Laith A. Jawad
Editor

Tigris and Euphrates Rivers: Their Environment from Headwaters to Mouth

Volume 1

Springer

Editor
Laith A. Jawad
School of Environmental and Animal Sciences
Unitec Institute of Technology
Auckland, New Zealand

Aquatic Ecology Series
ISBN 978-3-030-57569-4 ISBN 978-3-030-57570-0 (eBook)
https://doi.org/10.1007/978-3-030-57570-0

This Springer imprint is published by the registered company Springer Nature Switzerland AG.
The registered company address is: Gewerbestrasse 11, 6330 Cham, Switzerland

"Once Again to Zainab, my lovely wife"
To my daughters and granddaughters.....
You made me stronger, better and more
fulfilled than I could have ever imagined....
I love you all to the moon and back

Preface

It was a great book edited by Róska (1980), *Euphrates and Tigris, Mesopotamian Ecology and Destiny*. For a period of nearly 40 years, this book was ample in regard to the information that was incorporated into it as there was no flood of reports about the drainages of the twin rivers, Euphrates and Tigris, like what we have now.

Since I saw this book in the early 1980s, an idea came to me to produce a comprehensive book dealing with the drainages of the rivers of Euphrates and Tigris and the landscape they enclosed. This idea remained in my mind in spite of the two Gulf wars and the political turmoil that the region has confronted during the last 30 years. I carried with me through the countries that I settled in temporarily and permanently for the last 26 years the idea of editing a comprehensive publication about Euphrates–Tigris rivers until the opportunity came, nearly 2 and half years ago, when I sent the proposal of the book to Springer, and it has been accepted.

In planning the contents of the book, I took into consideration that of the book of Róska (1980) as a base and started to add and fill in the gaps those chapters that are missing and that contain information about subjects that have been forgotten in spite of being important to the region. I tried my best to make this book containing as much as possible the important aspects of life, the environment and the conservation of the Euphrates–Tigris rivers and their landscapes.

There are several reasons for choosing to edit a book about the Euphrates–Tigris rivers and the landscapes they encompassed. Among these are the resources of this region represented in water, oil, gas, minerals and agriculture. In addition, the historical aspects of the region and the social life of the people living in this area and the strategic geographic location, which is in the middle of the old world and accessible to all continents, are all factors that encourage me to edit such a book and make it available to the readers all over the world.

The fauna and the flora of the Mesopotamia are distinct and peculiar in their distribution and species content. Moreover, the invasion of the different species belonging to different plant and animal groups to this region made the study of these groups an important issue for most scientists.

The other important aspect that this book is dealing with is the issues of the environment and the conservation measures that are taking place at the present time and those that need to be planned in the future. The Mesopotamia is a land that faced over the long years several changes caused naturally and due to the activities of humans in this region for thousands of years. Conflicts of different types between different populations and ethnicities were evident in the historical parts of the region and are still in practice in the present time in some parts of Mesopotamia. Such conflicts and natural disasters have changed and still in the process of changing the resources of the area, which have a negative effect on the people living in the region.

The different chapters of this book were written by scholars who are experts in their field. Thus, they represent the best of their scientific work to educate readers about the status of the environment of the Euphrates–Tigris rivers and the inhabitants of their landscapes and show the important aspects of this region to the Middle East in particular and the world in general. It is important for the people living in Mesopotamia at least to be aware of their land and its resources and how to conserve such wealth as their ancestors in Sumer, Akkad and Assyria have done thousand years ago.

I would like to express my thanks to all the contributors of this book who agreed to share the pieces of their scientific work as chapters are written about different aspects of Mesopotamia in order make it available to the readers all over the world. Also, my sincere thanks should go to Springer that agreed to publish this book and make my dream a reality.

Auckland, New Zealand Laith A. Jawad

Contributors

Abed, Jasim	Iraq
Adriansen, Hanne Kirstine	Denmark
Ahmed, Sumaya M.	Iraq
Akıncı, Zeynep Sıla	Catalonia
Al-Edany, Taha Y.	Iraq
Al-Imarah, Faris J. M.	Iraq
Al-Janabi, Muhammad	Iraq
Al-Saad, Hamid T.	Iraq
Al-Sheikhly, Omar F.	Iraq
Al-Sulami, Amin	Iraq
Al-Taee, Asaad	Iraq
Ali, Atheer	Iraq
Atasağun, Sibel	Turkey
Audai M. Qasim	Iraq
Ayboğa, Ercan	Germany
Aydin, Gazel Burcu	Turkey
Bilgen, Arda	USA
Çamur-Elipek, Belgin	Turkey
Casellas, Antònia	Spain
Çetin, A. Kadri	Turkey
Dawood, Furat	Australia
Ergönül, M. Borga	Turkey
Esmaeili, Hamid Reza	Iran
Freyhof, Jörg	Germany
Genc, Ercument	Turkey
Ghaffari, Hanyeh	Iran
Glöer, Peter	Germany
Halihan, Todd	USA
Hassan, Fikrat M.	Iraq

Hayati, Dariush	Iran
Jajaei, Efat Salmani	Iran
Jawad, Laith A.	New Zealand
Jaweir, Haifa J.	Iraq
Jongerden, Joost	The Netherlands
Kaleli, M. Aydın	Turkey
Kalender, Leyla	Turkey
Kambouzia, Jafar	Iran
Kaya, Cüneyt	Turkey
Kaya, Dogukan	Turkey
Khalfallah, Myriam	Canada
Khayyun Amtair Rahi	Iraq
Koyuncu, Cafer Erkin	Turkey
Malekzadeh-Viayeh, Reza	Iran
Masseti, Marco	Italy
Maulood, Bahram K.	Iraq
Mehrabian, Ahmad Reza	Iran
Mohammed, Ruqayah	Iraq
Mostafavi, Hossein	Iran
Mousavi-Sabet, Hamed	Iran
Muhsin, Tawfik M.	Iraq
Nakayama, Mikiyasu	Japan
Öktener, Ahmet	Turkey
Pauly, Daniel	Canada
Pešić, Vladimir	Montenegro
Pourang, Nima	Iran
Safaei-Mahroo, Barbod	Iran
Sakamoto, Maiko	Japan
Saleh, Khalil I.	Iraq
Salewicz, Kazimierz A.	Austria
Salman, Nadir A.	Iraq
Scholz, Miklas	UK
Scurlock, JoAnn	USA
Secer, Faik Sertel	Turkey
Şen, Bülent	Turkey
Shafizade-Moghadam, Hossein	Iran
Sharif, Adel	UK
Solak, Cüneyt N.	Turkey
Sulaiman, Sadeq Oleiwi	Iraq
Tanaka, Yukio	USA
Taş-Divrik, Menekşe	Turkey
Teimori, Azad	Iran
Valinassab, Tooraj	Iran
Valizadeh, Naser	Iran

Waryani, Baradi	Pakistan
Wojtal, Agata	Poland
Yakhchali, Mohammad	Iran
Yaseen, Zaher Mundher	Vietnam
Yildirim, Pınar	Turkey
Yilmaz, Elif	Turkey
Yilmaz, Harun	Turkey
Zamanpoore, Mehrdad	Iran

Acknowledgements

First, the editor would like to thank each one of the authors, who contributed their time and expertise to this book. Without their support, this book would not have become a reality.

Second, the editor wishes to thank the staff of Springer DE in general about the efforts they put in publishing this book and in particular Éva Loerinczi for her distinguished service, valuable advice and suggestions that she delivered throughout the period of checking the text of this book.

Laith A. Jawad

Contents for Volume 1

1 **Introduction** 1
Laith A. Jawad

Part I Historical Perspectives

2 **Vertebrates of Upper Mesopotamia: Present Evidence and Archaeological Data** 13
Marco Masseti

3 **Fishing Gears and Methods: A Comparison of Ancient Mesopotamia and Other Ancient Worlds** 73
Laith A. Jawad

4 **Ichthyological Characteristics Available in the Fish Images Existed in the Art of the Ancient Mesopotamia** 85
Laith A. Jawad

5 **The Effectiveness of Ancient Mesopotamian Medical Practices: The Example of *šūšu*-Licorice** 127
JoAnn Scurlock

Part II The Abiotic Aspects of the Tigris–Euphrates River System

6 **Management of Water Resources Using Storage Reservoirs** 171
Kazimierz A. Salewicz, Maiko Sakamoto, and Mikiyasu Nakayama

7 **Estimation of Irrigated Agricultural Area and Water Consumption in Iraq** 205
Yukio Tanaka

8 **The Nature of Tigris–Euphrates Rivers Flow: Current Status and Future Prospective** 229
Zaher Mundher Yaseen, Sadeq Oleiwi Sulaiman, and Adel Sharif

9 Streamflow Alteration Impacts with Particular Reference to the Lower Zab River, Tributary of the Tigris River 243
Ruqayah Mohammed and Miklas Scholz

10 Ecohydrology in Iraq: Challenges and its Future Pathways 275
Laith A. Jawad

11 Oil Pollution in the Shatt Al-Arab River and its Estuary 1980–2018 285
Faris J. M. Al-Imarah and Hamid T. Al-Saad

12 Pesticides in the Waters, Sediments, and Biota of the Shatt Al-Arab River for the Period 1980–2017 299
Hamid T. Al-Saad and Faris J. M. Al-Imarah

13 Surface Water Salinity of the Euphrates, Tigris, and Shatt al-Arab Rivers 309
Khayyun Amtair Rahi and Todd Halihan

14 Renewable Energy for Water–Energy Nexus in Euphrates and Tigris River Basin: A Literature Review 337
Furat Dawood

15 Impacts of Dams on Aquatic Biodiversity, Fisheries, Fishes and Their Environment: Problems that Could Be Present in Iraq with Recommendations 369
Laith A. Jawad

16 Water, Politics and Dams in the Mesopotamia Basin of the Northern Middle East: How Turkey Instrumentalises the South-Eastern Anatolia Project for Political, Military and Strategic Interests 383
Joost Jongerden, Zeynep Sıla Akıncı, and Ercan Ayboğa

17 Dams and their Impacts on Fishes in Iran 401
Hamed Mousavi-Sabet

Part III The Biotic Aspects of the Tigris–Euphrates River System

18 Preliminary Review of the Aquatic Biodiversity in Al-Kahla River, Missan Province, Iraq 423
Laith A. Jawad

19 Freshwater Fish Biodiversity in Iraq: Importance, Threats, Status, and Conservation Challenges 479
Laith A. Jawad

20 **Effect of Climate Changes on the Freshwater Biodiversity in the Mesopotamian Plain: Recommendations for Avoidance and Plans for the Future** 499
Laith A. Jawad and Baradi Waryani

21 **River Corridors as a Refuge for Freshwater Biodiversity: Basic Information and Recommendations to the Policymakers for Possible Implications in Iraq** 515
Laith A. Jawad and Baradi Waryani

22 **Biodiversity of Fungi in Aquatic Environment of Iraq** 525
Tawfik M. Muhsin

23 **Potability of Drinking Water in Basra-Iraq** 541
Amin Al-Sulami and Asaad Al-Taee

24 **Algal Studies in Iraqi Inland Waters. A Review** 553
Bahram K. Maulood and Fikrat M. Hassan

25 **The Distribution of Epilitic Diatoms in the Turkish Part of the Tigris-Euphrates River Basin** 571
Cüneyt N. Solak, Agata Wojtal, Elif Yılmaz, M. Borga Ergönül, and A. Kadri Çetin

26 **Plant Biodiversity in Shatt Al-Arab Estuary and Ecological Variations** 621
Taha Y. Al-Edany

27 **Medicinal Plants of Shatt al-Arab River and Adjacent Area** 643
Taha Y. Al-Edany

28 **The Role of Plants as a Canopy in the Inland Waters: Basic Information for Application in Iraq** 663
Laith A. Jawad and Audai M. Qasim

29 **The Zooplankton Fauna of the Turkish Part of the Euphrates-Tigris River Basin** 679
Mehmet Borga Ergönül and Sibel Atasağun

30 **Kinds and Distribution of Icthyoplankton in Shatt Al-Arab River** 693
Sumaya M. Ahmed

31 **Rotifer Diversity in Iranian Waters: A Review** 705
Reza Malekzadeh-Viayeh

32 **Biodiversity of the Freshwater Amphipods in Iran** 743
Mehrdad Zamanpoore

33 **The Freshwater Molluscs of the Mesopotamian Plain** 763
Vladimir Pešić and Peter Glöer

34 **Freshwater Annelida of Iraq** 779
Haifa J. Jaweir

35 **A Critical Checklist of the Inland Fishes Native to the Euphrates and Tigris Drainages** 815
Jörg Freyhof, Cüneyt Kaya, and Atheer Ali

36 **Fish Fauna of Shatt al-Arab River, Basrah, Iraq: A More than Quarter a Century of Changes** 855
Laith A. Jawad, Abbas J. Al-Faisal, and Mustafa S. Faddagh Ziyadi

37 **The Common Carp, *Cyprinus carpio*: Effect on the Environment and the Indigenous Fish Species in Iraq** 877
Laith A. Jawad, Mustafa S. Faddagh Ziyadi, and Abbas J. Al-Faisal

38 **The Good and the Bad in Releasing the Grass Carp *Ctenopharyngodon idella* in the Freshwater System: Recommendations for the Policymakers in Iraq** 897
Laith A. Jawad and Falah Mutlak

39 **The Avifauna of Tigris and Euphrates River Basin** 913
Omar F. Al-Sheikhly

40 **The Feasible Approaches to Assist Migratory Birds Visiting the Southern Reaches of Mesopotamia** 937
Laith A. Jawad

41 **A Proposal for Establishing Bird Observatory Centre in the South of Iraq** 947
Laith A. Jawad

42 **The Potential Role of Waterbirds as a Vector in Dispersing Invertebrates and Plants in the South of Iraq** 951
Laith A. Jawad

43 **The Wild Mammals of the Tigris and Euphrates Rivers Basin** . . . 957
Omar F. Al-Sheikhly

44 **The Amphibians and Reptiles of Euphrates and Tigris Basin** 975
Barbod Safaei-Mahroo and Hanyeh Ghaffari

45 **A Preliminary Pictorial Guide to the Herpetofauna of Tigris and Euphrates River Basin** 1007
Omar F. Al-Sheikhly

46 Policy Guidance for Sustainable Aquaculture in the Inland Waters of Iraq 1041
Laith A. Jawad and Falah Mutlak

47 How Possible to Use the Desert Area in Iraq for Aquaculture Industry: Basic Facts and Recommondations 1047
Laith A. Jawad and Saad Mohammed Saleh Abdulsamad

48 Aquaculture Industry in Iraq: Past, Present, and Future Perspectives 1053
Nader A. Salman and Khalil I. Saleh

49 Towards an Improved Tilapia Farming in Iraq: Recommendations for Future Application 1077
Laith A. Jawad and Saad Mohammed Saleh Abdulsamad

50 Information and Recommendations of Aquaculture Stress and its Source in Hatcheries: The Iraqi Aquaculture Industry Investors 1089
Laith A. Jawad

51 Exotic and Invasive Freshwater Fishes in the Tigris-Euphrates River System 1103
Hamid Reza Esmaeili

Contents for Volume 2

Part IV Water Resources

52 The Ecology and Modelling of the Freshwater Ecosystems in Iran . . . 1143
Hossein Mostafavi, Ahmad Reza Mehrabian, Azad Teimori, Hossein Shafizade-Moghadam, and Jafar Kambouzia

53 Enhancing Rural Women's Participation in Fisheries in Iraq 1201
Laith A. Jawad

54 The Impact of Destructive Fishing Gear on the Fish Biodiversity in the Inland Waters of Iraq . . . 1207
Laith A. Jawad

55 Evaluating Variations in Fisheries by Means of Fishers' Information: Suggested Methodology to Improve Small-Scale Fisheries in Rivers in Iraq . . . 1225
Laith A. Jawad

56 A Possibility to Apply a Traditional Fisheries Enforcement Programme in the Inland Waters of Iraq . . . 1231
Laith A. Jawad

57 Market-Resource Relations and Fish Seller Livelihood as Seen in Inland Waters of Iraq . . . 1237
Laith A. Jawad and Audai M. Qasim

58 The Potential Impact of Deformities in Fishes upon Aquatic Production: Case of Iraq . . . 1253
Laith A. Jawad

59 The Impact of some Social Taboos on Fisheries in Iraq 1263
Laith A. Jawad

60 Inland Water Fishes and Fisheries in Iran 1277
Tooraj Valinassab and Hamid Reza Esmaeili

61 The Possibility of Introducing an Inland Fisheries Education in Iraq 1291
Laith A. Jawad

62 The Marine and Diadromous Fisheries of Iraq 1301
Myriam Khalfallah and Daniel Pauly

Part V Stress of the Environment of the Two Rivers

63 Fish Deformities in the Freshwater Fishes of Iraq: A Short Review and a Study Case on the Indian Catfish *Heteropneustes fossilis* 1311
Laith A. Jawad and Jasim Abed

64 The Phenomenon of Fluctuating Asymmetry: As Fish Welfare Indicator Represented by Case Study from the Freshwater Fishes of Iraq 1321
Laith A. Jawad and Jasim Abed

65 The Studies on Sediments Pollution by Different Types of Metals in Turkey 1339
Leyla Kalender

66 Macroplastic and Microplastic in the Freshwater Environment of Southern Iraq: Evidences Obtained from Freshwater Fish Species 1353
Laith A. Jawad

67 Heavy Metals in Freshwater Invertebrates of Iran: A Review on the Bioaccumulation and Effects 1375
Nima Pourang

68 Fish Parasites of Tigris and Euphrates River Systems 1389
Cafer Erkin Koyuncu, Harun Yilmaz, Faik Sertel Secer, Dogukan Kaya, and Ercument Genc

69 The Parasites of Fishes of the Euphrates and Tigris Rivers: Iraq and Turkey 1419
Ahmet Öktener

70 Ornamental Fishes: A Looming Danger for Inland Fish Diversity of Iraq 1445
Laith A. Jawad, Audai Qasim, and Muhammad Al-Janabi

Part VI The Health of the Human Community Inhabiting Freshwater Zones

71 Aquatic Snails as a Vector of Diseases to the Human in Iran 1459
Mohammad Yakhchali

72 Death by Drowning in Rivers in Iraq 1479
Laith A. Jawad

73 The Dangerous Catfish Species in the Freshwater System of Iraq: First Time Reports on Cases of Envenomation 1487
Laith A. Jawad

74 Fish Species of the Order Cypriniformes as a Source of Ichthyootoxin and Ichthyogallotoxin in Iraq: Cases Reports 1493
Laith A. Jawad

75 First Reports on Cases of Hallucinatory Fish Poisoning (Ichthyoallyeinotoxism) and Scombrotoxic Fish Poisoning in Iraq 1499
Laith A. Jawad

Part VII Conservation

76 Freshwater Management and Conservation in Iran: Past, Present, and Future 1507
Dariush Hayati and Naser Valizadeh

77 The Need of Biodiversity Conservation Strategies in Iraq: The State of Protected Areas 1535
Laith A. Jawad

78 Benthic Macroinvertebrates of the Tigris and Euphrates Rivers in Turkey 1539
Belgin Çamur-Elipek, Bülent Şen, Gazel Burcu Aydın, Menekşe Taş-Divrik, and Pınar Yıldırım

79 Freshwater Ecosystem Conservation in Iraq: Recommendations for Management 1561
Laith A. Jawad

80 Utilising Phenotypic Difference to Regulate Protection Value: A Scheme for Application of a Novel Approach to the Inland Water of Iraq 1565
Laith A. Jawad

81 Hatchery-Reared Fish Stocks Released into the Wild: A Conservation Problem as Seen in a Case Study from Iraq 1571
Laith A. Jawad

Part VIII Social Perspectives

82 Is the Glass Half Empty or Half Full? An Appraisal of the Four Decades of Turkey's Southeastern Anatolia Project (GAP) 1581
Arda Bilgen, Zeynep Sıla Akıncı, Antònia Casellas, and Joost Jongerden

Part IX Food Security

83 How Possible to Build Rice–Fish Farming in Iraq in order to Support Food Security Plan: Positive and the Negative Aspects 1601
Laith A. Jawad

84 The Importance of Non-commercial and Small-Sized Fish Species: A Proposal for an Additional Revenue to Iraq 1611
Laith A. Jawad

85 Sociocultural Aspects Influence Food Consumption Habits in Iraq: Management of Food Security 1625
Laith A. Jawad

86 A Preliminary Investigation of Determinants of Food Security in Rural Areas of Basrah Province, Iraq 1635
Laith A. Jawad

About the Editor

Laith A. Jawad obtained a degree (MSc) in fish taxonomy from the Zoology Department, University of Bristol, UK, in 1980. He continued as fish taxonomist at Basrah University, Iraq, where he worked for more than 20 years before he immigrated to New Zealand in 1997. During this time, he started the biochemical taxonomy of fishes of Iraq and published over 420 scientific papers and book reviews in leading scientific journals. He is the author and co-author of several textbooks in biology published in Arabic. Recently, he contributed five chapters to a book about coastal fishes, *Coastal Fishes: Habitat, Behavior and Conservation*, published by Nova Publishers, Canada, and authored a book about *Dangerous Fishes of the Eastern and Southern Arabian Peninsula* published by Springer in 2017. He served as fish biodiversity expert and consultant at the Ministry of Agriculture and Fisheries in Oman for the period 2008–2012 during which he co-authored two papers describing a new fish species from the Omani waters and reported over 80 fish species as a new record to the Omani waters. He authored a guide to the fishes of the southern coasts of Oman published by the Ministry of Agriculture and Fisheries in Oman in 2018. He also published over 90 papers on fish fauna of Oman, Iraq, Kuwait and Saudi Arabia. In 2013, he broadened his scientific contact and started to collaborate with over 60 scientists from more than 50 countries around the world in researches dealing with different aspects of fish taxonomy and ichthyology.

Chapter 1
Introduction

Laith A. Jawad

Euphrates and Tigris Rivers embrace between them the valley of what known as "Mesopotamia" or "the land between two rivers" (Fig. 1.1). These two rivers are among the long rivers in the Middle East and hold for thousands of years various civilizations that have left a significant mark in the human history.

In the different chapters of the present book, "*Tigris and Euphrates Rivers: Their environment from headwaters to mouth*", a comprehensive discussion for the different aspects of these two rivers, which include historical, hydrological, environmental, faunistic and floristic, conservation and management, and social. Therefore, in this introduction, only highlights will be shed on the main aspects of the Euphrates–Tigris River to draw the attention of the readers to the importance of these two rivers in human history.

1.1 There Were a Civilization

In the southern Mesopotamia, where there were isolated archaeological mounds that rise above this ancient plain are the remnants of early villages and cities that were at their maximum sizes during the third millennium B.C., which were extended over as much as 400 ha and contained as many as 80,000 people (Morozova 2005). This early knowledgeable civilization mainly depended upon the flow of the Tigris and Euphrates Rivers for its survival (Saggs 1988; Diakonoff 1991; Cole and Gasche 1998).

The Tigris–Euphrates delta was inhabited by farming societies maybe since early in the Holocene. Mesopotamian history is predictably separated into several periods.

L. A. Jawad (✉)
School of Environmental and Animal Sciences, Unitec Institute of Technology, Auckland, New Zealand

L. A. Jawad (ed.), *Tigris and Euphrates Rivers: Their Environment from Headwaters to Mouth*, Aquatic Ecology Series 11,
https://doi.org/10.1007/978-3-030-57570-0_1

Fig. 1.1 Map showing Mesopotamia (Courtesy of Goran tek-en, Germany. SVG development)

The world's first Sumer civilization, ran in the southern part of lower Mesopotamia, existed during the Sumerian period from about 3500 to 1600 B.C. Subsequently, the centres of power moved to the more northern parts of lower Mesopotamia, known as Akkad or Babylonia. Among the attainments of this early civilization are irrigation agriculture, flood-control systems, the world's first cities, codes of law, and cuneiform, the first writing system. Civilization has undergone a general decline during the middle Islamic period around 1000 A.D.

Lower Mesopotamia had numerous material benefits over neighbouring regions.

The irrigated areas had higher and more reliable agricultural yields, and goods were redistributed much more efficiently along a network of waterways (Algaze 2001).

Rivers were important transport and trade routes with the mountainous regions to the north. In exchange for agricultural products, such as barley, wheat, cereals, and wool, lower Mesopotamia received metals, building and semiprecious stones, gold, and timber (Mirsky 1982). Rivers were particularly valuable for the transport of highbulk cereals from city to city. Junctions in the channel network delivered ideal chances for early rulers of the region to employ control over downstream institutions (Morozova 2005). All of these factors either directly or indirectly originated from the fluvial geomorphology of the Mesopotamian plains (Wilkinson 2003).

As the first civilizations in Mesopotamia, the Sumerians, Babylonians, and Assyrians continued an economy chiefly based on agriculture. Meanwhile, water is indispensable for agriculture, these societies established complex irrigation networks and organized institutions in order to use water resources professionally. They created large-scale infrastructures for flood control and built barrages to regulate the flow of the Euphrates and Tigris Rivers. These civilizations also used these two rivers for the aims of navigation, transportation, and urban development (Dolatyar and Gray 2000). The significance of water for these early civilizations and their struggles for preserving an efficient water management system in the Mesopotamia was revealed in the Code of Hammurabi, which is accepted as the first written laws that regulated the use of water in the Euphrates–Tigris drainages.

The societies that consecutively occupied Mesopotamia enhanced the Mesopotamian culture as they introduced their traditional beliefs, practices, and customs. However, because of environmental conditions peculiar to Mesopotamia and the existence of an already complex civilization there, these newcomers eventually espoused the native culture, improved the inherited technical achievements in irrigation and agriculture, and advanced the social and civil institutions proper to a hydraulic civilization (Dolatyar and Gray 2000). As new ways of managing and scheming the natural forces of water were found, civilizations moved further away from the natural course of water resources. However, the social, economic, and political structure of these agrarian civilizations and their development depends upon a well-developed and well-maintained hydro-agricultural system.

The Mongol invasion in the thirteenth century destroyed much of these irrigation systems along the Euphrates and Tigris Rivers. In the following centuries, since the system became unusable by neglect and by the breakdown of central government administrations, Mesopotamia failed in terms of wealth and political importance (Hillel 1994).

1.2 Highlights on the Climate of Mesopotamia

Local climate continued semiarid until about 2000 B.C., when not only lower Mesopotamia but also the Euphrates catchment area to the north both became more arid (Aqrawi 2001). The period of dry climate was intermittent between about 5000 and 4000 B.C. by a wetter episode suggested by abundant organic-rich deposits dating to this period (Aqrawi 2001). Effect of small-scale climate variations on water supply and civilizations is well documented. For example, a period of slightly warmer climate from 1300 to 900 B.C. has been linked to reduced discharges, increased salinization of irrigated land, crop failures, famine, and hunger in both ancient Mesopotamia and Assyria (Neumann and Parpola 1987).

Bozkurt and Sen (2013) have studied the climatic changes in the Euphrates–Tigris Rivers drainages and suggested the following scenario for what will the climate be in the future. They proposed that all state simulations indicate surface temperature increases across the entire Euphrates–Tigris drainages. The increase is moderately bigger in the highlands in winter. Increase in annual surface temperature in the highlands ranges between 2.1 °C and 4.1 °C for 2041–2070, whereas it ranges between 2.6 °C and 6.1 °C for 2071–2099. Cold season temperature increase has the potential to greatly impact the regional hydrological cycle by reducing the snow cover and changing the seasonality of surface runoff.

In terms of precipitation, there is a general agreement between the simulations, which indicate a decrease in the highlands and northern parts of the basin and an increase in the southern parts. The results of Bozkurt and Sen (2013) confirm the findings of the previous studies that demonstrate similar spatial changes of precipitation in the Euphrates–Tigris drainages (Evans 2008; Onol and Semazzi 2009; Chenoweth et al. 2011). Precipitation in the highlands is projected to decrease by 33% under the higher emissions scenario by the end of the present century. Predictable changes in the annual evapotranspiration show normally statistically significant decreases in the drainages by the end of the century, which is most likely related to the projected decreases in precipitation.

A striking effect of warming could be seen on the snow water corresponding in the highlands of the drainages. It is also found that the greatest relative changes in the snow cover take place in the lower elevations, a result that is also reported by Ozdoğan (2011).

1.3 Groundwater Depletion and Drought, Two Crises to Confront

Water wealth is the uppermost need for the existence of all human beings and other living creatures and is the most vital basis for social and economic development. Essentially, the freshwater that can be easily used by human beings is found in lakes, swamps, and rivers and accounts for less than 1% of the total water (US Geological Survey 2017). As we all know, there has been a serious scarcity of water resources in the Middle East. Since the twentieth century, drinking water has been widely threatened in every Middle Eastern country. The water problems in the Middle East have been made worse by the influence of the global deficiency and rapidly growing populations in recent years. Moreover, the shortage of water resources and their uneven distributions create contradictions and conflicts between the countries of the Middle East (Allan 2002). Therefore, one of the most cared-about problems for Middle Eastern countries is how to utilize the all-out amount of river water resources (Food and Agriculture Organization 2017).

The Twin Rivers, Tigris and Euphrates are the two major rivers in the Middle East. The Tigris River is the place where mankind originated and is the birthplace of Ancient Babylon, which is one of the four ancient civilizations, and of Mesopotamian civilizations, which are the earliest civilizations in the world. There were approximately 20,000 km^2 of rich marshland in the Tigris–Euphrates drainages, and its average annual river discharge is approximately 30×109 m^3, with several hundred rare wild species living in the basin throughout the year and abundant fish and prawns. However, the area is becoming a desert because of the actions of many long years and endless war. Moreover, the onset of drought in 2007 (Integrated

Regional Information Networks 2010) made the aforesaid issue a greater challenge (Trigo et al. 2010). According to the recent report of the World Bank 2007, the average water depletion in half of the Middle Eastern countries is greater than the available precipitation, and approximately 85% of available precipitation is used for irrigation, which has made this area prone to desertification, especially in Iran, Iraq, Syria, and Jordan. Moreover, the Euphrates–Tigris drainage registers the second fastest rate of regional groundwater storage loss in the world after India. Voss et al. (2013) evaluated the groundwater trends in the Tigris–Euphrates western Iran region from 2003 to 2009 and showed that the region lost a volume of water that accounts for 63% of the total water storage change during the study period. On the other hand, Joodaki et al. (2014) estimated the groundwater in the Middle East and disclosed that the regions in the Middle East lost groundwater from 2003 to 2012. Furthermore, the anthropogenic influences to groundwater loss were also assessed by removing the natural variations of the groundwater predicted in the Middle East, which indicated that more than half of the groundwater loss in Iran may be accredited to human withdrawals.

1.4 International Law Perspective of the Waters of Euphrates–Tigris Rivers

The critical significance of water is impaired in arid and semi-arid regions such as the Middle East. In the case of Syria and Turkey, at the core of their political and strategic interaction lies the Euphrates and Tigris waters. Both rivers have seen the rise of ancient civilizations and the early development of irrigation practices dating back to the Sumerian and Akkadian periods (4000–5000 BC). Any fight over water can be said to exist when an actor feels inhibited in the attainment of national aims through the one-sided use of the resource by another actor (Daoudy 2009).

In order to assess past and potential agreements, an analytical negotiation framework is developed with a view to revealing the direct and indirect issues at stake and the coalition dynamics at work. The analysis starts theoretical and experimental links between water, negotiation (structure, process), power (asymmetries, coalition dynamics, strategies, development of alternatives), and security (direct/indirect interests such as national security, border security, territorial claims, economic development, and environmental concerns) (Daoudy 2009).

The joint negotiation and power viewpoints exert some motivating questions.

What is the weight of water sharing in the power diminuendos of the three key actors—Syria, Turkey, and Iraq? What have been and are the negotiation plans of the downstream riparians with regards to the powerful upstream neighbour? Given its apparent devastatingly greater share of geographic, political, and economic power (which will be examined later in the text), why has upstream Turkey agreed to a minimal allocation to downstream Syria? Perhaps illogically, the study accomplishes that asymmetries in power have favoured upstream/downstream interactions towards bilateral if not basin-wide arrangements. Primary undisclosed negotiation sources will serve as the main references (Daoudy 2009).

Negotiating power can also determine the dynamics taking place between respective riparians. Downstream or more vulnerable riparians can invert situations of power asymmetry by worsening the basin-dominant riparian's alternatives and thus reducing its degree of freedom. Syria's use of "issue-linkage" in its interactions with Turkey over water and security issues serves as the primary example. Regional instability is generally increased, but short-term collaboration over water may in fact be endorsed as bilateral agreements have been efficiently reached. One assumption that can be generalized from the analysis is that power asymmetries do not necessarily determine the results of negotiations. But the influence of power is inadequate as negotiation results are bilateral and temporary (Turkey and Syria, Syria and Iraq). The general distribution remains in the hand of upstream Turkey in the case of the Euphrates and Tigris rivers.

The countries contiguous the Tigris and Euphrates face technical, legal, and regional barriers to successful cooperation. In the Tigris–Euphrates drainage, data regarding streamflow, precipitation, evapotranspiration, water diversions, return flow, salinity, soil type, and other variables in relation to land resources, are very scarce, incomplete, and disputed at many locations. Furthermore, vital information linking to water and land resources of the region is not exchanged on a regular basis among the riparians. A diversity of different statistics concerning availability of irrigable land and soil water requirements in each riparian country are available depending on the origin of data and inclination of the experts.

The indeterminate political status, quest of short-term national interests, lack of regularized institutions, and incomplete information contributed significantly to the termination of any arrangement about dividing the waters of Euphrates–Tigris Rivers. The Arab countries have long blamed Turkey for disrespectful international water laws with regards to the Euphrates River. Iraq and Syria deliberate the river to be an international waterway that should be treated as an integrated entity by all users. However, a noteworthy legal obstacle is that Ankara regards the Euphrates as a transboundary river, which is under Turkey's exclusive sovereignty until it flows across the border (Daoudy 2009). According to Turkey, the Euphrates becomes an international river only after it joins the Tigris in lower Iraq to form the Shatt al-Arab, which then serves as the border between Iraq and Iran until it reaches the Arabian Gulf.

1.5 Towards Restoring of the Riverine Landscapes

Globally, rivers are the form of ecosystem best influenced by humans. This high level of effect has a number of motives. Rivers offer ecosystem goods and services that endure human societies; therefore, the history of influences on rivers is as long as human history; dating from early civilizations, e.g. the Nile in ancient Egypt, the Euphrates and Tigris in Mesopotamia, and the Yellow and Yangtze in China (Shaw

2003). Rivers occupy the lowest positions in landscapes; thus, collecting and assimilating impacts occurring over whole catchments (Naiman et al. 2002). Therefore, rivers are outstanding gauges of environmental alteration. Additionally, natural rivers have the size to contain a large diversity of habitats and species (Naiman et al. 2005). This is because they act as canals for varying amounts of water in more or less unstable channels, making them naturally dynamic (Leopold et al. 1995). Human-related impacts on natural as well as artificial river flows or channels, therefore, result in immediate ecosystem reactions.

Humans have only recently started reinstating or acclimatizing affected rivers (Jansson et al. 2007). Researchers are carefully following this development as shown by the increasing number of articles addressing river restoration ecology in the last two decades, and the billions of dollars spent annually on renovation (Bernhardt et al. 2005). However, well management among research efforts and better communication is needed for restoration efforts that are effective (Wohl et al. 2005).

Years of civil wars, international wars, and wars of withdrawal prove the sturdy association between natural resources and armed fight. Arguments over natural resources and their associated revenues can be among the motives that people go to war. Diamonds, timber, oil, water, and even bananas and charcoal can provide sources of financing to sustain conflict. In the case of the Mesopotamia that is rich in many natural resources such as water, plantation, and fertile soil and recently oil and gas have made this area a good reason for conflicts of different sorts including armed struggles. Struggles to negotiate an end to conflict progressively include natural resources. And conflicts related to natural resources are both more possible to revert than non-resource-related conflicts, and to relapse twice as fast.

Nearly after the end of any conflict, a window of chance opens for a conflict-affected country and the international community to establish security, rebuild, and consolidate peace—or risk conflict revert. This possibility also represents the chance to improve the management of natural resources and their incomes in ways that would otherwise be politically difficult to attain. The United Nations represented in the Peacebuilding Commission has started to recognize the significance of natural resources in post-conflict peacebuilding. In 2009, along with the United Nations Environment Programme, the commission published a pioneering report—*From Conflict to Peacebuilding: The Role of Natural Resources and the Environment*—that outlined the elementary ways in which natural resources donate to conflict and can be succeeded to support peacebuilding.

Given the difficulty of peacebuilding, practitioners and researchers alike are stressed to clear good practice. It is progressively obvious that natural resources must be involved as an initial issue; many questions remain, however, regarding opportunities, options, and trade-offs.

When decline has gotten states in which rivers are heavily negotiated, restoration might be the only option if rivers are again to provide useful environmental amenities. In a perfect world, one may want to restore rivers to what one observes as a "pristine state". Nevertheless, identification of such states needs knowledge about preceding human influences. In numerous cases, unawareness about historical land use and about the related effects on rivers makes it unbearable to discover the degree to which rivers have been changed from natural conditions and to accomplish what kinds of restoration are desired (Wohl 2004). Consequently, restoration efforts may set aims that are too committed or ones that are based on incorrect rules. Many rivers may give an untrue impress of naturalness while they have been meaningfully affected during the last few centuries by multiple human activities, e.g. flow regulation, to an extent that their present ecological integrity is heavily conceded (e.g. Wohl 2004).

So, in order to achieve successful river restoration, it is necessary to identify and target ecosystem states that would be able to interact with current surrounding landscapes, including other parts of the river, and that would be appreciated, or at least accepted, by human societies.

A number of questions, however, still lack conclusive answers. We do not always understand exactly why definite methods fail or succeed in specific cases, or which of a dozen related factors are the most important in defining the success of a peacebuilding effort. However, many distinct events linked to natural resources can be accepted now to develop the likelihood of long-term peace. By learning from peacebuilding experiences to date, we can evade reiterating the errors of the past and break the cycle of conflict that has come to characterize so many countries and in particular the Mesopotamia. We also hope that this undertaking represents a new way to understand and approach peacebuilding.

References

Algaze G (2001) Initial social complexity in southwestern Asia: the Mesopotamian advantage. Curr Anthropol 42:199–233

Allan JA (2002) Hydro-peace in the Middle East: why no water wars? A case study of the Jordan River basin. SAIS Rev 22(2):255–272

Aqrawi AA (2001) Stratigraphic signatures of climatic change during the Holocene evolution of the Tigris–Euphrates delta, lower Mesopotamia. Glob Planet Change 28(1–4):267–283

Bernhardt ES, Palmer MA, Allan JD, Alexander G, Barnas K, Brooks S, Carr J, Clayton S, Dahm C, Follstad-Shah J, Galat D, Gloss S, Goodwin P, Hart D, Hassett B, Jenkinson R, Katz S, Kondolf GM, Lake PS, Lave R, Meyer JL, O'Donnell TK, Pagano L, Powell B, Sudduth E (2005) Synthesizing U.S. river restoration efforts. Science 308:636–637

Bozkurt D, Sen OL (2013) Climate change impacts in the Euphrates–Tigris Basin based on different model and scenario simulations. J Hydrol 480:149–161

Chenoweth J, Hadjinicolaou P, Bruggeman A, Lelieveld J, Levin Z, Lange MA, Xoplaki E, Hadjikakou M (2011) Impact of climate change on the water resources of the eastern Mediterranean and Middle East region: modeled 21st century changes and implications. Water Resour Res 47:W06506

Cole SW, Gasche H (1998) Second- and first-millennium B.C. rivers in northern Babylonia. In: Gasche H, Tanret M (eds) Changing watercourses in Babylonia. Towards a reconstruction of the ancient environment in lower Mesopotamia. University of Chicago Press, Chicago, pp 1–158

Daoudy M (2009) Asymmetric power: negotiating water in the Euphrates and Tigris. Int Negot 14 (2):361–391

Diakonoff IM (1991) Early antiquity. University of Chicago Press, Chicago

Dolatyar M, Gray TS (2000) Water politics in the Middle East: a context for conflict or co-operation? St. Martin's Press, New York, p 121

Evans J (2008) Changes in water vapour transport and the production of precipitation in the eastern Fertile Crescent as a result of global warming. J Hydrometeorol 9:1390–1401
Food and Agriculture Organization (FAO) (2017) Irrigation. Water Development and Management Unit, FAO. http://www.fao.org/home/en/. Accessed 12 February 2019
Hillel D (1994) Rivers of Eden: the struggle for peace in the Middle East. New York, Oxford University Press, p 96
Integrated Regional Information Networks (2010) Syria: drought pushing millions into poverty. http://www.irinnews.org/Report/90442/SYRIA-Drought-pushing-millions-intopoverty. Accessed 9 January 2019
Jansson R, Nilsson C, Malmqvist B (2007) Restoring freshwater ecosystems in riverine landscapes: the roles of connectivity and recovery processes. Freshw Biol 52:589–596
Joodaki G, Wahr J, Swenson S (2014) Estimating the human contribution to groundwater depletion in the Middle East, from GRACE data, land surface models, and well observations. Water Resour Res 50:2679–2692
Leopold LB, Wolman MG, Miller JP (1995) Fluvial processes in geomorphology. Dover, New York
Mirsky A (1982) Influence of geologic factors on ancient Mesopotamian civilization. J Geol Educ 30:294–299
Morozova GS (2005) A review of Holocene avulsions of the Tigris and Euphrates rivers and possible effects on the evolution of civilizations in lower Mesopotamia. Geoarchaeology 20 (4):401–423
Naiman RJ, Bunn SE, Nilsson C, Petts GE, Pinay G, Thompson LC (2002) Legitimizing fluvial systems as users of water: an overview. Environ Manag 30:455–467
Naiman RJ, Décamps H, McClain ME (2005) Riparia: ecology, conservation and management of streamside communities. Elsevier, San Diego, CA
Neumann J, Parpola S (1987) Climatic change and the eleventh-tenth-century eclipse of Assyria and Babylonia. J Near East Stud 46:161–180
Onol B, Semazzi FHM (2009) Regionalization of climate change simulations over eastern Mediterranean. J Climatol 22:1944–1961
Ozdoğan M (2011) Climate change impacts on snow water availability in the Euphrates–Tigris basin. Hydrol Earth Syst Sci 15:2789–2803
Saggs HWF (1988) The greatness that was Babylon. A survey of the ancient civilization of the Tigris-Euphrates valley. Sidgwick & Jackson, London
Shaw I (2003) The Oxford history of ancient Egypt. Oxford University Press, Oxford, UK
Trigo RM, Gouveia CM, Barriopedro D (2010) The intense 2007–2009 drought in the Fertile Crescent: impact and associated atmospheric circulation. Agric For Meteorol 150:1245–1257
U.S. Geological Survey (2017) Earth's water distribution. USGS. https://water.usgs.gov/edu/earthwherewater.html. Accessed 12 February 2019
Voss KA, Famiglietti JS, Lo M, de Linage C, Rodell M, Swenson SC (2013) Groundwater depletion in the Middle East from GRACE with implications for transboundary water management in the Tigris-Euphrates-western Iran region. Water Resour Res 49:904–914
Wilkinson TJ (2003) Archaeological landscapes of the near east. University of Arizona Press, Tucson, AZ
Wohl EE (2004) Disconnected rivers. Yale University Press, New Haven, CT
Wohl EE, Angermeier PL, Bledsoe B, Kondolf GM, MacDonnell L, Merritt DM, Palmer MA, Poff NL, Tarboton D (2005) River restoration. Water Resour Res 41:1–12

Part I
Historical Perspectives

Chapter 2
Vertebrates of Upper Mesopotamia: Present Evidence and Archaeological Data

Marco Masseti

2.1 Introduction

The late Hellenistic king Antiochos I (69–34 B.C.)—who reigned over the kingdom of Commagene, founded north of Syria after the breakup of Alexander's empire—built his mausoleum on the top of the Nemrut Dağ, one of the highest peaks of the Eastern Taurus mountains in southeastern Anatolia (Fig. 2.1). Looking down from its privileged geographical location, this monument surveys the underlying progression of the sleepy floodplain of the large water bodies of northern Mesopotamia, term that means the "land between the rivers" in ancient Greek. These "rivers" referred to the Tigris and the Euphrates, locating the well-known Near Eastern alluvial plain which, since ancient times, had enlivened the desertic geography of the eastern Fertile Crescent, supporting the development of millenary civilizations, such as Sumerians, Akkadian, Babylonian, Assyrian, Parthians, Romans, and Muslims (Fig. 2.2). As far as is presently known, Mesopotamia also hosted the oldest permanent human settlements and the first evidence of the domestication of ungulates. The latter phenomenon seems, in fact, to have started in a few PPNB (Pre-Pottery Neolithic B) sites of southern Turkey, such as Nevali Çori, Göbekli Tepe, and Gürkütepe (Peters et al. 1999, 2005; cf. Schmidt 1999).

Medieval Muslim geographers (seventh to twelfth centuries AD) traditionally divided Mesopotamia into two areas. *Al-Sawad*, the "dark, black land," was the name used for southern Iraq, and refers to the strong contrast between the alluvial plain and the Arabian desert. In the middle and lower basin of the Tigris and Euphrates, the Iraqi marshlands are the most extensive wetland ecosystems in the

M. Masseti (✉)
Department of Biology, University of Florence, Florence, Italy

IUCN Deer Specialist Group, Gland, Switzerland
e-mail: marco.masseti@unifi.it

L. A. Jawad (ed.), *Tigris and Euphrates Rivers: Their Environment from Headwaters to Mouth*, Aquatic Ecology Series 11,
https://doi.org/10.1007/978-3-030-57570-0_2

Fig. 2.1 Partial view of the mausoleum of King Antiochos I (69–34 B.C.) of Commagene, on the top of the Nemrut Dağ, one of the highest peaks of the Eastern Taurus mountain range in southeastern Anatolia (photo by Marco Masseti)

Fig. 2.2 The lost Islamic city of Rasafa, the ancient Sergiopolis of the Romans, lies in the Syrian desert, south of the town of Raqqa, and the right bank of the Euphrates (photo by Marco Masseti)

Near East (Bedair et al. 2006). The *Jazīra* ("the island"), instead, coincides with the province of Upper Mesopotamia, geographically distributed throughout the territories of south-eastern Turkey, eastern Syria and northwestern Iraq Its principal towns are Al Hasakah, Qamishli, Raqqa, Deir ez-Zor and Harrân in the Syrian Arab

Republic, Sanlurfa, Mardin, and Diyarbakir in Turkey, and Mosul in Iraq. The Jazīra takes its name from the fact that the two main rivers, the Euphrates and Tigris, transform this part of Mesopotamia into almost an island (Hillenbrand 1985), as their sources in eastern Turkey are in close proximity. Upper Mesopotamia corresponds approximately to the territorial extension of what was the kingdom of Assyria, one of the main Near Eastern state, which lasted from perhaps as early as the twenty-fifth century B.C. up to 612–609 B.C. (Lloyd 1984).

2.2 Biogeographical Features

The geographic position of Upper Mesopotamia, in the land bridge between Eurasia and Africa, gives rise to a remarkable variety of bioclimatic and biogeographical conditions throughout its territory, permitting the coexistence of biological elements from the Mediterranean and Boreal regions with Irano-Turanian and Saharo-Sindian species (Guest 1966; Atallah 1977, 1978). As in other Near Eastern countries with ancient civilizations, the natural environment of this region and its vegetation have been degraded for at least several thousand years by fire, overgrazing, extensive cultivations, and the continuous development and transformation of human settlements (cf. Wirth 1971; Zohary 1973; Rösner and Schäbitz 1991). Thus, the distribution of biological elements has undergone since prehistorical times a process of redefinition of the natural resources and environment. The present study aims to give a review of the available knowledge of Upper Mesopotamian vertebrates, which is of even greater significance today in view of the situation of civil unrest that broke out in this Near Eastern region in the last years. In fact, as it is well known, very recently, almost the totality of the Jazīra has been dramatically upset by the civil war. Because of such a fact, all the data presented in this work were collected prior to the occurrence of recent local conflicts.

The natural environment of the Jazīra is characterized by a fertile steppe which, from northeastern Syria, extends east of the Euphrates beyond the Turkish border and to the south to Iraq, overcoming the isolated hill range of *Jabal Abdul Aziz*, toward and beyond the river Tigris. The southern boundary marks the traditional limit of rain-fed cultivation. There is a network of springs that feed the Euphrates through the Khabur river and other freshwater bodies of the area, such as the Balik and the Jaghjagha. Also, the Tigris is fed by several affluents such as the Ambar and the Oymataş, in Turkey, and the Great and Little Zab, in Iraq. This ecoregion is characterized by occasional smaller lakes, but not by extensive marsh-lake habitats. The summers are very hot with daily mean temperatures of about 40 °C in July and August. The subtropical climate is of the Mediterranean type, with an average annual precipitation between 250 and 500 mm (Evans 1994), and cold winters. The steppe vegetation includes *Pistacia*, *Prunus*, and *Rhamnus* scrub, with *Artemisia*, *Atriplex*, *Helianthemum*, and *Teucrium*. Many botanical endemics are confined to the isolated hills and *wadies* of the Upper Mesopotamian steppe. Several of the zoological species that inhabited this region were exterminated in recent historical times,

Fig. 2.3 Artistic reconstruction of an aspect of the *Jazīra*, which extends on the territories of Upper Mesopotamia, including the eastern territories of Syria, the southeastern edges of Anatolia, and northwestern Iraq. In the background of a *tell*, an artificial hill resulted from the repeated stratification of human settlements, a herd of the extinct Syrian onager, *Equus hemionus hemippus* (Geoffroy, 1855), is grazing together with few goitered gazelles, *Gazella subgutturosa* (Güldenstädt, 1780). A flock of pin-tailed sandgrouses, *Pterocles alchata* (L., 1766) is flying in the sky (drawing by Alessandro Mangione)

including the Syrian onager, *Equus hemionus hemippus* (Geoffroy, 1855), on Jabal Abdul Aziz in the 1930s (Misonne 1957; Harrison 1972) (Fig. 2.3). The Arabian or white oryx, *Oryx leucoryx* (Pallas, 1777), the species perhaps at the origin of the invention of the mythical unicorn, may have persisted in the most remote desert *wadies* until very recent times (Mountfort 1965; Masseti 2004) (Fig. 2.4). During the last decades, a program of reintroduction of this ungulate has been carried out in several of the countries of its ancient diffusion, such as Syria, Jordan, Israel, Saudi Arabia, the United Arab Emirates, and Oman (Jones 1988; Fletcher 2000; Serra et al. 2003a; Shalmon 2004; IUCN SSC Antelope Specialist Group 2017).

2.3 A Vanished World

Until relatively recent times, Upper Mesopotamia has been characterized by the occurrence of a very peculiar fauna, in many cases today extinct. Osteological data confirm the former occurrence of many of the zoological species represented in the ancient artistic productions, although many of them are no longer part of the extant local fauna.

Fig. 2.4 During the last decades, a program of reintroduction of the Arabian or white oryx, *Oryx leucoryx* (Pallas, 1777), has been carried out in several of the countries of its ancient diffusion, such as Syria, Jordan, Israel, Saudi Arabia, United Arab Emirates, and Oman (photo by Marco Masseti)

The archaeozoological and paleobotanic finds from the excavation of sites, such as Tell Abu Hureyra (Raqqa), Tell Hadidi (Tabqa), or Umm Dabaghiyah (Nineveh), allow reconstructing what had to be the natural characters of the Upper Mesopotamia floodplain, bordered by an environment of steppe vegetation. During the Epipaleolithic and early Neolithic of Tell Abu Hureyra (9000–7000 B.C.) the fauna was dominated by steppe herbivores, with abundant goitred gazelles, *Gazella subgutturosa* (Güldenstädt, 1780), onagers, the Arabian hare, *Lepus capensis* L., 1758, and rather rare mouflons, *Ovis orientalis* Gmelin, 1774, and wild goats, *Capra aegagrus* Erxleben, 1777 (see Legge and Rowley-Conwy 1986). Only in the later Neolithic do domestic sheep and goat become common, effectively replacing the gazelles (Legge 1975, 1977). In the Syrian Jazīra, the mammals of the river valley were represented at Tell Assouad (first half of the seventh millennium B.C.), by the bones of wild boar, *Sus scrofa* L., 1758, Mesopotamian fallow deer, *Dama dama mesopotamica* (Brooke, 1875), and red deer, *Cervus elaphus* L., 1758 (Helmer 1985) (Fig. 2.5). Also, the beaver, *Castor fiber* L., 1758, was a component of the floodplain fauna, apparently reaching in Upper Mesopotamia the southern limits of its Palaearctic geographic range.

Large carnivores were not uncommon in the region up to very recent historic times. The Asian or Indian lion, *Panthera leo persica* (Meyer, 1826), for example, survived in the gallery forests along the Upper Euphrates at least until the mid-nineteenth century (Masseti 2000, 2009a), as well as in southern and south-western Anatolia (Fellows 1841; Danford and Alston 1880; Kinnear 1920; Kumerloeve 1967; Kasparek 1986a; Kasparek and Kasparek 1990; Masseti and Mazza 2013) (Fig. 2.6). Remnant populations were reported from the region of Fethiye, in southern Turkey (Fellows 1841), on the southern bank of the river Esen (Koca Çay, south-western Anatolia) (Kasparek 1986a, b; Kasparek and Kasparek

Fig. 2.5 Stag, *Cervus elaphus* L., 1758, licking salt or drinking at a stream in an ivory bas-relief (*c.* 750–700 B.C.) from Khadatu (Arslan Tash), in the province of Aleppo about 30 km east of the Euphrates (photo by Marco Masseti)

Fig. 2.6 The Asian lion, *Panthera leo persica* (Meyer, 1826), survived in the gallery forests of Mesopotamia, at least, until the first half of the twentieth century (photo by Marco Masseti)

1990), as well as from the Euphrates valley (Danford and Alston 1880; Kinnear 1920; Kumerloeve 1967; Masseti 2000). Kasparek (1986a, b) noted that a small population of lions surely survived in southern Turkey at least until the mid-nineteenth century. It could still be found along the Tigris until 1918, and in Iran until 1957 (Schnitzler 2011; Masseti and Mazza 2013). According instead to Al-Sheikhly et al. (2015), the last representatives of the species were killed at the

Fig. 2.7 The so-called *Dying Lion*, wall relief from the palace of the Assyrian king Ashurbanipal which ruled Mesopotamian from 668 to 627 B.C. (photo by the British Museum, courtesy of the Trustees of the British Museum, London)

time of the military actions of the British forces (1916–1918). Among other things, the former occurrence of lions in Upper Mesopotamia is sumptuously testified by the subjects portrayed in the stone relieves from the palace of the king Ashurbanipal, at Nineveh (about 645 B.C.) (Strommenger and Hirmer 1963; Reade 1983; Masseti 2003) (Fig. 2.7). The latter represents the Assyrian monarch's hunts of his favorite prey. Among them lions are the most frequent subjects because at this time only royalty was allowed to kill this animal, considered as one of the greatest symbols of the hostile wildlife from which a Mesopotamian ruler was obliged to protect his land (Reade 1983). Mainly for this reason, killing lions was a meritorious activity. The former distribution of *P. leo* certainly included Greece, Asia Minor, the northern Arabian Peninsula, and Persia (Kinnear 1920; Harrison 1972; Masseti 2012; Masseti and Mazza 2013). Like other species, such as the wild ox, *Bos primigenius* Bojanus, 1827, which were employed in Assyrian royal hunts, even lions were not necessarily wild. Often they were brought to the hunting-grounds in cages, from which they were conveniently released one by one (Masseti 2003). There is a written document, for example, in which the writer asked anxiously what he should do about a lion trapped in his house: the lion was eventually caught in a cage and sent by boat to the town where the king was residing (Reade 1983).

Asiatic cheetahs, *Acinonyx jubatus venaticus* (Griffith, 1921), became apparently extinct in historical times too (Masseti 2009a) (Fig. 2.8). If any of these felids still survive in the western Near East, the area where they are most like to be found probably coincides with the remote desert tracks where the frontiers of Jordan, Iraq and Saudi Arabia meet (Corkill 1929; Harrison 1968; see also Masseti 1990). Its last report from the southern Iraqi desert is from near Busaiya (W Basra) (Al-Sheikhly et al. 2015). No recent record of the species exists from Anatolia. Yet there was a

Fig. 2.8 Asiatic cheetahs, *Acinonyx jubatus* (Schreber, 1775), probably still survive in remote desert tracks of the western Near East where the frontiers of Jordan, Iraq, and Saudi Arabia meet (photo by Anna M. De Marinis)

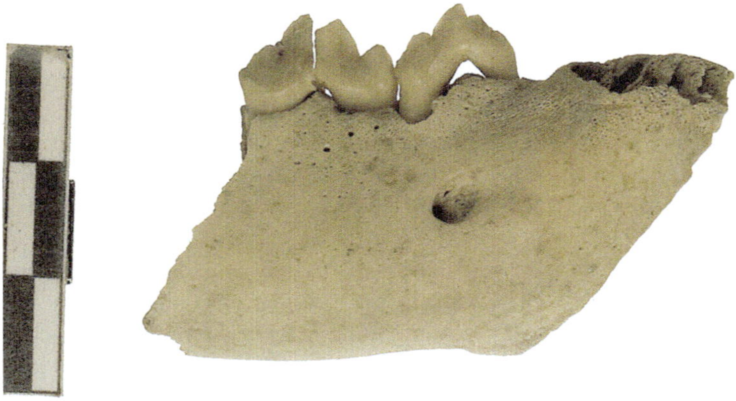

Fig. 2.9 Cheetah mandible found in the Early Bronze Age levels of Arslantepe (Malatya, southern Turkey) (photo by Giovanni Siracusano)

time when the felid was anything but rare in Mesopotamia, even in its northern parts, so much so that its bone remains were even found in proto-historic human settlements. In this regard, we can remember the discovery of a cheetah mandible in the Early Bronze Age levels of the archaeological site of Arslantepe (Malatya, southern Turkey) (Siracusano and Carlini 2010; Siracusano 2012) (Fig. 2.9). This finding is important both for zoological and for cultural reasons. Cheetahs have been used for hunting in the Near East and the Indian subcontinent since very ancient times. As far as is presently known, however, the oldest indirect evidence of the association between these felids and human beings comes from Italy, and more in particular from, the Etruscan tomb *Campana* (sixth century B.C.) of the necropolis of Veio

Fig. 2.10 Detail of the hunt of the wild horses, *Equus przewalskii* Poliakov, 1881, from the relief panels decorating the walls in the palace of Ashurbanipal (*c.* 645–635 B.C.) at Nineveh (photo by the British Museum, courtesy of the Trustees of the British Museum, London)

(Rome). In the latter artistic contest, ritual customs of oriental origin were apparently evoked.

It cannot be excluded that a wild—or feral—population of horses, *Equus przewalskii* Poliakov, 1881, survived in Upper Mesopotamia until fairly recent times, as it would be testified by some bas-reliefs still from the Ashurbanipal palace at Nineveh (Masseti 2003, 2018) (Fig. 2.10). In them, the depiction of some moments of hunting of perissodactyls, that are identical to *E. przewalskii* (Fig. 2.11), led to suppose an extension of the species distribution, still in the seventh century B.C., in today's northern Iraq. In any case, like the lions and the wild oxen employed in royal hunts, even the wild horses could not necessarily be wild, but imported alive for the king's amusements also from very far afield. Another representative of the Equidae family, eventually vanished from the land of Mesopotamia in very recent historical times, is the already mentioned Syrian onager which disappeared from its final refuge in the region of Jabal Abdul Aziz, in 1930s (Harrison 1972), the last herd being reported from the area of Jabal Sinjar in 1927 (Raswan 1935; Hatt 1959). Apropos this, Hatt (1959) observed that: "*Unfortunately, except for Xenophon's account of wild asses on the plains of the Euphrates during the campaign of 401 B.C., there are, so far as I know, no records of animals from about 500 B.C. to the middle of the nineteenth century A.D., when the reports of the Euphrates expedition (Ainsworth; Chesney), the archeologist Layard, and the travels of Lady Anne Blunt ushered in an era of new interest in this area and its animal life.*" The representation of Asian wild asses is, however, almost frequent in the artistic production of the geographic area comprised between the Levant and Mesopotamia at least since prehistory (see Masseti 2002). Hunting scenes involving

Fig. 2.11 The general description of the morphology of the equids portrayed in Ashurbanipal's hunt at Nineveh do not resemble asses (Masseti 2003): their limbs and ears are short like horses, while the tails are tufted just like those of the individuals of *E. przewalskii* shown in this photograph (photo by Ferdinando Ciani)

these equids are, for example, evoked in the wall paintings of Umm Dabaghiyah, in Iraq, dated to the seventh millennium B.C. (Cauvin 2000), and in the decoration of several western Near Eastern palaces and churches, as in the cases of the seventh-century mosaics from Dayr al-ʻAdas of the Bursa castle, south of Damascus, or the frescoes from Qaşr al-Hayr al-Gharbi, also in Syria (Schlumberger 1948; Schlumberger and Le Berre 1986; Fowden 2004). Other ancient artistic representations of Asian wild asses are known from the fifth century mosaic of the "personification of Ktisis" at the Beiteddine Palace (Lebanon), and the floor mosaics in the Byzantine church of Petra (Jordan) (Studer 2001), referred to the sixth century. Furthermore, the early Umayyad Qasr al-Amra (Hashemite Kingdom of Jordan), of the eighth century AD, is decorated with a magnificent scene in which onagers are being hunted being corralled in nests (Masseti 2015) (Fig. 2.12). Broadly speaking, it can be observed that osteological finds and archaeological documents place emphasis on Upper Mesopotamia, and more specifically the Jazīra, as a geographical area particularly congenial to onager hunting.

As already seen, also deer roamed once freely in the low Mesopotamian lands. Among them there was a peculiar form, the Persian fallow deer, *Dama dama mesopotamica* (Brooke, 1875), exclusive of the eastern Near East (Harrison and Bates 1991; Shalmon 2004; Masseti 2002), where its original range is reported east of the *Nur Dağlari*, the ancient *Amanus*; a mountain range of south-eastern Turkey which divides the coastal region of Cilicia from inland Syria, also apparently marking a biogeographic barrier (Masseti and Vernesi 2015) (Fig. 2.13). The other variety of fallow deer, the common fallow deer, *D. dama dama* (L., 1758), is instead traditionally regarded as naturally dispersed in the Mediterranean territories west of

Fig. 2.12 Detail of the onager hunt on the western wall of the great hall of Qasr al-Amra, located in the vicinity of the village of Azraq, in the desert of eastern Jordan (photo by Fabio Vianello)

Fig. 2.13 The Persian or Mesopotamian fallow deer, *Dama mesopotamica* (Brooke, 1875), is a subspecies exclusive of the eastern Near East, where its original range is reported east of the *Nur Dağlari*, a mountain range of south-eastern Turkey which divides the coastal region of Cilicia from inland Syria, also apparently marking a biogeographic barrier (photo by Marco Masseti)

the biogeographic barrier represented by the Amanus mountain (Masseti 2002; Masseti and Vernesi 2015). It has been artificially introduced, however, in many eastern areas, since very ancient times. A pair of cranial appendices of the latter subspecies have been discovered during the excavation of the Late Bronze Age site

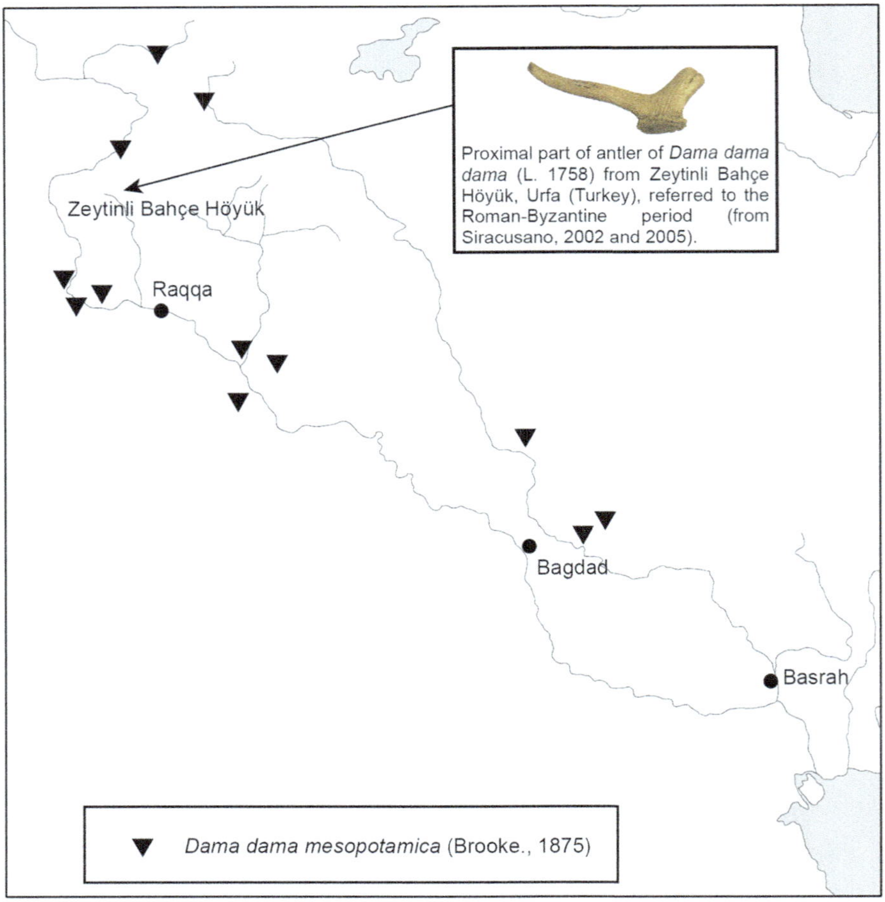

Fig. 2.14 Black triangles indicate the past and present distribution of the Mesopotamian fallow deer based on Masseti and Vernesi (2014&&). The documented ancient introduction of common fallow deer, *D. dama* (L., 1758) in the natural distributional range of *D. dama mespotamica* is also indicated. A proximal portion of an antler of the latter deer from Zeytinli Bahçe Höyük (Urfa/Adiyaman, southern Turkey) was referred to the Roman-Byzantine period (photo by Giovanni Siracusano)

of Tel Efshar, in Israel (Fergusson et al. 1985), while the proximal part of another antler was referred to the Roman-Byzantine period, from Zeytinli Bahçe Höyük, on the Euphrates left bank, south of the Taurus mountain chain, in the region of Şanliurfa/Adiyaman, in southern Turkey (Siracusano 2002, 2005) (Fig. 2.14). The Persian fallow deer has become comparatively rare throughout its range. Today, according to Werner et al. (2015), the total wild population does not exceed 250 adults, including wild living reintroduced individuals, such as those of the Judean Hills, in Israel. It is not known whether any individuals are remaining in the indigenous wild populations in the Iranian wildlife refuges of Dez and Karkeh.

2.4 The Beavers in the Khabur-Euphrates Basin

Up to 4000–5000 years ago, Eurasian beavers lived in the rivers from the mountain ranges of Eastern Anatolia reached the alluvial plains of Mesopotamia (Siracusano in press), allowing the diffusion of one of the southernmost population of the species in the whole Palaearctic biogeography up to very recent historical time.

The species was officially discovered by the British Expedition in Euphrates and Khabur of 1835. Ainsworth (1838), who was the surgeon of the expedition, reported the occurrence of these aquatic rodents where the route touched the already mentioned two rivers in modern Syria, not far from the Iraqi border. The occurrence of osteological remains of beavers in southern Anatolia and Upper Mesopotamia is documented by archeological evidence since, at least, the Paleolithic deposits of Shanidar Cave, in the Zagros Mountains (Iraq), that have been dated by radioactive carbon as 12,000 + 400 years old (Solecki 1957; Hatt 1959). Several authors, such as Patterson (1937) and Buitenhuis (1979), have reported bone fragments of this rodent from protohistoric and historical sites of this broad portion of the Near East, where the former distribution of the species was reviewed by Legge and Rowley-Conwy (1986). More recently, new data have further enriched our knowledge on the spread of beavers in Mesopotamia (Buitenhuis 1988, 1999; Becker 2005; Siracusano 2010), revealing the main concentration of their findings in the Near Eastern steppe between Turkey, Syria, and western Iran (Fig. 2.15). To these must also be added the most recent report of two osteological fragments of *C. fiber* (1 humerus and 1 mandible) which have been provided by the archaeological exploration of the Turkish site of Hirbemerdon Tepe (near Bismil) on the banks of the Tigris, and referred to the ancient Bronze Age (end of fourth—beginning of third millennium B.C.) (Remi Berton 2017, pers. com.) (Fig. 2.16). Thus, it is possible to track the past distribution of beaver in Mesopotamia from late Pleistocene onward (Table 2.1). The species may have continued to exist until the late nineteenth century in Anatolia and Syria if the reports of Danford and Alston (1880) are to be credited. It may have disappeared from Iraq much earlier as the result of deforestation and the general pressure of human population (Hatt 1959). Among the European travellers who occasionally reported the occurrence of beavers living in the Tigris–Euphrates basin during the nineteenth century, Legge and Rowley-Conwy (1986) recall Chesney and Ainsworth (1837), Byerly and Timbs (1838), Layard (1853), Hanney (1975), Helfer (1878), and Ainsworth (1888). According in particular to the last author, the rodents were found at Karkisha, near the confluence of the Khabur and the Euphrates: "*The Arabs brought us* [...] *the skin of a beaver for sale. They said that this animal, so valuable for its fur, was met with on the Khabur.*" Earlier literary sources have also been taken to suggest that these rodents lived in the Tigris–Euphrates basin into the historic period by Campbell Thompson (1926) and Brentjes (1964); a zoomorphic depiction on an orthostat from the site of Tell Halaf dated between 1000 and 800 B.C. gives a good representation of a beaver (Fig. 2.17) (von Oppenheim and Moortgat 1955; Brentjes 1964; see also Becker 2005). For the record, however, in his extensive survey of Near Eastern mammals, Harrison (1972) regards the majority

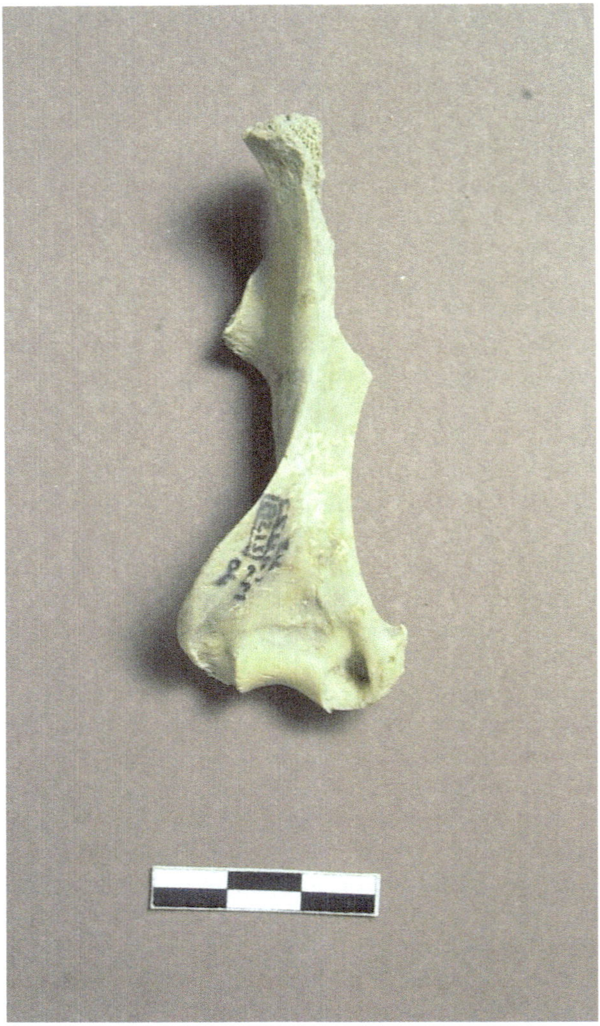

Fig. 2.15 Humerus of Eurasian beaver from the Ubaid-Chalcolithic levels of the site of Değirmentepe, Malatya (south-eastern Turkey) (photo by Giovanni Siracusano)

of these reports as "*far from satisfactory*," suggesting that otters were misidentified as beavers. Also according to Hatt (1959), several of the Mesopotamian beavers may possibly have been otters. In the last century, however, claims for the survival of the rodents were already restricted to the drainage of the river Ceyhan in southern Turkey (Kumerloeve 1967).

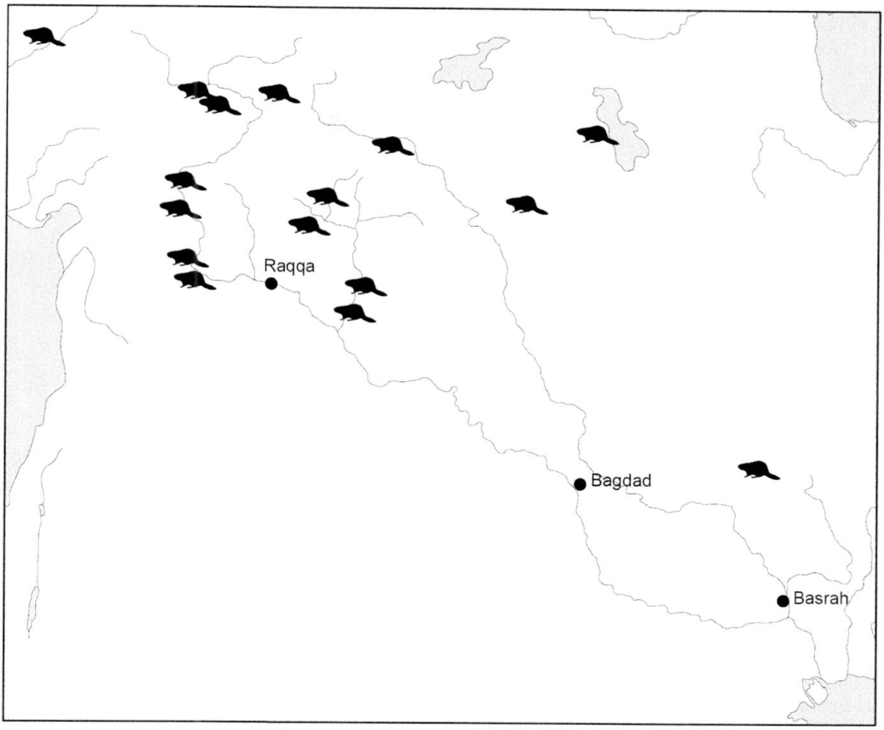

Fig. 2.16 Former occurrence of the Eurasian beaver, *Castor fiber* L., 1758, in Mesopotamia (data from Patterson 1937; Solecki 1957; Hatt 1959; Buitenhuis 1979; Legge and Rowley-Conwy 1986; Buitenhuis 1988, 1999; Becker 2005; Siracusano 2010, and Remi Berton, *in verbis*)

2.5 Extant Fauna. Mammals

Today the fauna of the Jazīra, even though it is very impoverished compared to the past both in the taxa and in the number of individuals, is still made up of numerous interesting species.

More than twenty species of Chiroptera have been recorded to date from Upper Mesopotamia (see Table 2.2). Among them, the occurrence of *Taphozous nudiventris magnus* Wettstein 1913, on the Syrian Euphrates is part of a northwestern dispersion of the Iraqi population, extending into southern Turkey (Shehab et al. 2004). The social vole, *Microtus socialis* Pallas, 1773) and the problematic *Microtus philistinus* Thomas, 1917—also considered in synonymy with *M. guentheri* (Danford and Alston 1880)—the Persian squirrel, *Sciurus anomalus* Gmelin, 1778, and the Indian crested porcupine, *Hystrix indica* Kerr, 1792, figure among the numerous rodents reported from the region (Hatt 1959; Shehab et al. 2004; Masseti 2016). The reporting, in particular, of a population of Persian squirrels in the surroundings of Deir ez-Zor raises interesting questions about the distribution of this species in eastern Syria, mainly associated with mixed and deciduous forests,

Table 2.1 Archaeological sites that provided osteological remains of Eurasian beaver, *Castor fiber* L., 1758, in Upper Mesopotamia and surrounding areas

Site	Chronology	Country	References
Cave Bisitun and Tepe Sarab	>35,000 years bp	Western Iran	Coon (1951)
Cave of Tamtama	Stone implements with no typical forms	Western Iran (Lake Urmia)	Coon (1951)
Shanidar Cave	12,000 ± 400 years bp (Palaeolithic deposits) and 10,600 ± 300 bp	North-eastern Iraq	Solecki (1957); Hatt (1959); Braidwood and Howe (1960); Perkins (1964)
Mureybet	Natufian (11000–10500 B. C.), Khiamian, (10500–9800 B.C.), and Pre Pottery Neolithic A (9800–8800 B.C.)	Syria, west bank of the river Euphrates	Cauvin (1977); Gourichon and Helmer (2004)
Jerf el Ahmar	Pre Pottery Neolithic A (PPNA)	Syria (Halula)	Stordeur (2000); Stordeur and Abbès (2002); Gourichon and Helmer (2004)
Belt cave	Uppermost Neolithic level	Northern Iran, southern shore of the Caspian Sea	Coon (1951)
Tell Abu Hureyra	7500–6000 B.C. (early and late Aceramic Neolithic)	Central Syria, Upper Euphrates	Moore (1975)
Değirmentepe	Ubaid-Chalcolithic levels	Southern Turkey (Malatya)	Siracusano (in press)
Tepe Sarab	6900 years B.C.	Western Iran	Protsch and Berger (1973)
Hirbemerdon Tepe	Bronze Age (end 4th-beginning 3rd millennium B.C.)	Turkey, near Bismil on the Tigris	Unpublished data (Remi Berton, pers. comm.)
Tell Bderi	3rd millennium B.C.	North-eastern Syria, eastern bank of the Khabur river	Becker (2005)
Tell Beydar	First half of the 3rd millennium B.C.	Southern Turkey, western bank of the river Khabur	Siracusano (2010)
Zeytinli Bahçe Hoyūk	2350–2200 years B.C.	Southern Turkey, eastern bank of the Euphrates	Siracusano (2010)
Arslantepe	Late Early Bronze Age	Southern Turkey (Malatya)	Bökönyi (1993)
Tell Halaf[a]	1000–800 B.C.	Northern Syria, province of Hassakè, opposite Ceylanpinar (Şanliurfa, Turkey)	von Oppenheim (1931); Legge and Rowley-Conwy (1986)
Tell Hadidi, west bank of the Euphrates	1800–1600 years B.C. (Middle Bronze Age 2)	Central Syria	Clason and Buitenhuis (1978); Buitenhuis (1979, 1988, 1999)

(continued)

Table 2.1 (continued)

Site	Chronology	Country	References
Alisar Hūyūk	1500–1200 years B.C. (Hittite period)	Central Turkey, along the Kızılırmak river	Patterson (1937)
Korucutepe I	Neo-Hittite levels,	Southern Turkey, Upper Euphrates	Boessnek and von der Driesch (1975)
Norsuntepe	Neo-Hittite levels	Southern Turkey, upper Euphrates	Boessnek and von der Driesch (1975)
Tell Sheikh Hamad	Middle and Late Assyrian period (13th century B.C.) and suburban areas (9th–6th century B.C.)	North-eastern Syria, eastern bank of the Khabur river	Becker (2005)

[a]Evidence for the historic presence of beavers comes from the site of Tell Halaf (southern Turkey), where this rodent appear carved on a stone stelae dated between 1000 and 800 B.C. (von Oppenheim and Moortgat 1955) gives a good representation of a beaver (Fig. 2.16, after von Oppenheim and Moortgat 1955)

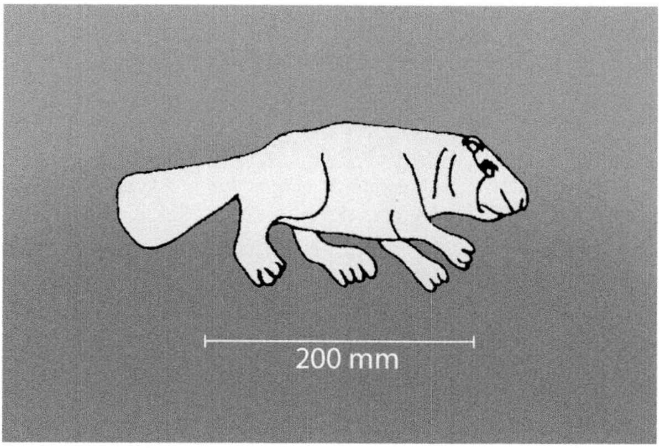

Fig. 2.17 Zoomorphic depiction of a beaver on an orthostat from the site of Tell Halaf, dated between 1000 and 800 B.C. (after von Oppenheim and Moortgat 1955)

avoiding strictly coniferous forests, but evidently much more rarely dispersed in steppe environments, such as those of eastern Syria. Recent records of the gray hamster, *Cricetulus migratorius* (Pallas, 1773), and the Balkan short-tailed mouse, *Mus macedonicus* Petrov & Ruzic, 1983, spread the range of both species eastwards, whereas the Asian garden dormouse, *Eliomys melanurus* (Wagner, 1839), appears to noteworthy extend its range further to the north (Shehab et al. 2004). Moreover, mammalian species unique to the wetlands of Lower Mesopotamia include the Bunn's short-tailed bandicoot rat, *Nesokia bunnii* (Khajuria, 1981), and the Mesopotamian gerbil, *Gerbillus mesopotamiae* Harrison, 1956 (cf. Stuart 2008). Together with the weasel, *Mustela nivalis* L., 1766, the marbled polecat, *Vormela peregusna* (Gueldenstaedt, 1770) (Fig. 2.18), is one of the most widespread small-sized carnivores. Thirteen species of large and medium-sized non-volant mammals are still

Table 2.2 Bats currently recorded from Upper Mesopotamia

English name	Scientific name	References
1. Geoffroy's trident leaf-nosed bat	*Asellia tridens* (É. Geoffroy, 1813)	Atallah and Harrison (1967); Nader and Kock (1983); Shehab et al. (2007); Omer et al. (2012)
2. Botta's serotine	*Eptesicus bottae* (Peters, 1869)	Shehab et al. (2004, 2007); Omer et al. (2012)
3. Anatolian serotine	*Eptesicus anatolicus* (Felten, 1971)	Omer et al. (2012)
4. Serotine	*Eptesicus serotinus* (Schreber, 1774)	Shehab et al. (2007)
5. Sind serotine	*Rhyneptesicus nasutus* (Dobson, 1877)	Omer et al. (2012)
6. Schreiber's bent-winged bat	*Miniopterus schreibersii* (Kuhl, 1817)	Wettstein (1913); Shehab et al. (2007)
7. Lesser mouse-eared myotis	*Myotis blythii* (Tomes, 1857)	Harrison and Lewis (1961)
8. Long-fingered bat	*Myotis capaccinii* (Bonaparte, 1837)	Shehab et al. (2004, 2007)
9. Geoffroy's bat	*Myotis emarginatus* (É. Geoffroy, 1806)	Niazi (1976); Benda (1996)
10. Greater Mouse-eared bat	*Myotis myotis* (Borkhausen, 1797)	Harrison and Lewis (1961); Nadachowski et al. (1990)
11. Hemprich's desert bat	*Otonycteris hemprichii* (Peters, 1859)	Atallah (1977); Shehab et al. (2004, 2007); Omer et al. (2012)
12. Kuhl's pipistrelle	*Pipistrellus kuhlii* (Kuhl, 1817)	Harrison and Bates (1991); Shehab et al. (2004, 2007); Omer et al. (2012)
13. Common pipistrelle	*Pipistrellus pipistrellus* (Schreber, 1774)	Benda et al. (2003)
14. Rüppell's pipistrelle	*Vansonia rueppellii* (Fischer, 1829)	Omer et al. (2012)
15. Gray Big-eared Bat	*Plecotus austriacus* (Fischer, 1829)	Trouessart and Kollman (1923)
16. Caucasian long-eared bat	*Plecotus macrobullaris* (Kuzjakin, 1965)	Shehab et al. (2007)
17. Lesser horse-shoe bat	*Rhinolophus ferrumequinum* (Schreber, 1774)	Shehab et al. (2007); Omer et al. (2012); Shehab and Mamkhair (2006)
18. Blasius's horseshoe bat	*Rhinolophus blasii* (Peters, 1867)	Wettstein (1913)
19. Mediterranean horseshoe bat	*Rhinolophus euryale* (Blasius, 1853)	Wettstein (1913); Trouessart and Kollman (1923); Omer et al. (2012); Shehab and Mamkhair (2006)
20. Lesser horse-shoe bat	*Rhinolophus hipposideros* (Bechstein, 1800)	Shehab et al. (2007); Omer et al. (2012)
21. Mehely's horseshoe bat	*Rhinolophus mehelyi* (Matschie, 1901)	Shehab et al. (2007)

(continued)

Table 2.2 (continued)

English name	Scientific name	References
22. European free-tailed bat	*Tadarida teniotis* (Rafinesque, 1814)	Shehab et al. (2007)
23. Egyptian fruit bat	*Rousettus aegyptiacus* (E. Geoffroy, 1810)	Shehab and Mamkhair (2004)
24. Naked-bellied tomb bat	*Taphozous nudiventris* (Cretzschmar, 1830)	Dobson (1878); Thomas (1915); Harrison (1964); Harrison and Bates (1991); Masseti (2001); Shehab et al. (2004, 2006, 2007); Omer et al. (2012); Masseti (2016)

Fig. 2.18 The marbled polecat, *Vormela peregusna* (Gueldenstaedt, 1770), is one of the most widespread small-sized carnivores of Upper Mesopotamia (photo by Gianluca Serra)

recorded from the Jazīra (Masseti 2001, 2004; Omer et al. 2012; Masseti 2016). However, there are some surprising absentees, such as the ratel or honey badger, *Mellivora capensis* (Schreber, 1776), which has been never reported from the Syrian Upper Mesopotamia but figures as widespread in the desert and arid steppes of central and southern Iraq, not however in the region of the Jazīra (Hatt 1959; Masseti 2009a; Omer et al. 2012). The last Syrian leopard, *Panthera pardus* (L., 1758), is reported to have been killed in 1963, west of the Euphrates valley, in the vicinity of the village of Bab Jannè (= "*the gate of paradise*"), Slonfeh, on the Alawit Mountains, about 20 km from the Turkish border (Masseti 2000), whereas it has been recently recorded in many localities of northern Iraq (Omer et al. 2012) (Fig. 2.19). Carnivores such as the wild cat, *Felis silvestris* (Schreber, 1775), the Asian jackal, *Canis aureus* L., 1758 (Fig. 2.20), the wolf, *Canis lupus*, *Canis lupus*

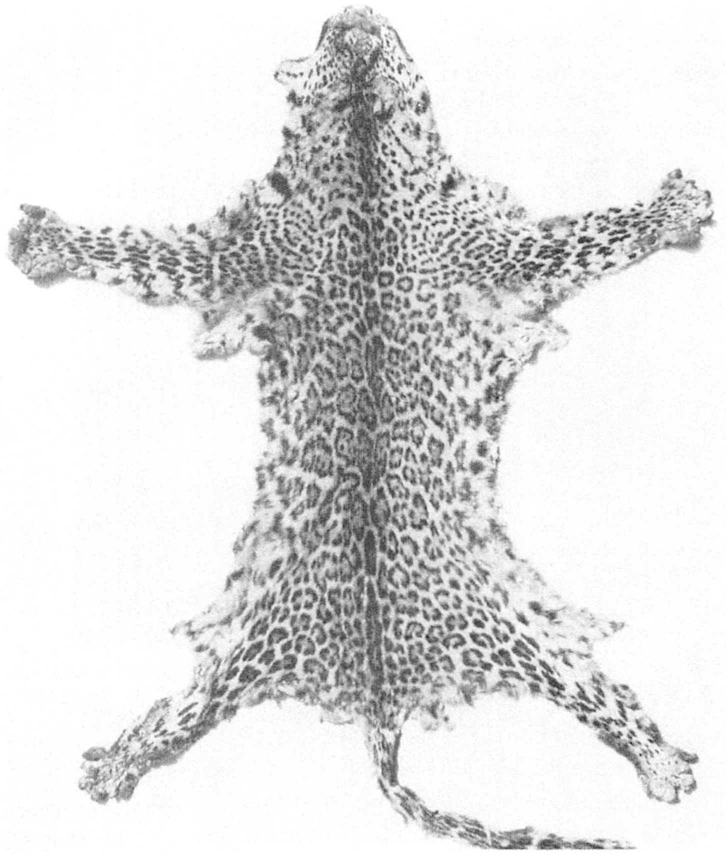

Fig. 2.19 The skin of one of the last Syrian leopard, *Panthera pardus* (L., 1758), an adult male killed at Nabi Yunes, Slonfeh, north-western Syria, in 1959, in the collection of the Museum Alexander König of Bonn (ZFMK n. 64.1171, see von Lehmann 1965)

L., 1758, the common fox, *Vulpes vulpes* (L., 1758), the Rűppel sand fox, *Vulpes rueppellii* (Schinz, 1825), and the striped hyena, *Hyaena hyaena* (l., 1758), still figure among the mammalian species most skilled at avoiding contacts with human beings, although they inhabit areas even densely settled by humans and sometimes behave as a commensal of man (Masseti 2009a, 2016). To all these must be added a felid particularly suitable for life in wetlands, the jungle cat, *Felis chaus* (Gueldenstaedt, 1776), whose world diffusion ranges from the Nile valley to the Indian subcontinent passing through Mesopotamia (Table 2.3).

There are other mammals the occurrence of which is still documented in the territories of Upper Mesopotamia, where their presence was much more widespread in former times. The already mentioned bas-reliefs of Ashurbanipal palace (seventh century B.C.), at Niniveh, undoubtedly describe, for example, a species of gazelle still present in Upper Mesopotamia, where its occurrence has also been confirmed by

Fig. 2.20 The Asian jackal, *Canis aureus* L., 1758, is among the mammalian species most skilled at avoiding contact with human beings, although it inhabits areas even densely settled by humans and behaves as a commensal of man (photo by Gianluca Serra)

archeological evidence (Masseti (2003) (Fig. 2.21). The ungulates represented refer to the already mentioned goitered gazelle, because of the accurate description of the morphology of the *taxon* (Masseti 2003). This is a rather heavily built gazelle, with indistinct flank and face stripes, and with the nose and sometimes the whole face largely white. Males have a goitered-like swelling in the mid-line of the throat, at least during the breeding season (Harrison and Bates 1991). The male horns are long and lyrate but in females they may be absent, especially in the variety *G. s. subgutturosa* Gueldenstaedt, 1780, distributed from former Soviet Union to the Levant (Kingswood and Kumamoto 1988) (Fig. 2.22). In southeast Asia, these subspecies are recorded from southwestern Anatolia, Syria, and northern Iraq (Harrison and Bates 1991; Masseti 2004). Among the gazelles that still inhabit the Near Eastern region, *G. s. subgutturosa* is the only variety showing hornless females, the same subspecies that is portrayed on the stone reliefs at Niniveh (Masseti 2003). In Upper Mesopotamia the southern range of the nominate form *G. s. subgutturosa* seems to overlap with the northern range of *G. s. marica* (Thomas, 1897), also known as Arabian sand gazelle (cf. Kingswood and Kumamoto 1988). North of the region of Deir ez Zor, in the very Upper Mesopotamia, gazelles of the subspecies *G. s. subgutturosa* are recorded up to Jabal al Bilas (north-eastern Syria) and Ceylanpınar (Şanlıurfa, south-eastern Turkey) (Kasparek 1986b; Masseti 2004), while the form *marica* has been reported already starting from the surroundings of Deir ez Zor (Masseti 2004). Although the distribution of gazelles in Iraq has been reported by Al-Sheikhly (2012), Omer et al. (2012) are of the opinion that further

Table 2.3 Birds recorded from the Jazīra in the present research (after Masseti 2016)

Species
1. Black-necked grebe, *Podiceps nigricollis* (Brehm, 1831)
2. White pelican, *Pelecanus onocrotalus* (L., 1758)
3. Pygmy cormorant, *Phalacrocorax pygmeus* (Pallas, 1773)
4. Bittern, *Botaurus stellaris* (L., 1758)
5. Little bittern, *Ixobrychus minutus* (L., 1766)
6. Squacco heron, *Ardeola ralloides* (Scopoli, 1769)
7. Night heron, *Nycticorax nycticorax* (L., 1758)
8. Cattle egret, *Bubulcus ibis* (L., 1758)
9. Great white egret, *Egretta alba* (L., 1758)
10. Little egret, *Egretta garzetta* (L., 1766)
11. Grey heron, *Ardea cinerea* (L., 1758)
12. Purple heron, *Ardea purpurea* (L., 1766)
13. White stork, *Ciconia ciconia* (L., 1758)
14. Black stork, *Ciconia nigra* (L., 1758)
15. Greater flamingo, *Phoenicopterus ruber* (L., 1758)
16. Shelduck, *Tadorna tadorna* (L., 1758)
17. Mallard, *Anas platyrhynchos* (L., 1758)
18. Teal, *Anas crecca* (L., 1758)
19. Garganey, *Anas querquedula* (L., 1758)
20. Pintail, *Anas acuta* (L., 1758)
21. Shoveler, *Anas clypeata* (L., 1758)
22. Marbled teal, *Marmaronetta angustirostris* (Ménétries, 1832)
23. Tufted duck, *Aythya fuligula* (L., 1758)
24. Pochard, *Aythya ferina* (L., 1758)
25. Ferruginous duck, *Aythys nyroca* (Gueldenstaedt, 1770)
26. Short-toed eagle, *Circaetus gallicus* (Gmelin, 1788)
27. Griffon vulture, *Gyps fulvus* (Hablizl, 1783)
28. Hen harrier, *Circus cyaneus* (L., 1766)
29. Pallid harrier, *Circus macrourus* (Gmelin, 1771)
30. Montagu's harrier, *Circus pygargus* (L., 1758)
31. Levant sparrowhawk, *Accipiter brevipes* (Severtzov, 1850)
32. Buzzard, *Buteo buteo* (L., 1758)
33. Long-legged buzzard, *Buteo rufinus* (Cretzschmar, 1827)
34. Honey buzzard, *Pernis apivorus* (L., 1758)
35. Booted eagle, *Hieraetus pennatus* (Gmelin, 1788)
36. Golden eagle, *Aquila chrysaetos* (L., 1758)
37. Pheasant, *Phasianus colchicus* (L., 1758)
38. Chukar, *Alectoris chukar* (J.E.Gray, 1830) [1]
39. Black francolin, *Francolinus francolinus* (L., 1766) [2]
40. Quail, *Coturnix coturnix* (L., 1758)
41. Water rail, *Rallus aquaticus* (L., 1758)
42. Moorhen, *Gallinula chloropus* (L., 1758)

(continued)

Table 2.3 (continued)

Species
43. Coot, *Fulica atra* (L., 1758)
44. Common crane, *Grus grus* (L., 1758) [3]
45. Houbara, *Chlamydotis undulata* (Jacquin, 1784)
46. Black-winged stilt, *Haematopus haematopus* (L., 1758)
47. Stone curlew, *Burhinus oedicnemus* (L., 1758)
48. Cream-coloured courser, *Cursorius cursor* (Latham. 1787)
49. Lapwing, *Vanellus vanellus* (L., 1758)
50. Spur-winged plover, *Hoplopterus spinosus* (L., 1758)
51. Curlew, *Numenius arquata* (L., 1758)
52. Common sandpiper, *Actitis hypoleucos* (L., 1758)
53. Black-headed gull, *Larus ridibundus* (L., 1766)
54. Pin-tailed sandgrouse, *Pterocles alchata* (L., 1766)
55. Black-bellied sandgrouse, *Pterocles orientalis* (L., 1758)
56. Turtle dove, *Streptopelia turtur* (L., 1758)
57. Cuckoo, *Cuculus canorus* (L., 1758)
58. Eagle owl, *Bubo bubo* (L., 1758)
59. Short-eared owl, *Asio flammeus* (Pontoppidan, 1763)
60. Barn owl, *Tyto alba* (Scopoli, 1769)
61. Hoopoe, *Upupa epops* (L., 1758)
62. Roller, *Coracias garrulus* (L., 1758)
63. Golden oriole, *Oriolus oriolus* (L., 1758)
64. Magpie, *Pica pica* (L., 1758)
65. Carrion crow, *Corvus corone cornix* (L., 1758)
66. Jackdaw, *Corvus monedula* (L., 1758)
67. Starling, *Sturnus vulgaris* (L., 1758)
68. Rose-colored starling, *Sturnus roseus* (L., 1758)

investigation is required to update the distribution of both the Arabian sand gazelle and Persian gazelle.

2.6 Birds and Reptiles

The present richness of birds of the *Jazīra* is evidently connected with that of the rest of Mesopotamia, and in particular, the Shatt al-Arab delta and marshes, in southern Iraq, which comprise one of the most important areas for waterfowl in the Near East, both in terms of number of birds and in terms of taxonomic diversity. Here, about 300 species have been recorded; and nearly half of them are wetland birds (Evans 1994). For the record, these marshes support almost the entire world population of the endemic Basra reed warbler, *Acrocephalus griseldis* (Hartlaub, 1891) (BirdLife International 2017), while the Iraq babbler, *Turdoides altirostris* (Hartert, 1909), is

Fig. 2.21 A herd of Persian gazelles, *Gazella subgutturosa subgutturosa* Gueldenstaedt, 1780, is represented on the bas-reliefs of Ashurbanipal palace (seventh century B.C.), at Niniveh (photo by the British Museum, courtesy of the Trustees of the British Museum, London)

Fig. 2.22 Unlike the Arabian sand gazelle, *G. subgutturosa marica* Thomas, 1897, the adult males of the Persian gazelle are characterized by long and lyrate horns, whereas in the females they may be absent (photo by Marco Masseti)

Fig. 2.23 The south Iraqi marshes were regarded as one of the only two known breeding sites in the Near East for sacred ibis, *Threskiornis aethiopicus* (Latham, 1790) (photo by Luigino Felcher)

also found in a few other areas of Iraq and south-western Iran (BirdLife International 2012). The Southern Iraq marshes are the only place where the African darter, *Aninga rufa* (Daudin, 1802), breeds in the Near East (Allouse 1953; Salim et al. 2009), while the Dead Sea sparrow, *Passer moabiticus* Tristram, 1864, is still regarded as a common species. Up to their drainage, which occurred between the 1950s and 1990s, these marshes were also regarded as one of the only two known breeding sites in the Near East for sacred ibis, *Threskiornis aethiopicus* (Latham, 1790) (Salim et al. 2009) (Fig. 2.23).

In the course of a research carried out several years ago in the Syrian Jazīra, about 70 species of birds have been reported, not restricted to wetland species but also including birds of prey, Passeriformes, and others (Masseti 2016). The given list is indicative of the birds of the area but certain occurrences need a more detailed comment (Table 2.4). The fact that, for example, wild geese are absent from this bird list does not mean that these birds do not belong to the avifauna recorded from the Jazīra. Wild geese of the genus *Anser*, for example, have been reported among the faunal remains since the end of the fourth millennium (Early Bronze Age) from several archaeological sites of the region, such as Zeytinli Bahçe (Urfa, south-eastern Turkey) (Siracusano 2005). It is well known that the wintering species of the area include many waterfowls, such as the ruddy shelduck, *Tadorna ferruginea* (Pallas, 1764), and the white-fronted goose, *Anser albifrons* (Scopoli, 1789), (Baumgart and Burkhard 1986; Evans 1994). Although there are not many records of the distribution of the marbled teal, *Marmaronetta angustirostris* Ménétries, 1832, in the Euphrates valley where it is comprised within the globally threatened species, it is locally regarded as a widespread breeding bird (Evans 1994; cf. Salim et al. 2009). The occurrence of common pheasants, *Phasianus colchicus* L., 1758, in Upper

Table 2.4 Archaeological evidence of elephants from Upper Mesopotamia between the first half of the second millennium and the eigth/seventh century B.C.

Site and locality	Chronology	Country	References
Arslantepe	Late Bronze Age	Turkey (Malatya)	Bökönyi (1985)
Değirmentepe	Iron Age	South-eastern Turkey (Malatya)	Siracusano (in press)
el Quitar			Buitenhuis (1988); Lister et al. (2013)
Maraş		Southern Turkey (Upper Euphrates)	Çakırlar and Ikram (2016)
Emar	Late Bronze Age	Syria	Gündem (2010); Gündem and Uerpmann (2003); Lister et al. (2013)
Zincirli		Southern Turkey	Lister et al. (2013)
Chagar Bazar	Late Bronze Age	North-eastern Syria (Al Hasaka)	Barnett (1982)
Tell Archana	Late Bronze—Iron Age	Syria	Çakırlar and Ikram (2016)
Tell Munbaqa	2nd millennium B.C.	Iraq (middle Euphrates	Boessneck et al. (1986); Fischer (2007)
Tell Sabi Abyad			
Tell Sheikh Hamad	7th century B.C.	North-eastern Syria	Becker (2005)
Tell Bderi	Early Bronze Age	North-eastern Syria	Becker (2005)
Nimrud			
Nuzi			
Babylon	*c.* 1800 B.C.	Iraq	Reuther (1926)
Haft Tepe	Middle Bronze Age	South-eastern Iran	Negahban (1979)

Mesopotamia is probably the result of a recent introduction (Masseti 2016). The species is considered as an extraneous element to the biogeography of the region, its most western homeland being comprised between the eastern shore of the Black Sea and the Caspian Sea (Zeuner 1963; Cramp 1980; Hill and Robertson 1988). The taxon was named after the ancient *Phasis* (present day Rion or Rioni), the main river of western Georgia which originates in the Caucasus mountains and flows west to the Black Sea. According to local people, also the black francolin, *Francolinus francolinus* (L., 1766), is not a species characteristic of the Jazīra, but was apparently introduced by the French possibly during their protectorate of Syria, between 1922 and 1941. It should not be forgotten, however, that this bird is regarded as autochthonous in remote coastal forests of north-eastern Turkey (Kasparek 1988). Not uncommon in the Jazīra, and in other areas of Syria and Iraq, is also the rose-colored starling, *Pastor roseus* L., 1758, a migrant characteristic of the steppe and open agricultural land from easternmost Europe across temperate southern Asia, (Masseti 2016).

It may be important to remember that along the Euphrates, near to the village of Birecik, in south-eastern Turkey at the border with Syria, there is one of the last nidification quarter of the Northern bald ibis, *Geronticus eremita* (L. 1758). Upper Mesopotamia is regarded, in fact, as the homeland of this species. Also the town of Raqqa, for example, was known to have hosted on its roof and in its walls an important colony of this bird still in the course of the second half of the nineteenth century (Mlíkovský 2012). Bals ibises were considered to have become extinct at the end of the 1980s, when the last survivors of the southeastern Anatolia colony—that of Birecik—where prevented from migrating and where held in semi-captivity (Masseti 1987; Kasparek 1992; Serra et al. 2008). A few years ago a relic breeding population of this bird was also discovered in the surroundings of Palmyra, in the Syrian desert (Serra et al. 2003b). The political unrest, which recently affected this territory, makes it fear even for the survival of the latter small colony.

Many species of reptiles and amphibians are also recorded from Upper Mesopotamia. Among them, the desert monitor, *Varanus griseus* (Daudin, 1802), was once regarded as one of the commonest lizards of the Near-Eastern steppes and deserts. However, it has been recorded from only a few localities of south-western Anatolia, in the last decades (Sindaco et al. 2000; Masseti 2016).

The Jazīra snakes include the Montpellier snake, *Malpolon monspessolanus* (Hermann, 1804), and at least two specific representatives of the genus *Natrix*: the grass snake, *Natrix natrix* (L., 1758), and the dice snake, *N. tessellata* (Laurenti, 1768) (Masseti 2016). Although the Montpellier snake belongs to the most African group of Psammophini, it is regarded as a circum-Mediterranean zoological element (De Haan 1997). In Asia Minor and Upper Mesopotamia, it presents a disjointed distribution, being dispersed in western, southern, and eastern Anatolia (see Sindaco et al. 2000). Among the most dangerous snakes in the region the desert black snake or black cobra, *Walterinnesia aegyptia* Lataste, 1887, can be mentioned; a highly venomous, medium-sized snake, which can grow to lengths of 1.3 m, completely black in color. The herpetofauna of Upper Mesopotamia also includes the beautiful Levantine viper or blunt-nosed viper, *Macrovipera lebetina* (L., 1758), a snake dispersed from the Near Eastern and western Asia mainland to northern India, parts of North Africa, and the Mediterranean island of Cyprus (Fig. 2.24).

2.7 Inside the Rivers and Along Their Shores

Mesopotamia also displays its own endemic zoological elements which, of course, cannot be found outside its boundaries, such as the already mentioned Bunn's short-tailed bandicoot rat, and the Mesopotamian gerbil (Petter 1961; Lay and Nadler 1975; Stuart 2008; Kock and Amori 2016).

Taxa indigenous to the Upper Mesopotamia water bodies include several sub-species of fish, most of which occur in the genera *Aphanius, Glyptothorax, Cobitis,*

Fig. 2.24 The Levantine viper or blunt-nosed viper, *Macrovipera lebetina* (L., 1758), is a snake dispersed from the near eastern and western Asia mainland to northern India, parts of North Africa, and the Mediterranean island of Cyprus ((courtesy Philipp Wagner, Forschung, and Artenschutz, Allwetterzoo Münster)

Orthrias, and *Schistura*. *Iranocypris* is a monotypic genus endemic to this region (Coad and Hales (2013). The autochthonous inland ichthyofauna is threatened by an increasing number of exotic fish receiving from different geographic realms including the Neotropical and Nearctic regions. Control of malaria and ornamental purposes, are the main reasons for these introductions. Recently, also a few individuals of the fish alligator gar, *Atractosteus spatula* (Lacepède, 1803), native to North America, were caught by local fishermen in the waters of Basrah (Mutlak et al. 2017), and in the lake of Marivan (Zarivar), a Tigris river tributary of Iran (Esmaeili et al. 2017). Although this is a Nearctic species, few notable sightings of it have been reported outside North America, including Turkmenistan (Salnikov 2010), Hong Kong, Singapore, and India. Iran is a new locality for this fish.

Perhaps, from time immemorial, even marine fish go up the Tigris and Euphrates. Written Arabic accounts from as early as 1263 A.D. hint at the presence of large aquatic predators in the Mesopotamian rivers, but it was not until very recent times that the Western World really took notice of such an occurrence (cf. Hunt 1951). Bull sharks, *Carcharhinus leucas* (Müller and Henle, 1839), for example, are known to frequent the Euphrates and Tigris as far inland as Baghdad (Coadt and Al-Hassan 1989). There have been even cases of the capture of two-meter sharks more than 200 km from the sea (Fig. 2.25). This type of fish is still today not uncommonly recorded and, according to Campbell (2007), it "... *is far better known to the Iraqi than is the crocodile*."

To the taxonomic class of reptiles would be referred the sightings of some other intriguing zoological species carried out in the course of the nineteenth century. Among the many legends that surround the sleepy flow of the Mesopotamian rivers,

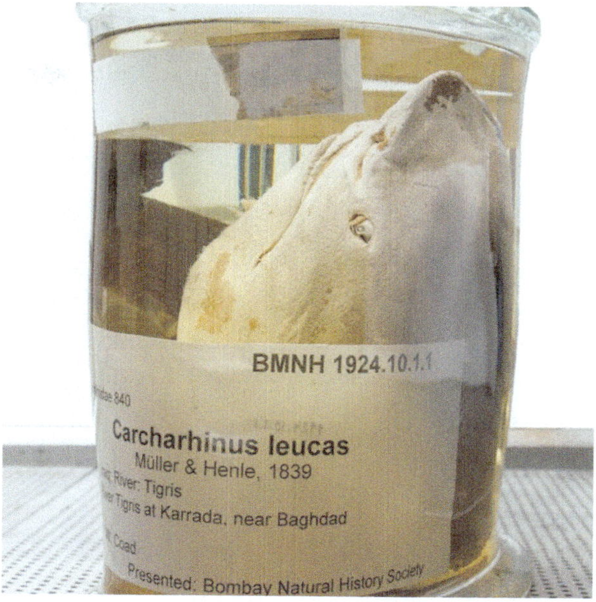

Fig. 2.25 Specimen of bull shark, *Carcharhinus leucas* (Müller and Henle, 1839), captured near Bagdad in 1924, and preserved in the collection of the Natural History Museum of London

there is also one reflecting the former, incredible, occurrence of crocodiles or similar reptiles The existence in the Upper Euphrates of this type of animals was described by Byerly and Timbs (1838) who referred an observation of Colonel Alexander George Chesney. As no specimen was ever captured, it was impossible to say whether the supposed loricate was: "*...a true crocodile, an alligator or a gavial*" (Byerly and Timbs 1838). Since then, however, it seems that the presence in Mesopotamia of crocodiles has never more been reported, as well as the assertion of Chesney never confirmed (Muzio 1925). According to Campbell (2007), the Iraqi word used for crocodile was *timsah*, of which, of course, there were none in Iraq. Therefore, it is not possible to understand if the legends about these reptiles originated from Africa, possibly as one told by indigenous people of the Black Continent themselves. Despite the fact that no representatives of the Crocodylidae taxonomic family have ever been reported from Mesopotamia, we must remember that Nile crocodiles, *Crocodylus niloticus* Laurenti, 1768, were dispersed in geographical areas not very far from here, in the swamps of the western Levant up to historical times (Ross and Magnusson 1989; King 1989; Thorbajarnarson et al. 1992; Kaplan 1993; Levin et al. 2009; de Gelder 2010). These reptiles became extinct in Palestine and western Syria only at the beginning of the twentieth century (Werner 1988; Ross 1989; Delfino et al. 2007).

If, even the former existence of crocodiles in Mesopotamia is strongly doubtful, on the other hand, the occurrence of another peculiar amphibian reptile, the Euphrates softshell turtle, also known as the Mesopotamian softshell turtle, *Rafetus euphraticus* (Daudin, 1801), is certain (Mobaraki and Mola 2011) (Fig. 2.26). This

Fig. 2.26 The Euphrates softshell turtle or Mesopotamian softshell turtle, *Rafetus euphraticus* (Daudin, 1801), is recognized as an autochthonous component of the zoogeography of Mesopotamia and its surroundings (photo by Şemsettin Turğa; courtesy Remi Berthon)

is a close relative of the Afrotropical softshell turtle, *Trionyx triunguis* (Forsskål, 1775), dispersed along the Nile valley up to south-eastern Anatolia, along the Levantine coast and rivers (Corsini-Foka and Masseti 2008; Masseti in press-a). With a distribution which extends from south-eastern Turkey to the north-western extent of the Persian Gulf, encompassing the rivers Euphrates and Tigris and their tributaries and other related water bodies in Syria, Iraq, and southwestern Iran (Khuzestan province) (Taskavak et al. 2016; Masseti 2016), the Euphrates softshell turtle is recognized as an autochthonous component of the zoogeography of Mesopotamia and its surroundings (Taşkavak 1998; Ghaffari et al. 2017; Ihlow et al. 2014). It is one of the least-known species of Trionychidae (Ghaffari et al. 2008), and its feeding habits are still imperfectly known. The species has generally been considered a carnivore, sometimes having been seen feeding on carcasses (Taskavak et al. 2016). The presence of large, flesh-eating turtles in the Tigris and Euphrates is commonly recorded by early travellers since the nineteenth century (Legge and Rowley-Conwy 1986). As far as 1842, for example, Ainsworth recorded several softshells feeding on the carcass of a wild ungulate: "*a number of Euphratic turtles tearing to pieces a stag*." Habitat destruction, pollution, and fisheries interactions (intentional killing) are the main threats to the survival of this species throughout its entire diffusion range (Ghaffari et al. 2008). Some of the most ancient evidences of human exploitation of the Mesopotamian softshell turtle have been provided by the archaeological excavation of the Late Bronze Age levels (1500–1068 B.C.) of the site of Kavuşan Hoyuk, located on the right bank of the river Tigris. Here, the faunal remains of the species cannot only be considered as consumption refuse (Berthon

2013, 2014), but also as elements of post-Assyrian funerary practices (Berthon et al. 2016). The unique burial finds from Kavuşan Hoyuk, where human beings are associated with Mesopotamian turtles, coupled with archaeological and textual records, underline the economic and symbolic significance of these animals for communities in prehistoric and early historical Mesopotamia.

2.8 Vanishing Wild Boars

Once the large carnivores were disappeared from the southernmost river banks of Mesopotamia, the only dangerous beast left in the marshes of the Shatt al-Arab, is the wild boar *Sus scrofa* L., 1758, the sole species still able to synthesize the unconscious archetype of all that is wild, savage, and dangerous. Despite the dramatic and progressive drainage of the southern Iraqi wetlands, together with the high level of hunting, this ungulate still represents one of the most widespread mammals of large dimensions in the southern marshes (Thesiger 1954, 1959; Harrison and Bates 1991; Bedair et al. 2006), where its habits were diffusely described by Thesiger (1954) who noted, among other things, that its hunt was until recently performed with the spear (cf. Meakin 1901; Masseti 2016) (Fig. 2.27).

According to Al-Sheikhly et al. (2015), wild boars were recorded from at least 33 sites throughout Iraq during 2013–2014, being abundant mainly along the Tigris and Euphrates and the southern marshes. More in particular, however, the species has long since disappeared from the middle Euphrates valley, where no report has

Fig. 2.27 Thesiger (1954) noted that, in the south Iraqi marshes, the wild boar hunt was performed with the spear

been registered between the years 1989 and 2004. Here, according to hunters at Deir ez-Zor, wild boars appear to have vanished many years before (Masseti 2004). Up to the end of the 1950s, these ungulates were instead reported to be widespread in most of Upper Mesopotamia, where they were common in every suitable spot (see Ainsworth 1838). According to Hatt (1959), for example, north of Baghdad, wild boars thrived in the maze of trails and rootings on the left bank of the Tigris, mainly in thickets and in fields where cover was adequate. In Syria, their occurrence is still confirmed in the Alawite mountains (east of Latakia), as well as in the hilly areas north of Aleppo (Zahoueh and Cheiko 1993), and are regarded as still rather common on the mountains of Bayr (Kassab), Jabal el Ansaryie, and along the Gharb plain (Masseti 2004).

According to Harrison and Bates (1991), the subspecies occurring in Syria have been referred to *S. s. libycus* (Gray, 1868), whereas Hatt (1959) was of the opinion that the form dispersed in Iraq is the bigger *S. s. attila* Thomas, 1912. Thesiger (1954) observed instead that the wild suids of the southern Mesopotamian marshes were the same as the Indian wild boar.

2.9 Mesopotamian Otters

As far as 1954, Thesiger noted that otters were widely hunted for their skins in the southern Iraqi marshes, and mentioned one person who shot 40 otters in the space of 2 months. No otters were, however, recorded during several surveys carried out between 1968 and 1979, and it seems likely that by that time the populations of these carnivores were becoming much depleted by the hunters (Bedair et al. 2006).

Traditionally, two species of otter are reported from the marshes of southern Mesopotamia: the Eurasian otter, *Lutra lutra* (L., 1758), and the Middle and Far Eastern smooth-coated otter, *Lutrogale perspicillata* (I. Geoffroy Saint-Hilaire, 1826). The geographical distribution, threats, and conservation status of both of these species have been the subjects of recent studies carried out in Iraq (Omer et al. 2012; Al-Sheikhly and Nader 2013). *L. lutra* is widespread in a huge areal which comprises most of the Palaearctic region, spanning from the western Iberian peninsula to Vietnam and Taiwan (Wozencraft 2005). The smooth-coated otter is instead found in Java, Sumatra and Borneo, northward to southwestern China and Vietnam, east through Nepal and Bhutan and India to Pakistan, excluding the Indus Valley (due to barrages and dams) (Wozencraft 2005). The Eurasian otter is still the most common otter recorded from northern, central, and western Iraq (Al-Sheikhly and Nader 2013). Here its habitat is restricted to densely vegetated banks of permanent rivers, stationary rain puddles, mountain streams, and reservoirs of Tigers and Euphrates basin, and marshes (see also Hatt 1959). Data collected between 1989 and 1998 indicate that a considerable population of these otters was still present in the region comprised between the Euphrates, its tributary the Khabur and Iraq (Kock et al. 1994; Masseti 2009a). Two stuffed specimens, captured in the surroundings of Deir-ez-Zor, confirmed, in particular, the occurrence of the species in the Syrian

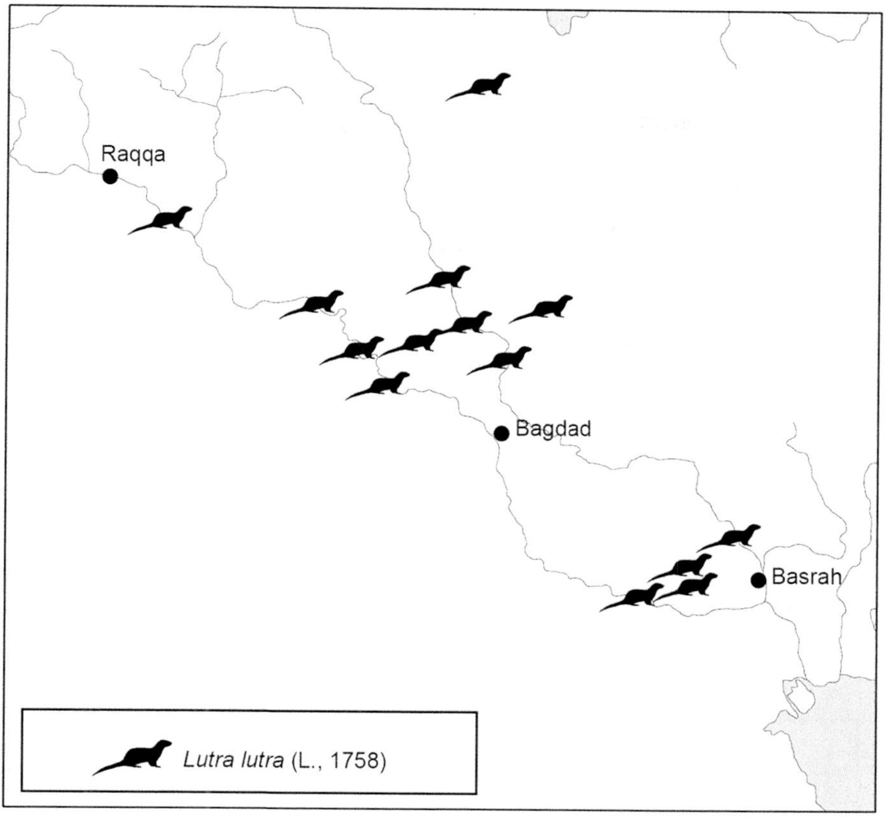

Fig. 2.28 Present distribution of the Eurasian otter, *Lutra lutra* (L., 1758), in Mesopotamia

Jazīra certainly up to the first half of the 1990s (Masseti 2001, 2004, 2009a). According to local people, there were numerous Eurasian otters in the Syrian Euphrates valley between the Iraqi border, Al Mayaddin and Doura Europos, but not in the north-westernmost area of Raqqa. Additional data were also reported from an island in the Euphrates at Doura Europos, from the site of Halabiyyeh and from Tell Sheikh, along the river Khabur, in the district of Deir-ez-Zor (Uhrin et al. 2000; Masseti 2009a) (Fig. 2.28).

As observed by Hatt (1959), the recognition of the smooth-coated otter in Mesopotamia is a matter of great interest. In Iraq, it has been described as the endemic subspecies *Lutrogale perspicillata maxwelli* Hayman, *1956* (Mason and Macdonald 1986), whose distribution was at first regarded as limited to the southern marshes (Harrison 1968; Harrison and Bates 1991; Wozencraft 2005; Karami et al. 2008). The occurrence of the species has been, however, recently documented also in north-eastern Iraq (Kurdistan Region, between Sulaymanya and the extreme North-East beyond the town of Erbil, from where 2 ♂♂ and 1♂ were recorded at a mountain river in TaqTaq (N 35°54′ E (Omer et al. 2012; Al-Sheikhly and Nader

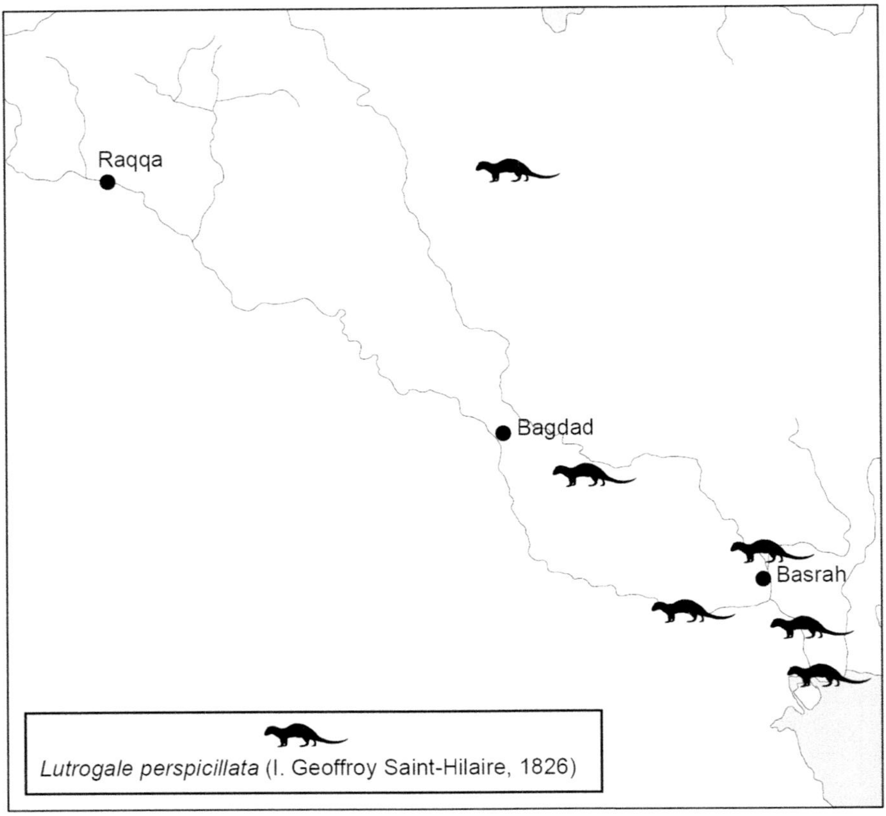

Fig. 2.29 Present distribution of the smooth-coated otter, *Lutrogale perspicillata* (I. Geoffroy Saint-Hilaire, 1826), in Mesopotamia

2013) (Fig. 2.29). In its Iraqi distribution, the smooth-coated otter prefers dense, tall reed beds, marshy lakes, and ponds. It was not possible, however, to obtain any information about its occurrence along the Syrian course of the Euphrates (Masseti 2009a). The current status of the isolated Iraqi population is uncertain. The fact that the carnivore inhabits a geographical portion of the Near East so far located from the remaining world areal of the species possibly indicates the range must once have been wider. However, one wonders if it cannot be the result of an ancient importation from abroad. Since immemorial time, smooth otters are used for commercial fishing in the Indian subcontinent. Today, in southern Bangladesh, these otters are still bred in captivity, trained, and used to chase fish into fishing nets (Feeroz et al. 2011).

2.10 Middle and Far Eastern Mongooses

Another medium-sized carnivore whose primary distribution has been described as limited to the Oriental Region, with the inclusion of eastern Arabia and southern Iran (Corbet 1978; Harrison and Bates 1991; Corbet and Hill 1992), the Indian gray mongoose, *Herpestes edwardsii* (É. Geoffroy Saint-Hilaire, 1818), was recently observed in northern Iraq. This record represents a large extension of the species range in Arabia (Al-Sheikhly and Mallon 2013; Al-Sheikhly et al. 2015). It confirms, after more than 100 years, the observation of Cheesman (1923), according to whom: "*. . . in May 28, 1917, I chased but failed to secure, a large mongoose beyond the oil fields at Maidan-i-Naptun. This might have been an Indian species . . .*". On the native occurrence of the Indian gray mongoose in Arabia there are, however, several serious doubts. There is, for example, evidence—dated since 2000 B.C. or just before—for the anthropochorous occurrence of the latter species and the small Indian mongoose, *Herpestes auropunctatus* (Hodgson, 1836), in the western Arabian peninsula and, more in particular, on the island of Bahrain (Uerpmann 1995). Moreover, by the ancient Sumerians the name of the mongoose refers to a god, which has control over the mice. According to Uerpmann (1995), the cult of this divinity is known from the early third millennium B.C., or probably already before. There is, however, no evidence of the primary distribution of these carnivores in Mesopotamia. Thus, there is no reason to exclude that they have been artificially imported from the Indus valley already in the course of the fourth millennium B.C. Who knows if the origin of the extant gray mongooses of northern Iraq can be somehow related to an ancient oriental introduction?

2.11 Aliens from the Indian Subcontinent and Beyond

There is considerable archaeological evidence for the circulation and trade of living biological elements, materials, and ideas between the Near East and the western Indian subcontinent. As we have just seen, cultural interaction between Mesopotamia and the Middle East seems to have been established since, at least, the fourth millennium B.C. Contacts with traders in the Arabian Gulf who brought goods from India and Arabia, by way of *Dilmun* (Bahrein), were also established since, at least, the third millennium B.C. (Leemans 1960; Lloyd 1984). Archaeological and textual documentation shows, for example, that merchants from Harappa, an important Bronze Age center of the Indus Valley civilization, were present in Sumer and Akkad, and a number of Indian animals were indeed brought to Mesopotamia as gifts or exotic goods from proto-history onward. Some of these evidences came from the period of approximately 2350–2150 B.C., named in southern Mesopotamia (Sumer) after the city of Agade (or Akkad), the Akkadian capital, whose Semitic monarchs united the region, bringing the rival Sumerian cities under their control by conquest. Agade was probably founded before the time of Sargon (r. ca. 2340–2285

BC), the dynasty's first king. In the Akkadian text *The Curse of Agade*, a story about the city Agade (dated to the Ur III Period, 2047–1750 B.C.), and how it was made wonderful by the goddess *Inanna*, several exotic animals were invoked to give a flavor on the cosmopolitan nature of the Akkadian capital:

> That monkeys, mighty elephants, water buffalo, exotic animals, as well as thoroughbred dogs, lions, mountain ibexes, and alum sheep with long wool would jostle each other in the public squares (Electronic Corpus of Sumerian Literature, "The Cursing of Agade", lines 21–24).

As it is easy to understand, all the animals mentioned in the latter text, including monkeys, elephants, buffaloes, and the *alum sheep with long wool*—perhaps a domestic breed characterized by a fleece of particularly fine hair—were exotic biological elements imported from the Middle East and not characteristic of Mesopotamia. Furthermore, an Ur III text describes a red dog originally from *Meluhha*, the Sumerian name of a prominent trading partner of the ancient Mesopotamian civilization, whose location in the Indus valley is still debated. The site was given in Mesopotamian literature as a source of god dust. The "red dog," which was given to king Hibbi-Sin as tribute from *Marhasi* (inland southwestern Iran) (McIntosh 2008), was probably not a domestic canid but a dhole, *Cuon alpinus* (Pallas, 1811), the Asiatic wild dog, once very widely distributed across India. Figurines of animals were also among the goods imported to Mesopotamia from the Indus valley. Also, the phenotypical characters of the representation of an Asian elephant, *Elephas maximus* L., 1758, in the Black Obelisk of Shalmaneser III (ninth century B.C.), together with monkeys with and without tail—here identified with the Akkadian word of *pagii*, a rare case where text and picture coincide (Andrew R. George, pers. comm.)—and their association on the other side of the monument with pictures of typical Indian mammals, such as the greater one-horned rhinoceros, *Rhinoceros unicornis* (L., 1758), and other ungulates, point to a Middle Eastern source of inspiration.

2.12 Deer of Mesopotamia

Curiously, the only deer that fall into the morphology of Middle Eastern species is portrayed in ancient Assyrian art, between the ninth and the eighth centuries B.C. (Masseti 2003). Yet, at least, four species occur today in the Near East and have been also reported from prehistorical times, on the basis of osteological evidence: the roe deer, *Capreolus capreolus* (L., 1758), the already mentioned red deer and common fallow deer, and the Persian or Mesopotamian fallow deer, *D. dama mesopotamica* (Uerpmann 1981, 1987; Harrison and Bates 1991).

In Mesopotamian art, one of the most interesting representations of deer is shown on the stone relieves from Sennacherib's palace, at Niniveh (about 700 B.C.) (Masseti 2003). Among the reeds of a marsh near the Assyrian town, appear certain animals that definitely inhabited this type of environment: a domestic sow with its

Fig. 2.30 Detail of a stone relieves from Sennacherib's palace, at Niniveh (about 700 B.C.) (photo by the British Museum, courtesy of the Trustees of the British Museum, London)

young, two hornless hinds and a stag, with well-developed but scantily pointed antlers (Fig. 2.30). Although the general aspect of this deer might resemble a small variety of *C. elaphus*, the shape of the antlers, the shortness of the limbs, and the rather large feet point to its identification with a tropical Oriental swamp-dwelling species, perhaps the thamin or Eld's deer, *Rucervus eldi* (McClevelend, 1842), the most endangered wild animal in Asia. The species, formerly distributed in the grassland–forest mosaics of the Indian subcontinent, adapted to its semiaquatic existence by developing elongated hooves and hard, hairless pasterns which assist in its movements in the morass and floating islands, and which are a distinguishing feature of this species (Israel and Sinclair 1988). This cervid is also called "brow-antlered deer," because the long first branch of the antlers, the brow tine, and the main beam form a continuous bow-shaped curve (Putman 1988). Further representations of deer of probable Middle Eastern origin are documented on the stone hunting scenes of Ashurbanipal, where a group of these ungulates is led by the beaters against a hunting net (Fig. 2.31). The morphology of these deer might be related to that of the Indian swamp deer or barasinga, *Rucervus duvaceli* (Cuvier, 1823), another endangered species, which was formerly recorded all along the base of the Himalayas, from upper Assam to Bhawalpur and Rohri in the upper Sind (Putman 1988). Like the thamin, this species too, adapted to the extensive Middle Eastern areas of marshy grassland, could have been imported into Mesopotamia

Fig. 2.31 Stone hunting scenes from the seventh century B.C. palace of Ashurbanipal, at Niniveh, where a group of Indian swamp deer or barasinga, *Rucervus duvaceli* (Cuvier, 1823), is led by the beaters against a hunting net (photo by the British Museum, courtesy of the Trustees of the British Museum, London)

from the Indian subcontinent by the Assyrian rulers, through the traders of the Arabian Gulf or via mainland Persia, and might have found convenient environmental conditions in the royal hunting parks, which were located along the Euphrates and Tigris shores (Masseti 2003).

As already seen, archaeological evidence furnishes enough documents to understand the long-term network established in commercial exchanges among Assyria and the nearest foreign countries. It may not be surprising that at some point in the recent natural history of the Near East the unexpected evidence of allochthones biological elements could attest to their artificial diffusion in the geographical range which concerns us: this evidence may in fact be a testimony of an imported species successful acclimatization (Masseti 2003). As far as is presently known, it is not possible to ascertain whether the Indian deer were physically present in Mesopotamia or their artistic depiction was merely the result of a well-developed artistic skill. Perhaps based on models imported already made from the Middle East. In any case, this illustrates how it is not always possible to document the past presence of a certain zoological species in a specific territory and/or a particular cultural context, solely on the basis of the data offered by archaeozoological research, especially when we are dealing with animals that were utilized by an absolute *elite* (cf. Masseti 2009b), such as the kings of Assyria.

2.13 Monkeys

Apart from the southern Arabian population of the sacred baboon, *Papio hamadryas* (L., 1758), native monkeys are today unknown in the Near East (see Harrison 1964), as well as in its fossil horizons (Fig. 2.32). In fact, other representatives of the Order Primates are completely absent from the Upper Pleistocene levels of this geographical portion of the Western Palaearctic; and, as far as is presently known, no bone of monkey has ever been yielded by the scientific exploration of any of the archaeological sites of Mesopotamia. Nonetheless, monkeys are sometimes evoked in local literature and art since very ancient times (McDermott 1938; van Buren 1939).

The awareness by the Near Eastern civilizations of the existence of animals like the primates is attested since the appearance of the earliest written documents, which dates back to the third millennium B.C. (Masseti in press-b). Among these, the epic of the hero Gilgamesh, king of Uruk, is a poem from Sumerian and Babylonian Mesopotamia that is often regarded as the earliest surviving great work of literature. Here, the *Cedar Forest*, the glorious realm of the gods of Mesopotamian mythology, is described. In the comment to a new tablet, discovered in 2011, Al-Rawi and George (2014) observe that, in the Babylonian literary imagination, the Cedar Forest was a dense jungle inhabited by exotic and noisy fauna (17–26). Among the other sounds of this wooded environment, the chatter of monkeys formed a symphony (or cacophony) that daily entertained the forest's potent guardian, the giant ▪ *umbaba*. In fact, as translated by Al-Rawi and George (2014), monkeys daily "… *bash out a rhythm in the presence of* ▪ *umbaba*." A propos this, Andrew R. George, in his oral paper contributed to the "Primates in Antiquity Symposium," held August 19, 2016, at Dartmouth College (New Hampshire, USA), emphasized the relationship between the musician monkeys of Gilgamesh Cedar Forest and the

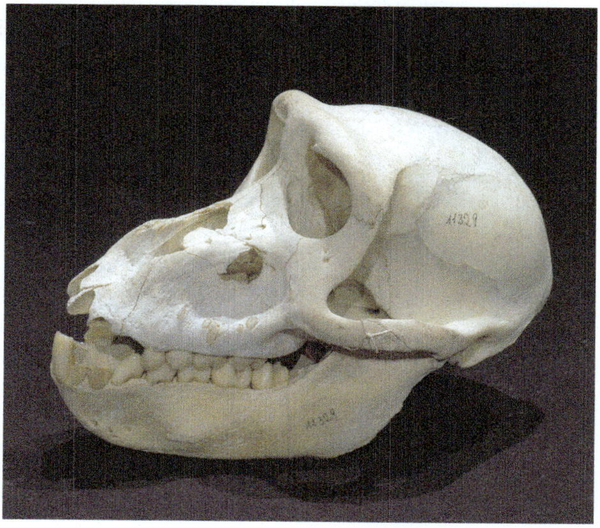

Fig. 2.32 Skull of a subadult female of the Arabian sacred baboon, *Papio hamadryas arabicus* (Thomas, 1900), the sole primate species still occurring in south-western Near East (photo by Saulo Bambi; courtesy of the Natural History Museum of the University of Florence, Zoological Section "La Specola")

many artistic evocations yielded by the archaeological exploration of ancient Mesopotamia. The translation of Al-Rawi and George (2014) gives us an unexpected description of the *"forest for the gods,"* mythologically located in a mountainous range which may have been perhaps inspired by Mount Lebanon, in the Levant, where the last centuries-old individuals of the Lebanon cedar, *Cedrus libani* A.Rich. 1823, are still preserved. No monkey, however, has ever inhabited any Near Eastern cedar forest, and, according to McDermott (1938), the presence of primates in the art of Mesopotamia indicates some exotic influence: "*The main source of this influence was Egypt—the close trade connections meant that specimens of the animals and more often artistic representations would be exported from Egypt to Mesopotamia as tribute, as presents, or as objects of trade. For example one of the tablets from Amarna lists gifts sent by Amenhotep IV (Ikhnaton, 1380-1362) to a Babylonian king—among these was a silver ape (cf. II Chronicles, 25, 27).*" But we cannot exclude that, possibly, primates have come into southern Mesopotamia from India, and that their trade passed throughout the eastern orographic chains. Sumerians believed, in fact, that monkeys came "*. . . from the east* [. . .] *In the year name of king Ibbi-Sin's twenty-third year, one hears of 'the monkeys coming from the mountains'. . .*" (Snell 2005). It is not easy to place geographically the location of these eastern mountains from where monkeys must have come. East of Mesopotamia, one can find the Zagros chain and the Iranian plateau. This means that, possibly, the monkeys have come in Mesopotamia from India, throughout these orographic systems; and so, Middle Eastern primates might have been regarded as elements of Gilgamesh's *Cedar Forest*. Also, the phenotypic characters of the monkeys evoked in many artistic representations of ancient Mesopotamia tell us of species of eastern origin, such as common Asian macaques of the species *Macaca mulatta* (Zimmermann, 1780). And very recently, osteological remains the latter species have been provided by the exploration of the Iranian site of Shahr-i Sokhta, in the province of Sistan and Baluchistan, dating to the third millennium B.C. (Minniti 2018, 2019).

There is even who, like van Buren (1939), says that several of the glazed frit monkey amulets of Akkadian Ur represent a species of anthropoid apes, and most likely a variety of gibbon of the Family Hylobatidae Gray, 1871. On the other hand, Houghton (1876-1877) suggests to identify some of the primates represented in Shalmaneser III Black Obelisk with Indian langurs of the genus *Semnopithecus* Desmarest, 1822 (Fig. 2.33), while van Buren (1939) is of the opinion that even northern plains gray langurs, *Semnopithecus entellus* (Dufresne, 1797), have been represented on many of the Mesopotamian artistic productions. Other authors, such as Hatt (1959) and Dunham (1985), believed to have further recognized several primates from the Middle to the Far East in the artistic artifacts of ancient Mesopotamia.

Fig. 2.33 The Indian elephant and the Asian monkeys of the Black Obelisk of Shalmaneser III, from Nimrud (about 825 B.C.) (photo by the British Museum, courtesy of the Trustees of the British Museum, London)

2.14 Syrian Elephants

Ivory is among the luxury materials most widespread in the ancient ornamental contexts of the southwestern Near East and, in particular, of Mesopotamia (see Mallowan 1978). This precious material was used for the decoration of furniture, such as chairs, tables, possibly beds, to ornament horse's bridles, to create boxes and cosmetic's containers, as holders for mirrors or fly swatters. The ancient world acquired its ivory either directly or through trade with Africa and the Middle East via the Levant, as attested by the Bronze Age shipwreck of Ulu Burun which had ivory as part of its cargo (Pulak 1996, 1998). Ivory, in fact, was obtained from Indian and African elephants and/or African or, even, Levantine hippopotamuses, *Hippopotamus amphibius* L., 1758 (Haas 1953; see Masseti 2003). There is some reason to assume that hippos were present at least in the coastal areas of the southern Levant, as indicated by the finding of subfossil remains of the species in local Bronze and Iron Age sites (Haas 1953; Uerpmann 1981, 1987). According, in fact, to Tchernov (1981, 1984a, 1991), the latter species was doomed to extinction in the western Near East in protohistoric/early historical time. We cannot exclude, however, that these bones might have been even imported there from the nearest African territories, such as the Nile Valley, where the species still occurred until the seventeenth century AD (Osborn and Helmy 1980).

It is generally believed that ivory was also accessible from the herds of elephants which were hunted in Syria between the first half of the second millennium and the ninth century, when they became extinct (cf. Miller 1986) (Fig. 2.34). In fact, according to pictorial, written, and osteological evidence, it seems that herds of elephants lived—possibly in a free-ranging state—in early historical times, in the

Fig. 2.34 Elephant tusks of supposed Syrian origin from the Minoan palace of Zakro (southern Crete), in the Archaeological Museum of Herakleion (photograph by Marco Masseti, courtesy of the Archaeological Museum of Herakleion, Greece)

region of north-west Syria, between the Oronte's Valley and the Khabur river (van Buren 1939; Arnold 1952; Hatt 1959; Brentjes 1969a; Drower 1973; Winter 1973; Hofmann 1974; Scullard 1974; Collon 1977; Corbet 1978; Clutton-Brock 1981; Barnett 1982; Bökönyi 1985; Miller 1986; Buitenhuis 1988, 1999; Houlihan 1996; Osborne and Osbornova 1998; Gabolde 2000; Masseti 2002, 2003). Several of the latter authors are of the opinion that the central range of the diffusion of the Mesopotamian proboscideans was the lost land of *Nij*, or *Neya*, where in 1464 B. C., the Egyptian pharaoh Tuthmosis I (1525–1512 B.C.) and his grandson Thutmosis III (1504–1450 B.C.) both took the opportunity to hunt elephants (Hatt 1959; Scullard 1974). The site of *Neya* was possibly located either in the *Gharb* plain, north of Hama (Drower 1973), or in the Euphrates region, not very far from the present site of Aleppo (Winter 1973; Barnett 1982). Also the Assyrian kings, Tiglathpileser I (1.1115–1.1102 B.C.), Adadnirari II (911–889 B.C.), Ashurnasirpal II (884–859 B.C.), and Shalmaneser III, all left accounts of elephants they killed or captured alive, presumably along the Euphrates (Hatt 1959).

Becker (2005) and Çakırlar and Ikram (2016) gave a summary of the available data on the distribution of protohistoric and early historic sites of the Near East with elephants remains from where it is possible to deduce that of the 15 sites considered in the study, at least ten are located within the Mesopotamian boundaries. More recently, elephant remains have been provided by the archaeological exploration of the Iron Age layers of Değirmentepe near Malatya, in south-eastern Turkey (Siracusano in press) (Fig. 2.35). Together with the findings of Arslantepe (Bökönyi 1985), from which it was only a few tens of kilometers away, the latter are the northernmost remains attributable to Mesopotamian proboscideans of historic times.

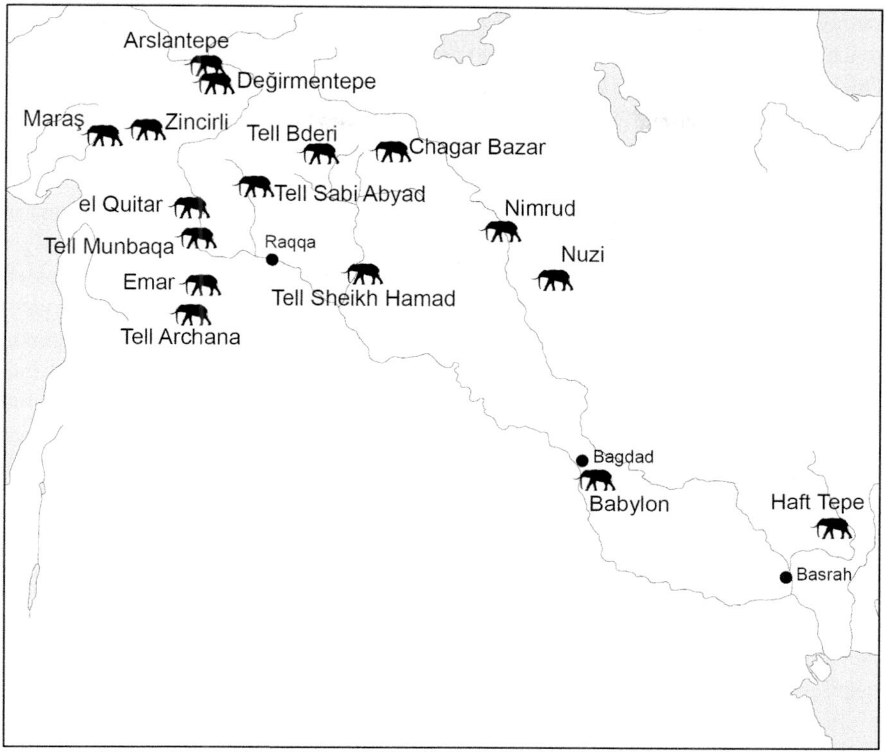

Fig. 2.35 Location of the protohistoric and early historic sites of Mesopotamia that provided osteological remains of elephants (data from Becker 2005, Çakırlar and Ikram 2016, and Siracusano in press)

Biogeographic considerations make us assume that the Syrian elephants must belong to the Asian species that is *Elephas maximus* L., 1758. As far as the end of the 1950s, Hatt (1959) was already of the opinion that the few teeth, which constituted the sole physical evidence of elephants, recovered until then from ancient Iraq, belonged to the same taxon. Teeth testifying the presence of *E. maximus* in Iraq in the late Pleistocene are also known (Hatt 1959; Al-Sheikhly et al. 2015).

The presence of elephants in north-west Syria during the late second and early first millennia B.C. coincided with a period of reduced human settlement and lower population density in the area comprised between the Oronte's valley and the Upper Mesopotamia (Miller 1986; cf. McClellan 1993). It is presumed that at this time the kind of woodland-savannah mosaic needed by elephants was found throughout much of the region. Miller (1986) argued that with the increasing demand for charcoal and fuel among sedentary communities of the Iron Age, forest resources declined to the point where elephant populations became extinct. Changes in metallurgy, politics, and patterns of rangeland management may also have contributed to the reduction of elephant populations to the point where they became vulnerable to overhunting (Miller 1986). The last pictorial evidence of the occurrence of proboscideans in Mesopotamia is carved on the already mentioned Black Obelisk of Shalmaneser III (859–824 B.C.), where a female Asian elephant is shown in the tribute of the land of *Muşri* to the Assyrian king (see Layard 1853). van Buren (1939) suggested that the so-called "Syrian elephants" were merely animals brought in tribute or used to stock royal hunting parks. Following van van Buren (1939), in view of the fact that no other representations of proboscideans have been found in Mesopotamian art, we can presume that the so-called "Syrian elephants" were merely animals brought in tribute or used to stock royal hunting parks (Masseti 2003). They did not represent endemism of the region, but the result of frequent contacts with the Indian subcontinent (Çakırlar and Ikram 2016; Siracusano in press). Recently, Çakırlar and Ikram (2016) formulated the hypothesis according to which, if one accepts that tamed elephants were known in the Indus Valley around 2500 B.C. (Clutton-Brock 2012), and considers the ethnographic record of elephant herding over long distances (Baker and Manwell 1983), together with the regular contacts between south-western Asia and the Indus Valley starting in the second half of the fourth millennium B.C., it is feasible to suggest that small groups of Asian elephants were transported to the Near East as part of the long-distance overland exchange.

Furthermore, Al-Rawi and George (2014) suggest that the creation of ▪ *umbaba*, the legendary guardian of the Cedar Forest of *Gilgameš*'s poem, was perhaps inspired by the observation of elephants. Thus, it cannot be ruled out that the source of inspiration for the invention of the mythical creature is to be referred precisely to the proboscideans who inhabited ancient Syria.

2.15 The Water Buffalo

Although there is no conclusive paleontological evidence that the Asian water buffalo, *Bubalus bubalis* (L., 1758), was indigenous to Mesopotamia, it cannot be denied that, today, this ungulate is among the most widespread domestic animals of the Euphrates and Tigris vallies, especially in its southernmost portion. The water buffalo is a coarsely robust breed, extensively used in the agricultural areas as a draft animal and for its milk, butter, cheese, and hides (Hatt 1959); it is well-adapted animal to swamps and areas subject to flooding (Abid and Fazaa 2007). These ungulates are largely immune to piroplasmosis, and in these territories perform better than the oxen, in terms not only of work, but also of meat and milk. Herding of water buffaloes is one of the traditional main economic activity in the life of the marsh Arabs of southern Iraq, the *Ma'dān* (Thesiger 1954, 1959; Roncalli and Mandel 1993; Kubba 2011) (Fig. 2.36). Thesiger (1954) wrote that the latter also used the animal for bride buying and for meat, although they usually slaughter only animals already near death. More than one century ago, Parish (1860) observed that although extremely common in the domestic state, these ungulates were more generally feral: "*... the abundant pastures affording such great facilities for it*

Fig. 2.36 Herding of water buffaloes, *Bubalus bubalis* (L., 1758), is one of the traditional main economic activities in the life of the marsh Arabs of southern Iraq, the *Ma'dān* (from Thesiger 1959)

increase. Those wild are a much greater in size than those domesticated, and have horns of enormous size. They are also endowed with an extraordinary amount of strength, and it is said can knock over a good-sized elephant. The horns and hides of this, as well as the oxen, form articles of commerce." Actually, domestic water buffaloes are to be found throughout the wetland areas of the Near East and particularly in the marshy, malaria-ridden territories, almost always kept in a free-ranging state (cf. Manetti 1921). Still in the nineteenth century, the Arabs of the Upper Euphrates employed the buffalo even in lion hunting (Blunt 1896). Until the end of the 1950s, these ungulates were still common in the river valleys from Basrah to Baghdad (Williamson 1949; Hatt 1959).

It seems that the water buffalo was domesticated independently in India and China from wild stock possibly before 6000 years ago. It has been introduced widely and used in domesticated herds for thousands of years across southern mainland Asia, Southeast Asia, and the Near East, and for at least 2000 years in far northeastern Africa and southern Europe, all areas where it is seen as naturalized and not invasive. Buffaloes were probably imported in the Near East not before the fourth millennium B.C. when these animals were introduced in Lower Mesopotamia following the arrival of nomadic human groups originating from India (cf. Lombard 1971). An alternative hypothesis is that the culturally advanced Uruk people, who came into the hills of Iraq from what is now central Turkey, already used the water buffalo as a domestic animal (Hatt 1959). The earliest evidence of the species in the Euphrates and Tigris plains consist of horn cores from the site of Grai Resh in the Sinjar hills, a village of Uruk culture, dated to around the first half of fourth millennium B.C. (Davis 1987). Some bones of water buffalo were also found

Fig. 2.37 Impression of a cylinder seal of the Uruk period, 3200–2700 B.C. (first half of the fourth millennium B.C.) (from Davis 1987)

in the post-Hittite (twelfth to seventh century B.C.) layers of the Turkish village of Boghazköy, built in the immediate vicinity of Hattusa, the ancient capital of the Hittite empire (Vogel 1952; Bökönyi 1974).

The importation of the domestic buffalo in Lower Mesopotamia is also documented by several artistic productions (Brentjes 1969b). The vivid depiction of this ungulate appears in a few Sumerian artifacts since the first half of the fourth millennium B.C. (Davis 1987), such as one cylinder seals of the Uruk period, 3200–2700 (first half of the fourth millennium B.C.) (Davis 1987) (Fig. 2.37), confirming its Mesopotamian occurrence, also documented as we have already seen by the contemporary discovery of osteological findings. Images of water buffaloes compare on Akkadian artistic productions of the late Agade period (2750–2100 B.C.), in another seal of Ibn-Sharrum, the scribe of king Sharkali-Sharri son of Naram-Sin, 2250 B.C. (Fig. 2.38), and are even mentioned in a few texts (McIntosh 2008). According to Hatt (1959), the species was frequently portrayed until about 2100 B.C. at which time it presumably became extinct. Failing to find representations of buffaloes after the third millennium B.C., Hilzheimer (1926) suggested that they may have disappeared from Mesopotamia by that time, whereas the occurrence on the Babylonian animal lists of about 1500 B.C. of a name translated as "water-ox," suggests to Oppenheim and Hartman (1945) knowledge of the water buffalo. The ungulate, however, was neither mentioned in literature, nor seen in artwork of the ancient Egyptians, Romans, or Greeks, to whom it was apparently unknown (Bökönyi 1974; Abid and Fazaa 2007). Thus, it cannot be excluded that these artiodactyls came already domesticated to Mesopotamia from the Indus valley, during the fourth millennium B.C., as the result of economic changes with the Harappa and Mohenjo-Daro cultures (which existed from about 3000 to

Fig. 2.38 Cylinder seal of Ibni-Sharrun, the scribe of king Sharkali-Sharri son of Naram-sin and king of Akkad, Mesopotamia (2340–2100 B.C.) (Paris, Louvre)

1500 B.C.). Later introduction was presumed by Smith (1827), Zeuner (1963), and Lombard (1971) who believed that these animals were imported into Persia, Mesopotamia, and northern Syria, where they became acclimatized in the marshes of the Orontes and, later on, even in Egypt, not before the early Middle Ages, possibly in the eighth century A.D. In any case, according to Postgate (2017), since water buffaloes were again introduced to Iraq during the early Islamic period, it is deceptive to treat the "marsh Arabs" way of life as inherited unbroken from prehistoric times.

The population of Mesopotamian water buffaloes suffered an evident and drastic decline due to the southern Iraqi marsh desiccation, an environmental disaster perpetrated by the previous Iraqi regime for political reasons over the years 1991–2003 (Alsaedy 2007). Drying reduced the number of these ungulates in the marshes, mainly due to their dependence on available water and reeds. Additionally, the buffalo's economic importance forced most breeders to leave the marshes and seek other wetland habitats far from the desiccated areas (Abid and Fazaa 2007). However, since 2003 many breeders were encouraged to return to the marshes after their re-flooding, driving their stock back from the villages around Baghdad where they had found a provisional shelter.

2.16 Concluding Remarks

"Middle-earth" between Eurasia and Africa, as well as between the two rivers Euphrates and Tigris, Mesopotamia has always represented a biogeographical crossroads between opposite natural entities, allowing the coexistence of different biogeographical realities, often also conflicting (cf. Atallah 1977, 1978; Serra et al.

2005). Among the species apparently extraneous to the zoogeography of the Near East, we can mention the African darter, and the sacred ibis, both Afrotropical elements that find in Mesopotamia one of the few breeding sites out of the Black Continent. The smooth-coated otter is instead a species primarily dispersed in most of the Indian subcontinent and southeastern Asia, with a curious, and very reduced, Near Eastern enclave of distribution. Another medium-sized carnivore characteristic of the Oriental zoogeographic region is the Indian gray mongoose, whose occurrence in northern Iraq has been recently confirmed. These species, and possibly others, must be added to those that, since prehistoric times, accompanied the movements of human communities, such as many of the domestic, commensal, and synanthropic vertebrates (cf. Tchernov 1984b; Tangri and Wyncoll 1989; Masseti 2002).

From immemorial time, Mesopotamia has hosted and suffered on its soil the plural succession of indigenous civilizations and the continuous passage of foreign cultures proceeding from afar. The consequences of this have inevitably reverberated on the primary composition of local environment and faunal balances that have undergone a prolonged alteration from prehistory onward, without apparent solution of continuity. We have seen that several zoological species have also been imported from abroad and that some of them have become naturalized in the new environment perhaps since antiquity, others have never done so. A few animals have been described as free-ranging in the Mesopotamian natural environment already in antiquity. In this regard, one of the most striking example we could recall is the elephant herd that apparently survived in Syria between the first half of the second millennium and the eigth century B.C. These Asian proboscideans were not endemic to the region, but had very likely an anthropochorous origin, being—as observed by Çakırlar and Ikram (2016)—the product of the power hungry Bronze Age elite in the Near Eastern region. Anthropochorous as well must be considered the appearance in the palaces of the Assyrian Mesopotamia of the barasinga imported from the Indian subcontinent or the wild horses of Central Asia. Another species extraneous to Mesopotamia is the bharal, or blue sheep, *Pseudois nayaur* (Hodgson, 1833), still dispersed in the mountain range of China, Tibet, and the Indian subcontinent (see Prater 1948; cf. Harris 2014). The bas-relief representation of this caprine decorates one of the walls of Ashurbanipal's throne room at Nimrud, about 865 B.C. (London, British Museum) (Fig. 2.39), already documenting the Assyrians' knowledge of this wild sheep. Elephants, Indian deer, wild horses, and blue sheep have disappeared long ago from the Near East but other Middle Eastern mammals are still reported from several sites in Upper Mesopotamia. It cannot be excluded, in fact, that even the smooth-coated otter may have been introduced for utilitarian purposes from India, already in early times. After all, we have already mentioned that, even today, human beings still use these otters for fishing in some parts of the Middle East.

The problem of the introduction of biological elements extraneous to the original ecosystems of Mesopotamia has increased over time, assuming a more and more worrying dimension. Even the deepest waters of the local rivers have not been spared from the importation of allochthonous biological elements and have undergone the invasion of fish of Neotropical and Nearctic origin, such as the voracious North American alligator gar. The invasion of ecosystems by exotic *taxa* is currently

Fig. 2.39 A protective genie carrying a bharal, or blue sheep, *Pseudois nayaur* (Hodgson, 1833), from Ashurbanipal's throne room at Nimrud, about 865 B.C. (courtesy British Museum, London). The species is still today characteristic of the Himalayan chain and of its mountainous surroundings

regarded as one of the most important causes of the loss of biodiversity. Today, in view of the vulnerability of what remains of the natural ecosystems, it is critically important to prevent further introductions. This, however, leaves the question of how to treat the allochthonous zoological populations of certified ancient anthropochorous origin, which instead deserve to be protected and considered in terms of a veritable "cultural heritage" (Masseti 2002, 2009c). Within the latter category we should, for example, comprise the smooth-coated otter and the water buffalo, both species of probable anthropochorous spread in Mesopotamia. The protection of these mammals and their study can provide an opportunity for testing a range of different evolutionary theories. Thus, the current challenge is how to use the specific knowledge to manage and conserve these anthropochorous populations. Their survival is significant not only in ecological but also in historical and archaeological terms (Masseti et al. 2008; Masseti 2009d). For these reasons too, their importance has to be considered on a par with that of a human artifact, as the dynamic testimony of ancient human intervention that is still available for our evaluation and our appreciation, with all the consequences that such an estimate brings with it.

Acknowledgments In the course of this project, I have been fortunate to have the help of many friends and colleagues. I am particularly grateful to Umberto Albarella, Department of

Archaeology, University of Sheffield (UK); Remy Berton, Muséum National d'Histoire Naturelle, Paris; Elena Maini, Department of History and Cultures-Alma Mater Studiorum, University of Bologna; Leonardo Salari, Department of Earth Sciences, University of Rome "La Sapienza"; and Giovanni Siracusano, Ancient Near East Zooarchaeology, Graduate School "Human Development in Landscapes," Christian-Albrechts-Universitat zu Kiel, Germany. Special thanks are due to Andrew R. George, SOAS University of London, for the enjoyable and stimulating discussions on the animals described in the ancient Mesopotamian poem of Gilgamesh.

References

Abid HS, Fazaa NA (2007) Water buffalo in the Iraqi marshes. Thi Qar and Missan Governorates. Nature Iraq status report. Nature Iraq Office, Sulaimani, Kurdistan. 29 pp

Ainsworth W (1838) Researches in Assyria, Babylonia, and Chaldaea: forming part of the labours of the Euphrates expedition. John W. Parker, London. 344 pp

Ainsworth WF (1842) Travels and researches in Asia minor, Mesopotamia, Chaldea and Armenia. Parker, London. 425 pp

Ainsworth WF (1888) A personal narrative of the Euphrates expedition, vol II. Kegan Paul, Teench, London. 492 pp

Allouse B (1953) Avifauna of Iraq. Al-Tafyudh Press, Baghdad. 522 pp

Al-Rawi FNH, George AR (2014) Back to the Cedar Forest: the beginning and end of Tablet V of the Standard Babylonian epic of Gilgameš. J Cuneiform Stud 66:69–90

Alsaedy K (2007) Iraqi Buffalo now. Ital J Anim Sci 6(suppl. 2):1234–1236

Al-Sheikhly OF (2012) Notes on antelopes in Iraq. Gunsletter 30(2):11–12

Al-Sheikhly OF, Mallon D (2013) The small Asian Mongoose *Herpestes javanicus* and the Indian Grey Mongoose *Herpestes edwardsii* in Iraq (Mammalia: Carnivora: Herpestidae). Zool Middle East 59(2):173–175

Al-Sheikhly OF, Nader IA (2013) The status of Iraq smooth-coated otter *Lutrogale perspicillata maxwelli* Hayman 1956 and Eurasian otter *Lutra lutra* Linnaeus 1758 in Iraq. IUCN Otter Spec Group Bull 30(1):18–30

Al-Sheikhly OF, Haba MK, Barbanera F, Csorba G, Harrison DL (2015) Checklist of the mammals of Iraq (Chordata: Mammalia). Bonn Zool Bull 64(1):33–58

Arnold R (1952) Das Verbreitungsgebiet der Elefanten zu Beginn der historischen Zeit. Zeitschrift für Säugetierkunde 17(2):73–82

Atallah SI (1977) Mammals of the Eastern Mediterranean region; their ecology, systematics and zoogeographical relationships. Säugetierkundliche Mitteilungen 26(1):241–316

Atallah SI (1978) Mammals of the Eastern Mediterranean region; their ecology, systematics and zoogeographical relationships. Säugetierkundliche Mitteilungen 25(4):1–50

Atallah SI, Harrison DL (1967) New records of rodents, bats and insectivores from the Arabian Peninsula. J Zool 153:311–319

Baker CMA, Manwell C (1983) Man and elephant: the 'dare theory' of domestication and the origin of breeds. Zeitschrift für Tierzuchtung und Zuchtungsbiologie 100:55–75

Barnett RD (1982) Ancient ivories in the Middle East. Quedem, vol 14. Hebrew University, Jerusalem, pp 54–62

Baumgart W, Burkhard S (1986) Ergebnisse ornithologisher Beobachtungen in der Syrischen Arabischen Republik. Teil 1: non-passeriformes. Mitt Zool Mus Berl 62 Suppl: AnnOrn 10:69–110

Becker C (2005) Small numbers, large potential—new prehistoric finds of elephant and beaver from the Khabur river/Syria. MUNIBE (Antropologia-Arkeologia) 57, Homenaje a Jésus Altuna :445–456

Bedair HM, Al Saad HT, Salman NA (2006) Iraq's southern marshes something special to be conserved; a case study. Marsh Bull 2(1):99–126
Benda P (1996) Distribution of Geoffroy's bat, *Myotis emarginatus* (Chiroptera: Vespertilionidae), in the Levant region. Folia Zoologica 45:193–199
Benda P, Hulva P, Andreas M, Uuhrin M (2003) Notes on the distribution of *Pipistrellus pipistrellus* complex in the Eastern Mediterranean: first records of *P. pipistrellus* for Syria and of *P. pygmaeus* for Turkey. Vespertilio 7:87–95
Berthon R (2013) New data on the exploitation of animal resources in the Upper Tigris river area (Turkey) during the second and first millennia BC. In De Cupere B, Linseele V, Hamilton-Dyer S (eds) Archaeozoology of the near East X. Proceedings of the tenth international symposium on the archaeozoology of South-Western Asia and adjacent areas. Ancient near Eastern studies, supplement 44, pp 145–162
Berthon R (2014) Small but varied. The role of rural settlements in the diversification of subsistence practices as evidenced in the Upper Tigris River area (Southeastern Turkey) during the second and first millennia BCE. J East Mediterr Archaeol Herit Stud 2(4):317–329
Berthon R, Erdal YS, Mashkour M, Kozbe G (2016) Buried with turtles: the symbolic role of the Euphrates soft-shelled turtle (*Rafetus euphraticus*) in Mesopotamia. Antiquity 90(349):111–125
BirdLife International (2012) *Turdoides altirostris*. IUCN red list of threatened species. Version 2013.2. International Union for Conservation of Nature. Accessed 26 Nov 2013
BirdLife International (2017) *Acrocephalus griseldis* (amended version of assessment). The IUCN red list of threatened species 2017: e.T22714757A118739069. https://doi.org/10.2305/IUCN.UK.2017-3.RLTS.T22714757A118739069.en
Blunt A (1896) Bedouin tribes of the Euphrates. Drallop, New York. 445 pp
Boessneck J, Peters J, von den Driesch A (1986) Tierknochenund molluskenfunde aus Munbaqa. Mitteilungen der Deutschen Orient-Gesellschaft zu Berlin 118:147–160
Boessnek J, von der Driesch A (1975) Tierknockenfunde von Korucutepe bei Elazig in Ostanatolien. In: van Loon MN (ed) Korucutepe I. Elsevier, Amsterdam, pp 1–220
Bökönyi S (1974) History of domestic mammals in Central and Eastern Europe. Akadémiai Kiadó, Budapest. 597 pp
Bökönyi S (1985) Subfossil elephant remains from Southwestern Asia. Paléorient 11(2):161–163
Bökönyi S (1993) Hunting in Arslantepe, Anatolia. In: Frangipane M, Hauptmann H, Liverani M, Matthiae P, Mellink M (eds) Between the rivers and over the mountains. Archaeologica, Anatolica et Mesopotamica, A. Palmieri dedicata. Università di Roma "La Sapienza", Roma, pp 341–360
Braidwood RJ, Howe B (1960) Prehistoric investigations in Iraqi Kurdistan. University of Chicago Press, Chicago. 184 pp
Brentjes B (1964) Der Syrische Biber. Zeitschrift fűr Jagdwiss 10:183–184
Brentjes B (1969a) Der syrische Elefant als Sudform des mammuts? Säugetierk Mitt 17:211–212
Brentjes B (1969b) Wasserbüffel in der Kulturen des Alten Orients. Zeitschrift für Sāugetierekunde 34:187–190
Buitenhuis H (1979) The faunal remains from Tell Hadidi. In: Kubasiewic M (ed) Archaeozoology I. Agricultural Academy, Szczecin (Poland), pp 164–175
Buitenhuis H (1988) Archeozoölogish onderzoek langs de Midden-Eufraat. Onderzoek van het faunamateriaal uit zes nederzettingen in Zuidoost-Turkije en Noord-Syrië daterend van ca. 10.000 BP tot 1400 AD. Wiskunde en Ntuurwetenschappen aan de Rijksuniversiteit Groningen. Doctoral thesis, 213 pp.
Buitenhuis H (1999) Archaeozoological aspects of Late Holocene economy and environment in the Near East. In: Bottema S, Entjes-Nieborg G, van Zeist W (eds) Man's role in the shaping of the Eastern Mediterranean landscape: Proceedings of the symposium on the impact of ancient man on the landscape of the E Med Region & the Near East: Groningen, 1989, pp 195–206
Byerly T, Timbs J (1838) The Euphrates expedition. In: The mirror of literature, amusement, and instruction, vol XXXI. J Limbird, London, pp 329–331

Çakırlar C, Ikram S (2016) When elephants battle, the grass suffers. Power, ivory and the Syrian elephant. Levant 48(2):167–183
Campbell CG (2007) Tales from the Arab tribes. The oral traditions among the great Arab tribes of southern Iraq. Routledge, London. 258 pp
Campbell Thompson R (1926) Assyrian Garidu = "Beaver". J Roy Asiatic Soc Lond 41:29–31
Cauvin J (1977) Les fouilles de Mureybet (1971-1974) et leur signification pour les origines de la sédentarisation au Proche-Orient. Annu Am Sch Orient Res 44:19–48
Cauvin J (2000) The birth of the gods and the origins of agriculture. Cambridge University Press, Cambridge. 259 pp
Cheesman RE (1923) Report on the mammals of Mesopotamia. In A survey of the fauna of Iraq. Mammals, birds, reptiles, etc. made by the members of the Mesopotamia Expeditionary Force "D" 1915-1919. The Bombay Natural History Society/Time Press, Bombay, pp 1–24
Chesney FR, Ainsworth W (1837) A general statement of the labours and proceedings of the expedition to the Euphrates, under the command of Col. Chesney, Royal Artillery, F.R.S. J Roy Geogr Soc VII(41):l–439
Clason AT, Buitenhuis H (1978) A preliminary report on the faunal remains of Nahr el Homr, Hadidi and Ta'as in the Tabqa Dam region in Syria. J Archaeol Sci 5:75–83
Clutton-Brock J (1981) Domesticated animals from early times. Heinemann-British Museum (Natural History), London, 208 pp
Clutton-Brock J (2012) Animals as domesticates. A world view through history. Michigan State University Press, East Lansing. 240 pp
Coad B, Hales J (2013) FEOW Freshwater Ecoregions of the World. 442: Upper Tigris & Euphrates. Copyright 2008 by The Nature Conservancy and World Wildlife Fund, Inc. All Rights Reserved
Coadt BW, Al-Hassan LA (1989) Freshwater shark attacks at Basrah, Iraq. Zool Middle East 3 (1):49–54
Collon D (1977) Ivory. Iraq 39:219–222
Coon C-S (1951) Cave explorations in Iran, 1949. Pennsylvania University Press (Museum monographs), Philadelphia. 126 pp
Corbet GB (1978) The mammals of the Palaearctic region. A taxonomic review. British Museum (Natural History)/Cornell University Press, London. 314 pp
Corbet GB, Hill JE (1992) The mammals of the Indomalayan region. Natural History Museum Publication/Oxford University Press, Oxford. 488 pp
Corkill NL (1929) On the occurrences of the cheetah (*Acinonyx jubatus*) in Iraq. J Bombay Nat Hist Soc 33:700–702
Corsini-Foka M, Masseti M (2008) On the oldest known record of the Nile Soft-shelled Turtle, *Trionyx triunguis* (Forskal, 1775), in the Eastern Aegean islands (Greece). Zool Middle East 43:108–110
Cramp S (1980) The birds of the Western Palearctic, vol 2. London, Oxford. 695 pp
Danford CG, Alston ER (1880) On the mammals of Asia minor', part II. In: Proceedings of the scientific meetings of the Zoological Society of London, 1880, pp 50–64
Davis SM (1987) The archaeology of animals. B.T. Batsford, London. 224 pp
de Gelder A (2010) Tracking and evaluating changes in vegetation and habitat structure in the En-Afeq Nature Reserve, Israel. Israel Nature and National Parks Protection Authority/ 2Wageningen University and Research Science and Management Division Dept. of Environmental Sciences, Wageningen. 109 pp
De Haan CC (1997) *Malpolon monspessolanus* (Hermann, 1804). In: Gasc J-P, Cabela A, Crnobrnja-Isailovoc J, Dolmen D, Grossenbacher K, Haffner P, Lescure J, Martens H, Matinez Rica J-P, Maurin H, Oliveira E, Sofianidou TS, Veirh M, Zuiderwijl A (eds) Atlas of amphibians and reptiles of Europe. Societas Europaea Herpetologicz/Museum National d'Histoire Naturelle, Paris, pp 366–367
Delfino M, Böhme M, Rook L (2007) First European evidence for transcontinental dispersal of *Crocodylus* (late Neogene of southern Italy). Zool J Linn Soc 149:293–307

Dobson GE (1878) Catalogue of the Chiroptera in the collection of the British Museum, London. 691 pp

Drower MS (1973) Syria *ca.* 1550-1400 B.C. Cambridge Ancient History, 3rd edn. Cambridge University, Cambridge, pp 417–525

Dunham S (1985) The monkey in the middle. Zeitschrift für Assyriologie 75:234–264

Esmaeili HR, Masoudi M, Chermahini MA, Esmaeili AH, Zarei F, Ebrahimi M (2017) Invasion of the neotropical and nearctic fishes to Iran. FishTaxa 2(3):126–133

Evans M (ed) (1994) Important bird areas in the Middle East. BirdLife, Cambridge. 410 pp

Feeroz MM, Begum S, Hasan MK (2011) Fishing with otters: a traditional conservation practice in Bangladesh. Proceedings of XIth International Otter Colloquium. IUCN Otter Spec Group Bull 28A:14–21

Fellows C (1841) An account of discoveries in Lycia, being a journal kept during a second excursion in Asia Minor (1840). John Murray, London. 621 pp

Fergusson WW, Porath Y, Paley S (1985) Late Bronze Period yields osteological evidence of *Dama dama* (Artiodactyla: Cervidae) from Israel and Arabia. Mammalia 49:209–214

Fischer E (2007) Ägyptische und ägyptisierende Elfenbeine aus Megiddo und Lachisch: Inschriftenfunde, Flaschen, Löffel. AOAT 47. Ugarit, Münster. 224 pp

Fletcher J (2000) Fletcher's game. A vet's life with Scotland's deer. Mercat, Edinburgh. 221 pp

Fowden G (2004) Quṣayr 'Amra. Art and the Umayyad Elite in Late Antique Syria. University of California Press, Berkeley. 390 pp

Gabolde M (2000) Les elephants de Niyi d'apres les sources egyptiennes. In: Beal J-C, Goyon J-C (eds) Des ivoires et des cornes dans les mondes anciens (Orient-Occident). Collection de l'Institut d'Archeologie et d'Histoire de l'Antiquite, Universite Lumiere-Lyon 2, Lyon, pp 129–140

Ghaffari H, Taskavak E, Karami M (2008) Conservation status of the Euphrates softshell turtle, *Rafetus euphraticus*, in Iran. Chelonian Conserv Biol 7(2):223–229

Ghaffari H, Taskavak E, Turkozan O, Mobaraki A (2017) *Rafetus euphraticus*. The IUCN Red list of threatened species 2017: e.T19070A1956551. https://doi.org/10.2305/IUCN.UK.2017-3.RLTS.T19070A1956551.en

Gourichon L, Helmer D (2004) Exploitation du petit gibier dans le Moyen Euphrate syrien du XIIe au IXe millénaires av. J.-C. In: Brugal J-P, Desse J (eds) Petits animaux et sociétés humaines. Du complément alimentaire aux ressources utilitaires XXIVe rencontres internationales d'archéologie et d'histoire d'Antibes. Éditions APDCA, Antibes, pp 415–433

Guest E (1966) The vegetation of Iraq and adjacent regions. In: Guest E, Al-Rawi A (eds) Flora of Iraq, vol 1. Ministry of Agriculture of the Republic of Iraq, Baghdad, pp 55–108

Gündem C (2010) Animal based subsistence economy of Emar during the Bronze Age. In Finkbeiner U, Saka F (eds) Emar after the Closure of the Tabqa Dam. The Syrian-German Çakırlar and Ikram 'When elephants battle, the grass suffers.' Power, ivory and the Syrian elephant Excavations 1996–2002. Volume I: Late Roman and Medieval cemeteries and environmental studies. Subartu, vol 25, pp 125–176

Gündem C, Uerpmann H-P (2003) Erste beobachtungen an den tierknochenfunden aus Emar (Syrien)—Grabungen bis 2002. Baghdader Mitteilungen 34:119–128

Haas G (1953) On the occurrence of *Hippopotamus* in the Iron Age of the coastal area of Israel (Tell Quasileh). BASOR 132:30–34

Hanney PW (1975) Rodents: their lives and habits. David and Charles, London. 2212 pp

Harris RB (2014) *Pseudois nayaur*. The IUCN red list of threatened species 2014: e.T61513537A64313015. https://doi.org/10.2305/IUCN.UK.2014-3.RLTS.T61513537A64313015.en.

Harrison DL (1964) The mammals of Arabia. I. Ernest Benn, London. 192 pp

Harrison DL (1968) The mammals of Arabia. II. Ernest Benn, London, pp 195–381

Harrison DL (1972) The mammals of Arabia. III. Ernest Benn, London. 670 pp

Harrison DL, Bates PJJ (1991) The mammals of Arabia. Harrison Zoological Museum, Sevenoaks. 354 pp

Harrison DL, Lewis RE (1961) The large mouse-eared bats of the Middle East, with a description of a new subspecies. J Mammal 42:372–380
Hatt RT (1959) Mammals of Iraq. Miscellaneous Publ Mus Zool Univ Mich 106:1–113
Helfer P (1878) Travels of Dr. und Mrs. He!fer in Syria, Mesopotamia, Burmah and other lands (narrated by Pauline, Countess Nostitz, formerly Mme. Helfer). Bentley, London. 428 pp
Helmer D (1985) Etude de la faune de Tell Assouad (Djezireh - Syrie) Sondage J. Cauvin. Cahiers de l'Euphrate 4:175–285
Hill D,Robertson P (1988) The pheasant. Ecology, management and conservation. Blackwell, Oxford, 281 pp
Hillenbrand C (1985) The history of the Jasira, 1100–1250: a short introduction. In Raby J (ed) The art of Syria and the Jasira 1100–1250. Oxford studies in Islamic art, vo 1, pp 176–177
Hilzheimer M (1926) Sägetierkunde und Archäologie. Zeitschr f Säugetierkunde 1:140–169
Hofmann I (1974) Die Artzugehorigkeit des syrischen Elefanten. Säugetierk Mitt 22(1):225–232
Houghton W (1876-1877) On the mammalia of the assyrian sculptures. Part II. Wild animals. Trans Soc Biblical Archaeol 5(33-64):319–383
Houlihan PF (1996) The animal world of the pharaohs. Thames and Hudson, London. 240 pp
Hunt RS (1951) The sharks of Ahwaz. J Roy Army Med Corp 97(2):79–85
Ihlow F, Ahmadzadeh F, Ghaffari H, Taskavak E, Hartmann T, Etzbauer C, Rödder D (2014) Assessment of genetic structure, habitat suitability and effectiveness of reserves for future conservation planning of the Euphrates Softshell Turtle *Rafetus euphraticus* (Daudin, 1802). Aquat Conserv Mar Freshw Ecosyst 24:831–840
Israel S, Sinclair T (1988) Indian wildlife. APA, Singapore. 363 pp
IUCN SSC Antelope Specialist Group (2017) *Oryx leucoryx*. The IUCN red list of threatened species 2017: e.T15569A50191626
Jones DM (1988) The Arabian oryx in captivity with particular reference to the herds in Arabia. In: Dixon A, Jones D (eds) Conservation and biology of desert antelopes. Christopher Helm, London, pp 47–57
Kaplan DY (1993) Biological diversity of Mediterranean landscapes in Israel through the development of nature reserves. Landsc Urban Plann 24(1–4):39–42
Karami M, Hutterer R, Benda P, Siahsarvie R, Kryštufek B (2008) Annotated check-list of the mammals of Iran. Lynx, n. s. 39(1):63–102
Kasparek M (1986a) On a historical occurrence of the lion, *Panthera leo*, in Turkey. Zool Middle East 1:9–10
Kasparek M (1986b) On a historical distribution and present situation of gazelles, *Gazella* spp., in Turkey. Zool Middle East 1:11–15
Kasparek M (1988) Zum ursprűnglichen Vorkommen und zur Wiedereinbűrgerung des Fasans, *Phasianus colchicus*, in der Tűrkey. Verh Orn Ges Bayern 24:725–735
Kasparek M (1992) Die Vögel der Tűrkey. Eine Übersicht. Max Kasoparek Verlag, Heidelberg. 128 pp
Kasparek A, Kasparek M (1990) Reiseführer Natur Türkei. BLV Verlagsgesellschaft mbH, München. 246 pp
King FW (1989) *Crocodylus niloticus* Laurenti 1768. In King FW, Burke RL (eds.) Crocodilian, tuatara, and turtle species of the world. A taxonomic and geographic reference. The Association of Systematics Collections, Washington, DC, pp 11–12
Kingswood SC, Kumamoto AT (1988) Research and management of Arabian Sand gazelle in the U.S.A. In: Dixon A, Jones D (eds) Conservation and biology of desert antelopes. Christopher Helm, London, pp 212–226
Kinnear NB (1920) The past and present distribution of the lion in southeastern Asia. J Bombay Nat Hist Soc 27:33–39
Kock D, Amori G (2016) *Gerbillus mesopotamiae*. (errata version published in 2017) The IUCN Red List of Threatened Species 2016: e.T9135A115091132. https://doi.org/10.2305/IUCN.UK.2016-3.RLTS.T9135A22463714.en

Kock D, Krupp F, Schneider W (1994) Einige Säugetiere aus dem Nahr al-Khabur-Gebiet, NE-Syrien. Säugetierkundliche Mitteilungen 35(4):177–183
Kubba S (2011) The Iraqi marshlands and the marsh Arabs. The Ma'dan, their culture and the environment. Ithaca, Reading, UK. 240 pp
Kumerloeve H (1967) Zur Verbreitung kleinasiatischer Raub-und Huftiere sowie einer Grossnager. Säugetierkundliche Mitteilungen 15:337–409
Lay DM, Nadler CF (1975) A study of *Gerbillus* (Rodentia: Muridae) east of the Euphrates River. Mammalia 39:423–445
Layard AH (1853) A popular account of discoveries at Nineveh. John Murray, London. 360 pp
Leemans WF (1960) Foreign trade in the old Babylonian period as revealed by texts from Southern Mesopotamia. Brill, Leiden. 196 pp
Legge AJ (1975) The fauna of Tell Abu Hureyra; preliminary analysis. Proc Prehist Soc 41:74–77
Legge AJ (1977) Origins of agriculture in the Near East. In: Megaw V (ed) Hunters, gatherers and first farmers beyond Europe. University Press, Leicester, pp 51–76
Legge AJ, Rowley-Conwy PA (1986) The beaver (*Castor fiber* L.) in the Tigris-Euphrates basin. J Archaeol Sci 13:469–476
Levin N, Elron E, Gasith A (2009) Decline of wetland ecosystems in the coastal plain of Israel during the 20th century: Implications for wetland conservation and management. Landsc Urban Plann 92(3–4):220–232
Lister AM, Dirks W, Assaf A, Chazan M, Goldberg P, Applbaum YH, Greenbaum N, Horwitz LK (2013) New fossil remains of Elephas from the southern Levant: implications for the evolutionary history of the Asian elephant. Palaeogeogr Palaeoclimatol Palaeoecol 386:119–130
Lloyd S (1984) The archaeology of Mesopotamia. Thames and Hudson, London. 251 pp
Lombard M (1971) L'Islam dans sa premiere grandeur (8e – 11e siècle). Flammarion, Paris. 280 pp
Mallowan MEL (1978) The Nimrud ivories. British Museum, London. 63 pp
Manetti C (1921) L'Anatolia meridionale. Istituto Agricolo Coloniale Italiano, Firenze. 313 pp
Mason CF, Macdonald SM (1986) Otters: ecology and conservation. Cambridge University Press, Cambridge. 236 pp
Masseti M (1987) A Birecik il centro per la protezione e lo studio dell'ibis eremita. *Atlante* (febbraio 1987). Istituto Geografico De Agostini, Milano, pp 96–97
Masseti M (1990) Fauna of Southern Jordan: notes on 22 endangered or extinct mammal and bird species. Studi per l'Ecologia del Quaternario 12:133–146
Masseti M (2000) Wild cats (Mammalia, Carnivora) of Anatolia. With some observations on the former and present occurrence of leopards in south-eastern Turkey and on the Greek island of Samos. Biogeographia 20:607–618
Masseti M (2001) A lost collection of birds and mammals from Northern Syria. In: Abstract book of the 13th international conference of the Society for the History of Natural History "Lost, stolen and strayed. The fate of missing natural history collection", 10–11 May 2001, Naturalis, Leiden, 22
Masseti M (2002) Uomini e (non solo) topi. Gli animali domestici e la fauna antropocora. Firenze University Press/Università di Firenze, Firenze. 337 pp
Masseti M (2003) Exploitation of fauna in the Near East during ancient historical times: observations on the mammalian species represented in Assyrian art. In: Albore Livadie C, Ortolani F (a cura di) Variazioni climatico-ambientali e impatto sull'uomo nell'area circum-mediterranea durante l'Olocene. Centro Universitario Europeo per i Beni Culturali, Ravello. Edipuglia, Bari, pp 375–390
Masseti M (2004) Artiodactyls of Syria. Zool Middle East 33:139–148
Masseti M (2009a) Carnivores of syria. In: Neubert E, Amr Z, Taiti S, Gümüs B (eds) Animal biodiversity in the Middle East. Proceedings of the first international congress: documenting, analysing and managing biodiversity in the Middle East, 20–23 October 2008, Aqaba, Jordan. ZooKeys 31:229–252. https://doi.org/10.3897/zookeys.31.170
Masseti M (2009b) In the gardens of Norman Palermo, Sicily (twelfth century A.D.). Anthropozoologica 44(2):7–34

Masseti M (2009c) Mammals of the Mediterranean islands: homogenisation and the loss of biodiversity. Mammalia 73:169–202
Masseti M (2009d) A possible approach to the "conservation" of the mammalian populations of ancient anthropochorous origin of the Mediterranean islands. Folia Zoologica 58(3):303–308
Masseti M (2012) Atlas of terrestrial mammals of the Ionian and Aegean islands. De Gruyter, Berlin. 302 pp
Masseti M (2015) The early 8th century A.D. zoomorphic iconography of the wall decorations in Qasr al-Amra, Hashemite Kingdom of Jordan. Anthropozoologica 50(2):69–85
Masseti M (2016) A lost collection (1989–1994) of reptiles, birds and mammals of the Syrian Jazira (Syrian Arab Republic). Arch Nat Hist 43(1):21–29
Masseti M (2018) Il cavallo, *Equus przewalskii* Poliakov, 1881, fra antichità e alto medioevo. In Camin L, Paolucci F (a cura di) A cavallo del tempo. L'arte di cavalcare dall'antichità al medioevo. Catalogo della mostra. Firenze, Limonaia del Giardino di Boboli, 25 giugno–14 ottobre 2018. Ministero dei beni e delle attività culturali e del turismo, Roma/Sillabe, Livorno, pp 152–176
Masseti M (in press-a) The Nile soft-shelled turtle or grey turtle, *Trionyx triunguis* (Forsskål, 1775): an Afrotropical-Mediterranean turtle. In: Leviton A et al (eds) Herpetofaunas of the islands of the Mediterranean Basin. Contributions to herpetology, 20
Masseti M (in press-b) Monkeys in the ancient Near East. From prehistory to the first Islamic caliphate. In: Veracini C, Diogo R, Wood B (eds) History of primatology. Taylor & Francis, Oxford
Masseti M, Mazza PPA (2013) Western European Quaternary lions: new working hypotheses. Biol J Linn Soc 109:66–77. https://doi.org/10.1111/bij.12032
Masseti M, Vernesi C (2015) Historic zoology of the European fallow deer, *Dama dama dama*: evidence from biogeography, archaeology and genetics. In: Baker K, Carden R, Madgwick R (eds) Deer and people. Windgather (Oxbow), Oxford, pp 13–22
Masseti M, Elena Pecchioli E, Vernesi C (2008) Phylogeography of the last surviving populations of Rhodian and Anatolian fallow deer (*Dama dama dama* L., 1758). Biol J Linn Soc 93 (4):835–844
McClellan TL (1993) La Siria settentrionale nella Tarda Eta del Bronzo. In: Rouault O, Masetti-Rouault MG (eds) L'Eufrate e il tempo. Electa, Milano, pp 85–86
McDermott WC (1938) The ape in antiquity. The Johns Hopkins Press, Baltimore. 338 pp
McIntosh J (2008) The ancient Indus Valley: new perspectives. ABC-CLIO, Santa Barbara, CA. 441 pp
Meakin B (1901) The land of the Moors. A comprehensive description. Swan Sonnenschein, London. 464 pp
Miller R (1986) Elephants, ivory, and charcoal: an ecological perspective. Bull Am Sch Orient Res 264:29–43
Minniti C (2018) Preliminary results from the study of animal remains at Shahr-i-Sokhta: new researches in 2017 season. In: Abstract book of the 11th international congress on the archaeology of the ancient Near East, 03–07 April 2018, Ludwig-Maximilians-Universität, Munich, 246
Minniti C (2019) New data on non-human primates from the ancient Near Est: the recent discovery of a rhesus macaque burial at Shahr-i Sokhta (Iran). Int J Osteoarchaeol 29(4):538–548
Misonne X (1957) Mammiferes de la Turquie sud-orientakle et du nord dr la Syrie. Mammalia 21:53–67
Mlíkovský J (2012) Ornithological records made by Jan Vilém Helfer in Syria in 1836. Sylvia 48:167–170
Mobaraki A, Mola A (2011) Mesopotamian soft shell turtle (*Raphetus euphraticus*), the strangest turtle of the Middle East. Wildl Middle East Newsl 5(4):6
Moore AMT (1975) The excavation of Tell Abu Hureyra in Syria: a preliminary report. Proc Prehist Soc 41:50–69

Mountfort G (1965) Portrait of a desert. The story of an expedition to Jordan. Collins, London. 192 pp

Mutlak F, Jawad L, Al-Faisal A (2017) *Atractosteus Spatula* (Actinopterygii: Lepisosteiformes: Lepisosteidae): a deliberate aquarium trade introduction incidence in the Shatt Al-Arab river, Basrah, Iraq. Acta Ichthyologica et Piscatoria 47(2):205–207

Muzio C (1925) Bacino Tigri-Eufrate. Mundus, Monografia 4ª. Casa Editrice Sonzogno, Milano, pp 1–28

Nadachowski A, Smielowski J, Rzebik-Kowalska B, Daoud A (1990) Mammals from the Near East in Polish collections. Acta Zoologica Cracoviensia 33(6):91–120

Nader IA, Kock D (1983) Notes on some bats from the Near East (Mammalia: Chiroptera). Zeitschrift für Säugetierkunde 48:1–9

Negahban EO (1979) Architecture of Haft Tepe. In: Kleiss W (ed) Akten des VII. Internationalen Kongresses für Iranische Kunst und Archäologie, München, 7–10 Sept 1976. Archäologische Mitteilungen aus Iran, Ergänzungsband 6:9–29

Niazi AD (1976) On the Mediterranean horseshow bat from Iraq. Bull Iraq Natl Hist Mus 7:167–176

Omer SA, Wronsk T, Alwas A, Elamin MH, Mohammed OB, Lerp H (2012) Evidence for persistence and a major range extension of the smooth-coated otter (*Lutrogale perspicillata maxwelli*; Mustelidae, Carnivora) in Iraq. Folia Zoologica 61(2):172–176

Oppenheim AL, Hartman LF (1945) the domestic animals of ancient Mesopotamia according to the thirteenth tablet of the series Har.ra=hubullû. Journ. Near East Stud 4:152–177

Osborn DJ, Helmy I (1980) The contemporary land mammals of Egypt (Including Sinai). Fieldiana Zool n.s. 5:1–579

Osborne DJ, Osbornova J (1998) The mammals of ancient Egypt. Aris & Phillips, Warmister. 213 pp

Parish W (1860) Diary of a journey with Sir Eyre Coote from Bussora to Aleppo in 1780 (?), from the Original MS. J R Geograph Soc Lond 30:198–211 (published by Wiley on behalf of The Royal Geographical Society with the Institute of British Geographers. https://doi.org/10.2307/1798302

Patterson B (1937) Animal Remains [from Alisan, central Anatolia]. Univ Chic Orient Inst Publ 30:294–309

Perkins D (1964) Prehistoric Fauna from Shanidar. Iraq. Science 144:1565–1566

Peters J, Helmer D, den Driesch A, Saña Segui M (1999) Early animal husbandry in the northern Levant. Palèorient 25(2):27–47

Peters J, von den Driesch A, Helmer D (2005) The upper Euphrates-Tigris basin: cradle of agro-pastoralism. In: Vigne J-D, Peters J, Helmer D (eds) The first steps of animal domestication. New archaeological approaches. Oxbow, Oxford, pp 96–124

Petter F (1961) Repartition geographique et ecologie des rongeurs desertiques (du Sahara occidental à l'Iran oriental). Mammalia 25(N° Special):1–222

Postgate N (2017) Early Mesopotamia: society and economy at the dawn of history. Taylor & Francis, London. 392 pp

Prater SH (1948) The book of Indian animals. Bombay Natural History Society/Oxford University Press, Oxford. 324 pp

Protsch R, Berger R (1973) Earliest radiocarbon dates for domesticated animals. Science 179:235–239

Pulak C (1996) A dendrochronological dating of the Uluburun ship. INA Q 23:12–13

Pulak C (1998) The Uluburun shipwreck: an overview. Int J Naut Archaeol 27(3):188–224

Putman R (1988) The natural history of deer. Christopher Helm, London. 191 pp

Raswan CR (1935) Black tents of Arabia; my life among the Bedouins. Little, Brown, Boston: xiii + 159

Reade J (1983) Assyrian sculpture. British Museum, London. 72 pp

Reuther O (1926) Die Innenstadt von Babylon (Merkes). J. C. Hinrichs, Leipzig. 276 pp

Roncalli M, Mandel M (1993) Il Tigri e l'Eufrate. I fiumi del paradiso. Edizioni San Paolo, Cinisello Balsamo, Milano. 312 pp
Rösner U, Schäbitz F (1991) Palynological evidence and sedimentological evidence for the historic environment of Khtouniye, Eastern Syrian Djezire. Paléorient 17(1):77–87
Ross CE (1989) Crocodiles and alligators. Facts on File, New York. 240 pp
Ross CA, Magnusson WE (1989) Living crocodilians. In: Ross CA (consulting ed) Crocodiles and alligators. Facts on File, New York, pp 58–73
Salim M, Porter R, Rubec C (2009) A summary of birds recorded in the marshes of southern Iraq, 2005–2008. BioRisk 3:205–219
Salnikov VB (2010) First finding of gar *Atractosteus* sp. (Actinopterygii, Lepisosteiformes, Lepisosteidae) in the Caspian Sea near the coast of Turkmenistan. Russ J Biol Invasion 1 (1):17–20
Schlumberger D (1948) Deux fresques omeyyades. Syria, Revue d'art oriental et d'archéologie 25:86–102
Schlumberger D, Le Berre M (1986) Qasr El-Heir El-Gharbi. Bibliothèque archéologique et historique 120:182–244
Schmidt K (1999) Frühe Tier-und Menschenbilder vom Göbekli Tepe – Kampagnen 1995-98. Ein kommentierter Katalog der Grossplastiik und der Reliefs. Istanbuler Mitteilungen 49:5–21
Schnitzler AE (2011) Past and present distribution of the North African–Asian lion subgroup: a review. Mammal Rev 41:220–243
Scullard HH (1974) The elephant in the Greek and the Roman world. Thames & Hudson, London. 288 pp
Serra G, Williamson D, Batello C (2003a) From indifference to awareness. Encountering biodiversity in the semi-arid rangelands Arab Republic. FAO, Rome. 47 pp
Serra G, Abdallah M, al Qaim G, Fayed T, Assaed A, Williamson D (2003b) Discovery of a relic breeding colony of Northern Bald Ibis *Geronticus eremita* in Syria: still in time to save the eastern population? Oryx 38:1–7
Serra G, Abdallah A, Assaed A, Al Qaim G, Abdallah AK (2005) A long-term bird survey of central Syrian desert (2000–2003). Part 1. Sandgrouse 27(1):9–23
Serra G, Scheisch AM, Al Qaim G (2008) Feeding ecology and behaviour of the last known surviving oriental Northern Bald Ibises, *Geronticus eremita* (Linnaeus, 1758), at their breeding quarters in Syria. Zool Middle East 43:55–68
Shalmon B (2004) Mammals. In: Dolev A, Perevolotsky A (eds) Vertebrates in Israel. Israel Nature and Parks Authority, Jerusalem, pp 211–289
Shehab AH, Mamkhair IH (2004) First record of the Egyptian fruit bat, *Rousettus aegyptiacus*, from Syria. Zool Middle East 33:73–78
Shehab AH, Mamkhair IH (2006) Notes on two horseshoe bats; *Rhinolophus ferrumequinum* and *Rhinolophus euryale* (Rhinolophidae, Chiroptera) from Northern Syria. Zool Middle East 39:25–30
Shehab A, Daud A, Kock D, Amr Z (2004) Small mammals recovered from owl pellets from Syria (Mammalia: Chiroptera, Rodentia). Zool Middle East 33(1):27–42
Shehab A, Mamkhair IH, Amr ZS (2006) First record of the lesser horseshoe bat, Rhinolophus hipposideros (Bechstein, 1800) (Rhinolophidae, Chiroptera) from Syria. Hystrix Ital J Mammal New Ser 17(2):161–166
Shehab A, Karatas A, Amr Z, Mamkhair I, Sözen M (2007) The distribution of bats (Mammalia: Chiroptera) in Syria. Vertebr Zool 57(1):103–132
Sindaco R, Venchi A, Carpaneto GM, Bologna MA (2000) The reptiles of Anatolia: a checklist and zoogeographical analysis. Biogeographia 21:441–554
Siracusano G (2002) Preliminary study of the fauna at Zeytinli Bahçe Hoyük. In: Frangipane M, Alvaro C, Balossi F, Siracusano G, Tuna N, Özturk J, Velibeyoglu J (eds) The 2000 Campaign at Zeytinli Bahçe Höyük Salvage Project of the Archaeological Heritage of the Ilisu and Carchemish Dam Reservoirs. Activities 2000. METU, Ankara

Siracusano G (2005) La fauna dei livelli preistorici di Zeytinli Bahçe (Urfa, Turchia). In: Malerba G, Visentini P (eds) Atti del 4° Convegno Nazionale di Archeozoologia (Pordenone, 13-15 novembre 2003). Quaderni del Museo Archeologico del Friuli Occidentale, vol 6, pp 205–213

Siracusano G (2010) Castori sull'Eufrate/*Beavers on the Euphrates*. In: Tagliacozzo A, Fiore I, Marconi S, Tecchiati U (eds) Atti del 5° Convegno Nazionale di Archeozoologia. Associazione Italiana di Archeozoologia. Museo Civico di Rovereto/Edizioni Osiride, Rovereto (Trento), pp 383–387

Siracusano G (2012) Il misterioso ghepardo (*Acynonix jubatus venatucus*) di Arslantepe (Anatolia centrale). In De Grossi Mazzorin J, Saccà D, Tozzi M (a cura di): Atti del 6° Convegno Nazionale di Archeozoologia. Associazione Italiana di Archeozoologia/Dipartimento di Scienze Archeologiche dell'Università di Pisa, pp 369–372

Siracusano G (in press) Zoologia fantastica o meraviglie della zoologia? Resti di animali che non ci sono, c'erano, ma non avrebbero dovuto esserci. Atti dell'8° Convegno Nazionale di Archeozoologia (Lecce, 2015): 13–18

Siracusano G, Carlini R (2010) The enigmatic discovering of "big cat" mandible in EBA level in the Eastern Anatolian site od Arslantepe (Malatya Turchia). In: Vigne J-D, Patou-Mathis M, Lefèvre C (eds) ICAZ 2010. 11th international conference of archaeozoology. Muséum National d'Histoire Naturelle, Paris, p 223

Smith CH (1827) Supplement to the order Ruminata. In: Griffith E (ed) Cuvier's animal Kingdom, vol IV. George B. Whittaker, London, pp 33–428

Snell DC (2005) A companion to the ancient near east. Blackwell, Maiden. 560 pp

Solecki RS (1957) Shanidar cave. Sci Am 197(5):58–65

Stordeur D (2000) Jerf el Ahmar et l'émergence du Néolithique au Proche-Orient. In: Guilaine J (ed) Premiers Paysans du monde. Naissances des agricultures. Éditions Errance, Paris, pp 31–60

Stordeur D, Abbès F (2002) Du PPNA au PPNB: mise en lumière d'une phase de transition à Jerf el Ahmar (Syrie). Bulletin de la Société préhistorique française 99(3):563–595

Strommenger E, Hirmer M (1963) Fūnf Jahrtausende Mesopotamien. Hirmer, Mūnchen. 178 pp

Stuart SN (2008) *Nesokia bunnii*. The IUCN Red List of Threatened Species 2008: e. T14660A4453417. https://doi.org/10.2305/IUCN.UK.2008.RLTS.T14660A4453417.en

Studer J (2001) Les mosaiques animalieres. In: Fiema ZT, Kanellopoulos C, Waliszewski T, Schick R (eds) The Petra Church. American Center of Oriental Research, Amman, pp 271–293

Tangri D, Wyncoll G (1989) Of mice and men: is the presence of commensal animals in archeological sites a positive correlate of sedentism? Paléorient 15(2):85–94

Taşkavak E (1998) Comparative morphology of the Euphrates soft-shelled turtle, *Rafetus euphraticus* (Daudin, 1802) (Reptilia, Testudines) in Southeastern Anatolia. Amphibia-Reptilia 19(3):281–292

Taskavak E, Atatür MK, Ghaffari H, Meylan PA (2016) *Rafetus euphraticus* (Daudin 1801). In: Rhodin AGJ, Iverson JB, van Dijk PP, Saumure RA, Buhlmann KA, Pritchard PCH, Mittermeier RA (eds) Euphrates softshell turtle. Conservation biology of turtles and tortoises: a compilation project of the IUCN/SSC Tortoise and Freshwater Turtle Specialist Group. Chelonian Research Foundation. 5. https://doi.org/10.3854/crm.5.098.euphraticus.v1.2016

Tchernov E (1981) The impact of the postglacial on the fauna of Southwest Asia. In: Frey W, Uerpmann H-P (eds) Beiträge zur Umweltgeschichte des Vorderen Orients. Beihefte zum Tübinger Atlas des Vorderen Orients, vol 8, pp 197–216

Tchernov E (1984a) Faunal turnover and extinction rate in the Levant. In: Martin PS, Klein RG (eds) Quaternary extinctions. The University of Arizona, Tucson, pp 528–552

Tchernov E (1984b) Commensal animals and human sedentism in the Middle East. In: Grigson C, Clutton-Brock J (eds) Animals and archaeology, vol 3. Early herders and their flocks. BAR International Series, vol 202, pp 91–115

Tchernov E (1991) The Middle Paleolithic mammalian sequence and its bearing on the origin of *Homo sapiens* in the Southern Levant. Cahiers de Paléoanthropologie. Editions du C.N.R.S, Paris, pp 77–88

Thesiger W (1954) The marshmen of Southern Iraq. Geogr J 120(3):272–281
Thesiger W (1959) The marsh Arabs. Penguin, London. 256 pp
Thomas O (1915) Notes on *Taphozous* and *Saccolaimus*. J Bombay Nat Hist Soc 24:57–63
Thorbajarnarson J, Messel H, King WF, Ross JP (1992) Crocodiles. An action plan for their conservation. IUCN/SSC Crocodile Specialist Group, Gland, Switzerland. 137 pp
Trouessart EL, Kollman M (1923) Mammals. In: Voyage Zoologique d'Henri Gadeau de Kerville en Syrie, vol 4, pp 59–64
Uerpmann H-P (1981) The major faunal areas of the Middle East during the late Pleistocene and early Holocene. In: Préhistoire du Levant. C.N.R.S., Paris, pp 99–106
Uerpmann H-P (1987) The Ancient Distribution of Ungulate in the Middle East. Beihefte zum Tübinger Atlas des vorderen Orients 27:13–171
Uerpmann M (1995) Early mongooses from Bahrain. In: Buitenhuis H, Uerpmann H-P (eds) Archaeozoology of the Near East II. Backhuys, Leiden, pp 64–71
Uhrin M, Benda P, Andreas M (2000) Správa o výskyte vydry riečnej (*Lutra lutra*) v Sýrii (Report on the occurrence of the Eurasian otter, *Lutra lutra*, in Syria). Vydra 9–10:53–54
van Buren ED (1939) The fauna of ancient Mesopotamia as represented in art. Pontificium Institutum Biblicum, Rome. 113 pp, XXIII pls.
Vogel R (1952) Reste von Jagd- und Haustieren. In: Bittel K, Naumann R (eds) Boghazköy – Hattusa. Wiss Veröff. D. Deuitsch. Orient-Ges., vol 63, pp 128–153
von Lehmann E (1965) Über die Säugetiere im Waldgebiet NW-Syriens. Sitzungsberichte der Gesellschaft Naturforschender Freunde zu Berlin (N.F.) 5(1):22–38
von Oppenheim MF (1931) Der Tell Halaf. Eine neue Kultur im ältesten Mesopotamien. Brockhaus, Leipzig. 588 pp
von Oppenheim MF, Moortgat A (1955) Tell Halaf. Walter de Gruyter, Berlin. 634 pp
Werner YL (1988) Herpetofaunal survey of Israel (1950-85), with comments on Sinai and Jordan and on zoogeographical heterogeneity. In: Yore-Toy Y, Tchernov E (eds) The zoogeography of Israel. W. Junk, Dordrecht, pp 355–387
Werner NY, Rabiei A, Saltz D, Daujat J, Baker K (2015) *Dama mesopotamica*. The IUCN Red List of Threatened Species 2015: e.T6232A97672550
Wettstein OV (1913) Wissenschaftliche Ergebnisse der Expedition nach Mesopotamien 1910. Die Chiroperernausbeute. Annalen Naturh Mus Wien 27:465–471
Williamson G (1949) Iraqi Livestock. Empire J Exp Agric 17:48–59
Winter I (1973) North Syria in the early first millennium B.C., with special reference to ivory carving. Doctoral dissertation. Columbia University. Ann Arbor, University Microfilm
Wirth E (1971) Syrien, eine geographische Landeskunde. Wissenschaftliche Buchgesellschaft, Darmstadt, 110 pp
Wozencraft WC (2005) Order Carnivora. In: Wilson DE, Reeder DM (eds) Mammal species of the world. A taxonomic and geographic reference. John Hopkins University Press, Baltimore. 743 pp
Zahoueh S, Cheiko M (1993) Etude écologique préliminaire de la région de Cyrrhus "Nabi-Houri" au nord d'Alep. Rapport de stage din d'étude. Faculté d'Agronomie, Universitè d'Alep, 92 pp
Zeuner FE (1963) A history of domesticated animals. Hutchinson, London. 560 pp
Zohary M (1973) Geobotanical foundations of the Middle East, vol I and II. CRC, Stuttgart-Amsterdam. 765 pp

Chapter 3
Fishing Gears and Methods: A Comparison of Ancient Mesopotamia and Other Ancient Worlds

Laith A. Jawad

Abstract Fishing is a vital economic issue of a large number of societies all the way through the world nowadays and has completed an important part in the life and survival of many prehistoric cultures. The physical environment of Mesopotamia Israel, undoubtedly could have contributed to the development of fishing communities through the usage of different fishing gears and methods. Human and fish have created a close relationship for a long time and it goes back to early history, when ancestors of human being developed the ability to use aquatic habitats.

Examples of the more significant current fishing gear were in common use in Ancient Mesopotamia. These included fish hooks and nets. The importance of fisheries in the economy of Ancient Mesopotamia is verified by the transportation of fish merchandises and the obligation of duties on fishing rights and in the obligation of a great number of the people in fish business. It is recommended that these basic fishing methods and gears extent in time, along many ways, from Mesopotamia to other parts of the world. At the end of this chapter, a comparison of the fishing gear and methods between the Ancient Mesopotamia and the other ancient regions of the world was given.

3.1 Introduction

Fisheries are a money making and active and method of obtaining food as professional fishermen catch what they utilize for their consumption (Jennings et al. 2001; Jawad 2006). Mesopotamia located in the west of the Asian continent (Almaca 1991). This region is designated by its exclusive topographical location. It is restricted by the Arabian Gulf and its northerly limits facing the southern marshes region (30° to 33° N, 45° to 48° E) (Fig. 3.1). The main wetland, Hawr al-Hammar,

L. A. Jawad (✉)
School of Environmental and Animal Sciences, Unitec Institute of Technology, Auckland, New Zealand

L. A. Jawad (ed.), *Tigris and Euphrates Rivers: Their Environment from Headwaters to Mouth*, Aquatic Ecology Series 11,
https://doi.org/10.1007/978-3-030-57570-0_3

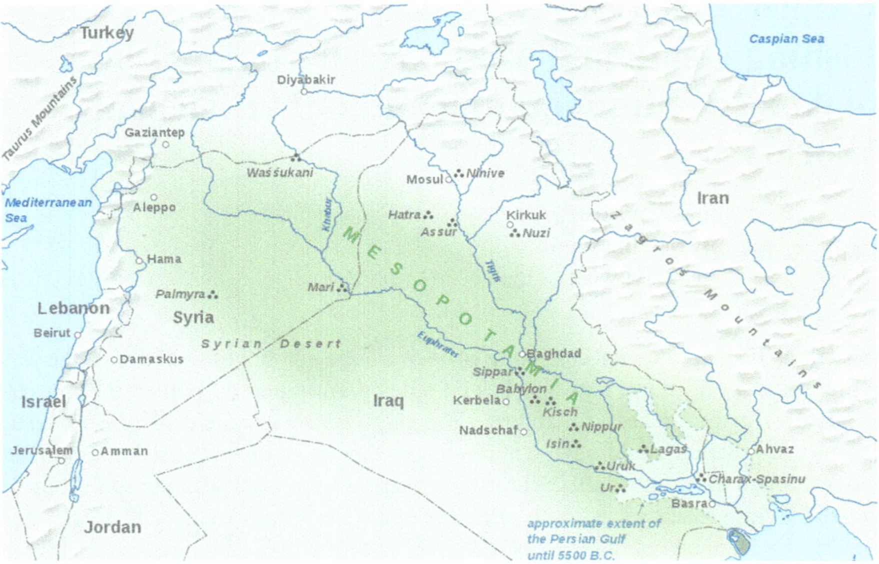

Fig. 3.1 Map of ancient Mesopotamia. Curtsey of Goran teken, CC BY-SA 4.0, https://commons.wikimedia.org/w/index.php?curid=30851043

is geologically new and related to 636 A.D. (Banister 1980; Berry et al. 1970; Rzóska 1980; Jawad 2006).

The Mesopotamian marshlands once formed the largest wetland ecosystem in the Middle East and Western Eurasia (Maltby 1994; Nicholson and Clark 2002). The Mesopotamian marshes are significant for economic, social, and biodiversity values signified by frequency of water flows, accumulation of nutrients and organic matter, and the production of commercially important vegetation and fish (USAID 2004). Marshes assist a variety of purposes for human and other ecosystems including acting as huge settling tanks for the silt deposited by the Tigris and Euphrates Rivers (Wilson 1925); acting as a natural filter for water and other pollutants in the two rivers (Partow 2001); acting as a natural sponge storing water during high river flow and releasing water during low flow; creating nursery grounds for fish and aquatic birds and refuges for terrestrial animals (USAID 2004; Jawad 2006); and being highly productive in vegetation cover used in several objectives. More important, they are considered the homebased of native human societies for thousands of years and are counted as the place of the mythical "Garden of Eden" (Eden Again Project 2003).

With the presence of a huge amount of freshwater in the Mesopotamian region, especially at its lower reaches, neither freshwater nor saltwater fish antiquity in this area has obtained the courtesy that they should be worthy (Jawad 2006).

The aim of this chapter is to offer a broad outline of the fishing gear and methods used in the lower region of Mesopotamia a number of issues associated with fish and comparing them with those used in the other ancient world regions.

3.2 Fish as a Commercial Commodity

The journey of fish as a commercial commodity has started since prehistory and continues to modern times. In this journey, human has caught fish for consumption and later as goods to exchange for valuable belongings and services (Potts 2012). The fishing technology has developed with the development of industry starting from the late 1700s and therefore man inventiveness and emergent of professional systems have directed the supply of fish for mass markets (Pitcher et al. 2013; Jawad 2006). This has created what is known latter as the commercialization of fish (Lam and Pitcher 2012).

In what way fisheries are achieved and ruled offers robust cultural and societal motives of individual performances that pride merchandises over supporting natural capitals and public relations. It is echoed in the present marketplace economy of fisheries, with worldwide export fishery merchandise now worth over US$129 billion (Potts 2012).

3.3 The Role of Fish Remains in Defining Fishing Gear and Sites

Potts (2012) talks about the role of fish remains as an indication for fishing gear and sites. Due to the importance and significance of this subject, I thought it worth given an overview of this section here in the present chapter highlighting those points regarding Lower Mesopotamian region.

Fish remains exist in the archaeological signs at locations of all periods all over the Near East. Nearly whole fish, or parts, containing fins, and particularly vertebrae, are able of enduring in archaeological sediments for thousands of years (Alperson-Afil et al. 2009; Potts 2012).

There are a possibility of under representation of a correct number of fish in the archaeological record due to several factors among them is the technique used in retrieving the bones. In some cases, the number of bones was very low that lead the archaeologists to believe that human settlements in the area were not ichthyofagi (Potts 2012) in spite of recovery of a large number of copper fish hooks, which indicates a considerable fishing activity went on the area (Sahrhage and Lundbeck 1992; von den Driesch 1993; Potts 2012). The other example of such incident can be drawn from a fishing area at Nippur in Mesopotamia, where six fragments of fish were recovered at Boessneck (1978) and Boessneck et al. (1984).

3.4 Fishing Gear and Practices

It is impossible to compare the results of the fishing efforts measured at present time based on sophisticated equipment with that of the ancient Mesopotamian region as in the latter era far simpler equipment was used, which can be considered to be misleading. Larger fish catch than the modern time as possible using small fishing efforts in the ancient time (Jawad 2006). This is due to the absence of fish stock depletion that became an icon of modern time (Jacobsen 2005).

It is still unknown how the early man trapped his fish (Potts 2012). The available information about the stone tool accumulation at Middle Pleistocene Gesher Benot Ya'aqov was dominated by grinders, scrapers, and borers (Alperson-Afil et al. 2009), all of which has no relation to catching fish. Furthermore, no evidence are there for any fishing tool that can fisherman gather fish with a bare hand (Jawad 2006; Potts 2012). Hook-and-line, nets, traps, and spears appeared at the start of the Neolithic period onward.

Fishhooks prepared from the pearl oyster shell (*Pinctada margaritifera*) are mainly characteristic of the locations on the Arabian Sea coast of Oman from the early fifth through the fourth millennium BC when they were replaced by copper fish hooks (Charpentier and Méry 1997). In some locations in the Arabian Gulf, metals such as copper or bronze were extensively used in hooks manufacturing starting with the bronze age (Potts 2000; Beech 2004; Moon 2005). Most of these hooks look like the present hook in use, with barbed end (Tallon 1987/2, in Potts 2012; Tallon 1987/1, in Potts 2012; Van Ess and Pedda 1992). Sahrhage and Lundbeck (1992) gave an illustration of two bronze fishing hooks drawn by Von Lorentz-König. They represent an early metal fishing hooks used in Mesopotamia as early as 5000 BC. One of these two metal fishing hooks looks like the English letter "J", with short and curved inward arm. The other hook also has the shape of the English letter "J", but the arm is curved inward further so the hook looks rounded rather than the letter "J." Both types of hooks have pointed short end provided with a triangular barb. The long arm of the hook is simply cylindrical and not flat. The long and the short arms of the two types of hooks are well separated from each other so to assist to penetrate the flesh of the fish properly. On the other hand, Woolley (1965) gave diagrams of two fishing hooks used during the Kassite Period (1595–1156 BC) in Babylonia. They differ from the other fishing hooks used in Mesopotamia in having curved long arm of the letter "J", with flattened end (Fig. 3.2). Such model of hooks is still in use in the southern marshes of Iraq (Jawad 2006). Sahrhage and Lundbeck (1992) have noted that in Assyria, fishing hooks were used, but with line and the rode was not known there.

Casting nets were in use by the sixth millennium BC on the coast of Oman (Charpentier 1996) and are in use at the present in Mesopotamia and considered one of the effective fish catching gear (Jawad 2006). These kinds of nets have not withstood in Arabia and do not appear in the archaeological excavations in spite of their existence was established by the presence of their large number. The plainest of these are made from stones, frequently of an equally even size in having a

Fig. 3.2 Fishing hooks form Mesopotamia. (a) about 3000 BC (Sahrhage and Lundbeck 1992); (b) about 2600 BC from Ur (Beech 2004)

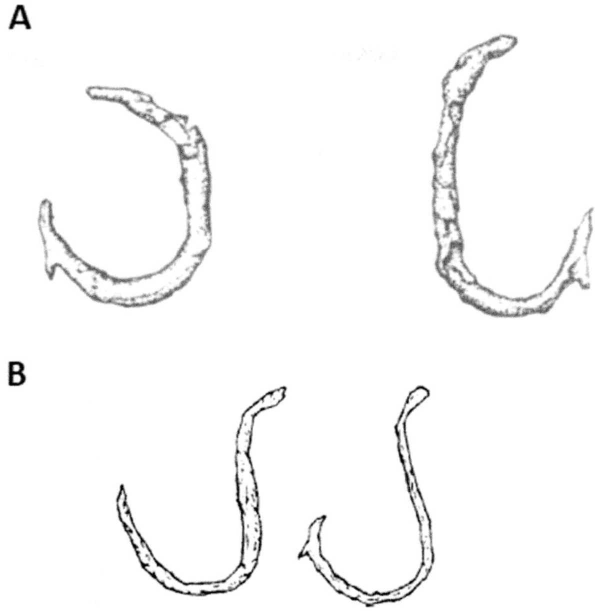

transverse furrow to support in binding them to the net (Charpentier and Méry 2008). It is interesting to note here that wooden floats are in use in the present time in the lower reaches of Mesopotamia by marsh Arabs, but instead of wood, parts of the date palms stem are used (Jawad 2006).

Although no identification of the Khafajah net fibers was published, the twist of the netting is clearly visible and one net—sinker still has a part of a net wrapped around it and tied off (Delougaz 1940). In the Ur III period ox tendons were used to make nets (Englund 2012), while in Hellenistic and Roman Galilee nets were made of flax (Hanson 1997).

Nets are generally classified as cast nets, requiring a single fisherman; surface gill nets, requiring 2–4 fishermen; and seine or dragnets, requiring 15–20 fishermen (Jawad 2006). These types are used in different environments for particular target species. In Oman, individual catches of 15–20 kilograms per throw using a cast net from the shore have been reported (Bekker-Nielsen 2005), however, this far surpasses the 2–3 kilograms evaluated for cast net method in southern Iraq (Jawad 2006; Potts 2012).

Catching fish using spears proposed an alternative to single hook-and-line fishing. In the latest past, fishing spikes were preferred by the Marsh Arabs of southern Iraq (Jawad 2006). Conferring to Philby (1959), their fishing is completed totally by tridents, comprising of a three-split pieces of metal secured to the end of a long reed, with which the fisherman, sat at the end of his boat and closely watching the depths below him as it swimming sluggishly by, shoves the water with a sharp, straight, downward stroke. An illustration displaying a fishing method, with a spear is available on an Early Dynastic (mid-third millennium BC) cylinder seal in Berlin

(Sahrhage 1999). What has been agreed as a three-split fishing spear shows uncommonly as a symbol on Mesopotamian cylinder seals from the Early Dynastic, Old Akkadian, Old Babylonian/Old Assyrian, and Neo-Assyrian eras (Black and Green 1992; Potts 2012).

The fish traps were perhaps utilized mostly, but, while current ones are often made of wire, ancient types were made of organic resources and therefore have left no hint in the archaeological findings (Potts 2012; Jawad 2006). The occurrence of interwoven reed fish catches has been expected in southern Mesopotamia, mostly due to their symbols on Egyptian wall reliefs (Sahrhage 1999). Jawad (2006) has shown that in the southern marshes of Iraq, certain types of traps are used to catch fish, which are considered as the creative work of the locals.

3.5 Similarities and Differences Between the Fishing Gears of Mesopotamia and the Other Ancient Regions of the World

The present section deals with the comparison of the fishing gears used in ancient Mesopotamia with those used in the other ancient world regions of the world (Jawad 2006). For evident reasons, ancient objects made from organic materials are the great unknowns of the past. Though, the technology available to us today enables us to obtain more information from the few fragments that are preserved. This method is more consistent than the traditional technique based on written sources and iconography, which remain, nevertheless, very important as well. The shortage of material data about artifacts such as fishing nets remains in enquiring contrast to the immense importance they had for everyday life in antiquity.

In the history of the ancient civilizations, hooks were maybe the only fishing artifacts that survive the concealed of thousands of years. As hooks are usually made of hard materials that resist decomposition such as bones, shells, and metals, they can be seen abundant in the archaeological sites where fishing activities took place beside the fish bones (Jawad 2006).

Although the evidence about fishing gears and methods in the Palaeolithic age are scarce, Sahrhage and Lundbeck (1992) mentioned that humans in that time used the bare hand to catch the fish. This could happen in rivers other than seas and after the fishes guided to a shallow area, where others can catch fish by their hands. This method of catching fish is in use at present time in the southern marshes of Iraq, with slight variation and it may be inherited from their ancestors of the Sumerian time. Jawad (2006) described three methods that are currently used by people living in the southern marshes of Iraq to catch fish. In those techniques, catching fish by hand is the main issue. The methods are Al-Shiah (Mud Dams) (Jawad 2006, Fig. 8), Al-Suwaise (Burning Method) (Jawad 2006, Figs. 9a and b), and Al-Tawamees or what is known as fishing by diving (Jawad 2006, Fig. 10), which involves selecting

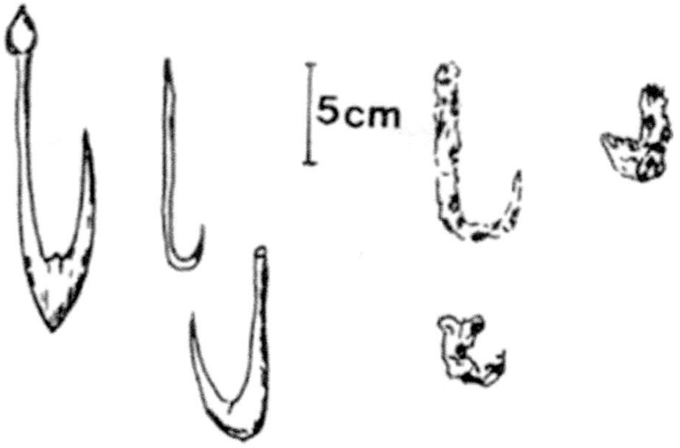

Fig. 3.3 Fishing hooks of the Mesolithic age

small reed floating islands for this kind of method of catching fish. The details of these methods are given by Jawad (2006).

The shape of the hooks that are used in fishing activities showed conservation over the times. The general "J" shape is dominant, but there are some variations in the length of the arms of the letter "J", distance between them, presence or absence of loop at the end of the long arm, and presence or absence and position of the barb (Potts 2012). The hooks that have been used in the Mesolithic age (15000–5000 BP) from Northern Europe showed differences from those used in Mesopotamia (Fig. 3.3) (Sahrhage and Lundbeck 1992; Beech 2004). In spite of having "J" shape, the long and the short arms of this letter was joined directly with an angle and not curved as in the hooks retrieved from Mesopotamia (3000–2600 BP). In addition, there is no barb at the end of the short arm of the letter "J" and this arm is thicker than the long arm (Sahrhage and Lundbeck (1992).

For the hooks of the Neolithic and the Bronze ages from Northern Europe, they look similar to those used in Mesopotamia, but they are different in being curved outward at the end of the long arm of the letter "J" (Fig. 3.4). Moreover, hooks with double curved ends have been used in these ages (Potts 2012).

Comparing the fishing hooks used during the Preceramic era in Peru with those used in Mesopotamia, it looks that the former had the long and the short arms of the letter "J" not parallel to each other as the short arm is displaced slightly outward and conspicuously short, absence of barb at the end of the short arm, the head of the long arm is either with a circle and a hole to fasten the rope or flattened (Fig. 3.5). On the other hand, the Phoenician fishing hooks that have been used 4000 BP were different from those of Mesopotamia in having the long arm of the letter "J" straight and flattened at the end, and significantly smaller short arm (Fig. 3.6).

The fishing hooks that have been used in Egypt during the Middle Empire (2180–1640 BC) look completely different from those used in the Mesopotamia 2000–3000 years before. The differences can be seen in: long arm of the letter "J" being curved, straight, or looped with hole; the short arm of the letter "J" being short, long, parallel, or not parallel to the long arm, equipped with barb or not; and the distance between the two arms of the letter "J" is either curved or straight. In

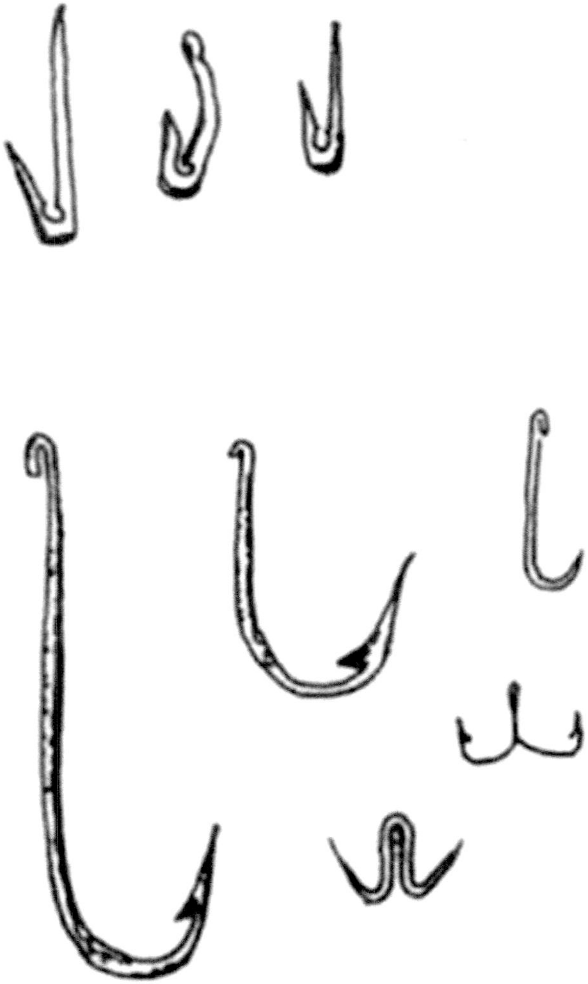

Fig. 3.4 Fishing hooks of the Neolithic age

Fig. 3.5 Fishing hooks of the preceramic era in Peru

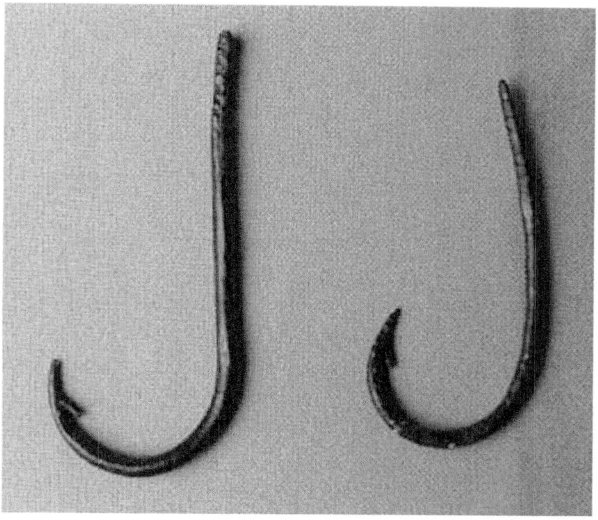

Fig. 3.6 Fishing hooks of the Phoenician time

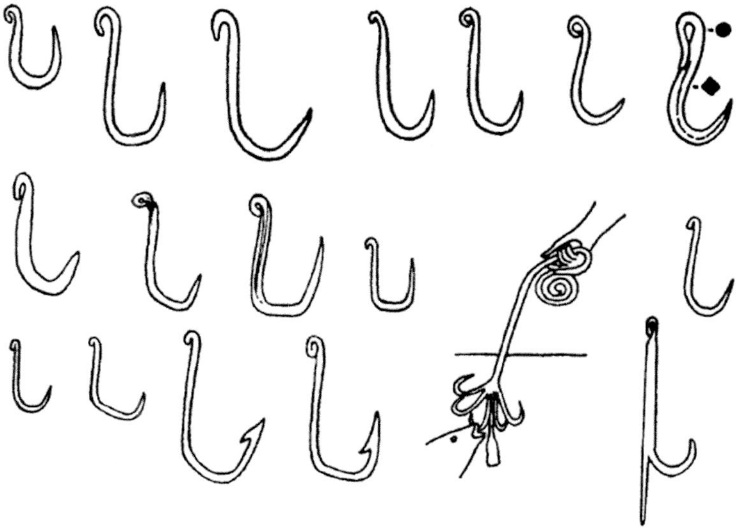

Fig. 3.7 Fishing hooks of Ancient Egypt

addition, hooks with two ends (one straight and the other curved) were also used in this period of time in Egypt (Fig. 3.7) (Aleem 1972).

Gaur (2004) described the fishing hook used by fishermen practiced fishing activities in Bet Dwarka Island, Gujarat, India during the Bronze Age (2600–1900 BC). He mentioned that Shinde and Thomas (1993) have considered the metallic fishhooks were probably developed by the Harappans, and were certainly superior to those from contemporary sites in Egypt and Mesopotamia.

Fig. 3.8 Headscarf of the men Marsh Arab showing the pattern of fishing net

Looking at the shape of the fishing hook that Gaur (2004) provided it does not differ from those used in Ur, Mesopotamia. In addition, Gaur (2004) did not specify in what aspect the fishing hooks of Harappan region were superior to those of Mesopotamia and Egypt. The fishing hooks have been used by fishermen in Mesopotamia thousands of years before those of Harappan region ((Sahrhage and Lundbeck 1992; Beech 2004).

Different from the other ancient world regions, Archaeological investigations did not show whether ancient Mesopotamian inhabitants were used to use fishing nets (Jawad 2006). It could be possible that they used sort of nets made of reed and since the materials are natural, then they perished over the years. Sociological evidence from the present life of the Marsh Arabs, the possible ancestors of the ancient Mesopotamians is available to show that the inhabitants of the lower reaches of Mesopotamia used sort of net in their fishing activities. This evidence can be seen in the pattern illustrated in the head scarf of men in Iraq in general and in the marsh area in particular (Jawad 2006). The pattern looks like a simple version of sprang technique of knot used in netting fishing net (Hald 1980). This was widely used in ancient Egypt. There are a few examples from the Neolithic in the north of Europe and Switzerland, but they are not very common (Rimantiene 1995). The edge of the headscarf is provided with wavy lines representing water (Fig. 3.8). This similarity needs to be investigated further in comparison with fishing artifacts collected from Mesopotamia. If this technique of nets is proved to be present in ancient Mesopotamia, then Egyptians were taken this knitting technique from Mesopotamia.

References

Aleem AA (1972) Fishing industry in ancient Egypt. Proc R Soc Edinb B Biol Sci *73*:333–343

Almaca C (1991) Evolutionary, biogeographical and taxonomic remarks on Mesopotamian species of *Barbus* s. s. Arquivos do Museu Bocage, Nova Serie 2(4):63–78

Alperson-Afil N, Sharon G, Kislev M, Melamed Y, Zohar I, Ashkenazi S, Rabinovich R, Biton R, Werker E, Hartman G, Feibel C (2009) Spatial organization of hominin activities at Gesher Benot Ya'aqov, Israel. Science 326(5960):1677–1680

Banister KE (1980) The fishes of the euphrates and tigris. In: Rzóska J (ed) Euphrates and tigris. Mesopotamian ecology and destiny. Dr. W. Junk Publisher, The Hague, pp 95–108

Beech MJ (ed) (2004) In the Land of the Ichthyophagi: Modelling fish exploitation in the Arabian Gulf and Gulf of Oman from the 5th millennium BC to the late Islamic period, vol 1217. British Archaeological Reports Limited, Oxford

Bekker-Nielsen T (2005) The technology and productivity of ancient sea fishing. In: Ancient fishing and fish processing in the black sea region, (black sea studies 2). Aarhus University Press, Aarhus, pp 83–95

Berry RW, Brophy GP, Naqash A (1970) Mineralogy of the suspended sediment in the Tigris, Euphrates, and Shatt al-Arab Rivers of Iraq, and the recent history of the Mesopotamian plain. J Sediment Petrol 40(1):131–139

Black J, Green A (1992) Gods, demons and symbols. University of Texas Press, Austin

Boessneck J (1977/1978) Tierknochenfunde aus Nippur. In: Gibson M (ed) Excavations at Nippur: twelfth season (oriental institute communications 23). The Oriental Institute, Chicago (IL), pp 153–187

Boessneck J, von den Driesch A, Steger U (1984) Tierknochenfunde der Ausgrabungen des Deutschen Archáologischen Instituts Bagdad in Uruk – Warka, Iraq. Baghdader Mitteilungen 15:149–189

Charpentier V (1996) Archaeology of the Eryhthraean Sea: craft specialization and resources optimization as part of the coastal economy on eastern coastlands of Oman during the 4th and 3rd millennium BC. In: G. Afanasev, S. Clueziou, J.R. Lukacs and M. Tosi (eds.), The prehistory of Asia and Oceania. 13th international congress of prehistoric and protohistoric sciences. Forli, Italy, pp 181–192

Charpentier V, Méry S (1997) Hameçons en nacre et limes en pierre d'Océanie et de l'Océan Indien: analyse d'une tendance. J de la Societé des Océanistes 2:147–156

Charpentier V, Méry S (2008, January) A neolithic settlement near the strait of Hormuz: Akab Island, United Arab Emirates. In: Proceedings of the seminar for Arabian studies. Archaeopress, Oxford, pp 117–136

Delougaz P (1940) The Temple oval at Khafajah. University of Chicago Press, Chicago

Eden Again Project (2003) Building a scientific basis for restoration of the Mesopotamian marshlands. Report of the Eden again project, The Iraq Foundation, Baghdad, Iraq

Englund RK (2012) Equivalency values and the command economy of the Ur III period in Mesopotamia. The Construction of Value in the Ancient World, Los Angeles, pp 427–458

Gaur AS (2004) A unique late bronze age copper fish-hook from bet Dwarka Island, Gujarat, west coast of India: evidence on the advance fishing technology in ancient India. Curr Sci 86 (4):512–514

Hald M (1980) Ancient Danish textiles from bogs and burials: a comparative study of costume and Iron age textiles, Copenhagen

Hanson KC (1997) The Galilean fishing economy and the Jesus tradition. Biblic Theol Bull 27 (3):99–111

Jacobsen ALL (2005) The reliability of fishing statistics as a source for catches and fish stocks in antiquity. Ancient fishing and fish processing in the Black Sea region. Aarhus University Press, Aarhus, pp 97–104

Jawad LA (2006) Fishing gear and methods of the lower Mesopotamian plain with reference to fishing management. Marina Mesopotamica 1(1):1–37

Jennings S, Kaiser MJ, Reynolds JD (2001) Marine fisheries ecology. Blackwell Science, Oxford, p 379

Lam ME, Pitcher TJ (2012) The ethical dimensions of fisheries. Curr Opin Environ Sustain 4 (3):364–373

Maltby E (ed) (1994) An environmental and ecological study of the marshlands of Mesopotamia. Draft consultative bulletin. Wetland Ecosystems Research Group, University of Exeter. Published by the AMAR Appeal Trust, London, p 224
Moon J (2005) Tools, weapons, utensils and ornaments. Bitumen artefacts and uses. In: Killick R, Moon J (eds) The early Dilmun settlement at Saar. (London-Bahrain archaeological expedition Saar excavation report 3). Archaeology International, Ludlow, pp 193–198
Nicholson, E. and Clark, P. (eds.) (2002). The Iraqi marshlands: a human and environmental study. London: The Amar Appeal International Charitable Foundation
Partow H (2001) The Mesopotamian marshlands: demise of an ecosystem. Early warning and technical assessment report. UNEP/DEWA/ATR. 01-3 rev. 1, division of early warning and assessment, United Nations environmental Programme, Nairobi, Kenya
Philby HSJ (1959) The eastern marshes of Mesopotamia. Geogr J 125:65–69
Pitcher TJ, Lam ME, Ainsworth C, Martindale A, Nakamura K, Perry RI, Ward T (2013) Improvements to Rapfish: a rapid evaluation technique for fisheries integrating ecological and human dimensions. J Fish Biol 83(4):865–889
Potts DT (2000) Ancient Magan, the secrets of tell Abraq. Trident Press, London
Potts DT (2012) Fish and fishing. In: A companion to the archaeology of the ancient near east, vol 1. Wiley-Blackwell, Oxford, pp 220–235
Rimantiene R (1995) Lietuva iki Kristaus, Vilnius
Rzóska J (ed) (1980) Euphrates and Tigris: Mesopotamian ecology and destiny. J. W. Junk, Holland, p 120
Sahrhage D (1999) Mesopotamia: fishing and fish cult in ancient Mesopotamia. Peter Lang GmbH, Frankfurt, p 241. (In German)
Sahrhage D, Lundbeck J (1992) A history of fishing. Springer-Verlag, Berlin, p 348
Shinde VS, Thomas E (1993) Man Environment, XVIII, 145–147
United States Agency for International Development, USAID (2004) The Iraq marshlands restoration program. www.usaid.gov/iraq
Van Ess M, Pedda F (1992) Uruk Kleinfunde II. Ausgrabungen in UrukWarka Endberichte Band 7. Mainz am Rhein
von den Driesch A (1993) Faunal remains from Habuba Kabira in Syria. In: Buitenhuis H, Clason AT (eds) Archaeozoology of the near east. Universal Book Services, Leiden, pp 52–59
Wilson AT (1925) The Delta of the Shatt al'Arab and proposals for dredging the Bar. Geogr J 65 (3):225–239
Woolley SL (1965) Kassite period and period of the Assyrian kings. Ur excavations VIII (London and Philadelphia, Joint publications of the British Museum and University of Pennsylvania)

Chapter 4
Ichthyological Characteristics Available in the Fish Images Existed in the Art of the Ancient Mesopotamia

Laith A. Jawad

Abstract Through the years since a human has inhabited earth, fish images and other fauna were depicted on the walls of the caves and other artefacts. With the appearance of great civilizations in the Near East, the fauna of this part of the world appeared in different artwork such as pottery, stone vessels, cylinder seals and reliefs dating too many different periods.

The ichthyological description of the fish images found in the art of ancient civilization is very rare and some workers have examined and described the fish images that appeared in them. No previous studies were on record about any ichthyological description of the fish images that appeared in the art of ancient Mesopotamia. In the present chapter, an ichthyological examination and description were given to the fish images of selected artefacts of a different period of ancient Mesopotamia and Egypt. At the end of the chapter, a comparison of the work of ancient Mesopotamian artists with that of the ancient Egyptians was given to envisage how the artists in both civilizations have developed the ichthyological perspective.

The present study has shown that (1) both ancient Mesopotamian and Egyptian artists have depicted the images of the fish species that live in their environment and those usually catch and eat. For Mesopotamia, the images of the species of the carp family, Cyprinidae have dominated the artwork, while in ancient Egypt, images of species of tilapia were overshadowed; (2) the credit should go to the ancient Mesopotamian for recording shark species in the freshwater environments worldwide; (3) there are still in use in the present time Iraq, some practices and rituals related to fish that have been used by ancient Mesopotamia; (4) colours are less used in artworks of Mesopotamia, but the Mesopotamian people were used colours in their artworks before the ancient Egyptian; (5) both ancient Mesopotamian and Egyptian artists were fully aware of some detailed ichthyological features of the fish and they depicted these features in different artefacts and (6) The ability to depict more detailed characters was developed for the artists of the two ancient civilizations through time as judged from the sequences of artworks examined.

L. A. Jawad (✉)
School of Environmental and Animal Sciences, Unitec Institute of Technology, Auckland, New Zealand

L. A. Jawad (ed.), *Tigris and Euphrates Rivers: Their Environment from Headwaters to Mouth*, Aquatic Ecology Series 11,
https://doi.org/10.1007/978-3-030-57570-0_4

4.1 Introduction

Art has a variety of extents and viewpoints. For example, it can be tackled as historic styles, such as by inferring the style in the historic context; as social history, by analyzing the social conditions by which different styles and forms appear; and as a symbol, such as interpreting it in its religious, ritual or symbolic settings (Gombrich 1995). Frothingham (1894) when investigating some theoretical issues of art, witnessed that in every period of civilization, the universal ideas that ruled the period were shown by multiple forms in the different interrelated arts from philosophy to painting. This concept may parallel the concept of Zeitgeist (Hegel 2009).

Fish images and other fauna of the Near Eastern fauna appeared in different artefacts and artwork pain such as pottery, stone vessels, cylinder seals, and reliefs dating to many different periods (Sahrhage 1999). Among the examples of such depiction are the images of fish appear on early third millennium painted pottery of so-called Susa D type from southwestern Iran (Amiet 1966); those on the early second millennium BC rock relief at Kurangun in Fars (Seidl 1986); and on cylinder seals of many different periods in the greater Mesopotamian region (von Osten 1934; van Buren 1948).

Figurines of fish were used in apotropaic rituals, and they sometimes appear on Neo-Assyrian reliefs, as protective spirits. As an example, in the Old Babylonian period, the figure of the fish—man or fish—centaur (Akkadian *kulull û*) appears in Mesopotamia (Wiggermann 1992); the Statues of fish–women (Akkadian *kuliltu*) in the Nabu temple of Nimrud are referred to in a Neo-Assyrian administrative text (Wiggermann 1992); and in the Kassite period (mid-second millennium BC) in southern Mesopotamia, the "fish-garbed figure" or fish—*apkallu*—a standing, bearded human male wearing the skin of a fish, its head mounted like a hat and its scaly body extending down over the shoulders and back, tail extending below the waist—appears, becoming more common in the art of the Neo-Assyrian period (Green 1986).

The ichthyological description of the fish images found in the art of ancient civilization is very rare. The work of Moyle and Moyle (1991) on Western art, with brief discussions of primitive and Asian art, takes in part the ichthyological issue of the fish images. Coad et al. (2000) described the ichthyological characteristics of the fish that appeared in the artwork of Rashid-al-Din Hamadani, thirteenth Century. In that painting, the Prophet Jonah has been patriated while the fish swallowing him. Moradi (2016) described in part the ichthyological concepts of the fish imagery in Iranian artwork. No previous studies were on record about any ichthyological description of the fish images that appeared in the art of the ancient Mesopotamia. The present chapter offers the first ichthyological description for fish images that appeared in the art of ancient Mesopotamia for the period 3000–539 BC. At the end of the chapter, a comparison of the work of ancient Mesopotamian artists with that of the ancient Egyptians were given to envisage how the artists in both civilizations have developed the ichthyological perspective.

4.2 Chronology of Mesopotamia and Egypt

In the present chapter, periods of the timeline of the ancient Mesopotamia and Egypt will be given. For those readers who are not familiar with such periods, a list of the eras that ancient Mesopotamia and Egypt passed through are given. All dates are B.C. and approximate and may subject to revision (Appendices 1 and 2) (Piccion (2019) and Hawass (2019), respectively.

4.3 Ichthyological Terms

Through the text of this chapter, ichthyological terms will be used. Following the term, a brief definition of each term will be given to aid those readers who are not familiar with the subject of ichthyology to know different parts of the fish body that the description of the fish images will contain. Illustration of the external major parts of the fish body is given in Fig. 4.1.

4.4 Ichthyological Description of the Fish Images

The fish images that appeared in the artworks of Ancient Mesopotamia were grouped according to their time period and arranged chronologically. The fish images were obtained from 7 periods extending from the Proto-Literate period (4000–2900 BC) to the Neo-Babylonian Period (626–539 BC). The total number of fish images described in the present chapter from Mesopotamia is 42, these are as follows: 3 images from the Proto-Literate Period (4000–2900 BC); 3 from the Early Dynastic Period (2900–2350 BC); 2 from Akkadian Empire Period (2282–2250 BC); 4 images from third Dynasty of Ur; 2 from the Old Assyrian Period (2025–1365); 16 from the

Fig. 4.1 Ichthyological terms of fish external morphology Courtesy of Laith Jawad

Fig. 4.2 Images of cylindrical seal (left) and stamp seal (right). Courtesy of Van Buren (1948)

Middle Babylonian Period (1570–689 BC); 1 from Neo-Assyrian Period (883–609 BC) and 2 images from the Neo-Babylonian Period (640–539 BC). From ancient Egypt, 9 images were examined and described.

Proto-Literate period 4000–2900 BC
Jemdet Nasr Period 3200–2900 BC

In this early period, the representation of the fish was very simple, but still the ancient Mesopotamian artist seems to know some morphological characteristics of the fish species that he/she is depicting on the clay tablets or seals. Van Buren (1948) in his work on the fish offerings in ancient Mesopotamia showed images of 2 cylindrical and one stamp seals. In these images, fish images were drawn. In the image of one of the cylindrical seal (Fig. 4.2, left image), images of two fishes were drawn. The outline of these images tells that the artist knows very well the shape of the fish (spindle shape) and the shape of the tail (caudal fin), where the two fishes in this seal look different species according to the shape of their tail. The fish appeared rising between the two men looks a member of the family Cyprinidae (carp family) judging from the shape of the body and the tail in spite of the absence of other fins and details on the head of the fish. The other fish appeared behind the man that looks busy offering or blessing a god, looks not a cyprinid fish due to the shape of the tail and the darkening of the ventral side of the fish body. This fish appeared having a very narrow caudal peduncle (the area where the tail is connected to the fish body) a character that not present in the fish species living in Euphrates–Tigris River drainages. Although there was one dorsal fin (fin found on the back of the fish body), but it is located much forward that it should be. In addition, there was a small line representing the eye of the fish, which was not in its normal position (further toward the mouth of the fish). Overall, this image represents fish that has no relationship to the fish species families living in Mesopotamia at the present time. In the image of the seal appeared on the right in Fig. 4.2, a large fish was drawn, with head directed upward. The outline of the fish body, the shape of the tail, the shape of the head and the position of the dorsal fin and the paired pelvic fins seemed all in agreement with the morphological characteristics of large-sized cyprinid fish species such as *Luciobarbus esocinus* (Biz) and *Luciobarbus xanthopterus* (Qattan). In this fish image, the artist illustrated perfectly the ichthyological characteristics of the caudal fin of the fish (Tail), where small lines were drawn representing the fin rays of

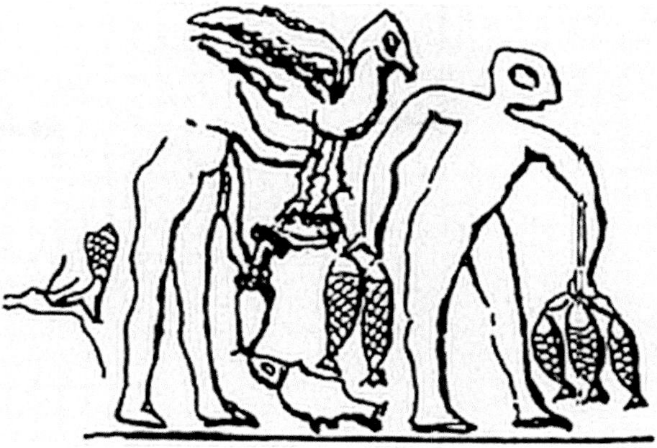

Fig. 4.3 Cylindrical seal with man carrying 5 small sized fishes (Courtesy of Van Buren (1948)

the tail (fine bony structures composing the fins of the fish) and the bifurcation of the tail, which is among the characteristic features of the members of the family Cyprinidae. The other fins, the dorsal and the pelvic fins, here the artist succeeded again in showing the correct ichthyological position of these two fins and in addition showed that the pelvic fins are paired. The choice of depicting an image of a large-sized cyprinid fish in cylindrical seal showing fish-offerings could represent how ancient Mesopotamians are keen to select the most valuable fish specimens and offered it to God as the two cyprinid species mentioned above are still having commercial importance due to their size and their taste.

In Fig. 4.3, Van Buren (1948) showed an image of another cylindrical seal with man carrying 5 small-sized fishes (2 in the right hand and 3 in the left hand). On the floor, another small-sized fish specimen showed lying back to the feet of the man. The process that man carrying the fish was fish-offerings as he presenting the fish to the god. The shape of the five fish specimens carried by the man in two bunches suggested they belong to the deep-bodied cyprinid species such as *Carasobarbus luteus* (Himri) or *Mesopotamichthys sharpeyi* (Buni), which are very common in the lower reaches of Mesopotamia. The artist in these images showed very well the scalation pattern on the fish body, which coincided with that of large-scale cyprinid species mentioned above. Although the head and the tail of the fishes were not drawn in detail, their shape looks in agreement of the cyprinid fish species. The fish specimen that lied on the floor at the foot of the man look different cyprinid species from those held in bunches. This fish had a more elongated body, with a smaller head and 2 dorsal fin. With latter characteristic, it is possible to conclude that this specimen is one of the mullet species (Family Mugillidae) and could well be *Planiliza abu*, a common freshwater mugillid fish species in Mesopotamia. Moreover, the artist does not draw scales on the fish body, this may due to the scales of the mugillid species are smaller than those of cyprinid fish species and hence he does not showed them. Another characteristic that may suggest this fish is a mullet species is the wide caudal peduncle, which is wide in *P. abu*. The way the fish specimens were grouped in bunches is very interesting as it is still in use in the present time

Fig. 4.4 Fish appeared in a stamp seal seems it has been picked up from the water by the two birds standing on its sides. Courtesy of Van Buren (1948)

Mesopotamia (Iraq). In this way, a green leaflet of date palm is used to thread through the mouth of the fish and then through the operculum (gill opening) and then to thread another specimen in the same way. Finally, the two ends of the date palm leaflet fastened together to enable to carry them. In the present Iraq, you can see such fish bunching is used by fishermen going around houses selling freshly caught fishes. It is also possible to note that the fish bunch with 3 fish specimens is heavier than that with the specimens as the man left hand is droopy down more than the right. Once again and through the fish images, it seems that fish-offerings are important and respective ritual among ancient Mesopotamians as the person who offers the fish chose the fat and heavy specimens to offer to god.

The fish appeared in Fig. 4.4 showed in a stamp seal seems it has been picked up from the water by the two birds standing on its sides. The artist has perfectly depicted the spindle shape of the fish with correct form of head and tail. The size of the eye and its position on the head are ichthyologically correct. I addition, the artist attempted to draw scales on the body of the fish. Here, it is possible to say that the artist has noted the size of the scales at the middle part is larger than any part of the fish body. This ichthyological fact is clear in the fish image shown in the seal. Other than the shape of the tail which is not bifurcated, the fish looks belong to the large-sized cyprinid species.

Early Dynastic Period (2900–2350 BC)
Sumerian civilization

Green and Black (1992) showed an image of a cylinder seal with Samaš, the sun god, in his anthropomorphized sacred boat (Fig. 4.5). This cylinder seal of the Akkadian Period from Ešnunna (modern Tell Asmar). Boat with a human head. The boats that are shown carrying people or gods by river or canal are on occasion

Fig. 4.5 Cylinder seal with Samaš, the sun god, in his anthropomorphised sacred boat. Courtesy of Green and Black (1992)

Fig. 4.6 Images of fish and sharks in a cylindrical seal. Courtesy of Green and Black (1992)

reduced with a prominent prow terminating in a human head, occasionally also with human torso and arms, with which the man-boat might actually row himself. In the image of another cylindrical seal (Fig. 4.6) by Green and Black (1992), the water god Ea and his two-faced minister god Usmû were shown. In this seal, Enki is represented as a seated god with a long beard, wearing a cap with many horns and a long, pleated robe. Streams of 60 water flow from his arms to the ground, sometimes with little fish swimming along with the flow.

Examining the fish images in Figs. 4.5 and 4.6 and judging from the shape of their body, head, pectoral fins and eye, it is possible to say that these images belong to a species of shark. The tail, which is a characteristic of the shark species, is known as heterocercal (the dorsal lobe is larger than the ventral lobe), but the artist in this image has reversed the two lobes. The scene in both cylindrical seals showed in Figs. 4.5 and 4.6 are in a freshwater habitat and the presence of a shark in such environment might raise a question of how such a marine creature can be illustrated to be present in the freshwater niches? The presence of species of sharks in the freshwater environment is not unusual event. In the modern history, the records of the sporadic presence of marine species in freshwater systems of Iraq goes back to the time of Günther (1874), who recorded the presence of the shark *Carcharhinus gangeticus* (=*C. leucas*) in the Tigris River near Baghdad. This record is a misidentification of a bull shark *Carcharhinus leucas,* as this species can tolerate pure freshwater and is often found hundreds of miles up rivers (A. Moore, personal communication). *Carcharhinus gangeticus* is a taxonomically difficult genus that may or may not be found in the waters of the Arabian Gulf (A. Moore, personal communication). On the other hand, Kennedy's (1937) list was significant in recording the presence of shark species *Carcharias lamia* Risso (could be misidentified for bull shark *Carcharhinus leucas*) in the Tigris River just north of Baghdad.

The images of shark appeared in the cylinder seal given by Green and Black (1992) (Figs. 4.5 and 4.6) can prove two issues, first, the presence of sharks in the freshwater water an event that happened nearly 6000 years ago and therefore, all documentation written in the modern history about the presence of cartilaginous fishes in the freshwater environment and their ability to adjust their physiological characters in relation to withstand the low level of salinity (Thorson et al. 1983) can be rendered confirmation to the simple observations of the ancient Mesopotamian depicted in the cylindrical seal nearly 6000 years ago. Second, the present cylindrical seals rendered any record that has been reported in the modern history to record for the first time the presence of shark species in the freshwater environment of Iraq (Hussain et al. 2012) as verification records.

The other fish image (Fig. 4.7) of this period is that shown in the The "Peace" side of the Standard of Ur, southern Iraq, about 2600–2400 BC. The "Peace" side of the Standard of Ur from a Royal tomb at Ur, southern Iraq, about 2600–2400 BC. Inlay of shell, red limestone and lapis lazuli in Bitumen on a hollow box. The main panels are known as "War" and "Peace". The "Peace" panel depicts animals, fish and other goods brought in procession to a banquet. Seated figures drink whilst a musician plays a lyre (Moradi 2017, from the British Museum's collection. Photo by CM Dixon/Print Collector/Getty Images). In this image, a man showed to carry two bunches of fishes, two fishes in each bunch. The general shape of the four fish specimens looks they belong to the deep body species of the family Cyprinidae. They might be *Carasobarbus luteus* (Himri) or *Mesopotamichthys sharpeyi* (Buni), which are very common in the lower reaches of Mesopotamia. In the drawing of the fishes, the artist has perfectly defines the cyprinid species through locating correctly the dorsal and ventral fins and in drawing large-sized scales on the fish body. The shapes of the head, eye and the tail are all in agreement with the general morphological characteristics of the two cyprinid species mentioned above. Colours of red

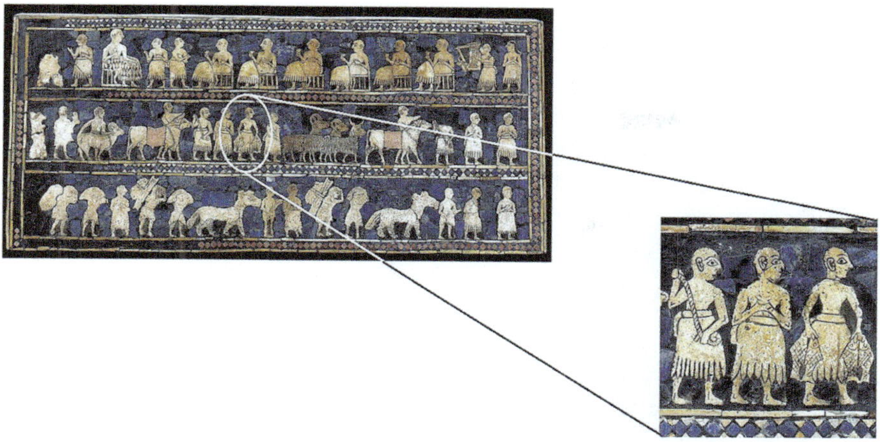

Fig. 4.7 The 'Peace' side of the Standard of Ur, southern Iraq, about 2600–2400 BC. Courtesy of British Museum's collection. Photo by CM Dixon/Print Collector/Getty Images

and blueish green are present in the image of the standard of Ur, which marks the early use of colour in the art of ancient Mesopotamia.

Akkadian Empire 2334–2154 BC
Sargonic period

Figure 4.8 showed a scene from the campaign of Asurbanipal against Te-umman, King of Susa. This is an adjoining relief from the palace of Asurbanipal. In this relief and in two locations, rivers or canals were shown. In both waterways, fish images were depicted. The artist in this relief has successfully described in drawing the basic ichthyological features of the fishes that have been drawn. The shape of the body, head and tail denoted that these fishes belong to the family Cyprinidae. It is very clear in these images that the tail of the fishes was bifurcated, a characteristic of the carp family members. Another basic character of the family Cyprinidae is the position of the pelvic fins at the middle of the body, a character that showed to be consistent in all fish images showed in this relief. In two fish images and for the first time, the artist has noted down the presence of the pectoral fins (shoulder fins). Also, an image of fish with a shallow body and small scales was also shown, which could be *Leuciscus vorax* (Shilq). In the lower waterway scene of the relief, dead men were shown and one of them was seen with a fish eating his head. This could be *L. vorax* as this species feed on fish and other aquatic animals.

Figure 4.9 showing a cylinder seal, with Enki seated on a throne, wearing a horned headdress with a flowing stream full of fish (The Trustees of the British Museum. https://www.britishmuseum.org/research/collection_online/collection_object_details.aspx?objectId=368705&partId=1&images=true&museumno=103317&page=1. Retrieved January 2019). In this seal, there were two fishes shown in front of Enki. They are not very well depicted as fishes though they look to belong to the family Cyprinidae judging from the position of the dorsal fin opposite the ventral fin and the general shape of the tail. The artist here did not pay

Fig. 4.8 Scene from the campaign of Asurbanipal against Te-umman, King of Susa. Courtesy of British Museum's collection. Photo by CM Dixon/Print Collector/Getty Images

attention to the detail of the fish image rather given more detail to the other characters in the seal, but he still can tell the group of the fishes they belong to.

Third Dynasty of Ur 2112–2004 BC
Sumerian Empire

In this period one image that contains fish is examined (Fig. 4.10). The image showed Apkallu (of fish) (Apkallu, Akkadian or Abgal, Sumerian are terms found in cuneiform inscriptions that in general mean either "wise" or "sage") next to the tree of life and under the symbol of Assur (Campbell 1991). The fish gown worn by

Fig. 4.9 Cylinder seal, with Enki seated on a throne, wearing a horned headdress with a flowing stream full of fish Courtesy of the Trustees of the British Museum. https://www.britishmuseum.org/research/collection_online/collection_object_details.aspx?objectId=368705&partId=1&images=true&museumno=103317&page=1

Fig. 4.10 Apkallu (of fish) standing next to the tree of life and under the symbol of Assur. Courtesy of Campbell (1991)

Apkallu depicting very well the large-sized cyprinid fish species such as *Luciobarbus esocinus* (Biz) and *Luciobarbus xanthopterus* (Qattan). These two species are among the fish species with high commercial value at the present time and it seems that they have the same value in ancient Mesopotamia due to having been selected to be worn by Apkallu. The artist here, gave an excellent description of the cyprinid fish species through the illustration of the fish. The body, head and tail

Fig. 4.11 A puradu-fish Apkallu appears to the left of the sacred tree, with twofish-men, apparently a merman and a mermaid, on the right. Courtesy of Campbell (1991)

shapes look exactly as those of cyprinid fish species. With the shape of the fish body, the artist knew how the dorsal and ventral contour of the fish body goes near the tail, which exactly what is present in the cyprinid fish species at the present time. In the head of the fish, the artist was careful in drawing and locating the eye on the side of the head of the fish. He did not forget to draw the gill cover behind the eye and a thick lips. The shape and the arrangement of the scales on the fish body were superior for an ancient artist like this. Not to forget, the artist knew very well where to put the dorsal fin, which in cyprinid it is unpaired and located near the mid-dorsal side of the body. The fish images reflect how the artist was precise in transferring the fish ichthyological characteristics to his artwork. It looks as though the artist had a good memory that helps him to draw such a detailed fish image.

In another image of this period (Fig. 4.11), a puradu-fish Apkallu appears to the left of the sacred tree, with two fish-men, apparently a merman and a mermaid, on the right (Campbell 1991). Here, the artist was not careful in depicting the ichthyological characteristics in the images of the fishes showed in his artwork. The fish coat Apkallu wearing look in general a cyprinid fish species judging from the unpaired dorsal fin, shape of the head and partly the shape and arrangement of the scales on the fish body. On the other hand, the lower parts of the body of the merman and a mermaid were better resemble the body of the members of the family Cyprinidae. Here, the artist, did not forget to put dorsal, pectoral and anal fins in their correct position on the fish body. In addition, he drew the tail of the fish with a degree of precision as he adds the fin-rays (fine bony elements supporting the fin of the fish) and did not forget to make the tail bifurcated as in the cyprinid fish species. Finally, the artist was successful in representing the shape and the arrangement of the scales on the fish body similar to those of the present time cyprinid fish species.

In Fig. 4.12, a scorpion-tailed bird-man showed, with a human upper torso, an avian body, and a scorpion tail (Green 1994; Dalley 2011). In this figure, the fish coat Apkallu wearing has been drawn as a sketch and without ichthyological detail

Fig. 4.12 A scorpion-tailed bird-man showed, with human upper torso, an avian body, and a scorpion tail. Courtesy of Dalley (2011)

for the fish, but it still can be seen that fish images represented members of the family Cyprinidae and the large-sized species of this family. Evidences for such a conclusion is based on shape of the body, tail and the position of the dorsal fin. Here, the artist did not care to add the shape and pattern of scales on the fish body.

In the Bas-relief (probably) of an Apkallu figure from the temple of Ninurta at Nimrud (Fig. 4.13) (Green and Black 1992). As an ichthyologist, the fish image depicted in this relief could be the best ichthyological representation of the ancient Mesopotamian artist to the fish. The striking ichthyological feature in this image is the shape and the arrangement of the scales on the fish body and how the size of the scales changes from small to large towards the tail of the fish, which is an ichthyological characteristic in most fish species. The other fish features that the artist had depicted in the image are the fine description of the fins in drawing the fin-rays of each fin, the presence of the pectoral fins, the position of all the fins on the fish body, shape of the head and the shape of the lips. The artist marked in the image a very specific ichthyological feature. He drew 3 curved lines below the mouth of the fish, which can also be considered as crown to Apkallu. These curved lines are the Branchiostegal rays (bony rays supporting the gill membranes behind the lower jaw). Certainly, the fish that the artist had in mind is belong to the family Cyprinidae and it could be one of the large-sized and high commercial value species such as *Luciobarbus esocinus* (Biz) and *Luciobarbus xanthopterus* (Qattan). It is not unusual for the artist to choose such kind of fish species as representing in his drawing a Wiseman of a society and he should choose a valuable fish species for this purpose.

Old Assyrian Period 2025–1365 BC

In Fig. 4.14, Ornan (2005) showed an Aramaic inscribed imagery conveyed on stamp seals Sin-in-the crescent may appear winged and sporting a stylized tree flanked by Apkallu, which, in Assyrian iconography, often accompany a stylized tree, topped by a winged disc. In these two seals, Apkallu figures were shown wearing the fish coat. Here, the artist did not give much attention to the

Fig. 4.13 The Bas-relief of an Apkallu figure from the temple of Ninurta at Nimrud. Courtesy of Green and Black 1992)

ichthyological detail in the fish images. The shape of the body and the tail were in agreement with those of the members of the family Cyprinidae. In Fig. 4.14, the scale pattern was clear on the fish body, while it is not so in the fish image showed above.

In another image from the same period (Fig. 4.15). A portrayal of human-shaped divinities discovered at Ashur, assigned to Sennacherib by an inscription, is rendered on a basalt basin found in the Ashur temple, adorned on its four corners with the images of lesser divinities holding the *Hegallu* jar with running water (Ornan 2005).

Fig. 4.14 An Aramaic-inscribed imagery conveyed on stamp seals Sin-in-the crescent may appear winged and sporting a stylized tree flanked by Apkallu. Courtesy of Ornan (2005)

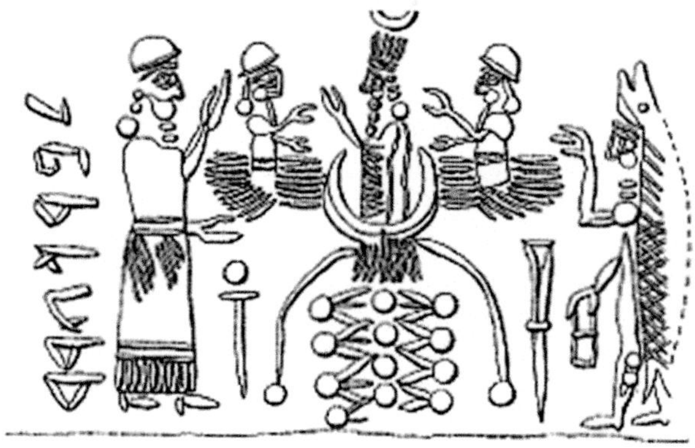

The image of the fishes on the coat that Apkallu was wearing carries slight ichthyological features, but still it is possible to say they belong to the family Cyprinidae. Here, the artist was not careful in adding the ichthyological detail to the fish images. Except for the scale pattern, shape of the head and the presence of the Branchiostegal rays below the mouth of the fish.

Middle Assyrian Period 1365–883 BC

Among the fish images that appeared in this period is Fig. 4.16 given by Ornan (2005). The image showing a Babylonian cylinder seal, depicting a worshipper before a large fish, which is an iconic depiction conveying fertility and abundance. The fish image showed in this seal looks arbitrary and the artist did not include the ichthyological features to the image. Except for the general shape of the fish body, which looks like a fish, other characters do not correspond to those of any fish

Fig. 4.15 A portrayal of human-shaped divinities discovered at Ashur, assigned to Sennacherib by an inscription. Courtesy of Ornan (2005)

species living in Mesopotamia at the present. To a lesser extent, the artist tried to bring the shape of the tail of the fish in the image so close to what is present in the members of the mullet, Family Mugilidae. The little lines drawn between the two lobes of the tail could represent the fin-rays of the caudal fin (tail fin). The other evidence for the fish in the image being a mullet, the presence of 2 dorsal fins.

Sun-dried clay figurines from this period of the goat-fish and the merman, probably from the city of Aššur were given by Green and Black (1992) (Fig. 4.17a and b). Although no details of their discovery are known, they will have been placed within a brick box buried in the foundations of a building to counter evil. The fish body of both the goatfish and the merman has some ichthyological characteristics showing that the artist wanted to say that the fish he depicted is one of the members of the common species of the carp family, Cyprinidae. The deep body of the two figurines, the presence of a dorsal fin and the shape of the tail suggested that the fish species could be *Mesopotamichthys sharpeyi* (Buni), a common cyprinid fish species in the lower Mesopotamian plain. The artist did not forget to include the fin-rays of the caudal fin (tail fin), which indicates that he is aware of the presence of these structures. The presence of the two figurines buried in the foundations of a building to keep away the evil eyes is a practice that presents people in Iraq are following suit, when they build a new house or when they move to a new house. People usually burry under the doorstep of the house some verses from the Quran wrapped in cloth and kept in a plastic bag. This is an evidence of the existence of an ancient Mesopotamian habit until the present time.

Green and Black (1992) displayed a Detail cylinder seal from this period showing the sun god lamaš, within his winged disc (supported by bull-men), stands on a horse, while a fish-garbed figure and worshipper (representing the owner of the seal) at the edge of the scene. In the fish image showed off the fish coat of the fish-garbed figure (Fig. 4.18), the ventral side of the fish is located on the back of the fish-garbed figure and the dorsal side on its abdomen. The artist here, did very well in matching

Fig. 4.16 A Babylonian cylinder seal, depicting a worshipper before a large fish, which is an iconic depiction conveying fertility and abundance. Courtesy of Ornan (2005)

Fig. 4.17 A sun-dried clay figurines. (**a**) the goat-fish; (**b**) the merman, probably from the city of Aššur. Courtesy of Green and Black (1992)

Fig. 4.18 Cylinder seal showing the sun god lamaš, within his winged disc (supported by bull-men). Courtesy of Green and Black (1992)

the fish image to one of the mullet species living at the present time in Mesopotamia. Judging from the elongated body shape, presence of two dorsal fins, shape of the head and the pattern of the scales on the body, it is possible to say that the fish could be a mullet species, Family Mugilidae. The lips and the operculum (cover of gills) were also noted in the fish image. In addition, the artist successfully added and in the correct position the pectoral fin, but it has been situated slightly backwards towards the tail instead of forward towards the head of the fish.

Figure 4.19 showing a swimming merman. Detail from a monumental stone relief from the royal palace of the Assyrian king Sargon (Green and Black 1992). The body of the merman looks exactly as shown in Fig. 4.17b, where the body of the merman is deep representing one of the deep-bodied species of the family Cyprinidae, with the pattern of the scales on the body of the fish and the shape of the tail.

The probable Syrian influence on Middle Assyrian divine imagery may be sustained by the later stela found at Terqa (Tell Ashara), reflecting Syrian traits both in its iconography and in the theme of its inscription, engraved by Tukulti-Ninurta II. In Fig. 4.20, the fish coat of Apkallu showed fish image drawn in different way from the previous fish images examined in this chapter. Here, the artist replaced the dorsal fins of the fish in 2 wavy movements of the fish. The edges of the caudal fin were absent, but the fin-rays supporting this fin were shown. The scale pattern was drawn uneven, where the pattern of the scales in the anterior (front) part of the fish body does not match that of the posterior (back) part. The eyes and lips were drawn correct and the fish looks to belong to the family Mugilidae judging from the shape of the body, head and the presence of two dorsal fins.

Fig. 4.19 A swimming merman. Details from a monumental stone relief from the royal palace of the Assyrian king Sargon. Courtesy of Green and Black (1992)

Fig. 4.20 The fish coat of Apkallu showed fish image. Courtesy of Ornan (2005)

Figure 4.21 showed a relief sculpture from Nineveh depicting fishermen swimming to catch fish (https://www.pinterest.com/pin/211131057061039/). There was one image of fish in Fig. 4.21, which look like a member of the family Cyprinidae judging from the shape of the tail and the presence of one dorsal fin located nearly at the middle of the fish body. The shape of the body, head and the mouth suggesting it is the piscivorous species *Leuciscus vorax*. Here, the ancient Mesopotamian artist and for the first time, showed the lateral line of the fish (line of scales with sensory holes running along the sides of the fish). This line is very clear in the fish image depicted in the present sculpture. The lateral line is an important ichthyological character used in the identification of the fish species. In *L. vorax*, this line extended backward from the head towards the tail through the lower half of the body. This way of extension of the lateral line has been shown exactly in the fish image depicted in the sculpture from Nineveh.

Fig. 4.21 A relief sculpture from Nineveh depicting fishermen swimming to catch fish Courtesy of Pinterest, https://www.pinterest.com/pin/211131057061039/

Fig. 4.22 Apkallu wearing a full-length fish cloak. Courtesy of Dalley (2011)

Dalley (2011) illustrated Apkallu wearing a full-length fish cloak (Fig. 4.22). The fish image looks like the large-sized species of the family Cyprinidae judging from the shape of the fish body, head and the pattern of the scales. However, the artist here did not give much attention to the ichthyological detail such adding the dorsal and other fins to the fish image.

Fig. 4.23 Puradu-fish Apkallu minister to an ill patient in bed. This plaque drawn by Faucher-Gudin, from a bronze plaque of which an engraving was published by Clermont-Ganneau. The original, which belonged to M. Péretié, was in the collection of M. de Clercq before it was acquired by the Louvre. http://www.gutenberg.org/files/17323/17323-h/17323-h.htm#linkBimage-0039%5B/caption%5D

Figure 4.23 showed detail from a drawing of a bronze plaque held in the Louvre. Puradu-fish Apkallu minister to an ill patient in bed. This plaque drawn by Faucher-Gudin, from a bronze plaque of which an engraving was published by Clermont-Ganneau.

The original, which belonged to M. Péretié, was in the collection of M. de Clercq before it was acquired by the Louvre. http://www.gutenberg.org/files/17323/17323-h/17323-h.htm#linkBimage-0039%5B/caption%5D (Wiggermann 2007). The image of the fish depicted in the coat of Apkallu has not been very well representing the ordinary image of fishes from the ichthyological point of view. Here, the artist gave a good presentation of the scale pattern on the fish body and the shape of the body and head to some extent are reasonable and look like those of the fish. The fish looks like one of the large-sized species of the family Cyprinidae. In Fig. 4.24 that showed Apkallu, a human figure wearing a fish-cloak. This figure was illustrated by Dalley (2011). As in Fig. 4.23, the fish image depicted in Apkallu fish coat has not been carefully drawn according to the ichthyological characteristics. The body and head shape look like fish and the presence of two dorsal fins might suggest that fish belong to the mullet family, Mugillidae.

Figure 4.25 given by Green and Black (1992) showing two figurines of the so-called fish-garbed figures, representing Apkallu. These figurines found together in a brick box buried in the foundations of the house of a priestly family at Aššur, probably dating to the reign of King Sargon (left) and to the reign of King Sennacherib (right). Telling from the shape of the fish body, head, tail and scale pattern that the fishes in both figurines belong to the family Cyprinidae. The size of the head denoted that they resemble large-sized cyprinid species. Here, the artist did not add much ichthyological detail may be the figurines are small in size, but he seemed to know the basic ichthyological characteristics of the family Cyprinidae judging from the position of the dorsal fin near the middle of the dorsal side of the fish body.

Linder (1986) wrote about the palace of Sargon 11, the ruler of the Assyrian Empire and how he produced the most worthy monuments of art by which the king

Fig. 4.24 Apkallu, a human figure wearing a fish-cloak. Courtesy of Dalley (2011)

Fig. 4.25 Two figurines of the so-called fish-garbed figures, representing Apkallu. Courtesy of Green and Black (1992)

demonstrated his concept of the empire. The wall reliefs highlight the activities of the ruler, by rejoicing his triumphs both over man and over nature. Among these memorials, the relief depicting lumber transport. In it, more than one achievement of the king is honoured: his large-scale building events that need huge shipments of wood; the construction and the use of sufficiently designed boats for the lumber transport; the employment from his vassal states of skilled manpower who were specialists in the nautical professions and thus introduced various innovations to suit the king's specific requirements.

In this relief, the fauna in its aquatic environment, is schematically embodied. It is therefore difficult to determine which species the artist wished to portray. Linder (1986) and after consulting marine biologists specialized in the Mediterranean environment suggested that the aquatic environment shown in the relief is riverine and not marine. This means that the scene was in Euphrates River and then Tigris River leading to Nineveh, where the palace of Sargon II was located. Linder (1986) did not comment on the images of fishes depicted in the relief and here a close examination for these images will be given.

Fish images were shown in three places of the relief of Sargon II palace (Fig. 4.26). Although it is difficult to determine the species of fishes in these images, it is possible to say that they belong to the members of the family Cyprinidae judging from the position of the pelvic fin, the shape of the body, head and the tail. In one fish image located at the bottom of the relief, the artist was successful in drawing the pectoral fin in its correct position. This is considered as an indication that the artist was aware of some ichthyological issues of the fish that he drawing.

Figures 4.27, 4.28 and 4.29 (Photo by CM Dixon/Print Collector) showed a series of the monuments of Nineveh including bas-reliefs from the palace of Sennacherib and bronzes from the ruins of Nimroud (Layard 1874). In these 3 reliefs, fish images were appeared, which might be drawn by the same artist. The images were very small and there is not much detail to see and determine the fish species or the group. On the other hand, it is possible to distinguish two kinds of species that belong to the family Cyprinidae based on the shape of the body and the position of the dorsal fin. The first to type is the broad bodied fish species such as *Carasobarbus luteus* (Himri) or *Mesopotamichthys sharpeyi* (Buni). The second type is the shallow and long-bodied species such as *Arabibarbus grypus* (Shaboot). In the top of Fig. 4.28, it showed a crab that attacking a small fish. This feeding habit of the crabs has been documented in the freshwater system of Iraq (Ali et al. 2007; Naser 2009)

In the bell-shaped stamp seal (Fig. 4.30) that showed a picture of Ea, the God of the deep, with the body of a fish and streams of water are discharging out of his shoulders (Rogers 1929), the body of Ea, which represented in fish body look like the body of cyprinid fish species judging from the pattern of the scales on the body and the position of the pectoral, anal and caudal fins. In spite that the fish image was very small, the artist was able to show some ichthyological characteristics some extent such as the correct position of the pectoral, dorsal and pelvic fins.

In the stone relief from the palace of King Sennacherib, a Phoenician warship was shown. King Sennacherib led an expedition against Phoenicia and Palestine in 702 BC. Assyrian (Porada 1945). In the image of this relief Fig. 4.31, images of a

Fig. 4.26 The relief of Sargon II palace. (Photo by CM Dixon/Print Collector)

few fishes were depicted. All these images demonstrated that the fishes belong to the deep-bodied species of the family Cyprinidae. This is based on the shape of the body, head, tail and the pattern of the scales. In addition, the artist was successful in locating the pelvic fins in their correct position.

Neo-Assyrian Period 883–609 BC

The relief shown in Fig. 4.32 shows cavalrymen leading their horses along a stream (Porada 1945). Layard (1874) published only one of his drawings of the room from which this fragment, and possibly the one below it, came, but he described the reliefs in this room as follows: "The Assyrian army was seen fording a broad river amidst wooded mountains". The sculptor had endeavoured to convey the idea of a valley by reversing the trees and mountains on one side of the stream. In this relief,

Fig. 4.27 A monument of Nineveh including bas-reliefs from the palace of Sennacherib and bronzes from the ruins of Nimroud. Courtesy of Layard (1874)

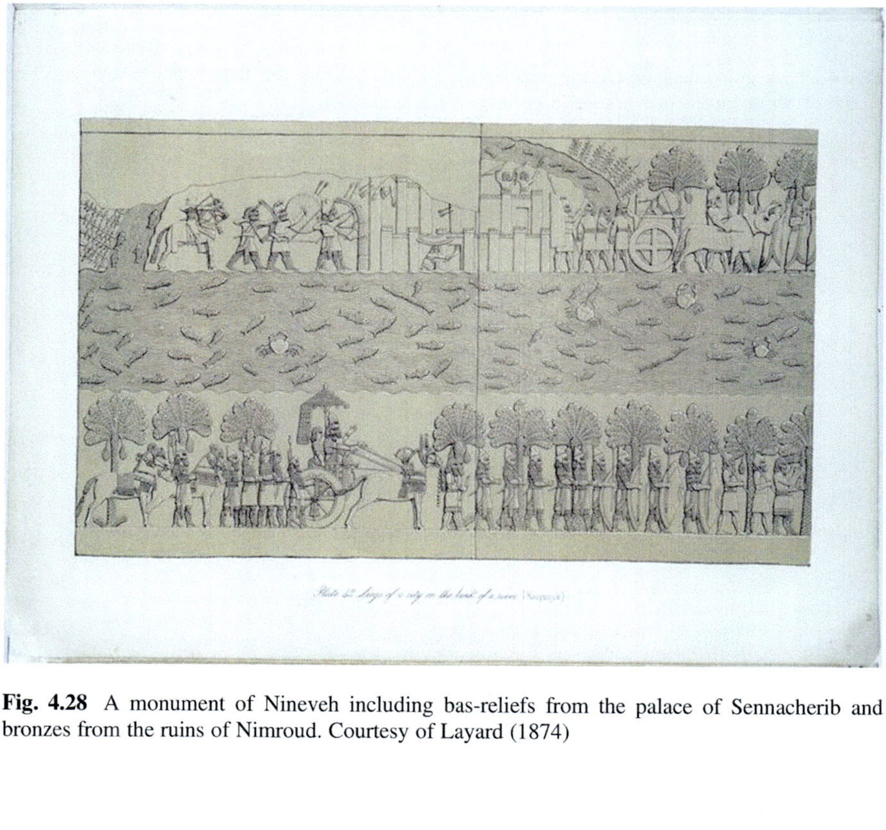

Fig. 4.28 A monument of Nineveh including bas-reliefs from the palace of Sennacherib and bronzes from the ruins of Nimroud. Courtesy of Layard (1874)

Fig. 4.29 A monument of Nineveh including bas-reliefs from the palace of Sennacherib and bronzes from the ruins of Nimroud. Courtesy of Layard (1874)

Fig. 4.30 The bell-shaped stamp seal showing picture of Ea, the God of the deep, with the body of a fish and streams of water are discharging out of his shoulders. Courtesy of Rogers (1929)

images of five fishes were depicted. Two images in the front of the cavalrymen and 3 between the man and the horse.

Looking at these images, it is possible to say that the artist has done a careful job in adding some detailed ichthyological features to the fish images that made it possible to recognize the fish family and to some extent even the species. The

Fig. 4.31 The stone relief from the palace of King Sennacherib, a Phoenician warship was shown. King Sennacherib led an expedition against Phoenicia and Palestine in 702 BC. Assyrian. Courtesy of Porada (1945)

Fig. 4.32 A cavalrymen leading their horses along a stream. Courtesy of Porada (1945)

shape of the body, head and tail suggested that the fishes depicted in this relief belonged to either *Carasobarbus luteus* (Himri) or *Mesopotamichthys sharpeyi* (Buni) (Family Cyprinidae). This is based on the position of the dorsal and pelvic fins in which the artist was carefully and correctly put in their positions. The size of scales in the two species mentioned above is large so the size of the scales shown on the body of the fishes in the images, especially that above the heads of the man and

Fig. 4.33 The relief of the Assyrian soldiers lead prisoners through a landscape of date palms, with a river beyond. Courtesy of Thomason (2001)

the horse. The artist was perfect in showing some ichthyological characteristics in the fish images. It looks that he was aware of these characters and therefore, depicted them clearly in the fish images. These characters are the fin-rays of the caudal fin (tail) and the lateral line on the side of the 3 fish images (2 at the bottom of the relief and 1 in the front of the cavalryman). Among other ichthyological features that the artist added to the fish images was the pectoral fin in its correct position.

Neo-Babylonian Period 625–539 BC

The relief of the Assyrian soldiers leads prisoners through a landscape of date palms, with a river beyond (Thomason 2001). At the far left, the heads of two clerks who are recording the details in a book and on a scroll. In this relief, three images of fishes were shown at the top of the relief and one at the middle (Fig. 4.33). The artist was very careful in making some ichthyological characters clear in these images. The species of the fishes in the images can be determined based on the shape of the body, head, tail and the scale pattern on the fish body. The artist did a good job in locating the pelvic and the dorsal fins of the fish in their correct positions. The artist was careful enough to draw the fin-rays of the dorsal fin, which can be counted as 8. This number is the characteristic of the species *Mesopotamichthyes sharpeyi* (Bunni), a common cyprinid fish species in the Euphrates–Tigris Rivers system.

Green and Black (1992) in Fig. 4.34 showed a Neo-Babylonian cast copper or bronze plaque, apparently used as a magical protection against evil spirits. Originally it probably hung in a wooden frame. Looking over the top is the god Pazuzu, was the king of the demons of the wind, brother of Humbaba and son of the god Hanbi. He also represented the southwestern wind, the bearer of storms and drought. The fish images were shown in two places of the plaque, the first at the centre, where Apkallu appeared attending a sick person and at the bottom of the plaque, where 4 fish images were shown. The fish image of the coat of Apkallu was drawn with precision by the artist so it resembled the large-sized species of the family Cyprinidae based on the shape of the body, head and the pattern of the scales. On the other hand, those fish

Fig. 4.34 A Neo-Babylonian cast copper or bronze plaque, apparently used as a magical protection against evil spirits. Courtesy of Green and Black (1992)

images shown at the bottom of the plaque were not drawn to show the ichthyological aspects of the fish except for the shape of the body and it is not possible to decide whether they are cyprinid or mugillid species.

4.5 Comparison of the Inclusion of the Ichthyological Features in the Fish Images Drawn by the Ancient Mesopotamian and Egyptian Artists

In this section, a comparison of the scientific abilities of the artists of both ancient Mesopotamia and Egypt have to add them to their work represented in fish images of different sorts that appeared in different artefacts across the thousands of years.

There is a sort of an agreement between the majority of Egyptologists on the outline of the chronology of Ancient Egypt. This scholarly covenant is the so-called Conventional Egyptian chronology, which consigns the beginning of the Old Kingdom in the twenty-seventh century BC, the beginning of the Middle Kingdom in the twenty-first century BC and the beginning of the New Kingdom in the mid-sixteenth century BC (Kitchen 1991). There are a number of "alternative chronologies" outside scholarly consensus that showed variation in the total period of both the Old and the New Kingdom. The early time of the Old Kingdom started in 2150–2040 BC, which coincides with the time of conquest of Akkad in 2150 BC within the Guti Interregnum. This makes the chronology of Ancient Mesopotamia started more than 2000 years before. These differences in time are important for the artists in both civilizations as with time the ability to depict the ichthyological features in the fish images will be developed.

Usually, artists draw natural history artefacts as they can see them in their environment surrounding them. This is not uncommon for the artists of the ancient civilizations of Mesopotamia and Egypt. As to the images of fishes, the Ancient Mesopotamian artists reacted with his environment and produced images of fish species that similar to what he usually see and caught in his daily life. Since the inland waters of Mesopotamia swarms with several species of the carp family, Cyprinidae, we expect that the fish images that the Mesopotamian artists will produce are different species of this family. This is true for the ancient Egyptian artists, who in his daily life see, caught and eat mainly tilapia species, family Cichlidae and depicted in his artworks.

The next issue to discuss here is the degree of precision in adding the ichthyological features to the fish images by artists from ancient Mesopotamia and Egypt. How this ability can be compared between the artists in both civilizations? In this section, an attempt will be made to answer this question after shedding light on the importance of fishes in the life of the ancient Egyptian and after comparing the usage of colours in the two civilizations.

Among the legends that the ancient Egyptians had was the tilapia legends. Amulets of the tilapia fish were perceived to possess fertility power as it was widely considered a symbol of regeneration and reproductive strength in this civilization. A description of the tilapia fish in Egyptian art mostly of the New Kingdom (roughly 1550–1070 BCE) follows, contrasted with scientific study of male tilapia virility. Although it is unlikely the ancient Egyptians were fully aware of all the particular tilapia traits, without doubt the species of Egyptian Tilapia (*Oreochromis niloticus* or *Tilapia niloticus*) was then and now famous for an unusual form of neoteny while rearing its young, which turned to be a scientific fact in the present time. Like modern fish aquaculture, Egyptians also farmed tilapia in closed ponds along the Nile, and this fish had its own special hieroglyph character as *in.t* (or '*int*) in texts (Betro 1996). In addition, the tilapia image was also worn as an amulet, not the least for pregnant women on necklaces or later seamed into coverings for the afterlife.

Chiriu et al. (2017) studied the usage of colours in ancient Mesopotamia and in particular in the life and culture of Sumerian. They answer a leading question: did Mesopotamian people apply some kind of colour to decorate their tablets or to put emphasis on selected words? The answer to this question came through the examination by Chiriu et al. (2017) of some administrative and literary Sumerian cuneiform tablets of mid-third Millennium B.C. from the site of Kish (central Mesopotamia, modern Iraq) were dug up in twentieth-century and stored at the Ashmolean Museum of the Oxford University. The results of their study showed that yellow, orange, red and white pigments have been detected and a possible identification has been proposed in this work. These results suggest that Sumerian people invented a new editorial style, to overcome the binary logic of engraving process and catch the reader's eye by decorating cuneiform tablets.

Besides of the meaning, the main result is that colours were applied by Sumerian scribe to decorate cuneiform tablets. Chiriu et al. (2017) suggested that Sumerian people should be given credit for the invention of a new editorial style, to overcome the binary logic of the engraving process and catch the eye of the reader by marking specific signs.

On the other hand, the ancient Egyptians had a great gratitude for life which is clearly portrayed through their art. Images of people enjoying themselves—whether in this life or the next—are as plentiful as those most often seen of the gods or funerary rituals. Egyptians decorated their homes, gardens, palaces, and tombs with inspiring works of art which echoed their gratitude for all that the gods had given them and accented these depictions with vibrant colours. Usually, the selected colours were not chosen randomly but each had a very specific symbolism for the Egyptians and were used to convey that significance (Mark 2017).

Color in ancient Egypt was used not only in truthful depictions of scenes from every life but to illustrate the heavenly kingdoms of the gods, the afterlife, and the stories and histories of the gods of the Egyptian pantheon. From the Old Kingdom until the country was annexed by Rome after 30 BCE, colour was an important component of every work of art fashioned by the Egyptians (Mark 2017).

The following is an ichthyological description of the fish images that appeared in the artistic work of ancient Egyptian through the Predynastic (3300–3000 BC) and up to the late period (305–30 BC), which covers around 3000 years. Figure 4.35 showing fish under Nebamun's boat. This damaged palette from the prehistoric Late Naqada II-III period, ca. 3300–3000 BCE, one of many (at least 35) early fish palettes, evidence for a long history of value of fish to Predynastic and Ancient Egypt (Hunt 2019). The fish images in this palette are considered among the most colourful ancient Egyptian artistic work in addition to the perfection of the ichthyological features of the fish images shown.

In this palette, images of incomplete tilapia fish (Family: Cichlidae) and a species of pufferfish, *Tetrodon* species (Family: Tetraodontidae). The Scandinavian ichthyologist Svein Fosså (Nilsen and Foss 2003) proposed that the image of tilapia in this very early palette is difficult to identify as tilapia because it is very small and likely damaged and is therefore uncertain. Closer examination showed a lack of a dorsal fin although there are detailed lines where the full dorsal fin should be and it also lacks a distal fin and its caudal (tail) fin if complete also appears wrong. It also shows some fragmentary breaks along its bottom and other anomalies that may result from damage. On the other hand, the ancient Egyptian artist was perfect in depicting some ichthyological features correctly. The fin rays of the caudal fin (tail) have been drawn precisely so that a branched and no-branched fin-rays were shown. This is a

Fig. 4.35 Fish under Nebamun's boat. This damaged palette from the prehistoric Late Naqada II-III period, ca. 3300–3000 BCE, one of many (at least 35) early fish palettes, evidence for a long history of value of fish to Predynastic and Ancient Egypt. Courtesy of Hunt (2019)

usual ichthyological character in teleosts fish, where the caudal fin has two types of fin-rays supporting it (Fricke 1983; Jawad and Jig 2017). The detailed structure of the fin-rays in all fins of the fish body has been noted carefully by the artist. The fin-ray of a teleost fish generally has cross bands along its length, which distinguish them from the spine that has no such feature. These cross bands were depicted in the fin-rays of all the fins of tilapia images shown in the palette. Also, the artist was precise in drawing the posterior edge of the anal fin and it curves behind the last anal fin-ray. On the other hand, the artist was not careful in adding the ichthyological characteristics to the fish image of the pufferfish shown at the bottom of the pallete. Here the general shape of the fish body is nearly similar to that of pufferfish, but the artist added scales on the upper part of the fish body, which not correct for the case of pufferfish as these fishes do not have scales on their body, but spines and spicules. The lower part of the fish body was drawn correct and inflated showing the distinctive character of the pufferfish species. It is not clear whether the artist added the 4 large teeth of the pufferfish that this group of fish characterized in. The species could be any species of the freshwater pufferfish that inhabits the Nile River such as lineatus puffer (*Tetraodon lineatus*), which is a tropical freshwater pufferfish found in the upper Nile, Chad, Senegal, Gambia, Geba, Volta and Turkana basins in Africa.

Aleem (1972) showed two images (Figs. 4.36 and 4.37) that depictedfishing activities in an aquatic environment in ancient Egypt in the period 2660–2180 BC. In Fig. 4.36, a diagram of two tilapia species specimens was speared by an ancient Egyptian fisherman from a boat. The shape of the body of the two fish, dorsal fin and the tail agree with the shape of the tilapia fish species. In Fig. 4.37, fishermen showed pulling a net in the Nile River. Here, several images of fish species appeared. Among those is the Nile pufferfish or lineatus puffer, *Tetraodon lineatus*. The artist was precise in drawing the lines on the upper side of the fish body to recognize this species. Other fish species are tilapia and species of catfishes. For the latter species, it is clear that the artist was careful in drawing the dorsal fin of these species erected and equipped with a hard spine. Usually, catfish erect its dorsal fin when the fish is in

Fig. 4.36 Fishing activities in an aquatic environment in ancient Egypt in the period 2660–2180 BC. Courtesy of Aleem (1972)

Fig. 4.37 Fishing activities in an aquatic environment in ancient Egypt in the period 2660–2180 BC. Courtesy of Aleem (1972)

danger. This is an indication that the artist was aware about this peculiar habit of this species of catfish. In the third image that Aleem (1972) has shown (Fig. 4.38, left), the artist has precisely added the barbels to the catfish showed under the boat in the water. This image was from the Middle Empire, Dynasty VII–XIV (2180–1640 BC).

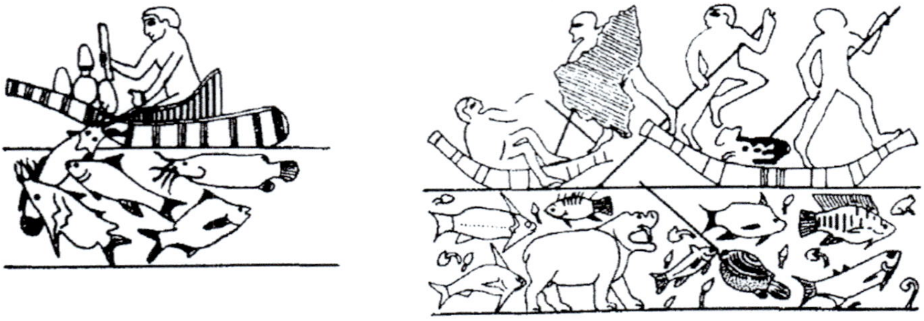

Fig. 4.38 Fishing activities in an aquatic environment in ancient Egypt in the period 2660–2180 BC showing catfish with barbels. Courtesy of Aleem (1972)

Fig. 4.39 An 18th Dynasty (1550–1307 BC) glowingly colored cosmetic bottle in the shape of a tilapia fish. Courtesy of Hunt (2019)

Hunt (2019) showed a beautiful striped glass object pictured in the lead photo (Fig. 4.39)—an 18th Dynasty (1550–1307 BC) glowingly collared cosmetic bottle in the shape of a tilapia fish. The colourful glass made this tilapia fish bottle a prized possession, and it was discovered under a floor in el-Amarna, Akhenaten's city that was soon abandoned during the following Ramesside dynasty (Kemp et al. 2013). This artefact is one of the most striking pieces on the British Museum's Egyptian collection of iconic glass objects and is both creative in design as well as anatomically perfect for a Nilotic ichthyologist (Brewer and Friedman 1989). The artist was very precise in adding the ichthyological features of the tilapia species. Starting with the fish shape, it is perfectly agreed that of the tilapia in being deep and slightly rectangular in shape. The head is somewhat elongated terminated with thick lips. All the fins in the correct ichthyological position, except for the pectoral and pelvic fins, where the former should slightly forward the latter. Finally, the position of the eye at the dorsal posterior corner of the head agrees with that of the tilapia species.

In an image of the central garden pool in Nebamun's tomb, Hunt (2019) showed a portrait that tilapia fish images were depicted (Fig. 4.40). This portrait was from the 18th Dynasty (1550–1307 BC) and kept at the British Museum. Four complete and one half tilapia fish images were shown in this figure. The positions of the fishes in

Fig. 4.40 The central garden pool in Nebamun's tomb showing a portrait that tilapia fish images were depicted. Courtesy of Hunt (2019)

both the top and the bottom of the figure were opposite to each other, which may represent that fish in swimming activity. Ichthyologic speaking, all the anatomical features were drawn perfectly and precisely by the artist, which include the shape of the body, head, tail and the fins. In addition, the artist added the fin-rays that support each fin as fine and careful details. Colours in the figure in general and the fish images, in particular, are excellent, especially the dark bands on the side of the fish body, which is considered among the characteristics of some tilapia species.

A dish from the New Kingdom period (1190–1070 BC) and during the XX-XXI dynasty is shown in Fig. 4.41 (left) by Hunt (2019). In this artefact, an outline sketch of two tilapia fishes appeared. Although, the fish images were not precisely drawn, but the artist was still able to depict the basic ichthyologic features of the tilapia species such as the shape of the body, head and the fins. The scales were replaced by dash lines, but they are still as in the natural.

In the same (Fig. 4.41, right), an Egyptian amulet was shown. The Egyptian amulets highlight the perceived protective value of the tilapia fish, especially to mother and children as a reflection of its careful parenting fish symbolic for protection. The Walters Museum commentary also notes that Egyptians thought of the tilapia as self-creating due to its discharge of its hatchlings from its mouth (Hunt 2019). In this figure, the shape of the fins and their locations; the shape of the scales and their pattern are correct and agree with those in nature, but the shape of the body and head look more rounded than rectangular as in the live tilapia fish. Here, some ichthyological features of the fish have been mist by the artist.

In another plate from the New Empire (1550–1295 BC, Dynasty XV-XX), sketch images of 4 tilapia fishes were depicted (Fig. 4.42). They look similar to those sketches showed in Fig. 4.41 (left). The exact ichthyologic features were there and in addition, the artist has added two more precise characters and these are the length

Fig. 4.41 A dish from the New Kingdom period (1190–1070 BC) and during the XX-XXI dynasty (left); an Egyptian amulet (right). Courtesy of Hunt (2019)

Fig. 4.42 A plate from the New Empire (1550–1295 BC, Dynasty XV-XX), sketch images of 4 tilapia fishes were depicted. Courtesy of Hunt (2019)

and shape of the pectoral fin. Here, the pectoral fin showed to be elongated with the dorsal fin-rays are longer than the ventral and the whole fin reaching the mid-side of the fish body, which comparatively agree with such characters in nature.

The oxyrhynchus fish, *Mormyrus kannume* (family: Mormyridae) was shown (Fig. 4.43). The artefact is an engraved neckless Ptelomic Period of the Xxi-XXXI Dynasty. In some regions of Egypt, this fish was highly regarded. The negative side of the oxyrhynchus fish that it is blamed for eating the penis of Osiris after his brother Seth disarticulated him and threw his body parts into the Nile River. The artefact of the Oxyrhynchus fish is an engraved necklace. The artist was successful in laying down the basic and fine ichthyologic characters of the oxyrhynchus fish, *Mormyrus kannume*. The shape of the body, head, tail and fins, both in number and in locations were all in agreement with those in nature. In nature, the body of this fish

Fig. 4.43 The oxyrhynchus fish, *Mormyrus kannume* (family: Mormyridae). Courtesy of Hunt (2019)

is covered by cycloid scales (Circular shaped scales without spines). The artist has shown this ichthyologic fact in his artefact as some engraving on the anterior dorsal side of the fish body just behind the gill opening.

From the above-mentioned ichthyological description of the fish images that appeared in the artworks of the ancient Egyptians and through the comparison of this work with that of the ancient Mesopotamia for nearly the same time periods, it is possible to conclude the followings: (1) the artworks of the ancient Egyptian artists have contained coloured more than that of the ancient Mesopotamia; (2) depicting ichthyological features were more precise in the art of the ancient Egyptian artist; (3) both ancient Mesopotamian and Egyptian artists have improved their perceptions of the ichthyologic features of the fish and depicted clearly as seen in their works in the later periods.

Appendix 1. Chronology of Mesopotamia

Ubaid Period		**5000–4000**
	4500	*Tell Hamoukar (urban centre, N. Mesopotamia)*
Proto-literate period		**4000–2900**
Early Uruk period	4000–3500	
	4000	*Trade contacts with N. Mesopotamia, Syria, Arabia*
	3750	*Founding of Uruk*
Late Uruk period	3500–3200	*Colonization of N. Mesopotamia*
	3300	*Invention of writing (Uruk)*
Jemdet Nasr period	3200–2900	*Initial trade contacts with Egypt*
Early dynastic period		**2900–2334**

(continued)

Sumerian civilization		
Akkadian empire		**2334–2154**
Sargonic period		
Guti interregnum		**2154–2112**
Conquest of Akkad	2154	
Liberation of Sumer	2120	
Third dynasty of Ur		**2112–2004**
Sumerian empire		
Fall of Ur (Elamites)	2004	
Isin-Larsa period		**2025–1887**
Dynasty of Larsa (Amorite)	2025–1863	
Dynasty of Isin (Sumero-Akkadian)	2017–1887	
Old Assyria		**2025–1365**
Amorites found Assyria	2025	
Old Babylonian conquest	1753	
Assyrian independence	1715	
Mitanni conquest	1500–1365	
Old Babylonian period		**1994–1595**
Babylon dynasty 1: Amorite		
Hammurabi	1792–1750	
Conquest of Assyria	1753	
Hittites sack Babylon	1595	
Dynasty of the sea-land		**1700–1570**
Middle Babylonia		**1570–689**
Dynasty 3: Kassite	1570–1225	
Assyrian domination	1225–1186	
Babylonian independence	1186–1157	
Elamite conquest	1157–1156	
Dynasties 4–9	1156–689	
Middle Assyria		**1365–883**
Absorption of Mitanni	1274	
Conquest of Babylon	1225–1186	
Neo-Assyria		**883–609**
Conquest of Babylon	689–625	
Conquest of Egypt	671–663	
Fall of Nineveh (Chaldeans)	609	
Neo-Babylonia		**625–539**
Dynasty 10: Chaldean		
Conquest of Assyria	609–539	
Fall of Babylon (Medes)	539	
Persian empire		**547–331**
Achaemenid dynasty		
Conquest of Troy	547	
Conquest of Ionia	546–478	

(continued)

Conquest of Babylon	539	
1st conquest of Egypt	525–404	
Conquest of Macedonia	492	
Sack of Babylon (Persians)	482	
1st Persian war	499–490	
2nd Persian war	480–479	
Liberation of Ionia	478	
Campaign of Xenophon	401	
2nd conquest of Egypt	343–332	
Macedonian invasion	334	
Liberation of Egypt	332	
Fall of Babylon (Macedonians)	331	
Macedonian empire		**334–305**
Reign of Alexander	334–323	
Era of the *Diadochi*	323–305	
Seleucid kingdom		**305–63**
Fall of Babylon (Parthians)	126	
Roman conquest	63	

Appendix 2. Chronology of Ancient Egypt (Hawass 2019)

First Intermediate Period: c. 2150–2040 BC

c.2150–2040	Dynasties 7–10	Collapse of central government country divided among local rulers famine and poverty

Middle Kingdom: c. 2040–1640 BC

c. 2040–1991	Dynasty II	Montuhotep II	Reunification of Egypt by Theban rulers
c. 1991–1783	Dynasty 12	Amenemhat ISenwosret IAmenemhat IISenwosret IISenwosret IIIAmenemhat IIIAmenemhat IVqueen Sobekneferu	Powerful central government; expansion into Nubia (Sudan) Capital at Lisht, near Memphis
c. 1783–0.1640	Dynasty 13		Rapid succession of rulers; country in decline

Second Intermediate Period: c. 1640–1550 BC

c. 1640–1580	Dynasty 14		Country divided with
c. 1585–1530	Dynasty 15 and 16		Asiatics ruling in the- Delta.
c. 1640–1550	Dynasty 17	Sekenenre Tao ISekenenre Tao IIKamose	Theban dynasty begins reunification process

New Kingdom: c. 1550–1070 BC

c.1550–1307	Dynasty I8	AhmoseAmenhotep ITuthmosis ITuthmosis IITuthmosis IIIQueen Hatshepsut Amenhotep IIIAkhenatenTutankhamunAyHoremheb,	Reunification and expulsion of Asiatics in North; annexation of Nubia in South. Period of greatest expansion and prosperity. Thebes (Luxor) became main residence.
c.1307–1196	Dynasty 19	Rameses ISeti IRameses IIMerneptahSiptahqueen Twosret	After glorious reign of Rameses II, prosperity threatened by incursions of 'sea peoples' in north. Residence in Delta.
c.1196–1070	Dynasty 20	SetnakhtRameses III–XI	Economic decline and weak kings ruling from the delta. Civil and workers' strikes. Royal tombs robbed.

Third Intermediate Period: c. 1070–712 BC

c. 1070–945	Dynasty 21	Smendes Siamun	Egypt in decline. Siamun*may* be the pharaoh who gave his daughter in marriage to Solomon
c. 945–712	Dynasty 22	Shoshenq IOsorkon IShoshenq II	'Shishak' of the bible. Egypt fragmented and politically divided.
c. 928–711	Dynasties 23–24	Osorkon IV	Egypt divided between local rulers.

Late Period: c.712–332 BC

c. 712–657	Dynasty 25	KashtaPiankhy (Py)ShabakaShebitkaTaharqaTantamani	Rulers from Kush (Sudan) united Egypt and started cultural revival. Threatened by Assyrians who invaded in 671, 667 and 663 BC. Last king fled south.
664–525	Dynasty 26	Psamtek INecho IIPsamtek II	Dynasty from sais in Delta. Defeated Kushite kings and continued rebuilding program after Assyrians left.
525–404	Dynasty 27	Cambyses	Egypt annexed into Persian empire.
404–343	Dynasties 28–30	AmyrtaiosNectanebo INectanebo II	Last native rulers of Egypt. Cultural renaissance and nationalism but political decline.
343–332	Dynasty 31	Artaxerxes III	Persian reconquest

Graeco-Roman Period: 332 BC—AD 642

331–304	Macedonian Dynasty	Alexander the Great	Macedonian rulers after death of Alexander in Babylon (323).
304–30	Ptolemaic dynasty	Ptolemy I–XV	Last ruler, Cleopatra VII, allied with Mark Anthony against Rome. Defeated at the Battle of Actium by Octavian

References

Aleem AA (1972) Fishing industry in ancient Egypt. Proc R Soc Edinb B Biol 73:333–343

Ali AH, Aziz NM, Hamza HA (2007) Abundance, occurrence, seasonal changes and species composition of macroinvertebrates in the restored Iraqi southern marshes. Marsh Bull 2 (1):80–95

Amiet P (1966) Elam. Auvers sur l'Oise

Betro MC (1996) Hieroglyphics: the writings of ancient Egypt. Abbeville, New York, p 114

Brewer D, Friedman R (1989) Fish and fishing in ancient Egypt. The natural history of Egypt, vol. II. Oxford & Warminster: Aris & Phillips, 1989 (Cairo: American University press, 1990).

Campbell J (1991) The masks of god. Vol. 3, occidental mythology. Penguin Group, New York

Chiriu D, Ricci PC, Carbonaro CM, Nadali D, Polcaro A, Collins P (2017) Raman identification of cuneiform tablet pigments: emphasis and colour technology in ancient Mesopotamian mid-third millennium. Heliyon 3(3):e00272

Coad BW, Atz GW, Keivany Y (2000) Fish imagery in art 82: Jonah and the fish. J Environ Biol Fish 57:9

Dalley K (2011) Exposed, uncovered, and declassified: lost Civilizations & Secrets of the past. Red Wheel/Weiser, Newburyport

Fricke R (1983) A method for counting caudal fin rays of actinopterygian fishes. Braunschweiger Naturkundliche Schriften 1:729–733

Frothingham AL (1894) The philosophy of art. Am J Archaeol Fine Arts 9:165–201

Gombrich EH (1995) The story of art. 16ths. Phaidon Press Ltd, London

Green A (1986) A note on the Assyrian "Goat-fish", "Fish-man" and "Fish-woman". Iraq 48:25–30

Green A (1994) "Mischwesen. B" Reallexikon der Assyriologie (RLA), 1994: 254–5. figure 15

Green A, Black J (1992) Gods, demons and symbols of ancient Mesopotamia: an illustrated dictionary. British Museum Press for the Trustees of the British Museum

Günther A (1874) A contribution to the fauna of the river Tigris. Ann Mag Nat Hist 14:36–37

Hawass Z (2019) Chronology of Ancient Egypt. http://www.guardians.net/hawass/chronology.htm. Retrieved 8 Jan 2019

Hegel G (2009) The philosophy of history. IAP, Scotts Valey, CA

Hunt P (2019) Ancient Egyptian tilapia story. http://www.electrummagazine.com/2012/09/ancient-egyptian-tilapia-fish-story/. Retrieved 14 January 2019

Hussain NA, Rasen AK, Al BY, Coad BW (2012) Bull shark occurrence *Carcharhinus leucas* (Valenciennes, 1839) at the inland waters of southern Iraq. J Univ Duhok 15(1):140–143

Jawad LA, Jig L (2017) Comparative osteology of the axial skeleton of the genus Pampus (family: Stromateidae, Perciformes). J Mar Biol Assoc U K 97(2):277–287

Kemp B, Stevens A, Dabbs G, Zabecki M, Rose J (2013) Life, death and beyond in Akhenaten's Egypt: excavating the south tombs cemetery at Amarna. Antiquity 87:64–78

Kennedy WP (1937) Some additions to the fauna of Iraq. J Bombay Nat Hist Soc 39:745–749

Kitchen KA (1991) The chronology of ancient Egypt. World Archaeol Chronologies 23:202

Layard AH (1874) Nineveh and Babylon: a narrative of a second expedition to Assyria during the years 1849, 1850, & 1851. J. Murray, London
Linder E (1986) The Khorsabad Wall relief: a Mediterranean seascape or river transport of timbers? J Am Orient Soc 106:273–281
Mark JJ (2017, January 08). Color in ancient Egypt. Ancient History Encyclopedia. Retrieved from https://www.ancient.eu/article/999/
Moradi Z (2016) Fish imagery in Iranian artwork. Iranian Journal of Ichthyology 2(4):244–261
Moradi Z (2017) Fish as a marvelous creature in myths and manuscripts: an overview. Iran J Ichthyol 4(3):188–219
Moyle PB, Moyle MA (1991) Introduction to fish imagery in art. Environ Biol Fish 31(1):5–23
Naser MD (2009) First record of the freshwater crab, Potamon Mesopotamicum Brandis, Storch & Türkay, 1998 (Decapoda, Brachyura, Potamidae) from the Al-Huwaizah marshes, Iraq. Crustaceana 82(12):1599–1602
Nilsen AJ, Foss SA (2003) Reef secrets. TFH-Microcosm
Ornan T (2005) The triumph of the symbol: pictorial representation of deities in Mesopotamia and the biblical image ban (Vol. 213). Saint-Paul
Piccion (2019) Mesopotamian chronology. http://piccionep.people.cofc.edu/graphics/mesop_chronol.html. Retrieved 8 January 2019
Porada E (1945) Reliefs from the palace of Sennacherib. Metrop Mus Art Bull 3:152–160
Rogers F (1929) Babylonian seal cylinders as a historical source, UIUC Master's Thesis, 1929.
Sahrhage D (1999) Fischfang und Fischkult im alten Mesopotamien (fish catching and cult figures in the ancient Mesopotamia). P. Lang, Frankfurt a/M.
Seidl U (1986) Die lamischen Felsreliefs von Kurangun und Naqsh-e Rustam. Dietrich Reimer Verlag, Berlin
Thomason AK (2001) Representations of the North Syrian landscape in neo-Assyrian art. Bull Am Sch Orient Res 323:63–96
Thorson TB, Brooks DR, Mayes MA (1983) The evolution of freshwater adaptation in stingrays. Natl Geogr Soc Res Rep 15:663–694
Van Buren ED (1948) Fish-offerings in ancient Mesopotamia. Iraq 10(2):101–121
von Osten HH (1934) Ancient oriental seals in the collection of Mr. Edward T. Newell, Chicago
Wiggermann FAM (1992) Mesopotamian protective spirits: the ritual texts. Styx, Groningen
Wiggermann FA (2007) Some demons of time and their functions in Mesopotamian iconography. Groneberg, B., Spieckermann. H.(Edd.). Die Welt der Götterbilder, BZaW 376:102–116

Chapter 5
The Effectiveness of Ancient Mesopotamian Medical Practices: The Example of *šūšu*-Licorice

JoAnn Scurlock

Although its roots go back into hoary antiquity, Ancient Mesopotamian medicine is best known from the Neo-Assyrian period (934–609 BCE). The language of common speech was Akkadian a Semitic language akin to Hebrew, Arabic and Aramaic, the latter of which eventually supplanted it. Texts were written on clay tablets in cuneiform script, a syllabic (not alphabetic) script borrowed from a much earlier culture with a completely different language. This language, Sumerian, had mostly one-syllable words which made it convenient to use as a sort of scientific shorthand, like our Latin with a similar cachet of hoary antiquity about it.

In the Neo-Assyrian period, there were two healing specialists who cooperated in the treatment of disease. One of these has left us very little direct documentation; his treatments are known to us primarily via a distinctive formatting in the texts drawn up by his colleague. His specialty was at the pharmacological end of the spectrum; he knew all there was to know about the 1000 odd medicinal plants and other substances that were administered in the form of bandages, salves, daubs, fumigants, potions, pills and enemas. Those not in palace employ will have operated out of a drug market or drug shop, the ancestor of the modern pharmacy. This specialist's only limitation was that he was not qualified to make diagnoses. He could, however, make up preparations for illnesses requiring diagnosis with the cooperation of his colleague. The archives of this colleague and the famous library of Aššurbanipal are our major source of information on Ancient Mesopotamian medicine. This man (and sometimes woman)'s specialty was at the diagnosis end of the spectrum, but he, too, treated patients when not too busy with numerous other duties. His centre of operation was a temple from which he emerged to make house calls. Both specialists used magic to help along their medicines but there was a difference: the plant specialist typically used spells; full-fledged magical rituals involving sacrifice to

J. Scurlock (✉)
Elmhurst College, Elmhurst, IL, USA
e-mail: r-beal@uchicago.edu

L. A. Jawad (ed.), *Tigris and Euphrates Rivers: Their Environment from Headwaters to Mouth*, Aquatic Ecology Series 11,
https://doi.org/10.1007/978-3-030-57570-0_5

divine patrons of medicine had to be performed by a person with access to a temple. These sacrificial rituals were alternatives to less expensive treatments for the same condition; medicine in ancient Mesopotamia was meant to be affordable..

In order to assess the efficacy of Ancient Mesopotamian medicine, two things are required. One of them is relatively easy, since we possess a good portion of a *Diagnostic and Prognostic Handbook*[1] and its accompanying *Therapeutic Handbook*, which together allow us to know, in many cases, for what symptom or disease or medical condition a particular treatment was designed. The other is more difficult, since we more often than not have no idea what the plant is that is being used in a given treatment.

The ancient Mesopotamian physician grouped the symptoms that he observed in his patients into symptom complexes and referred to them with an invented name (e.g. *bennu*), a descriptive name, (e.g."sick liver") or the name of an outside invisible attacking being thought to be causing the problem. Despite, or perhaps even due to willingness to attribute some diseases or conditions to what we would call supernatural causes (the ancestor of germ theory) and despite an extensive use of amulets in treatment, it is clear that the Ancient Mesopotamian physician practised in both diagnosis and treatment something we would recognise as medicine. This does not, however, tell us what we wish to know. It may be medicine, but was it effective medicine? How one might wonder, without microscopes and lab tests, could ancient Mesopotamian physicians have been able to discover truly effective treatments? Presumably, treated patients got better, but did the plants that were administered do more good than the accompanying magical ritual that enlisted the patient in his own cure?

Unfortunately, most of the Akkadian names for plants are still unidentified. So, to answer this question, we need to have a plant of secure identification, extensive use in traditional medicine and at least some pharmacological analysis and testing to which ancient Mesopotamian uses may be compared. The best candidate for this is *šūšu*, a plant which is more or less accepted on the basis of a very secure etymology[2] to be *Glycyrrhiza glabra* (licorice). It is also fairly well attested in the preserved texts, although there is no guarantee that all ancient Mesopotamian uses of the plant have come down to us.

As far as modern pharmacological studies are concerned, I was able to harvest from PubMed over 4000 references. These study the effects of the plant in vitro and in vivo on the usual laboratory rats and mice and human volunteers but also dogs, pigs, cats, Syrian golden hamsters, Mongolian gerbils, cattle, sheep, chickens, ducks, rabbits, pigeons and even fish of several varieties. Oh, and did I mention the cucumbers and sorghum? Lest anyone denigrate the trial and error method used by ancient Mesopotamians to discover plant uses, to note is that trial and error is the method still in use today with the difference that the modern in vitro and in vivo studies now supplemented by computer modelling make it possible to discover error

[1]See Scurlock (2014, 13–271).

[2]See CAD Š/III 385–386.

in a safe way. And how do modern researchers know where to look for suggestions? In study after study what do we hear? In Chinese/Turkish/Iranian/Indian traditional medicine, x plant has been used for y, so let us try it out and see if it works, which we generally assume it does or we would not be trying it out, the real interest being in whether it is worth being developed for commercial use and how exactly it does what it does.

Another caveat, modern studies cannot tell us the full range of licorice's potential uses, since only those diseases or conditions are worth the considerable cost of research and development where the current treatment has problems of toxicity, is prohibitively expensive, or is ineffective due to overuse of only one or two medicines to which the infectious agent has developed an immunity. Ancient Mesopotamian medicine had to worry about only one of these problems; there were no giant pharmaceutical companies, and the bitterly criticised tendency to have hundreds of different treatments for the same medical problem will have made it significantly more difficult for the infectious agent to protect itself from harm. Still, how effective could something eaten in Europe as a candy possibly be?

5.1 Ancient Mesopotamian Uses of *šūšu*-Licorice in Modern Perspective[3]

EARS:

Ancient Mesopotamian treatments cover a wide range of potential medical problems. For the category of eyes, ears, nose and mouth, *šūšu*-licorice is rarely attested. When it is, the problem to be treated is an ear infection, as might have been expected from the use of this plant for a variety of what we know to be infectious diseases affecting different parts of the body (see below).

OTITIS MEDIA[4]:

fumigants:

bitumen, ***šūšu*-licorice** and [...][5] are used [in a fumigant] for collection of [sweat in the ears], followed by a tampon with [...] and *kanaktu*-aromatic oil, a wash with [...] and a diet of hot things; treatment is for nine days (BAM 503 iii 69')

procedure lost:

powdered ***šūšu*-licorice** and [...] are used in a [...] for the right ear being heavy (BAM 503 iv 24)[6]

[3]The references used in the text passages are to copies of these texts; abbreviations are standard to Assyriology and may be found in the front of the *Chicago Assyrian Dictionary* (CAD). The line cited is the line in which the plant we are interested in actually appears. References to modern pharmacological studies are as cited in PubMed.

[4]See Scurlock and Andersen (2005, 204–205).

[5][...] indicates information lost due to damage to the clay tablet. A word within [] is a restoration.

[6]For the fuller context of this and the preceding reference, see Scurlock (2014, 367–387).

Modern experimental studies indicate antibacterial activity against one known cause of otitis media,[7] namely *Klebsiella pneumoniae* (Hosseinzadeh and Nassiri-Asi 2008, 711; Hosseinzadeh and Nassiri-Asi 2015, 1871).

FEVER:

An important use of *šūšu*-licorice in ancient Mesopotamia was for fever, particularly high fever. For *li'bu*-fever,[8] it it is recommended in a pharmacological text organised by plant use (BAM 1 i 43). This suggests that, in the salve for "curse" or "ghost" bloated insides or burning fever (BAM 199:2), it is the burning fever that is the specific target of the *šūšu*-licorice. Since however, *šūšu*-licorice was also used for gastrointestinal complaints, including both "curse" and "hand of ghost" (see below), the treatments for "internal fever" (fever with gastrointestinal symptoms)[9] may be directed against either the fever or the complaints, or both.

HIGH FEVER:

salves:

***šūšu*-licorice** is ground, (mixed) with oil and used in a salve for *li'bu*-fever (BAM 1 i 43)[10]

tarmuš. Imḫur-lim, imḫur-ešra, ḫašânu, atā'išu, ***šūšu*-licorice**, *maštakal, bīnu*-tamarisk seed, seed of *e'ru*-tree, seed of "lone plant", flax seed, *zalāqu*-stone, "white plant" (poplar resin), *imbû tâmti, kibrītu*-sulphur, *sīḫu*-wormwood, *argānu, barīrātu, kukru, burāšu*-juniper, *aprušu, ṣumlalû, suādu,* [. . .], *šupuḫru*-cedar, *šurmēnu*-cypress, myrrh,

ḫarmunu, "soiled rag" and charred human bone are (mixed) with oil and used in a salve for curse or ghost bloated insides or to keep away burning fever (BAM 199:2)[11]

baths:

a hole is dug out to the patient's specifications; the patient lies in it and is covered over with *anzaḫḫu*-frit, *šaršarru*-paste and *šammi bu'šāni*; after he comes out, he sits in a place for bathing and this grapevine root, *ḫašḫuru*-apple root, *armannu*-apricot root, *šallūru*-plum root, *šaššugu*-tree root, *šūnû*-chaste tree root, *amurdinnu*-rose root, *tublu*-tree root, *šurdu*-tree root, ***šūšu*-licorice** root, *kalbānu, sīḫu*-wormwood, *argānu, markuṣṣu, atā'išu, šammi bu'šāni* and *barīrātu* are (mixed) with water and used to bathe him for seven days for a fever that gets too high (Tsukimoto, Priests and Officials 199–200:13)[12]

FEVER WITH GI SYMPTOMS:

bandages:

***šūšu*-licorice leaves** are (mixed) with *isqūqu*-flour and *kasû* juice and used in a bandage for internal fever (BAM 579 i 1)

salves:

tarmuš. Imḫur-lim, imḫur-ešra, ḫašânu, atā'išu, ***šūšu*-licorice**, *maštakal, bīnu*-tamarisk seed, seed of *e'ru*-tree, seed of "lone plant", flax seed, *zalāqu*-stone, "white plant" (poplar resin),

---

[7]Harrison (1987, 584).

[8]See Scurlock and Andersen (2005, 29–32).

[9]See Scurlock and Andersen (2005, 129–131).

[10]For the fuller context, see Scurlock (2014, 273–280).

[11]This is restored from other passages; see Scurlock (2005 no. 187b).

[12]For the fuller context, see Scurlock (2014, 417–421).

imbû tâmti, *kibrītu*-sulphur, *sīḫu*-wormwood, *argānu*, *barīrātu*, *kukru*, *burāšu*-juniper, *aprušu*, *ṣumlalû*, *suādu*, [. . .], *šupuḫru*-cedar, *šurmēnu*-cypress, myrrh, *ḫarmunu*, "soiled rag" and charred human bone are (mixed) with oil and used in a salve for curse or ghost bloated insides (see below) or to keep away burning fever (BAM 199:2)[13]

potions:

ŠE.ḪAR leaves, ***šūšu*-licorice leaves** and 10 shekels of salt are poured into [...], left out overnight and used in a potion drunk on an empty stomach for internal fever (BAM 579 i 6)

In traditional North African medicine, extract of the root is used as a febrifuge (Boulos 1983, 123). In Chinese medicine, the flower is used, like the root, for fever (Duke and Ayensu 1985, 327). In Indian medicine, the root is considered cooling and is used in compound powders with sandal wood, madder, *Andropogon muricatus*, etc. for this purpose. With *Cydonia vulgaris* and *Andropogon muricatus* plus lesser quantities of camphor, saffron, cinnamon bark, *Cassia fistula* seeds, lettuce seeds, sandal, rose petal, water-melon seed and gum tragacanth, it is used for the pyrexia of phthisis. For "bilious fevers", it is used with coriander seeds, *Cyperus rotundus* and *gulancha* (Nadkarni 1957, 582, 583).

A traditional preparation that includes not only licorice but *Salix alba* (in Akkadian: *ṣarbatu*-poplar), one of the other plants used in Mesopotamian medicine with *šūšu*-licorice in the treatment of fever, was tested in aqueous extract on rabbits with yeast-induced pyrexia in a modern study. The rabbits were kept in air-conditioned comfort in an animal house in Karachi. Less appreciated was doubtless the digital thermometer that was inserted an exact 6 cm into their little recta to measure changes in temperature. The good news is that the preparation worked as well as the standard paracetamol. Even better, the rabbits who were deliberately given what should have been lethal doses of the preparation not only did not die, but they lost no hair or weight. On the contrary, tail erection indicated that they found the high dosage of the test product "stimulating" (Khan et al. 2017). It has been suggested that HMGB1[14] is implicated in febrile seizures, possibly contributing to their generation (Luo et al. 2014). If so, the administration of a plant known to bind HMGB1 (see below) should be of use in preventing the seizures that may develop with elevated temperature in young children and which may lead to epilepsy in later life.[15]

SKIN:

For skin eruptions, *šūšu*-licorice was recommended for pox diseases (*ašû*),[16] pointing abscesses,[17] and athlete's foot.[18] It main use was, however, for mycetoma

[13]This is restored from other passages; see Scurlock (2005 no. 187b).

[14]This is an abbreviation for High Mobility Group B1.

[15]Harrison (1987, 1924).

[16]See Scurlock and Andersen (2005, 224–226).

[17]See Scurlock and Andersen (2005, 65).

[18]See Scurlock and Andersen (2005, 80–81).

(*muruṣ kabbarti*), a fungal infection of the foot.[19] To note is that the plant is used by itself in some of these treatments and that the root is used for pox diseases, whereas the seeds and leaves are used for infections of the skin.

AŠÛ (POX DISEASES):
salves:

***šūšu*-licorice root**, [...] and *šimru*-fennel seed are ground in oil and used in a salve rubbed on the head and all the flesh for the head being full of *ašû* (AMT 1/3:14′).

potions:

***šūšu*-licorice root** is used in a potion drunk on an empty stomach for *ašû* afflicting the head (BAM 481:11')

procedure lost:

***šūšu*-licorice root** is ground, (mixed) with honey and used for MAŠ.TAB.BA afflicting the head (BAM 492:5')

"COCKSPUR" (POINTING ABSCESS):
bandages:

šarmadu, ***šūšu*-licorice leaves**, "fox grape" leaves, *urnû*-mint and [...] are ground together, softened in vinegar, beer and *kasû* juice, boiled in a *tamgussu* vessel, allowed to decoct in beer after it has cooked, daubed on the sore and bandaged over for "cockspur" that takes up a position on top of the sore (BAM 32:9'//BAM 417:9)[20]

ATHLETE'S FOOT:
salves:

***šūšu*-licorice oil** is used in a salve for sole of the foot being full of redness, accompanied by a wash with urine and a bath with *atā'išu*, [. . .], *nikiptu*, myrrh, KU.PAD salt, *Emesallim*-salt, [. . .], *Amanim*-salt, *agargarītu*-sulphur, *kibrītu*-sulphur, *zību*, *baluḫḫu*-aromatic, *ḫurātu*-madder, *ḫallūru*-chick pea, *kakku*-lentil, *kiššēnu*-bean, *nīnû*-mint, *kamunu*-cumin, [. . .] seed, *egemgirrû*-gingir, *bīnu*-tamarisk seed, [. . .], *puquttu*-thorn, *azupirānu*, *baluḫḫu* resin, "[ox] dung", [. . .], incense?, "white plant" (poplar resin), [. . .], *uḫḫūlu qarnānu*, *kukru*, [*kanaktu*] "fat", *adāru*-poplar seed, sesame residue, alum, [. . .], *ru'tītu*-sulphur, *rikibti arkabi*, *burāšu*-juniper and [. . .] (BAM 383:15)

daubs:

līparu-mulberry leaves, ***šūšu*-licorice leaves**, *e'ru*-tree leaves, *burāšu*-juniper and *kukru* are poured on the sore spots for itching eruptions on the feet due to burning *ruṭibtu*[21] (AMT 74/1 iii 5)

MURUṢ KABBARTI (MYCETOMA):
bandages:

amīlānu, "sunflower", *lišān kalbi*, *supālu*, *maštakal*, ***šūšu*-licorice leaves**, *burāšu*-juniper, *kukru*, *su'ādu*, *sīḫu*-wormwood, *argānu*, *barīrātu* and sesame residue are dried, crushed,

[19]See Scurlock and Andersen (2005, 78–80).

[20]For the fuller context, see Scurlock (2014, 438–440).

[21]See Scurlock and Andersen (2005, 210).

mixed with *isqūqu*-flour, decocted in oil and beer in a *tamgussu*-vessel, massaged into cloth and used in a bandage for *muruṣ kabbarti* (AMT 73/1 + 15/3 i 13)

[½ *qû*] of ***šūšu*-licorice seed**, ½ *qû* of sesame residue, [10] shekels of *baluḫḫu*-aromatic resin and 10 shekels of *suluppu*-dates are ground, decocted in beer in a *tamgussu*-vessel, massaged into cloth while hot and bandaged onto the patient's feet and the muscles of his neck left on three days for *muruṣ kabbarti* with black and white spots (AMT 73/1 + AMT 15/3 i 19)

fresh ***šūšu*-licorice** is ground and used in a bandage for *muruṣ kabbarti* in which the appearance of the sore spots is black (BAM 124 i 38)[22]

ašāgu-thorn leaves, *baltu*-thorn leaves, ***šūšu*-licorice leaves**, *qān šalāli* leaves, *ḫašḫuru*-apple leaves, *tittu*-fig leaves, *nurmû*-pomegranate leaves and dwarf palm leaves are dried, crushed, mixed with flour, decocted in a *tamgussu*-vessel and used in a bandage for *muruṣ kabbarti* in which the appearance of the sore spots is black (BAM 124 i 42)

[fresh] ***šūšu*-licorice leaves**, *buṣinnu* leaves and *lišān kalbi* leaves are crushed, decocted in a *tamgussu*-vessel, mixed with malt "dust" and flour and used in a bandage for *muruṣ kabbarti* in which the appearance of the sore spots is black (BAM 124 i 48)

sīḫu-wormwood, *argānu*, *barīrātu* and fresh ***šūšu*-licorice leaves** are ground, mixed with *isqūqu*-flour, decocted in *kasû* juice and used in a bandage left on for one day for *muruṣ kabarti* accompanied by stiff Achilles tendons (BAM 124 i 29//AMT 73/1+18/5 i 41)

daubs:

***šūšu*-licorice leaves** are ground and poured on for *muruṣ kabarti* (BAM 124 ii 41//AMT 74/1 ii 16).

In Chinese medicine, licorice is used for abscesses and burns. It is locally applied, mixed with honey, to boils, burns and sores. It is also used in topical dressings (Duke and Ayensu 1985, 327–328). In Ayurvedic medicine, *G. glabra* root is used for wounds, erysipelas, septic wounds and abscess (Biswas and Mukherjee, 2003, 28). In "traditional" European and Asian medicine, the fresh leaf is applied externally to wounds (Hosseinzadeh and Nassiri-Asi 2008, 711). In herbal medicine, it is used for skin eruptions, including dermatitis, eczema, pruritus and cysts (Hosseinzadeh and Nassiri-Asi 2008, 716).

The action is not simply emollient; modern experimental studies indicate that licorice leaves and roots have antibacterial activity against *Staphylococcus aureus* and a wide variety of bacterial, viral, fungal and protozoal diseases (Hosseinzadeh and Nassiri-Asi 2008, 711–712; Hosseinzadeh and Nassiri-Asi 2015, 1868–69, 1871). Included in the list of possible uses of this plant miracle drug would be treatments for malaria and leishmaniasis, the latter disease by selectively inhibiting fumarate reductase in the respiratory chain of the parasite (Hosseinzadeh and Nassiri-Asi 2015, 1869).

For the pox diseases, we are not (as has been claimed) simply treating the lesions themselves; experimental studies indicate antiviral activity against the cause of

[22]For the references quoted from BAM 124, see Scurlock (2014, 443–452).

chickenpox, Varicella zoster virus (Hosseinzadeh and Nassiri-Asi 2008, 712; Hosseinzadeh and Nassiri-Asi 2015, 1871). Licorice apparently works by inhibition of the penetration, uncoating or release of the virus particle (Tang and Eisenbrand 1992, 583). It should also be noted, although of no apparent relevance to ancient Mesopotamia, that *Glycyrrhiza* sp. are very promising as a treatment for HIV (Tang and Eisenbrand 1992, 583).

BONES/JOINTS:

The main use of *šūšu*-licorice in ancient Mesopotamia was to treat problems of the bones and joints and for arthritis.[23] In particular, there was a type of arthritis known to the ancient physician as *manzaz rabiṣi* ("emplacement of a *prosecutor-demon*"). The full set of symptoms describe fever, arthritis and "trembling". This last symptom would appear to be a reference to what we call Sydenham's chorea, a delayed manifestation of rheumatic fever.[24]

ARTHRITIS:

bandages:

[*kukkušu* flour(?)] of malt, gazelle dung and ***šūšu*-licorice leaves** are made into a dough with *kasû* juice, massaged into waterproof leather and used in a bandage for Achilles tendons so stiff the patient cannot walk about (AMT 70/3:8)

[*kukkušu* flour of malt?], ***šūšu*-licorice leaves**, "sunflower" leaves and gazelle dung are crushed, decocted in beer dregs and *kasû* juice in a *tamgussu*-vessel, massaged into cloth and used in a bandage for Achilles tendons so stiff the patient cannot walk about (AMT 68/1 r. 13)

kukkušu flour of malt and ***šūšu*-licorice leaves** are [crushed together], sifted, decocted in old beer and used in a bandage to relax feet so stiff the patient cannot walk after the spot has been rubbed with oil; the feet are expected to relax (AMT 68/1 r. 17)

[...], *maštakal*, *uḫḫūlu qarnānu*, ***šūšu*-licorice root**, *sikillu* root and *kasû* are crushed separately, sifted, decocted in *kasû* juice in a *diqāru*-bowl; roasted grain flour, emmer flour and *ḫašû*-thyme are sprinkled on; used in a bandage for the hands and/or feet, preceded by a massage, a bath with [...], *maštakal*, *uḫḫūlu qarnānu*, ***šūšu*-licorice root**, *sikillu* root and *kasû*, a salve with *imḫur-lim* and a steam bath (BAM 405:7')

saḫlû and ***šūšu*-licorice leaves** are crushed, sifted, decocted in a *dīqaru*-bowl in freshly opened beer, massaged into cloth and used in a bandage for Achilles tendons so stiff that the patient cannot walk on them, to soothe the tendons; the tendons are expected to be soothed and the feet to feel lighter (BAM 122 r. 5'//AMT 68/1 r. 9)[25]

baths:

[...], *maštakal*, *uḫḫūlu qarnānu*, ***šūšu*-licorice root**, *sikillu* root and *kasû* are (mixed) [with] beer, baked [in an oven] and used to continually rub and bathe his hands and/or feet,

[23]See Scurlock and Andersen (2005, 248–257).

[24]See Scurlock and Andersen (2005, 451).

[25]For the fuller context, see Scurlock (2014, 566–570).

followed by a salve with *imḫur-lim*, a steam bath and a bandage with [. . .], *maštakal*, *uḫḫūlu qarnānu*, ***šūšu*-licorice root**, *sikillu* root and *kasû* (BAM 405:7')

procedure lost:

[. . .], *imḫur-lim*, *imḫur-ešra*, *tarmuš*, ***šūšu*-licorice**, [. . .]-tree, *su'ādu*-chufa seed, [. . .], *kurkānu*-turmeric, *urnû*-mint, *nuṣabu*, [. . .], *ṣaṣumtu*, *kamkādu*, *kamantu*, *bīnu*-tamarisk seed, [. . .] seed, [. . .], *ašāgu*-thorn, *pillû*-mandragora, [. . .], *šibburrātu* and [. . .] are used for sore and swollen hips(?) (BAM 56 obv. 10')

RHEUMATISM:
baths:

***šūšu*-licorice**, *adāru*-poplar, *ḫaluppu*-wood, *šūnû*-chaste tree, *līparu*-mulberry, *mirišmara*, *lišān kalbi*, *uriyānu*, "eggplant" seed, *ḫašû*-thyme, "fox grape", *aktam*, flax seed and *elkulla* are dried, crushed, sifted, mixed with river water, baked in an oven and used to bathe the patient for inability to get up or stand and trembling of the hands and feet due to *manzaz rabiṣi* (AMT 69/2:7//AMT 70/7 i 13′).

In North African medicine, the roots are chewed for rheumatism (Boulos 1983, 123). Modern experimental studies have shown that the plant has anti-inflammatory properties (Hosseinzadeh and Nassiri-Asi 2015, 1870). Searching for an effective cure for rheumatoid arthritis, a chronic joint inflammation that causes structural damage to cartilage, bone and ligaments, modern researchers devised a series of tests to determine the usefulness of licorice. They were happy to report that treatment with plain or roasted licorice was able to prevent artificially induced mouse ear edema (an acute inflammation model) by four hours. In addition, the CIA mouse model for human rheumatoid arthritis (no, not waterboarding)[26] revealed that oral administration of licorice was able to inhibit paw swelling in the affected mice. Laboratory analysis showed that these preparations helped protect the joints by decreasing the levels of two proinflammatory, bone and cartilage destroying, cytokines by discouraging the spleen from producing them and, just for good measure, licorice prevented associated oxidative damage to the laboratory mice's liver and kidney tissues. Since plain and roasted licorice have differing chemical composition, the researchers were also able to tell to some extent, from the fact that the roasted variety worked better, which fractions were best suited to this application (Kim et al. 2010).

Part of the punch is that glycyrrhizin inhibits HMGB1 (see below), a notorious inflammation inducer now being implicated in the development of rheumatoid arthritis (Huang et al. 2016). This same trouble causer is now suspected of upregulating RAGE (it is after all an inflammatory agent) which may underlie the mechanistic basis of the progression of Parkinson's disease (Santoro et al. 2016). This progression can involve choreiform or choreoathetotic movements similar to those of Sydenham's chorea.[27] Since glycyrrhizin binds HMGB1, it should be a useful neuromodulator, helping to slow the progress of Parkinson's disease (Santoro

[26]CIA is an abbreviation for collagen-induced arthritis.

[27]Harrison (1987, 86, 2019).

et al. 2016). It may thus have been of more use than just for the arthritis in the ancient Mesopotamian example.

Nor is this the only licorice benefit. Modern experimental studies indicate antibacterial activity of licorice leaves and roots against *Staphylococcus aureus* and of glycyrrhizin against *Escherichia coli* (Hosseinzadeh and Nassiri-Asi 2015, 1871), both causes of acute bacterial arthritis.[28]

BONE FRACTURES:

A single text attests to the use in ancient Mesopotamia of *šūšu*-licorice to promote the healing of fractures.[29]

potions:

***šūšu*-licorice seed**, *hašû*-thyme seed and [...] are (mixed) with wine, left out overnight and used in a potion drunk on an empty stomach for any sort of fracture (*šimertu*) (BAM 158 iv 34)

Modern in vitro and in vivo tests on rats have concluded that isoliquiritigenin, a fraction of licorice, should be of use in preventing excessive bone resorption involved in such processes as postmenopausal osteoporosis and rheumatoid arthritis by interrupting the three MAPK cascades that drive the production of osteoclasts whose job it is to resorb old bone (Zhu et al. 2012). The formononetin contained in *Glycyrrhiza glabra* is a phytoestrogen and promises to help to prevent as well as to heal the bone fractures common in postmenopausal women (Kaczmarczyk-Sedlak et al. 2013). Similarly, as shown in in vitro, glabridin, an isoflavan purified from licorice root, has an oestrogen-like activity, encouraging osteoblasts to produce new bone and cartilage by increasing cell size, causing elevation of alkaline phosphatase activity, collagen content and osteocalcin secretion and decreasing the production of prostaglandin E and nitric oxide (Choi 2005).

LOW BACK PAIN:

Low back pain[30] of the less severe variety (*maškādu*) was also treated with *šūšu*-licorice.

baths:

***šūšu*-licorice**, [. . .] , *pillû*-mandragora leaves and [. . .] are baked in an oven and used in a bath for *maškādu* (K 2428+2548+6728 r. 4')

enemas:

erēnu-cedar, *šurmēnu*-cypress, "sweet reed", *asû*-myrtle, *ballukku*-aromatic, *šimšallu*-boxwood, myrrh, [. . .], *kukru*, *ṣumlalû*, *burāšu*-juniper, *su'ādu*-chufa, *baluḫḫu*-aromatic, *kurkānu*-turmeric, *nikiptu*, *sīḫu*-wormwood, *argānu*, *barīrātu*, ***šūšu*-licorice**, [. . .], *nīnû*-mint, *kasû*, *ḫašû*-thyme, *nuḫurtu*, *urnû*-mint, *tīyatu*, *azupīru*, *šumuttu*-vegetable, *samīdu*, lumps of malt, wheat groats and dates are boiled in beer and used in an enema for stabbing pain in the right or left side, preceded by a hot bandage with dates, crushed malt, yeast,

[28]Harrison (1987, 1463).

[29]See Scurlock and Andersen (2005, 247–248).

[30]See Scurlock and Andersen (2005, 257–258).

ballukku-aromatic, flax seed, *aktam* and winnowed beerwort and bath with *šūnû*-chaste tree infusion (BAM 3 iv 41)

procedure lost:

[. . .], *šammi bu'šāni*, *bīnu*-tamarisk, [. . .], ***šūšu*-licorice root** and [. . .] are used for *maškādu* (K 2428+2548+6728 r. 14')

[. . .] *errû*-coloquinth, *errû*-coloquinth kernel, *sīḫu*-wormwood, *argānu*, *barirātu*, [. . .], ***šūšu*-licorice**, ***šūšu*-licorice root**, male *pillû*-mandragora, [. . .], *kamkādu*, *kamantu*, *aprušu*, *bīnu*-tamarisk, [. . .], *šimru*-fennel, *šimru*-fennel root, *ṣaṣumtu*, *lišān kalbi*, [. . .], "fox grape", and [. . .] are used for stinging tongue, dizziness, [roaring ears], jabbing pain in the neck, hurting neck muscles, shifting inguinal regions (weakness in the hips), hurting breast, back and shoulders, [numb] arms, bloated insides, rumbling intestines, [sluggish fingers], gnawing pain in the feet and legs and [. . .] or sore hips or hurting lower back or constriction of DÚR.GIG (see below) or constriction of the urethra or [. . .] or *saḫḫu* in the kidneys or sick gallbladder or *amurriqānu*-jaundice or [. . .] or "hand" of curse or *maškādu* or *sagallu* or "hand of ghost" or [. . .] or if he has a *mukīl rēš lemutti* (headache) (AMT 22/2 obv. 12).

One of the common causes of low back pain is what we call sciatica. Licorice has long been known as a pain killer (see below), but this is not all that it does for this condition. Modern experimental studies show that licorice promotes sciatic nerve regeneration and functional repair (Jia et al. 2014).

HEART:

For the heart, we have two treatments with *šūšu*-licorice that, on the basis of the chest pain, shortness of breath, sleepiness, loss of appetite and the *birdu*-nodules, could be endocarditis or tuberculosis.[31]

ENDOCARDITIS?:

baths:

½ *qû* of *burāšu*-juniper, 10 shekels of *kukru*, 10 shekels of *ṣumlalû*, x *qû* of *qutru* seed, *ṣarbātu*-poplar leaves. *tittu*-fig leaves, *sirdu*-olive leaves, *nurmû*-pomegranate leaves, *ḫašḫuru*-apple leaves, [...], *baltu*-thorn leaves, *ašāgu*-thorn leaves, *errû*-coloquinth leaves, *tigilû*-melon and wild *tigilû*-melon leaves, *šūnû*-chaste tree leaves, ***šūšu*-licorice leaves**, potpourri, *baltu*-thorn root, *ašāgu*-thorn root, *uriyānu* and "swamp apple" are chopped and crushed together, (mixed) with beer, milk and well water; maltster's water is poured in, baked in an oven and used in a seven day bath for relaxed abdomen but sleepiness, internal fever, thirst, loss of appetite for bread and beer, *birdu*-nodules, shortness of breath, chest pain and impotence due to a curse (K 2426 i 15', cf. 25')

procedure lost:

bīnu-tamarisk, datepalm, *qān šalāli*, *šalluru*-plum leaves, *nurmû*-pomegranate leaves, *ḫašḫuru*-apple leaves, *adāru*-poplar leaves, *tittu*-fig leaves, [...] leaves, *e'ru*-tree leaves, *tullal* leaves, *sikillu* leaves, *baltu*-thorn leaves, *ašāgu*-thorn leaves, *sirdu*-olive leaves, *dulbu*-plane tree leaves, [...] leaves, *uriyānu* leaves, *buṣinnu* leaves, ***šūšu*-licorice leaves**, *šūnû*-chaste tree leaves, *errû*-coloquinth leaves, *sīḫu*-wormwood, [*argānu*, *barīrātu*], *kukru*, *burāšu*-juniper, *ṭūru*-aromatic, [...], *erēnu*-cedar and [...] are used for relaxed abdomen, sleepiness, internal fever, thirst, loss of appetite for bread and beer, *birdu*-nodules, shortness of breath, chest pain and impotence due to a curse (K 2426 i 25′).

---

[31] See Scurlock and Andersen 2005, 561–562.

If this is indeed endocarditis, modern experimental studies have shown the plant to have anti-inflammatory properties and to be of potential use in reducing myocardial inflammatory edema as shown by experimental myocardial damage in rats (Hosseinzadeh and Nassiri-Asi 2008, 711). Modern experimental studies also indicate antibacterial activity against some of the causes of endocarditis,[32] *Staphylococcus aureus*, *Enterococcus faecalis* and *Streptococcus mutans* (Hosseinzadeh and Nassiri-Asi 2015, 1871).

LUNGS:

For the lungs, the focus in attested ancient Mesopotamian treatments seems to be on secondary bacterial pneumonia in connection with *ṣētu* (enteric fever).[33] Since the plant is also used for just fever, one might suppose that this symptom was its target. However, *šūšu*-licorice was also separately used for *suʾālu*-cough[34] and *ḫaḫḫu* (bloody sputum).[35] Moreover, as we mentioned above, there is a treatment which could be for tuberculosis.

COUGH:

distillate daubs:

nīnû-mint is crushed; ***šūšu*-licorice root**, [...], *antaḫšum*-vegetable and *kukru* are mill ground and sifted; oil, beer and ghee are poured over it; poured into a bronze *dīqaru*-bowl; a porous *burzigallu*-vessel with a hole bored in it is put over top and the rim fluted with emmer dough; boiled over a fire; a reed straw is stuck in it; he is to suck it up and it is slapped while still hot on the lungs for *suʾālu*-cough; the patient is to suck up hot [honey and ghee]; to be used if he does not show improvement after another treatment (BAM 548 iv 7'÷BAM 552 iv 9')[36]

PHLEGM:

potions:

***šūšu*-licorice root** is ground, mixed with first quality beer and used in a potion for *ḫaḫḫu* (bloody sputum) (BAM 1 ii 43; CT 14.34 (80-7-19,356):2'; STT 92 ii 13)

SECONDARY PNEUMONIA:

bandages:

[...], *bīnu*-tamarisk leaves, *burāšu*-juniper, *kukru* and ***šūšu*-licorice leaves** are crushed together, sifted, mixed with sheep fat, massaged into leather and used to bandage the [breast] for difficulty breathing and diarrhoea due to *ṣētu*, preceded by a mouth rub with *ninû*-mint, a nasal rinse with alum, a lick with wild honey and a bath with *qiššû*-melon, *lišān kalbi* and *bīnu*-tamarisk infusion (SpTU 1.44:4)

---

[32] Harrison (1987, 972).

[33] See Scurlock and Andersen (2005, 52–61).

[34] See Scurlock and Andersen (2005, 178).

[35] See Scurlock and Andersen (2005, 42).

[36] For the fuller context, see Scurlock (2014, 465–469).

TUBERCULOSIS?:
baths:

½ *qû* of *burāšu*-juniper, 10 shekels of *kukru*, 10 shekels of *ṣumlalû*, x *qû* of *qutru* seed, *ṣarbātu*-poplar leaves. *tittu*-fig leaves, *sirdu*-olive leaves, *nurmû*-pomegranate leaves, *ḫašḫuru*-apple leaves, [...], *baltu*-thorn leaves, *ašāgu*-thorn leaves, *errû*-coloquinth leaves, *tigilû*-melon and wild *tigilû*-melon leaves, *šūnû*-chaste tree leaves, ***šūšu*-licorice leaves**, potpourri, *baltu*-thorn root, *ašāgu*-thorn root, *uriyānu* and "swamp apple" are chopped and crushed together, (mixed) with beer, milk and well water; maltster's water is poured in, baked in an oven and used in a seven day bath for relaxed abdomen but sleepiness, internal fever, thirst, loss of appetite for bread and beer, *birdu*-nodules, shortness of breath, chest pain and impotence due to a curse (K 2426 i 15', cf. 25')

procedure lost:

bīnu-tamarisk, datepalm, *qān šalāli*, *šalluru*-plum leaves, *nurmû*-pomegranate leaves, *ḫašḫuru*-apple leaves, *adāru*-poplar leaves, *tittu*-fig leaves, [...] leaves, *e'ru*-tree leaves, *tullal* leaves, *sikillu* leaves, *baltu*-thorn leaves, *ašāgu*-thorn leaves, *sirdu*-olive leaves, *dulbu*-plane tree leaves, [...] leaves, *uriyānu* leaves, *buṣinnu* leaves, ***šūšu*-licorice leaves**, *šūnû*-chaste tree leaves, *errû*-coloquinth leaves, *sīḫu*-wormwood, [*argānu*, *barīrātu*], *kukru*, *burāšu*-juniper, *ṭūru*-aromatic, [...], *erēnu*-cedar and [...] are used for relaxed abdomen, sleepiness, internal fever, thirst, loss of appetite for bread and beer, *birdu*-nodules, shortness of breath, chest pain and impotence due to a curse (K 2426 i 25′)

UNIDENTIFIED LUNG PROBLEMS:
bandages:

[...], *burāšu*-juniper, *kukru*, *ṭūru*-aromatic, *urnû*-mint leaves, ***šūšu*-licorice leaves**, [...] leaves and [...] are crushed together, sifted, mixed with sheep fat, massaged into leather and used in a bandage for lung problems, preceded by a mouth rub with [...] and *ḫašû*-thyme, a fumigation with alum and *erēnu*-cedar oil, a lick with [...] and wild honey and a bath(?) with "fox grape", *kasû* and *ḫašû*-thyme (AMT 54/1 r. 7')

***šūšu*-licorice leaves**, *šūnû*-tree leaves, *urnû*-mint [leaves], *ḫašḫuru*-apple leaves, winnowed potpourri and *uriyānu* leaves are mixed with first quality beer, redried, crushed, sifted, mixed with sheep fat and used in a breast and shoulders bandage for lung problems, preceded by a potion(?) with [...] and a *ša'iltu*-insect and a bath with hot *šūnû*-tree infusion (BAM 555 iii 63')

***šūšu*-licorice leaves** are used in a bandage for lung problems; alternated with another bandage with [...] in *kasû* juice, preceded by a potion with [...] (BAM 555 iii 68')

baths:

***šūšu*-licorice leaves**, *līparu*-mulberry leaves and *ḫašḫuru*-apple leaves are mixed with water and used to bathe the patient for lung problems, preceded by a mouth rub with [...], followed by a bandage with [...], *kukru*, wheat flour, "dove dung" (i.e. carob), gazelle droppings and [...] (AMT 24/5:9')

In North African medicine, the root extract is used for hoarseness of voice, cough and respiratory ailments and as an expectorant. Infusion of the root is used for cough due to its emolient, depurative and sweetening properties. The roots are chewed for throat troubles (Boulos 1983, 123). In Chinese medicine, the flower is an expectorant and used, like the root, for cough and difficulty in breathing. The root is antitussive, demulcent, emollient, expectorant and laxative. Licorice is used for cough, consumption, laryngitis, pharyngitis and sore throat. It is used in treating colds and sore throat (Duke and Ayensu 1985,

327–328). It is also used to treat bronchitis, tuberculosis and peptic ulcers (Huang 1993, 277).

In Indian medicine, the root is considered demulcent and expectorant. In Unani medicine, the root is used for diseases of the lungs as an expectorant and the extract is used to correct haemoptysis. The root in infusion, decoction, extract or lozenge is useful as a demulcent in inflammatory affection or irritable conditions of the bronchial tubes, as cough, hoarseness, sore throat and asthma. The inspissated juice is used in cough syrup. Mixed with lime juice and linseed, it is used as a valuable remedy for coughs and colds, feverishness, pain and distress of breathing. A confection called *Majoonai Soul* (decoction of licorice root, preserved grapes, white sugar, powdered chebulic and beleric myrobalans, cloves, nutmeg, round zedoary, cinnamon and a smaller amount of anisi fruit, emblic myrobalan and *Anethum sowa* made into a confection with the sugar and grapes) is a very useful expectorant in bronchitis. Lozenges made with extract of licorice root, cubebs, gum arabic, conium extract and sugar are useful in bronchial affections. For cough and asthma, compound pills were made up with licorice, acacia gum, black pepper, pellitory root, *gulacha*, saffron and sugar and for influenza with licorice, camphor, asafoetida and gum acacia (Nadkarni 1957, 582, 583, 584). In "traditional" European and Asian medicine, a root decoction is administered orally for cough. An aqueous extract of the stem is administered orally for tuberculosis. The root in decoction is administered orally for lung ailments (Hosseinzadeh and Nassiri-Asi 2008, 711).Coughing guinea pigs have benefitted from the antitussive properties of *G. glabra*. In addition, modern experimental studies have shown the plant to be potentially of use in obstructive pulmonary disease as a mucoregulator. It has been suggested that its cough relieving properties are due to mucilage in it or secretion produced by it covers the oral and throat mucosa, soothing irritability (Hosseinzadeh and Nassiri-Asi 2008, 715; Hosseinzadeh and Nassiri-Asi 2015, 1876–1877). Modern experimental studies have shown the plant to have anti-inflammatory properties. It is of potential use in inflammatory lung conditions (Hosseinzadeh and Nassiri-Asi 2008, 711) such as bronchitis and emphysema.[37]

Moreover, modern experimental studies indicate antibacterial activity against *Klebsiella pneumoniae* (Hosseinzadeh and Nassiri-Asi 2008, 711; Hosseinzadeh and Nassiri-Asi 2015, 1871) and antiviral activity against Influenza A and B virus (Hosseinzadeh and Nassiri-Asi 2008, 712; Hosseinzadeh and Nassiri-Asi 2015, 1868). In addition, the water extract of the root was effective against human respiratory syncytial virus in in vitro studies of human respiratory tract cells. It did the job by preventing viral attachment to and penetration into the attacked cells. Also, in a pattern repeated over and over again with licorice, the major benefit was not what the plant itself contributed to the situation but its ability to kick-start the body's own malfunctioning systems, in this case, persuading the attacked cells to protect themselves by secreting IFN-β which is a virus's nightmare, inhibiting not only penetration, but protein synthesis, genome replication and even virus assembly and releasing. Interestingly, the licorice effect was not produced by the major

[37] See Harrison (1987, 1089).

constituent of the root, glycyrrhizin; however, when glycyrrhizin hits the human gut, it is partially converted into 18 β-glycyrrhetinic acid by stomach bacteria and this fraction is effective against this as well as other viruses (Feng et al. 2013). If the tuberculosis reference is indeed referring to tuberculosis and not endocarditis as suggested as an alternative possibility above, glycyrrhetinic acid is "a major anti-tubercular agent" (Kalani et al. 2015).

GI TRACT:

GI complaints absorb a sizeable chunk of the ancient Mesopotamian therapeutic series. A correspondingly wide variety of gastrointestinal symptoms were treated with *šūšu*-licorice. Besides oesophogeal reflux,[38] intestinal worms,[39] both types of diarrhoea[40] and "hand of ghost"-alcoholic-ketoacidosis/dehydration,[41] there was severe indigestion (*kis libbi*),[42] "wind"[43] and "sick stomach".

ESOPHAGEAL REFLUX:
bandages:

kasû, *burāšu*-juniper, *kukru*, *ṣumalalû*, *labanātu*-incense, *lipāru*-mulberry leaves and ***šūšu*-licorice leaves** are ground together, mixed with sheep fat and used in a breast bandage for sore breast, esophogeal reflux, bile on belching and shortness of breath when speaking due to something wrong inside, preceded by a potion with *lišān kalbi* and a bath with *kasû* juice (BAM 575 iii 63)

INTESTINAL WORMS:
potions:

***šūšu*-licorice root** is ground, poured into oil and used [in a potion] for intestinal worms (*urbātu*); the patient is expected to expel the worm (BAM 159 ii 32)[44]

ENTEROTOXIC DIARRHOEA:
baths:

sīḫu, *argānu*-tree, *barirātu*, ***šūšu*-licorice**, *šūnû*-chaste tree, *kasû* and *aktam* are used in a bath for (a wasting) curse (BAM 156:17)[45]

ENTEROINVASIVE DIARRHOEA:
bandages:

līparu-mulberry leaves, ***šūšu*-licorice leaves**, *lišān kalbi* leaves, *ḫaḫḫinu*-thorn leaves, "crowfoot" leaves, *eʾru*-tree or *tittu*-fig leaves and *uzun lalî* leaves are crushed, made into a dough with beer, redried, crushed, mixed with sheep fat, massaged into leather and used in a bandage for blood being emitted from the anus like a woman with irregular bleeding due to

[38]See Scurlock and Andersen (2005, 132).
[39]See Scurlock and Andersen (2005, 82–83).
[40]See Scurlock and Andersen (2005, 51–52, 128–129).
[41]See Scurlock and Andersen (2005, 360–363, 501–503).
[42]See Scurlock and Andersen (2005, 131–132).
[43]See Scurlock and Andersen (2005, 124–125).
[44]For the fuller context, see Scurlock (2014, 495–498).
[45]For the fuller context, see Scurlock (2014, 329–333).

"flowing of the intestines" rather than DÚR.GIG, preceded by a potion with *kalû*-paste, alum and *nuḫurtu*, an anal suppository with powdered magnetic haematite and *erēnu*-cedar oil, a potion with *urnû*-mint and a bath with *baluḫḫu*-aromatic resin or just water (BAM 99:48// AMT 57/6 [+] AMT 43/1 iii 10')

"HAND" OF GHOST (ALCOHOLISM):
potions:

***šūšu*-licorice root** is used as a potion, salve or prophylactic against apparitions (SpTU 4 134:13b; cf. BAM 230:21)[46]

tarmuš, *imḫur-lim*, *imḫur-ešra*, *bīnu*-tamarisk seed, seed of *e'ru*-tree, *azallû*, *šumuttu*-vegetable, *kukru*, *burāšu*-juniper, *atā'išu*, alum, *šakirû* seed, *ardadillu* and ***šūšu*-licorice root** is (mixed) with beer and used in a potion for ghost hurting insides (BAM 161 iii 24'//BAM 165 i 7')[47]

***šūšu*-licorice root** is moistened with water, left out overnight, filtered and used in a potion drunk on an empty stomach for continually bloated insides and feet cramps (BAM 159 ii 17)

KIS LIBBI (SEVERE INDIGESTION):
potions:

pillû-mandragora root, ***šūšu*-licorice root**, *tarmuš*, *imḫur-lim*, *imḫur-ešra*, *tullal*, and *šakirû* are ground together, poured into beer, left out overnight and used in a potion drunk in the morning on an empty stomach for *su'ālu-cough* turning into indigestion (*kis libbi*) (BAM 574 i 2//BAM 573 i 9')

***šūšu*-licorice root** is (mixed) with water and used in a potion drunk on an empty stomach for indigestion (*kis libbi*) (BAM 574 i 8b)

male *pillû*-mandragora root, ***šūšu*-licorice root**, *imḫur-lim*, *imḫur-ešra tarmuš*, *maštakal* and *lišān kalbi* are ground, (mixed) with water and used in a potion for indigestion (*kis libbi*) (BAM 574 i 17)

***šūšu*-licorice root**, male *pillû*-mandragora root, *tarmuš*, *imḫur-lim*, *imḫur-ešra*, *maštakal* and *maštakal* seed are (mixed) with beer and used in a potion for indigestion (*kis libbi*) (BAM 574 ii 42)

"WIND":
bandages:

līparu-mulberry leaves and ***šūšu*-licorice leaves** are dried, crushed and sifted together, mixed with sheep fat, massaged into leather and used in a bandage for inability to keep down bread and beer, limp flesh (absence of febrile rash) and "wind" roiling about in the anus, preceded by a potion with ½ *qû* date juice, ½ *qû kasû* juice, ½ *qû* vinegar, 3 shekels of *nīnû*-mint, 2 shekels of pressed-out oil and 3 shekels of wild honey, a bath with *sīḫu*-wormwood, *argānu*, *barīrātu*, *kasû*, *kasû* leaves and *šūnû*-chaste tree seed in water and beer, a rub down in oil and a second potion with "fox grape" and *lišān kalbi* in beer (BAM 575 iv 53)

[46]For the text, see Scurlock (2005, no. 50b).

[47]For the text, see Scurlock (2005, no. 194).

SICK STOMACH:
bandages:

šūšu-licorice leaves, *buṣinnu* leaves, *šakirû* leaves, *lišān kalbi* leaves, *buāšu*-juniper and dates are ground, mixed with beer, redried, sifted, massaged into leather and used in a bandage for sick stomach; accompanied by emetic potions with *kikkirānu* in beer and male *pillû*-mandragora root in beer, a bath with honey and [...] in *šūnû*-chaste tree infusion and an enema with hot vinegar and oil (BAM 575 i 45)

potions:

ḫabigalbat, **šūšu-licorice seed**, *ḫašû*-thyme, male *šakirû* and *amīlānu*-mandrake are (mixed) with wine and used in a potion for stabbing pain (BAM 164:1)

rikibti arkabi, *tullal*, *qān šalāli* and **šūšu-licorice leaves** are (mixed) with *erēnu*-cedar resin and *kasû* juice and set out; honey, pressed-out oil, ghee and beer are poured on, left out overnight, filtered and used in a potion drunk on an empty stomach for bloated insides and continual mucus in the epigastrium (BAM 575 ii 46)

UNIDENTIFIED GI TRACT PROBLEMS:
procedure lost:

[...], **šūšu-licorice** [...], *amḫara*, [...], assorted aromatics, [...], *ašāgu*-thorn shoots, *baltu*-thorn shoots, [...] shoots, [...], [*sīḫu*], *argannu*, [*barīrātu*] and [...] are used for internal problems(?) (BAM 301:2')

[...], **šūšu-licorice** [...] and [...] are used for internal problems(?) (BAM 440:2')

In traditional North African medicine, extract of the root is used for gastritis and abdominal pains. A refreshing drink made from the root is used for gastric ulcer (Boulos 1983, 123). In Chinese medicine, the root is used for abdominal pain, diarrhoea, gastric imbalance, nausea, splenitis, swelling and ulcers (gastric, duodenal). Licorice is used for internal inflammations and ulcers (Duke and Ayensu 1985, 327–328). In Indian medicine, the root is considered a laxative. In Ayurvedic medicine, the root is used for ulcers. In Indian medicine generally, the root in infusion, decoction, extract or lozenge is useful as a demulcent in inflammatory affection or irritable conditions of the bowels, used as a tonic and mild laxative. Inspissated juice is made into laxative powders. It has been pronounced "wonderful" for relieving pain, discomfort and other symptoms caused by acid matter in the stomach. A compound powder of licorice root, fennel fruit, senna, sublimed sulphur and refined sugar is useful as a gentle laxative given to delicate patients (Nadkarni 1957, 582, 583). In "traditional" European and Asian medicine, a root decoction is administered orally for stomachache. An aqueous extract is administered by mouth for gastric ulcers. Aqueous extract of the root, administered orally, is a mild laxative (Hosseinzadeh and Nassiri-Asi 2008, 711).

For the stomach and bowels, modern experimental studies indicate antibacterial activity against *Escherichia coli* and *Staphylococcus aureus* (Hosseinzadeh and Nassiri-Asi 2008, 711; 2015, 1871). Modern experimental studies have shown the plant to be potentially of use in preventing ulcers, which it does in several ways, by preventing the adhesion of *Helicobacter pylori* to human stomach tissue (Wittschier et al. 2009), by attacking the microbe itself and through its anti-oxidant properties

(Hosseinzadeh and Nassiri-Asi 2015, 1877). Anti-adhesion sounds crazy, but a surprising number of bacteria (including *E. coli* and *Campylobacter jejuni*, a major cause of diarrhoea against which licorice root displays anti-adhesive properties) begin their nefarious work by attaching themselves to intestinal epithelial cells; because this therapy is not actually antibiotic, there is less worry about resistance to it being developed (Bensch et al. 2011).

Licorice also lowers gastric acidity, reduces pepsin activity and inhibits gastric secretion thus protecting against aspirin-induced ulceration (Huang 1993, 277; Tang and Eisenbrand 1992, 582). A traditional Turkish compound powder containing licorice was successful in controlling gastrointestinal bleeding in patients after more conventional haemostatic medicines had failed (Beyazit et al. 2011). Its effects on gastrointestinal motility are homeopathic, a small dose being spasmolytic possibly due to blockage of calcium channels and large doses being spasmogenic possibly due to the activation of muscarinic receptors (Hosseinzadeh and Nassiri-Asi 2015, 1877). Clinical studies have shown that oral licorice heals ulcers and is useful in other GI tract diseases such as dyspepsia (Raveendra et al. 2012). Served up with calcium carbonate, papain and apple cider vinegar, it has proved useful for gastro-esophogeal reflux in clinical trials (Brown et al. 2015).

LIVER AND GALL BLADDER:

There were also treatments in ancient Mesopotamia for the liver and gall bladder that employ *šūšu*-licorice. Most of these liver and gall bladder treatments were for the types of jaundice recognised by ancient Mesopotamians,[48] but a particularly interesting use was for hepatitis B.[49]

HEPATITIS B:

bandages:

lipāru-mulberry leaves, *tittu*-fig leaves, wild *bīnu*-tamarisk leaves, ***šūšu*-licorice leaves**, [...] leaves, *ḫašḫuru*-apple leaves and "fox grape" leaves are dried, crushed and sifted together, [made into a dough] with [. . .] water, redried, crushed, sifted and used in a bandage for continually hurting abdomen, churning(?) stomach, tenseness of the flesh(?) (febrile rash), bloating and stabbing pain(?) in the abdomen after jostling in a chariot, followed by a potion taken while holding the patient's tongue (BAM 397:5)

baths:

***šūšu*-licorice infusion** is used to bathe the patient for the liver afflicting him, stabbing pain in the abdomen, hurting abdomen, disorientation, churning stomach, tenseness of the flesh (febrile rash), bloated stomach and tenseness in the arms and legs due to sick liver, preceded by an abdominal bandage with *su'ādu*-chufa, *suḫḫuru*-aromatic and *ṣaṣuntu* in date juice , [...] and ghee, a rub with oil and a potion with pressed-out-oil (BAM 87:10)

---

[48]For the liver, the distinction is between *aḫḫāzu*-jaundice with bleeding and *amurriqānu*-jaundice with bloating or wasting. See Scurlock and Andersen 2005, 138–142. A further type, "gall" is fairly obviously jaundice due to biliary tree obstruction. To correct is Scurlock and Andersen 2005, 138.

[49]See Scurlock and Andersen 2005, 141.

OBSTRUCTIVE JAUNDICE:
potions:

***šūšu*-licorice peel** is (mixed) with beer and used in a potion for "gall" (BAM 578 i 23)[50]

***šūšu*-licorice root** is ground, (mixed) with oil and beer and used in a potion for "gall" (BAM 578 i 26)

AḪḪĀZU-JAUNDICE:
potions:

***šūšu*-licorice root** is ground, decocted in beer, left out overnight and used in a potion for *aḫḫāzu*-jaundice (BAM 578 iv 30)

***šūšu*-licorice root** is (mixed) with oil and beer and used in a potion for *aḫḫāzu*-jaundice (BAM 578 iv 41)

***šūšu*-licorice root** and *nurmû*-pomegranate root are poured into water, baked in an oven, taken out, filtered, allowed to cool and used in a potion for *aḫḫāzu*-jaundice (BAM 578 iv 42)

baths:

***šūšu*-licorice**, *šūnû*-chaste tree and 3 other [plants] are chopped together and used to bathe the patient for *aḫḫāzu*-jaundice (BAM 71:9')

AMURRIQĀNU-JAUNDICE:
potions:

***šūšu*-licorice root** is pulled up before sunrise[51] and minced; the juice is squeezed out and used in a potion for *amurriqānu*-jaundice (STT 92 ii 4; BAM 1 ii 62)

[1 shekel] of *tarmuš*, 1 shekel of ***šūšu*-licorice**, 1 shekel of *šurmēnu*-cypress, ½ shekel of burāšu-juniper, [...] and *karāšu*-leek are crushed, sifted, mixed with oil and beer and used in a potion for *amurriqānu*-jaundice; the patient is expected to vomit (BAM 52:86)

***šūšu*-licorice root** pulled before sunrise is dried, pulverised, decocted in beer, left out overnight and used in a potion drunk on an empty stomach for *amurriqānu*-jaundice(?) (BAM 159 ii 3)[52]

***šūšu*-licorice root** is soaked in milk, left out overnight, whisked into pressed-out oil and used in a potion for *amurriqānu*-jaundice (BAM 393:5)

***šūšu*-licorice root** is dried, ground and used in a potion for *amurriqānu*-jaundice (BM 38583:21')

***šūšu*-licorice root** is ground, (mixed) with oil and beer and used in a potion for *amurriqānu*-jaundice (BAM 578 iii 15)[53]

---

[50]For the fuller context of this and the following four references, see Scurlock (2014, 505–528).

[51]This practice sounds magical but will have helped avoid contamination of the root with ochratoxin A (Kalesi et al. 2013).

[52]For the fuller context, see Scurlock (2014, 528–531).

[53]For the fuller context of this and the following two references, see Scurlock (2014, 505–528).

***šūšu*-licorice root** used for (*amurriqānu*)-jaundice in the eyes (BAM 578 iii 25)

***šūšu*-licorice root** dried, ground, soaked in beer and used in a potion for *amurriqānu*-jaundice in the eyes (BAM 578 iv 18)

baths:

***šūšu*-licorice root** poured into water, baked in an oven and used to bathe the patient for *amurriqānu* (STT 92 ii 7)

***šūšu*-licorice root** is chopped, poured into water, baked in an oven and used to bathe the patient for *amurriqānu*-jaundice(?) (BAM 159 ii 6)[54]

In Chinese medicine, licorice root is used for jaundice and splenitis (Duke and Ayensu 1985, 327). In Unani medicine, the root is used for diseases of the liver (Nadkarni 1957, 582). For the liver and gall bladder, modern experimental studies indicate antibacterial activity against *Klebsiella pneumoniae* (Hosseinzadeh and Nassiri-Asi 2008, 711; Hosseinzadeh and Nassiri-Asi 2015, 1871), which can infect the biliary tract and peritoneal cavity.[55] Modern experimental studies also indicate antiviral activity against hepatitis A virus, hepatitis B virus and hepatitis C virus (Hosseinzadeh and Nassiri-Asi 2008, 712). In addition, modern experimental studies indicate that the plant has hepatoprotective properties that could make it useful in chronic hepatitis therapy (Hosseinzadeh and Nassiri-Asi 2015, 1870; cf. Huang 1993, 277). In Asia, glycyrrhizic acid is used intravenously for chronic hepatitis B and C and glycyrrhizin is used to treat patients with chronic liver damage and clinically as an antihepatitis agent (Hosseinzadeh and Nassiri-Asi 2008, 716). To note is that the ancient Mesopotamian treatments cited above avoid the GI tract, which is interesting, since oral administration has been observed in modern usage to severely hamper the efficiency of this particular treatment (Cohen 2005).

As for the ubiquitous jaundice, the children with infectious mononucleosis complicated liver impairment who were treated with licorice for two weeks not only showed improved cellular immune function, but had significantly reduced levels of bilirubin (apud Chigurupadi et al. 2016). Jaundice is caused by an excess of bilirubin. A common cause of this excess is overproduction beyond the capacity of the liver to process it. This, in turn, is as a result of haemolysis, the increased destruction of erythrocytes (red blood cells) in the reticuloendothelial system, which requires oxygen to perform.[56] Licorice is a well-known anti-oxidant and, in addition, in vitro studies of human erythrocytes obtained from a blood bank suggest that one of the fractions of *G. inflata*, licohalone A, can force erythrocytes to commit suicide (eryptosis), thus in principle preventing haemolysis (Egler and Lang 2015).[57] More prosaically, glycyrrhizin from *G. glabra* has been proven not only to reduce fatal

[54]For the fuller context, see Scurlock (2014, 528–531).

[55]Harrison (1987, 584).

[56]Harrison (1987, 184, 1320).

[57]This fraction is found only in *G. inflata*. However, this variety and *G. glabra* are practically interchangeable (Kondo et al. 2007) and the *glabridin* of *G. glabra* seems in some actions at least to be the equivalent of *G. inflata*'s licochalcones.

complications of subacute liver failure but also a marked decline in elevated bilirubin concentration in human patients with hepatitis. This was apparently accomplished by affecting serum enzymes in a way that encourages bilirubin clearance (Tewari et al. 2017: 8–9).

DÚR.GIG IN THE BOWELS:

Finally, there were treatments for an ancient Mesopotamian syndrome known as DÚR.GIG (literally "sick peritoneal region"). This syndrome subsumed chronic conditions that blocked or retarded the passage of urine or stool and could include ulcerative colitis as well as haemorrhoids (separately treated).[58]

ULCERATIVE COLITIS:

bandages:

kamantu seed, ***šūšu*-licorice leaves**, *līparu*-mulberry leaves, *maštakal* leaves, [...] leaves [...] and [...] are used in a bandage(?) for DÚR.GIG accompanied by needling pain, inability to keep down bread and water and mucus pouring out of the anus, preceded by potions with "fox grape" in beer, "white plant" in beer and [...] in wild honey, a bath with *šūnû*-chaste tree infusion and an enema with drawn wine in *kasû* juice (AMT 58/2:7)

HAEMORRHOIDS:

potions:

ḫašû-thyme, *nuḫurtu*, *atā'īšu*, [...], *ninû*-mint, *urnû*-mint, *tarmuš*, [*imḫur-lim*], *imḫur-ešra*, male *pillû* root, ***šūšu*-licorice root**, [...] and "sunflower" are ground, mixed with pressed-out oil and used in a potion drunk on an empty stomach to remove *umṣātu*-lesions (haemorrhoids) (AMT 17/5:5)

UNCERTAIN OR UNSPECIFIED:

anal suppositories:

[...], ***šūšu*-licorice**, riverbank silt and [...] are used in an anal suppository(?) for [DÚR.GIG] (AMT 43/2+ [BAM 7 pl. 18] iv 7')

procedure lost:

erēnu-cedar, *baltu*-thorn root, *ašāgu*-thorn root, *šūnû*-tree seed, ***šūšu*-licorice**, *aktam* root and *qutru* seed are used for DÚR.GIG (BAM 173:27)

***šūšu*-licorice** root and [...] are used for loss of motility due to [DÚR.GIG] (AMT 58/2:13)

In Chinese medicine, licorice root is used for haemorrhoids and swelling (Duke and Ayensu 1985, 327). The anti-inflammatory properties of licorice (Hosseinzadeh and Nassiri-Asi 2008, 711; Hosseinzadeh and Nassiri-Asi 2015, 1868, 1870) might have been expected to be useful in inflammatory bowel disease (both Crohn's disease and ulcerative colitis).[59] Indeed, rectally administered glycyrrhizic acid has shown significant protective effects against colitis in rats. Researchers discovered that treated mice lost less weight and had longer colons and significantly less severe

[58]See Scurlock and Andersen (2005, 150–153).

[59]Harrison (1987, 1277–90).

tissue damage than untreated mice. Laboratory analysis of the excised colons, bone marrow and spleens revealed that the benefits were due, inter alia, to the inhibition of HMGB1, thus decreasing the production of pro-inflammatory cytokines that play critical roles in the pathogenesis of colitis (Chen et al. 2017).

GENITO-URINARY TRACT:

Comparatively few conditions of the genito-urinary tract were treated with *šūšu*-licorice. DÚR.GIG (see above), constriction of the urethra, nephrotic syndrome[60] and urethritis complete the picture. More than half of the references appear to be infectious processes. These are the DÚR.GIG references, probably due to infection by *Schistosoma haematobium*,[61] and the two urethritis[62] references.

DÚR.GIG IN THE URINARY TRACT:

bandages:

kukru, *burāšu*-juniper, *erēnu*-cedar, *šurmēnu*-cypress, ***šūšu*-licorice leaves**, *šūnû*-chaste tree leaves, *līparu*-mulberry leaves, *ḫaluppu*-tree leaves, *tilamalû* leaves, [...] leaves, *ašāgu*-thorn [leaves], *baltu*-thorn leaves, [...] and *illūru*-berry are mixed with ox [fat], boiled over a fire and used in a bandage for DÚR.GIG with [...] and hurting lower back (STT 97 ii 22')

procedure lost:

[...] *errû*-coloquinth, *errû*-coloquinth kernel, *sīḫu*-wormwood, *argānu*, *barirātu*, [...], ***šūšu*-licorice**, ***šūšu*-licorice root**, male *pillû*-mandragora, [...], *kamkādu*, *kamantu*, *aprušu*, *bīnu*-tamarisk, [...], *šimru*-fennel, *šimru*-fennel root, *ṣaṣumtu*, *lišān kalbi*, [...], "fox grape" and [...] are used for stinging tongue, dizziness, [roaring ears], jabbing pain in the neck, hurting neck muscles, shifting inguinal regions (weakness in the hips), hurting breast, back and shoulders, [numb] arms, bloated insides, rumbling intestines, [sluggish fingers], gnawing pain in the feet and legs and [...] or sore hips or hurting lower back or constriction of DÚR.GIG or constriction of the urethra or [...] or *saḫḫu* in the kidneys or sick gall bladder or *amurriqānu*-jaundice or [...] or "hand" of curse or *maškādu* or *sagallu* or "hand of ghost" or [...] or if he has a *mukīl rēš lemutti* (headache) (AMT 22/2 obv. 12)

CONSTRICTION OF THE URETHRA:

potions:

[...], [2] shekels of *tarmuš*, 2 shekels of *imḫur-lim*, 2 shekels of *imḫur-ešra*, [2] shekels of "lone plant", 1 shekel of [...] seed, 2 shekels of [...], 2 shekels of [...], 2 shekels of [...], 2 shekels of *kurkānu*-turmeric, [2 shekels] of *nuṣabu*, 2 shekels of *arariânu*, 2 shekels of [...], [2 shekels] of *samānu*, 2 shekels of [...], 2 shekels of [...], 2 shekels of *paṭrānu*?, [...], 2 shekels of *errû*-coloquinth, 1 shekel of "white plant", 2 shekels of [...], 2 shekels of [...]-stone, 1 shekel of *tigilû*-melon, [...], [2] shekels of [...], 1 shekel of *šammi buʾšāni*, 2 shekels of [...], [...], 1 shekel of ***šūšu*-licorice**, 2 shekels of [...], [...], [...], 2 shekels of [...], [...], 1 shekel of [...], [...], 2 shekels of [...], 2 shekels of [...], [...], 2 shekels of *ṣumlalû*, 2 shekels of [...], 1 shekel of [...], 2 shekels of fragments of ostrich egg shell, [...], [2 shekels] of *nīnû*-mint, 2 shekels of [...], [...], [2 shekels] of *imbû tamtim*, [...], 2 shekels of *mūṣu*-stone, 2 shekels of [magnetic haematite, 2 shekels of white] *anzaḫḫu*-frit, 2 shekels of black *anzaḫḫu*-frit, 2 shekels of *pumice*, 2 shekels of *ḫarūbu*-carob pods, 2 shekels of

[60] See Scurlock and Andersen (2005, 106–107).

[61] See Scurlock and Andersen (2005, 109–110, 151, 253).

[62] See Scurlock and Andersen (2005, 89–93).

[...], 2 shekels of *biṣṣur atāne*-shell, 2 shekels of sea *biṣṣūru*-shell, 2 shekels of [*pallišu*-plant] stone, 2 shekels of *sāpinu*-plant stone, 2 shekels of *zalaqqu*-stone, 2 shekels of *ḫallūru*-chick peas, 2 shekels of [*kakku*-lentils], 2 shekels of *kiššēnu*-beans, 2 shekels of *šammi buʾšāni*, 2 shekels of *uḫḫūlu qarnānu*, 2 shekels of [...], 2 shekels of *kanaktu*-aromatic, 2 shekels of *abukkatu* resin, 2 shekels of *zibû*, 1/2 shekel of [...], 1/2 shekel of KÙ.PAD-salt, 1/2 shekel of *Amānu*-salt and 1/2 shekel of [*buḫru*]-salt are crushed, sifted and used in a potion(?) for constriction of the urethra, accompanied by a ritual and recitations (K 6493+6811+Bu 91 5-9,52 [BAM 7 pl. 11-12] obv. 18')

GONORRHEAL URETHRITIS:
gonorrheal urethritis:
bandages:

burāšu-juniper, *kukru*, *ṣumlalû*, *bīnu*-tamarisk leaves and ***šūšu*-licorice leaves** are ground together and used in a bandage on the head of the penis for the penis giving a stinging pain, "semen" flowing during urination, impotence, loss of libido and pus flowing constantly from the penis due to discharge (*mūṣu*), preceded by irrigation with *puquttu*-thorn and a potion and wash with *lišān kalbi* (BAM 112 i 24'//AMT 58/6:7)

līparu-mulberry [leaves], ***šūšu*-licorice leaves**, *ninû*-mint and *buṣinnu* leaves are ground together, boiled in sheep fat and milk and used in a bandage on the head of the penis for the penis giving a stinging pain, "semen" flowing during urination, impotence, loss of libido and pus flowing constantly from the penis due to discharge (*mūṣu*), preceded by irrigation with *urnû*-mint and a potion and wash with *sikillu* seed (BAM 112 i 30')

NEPHROTIC SYNDROME:
daubs:

***šūšu*-licorice oil** is used to rub the head before bandaging in a seven-day shaved head bandage with *kukru*, *burāšu*-juniper, *atāʾīšu*, *kamantu* seed, mill-ground *saḫlû*, *adāru*-poplar leaves, *kasû*, roasted grain flour, winnowed beerwort, *nikiptu*, and *šurdunû* in *kasû* juice for swollen head, eyes, feet and lips, positional vertigo, hurting hands and feet, tense limbs and numb body (Jastrow, TCPP obv. 41)

In North African medicine, extract of the root is used as a diuretic (Boulos 1983, 123). In "traditional" European and Asian medicine, the rhizome and root in infusion is administered orally for cystitis. The stem, administered orally, is used as a diuretic. The root in decoction is administered orally for kidney stones and ulcers (Hosseinzadeh and Nassiri-Asi 2008, 711). In Indian Medicine, the root is considered diuretic. In Unani medicine, the root is used for diseases of the bladder. In infusion, decoction, extract or lozenge, it is useful in catarrh of the genito-urinary tract such as dysuria and ardor urinae. A mixture of extracted juice of licorice roots and extracted juice of *Hermaphrodite amaranth* mixed with honey is considered a sovereign cure for all sorts of leucorrhoea and other uterine complaints (Nadkarni 1957, 582, 583–584).

Modern experimental studies indicate antibacterial activity against several kinds of bacteria implicated in urinary tract problems[63]: *Eschericia coli*, *Enterobacter aerogenes*, *Klebsiella pneumoniae*, *Staphylococcus aureus*, *Enterococcus faecalis* and *Streptococcus mutans* (Hosseinzadeh and Nassiri-Asi 2015, 1871). In addition,

[63]Harrison (1987, 549, 584, 1189–90).

modern in vitro studies of *Schistosoma mansoni*, a relative of *Schistosoma hematobium*, demonstrate that the root extract of *G. inflata* is an effective killer of these worms (Aleixo de Carvalho et al. 2015).

For the experiment, *Schistosoma mansoni* cercariae were harvested from Brazilian snails. These developed into adult worms inside hamsters, which were removed and incubated, two by two, in calf serum on a 24-well culture plate. The sexual fitness of the worms was measured by counting the number of eggs deposited per day in untreated and treated worms. At the maximum dose of extract used in the study, all worm pairs died within 24 hours (Aleixo de Carvalho et al. 2015).

What was discovered is that licorice interferes with the worms' ability to cause damage in the human body by inhibiting the production of the eggs. Adult schistosomes are essentially egg-laying machines. Charmingly paired for life, they swim about in the bloodstream of their victims leaving behind the results of their love-making in parts of the body specific to each type of worm, with *S. hematobium* favouring the bladder and ureters.[64] As usual with licorice, the effect was dose dependent, but 100% cessation in egg laying was achieved even without a lethal dose (Aleixo de Carvalho et al. 2015).

The way licorice kills is also interesting; this is by tampering with the worms' tegument whose ecto-enzymes are essential to its survival. If you make your home in human blood vessels, you need to be able not only to evade your host's defences but to keep your path clear of pesky blood clots.[65] It is also nice to be able to absorb nutrients, to metabolise lipids and cholesterol and to grow and repair damaged tissue. If you are a schistosome and your tegument comes to pieces, you are essentially out of luck. A fraction of licorice, licoflavone B, "massively" disintegrated the tegumental surface, with the males getting worse treatment than the females. Reassuringly, lethal doses of the licorice fraction did no harm to African green monkey kidney fibroblasts (Aleixo de Carvalho et al. 2015).

For urethritis, experimental studies indicate antiviral activity against Herpes simplex virus (Hosseinzadeh and Nassiri-Asi 2008, 711–712; Hosseinzadeh and Nassiri-Asi 2015, 1868–69), a common cause of urethritis. Perhaps, it would be good to test the plant against the gonococcal urethritis actually described in these references.

In addition, experimental studies have shown that licorice has the potential to prevent oxidative damage to the liver and kidneys, one of the causes of nephrotic syndrome (Kim et al. 2010). In managing the syndrome, *G. glabra* seems ideal,[66] since not only it is a promising treatment for hyperlipidemia (Birari et al. 2011), but three of its fractions show thrombin inhibitory effects in vitro (Tao et al. 2012) which should be of use in preventing thromboembolic complications that are reasonably common in nephrotic syndrome. If the *nikiptu* in the ancient Mesopotamian treatment for nephrotic syndrome is indeed *Euphorbia*, then it is a powerful diuretic, a bit

[64]Harrison (1987, 810).
[65]Harrison (1987, 810).
[66]Harrison (1987, 1177).

too powerful, in fact, and can be toxic to both liver and kidneys. It is therefore used in Chinese medicine to treat ascites in combination with *G. glabra*, which in vivo studies have shown to protect these organs. Depending on the dosage and relative proportions of the two plants, licorice can also antagonise the *Euphorbia* (Lin et al. 2016), preventing severe extracellular volume depletion, which can lead to acute renal failure. One has to be careful, however, since the wrong dose and proportion will actually make the *Euphorbia* more potent (Lin et al. 2016).

OBSTETRICS AND GYNAECOLOGY:

The attested use of *šūšu*-licorice in ancient Mesopotamia in obstetrics and gynaecology is also quite restricted. Apart from irregular bleeding, the attested references are to postpartum complications experienced by women who had just given birth, namely uterine atony and puerperal fever.[67]

IRREGULAR BLEEDING:
potions:

***šūšu*-licorice seed**, *nuḫurtu* and *burāšu*-juniper are ground, (mixed) with beer and used in a potion for irregular bleeding? (BAM 237 iv 4)[68]

PUERPERAL FEVER:
bandages:

1 *qû* of ***šūšu*-licorice flour**, 1 *qû* of sesame residue flour, 1 *qû* of malt flour, 1 *qû* of *kukru* flour and 1 *qû* of *burāšu*-juniper are made into a dough with *kasû* juice and used in a bandage over the canal for fever, dark spots on the body and internal suppuration following childbirth, preceded by a bath with gazelle dung from the early part of the year and a wash with oil and beer (BAM 240:62')[69]

baths:

***šūšu*-licorice**, *šūnû*-chaste tree, *aktam*, *kasû*, "sweet reed", *barīrātu* and *argānu* are boiled in water and used in a bath for internal fever, vomiting and retention of lochia after childbirth (BAM 240:58')

UTERINE ATONY:
baths:

erēnu-cedar, *šurmēnu*-cypress, *daprānu*-juniper, *kukru*, *burāšu*-juniper, *ṣumlalû*, *atā'išu*, *sīḫu*-wormwood, *argānu*, *barīrātu*, *kasû* and ***šūšu*-licorice seed** are boiled in water and beer in a *tamgussu*-vessel, filtered and used in a bath as part of a secondary treatment for distension and constipation due to her waters and her blood having gone back inside her, followed by a rubdown with oil (Lambert, *Iraq* 31 pl. 5:23)[70]

In North African medicine, extract of the root is used as an emmenagogue and to relax the uterine muscles. A decoction is boiled to facilitate the period due to the

[67]See Scurlock and Andersen (2005, 282).

[68]For the fuller context, see Scurlock (2014, 571–581).

[69]For the fuller context, see Scurlock (2014, 608–616).

[70]For the fuller context, see Scurlock (2014, 605–608).

presence of oestrogenic hormones in appreciable quantities (Boulos 1983, 123). In Indian medicine, the root is considered emmenagogue and so used in Unani medicine (Nadkarni 1957, 582). In "traditional" European and Asian Medicine, an aqueous extract of the rhizome is administered by mouth as a contraceptive (Hosseinzadeh and Nassiri-Asi 2008, 711).

Modern in vitro studies also show licorice root extract to have both oestrogenic and anti-oestrogenic (Boonmuen et al. 2016) **and** progestogenic and anti-progestogenic activity (Ahmed et al. 2014). Clinical studies show that licorice can induce regular ovulation and pregnancy in infertile hyperandrogenic patients (Hosseinzadeh and Nassiri-Asi 2008, 716). This is due to the fact that oestrogen stimulates the release of oxytocin, a nonapeptide that works on the myometrial and myoepithelial cells in the lining of the uterus, resulting in an increased force of contraction; progesterone, by contrast, inhibits uterine contraction.[71] Combining the two actions is what makes licorice a useful treatment for abnormal uterine bleeding. Moreover, it should also have been helpful in encouraging the descent of the lochia, a cause of what we call puerperal fever. Indeed, the American variety was once used in Texas to treat childbed fever and to help expel the placenta (Foster 2008).

Studies in vitro on the excised uteruses of virgin female Wistar rats indicate that isoliquiritigenin, a flavone from *G. glabra*, has a spasmolytic effect, relaxing uterine muscles (Shi et al. 2012). This finding was challenged on the grounds that the effect is species specific and would therefore not be applicable to humans. A follow-up study repeated the rat uterus analysis and compared the results to human tissues taken from 28–39-year-old women who had recently had a caesarian section. Glycyrrhetinic acid was found to significantly inhibit oxytocin-induced uterine contractions in both rat and human tissues, although the effect was indeed different in the two cases and in a way that favoured the use of licorice in humans (Sumi et al. 2014). Relaxation of the uterine muscles counteracts oxytocin-induced postpartum myometrial contractions. It may seem surprising that the same plant would both encourage and discourage uterine contractions. However, what you have with *G. glabra* is a complex of flavonoid components some of which agonize and some of which antagonise oestrogen, thus making it a selective oestrogen receptor modulator (Boonmuen et al. 2016), and the same is true for progesterone (Ahmed et al. 2014).

It is perhaps this modulatory effect that accounts for the Mesopotamian use of licorice to treat what we call uterine atony, which is, in modern medicine, treated with oxytocin but which can be caused by oxytocin augmentation of labour. Ancient Mesopotamian treatments differ from ancient Greek ones in preferring corrected to violent action. For example, you get the patient to vomit, and once the job is done, you stop the vomiting. In this case, you are using something that will produce uterine contractions that are neither violent nor prolonged and that you hope will correct the problem without harming the patient.

[71] Harrison (1987, 1820, 1826).

NEUROLOGY:

In ancient Mesopotamia, *šūšu*-licorice was used to treat pain, peripheral neuropathy, strokes and seizures. It also features in what can only be described as nervine tonics.

NERVINE TONIC:

baths:

tittu-fig leaves, *adāru*-poplar leaves, *ḫašḫuru*-apple leaves, [...] leaves, *nurmû*-pomegranate leaves, *karānu*-grape leaves, [...] leaves, *ḫaluppu*-tree leaves, *šaššugu*-tree leaves, [...] leaves, wild *bīnu*-tamarisk leaves, ***šūšu*-licorice leaves**, [*šūnû*-chaste tree] leaves, *ṣadānu* leaves, *buṣṣinu* leaves, [...] leaves, *ḫašānu* leaves, *sīḫu* leaves, *argānu* leaves, *barīrātu* leaves, *ṣaṣumtu* leaves, *kalbānu* leaves, *aktam* leaves, *līparu*-mulberry leaves, *e'ru*-tree leaves, *qan šalāli*-reed, *šimru*-fennel leaves, *maštakal* leaves, *erēnu*-cedar leaves, *šurmēnu*-cypress leaves, *asu*-myrtle leaves, *šimešallu* leaves, *daprānu*-juniper leaves, "sweet reed" leaves, *ṣumlalû* leaves, *balukku*-aromatic, assorted incense from wild plants. powdered malt, datepalm fronds, [...], *kasû*, *šitû* of *kasû*, [*baltu*]-thorn shoots and *ašāgu*-thorn shoots are boiled in river, well, canebrake, swamp and maltster's water, cows milk, cow/bull urine and *ḫīqu* made from first quality beer and used in a lotion of leaves for burning of *ṣētu*, *šibiṭ šāri*, *šimmatu*-numbness, *rimūtu* paralysis, "hand of ghost", "hand of curse", "hand" of mankind" or *kal murṣi* persisting despite pharmacology or medicine and not letting up; the patient's sore body is rubbed with ghee; *asu*-myrtle oil is poured over his head; the patient is bathed with it for seven days, morning and evening; *asû*'s lotion made from leaves (BAM 228:4)

procedure lost:

[...] *errû*-coloquinth, *errû*-coloquinth kernel, *sīḫu*-wormwood, *argānu*, *barirātu*, [...], ***šūšu*-licorice**, ***šūšu*-licorice root**, male *pillû*-mandragora, [...], *kamkādu*, *kamantu*, *aprušu*, *bīnu*-tamarisk, [...], *šimru*-fennel, *šimru*-fennel root, *ṣaṣumtu*, *lišān kalbi*, [...], "fox grape" and [...] are used for stinging tongue, dizziness, [roaring ears], jabbing pain in the neck, hurting neck muscles, shifting inguinal regions (weakness in the hips), hurting breast, back and shoulders, [numb] arms, bloated insides, rumbling intestines, [sluggish fingers], gnawing pain in the feet and legs and [...] or sore hips or hurting lower back or constriction of DÚR.GIG or constriction of the urethra or [...] or *saḫḫu* in the kidneys or sick gall bladder or *amurriqānu*-jaundice or [...] or "hand" of curse or *maškādu* or *sagallu* or "hand of ghost" or [...] or if he has a *mukīl rēš lemutti* (headache) (AMT 22/2 obv. 12)

PAIN[72]:

salves:

***šūšu*-licorice root**, [...], *ḫašû*-thyme and *atā'išu* are baked and used in a salve(?) for persistent "hand of ghost" applied with a copper [...]; accompanied by a salve with *tarmuš*, *imḫur-lim*, *imḫur-ešra* and [...] and a potion(?) with drawn wine (AMT 95/1:6')[73]

PERIPHERAL NEUROPATHY[74]:

bandages:

***šūšu*-licorice**, *šūnu*-chaste tree, *šimru*-fennel, *ḫašû*-thyme, *errû*-coloquinth, *kamantu* seed, *aktam* seed, *e'ru*-tree seed, *sīḫu*-wormwood, *argānu* and *barīrātu* are crushed and sifted together, decocted in beer in a *tamgussu*-vessel, sprinkled with emmer flour, massaged into

[72]See Scurlock and Andersen (2005, 284–289).

[73]For the text, see Scurlock (2005 no. 350).

[74]See Scurlock and Andersen (2005, 338–339).

leather and used hot in a bandage for lethargic and piercingly painful legs, followed by a potion and salve with *ḫašû*-thyme, *imḫur-lim* and *tarmuš* (BAM 152 iv 16//BAM 158 iii 28') [Aššur version]

[***šūšu*-licorice**] **root**, *šūnu*-chaste tree, [*šimru*-fennel], [. . .], *errû*-coloquinth, *maštakal* seed, *kamantu*, [*aktam* seed, *eʾru*-tree seed], either *azupīru* or *ḫašû*-thyme, *sīḫu*-wormwood, [*argānu*] and *barīrātu* are dried and crushed, decocted in milk and beer in a *tamgussu*-vessel, massaged into cloth and used in a bandage for lethargic and piercingly painful legs, followed by a potion and salve with [*ḫašû*-thyme, *imḫur-lim*], and *tarmuš* (AMT 70/7 i 3) [Nineveh version]

procedure lost:

***šūšu*-licorice leaves**, *baltu* –thorn leaves, *saḫlû* and *kamantu* and [. . .] are ground and used for feet that are feverish and lethargic so that he cannot raise them or walk about on them (AMT 69/7 ii 9).

In traditional North African medicine, a refreshing drink made from the root is used as an anti-spasmodic (Boulos 1983, 123). In Chinese medicine, the flower is used, like the root, for pain. It is used as a tonic and blood purifier (Duke and Ayensu 1985, 328). It is reported to have analgesic and anticonvulsant effects (Huang 1993, 277). In Unani medicine, the root is used as a nerve tonic (Nadkarni 1957, 582).

Traditional Chinese, Japanese and Korean medicine uses a compound powder of *glycyrrhiza* root and peony root known in Japan as shakuyakukanzoto as an anti-spasmodic for the treatment of skeletal muscle and intestinal cramps. Modern clinical studies have confirmed that SKT has antispasmodic properties, inducing a rapid and clinically significant improvement in muscle cramps arising from various causes including liver cirrhosis and diabetic peripheral neuropathy. Modern experimental studies suggest that the *glycyrrhiza* root in the preparation is doing the heavy lifting (with peony root helping to relieve the associated pain). All this is without the hepatotoxicity and central nervous system depression that are the bane of more conventional treatments (Lee et al. 2013).

Six fractions of the root worked together to relax smooth muscle in male Wistar rats whose left hind leg was subjected to electrical stimuli by a process of which the gruesome details are given in the article. One of the licorice fractions was busy at work in five minutes and pretty much done for the day at 30 minutes. Two more kicked in already in the first half hour, and two others joined later and lasted longer, with one still giving sporadic relief 60 minutes later. And all of this was without affecting motor coordination as tested by putting rats on a rotarod and seeing how long it took them to fall off with and without treatment with SKT (Lee et al. 2013).[75]

In another modern in vivo experiment, isoliquiritigenin, a flavone from *G. glabra* and the eager beaver of the other study, reduced the number of acetic acid-induced writhings of rats, a model for human visceral and inflammatory pain. The researchers tested to see whether the action was actually analgesic as opposed to purely anti-inflammatory. The result was that isoliquiritigenin increased reaction time to

[75]Only one of these fractions is found only in *G. uralensis*, the variety used in the experiment (Kondo et al. 2007).

hot-plate burned rat paws as measured by paw licking behaviour with an effect comparable to aspirin but at a much smaller dose. We are assured that the heat was shut off quickly so as to prevent permanent damage to the rats (Shi et al. 2012).

SEIZURES[76]:

Ancient Mesopotamian treatments for seizures are very frequently amulets, which sounds magical, but was actually a multiplex treatment involving alcohol-based potions and/or oil-based salves containing the same ingredients. Having the plants needed for said potion or salve already round the patient's neck made them readily available for emergency use.

amulets:

***šūšu*-licorice root**, male *pillû*-mandragora root, [...] seed, [...], *nīnu*-mint, *arzallû* and [...] are used in an amulet for AN.TA.ŠUB.BA (SpTU 2.48:15')

***šūšu*-licorice root**, male *pillû*-mandragora root, *ašqulālu*, gazelle dung, *imbû tamtim* and *nikiptu* are used in an amulet to remove AN.TA.ŠUB.BA (BAM 311:69')

murdudu, *ḫūratu*-madder root, ***šūšu*-licorice root**, "lone plant" root, *šumuttu*-vegetable, *lišān kalbi* and [...] are used in an amulet to remove AN.TA.ŠUB.BA (BAM 311:74')

murdudu, *baltu*-thorn root, ***šūšu*-licorice root**, [...], *uriyānu*, plant for "hand" of Ištar, [...], *lišān kalbi*, *ardadillu*, *maštakal* and [...] are used in an amulet for AN.TA.ŠUB.BA? (BAM 478:3')

***šūšu*-licorice root**, male *pillû*-mandragora root, *ašqulālu*, *agargarītu*-sulphur, [*imbû tamtim*] and *nikiptu* are used in an amulet for AN.TA.ŠUB.BA in an infant (STT 57:52)

For seizures including those in infants, modern experimental studies indicate antiviral activity against Herpes simplex virus (Hosseinzadeh and Nassiri-Asi 2008, 711–712; Hosseinzadeh and Nassiri-Asi 2015, 1868–69), which can cause encephalitis.[77] Modern experimental studies also indicate that the ethanol extract of *G. glabra* root has anticonvulsant effects in mice. Specifically, the extract was able to delay the onset of clonic convulsion and to reduce its duration with results comparable to diazepam, if in much larger doses (Chowdhury et al. 2013). How does it work? Given that oxidative stress is both the cause and consequence of epileptic seizures, licorice is an effective anti-oxidant. Curiously, licorice itself has very poor scavenging skills (González-Reyes et al. 2016). Instead, what it does is to stimulate the body's own antioxidant defence system to almost normal levels of activity, thus potentially preventing brain tissue damage due to free radicals (Chowdhury et al. 2013).

In another study, mice tested by a water maze test had significant cognitive impairment after an induced seizure of a type that is used as a model for human temporal lobe epilepsy. To the researchers' delight, the observed impairment was at least partially reversed by treatment with licorice flavonoids, decreasing "escape

[76]See Scurlock and Andersen (2005, 314–322).

[77]Harrison (1987, 694–5).

latency" and increasing the number of "target quadrant crossings" (Zeng et al. 2013). In addition, high-mobility group box 1 (HMGB1) "massively" released from damaged neurons has been implicated in causing delayed damage progresses after an epileptic seizure and contributing to the etiopathogenesis of future seizures by increasing neuronal excitement. Glycyrrhizin stops this process in its tracks, not only by binding HMGB1, but by suppressing neuronal death due to excitotoxicity (Luo et al. 2014).

STROKE:

It might surprise one who is under the mistaken impression that Mesopotamians were human salmon that turned grey and died shortly after thirty that there is ample evidence for strokes in ancient texts. In fact, a wide variety of strokes affecting various parts of the body are clearly described.[78]

bandages:

***šūšu*-licorice leaves** are made into a dough with *kasû* juice, decocted and used in a bandage for soothing the effects of a stroke (*mišittu*) and resulting paralysis (BAM 398:25)

***šūšu*-licorice leaves**, *šūnû*-chaste tree leaves, [...], *e'ru*-tree seed, *tigillû*-melon seed, *ḫašû*-thyme, *uriyānu*, [*sīḫu*-wormwood, *argānu*] are melded, boiled in a *diqāru*-bowl in oil and first quality beer and used in a bandage for soothing paralysis (AMT 52/5:14)

rubs:

saḫlû, *kasû*, ***šūšu*-licorice leaves** and "sunflower" are ground together, poured into water, baked in an oven and rubbed on the feet to soothe the effects of a stroke (*mišittu*) and resulting paralysis, followed by an oil rub (BAM 398:19)[79]

baths:

bīnu-tamarisk, GAB.LAM, *qān šalāli*, ***šūšu*-licorice**, *ḫašḫurru*-apple, *kirbān eqli*, *nurmû*-pomegranate, *adāru*-poplar, garden fruit, *mermer* and [...] sprouts are (mixed) with garden fruit, baked overnight in an oven and used to bathe the patient for stroke (*mišittu*) affecting the front part of the body; the patient is to be helped out of the bath and given a potion(?) with *e'ru*-tree seed in river mouth water followed by a salve(?) with gypsum and [...] (AMT 77/5+K 11127:10')

procedure lost:

***šūšu*-licorice seed**, *šakirû*, *ḫašû*-thyme and [...] are poured into [drawn] wine, left out overnight and used for stroke (*mišittu*) which "touches" the back (AMT 77/1 i 15//BAM 132:11'//BAM 133:7')

For strokes, probably the most interesting modern experimental study shows protection against brain damage in an artificially created middle cerebral artery occlusion. Injections of glycyrrhizic acid were effective both before and after the stroke, induced by anaesthetising male Sprague-Dawley rats and producing occlusion of the middle cerebral artery with a nylon suture. This was left in for 1 hour and then removed, allowing reperfusion for 12 hours to 14 days, after which the rats were

[78] See Scurlock and Andersen (2005, 327–330).

[79] For the fuller context, see Scurlock (2014, 561–566).

decapitated and their brains sectioned and thoroughly analysed. In the meantime, a variety of blood and neurological tests were performed, both on the test rats and a control sample. A nice thermal blanket and heat lamp kept rat rectal temperature at a comfortable 37 degrees centigrade and all was in accordance with proper animal research guidelines. There is no word about whether any of the rats filed a lawsuit against the researchers, but the treated rat's brains looked visibly healthier and treated rats were significantly more likely to struggle against being suspended by the tail, to be able to run along the floor in more or less straight lines and without falling on their sides and to more or less hang onto a balance beam, etc. Pretreatment by 30 minutes with GL actually suppressed the infarct (Kim et al. 2012).

Similarly positive results have been achieved with rats given artificially induced spinal cord injury. (Do not worry--the operation was performed under anaesthetic, the rats recovered on a heating pad under a warm towel and their little bladders were manually voided twice a day until they could recover bladder control.) Licorice root, it transpires, was helpful in ameliorating impaired movement due, not to the injury itself, but to the following degenerative process in which oxygen-derived free radicals have been implicated. Licorice not only inhibits the production of reactive oxygen species, but it also reduces the concentration of reactive nitrogen species that can actually smother cells. Amelioration in the affected rats was ascertained by measuring on a sliding scale the injured rats' ability to move their hind limbs, to stand unaided and to walk in a coordinated fashion (Genovese et al. 2009).

It was argued that the protective effects of glycyrrhizic acid in stroke were due to the chemical binding of HMGB1,[80] an endogenous danger signal, a sort of SOS released by neurons damaged by excitotoxicity in the ischaemic core. Unfortunately, HMGB1 proves its point by triggering inflammatory processes that result in damage to the brain in the surrounding regions. Not only does licorice bind HMGB1 but also has anti-excitotoxic and anti-oxidative effects in neurons (Kim et al. 2012). An earlier study (Kim et al. 2011) established that adding amino acids to the licorice preparation made for even better results. It has also been shown that glycyrrhetinic acid has promise as a possible preventative for future strokes in the affected patient due to its ability, when delivered orally, to inhibit factor Xa, thus preventing thrombosis without significantly affecting bleeding time (Jiang et al. 2014).

POISON:

Ancient Mesopotamians did not have the full range of chemical poisons known to us today. That did not, however, prevent people from being bitten or poisoned by swimming in the wrong river, eating the wrong food or nibbling on the wrong mushroom, and that favourite poison still voluntarily drunk to this day, namely alcohol.[81]

[80]This is an abbreviation for high mobility group protein B1.

[81]For details, see Scurlock and Andersen, 354–366.

NUMBNESS: (BITES AND STINGS)[82]

salves:

baltu-thorn shoots, *ašāgu*-thorn shoots, *qān šalāli* shoots, *šūnû*-chaste tree shoots, *bīnu*-tamarisk shoots, *bīnu*-tamarisk seed, *e'ru*-tree seed, *lišān kalbi* seed, *šimru*-fennel root, ***šūšu*-licorice root**, *ankinūtu*, *imbû tamtim*, *ašqulālu*, *šurdunû* seed, *azallû* and male *pillû*-mandragora root are [dried], crushed, sifted, (mixed) with oil and used in a salve for numbness (Rm. 265:15')[83]

fumigants:

baltu-thorn shoots, *ašāgu*-thorn shoots, *qān šalāli* shoots, *šūnû*-chaste tree shoots, *bīnu*-tamarisk shoots, *bīnu*-tamarisk seed, *e'ru*-tree seed, *lišān kalbi* seed, *šimru*-fennel root, ***šūšu*-licorice root**, *ankinūtu*, *imbû tamtim*, *ašqulālu*, *šurdunû* seed, *azallû* and male *pillû*-mandragora root are dried, crushed, sifted and used in a fumigant over coals for numbness of all the flesh (AMT 91/1:8)[84]

EXCESS SALIVATION (MUSHROOM POISONING)[85]:

potions:

15 grains of *tarmuš*, *atā'išu*, *maštakal*, [...], ***šūšu*-licorice root** and *imḫur-ešra* are ground together, (mixed) with beer and used in a potion for excess salivation, internal fever and fever to get the salivation to stop (Unger, AfK 1.24 obv. 4)

sikillu, *qān šalāli*, *burāšu*-juniper, *kanaktu*-aromatic, [...], ***šūšu*-licorice root** and *nikiptu* are ground together, (mixed) with wild honey and oil and used for spittle flowing and not to be stopped, preceded by a potion and aliment with *imḫur-lim*, *tarmuš* and *elikulla* (Unger, AfK 1.24 obv. 12)

sikillu, datepalm, [...], ***šūšu*-licorice root** and [...] fat are used for excess salivation, preceded by a potion with [*imḫur-lim*], *imḫur-ešra*, *elkulla* and [...] seed (BM 72246+72031+72008 ii 18)

BOTULISM (FOOD POISONING)[86]:

potions:

***šūšu*-licorice root** (root) and x shekels of *alluḫāru*-mineral are ground, decocted overnight in beer and used in a potion drunk on an empty stomach for cases where the patient eats bread and drinks beer and then his insides are unquiet; he gets progressively worse; and when he bathes, he is short of breath due to dirty substances having been introduced into his bread, beer and (bath) water, followed by a bath with *baluḫḫu*-aromatic resin, *bīnu*-tamarisk infusion and *maštakal* infusion (BAM 90 r. 6')--AMD 7.10.2

In Chinese medicine, the flower is considered to be alexipharmic and used, like the root, for difficulty in breathing (Duke and Ayensu 1985, 327). In Ayurvedic

[82]See Scurlock and Andersen (2005, 289–290).

[83]Neighboring treatments suggest that the numbness was due to "witchcraft".

[84]This appears alongside treatments for scorpion sting.

[85]See Scurlock and Andersen (2005, 356–357).

[86]The diagnosis is suggested by the specific association with intake of food and drink, the unquiet stomach, the difficulties in breathing and the rapid progression of the illness. Harrison (1987, 562).

medicine, the root is used against poisons, and in Unani medicine, the extract corrects purgatives (Nadkarni 1957, 582). In Indian medicine generally, the root is used in scorpion sting, which produces numbness (Nadkarni 1957, 582).

Licorice is a potent antitoxin, a property already noted in Chinese medical practice. Experiments have shown that glycyrrhetinic acid can lower the toxicity of strychnine, histamine, chloral hydrate, arsenate, snake venom, diphtheria toxin and tetanus toxin, inter alia. It can protect the heart against physostigmine and acetylcholine (Huang 1993, 277). On a more mundane level, experimental studies have shown the plant to have hepatoprotective properties against chemically induced liver damage in rats (Hosseinzadeh and Nassiri-Asi 2008, 712–713; Hosseinzadeh and Nassiri-Asi 2015, 1870, 1872). Some lucky male mice got to do some serious social drinking and an even luckier subset were protected from the resultant fatty liver by having licorice added to their liquid diet. Since the liver is encouraged to produce free radicals and cytokines by alcohol, the licorice benefit is due to its anti-inflammatory and anti-oxidant properties, as well as the suppression of the agent responsible for the deposition of fat in the liver (Jung et al. 2016). That human subjects can also be protected from liver damage due to binge drinking has also been established using volunteers who agreed to get drunk on vodka in overnight clinic visits for twelve days in a row (Chigurupadi et al. 2016). The plant is similarly protective of the heart and kidneys (Hosseinzadeh and Nassiri-Asi 2008, 713, 714–15; Hosseinzadeh and Nassiri-Asi 2015, 1875–1876; cf. Huang et al. 2016). In addition, experimental studies have shown the plant to be potentially of use in inhibiting mucus overproduction and protected lung epithelium in poisoned rats (Hosseinzadeh and Nassiri-Asi 2015, 1876).

"WITCHCRAFT":

The ancient Mesopotamian diagnostic category of "witchcraft" included not only cases of poisoning but also lung diseases that produced large amounts of phlegm as well as the whitish discharge of vaginitis (yeast infection).[87] The following are listings of plants that would be useful against some form of "witchcraft".

plant inventories:

[*ḫašû*-thyme], *tiyātu, atā'išu, kurkānu*-turmeric, *šibburātu* [...], *urnû*-mint, *nīnu*-mint, [...], azallû, *ṣaṣumtu*, "fox grape" [seed], *bīnu*-tamarisk seed, *e'ru*-tree seed, [...], "lone plant", *imḫur-lim, imḫur-ešra*, [...], *aktam, lišān kalbi, lišān kalbi* [seed], *saḫlû, kasû, šakirû* seed, *šumuttu*-vegetable, *pillû*-mandragora root, ***šūšu*-licorice root**, *uḫḫūlu qarnānu, ḫuluppu*-tree, *burāšu*-juniper, alum, *kalgukku*-clay, *šurdunnû* seed, *šammi ašî*, "white plant", "life plant", *kukru*, "sweet reed","swamp apple", *murdudû, sikillu, kirbān eqli*, silver gold rosette-blossom, gold rosette-blossom, stinking *ašāgu*-thorn, *azupīru, ardadillu, tarmuš* seed and *kamantu* are used for witchcraft (BAM 434 iii 63'//BAM 435 iii 6')--AMD 7.10.1

ḫašû-thyme, *tiyātu/nuḫurtu, atā'išu*, [...] fruit, *nīnu*-mint, *šibburātu, bīnu*-tamarisk, *maštakal, sikillu, šakirû*, canebrake reed, *qān šalāli*, datepalm, *baltu*-thorn leaves, *ašāgu*-thorn leaves, *pillû*-mandragora, ***šūšu*-licorice**, *lapat eqli*, [...], *ḫurātu*-madder, *murdudû, ardadillu, ḫašḫurru*-apple, *ašqulālu, merruta, lišān kalbi, ḫuluppu*-tree leaves, *imḫur-lim*,

[87]See Scurlock and Andersen (2005, 93, 503).

imḫur-ašna, *tarmuš*, *sīḫu*-wormwood, *argānu* and *barīrātu* are eaten separately to dispel *kišpu* lurking in garlic/onions (KUB 37 43 i 13'//KUB 37 44 i 26')

Modern *in vitro* studies suggest that the bark of *G. glabra* is more potent than two of the three antifungal agents currently used against Candidiasis (Sharma et al. 2016).[88]

ENDOCRINE DISORDERS:

ADDISON'S DISEASE[89]:Although one ancient Mesopotamian reference gives a full picture, the presenting complaint is often fulminating nausea, vomiting, diarrhoea and ill-defined abdominal pain, which may be so severe as to be mistaken for acute abdomen (Harrison 1987, 1770). This would appear to be what is referred to in the following treatment.

bandages:

kasû juice, [...], ***šūšu*-licorice leaves**, [...], *šūnû*-chaste tree leaves, *burāšu*-juniper and *kukru* are [...], massaged into leather and used in a bandage for sick stomach, stabbing pain, diarrhoea and vomiting mucus in the field, accompanied by an emetic potion with "white plant" in oil, and a bath with *maštakal*(?) and *lišān kalbi* (BAM 575 i 29)

In Chinese medicine, licorice root is used for Addison's disease and hyperthyroidism (Duke and Ayensu 1985, 327). In "traditional" European and Asian Medicine, an aqueous extract is administered by mouth for Addison's disease. Gan Cao is a drug that affects the adrenal cortex, stimulating it to increase mineralocorticoid secretion. It also potentiates and prolongs the action of cortisol. It is effective in the treatment of mild or moderate cases of Addison's disease (Huang 1993, 277; cf. Tang and Eisenbrand 1992, 582–583). There is a curious case of a woman with Addison's disease who was completely misdiagnosed and who should have died but managed to stay alive and be well for four years due to indulgence of cravings for European licorice sticks and "Dragon" soy sauce (Cooper et al. 2007).

GRAVES' DISEASE[90]:

Harrison (1987, 1769) notes that patients with idiopathic adrenal insufficiency (Addison's disease) are more than usually likely to also have Grave's disease. It is therefore hardly surprising that this endocrine disorder, too, turns up in ancient Mesopotamian texts.

potions:

imḫur-lim, *tarmuš*, [...], ***šūšu*-licorice seed**, *araru*(?), *šimru*-fennel and "swamp apple" are crushed, (mixed) with wild honey, ghee and oil, poured into beer, ground, left out overnight and used in a potion for fidgeting, feet full of *munû* type spots, impotence, staring eyes, itching, troubled words, forgetfulness and troubled heart due to *zikurudû* performed by

[88]For a discussion of Metabolic and Nutritional Diseases, see Scurlock and Andersen (2005, 155–164).

[89]See Scurlock and Andersen (2005, 160), but note that it is the heart and not the patient that is too low to speak; the translation there given should be corrected accordingly.

[90]See Scurlock and Andersen (2005, 161–162).

melting figurines or feeding them to a dog, followed by a stone charm with *ḫulālu*-chalcedony and a salve with *šurmēnu*-cypress oil (K 2351+:16)[91]

A preparation containing licorice root has long been used in Chinese medicine to control hyperthyroidism. *In vivo* studies in rats confirm an immunomodulatory effect of the preparation with the job of reducing the levels of thyroid hormone apparently accomplished by the synergistic use of licorice root with *Sargassum fusiforme* (Song et al. 2011).

5.2 What About Accidental Poisonings by Licorice?

The problem with all herbal medicines is that they are, well, medicines. What this means is that they have the potential to do real harm if used improperly. And the more sophisticated the plant action, the greater the potential for accidental poisoning. Modern herbal medicine has had to struggle with *Glycyrrhiza* species on precisely this issue. We are not simply talking about its tendency to produce allergic rashes in skin creams. Since we are dealing with a plant with significant metabolic effects, there is potential for literally life-threatening consequences of overdose (Hosseinzadeh and Nassiri-Asi 2015, 1880). These cases can be successfully treated in modern times, but this presupposes that the victim has access to a hospital.

A survey of modern literature suggests that this will have been significantly less of a problem for the ancient physician. Nobody in ancient Mesopotamia ate licorice candy. We do not mean the anise-flavoured American imitation but the European variety made from actual licorice. This is admittedly delicious, but eating too much of it accounts for most of the licorice-induced emergency room visits in Europe. Another modern use is in longevity licorice tea which is busy earning itself a reputation in rather the opposite direction. It should be noted that glycyrrhizin, the fraction of licorice that stimulates the adrenal cortex with sometimes fatal results, behaves surprisingly differently in patients with Addison's disease (primary adrenocortical deficiency) (Methlie et al. 2011), that is, in those who actually need it. Finally, there are likely to have been far fewer elderly women with essential hypertension taking potassium-excreting diuretics in ancient Mesopotamia, the major contraindications of the use of licorice in medicine (Yoshino et al. 2014). Moreover, the variety of licorice most readily available in Iraq today (*G. glabra*) has a lower concentration of three of the fractions, liquiritin, isoliquiritin and liquiritigenin, than the variety, *G. uralensis*, more typically used in Asian medicine, which will have made it safer to use (Kondo et al. 2007).

Modern herbal medicines are concentrates beyond the dreams of the ancient physician, which increases the chances of overdosing, as does the practice of using the root in processing cocoa and tobacco, not to mention putting it in just about every medicine. Ancient Mesopotamian physicians, as we know from late texts that give this information, measured actual dosages in grains (50 mg) and carats

[91]For the fuller context, see Scurlock (2014, 638–641).

(200 mg), not the shekels one sees alarmingly in the texts giving directions for the creation of compound medicines. They were also aware of the importance of correct dosage for individual patients, still a vexed problem in modern medicine.

The default in modern medicine is to use extracts or extracted fractions of herbal medicines singly, whereas most Mesopotamian treatments (as well as those of traditional Indian, Turkish, Iranian and Chinese, Japanese and Korean medicine) use decocted plant parts in combination. Ancient Mesopotamians have been bitterly criticised for this practice, but using a plant in combination with other plants has the potential for increasing its effectiveness and/or correcting unwanted side effects (Lee et al. 2013; Kim et al. 2011; Chigurupadi et al. 2016).

A fine example of synergy is licorice and silymarin. Both plants are effective at protecting the liver from chemically induced damage but, when used together, the effect is enhanced to the point of an almost 100% recovery of the treated rats (Rasool et al. 2014). For usages in chronic conditions in human subjects, where the danger of accidentally producing pseudohypercorticosteroidism is the greatest, there is a growing interest in chemically removing the unwanted fractions, producing deglycyrrhizinated licorice (Raveendra et al. 2012 [for dyspepsia]; Puram et al. 2013 [for peptic ulcer]). Where the fraction in question is actually wanted, an alternative is to combine the licorice with another plant that cancels out the undesired actions, possibilities suggested by plants used together with licorice in traditional Chinese medicine as for example Sophorae flavescentis (Shi et al. 2015).

Conversely, adding glycyrrhizin or glycyrrhetinic acid to current treatments for rheumatoid arthritis (NSAIDs and DMARDs) promises, it is argued, not only to increase their effectiveness, but to counteract the damage both sets of drugs do to the kidneys and cardiovascular system (DMARDs) or liver (NSAIDs) (Huang et al. 2016).

It was accidentally discovered that a Turkish compound powder containing licorice not only is effective in stopping GI bleeding (Beyazit et al. 2011) but also interferes with warfarin (coumadin), a synthetic imitation of an anticoagulant first discovered in mouldy clover. This is commonly used in medicine these days to treat/prevent blood clots. Before this plant-based drug came along, doctors used heparin, with potentially fatal results. Nonetheless, even by using warfarin in place of heparin, it is still difficult to control blood clots without causing bleeding (Jiang et al. 2014). Researchers engaged in testing their Ankaferd Blood Stopper in vivo found that, although the product was effective in shortening the duration and amount of bleeding in an amputated rat leg, it did its best work against warfarin-encouraged bleeding (Cipil et al. 2009). The researchers were apparently unaware of an earlier in vitro study of human liver cells and in vivo study of the livers of rats treated with warfarin and an aqueous extract of licorice root minced in that state-of-the-art scientific equipment, a Cuisinart CBT-500 W produced in East Windsor, New Jersey. This study suggested that activating PXR (pregnane X receptor), a xenobiotic orphan nuclear receptor, was what accounted for the clearance of warfarin and indeed the well-attested use of licorice as an antitoxin (Mu et al. 2006).

Drug–drug interactions are generally unwelcome, but perhaps not so much in this case, since licorice is itself well known as a potential preventer of blood clots (Tao et al. 2012; Jiang et al. 2014). You might say it was clearing out the competition. In any case, this would seem to represent another example, seen with its oestrogenic

and anti-oestrogenic (Boonmuen et al. 2016) **and** progestogenic and anti-progestogenic propensities (Ahmed et al. 2014), of the plant's modulatory effects, presumably reinforced in an anti-bleeding direction by the other plants included with it in the Turkish compound powder. It also suggests that the practice of using fractions rather than the whole root may not be the wisest choice with this plant and that dosages are of the essence since, as we have seen more than once, quite opposite effects may be obtained by using small and large doses.

So far, ancient Mesopotamian herbal medicine is getting an A for effectiveness. Still, there will have been uses for *šūšu*-licorice which the ancient physician was sophisticated enough to discover but which are still difficult to control. Classic for licorice in ancient Mesopotamia is the case of Grave's disease where we can actually demonstrate not only that the plant was doing its job, but that the ancient Mesopotamian physician succeeded in treating hyperthyroidism at the expense of giving some of his patients hypothyroidism.[92]

5.3 Conclusion

A review of traditional uses in various medical traditions confirms that any number of cultures, presumably acting more or less independently, decided that licorice was useful for a variety of diseases and medical conditions. What is absolutely astounding, however, is what modern pharmacological studies have to say about the effectiveness of these traditional treatments. The ability of ancient Mesopotamian physicians to sort through what are essentially chemicals used by plants to survive and thrive and to discover their medical uses without the benefit of modern tools such as the microscope, computer modelling and lab tests is little short of miraculous. One can only wonder what would have happened if there had been no one-two punch of Greek philosophy in the form of Greco-Roman humoral theory and the Christian intolerance of "demons" as causers of illness to set back literally by millennia the practice of medicine.

References

Mesopotamian Medicine

Scurlock JA (2005) Magico-medical means of treating ghost-induced illnesses in ancient mesopotamia. Ancient magic and divination III. Brill-Styx, Leiden

Scurlock JA (2014) Sourcebook for ancient mesopotamian medicine. writings from the ancient world 36. Society of Biblical Literature, Atlanta

[92]For details, see Scurlock (2014, 637–644).

Scurlock JA, Andersen BR (2005) Diagnoses in assyrian and babylonian medicine. University of Illinois Press, Urbana

Glycyrrhiza glabra/uralensis/inflata, *etc in "Traditional" Medicine*

Books

Boulos L (1983) Medicinal plants of North Africa. Reference Publications, Algonac
Duke JA, Ayensu ES (1985) Medicinal plants of China. Reference Publications, Algonac
Nadkarni KM (1957) Indian materia medica. Bombay, Popular Book Depot

Articles

Biswas TK, Mukherjee B (2003) Plant medicines of Indian origin for wound healing activity: a review. Int J Low Extrem Wounds 2:25–39
Steven Foster Group, Inc (2008) Licorice-Glycyrrhiza. http://www.Stevenfoster.com/education/monograph/licorice.html

Glycyrrhiza *sp. in Modern Medicine*

Books

Braunwald E, Isselbacher KJ, Petersdorf RG, Wilson JD, Martin JB, Fauci AS (1987) Harrison's principles of internal medicine, 11th edn. McGraw Hill, New York
Huang KC (1993) The pharmacology of chinese herbs. CRC Press, Boca Raton
Tang E, Eisenbrand G (1992) Chinese drugs of plant origin: chemistry, pharmacology, and use in traditional and modern medicine. Springer-Verlag, Berlin

Articles

Ahmed HM, Yeh JY, Tang YC, Chang WT, Ou BR (2014) Molecular screening of Chinese medicinal plants for progestogenic and anti-progestogenic activities. J Biosci. 39:453–461
Aleixo de Carvalho LS, Geraldo RV, de Moraes J, Silva Pinto PL, de Faria Pinto P, Pereira Odos S Jr, da Silva Filho AA (2015) Schistosomicidal activity and docking of *Schistosoma mansoni* ATPDase 1 with licoflavone B isolated from *Glycyrrhiza inflata* (Fabaceae). Exp Parasitol. 159:207–214
Bensch K, Tiralongo J, Schmidt K, Matthias A, Bone KM, Lehmann R, Tiralongo E (2011) Investigations into the antiadhesive activity of herbal extracts against *Camphylobacter jejuni*. Phytother Res. 25:1125–1132
Beyazit Y, Kekilli M, Haznedaroglu IC, Kayacetin E, Basaranoglu M (2011) Ankaferd hemostat in the management of gastrointestinal hemorrhages. World J Gastroenterol. 17:3692–3670

Birari RB, Gupta S, Mohan CG, Bhutani KK (2011) Antiobesity and lipid lowering effects of *Glycyrrhiza* chalcones: experimental and computational studies. Phytomedicine. 18:795–801
Boonmuen N, Gong P, Ali Z, Chittiboyina AG, Khan I, Doerge DR, Helferich WG, Carlson KE, Martin T, Piyachaturawat P, Katzenellenbogen JA, Katzenellenbogen BS (2016 Jan) Licorice root components in dietary supplements are selective estrogen receptor modulators with a spectrum of estrogenic and anti-estrogenic activities. Steroids. 105:42–49
Brown R, Sam CH, Green T, Wood S (2015) Effect of GutsyGum^{TM}, a novel gum, on subjective ratings of gastro-esophageal reflux following a refluxogenic meal. J Diet Suppl. 12(2):138–145
Chen X, Fang D, Li L, Chen L, Li Q, Gong F, Fang M (2017) Glycyrrhizin ameliorates experimental colitis through attenuating interleukin-17-producing T cell responses via regulating antigen-presenting cells. Immulol Res. 65:666–680
Chigurupadi H, Auddy B, Biyani M, Stoha SJ (2016) Hepatoprotective effects of a proprietary glycyrrhizin product during alcohol consumption: a randomized double-blind, placebo-controlled, crossover study. Phytother Res. 30:1943–1953
Choi EM (2005) The licorice root derived isoflavan glabridin increases the function of osteoblastic MC3T3-E1 cells. Biochem Pharmacol. 70:363–368
Chowdhury B, Bhattamisra SK, Das MC (2013) Anti-convulsant action and amelioration of oxidative stress by *Glycyrrhiza glabra* root extract in pentylenetetrazole-induced seizure in albino rats. Indian J Pharmacol. 45:40–43
Cipil HS, Kosar A, Kaya A, Uz B, Haznedaroglu IC, Goker H, Ozdemir O, Koroglu M, Kirazli S, Firat HC (2009) In vivo hemostatic effect of the medicinal plant extract Ankaferd Blood Stopper in rats preteated with warfarin. Clin Appl Thromb Hemost. 15:270–276
Cohen JI (2005) Licking latency with licorice. J Clin Invest. 115:591–593
Cooper H, Bhattacharya B, Verma V, McCulloch AJ, Smellie WS, Heald AH (2007) Liquorice and soy sauce, a life-saving concoction in a patient with Addison's disease. Ann Clin Biochem. 44:397–399
Egler J, Lang F (2015) Licochalcone a induced suicidal death of human erythrocytes. Cell Physiol Biochem. 37:2060–2070
Feng YC, Wang KC, Chiang LC, Shieh DE, Yen MH, San Chang J (2013) Water extract of licorice had anti-viral activity against human respiratory syncytial virus in human respiratory tract cell lines. J Ethnopharmacol. 148(2):466–473
Genovese T, Menegazzi M, Mazzon E, Crisafulli C, Di Paola R, Dal Bosco M, Zou Z, Suzuki H, Cuzzocrea S (2009) Glycyrrhizin reduces secondary inflammatory process after spinal cord compression injury in mice. Shock. 31:367–375
González-Reyes S, Santillán-Cigales JJ, Jiménez-Osorio AS, Pedraza-Chaverri J, Guevara-Guzmán R (2016) Glycyrrhizin ameliorates oxidative stress and inflammation in hippocampus and olfactory bulb in lithilum/pilocarpine-induced status epilepticus in rats. Epilepsy Res. 126:126–133
Hosseinzadeh H, Nassiri-Asi M (2008) Review of pharmacological effects of Glycyrrhiza sp. and its bioactive compounds. Phytother Res 22:709–724
Hosseinzadeh H, Nassiri-Asi M (2015) Pharmacological effects of *Glycyrrhiza* sp. and its bioactive constituents: update and review. Phytother Res. 29:1868–1886
Huang QC, Wang MJ, Chen XM, Yu WL, Chu YL, He XH, Huang RY (2016) Can active components of licorice, glycyrrhizin and glycyrrhetinic acid, lick rheumatoid arthritis? Oncotarget. 7(2):1193–1202
Jia YX, Li JR, Mao CY, Yin WT, Jiang RH (2014) Glycyrrhizin improves p75NTR-associated sciatic nerve regeneration in a BALB/c mouse model. Exp Ther Med. 7:1141–1146
Jiang L, Wang Q, Shen S, Xiao T, Li Y (2014) Discovery of glycyrrhetinic acid as an orally active, direct inhibitor of blood coagulation factor xa. Thromb Res. 133:501–506
Jung JC, Lee YH, Kim SH, Kim KJ, Kim KM, Oh S, Jung YS (2016) Hepatoprotective effect of licorice, the root of Glycyrrhiza uralensis Fischer, in alcohol-induced fatty liver disease. BMC Complement Altern Med. 16:19

Kaczmarczyk-Sedlak I, Wojnar W, Zych M, Ozimina-Kaminska E, Taranowicz J, Siwek A (2013) Effect of formononetin on mechanical properties and chemical composition of bones in rats with ovariectomy-induced osteoporosis. Evid Based Complement Alternat Med. 2013:457052

Kalani K, Chaturvedi V, Alam S, Khan F, Srivastava SK (2015) Anti-tubercular agents from *Glycyrrhiza glabra*. Curr Top Med Chem. 15:1043–1049

Kalesi M, Tabrizchi M, Sheikh-Zeinoddin M (2013) The effects of temperature and relative humidity on ochratoxin A formation in fresh liquorice root. Food Addit Contam Part A Chem Anal Control Risk Assess. 30:339–344

Khan MS, Hamid A, Akram M, Mustafa SB, Sami A, Shah SMA, Usmanghani K (2017) Antipyretic potential of herbal coded formulation (Pyrexol). Pak J Pharm Sci. 30:195–198

Kim KR, Jeong CK, Park KK, Choi JH, Park JH, Lim SS, Chung WY (2010) Anti-inflammatory effects of licorice and roasted licorice Extracts on TPA-Induced Acute Inflammation and collagen-induced arthritis in mice. J Biomed Biotechnol. 2010:709378

Kim SW, Lim CM, Lee HK, Lee JK (2011) The use of Stronger Neo-Minophagen C, a glycyrrhizin-containing preparation, in robust neuroprotection in the postischemic brain. Anat Cell Biol. 44:304–313

Kim SW, Jin Y, Shin JH, Kim ID, Lee HK, Park S, Han PL, Lee JK (2012) Glycyrrhizic acid affords robust neuroprotection in the postischemic brain via anti-inflammatory effect by inhibiting HMGB1 phosphorylation and secretion. Neurobiol Dis. 46:147–156

Kondo K, Shiba M, Nakamura R, Morota T, Shoyama Y (2007) Consitituent properties of licorices derived from *Glycyrrhiza uralensis*, *G. glabra* or *G. inflata* identified by genetic information. Biol Pharm Bull 30:1271–1277

Lee KK, Omiya Y, Yuzurihara M, Kase Y, Kobayashi H (2013) Antispasmodic effect of shakuyakukanzoto extract on experimental muscle cramps in vivo: role of the active constituents of Glycyrrhizae radix. J Ethnopharmacol. 145:286–293

Lin Y, Zhang Y, Shang E, Lai W, Zhu H, Fang Y, Qin Q, Zhao H, Lin N (2016) Effect of Glycyrrhiza on the Diuretic Function of *Euphorbia kansui*: An Ascites Mouse Model. Evid Based Complement Alternat Med. 2016:7620817

Luo L, Jin Y, Kim ID, Lee JK (2014) Glycyrrhizin suppresses HMGB1 inductions in the hippocampus and subsequent accumulation in serum of a kainic acid-induced seizure mouse model. Cell Mol Neurobiol. 34:987–997

Methlie P, Husebye EE, Hustad S, Lien EA, Løvås K (2011) Grapefruit juice and licorice increase cortisol availability in patients with Addison's Disease. Eur J Endocrinol 165:761–769

Mu Y, Zhang J, Zhang S, Zhou HH, Toma D, Ren S, Huang L, Yaramus M, Baum A, Venkataramanan R, Xie W (2006) Traditional Chinese medicines Wu Wei Zi (Schisandra chinensis Baill) and Can Gao (Glycyrrhiza uralensis Fisch) activate pregnane X receptor and increase warfarin clearance in rats. J Pharmacol Exp Ther. 316:1369–1377

Puram S, Suh HC, Kim SU, Bethapudi B, Joseph JA, Agarwal A, Kudiganti V (2013) Effect of gutgard in the management of *Helicobacter pylori*: a randomized double blind placebo controlled study. Evid Based Complement Alternat Med 2013:263805

Rasool M, Iqbal J, Malik A, Ramzan HS, Qureshi MS, Asif M, Qazi MH, Kamal MA, Chaudhary AG, Al-Qahtani MH, Gan SH, Karim S (2014) Hepatoprotective effects of *Silybum marianum* (Silymarin) and *Glycyrrhiza glabra* (Glycyrrhizin) in combination: a possible synergy. Evid Based Complement Alternat Med. 2014:641597

Raveendra KR, Jayachandra SV, Sushma KR, Allan JJ, Goudar KS, Shivaprasad HN, Venkateshwarlu K, Geetharani P, Sushma G, Agarwal A (2012) An extract of *Glycyrrhiza glabra* (GutGard) alleviates symptoms of functional dyspepsia: a randomized, double-blind, placebo-controlled study. Evid Based Complement Alternat Med. 2012:216970

Santoro M, Maetzler W, Stathakos P, Martin HL, Hobert MA, Rattay TW, Gasser T, Forrester JV, Berg D, Tracey KJ, Riedel G, Teismann P (2016) In-vivo evidence that high mobility group box 1 exerts deleterious effects in the 1-methyl-4-phenyl-1,2,3,6-tetrahydropyridene model and Parkinson's disease which can be attenuated by glycyrrhizin. Neurobiol Dis 91:59–68

Sharma H, Yunus GY, Agrawal R, Kalra M, Verma S, Bhattar S (2016) Antifungal efficacy of three medicinal plants *Glycyrrhiza glabra*, *Ficus religiosa*, and *Plantago major* against oral *Candida albicans*: a comparative analysis. Indian J Dent Res. 27:433–436

Shi Y, Wu D, Sun Z, Yang J, Chai H, Tang L, Guo Y (2012) Analgesic and uterine relaxant effects of Isoliquiritigenin, a flavone from *Glycyrrhiza glabra*. Phytother Res. 26:1410–1417

Shi L, Tang X, Dang X, Wang Q, Wang X, He P, Wang Q, Liu L, Liu X, Zhang Y (2015) Investigating herb-herb interactions: the potential attenuated toxicity mechanism of the combined use of Glycyrrhizae radix et rhizoma (Gancao) and Sophorae flavescentis radix (Kushen). J Ethnopharmacol. 165:243–250

Song XH, Zan RZ, Yu CH, Wang F (2011) Effects of modified Haizao Yuhu Decoction in experimental autoimmune thyroiditis rats. J Ethnopharmacol. 135:321–324

Sumi G, Yasuda K, Kanamori C, Kajimoto M, Nishigaki A, Tsuzuki T, Cho H, Okada H, Kanzaki H (2014) Two-step inhibitory effect of kanzo on oxytocin-induced and prostaglandin F2a-induced uterine myometrial contractions. J Nat Med. 68:550–560

Tao WW, Duan JA, Yang NY, Tang YP, Liu MZ, Qian YF (2012) Antithrombotic phenolic compounds from Glycyrrhiza uralensis. Filoterapia. 83:422–425

Tewari D, Mocan A, Parvanov ED, Sah AN, Nabavi SM, Huminiecki L, Ma ZF, Lee YY, Horbanczuk JO, Altanasov AG (2017) Ethnopharmacological approaches for therapy of jaundice: Part II. Highly used plant species from acanthaceae, euphorbiaceae, asteraceae, combretaceae and fabaceae families. Front Pharmacol. 8:690

Wittschier N, Faller G, Hensel A (2009) Aqueous extracts and polysaccharides from liquorice roots (*Glycyrrhiza glabra* L) inhibit adhesion of Helicobacter pylori to human gastric mucosa. J Ethnopharmacol. 7:218–223

Yoshino T, Yanagawa T, Watanabe K (2014) Risk factors for pseudoaldosteronism with rhabdomyolysis caused by consumption of drugs containing licorice and differences between incidence of these conditions in japan and other countries: case report and literature review. J Altern Complement Med. 20:516–520

Zeng LH, Zhang HD, Xu CJ, Bian YJ, Xu XJ, Xie QM, Zhang RH (2013) Neuroprotective effects of flavonoids extracted from licorice on kainate-induced seizure in mice through their antioxidant properties. J Zhejiang Univ Sci B. 14:1004–1012

Zhu L, Wei H, Wu Y, Yang S, Xiao L, Zhang J, Peng B (2012) Licorice isoliquiritigenin suppresses RANKL-induced osteoclastogenesis *in vitro* and prevents inflammatory bone loss *in vivo*. Int J Biochem Cell Biol. 44:1139–1152

Part II
The Abiotic Aspects of the Tigris–Euphrates River System

Chapter 6
Management of Water Resources Using Storage Reservoirs

Kazimierz A. Salewicz, Maiko Sakamoto, and Mikiyasu Nakayama

Abstract In this chapter, we are addressing complex problems regarding the operation of storage reservoirs in Iraq and relationships taking place between reservoir operators and agricultural water users during the water resources allocation process. These problems are addressed using various mathematical tools like simulation, multiple criteria optimization, and finally games theory. At first hydrologic characteristics of the river flow in both Euphrates and Tigris rivers were analyzed. Then, the properties of the traditional reservoir operation policy have been analyzed followed by the optimization analysis of the modified operation policy. In the following section of the chapter, the fundamental concepts of the game theory are presented along with the game theoretic analysis of the decision-making process taking place between reservoir operators and agricultural water users. Finally, conclusions drawn from the research conducted are presented.

Although the scope of our investigations covers quite a wide range of topics, there are still many questions open and they require further deep and comprehensive investigations. Such investigations are necessary for the well-being of the Iraqi people. It is our hope that the circumstances in Iraq will change soon, to allow for stable improvement of the overall situation, leading consequently to further peaceful development and prosperity.

Keywords Water resources · Storage reservoirs · Euphrates · Tigris · Iraq

---

K. A. Salewicz (✉)
Systems Analyst, Wien, Austria
e-mail: k.salewicz@gmail.com

M. Sakamoto · M. Nakayama
Graduate School of Frontier Sciences, The University of Tokyo, Tokyo, Japan
e-mail: m-sakamoto@k.u-tokyo.ac.jp; nakayama@k.u-tokyo.ac.jp

L. A. Jawad (ed.), *Tigris and Euphrates Rivers: Their Environment from Headwaters to Mouth*, Aquatic Ecology Series 11,
https://doi.org/10.1007/978-3-030-57570-0_6

6.1 Introduction

The intention of this chapter is to present a review of the research regarding the operation of storage reservoirs in Iraq, as a part of broader research dealing with Euphrates–Tigris Rivers.

The research project on management of the Euphrates–Tigris river system was mostly carried out within the framework of the research program "Sustainable basin-wide water policy in areas of rapid population growth." This program, led by Professor Kengo Sunada of the Yamanashi University, was conducted between 2003 and 2008 with funds provided by the CREST (Core Research for Evolutional Science and Technology) program of the Japan Science and Technology Agency (JST). Five times expert meetings were organized in Tokyo during the implementation of this research project with the participation of researchers from the basin countries of the Euphrates–Tigris river (Iraq, Syria, and Turkey) and a few non-basin countries (e.g., Japan, USA, Thailand). In addition to the expert meetings, two times an international symposia were organized in Tokyo for the purpose of disseminating the progress and findings out of the research project.

This study was also partly implemented within the framework of the Grants-in-Aid for Scientific Research (KAKENHI) of the Japan Society for the Promotion of Science. The program led by one of the authors, Mikiyasu Nakayama of the University of Tokyo, tried to develop a methodology to attain transparency in information and policy agenda among the stakeholders of shared water resources management. The numerical model developed in this study should be regarded as a useful tool to maintain transparency of information and policy agenda among basin countries of the Euphrates–Tigris Rivers.

Investigations reported in this chapter were initiated in very close cooperation with Professor Mukdat Ali from the Baghdad University, who provided information and data, who very actively participated in the research activities, and who directed research efforts to address practical problems. Unfortunately, any contact with Prof. M. Ali broke in December 2009 and since then we did not have any further communication with colleagues in Iraq.

As will be seen in the further sections of the chapter, the availability of water plays a crucial role for Iraq as a state and for its citizens and agriculture. From the scientific point of view, the scope of research topics dealing with water resources management in this country is almost unlimited. Unfortunately, since the US invasion of Iraq in 2003, the country has not achieved stability and is facing permanent turmoil and instability, with all negative consequences for its development, life of the citizens, and research. Because of the negative developments taking place in Iraq after 2003, our investigations are based on the data records until the year 2003, since the data regarding later years is either unavailable or unreliable.

While political instability in the river basin has been existing for more than a decade, namely since 2003 in Iraq, followed by Syria since 2011, rational and sustainable management of the Euphrates–Tigris river system should be one of the major prerequisites for post-conflict reconstruction of the region once the present turmoil is over. The importance of natural resources management in the post-conflict era, particularly of water resources, was shown by various case studies (Weinthal et al. 2014). This study should thus serve as a most important reference for the post-conflict days in the near future of the Euphrates–Tigris river basin.

6.2 Climatic and Hydrologic Conditions

Due to limited natural water resources and climatic conditions (hot desert climate or a hot semiarid climate to the northernmost part), Iraq is very strongly dependent upon the usage of water resources of two major rivers: Euphrates and Tigris. Average high temperatures are generally above 40 °C (104 °F) at low elevations during summer months (June, July, and August), while average low temperatures can drop to below 0 °C (32 °F) during the coldest month of the year during winter (The Washington Post 1999).

Tigris originates in the Armenian mountains of Turkey and has a total length of 1840 km (Ali 2005), of which 1400 km (76%) are in Iraq. The area of its river basin is equal to 253,000 km^2, out of which 126,000 km^2, that is 54%, belongs to Iraq. Up to 50% of its total discharge originates in Turkey. It is a source of water for approx. 14 million inhabitants.

Euphrates River originates from two tributaries in the Armenian mountains of eastern Turkey: Kara Su and Murat. The total area of the Euphrates river basin is estimated as 444,000 km^2, out of which 45%, that is 105,750 km^2 are in Iraq. Euphrates has no tributaries in Iraq and 97% of its flow comes from Turkey (88%) and Syria (9%). Euphrates provides water to approx. 12 million people.

Since in the case of both rivers Iraq is the most downstream country, as shown in Fig. 6.1, the inflows to Iraq are strongly dependent upon water management practices in the upstream countries, i.e., Syria and Turkey. Extensive usage of water resources in Turkey has a very significant impact on the availability of water in Iraq, especially due to the fact that existing agreements regarding water distribution between Turkey and Iraq are very vague and do not provide sufficient protection of Iraq's interests (Ali 2005).

To counterbalance these unfavorable (from the water user perspective) characteristics of the flow regime in both rivers, a number of engineering structures has been built (Ali 2006a) in Iraq (see Fig. 6.2). These structures are managed by the local operators reporting to regional Water Directorates and according to decision rules established by the Ministry of Water Resources (Ali 2005). The primary objective of these structures is to secure water supply for agricultural purposes. Availability of water is crucial for the existence of agriculture and for the ability of agriculture to produce a sufficient amount of crop.

It causes very high political and social pressure on institutions and facilities securing water supply (Ali 2006b) and strong conflicts between parties involved. Due to rather loose treatment of law regulations by farmers' community and insufficient legal regulations, farmers tend to take more water from rivers and

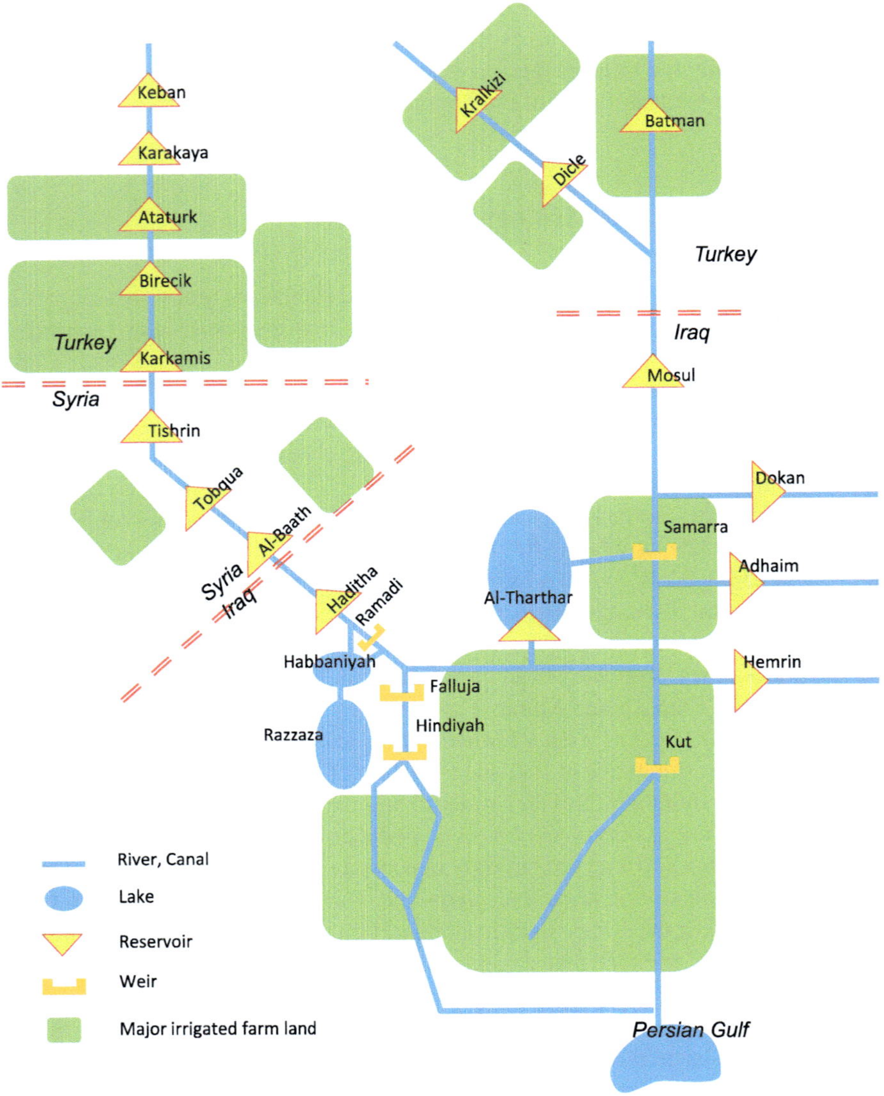

Fig. 6.1 Euphrates and Tigris River System (Tanaka 2011)

drainages, as allocated to them by respective water authorities (Ali 2006b). The real magnitude of illegal/unauthorized water withdrawals has not been assessed and analyzed before; therefore, in a course of our investigations, a model has been developed to calculate water flow balance along Euphrates and Tigris rivers (Salewicz and Tanaka 2006). The model calculated water balance and salt concentration along the course of the Euphrates and Tigris rivers separately. Flow values were computed at consecutive (subsequent) nodes representing river cross-sections

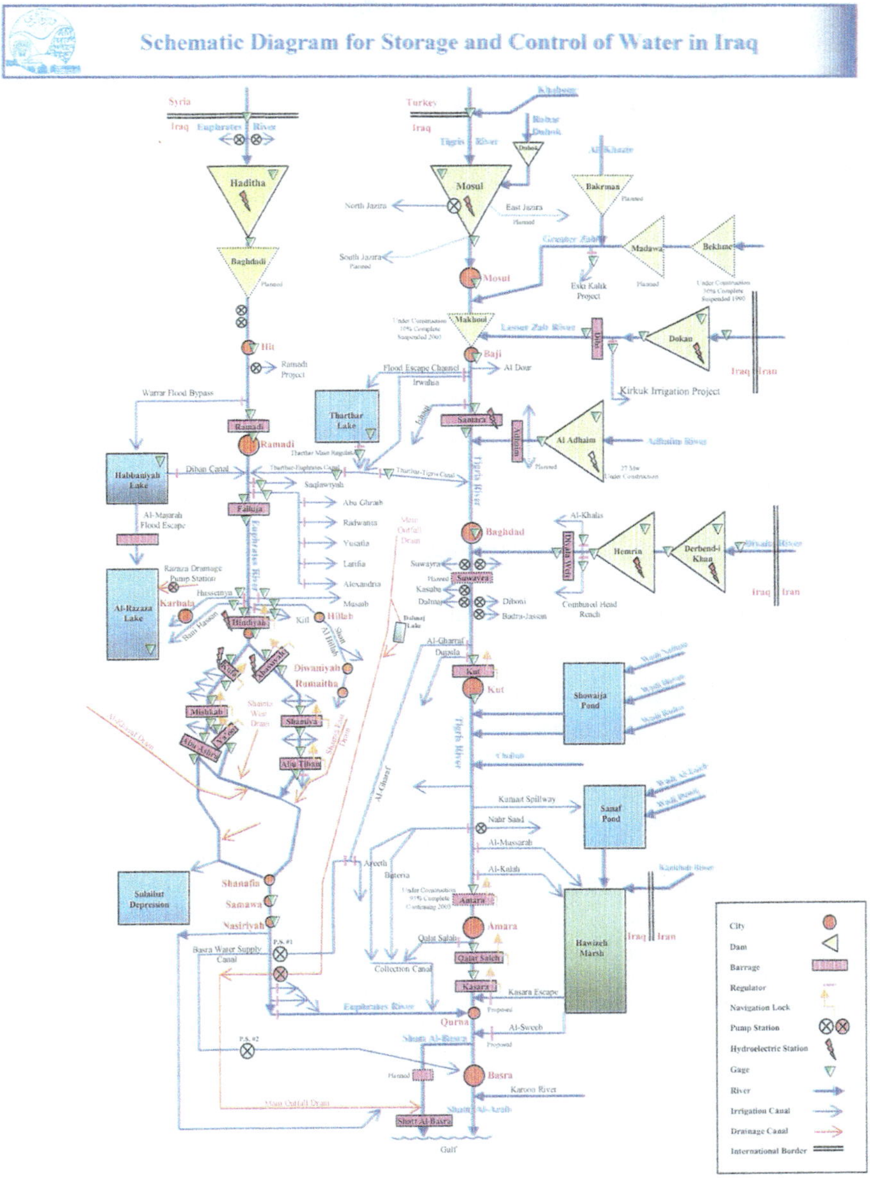

Fig. 6.2 Schematic diagram of Iraq storage and water control system (Ali 2006b)

or control profiles from upstream to downstream using mass balance equation for a given values (average monthly flows) of:

- Natural (unregulated) inflows to the system
- Withdrawals of water for irrigation and water transfer (decision variable)

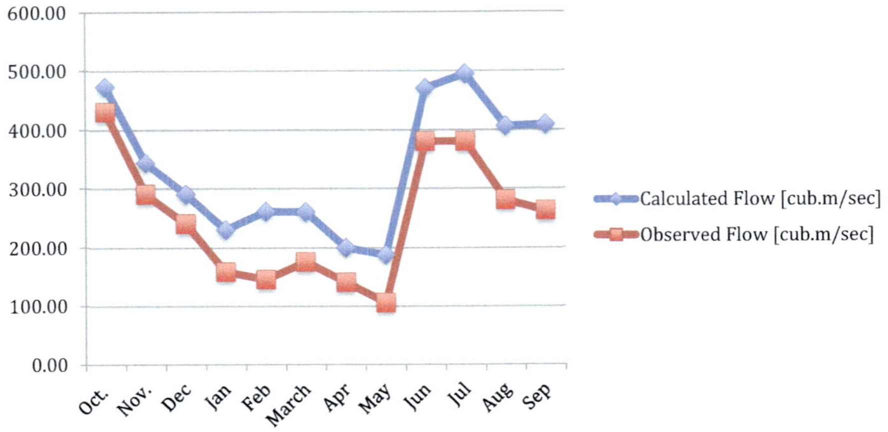

Fig. 6.3 Calculated and observed flows on Euphrates River at Hindiya Barrage

- Discharges from drains (model parameters related to land use and agricultural activities)
- Lateral inflow representing the cumulative effect of flow increase due to basin size increase

In parallel, saltwater concentration was computed using salt mass balance equations for the same profiles (nodes), where—in addition to already mentioned data—salt concentration values in considered streams of water (tributaries, withdrawals to water users, and discharges from water users) were taken into account.

The model used data about flows and salt concentration as obtained from Iraq and did not contain any estimated numerical parameters that could be used to adjust the computation process. Results of calculations performed along each of the rivers were compared with the flow and salt concentration values observed/measured at selected observation points. The model was validated using data associated with the hydrologic year 1998–1999, that is a year with the most complete hydrologic data record (Ali 2006b).

In the case of the Euphrates River, the first point, where the calculated values were compared with the real ones is located in Falluja, at 484 km from the border, which means 251 km from the starting point of the modeled river stretch. The accuracy of the computations regarding this stretch of the river appeared to be highly satisfactory: calculated flow values differed in average by 5, 83% from the observed values (i.e., calculated flows were generally lower than the real ones).

The second profile, where computed and observed flow and salt concentration values were compared was located at Hindiya Barrage (see Fig. 6.3), 619 km from the border (and 135 km south away from Falluja). The differences between computed and observed values of water flow appeared to be significantly higher: calculated flow values were bigger than the observed ones on average by 41, 52% per month.

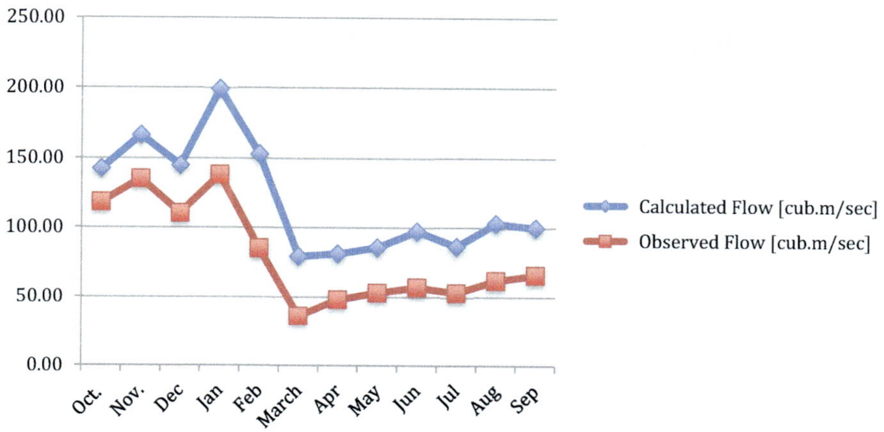

Fig. 6.4 Calculated and observed flows on Tigris River at the closing profile Qurna

In the case of the Tigris River, the calculations began for the Baji profile located 390 km from the border with Turkey. The first comparison between computed and measured values of flow and salt concentration was performed for the Samarra profile (located approx. 490 km from the Turkish border). The differences between calculated and measured values did not exceed a few percent. The next profile where the computed and observed values were compared is Baghdad, located at km 656. An average monthly error associated with the calculation of flow amounted to 47.69%.

The most significant differences between calculated and measured flow (exceeding even by 140% the observed value) could be observed during the vegetation period (July–September). The next comparison between computed and observed flow values was performed for Al-Kut Barrage (at 968 km, 308 km downstream of Baghdad).

Similar to results noted for Baghdad, the differences between calculated and observed flow values, in particular during the vegetation period, were very high (achieving 150% of the measured flows). A final comparison was performed for the closing profile at Qurna (at 1216 km), where the discrepancies between computed and measured flow values appeared to be even higher than at Al-Kut Barrage and for average monthly flow calculation the error amounted to 129.79% (see Fig. 6.4).

Consistently appearing significant differences between calculated and measured flow values, especially high during the vegetation period, support thesis formulated by Ali (2006b), that these differences can be contributed to illegal (not accounted for by the water authorities) withdrawals of water from the river by farmers. This particular aspect will be considered separately in detail in the following section of the chapter.

6.3 Traditional Operation of Major Reservoirs

The water potential of Euphrates and Tigris rivers depends mainly upon the type of the hydrologic year, where three types are distinguished: Dry, Normal, and Wet (Al-Faraj 2006) and results from natural (climatic) factors. Although the inflows to Iraq are influenced by reservoir operation policies in the upstream countries (Turkey, Syria), there is no cooperation between these countries and Iraq regarding release policies in the upstream countries. Since the sharing of hydrological data and information does not exist, the reservoir operation policy in Iraq is based on three scenarios representing each type of hydrologic year: Dry, Normal, and Wet.

The type of hydrologic year is determined based on the amount of water stored in the reservoirs at the beginning of the hydrologic year that is on October 1st of each year. The objectives of the reservoirs' operation are following:

1. Secure flood control
2. Meet water demands, primarily agricultural demands
3. Meet hydropower generation needs
4. Reach normal operation levels in the reservoirs at the end of the winter season (May 31st) and consequently meet as much as possible summer season requirements

Following recommendations of Iraqi experts (Ali 2006b), the subject of our studies were two reservoirs: Al-Tharthar reservoir (85 bln. cub. m of maximum storage) located along Tigris river together with much smaller Al-Habbaniya reservoir (3,76 bln. cub. m maximum storage) on the Euphrates river. These two reservoirs jointly create the system supplying water to major agricultural water users in the lower part of the Euphrates–Tigris river basin. The system has to meet three basic operational purposes:

1. Secure water supply to major agricultural users
2. Protect downstream region from spring floods caused by melting snow and high rainfalls in the upstream basin
3. Augment water flow in Euphrates river through balancing water transfer from Tigris to the Euphrates

The scheme of the case study system is shown in Fig. 6.5 and consist of the following major elements:

- Al-Tharthar and Al-Habbaniya reservoirs located along Tigris and Euphrates river, respectively
- Aggregated (understood as represented by summarized water demands) various agricultural water users in areas of Al-Dour, Al-Ishaki, Al-Saqlawiya, and Five Channels
- Water transfer channels: from Tigris to Al-Tharthar, from Euphrates to Al-Habbaniya, from Al-Tharthar to Tigris river, and from Al-Tharthar to Euphrates river

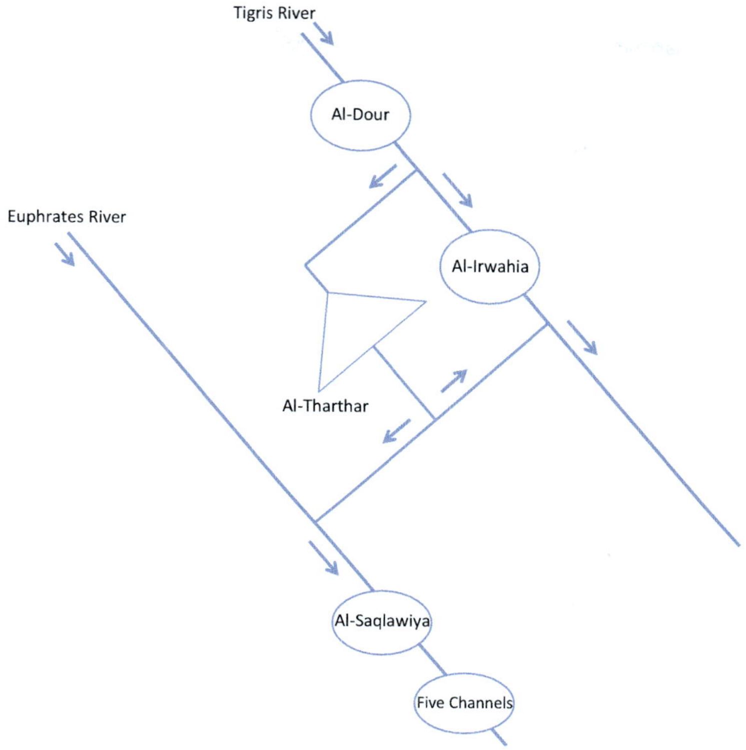

Fig. 6.5 Euphrates–Tigris case study system

Decision variables (average monthly values) to be made by respective reservoir operators are following:

- Water transfer from Tigris river to Al-Tharthar reservoir
- Water transfer from Euphrates river to Al-Habbaniya reservoir
- Release from Al-Tharthar
- Division of Al-Tharthar release between transfer to Tigris river and amount of water transfer to Euphrates river
- Release from Al-Habbaniya
- Water transfers to four considered aggregated agricultural water users

The approach adopted in Iraq to determine release from the reservoirs is based on a concept of Rule Curve (Loucks and Sigvaldason 1982), which associates release from the reservoir with the zone, in which the amount of water stored in the reservoir at the moment of decision-making belongs; and with the season (usually month) of the year. The concept of the Rule Curve is at best explained using two views:

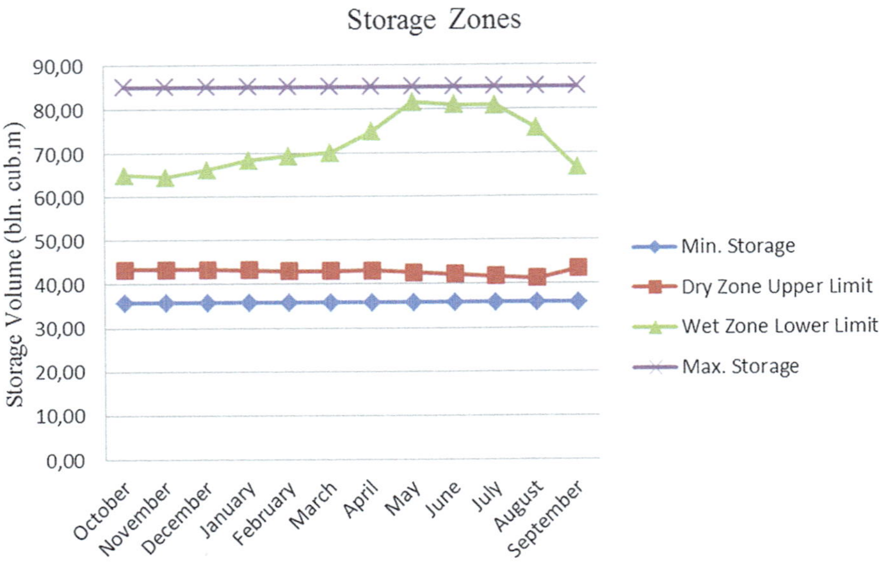

Fig. 6.6 Limits of Al-Tharthar storage zones

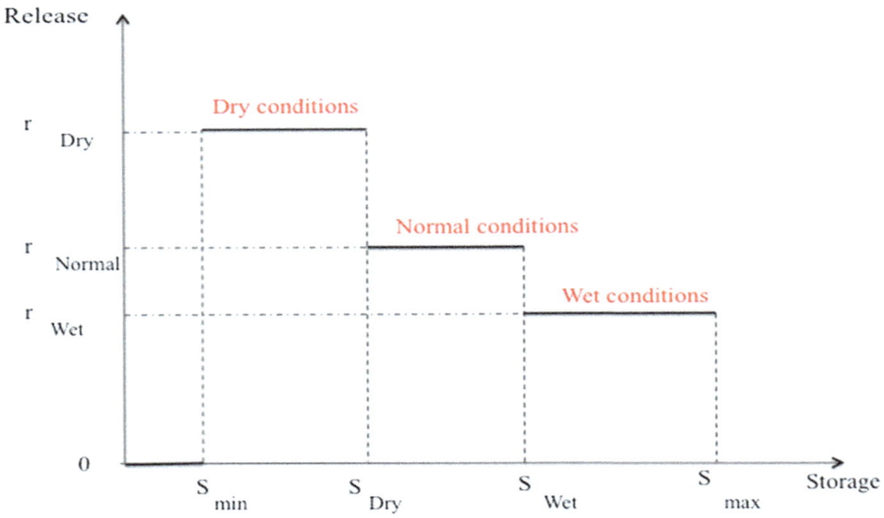

Fig. 6.7 Relationship between amount of water stored at the beginning of the months and amount of water released for each type of hydrological conditions (Dry–Normal–Wet) as determined at the beginning of the hydrologic year (October 1st)

1. Time-depending seasonal changes of the storage zones borders into which the storage volume of the reservoir has been divided (see Fig. 6.6)
2. Amount of monthly releases resulting from the volume of water stored in the reservoirs and determined for each month separately (see Fig. 6.7)

According to this concept, active storage of the reservoir is divided into three storage zones associated with Dry, Normal, and Wet type of hydrologic conditions (Al-Faraj 2006) in a course of the whole year. Figure 6.6 shows the changes of the storage zones border for the Al-Tharthar reservoir during the year. The same type of curve has been also determined for the Al-Habbaniya reservoir.

At the beginning of the hydrologic year, that is on October 1st, the reservoir operator determines the type of the forthcoming hydrologic year depending upon the amount of water stored at this day in the reservoir. If the amount of water in the reservoir is between minimum storage and dry zone upper limit, the coming hydrologic year is classified as Dry. If the amount of water in the reservoir lies between the upper limit of the dry zone and the lower limit of the wet zone, then the hydrologic year is considered to be Normal. Otherwise, the hydrologic year is considered to be Wet.

Then, every month a release from each of the reservoirs is determined using the curve that associates release value with the amount of water stored in the reservoir at the beginning of the particular months, as visualized in Fig. 6.7. S_{Dry} and S_{Wet} denote the upper limit of the dry zone and the lower limit of the wet zone, respectively.

Operation of the case study system performed using traditional operation policy has been simulated (Salewicz et al. 2014) for historical flow data record covering the period from October 1992 until September 2002. As an input to the case system, the following data of average monthly flows was used:

1. A river flow at upstream profiles at Haditha on Euphrates
2. River flow at Mosul on Tigris River
3. Inflows to Tigris from tributaries
4. Evaporation and precipitation time series for the same period of time as river flow data

All the data used (including reservoir operating rules) have been obtained from the local Iraqi experts.

Remarkably, it was not possible to obtain data and information as to by whom and how the traditional operation policy and its parameters had been derived.

The assessment of the simulation results has been performed by means of 34 criteria characterizing different results of the operation policy. These criteria encompass:

- Total release deficit calculated as the difference between target and actual release for Al-Tharthar and Al-Habbanniya reservoirs; and for 4 aggregated agricultural water users
- Maximum and minimum amount of water stored during simulation period in Al-Tharthar and Al-Habbaniya reservoir, respectively
- Maximum and minimum monthly release from Al-Tharthar and Al-Habbaniya
- Maximum and minimum monthly difference between target release and release both for Al-Tharthar and Al-Habbaniya

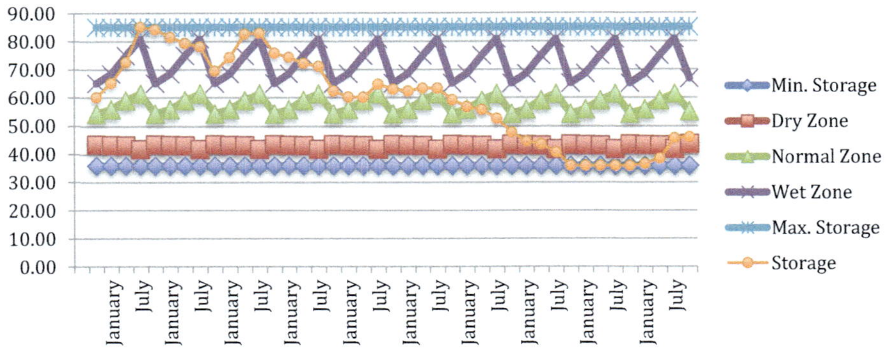

Fig. 6.8 Simulated Al-Tharthar reservoir storage

- Maximum and minimum monthly water transfer from Al-Tharthar
- Minimum monthly water supply to each aggregated agricultural water user
- Minimum yield factor (Tanaka 2011), which allowed to calculate the loss of harvest as a function of not met water demands, for each aggregated agricultural water user
- Final storage (at the end of simulation period) of Al-Tharthar and Al-Habbaniya reservoir
- Average monthly flow at Baghdad (Tigris) and Fallujah (Euphrates), respectively

The majority of above-mentioned criteria used to assess the performance of traditional reservoir operation policy is associated with meeting short-term operational objectives, such as meeting agricultural water demands; securing minimum water flow at selected river profiles or meeting target flows. Only one criterion, namely the value of the reservoir storage at the end of simulation (operation) period (= initial storage volume at the beginning of the next operation period) reflects the long-term objectives of the reservoir operator. Figure 6.8 shows the simulated trajectory of the Al-Tharthar reservoir together with the limits of storage zones associated with Dry, Normal, and Wet type of hydrologic year.

As simulation results demonstrate, in the case of both reservoirs, full range of active storage was used. The water supply of aggregated water users varied depending upon user location and overall hydrologic situation. Both upstream users, that is Al-Dour and Al- Ishaki did not suffer any water deficits over the simulation period, whereas downstream users like Al-Saqlawiya and Five Channels experienced water shortages in 4 out of 10 years. The worst situation took place in the hydrologic year 2000–2001, when the average water supply to these users was reduced down to 56% of annual demand. This, in turn, resulted in a reduction of the annual agricultural production to 94% of the planned harvest.

These results together with general dissatisfaction of the farmers with existing water allocation rules provided motivation to investigate possible adjustments of traditionally used operational policy to look for better fulfillment of operational goals.

6.4 Multiple Criteria Optimization of Parametric Reservoir Operating Rule

Closer analysis of the operating rule shown in Fig. 6.6 demonstrates its one specific feature, namely that reservoir releases determined using traditional operation policy, as shown in Fig. 6.7, change very strongly, even if the storage variation around zone border values (S_{Dry} and S_{Wet}) causing move from Dry to Normal zone (or from Normal to Wet zone) is minimal. In such cases, the release may surge or drop suddenly from one value to another, which in the case of water supply to agricultural users has no good physical justification. Therefore, to smooth (to eliminate sudden jumps of the release) the change of release value resulting from change of the storage zone, a modification of the operation policy has been proposed and linear interpolation of the release value has been proposed, as shown in Fig. 6.9.

Following above-mentioned idea of modifying the traditional operating policy, the operation of the system has been simulated applying the following modifications of this policy:

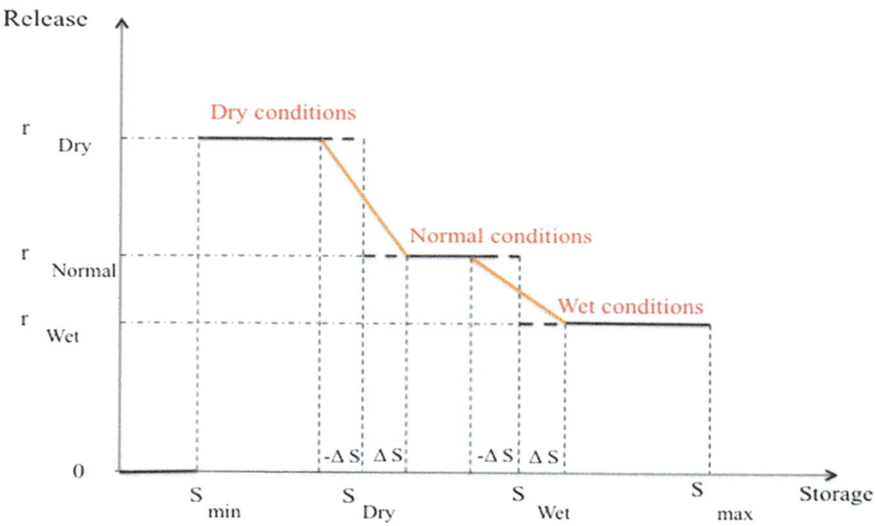

Fig. 6.9 Proposed modifications of the traditional operation policy

- With different values of parameter ΔS denoting the distance to the left and to the right of the storage zone border. Within the storage interval between S_{Dry}—ΔS and $S_{Dry} + \Delta S$ for the change between Dry and Normal storage zone (and analogously for the change between Normal and Wet zone) the value of the release has been calculated according to linear interpolation between release associated with Dry zone and release assigned to Normal zone (and analogously for the change between Normal and Wet zone)
- Determination of the type of the season (Dry–Normal–Wet) once a month (instead of determination of the type of the year performed once a year)
- Modification of the rule used to calculate transfer of the river flow into the reservoir during flood season by introducing flow threshold values and determination of floodwater transfer into reservoir proportionally to peak water flow

As in a case with traditional operating policy, the operation of the case study system has been simulated for numerous different variants of above-mentioned modifications introduced to the traditional operation policy. Results of every simulation run performed over 10 years long period have been evaluated not only with respect to 34 criteria characterizing various impacts of the operation policy but primarily with respect to their physical (practical) sense. Therefore, such variants of modified operation policy have been rejected from further analysis and investigation, which caused for instance depletion of the reservoir, strong deficits occurring at water users, and other, undesirable/unreasonable from a practical point of view, effects. This heuristic approach to the analysis of the simulation results allowed for separating from many, numerically correct solutions only these, which appeared viable with respect to fulfilling in a reasonable manner the overall objectives of the system operation: meeting water demands and secure operability of the reservoirs over a long period of time. Consequently, the sample of modified operation policy options has been limited to 49 out of more than 100 variants, which became subject of further, multiple criteria optimization analysis aimed at selection of the Pareto-optimal policy.

6.4.1 Multiple Criteria Analysis

Results of each simulation run containing values of 34 criteria can be seen as a discrete alternative to the operation policy. To choose Pareto-optimal solution out of considered 49 alternatives, preselected in a course of heuristic analysis of simulation results, the Web-based MCA application (Granat and Makowski 2009), (Granat et al. 2014), (Granat and Makowski 2009) (Wierzbicki et al. 2000) developed at the International Institute of Applied Systems Analysis in Laxenburg, Austria, was used.

The MCA application used in a course of the research allowed for analysis of a given set of discrete alternatives (= every of which being result of the simulation run); and each of the alternatives can be evaluated by more than one criterion associated with a specific attribute. The MCA supports in an interactive manner the user in analysis and evaluation of several alternatives to choose among them the best possible one. For each analyzed alternative, it is possible to develop several alternative instances (versions), each composed of selected attributes, that had been used for defining criteria.

It means, in particular, that for the problem alternative solutions consisting (like in a case considered here) of 34 attributes (criteria), a multiple criteria optimization can be performed for all criteria or for user-defined subset of criteria. Further, for each criterion, a type of optimization (maximization or minimization) can be chosen, and also priorities (relative importance) of criteria can be interactively set to reflect the decision-maker's preferences. These and other features embedded in the functionality of the MCA application make it an extremely useful and powerful optimization tool; however, it also requires a good understanding of the problem subject of the analysis.

Unlike single criteria optimization tools, MCA does not provide a solution in terms of quantified performance function value, but supports two types of analysis:

- Selection of the one (best) alternative for a given set of decision-maker's preferences expressed in terms of relative importance of attributes; or
- Ranking of the alternatives from the best to the worst according to predefined evaluation criteria

Consequently, the analysis performed with the application of MCA allowed the selection of the best alternative of operation policy for a given set of criteria (out of 34) and their preferences.

Figure 6.10 shows a sample MCA application screenshot with optimization results obtained for the full set of 34 criteria. A solid red triangle indicates the criterion value for the solution of the optimization, while blue dots indicate values of a specific criterion for other policy alternatives. Figure 6.11 shows a snapshot of the screen used to display results of the optimization run, namely ranking of the alternatives from the best to the worst one obtained for the same set of criteria, as the solution shown in Fig. 6.10. Optimization analysis has been performed for a full set of criteria (34) and for a number of subsets of this full set consisting of 34 criteria calculated during simulation.

Criteria reflecting water deficits or losses (like water deficit suffered by specific water user) have been subject of minimization, while criteria reflecting benefits or operational targets (like plant yield as a function of water supply level; user water supply; flow at the specific river cross-section; the amount of water stored in the reservoir; final storage of the reservoir at the end of simulation period) have been subject of maximization.

The optimizations have been performed for various groups of criteria (subsets chosen from the full set of 34 criteria). Each group (set) of criteria reflected specific, preferred objectives of the system operation. By selecting into the group criteria associated with minimization of water deficits, maximization of agricultural yield, and/or maximization of river flows, the short-term operational objectives had been reflected. When criteria associated with storage volumes had been chosen for optimization, this selection reflected a long-term objective, namely securing

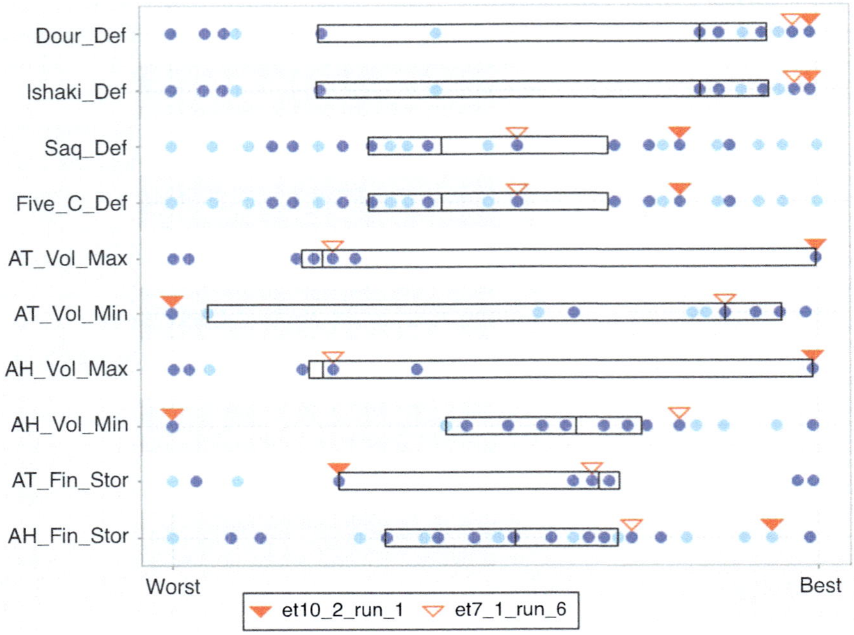

Fig. 6.10 Example MCA screenshot showing optimization results for given criteria listed along vertical line on the left side (like Dour_Def; Ishaki_Def, etc.) and given set of user priorities

operability of the reservoir over multiple years time period. The optimizations have been thus performed for the following groups (A–E) of criteria reflecting an ultimate objective to balance short-term operational goals like meeting agricultural water demands with long-term goals represented by the amount of water stored in the reservoir at the end of the simulation period:

(A) All 34 criteria. Optimal solution: traditional policy
(B) 20 criteria involving water supply deficits of agricultural users, releases from the reservoirs, maximum and minimum storage level attained during simulation period, agricultural yield factors, and final reservoir storage. Optimal solution: policy with interpolated release values for changing storage zones and monthly evaluation of the type of hydrologic status Dry–Normal–Wet
(C) 16 criteria involving water supply deficits of agricultural users, maximum and minimum storage level during simulation period, agricultural yield factors, and final reservoir storage. Optimal solution: either traditional policy or modified policy; result strongly depending upon small changes in priorities of individual criteria
(D) Water supply to the users and final reservoir storage (6 criteria). Optimal policy: traditional one
(E) Water supply to the users, agricultural yield factors, and final reservoir storage (10 criteria). Optimal solution: traditional one

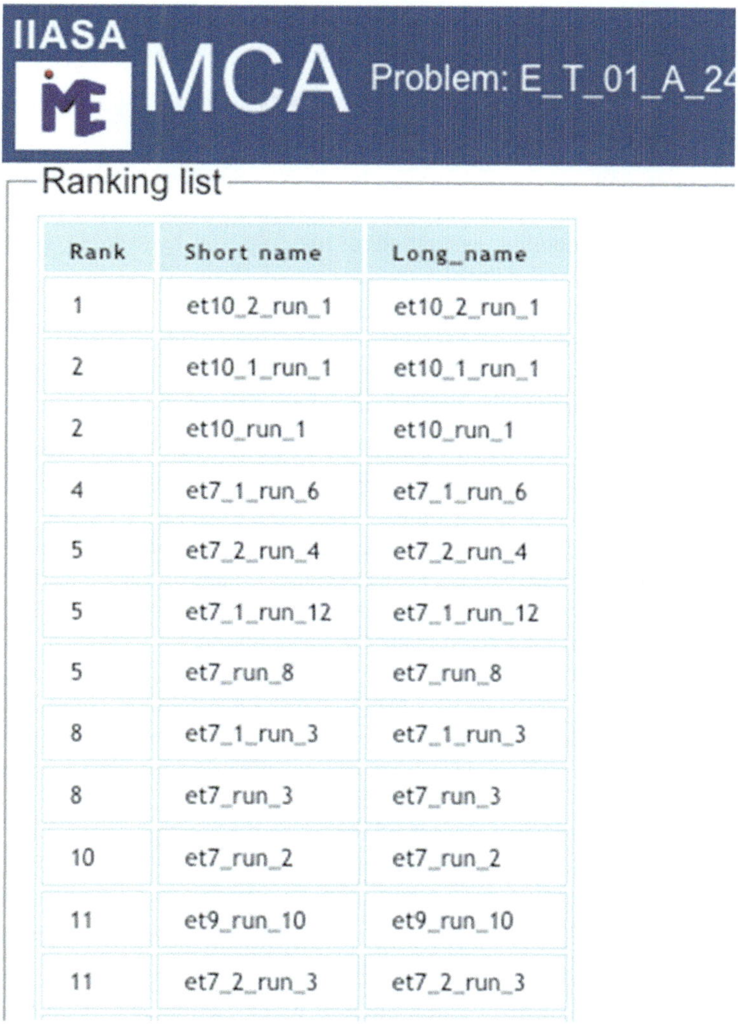

Rank	Short name	Long_name
1	et10_2_run_1	et10_2_run_1
2	et10_1_run_1	et10_1_run_1
2	et10_run_1	et10_run_1
4	et7_1_run_6	et7_1_run_6
5	et7_2_run_4	et7_2_run_4
5	et7_1_run_12	et7_1_run_12
5	et7_run_8	et7_run_8
8	et7_1_run_3	et7_1_run_3
8	et7_run_3	et7_run_3
10	et7_run_2	et7_run_2
11	et9_run_10	et9_run_10
11	et7_2_run_3	et7_2_run_3

Fig. 6.11 Screenshot showing MCA solution with the ranking of alternatives. Abbreviations appearing in the column "Short name" denote considered variants of the operation policy and et10_2_run_1 denotes traditional policy

The second group of criteria reflected the primary objective to maximize water supply to the users and consequently lead to maximization of agricultural production. This group contained the following criteria:

(F) Water supply deficits to the users (4 criteria). Optimal solution: traditional policy
(G) Agricultural yield factors (4 criteria). Optimal solution: slightly modified

traditional policy (adjusted releases from the reservoir to prevent depleting the reservoir, if storage is in a Dry zone).

Finally, the third group of criteria represented entirely long-term operational objectives and included:

(H) Maximum and minimum storage level attained during simulation period (4 criteria). Optimal solution: with interpolated release values for changing storage zones, monthly evaluation of the type of hydrologic status Dry–Normal–Wet and water transfer to the reservoir during flood period depending upon available free storage
(I) Maximum and minimum storage level during simulation period and final reservoir storage (6 criteria). Optimal solution: as above, but with different parameters of the linear approximation function of storage–release relationship

Results of the optimization (with equal priorities of criteria) were given in terms of the best alternative computed and as the ranking of remaining alternatives of the operation policy. Table 6.1 shows values of 12 selected criteria (out of 34) associated with operation policies chosen as optimal ones. As input to MCA optimization, the values of all 34 criteria associated with a specific variant of operation policy have been used. The values of criteria resulting from traditional policy referred to in Fig. 6.6 as et10_2_run_1 are given in the first row (optimal for variants (A), (D), (E), (F), (G)); the second row contains parameters of the optimal policy for the variant (B) and referred to in Fig. 6.6 as et7_1_run_4; row three contains parameters of the optimal policy for variant (C) and referred to et7_1_run_6; row four with parameters associated with variants (H) and (J), respectively, and referred to as et7_2_run5 and et7_2_run_12.

Results of the optimization performed for the above-mentioned groups of criteria appeared to be—at least to a certain extend—surprising. Traditional operation policy, which has been causing significant dissatisfaction of agricultural water users, appeared to be the best solution in following cases of optimization:

- The optimization has been performed for all 34 criteria.
- The set of criteria covered minimization of water user deficits; maximization of agricultural yield expressed in terms of minimum yield factor; and maximization of the final storage of the reservoirs, that is for 10 criteria.
- The set of criteria has been narrowed down to minimization of water user deficits only, and to maximization of the final storage volume of the reservoirs, that is for 6 criteria.
- Optimization was performed for criteria covering minimization of agricultural water users deficits (4 criteria), which reflects the most important operational criteria from the short-term, economic point of view.

These results show that the traditional operation policy is not only robust with respect to variations of the criteria used for the optimization but appeared to provide a reasonable compromise between two conflicting objectives: satisfying current water demands and allowing for the operability of the system in a long-term run.

Table 6.1 Optimal variants of operation policy obtained for chosen sets of criteria

Policy Description	Al-Tharthar Release Total Deficit (Target Release) (mln. cub. m) [MIN]	Al-Dour Total Deficit (Target Release) (mln. cub. m) [MIN. ABS ()]	Ishaki Total Deficit (Target Release) (mln. cub. m) [MIN. ABS ()]	Al-Saqlawiya Total Deficit (Target-Release) (mln. cub. m) [MIN.]	Five Channels Total Deficit (Target Release) (mln. cub. m) [MIN.]	Max volume of Al-Tharthar (bln. cub. m) [MAX.]	Min volume of Al-Tharthar (bln. cub. m) [MAX.]	Al-Tharthar final storage (bln. cub. m) [MAX.]
"Traditional" operational policy	5432.96	48.30	203.35	470.84	1623.93	85.00	35.82	46.05
Hydrologic status (dry–normal–wet) determined every month; storage zone border shift factors: 0%, 10%, 10%, and 3% of nominal border value.	−413,49	−10,60	−44,11	941,47	3247,13	71,84	41,60	53,38
Hydrologic status (dry–normal–wet) determined every month; storage zone border shift factors: 0%, 15%, 15%, and 3% of nominal border value.	5644.72	−6.39	−27.99	768.25	2649.64	72.17	40.86	53.21
Hydrologic status (dry–normal–wet) determined every month; water transfer to reservoir during	−3870.40	157.86	663.45	963.92	3324.65	85.00	41.63	59.25

(continued)

Table 6.1 (continued)

Policy Description	Al-Tharthar Release Total Deficit (Target Release) (mln. cub. m) [MIN]	Al-Dour Total Deficit (Target Release) (mln. cub. m) [MIN. ABS ()]	Ishaki Total Deficit (Target Release) (mln. cub. m) [MIN. ABS ()]	Al-Saqlawiya Total Deficit (Target-Release) (mln. cub. m) [MIN.]	Five Channels Total Deficit (Target Release) (mln. cub. m) [MIN.]	Max volume of Al-Tharthar (bln. cub. m) [MAX.]	Min volume of Al-Tharthar (bln. cub. m) [MAX.]	Al-Tharthar final storage (bln. cub. m) [MAX.]
flood depending upon free storage; storage zone border shift factors: 0%, 10%, 10%, and 3% of nominal border valuer.								
Hydrologic status (dry–Normal–wet) determined every month; water transfer to reservoir during flood depending upon free storage; storage zone border shift factors: 0%, 15%, 15%, and 0% of nominal border value.	−3609.32	154.88	650.85	1181.99	4076.81	85.00	41.42	58.86

This can be contributed to the fact that, due to the huge capacity of the reservoir, sharp changes of water supply associated with changes of the storage zone appear very seldom, which significantly reduced positive effect expected from modified policies involving linear interpolation of the release change resulting from storage zone changes.

Proposed various modifications of the traditional operational policy, as mentioned above, appeared to offer limited improvements of some criteria, but did not significantly contribute to the overall improvement of the operational results. This—to some extend surprising conclusion—can be interpreted as recognition of high professional qualifications of those unknown specialists, who developed traditional operation policy a couple of decades ago, when theoretical and computational engineering capabilities had been significantly less powerful than these days. In addition, considered modifications of the traditional operational policy did not provide significantly better results of operation, which—due to simplicity and transparency of the traditional policy—does not make in practice these modifications more attractive for decision-makers and water users.

6.5 Game Theoretic Approach

6.5.1 Bargaining as an Important Aspect in Water Resources Allocation in Iraq

As already mentioned, the uncertainty regarding the amount of water available in a downstream country like Iraq stems from two factors:

- Water consumption and operation of reservoirs in upstream countries, that is Turkey and Syria
- Natural hydrologic conditions in a downstream part of the basin

Both of these factors cannot be controlled and influenced by Iraqi water authorities, who have the task to secure water supply to communal and agricultural water users (farmers). Under these circumstances, governmental reservoir operators want agricultural farmers to use water in an efficient manner, whereas farmers want to obtain as much water as possible to maximize their benefit by expanding crop areas and extensively irrigate existing crops.

As explained in the previous sections of the chapter, water flow balance in the E–T river system shows discrepancies between real flows observed in the basin and "should be" water flow calculated based on valid water allocation rules.

According to traditional operation policy, the reservoir operator determines and announces to the farmers the type of a coming hydrologic year (Dry–Normal–Wet) based on the amount of water stored in the reservoir at the beginning of a hydrologic year (October 1). The announcement of the type of a coming hydrologic year is equivalent to the announcement of the water allocation rule, which will be applied by the reservoir operator during the coming year. It means that depending upon the type of the hydrologic year the amount of water available to the farmers for irrigation purposes will be determined based on the Rule Curve. At the same time, based on the type of coming hydrologic year announced by the reservoir operator, the farmers decide on the size of farming area and types of crop, which they are going to seed and cultivate during the coming year. Since the type of a coming hydrologic year is determined exclusively based on an amount of water stored in the reservoir on October 1 and without any consideration to meteorological conditions, the likelihood that in a coming year real rainfall pattern (and consequently amount of water available to farmers from precipitation) will be different from the one associated with announced type of hydrologic year is quite significant.

As we can see, the discrepancies between "should be" (ideal value of water flow calculated based on the mass balance equations at specific cross-sections) and real flow in the river result from uncertainty introduced by nature and from the water allocation process taking place as a sort of bargaining taking place between reservoir operator and farmers in the irrigated area.

In a course of this process, farmers—based on their experiences—may not follow the announcements of reservoir operators (type of the hydrologic year, amount of water allocated to them) to maximize their particular benefits. For the same reasons, the reservoir operator has an incentive to provide farmers with different information from the truth to influence their behavior in such a way, that they will make decisions consistent with his long-term objectives. Existing conflict of interests between farmers, who want to maximize their current profits and who disobey the allocation quotas established, and reservoir operator, who strives to secure the fulfillment of long-term objectives by setting limits on the amount of water allocated to farmers is one of the reasons why the water allocation rules have not worked as expected.

To investigate in more detail existing conflict of interests in a decision-making process associated with the allocation of water resources, a specific mathematical methodology, namely game theory, has been applied for the case study system.

Game theory is widely used to model such conflict situations and helps to understand their structures and possible consequences in a logical way. It is a powerful tool for analyzing ongoing conflicts and provides us with probable outcomes. There are various branches of the game theory, such as strategic-form games; extensive-form games; a graphical model for conflict resolution; cooperative games; and bargaining model. In our work, a noncooperative game in an extensive form has been applied to reflect the specifics of an interactive decision-making process taking place between reservoir operators and basin farmers and to examine the effectiveness of water allocation rule in the E–T River basin in Iraq.

6.5.2 Extensive-Form Game with Imperfect Information

This section provides a general, formal introduction to game theory in an extensive form (Gintis 2000), together with its fundamental concepts and essential assumptions.

An extensive-form game G consists of players making decisions, a set of payoffs resulting from the decisions made by players, and a game tree that models the sequence of players' decisions. A game tree includes nodes connected by branches, and each node corresponds to the decision made by a player. Each branch outgoing from the node represents a particular action that results from the decision made by the player at a given point of time, and the end of the branch determines the next point of play in the game. We require that the game tree has one node r called the root node, where the decision-making sequence starts; and a set T of nodes called terminal nodes, where the sequence of decision-making ends. We associate with each terminal node $t \in T$, and each player i, a payoff $\pi_i(t) \in R$ that maps decisions made by each player into the set R of real numbers.

In a course of the game, one is following the sequence of nodes, starting from the root node and going along branches until the terminal node is reached. In a game, any terminal node must be reached when starting from the root node. A player may know the particular node in the game tree, when it is his turn to move; or he may know the decisions that other players made beforehand, so that he may only know that he is at one of the several possible nodes. We call such a collection of nodes an information set.

Suppose each player $i = 1, \ldots, n$ chooses a strategy, that is options of decisions available for players, which could be different from a player to a player, s_i from a strategy set S_i; hence, $s_i \in S_i$. We call a combination of the strategy selection of all the players $i = 1, \ldots, n$ $s = (s_1, \ldots, s_n)$ a strategy profile for the game, and we define then the payoff function of player i, given strategy profile s, which maps strategy profile into real numbers as $\pi_i(s) : S \rightarrow R$, where R is a real number. A form of strategy profile defined by means of the probability distribution of selecting player decisions (behaviors) is called a mixed strategy; a mixed strategy has the form $\sigma_i = \left(p_i^1, \ldots, p_i^k\right)$, where the player i chooses his j-th strategy s_i^j with probability p_i^j, for $j = 1, \ldots, k$. and probabilities $p_i^1, \ldots, p_i^k$ are all non-negative and satisfy $\sum_{j=1}^{k} p_i^j = 1$.

If all p_i^j are zero except one, say $p_i^l = 1$, we say σ_i is a pure strategy. In an n-player game, where the player i has a pure strategy set S_i for $i = 1, \ldots, n$, a mixed strategy profile $\sigma = (\sigma_1, \ldots, \sigma_n)$ is the choice of a mixed strategy σ_i by each player. We define the payoffs of σ as follows:

Let $\pi_i(s_1, \ldots, s_n)$ be the payoff of player i when players use the pure strategy profile $(s_1, \ldots, s_n)$, and if s_i is a pure strategy for player i, let p_{s_i} be the weights of s_i. Then, we define the payoff function

$$\pi_i(\sigma) = \sum_{s_1 \in S_1} \cdots \sum_{s_n \in Sn} p_{s_1} p_{s_2} \cdots p_{s_n} \pi_i(s_1, \ldots, s_n) \tag{6.1}$$

as the expected payoff of player i for the mixed strategy profile σ.

Further on, we say a strategy profile $\sigma^* = (\sigma_1^*, \ldots, \sigma_n^*)$ is a Nash equilibrium if, for every player $i = 1, \ldots, n$ and every σ_i, we have payoff $\pi_i(\sigma^*) \geq \pi_i(\sigma_i, \sigma_{-i}^*)$ where player i chooses a strategy other than σ^* and all the other players, who are represented by $-i$, choose the strategy profile σ^*. In practical terms, it means that the Nash equilibrium is equivalent to the strategy profile, for which the payoff function is greater or equal to values of payoff function obtained for other strategy profiles. Choosing σ_i^* is at least as good for player i as choosing any other σ_i, given that the other players choose σ_{-i}^*. John Nash showed that every finite game has a Nash equilibrium in mixed strategy (Nash 1950).

The Nash equilibrium is important because we can know how players will play a game, if they are rational, that is when they behave in a rational manner. Rationality in the game theoretical context means that players always try to maximize their payoffs by selecting one strategy from their strategy set based on a given information.

A powerful way of finding a Nash equilibrium is to eliminate strategies bringing lower payoffs, that is dominated strategies. Suppose s_i and s'_i are two strategies for player i. We say s_i is strictly dominated by s'_i if, for every choice of strategies of the other players, i's payoff from choosing s'_i is strictly greater than i's payoff from choosing s_i. We say s_i is weakly dominated by s'_i if, for every choice of strategies of the other players, i's payoff from choosing s'_i is as least as great as i's payoff from choosing s_i. It is clear that a strictly dominated strategy s_i is never part of a Nash equilibrium since player i could do better by switching to another strategy that dominates s_i. We can, therefore, eliminate such strategies without eliminating any Nash equilibrium. We call this process the elimination of strictly dominated strategies, which in practical terms means the elimination of such player's strategies which are inferior with regard to payoff function values.

In an extensive-form game with perfect information (i.e., where each information set is represented by a single node), this procedure is formally called backward induction because we start the strategies elimination procedure at the end of the game (from the terminal node) and move backward, up to root node.

The backward-induction solution is called a subgame perfect Nash Equilibrium (Selten 1965) when at least one equilibrium can be obtained by eliminating strictly dominated strategies. Let us choose any terminal node $t \in T$ and find the parent node (that is the ascending node for the node t) of this terminal node, say node a. Suppose player i makes a decision at a and suppose i's highest payoff at a is attained at the node $t' \in T$. We erase all the branches from a so that a becomes a terminal node, and we attach the payoffs from t' to the new terminal node a. We also record i's move at a, so we can specify i's equilibrium strategy when we have finished the elimination. We repeat this procedure for all terminal nodes of the original game. This results in a truncated extensive-form game that is obtained from the original game tree. We repeat the process until the game tree has just one node branch going out from the root node. The strategies that are left after elimination at each node are a Nash equilibrium.

6.5.3 Modeling the Decision-Making Process

For the water resources allocation decision problem in the E–T case system, we have distinguished six players considered as decision-makers and further on referred to as Reservoir Operator, four aggregated water users representing Farmers around Al-Dour, Irwashia, Al-Saqlawiya, and Five Channels areas, and the Nature. The geographical relationship between these stakeholders is depicted in Fig. 6.5.

Noncooperative game theory in an extensive form has been chosen to model the sequence of decision-making steps made by the players, since it reflects at best the real course of the decision-making process taking place between Reservoir Operator and Farmers. The graphical representation of the game tree associated with the sequence of steps involved in a decision-making process is presented in Fig. 6.12.

The course of the decision-making process described in a form of a game tree begins with the Nature "deciding" upon the type of coming hydrologic year (Wet–Normal–Dry, each with a certain probability), which—in terms of game theory—represents strategy chosen by the Nature. In other words, the amount of river flow entering the basin in the coming operational year is determined at the first step of the game by the Nature. In the next step of the game tree, the Nature "decides" upon condition of storage by selecting with certain probabilities among Wet, Normal, and Dry.

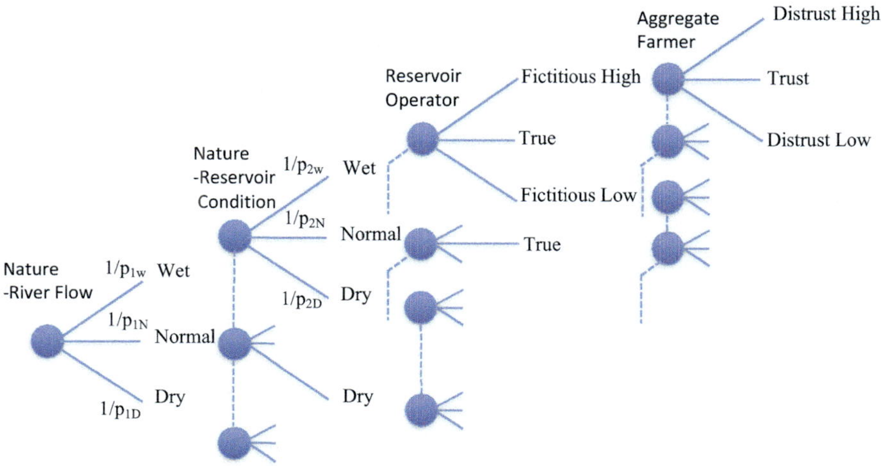

Fig. 6.12 A part of a decision tree (adapted from Sakamoto et al. 2013)

Following Nature's "decisions," the type of the hydrologic year to come (Wet, Normal, or Dry) is determined. This decision is made unilaterally by Reservoir Operator by applying the Rule Curve based on the storage volume on October 1st, which has been decided by the Nature in the previous stage of the game; therefore, it is not incorporated into the game tree at Reservoir Operator's decision stage.

The Reservoir Operator decides, however, about the announcement of the operating rule to be applied in a coming year to the farmers. He chooses one strategy from three available options: (a) he announces the true state of the reservoir selected at the previous decision stage (True), (b) he announces a fictitious higher state of reservoir (FH), or (c) he provides a fictitious lower state of reservoir (FL). For example, if the storage condition turns out to be Wet and the Reservoir Operator selects strategy FL, he announces that the operating rule NORMAL will be applied in the coming year.

Based on the Reservoir Operator's decision, the Farmers selects one strategy from the following three: Trust the announcement of Reservoir Operator Trust); Distrust the Operator and believe in a higher amount of water to be available—High (DH), or Distrust the Operator and expect a lower amount of water to be available—Low (DL).

By choosing the strategy Trust, the Farmers believe that the Reservoir Operator's announcement is true and decide upon the yearly agricultural plan accordingly. On the other hand, if the Farmers select DH or DL, they believe that the available amount of water will be respectively higher or lower than the amount announced by the Reservoir Operator and they cultivate their farmland according to their beliefs. For example, if the announcement by the Reservoir Operator is Dry and if the Farmers choose strategy DH, then they decide to cultivate an area appropriate to the amount of water that would be provided according to the operation rule Normal.

In the game, the operation of reservoirs is simulated for all possible combinations of decisions, and the real and fictitious occurrences of river flow within the hydrologic year are estimated. Each terminal node corresponds to an outcome representing one of the combinations of all the decisions. All the players except Nature obtain payoffs from each outcome, as the result of their decisions. The payoffs are calculated using the historical record of river flow data covering the time period 1992–2002, containing all types of river flow patterns (Dry, Normal, and Wet) and hydrologic years (Dry, Normal, and Wet).

6.5.4 Uncertainty Regarding Nature and Players' Decision

The Farmers do not know the real state of the reservoir and consequently, they do not know the true (real) decision of the Reservoir Operator. They know what Reservoir Operator communicates (announces) to them. It reflects a situation when the players do not necessarily know the decisions made by other players in advance. Extensive form games can incorporate this kind of uncertainty by employing the information set, as it was explained in the Subsection (2).

We incorporate information sets and associate uncertainty (probabilities) with the Reservoir Operator's decision made against Nature's decisions, and with the Farmers' decision made against Nature's and with the Reservoir Operator's decisions as shown in Fig. 6.12. The dotted lines appearing in Fig. 6.12 indicate which nodes belong to which information sets and, in other words, what uncertainty (expressed in terms of probability) the players are facing. The nodes representing the same situation for the players at the decision points are put into the same information set. For example, at the third stage of the game, when the Reservoir Operator makes a decision regarding the type of hydrologic year, the Reservoir Operator is unsure about the strategy that the Nature chose regarding future river flow pattern, although the state of the reservoir, that is the amount of water stored in the reservoir on October 1st is known to him. Therefore, the nodes corresponding to the same reservoir condition (=state of the reservoir) belong to the same information set, and we have three information sets, each corresponding to one of the possible reservoir states, that is Wet, Normal, or Dry. This is why the nodes labeled as "Dry" are connected with dotted lines. In this case, the Reservoir Operator can tell that the reservoir state belongs to the category Dry, but he cannot determine the accompanying river flow pattern. Please note that, for the sake of simplicity, not all the nodes and information sets reflecting state Normal and Wet are depicted in Fig. 6.12.

Figure 6.12 depicts the first and second stages of the game, where the Nature makes the decisions regarding the river flow pattern and the condition of reservoir storage, respectively.

In game theory, a probability distribution assigned to Nature's strategies is treated differently from the one associated with the behavior of remaining players. The probability associated with Nature's strategies is assumed as a priori values, while the probabilities associated with decisions of remaining players are deduced mathematically as a rational response to other players' possible decisions.

At the consecutive stage of the game, the Farmers select one strategy regarding their belief regarding the announcement of the Reservoir Operator. The information sets resulting from the decision made by the Reservoir Operator regarding the state of the reservoir are defined for Farmer's decisions in the same manner, as in a case of Reservoir Operator's decision.

For example, let us assume that the Reservoir Operator determined true release from the reservoir as Wet in accordance with the current amount of water stored in the reservoir and rule curve, and let us suppose that he decides to announce to the farmers the release policy as False Low (FL). Consequently, he announces to the farmers NORMAL policy. Similarly, let us assume that the true reservoir operating rule is identified by the Reservoir Operator as Normal, and then he decides to select announcement option True (T). Then, he will announce to Farmers policy NORMAL.

These two situations mean the same thing to the Farmers: they have to make a decision using the Reservoir Operator's announcement of NORMAL, although the true situation behind the Reservoir Operator's decision is not known to them and differs very significantly. To reflect such circumstances, the nodes representing the same condition as seen from Farmers' perspective are put into one information set and are connected with a dotted line. Consequently, there are three information sets corresponding to Wet, Normal, or Dry possible Reservoir Operator announcements to be made in the fourth stage of the game.

6.5.5 Modeling Players' Payoff

Let us describe now how the payoffs of farmers are calculated.

At the time of the decision-making by four aggregated Farmers depicted in Fig. 6.5, each of them makes his belief about the amount of water available for the coming year. Based on the decisions, the size of the crop area to be cultivated for the coming year is determined (a priori determined area). The true type of year will be revealed at the end of the year. The real payoff of the Farmer i is evaluated a posteriori using the function represented symbolically by Eq. (2):

$$\pi_F^i = \text{profit} - \text{cost} = f(A_i, a_i, a_i*) \tag{6.2}$$

where $\text{profit} = f(a_i*)$
$\text{cost} = f(a_i, A_i - a_i)$

A_i: maximum available farming area
a_i: farming area size decided a priori
a_i*: real (a posteriori) farming area size
$i = 1,\ldots,4$ (1 = Al-Dour, 2 = Irwashia, 3 = Al-Saqlawiya, 4 = Five Channels)

and where A_i is the maximum area that Farmer i possesses, a_i is the area that a Farmer decides to cultivate at the beginning of the hydrologic year, and a_i* is a posteriori area size that produces yields at the end of the year. Area a_i* is a result of the Farmer's decision as well as of the impact of the: (1) Nature's decision, namely the actual river flow caused by actual rainfall, intensity of sun radiation, and other climatic and agricultural factors, and (2) Reservoir Operator's decision, namely the actual amount of water allocated to the Farmers.

The profit and cost functions appearing in Eq. (6.2) are given by Eqs. (6.3) and (6.4), respectively. Note, that for simplification purposes and owing to the availability of data, economies of scale and nonlinear aspects were not considered here.

$$f(a_i*) = p \times g \times a_i* \tag{6.3}$$

$$f(a_i, A_i - a_i) = ca_i - (p \times g \times (A_i - a_i) - c(A_i - a_i)) \tag{6.4}$$

p: the price of wheat ($/ton)
g: production rate of wheat (ton/ha)
c: unit cultivation cost ($/ha)

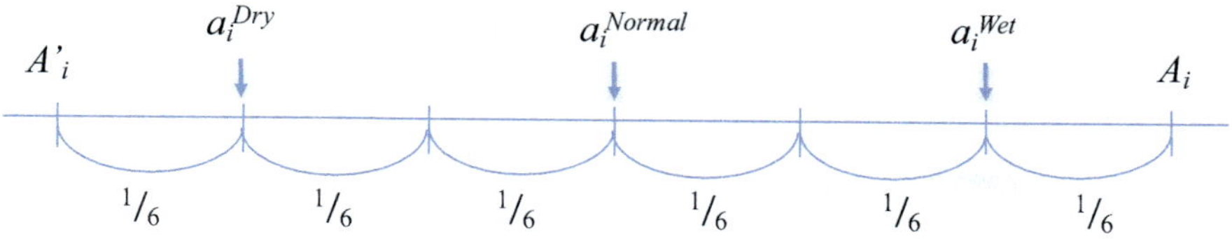

Fig. 6.13 Classification of type of year and decision variable for farmers

In Eq. (6.4), the first term reflects the Farmer's exact costs. Apart from this, we should consider another type of cost that is also involved in decision-making. It is assumed to be equivalent to the benefit that the Farmer could attain, if he would decide to cultivate the whole area available to him. The second term represents this kind of cost, which is called the opportunity cost in economics. The average values of parameters p, g, and c valid for this region are estimated based on the FAO database (FAO 2010), giving us price $p = 110$ \$/ton, production rate $g = 6.0$ ton/ha, and unit cultivation cost $c = 100$ \$/ha, respectively.

Concerning crop area a_i, the values of crop area were set for each Farmer i as a_i^{Dry}, a_i^{Normal}, and a_i^{Wet}, respectively, for the integration with the game model. Therefore, if Farmer i obtains announcement Dry and selects Trust as his policy, the value of a_i^{Dry} is applied to the following calculations. The range of a_i is estimated as the maximum a_i value (corresponding to A_i) minus the minimum a_i value. The values of a_i^{Dry}, a_i^{Normal}, and a_i^{Wet} are calculated as the minimum a_i plus one-sixth, three-sixths, and five-sixths of the range, respectively. Figure 6.13 shows the relationship between the types of the hydrologic year and the corresponding area sizes.

In Fig. 6.13, the maximum area size (A_i) and the minimum area size (A_i') was estimated for each Farmer through remote sensing data analysis, which was obtained from the results of the research described in Chapter "Estimation of irrigated agricultural land and water consumption in Iraq" of this book.

The area values are the following: (1) for Al-Dour, A_1 is 25,000 ha and A_1' is 4000 ha; (2) for Irwashia, A_2 is 9000 ha and A_2' is 100 ha; (3) for Al-Saqlawiya, A_3 is 20,000 ha and A_3' is 5000 ha; and (4) for Five Channels, A_4 is 32,000 ha and A_4' is 10,000 ha. The values of a_i representing a priori values of decision variables were calculated based on the following assumptions and estimated area parameters:

- For Al-Dour, $a_1^{\mathrm{Dry}} = 7500$ ha, $a_1^{\mathrm{Normal}} = 14{,}500$ ha, $a_1^{\mathrm{Wet}} = 21{,}500$ ha
- For Irwashia, $a_2^{\mathrm{Dry}} = 1580$ ha, $a_2^{\mathrm{Normal}} = 4540$ ha, $a_2^{\mathrm{Wet}} = 7500$ ha
- For Al-Saqlawiya, $a_3^{\mathrm{Dry}} = 7500$ ha, $a_3^{\mathrm{Normal}} = 12{,}500$ ha, $a_3^{\mathrm{Wet}} = 17{,}500$ ha
- For Five Channels, $a_4^{\mathrm{Dry}} = 13{,}600$ ha, $a_4^{\mathrm{Normal}} = 20{,}800$ ha, $a_4^{\mathrm{Wet}} = 28{,}000$ ha

The same values were used for a posteriori farming area a_i*, but the value for the true type of the year was applied. For example, if the Reservoir Operator identified the type of year as Dry and announced Dry by selecting strategy True, and if Farmer i selected Dry, consequently, the value of a_i^{Normal} was applied to a_i, and a_i^{Dry} was applied to a_i*.

It was assumed that the payoff function of the Reservoir Operator is given by Eq. (6.5) and his payoff is evaluated at the end of the hydrologic year, that is, for every year at the end of September.

$$\pi_R = w_{\mathrm{farmer}} \times \left(\frac{\sum_i w_i \pi_F^i}{\sum_i w_i \overline{\pi_F^i}} \right)^2 + w_{\mathrm{storage}} \times \left[1 - \left\{ \left(v_{\mathrm{final}} - v_{\mathrm{desired}} \right) / v_{\mathrm{active}} \right\}^2 \right] \quad (6.5)$$

w_{farmer} and w_{storage}: weighting coefficients for the farmer-related component and operator-related component respectively (note that $w_{\mathrm{farmer}} + w_{\mathrm{storage}} = 1$).
w_i: weight for Farmer i
π_F^i: payoff of Farmer i
$\overline{\pi_F^i}$: maximum value of payoff of Farmer i
v_{final}: final storage volume at the end of the simulation period
v_{desired}: target (desired) final storage volume
v_{active}: volume of active storage (= max_storage—dead_storage)

w_{farmer} and w_{storage} mean that the Reservoir Operator's objective is both to provide water to Farmers and to keep the storage condition at a certain level (v_{desired}). The payoff function of Farmer i, $\overline{\pi_F^i}$, is defined in the previous section. The role of the Reservoir Operator is to manage the basin so that the a posteriori evaluation of his performance, composed of a weighted sum of farmers' payoffs and storage volume, will be maximized.

The weights for the farmer- and operator-related components change depending on the objective of reservoir control. The relative weight for each Farmer's payoff is defined based on the average sizes of his crop area, estimated through the above-mentioned remote sensing data analysis as 10,000 ha for Al-dour, 3000 ha for Irwahia, 13,000 ha for Al-Saqlawiya, and 21,000 ha for Five Channels. Then, the resulting values of the weighting coefficients for each farmer were calculated as 0.21 for Al-dour, 0.06 for Irwahia, 0.28 for Al-Saqlawiya, and 0.45 for Five Channels.

6.5.6 Players' Rational Decisions Constituting Equilibrium

Following the formulation of the game, the effectiveness of the reservoir operation policy in the basin is examined through the application of the simulated river system data using game theoretical modeling.

First, the parameters w_{farmer}, w_{storage}, and v_{desired} in Eq. (6.5) need to be set in advance for the payoff calculation. The weighting coefficients of the farmer- and operator-related components (w_{farmer} and w_{storage}) should satisfy the conditions $0 \leq w_{\mathrm{farmer}}, w_{\mathrm{storage}} \leq 1$, and $w_{\mathrm{farmer}} + w_{\mathrm{storage}} = 1$. Accordingly, we assume three cases: (w_{farmer}, w_{storage}) = (0.3, 0.7), (0.5, 0.5), and (0.7, 0.3) reflecting higher priority of the farmer-related, short-term component over operator related, long-term component; equal priorities of both components; and higher priority of the operator-related component. For v_{desired}, we assume a boundary condition of $10 \leq v_{\text{desired}} \leq 90$ and three cases: 30, 50, and 70. Therefore, $3 \times 3 = 9$ sets of payoffs are calculated for each combination of strategies chosen by all the players.

Since there are seven decision-making stages in the decision tree (Nature has two stages, Reservoir Operator has one, and each of the four Farmers has one), and each stage consists of three possible outcomes, therefore the set of outcomes contains $3^7 = 2187$ combinations of possible strategy selections. For simplicity of computations, the four Farmers are first aggregated into two players, namely Upstream Farmer and Downstream Farmer, where Al-Dour and Irwahia are aggregated into Upstream Farmer, and Saqlawiya and Five Channels are aggregated into Downstream Farmer. After calculating the payoffs, we find that the Upstream Farmer and Downstream Farmer chooses the same strategy for any combinations of strategies of Nature and Reservoir Operator, or for any combinations of w_{farmer}, w_{storage}, and v_{desired} values considered in this study. Hence, Upstream Farmer and Downstream Farmer are aggregated into one so-called Aggregate Farmer.

After the game tree was truncated by eliminating the dominated strategies, it appeared that irrespective of Nature's decision regarding river flow type, it is the best (or at least not worse than the other choices) for Aggregate Farmer to select the strategy Distrust Low, when the announcement is WET or NORMAL, and the strategy Trust when the announcement is DRY. Regarding the Reservoir Operator, either True or Fictitious Low provide the maximum payoff, when the reservoir condition is Wet or Normal; and either True or Fictitious High provide the maximum payoff when the reservoir condition is Dry. This result applies to any strategy selection of Nature regarding river flow and any combination of w_{farmer}, w_{storage}, and v_{desired} values considered here.

The result reveals that mutual trust between the Reservoir Operator and Aggregate Farmer emerges only, when the reservoir condition is Dry; the Reservoir Operator selects True; and the Aggregate Farmer selects Trust. This applies to one-sixth of the total number of cases.

Sakamoto and Salewicz (2017) investigated the optimum reservoir condition for the Reservoir Operator for each of the 100 random patterns of possibilities assigned to the river flow types (Wet–Normal–Dry) and found that the state Normal as the reservoir condition is the most dominant strategy. Because the Normal is the most frequently emerging condition in reality, so the conclusion drawn from the river flow simulation based on the current operational rules as well as on interactive decision-making process between the reservoir operator and the farmers suggests that the traditional operational policy implemented at the Al-Tharthar reservoir for basin management is well designed.

6.6 Summary and Conclusions

The research described in previous sections of the chapter addressed, with different levels of detail, problems regarding the operation of reservoirs and mechanisms of distributing water among agricultural users.

The results demonstrated that:

1. The traditional operating policy appears to be quite efficient with regard to its ability to meet both short-term and long-term objectives.
2. The application of game theory has proven its capability to address and describe in a constructive manner the decision-making process taking place between the reservoir operator and agricultural water users (farmers).
3. The scope of the research and results obtained cover only a very small part of the plethora of problems, which Iraq is facing with regard to water resources management, agricultural water supply, water allocation rules, environmental aspects of water resources management, and many, many other fascinating topics.

Unfortunately, the current political situation in Iraq and around Iraq does not allow to undertake in a systematic and organized manner research efforts aimed at better understanding and resolving urgent problems facing water resources management in this country.

Managing the environment in a rational and sustainable manner is always problematic in post-conflict nations since such areas often lack good governance. Moreover, international societies tend to conduct their aid operations in a confusing fashion (Nakayama and Yamashiki 2013). As Bruch et al. (2016) suggest, post-conflict countries should effectively and equitably use natural resources to secure revenues from their extraction for the sake of job creation, infrastructure development, and provision of basic services for the citizens.

The use of natural resources needs to be carried out without leading to new conflicts, corruption, economic instability, or exceeding the management capacity of the government. There should be important principles to enhance post-conflict peacebuilding through rational management of natural resources.

The way and means of water resources management suggested in this study should be the basis for balanced development and management of the region, in a transparent manner in terms of information and policy agenda, in the future post-conflict days.

Acknowledgments The authors would like to express their appreciation and thanks to unknown colleagues from Iraq, who worked together with Professor Mukhdad Ali and who provided information and data that allowed us to conduct research reported in this chapter. It is our duty and pleasure to express particular appreciation to Professor Ali, without whom we would not be able to initiate and perform this research.

The work described in this chapter was supported by JSPS KAKENHI Grant Number JP15H02864, and by the Graduate School of Frontier Sciences of the University of Tokyo.

References

Al-Faraj F (2006) Visions on reservoir operation in Iraq. Internal report hydrological studies center, Directoriate General of the Water Resources Management, Baghdad, Iraq

Ali M (2005) Hydrology criteria of the Euphrates and Tigris Rivers. Unpublished Manuscript

Ali M (2006a) Irrigation schemes in Iraq. Unpublished Manuscript

Ali M (2006b) Personal communication

Bruch C, Muffett C, Nichols SS (2016) In: Bruch C, Muffett C, Nichols SS (eds) Part 4: local institutions and marginalized populations, governance, natural resources, and post-conflict Peacebuilding. Earthscan, London

FAO. (2010). "Agribusiness Handbook: Wheat Flour." http://www.fao.org/docrep/012/al376e/al376e.pdf April 6 2016

FAO (2019) Agribusiness handbook: wheat flour. http://www.fao.org/docrep/012/al376e/al376e.pdf April 6 2016

Gintis H (2000) Game theory evolving: a problem-centered introduction to modeling strategic interaction. Princeton University Press, Princeton

Granat J, Makowski M (2000) Interactive specification and analysis of aspiration-based preferences. European J Oper Res 122(2):469–485

Granat J, Makowski M (2009) Management of attainable tradeoffs between conflicting goals. J Comput 4(10):1033–1042. ISSN 1796-203X

Granat J, Makowski M, Ogryczak W (2014) Multiple criteria analysis of discrete alternatives with a simple preference specification: pairwise-outperformance-based approaches. European J Oper Res, draft available as IIASA report IR-09-24

Loucks DP, Sigvaldason OT (1982) Multiple-reservoir operation in North America in Z. Kaczmarek & J. Kindler (Eds.) The operation of multiple reservoir systems; international institute for applied systems analysis collaborative proceeding series, CP-82-S3, pp 1–105

Nakayama M, Yamashiki Y (2013) Governance in Timor-Leste: towards improvement in disaster and environment management. Asian J Environ Disaster Manage 5(4):299–306. https://doi.org/10.3850/S1793924013000084

Nash JF (2015) Equilibrium points in n-person games. Proc Natl Acad Sci U S A 1950 36(1):48–49

Sakamoto M, Salewicz KA (2017) Extensive-form game for examining mutual trust between a reservoir operator and agricultural water user. ASCE Journal of Water Resources Planning and Management (in press)

Sakamoto M, Salewicz KA, Tanaka Y, Nakayama M (2013) Game theoretical analysis of water allocation rule in lower basin of Euphrates and Tigris Rivers. IEEE International Conference on Systems, Man and Cybernetics, 2042–2049

Salewicz KA (2006) Development of water balance model for E-T River system. Unpublished Manuscript

Salewicz KA, Tanaka Y (2006) Euphrates-tigris water and salt concentration model – description. Internal Report, University of Tokyo; Tokyo, Japan

Salewicz KA, Sakamoto M, Nakayama M (2014) Multiple criteria optimization in analysis of conflict associated with Al-Tharthar reservoir operation. Proceedings of 2014 IEEE International Conference on Systems, Man, and Cybernetics; San Diego, CA, USA; October 5–8th, 2014; pp 2942–2947

Selten R (1965) Spieltheoretische Behandlung eines Oligopolmodells mit Nachfragentragheit. In Zeitschrift fur die gesamte Staatswissenschaft 12:201–324

Tanaka Y (2011) Land and water use analysis of Iraq for the cooperative transboundary water use of Euphrates and Tigris rivers. PhD Thesis. The University of Tokyo

The Washington Post (1999) Weather longterm historical data Baghdad, Iraq. http://www.washingtonpost.com/wp-srv/weather/longterm/historical/data/baghdad_iraq.htm Archived from the original on 29 June 2014

Weinthal E, Troell J, Nakayama M (2014) Water and post-conflict Peacebuilding (post-conflict Peacebuilding and natural resource management). Routledge, London

Wierzbicki A, Makowski M, Wessels J. (2000) Model-based decision support methodology with environmental applications. Kluwer Academic Publishers, Dordrecht, series: mathematical modeling and applications; ISBN 0–7923–6327-2; 2000

Chapter 7
Estimation of Irrigated Agricultural Area and Water Consumption in Iraq

Yukio Tanaka

Abstract The year-to-year irrigated area and water balance of central Iraq were estimated with a view to shed light to transboundary water dispute over Tigris and Euphrates Rivers caused partly by uncertainty of scientific data. Subpixel resolution technique based on Vegetation Fraction Model was applied to continuous AVHRR images to estimate actual irrigated area with Landsat images as verification/validation data. The water budget of target area was also analyzed to compare the evapotranspiration volume and estimated irrigated area. The estimated irrigated area was 6002 ± 1541 km^2 in spring harvest season and 4939 ± 726 km^2 in summer, which were substantially lower than the figures reported by public statistics. The water budget analysis also indicated that the evapotranspiration volume increases in proportion to the river inflow. It also suggested the possibility that substantial amount of evapotranspiration occurs outside of irrigated area. Both results imply that the transboundary water dispute over Tigris and Euphrates Rivers may be relaxed by expanding financial and technical assistance over the irrigation sector of Iraq.

7.1 Introduction

7.1.1 Transboundary Water Dispute Over Tigris and Euphrates Rivers

Cooperative and sustainable management of transboundary waters has long been a key focal pillar among broader global water agenda as has been stipulated in the Target 6.3 of the UN Sustainable Development Goals. The number of transboundary river basin is a variable figure as the global territory border continues to change, but there are as many as 286 transboundary river basins as of 2016, covering 42% of

Y. Tanaka (✉)
Water Global Practice, The World Bank, Washington, DC, USA
e-mail: ytanaka@worldbank.org

L. A. Jawad (ed.), *Tigris and Euphrates Rivers: Their Environment from Headwaters to Mouth*, Aquatic Ecology Series 11,
https://doi.org/10.1007/978-3-030-57570-0_7

total land surface area of the earth and accommodating 42% of global population (UNEP-DHI and UNEP 2015).

Transboundary water management embraces both opportunities and challenges. Channeling states and enabling water and other goods transport from state to state with its watercourses, transboundary rivers have potential to boost economic cooperation and shared prosperity among riparian states through international trade, climate change adaptation, food security, and so forth, provided appropriate institutional arrangement for river utilization exists among riparian states, as seen in the Danube River (Rieu-Clarke 2007). On the other hand, competitive use of limited river flow, or excessive discharge of polluted or flood water from upstream state to downstream, leads to deteriorated bilateral relationship among riparian states. One of the most critical cases was observed with the Jordan River, where the Six-Day War broke out in 1967 induced by Israel's aerial attack to Jordan's civil work site that aimed to divert the flow of the Jordan River to Jordanian territory (Hillel 1994; Allan 2001).

Tigris and Euphrates Rivers are not exceptional to transboundary water conflicts. Shared mainly by Turkey, Syria, and Iraq, and partly by Iran and Saudi Arabia, Tigris and Euphrates Rivers have one of the longest history of water utilization since the era of Mesopotamian civilization over more than 7000 years (Weiss 1978; Gleick 1988; Ludwig et al. 1993). During this long history, there had been water-related disputes from time to time, but none of them were basin-wide scale thanks to the ample flow of both rivers. It was when the riparian states started construction of large dams on the mainstream of rivers in late twentieth century that the issues scaled up from local to basin-wide (Naff 1984; Scheumann 1997).

The construction of large dams on the mainstream of Tigris and Euphrates drastically changed the flow regime of both rivers, and deteriorated the diplomatic relationship between Turkey, Syria, and Iraq. Particular instance was observed in 1974 when Syria started impounding the newly constructed Tabqa dam, causing decrease of flow volume of Euphrates into one-third of average years. Claiming augmented flow of Euphrates, Iraq mobilized its military power up to Syrian border until Saudi Arabia intervened for reconciliation and Syria agreed to release 40% of inflow (Starr 1991; Hillel 1994). Similar disputes continued among the riparian states, especially after Turkey announced a huge comprehensive development plan called Southeast Anatolia Project, also known as GAP, which ambitiously set a target of constructing 21 dams and reclamation of 1.6 million hectares of irrigated land (Dellapenna 1996).

One of the sources of dispute over Tigris and Euphrates Rivers lies in the discrepancy of hydrological data presented by different literatures/authorities. Turkish government in its official website claims its legitimacy to withdraw water, citing the river flow and water demand data that appear in academic literatures (Turkish Ministry of Foreign Affairs 1996). However, the river flow volume and water demand of each riparian state vary from literature to literature. Besides, most of major literatures, e.g., Kliot (1994), Bagis (1997), and Bilen (1994), derive flow volume data from Beaumont (1978), while Baumont (1978) refers to the data up to 1960s, which are not relevant to discuss the water disputes caused by water resources development after 1970s. Altinbilek (1997) conducts extensive analysis on the past, present, and future river discharge and water demand of Tigris and Euphrates, but these data are cited as "calculation by the author" and calculation process is not revealed to readers.

Theoretically, water demand volume of each riparian state should vary depending on the area actually irrigated, which also varies each year. However, most of the literatures assess the water demand based solely on average irrigated area—none of them shed light to the variability of irrigated area and water consumption volume, which is an essential information to discuss the appropriate water allocation among riparian states.

7.1.2 Objectives of This Chapter

With the uncertainty of scientific data discussed above in consideration, this chapter aims to delineate the land-water use situation of Iraq with particular focus on its year-to-year change. The study comprises of two components: first, it estimates the actual irrigated area of main cropped part of Iraq using images obtained from different satellite-borne sensors; second, the water balance of target area is analyzed so as to understand the water consumption, which is then compared with the irrigated area estimated in the first component.

7.2 Framework of the Study

7.2.1 Overview of Tigris and Euphrates Rivers

Tigris and Euphrates Rivers are often referred to as "twin rivers" for their geographic natures—although they have clear watershed boundary at their upstream in Turkish territory, these two rivers approach each other as they flow downstream, form a unified alluvial plain and marshland in Iraq, and eventually merge into Shatt al-Arab river at 200 km upstream of river mouth (Fig. 7.1). The share of river stretches and basin area by each riparian state is presented in Table 7.1. While the majority of Tigris River is dominated by Iraq both in terms of length and area, Euphrates river is shared fairly, evenly among three major riparians, namely Turkey, Syria, and Iraq.

Figure 7.2 presents the layout of major hydraulic structures and irrigation projects of the Tigris-Euphrates basin. As a consequence of massive water resources development projects by each state such as Turkey's Southeast Anatolia Project, both rivers are now regulated by a number of large dams.

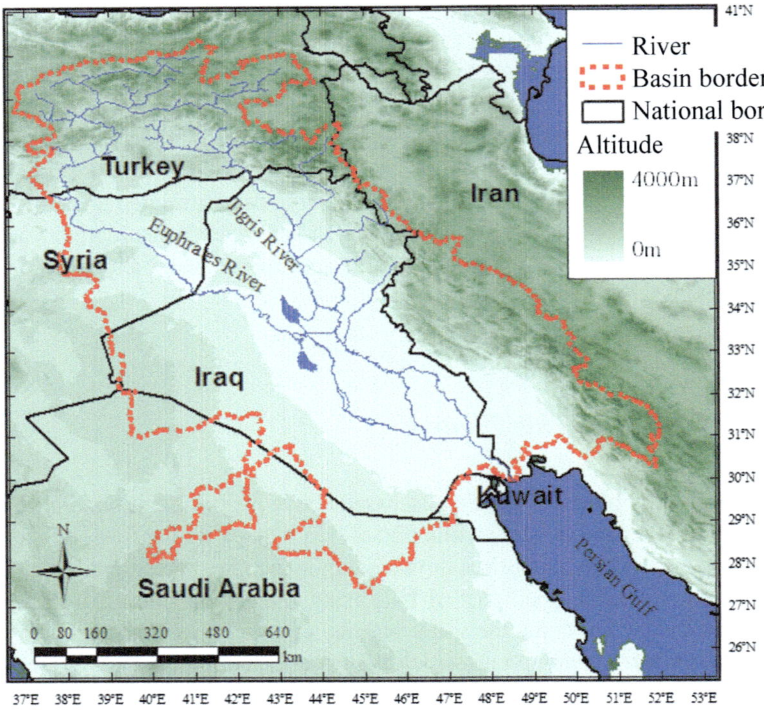

Fig. 7.1 Tigris and Euphrates Rivers basin

Table 7.1 Share of riparian state in terms of river length and basin area

	Euphrates				Tigris			
State	Length (km)	%	Area (km^2)	%	Length (km)	%	Area (km^2)	%
Turkey	1230	41	138,000	27.9	400	22	51,000	8.5
Syria	710	23	114,300	23.1	44	1	1000	0.2
Iraq	1060	36	177,000	35.7	1418	77	374,500	62.7
Iran	–	–	–	–	–	–	170,700	28.6
Saudi Arabia	–	–	66,000	13.3				
Total	3000	100	495,300	100.0	1850	100	597,200	100.0

Source: Kliot (1994), Altinbilek (2004); Global Drainage Basin Database, NIES

7.2.2 *Target AREA and Period of the Study*

As seen in Fig. 7.2, the main breadbasket of Iraq lies in the center to upper-south part of Iraq, down from Mosul along the Tigris and Hussaiba along the Euphrates, respectively. Therefore, this study will focus on the water balance in this region of the country (see the rectangular boundary with red-dashed line in Fig. 7.2).

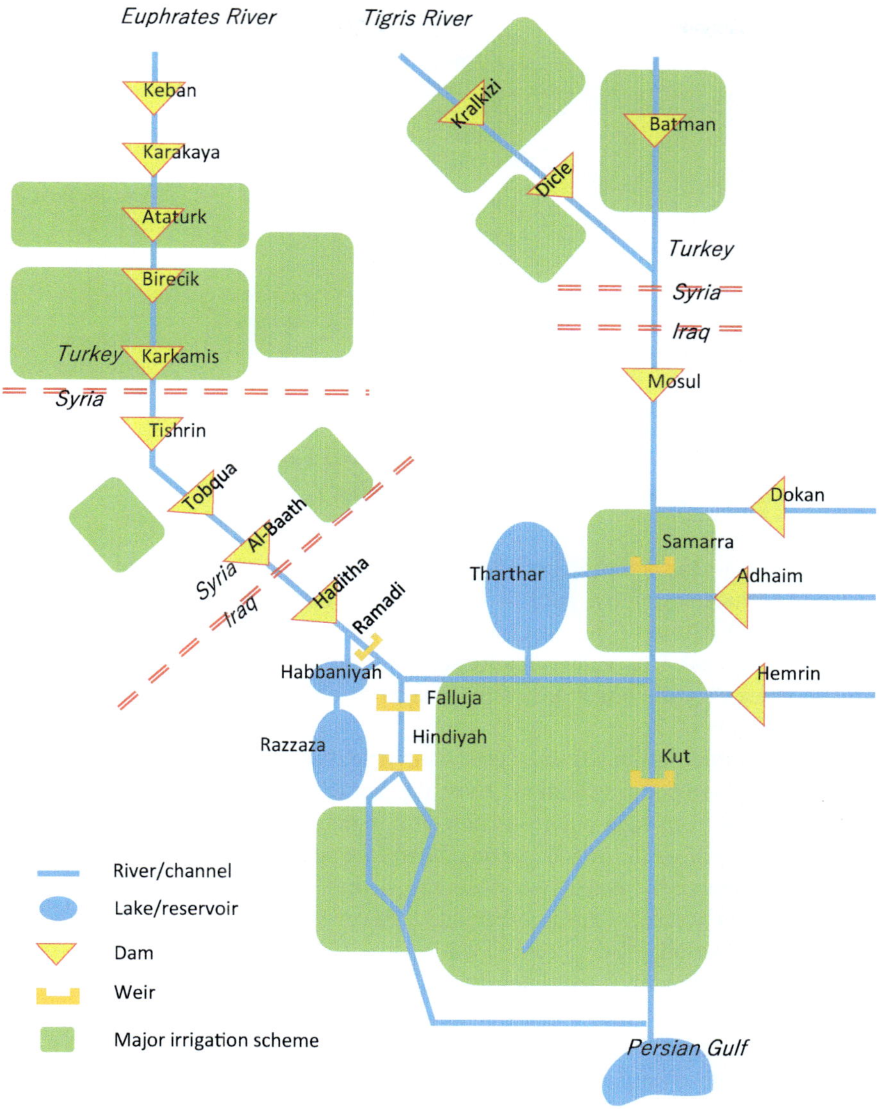

Fig. 7.2 Layout of major hydraulic structures and irrigation projects of the Tigris–Euphrates basin

The target period of analysis was set as 1981 to 2000 mainly due to the limitation of available river discharge data as well as satellite imageries.

Fig. 7.3 Schematic flow of analysis

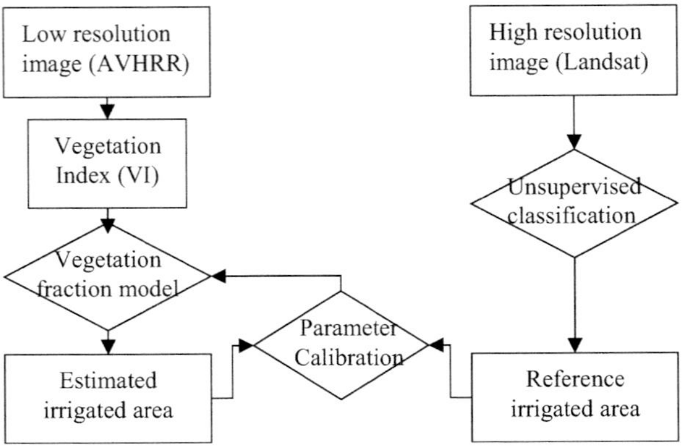

7.3 Estimation of Irrigated Agricultural Land of Iraq Using Satellite Remote Sensing Imageries

While year-on-year actual irrigated area is a necessary information to assess the water use efficiency, most publicly available satellite-derived land use and land cover (LULC) datasets[1] present LULC at a single point in time, or average LULC over certain period. Hence, this chapter aims to estimate the year-to-year variation of irrigated area over the target area using imageries obtained from different satellite-borne sensors.

The major trade-off in satellite-based earth observation is the one between spatial resolution and temporal resolution: given the fact the orbital speed of satellite does not vary much from each other, sensors with higher spatial resolution can capture narrower swath of earth surface at a time, therefore takes longer time to cover whole earth, resulting in less frequency of observing certain point of earth (or longer return period in other words). On the other hand, lower spatial resolution sensors have shorter return period.

To overcome this trade-off, intensive studies were dedicated to develop subpixel resolution techniques that enable to know the ratio of each land classification member (e.g., forest, bush, water body, etc.) within a single pixel of imagery with coarse spatial resolution (Adams et al. 1986; Inamura 1988; Smith et al. 1990; Gutman and Ignatov 1998).

This study applies the Vegetation Fraction model proposed by Gutman and Ignatov (1998) to develop a formula to obtain the subpixel content ratio of imageries from Advance Very High Resolution Radiometer (NOAA/AVHRR) using images of

[1]Some of the major works are provided by: Global Land Cover Characterization (GLCC) of USGS; Global Land Cover Network (GLCN) of FAO; and Global Land Cover Facility (GLCF) of University of Maryland. The Global Map of Irrigation Area (GMIA) of FAO specializes in mapping particularly the irrigated area.

Landsat Thematic Mapper (Landsat TM). The analysis process is outlined in Fig. 7.3.

7.3.1 *Materials and Methods*

Satellite Images Used
For AVHRR data, Pathfinder AVHRR Land (PAL) product is used instead of raw AVHRR images. PAL is produced as global 8-km grid with 10-day composite resampling technique to minimize the effect of clouds. It is also processed for atmospheric correction (James and Kalluri 1994). Channel 1 signal is used as visible red reflectance, and channel 2 as near-infrared for this study, respectively.

For Landsat data, images acquired by Landsat 5 and 7 are used for analysis. Digital signal value of each spectral band is converted into radiance first, then into reflectance value following the formula suggested in Landsat 7 Science Data User Handbook.

Vegetation Fraction Model
Normalized Difference Vegetation Index (NDVI) is the most common satellite-derived vegetation metrics. NDVI can be obtained with only two spectral band reflectance of satellite imagery, i.e., visible red and near-infrared (Tucker 1979). With its simplicity and intuitiveness, NDVI-based dataset product is widely disseminated by institute like NASA or USGS. However, it is also known that NDVI, the fraction of absorbed photosynthetically active radiation in definition, does not serve as direct indicator of vegetated area on ground (Wu 2014). Another known limitation is that NDVI is influenced by the surface reflectance from the soil, especially in arid land (Huete 1988).

Given these limitation, Gutman and Ignatov (1998) further developed discussion on the relationship between vegetation index and vegetated area assuming four different vegetation condition within a pixel, namely uniform full vegetation, dense uniform vegetation, nondense uniform vegetation, and variable density vegetation model. Under the dense/nondense vegetation condition, the relationship between vegetation index (*VI*; within the range of 0 to 1) and vegetation fraction (*VF*) is expressed as:

$$VI = VF \cdot VI_g + (1 - VF)VI_0 \tag{7.1}$$

where VI_0 and VI_g are the signals from bare soil and green vegetation. Given that, both VI_0 and VI_g are constants, this equation is translated to linear relationship between *VI* and *VF* under uniform vegetation condition. However, the comparison of Landsat and AVHRR imageries shows different *VF* against same value of *VI*, which implies the study area consists of multiple vegetation types. Although Iraq has several major crop types such as wheat, barley, rice, and maize, this study employs an assumption that these are categorized into two groups, namely high-density crops and low-density crops, where high-density crops have higher *VI* values when fully grown while low-density crops have lower *VI*. With this assumption incorporated, Eq. (7.1) is expressed as:

$$
\begin{gathered}
VI = VF_{\text{Low}} \cdot VI_{\text{Low}} + VF_{\text{High}} \cdot VI_{\text{High}} + \left(1 - VF_{\text{Low}} + VF_{\text{High}}\right) VI_0 \\
VF_{\text{High}} = \alpha \cdot VF \\
VF_{\text{Low}} = (1 - \alpha) \cdot VF
\end{gathered} \tag{7.2}
$$

VF_{Low} and VF_{High} are vegetation fractions of low-density crop and high-density crop, respectively, whereas VI_{Low} and VI_{High} are vegetation index values of low-density crop and high-density crop. α denotes the area ratio of high-density crop against total vegetation fraction of the pixel. Then, combination of Eq. (7.2) yields:

$$
VF = VF_{\text{Low}} + VF_{\text{High}} = \frac{VI - VI_0}{\alpha VI_{g\text{High}} + (1 - \alpha) VI_{g\text{Low}} - VI_0} \tag{7.3}
$$

VF and *VI* are variables that can be obtained from Landsat and AVHRR imageries, respectively, whereas $VI_{g\text{High}}$, $VI_{g\text{Low}}$, VI_0 are constant values, and α is unique to respective pixels. These constants are determined using Generalized Reduced Gradient (GRG) Method, one of nonlinear programming methods developed by Lasdon et al. (1978).

Vegetation Index

TSAVI (Transformed Sand Adjusted Vegetation Index) is used as satellite-derived vegetation index in this study, considering the influence of background soil reflectance and diversity of soil type in the study area (Baret et al. 1989). Comparative feature of different vegetation index is well presented in Baret and Guyot (1991). SAVI (Soil Adjusted Vegetation Index) also incorporates the influence of soil surface reflectance (Huete 1988), but it employs single constant to account for the influence of background soil. On the other hand, TSAVI applies different parameters per different soil types, which enables more precise estimation of plant activity over wide range of area covering different soil types.

TSAVI is presented as following:

$$
\text{TSAVI} = \frac{R_{\text{nir}} - aR_r - b}{aR_{\text{nir}} + R_r - ab} \tag{7.4}
$$

Where R_r and R_{nir} stand for the spectral reflectance measurements acquired in the red (visible) and near-infrared ranges, respectively; and a and b are parameters unique to each soil type.

For a given soil type, the red (R_r) and near-infrared (R_{nir}) reflectances are expressed as:

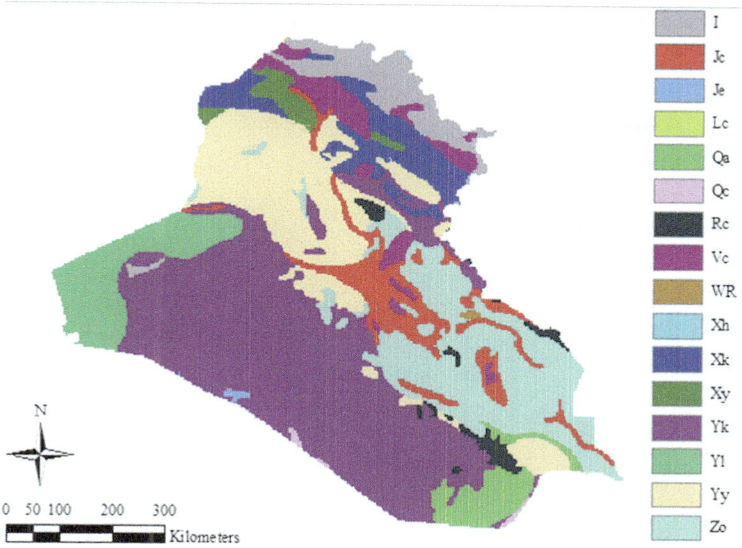

Fig. 7.4 Soil type distribution in Iraq (Source: FAO Digital Soil Map of the World)

$$R_{\text{nir}} = aR_r + b \tag{7.5}$$

According to the Digital Soil Map of the World (DSMW) presented by FAO, Iraq consists of 16 major soil types (Fig. 7.4). In order to determine the soil parameters for these 16 soil types, training sample reflectance data were collected through: (1) overlaying DSMW and global land cover classification imagery (Hansen et al. 1998); (2) picking up training pixels that are classified as bare soil on land cover classification imagery and also classified as 16 target soil types on DSMW; (3) collect red and near-infrared reflectance of AVHRR image as of March 2000 for the identified training pixels; and (4) plot the red and near-infrared reflectance per each soil type and modeled optimal parameters with linear regression.

Processing of Reference Irrigated Area with Landsat Imageries
The actual irrigated area, used to calibrate and validate the estimated vegetation fraction derived AVHRR imageries, was obtained through unsupervised classification of Landsat imageries during harvest seasons. Both the cropping calendar of Iraq and time-series NDVI trend suggest two harvest seasons, namely spring (March to April) and summer (August to September).

Sets of Landsat images with relatively good quality (with $\leq 10\%$ cloud coverage) that capture the harvest seasons are presented in Table 7.2. The sets of spring and summer in 2000 were used for the calibration of vegetation fraction model, while those of spring in 1999 and summer in 1992 were used for validation.

The Landsat imageries of respective seasons were classified into 20 classes using unsupervised ISODATA method (Irvin et al. 1997), then classes were merged into three classes, namely water body, bare land, and vegetated area. Theoretically,

Table 7.2 Landsat images used for analysis

Period	Path	Row	Date	Cloud cover
Summer 1992	167	38	8/18/1992	10%
	168	37	8/9/1992	10%
	168	38	8/18/1992	10%
	169	36	8/16/1992	0%
	169	37	8/16/1992	10%
Spring 1999	167	38	N/A	N/A
	168	37	3/14/1999	0%
	168	38	3/14/1999	0%
	169	36	3/5/1999	0%
	169	37	3/5/1999	0%
Spring 2000	167	38	3/9/2000	9%
	168	37	2/29/2000	0%
	168	38	3/16/2000	0%
	169	36	3/15/2000	10%
	169	37	3/15/2000	0%
Summer 2000	167	38	7/31/2000	0%
	168	37	8/31/2000	0%
	168	38	8/23/2000	0%
	169	36	8/14/2000	0%
	169	37	8/14/2000	0%

vegetated area can include nonfarm plants (e.g., street plants or wild plants) and nonirrigated farmland. However, considering extremely arid weather of the study area, plant growth without artificial watering is fairly negligible, therefore, the vegetated area obtained through image classification was regarded as reference irrigated area.

7.3.2 Results and Discussions

Determination of Soil Parameter and Calculation of TSAVI

Totally 1,084,864 pixels, equivalent to 133,933,827 km^2, were identified as training sample data to determine the soil parameter of 16 soil types. All the soil type could collect more than 300 pixels, which are enough for training sample. The results of linear regression are shown in Fig. 7.5. Eleven out of 16 soil types showed high (>0.9) R^2 value, while none of the soil showed R^2 value of less than 0.6.

Using the soil parameter obtained and reflectance of AVHRR images, TSAVI was calculated (Fig. 7.6). The value of TSAVI was around 0.1 on bare land area and 0.3 to 0.4 on irrigated farm area, which was quite reasonable.

Calibration and Validation of Vegetation Fraction Model

TSAVI image derived from AVHRR and reference irrigated area image from Landsat were related for the calibration period (spring and summer of 2000), and

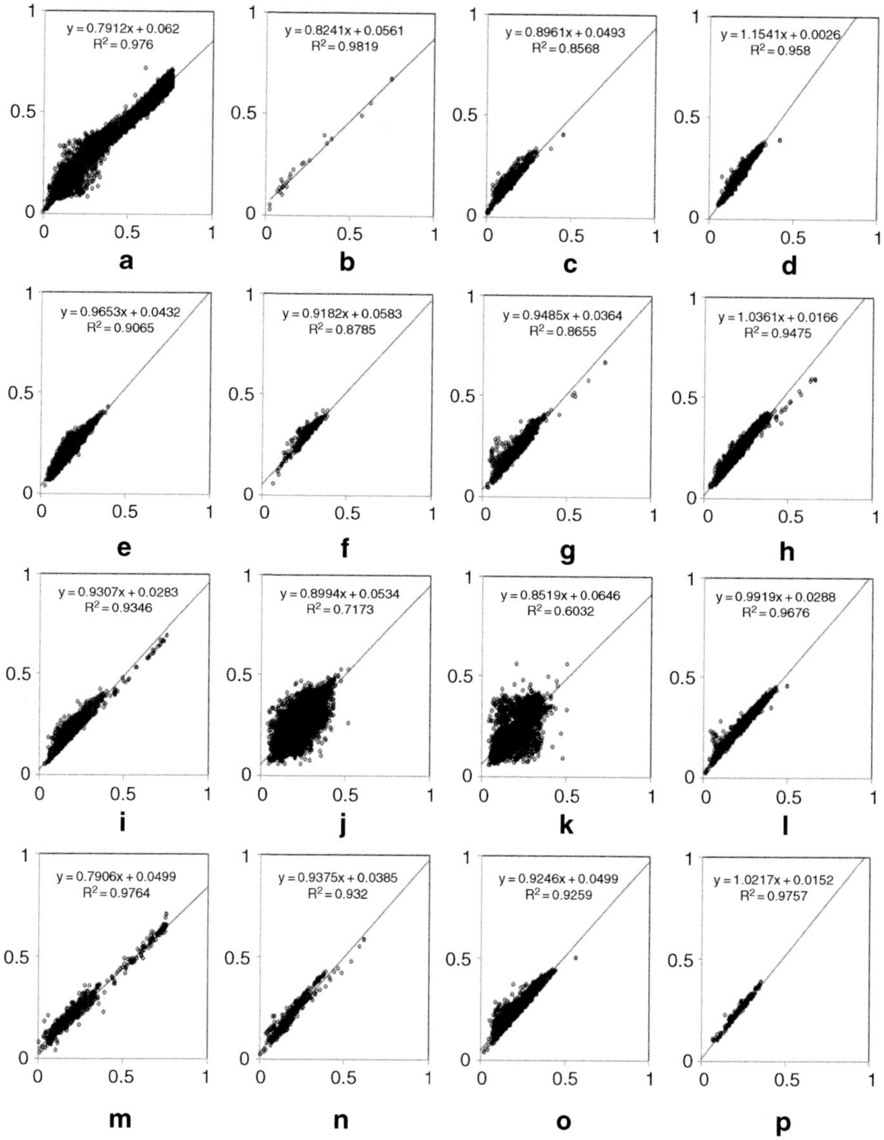

Fig. 7.5 Soil reflectance samples collected globally. Horizontal axis is the reflectance of visible red and the vertical axis is near-infrared

optimal values of VI_{gHigh}, VI_{gLow}, VI_0, and α were determined using GRG method, which enabled modeling the vegetation fraction with AVHRR image almost precisely (Fig. 7.7). The distribution of ratio of high-density crop area (α) is presented in Fig. 7.8. Relatively high ratio of high-density crop area was observed along Euphrates river and upstream of Tigris river, suggesting that the marsh zone in the

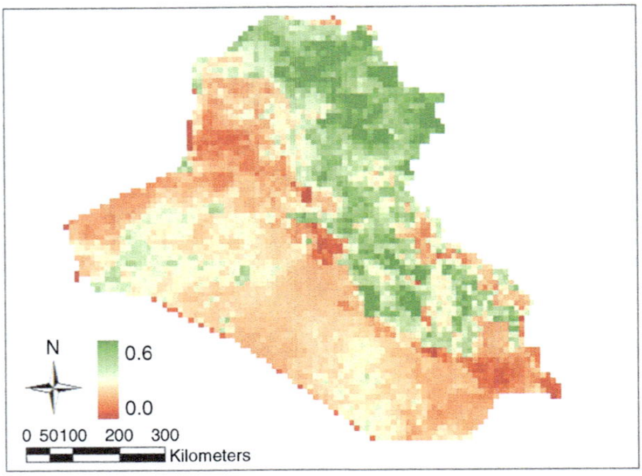

Fig. 7.6 Distribution of TSAVI value as of March 2000

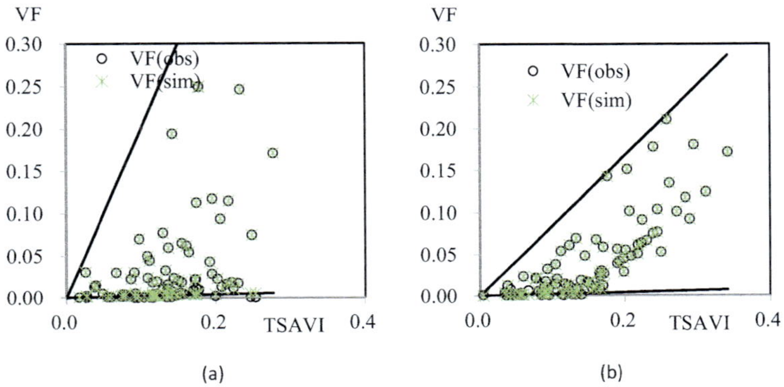

Fig. 7.7 Relation between TSAVI and Vegetation Fraction (observed and estimated)

downstream of Tigris is suffering from high salinity water due to inappropriate drainage management.

The calibrated vegetation fraction model was then validated with Landsat-derived irrigated area image of summer in 1992 and spring in 1999 (Fig. 7.9). The R^2 value was sufficiently high: 0.847 for summer and 0.838 for spring, respectively. Thus, it was concluded that the vegetation fraction model obtained is reliable enough to be applied for the estimation of irrigated area for the rest of study period (1982–2000).

Estimation of Irrigated Area in 1982–2000

The actual irrigated area estimated with the AVHRR-based vegetation fraction model is presented in Table 7.3.

Spring cropping showed higher irrigated area than summer. This coincides the fact that the major cropping season of Iraq is winter to spring with wheat and barley, while summer cropping, mainly paddy rice, is limited due to the limited water resources (Beaumont 1998; Kliot 1994; FAO 2003). On the other hand, spring irrigated area showed higher deviation than summer; particularly in 1984 and

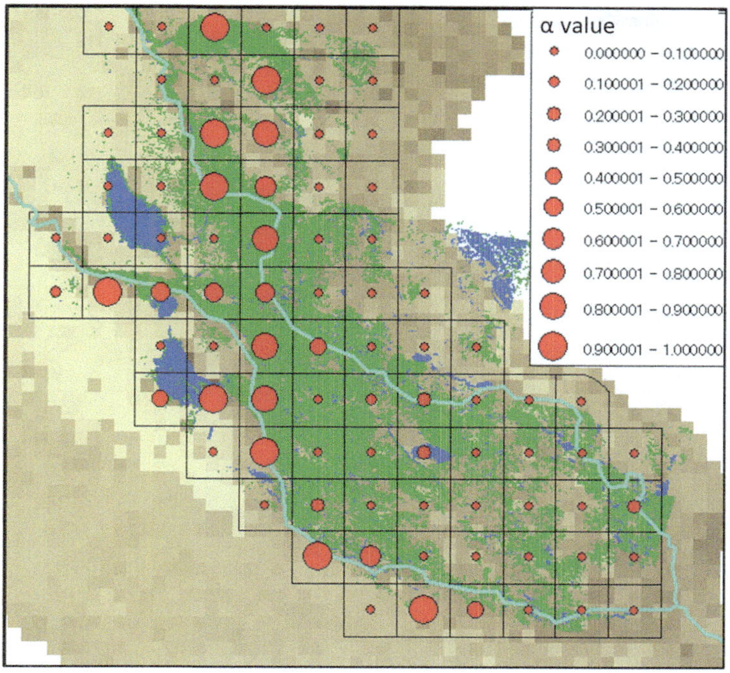

Fig. 7.8 Distribution of high-density crop ratio

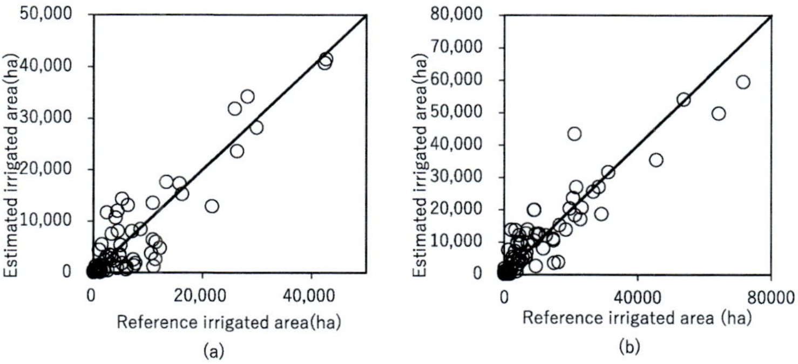

Fig. 7.9 Validation results of vegetation fraction model

1989, the spring irrigated area was even lower than the summer. This inconsistency may be attributed to higher dependency to rain-fed irrigation in spring cropping and unstable productivity thereby.

Substantial discrepancy was observed between estimated irrigated area of this study and the statistical cropped area provided by FAO and USDA throughout the study period (Fig. 7.10). The average area of this study was 1.10 million hectares, while that of FAO and USDA were 2.82 million hectares and 2.69 hectare, respectively. Possible reason for this gap is that the statistical data of FAO and USDA are

Table 7.3 Estimated irrigated area

Unit: 000 ha

Year	This study					Existing statistics	
	Irrigated area in spring		Irrigated area in summer		Total		
	AVHRR	Landsat	AVHRR	Landsat		FAO	USDA
1982	482		429		911	2.350	2.342
1983	422		429		851	2.470	2.472
1984	305		413		718	1.080	1.101
1985	609		442		1.051	2.970	2.971
1986	532		422		954	2.770	2.773
1987	478		451		929	1.930	2.263
1988	632		469		1.101	2.450	2.695
1989	404		476		880	1.620	1.955
1990	584		511		1.095	3.250	3.501
1991	575		497		1.072	3.340	4.312
1992	502		487	507	989	3.910	3.607
1993	644		603		1.247	4.580	3.247
1994	700		636		1.336	3.560	3.291
1995	864		548		1.412	3.160	3.272
1996	695		504		1.199	3.330	2.768
1997	692		578		1.270	2.760	2.430
1998	898		628		1.526	2.800	2.925
1999	793	809	449		1.242	2.710	1.740
2000	592	592	412	412	1.180	2.480	1.538
Average	600		494		1.103	2.817	2.695

Fig. 7.10 Trend of estimated irrigated area

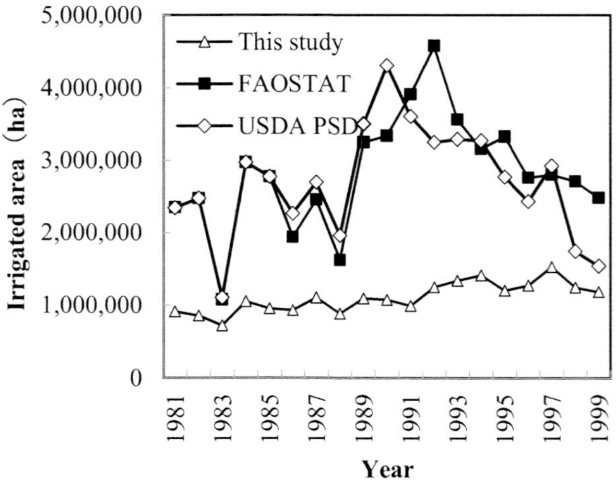

produced based on the reported data by local authorities, such as Ministry of Agriculture of Iraq in this case (USDA 2008). FAO (2003) also points out that Iraqi farmers tend to overclaim their cultivated area because farmer's distribution quota of fertilizer and pesticide by the government is determined based on the claimed cultivated area. Therefore, it is likely the statistical data of FAO and USDA are overestimated due to this overclaim. It is also noteworthy that the crop yield values calculated with FAO/USDA cropped area are extremely low, 0.8 ton/ha on average. Although both crop production data and cropped area data are provided by the same source, FAO and USDA, it is reasonable to consider production amount, which is available through the market, is relatively reliable than cropped area, the figure of which is sometimes not known even by farmers.

7.4 Water Budget Analysis of Iraq

7.4.1 Conceptualization of Water Balance of Study Area

This study employs lumped water balance model instead of distributed model considering the complex human intervention to the river flow and drainage. A general water balance equation of a certain terrestrial system is presented as:

$$\Delta ST = P + D_{\text{in}} + G_{\text{in}} - D_{\text{out}} - G_{\text{out}} - ET \tag{7.6}$$

Where P is precipitation, D_{in} and D_{out} are surface inflow and outflow, G_{in} and G_{out} are groundwater inflow and outflow, ET is evapotranspiration, and ΔST is change of storage. This section will elaborate modify this general equation to fit the situation of study area.

It is well known that the water storage can be regarded as stable state when seen in large spatial and temporal dimension (Tosaka 2006). Although the storage volume fluctuates in a short period due to seasonal rain, it recovers the same level at the beginning of each hydrological year. Therefore, this study regarded storage change is negligible when water balance is assessed based on each hydrological year. The study defined October as the beginning of hydrological year with consideration on the rainy season spans November to April.

Likewise, the groundwater flux ($G_{\text{in}} - G_{\text{out}}$) is regarded as negligible. The arid weather and extremely flat slope of the study area (as little as 1:10,000 or even less) imply very slow flux of groundwater underneath, while surface flow is fed by ample of snow-melt water. Therefore, it is reasonable to consider the groundwater is in relatively stable state when compared with surface water.

The evapotranspiration term of the water balance equation needs subdivision into different type of evaporation and transpiration happening in the study area, i.e., (1) evaporation from water body (ET_{lake}), (2) evapotranspiration from bare land (ET_{bare}), (3) evapotranspiration from irrigated farmland (ET_{farm}). ET_{lake}, dominantly accounted over large reservoirs like Tharthar and Habbaniyah, can be regarded as potential evaporation as sufficient water is fed from the huge water body. ET_{bare}, happening over nonirrigated area, can be regarded as offsetting the local precipitation. Historical rainfall data in Baghdad show the maximum monthly precipitation is no more than 30 mm, while potential evapotranspiration is no less than 80 mm; therefore, it is reasonable to assume virtually all the local precipitation is lost as evaporation with little contribution to the water balance of whole system.

With all these incorporated, the general water balance equation can be transformed as:

$$\int_0^T (P_{farm} + P_{lake} + D_{in} - D_{out} - ET_{farm} + ET_{lake})dt = 0 \tag{7.7}$$

where P_{farm} and P_{lake} are precipitations over farmland and water bodies, ET_{farm} and ET_{lake} are evapotranspirations from irrigated farm and water bodies. Since precipitation data, surface discharge data, and ET_{lake}, which can be estimated from temperature data, are known term, ET_{farm}, dominant factor of water consumption in Iraq, can be obtained as residual term of this equation.

7.4.2 *Materials and Methods*

River Discharge Data

River discharge data of Iraq publicly available are quite limited. Clawson et al. (1971) is one of the few available sources and is cited by various literatures, but it covers only 1960s (Smith et al. 2000; Isaev and Mikhailova 2009). Kavvas et al. (2011) models the water balance of Euphrates and Tigris River, but it is limited to 1937–1972 period when most large dams were not effective yet.

This study uses the discharge data obtained from Ministry of Water Resources of Iraq. This data cover monthly average discharge of 34 hydrological stations in the country. To verify the quality, the time-series discharge data at Hussaiba on Euphrates were cross-checked with those of same period at Youssef Pacha on upstream Euphrates within Syrian territory. They showed sufficiently high correlation ($R^2 = 0.75$). Among the 34 stations, the discharge of Mosul, Hussaiba, Greater Zab, Lesser Zab, Diyala, and Adhaim were used as inflow to the study area, while those of Nassiriyah, Amara, and Gharraf were used as outflow from the area (Fig. 7.11).

Precipitation Data

Global grid-base monthly climate data released by Climate Research Unit of East Anglia (CRU TS 3.0) were used as precipitation data. CRU data are based on 30,000 + meteorological stations worldwide, screen the error data, and resample into 0.5-degree geographical grid (Mitchell and Jones 2005).

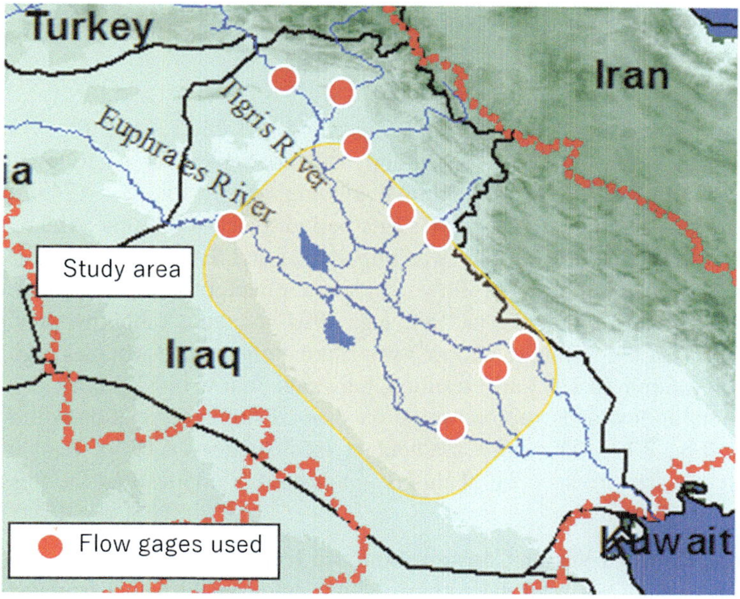

Fig. 7.11 Extent of study area and flow gages used

Potential Evaporation

Thornthwaite method is used to estimate the potential evapotranspiration (PET) depth from water bodies (Thornthwaite 1948). Being an empirical model, this method does not present physical connotation of evaporative phenomena, but with its convenience and reliance, it is widely used to date (Szilagyi et al. 1998; Xu and Singh 2002; Chen et al. 2005). The PET of i-th month (ET_i) is expressed as:

$$ET_i = 16\left(\frac{T_i}{I}\right)^a\left(\frac{N_i}{12}\right) \quad (7.8)$$

$$I = \sum_{i=1}^{12}\left(\frac{T_i}{5}\right)^{1.514} \quad a = \left(492390 + 17920I - 77.1I^2 + 0.675I^3\right) \times 10^{-6} \quad (7.9)$$

where T_i is the mean monthly temperature and N_i is the theoretical sunshine hours of i-th month.

CRU TS 3.0 was used as temperature input data. Since there are a number of small water bodies in the study area besides large reservoirs, the area of water bodies was determined by the land cover classification image derived from Landsat as of 2000.

7.4.3 Results and Discussions

Water Balance Analysis

The result of water balance analysis is presented in Table 7.4 and Fig. 7.12. Each hydrological year starts in October and ends in September of following year. Relatively low inflow, despite normal amount of precipitation (744 mm and 674 mm) is observed in 1989–1990. This period matches the time when the huge Ataturk Dam in Turkish territory was impounded and Syria and Iraq made protest against Turkey. Contrariwise, flow regulation effect of this huge dam can be observed after the completion in 1992, e.g., 1994 had remarkably low precipitation of 562 mm but the inflow (84.2 km^3) was almost on average whereas high precipitation of 946 mm in 1997 also resulted in average inflow of 85.8 km^3.

The total inflow and evapotranspiration showed strong correlation with R^2 value of 0.91 (Fig. 7.13). The slope gradient of 0.67 suggests 67% of total inflow is consumed in Iraq irrespective of inflow amount. This implies that farmers tend to save water in dry year.

Relations Between Water Consumption and Irrigated Area

The relation between total evapotranspiration amount and actual irrigated area is shown in Table 7.5. The amount of evapotranspiration per irrigated area is surprisingly high, ranging from 3250 mm to 20,200 mm per year, far above the potential

Table 7.4 Summary of water balance analysis

Unit: km^3

Year	D_{in}	P	D_{out}	ET_{lake}	ET_{farm}
1981	103.2	2.1	33.5	4.5	67.4
1982	70.8	2.0	33.2	4.2	35.3
1983	68.3	0.9	18.8	4.6	45.7
1984	82.7	3.1	25.2	4.5	56.2
1985	38.4	2.0	17.9	4.7	17.7
1986	98.6	1.8	19.9	4.8	75.7
1987	171.7	3.3	43.8	4.7	126.5
1988	49.0	1.4	31.9	5.1	13.4
1989	64.7	2.5	15.9	4.9	46.5
1990	46.3	2.1	15.5	4.8	28.2
1991	73.9	2.1	18.0	4.3	53.8
1992	104.8	3.3	20.0	4.6	83.5
1993	73.4	2.7	21.9	5.0	49.3
1994	113.3	3.7	28.6	5.0	83.4
1995	84.2	3.4	32.7	5.3	49.7
1996	80.6	2.8	21.7	4.5	57.2
1997	85.8	5.3	26.5	5.2	59.3
1998	47.0	2.6	14.6	5.7	29.3
1999	39.0	1.6	7.5	5.4	27.7
Average	78.7	2.6	23.5	4.8	52.9

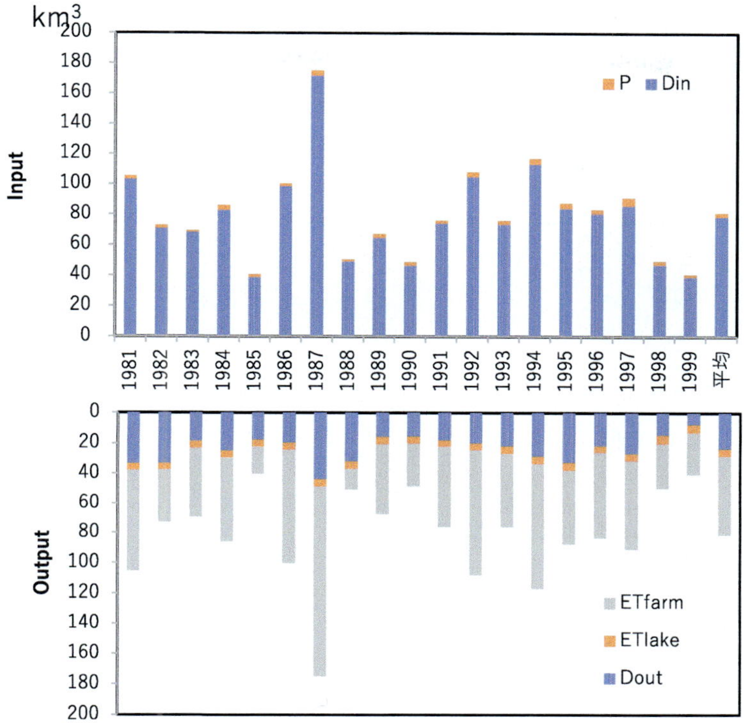

Fig. 7.12 Itemized breakdown of water balance

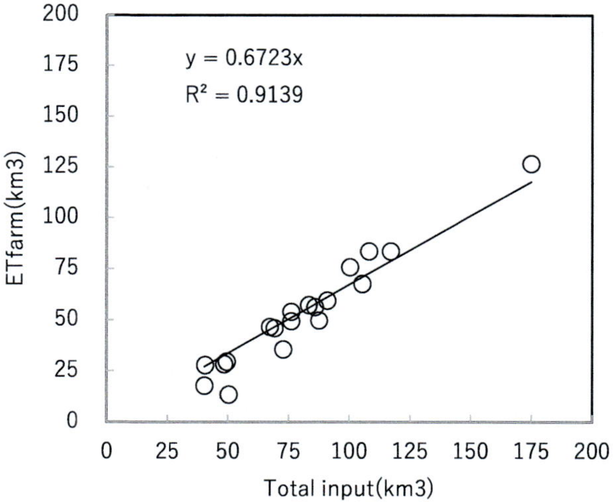

Fig. 7.13 Relation between total input and irrigation water use

Table 7.5 Relation of estimated evapotranspiration volume (ET_{farm}) and irrigated area

	ET_{farm} (km^3/yr)	Estimated spring irrigated area (000 ha)	ET per irrigated area (mm/yr)	Potential ET (mm/yr)	Minimum ET occurrence area (000 ha)
Year	A	B	A/B	C	A/C
1981	67	482	13.986	1.580	4.268
1982	35	422	8.374	1.565	2.258
1983	46	305	14.998	1.600	2.860
1984	56	609	9.223	1.632	3.442
1985	18	532	3.332	1.653	1.073
1986	76	478	15.829	1.768	4.280
1987	127	632	20.017	1.585	7.982
1988	13	404	3.323	1.827	735
1989	46	584	7.958	1.829	2.541
1990	28	575	4.902	1.676	1.682
1991	54	502	10.723	1.532	3.514
1992	83	644	12.959	1.627	5.129
1993	49	700	7.041	1.739	2.834
1994	83	864	9.656	1.764	4.729
1995	50	695	7.146	1.813	2.739
1996	57	692	8.267	1.599	3.577
1997	59	898	6.606	1.940	3.057
1998	29	793	3.697	2.104	1.394
1999	28	592	4.677	1.938	1.429
Average	53	600	9.090	1.725	3.133

evapotranspiration depth, which is around 2000 mm. This denotes that water is being lost in nonirrigated area too. If we assume all the evapotranspiration occurs at PET rate, the total area of evapotranspiration ranges from 735 thousand to 7.98 million hectares. Given the fact evapotranspiration at PET rate rarely happens unless the surface is always covered with water, it is reasonable to regard that the area that undergoes substantial evapotranspiration be even larger than this. These results indicate that large amount of water is lost for nonirrigation purposes, probably for conveyance loss and irrigation that does not result in harvest. On the other hand, in years with extremely high water loss, such as 1987 and 1994, excess water may be stored in dams as inter annual storage instead of onsite evapotranspiration.

The scatterplot of evapotranspiration per irrigated area against inflow volume showed moderate linear relation ($R^2 = 0.67$, Fig. 7.14). This suggests that larger amount of inflow does not necessarily lead to larger irrigated area, rather it leads to lower water use efficiency.

Interestingly, the irrigated area showed relatively high correlation ($R^2 = 0.76$) with the precipitation amount (Fig. 7.15). As seen in Fig. 7.12, contribution of rainfall to the total available water is quite limited, so relatively high amount of rainfall does not promise farmers of sufficient water for irrigation. However, given the limited accessibility and reliability of river flow information of Tigris and

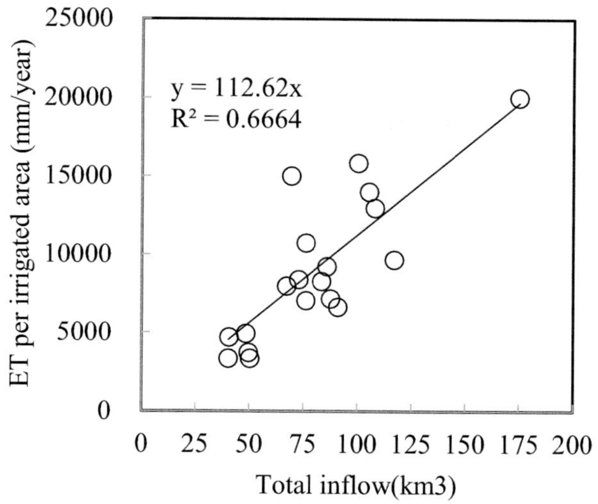

Fig. 7.14 ET per irrigated area against total inflow

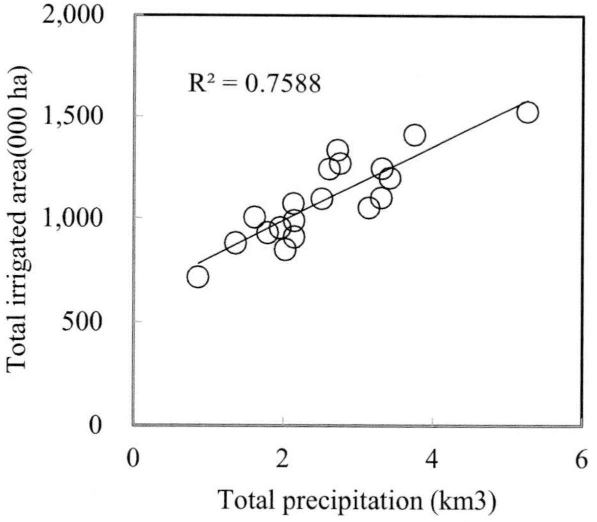

Fig. 7.15 Total irrigated area against total precipitation

Euphrates, it is likely that local farmers determine cultivation area of the year based on the rainfall amount.

7.5 Conclusions

This study analyzed the land-water use of Iraq, the lower most riparian state of Tigris and Euphrates River system which undergoes half-century-long transboundary water disputes. Given the unstable security situation which does not allow field survey, this study employed remotely available data, such as satellite imagery, statistical data, global dataset, as analysis material.

The land use analysis with satellite imagery revealed that the estimated irrigated area is much smaller than that of existing literatures or statistics. It also found that, out of two harvesting seasons, namely spring and summer, the irrigated area in summer tends to have less variance than summer, implying irrigation agriculture in summer enables stable cropping than rain-fed agriculture in spring.

The water budget analysis using multiple source of data, including river flow data of Iraq government and global climate data, revealed that the substantially high rate of evapotranspiration presumably occurs against the irrigated area estimated through the land use analysis. Also found was that about two-thirds of inflow water to Iraq is consumed within the study area. However, more inflow does not translate into more irrigated land or food production. Another finding was that the irrigated area is correlated with the precipitation volume, which does not have significant contribution to the overall water availability.

The limitation that this study does not involve on-site validation leaves questions as to the implication of respective findings. However, overall trend identified by this study, e.g., less-than-expected irrigated area and high rate of evapotranspiration, sheds strong signal of unfunctional irrigation facility and inefficient water management in the study area. This then can be translated into policy recommendation that the transboundary water dispute over Tigris and Euphrates Rivers may be relaxed by expanding financial and technical assistance over the irrigation sector of Iraq.

References

Adams JB, Smith MO, Johnson PE (1986) Spectral mixture modeling: a new analysis of rock and soil types at the viking lander 1 site. J Geophys Res AGU 91(B8):8098–8112

Allan JA (2001) The middle east water question: hydropolitics and the global economy. I.B. Tauris, London

Altinbilek D (1997) Water and land resources development in southeastern Turkey. J Water Resour Dev 13(3):311–332

Altinbilek D (2004) Development and management of the Euphrates-Tigris basin. Int J Water Res Dev 20(1):15–33

Bagis AI (1997) Turkey's hydropolitics of the Euphrates-Tigris basin. Water Resour Dev 13 (4):567–581

Baret F, Guyot G (1991) Potentials and limits of vegetation indices for LAI and APAR assessment. Remote Sens Environ 35(2–3):161–173

Baret F, Guyot G, Major DJ (1989) TSAVI: A vegetation index which minimizes soil brightness effects on LAI and APAR Estimation. IGARSS'89 and 12th Can Symp Remote Sensing, pp 1355–1358

Beaumont P (1978) The Euphrates river – an international problem of water resources development. Environ Conserv 5(1):35–44

Beaumont P (1998) Restructuring of water usage in the Tigris-Euphrates basin: the impact of modern water management policies. In: Coppock J, Miller JA (eds) Transformations of Middle Eastern natural environments legacies and lessons. Yale University, New Haven, Connecticut

Bilen O (1994) Prospects for technical cooperation in the Euphrates-Tigris basin. In: Biswas AK (ed) Int waters middle east from euphrates-tigris to nile. Oxford University Press, New York, pp 95–116
Chen DL, Gao G, Xu CY, Guo J, Ren GY (2005) Comparison of the Thornthwaite method and pan data with the standard Penman-Monteith estimates of reference evapotranspiration in China. Clim Res 28(2):123–132
Clawson M, Landsberg HH, Alexander LT (1971) The agricultural potential of the Middle East. American Elsevier Publishing, New York
Dellapenna JW (1996) The two rivers and the lands between: Mesopotamia and the international law of transboundary waters. BYU J Public Law 10:213–261
FAO (2003) Crop and food supply and nutrition assessment mission to Iraq. Spec Reports Alerts – ESC/Reports. FAO, Rome
Gleick PH (1988) The effects of future climatic changes on international water resources: the Colorado River, the United States, and Mexico. Policy Sci 21(1):23–39
Gutman G, Ignatov A (1998) The derivation of the green vegetation fraction from NOAA/AVHRR data for use in numerical weather prediction models. Int J Remote Sens 19(8):1533–1543
Hansen M, DeFries R, Townshend JRG, Sohlberg R (1998) UMD global land cover classification, 1 kilometer, 1.0. Department of Geography, University of Maryland, College Park
Hillel D (1994) Rivers of Eden: the struggle for water and the quest for peace in the Middle East. Oxford University Press, New York
Huete AR (1988) A soil-adjusted vegetation index (Savi). Remote Sens Environ 25(3):295–309
Inamura M (1988) Analysis of remotely sensed image data by means of category decomposition. Electron Commun Japan, Part 2 71(1):11–22
Irvin BJ, Ventura SJ, Slater BK (1997) Fuzzy and isodata classification of landform elements from digital terrain data in Pleasant Valley, Wisconsin. Geoderma 77:137–154. Elsevier
Isaev V, Mikhailova M (2009) The hydrography, evolution, and hydrological regime of the mouth area of the Shatt al-Arab River. Water Resour. MAIK Nauka/Interperiodica distributed exclusively by Springer Science+Business Media LLC 36(4):380–95
James ME, Kalluri SNV (1994) The pathfinder avhrr land data set - an improved coarse resolution data set for terrestrial monitoring. Int J Remote Sens 15(17):3347–3363
Kavvas ML, Chen ZQ, Anderson ML, Ohara N, Yoon JY, Xiang F (2011 Jan 1) A study of water balances over the Tigris–Euphrates watershed. Phys Chem Earth, Parts A/B/C Pergamon 36 (5–6):197–203
Kliot N (1994) Water resources and conflict in the middle east. Routledge, London
Lasdon LS, Waren AD, Jain A, Ratner M (1978) Design and testing of a generalized reduced gradient code for nonlinear programming. ACM Trans Math Softw ACM 4(1):34–50
Ludwig D, Hilborn R, Walters C (1993) Uncertainty, resource exploitation, and conservation - lessons from history. Science (80-) 260(5104):17
Mitchell TD, Jones PD (2005) An improved method of constructing a database of monthly climate observations and associated high-resolution grids. Int J Climatol 25(6):693–712
Naff T (1984) The Euphrates River and the Shatt Al-Arab. In: Naff T, Matson RC (eds) Water Middle East Confl or Coop. Westview Press, Boulder, pp 83–114
Rieu-Clarke AS (2007) Overview of Stakeholder Participation – What Current Practice and Future Challenges? – Case Study of the Danube Basin. Colo J Int Environ Law Policy 18(3):611–632
Scheumann W (1997) Conflicts on the Euphrates: An analysis of water and non-water issues. In: Scheumann W, Schiffler M (eds) Water middle east potential conflicts prospect coop. Springer, Berlin, pp 113–135
Smith MO, Ustin SL, Adams JB, Gillespie AR (1990) Vegetation in deserts: I. A regional measure of abundance from multispectral images. Remote Sens Environ 31(1):1–26
Smith RB, Foster J, Kouchoukos N, Gluhosky PA, Young R, Zhang J (2000) Hydrologic trends in the middle east: modeling and remote sensing, Aleppo
Starr JR (1991) Water wars. Foreign Policy 82:17

Szilagyi J, Rundquist DC, Gosselin DC, Parlange MB (1998) NDVI relationship to monthly evaporation. Geophys Res Lett 25(10):1753–1756

Thornthwaite CW (1948) An approach toward a rational classification of climate. Geogr Rev 38:55–94

Tosaka H (2006) Numerical hydrological analysis. University of Tokyo Press, Tokyo. (In Japanese)

Tucker CJ (1979) Red and photographic infrared linear combinations for monitoring vegetation. Remote Sens Environ 8:127–150

Turkish Ministry of Foreign Affairs (1996). Water issues between Turkey, Syria and Iraq [Internet]. Turkish Minist Foreign Aff. [cited 2018 Jan 31]. Available from: http://sam.gov.tr/wp-content/uploads/2012/01/WATER-ISSUES-BETWEEN-TURKEY-SYRIA-AND-IRAQ.pdf

UNEP-DHI and UNEP (2015) Transboundary river basins: status and trends, summary for policy makers. Nairobi

USDA (2008) USDA/FAS Production, Supply and Distribution Online Database. USDA

Weiss H (1978) Early irrigation agriculture. Science (80-) 1978(200):1377–1378

Wu W (2014 Jan 29) The Generalized Difference Vegetation Index (GDVI) for dryland characterization. Remote Sens Multidisciplinary Digital Publishing Institute 6(2):1211–1233

Xu CY, Singh VP (2002) Cross comparison of empirical equations for calculating potential evapotranspiration with data from Switzerland. Water Resour Manag 16(3):197–219

Chapter 8
The Nature of Tigris–Euphrates Rivers Flow: Current Status and Future Prospective

Zaher Mundher Yaseen, Sadeq Oleiwi Sulaiman, and Adel Sharif

Abstract The chapter emphasizes the understanding of the natural phenomena of the Tigris–Euphrates rivers and their water resources for the Iraq region. Iraq region is challenging a very disastrous water crisis owing to the water degradation of these two main water resources rivers. That is in parallel with human population's massive growth over the past decade. Tigris–Euphrates rivers embodies over 95% of the total Iraq nation water demand presented in the form of the domestic, agriculture, industrial, and several others. The deterioration status of these two rivers has become a life-threatening matter that required serious attention by the water resources engineers. In this phase of the book, a comprehensive review of the previous studies established on the water resources profit of these two rivers. In addition, the current status of the two rivers' flow pattern is studied in detail. Finally, the chapter briefed various practical solutions for the Tigris–Euphrates rivers sustainability and management.

Keywords Tigris–Euphrates rivers · Water resources · River engineering sustainability · Future prospective

Z. M. Yaseen (✉)
Sustainable Developments in Civil Engineering Research Group, Faculty of Civil, Ton Duc Thang University, Ho Chi Minh City, Vietnam

S. O. Sulaiman
Dams and Water Resources Department, College of Engineering, University of Anbar, Ramadi, Iraq

A. Sharif
Centre for Osmosis Research & Application, Chemical & Process Engineering Department, University of Surrey, Guildford, UK

L. A. Jawad (ed.), *Tigris and Euphrates Rivers: Their Environment from Headwaters to Mouth*, Aquatic Ecology Series 11,
https://doi.org/10.1007/978-3-030-57570-0_8

8.1 Introduction

Over the past few decades, scarcity of water resources has been a major obstacle in the Middle East and particularly in the Iraq region (Chenoweth et al. 2011; Sowers et al. 2011; Droogers et al. 2012). This is owing to numerous factors such as climatic changes and water demand increment. Going down history lane, the role of water in human life dates to the first civilization, but the initial effort toward water resource development dates commenced in the early 5500 BC in the Mesopotamia, made up of the basins of the Tigris–Euphrates rivers (Allan and Kliot 1994; Van De Mieroop 1997). The water from the Euphrates river served the water needs of the Sumerian and Babylonian civilizations (irrigation and household purposes) via canals (Antoniou et al. 2014). The Iraq Board of Development was established by the Kingdom of Iraq to initiate plans for the construction of development of systems for flood control and irrigation purposes in the 1950s. The major purpose of this plan is to protect the Iraqi capital city Baghdad from flooding. In 1959, the Dokan dam considered one of the earliest hydraulic structures was constructed on the Lesser Zab River (Hassan et al. 2017). Later on, several other dams were initiated in different locations for hydropower generation and irrigation purposes. To control flooding, a system comprising regulator dam was built in Ramadi city to store the water over Habbaniya Lake on the Euphrates river (Al-Heety 2011). Other systems such as regulators, Samarra dam, canal systems, and Lake Tharthar project were constructed on the Tigris River (Kibaroglu and Maden 2014).

The existence of these twin rivers in Iraq region conferred the country a rich status in terms of water resources until the 1970s (Beaumont 1998). The construction of water harvesting dams by Syria and Turkey along the Euphrates River after 1970s caused a decrease in the water flow and resulted in low water levels in some of the constructed new reservoirs (Freeman 2001; Zagonari and Rossi 2014). Consequently, the Iraqi Government initiated concise efforts toward building several water resources harvesting projects and especially in the west desert of Iraq (Al-Ansari and Knutsson 2011). This is explaining the reason, water resources development in Iraq reached its best period from 1970 to 1990. However, there was a decline in these developmental processes during the 1990s due to UN sanctions and the first Gulf War (Alnasrawi 2001). From 1960 to the early of the twenty-first century (about 5 decades), the supply and demand for water witnessed significant changes.

In 1977, the utilization of the Tigris–Euphrates rivers by the Turkish Government through the South-eastern Anatolia Project (GAP) (Unver 1997), has influenced these rivers' flow noticeably. The project constructed 19 hydraulic power plants and 22 multipurpose dams for land irrigation covers approximately 17,100 km^2. The storage capacity of this project was about 100 km^3, which is more than thrice the capacity of all the reservoirs in both Iraq and Syria. Of these dams, eight were proposed to be built on the Tigris river but only three were actually constructed (2 in 1997 and 1 in 1998). The annual water consumption by these irrigation projects within the GAP is about 22.5 km^3 when completed (Al-Ansari 2013). It should be noted, Tigris–Euphrates rivers are the main predominant freshwater resources for several nations and the major portion goes to Iraq region. Studying the rivers flow characteristics and trend are essential for different river engineering prospective. Based on this and for the best knowledge of the author, the current chapter is established.

8.2 The Hydrological Characteristic of the Tigris–Euphrates Basin

Throughout western Asia, the Tigris–Euphrates rivers are the largest regional rivers ever known (Cullen and DeMenocal 2000). Both are originating from the same Turkish region, though about 30 km apart. This region is notorious for its humid and cool climate, furnished with deep gorges and high mountains. These two great rivers flow separately from this region onto a wide and poorly drained valley (see Fig. 8.1). Halfway down their path, these rivers diverged for several kilometers apart but later rejoin about 160 km north of the Gulf near Qarmat Ali to form the Shatt al-Arab. Rocks and mountain gorges of Anatolia, together with the high plateaux of Syria and Iraq are the notable features of the upper basin of the rivers. At first, the rivers appear

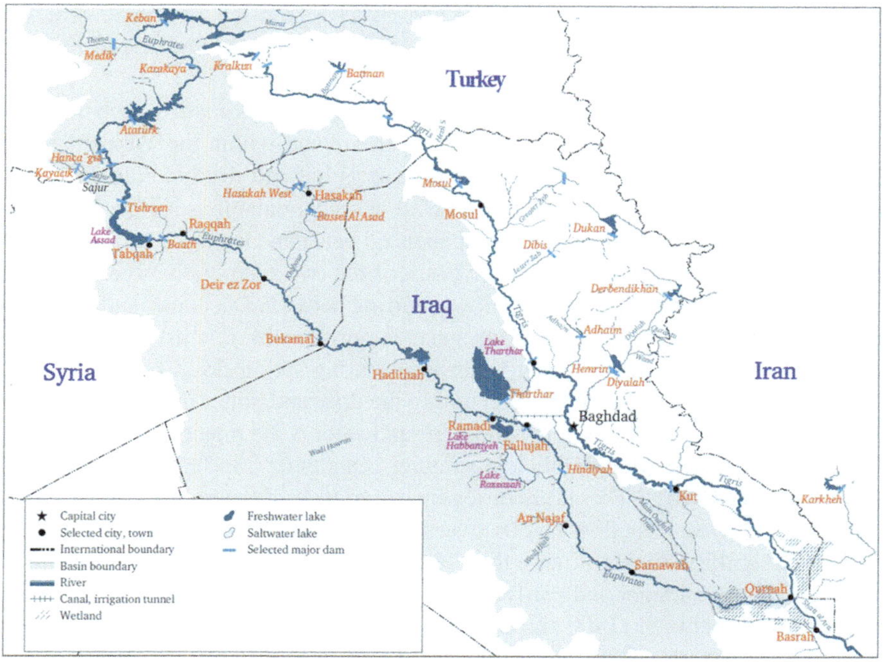

Fig. 8.1 The path of the Tigris–Euphrates rivers from source to mouth end including all the details of the hydraulic structures and the cities passing through

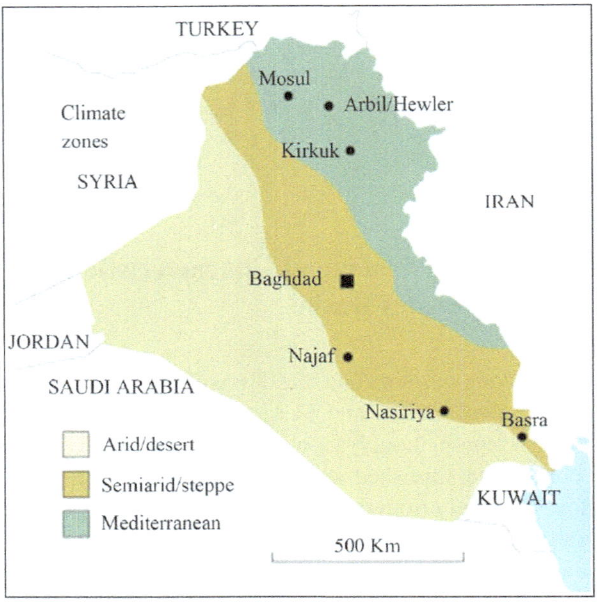

Fig. 8.2 The climatological characteristics of the Tigris–Euphrates rivers over Iraq region

to be parallel to each other, until they fall off the final limestone plateau onto the Mesopotamia plain. The climate of the two rivers region featured three different characteristics including arid, semiarid, and Mediterranean as exhibited in Fig. 8.2 (Issa and Al-Ansari 2014).

The catchment area and the Mediterranean have a similar climate though there are differences due to the mountainous nature of the region within the Turkish region (Yürekli 2015). During the summer, the climate is hot and dry but cold and rainy during the winter. There is a periodic snowfall in the mountainous region, but in the Mesopotamian basin, precipitation is experienced between October and May. Annually, the level of precipitation over the basin of the two rivers is between 100 and 1000 mm, with the heaviest precipitation occurring between December and February (Dezfuli et al. 2017) (see Fig. 8.3). In general, snow melting is usually witnessed around February, causing a higher water level during spring flows (from March to early June). There is a low-water level during the drier and hotter months of summer (July to October), but during this period, groundwater is the major source of runoff. On average, the range of monthly temperatures is about 6 °C in Jan and about 34 °C in July (Aqrawi 2001). The northern region experiences a lower temperature range throughout the year. The flow level between the flood season (during winter) and the spring are significantly high (more than 10 times) due to the rain and the melting of snow during the summer and partly during the autumn. The evaporation rate of the region limited between (1400 and 2000) mm as explained in Fig. 8.4. Finally, it is great to state that the Tigris–Euphrates water system is a highly remarkable water resources network that contributes to various nations and for several sectors (Fig. 8.5).

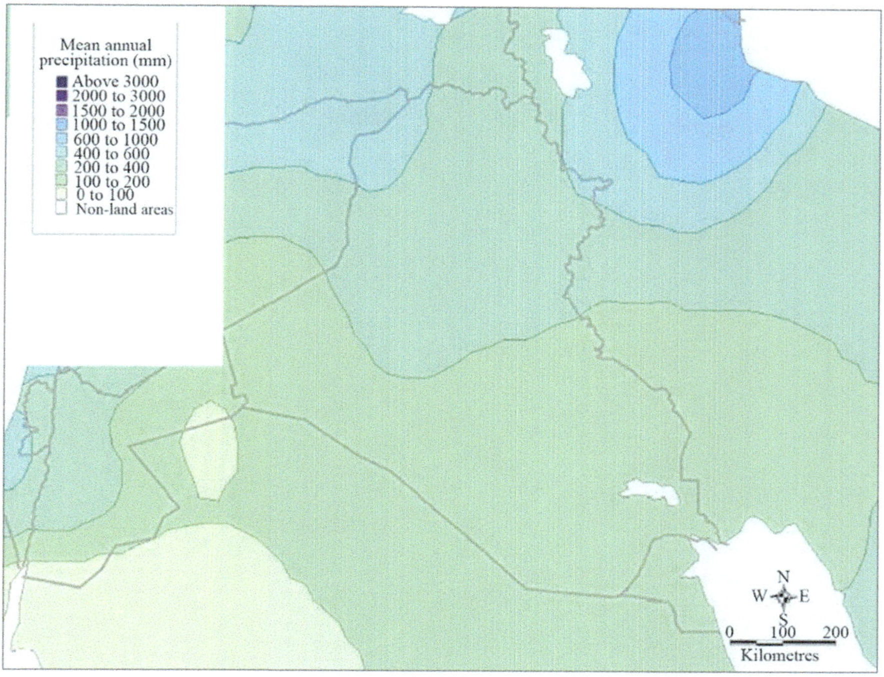

Fig. 8.3 The precipitation magnitudes over the Tigris–Euphrates rivers' basin

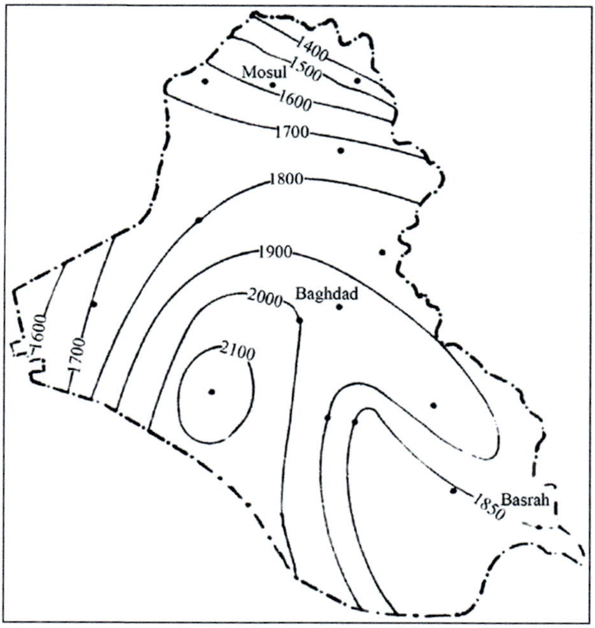

Fig. 8.4 The evaporation rate of the Tigris–Euphrates rivers

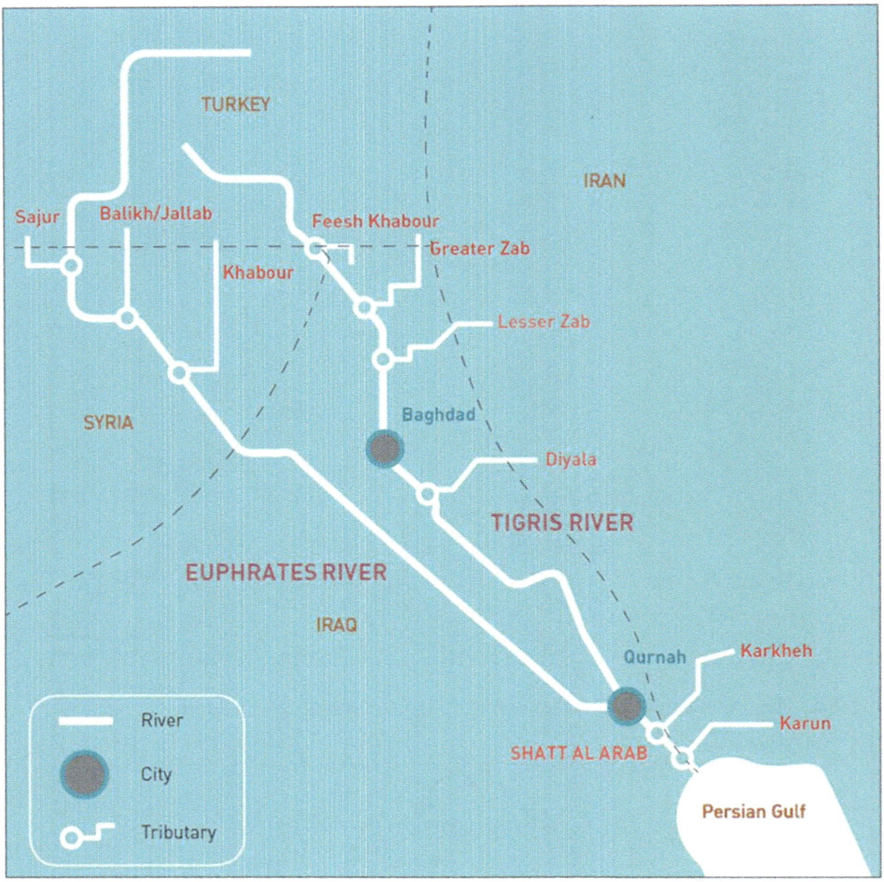

Fig. 8.5 Tigris–Euphrates rivers sketch basin system

8.2.1 Tigris River Sketch

The main source of water for the Tigris river is Hazar lake that elevated at 1150 ma.s.l and coordinated in the southeastern region of Turkey. The lake is surrounded by the Taurus mountain series with a height of over 3500 m. Tigris river is ever known as the second largest river in the western part of Asia. The small mountain lake of Jazar Golu in Turkey, located about 30 km north of the Euphrates catchment, is the source of the Tigris river headwaters. The route of the river is the hilly regions (southwestern portion) of the mountainous region that connects Turkey, Iran, and Iraq. The Tigris river maintains a direct flow toward Iraq and the Mesopotamian Plain, though it partly flows parallel to the Syrian border. About 400 km from the source, precisely at the Faish Al-Khabur village, the river crossed the Iraqi border. The river is fed by 8 major tributaries from the left bank (3 in Turkey and 5 in Iraq) before entering Iraq. The Iraqi tributaries flow from the Zagros mountains in the northeastern region of

Iraq, then, join the river body, before entering Baghdad. In the Diyarbakir area, the river channel is wide and shallow, but after joining the Batman tributary, it assumed a high velocity, became deep and narrow. Before the construction of the Mosul dam, the width of the river valley (floodplain) in the north of Mosul city to Faish Khabur was about 2–10 km, draining an area of about 375,000 km^2 covering Turkey, Syria, Iran, and Iraq. Out of the total flow length of 1862 km, only about 21% of the river flow length is in Turkish territory, while the remaining part is in Iraq. The river has a mean annual flow of 20–30 km^3 per year from Turkey before entering Iraq. When flowing through Iraq, all the named tributaries empty into the river, contributing an additional 25–29 km^3 of water per year. Before 1984, the river has a mean annual flow of 22.2 km^3 at Mosul city before dropping to 17.7 km^3. The river witnesses its annual hydrograph from Oct to Sept, while the maximum mean monthly discharge occurs in April. September is generally, the driest month in the region.

8.2.2 Euphrates River Sketch

The Turkish territories of Anatolia are the major water source for the Euphrates river, which represented the longest river in western Asia (Swearingen 1993). Muratsu and Furatsu are the main tributaries that formed the Euphrates river near Van lake (Allan and Kliot 1994). The tributaries met near Elazia city where it serves the region as the major source of water from the melted snow. The river, at an average slope of 2 m per km, flows 160 km southward of the Mediterranean before turning leftward to the southeast direction into Syria (flowing almost toward the Shatt al-Arab river). Upon entering the Syrian borders at Jarablis, the river is fed by two small tributaries (Al-Balikh and Al-Khabur), contributing to the volume of the Euphrates river (Beaumont 1998). The Euphrates enters the Iraqi territory at Hasaibah, having an annual flow of 28–30 km^3 per year at the Iraqi border. The river reached a giant alluvial delta at Ramadi, about 360 km from the Iraqi border, with an elevation of only 53 ma.s.l. From there onward, the river flows to the desert regions of Iraq where a reasonable portion of its water volume is lost to natural and man-made depressions and distributaries. There are several small tributaries that are mainly for irrigation purposes within the central and southern parts of Iraq. Within the Iraqi territory, the river is not fed by any tributary. Near Nasiriyah, the river bifurcates into several channels; some of the channels flow into the lake Al-Hammar marsh while the others rejoined Tigris at Qurna.

8.3 Tigris–Euphrates Rivers Flow Patterns

River flow pattern trend is one of the essential components related to river engineering comprehending. This is owing to the fact that river flow patterns correlated to various casual hydrological variables (e.g., evaporation, winds, temperature, humidity, and precipitation). The correlated variables might be also morphological

Fig. 8.6 The monitoring gauging stations exist on the Tigris–Euphrates rivers' basin

components including soil infiltration and groundwater level. In addition, human activities that signifies the major remarkable influence on the nature of river flow. Nevertheless, the noticeable climate change phenomena recognized recently. Due to all the mentioned reasons, it is highly necessitated and fundamental to study river flow characteristics over the decades. Tigris–Euphrates river are the main freshwater serves different nations and understand their flow trend is vastly magnificent for multiple water resources prospective. Before demonstrating the flow analysis of the two rivers, it is even worth to exhibit the monitoring gauging station over the two rivers. Figure 8.6 revealed all the monitoring gauging station was initiated on the Tigris–Euphrates rivers.

Tigris river comprises the biggest number of monitoring gauging stations owing to the numerous tributary rivers flowing toward the main river. However, two different coordinated stations located at the upstream and downstream (i.e., Mosul (T3) and Kut (T20)) were selected to be investigating the Tigris river flow trend (see Fig. 8.7). The studied time series river flow information covered a great sufficient

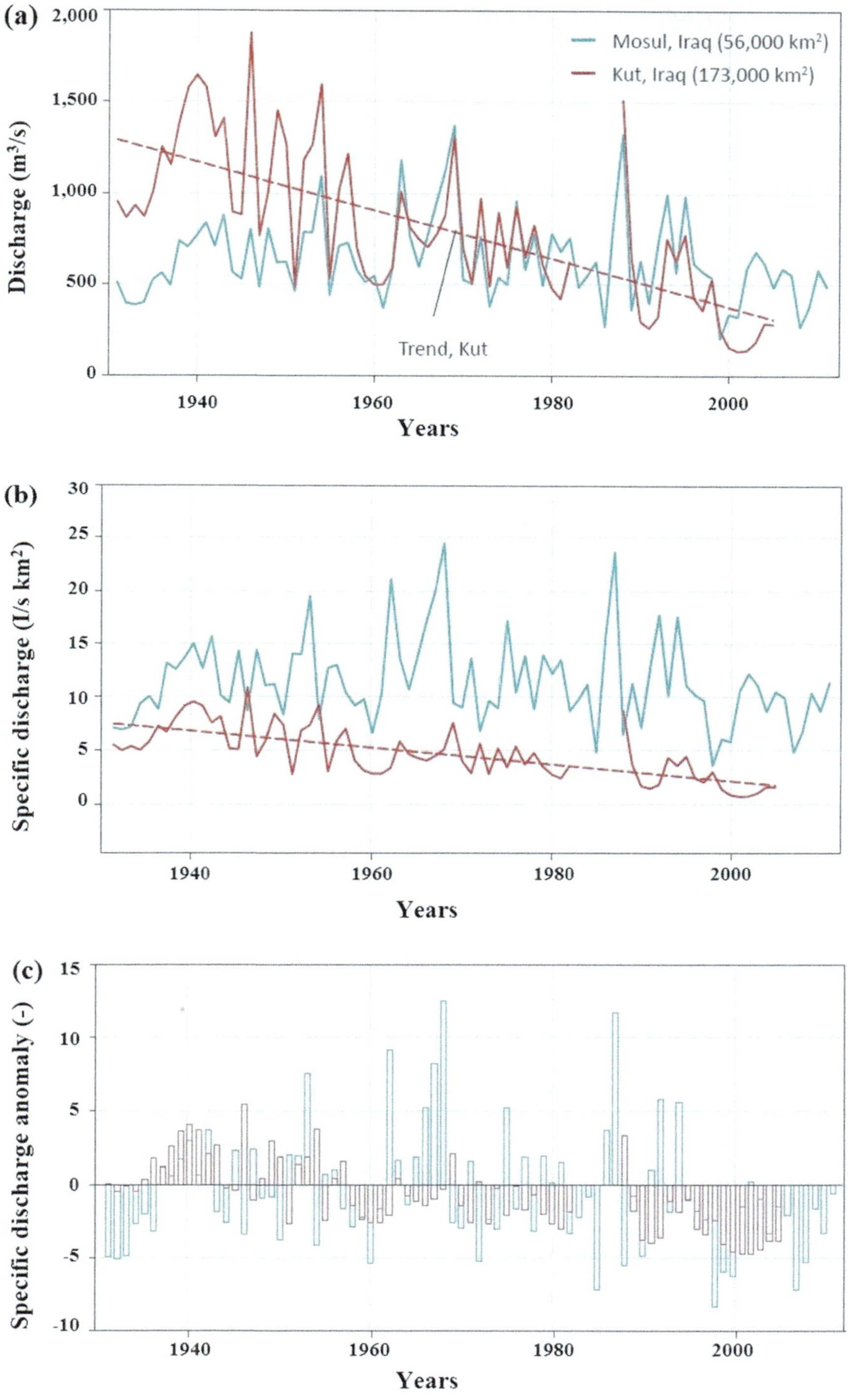

Fig. 8.7 The selected river flow gauging stations located on the Tigris river

period (1930–2011). The magnitude of the Tigris river discharge surpassed the amount noticed on the Euphrates river. The upstream station (T3) showed a consistent river flow pattern and hard to recognize and positive or negative trend. On the contrary with Kut station, the presented time series data demonstrated a remarkable negative vogue. Also, the data records of Kut station is mission over the last decade. This is due to poor management regarded with water resources ministry owing to the war situation.

Figure 8.8 exhibits the river flow pattern of the Euphrates river for different gauging stations (i.e., Jarablus (E1), Hussaybah (E2), Hit (E3), Hindiyah (E4) (see Fig. 8.6)) over nearly a century time-period start from 1930s until 2010. In general, the river flow mean discharge pattern shows a remarkable negative trend and particularly in the last three decades. Over the period before 1970, the river mean discharge approximately 30 BCM at E1 station. However, this discharge value dropped unto less than 26 BCM after the year 1974 until it reached below 23 BCM after 1990 (see Fig. 8.8a). The best reasonable explanation for this negative trend in the first place is due to the initiating of the GAP project in Turkey in addition to the several dams' construction by Syria and Iraq. The second reason is the massive changes in the climate variability and particularly over the drought periods. It is worth highlighting here, despite E1 station pattern defined the remarkable long-term trend, the other examined stations (e.g., E2, E3, and E4) indicated declining in the discharge magnitude after 1973. The last decade of the twenty-first century displayed a noticeable decline in the mean annual discharge volume and specifically at E2 station "Hussaybah." According to the Iraqi ministry of water resources, it has registered the lowest value at this station which is less than 250 m^3/s. On the other hand, Fig. 8.8b demonstrated the specific discharge over the same period at the four selected gauging stations. It should be noted that the downstream stations Hit (E3) and Hindiyah (E4) exhibited a lower discharge in comparison with the upstream stations (e.g., E1 and E2). Indeed, this is typically can be explained due to the hydrological characteristic of the region where no attributes supply the Euphrates river. The anomaly Euphrates river pattern is presented in Fig. 8.8c over the same available investigated period. Here, it can notice the water surplus and deficit over the long-term scale. Figure 8.8c is clearly indicated the wet and dry seasons that revealed the Euphrates regulations pattern. The investigated river time series information showed different prolonged drought events; however, the outstanding drought period located between 1983 and 1995 that episodic with wet event and particularly at the year of 1989. In general, the Euphrates river displayed a remarkable degradation in the average discharge and noticeably after the year 2000 owing to the massive dams' construction over the region.

8.4 Evaluation and Future Prospective

Based on the foregoing analysis, there were several conclusions abstracted from the past and current situations of the Tigris–Euphrates rivers. It is found that the problem of the whole Tigris–Euphrates basin is not only caused by external factors "Turkey

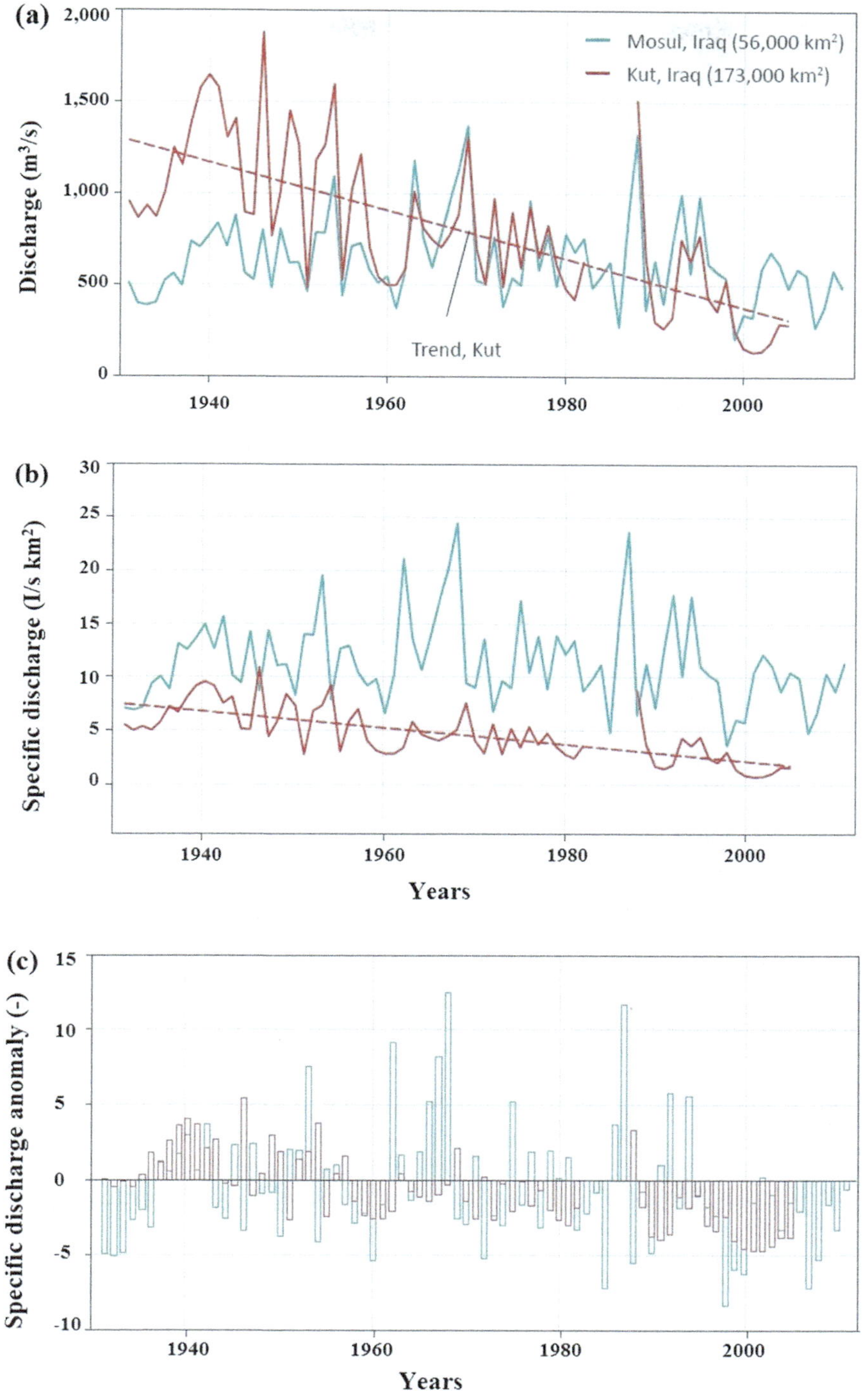

Fig. 8.8 The selected river flow gauging stations located on the Euphrates river

and Syria water controlling"; yet, there are several deficiencies and inadequate project planning and management over these two streams. Moreover, the past successive administrations did not consider the temporal development of the countries within the region and try to ensure water and food security in Iraqi based on the current developments. Additionally, this vital sector has suffered great negligence during the eighties and nineties of the last century. Most of the implemented water projects during this period have already been planned before. Many water hydraulic facilities have been established in bordering countries during this period; however, the Iraq region demonstrated no investment in water resources development. The successive Iraqi governments have had to deal with the issue of water security, and there has been a state of no binding international laws to regulate water resources between countries. Based on the above hypothesis and observations, there are several proposals suggested as follows:

1. Euphrates feeding channel: Based on the water projects comprehending over Iraq region, the southern part of Iraq has little or no water resources management projects (see Fig. 8.1). Owing to the high length of the Tigris river and its water resources availability over the Euphrates river, the Tigris river can provide the Euphrates river with sufficient amount of water through linking the two rivers by feeding water channel. The construction location of this channel best at the nearest coordinates where the two rivers are closest to each other. Particularly, south of the capital Baghdad at Kut city and north of the Musaib city on the Euphrates. The significance of this project can be highlighted as follows:

    (a) This channel will reinforce the expected shortage in the Euphrates river flow and especially during the drought periods and throughout the year.
    (b) Iraq will be able to benefit from the Tigris river in favor of the Euphrates river and greatly reduce its dependence on Turkey and Syria. In addition, it will guarantee Iraq's ability to negotiate politically in the future.
    (c) This project is extremely important for restoring water for the destructed marshes located in the southern part of Iraq.
    (d) Also, several side channels can be initiated on both sides of the proposed channel, which emphasizes the agriculture projects.

    The advantages that facilitate the implementation of this project are:

    (a) The length of this proposed channel does not exceed 45 kilometers.
    (b) The channel field location has excellent topography that assist the drilling and implementation.
    (c) Top all of that, the project can be implemented locally with manageable time and cost.

2. Shatt al-Arab's dam: The second proposal is constructing a dam at the joint of the Tigris and Euphrates rivers. The ideal benefit of this dam is to regulate the water passing to Shatt al-Arab and prevent the reverse flow from the Arabian Gulf at the northern city of Basra. In this way, water can be conserved from the salinity problems.

3. As a fact, Turkish government required to run the hydropower system for electric generation. As a result, the released water from this process required careful attention by Iraq government in which to store this water using small dams optimally designed and especially during winter season.
4. In addition to the stated strategic solutions, there are additional plans that can enhance Iraq's water capacity, such as increasing the storage capacity in the middle and southern regions and stressing on the development and management of more efficient irrigation and drainage systems with the use of modern technology.
5. Initiating desalination systems for water salinity. This is particularly with the south part of the Iraq region "the cities nearby the Arabian Gulf sea." Also, this plan can be established by taking advantages of the abundant gas and oil. Such a valuable source "gas" can be utilized for the desalination process without requiring energy expenditure. This plan can also be extended to include a pipeline to transfer seawater or desalinated water to the other parts of the middle of Iraq to meet water security requirements in the events of severe droughts in the region. The pipeline can also have the dual purpose of transferring freshwater from the middle to the south of Iraq during the times when plenty of freshwater is available in the region.
6. Last but not least, there must be a national plan and strategy to address the problem of water scarcity. The west desert of Iraq accommodates over 100 million cubic meters from the rainfall water. This requires a topographic study of the area and the possibility of constructing small dams and reservoirs for water harvesting as clearly reported by (Sayl et al. 2016).

References

Al-Ansari NA (2013) Management of water resources in Iraq: perspectives and prognoses. Engineering [Internet]. 5(8):667–684. Available from: http://www.scirp.org/journal/PaperInformation.aspx?PaperID=35541

Al-Ansari N, Knutsson S (2011) Toward prudent management of water resources in Iraq. J Adv Sci Eng Res 1:53–67

Al-Heety E (2011) a. Physico-Chemical Assessment of Euphrates River between Heet and Ramadi Cities, Iraq. J Water Resour Prot 3(November):812–823

Allan T, Kliot N. (1994) Water resources and conflict in the middle east. Geogr J [Internet]. 160 (3):337. Available from: http://www.jstor.org/stable/3059618?origin=crossref

Alnasrawi A (2001) Iraq: Economic sanctions and consequences, 1990–2000. Third World Q 22 (2):205–218

Antoniou G, Kathijotes N, Spyridakis DS, Angelakis AN (2014) Historical development of technologies for water resources management and rainwater harvesting in the Hellenic civilizations. Int Jo Water Res Develop 30(4):680–693

Aqrawi AAM (2001) Stratigraphic signatures of climatic change during the Holocene evolution of the Tigris-Euphrates delta, lower Mesopotamia. Glob Planet Change 28(1–2):267–283

Beaumont P (1998) Restructuring of water usage in the Tigris-Euphrates Basin: The impact of modern water management policies. Coppock, J and Miller, JA Transformation of Middle

Eastern natural environments: Legacies and lessons [Internet]. 113:168–186. Available from: http://environment.research.yale.edu/documents/downloads/0-9/103beaumont.pdf

Chenoweth J, Hadjinicolaou P, Bruggeman A, Lelieveld J, Levin Z, Lange MA, et al. (2011) Impact of climate change on the water resources of the eastern Mediterranean and Middle East region: Modeled 21st century changes and implications. Water Resour Res [Internet]. 47(6):W06506. Available from: https://doi.org/10.1029/2010WR010269

Cullen HM, DeMenocal PB (2000) North Atlantic influence on Tigris-Euphrates streamflow. Int J Climatol 20(8):853–863

Dezfuli AK, Zaitchik BF, Badr HS, Evans J, Peters-Lidard CD (2017) The role of low-level, terrain-induced jets in Rainfall variability in tigris–euphrates headwaters. J Hydrometeorol [Internet]. 18(3):819–835. Available from: http://journals.ametsoc.org/doi/10.1175/JHM-D-16-0165.1

Droogers P, Immerzeel WW, Terink W, Hoogeveen J, Bierkens MFP, Van Beek LPH et al (2012) Water resources trends in Middle East and North Africa towards 2050. Hydrol Earth Syst Sci 16 (9):3101–3114

Freeman K (2001) Water wars? Inequalities in the Tigris-Euphrates river basin. Geopolitics. 6 (2):127–140

Hassan R, Al-Ansari N, Ali AA, Ali SS, Knutsson S (2017) Bathymetry and siltation rate for Dokan Reservoir, Iraq. Lakes and Reservoirs: Res Manage 22(2):179–189

Issa I, Al-Ansari N (2014) Expected future of water resources within tigris-euphrates rivers Basin, Iraq. . . . of Water Resource and . . . [Internet]. (April):421–32. Available from: http://file.scirp.org/Html/2-9402125_45022.htm

Kibaroglu A, Maden TE (2014) An analysis of the causes of water crisis in the Euphrates-Tigris river basin. J Environ Stud Sci 4(4):347–353

Sayl KN, Muhammad NS, Yaseen ZM, El-shafie A (2016) Estimation the physical variables of rainwater harvesting system using integrated GIS-based remote sensing approach. Water Res Manage [Internet]. 30(9):3299–3313. Available from: https://doi.org/10.1007/s11269-016-1350-6

Sowers J, Vengosh A, Weinthal E (2011) Climate change, water resources, and the politics of adaptation in the Middle East and North Africa. Clim Chang 104(3–4):599–627

Swearingen WD (1993) The euphrates river and the southeast anatolia development project – Kolars, Jf, Mitchell,Wa. Geographical Rev 83(1):94–95

Unver IHO (1997) Southeastern Anatolia Project (GAP). Int J Water Res Devel 13(4):453–484

Van De Mieroop M (1997) The ancient mesopotamian city. Oxford University Press, Oxford

Yürekli K (2015) Impact of climate variability on precipitation in the Upper Euphrates-Tigris Rivers Basin of Southeast Turkey. Atmospheric Res 154:25–38

Zagonari F, Rossi C (2014) A negotiation support system for disputes between Iraq and Turkey over the Tigris-Euphrates basin. J Hydrol 514:65–84

Chapter 9
Streamflow Alteration Impacts with Particular Reference to the Lower Zab River, Tributary of the Tigris River

Ruqayah Mohammed and Miklas Scholz

Abstract Climate change and drought episode impact integrated with anthropogenic pressure have become an increasing concern for water resource managers, particularly in arid and semi-arid climatic zones. This chapter presents a comprehensive methodology to predict the prospective impact of such changes at a basin scale. The Lower Zab River Basin, northern Iraq, has been selected as a representative case study. The methodology has been achieved through estimation of drought severity and climate change impact during the human intervention periods to separate the influence of climatic abnormality and measure the hydrologic deviations as a result of streamflow regulation configurations. The Indicators of Hydrologic Alteration (IHA) method has been applied to quantify the hydrological alterations of numerous hydrological characteristics. The Hydrologiska Byråns Vattenbalansavdelning (The Water Balance Department of the Hydrological Bureau) hydrologic model was used to define the boundary conditions for the reservoir capacity yield model, which was applied to derive the reservoir capacity-yield-reliability relationships, comprising daily reservoir inflow from the basin with the size of 14,924 km^2 into a reservoir with the capacity of 6.80 Gm^3. Owing to the future precipitation reduction and potential evapotranspiration increase during the worst case scenario (−40% precipitation and +30% potential evapotranspiration), substantial reductions in the streamflow of between −56% and −58% are anticipated for the dry and wet seasons, respectively. Model simulations recommend that the

R. Mohammed
Civil Engineering Research Group, School of Computing, Science and Engineering, The University of Salford, Newton Building, Peel Park Campus, Salford, UK

M. Scholz (✉)
Civil Engineering Research Group, School of Computing, Science and Engineering, The University of Salford, Newton Building, Peel Park Campus, Salford, UK

Division of Water Resources Engineering, Department of Building and Environmental Technology, Faculty of Engineering, Lund University, Lund, Sweden

Department of Civil Engineering Science, School of Civil Engineering and the Built Environment, University of Johannesburg, Kingsway Campus, PO, Johannesburg, South Africa
e-mail: miklas.scholz@tvrl.lth.se

L. A. Jawad (ed.), *Tigris and Euphrates Rivers: Their Environment from Headwaters to Mouth*, Aquatic Ecology Series 11,
https://doi.org/10.1007/978-3-030-57570-0_9

reservoir reliability would generally decrease due to a decline in reservoir inflow. The study outcomes assist water resource managers and policymakers responsible for mitigating the effects of climate change.

Keywords Climate change · Anthropogenic intervention · Flow anomalies · Hydro-climatic change · Water resource management

9.1 Introduction

9.1.1 Background

Cyclical river flow component anomalies such as high and low flows resulting from natural (i.e. climate change) and man-made (i.e. river damming) climatic conditions have produced enormous concerns expressed by hydrologists due to negative consequences on riverine ecosystems (Suen 2010; Doll and Zhang 2010; Al-Faraj and Scholz 2014; Mittal et al. 2016). Climate variability would intensify the water resource stress such as the alteration in river flow seasonality (Vicuna and Dracup 2007; Minville et al. 2009) as well as reservoir capacity and performance (reliability, resilience, and vulnerability). Such change is likely to lead to a negative effect on water resource management.

For estimation of the natural and altered hydrologic systems, the Nature Conservancy (2009) has developed a simple tool, which is known as the Indicator for Hydrologic Alteration (IHA). Through calculating the hydrologic characteristics for the pre-alteration and post-alteration periods, the IHA can be utilised to assess how the natural flow regime would be altered due to river regulation such as dam construction. Literature exposes that considerable research work has been undertaken to assess the hydrologic anomalies of natural flow regimes (Yan et al. 2010; Gao et al. 2013; Sun and Feng 2013; Jiang et al. 2014; Wang et al. 2016). Yet, most research has focussed on the streamflow alteration from an ecological point of view (Suen 2010; Doll and Zhang 2010; Kim et al. 2011; Lee et al. 2014; Mittal et al. 2014; Stagl and Hattermann 2016) and not from a hydrological one. In addition, they did not consider incorporating anthropogenic intervention, climate change, and drought event impact within a comprehensive basin study. Therefore, the prime aim of this chapter is to evaluate the hydrological alterations of the streamflow due to anthropogenic intervention (i.e. dam construction) incorporated with climate change and drought phenomena. Based on the authors' best knowledge, this is a novel effort at exploring the anthropogenic intervention, climate change, and drought severity effects on basin hydrologic features.

9.1.2 *Rationale, Aim, Objectives, and Importance*

The impacts of climate change and anthropogenic interventions on a basin streamflow are subject to different climatic zones. Consequently, they should be investigated at a local (e.g. basin or sub-basin) scale. Recently, due to universal climate variability and local human-induced stresses, many areas have been affected by regular floods and droughts. Thus, evaluating the impact of such changes is important for understanding the mechanisms of hydrological responses in a basin, regional water resource organisation, and both flood and drought protection. The Lower Zab River Basin has witnessed substantial alterations in the meteorological and hydrological variables and human activities over the last few decades. Accordingly, this basin can be considered as an excellent case study for evaluating the hydrological influence of climate variability and man-made activities.

The current research aims to investigate the impact of climate change and anthropogenic intervention on the alteration of basin streamflow in addition to addressing multiple aspects of basin hydrology. The corresponding objectives are to (1) investigate the spatio-temporal characteristics of the meteorological data at monthly and annual timescales; (2) analyse the response of the potential evapotranspiration to the variations of mean air temperature; (3) assess temporal hydrological alteration of the basin related to the impact of anthropogenic intervention and climate change; (4) investigate the sensitivity of the contribution of the baseflow to the river discharge to the impact of anthropogenic and intervention climate change; and (5) develop the key reservoir Capacity-Yield-Reliability relationships for the basin.

9.2 Materials and Methods

9.2.1 *Illustrative Case Study Area*

The Lower (Lesser/Little) Zab River, which is the key stream of the Tigris River, is located near Erbil, north-eastern part of Iraq. The river originates from the Zagros Mountains in Iran and can be found at latitudes 36 °50′ N–35 °20′ N and longitudes 43 °25′ E–45°50′ E (Mohammed and Scholz 2016); see Fig. 9.1. The Lower Zab River flows about 370 km south-east and south-west through north-western Iran and northern Iraq before joining the Tigris River near Fateh city, which is located about 220 km north of Baghdad city (Mohammed et al. 2017a), with a total length of approximately 302 km, and about 80 km south of the Greater Zab River. The mean annual storage of the river at Dokan and Altun Kupri-Goma, which are the key flow stations, is about 6 billion cubic metres (BCM) and 7.8 BCM in this order (Mohammed et al. 2017b); Fig. 9.1. Dokan is the main dam that has been constructed within the upstream portion of the basin. The main function of the dam is to control the discharge of the river, store water for irrigation purposes, and provide hydroelectric power.

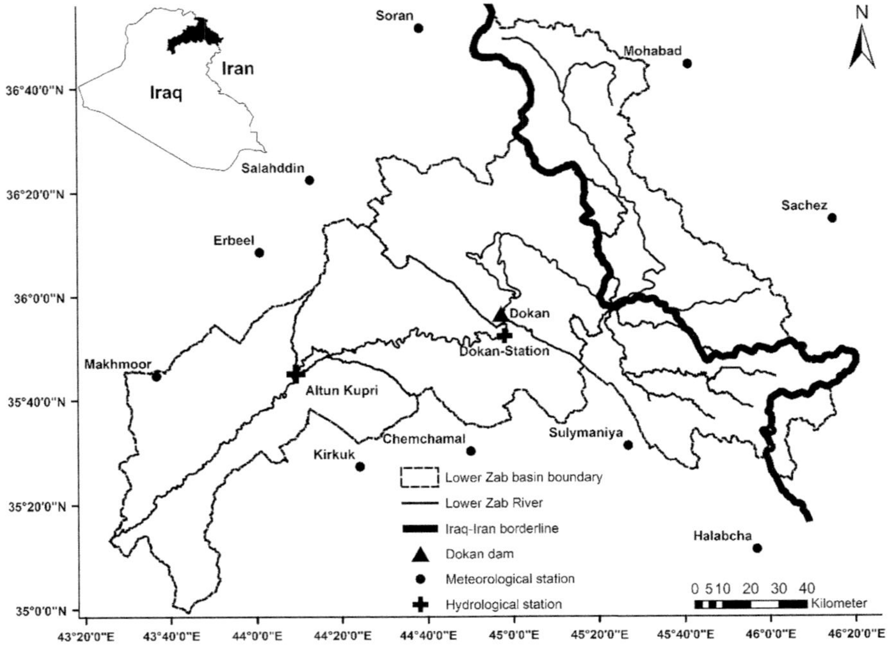

Fig. 9.1 Overview of the Lower Zab River Basin, located between Iraq and Iran, showing the hydro-climatologic station locations

There are a number of tributaries contributing to the river discharge such as the Banah and Qazlaga. The catchment area of the Lower Zab River Basin (LZRB) and its tributaries is approximately 20,605 km^2 with nearly 76% located in Iraq. Annual precipitation along the river decreases from $\geq$1000 mm in the Iranian Zagros to $\leq$200 mm at the confluence with the River Tigris. Mean temperatures follow the same gradient. The mountain valleys are usually subjected to colder winters than the corresponding foothill areas. However, summers in the latter are usually hotter (NOAA 2009).

The LZRB covers a range of relatively large watersheds and a wide range of climatic and hydrologic conditions. The upstream and downstream developments vary widely, which suggests a considerably wide range of uncertainties in climate change impacts on water resource availability that can be assessed. The magnitude of problems plaguing the LZRB has a great deal in common with other transboundary watersheds such as the Rhine, the Volta, and the Senegal, where the problems of shared water resource utilisation in a sustainable manner are expected to exacerbate under the collective impact of uncertainty surrounding climate change. Mohammed et al. (2017b) reported a severe drought over the basin during the water year 2007/2008 due to approximately 80% reduction in precipitation. The extended drought conditions seem to have had a disastrous impact on the lives of the people in the country. The limited access to water has led among others to the erosion of livelihoods and decrease in crop production (Mohammed and Scholz 2017b).

9.2.2 Data Availability, Collection, and Analysis

Daily meteorological data, which include precipitation as well as minimum and maximum air temperature, from ten stations with elevations extending from 319 m to 1536 m were gathered for the period from 1979/1980 to 2012/2013 (Table 9.1 and Fig. 9.1). Daily flow data at two key hydrometric stations, which are the Dokan (Latitude 35 ° 53′ 00″ N; Longitude 44 ° 58′ 00″ E) and Altun Kupri-Goma Zerdela (Latitude 35 ° 45′ 41″ N; Longitude 44 ° 08′ 52″ E) stations, were assessed (Fig. 9.1). The drainage area for the former is estimated to be 12,096 km^2 and data are available between 1931 and 2013, whereas the corresponding values for the latter are 8509 km^2 and between 1931 and 1993, respectively (Mohammed and Scholz 2017c).

Iraqi boundaries and the LZRB shape files have been downloaded from the Global Administrative Areas (GADM 2012) and the Global and Land Cover Facility (GLCF 2015) databases, respectively. GADM is a spatial database of the location of the world's administrative areas (or administrative boundaries), for use in GIS and similar software. GADM describes where these administrative areas are located, and for each area, it provides some attributes, such as the name and variant names, whereas GLCF is a centre for land cover discipline with a focus on research utilising remotely sensed satellite data and products to access land cover change for local to global systems in this order.

For the analysis and visualisation of the collected datasets, this study utilised many tools. ArcGIS 10.3 software has been applied for the hydro-climatic station

Table 9.1 Statistical properties of the meteorological variables after applying a non-parametric test for the decadal change

		T^d_m (°C)		P^e (mm)		PET^f (mm)	
Sub- basin	Station name	M–K[c]	*p*-value	M–K[c]	*p*-value	M–K[c]	*p*-value
US[a]	Sulymanya	0.358**	<0.01	–0.301**	<0.01	0.201	0.09
	Halabcha	0.572**	<0.01	–0.522**	<0.01	0.316**	<0.01
	Sachez	0.079	0.50	–0.328**	0.01	0.193	0.10
	Mahabad	0.603**	<0.01	–0.573**	<0.01	0.525**	<0.01
	Salahddin	0.452**	<0.01	–0.472**	<0.01	0.220	0.06
	Soran	0.380**	<0.01	–0.426**	<0.01	0.241*	0.05
DS[b]	Kirkuk	0.422**	<0.01	–0.553**	<0.01	0.420**	<0.01
	Makhmoor	0.462**	<0.01	–0.536**	<0.01	0.243	0.04
	Erbeel	0.351**	<0.01	–0.371**	<0.01	0.203	0.09
	Chemchamal	0.345**	<0.01	–0.412**	<0.01	0.139	0.24

[a]Upstream; [b]Downstream; [c]Mann–Kendall non-parametric test; [d]Mean air temperature; [e]Precipitation; and [f]Potential evapotranspiration

Note: Negative (–) and positive values indicate the decreasing and increasing trends, respectively

**Correlation is significant at the 0.01 level (2-tailed); and

*Correlation is significant at the 0.05 level (2-tailed)

site projections, Thiessen network calculations, and delineations of the basin. The statistical assessments for the daily hydro-climatic data, with trend test, monthly and annual value calculations, adjustments, and data gap filling were achieved by the Statistical Program for Social Sciences (SPSS 23). A one-way analysis of variance (ANOVA) test was performed using Daniel's XL Toolbox, which is an open-source Add-In for the common software tool Excel. This tool was also used to convert an excel graph to a picture with the required dimensions and resolutions. For the RDI index estimation, DrinC 1.5.73 software was applied.

The estimation of PET (mm) was undertaken according to the Food and Agriculture Organization Penman-Monteith standard procedure (Allen et al. 1998) applying the reference evapotranspiration ET_o 3.2 calculator. The Indicators of Hydrologic Alteration software (IHA 7.1) (The Nature Conservancy 2009) was utilised to evaluate the natural flow regime alteration that resulted from climate change linked to human-induced activities. For flow duration curve estimation and baseflow separation, HydroOffice (2015) (BFI + 3.0 and FDC) was used (https://hydrooffice.org/Downloads?Items=Software); this was achieved by applying the methodology that was developed by Mohammed and Scholz (2016).

A Fortran90 programme was developed and used for the application of the reservoir capacity yield (RCY) model. RS MINERVE 2.5 (2016) is a free downloaded software for the simulation of free surface runoff formation and propagation (Foehn et al. 2016). This software was used to run the Hydrologiska Byråns Vattenbalansavdelning (HBV) rainfall-runoff model (https://www.crealp.ch/down/rsm/install2/archives.html).Todetect the trend in the long-term hydro-climatic data, the distribution free Mann-Kendall (M-K) method was applied (Tabari and Taalaee 2011; Robaa and AL-Barazanji 2013). The method has the advantage of not assuming any data distribution and is similarly powerful to its parametric competitors (Mohammed et al. 2017a). Table 9.1 shows the M-K test results for the key meteorological variables.

9.2.3 Drought Identification

Drought is considered one of the leading water-related hazards (Giannikopoulou et al. 2014). For identifying and quantifying drought, there are many recommended approaches. Drought indices are considered as one of the most popular methods (Vangelis et al. 2013; Giannikopoulou et al. 2014; Mohammed and Scholz 2017a). A large number of meteorological drought indices with different intricacy have been utilised in various climatic conditions (Mohammed and Scholz 2017a). Latest drought trend studies (Sheffield et al. 2012; Vicente-Serrano et al. 2014) and drought scenarios under possible climate change predictions (e.g. Cook et al. 2014) depend on indices that take into consideration evapotranspiration and precipitation such as the reconnaissance drought index (RDI), which is considered in this study.

The RDI index is based on the ratio between two aggregated quantities of precipitation and potential evapotranspiration (Tsakiris and Vangelis 2005). The index can be expressed in alpha ($RDI_{\alpha k}$), normalised (RDI_n), and standard (RDI_{st}) forms. The RDI_{st} can be applied for drought severity calculations, whereas $RDI_{\alpha k}$ can be used as an aridity index. A positive RDI_{st} indicates a wet period, and a negative value is indicative of a dry period compared to the regional natural conditions. The $RDI_{\alpha k}$ is usually estimated using Eq. (9.1):

$$RDI_{\alpha_0^i} = \frac{\sum_{j=1}^{12} P_{ij}}{\sum_{j=1}^{12} PET_{ij}} \quad i = 1 \text{ to } N \text{ and } j = 1 \text{ to } 12 \tag{9.1}$$

where P_{ij} and PET_{ij} refer to the precipitation and the potential evapotranspiration of the jth month and the ith hydrological year, which in Iraq starts in October, and N is the total number of hydrological years for the equivalent climate.

The values of $RDI_{\alpha k}$ match both the lognormal and the gamma distributions for many locations at various timescales (Tigkas 2008). By utilising the gamma distribution, RDI_{st} can be estimated using Eq. (9.2):

$$RDI_{st}^{i} = \frac{y_i - \bar{y}}{\hat{\sigma}_y} \tag{9.2}$$

where y^i is $\ln(\alpha_{ki})$, $\bar{y}$ is its mathematical mean, and σ_y is the associated standard deviation.

Equation (9.3) is used for RDI_{st} evaluation concerning the gamma distribution (Tigkas 2008),

$$g\,(x) = \frac{1}{\beta^{\gamma} \times \Gamma(\gamma)} x^{\gamma-1} e^{\frac{x}{\beta}} \text{ for } x > 0, \tag{9.3}$$

where γ, β, and $\Gamma(\gamma)$ are the shape, the scale parameters, and the gamma function, respectively. The gamma probability distribution parameters γ and β can be estimated using Eqs. (9.4, 9.5 and 9.6) as follows:

$$\gamma = \frac{1}{4A}\left(1 + \sqrt{1 + \frac{4A}{3}}\right) \tag{9.4}$$

$$\beta = \frac{\bar{x}}{\gamma} \tag{9.5}$$

$$A = \ln\,(\bar{x}) - \frac{\sum \ln\,(x)}{N}. \tag{9.6}$$

When the cumulative precipitation datasets for the selected reference period contain zeros, the gamma function cannot be defined for $x = 0$. Therefore, a composite cumulative probability function, Eq. (9.7), might be utilised.

$$H(x) = q + (1 - q) \times G(x) \tag{9.7}$$

where q is the likelihood of zero precipitation and $G(x)$ is the gamma distribution cumulative probability. The probability of zero precipitation (q) can be computed by m/N, if m is zero in the αk time series. The gamma distribution cumulative probability $G(x)$ is replaced by the cumulative probability $H(x)$.

A positive value of RDI_{st} is related to a wet period. On the other hand, a negative value is indicative of a dry period in comparison to the natural conditions of the region. Drought severity phenomena increase when RDI_{st} values are minimal. The severity of drought can be classified into (extremely, severely, and moderately) dry, near normal, normal (moderately, very, and extremely), and wet classes based on the corresponding boundary RDI_{st} value ranges ≤ -2.00, -1.5 to -1.99, -1 to -1.49, 0.00 to -0.99, 0.99 to 0.00, 1.49 to 1.00, 1.99 to 1.5, and $\geq$ 2.00, respectively (Tigkas et al. 2012; Vangelis et al. 2013; Mohammed and Scholz 2017a).

9.2.4 *Hydrograph Analysis*

Particular activities that can affect dry weather may contain river flow regulations, where river discharge is controlled through infrastructure elements such as dams, weirs, or locks. Discharges from surface water storage structures for downstream stakeholders can make up the bulk of a river during dry times, reducing the groundwater contribution, which in turn decreases the dry weather index. Many researchers argue that dry weather assessment should be performed in unregulated reaches or, at least, the regulated basin area should be $\leq$10% of the basin area of the river gauge (Al-Faraj and Scholz 2014). This section explores the effects of weather variability and anthropogenic interventions on river flow alteration. The sensitivity of groundwater to such changes is subsequently assessed.

Mohammed and Scholz (2016) developed a generic methodology for BF separation from the daily average flow to the LZRB. They utilised two digital filtering algorithms (DFAs), which are Eckhardt (2005) and Chapman (1999).

The Eckhardt (2005) approach is used to achieve low pass filtering of the flow hydrograph to separate the BF using Eq. (9.8),

$$BF_t = \frac{(1 - BFI_{max}) \times \alpha \times BF_{t-1} + (1 - \alpha) \times BFI_{max} \times TF_t}{1 - \alpha \times BFI_{max}} \tag{9.8}$$

where BF (m^3/s) is the isolated baseflow, BFI is the baseflow index, which is the long-term ratio of BF to the total flow (TF), TF (m^3/s) is the total flow, α is the filter parameter, and t is the time step for $BF_t \leq TF_t$.

Chapman (1999) discussed the second recursive DFA that can be estimated by Eq. (9.9),

$$DF_t = \frac{3 \times \alpha^{-1}}{3 - \alpha} \times TF_{t-1} + \frac{2}{3 - \alpha} \times (TF_t - TF_{t-1}) \tag{9.9}$$

where DF (m^3/s) is the direct runoff, TF (m^3/s) is the total flow, α is the filter parameter, and t is known as the time step.

Two parameters are needed for the Eckhardt recursive method identification (Eckhardt 2005): (1) The recession constant stemming from the recession curve of the hydrograph valuation and (2) The BFI_{max} that cannot be measured but enhanced based on the results of other methods. As a starting point, BFI_{max} was taken to be 0.25 (Eckhardt 2005).

9.2.5 *Reservoir Capacity-Yield Simulation and Performance Indices*

As a first step for reservoir capacity-yield (RCY) evaluation, the Hydrologiska Byråns Vattenbalansavdelning (HBV) rainfall-runoff model was calibrated depending on the recorded dataset of the baseline period. The HBV model is an example of a semi-distributed conceptual rainfall-runoff model, simulating daily discharge depending on daily rainfall, temperature, and potential evapotranspiration as input. Subsequently, the RCY simulation was performed applying Eq. (9.10) (McMahon and Adeloye 2005),

$$S_{t+1} = S_t + R_t - D_t - \Delta E_t - L_t, \tag{9.10}$$

$$\text{Subject to } 0 \le S_{t+1} \le C_a \tag{9.11}$$

where S_{t+1} and S_t are reservoir storage volumes at the end and the beginning of a time period t, respectively; R_t is the inflow over period t; D_t is the actual water yield over period t; ΔE_t is the reservoir net evaporation loss within period t; L_t represents other losses; and C_a is the reservoir active capacity.

During the simulation procedures, the reservoir S_o is typically assumed to be full (McMahon and Mein 1978, 1986), and the downstream demand is usually considered as a specific fraction of the mean annual inflow. The usual period is 1 month, but any other period can be used. McMahon and Mein (1978) provided a complete method to calculate the reservoir operational probability of failure, which can be summarised as follows: (a) Assume that the reservoir is initially full ($S_o = C_a$); (b) Apply Eq. (9.11) month by month on the historical or generated monthly flows; (c) Plot (S_{t+1}) against time on a monthly timescale; and (d) Compute the OPOF by using Eq. (9.13).

The storage size estimated through behaviour analysis varies little with the starting month. In behaviour analysis, the vulnerability of the reservoir is estimated by ignoring the constraint on Eq. (9.12) and then choosing the maximum negative value of (S_{t+1}) during the analysis period, whereas the resilience is estimated by computing the maximum consecutive number of empty months during the analysis (Moy et al. 1986).

Following the simulation, three appropriate performance measures were assessed (McMahon and Adeloye 2005; McMahon et al. 2006), which are as follows:

1. The time-based reliability (R_e) can be defined as the percentage of the entire period under investigation during which a reservoir is capable of providing the full demand without any deficiencies, as indicated in Eq. (9.12),

$$R_e = 1 - \mathrm{OPOF} \tag{9.12}$$

   where R_e is the reservoir reliability(%) and OPOF(%) is the reservoir operational probability of failure, which is defined as the ratio of time units during which the reservoir is effectively empty to the total number of time units applied in the analysis, as shown in Eq. (9.13),

$$\mathrm{OPOF} = \frac{N_e}{N} \tag{9.13}$$

   where OPOF is the reservoir operational probability of failure (%), N_e is the time unit number during which the reservoir is empty, and N is the total number of time units in the streamflow time series.

   There is no limitation on the OPOF, but many studies consider 5% to be an acceptable limitation (McMahon and Adeloye 2005).
2. The resilience φ describes the reservoir's ability to recover from failure. Moy et al. (1986) defined it as the maximum number of consecutive periods of shortage that occur prior to recovery and can be expressed based on Eq. (9.14) as discussed previously (Fowler et al. 2003; Park and Kim 2014),

$$\varphi = \sum_{t=1}^{N} Y_t \tag{9.14}$$

   where Y_t is the number of the continuous shortage indicator, $Y_t = 1$ if there is a shortage in period (t), and $Y_t = 0$ otherwise, and $t = 1, 2, \ldots, N$, which is the total number of time units in the stream flow time series sequence of failure periods.
3. The vulnerability υ is a criterion to determine the significance of failure. Mathematically, it is expressed by Eq. (9.15) as shown previously (Fowler et al. 2003; Park and Kim 2014),

$$\upsilon = \max\left(Df_t\right) \tag{9.15}$$

   where υ is the system vulnerability and Df_t is a deficit at time t (McMahon and Adeloye 2005; McMahon et al. 2006).

9.2.6 Methodology Application

Figure 9.2 shows the methodology and visualises how the research objectives can be integrated. The proposed method can be summarised by the following six steps:

1. Hydro-climatic dataset analysis: The normality of hydro-climatic datasets was investigated using the Kolmogorov–Smirnov analysis as a first step before conducting change tests using statistical techniques. Depending on the results

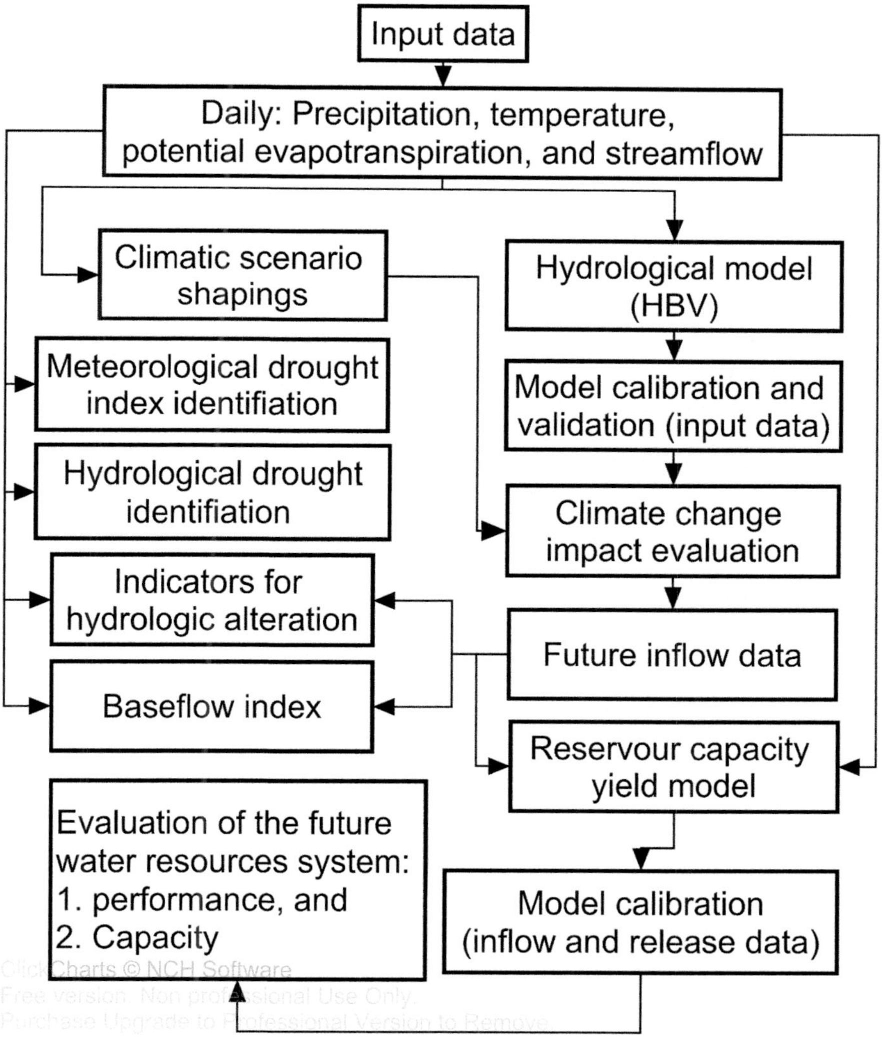

Fig. 9.2 The proposed methodology for assessing the impacts of human-induced climate change and drought events on the groundwater contribution to river flow

of these tests, most hydro-climatic data series applied in this research do not follow a normal distribution at a significance level p of 0.05. Regarding the non-normal distribution attributes of datasets, the widespread distribution-free non-parametric technique Mann–Kendall (M–K) analysis was applied to identify the variations in streamflow, precipitation, air temperature, and potential evapotranspiration time series in the LZRB. The M-K test was applied to identify sudden changes in the average level. Before applying non-parametric analysis, the influence of serial correlation in the dataset was eliminated. This research applied the trend-free pre-whitening (TFPW) method at 5% significant level (Oguntunde et al. 2011; Shahidian et al. 2012).

2. Hydrological year type's identification: To identify the wet and dry hydrological years, Yoo (2006) proposed that wet, dry, and normal hydrological years are defined as periods, which can be described by Eqs. (9.16, 9.17, and 9.18), respectively,

$$P_{av} > P_{av} + 0.75 \times \mathrm{SD} \tag{9.16}$$

$$P_{av} < P_{av} - 0.75 \times \mathrm{SD} \tag{9.17}$$

$$P_{av} - 0.75 \times \mathrm{SD} \leq P_{av} \geq P_{av} + 0.75 \times \mathrm{SD} \tag{9.18}$$

where P_{av} is the basin average precipitation and SD is the standard deviation.

3. Based on the approach by Mohammed and Scholz (2016), the outcomes of the FDC study have been combined with the results from Eq. (9.8) to gain the Eckhardt parameter α value after considering $\mathrm{BFI}_{max} = 0.25$ (Eckhardt, 2005) for perennial rivers with predominantly porous aquifers as a starting point. First of all, the long-term average annual fraction of TF from BF was estimated after obtaining the Q_{90} and Q_{50} numbers by applying the FDC method, connecting Eq. (9.8) with FDC (Al-Faraj and Scholz 2014). Considering $\alpha = 0.925$ as an initial value (Arnold and Allen 1999; Smakhtin 2001), the daily flow is filtered for various numbers of filter Eckhardt parameter α until the BFI is equivalent to the Q_{90}/Q_{50} ratio. Several BF time series were gained by applying the filtered Eckhardt parameter α value.
4. Climatic scenario: The delta perturbations in P (dP) of 0–40% (2% step) and PET perturbations (dPET) of 0–30% (2% step) were used for the streamflow alteration analysis. Twelve years (1988–2000), which are characterised by an average value of RDI_{st} close to zero, were applied for running the climate change scenarios. Although the delta perturbation method does not accommodate future alterations in the probability distribution of climatic characteristics and seasonality (and therefore the streamflow), it is, however, an effective technique in detecting tipping points at which a water storage structure such as a reservoir is expected to fail disastrously in providing water demand.
5. Streamflow simulation: The predicted streamflow time series by applying the HBV model was used as input to the RCY model. The corresponding outputs of the model are evaluated using the indicators of reliability, resilience, and vulnerability, with reference to the imposed demands.

6. Reservoir simulation: The RCY model has been utilised to develop the capacity-yield-reliability relationship. The first relation linked yield (%) with OPOF (%), and the second linked capacity (106 × m^3) with yield (%). These relations can be applied to test various adaptation strategies, whether they are structural or non-structural, against the range of different future scenarios to select the most effective adaptation measures.

9.3 Results and Discussion

9.3.1 Long-Term Hydro-Climatic Data Changes

Long-term trends in hydrological processes are potentially influenced by changing climate and anthropogenic intervention (Robaa and Al-Barazanji 2013; Shahidian et al. 2012; Mohammed et al. 2017a). Investigating such trends might support the identification of the starting points of the anthropogenic intervention. Yearly mean air temperature, precipitation, potential evapotranspiration, and streamflow data were analysed applying the M–K test to detect long-term trends for the period between 1979 and 2013. During the last 35 analysed years, the whole LZRB displayed a rising trend of mean air temperature with a maximum value of +0.67 °C for a decade, while a declining precipitation trend (Fig. 9.3) with a maximum decrease of 151 mm for a decade was noted. The LZRB yearly precipitation is around 720 mm. The maximum precipitation (1222 mm) was recorded for 1987/1988, while the corresponding minimum (250 mm) was assigned to 2007/2008 (Fig. 9.3).

The mean annual precipitation changed spatially from 56 mm at Kirkuk station to 1369 mm at Sulaymaniyah station. The upper basin had higher precipitation values than the lower one. An evident trend of air temperature increase during the last half century led to a significant increase in the potential evapotranspiration for the entire

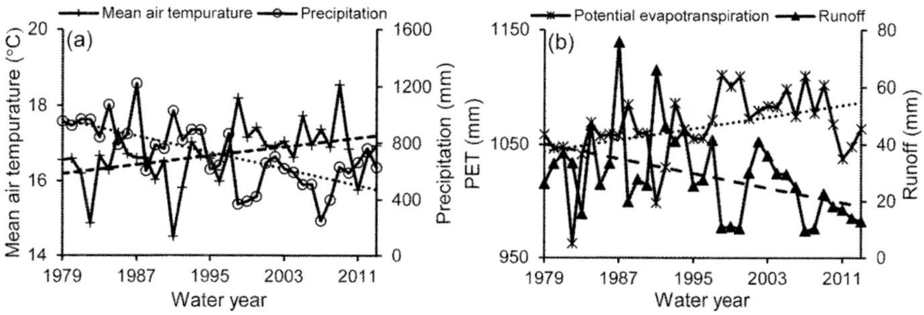

Fig. 9.3 Long-term trends for the key meteorological variables: (**a**) mean air temperature and precipitation and (**b**) potential evapotranspiration (PET) and the surface runoff over the Lower Zab River Basin for the time span from 1979 to 2013

Table 9.2 Statistical properties of the meteorological variables after applying a statistical test for the decadal change

Station name	Mean air temperature (°C)		Precipitation (mm)		Potential evapotranspiration (mm)	
	M–K[a]	*p*-value	M–K[a]	*p*-value	M–K[a]	*p*-value
Kirkuk	0.422**	<0.01	–0.553**	<0.01	0.420**	<0.01
Chemchamal	0.345**	<0.01	–0.412**	<0.01	0.139	0.24
Sulymanya	0.358**	<0.01	–0.301**	<0.01	0.201	0.09
Halabcha	0.572**	<0.01	–0.522**	<0.01	0.316**	<0.01
Makhmoor	0.462**	<0.01	–0.536**	<0.01	0.243	0.04
Salahddin	0.452**	<0.01	–0.472**	<0.01	0.220	0.06
Erbeel	0.351**	<0.01	–0.371**	<0.01	0.203	0.09
Soran	0.380**	<0.01	–0.426**	<0.01	0.241*	0.05
Mahabad	0.603**	<0.01	–0.573**	<0.01	0.525**	<0.01
Sachez	0.079	0.50	–0.328**	0.01	0.193	0.10

[a]Mann–Kendall non-parametric test
Note: Negative (–) and positive values indicate the decreasing and increasing trends, respectively
**Correlation is significant at the 0.01 level (2-tailed); and
*Correlation is significant at the 0.05 level (2-tailed)

LZRB. Based on the trend analysis (Table 9.2), the increase in PET rate was 39 mm per decade. With an average value of 1065.3 mm, the computed potential evapotranspiration for the basin changed from 962 mm in 1982/1983 to 1110 mm in 2007/2008 (Fig. 9.3). The obtained results indicate that the climate in the studied region is getting warmer and drier. The annual precipitation decreased. The yearly average air temperature increased, and the annual runoff depth decreased. These findings are largely in agreement with previous studies (Fadfil 2011; Robaa and AL-Barazanji 2013).

9.3.2 Drought Identification and Classification

For the purpose of drought occurrence detection, RDI (Tsakiris and Vangelis 2005) was calculated using the available data for precipitation and the estimated values of potential evapotranspiration. The index is classified as a meteorological drought index, because it depends essentially on meteorological parameters observed for various meteorological stations. Figure 9.4 presents the RDIst values as well as long-term basin average precipitation. These results indicate that a non-uniform cyclic pattern of drought and wet periods was observed for the RZRB concerning the studied time span. Evident seasonal droughts were recorded for 5 years during the examined record, particularly for 1998/1999, 1999/2000, 2000/2001, 2007/2008, and 2008/2009 (average RDI values −1.84, −1.67, −1.45, −2.91, and −1.53, respectively), which were also reported by many earlier studies such as Fadfil (2011) and UNESCO (2014).

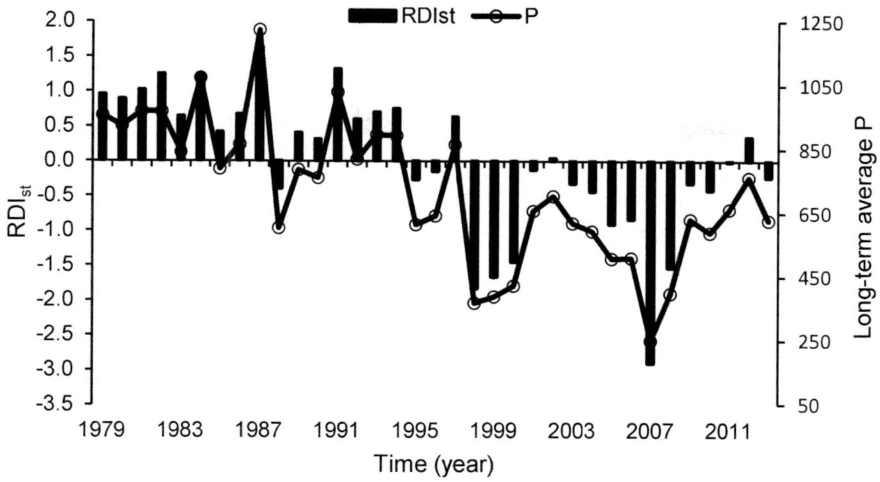

Fig. 9.4 The reconnaissance drought index (RDI) coupled with the long-term average precipitation (P) over the Lower Zab River Basin for the time span from 1979 to 2014

In general, drought usually happens at the beginning of the wet season, which is reflected by either a decline in the precipitation amount or a delay in the precipitation events. The RDI analysis shows that the drought severity over the studied basin has worsened significantly during the past 12 years. The RDI values computed between 1998 and 2011 show that critical droughts happened as the number of months with extended periods of precipitation shortage increased. In addition, the precipitation trend from the beginning of 2000 indicates that the region has been experiencing a reduction in rainfall and water resources in addition to an increase in drought periods.

9.3.3 Anthropogenic Intervention Impact

To assess the potential impact of the anthropogenic intervention on the streamflow and the contribution of groundwater to the baseflow of a basin, this study considered three time periods. Pre-alteration is the first period, covering the water years from 1931 to 1964. Post-damming is the second time period that spans from 1965 to 2013, whereas the hydrological years from 1931 to 2013 represent the third period, which is referred here as an integrated time period.

9.3.3.1 Streamflow

Figure 9.5 shows the anomalies of the median annual flow for the post-regulation condition (1965–2013) based on the long-term pre-alteration (1931–1964) median annual flow, which was estimated according to Eq. (9.19), basin average precipitation as well as wet and dry hydrological years,

$$\mathrm{LZR}_{\text{flow alteration}}\ (\%) = \frac{\mathrm{LZR}_{\text{altered flow}} - \mathrm{LZR}_{\text{unaltered flow}}}{\mathrm{LZR}_{\text{unaltered flow}}} \times 100 \qquad (9.19)$$

where $\mathrm{LZ}_{\mathrm{Raltered}}$ and $\mathrm{LZR}_{\mathrm{unaltered}}$ (m^3/s) are the median yearly changed discharge and the long-term median annual natural flow for the entirely unaltered condition in this order.

Figure 9.5 shows that the water year 1987 experienced a significant increase in the basin P_{av} of nearly 44% more than the normal year maximum threshold. The considerable increase in precipitation results in a substantial alteration in the streamflow to about 118%, with the corresponding annual mean flow volume of $2.31 \times 10^9\ m^3$ (Table 9.3). However, the opposite was recorded for the hydrological periods 1998–2001 and 2006–2008 (Table 9.3). These two periods experienced a steep decline in the basin average precipitation to nearly 40 and 60% in that order. The reduction in the basin average precipitation led to a substantial decline in the Lower Zab River flow by around 66, 77, and 79% with the corresponding yearly mean flow volumes of 0.35×10^9, 0.31×10^9, and $0.34 \times 10^9\ m^3$ for the period between 1998 and 2001. However, the hydrological period between 2006 and 2008 experienced approximately 52, 80, and 83% streamflow reductions with 0.76×10^9, 0.29×10^9, and $0.31 \times 10^9\ m^3$ annual mean flow volumes. Furthermore, between

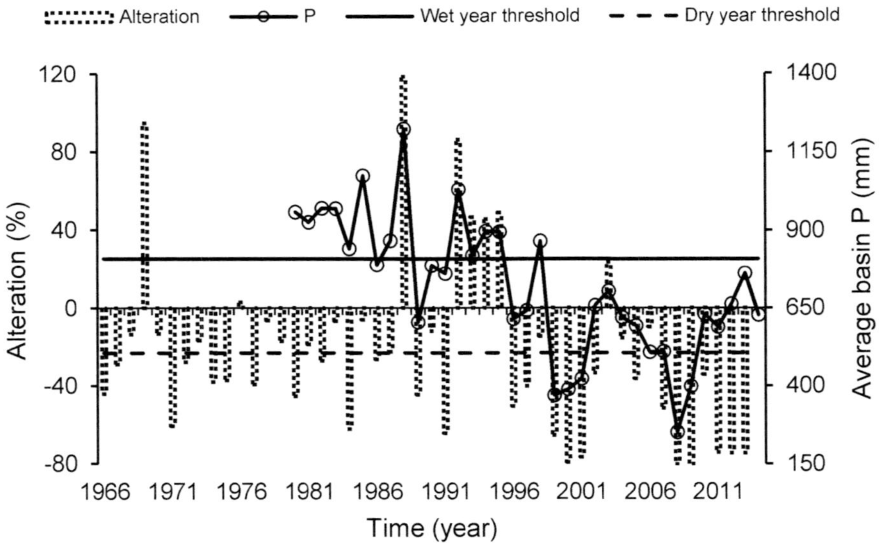

Fig. 9.5 Annual median anomaly for the period between 1966 and 2014 with both wet and dry year thresholds coupled with long-term average precipitation (P) over the study area for the time span between 1979 and 2014

Table 9.3 Annual river flows, storages, means of changed to unchanged storage ratios, and median anomalies between the periods from 1979–1980 to 1987–1988, and from 1998–1999 to 2008–2009

Hydrologic year	Flow (m^3/s)	Storage ($\times 10^9$ m^3)	Percentage change % Long-term mean yearly storage	Anomaly
Time-periods between 1979–1980 and 1987–1988				
1979–1980	148	4.68	0.79	–46
1980–1981	189	5.96	1.00	–19
1981–1882	210	6.61	1.11	–27
1982–1983	190	6.00	1.01	–7
1983–1984	88	2.79	0.47	–62
1984–1985	268	8.46	1.42	–6
1985–1986	148	4.66	0.78	–26
1986–1987	190	5.98	1.01	–23
1987–1988	436	13.75	2.31	+118
Time-periods between 1998–1999 and 2008–2009				
1998–1999	66	2.08	0.35	–66
1999–2000	65	2.04	0.34	–79
2000–2001	59	1.87	0.31	–77
2001–2002	171	5.40	0.91	–34
2002–2003	233	7.36	1.24	+24
2003–2004	205	6.46	1.09	–15
2004–2005	170	5.37	0.90	–37
2005–2006	170	5.38	0.90	–10
2006–2007	144	4.53	0.76	–52
2007–2008	54	1.70	0.29	–80

1991 and 2013, the normal flow regime has experienced a sharp reduction during which the flow alteration fluctuated between –75 and 86% with 0.31×10^9 and 1.24×10^9 m^3 maximum and minimum yearly average storage volumes in that order. Findings prove that climate change has negatively affected the Lower Zab River storage water availability.

The alteration in monthly median flows corresponding to the long-term natural flow regime and the three changed timescales and their related alteration are shown in Fig. 9.6. The alteration during the two altered time spans of the periods 1965–2013 and 1979–1987 could be considered relatively close to each other, particularly through the rainy months. The small variations can be assigned to the climate change impact, which was noticeable from the hydrological year 1998. The dramatic alteration during the non-rainy months was assigned to the effects of human-induced and climate change pressures in the upper part of the studied area, which in turn decreased the basin water storage availability.

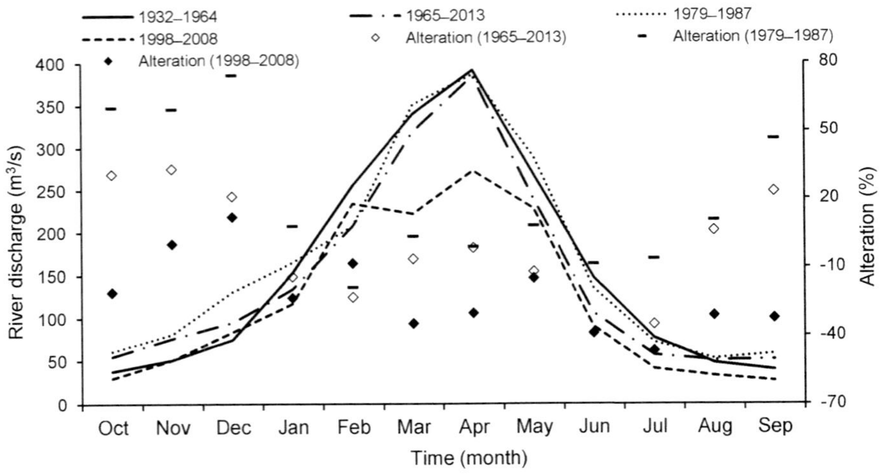

Fig. 9.6 The long-term median monthly flows of the pre-alteration and the three altered timescales coupled with their anomalies

9.3.3.2 Baseflow Contribution

Based on the outcome of the daily flow duration curve at Dokan location, the Q_{90} and Q_{50} discharges were 35 and 101, 31 and 100, and 33 and 100 m^3/s for the pre-alteration, post-alteration, and integrated time periods, respectively. Accordingly, the BFI, which is the Q_{90}/Q_{50} ratio, was nearly 35, 31, and 33%, respectively, representing volumes of water that the river might be gaining from sub-surface flow or other delayed shallow groundwater sources regarding the studied periods. This indicates that the BFI of the pre-alteration period was moderately greater than the values computed for the other two time periods. This difference can be attributed to the reduction in the groundwater involvement in the total Lower Zab River flow as a result of water being released from the reservoir during dry periods, which in turn reduces the BFI. Therefore, more consideration should be given to evaluate the aquifer characteristics and comprehend the aspects that might cause such alterations for improving the basin sub-surface resource management.

For the pre-alteration time period between 1931 and 1965 and taking into consideration Eckhardt's algorithm, outcomes reveal that the Eckhardt parameter α with a value of 0.982 produces a BFI that is the same as the one obtained from the FDC analysis. The yearly baseflow volumes over this period fluctuated between 0.632 billion cubic metres (bcm) recorded in 1958 and 2.826 bcm observed in 1953. The corresponding yearly baseflow depth varied between approximately 30 and 132 mm a year. The long-term yearly mean baseflow volume and the matching standard deviation were 1.478 and 0.512 bcm in this order (Fig. 9.7a and c). Whereas the annual baseflow magnitudes calculated by the Chapman algorithm over the same hydrological periods varied between 1.38 bcm recorded in 1958 and 5.71 bcm observed in 1953, the corresponding yearly base flow changed between

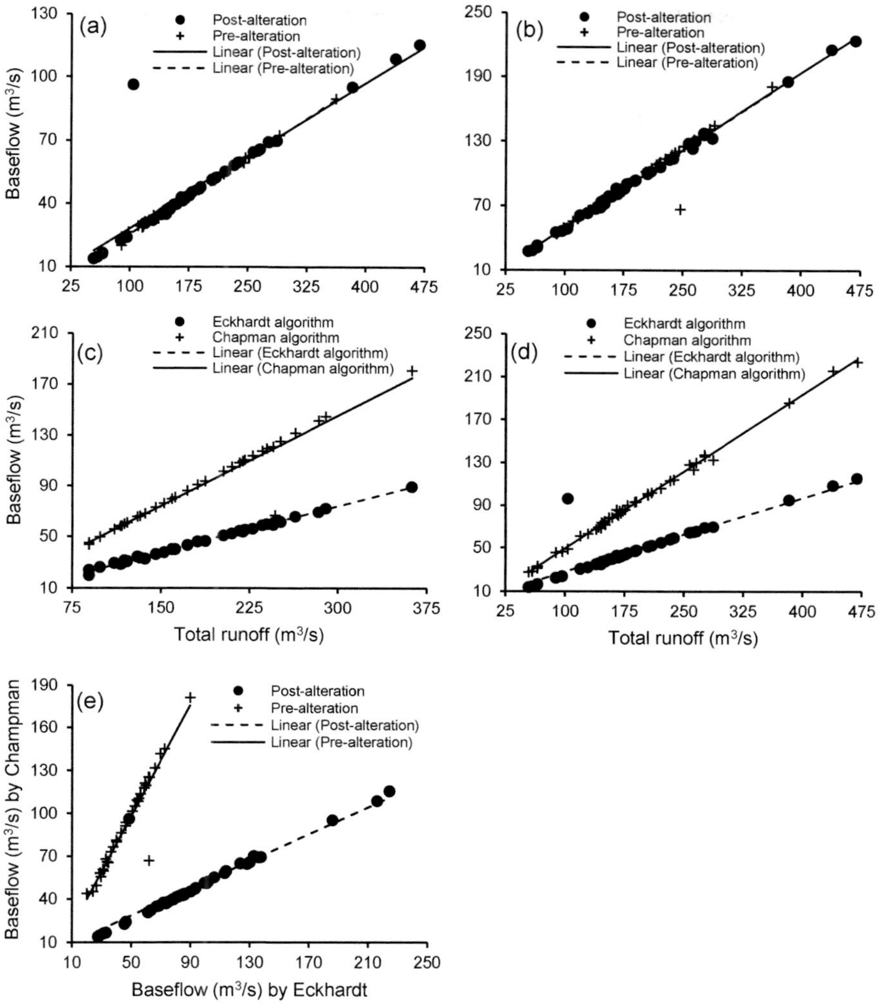

Fig. 9.7 The linear regression models for the relationships between the separated baseflow and the total runoff (**a**) using the Eckhardt method and (**b**) applying the Chapman recursive digital filtering algorithm for the pre-alteration and the post-alteration periods; (**c**) pre-alteration; and (**d**) post-alteration; and (**e**) the relationship between the two considered methods

approximately 65 mm a year and about 266 mm a year. The long-term yearly mean baseflow storage and the corresponding standard deviation were 2.91 and 1.05 bcm, respectively (Fig. 9.7b and d).

For the post-alteration period, and based on the outcome of the Eckhardt algorithm, the annual baseflow magnitudes vary between 0.437 bcm recorded in 2006 and 3.647 bcm observed in 1968. The matching yearly baseflow changed from nearly 21 mm to about 170 mm per annum. The long-term yearly mean baseflow storage and the corresponding standard deviation were 1.45 and 0.70 bcm,

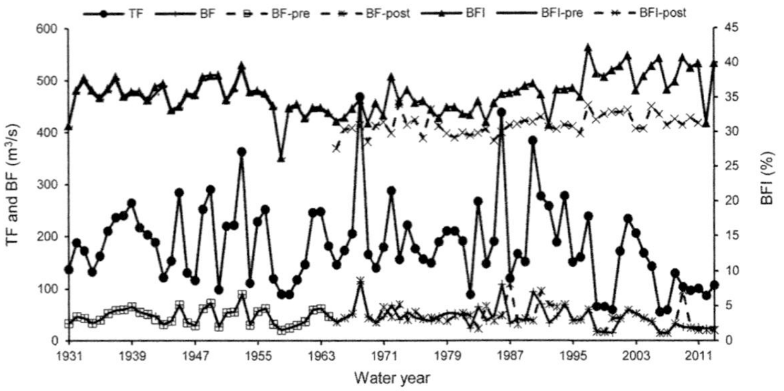

Fig. 9.8 Long-term monthly variation of total flow (TF), baseflow (BF), and baseflow index (BFI) at Dokan hydrometric station estimated by the Eckhardt filtering algorithm coupled with the flow duration curve (FDC) for the three studied periods

respectively (Fig. 9.7a and c). However, the annual baseflow magnitudes obtained from the Chapman algorithm varied from 0.871 bcm observed in 2006 to 7.077 bcm measured in 1968. The corresponding annual baseflow changed between approximately 40 mm/year and 330 mm/year. The long-term yearly mean baseflow volume and the conforming standard deviations were 2.88 and 1.31 bcm, respectively (Fig. 9.7b and d). Moreover, Fig. 9.7c and d characterise the baseflow, which is separated by the two considered filtering algorithms for the pre-alteration and post-alteration periods. The obtained results indicate that the two studied recurrence filtering algorithms are strongly correlated.

Furthermore, Fig. 9.8 shows the long-term variations of the total flow and the separated baseflow volumes coupled with BFI for the considered periods. The derived baseflow for all studied periods shows almost similar patterns. However, a noticeable variation has been observed for BFI for the post-alteration period, which is due to the water released from the river basin storage system.

9.3.4 *Climate Change Impact*

9.3.4.1 Baseflow Contribution

The detection of the effects of climate variability on the meteorological and hydrological variables over the study area can be considered as a substantial step in the analysis of the impact of climate change on basin water resource availability. The main consequence of climate change is that wet and dry years are characterised by high and low flows in that order.

Figure 9.9 shows the BFI and the time series for both wet and dry hydrological year thresholds. The hydrological year 1987 witnessed a significant ($p < 0.05$) increase in the basin average precipitation of about 44% more than the normal water

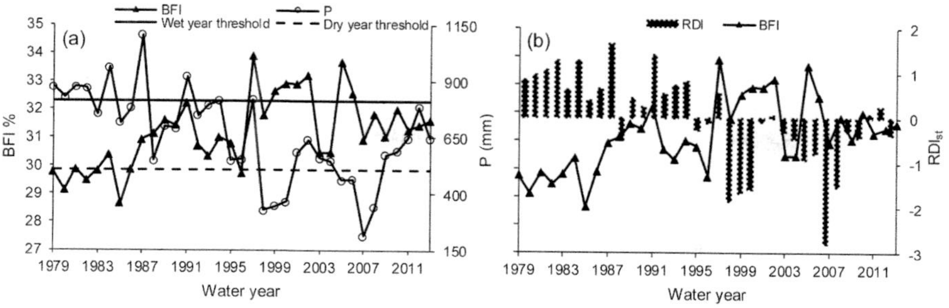

Fig. 9.9 (**a**) Long-term baseflow index (BFI) with both wet and dry year thresholds coupled with long-term average precipitation (P); (**b**) The reconnaissance drought index (RDI) coupled with the long-term BFI for the Lower Zab River Basin, for the time period between 1979 and 2014

year maximum threshold. This considerable increase in the amount of precipitation results in a dramatic change in the river flow of about 118% (Fig. 9.9a), which in turn reduces the contribution of the groundwater to the total river flow whereas the complete opposite was noticed for the hydrological periods between 1998 and 2001 and between 2006 and 2008. These two periods witnessed a sharp decline in the basin average precipitation to about 40 and 60%, respectively.

Figure 9.9b presents the RDI_{st} values calculated in addition to long-term average BFI and indicates that a non-uniform cyclic pattern of drought and wet periods was observed for the study area concerning the assessed time span. Based on the observed drought phenomena over the study area, the groundwater contribution was increased considerably by 30–35%.

Figure 9.10 shows that the basin potential to recharge the Lower Zab River with water from aquifers started to increase from April until reaching a maximum value by the end of June, remaining constant until the middle of August, and starting to decline slightly until the end of September. In addition, during the dry season, the BFI values revealed generally relatively high magnitudes of fluctuation, and they took relatively long time periods before reaching the peak point. Findings indicate that during the last few decades, climate change and drought phenomena have negatively affected the Lower Zab River storage water availability.

9.3.4.2 Streamflow and Reservoir Performance

Figure 9.11 shows the reservoir inflow magnitude and timing, which proves how climate change would strongly lead to a reduction in the reservoir inflow. In the worst case climatic scenario (−40% *P* and +30% PET), substantial variations in the inflow between −56% and −58% are anticipated for the dry and wet seasons, respectively. The inflow peak will decline, and there will be a marked shift in its magnitude, which can result in a dramatic effect on the basin water resource availability.

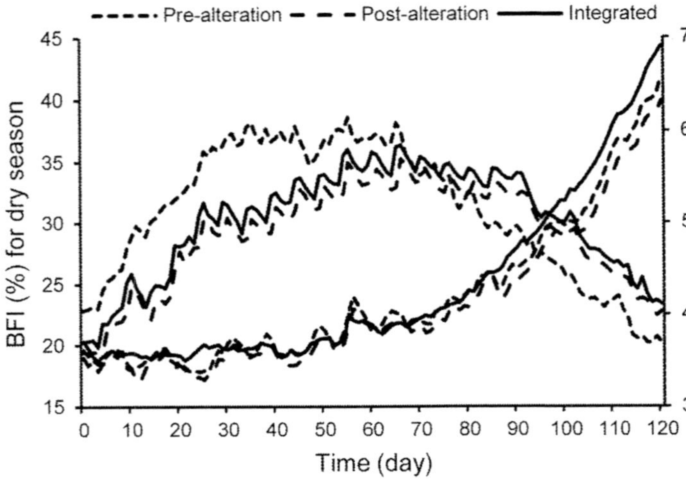

Fig. 9.10 Wet and dry seasonal variations of the baseflow index (BFI) at Dokan hydrometric station estimated by the Eckhardt filtering algorithm coupled with the flow duration curve for the three studied periods

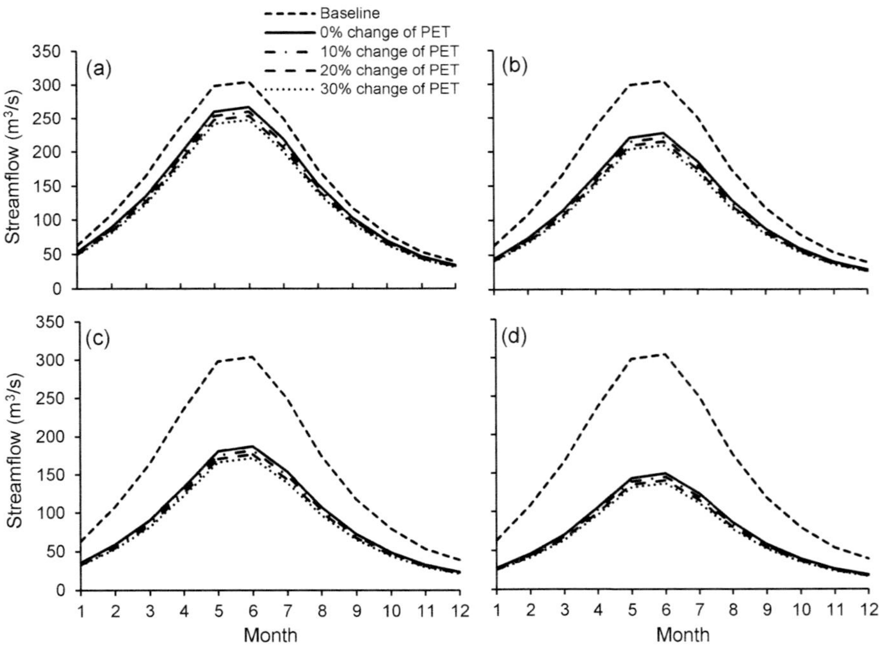

Fig. 9.11 Changes in the timing and magnitude of monthly average inflow to the Dokan reservoir under the delta perturbation climate change scenarios for (**a**) 10% increase in precipitation (P); (**b**) 20% increase in P; (**c**) 30% increase in P; and (**d**) 40% increase in P compared to the baseline values

The evaluation of the sensitivity of water supply performance on water scarcity brought about by climate change is important for water resource managers. For all considered scenarios, the future demand for water is assumed to continue without major change with respect to the baseline period. The climate change influence on the operation of the Dokan dam and its water system was assessed by entering modelled future inflows into the RCY model while maintaining the existing operating procedures. To simulate the operations of the Dokan reservoir, the HBV simulations of daily streamflow were applied to mimic water resources in the RCY model. The delta perturbations of climate change scenarios were applied. Then, the RCY model was run to assess the potential impacts of climate change on water resource availability and reservoir performance and capacity. After that, indicators were used for both the climate change scenarios and the baseline periods.

Moreover, Table 9.4 presents a summary of the potential performance of the reservoir. It is important to note that the reliability of the reservoir is generally high during the baseline scenario. The reliability would decline, while the resilience and vulnerability would increase as a result of precipitation reduction. However, Table 9.5 show that the reservoir volume required to meet the increase in water demand would decrease due to the decline of precipitation and runoff. For example, a 4% precipitation decrease can mean that the existing volume is too little (by as much as 29% for a water yield of 82%). Moreover, the uncertainty or the variability of the reservoir performance is characterised by the uncertainty Cv, i.e. the standard deviation divided by the mean. The uncertainty of the OPOF, resilience, vulnerability, and the required capacity varied between 0.78–0.07, 0.55–0.40, 0.87–0.47, and 0.46–0.10, respectively, which indicates that both uncertainty and variability increase as the basin becomes drier.

The so-called operational failure rates have been determined for each model run and climate change scenario, which led to an amount of OPOF/reliability for each unit time per scenario. The successive range of OPOF involves the possible range of climate change impacts upon the water resource system over the chosen period. The simulation results indicate that through combining adaptation measures, whether they are structural or non-structural, into water resource simulation (supply side), their effectiveness will be determined across a range of climate change and climate change variability by evaluating how each of the simulations react to a given adaptation strategy in terms of decrease (or otherwise) of OPOF and how much of the distribution is moved away from a pre-determined and undesirable level of hazard. The developed approaches can be used to evaluate how well an adaptation measure would work within the uncertainty of a climate change range and can, therefore, be used as a support tool for managing water resources. Based on the results of this representative case study, an adaptation operational approach can be inferred, where the policymakers, particularly in the semi-arid and arid regions, adjust the reservoir operating rules based on inflow estimations and the current state of the reservoir capacity at each specified period, which can result in a more effective and viable management of reservoirs.

Table 9.4 Summary of the water resources system performance under the collective impacts of precipitation (P) and potential evapotranspiration (PET) for a 82% yield

Considered time	Hydro-meteorological parameters	Values			
baseline	P (mm)	844.08			
	Inflow (m^3/s)	1009			
	Inflow$_{wet}$ (m^3/s)	2305			
	Inflow$_{dry}$ (m^3/s)	2249			
	Reliability %	99			
	Vurnlibility (10^6 m^3)	426			
	Resilience (month)	3			
% change of P	Hydro-meteorological parameters	% change of PET			
		0	10	20	30
0	P (mm)	844.08			
	PET (mm)	1009	1110	1211	1312
	Inflow (m^3/s)	1884	1832	1783	1737
	Infow$_{wet}$ (m^3/s)	1596	1550	1509	1470
	Inflow$_{dry}$ (m^3/s)	289	283	274	267
	Reliability %	92	91	90	88
	Vurnlibility (10^6 m^3)	426	427	428	428
	Resilience (month)	3	5	8	11
10	P (mm)	759.69			
	Inflow (m^3/s)	1623	1576	1531	1490
	Inflow$_{wet}$ (m^3/s)	1371	1331	1294	1260
	Inflow$_{dry}$ (m^3/s)	252	244	237	230
	Reliability %	79	77	74	72
	Vurnlibility (10^6 m^3)	426	427	428	429
	Resilience (month)	22	22	27	23
20	P (mm)	675.26			
	Inflow (m^3/s)	1369	1327	1287	1251
	Inflow$_{wet}$ (m^3/s)	1155	1120	1087	1057
	Inflow$_{dry}$ (m^3/s)	214	207	200	194
	Reliability %	63	58	56	53
	Vurnlibility (10^6 m^3)	430	431	432	433
	Resilience (month)	24	24	32	32
30	P (mm)	590.88			
	Inflow (m^3/s)	1123	1087	1053	1021
	Inflow$_{wet}$ (m^3/s)	946	916	888	862
	Inflow$_{dry}$ (m^3/s)	177	171	165	160
	Reliability %	44	40	37	36
	Vurnlibility (10^6 m^3)	435	436	436	436
	Resilience (month)	33	43	42	45
40	P (mm)	506.44			
	Inflow (m^3/s)	889	858	830	804
	Inflow$_{wet}$ (m^3/s)	748	722	699	677
	Inflow$_{dry}$ (m^3/s)	141	136	131	126

(continued)

Table 9.4 (continued)

Considered time	Hydro-meteorological parameters	Values			
	Reliability %	28	26	24	24
	Vurnlibility (10^6 m^3)	439	440	440	441
	Resilience (month)	61	82	94	73

Table 9.5 Statistical relationships between yield (Y; %) and operational probability of failure (OPOF, %) and reservoir capacity (C; 10^6 m^3) and yield for different reduction (%) in precipitation (P) and increase (%) in potential evapotranspiration (PET) using incremental climate change scenarios

%		Y (%)[a]				C (10^6 m^3)[b]			
P	PET	a_3	b_3	c_3	R^2	a_4	b_4	c_4	R^2
0	0	–0.018	1.51	71.36	0.99	6.41	–873.3	33,302	0.99
	10	–0.007	1.17	71.21	0.99	6.74	–571.3	10,968	0.99
	20	–0.008	1.15	69.68	0.99	5.63	–350.8	1302.6	0.99
	30	–0.005	1.02	68.93	0.99	2.25	226.54	–21,846	0.99
10	0	–0.004	0.93	65.25	0.99	–6.83	1727.5	–77,136	0.99
	10	–0.001	0.75	65.11	0.99	–7.03	1782.2	–81,386	0.99
	20	0.001	0.68	64.08	0.99	–7.06	1906.8	–88,811	0.99
	30	0.001	0.65	62.67	0.99	–4.82	1429.3	–69,695	0.99
20	0	0.004	0.37	61.92	0.99	–4.48	1224.8	–46,423	1.00
	10	0.006	0.24	62.95	0.99	–4.56	1251.1	–48,849	0.99
	20	0.007	0.08	64.32	0.99	–4.99	1339.9	–53,963	0.99
	30	0.001	–0.18	68.46	0.99	–5.52	1443.9	–59,918	0.99
30	0	0.017	–0.99	84.38	0.99	–2.99	905.8	–26,026	1.00
	10	0.019	–1.29	91.87	0.99	–3.57	1007.9	–31,067	0.99
	20	0.012	–1.59	98.98	0.99	–3.36	982	–30,995	1.00
	30	0.002	–1.98	109.48	0.99	–3.88	1081	–36,239	0.99
40	0	0.025	–2.01	111.42	0.99	–2.30	736.5	–13,283	0.99
	10	0.034	–3.48	156.36	0.99	–2.48	773	–15,447	0.99
	20	0.025	–2.20	105.70	0.99	–2.95	859	–19,765	1.00
	30	0.039	–4.31	181.91	0.99	–2.79	836.3	–19,458	1.00

[a]$= a_3 \times \text{OPOF}^2 (\%) + b_3 \times \text{OPOF} (\%) + c_3$; [b]$= a_4 \times Y^2 (\%) + b_4 \times Y (\%) + c_4$

9.4 Conclusions and Recommendations

A new methodology has been successfully applied to evaluate climate change, anthropogenic activities, and drought phenomenon impact on the streamflow alteration, baseflow (BF) involvement in the total flow (TF), and the flow to the basin storage system. The following case study-specific conclusions are drawn:

The region has already experienced increases in drought periods due to precipitation reductions after 1999. A non-uniform cyclical characteristic of wet and dry periods was witnessed for the considered period, and the seasonal droughts were recorded over the hydrologic years 1998/1999, 1999/2000, and 2000/2001. The hydrologic year 2007/2008 witnessed a severe drought, whereas the year 1987/1988 was characterised by a significant ($p < 0.05$) increase in the basin average precipitation. Almost 80% of the streamflow reduction was observed for the hydrologic years 1998–2002 and 2006–2008 as a consequence of a sharp drop in the basin average precipitation. A considerable alteration during the non-rainy months is attributed to the effect of human-induced activities and climate change pressure in the region, which led to a decrease in water resources in the basin.

The long-term mean annual BFI of the studied geographical region varied from 31, 33, and 35% for the pre-alteration, post-alteration, and integrated periods, respectively. The obtained results suggest that 31, 33, and 35% of the long-term discharge in the Lower Zab River Basin resulted from both the groundwater contribution and shallow sub-surface flow. The rather small variations in the baseflow contribution between the pre-alteration and the post-alteration period were linked to water released from the reservoir during the dry period to cover the downstream water requirements. Therefore, more consideration must be given to the assessment of the characteristics of aquifers and understanding the factors that might lead to such variations. This could lead to improvements in the management of groundwater resources within the basin. This is because it is critical to investigate the baseflow contribution to the total discharge as a result of the pressing requirement for logical use of water resources and strong competition among the riparian nations.

The average yearly baseflow index of the upper sub-basin was 0.14, which indicates that 14% of the long-standing LZRB flow can be sustained by groundwater flow. The hydrologic year 1987 witnessed a significant increase in the basin average precipitation of about 44% more than the normal hydrologic year maximum threshold. This resulted in a considerable change of the river flow to about 118%, in addition to the reduction in the groundwater contribution to the total flow of the river. Approximately 40 and 60% of the reduction in the average precipitation for the water years 1998–2001 and 2006–2008, respectively, have been reported. This caused substantial drops in the flow by between 52 and 83%, respectively, which led to reductions in the contribution of groundwater.

Over the water years 1998/1999, 1999/2000, 2000/2001, 2007/2008, and 2008/2009, a considerable increase in the contribution of sub-surface water was recorded, because of seasonal droughts. The matching mean RDI entries were −1.84, −1.67, −1.45, −2.91, and −1.53. Concerning the upstream sub-basin, findings show that the year 1997 had maximum annual BFI values of 0.42 and 0.53 estimated using Eqs. (9.8 and 9.9), respectively. Accordingly, 1997 can be considered as the driest year in the study period. The baseflow index values increased in April and peaked by the end of June. A gradual decrease has been recorded between August and September.

During the last few decades, there was a considerable alteration in the river flow in particular during the non-rainy months. This variation might be attributed to the influence of both river damming and climate variability pressure in the upper part of the case study area, which in turn decreased the basin storage system availability from both surface and underground sources. Research findings imply that river alteration and climate variability should be considered for future stream basin managing strategies to escape the temporal disparity between strategy and such changes.

The HBV model outputs recommend a critical decrease in the inflow to the Dokan reservoir due to a decline in the precipitation and increase in the potential evapotranspiration, which in turn lowers the reservoir capacity. For instance, a reduction of about 21% in streamflow is expected to result in 10% P reduction and 30% PET increase. Additionally, for the worst case scenario, a substantial decline in inflow of 56 and 58% was estimated for the dry and wet seasons in this order. The inflow maximum magnitude will decrease, and there will be a noticeable shift in their values, which can cause a dramatic effect on basin water resource availability.

A 4% precipitation decrease could indicate that the existing capacity is too little (by as much as 29% for a water yield of 82%). Variations in reservoir inflow will impact water utilisation purposes such as water supply, hydropower, irrigation, downstream water quality enhancement, and recreational uses. Accordingly, to moderate the negative hydrologic influences and to benefit from the positive effects, the potential climate change impact should be examined by water resource organisations in the future. Alterations in future monthly reservoir inflows require modifications of some of the reservoir management processes to use water more efficiently. If existing operational rules continue to be operated unchanged, projected climate change would further lead to a decline in the ability to supply water to all stakeholders. In addition, the research results show potential decreases in the water supply reliability and increases in the resilience to nearly 22% and 86%, respectively, due to dam inflow decreases. However, with higher mean temperatures and extended crop growing periods, it is expected that the demands for irrigation would also rise.

Any alteration that would improve the ability of a water resource system to accomplish the water needs for one sector would essentially require compromises from other stakeholders. Subsequently, it is essential to assess the monthly water supply to maximise the profit, which means that the supply system of a multi-purpose reservoir should have operating policies that can be adjusted to potential hydrological and climatic alterations. The inflow increase would provide the chance to reallocate preservation storage under a climate change scenario.

The need to estimate the existing water demand in the RCY model is a limitation. Despite the fact that water demand prediction is vital to evaluate water resource systems in the future, this chapter should be seen as an initial examination of the climate change impacts on the dam water supply reliability. In order to facilitate adaptation to climate change, basin managers ought to have a quantitative basis for establishing the adaptation strategies. Accordingly, results provide a basis for enabling future water resource system managers and planners of the Dokan reservoir and similar case studies, in particular, in arid and semi-arid regions to adapt to climate change.

Acknowledgement This research was financed by the Iraqi government via a Ph.D. scholarship for the lead author via Babylon University.

Disclosure Statement The authors declare that they have no conflict of interest.

Funding The researchers received funding from the Government of Iraq via Babylon University for the Ph.D. study of the lead author.

References

Al-Faraj F, Scholz M (2014) Incorporation of the flow duration curve method within digital filtering algorithms to estimate the baseflow contribution to total runoff. Water Resour Manag 28 (15):5477–5489

Allen RG, Pereira LS, Raes D, Smith M (1998) Crop evapotranspiration: Guidelines for computing crop water requirements. Food and Agriculture Organization (FAO) Irrigation and Drainage Paper 56, Rome, Italy

Arnold JG, Allen PM (1999) Automated methods for estimating baseflow and ground water recharge from streamflow records1. JAWRA J Am Water Resour Assoc 35(2):411–424. https://doi.org/10.1111/j.1752-1688.1999.tb03599.x

Chapman T (1999) A comparison of algorithms for streamflow recession and baseflow separation. Hydrol Process 13(5):701–714. https://doi.org/10.1002/(SICI)1099-1085(19990415)13:5<701:AID-HYP774>3.0.CO;2-2

Cook BI, Smerdon JE, Seager R, Coats S (2014) Global warming and 21st century drying. Climate Dyn 43(9–10):2607–2627. https://doi.org/10.1007/s00382-014-2075-y

Doll P, Zhang J (2010) Impact of climate change on freshwater ecosystems: a global-scale analysis of ecologically relevant river flow alterations. Hydrol Earth Syst Sci 14(5):783–799. https://doi.org/10.5194/hess-14-783-2010

Eckhardt K (2005) How to construct recursive digital filters for baseflow separation. Hydrol Process 19(2):507–515. https://doi.org/10.1002/hyp.5675

Fadfil MA (2011) Drought mapping using Geoinformation technology for some sites in the Iraqi Kurdistan region. Int J Digit Earth 4(3):239–257. https://doi.org/10.1080/17538947.2010.489971

Foehn A, García Hernández J, Roquier B, Paredes Arquiola J (2016) RS MINERVE – User's manual v2.6. RS MINERVE Group, Switzerland

Fowler HJ, Kilsby CG, O'Connell PE (2003) Modeling the impacts of climatic change and variability on the reliability, resilience, and vulnerability of a water resource system. Water Resour Res 39(8):1222. https://doi.org/10.1029/2002WR001778

GADM, Global Administrative Areas Database (2012) Boundaries without limits [Online] Available from http://www.gadm.org. Accessed 10 March 2015

Gao B, Yang D, Yang H (2013) Impact of the three gorges dam on flow regime in the middle and lower Yangtze River. Quat Int 304:43–50

Giannikopoulou AS, Kampragkou E, Gad FK, Kartalidis A, Assimacopoulos D (2014) Drought characterisation in Cyclades complex, Greece. Eur Water 47:31–43

GLCF, Global and Land Cover Facility (2015) Earth science data interface [Online] Available from http://www.landcover.org/data/srtm/. Accessed 05 March 2015

HydroOffice (2015) Download [Online] Available from https://hydrooffice.org/Downloads?Items=Software. Accessed 14 Dec 2015

Jiang L, Ban X, Wang X, Cai X (2014) Assessment of hydrologic alterations caused by the three gorges dam in the middle and lower reaches of Yangtze River, China. Water 6(5):1419–1434. https://doi.org/10.3390/w6051419
Kim BS, Kim BK, Kwon HH (2011) Assessment of the impact of climate change on the flow regime of the Han River basin using indicators of hydrologic alteration. Hydrol Process 25 (5):691–704. https://doi.org/10.1002/hyp.7856
Lee A, Cho S, Kang DK, Kim S (2014) Analysis of the effect of climate change on the Nakdong river stream flow using indicators of hydrological alteration. J Hydro Environ Res 8 (3):234–247. https://doi.org/10.1016/j.jher.2013.09.003
McMahon TA, Adeloye AJ (2005) Water resources yield. Water Resources Publications, Littleton, Colorado, USA
McMahon T, Mein RG (1978) Reservoir capacity and yield. Development in Water Science 9. Elsevier, Amsterdam
McMahon TA, Mein RG (1986) River and reservoir yield. Water Resources Publications, Ft, Collins, Colorado
McMahon TA, Adeloye AJ, Zhou SL (2006) Understanding performance measures of reservoirs. J Hydrol 324(1):359–382. https://doi.org/10.1016/j.jhydrol.2005.09.030
Minville M, Brissette F, Krau S, Leonte R (2009) Adaptation to climate change in the management of a Canadian water-resources system exploited for hydropower. Water Resour Manag 23 (14):2965–2986. https://doi.org/10.1007/s11269-009-9418-1
Mittal N, Mishra A, Singh R, Bhave AG, van der Valk M (2014) Flow regime alteration due to anthropogenic and climatic changes in the Kangsabati River, India. Ecohydrol Hydrobiol 14 (3):182–191. https://doi.org/10.1016/j.ecohyd.2014.06.002
Mittal N, Bhave AG, Mishra A, Singh R (2016) Impact of human intervention and climate change on natural flow regime. Water Resour Manag 30(2):685–699. https://doi.org/10.1007/s11269-015-1185-6
Mohammed R, Scholz M (2016) Impact of climate variability and streamflow alteration on groundwater contribution to the baseflow of the lower Zab River (Iran and Iraq). Environ Earth Sci 75:1392):1–1392)11. https://doi.org/10.1007/s12665-016-6205-1
Mohammed R, Scholz M (2017a) Impact of evapotranspiration formulations at various elevations on the reconnaissance drought index. Water Resour Manag 31:531–538. https://doi.org/10.1007/s11269-016-1546-9
Mohammed R, Scholz M (2017b) The reconnaissance drought index: a method for detecting regional arid climatic variability and potential drought risk. J Arid Environ 144:181–191. https://doi.org/10.1016/j.jaridenv.2017.03.014
Mohammed R, Scholz M (2017c) Adaptation strategy to mitigate the impact of climate change on water resources in arid and semi-arid regions: a case study. Water Resour Manag 31 (11):3557–3573. https://doi.org/10.1007/s11269-017-1685-7
Mohammed R, Scholz M, Nanekely MA, Mokhtari Y (2017a) Assessment of models predicting anthropogenic interventions and climate variability on surface runoff of the lower Zab River. Stoch Env Res Risk A 32:1–18. https://doi.org/10.1007/s00477-016-1375-7
Mohammed R, Scholz M, Zounemat-Kermani M (2017b) Temporal hydrologic alterations coupled with climate variability and drought for transboundary river basins. Water Resour Manag 31:1489–1502. https://doi.org/10.1007/s11269-017-1590-0
Moy WS, Cohon JL, ReVelle CS (1986) A programming model for analysis of the reliability, resilience, and vulnerability of a water supply reservoir. Water Resour Res 22(4):489–498. https://doi.org/10.1029/WR022i004p00489
NOAA (2009) National Oceanic and Atmospheric Administration Climate of Iraq. [Online] Available from https://www.ncdc.noaa.gov/oa/climate/afghan/iraq-narrative.html. Accessed 4 Nov 2015
Oguntunde PG, Abiodun BJ, Lischeid G (2011) Rainfall trends in Nigeria, 1901–2000. J Hydrol 411(3):207–218. https://doi.org/10.1016/j.jhydrol.2011.09.037

Park JY, Kim SJ (2014) Potential impacts of climate change on the reliability of water and hydropower supply from the multipurpose dam in South Korea. JAWRA J Am Water Resour Assoc 50(5):1273–1288. https://doi.org/10.1111/jawr.12190

Robaa SM, AL-Barazanji ZJ (2013) Trends of annual mean surface air temperature over Iraq. Nat Sci 11(12):138–145

RS MINERVE 2.5 Software (2016) Download [Online] Available from: https://www.crealp.ch/down/rsm/install2/archives.html. Accessed 15 June 2016

Shahidian S, Serralheiro RP, Serrano J, Teixeira J, Haie N, Santos F (2012) Hargreaves and other reduced-set methods for calculating evapotranspiration. In: Irmak A (ed) Evapotranspiration – remote sensing and modeling. InTech, Rijeka, Croatia, pp 59–80

Sheffield J, Wood EF, Roderick ML (2012) Little change in global drought over the past 60 years. Nature 491(7424):435–438. https://doi.org/10.1038/nature11575

Smakhtin VU (2001) Low flow hydrology: a review. J Hydrol 240(3–4):147–186. https://doi.org/10.1016/S0022-1694(00)00340-1

Stagl JC, Hattermann FF (2016) Impacts of climate change on riverine ecosystems: alterations of ecologically relevant flow dynamics in the Danube River and its major tributaries. Water 8 (12):566. https://doi.org/10.3390/w8120566

Suen JP (2010) Potential impacts to freshwater ecosystems caused by flow regime alteration under changing climate conditions in Taiwan. Hydrobiologia 649(1):115–128. https://doi.org/10.1007/s10750-010-0234-7

Sun T, Feng ML (2013) Multistage analysis of hydrologic alterations in the Yellow River, China. River Rese Appl 29(8):991–1003. https://doi.org/10.1002/rra.2586

Tabari H, Taalaee PH (2011) Analysis of trend in temperature data in arid and semi-arid regions of Iran. ISSN: 0921-8181. Glob Planet Change 72:1–10. https://doi.org/10.1016/j.gloplacha.2011.07.008

The Nature Conservancy (2009) Indicators of hydrologic alteration version 7.1 user's manual. The Nature Conservancy, June, 76

Tigkas D (2008) Drought characterisation and monitoring in regions of Greece. Eur Water 23 (24):29–39

Tigkas D, Vangelis H, Tsakiris G (2012) Drought and climatic change impact on streamflow in small watersheds. ISSN: 0048-9697. Sci Total Environ 440:33–41. https://doi.org/10.1016/j.scitotenv.2012.08.035

Tsakiris G, Vangelis H (2005) Establishing a drought index incorporation evapotranspiration. Eur Water 9(10):3–11

UNESCO (2014) United Nations educational, scientific and cultural organization. Integrated drought risk management-DRM national framework for Iraq. An analysis report. Retrieved from http://unesdoc.unesco.org/images/0022/002283/228343E.pdf UN-ESCWA and BGR (United Nations Economic and Social Commission for Western Asia; Bundesanstalt für Geowissenschaften und Rohstoffe, Inventory of Shared Water Resources in Western Asia, Beirut

Vangelis H, Tigkas D, Tsakiris G (2013) The effect of PET method on reconnaissance drought index (RDI) calculation. J Arid Environ 88:130–140. https://doi.org/10.1016/j.jaridenv.2012.07.020

Vicente-Serrano SM, Lopez-Moreno JI, Beguería S, Lorenzo-Lacruz J, Sanchez-Lorenzo A, García-Ruiz JM, Azorin-Molina C, Morán-Tejeda E, Revuelto J, Trigo R, Coelho F (2014) Evidence of increasing drought severity caused by temperature rise in southern Europe. Environ Res Lett 9(4):044001. https://doi.org/10.1088/1748-9326/9/4/044001

Vicuna S, Dracup JA (2007) The evolution of climate change impact studies on hydrology and water resources in California. Clim Chang 82(3):327–350. https://doi.org/10.1007/s10584-006-9207-2

Wang Y, Rhoads BL, Wang D (2016) Assessment of the flow regime alterations in the middle reach of the Yangtze River associated with dam construction: potential ecological implications. Hydrol Process 30(21):3949–3966. https://doi.org/10.1002/hyp.10921

Yan Y, Yang Z, Liu Q, Sun T (2010) Assessing effects of dam operation on flow regimes in the lower Yellow River. Procedia Environ Sci 2:507–516. https://doi.org/10.1016/j.proenv.2010.10.055
Yoo C (2006) Long term analysis of wet and dry years in Seoul, Korea. J Hydrol 318(1–4):24–36. https://doi.org/10.1016/j.jhydrol.2005.06.002

Chapter 10
Ecohydrology in Iraq: Challenges and its Future Pathways

Laith A. Jawad

Abstract There are several aspects that may affect the water supplies in Iraq. Such factors may lead to severe changes affecting the land and human societies. These differ from ecohydrological methodology for the aquatic habitat administration to gain extreme advantage of the aquatic assets, through attempts to recover the aquatic carrying capacity, endeavors to resolve the water discrepancy owing to land use change in the catchment region. An effort to link ecohydrology and adversity risk lessening, particularly the gradually approaching ecosystem tragedy, is emphasized insight of focusing on the emerging chances in Iraq.

10.1 Introduction

Ecohydrology, incorporating hydrology and ecology, is a vital ecotechnological speciality and offers energetic tackles that have been capable to resolve several habitat difficulties by very low-cost means (Jørgensen 2016). Lately, ecohydrology has been revealed as a scientific field, where some authors have emphasized its significance to the development of hydrology and ecology (Eagleson 2002; Grootjans et al. 1996; Nuttle 2002; Rodriguez-Iturbe 2000).

According to Zalewski et al. (1997), eco-hydrology can be defined as "the study of the functional relationship between hydrology and biota at catchment scale" and "a new method to attain maintainable control of water." To the others (Witte et al. 2004), ecohydrology indicates that this science is about associations of biota and the hydrological constituent of their habitat. Two major branches of "real" ecohydrology can be identified, one that is related to the vegetation structure and one that connected to biodiversity (Witte et al. 2004). The first branch clarifies geographical configurations and the dynamics in vegetation structure from hydrological procedures vice versa (e.g., Rietkerk et al. 2002; Rodriguez-Iturbe 2000; Rodriguez-Iturbe

L. A. Jawad (✉)
School of Environmental and Animal Sciences, Unitec Institute of Technology, Auckland, New Zealand

L. A. Jawad (ed.), *Tigris and Euphrates Rivers: Their Environment from Headwaters to Mouth*, Aquatic Ecology Series 11,
https://doi.org/10.1007/978-3-030-57570-0_10

et al. 2001; Van Breemen 1995; Van der Schaaf and Streefker 2002; Wainwright et al. 1999). The second branch of eco-hydrology is perhaps older than the first one. It contributes to the associations between hydrology and biodiversity, particularly the link between hydrology and the presence of individual plant species and plant groups. Baird and Wilby (1999) describe eco-hydrology as the study of plant–water associations and the hydrologic courses linked to plant development. Their explanation contains evapotranspiration investigations and research into the hydraulic opposition of water plants. On the other hand, Rodriguez-Iturbe (2000) and Rodriguez-Iturbe et al. (2001), Nuttle (2002) recognized ecohydrology as the sub-restraint used by the ecological and hydrologic sciences that fears with the consequences of hydrological procedures on the supply, structure and operation of bionetworks, and on the outcomes of biotic procedures on components of the water cycle.

Iraq, like many other Asian and the Middle Eastern states, is enduring a quick economic growth and it is predictable to be one of the chief economies in the Middle East. The oil industries have depreciated the soil and the rivers, and have augmented the damage to plant and animal lives. This sort of effect is similar to that happening in different parts of Asia (2009). Extensive upsurges in sediment packing to rivers and lakes have, consequently, deep inferences for the maintenance and administration of the natural habitats (Donohue and Molinos 2009).

Desertification is a major problem in Iraq, with climate limits, widespread removal of natural vegetation, extreme grazing, agricultural development in contrast to the natural environments, inappropriate irrigation systems and soil salinization have subsidized to land dilapidation, abridged water sources, and incomplete agricultural production in these regions, which assessed of 61 per cent of agricultural land in Iraq is vulnerable by salinity difficulties (Haktanir et al. 2004).

The whole Mesopotamia Basin would agonize severe desertification since the area only collects between 150 and 300 mm of rainfall annually but practices 1500–2500 mm of evaporation per year (Haktanir et al. 2004). The area also encounters severe temperatures and windstorms. Erosion has a chief impact on desertification. Population development and agricultural miss supervision have also impaired desertification difficulties in those regions. Population development will have increased 50% by 2025 in Iraq, Syria, and Turkey (Tigers and Euphrates Basin). Nearly there is no forest cover in Iraq in comparison with the neighboring countries like Turkey (26%), Lebanon (8%), and Syria (3%), but over half the forest area in those countries is ruined and unfruitful. With such a rate of desertification, there will be a significant rate of loss in the biodiversity. Similar case was observed in Indonesia (Navjot et al. 2004).

Iraq has profited from inland waters, which deliver source of water for millions of people daily, providing fishes as food, hydropower for electricity, and also for refreshment. Though, these purposes have altered due to anthropogenic activities upsetting mainly freshwater habitats (Alobaidy et al. 2010).

There are several ecological negative factors upsetting the lakes in Iraq (Toma 2013), which are comparable to those found in other parts of Asia (Haryani 2008). Worsening of water quality by different aspects of pollution represents the chief dilemma for Iraq (Toma 2013; Rahi and Halihan 2010; Jones et al. 2008). As a result, the lakes in Iraq including the man-made such as that near Dokan Dam, are shown to be extremely vulnerable due to the decreasing water quality level (Jawad 2003; Hassan et al. 2017). Similar events have been reported for lakes in Indonesia (Hargono and Pranoko 2003; Dam et al. 2001).

This chapter summarizes the concerns on water, habitat, and social development in Iraq and how ecohydrology can pick up the developing chances, notwithstanding the enormous contests postured by the fast expansion in the country. This will be headed by revealing for a short time the present ecohydrology studies in Iraq.

10.2 Matters on Water, Habitats, and Communal Growth in Iraq

Massive freshwater assets present in Iraq which are characterized by the occurrence of 924 km^2, with three major rivers with six tributaries, with the total bank areas of 283,110 km^2 (Al-Ansari and Knutsson 2011). Huge land changes have been in place since the 1990s of the last century, which include alterations in land use, urban and industrial expansions, and substructure networks. Such changes have caused in a noteworthy alteration of water flowing in Iraq.

The population of Iraq is about 20.4106 (1995) with a development rate of 3.6% (1980–1990). This proportion has plummeted since 1989 owing to harsh economic adversity. About 25% of the populations are living in rural regions. The population density ranges from 5 to 170 inhabitants/km^2. The lower number is in western desert and higher in the central part of the country (Al-Ansari and Knutsson 2011). The estimated population of Iraq in 2004 is 27,139,584 (Burnham et al. 2006).

10.3 The Existing Ecohydrology Uses and their Tests in Iraq

A restricted variety of methods and practices of ecohydrology have been employed in Iraq. These comprise some studies on numerous types of pollution, and Ecohydrology-based water administration situations for some aquatic areas (Salah et al. 2006). Such applications are already in use in other regions of the world (Chrismadha et al. 2011; Fakhrudin et al. 2009). In order to obtain similar results like that reached by other countries, such an application needs to be hosted in Iraq. These applications include the habitat engineering codes, particularly by familiarizing to the local aquatic environmental state. The purpose of applying such applications is to decrease the decline in the state of the aquatic habitats owing to the higher land usage ensuing in the sedimentation, release variation, and pollution. The second option is the ecohydrological engineering by coastline rearrangement and improvement of check dams to regulate the sedimentation and water level variations. The results of this option might increase water retention and the fishery productivity. The third option is these Ecohydrological methods to develop the aquatic carrying capacity in the lake's ecosystem. With this option, the fishery habitat will be threatened owing to the great water level variation. The countries that have used these three ecohydrological examples assessed an increase of 15% of the annual water spare in the reservoir through 240 ha of cultivation in the riverbank area (Ridwansyah and Aisyah 2006).

The overhead stated circumstances and methods demonstrate how significant is to comprehend the relationship between ecological and hydrological procedures. Their appreciative is the foundation to follow the Ecohydrology's main aim to progress a complete, problem-solving process by improving water means, biodiversity, and support the habitat facilities (Zalewski 2007, 2014). The cumulative worldwide amount of urbanization produces new hardships and new probabilities for running cities with different aspects of modern Wagner and Breil 2013). Obviously, in Iraq, both in country and in city zones, the idea of ecohydrology must be used through research and teaching. It is recognized that the plan of postgraduate education on ecohydrology is obligatory. It might be attained only by presenting the elementary thoughts and bases of ecohydrology into the present programs of natural means administration and habitat shelter plans.

10.4 Developing Chances

The opposing influence of destroying water class, such a distinct exercises from the previous part of the period in the more controlled fields were progressively joint and united. It is predictable that ecohydrology will have a growing influence on the attempts to decrease adversary environment factors that affect significantly water resources (Jørgensen 2016). A more integrative tactics of ecohydrology have appeared and delivered a new understanding into determining some crucial water linked difficulties. It made available an obvious outcome in improving the profits of economic growth in the state, which then would disregard its habitat constituents. It is known that ecohydrology might be competent to antedate the general and or else unavoidable undesirable tendencies in at least three chief conventional extents.

10.5 Ecohydrology and the Freshwater Fish Biodiversity in Iraq

Riverine ecosystems are among the natural systems that faced an intense human intervention during the last decades causing environmental damage and dilapidation and as a result, several fish species have become highly threatened, specifically in rivers where substantial plea is placed on freshwaters.

There may be some features that can cause the failure of the natural life in freshwaters, these are environment dilapidation and disintegration, presence of alien species, water path alterations, contamination, and worldwide changes in weather (Gibbs 2000; Saunders et al. 2002). Water contamination is the procedure that should be controlled by several states, but improvements in aquatic habitats are being restrained by environment lacks in freshwater bodies (Aarts et al. 2004). Fish considered to be the possible biomarkers of the condition of aquatic habitat in general (Schneiders et al. 1993), and fish are frequently the main component in environmental development (Schiemer 2000). There is one problem facing the freshwater management and that is the high diversity of fishes (Welcomme and Petr 2004). Consequently, maintenance programs want ways for retaining large river fishes, so they can extend possible conservation goals (Hoggarth et al. 1996; Sarkar and Bain 2007).

In Iraq, the chief rivers, Euphrates, Tigris, and Shatt al-Arab and the tributaries of Tigris and Shatt al-Arab Rivers supporting high level of biodiversity and present maintenance and nutritional refuge, but inappropriately there is a lack of investigation regarding these habitats from the standpoint of the management. The fish species living in this area are acknowledged to be extremely vulnerable owing to human activities pollution, environmental changes, and group mortality of fish have been conveyed (Jawad 2003, 2013).

To enlarge to the growth of basic awareness, studies of taxonomic variety or species wealth should be delivered, to abridge the strategies established by organisms in acclimatization to the changed habitats, and to concentrate on studying the species dispersal (Pouillly et al. 2006).

In Iraq, there are six main kinds of freshwater systems, these are: (a) rapid flowing water bodies such as rivers; (b) shallow ponds; (c) deep lakes; (d) open river; (f) wetlands (marshes); (e) lakes (natural and man-made). In the middle stretch of the rivers fish diversity is higher than any other water bodies. With the presence of the physical environment disparity, the dispersal of fishes in rivers and the habitat alteration could contribute considerably to the threat of the fish species.

Sheldon (1968) demonstrated that in running waters the number of fish species groups is linked with water depth. Several of the big fish species, with a protection significance such as *Barbus* displayed favorite to deeper water habitats in rivers or lakes. Diverse water environments exhibited an inclination toward certain fish groups depending mainly on depth and velocity of water current. The habitats of the large rivers differ from the smaller ones in nature of the flowing waters, amount of light infiltration to bed, the shape and size of the bottom, depth, discharge rates, and drenched area (Dunne and Leopold 1978; Cummins 1974). Depth is of the utmost influential ecological factor (Bain 1995).

The spatial distribution pattern of fish species in the rivers in Iraq may be linked to habitat topographies, disparity in fishing pressure, and contamination. Sarkar et al. (2010) established the noteworthy upsurge in species abundance, occurrence of several endangered species in the aquatic areas of the wildlife secured regions than the fished areas of a lotic water body. The configuration and livelihood of fish populations are intensely affected by environment construction. Nevertheless, in regard to freshwater fish species, there is substantial argument encircling the significance accredited to physical events such as flooding against competition and predation in constructing populations and amend species richness (Grossman et al. 1982).

The available studies specify that works on freshwater fishes in Iraq are inadequate and largely restricted to the fish biology and assemblages. Supply valuations are regularly missing and data on the ecology of most of the species are rare. Freshwater fish species found in tropical Asia use a large variety of food items from different sources (Wikramanayake and Moyle, 1989).

10.6 Ecohydrology and Fisheries Management in Iraq

Ecosystem "health" relies on freshwater administration. In various regions globally, and particularly in developing countries, water quality, and biota diversity have been considerably decreased (Zalewski 1998). This lessening in such areas has happened although the use of advanced environmental technology.

Among the significant issues in freshwater ecology and management is to recognize the resistance of this habitat to stress. The flexibility of the freshwater environment to changes dependents on hydrology, which has the power in changing the topography of the freshwater body's bank and bottom and in turn inflicting changes in the freshwater biota.

Biotechnology includes the use of enzymatic and immunological procedures that change problem. It also comprises alteration and regulation of biomatter flow at a higher level of biotic organization—the habitat. Therefore, supervising and organize of hydrological courses must generate a new solution for maintainable freshwater administration.

Fishes have been used as a device for decrease of indications of eutrophication in the freshwater systems around the world. In this application, ecohydrology method has been used in the plan of the renovation of the eutrophic water bodies. Water systems in Iraq are characterized by having a high density of plant and algae growths, especially in the areas where the water current is slow (Maulood et al. 1981; Al-Saadi and Al-Mousawi 1988; Richardson 2010; Richardson and Hussain 2006). The likelihood of incidents of algal bloom and a toxic type are highly likely. In case of toxic bloom, the possibility of passing the produced toxins by the algae to the people through the drinking water is not unexpected. The chlorination and boiling of water do not remove toxins. Only ozonization and active carbon considerably decreases the intensity, and then only to a level adjustable with the technique.

The rivers in Iraq with their different length are considered the chief source of phosphorus load in the freshwater system in Iraq as they drain agricultural lands from the north to the south of Iraq. The level of the phosphorous should be kerbed down so to eradicate toxic growth and attain a level of mezotrophy. It must be conceivable to accomplish such chief lessening event by merging the classic means and the new tactics. The classic methods encompass improvement of sewage treatment plants in the river catchment, while the new methods comprise ecosystem biotechnology. The adjustment of fish group structures to improving the filter-feeding constituent of the Zooplankton groups is organized in this method.

In Iraq and during the administration of the food web in freshwater habitats, the oligotrophic systems (good water quality and low production) the major part of energy is averted from primary producers (algae) to succeeding trophic stages (planktivorous and predatory fish) (Opuszynski 1980). The nutrients from dead algae are reused by the "microbial loop" and enduring high nutrient source upholds high intensities of algae, ensuing in decline of water quality. Benthic detritivores may be encouraged in the system as long as the bottom oxygen source of the water body is adequate.

Among the potentials to reinstate the normal status of water bodies in Iraq, water quality upgrading can be employed through using food network dynamics, controlling the fish admission for reproduction to the land/water ecotone concealed by land vegetation. Doing other projects at the same time such as the improvement of the predatory fish species such the cyprinid, *Luciobarbus vorax* should be functional, which powerfully decreases pelagic zooplanctivorous fish populations. In the advanced stages of the refurbishment, to alleviate good water clearness phases, when the predator fish species turn out to be less effective, it may be important to think through making of artificial littoral and/or island for development of these species. Such reinstatement can be defined as a reference to modifications of the energy flow procedure from detrital food network features of eutrophic environment to mesotrophic, high water quality, and high fish produce ecosystems.

10.7 Conclusions and Recommendations

A reasonable sum of information has been gathered and extra needs to enhance an ecohydrology school for a developing economy situation similar to that of Iraq, were recognized. The foundation of information contains the features of aquatic habitats, hydrological systems, (micro) climatic inconsistency, and the local living conditions.

Among the ecohydrological dilemmas in Iraq is the drought in spite of the presence of the three major rivers, tributaries, lakes (both natural and man-made), and wetlands (marshes). The global warming is among the factors that behind the drought problem in Iraq and this factor is hard to be resolved self-sufficiently or in short period of time. Therefore, regional or international cooperation is needed to face the drought problem. Nevertheless, there are other matters that can be resolved self-sufficiently in a comparatively short time. In such instances, Al-Ansari (2013) has few recommendations to the government of Iraq to take quick, practical, and secure action as quickly as possible. Among the recommendations put forward by Al-Ansari (2013), the followings are selected:

Implementation and design need to in place instantly. It should contain the following:

(a) Collaboration of manpower in the governmental and nongovernmental organizations.
(b) Using modernized techniques in irrigation.
(c) Adoption of a plan to restrict water consumption.
(d) Renovation of irrigation infrastructure.
(e) Agricultural and in particular irrigation training programs should be started in institutions.
(f) Treated wastewater plans should be introduced.
(g) Involving private sector in the process of enhancing irrigation.
(h) Southern marshes reinstatement.
(i) Reinstatement of public awareness programs.

References

Aarts BGW, Van den Brink FWB, Nienhuis PH (2004) Habitat loss as the main cause of the slow recovery of fish faunas of regulated large rivers in Europe: the transversal floodplain gradient. River Res Appl 20:3–23

Al-Ansari N (2013) Management of water resources in Iraq: perspectives and prognoses. Engineering 5(6):667–684

Al-Ansari N, Knutsson S (2011) Toward prudent management of water resources in Iraq. J Adv Sci Eng Res 1:53–67

Alobaidy AHMJ, Abid HS, Maulood BK (2010) Application of water quality index for assessment of Dokan lake ecosystem, Kurdistan region. Iraq J Water Res Prot 2(09):792

Al-Saadi HA, Al-Mousawi AH (1988) Some notes on the ecology of aquatic plants in the Al-Hammar marsh. Iraq Plant Ecol 75(3):131–133

Bain MB (1995) Habitat at the local scale: multivariate patterns for stream fishes. Bull Fr Peche Piscicult 337–339:165–177

Baird AJ, Wilby RL (1999) Eco-hydrology: plants and water in terrestrial and aquatic environments. Psychology Press

Burnham G, Lafta R, Doocy S, Roberts L (2006) Mortality after the 2003 invasion of Iraq: a cross-sectional cluster sample survey. Lancet 368(9545):1421–1428

Chrismadha T, Haryani GS, Fakhrudin M, Hehanussa PE (2011) The application of ecohydrology in lake management. In: Proc Nat. Sympos. Ecohydrology on Integrating Ecohydrological Principles for Good Water Governance (in Indonesian). Res. Centre for Limnology-LIPI, pp 25–44

Cummins KW (1974) Structure and function of stream ecosystems. Bioscience 24:631–641

Dam RA, Fluin J, Suparan P, van der Kaars S (2001) Palaeoenvironmental developments in the Lake Tondano area (N. Sulawesi, Indonesia) since 33,000 yr BP. Palaeogeogr Palaeoclimatol Palaeoecol 171(3):147–183

Donohue I, Molinos JG (2009) Impacts of increased sediment loads on the ecology of lakes. Biol Rev 84:517–531

Dunne T, Leopold LB (1978) Water in environmental planning. W. H. Freeman and Company, New York

Eagleson PS (2002) Ecohydrology: Darwinian expression of vegetation form and function. Cambridge University Press, Cambridge

Fakhrudin M, Haryani GS, Chrismadha T, Ridwansyah I (2009) Ecohydrology based Lake management: the case of lake Limboto, Gorontalo (in Indonesian). In: Proc. Natl. Conf. Lake in Indonesia, Denpasar, Ministry of Environment
Gibbs JP (2000) Wetland loss and biodiversity conservation. Conserv Biol 14(1):314–317
Grootjans A, Van Wirdum G, Kemmers R (1996) Ecohydrology in the Netherlands: principles of an application-driven interdisciplinary. Acta Botanica Neerlandica 45(4):491–516
Grossman GD, Moyle PB, Whittaker JO (1982) Stochasticity in structural and functional characteristics of an Indiana stream fish assemblage: a test of community theory. Am Nat 120:423–454
Haktanir K, Karaca A, Omar SM (2004) The prospects of the impact of desertification on Turkey, Lebanon, Syria and Iraq. In: Environmental challenges in the mediterranean 2000–2050. Springer, Dordrecht, pp 139–154
Hargono B, Pranoko W (2003) Problems and Constraints in the management plan of Lake Tempe, South Sulawesi, Indonesia. In: Proceedings of the Second Asia Pacific Training Workshop on Ecohydrology, LIPI/ UNESCO, 21–26 July, pp 47–58
Haryani GS (2008) Ecosystem based of inland water fishery management. In: Proceedings International Conference on Indonesian inland waters. Book 1: Keynote papers, Research Institute for Inland Fisheries, pp 77–84
Hassan R, Al-Ansari N, Ali AA, Ali SS, Knutsson S (2017) Bathymetry and siltation rate for Dokan reservoir, Iraq. Lakes Reserv Res Manage 22:179–189
Hoggarth DD, Cowan VJ, Halls AS, Aeron-Thomas M, McGregor JA, Garaway CA, Payne AI, Welcomme R (1996) Management guidelines for Asian floodplain river fisheries: Part 2: Summary of DFID research. Fisheries Technical Paper 384/2, F. Food and Agriculture Organization of the United Nations, Rome
Jawad L (2003) Impact of environmental change on the freshwater fish fauna of Iraq. Intern J Environ Stud 60(6):581–593
Jawad L (2013) Threatened freshwater fishes of Iraq, with remarks on their conservation status. Water Res Manage 3(2):27–36
Jones C, Sultan M, Yan E, Milewski A, Hussein M, Al-Dousari A, Al-Kaisy S, Becker R (2008) Hydrologic impacts of engineering projects on the Tigris- Euphrates system and its marshlands. J Hyd 353:59–75
Jørgensen SE (2016) Ecohydrology as an important concept and tool in environmental management. Ecohyd Hydrbiol 16(1):4–6
Maulood BK, Hinton GCF, Whitton BA, Al-Saadi HA (1981) On the algal ecology of the lowland Iraqi marshes. Hydrobiologia 80(3):269–276
Navjot S, Sodhi NS, Koh LP, Brook BW, Ng PKL (2004) Southeast Asian biodiversity: an impending disaster. Trends Ecol Evol 19:654–660
Nuttle WK (2002) Eco-hydrology's past and future in focus. Eos, Transactions AGU 83 (19):205–212
Opuszynski K (1980) The role of fisheries management in counteracting eutrophication processes. Dev Hydrobiol 2:263–269
Pouillly M, Barrera S, Rosales C (2006) Changes of taxonomic and trophic structure of fish assemblages along an environmental gradient in the upper Beni water shed (Bolivia). J Fish Biol 68:137–156
Rahi KA, Halihan T (2010) Changes in the salinity of the euphrates river system in Iraq. Reg Environ Change 10:27–35
Richardson CJ (2010) The status of Mesopotamian Marsh restoration in Iraq: A case study of transboundary water issues and internal water allocation problems. In: Towards new solutions in managing environmental crisis. Helsinki University, Helsinki, pp 59–72
Richardson CJ, Hussain NA (2006) Restoring the garden of Eden: an ecological assessment of the marshes of Iraq. AIBS Bull 56(6):477–489
Ridwansyah I, Aisyah S (2006). Hydrodynamic modeling in ecohydrology concept – integrating hydrological and limnological processes. In: Proc. Natl. Seminar on Limnology, Res. Centre for Limnology-LIPI, pp 1–15

Rietkerk M, Boerlijst MC, Van Langevelde F (2002) Self-organization of vegetation in arid ecosystems. Am Nat 160(4):524–530
Rodriguez-Iturbe I (2000) Ecohydrology: a hydrologic perspective of climate-soil vegetation dynamics. Water Reso Res 36(1):3–9
Rodriguez-Iturbe I, Porporato A, Laio F et al (2001) Plants in water-controlled ecosystems: active role in hydrologic processes and response to water stress. I. Scope and general outline. Advs Water Res 24(7):695–705
Salah AM, Al-Weshah R, Nelson EJ, Shammout MW (2006) Water resources management and modeling in Iraq, the Middle East, and the Nile Basin: towards knowledge transfer and capacity building. In: World environmental and water resource congress 2006: examining the confluence of environmental and water concerns, Omaha, pp 1–12
Sarkar UK, Bain MB (Mar. 2007) Priority habitats for the conservation of large river fishes in the Ganges river basin. Aquat Conserv 17:349–359
Sarkar UK, Gupta BK, Lakra WS (2010) Biodiversity, ecohydrology, threat status and conservation priority of the freshwater fishes of river Gomti, a tributary of river Ganga (India). Environmentalist 30(1):3–17
Saunders DL, Meeuwig JJ, Vincent ACJ (2002) Fresh water protected areas: strategies conservation. Conserv Biol 16(1):30–41
Schiemer F (2000) Fish as indicators for the assessment of the ecological integrity of large rivers. Hydrobiologia 422–423:178–271
Schneiders A, Verhaert E, Blust GD, Wils C, Bervoets L, Verheyen RF (1993) Towards an ecological assessment of watercourses. J Aquat Ecosyst Health 2:29–38
Sheldon AL (1968) Species diversity and longitudinal succession in stream fishes. Ecology 49:194–198
Toma JJ (2013) Limnological study of Dokan, Derbendikhan and Duhok lakes, Kurdistan region of Iraq. Open J Ecol 3(1):23
Van Breemen N (1995) How Sphagnum bogs down other plants. Trends Ecol Evol 10(7):270–275
Van der Schaaf S, Streefker JG (2002) Relationships between biotic and abiotic conditions. In: Schouten MGC (ed) Conservation and restoration of raised bogs: geological, hydrological and ecological studies. Dúchas, Dublin, pp 186–209
Wagner I, Breil P (2013) The role of ecohydrology in creating more resilient cities. Ecohydrol Hydrobiol 13:113–134
Wainwright J, Mulligan M, Thornes J (1999) Plants and water in drylands. In: Baird AJ, Wilby RL (eds) Ecohydrology. Routledge, London, pp 78–126
Welcomme R, Petr R (2004) Proceedings of the second international symposium on the management of large rivers for fisheries, vol 1. Food and Agriculture Organization of the United Nations, Regional Office for Asia and the Pacific, Bangkok
Wikramanayake ED, Moyle PB (1989) Ecological structure of tropical fish assemblages in wet-zone streams of Srilanka. J Zool (Lond) 218:503–526
Witte JPM, Meuleman JAM, Van der Schaaf S, Raterman B (2004) Eco-hydrology and biodiversity. In: Unsaturated zone modelling: Progress, challenges and applications. Kluwer, Dordrecht, NL, pp 301–329
Zalewski M (1998) Ecohydrology and fisheries management. Italian J Zool 65(S1):501–506
Zalewski M (2007) Ecohydrology – the use of water and ecosystem processes for healthy urban environments. In: International symposium on new directions in urban water management. UNESCO, Paris, pp 1–7
Zalewski M (2014) Ecohydrology and hydrologic engineering: regulation of hydrology-biota interactions for sustainability. J Hydrol Eng. https://doi.org/10.1061/(ASCE)HE.1943-5584.00009992014
Zalewski M, Janauer GA, Jolankai G (1997) Ecohydrology: a new paradigm for the sustainable use of aquatic resources. In: IHP-V Technical Documents in Hydrology no. 7. UNESCO, Paris

Chapter 11
Oil Pollution in the Shatt Al-Arab River and its Estuary 1980–2018

Faris J. M. Al-Imarah and Hamid T. Al-Saad

Abstract The Shatt Al-Arab river is the main river in southern Iraq used for supplying people in this area with water for different uses, drinking, irrigation, domestic, recreation, and fishing. For the last few decades, the Shatt Al-Arab river faced many problems of pollution, decrease in fresh water discharge, and increase in wastewater discharge by the river, in addition to that industrial, municipal, and irrigation, as well as agricultural wastes are main sources for contaminating the river water. The Shatt Al-Arab river is formed by the confluence of Tigris and Euphrates rivers at Qurnah. Due to increased activities of oil production in Basrah province, it is expected that the water of the Shatt Al-Arab river will deteriorate with oil and oil derivatives that enter into the river from different sources. Moreover, Shatt Al-Arab river faced many threats from oil pollution, transportation, and sinking vessels during the wars. The levels of petroleum hydrocarbons in the water, sediments, and biota from the Shatt Al-Arab river were recorded since 1984 and during the 1990s and extended through the 2000 years. Studies conducted by scientists from the Marine Science Centre of Basrah University revealed increased contamination in the water, sediments, and biota of the Shatt Al-Arab river with petroleum hydrocarbons.

Keywords Shatt Al-Arab river · Oil pollution · Contamination · Water · Sediments · Fishes

F. J. M. Al-Imarah (✉)
Department of Chemistry and Marine Environmental Pollution, Marine Science Center, Basrah, Iraq

H. T. Al-Saad
College of Marine Science, Basrah University, Basrah, Iraq

L. A. Jawad (ed.), *Tigris and Euphrates Rivers: Their Environment from Headwaters to Mouth*, Aquatic Ecology Series 11,
https://doi.org/10.1007/978-3-030-57570-0_11

11.1 Introduction

About 200 km long, 400–1500 m wide, combined effluent of Euphrates and Tigris below Qurna (Fig. 11.1) is the Shatt Al-Arab river to which the Karun contributes from Khorramshahr southward. The Shatt al-Arab river is influenced by tides in the Arabian Gulf, from about 5 m around Basra to about 3 m at its mouth, which contributes to the high level of dissolved salts in it. While, only 10% of the sediment deposited in the Shatt al-Arab river is brought by Tigris and Euphrates and 90% is carried south by Karun. At different times in the past, massive dredging operations have been undertaken in order to keep the Shatt al-Arab river navigable (Potts 2004).

11.2 Water Quality

Water is essential to human life and the health of the environment. Water has two dimensions that are closely linked: quantity and quality (http://www.environment.nsw.gov.au/water/waterqual.htm)

Fig. 11.1 Map of Iraq showing the conflict of the Euphrates and Tigris rivers to form the Shatt Al-Arab river

Water is never pure and is affected by community uses such as agriculture, urban and industrial, and recreation. Generally, the water quality of rivers is best in the headwaters, where rainfall is often abundant. Water quality frequently declines as rivers flow through regions where land and water use are intense and pollution from intensive agriculture, large towns, industry, and recreation areas increases.

Water levels in Iraq's rivers have rapidly decreased to less than a third of their normal capacity. Water levels may fall further in the coming years due to declining precipitation. Therefore, in Iraq, the quality of water used for drinking and agriculture is poor.

11.3 Water Pollution

Point sources or dispersed sources such as urban, agricultural, commercial, and industrial wastes, are main organic or inorganic water pollutants.

11.4 Water Pollution of the Shatt Al-Arab River

Different kinds of pollutants resulting from many sources enter the Shatt Al-Arab river and exert a great impact on the water quality of this river and its estuary, which eventually reflect upon living aquatic organisms. They comprise mainly as follows:

1. Sewage waters which contain both organic pollutants and nutrient salt enters the Shatt Al-Arab water without any treatment plant, that affect its water quality which leading to pollution.
2. Industrial wastes as a sources of pollutants.
3. In the agricultural fields of the Southern Iraqi, pest is being controlled by using pesticides. Measurable levels of DDT, endrine, and dieldrine were detected in water samples from the Shatt Al-Arab river, as well as in Hor Al-Hammar marsh that drains into the Euphrates river and subsequently reaches the Shatt al-Arab river.
4. Released metallic elements include essential ones (Co, Cu, Fe, Mg, Mn, and Zn) for the correct functioning of biochemical processes and nonessential elements include (Cd, Cr, Hg, Pb and so on) without any established biochemical function and represent as the more important contaminants in the aquatic environment (Viarengo 1985). Their toxic action is indicated by great affinity for amino acids and the SH groups of proteins (Hodson 1988).

   The main sources of metallic elements are sewage, waste water effluent, power plants cooling water discharge, auto emission, petroleum and petrochemical industry wastes, storm drain outfalls, and solid waste landfills (Bryan 1976; UNEP 1980).

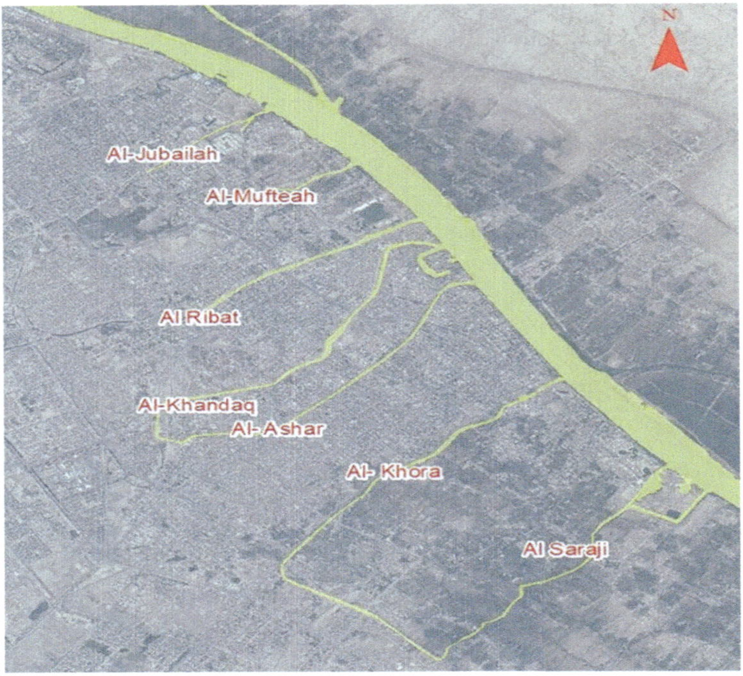

Fig. 11.2 Shatt Al-Arab River within Basrah City and its branches

5. Biological contamination: Microorganisms that affect human health greatly include pathogenic bacteria, pathogenic viruses, pathogenic protozoa (White 2015), and cyanobacteria. Bacteria that are currently targeted in water are general bacteria, total coliforms, and fecal coliforms (Inamori and Fujimoto n.d.).
6. Oil Pollution: The southern part of Iraq is exposed to oil pollution from different sources, mainly natural seepage, loading operations and transportations in which these major sources responsible for oil contamination in the sediments as well as the waters of the Shatt Al-Arab River, which empties into the North-West Arabian Gulf.

Oil spill in the aquatic systems could be identified by sight as well as sophisticated tools such as radar and optical satellites (Thomas 2017). Moreover, the levels of petroleum hydrocarbons could be measured and indicated as oil spill (Al-Saad et al. 1998).

Population of Basrah City depends upon the Shatt Al-Arab river for fresh water, even it is affected by many connected canals or branches namely: Al-Ashar, Al-Khandak, Al-Ribat, Al-Khorah, Al-Jubailah, Al-Mufteah, and Al-Saraji as shown in Fig. 11.2, which were used to be irrigation canals for the palm trees during the past, while now they are using for the discharge of waste water from the city of Basrah.

Since the first discovery of oil in the Middle East, which is back to 1908 from Massjed Suleiman (Iran), and transfer of oil towards the Arabian Gulf throughout Abadan City, the Basrah land and Shatt Al-Arab river were liable to pollution by oil (Kurt 2015). During the Iraqi–Iranian war (1980–1988), huge amounts of oil spilled in the Shatt Al-Arab waters due to military attaches, which led to sinking of many ship lounges in Shatt Al-Arab and its estuary at that time. Moreover, the northern part of the Shatt Al-Arab river is contaminated by waste water discharge through the Euphrates and Tigris rivers. During a study at 2005, a higher concentrations of PHCs were found near oil refinery terminals such as Mufteah at the middle sector of the Shatt Al-Arab River and Abadan at its southern part (Al-Fartossi 2013).

Early studies reported the contamination of waters of the Shatt Al-Arab river by petroleum hydrocarbons, Al-Saad (1983), DouAbul (1984), DouAbul and Al-Saad (1985), Al-Saad and Bedair (1989), Al-Saad (1995), Al-Saad (1998), Al-Timari et al. (2003), Awad (2004), Hantoush (2006), Hantoush (2007), DouAbul et al. (2014), and Farid et al. (2008).

It was estimated that this river transports about 48 tons of oil residues to the Arabian Gulf annually. The Arabian Gulf is also an extremely busy shipping line or oil transports, with accidental spilling being almost unavoidable. In combination, these sources provide a long-term input source of petroleum, some major spill, either unintentional or as consequence of military activities, has added occasional dramatic pulses of oil contamination to the long-term background.

11.5 Materials and Methods Usually Used for Oil Pollution

Oil pollution of water, sediments, and biota along the whole Shatt Al-Arab river starting from its confluence point of the Euphrates and Tigris Rivers at Qurnah down to its estuary at Al-Fao, have been investigated comprehensively by different researchers and organizations. Most of them revealed that the water of the Shatt Al-Arab River is contaminated to a certain extent with petroleum hydrocarbons of an oil origin. Most of the study stations chosen for sampling were close to the most crowded population along the Shatt Al-Arab River.

Petroleum hydrocarbons exist in water in two phases, dissolved and particulate. Water samples were collected by a water sampler in amber glass bottles, filtered on board by GFF filter paper type, filtrate water represents the dissolved petroleum hydrocarbons while retained materials in the GFF filter paper represent the particulate petroleum hydrocarbons.

Petroleum hydrocarbons were extracted from water samples according to UNEP (1993). On the other hand, the procedure of Goutx and Saliot (1980) was adopted for the extraction of petroleum hydrocarbons from sediments. For the extracts, the total petroleum hydrocarbons were estimated spectrofluorimetrically by UVF Shimadzu RF spectrofluorimeter fitted with the direct reading DR database unit and 1 cm path length quartz cell, excitation was at 310 nm and emission at 360 nm (Al-Saad 1995).

A calibration curve was done for Basrah light crude oil to calculate the concentrations of TPHs in each sample of water, sediment, and biota.

11.5.1 Petroleum Hydrocarbons in Water

11.5.1.1 PHC's in the Dissolved Phase of Water

Dissolved PHC's were solvent extraction according to UNEP (1993) in which the carbone tetrachloride solvent was used for extraction. Three-time portions of CCl_4 each of 50 ml were added to the filtrate and shook vigorously to disperse the solvent, shaking is repeated many times to ensure solvent dispersant and to each extract a small amount of anhydrous sodium sulfate was added to break any emulsion and to remove excess water. The combined extracts of CCl_4 were reduced in volume to less than 5 ml.

11.5.1.2 PHC's Extraction as Suspended Particulate Matter

The extraction procedure was based upon earlier study, Al-Imarah et al. (2016a, b). PHCs in GFF were extracted by organic solvents and determined by HPLC analysis.

11.5.1.3 PHC's in Sediments

The sediment samples were dried, grinded finely and sieved through a 63 μm mesh sieve, stored in glass containers until analysis. Twenty grams of sieved sediments were placed in cellulose thimble and the extraction of PHC's were adopted from Goutx and Saliot (1980).

11.5.1.4 PHC's in Fish Samples

Fishes were caught by net, the fish samples were wrapped in aluminum foil, stored in cool boxes, and frozen upon return to the laboratory of Marine Science Centre, and classified according to Fischer and Bianchi (1984) and Coad (2010).

The extraction of petroleum hydrocarbons from the fish samples were done according to the method of Grimalt and Oliver (1993).

Table 11.1 Concentrations of total petroleum hydrocarbons in water (μg/l) from the Shatt Al-Arab river, 1980–2017

Year	Level of PHC's μg/l	References
1980	5.6–14.2	Al-Saad (1983)
1984	12–86.7	DouAbul (1984)
1985	5.2–14.2	DouAbul and Al-Saad (1985)
1989	6.5–23.5	Al-Saad and Bedair (1989)
1995	3.97–38.29	Al-Saad (1995)
1998	1.3–35.0	Al-Saad (1998)
2003	2.5–47.0	Al-Timari et al. (2003)
2004	0.01–6.83	Awad (2004)
2005	2.247–50.232	Hantoush (2006)
2006	2.3–50.2	Hantoush (2007)
2010	5.67–9.48	Al-Imarah et al. (2010)
2011	1.077–11.816	Al-Fartossi (2013)
2012	18.20–30.81	DouAbul et al. (2014)
2014	5.18–37.58	Al-Hjuje et al. (2015a)
2015	8.124–22.98	Al-Imarah et al. (2017)
2016	3.09–30.87	Al-Saad et al. (2017)
2017	1.97–18.16	Al-Imarah et al. (2018)

Table 11.2 Concentrations of PHC's in the particulate phase of waters from the Shatt Al-Arab river 2013

Stations (The upper part of Shatt Al-Arab River)	Particulate petroleum hydrocarbons (μg/g)
Qurnah	55.752
Deer	62.294
Qarmah	77.46
Ashar	89.352
Abo Al-Khaseeb	109.562

11.6 Results and Discussion

Water of the Shatt Al-Arab river is liable to contain petroleum contaminants because of numerous operations. Table 11.1 shows the levels of total petroleum hydrocarbons in the dissolved fraction of waters from the Shatt Al-Arab river since 1980 together with current studies. Table 11.2 shows the levels of PHCs in the particulate fraction of water from the Shatt Al-Arab river.

The levels of the total petroleum hydrocarbons in the sediments of the Shatt Al-Arab river were recorded thoroughly by different researchers and these levels are listed in Table 11.3, while those that are reported in the muscles of fishes are listed in Table 11.4.

PHCs in the waters and sediments of the Shatt Al-Arab river is reduced from the northern part (Ali 2006) towards the southern part (Ali 2006; Ibraheem 2004).

As it is a tidal water body that receives fresh water from the Tigris and Euphrates rivers at its upstream boundary, the Shatt Al Arab river is the main source for all

Table 11.3 Ranges of the total petroleum hydrocarbons in the sediments of the Shatt Al-Arab river

Year	Range of TPHs (μg/g)	References
1980	2.6–44.0	DouAbul (1984)
1982	2.6–20.5	Al-Saad (1983)
1989	3.77–26.7	Al-Hamdi (1989)
1993	2.46–38.33	Al-Saad et al. (1995)
1997	0.108–37.02	Al-Khatib (1998)
2004	7.37–24.41	Ibraheem (2004)
2005	34.26–146.64	Ali (2006)
2010	7.37–24.81	Al-Imarah et al. (2010)
2011	28.463–63.1598	Al-Fartossi (2013)
2014	4.76–45.24	Al-Hjuje et al. (2015b)
2015	48.48–134.619	Al-Imarah et al. (2017)
2016	22,16–78.88	Al-Imarah et al. (2017)
2017	3.28–46.64	Al-Imarah et al. (2018)

Table 11.4 Concentrations of the total petroleum hydrocarbons in the muscles of fish from the Shatt A-Arab river

Year	Conc. TPH (μg/g)	References
1980	29.7–30.8	Al-Saad and Al-Asadi (1980)
1986	29.6–45.9	DouAbul et al. (1987)
1989	11.44–48.16	
1997	1.7–10.91	Al-Saad et al. (1997)
2001	7.97–35.73	Hantoush et al. (2001)
2012	5.695–6.695	Al-Khion (2012)
2014	2.785–13.32	Al-Saad et al. (2016)
2015	4.62–10.67	Al-Imarah et al. (2016a, b)
2017	0.407	Al-Imarah et al. (2017)

water treatment plants within Basrah governorate including the large and package treatment units. The river is exposed to different sources of pollution, it is affected by tides from the North-East of Arabian Gulf at its downstream boundary, and it is used as a disposal site for a portion of untreated sanitary sewage, which is discharged to it through the highly polluted lateral creek or branches namely Al-Ashar, Al-Khandaq, Al-Ribat, Al-Jubailah, Al-Khorah, Al-Mufteah, and Al-Saraji, as shown in Fig. 11.2, which were used to be irrigation canals for the palm trees during the past, while at the time being they are using for the discharge of waste water from the city of Basrah.

Therefore, the water of this river is of variable quality due to natural and man-made reasons. All of these pollutants are distributed in water, sediment, and biota, and in this review, the focus is upon oil pollution, the most effective pollutant in the environment of the Shatt Al-Arab river. Therefore, as it is affected mainly by oil pollutants, the water quality of the Shatt Al-Arab river was monitored during different sampling periods, since 1984 at six different stations along the river, together with sediments and fishes.

The sources of oil pollution in the Shatt Al–Arab river were assigned as oil refineries, rural runoff, electricity generating stations, sewage discharges, and river transportation activities, (DouAbul and Al–Saad 1985) in which sewage discharge

Table 11.5 Comparison of petroleum hydrocarbon residues in sediment samples from some regional and international locations

Location	Mean conc. µg/g	Reference studies
Coast of Kuwait	7.43–485.61	Metwally et al. (1997)
UAE	7	Sheridan (1999)
Saudi Arabia	13–540	Ehrhardt and Burns (1993)
Mediterranean	82–122	Ehrhardt and Burns (1993)
Arabian gulf	2.46–38.33	Al-Saad (1995)
Gulf of Aden	0.12–6.94	Heba et al. (2003)
Red sea	1.12–6.94	Heba and DouAbul (2000)
Aden coast	0.3–25.72	Bedair et al. (2006)
Shatt Al-Arab River	0.4–134.619	Present study

and urban runoff were the most significant sources of oil entering the Shatt Al–Arab river. Al–Saad (1995) reported that many aquatic organisms of the Shatt Al–Arab river including plants, algae, zooplankton, bacteria, and fish were capable to synthesize biogenic hydrocarbons.

During the Iraqi–Iranian war (1980–1988), huge amounts of different oil derivatives were spilled in the Shatt Al-Arab waters due to military attaches, which led to the sinking of many ship lounges in Shatt Al-Arab and its estuary at that time (Al-Fartossi 2013).

Moreover, the Shatt Al-Arab river is characterized by considerable inputs of different contaminants originated from the urbanized areas and industrial sites (Moyel and Hussain 2015). These pollutants may have severe impact on the water quality of the river and eventually upon living aquatic organisms (Farid et al. 2014).

According to a study for the total petroleum hydrocarbons, it was indicated that the sources of petroleum hydrocarbons in the muscles of fishes from the Shatt Al-Arab river are of two types namely pyrogenic and anthropogenic (Al-Imarah et al. 2017). Even though the detection of polluted petroleum hydrocarbons in aquatic and marine/organisms is a complicated task, since those organisms can also produce hydrocarbons (Al-Saad et al. 2008), the total amount of petroleum hydrocarbons detected in those organisms cannot be taken as an index for pollution by petroleum products only (Mottier et al. 2000).

In a comparison, the results reported in this review were contrasted with previous studies from some of the Gulf States and worldwide countries in which similar results were declared. Such studies indicated that the highest values were found near ports, industrial areas, and oil refineries, where boating activities, loading, and waste disposals are commonly practiced. The comparisons showed that the present data are much lower in magnitude (Table 11.5). Due to the pollution of the northwest part of the Arabian Gulf by petroleum hydrocarbons, it represents as another source for the pollution of the Shatt Al-Arab River with petroleum hydrocarbons (Al-Imarah et al. 2007).

11.7 Conclusion

The concentrations of PAHs measured in the Shatt Al-Arab river indicate that the contamination is considerable and comparable to those found in other places around the world. The comparison revealed the elevated petroleum hydrocarbon concentrations in the sediments of the Shatt Al-Arab river compared to those reported by Wang et al. (2011) and Zrafi et al. (2013).

Most studies conducted in the area of the Shatt Al-Arab river over the past years using high-quality monitoring techniques have observed a continuous pollution by petroleum hydrocarbons with an alternative change in their levels from north to south depending upon different factors, such as an increase of activities in the river, establishment of new wells for crude oil production on the land, discharge of pollutants by the Euphrates and Tigris rivers to the northern part of the Shatt Al-Arab River, as air pollutants during rain, and illegal transportation of crude oil through the river. Moreover, most of the studies revealed that the levels of petroleum hydrocarbons were comparable to those observed elsewhere. Consequently, the levels of petroleum hydrocarbons in the exchangeable and residual phases of the sediments are greater than in the dissolved and particulate phases of water, indicating that petroleum hydrocarbons in the Shatt Al-Arab river are transferred towards the Arabian Gulf mostly by sediments.

11.8 Recommendations

1. Fuel that is used for boating and fishing in the Shatt Al-Arab river should be regulated by certain laws to prevent the release of fuel to the water and deteriorate it.
2. As it is a very important source of water to be used for different purposes, it should be protected severely.
3. Rivers frequently act as conduits for pollutants by collecting and carrying wastewater from crowded cities, discharging them into the sea; therefore, artificial canals should be constructed alongside the Shatt Al-Arab river to carry and discharge waste water towards the Arabian Gulf.
4. Water levels in. Iraq's rivers have rapidly decreased to less than a third of their normal capacity. Water levels may fall further in the coming years due to declining precipitation. Therefore, in Iraq, the quality of water used for drinking and agriculture is poor; hence, a dam should be constructed on the Shatt Al-Arab river to keep losing water towards the Arabian Gulf.

References

Al-Fartossi FM (2013) The impact of maritime oil pollution in the marine environment: a case study of maritime oil pollution in the navigational channel of Shatt Al-Arab River. World Maritime University Dissertation

Al-Hamdi MM (1989) Hydrocarbon sources and vertical, distribution in sediment from Khor Al-Zubair, N.W. Arabian Gulf. M.Sc. Thesis, Basrah Univ., p 131

Al-Hjuje MM, Hussain NA, Al-Saad HT (2015a) Total petroleum hydrocarbons, n-alkanes, and Polycyclic aromatic hydrocarbons in water if Shatt Al-Arab river: part 1. GJBAHS 4(1):88–94

Al-Hjuje MM, Hussain NA, Al-Saad HT (2015b) Total petroleum hydrocarbons, n-alkanes, and polycyclic aromatic hydrocarbons in water if Shatt Al-Arab river: part 2. GJBAHS 4(1):95–100

Ali SAR (2006) Determination and Distribution of Petroleum Hydrocarbons in water and sediments from the Northern Part of Shatt Al-Arab River. Ph.D. Thesis, Coll. of Edu., Univ. of Basrah

Al-Imarah FJM, Hantoush AA, Nasir AM (2007) Petroleum hydrocarbons in water and sediments of northwest Arabian gulf 1980–2005. Aquat Ecosyst Health Manage 10(3):335–340

Al-Imarah FJM, Ali SA, Ali AA (2010) Temporal and special variations of petroleum hydrocarbons in water and sediments from Northern part of Shatt Al-Arab River, Iraq. Mesopot J Mar Sci 25 (1):65–74

Al-Imarah FJM, Nasir AM, AAK A-T (2016a) Concentrations of total petroleum hydrocarbons and chemical elements in fish muscles from North West of Arabian Gulf and Shatt Al-Arab. JPCBS 4(2):188–196

Al-Imarah FJM, Abdulrahman BA, Khoiedem AA (2016b) Monthly variations of particulate petroleum hydrocarbons in the waters of Shatt Al-Arab River/southern Iraq. Asian academic research J. Multidisciplinary 3(9):142–150

Al-Imarah FJM, Nasir AM, Al-Timari AAK, Al-Anbar LJM (2017) Levels of polycyclic aromatic hydrocarbons (PAHs) in muscles of commercial fishes from Iraqi water. Mesopotamian J Mar Sci 32(1):45–55

Al-Imarah FJM, Hantoosh AA, Al-Bidhani MFH, Abdul Nabi ZA, Abdulhadi ZM, Khwaidem AA (2018) Monitoring project for the determination of total petroleum hydrocarbons in water and sediments of Shatt Al-Arab river. Unpublished work

Al-Khatib FM (1998) Distribution of hydrocarbons compound and their sources in sediment cores from Shatt Al-Arab estuary and north-west Arabian gulf. M.Sc. Thesis, University of Basrah, College of Science, Biology Department, p 95

Al-Khion D (2012) Distribution of polycyclic nuclear compounds in Iraqi coast regions. Ph.D, Thesis, College of Agriculture, University of Basrah, p 171

Al-Saad HT (1983) A baseline study on the petroleum hydrocarbons pollution in Shatt Al-Arab River. M.Sc. Thesis, College of Science, Basrah University

Al-Saad HT (1995). Distribution and sources of hydrocarbons in Shatt Al-Arab Estuary and North-West Arabian Gulf. Ph.D. Thesis, Basra Univ., p 186

Al-Saad HT (1998) Seasonal distribution of dissolved and particulate hydrocarbons in Shatt AI-Arab estuary and north-west Arabian gulf. Mar Pollut Bull 36(10):850–855

Al-Saad HT, Al-Asadi MK (1980) Petroleum hydrocarbon concentrations in fishes from Shatt Al-Arab River. Marine Mesopotamica 4(2):233–242

Al-Saad HT, Bedair HM (1989) Hydrocarbons and chlorophyll a correlation in the waters of the Shatt Al-Arab River. Basrah Iraqi Marine Mesopotamica 4(1):117–126

Al-Saad HT, Abaychi JK, Shamshoom SM (1995) Hydrocarbons in the waters and sediments of Shatt Al-Arab estuary and north-west Arabian gulf. Marina Mesopotamica 10(2):393–410

Al-Saad HT, Shamshoom SM, Abaychi JK (1997) Hydrocarbons in fish from Shatt Al-Arab estuary and North West Arabian Gulf following the Gulf oil spill of 1991. Marine Mesopo 12 (2):213–231

Al-Saad HT, Shamshoom SM, Abayachi JK (1998) Seasonal distribution of dissolved and particulate hydrocarbons n Shatt Al-Arab Estuary and North West Arabian Gulf. Marine Pollut Bull 36(10):850–855

Al-Saad HT, DouAbul AA, Abd IM, Zukhair MK (2008) Uptake –release of pollution by *Hilsa ShadTenualosailisha* (Hamilton-Buchanan) fish collected from southern Iraq. Marine Mesopotamica 23(2):237–255
Al-Saad HT, Ali ST, Mahdi S, Alkafaji BY, Al-Hello A (2016) Polycyclic aromatic hydrocarbons (PAHs) in sediment samples from Euphrates River, Iraq. Int J Mar Sci 6(2):1–5
Al-Saad HT, Al-Timari AAK, Douabul AAZ, Hantoush AA, Nasir AM, Saleh SM (2017) Status of oil pollution in water and sediment from Shatt AlArab Estuary and North-West Arabian Gulf. Mesopot J Mar Sci 32(1):9–18
Al-Timari AAK, Hantoush AA, Nasir AM (2003) Petroleum hydrocarbons in southern Iraqi waters. Marina Mesopotamica 18(2):141–149
Awad N (2004) Determination of total petroleum hydrocarbons and heavy metals in waters and sediments from Shatt Al-Arab River. Marine Science Centre, Basrah
Bedair HM, Saeed MA, Al-Saad HT (2006) Status of oil pollution in sediment samples from the Aden Coast, Yemen. J Food Agric Environ 4(3&4):284–287
Bryan GW (1976) Metal concentration in the sea. Mar Poll Bull 7:353–356
Coad BW (2010) Fresh water fishes of Iraqand Al-Saad, H. T. (2014). Distribution, sources, and seasonal variations of hydrocarbons in Shatt Al-Arab River water. Journal of International Academic
DouAbul AAZ (1984) Petroleum residues in the water of the Shatt ai-Arab River and NW region of the Arabian Gulf. Environ Int 10(3):265–267
Douabul AAZ, Al-Saad HT (1985) Seasonal variations of oil residues in water of Shatt Al-Arab river Iraq. Water Air Soil Pollut 24:237–246
DouAbul AAZ, Abaychi JK, Al-Edanee TE, Ghani AA, Al-Saad HT (1987) Polynuclear aromatic hydrocarbons (PAHs) in fishes from the Arabian gulf. Bull Env Contam Toxicol 38:546–552
DouAbul AAZ, Al-Timari AA, Al-Saad HT, Hantoush AA, Nasir AM (2014) Total petroleum hydrocarbons in water and sediment of Shatt Al-Arab and Iraqi coastal (In Progress)
Ehrhardt M, Burns K (1993) Hydrocarbons and related photooxidation products in Saudi Arabian gulf coastal waters and hydrocarbons in underlying sediment and bioindicators bivalves. Mar Pollute Bull 27:187–199
Farid WA, Al-Saad HT, Al-Adhub AY (2008) Monitoring of hydrocarbons in Shatt Al-Arab River by using some species of mollusks. Marine Mesopotamica 23(2):305–321
Farid WA, Al-Eed AA, Shihab LA, Al-Saad HT (2014) Distribution, sources, and seasonal variations of hydrocarbons in Shatt Al-Arab River water. J Int Acad Res Multidiscip 2 (2):729–739
Fischer W, Bianchi G (1984) Species identification sheets for Fischer purpose, Western Indian ocean (fishing area 51). FAO, Rome
Goutx MA, Saliot A (1980) Relationship between dissolvedand particulate fatty acids, hydrocarbons, chlorophyll a and zooplankton biomass in Ville franche bay, Mediterranean Sea. Mar Chem 8:299–318
Grimalt JO, Oliver J (1993) Source in putelucidation in aquatic systems by factor and principal component analysis of molecular marker date. AnalChem Acta 278:159–176
Hantoush AA (2006) A study of oil pollution status in water and sediments of Shatt Al-Arab River South of Iraq. Ph.D Thesis, College of Science, University of Basrah, p 142 (In Arabic)
Hantoush AA (2007) A study of oil pollution status in water and sediments of Shatt Al-Arab River, South of Iraq. Ph.D. thesis, College of Science, University of Basrah, p 142.
Hantoush AA, Al-Imarah FJM, Abdullah SHB (2001) Seasonal variations of petroleum hydrocarbon concentrations in freshwater and marine fishes from Shatt Al-Arab River and NW Arabian gulf. Mar Mesop Spec Issue 16(2):441–451
Heba HM, DouAbul AAZ (2000) Concentration of petroleum hydrocarbons in fish, mussels and sediment samples from the Red Sea – Coast of Yemen. Bull Nat Inst Oceanogr Fish ARE 26:283–296
Heba HM, Saeed MA, Al-Shawafi N, Al-Saad HT (2003) Petroleum hydrocarbons and trace metals in mollusca Tivela ponderosa from the Gulf of Aden. J King Abdulaziz Univ Mar Sci 14:89–97

Hodson PV (1988) The effect of metal metabolism on uptake, disposition and toxicity in fish. Aqua Toxicol 11:3–18

Ibraheem SA (2004) Determination and distribution of total petroleum hydrocarbons, total organic carbon and nickel and vanadium metals in water and sediments from the southern sector of Shatt Al-Arab river, Iraq, Ph.D. Thesis Iraq: University of Basrah

Inamori Y, Fujimoto N (n.d.) Water Quality and Standards – Vol. II – Microbial/Biological 385 Contamination of Water ©Encyclopedia of Life Support Systems (EOLSS) Microbil/Biological 386 Contamination of Water. https://www.eolss.net/sample-chapters/C07/E2-19-05-03.pdf

Kurt B (2015) Discovery of oil and oil based environmental pollution in ottoman Iraq: the incident of Mohammarah (1913–1914). J Eurasina Stud IV(2):162–177

Metwally M, Al-Muzaini S, Jacob P, Bahloul M, Urushigawa Y, Sato S, Matsumura A (1997) Petroleum hydrocarbons and related heavy metals in the near–shore marine sediments of Kuwait. Environ Internat 23:115–121

Mottier P, Parisod V, Turesky RJ (2000) Quantitative determination of polycyclic aromatic hydrocarbons in barbecued meat sausages by gas chromatography coupled to mass spectrometry. J Agric Food Chem 48:1160–1166

Moyel MS, Hussain NA (2015) Water quality assessment of the Shatt Al-Arab River, Southern Iraq. J Coast Life Med 3(6):459–465

Potts DT (2004) Shatt Al-Arab River. Encyclopedia Iranica

Sheridan M (1999) Heavy metals in mangrove sediments of the United Arab Emirates shoreline (Arabian gulf). Water Air Soil Pollut 116:523–534

Thomas O (2017) Oil spill in the Persian Gulf. Skytruth. http://skytruth.org/2017/03/oil-spill-in-the-persian-gulf/

UNEP (1980) Report on land based sources of industrial pollution in Kuwait. Draft report, United Nations environment Programme. Kuwait action plan- Interm secretariat informal consultation on the implementation of group "B" projects

UNEP (1993) Gide lines for monitoring chemical contamination in the sea using marine organisms, reference methods for marine pollution studies, no. 20

Viarengo A (1985) Biochemical effects of trace metals. Mar Poll Bull 16:153–158

Wang C, Wang W, He S, Due J, Sun Z (2011) Sources and distribution of aliphatic and polycyclic aromatic hydrocarbons in Yellow River Delta nature reserve. China Appl Geochem 26:1330–1336

White S (2015) Water contamination: sources & solutions. Water quality products, contamination issues dominate water quality concerns worldwide World Maritime University

Zrafi I, Hizem L, Chalghmi H, Ghrabi A, Rouabhia M, Saidane-Mosbahi D (2013) Aliphatic and aromatic biomarkers for petroleum hydrocarbon investigation in marine sediment. Journal of Petroleum Science Research (JPSR), 2(4):146–155. Research for Multidisciplinary 2 (2):729–739

Chapter 12
Pesticides in the Waters, Sediments, and Biota of the Shatt Al-Arab River for the Period 1980–2017

Hamid T. Al-Saad and Faris J. M. Al-Imarah

Abstract Mesopotamia is predominantly an agricultural region and pest control is of great economic significance. Organochlorine pesticides have been used in Iraq and neighboring countries for more than five decades. Organochlorine pesticides enter the Shatt Al-Arab river through drainage from the farmlands of the Tigris and Euphrates basin, either adsorbed onto particulate matters or dissolved. Eventually, most contaminated particles settle to the Shatt Al-Arab sediments and only minor amounts are transported to the Gulf. Reports indicated that the Hor Al-Hammar lake and associated marshes in southern Iraq were sprayed with DDT and aldrin–dieldrin during the period 1950–1976, which have been indicated by the measurable levels of DDT, endrin, and dieldrin, which were encountered in different samples from the Shatt Al-Arab river and Hor Al-Hammar marsh. They transfer in the area either dissolved within a water column or a particular adsorbent, which settle down to the bottom or are transported to the Arabian Gulf. In another study, researchers observed high concentrations of $\sum$DDT residues, endrin, and dieldrin in the Indian shad (*Tenualosa ilisha*) muscles collected from the Shatt Al-Arab Estuary. Indian shad is a migratory fish species with a high fat content, which migrates between the northern Gulf waters and Iraq's marshes and river system for breeding and spawning.

Keywords Pesticides · Shatt Al-Arab River · Northern Arabian Gulf · GC · Analysis · Pollution

H. T. Al-Saad (✉)
College of Marine Sciences, University of Basrah, Basrah, Iraq

F. J. M. Al-Imarah
Marine Science Center, Basrah University, Basrah, Iraq

L. A. Jawad (ed.), *Tigris and Euphrates Rivers: Their Environment from Headwaters to Mouth*, Aquatic Ecology Series 11,
https://doi.org/10.1007/978-3-030-57570-0_12

12.1 Introduction

The term "pesticide" is a composite term that includes all chemicals that are used to kill or control pests. In agriculture, this includes herbicides (weeds), insecticides (insects), fungicides (fungi), nematocides (nematodes), and rodenticides (vertebrate poisons).

Agriculture is one of the few activities where pesticides as chemical compounds are used in which their ecological impacts in water are determined by their toxicity, persistence, degradation, and fate.

Due to the health effects that affect human beings, certain international organizations issued regulations and guidelines for using waters (UNEP 1993; WHO 1993). Pesticides are included in a broad range of organic micropollutants that have ecological impacts. Different categories of pesticides have different types of effects on living organisms.

Iraq is heavily dependent on the Euphrates and Tigris rivers as fresh waters for drinking, sanitation, and irrigation. The ecosystem that relays on the Tigris and Euphrates rivers is quite effective, especially the marshes of Iraq. Human activities caused severe salivation, which could cause serious water quality challenges in Iraqi cities.

The Tigris and Euphrates rivers are passing through many cities within Turkey, Syria, and Iraq, which receive different pollutants as waste, agricultural run-off, domestic, and commercial waters (Al Bomola 2010).

Tigris is joining with the Euphrates at Qurna south of Iraq and forming Shatt Al Arab, which flows 190 km before emptying into the Arabian Gulf (Isaev and Mikhailova 2009).

The major tributaries joining Tigris inside Iraq are Khabour River, near the Turkish-Iraqi border (Tusan), Greater Zab, at about 50 km downstream Mosul, Lesser Zab at about 220 km, upstream Baghdad, Adhaim river joins the Tigris river some 30 km downstream of Samarra, and Dyalah river that feeds into the Tigris below Baghdad.

The Shatt Al-Arab River is an important river as it is the only source of freshwater in the arid surroundings of southern Iraq. It is the prime freshwater source and pours about 5×10^9 m^3 nutrient-rich water into the Arabian Gulf each year (DouAbul and Al-Saad 1985).

Pesticides represented by organochlorines were recorded in fish tissues and water from different environmental sites (Ogunfowokan et al. 2012; Essumang et al. 2009; Malik et al. 2007).

Several factors affect the hydrological condition of the Shatt al-Arab river basin (Al-Mahmood 2009; Moyel 2014).

12.2 Methodology

Organochlorine pesticides were investigated in water, sediments, and fish samples from the Shatt Al-Arab river, its estuary, and adjacent Hor Al-Hammar marsh within the southern sector of Iraq.

12.2.1 *Study Area*

This study was focused on the aquatic system of southern Iraq represented by the Shatt Al-Arab river from its formation at Qurnah city down to Faw city and its estuary, as well as Hor Al-Hammar marsh, as shown in Fig. 12.1. Effective branches are leaches from western Hammar marsh represented by the Al-Shafy and Qarmat Ali rivers in addition to other branches that leach from the city of Basrah. These branches include as follows:

Fig. 12.1 Southern sector of Iraq showing Tigris, Euphrates, and Shatt Al-Arab rivers as well as Hor Al-Hammar Marsh to the west of Shatt Al-Arab

Saraji, Khora, Ashar, Khandek, Robat, Jubyla, and Shatt Al Turk all were characterized as agricultural canals. Sites of sampling were extended from Qurnah at the Euphrates-Tigris-Shatt Al-Arab delta (No. 1) to Al-Faw at Shatt Al-Arab estuary (No. 6) and along the whole Shatt Al-Arab river for sites 2–5, as shown in Fig. 12.1 (DouAbul et al. 1988).

12.2.2 Materials

Pesticides grade solvents acetone, methylene chloride, n-hexane, and acetonitrile were obtained commercially and were used as received. High-purity standard of pesticides (>99%) and their related compounds were used without further purification. Florisil PR, 60–100 mesh, and anhydrous granular sodium sulfate were purified by extraction with n-hexane for a minimum of 36 h in a soxhlet apparatus. Following cleanup extraction, they were oven-dried at 130 °C for 24 h prior to use. All glassware used throughout the investigation of pesticides in water, sediments, and biota was acetone and/or n-hexane-rinsed and then oven-baked at 300 °C for about 24 h prior to use.

12.2.3 Sample Collection

Within most studies for pesticide investigations, six sampling locations along the Shatt al- Arab river were selected (Abaychi and DouAbul 1985), Samples investigated were water, sediments, and fishes.

Internationally recognized guidelines were adopted for sampling (UNEP 1991) All samples of water, sediments, and biota were pretreated before transferring to the lab and kept cooled prior to analysis. Fish samples were cyprinid (*Barbus xanthopterus*; lean fish, residential species) and the Indian shad (*Tenualosa ilisha*; medium fatty, migratory species).

12.2.4 Extraction of Pesticides in Environmental Samples

12.2.4.1 Water Samples

Organochlorine pesticides in the dissolved phase of water were investigated following a procedure described by Dawson and Riley (1977). Gas chromatography with micro electron capture detector (GC-IECD) was used for the final detection of pesticides in the water samples.

12.2.4.2 Sediment Samples

For sediments, Soxhlet extraction was used for the extraction of pesticides in sediment samples. The final step was the analysis by gas chromatography.

12.2.4.3 Fish Samples

For fish samples, the extraction procedure employed for the edible portions of fishes was based upon the Draught Method of the Standing Committee of Analysis for the determination of organochlorine pesticides in fish tissues.

12.2.5 Analysis

All the samples were analyzed on Gas Chromatograph (Perkin Elmer Instruments, Auto System XL GC) equipped with an electron capture detector (ECD) and a Perkin Elmer wide-bore capillary column (PE-5 column) was used for the analysis of pesticides in all samples of water, sediments, and fishes as described in the literature (DouAbul et al. 1987; EPA 2007).

12.3 Results and Discussion

Routs for chlorinated pesticides entering the aquatic ecosystem are water run-off from agricultural areas and atmospheric fallout.

As the main class of toxic compounds, organochlorine pesticides have a relative chemical and biological stability (Al-Shawafi et al. 2009). They are soluble in liquids. During the period 1979–1991, the residues of organochlorine pesticides were investigated in water, sediments, and organisms collected from the Southern Iraqi ecosystem (DouAbul and Al-Timari 2014). The levels of organochlorine pesticides are shown in Table 12.1 (DouAbul et al. 1987; ROPME 1986). In the last decade, only two studies on organochlorine residues have been done. One of them in the Mesopotamian wetlands of southern Iraq in which the residues of p, p′-DDT have been detected in all examined samples of the studied area. This indicates its ability to persist under severe drying of previously exposed surface sediments (DouAbul et al. 2009) and the second by using the samples of water, sediment, fish, and shrimp from Hor Al-Hammar Marshes of Iraq, high concentration of chlordane in both water and sediments were shown (Al-Hilffi 2005).

Table 12.1 Concentrations (ng/l) of organochlorine pesticides in the dissolved phase of water from the Shatt Al-Arab river and Hor Al-Hammar Marsh

Site	Endrin	Aldrin	Dieldrin	∑DDT	Lindane	Heptachlor	References
Shatt Al-Arab	Nd	18–30	24–66	81–168	–	10–79	DouAbul et al. (1988)
Hor Al-Hammar	Nd-4.05	0.05–33.03	Nd-0.3	Nd-23.71	0.15–73.78	0.19–59.5	Al-Ali (2012)
Hor Al-Hammar				0.086	0.072	0.7	Latif et al. (2013)
Shatt Al-Arab	Nd-217	–	Nd-0.97	0.1–2560	–	Nd-679	DouAbul and Al-Timari (2014)

Table 12.2 Concentrations (ng/g) of organochlorine pesticides in the particulate phase of water from the Shatt Al-Arab river and Hor Al-Hammar Marsh

Site	Endrin	Aldrin	Dieldrin	∑DDT	Lindane	Heptachlor	References
Shatt Al-Arab	85–154	Nd	Nd	143–1221	–	68–186	DouAbul et al. (1988)
Hor Al-Hammar	Nd-30.94	Nd-8253	Nd-30.18	Nd-238.7	0.08–321.62	0.19–59.5	Al-Ali (2012)
Hor Al-Hammar				0.086	0.072	0.7	Latif et al. (2013)
Shatt Al-Arab	Nd-111	–	Nd-0.97	0.1–2560	–	Nd-679	DouAbul and Al-Timari (2014)

12.3.1 Organochlorine Pesticides in Water

The total concentrations of the selected pesticides are presented in Table 12.1 for the dissolved phase and Table 12.2 for the particulate phase of water from Shatt Al-Arab and Hor Al-Hammar marsh, the southern sector of Iraq. Certain pesticides recorded a high concentration in some locations close to a source input flow (Latif et al. 2013).

Al-Ali (2012) reported high levels for all selected pesticides in water samples from Hor Al-Hammar, on the other hand, DouAbul and Al-Timari (2014) reported high levels for these pesticides in water samples from the Shatt Al-Arab river in which they concluded that Hor Al-Hammar was the main source for pollution by pesticide in the waters of the Shatt Al-Arab river, especially ∑ DDT that has been banned in Iraq a long time ago but are used by marsh residence for agricultural requirements or fishing.

12.3.2 Organochlorine Pesticides in Sediments

The higher levels of pesticides were found in the sediments than water. The total organochlorine pesticide concentrations in the sediment of the Shatt Al-Arab river and Hor Al-Hammar marsh, the southern sector of Iraqi are presented in Table 12.3.

Table 12.3 Concentrations (ng/g) of organochlorine pesticides in the sediments from the Shatt Al-Arab river and Hor Al-Hammar Marsh

Site	Endrin	Aldrin	Dieldrin	∑DDT	Lindane	Heptachlor	References
Shatt Al-Arab	20	Nd	10–20	10–15	Nd	–	ROPME (1986)
Shatt Al-Arab	40	–	20	5	–	–	DouAbul et al. (1987)
Shatt Al-Arab	3–18	Nd-5	13–22	0.08624–220	–	Nd-24	DouAbul et al. (1988)
Hor A-Hammar	0.08	Nd	0.04	0.07	–	–	Al-Timari (1997)
Hor Al-Hammar	–	–	–	0.29–2.33	–	–	DouAbul et al. (2009)
Hor Al-Hammar	Nd-36-45	0.45–98.24	Nd-2.52	0.51–274.67	0.37–190	0.48–119.74	Al-Ali (2012)
Hor Al-Hammar	–	–	–	16.7	7.9	0.096	Latif et al. (2013)
Shatt Al-Arab	Nd-47	–	Nd-22	0.04–220	–	Nd-42	DouAbul and Al-Timari (2014)

During all studies presented in Table 12.3, it seems that the ∑DDT pesticides are the highest in concentrations in both the Shatt Al-Arab river and Hor Al-Hammar marsh compared with most of the studied pesticides. DouAbul et al. (2009) reported the values of 0.29–2.33 ng/g for ∑DDT, the only pesticide found in all investigated sediment samples from Hor Al-Hammar Marsh, which indicates their persistence under severe drying surface of the sediments.

12.3.3 Organochlorine Pesticides in Fishes

Fish samples can be considered as one of the most significant indicators in fresh water systems for the estimation of pesticide contamination. Organochlorine pesticides in fish tissues from the Shatt Al-Arab river and Hor Al-Hammar Marsh are listed in Table 12.4. The organochlorine concentration in fishes ranged from the below detection limit to 189 μg/kg being the lowest for Aldrin in Hor Al-Hammar marsh and the highest for ∑ DDT in the Shatt Al-Arab river.

12.4 Comparison with Worldwide Levels

For comparison, worldwide studies reported the presence of different kinds of pesticides with different levels of concentrations in water, sediment, and fishes. Most studies showed a higher concentration of pesticide residues in fish tissues

Table 12.4 Concentrations (μg/kg) of organochlorine pesticides in fish muscles from the Shatt Al-Arab rver and Hor Al-Hammar Marsh

Site	Endrine	Aldrin	Dieldrin	∑DDT	Lindane	Heptachlor	References
Shatt Al-Arab	16	–	2	13	–	3	DouAbul et al. (1987a)
Hor Al-Hammar	154	–	8	58	–	86	DouAbul et al. (1987a)
Shatt Al-Arab	3.90	–	2–32	16–189	–	–	DouAbul et al. (1987b)
Hor Al-Hammar	20	–	7	166	–	–	DouAbul et al. (1987b)
Hor Al-Hammar	Nd-3.31	0.05–25.79	Nd-0.35	0.02–61.71	0.23–17.25	0.95–50.63	Al-Ali (2012)
Shatt Al-Arab	Nd-154	–	Nd-32	1–189	–	Nd-6	DouAbul and Al-Timari (2014)

and sediment than that of water. Some physicochemical parameters viz. pH may have a direct influence on the solubility of these pesticides in river water.

Pesticide contamination of water bodies in other parts of the world has been shown by different workers. Erkmen et al. (2013) have shown total organochlorine pesticide concentrations in water and sediments, which ranged from 0.0014 to 0.0086 ppm and 0.0170 to 0.0391 ppm in Lake Manyas, Turkey. The total average pesticide residues in water samples from the four lagoons Chemu, Korle, Fosu, and Etsii are 2.6384 mg/l, 0.4992 mg/l, 0.3045 mg/l, and 1.3629 mg/l, respectively (Essumang et al. 2009).

In the Egyptian Mediterranean coast area, during the year 2010, the reported levels of pesticides (ng/g) in the surficial sediments were in the ranges as follows: PCBs (7.06–75.17), DDTs (4.89–29.93), Chloradane (0.12–0.99), and HCHs (3.20–26.71), (Dalia et al. 2013).

The presence of very high levels of pesticide residue in fish (Puntius sp.) ranging from 2.9599 to 5.0371 ppm indicates extensive bioaccumulation (Singh et al. 2015).

12.5 Conclusion

Certain pollution by most studies conducted for water, sediments, and fishes from southern Iraqi waterways was assigned certain levels of organochlorine pesticides such as with DDT and Aldrin-Dieldrin, which have been used heavily during the period 1950 to 1976 (DouAbul et al. 1987).

The observed concentrations are ∑DDT residues, endrin, and dieldrin in the muscles of Indian shad (*Tenualosa ilisha*) collected from the Shatt Al-Arab Estuary. Indian shad is a migratory fish species, which migrates between the northern Gulf waters and Iraq's marshes and Iraqi aquatic system for breeding and spawning.

Organochlorine pesticides have been used in Iraq and neighboring countries for more than five decades (Al-Omar et al. 1985). These compounds enter the aquatic environment by leaching directly to stream or groundwater, or by the erosion of contaminated soil. The measurable levels of DDT, endrin, and dieldrin were encountered in environmental samples from the Shatt al-Arab river, but invariably a significant incremental increase in their concentrations was observed in Hor al-Hammar Lake, a lake that drains into the Euphrates river and subsequently reaches the Shatt al-Arab river (DouAbul et al. 1987a). These findings were attributed to the fact that in Hor al-Hammar pesticides were applied, at various times and in various amounts, close to or over water causing fairly direct contamination.

References

Abaychi JK, DouAbul AAZ (1985) Trace metals in the Shatt Al-Arab River, Iraq. Water Res 19:457–462

Al Bomola AH (2010) Temporal and spatial changes in water quality of the Euphrates River – IRAQ. M.Sc. Thesis, Division of water resources engineering, Department of Building and Environmental Technology, Lund University

Al-Ali BS (2012) Level of some insecticide in water, sediment and some biota in east of hor Al-Hammar marshes, Iraq. PhD. Thesis, College of Agriculture, University of Basrah (in Arabic)

Al-Hilffi MAA (2005) Effect of pesticides in pollution of marsh environment of southern Iraq. Mesopotamian J Marine Sci 20(1):81–89

Al-Mahmood HK (2009) The monthly variations of discharge and effect that on a total dissolve suspended and salinity in Shatt al-Arab River (south of Iraq). Iraqi J Sci 50(3):355–368

Al-Omar M, Al-Bassomy M, Al-Ogaily N, Shebl DA (1985) Residue levels of organochlorine insecticides in lamb and beef from Baghdad. Bull Environ ContamToxicol 34:509–512

Al-Shwafi N, Al-trabeen K, Rasheed M (2009) Organochlorine pesticides and polychlorinated biphenyls carcinogens residual in some fish and Shell fish of Yemen. Jordan J Biol Sci 2 (1):23–28

Al-Timari AA (1997) Chlorinated hydrocarbons: pollutants or/and indicators of shrimp stock structure from Iraqi and Kuwaiti waters. Marina Mesopotamica 12(2):271–289

Dalia MS, Salem A, Khaled A, El-Memr A (2013) Assessment of pesticides and polychlorinated biphynels (PCBs) in sediments of the Egyption Mediterranian Coast, Egypt. J Aquatic Res 39:141–152

Dawson R, Riley JP (1977) Chlorine-containing pesticides and polychlorinated biphenyls in British coastal waters. Estuar and Coastal Mar Sci 4:55–69

DouAbul AAZ, Al-Saad HT (1985) Seasonal variations of oil residues in water of Shatt Al-Arab river, Iraq. Water Air and Soil Pollution 24(24):237–246

Douabul AAZ, Al-Timari AAK (2014) On the organochlorine pesticide residues in the marshes, Shatt Al-Arab and the Arabian gulf system. Mesopot J Mar Sci 29(2):97–114

DouAbul AAZ, Al-Saad HT, Al-Rekabi HN (1987) Residues of organochlorine pesticides in environmental samples from the Shatt al-Arab River, Iraq. Environ Pollut 43:175–187

DouAbul AAZ, Al-Omar M, Al-Obaidy SZ, Al-Ogaily N (1987a) Organochlorine pesticide residues in fish from the Shatt Al-Arab River, Iraq. Bull Environ Contam Toxicol 38:674–680

DouAbul AAZ, Al-Saad HT, Al-Rekabi HN (1987b) Residues of organochlorine pesticides in environment samples from Shatt Al-Arab River, Iraq. Environ Pollut 43:175–187

DouAbul AAZ, Al-Saad HT, Al-Timari AAK, Al-Rekabi HN (1988) Tigris–Euphrates Delta: a major Source of pesticides to the shatt Al-Arab River (Iraq). Arch Environ Contam Toxicol 17:405–418

DouAbul AAZ, Mohammed SS, Warner BG, Asada T (2009) Persistent DDE in the mesopotamina wetlands of Southern Iraq . Bull Enviro Contam Toxicol, Published on Line

EPA. U.S. Environmental Protection Agency (2007) Method 1699: pesticides in water, soil, sediment, biosolids, and tissue by HRGC/HRMS, 1200 Pennsylvania avenue NW, Washington, http://www.epa.gov/waterscience

Erkmen B, Yerli SV, Erkakan F, Kolankaya D (2013) Persistent organochlorine pesticide residues in water and sediment samples from Lake Manyas, Turkey. J Environ Biol 34:171–176

Essumang DK, Togoh GK, Chokky L (2009) Pesticide residue in the water and fish (lagoon tilapia) samples from lagoons in Ghana. Bull Chem Soc Ethiop 23:19–27

Isaev VA, Mikhailova MV (2009) The hydrology, evolution, and hydrological regime of the mouth area of the Shatt al-Arab River. Water Resour 36(4):380–395

Latif AKS, Ajmi RN, Zaki HF, Htait WA (2013) Investigation in the fate pesticides in water and sediments Iraqi marshland. Int J Envir Ecol Geol Engin 7(12):602–606

Malik A, Singh KP, Ojha P (2007) Residues of organochlorine pesticides in fish from the Gomti River, India. Bull Environ Contam Toxicol 78:335–340

Moyel MS (2014) Assessment of water quality of the Shatt al-Arab River, using multivariate statistical technique. Mesopotomia Environ J 1(1):39–46

Ogunfowokan AO, Oyekunle JAO, Torto N, Akanni MS (2012) A study on persistent organochlorine pesticide residues in fish tissues and water from an agricultural fish pond. Emirates J Food Agric 24:165–184

ROPME (1986) The 18 month marine pollution monitoring and research program in Iraq. Submitted by Marine Science Centre, Basrah Univ., Iraq, pp 52–72

Singh S, Bhutia D, Sarkar S, Rai BK, Pal J, Bhattacharjee S, Bahadur M (2015) Analyses of pesticide residues in water, sediment and fish tissue from river Deomoni flowing through the tea gardens of Terai region of West Bengal, India. Int J Fish Aquatic Stud 3(2):17–23

UNEP (1991) Sampling of selected marine organisms and sample preparation for the analysis of chlorinated hydrocarbons. Reference Methods for Marine Pollution Studies No. 12, Rev. 2, UNEP

UNEP (1993) Guide lines for monitoring chemical contamination in the sea using marine organisms, reference methods for marine pollution studies, no. 20

WHO (1993) Guidelines for drinking-water quality. In: Recommendations, vol 1, 2nd edn. World Health Organization, Geneva

Chapter 13
Surface Water Salinity of the Euphrates, Tigris, and Shatt al-Arab Rivers

Khayyun Amtair Rahi and Todd Halihan

Abstract Increasing salinity is a problem in many major river systems worldwide, especially in arid areas. The Euphrates, Tigris, and Shatt al-Arab river system is suffering from significant increases in salinity, especially in the lower reaches. Salinity increases coincide with the construction of dams on the upstream parts of the rivers. Salinities higher than 5000 ppm were recorded in the lower parts of the river system. The Shatt al-Arab is formed by the confluence of the Euphrates and Tigris rivers in southern Iraq. The salinity problem in the Shatt al-Arab is more severe than that in the two upstream rivers. Salinity values as high as half seawater salinity were observed in the Basra (Maqal) station in 2009. This chapter evaluates the salinity problem of the system composed of the three water courses. The salinity evolution, causes, and distribution are presented analytically. At the end of the chapter, salinity management and potential control options are discussed.

13.1 Introduction

Salinity is a measure of the total concentration of dissolved minerals in water and is often measured as Total Dissolved Solids (TDSs). TDS is commonly measured in the laboratory in milligram per liter (mg/l) or parts per million (ppm). In the field, the quantity is often measured electrically as Electrical Conductivity (EC) in units of milli- or micromhos (μmhos/cm) per centimeter or milli- or microsiemens per centimeter (μS/cm); mhos and siemens are equivalent. TDS is composed primarily of four major cations: Sodium (Na^+), Calcium (Ca^{2+}), Potassium (K^+), and Magnesium (Mg^{2+}) and four major anions: Chloride (Cl^-), Sulfate (SO_4^{2-}), Carbonate

K. A. Rahi (✉)
Mustansiriyah University, Baghdad, Iraq
e-mail: khayyun.rahi@okstate.edu

T. Halihan
Oklahoma State University, Stillwater, OK, USA
e-mail: todd.halihan@okstate.edu

L. A. Jawad (ed.), *Tigris and Euphrates Rivers: Their Environment from Headwaters to Mouth*, Aquatic Ecology Series 11,
https://doi.org/10.1007/978-3-030-57570-0_13

(CO_3^{2-}), and Bicarbonate (HCO^{3-}). Other significant constituents can include Nitrate (NO_3^-), Boron (B^{3+}), Iron (Fe^{3+}), Manganese (Mn^{2+}), and Fluoride (F^-) (CASS 2003). To convert EC to TDS, results are commonly multiplied by a linear conversion factor to obtain salinity (TDS) in ppm. In the laboratory, TDS can also be determined by an evaporation procedure.

In arid and semiarid zones, the salinization of water and soil resources is the most obvious effect of water quality degradation. Increasing salinity is a problem in many major rivers in the world (FAO 1976). As salinities pass 1000 mg/l, water becomes brackish and less potable for humans or crops (WHO 2004; IQBS 2001). Above 3000 mg/l, brackish water becomes unusable for most municipal or agricultural uses. Irrigation with these brackish waters adversely affects the production of most common crops (FAO 1976).

Salinization of surface water and groundwater occurs mainly due to a combination of natural and anthropogenic processes. In dryland environments, salts are deposited in the environment over time, stored in the subsurface, and then transported via shallow groundwater that discharges into rivers. Salinity may increase due to flow diversions, dam construction, irrigation processes, and the resulting decrease of natural discharge. The Amu Darya and Syr Darya rivers in Central Asia are a good example of these issues. These rivers had their discharges drop to zero due to the use of water for irrigation of cotton in the former Soviet Union (Weinthal 2002).

The Euphrates and Tigris rivers are suffering from high salinity, especially in the lower reaches. Noticeable salinity increases were recorded during the 1980s following the construction of dams on the upstream parts of the rivers (Rahi and Halihan 2010, 2018). The Tigris River salinity record shows values of 5205 and 2004 ppm, which were recorded in Amara in July 1989 and July 1994, respectively (MoI 1997). In the Lower Euphrates, the salinity in al-Nasiriyah is higher than 2000 ppm and reached as high as 5000 ppm for the last four decades (Fattah and Abdul Baki 1980; Rahi and Halihan 2010; CEB 2011a). More drastic increases in salinity are recorded in Shatt al-Arab in al-Basra. In 2009, the salinity of the Shatt has reached a record high of 18,500 ppm (CEB 2011b). This chapter presents a review of salinity evolution and analyses of salinity causes and distribution focused on the Tigris, Euphrates, and Shatt-al-Arab system. At the end of the chapter, a section will be devoted to the salinity management options for this region.

13.2 Tigris River

13.2.1 Tigris River Hydrology

The head of the Tigris River is located in southeastern Turkey on the southern side of the eastern Taurus Mountains (Altinbilek 2004). The Tigris River flows southeasterly until the confluence with the Euphrates at al Qurna. At this point, the two rivers form the Shatt al-Arab waterway, which discharges into the Arabian (Persian) Gulf.

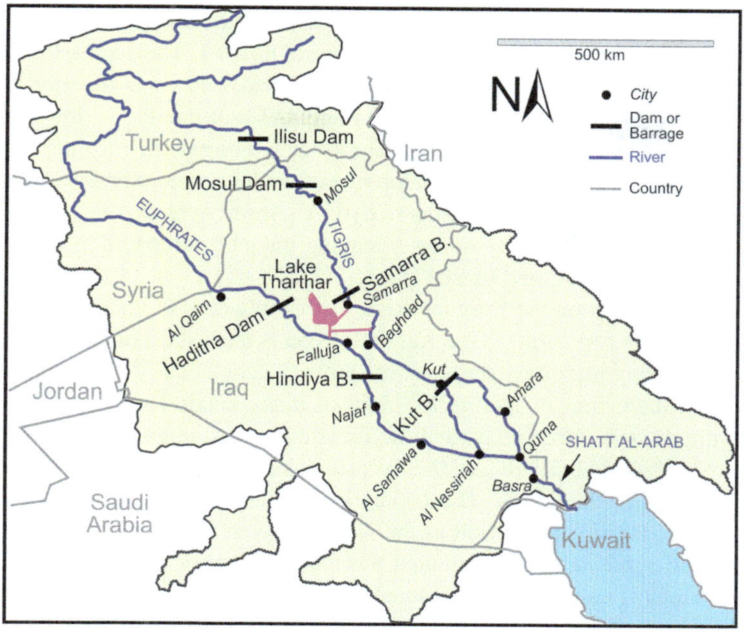

Fig. 13.1 Location map of the Euphrates, Tigris, and Shatt al-Arab Rivers (after Holmes 2010). Large dams and control structures (barrages) as well as major cities along the rivers are shown

Based on the course as defined, the Tigris is the second-largest river in western Asia (Fig. 13.1). The Tigris is approximately 1900 km in total length. The majority of the river is in Iraq, but 400 km of the headwaters are in Turkey and 32 km form the Turkey/Syria border.

The drainage basin of the Tigris River is 471,606 km^2 and extends across four countries: Turkey, Syria, Iraq, and Iran. About 12% of the basin area is in Turkey, 34% is in Iran, and a little more than half of the basin (54%) is located in Iraq (HARC 2015). Syria does not include a significant portion of the Tigris basin. Within Iraq, the Tigris is fed from the east by four major tributaries upstream of Baghdad: the Khabur, the Greater Zab, the Lesser Zab, and the al Udhaim. Downstream from Baghdad, the Diyala joins the Tigris as the fifth major tributary (Jalut et al. 2018).

Several large dams (>1 billion cubic meters or bcm of storage) have been constructed on the Tigris in Turkey. The Turkish retention capacity from large dams, both operative and under construction, is about 14 billion cubic meters (bcm). This reservoir capacity can retain the majority of the Tigris discharge generated in Turkey (UN-ESCWA 2013; Altinbilek 2004). The Ilisu Dam, located 65 km upstream from the Syrian-Turkish border, is the largest of these structures in Turkey (Fig. 13.1). The dam is intended for power generation and irrigation and can store 10.41 bcm in its reservoir (UN-ESCWA 2013).

Iraq has constructed several dams and barrages (regulators) on the river and its tributaries. The retention capacity of the Iraqi dams is 24 bcm, excluding Lake

Tharthar. Lake Tharthar (a natural depression) is located between the Euphrates and the Tigris and was originally used for flood control in 1954. The depression began to be used as a water supply reservoir in 1983. It is filled with water diverted from the Tigris. Outflow from the lake goes to the Euphrates and the Tigris downstream of Baghdad. The lake is functioning as a salinity generating device for the Euphrates and the Tigris due to the high evaporation rate of the lake, approximately three cubic kilometers each year (Rahi and Halihan 2010; Al-Ansari et al. 2014). Lake Tharthar function is to triple the salinity of water leaving the lake compared to the waters in the Tigris at Samarra (ICARDA 2012).

The Karkheh dam on the Karkheh River is the only dam in the Tigris basin with significant storage capabilities. The Karkheh River is the last tributary that flows into the Tigris, entering just above the town of Qurna (Vaghefi et al. 2013). In addition to the Karkheh dam, Iran is altering the flow of many small Tigris tributaries, with small hydraulic structures used for flow diverting and storage (Richardson 2016). The altered tributaries, which join the river downstream of Baghdad, result in lower discharge entering the Tigris. The tributaries include the Diyala, Galal Badra, al Chabab, Teeb, Dwaireeje, as well as the Karkheh river. Reduced flow from the last three tributaries has adversely affected the restoration process for the al-Hawizeh Marshes, which is one of the three ancient Mesopotamian marshes in Iraq (Richardson 2016).

The Tigris flow is further controlled by two primary head regulators in the system, the Kut and Samarra Barrages. The Kut Barrage was constructed in 1939 and the Samarra Barrage in 1954 (Fig. 13.1). The Samarra Barrage diverts water to Lake Tharthar and includes a separate irrigation canal that allows water to pass Lake Tharthar. The Kut Barrage sends water to the al Gharraf and the al Dujaila distributaries.

13.2.2 Salinity and the Alteration of the River Flow Regime

Changes in flow regime, especially the decrease in streamflow of natural rivers, will have a significant impact on stream salinity. The other major effect in these systems is irrigation systems causing salinization, leading to high salinity irrigation return flows. Historically, the Tigris waters were low salinity and suitable for all purposes (Buringh 1960; FAO 1976). The river salinity in Baghdad between 1924 and 1950 averaged at 260 ppm (FAO 1970). Mutlak et al. (1980) measured salinity of the river in Baghdad to be 500 ppm. At the time, no major flow control structures existed on the main river course. The changes in the flow that occurred in the river system are accompanied by an increase in salinity.

Large scale changes to the Tigris river began with plans for control structures developed early in the twentieth century with modifications continuing to present day (Haigh 1951; Harza Engineering Co. 1963). Two dams constructed on the Lower Zab and Diala in 1959 and 1961 were the first major modifications to the flow regime of the Tigris tributaries, respectively. On the main river course, major changes have started with the operation of Lake Tharthar (completed in the early 1980s). The full operation of Lake Tharthar as a water supply reservoir brought about major changes in the Tigris salinity in Baghdad and downstream. Rahi and Halihan (2018) indicate that salinity tripled in the lake compared to its inlet water salinity in Samarra. Following the Tharthar System, the Mosul Dam was put into operation in 1986. The structure greatly reduced the natural flow to the lower parts of the river without a noticeable increase in salinity.

Turkey's GAP Project (Southeast Anatolia Project 'Guneydogu Anadolu Projesi') (UNEP 2003) includes a series of dams with associated reservoirs on the upper reaches of both the Euphrates and Tigris rivers. The Ilisu Dam is the largest of these dams built for the GAP project and was put in operation during the second half of 2018. Research indicates that the discharge of the Tigris at the border between Turkey and Iraq would be decreased after the construction of the dam (UN-ESCWA 2013; Kolars 1994). The downstream impacts of the Ilisu and Cizre dams were evaluated, and this independent evaluation (PWA 2006) concluded: "The operation of the Ilisu Dam in combination with diversions from the future downstream Cizre project would probably significantly reduce summer flows in Syria and Iraq below historic levels. It is likely that a significant portion of the recommended minimum flow release from the Ilisu of 60 m^3/s during dry years would be diverted. It is even possible that with full implementation of the Ilisu/Cizre projects, during drought periods, all the summer flows could be diverted before it crossed the border." Flow reduction caused by the full operation of the Ilisu/Cizre scheme will further exacerbate the salinity problem downstream of the dams.

Agricultural irrigation return flow affects salinity in many arid streams, and these effects are evident in the Euphrates and Tigris basins. As an example, the increase in salinity in the lower reaches of the Euphrates is attributed primarily to irrigation return flow (Fattah and Abdul Baki 1980; Rahi and Halihan 2010). In the Tigris basin, irrigation practices occur in many locations that have poorly quantified and variable salinity. This results in uncertainty in the quantity and salinity of the irrigation return flows. Irrigation return flows in Turkey are not known to be a significant source of added salinity at this time. The site conditions exist; however, this may lead to salinity problems in the future (Odemis et al. 2010). Further downstream, it is evident that irrigation return flow has adversely affected water quality. The salinity increase in the southern parts of the river was attributed partly to the agricultural return flow coming from Iraq and Iran (ICARDA 2012; Al-Saady and Abdullah 2014).

13.2.3 Salinity of the Tigris River

13.2.3.1 Historical Review

The Tigris River waters are categorized by the Food and Agricultural Organization of the United Nation (FAO) during the early seventies as excellent for all purposes

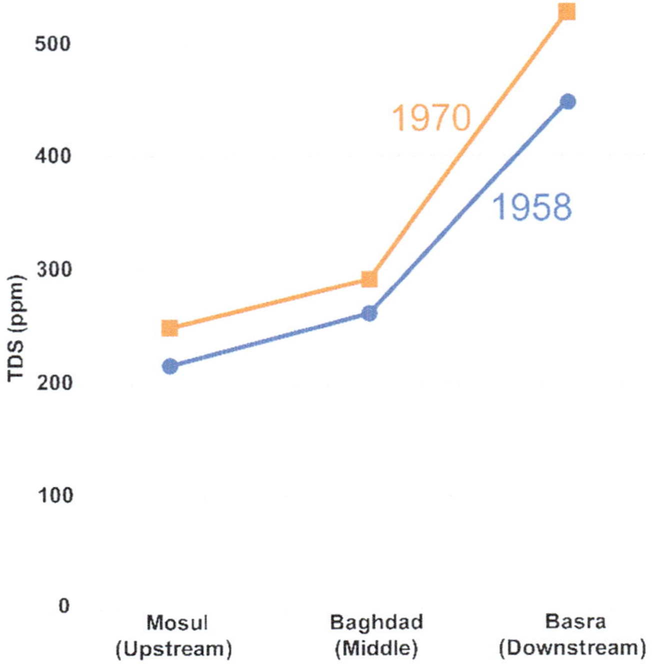

Fig. 13.2 Salinity distribution profile along the Tigris River for the year 1958 and the year 1970 when no major control structures existed in the upstream (Data from Al-Layla (1978)

(FAO 1976). FAO (1976) reported average salinity of 300 ppm of Tigris waters in Baghdad. FAO (1970) presented salinities along the river course from Mosul, in the North of Iraq, to Qurna, in the south. The salinities in Mosul, Baghdad, and Qurna were 260, 420, and 620 ppm, respectively. The same publication presents salinity data prior to 1950. The average monthly salinity recorded was 260 ppm. Data from Al-Layla (1978) show a salinity profile along the Tigris River for 1958 and 1970 (Fig. 13.2). The data from the north, middle, and south of Iraq show the same trend and close magnitude of salinity for the years available. The Mosul and Baghdad salinities were fairly close for both years and presented no limitations for domestic or agricultural usages. In the south, the salinity increased significantly, but remained fresh. The increase at Basra was most likely caused by drainage water from the marshes.

13.2.3.2 Recent Salinity Analysis

More recent data show that salinity of the Tigris upstream of Baghdad is in the range of 210–310 ppm (Fig. 13.3). These data compare salinities in Mosul, where most of its inflow comes from Turkey, and in Samarra, where flow from majority of Iraqi tributaries join the river. The average salinity for the years of record (2000–2010) is

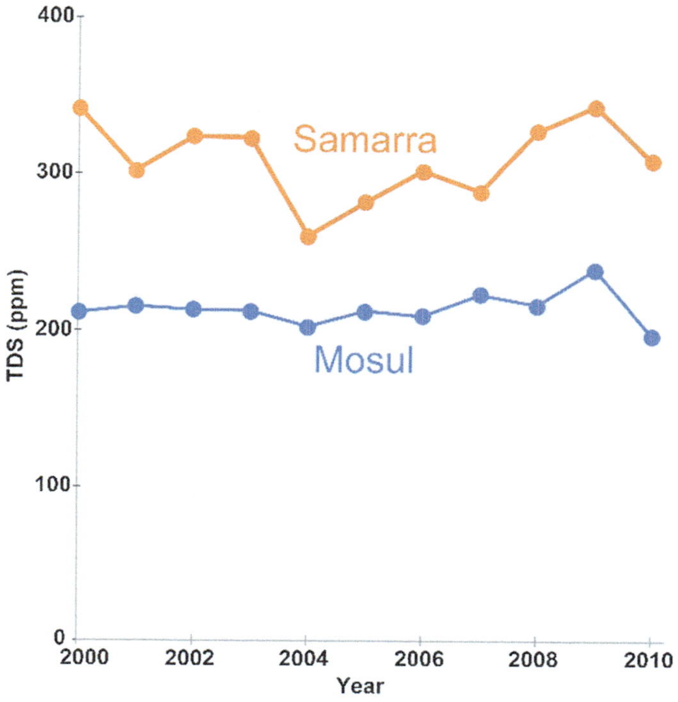

Fig. 13.3 Comparison of salinities as TDS (of the years 2000–2010) in Mosul (where most of its inflow comes from Turkey) and in Samarra (where flow from majority of Iraqi tributaries join the river)

210 ppm for Mosul and 310 ppm for Samarra. No salinity value has exceeded 400 ppm in either location. Even though the salinity in Samarra is higher than that of Mosul, the water is still of good quality and suitable for all purposes. The increase in salinity is probably due to the reduction in flow caused by upstream dams and evaporation. No significant agricultural return flow is known to discharge into Tigris in this reach of the river. Since salinity has not changed significantly from Mosul to Samarra, it will be considered similar for the discussion henceforward.

The salinity in Baghdad is at least twice as much compared to Samarra as obtained from more recent data. The average annual salinity in Baghdad was not less than 500 ppm during the available continuous record over the last three decades (1981–2010). The salinity of the river in Baghdad for the years 1979–2010, compared to the salinity in Samarra for the same period, illustrates the difference (Fig. 13.4). The average Baghdad salinity of about 648 ppm is more than twice that of Samarra (304 ppm). The increase in salinity of the Tigris in Baghdad occurred following the year 1983. In that year, the full operation of Lake Tharthar was completed. Most of the increase in salinity in Baghdad is attributed to the lake (Rahi and Halihan 2018). Much information about the salinity effects of Lake Tharthar is presented in a later section.

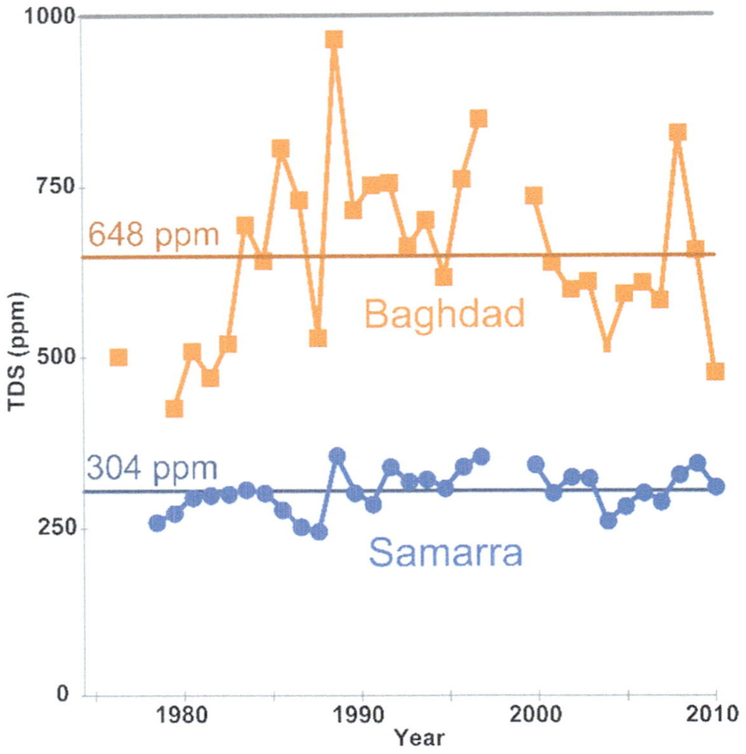

Fig. 13.4 Salinity values as TDS in Baghdad compared to salinities in Samarra (Data for the years 1979–2010, except 1998). Solid lines show average values of 304 ppm for Samarra and 648 ppm for Baghdad

13.2.3.3 Longitudinal Salinity Profile

The longitudinal salinity profile of the Tigris from Samarra to Amara shows significant changes in salinity after 1980 (Fig. 13.5). Amara is the last major city dependent on the Tigris for its water supply before the river meets the Euphrates in the town of Qurna. The figure presents two salinity profiles, one for the pre-1980 and the other for the post-1980 time period. The pre-1980 is the period when no major flood control structures were built on the upper reaches. The data for that period were obtained from published literature (FAO 1970, 1976; Buringh 1960).

The operation of major water storage structures upstream of Baghdad, namely, Lake Tharthar and Mosul Dam, started during the 1980s. The salinity profile for this period has not changed for decades upstream. However, minor changes may be associated with dry or wet years. Data for this period are from the unpublished records of the Iraqi Ministry of Water Resources. The salinity in Samarra, which is assumed to represent the natural state of the river, remained constant (about 300 ppm) throughout the covered time spans, while in Baghdad and Kut, salinity

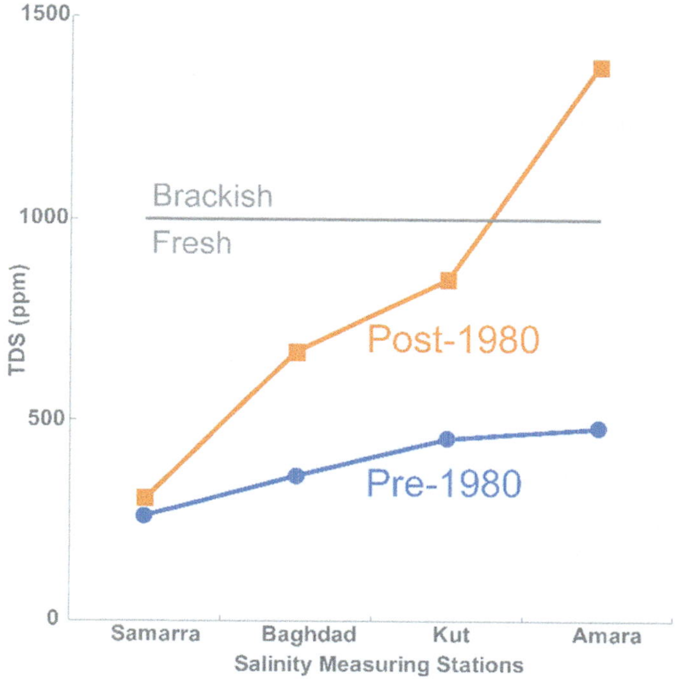

Fig. 13.5 Longitudinal salinity profile for the Tigris from Samarra to Amara for two periods. The pre-1980 time period represents salinity of "natural uncontrolled" flow (blue), and the post-1980 is the time period where major control structures in the upstream have started to operate (orange). The gray line represents the boundary between fresh and brackish salinities at 1000 ppm

has almost doubled during the last three decades. Nevertheless, Baghdad and Kut salinities have remained lower than 1000 ppm and the water is suitable for drinking and other uses.

The situation is changing significantly in the City of Amara. Salinity in Amara almost tripled in the post-1980 period compared to the pre-1980 period and quadrupled compared to the salinity in Samarra. Moreover, salinity in Amara has exceeded the allowable limit for drinking water for most of the time during the last four decades. Downstream, where the Tigris joins the Euphrates in the town of Qurna, the salinity quadrupled for the post-1980 relative to the predevelopment (Rahi and Halihan 2018). Further examination of Fig. 13.5 reveals that the post-1980 salinity rate of increase is getting steeper downstream of Kut. Rahi and Halihan (2018) have attributed this change to one or more of the following causes: inflow from al Shiwiaja marsh, inflow from the al-Shabaab tributary, or groundwater discharge. Most likely, all of these contributors intensify the rate of salinity increase.

As mentioned earlier, the increase in salinity downstream of Samarra is associated with the operation of the hydraulic control structures upstream. Research shows that the main contributor is Lake Tharthar (CEB 2011a; Rahi and Halihan 2018). While the increase of salt concentration in Samarra that could be linked to Mosul Dam and other smaller dams is about 30%, the increase in salt concentration in Baghdad is 100%. The salt load increase in Baghdad is caused by Lake Tharthar output, because the lake is the only major hydraulic control structure located (hydraulically) between Samarra and Baghdad. The lake is a massive natural gypsum doline depression that was converted to water management scheme and put in full operation in the early 1980s.

13.2.4 Lake Tharthar

Lake Tharthar is Located 120 km north-west of Baghdad between the Tigris and the Euphrates rivers (Fig. 13.6). Lake Tharthar is part of a western depression of the Mesopotamian tectonic zone (Jassim and Goff 2006). Other parts of the western depression are the Habbaniya and Razzaza lakes and the Bahr al Najaf depression. The surface area of the lake is 2710 km^2, and the highest level of storage is 65 m above mean sea level. The storage capacity of the lake at level 65 m is 85.6 billion cubic meters (bcm). The operational level range is 40–65 m, and the dead storage (below level 40 m) is 35.18 (bcm). The available (live) storage is about 50 bcm. The lake is part of a hydraulic structure system that includes the Samarra Barrage, the Tharthar intake regulator, the Tharthar outlet regulator, and the outlet regulators for the Euphrates and Tigris rivers.

The construction of the Tharthar System was conducted via two stages. The first was in 1956 when the Samarra Barrage and the Tharthar intake regulator were

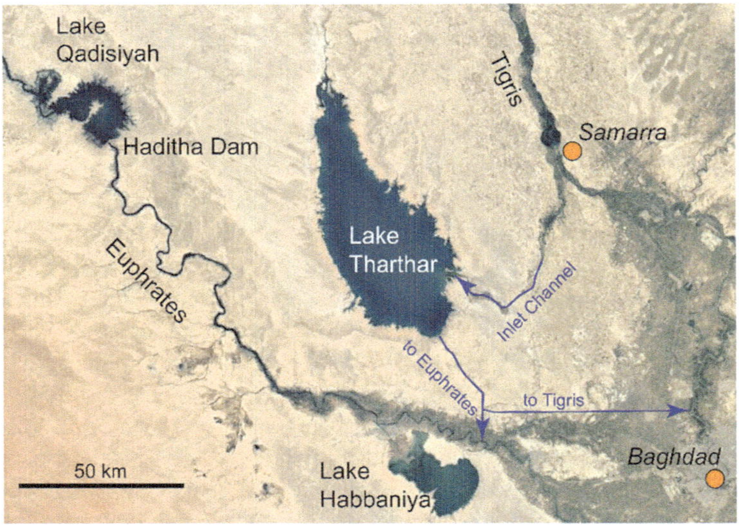

Fig. 13.6 Location map of the Lake Tharthar System (aerial photo from Google Earth). Inlet channel and outlet channels are highlighted

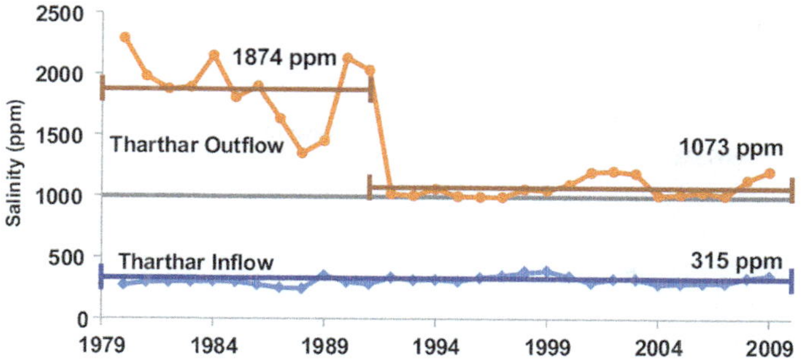

Fig. 13.7 Changes in the salinity in Lake Tharthar inflow and outflow. A gray line at 1000 ppm salinity indicates the boundary between fresh and brackish water based on WHO and IQBS drinking water salinity limits. Inflow average of 315 ppm is illustrated for the entire inflow record. Outflow averages shown are 1874 ppm prior to 1992 and 1073 ppm after 1992 (from Rahi and Halihan 2018)

completed. The lake was designated as a flood control reservoir for the Tigris River. The only outflow from the lake was via evaporation and possible outflow to groundwater. This mode of operation resulted in a significant increase in the lake salinity as the majority of water existed as vapor and salts concentrated from river water and the dissolution of gypsum in the local geology. It is possible that this salinity buildup is still contributing to the salinity of the outflow water.

The second stage was completed between 1976 and 1981. In 1976, the combined outlet regulator (maximum discharge 1100 cubic meters per second, cms) and the outlet regulator that supply water to the Euphrates River (capacity 600 cms) were completed. The Lake Tharthar water flowed to the Euphrates via a man-made channel and confluence with the river at a point upstream from al Fallujah.

The regulator that supplied water back to the Tigris River was completed in 1981 with maximum discharge of 500 cms. Salinity of the outflow from the lake is high (Fig. 13.6) and contributes to the high salinity of the Tigris in Baghdad and the lower reaches of the river. Rahi and Halihan (2018) discuss the salinization mechanism of the lake system to illustrate that it is still functioning as a salinization machine for the rivers (Fig. 13.7). The authors concluded that the high salinity of the lake may be caused by high evaporation rates, gypsiferous geologic formations beneath the lake, and groundwater inflow. Buringh (1960) indicated that the former Tharthar wadi and depression are underlain by gypsum mixed with salt deposits, but it is unclear how much each mechanism contributes to the overall lake salinity. Regardless of the mechanism(s), the lake is releasing water of high salinity to the Tigris River and negatively impacting the water supply of Baghdad and all users downstream. The salinity of the water released from the lake is more than three times that of the water feeding the lake (Fig. 13.7).

13.3 The Euphrates River

13.3.1 Euphrates River Hydrology

The Euphrates is defined as the longest river flowing through western Asia, originating in the Eastern Mountains of Turkey as a confluence of two streams; the Western Euphrates (locally called the Karasu) and the Eastern Euphrates (the Murat) form the headwaters of the river. The total length of the river from its origin, the head waters of Karasu in Turkey, to the town of Al Qurna in Iraq is 2786 km (UN-ESCWA and BGR 2013). The slope of the Euphrates averages 2 meters per kilometer over most of its course in Turkey and then crosses into Syria flowing southeast (Altinbilek 2004). The river flows for 680 km within Syria. The river then heads into northwestern Iraq at Al Qaim and continues flowing to the southeast. The Euphrates then joins with the Tigris at Al Qurna in southern Iraq to flow into the Shatt al-Arab waterway, which finally drains into the Arabian Gulf (Fig. 13.1).

The total drainage basin area is 440,000 km^2. The drainage basin area exists over four countries with approximately half of the basin located in Iraq (47%). Turkey (28%) and Syria (22%) each hold approximately a quarter of the basin. Saudi Arabia accounts for 3% of the basin area (UN-ESCWA and BGR 2013). The total annual streamflow of the Euphrates, entering Iraq, prior to dam construction (1932–1973) ranged between 15.3 bcm (in 1961) and a maximum of 63.4 bcm (in 1969) (Saleh 2010). Researchers reported different average annual flow for the river; Murakami (1995) reported an average annual flow of 33 bcm in the town of Hit, while Haigh (1951) reported a flow of 26.6 bcm for the years 1925–1946 in the same location. The mean annual flow of the Euphrates in Hit was calculated for this chapter using data from Saleh (2010) as 29.6 bcm (years of record 1932–1973). Upstream of Hit, no hydraulic structure was completed and put in operation prior to 1973.

At the present time, the discharge of the Euphrates is tightly controlled through a series of dams and hydraulic structures constructed by Turkey, Syria, and Iraq. Saudi Arabia does not utilize water resources from the Euphrates. The Turkish Keban dam and the Syrian Tabqa dam were completed on the Euphrates in 1975. In 1977, Turkey's GAP program was initiated (UNEP 2003). At least five major dams have been constructed on the Euphrates (Altinbilek 2004). The storage capacity of the Turkish reservoirs alone is greater than 90 bcm, approximately three times more than the mean annual flow of the river. The Syrian hydraulic construction included the development of three large dams. The storage capacity of these dams is 13.7 bcm.

As a result of these upstream activities, the Euphrates flow into Iraq was reduced to about 9.3 bcm for the water year 2008–2009 (CEB 2011a), which was a dry year. Comparing its streamflow with that of the lowest recorded (natural) dry year in 1961, the reduction in flow due to dams' construction is about 40 percent. In 1987, Turkey and Syria signed a protocol committing Turkey to release 500 cms over the Syrian border or about 16 bcm annually with no provision for water quality. After this, Syria and Iraq reached a water-sharing agreement, in which 58% of the waters will go to Iraq and the remainder going to Syria. UN-ESCWA and BGR, 2013, stated "in several instances, Turkey has limited these flows of water, bringing Syria and Turkey to the brink of war." Iraq, for its part, has constructed several structures along the Euphrates River course. Structures that are important for the Euphrates system are the Haditha Dam and the Lake Tharthar diversion scheme. The Lake Tharthar diversion scheme includes water transfer structures from the Tigris River to the Euphrates, through Lake Tharthar as discussed earlier (Fig. 13.6).

13.3.2 The Euphrates Salinity

13.3.2.1 History of Salinity Variation

Historically, the salinity of water in the Euphrates is low. At Al Qaim station, where the Euphrates River enters Iraq, a TDS of 467 ppm was cited by Hanna and Al Talbani (1970). FAO (1970) reported average salinity of 370 ppm in Fallujah (385 km from the Syrian Border) in the year 1950. Al-Hadithi (1978) presented salinity values at Al Fallujah station ranging from 420 to 710 ppm (recorded prior to 1973). A TDS of 525 ppm was recorded in the City of al Samawa in 1955 as reported by Rahi and Halihan (2010). Al Samawa is a southern Iraqi city, and it is located about 770 km from the Iraqi-Syrian Border.

Al-Layla (1978) published a salinity profile along the Euphrates River for two different years, 1958 and 1970. The author's data are reproduced here (Fig. 13.8). The salinity at Fallujah, representing the northern reach of the river, remained similar for the 2 years and was less than 400 ppm. A slight increase is noticed in salinity in Diwaniyah station at the middle reach of the river, compared to the Fallujah station, but values remain less than 500 ppm. The salinity increased in higher rates for the reach between the Diwaniyah and the Nasiriyah station at the south reach of the river. The salinity in Nasiriyah was higher than 500 ppm for the two recorded years. The Al-Layla (1978) data, however, show that the salinity level of the Euphrates along its flow course in Iraq was low and the water was suitable for drinking and irrigation. Currently, the Euphrates River salinity is highly altered. The Euphrates water is not suitable for domestic use over much of the river in Iraq and only slightly suitable for some crops in Samawa and further downstream.

In Turkey, a water quality evaluation of the Euphrates river (during 2002–2003) measured an average salinity of 237 ppm, a low fresh water value suitable for human consumption or crop productivity (UN-ESCWA and BGR 2013). The same authors reported a value of 248 at Jarablus near the Syrian-Turkish border. A TDS of 261 ppm was quoted by Scheumann (1993) at Keban in Turkey. The river water was classified, by the same author, as C2S1, which is suitable for irrigation and most other uses.

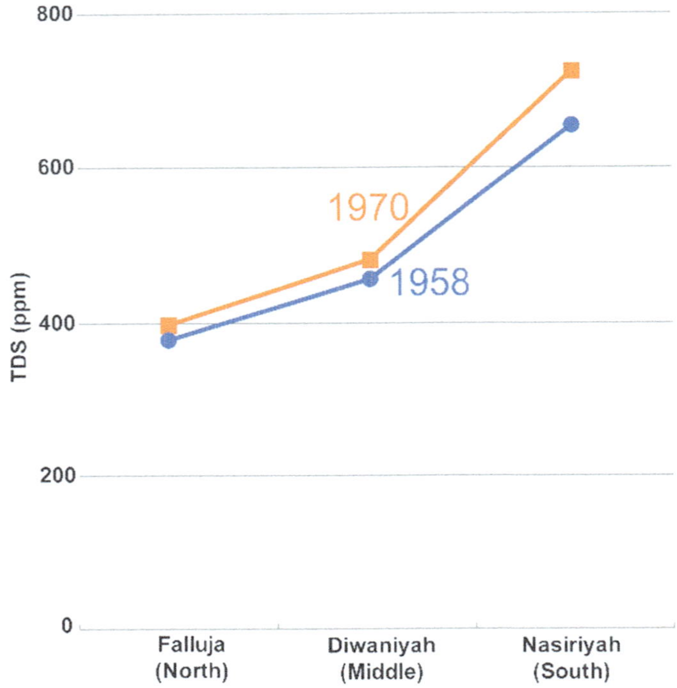

Fig. 13.8 Longitudinal salinity profile of the Euphrates for the year 1958 and the year 1970 when no control structure in the upstream portions of the river existed (data from Al-Layla 1978)

13.3.2.2 Recent Changes in Salinity

The salinity situation has changed dramatically following the construction of dams in the upper reaches of the Euphrates in Turkey, Syria, and Iraq. From the Turkish-Syria border to the Syria-Iraq border, the salinity of the Euphrates at least doubles (UN-ESCWA and BGR 2013).

A salinity profile of the Euphrates is available from al Qaim to al-Nasiriya (data from the water year 2000–2001) (Ali and Salewicz 2005). The salinity at al Qaim is about 1000 ppm, more than three times that of the river in Turkey; 1100 ppm in al Hindiya; 3000 ppm in al Samawa; and 4000 ppm in al-Nasiriyah. Rahi and Halihan (2010) compiled a salinity profile along the Euphrates River from the point where it enters Iraq (al Qaim) to the last major city (al-Nasiriyah) before the river joins the Tigris to form Shatt al-Arab. The profile compared two periods, pre-1970, where no dams were located on the river, and post-1980, where the river is highly regulated. The authors found that the salinity of the Euphrates had increased passing the 1000 ppm fresh water threshold from al-Kufa and all reaches downstream. The salinity in al Samawa is above 2000 ppm, and in al-Nasiriyah, it is above 3300 ppm. Rahi and Halihan (2010) presented measured TDS for the Euphrates reach that extends from al-Kufa to al-Nasiriyah. The TDS was 1100 ppm near Al-Kufa,

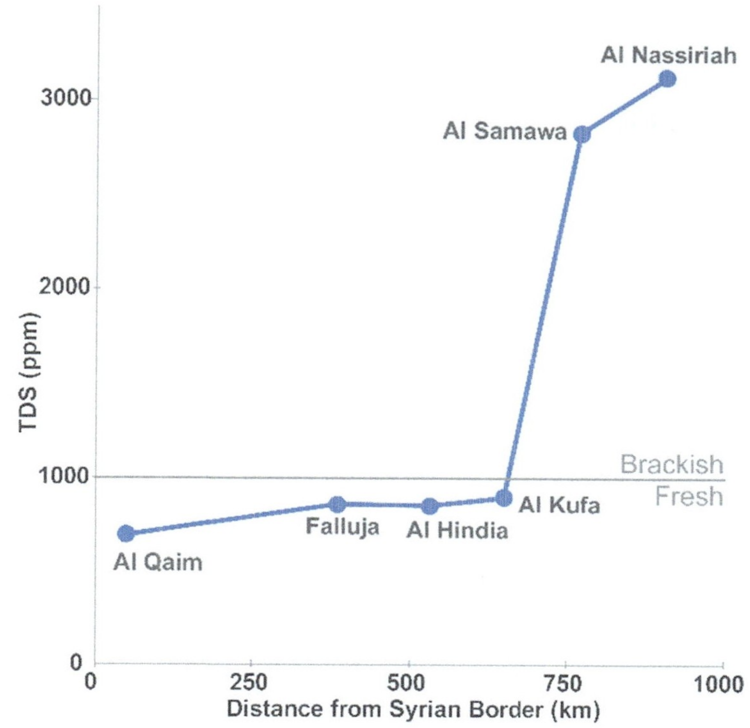

Fig. 13.9 Longitudinal salinity profile along the Euphrates River in Iraq for the years 2000–2015 (some years missing due to gaps in sampling at each location). The gray line illustrates the boundary for fresh water at 1000 ppm

increasing to 4000 ppm at al Samawa, and attained 5000 ppm at al-Nasiriyah. The authors indicate that the studied river reach receives irrigation return flow from at least four irrigation return flow drains along the stream.

While the Euphrates River salinity in Turkey remained within accepted limit, major changes occurred on the river reaches within Syria (Rahi and Halihan 2010; UN-ESCWA and BGR 2013). At the river reach that extends from the Iraqi-Syrian Border to al-Kufa, the water salinity more than tripled compared to its value in the Syrian-Turkey Border, but it is still below the threshold value set for drinking water by the Iraqi Bureau of Standards (Fig. 13.9). Salinity increases drastically in the lower parts of the Euphrates, the reach that starts just downstream from al-Kufa to al-Nasiriyah. A modern water salinity of around 3000 ppm exists for the cities of al Samawa and al-Nasiriyah (the years 2000–2015) (Fig. 13.9). Rahi and Halihan (2010) provided extensive review and analysis for the salinity within this part.

13.3.2.3 The Euphrates Salinity Profile within Iraq

Along the Euphrates River course within Iraq, the salinity profile differs in magnitude and to lesser extent in causes in comparison with the Tigris. However, the longitudinal salinity trend is the same for both rivers with salinity increases in the lower reaches. The salinity increases in al Fallujah relative to al Qaim by about 30% (from 580 to 840 ppm, Fig. 13.9). This increase is primarily due to the inflow from the Lake Tharthar system (Rahi and Halihan 2010). The increase in salinity in the Kufa to Nasiriyah reach was attributed to agricultural irrigation return flow (Fattah and Abdul Baki 1980; Rahi and Halihan 2010). In quantitative terms, salinity tripled in al Samawa compared to al-Kufa and quintupled compared to its value in al Qaim (Fig. 13.10).

The river salinity problem within Iraq is a multidimensional problem that requires additional studies to provide concrete solutions. Rahi and Halihan (2010) provided a review of the possible causes, some well-defined, some poorly understood, which

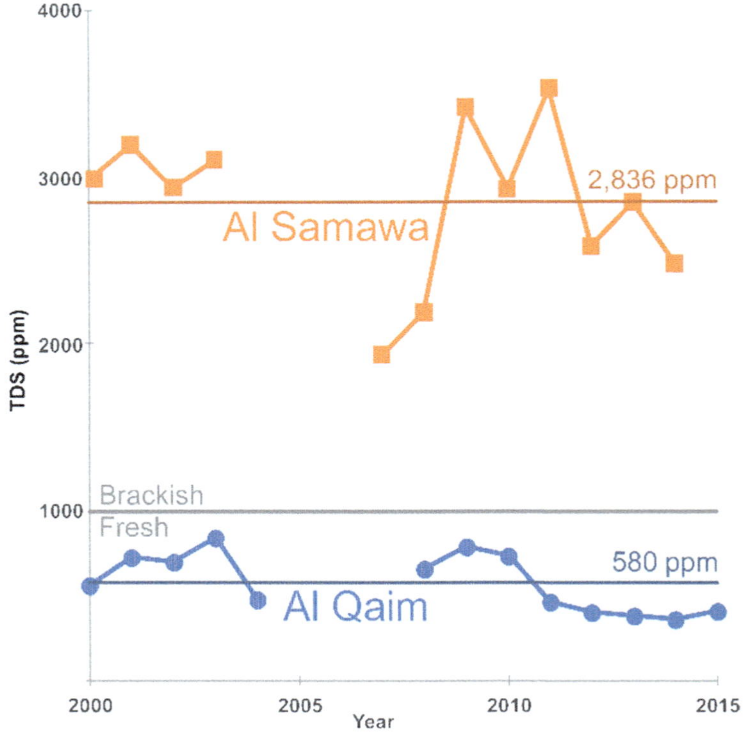

Fig. 13.10 Mean annual TDS at al Qaim (blue dots), near the Syria-Iraq border and al Samawa (orange squares), approximately 750 km downstream. Average at al Samawa is 2836 ppm and is shown with a solid orange line, and average at al Qaim is 580 ppm as shown with a solid blue line. Data are missing for the years 2005–2007 and 2004–2006 for al Qaim and al Samawa, respectively

may contribute to the salinization of the Euphrates waters. A summary of the author's analysis is presented in the next section.

13.3.3 The Euphrates Salinity Causes within Iraq

Four possible mechanisms for increasing the salinity of the Euphrates were investigated by Rahi and Halihan (2010). These mechanisms are summarized based on that work and included:

1. The decreased level of discharge entering Iraq and the increased salinity of the inflow waters.
2. The inflow coming from the Lake Tharthar system.
3. The irrigation return flow from Iraqi irrigation projects into the Euphrates.
4. Saline groundwater discharging from the Iraqi Western Desert (IWD) into the Euphrates.

The influence of each of the salinity mechanisms is evaluated in the following subsections. The analyses are ordered from upstream mechanisms to mechanisms that occur further downstream.

13.3.3.1 Decrease of Inflow from the Upper Euphrates

The total reservoir capacity of all the dams constructed on the upper Euphrates in Turkey as of 1997 was 90.9 bcm. When the GAP projects are completed in Turkey, the storage will be 94.8 bcm. This construction results in the capability to store about three times the average annual flow of the river of 29.6 bcm. In Syria and Iraq, the combined capacity of all dams along the Euphrates is 22.9 bcm. When added to the Turkish capacity, the gross reservoir storage capacity of all existing hydraulic structures on the Euphrates is 117.7 bcm. This provides the capacity to store approximately 4 years of the Euphrates flow.

Turkey and Syria negotiated a plan in 1987 to insure a minimum discharge of 500 m^3/s in the Euphrates as it entered Syria from Turkey. The water quality of the waters released into Syria was not specified (Beaumont 1996). The majority of water released from Turkey is irrigation return flow (Partow 2001; Topkaya 1998), which was evaluated to be approximately 700 ppm (Bilen, 1997). Other researchers (UN-ESCWA and BGR 2013) reported a salinity greater than 1000 mg/l, due largely to higher salinities of irrigation return flow. Additional impacts to water quality include pollution from upstream pesticide and fertilizer use influencing the water quality of the river. Data from the nearby Seyhan River basin show that salinity of the water used for irrigation was 225 ppm, while the irrigation return flow had increased to 1025 ppm (Scheumann 1993). It is anticipated that the salinity of the Euphrates water entering Iraq will eventually exceed the upper limit for domestic use.

The agreement reached between Syria and Iraq in 1990 decided that approximately 9.2 bcm per year should enter Iraq. Formally, the agreement was that Iraq was to have 58% of the Euphrates discharge that entered Syria during the year. This discharge is obtained from the primary channel of the Euphrates as it discharges from the Tabqa Dam. It also includes discharge from the Balikh and Khabur Rivers. The discharge from the Tabqa Dam is expected to be approximately 6.8 bcm/year with an anticipated salinity of 990 ppm. From the Khabur River, the discharge is approximately 1.4 bcm/year with a similar salinity. The discharge to Iraq from the Balikh River is approximately 1 bcm/year at 2000 ppm salinity. Based on these estimates, the predicted water quality for the Euphrates as it enters Iraq from Syria is at the fresh/brackish boundary of 1000 ppm when all the upstream irrigation schemes are included. After the GAP project is completed, Kolars (1994) estimated that the river flow into Iraq will be 6.6 bcm, which will primarily be irrigation return flow. Kolars (1994) did not estimate the salinity at this point. It is reasonable to assume the salinity of irrigation return flow to exceed 1000 ppm assigned as the upper limit for drinking water as further control structures are built upstream of Iraq.

13.3.3.2 Flow Diversion from the Tigris Through Lake Tharthar to the Euphrates

In the late 1970s, Iraq began to utilize the Lake Tharthar system to divert water from the Tigris to the Euphrates. Lake Tharthar water is a poor quality high salinity water even when compared to the present day degraded Euphrates river water quality. To quantify the effects of the Lake Tharthar system on the salinity of the Euphrates, a mass balance analysis was performed by Rahi and Halihan (2010). They estimated the total inflow entering Iraq to be 290 m^3/s. Iraq consumptive use from the Euphrates was estimated using values from Altinbilek (2004, Table 5), which was approximately 512 m^3/s. Therefore, Iraq needs to divert 222 m^3/s from Lake Tharthar to the Euphrates to maintain flow in the river. The salinity of water in Lake Tharthar was estimated as an average of 1500 ppm. Based on these figures, the estimated salinity of the Euphrates water downstream of the Lake Tharthar system was estimated to be approximately 1300 ppm. The 2000–2015 salinity, as measured, is about 900 in the al-Kufa station, which is located near Najaf downstream of the Lake Tharthar input.

13.3.3.3 Iraqi Irrigation Return Flow

Downstream from the city of al-Kufa, the Euphrates is used as an irrigation return flow drain carrying waters from several agricultural areas (Grego et al. 2004). Fattah and Abdul Baki (1980) studied irrigation return flow effects on water quality. Their results indicate "the influence of agricultural drainage disposal on the quality of the Euphrates River water between al Hindiya and al-Shinafiyah (a town located on the river just upstream from Al Samawa) is remarkably significant and may explain the major portion of the increase in concentration of TDS along that reach." In another portion of their paper, the authors stated that the major portion of the increase in concentration of TDS between Hit and al-Nasiriyah occurs between al-Hindiya and al-Shinafiyah, with a second large increase in salinity between al-Shinafiyah and al Samawa. Irrigation return flows in this reach of the river constitute approximately 70% of the discharge of the Euphrates. Rahi and Halihan (2010) have modeled the irrigation return flow, and their results confirmed the measured data as well as the finding of Fattah and Abdul Baki (1980). The authors used a mixing model to estimate the effects of irrigation return flow in the reach. The computed salinity of the Rahi and Halihan (2010) model was 2784 ppm, which is little less than the salinity value measured in al Samawa (Fig. 13.9).

13.3.3.4 Groundwater Discharge

The Euphrates River in Iraq may be a gaining stream with baseflow entering from the high salinity groundwater that flows through the aquifers in the Iraqi Western Desert (IWD). The IWD is bordering the Euphrates from the west as it travels through Iraq. These impacts due to groundwater baseflow are expected to impact the Euphrates that extends between al-Kufa and al-Nasiriyah.

The Iraqi Ministry of Irrigation completed hydrogeological characterization at two sites between al-Shinafiyah (just downstream from al-Kufa) and al Samawa (Ministry of Irrigation 2002). The data were later analyzed to assess groundwater/surface water interactions along the Euphrates. The study indicated that the area could be divided into a shallow unconfined aquifer with an underlying confined aquifer. The shallow unconfined aquifer discharges as baseflow to the Euphrates. No interaction could be found between the deeper confined aquifer and the river. The salinity of groundwater is quite high and ranges from 4400 to 9000 ppm. These two studies suggested that the Euphrates and the unconfined shallow aquifer are connected. This unconfined aquifer is local; it could not be linked to the massive aquifers of the IWD. The recharge source to this aquifer is most likely local irrigation water recharging the shallow system.

Estimates of the discharge from the aquifer to the river by Rahi and Halihan (2010) utilized a hydraulic conductivity of 10 m/day and a local groundwater gradient of 0.01. The estimated discharge per kilometer is approximately 0.007 m^3/s. At this discharge rate, groundwater would not constitute a major source of salinity. Rahi and Halihan (2010) concluded that the effect on the river water quality, which might be linked to groundwater baseflow is minor and is not strongly linked to the saline groundwater that flows from the IWD based on the available data.

Downstream of al Samawa, the significant increase of the salinity as the Euphrates flows to al-Nasiriyah is not well understood. Two possible causes have been suggested for the increase of salinity (Rahi and Halihan 2010). A link between the confined aquifer in the IWD and the river may exist along the reach, generating significant quantities of high salinity baseflow to the Euphrates, and/or the salinity comes from the dissolution of evaporative formations that are dominant in the area. Further studies would provide the data to test these two hypotheses. Electrical resistivity transects along these areas may provide significant insights due to the strong salinity signatures of the expected impacting waters.

13.4 Shatt al-Arab Waterway

13.4.1 Hydrology of Shatt al-Arab

The Shatt al-Arab River originates at the confluence of Tigris and Euphrates in the town of al-Qurna in southern Iraq and flows in a southeastern direction toward the Persian (or Arabian) Gulf (designated the Gulf hereafter). The salinity of the river was evaluated by Rahi (2018), and much of the discussion is paralleled here. Historically, the Shatt al-Arab functioned as a drain between the marshlands of lower Mesopotamia (southern marshes area of Iraq) and the Gulf.

The main marsh area is centered in the area north and west of the confluence of the Tigris and Euphrates. The area is divided into three major marshes: the al-Hammar Marshes, the Central Marshes, and the al-Hawizeh Marshes (Fig. 13.11). Prior to the 1990s, all three marshes drained into the Shatt al-Arab. The Al-Hawizeh marsh was draining through the al-Suwaib River. The al-Suwaib channel joins the left bank of the Shatt al-Arab approximately 5 km downstream from al-Qurna. The al-Shafi River and other small streams constituted the outlet of the Central Marshes to the Shatt al-Arab River. The al-Shafi River joins Shatt al-Arab about 50 km south of the confluence (not shown in Fig. 13.11). The al-Hammar marsh waters flow into Shatt al-Arab via the Garmat Ali River. Currently, the al-Hammar Marsh is being recharged by saline water from the Main Outfall Drain of Iraq (a primary irrigation return flow drain) and possibly from the Gulf through Khor Abdallah during high tides (Al-Gburi et al. 2017).

The Shatt al-Arab river is the smallest river of the three at about 192 km. The border between Iran and Iraq constitutes approximately half of the distance at 95 km (Partow 2001; Abdullah 2014). The Karun River, which is flowing to the southwest from the Zagros Mountains of Iran, discharges into the east bank of the river (Partow 2001). The Karun confluence is located about 40 km southeast of Basra (Partow 2001; Ghadiri and Ghadiri 2005; Abdullah 2014). The Garmat Ali River joins the Shatt al-Arab in Basra. It is the only major tributary from the west, and it drains the Hammar Marsh waters.

The Shatt al-Arab width increases from approximately 300 m near Qurna to about 700 m near Basra and more than 800 m near its mouth at the Gulf (UN-ESCWA and BGR 2013). The depth of the Shatt al-Arab is between 6 and 15 m, and the bed slope is 1–1.5 cm/km (Polservice 1980; Abdullah 2014). The UN-ESCWA and BGR (2013) estimated the basin size of the Shatt al-Arab to be 145,190 km^2, which included the Karkheh and Karun sub-basins. This area excludes the Euphrates and Tigris Basins upstream of Qurna. The basin definition varies as some calculations

Fig. 13.11 Location map and major features of the Shatt al-Arab River and vicinity (Modified from Fitzpatrick 2004)

evaluate the Tigris and Euphrates as part of a larger Shatt al-Arab basin and some refer to the river as the drain channel from the marshes with the Euphrates and Tigris as separate rivers that run to the marshes. Recently, the government of Iraq constructed a 129-km irrigation canal with its intake just south from the point where the Suwaib River meets the Shatt al-Arab.

The tides from the Arabian Gulf strongly influence much of the Shatt al-Arab. The semidiurnal and diurnal tides that affect the flow are important for agriculture productivity because they irrigate and drain agricultural areas adjacent to the riverbanks twice a day (Buringh 1960; Polservice). The relationship between the tidal magnitude during the cycles and the ultimate progression of tidal waters upstream in the Shatt al-Arab was studied by a consulting firm during the late 1970s (Polservice 1980). Results of their study confirm an inverse relationship between inland extent of tidal flows and the magnitude of discharge flowing into the Shatt al-Arab upstream (CEB 2011b; Polservice 1980).

13.4.2 *Salinity of Shatt al-Arab*

13.4.2.1 Salinity Issues for the Shatt al-Arab

The Shatt al-Arab has been impacted by higher salinity flows for much of the recorded data record. Research indicates that waters of Shatt al-Arab were generally fresh, except in the lower reaches, where it was brackish (Buringh 1960; Abdullah et al. 2016). Most of the early increases in salinity of the river were caused by the inflow from the marsh areas. In the marshes, salt concentration increased due to high evaporation rates as inflows decreased.

Large dams and reservoirs on the Euphrates and Tigris rivers have generated significant salinity increases in the Shatt al-Arab (UN-ESCWA and BGR 2013). The decrease in runoff and discharge in the upper river systems also increased salinity in the Shatt al-Arab (Rahi and Halihan 2010; UN-ESCWA and BGR). Inflow from the two rivers deceased in quantity and increased in salt concentration (Polservice 1980). The decrease of fresh water inflows to the Shatt al-Arab upstream allowed the saline wedge from the Arabian Gulf to migrate upstream (Polservice 1980; CEB 2011b). The fresh water inflows from the Tigris and the Euphrates have been reduced to historic low levels due to upstream hydraulic structures (UN-ESCWA and BGR, 2013). The modern inflow to the Shatt al-Arab is approximately 40 cms from the Tigris and approximately 12 cms from the Euphrates, with no flows entering from the Karun and Karkheh (Al-Mahmood 2015). This represents a decrease of ~96% from the previously reported discharge (UN-ESCWA and BGR 2013). The inflow from the Karun and Karkheh rivers has now been eliminated due to the Iranian use of the rivers via a range of hydraulic structures.

Marine water intrusion to the Shatt al-Arab is primarily driven by the tides. The tidally driven salinity advance in the river is inversely related to the amounts of fresh water discharge from the upstream rivers. The lower the discharges, the higher the tides are pushed upstream and the more coastal marine salinity migrates toward Qurna (the head of the Shatt al-Arab) (Polservice, 1980). There are limited measurements and studies of the tidal phenomenon and salinity variation available. Early work on the Karun and Karkheh rivers of Iran indicated that as their discharge was decreased, seawater intrusion may progress up the Shatt al-Arab (Harza 1963). The National Center for Water Resources Management concluded that the quality of Shatt al-Arab water is degraded not only by high salinity river and marine waters but also by oil spills from tankers as well. The center published a salinity value of 18,500 ppm in 2009 (CEB 2011b). Reported salinity values for the Shatt al-Arab at Seehan (south of Basra) were reported to be 2408 mg/L in 2011.

13.4.2.2 Salinity Analysis at Basra

Salinity increases in the Shatt al-Arab can be traced to dam construction upstream in the Euphrates River in the 1970s. However, field salinity data needed to clearly

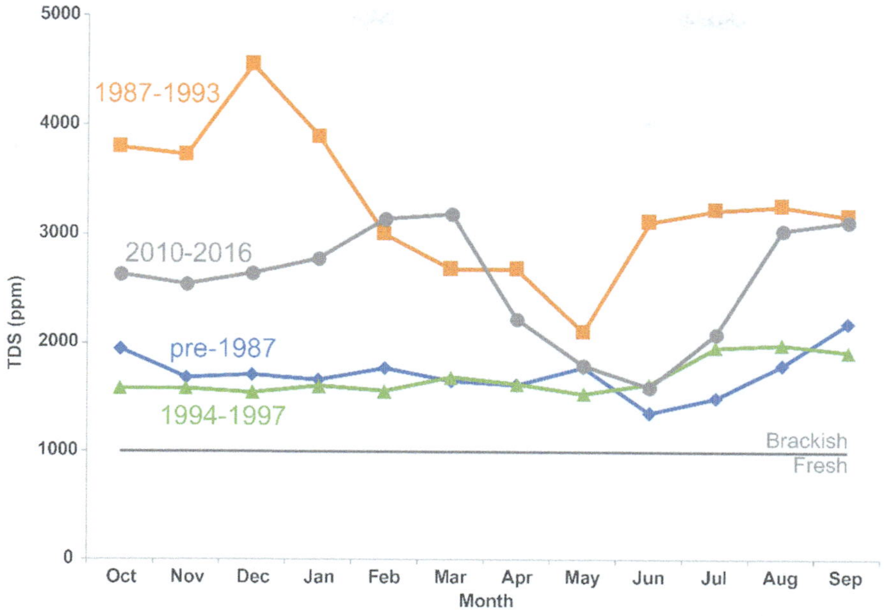

Fig. 13.12 Average TDS of the Shatt al-Arab at the Basra Station. All data are above the fresh water boundary indicated by the gray line. Note that the highest values (orange) occurred during a program of drying the marshes upstream of Basra

correlate the construction of large dams with salinity changes downstream are not available. Based on the available data, salinity in the city of Basra is analyzed for four different periods (Rahi 2018). Each period that was analyzed represents a change in the Tigris River system and/or the Mesopotamian marshes (Fig. 13.12).

The following salinity analysis for the Shatt al-Arab is from Rahi (2018) (Fig. 13.12). The pre-1987 period represents the time prior to operation of Mosul Dam, on the upstream portion of the Tigris River. The salinity is considered low relative to other periods, but still out of the range of fresh water. This high salinity was most likely caused by the high evaporation rates of the marshes that eventually drained to Shatt al-Arab (see Sect. 13.3). From 1987 to 2003, the flow to the lower parts of Tigris, the marshes, and to the Shatt al-Arab was limited by upstream users, first by the control structures upstream on the rivers and then by work to dry the marshes. The process of drying the marshes started in 1987 and ended in 1993. During those 7 years, the water supply to Shatt al-Arab was low quantity and poor quality. The waters of the Shatt al-Arab became high salinity during the entire year, which was the highest salinity period on record.

The most recent period of data is the salinity record for the years 2010–2016 (Fig. 13.12). In this interval, the Euphrates did not discharge to the Shatt al-Arab, and the Tigris discharged less than 50 cms to downstream. In recent years, most of the flow from the lower parts of the Euphrates and Tigris was diverted to restore the marshes. Salinity of the Shatt al-Arab is greater than 2000 ppm during the majority of the period. The most important conclusion for the salinity record of the Shatt al-Arab River is that it is above the standard limits for human consumption of 1000 ppm and is of limited use.

13.4.2.3 The Shatt al-Arab and the Marshes

The marshes are receiving their water supplies from the Tigris and the Euphrates rivers. Prior to 1974, most of the water was of low salinity; however, relatively long resident time and high evaporation rates in the marshes increase the salt concentration. Shatt al-Arab waters that were coming mainly from the marshes were of high salinity compared to those of the Euphrates and the Tigris (Buringh 1960). Following the planned drying of the marshes in the early 1990s, the inflows to Shatt al-Arab came primarily from direct flows from the Tigris and were relatively low salinity. The effects of those direct lower salinity flows are exhibited in the low salinity data from 1994 to 1997, which recorded relatively low salinity of Shatt al-Arab in Basra (Fig. 13.12). A significant contribution of these inflows comes from a river named al-Ezz, which was constructed as part of the marsh drying scheme (Fig. 13.11). The river collected water from the ends of two branches originating on the right bank of the Tigris just north of Amara. The al-Ezz River flows eastward, then southward, and ends in the Euphrates River five kilometers upstream of Qurna (Fig. 13.11). The relatively low-salinity waters of al-Ezz have a positive influence on the saline waters of the Euphrates, which in turn reduce the Shatt al-Arab salinity. The influence of lower salinity al-Ezz River water on the lowered salinity of the Shatt al-Arab is apparent in data at Qurna (Fig. 13.13).

13.5 Water Salinity Management Options

Water management measures are required to curtail and control water salinity increases. Advanced management schemes are practiced in countries such as the United States of America and Australia. In the United States, basin-wide salinity control and management programs have been implemented for the Colorado River (USBR 2013). The programs are conducted by a range of U.S. federal programs and state programs of the beneficiary states (Rahi 2018). The basin salinity is also supervised by the state-federal Colorado River Basin salinity control forum and the U.S. Environmental Protection Agency (U.S. EPA) (Morford 2014). The salinity control board implements best management practices for irrigation, erosion control, and reduction in natural saline spring discharge. Salinity control also regulates salinity releases at large dams. International treaty obligations between the United States and Mexico are met by keeping an average annual salinity of not more than 115 ppm ($\pm$30 ppm), greater than the average annual salinity at the Imperial Dam measuring station (USBR 2013; Morford 2014).

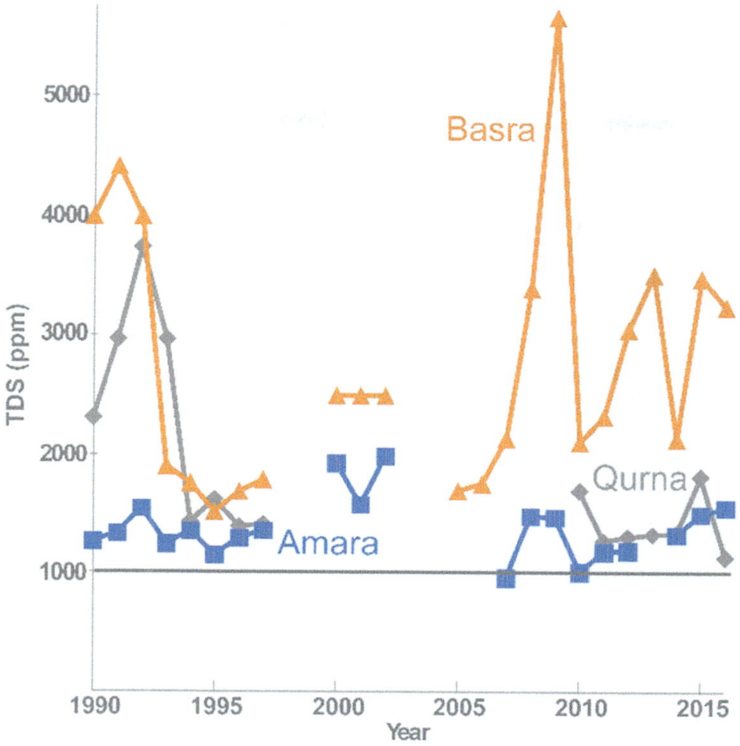

Fig. 13.13 Salinities of three stations (Amara, Qurna, and Basra compared to reflect the fresh water flow impact on salinity of Shatt al-Arab. The gray line indicates the fresh water/brackish water boundary at 1000 ppm

The South Australia Murray Darling Rivers system is another illustration of salinity control practices. The managing authorities at the rivers system have initiated a number of salinity control plans. The goal from the Murray Darling plans are to keep salinity at the city of Morgan, South Australia (well downstream, where it will affect water supplies for downstream users) at less than 800 μS/cm (about 512 ppm TDS) (BSM2030 2015). The Australian plan reached the salinity goal in 2010. The goal was achieved with a set of programs including salt interception schemes, changes in irrigation methods, and specific catchment programs (BSM2030 2015).

In the Euphrates, Tigris, and Shatt al-Arab River basins, incorporating Iran, Iraq, Saudi Arabia, Syria, and Turkey, no basin-wide management schemes are being practiced. Rahi and Halihan (2010, 2018) have assessed the salinity problems in both the Tigris and Euphrates rivers, evaluated the possible salinity control options, and concluded that the environmental flow is a suitable and needed low cost option. The minimum instream flow (MIF) or environmental flow concept "describes a fresh water flow (typically in-stream flow) that is maintained (or not allowed to be used for other, typically anthropogenic, purposes) solely for environmental reasons, to maintain the health and biodiversity of a particular water-related entity, such as a river, wetland, groundwater system, or estuary. For example, water may be extracted from a particular river for a particular industry. However, an environmental flow may be maintained down the river, not diverted to this industry, to maintain downstream river and/or estuarine ecosystems by allowing natural flows to progress through the system" (Peirson et al. 2002). For the Euphrates, an environmental flow rate is 178 cms for the annual minimum flow. This value is approximately a third of the historical minimum flows during the lowest flows of the Euphrates. This value is proposed as the minimum discharge needed to preserve the environment of the Euphrates (Rahi and Halihan 2010). For the Tigris, a minimum instream flow would be 185 cms. This value is approximately 15% of the mean historical flow of the river (Rahi and Halihan 2018). The combination of environmental flow of both rivers plus an estimated 100 cms from the Karun River is considered the environmental flow needed to preserve the integrity and biodiversity of Shatt al-Arab.

References

Abdullah SS (2014) Tide phenomena in the Shatt Al-Arab River, south of Iraq. J Arabian Gulf 42 (3&4):133–155

Abdullah AD, Gisen JIA, van der Zaag P, Savenije HHG, Karim UFA, Masih I, Popescu I (2016) Predicting the salt water intrusion in the Shatt al-Arab estuary using an analytical approach. Hydrol Earth Syst Sci 20:4031–4042. https://doi.org/10.5194/hess-20-4031-2016

Al-Ansari N, Ali A, Knutsson S (2014) Present conditions and future challenges of water resources problems in Iraq. J Water Res Pr 6(12):1066–1098. https://doi.org/10.4236/jwarp.2014.612102

Al-Gburi H, Al-Tawash B, Al-Lafta H (2017) Environmental assessment of Al-Hammar marsh, southern Iraq. Heliyon 3:1–26. https://doi.org/10.1016/j.heliyon.2017.e00256

Al-Hadithi AH (1978) Optimal utilization of the water resources of the Euphrates River in Iraq. Doctoral Dissertation. University of Arizona

Ali M, Salewicz KA (2005) Study on the salinity of Euphrates–Tigris Rivers, Iraq-model development status. Third expert meeting on Euphrates/Tigris River basin management. Japan Society for the Promotion of Science, the University of Tokyo

Al-Layla MA (1978) Effects of salinity on agriculture in Iraq, J. of the irrigation and drainage division. Am Soc Civil Eng 104(2):195–207

Al-Mahmood HK (2015) Hydrological changes in the lower part of Mesopotamian Basin. Iraq J Aquacult 12(1):47–70

Al-Saady YI, Abdullah EJ (2014) Water quality of Tigris River within Missan governorate eastern part of the Mesopotamia plain – Iraq, journal of Babylon university. Pure Appl Sci 9 (22):2489–2502

Altinbilek D (2004) Development and management of the Euphrates–Tigris basin. Int J Water Resour Dev 20(1):15–33. https://doi.org/10.1080/07900620310001635584

Basin Salinity Management 2030, (BSM2030) (2015) Published by the Murray–Darling Basin Ministerial Council, https://www.mdba.gov.au/sites/default/files/pubs/D16-34851-basin_salinity_management_strategy_BSM2030.pdf. Accessed 10 March 2016

Beaumont P (1996) Agricultural and environmental changes in the upper Euphrates catchment of Turkey and Syria and their political and economic implications. Appl Geogr 16(2):137–157

Bilen O (1997) Turkey and water issues in the Middle East. Southern Anatolia Project (GAP). Regional Development Administration. Ankara, Turkey

Buringh P (1960) Soils and soils conditions of Iraq. The Ministry of Agriculture, Baghdad

CASS (Central Arizona Salinity Study) (2003) Phase 1, report, technical appendix, p 309

CEB (Consultant Engineering Bureau) (2011a) Tigris and Euphrates Sampling (Final Report), College of Engineering, University of Baghdad, p 46, (Unpublished)

CEB (Consultant Engineering Bureau) (2011b) Shatt Al-Arab, final report (in Arabic), College of Engineering, University of Baghdad, p 101, (Unpublished)

FAO (Food and Agricultural Organization of the United Nations) (1970) Salinity Seminar Baghdad, Irrigation and Drainage Paper 7, Rome

FAO (Food and Agricultural Organization of the United Nations) (1976) Water quality for agriculture. Irrigation and drainage paper 29 Rev1. FAO, Rome

Fattah QN, Abdul Baki SJ (1980) Effects of drainage systems on water quality of major Iraqi rivers. Int Assoc Hydrol Sci Publ 130:265–270

Fitzpatrick RW (2004) Changes in soil and water characteristics of natural, drained and re-flooded soils in the Mesopotamian marshlands: implications for land management planning, Report from CSIRO Land and Water, p 81

Ghadiri H, Ghadiri M (2005) Marshlands of Mesopotamia and the Rivers that feed them, the 8th International River symposium, September 4–7, Brisbane, Australia, pp 1–15

Grego S, Micangeli A, Esposto S (2004) Water purification in the Middle East crisis: a survey on WTP and CU in Basra (Iraq) area within a research and development program. Desalination 165:73–79

Haigh FF (1951) Report on the control of the Rivers of Iraq and the utilization of their waters. Directorate general of irrigation

Hanna AB, Al Talbani K (1970) Evaluation of water quality in Iraq. First Technical Conference. Arab Federation of Agricultural Engineers, Khartoum, Sudan

Harza Engineering Company (1963) Hydrologic survey of Iraq: final report, Baghdad, Iraq

Holmes K (2010) Tigris and Euphrates River basin map. Annual precipitation in the river basin for 2010 (Iraq, Turkey, Iran, and Syria), scale 1:5,300,000

Houston Advanced Research Center (HARC) (2015) Report: Euphrates-Tigris Basin. http://www.harcresearch.org/sites/default/files/Project_Documents/Reports1-EuphratesTigris.pdf. Accessed 10 March 2016

ICARDA (International Center for Agricultural Research in the Dry Areas) (2012) Iraq salinity assessment, managing salinity in Iraq's agriculture, current state, causes, and impacts, an overview of the scope and scale of soil and water salinity in central and southern Iraq, report 1: situation analysis. ICARDA, Australia

IQBS (Iraqi Bureau for Standards) (2001) Iraqi quality standards for drinking water IQBS 417/2001

Jalut QH, Abbas NL, Mohammad AT (2018) Management of groundwater resources in the Al-Mansourieh zone in the Diyala River basin in eastern Iraq. Groundw Sustain Dev 6:79–86. https://doi.org/10.1016/j.gsd.2017.11.004

Jassim SZ, Goff JC (2006) "Geology of Iraq", Dolin, Prague, Czech Republic, p 352

Kolars J (1994) Problems of international river management: the case of the Euphrates. In: Biswas AK (ed) International waters of the Middle East – from Euphrates-Tigris to Nile. Oxford University Press, Oxford, pp 44–94

Ministry of Irrigation (MoI) (1997) (now Ministry of Water Resources), salinity data, unpublished, obtained

Ministry of Irrigation (MoI) (2002) (now Ministry of Water Resources), hydrogeological investigations of Al Atshan River (in Arabic). Baghdad, Iraq (Unpublished)

Morford S (2014) Salinity in the Colorado River Basin. Bureau of Reclamation, Phoenix, AZ

Murakami M (1995) Managing water for peace in the Middle East: alternative strategies. United Nations University Press, Tokyo, p 320

Mutlak SM, Salih BM, Tawfiq SJ (1980) Water Air Soil Pollut 13:9. https://doi.org/10.1007/BF02262520

Odemis B, Sangun MK, Evrendilek F (2010) Quantifying long-term changes in water quality and quantity of Euphrates and Tigris rivers, Turkey. Environ Monit Assess 170(1):475–490. https://doi.org/10.1007/s10661-009-1248-3

Partow H (2001) The Mesopotamian marshlands: demise of an ecosystem early warning and assessment technical report, UNEP/DEWA/Tr 01-3 rev. 1, division of early warning and assessment. United Nations environment program, Nairobi, Kenya
Peirson WL, Bishop K, Van Senden D, Horton PR, Adamantidis CA (2002) Environmental water requirements to maintain estuarine processes
Philip Williams and Associates (PWA) (2006) A review of the Hydrological and Geomorphic Impacts of the Proposed Ilisu Dam. Available from WEED, http://www2.weed-online.org/uploads/PWA_Ilisu_Report.pdf
Polservice (1980) Shatt Al-Arab project-feasibility report, VII, Ministry of Irrigation, Iraq, Unpublished report
Rahi KA (2018) Salinity management in the Shatt al-Arab river. Int J Eng Tech 7(4.20):128–133
Rahi KA, Halihan T (2010) Changes in the salinity of the Euphrates River system in Iraq. Reg Environ Change 10:27. https://doi.org/10.1007/s10113-009-0083-y
Rahi KA, Halihan T (2018) Salinity evolution of the Tigris River. Reg Environ Chang 18:2117. https://doi.org/10.1007/s10113-018-1344-4
Richardson CJ (2016) Mesopotamian marshes of Iraq. In: Finlayson CM et al (eds) The wetland book. Springer, Berlin. https://doi.org/10.1007/978-94-007-6173-5_70-1
Saleh DK (2010) Stream gage descriptions and streamflow statistics for sites in the Tigris River and Euphrates River basins, Iraq (no. 540). US Geological Survey
Scheumann W (1993) New irrigation scheme in Southeast Anatolia and in Northern Syria: more competition and conflict over the Euphrates. Q J Int Agric 32(3):240–259
Topkaya B (1998) Water resources in the Middle East: forthcoming problems and solutions for sustainable development of the region. Akdeniz University, Faculty of Engineering
UN-ESCWA and BGR (United Nations Economics and Social Commission for Western Asia; Bundesanstalt für Geowissenschaften und Rohstoffe) (2013) Chapter 3, tigris river basin. In: Inventory of shared water resources in Western Asia. UN-ESCWA and BGR, Beirut
UNEP (United Nations Environmental Program) (2003) Desk study on the environment of Iraq, Switzerland, 96 pp
USBR (United States Bureau of Reclamation) (2013) Quality of water, Colorado River basin, progress report no. 24, reclamation: managing rivers in the west. https://www.usbr.gov/uc/progact/salinity/pdfs/PR24final.pdf. Accessed 10 March 2016
Vaghefi SA, Mousavi SJ, Abbaspour KC, Srinivasan R, Yang H (2013) Analyses of the impact of climate change on water resources components, drought and wheat yield in semiarid regions: Karkheh River basin in Iran. Hydrol Process 28(4):2018–2032. https://doi.org/10.1002/hyp.9747
Weinthal E (2002) State-making and environmental, cooperation. MIT Press, Cambridge, MA, p 274
WHO (World Health Organization) (2004) International standards of drinking water. In: Recommendations, vol 1, 3rd edn. World Health Organization, Geneva, p 515

Chapter 14
Renewable Energy for Water–Energy Nexus in Euphrates and Tigris River Basin: A Literature Review

Furat Dawood

Abstract Understanding the interdependency of energy and water and the influencing factors as well as how this interrelation impacts the other essential sectors for the riparian states in the Euphrates and Tigris river basin. A literature review was conducted on the water stress, dispute and the renewable energy resources available in the basin. The water usage in the energy sector and the energy consumption in the water sector are reviewed. Nexus thinking and stakeholders' engagement approach were discussed to mitigate the water dispute in the basin. The asymmetry in power with the resources diversity in the riparian countries was promoted as an entry point for more collaboration and joint actions. The renewable energy resources availability was reviewed incomparable to the status of the renewable energy sources utilised, that already exist or are planned in the basin.

Keywords Euphrates and Tigris River Basin (ETRB) · Energy–water nexus · Renewable Energy (RE) in the ETRB · RE technologies in the water sector

14.1 Introduction

The Euphrates–Tigris River Basin (ETRB) has a total area of 879 790 km^2, distributed between Iraq (46%), Turkey (22%), the Islamic Republic of Iran (19%), the Syrian Arab Republic (11%), Saudi Arabia (1.9%), and Jordan (0.03%) (Lahn 2015; FAO 2009) as shown in Fig. 14.1.

The Euphrates–Tigris river systems flow from north to south (Salman et al. 2014). They originate from the highlands of Turkey and Iran where a number of tributaries join the Euphrates and Tigris in Iraq and Syria (UN-ESCWA and BGR 2013; FAO 2014; Lahn 2015; Jawad et al. 2017). The two rivers flow down the Mesopotamian plain, feeding the Mesopotamian Marshlands before merging near the city of

F. Dawood (✉)
College of Science, Health, Engineering and Education, Murdoch University, Canning Vale, WA, Australia

L. A. Jawad (ed.), *Tigris and Euphrates Rivers: Their Environment from Headwaters to Mouth*, Aquatic Ecology Series 11,
https://doi.org/10.1007/978-3-030-57570-0_14

Fig. 14.1 Euphrates–Tigris river basin geographic location; maps and image have been modified from FAO (2009), Kelly (2014), and ARENA (2018)

Al-Basra in southern Iraq (Jones et al. 2008), where they form the Shatt Al-Arab River. Shatt Al-Arab River runs for about 200 km and, empties in the Arabian Gulf (Persian Gulf) (Jawad et al. 2017). The Euphrates and Tigris regime depends heavily on winter rains and spring snowmelt in the Taurus and Zagros mountains (Gibson et al. 2016). The average annual precipitation in the ETRB is estimated at 335 mm (FAO 2009), although it varies along the basin area from 60 to 1500 mm within the basin (Daggupati et al. 2017). The annual mean discharge (natural flow) in normal weather conditions of the Euphrates is 29–35 billion cubic metres; whereas the Tigris provides an average of 52 billion cubic metres (Ahmmad 2010; Kibaroglu and Maden 2014; Zagonari and Rossi 2014).

Global warming and climate change have affected the ETRB with future projections indicating substantial reductions in the runoff of both the Euphrates and Tigris rivers (Ozdŏgan 2011; Kibaroglu and Maden 2014). Less water in the rivers, increased demand and rising populations further increase the stress on the ecosystems and water available for irrigation, food and energy production as well as domestic and industrial uses (Kibaroglu and Maden 2014; Zagonari and Rossi 2014). Water scarcity, dependence on trans-boundary water resources and the potential for continued disputes among the ETRB riparian countries pose many challenges (Kibaroglu and Maden 2014; Salman et al. 2014; Zagonari and Rossi 2014; FAO 2009). In response, Turkey, Iraq, and Syria have signed a memorandum of understanding in September 2009, to develop joint water flow-monitoring stations for more equitable apportionment and sustained water resources management (Salman et al. 2014).

Turkey's use of water has so far been limited, mainly due to hydropower generation and irrigation, which in the past was considered non-consumptive and not directly linked to water quality (FAO 2009). The water quality degrades downstream and riparian countries have complained about the quality of the water (Abdullah et al. 2016; FAO 2009). The return flow from irrigation causes water pollution and is expected to get worse as more land comes under irrigation (FAO 2009), which also increases the salinity of waters (Abdullah et al. 2016). Inflows from urban areas cause more anthropogenic pollution due to poor water treatment from the ineffective sewage treatment systems, causing solid waste to pile up on river banks (Kharrufa 2007, Jones et al. 2008; Abdullah et al. 2016; FAO 2009). Therefore, adopting aggressive policies to develop new agricultural projects, land use change, as well as the increase of energy consumption in the ETRB, without careful analysis of the environmental and hydrologic impacts, can have serious implications on water security, ecosystem deterioration, and the quality of water for human consumption (Jones et al. 2008; Abdullah et al. 2016; Daggupati et al. 2017).

Almost all forms of conventional energy harnessing, extracting and generating require water, whilst water treatment, distribution and irrigation water pumping consume energy (Rambo et al. 2017). The inter-dependencies between water and energy have been recognised and defined as Energy–Water Nexus (WEN), which is an analytical framework that treats water and energy use as an integrated system (IRENA 2015a, b; Lu and Chen 2016; Rambo et al. 2017). The management of WEN is closely linked to the carbon emissions that contribute to climate change (Jones et al. 2008). The Intergovernmental Panel on Climate Change (IPCC) predicts a pessimistic picture of the flows in the Tigris and Euphrates rivers (Janabi 2013). Climate change has become one of the most serious environmental issues due to its direct and indirect implications on communities, human prosperity and sociocultural factors, as well as the biodiversity of the natural ecosystem. Therefore, recognising water and energy interconnection provides a significant opportunity to save water and energy use, achieve sustainability and reduce the GHG emissions for the ETRB countries' benefit and in-line with the global agenda (Jones et al. 2008; Tan and Zhi 2016). The ETRB has reached a critical point due to the lack of awareness among citizens, the destructive impact of consecutive wars and the absence of effective rehabilitation plans (Bassem 2016). Therefore, "business as usual" is no longer a viable option in achieving sustainability and addressing environmental deterioration.

This chapter provides an intensive literature review for practical recommendations and future research directions about the water and energy nexus in the ETRB and offers avenues for further individual research.

14.2 Water–Energy Nexus

This section introduces the interlinkage between energy and water and discusses the implications of this nexus for the security of their respective sectors. The water and energy interdependencies have been recognised and defined as water treatment and distribution needs energy and similarly, producing energy utilized water (Yavuz

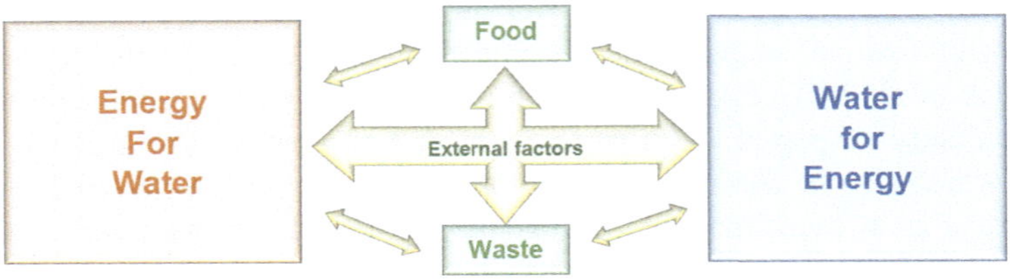

Fig. 14.2 Illustration of the Water–Energy Nexus with external factors

2015; Tan and Zhi 2016; Rambo et al. 2017). In spite of their close interdependence, most of the literature has often treated water and energy systems as separate systems over the entire life cycle (Rambo et al. 2017). Most of the literature has focused on unidirectional dependencies, such as the water use in energy or the energy use in water treatment (Tan and Zhi 2016).

The Water–Energy Nexus (WEN) has been developed as an analytical framework based on input–output analysis to depict and measure water and energy use as an integrated system [13, 14, 17, 19, 20]. However, WEN system can be influenced directly or indirectly by natural and/or anthropogenic factors, which include geography, climate change, government policies, technological advancement and socioeconomic development (Yavuz 2015; Lu and Chen 2016; Tan and Zhi 2016; Rambo et al. 2017; Zhang and Vesselinov 2017), as illustrated in Fig. 14.2.

According to the literature reviewed, the ETRB requires some special considerations due to the combination of high water stresses and a myriad of rapid changes in climate, socioeconomic and geopolitical conditions, as well as energy production and consumption (World Bank 2006; Daoudy 2008; Michel et al. 2012; Lahn 2015; Al-Ansari 2016; Kankal et al. 2016; Badran et al. 2017; Rambo et al. 2017). Therefore, the ETRB has been classified as a hot spot with the potential for continued disputes, at least in the immediate future (Daoudy 2008; Michel et al. 2012; Erickson 2013; Zagonari and Rossi 2014). Most of the literature has concluded that the ETRB must be dealt with as an integrated transboundary system, where water provides a useful point of entry to a nexus identification and analysis (World Bank 2006; Kibaroglu and Gürsoy 2015; Al-Ansari 2016). Achieving an optimal and sustainable resource allocation perspective needs nexus thinking in order to maximize the net socioeconomic benefits in the basin (Hansen 2012; Erickson 2013; Kucukmehmetoglu and Geymen 2014; Almusaed and Almssad 2015; Bilgen et al. 2015; IEA 2015a, b, 2016a, b, c, d, e).

Hence, the inefficient and unsustainable "business as usual" practices in the ETRB where, "up streamers use the water to produce more power, whilst down streamers use power to get more water" (Yavuz 2015) are no longer valid (Hansen 2012; Erickson 2013; Kucukmehmetoglu and Geymen 2014; Almusaed and Almssad 2015; Bilgen et al. 2015; IEA 2015a, b, 2016a, b, c, d, e; Yavuz 2015). Furthermore, the balance between the increasing supply and demand of water and energy, with obligations to reduce emissions and maintain sustainability (Global Agenda), is progressively becoming a more serious challenge for the riparian countries in the basin (Toklu 2013; Basaran et al. 2015; Dedinec et al. 2015; Najafi et al. 2015; IEA 2015a, b, 2016a, b, c, d, e; Rambo et al. 2017). In spite of the different national interests, power disparities and limited national capacities among the ETRB countries, these challenges have reached a critical point that cannot be dealt with in a unilateral and piecemeal manner (Daoudy 2008, Michel et al. 2012; Erickson 2013; Zagonari and Rossi 2014; Kibaroglu and Gürsoy 2015). In 1991, the former Secretary-General UN, Boutros Ghali said that "the next war in the Middle East will be over water, not politics" (Yavuz 2015). A nexus approach is envisioned in the literature to provide management strategies, hence decision makers have much to learn from nexus statistics and analysis (Kibaroglu and Gürsoy 2015; Larcom and van Gevelt 2017). The prerequisite for the transboundary political willingness of the ETRB countries to agree to a solution has not been studied extensively in the literature. However, it is important in developing and advancing the nexus thinking.

Hence, applying the nexus perspective to a transboundary basin helps river basin planners, managers and other stakeholders involved, to exploit synergies and mitigate trade-offs between different development opportunities whilst maintaining the overall sustainability in a shared basin. Literature looking at the ETRB through the lenses of the WEN showed that policy integration and coherence remains weak, particularly with the riparian states generally acting in a unilateral and uncoordinated manner (Siddiqi and Anadon 2011; Kibaroglu and Gürsoy 2015; Wasinger 2015; Yavuz 2015; Al-Ansari 2016; Badran et al. 2017; Daggupati et al. 2017; Rambo et al. 2017). The Joint Technical Committee is the only existing transboundary institution in the ETRB, but even they have not adopted any coherent objectives to govern transboundary water use and management (Kibaroglu and Gürsoy 2015; Wasinger 2015). Consequentially, the lack of a codified agreement or treaty that allocates Euphrates and Tigris waters among the riparians has led to a nexus that has not been studied extensively in the empirical literature (Lahn 2015; Wasinger 2015).

The identification and quantification of interconnections and interdependencies in the WEN will be discussed in the following section.

14.3 Quantifying Water–Energy Nexus

The quantitative evaluation of the water intensity in energy production and the energy intensity in the water utilisation chain highlights the strong environmental impacts on the ecosystem of the ETRB.

As water provides a point of entry to a nexus review in the transboundary basins, the next section will discuss water–energy correlations with a primary focus on calculating the water consumption of different energy productions.

14.3.1 Water Usage in Energy Production

Water is crucial for each stage of energy production, distribution and use, which has significant implications on energy security (UN-ESCWA 2013; UN-ESCWA 2015). Researchers have concluded that much of the attention has focused on how the availability of water impacts the different processes of energy production, rather than on how much energy the water sector uses (IEA 2016a, b, c, d, e).

Oil and gas production is a water-intensive sector, which after agriculture, is the largest water consumer in Iraq. Conversely, electricity generation systems (hydro-power) are the largest water consumers in Turkey and Syria. The water used in thermal power plants are predominantly for cooling as air is less effective than water in thermal efficiency (Chang et al. 2016; Rambo et al. 2017). The total water withdrawal for energy production is estimated to be 10–15% of the world's total water withdrawal (Chang et al. 2016; IEA 2016a, b, c, d, e).

14.3.1.1 Water in Hydropower Stations

Turkey has long been dependent on oil and gas imports, embarking on a programme with a particular emphasis on hydropower generation, like the lower Euphrates project in the late 1970s, which encompass the Tigris waters as well (Kibaroglu and Gürsoy 2015). Hence, 22 large dams and 19 hydropower plants were implemented in the South-eastern Anatolia Project (Güneydoğu Anadolu Projesi—GAP); which drained the river of 30 percent of its average annual flow and raised a red flag in Syria and Iraq (Kibaroglu and Gürsoy 2015; Wasinger 2015). Syria on the other hand started exploration for oil and gas in the early 1980s and launched the Euphrates Valley Project. Three dams were constructed with large hydropower plants; the multipurpose Tabqa Dam (Kibaroglu and Gürsoy 2015; Yıldız 2016), the Tishreen Dam and Baath dam in Raqqa (Erickson 2013).

Literature with Iran in focus reveals no accurate information about the hydro-electric schemes on the (15) Tigris river tributaries. Some reports stated that in March 2011, Iran signed a contract to build a 1,500 MW hydroelectric plant on the Bakhtiari tributary (Erickson 2013) followed by Khersan-3 dam project (IWPRD Co. 2018), which they flow directly into the Tigris river in eastern Iraq.

Iraq constructed 16 dams and eight hydroelectric power stations, seven plants of which, were on Tigris river and its tributaries (IEA 2015a, b, 2016a, b, c, d, e; UN-FAO 2018).

In spite of hydropower is considered as low carbon and renewable energy which offers significant environmental benefits, but it does not necessarily mean less water, as evapotranspiration (transpiration and evaporation) may cause water loss (Wu et al. 2009) and could, if not properly managed, exacerbate water stress (Chang et al. 2016; IEA 2016a, b, c, d, e).

14.3.1.2 Water in Thermal Power Stations

Thermal power plants withdraw a significant amount of water, mostly from surface water sources (rivers, lakes, and streams), to cool equipment since air is less effective than water in thermal efficiency for cooling, i.e. air is 1–7% efficient compared to water (Chang et al. 2016).

The steam utilisation and the cooling process are consuming water often through evaporative loss (Wang et al. 2017), and the much of it is returned to the water circle causes thermal pollution, which impacts the ecological system and diminishes the water quality (IEA 2016a, b, c, d, e). Hence, there is a real need for dry cooling technologies such as wet cooling towers, wet cooling ponds, dry cooling and hybrid systems that lower the water consumption, especially in the water-stressed areas (IEA 2015a, b).

Literature on the water usage of nuclear power stations revealed that they have similar water requirements compared to coal or thermal power plants (IEA 2015a, b; Tan and Zhi 2016).

Combined-cycle gas turbines (CCGT) have a higher thermal efficiency and some of the lowest rates of water withdrawals and consumption among thermal power plants, as they require less cooling (IEA 2016a, b, c, d, e). The efficiency of combined gas-fired turbines is around 55% (achievable under good operating conditions) compared to 31% for conventional gas-fired turbine plants (Kazem et al. 2014; IEA 2015a, b). Nevertheless, literature has revealed no accurate information about the current generation capacity of combined cycle power plants inside the ETRB boundaries (IEA 2015a, b). However, there is significant capacity in Turkey (Basaran et al. 2015; Kilickaplan et al. 2017), and Iran (Bahrami and Abbaszadeh 2013; Nejat et al. 2013; Najafi et al. 2015). There is no information of any large-scale CCGTs currently in Syria. On the other hand, Iraq is planning to convert over 9 GW of generators (originally commissioned as oil-fired plants) by 2020 to use natural gas as fuel. They are also planning to install 32 new GW CCGTs (around 40% of the total capacity) by 2035 (IEA 2015a, b; Edalati et al. 2017).

14.3.1.3 Water in Bioenergy

Bioenergy is derived from biomass to generate electricity, heat, and liquid fuels for transport. Biomass is often composed of agricultural products, forestry products, municipal waste and other hydrocarbon waste that is considered a low carbon or carbon neutral energy source. With that being said, it is important to note that low carbon does not necessarily mean less water consumption (IEA 2016a, b, c, d, e). Research on water consumption for bioenergy revealed that biomass feedstocks water requirement varies significantly from one region to another and that water consumption for biofuel production varies with processing technology (Gerbens-Leenes et al. 2009; Wu et al. 2009; Kim et al. 2015). For example, the classic trend shows:

- 10–17 litres of water are consumed to produce 1 litre of bioethanol. Life cycle consumption varies anywhere between 400 and 2600 litres of water per litre of bioethanol (Gerbens-Leenes et al. 2009; Kim et al. 2015), which makes it more water intensive than based on conventional oil (Kim et al. 2015).
- The water footprint for electricity production is estimated between 50 and 400 m^3 Water/GJ (Gerbens-Leenes et al. 2009).

However, research studies in the ETRB reveal that there is a potential for biogas production from the waste (landfill, municipal waste recycling and wastewater treatment), which needs attention to develop the planning and construction of a waste to electricity plants in the Basin (Knowles 2009; Shuokr Qarani et al. 2011; Dow 2013; Alnajjar 2015; Bilgen et al. 2015; Ozcan et al. 2015; Chabuk et al. 2016; Colleagues 2016; Karimi Alavijeh and Yaghmaei 2016; Al-Mossawi 2017; Bakhtiyari et al. 2017; Khishtandar et al. 2017; Toklu 2017).

The recent advancements in technology have expanded the potential in the biomass resources by including the agricultural residues, oilseeds and algae (Perona 2017; Sharma and Arya 2017; Toklu 2017; Yu et al. 2017, Yun et al. 2017; Vo et al. 2018). This allows for the sustainable development of the bioenergy industry without competing with the traditional agricultural industry for land resources that could otherwise be used for the production of food (Perona 2017; Toklu 2017). As things stand, the sensitivity of land use issues and the vital importance of food and water security impose significant constraints on the outlook for biomass to biofuels production in the ETRB riparian states (IEA 2015a, b).

14.3.1.4 Water in Fossil Energy Extraction and Distribution

Studies indicate that oil wells produce an average of 3–7 barrels of water for each barrel of oil, with the figure increasing for crude oil wells nearing the end of their productive lives (Jones et al. 1978; Veil et al. 2004; Kokal 2005). This produced water is very toxic, contaminated and polluted, which requires it to be heavily treated before any of it can be released back into the surface (Jones et al. 1978; Çakmakce et al. 2008; Hladik et al. 2014; Thiel et al. 2015). The white paper by the US Department of Energy in 2004 (Veil et al. 2004) stated that "the produced waters discharged from gas/condensate platforms are about ten times more toxic than the produced waters discharged from oil platforms". This issue has been reported in many areas in Iraq after the war as it has caused pollution and contaminated the surface water and soil, e.g. the recent reports on water pollution in the marshes area due to more oil and gas exploration in southern Iraq (Soaad Najy el Azzawy 2017).

Literature regarding oil and gas production in the ETRB riparian countries has revealed that Turkey has no big oil and gas reserves. Hence almost all of its oil and natural gas needs are imported (Toklu 2017). Iraq and Iran have big reserves of oil and gas and produce millions of barrels every day, whilst Syria had only started exploration for oil and gas in the early 1980s after launching the Euphrates Valley Project (Kibaroglu and Gürsoy 2015; Yıldız 2016).

On the other hand, the refinery process of the crude oil and the production of liquid fuels has been stated as a water-intensive industry or high life cycle water footprint (IEA 2015a, b), e.g. estimated a life cycle water consumed (water footprint) at a rate of 2.8–6.6 litres for each litre of gasoline produced from crude oil (Wu et al. 2009) and 13–16 from oil sands (Kim et al. 2015).

14.3.2 Energy Consumption in the Water Sector

All water services including purification, transport, sanitation and drinking water supply depend heavily on the energy available (IEA 2015a, b). The amount of energy consumption for water services depends on many factors such as water availability, topography, transport distances, water loss, efficiencies and level of treatment required (IEA 2015a, b). The World Energy Outlook report in 2016 by IEA provides the first systematic global estimate for the amount of energy used in the water services sector (IEA 2016a, b, c, d, e) revealed that "In 2014, some 4% of global electricity consumption was used to extract, distribute and treat water and wastewater, along with 50 million tonnes of oil equivalent of thermal energy, mostly diesel used for irrigation pumps and gas in desalination plants". This amount of energy consumption within the water sector is projected to double globally by 2040, mostly because of desalination projects that will account for 20% of water-related electricity demand (IEA 2015a, b, 2016a, b, c, d, e). The energy required in the water services in the Middle-East are projected to rise to 16% as higher levels of treatment grow, especially with the sharp rises in desalination capacity needed (IEA 2016a, b, c, d, e); however, the key issue identified here is that desalination is very costly (Rambo et al. 2017). As discussed in section (3.1.4), the produced water treatment in the oil and gas industry is an energy-intensive process, which introduces the need for cheap and clean energy. Hence, the renewable resources energy plays a crucial role in the socioeconomic development in and around the ETRB riparian states.

14.4 Renewable Energy in the ETRB

Regarding ETRB, most studies in the field of renewable energy resources have only focused on the resources for each country individually, not as a basin, due to the geopolitical and resource distribution diversity. In this section, a range of abundant renewable energy resources will be evaluated and mapped to demonstrate their viability and practicality in the ETRB riparian countries. The diversification of renewable energy resources can be considered as a strengthening element for the holistic basin energy system integration which has a direct effect on the water and food resources. Hence, sustainability in energy supply, especially electricity, requires more attention by the governments in the basin to ensure the diversification of energy sources other than fossil fuels. Literature has revealed that renewable energy sources, most of which are hydropower, could cause political disputes between the basin countries, especially as water has always played a big role in disputes and tensions within the ETRB (Erdem 2003; Yilmaz 2003; Hansen 2012; von Bogdandy and Wolfrum 2012; Kirschner and Tiroch 2012; KELLY 2014; Kibaroglu and Maden 2014; Zagonari and Rossi 2014; Lahn 2015; Wasinger 2015; Yavuz 2015; 2017 دارش).

Turkey has shown more investments in renewable energy resources like solar, wind and biomass energy (Yaniktepe et al. 2013; Basaran et al. 2015; Bilgen et al. 2015; Ozcan et al. 2015; Kilickaplan et al. 2017). A possible explanation for this might be that Turkey lacks fossil fuel resources compared to the other countries in the basin in particular Iraq and Iran. Despite that, most of the literature up till now did not draw a clear line between the basin area and the areas outside the boundaries for the ETRB from each riparian country. It reveals that solar, wind, geothermal and biomass energy are not being utilized sufficiently at present but that these energies could play an important role in the future of ETRB renewable energy. There is also great potential for utilising renewable energy resources inside the ETRB countries even outside the basin boundaries which can play a big role in the basin boundaries (Abed et al. 2014a, b).

The following different types renewable sources and technologies in the ETRB will be discussed in more details:

- Hydropower
- Solar energy

  Photovoltaic systems household scale to commercial scale
  Concentrated solar power (CSP) plants or salinity gradient solar ponds (SGSPs)
  Solar hot water for residential or industrial use
  Passive solar heating and daylighting in building design
  Space heating and cooling using solar heat pumps

- Wind energy

  Wind turbines for electricity generation
  Windmills for water pumping

- Bioenergy

  Biofuels
  Biopower
  Bioproducts

- Geothermal energy

  Geothermal Electricity Production from the earth's heat.
  Geothermal Direct Use of the hot water within the earth.
  Geothermal Heat Pumps by using the shallow ground to heat and cool buildings.

- Ocean, wave and tide
- Hydrogen and fuel cells

However, renewable sources intermittency and fluctuations such as oceanic, wind and solar introduce particular challenges to electricity grid stability and cause significant issues when trying to integrate these power generators into a grid system. Hence, the development of energy storage systems with robust policies and regulations is essential to prevent polluting the electricity grid and maintain the power quality and promote the renewable energy generation in the basin riparian countries (OECD 2009; Kaygusuz 2010; Ackerman and Fisher 2013; IEA 2015a, b; Amutha and Rajini 2016, Chang et al. 2016; IHA 2016; Murrant 2016; Badran et al. 2017; Kilickaplan et al. 2017).

On the other hand, there is a need for more research studies and investments to develop non-for electric utilisation of renewables like renewable energy technologies related to water. Renewables in water treatment, pumping and distribution can play a key role in addressing some of the most challenging aspects of the water–energy nexus. A range of practically applicable modern technologies will be discussed in the following sub-sections.

14.4.1 Renewables for Power; Resources, Current Trend and Potentials in the ETRB

In this sub-section, the renewable energy resources and their potentials in the ETRB are discussed and the role that they play in the EWN current and future trend.

14.4.1.1 Hydropower

The literature demonstrates that hydropower is considered the most developed renewable energy utilisation in the ETRB region where tens of dams and hydropower plants are installed (Erickson 2013; UN-ESCWA 2013; Al-Ansari 2016; Fawzi et al. 2016; IHA 2016; Kalender and Aytimur 2016; Kankal et al. 2016; Saeed et al. 2016; Yıldız 2016; Daggupati et al. 2017; Kilickaplan et al. 2017; IWRPD Co. 2018; FAO 2009). According to the International Hydropower Association (IHA) report in 2016 (IHA 2016), the installed hydropower capacity in the basin is 41,340 MW. The same report shows the asymmetric distribution of the water utilisation for power as Turkey has the lion share of 25,886 MW followed by Iran 11,196 MW then Iraq 2753 and Syria 1505 MW (IHA 2016). This shows the case of the water at the upstream is utilised to generate energy and the downstream countries using energy to get water which causing a conflict between the riparian countries (von Bogdandy and Wolfrum 2012; Erickson 2013; Kibaroglu and Maden 2014; Zagonari and Rossi 2014; Lahn 2015; Wasinger 2015; Yavuz 2015; Al-Ansari 2016; Yıldız 2016; Daggupati et al. 2017).

A literature review revealed that there is a lack of small-scale in-stream turbines to generate power that could serve remote and small communities. However, this needs further research to be verified.

14.4.1.2 Solar Energy

Literature in regard to the studies for individual countries and the global solar records, shown in Fig. 14.3, of the ETRB revealed that the solar energy resources are viable in the region of the basin and within the range of 1500–2400 KWh/m^2 per year (Kazem and Chaichan 2012; Maleh et al. 2012; Al-Nimr and Al-Shohani 2013; Hussein et al. 2013; Toklu 2013; Yaniktepe et al. 2013; Abed et al. 2014a, b; Saeed et al. 2016; WBG 2017a, b). The range of sunshine hours per day varies between winter and summer anywhere between 1970 and 2990 h of sunshine per year (Kazem and Chaichan 2012; Maleh et al. 2012; Hussein et al. 2013; Toklu 2013; Yaniktepe et al. 2013; Saeed et al. 2016; WBG 2017a, b).

In the last decade, the use of the solar energy for electricity power generation has progressed tremendously around the world due to the technology maturity and price reduction (IEA 2015a, b; Aghahosseini et al. 2018; Buttler and Spliethoff 2018; GRC 2018a, b) and can be considered cheaper than the traditional fossil fuel if all externalities are internalised, i.e. the pollution and its health and environmental impacts (Toklu 2013; Dai et al. 2018).

However, despite the strength of the solar resource in the basin, grid-connected solar electricity generation will remain a higher-cost option compared to fossil fuels, particularly in Iraq and Iran, resulting in fewer investments in the solar energy in the past and near future (IEA 2015a, b).

A series of IEA reports on Renewables and Waste for the years 2014 till 2016 revealed that the solar energy generation percentage out of the total energy for each country is unaccountable, except Turkey which has 3% energy generated by solar (IEA 2014a, b, c, d). This aligns with the solar energy installed capacity in the Arabian countries around the ETRB region, which is estimated to be less than 0.10%

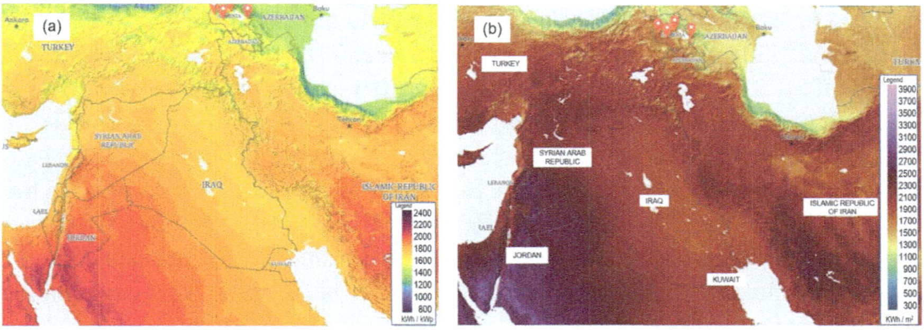

Fig. 14.3 Solar irradiance maps. (**a**) kWh/kWp per year; (**b**) KWh/m^2 per year; Credit to (IRENA 2018a, b)

Fig. 14.4 Examples of solar power plans in the ETRB region; (**a**) Hanwha Q CELLS and Kalyon Enerji win 1 GW solar tender in Turkey; Picture Credit: meteocotnrol/Phoenix Solar. (**b**) The Mohammed bin Rashid Al Maktoum Solar Park will have both PV and CSP components, picture credit: Server 2018. (**c**) Iran's last major solar power plant in the country's western province of Hamedan, picture credit: IPTV (2017)

of the total energy production in these countries (Badran et al. 2017). Some good studies were conducted on solar resources and the opportunities for harnessing that energy in the area surrounding ETRB area; they concluded that there is great potential for solar energy development. However, solar energy development requires immediate action and more investment to reduce the energy dependence on the fossil fuel and to mitigate the climate change (Bahrami and Abbaszadeh 2013; Gorjian and Ghobadian 2015), some examples are shown in Fig 14.4.

14.4.1.3 Wind Energy Resources

A literature review regarding wind energy resources in the ETRB and the surrounding areas is viable and needs to be considered for power generation or wind water pumping in order to reduce the electricity and fuel consumption. Again, there were no studies conducted regarding wind energy resources at the ETRB as a holistic system, rather some studies for the riparian countries individually, some references as examples are IEA (2016a, b, c, d, e), Saeed et al. (2016), Badran et al. (2017), Kilickaplan et al. (2017). The world bank group's renewable energy world atlas showed that the average wind speed in the ETRB is 5–7.5 m/s mostly in Iraq and Syria (250–500 W/m^2) and 3–6.6 m/s in Turkey and Iran (50–200 W/m^2) (WBG 2017a, b). Figure 14.5 shows the global estimation of the wind resources around the basin.

Some small-scale wind turbine in the remote areas does exist in Turkey and Iran, which they have been installed in the remote areas as off-grid stand-alone systems (IEA 2014a, b, c, d, 2016a, b, c, d, e). However, the literature reveals many studies regarding wind energy in the ETRB riparian countries. In Turkey, alone few studies have concluded the potential capacity is about 80 GW (Yaniktepe et al. 2013), and similar capacity in Syria (Amer and Hamed 2012), whilst there is a great potential in Iraq and Iran (Darwish and Sayigh 1988). Also, literature has revealed a great potential offshore the riparian countries which can be added to the basin wind energy resources and utilising potential (Abed et al. 2014a, b). Literature and the author

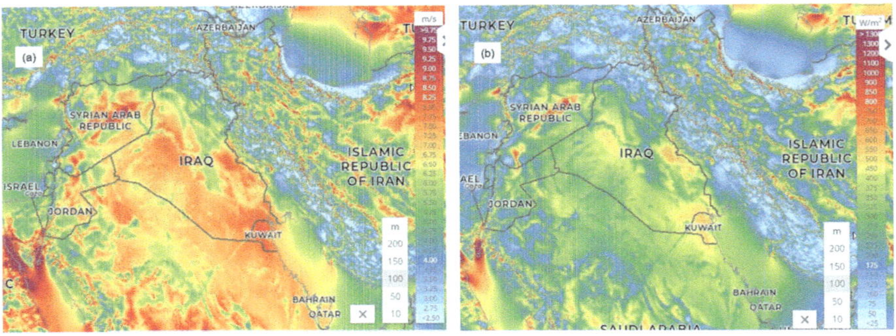

Fig. 14.5 Wind energy maps at 100 m above ground (a) Mean wind speed (m/s); (b) Mean power density (W/m2); Credit to (WBG 2017a, b)

concluded that the utilisation of the wind energy in the Basin requires more research studies and immediate action to support the growing need for clean energy to mitigate the climate change.

14.4.1.4 Biomass Energy Resources and Biofuels

The ETRB countries still rely heavily on the traditional use of solid biomass for cooking and space heating in particular in the remote or unelectrified areas, which may not be sustainable and can cause indoor air or greenhouse pollution (IEA 2015a, b; OECD 2015; Dawood et al. 2017). Biomass can be converted directly into liquid or gaseous fuels (biofuels) as a carbon-neutral energy as the carbon dioxide generated by combusting the biofuel is comparable to the amount of CO^2 that had been absorbed growing up the biomass (Badran et al. 2017; Perona 2017; Sharma and Arya 2017). The most common types of liquid biofuel are bioethanol and biodiesel, which are mostly used within transport sector (cars, trucks, buses, airplanes and trains) as direct fuel or blended with the fossil fuels (blending mandates) (OECD 2009, 2015; Badran et al. 2017; NREL 2018; World Bank Group 2017a, b). There are advanced biofuel technologies based on lignocellulosic biomass such as cellulosic-ethanol, biomass-to-liquids or synthetic gas and algae-based (bioreactors) biofuels (biological or chemical) catalysts (OECD 2015; Nelson et al. 2017). There has not yet evolved a single dominant technology or process strategy for biomass conversion into fuel (Badran et al. 2017; Perona 2017).

To the best of the author's knowledge, Turkey, for example has the highest biomass energy potential among the ETRB countries which is composed of the solid waste potential of 13.9 TW hth,a and solid biomass potential of 41.6 TW hth,a and biogas potential of 6.2 TW hth,a.; where currently Biomass waste and biogas power plant capacity reaches 2.8 GW (Ozcan et al. 2015; IEA 2016a, b, c, d, e; Kilickaplan et al. 2017). The world energy council website (WEC 2016), reported that "in the year 2005 Syria had municipal solid waste (4 million tonnes), wood

(0.5 million tonnes) and forestry and wood processing materials (0.2 million tonnes) available as bioenergy resources". The IEA report in 2015 on Iraq energy outlook reported 0% biomass primary energy demand by fuel and the biomass resource is moderate (IEA 2015a, b), and biomass energy is not being utilised (Kazem and Chaichan 2012; Mustafa and Mutlag 2013; Abed et al. 2014a, b).

The utilisation of the renewable biomass energy needs to develop the required regulations and policies to promote the biofuel utilisation which there has been a significant increase in the literature in this area (Basaran et al. 2015; Bilgen et al. 2015; Ozcan et al. 2015; Yuksel 2015; IEA 2016a, b, c, d, e; Tan and Zhi 2016; Kilickaplan et al. 2017).

However, biofuels production is a water-intensive process and land use change to produce crops either for food or biofuels, which may result endangering natural forests or food security (IEA 2015a, b; OECD 2015; Tan and Zhi 2016; Ziogou and Zachariadis 2016; Badran et al. 2017; Wang et al. 2017; Wichelns 2017).

However, recycling the green waste and biodegradable municipal waste mandates the collection of landfill gas by harnessing natural-rich resources like; food-waste biomass and sewage for power generation. This can generate a considerable amount of renewable bioenergy as well as saves water for non-potable reuse applications such as agricultural irrigation or industrial reuse (Peter Meisen 2011; IEA 2015a, b; Nelson et al. 2017).

14.4.1.5 Geothermal Energy Resources

Research reveals that there were fewer geothermal power plants in Turkey and Iran, than were outside the ETRB boundaries as shown in Fig. 14.6.

To the best of the author's knowledge, a small number of studies had been conducted in the region about the potential of geothermal energy in particular for electricity generation within and around the Basin; for example in Turkey (Toklu 2013; Yaniktepe et al. 2013; Melikoglu 2017), Iran (Bahrami and Abbaszadeh 2013; Soheil Porkhial 2013), Iraq (Abed et al. 2014a, b) and Syria (WEC 2016). The latest Geothermal Resources Council (GRC) bulletin report in Feb 2018 (GRC 2018a, b) has revealed that Turkey and Iran have just commissioned new geothermal power plants.

14.4.1.6 Ocean, Wave and Tide Resources

There has been little research on the ocean, wave and tide energy in the ETRB countries, despite that, there is some potential in the Mediterranean, black sea and Sea of Marmara. Most of the studies focus on the water salinity and desalination, e.g. (in Iraq) (Abdullah et al. 2016).

However, the literature reveals that there is significant potential for oceanic energy (shoreline, nearshore and offshore) along Turkey coastlines (Taşdemiroğlu 1991; Mustafa et al. 2010) and similar potential along Iran shores (Tavana 2009;

Fig. 14.6 Geothermal power and heat plants around the ETRB region; Credit to the Geothermal Resources Council (GRC) (GRC 2018a, b)

Saket and Etemad-Shahidi 2012). There was a rare research study (Mahmood et al. 2009) on the potential of the Breaker wind waves energy at the coast region in the extreme south of Iraq coastline and potential for tide energy utilisation.

14.4.1.7 Hydrogen and Fuel Cells

For decades, promoters of green energy have proclaimed the hydrogen economy referring to the vision of using hydrogen as a low carbon energy carrier or as a direct fuel replacing the polluting fossil fuels. Now the hydrogen economy is within sight and could finally become a reality (Walker et al. 2016; Apak et al. 2017; Gonernment of South Australia (GoSA) 2017a, b; Leonzio 2017; Perona 2017). Whilst some energy experts claim that hydrogen from renewable energy is proving in the market that this technology is already competitive, other experts think that it is still developing (Aouali et al. 2017; GoSA 2017a; Bailera et al. 2017; Esposito 2017; Leonzio 2017).

Hydrogen is attractive because when it is burned or reacted in a fuel cell to produce electricity, the only by-product is water (USDoE-FCO 2016; Gonernment of South Australia (GoSA) 2017a, b). In addition to transporting hydrogen or converting it to other gases (e.g. methane or ammonia), it may also be useful as a way to store or export renewable energy from intermittent sources Gonernment of South Australia (GoSA) 2017a, b. Hydrogen is an alternative to large-scale batteries or other storage systems like pumped hydro with a higher discharge depth and duration. Another possibility is to utilise hydrogen by blending a limited percentage of it with conventional fuel or natural gas without any required modification from the supply to the end user's side (Walker et al. 2016; Gonernment of South Australia (GoSA) 2017a, b). Despite the hydrogen merits and the extensive knowledge handling and distribution on a large scale, there are still some challenges mainly in storage technologies, which are still developing (Gutiérrez-Martín and Rodríguez-Antón 2016; Walker et al. 2016; Gonernment of South Australia (GoSA) 2017a, b). Literature lens on hydrogen shows that production technologies from renewable sources are mature and commercially available and the utilising technologies like fuel cells are also mature and reliable for large scale and commercial purposes (Kang and Holbrook 2015; Ferrero et al. 2016).

The literature reveals a lack of research studies of the deferent aspects of hydrogen economy development within the ETRB riparian states. However, some research studies in Turkey reveal the potential of hydrogen energy as a clean and sustainable source of energy that can be competitive to the conventional fuel price in the near future (Apak et al. 2017). The same paper concluded that the public response regarding hydrogen energy has been increasing and that the government should encourage the transition process. Other recent research studies modelled the solar–hydrogen hybrid system to evaluate the techno-economic viability and calculate the Levelized Cost of Electricity (LCE) over the system's lifetime of 25 years (Ozden and Tari 2016) and promoted the distributed generation and micro-grids (Uyar and Beşikci 2017). Whilst, in Iran there was more focus on hydrogen as energy storage in the renewable energy hybrid standalone systems and to power desalination plants from intermittent

renewable sources (Rezaei et al. 2018; Hakimi and Moghaddas-Tafreshi 2009; Shiroudi et al. 2013; Nasiri et al. 2015). Iraq increased awareness about pollution and the need for alternative cleaner fuel; a research study was conducted on the possibility of blending hydrogen to the gaseous fuel or liquid diesel (Kadhem et al. 2017). Literature and the author concluded that there is a lack of research studies in and around The ETRB countries.

14.4.1.8 Summary of the Renewables Status in the ETRB Riparian States

Table 14.1 summarizes the literature lens on the renewable power generation status in the ETRB.

14.4.2 Energy Storage Technologies

Energy storage is the key element to increase the renewable energy penetration percentage into the modern energy systems, improve the power quality and reduce the environmental impacts of the fossil fuel. Great amount of scientific efforts on different types of energy (e.g. electrical, mechanical and heat) and also, a wide

Table 14.1 Gross renewable resources in electricity generation

Country/Renewable energy source	Hydropower	Solar PV and thermal	Wind energy	Geothermal	Biofuels	Waste to energy	Tide wave ocean
Installed till 2016 *Generated in 2014*	*MW* *GWh*	*MW* *GWh*	*MW* *GWh*	*MW* *GWh*	*MW* *GWh*	*MW* *GWh*	*MW* *GWh*
Turkey (IEA 2016a, b, c, d, e; WEC 2016)	25900 40645	249 17	4690 8520	3510 2364	N/A 1082	N/A 104	N/A 0
Syria (IEA 2014a, b, c, d; WEC 2016)	1510 3000	N/A* 0	1 0	N/A 0	N/A 0	N/A 0	N/A 0
Iraq (IEA 2014a, b, c, d; WEC 2016)	2750 2931	N/A 0	N/A 0	N/A 0	N/A 0	N/A 0	N/A 0
Iran (IEA 2014a, b, c, d; WEC 2016)	11200 13862	17 0	117 358	81.5 N/A	N/A 47	N/A 0	N/A 0

N/A information not available

variety of storage technologies already reviewed in the literature (Aneke and Wang 2016; Farhadi and Mohammed 2016; Yu and Manthiram 2017; Baniassadi et al. 2018; Buttler and Spliethoff 2018; Lee et al. 2018; Robinius et al. 2018). These technologies have merits, challenges and different levels of technological maturity with many already proven for the commercial-scale application. Hence, the optimal energy storage system depends on sets of factors including geography, geopolitical, socioeconomic, resources, renewable energy technologies, cost-effectiveness and a mix of capacities and operation modes (Aneke and Wang 2016; Amirante et al. 2017).

Despite this, the pumped hydro energy storage has been used in the ETRB for decades (Johnson 2016), the topography limitations and the increase of water stress in the basin imposed the need for new distributed energy generation and storage in different scale of capacities as well as national scale of economy energy storage system in the riparian countries region. The new elegant innovations of power to gas, peer-to-peer (blockchain) energy trading and batteries (small and strata scale) need be developed in the region accordingly (Devlin et al. 2017; Aghahosseini et al. 2018; Robinius et al. 2018). Also, concentrated solar power plants or salinity gradient solar ponds are considered as a heat energy storage system that can help to overcome the intermittency and fluctuations in solar energy sources (Baniassadi et al. 2018). The literature on energy storage in the ETRB revealed that a great scientific effort has been done technically but not greatly coordinated with the governments and decision makers (Aghahosseini et al. 2018).

14.4.3 Renewables for Water Technologies

Renewable energy can be utilised for direct water services like treatment, distribution and irrigation to reduce the energy consumption in the water sector. As there are 6.5 to 7 million hectares used for irrigated agriculture in the ETRB distributed between the riparian states where 53% in Iraq followed by 18% in Iran, 15% in Turkey and 14% in Syria, which withdraw approximately 68 km^3 of water (FAO 2009). Hence, renewable energy can save a huge amount of energy in the agriculture sector, water treatment, distribution and wastewater recycling. The following subsections discuss different types of technologies to utilise renewable energy sources (mainly solar and wind) in water-related applications.

14.4.3.1 Renewable Energy in Waste Treatment

The sewage, industrial wastewater, agricultural return water, power stations, industrial factories, oil refineries, and research or hospital laboratories are currently a huge source of pollution in the ETRB in particular in Iraq and Syria (Al-Naseri et al. 2013; Awadh and Ahmed 2013, Fathi et al. 2013; Hussien et al. 2013; Topal and Arslan Topal 2016). However, if utilized appropriately, they can become a major potential

Fig. 14.7 Example of municipality waste; Source: Pictures copied from; Credit to (Khayamabshi 2016)

for efficient water reuse resource (Al-Mossawi 2014). Also, municipal solid waste which continuously increasing due to the rapid growth in population which poses a serious problem having adverse effects on environment and health; some example pictures from Iran are shown in Fig. 14.7. A similar if not worse situation has reported in Iraq and Syria, especially at the current status of war against terror (WHO 2018). The same World Health Organization fact sheet on Iraq (WHO 2018) situation stated that "Iraq is estimated to produce 31,000 tons of solid waste every day although estimated capacity to collect this waste is 4000 tons per day.

This gap in capacity is enormous and results in solid waste accumulating in streets or being dumped into natural depressions and empty lots". Whilst, Iraq has started its National Solid Waste Management Plan (NSWMP) development in 2007 by the collaboration of international waste management specialists, Kirkuk province is the first city in Iraq to benefit from solid waste management programme (Alnajjar 2016).

However, the waste to electricity by Landfill or Bio-digester technology with the right choice of electricity generation technology would lead to recycling the solid waste, generate energy (bioenergy) and improve the environment with by-product as a biofertilizer.

The literature on Turkey's situation reveals a different picture as waste to energy is much more in focus and these, development estimated some foreseen business opportunities of 9.5 billion Euro worth of solid waste investments with many landfills to gas projects already in business, as shown in Fig. 14.8.

The review of the Iran situation shows considerable development in this field and some commencing projects are reported (Khayamabshi 2016).

Fig. 14.8 Examples of municipality waste; pictures credit (TCN Co 2009; Yazgan 2013; WMW 2018)

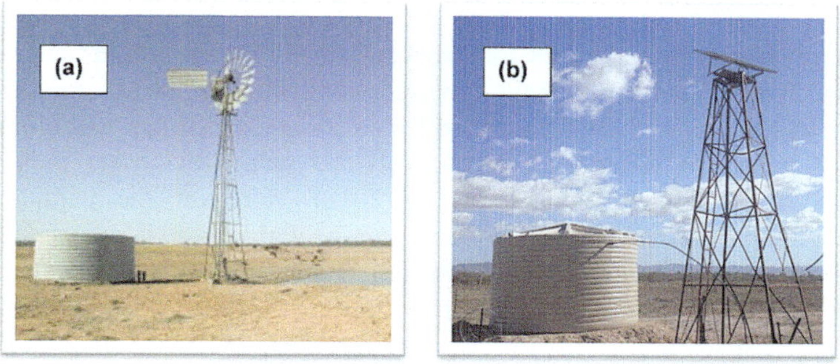

Fig. 14.9 Renewable energy in water pumping. (**a**) A traditional windmill irrigation system set-up; picture credit to Sontake and Kalamkar (2016); (**b**) Windmill upgrade to solar irrigation pump; picture credit to Mekhilef et al. (2013), Sontake and Kalamkar (2016)

14.4.3.2 Renewable Energy in Water Pumping and Conveyance

Solar photovoltaic and wind energy water pumping system has been a promising area of research for more than 50 years (Sontake and Kalamkar 2016). Solar PV irrigation systems potentially offer a cost-effective and reliable alternative to meeting growing energy demand to off-grid or gasoline-based pump sets; surface or underground for water pumping and conveyance whilst helping food production and sustaining livelihoods (IEA 2015a, b; Closas and Rap 2017). However, wind energy can be utilised in water pumping as a mechanical power to drive the pumps windmills) as shown in Fig. 14.9a or to generate electricity to drive electric motors for pumping (Sontake and Kalamkar 2016).

The main challenges were the initial cost and the intermittency of the renewable energy sources when continuous water flow is needed (Sontake and Kalamkar 2016). Hence, literature revealed that battery banks or water tanks had been used to compensate the renewable sources intermittency as illustrated in Fig. 14.9b. Whilst, government investment in subsidising the application of these systems can be the solution and create an incentive to increase the number of implementations (Sontake and Kalamkar 2016; Closas and Rap 2017; Niajalili et al. 2017).

In regard to the ETRB, the literature reveals that there have been lots of research studies, but very little has actually been done to implement these projects on a large scale due to the lack of awareness and government support (Senol 2012; Sontake and Kalamkar 2016; Niajalili et al. 2017). One of the research studies has calculated that a solar PV irrigation system can pay for itself and it will reach the conventional gasoline pumping system costs just in 9 years (Niajalili et al. 2017).

Literature and the author concluded that the application of a large number of small-scale irrigation systems based on renewable energy could save a huge amount of energy and reduce environmental pollution.

14.4.3.3 Renewable Energy for Heat

Heat energy is a major part of the total energy demand for households and industrial processes in the ETRB and globally. Therefore, utilising the solar thermal energy in the form of heat can reduce the demand for electricity, fossil fuel and biomass and results in producing clean heat (green heat) and reduce the water footprint and pollution (IEA 2015a, b; Baniassadi et al. 2018). Moreover, the literature shows that there are mature technologies for heat storage within insulated water or molten salt tanks over the ground or underground (Baniassadi et al. 2018). Also, the manufacturing technologies of direct or indirect heat collectors are well developed, commercially ready and highly efficient (Ramos et al. 2014; Shouman et al. 2015; IEA 2016a, b, c, d, e; Sharma et al. 2017; Suresh and Rao 2017; Farjana et al. 2018). Solar thermal energy can be utilised in many different domestic or industrial processes like solar dryers, steam production, space heating and heat pumps for air conditioning (Mekhilef et al. 2013; Pirasteh et al. 2014; Ramos et al. 2014). A literature review revealed some particular studies in the ETRB riparian countries for residential, agricultural and industrial solar thermal systems (Muneer et al. 2008; Kaygusuz 2010; Peter Meisen 2011; Bayrakcı and Koçar 2012; Kazem and Chaichan 2012; Bahrami and Abbaszadeh 2013; Abed et al. 2014a, b; Najafi et al. 2015; Soulayman and Sabbagh 2015; IEA 2016a, b, c, d; Badran et al. 2017).

14.5 Conclusion

The literature and the author concluded that the dispute over water and energy in the ETRB needs more joint work and should include all the riparian states. In 2008, Turkey, Iraq and Syria agreed to restart the Joint Trilateral Committee on water, which could be a start point to include all the stakeholders in the basin. The dispute ignited to restart this joint committee was originally caused by the fact that the upstream stakeholders utilise water to generate energy whilst the down streamers use more energy to compensate for the shortage of water. The asymmetry of political power and the resource diversity should be considered as an entry point for more collaboration among the riparian states.

The energy–water interdependence with all the other influencing factors and affected sectors like food, waste and pollution (nexus thinking) is promising a roadmap to mitigate the disputes in the ETRB. The literature revealed that there are viable renewable energy resources in the ETRB region, which can be utilised in the energy and water sectors to reduce the water–energy stress and mitigate the climate change.

This chapter reviewed the literature in regard to recognising and quantifying the interdependency between water and energy as well as the renewable energy resources availability and the utilisation status in the ETRB. The vital role of renewable energy and technologies in the energy and water sectors highlights the necessity of more research projects and implementations in the ETRB riparian states and offers avenues for further research studies.

References

Abdullah AD, Karim UFA, Masih I, Popescu I, Van der Zaag P (2016) Anthropogenic and tidal influences on salinity levels of the Shatt al-Arab River, Basra, Iraq. International Journal of River Basin Management 14(3):357–366

Abed FM, Al-Douri Y, Al-Shahery GMY (2014a) Review on the energy and renewable energy status in Iraq: The outlooks. Renewable and Sustainable Energy Reviews 39:816–827

Abed FM, Al-Douri Y, Al-Shahery GMY (2014b) Review on the energy and renewable energy status in Iraq: The outlooks. Renewable and Sustainable Energy Reviews 39(Supplement C):816–827

Ackerman F, Fisher J (2013) Is there a water–energy nexus in electricity generation? Long-term scenarios for the western United States. Energy Policy 59:235–241

Aghahosseini A, Bogdanov D, Ghorbani N, Breyer C (2018) Analysis of 100% renewable energy for Iran in 2030: integrating solar PV, wind energy and storage. International Journal of Environmental Science and Technology 15(1):17–36

Ahmmad YK (2010) Establishing a legal framework for the use and protection of Iraq's equitable right to the Tigris and the Euphrates River Basin. Doctor of Philosophy, University of Dundee

Al-Ansari N (2016) Hydro-politics of the tigris and euphrates basins. Engineering 8(3):33

Al-Mossawi MAJ (2014) Biological approach for recycling waste water in Iraq. Air, Soil and Water Research 7:111–115

Al-Mossawi MAJ (2017) Biological approach for recycling waste water in Iraq. Air Soil Water Res 7:ASWR.S17611

Almusaed A, Almssad A (2015) Building materials in eco-energy houses from Iraq and Iran. Case Studies in Construction Materials 2(Supplement C):42–54

Alnajjar AY (2015) Solid waste management in Iraq. Retrieved 17 March 2018, from https://www.bioenergyconsult.com/tag/solid-waste-management-in-iraq/

Alnajjar AY (2016) Solid waste management in Iraq, MENA

Al-Naseri SK, Mahdi ZM, Hashim MF (2013) Quality of water in hemodialysis centers in Baghdad, Iraq. Hemodialysis International 17(4):517–522

Al-Nimr M d A, Al-Shohani WAM (2013) Performance of photovoltaic module for different sites in Iraq. Arabian Journal for Science and Engineering 38(2):277–283

Amer ER, Hamed FA (2012) Reality and prospects of wind power in Syria. Energy Procedia 19:55–62

Amirante R, Cassone E, Distaso E, Tamburrano P (2017) Overview on recent developments in energy storage: mechanical, electrochemical and hydrogen technologies. Energy Conversion and Management 132:372–387
Amutha WM, Rajini V (2016) Cost benefit and technical analysis of rural electrification alternatives in southern India using HOMER. Renewable and Sustainable Energy Reviews 62:236–246
Aneke M, Wang M (2016) Energy storage technologies and real life applications – a state of the art review. Applied Energy 179:350–377
Aouali FZ, Becherif M, Ramadan HS, Emziane M, Khellaf A, Mohammedi K (2017) Analytical modelling and experimental validation of proton exchange membrane electrolyser for hydrogen production. International Journal of Hydrogen Energy 42(2):1366–1374
Apak S, Atay E, Tuncer G (2017) Renewable hydrogen energy and energy efficiency in Turkey in the 21st century. International Journal of Hydrogen Energy 42(4):2446–2452
Australian Renewable Energy Agency (ARENA) (2018) Bioenergy. Retrieved 16 March, 2018, from https://arena.gov.au/about/what-is-renewable-energy/bioenergy/
Awadh SM, Ahmed RM (2013) Hydrochemistry and pollution probability of selected sites along the Euphrates River, Western Iraq. Arabian Journal of Geosciences 6(7):2501–2518
Badran A, Murad S, Baydoun E, Daghir N (2017) Water, energy & food sustainability in the middle east the sustainability triangle. Springer International Publishing AG, Cham, Switzerland
Bahrami M, Abbaszadeh P (2013) An overview of renewable energies in Iran. Renewable and Sustainable Energy Reviews 24:198–208
Bailera M, Lisbona P, Romeo LM, Espatolero S (2017) Power to Gas projects review: Lab, pilot and demo plants for storing renewable energy and CO_2. Renewable and Sustainable Energy Reviews 69:292–312
Bakhtiyari Z, Yazdanpanah M, Forouzani M, Kazemi N (2017) Intention of agricultural professionals toward biofuels in Iran: Implications for energy security, society, and policy. Renewable and Sustainable Energy Reviews 69:341–349
Baniassadi A, Momen M, Amidpour M, Pourali O (2018) Modeling and design of solar heat integration in process industries with heat storage. Journal of Cleaner Production 170:522–534
Basaran ST, Dogru AO, Balcik FB, Ulugtekin NN, Goksel C, Sozen S (2015) Assessment of renewable energy potential and policy in Turkey – Toward the acquisition period in European Union. Environmental Science & Policy 46:82–94
Bassem W (2016) Iraq's environment faces crucial challenge. AlL-MONITOR Iraq pulse
Bayrakcı AG, Koçar G (2012) Utilization of renewable energies in Turkey's agriculture. Renewable and Sustainable Energy Reviews 16(1):618–633
Bilgen S, Keleş S, Sarıkaya İ, Kaygusuz K (2015) A perspective for potential and technology of bioenergy in Turkey: present case and future view. Renewable and Sustainable Energy Reviews 48:228–239
Buttler A, Spliethoff H (2018) Current status of water electrolysis for energy storage, grid balancing and sector coupling via power-to-gas and power-to-liquids: a review. Renewable and Sustainable Energy Reviews 82:2440–2454
Çakmakce M, Kayaalp N, Koyuncu I (2008) Desalination of produced water from oil production fields by membrane processes. Desalination 222(1):176–186
Chabuk A, Al-Ansari N, Hussain HM, Knutsson S, Pusch R (2016) Landfill site selection using geographic information system and analytical hierarchy process: a case study Al-Hillah Qadhaa, Babylon, Iraq. Waste Management & Research 34(5):427–437
Chang Y, Li G, Yao Y, Zhang L, Yu C (2016) Quantifying the water-energy-food nexus: current status and trends. Energies 9(2):65
Closas A, Rap E (2017) Solar-based groundwater pumping for irrigation: sustainability, policies, and limitations. Energy Policy 104:33–37
Colleagues MKAa (2016) Study Results from M.K. Alavijeh and Colleagues Update Understanding of Waste Management Research (Biochemical production of bioenergy from agricultural crops and residue in Iran). Energy Weekly News: 247
D. o. t. World Bank (2006) Iraq country water resource assistance strategy, World Bank

Daggupati P, Srinivasan R, Ahmadi M, Verma D (2017) Spatial and temporal patterns of precipitation and stream flow variations in Tigris-Euphrates river basin. Environmental Monitoring and Assessment 189(2):–50

Dai J, Wu S, Han G, Weinberg J, Xie X, Wu X, Song X, Jia B, Xue W, Yang Q (2018) Water-energy nexus: a review of methods and tools for macro-assessment. Appl Energy 210:393–408

Daoudy M (2008) Asymmetric Power:Negotiating water in the euphrates and tigris. International Negotiation 14(2009):359–389

Darwish ASK, Sayigh AAM (1988) Wind energy potential in Iraq. Journal of Wind Engineering and Industrial Aerodynamics 27(1):179–189

Dawood F, Urmee T, Shafiullah GM (2017) The renewable energy household lighting for Chibayish inhabitant's in Iraq. Renew Energy Environ Sustain 2:15

Dedinec A, Markovska N, Ristovski I, Velevski G, Gjorgjievska VT, Grncarovska TO, Zdraveva P (2015) Economic and environmental evaluation of climate change mitigation measures in the waste sector of developing countries. Journal of Cleaner Production 88(Supplement C):234–241

Department of Energy-Fuel Cell Office (USDoE-FCO) (2016) Pathways to commercial success. T. a. P. S. b. t. F. C. T. office. Fuel cell technologies office, Pacific Northwest National Laboratory

Devlin J, Li K, Higgins P, Foley A (2017) A multi vector energy analysis for interconnected power and gas systems. Applied Energy 192:315–328

Dow (2013) Basra waste treatment plant development - Iraq - construction project profile. Progressive Media Group, London

Edalati S, Ameri M, Iranmanesh M, Sadeghi Z (2017) Solar photovoltaic power plants in five top oil-producing countries in Middle East: a case study in Iran. Renewable and Sustainable Energy Reviews 69:1271–1280

Erdem M (2003) The Tigris-Euphrates rivers controversy and the role of international law

Erickson FL a EJ (2013) Strategic water iraq and security planning in the euphrates-tigris basin. Marine Corps University Press, Quantico, VA

Esposito DV (2017) Membraneless Electrolyzers for low-cost hydrogen production in a renewable energy future. Joule 1(4):651–658

FAO (2014) Report of the expert meeting on the review of fisheries and aquaculture activities in the Euphrates-Tigris basin: Erbil, Iraq, 11–12 November 2012. FAO fisheries and aquaculture report (1079): I, III, VII, VIII, 1–125

Farhadi M, Mohammed O (2016) Energy storage technologies for high-power applications. IEEE Transactions on Industry Applications 52(3):1953–1961

Farjana SH, Huda N, Mahmud MAP, Saidur R (2018) Solar process heat in industrial systems – a global review. Renewable and Sustainable Energy Reviews 82:2270–2286

Fathi RA, Matti LY, Al-Salih HS, Godbold D (2013) Environmental pollution by depleted uranium in Iraq with special reference to Mosul and possible effects on cancer and birth defect rates. Medicine, Conflict and Survival 29(1):7–25

Fawzi NAM, Goodwin KP, Mahdi BA, Stevens ML (2016) Effects of Mesopotamian Marsh (Iraq) desiccation on the cultural knowledge and livelihood of Marsh Arab women. Ecosystem Health and Sustainability 2(3):1–6

Ferrero D, Gamba M, Lanzini A, Santarelli M (2016) Power-to-gas hydrogen: techno- economic assessment of processes towards a multi-purpose energy carrier. Energy Procedia 101:50–57

Foof and Agriculture Organization (FAO) (2009) Euphrates-tigris basin. Retrieved 28 September 2017, from http://www.fao.org/nr/water/aquastat/basins/euphrates- tigris/index.stm#top

Geothermal Resources Council (GRC) (2018a) Bulletin. CA, USA, GRC

Geothermal Resources Council (GRC) (2018b) Geothermal energy news from around the world since 1972. T. G. R. C. Retrieved 5 March, 2018, from https://geothermal.org/

Gerbens-Leenes W, Hoekstra AY, van der Meer TH (2009) The water footprint of bioenergy. Proceedings of the National Academy of Sciences of the United States of America 106 (25):10219–10223

Gibson M, Lloyd SHF, Owen L (2016) Tigris-Euphrates river system. Encyclopædia Britannica, Chicago

Gonernment of South Australia (GoSA) (2017a) South Australian green hydrogen study. (Project No. A32334 | Rev 2): 132.

Gonernment of South Australia (GoSA) (2017b) A hydrogen roadmap for South Australia. Adelaide, South Australia Department of The premier and Cabinet

Gorjian S, Ghobadian B (2015) Solar thermal power plants: progress and prospects in Iran. Energy Procedia 75:533–539

Gutiérrez-Martín F, Rodríguez-Antón LM (2016) Power-to-SNG technology for energy storage at large scales. International Journal of Hydrogen Energy 41(42):19290–19303

Hakimi SM, Moghaddas-Tafreshi SM (2009) Optimal sizing of a stand-alone hybrid power system via particle swarm optimization for Kahnouj area in south-east of Iran. Renewable Energy 34 (7):1855–1862

Hansen R (2012) Civilization's drying cradle: water politics in the tigris-euphrates river basin. USA, United States Army War College

Hladik ML, Focazio MJ, Engle M (2014) Discharges of produced waters from oil and gas extraction via wastewater treatment plants are sources of disinfection by-products to receiving streams. Science of The Total Environment 466–467:1085–1093

Hussein BN, Essa IA, Hussein DJ (2013) Introducing a PV design program compatible with Iraq conditions. Energy Procedia 36(Supplement C):852–861

Hussien BM, Fayyadh AS, Hammed JM, Hamdani MAA (2013) Hydrochemical model and probable pollution in the water of Euphrates River (Qaem–Falluja). Arabian Journal of Geosciences 6(3):783–799

International Energy Agency (IEA) (2014a). Iran, Islamic Republic of renewables and waste for 2014. Retrieved 10 March 2018, from http://www.iea.org/statistics/statisticssearch/report/?country=IRAN&product=renewablesan dwaste&year=2014.

International Energy Agency (IEA) (2014b) Iraq: renewables and waste for 2014. Retrieved 10 March 2018, from http://www.iea.org/statistics/statisticssearch/report/?country=IRAQ&product=renewablesan dwaste&year=2014

International Energy Agency (IEA) (2014c) Syrian Arab Republic: renewables and waste for 2014. Retrieved 10 March 2018, from https://www.iea.org/statistics/statisticssearch/report/?country=Syria&product=Renewables andWaste

International Energy Agency (IEA) (2014d) Turkey: renewables and waste for 2014. Retrieved 10 March 2018, from http://www.iea.org/statistics/statisticssearch/report/?country=TURKEY&product=renewable sandwaste&year=2014.

International Energy Agency (IEA) (2015a) Iraq energy outlook. France, World energy outlook special report

International Energy Agency (IEA) (2015b) World energy outlook 2015, IEA

International Energy Agency (IEA) (2016a) Energy policies of IEA countries 2016 review Turkey

International Energy Agency (IEA) (2016b) Water energy nexus. World Energy Outlook 2016. France, IEA

International Energy Agency (IEA) (2016c) A delicate balance between water demand and the low-carbon energy transition. IEA reports

International Energy Agency (IEA) (2016d) Energy policies of IEA countries; 2016 review – Turkey

International Energy Agency (IEA) (2016e) World energy outlook 2016 – executive summary, IEA

International Hydropower Association (IHA) (2016) Hydropower status report

Iran Press TV (IPTV) (2017) Iran inaugurates its biggest solar plant ever. Retrieved 25 March, 2018, from http://www.presstv.com/Detail/2017/04/27/519635/Iran-inaugurates-its-biggest-solar-plant-ever

IWPRD Co. (2018) Kherasan 3. Retrieved 27 February 2018, from http://en.iwpco.ir/Khersan3/default.aspx.

Janabi H (2013) Climate Change Impact on Iraqi Water and Agriculture Sectors. MEES 56(12)

Jawad LA, Ligas A, Al-Janabi MIG (2017) Meristic character variability among populations of Silurus triostegus Heckel, 1843 from the Euphrates, Tigris, and Shatt al-Arab Rivers, Iraq. Archives of Polish Fisheries 25(1)

Johnson LB (2016) Sustainability of engineered rivers in arid lands: Euphrates-Tigris and Rio Grande/Bravo. School of Public Affairs, USA

Jones TJ, Neustadter EL, Whittingham KP (1978) Water-in-crude oil emulsion stability and emulsion destabilization by chemical demulsifiers

Jones C, Sultan M, Yan E, Milewski A, Hussein M, Al-Dousari A, Al-Kaisy S, Becker R (2008) Hydrologic impacts of engineering projects on the Tigris–Euphrates system and its marshlands. Journal of Hydrology 353(1-2):59–75

Kadhem J, Reza K, Ahmed W (2017) Alternative fuel use in Iraq: a way to reduce air pollution. EJERS, European Journal of Engineering Research and Science 2(5):20–30

Kalender L, Aytimur G (2016) REE geochemistry of Euphrates river, Turkey. Journal of Chemistry 2016:1–13

Kang DW, Holbrook JH (2015) Use of NH3 fuel to achieve deep greenhouse gas reductions from US transportation. Energy Reports 1:164–168

Kankal M, Nacar S, Uzlu E (2016) Status of hydropower and water resources in the Southeastern Anatolia Project (GAP) of Turkey. Energy Reports 2:123–128

Karimi Alavijeh M, Yaghmaei S (2016) Biochemical production of bioenergy from agricultural crops and residue in Iran. Waste Management 52:375–394

Kaygusuz K (2010) Sustainable energy, environmental and agricultural policies in Turkey. Energy Conversion and Management 51(5):1075–1084

Kazem HA, Chaichan MT (2012) Status and future prospects of renewable energy in Iraq. Renewable and Sustainable Energy Reviews 16(8):6007–6012

Kazem AA, Chaichan MT, Kazem HA (2014) Dust effect on photovoltaic utilization in Iraq: review article. Renewable and Sustainable Energy Reviews 37(Supplement C):734–749

Kelly M (2014) Land between the rivers: the water crisis in Iraq. Retrieved 18 November 2017, from https://eastofapreconceivedmiddle.wordpress.com/2014/09/24/iraq- week-5/

Kharrufa S (2007) Reduction of building waste in Baghdad Iraq. Building and Environment 42 (5):2053–2061

Khayamabshi E (2016) Municipality waste management in Iran. Tokyo, United Nations Industrial Development Organization

Khishtandar S, Zandieh M, Dorri B (2017) A multi criteria decision making framework for sustainability assessment of bioenergy production technologies with hesitant fuzzy linguistic term sets: The case of Iran. Renewable and Sustainable Energy Reviews 77:1130–1145

Kibaroglu A, Gürsoy SI (2015) Water–energy–food nexus in a transboundary context: the Euphrates–Tigris river basin as a case study. Water International 40(5-6):824–838

Kibaroglu A, Maden TE (2014) An analysis of the causes of water crisis in the Euphrates- Tigris river basin. Journal of Environmental Studies and Sciences 4(4):347–353

Kilickaplan A, Bogdanov D, Peker O, Caldera U, Aghahosseini A, Breyer C (2017) An energy transition pathway for Turkey to achieve 100% renewable energy powered electricity, desalination and non-energetic industrial gas demand sectors by 2050. Solar Energy 158:218–235

Kim HC, Wallington TJ, Mueller SA, Bras B, Guldberg T, Tejada F (2015) Life cycle water use of ford focus gasoline and ford focus electric vehicles. Journal of Industrial Ecology 20 (5):1122–1133

Kirschner AJ, Tiroch K (2012). The waters of Euphrates and Tigris: an international law perspective. Koninklijke Brill N.V., Max Planck Yearbook of United Nations Law.

Knowles JA (2009) National solid waste management plan for Iraq. Waste Management & Research 27(4):322–327

Kokal SL (2005) Crude oil emulsions: a state-of-the-art review

Kucukmehmetoglu M, Geymen A (2014) The significance and impacts of large investments over the determination of irrigated agricultural land use: The case of the Euphrates & Tigris River Basin. Land Use Policy 41:514–525

Larcom S, van Gevelt T (2017) Regulating the water-energy-food nexus: Interdependencies, transaction costs and procedural justice. Environmental Science & Policy 72:55–64
Lee B, Ko Y, Kwon G, Lee S, Ku K, Kim J, Kang K (2018) Exploiting biological systems: toward eco-friendly and high-efficiency rechargeable batteries. Joule 2(1):61–75
Leonzio G (2017) Design and feasibility analysis of a power-to-Gas plant in Germany. Journal of Cleaner Production 162:609–623
Lu Y, Chen B (2016) Energy-water nexus in urban industrial system. Energy Procedia 88:212–217
Mahmood AB, Al-Mahdi AA, Abdullah SS (2009) Breaker wind waves energy at Iraqi coastline. Mesop J Mar Sci 24(2):112–121
Maleh HAL, Tine HA, Naimeh W (2012) Environment & feasibility study to make use of solar energy in Syria. Energy Procedia 19:30–37
Mekhilef S, Faramarzi SZ, Saidur R, Salam Z (2013) The application of solar technologies for sustainable development of agricultural sector. Renewable and Sustainable Energy Reviews 18:583–594
Melikoglu M (2017) Geothermal energy in Turkey and around the World: a review of the literature and an analysis based on Turkey's Vision 2023 energy targets. Renewable and Sustainable Energy Reviews 76:485–492
Michel D, Pandya A, Hasnain SI, Sticklor R Panuganti S (2012) Water Challenges and Cooperative Response in the Middle East and North Africa. USA, The Brookings Project on U.S. Relations with the Islamic World
Muneer T, Asif M, Cizmecioglu Z, Ozturk HK (2008) Prospects for solar water heating within Turkish textile industry. Renewable and Sustainable Energy Reviews 12(3):807–823
Murrant D (2016) The water-energy nexus: quantifying the impact of water availability on future UK thermal power generation. Doctor of Philosophy, University of Birmingham
Mustafa SS, Mutlag AH (2013) Kirkuk municipal waste to electrical energy. International Journal of Electrical Power & Energy Systems 44(1):506–513
Mustafa S, Sulukan E, Uyar TS (2010) Wave energy and technical potential of Turkey. J Naval Sci Eng 6(2):34–50
Najafi G, Ghobadian B, Mamat R, Yusaf T, Azmi WH (2015) Solar energy in Iran: current state and outlook. Renewable and Sustainable Energy Reviews 49:931–942
Nasiri M, Ramazani Khorshid-Doust R, Bagheri Moghaddam N (2015) The status of the hydrogen and fuel cell innovation system in Iran. Renewable and Sustainable Energy Reviews 43:775–783
National Renewable Energy Laboratory (NREL) (2018) BioEnergy atlas. website Retrieved 6 March 2018, from https://maps.nrel.gov/bioenergyatlas/.
Nejat P, Morsoni AK, Jomehzadeh F, Behzad H, Saeed Vesali M, Majid MZA (2013) Iran's achievements in renewable energy during fourth development program in comparison with global trend. Renewable and Sustainable Energy Reviews 22:561–570
Nelson MJ, Nakhla G, Zhu J (2017) Fluidized-bed bioreactor applications for biological wastewater treatment: a review of research and developments. Engineering 3(3):330–342
Niajalili M, Mayeli P, Naghashzadegan M, Poshtiri AH (2017) Techno-economic feasibility of off-grid solar irrigation for a rice paddy in Guilan province in Iran: a case study. Solar Energy 150:546–557
Organisation for Economic Cooperation and Development (OECD) (2009) The economics of climate change mitigation. Paris, OECD
Organisation for Economic Cooperation and Development (OECD) (2015) Climate Change Mitigation. Paris, OECD
Ozcan M, Öztürk S, Oguz Y (2015) Potential evaluation of biomass-based energy sources for Turkey. Engineering Science and Technology, an International Journal 18(2):178–184
Ozden E, Tari I (2016) Energy–exergy and economic analyses of a hybrid solar–hydrogen renewable energy system in Ankara, Turkey. Applied Thermal Engineering 99:169–178
Ozdǒgan M (2011) Climate change impacts on snow water availability in the Euphrates-Tigris basin. Hydrol Earth Syst Sci 15:2789–2803

Perona JJ (2017) Biodiesel for the 21st century renewable energy economy. Energy Law J 38 (1):165–212
Peter Meisen JT (2011) The Water-Energy Nexus In The Jordan River Basin: The Potential For Building Peace Through Sustainability, Global Energy Network Institute (GENI).
Pirasteh G, Saidur R, Rahman SMA, Rahim NA (2014) A review on development of solar drying applications. Renewable and Sustainable Energy Reviews 31:133–148
Rambo KA, Warsinger DM, Shanbhogue SJ, Lienhard VJH, Ghoniem AF (2017) Water-energy nexus in Saudi Arabia. Energy Procedia 105:3837–3843
Ramos C, Ramirez R, Beltran J (2014) Potential assessment in Mexico for solar process heat applications in food and textile industries. Energy Procedia 49:1879–1884
Rezaei M, Mostafaeipour A, Qolipour M, Arabnia H-R (2018) Hydrogen production using wind energy from sea water: a case study on Southern and Northern coasts of Iran. Energy & Environment 29(3):0958305X1775005. https://doi.org/10.1177/0958305X17750052
Robinius M, Raje T, Nykamp S, Rott T, Müller M, Grube T, Katzenbach B, Küppers S, Stolten D (2018) Power-to-Gas: Electrolyzers as an alternative to network expansion – an example from a distribution system operator. Applied Energy 210:182–197
Saeed IM, Ramli AT, Saleh MA (2016) Assessment of sustainability in energy of Iraq, and achievable opportunities in the long run. Renewable and Sustainable Energy Reviews 58 (Supplement C):1207–1215
Saket A, Etemad-Shahidi A (2012) Wave energy potential along the northern coasts of the Gulf of Oman, Iran. Renewable Energy 40(1):90–97
Salman DA, Amer SA, Ward FA (2014) Water appropriation systems for adapting to water shortages in Iraq. JAWRA Journal of the American Water Resources Association 50 (5):1208–1225
Senol R (2012) An analysis of solar energy and irrigation systems in Turkey. Energy Policy 47:478–486
Server S (2018) Solar in the Middle East and North Africa: potential and development. Retrieved 24 March, 2018, from https://www.solarserver.com/solar-magazine/solar- report/solar-report/solar-in-the-middle-east-and-north-africa-potential-and- development.html
Sharma A, Arya SK (2017) Hydrogen from algal biomass: a review of production process. Biotechnol Rep (Amst) 15:63–69
Sharma AK, Sharma C, Mullick SC, Kandpal TC (2017) Solar industrial process heating: A review. Renewable and Sustainable Energy Reviews 78:124–137
Shiroudi A, Taklimi SRH, Mousavifar SA, Taghipour P (2013) Stand-alone PV- hydrogen energy system in Taleghan-Iran using HOMER software: optimization and techno- economic analysis. Environment, Development and Sustainability 15(5):1389–1402
Shouman ER, Sorour MH, Abulnour AG (2015) Economics of renewable energy for water desalination in developing countries. Eng Sci Technol Rev 5:1–5
Shuokr Qarani A, Hamidi Abdul A, Mohammed JKB, Mohd Suffian Y (2011) Appraisal of domestic solid waste generation, components, and the feasibility of recycling in Erbil, Iraq. Waste Management & Research 29(8):880–887
Siddiqi A, Anadon LD (2011) The water–energy nexus in Middle East and North Africa. Energy Policy 39(8):4529–4540
Soaad Najy el Azzawy (2017) Desertification, ArSco
Soheil Porkhial JN (2013) The potential use of geothermal energy in the persian gulf region. Research gate website.
Sontake VC, Kalamkar VR (2016) Solar photovoltaic water pumping system - a comprehensive review. Renewable and Sustainable Energy Reviews 59:1038–1067
Soulayman S, Sabbagh W (2015) Comment on 'Optimum tilt angle and orientation for solar collectors in Syria' by Skeiker, K. Energy Conversion and Management 89:1001–1002
Suresh NS, Rao BS (2017) Solar energy for process heating: a case study of select Indian industries. Journal of Cleaner Production 151:439–451

Tan C, Zhi Q (2016) The energy-water nexus: a literature review of the dependence of energy on water. Energy Procedia 88:277–284
Taşdemiroğlu E (1991) Wave-power potential along the coast of Turkey. Energy 16(4):779–782
Tavana HM (2009) Study of wave energy potential in Tonekabon (N Iran). 2009 1st International Conference on the Developments in Renewable Energy Technology (ICDRET)
TCN Co. (2009) Ankara landfill gas to energy project. Retrieved 20 March, 2018, from https://www.carbonneutral.com/interface/files/projects/pdfs/Ankara_Landfill_Gas_to_Energ y_Turkey_GS.pdf.
The International Renewable Energy Agency (IRENA) (2015a) Renewable energy in the water, energy & food nexus, T. I. R. E. A.
The International Renewable Energy Agency (IRENA) (2015b) Renewable energy in the water, energy & food nexus. UAE, Irena
The International Renewable Energy Agency (IRENA) (2018a) Global atlas – renewables. Retrieved 7 March 2018, from http://irena.org/globalatlas
The International Renewable Energy Agency (IRENA) (2018b) Global atlas. Retrieved 15 March, 2018, from http://www.irena.org/globalatlas
Thiel GP, Tow EW, Banchik LD, Chung HW, Lienhard JH (2015) Energy consumption in desalinating produced water from shale oil and gas extraction. Desalination 366:94–112
Toklu E (2013) Overview of potential and utilization of renewable energy sources in Turkey. Renewable Energy 50:456–463
Toklu E (2017) Biomass energy potential and utilization in Turkey. Renewable Energy 107:235–244
Topal M, Arslan Topal EI (2016) Investigation of tetracycline and degradation products in Euphrates river receiving outflows of trout farms. Aquaculture Research 47(12):3837–3844
UN-ESCWA (2015) The_water_energy_and_food_security_nexus_in_the_arab_region, Escwa Booklet
UN-ESCWA and BGR (United Nations Economic and Social Commission for Western Asia; Bundesanstalt für Geowissenschaften und Rohstoffe) (2013) Inventory of shared water resources in western Asia. Beirut., inventory of shared water resources in Western Asia: Chapter 1
UN-FAO (2018) Iraq. Retrieved 10 February, 2018, from http://www.fao.org/nr/water/aquastat/countries_regions/IRQ/.
Uyar TS, Beşikci D (2017) Integration of hydrogen energy systems into renewable energy systems for better design of 100% renewable energy communities. International Journal of Hydrogen Energy 42(4):2453–2456
Veil JA, Puder MG, Elcock D, Redweik RJ Jr (2004) A white paper describing produced water from production of crude oil, natural gas, and coal bed methane. United States, U.S. Department of Energy-Office of Scientific and Technical Information
Vo TTQ, Wall DM, Ring D, Rajendrana K, Murphy JD (2018) Techno- economic analysis of biogas upgrading via amine scrubber, carbon capture and ex-situ methanation. Applied Energy 212:1191–1202
von Bogdandy A, Wolfrum R (2012) Compilation of international treaties and other documents relative to the Euphrates and Tigris. Max Planck Institute for Comparative Public Law and International Law, Heidelberg
Walker SB, van Lanen D, Fowler M, Mukherjee U (2016) Economic analysis with respect to Power-to-Gas energy storage with consideration of various market mechanisms. International Journal of Hydrogen Energy 41(19):7754–7765
Wang S, Cao T, Chen B (2017) Water–energy Nexus in China's Electric Power System. Energy Procedia 105:3972–3977
Wasinger CE (2015) Peace be dammed? Water power and water politics in the Tigris-Euphrates Basin. Bowdoin Digital Commons, p 39.

Waste Management World (WMW) (2018) 16 GE engines for 3 Turkish landfill gas to energy projects. Retrieved 20 March, 2018, from https://waste-management-world.com/a/16-ge-engines-for-3-turkish-landfill-gas-to-energy-projects.
Wichelns D (2017) The water-energy-food nexus: Is the increasing attention warranted, from either a research or policy perspective? Environmental Science & Policy 69:113–123
World Bank Group (WBG) (2017a) 2017 the world bank group, the world bank. Retrieved 8 March 2018, 2018, from https://www.globalwindatlas.info/area/.
World Bank Group (WBG) (2017b) Energy data – global solar atlas." Retrieved 2 March 22018, 2018, from https://globalsolaratlas.info/area/Iraq.
World Energy Council WEC (2016) World energy resources. Retrieved 15 March, 2018, from https://www.worldenergy.org/data/resources/country/syria/biomass/.
World Health Organization (WHO) (2018) Environmental health – Iraq. Environmental health. Retrieved 20 March, 2018, from http://www.emro.who.int/irq/programmes/environmental-health.html.
Wu M, Mintz M, Wang M, Arora S (2009) Water consumption in the production of ethanol and petroleum gasoline. Environmental Management 44(5):981–997
Yaniktepe B, Savrun MM, Koroglu T (2013) Current status of wind energy and wind energy policy in Turkey. Energy Conversion and Management 72:103–110
Yavuz E (2015) Uncharted waters of the Middle East: re-visiting the Euphrates-Tigris conflict in troubled times. 1601731 M.S., Tufts University.
Yazgan M (2013) Development of sanitary landfills and waste-to-energy from landfill gas in Turkey. Turkey, NL EVD International
Yıldız D (2016) Natural Diminishing Trend of the Tigris and Euphrates Streamflows is Alarming for the Middle East Future. World Scientific News 47:279–297
Yilmaz M (2003) The war that never happened: the sharing of Euphrates –Tigris rivers' water between Turkey, Syria and Iraq. Master's Thesis, Naval Postgraduate School Monterey, CA
Yu X, Manthiram A (2017) Electrochemical energy storage with mediator-ion solid electrolytes. Joule 1(3):453–462
Yu N, Xing D, Li W, Yang Y, Li Z, Li Y, Ren N (2017) Electricity and methane production from soybean edible oil refinery wastewater using microbial electrochemical systems. International Journal of Hydrogen Energy 42(1):96–102
Yuksel I (2015) Water management for sustainable and clean energy in Turkey. Energy Reports 1:129–133
Yun Y-M, Sung S, Kang S, Kim M-S, Kim D-H (2017) Enrichment of hydrogenotrophic methanogens by means of gas recycle and its application in biogas upgrading. Energy 135:294–302
Zagonari F, Rossi C (2014) A negotiation support system for disputes between Iraq and Turkey over the Tigris-Euphrates basin. Journal of Hydrology 514:65–84
Zhang X, Vesselinov VV (2017) Integrated modeling approach for optimal management of water, energy and food security nexus. Advances in Water Resources 101:1–10
Ziogou I, Zachariadis T (2016) Quantifying the water–energy nexus in Greece. Int J Sustainable Energy 36(10):972–982
رشيد, ع. ا. ج. (2017). الموارد المائية في العراق. السليمانيه—العراق.

Chapter 15
Impacts of Dams on Aquatic Biodiversity, Fisheries, Fishes and Their Environment: Problems that Could Be Present in Iraq with Recommendations

Laith A. Jawad

Abstract Large dams can exceptionally disturb the environment and act as a non-alleviating issue in nature at several stages of construction. The environmental influences of building dams on some of the biodiversities, fisheries, fishes and their habitats, have been discussed in the present chapter. At the end of constructing a reservoir, fish species richness usually increases due to incorporation of close habitats, but richness decreases as reservoirs turn out to be old. Dams inspire release control, changing the seasonal cycles of floods. These influences are increased when dams are built in cascades.

Amongst the resolutions to overwhelm the effects of dams is dam elimination, which endures to assemble attention as a possible river renewal mean. By recurring riverine settings and sediment transport to previously held areas, riffle/pool sequences, gravel and cobble have recurred, along with increases in biotic diversity. Fish passageway has been an extra benefit of dam deletion. Dam removal is a significant other option for river restoration.

In this chapter, recommendations were put forward for the Iraqi governments to follow in order to assess the impacts of the dams on the freshwater habitats.

15.1 Introduction

The most important commodity for human being in nature is water. It is considered as one of the fundamental bases of development, economic prosperity and social well-being, and its shortage will cause many limitations. Water scarcities directly and indirectly affect the areas such as the control, storage and water supply, distribution types, transmission, factor productivity, planning and conservation

L. A. Jawad (✉)
School of Environmental and Animal Sciences, Unitec Institute of Technology, Auckland, New Zealand

L. A. Jawad (ed.), *Tigris and Euphrates Rivers: Their Environment from Headwaters to Mouth*, Aquatic Ecology Series 11,
https://doi.org/10.1007/978-3-030-57570-0_15

management. It is regarded to be built large and small dams and small irrigation networks of rivers in different countries to take advantage of the water potential to meet the needs and objectives such as drinking, agriculture, industry, generating non-polluting hydropower energy, flood control and improved water quality and quantity (Heidari et al. 2013).

Dams intersect streamflow, and produce hydrological changes along the integrated continuum of river ecosystems (Vannote et al. 1980) that eventually can be revealed in their associated fisheries. Both environments can be conducive to the establishment and maintenance of fish stocks appropriate for exploitation by fisheries (Jackson and Marmulla 2001).

Besides all the benefits of dams such as supply of drinking water, generation of hydroelectric power, increase of the water supply for irrigation, and recreational opportunities (McDonald et al. 2009), there are several environmental and sociological impacts that must be identified before construction of dams (Rashad and Ismail 2000). It is obvious that construction of dams has been escorted by negative environmental and social impacts such as influences on health through changes in water and food security (Brown et al. 2009). Amongst the important social negative influences of dam building is resettlement of local communities followed by lots of social disorder (Jackson and Sleigh 2000). Besides, dams attenuate floods, which may influence the ecology and agriculture dramatically, wherein the land is cultivated by soil moisture after floods recede (Tajziehchi and Karbassi 2011).

The aim of this chapter is to shed light on the adverse environmental issues that the dams might cause and suggest those unfavourable conditions to the decision-makers in Iraq in order to take the best action to meet and solve such incidences.

15.2 Ancient Mesopotamians and Irrigation Problems

A long time ago in the history and since humans have changed their style of life from nomadic and to settle down and live in groups, they started to look for a reliable and safe sources of water for their personal use and agriculture. The presence of water needs to be managed so humans need to protect their settlements and farmlands from flood (Garbrecht and Garbrecht 2004). Building simple reservoirs and diverting creeks and rivers were the first planned water management.

The management of irrigation has often been considered as vital to the understanding of early complex societies in Mesopotamia (Sanders and Marino 1970). The mass production was fulfilled through agricultural growth, which mainly depends on both small and large scale all irrigation schemes (Doolittle 1990). The history of irrigation in Iraq goes back to a period of nearly six thousand years. The region has been depicted by a semi-arid climate since the fourth millennium BC and annual precipitation rates amount to less than one hundred millimetres (Rost et al. 2011), and therefore, irrigation is essential for agriculture in that area. Irrigation was subsistent throughout this period for human communities and later became one of the economic foundations of states and empires (Pollock 1999).

In Iraq, the traditional irrigation scheme contains plans of construction of dams of different sizes to control the amount of water and the water flow within the canals (Rost et al. 2011). Similarly, records from ancient Mesopotamia showed many indications to the construction and maintenance of dams that their usage is not clear (Civil 1994). Therefore, archaeologists combine the traditional dam construction with that of the ancient designs in order to understand the aim behind the construction of dams by the ancient Mesopotamians.

The issue of irrigation and the maintenance of the irrigation structures were by low being enforced in the ancient Mesopotamia. In the codes of the Babylonian King Hammurabi (1711–1669 B.C.), there are 300 units, which apportioned with several portions of irrigation. One of the rules said: "if a man was careless in consolidation of the banks of his field and has not sustained his banks and then a break has happened in his bank and so he has let the waters remove (the soil on) the waterland, the man in whose bank the breach has happened shall substitute the corn which he has (caused to be) lost" (El-Yussif 1983).

One of the oldest irrigation system was implemented by the Assyrian King Sennachrib who ruled from 705 to 685 B.C. He implemented his irrigation system in his new capital Nineveh. The King Sennachrib irrigation system implies linking the River Gomel, a tributary of the Greater Zab River, with the Tigris by a canal 80 km long and by a watercourse. Water then kept in the stream called Khosar by building a small dam. The dam is undamaged to the present time. Afterwards, and throughout the Sasanian era (226–637 A.D.), two large waterway systems were built to upsurge the level of water in the Lower Diyala Plain as water was found insufficient for agricultures. One of these waterways is still running and active to the present time and known as Nahrawan canal. Once the Arab sovereignty started in Iraq, excessive care was given to irrigation systems such as those on the Nahrawan and its branches. In 842 A.D., the old Ishaki irrigation project was built.

Irrigation in Iraq entered a dark period since the invasion of Mongols and the destruction of Baghdad in 1258 A.D. (Ministry of Irrigation 1979). This depraved scenario has continued later on through the rules of the Mongols, Persians and then the Turks. The dark period of the irrigation history of Iraq resulted in a neglected field and agriculture, which are never fully recovered until the present time.

15.3 Impacts of Dams

The following set of impacts of dams is universal in nature and could happen in any country with slight variations depending on the habitat, geology and geomorphology of the landscape. No previous assessments of impacts of dams on the freshwater habitats were prepared. Therefore, the present selected general impacts are for the government of Iraq to adopt and assess the freshwater environment and follow the appropriate recommendations given at the end of this chapter to improve the status of the Iraqi freshwater habitats.

15.3.1 Ecological Impacts

Amongst the adverse measures that human have done to the Mother Nature are constructions of dams intruding freshwater water in rivers. As it has been mentioned earlier in this chapter, the benefits of the dams in general are: low-cost and effective power production, operative flood regulation, water resource, inundation management and entertaining aspects (Bednarek 2001).

Turning to the bad side of the dams, we can perceive that the existence of dams is challenging for numerous aquatic systems. Dams influence these habitats in several means: changing the normal cycle of flow, converting the biological and physical features of river channels and deltas and disintegrating the endurance of rivers (Stanford et al. 1996).

15.3.1.1 Impact on Water Temperature

Water temperature will be changed in the area behind the dam due to the slow flowing of the water. This changes will be within the water in the reservoir and downstream (Petts 1984; Yeager 1994). In addition, stratification in temperature will happen in the reservoir because the ecosystem of the reservoir is changed to something like that of the lake, which was categorised by greater surface areas and sluggish moving water. The upper layer of the reservoir (the epilimnion) warms and decreases in density, whilst cooler water endures on the bottom layer of the confinement (the hypolimnion). Gillilan and Brown (1997) have described what happened if water from the lower part of the dam was released and here I quote, "hypolimnetic releases are also often very low in dissolved oxygen (DO) due to a lack of mixing with well-oxygenated upper layers, no photosynthesis, and high biological oxygen demand. These low dissolved oxygen levels are insufficient to maintain the density of some organisms in the tail waters. Whether warm or cool water is released, changes in temperature downstream eliminate or shift the composition of species adapted to the natural water temperatures. If the dam releases warm, epilimnetic water, warm-water species often thrive downstream. For migrating cold-water fish, warm temperatures act as a thermal barrier to movement". The fish may otherwise select cooler tributaries, changing migration routes, and reducing the possibilities of reaching suitable spawning grounds".

The field observation (Kaeding and Zimmerman 1983), the laboratory experimentation (Berry 1988) and simulation experiments (Kaeding and Osmundson 1988) have shown that cold water released from the lower part of the dam has negative effects on the life-history of native fishes of the Colorado River basin. Clarkson and Childs (2000) have found that the opposing effects of low temperature on early life stage growth and development are comparable across the species they examined. They also concluded that early larval stages of some species enter cold coma when entering cold waters, with some mortalities. Berry (1988) suggested that slight changes in temperature could lead to other physiological or behavioral changes.

The results of entering cold coma in early stages of the life of fishes are rigorous. There are biological and physical effects that young fish individuals could face, amongst these are high rates of predation due to the inactive state (Countant et al. 1974), physical abrasion against the hard substances in the environment they live in, and complete coverage with bottom substances could happen, Clarkson and Childs (2000).

The low water temperature of the out water from the lower part of the dam is similar to that of winter months of the year. This means that the early life stage of fishes will face a rigorous development despite harmfully affecting many life-history constraints. Clarkson and Childs (2000) gave a good example on the effect of the cold water on fish larvae as they displayed postponements in change from larval stage to juvenile stage at lower temperatures, in a few circumstances spreading the larval stage through a whole season or further. In entering such lengthy alteration period and disclosure to cold water could couple with mortality due to food shortage (Papoulias and Minckley 1990), hydrological disruption (Robinson et al. 1998), predation (Ruppert et al. 1993) and parasitism (Clarkson et al. 1997).

15.3.1.2 Impact on Sediment Transport

Dam showed to obstruct sediment transport (Petts 1984; Kondolf 1997) as it dislocates the advance of sediment in rivers and alters the river's organisational environment (Kondolf 1997). This happened when the dam slows down velocity of the water in the river and the latter became unable to move rock, stone, and other large elements, which in turn causes in elevating the bottom of the upstream of the dam (Petts 1984; Fan and Springer 1990). Thinner units, such as sand and silt, sink nearer to the dam itself and can eventually fill the reservoir, restraining hydropower generation or water storage (Petts 1984). The crevices found between cobble and boulder will be filled with fine particles turning these habitats unfeasible to live in for many organisms. In doing so, a dramatic decrease in the biodiversity will result.

As to fish the changes in the sediments found in the substrate of the environment they live in can have a direct effect on the spawning of the fish species. Such results were obtained for salmonid species (Rabeini and Jacobson 1993). If the sediment particles increased in size, then fish might not be able to move them by the tail, hence this may delay the spawning process (Kondolf and Wolman 1993).

On the other hand, the fine sediments filling the basal part of the dam will retain nutrients and release clear water from the lower part of the dam. In this way, organisms near the lower part of the dam will have less or no nutrients in their food (Kondolf 1997), which in turn affect the growth of these organisms. A channel will be created from such an action, and enlarging its size leads to river's bank failures and riverside damage (Sear 1995). The newly altered environment might be hostile for certain creatures, creating variations in the biotic society configuration (Staggs and others 1995). The effect of a decrease in sediment transport also leads to damage of shoreline and environment. Otherwise, sediment involvements from tributaries below a dam remained still by the slower water velocities in the water originated from the lower part of the dam, which in turn could instigate deviances in environment.

15.3.1.3 Impact on Connectivity

Continuation in water masses and flow in river, which is known as "Connectivity" is a vital factor of approximately all features of an operative riverine system, comprising the upkeep of flow, water quality, temperature and sediment transportation (Ward and Stanford 1995). Such continuation is also imperative for allowing creatures to move through the river system. The presence of dams will break the strip of the river in severe way. Dams can separate inhabitants and habitations, generate physical and thermal obstacles for wandering and drifting stream organisms and upset communications between freshwater, terrestrial and coastal systems (Dynesius and Nilsson 1994; Stanford et al. 1996). For example, congested migration of diadromous fish has been an issue for numerous dammed rivers. With fragmented habitats, fish spent all the stored energy, whilst they are assembling above or below the dam.

15.3.1.4 Impact on Biogeochemical Cycles

With the presence of dams, the strip of the river will change into a reservoir. The later and as it has been mentioned above, the slow flow of water of the river induces fragment settling, turbidity declines and light transmissivity upsurges, improving primary production. Thus, from the upper part of the reservoir to the dam the river modifies from an allochthonous controlled system to a more lacustrine system, where autochthonous production of organic matter dominates.

Depletion of oxygen is a possible outcome of the changeover from a river to a lake. Cooke et al. (1993) and Cooke and Carlson (1989) have suggested that in the reservoir itself, the depletion of oxygen generates a decrease of nitrate, manganese hydroxides, iron hydroxides and sulphate. In contrast to low-dissolved oxygen points, supersaturated concentrations of oxygen and nitrogen can also happen due to reservoirs. Supersaturation below dams can originate from air and high oxygen concentrations that occur in the surface water of reservoirs with considerable primary production.

15.3.2 Impact on the Organisms

15.3.2.1 Impact on Plankton

Phytoplanktons are subtle aquatic organisms, and their configuration and richness can disclose the eutrophic state in a short time (Díaz-Pardo et al. 1998). Therefore, phytoplanktons are frequently employed to evaluate the influence of large-scale dams on the aquatic ecosystem (Lancelot et al. 2002). Damming not only cause alterations to the hydraulic settings of rivers but also imposes the differences of the phytoplankton structure and biomass at the same time. In the environment of the dams, frequently phytoplankton boosted owing to the lengthy hydraulic holding time, and the water quality in the reservoir was usually eutrophic (Kawara et al. 1998). Given the case of sorts of dams, the enlarged phytoplankton number can be ascribed to the joint effects of reduced flow velocity, increased water holding time and clarity, and decreased vertical mixing intensity (Sullivan et al. 2001). A significant reduction in the water flow usually happened after damming and with the increased retention time, phytoplankton can have an ample time for growth (Sullivan et al. 2001).

Zooplankton and ichthyoplankton may also face an impact from dams. The number of individuals of the zooplankton and fish larvae in the reservoir is significantly less than in the river. This circumstance perhaps is a consequence of the upsurge in water transparency due to retaining of held solids (Agostinho et al. 2004), making eggs and larvae more vulnerable to predation. Seemingly, this is one of the chief issues that clarify the decrease of migratory fish species below dams (Petrere 1996). Literally, Agostinho et al. (2002) stated that in an increased transparency of water, larvae of migratory and large-sized species are obviously preyed on, even by small fish species belonging to varied feeding habits (e.g. foragers, insectivores and planktophagues).

15.3.2.2 Impact on Macroinvertebrates

Habitat changes caused by dams can affect aquatic macroinvertebrates (Harding 1994). Such effects are significant because of the task that macroinvertebrates participate in the stream ecosystem function (Cummins and Klug 1979). In contrast to fish, aquatic insects have winged adult stages that allow for transfer between stream sections. Consequently, the exact effects of dams on aquatic macroinvertebrate groups can be mainly due to modification of environment, whilst influences on fish similarly contain migration impediment. The occurrence of a dam can modify many physical and chemical issues such as stream substrate, dissolved oxygen and water temperature (Waters 1995).

Bioenergetics investigations show a robust definite link between feeding rates and metabolism with temperature for both fish and insect communities (Wotton 1995). Augmented metabolic rate carries with it a need for increased levels of food quantity or quality in order to uphold development and survival proportions (Wotton 1994). Damming changes the normal flow of energy and resources, thus moving the physical and chemical features of downstream reaches and the biological society (Stanford and Ward 1979).

Temperature upsurges below dams have been given less consideration, and influences on the macroinvertebrate community have been mainly focused (Harding 1994; Wotton 1995). It has been proposed that effects on temperature are restricted to the area of stream straight below the dam and that temperatures rapidly levelled with the air (Brooker 1981). Dams modify temperature through different methods (e.g. reduced diel fluctuation) and in other seasons, nonetheless it is through the summer that temperatures reach a maximum. Issues distressing temperatures below dams comprise the size of the dam (specifically depth and surface area), endurance time and the discharge depth (Brooker 1981; Wotton 1995).

A durable connotation arises between environmental features of macroinvertebrate groupings, which suggests that disintegration and alteration (e.g. inundation, scouring and channelisation) of the riverine environment can have noteworthy consequences on biotic characteristics, containing failures in abundance and diversity of macroinvertebrates (Dynesius and Nilsson 1994).

Macroinvertebrate richness was at its lowermost at downstream action sites as a result of substrate roughening and diminished habitat diversity (Kondolf 1997; Camargo and Voelz 1998). Macroinvertebrate abundance depends on the presence of a mix of heterogeneous constituents of the soil that is found in the water flow (Waters 1995).

15.3.2.3 Impact on Fishes

Morphological deviations due to changes in habitat are prevalent in fishes, comprising intraspecific changes allocated to lentic and lotic conditions (Langerhans et al. 2003). Dams can affect the shape of the fish through changing its morphological traits. Several morphological test were introduced to detect such changes. Amongst those is the geometric morphometrics (GM) approach, which is considered a powerful tool that can complement other methods for stock identification (Cadrin 2013).

Haas et al. (2010) have studied the morphological differences that arise between reservoir and river populations of *Cyprinella venusta*, such as raising in body depth in reservoir inhabitants, parallel alterations perceived in populations of other species inhabiting naturally conflicting flow habitats (Langerhans 2008). Their results revealed that the average shape of reservoir populations varied in predictable way with reservoir size, suggesting that physical features of reservoirs may control evolutionary and ecological sets commending variations in the morphological features of occupant fish populations. They suggested that the discrepancy in body depth amongst river populations might account for the changes in the physical characteristics of reservoirs. They also suggested that the traits that have shown divergence are those traits that can have a direct impact on fitness (Langerhans and Reznick 2010) and intraspecific variation can be related to divergent selection caused by changes in habitats (Robinson and Wilson 1996). Adaptive responses to divergent selection could resemble the genetic difference and phenotypic flexibility only or combined together (West-Eberhard 1989). Flow regime differences between reservoirs and streams can result in divergent selection in fishes (Langerhans 2008).

Heidari et al. (2013) used this technique to show the effects of the construction of the Manjil and Tarik dams on the Sefidrud River, Iran on the body shape of *Capoeta gracilis* populations. They determined that the long-term separation of populations and interbreeding may cause variation in the morphology between populations, and offers a foundation for population difference (Dakin et al. 2007). In such cases where fish populations have become fragmented, Langerhans et al. (2003) suggested that characteristics of newly created habitats by constructing the dam could control evolutionary and ecological conditions driving alternations in the body shape of its resident fishes. Certain body proportions may show intraspecific variations. Heidari et al. (2013) have shown that the caudal peduncle and the body depth are the only two body measurements that showed differences between the populations in the Sefidrud River. Consequently, the changes in body shape may be associated with morphological adaptation to new habitats. For example, deeper body and caudal peduncle can be as a result of an adaptation for the rapid acceleration and movement, whereas, the more fusiform body (a more streamlined body) of the population could be an adaption to oppose the drag force (Nacua et al. 2010). Also, differences in the head region may indirectly be related to changes in food seeking habits reflecting the difference between feeding habits and availability of food resources (Langerhans et al. 2003). In addition to the fragmentation of the population of *C. gracilis*, Heidari et al. (2013) have suggested that the influence of the dam extended to the changes in the body shape of fishes and can be considered as main evolutionary drivers acting on aquatic biodiversity.

It is expected that species with simpler biological necessities or bigger reproductive flexibility thrive in settling in the new habitats, mostly in lacustrine and changeover zones. It is normal for fishes in new reservoirs or when fluctuations in the water level are recurrent to construct nests, lay large adhesive eggs, or grow some degree of parental care, may have impairments in recruiting young, essentially owing to the worsening of water quality, high predation impact and non-seasonal changes in the water level. Thus, aquatic macrophytes could reassure positive results on fish recruitment by proposing shelter, spawning grounds and feeding locations for small species (Casatti et al. 2003).

15.3.3 Impact on Fish Migration

Riverine fish migrate for many motives, usually linked with finishing the life cycle, looking for food, or to evade opposing environments (McDowall 1988). The function of dams in dropping the allocations of migratory fish has been well recognised (Petts 1984).

Fish passage services have been built mainly on the key stems of large rivers; nevertheless, fish usually use tributaries somewhat than main stems of large rivers to spawn. The influence of possible problems to fish migration into the spawning tributaries has infrequently been investigated (Croze and Larinier 2000). Additionally, latest telemetry studies have proved that fish of diverse species, will migrate over long distances within a river basin to reach their spawning grounds or for ontogenetic and trophic reasons (Parkinson et al. 1999). These views reinforce the implication of re-establishing the free circulation of fish through a river basin.

Fragmented inhabitants' upstream and downstream of dams frequently intricate self-governing population constructions (Almodóvar and Nicola 1999). Changes in population assembly between rivers and small lakes can reveal alterations in such features as energy supplies to reserve position, habitat type and food availability, as morphological reactions to ecological modification. The escaping of small numbers of fish from small sized water body into downstream riverine populations (Martinez et al. 1994) could be adequate to avert genetic departure.

15.3.4 Impact on Aquatic Biodiversity

The number of the freshwater species is under list (Stiassny 2002), lose species faster than terrestrial or marine biota (Sala et al. 2000), and face increasing human-originating influences (Arthington et al. 2010). Dams create a main danger to global freshwater species diversity (Vörösmarty et al. 2010); they lead to loss of inborn species but also to intrusion by exotic species, partially since alien species are probable to begin in altered or ruined freshwaters (Poff and Zimmerman 2010). The suggestions of dam impediment for local freshwater diversity have been inspected in small-scale investigations (Anderson et al. 2006), but a shortage of regional- or global-scale information has forced large-scale freshwater administration planning (Thieme et al. 2007).

15.4 Recommendations

The following recommendations have been created and put for different parts of the world, where dams have shown to generate impacts on the river systems. The dams in Iraq as in other parts of the world have shown impacts on the freshwater habitats, but such impacts that were observed in Iraq have never been observed. Therefore, the following recommendations are put forward to the government of Iraq, those are suitable to the freshwater habitats in Iraq can be followed.

1. Avoid areas of great biodiversity.
2. Avoid effects on areas of high productivity.
3. Encourage studies on biodiversity.

4. Do not obstruct migrations.
5. Preserve normality of flow.
6. Uphold release volume.
7. Endure water quality.
8. Look after exclusive species environments.
9. Halt increasing impressions of dams.
10. Include environmental staff early.
11. Control effects at old and new dams.
12. Remove leaks in dam and irrigation systems.

References

Agostinho A, Gomes LC, Fernandez DR, Suzuki HI (2002) Efficiency of fish ladders for Neotropical ichthyofauna. River Res Appl 18(3):299–306

Agostinho A, Bini LM, Gomes LC, Júlio HF Jr, Pavanelli CS, Agostinho CS (2004) Fish assemblages. In: Thomaz SM, Agostinho AA, Hahn NS (eds) The Upper Paraná River and its floodplain: physical aspects, ecology and conservation. Backhuys Publishers, Leiden, pp 223–246

Almodóvar A, Nicola G (1999) Effects of a small Hydropower Station upon Brown trout *Salmo trutta* L in the river Hoz Seca (Tagus Basin, Spain) one year after regulation. Regulated Rivers Res Manga 15(5):477–484

Anderson EP, Freeman MC, Pringle CM (2006) Ecological consequences of hydropower development in Central America: impacts of small dams and water diversion on Neotropical stream fish assemblages. River Res Appl 22:397–411

Arthington AH, Naiman RJ, McClain ME, Nilsson C (2010) Preserving the biodiversity and ecological services of rivers: new challenges and research opportunities. Freshw Biol 55:1–16

Bednarek AT (2001) Undamming rivers: a review of the ecological impacts of dam removal. Environ Manga 27:803–814

Berry JR (1988) Effects of cold shock on Colorado squawfish larvae. Southwest Nat 33:193–197

Brooker MP (1981) The impact of impoundments on the downstream fisheries and general ecology of rivers. Adv Appl Biol 6:91–152

Brown PH, Tullos D, Tilt B, Magee D, Wolf AT (2009) Modeling the costs and benefits of dam construction from a multidisciplinary perspective. J Environ Manag 90:S303–S311

Cadrin SX (2013) Morphometric landmarks. Stock identification methods, 2nd edn. Elsevier, London, pp 109–128

Camargo JA, Voelz NJ (1998) Biotic and abiotic changes along the recovery gradient of two impounded rivers with different impoundment use. Environ Monti Ass 50(2):143–158

Casatti L, Mendes HF, Ferreira KM (2003) Aquatic macrophytes as feeding site for small fishes in the Rosana Reservoir, Paranapanema River, Southeastern Brazil. Braz J Biol 63(2):213–222

Civil M (1994) The farmer's instructions. A sumerian agricultural manual. Aula Orientalis Supplementa 5. Barcelona

Clarkson RW, Childs MR (2000) Temperature effects of hypolimnial release dams on early life stages of Colorado River basin. Copeia 2:402–412

Clarkson RW, Robinson AT, Hoffnagle TL (1997) Asian tapeworm (*Bothriocephalus acheilognathi*) in native fishes from the little Colorado River, grand canyon. Arizona The Great Basin Nat 57:66–69

Cooke GD, Carlson RE (1989) Reservoir management for water quality and THM precursor control. AWWA Research Foundation, Denver

Cooke GD, Welch EB, Peterson SA, Newroth PR (1993) Restoration and management of lakes and reservoirs. Lewis Publishers, Boca Raton

Countant C, Ducharme H, FISHER J (1974) Effects of cold shock on vulnerability of juvenile channel catfish (*Ictalurus punctatus*) and largemouth bass (*Micropterus salmoides*) to predation. J Fish Res Bd Can 31:351–354

Croze O, Larinier M (2000) Libre circulation des poissons migrateurs et seuils en rivière dans le bassin RMC. Guide technique. Agence de l'Eau RMC. Conseil Supérieur de la Pêche, p 53

Cummins KW, Klug MJ (1979) Feeding ecology of stream invertebrates. Ann Rev Ecol Syst. 10:147–172

Dakin E, Porter BA, Freeman BJ, Long JM (2007) Genetic integrity of an isolated population of Shoal Bass (*Micropterus cataractae*) in the upper Chattahoochee River basin. Natural Resource Technical Report NPS/ NRWRD/NRTR—2007/366. National Park Service. Water Resources Division, Fort Collins, CO

Díaz-Pardo E, Vazquez G, López-López E (1998) The phytoplankton community as a bioindicator of health conditions of Atezca Lake. Mexico Aqu Ecosy Health Mana 1(3–4):257–266

Doolittle W (1990) Canal irrigation in prehistoric Mexico: the sequence of technological change. University of Texas Press, Austin

Dynesius M, Nilsson C (1994) Fragmentation and flow regulation of river systems in the northern third of the world. Science 266:753–762

El-Yussif F (1983) Condensed history of water resources developments in Mesopotamia. Water Intern 8(1):19–22

Fan S, Springer FE (1990) Major sedimentation issues and ongoing investigations at the Federal Energy Regulatory Commission. Pages 1015–1021 in Hydraulic engineering, Proceedings of the 1990 national conference of the American Society of Civil Engineers. American Society of Civil Engineers, New York

Garbrecht JD, Garbrecht GK (2004) Siltation behind dams in antiquity. In: Water resources and environmental history. American Society of Civil Engineers, Reston, pp 35–43

Gillilan DM, Brown TC (1997) Instream flow protection: seeking a balance in western water use, vol 417. Island Press, Washington, DC, p 2009

Haas TC, Blum MJ, Heins DC (2010) Morphological responses of a stream fish to water impoundment. Biol Lett 6(6):803–806

Harding JS (1994) Variations in benthic fauna between differing Lake outlet types in New Zealand. New Zealand journal of mar. Freshwater Res 21:417–427

Heidari A, Mousavi-Sabet H, Khoshkholgh M, Esmaeili HR, Eagderi S (2013) The impact of Manjil and Tarik dams (Sefidroud River, southern Caspian Sea basin) on morphological traits of Siah Mahi Capoeta gracilis (Pisces: Cyprinidae). Intern J Aqua Biol 1(4):195–201

Jackson DC, Marmulla G (2001) The influence of dams on river fisheries. FAO Fish Tech Pap 419:1–44

Jackson S, Sleigh A (2000) Resettlement for China's three gorges dam: socio-economic impact and institutional tensions. Communis Post-Commun Stud 33(2):223–241

Kaeding LR, Zimmerman MA (1983) Life history and ecology of the humpback chub in the little Colorado and Colorado Rivers of the grand canyon. Trans Am Fish Soc 112:577–594

Kaeding LR, Osmundson DB (1988) Interaction of slow growth and increased early-life mortality: an hypothesis on the decline of Colorado squawfish in the upstream regions of its historic range. Environ Biol Fish 22:287–298

Kawara O, Yura E, Fujii S, Matsumoto T (1998) A study on the role of hydraulic retention time in eutrophication of the Asahi River dam reservoir. Water Sci Technol 37(2):245–252

Kondolf GM (1997) PROFILE: hungry water: effects of dams and gravel mining on river channels. Environ Manag 21(4):533–551

Kondolf GM, Wolman MG (1993) The sizes of salmonid spawning gravels. Water Resour Res 29 (7):2275–2285

Lancelot C, Staneva J, Van Eeckhout D, Beckers J-M, Stanev E (2002) Modelling the Danube-influenced north-western continental shelf of the Black Sea: II. Ecosystem response to changes in nutrient delivery by the Danube River after its damming in 1972. Estuar Coast Shelf Sci 54:473–499

Langerhans RB (2008) Predictability of phenotypic differentiation across flow regimes in fishes. Integ Comp Biol 48(6):750–768

Langerhans RB, Reznick DN (2010) Ecology and evolution of swimming performance in fishes: predicting evolution with biomechanics. Fish locomotion: an eco-ethological perspective 200:248

Langerhans RB, Layman CA, Langerhans AK, Dewitt TJ (2003) Habitat-associated morphological divergence in two Neotropical fish species. Biol J Linn Soc 80(4):689–698

Martinez PJ, Chart TE, Trammell MA, Wullschleger JG, Bergersen EP (1994) Fish species composition before and after construction of a main stem reservoir on the White River, Colorado. Environ Biol Fish 40(3):227–239

McDonald K, Bosshard P, Brewer N (2009) Exporting dams: China's hydropower industry goes global. J Environ Mana 90:S294–S302

McDowall RM (1988) Diadromy in fishes. Timber Press, Portland

Nacua SS, Dorado EL, MAJ T, Demayo G (2010) Body shape variation between two populations of the white goby, *Glossogobius giuris*. Res J Fish Hydrobiol 5:44–51

Papoulias D, Minckley WI (1990) Food limited survival of larval razorback sucker, *Xyrauchen texanus*, in the laboratory. Environ Biol Fish 29:73–78

Parkinson D, Philippart JC, Baras E (1999) A preliminary investigation of spawning migrations of grayling in a small stream as determined by radio-tracking. J Fish Biol 55(1):172–182

Petrere M (1996) Fisheries in large tropical reservoirs in South America. Lakes & Reservoirs: Res Manage 2(1–2):111–133

Petts GE (1984) Impounded rivers: perspectives for ecological management. John Wiley, Chichester

Poff NL, Zimmerman JK (2010) Ecological responses to altered flow regimes: a literature review to inform the science and management of environmental lows. Freshw Biol 55(1):194–205

Pollock S (1999) Ancient Mesopotamia. The Eden that Never Was. Case studies in early societies. Cambridge University Press, Cambridge

Rabeini CF, Jacobson RB (1993) The importance of fluvial hydraulics to fish-habitat restoration in low-gradient alluvial streams. Freshw Biol 29(2):211–220

Rashad SM, Ismail MA (2000) Environmental-impact assessment of hydro-power in Egypt. Appl Energy 65(1):285–302

Robinson AT, Clarkson RW, Forrest RE (1998) Dispersal of larval fishes in a regulated river tributary. Trans Am Fish Soc 127:772–786

Robinson BW, Wilson DS (1996) Genetic variation and phenotypic plasticity in a trophically polymorphic population of pumpkinseed sunfish (Lepomis gibbosus). Evol Ecol 10:631–652

Rost S, Hamdani A, George S (2011) Traditional dam construction in modern Iraq: a possible analogy for ancient Mesopotamian irrigation practices. Iraq 73:201–220

Ruppert JB, Muth RT, Nesler TP (1993) Predation on fish larvae by adult red shiner, Yampa and Green Rivers, Colorado. Southwest Nat 38:397–399

Sala OE, Chapin FS, Armesto J, Berlow E, Bloomfiel J, Dirzo R, Huber-Sanwald E, Huenneke LF, Jackson RB, Kinzig A, Leemans R (2000) Global biodiversity scenarios for the year 2100. Science 287(5459):1770–1774

Sanders WT, Marino J (1970) New World prehistory: archaeology of the American Indian. Prentice-Hall, Englewood Cliffs

Stanford JA, Ward JV (1979) Dammed rivers of the world: symposium rationale. In: Ward JV, Stanford JA (eds) The ecology of regulated streams. Plenum, New York, pp 1–6

Stanford JA, Ward JV, Liss WJ, Frissell CA, Williams RN, Lichatowich JA, Coutant CC (1996) A general protocol for restoration of regulated rivers. Regul Rivers 12:391–413

Stiassny MLJ (2002) Conservation of freshwater fish biodiversity: the knowledge impediment. Verhandlungen der Gesellschaft für Ichthyol 3:7–18

Sullivan BK, Van Keuren D, Clancy M (2001) Timing and size of blooms of the ctenophore *Mnemiopsis leidyi* in relation to temperature in Narragansett Bay. RI Hydrobiologia 451 (1–3):113–120

Tajziehchi S, Karbassi AR (2011) Problems and challenges facing developing countries in order to execute the social impact assessment of dams—a review. Euro J Sci Res 56(4):489–495

Thieme M, Lehner B, Abell R, Hamilton SK, Kellndorfer J, Powell G, Riveros JC (2007) Freshwater conservation planning in data-poor areas: an example from a remote Amazonian basin (Madre de Dios River, Peru and Bolivia). Biol Conserv 135:484–501

Vannote RL, Minshall GW, Cummins KW, Sedell JR, Cushing CE (1980) The river 496 continuum concept. Can J Fish Aquat Sci 37:130–137

Vörösmarty CJ, McIntyre PB, Gessner MO, Dudgeon D, Prusevich A, Green P, Glidden S, Bunn SE, Sullivan CA, Liermann CR, Davies PM (2010) Global threats to human water security and river biodiversity. Nature 467(7315):555–561

Ward JV, Stanford JA (1995) The serial discontinuity concept: extending the model to floodplain rivers. Regul Rivers Res Manag 10:159–168

Waters TE (1995) Sediment in streams: sources, biological effects and control. American Fisheries Society, Bethesda, Maryland

West-Eberhard MJ (1989) Phenotypic plasticity and the origins of diversity. Ann Rev Ecol Syst 20 (1):249–278

Wotton RS (1994) Particulate and dissolved organic matter as food. In: Wotton RS (ed) The biology of particles in aquatic systems. Lewis, Boca Raton, FL

Wotton RS (1995) Temperature and lake-outlet communities. J Thermal Biol 20:121–125

Yeager BL (1994) Impacts of reservoirs on the aquatic environment of regulated rivers. Tennessee Valley Authority, Water Resources, Aquatic Biology Department, Norris, Tennessee. TVA/WR/AB-93/1

Chapter 16
Water, Politics and Dams in the Mesopotamia Basin of the Northern Middle East: How Turkey Instrumentalises the South-Eastern Anatolia Project for Political, Military and Strategic Interests

Joost Jongerden, Zeynep Sıla Akıncı, and Ercan Ayboğa

16.1 Introduction

Iran, Iraq and Turkey, and to a lesser extent Syria, are often said to be the water-rich exceptions in a broad, water-deficient northern Africa and Middle East region (Siddiqi and Anadon 2011: 13). Groundwater levels and water availability are falling in the wider Middle East region generally, and the future prospects are dim. According to NASA, the Levant region, comprising Cyprus, Israel, Jordan, Lebanon, Palestine, Syria and Turkey, is experiencing its worst drought for nine centuries.[1] Groundwater levels are decreasing as a result of a higher rate of extraction than recharging, while the quality of the water is at also risk due to environmental pollution (Tropp 2006). The river from which Jordan took its country name is beginning to run dry (Zurayk 2014). Iran, Iraq and Turkey, as well as Syria, are also experiencing increasing periods of drought and serious water crises (Kömüşcü et al. 2005; Barnes 2009; Douglas et al. 2010; Al-Ansari 2013; Gleick 2014; Lund 2014; Madani 2014). An increased intensity of drought is also reported in these countries (Kelleya et al. 2015; Ahmadi 2018;).

In the dominant water-crisis narrative, water shortage and drought are linked to population growth and climate change (Barnes 2009). In Iran, Iraq, Syria and

[1]nasa.gov/feature/goddard/2016/nasa-finds-drought-in-eastern-mediterranean-worst-of-past-900-years (All webpage citations accessed May 22, 2018.)

J. Jongerden (✉)
Rural Sociology, Wageningen University, Wageningen, The Netherlands
e-mail: joost.jongerden@wur.nl

Z. S. Akıncı
Department of Geography, Universitat Autònoma de Barcelona (UAB), Barcelona, Spain

E. Ayboğa
Keep Hasankeyf Alive, Batman, Turkey

L. A. Jawad (ed.), *Tigris and Euphrates Rivers: Their Environment from Headwaters to Mouth*, Aquatic Ecology Series 11,
https://doi.org/10.1007/978-3-030-57570-0_16

Turkey, however, the construction of dams is also playing a significant role, seriously aggravating the effects of low rainfall and change precipitation regime (Douglas et al. 2010). The drying of lakes, declining groundwater resources and salt storms, moreover, are increasingly seen as related to water management policies in general and dam construction in particular (Madani 2014: 316). In this context, an important case is the Tigris-Euphrates river basin, a vast, transboundary river system shared by Turkey, Syria, Iraq and also Iran (Voss et al. 2013). Intensive dam infrastructure development by Turkey in particular, where the rivers rise, is having a profound impact on water availability downstream. This is also the location of the non-state of Kurdistan, which immediately implies dynamic political dimensions to the natural resource issues.

Dams, as political and technological assemblies combining hydro-electrical and agricultural interventions with centralised bureaucracy, are conceived and utilised as material agents and pervasive symbols of modern state power (Mitchell 2002: 21). As an expression of 'modernity,' the construction of an intensive water infrastructure and the networks of irrigation canals and electricity cables that go with it, linking the periphery to the centre, peasants to markets and thus remote populations to a central administration, have been employed in the service of the nation-state (Mohamud and Verhoeven 2016). Additionally, through control over water resources, state bureaucracies are able to dictate flows and water availability in downstream states and make water part of a wider geopolitics.

As globally supported mega-projects for regional development, however, large-scale dam construction has been widely criticised by a range of NGOs and advocacy groups. International development agencies are now reluctant to promote such mega-projects, at least not without strong safeguards in place to ensure from the start that gains outweigh costs. A report World Bank (2012), for example, opened by emphasising the institution had 'directly or indirectly assisted only 3% of the dams in developing countries' and was now only financing 'about four dam projects a year—half the rate of the 1970s and 1980s.[2] Essentially, large-scale construction projects damming waterways to provide a range of water-supply, energy-production and other benefits have, as the report puts it, 'detrimental social and environmental impacts.' Previously regarded as inevitable 'side effects', these are now recognised as 'fundamental concerns'.

It is in this context, therefore, that we look at the construction of dams in the Mesopotamia basin and particularly the mainly Kurdish (Kurdistan) region of the northern Middle East, focussing on the South-eastern Anatolia Project (*Güneydoğu Anadolu Projesi*, GAP). Taking the case of GAP, the aim of this article is to show how the central, bureaucratic control of water resources has made the region and its people into objects of government. We present these dams as ordering mechanisms that have been employed as a means of geopolitics, causing and exacerbating drought and salinity issues rather than promoting rural development. The article is

[2]'[T]he World Bank pioneered the modelling of river basins and new methods of economic analysis of multipurpose projects in developing countries' (World Bank 2012).

composed of four main sections. First, it provides a background to dam construction in Turkey, Iran, Iraq and Syria and the structuring of Turkey's response to its 'Kurdish problem'. This is followed by a discussion of GAP as doing politics with water and a consideration of the Turkish state's forced migration, before widening the focus again to the broader Northern Middle East and Kurdistan region and the potential for violent, water-related international conflict.

16.2 Background

The first 'modern' dam in the Middle East was constructed in southern Iran. Construction of the Mohammad Reza Shah Dam on the River Dez was begun in 1959 and completed in 1963. Since the Islamic Revolution in Iran in 1979, dams have been built one after another. While in the period before 1979, 14 large dams were built in Iran, 541 dams have been completed in recent years, both large and small (Ayboğa and Akgün 2012: 1), while the construction of another 340 is under consideration (Madani 2014: 317). The dams are being constructed with reference to their role in rural development, turning rain-fed agricultural land into irrigated land in order to boost production, especially cash-crops, with urban electricity production and water supply also mentioned as important objectives. However, this dam building has come with adverse effects, also. The dams on the rivers feeding the Lake Urmia, the biggest lake in Iran, for example, have resulted in a shrinkage of the hypersaline lake, resulting in salt storms that threaten to destroy farmland around the lake (Ayboğa and Akgün 2012; Madani 2014).

In Syria, dam construction started with the coming to power of the Ba'ath party. Construction of the first dam began in 1968 and by 2001, Syria counted already 160 dams with another 28 planned. While the larger dams had hydro-electrical functions, most provided water for irrigation, linked to the government's vision to modernise agriculture through the promotion of irrigated agriculture (Barnes 2009). In Iraq, the first dam, the Ramadi-Habbaniya in Anbar province, was constructed to prevent downstream flooding and for irrigation purposes. By 1989, 11 more dams had been completed. The building of dams was halted after 1990, however, with the occupation of Kuwait and the start of a prolonged period of sanctions.

In Turkey, GAP in the southeast of the country was developed after the construction of the Keban Dam on the Euphrates in 1975. Initially aimed at the development of soil and water resources of the upper Euphrates (*Fırat*), the project expanded to the basin of the Tigris (*Dicle*) and Euphrates and became envisaged as a 'multi-sector and integrated regional development effort'.[3] Spread over nine provinces in the mainly Kurdish populated southeast of Turkey (Adıyaman, Batman, Diyarbakır,

[3]See: http://www.gap.gov.tr/en/latest-state-in-gap-page-47.html, date of access 17-2-2021.

Antep, Kilis, Siirt, Urfa, Mardin, and Şırnak),[4] also referred to officially as the South-eastern Anatolia Region or else as the Kurdistan region of Turkey or Northern Kurdistan. GAP has involved the construction of 22 dams, 19 hydro-electric power plants and hundreds of kilometres of irrigation canals. The GAP region covers more than 75,000 square kilometres, corresponding to almost 10% of the total land area of Turkey. The 1.8 million hectares of arable land served by the mega-project is about 20% of the total irrigable land in the country, and the population in the region is about 8.7 million people, just over 10% of the country's total.

The official aim of regional development plans in Turkey is stated as the elimination of economic disparities. However, the southeast along with the mainly Kurdish populated east of Turkey was a special concern, due not just to its economics but also its socio-cultural status as 'not Turk'. It was around 1990 that GAP was expanded into the field of development, against the background of the rise of the Kurdistan Workers Party (*Partiya Karkerên Kurdistan*, PKK). The PKK had emerged in the 1970s in the context of an absolute denial of Kurds and Kurdishness by the Turkish state, wherein there was no avenue of alternative (genuine) political expression for Kurds. The authorities had stood amazed at the massive support for the PKK, a 'problem' that it thus needed to address. Initially, we may distinguish between two responses. One was the security response, which involved the employment of the army and various other state military and police forces to combat and eradicate the movement, which first failed and then gained advantage but has never been fully successful. The other was an economic response. Through GAP, it was envisaged that, with the introduction of modern irrigation systems and variety in crop patterns, agricultural production would increase, peasants would modernise and integrate into national and international markets. In short, the Kurdish issue would be dissolved through economic progress, and a concern with 'rural development' emerged as a direct response to the 'threat' of the PKK.

16.3 Doing Politics with Dams: The Economic Response

When GAP became envisaged as a rural development project, the population in the region became its main object of governance. Rural development was not only supposed to contribute to added value and increased efficiencies for a market-oriented, capital-intensive agriculture, since it also incorporated a social development, but also supposed to contribute to abolition of the extended family and tribal institutions, regarded as pre-modern and outmoded, with the evolution instead of a modern lifestyle, characterised by nuclear families, city life and a reliance on (and thus loyalty to) the state rather than tribe for security and resources. Accomplished, therefore, thorough socio-economic reform involving the entry of capital into

[4]The Kurdish names of these provinces are in the same row: Semsûr, Elîh, Amed, Dîlok, Kîlîs, Sêrt, Rîha, Mêrdîn and Şirnex, respectively.

agriculture and urbanisation, the modern lifestyle was considered the best possible weapon against the persistence of Kurdish identity (Jongerden 2007: 112).

First, a relationship was drawn between an effective war against the PKK and knowledge about the people living in the insurgent areas. Knowledge about people was supposed to contribute to the practise of control over them. Thus, the transformation of the insurgent region and its population into legible (known) objects (Scott 1998) became growing interest of the authorities after 1989, and the task of turning the region and its population into objects of knowledge was mainly allocated to GAP (Özok 2004). Second, the simple promise of GAP was that it would create jobs for almost half of region's population (the official figure was 3.8 million jobs), bringing water to dry lands (1.8 million hectares) and improving agricultural production. Over time, this was extended with a notion of 'sustainable development' that would improve the region's transportation, education and health systems, boost tourism and generally promote rural and urban infrastructure (Benek 2009).

In the rationale for GAP, political problems were understood as deriving from economics in a very general way. Because the region was relatively poor, it was thought that dissatisfaction and disaffection were pervasive and the PKK was able to channel these sentiments to its own ends. Therefore, the logic went, by developing the Kurdish region economically, and the 'Kurdish issue' would take care of itself. GAP-employed social scientists argued that the project could stimulate and guide the development of a new lifestyle in the region, diminishing tribal relations and extended families and facilitating dependency on state institutions. GAP would not only help solve the economic problems of the region but also end conflict and provide order. Or, as then prime minister Tansu Çiller expressed in 1993, GAP would 'push the region into a growth process, as well as contribute to the establishment of societal peace and order nationwide' (GAP-BKİ 1993, as cited in Bilgen 2014: 17). Through GAP, that is, Kurds were to become Turks.

The Harran region south of Urfa city, bordering Syria, was treated as a model case for the state's rural development strategy. The immediate objective was to provide irrigation to the 150,000-hectare plain-land there (Kudat 1999), and the immediate result was a transformation of the agriculture there. With the introduction of irrigation, the land used for dry-farming crops, such as barley, lentils and chickpeas, decreased, while for water-intensive crops, like corn and sun flower, increased drastically, by close to 1500 and 7500%, respectively (Benek, 2009; Çelik and Gülersoy 2013). It was projected by GAP that cotton would occupy only 25% of the Harran plain in Urfa, but due to a combination of market demand, secure production, ease of management, storage and access to tools and machinery, it became the preferred crop of farmers and started to dominate the land. In 2002, cotton occupied some 85% of the land in Harran Plain and constituted 96% of total irrigated-land (Douglas et al. 2010). Around 2010, corn progressively replaced cotton in most irrigated land in the GAP region, mainly as a result of a fall in the price of cotton. The requirement of farmers to constantly adjust to a fluctuating market was not the only issue. A lack of infrastructure, such as drainage systems, and of extension services had strong adverse effects, while another issue concerned irrigation methods.

Farmers in Harran continued to use surface irrigation methods. Some scholars (Şahin et al. 2006; Bahçeci and Nacar 2009; Douglas et al.; 2010; Çelik and Gülersoy 2013) consider this a main cause of salination, although others hold an alternative view that the problem is rather 'the modernisation philosophy behind GAP' (Aygüney 2002, as cited in Bilgen 2017: 85), and the state's persistence in implementing modern irrigated agriculture without calculating social and environmental risks. The salination is a particular problem in Harran because the underground water sources there contain relatively high levels of salt. The end result, by 2005, was 15,000 hectares of salinity-affected areas in the plain and 40–-50,000 hectares under threat from a rising water-table alongside 450 tons of soil eroding every day (Douglas et al. 2010). This is not only crucial in terms of environmental degradation but also harmful to crop production and plant growth—clearly, in stark contrast to GAP's original promise to increase agricultural production and eliminate regional disparities.

It is not only Harran that is suffering from water-related problems. In the lower parts of the Karacadağ mountain chain, north of Harran, the farmers' demand for water increased especially after 2000, a result, among other things, of the shift to large-scale industrial and water-intensive farming (as well other factors, like decreasing precipitation since the end of the 1990s) and the availability of cheaper techniques for water wells with fuel driven motors. The groundwater level dropped over the course of a decade such that while farmers were previously able to pump water from a depth of 50 metres, today, they regularly have to go down 350–400 metres and sometimes even 500 metres. Karacadağ is not an exception. In this framework, it is important to mention that despite this adverse development, the State Hydraulic Works (*Devlet Su İşleri*, DSİ), which is responsible for water management in Turkey, did not show any concern about the drop in groundwater level. Thus, in their push to increase the harvest, farmers have been given free rein to exploit the groundwater without limits.

It is some 40 years after the start of the mega-project now, but GAP has not brought welfare to the Kurdish southeast as promised. Compared to other regions of Turkey, it has even regressed, being relatively poorer than it was in the 1980s. Except for Gaziantep, all the south-eastern provinces have fallen back in the economic strength rankings.[5] Meanwhile, the percentage of employed people in the GAP region has not increased significantly. Many people still migrate to the west of the country as seasonal workers every summer (see Sönmez 2017).

[5]See the figures provided by the Turkish Statistical Institute (*Türkiye İstatistikler Kurum*, TÜİK), no. 24580, September 25, 2017; at www.tuik.gov.tr/PdfGetir.do?id=24580

16.4 Doing Politics with Dams: The Security Response

Analysis of the rise of the PKK in the context of the security response led to the Turkish state's (re)definition of its Kurdish issue as a spatial problem. First, a direct relationship was made between effective combat of the PKK and a reduction of support for the guerrilla in the predominantly Kurdish countryside; then, a policy of forced evacuation was implemented, the simple removal of (1–3 million) people from the territory—who thus became internally displaced people (IDPs) and mostly resettled in cities, using family and other social networks to develop rural-urban 'tracks' (Jongerden 2007). The clearing of the countryside was intended to contribute to an isolation of the PKK, and it was in this context that the development of large infrastructure projects, such as dams or mines, was meant to help prevent village return and thence recovery of an environment hospitable to the insurgency. In 1993, in a top-secret letter written by then President Turgut Özal to Prime Minister Süleyman Demirel, the president stated that

> Starting with the most troubled zones, villages and hamlets in the mountains of the region should be gradually evacuated. (...) Security forces should immediately move in and establish complete control in such areas. To prevent the locals' return to the region, the building of a large number of dams in appropriate places is an alternative. (KHRP 2002: 118, 22)

According to estimations made by the Mesopotamia Ecology Movement (MEM), at least 400 villages emptied in the 1990s either have been or are planned to be flooded by dams. This would amount to 10% of all the evacuated villages.[6] It has not been possible to conduct proper research in the affected areas, however, and MEM suggests that this figure may be much higher.

At the time that Özal was referring to dams as a counter-insurgency measure, the DSİ had already started to plan further dozens of large and mid-size dams in the region. These were to be constructed with reference to socio-economic objectives, such as the production of energy and the development of agriculture by expanding the amount of land under irrigation, but many thought that the ulterior motive for the dam building was to permanently empty territory of its population as well as to flood possible hiding places for the PKK guerrilla and create watery barriers against their movement. Reasons given for this analysis included the elongational design, placement and siting of the dams (Jongerden 2010, Akıncı and Tan 2016). Dam building in the region thus became equally a project of construction (civil) engineering and social (demographic) engineering. Among the large dams built are three on the Zap River in Hakkari (Çolemêrg), eight on the Munzur River and its tributaries in Tunceli (Dersim) and five on the Peri River in Elazığ (Elazîz or Xarput)-Bingöl (Çewlîg)-Dersim, areas all noted for their high levels of PKK activity.[7]

---

[6]At www.mezopotamyaekoloji.org

[7]Turkish place names given in Kurdish in parentheses.

The construction of these water-supply/hydroelectric-cum-security dams took on a new dimension in 2007 when the Turkish government announced the imminent construction of eleven dams in the Hakkari and Şırnak (Şirnex) provinces along the border with Iraq and Iran (DSİ 2007: 142). In this case, the dams were planed neither for water supply (drinking or irrigation) nor for hydroelectric power production, but for security purposes only. The PKK guerrilla has had a strong presence in this mountainous area since the 1980s—it borders the main guerrilla camps in the north of Iraq, it has many caves and cross-border trails and, indeed, that is why most of its rural population was removed by the military during the 1990s. Thus, the DSİ Annual Report for 2007 stated that the eleven dams will be 'constructed for border security reasons'. According to various statements made by Turkish state authorities, the dams were to serve as a 'wall of water', with the sole purpose of making it difficult for PKK fighters to penetrate Turkey's borders in some areas and to limit their movement and hiding capacity in the greater region.

When questions were raised in parliament on the matter in 2012, the Minister of Forestry and Hydraulic Affairs denied the construction of dams as ongoing under the heading of 'security dams':

> There is no such dam built under my Ministry under the name of 'security dam'. The water constructions, which are planned for drinking water, irrigation, energy production, and storm mitigation, are technically called 'litigated water embankments (dams)'. All the project phases of the mentioned facilities, including the Environmental Impact Assessment report, are built in accordance with legislation.[8]

Manifestly, this was in clear contradiction of what the state's water management organisation had stated in 2007. Since then, the plans for the most of these dams have included equipping them as hydro-electric power plant, giving cover for the claim that they are not being built for security reasons. In 2019, five of these dams have been completed, and the construction of the other six continues.

16.5 Forced Migration

Just as territory was evacuated as a part of the military response to the PKK insurgency, so also has the dam-building involved in the economic response to the 'Kurdish issue' resulted in mass population displacement. It is estimated that hundreds of thousands of people were uprooted from their homes and lost their livelihoods during the GAP development. A precise figure for the number of dam-induced displacements is difficult to ascertain due to contradicting totals given in various publications. According to the regional development administration GAP-BKİ (1994: 2, 66–77), precisely 181,210 people from 336 villages were to be displaced due to dams constructed as part of the project where dams either had been

[8]Response to Şırnak parliamentary deputy Hasip Kaplan's parliamentary question BDP 7/5848 (June 1, 2012); at www2.tbmm.gov.tr/d24/7/7-5848sgc.pdf

completed, were under construction or in the planning stage. However, they did not include the Keban dam. Çakırca (2015: 557) gives the number of settlements affected by the Keban dam as five cities, nine counties, and 258 villages, without giving the populations (or number of households).

Özkalaycı and İçten (2005, as cited in Şatıroğlu and İrge 2012: 54) draw attention to the numbers given by DSİ, of 355,000 people displaced as a result of the construction of 198 large-scale dams up until 2005 in Turkey. It is stressed that, however, these figures are probably much lower than the reality. Taşlı (2012), meanwhile, puts the total number of displaced people due to GAP at 300,000, highlighting 100,000 as a result of Ataturk Dam alone. Again, though, this total does not include the Keban dam, as well as some other GAP dams, and if we add the numbers of people displaced by those, then, according to MEM, somewhere around 400,000 people have been displaced so far.

The Ilısu dam takes its name from the village of Ilısu (Germav), about 200 kilometres east of Mount Karacadağ, situated on the bank of the River Tigris and in the middle of the land to be flooded. This dam, which Turkey started filling at the end of 2019, covers an area of 313 square kilometres and swallows, completely or in part, 199 settlements. The inhabitants of those places—some 55,000–100,000 people have lost their livelihoods and mostly join the caravan of internally displaced Kurds in Turkey.[9] The ancient city of Hasankeyf, and the more than 300 historical sites around it, are among the places flooded, despite the fact that it has protected status (Eberlein et al. 2010). With its ancient cave dwellings, reckoned as one of the world's oldest continuously inhabited settlements and home to several cultures over the millennia, effectively a huge, living, open-air medieval museum of antiquities, the cultural heritage value of this site is unique and its picturesque setting made it a popular international tourist destination. Hasankeyf, standing in the way of the Ilısu dam, has thus prompted a protest movement and stands as emblematic of state intransigence. It is far from alone, however, with the resettlement of displaced populations and land consolidation also being controversial issues in the region.

First, resettlement of the population displaced by the dams comprises another type of forced migration, since the resettlement locations offered are situated in the western parts of country (İzmir, Aydın, etc.). Of course, this is questionable if the main object of the exercise is regional development—but it is entirely in keeping with the logic of an economic response to the Kurdish issue. Second, only land-owners holding a land title can benefit from the expropriation payment, and landless farmers are losing their livelihoods without compensation, support or redress of any form, financial or other. This is particularly pertinent since land ownership distribution in the region is quite disproportionate, with 8% of farmers holding half of all

[9]The DSİ puts the IDP number at 15,000 and the number of people who will be affected at 55,000, while a study by a former World Bank expert includes 23,000 covers people displaced in the 1990s, thus estimating the number of affected people to be 78,000 (Eberlein et al. 2010: 296); however, this does not include some 3000 nomadic families that use the Tigris River there for their animals and thus makes a total in the vicinity of 100,000 affected people.

lands, around two-fifths having just five hectares per person and about the same proportion again being landless (Taşlı 2012; Yadırgı 2017; Bilgen 2017).

For those receiving compensation, there are also issues, since the reimbursement is primarily monetary. This is contrary to recommendation of the World Bank and Organisation for Economic Co-operation and Development (OECD), which advocates land-for-land compensation, since cash-for-land compensation is generally associated with a loss of livelihood. Moreover, the amount of compensation is low, not only because of the small size of plots owned by the vast majority of local households but also because the compensation rates do not reflect market price. Further to the injustice is the practical effect that many families cannot afford to buy similar amounts of land elsewhere and one-off payments are commonly consumed by relocation costs or go unpaid.

The stance towards compensation payment taken by the GAP authority supplies a final twist to this sorry tale. In a report made by GAP (2016: 53–54), the land consolidation of almost 2.5 million hectares was proudly announced as the largest in the world—a claim followed by the highlighting of 2.5 billion Turkish Lira savings made by *not* paying compensation for lands where canals are to pass through, as per the regulations. Again, beyond the moral issue of justice, it seems somewhat inappropriate to apply such regulations to a region suffering from the economic disparities that GAP is supposed to address. As an economic response to the Kurdish problem, indeed, it seems more likely to be counterproductive, affecting an economic oppression that will only store up problems for the future. Regardless, the end result is certainly that many people do lose their livelihoods—the state shows no interest in investigating to quantify this, let alone ameliorate it, but if the numbers affected must run to the order of several tens of thousands of people. Quite commonly, the previous owners of small parcels of land are impoverished, turned into labourers or crop-sharers (Akça et al. 2013)—or, of course, they simply become destitute and head for the city slums or worse.[10]

16.6 The Northern Middle East

As indicated, there are major water supply issues in the three other states with lands situated in the Mesopotamia basin. These include the 'indirect' displacement effects of dam construction. The Ba'ath party's promotion of water-intensive agriculture (Barnes 2009), enabled mainly through the construction of dozens of dams, was followed by a multi-year period of drought and failed harvests. This led to a huge displacement of farmers in Syria and is thought to have contributed to the outbreak

[10]Notably, there is only one dam, at Birecik, in the eastern part of the region (between Şanlıurfa and Gaziantep), for which a 'resettlement and rehabilitation project has been planned and implemented in order to minimise the effects of displacement on the affected communities' (with mixed results), as Çiğdem Kurt (2013, i) notes in her doctoral thesis; as she also notes, Birecik is mainly Turkish-speaking, i.e. not Kurdish.

of the (civil) war there (Gleick 2014). An estimated 800,000 people lost their livelihoods, among them thousands of rural families (Lund 2014). In Iraq, dam construction and hydrological engineering have been referred to as 'fundamental instruments of the destruction of the southern marshlands and the communities that lived there', displacing up to 100,000 people (Fawcett and Tanner 2002: 30–31).

On the Iranian side of the border, dam construction is equally threatening the marshes and displacing its inhabitants. A United Nations Environment Programme study concluded that because of 'the cumulative impacts of dam construction upstream and intensive drainage schemes in and around the marshlands... [i]n less than a decade, one of the world's largest and most significant wetland ecosystems has completely collapsed' (UNEP 2001: 29). Iran's planned construction of a dam in Iranian Kurdistan is expected to have serious downstream effects, in particular for those on the Iraqi side of the state boundary, which may have successive effects on populations in central and southern Iraq. As expressed by Chomani and Bijnens (2016: 18), 'When Iran finishes the construction of the Daryan Dam and begins diverting the water of the Şirwan River, the KRI [the de facto autonomous Kurdistan Region of Iraq] will face a serious water crisis in Halabja Governorate... [which] will force the KRG [Kurdistan Regional Government] to reconsider the amount of water flows it releases to other parts of Iraq'.

16.7 Transnational Conflicts

In 1984, the World Bank refused to provide financial support to GAP because of its expected adverse impacts on Iraq and Syria and violation of resettlement guidelines.[11] It later granted two funds totalling 650,000 USD for the regional development of urban and rural infrastructure in 1997 and 1998.[12] The World Bank justified this decision, among other reasons, by the lack of any existing mutual agreement among the three states on the share of the Euphrates and Tigris rivers. In the last 20 years, the three states have met several times in order to discuss dam projects on the two rivers, but no mutual agreement covering the whole basin has yet been forthcoming. This is mainly caused by the stance taken by Turkey, which has not ratified the UN Convention on the Non-Navigational Use of International Watercourses from 1997, which has become the international standard for agreements on transnational rivers.[13]

Shortly after its foundation, the Republic of Turkey indicated its intention to construct large dams on the Euphrates and Tigris as a means of gaining regional

[11]According to a consultant on resettlement for the World Bank, speaking to the author Ercan Ayboğa in a meeting in 2006, in Batman. See also: https://www.rivernet.org/turquie/ilisu.htm, date of access February 12, 2021

[12]See gap.gov.tr/en/internationally-funded-projects-page-15.html

[13]At legal.un.org/ilc/texts/instruments/english/conventions/8_3_1997.pdf

power in the Middle East. It has only been with the recent construction of large dams on the Euphrates, however, that it has acquired the tools for this (following the collapse of the Ottoman Empire, the new Turkish state had no method by which to exert such power, and after WWII, it went through many crisis). Particularly after 1990, the Turkish state authorities have felt able to engage in regional power politics, a period when the conflict with the PKK escalated. Turkey's regional policy is linked very directly to the unsolved Kurdish question, as it is also for the country's three Middle Eastern neighbours, Syria, Iraq and Iran, where the Kurdistan is situated.

The first big water dispute occurred in 1975. Syria and Iraq had come to the brink of war when the building of both the Keban Dam in Turkey and the Tabqa Dam in Syria combined with a drought to create serious problems in Iraq. In 1989, two Syrian MIG fighter jets entered Turkish airspace and shot down a survey plane belonging to the state Land Registry Directorate. Turkey responded by announcing that it would cut off the flow of water to Syria from the Euphrates (Lawson 2013: 126). Turkey mobilised its forces in January 1990, when it did cut supply, allegedly for a test to fill the reservoir for the Atatürk Dam, reducing the flow of water into Syria and Iraq by 75%. Iraq had threatened to bomb the dam and so Turkey threatened to end the water flow to Syria and Iraq.

With the construction of the Atatürk Dam, several high-ranking Turkish politicians linked the water issue to Syria's dealing with the PKK. In 1992, Turkey's President Özal stated that his country would not inflict damage on Iraq and Syria if they cooperated in Turkey's fight against the PKK. In the same year, Prime Minister Demirel stated that it was 'impossible to engage in negotiations over water while allowing terrorism'. Foreign Minister Hikmet Çetin added, 'The water issue should not be thought of as so important. If we have good relations with one another, we will not cause problems.' The water issue was thus used to press Syria to end its tacit support of, or at least non-intervention in, PKK activities in its border region with Turkey, a strategy that eventually contributed to PKK leader Abdullah Öcalan's leaving the country and being captured (in Kenya in 1999).

Although there are several areas of water-related conflict in the Middle East, including the Jordan-Yarmuk and Nile rivers, which involve several countries, the fiercest conflicts have occurred over the Euphrates and Tigris, between Iraq, Syria and Turkey. Iraq and Syria have filed numerous claims over the years accusing Turkey of causing a water shortage. Although Iraq has constructed seven dams in the Euphrates basin and sixteen dams in the Tigris so as to more effectively utilise its fresh-water resources, water availability in Iraq has been heavily affected by the upstream construction in Turkey, and Syria also, to a lesser extent (Al-Ansari and Knutsson 2011; Al-Ansari 2013).

In the midst of a severe water-supply shortage in early 2009, on May 12, the Iraqi parliament pressed its government to demand a greater share of water from its neighbour. The parliament claimed that as a result of Turkey's infrastructure work in the river basin of the Euphrates and Tigris, spring water reserves in Iraq had dropped to a total of 11 billion cubic metres, compared to 40 billion cubic metres 3 years previously. According to Iraqi experts, rainfall had not been below normal levels, and the shortages had been created by Turkey, which was cutting off water flows while filling its newly constructed dams on the Euphrates.[14]

Water shortage in Iraq is also caused by Iran's recent widespread dam construction. There are several rivers flowing from western Iran into the Tigris in Iraq, contributing over 40% of the Tigris capacity (the contribution to the Tigris basin from Turkish state territory is around 55%). Grave impacts of the Iranian dams have been experienced and discussed since 2010 (Iran itself has a decades-long history of disputes with Afghanistan and Azerbaijan because of their dam construction). Since Iran as state is much stronger and Iraq is partly dependent on it, however, the Iraqi government has not raised the issue of water-flow reductions in this case (which has started to change, with the Iraqi parliament discussing the water cut by Iran after criticism of Iran from civil society organisations and media outlets).

Overall, the shortage of water in Iraq is leading to environmental disaster (including a draining and salination of the marshes in the south of the country) and displacement of its peasant population (Jongerden 2010). Dams on the Euphrates (at Tishrin, Tabqa and Baath) have already not been enough to produce the usual amount of electricity and supply farmers with irrigation water for decades. Farmers in Iraq's south say they face serious difficulties, driving many to despair. Alewi al-Shimmari, a rice farmer living in Diwaniyah, south of Baghdad, says 'More than 50% of families working as farmers have left their villages and gone to the city.' Al-Shimmari used to grow rice throughout his 40-hectare farm, but the drought has reduced this to just five hectares. 'Lands that once were green farms are now turned to desert,' he notes.

Syria has also accused Turkey of effectively releasing contaminated water, arguing that water cut-backs had effectively deposited dangerously increased concentrations of faecal matter on irrigated fruits and vegetables, leading to an acute outbreak of cholera. Contamination and increased levels of salinity in the rivers, it is argued, have also contributed to a sharp reduction in fishing, an important source of food and livelihoods. Turkey acknowledged the need for water of its neighbours, yet maintained that it could not 'allow our own water and energy management to run into problems' (Jongerden 2010).

Now, there is a new, mainly Kurdish actor on the scene, the Democratic Federation of Northern Syria (DFNS), in its hostility to which, Turkey uses the GAP dams as a weapon in the most direct way to date. The DFNS is a multi-ethnic and democratic administrative entity in the northern part of Syria, in which Kurds play the leading role.[15] Linked by political philosophy to the PKK, raising the spectre for Turkey of a 'threat' along its southern border (through contiguous territory with the KRI) and considered by Turkey an obstacle to its developing influence in northern Syria, the DFNS is faced with constant aggression (including, in March 2018, the military occupation of Afrîn). Thus, according to the Cizîre canton administration in

---

[14]reuters.com/article/us-iraq-water/iraq-parliament-demands-more-water-from-neighbors-idUSTRE54B3ZA20090512

[15]theregion.org/article/13413-the-euphrates-turkey-039-s-tool-of-destabilisation-of-rojava

the DFNS and many other statements and reports,[16] in the spring and summer of 2017, Turkey released only the half of the contractually committed Euphrates flow rate (250 rather than 500 m^3/sec).[17]

According to a previous NATO conflict scenario, predating the most recent turmoil, Syria and Iraq execute a joint invasion of Turkey. In this scenario, the invasion occurs against the background of a severe three-year drought in Iraq and Syria, ascribed to Turkey's water policies, and an unstable political situation in the region. According to an Uppsala Model UN scenario, meanwhile, Turkey and Iraq come to the brink of war after a failed attempt by an illegal organisation from Iraq to explode one of Turkey's dams. Iraq condemns the assault but accuses Turkey of denying the country access to the water. Turkey blames the Iraqi government for the attack, demands the arrest of those responsible and threatens to cut water supplies completely if Iraq does not comply. Forces are mobilised and war looms. These scenarios clearly indicate the potential for armed conflict over water in this part of the Middle East.

16.8 Conclusions

Understanding dams as political-technical ordering mechanisms, the objective of this chapter has been to show how central and bureaucratic control over water resources through the construction of large infrastructural projects has transformed the (mostly Kurdistan) Mesopotamia basin of the northern Middle East region and its peoples into objects of government. We have taken GAP in the mainly Kurdish, south-eastern region of Turkey as a case study, discussing its engagement as a means for geopolitics within the field of rural development. This is related to debates in the wider literature on the employment of large water infrastructures in the production of nation-state identities and to demonstrate the strength of the modern state, as well as on the politics of infrastructure in general (Winner 1980; Easterling 2016).

Through its engagement in water-for-agriculture and a so-called modernisation of agricultural practices and lifestyles in the region, GAP was supposed to counter the influence of the PKK over the Turkey's Kurdish population and address the state's Kurdish issue at its socio-economic and cultural roots. In the official discourse, Kurdish identity was conceived as the outcome of economic underdevelopment and social backwardness and so it was assumed that rural development was a means through which the issue could be resolved. However, the introduction of irrigated

[16]E.g. anfenglish.com/rojava/the-turkish-state-uses-the-water-of-euphrates-as-a-threat-26014

[17]Turkey guaranteed to Syria a minimum, year-long flow of 500 m^3/sec from the Euphrates in 1987, through an informal protocol agreed when the construction of the Atatürk Dam was ongoing; the average flow over the year before construction of the large dams and precipitation decrease had been around 800 m^3/sec. Prior to the war in Syria, Turkey had violated this agreement few times, but only in a limited way (except for one-month cut linked to filling the Atatürk Dam Reservoir). See fao.org/nr/water/aquastat/countries_regions/Profile_segments/SYR-IntIss_eng.stm

agriculture in the region did not bring the envisaged prosperity and actually confronted farmers with new problems. Although water was provided, the development of drainage systems did not keep pace and agricultural extension services were poor, which contributed to a rapid salination of the productive land in the region.

This rural development practice could already be understood as a form of geopolitics—the exercise of power through control over geographical space. Historically, the concept has been strongly connected to the project of colonisation that legitimised territorial expansion and acquisition. Today, in the case of GAP, the control over water resources through the development of large infrastructural projects is being used as a tool of control and coercion of people and land. This includes the development of security dams, functioning for counter-insurgency as their principal objective, both directly by hampering the possibility of guerrilla movement by the creation of obstacles in the landscape and indirectly by means of demographic engineering. The destruction of deeply rooted social bonds and effective relocation of the local people in cities are thought to contribute to integration into Turkish society. In fact, this is a project of cultural assimilation grounded on eradication.

Dam construction generally has massively impacted the Mesopotamia basin, upriver states and central authorities generally employing their access to power as tools of political leverage. In this context, GAP has wider political implications vis à vis international relations. In the Turkish-Kurdish context, moreover, control over water is also used to destabilise the DFNS. It becomes weaponised in the state war on Kurdistan as an alternative territory.

Thus, the spoils of drought, war, internal displacement and the destruction of cultural heritage are not collateral damage of imperfect design, but the calculated and even intended objectives of dam construction.

Bibliography

Ahmadi D (2018) Iran drought here to stay. Financial Tribune, January 16, 2018

Akça E, Fujikura R, Sabbağ Ç (2013) Atatürk dam resettlement process: increased disparity resulting from insufficient financial compensation. Int J Water Resour Dev 29:101–108

Akıncı ZS, Tan P (2016) Waterdams as dispossession: ecology, security, colonization. In: Graham J, Blanchfield C, Anderson A, Carver J, Moore J (eds) Climates: architecture and the planetary imaginary. Lars Müller Publishers, Zürich

Al-Ansari NA (2013) Management of water resources in Iraq: perspectives and prognoses. Engineering 5(8):667–684

Al-Ansari N, Knutsson S (2011) Toward prudent management of water resources in Iraq. J Adv Sci Eng Res 1:53–67

Ayboğa E, Akgün İ (2012) Iran's dam policy and the case of Lake Urmia. At ekopotamya.net/index.php/2012/07/irans-dam-policy-and-the-case-of-the-lake-urmia/

Aygüney, N. (2002). Burdens of 'development' in Southeastern Turkey: Salinization and socio-cultural disruption (Master's thesis, Lund University, Sweden).

Bahçeci, İ and Nacar, A. S (2009). Subsurface drainage and salt leaching in irrigated land in south-east Turkey. Irrigation and Drainage 58: 346–356.

Barnes J (2009) Managing the waters of Ba'ath country: the politics of water scarcity in Syria. Geopolitics 14:510–530

Benek S (2009) Ortaya çıkışı, gelişme seyri ve bölgeye etkileri bakımından Güneydoğu Anadolu Projesi. Ankara Üniversitesi SBF Dergisi 64(3):46–71
Bilgen A (2014) A static nexus or a dynamic network? Rethinking the security-development relationship within the context of southeastern Anatolia project. Res Turkey III(4):12–23
Bilgen A (2017) Demystifying the (post-) politics of Southeastern Anatolia Project (GAP): An analysis of the what, why and how of GAP and the operation of development in Turkey from a critical perspective. PhD Thesis Wheinischen Friedrich-Wilhelms-Universitat, Bonn
Çakırca D (2015). Keban Barajı ile neler kaybettik. *İnşaat Mühendisleri Odası, 4.* Su Yapıları Sempozyumu Kitabı, pp 550–561
Çelik MA, Gülersoy AE (2013) Güneydoğu Anadolu Projesi'nin (GAP) Harran Ovası Tarımsal Yapısında Meydana Getirdiği Değişimlerin Uzaktan Algılama ile İncelenmesi. Uluslararası Sosyal Araştırmalar Dergisi 6(28):46–54
Chomani K, Bijnens T (2016) The impact of the Daryan Dam on the Kurdistan Region of Iraq (save the Tigris and Iraqi marshes campaign). At waterkeepersiraq.org
Douglas EM, Kimbrunner JF, Özertan G, Simpson KE, Terkla DG, Wisser D (2010) Sustainable use of irrigation water: the case of Turkey. Paper prepared for presentation at the 13th Biennial Conference of the International Association for the study of the Commons (IASC), Hyderabad, India
DSİ (2007) Dsi Genel Müdürlüğü 2007 Yılı Faaliyet Raporu (T.C. Çevre ve Orman Bakanlığı Devlet Su İşleri Genel Müdürlüğü). DSİ: Ankara. At dsi.gov.tr
Easterling K (2016) Extrastatecraft: the power of infrastructure space. London, Verso
Eberlein C, Drillisch H, Ayboğa E, Wenidoppler T (2010) The Ilisu dam in Turkey and the role of export credit agencies and NGO networks. Water Altern 3:291–312
Fawcett J, Tanner V (2002) The internally displaced people of Iraq. The Brookings Institution-SAIS Project on Internal Displacement, Washington DC
GAP-Bölge Kalkınma İdaresi (1993) GAP: 1992 Durum Raporu. GAP-BKİ, Ankara
GAP Bölge Kalkınma Dairesi (2016) Güneydoğu Anadolu Projesi Son Durum. At yayin.gap.gov.tr
GAP Bölge Kalkınma İdaresi (1994) GAP Bölgesi Baraj Göl Aynasında Kalacak Yörelerde İstihdam ve Yeniden Yerleştirme Sorunları Araştırması. Sosyoloji Derneği, Ankara
Gleick PH (2014) Water, drought, climate change, and conflict in Syria. Water Climate Soc 6:331–340
Jongerden J (2007) The settlement issue in Turkey and the Kurds: an analysis of spatial policies, modernity and war. Leiden and Boston, Brill
Jongerden J (2010) Dams and politics in Turkey: utilizing water, developing conflict. Middle East Policy, XVII, pp 137–143
Kelleya, Colin P., Shahrzad Mohtadib, Mark A. Canec, Richard Seagerc, and Yochanan Kushnir. 2015. 'Climate change in the Fertile Crescent and implications of the recent Syrian drought', 112: pp. 3241–46
KHRP (2002) Internally displaced persons: the Kurds in Turkey. London, Kurdish Human Rights Project
Kömüşcü AU, Erkan A, Turgu E (2005) Normallestirilmiş yağış indeksi metodu (SPI) ile Turkiye'de kuraklık oluşumunun coğrafik analizi. Turkish State Meteorological Service. At meteor.gov.tr/2005/arastirma/dogalafetler/kuraklik.htm
Kudat A (1999) Şanlıurfa and Harran Plains on-farm and village development project. Social development notes, World Bank, Washington, DC, pp 1–4. At documents.worldbank.org
Kurt, Ç (2013) The impact of the Southeast Anatolia project (GAP) on displaced families: household livelihoods and gender relations. PhD thesis, University of Newcastle, UK
Lawson FH (2013) Global security watch. Santa Barbara, Praeger
Lund A (2014) Drought, corruption, and war: Syria's agricultural crisis. Carnegie Middle East Center. At carnegie-mec.org/diwan/55376
Madani K (2014) Water management in Iran: what is causing the looming crisis? J Environ Stud Sci 4(4):315–328

Mitchell T (2002) Rule of experts, Egypt, techno-politics, modernity. Berkeley, University of California Press

Mohamud M, Verhoeven H (2016) Re-engineering the state, awakening the nation: dams, islamist modernity and nationalist politics in Sudan. Water Altern 9:182–202

OECD (2008) Environmental performance reviews: Turkey 2008. Conclusion and recommendations. At oecd.org

Özok N (2004) Social development as a governmental strategy in the Southern Anatolia project. M. Sc. Thesis (Bogazici University: Istanbul)

Özkalaycı Z, İçten H (2005) Yeniden Yerleşim Planlamaları ve Devlet Su İşleri Genel Müdürlüğü'ndeki Uygulamaları. Ankara: TMMOB Harita ve Kadastro Mühendisleri Odası 10. Türkiye Harita Bilimsel ve Teknik Kurultayı

Şahin, R., Yılmaz, N., Işık, E. (2006). Irrigation and environment. In International Congress on River Basin Management (pp. 425–437). Ankara, Turkey.

Şatıroğlu, İrge K (2012) 'Zorunlu göç ve yeniden yerleşimin toplumsal çevre üzerindeki etkisi: tahtalı barajı örneği' PhD thesis. Ankara üniversitesi sosyal bilimler enstitüsü sosyal çevre bilimleri anabilim dalı

Scott JC (1998) Seeing like a state: how certain schemes to improve the human condition have failed. New Haven, Yale University Press

Siddiqi A, Anadon LD (2011) The water-energy nexus in Middle East and North Africa. Energy Policy 39:4529–4540

Sönmez M (2017). Güneydoğu'da Yeniden Yoksullaşma. At mustafasonmez.net/?p=6046

Taşlı İ (2012). Güneydoğu Anadolu Projesine Eleştirel bir Bakış. At www.ekopotamya.net

Tropp H (2006). Water scarcity challenges in the Middle East and North Africa (MENA). human development report, Stockholm International Water Institute. At undp.org

UNEP (2001) The Mesopotamian marshlands: demise of an ecosystem. Nairobi, United Nations Environment Programme

Voss KA, Famiglietti JS, Lo MH, de Linage C, Rodell M, Swenson SC (2013) Groundwater depletion in the Middle East from GRACE with implications for transboundary water management in the Tigris-Euphrates-Western Iran region. Water Resour Res 49:904–914

Winner L (1980) Do artifacts have politics? Deadulus 109:121–136

World Bank (2012). World Bank lending for large dams: a preliminary review of impacts. Independent Evaluation Group for the World Bank Group, Washington, DC. At worldbank.org

Yadırgı V (2017) The political economy of the Kurds of Turkey. From the ottoman empire to the Turkish republic. Cambridge University Press, New York

Zurayk R (2014) The fatal synergy between war and drought in the eastern Mediterranean. J Agric Food Sys Community Dev 4:9–13

Chapter 17
Dams and their Impacts on Fishes in Iran

Hamed Mousavi-Sabet

Abstract Various impacts of dams on aquatic organisms have been well documented in the recent past. The construction of a dam can make significant changes in the ecosystem of a river and particularly affect fish communities including obstruct the upward migration, reduce the genetic diversity and affect morphology, reproduction and growth indices, etc. After China and Turkey, Iran is the third country in dam construction in the world. Construction of dams is essential for socio-economic development in an arid country like Iran. Dams construction started since 1950s in Iran, which to date there are more than 600 (big and small) constructed dams reported from the country. Despite various benefits, construction of dams has also many assessable negative environmental impacts particularly on fishes. About 300 fish species ($\cong$ 100 endemics) listed from water basins of Iran, which the fauna is mostly affected by the dams. The damming and its effects on fish populations in Iran is reviewed. As a result, almost no environmental considerations (in relation with ichthyofauna) have been observed in the damming, such that these dams normally designed and constructed without any fishways or fish ladders. Therefore, the dams mainly could (1) change aquatic ecosystems, (2) make limitation for downstream dispersal and blocking the upstream migration altogether, (3) periodically dried out downstream, (4) produce morphologically/genetically different populations in up and downstream and (5) affect the growth indices and reproductive characteristics, etc. Also, the introduction of non-native/invasion fishes into the reservoirs is an important threat for native fish communities. The diadromous species within the families Acipenseridae, Cyprinidae, Gobiidae, Petromyzontidae and Salmonidae were negatively affected by dams, whereas some native cyprinids and loaches those threatened by the drought were positively associated with the damming, where they could survive themselves in the reservoirs. It can be concluded that dams in Iran have negatively affected native – especially migratory – fish species by

H. Mousavi-Sabet (✉)
Department of Fisheries, Faculty of Natural Resource, University of Guilan, Rasht, Iran

The Caspian Sea Basin Research Center, University of Guilan, Rasht, Iran
e-mail: mousavi-sabet@guilan.ac.ir

L. A. Jawad (ed.), *Tigris and Euphrates Rivers: Their Environment from Headwaters to Mouth*, Aquatic Ecology Series 11,
https://doi.org/10.1007/978-3-030-57570-0_17

blocking their migration routes, whereas favouring non-native/invasion fish species, or altering existing aquatic habitats. Conservation programmes favouring native/endemic fishes in the constructed dams are strongly recommended, and fishways should be mandatory in the under construction/planned dams.

17.1 Brief Overview of Dam Impacts

Dam construction process, the constructed dam and the reservoir behind the dam have several impacts on the area, including social, economic, geophysical, hydrological, water quality, climate, ecological, flora and fauna, etc. From all these impacts which related to each other very closely, the dam impacts on aquatic ecosystems and fishes seem to be the most important one from ecological view. As Han et al. (2008) well reviewed, dams are a major threat to aquatic habitats in the world (Saunders and Hobbs 1991; Dynesis and Nilsson 1994), which fragmentation of freshwater ecosystems is an obvious negative impact of damming (Dudgeon 2000; Joy and Death 2001; Park et al. 2003; Katano et al. 2006). Fishes are affected directly by dams in terms of blocking their migration routes, missing the spawning grounds by inundation within the reservoirs, irregular water releases and periodic inundation or drying-out of spawning habitats in downstream of the dam (Horváth and Municio 1998). Dams have significant impacts on fish communities (Joy and Death 2001; Park et al. 2003) and species richness (Holmquist et al. 1998; Fukushima 2005), particularly the diadromous species (Joy and Death 2001). The diadromous fishes need to migrate between freshwater and saltwater ecosystems at least once during their life cycles. Therefore, the diadromous species are most sensitive to the negative effects of damming, which the dams may block their migration routes altogether (Dauble and Watson 1997; Dauble and Geist 2000; Kareiva et al. 2000; Joy and Death 2001; Morita and Yamamoto 2001; Morita and Yokota 2002; Dauble et al. 2003). In addition, fish populations can be affected indirectly (to a different level depending on species) by modification in physico-chemical water parameters, e.g. velocity, temperature and water quality (Horváth and Municio 1998). The next important potential impacts of damming will appear by shifting from a lotic aquatic ecosystem to a lentic habitat by the constructed reservoir behind the dam, especially when it comes with the introduction of non-native/invasion fishes into the reservoirs (Martinez et al. 1994). This habitat changing caused by damming can change the fish communities, population structures and areas utilized by a particular species. Also, probably non-native fish species will introduce to the constructed reservoir by several reasons (Mousavi-Sabet and Eagderi 2014, 2016; Esmaeili et al. 2017; Mousavi-Sabet 2018). These changes in the ecosystem which affected by damming can lead to local extinction of numerous populations of native freshwater fishes, even without any radical changes in hydrologic or thermal properties (McKinney 2001; Maezono and Miyashita 2003; Gido et al. 2004; Marchetti et al. 2006). Dams in general, especially large ones, have blocked not only the upstream routs (due to habitat fragmentation and reservoir construction) but also affect negatively the downstream reaches. Large dams also change the water flow, sediment transport, thermal conditions and chemical cycling, and such modifications can destruct downstream habitat for riparian organisms and fishes (De Merona and Albert 1999; Paragamian 2002; Poff and Hart 2002; Preece and Jones 2002). In addition to these local effects, large dams also may affect the whole habitat of a river drainage.

17.2 Damming History and Dams in Iran

The damming activities started in Iran in the 1950s. About 14 large dams were constructed with the help of international engineers and advisors during two decades before the Islamic Revolution (1978). After the revolution, dam building in Iran was significantly increased, by more than 200 contracting companies, 70 consultant firms and 30 corporations as well as hundreds of hydroelectric manufacturing units having been established inside of Iran in less than three decades (IME 2006). The country was constructing 85 small/large dams in 2007, when about 500 dams (180 large and 320 small dams) were already constructed since 1979 (Iran-Daily 2007). In 2010, Iran has 588 constructed dams (big and small), with 137 more under construction and 546 planned (Press TV 2010). According to the latest available official list, there are 611 constructed dams (big and small), with 146 more under construction and 520 planned (WRM 2019). For more details regarding numbers of constructed dams in different Iranian water basins, see Table 17.1.

17.3 How Did Dams Change Aquatic Ecosystems in Iran?

The environmental effects of dams (particularly large ones) are numerous and varied, including direct impacts to the biological, ecological, chemical and physical properties of rivers. According to the guidelines of the International Committee on Large Dams, large dams affect the region and ecosystem by physical, chemical, biological, ecological, health, economical, social and pollution effects (Heydari et al. 2013). In terms of biological impacts, large dams can reduce nutrient concentrations in downstream, also planktonic bloom, plant growth, species extinction, fish spawning areas limitation, thermal stratification, production of new species or populations, decline in fish populations, various modifications in water quality, increased opacity of water, release of toxic substances (pesticides, toxic metals, etc.), increasing the concentration of pollutants in the intake water in periods of water shortage, deterioration of vegetation and greenhouse gas emissions. Also, increasing of BOD in water (at first), formation of anaerobic degradation environment, formation of dark and funky environment, the exceeding of phytoplankton, growing of macro-flora in the water, increasing plant, increasing evaporation and

Table 17.1 Numbers of dams in Iran by basin (according to WRM 2019)

Basin	Planned	Under construction	Constructed	Total
Bedjestan	8	1	33	42
Caspian	157	40	186	383
Esfahan	6	0	6	12
Hormuz	1	1	27	29
Jaz-Murian	20	1	9	30
Kavir	12	4	58	74
Kerman-Na'in	7	1	6	14
Kor-Maharlu	3	1	3	7
Lut	12	3	26	41
Makran	12	4	29	45
Mashkid	7	0	10	17
Namak	50	6	64	120
Persis	5	7	5	17
Sirjan	8	2	6	16
Sistan	2	2	8	12
Tedzhen	15	3	22	40
Tigris	156	61	58	275
Urmia	39	9	55	103
Total	520	146	611	1277

transpiration, and making a barrier to stop fish from passing, can be listed as ecological impacts (Heydari et al. 2013; Manouchehri and Mahmoodian 2002).

Ecosystem fragmentation: One of the most negative impacts of dams is fragmentation of river ecosystem into upstream and downstream (Fig. 17.1), which in particular affects fish movements. In Iran, downstream of dams are normally dried out in small rivers (periodically in large ones) (Fig. 17.2). Also, the slow water flow in downstream of large rivers has created an opportunity to harvest sand (Fig. 17.3). Therefore, we can see a critical destruction of fish habitats in upstream due to sedimentation and in downstream due to the harvest of sand.

Negative impacts on natural water bodies: Heydari et al. (2013) reviewed the environmental impact of some large dams in Iran and stated that Sivand Dam in Kor River basin threats the Tashak and Bakhtegan lakes, where known as the winter habitat for birds which migrate from Siberia and Russia to Iran; Gotvand Dam which is located on Karun River in southwestern of Iran, one of the major challenges created regarding the dam is the discussion of domes and salt veins around the dam inundation area which is underwater after dam inundation and may lead to too much salt in the water in the downstream area of the dam, which the salinization can threats the freshwater fish populations; and Zhaveh, Dariyan, Azad, Kanisib, Zangabad, Shivashan and Sardasht dams have several impacts on their regions.

Due to climate change and drought together with the damming and poor water management impacts on aquatic ecosystems in Iran, a critical/significant drop in both lentic and underground water levels has been occurred in recent decades. In case of

Fig. 17.1 Manjil Dam (above) and Tarik Dam (below, behind the bridge) on Sefid River, fragmented the river ecosystem into up- and downstream and blocking migration routes

Fig. 17.2 Ashar Dam in Mashkid basin. The dam and its reservoir (right) and the completely dried out downstream (left)

the Lake Urmia in northwest of Iran, the third largest saltwater lake around the earth, more than 70% of the lake surface areas have already dried due to poor water management and construction of 48+ dams (Garousi et al. 2013). However, similar dams' impacts are assessable for other lakes/wetlands in Iranian water basins, e.g. Hamun-e Puzak, Hamun-e Saberi and Hamun-e Hirmand in Sistan basin (dam construction in Afghanistan), Gavkhuni Wetland in Esfahan basin (dam construction

Fig. 17.3 Harvesting sand from riverbed. Above: in Mazandaran Province's rivers, in the southern Caspian Sea basin (photo from www.pana.ir), below: Ilam Province, in the Tigris River basin

on Zayandeh River), Jazmurian Wetland in Jazmurian basin, Gandoman (Fig. 17.4), and Shadegan Wetlands in the Tigris River basin, etc.

Generally, main impacts of dams on aquatic ecosystems in Iran can be summarised as follows: (1) fragmentation of river ecosystem, (2) transformation upstream of the dam from a free-flowing river ecosystem (lotic) to an artificial slack-water reservoir habitat (lentic) (Fig. 17.5), (3) periodic drying in downstream of the dam, (4) impact on water level and salinization of natural water bodies in the basin, (5) habitat destruction by sedimentation in upstream of the dam, and sand harvesting in the downstream and (6) effects on diversity and richness of fauna and flora, etc.

Fig. 17.4 The Gandoman Wetland's ecosystem changing to meadow

Fig. 17.5 The artificial slack-water reservoir habitat, behind Shahid-Rajaee Dam, on Tajan River, in the southern Caspian Sea basin

17.4 Freshwater Fishes of Iran

The territory of Iran is important from a zoogeographical point of view, as it straddles several major ecoregions of the world including the Palaearctic, Ethiopian and Oriental realms (Nalbant and Bianco 1998; Coad 1998; Esmaeili et al. 2020) as well as having some exotic elements from the Nearctic and Neotropical realms (Mousavi-Sabet and Eagderi 2014, 2016; Esmaeili et al. 2017; Mousavi-Sabet 2018). Also, western portion of Iran is parts of Irano-Anatolian hot spot with high biodiversity especially freshwater fish diversity (Jouladeh-Roudbar et al. 2015). The country features containing 19 basins and three main climatic zones including arid and semi-arid regions of the interior and far south, Mediterranean climate and humid and semi-humid regions. In addition, Iran's plateau with a vast desert located in the central areas and two mountain ranges, Zagros in the west and Alborz in the north, comprise a significant portion of its territory. The complex and varied climates, topography, geological formations and anthropological management of natural resources have led to a varied and unique biological diversity. Many authors were interested to review the biodiversity of Iranian freshwater fishes. According to the latest checklist (Esmaeili et al. 2018), freshwater fishes of Iran comprise 297 species in 109 genera, 30 families, 24 orders and 3 classes. However, presence of 23 reported species in Iranian waters needs confirmation by specimens. The most diverse order is the Cypriniformes with 176 species (59.3%), followed by Gobiiformes with 42 species (14.1%), Cyprinodontiformes (19 species, 6.4%), Clupeiformes (11 species, 3.7%), Salmoniformes (7 species, 2.36%), Acipenseriformes, Siluriformes, Mugiliformes and Cichliformes each with 6 species (2.02%), Perciformes (3 species, 1.01%), Scorpaeniformes (2 species, 0.67%) and 13 other orders each with one species (0.34%). The most diverse family is Cyprinidae with 123 reported species (42.7%) followed by Nemacheilidae with 46 species (16%), Gobiidae (42 reported species, 14.6%), Aphaniidae (15, reported species, 5.05%), Clupeidae (11 reported species, 3.70%), Cobitidae and Salmonidae (each with 7 reported species, 2.36%), Acipenseridae, Cichlidae and Mugilidae (each with 6 reported species, 2.02%), Poeciliidae (with 4 reported species, 1.35%), Percidae (with 3 reported species, 1.01%) and Gasterosteidae, Siluridae and Sisoridae (each with 2 reported species, 0.67%). Fifteen families have 1 reported species (0.34% each). Iran comprises 95 endemic species (32% of total reported species) in 7 families (Fig. 17.4): Cyprinidae (45, 47.37%), Nemacheilidae (29, 30.53%), Aphaniidae (12, 12.63%), Cichlidae (3, 3.16%), Cobitidae (4, 4.21%), Gobiidae (1, 1.05%) and Sisoridae 1 (1, 1.05%). Twenty-nine exotic species in 11 families are listed from Iranian basins. Cyprinidae with 10 species (34.50% of the total exotic species) is ranked first followed by the Salmonidae and Poeciliidae (each with 4 species (13.8%), Cichlidae (with 3 species, 10.3%), Mugilidae (2 species, 6.9%) and 6 families (Anguillidae, Gasterosteidae, Gobiidae, Heteropneustidae, Lepisosteidae and Serrasalmidae) each with only one species or 3.45%.

17.5 Artificial Reservoirs, Suitable Habitats for Exotic Species

As mentioned above, transformation upstream of a dam from a lotic ecosystem to an artificial lentic habitat is one of the most important ecological impacts of the dam. Since most of the natural aquatic habitats critical for freshwater fish survival have been negatively affected by human activities, the effectiveness of artificial structures in providing new and suitable habitats for fishes is interesting to investigate (Zeng et al. 2017). As well discussed by Zeng et al., biodiversity changes resulting from dam construction may increase susceptibility to species invasion in a number of ways (Chapin et al. 2000). The human introductions of alien/exotic fishes are deemed to be the second largest cause of species extinction after the habitat destruction (Vitousek et al. 1997; Xie et al. 2001; Zeng et al. 2017). An impressive amount of autochthonous species has been extirpated throughout the world following the introduction of alien/exotic fish species into their habitats (Lever 1996; Taylor et al. 1984). Exotic fishes can affect local ecosystems in various ways (Simon and Townsend 2003). They can prey on or compete with native fishes for food and other resources, which may lead to dramatic declines in the abundance of some local species, and even the extinction of some species in extreme circumstances (Dudgeon and Smith 2006; Raghavan et al. 2008; Zhang et al. 1997; Bhatt and Pandit, 2015). Introductions of alien/exotic fish species and extirpations of native fishes in freshwater may alter fish community dynamics in a given space or over a length of time, thus resulting with overall community homogeneity (McKinney and Lockwood 1999; Olden 2006) or heterogeneity (Taylor 2010). However, despite the negative effects of alien/exotic fish species, they have been introduced into many aquatic ecosystems in large numbers, both unintentionally and intentionally, for various reasons (Welcomme 1988). Aquaculture development was the main motivation for most of these introductions.

In Iran, the artificial reservoirs could be considered as stable ecosystems to conservation of native fishes. Despite this good potential to conserve endemic/native species which are mostly threatened by the drought in the country, unfortunately the artificial reservoirs became suitable habitats for exotic species (Fig. 17.6). Five exotic species containing, *Oncorhynchus mykiss*, *Ctenopharyngodon idella*, *Hypophthalmichthys molitrix*, *Hypophthalmichthys nobilis* and *Cyprinus carpio* transplanted by Iranian Fisheries Organization (Shilat) for aquaculture purposes throughout Iran. Additionally, *Alburnus hohenackeri*, *Carassius auratus*, *Carassius gibelio*, *Hemiculter leucisculus*, *Pseudorasbora parva*, *Rhinogobius lindbergi* and *Gambusia holbrooki* (Fig. 17.7) transferred with them inadvertently to the reservoirs (Jouladeh-Roudbar et al. 2015). Similar to rainbow trout in Iran, other piscivorous species such as largemouth bass (*Micropterus salmoides*) and bluegill (*Lepomis macrochirus*) have been extensively introduced into reservoirs worldwide, mainly for sport fishing (Murakami and Washitani 2002; Taylor and Weyl 2017). For example, *Oreochromis niloticus* was a typical alien/invasion species introduced to many aquatic ecosystems for aquaculture purposes, which established and severely

Fig. 17.6 A very small reservoir behind a small dam in Mashkid River basin, which could be an enough good habitat for the drought-affected native fishes in the region

threatened the survival of native fishes, as well as the structure and function of the freshwater ecosystem (Tan 2012). There is also a serious warning about the introduction of the invasive Tilapia into the Iranian water basins (Khaefi et al. 2014; Roozbhfar et al. 2014; Valikhani et al. 2016).

17.6 Dams Impact on Fish Diversity and Richness in Iran

As discussed, construction of large dams blocks the seasonal migration routes of diadromous fish species which contribute to the heterogeneity of biota (Freeman et al. 2002). The exploitation of these dams will cause environmental modifications, including flow variation, depth, dissolved oxygen and temperature fluctuation (Liu 1992; Salazar 2000). Furthermore, construction of dams will simplify the natural physical structure of the river with the loss of heterogeneous habitats such as lakes and riffles. All these changes can affect the fish diversity and richness in the river impacted by dam. Regarding Iranian ichthyofauna, dams' effects are in favour of large cyprinids and catfish (e.g. the genera *Capoeta*, *Luciobarbus* and *Silurus*, Fig. 17.8) those prefer the lentic ecosystems. These species reach to their maximum sizes and/or produce the predominant populations in the reservoirs. Normally, downstream of dams are not good habitats for any fish group, due to shallow and polluted water, destructed habitat, periodic drying, sedimentation by opening dam gates (Fig. 17.9) and easy access for local fisherman. In case of reservoirs in the

Fig. 17.7 Some of the most distributed exotic species in Iran: *Carassius gibelio* (above), *Pseudorasbora parva* (centre) and *Rhinogobius lindbergi* (below)

southern Caspian Sea, Hari River and the Tigris River basins, the predominant cyprinids are belonging to the native genera *Capoeta*, *Cyprinus* (native in the Caspian Sea basin, non-native in other basins), *Luciobarbus* and *Squalius*, whereas predominant exotics are within the genera *Ctenopharyngodon*, *Hypophthalmichthys*, *Carassius*, *Hemiculter* and *Pseudorasbora*. In these basins, it seems that the native genera *Alburnus*, *Alburnoides* and *Barbus* are negatively affected by dams in term of

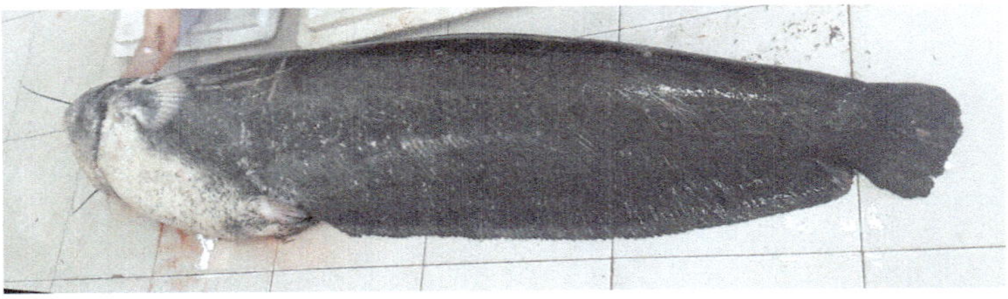

Fig. 17.8 *Silurus glanis*, about 200 cm in TL, caught by local fisherman from Aras Dam on Aras River, in the southwestern Caspian Sea basin

Fig. 17.9 Sedimentation and habitat destruction due to opening Manjil dam's gates in Sefid River, in the southern Caspian Sea basin

diversity and richness. Other small species such as small cyprinids, loaches and gobies need further investigation to assessing the dam impacts.

17.7 Dams Impacts on Migratory Fishes in Iran

As well reviewed by Ferguson et al. (2011), in large rivers, dams are obstacles to fishes that require movement to complete their life cycle (Larinier 2001; Winter and Van Densen 2001; Zigler et al. 2004; Belliard et al. 2018), which can cause decreasing population trends (Parrish et al. 1998; Jackson and Marmulla 2001; Valbo-Jørgensen et al. 2008). Dams can reduce species diversity and catch per unit effort of fisheries for short- and long-distance migratory species (Fernandes et al. 2009; Belliard et al. 2018). Providing passage (fishway) for migratory fish species at dams is critical for maintaining the viability of potamodromous and

Table 17.2 Some of the important migratory fish species of Iran

Diadromous				
Anadromous		Catadromous	Amphidromous	Potamodromous
Acipenser spp.	*Huso huso*	*Anguilla anguilla*	*Atherina caspia*	*Barbus* spp.
Alburnus chalcoides	*Rutilus kutum*		*Carcharhinus leucas*	*Capoeta* spp.
Alosa kessleri	*Salmo caspius*		*Ponticola gorlap*	*Chondrostoma* spp.
Caspiomyzon wagneri	*Tenualosa ilisha*			*Luciobarbus* spp.
Cyprinus carpio	*Vimba persa*			*Salmo trutta*

diadromous fish populations (Lucas and Baras 2001). The real anadromous species in Iran are mainly restricted to the Caspian Sea and Tigris basins, where the big rivers flow (Table 17.2). The genera *Acipenser, Alburnus, Alosa, Caspiomyzon, Cyprinus, Huso, Rutilus, Salmo* and *Vimba* in the southern Caspian Sea basin and the genus *Tenualosa* in the Tigris River basin are the main anadromous species, and the exotic *Anguila anguila* is the only catadromous species in Iran. There are available reports that show a critical decline in anadromous species stocks in recent decades. Just in one case, the richness of sturgeons (family Acipenseridae) decreased from 14,959 tons (in 1991) to 23 tons (in 2016), >98% in a 25-year period. Certainly, the damming (inaccessibility to spawning site and habitat destruction) along with overfishing is the main reasons for this decline. Despite this huge decline, the remaining wild sturgeon populations in the Caspian Sea especially in Iran are mainly recouped by hatchery-bred stocks, since dams have blocked natural spawning migration routes in the northern Caspian Sea (Anvarifar et al. 2011). Furthermore, Holcik (1999) stated that while dramatic declines in migratory fishes such as the members of families Petromyzontidae, Acipenseridae, Salmonidae and Clupeidae are well known in European rivers, other fishes, the so-called resident or nonmigratory fishes which perform in-stream movements, require attention.

17.8 Morphologic and Genetic Differentiation in Fishes Caused by Dams

River fragmentation due to damming probably changes the genetic diversity within populations and genetic differentiation between populations (Yamamoto et al. 2004). It has been also suggested that the fragmentation of river ecosystems can change the migration patterns among fish populations (Horváth and Municio 1998; Craig 2001; Jager et al. 2001), producing 'genetic stocks' that are reproductively isolated units and are genetically different from other stocks (Carvalho and Hauser 1994). Dakin et al. (2007) showed that the Morgan-Falls dam on Chattahoochee

River has reduced the genetic diversity in the population of shoal bass, *Micropterus cataractae*, especially in the upstream population compared to the downstream population, and the morphological differences between the two populations were highly significant. Yamamoto et al. (2006, 2016) also reported similar results in white-spotted charr, *Salvelinus leucomaenis*, populations 20 years after the dam construction. Morphologically and genetically differentiations in cyprinids populations in different Iranian water basins have been reported by various studied impacts (Mousavi-Sabet and Anvarifar 2013; Khataminejad et al. 2013; Kohestan-Eskandari et al. 2013, 2014; Paknejad et al. 2014; Vatandoust et al. 2014a, b, 2015). Anvarifar et al. (2013) evaluated the effects of the Shahid-Rajaee dam, 14 years after its construction on Tajan River in the southern Caspian Sea basin, on morphometric and genetic divergence between up- and downstream populations of *Capoeta capoeta gracilis*. They stated that there are 2 morphologically and genetically distinct populations of *C. c. gracilis*, probably due to limited downstream dispersal and elimination of upstream migration. Also, Azizi et al. (2015) showed similar morphological differences in *Alburnoides eichwaldii* (now *Alburnoides tabarestanensis*) populations in the same river caused by the same dam. Heidari et al. (2013, 2014) examined the hypothesis and morphological differences among the three populations of *Capoeta gracilis* in up- and downstream of Manjil and Tarik dams in Sefid River in the southern Caspian Sea basin. Their obtained results indicated the body shape differences in the populations of *Capoeta gracilis* in Sefid River across the Manjil and Tarik dams probably due to the dam construction, showing anthropogenic transformation of rivers influences body shape in an aquatic organism. The long-term reproductive isolation of fish populations and interbreeding can lead to morphometric variations between populations, and these morphometric variations can provide a basis for population differentiation.

17.9 Dam Impacts on Fish Population Structures and Biological Characteristics

Dam impacts on biological indices of fishes in up- and downstream are barely known in Iran. However, as mentioned above, the change in ecosystem types from lotic to lentic is a significant effective factor which affects biological indices. One of the most studied biological factors that are significantly related to the habitat conditions is the length-weight relationship (LWR). The LWRs of freshwater fishes are useful in determining weight and biomass when only length measurements are available and are required for fishery management and conservation (Mousavi-Sabet et al. 2017). When we examined growth indices in cyprinids (*Capoeta razii*) caught from up- and downstream of Manjil and Tarik dams (on Sefid River, in the southern Caspian Sea basin), it was clear that the studied growth indices including length-weight (LWR) and length-length (LLRs) relationships were significantly affected by dams (Mousavi-Sabet et al. 2018). As our results indicate, despite positive allometric growth in the population caught from dam reservoir, other studied populations have negative allometric growth. Also, the averages of recorded length and weight in dam lake population were up to six times more than the ranges in downstream population. It was clear that the caught specimens from dam reservoir were larger in size in comparison with those live in downstream in the same age; therefore, the number of eggs and other reproductive indices were significantly different comparing downstream population. Further studies are suggested to evaluate dams impacts on fishes' biological factors.

17.10 What Can We do in Regard to the Irresponsible Damming?

As a conclusion, it is accepted that the construction of dams is necessary for socio-economic development in an arid country like Iran, but the absence of any environmental considerations in relation with ichthyofauna cannot be ignored. We believe despite various benefits, dam construction has resulted in many negative environmental impacts particularly on fishes. There are more than 300 fish species (about 100 endemics) in our country, which are mostly affected by damming and need urgent attention. It is concluded that dams mainly could change aquatic ecosystems, limited downstream dispersal and blocking migration routes, periodic drying, affect fishes in term of morphology, genetic, growth and reproductive indices, etc. Also, non-native fishes should be considered as an important threat for native fish communities in reservoirs. However, further studies are suggested to evaluate dams impacts on fishes' in Iran. And finally, conservation programmes favouring native/endemic fishes in the constructed dams are strongly recommended, and also, fishways should be mandatory in the under construction/planned dams.

References

Anvarifar H, Khyabani A, Farahmand H, Vatandoust S, Anvarifar H, Jahageerdar S (2011) Detection of morphometric differentiation between isolated up- and downstream populations of Siah Mahi (*Capoeta capoeta gracilis*) (Pisces: Cyprinidae) in the Tajan River (Iran). Hydrobiologia 673:41–52

AnvariFar H, Farahmand H, Silva DM, Bastos RP, AnvariFar H (2013) Fourteen years after the Shahid-Rajaei dam construction: an evaluation of the morphometric and genetic differentiation between isolated up- and downstream populations of *Capoeta Capoeta gracilis* (Siah Mahi, Pisces: Cyprinidae) in the Tajan River (Iran). Genet Mol Res 12:3465–3478

Azizi F, Anvarifar H, Mousavi-Sabet H (2015) Morphological differentiation between isolated populations of Caspian spirlin (*Alburnoides eichwaldii*) (Pisces: Cyprinidae) affected by dam. Int J Anim Biol 1(2):28–37

Belliard J, Beslagic S, Delaigue O, Tales E (2018) Reconstructing long-term trajectories of fish assemblages using historical data: the Seine River basin (France) during the last two centuries. Environ Sci Pollut Res 25:23430–23450. https://doi.org/10.1007/s11356-016-7095-1

Bhatt JP, Pandit MK (2015) Endangered Golden mahseer *Tor putitora* Hamilton: a review of natural history. Rev Fish Biol Fish 26(1):25–38. https://doi.org/10.1007/s11160-015-9409-7
Carvalho GR, Hauser L (1994) Molecular genetics and the stock concept in fisheries. Rev Fish Biol Fish 4:326–350
Chapin FS, Zavaleta ES, Eviner VT, Naylor R, Vitousek PM, Reynolds HL et al (2000) Consequences of changing biodiversity. Nature 405(6783):234–242
Coad BW (1998) Systematic biodiversity in the freshwater fishes of Iran. Italian J Zool 65 (Supplement):101–108
Craig JF (2001) Large dams and freshwater fish biodiversity. World commission on dams, prepared for thematic review II.1: dams, ecosystem functions and environmental restoration, p 59
Dakin EE, Porter BA, Freeman BJ, Long JM (2007) Genetic integrity of an isolated population of shoal bass (*Micropterus cataractae*) in the upper Chattahoochee River basin. Natural resource technical report NPS/NRWRD/NRTR—2007/366. Fort Collins, CO., National Park Service. Water Resources Division
Dauble DD, Geist AR (2000) Comparison of mainstream spawning habitats for two populations of fall Chinook salmon in the Columbia River basin. Regul Rivers Res Manage 1634:53–61
Dauble DD, Watson DG (1997) Status of fall Chinook salmon populations in the mid-Columbia River, 1948–1992. N Am J Fish Manage 17:283–300
Dauble DD, Hanrahan TP, Geist DR, Parsley MJ (2003) Impacts of the Columbia River hydroelectric system on main-stem habitats of fall Chinook salmon. N Am J Fish Manage 23:641–659
De Merona B, Albert P (1999) Ecological monitoring of fish assemblages downstream of a hydroelectric dam in French Guiana (South America). Regul Rivers Res Manage 15:339–351
Dudgeon D (2000) The ecology of tropical Asian rivers and streams in relation to biodiversity conservation. Annu Rev Ecol Syst 31:239–263
Dudgeon D, Smith REW (2006) Exotic species, fisheries and conservation of freshwater biodiversity in tropical Asia: the case of the Sepik River, Papua New Guinea. Aquat Conserv Mar Freshwat Ecosyst 16:203–215
Dynesis M, Nilsson C (1994) Fragmentations and flow regulation of river systems in the northern third of the world. Science 266:753–762
Esmaeili HR, Masoudi M, Amini CM, Esmaeili AH, Zarei F, Ebrahimi M (2017) Invasion of the Neotropical and Nearctic fishes to Iran. Fish Taxa 2(3):126–133
Esmaeili HR, Sayyadzadeh G, Eagderi S, Abbasi K (2018) Checklist of freshwater fishes of Iran. Fish Taxa 3(3):1–95
Esmaeili HR, Sayyadzadeh G, Zarei F, Kafaei S, Coad BW (2020) Phylogeographic pattern and population structure of the Persian stone loach, *Oxynoemacheilus persa* (Heckel 1847) (family: Nemacheilidae) in southern Iran with implications for conservation. Environ Biol Fish 103:77–88. https://doi.org/10.1007/s10641-019-00934-y
Ferguson JW, Healey M, Dugan P, Barlow C (2011) Potential effects of dams on migratory fish in the Mekong River: lessons from Salmon in the Fraser and Columbia Rivers. Environ Manag 47:141–159. https://doi.org/10.1007/s00267-010-9563-6
Fernandes R, Agostinho A, Ferreira E, Pavanelli C, Suzuki H, Lima D, Gomes L (2009) Effects of the hydrological regime on the ichthyofauna of riverine environments of the upper Parana' river floodplain. Brazilian J Biol 69(2, Suppl):669–680
Freeman R, Bowerman W, Grubb T, Bath A, Dawson G, Ennis K, Giesy J (2002) Opening rivers to Trojan fish: the ecological dilemma of dam removal in the Great Lakes. Conserv Pract 3:35–40
Fukushima M (2005) The dam-related decline of freshwater fish diversity: analysis of the data collected from Hokkaido during the last half century. Ecol Soc Jpn 55:349–357. (in Japanese)
Garousi V, Najafi A, Samadi A, Rasouli K, Khanaliloo B (2013) Environmental crisis in Lake Urmia, Iran: a systematic review of causes, negative consequences and possible solutions. Proceedings of the 6th International Perspective on Water Resources & the Environment (IPWE) Conference at: Izmir, Turkey. https://doi.org/10.13140/RG.2.1.4737.0088
Gido KB, Schaefer JF, Pigg J (2004) Patterns of fish invasions in the Great Plains of North America. Biol Conserv 118:121–131

Han M, Fukushima M, Kameyama S, Fukushima T, Matsushita B (2008) How do dams affect freshwater fish distributions in Japan? Statistical analysis of native and nonnative species with various life histories. Ecol Res 23:735. https://doi.org/10.1007/s11284-007-0432-6

Heidari A, Mousavi-Sabet H, Khoshkholgh M, Esmaeili HR, Eagderi S (2013) The impact of Manjil and Tarik dams (Sefidroud River, southern Caspian Sea basin) on morphological traits of Siah Mahi *Capoeta gracilis* (Pisces: Cyprinidae). Int J Aquat Biol 1(4):195–201

Heidari A, Khoshkholgh M, Mousavi-Sabet H (2014) Tracing the effects of Sefidrud dams on *Capoeta gracilis* (Cyprinidae) populations using truss distances in southern Caspian Sea basin. Iranian J Ichthyol 1(2):106–113

Heydari M, Othman F, Noori M (2013) A review of the environmental impact of large dams in Iran. Int J Adv Civil Struct Environ Eng 1(1):1–4

Holcik J (1999) The impact of stream regulations upon the fish Fauna and Measures to prevent it, p. 13. Abstract in: Stomboudi, M. T., M. Kottelat & R. Barbieri (eds), Workshop on Mediterranean stream fish ecology and conservation, Rhodes, Hellas, 1–3 November 1999

Holmquist JG, Schmidt-Gengenbach JM, Yoshioka BB (1998) High dams and marine-freshwater linkage: effects on native and introduced fauna in the Caribbean. Conserv Biol 12:621–630

Horváth E, Municio MAT (1998) Impacts of dams on fish fauna. 2nd Int. PhD symposium in civil engineering, Budapest. 1-8

IME (Iran Minister of Energy) (2006) The eight-year performance report of the Ministry of Energy during 1998–2006. p. 97. [in Persian]

Iran-Daily - Domestic Economy (2007) 88 major dams under construction. Available at: https://web.archive.org/web/20070702111417/http://iran-daily.com/1386/2877/html/economy.htm#s237918

Jackson D, Marmulla G (2001) The influence of dams on river fisheries. In: Marmulla, G (ed) Dams, fish and fisheries. Opportunities, challenges and conflict resolution. FAO fisheries technical paper no. 419, Rome, pp 1–44

Jager HI, Chandler JA, Lepla KB, Winkle WV (2001) A theoretical study of river fragmentation by dams and its effect on white sturgeon populations. Environ Biol Fish 60:347–361

Jouladeh-Roudbar A, Vatandoust S, Eagderi S, Jafari-Kenari S, Mousavi-Sabet H (2015) Freshwater fishes of Iran; an updated checklist. AACL Bioflux 8(6):855–909

Joy MK, Death RG (2001) Control of freshwater fish and crayfish community structure in Taranaki, New Zealand: dams, diadromy or habitat structure? Freshw Biol 46:417–429

Kareiva P, Marvier M, McClure M (2000) Recovery and management options for spring/summer Chinook salmon in the Columbia River basin. Science 290:977–979

Katano O, Nakamura T, Abe S, Yamamoto S, Baba Y (2006) Comparison of fish communities between above- and below dam sections of small streams: barrier effect to diadromous fishes. J Fish Biol 68:767–770

Khaefi R, Esmaeili HR, Zareian H, Babaei S (2014) The first record of the redbelly tilapia, *Tilapia zillii* Gervais, 1848 in freshwaters of Iran. Turkish J Zool 38:96–98

Khataminejad S, Mousavi-Sabet H, Sattari M, Vatandoust S, Eagderi S (2013) A comparative study on body shape of the genus *Alburnus* (Rafinesque, 1820) in Iran, using geometric morphometric analysis. Caspian J Environ Sci 11:205–215

Kohestan-Eskandari S, AnvariFar H, Mousavi-Sabet H (2013) Detection of morphometric differentiation of *Liza aurata* (Pisces: Mugilidae) in southeastern of the Caspian Sea, Iran. Our Nature 11(2):126–137

Kohestan-Eskandari S, AnvariFar H, Mousavi-Sabet H, Yousefi M, Khanzade M (2014) A morphology-based hypothesis for homeward migration success and population differentiation in the anadromous kutum *Rutilus kutum* (Pisces: Cyprinidae) along the southern Caspian Sea, Iran. Folia Zool 63(3):151–160

Larinier M (2001) Environmental issues, dams and fish migration. In: Marmulla, G (ed) Dams, fish and fisheries. Opportunities, challenges and conflict resolution. FAO fisheries technical paper no. 419, Rome, pp 45–90

Lever C (1996) Naturalized fishes of the world. Academic Press, London, p 408

Liu J (1992) Fish resources of the Yangtze River basin and the tactics for their conservation. Res Environ Yangtze Basin 1:17–23

Lucas M, Baras E (2001) Migration of freshwater fishes. Blackwell Science, Oxford, p 420

Maezono Y, Miyashita T (2003) Community-level impacts induced by introduced largemouth bass and bluegill in farm ponds in Japan. Biol Conserv 109:111–121

Manouchehri GR, Mahmoodian SA (2002) Environmental impacts of dams constructed in Iran. Int J Water Resour Dev 18:179–182. https://doi.org/10.1080/07900620220121738

Marchetti MP, Lockwood JL, Light T (2006) Effects of urbanization on California's fish diversity: differentiation, homogenization and the influence of spatial scale. Biol Conserv 127:310–318

Martinez PJ, Chart TE, Trammel MA (1994) Fish species composition before and after construction of a main stem reservoir on the White River, Colorado. Environ Biol Fish 40:227–239

McKinney ML (2001) Effects of human population, area, and time on non-native plant and fish diversity in the United States. Biol Conserv 100:243–252

McKinney ML, Lockwood JL (1999) Biotic homogenization: a few winners replacing many losers in the next mass extinction. Trends Ecol Evol 14:450–453

Morita K, Yamamoto S (2001) Effects of habitat fragmentation by damming on the persistence of stream-dwelling char populations. Conserv Biol 16:1318–1323

Morita K, Yokota A (2002) Population viability of stream-resident salmonids after habitat fragmentation: a case study with white spotted char (*Salvelinus leucomaenis*) by an individual based model. Ecol Model 155:85–94

Mousavi-Sabet H (2018) Range extension of an exotic sailfin molly *Poecilia latipinna* (Lesueur, 1821) in Iran. Poec Res 8(1):18–23

Mousavi-Sabet H, Anvarifar H (2013) Landmark-based morphometric variation between *Cobitis keyvani* and *Cobitis faridpaki* (Pisces: Cobitidae), with new habitat for *C. faridpaki* in the southern Caspian Sea basin. Folia Zool 62(3):167–175

Mousavi-Sabet H, Eagderi S (2014) First record of *Poecilia reticulata* Peters, 1859 (Cyprinodontiformes, Poeciliidae) from natural freshwaters of Iran. Poeciliid Res 4(1):19–23

Mousavi-Sabet H, Eagderi S (2016) First record of the convict cichlid, *Amatitlania nigrofasciata* (Günther, 1867) (Teleostei: Cichlidae) from the Namak Lake basin, Iran. Iranian J Ichthyol 3 (1):25–30

Mousavi-Sabet H, Heidari A, Vatandoust S (2017) Length–weight and length–length relationships for nine species of the genus *Paracobitis* (Nemacheiliidae) in Iran. J Appl Ichthyol 33:617–619. https://doi.org/10.1111/jai.13278

Mousavi-Sabet H, Heidari A, Salehi M (2018) How did dams affect length-weight and length-length relationships of *Capoeta razii* (Cyprinidae) in Sefid River, the southern Caspian Sea basin? Int J Aquat Biol 6(5):274–280. https://doi.org/10.22034/ijab.v6i5.534

Murakami O, Washitani I (2002) Handbook of alien species in Japan. Chijin-shokan, Tokyo

Nalbant TT, Bianco PG (1998) The loaches of Iran and adjacent regions with description of six new species (Cobitoidea). Italian J Zool 65(Supplement):109–123

Olden J (2006) Biotic homogenization: a new research agenda for conservation biogeography. J Biogeogr 33:2027–2039

Paknejad S, Heidari A, Mousavi-Sabet H (2014) Morphological variation of shad fish *Alosa brashnicowi* (Teleostei, Clupeidae) populations along the southern Caspian Sea coasts, using a truss system. Int Aquat Res 2(6):330–336

Paragamian VL (2002) Changes in the species composition of the fish community in a reach of the Kootenai River, Idaho, after construction of Libby dam. J Freshw Ecol 17:375–383

Park YS, Chang J, Lek S, Cao W, Brosse S (2003) Conservation strategies for endemic fish species threatened by the three gorges dam. Conserv Biol 17:1748–1758

Parrish D, Behnke R, Gephard S, McCormick S, Reeves G (1998) Why aren't there more Atlantic salmon (*Salmo salar*)? Can J Fish Aquat Sci 55(Suppl. 1):281–287

Poff NL, Hart DD (2002) How dams vary and why it matters for the emerging science of dam removal. Bioscience 52:659–668

Preece RM, Jones HA (2002) The effect of Keepit dam on the temperature regime of the Namoi River, Australia. River Res Appl 18:397–414

Press TV (2010) Iran's environment in danger. Available at: https://web.archive.org/web/20120312074439/http://presstv.com/program/152328.html

Raghavan R, Prasad G, Anvar-Ali PH, Pereira B (2008) Exotic fish species in a global biodiversity hot spot: observations from river Chalakudy, part of Western Ghats, Kerala, India. Biol Invasions 10:37–40

Roozbhfar R, Dehestani-Esfandabadi M, Roozbehfar S (2014) First record of the redbelly tilapia, *Tilapia zillii* Gervais, 1848 in Iran. J Appl Ichthyol 30(5):1045–1046

Salazar JG (2000) Damming the child of the ocean: the three gorges project. J Environ Dev 9:160–174

Saunders DA, Hobbs RJ (1991) Biological consequences of ecosystem fragmentation: a review. Conserv Biol 5:18–32

Simon KS, Townsend CR (2003) Impacts of freshwater invaders at different levels of ecological organisation, with emphasis on salmonids and ecosystem consequences. Freshw Biol 48:982–994

Tan ZY (2012) The study of Tilapia industry development situation in Guangxi. PhD Thesis, Guangxi University, Nanning, China

Taylor EB (2010) Changes in taxonomy and species distributions and their influence on estimates of faunal homogenization and differentiation in freshwater fishes. Divers Distrib 16:676–689

Taylor GC, Weyl OLF (2017) Age, growth and reproduction of non-native largemouth bass *Micropterus salmoides* (Lacépède, 1802) populations in two temperate African impoundments. J Appl Ichthyol 33:767–775. https://doi.org/10.1111/jai.13353

Taylor JN, Courtenay WR, McCann JA (1984) Known impacts of exotic fishes in the continental United States. Distribution, biology and management of exotic fishes. John Hopkins University Press, Baltimore, MD, pp 322–373

Valbo-Jørgensen J, Marmulla G, Welcomme RL (2008) Migratory fish stocks in Transboundary basins — implications for governance, management and research. In: Lagutov V (ed) Rescue of Sturgeon Species in the Ural River basin. NATO science for peace and security series C: environmental security. Springer, Dordrecht. https://doi.org/10.1007/978-1-4020-

Valikhani H, Abdoli A, Kiabi BH, Nejat F (2016) First record and distribution of the blue tilapia, *Oreochromis aureus* (Steindachner, 1864) (Perciformes: Cichlidae) in inland waters of Iran. Iran J Ichthyol 3(1):19–24

Vatandoust S, Abdoli A, Anvarifar H, Mousavi-Sabet H (2014a) Morphometric and meristic characteristics and morphological differentiation among five populations of Brown trout *Salmo trutta fario* (Pisces: Salmonidae) along the southern Caspian Sea basin. Int J Zool Res 3(2):56–65

Vatandoust S, Nejati M, Anvarifar H, Mousavi-Sabet H (2014b) Morphological differentiation of *Vimba persa* (Pisces: Cyprinidae) along the southern Caspian Sea Basin, Iran. Int J Zool Res 3 (2):94–102

Vatandoust S, Mousavi-Sabet H, Razeghi-Mansour M, AnvariFar H, Heidari A (2015) Morphometric variation of the endangered Caspian lamprey, *Caspiomyzon wagneri* (Pisces: Petromyzontidae), from migrating stocks of two rivers along the southern Caspian Sea. Zool Stud 56:1–9

Vitousek PM, Mooney HA, Lubchenco J, Melillo JM (1997) Human domination of Earth's ecosystems. Science 277:494–499

Welcomme RL (1988) International introductions of inland aquatic species. FAO, UN, Rome

Winter H, Van Densen W (2001) Assessing the opportunities for upstream migration of non-salmonid fishes in the weir regulated river Vecht. Fish Manag Ecol 8:513–532

WRM (Iran Water Resources Management Company) (2019) Dams of Iran. Available at: http://daminfo.wrm.ir

Xie Y, Li Z, Gregg WP, Li D (2001) Invasive species in China—an overview. Biodivers Conserv 10:1317–1341

Yamamoto S, Morita K, Koizumi I, Maekawa K (2004) Genetic differentiation of white-spotted charr (*Salvelinus leucomaenis*) populations after habitat fragmentation: spatial–temporal changes in gene frequencies. Conserv Genet 5:529–538

Yamamoto S, Maekawa K, Tamate T, Koizumi I, Hasegawa K, Kubota H (2006) Genetic evaluation of translocation in artificially isolated populations of white-spotted charr (*Salvelinus leucomaenis*). Fish Res 78:352–358

Yamamoto S, Kubota H, Hasegawa K, Nakamura T (2016) Census and effective population sizes of white-spotted charr (*Salvelinus leucomaenis*) in a fragmented landscape. Ecol Freshw Fish 25:612–621. https://doi.org/10.1111/eff.12239

Zeng L, Zhou L, Guo D-L, Fu D-H, Xu P, Zeng S, Tang Q-D, Chen A-L, Chen F-Q, Luo Y, Li G-F (2017) Ecological effects of dams, alien fish, and physiochemical environmental factors on homogeneity/heterogeneity of fish community in four tributaries of the Pearl River in China. Ecol Evol 7:3904–3915

Zhang G, Cao W, Chen Y (1997) Effects of fish stocking on lake ecosystems in China. Acta Hydrobiol Sin 21:271–280

Zigler S, Dewey M, Knights B, Runstrom A, Steingraeber M (2004) Hydrologic and hydraulic factors affecting passage of paddlefish through dams in the upper Mississippi River. Trans Am Fish Soc 133:160–172

Part III
The Biotic Aspects of the Tigris–Euphrates River System

Chapter 18
Preliminary Review of the Aquatic Biodiversity in Al-Kahla River, Missan Province, Iraq

Laith A. Jawad

Abstract Al-Kahla River is situated in the southeastern part of the Euphrates-Tigris River Basin near the province of Missan. This river is one of two branches of the Tigris River, which also called Gate of Al-Kahla, respectively. The flows in the Nahr happened between October and January and the highest flows occurred between February and July. Towards the southern part of this river, the Nahr Al-Kahla River divides into two smaller rivers—the Nahr az Zubayr (also referred to as Al-Zubair or Nahr Abu Kilaf) and the Nahr al Akrah (also known as Um al Toos or Hischa).

Aquatic biodiversity in Iraq is not well investigated and needs years to study and surveys in order to put the baseline of the biological components of the rivers. Apart from the marsh areas which their biodiversity has a certain degree of coverage, the biodiversity of the branches of the Tigris River at its lower reaches has never being investigated or surveyed properly. Therefore, no integrated publication on any plant or animal groups is available of these river's branches.

The present chapter puts in the hand of the readers a preliminary floristic and faunistic survey of one of the rivers, Al-Kahla River, at the lower reaches of Tigris River in Iraq. This chapter will be the baseline study for any future survey or investigation about in the area.

18.1 Introduction

Iraq has a unique geographical position at the heart of the Middle East, between latitudes 29°27′ and 37°23’ North and longitudes 38°42′ and 48°23′ East. From the east, it is bounded by Islamic Republic of Iran, Turkey in the north, Syria in the northwest, Jordan and Saudi Arabia in the west and southwest, and Kuwait and the Arabian Gulf in the south.

L. A. Jawad (✉)
School of Environmental and Animal Sciences, Unitec Institute of Technology, Auckland, New Zealand

L. A. Jawad (ed.), *Tigris and Euphrates Rivers: Their Environment from Headwaters to Mouth*, Aquatic Ecology Series 11,
https://doi.org/10.1007/978-3-030-57570-0_18

Running from the utmost north to its southern borders, two great rivers, the Tigris and the Euphrates, both of which rise in the eastern mountains of Turkey and enter Iraq along its northwestern borders. After flowing for some 1200 km through Iraq, Tigris and Euphrates converge at Qarmat Ali, just north of Basrah City to form tidal Shatt al-Arab River, which flows some 110 km to enter the Arabian Gulf.

Unlike Euphrates River, Tigris River receives four main tributaries, the Khabour, Great Zab, Little Zab and Diyala. The latter river rises in the mountains of eastern Turkey and northwestern Iran and flows in southwestern direction until they meet the Tigris River. A fifth tributary, Al-Authaim, which is considered a seasonal river rising in the highlands of northern Iraq, also flows into Tigris River.

Maysan province lies in the southeastern part of Iraq, within the Precipitation Valley. The total area of the province is 16,072 square kilometres (km^2), which represents approximately 4% of the total land area of Iraq. The geographical nature of the province is divided between land and water, where the marshlands form more than 40% of the total area extending between its southern and southeastern parts as well as the southwestern part.

At Al Amarah, the Tigris River splits into three channels. These are (1) Main Tigris River that flows towards Basrah in the south; (2) Musharrah River that flows towards the east; and (3) Nahr Al-Kahla River (also sometimes referred to as Chahala River) that flows towards the southeast (MacDonald 2010).

Further southeast, Al-Kahla River splits into two smaller rivers, Al-Zubair River and Akrah River. The width of Al-Zubair River varies between 50 and 150 m while the width of Akrah River is around 65 m (MacDonald 2010). The third main watercourse is the Al-A'waj Canal, a relatively large off-take from Al-Zubair River. Several smaller canals and ditches are found within the main Al-Kahla River that fed predominantly fed by the larger rivers. Most of these canals and ditches have been used for several human usages such as irrigation and fishing both past and present.

In the southern marshes of the River Tigris in general and in Al-Kahla River district are in particular, some permanent and seasonal ponds, lakes and marshes are present. Hawr Al Hakkah and the Umm Al-Niaj are the largest lakes present in the area. The latter is also known as Hawizeh Marshes. Al-Hawizeh marsh groups are fed by the Musharrah and Al-Kahla Rivers. The aquatic environment of these three branches of the Tigris River is similar to that of the marsh area and more specifically to Al-Huwaiza Marsh (Ali and Yousif 2010).

Freshwater makes up only 0.01% of the World's water and covers only 0.8% of the Earth's surface, yet this tiny fraction of global water supports at least 100,000 species out of approximately 1.75 million—almost 6%. Not surprisingly, considering their landscape position and value as a natural resource, freshwaters are experiencing declines in biodiversity far greater than those in the most affected terrestrial ecosystems.

Freshwater ecosystems may well be the most endangered ecosystems in the world. Declines in biodiversity are far greater in freshwaters than in the most affected terrestrial ecosystems (Sala et al. 2000). What makes freshwater habitats and the biodiversity that they support especially vulnerable to human activities and environmental change? The main reason is the disproportionate richness of inland waters as a habitat for plants and animals. Over 10,000 fish species live in freshwater (Lundberg et al. 2000), approximately 40% of global fish diversity and one-quarter of global vertebrate diversity. When amphibians, aquatic reptiles and mammals are added to this freshwater fish total, it becomes clear that as much as one-third of all vertebrate species are confined to freshwater (Gleick 1996).

Various anthropogenic activities threaten the biodiversity of rivers and their associated wetlands at global and regional scales and may well impair or significantly reduce the ecosystem services those rivers and wetlands provide (Dudgeon 1999). These threats can be placed in four categories: flow alteration or regulation (including impoundment by dams, water extraction for irrigation, and so on); pollution; drainage-basin alteration (especially deforestation); and overharvesting (mainly of fishes). Researchers, increasingly aware of the potential impact of hydrologic alterations on biodiversity, have begun to focus on the environmental and social consequences of large-scale hydroelectric development (Rosenberg et al. 1997).

Habitat degradation is brought about by an array of interacting factors. It may involve direct effects on the aquatic environment or indirect impacts that result from changes within the drainage basin. Flow modifications due to transferring huge amount of water are ubiquitous in running waters (Nilsson et al. 2005).

Water regimes influence aquatic biodiversity via several, inter-related mechanisms operating over a range of spatial and temporal scales. The maintenance of natural variability in flows and water levels is therefore essential to underpin conservation strategies for freshwater biodiversity and habitats. This requires establishing a hydrological regime that mimics natural variability in flows and water levels rather than focusing on minimum levels only.

Biodiversity of Al-Kahla River is poorly studied, but it is more or less similar to that of the neighbouring rivers and marshes found in the lower reaches of Tigris River. There are several threatened or nearly threatened aquatic species that need to be taking into consideration before any hydrological activity starts in the area.

The idea of transferring water from Al-Kahla River for oil field requirements will have a dramatic effect on the biota of the river. Such changes will start on the level of lower taxonomic groups and find its way up the pyramid of taxonomic hierarchy reaching the top-level organism in the ecosystem. In another words, the total biodiversity of the river will be changed if it is not damaged completely. To explain such effects on the riverine environment of Al-Kahla River, it is important to show the effect on each taxonomic group separately.

The purpose of this study is to give the status of the existing aquatic biodiversity of Al-Kahla River which in turn supports any baseline study that might set to study the environment of the area in general. In giving the status of the river in question, it will (1) provide a rapid overview and preliminary study to the aquatic biodiversity of Al-Kahla River; (2) list the species of the aquatic flora and fauna that recorded from Al-Kahla River environment; (3) highlight the role and significance of each taxonomic group in the riverine environment in question; (4) point out the possible effect of the project activity on the status of the biodiversity of Al-Kahla River in general.

18.2 The Biodiversity of Aquatic Micro-organisms of Al-Kahla River

18.2.1 Fungi Groups

Aquatic fungi or what usually known as water moulds represent a unique group that is well adapted the fresh aquatic environments (Johnson and Sparrow 1962; Dick 1973). There are two groups of aquatic fungi, freshwater which the majority are belonging to the class Oomycetes and marine fungi which are adapted to live in saline waters and they mainly are members of the groups Ascomycetes and Basidiomycetes (Kohlmeyer and Kohlmeyer 1979).

Aquatic fungi showed wide distribution in different water bodies such as lakes, rivers, pond, estuaries and seas worldwide (Johnson et al. 2004). They play a major role in biodegradation in aquatic habitats and some species causing fish diseases and others causing plant diseases (Alexopoulos et al. 1996).

Rivers and marshes provide suitable environment for the fungi to grow. In the aquatic macrophytes, *Cyperus*, *Phragmites* and *Typha* serve as good substrates for fungi (Wong et al. 1998). In addition, aquatic mycobiota inhabits sediments of rivers and different freshwater bodies (Ulfig et al. 1998).

The list of aquatic fungi that recorded from the Al-kahla River is given in Table 18.1. The total number of species reported to present in this river is 36 belonging to 10 genera contained in 5 families. The largest family is Saprolegniaceae with 28 species belonging to 4 genera.

Dictyuchus monosporus and *Saprolegnia ferax* among other species were the most abundant in aquatic habitat in the lower reaches of Tigris River (Mahmood et al. 2007). Seasonal variation in the occurrence of some species like members of the genera *Calyptralegnia* and *Saprolegnia* is evident in the southern Iraqi rivers (Rattan et al. 1980; Yousif and Harry 2008). The maximum occurrence of these fungi was in winter and minimum in summer. This fluctuation pattern was attributed to the temperature as a major factor affecting the growth and reproduction of aquatic fungi.

The generic composition of the aquatic fungi in the sediment of the rivers and marshes in the south of Iraq is more or less similar to that of Shatt al-Arab River and North-west Arabian Gulf (Abdullah et al. 2000), that of river sediments in Nagasaki, Japan (Ueda 1980), and that of the Neuse-Newport estuarine system in coastal North Carolina, USA (Bourt and Johnson 1962). There is also similarity with the fungal community inhabiting mud in the tidal zone of Khor Al-Zubair canal, Southern Iraq (Abdullah et al. 2007).

Table 18.2 shows the total number of species of aquatic fungi reported to present in the sediment of Al-Kahla River is 49 belonging to 27 genera contained in 13 families. The largest family is Trichocomaceae with 16 species belonging to 7 genera.

The aquatic fungi found in sediments listed in Table 18.3 can be divided into four groups according to their occurrence; high-frequency species such as *Aspergillus*

Table 18.1 List of aquatic fungi presents in River Al-Kahla

Scientific name	Family
Jahnula bipileata	Aliquandostipitaceae
Aniptodera fusiformis	Halosphaeriaceae
Nais inornata	Halosphaeriaceae
Clavatospora bulbosa	Hypocreomycetidae
Achlya ambisexualis	Saprolegniaceae
Achlya debaryana	Saprolegniaceae
Achlya dubia	Saprolegniaceae
Achlya klebsiana	Saprolegniaceae
Achlya oblongata var. gigantica	Saprolegniaceae
Achlya polyandra	Saprolegniaceae
Achlya prolifera	Saprolegniaceae
Achlya proliferoides	Saprolegniaceae
Achlya racemosa	Saprolegniaceae
Achlya treleaseana	Saprolegniaceae
Achlya aff. caroliniana	Saprolegniaceae
Dictyuchus achlyoides	Saprolegniaceae
Dictyuchus carpophorous	Saprolegniaceae
Dictyuchus magnusii	Saprolegniaceae
Dictyuchus monosporus	Saprolegniaceae
Dictyuchus pseudodictyon	Saprolegniaceae
Dictyuchus pseudoachlyoides	Saprolegniaceae
Calyptralegnia ripariensis	Saprolegniaceae
Saprolegnia anisospora	Saprolegniaceae
Saprolegnia australis	Saprolegniaceae
Saprolegnia diclina	Saprolegniaceae
Saprolegnia delica	Saprolegniaceae
*Saprolegnia f*erax	Saprolegniaceae
Saprolegnia hypogyna	Saprolegniaceae
Saprolegnia litoralis	Saprolegniaceae
Saprolegnia parasitica	Saprolegniaceae
Saprolegnia terrestris	Saprolegniaceae
Saprolegnia uliginosa	Saprolegniaceae
Corollospora pseudopulchella	Sordariomycetes
Corollospora pulchella	Sordariomycetes
Olpidium bornovanus	Spizellomycetaceae
Olpidium brassicae	Spizellomycetaceae
Total number of species	36
Total number of genera	10
Total number of families	5

Table 18.2 List of aquatic fungi found in the river sediments

Scientific name	Family
Chaetomium atrobrunneum	Chaetomiaceae
Chaetomium globosum	Chaetomiaceae
Humicola grisea	Chaetomiaceae
Thielavia terricola	Chaetomiaceae
Phoma sp.	Gnomoniaceae
Cladophialophora bantiana	Herpotrichiellaceae
Cladosporium herbarum	Herpotrichiellaceae
Cladosporium spongiosum	Herpotrichiellaceae
Acremonium curvulum	Hypocreaceae
Acremonium kiliense	Hypocreaceae
Acremonium reseogriseum	Hypocreaceae
Trichoderma harzianum	Hypocreaceae
Graphium putredinis	Microascaceae
Microascus cinereus	Microascaceae
Microascus trigonosporus	Microascaceae
Pseudallescheria ellipsoideum	Microascaceae
Pseudoeurotium multisporum	Microascaceae
Trichurus spiralis	Microascaceae
Absidia corymbifera	Mucoraceae
Mucor hiemalis	Mucoraceae
Rhizopus sp.	Mucoraceae
Fusarium moniliforme	Nectriaceae
Fusarium oxysporum	Nectriaceae
Stachybotrys atra	Niessliaceae
Stachybotrys oenanthes	Niessliaceae
Chrysosporium merdarium	Onygenaceae
Alternaria alternata	Pleosporaceae
Alternaria chlamydospora	Pleosporaceae
Curvularia lunata	Pleosporaceae
Ulocladium chlamydosporum	Pleosporaceae
Preussia dispersa	Sporormiaceae
Preussia nigra	Sporormiaceae
Sordaria fimicola	Sordariaceae
Aspergillus candidus	Trichocomaceae
Aspergillus clavatus	Trichocomaceae
Aspergillus flavus	Trichocomaceae
Aspergillus fumigatus	Trichocomaceae
Aspergillus niger	Trichocomaceae
Aspergillus terreus	Trichocomaceae
Aspergillus versicolor	Trichocomaceae
Aspergillus wentii	Trichocomaceae
Byssochlamys nivea	Trichocomaceae

(continued)

Table 18.2 (continued)

Scientific name	Family
Emericella nidulans var. *nidulans*	Trichocomaceae
Eurotium cristatus	Trichocomaceae
Neosartorya sp.	Trichocomaceae
Paecilomyces variotii	Trichocomaceae
Talaromyces flavus	Trichocomaceae
Talaromyces stipitatus	Trichocomaceae
Talaromyces trachyspermus	Trichocomaceae
Total number of species	49
Total number of genera	27
Total number of families	13

terreus, A. niger and *Acremonium kiliense*; moderate frequency species, i.e. *Graphium putredinis, Preussia dispersa, Aspergillus fumigatus, Dichotomomyces ceipii* and *Rhizopus* sp.. The remaining species in the list are either designated with low- or very-low-frequency species.

The mere isolation of a fungus from a certain habitat does not mean that such a fungus has a certain activity in that habitat; species with low and very low occurrence are well known either soil-inhabiting fungi or associated with aquatic macrophytes (Domsch et al. 1980).

The high frequency of aquatic fungi present in Al-Kahla River in particular and in rivers in the lower reaches of Tigris River has led to the suggestion by several authors (Ito and Nakagiri 1997) that these species may live in active form in the mud and they expect that these species may contribute to the microbial activity in mud. Al-Dossari et al. (2001) reported the capability of *Aspergillus terreus* and *Acremonium killense* strains isolated from surface sediments of Shatt Al-Arab River in southern Iraq in degradation of a mixture of five polycyclic aromatic hydrocarbons in laboratory. More recently, Al-Dossari (2008) reported high ability for degradation of crude oil by *Aspergillus niger, A.terreus, Paecilomyces* sp., and *Acremonium* sp., isolated from sediments of southern marshes of Iraq.

18.2.1.1 The Role of Fungi in the Ecosystem

Fungi play an important role in aquatic ecosystem in the degradation of dead plant materials as well as materials from animal origin such as insect skeleton, fish scales and hair (Wong et al. 1998). Moreover, fungi and other microorganisms have the ability to degrade several pollutants including crude oil in the aquatic ecosystem and utilize them as nutrient sources (Davis and Westlake 1979). They may also metabolize such pollutants to substrates with low harmful effect on the environment (Boonchan et al. 2000).

In the lower reaches of Tigris River, changes in the environmental factors such as the level of water salinity might cause a decline in the bloom of certain aquatic fungi

Table 18.3 List of micro-algae in the Al-Kahla River

Scientific name	Family
Division: Chlorophyta	
Characium ambiguum	Characiaceae
Characium ornithocephalum	Characiaceae
Cladophora glomerata	Cladophoraceae
Cosmarium subtumidium	Cladophoraceae
Euastrum dub	Cladophoraceae
Oedogonium sociale	Oedogoniaceae
Scenedesmus arcuatus	Scenedesmaceae
Scenedesmus bijuga	Scenedesmaceae
Spirogyra nitida	Scenedesmaceae
Spirogyra scrobiculata	Zygnemataceae
Division: Rhodophyta	
Compsopogon sp.	Compsopogonaceae
Division: Chrysophyta	
Characiopsis spinifer	Botryochloridaceae
Achnanthes brevipes	Achnanthaceae
Achnanthes hungarica	Achnanthaceae
Achnanthes microcephala	Achnanthaceae
Achnanthes minutissima	Achnanthaceae
Amphora coffeaeformis	Achnanthaceae
Amphora commutata	Achnanthaceae
Amphora mexicana var. *major*	Achnanthaceae
Anomoeoneis exilis	Achnanthaceae
Cocconeis pediculus	Achnanthaceae
Cocconeis placentula	Achnanthaceae
Cocconeis placentula var. *euglypta*	Achnanthaceae
Cocconeis placentula var. *lineata*	Achnanthaceae
Gomphonema lanceolatum	Achnanthaceae
Gomphonema olivaceum	Achnanthaceae
Gomphonema parvulum	Achnanthaceae
Gomphonema sphaerophorum	Achnanthaceae
Navicula cincta	Achnanthaceae
Navicula cryptocephala	Achnanthaceae
Navicula cuspidata	Achnanthaceae
Navicula inflata	Achnanthaceae
Navicula radiosa	Achnanthaceae
Navicula radiosa var. *tenella*	Achnanthaceae
Nitzschia amphibia	Achnanthaceae
Nitzschia frustulum	Achnanthaceae
Nitzschia palea	Achnanthaceae
Rhoicosphenia curvata	Achnanthaceae
Rhopalodia gibba	Achnanthaceae

(continued)

Table 18.3 (continued)

Scientific name	Family
Synedra affinis	Achnanthaceae
Synedra fasciculata	Achnanthaceae
Synedra ulna	Achnanthaceae
Synedra ulna var. *biceps*	Achnanthaceae
Division: Cyanophyta	
Chroococcus dispersus	Chroococcaceae
Chroococcus limneticus	Chroococcaceae
Chroococcus turgidus	Chroococcaceae
Gomphosphaeria lacustris	Chroococcaceae
Lyngbya majuscula	Oscillatoriaceae
Merismopedia glauca	Oscillatoriaceae
Merismopedia tenuissima	Oscillatoriaceae
Oscillatoria princeps	Oscillatoriaceae

such as the members of the genus *Saprolegnia* where they showed significant decline in their frequency and dominance (Mazin et al. 2011) as compared with a previous study by Ministry of Agriculture, Iraq (1977). In Shatt al Arab River, there is an inverse correlation between the number of Saprolegnoid fungi and phytoplankton numbers (Al-Saddi et al. 1979). Such correlation might cause unbalancing in the total number of phytoplankton which might lead to direct effect on the food of the filter feeder fish species that live in this water body and results in changing the fish species composition in the affected area. Although this kind of study has not been performed in Al-Kahla River and since this river is not far away from Shatt al-Arab River and lies in the same geographical region, there is no reason for such effect to be absent in Al-Kahla River environment.

18.2.1.2 Threats and Sensitivities

The movement in river water masses and waves resulted from the activity of transferring huge quantity of water from the river will disrubted the species of aquatic fungi and stop their function in the aquatic environment as major degraders of dead plant materials as well as materials from animal origin such as insect skeleton, fish scales and hair. Disturbance of aquatic fungi and other microorganisms will also affect the degradation of several pollutants including crude oil in the aquatic ecosystem and stop their reutilization as nutrients.

The activity of the project will result in change of the water salinity that in turn cause a decline in the bloom of certain aquatic fungi such as the members of the genus *Saprolegnia*.

18.2.2 Flora of Micro-Algae

The study of algae in aquatic ecosystems has a great importance since it is the base level in the food chain. Aquatic ecosystems are dynamic with several biotic and abiotic variables changing in space and time. Studies have shown that rivers and marshes are suitable areas for the growth of several types of algae and higher aquatic plants. The rivers in the lower Mesopotamia seem especially suitable for growth of algae so that they diversify widely due to the slow flow of the water attributable to low gradients and suitable nutrient concentrations and temperatures (Yaaqub 1992). Therefore, these algae have been widely used for water quality monitoring, and as they are primary producers, they are easily affected by physical and chemical variations in their environment (Bartram and Balance 1996).

Micro-algae in Al-Kahla River fall in four divisions; these are Chlorophyta, Chrysophyta, Rhodophyta and Cyanophyta. Species wise, the largest of these four divisions is Chrysophyta with 32 species belonging to 11 genera. The other three divisions contain 10, 1 and 8 for Chlorophyta, Rhodophyta and Cyanophyta, respectively. On the other hand, Chlorophyta shown to embraces 5 families and both Chrysophyta and Cyanophyta with 2 families each. The domination of the Chrysophyta division is similar to what happened in the nearby marshes (Al-Mashhadani 2000). The majority of the micro-algae species were of benthic origin. This is a common feature of rivers and streams, and it is due to current turbulence (Al-Saadi et al. 1996).

Many environmental factors could affect the distribution, diversity and abundance of algae in the rivers and marshes. Such factors are light, temperature, type and growth phase of host plant, depth, nutrients, etc. (Michael et al. 2008). Some studies focused on the qualitative and quantitative changes in population dynamics of algae on different host macrophytes in different aquatic systems (Kassim et al. 2000).

18.2.2.1 The Role of Micro-Algae in the Ecosystem

Mitsch and Gosselink (2000) mentioned the importance of the contributions of micro-algae in non-tidal freshwater system. Other authors were explaining the ecological importance of these organisms in aquatic systems (Simkhada et al. 2006).

Planktonic algae are sensitive to environmental changes and comprise a necessary component of a useful monitoring programme (Schindler 1987). Arguably, all components of a water body function are influenced in major ways by the dynamics of phytoplankton and zooplankton. Phytoplankton are the primary producers whereas zooplankton is the central trophic link between primary producers and fish (Tatrai et al. 1997).

Micro-algae have been used as indicators of nutrient conditions as well as pollution in water (APHA 2005). Trophic status of any water body may be indicated by different phytoplankton taxa or a number of ratios between different groups (Hutchinson 1967). Diatoms are one of the major micro-algae in the river ecosystem, and they could be used for monitoring environmental changes such as those changes in trophic status of water bodies (Dixit et al. 1992). There should a monitor programme for the following factors like light penetration, temperature, pH, water flow, nutrient levels and land use, in particular for cattle grazing in order to not affecting the role of the micro-algae in the freshwater ecosystem.

Planktonic algae are the first response assemblage of organisms directly affected by water quality conditions (Lacouture et al. 2006). They are also the basis of food web in aquatic environments. Therefore, it is important to monitor such organisms in nature.

Epiphytic algal community appear an important fraction of the primary production of water body and are important as shelter and food for many invertebrates and fish (Cattaneo and Kalff 1987). The relationship between epiphytic algae and their host macrophytes is poorly understood (Morin 1986). The epiphytic algae may reduce growth and production of aquatic plant (Takashi et al. 2004), in plants without epiphytic algae, the net photosynthesis was significantly higher than in plants with epiphytic algae. Epiphytic owing to their close association with aquatic macrophytes may utilize dissolved organic matter products released by their hosts in freshwater habitats (Eminson 1978).

18.2.2.2 Threats and Sensitivities

The abundance of the planktonic algae will be affected by the project activity in the area as they sensitive to any environmental changes, therefore, their role as useful monitoring programme will be hindered. The disturbance in the dynamics of micro-algae as primary producers will affect the function of all the components of the water body.

Their role as indicators of nutrient conditions as well as pollution will be cancelled due to the huge loss in their abundance. The water movement resulted from the project activity will increase water turbidity, decreasing oxygen level, changing pH and affecting water flow. Such changes will make monitoring programme impossible.

Due to the effects mentioned above on the microalgae, the relationships between these organisms and their host macrophytes on one side and micro-algae and invertebrate group on the other will be break down.

18.2.3 Zooplankton

Much of the early literature on zooplankton in running waters is reviewed by Bryan (1976). The author refers to many early records of rotifers collected from the main water channel of rivers and streams (Lauterbom 1893) and many other rivers in Europe (Zacharias 1898), and the Missouri (Berner 1951).

In an aquatic ecosystem, interaction occurs between living and non-living components. Environmental factors comprising physical and chemical components have been reported in several studies to have a great influence on the well-being of aquatic species, plankton inclusive (Okogwu and Ugwumba 2006).

The earliest study related to zooplankton in the southern rivers and marshes of Iraq is that of Gurney (1921). In this preliminary study, the author dealt with the crustacean group in the major rivers and marshes and managed to list number of microscopic invertebrate organisms including 13 species of cladocera and three species of rotifer in this region. Since then, several works have published on this group (Khalaf and Smirrnov 1976; Al-Saboonchi et al. 1986; Al-Qarooni 2005; Ajeel et al. 2006).

There are four main taxonomic groups of zooplankton in Al-Kahla River, and these are Cladocera with 39 species belonging to 18 genera and 6 families, Copepoda with 8 species belonging to 5 genera and 2 families, Amphipoda with one species and Rotifera with 46 species belonging to 24 genera and 14 families (Table 18.4). It is obvious that Rotifera is the richest group species wise followed by Cladocera. Amphipods are represented by only one species. This is unusual for such group, but such low record is certainly due to lack of regular and efficient aquatic surveys in the area.

In addition to the above-mentioned zooplankton groups, few other groups of microscopic invertebrates are also present in Al-kahla River these are free-living Nematodes, Ostracods, Harpacticoida, Acari and Calanoida. Their numbers are far below of those mentioned in the main four groups.

Qualitative and quantitative distribution of zooplankton in the freshwater systems is under the effect of some environmental factors such as temperature and pH which are shown to have direct effect on zooplankton. On the other hand, variations in zooplankton density are related to changes in electrical conductivity (EC) and Secchi Disc transparency (SDT).

18.2.3.1 Role and Importance of Zooplankton in the Freshwater Ecosystem

The production of many smaller zooplanktonic species is very abundant and diversified group (Ajeel 1998). In an aquatic ecosystem, zooplankton form the microscopic animals (Redmond 2008) that play an important role in an aquatic food chain as they are largely consumed by fishes and other higher organisms in food chain.

Zooplanktons are important component in channeling energy and nutrients from algae and bacteria to fish. Because they are highly productive and important in fish diets, an improved understanding of zooplankton production and growth can be applied to increase fish production in both aquaculture facilities and the wild.

Secondary production may be defined as rate of energy storage at consumer trophic levels (Odum 1971), or rate of production or rate of elaboration of living matter through the interaction between the organism and the space where it lives (Wheaton 1979), or the amount of living matter elaborated per unit area per unit time

Table 18.4 List of zooplankton in the Al-Kahla River

Scientific name	Family
Taxonomic group: Cladocera	
Bosmina coregoni	Bosminidae
Bosmina longirostris	Bosminidae
Alona affinis	Chydoridae
Alona circumfimbriata	Chydoridae
Alona costata	Chydoridae
Alona guttata	Chydoridae
Alona rectangula	Chydoridae
Camptocercus rectirostris	Chydoridae
Camptocercus uncinatus	Chydoridae
Chydorus barrois	Chydoridae
Chydorus eurynatus	Chydoridae
Chydorus sphaericus	Chydoridae
Chydorus ventricosus	Chydoridae
Indialona macronyx	Chydoridae
Dunhevedia crassa	Chydoridae
Eurycerus glacialis	Chydoridae
Pleuroxus aduncus	Chydoridae
Pleuroxus denticulatus	Chydoridae
Pleuroxus similis	Chydoridae
Pxyurella singalensis	Chydoridae
Ceriodaphnia cornuta	Daphniidae
Ceriodaphnia reticulate	Daphniidae
Daphnia dubia	Daphniidae
Daphnia longispina	Daphniidae
Daphnia lumholtzi	Daphniidae
Daphnia magna	Daphniidae
Daphnia pulex	Daphniidae
Simocephalus expinosus	Daphniidae
Scapholeberis kingi	Daphniidae
Simocephalus mucronata	Daphniidae
Simocephalus vetulus	Daphniidae
Echinisca rosea	Macrothricidae
Ilyocryptus sordidus	Macrothricidae
Ilyocryptus spinifer	Macrothricidae
Macrothrix spinosa	Macrothricidae
Macrothrix rosea	Macrothricidae
Moina braciata	Moinidae
Moina rectirostris	Moinidae
Diaphanosoma brachyurum	Sididae
Latonopis occidentalis	Sididae

(continued)

Table 18.4 (continued)

Scientific name	Family
Taxonomic group: Copepoda	
Canthocamptus staphylinus	Canthocamptidae
Acanthodiaptomus denticornis	Diaptomidae
Acanthodiaptomus salinus	Diaptomidae
Diaptomus blanci	Diaptomidae
Diaptomus chevreuxi	Diaptomidae
Diaptomus vulgaris	Diaptomidae
Eudiaptomus vulgaris	Diaptomidae
Phyllodiaptomus irakiensis	Diaptomidae
Cyclops sp.	Cyclopidae
Eucyclops agilis	Cyclopidae
Macrocyclops albidus	Cyclopidae
Mesocyclops leukarti	Cyclopidae
Paracyclops fimbriatus	Cyclopidae
Taxonomic group: Amphipoda	
Paryhale basrensis	Hyalidae
Taxonomic group: Rotifera	
Brachionus angulars	Brachionidae
Brachionus calcyflorus	Brachionidae
Brachionus quadridentatus	Brachionidae
Brachionus urceolaris	Brachionidae
Keratella cochleeris	Brachionidae
Keratella hiemalis	Brachionidae
Keratella valga	Brachionidae
Keratella quadrata	Brachionidae
Keratella tropica	Brachionidae
Notholca acuminate	Brachionidae
Notholca squamula	Brachionidae
Platuias quadricornis	Brachionidae
Squatinella mutica	Brachionidae
Dipleuchlanis propatula	Euchlanidae
Euchlanis dilatata	Euchlanidae
Manfredium eudactylotum	Euchlanidae
Camptocercus rectirostris	Eurycercidae
Cephalodella auriculata	Eurycercidae
Gastropus stylifer	Gastropodidae
Lecane depressa	Lecanidae
Lecane elasme	Lecanidae
Lecane luna	Lecanidae
Lecane ohioensis	Lecanidae
Monostyla bulla	Lecanidae
Monostyla closterocerca	Lecanidae

(continued)

Table 18.4 (continued)

Scientific name	Family
Monostyla lunaris	Lecanidae
Monostyla quadridentata	Lecanidae
Colurella adriatica	Lepadellidae
Colurella bicuspidata	Lepadellidae
Colurella obtusa	Lepadellidae
Lepadella ovalis	Lepadellidae
Lepadella patella	Lepadellidae
Mytilina mucronata	Mytilinidae
Mytilina ventralis	Mytilinidae
Cephalodella exigua	Notommatidae
Cephalodella mucronata	Notommatidae
Philodinavus paradoxus	Philodinavidae
Scaridium longicaudum	Scaridiidae
Polyarthra ramata	Synchaetidae
Testudinella patina	Testudinellidae
Trichocerca longiseta	Trichocercidae
Trichocerca porcelus	Trichocercidae
Trichocerca cylindrical	Trichocercidae
Macrochaetus subquadratus	Trichotriidae
Trichotria pocillum	Trichotriidae
Trichotria tetractis	Trichotriidae

(Downing 1984). Therefore, the organism's success in an environment might be a function of its ability to fix and retain energy (Lindeman 1942). Moreover, secondary production is not a distinct entity by itself rather it is part of a larger scheme of the movement of material through the ecosystem, and this is based on the activities of individuals and populations of animals (Edmondson 1974). Furthermore, it is one of the most important components of the energy budget.

Under favourable environmental conditions, zooplanktons such as Cladocera may be monocyclic or dicyclic (Pennak 1978). Cladocerans such as *Daphnia pulex* are also reported to be rare in rapid streams and grossly polluted waters (Pennak 1978). Additionally, some species, particularly rotifers, could show unaccountable disappearance for certain period of time under harsh conditions although they are cosmopolitan in distribution (Pennak 1978).

If the environmental conditions continue to be unfavourable, consequently, zooplanktons will be unable to maintain footing and therefore develop a stable community in a lotic environment, which is fundamental for reproduction success.

The turbulence of water due to any mechanical activities will lead to low transparency level which in turn limits light penetration and cause lowering the diversity of zooplankton species.

Threats and Sensitivities

The number and abundance of zooplankton will be directly affected by changes in phytoplankton. These changes will transmit to the zooplankton feeders and the aquatic food chain and the secondary production will be disrubted.

With the great deficiency in zooplankton abundance, their role in channeling energy and nutrients from algae and bacteria to fish will be affected. Their role in fish diets and increase fish production in both aquaculture facilities and the wild will not be possible.

If the environmental conditions continue to be unfavourable, consequently, zooplanktons will be unable to maintain footing and therefore develop a stable community in a lotic environment, which is fundamental for reproduction success.

The turbulence of water due to any mechanical activities will lead to low transparency level which in turn limits light penetration and cause lowering the diversity of zooplankton species.

18.2.4 Meiofauna

Meiofauna organisms are micro-benthic creatures that live in the bed of the river or any aquatic system. Several studies have shown that meiofauna can no longer be considered as holobenthic infauna existing only among sediment particles (Chandler and Fleeger 1983). This is especially true in muddy substrates where fauna lack adhesive capabilities, and where sediments and their fauna are easily suspended by water movement. Harpacticoid copepods seem particularly vulnerable to suspension by tidal currents as several species are active on sediment surfaces (Palmer 1984) or are epibenthic or hyperbenthic, i.e. associated with near-bottom water and superficial sediments (Sibert 1981).

In addition to different groups of worms, the meiofauna also includes different developmental stages of aquatic insects and mollusca. In Table 18.5, the taxonomic

Table 18.5 List of meiofauna in the Al-Kahla River

Full scientific, genus or family names	Higher taxonomic group
Alaimus sp.	Nematoda
Triolopus sp.	Nematoda
Dorylaimus sp.	Nematoda
Aeolosomatidae	Oligochaeta
Lumbricidae	Oligochaeta
Enchytraeidae	Oligochaeta
Naididae	Oligochaeta
Tubicidae	Oligochaeta
Lumbriculidae	Oligochaeta
Pristina macrochaeta	Naididae, Oligochaeta
Pristina longiseta	Naididae, Oligochaeta

groups of worms are only shown. The aquatic insects and mollusc composition of meiofauna are included in the invertebrate section.

Representation of the members of the Oligochaeta is better than those of the Nematoda. It is clear from Table 18.5 that there are not many of meiofauna species are reported from the freshwater systems of Iraq. This is due to the lack of invertebrate surveys in the area. More species of meiofauna will be recorded from the lower reaches of Tigris River once scientific investigations start.

18.2.4.1 Significance of Meiofauna in Al-Kahla River Ecosystem

Permanent and temporal members of benthic communities in general and meiofauna in particular are important food sources for many commercial fishes and other organisms (Barnes and Hughes 1992). Meiofauna are also important indicators for monitoring and assessing of pollution effects, especially oil pollution, on the ecosystem (Warwick 1993).

Biological response and ecological indices of all benthic communities including meiofauna have been used to document pollution effects; for example, in unusual circumstances of high pollution, greater mortality of foraminifera and shell abnormalities (Zare-maivan 2011) and changes in secondary production and transferring of energy through food chains have been observed (Nabavi 2001).

Moreover, meiofauna may also enhance the use of detritus by macrofauna (Valiela 1995). The presence of meiofauna in the aquatic ecosystem will enhance the degradation of roots. The higher bacterial degradation under anoxic conditions (without cycloheximide) may result from the presence of anaerobic flagellates that could enhance the degradation when compared to the aerobic microbial degradation for the same period of time (Lillebø et al. 1999). Other studies showed that polychaetes can incorporate significantly more carbon from detritus of some aquatic macrophytes when meiofauna is present (Valiela 1995). Although bacteria are the first catabolizers of organic matter, benthic macrofauna feeds on detritus formed by bacterial decomposition activity and both macrofauna and meiofauna may feed directly on bacteria (Enriques et al. 1993).

The movement of meiofauna is known as bioturbation. Such movement is important to life of the residents of any aquatic ecosystem as stimulation of nutrient generation will occur. In such process, oxygenation and mineralization will be increase and alters the sediment geochemistry (Tamaki and Ingole 1993). In addition, bioturbation enhances the exchange of important constituents such as carbon, nitrogen, sulphur, phosphate and silica which increase the productivity in the sediment and water column (Ingole 2005).

Threats and Sensitivities

The number and abundance of zooplankton will be directly affected by changes in phytoplankton. These changes will transmit to the zooplankton feeders and the aquatic food chain and the secondary production will be disrubted.

With the great deficiency in zooplankton abundance, their role in channeling energy and nutrients from algae and bacteria to fish will be affected. Their role in fish diets and increase fish production in both aquaculture facilities and the wild will not be possible.

If the environmental conditions continue to be unfavourable, consequently, zooplanktons will be unable to maintain footing and therefore develop a stable community in a lotic environment, which is fundamental for reproduction success.

The turbulence of water due to any mechanical activities will lead to low transparency level which in turn limits light penetration and cause lowering the diversity of zooplankton species.

18.3 Mosses and Ferns in the River Al-Kahla

No studies are available on mosses and ferns in Al-Kahla River in particular and in Iraqi inland waters. The records of few species of these two groups included through aquatic ecology surveys in the area. The only species seem to be present of these two groups are the moss, *Ricciocarpas natans* (L.), and the fern, *Salvinia natans* (L.). Both of these two species have a floating habit.

18.4 Flora of Higher Aquatic Plants

The presence of macrophytes is one of the defining features of rivers and wetlands in Iraq. The aquatic plants in general play a vital role in the aquatic ecosystems through their effect on the structure and function of aquatic ecosystems to alter water movement regimes (Van der Valk 2006). They are important in providing shelter, food for many levels of wildlife, and at the same time produce oxygen needed by animals (Chambers et al. 2008). In addition, they have great influence in the ecosystem by taking up nutrient, releasing dissolved organic matter, increasing sedimentation and improving water clarity (Scheffer et al. 1993).

Aquatic macrophytes constitute a major component of species richness (Wetzel 2001); hence, changes in freshwater macrophyte diversity also affect associated organisms as well as the single macrophyte species (Heino and Toivonen 2008). Aside of changes in plant biodiversity due to direct human disturbances water quality degradation also heavily impacts on plant populations. Therefore, the aquatic macrophytes are considered to be keystone elements in freshwater ecosystems (Abdullhasan 2009).

Seasonal variations in water flow level and severe water shortages are considered among the reasons that affect the biodiversity and macrophyte communities in the rivers and wetlands connected to these rivers. Therefore, in situ preservation through the establishment of a natural reserve as a conservation mechanism has shown to be one of the most effective and least expensive means to protect biodiversity (Bowes 1999).

The aquatic plant community is divided into three groups:

(a) Emergents: the members of this group are characterized with fast re-establishment in the new areas. *Phragmites australis* (Reed), *Typha domingensis* (Cattail) and *Schonemoplectus litoralis* (Sedge) are the common species in this group.
(b) Submergents: a submergent plant or submergent vegetation is a plant that is completely beneath the surface of water. Most submergent plants are firmly rooted in the soil. *Hydrilla verticillata* is the common and dominant species in the rivers of the lower reaches of Tigris River. The species of *Ceratophyllum* and *Potamogeton* are also among members of this group.
(c) Floating leaved-plants: aquatic plants rooted in sediment with leaves floating on the surface. *Nymphaea alba*, with large cordate leaves and white flowers, is the common and dominant aquatic plants in the southern Iraq waterways.

The species of macrophytes present in Al-Kahla River are listed in Table 18.6. This list is compiled from the literature as there is no checklist of aquatic plants is on records (Jones et al. 2010; Mohammed et al. 2011; Ali 2011; Yamakuta 2012). In the River Kahla, there are 79 species of macrophytes belonging to 46 genera and contained in 34 families. Species number wise, the largest genus is *Cyperus* with 8 species followed by *Potamogeton* with 7 species. As to the families, the largest is Cyperaceae with 13 species and Potamogetonaceae with 7 species.

Among the macrophytes living in the River Al-Kahla, the species *Phragmites australis* which shown to be an effective and successful invasive plant, sexually reproducing by seeds and expanding through rhizomes and having the ability to rapidly recover after damage to aboveground growth (Meyerson et al. 2000). The plant often forms monocultures with very densely vegetation cover under strenuous conditions (Al-Hilli et al. 2009). Other aquatic plant species living in River Al-Kahla such as *Typha domingensis* and *Schoenoplectus litoralis* which are more sensitive to salt (Abd El-Ghani 2000) showed restricted growth in the form of small patches where water was still present.

Saltbush, *Atriplex leucoclada,* and salt cedar, *Tamarix ramosissima,* are also present. The drastic changes that occurred during the last two decades in the area have led to an increase in soil salinity that will create a favourable environment for establishing such saltbush species (Al-Mayaha et al. 2012).

Invasive species of macrophytes such as *Hydrilla verticillata* has been observed in the waters of the lower reaches of Tigris River and in Al-Kahla River. This species may tend to overpopulate water resources forming dense canopy (Langeland 1990) and can regenerate from small cuttings. It competes with native species, and when die, it releases nutrients that and create poor aesthetics (Walley 2007).

Table 18.6 List of macrophytes reported from the waters of Al-Kahla River

Scientific name	Family
Alisma laniceolatum	Alismataceae
Alisma plantagoaquatica	Alismataceae
Damasonium alisma	Alismataceae
Sagittaria sagitifolia	Alismataceae
Alternanthera sessilis	Alternanthaceae
Cynancum acutum	Apocynaceae
Lemna gibba	Araceae
Lemna minor	Araceae
Lemna perpusilla	Araceae
Lemna trisulca	Araceae
Oxystelma esculentum	Asclepiadaceae
Aster tripolium	Asteraceae
Nasturtium officenale	Brassicaceae
Rorippa amphibian	Brassicaceae
Butomus umbellatus	Butomaceae
Ceratophyllum demersum	Ceratophyllaceae
Ceratopteris thalicroides	Ceratophyllaceae
Sonchus maritimus	Compositae
Bolboschoenus maritimus	Cyperaceae
Cladium mariscus	Cyperaceae
Cyperus corymbosus	Cyperaceae
Cyperus difformis	Cyperaceae
Cyperus iria	Cyperaceae
Cyperus lavegatus	Cyperaceae
Cyperus longus	Cyperaceae
Cyperus malaccensis	Cyperaceae
Cyperus michelians	Cyperaceae
Cyperus rotundus	Cyperaceae
Fimbristylis bisumbillata	Cyperaceae
Fimbristylis littoralis	Cyperaceae
Fimbristylis sieberiana	Cyperaceae
Bergia ammannioides	Elatinaceae
Bergia capensis	Elatinaceae
Schenoplectus litoralis	Hemerocallidaceae
Schenoplectus maritimus	Hemerocallidaceae
Schenoplectus triquater	Hemerocallidaceae
Myriophyllum spicatum	Haloragaceae
Myriophyllum verticillatum	Haloragaceae
Ottelia alismoides	Hydrocharitaceae
Juncus acutus	Juncaceae
Juncus articulates	Juncaceae
Juncus rigidus	Juncaceae

(continued)

Table 18.6 (continued)

Scientific name	Family
Lycopus europaeus	Lamiaceae
Mentha aquatica	Lamiaceae
Marsilea capensis	Marsileaceae
Nymphoides indica	Menyanthaceae
Nymphoides petata	Menyanthaceae
Najas graminea	Najadaceae
Najas marina	Najadaceae
Najas minor	Najadaceae
Nymphaea alba	Nymphaeaceae
Ludwigia repens	Onagraceae
Peplidium maritimum	Phrymaceae
Baccopa monniera	Plantaginaceae
Arundo donax	Poaceae
Diplache fusca	Poaceae
Echinoeloa crasszalli	Poaceae
Paspalum paspaloides	Poaceae
Phragmites australis	Poaceae
Polygonum amphibium	*Polygonaceae*
Polygonum lapathifolium	*Polygonaceae*
Polygonum persicaria	*Polygonaceae*
Polygonum salicifolium	*Polygonaceae*
Polypogon monspeliensis	*Polygonaceae*
Potamogeton berchteldii	Potamogetonaceae
Potamogeton crispus	Potamogetonaceae
Potamogeton lucens	Potamogetonaceae
Potamogeton nodosus	Potamogetonaceae
Potamogeton pectenatus	Potamogetonaceae
Potamogeton perfoliatus	Potamogetonaceae
Potamogeton pusillus	Potamogetonaceae
Samolus valerandi	Primulaceae
Rannunculus sphaerospermus	*Rannunculaceae*
Rannunculus trichuphyllus	*Rannunculaceae*
Ruppia maritima	Ruppiaceae
Salvinia natans	Salviniaceae
Thelypteris palustris	Thelypteridaceae
Sparganium erectum	Typhaceae
Paniam repens	Verbenaceae
Phyla nodiflora	Verbenaceae
Total number of species	79
Total number of genera	46
Total number of families	34

The absence or loss of many important native submerged species will certainly have a direct impact on the whole productivity of the rivers in Iraq such as shrimps, fish and birds (including migratory species). Local people have the habit of using most of the aquatic plants as food, especially during spring when such vegetation flourish (Al-Hilli 1977). On the other hand, aquatic plants are considered important to the native fish species such as *Cyprinus carpio* whose stomach content has shown to constitute 18.9% of aquatic plants (Al-Kanaani 1989).

18.4.1 Role of Aquatic Macrophytes in the Riverine Environment

Macrophytes colonize many different types of aquatic ecosystems, such as lakes, reservoirs, wetlands, streams, rivers, marine environments and even rapids and falls.

Among the roles of macrophytes in the riverine environment is the nutrient cycling, for example through transference of chemical elements from sediment to water, by both active and passive processes (Camargo et al. 2003). Limiting nutrients released by macrophytes, like phosphorus and nitrogen, are rapidly used by micro-algae and bacteria (which also use organic carbon released by macrophytes); these microorganisms may be free-living or attached to macrophyte surfaces and their detritus (Stets and Cotner 2008). In addition, several species of macrophytes produce an elevated percentage of refractory matter (basically fibrous material) that is relatively slow to decompose (Bianchini 2003), and thus, they also contribute to a return of carbon to sediment. Macrophytes may also influence nutrient cycling in two other ways: retention of solids and nutrients by their submersed roots and leaves (Pott and Pott 2003).

Owing to their high rate of biomass production, macrophytes have primarily been characterized as an important food resource for aquatic organisms, providing both living (grazing food webs) and dead organic matter (detritivorous food webs). It is true that macrophytes may represent an important source of organic matter for aquatic herbivores and detritivores in some ecosystems (Poi de Neife and Casco 2003).

The effect of macrophytes on populations and communities has been widely demonstrated for a variety of organisms, such as micro- and macroinvertebrates (e.g. Takeda et al. 2003), fish (Theel et al. 2008) and waterbirds (Klaassen and Nolet 2007).

The importance of macrophytes as refugia (via an increase in habitat complexity) has been suggested for invertebrates and fish in field studies (Agostinho et al. 2007a, b). Possibilities for shelter increase with such structural complexity and in the same way decrease visual contact among predators and their prey (Dibble et al. 1996). Consequently, with the presence of macrophytes, we expect higher diversity of invertebrates and small fish on plants with intermediate levels of complexity (Agostinho et al. 2007b).

Rooted submersed and floating as well as emergent macrophytes exhibit their role in bioaccumulation and filtration mostly in shallow water, namely littoral zone of rivers, canals and lakes and in slow streams (Bunn et al. 1998). An evident nutrient and heavy metal bioconcentration ability significantly enlarge the possibilities of utilization of aquatic plants not only for bioindication, but, also, for the purification of water, substratum and littoral zone. Utilization of aquatic plants for a biological clean-up technique in various polluted ecosystems is highly acceptable due to their high biomass production resulting in a high uptake of macronutrients (N, P, K, S) and heavy metals. In addition, the ability of most of aquatic plants to highly tolerate a specific metal is of a great importance (Matagi et al. 1998).

Rooted macrophytes serve as a living link between the sediment, water and (sometimes) atmosphere in wetlands, lakes and rivers. Recent studies also suggest that macrophytes play a central role in rivers, lakes and marshes which can have two possible stable equilibria: a clear-water state that is dominated by aquatic macrophytes and a turbid-water state that is dominated by phytoplankton (Jeppesen et al. 1998). Macrophytes maintain the clear-water state by a variety of mechanisms (e.g. stabilizing sediments, promoting zooplankton populations) whose relative importance is probably variable (Madsen et al. 2001).

18.4.1.1 Threats and Sensitivities

The aquatic macrophytes will be among the most affected by the activities of the project in Al-Kahla River. Their roles in several aspects of the ecology and biology of the riverine ecosystem will be hindered due to project activities. They will not play the role of important habitat structures and not be able to be highly influential on the composition of the associated fauna and influence interspecific relationships. Due to their reduction in abundance and complexity, there will be a decrease in animal abundance, richness and diversity. They will fail in providing the mechanisms that involve the availability of habitat, which increases the possibility of available food and consequently attracts other organisms.

18.5 Invertebrate Fauna

River waters are frequently referred to as highly variable in space and time (Minshall et al. 1983). Consequently, it is not surprising to find that riverine landscapes are physically complex. Organisms use different parts of the habitat to varying degrees; some for foraging or hiding from enemies, others during dispersal.

The spatial distribution of invertebrates in the riverine landscape is determined not only by habitat conditions, but also to a large extent by active movements of the invertebrates. In streams, movement of invertebrates takes place in many other forms as well. The needs of feeding, mating and dispersal are manifested within—as well as between different parts of the river system.

Through movements associated with these needs, particularly when they take place across system boundaries, invertebrates are involved in spatially separated ecological processes, which in this way become linked to each other. The importance of such linkages in the riverine landscape is, no doubt, considerable (e.g. Nakano and Murakami 2001). In addition to the direct movements of individuals, invertebrates affect the exchange and flow of nutrients and organic matter with quantitatively considerable consequences. Riverine invertebrates are involved to a greater or lesser extent in dispersal taking place over smaller or larger distances, to find food or mates, to colonize other areas within the same or different streams, or to survive adverse conditions.

The main channel of the river is usually fast flowing and unstable with relatively few species tolerant of a shifting substratum and a predominance of lotic species. In the side channels, diversity as well as production tends to increase (Marchese and Ezcurra de Drago 1992), and invertebrates with a lentic mode of life are more common.

In addition to natural factors, various anthropogenic disturbances could affect invertebrate species composition, influencing the composition in a similar manner. The impact of construction work, such as bridges, may cause such changes. Human activity in general has caused disturbance in streams and rivers all over the world to such.

The information on the invertebrate fauna of the rivers in the lower reaches of the Tigris River is fragmentary as it has not been studied extensively. The main taxonomic groups in the Al-Kahla River as well as in the other rivers in southern Iraq are Molluscs (including gastropods and bivalves) and arthropods (including isopods, amphipods and insects—particularly the dragonflies and beetles). Therefore, the report at hand will discuss the biodiversity of these two groups in detail.

There are five major invertebrate groups in the water of Al-Kahla River these are Ostracoda with 6 species belonging to 6 genera, Malacostraca with one species, Decapoda with 26 species belonging to 17 genera and 6 families, Insecta with total number of species 35 belonging to 24 genera contained in 12 families and, finally, Mollusca with total number of species of 41 belonging to 22 genera contained in 18 families. For Insect group, Odonata is represented by 28 species belonging to 17 genera and contained in 6 families. The other insect groups are represented by 7 species belonging to 7 genera and 6 families. For molluscs, Gastropod is represented by 32 species belonging to 16 genera and contained in 14 families. For bivalve, there are 9 species belonging to 6 genera and contained in 4 families (Table 18.7).

It is clear from Table 18.7 that gastropods group holds the highest number of species (32) followed by the Odonata (28). In spite of the high number of molluscs species and insects with 41 and 33 species, respectively, there are more species to be discovered from these two groups once a comprehensive scientific investigation starts.

Table 18.7 Species list of major invertebrate groups that present in the Al-Kahla River

Scientific name	Family/higher taxonomic group
Taxonomic group: Ostracoda	
Cypria sp.	Crustacea
Physocypria sp.	Crustacea
Eucypria sp.	Crustacea
Cyprinotus sp.	Crustacea
Herpetocypris sp.	Crustacea
Ilyocypris sp.	Crustacea
Taxonomic group: Malacostraca	
Syncaris sp.	Atiydae
Taxonomic group: Decapoda	
Ataephyra mesopotamica	Atiydae
Caridinia baboulti basrensis	Atiydae
Macrobrachium equidens	Palaemonidae
Macrobrachium rude	Palaemonidae
Elamenopsis kempi	Hymenosomatidae
Seasarma boulengeri	Seasarmidae
Taxonomic group: Insecta Order: Odonata	
Aeshna mixta	Aeshnidae
Anax ephippiger	Aeshnidae
Anax parthenope	Aeshnidae
Calopteryx splendes	Calopterygidae
Ischnura evansi	Coenagrionidae
Ischnura fountaineae	Coenagrionidae
Ischnura senegalensis	Coenagrionidae
Anormagomphus kiritshenkoi	Gomphidae
Gomphoidus sp.	Gomphidae
Gomphus kinzelbachi	Gomphidae
Lindenia tetraphylla	Gomphidae
Onychogomphus flexuosus	Gomphidae
Sympecma paedisca	Lestidae
Brachythemis fuscopalliata	Libellulidae
Crocothemis erythraea	Libellulidae
Crocothemis servilia	Libellulidae
Diplacodes lefebvrii	Libellulidae
Libellula pontica	Libellulidae
Orthetrum sabina	Libellulidae
Orthetrum taeniolatum	Libellulidae
Orthetrum trinacria	Libellulidae
Pantala flavescens	Libellulidae
Selysiothemis nigra	Libellulidae
Sympetrum arenicolo	Libellulidae
Sympetrum fonscolombii	Libellulidae

(continued)

Table 18.7 (continued)

Scientific name	Family/higher taxonomic group
Sympetrum striolatum	Libellulidae
Trithemis annulata	Libellulidae
Trithemis festiva	Libellulidae
Other insect taxonomic groups	
Melaoplus sp.	Acrididae
Agrion splendens	Agriidae
Belostoma sp.	Belostomatidae
Lethocerus sp.	Belostomatidae
Anisoptera sp.	Dipterocarpaceae
Gryllotalpa sp.	Gryllotalpidae
Anabrus sp.	Tettigoniidae
Taxonomic group: Mollusca Gastropoda	
Amnicola (Alocinna) ejecta	Amicolidae
Bithynia badiella	Bythyniidae
Cerithidea (Cerithideopsilla) cingulate	Cerithidae
Lymnaea auricularia	Lymnaeidae
Lymnaea canalifera	Lymnaeidae
Lymnaea gedrosiana	Lymnaeidae
Lymnaea lagotis	Lymnaeidae
Lymnaea natalensis	Lymnaeidae
Lymnaea tenera	Lymnaeidae
Melanopsis nodosa	Melanopsidae
Melanopsis praemorsa	Melanopsidae
Neritina (Dostia) schlaeflii	Neritidae
Neritina (Dostia) violacea	Neritidae
Theodoxus (N.) euphraticus	Neritidae
Theodoxus (N.) jordani	Neritidae
Theodoxus (N.) macrii	Neritidae
Theodoxus mesopotamicus	Neritidae
Cleopatra bulimoides	Paludomidae
Physa acuta	Physidae
Bulinus contortus	Planorbidae
Bulinus truncatus	Planorbidae
Gyraulus albus	Planorbidae
Gyraulus convexiusclus	Planorbidae
Gyraulus costulatus	Planorbidae
Gyraulus intermixtus	Planorbidae
Potamides conicus	Pomtididae
Tricula palmyrae	Pomatiopsidae
Stenothyra iraqensis	Stenothyridae
Melanoides nodosum	Thiaridae
Melanoides tuberculata	Thiaridae

(continued)

Table 18.7 (continued)

Scientific name	Family/higher taxonomic group
Bellamya bengalensis	Viviparidae
Bellamya unicolor	Viviparidae
Bivalvia	
Corbicula cor	Corbiculidae
Corbicula fluminalis	Corbiculidae
Corbicula fluminea	Corbiculidae
Corbicula tigridis	Corbiculidae
Saccostrea cucculata	Ostreidae
Theora mesopotamica	Tellinoidea
Unio tigrids	Unionidae
Pseudodontopsis euphraticus	Unionidae
Anodonta (Anodonta) vescoiana	Unionidae

18.5.1 Role of Invertebrates in the Freshwater System

Aquatic invertebrates are involved in many different processes in riverine ecosystems (e.g. Wallace and Webster 1996). For example, they contribute significantly to nutrient cycling and the turnover of organic material, whether produced within the system or entering from the riparian zone. As a resource for grazing, shredding, deposit- and suspension-feeding macroinvertebrates, there is the biofilm, consisting of algae, fungi, bacteria and protozoans, along with dissolved and particulate organic matter of highly variable quantity and quality. Predators, including fish, depend substantially on macroinvertebrates, and it has been proposed that most of the latter's production may go to support predators (Huryn and Wallace 2000).

Most stream insects leave the water for mating, feeding and dispersal and may then be significant in the surrounding ecosystem, primarily as food for birds, spiders and other predators (Jackson and Resh 1989). However, they may be also significant in other processes, such as pollination and blood-feeding, thus negatively affecting birds and mammals, including humans (Crosskey 1990).

18.5.2 Endemic and Threatened Invertebrate Species in Al-Kahla River

There are four species of dragonflies that reported in the Red List of IUCN (2010) as near-threatened, vulnerable or data deficient. These are *Gomphus kinzelbachi*, *Brachythemis fuscopalliata*, *Libellula pontica* and *Anormogomphus kiritshenkoi*. *Gomphus kinzelbachi* is globally data deficient species and only known from Iraq and Iran, and therefore, it considered a regional endemic.

With the near-complete destruction of the marshes, any such species would now be critically endangered (Evans 2002).

Aquatic invertebrates are generally vulnerable to fluctuations of water level, changes in salinity and prolonged dryness of the substratum. Dragonflies require water for most of the year. Therefore, drainage and habitat alteration are the main threats to them.

18.5.2.1 Threats and Sensitivities

The reduction in the abundance of invertebrates in the riverine environment will in turn affect the food chain that involves algae, fungi, bacteria and protozoans. As a result, fish will have less food resources especially the predators.

Changing in the river's environment due to the project activity will disturb the ecological niches of stream insects and involve in their daily activity, and they may leave the area for more settled environment. In doing so, the stream insect biodiversity will be decreased, and in turn, the diversity of birds, spiders and other predators will be affected as these insects usually leave the water for mating, feeding and dispersal and may then be significant in the surrounding ecosystem. Their absence from the riverine environment might also affect other biological important processes such as pollination.

The effect of the project activity on the biodiversity of the river in particular and the invertebrates in particular will also involve the near-threatened insect species such as those of dragonflies that reported in the Red List of IUCN (2010).

18.6 The Vertebrate Fauna of Al-Kahla River

18.6.1 Ichthyofauna

The history of the study of the fish fauna of Iraq goes back to the time of Sumer, Babylon and Assyria (Saggs 1962). Ancient Mesopotamians succeeded in identifying and naming several freshwater and marine species, which were recorded on clay tablets (Landsberger 1962). However, no further records are then available until the nineteenth century, when Günther (1874) reported on the fauna of the River Tigris. Few works on fishes in the Middle East were published in the first half of the eighteenth century that contain descriptions of a number of species later reported from Iraq. The authors of these works did not collect the specimens from Iraqi waters, and thus, they are considered out with the scope of this study. During the 137 years since the publication of Günther (1874), a large number of publications have dealt with several aspects of Iraqi fish taxonomy such as description of new species, range extensions and checklists of species of both freshwater and marine species.

18.6.1.1 Freshwater Fish Species

The total number of the freshwater fish species living in Al-Kahla River is 33 belonging to 25 genera contained in 12 families. Members of the family Cyprinidae dominate this river as they did in the remaining Tigris-Euphrates rivers system. There are 18 species belonging to 12 genera in this family. The genus *Barbus* represents the largest genus of the family Cyprinidae with 5 species (Table 18.8).

Table 18.8 List of freshwater fish species present in Al-Kahla River

Scientific name	Family
Acanthobrama marmid	Cyprinidae
Alburnoides bipunctatus	Cyprinidae
Alburnus caeruleus	Cyprinidae
Alburnus mossulensis	Cyprinidae
Aspius vorax	Cyprinidae
Barbus (Barbus) lacerta	Cyprinidae
Barbus (Carasobarbus) luteus	Cyprinidae
Barbus (Mesopotamichthys) sharpeyi	Cyprinidae
Barbus (Luciobarbus) subquincunciatus	Cyprinidae
Barbus (Luciobarbus) xanthopterus	Cyprinidae
Carassius auratus	Cyprinidae
Ctenopharyngodon idella	Cyprinidae
Cyprinus carpio	Cyprinidae
Garra rufa	Cyprinidae
Garra variabilis	Cyprinidae
Hemiculter leucisculus	Cyprinidae
Hypophthalmichthys molitrix	Cyprinidae
Hypophthalmichthys nobilis	Cyprinidae
Cobitis taenia	Cobitidae
Paracobitis malapterura	Balitoridae
Glyptothorax kurdistanicus	Sisoridae
Silurus triostegus	Siluridae
Heteropneustes fossilis	Heteropneustidae
Mystus pelusius	Bagridae
Liza abu	Mugilidae
Aphanius dispar	Cyprinodontidae
Aphanius mento	Cyprinodontidae
Aphanius mesopotamicus	Cyprinodontidae
Gambusia holbrooki	Poeciliidae
Poecilia latipinna	Poeciliidae
Mastacembelus mastacembelus	Mastacembelidae
Oreochromis aureus	Cichlidae
Oreochromis niloticus	Cichlidae
Tilapia zillii	Cichlidae

Exotic Fish Species in the Marshes

There are 12 exotic freshwater species in the Tigris-Euphrates system that includes Al-Kahla River. These fish species have been introduced during the second half of the twentieth century (Coad 1996). Few but specific data are available on the effect of these exotic species on native fish species living in the southern reaches of Tigris River. The Common Carp *Cyprinus carpio* was introduced between 1960 and 1972 to Iraq. It has been threatening at least three native species (*Barbus sharpeyi, Barbus grypus* and *Barbus xanthopterus*) since the 1970s. This has reportedly been caused by a sharp increase in its number and increased benthic food competition (Al-Kanaani 1989, Jawad 2003). Native species have reportedly also become rare after being outcompeted by other introduced species, e.g. *Barbus sharpeyi* by *Ctenopharyngodon idella* (Richardson 2008; Barak and Mohamed 1983; Jasim 1988).

The Stinging Catfish *Heteropneustes fossilis* was probably introduced in the 1950s as a biological control agent against *Bulinus truncatus*, the snail that acts as an intermediate host for *Schistosoma haematobium*. It competes with the endemic *Barbus sharpeyi*, but competition is not as strong as with the Common Carp. *Heteropneustes fossilis* is poisonous, with records of fatalities among inhabitants of rivers in southern Iraq (Coad 1996).

The Mosquito Fish *Gambusia holbrooki/affinis* was introduced to Iraq at an unknown date as a biological control agent against mosquito vectors of malaria. This reportedly caused devastation among native fishes, as the species also feed on fish eggs (Jawad 2003).

Coad (1996) reported that the Gold Fish *Carassius auratus* may hybridize with the endemic *Barbus sharpeyi*. This may dilute the original gene pool of the latter species or reduce net fecundity, if sterile hybrids are produced (Table 18.9).

Table 18.9 List of exotic fish species in Al-Kahla River

Scientific name	Family
Ctenopharyngodon idella	Cyprinidae
Cyprinus carpio	Cyprinidae
Carassius carassius	Cyprinidae
Hemiculter leucisculus	Cyprinidae
Hypophthalmichthys molitrix	Cyprinidae
Hypophthalmichthys nobilis	Cyprinidae
Heteropneustes fossilis	Heteropneustidae
Gambusia holbrooki	Poeciliidae
Poecilia latipinna	Poeciliidae
Oreochromis aureus	Cichlidae
Oreochromis niloticus	Cichlidae
Tilapia zillii	Cichlidae

Table 18.10 List of marine fish species possibly present in Al-Kahla River

Scientific name	Family
Thryssa hamiltonii	Engraulidae
Thryssa mystax	Engraulidae
Tenualosa ilisha	Clupeidae
Liza klunzingeri	Mugilidae
Rhynchorhamphus georgi	Hemirhamphidae
Acanthopagrus latus	Sparidae
Bathygobius fuscus	Gobiidae

Marine Fish Species

The fish species listed in Table 18.10 are the marine species that are recorded to be present in Al-Kahla River. Some of the marine fish species shown to travel up the Tigris and Euphrates Rivers to varying degrees but dams and water diversion schemes prevent their northward movement.

More marine species are reported from the marshes near to Al-Kahla River. The aquatic environment of these marshes and especially Al-Huwaiza Marsh which feed by Al-Kahla River is more or less similar to that of Al-Kahla River. Therefore, there is no reason why more marine species can invade this waterway. Among the reasons stopped reporting, their presence in this river is the severe lack of ichthyological investigations in the area which for sure will aid in turning up more marine species.

18.6.2 *The Use of Inland Water System as Necessary Grounds for Fish*

The importance of the rivers in the lower reaches of Tigris-Euphrates Rivers system as nursery grounds to the marine species of the Arabian Gulf is documented for several rivers and marshes in the area including Al-Kahla River and its adjacent Al-Huwaiza Marsh. Some species utilize deep water river channels for migration. Marine fish and invertebrates like certain species of shrimp migrated into the southern marshes and rivers. For conservation, the marine fish species and penaeid shrimp that migrate up the stream need to have favourable water conditions such as temperature, flow and salinity in order to allow uninterrupted migration corridors, spawning substrate and water conditions (Salman and Bishop 1990).

18.6.3 *Human Use of Fish*

Cyprinids have long been popular food fishes in Eurasia in general and in the Mesopotamia in particular. In Iraq and in the lower reaches of Tigris River, there are 12 cyprinid species of economic importance. Among this number, there are

5 *Barbus* species. In addition, there are several fish farms in different parts of Iraq that culture several cyprinid species. This aquaculture activity will fulfil the huge demand for fish consumption in Iraq in general and local people in the southern Iraq as fish meal represents their main food. Iraqi people have several recipes where fish enter as the main component. For example, the marsh Arabs used historical method of eating fish. This method is drying the fish in the sun after salting. The Sumerians, the ancestors of the marsh Arabs, used to practice this way of fish processing around 5000 years B.C.

Another major industry involving cyprinids is the aquarium trade. Although most of colourful aquarium cyprinid species come from South East Asia, there are several cyprinids species that used in this trade are living in Iraq. Among these species are *Carassius auratus* and *Cyprinus carpio*.

The silurid fish, *Silurus triostegus*, is another common freshwater fish species which is common in the rivers and marshes of the south of Iraq. There is huge potential of economic use of this species in Iraq. Although the Shiat Muslem that represents the majority of population in southern Iraq do not eat this fish or specific religious reasons, this fish is usually caught from the rivers and marshes of southern Iraq and sell in other Iraqi provinces where it consumed in large quantities.

Fisheries in the river and marshes in southern Iraq are crucial to the life of the local people as it is one of their main economic foundations of the traditional culture (Tkachenko 2002).

Over the previous years, there was significant inland fish catch of Iraq with more than 60% coming from the rivers and marshes in southern Iraq (Partow 2001). These catches will tend to decline if serious changes in the resource base and reduction in the activity of local people are introduced.

Following the re-flooding of the nearby Marshes since 2003, fisheries have returned to some areas of the marshes and flourished in the rivers on the other hand. Therefore, new destruction to the aquatic environment will not be tolerated anymore by the living organisms. *Barbus sharpeyi* was still caught but at much-reduced numbers and size, while the introduced cyprinid *Carassius carassius* comprised 46% of the summer 2004 catch in Al-Kahla River and the adjacent Al-Huwaiza Marshes, respectively. The catfish *Silurus triostegus* also increased in relative catch (up to 60% weight) (Development Alternatives Inc. 2004). The rivers and marshes of the southern Iraq are also crucial for fisheries elsewhere. Several marine fish species of great economic importance in the Arabian Gulf are dependent on the estuarine systems and marshes, either for spawning like the Clupeidae (Hussain and Jabir 1994) or for feeding, such as the Mugilidae and Sparidae (Hussain and Hamza 1987).

18.6.4 Threats and Sensitivities

With the start of the project activities in Al-Kahla River area, clear reduction in the biota that take the river as home to live in was observed. This significant reduction in the biota will change the nature of the environment of the river and make it impossible for fish (both marine and freshwater) and shrimps to take as nursery grounds.

Fish living in Al-Kahla River represents the major food source for the locals. They depend on this type of food for their daily consumption. With the destruction of the food resources that the fish depend on will have a direct effect on the abundance of the fishes in the area. Fish will move out of the river's environment looking for area rich in the food they usually feed on. This is an important social issue as locals might take it further to the local authority to find solution as the activity of the project will destroy their fishing grounds.

18.7 Herpetofauna of the Al-Kahla River

Amphibians and reptiles constitute an important and diverse fauna associated with both isolated wetlands and stream or river floodplains. Amphibians use a wide range of terrestrial habitats adjacent to wetlands and streams. Most of these habitats are related to foraging, refuge, or overwintering sites and typically consist of leaf litter, coarse woody debris, and small mammal burrows. Reptiles, such as turtles and aquatic snakes, often live and forage in aquatic habitats most of the year but immigrates to upland habitats to nest or overwinter (Bodie 2001).

The herpetofauna is poorly studied. So far, about 96 species of reptiles and amphibians have been recorded from Iraq (In-den-Bosch 2003), but only a relatively small proportion of them occur in the rivers ad marshes of southern Iraq. The country's herpetofauna was studied extensively during the 1920s when British troops were in Iraq (Boulenger 1918, 1919, 1920a, b; Corkill 1932; Angel 1936; Schmidt 1939). Earlier studies were summarized by Allouse (1955) and Mahdi and Georg (1969). The few scattered studied that were published in the mid-twentieth Century (Haas 1952; Reed and Marx 1959; Haas and Werner 1969) added little to the understanding of the herpetofauna of Iraq, but created some confusion themselves. The last significant series of articles on Iraqi herpetofauna were published in the early 1960s (Khalaf 1960 1961) and include a general account, again without localities (Khalaf 1959).

So far, in Al-Kahla River area, amphibian is represented by 6 species belonging to 4 genera and contained in 3 families. On the other hand, reptilian showed little more biodiversity variation than amphibian and shown to be represented by 8 species belonging to 15 genera contained in 8 families.

The information about the biodiversity of amphibian and reptiles in Iraq in general and in the river environment is very scarce; therefore, further herpetological investigations are needed as it may yield additional species.

The Euphrates Soft-shell Turtle *Rafetus euphraticus* is reported by the IUCN Red List of threatened species (IUCN 2010) as endangered species. In addition, the common tree frog, *Hyla arborea,* is reported to have near-threatened species.

Table 18.11 List of Amphibian and reptiles present in the area of Al-Kahla River

Scientific name	Family
Amphibia	
Bufo viridis	Bufonidae
Bufo viridis	Bufonidae
Hyla arborea	Hylidae
Hyla savignyi	Hylidae
Pelophylax ridibunda	Ranidae
Rana esculenta	Ranidae
Reptalia	
Clemmys caspia	Geoemydidae
Cyrtopodion scaber	Gekkonidae
Cyrtopodion heterocercum	Gekkonidae
Stenodactylus doriae	Gekkonidae
Stenodactylus sleveni	Gekkonidae
Bunopus tuberculatus	Gekkonidae
Asaccus elisae	Gekkonidae
Hemidactylus flaviviridis	Gekkonidae
Hemidactylus persicus	Gekkonidae
Ophisops elegans	Lacertidae
Mabuya aurata septemtaeniata	Scincidae
Trachylepis vittata	Scincidae
Eryx jaculus	Boidae
Platyceps ventromaculatus	Colubirdae
Natrix tessellata	Colubirdae
Rafetus euphraticus	Trionychidae
Mauremys caspica	Bataguridae
Varanus griseus	Varanidae

Among the causes that contribute to the threat of the herpetofauna in the aquatic environments in southern Iraq are (1) Destruction of the Mesopotamian marshes and consequent degradation of the environment of the neighbouring rivers; (2) Destruction of forest cover; (3) Desertification and drought; (4) Pollution and contamination; (5) Military Operations and Insurgency Conflicts; and (6) Irrigation and Soil Salinization (Table 18.11).

18.7.1 Role of Amphibian and Reptilian Species in the Aquatic Environment

The ecological roles of amphibian and reptiles in the river's environment as potential regulators of decomposer populations in ecosystems and ecosystem processes associated with detritus-litter food webs are already documented Konishin et al. 2001).

Some researchers have suggested that amphibian in particular and reptilian in general may provide an important indirect regulatory role in the processing of detritus-litter by ingestion of detritivore prey (Stebbinsm and Cohen 1995). Amphibian usually feed on invertebrates, which themselves feed on the bacteria and fungi in the river floor microflora, would promote a more rapid rate of leaf litter decomposition.

Amphibian and reptilian live in the river environment may play a dispersal vector during their movement in the area. Amphibian may transport food items such as molluscs and shrimp eggs originating from the river's environment out to another adjacent water body. In this way, they contribute in the dispersal of the invertebrate species between adjacent water bodies.

Amphibian also may play an important role in the riparian ecotone via processes of chemical transformation. They are reported to oxidize ingested aromatic hydrocarbons followed by conjugations to glucuronides and organic sulphates (National Research Council 1981). Given their moderate lifespans and high numbers in ecosystems, amphibian may be a critical food-web link in the bioaccumulation of persistent chemicals such as mercury and PCBs. They play a good toxicological role as elemental sinks, chemical transformers, and cross-links of organic molecules and heavy metal ions between aquatic and terrestrial environments (Sparling et al. 2000).

18.7.1.1 Threats and Sensitivities

The reduction in the amphibian and reptilian abundance will affect the dispersal process from the riverine environment to another water body. Those two animal groups play a dispersal vector during their movement in the area. Amphibian may transport food items such as molluscs and shrimp eggs originating from the river's environment out to another adjacent water body.

As a result of project activity, amphibian role as chemical transformers will be affected. They are reported to oxidize ingested aromatic hydrocarbons followed by conjugations to glucuronides and organic sulphates.

18.8 Aquatic Avifauna of Al-Kahla River

Numerous studies have been carried out on the avifauna in urban landscapes worldwide (Marzluff et al. 2001). This includes studies addressing species richness and its relationship to their living area (Park and Lee 2000) and studies on birds in riparian systems of urban areas, and ponds and lakes in cities (Trauti and Hostetler 2003).

Rivers and wetlands support and maintain a diverse community of birds (Duncan et al. 1999). Man has been aware of the link between birds and wetlands for thousands of years. These habitats are useful for birds for breeding, nesting and rearing of young (Acuna et al. 1994). Many wetlands such as those present in the southern Iraq are stopovers for migratory birds. Declining number of wetland-associated birds is partly attributed to the loss and destruction of wetlands (Mads et al. 2002).

The birds of Iraq have long investigational history. It goes back to the year 1886 when Sharp published his work on the bird of Iraq in general. Latter and during the nineteenth century, large number of publications have been appeared and concentrate on the taxonomic feature of avifauna in Iraq. Among these studies are Sassi (1912), Tomlinson (1916), Ticehurst (1920a, b), Chapman and McGeoch (1956), Harrison (1959), Marchant (1963a, b, c), Kainady (1976a, b) and Mahdi (1982). The most comprehensive accounts of Iraqi birds were published by Allouse (1953) and Moore and Boswell (1956–1957). They stated that the total number of bird species recorded in Iraq was 375, of which 134 were aquatic. Recently, Salim and Porter (2006) published 'A Field Guide to the Birds of Iraq', which covers 387 species of birds, and Porter and Salim (2010) compiled an up-to-date, critically revised checklist of the birds of Iraq.

The waterbirds in Iraq have been studied extensively by several workers, e.g. Georg Kainady and Vielliard (1968, 1970), Koning and Dijksen (1973), Carp (1975a, b, 1980), Scott and Carp (1982) and Scott (1995). These studies provide the best information about wintering avifauna and indicate that the total wintering population of waterfowl in Iraq in the 1960s and 1970s may have been several million birds, with the southern rivers and marshes being the main centre of distribution. The survey in 1979 on the water birds of Iraq has reported over 475,000 waterfowl belonging to 81 species (Scott 1995).

The southern rivers of Iraq and the marsh areas were put in the context of bird migration throughout western Asia as they are important major wintering and resting areas for migrating waterbirds in western Eurasia (Scott and Evans 1994). Georg and Savage (1968) considered Al-Huwaiza marsh and its adjacent rivers including Al-Kahla River among the possible habitats that inhabited by two-third of the wintering waterfowl of the Middle East.

The total number of aquatic bird species visiting the vicinity of Al-Kahla River is 53. These species are belonging to 38 genera contained in 7 orders. The order Charadriiformes is the largest order of the seven orders of aquatic birds in the area with 17 species belonging to 7 genera following by Ciconiiformes with 15 species belonging to 11 genera and Ralliformes with 9 species and 8 genera. The remaining three orders are represented by lower number of species, Podiciformes with 4 species belonging to 3 genera; Pelecaniformes with 3 species belonging to 3 genera; and Coraciiformes with 2 species belonging to 2 genera (Table 18.12).

Among the aquatic bird species that has been stopped since 1980s is the Sacred Ibis, *Threskiornis aethiopicus* (Scott 1995). Recently, 27 individuals of this species have been discovered in the vicinity of lower reaches of Al-Kahla River in area adjacent to Al-Huwaiza Marsh (Abed 2007).

Other than the aquatic bird fauna that recorded in the vicinity of Al-Kahla River, there are number of bird species that reported to pay visit to the area of Al-Kahla River. The total number of species of these non-aquatic birds is 13 belonging to 18 genera contained in 5 orders. The largest order is Passeriformes with 10 species

Table 18.12 List of aquatic bird species recorded in the vicinity of Al-Kahla River

Scientific name	Common English name
Order: Anseriformes	
Anser anser	Greylag goose
Aythya fuligula	Tufted duck
Gallinula chloropus	Moorhen
Porzana porzana	Spotted crake
Order: Charadriiformes	
Alceda atthis	Common kingfisher
Childonias hybridus	Whiskered tern
Childonias leucopterus	White-winged black tern
Gallinago gallinago	Common snip
Larus argentatus	Herring gull
Larus cnus	Common gull
Larus genei	Slender-billed gull
Larus icthyaetus	Great black-headed gull
Larus minutus	Little gull
Larus ridibundus	Black-headed gull
Limosa limosa	Black-tailed godwit
Sterna albifrons	Little tern
Sterna hirundo	Common tern
Sterna repressa	White-cheeked tern
Tringa nebularia	Greenshank
Tringa ochropus	Green sandpiper
Tringa stagnatilis	Marsh sandpiper
Order: Ciconiiformes	
Anas crecca	Teal
Anas penelope	Wigeon
Anas platyrhynchos	Mallard
Anser anser	Greylag goose
Ardea cinerea	Grey heron
Ardea purpurea	Purple heron
Ardeola ralliodes	Squacco heron
Bubulcus ibis	Cattle egret
Coconia coconia	White stork
Egretta alba	Great white heron
Egretta garzetta	Little egret
Nycticorax nycticorax	Night heron
Platalea leucorodia	Spoonbill
Plegadis falcinellus	Glossy ibis
Threskiornis aethiopicus	Sacred ibis
Order: Coraciiformes	
Halcyon smyrnensis	White-breasted kingfisher
Ceryl rudis	Pied kingfisher

(continued)

Table 18.12 (continued)

Scientific name	Common English name
Order: Pelecaniformes	
Anhinga rufa	Darter
Botaurus stellaris	Bittern
Lxobrychus minutus	Little bittern
Order: Podiciformes	
Phalacrocorax carbo	Cormorant
Phalacrocorax pygmeus	Pygmy cormorant
Podiceps cristatus	Crested grebe
Tachbaptus ruficollis	Little grebe
Order: Ralliformes	
Chetusia lecura	White-tailed
Calidris minuta	Little stint
Calidris alpina	Dunlin
Charadrius alexandrius	Kentish plover
Fulica atra	Coot
Himantopus himantopus	Black-winged stilt
Porophyrio porophyrio	Purple gallinule
Recurvirostra avosetta	Avocet
Vanellus indicus	Red-wattled plover

belonging to 9 genera followed by Falconiformes with 5 species belonging to 3 genera. Each of the orders Columbiformes and Coraciiformes has 3 species belonging to 3 genera while Galliformes showed to have 2 species belonging to 2 genera (Table 18.13).

This no-aquatic bird group shares the environment of Al-Kahla River with the aquatic avifauna living in the same area. Both groups are integrated with each other to form a balanced ecosystem as they share the same natural resources.

The IUCN database lists 404 species of birds that are believed to exist in Iraq, of which 14 are included in the global list of threatened species (Table 18.14). The two endemic birds of Iraq, the Basra reed warbler, *Acrocephalus griseldis*, the Iraq babbler, *Turdoides altirostris,* are southern Iraq Rivers and marshland birds and are found primarily in this area. Two of these species, the Iraq babbler and Basra reed warbler, are known to breed only in the lower reaches of Mesopotamia.

There are seven aquatic bird species/subspecies that are endemic to the southern rivers and marshes of Iraq. The IUCN status of these species is shown in Table 18.15. The endemic subspecies in southern Iraq Rivers and marshes can be regarded as examples of 'evolution in progress'. These populations of southern Iraq aquatic birds should be considered of similar conservation value as the populations of globally threatened species because (1) the evolutionary process itself is increasingly seen as a target of conservation management, and (2) the possibility that the rivers and marshes of southern Iraq have given rise to several examples of ongoing bird evolution highlights the unique quality of this area. These species also

Table 18.13 List of birds other than aquatic species recorded in the vicinity of Al-Kahla River

Scientific name	English Common name
Order: Columbiformes	
Columba livia	Rock dove
Pterocles alchata	Pin-tailed sandgrouse
Streptopelia decaocto	Collared dove
Order: Coraciiformes	
Coracias benghalensis	Indian roller
Merops superciliosus	Blue-cheeked bee-eater
Upupa epops	Hoopoe
Order: Falconiformes	
Accipter nisus	Sparrowhawk
Aquila chrysaetos	Golden eagle
Aquila nipalensis	Steppe eagle
Circus aeruginosus	Marsh harrier
Circus cyaneus	Hen harrier
Order: Galliformes	
Ammoperdix griseogularis	See-see partridge
Francolinus francolinus	Black francolin
Order: Passeriformes	
Ammomanes deserti	Desert lark
Corvus corax	Raven
Corvus corone cornix	Hooded crow
Erithacus rubecula	Robin
Galerida cristata	Crested lark
Mtacilla alba	White wagtail
Passer domesticus	House sparrow
Pycnonotus leucogenys	White-cheeked bulbul
Sturnus vulgaris	Starling
Turdoides altirostris	Iraq babbler

contribute to the status of the lower reaches of Tigris and Euphrates Rivers as one of only a few global high-priority centres of bird endemism (Bird Life-International 2010).

18.8.1 The Role of Aquatic Birds in the River Environment

Avifauna plays important roles in the river's environment, and here, these roles are briefly discussed according to their importance to the ecosystem.

Consumption of the energy at the higher levels of food chain makes birds useful biological indicators (Furners et al. 1993). Wetland bird communities are important biological indicators.

Table 18.14 List of threatened bird species recorded in the vicinity of Al-Kahla River

Scientific name	English common name	Biological status in the area	IUCN status
Acrocephalus griseldis	Basrah Reed warbler	Breeding	EN
Anser erythropus	Lesser white-fronted goose	Wintering	VU
Aquila clanga	Greater spotted eagle	Wintering	VU
Aquila heliaca	Eastern Imperial eagle	Wintering	VU
Chlamydotis macqueenii	Macqueen's bustard	Breeding	VU
Falco naumannii	Lesser kestrel	Passage	VU
Marmaronetta angustirostris	Marbled teal	Breeding and wintering	VU
Neophron percnopterus	Egyptian vulture	Passage	EN
Pelecanus crispus	Dalmatian pelican	Wintering	VU
Branta ruficollis	Red-breasted goose	Rare winter vagrant	EN
Falco cherrug	Saker falcon	Scarce winter visitor	EN
Haliaeetus leucoryphus	Pallas's fish-eagle	Scarce winter visitor	VU
Numenius tenuirostris	Slender-billed curlew	Regular winter visitor	CR
Oxyura leucocephala	White-headed duck	Rare winter visitor4	EN
Vanellus gregarious	Sociable lapwing	Passage	CR

Table 18.15 List of endemic aquatic bird species/ subspecies in the vicinity of Al-Kahla River

Scientific name	English common name	IUCN status
Anhinga rufa chanteri	African (Levant) darter	Species LC, subspecies possibly CR
Corvus corone capellanus	Hooded crow	Species LC, subspecies possibly LC
Francolinus francolinus arabistanicus	Black francolin	Species LC, subspecies possibly NT
Hypocolius ampelinus	Grey Hypocolius	Regional endemic LC
Pycnotus leucotis mesopotamiae	White-eared bulbul	Species LC, subspecies possibly LC
Tachybaptus ruficollis iraqensis	Little grebe	Species LC, subspecies possibly VU
Turdoides altirostris	Iraq babbler	Species LC

There are many factors affecting the relationship between wetlands characteristics and birds. These include the availability of habitats and healthy water, food, shelter and predators.

Birds are responsible for certain degree about the trophic state of a water body by importing nutrients. One of the very important functions that the wetlands perform is to provide suitable site for the breeding of the resident as well as a wintering ground for short- and long-distance migratory water birds. Palmgren (1936) related the trophic state of lakes and rivers (e.g. eutrophic, oligotrophic) to bird communities (species composition) and their abundance.

The capacity of aquatic plants and invertebrates to colonize new habitats and distribute themselves over large geographic ranges has long fascinated naturalists

(Lyell 1832). Such ubiquity has often been ascribed to the potential for waterfowl to transport propagules of such organisms (Avise 2000). Waterbirds can carry plant and animal propagules both externally (attached to plumage or feet) and internally (after surviving digestion; Figuerola and Green 2002). As waterfowl is important vectors for invertebrate dispersal, waterfowl movements should be a better predictor of invertebrate genetic structure than geographical distances between populations (Akihisa and Satoshi 2001).

Many aquatic plants and invertebrates lack the mobility necessary to travel directly from one catchment to another, to colonize new areas and to disperse to neighbouring water bodies in different catchments. Despite this apparent isolation of freshwater habitats, many aquatic plant and invertebrate species have widespread distributions (Brown and Gibson 1983; WCMC 1998), spanning several continents in some cases.

Many waterbirds undertake long migratory journeys from their breeding areas at extreme latitudes towards more temperate areas during the winter (Del Hoyo et al. 1992). *Daphnia laevis* reflects the major waterfowl (ducks, geese and swans) flyways in North America, with higher similarity between populations in a North-South than an East-West direction (Taylor et al. 1998). Similarly, the genetic distribution of the bryozoan *Cristatella mucedo* in northern Europe follows the major waterfowl flyway in this region (Freeland et al. 2000).

Aquatic birds are responsible for level of nutrients in the rivers and marshes where they live. Post (1998) demonstrated that nutrient loading is proportional to bird densities and that birdborne nutrients could account for high percentage of the nitrogen and the phosphorus input to these systems. Chlorophyll concentrations might be affected by the addition of nitrogen and phosphorus in guano.

Stable isotope results from fish and crayfish demonstrate that the nitrogen transported by aquatic birds is incorporated in the pond food webs. It provides a cumulative indicator that corresponds with the modelling results—birds nearly double the nitrogen load to refuge ponds where high roosting densities develop (Post 1998).

18.8.1.1 Threats and Sensitivities

With the reduction of food items as a result of project activities, the number of aquatic birds will reduce accordingly. This reduction in number of aquatic birds will be coinciding with the brake of the relationship between wetlands characteristics and birds.

As a result of what will happen in no. 24, riverine environment will not be suitable breeding and wintering sites as the trophic state of the river's environment and the bird communities will be affected.

With the reduction of the number of birds in the area, the process of transporting aquatic plan and invertebrate's propagules to a new environment will be very slow or even stop completely.

The level of nutrients in the rivers and marshes where the aquatic birds live will be disrubted as the result of reduction in bird's number. That nutrient loading is proportional to bird densities and that birdborne nutrients could account for high percentage of the nitrogen and the phosphorus input to these systems.

18.9 Mammals Living in the Vicinity of Al-Kahla River

The mammal fauna of Iraq in general and that found in the vicinity of the rivers and marshes in southern Iraq in particular are poorly studied. The present studies showed that the total number of mammal species in Iraq is 74. This number is quite low if it compares with the adjacent countries like Iran and Turkey. Certainly, the future regular and integrated mammal investigations will add more species to the present check-list.

Similarly, the mammal species that inhabit the rivers and marsh environment are in the same position of the terrestrial mammal species in having severe short in the studies that related to their taxonomy and biodiversity.

The list of mammal species that recorded to be present in the environment of the rivers and marshes in the southern Iraq is shown in Table 18.16. The list is based mainly on the work of Harrison and Bates (1991) and Scott (1995).

The total number of mammal species that living in the vicinity of Al-kahla River is 38 belonging to 31 genera and contained in 6 orders. The largest order in the area is Varnivora with 11 species belonging to 9 genera followed by Rodentia with 11 species belonging to 8 genera. The orders Chiroptera and Insectivora have 8 and 5 species belonging to 6 and 4 genera, respectively. There are only 2 species of Artiodactyla belonging to 2 genera while order Lagomorpha contains one species only.

Due to the heavy demand on the skin of the two species of otter, the Common Otter *Lutra lutra* and the Smooth-coated Otter (or Maxwell's Smooth-coated Otter) *Lutrogale perspicillata maxwelli,* the two species became extremely rare in the lower reaches of Tigris-Euphrates River systems.

The IUCN Red List of threatened species contains seven species that are known to occur in the rivers and marshes in the southern Iraq and considered globally threatened. Three of them are near-threatened (NT), one is endangered (EN), and three are vulnerable (VU) (Table 18.17).

18.9.1 The Role of Mammals in the River Environment

Interestingly, no mammals appear unique to any aquatic ecosystem, given that each species has been reported to be distributed extensively in or along freshwater inland marshes or along the borders of small ponds, or streams (Daiber 1982). However, nearly all of these species benefit from the rich aquatic and terrestrial food found the

Table 18.16 List of mammal species that recorded in the vicinity of Al-Kahla River

Scientific name	Common name	IUCN status
Order Insectivora		
Hemiechinus auritus	Long-eared hedgehog	LC
Paraechinus aethiopicus	Ethiopian hedgehog	LC
Crocidura suaveolens	Lesser white-toothed shrew	LC
Suncus murinus	Asian house shrew	LC
Suncus etruscus	Pygmy white-toothed shrew	LC
Order Chiroptera		
Rhinopoma hardwicke	Lesser mouse-tailed bat	LC
Taphozous nudiventris	Naked-rumped tomb bat	LC
Eptesicus bottae	Botta's Serotine	LC
Eptesicus nasutus	Sind Serotine bat	LC
Pipistrellus kuhlii	Kuhl's Pipistrelle	LC
Pipistrellus rueppellii	Rüppel's Pipistrelle	LC
Otonycteris hemprichii	Desert long-eared bat	LC
Myotis capaccinii	Long-fingered bat	VU
Order Carnivora		
Canis aureus	Golden jackal	LC
Canis lupus	Grey wolf	LC
Vulpes vulpes	Red fox	LC
Lutrogale *perspicillata maxwelli*	Smooth-coated otter	VU
Lutra lutra	Eurasian otter	NT
Herpestes javanicus	Small Indian mongoose	LC
Mellivora capensis	Honey badger	LC
Hyaena hyaena	Striped Hyaena	NT
Felis silvestris	Wild cat	LC
Felis chaus	Jungle cat	LC
Caracal caracal	Caracal	LC
Order Artiodactyla		
Gazella subgutturosa	Goitered gazelle	VU
Sus scrofa	Eurasian wild pig	LC
Order Lagomorpha		
Lepus capensis	Cape hare	LC
Order Rodentia		
Hystrix indica	Indian crested porcupine	LC
Allactaga euphratica	Euphrates jerboa	NT
Jaculus jaculus	Lesser Egyptian jerboa	LC
Gerbillus mesopotamicus	Harrison's gerbil	NE
Gerbillus cheesmani	Cheesman's gerbil	LC
Tatera indica	Indian gerbil	LC
Meriones crassus	Sundevall's Jird	LC
Nesokia bunnii	Bunn's short-tailed bandicoot rat	EN
Nesokia indica	Short-tailed bandicoot rat	LC
Rattus rattus	Black rat	LC
Rattus norvegicus	Brown rat	LC

Table 18.17 List of endangered mammal species reported to live in the vicinity of Al-Kahla River

Scientific name	Common name	IUCN status
Order: Chiroptera		
Myotis capaccinii	Long-fingered bat	VU
Order: Carnivora		
Lutrogale perspicillata maxwelli	Smooth-coated otter	VU
Lutra lutra	Eurasian otter	NT
Hyaena hyaena	Striped hyena	NT
Order: Artiodactyla		
Gazella subgutturosa	Goitered gazelle	VU
Order: Rodentia		
Allactaga euphratica	Euphrates jerboa	NT
Nesokia bunnii	Bunn's short-tailed bandicoot rat	EN

banks of these areas. As is typical of most small mammal populations, local populations increase when living conditions are ideal and decline when they are severe, resulting in great fluctuations in numbers from year to year and even from season to season (Webster et al. 1985).

The significant of mammals living in or near the aquatic environment can be seen through the following main aspects:

1. On the banks of rivers and marshes, large number of small mammals uses their long claws and strong forefeet to turn plant litter and dig shallow scratchings, pits and/or holes when foraging for a variety of foods that include fruits, seeds, roots, insects, fungi, tubers and invertebrates (Strahan 1995). During the foraging process, they mix organic matter into the soil, spread mycorrhizal fungi and seeds and improve conditions for water collection, absorption and seed germination (Garkaklis et al. 1998). These diggings and scratchings therefore represent a substantial, ongoing importance to the soil. The action of their digging moves the soil under and behind the animal often resulting in the soil being turned over and organic matter broken into smaller particles and mixed into the soil (Soule and Piper 1992). Intermixed organic matter provides substrate for a wide range of soil biota, including bacteria, fungi, actinomycetes, nematodes, algae, protozoa and viruses (Killham 1994) and is the first step in soil building. In the absence of this substrate, soil biota is dramatically reduced in number and variety (Jordon 1998). This process may be particularly important in drier areas where a soil crust can form underneath plant litter which could prevent its utilization by soil biota. This intermixing of organic matter with the soil is important for nutrient cycling (Andersen and MacMahon 1985). The breaking up of organic matter and mixing into the soil also has the potential to reduce the mass of readily combustible plant material accumulating on the banks.
2. The other significant importance of digging of small mammals is to increase and maintain the biodiversity of other animal groups living in the same environment. There is a relationship between the successional stages of biological soil crusts

and the distribution of lizards, geckos and diurnal lizards as these organisms strongly preferred the fragile crust and the distributions of burrows of these species are affected by soil surface characteristics. Soil crusts may thus play an important role in determining burrow location. If the crust left untouched for several years, these fragile crusts converted into harder and thicker crusts due to successional processes. When fully developed, the hard crusts are avoided by the lizards as they are probably too hard to dig into.

3. Some small mammals may also help maintain the balance of trees, shrubs and grasses in the river's bank environment (Noble 1993) by browsing on seeds and new seedlings that germinate after rain.
4. Depending on the season, herbivores may act as predators (by removing plants or seeds) or as parasites (by partially reducing plant biomass), or they may promote some form of mutualistic association (by distributing seed and plant fragments to favourable growth environments (Crawley 1983). Feeding method and seasonal variations in food chemistry all strongly influence riparian community structure, plant physiognomy, competition, soil development and propagule dispersal. The influence of herbivores on plants should seem self-evident because the ecological effects and consequences of feeding on the plant community are well known (DeAngelis et al. 1973).
5. Metal concentrations in small mammals can provide information regarding potential for bioaccumulation, within the ecosystem and food chain transport of these contaminants (Johnson et al. 1978). Stable isotope analysis of small mammal tissues can provide important information on these processes because, when properly applied, this technique can produce estimates of trophic position within complex trophic webs of ecological communities (Post 2002).
6. Small mammals are important animals in riverine ecosystems, because of their feeding and burrowing activities and their role in food webs (Wijnhoven et al. 2005). They are important prey items for several endangered or protected birds of prey and carnivorous mammals (Leuven et al. 2005), for which many aquatic areas are assumed to function as important conservation areas (De Nooij et al. 2004). The distribution of small mammal species in the riverine environment is very important as they considered as predators predominantly forage in areas with abundant prey. Riverine landscapes are dynamic, and biologically and spatially complex, characterized by a successional landscape mosaic with high habitat heterogeneity (Robinson et al. 2002).
7. The large mammal species in the riverine environment play an important role as they considered as the source of nutrients to the aquatic environment through their faeces that accumulate on the banks and sweep to the water of the river after the first rainfall in the area.

18.9.1.1 Threats and Sensitivities

Reduction in the aquatic biota of Al-Kahla River as a result of project activities will in turn reduce the number of small mammals that live on the bank of the river. Such reduction will reduce their normal activities in digging a scratching the soil of the bank searching for food. This activity will decrease the mix organic matter into the soil, spread mycorrhizal fungi and seeds and make less improve for conditions of water collection, absorption and seed germination. The reduction in digging activity will reduce the biodiversity of other animal groups living in the same environment.

Among the effect of the activity of the project in the area and due to reduction in small mammal number, the ecological effects and consequences of feeding on the plant community will be disrubted.

When the feeding habit of the small mammal in the riverine environment changes the number of prey that those mammals used to feed on will increase. Such imbalance in the ecosystem will affect other biota in other trophic levels of Al-Kahla River.

The source of nutrient that originates from the faeces of large mammal species that live in the riverine environment play will be reduced or stop completely. Such changes will have dramatic changes in the aquatic environment of the river in question.

18.10 Factors Affecting the Aquatic Biodiversity of Al-Kahla River Other than Changing Water Masses

As with other industrial projects, there will be direct and indirect effects of the project on the biota of aquatic environment and any petrol-related project in the area of the Al-Kahla River. There are different kinds of such effect; among the most important effect is the pollution in its different types.

Potential Contamination Sources in the river environment may include but are not limited to: the existing leaks and spills associated with the existing equipment present in the area, tanks that used for fuel and chemical storage, pipes and valves. The possible contaminants that seem to have a direct effect on the river environment biota include hydrocarbons such as fuels, oils and lubricants, acids, polymers, glycols, methanol, corrosion inhibitors and biocides. Besides the above-mentioned contaminants, wastewaters such as ‘produced water’ and general waters such as run-off and sewage are also listed under the possible materials that affect biodiversity of the river environment. It is obvious that all these may contain elevated concentrations of hydrocarbons, heavy metals, natural salts, sulphur (MacDonald 2010). The effects of the contaminants mentioned above have a pronounced effect on different life stages of the biota of Al-Kahla River. The discussion of such effect is beyond the scope of the present review. Therefore, a separate review is recommended to cover such effects.

References

Abd El-Ghani MM (2000) Vegetation composition of Egyptian inland saltmarshes. Bot Bull Acad Sin 41:305–314

Abdullah SK, Al-Dossari MA, Al-Saad HT (2000) A mycofloral study on aquatic sediments of Shatt Al-Arab estuary and north-west Arabian Gulf. Basrah J Sci 18:1–14

Abdullah SK, Al-Saadoon AH, Al-Salihy, M.H. (2007) Fungi from the tidal zone of Khawr Al-Zubair canal, southern Iraq. Marsh Bull 2:18–31

Abdullhasan NA (2009) Habitat mapping project of the proposed Iraqi marshlands National Sanctuary area. BioRisk 3:55–68

Abed JM (2007) Status of water birds in restored southern Iraqi marshes. Marsh Bulletin 2(1):64–79

Acuna R, Contras F, Kerekes J (1994) Aquatic bird densities in two coastal lagoon systems in Chiapas state, Mexico, a preliminary assessment. Hydrobiologia 279(280):101–106

Agostinho AA, Gomes LC, Pelicice FM (2007a) Ecologia e manejo de recursos pesqueiros em reservatórios do Brasil. Eduem, Maringá, p 501

Agostinho AA, Thomaz SM, Gomes LC, Baltar S (2007b) Influence of the macrophyte *Eichhornia azurea* on fish assemblage of the upper Paraná River floodplain (Brazil). Aquat Ecol 41:611–619

Ajeel SG (1998) Population dynamics and bioenergetics of two Cladocerans *Simocephalus vetulus* (Muller) and *Daphnia magna* (Straus) in Basrah with a reference to zooplankton. PhD Thesis, Basrah University, 154p

Ajeel SG, Ali MH, Salman SD (2006) Secondary production of the Cladoceran *Simocephalus vetulus* in a temporary Pool at Basrah. Iraq Marina Mesopotamica Online 1:38–47

Akihisa H, Satoshi M (2001) Habitat use and diversity of water birds in a coastal lagoon around Lake Biwa, Japan. Ecol Res 16:543–553

Al-Dossari MA (2008) Isolation and identification of some fungi from sediments of Southern marshes of Iraq and their ability in biodegradation of crude oil *in vitro*. PhD thesis, Basrah University 116pp

Al-Dossari MA, Al-Saad HT, Abdullah SK (2001) The biodegradation ability of some fungi isolated from sediments of Shatt Al-Arab estuary for some polycyclic aromatic compounds. Marina Mesopotamica 16:191–207

Alexopoulos CJ, Mims CW, Blackwell M (1996) Introductory mycology. Wiley, Hoboken, NJ, 869 pp

Al-Hilli MR (1977) Studies on the plant ecology of the Ahwar region in southern Iraq PhD Thesis, Faculty of Science University of Cario, Egypt, 460p

Al-Hilli MRA, Warner BG, Asada T, Douabul A (2009) An assessment of vegetation and environmetal controls in the 1970s of the Mesopotamian wetlands of southern Iraq. Wetl Ecol Manag 17:207–223

Ali HH (2011) Macrophytes of the lower reaches of Mesopotamia. M.Sc. Thesis, University of Ammarah, Iraq, 380p

Ali HY, Yousif SL (2010) Study on the topography of the Tigris River branches in lower reaches of Mesopotamia. Report to the Ministry of Agriculture, 230pp

Al-Kanaani SM (1989) Diet overlap among the common carp *Cyprinus carpio* L. and three native species in Al-Hammar marshes, southern Iraq. M.Sc. Thesis, Basrah University, Iraq (In Arabic)

Allouse BE (1953) The avifauna of Iraq. Al-Tafayyudh Press, Baghdad

Allouse BE (1955) A bibliography of the vertebrate fauna of Iraq and neighbouring countries. III Reptiles and amphibians. Iraq Natural History Museum publications no. 6. Baghdad, Iraq Natural History Museum. 5: 23 pp

Al-Mashhadani GU (2000) Study on the distribution of microalgae in the marsh area of the southern Iraq. J Sci 6:43–65

Al-Mayaha ARA, Al-Abbawya DAH, Widad MT, Al-Assadia WMT, Al-Saadia SAA, Al-Edanyb TY (2012) Status of aquatic Macrophytes in Saffia Nature Reserve, south of Huwaiza Marsh, Iraq. Marsh Bull 7:1–16

Al-Qarooni EHM (2005) Study of seasonal abundance of some aquatic invertebrates in southern Iraqi marshes. M.Sc. Thesis, Basrah University, 97p
Al-Saadi HA, Al-Lami AA, Kasim TI (1996) Algal ecology and composition in the Garmat Ali River, Iraq. Regul Rivers 12:27–38
Al-Saboonchi AA, Barak NA, Mohammed AM (1986) Zooplankton of Garma Marshes, Iraq. J Biol Sci Res Centre 17:33–40
Al-Saddi HA, Rattan SS, Muhsin TM, Hameed AH (1979) Possible relationship between phytoplankton numbers and saprolegnoid in Shatt Al-Arab River near Basrah, Iraq. Hydrobiologia 63:57–62
Andersen DC, MacMahon JA (1985) Plant succession following the Mount St. Helens volcanic eruption: facilitation by a burrowing rodent, *Thomomys talpoides*. Am Midl Nat 114:62–69
Angel F (1936) Reptiles et batraciens de Syrie et de Mesopotamie recoltes par M. P Pallary. Bull Inst Egypt 18:107–116
APHA, WWA, WEF (2005) Standard methods for examination of water and waste water, 21st edn. American Public Health Association, Washington, DC
Avise JC (2000) Phylogeography. Harvard University Press, Cambridge, MA
Barak NAA, Mohamed ARM (1983) Biological study of the cyprinid fish, Barbus luteus (Heckel). J Biol Sci Res 14:53–70
Barnes RSK, Hughes RN (1992) An introduction to marine ecology. Blackwell Scientific Publication, Oxford, 296 p
Bartram J, Balance R (eds) (1996) Water quality monitoring: practical guide to the design and implementation of freshwater quality studies and monitoring programmes. E. and F.N. Spon Co., London
Berner LM (1951) Limnology of the lower Missouri River. Ecology 32:1–366
Bianchini I (2003) Modelos de crescimento e decomposição de macrófitas aquáticas. In: Thomaz SM, Bini LM (eds) Ecologia e Manejo de Macrófitas Aquáticas. Eduem, Maringá, pp 85–126
Bird Life-International (2010) World Bird Database. http://www.birdlife.org/datazone. Retrieved 1 October 2010
Bodie JR (2001) Stream and riparian management for freshwater turtles. J Environ Manag 62:443–455
Boonchan S, Britz ML, Stanles GA (2000) Degradation and mineralization of high molecular weight polycyclic aromatic hydrocarbons by defined fungal-bacterial cultures. Appl Environ Microbiol 66:1007–1019
Boulenger GA (1918) Description of a new lizard of the genus *Acanthodactylus* from Mesopotamia. J Bombay Nat Hist Soc 25:373–374
Boulenger GA (1919) On a new variety of *Acanthodactylus boskianus* Daud from the Euphrates. Ann Mag Nat Hist 9:549–550
Boulenger GA (1920a) A list of lizards from mesopotamia. J Bombay Nat Hist Soc 27:351–352
Boulenger GA (1920b) A list of snakes from mesopotamia. J Bombay Nat Hist Soc 27:347–350
Bourt SY, Johnson TW (1962) Some biological observations on fungi in estuarine sediments. Mycologia 54:181–193
Bowes B (1999) A color atlas of plant propagation and conservation. Manson Publishing, London, pp 177–186
Brown JH, Gibson AC (1983) Biogeography. C. V. Mosby Company, St Louis, MO
Bryan JR (1976) A Winkler's procedure for oxygen measurement. J Exp Microbial Ecol 21:191–197
Bunn SE, Davies PM, Kellaway DM, Prosser IP (1998) Influence of invasive macrophytes on channel morphology and hydrology in an open tropical lowland stream, and potential control by riparian shading. Freshw Biol 39:171–178
Camargo AFM, Pezzato MM, Henrysilva GG (2003) Fatores limitantes à produção primária de macrófitas aquáticas. In: Thomaz SM, Bini LM (eds) Ecologia e Manejo de Macrófitas Aquáticas. Eduem, Maringá, pp 59–83

Carp E (1975a) Report on a mission to Iraq: January/February 1975 (unpublished). IWRB, Slimbridge, UK
Carp E (1975b) Waterfowl counts in Iraq. IWRB Bull 39(40):51–55
Carp E (1980) A directory of Western Palaearctic wetlands. Nairobi/Kenya and Gland/Switzerland, UNEP and IUCN
Cattaneo A, Kalff J (1987) Seasonal changes in the epiphytic community of natural and artificial maophyes in lakes memphremagog. Hydrobiologia 60:135–144
Chambers AP, Lacoul P, Murphy KJ, Thomaz SM (2008) Global diversity of aquatic macrophytes in freshwater. Hydrobiology 595:9–26
Chandler GT, Fleeger JW (1983) Meiofaunal colonization of azoic estuarine sediment in Louisiana: mechanisms of dispersal. J Exp Mar Biol Ecol 69:175–188
Chapman EA, McGeoch JM (1956) Recent field observations from Iraq. Ibis 98:577–594
Coad BW (1996) Zoogeography of the fishes of the Tigris-Euphrates basin. Zool Middle East 13:51–70
Corkill NL (1932) The snakes of Iraq. J Bombay Nat Hist Soc 35:552–572
Crawley MJ (1983) Herbivory: the dynamics of animal-plant interactions. Blackwell, Oxford
Crosskey RW (1990) The natural history of blackflies. John Wiley, Chichester
Daiber FC (1982) Animals of the tidal marsh. Van Nostrand Reinhold, New York
Davis JS, Westlake DW (1979) Crude oil utilization by fungi. Can J Microbiol 25:146–156
De Nooij RJW, Lenders HJR, Leuven RSE, de Blust G, Geilen N, Goldschmidt B, Muller S, Poudevigne I, Nienhuis PH (2004) BIO-SAFE: assessing the impact of physical reconstruction on protected and endangered species. River Res Appl 20:299–313
DeAngelis DL, Mulholland PJ, Palumbo AV, Steinman AD, Huston MA, Egerton FN (1973) Changing concepts of the balance of nature. Q Rev Biol 48:322–350
Del Hoyo J, Elliott A, Sargatal J (1992) Handbook of the birds of the world. Ostrich to Ducks, vol 1. Lynx Edicions, Barcelona
Development Alternatives Inc (2004) Iraq marshlands restoration program action plan. DAI for USAID, Washington, DC
Dibble ED, Killgore KJ, Dick GO (1996) Measurement of plant architecture in seven aquatic plants. J Freshw Ecol 11:311–318
Dick MW (1973) Saprolegniales. In: Ainsworth GC, Sussman AS (eds) The fungi, vol 4. Academic Press, New York, pp 113–144
Dixit SS, Smol JP, Kingston JC, Charles DF (1992) Diatoms: powerful indicators of environmental change. Environ Sci Technol 26:23–33
Domsch KH, Gams W, Anderson TH (1980) Compendium of soil fungi, vol 1. Academic Press, London, 859 pp
Downing JA (1984) Assessment of secondary production: the first step. In: Downing JA, Rigler FH (eds) A manual on method for the assessment of secondary production in freshwater, IBP Hand book No. 17. Blackwell, Oxford, pp 1–18
Dudgeon D (1999) Tropical Asian streams: zoobenthos, ecology and conservation. Hong Kong University Press, Hong Kong
Duncan P, Hewison AJM, Houte S, Rosoux R, Tournebize T, Dubs F, Burel F, Bretagnolle V (1999) Long-term changes in agricultural practices and wildfowling in an internationally important wetland, and their effects on the guild of wintering ducks. J Appl Ecol 36:11–23
Edmondson WT (1974) Secondary production. Mitt Int Ver Theor Angew Limnol 20:229–272
Eminson DF (1978) A comparison of diatom epiphytes, their diversity and density, attached to Myriophyllum spicatum L. in Norfolk dykes and broads. Br Phycol J 13:57–64
Enriques E, Duarte CM, Sand-Jensen K (1993) Patterns in decomposition rates among photosynthetic organisms: the importance of detritus C:N:P content. Oecologia 94:457–471
Evans MI (2002) The ecosystem. In: Nicholson E, Clark P (eds) The Iraqi marshlands: a human and environmental study. Politico's Publishing, London, pp 201–222
Figuerola J, Green AJ (2002) Dispersal of aquatic organisms by waterbirds: a review of past research and priorities for future studies. Freshw Biol 47:483–494

Freeland JR, Noble LR, Okamura B (2000) Genetic diversity of north American populations of *Cristatella mucedo*, inferred from microsatellite and mitochondrial DNA. Mol Ecol 9:1375–1389

Furners RW, Greenwood JJD, Jarvis PJ (1993) Can birds be used to monitor the environment? In: Furners INRW, Greenwood JJD (eds) Birds as monitors of environmental change. Chapman & Hall, London, pp 1–41

Garkaklis MJ, Bradley JS, Wooller RD (1998) The effects of Woylie (*Bettongia penicillata*) foraging on soil water repellency and water infiltration in heavy textured soils in southwestern Australia. Aust J Ecol 23:492–496

Georg Kainady PV, Vielliard J (1968) Waterfowl counts in Iraq: 21 December 1967 - 18 January 1968. Wetlands International, Wageningen, NL

Georg Kainady PV, Vielliard J (1970) Midwinter observations on birds of central and South Iraq. Bull Iraq Nat Hist Mus 4:61–85

Georg PV, Savage CDW (1968) Status of the main waterfowl resorts in Iraq. International regional meeting on conservation of wildfowl resources. Nauka, Leningrad

Gleick PH (1996) Water resources. In: Schneider SH (ed) Encyclopedia of climate and weather. Oxford University Press, New York, pp 817–823

Günther A (1874) A contribution to the fauna of the River Tigris. Ann Mag Nat Hist 14:36–37

Gurney R (1921) Freshwater crustacean collected by Dr. P.A. Buxton in Mesopotamia and Persia. J Bombay Nat Hist Soc 27:835–844

Haas G (1952) Two collections of reptiles from Iraq, with descriptions of two new forms. Copeia 1952:20–22

Haas G, Werner YL (1969) Lizards and snakes from southwestern Asia, collected by Henry Field. Bull Mus Comp Zool 138:327–406

Harrison JM (1959) Notes on a collection of birds made in Iraq by F/Lt. David L. Harrison. Bull Br Ornithol Club 79(31–36):49–50

Harrison DL, Bates PJJ (1991) The mammals of Arabia. Harrison Zoological Museum, Sevenoaks, UK

Heino J, Toivonen H (2008) Aquatic plant biodiversity at high latitudes: patterns of richness and rarity in Finnish freshwater macrophytes. Boreal Environ Res 13:1–14

Huryn AD, Wallace JB (2000) Life history and production of stream insects. Annu Rev Entomol 45:83–110

Hussain NA, Hamza HA (1987) Some biological aspects of the freshwater population of the Shanag *Acanthopagrus latus* (Houtuyn) in the Shatt Al-Arab River, Iraq. Marina Mesopotamica 2:29–40

Hussain NA, Jabir HA (1994) On the biology of sbour *Tenualosa ilisha* (Hamilton) in the Shatt Al-Arab River, south of Iraq, with notes on their distribution in the Arabian Gulf. Marina Mesopotamica 9:115–139

Hutchinson GE (1967) A treatise on limnology (Vol. II): Introduction to lake biology and the limnoplankton. Wiley, New York

In-den-Bosch HAJ (2003) Iraqi herpetology: an introductory checklist. Podarcis 4:53–72

Ingole B (2005) Indian Ocean coast, coastal ecology. Western Washington University, WA, U.S.A. Encyclopaedia of coastal science. Springer, Dordrecht, Netherlands, pp 546–554

Ito T, Nakagiri A (1997) A mycofloral study on mangrove mud in Okinawa, Japan. IFO Res Commun 18:32–39

IUCN (2010) The IUCN red list of threatened species. www.iucnredlist.org. Retrieved 2 June 2010

Jackson JK, Resh VH (1989) Distribution and abundance of adult aquatic insects in the forest adjacent to a northern California stream. Environ Entomol 18:278–283

Jasim, A. A. W. 1988. The reproductive biology of *Barbus sharpeyi* in Hammar marsh, Iraq, M. Sc. Basrah, Basrah University

Jawad LA (2003) Impact of environmental change on the freshwater fish fauna of Iraq. Int J Environ Stud 60:581–593

Jeppesen E, Søndergaard M, Søndergaard M, Christoffersen K (eds) (1998) The structuring role of submerged Macrophytes in lakes. Ecol Ser 131:423

Johnson TW, Sparrow FK (1962) Fungi in oceans and estuaries. Journal of Cramer Publication, Weinheim, 668 pp

Johnson MS, Roberts RD, Hutton M, Inskip MJ (1978) Distribution of lead, zinc and cadmium in small mammals from polluted environments. Oikos 30:153–159

Johnson TW, Seymour RL, Padgett DE (2004) Biology and systematics of the Saprolegniaceae. Academic Press, New York

Jones S, Salim M, Hani T (2010) Botanical survey in the southern Iraq. Report to the Ministry of Agriculture, Iraq, 670pp

Jordon CF (1998) Working with nature: resource management for sustainability. Harwood Academic, Amsterdam

Kainady PVG (1976a) First positive breeding record of Acrocephalus arundinaceus Eurasian Great Reed Warbler for Iraq. Bull Basrah Nat Hist Mus 3:101–105

Kainady PVG (1976b) The Indian Pygmy Goose, *Nettapus coromandelianus*, in Basrah, Iraq. Bull Basrah Nat Hist Mus 3:107–109

Kassim TI, Sabri AW, Al-Lami AA (2000) Ecological study on epiphytic algae community in the River Tigris at Sammarra impoundment. Sci J Iraqi Atomic Energy Commission 2:33–51

Khalaf KT (1959) Reptiles of Iraq with some notes on the amphibians. Baghdad

Khalaf KT (1960) Notes on a collection of lizards and snakes from Iraq. Publ Iraq Nat Hist Mus 18:12–18

Khalaf KT (1961) Some new records of lizards from Iraq. Bull Iraq Nat Hist Mus 1:1–2

Khalaf AN, Smirrnov MN (1976) On littoral Cladocera of Iraq. Hydrobiologia 51:91–94

Killham K (1994) Soil ecology. Cambridge University Press, Cambridge

Klaassen M, Nolet BA (2007) The role of herbivorous water birds in aquatic systems through interactions with aquatic macrophytes, with special reference to the Bewick's swan-fennel pondweed system. Hydrobiologia 584:205–213

Kohlmeyer J, Kohlmeyer E (1979) Marine mycology: the higher fungi. Academic Press, New York, 690 pp

Koning FJ, Dijksen LJ (1973) IWRB mission to Iraq and Syria, December 1972. IWRB Bull 35:57–62

Konishin M, Nakano S, Iwata T (2001) Trophic cascading effects of predatory fish on leaf litter processing in a Japanese stream. Ecol Res 16:415–422

Lacouture RV, Johnson JM, Buchanan C, Marshall HG (2006) Phytoplankton index of biotic integrity for Chesapeake Bay and its tidal tributaries. Estuar Coasts 29:598–616

Landsberger B (1962) The fauna of ancient Mesopotamia. MSL VIII/2, second part. Pontificium Institutum Biblicum, Roma 204, Piazza Pilotta 35. Publié avec le concourse financier de l'UNESCO et sous less auspices du Conseil International de la Philosophie et des Sciences Humaines, 180pp

Langeland KA (1990) Hydrilla (*Hydrilla verticillata* (L.F.) Royle), a continuing problem in Florida waters. Circular 884. Cooperative Extension Service, IFAS, University of Florida, Gainesville, FL

Lauterbom R (1893) Beitrage zur Rotatorienfauna des Rheins und seiner Altwasser. Zool lb Abt Syst 7:254–273

Leuven RSEW, Wijnhoven S, Kooistra L, de Nooij RJW, Huijbregts MAJ (2005) Toxicological constraints for rehabilitation of riverine habitats: a case study for metal contamination of floodplain soils along the Rhine. Arch Hydrobiol 155:657–676

Lillebø AI, Flindt MR, Pardal MA, Marques JC (1999) The effect of macrofauna, meiofauna and microfauna on the degradation of *Spartina maritima* detritus from a salt marsh area. Acta Oecol 20:249–258

Lindeman RL (1942) The trophic-dynamic aspect of ecology. Ecology 23:399–418

Lundberg G, Kottelat M, Smith GR, Stiassny MLJ, Gill AC (2000) So many fishes, so little time: an overview of recent ichthyological discovery in continental waters. Ann Missouri Bot Gard 87:26–62
Lyell C (1832) Principles of geology, vol 2. J. Murray, London
MacDonald M (2010) Halfaya oil field ESIA environmental baseline study. Mott MacDonald, Croydon, p 142
Mads C, Forchhammer EP, Nils CHRS (2002) North Atlantic oscillation timing of long- and short-distance migration. J Anim Ecol 71:1002–1014
Madsen JD, Chambers PA, James WF, Koch EW, Westlake DF (2001) The interaction between water movement, sediment dynamics and submersed macrophtes. Hydrobiologia 444:71–84
Mahdi S (1982) Aquatic birds of Iraq. Iraq Natural History Museum, Baghdad
Mahdi N, Georg PV (1969) A systematic list of vertebrates of Iraq. Iraq Natural History Museum, Baghdad
Mahmood FH, Hussain MS, Salim TL (2007) Mycological survey in the lower reaches of Tigris River. J Ministry Environ 3:56–89
Marchant S (1963a) Migration in Iraq. Ibis 105:369–417
Marchant S (1963b) Notes on the winter status of certain species in Iraq. Ardea 51:237–243
Marchant S (1963c) The breeding of some Iraqi birds. Ibis 105:516–557
Marchese M, Ezcurra, de Drago I (1992) Benthos of the lotic environments in the middle Paranà river system: transverse zonation. Hydrobiologia 237:1–13
Marzluff JM, Bowman R, Donnelly R (2001) Avian ecology and conservation in an urbanizing world. Kluwer Academic Publishers, Boston, MA
Matagi SV, Swai D, Mugabe R (1998) A review of heavy metal removal mechanisms in wetlands. Afr J Trop Hydrobiol Fish 8:23–35
Mazin GH, Hilly JK, Rasheed TT (2011) Study on the effect of environmental factors on the biology of some aquatic fungi in the lower reaches of Tigris River. Sci J Diyala 4:67–89
Meyerson LA, Saltonstall K, Windam L, Kiviat E, Findlay S (2000) A comparison of *Phragmites australis* in freshwater and brackish marsh environments in North America wetlands. Ecol Manag 8:89–103
Michael TS, Shin HW, Hanna R, Spafford DC (2008) A review of epiphytic community development: surface interactions and settlement on seagrass. J Environ Biol 29:629–638
Ministry of Agriculture (1977) The environment of the aquatic fungi in the lower reaches of Mesopotamia. Report of the environmental committee, Ministry of Agriculture, Iraq, 456pp
Minshall GW, Petersen RC, Cummins KW, Bott TL, Sedell JR, Cushing CE, Vannote RL (1983) Interbiome comparison of stream ecosystem dynamics. Ecol Monogr 53:1–25
Mitsch WJ, Gosselink JG (2000) Wetlands, 3rd edn. Wiley, Hoboken, NJ
Mohammed M, Faris H, Andy J (2011) Ecological and taxonomical studies of the aquatic plants in the rivers souther of Iraq. Report to the Agriculture Office, Ammarah Province, Iraq, 560 pp
Moore HJ, Boswell C (1956–1957) Field observations on the birds of Iraq (Parts I and II 1956, Part III 1957). Iraq Natural History of Museum Publication No. 9, 10, 12. Baghdad, Iraq Natural History Museum, 10 pp
Morin JO (1986) Initial colonization of periphyton on natural and artificial species of *Myriophyllum heterophyllum* Mivhx. Fresh Water Biol 16:685–749
Nabavi SMB (2001) Use of diversity indices as coastal pollution indicators in Khure Musa (Mahsahr creeks). Sci Q J 32:57–61
Nakano S, Murakami M (2001) Reciprocal subsidies: dynamic interdependence between terrestrial and aquatic food webs. Proc Natl Acad Sci USA 98:166–170
National Research Council (1981) Testing for effects of chemicals on ecosystems. National Academic Press, Washington, DC, 103 pp
Nilsson C, Reidy CA, Dynesius M, Revenga C (2005) Fragmentation and flow regulation of the world's large river systems. Science:2345–2347
Noble JC (1993) Relict surface-soil features in semi-arid mulga (*Acacia aneura*) woodlands. Rangelands J 15:48–70

Odum EP (1971) Fundamentals of ecology, 3rd edn. W. B. Saunders, London, 574 pp
Okogwu OI, Ugwumba OA (2006) The zooplankton and environmental characteristics of Ologe lagoon, south West Nigeria. Zoologist 1:86–92
Palmer MA (1984) Invertebrate drift: behavioral experiments with intertidal meiobenthos. Mar Behav Physiol 10:235–253
Palmgren P (1936) Uber die Vogelfauna der Binnengewasser Alands. Acta Hydrobiol 17:1–59
Park CR, Lee WS (2000) Relationship between species composition and area in breeding birds of urban woods in Seoul, Korea. Landscape Urban Plan 51:29–36
Partow H (2001) The mesopotamian marshlands: demise of an ecosystem. UNEO/DEWA TG.01-3. Nairobi, UNEP, Division of Early Warning and Assessment
Pennak RW (1978) Freshwater invertebrates of the United States. Wiley, Hoboken, NJ, 803pp
Poi de Neife AS, Casco SL (2003) Biological agents that accelerate winter decay of *Eichhornia crassipes* Mart. Solms. in northeastern Argentina. In: Thomaz SM, Bini LM (eds) Ecologia e Manejo de Macrófitas Aquáticas. Eduem, Maringá, pp 127–144
Porter RF, Salim M (2010) A provisional checklist of the birds of Iraq. Marsh Bull 5:56–95
Post DM (1998) The role of migratory waterfowl as nutrient vectors in managed wetlands. Conserv Biol 12:910–920
Post DM (2002) Using stable isotopes to estimate trophic position: models, methods, and assumptions. Ecology 83:703–718
Pott VJ, Pott A (2003) Dinâmica da vegetação aquática do Pantanal. In: Thomaz SM, Bini LM (eds) Ecologia e Manejo de Macrófitas Aquáticas. Eduem, Maringá, pp 145–162
Rattan SS, Muhsin TM, Ismail AS (1980) Notes on the occurrence and seasonal periodicity of Saprolegniaceae in Shatt Al-Arab River, Basrah, Iraq. Kavaka 8:41–48
Redmond WA (2008) Lead. Microsoft® Encarta® 2007 [DVD]. Microsoft Corporation, 2007
Reed CA, Marx H (1959) A herpetological collection from northeastern Iraq. Trans Kans Acad Sci 62:91–122
Richardson CJ (2008) Wetlands of mass destruction: can the garden of Eden be fully restored? Natl Wetlands Newsl 30:2–7
Robinson CT, Tockner K, Ward JV (2002) The fauna of dynamic riverine landscapes. Freshw Biol 47:661–677
Rosenberg DM, Berkes F, Bodaly RA, Hecky RE, Kelly CA, Rudd JWM (1997) Large-scale impacts of hydroelectric development. Environ Rev 5:27–54
Saggs HWF (1962) The greatness that was Babylon. Sidgwick & Jackson, London, 562pp
Sala OE, Chapin FS, Armesto JJ, Berlow R, Bloomfield J, Dirzo R, Huber-Sanwald E, Huenneke LF, Jackson RB, Kinzig A, Leemans R, Lodge D, Mooney HA, Oesterheld M, Poff NL, Sykes MT, Walker BH, Walker M, Wall DH (2000) Global biodiversity scenarios for the year 2100. Science 287:1770–1774
Salim M, Porter RF (2006) Birds of Iraq (in Arabic). Amman, Nature Iraq and BirdLife International
Salman SD, Bishop JM (1990) *Exopalaemon styliferus* (H. Milne Edwards, 1840) in the northern Arabian Gulf and the inland waters of Iraq (Decapoda, Caridea, Palaemonidae). Crustaceana 59:281–288
Sassi M (1912) Liste von Vogelbalgen aus Mesopotamien, vol 26. Annals and of Natural History Hofmus, Wien, pp 116–119
Scheffer M, Hosper SH, Meijer ML, Moss B, Jeppesen E (1993) Alternative equilibria in shallow lakes. Trends Ecol Evol 8:275–279
Schindler DW (1987) Detecting ecosystem response to anthropogenic stress. Can J Fish Aquat Sci 44:6–25
Schmidt KP (1939) Reptiles and amphibians from southwestern Asia. Chicago Field Mus Nat Hist Zool Ser 24:49–92
Scott DA (ed) (1995) A directory of wetlands in the Middle East. IUCN, Gland, Switzerland
Scott DA, Carp E (1982) A midwinter survey of wetlands in mesopotamia. Sandgrouse 4:60–76

Scott DA, Evans MI (1994) Wildlife. An environmental and ecological survey of the marshlands of Mesopotamia E. AMAR Appeal Trust, Maltby

Sibert JR (1981) Intertidal hyperbenthic populations in the Nanaimo estuary. Mar Biol 46:259–265

Simkhada BM, Jutter I, Chimonides J (2006) Diatoms in lowland ponds of Koshi Tappu, Eastern Nepal relationships with chemical and habitat characteristics. Int Rev Ges Hydrobiol 91:574–593

Soule JD, Piper JK (1992) Farming in nature's image: an ecological approach to agriculture. Island Press, Washington, DC

Sparling DW, Linder G, Bishop CA (2000) Ecotoxicology of amphibians and reptiles. Pensacola, Florida Society of Environmental Toxicology and Chemistry, 904 pp

Stebbinsm RC, Cohen N (1995) A natural history of amphibians. Princeton University Press, Princeton, NJ, 316 pp

Stets EG, Cotner JB (2008) Littoral zones as sources of biodegradable dissolved organic carbon in lakes. Can J Fish Aquat Sci 65:2454–2460

Strahan R (1995) The mammals of Australia. Reed Books, Chatswood

Takashi A, Sultana M, Jagath Manatunge J, Fujino T (2004) The effect of epiphytic algae on the growth and production of Potamogeton perfoliatus L. in two light conditions. Environ Exp Bot 52:225–238

Takeda AM, Souza-ranco GM, Melo SM, Monkolski A (2003) Invertebrados associados às macrófitas aquáticas da planície de inundação do altorio Paraná (Brasil). In: Thomaz SM, Bini LM (eds) Ecologia e Manejo de Macrófitas Aquáticas. Eduem, Maringá

Tamaki A, Ingole B (1993) Distribution of juvenile and adult ghost shrimps *Callianassa japonica* Ortmann (Thalassinidae), on intertidal sand flat: intraspecific facilitation as a possible pattern-generating factor. J Crustac Biol 13:175–183

Tatrai I, Olah J, Paulovits G, Matyas K, Kawiecka BI, Jozsa V, Pekar F (1997) Biomass dependent interactions in pond ecosystems: responses of lower trophic levels to fish manipulation. Hydrobiologia 345:117–129

Taylor DJ, Finston TL, Hebert PDN (1998) Biogeography of a widespread freshwater crustacean: pseudocongruence and cryptic endemism in the North American *Daphnia laevis* complex. Evolution 52:1648–1670

Theel HJ, Dibbie ED, Madsen JD (2008) Differential influence of a monotypic and diverse native aquatic plant bed on a macroinvertebrate assemblage; an experimental implication of exotic plant induced habitat. Hydrobiologia 600:77–87

Ticehurst CB (1920a) A new species of reed warbler from the marshes of lower Mesopotamia (Acrocephalus babylonicus). Bull B.O.C 41:12–13

Ticehurst CB (1920b) Remarks on the avifauna of Mesopotamia. Bull B.O.C 41:53–54

Tkachenko A (2002) The economy of the Iraq marshes in the 1990s. In: Nicholson E, Clark P (eds) The Iraqi marshlands: a human and environmental study. Politico's Publishing, London, pp 36–63

Tomlinson AG (1916) Notes on the birds of Mesopotamia. J Bombay Nat Hist Soc 24:825–829

Trauti AH, Hostetler ME (2003) Urban lakes and waterbirds: effects of development on avian behavior. Waterbirds 26:290–302

Ueda S (1980) A mycofloral study on river sediments in Nagasaki, Japan. Trans Mycol Soc Jpn 21:495–504

Ulfig KG, Cano J, Vidal P, Figueras MJ, Lukasik W (1998) A preliminary study of the occurrence of actidione resistant fungi in sediments of Catalonian river mouth (Spain) 1. Keratinophilic fungi and related onygenales. Mycopathologia 141:143–151

Valiela I (1995) Marine ecological processes, 2nd edn. Springer, New York, pp 385–461

Van der Valk AG (2006) The biology of freshwater wetlands. Oxford University Press, Oxford, p 173

Wallace BJ, Webster JR (1996) The role of macroinvertebrates in stream ecosystem function. Annu Rev Entomol 41:115–139

Walley RC (2007) Environmental factors affecting the distribution of native and invasive aquatic plants in the Atchafalaya River basin, Louisiana, USA. M.Sc Thesis, 78p
Warwick RM (1993) Environmental impact studies on marine communities: pragmatical considerations. Aust J Ecol 18:63–87
WCMC, 1998. Freshwater biodiversity: A preliminary global assessment. World Conservation Press, Cambridge
Webster WD, Parnell JF, Briggs WC (1985) Mammals of the Carolinas, Virginia, and Maryland. University of North Carolina Press, Chapel Hill, NC
Wetzel RG (2001) Limnology, 3rd edn. Academic Press, San Diego, p 1006
Wheaton B (1979) Rivers, lakes and marshes. Hadder and Stoughton Ltd, London, p 371
Wijnhoven S, van der Velde G, Leuven RSW, Smits AJM (2005) Flooding ecology of voles, mice and shrews: the importance of geomorphological and vegetational heterogeneity in river floodplains. Acta Theriol 50:453–473
Wong MKM, Goh TK, Hodgkiss J, Hyde DK, Ranghoo VM, Tsui CKM, Wong WSW, Yeuen TK (1998) Role of fungi in freshwater ecosystems. Biodivers Conserv 7:1187–1206
Yaaqub A (1992) Algae of southern Iraqi marshes. In: Proceedings of the Forum on Southern Iraqi Marshes Environment. December 20–21, 1992. Basrah, Iraq (In Arabic)
Yamakuta YY (2012) Preliminary report on the macrophytes in the branches of Tigris River. Intrim Report, Ministry of Agriculture, Iraq, 560 pp
Yousif UU, Harry MN (2008) Some aquatic fungi in the southern of Iraq. Report to the Ministry of Agriculture, 38 pp
Zacharias O (1898) Das Potamoplankton. Zool Anz 21:4148
Zare-maivan H (2011) Determining heavy metals, organic pollutants and agricultural insecticides in Arvand Rood, Hafar and Bahmanshir Rivers. Report of Department of Environment, Iran

Chapter 19
Freshwater Fish Biodiversity in Iraq: Importance, Threats, Status, and Conservation Challenges

Laith A. Jawad

Abstract Freshwater biodiversity in general and freshwater fish biodiversity in particular are among the hot global issues that gained attention in an ongoing engagement.

It is clear that the inland waters and freshwater biodiversity are very much treasured natural resources that human cannot depend on as they are important in economic, cultural, aesthetic, scientific, and educational terms. Therefore, their conservation and management are important and prerequisite state to the interests of all humans, nations, and governments. However, this valuable natural treasure is in disaster. Fresh waters are facing failures in biodiversity far greater than the terrestrial ecosystems, and if human activities continue to have the same pace, then this biodiversity will vanish earlier than expected.

In the present chapter, documentation on the freshwater fish biodiversity of Iraq is given in the frame of species richness, taxonomy and nomenclature, how valuable the freshwater fishes of Iraq are, endangered species, threat and challenges, and conservation. The interactions of all these issues have led to a decline in the population of a certain number of freshwater species. Also, a clear range reduction of other species has been observed.

As in many other world countries, the freshwater fishes of Iraq are facing a strong competition on both food resources and habitats from introduced species such as different species of carp and tilapia. Therefore, measures of protection of freshwater biodiversity are perhaps the eventual need for the conservation of this fish group in spite of the difficulties that such issues might confront the influence of the upstream drainage network, the surrounding land, the riparian zone, and—in the case of migrating aquatic fauna—downstream reaches.

Globally, these fundamentals are very hard to bring to reality. Immediate action is needed to create large protected areas that include rivers and lakes and balance between human needs and the freshwater biodiversity should be in place to keep the natural ecological balance. Therefore, the absence and presence of the species should

L. A. Jawad (✉)
School of Environmental and Animal Sciences, Unitec Institute of Technology, Auckland, New Zealand

L. A. Jawad (ed.), *Tigris and Euphrates Rivers: Their Environment from Headwaters to Mouth*, Aquatic Ecology Series 11,
https://doi.org/10.1007/978-3-030-57570-0_19

be checked continuously and the invasive species should be reported, especially those very hazardous to the native fish fauna and for the habitats. Distinguishing this need will necessitate acceptance of a new pattern for biodiversity safety and freshwater ecosystem.

19.1 Introduction

On the surface of our mother earth, freshwater ecosystems offer habitats for around 10% of the world's known species, which include a quarter of all vertebrates, and supply humans with several aspects of resources (Strayer and Dudgeon 2010). The biodiversity of the freshwater habitats constitutes an essential component on earth, with a species richness higher than terrestrial and marine ecosystems combined (Gleick 1996). The usefulness of the freshwater systems is innumerable to both human society and other organisms. They provide food of different types and water for drinking in addition to other ecological benefits such as nutrient cycling, flood control, and water filtration.

With the high diversity and with all the advantages to human, the biodiversity of the freshwater systems remains under continuous threat due to human activities, which can be categorized as follows: overexploitation, water pollution, flow modification, destruction or degradation of habitat, and invasion by exotic species. In addition to all are the global scale environmental changes (Dudgeon et al. 2006). These globally mounting pressures have led to freshwater biodiversity dropping into a state of catastrophe (Vorosmarty et al. 2010), causing a decrease greater than is seen in either terrestrial or marine kingdoms with freshwater species populations lessening on average by 76% since 1970 (Strayer and Dudgeon 2010).

Conservation and management are serious if man's increasing demand for water continues relentless and the species losses continue at current rates then, the chance of preserving much of the remaining biodiversity in freshwater will be diminishing.

In general, the information of the total freshwater diversity is tragically incomplete, in particular between invertebrates and microbes, especially in tropical latitudes that back most of the species of the world. Vertebrates also are not completely known, including well-known taxa, such as fish (Stiassny 2002).

The most influential factor that leads to decline in aquatic biodiversity is water shortage. This problem is affecting human life in the arid zones of the world. In these areas, governments of many countries are seeking critical plans to develop new agricultural societies without cautious analysis of the environmental and hydrologic impacts of these projects. Examples of such problems are present in countries of the Middle East like Iraq, where destructive water administration plans by Iraq and its neighbours have created severe change to the Tigris and Euphrates watershed and the Mesopotamian Marshes downstream in southern Iraq (Jones et al. 2008).

One of the causes that caused shortage in water supply is the dispute between countries situated on the same river/s. The Euphrates–Tigris and its tributaries aided as the cradle for many civilizations that grow in Mesopotamia, 'the land between two rivers'. Historians have noted the development of the Mesopotamian civilizations, with the oldest dating back to 10,000 BC. The well-known civilizations in Mesopotamia (Iraq) were those of the Sumerians, Acadians, Babylonians, and Assyrians, who organized efficient hydraulic civilizations that supported some 20 million inhabitants at their peak and were based on a well-maintained irrigation and flood control system. The history of water-related disagreements in the Middle East, especially in the Euphrates–Tigris basin, goes back 6000 years and is described in many myths, legends, and historical accounts that have survived from earlier times. These disputes range from struggles over reach to satisfactory water resources to intentional attacks on water supply systems during wars (Gleick 1996). The fall of the Ottoman Empire brought in new borders together with a new trans-boundary river basin in the Middle East. The rise of hydro-political problems among the riparian of the Euphrates–Tigris goes back no more than 50 years (Altinbilek 2004).

The Euphrates–Tigris Rivers originated and fed from the mountains at the north and eastern Turkey, Iraq, and Iran. The Euphrates River (the pure river) extends nearly 3000 km, and it is the longest river in western Asia (Altinbilek 2004). This river originates exactly from area near Mount Ararat at heights of around 4500 m near Lake Van and then drops on average 2 m/km of length in Turkey and then crosses into Syria flowing south-east. Its journey in Syria covers 680 km before it enters the Iraqi borders at Al Qaim. Once it is in the Iraqi lands, it passes through alluvial delta and then goes through the desert of Iraq loosing water quantities in some desert areas. In reaching the City of Nasiriyah, the river becomes a net of channels, some of which drain into the shallow Al-Hammar Marsh and the rest join the Tigris River at Qurna.

The other twin of Euphrates River, the Tigris (The Akkadian name means the river with high banks), is considered the second-largest river in western Asia. This river initiates near Lake Hazar (elevation 1150 m) in eastern Turkey. Within Turkey, several tributaries join this river and more tributaries link to Tigris River in Iraq originated from Iran. At the southern Mesopotamia, in Qurna, the two rivers meet each other and form another river which is called Shatt-al-Arab, with a kilometre wide and 190 km long. The tributaries originated from Iran are the greater and lesser Zab, Diyalah and Kharun rivers. In addition, the River Kharkeh emanating from Iran flows into the Mesopotamian marshes of southern Iraq.

The Euphrates–Tigris river system is known for its irregularity in its hydrological regime of flow both between and within years. Periodic floods at the lower reaches of Mesopotamia may occurred with snow-melt in spring. Variation in the minimum and the maximum values of monthly flows of the two rivers. In the previous decades, the concentration of discharge over the months of April and May can cause not only extensive spring flooding, inundating large areas, but also the loss of much-needed water required for irrigation and power generation purposes during the summer season.

Biodiversity in general and aquatic biodiversity in particular are well studied in Iraq; the bird group is the best studied among the vertebrate animals (Rubec et al. 2009) and comes next are the freshwater fishes (Coad 1991, 2010). The biodiversity investigations in Iraq are characterized in being unorganized and not integrated in different parts of Iraq. Also, there are no concluding reports disclosing the status of the biodiversity in general. What is present at the moment are reports on range extensions and checklists of the species, in which the authors do not take their studies further and write a conclusion to their findings.

This article investigates why the transfer of knowledge to conservation action has, in the case of freshwater biodiversity in Iraq, been largely ineffective. The insolvency is due to the special topographies of freshwater habitats—and the biodiversity they support—that make them especially susceptible to human activities. In this chapter, clarification of why freshwater biodiversity is of outstanding global importance is given and possible instances at which humans have produced rapid and significant declines in freshwater species and habitats are described.

19.2 Iraq Freshwater Riches

Freshwater ecosystems may well be the most threatened ecosystem in the world. Drops in biodiversity are considerably greater in fresh waters than in the most affected terrestrial ecosystems (Sala et al. 2000). The uneven richness of inland waters as a habitat for plants and animals makes freshwater habitats and the biodiversity that they maintain especially susceptible to human activities and environmental change. Examples of such an unevenness can be seen in freshwater fish species. Over 10,000 fish species live in fresh water (Lundberg et al. 2000), around 40% of global fish diversity and one quarter of global vertebrate diversity. By adding amphibians, aquatic reptiles (crocodiles and turtles) and mammals (otters, river dolphins, and platypus) to this freshwater-fish total, it becomes obvious that as much as one third of all vertebrate species are confined to fresh water. However, surface freshwater habitats contain only around 0.01% of the world's water and cover only about 0.8% of the Earth's surface (Gleick 1996).

Freshwater system of Iraq provides a wide variety of ecosystem services, including water, food, and income. For example, at the southern marsh areas, reeds of different species are used as building materials and constitute a key source of income in the area. The drying of the southern marshes of Iraq hindered the activities such as fishing and reed harvesting, which contributed to the economy of local communities, as well as losing the potential opportunities from ecotourism. Such results are similar to that happened in the central marshes of Turkey (Karadeniz et al. 2009). Many species of waterfowl are hunted in the areas near the freshwater system in Iraq and considered as an important socio-economic activity across the region.

However, traditional fisheries are being replaced by the harvesting of the non-native and invasive carp species as it generates a considerable income for local communities. The dependence of human populations on healthy freshwater ecosystems is no more evident than in Iraq.

The number of aquatic groups in Iraq is too many that the scope of this chapter cannot accommodate with. In the major biodiversity assessments, the focus is usually made on a limited range of taxonomic groups, most frequently including those groups that deliver noticeable benefits to humans through direct consumption, or the more compelling groups, such as mammals and birds. In the case of aquatic biodiversity, it is the birds and fishes that have received most consideration, but it is vital to gather evidence about other constituents of the food web that are imperative to the upkeep of healthy functioning of freshwater ecosystems. Therefore, attention will be given for the fish group only as the biodiversity of the other major aquatic groups will be dealt in separate chapters of this book.

Fishes are known for their wide spectrum in regards to trophic levels and ecological roles, so data on their distributions and conservation status will offer a useful signal of the general status of the associated freshwater ecosystems.

19.2.1 Fishes

Fishes are considered one of the most important freshwater products at an international measure (Dudgeon et al. 2006). FAO has given an estimation of more than 6% of the world's annual animal protein supplies for humans is that of fishes (FAO 2007) and can be considered as a food security and employment too (Dugan et al. 2010). Globally, there are nearly 13,000 freshwater fish species or about 15,000 species if brackish water species are encompassed (Lévêque et al. 2008).

19.2.1.1 Freshwater Fish Diversity

The freshwater fish group of Iraq is a part of that of the Eastern Mediterranean region that comprises Turkey, the Levant, the southern Caucasus, and Mesopotamia. This geographical zone overlays with three biodiversity hotspots (Myers et al. 2000): the Mediterranean Basin, Irano-Anatolian, and the Caucasus, and incorporates 14 freshwater ecoregions (Abell et al. 2008). According to Freyhof and Brooks (2011) and Darwall et al. (2011), there are 322 species of freshwater fishes present in the Eastern Mediterranean region, two thirds (66.8%/215 species) of which are endemic to the region.

The ichthyological history of the fish fauna of Iraq takes us back to the time of Sumer, Babylon, and Assyria (Saggs 1962). The people of ancient Mesopotamia have succeeded in identifying and naming several fish species, which they recorded on clay tablets (Landsberger 1962). In the modern history of Iraq, probably Heckel (1843) is considered the first who made an ichthyological survey in the Mesopotamia and succeeded in describing 17 freshwater fish species from the Tigris River at Mosul City, northern Iraq. The work of Hasselquit (1722–1752) and Russell (1742–1753) (Cited in Coad 2010) cannot be considered the start of the ichthyological investigations in Iraq as those authors did not collect the specimens from Iraqi waters in spite of the fact that the species they described were actually present in Iraq later on.

During the 173 years since the publication of Heckel (1843), the journey of the fish study has created a large number of publications dealt with several aspects of Iraqi fish taxonomy, such as descriptions of new species, range extensions, and checklists of both freshwater and marine species.

The number of families and species allocated by some authors in the 20th century is very high and does not truly represent the ichthyological fauna of Iraq; the number of families stated by Khalaf (1961), Mahdi (1962), Mahdi and George (1969), Al-Nasiri and Hoda (1975), and Al-Daham (1977) cannot be relied upon as they represent compilation from other works, rather than collected in the field. The work of Coad (2010) is reasonable and the number of freshwater taxa can be used for future biodiversity studies of Iraq's fish fauna. Coad (2010) probably is the first in the modern ichthyological history of Iraq to resolve all the synonym and rejected taxa issues tracing to several authors over a long period of time. The record of the number of freshwater fish species increased dramatically during the period 1937–1969, then dropped significantly. Jawad et al. (2012) suggested that since 1975 and until 2012, only four additions to the list of freshwater fish species have been made. Recently, two more freshwater fish species, *Oxynoemacheilus gyndes* and *Eidinemacheilus proudlovei*, were described from the northern part of Iraq (Freyhof et al. 2016; Freyhof and Abdullah 2017). The availability of marine specimens and the large number of species in the marine environment of Iraq made it possible for this component of the fauna to be better documented than the freshwater component.

19.2.1.2 Discrepancies in the Nomenclature of the Freshwater Fish of Iraq Over Time

Within the freshwater fish fauna, some generic and specific names were used only once by the authors who erected them and never appeared again in the literature. For example, the genus *Barynotus* and the species *Barbus faoensis* (= *B. sharpeyi*) appeared in the works of Günther (1868, 1874). The genus *Tor* was erected by Karaman in (1971) and used by Misra (1947), and *Girardinus fosciloides* (= *Poecilia latipinna*) was used by Kennedy (1937). In 1874, Günther used the name *Capoeta trutta*, but since then all the authors who have dealt with this species used *Varicorhinus trutta* instead, which is the junior synonym of *C. trutta*. Recently, Coad (2010) used *C. trutta* and considered *Varicorhinus* as a synonym.

Some other scientific names were used for a certain period of time, then disappeared. For example, *Alburnus sheitan* was first used by Misra (1947) and then by Khalaf (1961) and Mahdi (1962). Although *Saccobranchus fossilis* is an old combination, it was used by Mahdi (1962), Mahdi and George (1969), and Al-Nasiri and Hoda (1975). Khalaf (1961) used the new combination *Heteropneustes fossilis*. The inconsistency in the usage of the common name of the Indian stinging catfish reflects the lack of the authors' knowledge of ongoing changes in the taxonomic nomenclature of the native species. Scientific records of the introduced species (such as some cyprinid species) did not make their appearance until Al-Nasiri and Hoda

(1976) mentioned them in their work. This year could mark the start of invasion of the freshwater system of Iraq by species such as the common carp, *Cyprinus carpio*, and the silver carp, *Hypophthalmichthys molitrix*.

19.2.1.3 Competition with the Native Species

Comparing lists from the 1960s and 1970s with those of the 1980s, it is clear that the number of *Barbus* species has dropped dramatically in the Shatt al-Arab River and the marsh areas, as they were replaced by introduced species such as *Cyprinus carpio*, *Carassius auratus*, *Ctenopharyngodon idella*, and *Hypophthalmichthys molitrix* (Khalaf 1961; Mahdi 1962; Mahdi and George 1969; Al-Nasiri and Hoda 1975, 1976; Al-Hassan et al. 1989). Such major replacements in the species composition are mainly due to significant changes in the environment (Jawad 2003, 2006).

19.2.1.4 Range Extension

The records of range extension of the fish species in Iraq over a period of 80 years are listed in Table 19.1. It is clear from this table that there is a sporadic riverine and marine presence of fish species in central and southern parts of Iraq. In addition to these two movements, there is also evidence of invasion of rivers and streams by aquaculture species.

Records of the sporadic presence of marine species in freshwater systems of Iraq go back to the time of Günther (1874), who recorded the presence of the shark *Carcharhinus gangeticus* (=*C. leucas*) in the Tigris River near Baghdad. Conversely, records of freshwater species in the marine environment started when Günther (1868) described *B. faoensis* (= *B. sharpeyi*) from marine water in the vicinity of Fao City in southern Iraq. The record of this principal freshwater fish species has not been confirmed, whether it was collected from the marine environment or obtained from a fish market.

The sporadic riverine presence of marine species can be explained by significant changes in both estuarine and freshwater systems, enabling marine species to extend their range northward towards the freshwater system of Mesopotamia (Al-Saad et al. 2009; Sultan and Ahmed 2009). As to the invasion of marine waters by a few freshwater species, this can be explained by their salinity tolerance and ability to survive in estuarine marginal marine environments (Sultan and Ahmed 2009). The presence of escaped aquaculture species in rivers and streams (Al-Hassan 1994; Coad and Hussain 2007) may be related to extensive fish culture activities in Iran. The record of the shark *Carcharias gangeticus* by Günther (1874) from the Tigris River in the vicinity of Baghdad City is a misidentification of a bull shark *Carcharhinus leucas*, as this species can tolerate pure freshwater and is often found hundreds of miles up rivers (A. Moore, personal communication).

Table 19.1 Chronological list of the number of the newly recorded fish species in the Iraqi fresh and marine waters together with their locality

Published work	Number of new records of species	Locality
Kennedy (1937)	1	Tigris River at Amara and Baghdad Cities
	1	Tigris River at Baghdad City
Hora and Misra (1943)	1	Shatt al-Arab River at Fao City
Misra (1947)	7	Al-Hammar Marsh
Menon (1956)	1	Habbaniyah Lake
	2	Al-Hammar Marsh
	1	Shatt al-Arab River
Mahdi (1971)	7	Shatt al-Arab River
Al-Daham (1974)	2	Khor al-Umia
Al-Daham (1975)	1	Marine waters of Iraq
Al-Daham (1976)	1	Marine waters of Iraq
Nader and Jawdat (1977)	19	(10 records) Marine and (9 records) marsh areas
Al-Rawi et al. (1978)	7	Lesser Zab River
Shafi et al. (1980)	1	Derbendikhan Dam Lake
Al-Hassan and Naama (1986)	4	Euphrates River at Nasiria City
Al-Hassan and Al-Badri (1986)	2	Khor al-Zubair
	1	Khor Abdullah
	1	Shatt al-Arab River
Al-Hassan and Muhsin (1986)	2	Khor al-Zubair
Al-Hassan and Miller (1987)	1	Khor Abdullah
Mohammed et al. (1993)	1	South of Fao City
Al-Hassan (1994)	1	Shatt al-Arab River
Coad and Hussain (2007)	1	Huwaeza Marsh
Saleh (2007)	1	Euphrates River
Abd and Abd (2009)	1	Huwaeza Marsh
Jawad et al. (2009)	1	Tigris River
Mutlak and Al-Faisal (2009)	1	Man-made river (third river)
Jawad et al. (2009)	1	Tigris River
Jawad et al. (2010)	1	Marine waters of Iraq
Al-Mukhtar et al. (2011)	1	Marine waters of Iraq
Jawad et al. (2012)	1	Shatt al-Arab River
Al-Zaidy (2013)	1	Southern marshes
Jawad and Al-Badri (2014)	3	Marine waters of Iraq
Sadek and Jawad (2014)	3	Marine waters of Iraq

(continued)

Table 19.1 (continued)

Published work	Number of new records of species	Locality
Jawad and Hussain (2014)	3	Marine waters of Iraq
Jawad et al. (2014a–c)	1	Marine waters of Iraq
Jawad et al. (2014a–c)	2	Marine waters of Iraq
Al-Badri and Jawad (2014)	2	Marine waters of Iraq
Jawad et al. (2014a–c)	3	Marine waters of Iraq
Al-Faisal and Mutlak (2014)	1	Shatt al-Arab River
Al-Faisal and Mutlak (2015)	1	Shatt al-Arab River
Ziyadi et al. (2015)	1	Lake
Jawad and Al-Badri (2015)	3	Marine waters of Iraq
Jawad (2015)	4	Marine waters of Iraq
Jawad et al. (2016a, b)	1	Southern marshes
Jawad et al. (2016a, b)	1	Shatt al-Arab River
Mohamed et al. (2016)	1	Shatt al-Arab River
Al-Daraji et al. (2017)	1	Marine waters of Iraq
Mutlak et al. (2017)	1	Shatt al-Arab River
Ziyadi et al. (2018)	1	Marine waters of Iraq

G. gangeticus is a taxonomically difficult genus that may or may not be found in the waters of the Arabian Gulf (A. Moore, personal communication).

The record of *Hilsa ilisha* (= *Tenualosa ilisha*) by Menon (1956) from the natural Lake Habbaniyah (80 km west of Baghdad) is clearly a mistake in the labelling of the specimen. This assumption is based on the following: (1) Menon (1956) did not collect the fish specimens by himself, they were sent to him from Iraq; (2) the northernmost record for this species is from the northern reaches of Hammar Marsh, south of Thiqar Province (Al-Hassan 1993); (3) no marine species was recorded from the Euphrates water north of Thiqar Province; (4) the hydrological parameters of Lake Habbaniyah are not the environmental factors suitable for the riverine presence of *T. ilisha*; and (5) finally food items available at Hammar Marsh are completely different from those of Lake Habbaniyah. Kennedy (1937) recorded the aquarium fish *Girardinus fosciloides* (= *Poecilia latipinna*) from the Tigris River at Baghdad. He misspelled *fosciloides* as *poeciloides* and *Girardinus* (= *Limia poeciloides*) is a synonym of *Poecilia latipinna*. This species is of South American origin and definitely an escapee from aquarium culture. Recently, Coad (2010) reported the presence of this species in the Shatt al-Arab River and attributed its presence to escapes from commercial aquarium fish activity in southern Iraq.

Further specimens of this species were collected by the present author from Qarmat Ali (north of Basrah), the Shatt al-Arab River (at Basrah), and from different branches of the Shatt al-Arab River.

Over the period of 173 years, there were 108 cases of range extension that demonstrate one of the following forms of ecological distribution: sporadic riverine and marine presence, aquaculture escape, and freshwater new records. There are four records from the Tigris, five from the Euphrates, 17 from the Shatt al-Arab, one from the Lesser Zab River, one from man-made river at south of Iraq, 22 from the marsh areas, one from natural lake, two from man-made lakes, and 49 from the marine waters of Iraq (Table 19.1).

19.2.1.5 Endangered Freshwater Fishes of Iraq

The threatened freshwater fish fauna of Iraq show difficulties ensuing chiefly from habitat alteration by man. The evolutionary history of the fauna has left it especially subtle to biotic interactions. In addition, many forms are of such limited distribution that the whole taxon can confront obliteration by minor alarms. Freshwater fishes of Iraq are confronting the same general types of ecological difficulties that are causing destructions throughout the world. The interaction of economics with apparent value in society has led us into the numerous ecological difficulties challenging us today. There is some indication to propose that society is making some initial efforts to slow the rate of extinction. Possibly this is occurring because the assumptions of ecologists, philosophers, and theologians regarding the relationship of man and environment are to a degree being translated into legislation as well as into conventional wisdom.

Groups of the Threatened Fishes of Iraq

Jawad (2013) has evaluated the conservation status of the freshwater fishes of Iraq and grouped those endangered species as shown in Table 19.2. In this table, the conservation status of the twenty freshwater fish species of Iraq using IUCN Red List categories and criteria is shown (IUCN 2001). There are three IUCN categories recognized for the twenty freshwater fish species of Iraq. There are three species assigned to the regionally extinct in the wild category (RE), five species to the critically endangered criteria (CR), and twelve species for the vulnerable criteria (VU) (Table 19.1). The fish species in the aforementioned three categories are confined to one or all the following states: (1) a certain area, (2) fragmentation of the population, (3) dramatically reduced in population size, and (4) subject to heavy fishing all year round.

The distribution of threatened freshwater fishes of Iraq that deals with in this chapter is in the Tigris-Euphrates River system of Iran, Turkey, Syria, and Iraq. This distribution is based on Jawad (2013). There are significant differences between the past and present time distributions. In the past, these species used to swarm the

Table 19.2 IUCN Red List status of the threatened freshwater fishes of Iraq

Species	Threatened categories	Red list criteria	Decline rate over years
Luciobarbus esocinus (Heckel, 1843)	RE	–	>90% over 35 years
Barbus rajanorum (Heckel, 1843)	RE	–	>90% over 45 years
Barbus subquincunciatus (Gunther, 1868)	RE	–	>90% over 35 years
Oxynoemachelius argyrogramma (Heckel, 1847)	CR	A2ace	>30% over 45 years
Oxynomachelius frenatus (Heckel, 1843)	CR	A2ace	>30% over 45 years
Barbus grypus (Heckel, 1843)	CR	CRA2CDE	>50% over 25 years
Barbus lacerta (Heckel, 1843)	CR	A2ACE	>75% over 30 years
Hemigrammocapoeta elegans (Gunther, 1868)	CR	A2ACE	>75% over 30 years
Barbus barbulus (Heckel, 1847)	VU	A1A+1C	>30% over 30 year
Luciobarbus kersin (Heckel, 1843)	VU	A1A+1C	>30% over 35 years
Luciobarbus xanthopterus (Heckel, 1843)	VU	A2A+3A +1C	>305 over 25 year
Barilius mesopotamicus (Berg, 1932)	VU	A2a	>30% over 35 years
Capoeta barroisi (Lortet, 1894)	VU	A2a	>30% over 40 years
Capoeta damascina (Valenciennes, 1812)	VU	A2a	>30% over 45 years
Capoeta trutta (Heckel, 1843)	VU	A2a	>30% over 45 years
Chondrostoma regium (Heckel, 1843)	VU	A2a	>30% over 45 years
Cyprinion kais (Heckel, 1843)	VU	A2a	>30% over 35 years
Cyprinion macrostomum (Heckel, 1843)	VU	A2a	>30% over 35 years
Squalius cephalus (Linnaeus, 1758)	VU	A2a	>30% over 45 years
Squalius lepidus (Heckel, 1843)	VU	A2a	>30% over 45 years

Tigris-Euphrates rivers system, Shatt al-Arab River, the marsh areas, and the natural and man-made lakes (Table 19.3). Some of the species occur only in the upper reaches of the Tigris River, whereas others are confined to its small tributaries or to the natural and man-made lakes (Table 19.3). Several factors that are considered

Table 19.3 Distribution of the threatened freshwater fishes of Iraq

Species	Past distribution	Current distribution
Barbus esocinus (Heckel, 1843)	Tigris-Euphrates Rivers system, Shatt al-Arab River, marsh areas, natural and man-made lakes	Victory Palace fish lakes, Baghdad
Barbus rajanorum (Heckel, 1843)	=	Confined to Dokan Dam Lake
Barbus subquincunciatus (Gunther, 1868)	Upper reaches of Tigris and Euphrates Rivers, Lesser and Greater Zab Rivers, Diyala River	Confined to Derbendikhan Dam Lake
Barbatula argyrogramma (Heckel, 1849)	Diyala River, Lesser Zab River, Dokan Dam Lake	Confined to Diyala River
Barbatula frenata (Heckel, 1843)	Lesser Zab river, Khalis River, Habbaniyah Lake	Confined to Khalis River
Barbus grypus (Heckel, 1843)	Tigris-Euphrates Rivers system, Shatt al-Arab River, marsh areas, natural and man-made lakes	Upper reaches of Tigris and Euphrates Rivers, Greater and Lesser Zab Rivers, natural and man-made lakes
Barbus lacerta (Heckel, 1843)	Upper Tigris River Tributaries, Diyala River, natural and man-made lakes	Greater and Lesser Zab Rivers
Hemigrammocapoeta elegans (Gunther, 1868)	Tigris River Basin	Greater Zab river
Barbus barbulus (Heckel, 1847)	Tigris-Euphrates Rivers system, Shatt al-Arab River, marsh areas, natural and man-made lakes	Confined to the Lesser Zab River
Barbus kersin (Heckel, 1843)	=	Confined to Adhaim and Diyala Rivers
Barbus xanthopterus (Heckel, 1843)	=	Tigris-Euphrates Rivers system, Shatt al-Arab River, marsh areas, natural and man-made lakes
Barilius mesopotamicus (Berg, 1932)	Greater Zab River	Greater Zab River
Capoeta barroisi Lortet in (Barroisi, 1894)	Tigris-Euphrates Rivers system, natural and man-made lakes	Confined to Khalis River
Capoeta damascina (Valenciennes in Cuvier and Valenciennes, 1812)	=	Confined to Diyala River and its branches
Capoeta trutta (Heckel, 1843)	=	Greater and Lesser Zab Rivers
Chondrostoma regium (Heckel, 1843)	=	=
Cyprinion kais (Heckel, 1843)	Greater and Lesser Zab Rivers, Diyala River, Upper Euphrates River	Confined to the upper Tigris River north of Mosul City, Greater Zab River
Cyprinion macrostomum (Heckel 1843)	Tigris-Euphrates Rivers system, natural and man-made lakes	=

(continued)

Table 19.3 (continued)

Species	Past distribution	Current distribution
Squalius cephalus (Linnaeus, 1758)	=	Lesser Zab River
Squalius lepidus (Heckel, 1843)	Tigris River Tributaries	=

vital in the determination of the restriction of the fish species to a certain area are as follows: (1) the preference of the species to the environmental conditions present in the area, (2) absent or fewer competitors, (3) absent or fewer enemies, and (4) pollution. Ten of the threatened species in question seem to be confined to the Greater and Lesser Zab rivers, the main two tributaries of Tigris River, whereas only two species, *Barbus subquincunciatus* and *Barbus rajanorum*, have refuge at the man-made dam lakes of Dokan and Derbendikhan, respectively. Other threatened species have retreated to other Tigris River tributaries at the eastern part of Iraq. The change in the distribution of these species started in the late 1950s and early 1960s, and since then an increased degradation and fragmentation is evident in their population (Maza 2005).

The twenty threatened fish species belong to the two families, Cyprinidae and Nemacheilidae of the order Cypriniformes. Family Cyprinidae has the larger number of the threatened species (18), whereas only two species belong to Nemacheilidae. The 18 cyprinid species belong to eight genera and five genera belonging to the genus *Barbus*. Genera with two and three species are *Luciobarbus, Squalius, Cyprinion,* and *Capoeta*, and *Hemigrammacapoeta*, *Barilius*, and *Chondrostoma* are monotypic genera. Genus *Barbus* has two regionally extinct wild species (RE), two critically endangered species (CR), and one vulnerable species (VU).

19.2.1.6 Threats

The main threats to the freshwater fish species of Iraq are the following: (1) habitat loss, degradation, and deteriorating level of water quality; (2) pollution of water and sediment: pollution resulting from increased oil production activities; (3) improper use of insecticides and chemical fertilizers; (4) competition for food and habitats: increased competition resulting from the introduction of different species of carp; (5) the presence of predators of eggs, larvae, and young: predators of the native fish species of *S. glanis*. The effect of predators became serious when the government of Iraq, in an attempt to control malaria transmitting mosquito larvae, mistakenly introduced *Heteropneustes fossilis*; (6) pathological factors: pathological outbreaks are common events for the native species from different parts of Iraq. Such events weaken individuals and reduce their reproductive capability; (7) over fishing and illegal fishing: the efforts of the Department of Fisheries at the Ministry of Agriculture in Baghdad towards reducing illegal fishing activities by legislating new rules were too limited and insufficient to protect the native freshwater fish populations;

(8) the failure of the government of Iraq to release captive-bred individuals to enhance the wild population: chaos in the fishing administration was a characteristic feature due to the ongoing political instability in Iraq; (9) freshwater aquaculture: aquaculture emerged as a national industry in Iraq in the 1960s. Several species introduced for aquaculture purposes found their way to the rivers and streams accidentally through mishandling; (10) utilization: most of the threatened species are commercially important. Their flesh is very delicious in general and species such as *L. esocinus*, *L. xanthopterus*, and *Barbus grypus* sell for high prices. Thus, there is heavy public demand for those fish species, which encourages anglers to increase fishing intensity for higher profit; (11) intrinsic factors: intrinsic factors seem to affect several biological issues of the threatened fish species; (12) human disturbances: disturbance could be in a form of new roads or fire which directly affect the ecosystem; and (13) flow modification: modification of flow is one of the five interacting categories that create threats to global freshwater biodiversity in general. In the Mesopotamian drainage, several dams are found on both Tigris and Euphrates Rivers in Turkey, Syria, and Iraq.

19.2.1.7 The Advantageous the Iraq Freshwater Fish Biodiversity

In general, freshwater fish biodiversity may contribute to the broad variety of valuable goods and services for human societies (Covich et al. 2004). The fishes have a direct contribution to economic productivity and contribute in other components such as insurance value in light of unexpected events, value as a storehouse of genetic information, and value in supporting the provision of ecosystem services (e.g. clearing water) (Heal 2000). To date, no estimation has been done for the full value of the freshwater biodiversity in general and Iraqi freshwater biodiversity in particular through assessing each component mentioned above.

This is true in the case of Iraq with the political and environmental chaos that took place for a long time. Of particular concern is the decline in populations of some freshwater fish species in Iraq to a level whereby they have become so scarce that their ecological roles have degraded to an extent where they might as well be extinct.

It may be possible to meet human needs for water without loss of most inland-water species including fishes, but this will require implementation of environmental water allocations that allow natural patterns of flow variability and include a range of flows, not just a minimum level of flow (Bunn and Arthington 2002). At the present time, scientists can neither estimate the quantities of water that can be extracted nor the temporal changes in flow that can be tolerated by freshwater taxa in general and in fish in particular (Dudgeon et al. 2006). Maintenance of the freshwater biodiversity is considered as a critical task and scientists have to examine whether the water use or ecosystem modifications are sustainable and address all use of freshwater organisms as bio-indicators of habitat condition (Karr and Chu 1999).

19.2.1.8 Conservation Measures

Inland waters constitute valuable natural resources in economic cultural, aesthetic, scientific, and educational terms. Their conservation and management are critical to the interest of all nations and governments. With regards to policy-based actions, there are several issues needed to be implemented and put into action in the immediate future. Among these are (1) development and implementation of management plans; (2) development and implementation of legislation at the national level; and (3) community management, including governance, resource stewardship, and livelihood alternatives.

Communication and education issues are as follows: (1) formal education starting from the primary school level, (2) awareness programs especially for the illiterate people in the rural and marsh areas, and (3) capacity-building and training at the secondary school and tertiary levels to have environmentally well-educated personnel to serve as fishery inspectors or community educators. Several issues are in need of consideration in regards to habitat and site-based actions. They are maintenance and conservation, restoration, and identification of protected areas. No protected areas have yet been recognized in Iraq. In order to recognize a protected area, it is important to establish, manage, expand, and look for community initiatives.

Among conservation actions, species-based issues arise as a major step to be considered by Iraqi authorities to limit the influence of the introduced species on the native freshwater fish fauna. To achieve such goals, the Iraqi government needs to (1) place benign introduction among their priorities; (2) harvest and trade management should be clearly considered in the fishing policy; (3) public awareness should take a vital role in promoting these issues; and (4) the necessity to establish a genome resource bank in Iraq.

If tendencies in human demands for water remain unaltered and species losses continue at current rates, the opportunity to conserve much of the remaining fish biodiversity in fresh water will vanish before the water from the surface of the earth will disappear. Such occasion costs will be overstated by a noteworthy loss in choice values of species yet unknown for human use. In addition, these fundamental ecological and possible financial losses may well be permanent. Significantly, actual conservation measures will be needed to make a major change in approach towards freshwater biodiversity and ecosystem administration, including general acknowledgement of the catchment as the focal management unit, and greater recognition of the trade-offs between species conservation, overall ecosystem reliability, and the provision of goods and services to humans.

References

Abell R, Thieme M, Revenga C, Bryer M, Kottelat M, Bogutskaya N, Coad B, Mandrak N, Contreras-Balderas S, Bussing W, Stiassny MLJ, Skelton P, Allen GR, Unmack P, Naseka A, Ng R, Sindorf N, Robertson J, Armijo E, Higgins J, Heibel TJ, Wikramanayake E, Olson D,

Lopez HL, Reis RED, Lundberg JG, Sabaj Perez MH, Petry P (2008) Freshwater ecoregions of the world: a new map of biogeographic units for freshwater biodiversity conservation. BioScience 58:403–414
Al-Badri ME, Jawad LA (2014) New fish records from the marine waters of Iraq. Cah Biol Mar 55:431–436
Al-Daham NK (1974) Two new records of sharks from Iraqi territorial waters with their descriptions. Bull Basrah Nat Hist Mus 1(1):99–111
Al-Daham NK (1975) The first record of the dragon fish *Pegasus natans* Linnaeus from the Arab Gulf with notes on its distribution. Bull Basrah Nat Hist Mus 2:55–57
Al-Daham NK (1976) The stingray *Taeinura melanospilos* a new record from the Arab Gulf. Bull Basrah Nat Hist Mus 3:117–118
Al-Daham NK (1977) Fishes of Iraq and the Arab Gulf, vol. 1. Squaliformes to Atherniformes. Centre for Arab Gulf Studies Publications No. 9 (In Arabic with Latin names)
Al-Daraji SA, Jawad LA, Al-Faisal AJ, Taha A (2017) Second appearance of the burrowing goby Trypauchen vagina (Bloch & Schneider, 1801) in the marine waters of Iraq. Cah Biol Mar 58:229–232
Al-Faisal AJ, Mutlak FM (2014) First record of the Nile tilapia *Oreochromis aureus* (Linnaeus, 1758), from the Shatt Al-Arab River, South Iraq. Mesopotamian J Mar Sci 29:45–50
Al-Faisal AJ, Mutlak FM (2015) First record of the Nile tilapia *Oreochromis niloticus* (Linnaeus, 1758), from the Shatt Al-Arab River, Southern Iraq. Int J Mar Sci 5:1–3
Al-Hassan LAJ, Al-Badri MEH (1986) First record of some fishes from Khor al-Zubair, Khor Abdullah and Shatt al-Arab, Basrah, Iraq. Cybium 10(3):295–297
Al-Hassan LAJ, Miller PJ (1987) *Rhinogobius brunneus* (Gobiidae) in the Arabian Gulf. Jpn J Ichthyol 33(4):405–408
Al-Hassan LAJ, Muhsin KA (1986) The presence of *Barbus luteus* and *Heteropneustesfossilis* in Khor al-Zubair, North-west of the Arabian Gulf. Zool Middle East 1:116–118
Al-Hassan LAJ, Naama A (1986) New record for some Arab Gulf fishes in the freshwater systems of Iraq. Bull Basrah Nat Hist Mus 6:35–44
Al-Hassan LAJ (1993) Additional synopsis of biological data on *Tenualosa ilisha*. Marina Mesopotamica suppl no 2:1–23
Al-Hassan LAJ (1994) The silver carp, *Hypophthalmichthys molitrix* (Cyprinidae) in the Shatt al-Arab River, Iraq. Cybium 18:204
Al-Hassan LAJ, Hussain NA, Soud KD (1989) A preliminary annotated check-list of the fishes of Shatt al-Arab River, Basrah, Iraq. Polskie Archiwum Hydrobiologii 36(2):283–288
Al-Mukhtar MA, Jawad LA, Al-Faisal AJ, Mustafa F (2011) First record of Dotted grouper Epinephelus epistictus (Temminck & Schlegel, 1842) (Pisces: Serranidae) for the fish fauna of Iraq. Zool Middle East 54:136–137
Al-Nasiri SK, Hoda SMS (1975) Survey of fish fauna of Shatt al-Arab (from Abu al-Khasib to Karmat Ali). Bull Basrah Nat Hist Mus 2:36–46
Al-Nasiri, S. K. and S. M.S. Hoda (1976). A guide to the freshwater fishes of Iraq. Basrah Natural History Museum Publication No. 1, 124 pp
Al-Rawi AH, Al-Obaidi S, Jawdat SZ (1978) A list of fishes collected from the Little Zab River in Iraq: 1. Bulletin of the Biological Research Centre. Baghdad 10:3–12
Al-Saad HT, Al-Hello MA, Al-Taein AM, Douabul AZ (2009). Environmental study on the marshes of southern Iraq. In: 3rd Marsh conference, University of Basrah, (Abstract)
Altinbilek D (2004) Development and management of the Euphrates–Tigris basin. Int J Water Resour Dev 20(1):15–33
Al-Zaidy KJ (2013) First record of *Tilapia zillii* Gewais, 1848 in Al-Delmj marsh west Al-Diwania city middle of Iraq. Diyala Agric Sci J 5:9–16
Bunn SE, Arthington AH (2002) Basic principles and ecological consequences of altered flow regimes for aquatic biodiversity. Environ Manag 30:492–507
Coad BW (1991) Fishes of the Tigris-Euphrates Basin: a critical check-list. Syllogeus No. 68

Coad BW (2010) Freshwater fishes of Iraq. Pensoft Series Faunistica No. 93. Pensoft, Moscow, 274 pp

Coad BW, Hussain NA (2007) First record of the exotic species *Hemiculter leucisculus* (Actinopterygii: Cyprinidae) in Iraq. Zool Middle East 40:107–109

Covich AP, Ewel KG, Hall RO, Giller PE, Goedkoop W, Merrit DM (2004) Ecosystem services provided by freshwater benthos. In: Wall DH (ed) Sustaining biodiversity and ecosystem services in soil and sediments, pp 45–72

Darwall WRT, Smith KG, Allen DJ, Holland R, Harrison I, Brooks E (eds) (2011) The diversity of life in African freshwaters: under water, under threat. An analysis of the status and distribution of freshwater species throughout mainland Africa. IUCN, Gland, Switzerland

Dudgeon D, Arthington AH, Gessner MO, Kawabata Z-I, Knowler DJ, Leveque C, Naiman RJ, Prieur-Richard A-H, Soto D, Stiassny MLJ, Sullivan CA (2006) Freshwater biodiversity: importance, threats, status and conservation challenges. Biol Rev 81:163–182

Dugan P, Delaporte A, Andrew N, O'Keefe M, Welcomme R (2010) Blue harvest: inland fisheries as an ecosystem service. Penang, Malaysia

FAO (2007) *The state of world aquaculture and fisheries 2006*. Fisheries and Aquaculture Department, Food and Agriculture Organization of the United Nations, Rome, Italy

Freyhof J, Abdullah YS (2017) Two new species of *Oxynoemacheilus* from the Tigris drainage in Iraqi Kurdistan (Teleostei: Nemacheilidae). Zootaxa 4238

Freyhof J, Brooks E (2011) European red list of freshwater fishes. Publications Office of the European Union, Luxembourg

Freyhof J, Abdullah YS, Ararat K, Ibrahim H, Geiger MF (2016) Eidinemacheilus proudlovei, a new subterranean loach from Iraqi Kurdistan (Teleostei; Nemacheilidae). Zootaxa 4173(2016)

Gleick PH (1996) Water resources. In: Schneider SH (ed) Encyclopaedia of climate and weather. Oxford University Press, New York, pp 817–823

Günther A (1868) Catalogue of the fishes in the British Museum, vol 7. Taylor & Francis, London, pp 1–512

Günther A (1874) A contribution to the fauna of the River Tigris. Ann Mag Nat Hist 14:36–37

Heal GM (2000) Nature and the marketplace: capturing the value of ecosystem services. Island Press, Washington, DC

Heckel JJ (1843) Abbildungen und Beschreibungen der Fische Syriens nebst einer neuen Classification und Characteristik sämmtlicher Gattungen der Cyprinen. Stuttgart, 109 pp

Hora SL, Misra KS (1943) On a small collection of fish from Iraq. J Asiat Soc Bengal 9(1):1–15

IUCN (2001) IUCN red list categories: Version 3.1. Prepared by the IUCN Species Survival Commission. IUCN, Gland, Switzerland and Cambridge, UK

Jawad LA (2003) Impact of environmental changes on the freshwater fish fauna of Iraq. Int J Environ Stud 60(6):581–593

Jawad LA (2006) Fishing gear and methods of the Lower Mesopotamian Plain with reference to fisheries management. Marina Mesopotamica 1(1):1–39

Jawad L (2013) Threatened freshwater fishes of Iraq, with remarks on their conservation status. Water Res Manag 3(2):27–36

Jawad LA (2015) Four new records of fishes from the Arabian Gulf coast of Iraq. Bol Inst Pesca São Paulo 41(4):1033–1042

Jawad LA, Al-Badri ME (2014) *Lophiomus setigerus* (Vahl, 1797), *Nemipterus zysron* (Bleeker, 1856), and *Parascolopsis eriomma* (Jordan & Richardson, 1909) (Osteichthyes: Lophiidae and Nemipteridae) in the marine waters of Iraq. Zool Middle East 60(2):186–188

Jawad L, Al-Badri M (2015) *Bodianus macrognathos* (Teleostei: Labridae), *Coris nigrotaenia* (Teleostei: Labridae) and *Bothus pantherinus* (Teleostei: Bothidae) in the Iraqi marine waters. North West J Zool 11(2):347–350

Jawad LA, Hussain S (2014) First record of *Antennarius indicus* (Pisces: Batrachoidiformes: Antennaridae), *Equulites elongatus* (Pisces: Perciformes: Leiognathidae) and second record of *Cheilinus lunulatus* (Pisces: Perciformes: Labridae) from the Marine Waters of Iraq. Int J Mar Sci 4(40):1–5

Jawad LA, Hussein SA, Fahad KK (2009) *Glyptothorax kurdistanicus* (Berg, 1931) in the lower reaches of Tigris River, Iraq. J Appl Ichthyol 25:779–781

Jawad LA, Hussein SA, Bulbil F (2010) The rare Ranzania laevis (Pennant, 1776) (Tetraodontiformes: Molidae) in the marine waters of Iraq. J Appl Ichthyol. https://doi.org/10.1111/j.1439-0426.2010.01609.x

Jawad LA, Al-Faisal A, Al-Mukhtar M (2012) Presence of the non-indigenous fish species, Gibel carp, *Carassius gibelio* (Family: Cyprinidae) in Basrah Province freshwater systems, Iraq. Water Res Manag 2(4):41–44

Jawad LA, Al-Badri ME, Fricke R (2014a) New records of thicklips and grunts from the marine waters of Iraq (Teleostei: Haemulidae). J Ocean Sci Found 12:18–24

Jawad LA, Al-Mukhtar MA, Faddagh MS (2014b) Confirmation of the presence of *Heniochus acuminatus* (Linnaeus, 1758) (Chaetodontidae) and *Pomacanthus maculosus* (Forsskål, 1775) (Pomacanthidae) in Iraqi marine waters, Arabian Gulf. Arxius de Miscel·lània Zoològica 12:124–129

Jawad LA, Al-Mukhtar MA, Al-Faisal AJ, Hammed T (2014c) New record of Neobythites steatiticus Alcock, 1894 (Actinopterygii: Ophidiidae) from the marine waters of Iraq. Int J Mar Sci 4(32):1–3

Jawad LA, Al-Mukhtar MA, Subr SH (2016a) Occurrence of population of *Photopectoralis bindus* (Valenciennes, 1835) in Shatt al-Arab River, Basrah, Iraq. Int J Mar Sci 6:1–4

Jawad LA, Nasir NA, Al-Faisal AJ, Sultan QM, Mahlhal MS (2016b) Day's Goby, *Acentrogobius dayi* Koumans, 1941 (Pisces: Gobiidae) in the Southern Marshes, Thiqar, Southwest Baghdad, Iraq. Int J Mar Sci 6:1–5

Jones C, Sultan M, Yan E, Milewski A, Hussein M, Al-Dousari A, Al-Kaisy S, Becker R (2008) Hydrologic impacts of engineering projects on the Tigris–Euphrates system and its marshlands. J Hydrol 353(1-2):59–75

Karadeniz N, Tiril A, Baylan E (2009) Wetland management in Turkey: problems, achievements and perspectives. Afr J Agric Res 4(11):1106–1119

Karr JR, Chu EW (1999) Restoring life in running waters: better biological monitoring. Island Press, Washington, DC

Kennedy WP (1937) Some additions to the fauna of Iraq. J Bombay Nat Hist Soc 39:745–749

Khalaf KT (1961) The marine and freshwater fishes of Iraq. Al-Rabitta Press, Baghdad. 164 pp

Landsberger B (1962) The fauna of ancient Mesopotamia. MSL VIII/2, second part. Pontificium Institutum Biblicum, Roma 204, Piazza Pilotta 35. Publié avec le concourse financier de l'UNESCO et sous less auspices du Conseil International de la Philosophie et des Sciences Humaines, 180 pp

Lévêque C, Oberdorff T, Paugy D, Stiassny MLJ, Tedesco PA (2008) Global diversity of fish (Pisces) in freshwater. In: Balian EV, Lévêque C, Segers H, Martens K (eds) The freshwater animal diversity assessment, Hydrobiologia, vol 595. Springer, Dordrecht, pp 545–567

Lundberg G, Kottelat M, Smith GR, Stiassny MLJ, Gill AC (2000) So many fishes, so little time: an overview of recent ichthyological discovery in continental waters. Ann Mo Bot Gard 87:26–62

Mahdi N (1962) Fishes of Iraq. Ministry of Education, Baghdad. 82 pp

Mahdi N (1971) Additions to the marine fish fauna of Iraq. Iraq Natural History Museum, Special Publication No. 28: 47pp

Mahdi N, George PV (1969) A systematic list of the vertebrates of Iraq. Iraq Nat Hist Mus Publ Baghdad 26:1–104

Maza HY (2005) Fisheries catch records. J Fish Tech 73:112–125

Menon MAS (1956) On a third collection of fish from Iraq. Rec Indian Mus 54:139–157

Misra KS (1947) On a second collection of fish from Iraq. Rec Indian Mus 45:115–127

Mohammed ARM, Al-Hassan LAJ, Ali TS (1993) The presence of a cyprinid fish, *Barbus luteus* in marine waters of Iraq. Arquivos do Museu Bocage 2(25):415–416

Mohamed ARM, Abood AN, Jawad LA (2016) Presence of *Carasobarbus sublimus* (Coad & Najafpour, 1997) in the upper reaches of Shatt al-Arab River, Basrah, Iraq. Zool Ecol. https://doi.org/10.1080/21658005.2016.1272833

Mutlak FM, Al-Faisal AJ (2009) A new record of two exotic cichlids fish *Oreochromis aureus* (Steindacher, 1864) and *Tilapia zilli* (Gervais, 1848) from south of the main outfall drain in Basrah city. Mesopotamian J Mar Sci 24(2):160–170

Mutlak F, Jawad L, AL-Faisal A (2017) *Atractosteus spatula* (Actinopterygii: Lepisosteiformes: Lepisosteidae): a deliberate aquarium trade introduction incidence in the Shatt Al-Arab River, Basrah, Iraq. Acta Ichthyologica et Piscatoria 47(2):205–207

Myers N, Mittermeier RA, Mittermeier CG, de Fonseca GAB, Kent J (2000) Biodiversity hotspots for conservation priorities. Nature 403:853–858

Nader IA, Jawdat S (1977) First records of fish from Iraqi waters. Bull Fac Sci Riyadh Univ 8:429–443

Rubec C, Alwash A, Bachmann A (2009) The key biodiversity areas project in Iraq: objectives and scope 2004–2008. In: Krupp F, Musselman L J, Kotb M M A, Weidig I (eds) Environment, biodiversity and conservation in the Middle East. Proceedings of the first Middle Eastern biodiversity congress, Aqaba, Jordan, 20–23 Oct 2008. BioRisk 3:39–53

Saggs HWF (1962) The greatness that was Babylon. Sidgwick and Jackson, London. 562 pp

Sala OE, Chapin FS, Armesto JJ, Berlow R, Bloomfield J, Dirzo R, Huber-Sanwald E, Huenneke LF, Jackson RB, Kinzig A, Leemans R, Lodge D, Mooney HA, Oesterheld M, Poff NL, Sykes MT, Walker BH, Walker M, Wall DH (2000) Global biodiversity scenarios for the year 2100. Science 287:1770–1774

Saleh KI (2007) First record of cichlid fish *Tilapia zilli* in the Euphrates River near Musaib City, Centre of Iraq (Abstract). 2nd Fisheries Conference, Basrah University, March 1–4, 2007

Shafi M, Shalli MR, Saeed BM (1980) *Capoeta pestai* (Pietschmann) (Cyprinidae) a new record for Iraqi waters. Bull Nat Hist Res Cent Baghdad 7(4):173–175

Strayer DL, Dudgeon D (2010) Freshwater biodiversity conservation: recent progress and future challenges. J N Am Benthol Soc 29(1):344–358

Stiassny MLJ (2002) Conservation of freshwater fish biodiversity: the knowledge impediment. Verhandlungen der Gesellschaft für Ichthyologie 3:7–18

Sultan FAM, Ahmed SM (2009) Effect of salinity acclimation on growth in *Acanthopagrus latus* (Houttyn, 1782) juveniles. In: The 6th scientific conference of fisheries resources, 3–4 Mar 2009, Basrah, Iraq (Abstract)

Vorosmarty CJ, McIntyre PB, Gessner MO, Dudgeon D, Prusevich A, Green P, Glidden S, Bunn SE, Sullivan CA, Liermann CR, Davies PM (2010) Global threats to human water security and river biodiversity. Nature 467:555–561

Ziyadi MSF, Jawad LA, Al-Mukhtar MA, Pohl T (2015) Day's goby, Acentrogobius dayi Koumans, 1941 (Pisces: Gobiidae) in the desert Sawa Lake, south-west Baghdad, Iraq. Mar Biodivers Rec 8:e148. Published online. Page 1 of 5. https://doi.org/10.1017/S1755267215001220

Ziyadi MSF, Jawad LA, Al-Mukhtar MA (2018) *Halicampus zavorensis* Dawson, 1984 (Syngnathidae): new record for Iraqi marine waters and for the Arabian Gulf area. Cah Biol Mar 59:121–126

Chapter 20
Effect of Climate Changes on the Freshwater Biodiversity in the Mesopotamian Plain: Recommendations for Avoidance and Plans for the Future

Laith A. Jawad and Baradi Waryani

Abstract Freshwater ecosystems offer a special set of goods and services valued highly by the public that are inseparably linked to their flow and the interaction of these systems with the landscape. However, most freshwater ecosystems are within the anthropogenic activities that can create different types of stresses such as development, dams, or extractive uses. The climatic changes will be an additional impact to the freshwater ecosystems as these changes have the potential to modify rainfall, temperature, run-off patterns, and to disrupt biological communities and sever ecological linkages. In order to protect these valuable ecosystems, management must put a plan to be based on within environment resources. Saving rivers and riparian corridors is a priority in such plans, and this needs partnerships among multiple associates in the particular river basins. Applying renovation schemes positively can be used to save present resources so that exclusive reactive restoration to repair damage associated with a changing climate is curtailed. Freshwater biodiversity comes first in those plans so as to provide diversifying and replicating habitats of special importance.

20.1 Introduction

Changes that are happening on the global scales have significant effects on the biodiversity (Sala et al. 2000). Such changes may take different forms of anthropogenic influences like land-use alterations, nitrogen deposition, and invasions of

L. A. Jawad (✉)
School of Environmental and Animal Sciences, Unitec Institute of Technology, Auckland, New Zealand

B. Waryani
Department of Fresh Water Biology and Fisheries, University of Sindh, Jamshoro, Sindh, Pakistan

L. A. Jawad (ed.), *Tigris and Euphrates Rivers: Their Environment from Headwaters to Mouth*, Aquatic Ecology Series 11,
https://doi.org/10.1007/978-3-030-57570-0_20

exotic species; much recent interest has been directed at climate change (Parmesan 2006). During the history of the earth, substantial climate changes have been evident on different rate and magnitude, but what is happening during the present time cannot predict its consequence in the future (IPCC 2001). As with both terrestrial and marine environment, aquatic ecosystems are as susceptible to changes that happened globally. Sala et al. (2000) considered lentic (i.e. lakes and ponds) and lotic (i.e. streams and rivers) ecosystems to be most subtle to land-use change, exotic species, and climate change in a global-scale valuation. Variation in the effects of environment of the aquatic ecosystems could be happened between regions and high levels, within the aquatic habitats at high latitudes being more intensely endangered by climate alteration than by others in the lower latitudes.

For the freshwater ecosystem and for rivers in specific, climate variation is not the only hazard as most of these ecosystems are within the range of the anthropogenic activities that include agriculture, urbanisation, or suburban development or affected by dams. Consequently, management sets for climate modification often overlap with and include activities that influence human use of the land or water. By the growing effect by means of land by human, the ability of a river to bring expected habitats goods and facilities in the future will be influenced increasingly by how it is controlled.

The present chapter will give a short review on the impacts of climate fluctuations on the freshwater aquatic ecosystem and how these changes affect in particular the distribution of the species in combination with the other anthropogenic influences. The recommendations will be addressed to the policymakers in the countries located in the Mesopotamian plane to take a prompt action to save this vulnerable environment.

20.2 Historical Record of the Climatic Changes in the Mesopotamian Plain

In the recorded history of humanity, probably the first drastic climatic changes happened in the Mesopotamian plain are that occurred near 4200 B. P. during the reign of Sargon of Akkad, the king that succeeded in establishing the world's first unified empire in Mesopotamia connecting the isolated agricultural neighbourhoods of northern Mesopotamia with the compound city-states in the south (Cullen et al. 2000). One of the factors that lead to the success of this empire was the fertile, rain-fed agricultural production of the wide, northern Mesopotamian plains. The Akkadian Empire has collapsed 100 years later near 4200 B. P. (Weiss et al. 1993) due to a chief environmental changes that linked to collapse of the empire according to the archaeological investigations at Tell Leilan in northeast Syria told. Weiss et al. (1993) has suggested that such environmental changes could reflect an unexpected beginning of an arid circumstance, which may have subsidised to the perceived failure. The geological evidences taken from the soil may indicate restricted occurrences dissimilar to larger scale local aridification.

Yoffee and Cowgill (1988) thought that the Akkadian collapse had been formerly attributed to human factors such as invasion and political imbalance, but Sandweiss et al. (1999) believe the Akkadians social collapse was as a result of severe climatic changes. In both cases, the reasons may be resolved by the study of the record of Holocene palaeoclimatic disparities in Mesopotamia as conserved in a marine sediment core from the Gulf of Oman (Cullen et al. 2000).

The northern part of Mesopotamia is characterised in having climate that is typically semi-arid, with strong seasonality in both precipitation and temperature (Cullen et al. 2000). In this area, winters are cool and wet (100–300 mm/year), while summers are hot, very dry, and obvious by a continued northwest wind recognised locally as the shamal. Such an atmosphere will be a good foundation of mineral dust, with numerous regions stating over 200 days/year of perceptibility diminishing due to dust (Pye 1987). Such minerals will be conveyed to the Arabian Gulf and Gulf of Oman with an assessed bulk fluidity of 100×106 tons/year (Pye 1987) according to the examination of satellite images and marine sediments (Al-Bakri et al. 1984; Al-Ghadban 1990; Sirocko and Sarnthein 1989).

The mineralogical and geochemical evidences obtained from Gulf of Oman have shown the onset of arid conditions in Mesopotamia has occurred near 4025 ± 150 B. P. Claussen et al. (1999) have suggested that these aridification incidents were short-lived, continue for only a few centuries, and could be resulted from the significant variations in ocean-atmosphere-vegetation border settings. The climatic change event happened near 4025 B. P. was one of the largest incidents with large amplitude during the Holocene (Cullen et al. 2000).

In the Middle East, there are other palaeoclimate records that document arid circumstances start about 4000–4200 B. P. such as that happened in Lake Van of eastern Turkey that situated at the upper region of the Tigris and Euphrates Rivers (Lemcke and Sturm 1997). The other examples are from the cave deposits in Israel (Bar-Matthews et al. 1997) and the Dead Sea level that dropped sharply by ~100 m during this period (Frumkin 1991). Evidences from the archaeological and the regional palaeoclimate information indicate that unexpected climate change could play a vital role in contributing to the downfall of these erudite societies which had ruled Mesopotamia. With such climatic changes, the Akkadians had applied high-level grain-keeping and water-organisation knowledge to upkeep themselves versus seasonal disparities in rainfall (Weiss et al. 1993).

20.3 The Current Climate of the Mesopotamian Plain

The climate of the Mesopotamian plain is chiefly governed by wide variety of circulation patterns, teleconnections, and regional topography. Highlands of this plain show the attributes of cold-continental climate, while lowlands are classified as hot desert climate and hot semi-arid climate according to Koppen-Geiger climate classification (Bozkurt and Sen 2013). The influence of masses of air is covering both the lowlands and highlands of the plain in summer time, whereas highlands are under the effects of polar air masses in winter (Bozkurt et al. (2012).

In Iraq, the system of rainfall is following that of the Mediterranean region, where rainfall happens almost in winter, autumn, and spring and vanishes in summer. According to the annual rainfall factor, it is possible to define three rainfall regions in these areas: northern region, middle region, and southern region (Shubbar 1999). There is variation in the amount of rainfall that each part of Iraq receives. It ranges between 50 mm/year in the SW to 1200 mm/year in the NE (Al-Timimi and Al-Jiboori 2013).

The rain that falls on Iraq is a result of the cyclones originating from the Atlantic Ocean and moving across Iraq from the west, with variation in the number of these cyclones according to the seasons, months, and places over which they are passing. Normally, there are more cyclones in the winter and decreasing in the autumn and lastly disappear altogether in the summer. South part of Iraq evident more cyclones moving than the northern part, with higher amount of rain than the south due to the precipitations in the north are orographic as much as it is cyclonic (Hassan and Mashkour 1972).

20.4 Predictable Designs of Climate Alteration and Freshwater Habitats

Among the unbelievable climatic variations that happened during the last few decades is the increased in the earth surface temperatures by 0.6 °C. The global temperatures may keep rising if the anthropogenic emissions of the carbon dioxide gas are not restricted. With the continue rising in the temperature of the earth, utmost land areas are warming extra quickly than the global normal, and there also will be clear regional changes (IPCC 2002). Therefore, high-latitude areas could be experiencing more warming than other regions on earth (ACIA 2005). As a result, growing temperature will have profound effect on the ice cover, which in turn has intense impacts on freshwater inhabitants (Magnuson et al. 2000).

The climate change effects have an impact on the precipitation patterns. In high-latitude areas, rainfall is probable to upsurge in both summer and winter, with augmented year-to-year difference in amount of rain (IPCC 2002). Precipitation will not affect the hydrology of the region, but also influences the features of freshwater habitats, and also on the inhabiting these environments (Poff 1992).

The Eastern Mediterranean region and the Mesopotamia are under the effect of the North Atlantic Oscillation (NAO). This is a dominant style of Atlantic sector climate alterability (Eshel et al. 2000). Similar to the El Niño Southern Oscillation (ENSO) (Cane et al. 1994), the NAO has been revealed to impact both marine and terrestrial environments (Beniston 1997). Different from ENSO, there is still substantial uncertainty as to whether the NAO is a coupled ocean-atmosphere process.

Coupled, or two-way interactions between the ocean and atmosphere suggest the potential for a degree of expectedness. Predictability exists primarily within the long-term memory of the ocean and provides the opportunity for long-range climate forecasting (Mason et al. 1999).

The influence of the NAO on interannual to decadal precipitation variability in the Middle East has been studied by Cullen et al. (2002). Their results have shown that the accepted signature of the NAO, with enhanced precipitation in Western Europe during high index periods, and in the eastern Mediterranean during low index periods, is obvious (Cullen and deMenocal 2000; Eshel et al. 2000). There were noteworthy precipitation spreads through the northern Mediterranean approximately to the Caspian Sea with an improved secondary maximum sideways the west coast of Spain and Portugal and over Turkey (Cullen et al. 2002).

The results of Cullen et al. (2002) have also showed the presence of a relationship between the eastern Mediterranean and Atlantic sector due to the regulation of the NAO to the Atlantic heat and moisture fluxes into the Mediterranean region (Turkes 1996). These types of cyclones that NAO transfer to the Middle East will have a dominant effect on the source of rainfall and river run-off; therefore, NAO-related changes in Atlantic westerly heat/moisture transport and Atlantic/Mediterranean SSTs can be predicted to impact Middle Eastern climate.

Ecosystems types may show different levels of climate change effects (Poff et al. 2002). Accordingly, small streams could show strong impact than larger rivers owing to the similar associations between air temperature and water temperature of trivial streams. Those little streams are likewise extra susceptible to low flows and floods originating from predicted changes in precipitation. In the same way, little ponds may experience more thermal influence than larger water bodies.

20.5 Climate Change Impacts on Water Resources in the Mesopotamian Plain

It has been noted that in the Middle East the water resources are under a heavy and increasing stress. Therefore, any changes in climatic patterns that would increase temperatures and reduce rainfall would greatly intensify current problems. The general method used for evaluating the influences on hydrologic systems is to gain climate information that represents past data to use them as input to basin hydrologic models (Bou-Zeid and El-Fadel 2002).

Studies performed by Bou-Zeid and El-Fadel (2002) have shown that temperature changes during the winter months (January, February, and March) and the summer months (June, July, and August) in addition to the rainfall change during the wet season (October to April). With the high temperature during summer months to up to 2.1 °C, the areas at the Mediterranean coast like Lebanon, Israel, and coastal Syria would be the least affected. On the other hand, the effects on the groundwater aquifers in these areas are higher due to the increased seawater intrusion because of the higher sea levels.

Due to the high evaporation in the area, the demand for water to be used for irrigation will be increased accordingly. Bou-Zeid and El-Fadel (2002) have suggested that both increased in temperature, evapotranspiration, and constant precipitation are likely to happen to cause desertification. In winter months, slight increase in temperature occurred, which boosts evapotranspiration and reduces potential groundwater recharge. Such scenario of changes will most likely lead to climate change in the region.

20.5.1 Effects of Climate Change on Flow of Rivers

According to the FAO (2009), the normal yearly release rate of the Tigris-Euphrates is ranging from as low as 30 km^3 to as much as 84 km^3. Chenoweth et al. (2011) suggested that the usual yearly Tigris-Euphrates river release might reduce by 9.5% by the 2040–2069 period. The decrease in the amount of water is highest in Turkey at 12%, while it is only 4% in Iraq. There is an additional decline in the amount of water of the in river discharge by 2070–2099; nevertheless, the reduction is less than 1%. It has been calculated by Chenoweth et al. (2011) that at the end of the present century there will be about 10% decrease in overall river basin discharge due to the changes in the climate.

Similar to the case of Tigris-Euphrates river basin, there is annual variation in the amount of water in the Jordan River. At the closing of the century, Chenoweth et al. (2011) have suggested that accessible water resources in this river will be reduced to about 22% fall by 2040–2069 and a 30% fall by 2070–2099.

20.5.2 Expected Changes to the Rainfall and Internal Water Resources in the Mesopotamian Plain

Chenoweth et al. (2011) proposed that the average yearly rainfall in the area will descent from 491 mm in the 1961–1990 period to 443 mm in the 2040–2069, with a general decrease in amount of rain of 10%. Their results have shown that some of the countries such as Armenia, Bahrain, Georgia, Kuwait, and Qatar have experienced an annual in precipitation. They argued that Bahrain, Kuwait, and Qatar in spite of having an increase in the annual precipitation, but it is unimportant owing of their current absence of water supplies.

In general, inner water capitals alteration in line with rainfall through the area in general and the inside water assets will decline in their amount from 464 to 419 km^3 in 2040–2069 relative to 1961–1990.

There will be further small decrease in the amount of rain in the region at the end of the period of 2070–2099 starting from the mid-century Chenoweth et al. (2011). The results given by Chenoweth et al. (2011) about the amount of rain that expected to be in the region are very frightening. They showed a decline in the amount rain that the countries of the region having will decline from 443 mm in the period 2040–2069 to 436 mm in 2070–2099, therefore providing a regular more reduction in each country of 1.5% comparative to mid-century and 11% relative to 1961–1990.

20.6 Anthropogenic Impacts and the Climate Changes

In this section, specific highlighting is given to the following anthropogenic stressors: acidification, eutrophication, land-use change, and exotic species, keeping in mind that other sorts of impacts are present in the nature besides those mentioned above that can bring about alterations in the freshwater biodiversity.

20.6.1 Acidification

It has been documented that there is an interaction between climate changes and acidification. In the freshwater ecosystems, acidification in streams can be enhanced due to climate warming, which in turn have negative effects in the process of recovery of acidified lakes (Schindler 1997), but opposite opinions have been succumbed (Schindler et al. 1996). Therefore, it is problematic to conclude the inclusive outcomes of climate modification for the acidity of freshwater habitats, since regional factors other than climate changes might bring an increase in the alkalinity of the water of the lakes such as the effect of the run-off through the catchments, which may also lead to an increase in the alkalinity of the water of the lake (Schindler 1997). Such an increase in the acidity will have an inverse effect on the freshwater biodiversity (Giller and Malmqvist 1998) through conflicting consequences on seasonal fluctuations in the acidic and neutral streams (Durance and Ormerod 2007).

20.6.2 Eutrophication

Amount of nutrients entering the freshwater system can easily be affected by climate changes. Schindler (1997) found that climate warming could cause simultaneous decrease in the level of phosphorus concentration of the freshwater body like lakes water. This decrease will result in changes in phytoplankton communities. With the case of numerous freshwater habitats which are eutrophic owing to either natural or anthropogenic reasons, or augmented nutrient levels in such systems might frequently lead to decreases in biodiversity.

It has been expected in the case of the cold lakes to have an intensified algal production and biomass with both warming climate and increased nutrient contributions. In non-boreal water bodies, streams and lakes differ broadly in nutrient intensities, ranging from oligotrophic to eutrophic. In those areas, the intensified nutrient production in connotation with climate modification is having a clear effect on the biodiversity (Heino et al. 2009). An increase in some organism groups may happen in the oligotrophic lakes. Tammi et al. (1999) and Jeppesen et al. (2000) gave an indirect support for these suggestions based on examples drawn from fish group, when fish assemblage structure characteristically varies beside a gradient from oligotrophic to eutrophic lakes.

20.6.3 Land Cover Alterations

Land cover is one of the important factors in the settlement of the freshwater ecosystem. With climate changes, this cover will have a drastic effect, which in turn leads to a direct change in the riparian zones. The ecological changes might take several forms among those are the natural results of changes in terrestrial vegetation and the human-induced modifications of the vegetation (Saetersdal et al. 1998). In the cold region of the earth, forests are usually dominated by coniferous trees and such dominance may change some other species (Kellomäki et al. 2005) and might substitute by broadleaved deciduous trees found in the forests in the further southerly areas (Sykes and Prentice 1995). Freshwater system occurs in such forest will be impacted by the change in the dominance of the species of the trees of the forest. With changes in the species of the trees, invertebrates consumers community will be changed too, which in turn has significant changes on the riparian vegetation (France 1995).

All these patterns of variations in land usage would obviously have straight and unintended effects on biodiversity in freshwater habitats, since variations in the land expenditure at the bank level are shown in the physical and chemical features of these habitats (Allan 2004). For example, if the agricultural land practices have increased the consequence it is likely to increased loads of nutrients to freshwater ecosystems, which lead to a change in water quality, habitat characteristics, and biodiversity (Sponseller et al. 2001). In general, the more agricultural land in the freshwater catchments, the less the freshwater biodiversity due to the increased amounts of nutrient run-off (Allan 2004).

The other effect of removing plant cover and change land for agriculture is the increase in the climate warming on the upper part of the stream environment. Therefore, algae might be the main organism group to be influenced by intensified light and temperature after the elimination of riparian vegetation, which cause an increase algal biomass that can alter the society configuration (Allan 2004). Accordingly, the community of the macroinvertebrates that depend on their food on the other invertebrate animals will change as their main food items have been replaced by grazers and decreased in the abundance of shredders (Delong and Brusven 1998).

20.6.4 Alien Species

There are a three ways that climate changes can be associated to the possible invasion of an exotic species, these are (1) growing invisibility of ecosystems, (2) the impacts of climatic settings on resident species, and (3) growing the intrusive possible of alien species (Thuiller et al. 2007). For the cold freshwater system, the species present are determined by the minimum water temperature during winter. Any increase in water temperature due to any factors will give a way to the exotic fish species to invade northern areas, where they find suitable habitats to breed in, with remarkable impacts on natural organisms, biotic societies, and ecosystem procedures (Rahel and Olden 2008). Such stressors could be in the form of predation, struggle, and disperse of parasites and illnesses to which species native to cold freshwater habitats are not altered (Wrona et al. 2006).

20.7 Species Adaptation to Climate Changes in Freshwater Biodiversity and Conservation

The adaptation and the distribution of the freshwater organisms can be significantly affected by the climatic changes. Therefore, organisms need to take new steps in their lives to adapt to such changes as much as possible so they can survive.

In the case of the freshwater inhabitants, the utmost encounters of climate alteration acclimatisation will be in association with the level to which the aquatic ecosystems are diminished, the capabilities of species to transfer to higher latitudes or altitudes, and potentials to overcome the impacts of augmented separation of aquatic ecosystems owing to human actions (Poff et al. 2002). Therefore, in order to save freshwater biodiversity, man should assist nature to facilitate the adaptation of the species in its environment with the changes in the climate. Such anthropogenic assistant is known as "The planned adaptation."

The planned adaptation should be in association with the upkeep of biodiversity in the versus the climate alteration. This includes formation of sheltered areas, habitat organisation, scheduling of dispersion hallways for various types of species (Hannah et al. 2002), and dropping influences of human activities like pollution, habitat devastation, and alien species intrusion on aquatic habitats (Poff et al. 2002).

The issue of the planned climate alteration acclimatisation composed of a varied collection of themes, but the subsequent discussion will focus on some acclimatisation events and study these in the contextual of the preservation of freshwater biodiversity.

20.7.1 Protected Areas

Usually, systems of sheltered regions offer the maximum effective method of keeping biodiversity against the climate modification. Several findings have gone through checking how plausible the preserved regions are to uphold biomes and species (Scott et al. 2002), in addition to other researches that targeted the development of approaches that might support in picking preserved regions regarding the current and future deviations of species' distribution (Pyke and Fischer 2005). To avoid species pooling in the protected areas, these spaces should be large enough to have different habitats to accommodate wide variety of regional climatic conditions (Pöyry and Toivonen 2005).

Problems may arise in the establishment of the protected areas and even these places are designed perfectly and according to the ecological needs of the species. It may be the freshwater biodiversity is still not fully sheltered protected in the new protected areas network. The new protected areas need to be available to accommodate any shifts in freshwater species' ranges in the future. To be such efficient, these places must obviously be grounded on the features of freshwater habitats and the necessities of freshwater inhabitants (Toivonen et al. 2004; Abell et al. 2007). In addition, the design of the protected areas should have a combination of information originated from the freshwater and terrestrial biodiversity in order to achieve the maximum productive possibility for saving comprehensive biodiversity (Abell 2002).

20.7.2 Large and Environmentally Heterogeneous Areas

In choosing large protecting area, this means that such an area will be characterised in having wide range of habitats that are suitable for different species of organisms and imperative significance for the working of freshwater habitats (Dudgeon et al. 2006). More variable protected areas such as containing mountainous and lowland areas will assist freshwater inhabitants to discover appropriate temperature settings subsequent climate alteration.

20.7.3 Species Dispersal Paths

When climatic changes set in area, species need to find way out of the area to survive. These ways or what are known as "dispersal paths" or "dispersal corridors" are very important for freshwater inhabitants that depend on the freshwater habitats for fruitful spreading among aquatic areas. In some cases, these corridors and paths are not present in the area where climatic changes happened or they have been obstructed by anthropogenic activities such as dams, the species then will have no choice but to not respond to the climatic changes and this means an inevitable extension (Malmqvist and Rundle 2002). With organisms other than fish such as insects and terrestrial species, the problem might be simpler as these organisms can fly or move overland and find another suitable habitats.

20.7.4 Environment Restoration and Controlling

It is important to keep restoring and managing the protected regions in order to continue being naturally comparable ecosystems so the populations of species will be viable (Suffling and Scott 2002). These efforts can be listed as follows: (1) static such in case of sustained up keeping of existing biodiversity using a present sheltered regions; (2) passive as in permitting ecological reactions to climate modification to mother nature deprived of human interference); and (3) flexible activities directing at the enlargement of aptitudes of existing societies and species to acclimatise to climate alteration. As to the complexity at the freshwater biodiversity, it might as well to combine two or more efforts and apply them at the same time. The other problem is the financial support as most of the conservation projects aim to protect a restricted number of organisms and it will be difficult to choose the species need to be protected (Suffling and Scott 2002).

20.8 Suggested Recommendations for Management of the Climatic Changes Effects on the Freshwater Ecosystem in the Mesopotamian Plain

Freshwater ecosystems and in particular rivers are progressively at danger owing to land-use variations, population development, pollution releases, flow-system changes due to building dams and extreme groundwater siphoning. Predictable climate modification might augment to and upsurge these hazards throughout its option to alter rainfall quantity, temperature, and overflow designs, besides to disturb of the biological groups. Consequently, an exceptional resources requisite to be generated for such a venture to make accessible the tools, training, and financial assets to backing energetic administration steps.

Restoring of a freshwater ecosystem offers shelter for river habitat in addition to decrease the influences of both floods and droughts on people that be influenced by on rivers for their water. Also, providing watersheds for the protected areas will provide suitable habitats for the threatened species. For species of special attention, cautious observing of group sizes and their health can be attentive directors to take movements afore species are at danger.

The following recommendations will ensure a successful development of plans for the freshwater ecosystem. These recommendations are generic in nature and can fit most countries in the Mesopotamian plane.

1. Improving water control abilities.
2. Creating manpower to help in the technical issues through the education process of the local managers.
3. Assigning more river paths as protected measures to restore lands nearby to rivers.
4. Implementing plans to combine use of both groundwater and surface water.
5. Creating plans for the restoration and renovation of the maximum susceptible river parts.
6. Distinct consideration should be given to keeping, repositioning, or generating habitats of different ecological nature that provides different needs for the species.

References

Abell RA (2002) Conservation biology for the biodiversity crisis: a freshwater follow-up. Conserv Biol 16:1435–1437

Abell RA, Allan JD, Lehner B (2007) Unlocking the potential of protected areas for freshwaters. Biol Conserv 134:48–63

ACIA (2005) Arctic climate impact assessment—scientific report. Cambridge University Press, Cambridge. http://www.acia.uaf.edu/pages/scientific.htm

Al-Bakri D, Khalaf F, Al-Ghadban A (1984) Mineralogy, genesis, and sources of surficial sediments in the Kuwait marine environment, northern Arabian Gulf. J Sediment Petrol 54:1266–1279

Al-Ghadban AN (1990) Holocene sediments in a shallow bay, southern coast of Kuwait, Arabian Gulf. Mar Geol 92:237–254

Allan JD (2004) Landscapes and riverscapes: the influence of land use on stream ecosystems. Annu Rev Ecol Evol Syst 35:257–284

Al-Timimi YK, Al-Jiboori MH (2013) Assessment of spatial and temporal drought in Iraq during the period 1980–2010. Int J Energy Environ 4(2):291–302

Bar-Matthews M, Ayalon A, Kaufman A (1997) Late quaternary paleoclimate in the eastern Mediterranean region from stable isotope analysis of speleothems at Soreq Cave, Israel. Quat Res 47:155–168

Beniston M (1997) Variations of snow depth and duration in the Swiss Alps over the last 50 years: links to changes in large-scale climatic forcings. Clim Change 36:281–300

Bou-Zeid E, El-Fadel M (2002) Climate change and water resources in Lebanon and the Middle East. J Water Resour Plann Manag 128(5):343–355

Bozkurt D, Sen OL (2013) Climate change impacts in the Euphrates–Tigris Basin based on different model and scenario simulations. J Hydrol 480:149–161

Bozkurt D, Turuncoglu U, Sen OL, Onol B, Dalfes HN (2012) Downscaled simulations of the ECHAM5, CCSM3 and HadCM3 global models for the eastern Mediterranean-Black Sea region: evaluation of the reference period. Clim Dyn 39:207–225

Cane M, Eshel G, Buckland R (1994) Forecasting Zimbabwean maize yield using Eastern Equatorial Pacific Sea surface temperature. Nature 370:204–207

Chenoweth J, Hadjinicolaou P, Bruggeman A, Lelieveld J, Levin Z, Lange MA, Xoplaki E, Hadjikakou M (2011) Impact of climate change on the water resources of the eastern

Mediterranean and Middle East region: modeled 21st century changes and implications. Water Resour Res 47(6)
Claussen M, Kubatzki C, Brovkin V, Ganopolski A, Hoelzmann P, Pachur HJ (1999) Simulation of an abrupt change in Saharan vegetation in the mid-Holocene. Geophys Res Lett 26:2037–2030
Cullen HM, deMenocal P (2000) North Atlantic influence on Tigris-euphrates streamflow. Int J Clim 20:853–863
Cullen HM, deMenocal PB, Hemming S, Hemming G, Brown FH, Guilderson T, Sirocko F (2000) Climate change and the collapse of the Akkadian empire: evidence from the deep sea. Geology 28(4):379–382
Cullen HM, Kaplan A, Arkin PA (2002) Impact of the North Atlantic oscillation on middle Eastern climate and streamflow. Clim Change 55(3):315–338
Delong MD, Brusven MA (1998) Macroinvertebrate community structure along the longitudinal gradient of an agriculturally impacted stream. Environ Manag 22:445–457
Dudgeon D, Arthington AH, Gessner MO, Kawabata Z-I, Knowler DJ, Leveque C, Naiman RJ, Prieurrichard A-H, Soto D, Stiassny LJ, Sullivan CA (2006) Freshwater biodiversity: importance, threats, status and conservation. Biol Rev 81:163–182
Durance I, Ormerod SJ (2007) Climate change effects on upland stream macroinvertebrates over a 25-year period. Glob Change Biol 13:942–957
Eshel G, Farrell CM, Farrell B (2000) Forecasting Eastern Mediterranean Droughts. Mon Weather Rev 128:3618–3630
Food and Agriculture Organization (FAO) (2009) Irrigation in the Middle East region in figures: AQUASTAT survey—2008. FAO Water Rep. 34, Frenken K (ed), Rome
France RL (1995) Macroinvertebrate standing crop in littoral regions of allochthonous detritus accumulation: implications for forest management. Biol Conserv 71:35–39
Frumkin A (1991) The Holocene climatic record of the salt caves of Mount Sedom, Israel. The Holocene 1:191–200
Giller PS, Malmqvist B (1998) The biology of streams and rivers. Oxford University Press, Oxford
Hannah L, Midgley GF, Lovejoy T, Bond WJ, Bush M, Lovett JC, Scott D, Woodward FI (2002) Conservation of biodiversity in a changing climate. Conserv Biol 16:264–268
Hassan AH, Mashkour M (1972) Mean annual precipitation in Iraq. J Geol Soc Iraq 5:119–127
Heino J, Virkkala R, Toivonen H (2009) Climate change and freshwater biodiversity: detected patterns, future trends and adaptations in northern regions. Biol Rev 84(1):39–54
IPCC (2001) Climate change 2001: the scientific basis. Intergovernmental Panel on Climate Change: Working Group I. http://www.grida.no/climate/ipcc_tar/wg1/index.htm
IPCC (2002) Climate change and biodiversity. IPCC technical Paper V. Intergovernmental Panel on Climate Change. http://www.ipcc.ch/pub/tpbiodiv.pdf
Jeppesen E, Jensen JP, Søndergaard M, Lauridsen T, Landkildehus F (2000) Trophic structure, species richness and biodiversity in Danish lakes: changes along a phosphorus gradient. Freshw Biol 45:201–218
Kellomäki S, Strandman H, Nuutinen T, Peltola H, Korhonen KT, VA ¨ Isänen, H. (2005) Adaptation of forest ecosystems, forests and forestry to climate change. Finn Environ Inst Mimeogr 334:1–44
Lemcke G, Sturm M (1997) d18O and trace element measurements as proxy for the reconstruction of climate changes at Lake Van (Turkey): preliminary results. In: Dalfes HN et al (eds) Third millennium B.C. climate change and old world collapse (vol 49). Springer, Berlin, pp 178–196
Magnuson JJ, Robertson DM, Benson BJ, Wynne RH, Livingstone DM, Arai RA, Barry RG, Card V, Kuusisto E, Granin NG, Prowse TD, Stewart KM, Vuglinski VS (2000) Historical trends in lake and river ice cover in the Northern Hemisphere. Science 289:1743–1746
Malmqvist B, Rundle S (2002) Threats to the running water ecosystems of the world. Environ Conserv 29:134–153
Mason S, Goddard L, Graham N, Yulaeva E, Sun L, Arkin P (1999) The IRI seasonal climate prediction system and the 1997/1998 El Niño Event. Bull Am Meteorol Soc 80:1853–1873

Parmesan C (2006) Ecological and evolutionary responses to recent climate change. Annu Rev Ecol Evol Syst 37:637–669
Poff NL (1992) Regional hydrologic response to climate change: an ecological perspective. In: Firth P, Fisher SG (eds) Global climate change and freshwater ecosystems. Springer, New York, pp 88–115
Poff NL, Brinson MM, Day JW Jr (2002) Aquatic ecosystems and global climate change: potential impacts on inland freshwater and coastal wetland ecosystems in the United States. Pew Center on Global Climate Change, Arlington, VA, p 56
Pöyry J, Toivonen H (2005) Climate change adaptation and biological diversity. Finn Environ Inst Mimeogr 333:1–46
Pye K (1987) Aeolian dust and dust deposits. Academic, London. 334 p
Pyke CR, Fischer DT (2005) Selection of bioclimatically representative biological reserve systems under climate change. Biol Conserv 121:429–441
Rahel FJ, Olden JD (2008) Assessing the effects of climate change on aquatic invasive species. Conserv Biol 22:521–533
Saetersdal M, Birks HJB, Peglar SM (1998) Predicting changes in Fennoscandia vascular-plant species richness as a result of future climatic change. J Biogeogr 25:111–122
Sala OE, Chapin FS III, Armesto JJ, Berlow E, Bloomfield J, Dirzo R, Huber-Sanwald E, Huenneke LF, Jackson RB, Kinzig A, Leemans R, Lodge DM, Mooney HA, Oesterheld M, Poff NL, Sykes MT, Walker BH, Walker M, Wall DH (2000) Global biodiversity scenarios for the year 2100. Science 287:1770–1774
Sandweiss DH, Maasch KA, Anderson DG (1999) Transitions in the mid-Holocene. Science 283:499–500
Schindler DW (1997) Widespread effects of climate warming on freshwater ecosystems in North America. Hydrol Process 11:1043–1067
Schindler DW, Bayley SE, Parker BR, Beaty KG, Cruikshank DA, FEE EJ, Schindler EU, Stainton MP (1996) The effects of climate warming on the properties of boreal lakes and streams at the Experimental Lakes Area, Northwestern Ontario. Limnol Oceanogr 41:1004–1017
Scott D, Malcolm JR, Lemieux C (2002) Climate change and modelled biome representation in Canada's national parks: implications for system planning and park mandates. Glob Ecol Biogeogr 11:475–484
Shubbar RM (1999) Climate variation indices of Iraq. M.S.C. Thesis, Collage of Science, AL-Mustansiriyah University, Baghdad, Iraq
Sirocko F, Sarnthein M (1989) Wind-borne deposits in the northwestern Indian Ocean: record of Holocene sediments versus modern satellite data. In: Leinen M, Sarnthein M (eds) Paleoclimatology and paleometeorology: modern and past patterns of global atmospheric transport. Kluwer, Berlin, pp 401–433
Sponseller RA, Benfield EF, Valett HM (2001) Relationships between land use, spatial scale and stream macroinvertebrate communities. Freshw Biol 46:1409–14024
Suffling R, Scott D (2002) Assessment of climate change effects on Canada's national park system. Environ Monit Assess 74:117–139
Sykes MT, Prentice IC (1995) Boreal forest futures: modelling the controls on tree species range limits and transient responses to climate change. Water Air Soil Pollut 82:415–428
Tammi J, Lappalainen A, Mannio J, Rask M, Vuorenmaa J (1999) Effects of eutrophication on fish and fisheries in Finnish lakes: a survey based on random sampling. Fish Manag Ecol 6:173–186
Thuiller W, Richardson DM, Midgley GF (2007) Will climate change promote alien plant invasions? In: Nentwig W (ed) Biological invasions, ecological studies., Vol, vol 193. Springer, Berlin, pp 197–211
Toivonen H, Leikola N, Kallio M (2004) Sisävesien suojelualueverkon edustavuuden arviointia. Suomen Ympäristö 713:1–113
Turkes M (1996) Meteorological drought in Turkey: a historical perspective 1930–1993. Drought Netw News 8:17–21

Weiss H, Courty M, Wetterstrom W, Guichard F, Senior L, Meadow R, Curnow A (1993) The genesis and collapse of 3rd millennium North Mesopotamian civilization. Science 261:994–1004

Wrona FJ, Prowse TD, Reist JD, Hobbie JE, Levesque LMJ, Vincent WF (2006) Climate change effects on aquatic biota, ecosystem structure and function. Ambio 35:359–369

Yoffee N, Cowgill W (1988) The collapse of ancient Mesopotamian states and civilizations. University of Arizona Press, Tucson, AZ. 241 p

Chapter 21
River Corridors as a Refuge for Freshwater Biodiversity: Basic Information and Recommendations to the Policymakers for Possible Implications in Iraq

Laith A. Jawad and Baradi Waryani

Abstract Rivers provide a special group of benefits and services highly appreciated by the public that are intimately linked to their flow and its interaction with the landscape. Protection of a greater number of rivers and corridors is vital. This will need partnerships among multiple cohorts in the respective river basins and wise land-use planning to minimize additional development in watersheds with valued rivers. Ensuring environmental flows by purchasing or leasing water rights and/or altering reservoir release patterns will be needed for many rivers. Implementing restoration projects can be used to protect existing resources so that expensive reactive restoration to repair damage associated with any changing is minimized. Special attention should be given to diversifying and replicating habitats of special importance and to monitoring populations at high risk or of special value so that management interventions can occur if the risks to habitats or species increase significantly over time. The set of recommendations given will be considered the first step into the river management in Iraq, and policymakers should follow in order to conserve the freshwater biodiversity in Iraq.

21.1 Introduction

It has been known for a long time that rivers are characterized with high ecological value, but in contrary, they are highly endangered ecosystems (Palmer et al. 2010). Waterways that created for the sake of human transportation have the same characteristics (Wolter and Vilcinskas 1997).

L. A. Jawad (✉)
School of Environmental and Animal Sciences, Unitec Institute of Technology, Auckland, New Zealand

B. Waryani
Department of Fresh Water Biology and Fisheries, University of Sindh, Jamshoro, Sindh, Pakistan

L. A. Jawad (ed.), *Tigris and Euphrates Rivers: Their Environment from Headwaters to Mouth*, Aquatic Ecology Series 11,
https://doi.org/10.1007/978-3-030-57570-0_21

Without any modifications to the rivers, flood control will become a serious matter as the case at the end of the last century, where 77% of the rivers in Europe, the Commonwealth of Independent States (CIS) and North America were seriously modified (Cowx and Welcomme 1998). The floodplains are considered as ecological corridors for species spreading and they provide home space, which results in remarkably high levels of biodiversity (Naiman and Decamps 1997). Vice versa, biodiversity is an important teamster of ecosystem functionality (MEA 2005; Naeem et al. 1994), and therefore, the functional biodiversity method has received increasing thoughtfulness during the last few years (Petchey and Gaston 2006). Ecosystem functioning is rather determined by the traits and characteristics of species than by mere species numbers (Díaz and Cabido 2001). In functional diversity research, the range and value of those traits (e.g. ability to fix nitrogen, growth form, and dispersal mode) are studied and used as a measure of biodiversity.

Biodiversity and functionality of riparian systems are intensely affected by human activities, which change the river body itself, such as hydromorphological changes, river impoundment, and water management (Naiman and Decamps 1997). In addition, change and intensification of human land use influence riparian systems (Méndez-Toribio et al. 2014). Together, these events caused the above-mentioned loss in floodplain space and functionality. Such changes have drawn the attention of the policymakers, starting with the Ramsar Convention on Wetlands (United Nations 1971), the European Habitats Directive (92/43/EEC, The Council of the European Communities 1992), the Convention on Biodiversity (CBD), and the Millennium Ecosystem Assessment (MEA 2005).

The similarity between artificial water bodies and rivers is simply trivial (Hatcher et al. 1999). Canals have a regular structure, a negligible flow velocity, and regulated water levels, with lack of the dynamics of the rivers (Willby et al. 2001). With such differences, they cannot deliver the same functions like riparian systems with respect to nutrient cycling or flood retention. Still they provide habitat space, increase the connectivity within a landscape, and thus might serve as migration corridors (Jesus Casas et al. 2011). Fish species might use canals as secondary habitats (Waltham and Connolly 2007; Wolter 2001) also invertebrate (Grumiaux and Dhainaut-Courtois 1996) for survival when their natural habitat is spoilt or vanished, as described (Gómez and Araujo 2008).

The aim of this chapter is to give basic information about the river corridors, their benefit, and their protection. At the end of the chapter, a set of recommendations have been given in order to be taken by the policymakers in Iraq to create river corridors and initiate river management for the conservation of the freshwater biodiversity.

21.2 The Contribution of Streams and Springs to Biodiversity in River Networks

The diversity of life in the streams and springs subsidizes to the biodiversity of a river system and its riparian network. Small streams differ widely in physical, chemical, and biotic attributes, thus providing habitats for a range of unique species. Springs species include permanent residents as well as migrants that travel to headwaters at particular seasons or life stages. Movement by migrant's links springs with downstream and terrestrial ecosystems, as do spreads such as emerging and drifting insects.

Even intermittent streams may support rich and distinctive biological communities, in part because of the expectedness of dry periods. The effect of springs on downstream systems emerges from their subsidizes that meet unique habitat requirements of residents and migrants by offering a refuge from temperature and flow extremes, competitors, predators, and introduced species; serving as a source of colonists; providing spawning sites and rearing areas; being a rich source of food; and creating migration corridors throughout the landscape. Degradation and loss of headwaters and their connectivity to ecosystems downstream endanger the biological integrity of entire river networks.

The biota living in streams and springs can be divided into five main groups: (1) species that are exclusive to these small ecosystems; (2) species that are found in these and larger streams, although their abundance may vary with stream size; (3) species that move into streams seasonally as the spring network expands and contracts or as downstream conditions grow less favourable; (4) species that spend most of their lives in downstream ecosystems, but require headwaters at particular life-history stages (e.g. for spawning or nursery areas); and (5) species that live around but not in headwater streams, requiring the moist habitat they provide or feeding on the products of headwaters (e.g. benthic, emerging, or drifting insects) (Meyer et al. 2007).

Streams and springs should always be connected to water bodies downstream. Such connectivity is important for the maintenance of species diversity in downstream ecosystems (Labbe and Fausch 2000). Among the ways that small streams maintain diversity in the river network is by providing a source of colonists for recovery of downstream systems following disturbance (Progar and Moldenke 2002). Small streams also provide movement corridors for plants and animals across the landscape. Their riparian zones provide cooler conditions than those found in the uplands (e.g. Richardson et al. 2005). The flight paths of adult aquatic insects are intense along streams and riparian zones, which serve as dispersal corridors (Petersen et al. 2004).

In spite of their sole assistances to and significance in preserving the diversity and functional integrity of entire river systems, small streams are continually under danger by human activity (Meyer and Wallace 2001). Threats may originate from groundwater extraction, which threatening species associated with small springs (Hubbs 1995), has caused tributaries of Kansas streams for example to go dry, resulting in the extirpation of 16 species from the river system (Cross and Moss 1987). Other land activities such as agriculture, urbanization degrade, and eliminate headwater habitats can also have an impact on small streams (Meyer and Wallace 2001). The increasing impact of damaged headwaters subsidizes to the loss of ecological integrity in ecosystems downstream. Small streams are thus a vital part of the biological integrity of our nation's waterways. Deprivation of headwater habitats and loss of their connections to larger streams have negative consequences not only for inhabitants of small streams but also for the diversity of downstream and riparian ecosystems.

21.3 River Corridors

A river corridor can be generally defined as the river and river channel together with their associated wildlife and the adjacent riparian ecosystem. The linear character of a river corridor is a vital aspect of its value for conservation, since river corridors link the uplands with the lowlands and may also link otherwise isolated habitats and populations (Gardiner 1991). The corridor, or linear subdivisions of the corridor (known as stretches), may thus be regarded as management parts. Since riparian vegetation frequently grades into adjacent habitats, the river corridor is a theoretical rather than a physical feature. The apparent width of the corridor is generally random because of this gradation and also varies along its length according to the form of the landscape and the nature of the adjacent vegetation (Budd et al. 1987).

Due to the complexity of biophysical interactions within river corridors, suitable classification systems that identify corridor units according to their character and sensitivity to management should be created. However, any classification system should take into consideration that the river corridors can be subdivided into units by clearly defined boundaries. The gradually varying nature of river environments both along and across their corridors tends to prevent the identification of clear boundaries, so a clash exists between describing the river corridor environment in a clear and actual way and introducing boundaries which are to some extent arbitrary.

The classification of river corridors may simply aim to describe what is there in a brief and actual form. Here, classification suggests the allocation of river units to a series of classes, whereas indexation can infer the estimation of the value of either a continuous or discrete index for each unit. Such classifications depend the collection of data on some aspect or aspects of the river corridor environment at an appropriate spatial scale or scales. Among the classifications, there are criteria about judgements of the value of particular river corridor units or in assessing their sensitivity to management.

The following are characteristics of any classification of the river's corridors need to be in order to support conservation and management decision-making: (1) include as many features of river channel and corridor form, process, and ecology as possible, so linking to the range of disciplines concerned with rivers; (2) include information on the dynamics of the fluvial system, so integrating a temporal viewpoint; (3) be capable of application at a range of spatial scales; (4) give reliable and reproducible results.

21.3.1 Planning to Build River's Corridor

The main aim of building river's corridors is for the persistence of biodiversity (Rouget et al. 2003). Also within the aim in designing corridors is to represent the biological gradients within each biogeographically distinct water catchment. Corridor design, therefore, focused primarily on ensuring biodiversity persistence (i.e. the long-term maintenance of ecological and evolutionary processes), on which the conservation assessment is founded.

The design of the corridor must contain the following four key functions in order to (1) maintain ecological processes (gradients) to enable movement of biota over ecological and evolutionary time scales; (2) ensure habitat retention and connectivity; (3) maximize wildlife habitat suitability; and (4) represent biodiversity pattern (to integrate biodiversity persistence and representation).

Corridors were developed in three phases: phase 1, primarily forced by biological process considerations, identified the core area of the corridor (referred to as "conservation paths"); phase 2 expanded the core area to develop representation of habitats and the persistence of processes; and phase 3 further expanded corridors into areas of high irreplaceability value for biodiversity pattern.

Phase 1: Recognition of the Area to Be Ecologically Assessed
The area that needs to be conserved should include upland-lowland and climatic gradients operating at a macroscale, which should show the following importance: (1) run through thicket vegetation types; (2) are not in transformed habitats (urban areas excluded from the analysis); (3) run through habitats highly suitable for wildlife; (4) include other process components (i.e. riverine corridors, biome interfaces, sand movement corridors); (5) tie protected areas; and (6) are not in areas likely to be transformed in future.

On the other hand, some criteria need to set in the process of designing the corridors that ensure the functionality of the selected areas. These relate to (1) the presence of thicket vegetation and its condition; (2) the occurrence of process components; (3) the degree of suitability of wildlife habitat; (4) the location of protected areas; and (5) future land-use pressures.

The cost of each criterion to develop is depended on the nature of the criterion itself, where criteria of higher rank override lower rank criteria (i.e. intact habitat was always more suitable than transformed or degraded habitat irrespective of wildlife habitat suitability).

An area of about 1 km wide needs to be reserved within single primary water catchments by attaching them to major river mouths and ending them at the northern margin. River mouths were selected because of key ecological processes associated with their estuaries and wetlands (Heydorn and Tinley 1980). Based on the landscape suitability surface, least-cost surface analysis identified the best option to link start and end points. Urban areas, including rural settlements, were excluded (i.e. the paths could not traverse urban areas).

Phase 2: Enlarging Conservation Path Along the Corridors
The area besides the corridors is known as "conservation paths." This area is recommended to be increased toward the corridor 1 km wide. It denotes a nearly optimal location and the bare minimum extent for conserving processes along upland-lowland and macroclimatic gradients. The reason of expanding such an area is to (1) buffer the conservation path; (2) include fixed process substitutes; (3) achieve targets for vegetation types; (4) select areas highly suitable for wildlife; and (5) include existing protected areas. The expansion was attuned to evade areas threatened by future land-use pressures

Phase 3: Enlarging the Corridors
The final step in designing the river's corridor is to explore the extent to which corridors could be expanded to capture areas of high irreplaceability for biodiversity. Irreplaceability values need to be recalculated starting from the current configuration of corridors (from stage 2). Considered the contribution of corridors to targets for biodiversity features should be accounted for assuming that each of the corridors was afforded conservation management relatively consistent with that of protected areas. The identified planning units were important for achieving remaining biodiversity targets.

21.3.2 How to Evaluate the Effectiveness of Corridor

In order to check the competence of the design of the corridor, it is always a good idea to compare the design with those of established and well-known corridors worldwide and in environment similar to that the corridor needs to be built in and compare the criteria: extent of natural area, thicket representation, achievement of pattern targets (vegetation types), achievement of process targets, and avoidance of land-use pressures (i.e. implementation constraints), and linkages to protected areas (implementation opportunities) (Rouget et al. 2006).

21.4 Public Awareness of River Corridors

It is an essential practice to make the people who live or own in the area where the river corridors will be built. Such move is different from the previous steps that used to be. The public input has regularly depended upon discussions involving formal meetings with interest groups and local politicians, but the means of public consultation are changing. With direct consultation with the public who are concerned with river corridors, the organization building these corridors now relies less on public meetings and increasingly more on a direct approach and involvement of the public.

The project of building river corridors will face huge problems if the owners of the areas or living in the areas where the corridors designed to be built in not have been consulted before starting the project. In Iraq, most of the areas where corridors need to be built in are either owned by tribes of people belonging to certain tribes living in the area. Dealing with tribe is major dilemma and settling such problem will need time and money. In the urban areas of Iraq, specialized research companies can do such survey and obtain the response of the public about the river corridor project.

House and Fordham (1997) have shown that there are an enlarged amount of research programmes in the area of public approaches to rivers and to river management. In order to get a direct response of the public, House and Fordham (1997) have suggested to use questionnaire surveys and semi-structured interviews with local residents living within a proposed project area or river catchment, which have also been mentioned by others (Tapsell and Tunstall 1994; Sawyer and Fordham 1994).

Among the issues that the public usually raise are the environmental factors in the management of rivers in general (Witherspoon 1994) and flood hazard management in particular. The public may not support the project of the river management if no sign of environmental sensitivity has been included in the project (Tunstall et al. 1994).

In order to reach a compromise in the public awareness about the river corridor, social surveys can form part of a public education programme, as participation in such interviews invariably leaves the participants both better informed, as information is often provided during the questionnaire survey, and more reflective than before, as few people will previously have spent an hour or more deliberating on the issues raised by the survey.

21.5 Protection of River Corridors

The lands contiguous to river channels are important to aquatic and terrestrial ecosystems (Smith et al. 2008), offer vital ecosystem amenities (Postel and Carpenter 1997), and are socially important (Millennium Ecosystem Assessment 2005). The worth of river corridors to both aquatic ecosystems and public safety has led the governments and organizations looking after the river corridors in the world to invest substantial time and capitals to establish a river corridor protection programme over the past decade. This approach goes beyond the customary view of buffers as land use setbacks to maintain water quality. For such a protection programme, geomorphic-based river corridors are being established to maintain natural channel form and functions, as well as serious ecosystem services such as flood and erosion hazard mitigation.

According to several research performed in the field of river management and in particular in protecting river corridors (Kline and Cahoon 2010), there are two chief mechanisms for such protection process: (1) state and municipal land-use restrictions on development within defined fluvial erosion hazard areas; and (2) a programme to support the purchase of development and channel management rights in river corridor conservation easements. The purpose of establishing this easement is to give the river the space to re-establish a natural slope, meander pattern, and flood-plain connection.

21.6 Recommendations to the Policymakers in Iraq About Creating River Corridors in Their Country

In this chapter, the author has aimed to deliver to the reader a programme for protecting river corridors. The following recommendations were composed from researches such as Fischer and Fischenich (2000) and Wenger and Fowler (2000) on the issue of river management and modified to fit the environment of Iraq.

1. Initiate a public information campaign to explain the benefits of river corridors.
2. Identify critical river corridor areas in which existing land uses may pose a threat to water quality. Such areas include cattle watering spots, areas where chicken waste is applied to fields, older homes with septic drain fields, etc.
3. Identify high-priority wildlife habitat areas, historic or prehistoric sites, and other.
4. Create limits on waterproof surfaces to control run-off.
5. Properly impose erosion and sedimentation control statutes.
6. Adjust the jurisdiction's existing flood damage prevention ordinance to include importance of the importance of limiting floodplain development for purposes of flood storage, water quality protection, and wildlife habitat preservation. Prohibit activities in the floodplain that could directly threaten water quality, including application of fertilizers and pesticides, siting of animal waste lagoons, and disposal of hazardous materials, including motor oil.
7. Establish a turbidity standard to monitor erosion and sedimentation control and river corridor effectiveness in the area selected.
8. Think at a watershed scale when planning for or managing corridors. Many species that primarily use upland habitats may, at some stage of their life cycle, need to use corridors for habitat, movements, or dispersal.
9. Continuous corridors are better than fragmented corridors.
10. Wider corridors are better than narrow corridors.
11. Riparian corridors are more valuable than other types of corridors because of habitat heterogeneity, and availability of food and water.
12. Several corridor connections are better than a single connection.
13. Structurally diverse corridors are better than structurally simple corridors.

14. Native vegetation in corridors is better than non-native vegetation.
15. Practice ecological management of corridors; burn, flood, open canopy, etc., if it mimics naturally occurring historical disturbance processes.

References

Budd WW, Cohen PL, Saunders PR, Steiner FR (1987) Stream corridor management in the Pacific Northwest: I. Determination of stream corridor widths. Environ Manag 11:587–597

Cowx IG, Welcomme RL (1998) Rehabilitation of rivers for fish. Fishing News Books, Oxford, UK, pp 260

Cross FB, Moss RE (1987) Historic changes in fish communities and aquatic habitats in plains streams of Kansas. In: Matthews WJ, Heins DC (eds) Community and evolutionary ecology of North American stream fishes. University of Oklahoma Press, Norman, OK, pp 155–165

Díaz S, Cabido M (2001) Vive la difference: plant functional diversity matters to ecosystem processes. Trends Ecol Evol 16:646–655

Fischer RA, Fischenich JC (2000) Design recommendations for riparian corridors and vegetated buffer strips (No. ERDC-TN-EMRRP-SR-24). Army engineer waterways experiment station Vicksburg MS engineer research and development center.

Gardiner JL (1991) Influences on the development of river catchment planning in the Thames Basin. PhD thesis, University of Southampton

Gómez I, Araujo R (2008) Channels and ditches as the last shelter for freshwater mussels: the case of Margaritifera auricularia and other naiads inhabiting the mid Ebro River Basin, Spain. Aquat Conserv Mar Freshw Ecosyst 18:658–670

Grumiaux F, Dhainaut-Courtois N (1996) Benthic macroinvertebrate communities of the canal a Grand Gabarit, Aa river, and the canalized Aa River in the north of France. J Freshw Ecol 11:131–138

Hatcher D, Eaton J, Gibson M, Leah R (1999) Methodologies for surveying plant communities in artificial channels. Hydrobiologia 415:87–91

Heydorn AEF, Tinley KL (1980) Estuaries of the Cape. Part 1. Synopsis of the Cape coast. Natural features, dynamics and utilization. Research report N-380. Council for Scientific and Industrial Research, Stellenbosch, South Africa

House M, Fordham M (1997) Public perceptions of river corridors and attitudes towards river works. Landsc Res 22(1):25–44

Hubbs C (1995) Springs and spring runs as unique aquatic systems. Copeia 1995(4):989–991

Jesus Casas J, Sanchez-Oliver JS, Sanz A, Furné M, Trenzado C, Melchor J et al (2011) The paradox of the conservation of an endangered fish species in a Mediterranean region under agricultural intensification. Biol Conserv 144:253–262

Kline M, Cahoon B (2010) Protecting river corridors in Vermont. J Am Water Resour Assoc (JAWRA) 1–10. https://doi.org/10.1111/j.1752-1688.2010.00417.x

Labbe TR, Fausch KD (2000) Dynamics of intermittent stream habitat regulate persistence of a threatened fish at multiple scales. Ecol Appl 10:1774–1791

MEA (Millennium Ecosystem Assessment) (2005) Ecosystems and human well-being: biodiversity synthesis. World Resources Institute, Washington, DC

Méndez-Toribio M, Zermeño-Hernández I, Ibarra-Manríquez G (2014) Effect of land use on the structure and diversity of riparian vegetation in the Duero river watershed in Michoacén, Mexico. Plant Ecol 215:285–296

Meyer JL, Wallace JB (2001) Lost linkages and lotic ecology rediscovering small streams. In: Huntly NJ, Levin S (eds) Ecology: achievement and challenge. M.C. Press, Blackwell, Malden, MA, pp 295–317

Meyer JL, Strayer DL, Wallace JB, Eggert SL, Helfman GS, Leonard NE (2007) The contribution of headwater streams to biodiversity in river networks. JAWRA J Am Water Resour Assoc 43 (1):86–103

Millennium Ecosystem Assessment (2005) Ecosystems and human well-being: wetlands and water synthesis. World Resource Institute, Washington, DC
Naeem S, Thompson J, Lawler SP, Lawton JH, Woodfin RM (1994) Declining biodiversity can alter the performance of ecosystems. Nature 368:734–737
Naiman RJ, Decamps H (1997) The ecology of interfaces: Riparian zones. Annu Rev Ecol Syst 28:621–658
Palmer MA, Menninger HL, Bernhardt E (2010) River restoration, habitat heterogeneity and biodiversity: a failure of theory or practice? Freshw Biol 55:205–222
Petchey O, Gaston KJ (2006) Functional diversity: back to basics and looking forward. Ecol Lett 9:741–758
Petersen I, Masters Z, Hildrew AG, Ormerod SJ (2004) Dispersal of adult aquatic insects in catchments of differing land use. J Appl Ecol 41:934–950
Postel S, Carpenter S (1997) Freshwater ecosystem services. In: Daily G (ed) Nature's services. Island Press, Washington, DC, pp 195–214
Progar RA, Moldenke AR (2002) Insect production from temporary and perennially flowing headwater streams in Western Oregon. J Freshw Ecol 17:391–407
Richardson JS, Naiman RJ, Swanson FJ, Hibbs DE (2005) Riparian communities associated with Pacific Northwest headwater streams: assemblages, processes and uniqueness. J Am Water Resour Assoc 41(4):935–948
Rouget M, Cowling RM, Pressey RL, Richardson DM (2003) Identifying spatial components of ecological and evolutionary processes for regional conservation planning in the Cape Floristic Region, South Africa. Divers Distrib 9:191–210
Rouget M, Cowling RM, Lombard AT, Knight AT, Kerley GI (2006) Designing large-scale conservation corridors for pattern and process. Conserv Biol 20(2):549–561
Sawyer JA, Fordham MH (1994) Upper Kennet Public Perception Study. Final report to the National Rivers Authority (Enfield, Flood Hazard Research Centre)
Smith MP, Schiff R, Opperman J (2008) Riparian zones: they aren't just for buffers any more. Water Resour IMPACT 10(3):6–8
Tapsell SM, Tunstall SM (1994) River mole flood alleviation scheme survey. Final report to the National Rivers Authority, Thames region (Enfield, Flood Hazard Research Centre)
The Council of the European Communities (1992) Council Directive 92/43/EEC of 21 May 1992 on the conservation of natural habitats and of wild fauna and flora, consolidated version 01.01.2007
Tunstall S, Fordham M, Glenn C (1994) Eton Wick: a survey of residents' perceptions of flood risk and flood alleviation schemes, final report. Final report to the National Rivers Authority Thames Region (London, UK, Middlesex University, Flood Hazard Research Centre)
United Nations (1971) Convention on Wetlands of International Importance especially as Water-fowl Habitat Ramsar (Iran). UN Treaty Series 14583. As amended by the Paris Protocol, 1982 and Regina Amendments, 1987
Waltham NJ, Connolly RM (2007) Artificial waterway design affects fish assemblages in urban estuaries. J Fish Biol 71:1613–1629
Wenger SJ, Fowler L (2000) Protecting stream and river corridors: creating effective local riparian buffer ordinances. Carl Vinson Institute of Government. The University of Georgia. 68pp
Willby NJ, Pygott JR, Eaton JW (2001) Inter-relationships between standing crop, biodiversity and trait attributes of hydrophytic vegetation in artificial waterways. Freshw Biol 46:883–902
Witherspoon S (1994) The greening of Britain: romance and rationality. In: Jowell C, Curtice J, Brook L, Ahrendt D (eds) British social attitudes survey: the 11th report. Aldershot, Dartmouth
Wolter C (2001) Conservation of fish species diversity in navigable waterways. Landsc Urban Plan 53:135–144
Wolter C, Vilcinskas A (1997) Characterization of the typical fish community of inland waterways of the north-eastern lowlands in Germany. Regul Rivers Res Manag 13:335–343

Chapter 22
Biodiversity of Fungi in Aquatic Environment of Iraq

Tawfik M. Muhsin

Abstract In this chapter, an estimation of fungal biodiversity in aquatic environment of Iraq was based on past and updated reports from the literature. The list of fungal species for commonly isolated species either directly from water samples or from submerged substrates in freshwater and brackish water habitats is provided. Most of the studies carried out on fungi in aquatic habitats of Iraq were focused on fungi which belong to the phyla Oomycota, Ascomycota and Deuteromycota. The greatest number of species comprised the Oomycota mainly belongs to the family Saprolegniaceae with most common occurred genera *Achlya* and *Saprolegnia*. However, majority of fungi isolated from submerged substrates belonged to Ascomycota and Deuteromycota. The fungal diversity in waters of this region is influenced mainly by the temperature and salinity factors. It is interesting to state from the data available that several species of marine fungi are becoming inhabitants of freshwater and brackish water of Iraq. Nevertheless, high species diversity of freshwater fungi was remarkable during winter season. It can be predicted that a low diversity of freshwater fungal species in summer is due to the elevation of salinity of these waters bodies when comparing the data of the past and recent dates of reports. Also, the relationship between the biodiversity of aquatic fungi in Iraq with some environmental factors over different periods of time is elucidated.

Keywords Biodiversity · Ecological factors · Freshwater fungi · Marine fungi

22.1 Introduction

Fungi are common in aquatic environments, and their major role along with bacteria is the decomposition process and recycling of nutrients in nature. Among the fungal groups, aquatic fungi (water moulds) are commonly occurred and adapted

T. M. Muhsin (✉)
Department of Biology, College of Education for Pure Sciences, University of Basrah, Basrah, Iraq

L. A. Jawad (ed.), *Tigris and Euphrates Rivers: Their Environment from Headwaters to Mouth*, Aquatic Ecology Series 11,
https://doi.org/10.1007/978-3-030-57570-0_22

to thrive well, grow and reproduce submerged in water of various aquatic ecosystems including rivers, lakes, ponds, marshes, estuaries, seas and oceans. They are mostly saprophytic; however, some species are parasitic on algae and fish causing fish diseases (Willoughby 1994). Among the aquatic fungi, the saprolegnoid members are the most common in freshwater habitats and belong to the family Saprolegniaceae (Oomycota) which includes about 150 well-known species mainly belong to the genera *Saprolegnia* and *Achlya* (Johnson et al. 2002), while other members of aquatic fungi are living in brackish and marine habitats and mostly belong to Ascomycota and Basidiomycota. Although noteworthy studies have been carried out on the taxonomy of aquatic fungi, nonetheless, the most extensive pioneer works have been conducted by Coker (1923), Johnson (1956), Seymour (1970), Kohlmeyer and Kohlmeyer (1979) and Johnson et al. (2002). Nonetheless, a controversial taxonomic disposition of freshwater fungi is in debate and traditionally is filed under the division Oomycota (Alexopoulos et al. 1996); however, recently based on the phylogenic approach using DNA sequencing, this group of organisms has been transferred into the kingdom Chromalveolata (Phylum Straminipila) (Keeling et al. 2005; Webster and Weber 2007; Weber 2009). From the ecological point of view, many studies have been conducted on fungi in freshwater ecosystems over the world (see Shaerer et al. 2007). It has been stated that the greatest biodiversity for fungal groups occurs in temperate and Asian tropical areas. However, the occurrence of some fungal species overlaps among terrestrial and freshwater habitats. The biodiversity pattern of fungi in aquatic systems has been correlated with some environmental factors (Thuiller 2007). This chapter deals exclusively with the biodiversity of fungi in aquatic environment of Iraq and to estimate the biodiversity in relation with some ecological factors over different periods of time and a list of fungal species recorded in waters of this region is provided.

22.2 Taxonomic Studies on Aquatic Fungi of Iraq

A number of taxonomic studies on aquatic fungi of freshwater in Iraq have been conducted by using water sampling techniques from different locations of Shatt Al-Arab river southern Iraq (Muhsin 1979). The most common technique used for isolation of aquatic fungi is the baiting method by using selected natural substrates (baits) such as sesame seeds placed into Petri dishes containing water samples. Water samples with baits were incubated at 18–20 °C; the fungal growth colonized baits can be observed in few days. The identification of aquatic fungi basically depends on the morphological characteristics of somatic and reproductive fungal structures including hypha, zoosporangia, oogonia and antheridia (Fig. 22.1) and according to Johnson (1956), Seymour (1970) and Johnson et al. (2002). Based upon the studies carried on aquatic fungi in Iraq, twenty species were recorded from freshwaters of Shatt Al-Arab river in Basrah, Southern Iraq (Muhsin 1979) (Table 22.1). Out of them, four species belong to the genus *Dictyuchus* and two species of the genus *Calyptralegnia* were reported and the description, illustrations

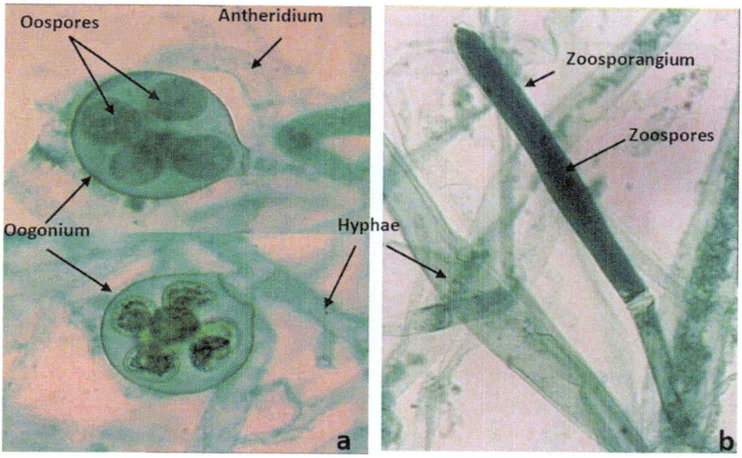

Fig. 22.1 Reproductive structures of Saprolegnoid freshwater fungi, (**a**) sexual structures (oogonia containing oospores and antheridia) and (**b**) asexual structures (zoosporangia containing zoospores) (×40 magnification)

and taxonomic key were provided (Rattan et al. 1978). This study revealed that *D. monosporus* was the most abundant in aquatic habitat of Iraq, while the members of *Calyptralegnia* occurred in spring and autumn only. Later on, a new species *C. basraensis* Muhsin was reported and described by Muhsin (1984). On the other hand, five species of the genus *Saprolegnia* were reported from Shatt Al-Arab waters (Ismail et al. 1979). *S. parasitica* and *S. ferax* are the most dominant species in these waters. Within the genus *Achlya*, 11 species were reported from the same studied locations and a detailed description, illustrations and a proposed key for these species were provided (Muhsin et al. 1984). However, *Achlya polyandra* was the most dominant species among this genus in waters. More aquatic fungal species including five species of *Saprolegnia* and two species of *Achlya* were newly recorded from water bodies of Iraq (Muhsin and El-Habeeb 1999). Only two species of *Olpidium* were reported by Abdullah et al. (2003). In regard to the marine fungi from Iraqi waters, a total of 33 species are reported (Table 22.2), out of them 11 species were reported from brackish and saline waters including four species of Ascomycota, five species of Deuteromycota and two species of Coelomycetes (Abdulkadir and Muhsin 1991). From submerged wood, ten fungal species including seven species belong to Ascomycota and three species of Deuteromycota were reported and full taxonomic description was given (Muhsin and Khalaf 2002). More recently, Al-Saadoon and Al-Dossary (2014) have reported 67 species of freshwater and marine fungi isolated from different types of submerged plant substrates from various locations of water bodies in Iraq. Additional aquatic fungal species have been isolated from water bodies in Basrah; however, no taxonomic description has been given, and these include *Aphanomyces laevis, Calyptelegnia achlyoides, Pythium debaryanum, Saprolegnia turfosa* and unidentified taxon (Muhsin et al. 1993).

Table 22.1 Aquatic fungi recoded from aquatic environment in Southern Iraq

Fungal species	References
Oomycota	
Achlya ambisexualis Raper	Muhsin et al. (1984)
A. debaryana Humphrey	Muhsin et al. (1984)
A. dubia Coker	Muhsin et al. (1984)
A. klebsiana Pieters	Muhsin et al. (1984)
A. oblongata var. gigantica Forbes	Muhsin et al. (1984)
A. polyandra Hildebrand	Muhsin et al. (1984)
A. Prolifera Nees	Muhsin et al. (1984)
A. Proliferoides Coker	Muhsin et al. (1984)
A. racemosa Hildebrand	Muhsin et al. (1984)
A. treleaseana (Humphrey) Kaufmann	Muhsin et al. (1984)
A. aff. caroliniana Coker	Muhsin et al. (1984)
Dictyuchus achlyoides (Coker and Couch) Coker	Rattan et al. (1978)
D. carpophorus Zopf	Rattan et al. (1978)
D. magnusii Lindst.	Muhsin (1979)
D. monosporus Leitgeb	Rattan et al. (1978)
D. pseudodictyon Coker and Braxton	Muhsin (1979)
D. pseudoachlyoides Beneke	Rattan et al. (1978)
Calyptralegnia basraensis	Muhsin (1994)
C. ripariensis Hohnk	Rattan et al. (1978)
Saprolegnia anisospora de Bary	Ismail et al. (1979)
S. australis Elliot	Muhsin and El-Habeeb (1999)
S. declina Humphrey	Ismail et al. (1979)
S. delica Coker	Muhsin and El-Habeeb (1999)
*S. f*erax (Gruith) Thuret	Ismail et al. (1979)
S. hypogyna (Pringsheim) de Bary	Muhsin and El-Habeeb (1999)
S. litoralis Coker	Muhsin and El-Habeeb (1999)
S. parasitica Coker	Ismail et al. (1979)
S. terrestris Cookson ex Seymour	Ismail et al. (1979)
S. uliginosa Johannes	Muhsin and El-Habeeb (1999)
Chytridiomycota	
Olpidium bornovanus (Sahityama) Karling	Abdullah et al. (2003)
Olpidium brassicae (Wor.) Dang	Abdullah et al. (2003)

22.3 Ecological Studies on Aquatic Fungi

Ecological studies of aquatic fungi were carried out by many researchers over the world to elucidate the effects of various environmental factors on the fungal biodiversity in certain aquatic habitat. It has been reported that water temperature plays a major role on population dynamics including occurrence, frequency and seasonal variation of fungal assemblages inhabiting aquatic habitats (Johnson et al. 2002). However, analysis of these parameters needs an appropriate methodology and

Table 22.2 Marine fungi recorded from different aquatic habitats in Iraq

Fungal species	References
Arxiomyces zubairiensis Abdullah and Al-Saadon	Abdullah and Al-Saadoon (1994b)
Aniptodera chesepeakensis Shearer and Miller	Abdullah and Abdulkadir (1987)
A. mauritianicenis Hyde, Ho and Tsu	Al-Saadoon and Abdullah (2001)
A. palmicola Hyde, Ho and Tsu	Al-Saadoon and Abdullah (2001)
Basramyces marinus Abdullah, Abdulkadir and Goos	Abdullah et al. (1989)
Camarosporium roumeguerii Saccardo	Abdulkadir and Muhsin (1991)
Cirrenalia macrocephala (Kohlm.) Meyers and Moore	Abdulkadir and Muhsin (1991)
Clavatospora bulbosa (Anastas.) Nakagi and Tubaki	Muhsin and Khalaf (2002)
Coniothyrium obiones Jaap	Abdulkadir and Muhsin (1991)
Corollospora maritima Werdermann	Abdulkadir and Muhsin (1991)
C. mesopotamica Al-Saadoon	Al-Saadoon (2006)
Didymosphaeria enalia Kohlm	Abdulkadir and Muhsin (1991)
D. futilis (BerkandBr.) Rehn.	Abdullah and Abdulkadir (1987)
Leptosphaeria albopunctata (Westendorp) Saccardo	Abdullah and Abdulkadir (1987)
L. oremaris Linder	Abdullah and Abdulkadir (1987)
Lulworthia grandispora Meyers	Muhsin and Khalaf (2002)
L. medusa (Ellis et Everh) Cribb et Cribb	Muhsin and Khalaf (2002)
Mycosphaerella neumatophora Kohlm.	Muhsin and Khalaf (2002)
Nais inornata Kohlm.	Muhsin and Khalaf (2002)
Ophiobolus australiensis Johnson and Sparrow	Abdulkadir and Muhsin (1991)
Periconia prolifera Anastas	Muhsin and Khalaf (2002)
Pleospora gaudefroyi Patouillard	Abdulkadir and Muhsin (1991)
P. herbarum (Fr.) Rabenh.	Abdullah and Abdulkadir (1987)
Preusia dispersa (Clun) Cain	Abdullah and Abdulkadir (1987)
Preussia aqulirostrata Guarro, Abdullah, Gene and Al-Saadoon	Guarro et al. (1997b)
Pseudallescheria desertorum (Arx et Mustafa) McGinnis	Muhsin and Khalaf (2002)
Pyrenophora typaecola (Cke.) Mull.	Abdullah and Abdulkadir (1987)
Savoryella lignicola Jones and Eaton	Al-Saadoon and Abdullah (2001)
Sphaerulina oraemaris Linder	Muhsin and Khalaf (2002)
Syspastospora tetraspora Abdullah and Al-Saadon	Abdullah and Al-Saadoon (1994a)
Trichocladium linderi Crane and Shearer	Abdulkadir and Muhsin (1991)
Zalerion varium Anastas.	Abdulkadir and Muhsin (1991)
Zopfiella submerse Guarro, Al-Saadoon, Gene and Abdullah	Guarro et al. (1997a)

accuracy in sampling method and statistical analysis applied. On the other hand, the environmental factors including pH, salinity and dissolved oxygen are also important parameters to be correlated with the fungal population dynamic structures and biodiversity (Johnson et al. 2002). Nevertheless, the temperature is the major factor influencing the growth, reproduction and subsequently the population dynamics and diversity of fungi in aquatic environment. Also, the temperature has an impact on the spores germination, hyphal growth and production of sexual organs represented by the oogonia and antheridia (Dick 1973; Johnson et al. 2002). Dissolved oxygen is also another factor affecting the distribution and diversity of fungi in aquatic habitats; therefore, aquatic fungi vary greatly with the respect of their requirement for dissolved oxygen. Low oxygen may inhibit the growth of freshwater fungi mainly the species which belong to the Oomycota (Dick 1973). Other environmental factors like pH also have an impact on the occurrence and distribution of aquatic fungi reported that aquatic fungi can be divided into neutral species (thrive well at pH 5.6–7.4), acidic species (grow at pH below 5.2) and alkaline species (grow at pH above 7.8). Generally, pH of waters of Iraq ranges between7.0 and 8.2; however, from the data available, it is difficult to correlate between this factor and the biodiversity of aquatic fungi in this region. Nonetheless, salinity seems to be the limiting factor influencing the diversity of fungi in aquatic environment. Accordingly, fungal species which live in freshwater are known as freshwater fungi and those are mostly belong to Oomycota while other groups of fungi which live in saline waters are so-called marine fungi and mainly belong to Ascomycotina and Basidiomycotina.

22.4 Biodiversity of Aquatic Fungi of Iraq

Fungi are among the most important organisms in the world, not only because of their important roles in ecosystem but also because of their influence on human activities (Mueller et al. 2004). Aquatic fungal biodiversity is defined as the number and variability of living fungal species in certain aquatic ecosystem; however, many factors affect the loss of biodiversity such as habitat change, climate change, invasive species, over-exploitation and pollution. Nonetheless, water bodies in Iraq are characterized by a high fluctuation of temperature over the year. High water temperature is often recorded in summer (30–32 °C) and low temperature in winter (10–11 °C) seasons (Ismail et al. 1979). It has been noticed that the most species of Saprolegnoid fungi in waters of Iraq are sexually reproduced by forming oogonia and antheridia at low water temperature in winter season; on contrary, asexual reproduction by formation of zoosporangia and zoospores was dominated at high temperature in summer season (Muhsin 1979). The diversity of aquatic fungi in waters, based on collective data from past and recent studies, is greatly affected by water temperature. Generally, high total number of fungal species was found in winter season (December-February) and low total fungal species in summer (June-August) (Fig. 22.2). Also, the occurrence and frequency of aquatic fungal species are affected by the fluctuation of water temperature (Rattan et al. 1980). Among the recorded species of fresh aquatic fungi in waters of Iraq, *Saprolegnia parasitica*, *S. ferax* and *Dictyuchus monosporus* were the most dominant fungi in Shatt Al-Arab

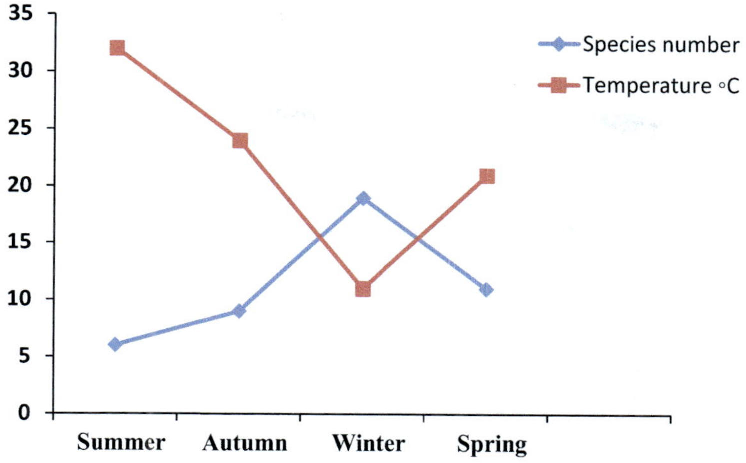

Fig. 22.2 Diversity of aquatic fungal species in relation to water temperature over seasons of Iraq

Table 22.3 Total number of species, total number of isolates and species diversity values (Dv) using Simpson's index of aquatic fungi over three sampling periods

Sampling periods	Total of species	Total of isolates	Simpson's index (Dv)
1976–1977	20	220	0.72
1989–1990	12	122	0.61
2008–2009	4	60	0.37

river southern Iraq (Muhsin 1979; Rattan et al. 1980). There seems to be a common occurred species of freshwater fungi in different water habitats in Iraq represented by *S. parasitica, S. ferax* and *D. monosporus*. These species were reported from southern Iraq (Shatt Al-Arab river, Basrah) (Rattan et al. 1978; Ismail et al. 1979; Muhsin et al. 1993; Muhsin et al. 2011), from middle of Iraq (Tigris river, Baghdad) (Rasheed and Hassan 2013) and from north of Iraq (Sarchnar lake, Sulaymaniyah) (Farkha and Abdulrahman 2011). A survey of aquatic fungi in Shatt Al-Arab River showed that the diversity of Saprolegnoid fungi is drastically declined over collecting periods of time according to studies of Muhsin (1979), Muhsin et al. (1993) and Muhsin (2011). In order to estimate the biodiversity values of aquatic fungi in Iraqi waters, Simpson's index (Simpson 1949) was applied. It appeared that aquatic fungal diversity is higher (0.72) in waters during the year (1976–1977) than the diversity in (1989–1990) and in (2008–2009) (Table 22.3). Changes in the fungal biodiversity over time can be related to the changes of water environmental factors. This may explain that changing in the ecological factors of the aquatic environment of these waters may be due to the increase of water salinity levels (from 0.7‰ in the year 1977 to 2.5‰ in 1990 and to 7.3‰ in the year 2009). An inverse relationship between the fungal species diversity and water salinity was observed (Fig. 22.3). On

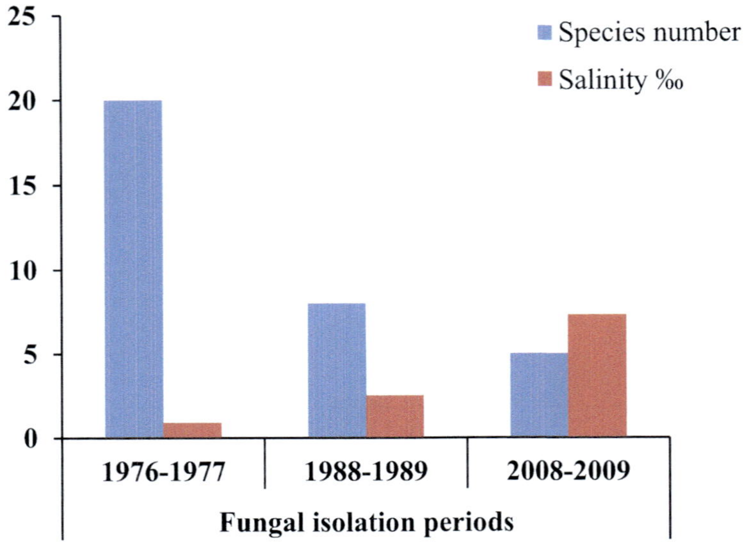

Fig. 22.3 Correlation between the diversity of fungi and salinity in aquatic environment of Iraq over three sampling periods

the other hand, the seasonal periodicity of Saprolegnoid fungi in Iraqi waters has been studied by Rattan et al. (1980); a maximum occurrence and frequency of these fungi were in winter and minimum in summer. This fluctuation pattern based upon the fungal number of species was attributed to the water temperature fluctuation over the year (Fig. 22.4). Twenty species of aquatic freshwater fungi were reported from waters of Shatt Al-Arab river (Sothern Iraq) (Table 22.1). Other studies showed that water samples from Tigris river (near Baghdad, middle of Iraq) revealed 11 species of Saprolegniaceae were reported (Butty 2008) and nine species including five species of *Achlya,* four species of *Saprolegnia* and one species of each of *Dichtyuchus* (Rasheed and Hassan 2013) (Table 22.4), while from waters of Sarchnar lake in Sulaymaniyah (northern Iraq), three species of each of the genus *Achlya* and *Saprolegnia* have been reported (Farkha and Abdulrahman 2011). Moreover, it is well known that pollution of different sources is generally increased during the last decades in aquatic environment over the world (Agarwal 2005). Pollutants such as heavy metals have been reported in waters of Iraq (Abdullah 2013); however, so far there is no report regarding the effects of pollutants on the biodiversity of fungi in these aquatic habitats. It can be stated that an increase of pollution of waters may affect the diversity of aquatic fungi in Iraqi waters; nonetheless, at this point, there is no evidence to correlate between the pollutants levels and fungal biodiversity and needs further investigations.

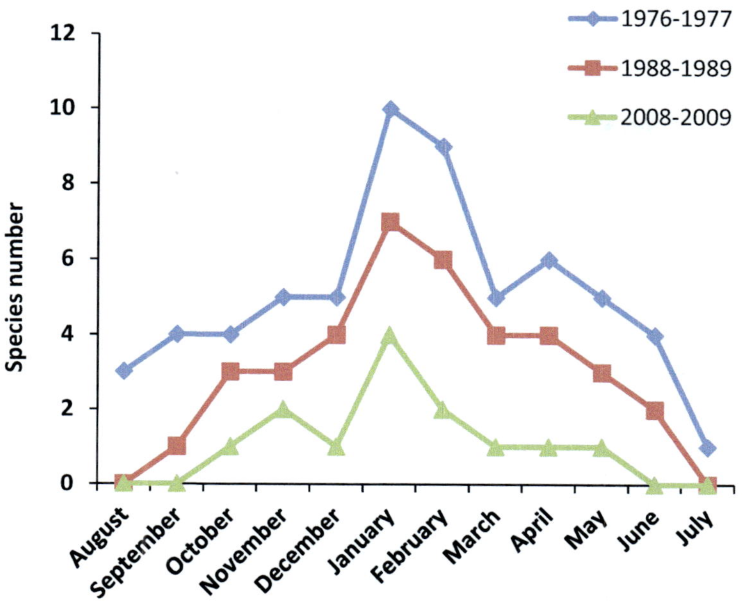

Fig. 22.4 Comparison of fungal diversity in Shatt Al-Arab River south of Iraq over three sampling periods

22.5 Fungi from Submerged Substrates in Aquatic Habitats of Iraq

A large amount of falling plant leaves and wood are available in aquatic habitats as prime source of carbon used by fungi (Barlocher and Kendrick 1974). From submerged plant substrates, many fungal species have been reported (Shaerer et al. 2007). In Iraq, a number of works have been done on fungi of submerged plants. Abdullah and Abdulkadir (1987) reported some freshwater and marine species from southern marshes of Iraq. A new marine species has been designated from marsh water habitat by Abdullah et al. (1989). Also, Abdulkadir and Muhsin (1991) have reported 11 marine species from submerged plants in waters southern Iraq (Table 22.5). More marine fungal species have been also isolated from submerged wood in various water bodies in Basrah and ten species including seven species belong to Ascomycota and three species of Deuteromycota were recorded and their description and illustrations are given (Muhsin and Khalaf 2002). In addition, two new species *Syspastospra tetraspora* (Abdullah and Al-Saadoon 1994a) and *Arxiomyces zubairiensis* (Abdullah and Al-Saadoon 1994b) were isolated from decaying culms of *Phragmitis* and *Arundo* submerged in waters of Khawr Al-Zubair estuary. Some ascomycetous marine fungi have been also reported as new species (Guarro et al. 1996, 1997a, b). Recently, Al-Saadoon and Al-Dossary (2014) have reported 67 species of freshwater and marine fungi isolated from

Table 22.4 Aquatic fungal species reported from Tigris River and Sarchnar Lake in Iraq

Tigris River (Middle of Iraq)	References	Sarchnar Lake (North of Iraq)	References
Achlya carolina	Farkha and Abdulrahman (2011)	*Achlya americana*	Rasheed and Hassan (2013)
A. debaryana	Farkha and Abdulrahman (2011)	*A. apiculata*	Rasheed and Hassan (2013)
A. dubia	Farkha and Abdulrahman (2011)	*A. racemosa*	Rasheed and Hassan (2013)
A. Klebsiana	Farkha and Abdulrahman (2011)	*Saprolegnia ferax*	Rasheed and Hassan (2013)
A. prolifera	Butty (2008)	*S. litoralis*	Rasheed and Hassan (2013)
A. polyandra	Butty (2008)	*S. parasitica*	Rasheed and Hassan (2013)
A. americana	Butty (2008)		
A. racemosa	Butty (2008)		
A. treleaseana	Butty (2008)		
A. debaryana	Butty (2008)		
A. ycologica	Butty (2008)		
Calyptrlegnia ripariensis	Butty (2008)		
Dictyuchus monosporus	Farkha and Abdulrahman (2011)		
D. magnusi	Butty (2008)		
Saprolegnia anisospora	Farkha and Abdulrahman (2011)		
S. diclina	Farkha and Abdulrahman (2011)		
Saprolegnia ferax	Butty (2008)		
S. uliginosa	Butty (2008)		
S. terrestris	Butty (2008)		
S. parasitic	Farkha and Abdulrahman (2011)		

different types of plant substrates from various locations of water bodies in Iraq. These represent 46 species (32 genera) of Ascomycota, 19 species (19 genera) of Hyphomycetes and two species (2 genera) of Coelomycetes (Table 22.6). It is worth mentioning that some marine fungal species became inhabitant in freshwater of Shatt Al-Arab waters.

22.6 Transient Fungi in Aquatic Habitats of Iraq

The fungal population in water ecosystem either representing the true aquatic fungi (residents) or those are washed or blown from soils into waters (transients). Most of the recorded transient fungal species from waters of Iraq belong to the Phylum Deuteromycota (hyphomycetes). From ground waters in Basra (Southern Iraq),

Table 22.5 Fungal species reported from submerged plant substrates in brackish and saline waters in Iraq

Fungal species	Substrates
Drechslera halodes	*Cyperus rotundus*
Basramyces marinus	=
Didymosphaearia analia	*Arundo donax*
Zalerion varium	=
Camarosporium roumeguerii	*Salsola baryosma*
Cirranalia macrocephala	*Phoenix dactylifera*
Corollospora maritime	=
Ophiobolus australiensis	=
Coniothyrium obions	*Tamarix aphylla*
Pleospora gaudefroyi	*Suaeda vermiculata*
Trichocladium linderi	*Typha australis*

From Abdulkadir and Muhsin (1991)

twenty fungal species were reported with variable occurrence percentages (Table 22.7). The frequency percentage of the isolated fungi from ground waters was high during winter season and low during summer season. The members of *Aspergillus* were highly frequent (31.2%) followed by *Penicillium* (22.8%) and *Fusarium* (18.7%). The increase of fungal population in ground waters during winter has been related to the washed fungal propagules from soil into waters due to heavy rain usually occurred in winter. Overall, the diversity of fungi in ground waters of Iraq is mainly affected by the water salinity; an inverse correlation between Na and Cl ions concentration and number of fungal species were detected. Another studies on Shatt Al Arab waters reported 40 species belong to 20 genera of Deuteromycota were recorded as cellulolytic fungi (Table 22.8) which are capable of degrading cellulose substrates in aquatic environment. In this study, the diversity of cellulolytic fungi was correlated with water temperature during the sampling period. More likely, high fungal diversity (17 species) was in winter when the water temperature is low (12 °C) while a low fungal diversity (3 species) in summer with high water temperature reached 32 °C. Highest occurrence percentages (100%) of cellulolytic fungi was represented by the species of the genera *Aspergillus*, *Cephalosporium*, *Fusarium*, *Oidendron* and *Trichoderma*. The role of cellulolytic fungi in aquatic habitats is the biodegradation of plant substrates which are composed of cellulose mainly. The degradation process of plant substrates by aquatic fungi is attributed to the capability of fungi of producing cellulose enzymatic activity.

Table 22.6 List of fungi isolated from submerged plant substrates in water habitat in Iraq

Ascomycota	Deuteromycota (Hyphomycetes)
Aniptodera chesapeakensis	*Alternaria alternata*
A. fusiformis	*Aureobasidium pullulans*
A. muritaniensis	*Bactrodesium linderi*
A. palmicoda	*Beltrania rhombic*
Arxiomyces campanulatus	*Cirrenalia macrocephala*
A. zubairiensis	*Clavatospora bulbos*
Canariomyces notabilis	*Cumulospora marina*
Chaetomium globosum	*Cylindrocladium camelliae*
Coniochaeta saccardoi	*Dendryphiella arenaria*
Corollspora maritime	*Exserohilum rostratum*
C. pseuodopulchella	*Halenospora varia*
Decorospora gaudefroy	*Halosigmodea parvula*
Didmosphaeria fulis	*Moromyces varius*
Jahnula bipileala	*Monodictys pelagica*
Kirachsteiniothelia maritime	*Periconia prolific*[a]
Leptosphaeria agnita	*Stachybotrys atra*
Lignicola laevis	*Trichoderma alopallonellum*
Lulworthia grandispora[a]	*Virganella atra*[a]
L. medusa[a]	*Zygosporum masoni*[a]
Marinosphaeria mangrovei	
Monosporoasus eutypoides	
Mycosphaerella pneumatophora[a]	Deuteromycota (Coelomycetes)
Nais aquatic	*Camarosporium roumeguerii*
N. inormata[a]	*Coniothyrium obions*
Natanlispora refoquens	
Ophiopolus australensis	
Phaeosphaeria albpunchata	
P. oraemaris	
P. typhanus	
Pleaospora herbarum	
Podospora dolichpodalis	
P. inquinata	
Perussia aqulirostata	
P. dispersa	
Pseudoallescharia desertorum[a]	
Pseudohaloneteria phialidica	
Pseudolignicola siamnsis	
Pyreophora typhaecola	
Savoyella lignicola	
Sphaerulina orae-maris[a]	
Syspastospora tetraspora	
Verruculina enalia	

(continued)

Table 22.6 (continued)

Ascomycota	Deuteromycota (Hyphomycetes)
Zopfiella cephaliothecoidea	
Z. karachiensis	
Z. latpes	
Z. submerse	

From Al-Saadoon and Al-Dossary (2014)
[a]Species are reported from submerged wood

Table 22.7 Occurrence % of twenty transient fungal species isolated from ground waters in Southern Iraq

Fungal species	Occurrence %
Alternaria alternata	50
Aspergillus fumigatus	100
A. niger	75
A. terreus	50
Aureobasidium pullulans	25
Cladosporium cladosporioides	50
C. herbarum	100
Drechslera australiensis	25
Fusarium moniliforme	25
F. oxysporum	100
Geotrichum candidum	25
Humicola grisea	25
Mucor hiemalis	50
Paecilomyces variotii	50
Penicillium notatum	100
Pithomyces sacchari	25
Stachybotrys atra	25
Thermomyces lanuginosus	25
Trichoderma viride	50
Ulocladium botrytis	75

22.7 Conclusions

Based on the collected data from different studies carried out on aquatic fungi in waters of Iraq, it appeared that most of fungal species inhabiting the freshwater habitats belong to the family Saprolegniaceae (Oomycota) while in brackish and saline waters the most abundant species belong to Ascomycota followed by Deuteromycota. Due to increase of water salinity during the last decade, some marine fungal species become inhabitants in Iraqi water bodies and there is a general decline in the biodiversity of freshwater fungi.

Table 22.8 Occurrence % of cellulolytic fungi from waters of Shatt Al-Arab River Southern Iraq

Genera	Occurrence %
Alternaria (3 species)	40
Aspergillus (6 species)	100
Cladosporium (3 species)	30
Cephalosporium (3 species)	100
Fusarium (1 species)	100
Geotrichum (1 species)	40
Gliocladium (1 species)	20
Humicola (2 species)	30
Mucor (2 species)	60
Oidendron (3 species)	90
Paecilomyces (1 species)	40
Penicillium (3 species)	80
Phoma (1 species)	20
Sepedonium (1 species)	20
Stachybotrys (2 species)	30
Stemphylium (1 species)	20
Trichoderma (1 species)	100
Trimmatostroma (1 species)	20
Ulocladium (3 species)	40
Verticillium (1 species)	50

References

Abdulkadir MA, Muhsin TM (1991) Marine fungi from Iraq. Marin Mesopotam 6:193–207

Abdullah EJ (2013) Quality assessment for Shatt Al-Arab River using heavy metal pollution index and metal index. J Environ Earth Sci 3:114–120

Abdullah SK, Abdulkadir MA (1987) Freshwater and marine ascomycotina from the southern marshes of Iraq. Marin Mesopotam 2:65–74

Abdullah SK, Al-Saadoon AH (1994a) *Syspastospora tetraspora* sp. nov., a new ascomycetes from Khawr Al-Zubair estuary, Southern Iraq. Marin Mesopotam 9:83–89

Abdullah SK, Al-Saadoon AH (1994b) *Arxiomyces zubairiensis* sp. nov. from the Khawr Al-Zubair estuary, Southern Iraq. Marin Mesopotam 9:245–250

Abdullah SK, Abdulkaddir MA, Goos RD (1989) *Basramyces marinus* (nom. nov.) Hyphomycets from the Southern marshes of Iraq. Int. J Mycol Lichenol 4:181–186

Abdullah SK, Abbas AF, Hammodi KJ (2003) A study of the *Olpidium* species in Basrah region. Iraqi J Agric 8:36–44

Agarwal SK (2005) Water pollution. APH, India

Alexopoulos CJ, Mims CW, Blackwell M (1996) Introductory mycology. Wiley, New York. 869 pp

Al-Saadoon AH (2006) A new arenicolous species of *Corollospora* from Iraq. Marsh Bull 2:134–139

Al-Saadoon AH, Abdullah SK (2001) Some interesting ascomycetes from Iraq. Iraqi J Biol 1:125–134

Al-Saadoon AH, Al-Dossary MN (2014) Fungi from submerged plant debris in aquatic habitats in Iraq. Int J Biodivers Conserv 6:468–487

Barlocher F, Kendrick B (1974) Dynamics of fungal populations on leaves in a stream. J Ecol 62:761–91

Butty AN (2008) Isolation of Saprolegniaceae from Tigris River and notes on their occurrence in winter season. J Environ Sci Health 24:505–513

Coker WC (1923) The Saprolegniaceae with notes on the other water molds. Univ North Carolina Press, Chapel Hill. 201pp

Dick MW (1973) Saprolegniales. In: Ainsworth GC, Sussman AS (eds) The fungi. Academic, San Diego, p 113

Farkha TK, Abdulrahman AS (2011) An ecological study on the occurrence and distribution of aquatic fungi in Sarchnar water spring within Sulaimani Province / Kurdistan Region of Iraq. J Al-Nahrain Univ 14:137–145

Guarro J, Abdullah SK, Al-Saadoon AH, Gene J (1996) A new *Zopfiella* (Lasiosphaeriaceae) from Iraq. Mycotaxon 54:197–202

Guarro J, Abdullah SK, Gene J, Al-Saadoon AH (1997a) A new species of *Preussia* from submerged plant debris. Mycol Res 101:305–308

Guarro J, Al-Saadoon AH, Gene J, Abdullah SK (1997b) Two new cleistothecial ascomycetes from Iraq. Mycologia 89:955–961

Ismail AS, Rattan SS, Muhsin TM (1979) Aquatic fungi of Iraq: species of *Saprolegnia*. Hydrobiologia 65:83–93

Johnson TW (1956) The genus *Achlya*: morphology and taxonomy. University of Michigan Press, 180 pp

Johnson TW, Seymour RL, Padgett DE (2002) Biology and systematics of the Saprolegniaceae. Academic, New York

Keeling PJ, Burger G, Durnford DG, Lang BF, Lee RW, Pearlman RE, Roger AJ, Gray M (2005) The tree of eukaryotes. Trends Ecol Evol 20:670–676

Kohlmeyer J, Kohlmeyer E (1979) Marine mycology: the higher fungi. Academic, New York. 690 pp

Mueller GM, Bills GF, Foster MS (2004) Biodiversity of fungi: inventory and monitoring methods. Academic, London

Muhsin TM (1979) Studies on the Saprolegniaceae of Shatt Al-Arab River near Basrah, Iraq. MSc. Thesis, University of Basrah, Iraq. 120 pp

Muhsin TM (1994) *Calyptralegnia basraensis* sp. nov. (Saprolegniaceae) from Iraq. Pol Arch Hydrobiol 41:415–418

Muhsin TM, El-Habeeb EK (1999) Aquatic fungi from Iraq: new records of Saprolegniaceae. J Basrah Res 22:77–86

Muhsin TM, Khalaf KT (2002) Fungi from submerged wood southern Iraq. Iraqi J Biol 2:455–463

Muhsin TM, Rattan SS, Ismail AS (1984) Aquatic fungi of Iraq: species of *Achlya*. Sydow Annal Mycol 37:224–237

Muhsin TM, El-Habeeb EK, Athbi AM (1993) Spatial and temporal occurrence and frequency of aquatic fungi in Shatt Al-Arab river, Basrah, Iraq. Marin Mesopotam 8:105–117

Muhsin TM, Al-Duboon AA, Bader SK (2011) Ecology of water molds in Shatt Al-Arab River and fish aquaria of Basra University. J Basra Res 37:68–82

Rasheed KA, Hassan KF (2013) Fungal biodiversity in the freshwater environment of Iraq. Assi Univ Bull Environ Res 16:12–18

Rattan SS, Muhsin TM, Ismail AS (1978) Aquatic fungi of Iraq: species of *Dictyuchus* and *Calyptralegnia*. Sydow Annal Mycol 31:1–6

Rattan SS, Muhsin TM, Ismail AS (1980) Notes on the occurrence and seasonal periodicity of Saprolegniaceae in Shatt Al-Arab river, Basrah, Iraq. Kavaka 8:41–48

Seymour RL (1970) The genus *Saprolegnia*. Nov Hedwig 10:1–12

Shaerer CA, Descal E, Kohlemyer B, Kohlemyer J, Marvanova L, Padgett D, Porter D, Raja H, Schmit JP, Thorton HA, Voglymayr H (2007) Fungal biodiversity in aquatic habitats. Biodivers Conserv 16:49–67

Simpson E (1949) Measurement of diversity. Nature 163:688

Thuiller W (2007) Biodiversity: climate change and the ecologist. Nature 448:550–552
Weber RWS (2009) Recent developments in the molecular taxonomy of fungi. In: Anke A, Weber D (eds) The mycota, physiology and genetics. Springer, pp 1–15
Webster J, Weber RWS (2007) Introduction to fungi. Cambridge Univ Press, 841 pp
Willoughby LG (1994) Fungi and fish diseases. Pisces. pp 57

Chapter 23
Potability of Drinking Water in Basra-Iraq

Amin Al-Sulami and Asaad Al-Taee

Abstract Since the 1980s, there has been a general marked deterioration in water quality in Iraq, reflecting the environmental degradation of the country caused by successive armed conflicts. In major cities as Basra, the destruction of infrastructure has damaged sewerage and water supply systems, preexisting inadequate wastewater treatment works, many pumping stations, and related engineering infrastructure.

Certainly, this damage has resulted in massive pollution of watercourses and rivers. The environmental degradation has been exacerbated by an unreliable supply of electricity, resulting in additional serious pollution problems. This has been compounded by widespread corruption and mismanagement. In consequence, the citizens no longer have access to safe piped-drinking water, meeting internationally accepted standards of water quality.

In order to assist, the local government undertakes a water service infrastructure rehabilitation and renewal program and it will be necessary to implement a three-phase program of scientific and technological research. A comprehensive survey of drinking water quality should be undertaken to determine where there are significant water treatment problems; laboratory and pilot-scale research should then be undertaken to determine the most appropriate water treatment solutions to overcome these problems and through technology transfer, the outcomes of this research will be applied to infrastructure design and construction.

23.1 Introduction

The objective of this study is to review the status of drinking water in Basra Governorate and advance sustainable solutions to restore drinking water quality to internationally acceptable standards. Also, to develop science and technology,

---

A. Al-Sulami (✉)
Basrah University, Basrah, Iraq

A. Al-Taee
Marine Science Centre, Basrah University, Basrah, Iraq

L. A. Jawad (ed.), *Tigris and Euphrates Rivers: Their Environment from Headwaters to Mouth*, Aquatic Ecology Series 11,
https://doi.org/10.1007/978-3-030-57570-0_23

expertise in water supply and treatment to international standards through the establishment of water research center in Basra.

This study is of national significance having country wide implications for improved health and lifestyle (Murthy, 2011). It will review the status of the drinking water by looking into the infrastructures and the quantity and quality of piped water as provided by local authorities and revealed via academic research. Eventually, it will provide information on water treatment problems that need attention and suggest the most appropriate water treatment solutions to overcome these problems.

23.2 Water Supplies to Basrah

The Basra governorate (Fig. 23.1), where the Basra city is located, has a population of about four million, its water supply from three sources, Shatt Al-Arab River, Tigris River, and Bada'a. Bada'a is a lake that receives its water from Tigris through the Gharraf River and lies approximately 250 km north of Basra, fresh water is pumped through an open canal from Bada'a to the Basra city. While, Um Qasr is supplied from groundwater.

According to the Directorate of Basra Water, the water is treated in 312 stations of submerged main and main water from these only 21 treatment works (Table 23.1) and distributed through approximately 13,500 km pipe network. An average of 442,000,000 l of drinking water is pumped each day, providing theoretically 147.5 l for each citizen (about 330 l/citizen/day in 1970), (UNCT 2010; World Bank 2007). Several quarters and villages are either cut off or have interrupted supply and in all cases, household electric pumps are required, adding financial burden and expenditure of electricity. This is because water normality does not reach the top level. Also, it is used to lift water to fill up over-roof storage tanks for various daily usages.

Iraq faced difficulties in meeting the target of 91% of households using a safe drinking water supply by 2015.

1. Currently, 20% of households in Iraq use an unsafe drinking water source.
2. Furthermore, 16% have daily problems with supply.
3. Leaking sewage pipes and septic tanks contaminate the drinking water network with wastewater.
4. 80% of Iraqis do not treat their water before drinking.

23.3 Chemical Analysis

In 2004, the Iraqi company Al-Khebra implemented the project called "Salinity balance of tidal flow Shatt Al-Arab waterway, Basra city" (Al-Khebra Bureau 2004) in collaboration with the environmental Czech company GEO test Brno, a.s. Analyses included the determination of significant physical and chemical

Fig. 23.1 Map of Basra Governorate

parameters and the concentration of toxicologically important trace elements in water samples

They drew attention to the urgency of the problem of the high mineralization (salinity) in the Shatt Al-Arab watercourse. The situation is extremely unsatisfactory with salinity 1750–4000 mg/l and currently exceeds 4000 mg/l. The current World Health Organization Guidelines for Drinking Water Quality (WHO 2017) state the following:

Table 23.1 overview of main water treatment stations in Basra 2018 (Basra water Directorate)

Project name	Disinfection	Water quantity produced m^3/day	A source of water
Basra Unified	Chlorine	60,000	Shatt al-Arab+ BW[a]
Al-Jubaila	=	16,000	= = =
Al-Bradhaya No.1	=	16,000	= = =
Al-Bradhaya No.2	=	16,000	= = =
Shatt al-Arab	=	17,000	= = =
Khor al-Zubair	=	24,000	= = =
Al-Shuaiba	=	12,000	BW
Al-Qurna Unified	=	18,000	Tigris River
Old Medaina	=	8000	= =
Al-Nashwa	=	6000	= =
Al-Hartha 25 million camps	=	80,000	Shatt al-Arab+ BW
Abe-Sukhair camps	=	85,000	BW
Garma camps No.1	=	24,000	Shatt al-Arab+ BW
Garma camps No.2	=	500	Shatt al-Arab+ BW
Al-Labbani	=	8000	Shatt al-Arab+ BW
Sehan	=	14,000	Shatt al-Arab
Muhella	=	8,000	Shatt al-Arab+ BW
Al-Shuaiba camps No.1	=	8000	BW
Al-Shuaiba camps No.2	=	2000	BW
Al-Jubaila camps	=	12,000	Shatt al-Arab+ BW
Al-Rubat camps	=	8000	Shatt al-Arab+ BW

Total = 442,500 m^3/day
[a]BW: Bada'a Water Canal

23.3.1 Total Dissolved Solids

The palatability of water with a TDS level of 500 mg/l is considered to be good; drinking water becomes significantly and increasingly unpalatable at TDS levels greater than about 1000 mg/l. The presence of high levels of TDS may also be objectionable to consumers, owing to excessive scaling in water pipes, heaters, boilers, and household appliances.

23.3.2 Bacteriological Studies

Our interest in studying fresh water and piped drinking water started in 1978, when we evaluated the presence of enteric bacteria in the Shatt Al-Arab River and Al-Ashar creek, emanating from it through the city center (Ali and Al-Sulami 1978). More than 20 species of enteric bacteria were identified, including fecal coliforms.

In a consequent study, Al-Sulami and Al-Bakri (1988) isolated four species of *Proteus* from Mugil fishes (*Liza abu*) obtained from the Shatt Al-Arab River. In 1991, the bacteriological suitability of ground water from the Basra western desert was evaluated. Only 20% of wells were within the internationally accepted limits (Al-Sulami and Yaseen 1991).

In order to verify whether Basra drinking water is up to international standard (WHO 1993) up to 2012, many extensive studies have been undertaken on drinking water quality, Al-Taee (2001); Al-Taee and Shamshoom (2001); Al-Taee (2004) and Al-Taee (2005) study on the bacteriological quality of drinking water in different districts in Basra, whose sources are from three water treatment works as Al-Bradhaaya, Al-Maakel, and Al-Hartha. The results of these studies show that water leaving the water treatment works is of a quality unacceptable for drinking water purposes (Table 23.2).

Further studies (Al-Taee et al. 2008; Al-Sulami et al. 2012a, b, 2013a, b, 2014) revealed that drinking water from all stages of treatment from most districts in Basra harbor fecal indicators and several kinds of pathogenic bacteria (Tables 23.3, 23.4, and 23.5) (Fig. 23.2). This water clearly violated the international standards (WHO 2002).

The WHO Guidelines states as follows:

"The most common and widespread health risk associated with drinking water is microbial contamination, the consequences of which mean that its control must always be of paramount importance." Clearly, the water treatment systems at these waterworks are ineffective in providing safe disinfection. New treatment and control measures are required to restore water quality to internationally accepted standards and protect the public from waterborne diseases.

Also, in 2005, limited chemical analysis was undertaken on a number of these water supplies (Table 23.6). The results of this analysis also highlighted the inadequacy of the water treatment systems in terms of suspended solids and turbidity removal. The very high salinity of some of the samples is also a cause for concern. A much more comprehensive analysis is needed to determine the extent of possible pollutants such as heavy metals, organic contaminants, and sewage.

Table 23.2 Identification of bacterial isolates from different districts in the Basra city during 2001

Identified bacteria	Isolates no./CFU
Salmonella Spp.	0–106
Vibrio cholerae	0–1213
Escherichia coli	0–6
Enterobacter agglomerans	0–3
Ent. Taylorae	0–2
Ent. Gergoviae	0–4
Ent. Asburiae	0–3
Ent. Sakazaki	0–2
Klebseilla ornithinolytica	0–2
K. pneumonia	0–4
Hafnia alvei biogroup 1	0–3
Proteus mirabilis	0–2
Rahnella aquatilis	0–1
Serratia fonticola	0–3
Serratia odorifera biogroups 2	0–4
Citrobacter diversus	0–2
Citrobacter frundii	0–2
Kluyvera ascorbata	0–1
Buttiauxella agrestis	0–2
Budvicia aquatic	0–2
Pseudomonas euroginosa	0–15
Bacillus spp.	0–60
Total non-enteric bacteria	0–15
Aspergillus terrus	0–1
Aspergillus flavus	0–3
Penicillium sp.	0–2

CFU, colony forming units

23.4 Conclusion

Examination of the results described above shows a profound contravention of International Standards (WHO 1998; EPA 2018) for drinking water quality. Also, there are indications of the presence of pathogens in Basra drinking water.

It is clear that Basra's surface and drinking water are fastly deteriorating with time. It represents a crisis at certain points in the 80s of the last century and culminates to the catastrophic situation at the present time, several reasons can be cited to visualize trends for worsening water quality. At the top of them is the ignorance of the local and central authorities in managing the situation, compounded by their widespread corruption. Also, the dwindling supply of fresh water from Turkey and Iran is decisive in exacerbating the problem. Again, the failure of local authorities to establish sewage treatment plants and the indifference on the part of citizens, in general, on how well they utilize and customize what is available on water resources.

Table 23.3 average of *Mycobacterium, Helicobacter pylori*, total and fecal coliform from different quarters of Basrah governorate (CFU/100 ml) during 2006–2007

Quarters	TC	FC	*Mycobacterium avium complex* and other Mycobacteria	*Helicobacter pylori*
Al-Zubair	207.727	254.090	41	1
Old-Basra	236.071	177.241	15	2
Al-Ashar	415.416	392.692	11	–
Garmat-Ali	44.736	13.684	36	–
Shatt-al-Alarab	113	76	10	–
Al-Hussain Q.	106	124.545	10	–
Al-Jamiayat	16	24	3	4
Al-Maaqal	81	48	8	3
Al-Hakeemya	60	108.888	18	–
Al-Asmaaee	13.73	10	Nil	–
Al-junaina	10	0	13	–
Al-tuwaisa	322	302	3	–
5 Miles	5	92.3	7	–
Al-abela	6.666	30	2	–
Al-Jubaila	525	20	5	–
Al-guzaiza	186.666	56.666	Nil	–
Al-Eskan	196.666	206.666	Nil	–
Al-qibla	1500	100	48	–
Al-Jazair	380	200	Nil	–
Al-Qurna	0	2	ND	–
Abu-AlKhaseeb	1000	1300	ND	–

One can advance several solutions to this hazardous endemic problem after critical scientific evaluation, including desalination plants, harnessing sun energy for the evaporation of raw water, recycling used water, and filtration.

Table 23.4 Average of the logarithmic number of *Legionella pneumophila* and fecal coliform (CFU/100 ml) isolated from different plants in the Basrah governorate during 2008–2009

	Raw water		Precipitation tankers		Filtration tankers		Water coming out of the plant	
Sanitation plant	*L. pneumophila*	FC	*L. pneumophila*	FC	*L. pneumophila*	FC	*L. pneumophila*	FC
Al-Baradhiya	3.75257	4	3.6808	4	3.7456	3.54	1	3.67
Al-Jubaila I	4	4	3.6731	4	3.2385	2.46	3.1808	2.42
Al-Jubaila II	4	4	2.4471	3.68	3.5536	4	2.4146	3.68
Al-Ru bat	1.6020	3.18	2.0965	2.82	1.1412	2.34	0	1.73
Al-Fayhaa	2.4842	2.93	2.3520	2.91	ND	ND	1	2.47
Al-Lab bani	1.8740	3.06	2.0791	2.91	ND	ND	0	2.20
Al-Abbas	3.5449	4	1.9030	2.34	1.7781	2.77	0	0
Al-Ma'aqal	3.7463	3.49	3.6989	3.40	1.7385	3.02	1.4515	0
Hamdan	1.3890	2.74	0	1.30	0	1.60	0	1.53
Shatt Al-Arab	1.7235	2.24	1.6989	2.57	0	0	0	1.47
Uwasain	1.3010	2.94	1.9030	2.93	ND	ND	0	1.38
Muhaila	0	2.55	1.8740	2.44	0	1.12	0	0
Muhajran	1.9030	3.08	0	0	0	0	0	0
Average	2.40	3.24	2.10	2.71	1.51	2.08	0.61	1.58

Table 23.5 Average of the logarithmic number of *Aeromonas* bacteria and fecal coliform (CFU/100 ml) isolated from different plants in the Basra governorate during 2008–2009

Sanitation plant	Raw water		Precipitation tankers		Filtration tankers		Water coming out of the plant	
	Aeromonas	FC	*Aeromonas*	FC	*Aeromonas*	FC	*Aeromonas*	FC
Al-Baradhiya	4	4	3.67	4	2.64	3.54	1.47	3.67
Al-Jubaila I	4	4	4	4	4	2.46	2.32	2.42
Al-Jubaila II	4	4	4	3.68	4	4	4	3.68
Al-Rabat	2.55	3.18	1.30	2.82	1.02	2.34	0	1.73
Al-Fayhaa	4	2.93	3.69	2.91	ND	ND	1	2.47
Al-Labani	3.57	3.06	1.87	2.91	ND	ND	2.33	2.20
Al-Abbas	4	4	2	2.34	2.68	2.77	0	0
Al-Ma'aqal	3.07	3.49	4	3.40	2.48	3.02	0	0
Hamdan	0	2.74	1.38	1.30	1.30	1.60	0	1.53
Shatt Al-Arab	1.84	2.24	2.34	2.57	1.38	0	1	1.47
Uwasain	2.75	2.94	2.73	2.93	ND	ND	0	1.38
Muhaila	1.30	2.55	2.83	2.44	1.01	1.12	0	0
Muhajran	2.47	3.08	1.87	0	1.38	0	0	0
Average	2.89	3.24	2.74	2.71	2.19	2.08	0.93	1.58

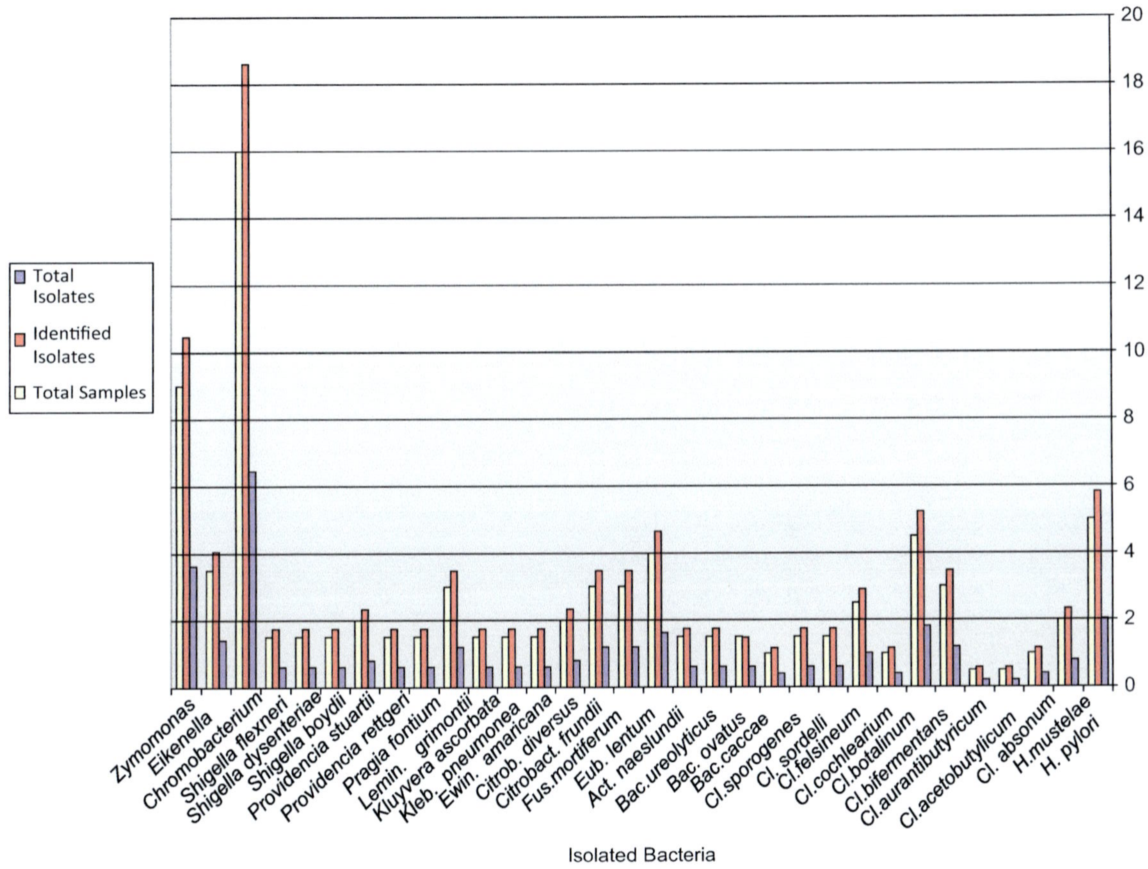

Fig. 23.2 The isolated bacteria and their percentage according to the number of total isolates, identified isolates, and total samples in 2006–2007

Table 23.6 Physiochemical parameters of drinking water from 10 stations in Basra, 2005

Treatment station	PH	TDS (mg/l)	TSS (mg/l)	EC (dS/m)	Tur (NTU)	Mg (mg/l)	Ca (mg/l)	HCO_3 (mg/l)	Cl (mg/l)	SO_4 (mg/l)
Al-Qurna	7.3	1280	16	2.22	22.3	165.24	152	317.2	710	362.59
Al-Medaina	7.5	834	5	2.95	2.47	140.94	160	244	710	199.68
Shatt Al-Arab	7.4	954	10	1.94	6.6	121.5	112	268.4	639	225.96
Abu Al-Khaseeb	7.5	2508	1	2.69	5.25	106.92	152	256.2	639	73.56
Al-Jama'ayat	6.5	916	17	1.48	2.69	87.48	96	244	497	278.5
Safwan	6.4	1122	13	2.12	0.6	–	280	195.2	639	178.66
5-Miles	7.2	–	–	2.0	6.47	–	304	231.8	710	278.15
Al-Junaina	7.2	1234	34	1.96	27.42	116.15	152	317.2	461.5	270.48
Al-Hartha	7.3	826	47	2.12	6.47	–	296	219.6	461.5	252.23
Al-Ma'akel	6.9	1100	31	1.75	3.48	–	216	170.8	639	168.15

References

Ali ZM, Al-Sulami AA (1978) Preliminary studies on Enterobacteriaceae in Shatt-Al-Arab River and Al-Ashar creek. J Asiatic Soc Bangladesh (SC) 3(2):27–37

Al-Khebra Bureau (2004) Salinity balance of tidal flow Shatt Al-Arab waterway, Basra city

Al-Sulami AA, Al-Bakri G (1988) Biochemical variation of Proteus strains isolated from Mugil fishes (*Liza abu*). Basrah J Agric Sci 2(1-2):267–274

Al-Sulami AA, Yaseen HA (1991) Bacteriological suitability of water from Basrah wells for drinking water. Sci Technol 24(2):89–93

Al-Sulami AA, Al-Taee AMR, Wida'a QH (2012a) Isolation and identification of *Mycobacterium avium* complex and other nontuberculosis mycobacteria from drinking-water in Basra governorate, Iraq. East Mediterr Health J 18(3):274–278

Al-Sulami AA, Al-Taee AMR, AL-Rubyai AY (2012b) Recovery of *Helicobacter pylori* from sewage in Basrah Governorate and its confirmation by 16S rRNA. Air Water Born Dis 1(5):1–4

Al-Sulami AA, Al-Taee AMR, Yehyazarian AA (2013a) The effect of Aeromonas spp. on the growth of *Legionella pneumophila* in vitro. Air Water Born Dis 2:1–4

Al-Sulami AA, Al-Taee AMR, Yehyazarian AA (2013b) Isolation and identification of *Legionella pneumophila* from drinking water in Basra governorate, Iraq. East Mediterr Health J 19 (11):936–941

Al-Sulami AA, Al-Taee AMR, Yehyazarian AA (2014) Isolation and identification of *Aeromonas* spp. From drinking water in Basrah governorate. IJDR 4(8):1692–1697

Al-Taee AMR (2001) Assessment of water quality due to microbial growth in drinking water distribution systems in Basrah city. Marina Mesopotamica 16(1):37–46

Al-Taee AMR (2004) Prevalence of pathogenic vibrios in aquatic environment and drinking water. Marina Mesopotamica 19(1):1–5

Al-Taee AM (2005) Scanning the bacteria in soil and water of Basrah city from Qurna to Fao. A project with CPI

Al-Taee AMR, Shamshoom SM (2001) Salmonella detection in drinking water using new incubation regime. Marina Mesopotamica 16(2):143–148

Al-Taee AMR, Al-Sulami AA, Jumaa MG (2008) Isolation and identification of *Helicobacter pylori* and other pathogens from drinking water in Basrah city II Isolation and identification of pathogenic bacteria. Conf Mar Sci Center:1–16. (In Arabic)

EPA (2018) Edition of the drinking water standards and health advisories. Office of Water U.S. Environmental Protection Agency, Washington, DC

Murthy SL (2011) Iraq's constitutional mandate to justly distribute water: the implications of federalism, Islam, International Law and Human Rights. Publication forthcoming in the George Washington International Law Review

UNCT (2010) Government of Iraq, The Millennium Development Goals in Iraq

WHO (1993) Guidelines for drinking-water quality, 2nd edn. Geneva

WHO (1998) Guidelines for safe recreational-water environments. Coastal and fresh-waters (Draft for consultation. WHO/EOS/98.14.), vol 1. World Health Organization, Geneva

WHO (2002) Guidelines for drinking-water quality. Addendum microbiological agents in drinking water, 2nd edn. Geneva

WHO (2017) Guidelines for drinking-water quality. Incorporating the first addendum, 4th edn. Geneva

World Bank/COSIT/KRSO IHSES (2007)

Chapter 24
Algal Studies in Iraqi Inland Waters. A Review

Bahram K. Maulood and Fikrat M. Hassan

Abstract Algal studies in Iraq and Kurdistan may go back to the beginning of this century (1910), it did not flourish until establishing the two universities (Basra and Sulaimaniyah) in south and the north of the country, respectively.

A review of algal study shows the presence of more than 2700 identified algal taxon in Iraq. This investigation reveals the need for rechecking the identification and nomenclature of most of the taxon, particularly the diatoms. The existing gap of knowledge of algae in this part of the world have already been reduced; however, still there is a much more need to be carried out by scientists in order to fulfill or at least reduce the gap of knowledge in this respect in this part of the world. However, physiological and biochemical studies on this sort of flora are quite rear in Iraq; therefore, one may regard this country to be a virgin yard for such investigation in parallel with more accurate and detail ecological and phycological studies. Quite many other references, particularly M.Sc. Thesis on such subjects, have not been included in this review, only attention was given to the most significant publication, whereas others have been excluded in this study.

24.1 Introduction

Phytoplankton is an important tool for understanding the maintenance of species diversity (Stomp et al. 2011). Actually, species richness can be a simple measure to quantify and express the complexity of an area (Nabout et al. 2007). According to McCann (2000), the idea of the influence of species richness over the dynamics and

B. K. Maulood (✉)
Department of Biology, College of Science, University of Salahaddin, Erbil, Kurdistan Region, Iraq

Hawler Botanical Garden, Erbil, Kurdistan Region, Iraq

F. M. Hassan
Department of Biology, College of Science for Women, University of Baghdad, Baghdad, Iraq
e-mail: fikrat@csw.uobaghdad.edu.iq

L. A. Jawad (ed.), *Tigris and Euphrates Rivers: Their Environment from Headwaters to Mouth*, Aquatic Ecology Series 11,
https://doi.org/10.1007/978-3-030-57570-0_24

functioning of the community has become a well-known phenomenon, especially in relation to stability and productivity of an aquatic ecosystem, which is essential to understand factors that lead to the phytoplankton richness.

It is a well-known fact that Iraq is situated within a most fertile land of Mesopotamian, its territory is quite rich with inland water bodies. In the north, apart from Tigris and Euphrates, too many numbers of springs, water channels, and impoundments exist which do support various aquatic life, never the less in the middle of Iraq apart from the twin rivers and their branches, the presence of Tharthar impoundment and various irrigation channels provides enough habitat for aquatic life. However, the Arab marshes in the south with Shatt al-Arab do support and supply environment for brackish life (Maulood et al. 1979a, b, 1981; Al-Saboonchi et al. 2011; Al-Mayah et al. 2014) and finally the Gulf represents the marine habitat within the Iraqi territory.

Algae (Benthic and planktonic) as the main producers in the aquatic environment will be the present report to be dealt with. It is, in fact, the first step within the food chain in the aquatic ecosystem. Actually, nannoplankton, tychoplankton, phytoplankton, benthos, attached algae, seaweed, micro plankton, and macro plankton constitute the producers of the aquatic ecosystem and well be the concern of the present chapter.

24.2 Brief History of Algal Studies in Iraq

Actually, the scope of the primary producer in the inland water systems, in Iraq, generally and mainly is confined to algae only. Their role in the environment, their biomass and total number, taxonomy, their habitat and morphology, their group classification, and finally their contribution to the aquatic ecosystem as pollution indicators will be an attempt to be dealt with in the present report. The role of phytoplankton in general and in more detail information of its productivity as well as its ecology may be found in (Boney 1975 and Round 1975). However, the present attempt will be a review to most published algal studies in Iraq.

Such studies actually did not start comprehensively till the establishment of Basrah University in the south and Sulaimaniyah University in the Kurdistan part of Iraq. In contrast, investigation and research on algae in the middle of Iraq were not initiated comprehensively until almost one decade later, apart from Al-Kaisi (1964).

The role of Al-Saadi, Hadi, Maulood, Hinton, and Al-Mousawi in progressing such studies in Iraq (north, south, and middle of Iraq) is admirable and almost all other new generations of scientists in the country in this respect were under their supervision or cooperation without any exception. Nevertheless, the role of phycologists in Iraq in comparison to neighboring countries is distinctive as they continuously follow up and they covered most aquatic spots within Iraq for various aspects of algal studies.

The absence of an algal society in Iraq or any research center specific to algae or limnology made the publication in this subject to be scattered, dispersed, and disorganized scientifically. Nevertheless, the outcome was the publication of a vast number of papers with different slandered incomplete descriptions and photographs or illustrations of various species, which made one to doubt the identification. Therefore, the need for a consistent review of all or most papers is urgent.

It is clear that not only different check lists of algae have been published during the last 50 years but also few check lists specifically for definite area have been also performed. Actually, the background knowledge for this attempt is so promising in one hand, whereas it reflects the real fact that still knowledge and information about algae in Iraq have been so confusing on the other hand.

In spite of quite many valuable published papers on lentic and lotic primary producers (algae), it is clear that the total number of recorded and identified algae has increased from around 1500 to more than 2600 with Chrysophyta species make up more than 46% of the total flora, whereas Rhodophyta and Charophyta constitute the least number of taxon, as each represented less than 0.5% of the total flora (Saadulla and Maulood 1993; Maulood et al. 2013). Still the need for reviewing phytoplankton or algal studies in Iraq, in general, becomes a real need as it has been declared recently by that an urgent need for the reestablishment of the names of known algae in Iraq, particularly diatoms, is the request (Prof. Al-Handal personal communication). However, many published names of quite a few diatoms and dinoflagellate are confusing and need reconsideration in accordance to most recent taxonomical bases and classification (Al-Handal 1988; Al-Handal and Abdulla 2010; Al-Handal et al. 1989; Al-Saboonchi et al. 1982; Hadi et al. 1984; Islam 1982; Islam and Hameed 1985; Pankow et al. 1979; Saadulla and Maulood 1993; Al-Hassany and Hassan 2014, 2015; Hassan et al. 2012a, b; Al-Handal et al. 2018; Toma et al. 2018).

Laith Jawad, the editor of this book, has invited us with many thanks to write an essay on algal studies in Iraq. In fact, it was not an easy task, but we did our best to forward the present article in such a way to involve almost all published papers, actually, the attempt is in such a way to exclude the unpublished thesis. Undouble this will encourage to much more comprehensive reviews in the near future.

In spite of appearance of more than hundred M.Sc. and Ph.D thesis during the last 50 years on various aspects of algae from different universities and institutes in Iraq, still much more areas remain to be fulfilled in various aspects of algal studies. In the present review, the attempt is to exclude the results of most of unpublished projects and most of unconfirmed papers. The first attempt of gathering various scattered publications of algae in the area done by Hinton and Maulood (1983), when they listed the first check list of algae in the area that included 1296 species and subsequently the check list published by Islam and Hameed (1985) includes 1328 species where ten years later the second check list published included 1900 species by Maulood et al. (1993), then after ten years after the second check list published by Maulood and Toma (2004) included 2312 species. The present check list adds (334) species. Nevertheless, it is evident that more than 2646 taxa of algal species have been reported so far in accordance to the last published check list of algae in Iraq (Maulood et al. 2013). Rechecking the names and identification is going to become a necessary goal and target of algologist in Iraq. As doubt is rising on their identification and naming of algae, particularly diatoms, rechecking names and identification is going to become one of the aims of the present and future phycologists (local and others).

The present review is planned in such a way to forward the history of algal studies in Iraq, which involve their biomass and quantities in various habitat, their systematic classification and species number, their role in the environment and pollution, their periodicity, and seasonal variation. After this focus will be on the main dominant group that makes up most of the bulk or the producers. As a matter of fact, in most aquatic ecosystems in Iraq, diatoms are the dominant algal group that comes in accordance to other parts of the world. Apart from standing water systems where central diatoms make up the main bulk of biomass, the dominant group of standing crops in almost all running water bodies is acentric diatoms (Pennales). In general, blue-green algae and Chlorophyta dominate small ditches of water bodies in Iraq, particularly in summer. Blooming of blue-green algae or dinoflagellate has been also reported in quite many aquatic ecosystems in the country, particularly in the polluted areas, nearby towns and populated sides, where even the color of water changes to blue or greenish because of blue-green algal dominance. In contrast, in fast running water ecosystems in Iraqi Kurdistan, the dominance of Rhodophyta particularly Batrachospermum will be evident.

24.3 The Present Status of Algal Studies in Iraq

Iraq is one of the richest countries of the middle East for water resources, as it has a huge number of surface and underground water systems. The fresh water in the country (lentic and lotic) environment includes two rivers (Tigris and Euphrates) and their main five tributaries, which are situated on the east side of Tigris, ponds, lakes, impoundments, springs and ditches, whereas Shatt al-Arab and east and west marshes of Tigris and Euphrates that are represented mainly by Al-Hammar and Chibiyesh on Euphrates and Hawizah marshes on Tigris represent the brackish water of Iraq, which include oligohaline and mesohaline water system in Iraq. Finally, the gulf water represents the only marine water system in Iraq. Therefore, it becomes quite evident that dealing with any components of such an ecosystem will not be an easy task, since quite many publications and academic theses have been produced, particularly after establishing different universities in different parts of Iraq. Meanwhile, the role of Basra University in the south and Sulaimaniyah university in the north, since 1974, should always be referred to.

Many algal studies have been already conducted on lentic and lotic water bodies in Iraq, meanwhile, there are quite many areas and spots that have not been yet investigated around the country. The first comprehensive study on phytoplankton in Iraq may be that of PhD thesis of (Al-Kaisi 1964) from the University of Wales UK, who performed his project on algae of the water system in Iraq (Razzazah and Habbaniyah). This was followed by a number of preliminary phycological studies in the center of Iraq (Al-Kaisi 1970, 1974 and 1976). (Maulood et al. 2013) declared that the first and pioneer study on algae in the water ecosystem of Iraq and Kurdistan region was conducted by Handel-Mezzeti in 1910. However, although this survey was about flowering plants mainly around the west of Tigris, still the paper includes a list of identified algae in this area. In fact, this paper is the first published background information on algae in Mesopotamia and Kurdistan, nevertheless, the results of this survey were not published until 1942 (Kolbe and Krieger 1942).

Nevertheless, few decades later (Abdin et al. 1957) reported a list of algae that was found from and around Baghdad and the study was quite preliminary and most taxa have been identified till the genus level only. Whereas, (Al-Kaisi 1970, 1974, 1976) performed an algological study in the mid and south of Iraq as well as on the rice field of the southeastern part of Iraq. In fact, a contribution to the genus Cyclotella has been revealed by him. This paper followed his PhD thesis on phytoplankton in Iraq, which was confined to Habbaniyah and Razzazah lakes in 1964. In contrast, in the north of Iraq, Al-Hamed (1976) reported algal flora in the Dokan lake that contains the presence of only 23 taxa, 17 of which belong to Chlorophyta, whereas blue-green algae and Pyrophyta were represented by 4 and 2 taxa, respectively. However, the identification here also did not extend to the species level. Nevertheless, Hirano (1973) on his paper under the title of freshwater algae from Mesopotamia showed that diatoms were dominating algal flora representing more than 115 diatom species; however, few species of green algae were also included. His study was confined to the water systems along the border between Iran and Iraq only. Nevertheless, his study was concentrated on diatom flora and did not mention any other groups of algae that were performed other than a few green algae.

Al-Lami et al. (1998) recorded 55 phytoplankton in the lake during autumn 1992, these algae belonging to diatoms (65%), green (21.8%), and blue green (9.7%). A remarkable presence of some diatomic species was noticed such as Nitzschia (9 spp.), Cyclotella (6 spp.), and Navicula (6 spp.). Also, recorded spatial variation in phytoplankton quantities study, they reported that the algae Chlorella sp. encompasses the majority of the total number of phytoplankton at the northern part of the lake, while at the southern part of the *Cyclotella ocellata* was widely held of the total number of phytoplankton. Only two dinoflagellates were recorded in the northern part of the lake, and listed them as generic (*Ceratium* and *Glenodinium*).

Kassim et al. (2006) studied the distribution of phytoplankton in the water column of the lake during 1996–1997 at the mid of the lake; in this study, 116 taxa were identified in which 70 taxa belonged to diatoms (60 green, 30 tax, 25.8%), blue-green (12 taxa, 10.3%). Some species recorded a high number of species such as Nitzschia (species), Navicula (7spp.), Cymbella (species), and Cyclotella (species). Four species of diatoms were considered as the highest in cell number and these species are *Aulacoseira granulate*, *Cyclotella ocellata*, *Navicula cryptocephala,* and *Nitzschia palea*. The study revealed that the distribution of phytoplankton varied among different depths, and they found only *Arthrospria spiruliodes*, *Chroococcus minor*, *Pediastrum boryanum*, Amphora sp., *Gomphonies olivacea*, *Navicula* sp., and Rhopalodia gibba at the surface water. While some others are found only at 10 meters depth such as: *Tetraedron caudatum*, *Surirella biserata,* and Surirella sp.

Al-Ghafily and Al-Tamimi (2009) recorded in the study diurnal variation of phytoplankton in the lake during 2002, 20 taxa belonged to diatoms (species), blue-green (seven species), green (three species), and *Euglenophyceae* (one species). They found that four genera such as *Cyclotella*, *Navicula*, *Botryococcus,* and *Oscillatoria* were present at all study projects.

Algae in both the lentic water ecosystem as well as in the lotic ecosystem in Iraq that involve (impoundments, reservoirs, lakes, springs, marshes, ponds, dams as well as rivers, tributaries, water channels, river branches, drainage, and irrigation channels) have been considered during the present review nevertheless the review extent also to the algal flora of Shatt Al-Arab and the Gulf water also. All aspects of the algal study have been covered to a certain extend. Detailed information about all these aquatic ecosystems may be found in (Al-Kaisi 1964; Al-Hamed 1966, 1976; Maulood et al. 1979a, b; Pankow et al. 1979; Hinton and Maulood 1979; Hinton and Maulood 1979a; Hinton and Maulood 1979b; Shaban 1980; Maulood et al. 1980; Al-Saadi et al. 1981; Al-Saboonchi et al. 1982; Kassim 1986; Maulood and Al-Mousawi 1989; Saadulla 1989; Ismail 1989; Kassim and Al-Saadi 1994; Al-Barzingy 1995; Toma 2000; Al-Nakashabandi 2002; Kassim et al. 2006; Al-Saadi et al. 2008; Toma 2011a, b; Bilbas 2014; AL-Handal et al. 2014; Ahmad 2016), more details on the location, origin, surface area, the depth, salt content and, other characteristics of these lakes and reservoirs have been previously described by Rzoska (1980) and Maulood et al. (2013) Islam and Hameed (1985).

Al-Kaisi (1964) studied phytoplankton in Habbaniyah and Abu Dibbis lakes in the central part of Iraq. Nevertheless, he is the first who estimated phytoplankton composition and biomass of algae in Iraq. The seasonal variation of phytoplankton was also referred throughout his investigation. The periodicity and the dominancy of different algal groups were also considered throughout his dissertation. His study may be referred to as a pioneer comprehensive study on algae in Iraq that dealt with habitat, description, taxonomy, and ecology of more than 400 species of diatoms and non-diatom algal taxon.

Dominancy of diatoms among the standing crop of the two lakes was well documented throughout the study. They were represented mainly by Cyclotella and Melosira. The spring and autumn peaks of biomass have been illustrated throughout the research. Blooms of Cyanophyta have been recorded and referred to in the thesis. A sudden increase of blue-green algae in the area was noticed. He showed that blue-green algae are mainly represented by *Aphanizomenon* sp., *Microcystis* sp., *Merismopedia* sp., and Anabaena spp. The large number of Cyanophyta was more evident in summer whereas Dinophyceae were represented mainly by *Ceratium*.

The brief and preliminary study of Al-Hamed (1966) showed the spring (April) and autumn (October) peak of phytoplankton in the Dokan lake in the north of Iraq as well as two minima of phytoplankton in winter and in late spring without referring to the composition of algal groups, in fact 1976, he referred to the composition of plankton in the Dokan lake and referred to the presence of 23 algal taxa as follows

17, 4, 2 green algae, blue-green, and golden green, respectively, from the Dokan lake (Al-Hamed 1976).

Al-Kaisi (1974) reported the distribution of algae along the road between Basrah and Baghdad; moreover, he recorded 289 species of algae in the rice field of Mesopotamian marshes, a brief description and discussion of Cyclotella species have been forwarded (Al-Kaisi 1974 and 1976).

In fact, detailed and comprehensive studies on Iraqi inland water in respect to algae and phytoplankton did not start until establishing the university of Basra in the south and Sulaimaniyah in the north. Nevertheless, quite few comprehensive studies on phytoplankton in the Dokan Lake were performed in respect to the type of algae, its total number, productivity, chlorophyll content, and biomass (Shaban 1980; Toma 2000; Bilbas 2014). They reported the presence of more than 400 taxa of algae in the impoundment with dominancy of Cyclotella and Melosira, presence of Anabaena, Desmids, Dinobryon, and Scenedesmus were indicated. Whereas, Duhok impoundment has been studied in detail by Al-Nakashabandi (2002) he showed that the total number of phytoplankton is approaching 1,000,000 cells per liter with dominancy of diatoms and presence of dinoflagellate, blue-green algae, and Chlorophyta was also evident (Al-Nakashabandi 2002).

Running water that involves upper and Lower Zab, Khabour, Sirwan, Diyala as well as Tigris, and Euphrates beside the irrigation drainage channels in middle and south such as Saqualawiyiah, Yousefia, Hindiya, and Mosiab were been dealt with (Abdul-Jabbar 1981; Khamese 1979; Al-Nimma 1982; Al-Nimma and Maulood 1992; Al-Shahari 2006, and others) and the presence of large number of various types of algae representing almost all groups of known algae apart from brown algae, which also have been reported to be present in the marshes by al-Miah 2003.

The marshes, Shatt Al-Arab and its different channels have been dealt with by (Hinton and Maulood 1980, 1982; Al-Handal 1988; Kassim 1986, Maulood et al. 2013; Hassan et al. 2012a, b; Al-Hassany and Hassan 2014, 2015; Al-Gaff 2016), and many others. The diatom and non-diatoms flora of such an ecosystem were summarized in the recent check list of algae in Iraq (Maulood et al. 2013).

Hadi et al. (1984), Al-Mousawi et al. (1994), and Al-Saadi et al. (1996) are the main contributors of algal flora in such an ecosystem and the presence of *Thalassiosira* and *Sceletonema* as well as Fragilaria, Nitzschia, Synedra, and Cymbella was found to be the main components of algal flora of this ecosystem. Al-Handal et al. (2018) on their study documented 96 diatom taxa belonging to 33 genera, 18 of which were not previously known from the Arabian (Persian) Gulf including six taxa of Amphora and five of *Campylodiscus*. The most common genera were *Amphora* with 17 taxa, *Nitzschia* with 10 taxa, and Navicula with nine taxa. Most notably is the very rare occurrence of Mastogloia, which has been previously documented as species in coral reef habitats. Similarly, *araphid* and *monoraphid* species were uncommon in these diatom communities. The large species of the epipelic diatom Trachyneis are well represented and appeared commonly in all samples examined. Meanwhile, they also reported that the most common genera were Amphora with 17 species, *Nitzschia* with 10 species, and Navicula with nine species. However, none of these taxa were "common" in this flora. One of the characteristic diatom assemblages of coral reefs is the diversified occurrence of Mastogloia, which generally prefers warm water. This might be attributed to the unfavorable conditions prevailing in Basra corals, particularly the type of muddy substrate. Also, noteworthy is the occurrence of several species of the relatively large diatom Trachyneis. All the taxa of this genus were common in all samples examined. These species are considered as marine but have been previously found thriving in the brackish water of the Shatt Al-Arab estuary, Southern Iraq. Most of Trachyneis taxa are epipelic and the silty muddy sediment around Basra corals seem to provide favorable habitats for them.

More investigations will likely reveal greater specific diversity, and additional sampling events during different seasons will enhance our understanding of the diatom communities living on, around, and drifting into this remarkable reef.

In general, it can be easily pointed out that the main common macrophytes algae in Iraqi inland waters are Chara and Nitella, whereas Batrachospermum is also common in cold water. Whereas common micro flora of Iraqi inland water are in respect to diatoms Navicula, Diatoma, Cymbella, Cocconeis, Gomphonema, and Cyclotella, whereas the non-diatoms micro flora are Cosmarium, Scendesmus, Oscillatoria, Chroococcus, Vaucheria, Mougeotia, Spirogyra, Zygnema, Pediastrum, and Peridinium. Whereas, Peridinium, Ceratium, and Dinobryon may also be found in the polluted inland water.

24.4 Remarks and Conclusion

Similar to the flora of other inland water systems in different parts of the world, Iraqi inland water is also characterized with the dominancy of diatoms in general. Diatoms belonging to the central in standing water systems and acentric diatoms (Pennales) make up the main bulk of standing crop in running water systems in general. However, the blooms of blue-green algae have also been recorded, particularly Anabaena sp. inn impoundment and lakes during spring and summer. In contrast, Green algae (*Spirogyra* sp., *Zygnema* sp., *Cladophora* sp., *Stigeoclonium* sp., and *Desmids* flourish different ditches, ponds, and small water channels in summer. The presence of Dinoflagellate and blue-green algae in the polluted water system has been recorded. Red algae on the other hand was found in cold running water in the north.

Still knowledge and information about algae in Iraq was still so confusing that the need for reviewing phytoplankton or algal studies in Iraq, in general, becomes a real need as it has been declared recently by Al-Handal (2018) that an urgent need for the reestablishment of the names of known algae in Iraq, particularly diatoms, is the request. However, many published names of quite a few diatoms and dinoflagellate are confusing and need reconsideration in accordance to most recent taxonomical bases and classification. This is certainly true with respect to diatoms.

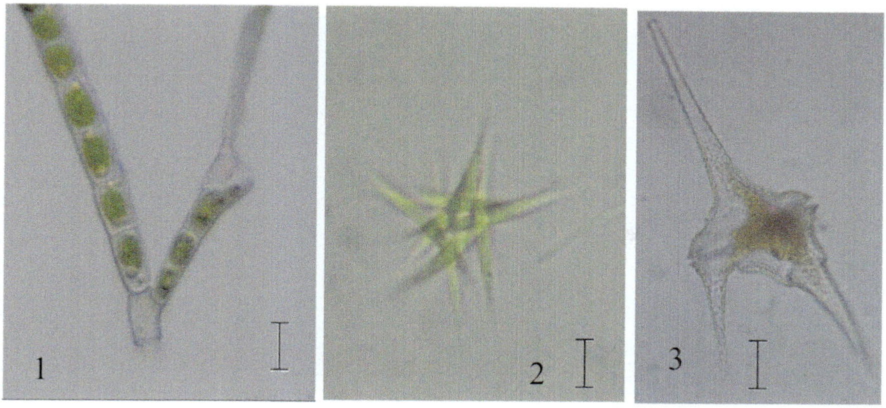

Plate 24.1 Lotic system. Each par= 10μm. (1) Bulbochaete sp. (2) Ankistrodesmus falcatus (Cord.) Ralfs. (3) Ceratium hirundinella (Müll.) Dujardin

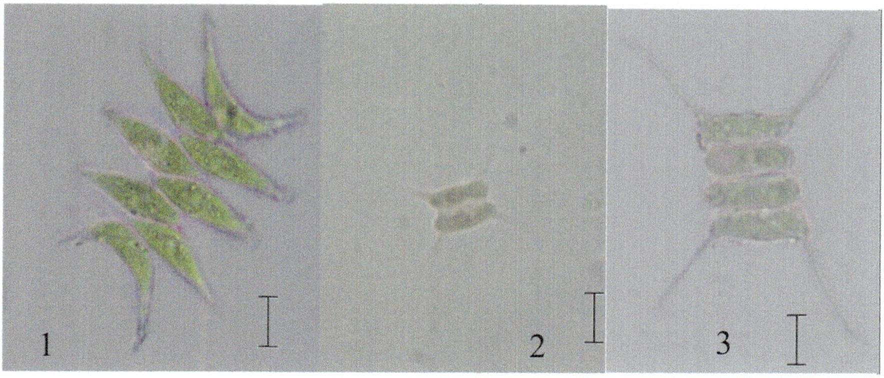

Plate 24.2 Each par= 10μm. (1) Scenedsmusacuminatus(Lag.) Chodat. (2) S. intermedius (Chodat.) (3) S. opoliensis (P.) Richter

Rechecking the names and identification is going to become a necessary goal and target of algologists in Iraq. As doubt is increasing on their identification and naming of algae, particularly diatoms, rechecking names and identification is going to become one of the aims of the present and future phycologists (local and others).

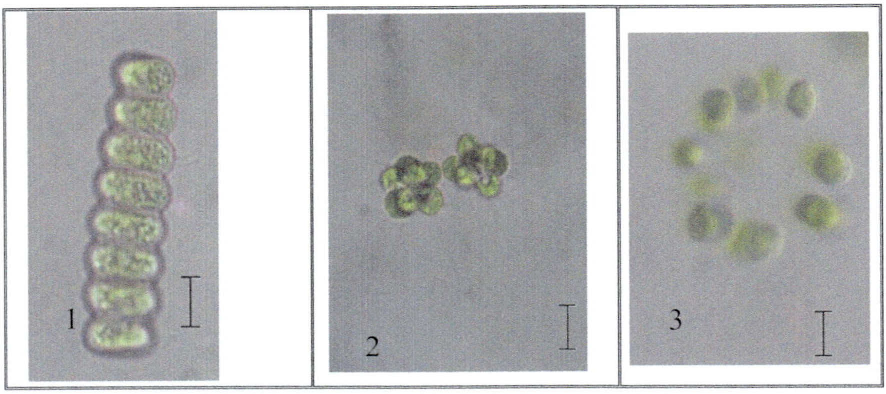

Plate 24.3 Each par= 10μm. (1) S. bijuga (Turp.) Lagerheim. (2) Micractinium pusillum Fresenius. (3)Dictyosphaerium ehrenbergianum Naegeli

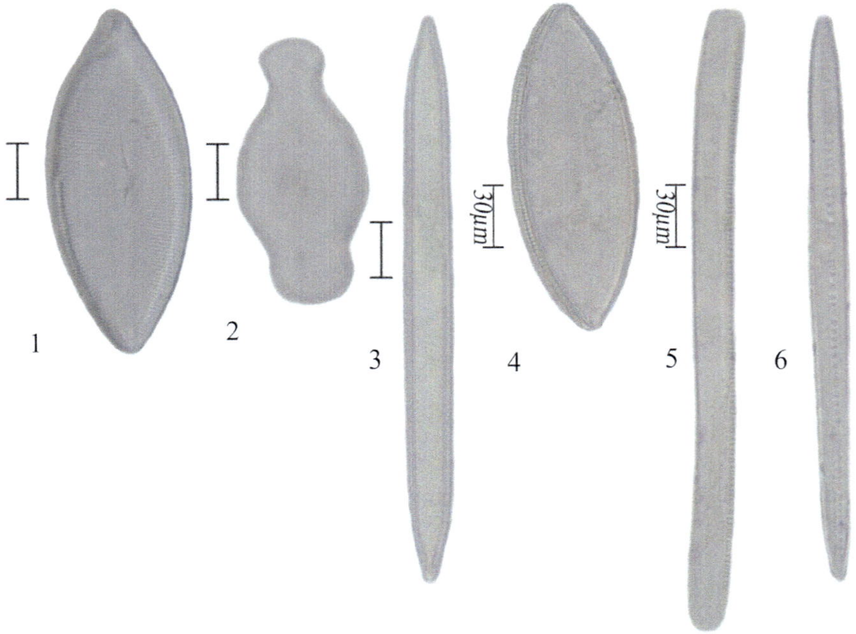

Plate 24.4 Each par= 10μm. (1) Calonies permagna Bailey. (2) Didymosphenia geminate (lyngb.) Schmidt. (3) Syndra ulna var. oxyrhynchus Kützing. (4) Nitzschia circumsuta (Bailey) Grunow. (5) Nitzschia sigmoidea (Ehr.) W.Smith. (6) Bacillaria paradoxa Gmelin

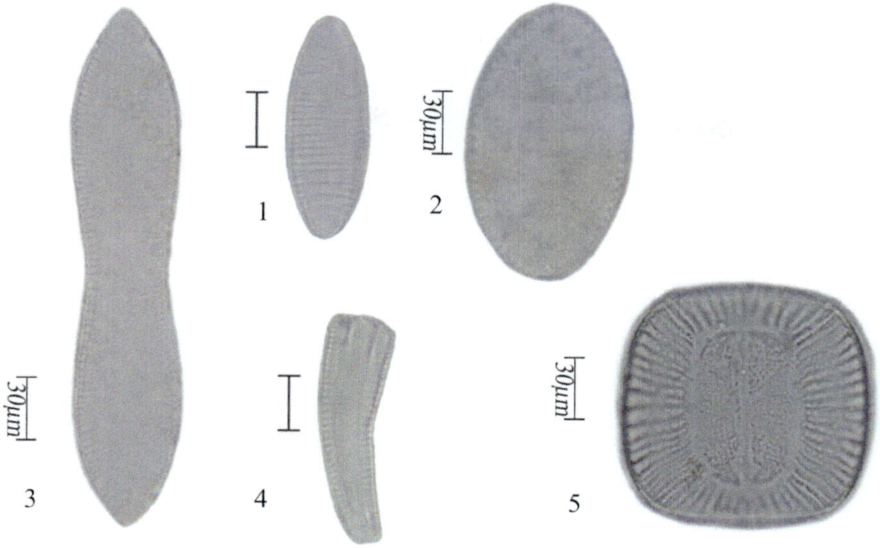

Plate 24.5 (1) Diatoma vulgare var. brevis Grunow. (2) Cymatopleura elliptica (Br?b.) Smith. (3) C. solea (Br?b) Smith. (4) Rhoicosphenia curvata (Kütz.) Grunow. (5) Campylodiscus clypeus Ehrenberg

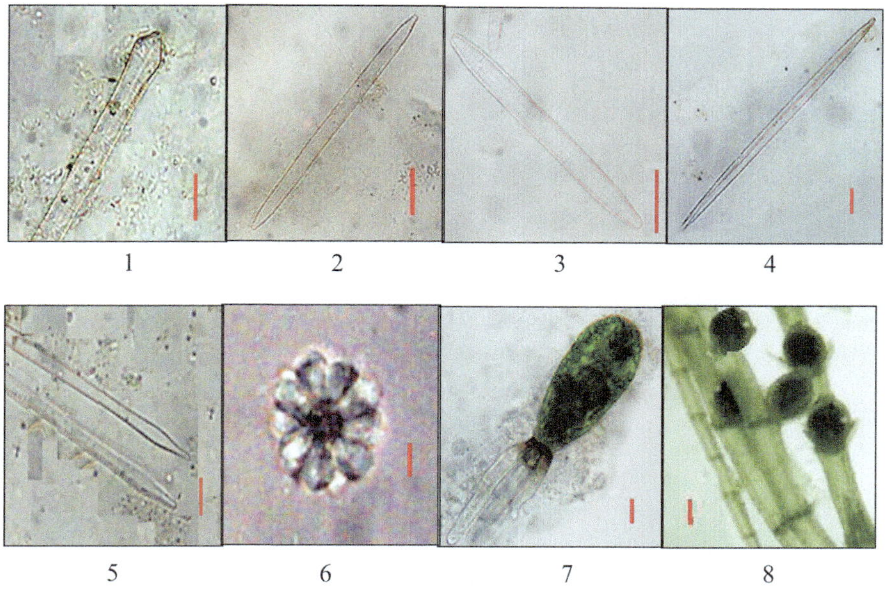

Plate 24.6 (lentic) Each par= 10μm. (1) Synedra capitata. (2) S. fasciculata. (3) S.pulchella. (4) S. ulna. (5) S. ulna var.balatonis. (6) Synura uvella. (7) Botrydium granulatum. (8) Chara sp.

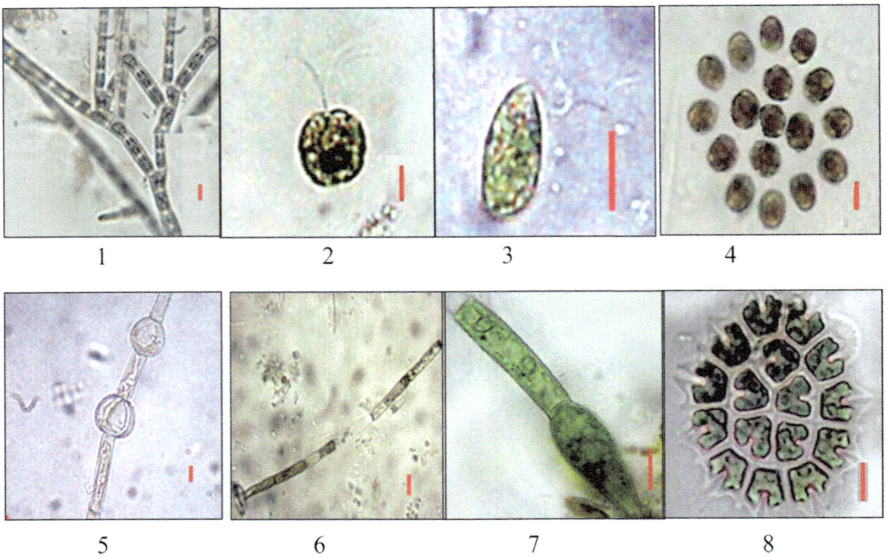

Plate 24.7 Each par= 10μm. (1) Stigeoclonium lubricum. (2) Chlamydomonas polypyreoideum. (3) C.snowiae. (4) Pandorina morum. (5) Oedogonium cardiacum. (6) O. gallicum. (7) O. plurisporum. (8) Pediastrum biradiatum var. emarginatum

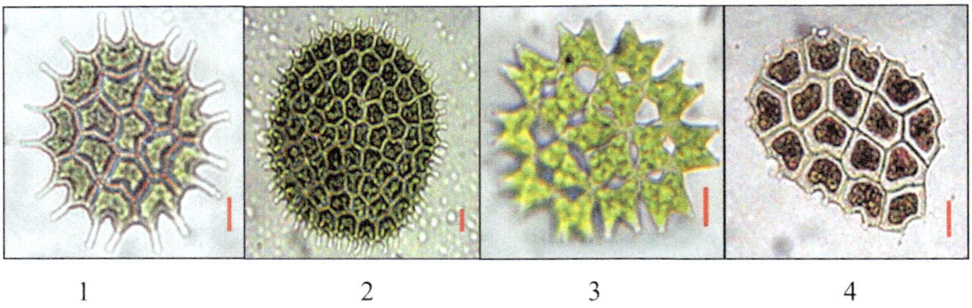

Plate 24.8 (1) P. boryanum. (2) P.boryanum var. undulatum. (3) P. duplex. (4) P. duplex var. rotundatum

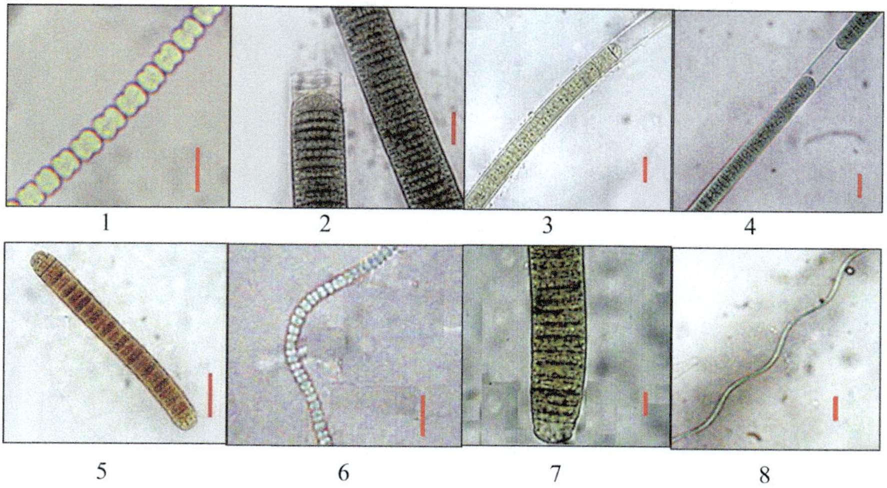

Plate 24.9 Each par= 10μm. (1) Nostoc linckia. (2) Lyngbya aestuarii. (3) L. martensiana. (4) L. taylorii . (5) Oscillatoria curviceps. (6) O. lacustris. (7) O. limosa. (8) O.minima

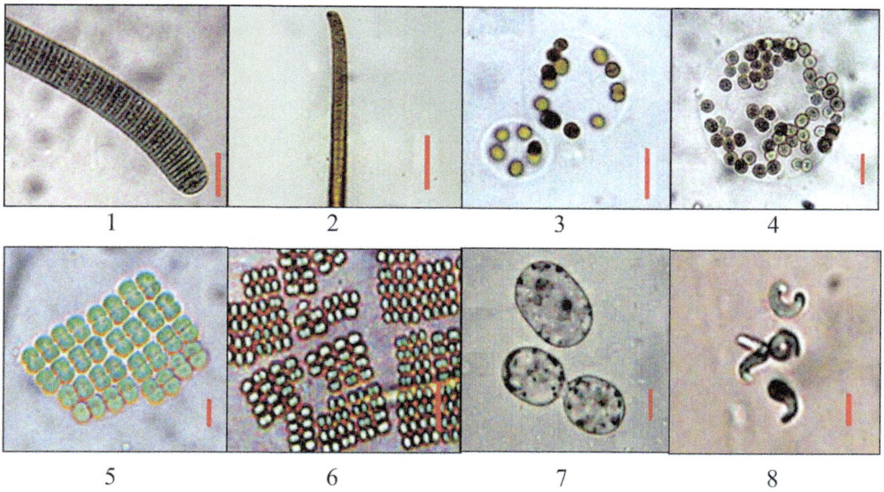

Plate 24.10 Each par= 10μm. (1)Oscillatoria sancta. (2) O.tenuis. (3) Aphanocapsa grevillei. (4) A.rivularis. (5) Merismopediaglauca. (6) M. punctata. (7) Synechococcus aeruginosus. (8) Rhabdo dermasigmoideavar.minor

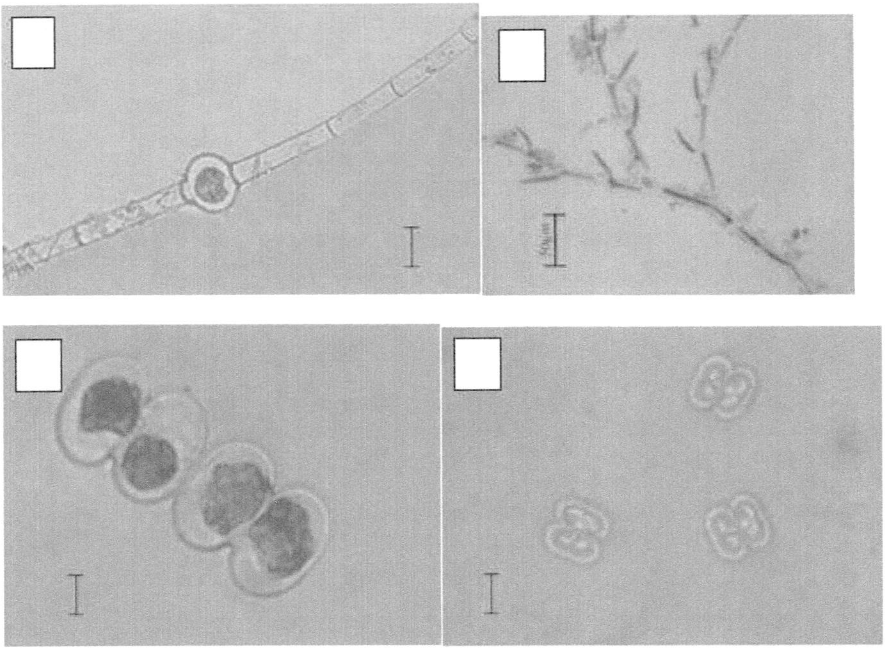

Plate 24.11 (Marsh) Each scale = 10 μm unless otherwise mentioned. (1) Oedogonium sociale. (2) Cladophora glomerata. (3) Cosmarium subtumidium. (4) Cosmocladium sp.

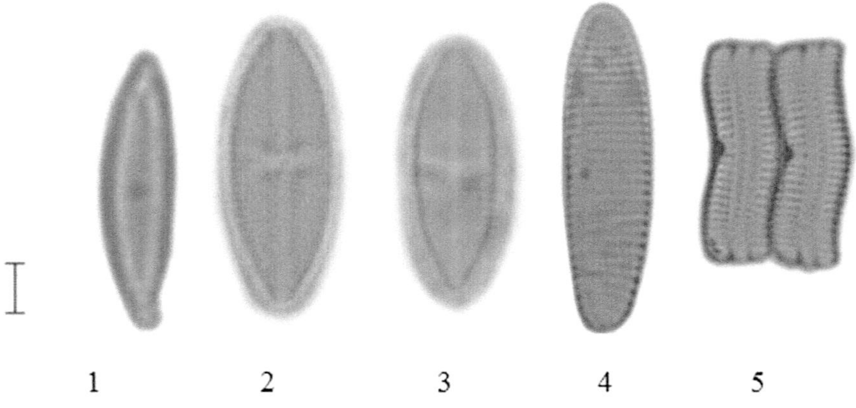

Plate 24.12 Each scale = 10 μm unless otherwise mentioned. Figures 1 and (2) Achnanthes brevipesvar intermedia. Figures 3 and (4) Achnanthes hungarica. (5) Achnanthes microcephala

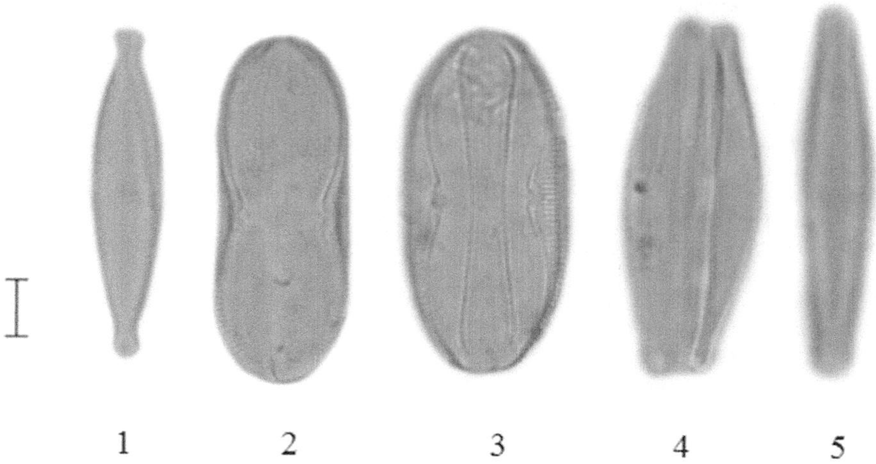

Plate 24.13 Each scale = 10 μm unless otherwise mentioned. (1) Achnanthes minutissima. (2) Amphora coffeaeformis. (3) Amphora commutate. (4) Amphora mavicanavarmajor. (5) Anomoeoneis exilis

Acknowledgements Many thanks to Hawler Botanical Garden and Assistant professor Janan Jabbar Toma M.Sc. in Phycology for assisting and helping in the preparation of the many scripts.

References

Abdin G, Al-Kaisi K, Naib F (1957) Some observations on the algal flora in and around Baghdad. Bull Coll Arts Sci Baghdad 2:21–43

Abdul-Jabbar RA (1981) Ecological study on lesser Zab.M.Sc. Thesis, University of Sulaimaniyah, Iraq

Ahmad AQ (2016) Phycolimnological study on some springs around Safeen Mountain area. M.Sc thesis. University of Salahaddin. College of Sciences, Erbil, Iraq

Al-Barzingy YOM (1995) Phycological study within Arbil province. M.Sc. Thesis. Univ. of Salahaddin-Arbil

AL-Ghafily AA, AL-Tamimi AA (2009) Diurnal variation of phytoplankton and related ecological parameters of one location in Habbaniya Lake, Iraq. Ibn Al-Haitham J Pure Appl Sci 22 (2):72–82

Al-Gaff HAA (2016) Impact of Hadithah dam and some pollution sources on phytoplankton community composition and its environment in northern part of Euphrates River, Al-Anbar Province. Ph.D thesis, College of Science, University of Al-Anbar, Iraq

Al-Hamed MI (1966) Limnological studies on the inland water of Iraq. Bull Iraqi Nat Hist Mus 3 (5):1–22

Al-Hamed MI (1976) Limnological investigation on the Dokan lake of Iraq. Bull Iraqi Nat Hist Mus 3(7):34–50

Al-Handal AY (1988) New addition to the algal flora of Iraq. J University Kuwait Sci 18:97–103

Al-Handal AY, Abdulla DS (2010) Diatoms from the restored Mesopotamian marshes of south of Iraq. Algol Stud 133:65–103

Al-Handal AY, Al-Shaheen M (2019) Diatoms in the wetlands of Southern Iraq. Bibil Diatomol 67, p 252
Al-Handal AY, Al-Assa AS, Al-Mukhtar MA (1989) Occurrence of some filamentous algae in the river Shatt al-Arab, Iraq. Marine Mesopotamica J Mar Sci Cent Univ Basrah 4(1):67–81
AL-Handal AY, Gobin CR, Abdulla D, Ali MH (2014) Cocconeis sawensis sp. nov. (Bacillariophyceae) from a saline lake (Sawa Lake), South Iraq: comparison with allied taxa. J Phytotaxa 181(4):216–228. Magnolia Press
Al-Handal AY, Thomas EW, Pennesi C (2018) Marine benthic diatoms in the newly discovered coral reef, off Basra coast, southern Iraq. J Phytotaxa 372(2):111–152
Al-Hassany JS, Hassan FM (2014) Taxonomic study of epiphytic diatoms on aquatic plants from Al-Hawizah marshes southern of Iraq. Asian J Natural Appl Sci 3(1):1–11
Al-Hassany JS, Hassan FM (2015) Descriptive study of some epiphytic algae (non diatoms) after restoration of Mesopotamian marshes southern of Iraq. Mesop Environ J 1(2):96–108
Al-Kaisi KA (1964) Studies on algae of a water system in Iraq. Ph.D. thesis. Univ. Collage of North WalesBagor, UK, p 182
Al-Kaisi KA (1970) Introductory study on the algae of mid and South Iraq. Bull Sci, Baghdad 11:45–80
Al-Kaisi KA (1974) The genus Cyclotella Ktz. From some aquatic habitats in Iraq. Bull Sci Baghdad 15:21–40
Al-Kaisi KA (1976) Contribution to the algal flora of rice fields of southeastern Iraq. Nova Hedwigia 27:813–827
Al-Lami AA (1986) Ecological study on phytoplankton on some marsh regions southern Iraq. M. Sc. Thesis, College of Science, University of Basrah
Al-Lami AA, Al-Saadi HA, Kassim TI (1998) Limnological features of Qadisia Lake, north-West Iraq. Al Mustansiryia J Sci 9(2):59–66
Al-Mayah AA, Al-Hilli MR, Hassan FM (2014) Marsh Flora of southern Iraq Befor Desication. Al-baser Company. S.a.R.I Printing and publishing, p 195
Al-Mousawi AH, Al-Saadi HA, Hassan FM (1994) Spatial and seasonal variation of phytoplankton population and related environment in Al-Hammar marsh. Iraq Basra J Sci 12:10–20
Al-Nakashabandi IYR (2002).A phycolimnological study on Duhok impoundment and its main watershed. Ph.D thesis, University of Dohuk
Al-Nimma BA (1982) A studies on the limnology of the Tigris and Euphrates river. M.Sc. Thesis University of Salashaddin, Erbil, Iraq
Al-Nimma BA, Maulood BK (1992) A study on the distribution of algae in some thermal springs in Nineveh province. Iraq J Ibn Al-Haitham 3:33–43
Al-Saadi HA, Antoine SE, Naural Islam AKM (1981) Limnological investigations in Al-Hammar marsh in southern Iraq. Nova Hedwigia 35:157–166
Al-Saadi HA, Al-Lami AA, Kassim TI (1996) Algal ecology and composition in the Garmat-Ali River, Iraq. J Res Manage 12:27–38
Al-Saadi HA, Hassan FM, Alkam FM (2008) Phytoplankton and related nutrient to Sawa Lake, Iraq. J Dohuk Univ 11(1):67–76
Al-Saboonchi A, Mohamed ARM, Alobaidy AHMJ, Abid HS, Maulood BK (2011) On the current and restoration conditions of the southern Iraqi marshes: application of the CCME WQI on East Hammar marsh. J Environ Prot 2(3):316–322
Al-Saboonchi AA, Mohammed ARM, Burak NA (1982) A study of phytoplankton in the Garma marshes. Iraq Iraqi of Marine Sci 1(1):67–78
Al-Shahari YJI(2006) A Phycological study on some water system within Mosul city. Ph.D. Thesis, Mosul University, Mosul, Iraq
Bilbas AHA (2014) Ecosystem health assessment of Dokan Lake, Kurdistan region of Iraq. Ph.D thesis University of Salahaddin, Erbil, Iraq
Boney AD (1975) Phytoplankton. University Press, Glasgow
Hadi RAM, Azhar A-S, Yousif Haroon AK (1984) Diatoms of the Shatt Al-Arab River, Iraq. Nova Hedwigia 39:517–557

Hassan FM, Toma JJ, Ismail AM, Al-Hassany JS, Hadi RMA, Maulood BK (2012a) A contribution to algal flora in Baghdad area, Iraq. J Adv Lab Res Biol 3(2):77–83
Hassan FM, Hadi RA, Kassim TI, Al-Hassany JS (2012b) Systematic study of epiphytic algal after restoration of Al-Hawizah marshes, southern of Iraq. Int J Aquatic Science 3(1):37–57
Hinton GCF, Maulood BK (1979) Some diatoms from brackish water habitat in southern Iraq. Nova Hedwigia
Hinton GCF, Maulood BK (1979a) Some diatoms from brackish water habitat in southern Iraq. Nova Hedwigia
Hinton GCF, Maulood BK (1979b) Fresh water diatoms from Sulaimaniyah, Iraq. Nova Hedwigia Band 31(1+2):449–466
Hinton GCF, Maulood BK (1980) Some diatoms from brackish water habitate in southern Iraq. Nova Hedwigia 31:475–486
Hinton GCF, Maulood BK (1982) Contribution to algal flora of Iraq, the non-diatom flora of the southern marshes. Nova Hedwigia 37:49–63
Hinton GCF, Maulood BK (1983) Check list of the algae from inland water in Iraq. J Univ of Kuwait 10:191–265
Hirano M (1973) Freshwater algae from Mesopotamia. Center Biol Lab Kyoto Univ 24:105–119
Islam AAKM (1982) Marsh algae from southern Iraq. Int Revue Ges Hydrobiol 67:245–260
Islam AAKM, Hameed HA(1985) Check list of algae with a note on the limnological and oceanographic studies in Iraq (1942–1982). Asiatic Society of Bangladesh, Dhoka-2, Bangladesh, p 63
Ismail AM(1989) An ecological comparative study between the lake of Baghdad tourist island and river Tigris in Baghdad. M.Sc. Thesis Baghdad University
Kassim TI (1986) An ecological study on the benthic algae in some marsh area southern Iraq. M.Sc. Thesis. Univ. of Basrah
Kassim TI, Al-Saadi HA (1994) On the seasonal variation of the epipelic algae in marsh area (southern Iraq). Acta Hydrobiologia 36(2):191–200
Kassim T, Al-Saadi H, Farhan R (2006) Vertical distribution of phytoplankton in Habbaniyah lake, Iraq. Marsh Bulletin 1(1):19–31
Khamese HS (1979) An ecological study on water pollution in Tangero valley. M.Sc. Thesis University of Sulaimaniyah, Iraq
Kolbe RW, Krieger W (1942) Swsswasserlagen aus Mesopotamian and Kurdistan. Ber Ges 60:336–355
Maulood BK, Al-Mousawi AHA (1989) Limnological investigation on Sawa lake. Iraq J Agric Sci 2(1,2):113–122
Maulood BK, Toma JJ (2004) Check list of algae in Iraq. Sci J Babylon Univ 9(3):1–71
Maulood BK, Hinton GCH, Kamees HS, Saleh FAK, Shaban AA, Shahwani SMH (1979a) An ecological survey of some aquatic ecosystem in southern Iraq. J Trop Ecol 20(1):27–40
Maulood BK, Hinton GCF, Kamees HS, Saleh FAK, Shaban AA, Al-Shahwani SMH (1979b) An ecological study of some aquatic ecosystem in southern Iraq. Trop Ecol 20(1):27–40
Maulood BK, Hinton GCF, Al-Dosky HS (1980) A study on the blue green algae of Arbil province, Iraq. Zanco (Sci J. Univ. Sulaimaniyah, Iraq). Series A6(2):67–90
Maulood BK, Hinton GCF, Whitton BA, Al-Saadi HA (1981) On the algal ecology of the lowland Iraqi marshes. J Hydrobiologia 80:269–276
Maulood BK, Hadi RMA, Saadalla HAA, Kassim TI, Al-Lami AA (1993) Check-List of algae in Iraq. Marina Mesopotamia Suppl. (1):1–128
Maulood BK, Hassan FM, Al-Lami AA, Toma JJ, Ismail AM (2013) Check list of algal Flora in Iraq. Ministry of Environment of Iraq, Baghdad
McCann KS (2000) The diversity stability debate. Nature 405(6783):228–233. Pmid:10821283. https://doi.org/10.1038/35012234
Nabout JC, Nogueira IS, Oliveira LG, Morais RR (2007) Phytoplankton diversity (alpha, beta, and gamma) from the Araguaia River tropical floodplain lakes (Central Brazil). Hydrobiologia 575 (1):455–461. https://doi.org/10.1007/s10750-006-0393-8

Pankow H, Al-Saadi HA, Hug MF, Hadi RAM (1979) On the algal flora of marshes near Quarna (southern Iraq). Willedenowia 8(3):493–506
Round AB (1975) The biology of algae, 2nd edn. Cambridge University Press, Cambridge
Rzoska J (1980) Euphrates and Tigris, Mesopotamian ecology and destiny. In: Junk W (ed) Monographiae Biologica, vol 38. The Hague-Boston, London, p 109
Saadulla HAA (1989) Ecological study on the effect of Saqlawiya irrigation channel on Tigris River within Baghdad. M.Sc. Thesis. Baghdad University
Saadulla HAA, Maulood BK (1993) Limnological study on Al-Khair canal, Baghdad with special references to algal flora 4(1):20–32
Shaban AAG (1980) An ecological study on phytoplankton in Dokan lake. M.Sc. Thesis, Univ. of Sulaimaniyah
Stomp M, Huisman J, Mittelbach GG, Litchman E, Klausmeier C (2011) Large-scale biodiversity patterns in freshwater phytoplankton. Ecology 92(11):2096–2107. Pmid: 22164834. https://doi.org/10.1890/10-1023.1
Toma JJ (2000) Limnological study of Dokan lake, Kurdistan region of Iraq. M.Sc. Thesis, Erbil, Iraq
Toma JJ (2011a) Limnological study of Dokan lake, Kurdistan region of Iraq. J Environ Sci 6:1–12
Toma JJ (2011b) Physical and chemical and algal composition of Derbendikhan lake. Iraq J Curr World Environ 6(1):17–27
Toma JJ, Shekha YA, Al-Barzingy YO (2018) An ecological assessment for water quality of some water bodies in Koysenjaq-Erbil. Iraq J Al-Nahrain Univ 21(2):119–129

Chapter 25
The Distribution of Epilitic Diatoms in the Turkish Part of the Tigris-Euphrates River Basin

Cüneyt N. Solak, Agata Wojtal, Elif Yılmaz, M. Borga Ergönül, and A. Kadri Çetin

Abstract Tigris and Euphrates rivers are transboundary rivers originated in Southeastern Anatolia region of Turkey and flow into Persian Gulf. These two river systems are one of the major drainage basins in southwestern Asia. Benthic diatoms are well environmental indicators in biomonitoring studies because of reacting rapidly to disturbance of water, e.g., physicochemical conditions of water or to pollution-affected catchment area. The aim of the present study is to reveal the distribution of diatoms in Tigris-Euphrates river catchment. For this purpose, diatom samples, which were collected from 20 different sampling locations in 2014, were investigated in the study. As a result, a total of 153 epilitic diatoms were identified and 32 of them were new record for Turksh Freshwater Diatom Flora.

Keywords Diatom · Euphrates · New records · Tigris · Turkey

25.1 Introduction

Tigris and Euphrates rivers are transboundary rivers originated in Southeastern Anatolia region of Turkey and flow into Persian Gulf. These two river systems are one of the major drainage basins in southwestern Asia (Hassan et al. 2010). The

C. N. Solak (✉) · E. Yılmaz
Dumlupınar University, Science & Art Faculty, Department of Biology, Kütahya, Turkey

A. Wojtal
Institute of Nature Conservation of the Polish Academy of Sciences, Kraków, Poland
e-mail: wojtal@iop.krakow.pl

M. B. Ergönül
Ankara Universit, Faculty of Science, Department of Biologyy, Ankara, Turkey

A. K. Çetin
Fırat University, Faculty of Science, Department of Biology, Elazığ, Turkey

L. A. Jawad (ed.), *Tigris and Euphrates Rivers: Their Environment from Headwaters to Mouth*, Aquatic Ecology Series 11,
https://doi.org/10.1007/978-3-030-57570-0_25

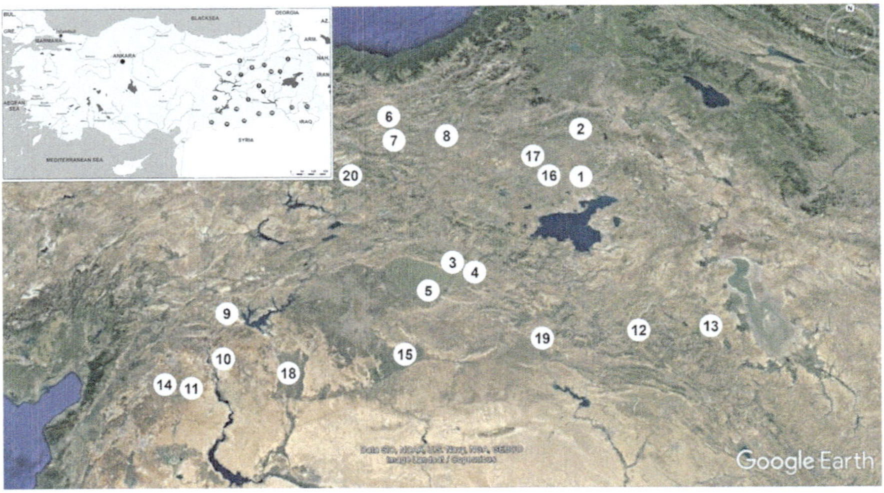

Fig. 25.1 Map showing the sampling locations in the Turkish part of the Tigris-Euphrates River Basin

Euphrates (Fırat—catchment area: 127304 km^2) and Tigris River (Dicle—catchment area of 57,614 km^2) are the largest basin in Turkey. Euphrates River begins in the highlands northeast of the Anti-Taurid Mountains (>3000 m), and the river forms a large 'C' before entering Syria. The Tigris River (Dicle), draining southeastern Anatolia, is the other largest river, the basin drains a mountainous area (>3000 m asl), and the river enters Iraq at ~300 m asl. Karasu and Murat Rivers are the main branches of Euphrates River, while the Batman, Garzan, Botan, and Hezil Rivers are the main branches of Tigris River in Anatolia. The rivers flow into the Shatt al-Arab in southern Iraq before entering the Persian Gulf (Akbulut et al. 2009a) (Fig. 25.1).

Benthic diatoms are well environmental indicators in biomonitoring studies because of reacting rapidly to disturbance of water, e.g., physicochemical conditions of water or to pollution-affected catchment area. So, the organisms are recommended by the Water Framework Directives (Water Framework Directive, European Parliament 2000 directive, 2000/60/EC) (Ács et al. 2004). Although there are partial data on the diatom flora of some rivers and lakes located in the basin (Altuner and Gürdüz 1991, 1996; Şen et al. 2001 Çetin and Yıldırım 2007; Pala-Toprak 2007; Pala-Toprak and Çaglar 2008; Yılmaz and Çağlar 2011; Koçer and Şen 2012), the diatom flora of several water bodies in the basin is still lacking. Thus, the aim of the present study is to reveal the distribution of diatoms in Tigris-Euphrates river catchment. For this purpose, diatom samples, which were collected from 20 different sampling locations in 2014, were investigated.

25.2 Materials and Methods

25.2.1 *Sample Collection*

The name and the map of the sampling locations are shown in Table 25.1 and Fig. 25.1. The samples were collected by brushing the submerged stones in the water bodies. The samples were preserved in ethanol.

25.2.2 *Sample Processing, Observation, and Identification*

Samples were boiled in H_2O_2 and HCl to remove the organic matter from frustules. After washing three times in distilled water, the material was air-dried on cover glasses and mounted in Naphrax solution. Diatoms were observed using a Nikon Ci Light Microscope (LM). The diatoms were identified according to Hofmann et al. (2011); Krammer and Lange-Bertalot (1998, 1991); Lange-Bertalot (2001); Levkov (2009); Wojtal (2009); and Wojtal and Kwandrans (2006). The dimensions (length, width, and number of striae/10 μm) of each taxon were based on measurements done in this study. The distribution of the epilitic diatoms recorded was evaluated according to Solak and Wojtal (2012), Solak et al. (2012, 2017), and Gönülol (2017).

25.2.3 *Diatom Abundance*

Approximately, 400 valves were counted for each sample. Description of the abundance of diatoms was categorized as 5 levels. If the abundance value was less than 1%, it was signed as "+"; if the value was between 1 and 5%, it was signed as "+ +"; if the value was between 5 and 15%, it was signed as "+++"; if the value was between 15 and 50%, it was signed as "o"; and if the value was higher than 50%, it was signed as "oo".

25.3 Results

A total of 153 epilitic diatoms were recorded from 20 different sampling locations in the Turkish part of the Tigris-Euphrates river basin (Table 25.2). Among the species found, *Amphora alpestris* Levkov, *A. copulata* (Kützing) Schoeman and Archibald, *Amphora liriope* Nagumo, *A. lange-bertalotii* var. *tenuis* Levkov and Metzeltin, *Brachysira neglectiformis* Lange-Bertalot, *Cymbella excisa* var. *procera* Krammer, *C. excisa* var. *subcapitata* Krammer, *C. exigua* Krammer, *C. peraffinis* Tynni, *Cymbopleura lange-bertalotii* Krammer, *C. vrana* Krammer, *Encyonopsis*

Table 25.1 Sampling locations in the Turkish part of the Tigris-Euphrates river basin

Station	River name	Province	District	Coordinates	
TE1	Karasu Stream	AĞRI	Patnos	42° 59′ 43.000″ E	39° 08′ 25.000″ N
TE2	Murat River	AĞRI	Merkez	43° 03′ 34.000″ E	39° 42′ 35.000″ N
TE3	Batman River	BATMAN	Merkez	41° 10′ 31.000″ E	38° 02′ 05.000″ N
TE4	Garzan stream	BATMAN	Beşiri	41° 20′ 45.410″ E	37° 57′ 49.620″ N
TE5	Dicle River	DİYARBAKIR	Bismil	40° 39′ 34.000″ E	37° 50′ 32.000″ N
TE6	Karasu River	ERZİNCAN	Çayırlı	40° 08′ 01.000″ E	39° 50′ 04.000″ N
TE7	Tuzla stream	ERZİNCAN	Tercan	40° 15′ 52.000″ E	39° 43′ 51.000″ N
TE8	Gördelli stream	ERZURUM	Çat	41° 00′ 33.000″ E	39° 40′ 39.000″ N
TE9	Karasu River	GAZİANTEP	Araban	37° 53′ 38.000″ E	37° 26′ 02.000″ N
TE10	Bozalioğlu stream	GAZİANTEP	Kargamış	37° 59′ 26.000″ E	36° 56′ 57.000″ N
TE11	Sacırsuyu stream	GAZİANTEP	Oğuzeli	37° 40′ 28.000″ E	36° 45′ 02.000″ N
TE12	Çığlı (zap) stream	HAKKÂRİ	Çukurca	43° 35′ 05.000″ E	37° 16′ 15.000″ N
TE13	Şemdinli stream	HAKKÂRİ	Şemdinli	44° 34′ 12.000″ E	37° 17′ 54.000″ N
TE14	Sinnep stream	KİLİS	Merkez	37° 14′ 01.000″ E	36° 43′ 05.000″ N
TE15	Altınsuyu stream	MARDİN	Kızıltepe	40° 25′ 38.000″ E	37° 02′ 05.000″ N
TE16	Ağbu stream	MUŞ	Malazgirt	42° 30′ 39.000″ E	39° 09′ 10.000″ N
TE17	Kocasu River	MUŞ	Malazgirt	42° 15′ 50.000″ E	39° 22′ 09.000″ N
TE18	Hut stream	SANLIURFA	Akçakale	38° 54′ 52.000″ E	36° 49′ 48.000″ N
TE19	Yukarısaksan stream	ŞIRNAK	Silopi	42° 20′ 56.000″ E	37° 12′ 59.000″ N
TE20	Munzur River	TUNCELİ	Merkez	39° 32′ 24.000″ E	39° 06′ 34.000″ N

falaisensis (Grunow) Krammer, *E. krammeri* Reichardt, *E. thumensis* Krammer, *Gomphonema aequale* Gregory, *G. exilissimum* (Grunow) Lange-Bertalot and E. Reichardt, *G. jadwigiae* Lange-Bertalot and Reichardt, *Gyrosigma kuetzingii* (Grunow) Cleve, *Luticola hlubikovae* Levkov, Metzeltin and Pavlov, *Navicula exilis* Kützing, *N. lacuum* Lange-Bertalot et al., *N. lundii* Reichardt, *N. simulata* Manguin,

Table 25.2 The distribution and % abundance of diatoms in sampling stations

	TE 1	TE 2	TE 3	TE 4	TE5	TE 6	TE 7	TE 8	TE 9	TE 10	TE 11	TE 12	TE 13	TE 14	TE 15	TE 16	TE 17	TE 18	TE 19	TE 20
Achnanthidium biasolettianum			++	++	o	+++														
Amphora alpestris																	+			
Amphora copulata		+				++						++								
Amphora liriope		+																		
Amphora ovalis						+						+								
Amphora pediculus	+	+++	++	++		++	+++	+++	++			+		++	++	+	+	++	+	
Amphora lange-bertalotii var. *tenuis*																		+		
Aulacoseira granulata																				++
Brachysira neglectiformis		+																		
Cocconeis pediculus	++	+++	+	+				++			+	++	+	++	+	+				
Caloneis amphisbaena																	++			
Caloneis macedonica																				+
Craticula ambigua										+	+									
Craticula accomoda								+		++										+
Craticula cuspidata											+++									
Craticula minusculoides										+++										
Craticula molestiformis										+										
Craticula subminuscula			+				++			o				+						

(continued)

Table 25.2 (continued)

	TE 1	TE 2	TE 3	TE 4	TE5	TE 6	TE 7	TE 8	TE 9	TE 10	TE 11	TE 12	TE 13	TE 14	TE 15	TE 16	TE 17	TE 18	TE 19	TE 20
Cyclostephanos invisitatus														+						
Cyclotella meneghiniana				oo							oo			o			+			
Cymatopleura elliptica																		+		
Cymatopleura solea																	+			
Cymatopleura solea var. apiculata																				+
Cymbella compacta															++	++				
Cymbella excisa	o	++	oo	++ +	o	oo			o		+	oo	oo	++ +	++	++	+	o	++ +	++ +
Cymbella excisa var. *procera*	++ +	+							++									++		
Cymbella excisa var. *Subcapitata*			+																	
Cymbella exigua			+																	
Cymbella lange-bertalotii	++ +		++									++			++ +	++ +		++		
Cymbella parva												+								
Cymbella peraffinis																	+			
Cymbella tumida															++ +					
Cymbopleura amphicephala	++ +				+	+						+								++
Cymbopleura lange-bertalotii						+														
Cymbopleura vrana					+															

Diatoma moniliformis	++ +		++ +		+++										++					++
Diatoma vulgaris											+			+	++	oo	+	+	o	
Didymosphenia geminata																		+		
Diploneis separanda											+									
Discostella stelligera var. *tenuis*														+						
Encyonema auerswaldii												+								+
Encyonema minutum												+			++	++		++ +		++ +
Encyonema reichardtii																			+	
Encyonema silesiacum	+														++			++ +	++	+
Encyonema ventricosum		++ +	+												++	++				
Encyonopsis cesatii																			+	
Encyonopsis falaisensis		+																		
Encyonopsis krammeri		+																		
Encyonopsis microcephala					+							+							++	
Encyonopsis minuta	+	+		++	o	++ +			++			++ +	++	++ +						
Encyonopsis subminuta			++		+++															
Encyonopsis thumensis						+														
Eunotia pectinalis																			+	
Fallacia lenzii					+															
Fallacia pygmaea											++									
Fistulifera saprophila		+					+	+		++										

(continued)

Table 25.2 (continued)

	TE 1	TE 2	TE 3	TE 4	TE5	TE 6	TE 7	TE 8	TE 9	TE 10	TE 11	TE 12	TE 13	TE 14	TE 15	TE 16	TE 17	TE 18	TE 19	TE 20
Fragilaria austriaca			++ +																	
Gomphonema aequale								+												
Gomphonema exilissimum		++							+											++
Gomphonema jadwigiae		+																		
Gomphonema micropus		+																		
Gomphonema olivaceum	++		++ +									+				+				+
Gomphonema parvulum		++					+	++	+	++ +	++		+							
Gomphonema pumilum		++							++ +				+					++		
Gomphonema tergestinum			+											+						
Grunowia tabellaria				+								++						+		
Gyrosigma kuetzingii				+																++
Halamphora montana																			+	
Hannaea arcus															++	+		++ +		
Hippodonta capitata																				+
Lindavia balatonis																				++
Luticola hlubikovae										+										
Luticola goeppertiana							o				+						+			
Mayamaea permitis		+					++ +													

Melosira varians				++										++ +		++	o			++ +
Navicula antonii																		++		
Navicula capitatoradiata	+	++ +		++	+++				++			+	++ +	++ +	o	+	+	++ +	oo	
Navicula cincta																	+			
Navicula cryptotenella	++	+	+		++			o					++ +	+	+					+
Navicula cryptotenelloides					++		+	++ +			+		++							
Navicula erifuga								++					++							
Navicula exilis																	+			
Navicula gregaria	+	++			+												o		++	
Navicula kotschyi					+															+
Navucla lacuum																	+			
Navicula lanceolata															+					
Navicula libonensis																	+			
Navicula lundii							+													
Navicula moskalii					+	+												++ +		
Navicula novaesiberica												++				++		+		
Navicula oblonga																	+			
Navicula radiosa		+							+								+	+		
Navicula reichardtiana	++	++ +	++						+											
Navicula rostellata									+				++							+
Navicula simulata								++ +					++	+						

(continued)

Table 25.2 (continued)

	TE 1	TE 2	TE 3	TE 4	TE5	TE 6	TE 7	TE 8	TE 9	TE 10	TE 11	TE 12	TE 13	TE 14	TE 15	TE 16	TE 17	TE 18	TE 19	TE 20
Navicula tripunctata	++	+		++			+	o	++		++			++		++ +		++	++ +	+
Navicula trivialis					+						++									
Navicula vandamii		+																	+	
Navicula vandamii var. *mertens*		+																		
Navicula veneta		+		+											+		o			++
Navicula wildii																	+			
Navigeia decussis						+												++		
Neidiopsis binodiforme																		++		
Nitzschia acicularis																	+			
Nitzschia amphibia				++		++ +	++ +	+			++	+								
Nitzschia capitellata		++			+			+									+			o
Nitzschia clausii														++						
Nitzschia communis																	++			
Nitzschia denticula																		++		
Nitzschia dissipata	++						++								o	++			++	+
Nitzschia dissipata var. media	++	+	++	++	++		++	++					++		++ +				++	+
Nitzschia dissipata ssp. oligotrophenta					+															
Nitzschia filiformis																				+
Nitzschia fonticola	++	++										+				++	++		++	
Nitzschia frustulum	+	+		++		+														++

Nitzschia heufleriana	++													+	++	++ +				++
Nitzschia inconspicua		++		++			o	+	oo	+				++			+			+
Nitzschia intermedia				++									+	++						
Nitzschia linearis	++	++			+										+	+	++	+		++
Nitzschia palea										o	++									
Nitzschia pumila						+														
Nitzschia recta																++		+		
Nitzschia rosenstockii								+												
Nitzschia sociabilis																+				
Nitzschia sublinearis					+								+		+	+				
Nitzschia subtilis															++					
Nitzschia umbonata										++ +	o									
Pantocsekiella delicatula			++	++										++ +						
Pantocsekiella iranica														++						
Pantocsekiella ocellata			++																	++ +
Pinnularia brebissonii	+																			+
Planothidium lanceolatum		+															++			
Pseudostaurosira brevistriata																	+			
Reimeria sinuata		++ +	++	+					+			+			++	++		++	+	
Reimeria uniseriata	+											+								+
Rhoicosphenia abbreviata	+	++		+				++			+							+		

(continued)

Table 25.2 (continued)

	TE 1	TE 2	TE 3	TE 4	TE5	TE 6	TE 7	TE 8	TE 9	TE 10	TE 11	TE 12	TE 13	TE 14	TE 15	TE 16	TE 17	TE 18	TE 19	TE 20
Sellaphora bacillum																		+		
Sellaphora pupula																		+	+	
Sellaphora seminulum								+			+									
Stauroneis separanda																		+		+
Stauroneis tackei																	++			
Staurosirella pinnata																				+
Stephanodiscus balatonis																			+	
Stephanodiscus neoastreae																				++
Surirella angusta		+		+			+										+	+		
Surirella brebissonii								+												
Surirella minuta		++ +					++							+	++	++				
Tabularia fasciculata																	+			
Tryblionella apiculata				+	+++		+							+			++			
Tryblionella hungarica																	++			+
Ulnaria acus												++							+	
Ulnaria biceps	+				++															++
Ulnaria ulna	+										+	++							++	

+: < 1%, ++: 1% ≤ x > 5%, +++: 5% ≤ x > 15%, o: 15% ≤ x > 50%, oo: ≥ 50%

N. vandamii var. *vandamii* Schoeman and Archibald, *N. vandamii* var. *mertens* Lange-Bertalot, *N. wildii* Lange-Bertalot, *Nitzschia denticula* Grunow, *N. dissipata* ssp. *oligotrophenta* Lange-Bertalot, *N. pumila* Hustedt, *N. rosenstockii* Lange-Bertalot, *N. subtilis* (Kützing) Grunow, and *Stauroneis separanda* Lange-Bertalot and Werum were new records for the Turkish freshwater diatom flora.

Achnanthidium biasolettianum (Grunow) Bukhtiyarova Pl.1-Fig. 1

Basionym. *Achnanthes biasolettiana* Grunow.

Ref. Krammer and Lange-Bertalot 1991b (p.62-pl.36:1–31).

Dimensions of studied samples: Valves 26.8–31.3 μm length and 4.7–5.2 μm width, 20–22 striae in 10 μm.

Distribution among the sampling locations: TE3, TE4, TE5, and TE6.

Distribution in Turkey: Common.

Amphora alpestris Levkov Pl.1-Fig. 2

Ref.: Levkov 2009 (p.34-pl.51:1–14).

Dimensions of studied samples: Valves 23.9 μm length and 5.7 μm width, 15 striae in 10 μm.

Distribution among the sampling locations: TE17.

Distribution in Turkey: NEW RECORD.

Amphora copulata (Kützing) Schoeman and Archibald Pl.1-Fig. 3

Basionym. *Frustulia copulata* Kützing.

Ref. Levkov 2009 (p.49-pl.46:13–23, 59:1–13); Hofmann et al. 2011 (p.95-pl.91:4–8).

Dimensions of studied samples: Valves 16.6–20.4 μm length and 4.8–6.5 μm width, 13–14 striae in 10 μm.

Distribution among the sampling locations: TE2, TE6, and TE12.

Distribution in Turkey: NEW RECORD.

Amphora liriope Nagumo Pl.1-Fig. 4

Ref.: Levkov 2009 (p.75-pl.43:14–22).

Dimensions of studied samples: Valves 26.3 μm length and 5.6 μm width, 16 striae in 10 μm.

Distribution among the sampling locations: TE2.

Distribution in Turkey: NEW RECORD.

Amphora ovalis (Kützing) Kützing Pl.1-Fig. 5

Basionym. *Frustulia ovalis* Kützing.

Ref. Levkov 2009 (p.96-pl.1:1–5); Hofmann et al. 2011 (p.98-pl.90:1–3).

Dimensions of studied samples: Valves 42.7–50.3 μm length and 10.4–12.5 μm breadth, 11 striae in 10 μm.

Distribution among the sampling locations: TE6 and TE12.

Distribution in Turkey: Common.

Amphora pediculus (Kützing) Grunow Pl.1-Fig. 6

Basionym. *Cymbella pediculus* Kützing.

Ref. Levkov 2009 (p.101-pl.55:31–34, 78:40–47); Hofmann et al. 2011 (p.98-pl.91:29–33).

Dimensions of studied samples: Valves 10.2–14.1 μm length and 2.9–3.1 μm width, 16–17 striae in 10 μm.

Distribution among the sampling locations: TE1-TE4, TE6-TE9, TE12, and TE14-TE19.

Distribution in Turkey: Common.

Amphora lange-bertalotii var. ***tenuis*** Levkov and Metzeltin Pl.1-Fig. 7

Ref. Levkov 2009 (p.73-pl.53:1–12); Hofmann et al. 2011 (p.95-pl.90:10–12).

Dimensions of studied samples: Valves 35.2 μm length and 7.0 μm width, 14 striae in 10 μm.

Distribution among the sampling locations: TE18.

Distribution in Turkey: NEW RECORD.

Aulacoseira granulata (Ehrenberg) Simonsen Pl.1-Fig. 8

Basionym. *Gaillonella granulata* Ehrenberg.

Ref.: Bey and Ector 2013 (Vol.1-p.12).

Dimensions of studied samples: Valves mantle depth 14.5–15.9 μm and 12–13 interstriae in 10 μm.

Distribution among the sampling locations: TE20.

Distribution in Turkey: Common.

Brachysira neglectiformis Lange-Bertalot Pl.1-Fig. 9

Ref.: Hofmann et al. 2011 (p.110-pl.59:1–5).

Dimensions of studied samples: Valves 24.0 μm length and 5.1 μm width.

Distribution among the sampling locations: TE2.

Distribution in Turkey: NEW RECORD.

Cocconeis pediculus Ehrenberg Pl.1-Fig. 10

Ref.: Krammer and Lange-Bertalot 1991b (p.89-pl.55:1–8, 57:1–4).

Dimensions of studied samples: Valves 18.9–27.7 μm length and 15.4–21.1 μm width, 16–19 striae in 10 μm.

Distribution among the sampling locations: TE1-TE4, TE8, and TE11-TE16.

Distribution in Turkey: Common.

Caloneis amphisbaena (Bory) Cleve Pl.1-Fig. 11

Basionym. *Navicula amphisbaena* Bory.

Ref.: Krammer and Lange-Bertalot 1986 (p.385-pl.168:4,5); Hofmann et al. 2011 (p.116-pl.69:1–4).

Dimensions of studied samples: Valves 73.7–78.3 μm length and 24.3–26.1 μm width, 16–17 striae in 10 μm.

Distribution among the sampling locations: TE17.

Distribution in Turkey: Common.

Caloneis macedonica Hustedt Pl.1-Fig. 12

Ref.: Krammer and Lange-Bertalot 1986 (p.395-pl.175:7–9); Bey and Ector 2013 (Vol.3, p.366).

Dimensions of studied samples: Valves 37.4 μm length and 7.9 μm width, 21 striae in 10 μm.

Distribution among the sampling locations: TE20.

Distribution in Turkey: It was found only in Inner Anatolia by Baykal et al. (2009).

Craticula accomoda (Hustedt) Mann Pl.1-Fig. 13

Basionym. *Navicula accomoda* Hustedt.

Ref.: Lange-Bertalot 2001 (p.108-pl.93:1–6).

Dimensions of studied samples: Valves 20.7–22.3 μm length and 6.6–7.1 μm width, 23–25 striae in 10 μm.

Distribution among the sampling locations: TE8, TE10, and TE20.

Distribution in Turkey: Common.

Craticula ambigua (Ehrenberg) Mann Pl.1-Fig. 14

Basionym. *Navicula ambigua* Ehrenberg.

Ref.: Lange-Bertalot 2001 (p.109-pl.82:4–8, 83:3,4).

Dimensions of studied samples: Valves 60.7–65.4 μm length and 17.2–18.4 μm width, 17–18 striae in 10 μm.

Distribution among the sampling locations: TE10 and TE11.

Distribution in Turkey: Common.

Craticula cuspidata (Kützing) Mann Pl.1-Fig. 15

Basionym. *Frustulia cuspidata* Kützing.

Ref.: Lange-Bertalot 2001 (p.111-pl.82:1–3, 83:1,2).

Dimensions of studied samples: Valves 96.3–101.5 μm length and 24.7–25.4 μm width, 14 striae in 10 μm.

Distribution among the sampling locations: TE11.

Distribution in Turkey: Common.

Craticula minusculoides (Hustedt) Lange-Bertalot Pl.1-Fig. 16

Basionym. *Navicula minusculoides* Hustedt.

Ref.: Lange-Bertalot 2001 (p.115-pl.93:7,15).

Dimensions of studied samples: Valves 13.2–14.7 μm length and 4.5–4.6 μm width, 26 striae in 10 μm.

Distribution among the sampling locations: TE10.

Distribution in Turkey: It was found only in Western Anatolia by Albay and Aykulu (1994) and Çelekli (2006).

Craticula molestiformis (Hustedt) Mayama Pl.1-Fig. 17

Basionym. *Navicula molestiformis* Hustedt.

Ref.: Lange-Bertalot 2001 (p.116-pl.93:19–28); Hofmann et al. 2011 (p.140-pl.45:16–20).

Dimensions of studied samples: Valves 11.4–11.7 μm length and 3.8–3.9 μm width, 28 striae in 10 μm.

Distribution among the sampling locations: TE10.

Distribution in Turkey: This was newly recorded in Western Anatolia (Solak et al. 2016a).

Craticula subminuscula (Hustedt) Mayama Pl.1-Fig. 18

Basionym. *Navicula subminuscula* Manguin.

Ref. Krammer and Lange-Bertalot 1986 (p.223-pl.76:21–26); Hofmann et al. 2011 (p.202-pl.42:45–50; Wetzel et al. 2015 (p.229).

Dimensions of studied samples: Valves 9.2–10.5 μm length and 4.2–4.8 μm width, 15–16 striae in 10 μm.

Distribution among the sampling locations: TE3, TE7, TE10, and TE14.

Distribution in Turkey: Common.

Cyclostephanos invisitatus (Hohn and Hellermann) Theriot et al. Pl.1-Fig. 19

Basionym. *Stephanodiscus invisitatus* Hohn and Hellerman.

Ref. Krammer and Lange-Bertalot 1991a (p.63-pl.67:3, 4), Håkansson 2002 (p.67-Figs. 221–225), Wojtal and Kwandrans 2006 (p.198-pl.15:9, 16:12–14,17), Kiss et al. 2012 (p.331-Fig. 10:D-F), Bey and Ector 2013 (Vol.1, p.22), Houk et al. 2014 (p.49-pl.160:1–8).

Dimensions of studied samples: Valve: diameter 9.1–10.4 μm and 12–14 interstriae in 10 μm.

Distribution among the sampling locations: TE14.

Distribution in Turkey: This was newly recorded in Western Anatolia (Solak et al. 2016a).

Cyclotella meneghiniana Kützing Pl.1-Fig. 20

Ref. Krammer and Lange-Bertalot 1991a (p.44-pl.44:1–10), Håkansson 2002 (p.79-Figs. 263–268), Wojtal and Kwandrans 2006 (p.186-pl.4:18–21, 7:1–13, 9:1–8, 10:1–5), Kiss et al. 2012 (p.337-Fig. 14:A-C), Bey and Ector 2013 (Vol.1, p.30), Cavalcante et al. 2013 (p.243-pl.8:A-O), Houk et al. 2010 (p.16-pl.143:1–15).

Dimensions of studied samples: Valve: diameter 11.3–16.7 μm and 8–10 interstriae in 10 μm.

Distribution among the sampling locations: TE4, TE11, TE14, and TE16.

Distribution in Turkey: Common.

Cymatopleura elliptica (Brébisson) Smith Pl.2-Fig. 1

Basionym.*Surirella elliptica* Brébisson.

Ref.: Hofmann et al. 2011 (p.143-pl.123:1–3).

Dimensions of studied samples: Valves 80.8 μm length and 17.6 μm width, 4 fibulae in 10 μm.

Distribution among the sampling locations: TE18.

Distribution in Turkey: Common.

Cymatopleura solea (Brébisson) Smith Pl.2-Fig. 2

Basionym. *Surirella solea* Brébisson.

Ref.: Hofmann et al. 2011 (p.144-pl.124:1–4).

Dimensions of studied samples: Valves 78.6 μm length and 53.3 μm width, 8 fibulae in 10 μm.

Distribution among the sampling locations: TE17.

Distribution in Turkey: Common.

Cymatopleura solea *var.* ***apiculata*** (Smith) Ralfs Pl.2-Fig. 3

Basionym. *Cymatopleura apiculata* W.Smith.

Ref.: Hofmann et al. 2011 (p.144-pl.125:1–5).

Dimensions of studied samples: Valves 41.6 μm length and 12.8 μm width, 8 fibulae in 10 μm.

Distribution among the sampling locations: TE20.

Distribution in Turkey: Common.

Cymbella compacta Østrup Pl.2-Fig. 4

Ref.: Krammer 2002 (p.150-pl.173:1–8, 174:1–15; 175:1–7); Hofmann et al. 2011 (p.148-pl.79:6–10).

Dimensions of studied samples: Valves 65.4–69.2 μm length and 13.7–14.1 μm breadth, maximum length/breadth ratios about 5.0, and 11–12 striae in 10 μm.

Distribution among the sampling locations: TE15 and TE16.

Distribution in Turkey: Common.

Cymbella excisa Kützing Pl.2-Fig. 5

Ref.: Krammer 2002 (p.26-pl.8:1–26, 9:19–25); Hofmann et al. 2011 (p.150-pl.77:23–28).

Dimensions of studied samples: Valves 21.2–26.8 μm length and 7.3–7.9 μm width, 11–12 striae in 10 μm.

Distribution among the sampling locations: TE1-TE6, TE9, and TE11-TE20.

Distribution in Turkey: This was newly recorded in Western Anatolia (Solak et al. 2016b).

Cymbella excisa var. ***procera*** Krammer Pl.2-Fig. 6

Ref.: Krammer 2002 (p.28-pl.9:1–7, 10:10–13).

Dimensions of studied samples: Valves 23.4–27.2 μm length and 7.9–8.4 μm width, 11–12 striae in 10 μm.

Distribution among the sampling locations: TE1, TE2, TE9, and TE18.

Distribution in Turkey: NEW RECORD.

Cymbella excisa var. ***subcapitata*** Krammer Pl.2-Fig. 7

Ref.: Krammer 2002 (p.28-pl.10:1–9,14,18).

Dimensions of studied samples: Valves 26.2–29.3 μm length and 7.6–7.8 μm width, 11 striae in 10 μm.

Distribution among the sampling locations: TE3.

Distribution in Turkey: NEW RECORD.

Cymbella exigua Krammer Pl.2-Fig. 8

Ref.: Krammer 2002 (p.30-pl.10:19–24).

Dimensions of studied samples: Valves 29.4 μm length and 8.3 μm width, 13 striae in 10 μm.

Distribution among the sampling locations: TE3.

Distribution in Turkey: NEW RECORD.

Cymbella lange-bertalotii Krammer Pl.3-Fig. 1

Ref.: Krammer 2002 (p.152-pl.179:1–6, 180:1–5).

Dimensions of studied samples: Valves 67.2–72.4 μm length and 13.4 μm width, 12 striae in 10 μm.

Distribution among the sampling locations: TE1, TE3, TE12, TE15, TE16, and TE18.

Distribution in Turkey: This was newly recorded in Inner Anatolia by Barinova et al. (2014).

Cymbella parva (W.Smith) Kirchner Pl.2-Fig. 9

Basionym. *Cocconema parvum* W. Smith.

Ref.: Krammer 2002 (p.35-pl.6:1–19).

Dimensions of studied samples: Valves 40.1 μm length and 9.7 μm width, 10 striae in 10 μm.

Distribution among the sampling locations: TE12.

Distribution in Turkey: Common.

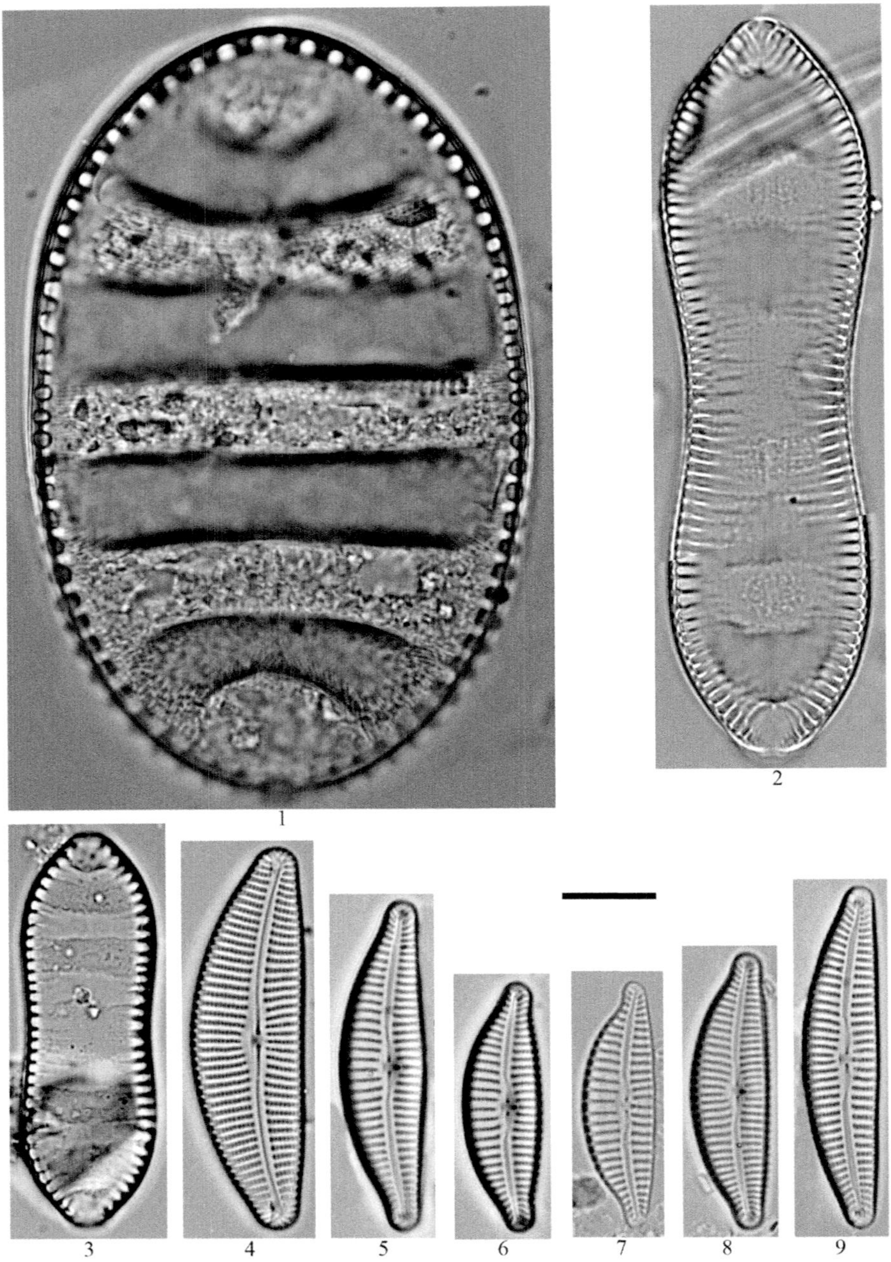

Plate 25.1 1-*Cymatopleura elliptica*, 2-*C. solea*, 3-*C. solea* var. *apiculata*, 4-*Cymbella compacta*, 5-*C. excisa*, 6-*C. excisa* var. *procera*, 7-*C. excisa* var. *subcapitata*, 8-*C. exigua*, and 9-*C. parva*. Scale bar: 10 μm.

Cymbella peraffinis Tynni Pl.3-Fig. 2

Ref.: Krammer 2002 (p.32-pl.14:1–7).

Dimensions of studied samples: Valves 27.9 μm length and 7.3 μm width, 10 striae in 10 μm.

Distribution among the sampling locations: TE17.

Distribution in Turkey: NEW RECORD.

Cymbella tumida (Brébisson) Van Heurck Pl.3-Fig. 3

Basionym. *Cocconema tumidum* Brébisson.

Ref.: Krammer 2002 (p.141-pl.162:1–8).

Dimensions of studied samples: Valves 42.4–67.2 μm length and 15.3–16.8 μm breadth, 11–12 striae in 10 μm.

Distribution among the sampling locations: TE15.

Distribution in Turkey: Common.

Cymbopleura amphicephala (Nägeli) Krammer Pl.3-Fig. 4

Basionym. *Cymbella amphicephala* Naegeli ex Kützing.

Ref.: Krammer 2003 (p.70-pl.91:1–18); Hofmann et al. .

Dimensions of studied samples: Valves 24.3–25.9 μm length and 7.8–8.2 μm width, 11–12 striae in 10 μm.

Distribution among the sampling locations: TE1, TE5, TE6, TE12, and TE20.

Distribution in Turkey: Common.

Cymbopleura lange-bertalotii Krammer Pl.3-Fig. 5

Ref.: Krammer 2003 (p.75-pl.99:1–4).

Dimensions of studied samples: Valves 32.5 μm length and 10.1 μm width, 10 striae in 10 μm.

Distribution among the sampling locations: TE6.

Distribution in Turkey: NEW RECORD.

Cymbopleura vrana Krammer Pl.3-Fig. 6

Ref.: Krammer 2003 (p.67-pl.93:9–18).

Dimensions of studied samples: Valves 35.0 μm length and 10.2 μm width, 14 striae in 10 μm.

Distribution among the sampling locations: TE5.

Distribution in Turkey: NEW RECORD.

Diatoma moniliformis (Kützing) D.M.Williams Pl.3-Fig. 7

Basionym. *Diatoma tenue* var. *moniliforme* Kützing.

Ref.: Krammer and Lange-Bertalot 1991a (p.98-pl.96:11–21); Hofmann et al. 2011 (p.174-pl.2:11–15).

Dimensions of studied samples: Valves 19.3–27.8 μm length, 4.8–5.2 μm width, and 5–6 transapical bars in 10 μm.

Distribution among the sampling locations: TE1, TE3, TE5, TE15, and TE20.

Distribution in Turkey: Common.

Diatoma vulgaris Bory Pl.3-Fig. 8

Ref.: Hofmann et al. 2011 (p.175-pl.3:20–25).

Dimensions of studied samples: Valves 32.4–47.7 μm length, 13.9–15.8 μm width, and 7–8 transapical bars in 10 μm.

Distribution among the sampling locations: TE11 and TE14-TE19.

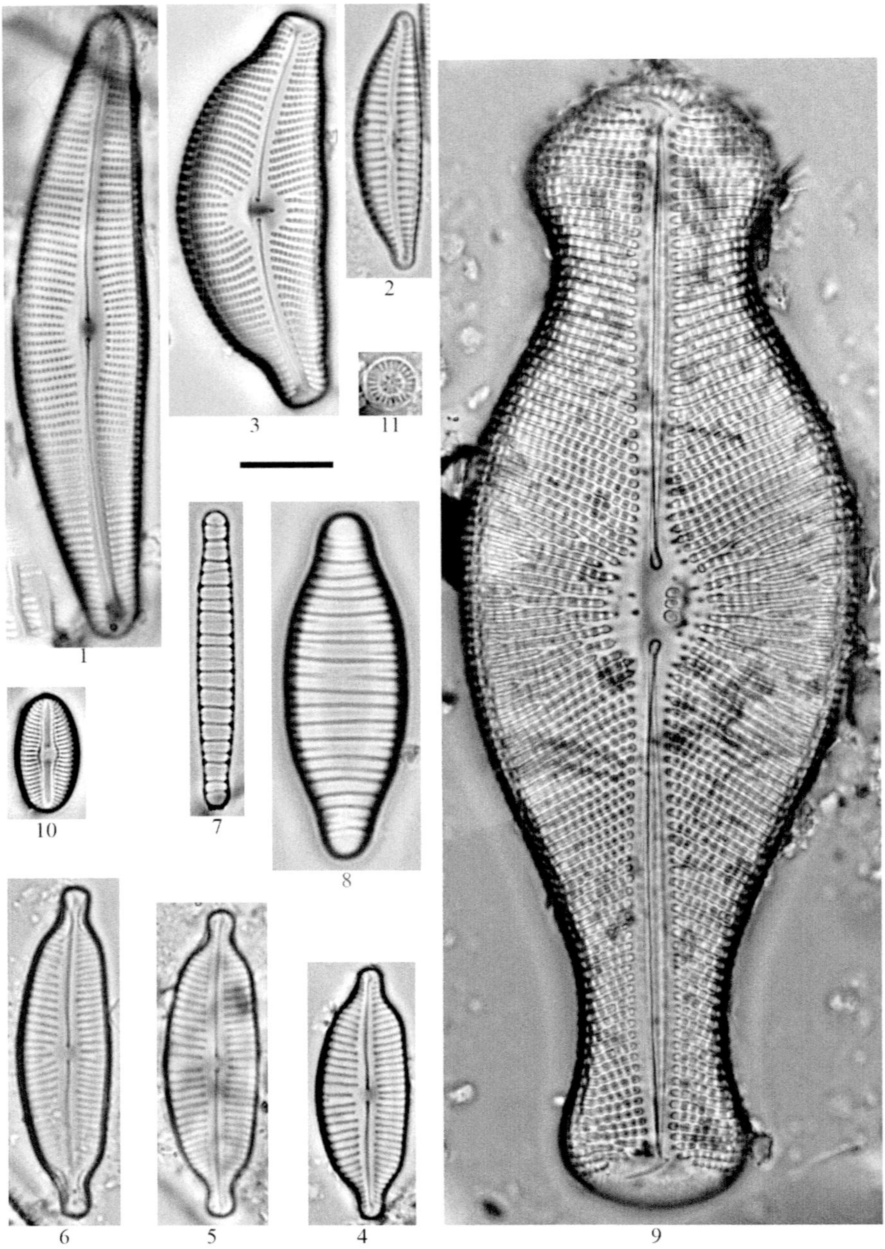

Plate 25.2 1-*C. lange-bertalotii,* 2-*C. peraffinis,* 3-*C. tumida*, 4-*Cymbopleura amphicephala,* 5-*C. lange-bertalotii,* 6-*C. vrana,* 7-*Diatoma moniliformis,* 8-*D. vulgaris*, 9-*Didymosphenia geminata*, 10-*Diploneis separanda,* and 11-*Discostella stelligera* var. *tenuis*. Scale bar: 10 μm.

Distribution in Turkey: Common.

Didymosphenia geminata (Lyngbye) M.Schmidt Pl.3-Fig. 9

Basionym. *Echinella geminata* Lyngbye.

Ref.: Hofmann et al. 2011 (p.176-pl.100:1).

Dimensions of studied samples: Valves 121.2 μm length, 41.0 μm width, and striae in 10 μm.

Distribution among the sampling locations: TE18.

Distribution in Turkey: Common.

Diploneis separanda Lange-Bertalot Pl.3-Fig. 10

Ref.: Hofmann et al. 2011 (p.183-pl.66:11–15).

Dimensions of studied samples: Valves 12.3–14.9 μm length and 6.2–7.2 μm width, 19–21 striae in 10 μm.

Distribution among the sampling locations: TE11.

Distribution in Turkey: This was newly recorded in Western Anatolia (Yılmaz 2017).

Discostella stelligera var. ***tenuis*** (Hustedt) Houk and Klee Pl.3-Fig. 11

Basionym. *Cyclotella stelligera* var. *tenuis* Hustedt.

Ref. Houk et al. 2010 (p.47, Fig. 307:1–8).

Dimensions of studied samples: Valve 5.3–8.9 μm diameter and 12–16 interstriae in 10 μm.

Distribution among the sampling locations: TE14.

Distribution in Turkey: This was newly recorded in Thrace Region (Solak et al. 2018).

***Encyonema auerswaldii* Rabenhorst** Pl.4-Fig. 1

Ref.: Krammer 1997a (p.117-pl.66:1–6); Hofmann et al. 2011 (p.190-pl.86:1–4).

Dimensions of studied samples: Valves 25.6–26.1 μm length and 10.4 μm breadth, 10 striae in 10 μm.

Distribution among the sampling locations: TE12.

Distribution in Turkey: Common.

Encyonema minutum (Hilse) Mann Pl.4-Fig. 2

Basionym. *Cymbella minuta* Hilse.

Ref.: Krammer 1997a (p.53-pl.6:19–27, 25:1–19); Hofmann et al. 2011 (p.188-pl.87:33–40).

Dimensions of studied samples: Valves 10.5–11.3 μm length and 4.6–5.0 μm breadth, 16–17 striae in 10 μm.

Distribution among the sampling locations: TE12, TE15, TE16, TE18, and TE20.

Distribution in Turkey: Common.

Encyonema reichardtii (Krammer) Mann Pl.4-Fig. 3

Basionym. *Cymbella reichardtii* K. Krammer.

Ref.: Krammer 1997a (p.53-pl.6:19–27, 25:1–19); Ref.: Hofmann et al. 2011 (p.191-pl.87:23–29).

Dimensions of studied samples: Valves 11.4 μm length and 3.7 μm breadth, 21 striae in 10 μm.

Distribution among the sampling locations: TE19.

Distribution in Turkey: This was newly recorded in Western Anatolia (Solak et al. 2016b).

Encyonema silesiacum (Bleisch) D.G.Mann Pl.4-Fig. 4

Basionym. *Cymbella silesiaca* Bleisch.

Ref.: Krammer 1997a (p.72-pl.4:1–18); Ref.: Hofmann et al. 2011 (p.192-pl.87:11–17).

Dimensions of studied samples: Valves 30.1–36.7 μm length and 9.2–10.7 μm breadth, 13–14 striae in 10 μm.

Distribution among the sampling locations: TE1, TE15, and TE18-TE20.

Distribution in Turkey: Common.

Encyonema ventricosum (C.Agardh) Grunow Pl.4-Fig. 5

Basionym. *Frustulia ventricosa* C.Agardh.

Ref.: Krammer 1997a (p.98-pl.6:5–17); Ref.: Hofmann et al. 2011 (p.192-pl.87:18–22).

Dimensions of studied samples: Valves 17.5–20.1 μm length and 6.2–7.0 μm breadth, 12–14 striae in 10 μm.

Distribution among the sampling locations: TE2, TE3, TE15, and TE16.

Distribution in Turkey: It was found only in Inner Anatolia by Demir et al. (2014), Morkoyunlu-Yüce and Ertan (2014), and Morkoyunlu-Yüce et al. (2015).

Encyonopsis cesatii (Rabenhorst) Krammer Pl.4-Fig. 6

Basionym. *Navicula cesatii* Rabenhorst.

Ref.: Krammer 1997b (p.152-pl.182:1–13); Ref.: Hofmann et al. 2011 (p.195-pl.89:1–11).

Dimensions of studied samples: Valves 27.4–34.3 μm length and 6.2–6.7 μm breadth, 19–20 striae in 10 μm.

Distribution among the sampling locations: TE19.

Distribution in Turkey: It was found only in Inner Anatolia by Baykal and Açıkgöz (2004) and Blacksea Region by Baytut and Gönülol (2016).

Encyonopsis falaisensis (Grunow) Krammer Pl.4-Fig. 7

Ref.: Krammer 1997b (p.116-pl.163:1–5).

Dimensions of studied samples: Valves 28.4 μm length and 5.4 μm breadth, 18 striae in 10 μm.

Distribution among the sampling locations: TE2.

Distribution in Turkey: NEW RECORD.

Encyonopsis krammeri Reichardt Pl.4-Fig. 8 Ref.: Krammer 1997b (p.99-pl.145:1–18).

Dimensions of studied samples: Valves 17.2 μm length and 3.2 μm breadth.

Distribution among the sampling locations: TE2.

Distribution in Turkey: NEW RECORD.

Encyonopsis microcephala (Grunow) Krammer Pl.4-Fig. 9

Basionym. *Cymbella microcephala* Grunow.

Ref.: Krammer 1997b (p.91-pl.143:1,4,5,8–26); Ref.: Hofmann et al. 2011 (p.197-pl.89:35–39).

Dimensions of studied samples: Valves 11.8–15.4 μm length and 2.9–3.6 μm breadth, 24 striae in 10 μm.

Distribution among the sampling locations: TE5, TE12, and TE19.
Distribution in Turkey: Common.
Encyonopsis minuta Krammer and E.Reichardt Pl.4-Fig. 10
Ref.: Krammer 1997b (p.95-pl.143a:1–29); Ref.: Hofmann et al. 2011 (p.198-pl.87:25–34).
Dimensions of studied samples: Valves 10.5–11.1 μm length and 3.4–3.6 μm breadth, 23 striae in 10 μm.
Distribution among the sampling locations: TE1, TE2, TE4-TE6, TE9, and TE12-TE14.
Distribution in Turkey: This was newly recorded in Western Anatolia (Solak et al. 2016b).
Encyonopsis subminuta Krammer and E.Reichardt Pl.4-Fig. 11
Ref.: Krammer 1997b (p.96-pl.144:1–11); Ref.: Hofmann et al. 2011 (p.191-pl.87:23–29).
Dimensions of studied samples: Valves 12.4–15.1 μm length and 3.5–3.9 μm breadth, 24–25 striae in 10 μm.
Distribution among the sampling locations: TE3 and TE5.
Distribution in Turkey: This was newly recorded in Western Anatolia (Solak et al. 2016b).
Encyonopsis thumensis Krammer Pl.4-Fig. 12
Ref.: Krammer 1997b p.103-pl.158:16–22
Dimensions of studied samples: Valves 13.5–15.9 μm length and 3.3–3.5 μm breadth, 22–24 striae in 10 μm.
Distribution among the sampling locations: TE6.
Distribution in Turkey: NEW RECORD.
Eunotia pectinalis (Kützing) Rabenhorst Pl.4-Fig. 13
Basionym. *Himantidium pectinalis* Kützing.
Ref.: Hofmann et al. 2011 (p.234-pl.12:1–6); Lange-Bertalot et al. 2011 (p.193-pl.166:1–13).
Dimensions of studied samples: Valves 132.2 μm length and 11.5 μm breadth, 9 striae in 10 μm.
Distribution among the sampling locations: TE19.
Distribution in Turkey: Common.
Fallacia lenzii (Hustedt) Lange-Bertalot Pl.4-Fig. 14
Basionym. *Navicula lenzii* Hustedt.
Ref.: Werum and Lange-Bertalot 2004 (p.159-pl.33:1,5); Hofmann et al. 2011 (p.243-pl.46:3–7).
Dimensions of studied samples: Valves 23.2–15.5 μm length and 5.4–5.8 μm breadth.
Distribution among the sampling locations: TE5.
Distribution in Turkey: This was newly recorded in Western Anatolia (Yılmaz 2017).
Fallacia pygmaea (Kützing) Stickle and Mann Pl.4-Fig. 15
Basionym. *Navicula pygmaea* Kützing.
Ref.: Hofmann et al. 2011 (p.245-pl.46:31–34).

Dimensions of studied samples: Valves 21.4–30.2 μm length and 10.7–11.4 μm breadth, 24–25 striae in 10 μm.

Distribution among the sampling locations: TE11.

Distribution in Turkey: Common.

Fistulifera saprophila (Lange-Bertalot and Bonik) Lange-Bertalot Pl.4-Fig. 16

Basionym. *Navicula saprophila* H. Lange-Bertalot and K. Bonik.

Ref. Krammer and Lange-Bertalot 1986 (p.207-pl.74:34, 35); Hofmann et al. 2011 (p.248-pl.49:6,7).

Dimensions of studied samples: Valves 6.7–7.2 μm length and 3.7–3.8 μm breadth.

Distribution among the sampling locations: TE2, TE7, TE8, and TE10.

Distribution in Turkey: This was newly recorded in Western Anatolia (Solak et al. 2016a).

Fragilaria austriaca (Grunow) Lange-Bertalot Pl.4-Fig. 17

Basionym. *Synedra austriaca* Grunow.

Ref.: Krammer and Lange-Bertalot 1991a (p.126-pl.106:21–24); Hofmann et al. 2011 (p.257-pl.7:31–35).

Dimensions of studied samples: Valves 22.1–25.1 μm length and 3.4–3.6 μm breadth, 17–18 striae in 10 μm.

Distribution among the sampling locations: TE3.

Distribution in Turkey: Common.

Gomphonema aequale Gregory Pl.4-Fig. 18

Ref.: Levkov et al. 2016 (p.23-pl.91:1–42).

Dimensions of studied samples: Valves 22.3 μm length and 6.8 μm breadth, 7 striae in 10 μm.

Distribution among the sampling locations: TE8.

Distribution in Turkey: NEW RECORD.

Gomphonema exilissimum (Grunow) Lange-Bertalot and E.Reichardt Pl.4-Fig. 19

Basionym. *Gomphonema parvulum* var. *exilissimum* Grunow.

Ref.: Levkov et al. 2016 (p.49-pl.127:1–33, 128:1–41); Hofmann et al. 2011 (p.303-pl.99:11–15).

Dimensions of studied samples: Valves 15.0–26.1 μm length and 5.5–5.9 μm breadth, 11–13 striae in 10 μm.

Distribution among the sampling locations: TE2, TE9, and TE20.

Distribution in Turkey: NEW RECORD.

Gomphonema jadwigiae Lange-Bertalot and Reichardt Pl.4-Fig. 20

Ref.: Levkov et al. 2016 p.67-pl.81:1–30.

Dimensions of studied samples: Valves 20.2 μm length and 6.1 μm breadth, 13 striae in 10 μm.

Distribution among the sampling locations: TE2.

Distribution in Turkey: NEW RECORD.

Gomphonema micropus Kützing Pl.4-Fig. 21

Ref.: Levkov et al. 2016 (p.83-pl.96:21–40, 97:1–37); Hofmann et al. 2011 (p.305-pl.98:21–25).

Dimensions of studied samples: Valves 24.9–37.8 μm length and 6.9–8.7 μm breadth, 12–15 striae in 10 μm.

Distribution among the sampling locations: TE2.

Distribution in Turkey: Common.

Gomphonema olivaceum (Hornemann) Brébisson Pl.4-Fig. 22

Basionym. *Ulva olivacea* Hornemann.

Ref.: Levkov et al. 2016 (p.91-pl.184:1–47, 185:1–36, 186:1–38); Hofmann et al. 2011 (p.310-pl.95:1–6).

Dimensions of studied samples: Valves 17.8–28.9 μm length and 6.7–7.5 μm breadth, 10–12 striae in 10 μm.

Distribution among the sampling locations: TE1, TE3, TE12, TE16, and TE20.

Distribution in Turkey: Common.

Gomphonema parvulum (Kützing) Kützing Pl.4-Fig. 23

Basionym. *Sphenella parvula* Kützing.

Ref.: Levkov et al. 2016 (p.98-pl.102:1–38, 103:1–18; 104:1–24); Hofmann et al. 2011 (p.312-pl.99:1–5).

Dimensions of studied samples: Valves 13.7–17.2 μm length and 6.0–7.2 μm breadth, 15–16 striae in 10 μm.

Distribution among the sampling locations: TE2, TE7-TE11, and TE13.

Distribution in Turkey: Common.

Gomphonema pumilum (Grunow) E.Reichardt and Lange-Bertalot Pl.4-Fig. 24

Basionym. *Gomphonema intricatum* var. *pumilum* Grunow.

Ref.: Levkov et al. 2016 (p.110-pl.151:26–52); Hofmann et al. 2011 (p. 315, Fig. 97: 10–14).

Dimensions of studied samples: Valves 24.7–29.2 μm length and 4.9–5.3 μm breadth, 11–13 striae in 10 μm.

Distribution among the sampling locations: TE2, TE9, TE13, and TE18.

Distribution in Turkey: This was newly recorded in Western Anatolia (Solak et al. 2016a).

Gomphonema tergestinum (Grunow) Fricke Pl.4-Fig. 25

Basionym. Gomphonema semiapertum var. tergestina Grunow.

Ref.: Levkov et al. 2016 (p.128-pl.163:1–31, 164:1–32); Hofmann et al. 2011 (p.317-pl.96:22–26).

Dimensions of studied samples: Valves 15.4–18.9 μm length and 4.3–5.7 μm breadth, 12–15 striae in 10 μm.

Distribution among the sampling locations: TE3 and TE14.

Distribution in Turkey: It was found only in Inner Anatolia and Blacksea (Solak et al. 2016a).

Grunowia tabellaria (Grunow) Rabenhorst Pl.4-Fig. 26

Basionym. *Denticula tabellaria* Grunow.

Ref.: Krammer and Lange-Bertalot 1988 (p.53-pl.39:10–13); Hofmann et al. 2011 (p.464-pl.117:1–5); Bąk et al. 2012 (p.260-pl.76).

Dimensions of studied samples: Valves 15.7–21.4 μm length and 7.0–8.7 μm breadth, 22–24 striae and 5–6 fibulae in 10 μm.

Distribution among the sampling locations: TE4, TE12, and TE18.

Distribution in Turkey: Common.

Gyrosigma kuetzingii (Grunow) Cleve Pl.4-Fig. 27

Basionym. *Pleurosigma kuetzingii* Grunow.

Ref.: Hofmann et al. 2011 (p.324-pl.63:1–3).

Dimensions of studied samples: Valves 97.1–112.4 μm length and 13.5–15.3 μm breadth, 17–18 striae in 10 μm.

Distribution among the sampling locations: TE4 and TE20.

Distribution in Turkey: NEW RECORD.

Halamphora montana (Krasske) Levkov Pl.4-Fig. 28

Basionym. *Amphora montana* Krasske.

Ref.: Levkov 2009 (p.207-pl.93:10–19, 26–45); Hofmann et al. 2011 (p.328-pl.92:12–14).

Dimensions of studied samples: Valves 15.5–16.4 μm length and 3.5–3.7 μm breadth.

Distribution among the sampling locations: TE19.

Distribution in Turkey: Common.

Hannaea arcus (Ehrenberg) R.M.Patrick Pl.4-Fig. 29

Basionym. *Navicula arcus* Ehrenberg.

Ref.: Hofmann et al. 2011 (p.332-pl.4:8–12).

Dimensions of studied samples: Valves 35.3–29.7 μm length and 5.7–6.3 μm breadth, 15–16 striae in 10 μm.

Distribution among the sampling locations: TE15, TE16, and TE18.

Distribution in Turkey: Common.

Hippodonta capitata (Ehrenberg) Lange-Bertalot et al. Pl.4-Fig. 30

Basionym. *Navicula capitata* Ehrenberg.

Ref.: Lange-Bertalot 2001 (p.98-pl.75:1–6); Hofmann et al. 2011 (p.335-pl.51:1–6).

Dimensions of studied samples: Valves 23.4–25.5 μm length and 6.9–7.2 μm breadth, 9 striae in 10 μm.

Distribution among the sampling locations: TE20.

Distribution in Turkey: Common.

Lindavia balatonis (Pantocsek) T.Nakov et al. Pl.4-Fig. 31

Basionym. *Cyclotella balatonis* Pantocsek.

Ref. Budzyńska and Wojtal 2011 (p.512-Fig. 1–22); Kiss et al. 2012 (p.340-pl.15:D-F); Bey and Ector 2013 (Vol.1, p.56); Houk et al. 2010 (p.39-pl.269:1–11).

Dimensions of studied samples: Valve 13.6–16.5 μm diameter and 15–17 interstriae in 10 μm.

Distribution among the sampling locations: TE20.

Distribution in Turkey: This is a rare species in Western Anatolia (Solak et al. 2018).

Luticola hlubikovae Levkov, Metzeltin, and Pavlov Pl.4-Fig. 32

Ref.: Levkov et al. 2013 (p.130-pl.55:18–29).

Dimensions of studied samples: Valve 19.6–25.1 μm diameter and 7.4–7.7 μm breadth, 19–21 striae in 10 μm.

Distribution among the sampling locations: TE10.

Distribution in Turkey: NEW RECORD.

Luticola goeppertiana (Bleisch) D.G.Mann Pl.4-Fig. 33

Basionym. *Stauroneis goeppertiana* Bleisch.

Ref.: Levkov et al. 2013 (p.122-pl.55:1–17, 56:1–7, 58:1–19); Hofmann et al. 2011 (p.346-pl.45:22–26).

Dimensions of studied samples: Valves 25.4–28.3 μm length and 3.1–3.2 μm breadth.

Distribution among the sampling locations: TE7, TE11, and TE17.

Distribution in Turkey: It was found only in Inner Anatolia by Baykal and Açıkgöz (2004), Varol and Şen (2014), and Black sea Region by Taş and Yılmaz (2015).

Mayamaea permitis (Hustedt) K.Bruder and Medlin Pl.4-Fig. 34

Basionym. *Navicula permitis* Hustedt.

Ref. Lange-Bertalot 2001 (p. 136, Fig. 104: 7–13); Hofmann et al. 2011 (p.354-pl.49:13–19).

Dimensions of studied samples: Valves 7.0–7.3 μm length and 7.2–7.9 μm breadth, 17–18 striae in 10 μm.

Distribution among the sampling locations: TE2 and TE7.

Distribution in Turkey: This was newly recorded in Western Anatolia (Solak et al. 2016a).

Melosira varians C.Agardh Pl.5-Fig. 1

Ref. Krammer and Lange-Bertalot 1991a (p.7-pl.4:1–8), Wojtal 2009 (p.238-pl.1:1–4), Bey and Ector 2013 (Vol.1, p.48), Cavalcante et al. 2013 (p.246-pl.11:A).

Dimensions of studied samples: Valve diameter 19.3–23.5 μm.

Distribution among the sampling locations: TE4, TE14, TE16, TE17, and TE20.

Distribution in Turkey: Common.

Navicula antonii Lange-Bertalot Pl.5-Fig. 2

Ref. Rumrich et al. 2000 (p.155-pl.46:18–21; Lange-Bertalot 2001 (p.15-pl.13:1–15, 28:6); Hofmann et al. 2011 (p.373-pl.32:11–15).

Dimensions of studied samples: Valves 12.7–18.4 μm length and 5.9–6.2 μm breadth and 14–15 striae in 10 μm.

Distribution among the sampling locations:

Distribution in Turkey: This was newly recorded in Western Anatolia (Solak et al. 2016a).

Navicula capitatoradiata H.Germain Pl.5-Fig. 3

Ref. Lange-Bertalot 2001 (p.22-pl.29:15–20); Hofmann et al. 2011 (p.374-pl.36:28–34).

Dimensions of studied samples: Valves 30.4–34.7 μm length and 7.3–7.7 μm breadth, 14–15 striae in 10 μm.

Distribution among the sampling locations: TE1, TE2, TE4, TE5, TE9, and TE12-TE19.

Distribution in Turkey: Common.

Navicula cincta (Ehrenberg) Ralfs Pl.5-Fig. 4

Basionym. *Pinnularia cincta* Ehrenberg.

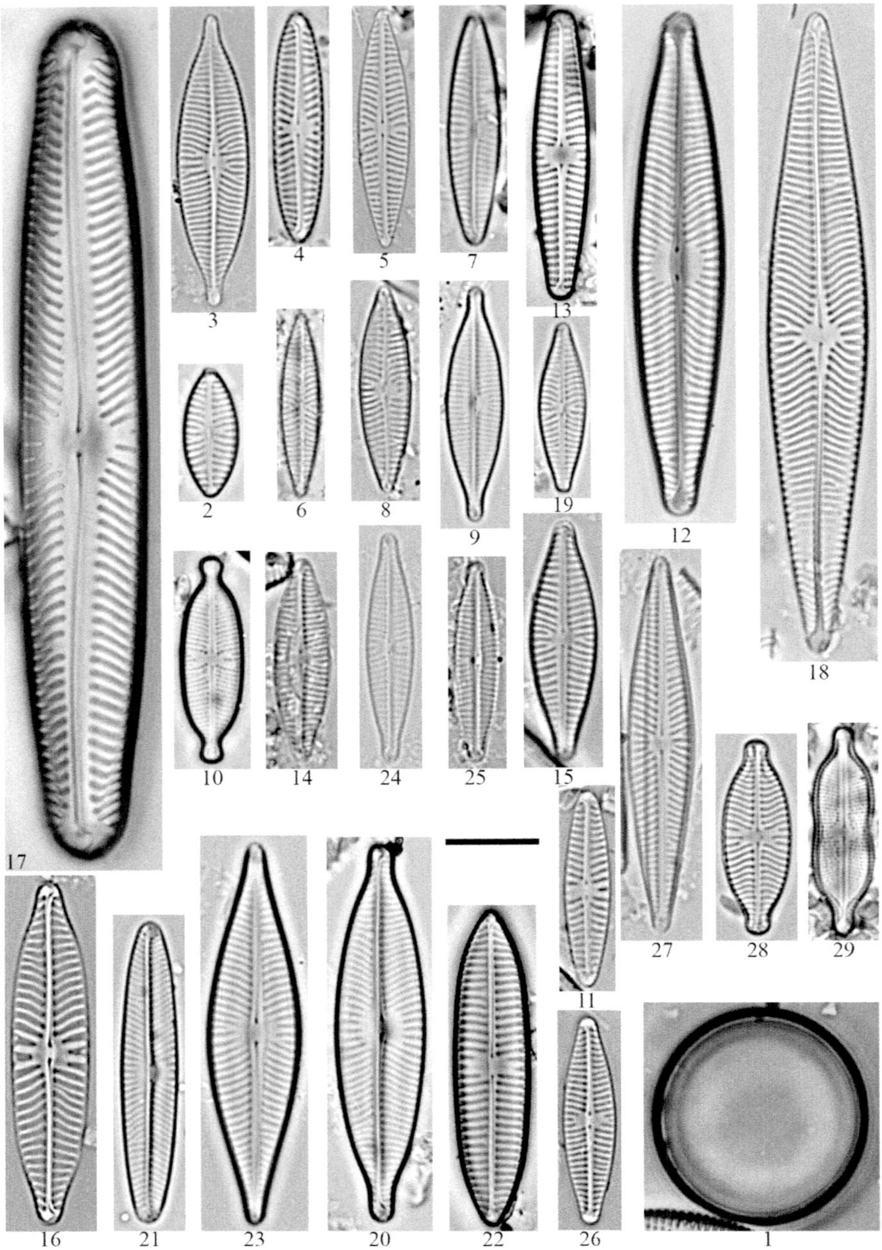

Plate 25.3 1-*Melosira varians*, 2-*Navicula antonii,* 3-*N. capitatoradiata*, 4-*N. cincta*, 5-*N. cryptotenella*, 6-*N. cryptotenelloides,* 7-*N. erifuga,* 8-*N. exilis*, 9-*N. gregaria*, 10-*N. kotschyi*, 11-*N. lacuum*, 12-*N. lanceolata*, 13-*N. libonensis,* 14-*N. lundii,* 15-*N. moskalii,* 16-*N. novaesiberica,* 17-*N. oblonga*, 18-*N. radiosa*, 19-*N. reichardtiana,* 20-*N. rostellata*, 21-*N. simulata*, 22-*N. tripunctata*, 23-*N. trivialis,* 24- *Navicula vandamii,* 25- *Navicula vandamii* var. *mertens,* 26-*N. veneta,* 27-*N. wildii,* 28- *Navigeia decussis,* and 29-*Neidiomorpha binodiformis*. Scale bar: 10 µm.

Ref.: Lange-Bertalot 2001 (p.26-pl.41:1–29); Hofmann et al. 2011 (p.376-pl.32:31–36).

Dimensions of studied samples: Valves 24.7–25.3 μm length and 5.2 μm breadth, 11 striae in 10 μm.

Distribution among the sampling locations: TE17.

Distribution in Turkey: Common.

Navicula cryptotenella Lange-Bertalot Pl.5-Fig. 5

Ref. Krammer and Lange-Bertalot 1986 (p.62-pl.18:22–23, 19:1–10, 27:1,4); Lange-Bertalot 2001 (p.28-pl.26:17–32, 27:19–22, 28:3, 69:5); Hofmann et al. 2011 (p.378-pl.32:1–5).

Dimensions of studied samples: Valves 22.4–27.9 μm length and 5.4–5.6 μm breadth, 14–15 striae in 10 μm.

Distribution among the sampling locations: TE1-TE3, TE5, TE8, TE13-TE15, and TE20.

Distribution in Turkey: Common.

Navicula cryptotenelloides Lange-Bertalot Pl.5-Fig. 6

Ref. Lange-Bertalot 2001 (p.29-pl.26:8–16); Hofmann et al. 2011 (p.379-pl.32:6–10).

Dimensions of studied samples: Valves 18.7–21.4 μm length and 3.6–4.2 μm breadth, 18 striae in 10 μm.

Distribution among the sampling locations: TE5, TE7, TE8, TE11, and TE13.

Distribution in Turkey: It was found only in Western Anatolia (Solak et al. 2016a).

Navicula erifuga Lange-Bertalot Pl.5-Fig. 8

Ref.: Lange-Bertalot 2001 (p.28-pl.26:17–32); Hofmann et al. 2011 (p.382-pl.38:12–16).

Dimensions of studied samples: Valves 24.9–26.7 μm length and 5.5–6.2 μm breadth, 13 striae in 10 μm.

Distribution among the sampling locations: TE8 and TE13.

Distribution in Turkey: It was found only in Inner Anatolia by Karacaoğlu et al. (2004) and Dalkıran et al. (2016).

Navicula exilis Kützing Pl.5-Fig. 8

Ref.: Hofmann et al. 2011 (p.383-pl.31:1–5).

Dimensions of studied samples: Valves 22.5 μm length and 5.5 μm breadth, 15 striae in 10 μm.

Distribution among the sampling locations:

Distribution in Turkey: NEW RECORD.

Navicula gregaria Donkin Pl.5-Fig. 9

Ref.: Lange-Bertalot 2001 (p.87-pl.39:15–22); Hofmann et al. 2011 (p.384-pl.39:5–11).

Dimensions of studied samples: Valves 22.4–26.7 μm length and 6.2–7.4 μm breadth, 18–20 striae in 10 μm.

Distribution among the sampling locations: TE1, TE2, TE5, TE17, and TE19.

Distribution in Turkey: Common.

Navicula kotschyi Grunow Pl.5-Fig. 10

Ref. Krammer and Lange-Bertalot 1986 (p.169-pl.60:10–15); Hofmann et al. 2011 (p.386-pl.40:38–41).

Dimensions of studied samples: Valves 21.5–24.7 μm length and 6.0–6.4 μm breadth, 22–24 striae in 10 μm.

Distribution among the sampling locations: TE5 and TE20.

Distribution in Turkey: It was found only in Inner Anatolia and Northern Anatolia by Solak et al. 2016a).

Navicula lacuum Lange-Bertalot et al. Pl.5-Fig. 11

Ref.: Hofmann et al. 2011 (p.387-pl.32:37–43).

Dimensions of studied samples: Valves 20.5–22.3 μm length and 4.8–5.1 μm breadth, 11–13 striae in 10 μm.

Distribution among the sampling locations: TE17.

Distribution in Turkey: NEW RECORD.

Navicula lanceolata Ehrenberg Pl.5-Fig. 12

Ref.: Lange-Bertalot 2001 (p.85-pl.38:8–18); Hofmann et al. 2011 (p.387-pl.38:20–26).

Dimensions of studied samples: Valves 50.1–55.3 μm length and 9.5–10.1 μm breadth, 10–11 striae in 10 μm.

Distribution among the sampling locations: TE15.

Distribution in Turkey: Common.

Navicula libonensis Schoeman Pl.5-Fig. 13

Ref. Lange-Bertalot 2001 (p.45-pl.43:7–14); Hofmann et al. 2011 (p.388-pl.29:20–24).

Dimensions of studied samples: Valves 28.4–35.3 μm length and 6.0–7.1 μm breadth, 12 striae in 10 μm.

Distribution among the sampling locations:

Distribution in Turkey: It was found only in Northern Anatolia by Tunca et al. (2014).

Navicula lundii Reichardt Pl.5-Fig. 14

Ref.: Lange-Bertalot 2001 p.46-pl.22:17–24

Dimensions of studied samples: Valves 22.2 μm length and 5.9 μm breadth, 16 striae in 10 μm.

Distribution among the sampling locations: TE7.

Distribution in Turkey: NEW RECORD.

Navicula moskalii Metzeltin et al. Pl.5-Fig. 15

Ref.: Lange-Bertalot 2001 (p.50-pl.14:1–14); Hofmann et al. 2011 (p.389-pl.34:9–13).

Dimensions of studied samples: Valves 25.2—26.1 μm length and 7.2–7.8 μm breadth, 12–13 striae in 10 μm.

Distribution among the sampling locations: TE5, TE6, and TE18.

Distribution in Turkey: This was newly recorded in Western Anatolia (Yılmaz 2017).

Navicula novaesiberica Lange-Bertalot Pl.5-Fig. 16

Ref. Lange-Bertalot 2001 (p.89-pl.36:10–14); Hofmann et al. 2011 (p.408-pl.39:21–23).

Dimensions of studied samples: Valves 35.1–44.3 μm length and 8.1–9.2 μm breadth, 9–10 striae in 10 μm.

Distribution among the sampling locations: TE12, TE16, and TE18.

Distribution in Turkey: This was newly recorded in Western Anatolia (Solak et al. 2016a).

Navicula oblonga (Kützing) Kützing Pl.5-Fig. 17

Basionym. *Frustulia oblonga* Kützing.

Ref.: Lange-Bertalot 2001 (p.51-pl.6:12–14); Hofmann et al. 2011 (p.390-pl.34:1–3).

Dimensions of studied samples: Valves 86.4–91.6 μm length and 12.8–14.1 μm breadth, 7–8 striae in 10 μm.

Distribution among the sampling locations: TE17.

Distribution in Turkey: Common.

Navicula radiosa Kützing Pl.5-Fig. 18

Ref.: Lange-Bertalot 2001 (p.59-pl.8:1–7), Hofmann et al. 2011 (p.394-pl.35:1–5).

Dimensions of studied samples: Valves 65.7–70.9 μm length and 9.9–11.0 μm breadth, 10 striae in 10 μm.

Distribution among the sampling locations: TE2, TE9, TE17, and TE18.

Distribution in Turkey: Common.

Navicula reichardtiana Lange-Bertalot Pl.5-Fig. 19

Ref. Lange-Bertalot 2001 (p.63-pl.13:25–35); Hofmann et al. 2011 (p.395-pl.31:29–33).

Dimensions of studied samples: Valves 15.4–18.2 μm length and 4.8–5.1 μm breadth, 16–17 striae in 10 μm.

Distribution among the sampling locations: TE1-TE3 and TE9.

Distribution in Turkey: It was found in only Inner Anatolia by Atıcı and Obalı (2010).

Navicula rostellata Kützing Pl.5-Fig. 20

Ref.: Lange-Bertalot 2001 (p.91-pl.35:1–6), Hofmann et al. 2011 (p.397-pl.37:10–14).

Dimensions of studied samples: Valves 38.7–41.7 μm length and 9.0–9.5 μm breadth, 12–13 striae in 10 μm.

Distribution among the sampling locations: TE9, TE13, and TE20.

Distribution in Turkey: Common.

Navicula simulata Manguin Pl.5-Fig. 21

Ref.: Hofmann et al. 2011 (p.400-pl.38:6–11).

Dimensions of studied samples: Valves 32.1–34.5 μm length and 6.1–6.5 μm breadth, 14 striae in 10 μm.

Distribution among the sampling locations: TE8, TE13, and TE14.

Distribution in Turkey: NEW RECORD.

Navicula tripunctata (O.F.Müller) Bory Pl.5-Fig. 22

Basionym. *Vibrio tripunctatus* O.F. Müller.

Ref.: Lange-Bertalot 2001 (p.73-pl.1:1–8), Hofmann et al. 2011 (p.403-pl.35:11–16).

Dimensions of studied samples: Valves 32.5–47.4 μm length and 8.1–8.4 μm breadth, 11 striae in 10 μm.

Distribution among the sampling locations: TE1, TE2, TE4, TE7-TE9, TE11, TE14, TE16, and TE18-TE20.

Distribution in Turkey: Common.

Navicula trivialis Lange-Bertalot Pl.5-Fig. 23

Ref.: Lange-Bertalot 2001 (p.73-pl.29:1–6), Hofmann et al. 2011 (p.403-pl.35:11–15).

Dimensions of studied samples: Valves 37.4–42.3 μm length and 9.4–10.5 μm breadth, 12 striae in 10 μm.

Distribution among the sampling locations: TE5 and TE11.

Distribution in Turkey: Common.

Navicula vandamii var. ***vandamii*** Schoeman and Archibald Pl.5-Fig. 24

Ref.: Lange-Bertalot 2001 (p.93-pl.37:16–23).

Dimensions of studied samples: Valves 24.9–25.4 μm length and 5.0–5.1 μm breadth, 16 striae in 10 μm.

Distribution among the sampling locations: TE2 and TE19.

Distribution in Turkey: NEW RECORD.

Navicula vandamii var. ***mertens*** Lange-Bertalot Pl.5-Fig. 25

Ref.: Lange-Bertalot 2001 (p.94-pl.38:1–7).

Dimensions of studied samples: Valves 21.8 μm length and 4.6 μm breadth, 18 striae in 10 μm.

Distribution among the sampling locations:

Distribution in Turkey: NEW RECORD.

Navicula veneta Kützing Pl.5-Fig. 26

Ref.: Lange-Bertalot 2001 (p.78-pl.14:23–30), Hofmann et al. 2011 (p.406-pl.31:44–48).

Dimensions of studied samples: Valves 17.9–22.4 μm length and 5.0 μm breadth, 15–16 striae in 10 μm.

Distribution among the sampling locations: TE2, TE4, TE15, TE17, and TE20.

Distribution in Turkey: Common.

Navicula wildii Lange-Bertalot Pl.5-Fig. 27

Ref.: Hofmann et al. 2011 (p.410–35:6–10).

Dimensions of studied samples: Valves 40.7 μm length and 6.8 μm breadth, 13 striae in 10 μm.

Distribution among the sampling locations:

Distribution in Turkey: NEW RECORD.

Navigeia decussis (Østrup) Bukhtiyarova Pl.5-Fig. 28

Basionym. *Navicula decussis* Østrup.

Ref.: Lange-Bertalot 2001 (p.123-pl.95:1–17), Hofmann et al. 2011 (p.284-pl.51:40–44).

Dimensions of studied samples: Valves 20.4–22.1 μm length and 6.8–7.3 μm breadth, 15–16 striae in 10 μm.

Distribution among the sampling locations: TE6 and TE18.

Distribution in Turkey: Common.

Neidiomorpha binodiformis (Krammer) Cantonati et al. Pl.5-Fig. 29

Basionym. *Neidium binodiforme* Krammer.

Ref.: Cantonati et al. 2010 (p.197-Figs.10–16); Hofmann et al. 2011 (p.417-pl.53:23–25).

Dimensions of studied samples: Valves 20.4–22.7 μm length and 6.5–7.1 μm breadth, 25–26 striae in 10 μm.

Distribution among the sampling locations: TE18.

Distribution in Turkey: It was found only in Black sea by Soylu et al. 2010 and Inner Anatolia by Varol and Şen (2014).

Nitzschia acicularis (Kützing) W.Smith Pl.6-Fig. 1

Basionym. *Synedra acicularis* Kützing.

Ref.: Krammer and Lange-Bertalot 1988 (p.123-pl.85:1–4); Hofmann et al. 2011 (p.123-pl.107:20–24).

Dimensions of studied samples: Valves 46.7–53.4 μm length and 3.2–4.8 μm breadth, 17–18 fibulae in 10 μm.

Distribution among the sampling locations: TE17.

Distribution in Turkey: Common.

Nitzschia amphibia Grunow Pl.6-Fig. 2

Ref.: Hofmann et al. 2011 (p.433-pl.117:9–15).

Dimensions of studied samples: Valves 16.4–27.8 μm length and 4.4–4.9 μm breadth, 16–17 striae and 8–9 fibulae in 10 μm.

Distribution among the sampling locations: TE4, TE6-TE8, TE11, and TE12.

Distribution in Turkey: Common.

Nitzschia capitellata Hustedt Pl.6-Fig. 3

Ref.: Krammer and Lange-Bertalot 1988 (p.88-pl.62:1–12); Hofmann et al. 2011 (p.438-pl.113:11–16).

Dimensions of studied samples: Valves 26.3–32.8 μm length and 4.7–5.1 μm breadth, 12–13 fibulae in 10 μm.

Distribution among the sampling locations: TE2, TE5, TE8, TE17, and TE20.

Distribution in Turkey: Common.

Nitzschia clausii Hantzsch Pl.6-Fig. 4

Ref.:Krammer and Lange-Bertalot 1988 (p.27-pl.19:1–6); Hofmann et al. 2011 (p.438-pl.116:15–18).

Dimensions of studied samples: Valves 35.4–39.7 μm length and 4.0–4.4 μm breadth, 12 fibulae in 10 μm.

Distribution among the sampling locations: TE14.

Distribution in Turkey: Common.

Nitzschia communis Rabenhorst Pl.6-Fig. 5

Ref.: Krammer and Lange-Bertalot 1988 (p.110-pl.79:1–6); Hofmann et al. 2011 (p.439-pl.112:1–5).

Dimensions of studied samples: Valves 18.9–23.1 μm length and 4.0–4.8 μm breadth, 11–13 fibulae in 10 μm.

Distribution among the sampling locations: TE17.

Distribution in Turkey: It was found only in Eastern Anatolia by Öztığ (1957) and Gessner (1957) and Inner Anatolia by Baykal (2006).

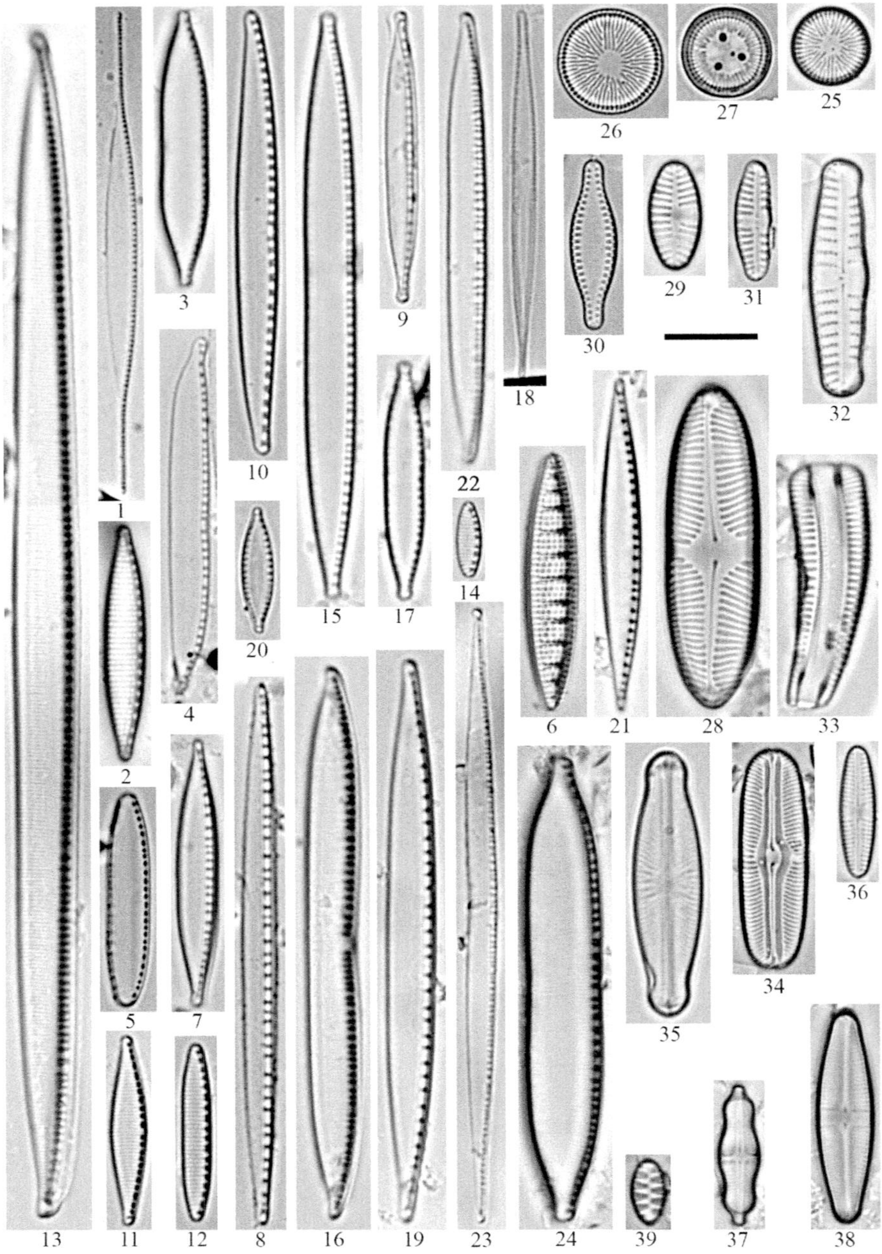

Plate 25.4 1-*Nitzschia acicularis*, 2-*N. amphibia*, 3-*N. capitellata*, 4-*N. clausii*, 5-*N. communis*, 6-*N. denticula*, 7-*N. dissipata*, 8-*N. dissipata* var. *media*, 9-*N. dissipata* ssp. *oligotrophenta*, 10-*N. filiformis*, 11-*N. fonticola*, 12-*N. frustulum*, 13-*N. heufleriana*, 14-*N. inconspicua*, 15-*N. intermedia*, 16-*N. linearis*, 17-*N. palea*, 18-*N. pumila*, 19-*N. recta*, 20-*N. rosenstockii*, 21-*N. sociabilis*, 22-*N. sublinearis*, 23-*N. subtilis*, 24-*N. umbonata*, 25-Pantocsekiella delicatula, 26-*P. iranica*, 27-*P. ocellata*, 28-*Pinnularia brebissonii*, 29-*Planothidium lanceolatum*,

Nitzschia denticula Grunow Pl.6-Fig. 6

Ref.: Krammer and Lange-Bertalot 1988 (p.143-pl.99:11–23); Hofmann et al. 2011 (p.441-pl.26–31).

Dimensions of studied samples: Valves 23.7–28.6 µm length and 4.9–5.3 µm breadth, 5–6 fibulae in 10 µm.

Distribution among the sampling locations: TE18.

Distribution in Turkey: NEW RECORD.

Nitzschia dissipata (Kützing) Rabenhorst Pl.6-Fig. 7

Basionym. *Synedra dissipata* Kützing.

Ref.: Hofmann et al. 2011 (p.441-pl.109:8–13).

Dimensions of studied samples: Valves 23.4–29.1 µm length and 4.3–4.9 µm breadth, 10–11 fibulae in 10 µm.

Distribution among the sampling locations: TE1, TE7, TE15, TE16, TE19, and TE20.

Distribution in Turkey: Common.

Nitzschia dissipata** var. **media (Hantzsch) Grunow Pl.6-Fig. 8

Ref.: Krammer and Lange-Bertalot 1988 (p.19-pl.11:8–14); Hofmann et al. 2011 (p.441:14–18).

Dimensions of studied samples: Valves 43.7–62.4 µm length and 3.6–4.8 µm breadth, 7–9 fibulae in 10 µm.

Distribution among the sampling locations: TE1-TE5, TE7, TE8, TE13, TE15, TE19, and TE20.

Distribution in Turkey: Common.

Nitzschia dissipata** ssp. **oligotrophenta Lange-Bertalot Pl.6-Fig. 9

Ref. Hofmann et al. 2011 (p.453:3–7).

Dimensions of studied samples: Valves 27.4–32.4 µm length and 3.3–3.8 µm breadth, 11–12 fibulae in 10 µm.

Distribution among the sampling locations: TE5.

Distribution in Turkey: NEW RECORD.

Nitzschia filiformis (W.Smith) Van Heurck Pl.6-Fig. 10

Basionym. *Homoecladia filiformis* W. Smith.

Ref. Hofmann et al. 2011 (p.443-pl.116:1–6).

Dimensions of studied samples: Valves 38.4–47.6 µm length and 4.4–4.9 µm breadth, 9–10 fibulae in 10 µm.

Distribution among the sampling locations: TE20.

Distribution in Turkey: Common.

Nitzschia fonticola Grunow Pl.6-Fig. 11

Ref.: Krammer and Lange-Bertalot 1988 (p.103, Fig. 75:1–22); Hofmann et al. 2011 (p.444-pl.108:9–15).

Plate 25.4 (continued) 30-*Pseudostaurosira brevistriata*, 31-*Reimeria sinuata*, 32-*R. uniseriata*, 33-*Rhoicosphenia abbreviata*, 34-*Sellaphora bacillum*, 35-*Sellaphora pupula*, 36-*Sellaphora seminulum*, 37-*Stauroneis separanda*, 38-*Stauroneis tackei*, and 39-*Staurosirella pinnata*. Scale bar: 10 µm.

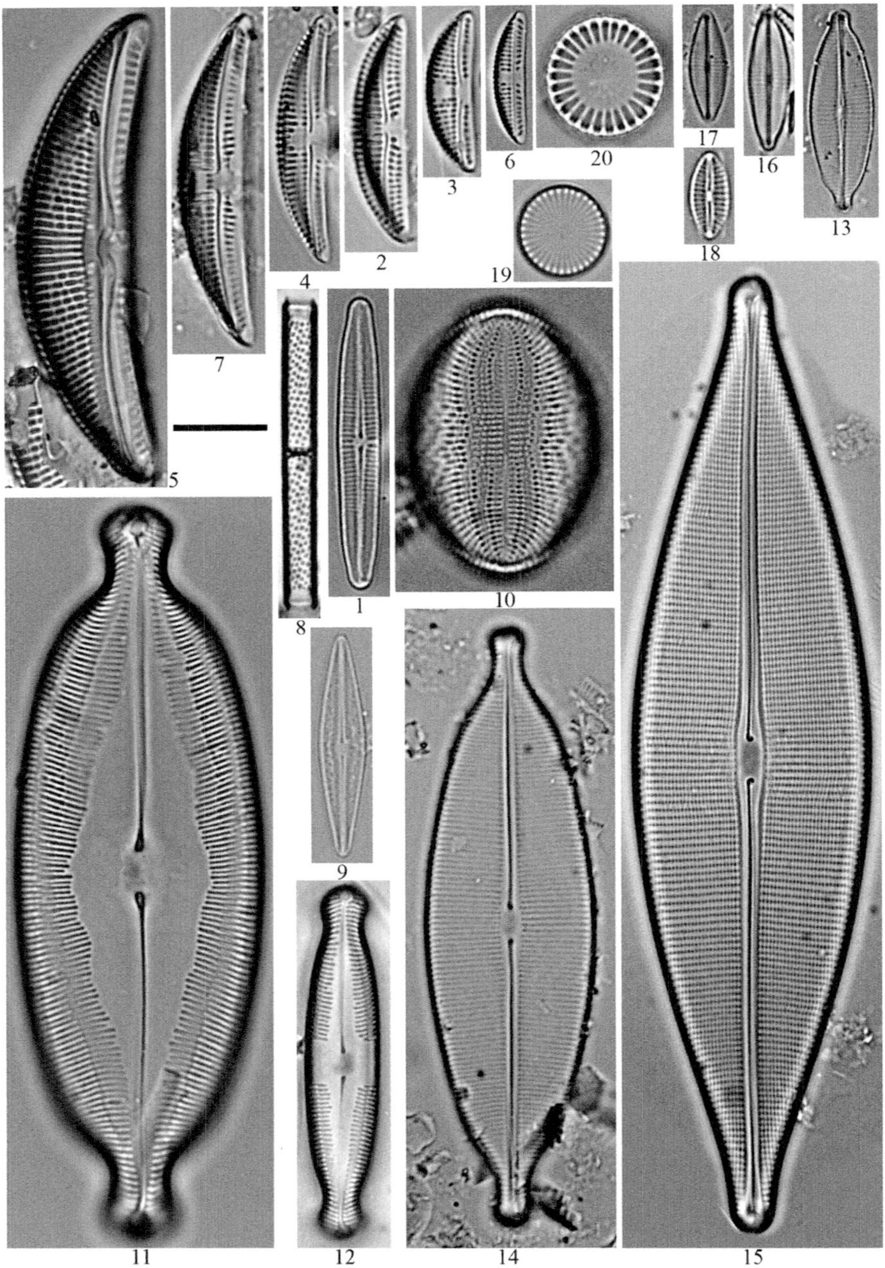

Plate 25.5 1-*Achnanthidium biasolettianum*, 2-*Amphora alpestris*, 3-*A. copulata*, 4-*A. liriope*, 5-*A. ovalis*, 6-*A. pediculus*, 7-*A. lange-bertalotii* var. *tenuis*, 8-*Aulacoseira granulata*, 9-*Brachysira neglectiformis*, 10-*Cocconeis pediculus*, 11-*Caloneis amphisbaena*, 12-*C. macedonica*, 13-*Craticula accomoda*, 14-*C. ambigua*, 15-*C. cuspidata*, 16- *C. minusculoides*, 17-*C. molestiformis*, 18-*C. subminuscula*, 19-*Cyclostephanos invisitatus*, and 20-*Cyclotella meneghiniana*. Scale bar: 10 μm.

Dimensions of studied samples: Valves 19.2–25.7 µm length and 3.6–4.0 µm breadth, 27–28 striae and 10–12 fibulae in 10 µm.

Distribution among the sampling locations: TE1, TE2, TE12, TE16, TE17, and TE19.

Distribution in Turkey: Common.

Nitzschia frustulum (Kützing) Grunow Pl.6-Fig. 12

Basionym. *Synedra frustulum* Kützing.

Ref. Hofmann et al. 2011 (p.445-pl.112:28–34).

Dimensions of studied samples: Valves 17.5–28.7 µm length and 3.2–3.3 µm breadth, 23–24 striae and 10–11 fibulae in 10 µm.

Distribution among the sampling locations: TE1, TE2, TE4, TE6, and TE20.

Distribution in Turkey: Common.

Nitzschia heufleriana Grunow Pl.6-Fig. 13

Ref.: Krammer and Lange-Bertalot 1988 (p.22-pl.13:1–5); Hofmann et al. 2011 (p.448-pl.109:19–23).

Dimensions of studied samples: Valve 113.8–131.4 µm length and 6.2–7.3 µm breadth, 9–10 fibulae and 25–26 striae in 10 µm.

Distribution among the sampling locations: TE1, TE14-TE16, and TE20.

Distribution in Turkey: Common.

Nitzschia inconspicua Grunow Pl.6-Fig. 14

Ref.: Krammer and Lange-Bertalot 1988 (p.95-pl.69:1–13); Hofmann et al. 2011 (p.446-pl.112:35–40).

Dimensions of studied samples: Valves 7.2–9.5 µm length and 2.5–2.9 µm breadth, 9–10 fibulae in 10 µm.

Distribution among the sampling locations: TE2, TE4, TE-TE10, TE14, TE17, and TE20.

Distribution in Turkey: Common.

Nitzschia intermedia Hantzsch Pl.6-Fig. 15

Ref.: Hofmann et al. 2011 (p.449-pl.107:1–8).

Dimensions of studied samples: Valves 58.4–63.1 µm length and 5.2–5.4 µm breadth, 11 fibulae in 10 µm.

Distribution among the sampling locations: TE4, TE13, and TE14.

Distribution in Turkey: Common.

Nitzschia linearis W.Smith Pl.6-Fig. 16

Ref.: Krammer and Lange-Bertalot 1988 (p.69-pl.55:1–4); Hofmann et al. 2011 (p.452-pl.106:1–3).

Dimensions of studied samples: Valves 59.0–71.4 µm length and 5.4–5.8 µm breadth, 11–12 fibulae in 10 µm.

Distribution among the sampling locations: TE, TE2, TE5TE15-TE18, and TE20.

Distribution in Turkey: Common.

Nitzschia palea (Kützing) W.Smith Pl.6-Fig. 17

Basionym. *Synedra palea* Kützing.

Ref.: Krammer and Lange-Bertalot 1988 (p.85-pl.59:1–10); Hofmann et al. 2011 (p.454-pl.111:1–9).

Dimensions of studied samples: Valves 22.7–38.4 μm length and 3.8–4.7 μm breadth, 11–13 fibulae in 10 μm.

Distribution among the sampling locations: TE10 and TE11.

Distribution in Turkey: Common.

Nitzschia pumila Hustedt Pl.6-Fig. 18 Ref.: Krammer and Lange-Bertalot 1998 (p.115-pl.81:14–15).

Dimensions of studied samples: Valves 39.6 μm length and 2.6 μm breadth, 20 fibulae in 10 μm.

Distribution among the sampling locations: TE6.

Distribution in Turkey: NEW RECORD.

Nitzschia recta Hantzsch ex Rabenhorst Pl.6-Fig. 19

Ref.: Krammer and Lange-Bertalot 1988 (p.20-pl.12:1–11).

Dimensions of studied samples: Valves 49.2–58.4 μm length and 4.9–5.4 μm breadth, 6–7 fibulae in 10 μm.

Distribution among the sampling locations: TE16 and TE18.

Distribution in Turkey: Common.

Nitzschia rosenstockii Lange-Bertalot Pl.6-Fig. 20 Ref.: Krammer and Lange-Bertalot 1998 (p.116-pl.81:17-20A).

Dimensions of studied samples: Valves 13.6 μm length and 3.6 μm breadth, 15 fibulae in 10 μm.

Distribution among the sampling locations: TE8.

Distribution in Turkey: NEW RECORD.

Nitzschia sociabilis Hustedt Pl.6-Fig. 21

Ref.: Krammer and Lange-Bertalot 1988 (p.83-pl.83:1–9).

Dimensions of studied samples: Valves 25.7–36.1 μm length and 3.9–4.5 μm breadth, 9–10 fibulae in 10 μm.

Distribution among the sampling locations: TE16.

Distribution in Turkey: Common.

Nitzschia sublinearis Hustedt Pl.6-Fig. 22

Ref. Hofmann et al. 2011 (p.463-pl.107:15–19).

Dimensions of studied samples: Valve 48.4 μm length and 4.1 μm breadth, 14 fibulae in 10 μm.

Distribution among the sampling locations: TE5, TE13, TE15, and TE16.

Distribution in Turkey: Common.

Nitzschia subtilis (Kützing) Grunow Pl.6-Fig. 23

Basionym. *Synedra subtilis* Kützing.

Ref.: Krammer and Lange-Bertalot 1988 (p.70-pl.55:7–10).

Dimensions of studied samples: Valve 66.1 μm length and 3.2 μm breadth, 17 fibulae in 10 μm.

Distribution among the sampling locations: TE15.

Distribution in Turkey: NEW RECORD.

Nitzschia umbonata (Ehrenberg) Lange-Bertalot Pl.6-Fig. 24

Basionym. *Navicula umbonata* Ehrenberg.

Ref.: Krammer and Lange-Bertalot 1988 (p.65-pl.51:1-6A); Hofmann et al. 2011 (p.467-pl.105:4–8).

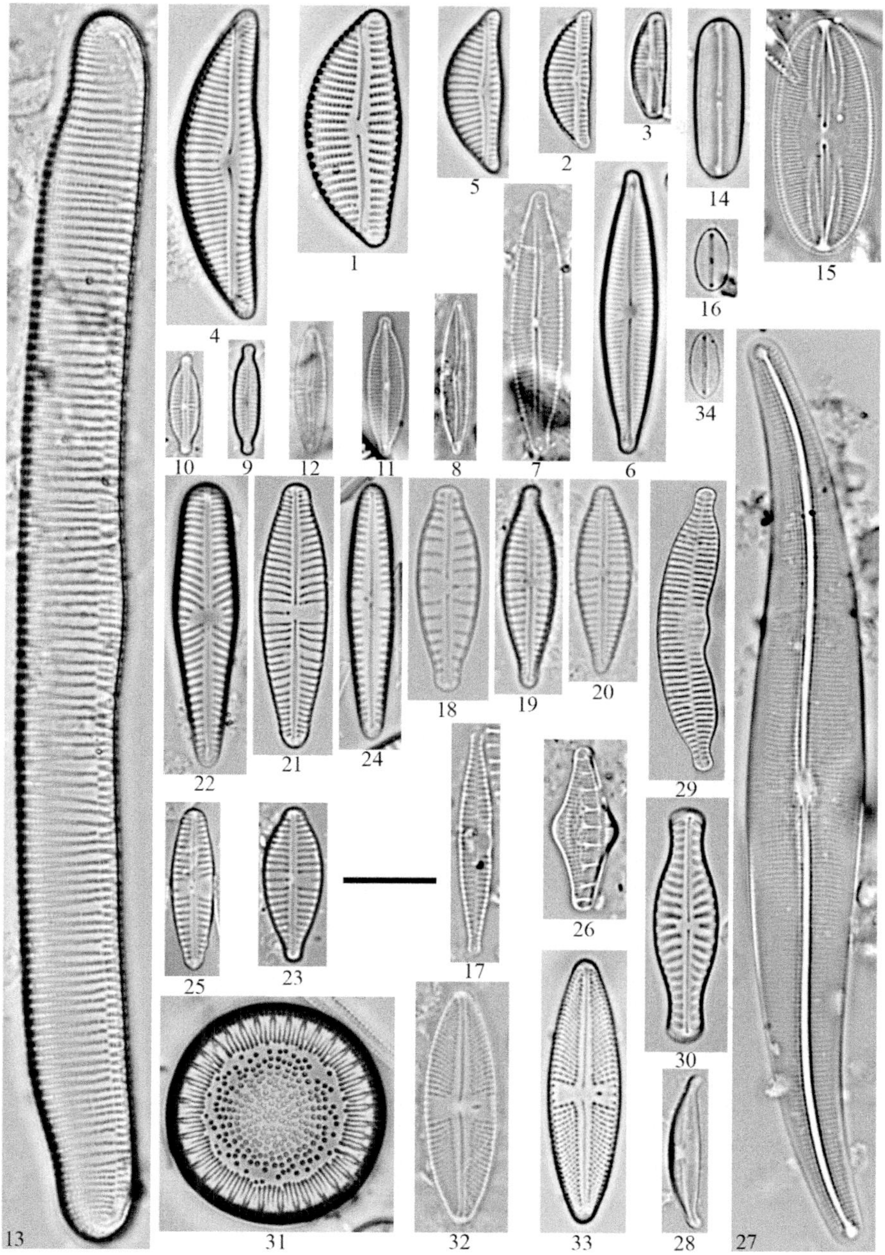

Plate 25.6 1-*Encyonema auerswaldii*, 2-*E. minutum*, 3-*E. reichardtii*, 4-*E. silesiacum*, 5-*E. ventricosum*, 6-*Encyonopsis cesatii*, 7-*E. falaisensis*, 8-*E. krammerii*, 9-*E. microcephala*, 10-*E. minuta*, 11-*E. subminuta*, 12-*E. thumensis*, 13-*Eunotia pectinalis*, 14-*Fallacia lenzii*, 15-*F. pygmaea*, 16-*Fistulifera saprophila*, 17-*Fragilaria austriaca*, 18-*Gomphonema aequale*, 19-*G. exilissimum*, 20-*G. jadwigiae*, 21-*G. micropus*, 22-*G. olivaceum*, 23-*G. parvulum*, 24-*G. pumilum*, 25-*G. tergestinum*, 26-*Grunowia tabellaria*, 27-*Gyrosigma kuetzingii*, 28-*Halamphora montana*, 29-*Hannaea arcus*, 30-*Hippodonta capitata*, 31- *Lindavia balatonis*, 32-*Luticola hlubikovae*, 33-*L. goeppertiana*, and 34-*Mayamaea permitis*. Scale bar: 10 μm.

Dimensions of studied samples: Valves 48.4–63.2 μm length and 7.7–8.0 μm breadth, 7–9 fibulae in 10 μm.

Distribution among the sampling locations: TE10 and TE11.

Distribution in Turkey: Common.

Pantocsekiella delicatula (Hustedt) K.T.Kiss and E.Ács Pl.6-Fig. 25

Basionym. *Cyclotella delicatula* Hustedt.

Ref. Krammer and Lange-Bertalot 1991b (p.51-pl.52:3), Wojtal and Kwandrans 2006 (pl.7:14–19, 8: 1–7), Houk et al. 2010 (p.32-pl.228:1–19).

Dimensions of studied samples: Valve 8.4–16.7 μm diameter,18–21 interstriae in 10 μm.

Distribution among the sampling locations: TE3, TE4, and TE14.

Distribution in Turkey: This was newly recorded in Western Anatolia (Solak et al. 2018).

Pantocsekiella iranica (T.Nejadsattari et al.) K.T.Kiss et al. Pl.6-Fig. 26

Basionym. *Cyclotella iranica* T. Nejadsattari et al.

Ref. Kheiri et al. 2013 (p.37, Fig. 2–10).

Dimensions of studied samples: Valve 10.7–13.5 μm diameter, 16–18 interstriae in 10 μm.

Distribution among the sampling locations: TE14.

Distribution in Turkey: This was newly recorded in Western Anatolia (Solak et al. 2018).

Pantocsekiella ocellata (Pantocsek) K.T.Kiss and E.Ács Pl.6-Fig. 27

Basionym. *Cyclotella ocellata* Pantocsek.

Ref. Krammer and Lange-Bertalot 1991b (p.51-pl.50:1–11), Håkansson 2002 (p.85-Figs.309–318), Wojtal and Kwandrans 2006 (p.188-pl.7:26–27), Kiss et al. 2012 (p.339-pl.15:A-C), Bey and Ector 2013 (Vol.1, p.32) Houk et al. 2010 (p.26-pl.197:1–10).

Dimensions of studied samples: Valve 9.5–13.4 μm diameter, 15–20 interstriae in 10 μm.

Distribution among the sampling locations: TE3 and TE20.

Distribution in Turkey: Common.

Pinnularia brebissonii (Kützing) Rabenhorst Pl.6-Fig. 28

Basionym. *Navicula brebissonii* Kützing.

Dimensions of studied samples: Valves 34.7–38.9 μm length and 9.8–11.2- μm width, 13–14 striae in 10 μm.

Distribution among the sampling locations: TE1 and TE20.

Distribution in Turkey: Common.

Planothidium lanceolatum (Brébisson ex Kützing) Lange-Bertalot Pl.6-Fig. 29

Basionym. *Achnanthidium lanceolatum* Brébisson ex Kützing.

Ref.: Krammer and Lange-Bertalot 1991b (p.75-pl.41:1–8), Hofmann et al. 2011 (p.510-pl.24:41–47).

Dimensions of studied samples: Valves 10.7–16.4 μm length and 5.7–6.4 μm width, 12–14 striae in 10 μm.

Distribution among the sampling locations: TE2 and TE17.

Distribution in Turkey: Common.

Pseudostaurosira brevistriata (Grunow) D.M.Williams and Round Pl.6-Fig. 30

Basionym. *Fragilaria brevistriata* Grunow.

Ref.: Krammer and Lange-Bertalot 1991a (p.162-pl.130:9–16), Hofmann et al. 2011 (p.258-pl.9:25–29).

Dimensions of studied samples: Valves 14.7–18.5 μm length, 4.5–5.1 μm width, 14–15 striae in 10 μm.

Distribution among the sampling locations: TE17.

Distribution in Turkey: Common.

Reimeria sinuata (W.Gregory) Kociolek and Stoermer Pl.6-Fig. 31

Basionym. *Cymbella sinuata* Gregory.

Ref.: Krammer and Lange-Bertalot 1986 (p.341-pl.148:10–17), Hofmann et al. 2011 (p.526-pl.89:50–61).

Dimensions of studied samples: Valves 12.4–15.7 μm length and 3.9–4.6 μm breadth, 11–12 striae in 10 μm.

Distribution among the sampling locations: TE2-TE4, TE9, TE12, TE15, TE16, TE18, and TE19.

Distribution in Turkey: Common.

Reimeria uniseriata S.E.Sala, J.M.Guerrero and M.E.Ferrario Pl.6-Fig. 32

Ref.: Bąk et al. 2012 (p.297-pl.62), Bey and Ector 2013 (Vol.5, p.970).

Dimensions of studied samples: Valves 22.9–26.1 μm length and 5.9–6.3 μm breadth, 9–10 striae in 10 μm.

Distribution among the sampling locations: TE1, TE12, and TE20.

Distribution in Turkey: Rare species in Turkish inland waters (Solak et al. 2016b).

Rhoicosphenia abbreviata (C.Agardh) Lange-Bertalot Pl.6-Fig. 33

Basionym. *Gomphonema abbreviatum* Agardh.

Ref.: Krammer and Lange-Bertalot 1986 (p.381-pl.91:20–28); Hofmann et al. 2011 (p.527-pl.18:42–48).

Dimensions of studied samples: Valves 26.5–28.5 μm length and 7.9–8.4 μm breadth, 14–16 striae in 10 μm.

Distribution among the sampling locations: TE1, TE2, TE4, TE8, TE11, and TE18.

Distribution in Turkey: Common.

Sellaphora bacillum (Ehrenberg) D.G.Mann Pl.6-Fig. 34

Basionym. *Navicula bacillum* Ehrenberg.

Ref.: Krammer and Lange-Bertalot 1986 (p.187-pl.67:2–4); Hofmann et al. 2011 (p.532-pl.41:15–20).

Dimensions of studied samples: Valves 23.5 μm length and 7.3 μm breadth, 20 striae in 10 μm.

Distribution among the sampling locations: TE18.

Distribution in Turkey: Common.

Sellaphora pupula (Kützing) Mereschkowsky Pl.6-Fig. 35

Basionym. *Navicula pupula* Kützing.

Ref.: Krammer and Lange-Bertalot 1986 (p.187-pl.68:1–21); Hofmann et al. 2011 (p.536-pl.41:1,2).

Dimensions of studied samples: Valves 18.7–29.2 μm length and 6.8–7.6 μm breadth, 20–21 striae in 10 μm.

Distribution among the sampling locations: TE18 and TE19.

Distribution in Turkey: Common.

Sellaphora seminulum (Grunow) D.G.Mann Pl.6-Fig. 36

Basionym. *Navicula seminulum* Grunow.

Ref. Hofmann et al. (2011, as *Sellaphora joubaudii*) (p.533 - pl.42: 27–31); Wetzel et al. 2015 p.215-Figs. 158–179, 295–301).

Dimensions of studied samples: Valves 12.4–14.3 μm length and 3.5–3.7 μm breadth, 16–17 striae in 10 μm.

Distribution among the sampling locations: TE8 and TE11.

Distribution in Turkey:: It was found only in Inner Anatolia by Baykal and Açıkgöz (2004).

Stauroneis separanda Lange-Bertalot and Werum Pl.6-Fig. 37

Ref.: Hofmann et al. 2011 (p.547-pl.58:4–7); Bey and Ector 2013 (Vol.4, p.740).

Dimensions of studied samples: Valves 13.5–16.0 μm length and 4.6–4.9 μm breadth, 25 striae in 10 μm.

Distribution among the sampling locations: TE18 and TE20.

Distribution in Turkey: NEW RECORD.

Stauroneis tackei (Hustedt) Krammer and Lange-Bertalot Pl.6-Fig. 38

Basionym. *Navicula tackei* Hustedt.

Bey and Ector 2013 (Vol.4, p.747).

Dimensions of studied samples: Valves 18.9–22.6 μm length and 5.9–6.5 μm width, 24–25 striae in 10 μm.

Distribution among the sampling locations: TE17.

Distribution in Turkey: This was newly recorded in Western Anatolia (Yılmaz 2017).

Staurosirella pinnata (Ehrenberg) D.M.Williams and Round Pl.6-Fig. 39

Ref.: Krammer and Lange-Bertalot 1991a (p.156-pl.133:1–18); Hofmann et al. 2011 (p.265-pl.10:41–46).

Dimensions of studied samples: Valves 7.1–9.2 μm length and 3.6–4.2 μm width, 11–12 striae in 10 μm.

Distribution among the sampling locations: TE20.

Distribution in Turkey: Common.

Stephanodiscus balatonis Pantocsek Pl.7-Fig. 1

Ref. Houk et al. 2014 (p.33-pl.29:1–8).

Dimensions of studied samples: Valve 13.6–16.5 μm diameter, 10 interstriae in 10 μm.

Distribution among the sampling locations: TE19.

Distribution in Turkey: This was newly recorded in Inner Anatolia (Solak et al. 2018).

Stephanodiscus neoastreae Håkansson and Hickel Pl.7-Fig. 2

Ref. Krammer and Lange-Bertalot 1991 (p.68-pl.69:3, 71:3a-5b), Kiss et al. 2012 (p.348-pl.20:D-F), Bey and Ector 2013 (Vol.1, p.68), Houk et al. 2014 (p.31-pl.89:1–6).

Dimensions of studied samples: Valve 15.9–17.2 μm diameter, 9–10 interstriae in 10 μm.

Distribution among the sampling locations: TE20.

Distribution in Turkey: This was newly recorded in Inner Anatolia (Solak et al. 2018).

Surirella angusta Kützing Pl.7-Fig. 3

Ref.: Krammer and Lange-Bertalot 1988 (p.187-pl.133:6–13); Hofmann et al. 2011 (p.554-pl.131:1–5).

Dimensions of studied samples: Valves 31.2–35.7 μm length and 7.9–8.3 μm breadth, 6–7 fibulae in 10 μm.

Distribution among the sampling locations: TE2, TE4, TE7, TE17, and TE18.

Distribution in Turkey: Common.

Surirella brebissonii Krammer and Lange-Bertalot Pl.7-Fig. 4

Ref.: Hofmann et al. 2011 (p.556-pl.130:11–16).

Dimensions of studied samples: Valves 28.5 μm length and 17.3 μm breadth, 5 fibulae in 10 μm.

Distribution among the sampling locations: TE8.

Distribution in Turkey: Common.

Surirella minuta Brébisson ex Kützing Pl.7-Fig. 5

Ref.: Hofmann et al. 2011 (p.558-pl.131:6–12).

Dimensions of studied samples: Valves 17.2–23.4 μm length and 9.7–10.3 μm breadth, 7–8 fibulae in 10 μm.

Distribution among the sampling locations: TE2, TE7, and TE14-TE16.

Distribution in Turkey: Common.

Tabularia fasciculata (C.Agardh) D.M.Williams and Round Pl.7-Fig. 6

Basionym. *Diatoma fasciculata* C. Agardh.

Ref.: Hofmann et al. 2011 (p.564-pl.4:3–7).

Dimensions of studied samples: Valves 51.9 μm length and 3.1 μm breadth, 18 striae in 10 μm.

Distribution among the sampling locations: TE17.

Distribution in Turkey: Common.

Tryblionella apiculata W.Gregory Pl.7-Fig. 7

Ref.: Krammer and Lange-Bertalot 1988 (p.43-pl.35:1–6); Hofmann et al. (p.439-pl.104:18–22).

Dimensions of studied samples: Valves 28.4–37.4 μm length and 5.4–6.1 μm breadth, 15–16 fibulae in 10 μm.

Distribution among the sampling locations: TE4, TE5, TE7, TE14, and TE17.

Distribution in Turkey: Common.

Tryblionella hungarica (Grunow) Frenguelli Pl.7-Fig. 8

Basionym. *Nitzschia hungarica* Grunow.

Ref.: Krammer and Lange-Bertalot 1988 (p.37-pl.34:1–3); Hofmann et al. (p.449-pl.104:3–7).

Dimensions of studied samples: Valves 49.4–68.3 μm length and 7.2–7.7 μm breadth, 18 striae in 10 μm.

Distribution among the sampling locations: TE17 and TE20.

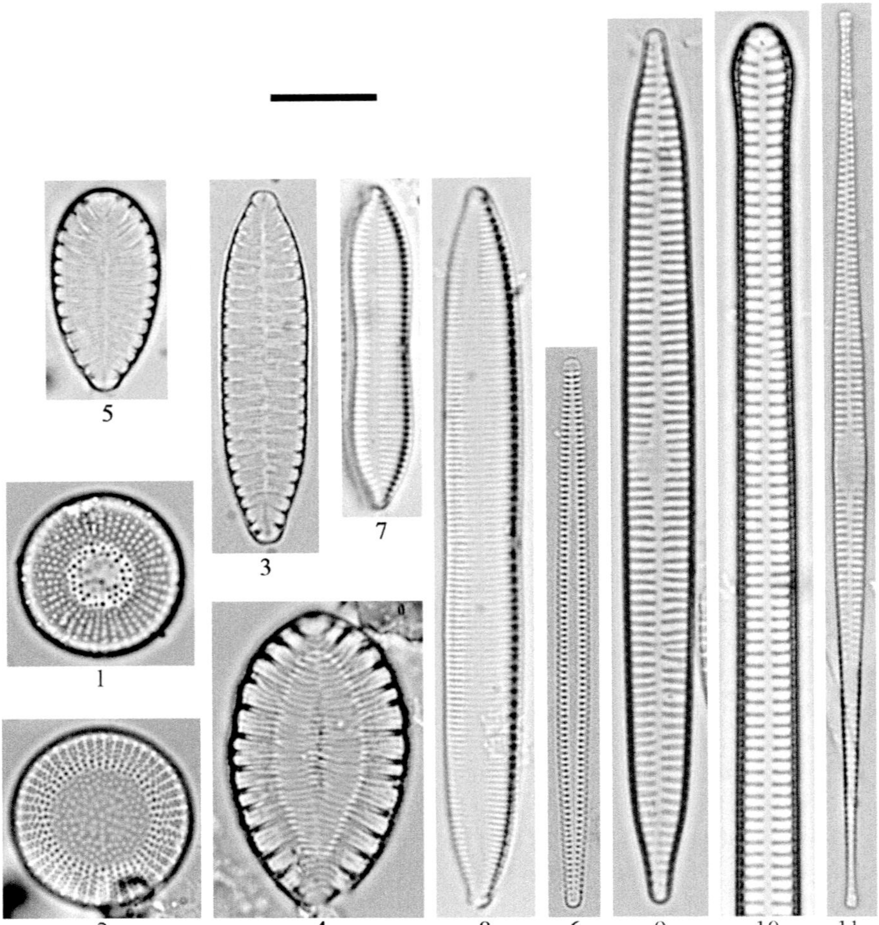

Plate 25.7 1-*Stephanodiscus balatonis*, 2-*S. neoastreae*, 3-*Surirella angusta*, 4-*S. brebissonii*, 5-*S. minuta*, 6-*Tabularia fasciculata*, 7-*Tryblionella apiculata*, 8-*T. hungarica*, 9-*Ulnaria acus*, 10-*U. biceps*, and 11-*U. ulna*. Scale bar: 10 µm.

Distribution in Turkey: Common.

Ulnaria acus (Kützing) Aboal Pl.7-Fig. 9

Basionym. *Synedra acus* Kützing.

Ref.: Krammer and Lange-Bertalot 1991a (p.134-pl.117:8–14); Bey and Ector 2013 (Vol.2, p.282).

Dimensions of studied samples: Valves 104.1–112.1 µm length and 4.2–4.4 µm width, 11–12 striae in 10 µm.

Distribution among the sampling locations: TE12 and TE19.
Distribution in Turkey: Common.
Ulnaria biceps (Kützing) Compère Pl.7-Fig. 10
Basionym. *Synedra biceps* Kützing.
Ref.: Krammer and Lange-Bertalot 1991a (p.146-pl.121:1–5); Bey and Ector 2013 (Vol.2, p.284).
Dimensions of studied samples: Valves 170.4 μm length and 6.2 μm width, 10 striae in 10 μm.
Distribution among the sampling locations: TE1, TE5, and TE20.
Distribution in Turkey: Common.
Ulnaria ulna (Nitzsch) Compère Pl.7-Fig. 11
Basionym. *Bacillaria ulna* Nitzsch.
Ref.: Krammer and Lange-Bertalot 1991a (p.143-pl.122:1–8); Hofmann et al. 2011 (p.276-pl.5:6–11).
Dimensions of studied samples: Valves 87.4–104.3 μm length and 6.1–7.9 μm width, 11–13 striae in 10 μm.
Distribution among the sampling locations: TE1, TE11, TE12, and TE19.
Distribution in Turkey: Common.

25.4 Discussion

The first study on the diatom composition in the basin was done by Ehrenberg (1844) in Karasu River, which is one of the main tributaries of Euphrates River. After the work, it is about 11.6% of total studies in Turkey by different research groups in this region. However, there are more comprehensive works in Euphrates catchment according to the diatoms (Table 25.3).

Amphora pediculus, Cocconeis pediculus, Cymbella excisa, Navicula capitatoradiata, Navicula tripunctata, and *Nitzschia dissipata* var. *media* were the most abundant species in the sampling stations. These taxa are also very common in other river systems in Turkey (Solak et al. 2012). *Cymbella affinis* Kützing is one of the most common species in Turkish flora. The species were revised by Krammer (2002), and the taxa was split out several new species like *Cymbella excisa*, which was also recorded from running waters in Kütahya. Two varieties of this species (*Cymbella excisa* var. *procera* and *C. excisa* var. *subcapitata*) were recorded in this study. *Cymbella excisa* var. *subcapitata* was widely distributed and abundant in Europe and North America. *Amphora* genus was represented with six taxa; *A. ovalis* and *A. pediculus* are very common species in Turkish inland waters, while *Amphora alpestris, A. copulata, A. liriope*, and *A. lange-bertalotii* var. *tenuis* were new records for the flora. *A. alpestris, A. lange-bertalotii* var. *tenuis*, and *A. copulata* are confused each other; however, striae density and shape of central area are distinctive features to differ the taxa (Levkov 2009). *Encyonopsis* genus is one of the less representative taxa in Turkish flora and represented with 7 species in Turkey. Among them, *E. cesatii* is only found in Inner Anatolia and Black sea Regions.

Table 25.3 List of literature comprising diatom data for rivers, lakes, and reservoirs in the basin

River Basin/ Size	Rivers/Lakes/Reservoirs (References)
Tigris	**Rivers** Kırkgözeler (Şen and Aksakal 1988) **Lakes/reservoirs** Devegeçidi Dam Lake (Baykal et al. 2004), Dicle River Basin Wetlands (Açıkgöz et al. 2007, Varol and Şen 2017)
Euphrates	**Rivers** Karasu River (Ehrenberg 1844, Altuner and Gürbüz 1988, 1989, 1990, 1991; Gürbüz and Kivrak 2002; Gürbüz and Ertuğrul 2003), Selli River (Şen et al. 1995), Keban River (Aksın et al. 1999), Cip River (Yavuz and Çetin 2000, 2001), Kürk River (Yildirim et al. 2003), Peri River (Pala-Toprak and Çaglar 2008; Yildirim and Çetin 2009), Kemaliye (Akbulut et al. 2009b), Kozluk Creek (Pala-Toprak et al. 2017) **Lakes/reservoirs** Hazar Lake (Şen 1988, Şen et al. 1999, Maraşlıoğlu and Soylu 2018), Tercan Dam Reservoir (Altuner and Gürbüz 1990, 1994, 1996), Keban Dam Reservoir (Akbay et al. 1999, Çetin and Şen 1998, 2006, Pala-Toprak and Çaglar 2006; Pala-Toprak 2007), Deli Çermik Termal Pond (Altuner and Pabuçcu 1993, 1994), Palandöken Pond (Gürbüz 2000; Gürbüz and Altuner 2000), Sürgü Dam Reservoir (Çetin and Yıldırım 2000, 2003), Tadım Pond (Şen et al. 2001), Kuzgun Dam Reservoir (Gürbüz and Kivrak 2003), Orduzu Dam Reservoir (Çetin and Şen 2004), Özlüce Dam Reservoir (Şen et al. 2005), Suluçayır Pond (Özer and Pala-Toprak 2009)

However, the species is also common in Kütahya flowing waters (unpublished data); *E. microcephala* is common for the flora. *E. eifeliana, E. lanceola, E. moseri, E. minuta*, and *E. subminuta* were newly recorded from Western Anatolia (Barinova et al. 2014, Solak et al. 2016b) and Black sea Region (Kara and Şahin 2001). Additionally, *E. falaisensis, E. krammeri*, and *E. thumensis* were new record for the flora. *Navicula* genus is the most representative group in the study. The members of the genus consist of 17.5% of total diatom taxa. Among the taxa, *N. capitatoradiata, N. cincta, N. cryptotenella, N. gregaria, N. lanceolata, N. oblonga, N. radiosa, N. rostellata, N. tripunctata, N trivialis*, and *N. veneta* were the most common taxa, while, *N. exilis, N. lacuum, N. lundii, N. simulata, N. vandamii, N. vandamii* var. *mertens*, and *N. wildii* were new record for the flora.

Acknowledgement The authors would like to thank to Dr. Ali Salur. The samples were collected by him

References

Açıkgöz İ, Baykal T, Akbulut A, Udoh A, Yıldız K, Şen B (2007) Güneydoğu Anadolu Bölgesindeki (Dicle Havzası) Önemli Sulak Alanların Alg Florasında Gözlenen Bol ve Yaygın Türlerin Dağılımı. XIV. Ulusal Su Ürünleri Sempozyumu. 4–7 Eylül 2007, Muğla

Ács É, Szabó K, Tóth B, Kiss KT (2004) Investigation of benthic algal communities, especially diatoms of some Hungarian streams in connection with reference conditions of the water framework directives. Acta Bot Hungar 46(3–4):255–277

Akbay N, Anul N, Yerli S, Soyupak S, Yurteri C (1999) Seasonal distribution of phytoplankton in the Keban dam reservoir. J Plankton Res 21(4):771–787

Akbulut NE, Bayarı S, Akbulut A, Şahin Y (2009a) Rivers of Turkey. In: Tockner K, Robinson CT, Uehlinger U (eds) Rivers of Europe. Elsevier, Oxford, pp 643–672

Akbulut A, Demirsoy A, Akbulut N, Arslan N, Özgür ME, Darilmaz M (2009b) Kemaliye ve Çevresinin Sucul Canlılar Açısından Değerlendirilmesi. 5.Geleneksel Ulusal Su Ürünleri Bilimsel ve Kültürel Platformu. 31 Mayıs-1 Haziran 2008, Erzincan

Aksin M, Çetin K, Yildirim V (1999) Keban Çayı (Elazığ – Turkey) algleri. F Ü Fen ve Müh Biliml Derg 11(1):59–65

Albay M, Aykulu G (1994) Algal features of Göksu stream (Istanbul) ll. Epipelic algae Istanbul University. J Fish Aquat Prod 8:119–128

Altuner Z, Gürbüz H (1988) Karasu Nehrinin epilitik diyatomeleri. IX. Ulusal Biyoloji Kongresi: 223–230. 21–23rd Eylül 1988, Sivas

Altuner Z, Gürbüz H (1989) Karasu (Fırat) Nehri fitoplanktonu üzerinde bir arastırma. İ Ü Su Ürünleri Derg 3(1–2):151–176

Altuner Z, Gürbüz H (1990). Tercan Baraj Gölü fitoplanktonları üzerine bir çalışma. X. Ulusal Biyoloji Kongresi: 131–140. 18–20 Temmuz 1990, Erzurum

Altuner Z, Gürbüz H (1991) Karasu (Fırat) Nehri epipelik alg florası üzerine bir arastırma. Türk Bot Derg 15(3):253–267

Altuner Z, Pabuçcu K (1993) Köprüköy Deli Çermik alg florası-I. İ Ü Su Ürünleri Derg. 7:77–90

Altuner Z, Pabuçcu K (1994) Köprüköy Deli Çermik alg florası-II. İ Ü Su Ürünleri Derg 8:95–114

Atıcı T, Obalı O (2010) The diatoms of Asartepe dam Lake (Ankara), with environmental and some physicochemical properties. Turk J Bot 34:541–548

Bąk M, Witkowski A, Żelazna-Wieczorek J, Wojtal AZ, Szczepocka E, Szulc K, Szulc B (2012) Klucz do Oznaczania Okrzemek w Fitobentosie na Potrzeby Oceny Stanu Ekologicznego wód Powierzchniowych w Polsce. Biblioteka Monitoringu Środowiska, Warszawa

Barinova S, Romanov R, Solak CN (2014) New record of *Chara hispida* (L.) Hartm. (Streptophyta: Charophyceae, Charales) from the Işıklı Lake (Turkey) and critical checklist of Turkish Charophytes. Natural Res Conserv 2(3):33–42

Baykal T (2006) Çamlıdere Baraj Gölü diyatomeleri. İ Ü Su ürünleri Dergisi 20:45–63

Baykal T, Açıkgöz İ (2004) Hirfanlı Baraj Gölü algleri. Gazi Üniversitesi Kırşehir Eğitim Fakültesi 5(2):115–136

Baykal T, Açikgöz İ, Yildiz K, Bekleyen A (2004) A study on algae in Devegeçidi Dam Lake. Türk Bot Derg 28:457–472

Baykal T, Akbulut A, Açıkgöz İ, Udoh AU, Yıldız K, Şen B (2009) New records for the freshwater algae of Turkey. Turk J Bot 33:141–152

Baytut Ö, Gönülol A (2016) Phytoplankton distribution and variation along a freshwater-marine transition zone (Kızılırmak River) in the Black Sea. Oceanol Hydrobiol Stud 45(4):453–465

Bey MY, Ector L (2013) Atlas des diatomées des cours d'eau de la région Rhône-Alpes, p 1182

Budzyňska A, Wojtal AZ (2011) The centric diatom *Puncticulata balatonis* (Pantocsek) Wojtal et Budzyňska, comb. nov., in the plankton of eutrophic-hypertrophic Rusalka Lake (Western Poland). Nova Hedwigia 93:509–524

Cantonati M, Lange-Bertalot H, Angeli N (2010) *Neidiomorpha* gen. Nov. (Bacillariophyta): a new freshwater diatom genus separated from *Neidium* Pfitzer. Bot Stud 51:195–202

Cavalcante KP, Tremarin PI, Ludwig TAV (2013) Taxonomic studies of centric diatoms (Diatomeae): unusual nanoplanktonic forms and new records for Brazil. Acta Bot Bras 27:237–251

Çelekli A (2006) Net diatom (Bacillariophyceae) flora of Lake Gölköy (Bolu). Turk J Bot 30:359–374

Çetin AK, Şen B (1998) Diatoms (Bacillariophyta) in the phytoplankton of Keban reservoir and their seasonal variations. Türk Bot Derg 22:25–33
Çetin AK, Şen B (2004) Seasonal distribution of phytoplankton in Orduzu dam Lake (Malatya, Turkey). Türk Bot Derg 28:279–285
Çetin AK, Şen B (2006) Benthic diatom flora (Bacillariophyta) of Keban reservoir (Elazig, Turkey). Int J Algae 8(2):162–173
Çetin AK, Yıldırım V (2000) Species composition and seasonal variation of the phytoplankton in Sürgü reservoir (Malatya, Turkey). Acta Hydrobiol 42:21–28
Çetin AK, Yıldırım V (2003) Epilithic and epiphytic diatoms of the Surgu reservoir (Malatya, Turkey). Int J Algae 5(1):37–45
Çetin AK, Yıldırım V (2007) Distribution and occurrence of the diatom community in Göksu stream, Adiyaman. Turkey Fresen Environm Bull 16:555–560
Dalkıran N, Zünbülgil B, Karacaoğlu D, Dere Ş (2016) Uluabatgölü epifitik diyatomelerinin uzun dönemdeki değişimleri. LimnoFish 2(3):153–163
Demir AN, Fakıoğlu Ö, Dural B (2014) Phytoplankton functional groups provide a quality assesment method by the Q assemblage index in Lake Mogan (Turkey). Turk J Bot 38:169–179
Ehrenberg CG (1844) Untersuchungen über die kleinsten Lebensformen im Quellenlande des Euphrats und Araxes, so wie über eine an neuen Formen sehr reiche, marine Tripelbildung von den Bermuda-Inseln. Ber Bekanntm Verh Königl Preuss Akad Wiss Berlin 1844:253–275
Gessner F (1957) Van See zur limnologie des großen Soda - Sees in Ostananatolien (Turkei). Arch Hydrobiol 53:1–22
Gönülol A (2017) Turkish algae electronic publication, Samsun, Turkey. http://turkiyealgleri.omu.edu.tr
Gürbüz H (2000) Palandöken Göleti bentik alg florası üzerinde kalitatif ve kantitatif bir arastırma. Türk Biyol Derg 24:31–48
Gürbüz H, Altuner Z (2000) A quantitative and qualitative study on the phytoplankton of Palandöken pond. Türk Biyol Derg 24:13–30
Gürbüz H, Ertuğrul M (2003) Determination of medium elements in algae of Karasu (Fırat) river by EDXRF using an 241Am excitation source. Instrumentatıon Sci. Tech 31(2):189–196
Gürbüz H, Kivrak E (2002) Use of epilithic diatoms to evaluate water quality in the Karasu River of Turkey. J Environm Biol 23(3):239–246
Gürbüz H, Kivrak E (2003) Seasonal variations of benthic algae of Kuzgun dam reservoir and their relationship to environmental factors. Fresen Environm Bull 12(9):1025–1032
Håkansson H (2002) A compilation and evaluation of species in the general *Stephanodiscus, Cyclostephanos* and *Cyclotella* with a new genus in the family Stephanodiscaceae. Diatom Res 17:1–139
Hassan FM, Maysoon MS, Awaz BM (2010) A limnlogical study in Euphrates River from Al-Hindiya barrage to Al- Kifil city, Iraq. Basrah J Sci 28(2):273–288
Hofmann G, Werum M, Lange-Bertalot H (2011) Diatomeen im Süßwasser-Benthos von Mitteleuropa. Bestimmungsflora Kieselalgen für die ökologische Praxis. Über 700 der häufigsten Arten und ihre Ökologie. A.R.G. Gantner Verlag K.G, Rugell, LI, p 908
Houk V, Klee R, Tanaka H (2010) Atlas of freshwater centric diatoms with a brief key and descriptions. Part III. Stephanodiscaceae a: *Cyclotella, Tertiarius, Discostella*. Fottea 10 (Supplement):1–498
Houk V, Klee R, Tanaka H (2014) Atlas of freshwater centric diatoms with a brief key and descriptions. Part IV. Stephanodiscaceae B. Fottea 14(Supplement):1–530
Kara H, Şahin B (2001) Epipelik and epilithic algae of Değirmendere River (Trabzon-Turkey). Turk J Bot 25:177–186
Karacaoğlu D, Dere Ş, Dalkıran N (2004) A taxonomic study on the phytoplankton of Lake Ulubat (Bursa). Turk J Bot 28:473–485
Kheiri S, Nejadsattari T, Asri Y, Hamdi SMM, Spaulding S, Edlund MB (2013) Cyclotella *iranica* sp. nov. (Bacillariophyta: Coscinodiscophyceae), a new diatom from the Karaj River, Iran. Phytotaxa 104:35–42

Kiss KT, Klee R, Ector L, Ács É (2012) Centric diatoms of large rivers and tributaries in Hungary. - morphology and biogeographic distribution. Acta Bot Croat 71:311–363
Koçer MA, Şen B (2012) The seasonal succession of diatoms in phytoplankton of a soda Lake (lake Hazar, Turkey). Turk J Bot 36:738–746
Krammer K (1997a) Die cymbelloiden Diatomeen. Eine Monographie der weltweit bekannten Taxa. Teil 1. Allgemeines und *Encyonema* part. Bibl Diatomol 36:1–382
Krammer K (1997b) Die cymbelloiden Diatomeen. Eine Monographie der weltweit bekannten Taxa. Teil 2. Encyonema part., *Encyonopsis* und *Cymbellopsis*. Bibl Diatomol 37:1–469
Krammer K (2002) Diatoms of the European inland waters and comparable habitats: *Cymbella*. Diatoms Europe 3:1–584
Krammer K (2003) Diatoms of the European inland waters and comparable habitats: *Cymbopleura, Delicata, Navicymbula, Gomphocymbellopsis, Afrocymbella*. Diatoms Europe 4:1–530
Krammer K, Lange-Bertalot H (1986–1991b) In: Ettl H, Gerloff J, Heynig H, Mollenhauer D (eds) Süsswasserflora von Mitteleuropa: Bacillariophyceae. Gustav Fischer Verlag, Stuttgart
Lange-Bertalot H (2001) Diatoms of the European inland waters and comparable habitats: *Navicula* sensu stricto. Diatoms Europe 2:1–526
Lange-Bertalot H, Bąk M, Witkowski A, Tagliaventi N (2011) Diatoms of the European inland waters and comparable habitats: *Eunotia* and some related genera. Diatoms Europe 6:1–747
Levkov Z (2009) Diatoms of the European inland waters and comparable habitats: *Amphora* sensu stricto. Diatoms Europe 5:1–916
Levkov Z, Metzeltin D, Pavlov A (2013) Diatoms of the European inland waters and comparable habitats: *Luticola* and *Luticopsis*. Diatoms Europe 7:1–698
Levkov Z, Mitić-Kopanja D, Reichardt E (2016) The genus Gomphonema from the republic of Macedonia. Diatoms Europe 8:1–552
Maraşlıoğlu F, Soylu EN (2018) New diatom records for Turkish freshwater algal flora from Lakes Ladik (Samsun, Turkey) and Hazar (Elazığ, Turkey). Turk J Fish Aquatic Sci 18(3):463–474
Morkoyunlu-Yüce A, Ertan ÖO (2014) Kovada channel phytoplankton (Isparta-Turkey). J Biol Environ Sci 8(24):165–171
Morkoyunlu-Yüce A, Ertan ÖO, Yıldırım MZ (2015) Epiphytic and epilithic diatoms in dam lakes (Euphrates – Turkey). Yunus Araştırma Bülteni 3:45–51
Özer G, Pala-Toprak G (2009) Suluçayır Düzü (Sivrice/Elazığ)'nde bulunan bir gölet (TMİ12)'in epipsammik diyatomeleri ve mevsimsel değişimleri. F Ü Fen Biliml Derg 31(1):79–88
Öztığ F (1957) Uber einige diatomee-formen aus Vansee, Revue de la Faculte des Sciences del Universite d'Istanbul Botanique. Ser B 21:267–270
Pala-Toprak G (2007) Keban Baraj Gölü Gülüşkür Kesimindeki Planktonik Algler ve Mevsimsel Değişimleri II-Bacillariophyta, Science and Eng. J Fırat Univ 19(1):23–32
Pala-Toprak G, Çaglar M (2006) Keban Baraj Gölü epilitik diyatomeleri ve mevsimsel degisimleri. F Ü Fen Müh Biliml Derg 18(3):323–329
Pala-Toprak G, Çaglar M (2008) Peri Çayı (Tunceli-Türkiye) epilitik diyatomeleri ve mevsimsel degisimleri. F Ü Fen Müh Biliml Derg 20(4):557–562
Pala-Toprak G, Çağlar M, Selamoğlu Z (2017) Study on epilithic diatoms in the Kozluk Creek (Arapgir-Malatya, Turkey). Iranian J Fish Sci 16(1):441–450
Rumrich U, Lange-Bertalot H, Rumrich M (2000) Diatomeen der Anden. Von Venezuela bis Patagonien/Feuerland. Iconographia Diatomologica 9:1–649
Şen B (1988) Hazar Gölü (Elazığ) alg florası ve mevsimsel değişimleri üzerine gözlemler. Kısım I: Littoral bölge. IX. Ulusal Biyoloji Kongresi: 289–298. 21–23rd Eylül 1988, Sivas
Şen B, Aksakal M (1988) Kırkgözeler'de (Elazıg) alg populasyonlarının *Potamogeton* sp. ve *Nasturtium officinale* üzerindeki mevsimsel yogunlukları ve degisimleri. IX. Ulusal Biyoloji Kongresi: 485–494. 21–23 Eylül 2000, Sivas
Şen B, Topkaya B, Alp MT, Özrenk F (1995) Organik madde ile kirlenen bir Çay (Selli Çayı, Elazığ) içindeki kirlilik ve algler üzerine bir araştırma. II. Ulusal Ekoloji ve Çevre Kongresi: 599–610. 11–13th Eylül 1995, Ankara

Şen B, Alp MT, Özrenk F, Ercan T, Yildirim V (1999) A study on the amonths of plant nutrients and organic matter carried into Lake Hazar (Elazığ, Turkey). Fresen Environm Bull 8:272–279

Şen B, Çaglar M, Pala-Toprak G (2001) Tadım Göleti (Elazıg) diyatomeleri ve yıl içindeki dagılımları. F Ü Fen Müh Biliml Derg 13(2):255–261

Şen B, Toprak-Pala G, Çağlar M (2005) Özlüce Baraj Gölü (Kiğı/Bingöl) Epilitik Diyatomeleri ve Mevsimsel Değişimleri. F Ü Fen ve Mühendislik Bilimleri Dergisi 17(2):310–318

Solak CN, Wojtal AZ (2012) Diatoms in springs and streams of Türkmen Mountain (Sakarya river basin) common in Turkish inland waters. Polish Botanical J 57(2):375–425

Solak CN, Ector L, Wojtal AZ, Ács É, Morales EA (2012) A review of investigations on diatoms (Bacillariophyta) in Turkish inland waters. Nova Hedwig Beih 141:431–462

Solak CN, Kulikovskiy M, Kaleli A, Gönülol A (2016a) Rare and new records for diatoms of Turkey from Kütahya running waters. Oceanol Hydrobiol Stud 45(4):564–587

Solak CN, Kaleli A, Baytut Ö (2016b) The distribution of Cymbelloid diatoms in Yalova running waters. Turkish J Fish Aquat Sci 16(4):953–959

Solak CN, Kulikovskiy M, Kiss TK, Kaleli A, Kociolek JP, Ács É (2018) The distribution of centric diatoms in different river catchments in the Anatolian peninsula, Turkey. Turkish J Botany 42 (1):100–122

Soylu EN, Maraşlıoğlu F, Gönülol A (2010) Gıcı Gölü (Samsun-Bafra) epipelik algleri ve mevsimsel değişimi. J FisheriesSciences.com 4(4):437–445

Taş B, Yılmaz Ö (2015) Cimil Deresinin (Rize, Türkiye) epilitik alg çeşitliliği. Türk Tarım – Gıda Bilim ve Teknolojisi Dergisi 3(10):826–833

Tunca H, Ongun-Sevindik T, Bal DN, Arabacı S (2014) Community structure of epiphytic algae on different macrophytes ar Acarlar floodplain forest (northern Turkey). Chinese J Osean and Limno 32(4):845–857

Varol M, Şen B (2014) Dicle nehrinin planktonik alg florası. J FisheriesSciences.com 8(4):252–264

Varol M, Şen B (2017) Abiotic factors controlling the seasonal and spatial patterns of phytoplankton community in the Tigris River, Turkey. River Res Applic 2017:1–11. https://doi.org/10.1002/rra.3223

Werum M, Lange-Bertalot H (2004) Diatoms in springs from Central Europe and elsewhere under the influence of hydrologeology and anthropogenic impacts. Iconographia Diatomologica 13:1–417

Wetzel CE, Ector L, Van de Vijver B, Compère P, Mann DG (2015) Morphology, typification and critical analysis of some ecologically important small naviculoid species (Bacillariophyta). Fottea 15:203–234

Wojtal AZ (2009). The diatoms of Kobylanka stream near Kraków (Wyżyna Krakowsko-Częstochowska, S Poland). Polish Bot J 54:129–330

Wojtal AZ, Kwandrans J (2006) Diatoms of the Wyżyna Krakowsko Częstochowska upland (S Poland) – Coscinodiscophyceae (Thalassiosirophycidae). Polish Botanical J 51:177–207

Yavuz OG, Çetin AK (2000) Cip Çayı (Elazığ, Turkey) pelajik bölge algleri ve mevsimsel değişimleri. F Ü Fen Müh Biliml Derg 12(2):9–14

Yavuz OG, Çetin AK (2001) Cip Çayı (Elazığ – Türkiye) Epilitik, Epifitik ve Epipelik Alg Florası. F Ü Fen Müh Biliml Derg 13(2):9–14

Yildirim V, Çetin AK (2009) Periphytic diatom assemblages from Cıp stream, Turkey: an examinatıon of community relationships and habitat references. Fresen Environm Bull 18 (9):1546–1550

Yildirim V, Şen B, Çetin AK, Alp MT (2003) Hazar Gölü'ne dökülen Kürk Çayı'nın (Elazıg) epipelik diyatome florası. F Ü Fen Müh Biliml Derg 15(3):329–336

Yılmaz E (2017) The benthic naviculoid diatoms of Yedigöller District (Bolu). Dumlupınar University Institute of Science. Master Thesis [in Turkish]

Chapter 26
Plant Biodiversity in Shatt Al-Arab Estuary and Ecological Variations

Taha Y. Al-Edany

Abstract Shatt al-Arab River is the only freshwater source in Basrah Governorate. It is formed by the confluence of rivers Tigris and Euphrates at Qurna Town, 70 km to the north of Basrah City.

The river is 204 km long. Its lowest 84 km form the border between Iraq and Iran. Many aquatic plants either submerged, floating, or emergent grow in different sites of the river; most of them are widespread or cosmopolitan, but few are restricted to the river among Iraqi water bodies. Several environmental variations had taken place in Shatt al-Arab and the area. There is a deterioration in the quantity and quality of water in the river due to the large demand in the upstream of the rivers Tigris and Euphrates and the adjacent countries from where water comes. Tens of dams were also constructed. Millions of people living around Shatt al-Arab, untreated domestic wastes, petroleum pollution, pesticides, and intrusion of seawater are highly affecting the biodiversity of aquatic plants in the river. Many plant species became rare or totally disappeared and some others are newly recorded. Few plant species of the river are nonflowering, five are macroalgae, four of which belong to Characeae and three are Bryophytes. The majority are flowering plants; 11 families and 15 species are dicotyledons, and each family is represented by one or two species except Ranunculaceae (3 species). While there are 9 families and 27 species of monocotyledons, some of its families are represented by considerable species number such as Cyperaceae (9 species), Gramineae (5 species), Potamogetonaceae (4 species), and Juncaceae (3 species). Other families are represented by one or two species.

26.1 Introduction

Aquatic macrophytes are essential components of the natural ecosystem that anchor in and stabilize the riverbank, provide shelter, habitat and food for fishes and other animals, support places for fish spawning and feeding. Some aquatic plants can be

T. Y. Al-Edany (✉)
College of Agriculture, University of Basrah, Basrah, Iraq

L. A. Jawad (ed.), *Tigris and Euphrates Rivers: Their Environment from Headwaters to Mouth*, Aquatic Ecology Series 11,
https://doi.org/10.1007/978-3-030-57570-0_26

beneficial as food for man and birds or maybe medicinally important. Some have esthetic significance (Haslam 1978). They are more productive than phytoplanktons under considerable conditions and useful to oxygenate water and produce organic matter to the food web and provide important nutrients (Al-Saadi and Al-Mousawi 1984).

The high uptake capacity of mineral elements by several aquatic plants (hyperaccumulation) has stimulated some proposals for using higher aquatic plants in the tertiary treatment for domestic or industrial wastes.

Some of these plants are very common, widespread, tolerant for many ecological changes, for example *Phragmites australis*, *Typhadomingensis*, and *Schoenoplectuslitoralis* and to some extent *Salvinianatans*, *Lemna* spp., *Bacapamonnieria,* and *Ceratophyllum demersum* (Al-Hilli et al. 2008). Other macrophytes tend to be less spread and grow in nearly restricted areas.

Phytogeographically most areas of Turkey, Iran, and part of the North-Eastern Iraq are located within the Irano-Anatolian area which is one of five subregions of the Irano-Turanin Region. This subregion is one of the richest areas in the world in species and genera as well as it is the most active in the formation of species. Most hydrophytes in Iraq are interregional plants and of boreal origin that water habitat is highly reducing the effect of high temperature (Guest 1966).

Syria lies within the Irano-Turanin and the Mediterranean Regions where there is a wealth of species. So, most of the plant species of Shatt al-Arab (and Iraq) are represented in these countries also.

Most hydrophytes are less affected by climate change and are almost cosmopolitan in distribution, and tolerate different habitat conditions. Some others tend to be more affected directly or indirectly according to the change in many environmental factors. So, some species became rare, very rare, highly reduced, or disappeared. On the other hand, some plants are newly recorded in recent years or show more dispersal than before.

According to the accelerated deterioration of the habitat and quality and quantity of water in Shatt al-Arab River and the reduction in the number of individuals and species or disappearance of some plant species which may continue to years or decades, it is important to determine these plants, the sites where they grow and live, their past and present status, and their importance to the environment and life.

26.2 History of Shatt al-Arab River

The river system of Tigris, Euphrates, and Shatt al-Arab consists of the greatest water resource in Western Asia. Shatt al-Arab River is considered as the only source of freshwater to Basrah city. Numerous canals and creeks branched of this river are for irrigation of palm orchards along the river (Al-Fartoosi 2013). Al-Faidhi (1965) mentioned 635 named rivers forking out of Shatt al-Arab River comprising 470 rivers branching from the western side and 165 from the eastern side.

Shatt al-Arab is formed by the meeting of Tigris and Euphrates Rivers near Qurna, 70 km to the north of Basrah city. After covering a distance of 120 km, it enters the border area between Iraq and Iran to a distance of 84 km till it discharges into the Arab Gulf at Fao city, with a total length of 204 km. The river widens and contracts over its journey from 250 to 300 m near Qurna to almost 500 m near the city of Basrah and up to 2250 m near its mouth to the south of Fao. The narrowest part is to the south of Karun junction with the river (Al-Mayah et al. 2016a).

Shatt al-Arab Delta is classified as estuarine, affected by tides and waves; its riverine entrance widens toward the Gulf and river sediments accumulate into shallow and narrow part of this Gulf.

Depth of Shatt al-Arab river is variable ranging from 4 m at Qurna where Tigris and Euphrates meet, 8.2 m at Dair, 9.2 m at Hartha, 15 m at the city of Basrah (24 m near the port of Maqil) (Al-Khafaji 2010) and 7 m at the mouth of the river. Elevation of the area with Basrah city is about 2 m above the sea level, with a decline of 1.5 cm/km along the river from meeting to the mouth with a total height difference of 3 m. Hydrology of Shatt al-Arab depends on that of rivers Tigris, Euphrates, and Karun in addition to the Arab Gulf water that mix with the water of Shatt al-Arab during high tide.

Geologically, the Shatt al-Arab river is relatively recently formed and is one of the Earth's newest characteristics created by the retreat of the shoreline of the Gulf (Al-Fartoosi 2013).

The past form of the Arab Gulf at the beginning of Holocene (12,000 years ago) was not as present where the Tigris and Euphrates rivers flowing through a marshland and draining into the Arab Sea at the Strait of Hurmuz. While Al-Hamad et al. (2015) referred that the Shatt al-Arab River may have been formed during 2000–1600 years before the present time. Since before 2000 years, there were Tigris and Euphrates rivers discharging separately in the Arab Gulf, the Shatt al-Arab River has been formed in the last 1500 years.

26.3 Geomorphology of Shatt al-Arab

To study the situation of geomorphology of any river in the world, it is important to recognize its life cycle of development which is divided into three phases depending on current speed, river decline, and channel breadth (Al-Whaely 2009 and Al-Hamad 2017) namely:

1. Instractive stage which is characterized by its high discharge and increase of river slope.
2. Constractive stage where discharge and decline are equal, and the breadth of the channel increases.
3. Perstractive stage, which is the last phase, where the sedimentation rate is higher than erosion. Shatt al-Arab is classified now as in the final stage especially after the decrease of Karun River discharge to about zero most of the year.

Palmer et al. (2008) stated that flow variability over time and space is an essential property of natural rivers and their flow system. It defines geomorphic adjustment and biotic composition.

Shatt al-Arab Estuary is now under huge stress due to vast water utilization and elevated salinity which increased toward the mouth. Degradation of water quality of the river started in the middle of last century due to the construction of some factories on the riverside.

26.4 Salinity in Shatt al-Arab River

Salinity is affected by the volume and freshness of water imported and exported in specific period of time as well as temperature, sun brightness, amount of precipitation, evaporation, and quality of area where the river cross.

Around the world, salinity of rivers increases annually in recent period and this salinity increases downstream of the rivers.

The salinity of Shatt al-Arab may be higher than other rivers of the world. Although seawater intrusion is considered as the main origin of natural salinity in the Shatt al-Arab river, Tigris, Euphrates, Karun, Kerkheh, and Swaib Rivers and drainage water coming from creeks and branches during low tide may also be other sources, making the river water unfit for human use and unacceptable for irrigation purposes (Abdullah et al. 2016).

The main creaks of the Shatt al-Arab river are located in urban places where millions of people live; thus, their utilization is changed from irrigation and navigation to be store for sewage and wastewater.

Many authors (Table 26.1) stated the changes in salinity of Shatt al-Arab during a period of about four decades in different sites and times of high and low tides from confluence to the mouth of the estuary.

Salinity resulted in a reverse effect on biodiversity in the area leading to disorder in the ecological system where marine animals and plants were present in the inland

Table 26.1 Changes of salinity in the Shatt al-Arab river over four decades at different sites, times of the year, and periods of high and low tides (After Hameed et al. 2013 with additions).

Salinity (ppt)	References
0.74–1.04	Huq et al. (1978)
0.29–18.9	Saad et al. (1978)
0.83–3.6	Al-Aubaidy and Al-Hello (1996)
0.29–0.78	Al-Hello (2001)
0.89–2.7	Al-Essa (2004)
1.3–2.1	Al-Asadi et al. (2006)
1.02–35.66	Directorate of Basrah Environment (2008–2009)
0.6–40.4	Hameed et al. (2013)
0.75–6.19	Mohamed et al. (2015)

water of Shatt al-Arab, the habitat that was not suitable for their life before (Khalaf and Zayed 2009).

26.5 Climate Changes and Challenges

Climate changes and drought are the main risks facing our area. These changes cause a shortage in water supplies (Yaseen et al. 2016). Shatt al-Arab is now facing very serious ecological risks because of various human activities in the countries participating in rivers in increasing water requirement and decreasing water quality.

There are extensive water withdrawals and diversions in the upstream where there are 30 operational dams and 20 planned or under construction (Al-Abaichy and Al-Khaddar 2010).

The South Eastern Anatolian Project (GAP) is now completed and will soon lead to raise salinity and cause critical changes in the ecosystem, regime of river flow, and environmental and geomorphic impacts (Kordic and Milankovic 2016).

The physical and chemical properties of natural waters are considered as the most important factors affecting aquatic life on all its sides such as temperature, dissolved oxygen, acidity, and content of carbon dioxide, chlorine, nutrients, and hardiness. Some parameters in Shatt al-Arab are summarized in Table 26.2.

Direct evaporation of surface water of river leaves salt residues behind. Dissolved oxygen is now low in Shatt al-Arab (Al-Fartoosi 2013). Precipitation is much lower than evapotranspiration; the ratio may reach 1:22 or less (Guest 1966; Al-Mayah et al. 2016a).

Shatt al-Arab region has an arid climate with a hot, dry, long, dead summer and short mold winter. Large differences in temperature can be found every month as well as between day and night.

Table 26.2 Changes in some of the elements and nutrient ions in the Shatt al-Arab River in different sites and periods of tide during three decades

Reference Parameter (ppm)	Rezk and Al-Edany (1980)	Al-Imara et al. (2001)	Al-Essa (2004)	Hassan et al. (2011)
Total residue	959.75–1490.00	–	–	2019.09–29224.00
Chlorine	198.2–405.33	85.30–2410.00	–	715.09–1030.83
Sodium	131.50–201.71	800.0–2825.00	–	–
Potassium	3.26–4.53	70.00–200.00	–	–
Calcium	126.58–164.10	120.00–480.00	50.00–270.00	–
Magnesium	88.40–120.30	136.20–343.00	5.50–11.50	–
Phosphorus	0.08–0.43	0.15–4.07	0.20–12.50	–
Nitrate nitrogen	0.93–1.75	42.90–317.60	0.10–4.30	–
Sulfate	397.73–542.25	19.20–62.80	–	–

Mean minimum temperature is recorded in January, and it was about 8.4 °C and mean maximum in July reaching 45 °C, and the absolute minimum may reach 0 °C or less but the absolute maximum may be more than 50 °C (Hussein et al. 1991).

The summer of 2017 recorded standard high levels of temperature when 12 days of July and 15 days of August recorded temperatures more than 50 °C, while the highest was 52.8 °C on July 11th (Basrah Airport Meteorological Station).

The Shatt al-Arab River witnesses a range of hydrological changes due to environmental changes but the largest ones were attended at the last three decades.

26.6 Pollution in Shatt al-Arab River

Shatt al-Arab River is contaminated with many materials such as domestic sewage, petroleum hydrocarbons, agricultural drainage, and organic matter thrown in the river or from decaying aquatic organisms, heavy metals, or suspended particles.

These pollutants enter the river either indirectly from the rivers Tigris and Euphrates and tributaries as well as the main creeks and branches of Ashar, Khandaq, Robat, and Khora, or directly by pollutants produced inside Basrah Province (Abdullah 2016), such as human activities, ammunitions and about 280 ships dumping into the river in the period of 1980–1988 during the Iraq–Iran war (Hassan et al. 2011). Some drowned ships may contaminate the river water by their load of oil, chemicals, inflammable or explosive materials.

Petroleum hydrocarbons may come through transporting of oil, oil refineries, power stations, oil seepage from fields, and municipal wastes. Al-Saad et al. (2016) stated the concentrations of these pollutants through three decades along the river (Table 26.3).

Water of Shatt al-Arab river contains some concentrations of heavy metals such as Fe (1.8 $\mu gm.l^{-1}$), Mn (3.4 $\mu gm.l^{-1}$), Ni (1.3 $\mu gm.l^{-1}$), and Zn (7.6 $\mu gm.l^{-1}$).

Table 26.3 Changes in the concentrations of petroleum hydrocarbons in the Shatt al-Arab River over three decades in different sites, times of the year, and periods of tides (after Al-Saad et al. 2016)

Petroleum hydrocarbons (nanometer per liter)	Reference
5.2–14.2	Al-Saadi and Al-Mayah (1983) and Hantoosh (2006) cited in Al-Saad et al. (2016)
3.08–14.37	Al-Saad (1995)
3.1–21.0	Al-Saad et al. (1999)
21–55	Al-Timari (2000)
0.007–13.36	Al-Saadoon (2002)
2.247–50.230	Al- Saad et al. (2016)
3.76–16.76	Al-Imarah et al. (2007)
7024–17.39	Al-Saad et al. (2016)

Table 26.4 Heavy metal content of some aquatic submerged plants in Shatt al-Arab river (μgm. l^{-1}) (After Hussein et al. 1991)

Species	Heavy Cd	Cu	Fe	Mn	Ni	Pb	Zn
Ceratophyllum demersum	0.38	7.3	23.9	2.132	122.90	1.13	–
Vallisneria spiralis	1.1	4.9	10.7	2.112	78.20	0.063	31.2
Potamogeton lucens	0.12	5.5	13.1	28.060	106.71	0.08	18.1

Table 26.5 Amount of chlorinated pesticides in two sites in Shatt al-Arab river (nanogram.l^{-1}) (After Hussein et al. 1991)

Pesticide	Site in Shatt al-Arab river: Garmat Ali	Abul-Khasib
DDT	150	83
Aldrin	18	30
Dieldrin	24	16
Cic-chloridane	37	38
Trans-chloridane	24	12

Some aquatic plants contain amounts of heavy metals; it was recorded that *Ceratophyllum demersum* concentrates copper and iron while *Vallisneria spiralis* accumulates cadmium, zinc, and lead (Table 26.4).

Considerable amounts of chlorinated pesticides were recorded at Garmat Ali and Abul-Khasib (Table 26.5). It is thought that the rivers of Tigris and Euphrates are their origin.

26.7 Plants of Shatt al-Arab River and Its Branches

Aquatic plants are growing and distributed along the banks of Shatt al-Arab and its branches in various places. Some are submerged, others are floating or emergent. They belong to several taxa, some are nonflowering, and most are flowering belonging either to monocots or dicots.

Although many of them are cosmopolitan and widespread in Iraq, some are more distributed in the Middle and Southern Iraq. It seems that the biodiversity of aquatic plants increases southward in marshes and the Shatt al-Arab River. Few species are restricted to this river and few others are restricted to the northern area in their distribution.

Some invasive plants had been recorded in the recent decades and years such as *Eichhornia crassipes* which was introduced to some special lakes in the 1980sand reached Tigris River near Baghdad and then spread southwards to Kut, North Amara, Nasiriyah, and Gharraf. It is not found in the Shatt al-Arab River probably due to the salinity of the water. In 2004, *Hydrilla verticillata* was discovered in Abuzereg Marsh Southern Nasiriyah, and then arrived to Shatt al-Arab through Garmat Ali Canal up to Ashar and Abul-Khasib.

The last discovered species is the fern *Azolla filiculoides* in some river branches at Hartha, 20 km to the north of Basrah by Al-Mayah et al. (2016b). Many other authors dealt with the aquatic plants in the river from taxonomical or ecological sides such as Townsend and Guest (1966, 1968, 1974, 1980, 1985), Rezk and Al-Edany (1980, 1981), Al-Edany and Al-Mousawi (1986), Al-Mayah and Al-Hmaiem (1991), Al-Saadi et al. (1996) and Al-Asadi (2009), while some others dealt with the river limnology and chemical composition of water and sediments including Al-Essa (1981), Antoine and Al-Saadi (1982), Al-Asadi (1983), Jawad (1994), Ali and Abdullah (1997), Al-Suwaige (1999), Hussein and Attee (2000a, b), Al-Shawi (2006), Aziz et al. (2006), Al-Asadi and Yaseen (2013), Ghazanfar and Edmondson (2013), and Muhamed et al. (2015).

Algae

Fam. Characeae

1. *Chara canensis* Des and Lois

Annual macroalgae with higher macrophytes appearance. Submerged in calcareous shallow ponds. Main axis with dense spines.

Flowering: Spring.

Habitat: Garmat Ali.

Importance: environmental indicators.

Distribution: widespread.

Status: occasional.

2. *Chara globularis* Thui

Freshwater annual submerged algae in rivers and ponds. Main axis with rudimentary spines.

Flowering: Spring.

Habitat: Shatt al-Arab near Tannuma.

Distribution: widespread.

Status: occasional.

3. *Chara vulgaris* L.

Annual macroalgae with variable shapes. Sometimes similar to *Ceratophyllum*, submerged in fresh or brackish water.

Flowering: Spring.

Habitat: Shatt al-Arab River.

Importance: fodder for birds and fishes, environmental indicator.

Distribution: widespread.

Status: occasional.

***Nitella tenuissima* (Desf.) Kts.**

Differ from *Chara* by the presence of corona with 10 cells in two tires of 5 cells each, submerged in freshwater.

Flowering: Spring.

Habitat: Salihiyah near Shatt al-Arab River.

Distribution: Southern Iraq.

Status: very rare.

Fam. Ulvaceae

***Enteromorpha intestinalis* (L.) Link**

Green, annual, tubular, long, anchored, floating, intestinal-like, in polluted saline or brackish water.

Flowering: Spring.

Habitat: Garmat Ali.

Distribution: Marshes and Khor al-Zubair.

Status: rare.

Bryophyta

Fam. Fontinalaceae

Fontinalis antipyretica **L.**

Perennial moss, reproducing by spores. Stem triangular branched. Leaves small, rigid.

Flowering: Winter.

Habitat: wet and shaded places, in Shatt Al-Arab at Ashar.

Distribution: all of Iraq.

Status: rare.

Fam. Salviniaceae

Azolla filiculoides **Lam.**

A small free-floating fern, 3–7 mm long, roots simple, numerous, unbranched, hang down from beneath the rhizomes, up to 25 mm long. Stem a horizontal rhizome lies under water surface, yellowish or olive green.

Flowering: unknown.

Habitat: still and slowly flowing water in canals and river branches or in brackish water in Hartha (20 km north of Basrah).

Importance: biofertilizer for rice paddy.

Distribution: Sulaiymaniyah and Amara.

Status: rare, newly recorded.

Salvinia natans **(L.) All.**

Free-floating moss, missing true stem and roots. Leaves simple, small. Plant appearance dissimilar to mosses.

Flowering: April–December.

Habitat: still and slowly flowing water in river branches at Qurna, Garmat Ali, Tannuma, Maqil, and Ashar.

Importance: good fodder to wild duck and foul.

Distribution: Samawa, Nasiriyah, Amara.

Status: common.

Angiospermae (Flowering plants)

Dicotyledoneae

Fam. Amaranthaceae

Alternanthera sessilis **(L.) R. Br. ex DC.**

Perennial, ascending or prostrate. Stem 0.1–1 m long, green or purplish. Leaves variable in shape and size, opposite and nearly sessile, linear lanceolate to oblong. Inflorescence sessile, flowers bisexual, petals white.

Flowering and fruiting: December–March.

Habitat: Garmat Ali, Jezira, Tannuma.

Importance: fodder for sheep.

Distribution: Basrah only.

Status: rare.

Fam. Asteraceae (Compositae)

***Ecliptaalba* (L.) Hausskn.**

Annual plant. Stem rigid, 15–20 cm tall. Leaves opposite. Flowers white.

Flowering: March–May.

Habitat: muddy tidal bank of Shatt al-Arab, creeks, canals, and fields in Qurna, Garmat Ali, and Fao.

Importance: medicinal.

Distribution: Baghdad, Kut, Hilla.

Status: rare.

Fam. Ceratophyllaceae

***Ceratophyllum demersum* L.**

Entirely submerged, perennial, anchored, rootless dark green up to 100 cm long. Leaves in whorls, more crowded toward the tip, once or twice forked. Flowers not clear, nutlet small and compressed with terminal spine and two basal spines.

Flowering and fruiting: Summer.

Habitat: on the bank of Shatt al-Arab River and creeks in Garmat Ali, Tannuma, and Abul-Khasib.

Importance: shelter for fishes, good fodder for birds, medicinally important, ecological indicator.

Distribution: widespread in Iraq.

Status: rare in Shatt al-Arab.

Fam. Haloragaceae

***Myriophyllum spicatum* L.**

Perennial, submerged, leaves in whorls, pinnate inflorescence emergent. Flower in interrupted spike. Upper (male) flowers often closer together than lower (female) flowers or more than hermaphrodite flowers. Fruit subglobose.

Flowering and fruiting: October–May.

Habitat: Shatt al-Arab River and creeks in Hartha and Abul-Khasib.

Distribution: North, Central, and Southern Iraq.

Importance: shelter for planktons and fishes.

Status: occasional.

***Myriophyllum verticillatum* L.**

In vegetative aspect, it is similar to *M. spicatum*. Hermaphrodite flowers more than in *M. spicatum*. Leaver submerged, in whorls, pinnate. Inflorescence emerged. Fruit subglobular.

Flowering and fruiting: Spring and summer.

Habitat: bank of river at Abul-Khasib.

Distribution: Mosul, Nasririyah, and Amara.

Importance: shelter for planktons and fishes.

Status: disappeared.

Fam. Lamiaceae

***Mentha longifolia* (L.) Huds.**

Perennial herb, 60–100 cm long with creeping rhizomes. Leaves prolonged to lanceolate, petiolate, and aromatic. Flowers numerous in cylindrical spike.

Flowering and fruiting: Spring.

Habitat: muddy tidal flats of creaks at Abul-Khasib, Siba, and Fao.

Distribution: Mosul, Sulaimaniyah, Rawanduz.

Importance: Eaten fresh or dry.

Status: very rare.

Fam. Menianthaceae

***Nymphoides indica* (L.) O. Ktze**

Floating herb. Stems similar to petioles. Root projected from lower side of nodes. Leaves simple, alternate, circular.

Flowering and fruiting: March–June.

Habitat: Shatt al-Arab branches and creeks at Hartha.

Importance: good forage, medicinal properties.

Distribution: Lower Mesopotamia.

Status: very rare—disappeared.

Fam. Onagraceae

***Ludwigia stolonifera* (Guill and Per.) Raven**

Stem spreading, creeping or ascending, rooting at nodes. Leaves lanceolate, flowers axillary, petals bright yellow. Capsule dark brown, shining, pediculate.

Flowering: Winter–Summer.

Habitat: shallow water and river banks at Tannuma.

Distribution: Amara and Nasiriyah.

Status: disappeared.

Fam. Polygonaceae

***Polygonum lapathiolifolia* (L.) S.F. Gray**

Perennial, stem creeping or ascending, leaves sessile, lanceolate with ochrea, margin serrate.

Flowering and fruiting: February–April.

Habitat: on the banks of creeks in Abul-Khasib and Siba.

Importance: good fodder for cattle.

Distribution: Kut, Basrah.

Status: very rare.

Fam. Ranunculaceae

1. *Ranunculus sceleratus* L.

Annual herb, erect, smooth, 20–60 cm tall. Leaves three-lobed. Inflorescence dense. Flowers small, petals pale yellow.

Flowering and fruiting: February–April.

Habitat: in date palm orchards in Basrah and Abul-Khasib, common on the tidal mud of Shatt al-Arab in Siba.

Distribution: Amara.

Status: occasional.

2. *Ranunculus sphaerospermus* Boiss. et Blanche

Annual herb. Laminate floating leaves absent, submerged leaves with numerous capillary segments. Flowers solitary, axillary, sepals ovate, petals white broadly cuneate- obovate.

Flowering and fruiting: March–May.

Habitat: river banks in Qurna, between Qurna and Basrah, Garmat Ali, Basrah.

Distribution: Ramadi, Babylon, Samawa, Diyala, Amara.

Status: very rare.

3. *Ranunculus trichophyllus* var. *ronii* (Lagger) Pikli

Annual or perennial. Laminate floating leaves absent, submerged leaves with numerous capillary segments. Flowers solitary, petals white.

Flowering and fruiting: March–May.

Habit at: banks of Shatt al-Arab and creeks in Qurna, Garmat Ali, Maqil, Kibasi, Southern Basrah.

Distribution: Ramadi, Babylon, Baghdad, Amara.

Status: very rare.

Fam. Scrophulariaceae

***Bacopa monnieria* (L.) Hayata et Matsum**

Perennial fleshy herb. Stem long creeping, up to 30 cm long. Leaves opposite, sessile, ovate. Flowers solitary, petals pale violet. Capsule 2–3 mm long, ovate.

Flowering and fruiting: March–May.

Habitat: Shatt al-Arab banks, creeks and canals, common in Garmat Ali, Tannuma, and Siba.

Importance: medicinal.

Distribution: Ramadi, Baghdad, Hilla, Diwaniyah.

Status: common.

Fam. Verbenaceae

***Phyla nodiflora* (L.) Greene**

Perennial herb, stem creeping, rooting at nodes. Leaves ovate serrate. Inflorescence dense, flowers pink.

Flowering and fruiting: April–November.

Habitat: wet places on banks of rivers, creeks and irrigation canals in Qurna, Hartha, Garmat Ali, Maqil, Tannuma, and Fao.

Distribution: Central and Southern Iraq.

Status: occasional.

***Verbena officinalis* L.**

Annual up to 100 cm tall. Leaves elongate, serrate, or lobate. Flowers small.

Flowering and fruiting: April–June.

Habitat: bank of Shatt al-Arab River, creeks and canals in Abul-Khasib, Siba, and Fao.

Importance: medicinal.

Distribution: Sulaimaniyah, Dohuk, Kut.

Status: very rare.

Monocotyledoneae

Fam. Cyperaceae

1. *Cyperus alopecuroides* Rottb.

Perennial or annual with fleshy roots. Rhizomes short, thick, woody, culms solitary, 90–150 cm tall, trigonous, leafy. Leaves shorter than stems. Inflorescence once or twice anthelate spikes digitate, oblong to cylindrical. Achene ellipsoid.

Flowering and fruiting: March–May.

Habitat: in marshes and tidal estuary, on mud in Basrah.

Importance: environmentally important.

Distribution: Amara.

Status: occasional.

2. *Cyperus corymbosus* Rottb.

Creeping perennial. Root succulent. Rhizome looping, woody, covered with fibrous scales. Stem 60–120 m. Leaves basal, bladeless. Spikelets up to 25 * 1.25 mm. quadrangular, red-brown. Achene ellipsoid. Ripe achenes are very rare.

Flowering and fruiting: December–January.

Habitat: Garmat Ali, Sindbad Island, Maqil, Khora Creek.

Distribution: restricted to Basrah district.

Status: rare.

3. *Cyperus difformis* L.

Weak, slender annual herb with fibrous reddish roots. Stem triquertus. Leaves few, basal, short, soon withering. Inflorescence once or twice anthelate. Spikes spherical, 1 cm across, very dense. Achene ellipsoid sharply triquertus.

Flowering and fruiting: August–October.

Habitat: bank of ditches in Basrah.

Distribution: Dohuk, Diyala, Hilla, Nasiriyah, Amara.

Status: occasional.

4. *Cyperus malaccensis* Lam.

Rhizomatous perennial. Root fibrous. Rhizome thick, reddish, clothed with triangular red-brown scales. Stem 50–150 cm trigonous. Leaf sheaths long. Inflorescence capitate, spikelet linear, 10–30 cm long, erect. Glumes oblong. Achene ellipsoid, 1.5 * 0.5 mm, dark brown.

Flowering and fruiting: October.

Habitat: banks of the tidal estuary in Garmat Ali, Maqil, Ashar, Abul-Khasib, Siba, and Fao.

Importance: fodder plant.

Distribution: In Basrah district only.

Status: common.

***4. Cyperus rotundus* L.**

Creeping perennial with woody roots. Rhizomes rigid, tuber. Stem slender triquertus 15–40 cm tall. Leaves basal, crowded, and membranous. Spikes numerous. Achenes rarely produced.

Flowering and fruiting: May–August.

Habitat: damp places, canals, creeks, and bank of Shatt al-Arab River from Qurna to Fao.

Importance: tubers of medicinal uses.

Distribution: widespread in Iraq in Arbil, Mosul, Diyala, Ramadi, Baghdad, Hilla, Diwaniya, and Amara.

Status: common.

Fimbristylis bisumbellata **(Forssk.) Bub.**

Annual, fibrous-rooted stem 5–25 cm trigonous or hexagonal. Blade shorter or longer than stem. Inflorescence 2–3 anthelate, loose with many solitary spikelets. Nut obovate, 0.5 * 0.45 mm.

Flowering and Fruiting: August–January.

Habitat: bank of Shatt al-Arab near Basrah.

Importance: grazed by livestock.

Distribution: from north to south of Iraq, very common in the irrigated Lower Mesopotamia.

Status: occasional in Shatt al-Arab.

Fimbris tylissieberiana **Kunth**

Perennial with coarse fibrous roots. Rhizomes short, woody. Stems crowded, 15–55 cm, stiff. Blades of upper leaves up to 25 cm. Inflorescence anthelate with 10 spikelets. Achene broadly obovate, pale dark.

Flowering and fruiting: June–October.

Habitat: bank of Shatt al-Arab between Basrah and Fao.

Importance: grazed by sheep and water buffaloes.

Distribution: Salaimaniyah, Samawa, Amara.

Status: very rare.

Schcenoplectus litoralis **(Schrad.) Palla**

Perennial rhizomatous with wiry roots. Rhizome thick producing leafy, submerged shoots. Stem erect, 100 cm tall, trigonous. Emergent leaves much reduced, 3–10 cm. Inflorescence terminal, spikelets many. Achene broadly obovoid, dark brown to black.

Flowering and fruiting: March–September.

Habitat: along Shatt al-Arab river and creeks from Qurna to Fao.

Importance: environmental, industrial, and forage.

Distribution: widespread in Iraq, along Tigris and Euphrates and their tributaries from Dohuk and Sulaimaniyah to Amara.

Status: common.

Schocnoplectus triqueter **(L.) Palla**

Rhizomatous, glabrous, perennial. Stem 100 cm long, triangular. Leaves few, sheaths long red-brown blade short often 2 cm. Inflorescence corymbose, spikelets ovoid 8–15 mm long, red-brown. Achene broadly obovoid. 2.25 * 1.5 mm, pale to mid-brown.

Flowering and fruiting: March–November.

Habitat: ditches among date palm trees near tidal river mouth on lightly flooded mudflats along river edge in Basrah and Siba on Shatt al-Arab.

Distribution: restricted to Basrah District.

Status: rare.

Fam. Hydrocharitaceae

Hydrilla verticillata **(L. f.) Royle**

Invader plant, perennial, submerged, stem long. Leaves small, membranous. Plant reproducing sexually and vegetatively.

Flowering and fruiting: unknown.

Habitat: Hartha, Maqil, Ashar.

Importance: ecologically dangerous, severely shading endemic submerged plants, preventing fishing.

Distribution: Southern Iraq.

Status: common.

***Vallisneria spiralis* L.**

A caulus, stoloniferous submerged herb. Leaves ribbon-like, 30 cm long. Male spathes short pedunculated. Female pedunculated, becoming spirally twisted after anthesis.

Flowering and fruiting: February–August.

Habitat: Shatt al-Arab bank in Garmat Ali, Tannuma, Ashar, Salihiyah, Hamdan, Abul-Khasib.

Importance: ecological indicator, oxygenating water, shelter for fish spawning.

Distribution: along Tigris River from Mosul to Amara.

Status: very rare in Shatt al-Arab.

Fam. Juncaceae

Juncus acutus* L. subsp. *acutus

Perennial, rhizome reduced, branching from the stem base. Stems erect, rigid, 50–150 cm tall. Inflorescence with two leafy bracts, dense, spherical, or elongated. Capsule 4–6 mm, spherical.

Flowering: March–April, fruiting: April–August.

Habitat: marshy places, near irrigation channels and bank of Shatt al-Arab, in Dair 40 km north of Basrah, Hartha, Abul-Khasib. Siba and Fao.

Importance: stems for making baskets and chair seats, seeds of laxative effect.

Distribution: Baghdad and Diyala.

Status: common.

***Juncus articulatus* L.**

Perennial rush with creeping rhizome. Stem 5–60 cm long from nodes. Inflorescence wide with 5–20 heads, heads 5–15 flowered. Flowers without bracteoles. Capsule trigonous, ovoid. Seed without seed coat, 0.5–0.6 mm, ovoid.

Flowering and fruiting: January–December.

Habitat: date gardens, along the bank of tidal estuary from Maqil, Basrah to Fao and Robat Creek.

Distribution: Dohuk, Arbil, Sulaimaniyah, Diyala, Ramadi.

Status: common.

***Juncus rigidus* Desf.**

Mat forming perennial with a thick creeping rhizome. Stems not septate. Inflorescence with two leafy bracts, lax, usually elongated, 5–35 cm long. Flowers in few flowered heads. Capsule 3.5–5 mm. Seeds 0.9–1.5 mm.

Flowering: January–June, fruiting: May–December.

Habitat: near margin of Shatt al-Arab and creaks in Basrah, 50 km S.E. of Basrah, Siba, and 10 km N of Fao.

Distribution: along Tigris and Euphrates Rivers.

Status: common.

Fam. Lemnaceae

***Lemna gibba* L.**

Thallus free-floating, small broadly ovate. Leaves yellow-green. Daughter thali budding laterally covering water surface.

Flowering and fruiting: rarely flowering or fruiting.

Habitat: common in stand or slowly flowing water, in Shatt al-Arab and creeks in Qurna, Hartha, Garmat Ali, Maqil, and Tannuma.

Importance: environmental indicator, fodder for birds.

Distribution: Qurna, Garmat Ali, Ashar.

Status: widespread.

Fam. Poaceae (Gramineae)

***Diplachne fusca (*L.) P. Beauv.**

Perennial with branched culms, up to 100 cm tall. Leaf blades 20 cm long. Panicle 15–30 cm long, spikelets 6–10 mm long.

Flowering and fruiting: March–November.

Habitat: canal banks and mudflats of intertidal area in Qurna, Garmat Ali, and Basrah District.

Importance: suitable forage for buffaloes.

Status: occasional.

***Panicum repens* L.**

Perennial with long wide-spreading rhizomes. Culms erect up to 60 cm long. Leaves linear. Panicle open 6–10 cm tall, spikelets about 2.5 cm long.

Flowering and fruiting: March–December.

Habitat: bank of Shatt al-Arab and creeks in the intertidal zone and mudflats in Hartha, Garmat Ali, Jezira, Tannuma.

Importance: good fodder for buffaloes.

Distribution: Hilla to Nasiriyah and Kut to Amara.

Status: occasional.

***Paspalum paspaloides* (Michx) Scribn.**

Perennial with widely spreading stolons and rhizomes. Culms up to 40 cm long. Leaf blade flat up to 12 cm long. Inflorescence in pair of false racemes, spikelets oblong.

Flowering and fruiting: June–January.

Habitat: along ditches and intertidal zone in Garmat Ali and Jezira.

Distribution: Kirkuk, Baghdad, Hilla, Diwaniyah, Nasiriya, Amara.

Status: common.

***Phragmites australis (*Cav.) Trin ex Steudel**

A stout rhizome—perennial, up to 5 m tall or more, culms simple rarely branched. Leaf blades flat, 20–60 cm. Panicle plume-like, up to 30 cm long, much branched.

Flowering and fruiting: May–December.

Habitat: very common on the bank of Shatt al-Arab and creeks.

Importance: environmental, fodder.

Distribution: along Tigris and Euphrates Rivers.

Status: very common.

***Saccharum ravennae (*L.) Murr.**

Culms caespitose, up to 3 m tall, forming dense tufts. Lower leaves up to 1 m long. Panicle silky, erect, very dense, 20–50 cm long.

Flowering and fruiting: September–October.

Habitat: Khora Canal and bank of Salihiyah River.

Importance: of little value as fodder plant.

Distribution: North Iraq.

Status: disappeared—very rare.

Fam. Potamogetonaceae

***Potamogeton crispus* L.**

Submerged, perennial. Stem compressed. Leaves linear oblong 10 cm long, 5–10 mm broad, undulate or crisped. Spike erect, flowers wind pollinated. Drupelet 4–5 mm long.

Flowering: March, fruiting: April.

Habitat: along the bank of Shatt al-Arab in Qurna, Garmat Ali, Maqil, Basrah, Abul-Khasib, Siba, and Fao.

Importance: ecological indicator, fodder for birds and fishes.

Distribution: mainly in Basrah but found south of Amara also.

Status: common.

***Potamogeton lucens* L.**

Perennial, rhizomatous, submerged plant. Stem terete up to 100 cm long. Leaves elliptic to lanceolate, 10–20 cm long. Short petiolate. Spike emergent, flowers wind pollinated. Drupelets 3–4 mm long.

Flowering: March–April, fruiting: March–August.

Habitat: fresh water in Shatt al-Arab branches and canals in Qurna and Garmat Ali.

Importance: fodder for birds.

Distribution: Sulaimaniyah, Mosul, Arbil, Nasiriyah, Amara.

Status: rare.

***Potamogeton nodosus* Poir**

Perennial, leaves floating or submerged, long peteolated, floating leaves elliptic 3–12 mm long, submerged leaves lanceolate. Flowers wind pollinated. Drupelets 3.5–4 cm long.

Flowering: February–August, fruiting: April–October.

Habitat: bank of Shatt al-Arab in Garmat Ali, between Basrah and Abul-Khasib, Siba and Fao.

Distribution: widespread in Iraq in Sulaimaniyah, Arbil, Nasiriyah, and Amara.

Status: rare in Shatt al-Arab.

***Potamogeton pectinatus* L.**

Rhizomatous perennial. Stem slender, highly branching. Leaves submerged, narrowly linear to filiform. Flowers water pollinated. Drupelets 3–5 mm long.

Flowering: February–September, fruiting: June–September.

Habitat: shallow brackish water in creeks in Hartha, Garmat Ali, Maqil, Ashar, and Abul-Khasib.

Importance: environmental indicator, fodder for birds.

Distribution: Kirkuk, Ramadi, Baghdad, Nasiriyah.

Status: occasional.

Fam. Ruppiaceae

***Ruppia maritima* L.**

Perennial submerged plant. Stems filiform, much branched. Leaves opposite or alternate, filiform, pale green.

Flowering and fruiting: March–November.

Habitat: brackish or saltwater in Shatt al-Arab, pools and creeks in Qurna, Garmat Ali, Abul-Khasib, and Siba.

Distribution: Najaf, Samawa, and Nasiriyah.

Status: occasional.

Fam. Typhaceae

***Typha domingensis* Pers.**

Erect rhizomatous plant. Flowering stem 1.5–3 m high. Leaves overlapping stem, blade linear. Male inflorescence at the top, separated from female inflorescence by a gap.

Flowering: April–June, fruiting: September–December.

Habitat: river bank along Shatt al-Arab in Qurna, Garmat Ali, Abul-Khasib, and Fao. Sometimes interfere with *Phragmites*.

Importance: environmental, industrial, and food (pollen are eaten).

Distribution: Dohuk, Mosul, Baghdad, Ramadi, Amara.

Status: very common.

Fam. Zennichelliaceae

***Zennichellia palustris* L.**

Submerged plant with wiry multi-branched rhizome. Leaves narrow 2–6 cm long. Female flowers with 2–9 carpels. Achene 3–6 mm long.

Flowering and fruiting: January–November.

Habitat: fresh and brackish water in Shatt al-Arab and canals in Garmat Ali, Maqil, Abul-Khasib, Ashar, between Siba and Fao.

Distribution: Sulaimaniyah, Kirkuk, Baghdad, Hilla, Kut, Amara, Nasririyah.

Status: rare.

References

Abdullah AD (2016) Modelling approaches to understand salinity variations in a highly dynamic tidal river, the case of Shatt al-Arab river. Ph.D. Thesis. Delft University, Netherland, p 186

Abdullah AD, Gisen JIA, Van der Zaag P, Savenije HHG, Karim UFA, Masih I, Popescu I (2016) Prediction the salt water intrusion in the Shatt al-Arab Estuary using an analytical approach. Hydrol Earth Syst Sci 20:4031–4042

Al-Abaichy J, Al-Khaddar R (2010) Study of Shatt al- Arab River: the latest environmental disaster. report prepared to Iraq Institution for Economic Reform (IIER). (unpublished)

Al-Asadi MK (1983) System of nutrient distribution in Shatt al-Arab River and some of its branches at Basrah City. M.Sc. Thesis. Basrah University, p 163

Al-Asadi MS, Salman NA, Mahdi AA (2006) Effect of waste discharges on nutrients content and growth of Chlorella sp. from Shatt Al-Arab River. J King Abdulaziz Univ Mar Sci 17:89–101

Al-Asadi WM (2009) Morphological and ecological study on the stranger aquatic plant *Hydrillaverticillata* (L.f.) Royle. M.Sc. Thesis. Department of Biology, College of Science, University of Basrah p 159

Al-Asadi KA, Yaseen BR (2013) The changes in the salinity of the Shatt al-Arab for the years 2008–2009 threating the life in the geographical territory. Proc. Water Resources in Iraq and the reality problems and optimizing planning for investment. College of Literature. Dhi-Kar University, Iraq

Al-Aubaidy A, Al-Hello A (1996) Monthly variations of the nutrients and other related factors in Shatt Al-Arab River, Basrah, Iraq. Marina Mesopotamica 11(2):365–383

Al-Edany TY, Al-Mousawi AH (1986) Regional variation in the chemical composition of soil and dominant aquatic plants in Shatt al-Arab, Iraq. Bangladesh J Bot 15(1):87–94

Al-Essa SA (1981) Limnological study of Shatt al-Arab River and Al-KhandakCreek. M.Sc. Thesis. University of Basrah, p 120

Al-Essa SA (2004) Ecological study of the aquatic plants and epiphytic algae in Shatt Al-Arab River (Doctoral dissertation, Ph.D. thesis, University of Basrah)

Al-Faidhi S (1965) The great Basrah. Dar al-Tadhamin, Baghdad, p 80

Al-Fartoosi FM (2013) The impact of maritime oil pollution in the marine environment case study of maritime oil pollution in the navigational channel of Shatt al-Arab. M.Sc. Thesis, World Maritime University, Malmo, Sweden, p 54

Al-Hamad SS (2017) Geological history of Shatt al- Arab River, south of Iraq. M.Sc. Thesis. College of Science, University of Basrah, p 121

Al-Hamad SS, Albadran BN, Pournelle JR (2015) Geological history of Shatt al-Arab River, south of Iraq. Intern J Sci Res 1(6):2029–2039

Al-Hello A (2001) Some characters of Shatt Al-Arab River water and its suitability for different uses. Marina Mesopotamica 16:295–308

Al-Hilli MR, Warner BG, Asada T, Douabul A (2008) Assessment of vegetation and environmental controls in the 1970s of the Mesopotamian wetlands of southern Iraq. Wetl Ecol Manag 2008. https://doi.org/10.1007/s11273-008-9099-1

Ali HH, Abdullah DS (1997) Energy values of aquatic plants from the Shatt al-Arab habitat, Basrah. Iraq Marina Mesopotamica 12(1):1–6

Al-Imara FJM, Alaiwi YJ, Muanis FS (2001) Monthly variations of nutrients and chlorophyll a in water of Shatt al-Arab River. Mesopatamian J. Marine Sci 16(1):347–357

Al-Imarah FJ, Hantoush AA, Nasir AM (2007) Petroleum hydrocarbons in water and sediments of N. W. Arabian Gulf, 1980 – 2005. Aquat Ecosyst Health Manage 10:335–340

Al-Khafaji SN (2010) Changes in Shatt al-Arab River bed and its effects on Iraqi land. College of Arts J 93:435–465

Al-Mayah AA, Al-Hmaiem FI (1991) Aquatic plants and algae. Dar Al-Hikma, Basrah University, Basrah, p 735

Al-Mayah AA, Al-Edany TY, Al-Asadi WT (2016a) Ecology and Flora of Basrah. Jeekor Printing, Beirut, Lebanon, p 686

Al-Mayah AA, Al-Saadi SM, Abdullah JN (2016b) A new generic record (*Azolla*, Salviniaceae) to the aquatic pteridoflora of Iraq. Indian J Appl Sci 6(2):21–23

Al-Saad HT (1983) A baseline study on petroleum hydrocarbons pollution in Shatt Al-Arab River (Doctoral dissertation, M.Sc. thesis, Basrah Univ)

Al-Saad HT (1995) Distribution and sources of hydrocarbons in Shatt Al-Arab Estuary and North-West Arabian Gulf. Ph.D. thesis, Basra Univ., 186p

Al-Saad HT, Farid WA, Al-Adhub AY (1999) Distribution and seasonal variations of n-alkanes in some species of molluscs from Shatt Al-Arab River. Mesopot J Mar Sci 20(2):182–196

Al-Saad HT, Al-Khion DD, Majeed ES, Hantoush AA, Saleh SM, Alhello AA (2016) Environmental assessment of polycyclic aromatic hydrocarbons pollutants in the waters of the side branches of the Shatt Al-Arab river. Iraqi J Aquac 13(1)

Al-Saadi HA, Al-Mayah AA (1983) Aquatic plants of Iraq. Basrah University Press, Basrah, p 192

Al-Saadi HA, Al-Mousawi AH (1984) On the chemical composition of aquatic plants in Shatt al-Arab near Basrah, Iraq. Bangladesh J Bot 13(2):137–146
Al-Saadi HA, Al-Edany TY, Neama JD (1996) On the distribution and ecology of aquatic plants in the Shatt al-Arab River. Iraq. Marina Mesopotamica 11(1):49–62
Al-Saadoon WJ (2002) Determination and distribution of total petroleum hydrocarbon and trace metals in water and sediments from Shatt AlBasrah and Khor-Al-Zubair, Southern Iraq. Ph.D. thesis, College of Educ., Basrah Univ., 151pp
Al-Shawi TTM (2006) Comparative study of some physio-chemical characteristics for northern Al-Hammar marsh waters before destroyed and after rehabitation 2004. Marsh Bulletin 1 (2):127–133
Al-Suwaige AR (1999) Comparative limnological study of Shatt al-Arab estuary and Khora Canal. M.Sc. Thesis Basrah University, p 161
Al-Timari AA (2000) Oil pollution in Shatt Al-Arab water: studying the monthly variations of polycyclic aromatic hydrocarbons (PAHs). J Mar Mesopotamia 15:535–548
Al-Whaely OQ (2009) Sedimentology and geomorphology of Shatt al- Arab Islands between Basrah andSiba Towns. M.Sc. Thesis, College of Science, University of Basrah, p 68
Antoine SE, Al-Saadi HA (1982) Limnological studies on the polluted Ashar canal and Shatt al-Arab river at Basrah. Iraq Int Rev Ges Hydrobiol 67(3):405–418
Aziz NM, Al-Adhub AHY, Al-Imarah FJM (2006) *Phragmitesaustratis*and*Typhadomengensis* as bio accumulators and biomonitors of trace metals along Shatt al-Basrah Canal, south of Hammar marshes. Marsh Bulletin 1(2):172–183
Directorate of Basrah Environment (2008–2009) Routine water evaluation of Shatt al-Arab River, Basrah. Technical Report no. 3. 23p
Ghazanfar, S.A., Edmondson, J.R. Flora of Iraq. Vol. 5, Part TwoMinistry of Agriculture, Baghdad. 2013. 349 pp.
Guest E (1966) Flora of Iraq, vol 1. Ministry of Agriculture, Baghdad, p 213
Hameed HA, Ali MH, Aljorany YS, Hassan WF, Al-Hello A (2013) Assessing changes in seawater intrusion and water quality of the Shatt Al-Arab River, Iraq. Annales de Limnologie-Int J Limnol 49(3):199–206. EDP Sciences
Hantoush AA (2006) A study of oil pollution status in water and sediments of Shatt Al-Arab River-south of Iraq (Doctoral dissertation, Ph.D. thesis, College of Science, University of Basrah)
Haslam SM (1978) River plants. Cambridge University Press, Cambridge, 396 pp
Hassan WF, Hassan IF, Jasim AH (2011) The effect of industrial effluents polluting water near their discharging in Basrah governorate. Iraq Basrah J Res 37(1):21–32
Huq MF, Al-Saadi HA, Hameed HA (1978) Phytoplankton ecology of Shatt al-Arab River at Basrah, Iraq: with 2 figures in the text. Internationale Vereinigung für theoretische und angewandte Limnologie: Verhandlungen 20(3):1552–1556
Hussein SA, Attee RS (2000a) Comparative studies on limnological features of the Shatt al-Arab Estuary and MhejranCanal.I. Monthly variations in abiotic factors. Basrah J Agric Sci 13 (1):49–59
Hussein SA, Attee RS (2000b) Comparative studies on limnological feature of the Shatt al-Arab Estuary and MhejranCanal II Monthly variations of nutrients. Basrah J Agric Sci 13(2):53–61
Hussein NA, Al-Najjar HHK, Al-Saad HT, Yousif OH, Al-Saboonchi AA (1991) Shatt al-Arab, fundamental scientific studies. Marine Science Center Publications, University of Basrah, p 392
Jawad AM (1994) Study of some chemical and physical indices of water of Shatt al-Arab River at Basrah City. Mesopotamian J Marine Sci 9(2):377–396
Khalaf SK, Zayed JA (2009) Shatt al-Arab salanization, status and possible managements. Ministry of Environment, p 12.
Kordic M, Milankovic M (2016) Analysis of the situation of the Euphrates-Tigris- Shatt al-Arab River system and possible solutions of the devastation of the ecosystem. Geol Geophys Environ 42(1):83–84
Mohamed ARM, Hussein SA, Lazem LF (2015) Spatiotemporal variability of fish assemblage in the Shatt Al-Arab River, Iraq. J Coast Life Med 3(1):27–34

Palmer AM, Liermann CAR, Nilson C, Flörke M, Alcamo T, Lake PS, Bond N (2008) Climate change and the worlds river basin anticipating management options. Front. Ecol. Environ, p 18. www.frontier sinecology.org
Rezk MR, Al-Edany TY (1980) Ecology of *Phragmitesaustratis* (Cav.) Trin ex Steud. In Shatt al-Arab, Iraq. Part I. Seasonal and regional variations in the chemical constituents of water and soil in reed beds. Pol. Arch. Hydrobiol 27(3):349–358
Rezk MR, Al-Edany TY (1981) Ecology of *Phragmitesaustratis* (Cav.) Trin. exSteud. in Shatt al-Arab, Iraq. Part II. Reed growth as protected by the chemical constituents of its beds. Pol Arch Hydrobiol 28(1):19–31
Saad MA (1978) Seasonal variations of some physicochemical conditions of Shatt al-Arab estuary, Iraq. Estuar Coast Mar Sci 6(5):503–513
Townsend CC, Guest E (1966) Flora of Iraq, vol 2. Ministry of Agriculture, Baghdad, p 184
Townsend CC, Guest E (1968) Flora of Iraq, vol 9. Ministry of Agriculture, Baghdad, p 588
Townsend CC, Guest E (1974) Flora of Iraq, vol 3. Ministry of Agriculture, Baghdad, p 662
Townsend CC, Guest E (1980) Flora of Iraq, vol 4. Ministry of Agriculture, Baghdad, p 1199
Townsend CC, Guest E (1985) Flora of Iraq, vol 8. Ministry of Agriculture, Baghdad, p 440
Yaseen BR, Al-Asady KA, Kazem AA, Chaichan MT (2016) Environmental impacts of salt tide in Shatt al-Arab River, Basrah. Iraqi J Environ Sci Toxico Food Technol 109(1):35–43

Chapter 27
Medicinal Plants of Shatt al-Arab River and Adjacent Area

Taha Y. Al-Edany

Summary Plants have an important role in the lives of humans and other living organisms. They are important for food, fodder, dwelling, clothing, and therapeutics. Humans along periods and times obtained medicines from plants, animals, and minerals to treat different diseases. Chinese, Indians, Egyptians, Babylonians, Assyrians, Akkadians, Sumerians, Greeks, and Romanians had many records and descriptions about hundreds or thousands of medicinal plants and drugs. After that Arabs wrote excellent works for amelioration of human sufferings due to ailments. Ultimately Europeans develop the preparations of drugs and separate the active principles as pure materials.

Although most medicinal plants are growing in dry arid zones of Iraq, aquatic and wetland plants have considerable amounts of active principles used in traditional folk and modern medicine. Sometimes the whole plant is used, or only one or more parts have therapeutic importance. These plants faced catastrophic changes in their environment with alteration in physical and chemical properties and shortage in the quantity of water supply which is reflected on the vegetation composition where many species became rare or totally disappeared, while few are newly recorded.

27.1 Introduction

Plants are a basic source for human and other living organism's lives. Humans recognized and classified several plants and wild herbs in their wide environment depending on their morphological characters, where they found plenty of plants for food, some other fodder, dwelling, clothing, heating, and medicine.

Medicines were mainly obtained from plants, animals, or mineral origins. Although active principles obtainable for drugs are present in favorable concentrations in plants growing in the dry arid regions of Iraq, aquatic and wetland plants

T. Y. Al-Edany (✉)
College of Agriculture, University of Basrah, Basrah, Iraq

L. A. Jawad (ed.), *Tigris and Euphrates Rivers: Their Environment from Headwaters to Mouth*, Aquatic Ecology Series 11,
https://doi.org/10.1007/978-3-030-57570-0_27

have also considerable amounts of these principles and are used in popular medicine in many parts of the world.

This issue aimed to discuss the pharmaceutical uses and the high value of the aquatic and moist land weedy plants in Shatt al-Arab River and adjacent area in marshes and cultivated farms, to make knowledge about it for people in order to encourage them to take care, maintain, or even cultivate by a way or another, that natural plants are still economically significant and have a considerable value as potential source of drug.

27.2 Historical Review

In ancient China (5000–4000 B.C.), ephedra or the drug ephedrine was used to treat asthma, as well as the leaves of tea which was used as stimulant.

Egyptians had records including names of medicinal plants and prescriptions about the use of fenugreek, castor oil, colocynth, and other plants discovered in their tombs.

Greek and Romanians described the use of hundreds of medicinal plants.

In Iraq, the first signals about drugs appeared during the Akkadian period (2370–2160 B.C.), while the effect of drugs had a significant impact on the general living in the Sumerian literature.

Also, pharmaceutical preparations such as decoctions, poultices, salves, and coryllium were prepared. The ancient references stated in that time there was a rank of people preparing drugs and cosmetics known as (Pasisu), placed in certain streets in the capital Babylon. Some clay slates showed medicinal prescriptions were they divided into three fields, the first field contain the herb name, the second one include ailment state, while the third one refers to the amount of drug or the dose if taken before or after meal. Babylonians also recognized about 250 botanical and 120 mineral drugs.

Assyrians added more records to the Babylonian pharmaceutical knowledge.

Arabs wrote excellent works for the amelioration of human sufferings due to ailments. Arab medicine is rich in bright names of herbalists such as Ibn Sina, Jabber Ibn Hayan, Ibn Al-Baitar, Al-Razi, and others who search, study, and translate the Greek, Rumanian, Indian, and Persian books of drugs. In their books, they described thousands of medicinal plants and prescriptions for different diseases that attack humans in their areas. Some of their books were translated into Latin and French.

Ultimately "the Europeans" who made use of encyclopedic works and developed the preparations of drugs to powder or extracts and separate the active principles of medicinal plants as pure or crystalline materials manufactured as tablets, syringes, salves, etc.

27.3 The Ecology of Medicinal Plants

Medicinal plants belong to various groups and families, grow in different habitats, some are herbs, others are shrubs or trees. Some are wild, weeds or cultivated, native, endemic, or introduced.

Habitat differing between cultivated fields, waste areas, along irrigation canals, sides of ditches, on mudflats, riverine thickets, in shallow pools, edge of saline marshes, or weeds in rice and other crop fields. Few species are floating or submerged in fresh or brackish water. In some species the whole plant is used, in others only leaves, stems, flowers (or inflorescence), fruits, seeds, roots, or rhizomes have therapeutic importance.

Aquatic herbs, in general, showed a decrease in the number of species and individuals in the last decades. The scarcity is ascribed to the desiccation of marshes and the shortage in water supply from the river head which leads to change in physical and chemical properties of the water in the Shatt al-Arab River, side rivers, and branches. This is reflected on the vegetation composition. So many species became rare or totally disappeared, while few are newly discovered.

Table 27.1 shows the form of life and the plant parts medicinally used for each species.

27.4 Description and Therapeutic Value of Plants

Pteridophytes.

Family Marsileaceae.

***Marsilea capensis* A. Br.**

A fern with creeping slender horizontal rhizomes. Fronds one at a node. Internodes in water form up to 4 cm long, much contracted in land-forms, stipes slender, in water forms 20 cm long. Sporangia is about 6 cm long. Common in shallow water and mud banks in North Qurna.

Medicinal uses.

Leaf juice drunk for snakebite, dermatitis, diuretic, febrifuge, refrigerant, and ulcers, and decoctions for fever and swellings. The whole plant is used for abscesses, backache, and boils.

Family: Adiantaceae.

***Adiantum capillus-veneris* L.**

Short-creeping fern with rhizomes covered with brown scales. Fronds up to 40 cm long, blades lax and drooping, bipinnate in adult fronds; the indusial flaps becoming membranous and brownish at maturity. Occasionally on the lower plain.

Medicinal uses:

Herbpectoral, expectorant, catarrhal, diuretic, emmenagogue, tonic, soothing, astringent, emollient, used in cold and pulmonary catarrh. Powder is used as alleviant in kidney complaints. Leaves in asthma and alopecia. Leaves and rhizomes

Table 27.1 Habitat and the parts of plant used in medicinal plants

Scientific name	Habitat				Parts of the plant used							
	A	M	R	W	F	H	J	L	RR	SF	ST	WP
Adiantum capillus-veneris L.		+				+		+				
Alhagi graecorum Boiss.				+		+					+	
Ammi majus L.				+		+				+		
Anagallis arvensis L.				+		+						
Arundo donax L.		+	+						+			
Bacopa monnieria (L.) Hayata *et* Matsum	+	+	+					+				
Ceratophyllum demersum L.	+	+									+	
Ceratopteris thalictroides (L.) Brongn.	+	+				+						
Cichorium intybus L.				+		+		+	+			
Convolvulus arvensis L.				+		+			+			
Cressa cretica L.				+							+	
Cuscuta chinensis Lam.				+						+		
Cuscuta planiflora Ten.				+		+				+		
Cynodon dactylon (L.) Pers.		+	+	+		+	+		+			
Cyperus rotundus L.		+	+	+					+			
Digitaria sanguinalis (L.) Scop.		+		+		+						
Echinocola crus-galli (L.) P. Beauv.			+			+			+			+
Eclipta alba (L.) Hausskn.		+		+			+				+	
Euphorbia helioscopia L.		+		+			+	+	+	+	+	
Euphorbia hirta L.				+			+					
Gallium aparine L.				+		+						
Glycerrhiza glabra L.			+	+			+		+			
Imperata cylindrica (L.) P. Beauv.			+	+	+				+			
Lactuca serriola L.				+		+				+		
Lawsonia inermis L.[a]					+			+				
Lemna minor L.	+					+						
Lolium temulentum L.				+		+				+		
Ludwigia stolonifera (Guill and Perr.) Raven	+					+				+		
Lycopus europaeus L.	+	+	+			+						
Malva neglecta Wallr.				+	+			+				
Malva parviflora L.				+	+	+		+		+		
Malva sylvestris L.				+	+			+				
Marsilea capensis A. Br.	+	+						+			+	
Medicago sativa L.[a]								+		+		
Melilotus indica (L.) All.				+								
Mentha longifolia (L.) Huds.		+	+			+						
Nasturtium officinale R.Br.	+					+	+				+	
Nymphoides peltata (Gmel) O. Ktze	+					+						
Phyla nodiflora (L.) Greene			+			+						

(continued)

Table 27.1 (continued)

Scientific name	Habitat				Parts of the plant used							
	A	M	R	W	F	H	J	L	RR	SF	ST	WP
Phoenix dactylifera L.[a]					+					+		
Phragmites australis (Cav.) Trin. ex Steudel	+	+	+	+	+	+		+	+			+
Plantago lanceolata L.				+	+			+			+	
Plantago major L.				+	+	+		+				
Polygonum aviculare L.				+			+					
Portulaca oleracea L.				+		+	+			+		
Potamogeton crispus L.	+					+						
Potamogeton nodosus Poir.	+					+						
Potamogeton pectinatus L.	+					+						
Ranunculus sceleratus L.	+	+	+			+		+	+	+		
Solanum nigrum L.				+		+		+		+		+
Sonchus oleraceus L.		+		+			+	+			+	
Spirodela polyrrhiza (L.) Schleid.	+						+				+	
Stellaria media (L.) Vill.				+			+				+	
Trigonella foenum-graecum L.[a]										+		
Typha domingensis Pers.	+	+	+		+			+	+			
Vallisneria spiralis L.	+					+						
Verbena officinalis L.		+		+	+	+		+	+			

A Aquatic, *M* Moist Places, *R* Riverine, *W* Weed, *F* Flowers, *H* Herb, *J* Juice or sap, *L* Leaves, *RR* Roots or rhizomes, *SF* Seed or fruit, *ST* Stem, *WP* Whole plant
[a]Cultivated

in chronic cough and cold. Root treats grippe, stops bleeding, rheumatism, arthritis, dysentery, and breast inflammation. Syrup made of fronds is purgative and in catarrhal infection. Large doses acting as emmtic.Infusion for tea to heal the bowels as mild purgative.

***Ceratopteris thalictroides* (L.) Brongn.**

The plant is glabrous. Sterile fronds up to 45 cm or more, flaccid and fleshy;

Fertile fronds are produced on adult plants up to 60 cm, rarely 100 cm. Sporangia scattered, irregular along the veins. Common in ditches along swampy places north of Basrah.

Medicinal uses:

The fronds being pounded and used as skin poultice. It is also used as tonic or styptic.

Angiospermae (Flowering Plants).

Class: Dicotyledones.

Family: Apiaceae (Umbelliferae).

***Ammi majus* L.**

Annual herb, up to 80 cm high. Leaves pinnately lobed. Flowers greenish white in compound umble. Fruit brown.

Medicinal uses:

Herb is said to be poisonous due to its content of furo-cumarine, ammoidin, amidin, and majudin. Seeds are used in the treatment of leucoderma patches, used orally, or as 1% liniment. Of the coumarines present ammoidine is considered effective in skin disorders. Fruits are also medicinally important.

Family: Asteraceae (Compositae).

***Cichorium intybus* L.**

Perennial herb reproducing by seeds. It is a weed growing in cultivated fields and waste areas. Stem erect, 30–100 cm high, branching from the base. Leaves hirsute, lower leaves with long, acute, terminal segments, and upper leaves are lanceolate. Lateral heads sessile, while terminal heads peduncle. Seeds gray or light brown, oblong, or curved shape.

Medicinal uses:

Herb is diuretic, stomachic, aperients, sedative, tonic, vomiting or given in fever, diarrhea, dilation of spleen, disorders of kidney, liver, urinary canal, and stomach. Root is laxative, diuretic, sudorific, depurative, diaphoretic, tonic, cholagogue, febrifuge, and against typhoid fever. Flower heads are diuretic and carminative. Fruit is cholagogue.

***Eclipta alba* (L.) Hausskn.**

Annual herb. Leaves opposite. Flowers white, inflorescence involucres, solitary, and pedunculated. Fruit sypsela. The plant grows in river banks and irrigation canals.

Medicinal uses:

Herb is tonic, deobstruent in hepatic and spleen enlargement and skin diseases, hwmostatic, antihemorrhagic, for dysentery, epistaxis, and neurasthenia. Combination of the herb juice and some aromatic ingredients is used for catarrhal and jaundice. Leaf juice is used for premature graying of hair and for teething. The fresh plant is considered anodyne and absorbent.

***Lactuca serriola* L.**

Annual or biennial, 30–90 cm high. Stem erect. Simple, leafy, terminating into an elongated pyramidal panicle. Lower leaves ovate. Heads 8–15 mm long on very short pedicels, florets yellow. Achene elliptic, blackish, terminating in a white beak as long as or longer than achene.

Medicinal uses:

Herb is diuretic, antispasmodic, sedative, emollient, anticolic, and toxic. Decoction is beneficial for insomnia. Seed powder is used against cough, oil of seeds has hypnotic and antipyretic effects.

***Sonchus oleraceus* L.**

Annual herb reproducing by seeds. Stem erect, stout, hollow. Leaves sessile, glabrous, and toothed. Head many-flowered becomes white and fluffy in fruits. Flowers small, yellow. Achene brown, pappus longer than achenes, soft, white.

Medicinal uses:

Herb is an insecticide, treats asthma, bronchitis, cough, ophthalmia, insomnia, pertussis, and swellings. Plant juice a powerful hydrogogue, cathartic and anticancer for breast, skin, and stomach. Injections of extracts damage sarcoma cells. It is also used for abscesses, jaundice, morphisms, piles, snakebite, sores, stomach, tenesmus, and wounds. Leaves are tonic purgative, diuretic, and antipyretic.

Family: Brassicaceae (Cruciferae).

***Nasturtium officinale* R. Br.**

Perennial herb. Stem glabrous. Leaves imparipinnate, leaflets 3–9 and 2 mm long. The plant grows in running water. Rare in southern Iraq.

Medicinal uses.

Plant juice linimetic for burns and hair growth. Internally it is used as antiscorbutic, depurent, against rheumatism, kidney ailment, and tuberculosis. It is also an expectorant, digestive, stimulant, diuretic, and emmenagogue.

Family: Caryophyllaceae.

***Stellaria media* (L.) Vill.**

Annual or overwintering herb. Stems 10–40 nm. Long, decumbent, or ascending, diffusely branched from the base. Leaves ovate, the lower ones 5–20 mm long, the upper ones up to 25 mm. Petals white, not exceeding the sepals. Capsule ovoid-oblong.

Medicinal uses:

Herb is postpartum depurative, emmenagogue, lactogogue, promote circulation, treats mucus disorder, antitoxic, diuretic, expectorant, laxative, demulcent, refrigerant, very good blood purifier, spasm against itching and piles. Herb juice heals heartbeats, rheumatism, vulnerary, and laxative. Fruit is good for urinary complaints, treating skin disease and fever.

Family: Ceratophyllaceae.

***Ceratophyllum demersum* L.**

Entirely submerged perennial herb, rootless but anchored, appearing branched, 30–150 cm long. Nutlet compressed, ellipsoid, 4–5 mm long with one terminal and two basal opposite spines, 2–5 mm long. Common in still water.

Medicinal uses:

Although it is reported therapeutically inactive, but it has oleaginous, bitter, and fragrant principles. The herb purifies water around.

Family: Convolvulaceae.

***Cressa cretica* L.**

Suffrutescent. Stems slender, terete, woody, 30–40 cm long. Leaves sessile ovate–lanceolate, 5–6 mm long. Flowers white, aggregated in dense spikes at the end of branches. Capsule ovoid, 2.5–5 mm long. Seeds ovoid, glabrous.

Medicinal uses:

All parts are used. It is used as alterative, stomachic, tonic, and aphrodisiac.

***Convolvulus arvensis* L.**

Perennial herb, 30–60 cm high, glabrous, or pubescent. Stems prostrate or climbing. Lower leaves on a petiole shorter than the blade, the blades sagittate, or hastate, 15–30 mm long. Flowers solitary or 2–3 together. Corolla white or pink, 20–30 mm long. Ovary glabrous. Fruit capsule.

Medicinal uses:

Herb is laxative, cholagogue, vulnerary, healing wounds. Roots cathartic and laxative. Flowers laxative, febrifuge vulnerary. Resinous substances have an acrid taste and cathartic in action.

Family: Cuscutaceae.

***Cuscuta chinensis* Lam.**

Plant is parasitic. Stem is slender. Flowers 2–3.5 mm long, short pediculate, in dense clusters. Calyx loose, inclosing the corolla. Corolla sub-globular. Capsule depressed-globose. Plant is seen on *Alhagi* and some other plants.

Medicinal uses:

The seeds are used as a tonic, diaphoretic, and demulcent.

***Cuscuta planiflora* Ten.**

Stems are very slender. Flowers are 1.5–2 mm long, sessile, in dense, globose, sessile clusters, 5–8 mm in diameter. Calyx enclosing the corolla, campanulate-globose, pale pink to whitish. Capsule depressed globose.

Medicinal uses:

Stems are used for obesity. Infusion of seeds for catarrh.

Family: Euphorbiaceae.

***Euphorbia helioscopia* L.**

An annual weed, reproducing by seeds, growing in cultivated fields, riverside, and damp places. Plant erect or ascending, glabrous, 10–50 cm high. Stem branching from the base. Leaves glabrous, spathulate, 1–3 cm long. Umble 5-rayed. Floral leaves elliptical or round, 1–2 cm long. Capsule glabrous, globular, and trisulcate. Seeds ovoid, brown, and net-veined.

Medicinal uses:

Plant is toxic and used as hydrogogue cathartic. Roots have anthelmintic properties, vesicatory, and laxative. Leaves vermifuge, toxic, used in fever, dropsy, and in malaria. Seeds are given in cholera, rheumatism, and neuralgia. Oil of seeds has purgative properties. Plant juice is used as a liniment in neuralgia, rheumatism, and applied to warts. Latexes highly irritating the mucus membrane and toxic to mammals and fishes.

***Euphorbia hirta* L.**

Hispid erect ascending herb, usually with simple stem. Leaves rhombic—ovoid to rhombic—lanceolate, petiole 1–2 mm long. Cyathia aggregated together into dense clusters. Cyathial involucres 1 mm long. Capsule 1.3 mm across. Seeds 0.5 mm quadrangular, grayish, and ecarunculate.

Medicinal uses:

Root is administered to allay vomiting, antiasthmatic, pectoral, and as a remedy for cough, bronchial and pulmonary disorders, it affords in paroxysmal asthma. Latex anthelmintic, expectorant, diuretic, protect mucous membranes and heal wounds.

Family: Fabaceae (Leguminosae).

***Alhagi graecorum* Boiss.**

Erect, bushy perennial subshrub, 25–100 cm high. Stem branched, densely armed with stout spines, up to 5 cm long. Leaves simple, ovate, and obtuse, up to 20 mm long. Inflorescence axillary, 1–3 flowered. Corolla rose to reddish. Pod linear, 10–35 mm, glabrous at maturity. Seeds 2.5 mm, subspherical to rhomboidal, brown, and smooth.

Medicinal uses:

Manna, the exudates from stem, is mild laxative and expectorant, arthritis used in difficult respiration. All parts heal hemorrhoids. Syrup is used to alleviate leukoma and impairment of vision, employing it as eye salve, nose drops to assuage chronic migraine.

***Glycerrhiza glabra* L.**

Perennial, frequently glandular herb, with imparipinnate leaves. Inflorescence a lax to subcapitate axillary raceme or spike. Flowers bluish, white, or yellow. Calyx tubular, deeply divided. Corolla papilionacious. Pod ovate-oblong or linear oblong, late dehiscent. Seeds estrophiolate.

Medicinal uses:

Roots and rhizomes are a flavoring agent, demulcent, expectoral, tonic, gentle laxative, remove irritating effects of acids, soothing sore throats, coughs, heartburns, ulcer, colic, asthma, thrust, and diseases of liver, stomach, and intestine. Extract is used to sweet medicinal preparations. Useful in the treatment of Addison's disease.

***Medicago sativa* L.**

Erect or ascending perennial herb, 30–90 cm much branched and bushy. Stems and branches are pale green. Leaflets narrowly oblong-ovate. Raceme rather dense, oblong, to 4 cm long, with 10–25 flowers. Pod spiral with 2–3 lax coils. Seeds ovoid to deltoid, brownish, 2–2.5 mm long, smooth.

Medicinal uses.

Fattening, brain tonic, heart tonic, antiscorbutic, stypic. Seeds possess emmenagogue lactogenic properties, used as a poultice for boils, claimed to be of value in arthritis, or eaten fresh to soothe diabetes.

***Melilotus indica* (L.) All.**

Erect, simple, or much branched annual herb, 20–40 cm. Stems and branches slender, wiry. Leaflets of lower leaves shortly cuneate, obovate, of the upper leaves lengthening to ovate or oblong, 8–20 mm long. Raceme slender, many flowered. Corolla yellow, small. Pod globular-ovoid, 2–3 mm. Seeds ovoid, yellowish to chestnut, solitary, or rarely 2 in each pod.

Medicinal uses:

Emollient, discontent, and used as fomentation against swellings. It may possess narcotic and astringent properties. It is said to be demulcent, carminative, soothing colic, hematostatic, and helpful in vulnerary. It is sometimes eaten fresh to soothe diabetes. Seeds are used in bowels complaints and infantile diarrhea.

***Trigonella foenum-graecum* L.**

Erect annual herb, 15–30 cm high, glabrous. Leaf trifoliate, leaflets broadly cuneate-oblong. Flowers creamy, solitary, or in pairs. Corolla about twice as long as the calyx. Pod large, erect, 6–10 cm long, seeds subtetragonous-ovoid, purplish-brown to yellowish-brown.

Medicinal uses:

Seeds demulcent and emollient. The ground drug is employed for the preparation of poultices, enemata, and plasters or to treat allergies, anemia, coughs, digestion, emphysema, migraine, mucous membranes, sore throat, inflammation, and ulcer lungs. Oil of seeds is aphrodisiac, galactagogue, diuretic, emmenagogue, laxative, healing constipation, tonsil, and respiration. Pounded seeds are important in vulnerary, arthritis, and diabetes.

Family: Lamiaceae (Labiatae).

***Lycopus europaeus* L.**

Perennial herb, 30–60 cm high, glabrous, or minutely pubescent. Stem erect, tetragonal. Leaves long, lanceolate, 10 cm long, with numerous triangular lobes. Inflorescence verticillate, many flowered. Calyx 3 mm long. Corolla white with some purple dots on the lower lip. Fruit nutlet. Found in channels and ditches.

Medicinal uses.

Herb is astringent and febrifuge. It contains essential oil, malic acid, tannin, coloring matter, and calcium oxalate.

***Mentha longifolia* (L.) Huds.**

Perennial herb, 60–100 cm high, with creeping rhizome. Stem erect, simple or branched above. Leaves short petioled, 3–8 cm long, 1–3 cm wide. Flowers in many flowered whorls forming terminal cylindrical spikes, 4–10 cm long. Corolla lilac, hairy outside, glabrous within. Found on the canal banks and side of ditches.

Medicinal uses:

Herb carminative, antiseptic, stimulant. Leaves are used against rheumatic pains. Plant decoction is given in fever, apoplexy, bowels and bile disorders, respiratory system, asthma, cough. It is also respected as cholagogue, anti-inflammatory, diuretic, corroborant for liver, mastitis, cold, and arthritis. It is said that the mint odor of the herb drive away snakes.

Family: Malvaceae.

***Malva parviflora* L.**

Annual herb, 10–25 cm high, erect, or with ascending branches. Leaves reniform or orbicular, 2.5–10 cm in diameter. Calyx moderately to strongly accrescent in fruit. Petals obovate, 4.5 mm Seeds 1.4–1.75.

Medicinal uses:

Herb is emollient and pectoral. Decoction of leaves is taenicide. Lotion is given is applied for bruises, hoarseness, bronchitis, and fever. Leaves are used as poultice for wounds and swellings. Seeds are said to possess demulcent effects and given in cough, ulcers of the bladder.

***Malva sylvestris* L.**

Coarse perennial herb, erect, decumbent, or ascending, 30–90 cm high, frequently much branched, and bushy. Leaves orbicular-reniform. The basal 5–10 cm in diameter, crenate lobed, stem leaves are deeply lobed. Flowers form leafy racimes, pink to rosy-purple with darker veins. Petals 18–3 mm, 3–4 times as longer as the calyx or more. Carpels 2.5 mm in diameter. Seeds dull brownish, 2 mm.

Medicinal uses.

Herb soothes inflammation of mouth and throat, helpful for earaches. Decoction of stem and leaves resolving for tonsils and mouth inflammation, cough, and diarrhea. Leaves are cathartic, diuretic, and anti-inflammatory for renal ulcers and dermatitis.

***Malva neglecta* Wallr.**

Annual herb, 12–40 cm high, with numerous decumbent branches. Leaves orbicular-reform in outline, 3–7 cm in diameter. Flowers in axillary fascicles, white or lilac with pinkish veins. Petals obovate, much exceeding the calyx. Carpels 2.5 mm in diameter. Seeds brown, minutely rugulose.

Medicinal uses:

Leaf laxative, emollient in catarrh of respiratory organs, and intestine.

Flowers pectoral, expectorant, laxative, refrigerant, diuretic, emollient. Flowers of *Malva sylvestris* and leaves of *M. neglecta* are particularly suitable for coughs and colds.

Family: Lythraceae.

***Lawsonia inermis* L.**

A small cultivated shrub with opposite elliptic to obovate leaves. Flowers 4-merous, white, in terminal panicles, fragrant. Capsule 5 mm in diameter. It is believed to be indigenous in South Western Asia. It is cultivated in Central and Southern Iraq. It is said that (hinna) of Southern Basrah and Shatt al Arab is the best.

Medicinal uses:

Chiefly used as coloring agent for hair, skin, and nails. Medicinally used as a coolant, vulnerary against sunburn, depurative, itchy skin, and prevent hair drop. Flowers are soporific. Extract is bactericidal, antitumor, alterative, anthelmintic, astringent, collyrium, deodorant, and emmenagogue, fungicidal, sedative and vulnerary. Folk remedies for boils, bronchitis, bruises, burns, calculous, headache, herpes, hysteria, inflammations, jaundice, skin sores, leprosy, whitlow, and other uses.

Family: Menianthaceae.

***Nymphoides peltata* (Gmel.) O. Ktze.**

Floating leaved herb. Leaves broadly ovate to orbicular, bright green, 2.5–10 long, margin sinuate, petiole of inflorescence leaves 2–10 cm. Fascicles 2–5 flowered. Corolla lobes bright yellow,20–25 mm in diameter. Capsule narrowly ovoid, up to 25 mm. Seeds are ovate and flat.

Medicinal uses:

Herb is a diuretic, febrifuge, and refrigerant. It is said to have lactogenic properties when buffaloes feed on, Bruised plant for burns, fever, rodent ulcers, snakebites, and swellings.

Family: Onagraceae.

***Ludwigia stolonifera* (Guill & Perr.) R. H. Raven.**

Aquatic herb of shallow water and moist soils. Stem spreading, prostrate, or ascending, up to 1 m or longer, rooting at nodes, with tufts of white aerenchyma from submerged stems. Leaves narrowly lanceolate to narrowly elliptic, 2–9 cm. Flowers axillary. Sepals 5–10 mm. Petals 5, bright yellow, 8–10 mm long. Capsule 15–27 mm, dark brown, shining. Seeds 1.1–1.2 mm.

Medicinal uses:

Herb plaster or poultice for ulcers and skin complaints. Pounded plant for boils, rheumatism, and syphilis. Decoction for abscesses, boils, infection of urinary tract and snakebite. Antidote for infections, intoxication, mycosis, eye, and throat infections.

Family: Plantaginaceae.

***Plantago lanceolata* L.**

Perennial herb, acaulous, 60 cm high. Leaves alternate in basal rosette, lanceolate to lanceolate-oblong, usually petulate. Peduncles 25–50 cm, usually erect. Spike globose to cylindrical, dense. Calyx scarious except midrib. Corolla glabrous, lobes ovate-lancolate. Capsule 2 seeded. Seed ovoid, smooth.

Medicinal uses:

Stems, leaves, and flowers treat chest diseases, good appetizer, blood purifier, refrigerant, diuretic, deobstruent, mildly astringent, stimulant for sores. Leaves whole or bruised used as a poultice, healing tuberculosis, asthma, lung, gum, spleen, epilepsy, dyspepsia, diarrhea, decoagulation of blood, and respiratory tract diseases. Infusion of leaves is vulnerary, anthelmintic, and helpful in whooping cough, inflammation of bladder, and soothing toothache.

***Plantago major* L.**

Perennial herb, acaulous, 20–40 cm tall with short rootstock and many fibrous roots. Leaves large, elliptic-ovate to orbicular-ovate, 10–25 cm. Spike dense or lax, linear, or linear-cylindrical, 5–30 cm. Sepals ovate-elliptic to orbicular. Corolla greenish or yellowish. Capsule globose or conical. Seeds angular smooth.

Medicinal uses:

Stems, leaves, and flowers are used as blood purifiers and treat chest diseases. The herb is one of the best remedies for all cuts, skin infections, and chronic skin problems, work well in a score of ailments as diarrhea, stomach troubles and ulcer, hemorrhoids, and inflammations. Seeds are used as poultice.

Family: Polygonaceae.

***Polygonum aviculare* L.**

Annual or perennial herb, 30–60 cm high, branching from the base, leafy along the length with elongate internodes. Ochrea short, silvery brownish at the base. Flowers 3–6 together, on a very short pedicel. Perigon herbaceous, greenish, or reddish. Achene 2–4 mm long.

Medicinal uses:

Herb is pounded as poultice for sores and wounds, and active to treat constipation, hemorrhage, and dysentery. It is diuretic, expectorant, astringent, hemorrhoid, expectorant, pectoral, styptic, hemostatic, antiperiodic, tonic, vulnerary, and antiseptic, used in kidney diseases. Decoction of the herb is helpful in diarrhea and tuberculosis. It can be used as gargling for tonsils.

Family: Portulacaceae.

***Portulaca oleracea* L.**

Annual herb, 10–20 cm, glabrous, succulent. Stem prostrate or ascending, branching from the base. Leaves spirally arranged, ovate-oblong, 1–2 cm long. Flowers 1–3, terminal, or sessile. Sepals are unequal. Petals free yellow, soon falling. Capsule ovoid-globose. Seeds brownish-black, shining.

Medicinal uses:

Herb is used as antiscorbutic, cooling, diuretic, anthelmintic, antiemetic, hemorrhage. Herb juice is hemorrhoid and used against burning sensation of heat. Seeds are vermifuge.

Family: Primulaceae.

***Anagallis arvensis* L.**

Annual herb, 6–30 cm, glabrous. Stem procumbent or ascending, quadrangular. Leaves opposite or whorled, sessile, ovate to triangular ovate. 15–25 mm long, dotted with black glands beneath. Flowers solitary, on long peduncles. Calyx—teeth narrow—lanceolate, 4–5 mm long. Corolla rotates, up to 12 mm in diameter, blue, or orange. Capsule 3–4 mm in diameter.

Medicinal uses:

Herb is used for cerebral affection or employed externally for ill-conditioned ulcer, sometimes used to intoxicate fishes. It has poisonous effects on dogs, rabbits, and sheep.

Family: Ranunculaceae.

***Ranunculus sceleratus* L.**

Erect or ascending, annual, 25–60 cm, with numerous erect or divaricate branches from near the base upward. Stem and branches succulent, glabrous. Basal leaves long petiolate, palmately trilobed, upper leaves progressively less divided, and sessile. Flowers many, in axillary cymes. Sepals 3–4 mm, oblong-ovate. Petals pale yellow, oblong. Achenes numerous, very small, 1–1.25 mm. Found on mud by the river bank.

Medicinal uses:

Herb treats tuberculosis, adenopathy, malaria, arthralgia, antitoxic, anti-inflammatory, and useful as emmenagogue and galactogogue. Leaf ant rheumatic, case dermatitis. Bruised leaves cause skin blisters. Leaf juice removes warts, used together with bruised root as counter-irritant to alleviate headache. Root ant rheumatic, classified as an active agent against KB tumor. Seeds tonic, for cold, general debility, rheumatism, and spermatorrhoea.

Family: Rubiaceae.

***Galium aparine* L.**

Annual herb, up to 180 cm long, with flaccid stout quadrangular stem. Leaves up to 60 mm. Corolla 1.5–2 mm in diameter, always whitish, mericarp 2–5 mm long.

Medicinal use:

Herb diuretic, depurative, antimicrobial, aperient, atraumatic, alterative, tonic, astringent, sudorific, antispasmodic, antiscorbutic, anticancer, and refrigerant. Applied for hepatoma, leukemia, cancer of breast, glands, throat, and tongue. Juice is a blood purifier and heals psoriasis.

Family: Scrophulariaceae.

***Bacopa monnieria* (L.) Hayata et Matsum.**

Perennial herb, 10–30 cm, glabrous, flesh. Stem creeping, rooting at the axils. Leaves opposite, sessile, obovate-cuneiform, 1–1.5 cm long, the upper ones gradually smaller. Flowers solitary in the axils, on 5–10 mm long pedicels, corolla pale violet, 5–6 mm long. Capsule 2–3 mm long, ovate, and acute.

Medicinal uses:

Herb nervine, tonic, beneficial in epilepsy, insanity, epilepsy, and hysteria. Powdered leaves manage cases of asthenia, nervous breakdown, and other random conditions. The plant is a direct cardiac tonic.

Family: Solanaceae.

***Solanum nigrum* L.**

Annual, 30–60 cm, glabrous or pubescent. Stem branched from the base. Leaves ovat-deltoid, entire, or shallowly sinuate-dentate. Flowers 3–6 flowered, umbellate, and short peduncled. Corolla white, pubescent, twice as long as the calyx. Berries globose, about 8 mm diameter, black.

Medicinal uses:

Herb diuretic, antibacterial, febrifuge, treat mastitis, cervicitis, chronic bronchitis, dysentery, the entire plant is decocted for abscesses, cancer of cervix inflammation, leucorrhea, heavy female discharges, and sore throats. Leaf, root, and stalk are used for cancerous sores, wounds leukoderma, nausea, purging, nervous disturbance and plant juice for hydrogogue, cathartic, diuretic, and alterative. Extract decreases cardiac activity, reduces blood pressure, and is vasodilator, analgesic, and antispasmodic. Young shoots are eaten as pot herb considered a tonic for virility in men and for dysmenorrhea in women, for dysentery, sore throat, and whitlow.

Family: Verbenaceae.

***Phyla nodiflora* (L.) Greene.**

Trailing perennial herb, 12–60 cm, branches angular. Leaves cuneate-obovate, 7–30 cm long with mostly 7–9 coarse, sharp teeth around the upper margin. Spike ovoid, dense, becoming cylindrical, up to 2.5 cm long in fruit. Calyx 1.25 mm, membranous, and bilobed. Corolla 2.25–2.5 mm, pink, lobes included. Drupe roundish, 1.25–1.75 mm, yellowish.

Medicinal uses:

Plant used for hookworm, said to be coolant, diuretic, febrifugal, stopping of bowels and pain in knee joints, poultice for boils, swollen cervical glands, erysipelas, and indolent ulcers. Juice of the plant is used for children's dyspepsia, gonorrhea, emmenagogue, bronchitis, and respiratory disease. Alcoholic extract shows antibacterial activity against *Escherichia coli*. Plant is also used for adenopathy, arthritis, fever diaper rash, neuralgia, and vertigo. Decoction of inflorescence with vinegar is deodorant.

***Verbena officinalis* L.**

Perennial herb, 30–100 cm with numerous tough stems and woody rootstock. Leaves lanceolate or lanceolate-oblong in outline, 2–11 cm, pinnatifid. Spikes terminal on the stem and branches, at first dense but elongating and lax in fruit. Calyx 2–2.5 cm. Corolla 4–5 mm, pale lilac with yellowish throat, but deeper pink or the outer surface. Nutlets 1.75–2.0 mm, shining, and brown.

Medicinal uses:

Plant de-coagulant, detoxicant, corroborant, antiscorbutic, tonic, antispasmodic, antibacterial, antitoxic, anti-inflammatory, febrifuge, detersive, nervine, diuretic, emmenagogue, emetic, sudorific, used for amenorrhea, anemia, ascites, cirrhosis, dropsy, dyspepsia, edema, fever, hepatitis, inflammation, malaria, mastitis, nephritis, parturition, urinary tract infection, whooping cough, and heartthrob. Leaf for congestion, dropsy, and constipation. Extract of leaves is antibacterial, hepatitis, against cold, fomentation for sorts and wounds. Decoction is helpful in throat and tonsil inflammation, headache, and psychological troubles.

Class: Monocotyledones.

Family: Arecaceae (Palmae).

***Phoenix dactylifera* L.**

Cultivated, solitary with massive erect trunk, ultimately 25 more tall, 40–50 cm diameter, trunk dull brown. Crown comprising about 50 or more fresh leaves, leaf length to 4.5 m, apparent petiole 50–100 cm, long, armed with acanthophylls to 15 cm long, leaflets to 40x2 cm, 80 or more on each side of the rachis. Staminate inflorescence with the peduncle up to 60 cm, pistillate inflorescence similar but greatly elongating after anthesis. Pistillate flower globose. Fruit very variable, 4–7 cm, green, yellow, orange-brown to deep chestnut. Seed variable to 2.4 cm.

Medicinal uses:

Pollen is antifebrile, purgative, hemostatic, emmenagogue, and tonic, helpful in thrust, inflammation, tenesmus, asthma, cough, and aphrodisiac. Fruit demulcent, expectorant, nutrient, laxative, aphrodisiac, useful in asthma, chest complaints, fever, and gonorrhea.

.

Family: Cyperaceae.

***Cyperus rotundus* L.**

Creeping perennial herb with wooly roots. Rhizomes are brown. Stem rigid 15–40 cm triquertous. Leaves several, basal, and crowded. Inflorescence one or twice anthelate, rays 3–9 cm, spike ovoid, lax, spikelets linear-lanceolate, 1–3 cm. Achene ellipsoid, 1.5 cm long, red, or dark brown.

Medicinal uses:

Tubers are astringent, diaphoretic, carminative, deodorant, given in disorder of stomach and irritation of bowels, used for fever, relapsing fever, long-standing fever, chronic diarrhea, dyspepsia, cases of dysentery, gastric, and intestinal disorders. Infusion or decoction of the tubers is given in fever, diarrhea, dysentery, dyspepsia, vomiting, and cholera.

Family: Hydrocharitaceae.

***Vallisneria spiralis* L.**

Acaulous herb, stoloniferous. Leaves rosulate, ribbon-like up to 30 cm or more cm long, 1 cm broad, 5–7 nerved. Male spathes short, pedunculate. Female peduncles becoming spirally twisted after anthesis.

Medicinal uses:

Plant for leucorrhea, made into a tea with sesame to improve the appetite, demulcent, refrigerant, stomachic, and aperitif.

Family: Lemnaceae.

***Lemna minor* L.**

Free-floating aquatic.Thallus dark green, flat, oblong-ovate, 2.5–6 mm long, 1.5–4 mm wide. Root solitary, arising along a shallow groove. Floral pocket lateral, spathe open at the top. Staminate flowers 2 each with a single stamen. Pistillate flower is solitary. Seed solitary, not ribbed, and reticulate.

Medicinal uses.

Herb cooling, diuretic, astringent, alterative, antiscorbutic, used for circulation, measles, swollen feet, considered depurative, and soporific.

***Spirodela polyrrhiza* (L.) Schleid.**

Frond flat, rounded, rather thick, about 5–7 mm across, brownish-red beneath, 5–11 nerved, with a cluster of fine rootlets beneath. Stamens 2, anthers 4-roomed. Ovary with 2 or more ovules, but only one developing into seed. In stagnant water.

Medicinal uses:

Plant carminative, diaphoretic, and diuretic. Whole plant decocted for cold, fever, hemorrhagic, purpura, inadequate measles rash, nephritis, and urticaria.

Family: Poaceae (Gramineae).

***Arundo donax* L.**

A perennial with woody rhizomes and with very stout, erect culms up to 4 m in height, simple or shortly branched from the nodes. Leaf blades 5–6 cm, cordate at the base up to 60 cm long. Inflorescence a plume-like panicle, much branched.

Medicinal uses:

The defatted ethanolic extract of the rhizomes produces hypotensive and anti-spasmodic effects against histamine, serotonin, and acetylcholine-induced spasms. Graminine and donaxine started to increase blood pressure when given in small doses, but reduce the pressure in higher doses. Rootstock contains essential oil and bitter principle, which can be used medicinally.

***Cynodon dactylon* (L.) Pers.**

Perennial herb spreading in all directions by rhizomes and stolons, sending up erect, leafy shoots up to 30 cm tall. Spikes digitate 2–4 cm long. Leaf-blades flat, lanceolate up to 6 cm long. Spikelets crowded, small, and secund.

Medicinal uses:

The grass is used in dropsy and anasarca. Juice of the plant is astringent and applied to fresh cuts and wound to stop bleeding. It is also prescribed in hysteria, epilepsy, insanity, chronic diarrhea, dysentery, and catarrh. Decocted juice is helpful in calculous. Decoction of roots is given as diuretic, in dropsy, and in secondary syphilis. An infusion of the roots is given for stopping bleeding of piles. Rhizome in genital-urinary disorder.

***Digitaria sanguinalis* (L.) Scop.**

Annual grass. Culms in favorable habitats up to 90 cm tall. Leaf blades linear-acuminate, up to 15 cm long, 8 mm wide. Racemes 5–12 cm long.

Medicinal uses:

Plant decoction is used for gonorrhea, sclerosis of the breast, cataracts, and debility. It is said to be emetic.

***Echinocloa crus-galli* (L.) P. Beauv.**

A tufted annual with erect or spreading culms up to 1 m tall, stout, succulent, and glabrous at nodes. Leaf blades flat, up to 30 cm long. Panicle of spaced racemes, either solitary or two or three together, spikelets secund, 3–4 mm long. A very common weed in rice fields.

Medicinal uses:

Plant is tonic and used in the diseases of spleen and for checking hemorrhage. Stem and shoot applied as a styptic to wounds. Root pounded and applied to wounds.

***Imperata cylindrica* (L.) P. Beauv.**

Perennial herb, calms up to 120 cm tall erect or geniculate from the base, rhizome long, thick, succulent, scaly. Leaf blades stiffly erect. Panicle very dense, erect, cylindrical up to 20 cm long Spikelets 4–5 mm long.

Medicinal uses.

Plant is used for drug withdrawal. Its decoction is used against diarrhea, dysentery, and gonorrhea. Flowers for hemorrhage wounds, hematophysis, epistaxis, hematotensis, and hematuria. Decoction for nephritic edema, high fever, thirst, and urinary tract infection. Fruiting spike is given as a sedative. Root is considered as astringent, antifebrile, antivirus, diuretic, hemostat, restorative and tonic, helpful for asthma, cancer, dropsy, epistaxis, fever, hematuria, jaundice, and nephritis. Root bark for diarrhea, gonorrhea, febrifuge, restorative, and tonic. Rhizome as restorative, tonic, and antipyretic.

***Loliunm temulentum* L.**

Annual weed with tufted, rarely solitary, stiff culms, up to 60 cm tall. Leaf blades linear-acuminate, up to 30 cm long. Spikelets erect, 10–30 cm long, rigid. Upper glumes longer than the spikelets, lemmas usually wound, becoming swollen and hard at maturity, 6–8 mm long.

Medicinal uses:

Plant decoction is hemostatic, diuretic. Seeds are harmful with narcotic effect, toxic, and helpful as a poultice for dermatitis.

***Phragmites australis* (Cav.) Trin. ex Steudel.**

A stout, rhizomatous, perennial grass, up to 4 m tall, culms simple, rigid, smooth, and glabrous. Leaf blades flat, 20–60 cm long. Panicle plume-like, erect, purplish, or brownish finally silvery, up to 30 cm. Long much branched. Spikelets widely gaping, 2–6 flowered.

Medicinal uses:

Herb is diaphoretic, diuretic, and said to be applied for leukemia. Leaves for bronchitis, cholera, fish, and shrimp poisoning, ashes are styptic. Shoot is antidotal, antiemetic, antipyretic, refrigerant, diuretic, febrifuge, sialagogue, arthritis, cough, earache, fever, hematuria, hiscups, nausea, pulmonary abscesses, sore throat, sun stroke, and toothache.

Family: Potamogetonaceae.

***Potamogeton nodosus* Poir.**

Perennial herb, stem terete. Leaves floating or submerged. Floating leaves opaque and coriaceous, elliptic or ovate-elliptic, 3–12 cm long, submerged leaves translucent, and lanceolate. Flowers wind-pollinated. Druplets 3.5–4 mm long.

Medicinal uses:

Plant is febrifuge and resolving.

***Potamogeton pectinatus* L.**

Perennial herb, stems slender, and terete. Leaves submerged, narrowly linear to filiform, under 2 mm broad, entire, sessile, acute, sheath united with the leaf base, and convolute. Flowers water-pollinated. Drupelets 3–5 mm long.

Medicinal uses:

Decoction is used to soothe feverish liver.

***Potamogeton crispus* L.**

Perennial herb, stem compressed. Leaves submerged, translucent, linear to linear-oblong, 5–10 mm broad, sessile, 3–5 nerved, and usually undulate. Flowers wind-pollinated. Drupelets long-beaked, connate at the base, and 4–5 mm long.

Medicinal uses.

Extract of the herb are antibacterial, healing hepatic cells, anticancer of ovary, and have protective effect on chronic disease, diabetes, cardiac, and bowel disease as well as against AIDS virus infection.

Family: Typhaceae.

***Typha domingensis* Pers.**

Perennial herb, rhizome stout, 7–10 mm wide, Flowering stem 1–3 m tall. Leaves overlapping the stem broad sheathing base. Male and Female inflorescences separated by a gap of mostly 1–3 cm, male spike 17–35 cm long, female spike 18–23 cm at maturity. Fruit 1.5 mm, yellow, linear-cylindrical.

Medicinal uses:

Leaf is diuretic. Pollen are astringent, desiccant, diuretic, and hemostat, vulnerary.

Recommended literature

Al-Hity M (1989) Pioneer pharmacologists in history. Studies about the Arab sciences history. Baghdad University, Iraq, pp 75–87

Ali KI, Al-Baldawi MH, Al-Kassie GA, Jamel YJ (2010) Drug plants. Baghdad, p 110. (In Arabic)

Al-Mayah AA (2013) Medicinal plants and herbal therapy. Basrah University, Iraq, p 358. (In Arabic)

Al-Rawi A, Chakravarty HL (1964) Medicinal plants of Iraq. Ministry of Agriculture, Baghdad, Iraq, p 109

Al-Saadi HA, Al-Mayah AA (1983) Aquatic plants of Iraq. Basrah University, Iraq, p 192

Al-Shahat NA (1992) Aromatic plants. Cairo, Egypt, p 473. (In Arabic)

Ashour A (1985) Therapy by herbs. Quran Bookshop, Cairo, p 216. (In Arabic)

Chakravarty HL (1976) Plant wealth of Iraq. Ministry of Agriculture, Baghdad, Iraq, p 505

Lupoae P, Cristea V, Borda D, Lupoae M, Gurau G, Dicina RM (2015) Phytochemical screening: antioxidant and antibacterial properties of Potamogeton species in order to obtain valuable feed additives. J Oleo Sci 64(10):111–1123

Duke JA, Ayensu ES (1985) Medicinal plants of China. Reference Publications, Michigan, p 705

Dutta SC, Banerjee AK (1978) Useful weeds of West Bengal rice fields. Econ Bot 32:297–310

Ghazanfar SA, Edmondson JR (2013) Flora of Iraq. vol. 5, Part 2. Ministry of Agriculture, Baghdad. 696 p 349

Ghazanfar SA, Edmondson JR (2016) Flora of Iraq. vol. 5, Part 1. Ministry of Agriculture, Baghdad, p 285

Guest E (1966) Flora of Iraq, vol 1. Ministry of Agriculture, Baghdad, p 213

Holm LR, Doll J, Holm E, Pancho J, Herberger J (1997) World weeds. Natural histories and distribution. John Wiley & Sons, New York, p 1129

Husain FT (1979) Medicinal plants. Arabic House Publishers, Libya, p 557. (In Arabic)

Kadhim SH (1986) Medicinal herbs. Ministry of Information, Baghdad, Iraq, p 230

Li TS (2006) Taiwanese native medicinal plants, phytopharmacology and therapeutic value. Taylor & Francis Group, New York, p 328

Padma S, Deepika S (2013) Explotation of weed plants as beneficial purpose. J Biol Sci Opinion 1 (2):123–127
Qudama A (1985) Dictionary of food and herbal therapy. Dar Al-Nafaes, Beirut, p 806. (In Arabic)
Rechinger KH (1964) Flora of lowland Iraq. Verlag Von Cramer, Weinheim, p 746
Ruwaiha A (1983) Herbal therapy. Dar al-Qalam, Beirut, p 559. (In Arabic)
Saeed OA (1981) On the way to the herbalist. Azal Publishers, Beirut, p 100. (In Arabic)
Townsend CC, Guest E (1966) Flora of Iraq, vol 2. Ministry of Agriculture, Baghdad, p 184
Townsend CC, Guest E (1968) Flora of Iraq, vol 9. Ministry of Agriculture, Baghdad, p 588
Townsend CC, Guest E (1974) Flora of Iraq, vol 3. Ministry of Agriculture, Baghdad, p 662
Townsend CC, Guest E (1980) Flora of Iraq, vol 4. Ministry of Agriculture, Baghdad, p 1199
Townsend CC, Guest E (1985) Flora of Iraq, vol 8. Ministry of Agriculture, Baghdad, p 440

Chapter 28
The Role of Plants as a Canopy in the Inland Waters: Basic Information for Application in Iraq

Laith A. Jawad and Audai M. Qasim

Abstract The water temperature of stream can limit the growth and endurance of aquatic biota, while covering of the plantation on the stream catchments plays a vital role in inhibiting river warming. Water temperature directly responded to air temperature and cloudiness. Removal of plantation especially those with large leaves can upsurge solar radiation in the stream catchment area as well as wind speed and disclosure to air, which can increase during air in summer, soil, and stream temperatures and decrease in relative humidity. Stream temperature upsurges and subsequent removal of plantation are not only chiefly organized by variations in separation but also governed by stream hydrology and channel shape.

Recently, eco-environment degradation processes and changes have been detected in Iraq. Examples on such great changes in the environment can be seen in the environment of Basrah City, south of Iraq. Observed changes are mainly desertification, secondary salinization, urbanization, vegetation degradation, and loss of wetlands.

Recommendations to restore the streams and their catchments were suggested in order to be followed in any developmental program that will be taken by the government.

28.1 Introduction

Water temperature in streams and springs is considered a limiting factor for a number of aquatic organisms such as fish and invertebrates (Logez and Pont 2013; Pletterbauer et al. 2014). The losses and gain in water temperature in small freshwater body depend on the quantity of stream release, topography, and atmospheric settings along the river way for example (Evans et al. 1998; Caissie 2006;

L. A. Jawad (✉)
School of Environmental and Animal Sciences, Unitec Institute of Technology, Auckland, New Zealand

A. M. Qasim
Department of marine vertebrates, Marine Science Centre, University of Basrah, Basrah, Iraq

L. A. Jawad (ed.), *Tigris and Euphrates Rivers: Their Environment from Headwaters to Mouth*, Aquatic Ecology Series 11,
https://doi.org/10.1007/978-3-030-57570-0_28

Pletterbauer et al. 2014). Also, another set of characters may affect water temperature of streams such as upstream water temperature and discharge, bed heat conduction, and solar radiation (incident, reflected) (Caissie 2006; Logez and Pont 2013). Air temperature and short-wave radiation are deliberated as main impelling issues on water temperatures of streams (Evans et al. 1998).

The studies that have dealt with the effect of riparian vegetation on river water temperature have recognized certain parameters to measure in order to assess whether the vegetation cover is enough to keep the water temperature of stream in its normal range (Moore et al. 2005; Gomi et al. 2006; Caissie 2006; Davies-Colley et al. 2009). Among these important parameters are the buffer width, vegetation height, and its density. On the other hand, the height and density of the river bank vegetation have an equal or even greater influence on water temperature (Evans et al. 1998; Gomi et al. 2006; DeWalle 2010), and Davies-Colley et al. (2009) and DeWalle (2008) have shown that the ratio of canopy height to stream width has specific high effect on stream light disclosure and is consequently particularly applicable (Logez and Pont 2013).

Changes in water temperature of small rivers and streams are somewhat quick (Moore et al. 2005; Caissie 2006; Davies-Colley et al. 2009; DeWalle 2010), and river bank vegetation effect can be great (Poole and Berman 2001; Gomi et al. 2006). Changes in water temperature vary with the size of the river. Shading could be dissimilar for greater rivers where water temperature variations are normally lower (Poole and Berman 2001; Moore et al. 2005; DeWalle 2010).

In Iraq, deforestation is a main problem and riparian areas along the rivers, especially the small streams, are decreasing substantially (Haktanir et al. 2004). Therefore, this chapter aims to review short studies performed on the effects of river bank vegetation on river water temperature during summer and how to manage the presence of vegetation along the sides of the rivers to keep water temperatures of these water bodies in its normal range. At the end of the chapter, recommendations have been given about managing the riparian areas, and so the policy makers in Iraq can take up.

28.2 The Role of Stream Bank Plantation in Improving and Protecting Chemical Water Quality

Worldwide, protecting sources of clean water and refining the chemical value of ruined waters for both human drinking and habitat health have become vital goals (Baker et al. 2006; Arthurton et al. 2007). One approach that can attain these aims is administration of vegetation on the stream bank. It is well recognized that vegetated riparian areas can powerfully impact the chemical composition of contiguous streams, chiefly through the elimination of nutrients in runoff from agricultural uplands (Baker et al. 2006; Arthurton et al. 2007). Vegetation refurbishment and management in the zone adjacent to streams are therefore widely proposed and endorsed in agricultural zones to recover chemical water quality in rivers (NRC 2002; Baker et al. 2006).

Dosskey (2001) and Arthurton et al. (2007) have suggested that the level and time frame of water class enhancement rely on the sort of contaminant and the methods that impact it such as (1) site situations that govern how significant each procedure is and (2) the extent of dilapidation in these procedures that happened prior to refurbishment. An understanding of these fundamental processes is vital for successfully using vegetation state as a marker of water quality protection and for precisely evaluating outlooks for water quality enhancement through renovation of permanent vegetation (Dosskey 2001; NRC 2002; Baker et al. 2006; Arthurton et al. 2007).

28.3 The Date Palm: THE Main Canopy Vegetation in Stream Area in Iraq

In Iraq, date palms are usually planted near the banks of rivers and streams in large quantities. Therefore, their long leaves form an excellent canopy and create a decent shade to the rivers and streams nearby (Fig. 28.1).

The date palm (*Phoenix dactylifera* L.) is deliberated a figure of life in the Middle East, since it bears high temperatures, drought, and salinity more than many other fruit produce plant species. In the long history of the human civilization, date palm is considered as one of the oldest trees from which man has obtained benefit (Zohary and Hopf 2000). It has been suggested that the cultivation of the date palm goes as far back as 4000 B.C. in southern Iraq. Evidences on using dates as food were obtained from prehistoric Egypt (Barreveld 1993; Zohary and Hopf 2000). Relics of dates have been discovered on a number of Neolithic sites, chiefly in Syria and Egypt. This indicates that they were being consumed by man as much as 7000–8000 years ago (Lunde 1978; Barreveld 1993).

The date palm is counted a renewable natural resource because it can be substituted in a comparatively short period of time or used through upkeep attempts without running down (Lunde 1978; Barreveld 1993; University of Delaware 2004). The date palm maintains its value for growers as it gives a wide range of harvests and facilities, including many provisions of life. The date, the primary product of the palm, is rich in protein, vitamins, and mineral salts. That is why it signifies a vital constituent of food for the farmer himself and his animals. All subordinate goods of the palm result from annual clipping and have indispensable usages for the farmer (Lunde1978) (Fig. 28.2).

Fig. 28.1 Date palm trees used as canopy in Iraq. (**a**) Tall date palm trees; (**b**) other plantations are usually cultivated in the shadow of the date palm trees; (**c**) cultivation of date palm close to each other

Fig. 28.2 Pieces of mat made of dry leaves of date palm

28.4 Socioeconomic and Traditional Importance of Dates

Date palm is socioeconomically and conventionally important for native societies where the culture succeeds (Barreveld 1993; Jain et al. 2011). Formation of date palm meadows assisted nomad groups in the past to live and start societies and start agriculture. These groups became a center for marketing/trading supplies and animal and other products. Dates were and still are among the most imperative and appreciated products for trading at these societies. They are respected for their food and feed usages (Jain et al. 2011).

A completely new business has also been established in recent years around date palm and dates. For instance, workshops with more or less complicated ways of pitting, penetrating, crushing, and sieving dates provide a significant number of local jobs. Excess production not eaten fresh frequently is changed by natives into paste, spread, powder (used locally as a sugar), jam, jelly, juice, sirup, vinegar, and alcohol (Jain et al. 2011). Unacquainted dates do not go discarded. They are frequently dehydrated, crashed, and mixed with grains and straw to become an appreciated feed for domestic animals.

The date palm tree can give more than their fruit to the benefited human being. Young leaves and terminal buds are sometimes prepared as vegetables. Seeds are also often roasted and crushed (Barreveld 1993; El Hadrami et al. 2011). The stems of the date palm tree can be used for building houses and bridges over small streams. The leaves are also included in the building house sectors in southern of Iraq (Fig. 28.3) (El Hadrami et al. 2011). Date palm leaves are frequently included in making mats, screens, baskets, crates, and fans or for religious devotions (Fig. 28.4). The pulp of the date palm tree can be eaten fresh as a fruit (Fig. 28.5).

Medically, dates have been prescribed for several diseases and sicknesses. The fruit is rich in tannins, making it a good severe cure for intestinal distresses (El Hadrami et al. 2011). Preparations such as infusions, decoctions, sirups, and pastes are often administered against colds, sore throat, and bronchial cough. They are also taken to release fever, liver and abdominal aches, cystitis, gonorrhoea, and edema. The roots are used to treat toothache, and the pollen is appreciated for its estrogenic compound, estrone (El Hadrami and Al-Khayri 2012).

28.5 Degradation of Date Palm Trees in the Arab Countries: Indicators and Causes

The state of Date palms in Iraq is similar to that in the remaining Middle east countries that are facing hazard from diseases, pests, habitat variations, and socio-economic causes (FAO 1982).

Over the last decade, yield of date palms has deteriorated in the traditional farming zones that pests and diseases have induced important impacts on date production in the Arab countries for example (FAO 1982; Malumphy and Moran 2009). Unfortunately, pests and diseases spread increasingly with the development of trade and travel in the globalizing world system (UN Press Release 2004). Additionally, habitat effects, such as drought and salt and conventional cultural methods, are common difficulties in the areas and put more pressures on those already present (Oihabi 2001; Malumphy and Moran 2009).

The other cause that made a reduction in the yield volume of date palms and the dilapidation of the feature of production itself is the loss of manpower (Fig. 28.6). The workers, who used to work at the date palm fields in the countries where dates are usually produced, have left to the urban areas to increase their income (de Haas 1998; Oihabi 2001; Malumphy and Moran 2009). The needed agricultural events such as soil grounding, selection, and planting out of palm offshoots are considered hard jobs that are suitable only for young people. With the loss of the young manpower, the maintenance of date palms is generally reduced and resulted in field having densely wild grown date palms and consequently reducing harvests (de Haas 1998; Oihabi 2001).

In the last half of century, date palm trees have been uncovered to deprivation owing to widespread misuse ensuing from the upsurge in both the human society and the number of domestic animals (de Haas 1998). Hundreds of date palm fields have been converted to urban sites, where date palm trees have been cleared to prepare the

Fig. 28.3 Stems of date palm tree used for house roofing. (**a**) Date palm trunks prepared for roofing process; (**b**) date palm trunks after cutting; (**c**) completed roof

Fig. 28.4 Handmade artifacts made from dry leaves of date palm trees

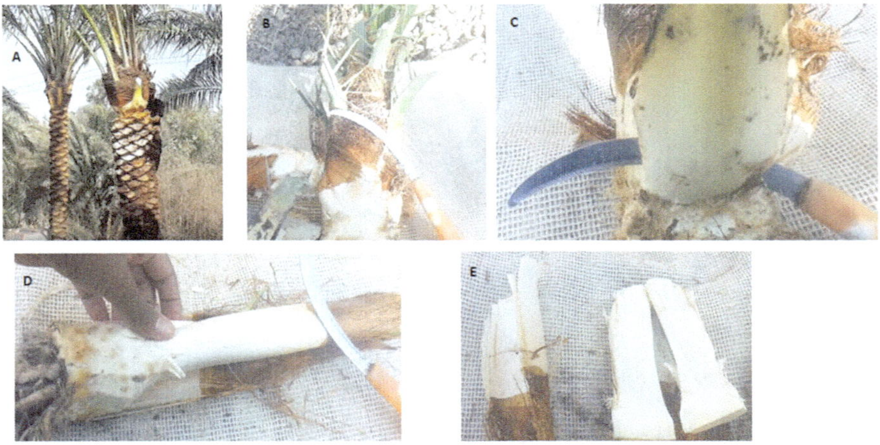

Fig. 28.5 Steps to extract the pulp of the date palm trees. (**a**) Cutting the head of the date palm tree; (**b**) removing the hard layers of plant tissues, (**c**) removing the soft plant tissues, and (**d**) exposing and removing the pulp

Fig. 28.6 Neglected date palm trees showing loss of leaves

land for building houses (Fig. 28.7). Effects of war in the regions where date palms are usually grown were evident in reducing the number of these trees dramatically. 8 years spent in the first Gulf War (Iraq-Iran war) has produced hundreds of date palms that are classified as nothing but dead plant remains (Fig. 28.8).

Among the common reasons of date palm dilapidation in the Arab countries are the lack of vegetation canopy. In some regions and owing to high grazing and harvesting of wood for fuel, soil erosion is created. With the augmented wind velocity and storm frequency, an increase in the levels of hovering dust in the air will increase. The other factor is the decreased infiltration of water into soil due to the increase in runoff and floods and reduced water table levels in some areas.

28.6 Wildfire as an Ecological Factor for Streams

One of the most worst ecological factors, which are natural disturbances, is the wildfire (Agee 1998). The effect of this factor has been investigated on the vegetation, soils, water yield of diverse areas, and numerous forest kinds (Gresswell 1999; Agee 1998; Brown 2000; Arno and Allison-Bunnell 2002). The area adjoining the stream usually experienced different natural disturbances such as fire, flooding, debris flows, and landslides (Naiman et al. 1993).

Fig. 28.7 Date palm field has been prepared for house building after removal of the trees

Fig. 28.8 Effects of the Iraq-Iran war on the date palms in the southern of Iraq. Images showing burned trees and others without heads

Several aspects need to be understood in order to encompass the behavior of fire in the area adjacent to streams (Agee 1998; Brown 2000; Arno and Allison-Bunnell 2002). Among these issues are understanding the mechanism of spread of fire spread and burn patterns, fuel loads and spreading, fuel chemistry and flammability, and fuel moisture (Naiman et al. 1993; Patten 1998). It is not possible to separate human activities from the effects and properties of fire. For instance, water usage and administration and man-made disturbances (Gresswell 1999; Brown 2000; Arno and Allison-Bunnell 2002) are among the main anthropogenic factors on fire behavior. To manage such an effect, the urban-rural line needs to manage first (Agee 1998; Gresswell 1999; Kauffman 2001). Moreover, in addition, foretelling prototypes of fire performance under diverse situations of land and water usage and global climate modification need to be considered and put in contingency plans (Patten 1998; Kauffman 2001; Arno and Allison-Bunnell 2002).

28.7 Effect of Fire on the Soil of the Stream: Case of the Southern Marshes of Iraq

Soon after the genocide and the ecocide that Saddam Hussein has committed against Marsh Arabs in 1991, he committed another crime. To make sure that no one will come out of the marsh area alive, he ordered his forces to burn the reeds in the marshes. The author has witnessed this burning reed crime. Saddam Has ordered members of his popular army to be present in the marsh area and provided them with all the tools that enable them to burn completely the reed forest of the marshes. The smoke of the fire extends to over 100 Km south of the marsh area and covered the sky of Basrah city. The smoke was too thick that many people who have problem in breathing have been taken to the hospital.

The reason behind drying the marshes and burning the reeds is to deny the rebels a place to hide. The government troops found it problematic to combat in the marshes as their heavy equipment and vehicles could not move through such territory. As a result, the resolution was to strengthen and enlarge the strategies to empty the marshes completely from water. This strategy and burning the reeds were completely and quickly executed by Iraqi administration and the Marsh Arabs, and the Shi'ite troops were killed, seized, or driven out and into expatriate camps mostly inside Iran (UNEP 2003). Chemical arms, weaponry, and minefields were involved to take out the remaining marsh people. Prevalent mass killings and forced replacements meant that by the year 2000, only 20,000 marsh Arabs out of a total population of 500,000 endured in their customary regions (Carpenter and Ozernoy 2003; UN 2002).

The marshes were dry when the fire has set in the reeds, and therefore, the heat resulting from the huge fire has affected the soil of the marshes (Fig. 28.8). Fitzpatrick (2004) in his report on the changes in the soil of the marshes has shown that soil of the marshes has been completely devastated. To let the readers many aspects of this report, here I quote Fitzpatrick (2004), "The severely burned locations in the marsh area were designated by enflamed surface soils, loss of organic matter, and occasionally white to gray ash on top of the soil. The inflamed surface soil layer fluctuated in thickness from 1 to 80 cm and was caused by a darkened soil layer 1–15 cm thick. Biomass burning has had a profound effect on the functioning of these soils. Both the direct impacts of fire and also the overall variations to the habitat in a post-fire state have led to short-, medium-, and long-term deviations in the soil (Fitzpatrick (2004). These are linked to soil performance in the physical, biological, and chemical aspects and also contain variations to combined stability, pore size distribution, water repellence, and runoff response. High temperature burning (>500 degrees centigrade) of the dried marshland soils has irreversibly destroyed the original soil components (e.g., organic matter, iron pyrite, and layer silicates) and formed high concentrations of magnetic cemented/ceramic-like gravel (>60%) in the upper 1–50 cm (Fitzpatrick 2004). These fragments have significant inferences for chemical and physical procedures in these soils because they increase permeability and provide a physical restriction to the root growth of sensitive plants".

Fitzpatrick (2004) also noted that the dried reeds were up to 2–3 m high when the burning occurred. The fuel load was enormous to burn such dry reed. Therefore, the fires would cause temperatures to exceed 300 degrees centigrade (i.e., as though in a kiln or oven). Topsoil between 15 and 50 cm transformed irreversibly into ceramic bricks or hard cemented (fused) ceramic-like porous fragments (Carpenter and Ozernoy 2003).

28.8 Decreasing Plant Cover Due to the Environmental Changes: Case of Southern of Iraq

In south of Iraq, the decreasing plan cover is forming a serious problem in recent years. There is no precise indication about the area of plant concealment in Iraq (Jabbar 2001; Hadeel et al. 2010; Al-saadi 2002). The land of Iraq has experienced numerous changes over the past few years (UNEP 2001). The agricultural land in Iraq is about 12%. Most of which is in the region of the Tigris and Euphrates Rivers (Al-saadi 2002; Jabbar 2001).

In their study, Hadeel et al. (2010) have shown that voluminous changes in the land at Basrah City, south of Iraq, have been performed during the last decade. The urban area and sand land have increased, with a significant decrease in farmland (Al-saadi 2002; Hadeel et al. 2010). The reasons behind such changes could be the expansion of new network and huge development in military aspect, which transformed huge agricultural lands to build up these areas. In addition, several other factors may be considered vital for such environmental changes. Jabbar (2001) and Hadeel et al. (2010) have identified four important elements, which are as follows: (1) Basrah City has been evolving fast under the habitat of rapid growth of the southern part of Iraq. For instance, the rural development of the marshland and the rural-urban zone of Safwan-Zubair, south of the city, have increased rapidly; (2) continuous increase in number of inhabitants and housing stresses enhanced the real estate expansion on the fringes of Basrah Province, which transformed agricultural lands to housing land; (3) traffic, water, power, natural gas, and other numerous substructures are some other lashing powers of urban growth and extension (Hadeel et al. 2010).

28.9 Why Restorations are Needed

Palmer et al. (2007) and Arthurton et al. (2007) have suggested an urge for more restorations and proposed the following reasons: (1) yet where human societies are not mounting, water withdrawals and human substructure on the land endure to upsurge intensely Arthurton et al. (2007); (2) assumed upcoming weather plans, many rivers will face new flow and sediment rules (Poff et al. 2002; Palmer et al. 2007).

Palmer et al. (2007) have also suggested that renovation along with sensible administration movements may ease the predictable influences and possibly even withstand a river ecosystem's capability to react to and be tough in the meet water usage and climate modification.

References

Agee JK (1998) The landscape ecology of western forest fire regimes. Northwest Sci 72:24–34. (Special Issue)

Al-Saadi KM (2002) Use of remote sensing data for detection the desertification in dry climate areas. First Iraqi Scientific Symposium in Meteorology and Seismic (Dryness in Iraq). Baghdad, pp. 26–27

Arno SF, Allison-Bunnell S (2002) Flames in our forest: disaster or renewal? Island Press, Washington, DC, p 227

Arthurton R, Barker S, Rast W, Huber M (Coordinators) (2007) Water. In: Global environment outlook: environment for development. United Nations Environmental Programme, Nairobi, Kenya, Chapter 4, pp 115–156. http://www.unep.org/geo/ geo4/report/04_Water.pdf. Accessed June 2009

Baker JL, Helmers MJ, Laflen JM (2006) Water management practices, rain-fed cropland. In: Schnepf M, Cox C (eds) Environmental benefits of conservation on cropland: the status of our knowledge. Soil and Water Conservation Society, Ankeny, Iowa, pp 89–130

Barreveld WH (1993) Date palm products. Agricultural services bulletin no. 101, food and agriculture organization of the United Nations. FAO, Rome

Brown JK (2000) Introduction and fire regimes. In: Brown JK, Smith JK (eds) Wildland fire and ecosystems: effects of fire on Flora. Gen. Tech. Rep. RMRS-GTR-412-vol 2. USDA Forest Service, Ogden, UT, pp 1–8

Caissie D (2006) The thermal regime of rivers: a review. Freshw Biol 51:1389–1406

Carpenter B, Ozernoy I (2003) Water world: marshes in Iraq in process of being restored. U.S. News & World Report

Davies-Colley RJ, Meleason MA, Hall RMJ, Rutherford JC (2009) Modelling the time course of shade, temperature, and wood recovery in streams with riparian forest restoration. N Z J Mar Freshw Res 43:673–688

de Haas H (1998) Socio-economic transformations and oasis agriculture in southern Morocco. In: de Haan L, Blaikie P (eds) Looking at maps in the dark: directions for geographical research in land management and sustainable development in rural and urban environments of the third world. KNAG/FRW UvA, Utrecht/Amsterdam, pp 65–78

DeWalle DR (2008) Guidelines for riparian vegetative shade restoration based upon a theoretical shaded-stream model. J Am Water Resour Assoc 44:1373–1387

DeWalle DR (2010) Modeling stream shade: riparian buffer height and density as important as buffer width. J Am Water Resour Assoc 46:323–333

Dosskey MG (2001) Toward quantifying water pollution abatement in response to installing buffers on crop land. Environ Manag 28(5):577–598
El Hadrami A, Al-Khayri JM (2012) Socioeconomic and traditional importance of date palm. Emirates J Food Agric 24(5):371
El Hadrami A, Daayf F, El Hadrami I (2011) Secondary metabolites of date palm. In: Jain SM, Al-Khayri JM, Johnson DV (eds) Date palm biotechnology. Springer, Netherlands, pp 653–674
Evans EC, McGregor GR, Petts GE (1998) River energy budgets with special reference to river bed processes. Hydrol Process 12:575–595
FAO (1982) Plant production and protection: paper no. 35. Date production and protection. FAO, Rome, p 294
Fitzpatrick RW (2004) Changes in soil and water characteristics of natural, drained and re-flooded soils in the Mesopotamian marshlands: Implications for land management planning. CSIRO land and water: client report
Gomi T, Moore RD, Dhakal AS (2006) Headwater stream temperature response to clear-cut harvesting with different riparian treatments, coastal British Columbia, Canada. Water Resour Res 42:W08437
Gresswell RE (1999) Fire and aquatic ecosystems in forested biomes of North America. Trans Am Fish Soc 128:193–221
Hadeel AS, Jabbar MT, Chen X (2010) Environmental change monitoring in the arid and semi-arid regions: a case study Al-Basrah Province, Iraq. Environ Monit Assess 167(1–4):371–385
Haktanir K, Karaca A, Omar SM (2004) The prospects of the impact of desertification on Turkey, Lebanon, Syria and Iraq. In: Environmental Challenges in the Mediterranean 2000–2050. Springer, Dordrecht, pp 139–154
Jabbar MT (2001) Soil loss by wind erosion for three different textured soil treated with polyacrylamide and crude oil, Iraq. J China Univ Geosci 2:113–116
Jain SM, Al-Khayri JM, Johnson DV (eds) (2011) Date palm biotechnology. Springer, Netherlands
Kauffman JB (2001) Workshop on the multiple influences of riparian/stream ecosystems on fires in western forest landscapes. Report submitted to Rocky Mountain Research Station, Stream Systems Technology Center, July 2001
Logez M, Pont D (2013) Global warming and potential shift in reference conditions: the case of functional fish-based metrics. Hydrobiologia 704:417–436
Lunde P (1978, March/April) A history of dates. Saudi Aramco World 29(2):176–179
Malumphy C, Moran H (2009, November) Plant Pest factsheet: red palm weevil-*Rhynchophorus ferrugineus*. The food and environment research agency (Fera). Sand Hutton, New York
Moore RD, Spittlehouse DL, Story A (2005) Riparian microclimate and stream temperature response to forest harvesting: a review. J Am Water Resour Assoc 41:813–834
Naiman RJ, Décamps H, Pollock M (1993) The role of riparian corridors in maintaining regional biodiversity. Ecol Appl 3:209–212
NRC (National Research Council) (2002) Riparian areas: functions and strategies for management. National Academy Press, Washington, DC, p 428
Oihabi A (2001, March) Date palm in North Africa: trumps and problems. In: The second international conference on date palms, Department of Arid Land Agriculture and Department of agriculture and livestock College of Food Systems, UAE university. UAE, Ai-Ain, pp 25–27
Palmer M, Allan JD, Meyer J, Bernhardt ES (2007) River restoration in the twenty-first century: data and experiential knowledge to inform future efforts. Restor Ecol 15(3):472–481
Patten DT (1998) Riparian ecosystems of semi-arid North America: diversity and human impacts. Wetlands 18:498–512
Pletterbauer F, Melcher A, Ferreira T, Schmutz S (2014) Impact of climate change on the structure of fish assemblages in European rivers. Hydrobiologia 744:235–254
Poff NL, Brinson MM, Day JW (2002) Aquatic ecosystems and global climate change: potential impacts on inland freshwater and coastal wetland ecosystems. Prepared for the Pew Center on Global Climate Change, Arlington, VA. Available from www.pewclimate.org/projects/aquatic.clim Accessed 20 June 2007

Poole GC, Berman CH (2001) An ecological perspective on in-stream temperature: natural heat dynamics and mechanisms of human-causedthermal degradation. Environ Manag 27(6):787–802

UN Press Release (2004) Global Date Palm Production at Risk due to pests, diseases, SAG/276. Food and Agriculture Organization of the United Nations (FAO), Rome, Italy. http://www.un.org/News/Press/docs/2004/sag276.doc.htm

UNEP (2001) The Mesopotamian marshlands: the demise of an ecosystem. Early warning and technical assessment report, UNEP/DEWA/TR.01–3. http://www.grid.unep.ch/activities/ sustainable/tigris/mesopotamia.pdf. Accessed May 2003

United Nations (2002) The demise of Mesopotamian marshlands. UN Chronicle, 34. UN, New York

United Nations Environment Programme (2003) Environment in Iraq: UNEP progress report. UNEP, Geneva

University of Delaware (2004) Renewable natural resources. University of Delaware Cooperative Extension, College of Agriculture and Natural Resources, Newark

Zohary D, Hopf M (2000) Domestication of plants in the Old World, 3rd edn. Oxford University Press, Oxford, UK

Chapter 29
The Zooplankton Fauna of the Turkish Part of the Euphrates-Tigris River Basin

Mehmet Borga Ergönül and Sibel Atasağun

Abstract There is growing interest in the zooplankton species composition of inland water bodies in Turkey. Although there are partial data from several water bodies located in the Turkish part of the Euphrates-Tigris river basin, a complete dataset is missing. In this study, studies on the zooplankton fauna of reservoirs, lakes and running waters in the basin are reviewed. A total of 256 zooplankton species, including 37 cladocerans, 12 copepods and 207 rotifers, have been recorded, which correspond to 38.5% of the total zooplankton fauna in Turkey.

Keywords Zooplankton · Rotifera · Cladocera · Copepoda · Anatolia · Tigris · Euphrates

29.1 Introduction

Zooplankton constitute a major component of the food webs in aquatic habitats (Havens 2002). Due to their high reproduction and assimilation rates, they play crucial roles in energy flow and nutrient cycling (Saler and Şen 2002b). Zooplankton species composition can be also used as an indicator of water quality and pollution (Ren et al. 2011; Garcia-Chicote et al. 2017). Several zooplankton species such as *Daphnia magna* (Hoang et al. 2007; Altındağ et al. 2008; Ergönül et al. 2012) and *Artemia salina* (Rajabi et al. 2015) are accepted as reliable laboratory models for toxicity studies.

In freshwater habitats, zooplankton are represented by three major groups: Cladocera, Copepoda and Rotifera (Bekleyen 2003). Rotifers are the dominant group in freshwater habitats. Several researchers investigated the species composition and seasonal fluctuations in freshwaters in Turkey (Altındağ et al. 2008; Yıldız et al. 2007; Ergönül et al. 2016). However, there are limited data on the zooplankton

M. B. Ergönül (✉) · S. Atasağun
Faculty of Science, Department of Biology, Ankara University, Ankara, Turkey
e-mail: borga@science.ankara.edu.tr

L. A. Jawad (ed.), *Tigris and Euphrates Rivers: Their Environment from Headwaters to Mouth*, Aquatic Ecology Series 11,
https://doi.org/10.1007/978-3-030-57570-0_29

species composition at a basin level (Saler et al. 2011a; Ergönül et al. 2015), especially from the eastern and southeastern part of Turkey (Bulut and Saler 2014a, b). Euphrates-Tigris river basin is a transboundary basin and 22% of the basin is located in Anatolia (Lehner et al. 2008), which corresponds to 25% of the country. Although there are a few studies dealing with the zooplankton fauna of the basin from Iraq, Iran and Syria (Abdulwahab and Rabee 2015; Merza 2017), there is limited information on the Turkish part of the basin. Thus, the aim of this study is to present a list of zooplankton species recorded in the Turkish part of the Euphrates-Tigris river basin.

29.2 Materials and Methods

A total of 34 studies, which were conducted at 35 different sampling locations located in the Turkish part of the Euphrates-Tigris river basin, were reviewed. Eleven of these of studies were carried out on running waters and the remaining were from reservoirs and lakes. The sampling locations for zooplankton species composition of water bodies located in the basin are illustrated in the map (Fig. 29.1).

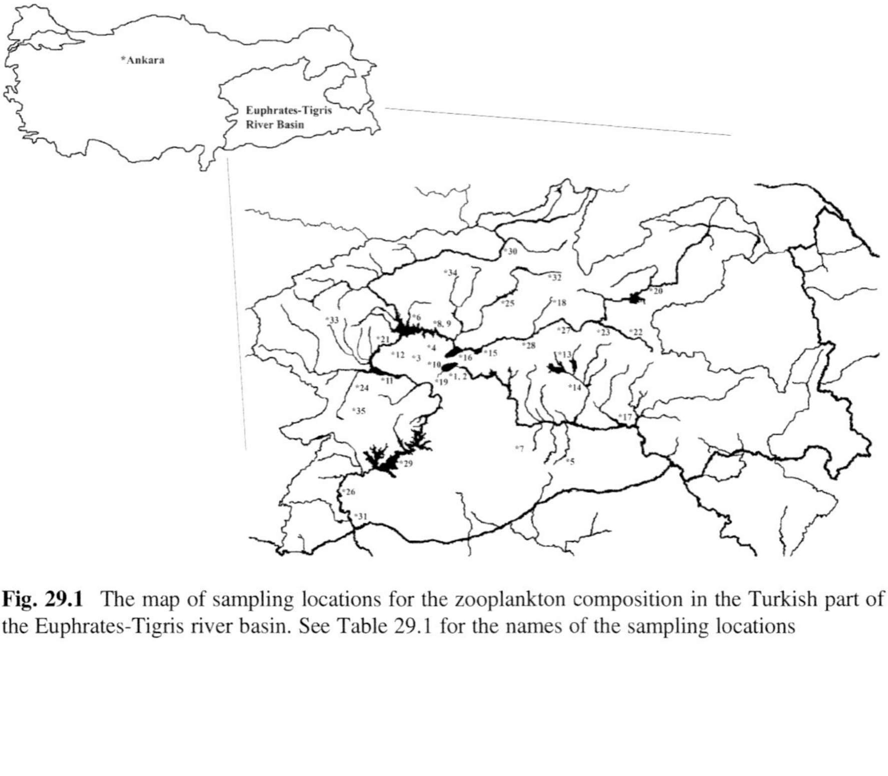

Fig. 29.1 The map of sampling locations for the zooplankton composition in the Turkish part of the Euphrates-Tigris river basin. See Table 29.1 for the names of the sampling locations

29.3 Results and Discussion

A total of 256 zooplankton species, including 37 cladocerans, 12 copepods and 207 rotifers, have been recorded from the Turkish part of the Euphrates-Tigris river basin. The list of species and the sampling localities are shown in Table 29.1.

Table 29.1 The zooplankton species recorded from the Turkish part of Euphrates-Tigris River Basin

	Sampling localities
Cladocera	
Alona costata	14
Alona guttata	29
Alona quadrangularis	13
Alona rectangula	5, 7, 11, 13, 14, 21, 22, 23, 24, 26, 31, 35
Biapertura affinis	5, 7
Bosmina coregoni	35
Bosmina longirostris	5, 7, 10, 11, 18, 20, 21, 22, 23, 24, 25, 26, 27, 28, 29, 30, 31, 32, 34, 35
Ceriodaphnia dubia	5
Ceriodaphnia megops	5
Ceriodaphnia pulchella	13, 14, 26, 31
Ceriodaphnia quadrangula	29, 35
Ceriodaphnia reticulata	1, 5, 10, 18, 19, 20, 21, 22, 24, 25, 26, 28, 31, 32, 34, 35
Chydorus sphaericus	5, 7, 11, 13, 14, 23, 24, 25, 26, 27, 28, 31, 32, 34, 35
Coronatella rectangulata	33
Cornigerius lacustris	1, 10
Daphnia birgei	24
Daphnia cucullata	5, 7, 11, 13, 14, 21, 22, 23, 24, 28 30, 31, 35
Daphnia galeata	13, 14, 26, 35
Daphnia longispina	10, 11, 19, 21, 23, 24, 25, 26, 29, 30, 31, 32, 34, 35
Daphnia magna	21, 35
Daphnia pulex	21
Diaphanosoma birgei	26, 27, 31, 35
Diaphanosoma brachyurum	29
Diaphanosoma lacustris	1, 5, 7, 10, 13, 14, 23, 28, 30, 32, 35
Diaphanosoma orghidani	5, 7
Disparalona rostrata	5, 7, 14, 23, 24, 26, 27
Ilyocryptus sordidus	14
Leptodora kindtii	7, 13, 20, 23, 25, 32
Leydigia leydigi	7, 14, 22, 23, 24, 26, 31, 32
Macrothrix groenlandica	13, 14
Macrothrix laticornis	5, 7, 13, 14, 34
Moina micrura	5, 7, 14, 31
Moina macrocopa	30

(continued)

Table 29.1 (continued)

	Sampling localities
Pleuroxus aduncus	14, 28, 34
Sida crystallina	35
Simocephalus expinosus	14
Simocephalus vetulus	5, 25
Copepoda	
Acanthocyclops robustrus	23
Acanthodiaptomus denticornis	1, 10, 13, 14, 19, 20, 21, 22, 23, 24, 25, 26, 28, 30, 31
Cyclops strenuus	24, 26, 31, 35
Cyclops vicinus	1, 7, 10, 13, 14, 18, 19, 20, 21, 22, 23, 24, 25, 26, 28, 30, 31, 32, 34, 35
Diacyclops bicuspidatus	24, 26
Eucyclops serrulatus	5, 7
Macrocyclops albidus	13, 23, 24, 31
Megacyclops viridis	26
Nitocra hibernica	24, 26, 31, 35
Sinodiaptomus sarsi	7
Thermocyclops crassus	7, 26
Thermocyclops dybowskii	5, 7, 13
Rotifera	
Anuraeopsis fissa	13, 14, 17, 29
Ascomorpha ecaudis	13, 14, 17, 23, 29, 35
Ascomorpha ovalis	5, 13, 14, 17, 21, 22, 29
Ascomorpha saltans	2, 8, 9, 13, 15, 17, 18, 21, 22, 24, 25, 26, 31, 32, 34, 35
Aspelta angusta	14, 17
Aspelta aper	17
Aspelta curvidactyla	14, 17
Aspelta psitta	17
Asplanchna brightwellii	11, 14, 17
Asplanchna priodonta	4, 6, 8, 9, 11, 12, 13, 14, 15, 16, 17, 20, 21, 22, 23, 24, 25, 26, 27, 28, 29, 31, 34, 35
Asplanchna sieboldi	5, 8, 9, 22, 25, 31, 34, 35
Asplanchnopus hyalinus	15, 17
Brachionus angularis	2, 4, 8, 9, 14, 17, 19, 21, 22, 23, 24, 26, 27, 28, 30, 31, 32, 34, 35
Brachionus bennini	17
Brachionus bidentatus	14, 17, 26
Brachionus budapestinensis	14, 17
Brachionus calyciflorus	2, 6, 14, 15, 16, 17, 21, 30, 31, 35
Brachionus caudatus	5, 24, 26
Brachionus diversicornis	17
Brachionus durgae	17
Brachionus falcatus	5, 17
Brachionus leydigi	14, 17
Brachionus murphyi	17

(continued)

Table 29.1 (continued)

	Sampling localities
Brachionus nilsoni	17
Brachionus plicatilis	2, 4, 15
Brachionus quadridentatus	2, 4, 14, 17, 19, 20, 24, 28, 29, 31, 34, 35
Brachionus rubens	14, 17
Brachinous sericus	17
Brachionus sessilis	13
Brachionus urceolaris	2, 3, 6, 12, 14, 15, 16, 17, 30
Cephalodella auriculata	13, 14, 17, 27, 34
Cephalodella catellina	14, 17, 28, 30, 34, 35
Cephalodella forficata	14, 17
Cephalodella forficula	4, 8, 9, 14, 15, 16, 17, 21, 23, 28, 29, 30, 31
Cephalodella gibba	2, 3, 4, 5, 8, 9, 13, 14, 15, 16, 17, 19, 20, 21, 22, 23, 25, 26, 27, 28, 29, 30, 31, 32, 34
Cephalodella gigantea	17
Cephalodella hollowdayi	14
Cephalodella hoodii	14, 17
Cephalodella megalocephala	14, 17
Cephalodella misgurnus	14, 17
Cephalodella obvia	17
Cephalodella stenroosi	17
Cephalodella sterea	14, 17
Cephalodella tenuiseta	17
Cephalodella theodora	14, 17
Cephalodella tinca	17
Cephalodella ungulata	14, 17
Cephalodella ventripes	14, 17, 27, 29, 30, 34
Collotheca campanulata	13
Collotheca mutabilis	5, 13, 14, 17
Collotheca ornata	5, 17
Colurella adriatica	2, 14, 15, 17, 23, 24, 26, 27, 28, 29, 31, 34
Colurella colurus	14, 17, 26, 27, 29, 31
Colurella obtusa	21, 29, 30, 34
Colurella uncinata	4, 6, 8, 9, 12, 15, 16, 17, 29
Conochilus coenabasis	29
Conochilus dossuarius	5, 13
Conochilus hippocrepis	23
Colurella uncinata	14
Conochilus unicornis	3, 13, 14, 17, 23
Dicranophoroides caudatus	14, 17
Dicranophorus aspondus	14, 17
Dicranophorus epicharis	14, 17
Dicranophorus forcipatus	14, 17
Dicranophorus luetkeni	17

(continued)

Table 29.1 (continued)

	Sampling localities
Donneria sudzukii	17
Encentrum incisum	13
Encentrum martes	14, 17
Encentrum putorius	14, 17
Encentrum saundersiae	14, 17, 34
Encentrum uncinatum	13, 14, 17
Eosphora ehrenbergi	17
Eosphora thoides	17
Epiphanes brachionus	14, 17
Epiphanes macroura	17
Epiphanes senta	8, 9, 15, 17, 30
Euchlanis deflexa	14, 17
Euchlanis dilatata	17, 21, 28, 30, 34
Euchlanis incisa	14, 17
Euchlanis lyra	14, 17
Euchlanis dilatata	5, 13, 22, 23, 24, 26, 27, 29, 31, 32
Filinia branchiata	14, 17
Filinia longiseta	2, 5, 6, 8, 9, 11, 13, 14, 16, 17, 19, 21, 22, 23, 26, 28, 29, 30, 35
Filinia opoliensis	17, 24, 31
Filinia terminalis	2, 4, 6, 12, 13, 14, 15, 16, 17, 23, 27, 30
Gastropus stylifer	13, 14, 17, 23
Habrotrocha sp.	11
Hexarthra fennica	2, 8, 9, 14, 17, 24, 26, 29, 30
Hearthra intermedia	13, 14, 17, 24, 26
Hexarthra mira	8, 9, 21, 24, 35
Hexarthra oxyuris	29
Itura aurita	14, 17
Kellicottia longispina	2, 3, 4, 6, 8, 9, 11, 12, 13, 14, 15, 16, 17, 19, 20, 21, 22, 23, 24, 25, 26, 31, 32, 34, 35
Keratella cochlearis	3, 5, 6, 8, 9, 11, 12, 13, 14, 15, 16, 17, 18, 19, 20, 21, 22, 23, 24, 25, 26, 27, 28, 29, 30, 31, 32, 34, 35
Keratella hispida	22
Keratella quadrata	3, 5, 8, 9, 11, 12, 14, 15, 16, 17, 18, 20, 21, 22, 24, 25, 26, 27, 28, 30, 31, 32, 34, 35
Keratella tecta	13, 14, 17, 21, 22, 23, 24, 26, 27, 28, 30, 31, 34
Keratella tropica	5, 11, 14, 17, 24, 26, 31, 32
Keratella valga	14, 17, 24, 26, 31
Lecane aculeata	17
Lecane arcuata	17
Lecane aspasia	14, 17
Lecane bifastigata	14, 17
Lecane bulla	14, 17, 24, 26, 27, 29, 31, 34
Lecane closterocerca	13, 14, 17, 23, 26, 28, 29, 30

(continued)

Table 29.1 (continued)

	Sampling localities
Lecane curvicornis	29
Lecane cornuta	15, 17, 34
Lecane decipiens	14, 17
Lecane flexis	13, 14, 17
Lecane furcate	17, 29
Lecane hamate	14, 17, 29
Lecane hastate	14, 17, 29
Lecane homemanni	17
Lecane inopinata	17
Lecane ivli	17
Lecane leonita	17
Lecane ludwigi	14, 17
Lecane luna	2, 4, 5, 6, 13, 15, 16, 17, 21, 23, 24, 25, 26, 27, 28, 29, 32, 34
Lecane lunaris	3, 12, 13, 14, 15, 16, 17, 21, 23, 27, 28, 29, 30, 34
Lecane margalefi	14, 17
Lecane nana	14, 17
Levane papuana	14, 17
Lecane pyriformis	6, 15, 16, 17
Lecane quadridentata	11, 14, 17
Lecane scutata	5, 13, 14, 17
Lecane shieli	14, 17
Lecane stenroosi	14, 17
Lecane stichaea	14, 17
Lecane tenuiseta	17
Lecane thienemanni	14, 17, 29
Lepadella acuminata	5, 13, 14, 17
Lepadella costata	6, 14, 15, 17
Lepadella ehrenbergii	17
Lepadella latusinus	17
Lepadella ovalis	3, 4, 6, 15, 16, 17, 18, 21, 25, 27, 30, 32, 34
Lepadella patella	11, 13, 14, 15, 17, 27, 28, 29, 30, 31, 34
Lepadella quadricarinata	13, 14, 17
Lepadella quinquecostata	17
Lepadella rhomboids	11, 14, 17
Lindia torulosa	14, 17
Lopocharis salpina	13, 14, 17, 29
Monommata arndti	5, 14, 17
Monommata dentate	29
Mytilina bisulcata	14, 17
Mytilina mucronata	30
Mytilina trigona	6, 24, 26
Mytilina unguipes	17
Mytilina ventralis	14, 17, 31

(continued)

Table 29.1 (continued)

	Sampling localities
Notholca acuminata	2, 18, 25, 28, 30, 32
Notholca squamula	2, 3, 4, 5, 6, 8, 9, 12, 13, 14, 15, 16, 17, 18, 19, 20, 21, 22, 24, 25, 26, 27, 28, 30, 31, 32, 34
Notommata aurita	14, 17
Notommata codonella	17
Notommata copeus	17
Notommata crytopus	14, 17
Notommata glyphura	14, 17, 27
Notommata pachyura	17
Notommata pseudocerberus	14, 17
Notommata tripus	17
Octotrocha speciosa	14, 17
Plationus patulus	14, 17
Platyias quadricornis	14, 17
Philodina megalotrocha	29
Philodina roseola	4, 12, 15, 18, 24, 26, 28, 30, 34
Pleurotrocha petromyzon	13, 14, 17
Pleurotrocha sigmoidea	17
Polyarthra dolichoptera	5, 11, 13, 14, 15, 17, 19, 21, 22, 23, 24, 25, 26, 27, 28, 29, 30, 31, 32, 35
Polyarthra remata	8, 9, 20
Polyarthra vulgaris	3, 5, 6, 9, 11, 12, 13, 14, 15, 16, 17, 18, 20, 29
Pompholyx complanata	5, 13, 14, 17
Pompholyx sulcata	13, 14, 17, 24, 26, 30, 31
Proales fallaciosa	14, 17, 27
Proales sigmoidea	14
Proales Theodora	14, 17
Resticula melandocus	17
Resticula nyssa	13, 14, 17
Rotatoria neptunia	4, 24, 26, 32
Rotatoria rotatoria	26, 28
Scaridium longicaudum	14, 17
Stephanoceros fimbriatus	14, 17
Synchaeta kitina	13, 17
Synchaeta oblonga	5, 11, 13, 14, 17, 23, 27, 29, 32, 34, 35
Synchaeta pectinata	3, 6, 8, 9, 13, 14, 15, 16, 18, 20, 21, 22, 23, 25, 26, 27, 28, 29, 30, 31, 32, 34, 35
Synchaeta stylata	13, 14, 17, 24, 26
Synchaeta verrucosa	35
Taphrocampa selenura	14, 17
Testudinella patina	8, 9, 14, 24, 26, 30, 31
Testudinella mucronata	5
Testudinella truncata	5

(continued)

Table 29.1 (continued)

	Sampling localities
Tirchocerca bicristata	14, 17
Trichocerca braziliensis	17
Trichocerca capucina	13, 14, 17, 21, 23, 25, 26, 28, 29, 30, 32
Trichocerca cylindrica	5, 8, 9
Trichocerca heterodactyla	13
Trichocerca insignis	14
Trichocerca longiseta	14, 17, 31
Trichocerca porcellus	17
Trichocerca pusilla	13, 14, 17
Trichocerca rattus	14, 17
Trichocerca similis	5, 13, 14, 17, 24, 26, 28, 29, 30, 31
Trichocerca taurocephala	14, 17
Trichocerca tenuior	13, 14, 17
Trichotria curta	17
Trichotria pocillum	14, 17, 30
Trichotria tetractis	6, 13, 14, 15, 16, 17, 21, 25, 27, 30, 32, 34
Tripleuchalanis plicata	14, 17
Wolga spinifera	14, 17
Wulfertia kivuensis	17

1. Hazar Lake—Tellioğlu and Şen 2001	19. Kürk Stream—Saler et al. 2011a
2. Hazar Lake—Tellioğlu and Şen 2002	20. Peri Stream—Saler et al. 2011b
3. Tadım Pond—Saler and Şen 2002a	21. Görgüşan Stream—İpek and Saler 2012
4. Cip Dam Lake—Saler and Şen 2002b	22. Ohi Stream—İpek and Saler 2013a, b
5. Göksu Dam Lake—Bekleyen 2003	23. Kalecik Dam Lake—Bulut and Saler 2013
6. Keban Dam Lake (Çemişgezek Region)—Saler 2004	24. Sürgü Dam Lake—İpek and Saler 2013a, b
7. Devegeçidi Dam Lake—Bekleyen 2006	25. Uzunçayır Dam Lake—Saler et al. 2014
8. Keban Dam Lake (Pertek Region)—Tellioğlu and Akman 2007a	26. Hancağız Dam Lake—Saler and Alış 2014
9. Keban Dam Lake (Pertek Region)—Tellioğlu and Akman 2007b	27. Beyhan Dam Lake—Bulut and Saler 2014a
10. TMI-12 Pond—Saler and Arslan 2007	28. Murat River—Bulut and Saler 2014b
11. Karakaya Dam Lake—Özhan 2007	29. Atatürk Dam Lake—Kaya 2014
12. Kepektaş Dam Lake—Saler 2009	30. Karasu River—Saler et al. 2015
13. Kral Kızı and Dicle Dam Lake—Gökot 2009	31. Karkamış Dam Lake—İpek and Saler 2016a
14. Tigris River—Gökot 2009	32. Özlüce Dam Lake—İpek and Saler 2016b
15. Keban Dam Lake (Gülüşkür Region)—Saler and Sen 2010	33. Tohma Stream—Saler and İpek 2016
16. Keban Dam Lake (Ova Region) Saler 2010	34. Tahar Stream—Öcalan and Saler 2016
17. Tigris River—Bekleyen et al. 2011	35. Boztepe Recai Kutan Dam Lake—Saler et al. 2017
18. Munzur River—Saler 2011	

Table 29.2 Number of zooplankton species recorded from reservoirs/lakes and running waters in the Turkish part of the Euphrates-Tigris river basin

Zooplankton groups	Reservoirs/lakes	Running waters	Total
Cladocera	30	16	37
Copepoda	12	2	12
Rotifera	109	186	207
Total	151	204	256

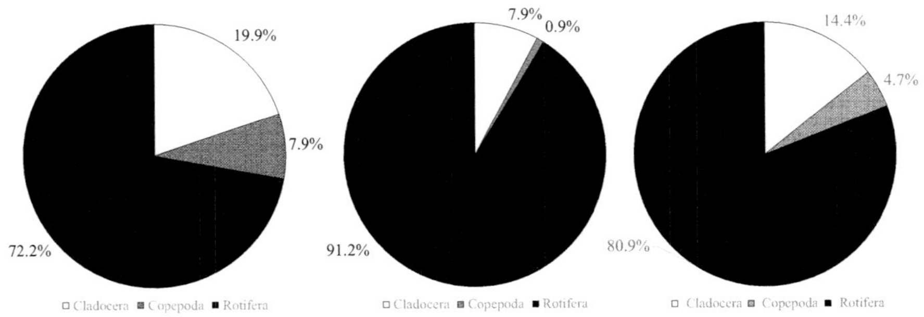

Fig. 29.2 The percentage of the number of species recorded from reservoirs/lakes, running waters and whole basin (Turkish part of the Euphrates-Tigris river basin) for each zooplankton group

The number of species recorded in running waters and reservoirs/lakes is shown in Table 29.2. The number of cladoceran and copepod species recorded from reservoirs/lakes was higher than the running waters. However, the number of rotifer species was higher in the running waters.

The relative percentage of zooplankton groups recorded from reservoirs/lakes, running waters and whole basin (Turkish part) is shown in Fig. 29.2.

The most common cladoceran species was *Bosmina longirostris*, which has been recorded in 20 out of 35 different sampling locations in the basin. The most common copepod and rotifer species were *Cyclops vicinus* and *Keratella cochlearis*, which were recorded in 20 and 29 different sampling locations, respectively. The genus *Daphnia* was represented with 6 species, *Cyclops* and *Thermocyclops* with 2 species each; *Lecane* was represented with 31 species among Cladocera, Copepoda and Rotifera, respectively.

The zooplankton fauna of Turkish inland waters includes 661 species, with 103 belonging to Cladocera, 141 to Copepoda and 417 to Rotifera (Table 29.3). In this study, 37 Cladocera species have been recorded corresponding to 35.9% of the total Cladocerans, 8.5% of the copepods and 49.6% of the rotifers. Although there is sufficient information on the zooplankton composition of individual water bodies (Altındağ et al. 2008; Yıldız et al. 2007; Ergönül et al. 2016), there are limited data at a basin level. According to Bulut and Saler (2014a, b), a total of 185 zooplankton species have been recorded from the Eastern and Southeastern Anatolian waters, which include Euphrates-Tigris river basin.

Table 29.3 Number of species recorded from Turkish inland waters and Turkish part of the Euphrates-Tigris river basin

Zooplankton groups	Turkish inland waters (Ustaoğlu 2015)	Eastern and Southeastern Anatolia (Bulut and Saler 2014a, b)	Turkish part of the Euphrates-Tigris river basin (this study)
Cladocera	103	31	37 (35.9%)
Copepoda	141	16	12 (8.5%)
Rotifera	417	138	207 (49.6%)
Total	661	185	255 (38.5%)

Numbers in parentheses indicate the percentage recorded in basin compared to whole Turkish inland waters

Since the zooplankton fauna of the water bodies located in the Eastern and Southeastern region of Turkey received little attention (Bekleyen et al. 2011), it is quite possible to find new zooplankton species records in the region. Thus, future studies should focus on the region using the trophi morphology for rotifers to ensure the identification.

References

Abdulwahab S, Rabee AM (2015) Ecological factors affecting the distribution of the zooplankton community in Tigris River at Baghdad region, Iraq. Egypt J Aquat Res 41:187–196

Altındağ A, Ergönül MB, Yigit S, Baykan Ö (2008) The acute toxicity of lead nitrate on *Daphnia magna* Straus. Afr J Biotechnol 7(23):4298–4300

Bekleyen A (2003) A taxonomical study on the zooplankton of Göksu Dam Lake (Diyarbakır). Turk J Zool 27:95–100

Bekleyen A (2006) Devegeçidi Baraj Gölü'nün (Diyarbakır) Cladocera ve Copepoda (Crustacea) Türleri. Anadolu Üniversitesi Fen-Edebiyat Dergisi 23(3–4):413–415

Bekleyen A, Gökot B, Varol M (2011) Thirty-four new records and the diversity of the Rotifera in the Turkish part of the Tigris River watershed, with remarks on biogeographically interesting taxa. Sci Res Essays 6(30):6270–6284

Bulut H, Saler S (2013) Kalecik Baraj Gölü (Elazığ-Türkiye) Zooplanktonu. Fırat Üniversitesi Fen Bilimleri Dergisi 25(2):99–103

Bulut H, Saler S (2014a) Zooplankton of Beyhan Dam Lake (Elazığ-Turkey). Turk J Sci Technol 9 (1):23–28

Bulut H, Saler S (2014b) Murat Nehri'nin (Elazığ Palu İlçe Merkezi Sınırları İçindeki Bölümün'de) Zooplanktonu ve Değişimi. Türk Tarım-Gıda Bilim ve Teknoloji Dergisi 2(1):13–17

Ergönül MB, Atasağun S, Beşer T (2012) The acute toxicity of zinc chloride on *Daphnia magna* straus. Gazi Univ J Sci 25(2):313–316

Ergönül MB, Erdoğan S, Altındağ A, Atasağun S (2015) Rotifer Fauna of Gediz River Basin, Turkey. World J Fish Mar Sci 7(1):69–73

Ergönül MB, Erdoğan S, Altındağ A, Atasağun S (2016) Rotifera and Cladocera fauna of several lakes from Central Anatolia, Marmara and Western Black Sea region of Turkey. Turk J Zool 40:141–146

Garcia-Chicote J, Armengol X, Rojo C (2017) Zooplankton abundance: a neglected key element in the evaluation of reservoir water quality. Limnol Ecol Manag Inland Waters. https://doi.org/10.1016/j.limno.2017.11.004

Gökot B (2009) Dicle Nehri zooplanktonunun mevsimsel değişimi üzerine bir araştırma. Doktora Tezi. Dicle Üniversitesi
Havens KE (2002) Zooplankton structure and potential food web ınteractions in the plankton of a subtropical chain-of-lakes. Sci World J 2:926–942
Hoang TC, Tomasso JR, Klaine SJ (2007) An integrated model describing the toxic responses of *Daphnia magna* to pulsed exposures of three metals. Environ Toxicol Chem 26(1):132–138
İpek N, Saler S (2012) Görgüşan ve Geban Deresi (Elazığ-Türkiye) Zooplanktonu. J Fisheriessciences.com 6(2):155–163
İpek AN, Saler S (2013a) Zooplankton of Sürgün Dam Lake (Malatya-Turkey). BEÜ Fen Bilimleri Dergisi 2(1):39–43
İpek N, Saler S (2013b) Zooplankton community structure of Ohi Stream (Elazığ-Turkey). J Fisheriessciences.com 7(1):83–88
İpek AN, Saler S (2016a) Zooplankton composition of Karkamış Dam Lake (Gaziantep-Turkey). Biherian Biol 10(2):86–89
İpek AN, Saler S (2016b) Zooplankton fauna of Özlüce dam Lake (Bingöl-Turkey). Bitlis Eren Univ J Sci 5(1):86–90
Kaya İ (2014) Atatürk Baraj Gölü set ile tünel girişi arası bölgenin zooplankton faunası. Yüksek Lisans Tezi. Harran Üniversitesi
Lehner B, Verdin K, Jarvis A (2008) New global hydrography derived from spaceborne elevation data. EOS Trans Am Geophys Union 89(10):93–94
Merza FA (2017) Checklist of Cladocera and Copepoda Taxa in Iraqi waters. Int J Sci Eng Res 8 (4):134–151
Öcalan A, Saler S (2016) Tahar Çayı (Tunceli) zooplanktonu. Fırat Üniv Müh Bil Dergisi 28 (2):1–10
Özhan D (2007) Karakaya Baraj Gölü su kalitesinin zooplankton kompozisyonu ile değerlendirilmesi. Yüksek lisans tezi. İnönü Üniversitesi
Rajabi S, Ramazani A, Hamidi M, Naji T (2015) *Artemia salina* as a model organism in toxicity assessment of nanoparticles. DARU J Pharm Sci 23:20
Ren L, Zhang Z, Zeng X, Ma Y, Zeng Y, Zhou C (2011) Community structure of zooplankton and water quality assessment of Jialing River in Nan Chong. Procedia Environ Sci 10:1321–1326
Saler S (2004) Observation on the seasonal variation of Rotifera fauna of Keban Dam Lake (Çemişgezek Region). F Ü Fen ve Mühendislik Bilimleri Dergisi 16(4):695–701
Saler S (2009) Rotifers of Kepektas Dam Lake (Elazığ-Turkey). Iran J Sci Technol 33(A1):121–126
Saler S (2010) Fluctuations in Rotifera fauna of Keban Dam Lake Ova Region (Elazıg-Turkey) during spring. Turk J Sci Technol 5(2):53–58
Saler S (2011) Zooplankton of Munzur River (Tunceli, Turkey). J Anim Vet Adv 10(2):192–194
Saler S, Alış N (2014) Zooplankton of Hancağız Dam Lake (Gaziantep-Turkey). J Surv Fish Sci 1 (1):36–45
Saler S, Arslan S (2007) TMİ 12 Göleti (Elazığ) Copepoda ve Cladocera (Crustacea) faunası. E Ü Su Ürünleri Dergisi 24(1–2):191–193
Saler S, İpek NA (2016) Zooplankton composition of Tohma Stream (Malatya-Turkey). J Aquacul Eng Fish Res 2(1):30–35
Saler S, Şen D (2002a) A taxonomical study on the Rotifera fauna of Tadım Pond (Elazığ). EÜ Su Ürünleri Dergisi 19(3–4):497–500
Saler S, Şen D (2002b) Seasonal variation of Rotifera fauna of Cip Dam Lake (Elazığ-Turkey). Pak J Biol Sci 5(11):1274–1276
Saler S, Sen B (2010) Long term changes in Rotifera fauna of Gülüşkür Bay (Keban Dam Lake Elazig Turkey). J Anim Vet Adv 9(14):1909–1912
Saler S, İpek N, Aslan S (2011a) Kürk Çayı (Elazığ-Türkiye) zooplanktonu. J Fisheriessciences.com 5(3):219–225
Saler S, Eroğlu M, Bulut H (2011b) Peri Çayı (Tunceli-Türkiye) zooplanktonu. Ecol Life Sci 6 (2):14–20

Saler S, Bulut H, Baysal N (2014) Zooplankton of Uzunçayır Dam Lake (Tunceli-Turkey). J Fisheriessciences.com 8(1):1–7

Saler S, Bulut H, Birici N, Tepe R, Alpaslan K (2015) Karasu Nehri Erzincan nin zooplanktonu. Eğirdir Su Ürünleri Fakültesi Dergisi 1(1):10–16

Saler S, Alpaslan K, Karakaya G, Gündüz F (2017) Boztepe Recai Kutan Baraj Gölü (Malatya-Türkiye) zooplankton faunası. Ege Üniversitesi Su ürünleri Dergisi 34(3):261–267

Tellioğlu A, Akman F (2007a) A taxonomical study on the Rotifera fauna in Pertek region of Keban Dam Lake. EÜ Su Ürünleri Dergisi 24(1–2):135–136

Tellioğlu A, Akman F (2007b) Seasonal variation in the species diversity of planktonic rotifers in the Pertek region of Keban dam Lake, Turkey. J Freshwater Ecol 22(4):609–615

Tellioğlu A, Şen D (2001) Hazar Gölü (Elazığ) Copepoda ve Cladocera faunasının mevsimsel dağılımı. G.Ü Gazi Eğitim Fakültesi dergisi 21(2):7–18

Tellioğlu A, Şen D (2002) Hazar Gölü (Elazığ) Rotifer faunasının taksonomik yönden incelenmesi. EÜ Su Ürünleri Dergisi 19(1–2):205–207

Ustaoğlu MR (2015) An updated Zooplankton biodiversity of Turkish Inland Waters. J Limnol Freshw Fish Res 1(3):151–159

Yıldız S, Altındağ A, Ergönül MB (2007) The seasonal fluctuations in the zooplankton community of an eutrophic lake; Lake Marmara (Manisa, Turkey). Turk J Zool 31:121–126

Chapter 30
Kinds and Distribution of Icthyoplankton in Shatt Al-Arab River

Sumaya M. Ahmed

Abstract The icthyoplankton community structure in Shatt Al-Arab River is classified into two categories: fresh water fish larvae dominating in the upper reaches of the river and marine fish larvae dominating in the lower reaches of the river. The larvae of Mugilidae and Cyprinidae dominate the icthyoplankton. The abundance and distribution of icthyoplankton in Shatt Al-Arab River coincide with increasing water temperature and blooming in phytoplankton and zooplankton during summer season. Most fish larvae in Shatt Al-Arab River were recorded during spring and summer. However, no fish larvae were recorded during autumn and winter.

30.1 Introduction

30.1.1 Shatt Al-Arab River

Shatt Al-Arab River formed after the confluence of the Tigris and Euphrates at Qurna. It flows southeastern about 196 Km to open in Arabian Gulf (Total length: 169 km, width: 400–1500 m, and depth of channel: 6–24 m) (Al-Wuhaily 2009). Shatt Al-Arab River is affected by diurnal and semidiurnal high tide (Abaychi and Al-Obaidy 1987).

Shatt Al-Arab River is a unique euryhaline environment, and its water is a mixing of fresh water of Tigris and Euphrates rivers with the salty water of the Arabian Gulf, besides its high primary and secondary productivity, which is characterized with two peaks; first peak during autumn and the second during spring (Al-Imara and Jawad 1994) with different species of aquatic plants, which is a good source as shelter, spawning sites and feeding for fish.

S. M. Ahmed (✉)
Department of Fisheries and Marine Resources, College of Agriculture, University of Basrah, Basrah, Iraq

L. A. Jawad (ed.), *Tigris and Euphrates Rivers: Their Environment from Headwaters to Mouth*, Aquatic Ecology Series 11,
https://doi.org/10.1007/978-3-030-57570-0_30

Shatt Al-Arab River is the main source of fresh water to Arabian Gulf and therefore plays major role in maintaining the ecological balance of marine habitats in the northern Arabian Gulf.

Water temperature in Shatt Al-Arab River ranges between 12 °C in December and 38 °C in July, while salinity ranges between 1.03‰ and 3.8‰ in upper reaches and 9.3–36‰ in the lower reaches. Transparency ranges between 38.3 cm in August and 72.3 cm in May (Al-Okailee 2016). Mohamed et al. (2012b) recorded salinity gradient in Shatt Al Arab River with significant differences among sites and ranged between 0.8 and 7.6 ppt.

30.2 Community Structure of Icthyoplankton in Shatt Al-Arab River

Larval fish distribution may provide insight into the factors influencing recruitment dynamics such as the location and suitability of spawning ground (Donald 1997).

Study of icthyoplankton distribution can be a good tool to understand adult spawning strategy and identify the time of recruitment (Cushing 1990); besides, the analysis of fish community structure is wildly considered as an integrated indicator of the ecological states of water bodies.

The icthyoplankton surveys in Shatt Al-Arab River are scarce and limited to that of Al-Okailee (2001) who determined the abundance and distribution of fish egg and larvae in Shatt Al-Arab estuary. Al-Okailee (2010) determined the spatial and seasonal composition of ichthyoplankton and trophic relationship in the upper reaches of Shatt Al-Arab River.

The aim of this work is to determine the structure of ichthyoplankton communities in Shatt Al-Arab River and then show the importance of Shatt Al-Arab River as spawning, shelter, and nursery site for many fish species.

30.3 Specific Composition of Fish Larvae in Shatt Al-Arab River

The fishes of Shatt Al-Arab River include 32 species from which 12 species are marine source (Hussain et al. 1989), while Younis et al. (2001) recorded 25 fish species. However, Lazem (2009) recorded 26 species of which 9 are marine.

Icthyoplankton community structure in Shatt Al-Arab River differs between upper reaches and lower reaches of the river, because the lower reaches of Shatt Al-Arab are affected by salty marine water from Arabian Gulf, and so the salinity is fluctuated daily in the lower reaches; therefore, the structure of icthyoplankton is also affected. So the euryhaline species of fish larvae can use the lower reaches of Shatt Al-Arab River as nursery area.

Table 30.1 Families of fish larvae from the upper reaches of Shatt Al-Arab River

Family	Percentage
Mugilidae	41.69
Cyprinodontidae	19.95
Cyprinidae	18.28
Cichlidae	10.25
Clupeidae	9.81

Table 30.2 families of fish larvae from the lower reaches of Shatt Al-Arab River

Family	Percentage
Gobiidae	38.02
Engraulidae	29.35
Clupeidae	17.71
Sciaenidae	10.26
Soleidae	1.9
Cynoglossidae	1.54
Polynemidae	0.83
Syngnathidae	0.27

Most fish larvae in Shatt Al-Arab River were recorded during spring and summer. However, no fish larvae were recorded during autumn and winter (October–January) (Al-Okailee 2010). Mean abundance of fish larvae was estimated to be 2028 larvae/ 10 m^2 (Al-Okailee 2010).

Spawning activity of the fish in Shatt Al-Arab River is supposed to be during March–October, while no spawning activity was recorded during winter (November–February (Mohamed et al. 2015).

Al-Okailee (2016) recorded 13 larval fish families in Shatt Al-Arab River, of which five families were recorded from upper reaches and eight families from lower reaches (Tables 30.1 and 30.2). It is clear that fish larvae in the lower reaches are related to marine species, which tolerate the fluctuation in salinity.

In the upper reaches of Shatt Al-Arab, five families of fish larvae were recorded, represented by Cyprinidae (*Cyprinus carpio, Carassius auratus, and Alburnus mossulensis)*, Mugilidae (*Liza abu*), Cyprinodontidae (*Aphanius* sp.), Clupeidae (*Tenualosa ilisha*), and Hemiramphidae (*Hemiramphus* sp.) (Al-Okailee 2010). *Liza abu* larvae represented 62.96% of the total fish larvae collected.

Five families of fish larvae: Mugilidae, Cyprinodontidae, Cyprinidae, Cichlidae, and Clupeidae were identified from upper reaches of Shatt Al-Arab River. While in the lower reaches, eight larval fish families were identified: Gobiidae, Engraulidae, Sciaenidae, Soleidae, Clupeidae, Polynemidae, Bothidae, and Cynoglossidae, which are marine fish.

Most fish larvae from Shatt Al Arab River used aquatic plants as shelter, which is clear from total abundance of fish larvae, which collected by plankton net was 2048 larvae/10 m^2, while that collected from aquatic plants was estimated to be 97,800 larvae/10 m^2 (Al-Okailee 2010).

Many researchers reported that several marine fish species move to the lower and upper reaches of Shatt Al-Arab River (Hussain et al. 1989; Lazem 2009).

Al-Okailee (2010) noticed that larvae of Mugilidae, Cyprinodontidae, Cyprinidae, and Clupeidae appeared during March to September in the upper reaches of Shatt Al-Arab.

Tenualosa ilisha **Larvae**

It has been known for a long time that the river shad *T. ilisha* lives in the sea for most of its life and migrates to rivers for spawning. After egg laying and hatching of larvae, these fish go back to marine habitat for further growth.

T. ilisha is an anadromous Clupeidae that migrates from Arabian Gulf toward freshwater rivers for spawning. The normal habitat of *T. ilisha* is the lower regions of estuaries and offshore areas; during breeding season, they ascend the rivers and after spawning return to the original habitat. In Iraq, it migrates in Shatt Al-Arab River and surrounding marshes up to 100 km (Al-Mukhtar et al. 2016).

T. ilisha larvae compromise 8.57% of the total larvae collected from Shatt Al Arab River (Mohamed et al. 2012a, b) during 6 months extending from March to September with peak of abundance (880 larvae/10 m^2) during August.

It appears that water temperature affects the occurrence of *T. ilisha* in Shatt Al-Arab River as peak of abundance of *T. ilisha* larvae was corresponding to the highest water temperature during August; however, the larvae disappear from Shatt Al-Arab during winter (Mohamed et al. 2012b).

The length of *T. ilisha* larvae in Shatt Al-Arab River ranged between 5.0 and 18.0 mm (Al-Okailee 2010). The presence of these larvae indicates that this river is one of the spawning grounds for this species.

T. ilisha larvae are present along Shatt Al-Arab River, but the highest number and peak of abundance are found in the shallower and denser vegetation regions as the water depth does not exceed 1.3 m possibly to avoid drifting by current (Al-Okailee 2010).

Al-Noor (1998) pointed out that *T. ilisha* larva distributed along the banks of Shatt Al-Arab River within the area extended from 30 to 120 km from the estuary. These banks are characterized by slow current and thick growth of aquatic plants.

Water temperature and salinity showed significant positive correlation with the number of *T. ilisha* larvae, while transparency exhibited significant negative correlation with the number of larvae (Mohamed et al. 2012b).

The occurrence of *T. ilisha* larvae in the upper reaches of Shatt Al-Arab River can consider this area as one of the main sites of spawning of *T. ilisha* (Mohamed et al. 2012b); however, no egg of *T. ilisha* was collected.

Spawning season of *T. ilisha* in Iraq was estimated to be during March to August coinciding with the flooding season of Tigris and Euphrates rivers (Mohamed et al. 2012b).

Al-Mukhtar et al. (2016) reported that the spawning season of *T. ilisha* in Shatt Al-Arab River extended from March–November with two distinct peaks. The same result was reported by Al-Hassan (1999) who referred to the long spawning season of *T. ilisha* in Shatt Al-Arab River with two peaks: first peak during March–May and

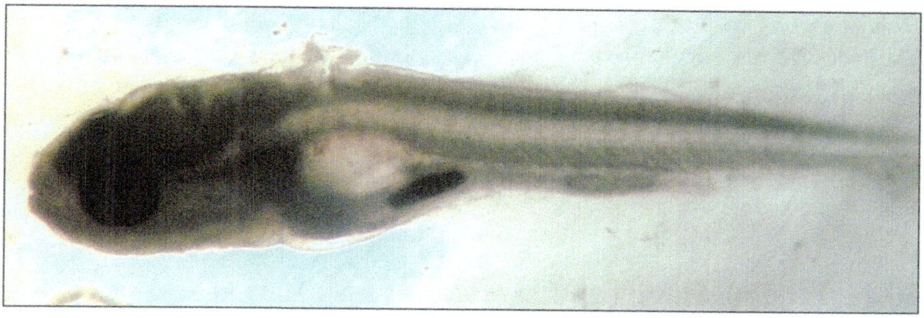

Fig. 30.1 *Liza abu* larvae

the second peak from July–August. However, Al-Noor (1998) found that spawning season of *T. ilisha* in Shatt Al-Arab River extended from March–October.

The young of the year of *T. Ilisha* was reported by Hussain et al. (1997) who collected young of hilsa from shallow banks in the northern parts of Shatt Al-Arab River.

Aphanius sp. Larvae

Very common larvae are present in the upper reaches of Shatt Al-Arab River with percentage of 19.95% of total fish larvae collected (Al-Okailee 2010).

The larvae of *Aphanius* sp. appear during April–July with peak of abundance during April. Most *Aphanius* larvae prefer aggregation between aquatic plants and shallow habitat.

L. abu Larvae

It is the most abundant fish larva in Shatt Al-Arab River with percentage of 62.96% of the total larvae collected (Fig. 30.1) (Al-Okailee 2010). These larvae appear during February–May with peak of abundance during March.

Liza abu larvae also dominate the icthyoplankton in the marshes of southern Iraq (Hussain et al. 2013).

Cyprinus carpio Larvae

Representing 5.91% of total larvae collected, these larvae appear during February–May with peak of abundance during March.

Carassius auratus Larvae

The most abundant cyprinid fish larvae in Shatt Al-Arab River represent 11.5% of the total larvae collected. These larvae appear during February–May with peak of abundance during March.

Alburnus mossulensis Larvae

These larvae appear during February–July with peak of abundance during February, Which represent 1.01% of total larvae collected.

***Hemiramphus* sp. Larvae**

These larvae are euryhaline species, very common in the north western parts of Arabian Gulf as adult. These larvae appear during August only with percentage of 0.07% of total larvae collected. Al-Nasiri and Shamsul-Huda (1975) first found *Hemiramphus* larvae from Shatt Al-Arab River.

30.4 Feeding Habits and Strategy of Fish Larvae

Diet analysis of fish allows us to understand their feeding strategy and indirectly indicate community energy flow (Ramirez-Luna et al. 2008).

The strategy and food composition of some fish larvae in Shatt Al-Arab River were studied by Al-Okailee (2010) and found that the larvae of *C. carpio* are not specialized in feeding, consume mainly diatoms followed by copepod and organic detritus. However, in adult stage *C. carpio* consumes Crustacea, Mollusca, diatoms, and detritus.

Carassius auratus larvae consume diatoms and detritus mainly followed by filamentous algae and copepod. This larva is not specialized in feeding, while in its adult stage, *C. auratus* consumes algae, detritus Crustacea, aquatic plants, and aquatic insects.

The larvae of *Alburnus mossulensis* consume copepods (larval and adult stages), Mollusk egg, diatoms, *Daphnia*, and organic detritus. These larvae are not specialized in feeding. However, Hussain et al. (1997) find that the adult of *A. mossulensis* consumes copepods, Mollusk, Crustacea, and aquatic insects.

Ahmed (2011) analyzed the gut content of three cyprinid fish larvae (*C. carpio, C. auratus*, and *A. mossulensis*) from southern marshes of Iraq and found that these larvae consume mainly zooplankton dominated by copepod both adult and larval stages followed by Cladocera, Rotifers, aquatic insects, and Ostracoda, and the food content of plant origin also exists but with less important.

The larvae of *Liza abu* consume copepod (adult and larval stage), mollusk egg, diatoms, *Daphnia*, and detritus. These larvae are not specialized in feeding. However, Ahmed (2004) found that *Liza carinata* (Mugilidae) from northern parts of Arabian Gulf consumes mainly diatoms in the smallest larvae and then shifts in feeding to consume copepods (both adult and larval stages) in the larger larvae. However, in the adult stage, it consumes detritus, algae, aquatic plants, and diatoms.

The larvae of *Aphanius* sp. consume Copepoda (adult and larval stages) mainly and with less important egg of mollusk, diatoms, *Daphnia*, and detritus. These larvae are nonspecialized feeders.

Larvae of *T. ilisha* consume copepods (adult and larval stage), *Daphnia*, and detritus. These larvae are not specialized in feeding. However, the juveniles of *T. ilisha* consume mainly algae, diatoms, and zooplankton, while in adult stage, it stops feeding during its spawning migration to freshwater (Al-Noor 1998; Lazem 2009).

The larvae of *Ilisha megaloptera* (Clupeidae) from northern parts of Arabian Gulf consume mainly diatoms, while shifting in feeding occurs in juvenile stage to zooplankton (Ahmed et al. 2005).

30.5 Fish Eggs in Shatt Al-Arab River

Fish eggs are classified into two categories: planktonic and demersal, nearly all freshwater fish lay demersal eggs, which are generally larger than 1 mm. The demersal eggs are frequently adhesive (Kendal et al. 1983).

Fish eggs from Shatt Al-Arab River were mostly collected during spring and summer seasons (February–July) with peak of abundance during March and disappear during August–January (Al-Okailee 2010).

Fish egg in Shatt Al-Arab River divided into three families: Cyprinidae, Mugilidae, and Cyprinodontidae. The egg of *Liza abu* (Mugilidae) recorded the higher percentage (48.2%) among the total fish egg, followed by egg of Cyprinidae (42.1%) (Al-Okailee 2010).

30.5.1 Fish Egg Distribution

Fish egg in Shatt Al-Arab River was collected from water by plankton net and from aquatic plant. Most fresh water fish lays adhesive egg as a mass lying on aquatic plants in shallow water to avoid drifting by water current. Most eggs were collected from aquatic plants. In March, egg abundance was 268 egg/10 m^2 by plankton net, while it was 82,280 egg/10 m^2 during March also from aquatic plants (Al-Okailee 2010).

Cyprinid eggs appear during February–May with peak of abundance during March. Most cyprinid eggs were collected from aquatic plants.

Mugilidae eggs appear during February–May with peak of abundance during March. Most mugilid eggs were collected from aquatic plants.

Cyprinodontidae eggs appear during April–July with peak of abundance during April. Most eggs were collected from aquatic plants.

30.6 Discussion

Shatt Al-Arab River and adjacent marshes represent a specific spawning, feeding, and nursery habitat for many fresh water and marine fish.

So the presence of fish egg and larvae indicates the important of Shatt Al-Arab River as spawning, nursery, and feeding ground.

The emergence of fish larvae in Shatt Al-Arab River during hot season reveals that it coincides with increasing water temperature, which stimulates maturation of fish gonad and egg and larval growth (Walter 2007). Besides, larval growth coincides with increasing primary and secondary productivity (Ara et al. (2009).

The increasing water salinity of Shatt Al-Arab River in recent years affects the abundance and presence of adult fish and their larvae, which is evident from the presence of marine species of fish (Mohamed et al. 2015); besides, the reduction in Shatt Al-Arab River flow is likely to impact fish recruitment pattern, causing change in fish stock size, which is affected by changing the nursery and spawning environment (Ben Hasan 2016).

Increasing salinity also affects the icthyoplankton community structure in Shatt Al-Arab River, which is clear from the kinds of larval families in the upper and lower reaches, as marine larval fish families are dominated in the lower reaches, while the freshwater larval fish families are dominated in the upper reaches.

It is clear that the lower reaches of Shatt Al-Arab River are affected by both freshwater and salt water from marine habitat and also the high salinity and turbidity, which prevail the lower reaches that attract many euryhaline species (Mohamed et al. 1993).

Al-Hassan and Hussain (1985) showed that hydrological parameters influence the penetration of marine species to Shatt Al-Arab River.

The aquatic plants and shallow water also represent good factors for distribution of fish larvae. Overton and Rulifson (2007) referred that most freshwater fish larvae prefer shallow water with aquatic plant cover as shelter. Al-Noor (1998) recorded that *T. ilisha* larvae used shallow water with dens cover of aquatic plants and slow water current as spawning sites; also, *C. carpio* is known to use aquatic plants for egg laying and shelter for larvae (Whitener 2004).

The majority of fish larvae in Shatt Al-Arab River were collected from aquatic plants rather than by plankton net, which means that these larvae use aquatic plants as shelter from predation and avoid drifting by water current.

Most fish eggs appear during spring and early summer (February–May), while fish larvae mostly appear during spring till late summer (February–September), which coincides with peak of abundance of adult fish (Mohamed et al. 2015).

Mohamed et al. (2012b) listed 19 fish families as adult from Shatt Al-Arab River, with Cyprinidae being the dominated fish family represented by 10 species. Furthermore, Lazem (2009) listed 26 species of adult fish families from Shatt Al-Arab, from which 17 freshwater fish families and 9 families are marine.

The previous studies showed that *T. ilisha* fish enters Shatt Al-Arab River from Arabian Gulf for spawning from March–August, which coincides with seasonal flood of Tigris and Euphrates.

Mohamed et al. (2015) reported that number of fish increases in summer and autumn and sharply decreases during winter, with *Tenualosa ilisha* being the most abundant species followed by *Carassius auratus* and *Liza klunzingeri.*

Exotic species in Shatt Al-Arab were also recorded by Al-Faisal and Mutlak (2014), represented by species belonging to five families: Cichlidae, Cyprinidae,

Heteropneustidae, Pangasiidae, and Poeciliidae and referred that some exotic freshwater species were introduced to Iraq for improving fishery potential.

The ichthyofauna of Shatt Al-Arab River is dominated by Mugilidae and Cyprinidae (Hussain et al. 1997).

58 fish species belong to 46 genera, 27 families, and 16 native, 10 elian, and 32 marine species; Cyprinidae is the dominant family (15 species and 10 genera) (Mohamed et al. 2015). For that, Mohamed et al. (2015) divided the fish assemblages of Shatt Al-Arab River into three ecological fish guilds: common species, seasonal species, and occasional species.

The feeding habits and strategy give a good idea about relationship between fish (Luna et al. 2008). We see that the gut content of fish larvae in Shatt Al-Arab river consists mainly of diatoms, algae, copepods, daphnia, egg of mollusk, and organic detritus, which means that they are not specialized in feeding and they consume both phytoplankton and zooplankton. This shows that the availability of food resources coincides with abundance and emergence of fish larvae in Shatt Al-Arab River, which coincides with increase in water temperature and gives best environmental conditions for high growth rates of fish larvae (Ara et al. 2009).

The nonspecialized feeding means that these larvae are generalist feeders, and this strategy results in lower competition and allows larvae to coexist in relatively high density in its habitat (Ahmed 2011).

The sufficient food availability in Shatt Al-Arab River results in less direct competition for food resources, which makes Shatt Al-Arab river a suitable nursery and feeding ground for many fish larvae.

Al-Dubaikel (2016) finds that feeding intensity of fish in Shatt Al-Arab river was higher in spring and autumn, indicating appropriate water temperature and availability of food resources.

Larvae of local and endemic freshwater fishes (*Barbus sharpeyi, Aspius vorax,* and *Barbus luteus*) were absent in plankton samples, which may be due to migration of adult to favorite area for spawning.

Lower reaches being more diverse are comparable to the upper reaches, which play a role as feeding and nursery ground for fish larvae of fresh water and estuarine species.

The previous studies showed that Shatt Al-Arab River (mainly upper reaches) is a good spawning, feeding, and nursery ground for some fish.

We can conclude that Shatt Al-Arab River plays a vital role in fish conservation and management. Besides, the upper reaches of Shatt Al-Arab River can be considered as a favorable spawning ground for some freshwater fish species, while lower reaches of Shatt Al-Arab River are less important as spawning site for fish maybe due to fluctuation in salinity. However, the lower reaches of Shatt Al-Arab are more diverse than the upper reaches, most likely because Lower reaches of Shatt Al-Arab consider rich in nutrient and food resources.

References

Abaychi JK, Al-Obaidy SZ (1987) Concentration of trace metals in aquatic plants from Shatt Al-Arab River. Iraq J Biol Sci Res 8(2):123–129

Ahmed SM (2004) Selective feeding by the larvae of Golden mullet (*Liza carinata*: Mugilidae) from Khor Abdulla/North Western Arabian gulf. Basrah J Agric Sci 17(2):31–38

Ahmed SM (2011) Note on feeding relationships of three species of cyprinid fish larvae in Al-Huwaiza marsh, southern Iraq. Mesopotamian J Mar Sci 26(1):35–46

Ahmed SM, Al-Okailee MT, Hussain NA (2005) Food and feeding activity of *Ilisha megaloptera* larvae in Shatt Al-Arab estuary. Iraq J Aqua 2:93–99

Al-Dubaikel AY (2016) Analysis of natural food composition of fishes in Shatt Al-Arab River, southern Iraq. Jordan J Biol Sci 9(2):89–96

Al-Faisal AJ, Mutlak FM (2014) Exotic freshwater fishes in the southern Iraq. Marsh Bull 9(1):65–78

Al-Hassan LJ (1999) Shad of Shatt Al-Arab River, Iraq. Shad J 4:1–4

Al-Hassan LJ, Hussain NA (1985) Hydrological parameters influencing the penetration of Arabian Gulf fishes into the Shatt Al Arab river, Kaq. Cybium (Paris) 9(1):7–16

Al-Imara FJM, Jawad AM (1994) Physico-chemical parameters of southern Iraqi waters, north west Arabian gulf. Mar Mesopotamica 9:1):1–1)12

Al-Mukhtar MA, Jasim W, Mutlak F (2016) Reproductive biology of Hilsa shad *Tenaulosa ilisha* (Teleostei: Clupeidae) during spawning migration in the Shatt Al-Arab River and southern Al-Hammar marsh, Basra, Iraq. J Fish Aquat Sci 11(1):43–55

Al-Nasiri SK, Shamsul-Hoda SM (1975) Some notes on the larvae and juveniles of *Hemirhamphus xanthopterus* in Iraqi waters. BullBasrah Nat Hist Mus 3:3–12

Al-Noor SSH (1998) Reproductive biology of *Tenualosa ilisha* in Shatt Al-Arab River. PhD Thesis (In Arabic). College of Agriculture. University of Basrah Iraq

Al-Okailee MT (2001) Abundance and distribution of fish eggs and larvae in Shatt Al-Arab estuary in North West Arabian Gulf. M.Sc. Thesis. University of Basrah, Iraq. (In Arabic)

Al-Okailee MT (2010) The spatial and seasonal composition of some icthyoplankton and trophic relationships in north part of Shatt Al-Arab River. PhD Thesis. College of Agriculture, Basrah University, Iraq (In Arabic)

Al-Okailee MT (2016) Comparative study of fish larval communities in north and south parts of Shatt Al-Arab River. Iraqi J Aquacult 13(1):86–101

Al-Wuhaily OQK (2009) Sedimentological and geomorphological study of Shatt Al-Arab Islands between Basrah and Saybah. M.Sc. Thesis (In Arabic). College of Science. University of Basrah, Iraq

Ara R, Arshad A, Amrullah NASM, Dand S, NorAzwady AA, Mazlan AG (2009) Feeding habits and temporal variation of diet composition of fish larvae (Osteichthyes:Sparidae) in the Sungai Pulai Sea grass bed, Johore, Peninsula. Malaysia J Biol Sci 9(2):445–451

Ben Hasan A (2016) Fisheries in change environment: the impact of the reduction in Shatt Al-Arab flow on near shore fish stock in the northern Persian gulf. M.Sc Thesis. University of British Columbia (Vancouver)

Cushing D (1990) Plankton production and year-class strength in fish population: an update of the match/mismatch hypothesis. Adv Mar Biol 26:249–294

Donald DB (1997) Relationship between year-class strength for goldeyes and selected environmental variables duringthe first year of life. Trans Am Fish Soc 1(26):361–368

Hussain NA, Ali TS, Saud KD (1989) Seasonal fluctuations and composition of fish assemblage in the Shatt Al-Arab River at Basrah, Iraq. J Biol Sci Res 20(1):139–150

Hussain NA, Younis KH, Yousif UH (1997) The composition of small fish assemblages in the river Shatt Al-Arab near Basrah, Iraq. Acta Hydrobiol 39:29–37

Hussain NA, Ahmed SM, Lazem LF (2013) Occurrence and abundance of fish larvae in the restored Mesopotamian marshlands, southern Iraq. JKAU Mar Sci 24(2):15–27

Kendal AW, Ahlstrom EH, Moser HG (1983) Early life history stages of fishes and their characters. In: Ontogeny and systematic of fishes. International symposium, California, 11–22
Lazem LF (2009) Composition structure of fish community assemblage and their connection to the prevailing abiotic conditions in Garamat Ali River, southern Iraq. M.Sc. Thesis (In Arabic). College of Agriculture. University of Basrah, Iraq
Luna VR, Navia AF, Rubio EA (2008) Food habits and feeding ecology of an estuarine fish assemblage of northern pacific coast of Ecuador. Am J Aquat Sci 3(3):361–372
Mohamed ARM, Al- Hassan LAJ, Ali TS (1993) The presenceof Cyprinid fish Barbus lutes in marine water of Iraq. Nova Serie 2(25):415–416
Mohamed ARM, Resen AK, Taher MT (2012a) Longitudinal patterns of fish community structure in the Shatt Al-Arab River, Iraq. Basrah J Sci 30(2):65–86
Mohamed ARA, Ahmed SM, Al-Okailee MT (2012b) Variations in occurrence, abundance and diet of Hilsa, *Tenualosa ilisha* larvae in the north of Shatt Al-Arab River, Basrah, Iraq. Basrah J Agric Sci 25(2):40–52
Mohamed ARA, Hussein SA, Lazim LF (2015) Spatiotemporal variability of fish assemblage in Shatt Al-Arab River, Iraq. J Coast Life Med 3(1):27–34
Overton AS, Rulifson RA (2007) Evaluation of plankton surface pushnet and oblique tows for comparing the catch of diadromous larval fish. Fish Res 80:99–104
Ramirez-Luna V, Navia AF, Rubio EA (2008) Food habits and feeding ecology of an estuarine fish assemblage of northern Pacific coast of equator. Pan-Am J Aquat Sci 3(3):361–372
Walter C (2007) Temperature influence of the fish assemblage structure in large lowland river, the lower Oder River, Germany. Ecol Freshw Fish 16:493–503
Whitener K (2004) Used of restored Central California flood plain by larvae of native and alien fishes. Am Fish Soc Symp 39:125–140
Younis KH, Hussain NA, Yousif UH (2001) Food and diet overlap of small fish assemblage in the upper reaches of Shatt Al-Arab River, Iraq. Mar Mesopotamica 16(1):129–139

Chapter 31
Rotifer Diversity in Iranian Waters: A Review

Reza Malekzadeh-Viayeh

Abstract Iran is the second largest country in the Middle East with a significant contribution to the regional and global biodiversity. While the country is considered as a home to a wide range of flora and fauna, much of its biological assets has not yet been apprehended. Rotifers, the members of phylum Rotifera, comprise some 2000 species of minute aquatic invertebrates. They have increasingly attracted efforts from the global taxonomists and limnologists not only due to their crucial role in aquatic food webs, but also for their growing applications in fundamental biological research and aquaculture industry. The oldest available records of Iranian rotifers date back to almost 70 years ago. Notwithstanding that the task to discover the aquatic invertebrates, such as rotifers, of all Iranian waters is far from complete, the last decade has witnessed an expanding quantity of research and publications on their diversity. In this chapter, a review of the surveys on phylum Rotifera in Iran has been provided, including a checklist of the identified rotifer taxa, based on almost all available data. A total of 366 rotifer species and subspecies of 69 genera under 27 families have so far been identified from Iranian waters, most of which (95%) belong to the subclass Monogononta and the remaining (18 species of 8 genera in 4 families) to Bdelloidea. Biogeography of the rotifers, a comparison of the regional rotifer diversity, and prospects for future studies have also been discussed.

Keywords Rotifers · Species diversity · Iranian waters

31.1 Introduction

Estimation of global biodiversity has increasingly been granted not only due to its intrinsic scientific value but also for the vital contribution it has to human well-being and food safety. Several examples of biodiversity applications in modern societies

---

R. Malekzadeh-Viayeh (✉)
Urmia University, Urmia, Iran

L. A. Jawad (ed.), *Tigris and Euphrates Rivers: Their Environment from Headwaters to Mouth*, Aquatic Ecology Series 11,
https://doi.org/10.1007/978-3-030-57570-0_31

exist, including the use of natural products for pharmaceutical development or extraction of marine resources. In addition, biodiversity provides ecosystems' health by supporting their stability, functionality and sustainability. These healthy ecosystems can, in turn, facilitate various crucial functions, such as recycling of organic waste, carbon sequestration, oxygen production, protection of water quality, mitigation of the effects of pollutants, and improve human and animal health (Clark et al. 2014; Jowkar et al. 2016). While the diversity of several Iranian terrestrial animal and plant groups (Ghasemi et al. 2012; Farashi and Shariati-Najafabadi 2017), fishes (Coad 1998, 2006; Esmaeili et al. 2017), and aquatic macroinvertebrates (Sharifinia 2015; Gerami et al. 2016) has been documented, not much is known about microinvertebrate fauna of the country.

Rotifers (phylum Rotifera) are a group of microscopic and primary freshwater invertebrates that occur in almost all types of habitat, from large permanent lakes to small temporary puddles and interstitial and capillary waters (Segers 2008). They are the dominant zooplankton group in inland waters and play a major role in energy transfer and nutrient cycling (Malekzadeh-Viayeh and Špoljar 2012). Due to their short life cycles, rotifers react quickly to changes in environmental conditions. Hence, their species composition and abundance may be used as biological indicators reflecting the alterations in water quality (Gutkowska et al. 2013). Rotifers offer interesting challenges to developmental and cell biologists because of their remarkable combination of development and plasticity of forms in response to environmental variations. In addition, they present superb research materials for many problems of current interest to ecologists, including population dynamics and genetics, resource partitioning, and adaptation of form, function, and life history (Ruttner-Kolisko 1974). Rotifers have also been used as primary live foods in the aquaculture of a variety of fish and crustaceans (Lubzens et al. 1989), the demand for whom has gradually been increased with the flourishing aquaculture practices worldwide (Suantika et al. 2003).

While rotifers have first been described in the 1600s, they have attracted much attention of the global naturalists since the twentieth century to disclose their diverse biological attributes. The expanded applications of rotifers in various scientific investigations as well as in biotechnology and aquaculture have promoted studies on their biodiversity, biogeography, ecology and conservation. Scientists have become more interested than ever to explore and describe rotifers from all over the world. This has led to the isolation and introduction of hundreds of these microinvertebrates from different continents (Murray 1913; de Ridder 1986; Shiel 1995; Pejler 1998; Modenutti 1998; Sarma 1999; Duggan et al. 2001; Jersabek 2003; Thorp and Covich 2009; Athibai et al. 2013; Kriska 2014), exceeding the number of described rotifer taxa to over 2000 (Smith 2001; Segers 2008), which is still undoubtedly much less than their actual number. The fact that new species are currently being discovered from various parts of the globe suggests that the taxonomic studies on rotifers are far from complete in spite of their 300-year-old history (Sarma 2006). Classically, three groups/subclasses are recognized within the phylum Rotifera: Monogononta, Bdelloidea and Seisonacea with 1570, 461 and 3 species, respectively, whereas molecular studies indicate that the exclusively endoparasitic Acanthocephala are also rotifers (Fontaneto et al. 2008). Among all other living organisms, rotifers are common in freshwater and marine habitats, but some species also live in such extreme environments as damp soil or on mosses (Fontaneto and De Smet 2015). Most species of the rotifers are cosmopolitan, but there also are biotope-specific or endemic taxa whose distribution is restricted to certain biogeography or ecological niche (Segers 2008). While rate of the rotifer endemism is still unclear, enhanced research efforts may find additional endemic taxa from different geographical areas (Segers and De Smet 2008).

In order to contribute to the knowledge of regional biodiversity, a review of the surveys on Iranian rotifers is herein provided which includes a first-ever and nearly-complete checklist of the identified rotifer taxa from the country using the available data. Biogeographical distribution of the reported rotifers, a comparison of the regional rotifer diversity, and general remarks on the performed investigations, flaws, and perspectives are discussed.

31.2 Geography, Climate and Hydrology of Iranian Plateau

Although the country presently known as 'Iran' is just part of the vast ancient Persian Empire territory, covering an area of about 1,648,000 km^2, it even now is one of the largest countries not only in the region but also on a global scale (*i.e.*, the second largest country in the Middle East and 18th in the world) (Jowkar et al. 2016; Rezaei et al. 2017). The country has a complex topography (Fallah et al. 2017), and more than half of its area is covered by mountains and deserts (Mirzaie-Nadowshan 2009). Iranian Plateau, encompassing Iran and parts of a number of neighboring countries, has an old formation history and connects to Mesopotamia to the west where the leading Elamite Civilization was established and was a historical start point for agriculture development by human (Derenko et al. 2013).

Iran is located between arid or continental climate zone of Central and West Asia and the Mediterranean climate zone, and, therefore, generally has a four-season climate influenced by the surrounding water basins, Caspian, Aral and Black seas in the north, Indian Ocean, Persian Gulf and Oman Sea in the south, Atlantic Ocean in the west, and Mediterranean Sea in northwest, as well as by mountain ranges, highlands and deserts. However, the main meteorological attributes, *e.g.*, length of the seasons, temperature, precipitation, humidity and air pressure, may differ greatly in different regions of the country (Ghorbani 2013; Fallah et al. 2017). Temperature can vary between −20 and +50 °C, while precipitation fluctuates from less than 50 mm to over 1000 mm per year (Madani et al. 2016). The Mediterranean climate governs large part of Iran, which is affected by changes in Westerly activities. Thus, except for the western parts and the northern coastal areas, Iran's climate is mainly arid and semi-arid (Fallah et al. 2017). About 75% of the total land area of Iran is dominated by arid or semi-arid climate with annual precipitation rates from ~350 mm to less than 50 mm (Kehl 2009). The average annual precipitation across the country is estimated to be near 250 mm, about one-third of the global annual precipitation (Mirzaie-Nadowshan 2009).

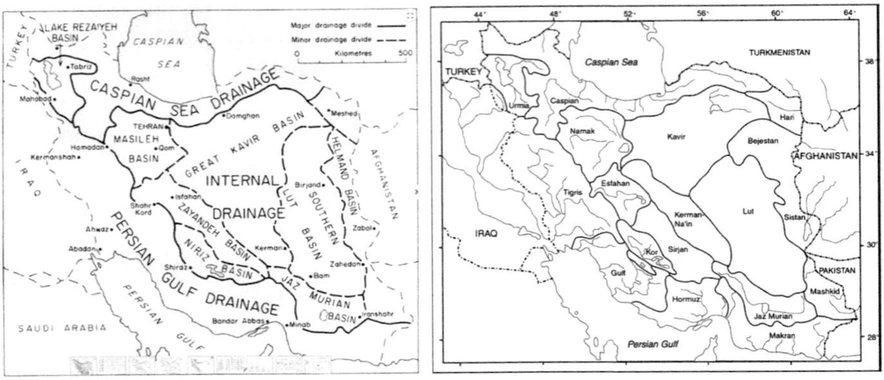

A B

Fig. 31.1 Described drainage basins for Iranian Plateau by Beaumont (1982) (**a**) and Coad (2006) (**b**)

Hydrologically, Iran is bound with massive bodies of water to its north (Caspian Sea) and south (Persian Gulf and Sea of Oman). With an estimated area of over 390,000 km^2, Caspian Sea is the largest landlocked waterbody in the globe with extended coasts in five countries of Iran, Russia, Kazakhstan, Turkmenistan and Azerbaijan. It once was a part of the ancient Tethys Sea, being completely isolated roughly 1 million year ago. However, it has retained a wealthy collection of fauna of marine origin that currently lives in its brackish water with salinities ranging from 10 to 13 ppt on average. Having its surface water located 22 m below the average sea level, Caspian Sea contains 40% of the world inland waters and its maximum depth is about 1000 m with a mean depth of 210 m. Some 130 rivers of various sizes drain into the sea, of which the northern rivers of Volga and Ural constitute the main (over 90%) water input (Birshtain et al. 1968; Kostianoy and Kosarev 2005; Asian Development Bank 2010; Jamshidi and Abu Bakar 2012). Persian Gulf and Sea of Oman are the only actual marine ecosystems in Iranian territory. Persian Gulf is a semi-closed projection of the Indian Ocean connecting to the oceanic waters through the Strait of Hormuz and Sea of Oman. It is located in one of the hottest regions in the world and is characterized by its harsh environmental conditions where the air temperature can rise to over 50 °C in summer, elevating the water salinity to 45 ppt (Moradi et al. 2014). Persian Gulf is surrounded by the Iranian shorelines in the north and those of several Arabic states in the south and southwest, which share the rich natural resources of the Gulf including its fauna and flora.

Iranian inland waters belong to several drainage basins whose number is defined to be up to 19 (Coad 1998, 2006). Beaumont (1982) specified 6 water basins for Iranian Plateau; among them, the largest is the Great Kavir (Kavir-e-Markazi) basin which covers about one-third of the total country area (Fig. 31.1). While the lakes in Iran have diverse formation history, many of them are the remnants of the Tethys Sea, parts of which turned into swamps and parts still exist. The most important existing lakes are Urmia (Orumieh/formerly, Rezāieh) Lake, Hāmoon, Qom Namak,

Bakhtegān, Gāvkhooni, Jāzmooriān, Zarivār, Mahārloo, Parishān, and Tār. Most of the major Iranian rivers run in two principal basins: Caspian Sea and Persian Gulf basins. The main rivers flowing ultimately into the Caspian Sea (Aras, Sefid Rud, Chālus, Harāz, Sehezār, Bābol, Talār, Tajan, Gorgān, Atrak, Qarasoo, and Nekā) originate from the northern slopes of the Alborz Mountains. The Zāgros Mountains accommodate the main headsprings of the rivers running into the Persian Gulf and Oman sea watersheds. The major rivers of the latter basin are Arvand Rud, Gāmāsb, Kāroon, Jarrāhi, Zohreh, Dālaki, Mend, Shoor, Mināb, Mehrān and Naband (Sharifinia 2015). There also is a vastly-extended network of rivers in Iran, most of which are seasonally filled with water. Almost all water systems and resources are part of or in connection with the main water basins flowing eventually into the Caspian Sea in the north, Persian Gulf and Oman Sea in the south, and several confined basins in south, center, and east parts of the country (Ghorbani 2013). Iran shares several water basins with its neighboring countries. Based on a report by the Iranian Parliament Research Center, there are 89 rivers in Iran connected to the regional water drainages. Of these, 17 are shared with the adjacent countries, 4 are inflowing, and 68 are outflowing. Among the surrounding countries, Iraq has the highest topographic and hydrologic affinity with Iran and a number of 18 rivers are flowing into Iraq through the water basins in West Iran (http://rc.majlis.ir/en). The main transboundary rivers with considerable water flow through or into Iran are Aras, Āstārāchāy, Tajan (Tejen), Atrak, Sārisoo, Qarasoo, Zāb, Sirvān, Arvand Rud (Shatt al-Arab), Helmand (Hirmand), Harirud, Farāh and Khāsh. Ruttner-Kolisko (1980), based on their hydrological and chemical characteristics, classified Iranian inland waters in four categories: (1) exorheic rivers, reservoirs and swamps, (2) high alpine waters, (3) gradually–evaporating endorheic rivers and lakes, and (4) Astatic (unstable/temporary) desert springs and ponds.

Given that the climate of Iran has historically been regarded as semi-arid, the extended dry seasons in the country, chiefly in the last decade, have become a key concern. Drying lakes and rivers, declining groundwater levels, land subsidence, water quality deterioration, desertification, soil erosion, and dust and salt storms are the modern problems of a nation who was once one of the world's pioneers in sustainable water management (Madani et al. 2016). The annual precipitation is now less than the potential annual loss of water through evapotranspiration. Low precipitation and high evaporation rates, intense solar radiation and winds transferring the dry air masses, and mountain ranges, which prevent the transfer of moisture-laden air masses from Caspian and Mediterranean seas, are known as the main causes of the land dryness (Kehl 2009). These trends have more adverse effects when aquatic lives come into account. Several small to large lakes and rivers have been partly or completely dried out in the country. According to a recent study by the Iranian Parliament Research Center, 31 out of the 80 major Iranian lakes and wetlands, which are distinguished globally, have lost 60 to 100% of their water (http://rc.majlis.ir/en). Other than the global climate change, the state water mismanagement is blamed for its contribution to such losses (Madani et al. 2016). While impacts of the drought on larger residents of the inland water systems are noticeable, *e.g.*, mass mortality of fishes in Hāmoon Wetland in East and Zāyandeh-Rud River in Central

Iran, it is not clear to what extent the small aquatic invertebrates have been affected by this phenomenon.

31.3 A Biodiversity Hotspot

Iran contributes a significant share to the regional biological resources. Owing to its diverse geomorphological, topographical and climatic conditions, it is predictable that Iranian Plateau hosts a substantial blend of plant and animal taxa with high endemism rate. This is supported by the fact that the Plateau has historically acted as a bridge to connect three continents of Asia, Europe and Africa, resulting in its occupation by the European and Asian flora and fauna (Ghorbani 2013). Iran is located in the Palearctic realm at the crossroads of four biogeographical regions: the Euro-Siberian, the Irano-Touranian, the Nubo-Sindian and the Saharo-Arabian regions. The variety of landscapes arising from both this unique intersection of biogeographies and the physical and evolutionary processes operating across ecosystems and organisms have produced a diverse selection of flora and fauna. According to a recent estimation by the World Bank and Global Environment Facility, Iranian biodiversity has the highest economic value across Western Asia and the Middle East (Jowkar et al. 2016). The country's biodiversity hotspots are mostly distributed in north and west of Iran along the Alborz and Zāgros mountain ranges and are parts of the Irano-Anatolian biodiversity hotspot, which is designated as the 20th global hotspot region (Farashi and Shariati-Najafabadi 2017).

31.4 Rotifer Survey in Iran

Considering that scant records of field studies on Iranian water systems are available dating back to early and the mid-1800s (Marcet 1819; Abich 1856), Iranian Plateau remained relatively unexplored limnologically until the middle of the twentieth century. Heinz Löffler, a German limnologist, in a series of publications starting in 1949, described the environmental features of a number of Iranian water bodies and reported several fish, invertebrate and algal taxa, including 79 species and subspecies of rotifers from different locations in Iran: Hāmoon Lake in Sistān and Baluchestān Province, Lut Desert water basin in East Iran, Niriz (Bakhtegān) and Parishān Lakes in Fārs Province, Zāyandeh Rud water basin in Isfahān (Esfahān) Province, Zarivār Lake in Kurdistān (Kurdestān) Province, Urmia Lake in West Azarbaijān Province, and Anzali Wetland in Guilān (Gilān) Province (Löffler 1949/1950, 1953, 1961, 1981). His educational background and experience along with the adorable enthusiasm aided him to provide some of the exemplified technical and yet-valid reports with detailed characterization of the examined water bodies and many of the identified organisms. Over almost the same time period, an atlas of the Caspian Sea invertebrates, including the names and descriptions of 32 rotifer taxa, was published by Birshtain et al. (1968). They described the rotifers of the Caspian basin to be of three categories: (1) pre-estuarine, which accidentally enter into the coastal waters from the flowing rivers and are occasional members of the Caspian fauna, (2) euryhaline or eurybioant taxa, which are freshwater residents, but live in shallow, less saline coastal waters within the aquatic plants, and (3) stenohaline and less-frequently, euryhaline species living in saline waters. They also suggested that the number of rotifer taxa in the sea exceeds 300 species. While the reported taxa belonged to different parts of the sea and the exact sampling sites in the southern (Iranian) waters have not been delineated, they all are included in the list of Iranian rotifers collected in this review. The other pioneering studies on Iranian water systems were performed by two Austrian scientists, namely, Anton W. Ruttner and Agnes Ruttner-Kolisko. While their main focus seems to have been on investigating the geology and hydrology of some Iranian inland water basins (Ruttner-Kolisko 1966; Ruttner and Ruttner-Kolisko 1972, 1973), Ruttner-Kolisko (1980) reported the identification of about 100 rotifer species from Iranian salt lakes and desert pools by Löffler and herself. Although in her short article, she has briefly discussed on the ecological preferences of some rotifer species, she has not provided information on the identity of her rotifer inventories and their sampling locations. During the next two decades, no published report on Iranian rotifers was released. However, this does not necessarily imply that they have not been inspected by the native or overseas researchers. After a period of 'dormancy', research on the rotifers has been accelerated at the beginning of the twenty-first century, resulting in a surge in the publications by which a considerable number of Iranian rotifers were introduced. These accelerated efforts were mainly due to (1) a provoked general interest in Iranian biologists to investigate new and less-known subjects and present and publish their findings in international journals and communities, (2) the global trend of studying on more specific topics or group of organisms such as the phylum Rotifera, (3) a recent nationwide motivation to expand rotifer applications in aquaculture, and (4) the swift spread of Internet globally, which allowed the scientists to share their observations and experiences.

Anzali Wetland, an outstanding ecological heritage and one of the 25 Iranian wetlands of international importance (Ramsar Sites), and its adjacent water systems in Southern Caspian Sea are relatively extensively studied, mainly due to their proximity to one of the well-established fisheries research institute in the region. Most of these studies witnessed a considerably higher diversity and abundance of rotifers compared to the other zooplankton groups. Sabkara and Makaremi (2004) reported 50 genera of various zooplanktonic groups, including 26 rotifer genera, from Anzali Wetland. Later, they isolated 90 species of rotifers belonging to 36 genera from the wetland, of which 42 species of 13 genera were from the family Brachionidae (Sabkara and Makaremi 2009). Fallahi-Kaporchali et al. (2015) reported 30 rotifer genera from Anzali Wetland and the estuarine and coastal waters of Southern Caspian Sea. Fallahi and Sabkara (2015) noted that Rotifera was the most abundant zooplanktonic group in the wetland, comprising 31 out of the 60 identified genera of zooplanktons. However, the rotifers were not specified in their report. Golmarvi et al. (2017) reached a similar conclusion while studying the zooplankton of the wetland, noticing that out of the 61 zooplankton species, 30 belonged to the phylum Rotifera. Sabkara and Makaremi (2018) and Sabkara (2019) reported 89 species of 37 genera of rotifers from the wetland. In a recent survey, a total of 29 rotifer genera of 17 families have been identified from Anzali Wetland by Alavi et al. (2019). According to the latter study, Brachionidae and *Brachionus* were the most diverse rotifer taxa in the wetland, represented by 13 genera and 8 species, respectively. Rowshan-Tabari et al. (2003) investigated zooplanktons across the Iranian coastal line of the Caspian Sea at various depths, 10–100 m, and found Rotifera as the third most abundant zooplanktonic group (11%) after Cladocera (54%) and Copepoda (15%). Sabkara et al. (2006) investigated the zooplanktons of Kargānrud River flowing into the Caspian Sea and introduced *Keratella*, *Philodina*, *Synchaeta* and *Cephalodella* as the most abundant rotifer genera in the river comprising up to 17% of the total annual zooplankton abundance. Ghane-Sasansaraie (2007) by performing a physico-chemical and biological investigation on three major rivers in the Caspian Sea basin showed that rotifers from 11 genera were residing in the rivers. Khodaparast et al. (2012) recorded a total of 24 zooplankton species, including 9 rotifers, in the water column of three sampling sites at South Caspian Sea. Bagheri et al. (2013, 2014) reported the collective observations of their comprehensive study on zooplanktons of southwestern coast of the Caspian Sea during the years 1996–2010, in which 16 rotifer taxa were identified. Rowshan-Tabari et al. (2014) searched a broader area of the Southern Caspian coast for the zooplanktons, elapsing all three coastal provinces of Gilān, Māzandarān and Golestān. Of the 22 identified zooplankton species in their study, rotifers had the highest diversity (9 taxa) while were the second most abundant group after the copepods. According to their report, *Asplanchna priodonta* was the most abundant zooplankton in winter, reaching a density of over 3500 ind/m^3.

Studies on the rotifer diversity of other parts of the country have been sporadic. Sabkara and Makaremi (2003) investigated the zooplankton communities of Māku Dam Lake in Northwest Iran. They identified 17 rotifer taxa and found that Rotifera was the most abundant group in the lake, comprising two-third of the annual zooplankton population. Shayestehfar et al. (2008) introduced 13 rotifer taxa from Kor River in Fārs Province, South Iran. Mohammadzadeh et al. (2009) reported 10 rotifer genera from Amirkolāyeh Wetland in Northern Iran. Sabkara and Makaremi (2011) isolated 13 rotifer genera from Shoorābil Lake in Ardabil, Northwest Iran. Shayestehfar and Abdovis (2011) reported 6 rotifer species from a section of Kārun River running through Ahvāz City in Southwest Iran. They also observed rotifer population densities of up to 96 ind/l in the river. Jafari et al. (2011) conducted an ecological survey on Harāz River flowing in Māzandarān Province, North Iran, by which introduced 25 rotifer species with densities of 25–470 ind/m^3. Salavatian et al. (2011) identified 15 genera of zooplanktons, including 8 rotifer genera, while investigating Lār Dam Lake in Māzandarān Province. Papahn-Shooshtari et al. (2012) reported 4 families of Rotifera from Hoorolazim Wetlands, a complex of connected bodies of water on Iran-Iraq border. Sabkara and Makaremi (2013) documented the results of an earlier study on the zooplankton of Aras Dam Lake in Northwest Iran, in which 11 rotifer genera had been identified. The lake was also investigated by Mohsenpour-Azari (2017) who observed 19 rotifer taxa in its water. Farashi et al. (2014) found four rotifer genera, *Lecane*, *Brachionus*, *Trichocerca* and *Philodina*, in a cave stream, a tributary of the Dez River, in Lorestān Province, West Iran. The cave ecosystem was characterized by low water transparency, plankton abundance and concentrations of nutrients, while rotifers were the most abundant zooplankton therein. Ansari (2013) and Abbasi (2016) identified 25 and 19 rotifer species, respectively, from Shāzand suburb near Arāk City, Central Iran. Ansari et al. (2013, 2014) performed morphological analyses to discriminate inter- and intrapopulation variations in the rotifers *Lepadella patella*, *Brachionus urceolaris* and *B. quadridentatus* from Arāk County. Fathi et al. (2015) reported the predominance of rotifers of the genera *Keratella*, *Monostyla* (*Lecane*) and *Polyarthra* with abundances from 9000 to over 150,000 ind/m^3 in Choghākhor Wetland in Chāhārmahāl and Bakhtiāri Province, Central Iran. They witnessed the existence of a total of 16 rotifer genera in the wetland. Mohammadi and Reihan-Reshteh (2015) reported 23 monogonont rotifers from Harāz and Darakeh Rivers and Chitgar (Persian Gulf Martyrs) artificial lake in Tehran Province. Bagheri et al. (2017) found 37 zooplanktonic groups in Chitgar Lake in Tehran City, capital of Iran, among which Rotifera was the most diverse one represented by 20 genera. The lake is fed by Kan River originating from Mount Tochāl in Alborz Mountain Range. In a later study, however, Bagheri et al. (2018) only observed 11 rotifer genera in Kan River. Salavatian et al. (2016) reported the zooplankton of Arasbārān Dam Lake in Northwest Iran, including a list of 19 identified rotifer genera. Sinaei et al. (2017) in their study on the plankton of Sarbāz River in East Iran reported the genus *Brachionus* as the only rotifer taxon detected in the river. Ebrahimi-Dorche et al. (2018) identified 4 rotifer genera among the planktons of Zāyandehrud Dam Lake in central Iran.

During the years 2008 to 2010, three research projects were designed and performed at *Artemia* and Aquaculture Research Institute of Urmia University aiming at the discovery of biodiversity, ecological preferences, phylogenetics and aquaculture potential of the Iranian rotifers. Extensive sampling effort was invested across more than 40 water bodies in two provinces of West Azarbaijān, in Northwest, and Hormozgān, in South of Iran, during four seasons. As a result, more than 110 planktonic and epiphytic rotifer species and subspecies belonging to 41 genera and 22 families, including 61 new records from Iran, were identified and reported in the final project report and several peer-reviewed publications (Malekzadeh-Viayeh 2010a, b; Khaleqsefat et al. 2011; Malekzadeh-Viayeh and Špoljar 2012; Malekzadeh-Viayeh et al. 2014; Mills et al. 2017). Phylum Rotifera was also a focus of research by Faculty of Science of Tehrān University, through which the diversity of the local rotifer fauna was further explored. Hakimzadeh-Khoei et al. (2011) and Kordbacheh and Rahimian (2012) recorded 115 and 113 rotifer taxa, respectively, from various freshwater systems in Tehrān Province and its neighboring areas. Reihan-Reshteh and Rahimian (2014) reported 66 rotifer species from four major rivers of Kārun (Kāroon), Karkhe, Dez and Jarrāhi-Maroon, Shādegān

Wetland, and a number of small-to-large natural or artificial ponds in Khuzestān (Khoozestān) Province. Shādegān Wetland was earlier investigated by Kholfe-Nilsaz (2009) who found that Rotifera, with 12 out of the 24 genera, was the most diverse zooplanktonic group in the wetland, and among the rotifers, *Brachionus* and *Asplanchna* were the most abundant genera.

Similar to other planktonic groups, rotifers have also been studied out of their natural habitats, in small artificial ponds and water reservoirs. Mahdizadeh et al. (2006) recorded 14 rotifer genera from the earthen fish pond in Gilān Province. They have also observed significantly higher abundances of the rotifers in different seasons compared to the other zooplankton groups. Kamali-Sansiqi et al. (2014) identified the members of eight rotifer genera of *Adineta*, *Asplanchna*, *Brachionus*, *Gastropus*, *Keratella*, *Philodina*, *Polyarthra* and *Rotaria* in the earthen fish ponds of Gonbad-e-Kāvus in Golestān Province, Northeast Iran. While at least part of the rotifer communities in fish ponds may have originated in a different geographical location (*e.g.*, they may have been transferred alive or as resting eggs by water, fish larvae or organic fertilizers), they are herein considered to belong to where they are found.

To provide a more thorough inventory of the Iranian rotifers, the native species used in laboratory experiments are also included in the inventory presented in this chapter. Ahmadifard (2007) studied growth and body compositions of *B. calyciflorus* isolated from Anzali Wetland after being fed with different algal species. Farhadian et al. (2015) performed laboratory culture experiments on *Euchlanis dilatata* isolated from Hannā Dammed Lake in Semirom, Isfahān, using diverse feed items. Seyyedi-Abalvan et al. (2015) examined the effects of different nickel concentrations on population growth of *B. calyciflorus* sampled from Chāh-Nimeh reservoirs in Zābol, East Iran. Rufchaie et al. (2015) fed Persian sturgeon (*Acipenser persicus*) with vitamin C-enriched *B. calyciflorus* of Anzali Wetland to compare the nutritional value of different live foods by measuring the growth indexes of the fish.

While zooplankton communities of the Persian Gulf and Sea of Oman have recently been under investigation, the priority has been granted to the estimation of diversity and abundance of macrozooplanktons (Fazeli et al. 2013; Farhadian and Pouladi 2014; Abedi et al. 2014; Abedi 2015; Mokhayer et al. 2017). Pouladi et al. (2013) observed *Brachionus* rotifers among the zooplankton of Helle River estuary where the river meets the Persian Gulf, albeit only in their spring sampling. Recently, Khafaeizadeh et al. (2017) isolated 12 planktonic rotifer species from Bahmanshir River and its estuary on the northwestern coast of the gulf. Apart from less than a handful of records, almost nothing is known about the Iranian marine rotifer fauna. The distribution of locations where Iranian rotifers have been found and reported from is shown in Fig. 31.2.

Nearly all studies on Iranian rotifers have been performed based solely on their morphology. To contribute to the universal efforts addressing the question of cryptic speciation and puzzling taxonomy in *Brachionus plicatilis* species complex, we analyzed 15 clonal cultures of the putative members of this complex sampled from different localities in Iran using both their morphological and genetic (cytochrome c oxidase subunit 1-*COI*-DNA fragment) attributes (Malekzadeh-Viayeh et al. 2014).

Fig. 31.2 Map of Iran and its provincial districts. Red solid squares are to show the number of reports on Iranian rotifers in each district/province. Solid circles provide a rough comparison of the number of rotifer taxa reported from each district, where each circle is representative of about five taxa

Morphological analyses, based on nine linear measurements of the lorica, discriminated the 9 *Brachionus* rotifer strains whose body dimensions were available (Fig. 31.3), and the molecular comparisons illustrated that the rotifers belonged to four distinct entities: *B. plicatilis* senso stricto, s.s., *Brachionus* 'Austria', *Brachionus* 'Tiscar' and a putative native Iranian form (Fig. 31.4). However, the subsequent study by Mills et al. (2017) employing a nuclear gene region, the internal transcribed spacer subunit 1 (ITS1), in addition to the mitochondrial *COI* gene fragment discovered that none of the *Brachionus* rotifers from Iran could be regarded as a distinct or endemic species, as they constructed sister groups with those from other parts of the world.

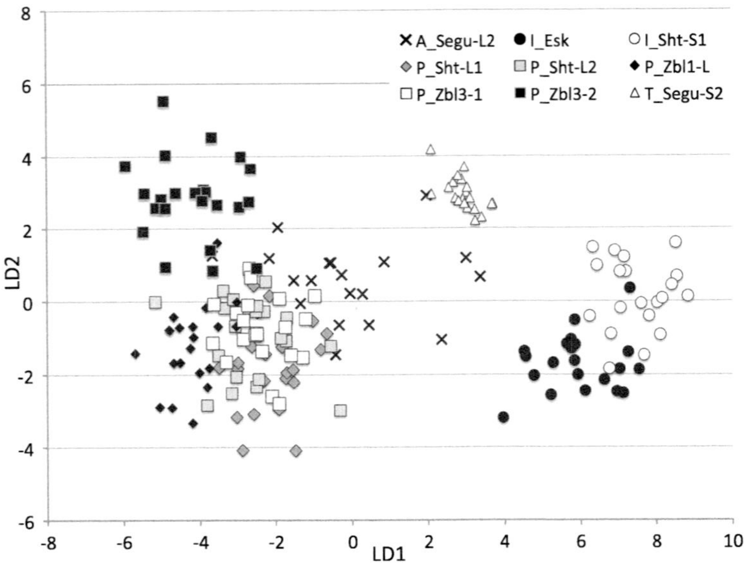

Fig. 31.3 The scatter plot resulted from a discrimination analysis based on the linear dimensions of nine *Brachionus* rotifer strains from Iran, denoted by different geometric shapes (Malekzadeh-Viayeh et al. 2014)

31.5 Notes on Iranian and Regional Rotifer Diversity

According to the existing literature, a total of 366 rotifer species and subspecies belonging to 69 genera and 27 families have so far been identified in Iranian waters. These exclude several rotifer taxa, which have been reported at genus or higher taxonomic levels. Regardless of sources of the records, all scientific and authority names were checked against the recently-updated global rotifer checklists and refined to omit the spelling errors and invalid taxa. A great majority of the rotifers (95.08%) belong to the subclass Monogononta, whereas 18 species (4.91%) representing 8 genera of 4 families are from the subclass Bdelloidea (Tables 31.1 and 31.2). The most diverse family is Brachionidae (58 species and subspecies, 15.85%), followed by Notommatidae (49 species and subspecies, 13.39%) and Lecanidae (44 species and subspecies, 12.02%). The families represented by the highest number of genera are Notommatidae (9 genera), Brachionidae (7 genera), and Euchlanidae, Flosculariidae and Philodinidae (5 genera each). *Lecane* is the most diverse genus (with 44 species and subspecies), followed by *Brachionus* (30 species and subspecies), *Cephalodella* (24 species and subspecies) and *Trichocerca* (23 species and subspecies). Several other rotifer families are present by only a single genus and species. The most widespread species with the highest distribution ranges are *Lecane lunaris* (found in 11 locations/water bodies), *Brachionus plicatilis* and *B. quadridentatus* (10 locations each), *L. bulla*, *L. luna* and *B. urceolaris* (9 locations each), and *B. calyciflorus*, *B. angularis*, *Keratella*

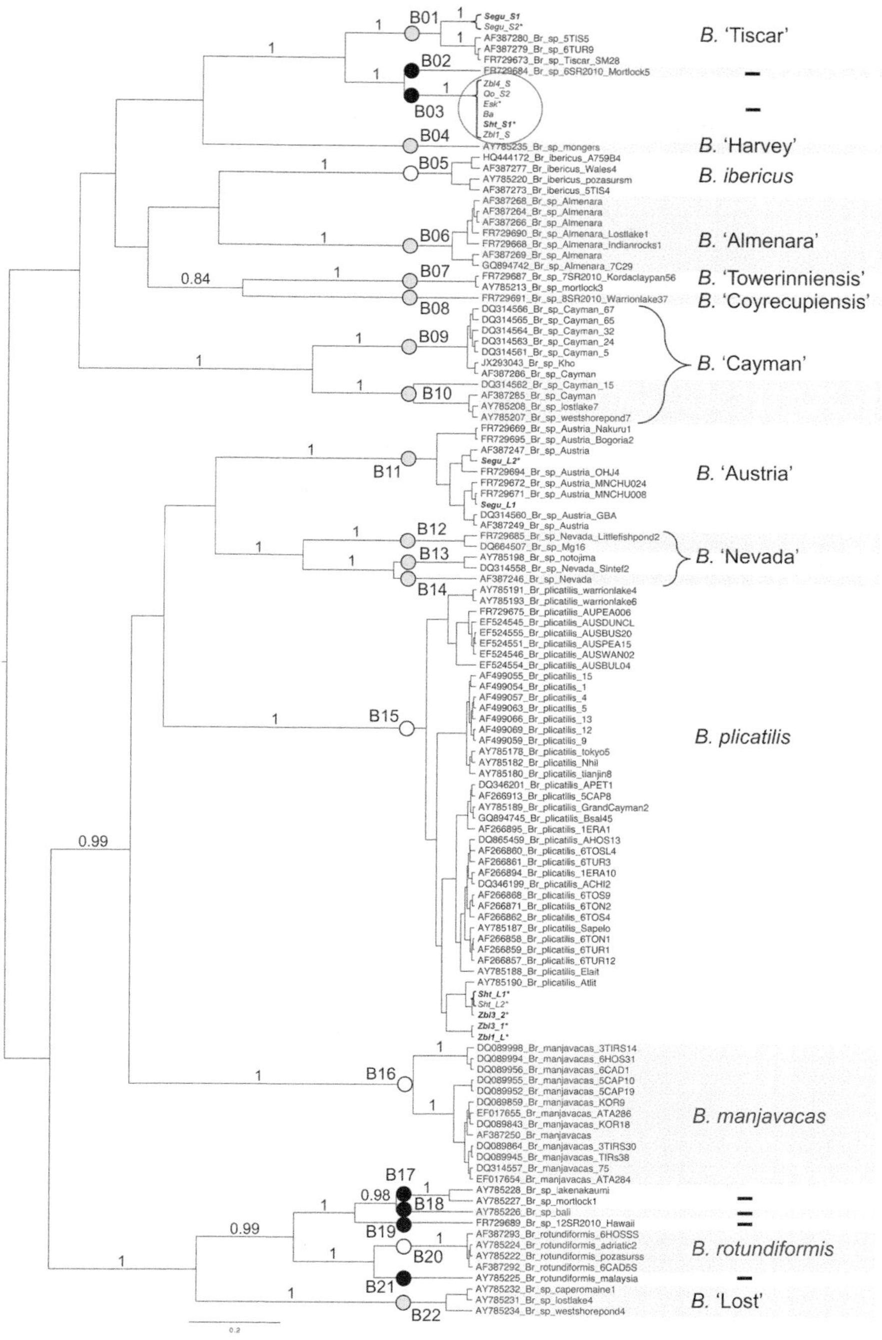

Fig. 31.4 Phylogenetic tree constructed based on the comparisons of the cytochrome c oxidize subunit 1 (*COI*) gene region of several members of the *Brachionus plicatilis* species complex from all over the world. The *Brachionus* rotifer strains from Iran made a distinct clade (shown in red circle) based on the *COI* phylogeny (Malekzadeh-Viayeh et al. 2014)

Table 31.1 Name and characteristics of the locations where the Iranian rotifers have been sampled from and their corresponding references

Site no.	Location/province	Climate	Position	Source
1	Caspian Sea	Humid continental	North	Birshtain et al. (1968), Rowshan-Tabari et al. (2003, 2014), Segers and De Smet (2008), Khodaparast et al. (2012), Bagheri et al. (2013, 2014), Fallahi-Kaporchali et al. (2015), Sabkara (2019), https://www.zin.ru
2	a. Anzali Wetland; Rivers in South Caspian Sea basin (b. Kargānrud, c. Haviq and d. Shafārud), e. Earthen fish ponds, and f. Amirkolāyeh Wetland, Gilān	Humid continental	North	Löffler (1961), Sabkara and Makaremi (2004, 2009, 2018), Sabkara et al. (2006), Mahdizadeh et al. (2006), Ghane-Sasansaraie 2007), Ahmadifard 2007), Mohammadzadeh et al. (2009), Fallahi-Kaporchali et al. (2015), Fallahi and Sabkara (2015), Rufchaie et al. (2015), Golmarvi et al. (2017), Sabkara (2019), Alavi et al. (2019)
3	a. Harāz River; b. Lār Dam Lake, Māzandarān	Humid continental	North	Jafari et al. (2011), Salavatian et al. (2011)
4	Golestān	Humid continental	North	Kamali-Sansiqi et al. (2014)
5	Tehrān	Moderate winters and hot summers, semi-arid steppe to alpine climate	North	Hakimzadeh-Khoei et al. (2011), Kordbacheh and Rahimian (2012), Mohammadi and Reihan-Reshteh (2015), Bagheri et al. (2017, 2018)
6	West Azarbaijān	Cool-temperate to cold/ Mediterranean, semi-arid	North-West	Löffler (1949, 1953, 1961), Sabkara and Makaremi (2003), Malekzadeh-Viayeh (2010a,b), Khaleqsefat et al. (2011), Malekzadeh-Viayeh and Špoljar (2012), Sabkara and Makaremi

(continued)

Table 31.1 (continued)

Site no.	Location/province	Climate	Position	Source
				(2013), Malekzadeh-Viayeh et al. (2014), Mohsenpour-Azari (2017)
7	Arasbārān Dam Lake, East Azarbaijān	Cool-temperate to cold/semi-arid	North-west	Salavatian et al. (2016)
8	Shoorābil Lake, Ardabil	Cool-temperate to chilly/semi-arid	North-west	Sabkara and Makaremi (2011)
9	Zarivār Lake, Kurdestān	Moderate to cold, semi-arid to dry	West	Löffler 1961
10	Arāk and Shāzand Suburbs, Markazi	Moderate, with quite chilly summers and icy winters, desert, semi-desert, and moderate to cold mountainous	Center	Ansari (2013), Ansari et al. (2013), Abbasi (2016)
11	Dez River, Lorestān	Moderate to cold, subhumid continental	West	Farashi et al. (2014)
12	Khuzestān	Very hot to moderate and dry to semi-arid	South-West	Kholfe-Nilsaz (2009), Malekzadeh-Viayeh (2010a), Shayestehfar and Abdovis (2011), Papahn-Shooshtari et al. (2012), Reihan-Reshteh and Rahimian (2014), Khafaeizadeh et al. (2017)
13	Choghākhor Wetland, Chāhārmahāl and Bakhtiāry	Moderate to very cold and wet (rain and snow)	South-West	Fathi et al. (2015)
14	Zāyandeh Rud River basin; Gāvkhooni Wetland; Hanna Dam Lake, Isfahān	Moderate, dry	Center	Löffler (1961), Farhadian et al. (2015), Ebrahimi-Dorche et al. (2018)
15	Bakhtegān and Parishān Lakes; Kor River, Fārs	Temperate, semi-arid	South-West	Löffler (1949, 1961, 1981), Shayestehfar et al. (2008)
16	Lut Desert water basin, Kermān	Hot and dry	South-East	Löffler (1961)
17	Hāmoon Lake; Sarbāz River, Sistān and Baluchestān	Hot to moderate, arid	South-East	Löffler (1961), Seyyedi-Abalvan et al. (2015), Sinaei et al. (2017)
18	Hormozgān	Very hot to moderate, humid-subtropical	South	Malekzadeh-Viayeh (2010a, b)

Table 31.2 Checklist of Iranian rotifers and their distribution across the country

Rotifer taxa	Location
Brachionidae Harring, 1838	
Brachionus angularis Gosse, 1851	1, 2a, 3a, 5, 6, 10, 12, 15
Brachionus angularis angularis Gosse, 1851	1
Brachionus angularis bidens Plate, 1886	1
Brachionus bidentatus Anderson, 1889	5, 6, 15
Brachionus calyciflorus Pallas, 1766	1, 2a, 3a, 5, 6, 12, 17, 18
Brachionus calyciflorus amphiceros Ehrenberg, 1838	1
Brachionus falcatus Zacharias, 1898	2a
Brachionus forficula Wierzejski, 1891	1
Brachionus budapestinensis Daday, 1885	1, 2a
Brachionus diversicornis Daday, 1883	1, 2a, 5, 12, 15
Brachionus bennini Leissling, 1924	6
Brachionus havanaensis Rousselet, 1911	12, 15
Brachionus leydigii Cohn, 1862	1, 5, 6, 12
Brachionus plicatilis Müller, 1786	1, 2a, 5, 6, 12, 14, 15, 16, 17, 18
Brachionus plicatilis plicatilis Müller, 1786	1, 12
Brachionus plicatilis decemcornis Fadeev, 1925	1
Brachionus asplanchnoidis Charin, 1947	6
Brachionus 'Tiscar'	6
Brachionus quadridentatus Hermann, 1783	1, 2a, 3a, 5, 6, 10, 12, 15, 17, 18
Brachionus quadridentatus quadridentatus Hermann, 1783	1, 12
Brachionus rotundiformis Tschugunoff, 1921	1, 12
Brachionus cf. *rotundiformis*	12, 18
Brachionus nilsoni Ahlstrom, 1940	1, 6
Brachionus ibericus Ciros-Pérez, Gómez & Serra, 2001	18
Brachionus rubens Ehrenberg, 1838	2a, 5
Brachionus urceolaris Müller, 1773	1, 2a, 5, 6, 10, 12, 15, 17, 18
Brachionus urceolaris urceolaris Müller, 1773	12
Brachionus variabilis Hempel, 1896	2a, 5
Brachionus sericus Rousselet, 1907	12
Brachionus sp.	1, 2a, 2d, 2e, 4, 6, 7, 8, 11, 12, 13, 15
Keratella cochlearis (Gosse, 1851)	1, 2a, 5, 6, 10, 15
Keratella cochlearis cochlearis (Gosse, 1851)	8
Keratella tropica (Apstein, 1907)	1, 2a, 5, 6, 12, 15, 17, 18
Keratella lenzi Hauer, 1953	6
Keratella irregularis (Lauterborn, 1898)	6
Keratella procurva (Thorpe, 1891)	16, 17
Keratella quadrata (Müller, 1786)	1, 2a, 3a, 5, 6, 9, 15
Keratella quadrata quadrata (Müller, 1786)	12
Keratella tecta (Gosse, 1851)	6, 10, 12
Keratella f. *tecta*	6

(continued)

Table 31.2 (continued)

Rotifer taxa	Location
Keratella serrulata (Ehrenberg, 1838)	18
Keratella testudo (Ehrenberg, 1832)	17
Keratella valga (Ehrenberg, 1834)	1, 2a, 5, 12, 15
Keratella sp.	1, 2a, 2b, 2c, 2d, 2e, 3b, 4, 5, 6, 7, 8, 12, 13, 14
Notholca acuminata (Ehrenberg, 1832)	1, 2a, 3b, 5, 6, 9
Notholca labis Gosse, 1887	1, 5
Notholca salina Focke, 1961	6
Notholca salina salina Focke, 1961	12
Notholca squamula (Müller, 1786)	1, 2a, 5, 6, 9, 14, 15
Notholca striata (Müller, 1786)	5
Notholca psammarina Buchholz & Rühmann, 1956	5
Notholca foliacea (Ehrenberg, 1838)	3a
Notholca caudata Carlin, 1943	1, 6
Notholca cinetura Skorikov, 1914	1
Notholca sp.	2a, 6, 8, 13
Kellicottia longispina (Kellicott, 1879)	1, 3a, 7
Platyias quadricornis (Ehrenberg, 1832)v	1, 2a, 3a, 5, 6, 9, 12, 15
Platyias sp.	2a, 2f
Plationus patulus (Müller, 1786)	1, 2a
Anuraeopsis fissa (Gosse, 1851)	2a, 6
Anuraeopsis navicula Rousselet, 1911	2a
Anuraeopsis sp.	2a, 2e, 2f, 5, 7, 13
Notommatidae Hudson & Gosse, 1886	
Notommata aurita (Müller, 1786)	1, 5, 17
Notommata brachyota Ehrenberg, 1832	6
Notommata copeus Ehrenberg, 1834	1, 5, 12, 15
Notommata diasema Myers, 1936	6
Notommata glyphura Wulfert, 1935	5, 6
Notommata pygmaea Harring & Myers, 1922	6
Notommata tripus Ehrenberg, 1838	8
Notommata pseudocerberus de Beauchamp, 1908	1
Notommata sp.	2a, 6, 17
Cephalodella catellina (Müller, 1786)	2a, 5, 6, 12, 18
Cephalodella forficata (Ehrenberg, 1832)	6
Cephalodella physalis Myers, 1924	12
Cephalodella forficula (Ehrenberg, 1838)	1, 5, 6, 10, 12
Cephalodella gibba (Ehrenberg, 1830)	2a, 3a, 5, 6, 12, 18
Cephalodella gibboides Wulfert, 1951	6
Cephalodella gracilis (Ehrenberg, 1830)	10
Cephalodella gracilis gracilis (Ehrenberg, 1830)	6
Cephalodella inquilina Myers, 1924	8, 10
Cephalodella lepida Myers, 1934	6

(continued)

Table 31.2 (continued)

Rotifer taxa	Location
Cephalodella maior (Zavadovsky, 1926)	5, 12
Cephalodella misgurnus Wulfert, 1937	5
Cephalodella cf. *mus* Wulfert, 1956	6
Cephalodella obvia Donner, 1951	6
Cephalodella plicata Myers, 1924	5
Cephalodella stenroosi Wulfert, 1937	5, 6
Cephalodella sterea (Gosse, 1887)	6
Cephalodella tincaformis Koste, 1992	6
Cephalodella ventripes (Dixon-Nuttall, 1901)	2a, 5, 6
Cephalodella vittata Kutikova, 1985	6
Cephalodella anebodica Bērziņš, 1976	5
Cephalodella belone Myers, 1924	10
Cephalodella compressa Myers, 1924	10
Cephalodella pachyodon Wulfert, 1937	2a
Cephalodella sp.	2a, 2b, 2c, 2d, 2e, 3a, 5, 6, 7, 8, 9
Eosphora anthadis Harring & Myers, 1922	6
Eosphora cf. *anthadis*	12
Eosphora ehrenbergi Weber & Montet, 1918	6
Eosphora najas Ehrenberg, 1830	6, 12, 17
Eosphora therina Harring & Myers, 1922	18
Eothinia elongata (Ehrenberg, 1832)	6
Monommata actices Myers, 1930	5, 6
Monommata longiseta (Müller, 1786)	5
Monommata grandis Tessin, 1890	2a
Monommata sp.	2a, 5
Resticula melandocus (Gosse, 1887)	6, 12
Resticula nyssa Harring & Myers, 1924	6
Pleurotrocha atlantica Myers, 1936	12, 18
Pleurotrocha cf. *atlantica*	18
Pleurotrocha petromyzon Ehrenberg, 1830	5, 12
Pseudoharringia similis Fadeev, 1925	1
Taphrocampa annulosa Gosse, 1851	12
Drilophaga judayi Harring & Myers, 1922	10
Lecanidae Bartos, 1959	
Lecane aculeata (Jakubski, 1912)	2a, 12
Lecane bifastigata Hauer, 1938	12
Lecane bulla (Gosse, 1851)	1, 2a, 3a, 5, 6, 9, 14, 17, 18
Lecane bulla bulla (Gosse, 1851)	12
Lecane closterocerca (Schmarda, 1859)	1, 5, 6, 9, 10, 12
Lecane elsa Hauer, 1931	5
Lecane cf. *glypta* Harring & Myers, 1926	17
Lecane flexilis (Gosse, 1886)	2a, 6

(continued)

Table 31.2 (continued)

Rotifer taxa	Location
Lecane depressa (Bryce, 1891)	6
Lecane furcata (Murray, 1913)	5, 6, 9
Lecane grandis (Murray, 1913)	2a, 5, 12
Lecane hamata (Stokes, 1896)	1, 2a, 5, 6, 9, 12, 15
Lecane hastata (Murray, 1913)	6, 10, 18
Lecane inconspicua Segers & Dumont, 1993	8
Lecane lamellata (Daday, 1893)	5, 6, 12
Lecane leontina (Turner, 1892)	5
Lecane ludwigii (Eckstein, 1883)	1, 2a, 9
Lecane luna (Müller, 1776)	1, 2a, 5, 6, 9, 12, 14, 15, 17
Lecane lunaris (Ehrenberg, 1832)	1, 3a, 5, 6, 9, 10, 12, 14 15, 17, 18
Lecane nana (Murray, 1913)	5
Lecane obtusa (Murray, 1913)	5, 9, 15
Lecane opias (Harring & Myers, 1926)	5
Lecane arcuata (Bryce, 1891)	5
Lecane venusta Harring & Myers, 1926	5
Lecane papuana (Murray, 1913)	6, 12
Lecane plesia Myers, 1936	12
Lecane pumila (Rousselet, 1906)	6
Lecane punctata (Murray, 1913)	6, 12
Lecane pyriformis (Daday, 1905)	5, 10, 12
Lecane quadridentata (Ehrenberg, 1830)	1, 2a, 5, 6, 9, 14, 15
Lecane rhenana Hauer, 1929	1
Lecane scutata (Harring & Myers, 1926)	5, 12
Lecane stenroosi (Meissner, 1908)	1, 12, 15, 18
Lecane sympoda Hauer, 1929	5, 9
Lecane tenuiseta Harring, 1914	9
Lecane thalera (Harring & Myers, 1926)	2a, 6, 12, 15, 17, 18
Lecane curvicornis (Murray, 1913)	2a
Lecane cornuta (Müller, 1786)	1, 2a
Lecane crepida Harring, 1914	1
Lecane crenata (Harring, 1913)	1
Lecane ungulata (Gosse, 1887)	1, 12
Lecane broaensis Segers & Dumont, 1995	10
Lecane inermis (Bryce, 1892)	10
Lecane paradoxa (Steinecke, 1916)	10
Lecane sp.	1, 2a, 2c, 2e, 2f, 5, 6, 7, 11, 12, 13
Trichocercidae Harring, 1913	
Trichocerca cavia (Gosse, 1886)	6, 14, 15
Trichocerca caspica (Tschugunoff, 1921)	1, 2a
Trichocerca intermedia (Stenroos, 1898)	1
Trichocerca bicristata (Gosse, 1887)	1

(continued)

Table 31.2 (continued)

Rotifer taxa	Location
Trichocerca capucina (Wierzejski & Zacharias, 1893)	1
Trichocerca musculus (Hauer, 1936)	1
Trichocerca similis (Wierzejski, 1893)	1, 6, 10
Trichocerca tigris (Müller, 1786)	1, 2a
Trichocerca bidens (Lucks, 1912)	6
Trichocerca agnatha Wulfert, 1939	6
Trichocerca cylindrica (Imhof, 1891)	1, 5
Trichocerca elongata (Gosse, 1886)	1, 2a, 3a, 5, 6
Trichocerca heterodactyla (Tschugunoff, 1921)	1
Trichocerca longiseta (Schrank, 1802)	1, 2a, 5, 6, 9, 15
Trichocerca myersi (Hauer, 1931)	5
Trichocerca dixonnuttalli (Jennings, 1903)	9, 15
Trichocerca porcellus (Gosse, 1851)	2a, 5, 6, 12
Trichocerca pusilla (Jennings, 1903)	1, 2a, 5, 6, 9, 12, 15
Trichocerca rattus (Müller, 1776)	1, 2a, 6, 9, 15
Trichocerca stylata (Gosse, 1851)	1, 2a, 5, 16
Trichocerca tenuior (Gosse, 1886)	5
Trichocerca weberi (Jennings, 1903)	1, 6, 9, 15
Trichocerca obtusidens (Olofsson, 1918)	2a
Trichocerca sp.	1, 2a, 2d, 2e, 5, 6, 7, 8, 17
Lepadellidae Harring, 1913	
Lepadella acuminata (Ehrenberg, 1834)	1, 6, 10
Lepadella biloba Hauer, 1958	10
Lepadella (*Lepadella*) *biloba* Hauer, 1958	5
Lepadella (*Lepadella*) *costata* Wulfert, 1940	5
Lepadella heterostyla (Murray, 1913)	2a, 9, 15
Lepadella (*Lepadella*) *eurysterna* Myers, 1942	5
Lepadella ovalis (Müller, 1786)	1, 2a, 3a, 5, 6, 10
Lepadella (*Lepadella*) *ovalis* (Müller, 1786)	5
Lepadella patella (Müller, 1773)	1, 2a, 5, 6, 9, 10, 14, 15
Lepadella (*Lepadella*) *patella* (Müller, 1773)	5
Lepadella patella patella (Müller, 1786)	12
Lepadella cf. *discoidea* Segers, 1993	6
Lepadella (*Lepadella*) *punctata* Wulfert, 1939	5
Lepadella quadricarinata (Stenroos, 1898)	5, 12
Lepadella (*Lepadella*) *quadricarinata* (Stenroos, 1898)	5
Lepadella triptera (Ehrenberg, 1830)	1, 5, 6, 9, 15
Lepadella (*Lepadella*) *triptera* (Ehrenberg, 1830)	5
Lepadella neglecta Segers & Dumont, 1995	10
Lepadella (*Lepadella*) *apsida* Harring, 1916	2a
Lepadella (*Heterolepadella*) *ehrenbergii* (Perty, 1850)	2a
Lepadella (*Lepadella*) *hyalina* Smirnov, 1927	2a

(continued)

Table 31.2 (continued)

Rotifer taxa	Location
Lepadella sp.	2a, 2d, 2f, 5, 7, 8, 13
Colurella adriatica Ehrenberg, 1831	1, 2a, 3a, 5, 9, 12, 14, 15
Colurella colurus (Ehrenberg, 1830)	1, 5, 6, 10
Colurella obtusa (Gosse, 1886)	2a, 5, 6, 10
Colurella sanoamuangae Chittapun, Pholpunthin & Segers, 1999	12
Colurella uncinata (Müller, 1773)	1, 3a, 5, 6
Colurella uncinata bicuspidata (Ehrenberg, 1832)	6, 12
Colurella cf. *sinistra* Carlin, 1939	6
Colurella geophila Donner, 1951	2a
Colurella sp.	2a, 2c, 5, 6, 7
Squatinella lamellaris (Müller, 1786)	1, 2a, 12
Squatinella rostrum (Schmarda, 1846)	1, 2a, 5, 6, 9, 12, 15
Squatinella sp.	2a
Synchaetidae Hudson & Gosse, 1886	
Synchaeta cecilia Rousselet, 1902	1
Synchaeta cecilia f. *fusipes* Buchholz, 1954	1
Synchaeta grandis Zacharias, 1893	1
Synchaeta littoralis Rousselet, 1902	1, 6
Synchaeta tremula (Müller, 1786)	1, 6
Synchaeta neapolitana Rousselet, 1902	1
Synchaeta kitina Rousselet, 1902	1
Synchaeta longipes Gosse, 1887	1
Synchaeta baltica Ehrenberg, 1834	1
Synchaeta oblonga Ehrenberg, 1832	1, 2a, 5, 6
Synchaeta pectinata Ehrenberg, 1832	1, 2a, 5, 6, 9, 15, 17
Synchaeta stylata Wierzejski, 1893	1, 2a, 5
Synchaeta vorax Rousselet, 1902	1, 2a, 5
Synchaeta johanseni Harring, 1921	2a
Synchaeta sp.	1, 2a, 2b, 2d, 2e, 2f, 3b, 5, 6, 7, 8
Polyarthra dolichoptera Idelson, 1925	1, 2a, 3b, 5, 6
Polyarthra euryptera Wierzejski, 1891	1, 6
Polyarthra longiremis Carlin, 1943	1
Polyarthra luminosa Kutikova, 1962	1
Polyarthra remata Skorikov, 1896	1, 5, 6, 10
Polyarthra vulgaris Carlin, 1943	1, 2a, 3a, 5, 6, 9, 15, 17
Polyarthra major Burckhardt, 1900	1, 3a
Polyarthra minor Voigt, 1904	1, 6
Polyarthra sp.	2a, 2e, 4, 5, 6, 7, 8, 12, 13, 14
Ploesoma lenticulare Herrick, 1885	1
Ploesoma truncatum (Levander, 1894)	1
Ploesoma hudsoni (Imhof, 1891)	1
Ploesoma sp.	12

(continued)

Table 31.2 (continued)

Rotifer taxa	Location
Dicranophoridae Harring, 1913	
Dicranophorus forcipatus (Müller, 1786)	5, 6, 10
Dicranophorus luetkeni (Bergendal, 1892)	5
Dicranophorus epicharis Harring & Myers, 1928	6
Dicranophorus grandis (Ehrenberg, 1832)	1
Dicranophorus dolerus Harring & Myers, 1928	5
Dicranophorus sp.	2a
Dicranophoroides caudatus (Ehrenberg, 1834)	1, 5, 9, 12, 15
Encentrum uncinatum (Milne, 1886)	1, 3a
Encentrum saundersiae (Hudson, 1885)	2a, 6
Encentrum cf. *algente* Harring, 1921	6
Encentrum putorius Wulfert, 1936	12
Encentrum lutra Wulfert, 1936	5
Encentrum orthodactylum Wulfert, 1936	5
Encentrum cf. *putorius*	5
Encentrum ussuriensis De Smet & Chernyshev, 2006	10
Encentrum walterkostei Jersabek, 1994	10
Encentrum sp.	2a
Paradicranophorus hudsoni (Glascott, 1893)	1
Paradicranophorus aculeatus (Neizvestnova-Zhadina, 1935)	12
Aspelta angusta Harring & Myers, 1928	12
Hexarthridae Bartos, 1959	
Hexarthra fennica (Levander, 1892)	2a, 5, 6, 15, 16, 17
Hexarthra intermedia (Wiszniewski, 1929)	5, 9, 15, 17
Hexarthra mira (Hudson, 1871)	1, 2a, 5, 6, 9, 15, 17
Hexarthra oxyuris (Zernov, 1903)	1, 5
Hexarthra bulgarica (Wiszniewski, 1933)	6
Hexarthra jenkinae (de Beauchamp, 1932)	6
Hexarthra polyodonta (Hauer, 1957)	5
Hexarthra sp.	2a, 5, 6, 13
Asplanchnidae Eckstein, 1883	
Asplanchna brightwellii Gosse, 1850	1, 2a, 3b, 5, 6, 9, 12, 15
Asplanchna girodi Guerne, 1888	8
Asplanchna priodonta Gosse, 1850	1, 2a, 3a, 5, 6, 9, 12, 15
Asplanchna sieboldii (Leydig, 1854)	1, 6, 17
Asplanchna herricki Guerne, 1888	1
Asplanchna sp.	1, 2a, 2e, 2f, 4, 5, 6, 7, 8, 12, 13, 14
Asplanchnopus multiceps (Schrank, 1793)	1
Asplanchnopus sp.	1
Harringia rousseleti Beauchamp, 1911	15
Harringia eupoda (Gosse, 1887)	1
Euchlanidae Ehrenberg, 1838	

(continued)

Table 31.2 (continued)

Rotifer taxa	Location
Euchlanis dilatata Ehrenberg, 1832	1, 2a, 5, 6, 14, 16, 17, 18
Euchlanis dilatata dilatata Ehrenberg, 1832	12
Euchlanis incisa Carlin, 1939	1, 5
Euchlanis lyra Hudson, 1886	6, 12
Euchlanis parva Rousselet, 1892	9, 15, 18
Euchlanis pyriformis Gosse, 1851	1
Euchlanis calpidia (Myers, 1930)	1
Euchlanis alata Voronkov, 1912	1
Euchlanis arenosa Myers, 1936	1
Euchlanis deflexa (Gosse, 1851)	1, 3a
Euchlanis oropha Gosse, 1887	1
Euchlanis triquetra Ehrenberg, 1838	1
Euchlanis sp.	2d, 2f, 5, 12, 13
Beauchampiella eudactylota (Gosse, 1886)	1, 5, 15
Tripleuchlanis plicata (Levander, 1894)	1, 12
Dipleuchlanis propatula (Gosse, 1886)	1, 6
Diplois daviesiae Gosse, 1886	6
Flosculariidae Ehrenberg, 1838	
Ptygura furcillata (Kellicott, 1889)	5
Ptygura melicerta Ehrenberg, 1832	6
Ptygura tridorsicornis Summerfield-Wright, 1957	6
Sinantherina socialis (Linnaeus, 1758)	1
Sinantherina semibullata (Thorpe, 1893)	2a, 5
Sinantherina sp.	3a
Floscularia ringens (Linnaeus, 1758)	1, 5
Floscularia sp.	1
Lacinularia flosculosa (Müller, 1773)	1
Limnias melicerta Weisse, 1848	12, 15
Testudinellidae Harring, 1913	
Testudinella incisa (Ternetz, 1892)	5
Testudinella mucronata (Gosse, 1886)	5
Testudinella parva (Ternetz, 1892)	5
Testudinella patina (Hermann, 1783)	1, 2a, 5, 6, 9, 14, 15, 17
Testudinella truncata (Gosse, 1886)	2a
Testudinella sp.	2f
Pompholyx complanata Gosse, 1851	1
Pompholyx sulcata Hudson, 1885	1, 2a
Pompholyx sp.	2a, 5, 7, 8, 13
Mytilinidae Harring, 1913	
Mytilina mucronata (Müller, 1773)	1, 2a, 6, 9, 15
Mytilina ventralis (Ehrenberg, 1830)	1, 2a, 6, 9, 15
Mytilina ventralis brevispina (Ehrenberg, 1830)	9, 12, 14, 18

(continued)

Table 31.2 (continued)

Rotifer taxa	Location
Mytilina sp.	2a
Lophocharis oxysternon (Gosse, 1851)	1, 2a, 12
Lophocharis salpina (Ehrenberg, 1834)	3a, 5, 6, 9, 12, 18
Lophocharis naias Wulfert, 1942	5
Trichotriidae Harring, 1913	
Trichotria pocillum (Müller, 1776)	1, 2a, 5, 6, 12
Trichotria tetractis (Ehrenberg, 1830)	1, 2a, 3a, 5, 6, 9, 15
Trichotria tetractis similis (Stenroos, 1898)	5, 12, 18
Trichotria curta (Skorikov, 1914)	1
Macrochaetus altamirai (Arévalo, 1918)	12, 17
Macrochaetus collinsii (Gosse, 1867)	12
Macrochaetus subquadratus Perty, 1850	2a
Macrochaetus sp.	1, 2f, 5
Wolga spinifera (Western, 1894)	1
Trochosphaeridae Harring, 1913	
Filinia limnetica (Zacharias, 1893)	1, 5
Filinia brachiata (Rousselet, 1901)	6
Filinia longiseta (Ehrenberg, 1834)	1, 2a, 6, 15, 16, 17
Filinia terminalis (Plate, 1886)	1, 5, 6
Filinia cornuta (Weisse, 1847)	2a
Filinia sp.	1, 2a, 2e, 5, 6, 7, 8, 13
Collothecidae Harring, 1913	
Collotheca libera (Zacharias, 1894)	1
Collotheca campanulata (Dobie, 1849)	1, 2a
Collotheca mutabilis (Hudson, 1885)	1
Collotheca atrochoides (Wierzejski, 1893)	1
Collotheca heptabrachiata (Schoch, 1869)	5
Collotheca ornata (Ehrenberg, 1832)	1, 5, 6
Collotheca pelagica (Rousselet, 1893)	1, 5
Collotheca coronetta (Cubitt, 1869)	2a
Collotheca sp.	3b, 5, 7, 9, 13
Stephanoceros fimbriatus (Goldfusz, 1820)	1, 5
Proalidae Harring & Myers, 1924	
Proales reinhardti (Ehrenberg, 1834)	1
Proales minima (Montet, 1915)	5
Proales theodora (Gosse, 1887)	5, 12
Proales fallaciosa Wulfert, 1937	10
Proales decipiens (Ehrenberg, 1832)	2a
Proales sp.	2a
Conochilidae Harring, 1913	
Conochilus (*Conochiloides*) *dossuarius* Hudson, 1885	5
Conochilus hippocrepis (Schrank, 1803)	1

(continued)

Table 31.2 (continued)

Rotifer taxa	Location
Conochilus (*Conochilus*) *hippocrepis* (Schrank, 1803)	5
Conochilus unicornis Rousselet, 1892	1, 9, 17
Conochilus (*Conochiloides*) *coenobasis* (Skorikov, 1914)	1
Conochilus (*Conochiloides*) sp.	13
Epiphanidae Harring, 1913	
Epiphanes senta (Müller, 1773)	1, 5, 6, 10
Epiphanes brachionus (Ehrenberg, 1837)	2a
Epiphanes sp.	1, 2e
Proalides tentaculatus de Beauchamp, 1907	2a, 6
Proalides sp.	2a, 7, 8
Rhinoglena frontalis Ehrenberg, 1853	2a
Rhinoglena sp.	2a
Scaridiidae Manfredi, 1927	
Scaridium longicaudum (Müller, 1786)	1, 2a, 3a, 5, 6, 14, 15
Scaridium sp.	2a, 7
Gastropodidae Harring, 1913	
Gastropus stylifer Imhof, 1891	1
Gastropus minor (Rousselet, 1892)	1, 6
Gastropus hyptopus (Ehrenberg, 1838)	2a
Gastropus sp.	4
Ascomorpha ecaudis Perty, 1850	3a
Ascomorpha agilis Zacharias, 1893	5
Ascomorpha sp.	2a, 2e, 3b, 5, 7, 12, 13, 14
Lindiidae Harring & Myers, 1924	
Lindia (*Lindia*) *truncata* (Jennings, 1894)	6, 18
Lindia (*Lindia*) *janickii* Wiszniewski, 1934	6, 18
Lindia (*Lindia*) *torulosa* Dujardin, 1841	12
Ituridae Sudzuki, 1964	
Itura aurita (Ehrenberg, 1830)	1, 6
Itura chamadis Harring & Myers, 1928	2a
Itura viridis (Stenroos, 1898)	2a
Philodinidae Ehrenberg, 1838	
Philodina roseola Ehrenberg, 1832	5, 6, 10, 12, 15, 18
Philodina erythrophthalma Ehrenberg, 1830	2a
Philodina citrina Ehrenberg, 1832	5, 6
Philodina megalotrocha Ehrenberg, 1832	6
Philodina sp.	1, 2a, 2b, 2d, 2e, 2f, 3b, 4, 5, 6, 7, 11
Rotaria tardigrada (Ehrenberg, 1830)	5
Rotaria neptunia (Ehrenberg, 1830)	2a, 5, 6, 16, 17
Rotaria neptunoida Harring, 1913	6
Rotaria macrura (Ehrenberg, 1832)	5
Rotaria rotatoria (Pallas, 1766)	5, 6, 18

(continued)

Table 31.2 (continued)

Rotifer taxa	Location
Rotaria socialis (Kellicott, 1888)	5
Rotaria citrina (Ehrenberg, 1838)	10
Rotaria sp.	2a, 2c, 2d, 2e, 4, 5, 7, 8, 12
Dissotrocha macrostyla (Ehrenberg, 1838)	5
Dissotrocha aculeata (Ehrenberg, 1832)	2a, 6
Embata laticeps (Murray, 1905)	5
Macrotrachela quadricornifera Milne, 1886	5
Habrotrochidae Bryce, 1910	
Habrotrocha constricta (Dujardin, 1841)	5
Habrotrocha sp.	6
Philodinavidae Harring, 1913	
Philodinavus paradoxus (Muray, 1905)	5
Adinetidae Hudson & Gosse, 1886	
Adineta vaga (Davis, 1873)	10
Adineta sp.	4, 6
Bdelloidea (unspecified taxa)	9, 14, 15, 17

The scientific names are updated according to the global rotifer checklist provided by Segers (2007), Rotifer World Catalog (http://rotifera.hausdernatur.at/), and World Register of Marine Species (WORMS) (http://www.marinespecies.org). For *Brachionus plicatilis* species complex, the names are adapted to the recent revisions by Mills et al. (2017). Based on such considerations, some of the reported taxa with invalid scientific names are excluded. Numbers refer to the locations described in Table 31.1

tropica, *Platyias quadricornis*, *Lepadella patella*, *Colurella adriatica*, *Polyarthra vulgaris*, *Asplanchna brightwellii*, *A. priodonta*, *Euchlanis dilatata* and *Testudinella patina* (8 locations each). A large number of recorded rotifers (203 species and subspecies and a number of those identified at genus level) were each found only once or in a single location or water body. These include 81 taxa observed only in Caspian Sea water basin, 44 in Tehrān and 43 in West Azarbaijān provinces. More studies are required to elucidate to what degree the distribution of Iranian rotifer fauna is geographic-specific or is related to the environmental or climatic conditions. The rotifers have been sampled from various aquatic ecosystems comprising the small and large rivers, lakes, wetlands, temporary and permanent lagoons, pools, sinkholes and artificial reservoirs. The largest investigated water body is the Caspian Sea. Considering that the rotifers of some water systems have been reported at genus or higher levels (*e.g.*, 10 rotifer genera from Amirkolāyeh Wetland—Mohammadzadeh et al. 2009, 12 genera from Shādegān Wetland—Kholfe-Nilsaz 2009, 13 genera from Shoorābil Lake—Sabkara and Makaremi 2011, 8 genera from Lār Dam Lake—Salavatian et al. 2011, 4 rotifer families from Hoorolazim Wetlands—Papahn-Shooshtari et al. 2012, 11 genera from Aras Dam Lake—Sabkara and Makaremi 2013, 16 genera from Choghākhor Wetland—Fathi et al. 2015, and 19 genera from Arasbārān Dam Lake—Salavatian et al. 2016), these bodies of water seem to have a wealth of rotifer diversity that may substantially increase the number of Iranian rotifers.

While the Iranian provinces have been formed based mainly on the administrative or cultural considerations, many of them represent specific geography and climate. On the other hand, there may be similar conditions in distinct provinces. Similarly, the natural watershed boundaries of the country coincide neither with the administrative divisions nor with the climatic zones. As a result, shaping a robust framework for species distribution and biogeography based on the provincial divisions cannot be reasonable. Notwithstanding this, the provincial borders are used in order to locate the study sites and to provide a rough comparison of the extent of the studies countrywide. With regard to this, the highest rotifer diversity (234 taxa) has so far been reported from the northern provinces of Gilān, Māzandarān and Golestān, situated in the water basin of the Caspian Sea. These are followed by Tehrān and West Azarbaijān as the next most-intensively-investigated provinces with the highest numbers of rotifers reported (164 and 151 taxa, respectively). Although a considerable amount of data has so far been gathered on Iranian rotifer diversity in some locations, a vast portion of the country has not yet been searched for these ubiquitous creatures; that is, rotifers have been reported from only 17 out of the 31 provinces in Iran. For eight provinces, only a single report is available and two of these reports are older than half a century (Table 31.1 and Fig. 31.2). The investigated areas are located in varying latitudes with diverse climatic conditions. However, as can be seen in Fig. 31.2, they are mostly distributed in northern and western parts of the country where moderate to cold weather and semidry to humid climatic conditions are dominant and the average annual precipitation is higher than that of the eastern and central areas. Thus, while higher rotifer records from some parts of the country are primarily a consequence of more study efforts, they also reflect the suitability of environmental conditions and availability of aquatic habitats for diversification and colonization of the rotifers in those territories. Not unexpectedly, the rotifer-rich regions coincide with the distribution of major biodiversity hotspots in Iran (Farashi and Shariati-Najafabadi 2017). Nevertheless, given that the intensity of studies has also been very different in the investigated locations, the current list of Iranian rotifer taxa is undoubtedly very incomplete.

As the country is located at the crossroad of various biogeographies with diverse climatic conditions (Jowkar et al. 2016), it is no surprise that the representatives of various biogeographies are present in the rotifer inventories from Iran. While vast majority of rotifer taxa observed in Iran are cosmopolitan in nature, the taxa reported exclusively from single biogeographies across the globe (Segers 2007) have also been found in Iranian waters. These include the Nearctic (*Aspelta chorista*, *Cephalodella ablusa*, *C. abstruse*, *C. angusta*, *C. asarcia*, *C. astricta*, *C. collactea*, *C. compressa*, *C. conjuncta*, *C. compressa*, *C. conjuncta*, *C. derbyi*, *C. dixonnuttalli*, *C. eunoma*, *C. montana*, *C. mucosa*, *Euchlanis arenosa* and *Notommata pygmaea*), Neotropical (*C. tincaformis*, *Lecane broaensis* and *Lepadella* (*Lepadella*) *neglecta*) and Palearctic (*Brachionus asplanchnoides*, *B. ibericus*, *C. gibboides*, *C. mus*, *C. vittata*, *Encentrum walterkostei*, *L. inconspicua*, *L. paradoxa*, *Lepadella* (*Lepadella*) *costata*, *Notholca cinetura*, *Paradicranophorus aculeatus*, *Pseudoharringia similis*, *Ptygura tridorsicornis*, *Synchaeta cecilia* and *Trichotria curta*) rotifers. Nevertheless, no exclusively Australian, Antarctic, African, or Oriental rotifers have so far been reported from Iran (Table 31.2). Reihan-Reshteh and Rahimian (2014) suggested that multi-original nature of several Iranian rotifers is a result of the existence of natural bridges in the region which has facilitated the cross of organisms between the biogeographies.

From a regional perspective, except for India and Turkey, studies on rotifer biodiversity of the neighboring countries have been rather occasional. Because several water basins extend into and are shared among the adjacent countries, it would be beneficial to review the history of regional studies and the resulting rotifer inventories. A very early study on the plankton of the northern Caspian Sea was carried out by Tschugunoff (1921) who identified 92 zooplankton species, including 42 rotifers. Salahova et al. (2014) reported 20 rotifer species from 6 sampling sites on the Caspian Sea coasts in Azerbaijan Republic. *Trichocerca* (4 species), *Brachionus* (4 species) and *Synchaeta* (3 species) were the most abundant genera. They also cited a preceding survey, according to which, about 190 mesoplankton species live in the Caspian Sea, and among these, some 82 are in southern part of the Sea, with Rotifera being the most diverse mesoplankton group. A list of 145 rotifer species from Caspian Sea basin has been provided by Caspian Sea Biodiversity Project and almost all of them are from the northern freshwaters, mainly the Volga River territory (https://www.zin.ru/projects/caspdiv/caspian). Several rotifer species sampled in the south Caspian Sea are also freshwater residents originating from the associated water bodies, such as Anzali Wetland (Bagheri et al. 2013). While such species as *Trichocerca caspica*, *Trichocerca* (*Diurella*) *heterodactyla* and *Keratella tropica* f. *taurocephala* are marked as endemic to Caspian Sea (Segers and De Smet 2008, https://www.zin.ru), questions are raised over their taxonomic validity and endemicity. For instance, *T. caspica* and *T. heterodactyla* are deemed to be the erroneous alternatives of *T. marina* and *T. dixonnutalli*, respectively (Dr. H. Segers pers.com). It is assumed that more efforts have been done by the former Soviet Union and Russian scientists to evaluate the Caspian biological resources. However, as many of them have not been publicized or reported in English, they cannot be accessed in the online literature or universal data banks. Notwithstanding that a substantial fraction of the Caspian Sea invertebrates are yet to be discovered, a decline in their diversity and density has been noted. Climate as well as the hydrological changes, environmental degradation, and the invasive species are regarded as the main causes of the alteration in zooplankton communities in the south Caspian Sea (Bagheri et al. 2013, 2014; Rowshan-Tabari et al. 2014). India is a rather well-studied country in the region for its rotifer fauna. The first report on Indian rotifers belongs to the year 1889 (Sinha 2014). Since then, several studies have been conducted on their diversity (*e.g.*, Dhanapathi 1974; Arora and Mehra 2003; Sharma and Sharma 2005, 2011; Siddiqi and Karuthapandi 2013; Sharma 2017), resulting in the discovery of more than 490 rotifer species; which represents India as the host for the highest rotifer diversity in South-West Asia (Sinha 2014). While the records of rotifers for as old as the year 1903 are available from Turkey, the new surge of research on its rotifer fauna started in the 1980s (Dumont and De Ridder 1987;

Bekleyen 2001; Kaya and Altındağ 2007a,b; Kaya et al. 2007; Bekleyen et al. 2011; Ustaoğlu et al. 2012). The current list of rotifers from Turkey comprises 417 taxa (Ustaoğlu 2015). Considering that these taxa are recorded in 263 published literature, which are far more than the published reports of Iranian rotifers, if a comparison of rotifer discovery per sampling effort would be reasonable, rotifer diversity in Iran may be relatively higher than that of Turkey. A survey of water bodies in several locations in Turkey during the years 1992–1999 led to the detection of only 71 rotifer species (Altındağ and Yiğit 2001). Özdemir Mis and Ustaoğlu (2017) isolated 63 rotifer taxa by studying 59 high-mountain lakes in Northeast Turkey. Interestingly, the four most diverse rotifer genera in Iran and Turkey are the same and despite the much higher total rotifer records for Turkey, the number of species of its first four genera is similar to those for their Iranian congeners (*i.e.*, *Lecane* with 51, *Cephalodella* with 29, *Trichocerca* with 27 and *Brachionus* with 20 species and subspecies) (Ustaoğlu 2015). A further notable point in comparisons of the rotifer diversity in Iran and Turkey is that a higher diversity of *Brachionus* rotifers has been observed in the former country. There was no record of Rotifera from Arabian Peninsula until the late twentieth century. Segers and Dumont (1993) identified 118 monogonont rotifers from five Arabian states in south Persian Gulf. Similar to that for Turkey, the three most specious rotifer genera in the Peninsula and Iran are the same. While geographical proximity might explain such similarities, despite the diverse climates in most Iranian territories, Arabian Peninsula has almost constant climatic conditions year-round with high temperature and evaporation rates and most likely, the aquatic habitats with higher water salinities. According to the discussion provided by Aloufi and Obuid-Allah (2014), in spite of a considerable body of studies taken place in Saudi Arabian waters, they support a rather low zooplankton, including rotifer, diversity. Research on the rotifers in Iraq started more recently. Rabee (2010) and Abd Al-Rezzaq et al. (2014) reported 32 and 128 rotifer taxa from Iraqi waters, respectively. Abdulwahab and Rabee (2015) found 65 rotifer taxa from the Tigris River. Ahmed and Ghazi (2014) and Hammadi et al. (2016) identified 26 and 99 rotifer species, respectively, from Shatt al-Arab, a conjoined river of Tigris and Euphrates flowing into the Persian Gulf. Rotifers have also been considered in some ecological studies in Iraq (Al-Saboonchi et al. 2012; Salman et al. 2014). The external and internal conflicts during the last two decades may have been a reason that only a limited part of Iraq has been limnologically investigated. Reihan-Reshteh and Rahimian (2014) found a high similarity between the rotifer fauna of Khuzestān Province, Southwest Iran, and those of the Arabian waters and considered it as a consequence of the geological events connecting South Iranian Plateau to Arabian Peninsula. Reports on rotifers from Pakistan are scant, but its rotifer diversity appears to be worth noticing (Baloch and Soomro 2004; Sahato and Lashari 2004; Sulehria and Malik 2013; Hussain et al. 2016). While a majority of reports on regional rotifer diversity have been published in domestic or mediocre journals and are not backed up by detailed illustrations or proof of the identifications, they can still reflect the existence of a wide variety of rotifers in this part of the world, which are worthy of being accounted in global biodiversity estimations. Regardless of extent of the studies, a brief comparison of the rotifer diversity in the Near- and Middle Eastern countries testifies the high species richness of the Iranian rotifer fauna. It should also be noted that despite the timely literature review to provide the most complete list of the rotifers, there might still exist more records under university theses or locally-provided project reports which have not been inseminated. Thus, the number of identified rotifer species from Iran could exceed the present counts.

Considering that there may be more studies underway on Iranian rotifers, based on the current data, only a small proportion of the country's area has been searched for the rotifers and many pristine areas exist which are yet to be studied (Fig. 31.2). Thus, Iran can be considered as a virgin source and a potential hotspot of rotifer biodiversity, and it is predictable that many more rotifer taxa would be discovered and introduced from the country and its conjugated water basins in the future. Furthermore, Iranian marine rotifers, *i.e.*, those living in Persian Gulf and Sea of Oman, are left to be investigated and identified, holding promise for the observation of higher and exclusive rotifer diversity.

Despite acceleration of the research on Iranian rotifers in recent years, several factors may have contributed to the overall scarcity of studies and information. They include the limited number of skilled scientists, as well as difficulties to inspect small-sized and highly diverse rotifer groups. These are more effective when the more complex forms are to be examined. For instance, bdelloid rotifers can, at present, be identified only while alive and need to be examined during feeding and creeping (Segers 2008). This might be a reason why not many of them are among the rotifers recorded from all over the world, including Iran. Furthermore, while a considerable share of the introduced taxa are reported at genus or higher levels, it is not clear whether all of those at species or subspecies level have been identified accurately, identified, as no complementary approach, *e.g.*, using electron microscopy and molecular techniques, has been applied to confirm the identities. Thus, although this review discloses a fairly high diversity of rotifers discovered from Iranian mainland waters, the number of recorded rotifer species should be considered with caution, as rotifer identification at species level is mostly confusing and for many rotifer species requires detailed infrastructural examination of body sections such as their trophi (Segers 2008), assessment of reproductive isolation (Snell 1989), and application of genetic markers (Fontaneto 2014); neither of which has been provided in the records of Iranian rotifers, except for few morphological and molecular studies (Malekzadeh-Viayeh 2010b; Kordbacheh and Rahimian 2012; Malekzadeh-Viayeh et al. 2014; Mills et al. 2017). Moreover, it has been proven that for many rotifer species, several cryptic taxa exist within a single species (Leasi and Norenburg 2014). Therefore, future investigations on rotifer diversity in Iran must be assisted by precise ultrastructural and molecular inspections to provide an actual estimation of the diversity. Further studies should also address the relationships between environmental variables and the rotifer diversity and the question of possible rotifer dispersal across the country naturally or by human, birds or other animal vectors, as well as the historical affinity of the rotifers from water basins of the neighboring countries through genetic analyses.

A notable obstacle for further rotifer surveys in Iran is the consistent long-term drought, which has impacted the country during the last decade, shrinking the large and completely drying the small water bodies. Rotifers are among the few animal entities that produce dormant/resting eggs during their life cycle. These eggs can retain their viability for several decades if embedded within the sediments in oxygen-free conditions (Hairston 1996; Brendonck and De Meester 2003). Maximum age of the hatchable zooplankton eggs is estimated to be 332 year (Hairston et al. 1995). Piscia et al. (2012) found that *B. calyciflorus* resting eggs as old as ca. 100 years were still viable. Thus, optimal rotifer survey for a more thorough recording of Iranian rotifer diversity may require recovery of the dormant eggs from the dried water-systems beds and their identification following the production of laboratory cultures.

References

Abbasi M (2016) A faunistic study of Rotifera from Shazand, Arak, Iran. MSc Thesis in Animal Biosystematics. Arak University, 130pp (in Farsi)

Abd Al-Rezzaq AJ, Salman SD, Saleh MM (2014) Rotifera of the Al-Hilla River, Iraq. Int J Adv Biol Res 4(2):158–164

Abdulwahab S, Rabee AM (2015) Ecological factors affecting the distribution of the zooplankton community in the Tigris River at Baghdad region, Iraq. Egypt J Aquat Res 41(2):187–196

Abedi E (2015) The zooplankton of Bardestan Creek in the Persian Gulf. J Pers Gulf (Mar Sci) 6 (20):21–28

Abedi E, Arebi E, Sepahvand V et al (2014) Zooplankton community composition in the north-western part of the Persian Gulf during PG-GOOS oceanographic cruise. J Pers Gulf (Mar Sci) 5 (18):17–26

Abich H (1856) Vergleichend chemische Untersuchugen des Wassers des Kaspischen Meeres, Urmia- und Vansees. Mem Acad St Petersb 6:9

Ahmadifard N (2007) Effects of different algal types and concentrations on growth and fatty acids contents of the freshwater rotifer, *Brachionus calyciflorus*. MSc Thesis, Tarbiat Modarres University, Tehran, Iran (in Farsi)

Ahmed HK, Ghazi AHH (2014) Rotifer diversity at Shatt Al-Arab River and Shatt Al-Basrah canal, south of Iraq, during the abnormal rising of water salinity. Mesopot J Mar Sci 29(2):145–154

Alavi N, Bani A, Bagheri S (2019) Species inventory of Rotatoria in the Anzali wetland, Iran. Iran J Fish Sci 18(4):1060–1067

Aloufi ABA, Obuid-Allah AH (2014) Zooplankton communities inhabiting inland waters of Al-Madinah Almonawwarah region. Saudi Arabia Int J Adv Res 2(7):1137

Al-Saboonchi A, Abid HS, Alobaidy AMJ, Maulood BK (2012) Assessment of environmental changes in the Iraqi marshes by index of biological integrity. J Environ Prot 3:681–688

Altındağ A, Yiğit S (2001) A short list of rotifers from Turkey. Zool Middle East 22(1):129–132

Ansari S (2013) A faunistic survey of the rotifers of Arak, Iran. MSc thesis in Animal Biosystematics. Arak University, 170pp (in Farsi)

Ansari S, Shayestehfar A, Talebi SM (2013) Population variations of the species of *Lepadella patella* based on phenotype plasticity in Arak County, Iran. Sci Vis 13(4):162–168

Ansari S, Shayestehfar A, Talebi SM (2014) The inter and intrapopulation variations in two species of genus *Brachionus* (Rotifera), in Iran: an ecological review. Eur J Exp Biol 4(1):420–430

Arora J, Mehra NK (2003) Species diversity of planktonic and epiphytic rotifers in the backwaters of the Delhi segment of the Yamuna River, with remarks on new records from India. Zool Stud 42(2):239–247

Asian Development Bank (2010) Central Asia atlas of natural resources, 211 pp

Athibai S, Segers H, Sanoamuang L (2013) Diversity and distribution of Brachionidae (Rotifera) in Thailand, with a key to the species. J Limnol 72(2s):345–360

Bagheri S, Sabkara J, Mirzajani A, Khodaparast H, Yosefzad E, Yeok FS (2013) List of zooplankton taxa in the Caspian Sea waters of Iran. J Mar Biol. https://doi.org/10.1155/2013/134263
Bagheri S, Niermann U, Mansor M, Yeok FS (2014) Biodiversity, distribution and abundance of zooplankton in the Iranian waters of the Caspian Sea off Anzali during 1996–2010. J Mar Biol Assoc U K 94(1):129–140
Bagheri S, Sabkara J, Yousefzad E, Zahmatkesh Y (2017) Ecological study of zooplankton communities in the Persian Gulf Martyrs Lake (Chitgar-Tehran) and the first report of the freshwater jellyfish *Craspedacusta* sp. (Cnidaria, Limnomedusae) in Iran. Sci J Iran Fish 25 (5):113–128. (in Farsi)
Bagheri S, Sabkara J, Makaremi M (2018) Ecological study of zooplankton in Kan River. J Mar Sci Technol Res 13(1):41–52. (in Farsi)
Baloch WA, Soomro AN (2004) Planktonic rotifers occurring in the adjacent areas of Hyderabad. Pak J Zool 36(4):319–322
Beaumont P (1982) The hydrology and water resources of the Iranian plateau. Encyclopedia Iranica I/1, pp 28–39. Available from http://www.iranicaonline.org/articles/ab-iii-the-hydrology-and-water-resources-of-the-iranian-plateau
Bekleyen A (2001) A taxonomical study on the Rotifera fauna of Devegeçidi Dam Lake (Diyarbakır-Turkey). Turk J Zool 25:251–255
Bekleyen A, Gökot B, Varol M (2011) Thirty-four new records and the diversity of the Rotifera in the Turkish part of the Tigris River watershed, with remarks on biogeo-graphically interesting taxa. Sci Res Essays 6(30):6270–6284
Birshtain JA, Vinogradov LT, Kondakov NN et al. (1968) The atlas of Caspian Sea invertebrates. Translated by Dlinad L. and Nazari F. IFRO Publications, 610 pp (In Farsi)
Brendonck L, De Meester L (2003) Egg banks in freshwater zooplankton: evolutionary and ecological archives in the sediment. Hydrobiologia 491:65–84
Clark NE, Lovell R, Wheeler BW et al (2014) Biodiversity, cultural pathways, and human health: a framework. Trends Ecol Evol 29(4):198–204
Coad BW (1998) Systematic biodiversity in the freshwater fishes of Iran. Ital J Zool 65 (S1):101–108
Coad BW (2006) Endemicity in the freshwater fishes of Iran. Iran J Anim Biosyst 1(1):1–13
De Ridder M (1986) Annotated checklist of nonmarine Rotifera from African inland waters. Koninklijk Museum voor Midden Afrika, Tervuren, Zoologische Documentatie 21:123
Derenko M, Malyarchuk B, Bahmanimehr A et al (2013) Complete mitochondrial DNA diversity in Iranians. PLoS One 8(11):e80673. https://doi.org/10.1371/journal.pone.0080673
Dhanapathi MVSSS (1974) Rotifera from Andhra Pradesh, India. Hydrobiologia 45:357–372
Duggan IC, Green JD, Shiel RJ (2001) Distribution of rotifers in North Island, New Zealand, and their potential use as bioindicators of lake trophic state. Hydrobiologia 446/447:155–164
Dumont HJ, De Ridder M (1987) Rotifers from Turkey. Hydrobiologia 147(1):65–73
Ebrahimi-Dorche E, Zare-Shahraki M, Farhadian O, Keivany Y (2018) Seasonal variations of plankton structure as bioindicators in Zayandehrud Dam Lake, Iran. Limnol Rev 18(4):157–165
Esmaeili HR, Mehraban HR, Abbasi K et al (2017) Review and updated checklist of freshwater fishes of Iran: taxonomy, distribution and conservation status. Iran J Ichthyol 4(Suppl. 1):1–114
Fallah B, Sodoudi S, Russo E et al (2017) Towards modeling the regional rainfall changes over Iran due to the climate forcing of the past 6000 years. Quat Int 429:119–128
Fallahi M, Sabkara J (2015) The study of zooplankton population structure in Anzali wetland. Iran Sci Fish J 24(2):29–42. (In Farsi)
Fallahi-Kaporchali M, Shapoori M, Mansouri S (2015) The identification and study of density of Rotatoria in Anzali wetland in comparison with estuary region and Caspian Sea. J Wetl Ecobiol 7(3):21–32
Farashi A, Shariati-Najafabadi M (2017) Biodiversity hotspots and conservation gaps in Iran. J Nat Conserv 39:37–57

Farashi A, Kaboli M, Rezaei HR et al (2014) Plankton composition and environmental parameters in the habitat of the Iranian cave barb (*Iranocypris typhlops*) in Iran. Anim Biodivers Conserv 37 (1):13–21

Farhadian O, Pouladi M (2014) Seasonal changes in the abundance and biomass of zooplankton from shallow mudflat river-estuarine system in Persian Gulf. Braz J Aquat Sci Technol 18 (2):19–29

Farhadian O, Daghighi L, Ebrahimi-Dorche E (2015) Growth and production of freshwater rotifer *Euchlanis dilatata* (Monogononta: Euchlanidae) fed on different diets. J Anim Res 28 (2):199–209. (in Farsi)

Fathi P, Ebrahimi E, Farhadian O et al (2015) Seasonal trophy assessment in Choghakhor wetland based on zooplankton communities. J Anim Res 28(3):371–382. (in Farsi)

Fazeli N, Savari A, Nabavi MB, Zare R (2013) Seasonal variation of zooplankton abundance, composition and biomass in the Chabahar Bay, Oman Sea. Int J Aquat Biol 1(6):294–305

Fontaneto D (2014) Molecular phylogenies as a tool to understand diversity in rotifers. Int Rev Hydrobiol 99(1–2):178–187

Fontaneto D, De Smet W (2015) Rotifera. In: Schmidt-Rhaesa A (ed) Handbook of zoology, gastrotricha, cycloneuralia and gnathifera, Vol 3: Gastrotricha and Gnathifera. de Gruyter, Berlin, pp 217–300

Fontaneto D, De Smet WH, Melone G (2008) Identification key to the genera of marine rotifers worldwide. Meiofauna Marina 16:75–99

Gerami MH, Patimar R, Negarestan H, Jararian H, Mortazavi MS (2016) Temporal variability in macroinvertebrates diversity patterns and their relation with environmental factors. Biodiversitas 17(1):36–43

Ghane-Sasansaraie A (2007) Limnological survey of three major rivers in southern coast of Caspian Sea in Guilan Province (Haviq, Karganrud and Shafarud rivers) with an emphasize on the pollutants. Final Project Report. Iranian Fisheries Science Research Institute, 158pp (In Farsi)

Ghasemi S, Mola-Hoveizeh N, Zakaria M et al (2012) Relative abundance and diversity of waterbirds in a Persian Gulf mangrove forest. Iran Trop Zool 25(1):39–53

Ghorbani M (2013) Nature of Iran and its climate, Chapter 1. In: The economic geology of Iran: mineral deposits and natural resources. Springer, New York, pp 1–44

Golmarvi D, Fallahi-Kapourchali M, Mashinchian-Moradi A et al (2017) Influence of physico-chemical factors, zooplankton species biodiversity and seasonal abundance in Anzali International Wetland. Iran Open J Mar Sci 7:91–99

Gutkowska A, Paturej E, Kowalska E (2013) Rotifer trophic state indices as ecosystem indicators in brackish coastal waters. Oceanologia 55(4):887–899

Hairston NG Jr (1996) Zooplankton egg banks as biotic reservoirs in changing environments. Limnol Oceanogr 41(5):1087–1092

Hairston NG Jr, Van Brunt RA, Kearns CM, Engstrom DR (1995) Age and survivorship of diapausing eggs in a sediment egg bank. Ecology 76(6):1706–1711

Hakimzadeh-Khoei M, Kaya M, Altindag A (2011) New records of rotifers from Iran with biogeographic considerations. Turk J Zool 35:395–402

Hammadi NS, Salman DS, Al-Essa SA (2016) Diversity of Rotifera in the Shatt Al-Arab region, south of Iraq. J Fish Sci

Hussain A, Sulehria AQ, Ejaz M, Maqbool A (2016) Population dynamics of rotifers in the floodplain of River Ravi, Pakistan. Pak J Zool 48(1):215–225

Jafari N, Nabavi SM, Akhavan M (2011) Ecological investigation of zooplankton abundance in the river Haraz, Northeast Iran: impact of environmental variables. Arch Biol Sci Belgrade 63 (3):785–798

Jamshidi S, Abu Bakar NB (2012) Seasonal variations in temperature, salinity and density in the southern coastal waters of the Caspian Sea. Oceanology 52(3):380–396

Jersabek CD (2003) Freshwater Rotifera (Monogononta) from Hawai'i—a preliminary checklist. In: Evenhuis NL, Eldredge LG (eds) Records of the Hawaii Biological Survey for 2001–2002 - part II: notes, Bishop museum occasional papers, vol 74. Springer, Dordrecht, pp 46–72

Jowkar H, Ostrowski S, Tahbaz M, Zahler P (2016) The conservation of biodiversity in Iran: threats, challenges and hopes. Iran Stud 49(6):1065–1077

Kamali-Sansiqi M, Mousavi-Nadooshan R, Qelichi A (2014) Diversity and abundance of rotifers in earthen fish ponds of Gonbade Kavous suburb, East Golestan, Iran. J Aquat Ecol 4(1):7–17. (in Farsi)

Kaya M, Altındağ A (2007a) A taxonomic study on the families Lepadellidae and Trichocercidae (Rotifera: Monogononta) of Turkey. Chin J Oceanol Limnol 25(4):423–426

Kaya M, Altındağ A (2007b) Brachionidae (Rotifera: Monogononta) species from Turkey. Asian J Anim Sci 1:40–47

Kaya M, Yiğit S, Altındağ A (2007) Rotifers in Turkish inland waters. Zool Middle East 40:71–76

Kehl M (2009) Quaternary climate change in Iran - the state of knowledge. Erdkunde 63(1):1–17

Khafaeizadeh K, Sakhaei N, Doostshenas B, Ghanemi K, Zolqaranein H (2017) Biodiversity of planktonic rotifers along a horizontal salinity gradient in the Bahmanshir River estuary (northern Persian gulf). J Anim Environ 9(3):401–410. (in Farsi)

Khaleqsefat E, Pashaei-Rad S, Malekzadeh-Viayeh R (2011) Lecanid rotifers (Rotifera: Monogononta: Lecanidae) from Iran. Turk J Zool 35:49–55

Khodaparast N, Roohi A, Shahpoori M, Rowshan-Tabari M, Rahmati R, Nasrollahzadeh-Saravi H, Vahedi F (2012) A study on the species composition and population structure of zooplanktons at various depths of South Caspian Sea following the invasion by the Ctenophore *Mnemiopsis leidyi*. J Mar Biol 4(15):12. (in Farsi)

Kholfe-Nilsaz M (2009) Planktonic frequency and biodiversity survey of Shadegan wetland and determination of its trophic status. J Mar Biol 1(3):1–13

Kordbacheh A, Rahimian H (2012) Annotated checklist of rotifers of Tehran Province, Iran, with notes on new records. Prog Biol Sci 2:59–67

Kostianoy AG, Kosarev AN (2005) The Caspian Sea environment. Springer, New York, 277pp

Kriska G (2014) Freshwater invertebrates in Central Europe: a field guide. Springer, New York, 410pp

Leasi F, Norenburg JL (2014) The necessity of DNA taxonomy to reveal cryptic diversity and spatial distribution of meiofauna, with a focus on Nemertea. PLoS One 9(8):e104385. https://doi.org/10.1371/journal.pone.0104385

Löffler H (1949/1950) Limnologische Beobachtungen an Iranischen Binnengewassern. II. Teil: systematische beschreibung der rotatorien und crustaceen aus den untersuchten gewassern

Löffler H (1953) Limnologische Ergebnisse der österreichischen Iranexpedition 1949/50. Naturwissenschaft Rundsch 2:64–68

Löffler H (1961) Beitrage zur kenntnis der Iranischen binnengewasser II. Regional-limnologische studie mit besonderer Berucksichtigung der crustaceenfauna. Int Revue Ges Hydrobiol 46:309–406

Löffler H (1981) The winter condition of Lake Niriz in southern Iran. Verh Internat Verein Limnol 21:528–534

Lubzens E, Tandler A, Minkoff G (1989) Rotifers as food in aquaculture. Hydrobiologia 186/187:387–400

Madani K, Agha KA, Mirchi A (2016) Iran's socio-economic drought: challenges of a water-bankrupt nation. Iran Stud 49(6):997–1016

Mahdizadeh GR, Ahmadi MR, Saberi H, Kiabi B, Vosougi GH (2006) A study on the distribution and abundance of zooplankton in the fish earthen ponds of Lakān District, Gilān. J Mar Sci Technol 5(3–4):77–85. (in Farsi)

Malekzadeh-Viayeh R (2010a) An overview of rotifers of the family Notommatidae (Rotifera: Monogononta: Ploima) from Iran. Caspian J Env Sci 8:127–139

Malekzadeh-Viayeh R (2010b) Biodiversity and ecology of rotifers from West Azarbaijan and Hormozgan provinces, Iran. Final Project Report. Artemia and Aquatic Research Institute, Urmia University, 273pp

Malekzadeh-Viayeh R, Špoljar M (2012) Structure of rotifer assemblages in shallow waterbodies of semi-arid Northwest Iran differing in salinity and vegetation cover. Hydrobiologia 686:73–89

Malekzadeh-Viayeh R, Pak-Tarmani R, Rostamkhani N, Fontaneto D (2014) Diversity of the rotifer *Brachionus plicatilis* species complex (Rotifera: Monogononta) in Iran through integrative taxonomy. Zool J Linnean Soc 170:233–244

Marcet A (1819) Lake Ourmia, or Urumea, in Persia. Ann Philos 14:150–151

Mills S, Alcántara-Rodríguez JA, Ciros-Pérez J et al (2017) Fifteen species in one: deciphering the *Brachionus plicatilis* species complex (Rotifera, Monogononta) through DNA taxonomy. Hydrobiologia 796(1):39–58

Mirzaie-Nadowshan H (2009) Climate variation causes genetic variation in Iranian rangeland plant species. In: Kharazipour AR, Schöpper C, Müller C, Euring M (eds) Review of forests, wood products and wood biotechnology of Iran and Germany - Part III. Universitätsverlag Göttingen, Göttingen, pp 149–161

Modenutti BE (1998) Planktonic rotifers of Samborombón River basin (Argentina). Hydrobiologia 387:259–265

Mohammadi M, Reihan-Reshteh R (2015) A faunistic study on the rotifers of Tehran, Iran with some new records. J Anim Environ 7(2):143–152. (in Farsi)

Mohammadzadeh M, Nezami-Balouchi SA, Keivan A, Khara H (2009) Diversity and spatio-temporal abundance of various zooplanktonic groups in Amirkolayeh wetland, Lahijan, Iran. J Biol Sci Lahijan Azad Univ 3(2):61–69. (in Farsi)

Mohsenpour-Azari A (2017) Effects of environmental factors on growth parameters of the Aras freshwater crayfish. Final Project Report, Iranian Fisheries Science Research Institute, 105pp (in Farsi)

Mokhayer Z, Mousavi-Nadushan R, Rabbaniha M, Fatemi MR, Jamili S (2017) Community composition and diversity of zooplankton in the northwest Persian gulf. Iran J Fish Sci 16 (2):722–732

Moradi M, Malekzadeh-Viayeh R, Eshaghi-Rad J (2014) Biodiversity of scleractinian corals in the reefs of Qeshm and Larak Islands of the Persian Gulf, in association with environmental variables. J Mar Biol Assoc U K 94(5):907–916

Murray J (1913) South American Rotifera, Part II. J Microscopy 33(4):341–362

Özdemir Mis D, Ustaoğlu MR (2017) Distribution of rotifers of high mountain lakes in the eastern Black Sea range of Turkey. Turk J Zool 41:674–685

Papahn-Shooshtari F, Dehqan S, Baqeri R (2012) A study of the plankton composition and abundance in Hoorolazim Wetland in Khuzestan Province. J Wetl Ecobiol 3(10):6. (in Farsi)

Pejler B (1998) History of rotifer research in northern Europe. Hydrobiologia 387:1–8

Piscia R, Guilizzoni P, Fontaneto D et al (2012) Dynamics of rotifer and cladoceran resting stages during copper pollution and recovery in a subalpine lake. Int J Limnol 48:151–160

Pouladi M, Farhadian O, Vazirizadeh A (2013) Zooplankton composition, abundance and biomass in Helle River estuary, Persian gulf coast in Bushehr Province. J Fish (Iran J Natl Res) 66 (3):255–270. (in Farsi)

Rabee AM (2010) The effect of Al-Tharthar-Euphrates canal on the quantitative and qualitative composition of zooplankton in Euphrates River. J Al-Nahrain Univ 13(3):120–128

Reihan-Reshteh R, Rahimian H (2014) Rotifers of Southwest Iran: a faunistic and biogeographical study. Turk J Zool 38:525–537

Rezaei S, Dana LP, Ramadani V (2017) Iranian entrepreneurship: deciphering the entrepreneurial in Iran and in the Iranian diaspora. Springer, Cham, 374pp

Rowshan-Tabari M, Takmilian K, Sabkara J, Roohi A, Rostamian MT (2003) A study on the distribution of zooplankton in South Caspian Sea. Iran Sci Fish J 3:83–96. (in Farsi)

Rowshan-Tabari M, Fatemi SMR, Pourgholam R, Mousavi-Nadoushan R (2014) Latest diversity trend and seasonal abundance of population of zooplankton (holoplankton) communities in the South Caspian Sea, Iran. Iran J Fish Sci 13(2):437–448

Rufchaie R, Fallahi-Kaporchali M, Parvaneh D, Chubian F (2015) Effects of enriched freshwater rotifer with vitamin C on growth indices of Persian sturgeon (*Acipenser persicus*) larvae rearing. Iran Sci Fish J 24(1):139–144. (in Farsi)

Ruttner WA, Ruttner-Kolisko A (1972) Some data on the hydrology of the Tabas-Shirgesht-Ozbak-kuh area (East Iran). Jahrb Geol J BA (Wien) 115:1–48
Ruttner WA, Ruttner-Kolisko A (1973) The chemistry of springs in relation to the geology in an arid region of the Middle East (Khurasan, Iran). Verh Jnternat Verein J Limnol 18:1751–1752
Ruttner-Kolisko A (1966) The influence of climatic and edaphic factors on small astatic waters in the east Persian salt desert. Verh Internat Verein Limnol 16:524–531
Ruttner-Kolisko A (1974) Plankton rotifers: biology and taxonomy. English translation of Die Binnengewasser 26, part 1, 146 pp
Ruttner-Kolisko A (1980) Rädertiere aus Iran und ihre Beziehung zum Chemismus des Wassers. Jber Biol Station Lunz 3:155–162
Sabkara J (2019) Diversity and distribution of various zooplanktonic groups in Anzali wetland and the Caspian Sea coastal waters. Q J Ecol Water Resour 2:61–76. (in Farsi)
Sabkara J, Makaremi M (2003) A study on the diversity and density of zooplankton in Maku Dam Lake. Iran Sci Fish J 12(2):29–46. (In Farsi)
Sabkara J, Makaremi M (2004) Abundance and distribution pattern of planktons in Anzali lagoon. Iran Sci Fish J 13:87. (In Farsi)
Sabkara J, Makaremi M (2009) Diversity and distribution of rotifers of the family Brachionidae from Anzali wetland. In: The First National Symposium on Iranian Wetlands, Ahvaz, Iran (in Farsi)
Sabkara J, Makaremi M (2011) Diversity and abundance of plankton and their role in Shoorābil Lake ecosystem. J Biol Sci Lahijān Azad Univ 5(1):31–46. (in Farsi)
Sabkara J, Makaremi M (2013) Diversity and abundance of planktons and their effects on fish culture in Aras Dam Lake. J Aquacult Dev 7(2):41–59. (in Farsi)
Sabkara J, Makaremi M (2018) Morphology, diversity and distribution of the phylum Rotifera (Rotatoria) in different parts of Anzali wetland. J Biol Biotechnol Sci Stud 4(1):38–54. (in Farsi)
Sabkara J, Makaremi M, Mohammadjani T (2006) Density and distribution of the plankton in Karganrud River. Pajouhesh Sazandegi 73:65–73. (in Farsi)
Sahato GA, Lashari KH (2004) Diversity and seasonal occurrence of planktonic rotifers in Keenjhar Lake district Thatta Sindh, Pakistan. Pak J Biotechnol 1(2):28–32
Salahova SZ, Topchiyeva SA, Alakbarov IK (2014) Zooplankton community of the coastal zone of Azerbaijan sector of the Caspian Sea. J Appl Environ Biol Sci 4(5):141–146
Salavatian M, Abdollahpour-Biria H, Nezami-Baloochi S et al (2011) Temporal and spatial variations in density and distribution of zooplankton communities in Lār dam lake. J Biol Sci Lahijan Azad Univ 5(1):47–59. (in Farsi)
Salavatian M, Abedini A, Sabkara J (2016) Diversity and abundance of zooplankton in Arasbārān Dam Lake. J New Technol Aquacult Dev 10(1):1–10. (in Farsi)
Salman SD, Abbas MF, Ghazi AHM et al (2014) Seasonal changes in zooplankton communities in the re-flooded Mesopotamian wetlands, Iraq. J Freshwater Ecol 29(3):397–412
Sarma SSS (1999) Checklist of rotifers (Rotifera) from Mexico. Environ Ecol 17:978–983
Sarma SSS (2006) International Rotifer Symposia: prospects and retrospects from Rotifera XI. Saline Syst 2:6. https://doi.org/10.1186/1746-1448-2-6
Segers H (2007) Annotated checklist of the rotifers (phylum Rotifera), with notes on nomenclature, taxonomy and distribution. Zootaxa 1564:1–104
Segers H (2008) Global diversity of rotifers (phylum Rotifera) in freshwater. Hydrobiologia 595:49–59
Segers H, De Smet WH (2008) Diversity and endemism in Rotifera: a review, and *Keratella* Bory de St Vincent. Biodivers Conserv 17(2):303–316
Segers H, Dumont HJ (1993) Rotifers from Arabia, with description of two new species. Fauna Saudi Arabia 13:3–26
Seyyedi-Abalvan M, Ghaffari M, Gharaei A (2015) The effect of nickel (Ni) on the growth and population density of freshwater rotifer *Brachionus calyciflorus*. J Anim Res 28(2):223–232. (in Farsi)

Sharifinia M (2015) Macroinvertebrates of the Iranian running waters: a review. Acta Limnol Bras 27(4):356–369

Sharma BK (2017) First report of freshwater rotifers (Rotifera: Eurotatoria) from south Andaman, India: composition and interesting elements. J Asia Pacific Biodivers 10(2):261–266

Sharma BK, Sharma S (2005) Biodiversity of freshwater rotifers (Rotifera, Eurotatoria) from North-Eastern India. Zoosyst Evol 81(1). https://doi.org/10.1002/mmnz.200310002

Sharma BK, Sharma S (2011) Faunal diversity of rotifers (Rotifera: Eurotatoria) of Nokrek Biosphere Reserve, Meghalaya, India. J Threat Taxa 3(2):1535–1541

Shayestehfar A, Abdovis S (2011) Diurnal fluctuations in population density of rotifera in relation to some physical and chemical parameters from Karun River, Khuzestan Province, Iran. Agric Sci Res J 1:272–276

Shayestehfar A, Soleimani M, Mousavi SN, Shirazi F (2008) Ecological study of rotifers from Kor River, Fars, Iran. J Environ Biol 29:715–720

Shiel RJ (1995) A guide to identification of rotifers, cladocerans and copepods from Australian inland waters. C.R.C.F.E. Identification guide, 3, Murray-Darling Freshwater Research Centre, Albury

Siddiqi SZ, Karuthapandi M (2013) A report on Lecanidae (Rotifera: Monogononta) from Andhra Pradesh, India, including six new distribution records with notes on their contemporary taxonomic nomenclature. J Threat Taxa 5(11):4556–4561

Sinaei M, Ashrafali H, Blooki M (2017) The first report on the identification of planktonic communities from Sarbaz River in Sistan and Baluchestan Province. J Environ Sci Technol 19(4):617–627. (in Farsi)

Sinha B (2014) Annotated checklist of Indian Rotifera. ZSI e-publication. Zoological Survey of India, Kolkata, 36 pp

Smith DG (2001) Rotifera (wheel animals), Chapter 7. In: Smith DG (ed) Pennak's freshwater invertebrates of the United States. Wiley, New York, pp 129–190

Snell TW (1989) Systematics, reproductive isolation and species boundaries in monogonont rotifers. Hydrobiologia 186(1):299–310

Suantika G, Dhert P, Sweetman E et al (2003) Technical and economical feasibility of a rotifer recirculation system. Aquaculture 227:173–189

Sulehria AQ, Malik MA (2013) Diversity indices of pelagic rotifers in camp Balloki Water Park, Lahore, Pakistan. Turk J Zool 37:699–705

Thorp J, Covich A (2009) Ecology and classification of North American freshwater invertebrates, 3rd edn. Academic Press, New York. 1021pp

Tschugunoff NL (1921) Über das Plankton des nördlichen Teiles des Kaspisees. Arb Wolgastat 8:107–162

Ustaoğlu MR (2015) An updated zooplankton biodiversity of Turkish inland waters. J Limnol Freshwater Fish Res 1(3):151–159

Ustaoğlu MR, Altindağ A, Kaya M et al (2012) A checklist of Turkish Rotifera. Turk J Zool 36:607–622

Chapter 32
Biodiversity of the Freshwater Amphipods in Iran

Mehrdad Zamanpoore

32.1 Introduction

Biological diversity of organisms in any given region is affected by many factors particularly the ecology and special geography of that region. The country of Iran lies in the dry belt of Asia, as about 80% of its area has a dry or semidry climate (Breckle 2002). Extensive parts of these areas are actually hot deserts of the Central Plain (Fig. 32.1). The prompt image that comes to any mind, when talking about Iran, is a through a dry desert with none or little amount of surface waters. As a result, when inland aquatic fauna is discussed, the logical proposition would be that one should not expect high biodiversity in Iran as it seems that there are not many aquatic habitats available. As a matter of fact, this is not true. Except for the green margin of South Caspian shore, which has a wet Mediterranean climate, a considerable percentage of lands covering Iran is mountainous, mainly in two great mountain chains of the Zagros and the Alborz (Fig. 32.1), with numerous sources of surface waters from small springs and brooks to long rivers (e.g., Hirmand, Harirud, Aras, etc.) and their endpoint wetlands. There are 24 recorded Ramsar sites in the country with 1,486,438 hectares of area (The Ramsar Convention 2018), and still many others are in the line to be accepted in the list. These water bodies are habitats to a diverse range of aquafauna from microscopic planktons to macroinvertebrates and fishes.

Drastic differences of elevation among lands have caused high climatic disparity among different parts of Iran, which in turn causes variation between populations of the biota. Recent research studies illustrate this variation by the rich diversity in alpine and subalpine flora (Noroozi et al. 2008) and high endemism up to 85% in

M. Zamanpoore (✉)
Department of Aquatics and Fisheries, Fars Agricultural and Natural Resources Research Center, Agricultural Research, Education and Extension Organization (AREEO), Shiraz, Iran

L. A. Jawad (ed.), *Tigris and Euphrates Rivers: Their Environment from Headwaters to Mouth*, Aquatic Ecology Series 11,
https://doi.org/10.1007/978-3-030-57570-0_32

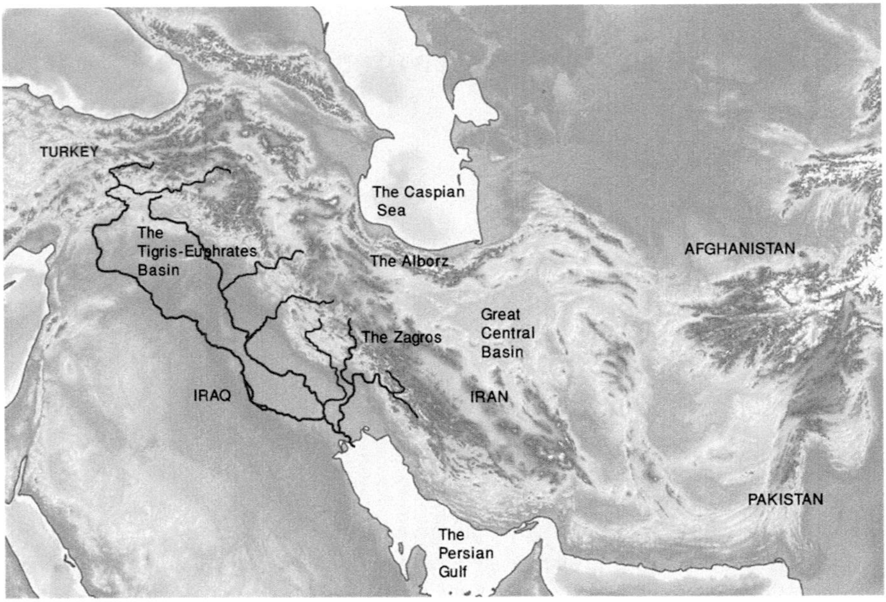

Fig. 32.1 Topographic map of Iran and surrounding regions, showing main channels of the Tigris-Euphrates river system and their reaches inside Iran

mountain plants (Akhani 2006). Other research studies on vertebrate animals in the Zagros Mountains show great diversity, for instance, in reptiles with 36 endemic species belonging to 17 genera and seven families (Gholamifard 2011), including snakes (Nilson and Rastegar-Pouyani 2007), agamid (Macey et al. 1998), and gekkonid (Rastegar-Pouyani 2006) lizards. Ahmadzadeh and Kheyrandish (2006) reported 20 species from 13 genera of lizards from only the northwest of Iran.

Biodiversity of aquatic fauna is specifically high in Iran: more than 161 freshwater fish species belonging to 20 genera and 6 families, with 33 of them being endemics (Coad 2006). Fewer works can be found on invertebrates as a whole, and fewer data are available on their biodiversity. Valuable attempts have begun in recent years to collect information on some of the invertebrate taxa, but there are still many groups waiting for attention from scientists.

A number of concise works have been published on aquatic macroinvertebrates in Iran, many of them on Coleoptera, with 67 species reported from only one family, Hydraenidae (Darilmaz et al. 2018). From Odonata, 98 species are reported from Iran in which almost all of them were from the Alborz and the Zagros Mountain ranges and mostly from the Zagros (Heidari and Dumont 2002; Sadeghi and Mohammadalizadeh 2009). Three species of freshwater shrimps were reported from Southern Zagros surface waters (Gorgin 1996). In the quite recent research studies of speleobiology, 89 invertebrate species including 36 species of chelicerates, 15 species of crustaceans, 34 species of hexapods, and 4 species of myriapods have been recorded from underground waters in caves of the west and northwest of Iran (Malek-Hosseini and Zamani 2017).

Most focused works on amphipods have just begun a few years ago; however, these works show the significant biodiversity of the taxon in this region. The first record of freshwater amphipods appeared in Karaman (1934) describing *Gammarus pulex persicus* from Northwest Iran, which was later proved to be a synonym of *G. komareki*. The next records are those of Birstein (1945) reporting *G. lacustris* and Löffler (1956) with a short description of an amphipod, both from the Northwest. Ruffo (1979) worked on a single specimen as well, described a new genus (and new species) from the Central Zagros, and Mateus and Mateus (1990) examined a large collection of amphipods from central and southern Zagros and described several new species. Karaman (1998) identified a well-dwelling niphargid from north of Iran, and Stock et al. (1998) identified and described a handful of new species in an intensive collection from all over the Caspian Sea shore and some other localities in central Iran. From this time, the tendency for collecting and examining amphipods got accelerated, which appeared in a series of original publications such as Khalaji-Pirbalouty and Sari (2004), Khalaji-Pirbalouty and Sari (2006), Zamanpoore et al. (2009), Zamanpoore et al. (2010), Hekmatara et al. (2011), and Semsar-Katouzian et al. (2016) for *Gammarus* species, and Hekmatara et al. (2013), Esmaeili-Rineh and Sari (2013), Esmaeili-Rineh et al. (2015), Esmaeili-Rineh et al. (2016), (2017a, b), and Mamaghani-Shishvan et al. (2017) for *Niphargus* species. More interestingly, recent research examined *Gammarus* species in the Northern Zagros using molecular methods and analysis of genetic markers and suggested 42 species among five previously recognized morphospecies (Katouzian et al. 2016). Apart from any judgment about under–/overestimations, the most significant perception of these findings is a realization of the magnitude of biodiversity among the amphipod taxa.

These investigations stepped up our knowledge of amphipods in epigean and hypogean ecosystems in Iran. Extensive areas among the Zagros and the Alborz have been searched, and the resulting information on newly described species gave a better understanding of their biodiversity in the whole country. Despite this, our current knowledge is sometimes mixed with issues of obscurity, like lack of precision in identification (as in the case of repeatedly making synonyms in Mateus and Mateus 1990), ignoring the existing data (Mateus and Mateus 1990, was seemingly hidden from researchers in the early stages), and failure in taxonomic expertise and accuracy, which resulted in reporting species (e.g., in Sharifinia 2015) that have not ever seen even in the Palearctic (Ruiz et al. 2011). Many locations have not been fully investigated (especially for hypogean species), apparently due to the difficulty of sampling excursions to alpine areas and the caves in high elevations. There are also scant amphipod samples collected from mountainous regions of medial East and south East of Iran as well.

Ecological importance of amphipods in aquatic ecosystems has been researched and documented very frequently (Gerhardt et al. 2011; Nelson 2011). They have absolutely diverse modes of feeding: herbivores, detritivores, carnivores, or omnivores. Detritivores help the relatively weak food chain of alpine spring-brooks recirculate; by this mode of feeding, amphipods pass matter through the food chain, produce animal biomass, provide stability for larger hunting components, and cleanse their ecosystem. As well as most plecopterans, trichopterans, some tipulid dipterans, and gammaridean amphipods are shredders, which can specify them as fundamental elements of the spring and river ecosystems (Graça 2001; Tachet et al. 1987). One other essential effect of amphipods is regulation of periphyton algae densities (Friberg and Jacobson 1994). Even subterranean taxa are eminently dependent on the organic debris coming from the surface, hence taking part in controlling and regulating the amount of organic matter in hypogean environments.

Pollutions on the other hand may greatly influence amphipod survival. Increase of human population in recent decades has expanded rural and urban settlements as well as agricultural lands, which can directly and indirectly affect natural wildlife habitats, especially in freshwater ecosystems that are of the most susceptible systems against disturbances (Graça 2001). As a consequence, many of the springs and low order rivers have lost many of their biota, and no amphipods can be found in places where they had inhabited before. Climatic changes, which appeared as a severe drought in the last decade, exerted even more pressure on aquatic ecosystems, so that many of them, including spring-brooks and rivers of first and second order, are thoroughly dried.

Ecosystems of western Iran, from northwest to southwest of the Zagros, lie in the catchments area of Arvandrud or the Tigris-Euphrates (Fig. 32.1). This status of hydrological relation connection, which creates a unidirectional network between west of Iran and east of Tigris-Euphrates catchments area in Iraq, establishes an influential connection between aquatic habitats and dispersal routes of the two sides, which, due to the nature of dispersal processes, is bidirectional. Appreciation of the distribution patterns of the biota in every part of the Tigris-Euphrates territory can provide a better understanding of the past and future patterns of the distribution of populations and communities in both upstream and downstream habitats of this complicated network of diverse aquatic systems. Freshwater amphipods of Iran have received much attention than many of other invertebrate taxa, and a thorough collection of research studies on their taxonomy and biogeography has been published.

In this chapter, I will demonstrate biodiversity of this group of freshwater crustaceans in Iran. This recognition can be used as a model for understanding macroinvertebrate biodiversity in Tigris-Euphrates catchment area and provide a basis for unraveling their distribution patterns and processes in evolutionary backgrounds.

32.2 The Iranian Plateau and the Tigris-Euphrates Basin

Iran is geographically on the Iranian tectonic plate in western and central Asia. In west and south of this plate lies the Arabian Plate, which hosts some parts of the Zagros. The main catchment of Iran is the Central Plain, a closed watershed basin covering more than half of the country's total area. Besides the catchment of the Caspian Sea in north and Persian Gulf and the Sea of Oman in south, other major catchment areas include the catchment area of Urmia Lake in northwest, Eastern Territories in east, and Karakum in northeast (Office of Standards and Technical Criteria 2004).

Invasive parts of the lands in Iran are mountainous, with elongated chains of mountains encircling the majority of the country in the Central plain. All over the north is the great wall of the Alborz, and all over the west and south lies the outspread network of mountains called the Zagros (Fig. 32.1). The Zagros separates the Central Plain Basin and Urmia Lake Basin from the Persian Gulf and the Sea of Oman in south, as well as the Tigris-Euphrates Basin in west, while the Alborz separates the Central Plain from the Caspian Sea Basin. The two mountain chains partially join in the northwest corner and connect with the Pontic Mountain in the northern course and with Taurus in the southern course, both in Turkey, to form the southern margin of the Eurasian Plate together with the Rhodopes, Alps, and the Atlas (Petrov 1955).

The Zagros, ca. 1200 km long and 200–300 km wide, is extended all over the eastern boundaries of the Tigris-Euphrates Basin from southeast to northeast (Tatar et al. 2004). The most elevated point reaches ca. 4500 m asl, and its main source of water is winter snow (Bozkurt and Sen 2011). The Pontic and the Taurus merge in the north and make Anti-Taurus with an average elevation of ca. 3000 m asl, producing headwaters of the region's large rivers: Aras in the east flowing toward the Caspian Sea and Euphrates and Tigris in the south (Dyson 1968). Tigris receives many of the large reaches of the Zagros as well (e.g., Karkhe, Karun, Dez, Sirvan, Dialeh, etc.). The whole catchment area of the Tigris-Euphrates Basin exceeds 900,000 km2, which passes throughout the Mesopotamian plains (UN-ESCWA and BGR 2013). The rivers join in southern parts to make Arvandrud or Shatt al Arab end up in the Persian Gulf.

Availability of water is crucial for the formation and stability of aquatic ecosystems and for dispersal and establishment of living organisms. The northwest part of the Iranian Plateau is more elevated and receives more precipitation than the Anatolian Plateau (Bozkurt and Sen 2011). Despite the low precipitation in eastern plains of the Central Basin, high elevated terrains of the Zagros, Pontic, and Taurus Mts. have annual precipitation records of more than 1000 mm (Alex 1985). This is not even the long-term image of the water budget of the region. Snowline has been recorded somewhere between 3800 and 5000 m in the twentieth century (Bobek 1953; Ehlers 1980), while it was estimated to be 600 to 1100 meters lower in the Pleistocene, due to the more extent glaciation, resulting in lower temperatures and higher precipitations than the contemporary time (Bobek 1937; Ferrigno 1991). Therefore, it can be imagined that situations were well adjusted for cold-dwelling aquatic animals in all the upstream sources of water in Tigris-Euphrates Basin from the Pleistocene epoch, the fact that can explain to some extent their rich biodiversity at recent times.

32.3 The Amphipod Fauna

Amphipoda (Arthropod: Crustacea) is a large order of mostly aquatic malacostracans with more than 10,000 described species (Horton et al. 2018). They lack the common crustacean carapace, and their body is laterally compressed, with their lengths averaging 5 to 15 mm (Väinölä et al. 2008). A majority of amphipods (more than 80%) are marine, but there are many species (53 families, 293 genera) living in various freshwater habitats especially in cold Springs (Väinölä et al. 2008) and with a high degree of biodiversity in hypogean habitats (Sket 1999). All of the amphipod species in inland waters of Iran belong to the Gammaridea, a large suborder with 41 (Hou and Sket 2016) families.

Amphipods of Iran's inland waters are found in a variety of geographical regions and diverse types of habitats. However, most populations of the known species have been found in alpine spring-brooks and streams and occasionally in running waters in lowlands of mountain outskirts. Various species of amphipods are reported from both mountain chains of the Alborz, from east to west, and the Zagros, from southeast to northwest, mostly of the genus *Gammarus* (from surface waters) and the genus *Niphargus* (from hypogean habitats or cave water systems).

In this chapter, only 'freshwater' amphipods are presented and reviewed. A few species inhabiting both brackish waters of the Caspian Sea and their freshwater river reaches are included as well, and species of open waters of the Caspian Sea as well as those from the Persian Gulf and the Sea of Oman are excluded. In this context, recorded amphipods of Iran are composed of four families, seven genera, and 41 species (Table 32.1). Quite newly, a specimen of the Family Crangonyctidae has been found within the country. It is proved to be a new species of *Crangonyx* and is under description.

The species inventory and distribution range of each species (or its presently known distribution status) as well as a brief note on their ecology (when provided in the original sources) is presented.

Class Malacostraca Laterille, 1802
Order Amphipoda Latreille, 1816
Suborder Gammaridea Latreille, 1802
Family Gammaridae Latreille, 1802
Genus *Gammarus* Fabricius, 1775

Table 32.1 Recorded families and genera of freshwater amphipods (Gammaridea) in Iran

Families	Genera	Number of species
Gammaridae	*Gammarus*	18
	Obesogammarus	2
	Turcogammarus	1
	Phreatomelita	1
Niphargidae	*Niphargus*	17
Pontogammaridae	*Pontogammarus*	1
Crangonyctidae	*Crangonyx*	1
Total amphipods		41

Family Gammaridae includes 52 genera (Hou and Sket 2016), and *Gammarus* is the most diverse epigean genus in this family. Separation of *Gammarus* is estimated to happen 32 mya (Hou and Sket 2016). It is well spread in inland waters of the Nearctic and Palearctic, especially in the mountainous regions of the Mediterranean and the Near East (Väinölä et al. 2008). Occurrence of the genus *Gammarus* in the Palearctic is currently known to be centered in Europe, but the range of the genus is extended to eastern borders of the Palearctic to the Far East (Hou et al. 2014). It has been shown that there are 18 valid freshwater *Gammarus* species in Iran (Zamanpoore et al. 2011), while two new subspecies were recently described (Özbek and Rasouli 2014; Semsar-Kazerooni et al. 2016).

Gammarus anodon Stock, Mirzajani, Vonk, Naderi & Kiabi 1998.

Distribution. Endemic species, restricted to the type locality in Hasheelan wetland, Kermanshah province, north-west Zagros (Stock et al. 1998).

Gammarus bakhteyaricus Khalaji-Pirbalouty and Sari 2004.

Distribution. Endemic to a few localities in the Central Zagros: Chaharmahal va Bakhtiari (Khalaji-Pirbalouty and Sari 2004) and Esfahan (Ebrahimnezhad et al. 2005) Provinces. The habitats are small springs in high mountains of ca. 2300 m above sea level with slow water currents (Khalaji-Pirbalouty and Sari 2004).

Gammarus balouchi Khalaji-Pirbalouty and Sari 2006.

Distribution. Endemic in just one locality in Chaharmahal va Bakhtiari Province, the Central Zagros in high altitudes of ca. 2100 m asl. The habitat is recorded as a waterfall with aquatic macrophytes with a gravel-covered stream bed (Khalaji-Pirbalouty and Sari 2006).

Gammarus crinicaudatus Stock, Mirzajani, Vonk, Naderi & Kiabi 1998.

Distribution. Several localities in Southern Zagros, Fars Province, inhabiting springs and small streams in altitudes of 712 m to 2247 m asl (Stock et al. 1998; Zamanpoore et al. 2011). Mean water temperature was 16.2 with a maximum of 22 °C, with exception of winter temperatures of 1 °C. Electric conductivity is recorded from 150 to 810 μS/cm in different seasons (Zamanpoore et al. 2011).

Gammarus hegmatanensis Hekmatara, Sari & Heidari Baladehi 2011.

Distribution. Narrowly endemic in Northern Zagros, close to the capital Hamadan, Hamadan Province, in altitudes of 1677–2109 m asl (Hekmatara et al. 2011).

Gammarus komareki Schäferna 1922.

Distribution. This is a widely distributed species, ranging from Eastern Europe to east of Iran on a west-east axis. In Iran, it can most frequently be seen in Alborz Mountains (Alizade-Eghtedar and Sari 2007; Stock et al. 1998), but many of its populations have been recorded in the Zagros Mountains as well (Ebrahimnezhad et al. 2005; Khalaji-Pirbalouty and Sari 2006; Stock et al. 1998; Zamanpoore et al. 2011). It is found in cold water (ca. 12 °C) habitats with slow currents among aquatic macrophytes (Karaman and Pinkster 1977), at elevations ca. 2200 m asl (Khalaji-Pirbalouty and Sari 2004; Zamanpoore et al. 2011).

G. komareki aznavensis Özbek and Rasouli 2014.

Distribution. This subspecies is endemic to Aznav County, Ardabil Province, northwest of Iran. It was found in a habitat with maximum depth of 130 cm, water temperature of 11 °C, and electrical conductivity of 310 μS in a slow running spring water with dense submersed macrophytes (Özbek and Rasouli 2014).

Gammarus lacustris G.O. Sars, 1863.

Distribution. This widely distributed cosmopolitan species has a restricted northwest distribution in Iran in west of the Alborz, generally in stagnant waters and lakes like Ne′ur Lake with temperatures below 20 °C (Alizade-Eghtedar and Sari 2007; Birstein 1945; Stock et al. 1998; Zamanpoore et al. 2011).

Gammarus lobifer Stock, Mirzajani, Vonk, Naderi & Kiabi 1998.

Distribution. This is an endemic species confined to a local area in lower central Zagros Yasooj, Kohgiluye va Buierahmad Province, found only in three cold springs (Stock et al. 1998).

Gammarus loeffleri Zamanpoore, Grabowski, Poeckl & Schiemer Zamanpoore et al. 2010.

Distribution. *G. loeffleri* is endemic to a wide area in east and south east of Fars province, Southern Zagros, where water has naturally higher salinities and temperatures. Its habitats include springs at altitudes from as low as 1100 m asl, lower than many other Zagros species. Water temperatures were recorded up to 24 °C, and the electric conductivity was measured from up to 1000 μS/cm (Zamanpoore et al. 2010).

Gammarus lordeganensis Khalaji-Pirbalouty and Sari 2004.

Distribution. Apparently an endemic, this species is reported from just two springs in the Central Zagros, Chaharmahal va Bakhtiari Province, which are cold (ca. 12 °C) mountain springs in altitudes of some 1600 m asl (Khalaji-Pirbalouty and Sari 2004).

Gammarus paricrenatus Stock, Mirzajani, Vonk, Naderi & Kiabi 1998.

Distribution. This species is also found in just a single locality, in cold waters of Goree Gol wetland in west of Alborz, Tabriz, East Azarbaijan Province (Alizade-Eghtedar and Sari 2007; Stock et al. 1998).

Gammarus parthicus Stock, Mirzajani, Vonk, Naderi & Kiabi 1998.

Distribution is recorded from Ardebil, in northwest (Stock et al. 1998; Zamanpoore et al. 2011), Esfahan (Zamanpoore et al. 2011) in Central Basin, north of Central Zagros, and in Khorasan Razavi Province, far east of Alborz (Zamanpoore et al. 2011). Water temperature was 13 °C, and pH was 7 (Stock et al. 1998).

Gammarus pretzmanni Mateus and Mateus 1990.

The species is found in Markazi Province as well as several localities from northern and central Zagros (Ebrahimnezhad et al. 2005; Khalaji-Pirbalouty and Sari 2004; Mateus and Mateus 1990; Stock et al. 1998).

Gammarus pseudosyriacus Karaman and Pinkster 1977.

This species is distributed in a vast area of the northern, central, and southern Zagros in its eastern fronts, which lies in the catchments of the Central Basin (Ebrahimnezhad et al. 2005; Khalaji-Pirbalouty and Sari 2004; Mateus and Mateus 1990; Zamanpoore et al. 2011). *Gammarus pseudosyriacus* is one of the two most widely distributed gammaridan species in Iran. Their southern range is supposed to be expanded southward to the province of Kerman (Zamanpoore et al. 2011). They live in habitats with a wide range of temperatures (5–34 °C) and salinity (120 to 1015 μS/cm). In many occasions, they are found in lowland desert-like regions with altitude of 905 m asl, as well as in alpine regions in 2600 m asl (Karaman and Pinkster 1977; Khalaji-Pirbalouty and Sari 2004; Zamanpoore et al. 2011).

G. p. issatisi Semsar-Kazerooni, Zamanpoore and Sadeghi 2016.

The subspecies has been described from Qanat e Hojjatabad, Yazd province, south of Iran, separated from the main Zagros populations of *G. p. pseudosyriacus* by a dry desert plain (Semsar-Kazerooni et al. 2016).

Gammarus sepidannus Zamanpoore, Poeckl, Grabowski & Schiemer 2009.

This endemic species is described from only a limited area in Sepidan, Fars Province, in the upper reaches of the Zohre River, south of the Central Zagros. Habitats are a few small cold (8–13 °C) alpine springs with very low conductivities (180 to 410 μS/cm), 2053 m to 2350 m above sea level. Springs were surrounded by snow cover throughout the whole winter (Zamanpoore et al. 2009).

Gammarus shirazinus Zamanpoore, Grabowski, Poeckl & Schiemer 2010.

Distribution is almost exclusively restricted to Maharlu Lake Basin in southern Zagros in Shiraz Plain (1460 m asl). Springs are relatively highly saline (conductivities of up to 1800 μS/cm), and the highest water temperatures were recorded as 25 °C (Zamanpoore et al. 2010).

Gammarus sirvannus Hekmatara, Sari & Heidari Baladehi 2011.

This species was collected from only a single spot in Kermanshah Province, northwest of the Zagros, at the altitude of ca. 700 m asl. The Sirvan river in which the species was found is the habitat of two other *Gammarus* species, *G. parthicus* and *G. anodon*. Interestingly, the river is on the western outskirt of the Zagros, joining the Tigris River at its end (Hekmatara et al. 2011).

Gammarus zagrosensis Zamanpoore, Poeckl, Grabowski & Schiemer 2009.

Another endemic species inhabits the same geographic range as *G. sepidannus* in cold regions of lower Central Zagros (Zamanpoore et al. 2009).

Genus *Obesogammarus* Stock 1974.

There are two species of the genus *Obesogammarus* recorded from Iran up to now.

Obesogammarus acuminatus Stock, Mirzajani, Vonk, Naderi and Kiabi 1998.

Distribution. This species is recorded from different localities and habitats along the southern margins of the Caspian Sea, including Anzali Bay (a fresh water lagoon), Sefidrud River and estuary, and Ameerkelayeh Wetland, Gilan Province, and in the brackish waters of the Caspian Sea in Sari, Mazandaran Province (Stock et al. 1998).

Obesogammarus crassus (Sars 1894).

Distribution. Recorded from Khazarabad, Sari, Mazandaran Province, few kilometers away from the Caspian seashore. It was recorded from the brackish water of the sea as well (Stock et al. 1998).

***Genus Turcogammarus* Karaman & Barnard, 1979.**

Turcogammarus turcarum (Stock 1974).

Synonymized name: Obesogammarus turcarum Stock 1974 (Horton et al. 2018).

Distribution. The species is recorded from Bor Alaan spring and somewhere near to Aras Reservoir in far north-west of the country, West Azerbaijan province (Stock et al. 1998).

Genus *Phreatomelita* Ruffo 1979**.**

Phreatomelita paceae Ruffo 1979.

Distribution. Shahrekord, Chaharmahal va Bakhtiari Province, Midwest of Iran (Ruffo 1979).

Family Niphargidae Bousfield 1977Genus *Niphargus* Schiödte 1849.

Niphargus comprises the largest genus among freshwater amphipods with some 350 described species up to the recent time (Väinölä et al. 2008). Excluding a few surface species, a majority of species are stygobiotic, living in subterranean streams, fissure systems, wells, and springs (Fišer et al. 2009). The most significant feature of this group of animals is the lack of eyes. Generally, niphargids show more elongated bodies compared to gammarids, despite their absolute variability of body length, which ranges from 2–35 cm (Väinölä et al. 2008).

Rapidly growing attention to collection and taxonomy of groundwater amphipod revealed a significant diversity of *Niphargus* in Iran, which resulted in identification of 16 new endemic species in just the last 5 years (including the last one in press), which makes a total species number of 17 with addition of formerly recorded *N. valachicus*.

N. alisadri, Esmaeili-Rineh and Sari 2013.

Distribution. This species was collected from Alisadr Cave, Hamadan Province, in Northwest Iran (Esmaeili-Rineh and Sari 2013).

N. bisitunicus, Esmaeili-Rineh, Sari & Fišer 2015.

Distribution. Sarab e Bisotun, 25 Km west of Kermanshah City, Kermanshah Province, West Iran (Esmaeili-Rineh et al. 2015).

N. borisi Esmaeili-Rineh, Sari & Fišer 2015.

Distribution. Belgheis (Belqais) spring, 4 Km to Choram City, Kohgiluyeh va Buierahmad Province, west Iran (Esmaeili-Rineh et al. 2015).

N. daniali Esmaeili-Rineh and Sari 2013.

Distribution. Danial cave (ca. 20 km south of Tonekabon City, Mazandaran Province, north of Iran (Esmaeili-Rineh and Sari 2013). Locality is derived from the coordinates in the original report.

N. darvishi Esmaeili-Rineh, Sari & Fišer 2015.

Distribution. Dimeh spring, Koohrang city, Chaharmahal va Bakhtiari Province, west of Iran (Esmaeili-Rineh et al. 2015).

N. hakani Esmaeili-Rineh, Mirghaffari & Sharifi 2017a, b.

Distribution. Khedergoli spring, Razan City, Hamadan Province, Northwest of Iran (Esmaeili-Rineh et al. 2017a, b).

N. keeleri Zamanpoore and Bakhshi 2020.

Distribution. Keeler Cave, Yasooj, Kohgiluye Va Buierahmad Province, west of Iran (Zamanpoore et al. 2020).

N. khayyami Hekmatara, Zakšek, Heidari Baladehi & Fišer 2013.

Distribution. Ghoori-Ghaleh cave, Ravansar-Paveh, Kermanshah Province, West Iran (Hekmatara et al. 2013).

N. khwarizmi Hekmatara, Zakšek, Heidari Baladehi & Fišer 2013.

Distribution. Cheshmeh Kahriz (a Qanat), Kharvana village, 141 km north of Tabriz, East Azarbaijan Province, northwest of Iran (Hekmatara et al. 2013).

N. kermanshahi Esmaeili-Rineh, Heidari, Fišer & Akmali 2016.

Distribution. Kangarshah Spring, Sahneh City, Kermanshah Province, west Iran (Esmaeili-Rineh et al. 2016).

N. kurdistanensis Mamaghani-Shishvan, Esmaeili-Rineh & Fišer 2017.

Distribution. Shoei Cave, Baneh City and Darvish Olya Cave, Marivan City, Kurdistan Province, northwest of Iran (Mamaghani-Shishvan et al. 2017).

N. sharifi Esmaeili-Rineh, Sari & Fišer 2015.

Distribution. Sarab e Robat, Khorramabad City, Lorestan Province, west of Iran (Esmaeili-Rineh et al. 2015).

N. valachicus Dobreanu & Manolache 1933.

Distribution. Ghaem Shahr, Mazandaran Province, North Iran (Karaman 1998).

N. persicus Esmaeili-Rineh, Sari, Fišer & Bargrizaneh2017a, b.

Distribution. Tirebagh Spring, 60 km north of Marvdasht City, Fars Province, South Iran (Esmaeili-Rineh et al. 2017a, b). Details of the locality are partly derived from the coordinates in the original report.

N. hosseiniei Esmaeili-Rineh, Sari, Fišer & Bargrizaneh2017a, b.

Distribution. The species was found in three localities in the west and north-west of Iran: Nojivaran spring and Ghoori Ghale Cave, Kermanshah Province, and Bor Alaan spring, West Azarbayejan Province (Esmaeili-Rineh et al. 2017a, b), west and northwest Iran.

N. ilamensis Esmaeili-Rineh, Sari, Fišer & Bargrizaneh 2017a, b.

Distribution. Sarab-e-Moord and Sarab e Kanipahn, Ilam Province, West Iran (Esmaeili-Rineh et al. 2017a, b).

N. sohrevardensis Esmaeili-Rineh, Sari, Fišer & Bargrizaneh 2017a, b.

Distribution. This species has an exceptional expanded distribution, in two localities with an aerial distance of ca. 500 km to each other, Sohrevard Spring, 110 Km in west of Zanjan City (coordinates of the spring in the original report lies inside the city of Zanjan), Zanjan Province, Iran, and Razbashi spring, Lorestan Province (Esmaeili-Rineh et al. 2017a, b), northwest Iran.

Family Pontogammaridae Bousfield 1977.

Genus *Pontogammarus* Sowinsky 1904.

Pontogammarus maeoticus (Sowinsky 1894).

Distribution. This species is collected from a long distance from Astara in west, to Babolsar in the middle of the eastern half of the Caspian Sea margins (Stock et al. 1998). I include this species here because Stock et al. (1998) mentioned that it "penetrates" into the upstream rivers. Nevertheless, all of their samples were taken from seashore, and none of them are actually recorded from freshwaters far from the sea. One exception is a fresh water locality in Kashmar in the Central Basin, with a distance of at least 700 km from the nearest other recorded points in the Caspian Sea shore. Based on the general distribution pattern of amphipods in the country, especially that of gammarid species in Alborz Region, this can obviously be regarded as a mistaken record or false identification.

Family Crangonyctidae Bousfield 1973.
Genus *Crangonyx* Bate 1859.
Crangonyx sp.
A new cave amphipod sample has just been collected by Y. Bakhshi, PhD candidate in University of Shiraz, from Southern Iran, which is identified by the author as a crangonyctid, a new record for the whole Middle East region. The description is under preparation.

32.4 Amphipod Biogeography in Iran

Distribution of ancestral amphipods in the Zagros and Alborz and their subsequent endemism were affected by several natural phenomena and processes. Specific mechanisms of orogenesis have resulted in a branched morphology of the mountains, especially in the Zagros, a linear belt of asymmetrically folded mountain chain forming a series of ranges inside it (Tatar et al. 2004). As a result, numerous valleys and isolated plains are formed as subbasin catchments. Isolation of catchment areas is crucial for the establishment of differentiation among aquatic populations. Even more, mountainous relief favors breakup of species into isolated populations in the high altitudes (Löffler 1984). On the other hand, permanent and stable surface waters are sparse through the region and most wetlands are small and/or just ephemeral. As a consequence, sporadic aquatic ecosystems were formed in springs and artificial ponds and canals as exclusive natural refuges for aquatic organisms (Coad 2006). Löffler (1984) states that the amount of variation among populations is higher in limnic ecosystems, probably due to recurrent hybridizations. These features altogether might have led to endemism, as climatic differences and geographical isolation helped divergent evolution to increase biodiversity in this peculiar region (Coad 2006; Macey et al. 1998; Zamanpoore et al. 2011). Coleoptera, Odonata, fishes, and amphipods are examples of aquatic taxa who received much extensive and systematic attention, and this high biodiversity is particularly noticeable among them (Coad 2006; Darilmaz et al. 2018; Katouzian et al. 2016; Sadeghi and Mohammadalizadeh 2009; Zamanpoore et al. 2011).

Based on the total data gathered from extensive parts of the country on taxonomy and distribution of the genus *Gammarus* in inland waters of Iran, 15 species are known to be narrowly localized, most probably indigenous endemics, while three species *G. lacustris*, *G. komareki*, and *G. pseudosyriacus* have broader distributions. Many of the narrowly endemic species were recorded from just a very few localities, sometimes from only a single spring, e.g., *G. anodon*, *G.lobifer*, *G. paricrenatus* (Stock et al. 1998), and *G. baloutchi* (Khalaji-Pirbalouty and Sari 2006). On the other hand, *G. lacustris* has a Holarctic distribution and can be easily regarded as a cosmopolitan species, *G. komareki* is endemic to a rather larger-scale area of the Balkan Peninsula, Asia Minor, and Alborz Region, and *G. pseudosyriacus* is an extensively endemic of Asia Minor and the Zagros Region (Zamanpoore et al. 2011).

Among the Zagros endemics, species, which occupied water bodies in outskirts of the mountains or in more extensive plains, obviously show broader distributions compared to those in narrow valleys and have been extended even to more than one watershed basin (e.g., *G. crinicaudatus* and *G. loeffleri* which are spread in far eastern end-branches of the Zagros (Zamanpoore et al. 2011)). Despite this, the latter three species have occupied various geographical locations including both high altitudes and lowland doorsteps of the mountains, so that their broad distribution must probably have been acquired through different processes from those of the Zagros endemics.

There is an interesting point in the distribution pattern of *G. komareki* and *G. pseudosyriacus*. *G. komareki* is recorded from all the Alborz Mountains, east to west (Stock et al. 1998). Outside Iran, it has been recorded from Lesser Caucasus (Armenia and Azerbaijan), Eastern Taurus, Pontic and Koroglu Mts. (Turkey), to Rhodope Mts. (Greece and Bulgaria) (Karaman and Pinkster 1977; Khalaji-Pirbalouty and Sari 2004; Özbek et al. 2009; Stock et al. 1998). On the more southern part of the geography of the region, the range of *G. pseudosyriacus* includes the Lebanon Mountains in Lebanon down to its southern areas, Anti-Lebanon Mts. in Syria, and Western Taurus Mts. in Turkey (Karaman and Pinkster 1977; Özbek and Ustaoğlu 2006; Özbek et al. 2009) to join its other populations on entire Zagros Mountains (Zamanpoore et al. 2010). Regarding the two northern (Pontic) and southern (Taurus) routes of mountain chains in Turkey toward west of the Eurasian Plate, it can be obviously seen that one species (*G. komareki*) chose the northern course, while the other (*G. pseudosyriacus*) went through the southern course.

It is well demonstrated that many living organisms have used long Mountain ranges as climatic bridges (Löffler 1984). It was possible for the two species to use their own chain of mountains as a bridge or corridor for passing thousands of kilometers in their dispersal route.

Recently, *G. lacustris* has been reported from a few locations in the western Alborz close to northwest borders of Iran in some lakes and reservoirs and a spring (Stock et al. 1998). It is also distributed in epigean freshwaters of Turkey, suggesting its possible route of dispersal as proposed for *G. komareki*. The affinity of the species to inhabit lacustrine (Karaman and Pinkster 1977) rather than riverine situations might be one possible explanation of its limited distribution in Iran.

Obesogammarus acuminatus is endemic to southern margins of the Caspian Sea (Stock et al. 1998). *Turcogammarus turcarum* is a Pontocaspian amphipod with a wider range in Russia (Stock 1974) and west Asia including Turkey (Özbek and Ustaoğlu 2006), and *Obesogammarus crassus*, another Pontocaspian species, has the broadest range from Eastern Europe, Lithuania, Poland, and Germany to Eastern Russia (Dobson 2012).

The monospecific endemic genus (and species) *Phreatomelita paceae* is closely related to some Indo-Pacific genera, which are also recorded in subterranean fresh-water, and to the amphipods of the Hadza group, as defined by Stock (1977). This new genus is especially closely linked with *Psammoniphargus*, Ruffo 1956, mono-specific and endemic genus of the island of Réunion (Mascarenhas Archipelago, Indian Ocean), to the genus *Paraniphargus* Tattersal 1925, the Andaman islands

(Bay of Bengal) and Java, and to the genus *Galapsiellus* Barnard 1976 known for the Galapagos archipelago.

Prior to very recent years, it was suggested that the species richness of *Niphargus* downturns from Western Balkans toward the Middle East (Zagmajster et al. 2014). However, it is clear now that this genus is well distributed in the latter region, more diverse than it was once realized (Fišer et al. 2009), indicating a tremendous proportion of endemism in karstic hydrological systems (Fišer et al. 2005). It has been shown that Western Palearctic is the origin and center of distribution range for hypogean *Niphargus*, the largest freshwater amphipod genus (Väinölä et al. 2008), from where a large-scale wide-ranging radiation toward Southern and Eastern Europe occurred around 30 MYA (McInerney et al. 2014). These Eastern boundaries of the genus range are supposed to be under repeated independent colonization, the last of which split in Miocene, making the clade which radiated progressively to the Zagros Mountains (Esmaeili-Rineh et al. 2015).

Of the 17 niphargid species in Iran, only one (*N. valachicus*) has a broad distribution range, a Pontocaspian species recorded from Bulgaria, Croatia, Hungary, Romania, Serbia and Montenegro, Slovakia, and Slovenia (Sket 1996). Thirteen species are narrow endemics of phreatic waters of single caves, while a wider distribution is seen in *N. hosseiniei* with three recorded localities in west and northwest provinces of Kermanshah and West Azarbayejan, *N. sohrevardensis* recorded from two localities as far as ca. 500 km in Zanjan and Lorestan Provinces (Esmaeili-Rineh et al. 2017a, b), and *N. kurdistanensis* from two localities in Baneh and Marivan, both in Kurdistan Province (Mamaghani-Shishvan et al. 2017).

32.5 Conclusions and Future Research Priorities

Extensive mountains of Iran show a great degree of biodiversity, and the recent data from conservation status of these regions convinced everyone that they well fit into the conception of biodiversity hotspots (Mittermeier et al. 2011), regions with exceptional level of endemism (i.e., at least 1500 vascular endemic plants), which are undergoing exceptional habitat loss (i.e., at least 70% of its original vegetation), as defined by Myers et al. (2000) and Mittermeier et al. (2011). Up to now, 41 amphipod species have been recorded or described from freshwater habitats throughout Iran. They constitute families Gammaridae (*Gammarus*, 18 sp. and 2 ssp., *Obesogammarus*, 2 sp., *Turcogammarus*, 1 sp., and *Phreatomelita*, 1 sp.), Niphargidae (*Niphargus*, 17 sp.), Pontogammaridae (*Pontogammarus*, 1 sp.), and the under description Crangonyctidae (*Crangonyx* sp.). These data are results of a relatively short time (mostly from the last 20 years), which makes their magnitude of biodiversity more noticeable.

Restricted aquatic habitats of amphipods in sparse freshwater sources make their populations potentially much vulnerable to any natural and human-induced environmental change (Conlan 1994). The drastic increase in human population in Iran through the last 50 years was proceeded by huge "developments" with all their various potentials for inserting high pressures on aquatic life: pollution of surface and groundwater, expansion of rural and urban areas, new road constructions, overexploitation of surface and ground waters by the ever-growing agriculture, and trans-basin water diversions. Several species of exotic hunters are new threats to all aquatic taxa (Coad 2006; Coad and Abdoli 1993). Increased need of the nation for food supplies, in an easy, fast, and guaranteed way of production, opened doors for introduction of exotic aquaculture fishes directly to natural waters or indirectly via fishes escaped from fish farms. Due to the fact that most of the listed amphipod species are narrow endemics (i.e., have only limited local distributions), everyone can expect that they are close to the threat of extinction, as reported for similar situations elsewhere (Sket 1999). As realization of the extent of disturbance in natural aquatic habitats in the Zagros and Alborz Mountains, besides the notable endemism in amphipods is cleared up, it gets more convenient to recognize the region as a biodiversity hotspot. These mountains and basins of the Irano-Anatolian biogeographic region host a lot of local centers of endemism and are actually accepted as biodiversity hotspots (Mittermeier et al. 2011).

Reflective conservation plans have not yet been started, but valuable scientific effort has begun using molecular genetic methods to predict genetic diversity loss among the amphipod gene pool (Esmaeili-Rineh et al. 2017b; Katouzian et al. 2016; Mamaghani-Shishvan et al. 2017). Still, these works cover only a small fraction of the whole amphipod taxa, which need an overall and holistic look into their phylogeny and phylogeography.

There are still spacious gaps in the knowledge of freshwater amphipods of Iran, even in the first steps of collection and morphological identification. None of the springs and streams in eastern mountains, in the great province of Sistan va Baluchestan, have been visited for sampling amphipods, and areas with no data on subterranean amphipods are even bigger; the prospect of the actual biodiversity is much promising as they have yet been collected mostly from northwest of the country.

Another priority for understanding amphipod biodiversity in Iran, as a country which shares part of the Tigris-Euphrates basin, is precise investigation for amphipods in freshwaters of Iraq. There is currently a big gap in the distribution range of the whole taxon from the Zagros to the eastern margins of the Mediterranean coast, with exceptional records of two Niphargids (*Niphargus itus* Karaman 1986 and *N. nadarini* Alouf 1972). Many of the amphipods recorded from northwest of Iran live in headwater reaches of the Tigris. Whether or not any of them could migrate downstream and establish populations would be illuminative progress in understanding patterns of amphipod biogeography in one of the most fascinating ecoregions of the world.

References

Ahmadzadeh F, Kheyrandish A (2006) Lizard biodiversity in northwestern Iran. Environ Sci 11:43–54

Akhani H (2006) Flora Iranica: facts and figures and a list of publications by K. H. Rechinger on Iran and adjacent areas. Rostaniha 7(suppl. 2):19–61

Alex M (1985) Klimadaten ausgewählter Stationen des Vorderen Orients. Beihefte zum Tübinger Atlas des Vorderen Orients. Reihe A, Nr. 14. Dr. Ludwig Reichert Verlag, Wiesbaden

Alizade-Eghtedar H, Sari AR (2007) Dojurpayan e Ponto-Caspian va goruhe guneii e *Gammarus pulex* (Crustacea: Amphipoda) dar Azarbaijan e Iran. Majalle ye Olum e Daneshgahe Tehran 33 (4):35–47

Birstein JA (1945) Zametka o presnovodnykh visshikh rakoobraznykh Turkmanii i Irana. Uchenye Zapiski Moskovskogo Gosudarstvennogo Universiteta 83:151–164

Bobek H (1937) Die Rolle der Eiszeit in Nordwestiran. Zeitschr Gletscherk 25:130–183

Bobek H (1953) Zur eiszeitlichen Vergletscherung des Alborzgebirges. Nordiran Carinthia 2:97–104

Bozkurt D, Sen OL (2011) Hydrological response of past and future climate changes in the Euphrates-Tigris Basin. Geophys Res Abstr 13:EGU2011–EG11072

Breckle SW (2002) Walter's vegetation of the earth. The ecological systems of the geo-biosphere. Springer, Heidelberg, p 527

Coad BW (2006) Endemicity in the freshwater fishes of Iran. Iranian J Animal Biosystematics. 1 (1):1–13

Coad BW, Abdoli A (1993) Exotic fish species in the fresh waters of Iran. Zool Middle East 9:65–80

Conlan KE (1994) Amphipod crustaceans and environmental disturbance: a review. J Nat Hist 28:519–554

Darilmaz MC, Ostovan H, Ghahari H (2018) Annotated checklist of the Hydraenidae (Coleoptera: Staphylinoidea) recorded from Iran. Int J Freshwater Entomol 38(4):239–253

Dobson M (2012) Identifying invasive freshwater shrimps and isopods. UK, Freshwater Biological Association, p 30. https://secure.fera.defra.gov.uk/nonnativespecies/news/index.cfm?id=77

Dyson RH (1968) The archaeological evidence of the second millennium B.C. on the Persian plateau. Cambridge University Press

Ebrahimnezhad M, Hosseini L, Sari A (2005) Collecting and identification of *Gammarus* species of the Zayandeh-Rood River. Majalle ye Zistshenasi e Iran 18(3):218–227

Ehlers E (1980) Iran: Grundzüge einer geographischen Landeskunde. Wissenschaftliche Buchgesellschaft, Darmstadt

Esmaeili-Rineh S, Heidari F, Fišer C, Akmali V (2016) Description of new endemic species of the genus *Niphargus* Schiödte, 1849 (Amphipoda: Niphargidae) from a karst spring in Zagros Mountains in Iran. Zootaxa 4126:338–350. https://doi.org/10.11646/zootaxa.4126.3.2

Esmaeili-Rineh S, Mirghaffari SA, Sharifi M (2017a) The description of a new species of *Niphargus* from Iran based on morphological and molecular data. Subterr Biol 22:43–58

Esmaeili-Rineh S, Sari A (2013) Two new species of *Niphargus* Schiödte, 1849 (Crustacea: Amphipoda: Niphargidae) from two caves in Iran. J Nat Hist 47:2649–2669. https://doi.org/10.1080/00222933.2013.802041

Esmaeili-Rineh S, Sari A, Fišer C (2015) Making future taxonomy of *Niphargus* (Crustacea: Amphipoda: Niphargidae) in the Middle East easier: DELTA database of Middle East species with description of four new species from Iran. Zootaxa 4020(3):401–430

Esmaeili-Rineh S, Sari A, Fišer C, Bargrizaneh Z (2017b) Completion of molecular taxonomy: description of four amphipod species (Crustacea: Amphipoda: Niphargidae) from Iran and release of database for morphological taxonomy. Zool Anz:1–23. https://doi.org/10.1016/j.jcz.2017.04.009

Ferrigno JG (1991) Glaciers of the Middle East and Africa – glaciers of Iran. In: Williams RS, Ferrigno JG (eds) Satellite image atlas of glaciers of the world. U.S. Geological Survey professional paper 1386-G. United States Government Printing Office, Washington, pp 31–47

Fišer C, Sket B, Trontelj P (2005) *Niphargus* homepage. 2005. URL http://www.bf.uni-lj.si/bi/zoologija/cene_fiser/niphargus

Fišer C, Trontelj P, Luštrik R, Sket B (2009) Toward a unified taxonomy of *Niphargus* (Crustacea: Amphipoda): a review of morphological variability. Zootaxa 2061:1–22

Friberg N, Jacobson D (1994) Feeding plasticity of detritivore shredders. Freshwat Biol 32 (1):133–142

Gerhardt A, Bloor M, Mills CL (2011) *Gammarus*: important taxon in freshwater and marine changing environments. Int J Zool 524276:1–2. https://doi.org/10.1155/2011/524276

Gholamifard A (2011) Endemism in the reptile fauna of Iran. Iranian J Animal Biosystematics 7 (1):13–29

Gorgin S (1996) The first record of two species of freshwater shrimps Decapoda, Caridea, Atyidae from Iran. Crustaceana (Leiden) 69(5):662–668

Graça MA (2001) The role of invertebrates on leaf litter decomposition in streams-a review. Int Rev Hydrobiol 86:383–393

Heidari H, Dumont HJ (2002) An annotated checklist of Odonata of Iran. Zool Middle East. 26:133–150

Hekmatara M, Sari A, Heidari BM (2011) Two new *Gammarus* species (Crustacea: Amphipoda: Gammaridae) from Zagros Mountains. Iran. Zootaxa. 2894:39–57

Hekmatara M, Zakšek V, Heidari Baladehi M, Fišer C (2013) Two new species of *Niphargus* (Crustacea: Amphipoda) from Iran. J Nat Hist 47:1421–1449. https://doi.org/10.1080/00222933.2012.743616

Horton T, Lowry J, De Broyer C, Bellan-Santini D, Coleman CO, Corbari L, Daneliya M, Dauvin J-C, Fišer C, Gasca R, Grabowski M, Guerra-García JM, Hendrycks E, Hughes L, Jaume D, Jazdzewski K, Kim YH, King R, Krapp-Schickel T, LeCroy S, Lörz A-N, Mamos T, Senna AR, Serejo C, Sket B, Souza-Filho JF, Tandberg AH, Thomas J, Thurston M, Vader W, Väinölä R, Vonk R, White K, Zeidler W (2018) World Amphipoda Database. 2018. http://www.marinespecies.org/amphipoda Accessed on 2018-03-24

Hou Z, Li J, Li S (2014) Diversification of low dispersal crustaceans through mountain uplift: a case study of Gammarus (Amphipoda: Gammaridae) with descriptions of four novel species. Zool J Linnean Soc 170(4):591–633. https://doi.org/10.1111/zoj12119

Hou Z, Sket B (2016) A review of Gammaridae (Crustacea: Amphipoda): the family extent, its evolutionary history, and taxonomic redefinition of genera. Zool J Linnean Soc 176:323–348. https://doi.org/10.1111/zoj.12318

Karaman S (1934) Über asiatische Süsswassergammariden. Zool Anz 106(5/6):127–134

Karaman GS (1998) First discovery of the family Niphargidae (Gammaridea) in Iran (contribution to the knowledge of the Amphipoda 243). Glas Od Pr Glasnik Odjeljenja Prirodnih Nauka–Crnog Akad Nauka Umjet 12:9–22

Karaman GS, Pinkster S (1977) Fresh water *Gammarus* species from Europe, North Africa and adjacent regions of Asia (Crustacea, Amphipoda), part 1. *Gammarus pulex*-group and related species. Bijdr Dierk 47(1):1–97

Katouzian AR, Sari A, Macher JN, Weiss M, Saboori A, Leese F (2016) Drastic underestimation of amphipod biodiversity in the endangered Irano-Anatolian and Caucasus biodiversity hotspots. Sci Rep 6:22507. https://doi.org/10.1038/srep22507

Khalaji-Pirbalouty V, Sari A (2004) Biogeography of amphipods (Crustacea: Amphipoda: Gammaridae) from the central Zagros Mountains, Iran, with descriptions of two new species. J Nat Hist 38:2425–2445

Khalaji-Pirbalouty V, Sari A (2006) A description of *Gammarus baloutchi* n. sp. (Amphipoda: Gammaridae) from Iran, based on light and electron microscopy. Zool Meded 801(6):91–100

Löffler H (1956) Limnologische beobachtungen an Iranischen binnengewässern II. Teil: Systematische beschreibung der rotatorien und crustaceen aus den untersuchten gewässern. Hydrobiologia 8(3–4):239–278

Löffler H (1984) The importance of mountains for animal distribution, species speciation, and Faunistic evolution (with special attention to inland waters). Mt Res Dev 4(4):299–304

Macey JR, Schulte JA, Ananjeva NB, Larson A, Rastegar-Pouyani N, Shammakov SM, Papenfuss TJ (1998) Phylogenetic relationships among agamid lizards of the *Laudakia caucasia* species group: testing hypotheses of biogeographic fragmentation and an area Cladogram for the Iranian plateau. Mol Phylogenet Evol 10(1):118–131

Malek-Hosseini MJ, Zamani A (2017) A checklist of subterranean arthropods of Iran. Subterr Biol. 21:19–46

Mamaghani-Shishvan M, Esmaeili-Rineh S, Fišer C (2017) An integrated morphological and molecular approach to a new species description of amphipods in the Niphargidae from two caves in west of Iran. Zool Stud 56:33

Mateus PA, Mateus E (1990) Etude d'une collection d'amphipodes specialement du sud-ouest asiatique du Museum d'Histiore Naturelle de Vienne (Autriche). Ann Naturhist Mus Wien 91 (B):273–331

McInerney CE, Maurice L, Robertson AL, Knight LRFD, Arnscheidt J, Venditti C, Dooley JSG, Mathers T, Matthijs S, Erikkson K, Proudlove GS, Häfling B (2014) The ancient Britons: groundwater fauna survived extreme climate changes over tens of millions of years across NW Europe. Mol Ecol 23:1153–1166

Mittermeier RA, Turner WR, Larsen FW, Brooks TM, Gascon C (2011) Global biodiversity conservation: the critical role of hotspots. In: Zachos FE, Habel JC (eds) Biodiversity hotspots: distribution and protection of conservation priority areas. Springer, Heidelberg, pp 3–23

Myers N, Mittermeier RA, Mittermeier CG, da Fonseca GA, Kent J (2000) Biodiversity hotspots for conservation priorities. Nature 403:853–858. https://doi.org/10.1038/35002501

Nelson D (2011) *Gammarus*-microbial interactions: a review. Int J Zool 295026:1–6. https://doi.org/10.1155/2011/295026

Nilson G, Rastegar-Pouyani N (2007) *Walterinnesia aegyptia* Lataste, 1887 (Ophidia: Elapidae) and the status of *Naja morgani* Mocquard 1905. Russian J Herpetol 14(1):7–14

Noroozi J, Akhani H, Breckle SW (2008) Biodiversity and phytogeography of the alpine flora of Iran. Biodivers Conserv 17:493–521

Office of Standards and Technical Criteria (2004) Dastoore amal va zavabete taghsimbandi va kodgozari hozehaye abriz va mahdoodehaye motaleati dar sathe keshvar. Iranian Water Resources Management Company

Özbek M, Balik S, Topkara ET (2009) Contribution to the knowledge on the distribution of malacostraca (Crustacea) species of central and southern Anatolia, with some ecological notes. Turk J Zool. 33(1):47–55

Özbek M, Rasouli H (2014) *Gammarus komareki aznavensis* subsp. nov., a new amphipod subspecies from Iran (Amphipoda: Gammaridae). Turk J Zool 38(3):326–333. https://doi.org/10.3906/zoo-1306-1

Özbek M, Ustaoğlu MR (2006) Check-list of malacostraca (Crustacea) species of Turkish inland waters. E.U. J Fish Aquatic Sci 23(1–2):229–234

Petrov MP (1955) Iran: Fiziko-geograficheskii ocherk. Moscow

Rastegar-Pouyani N (2006) Systematics of the genus *Asaccus* (Sauria: Gekkonidae) on the Zagros Mountains, Iran. In: Vences M, Köhler J, Ziegler T, Böhme W (eds) Herpetologia Bonnensis II. Proceedings of the 13th congress of the Societas Europaea Herpetologica, pp 115–116

Ruffo S (1979) Descrzione di gue nouvi anfipodi anoftalmi dell'Iran e dell Madagascar (*Phreatmelita paceae* n. gen. n. sp., *Dussartiella madegassa* n. gen. n. sp.). Boll Mus Civ St Nat Verona I:419–440

Ruiz GM, Fofonoff PW, Steves B, Foss SF, Shiba SN (2011) Marine invasion history and vector analysis of California: a hotspot for western North America. Divers Distributions 17:362–373

Sadeghi S, Mohammadalizadeh J (2009) Additions to the Odonata Fauna of Iran. Iranian J Sci Technol 33(A4):355–359

Sars GO (1894) Crustacea Caspia. Part III. Amphipoda. Gammaridae (1st part.). Bull Acad Imper Sci St Petersbourg 1(5):343–378

Semsar-Kazerooni M, Zamanpoore M, Sadeghi S (2016) Redescription of *Gammarus pseudosyriacus* (Karaman & Pinkster, 1977) and description of a new subspecies from southern Iran (Crustacea, Amphipoda, Gammaridae). ZooKeys 598:57–73

Sharifinia M (2015) Macroinvertebrates of the Iranian running waters: a review. Acta Limnol Bras 27(4):356–369

Sket B (1996) *Niphargus valachicus*. The IUCN Red List of Threatened Species 1996. e.T14805A4461365. doi: https://doi.org/10.2305/IUCN.UK.1996.RLTS.T14805A4461365.en

Sket B (1999) The nature of biodiversity in hypogean waters and how it is endangered. Biodivers Conserv 8:1319–1338

Stock JH (1974) The systematics of certain Ponto-Caspian Gammaridae (Crustacea, Amphipoda). Mitt Hamb Zool Mus Inst 70:75–95

Stock JH (1977) The taxonomy and zoogeography of the hadziid amphipoda, with emphasis on the west Indian taxa. Studies on the Fauna of Curaçao and other Caribbean Islands 55(1):1–130

Stock JH, Mirzajani AR, Vonk R, Naderi S, Kiabi BH (1998) Limnic and brackish water Amphipoda (Crustacea) from Iran. Beaufortia 48(9):173–234

Tachet H, Bournaud M, Richoux P (1987) Introduction à l'étude des macroinvertébrés des eaux douces (systématique élémentaire et aperçu écologique). Université Lyon, Lyon, France, p 155

Tatar M, Hatzfeld D, Ghafory-Ashtiany M (2004) Tectonics of the central Zagros (Iran) deduced from microearthquake seismicity. Geophys J Int 156(2):255–266. https://doi.org/10.1111/j.1365-246X.2003.02145.x

The Ramsar Convention (2018) The List of Wetlands of International Importance. pp. 57. https://www.ramsar.org/sites/default/files/documents/. . ./sitelist.pdf

UN-ESCWA and BGR (United Nations Economic and Social Commission for Western Asia; Bundesanstalt für Geowissenschaften und Rohstoffe). Inventory of Shared Water Resources in Western Asia. Beirut. (2013)

Väinölä R, Witt JDS, Grabowski M, Bradbury JH, Jazdzewski K, Sket B (2008) Global diversity of amphipods (Amphipoda; Crustacea) in freshwater. Hydrobiologia 595:241–255

Zagmajster M, Eme D, Fišer C, Galassi D, Marmornier P, Stoch F, Cornu JF, Malard F (2014) Geographic variation in range size and beta diversity of groundwater crustaceans: insights from habitats with low thermal seasonality. Glob Ecol Biogeogr 23:1135–1145

Zamanpoore M, Grabowski M, Poeckl M, Schiemer F (2010) Two new *Gammarus* species (Crustacea, Amphipoda) from warm springs in the south-east pre-alpine area of the Zagros, Iran: habitats with physiological challenges. Zootaxa 2546:31–51

Zamanpoore M, Poeckl M, Grabowski M, Schiemer F (2009) Two new sympatric species of freshwater *Gammarus* (Crustacea: Amphipoda) from southern Zagros region. Iran Zootaxa 2136:21–39

Zamanpoore M, Poeckl M, Grabowski M, Schiemer F (2011) Taxonomic review of freshwater*Gammarus* (Crustacea: Amphipoda) from Iran. Zootaxa 3140:1–14

Zamanpoore M, Bakhshi Y, Sadeghi S, Malek-Hosseini MJ (2020) *Niphargus keeleri* n. sp., a new cave Amphipod (Arthropoda: Crustacea) from the Zagros Mountains, Iran. J Nat Hist 53(43–44):2621–2635

Chapter 33
The Freshwater Molluscs of the Mesopotamian Plain

Vladimir Pešić and Peter Glöer

Abstract The present paper revises diversity, endemism and threats for freshwater molluscs that inhabit the plains between the Tigris and Euphrates rivers. A systematic account of the extant freshwater mollusc fauna of Iraq includes 35 mollusc species, twenty species of Gastropoda, and five species of Bivalvia. The recent malacofauna in the Mesopotamian plain is the mix of the species that evolved from the saline tolerant freshwater taxa in the lakes that existed in former times and the geologically recent immigrants that either reached Mesopotamian plain from the Palaearctic or from the east.

The gastropods of the family Melanopsidae dominate in the gastropod assemblage, while the Asian clam *Corbicula fluminea* is the most common mussel in Tigris and Euphrates basins. The Mesopotamian plain harbours a few endemic species of freshwater molluscs.

Freshwater molluscs are affected by multiple threats: water abstraction, natural system modification by constructing large dams, draining of the lower basin marshes and impact of invasive species being found as most impacting. About 29% of Iraqi mollusc fauna is non-native, indicating that malacofauna of the Mesopotamian is among the most threatened in the world by invasive species.

Keywords Mesopotamia · Molluscs · Diversity · Endemism · Conservation

V. Pešić (✉)
Department of Biology, University of Montenegro, Podgorica, Montenegro

P. Glöer
Biodiversity Research Lab, Hetlingen, Germany
e-mail: gloeer@malaco.de

L. A. Jawad (ed.), *Tigris and Euphrates Rivers: Their Environment from Headwaters to Mouth*, Aquatic Ecology Series 11,
https://doi.org/10.1007/978-3-030-57570-0_33

33.1 Introduction

In the ancient Mesopotamia, molluscs were widely used, mainly as a source for food, but their shells were also exploited as a valuable raw material, employed in manufacture inlays, personal ornaments, seals and utensils (MacKay 1999). Moreover, the shells played role in religious life (Aynard 1966) and in specific socio-ritual functions (Gensheimer 1984). Freshwater molluscs played much less important role than marine ones, and the mussels of genus *Unio* were the most used freshwater molluscs. These mussels were used as food, whilst their shells were used as personal ornaments or containers (MacKay 1999). Some freshwater molluscs were found in bitumen stopper: for example, the bitumen stands for ostrich-egg cups at Kish, which were inlaid with pieces of *Anodonta* mussels in Early Dynastic III (Mackay 1999).

Much of the knowledge of the freshwater molluscan fauna for the Mesopotamian region is fragmented. The first paper about freshwater molluscs of Mesopotamia has been published in 1874 by Eduard von Martens who reported molluscs collected by the botanist Prof. Hausknecht from his expedition to Persia (Allouse 1956). In the same year, Mousson (1874) published new data on freshwater molluscs of Mesopotamia based on the material collected by Dr. Schlaefli. He listed 19 species of freshwater gastropods and eight bivalve species from Mesopotamia.

The most comprehensive papers on Mesopotamian malacofauna were published by Annandale (1918, 1920) and Prashad (1921) who listed 14 gastropod species. Germain (1921a, b, 1922, 1921-1924) and Pallary (1929, 1939) studied malacofauna of Syria. Over the following century, a number of subsequent works (Najim 1950, 1959, 1960; Soyer 1961) contributed to our knowledge of the molluscs of Mesopotamia and the most recent summary was provided by Ahmed (1975) who provided a complete checklist of species living in Iraq. Plaziat and Younis (2005) synthesised the data on the Quaternary freshwater malacofauna of lower Mesopotamia. This region was almost destroyed as a result of the draining of the lower basin marshes in the 1990s (Partow 2001).

In the last decade of the twenty-first century, a number of publications were published (Abdul-Sahib and Abdul-Sahib 2005, 2008; Naser 2006, 2010; Glöer and Naser 2007, 2008; Glöer et al. 2008; Naser and Son 2009; Haase and Wilke 2010; Mohammad 2014; Al-Fanharawi and Ibrahim 2014; Al-Waaly et al. 2014; Salman and Nassar 2014), increasing our knowledge on the Mesopotamian malacofauna. Most of these papers include faunistical information, including description of new species (Glöer and Naser 2008, 2013; Glöer et al. 2008). On the other hand, the current scientific literature on the ecology of freshwater malacofauna of Mesopotamia is still scant (Harris 1965) but with evident progress in the last few years (Mohammad 2014; Al-Fanharawi and Ibrahim 2014; Salman and Nassar 2014). Most of the published ecological works were devoted to *Bulinus truncatus* (the intermediate host for *Schistosoma haematobium*) and *Radix auricularia* (the intermediate host of *Fasciola gigantica*) (Watson 1958; Najarian 1961; Al-Asadi 2011; Al-Yaquob 2011).

The aim of this study is to synthesise all accumulated data concerning the species content, distribution and taxonomy of the freshwater molluscs that inhabit the plains between the Tigris and Euphrates rivers. This region is covered with alluvial sediments of Pleistocene and Holocene and represents a complex of shallow freshwater lakes, swamps, marshes and seasonally inundated plains.

33.2 Distribution and Species Richness of Molluscs of Mesopotamia

A total of 35 mollusc species, twenty species of Gastropoda and five species of Bivalvia have been included into annotated list of the freshwater molluscs of Iraq (Table 33.1). The fauna of freshwater molluscs of Iraq is taxonomically impoverished as compared to the fauna of neighbouring countries: Turkey (Yıldırım et al. 2006) and Iran (Glöer and Pešić 2012). It is worth mentioning that the compiling of the species list for the Mesopotamian region has been a difficult task and it is likely that a number of species that live in Syrian, Turkish and Iranian parts of the catchment may have been missed. For example, Pallary (1929, 1939) listed *Pyrgula syriaca* and *P. euphratica* for Mesopotamia, but these species have not been found by the later authors. It is possible that some of the 'older'species are valid taxa, but we cannot exclude that they are apparent synonyms of other species, which in the future require a thorough revision of this list from both nomenclatorial and taxonomic points of view. The taxonomy of some groups especially Melanopsidae and Planorbidae (see discussion below under these families) is either presently under revision or is in urgent need of revision. Moreover, it is very likely that new species will be discovered, particularly by applying molecular techniques.

The current molluscan fauna of Mesopotamian region is the result of spectacular changes in the evolution of the whole region, which are present in the region over the millions of years and primarily were shaped by the existence of vast lakes with marked salinity gradients and fluctuations and the possibility of migration among these ecosystems (Werner et al. 2007; Wesselingh 2007). From a biogeographic point of view, the Mesopotamian freshwater fauna belongs to Mediterranean palaearctic domain (Plaziat and Younis 2005). The Mesopotamian plain was formed at the end of Zagros orogeny in Miocene–Pliocene and the modern malacological fauna is a small relict of the fauna formed at that time (Seddon et al. 2014). The higher richness in some families like Neritidae and Melanopsidae is directly linked to the evolutionary radiation of saline tolerant freshwater taxa in the lakes that existed in former times (Seddon et al. 2014). On the other hand, a large part of the recent malacofauna consists of geologically recent immigrants that either reached Mesopotamian plain from the Palaearctic (e.g. *Bithynia* and *Radix)* or from the east (e.g. *Bellamya* and *Corbicula*).

Table 33.1 List of mollusc species occurring in Iraq. Endemic species are marked by an asterisk

Taxa	Red List Category
Neritidae Lamarck, 1809	
Neritina schlaeflii Mousson 1874	
Neritina mesopotamica Martens, 1874	
Neritina euphratica (Mousson 1874)	Least concern
Theodoxus jordani (Sowerby, 1832)	
Viviparidae J.E. Gray, 1847	
Bellamya bengalensis (Lamarck, 1822)	
Melanopsidae H. & A. Adams, 1854	
Melanoides tuberculatus O.F. Müller, 1774	Least concern
Melanopsis costata (Olivier, 1804)	Least concern
Melanopsis subtingitana Annandale 1918	
Melanopsis nodosa Férussac, 1823	Least concern
Melanopsis buccinoidea (Olivier, 1801)	Least concern
Bithyniidae Troschel, 1857	
Bithynia badiella (Küster, 1853) (?)	
Bithynia iraqensis Pallary 1939	
* *Bithynia hareerensis* Glöer and Naser 2008	
* *Bithynia ejecta* Mousson 1874	
Hydrobiidae Troschel, 1857	
Hydrobia lactea (Küster, 1852)	Near threatened
Ecrobia grimmi (Clessin & Dybowski, 1888)	
* *Assiminea mesopotamica* Glöer & Naser 2007	Data deficient
* *Assiminea zubairensis* Glöer and Naser 2013	
Potamopyrgus antipodarum (J.E. Gray, 1843)	
Lymnaeidae Rafinesque, 1815	
Radix auricularia (Linnaeus, 1758)	Least concern
Radix lagotis (Schrank, 1803)	Data deficient
Galba truncatula (O.F. Müller, 1774)	Least concern
Planorbidae Rafinesque, 1815	
Planorbis intermixtus Mousson 1874	
Gyraulus euphraticus (Mousson 1874)	Least concern
* *Gyraulus huwaizahensis* Glöer and Naser 2008	Data deficient
Gyraulus convexiusculus (Hutton, 1849)	Least concern
Gyraulus chinensis (Dunker, 1848)	Least concern
Bulinus truncatus (Audouin, 1827)	Least concern
Ferrissia californica (Rowell, 1863)	Least concern
Physidae Fitzinger, 1833	
Physa acuta (Draparnaud, 1805)	Least concern
Bivalvia	
Unionidae Rafinesque, 1820	
Unio tigridis Bourguignat, 1852	Least concern
Pseudodontopsis euphraticus (Bourguignat, 1852)	

(continued)

Table 33.1 (continued)

Taxa	Red List Category
Anodonta vescoiana Bourguignat, 1857	Near threatened
Corbiculidae J.E. Gray, 1847	
Corbicula fluminea (O.F. Müller, 1774)	Least concern
Dreissenidae J.E. Gray, 1840	
Dreissena siouffi Locard, 1893	

33.3 Diversity of Gastropoda

Altogether, there are 20 gastropod species belonging to eight families that have been recorded so far from Iraq (Table 33.1). The family Neritidae includes four species, three species of the genus *Neritina* and the widespread *Theodoxus jordani*, which inhabits lakes, marsh channels, fluviatile channels and estuaries (Plaziat and Younis 2005; Abdul-Sahib and Abdul-Sahib 2008). Molecular analysis revealed that the population of *Th. jordani* from its *locus typicus* (Jordan River) is genetically conspecific with the population from Shatt-Al Arab (Katharina Kurzrock pers. comm., FREDIE-Project). *Neritina* species predominantly inhabits brackish waters. Plaziat and Younis (2005) mentioned that *Neritina schlaeflii* forages in the muddy intertidal zone near the mouth of the Shatt al Arab where the salinity attained 2‰. Most *Neritina* species seem to be restricted to the Shatt al-Arab, i.e. the joined downstream part of the Tigris and Euphrates between Basrah and the Persian Gulf in Iraq, but some of the species such as *N. euphratica and N. mesopotamica* were also found in Khuzestan province in Iran (Glöer and Pešić 2012) (Figs. 33.1 and 33.2).

The family Viviparidae includes one species, *Bellamya bengalensis*. This species is a recent invader (Soyer 1961) and is common in the lower Euphrates, Tigris and Shatt Al-Arab Rivers (Nesemann 2007; Al-Fanharawi and Ibrahim 2014; Salman and Nassar 2014). It lives in fresh water on the bottom muds of ponds, marshes and marsh channels but can also be found in estuarine shore deposits (Plaziat and Younis 2005). The specimens of *Bellamya bengalensis* (as well as the species of the genus *Radix)* that inhabit habitats with high salinity in southern part are smaller than those that live in the central part of Mesopotamia (Al-Waaly et al. 2014).

The gastropods of the family Melanopsidae are the most common snails in Euphrates and Tigris drainage, but systematic status of many species is still unclear. For example, Harris (1965) listed 20 taxa of *Melanopsis*, but Plaziat and Younis (2005) lumped all these taxa together to *M. praemorsa*, indicating that further study is necessary to clarify taxonomic status of Mesopotamian melanopsid populations. In this paper, we list six species of Melanopsidae for the Tigris and Euphrates basin based on their conchological characteristics. Fossil specimens of the melanopsid species have been reported from the Euphrates valley in Syria (Freedman and Lundquist 1977). Most melanopsid snails are widespread and only *Melanopsis subtingitana* has a more restricted distribution in central and lower Euphrates (Naser 2006; Salman and Nassar 2014), Al-Hammar marshes and the Garmat-Ali, the canalised river that drains later marshes in the southern part of Mesopotamia

Fig. 33.1 Gastropod species occurring in the Mesopotamian plain. (**a**) *Neritina euphratica* (syntype); (**b**) *Neritina mesopotamica* (syntype ZMZ528916, photo Eike Neubert); (**c**) *Neritina schlaeflii*; (**d**) *Theodoxus jordani*; (**e**) *Radix lagotis*; (**f**) *R. auricularia*; (**g**) *Melanopsis costata*; (**h**) *M. nodosa*; (**i**) *M. subtingitana*; (**k**) *M. buccinoidea*; (**l**) *Assiminea mesopotamica*; (**m**) *A. zubairensis*; (**n**) *Melanoides tuberculatus*; (**o**) *Ecrobia grimmi*. (Photos by Peter Glöer)

Fig. 33.2 Gastropod species occurring in the Mesopotamian plain. (**a**) *Potamopyrgus antipodarum*; (**b**) *Galba truncatula* (topotype); (**c**) *Physa acuta*; (**d**) *Bithynia ejecta* (Syntype ZMZ524006, Samava); (**e**) *Bithynia hareerensis*; (**f**) *Bellamya bengalensis*; (**g**) *Bulinus truncatus*; (**h**) *Gyraulus huwaizahensis*; (**i**) *Gyraulus convexiusculus*; (**k**) *Gyraulus chinensis*; (**l**) *Gyraulus euphraticus*; (**m**) *Planorbis intermixtus*. (Photos by Peter Glöer)

(Naser 2006). Most species are ubiquitous and inhabit diverse habitats including springs and pools, irrigation canals, ponds, streams, swamps, ditches and rivers. Some species, i.e. *Melanoides tuberculatus* and *Melanopsis costata*, are resistant to severe drought (Dzikowski et al. 2003) and most species of the family tolerate salinities. For example, *M. costata* has been recorded to withstand salinity of almost 0.4 psu (Falniowski et al. 2002), while *M. tuberculatus* can tolerate salinities of up to 23 psu (Plaziat and Younis 2005).

The euryhaline *Melanoides tuberculatus* is widely distributed in mountain streams in the northern part of Iraq and the rivers, pools, lakes and ditches in the southern part of the country (Al-Waaly et al. 2014). *Melanopsis nodata* is widely distributed in Tigris and Euphrates drainage (Al-Fanharawi and Ibrahim 2014; Salman and Nassar 2014), but more common in the lower Mesopotamia (Naser 2006) reaching an abundance of 245.45 ind./m^2 in the lower part of Euphrates (Al-Fanharawi and Ibrahim 2014). *Melanopsis costata* is known from Euphrates River but also from the agriculture canals of Khuzestan Province in southwest Iran (Farahnak et al. 2006). Haller et al. (2005]) showed that the latter species hybridises with *M. buccinoidea* over the last 1.5 Myr. *Melanopsis buccinoidea*, widely distributed in artesian springs and irrigation channels of Ain Al-Tamura area (Mohammad 2014) and the upper part of Euphrates (Heller et al. 2005), seems to be absent from the lower Mesopotamia. On the other hand, the smooth shelled *Melanopsis* populations from Iranian Khuzestan and Bushehr provinces belong to *Melanopsis doriae* (see Glöer and Pešić 2012), whose presence in Iraq is still not confirmed.

The Bithyniidae include four species known from Mesopotamia. Occurrence of some species in Mesopotamia like *Bithynia badiella* (Annandale 1918) is questionable. The genus seems to be more common in the northern part of the catchment and Al-Waaly et al. (2014) failed to find bithyniid species in central and southern part of Iraq. *Bithynia iraqensis* was described from Gireza spring (Qalkand) in the northern part of Iraq (Pallary 1939). Prashad (1921) reported *Bithynia ejecta* from the Euphrates at Nasiriyah and at Fallujah. Glöer and Naser (2008) described *B. hareerensis* from Garmat Ali River, a naturally regulated river that drains the Al-Hammar marsh.

The Hydrobiid snails are neither abundant nor diverse in the water bodies in the Mesopotamian plain. Taxonomic status of some species like *Hydrobia lactea* (originally described from Tigris near Mosul) is unresolved and requires application of molecular techniques (Neubert and Amr 2012). Most hydrobiid species like those belonging to *Assiminea* and *Ecrobia* show tolerance to brackish waters. Two endemic *Assiminea* species recently were described for Shatt Al-Arab-Fao region (Glöer and Naser 2008, 2013). *Assiminea zubairensis* lives where salinity may exceed 37 psu, while *A. mesopotamica* occurs in regions of lower salinity (Glöer and Naser 2013). Two hydrobiid species *Ecrobia grimmi and Potamopyrgus antipodarum* are recent invaders. Based on the molecular results, Haase and Wilke (2010) hypothesised that population of *Ecrobia grimmi* from the mixomesohaline Lake Sawa (Iraq) was possibly transported by migrating birds from the Caspian Sea. However, it should be taken into account that there is no direct evidence for hydrobiids travelling with birds and is possible that the lake has not been colonised directly from the Caspian Sea. *Potamopyrgus antipodarum* was found first time in 2008 collected from the banks of Garmat Ali River, part of Shatt Al-Arab (Naser and Son 2009).

Several studies demonstrated the presence of exceptionally rich hydrobiid fauna in the springs and streams in Eastern Anatolia, which belongs to Euphrates drainage (Schütt and Şeşen 1993; Yıldırım et al. 2006; Sahin et al. 2012). Twelve hydrobiid genera, i.e. *Islamia, Sadleriana, Pseudamnicola, Belgrandiella, Sheitanokok, Hydrobia, Ventrosia, Pyrgorientalia, Bythinella, Kirelia, Anadoludamnicola* and *Sivasi*, have been reported from Eastern Anatolia (Sahin et al. 2012). The rich hydrobiid fauna can also be expected in the springs and streams on the western slopes of Zagros Mountains of Iran, which belongs to the Tigris drainage (Glöer and Pešić 2009; Delicado et al. 2016). Glöer and Pešić (2009) described two species *Intermaria zagrosensis* and *I. kermanshahensis* from the springs and streams in Kermanshah province in Iran.

The diversity of lymnaeid snails in Mesopotamia is low. The family includes the common and widespread *Galba truncatula* (Najim 1959; Dautzenberg 1894) and two species of the genus *Radix*, i.e. *R. auricularia* and *R. lagotis*, which are rather common in Mesopotamian plain especially in the waters rich in vegetation. *Radix auricularia* is one of the most dominant species in Shatt Al-Arab (Ahmed 1975; Plaziat and Younis 2005). Taxonomic status of Mesopotamian populations of *Radix lagotis* should be verified by the application of molecular methods (Schniebs et al. 2015).

The seven species of Planorbidae have been recorded from the Mesopotamian plain so far, but systematic of this group requires further study. Occurrence of some species in Mesopotamia, like *Gyraulus ehrenbergi* (Najim 1959), an African species, is probably a misidentification. *Planorbis intermixtus*, a common species in Iran (Glöer and Pešić 2012), is not very abundant in southern Mesopotamia and usually is found in the subaqueous portion of the rooted vegetation of lakes and swamps (Plaziat and Younis 2005). Records of Annandale (1920) are questionable: his specimen of *P. intermixtus* from the river deposits of Lower Mesopotamia has a diameter of about 4 mm, while the typical *P. intermixtus* is a larger species, with a diameter of about 6.5-8.6 mm. *Gyraulus euphraticus* is a common species in the Euphrates floodplain. Only one planorbid species have a restricted distribution: *Gyraulus huwaizahensis* described from the Al Huwaizah marshes in the upper Tigris-Euphrates delta in Iraq (Glöer and Naser 2008). Three species *Ferrissia californica, Gyraulus convexiusculus* and *G. chinensis* are invaders. Two later *Gyraulus* species are native to south Asia: *G. convexiusculus* has been reported from Euphrates (Annandale 1918, 1920) and Shatt Al-Arab and the Garmat-Ali River (Ahmed 1975), while *G. chinensis* has been found in the region of lower Euphrates near Basra and in Al-Hammar Marshes (Glöer and Naser unpublished). *Ferrissia californica*, a North American cryptic invader, was recently reported by Marrone et al. (2014) from Tigris River at Qurna.

Bulinus truncatus is one of the best studied gastropod species (Watson 1958; Najim 1959; Al-Waaly et al. 2014). This species occurs in marshes, swamps, irrigation canals and drains in central Iraq along the Tigris and Euphrates rivers; in the northern part of the country, it is extinct and it does not occur in southern Mesopotamia (Watson 1958). Najarian (1961) found *B. truncatus* in the canals of central Iraq. Nowadays, this species could not be found in Iraq, which is the result of the intense chemical control campaign against *Bulinus truncatus* over the last five decades (Al-Waaly et al. 2014).

The family Physidae includes one species *Physa acuta*, which is one recent invader, first time discovered in Addawoudy near Bagdad city (Najim 1959). Today, this species, whose impact on the indigenous taxa is well documented (Naser et al. 2011), is a common gastropod in Iraq (Al-Waaly et al. 2014).

33.4 Diversity of Bivalvia

Five species included in our checklist belong to three families and 4 genera of Bivalvia. *Unio tigridis* is widely distributed mussel species in most of the Tigris-Euphrates catchment (Falkner 1994; Ahmed 1975; Al-Mahdawi and Al-Dulaimi 2004; Al-Fanharawi and Ibrahim 2014). According to Plaziat and Younis (2005), it is the most common unionid species in Lower Mesopotamia. Naser (2010) mentioned that *Unio tigridis*, *Pseudodontopsis euphraticus* and *Anodonta vescoiana* are rare in Shatt Al-Arab. *Pseudodontopsis euphraticus* is reported from Tigris near Bagdad (Mousson 1874) and marshes near Qurna and Shatt Al-Arab (Ahmed 1975; Plaziat and Younis 2005), burrowing in the bottom muds. *Anodonta vescoiana* (Fig. 33.3) is present over the Tigris and Euphrates basin (Mousson 1874; Falkner 1994), with the recent records from the lake Hammar wetlands near the coast

Fig. 33.3 *Anodonta vescoiana*. Assad dam near Halawa (leg. J. Boessneck & A. von den Driesch 1985). Shell set with a colony of *Dreissena siouffi* by which the *Anodonta* possibly died. Length of the shell is 13.8 cm (Photo by G. Falkner)

(Abdul-Sahib and Abdul-Sahib 2009; Plaziat and Younis 2005). This species, less frequent than other unionid species, is indicative for the lacustrine environment (Plaziat and Younis 2005).

The most common mussel in Tigris and Euphrates basins is the Asian clam *Corbicula fluminea*, which is native to southern and eastern Asia. The taxonomic status of this species is still unclear, and according to Korniushin's (2004) molecular analysis, *C. fluminea* and *C. fluminalis*, which in the past have often been confused with each other, are parthenogenetic clones of the same species. In this list, we consider them as morphotypes, rather than as separate species. Both the morphotypes of Asian clams occurred in Tigris and Euphrates catchment: *C. fluminea* (Mousson 1874; Ahmed 1975; Al-Fanharawi and Ibrahim 2014) and *C. fluminalis* (Ahmed 1975; Plaziat and Younis 2005). This species dominates in running waters reaching an abundance of 572.7 ind./m^2 in the lower part of Euphrates (Al-Fanharawi and Ibrahim 2014), as well as in many lakes and marshes (Plaziat and Younis 2005).

The family Dreissenidae includes *Dreissena siouffi*, endemic of Euphrates River (Falkner 1994; Schütt and Şeşen 2007). Recently, this species was found in the upper part of Euphrates in the Birecik dam area in Turkey (Ekin et al. 2008). The invasive zebra mussel (*Dreissena polymorpha*) has been reported from the upper part of Euphrates, causing technical and economic damage in Atatürk dam and hydropower plants built on the Euphrates River since 1997 (Bobat et al. 2004).

33.5 Major Threats to Molluscan Fauna

There are multiple drivers of threat to freshwater molluscs in the Mesopotamian plain: (1) water abstraction for domestic supplies and agriculture, (2) the large dams in the upper reaches of the Euphrates and Tigris Rivers in Turkey and Syria, and on the Karun and Karkheh Rivers in Iran (Partow 2001), strongly impacting downstream water use (Beaumont 1996), and (3) the projects of draining of the lower basin marshes (Seddon et al. 2014). The draining of the lower basin marshes in the second half of the twentieth century and particularly in the 1990s after the second Gulf War 1991 made the iconic Iraqi marshlands almost disappear: for example, once 120 km long Lake Hammar practically disappeared between 1992 and 1994 (Munro and Touron 1997). The loss of wetlands consequently led to negative impact on the freshwater mollusc populations. Moreover, this fauna is highly susceptible to climatic changes. The climate change in the Eastern Mediterranean region is predicted to become dryer and warmer, with a particular increase in the frequency of hot summer days and high temperature events (Seddon et al. 2014). Possibly, these changes will lead to salinisation and eutrophication of the waters in the marshes affecting mollusc assemblages. This process is significantly impaired by the impact of invasive species and habitat alteration.

The mollusc assemblage in the Tigris and Euphrates basin is threated by the alarming number of invasive species, which impacts autochthonous species. About

29% of Iraqi mollusc fauna (seven gastropod and one species of Bivalvia) is non-native, which is probably one of the largest shares of non-native species in the fauna of the molluscs of one country. The first invasive gastropod species *Bellamya bengalensis* has been introduced from India to Iraq between 1929 and 1939 (Pallary 1939), followed by the findings of *Physa acuta* in the 1950s (Najim 1959). The recent invaders include the New Zealand species *Potamopyrgus antipodarum* (Naser and Son 2009), hydrobiid *Ecrobia grimmi* (Haase and Wilke 2010) possibly introduced from Caspian Sea and the North American cryptic invader *Ferrissia californica* possibly introduced via Europe (Marrone et al. 2014).

The characteristic of the mollusc fauna of Mesopotamian plain is that most species are common. Only five species known from the type localities can be considered as possibly locally endemic (Table 33.1), indicating that this region does not belong to the hot spots of gastropod diversity at the global scale (Strong et al. 2008).

Acknowledgements We would like to thank Eike Neubert (Bern) for the photo of *Bithynia ejecta* and to M. K. Mohammad (Baghdad) and M. D. Naser (Basra) for providing samples from Iraq.

References

Abdul-Sahib IM, Abdul-Sahib EM (2005) The abundance and distribution of Mollusca in common waters of southern region of Iraq. Dept. Marine Biology, Marine Science Centre, University of Basrah; 7 pp.

Abdul-Sahib IM, Abdul-Sahib EM (2008) The abundance and distribution of Molluscan in common waters of southern region of Iraq. Al-Qadisyah Journal for pure science 13(3):1–7

Abdul-Sahib IM, Abdul-Sahib EM (2009) A new record of the freshwater clam, Anodonta vescoiana Bourguignat, 1857 (Mollusca: Bivalvia) from Al-Ezz River, Iraqi Marshes. Mesopot J Mar Sci 24:7–12

Ahmed MM (1975) Systematic study on Mollusca from the Arabian gulf and Shatt al-Arab. Iraq. Center for Arab Gulf Studies University Basrah, 78 pp

Al-Asadi SAM (2011) Effect of calcium chloride on hatching, growth and survival of snails *Lymnaea auricularia* – Intermediat host of *Fasciola gigantica*. Marsh Bulletin 6:125–133

Al-Fanharawi AA, Ibrahim SS (2014) Environmental study of the benthic mollusks in Euphrates River at Samawa City, Iraq. Int J Sci Res (Raipur) 3:1955–1960

Allouse BE (1956) A bibliography on the invertebrate Fauna of Iraq and Neighbouring countries. I. Mollusca. Iraq National History Museum. 38 pp.

Al-Mahdawi M, Al-Dulaimi M (2004) Reproductive behavior of freshwater clams *Unio tigridus* (Bou.) and *Pseudodontopsis euphraticus* (Bou.) in Al-Habbanyah Lake-Iraq. J King Abdulaziz Univ Marine Sci 15(1):3–20

Al-Waaly ABM, Mohammad MK, Al-Miali HM (2014) Freshwater snails diversity in the middle and south regions of Iraq. Advances in Bioresearch 5(3):166–171

Al-Yaquob AJ (2011) Effect of salinity and pH on hatchability and survival of the snails *Lymnaea auricularia*. Marsh Bulletin. 6:62–72

Annandale N (1918) Freshwater shells from Mesopotamia. Records of the Indian Museum. 15:159–170

Annandale N (1920) Report on the freshwater gastropod mollusks of lower Mesopotamia. Part II. The family Planorbidae. Records of the Indian Museum 18(3):147–148

Aynard JM (1966) Coquillages Mésopotamiens. Syria XLIII. p. 21–37
Beaumont P (1996) Agricultural and environmental changes in the upper Euphrates catchment of Turkey and Syria and their political and economic implications. Appl Geogr 16:137–157
Bobat A, Hengirmen MO, Zapletal W (2004) Zebra mussel and fouling problems in the Euphrates basin. Tur J Zool 28:161–177
Dautzenberg P (1894) Liste des mollusques terrestres et fluviatiles recueillis par M. Th. Barrois en Palestine et en Syrie. Revue biologique du Nord de la France 6:329–353
Delicado D, Pešić V, Glöer P (2016) Unraveling a new lineage of Hydrobiidae genera (Caenogastropoda: Truncatelloidea) from the Ponto-Caspian region. Eur J Taxon 208:1–29
Dzikowski R, Diamant A, Paperna I (2003) Trematode metacercariae of fishes as sentinels for a changing limnological environment. Dis Aquat Org 55:145–150
Ekin I, Bashan M, Şeşen R (2008) *Dreissena siouffi* (Locard 1893) (Bivalvia: Dreissenidae) collected from Euphrates and its fatty a cid composition. Firat Univ J Sci Engg 20:243–250
Falkner G (1994) Systematik vorderorientalischer Najaden als Vorstudie zur Bearbeitung archäologischer Funde. Forschungen und Berichte zur Vor- und Frühgeshcichte in Baden-Württemberg 53:135–161
Falniowski A, Heller J, Mazan-Mamczarz K, Szarowska M (2002) Genetic structure of populations of the closely related species of *Melanopsis* (Gastropoda: Cerithiacea) in Israel. J Zool Syst Evol Res 40:92–104
Farahnak A, Vafaie-Darian R, Mobedi I (2006) A faunistic survey of Cercariae from fresh water snails: *Melanopsis* spp. and their role in disease transmission. *Iran J Public Health* 35(4):70–74
Freedman DN, Lundquist JM (1977) Archeological reports from the Tabqa dam project: Euphrates Valley, Syria. Annu Amer Sch orient Res 44:1–177
Gensheimer TR (1984) The role of shell in Mesopotamia: evidence for trade exchange with Oman and the Indus Valley. Paleorient 10:65–73
Germain L (1921a) Mollusques terrestres et fluviatiles de Syrie. Voyage zoologique d'Henri de Kerville En Syrie 1:1–523
Germain L (1921b) Catalogue of the Planorbidae in the Indian museum. Rec Indian Museum 21:81–210. pl. 1-4
Germain L (1922) Mollusques terrestres et fluviatiles de Syrie. Voyage zoologique d'Henri de Kerville En Syrie. 2:1–243 + 22 pl.
Glöer P, Naser MD (2007) Gyraulus huwaizahensis n. sp. – a new species from Mesopotamia, Iraq (Mollusca: Gastropoda: Planorbidae). Mollusca 25:147–152
Glöer P, Naser MD (2008) *Bithynia hareerensis* n. sp., a new *Bithynia* from Mesopotamia (Iraq) (Gastropoda: Bithyniidae). Mollusca 26:159–162
Glöer P, Naser MD (2013) *Assiminea zubairensis* n. sp., a new *species* from Iraq (Mollusca: Gastropoda: Assiminiidae). J Conchol 41:305–307
Glöer P, Naser MD, Yasser AG (2008) *Assiminea mesopotamica* n. sp., a new *species* from Mesopotamia, Iraq (Mollusca: Gastropoda: Assiminiidae). Mollusca 25:3–6
Glöer P, Pešić V (2009) New freshwater gastropod species of the Iran (Gastropoda: Stenothyridae, Bithyniidae, Hydrobiidae). Mollusca 27(1):33–39
Glöer P, Pešić V (2012) The freshwater snails (Gastropoda) of Iran, with descriptions of two new genera and eight new species. ZooKeys 219:11–61
Haase M, Naser MD, Wilke T (2010) *Ecrobia grimmi* in brackish Lake Sawa, Iraq: indirect evidence for long-distance dispersal of hydrobiid gastropods (Caenogastropoda: Rissooidea) by birds. J Molluscan Stud 76:101–105
Harris SA (1965) Ecology of the freshwater mollusca of Iraq. Can J Zool 43:509–526
Heller J, Mordan P, Ben-Ami F, Siva N (2005) Conchometrics, systematics and distribution of Melanopsis (Mollusca: Gastropoda) in the Levant. Zool J Linnean Soc 144:229–260
Korniushin A (2004) A revision of some Asian and African freshwater clams assigned to *Corbicula fluminalis* (Müller, 1774)(Mollusca: Bivalvia: Corbiculidae) with a review of anatomical and reproductive features based on museum collections. Hydrobiologia 529:251–270

MacKay E (1999) A sumerian palace and the "A" Cemetery. The field museum of natural history, Chicago, USA Moorey PRS. Ancient mesopotamian material and industries. The archaelogical evidence. Eisenbrauns, Winona Lake, Indiana; p. 1–415

Marrone F, Naser MD, Amaal GY, Sacco F, Arculeo M (2014) First record of the north American cryptic invader *Ferrissia fragilis* (Tryon, 1863) (Mollusca: Gastropoda: Planorbidae) in the Middle East. Zool Middle East 60(1):39–45

Mohammad MK (2014) Ecology of the freshwater snail *Melanopsis buccinoidea* (Olivier, 1801) in Ain Al-Tamur, Kerbala Province. Int J Curr Microbiol Appl Sci 3(2):390–394

Mousson MA (1874) Coquilles terrestres et fluviatiles recueillies par M. le Dr Alex. Schlaefli en Orient. Journal de conchyliologie 22:5–60

Munro DC, Touron H (1997) The estimation of marshland degradation in southern Iraq using multitemporal Landsat TM images. Int J Remote Sens 18(7):1597–1606

Najarian HH (1961) Biological studies on the snail *Bulinus truncatus*, in Central Iraq. Bull Wld Hlth Org 25:435–446

Najim AT (1959) Notes on the distribution of some mollusks in Iraq. Proc Malac Soc Lond. 33:159–163

Najim AT (1960) Further notes on the distribiution of mollusks in Iraq with the first record of Physa. Proc Malac Soc Lond 33:299–301

Naser MD (2006) Some notes on Melanopsidae (Gastropoda) of Mesopotamia. Iraq Aqua J 2:85–90

Naser M. (2010) The diversity of the Molluscs of Shatt Al- Arab, Basrah, Iraq. In: Pešić V, editor. The book of abstracts and Programme, IV international symposium of ecologists of Montenegro. Budva, 2010. p. 63

Naser MD, Son MO (2009) First record of the New Zealand mud snail *Potamopyrgus antipodarum* (gray 1843) from Iraq: the start of expansion to Western Asia? Aquat Invasions 4(2):369–372

Naser MD, Son MO, Yasser AG (2011) Assessing the risks of invasions of aquatic invertebrates in the Shatt Al-Arab River. Russ J Biol Invasions 2:120–125

Neubert E, Amr ZSS (2012) Hydrobia lactea. The IUCN Red List of Threatened Species. e. T189165A14211701. 2012. https://doi.org/10.2305/IUCN.UK.2012.RLTS. T189165A14211701.en

Pallary P (1929) Première Addition à la Faune Malacologique de la Syrie. Mémoires a l'Institut d'Égypte 39(6):43. pp, 3 Taf

Pallary P (1939) Deuxième Addition à la Faune Malacologique de la Syrie. Mémoires a l'Institut d'Égypte 39(8):141. pp, 7 Taf

Partow H (2001) The Mesopotamian marshlands: demise of an ecosystem. UNEO/DEWA TG.01-3. Nairobi, UNEP, division of early warning and. Assessment

Plaziat J-C, Younis WR (2005) The modern environments of mollusca in southern Mesopotamia, Iraq: A guide to paleogeographical reconstructions of Quaternary fluvial, palustrine and marine deposits. Carnets de Geologie. p.1–18

Prashad B (1921) Report on the freshwater gastropod mollusks of lower Mesopotamia. Part III. The families Neritidae, Hydrobiidae and Melaniidae. Rec Indian Museum 18:215–227

Sahin SK, Koca SB, Yildirim MZ (2012) *New genera Anatolidamnicola* and *Sivasi* (*Gastropoda*: *Hydrobiidae*) from *Sivas* and *Malatya* (*Turkey*). Acta Zool Bulg 64(4):341–346

Salman JM, Nassar AJ (2014) Variation of some Physico-chemical parameters and biodiversity of gastropods in Euphrates River, Iraq. Int J Environ Sci Develop 5(3):328–331

Schniebs K, Georgiev D, Glöer P, Hundsdoerfer AK (2015) A molecular genetic evidence of the occurrence of the freshwater snail *Radix lagotis* (Schrank, 1803) (Gastropoda: Lymnaeidae) in Bulgaria. Ecol Montenegrina 3:29–39

Schütt H, Şeşen R (1993) *Pseudamnicola* species and other freshwater gastropods (Mollusca, Gastropoda) from East Anatolia (Turkey), the Ukraine and the Lebanon. Basteria 57:161–171

Schütt H, Şeşen R (2007) The freshwater mussel Dreissena siouffi (Locard) in the river Euphrates. Triton 16:1–4

Seddon MB, Kebapçi Ü, Lopes-Lima M, van Damme D, Smith KG (2014) Chapter 4. Freshwater mollusks. In: Smith KG, Barrios V, Darwall WRT, Numa C (eds) The status and distribution of freshwater biodiversity in the eastern Mediterranean. The IUCN Red List of Threatened Species, pp 43–56

Soyer R (1961) Contribution à la malacology de la Mésopotamie (Irak). Cahiers des Naturalistes, Paris. (n.s.), t. 17:65–68

Strong EE, Gargominy O, Ponder WF, Bouchet PH (2008) Global diversity of gastropods (Gastropoda; Mollusca) in freshwater. Hydrobiologia 595:149–166

Watson JM (1958) Ecology and distribution of *Bulinus truncatus* in the Middle East. Bull. Org. mond. Santé, Bull Wld Hlth Org

Werner E, Piller W, Harzhauser M, Mandic O (2007) Miocene central Paratethys stratigraphy – current status and future directions. Stratigraphy 4(2/3):151–168

Wesselingh F (2007) Long-lived Lake Molluscs as island faunas: a bivalve perspective. In: Renema W (ed) Biogeography, time, and place: distributions, barriers, and islands. Germany, Springer Verlag, pp 275–314

Yıldırım MZ, Bahadir Koca S, Kebapçı Ü (2006) Supplement to the Prosobranchia (Mollusca: Gastropoda) Fauna of fresh and brackish waters of Turkey. Tr J Zool 30:197–204

Chapter 34
Freshwater Annelida of Iraq

Haifa J. Jaweir

The Annelida is a large phylum, with over than 22,000 species, including ragworms, earthworms, and leeches. They are adapted to various environments. Some species exist in and have adapted to marine environments, others in fresh water, and yet others in moist terrestrial environments. Freshwater Annelids are classified into the following subgroups:

34.1 Class Polychaetes

The Polychaeta are the largest and the most diverse of the Annelid groups, which include about 12,000. Most species are marine, but some have adapted to brackish or even freshwater and a very few are terrestrial. The number of freshwater species in the world is quite small when compared to marine species. The largest numbers of freshwater species are of the families Nereidae & Sabellidae due to their ability to withstand extreme salinity changes (Glasby and Timm 2008).

In Iraq, two species of family Nereidae were identified in Shatt Al-Arab, *Dendronereides heteropoda*, Southern, 1921 (Fauvel, 1953), and *Namalycastis indica*, Southern, 1921 (Jaweir 1987). Both species were also recorded in the Southern marshes of Iraq by Sabtie (2009).

Polychaeta are characterized by the presence of multiple chaetae per segment. They have parapodia that function as limbs. The head has short sensory projections, including palps and tentacles, in addition to the nuchal organs that are thought to be chemosensors. The diet consists of minute aquatic plants and animals in some species, while others are purely carnivorous. The sexes are usually separate; sperm

H. J. Jaweir (✉)
Department of Biology, Coll. of Science for Women, University of Baghdad, Baghdad, Iraq

L. A. Jawad (ed.), *Tigris and Euphrates Rivers: Their Environment from Headwaters to Mouth*, Aquatic Ecology Series 11,
https://doi.org/10.1007/978-3-030-57570-0_34

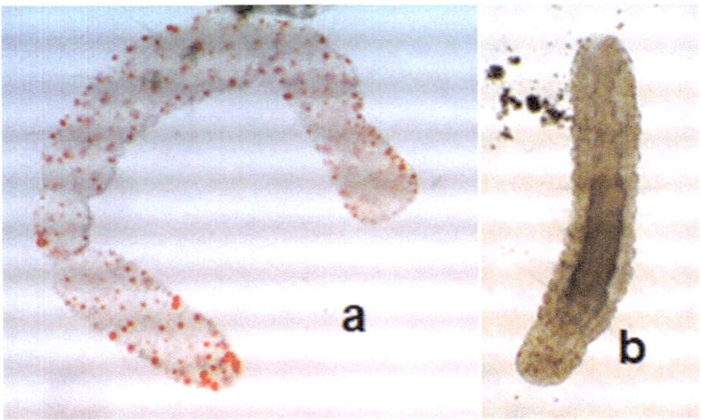

Plate 34.1 (**a**) *Aeolosoma quarternarium*; (**b**) *A.leidyi*

and eggs are discharged directly into the water, where fertilization occurs. The fertilized eggs developed to ciliated and free-swimming larvae, the trochophores. Some freshwater polychaete species are sequential hermaphroditic, i.e., the gonad functions first as one sex and then way to the other, mostly protendric (male-to-female); other species reproduce by budding, in which a portion of an adult's body breaks away to form a new individual (Architomy) (Rodriguez and Reynoldson 2011).

Species of family, Aeolosomatidae, may be treated as a separate class, Aphanononeura, according to the molecular data (Timm 1981; Parker 1982). They are previously classified as oligochaetes, usually sampled with freshwater oligochaetes. Timm (2009) classified this group as Polychaeta. They are small transparent worms with colored droplets and have simple hair setae in both dorsal and ventral bundles (Plate 34.1). They are hermaphrodite with ventral copulatory gland in some segment. In Iraq, four species of Aeolosoma were recorded (Table 34.1).

34.2 Class Clitellates

These have few or no chaetae per segment and no nuchal organs or parapodia. However, they have a ring-shaped reproductive organ around their bodies, the Clitellum, which produces a cocoon that stores and nourishes fertilized eggs until they hatch. The clitellates are subdivided into.

34.2.1 Subclass Hirudinea (Leeches)

Most leeches are freshwater animals, but many terrestrial and marine species occur. Freshwater leeches prefer to live in still or slowly flowing waters, but specimens have been collected from fast flowing streams. Some species are considered amphibious as they have been observed in both terrestrial and aquatic habitats. Marine

Table 34.1 List of aquatic oligochaetes of Iraq

Family	species
Aeolosomatidae	*Aeolosoma quarternarium* Ehernberg, 1828
	Aeolosoma leidyi Cragin, 1887
	Aeolosoma Variegatum Vejdovsky 1884
	Aeolosoma hemprichi Ehrenberg 1828
Naidiaea Naidinae	*Chaetogaster cristallinus* Vejdovsky, 1883
	Chaetogaster diaphanous (Gruithuisen, 1828
	Chaetogaster diastrophus (Gruithuisen, 1828)
	Chaetogaster langi Bretscher, 1869
	Chaetogaster limnaei Von Baer, 1827
	Dero (Aulophorus) furcatus(Müller, 1773)
	Dero (Aulophorus) indicus Naidu, 1962
	Dero (Dero) obtusa d'Udekem, 1855
	Dero (Dero) digitata (Müller, 1774)
	Dero (Dero) dorsalis Ferronnière, 1899
	Dero (Dero) evelinae, Marcus 1943
	Dero (Dero) nivea Aiyer, 1929
	Homochaeta naidina Bretscher, 1896
	Nais christinae Kasprzak, 1973
	Nais communis Piguet, 1906
	Nais elinguis Müller, 1773
	Nais pardalis Piguet, 1906
	Nais simplex Piguet, 1906
	Nais varaibilis Piguet, 1906
	Nais pseudobtusa Piguet, 1906
	Paranais litoralis (Müller, 1784)
	Paranais simplex Hrabe, 1936
	Paranais frici Müller, Müller, Hrabe, 1941
	Ophidonais serpentine (Müller, 1773)
	Allonais gwaliorensis (Stephenson, 1920)
	Allonais inaequalis (Stephenson, 1920)
	Stephensoniana trivandrana (Aiyer, 1929)
	Stylaria lacustris (Linnaeus, 1767)
	Stylaria fossularis Leidy, 1852
	Specaria josinae (Vejdovsky, 1883)
	Slavina appendiculata (d'Udekem, 1855)
	Stylaria lacustris (Linnaeus, 1767)
	Stylaria fossularis
Naididae Pristininae	*Pristina macrochaeta* Stephenson, 1931
	Pristina longiseta Ehrenberg, 1828
	Pritsina aequiseta Bourne, 1891
	Pristina foreli (Piguet, 1906)
	Pristina biserrata Chen 1940

(continued)

Table 34.1 (continued)

Family	species
	Pristinella jenkinae (Stephenson. 1931)
	Pristinella sima (Marcus, 1944)
	Pristinella osborni (Walton, 1906)
Naididae Tubificinae	*Aulodrilus pigueti* Kowalewski, 1914
	Embolocephalus velutinus (Grube, 1879)
	Limnodrilus claparedeianus Ratzel, 1869
	Limnodrilus hoffmeisteri Claparede, 1862
	Limnodrilus profundicola (Verrill, 1871)
	Limnodrilus udekemianus Claparede, 1862
	Limnodrilus cervix Brinkhurst, 1963
	Monopylephorus irroratus (Verrill, 1873)
	Potamothrix bavaricus (Oschmann, 1913)
	Potamothrix hammoniensis (Michaelsen, 1901)
	Psammoryctides barbata (Grube, 1861)
	Psammoryctides moravicus (Hrabe, 1934)
	Tubifex tubifex (Müller, 1774)
	Tubificoides benedeni (d'Udekem, 1855)
Branchiurinae	*Branchiura sowerbyi* Beddard, 1892
Lumbriculidae	*Lumbriculus variegatus* Müller, 1774
Lumbricidae	*Eiseniella tetraedra* (Savigny, 1826)

species are mostly blood-sucking parasites, mainly on fish, while most freshwater species are predators. They have suckers at both ends of their bodies.

Leeches are bilaterally symmetrical, with thick muscular bodies. Usually, they are dorsoventrally flattened and segmented, though the segments are not often seen. Some leeches are long and worm-like, and others are pear-shaped and broad. Most can vary considerably in shape both between the elongated and contracted state and between the starved and full condition. The body tapers toward the head and has a small oral sucker surrounding the mouth and a larger caudal sucker at the rear end, except the marine fish parasites, Pisciolidae, which have a larger oral sucker. The anus is on the dorsal surface just in front of the rear sucker. In Euhirudinea the body consist of 32–34 segment, and each segment is divided to externally 2–16 annuli, while Acanthobdellida (a small group of fish leeches) have 29, but counting is difficult because four to six segments are included in an front sucker and seven in a rear sucker, while the remaining segments are secondarily annulated to give two to five apparent segments per internal septum (internal membrane). Unlike other annelids, leeches do not have parapodia or chaetae (except for Acanthobdellida). Leeches usually have three jaws and make a Y-shaped incision. Their Sizes range between 7 and 300 mm long when extended.

There are around 500 species of leeches worldwide. These are divided into two major infraclasses

1. Euhirudinea: the 'true' leeches—marine, freshwater and terrestrial—which have suckers at both ends and lack chaetae (bristles). They have 32 internal segments when mature. The Euhirudinea is further divided into two orders:
   (a) Rhynchobdellida: jawless marine and freshwater leeches with a protrusible proboscis and true vascular system.
   (b) Arynchobdellida: jawed and jawless freshwater and terrestrial leeches with a nonprotrusible muscular pharynx and a hemocoelomic system.
2. Acanthobdellida: a small northern hemisphere infraclass ectoparasitic on salmoniid fish, which lacks an anterior sucker and retain chaetae.

The first information on the leeches in Iraq was given by Herzog (1969) who recorded *Piscicola* sp. from *Barbus schejch* in different regions of Iraq. Two species were reported from Greater Zab River, namely, *Fadejewobdella quinqueannulata* (Ali 2007b) and *Dina lineata* (Ali and Jaweir 2013), other species reported in Iraq were, namely, *Hemiclepsis marginata* from *Barbus sharpeyi* (Kalifa 1985), *Cystibranchus mastacembeli* from *Mastacembelus simach* collected from Tigris river in Mosul city (Rahemo 1989), *Piscicola geometra* from *Aspius vorax* (Mhaisen et al. 1997), and *Cystibranchus mammillatus* from *Mastacembelus mastacembelus* from Greater Zab River in Kurdistan region (Bashê 2008) and *Glossiphonia heteroclita* by Al-Raie (2014).

34.2.2 Subclass Oligochaetes

Oligochaetes have a sticky pad in the roof of the mouth. Most are burrowers that feed on wholly or partly decomposed organic materials.

Oligochaetous worms are predominantly aquatic and terrestrial class, though a few species are marine. The aquatic oligochaetes are divided to two superorders,

Superorder: Megadrili are relatively robust worms related to earthworms.

Superorder: Microdrili are mostly smaller, thin-bodied worms without any close terrestrial relatives.

Aquatic oligochaetes are usually small, ranging from 1 mm to a few centimeters in length and represent about third of almost valid species described to date; in addition, 4 of the 14 described megadrile families include species that occur in aquatic habitats. The microdriles are composed of 13 families, which are fully aquatic, with the exception of the primarily terrestrial family Enchytraeidae. About 1700 valid species of aquatic oligochaetes are known to date, of these about 1100 are freshwater. The most specious group is the tubificid worms (Naididae: Tubificinae) with more than 1000 described species including 582 that are considered as freshwater inhabitants. More than 60 species of megadriles are also considered aquatic. Three families in the orders Tubificida and Lumbriculida are common in freshwater, Naididae, Lumbriculidae, and Enchytraeidae, in addition to the minor families, Pavidrilidae; Dorydrillidae; Phreodrilidae, and propappidae. (Martin et al. 2008).

Traditionally, 'Naididae' has been treated as a family, but phylogenetic analysis based on 18S rDNA sequences supports the premise that all members of the former Clitellate family Naididae are phylogenetically nested within the former family Tubificidae (Brinkhurst and Jamieson 1971). Erséus and Gustarsson (2002) had proposed that these taxa together should be regarded as a single family to ovoid this paraphyly of Tubificidae, suggesting that all Niadides should be classified as members of Tubificidae, and for this reason, Erséus et al. (2005) submitted an application to the ICZN commission requesting to use its power to give precedence to Tubificidae, Vejdovesky 1876 over Naididae, Ehrenberg 1828, but the members of ICZN voted against it, stating that the use of the former name (Tubificidae) was not so great as to justify reversal of procedure and that no confusion with result from adherence to priority. Since the family-group name Naididae is older than Tubificidae, and thus, according to the International Code of Zoological Nomenclature, it should have precedence over the latter.

Among the family Naididae, Tubificid worms (subfamily Tubificinae) probably are the best known of the freshwater oligochaetes. They are most commonly found in soft sediments rich in organic matter, and several species characteristically live in sites that receive organic pollution. Like all aquatic oligochaetes, tubificids respire cutaneously, but a unique feature of this family is that some species can tolerate anoxic conditions. Most tubificids are deposit feeders, subsisting on organic detritus and its associated microflora. A few of the tubificids reproduce predominantly by fragmentation (e.g., *Aulodrilus* spp. and *Tasserkidrilus harmani*), but most are sexually reproducing hermaphrodites.

Naidid worms (subfamily Naidinae) are an ecologically diverse group of worms common in both running and standing waters. Many naidids are sediment dwellers, like the tubificids, but other species are characteristically found associated with aquatic organism, such as mollusks (Gorni and Alves 2006), macrophytes (Alves and Gorni 2007), bryophytes (Corbi et al. 2005; Gorni and Alves 2007), and sponges (Gorni and Alves 2008). They include detritivores (e.g., *Specaria josinae*), algivores (e.g., *Amphichaeta americana*), carnivores (e.g., *Chaetogaster diaphanus*), and even a parasite of snails (*Chaetogaster limnaei*). Sexual reproduction is rare in most species. Reproduction occurs predominantly by paratomy.

Timm (2009) excluded genus *Pristina* from previous family Naididae as recommended by Envall et al. (2008), arranged them in a separate family named as family Pristinidae, which are very small worms breeding mostly by budding (paratomy) like Niadidae, Sexually mature individuals, which occur seldom, reveal forward position of reproductive system, with male pores and clitellum mostly in VIII and spermathecae in VII, living on bottom surface and water plants, but never swimming, And Dorsal setae in *Pristina* species beginning in segment II as in Tubificid.

Enchytraeids are semiaquatic worms, common in marginal aquatic habitats as marshes, small streams, springs, and interstitial waters along the margins of streams, and they are found occasionally in the sediments of lakes and rivers as well. Two unidentified species belonging to genus *Achaeta* and *Enchytraes* were recorded from wet soil in Iraq (Jaweir and Albayati 2016).

Aquatic oligochaetes play crucial roles in organic matter dynamics and trophic energy transfer in river ecosystem. As a group, they are the primary food source for most stream and river fishes. Their taxonomic, habitat, and life-history diversity ensure that an array of food types is available to many fish species over the entire annual cycle. They also conduct the less apparent but no less important work of decomposing leaf litter and small particles of organic debris on the stream bottom or in the water column and of grazing stream algae, fungi, and bacteria. Benthic macroinvertebrates, in general, and aquatic oligochaetes, in particular, are of importance to aquatic biologists because they can indicate pollution effects on the river ecosystem, and environmental quality has been assessed with respect to oligochaete species ecology (Rodriguez and Reynoldson 2011).

Aquatic Oligochaeta diversity in the Middle East is still not well determined, although studies in Turkey‘s fauna have increased in the last 20 years, and about 150 oligochaetes and 6 Aphanoneura species have been reported up to date (Arslan and Margan 2018). Arslan and Sahin (2003a, b) recorded nine species of Naididae-Naidinae, including *Ophidonais, serpentina; Stylaria lacustris; Nais variabilis; N. pardalis; Pristina longiseta, P. foreli; P. proboscidea*; *P. aequiseta*, and *Pristinella jenkinae.*, collected from Sakarya river, while Tas et al. (2004) referred to 37 species recorded from different Turkish thrace, among them, the naidid worm was the most dominant, including 27 species. Yildiz and Balik (2005) recorded 24 species of Tubificinae; 14 species of Naidinae; 3 species of Enchytraeidae; and 2 species of Lumbriculidae, in addition to one species of Haplotaxidae. In Kovado lake, Arslan and Sahin (2006) collected 37 species of macrobenthic invertebrates including 25 species of Oligochaeta; they referred to the high frequencies of *Tubifex tubifex; Limnodrilus hoffmeisteri; Potamothrix hammoniensis; and Nais communis.* A total of 17 species of Tubifinae; 28 species of Naidinae; 3 species of Enchytraeidae; and 2 species of Lumbriculidae were also recorded in Yuvarlak stream by Yildiz et al. (2007); among them, the species *Tubifex newaensis; Nais alpine; and N. behning* were considered as new records to Turkish fauna. Arslan and Ilhan (2010) referred to 25 species of aquatic oligochaete collected from Porsuk stream, which is a branch of Sakarya River, while Yildiz and Ahiska (2010) collected samples of aquatic oligochaetes from 39 lakes east to Black Sea in Turkey and referred to the first time recording of *Stylodrilus heringianus. Nais stolci* was also added to the Turkish fauna by Yildiz and Ahiska (2010).

Tubifex tubifex, Limnodrilus claparedeianus, L. hoffmeisteri, Potamothrix hammoniensis, P. bedoti, Branchiura sowerbyi, Nais pardalis, Ophidonais serpentina, Dero digitata, Stylaria lacustris, Slavina appendiculata, and Mesenchytraeus sp. were identified by Nazarhaghighia et al. (2014) in the Anzali International Wetland, north-western Iran.

In Iraq, aquatic oligochaete studies were started after 2000. Before that time, this group was only indicated as a port of benthic invertebrate’s community (Al-Lami et al. 1997; Kassim et al. 1997). Ali (2007a) referred to 8 species of aquatic oligochaetes recording within the benthic macroinvertebrates in the middle sector of Greater Zab River, while Nashaat (2010) referred to the presence of *Limnodrilus hoffmeisteri* and *Branchiura sowerbyi* in Tigris River. In the Iraqi southern marshes,

Jaweir et al. (2012) studied the aquatic oligochaete communities in three southern marshes in Iraq, including Al- Hawieza, Al-Chibayish, and Al-Hammar, and their results revealed 10 species belonging to 14 genera and 4 families. Naidinae was represented by 8 species; Tubificinae 8 species; and Lumbriculidae 2 species; in addition to one species belonging to family Lumbricidae, with *Limnodrilus spp.* being the most abundant. Three new record species were added to the Iraqi fauna by Jaweir (2011) collected from Al-Hawiezah marshes. Two new species were also recorded by Al-Abbad (2010), including *Pristina proboscidea* and *P. aequiseta;* in addition, other two species were recorded by Al-Abbad and Al-Mayah (2010) as new records for Iraqi fauna, including *P. longiseta* and *P. macrochaeta;* Al-Abbad (2012) referred to 11 species of Naidinae, which are also recorded in the marshes. In Tigris River and some water surface within Baghdad city, 23 species of Naidinae were recorded by Jaweir and Rhadi (2012) in addition to 12 species of Tubificinae by Jaweir and Alwan (2013). Jaweir and Al-Janabi (2012) identified 19 species in the middle sector of Euphrates River at Al-Massayab city. Jaweir (2014) prepared a checklist of 60 species of aquatic oligochaetes recording in Iraq. Aquatic oligochaete community Al-Delmage Lake, Iraq, was studied by Jaweir and Al-Sarai (2016), and their results revealed the presence of 11 species.

34.3 Sampling Techniques of Aquatic Oligochaete

Sampling freshwater oligochaetes for identification is recommended in spring, where the most individuals are sexually mature. Naididae, Lumbriculidae, and other oligochaetes, which live in the sediments, can be collected with other macroinvertebrates when sampling freshwater zoobenthos, using various net or bottom sampler, grabs, or dredges. Large worm can easily sort from sieving residue in the field by spreading the sediment on a white tray. A magnification hand lens may be convenient for this purpose. The smaller meiobenthos forms, like Naidinae and Pristininae, can get lost in macroscopical sorting, but they are well visible under a low power dissecting microscope. Small worms may be obtained by using the wet-funnel method recommended by Nielson and Christensen (1959) for sorting Enchytraeidae and other small oligochaetes. A soil sample of about 2 cm thick is replaced up wise down on wire gauze in a water-filled funnel with rubber tubing, closed by screw clip, and attached to the stalk. 40–60 watt bulb is placed above the funnel actually touching the sample. After ½- 1 hour, a large proportion of the worms present in the sample will have moved downward through the sediment and fallen into the lower part of the funnel. By opening the clamp and letting out a little of the water, they may be collected in a Petri dish (the upper layer of the water has warmed up considerably, and if it is emptied into dish, the worm will be damaged). Petri dish is then left in a cool place (5–18^0 C) for 12–24 hours. After this period, the worms will have emptied the gut they contained, which might interfere with subsequent examination.

34.4 Preservation and Mounting

Some oligochaetes are identified, which are most effective alive, when all tissues are more or less transparent, and the internal organs can be easily observed, especially for Enchytraeidae, which should be studied alive ever possible as they do not have many external identification characteristics (Timm 2009). Worms can be easily anesthetized by drops of carbonized drinking water.

The worms may be killed and preserved in 70% alcohol. *Branchiura sowerbyi* and some other species tend to fragments in alcohol, and so this must be narcotized in 5% magnesium chloride before preservation. Preserved worms are stored in 70% alcohol or 4% formalin. It is recommended to change alcohol repeatedly, especially when the mass of worm is large. Before identification, the preserved specimens should be transferred to 30% alcohol and then to water and placed on slide; add a few drops of mounting fluid (Brinkhurst 1963).

In temporary mount, glycerin can used as clearing media (Nielsen and Christensen 1959; Timm 2009). It is reasonable to soak worms in glycerin for several hours or days before mounting under a cover slide. As glycerin makes the worm soft, it is possible to flattened them by exerting slight pressure on the cover slides. For better exposition of setae and some other details, Amman's lactophenol was recommended by Brinkhurst (1963), in which phenol (carbolic acid) and lactic acid are added to glycerol. This mounting fluid is prepared as follows: Carbolic acid, 400 g; Lactic acid, 400 ml; Glycerol, 800 ml; and water, 400 ml. The worms should be covered with cover slip and left in this fluid at least for several hours before examination.

For permanent preparation, Amman's lactophenol can be replaced by polyvinyl lactophenol (Brinkhurst, 1963). When this has dried sufficiently, the preparation should be ringed, preferably by glycol. Permanent mount can also be made by Canada balsam or other synthetic resin (Beylich 2005); Balsam makes chaetae even slightly transparent; on the other hand, several internal organs can be well visible. Timm (2009) recommended to keep the specimens for some days or weeks under a cover glass in glycerin, before mounting them permanently in balsam. This will turn them flat and will avoid breaking of the worm's body and chaetae during transition through ethanol and xylene to balsam where they become relatively brittle.

The megadrilus worms, or the earthworms, can be identified without any mounting. Mature specimens of Glossoscolecidae and Lumbricidae may be identified using only a low-power stereoscopic microscope.

34.5 Identification Criteria

The characters on which oligochaetes families are defined involve the position and arrangement of the reproductive organs. Specimens in reproductive phase show a clitellum around the genital segments. Identification of nonreproductive specimens to family level can be difficult or impossible although certain taxa can be recognized by overall body shape and size or by distinctive gills or setae.

Identification of aquatic Oligophages is started by establishing the anterior and the posterior end. The anterior end is usually thicker, begins with prostomium, or preoral lobe, which is homologous to the upper hemisphere of the trochophore larva in polychaetes, and has a tactile function. The mouth is positioned directly under prostomium, on the ventral side of the first segment, peristomium, or the buccal segment. The prostomium is variable in different groups, and it may be:

1. Prolobate, which is separated from the peristomium by a simple transverse furrow, which resembles the intersegmental groove.
2. Zygolobate, when fused with peristomium.
3. Epilobate, when it protrudes deep into peristomium.
4. Tanylobate, when its posterior margin approaches segment II.

The prostomium is occasionally elongated and produced as proboscis in some species of Naididae, (*Stylaria*, and some *pristina spp.*) and also Lumbriculidae (some *Rhynchelmis spp.).* In Tubificidae, the prostomium is rounded or triangular, prolobate or zygolobate, no proboscis is found in Tubificid Worms (Plate 34.2). Dorydrillidae has a rounded prostomium, and *Propappus volki* (propappidae) has a

Plate 34.2 Types of Prostomium. (**a**) proboscis of *Stylaria;* proboscis of *Rhynchelmis;* (**c**) rounded prostomium of *Stylodrilus;* (**d**) prostomium of *Tubifex;* (**e**) prostomium of **Lumbriculus**

short proboscis. Prostomium of *Haplotaxis gordiodes* has an elongated transversal groove. Eyes may occur at the base of prostomium in Naididae.

The first segment "peristomium" is devoid of setae, which are otherwise arranged in four bundles on each segment, two dorsal-lateral, termed dorsal, and two ventral-laterals, termed ventral. Segment II is the first setae-bearing segment. The total number of segments is usually variable since they can easily grow, get lost, and regenerate again in the tail part. In aquatic oligochaetes, which reproduce in an asexual way by paratomy, the number of segments before the budding zone can be species specific.

34.6 The Setae or Chaetae

The most important feature to examine is the setae, which are uniform in some oligochaetes, while in other, they are different in the dorsal and ventral bundles. The shape, number, and size of setae are not identical within a single specimen; however, the uniformity in these differences constitutes an essential character of species, for that reason, to determine the various sorts of setae in the specimen; dorsal and ventral bundles from several regions of the body should be examined.

In general, the setae in aquatic oligochaetes are of two types:

1. Hair setae, characterized by elongated, slender form, and the absence of nodule. They are found in the dorsal bundle of most Naididae. Hair setae may be serrate or smooth.
2. Sigmoid setae or crotchets, which are found in both dorsal and ventral bundles; they consist of S-shaped shaft, which has a thickened region at some point, and the nodules, which may be absent in several oligochaetes (most Enchytraeidae, Lumbricidae, and others). Distal end of crotchet may be simple pointed (Plate 34.3a), rounded, or bifid.

Bifid setae terminate into upper and lower tooth, occurring on the convex and concave side of the seta, respectively (Plate 34.3b&d). The relative length of teeth varies and is species specific. Bifid setae are found both in dorsal and ventral bundles of Naididae, Lumbricidae, and others.

Bifid setae may be ornamented especially in dorsal bundles, by a series of fine intermediate teeth, and so they are called pectinate setae (Plate 34.3e). In Palmate setae, (plate 34.3f&g) the intermediate teeth are replaced by a continuous thin membrane extending between the two main teeth. Sometimes, the distal end of setae with reduced teeth is broad and is referred to as oar-shaped setae. Pectinate, palmate, and oar-shaped (plate 34.4d) are characterized setae of Tubificidae and occur only in dorsal bundles.

Needle setae are long dorsal setae found in family Naididae, subfamily Naidinae, accompanied by hair setae, often lack a node, or have only inconspicuous one, and hence, they closely approach in their forms hair setae (Plates 34.4a & b). Needle setae may be straight, simple pointed, or bifid. In some species, some intermediate

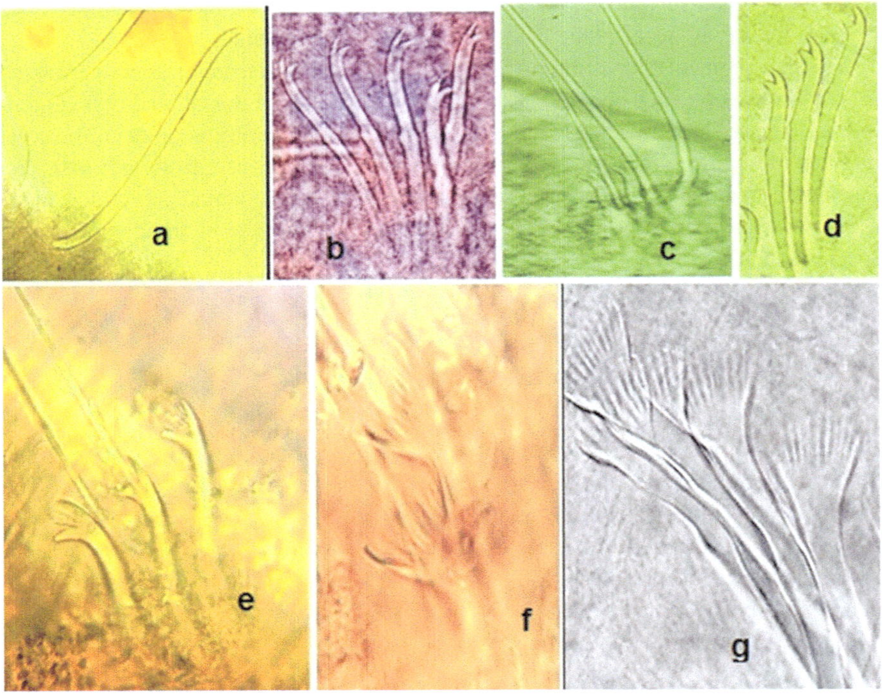

Plate 34.3 Types of setae. (**a**) simple pointed; (**b&d**), bifid setae; (**c**) hair setae; (**e**) pectinate setae; (**f&g**), palmate setae

teeth can be visible between the lateral teeth as in *Allonais*. Only one straight, stick-shaped seta is found in the dorsal bundle of *Ophidonais serpentine* (Plate 34.4c).

The presence of giant setae on some segments of some species of Naididae and Enchytraeidae is noteworthy. These setae are sharply distinguished from others by their thickness and length, and they may be either single pointed, bifid, or even hair.

Genital setae are the setae carried by segments in the region of genital pore and help in the process of copulation. They are usually simple pointed, with an important modification. They are distally hollowed or grooved. The genital setae include the spermathecal setae (Plate 34.5), arranged around the spermathecal opening, and penial setae (Plate 34.6), arranged around the male genital pores. Genital setae in Naididae are represented by penial seta only, with the exception of *Pristina longiseta*, in which the penial setae are replaced by spermathecal in segment VI. Some species of Tubificidae have spermathecal setae in segment X (e.g., *Potamothrix spp. and Psammoryctides spp.*), and others have modified penial setae hidden in body wall, near male pore, in segment XI. Penial setae and male pores can occur in VII in some species (*Aulodrilus pigueti*), while in *A. pluriseta*, no modified penial setae present, but male pore also in VII. Both spermathecal and penial setae are found in *Haber speciosus*. No modified genital setae are found in Lumbiculidae and Enchytraeidae.

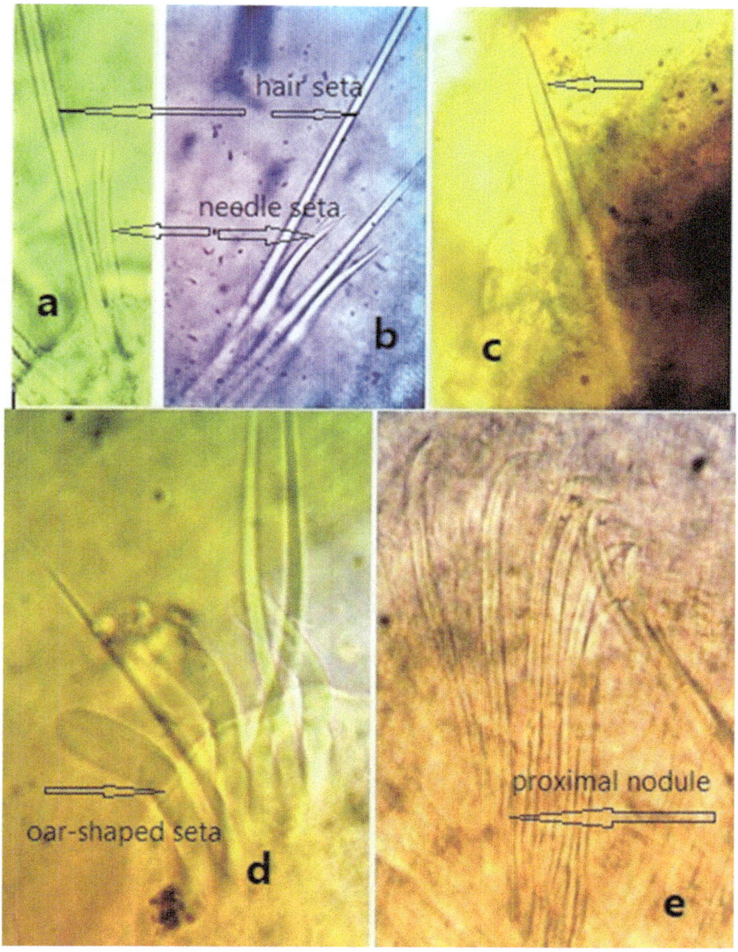

Plate 34.4 Type of setae. (**a**) Hair &Needle seta of *Nais elinguis*; (**b**) Hair &Needle seta of *Allonais inaequalis;* (**c**) Dorsal seta of *Ophidonais serpentine*; (**d**) Oar-shaped seta of *Aulodrilus pigueti*; (**e**) setae of *Stylaria lacustris*

The number of setae in each bundle is an important taxonomic characteristic. The two setae are present per bundle characteristic of Lumbricidae and Lumbriculidae, while in Naididae, the number of setae in bundle is variable and is constant in species. It may range between 2 and 20 setae per bundle. The setae in each bundle are generally arranged in a line, which is perpendicular to the main axis of the body and less often obliquely situated.

Genital organs are very important for aquatic oligochaete identification. Many species cannot be identified even to genus of family level when immature. All aquatic oligochaetes are hermaphrodite, the pairs testes, ovaries, genital ducts, and spermathecae are concentrated in few segments usually in V-VII in most Naididae

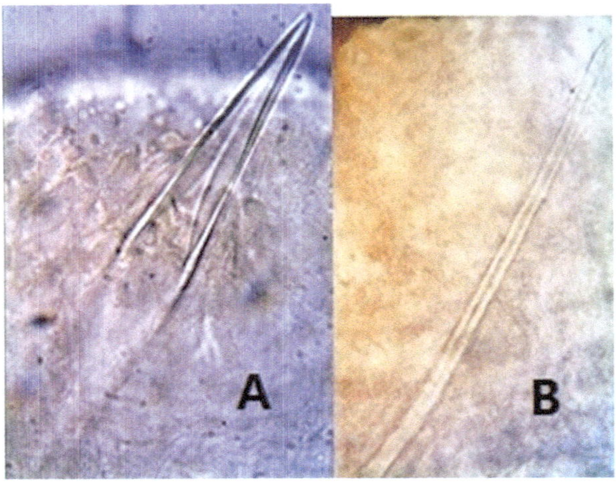

Plate 34.5 Type of spermathecal setae. (**a**) Spermathecal seta of *Pomatothrix spp.;* (**b**) Spermathecal seta of *Psammorhyctides spp.*

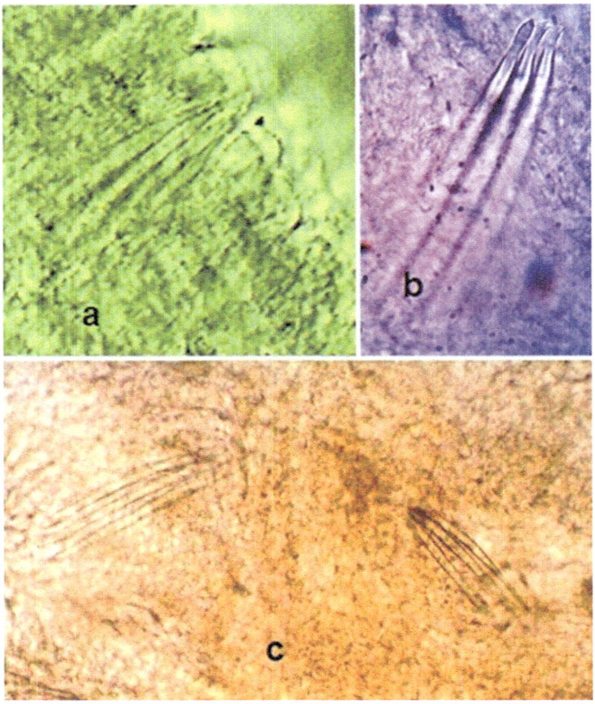

Plate 34.6 Type of pineal setae. (**a**) Pineal setae of *Rhyacodrilus*. (**b**) Pineal setae of *Nias;*(**c**) Pineal setal dundles of *Ophidonais*

and X-XIII in Tubificidae, Lumbriculidae, and Enchytraeidae (Enchytraeidae have spermathecae lie in V). Their exact location is of great taxonomic value.

The vas deferens, or the sperm duct, can end with penis, hidden in a special penial pouch. It is a thickened distal part of the male gonoduct with a special musculature

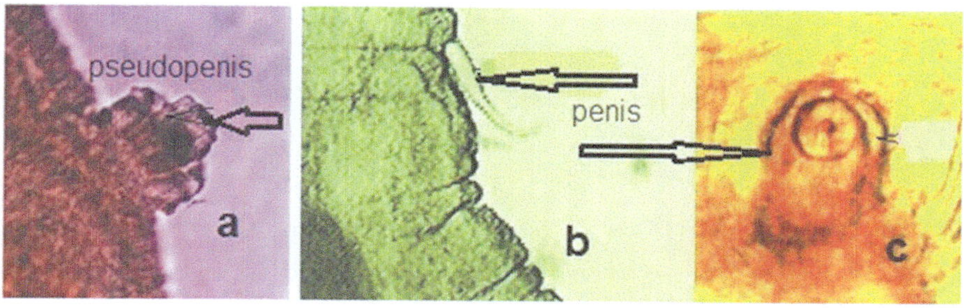

Plate 34.7 (**a**) Pseudopenis of *Monopylephorus*; (**b**) External penis of ***Stylodrilus heringianus;*** (**c**) Tub-shape penis sheath

and chitinous cover. When the distal end of the atrium itself is protrusible, it is called pseudopenis (Plate 34.7a). Penis and pseudopenis can be found in some species of Lumbriculidae and few species of Tubificinae and absent in Naidinae. Two external transparent penises, well visible in mature *Stylodrilus heringianus* (Plate 34.7b), are directed backward in X. While in *S. brachystylus*, mature individuals have a large internal penis, sitting on muscular bulbs, sometimes, their tips slightly protrude from penial sac in X. A true penis is rare among Enchytraeidae, but there is a copulatory apparatus called penial bulbi, which is a glandular thickening of the muscles of the body wall near the distal part of the vas efferent, which passes through the bulbus in form of a thin tube. The penial bulbus is sometimes embedded in the epidermal fold of the body or pseudopenial sac.

Penis sheaths of some Tubificid worms are variable in shape and size in different species and valuable in recognizing some species. It may be cylindrical, tub-shaped, or cone shaped (Plate 34.7c). In cylindrical type, the ratio of its length to its proximal width is important to differentiate between different species of *Limnodrilus* (Plate 34.8).

The Clitellum In mature oligochaetes, during a period of sexual reproduction, a certain section of the epidermis swell, frequently differs in color from adjacent segments and is refer to as “Clitellum”. It is glandular in nature, which exudes a secretion that forms the membrane and the nutritive fluid of egg cocoons.

The clitellum is very uniform in all typical oligochaetes. It includes segments that bear male and female genital pores, which indicates the intraclitellar position of the pore. It occupies few segment, in Naidinae—Naididae (IV-VII) and Tubificinae-Naididae (IX-XI), with the exception of *Aulodrilus pigueti* (VII-VIII) and *Rhyacodrilus coccineus* (½ X- ½ XIII). In Lumbriculidae, clitellum, one cell thick, often covers many segments, including those with male and female pores, while Enchytraeidae have clitellum in few segments in the region of the gonads. Testes and ovaries are one pair of each, in XI & XII (plate 34.9). The clitellum is a temporary in the majority of aquatic oligochaetes, and it disappears when the mating season ends and reappears when mating begins.

Body surface is also essential for identification, which can be smooth or more or less beset by small papillae, sometimes also with adhering fine detritus.

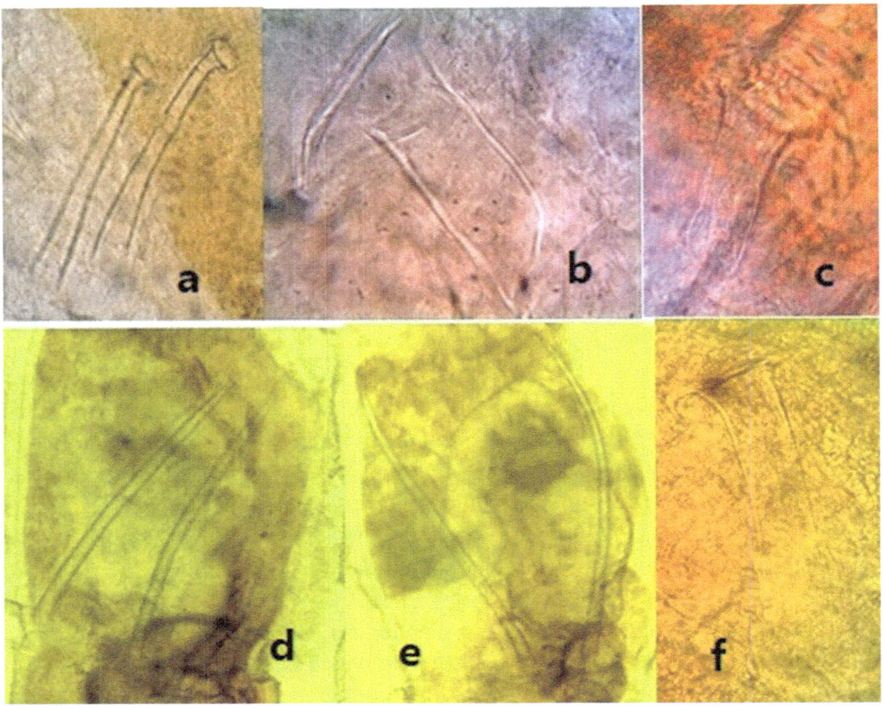

Plate 34.8 Cylindrical type penis sheath of *Limnodrilus spp.* (**a**) *L. hoffmeisteri;* (**b**) *L. silvani*; (c) *L. udekimianus;* (**d**) *L. claparedeianus;* (**e**) *L. cervix;* (**f**) *L. hoffmeisteri (short)*

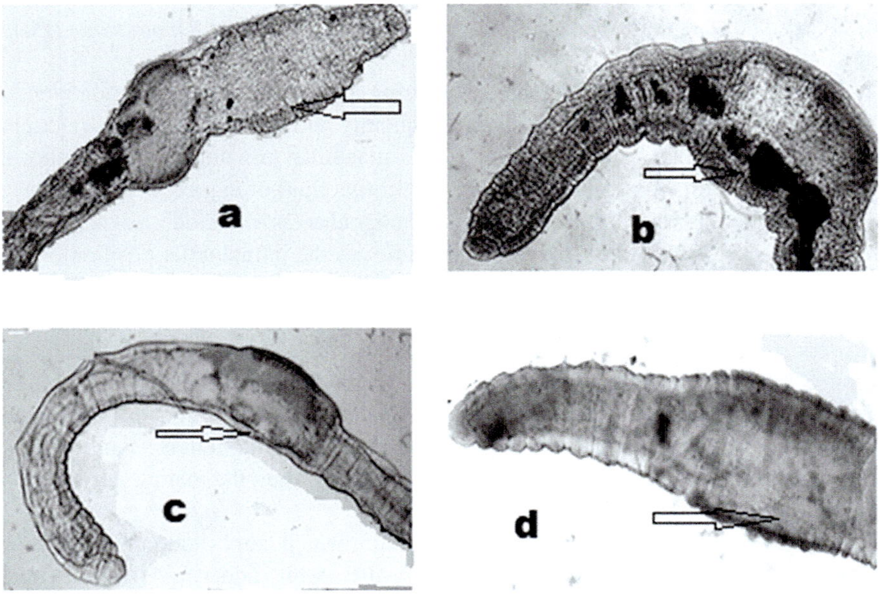

Plate 34.9 type of Clitella. (**a**) Naidinae; (**b**) Tubificinae; (**c**) Enchytraeidae; (**d**) lumbriculidae

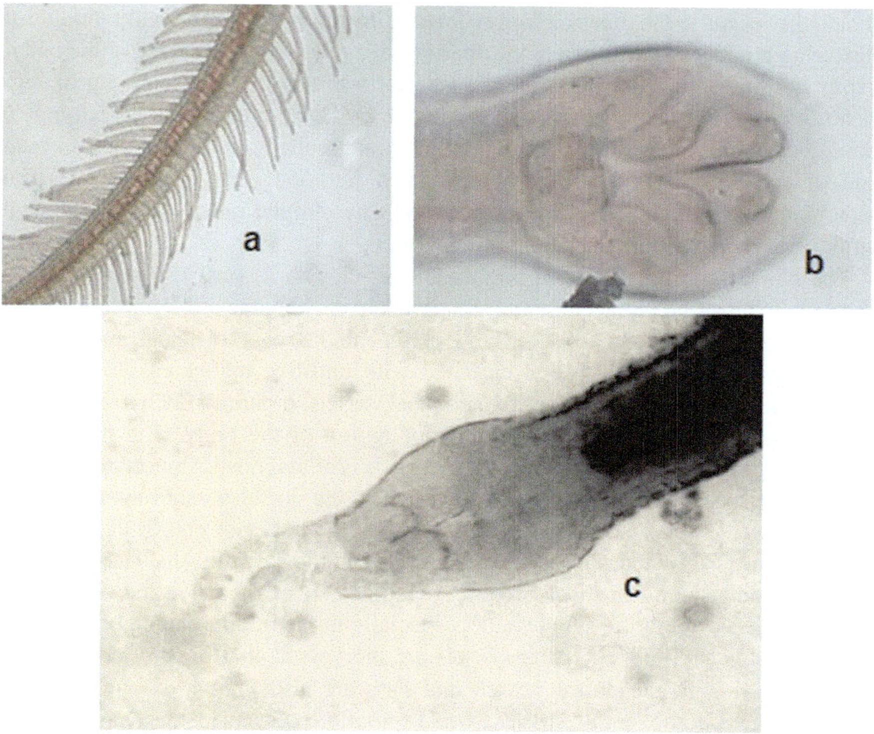

Plate 34.10 Types of gills. (**a**) Posterior gills of *Branchiura sowerbyi;* (**b**) Posterior gills of *Dero (Dero)spp.;* (**c**) Posterior brachial fossa of *Dero(Aulophorus) spp.*

Intersegmental furrows can be either deep or poorly expressed in preserved specimen or sometimes accompanied by secondary annuli.

Gills are one of the external characters essential for identification. They are not common among aquatic oligochaetes, but are highly characteristic when present. *Branchiura sowerbyi* (Plate 34.10a) is the only Tubificid worm, which has posterior fingerlike gills of variable length and one dorsal and one ventral per segment in tail region. In *Branchiodrilus* (Naididae), the digitiform gill filaments are situated in anterior dorsal. *Dero (Dero) spp.* (Plate 34.10b) have a posterior end with branchial disk or fossa, surrounding anus and bearing leaf-like gills, while in *Dero (Aulophorus) spp*, the posterior border of brachial fossa is projected into two palps (Plate 34.10c).

There are some internal criteria essential for identification of some aquatic oligochaetes. In the digestion system, the connection between esophagus and intestine, the various types of glands, alongside the esophagus, could be of great taxonomic importance. The esophagus posteriorly enlarges abruptly or gradually to form the digestive tract, which contains chloragogens through its length. The pharynx and esophagus are provided with various types of glands, and the most important one, as taxonomic criteria, is the septal glands, which are multicellar glands that occur in the anterior region of the alimentary canal. The bulky bodies of these glands adhere to the septa in the pharyngeal and esophageal region of the body. Extremely thin intracellar ducts extend anteriorly through several pharyngeal segments. Peptonephridia, in some representative of Enchytraeidae, is a pharyngeal gland resembling the nephridia externally, it is long-coiled tube, which lies on each side of the esophagus or under it and open into the pharyngeal cavity. The peptonephridia participate in digestion and bear no morphological or physiological similarities to nephridia (Chekanovskyana 1963).

The blood vascular system in oligochaetes is closed. It consists of an intestinal sinus (which may be replaced by intestinal plexus in Naididae and Tubificidae) and dorsal, ventral, and circular vessels. A peripheral network of small vessels and capillaries is also present. Blood enriched with nutritive matter passes from the intestinal sinus or intestinal plexus to the dorsal vessels. In family Enchytraeidae, the dorsal vessel is present only in the anterior region of the body. The number of segments through which the dorsal vessel extends is of diagnostic importance both at generic and specific levels. In other oligochaetes, the dorsal vessel runs above the alimentary canal and extends from the anterior to the posterior end of the body. In Naididae and Tubificidae, the dorsal vessel is slightly laterally displaced, almost toward the left side. The displacement sometimes so marked that the dorsal and ventral vessels actually lie side by side, especially in the middle region of the body (Chekanovskyana 1963). Circular vessels are the vessels that directly connect the ventral vessel with the dorsal vessel, and they are also known as commissural, transverse, and dorsoventral vessels. Great variation is seen in the character and position of these vessels. The distinguishing characters are of taxonomic significance. A reduction in the transverse connection is very common in aquatic oligochaetes. In family Naididae, either the postlarval segments are completely devoid of circular vessels or the number present is very small. The presence of two circular vessels in the anterior region of the body is characteristic of family Enchytraeidae. A considerably greater number of circular vessels occur in the Tubificidae and Lumbricidae, but not in every segment. A distinctive characteristic of circular vessels is observed in the family Lumbriculidae, which is also characterized by the presence of blind processes on the dorsal vessel, which are often short and bear digitiform lateral outgrowth.

Large, roundish coelomocytes are abundant in the body cavity of some groups, particularly used in recognition of subfamily Rhyacodrilinae (Naididae).

Nephridia in aquatic oligochaetes consist of anteseptal portion, which lies anterior to the septum, and the postseptal portion, which lies posterior to the septum and extends through two successive segments. The antiseptic part consists of ciliated funnel communicating with the celom, while postseptal part consists of coiled tube of variable length, which forms a number of loops. The loops may cohere partially or completely. In the latter case, the nephridia said to be "compact".

All representatives of Naididae and Enchytraeidae have characteristically small funnel 1–4 ciliated cells, while a majority of Lumbriculidae and Lumbricidae have a large funnel, lined with ciliated epithelium. One of the morphological criteria for determining the boundary between the larval and post larval body of oligochaetes is the position of the first pair of nephridia in the body. There is a total reduction of nephridia in some marine oligochaetes as in marine *Paranais spp.* This is probably associated with their decreased osmoregulatory function.

34.7 Family Naididae

34.7.1 Subfamily Naidinae

Species of subfamily naidinae are small, transparent worms, usually less than 2 cm long; Prostomium, with or without proboscis; eye present or absent; Clitellum in a few segments in the region of gonads; testes and ovaries, one pair each, in IV-V, V-VI, or VII-VIII; male efferent apparatus paired; funnels in segment with testes, atria in segment with ovaries, usually opening apart. No penes, penial setae are often present; often diffuse prostate gland cells on vasa deferentia or ateria; spermathecae in segment bearing testes, usually an unpaired sperm-sac and ovisac formed. Asexual reproduction by paratomy (budding), forming a chain of individuals, mature individuals occur seldom. Cosmopolitan, living mostly on bottom surface and aquatic plants, many species are able to swim. In Iraq, about 32 species were already identified (Table 34.1).

Most Naidinae species can be easily identified to the species level by their locomotory chaetae. Dorsal chaetae entirely lacking in *Chaetegaster spp.* (Plate 34.11); In *Pseudochaetogaster longemer,* Lafont, 1981, the species which are closely related to *C. diastrophus*, bearing dorsal bundles with two bifid chaetae from VI on, while in *C. palustris*, long sensory hairs occurs on both prostomium and body segments (Timm 2009).

In all other Naidid worms, dorsal chaetae are present with or without hair chaetae. The beginning of dorsal chaetae is used to differentiate between genera. They begin in segment II, in *Homochaeta* & *Pristina*; III in *Amphichaeta;* or further backward in, IV, V, or VI. In *Haemonais waldogeli,* dorsal bundles beginning from XVIII-XX

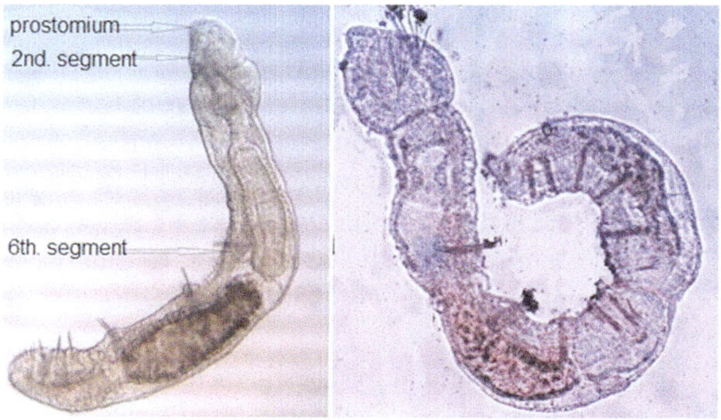

Plate 34.11 *Chaetogaster spp.*

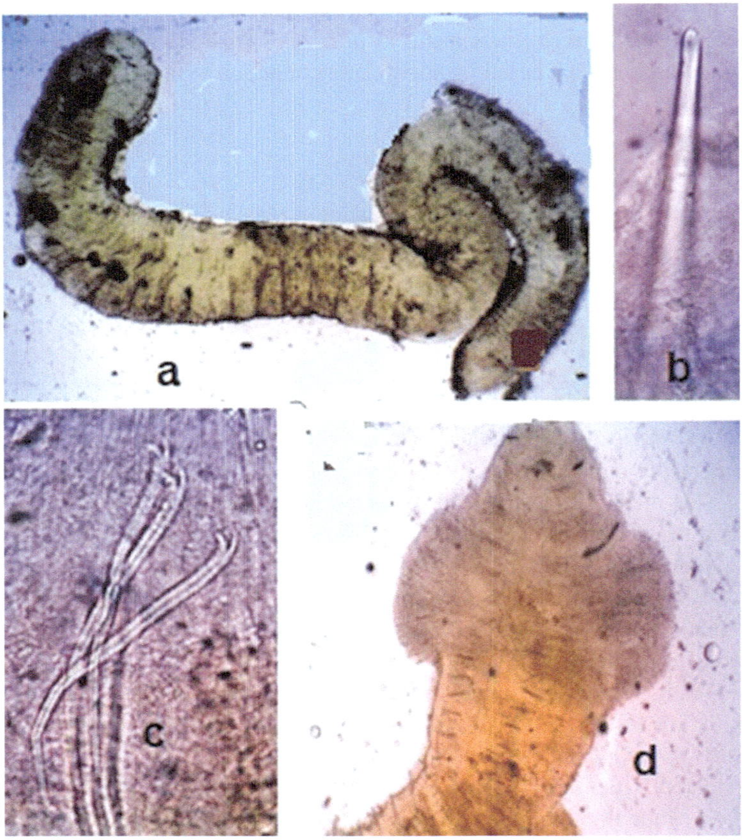

Plate 34.12 ***Ophidonais serpentine.*** (**a**) whole Asexual worm; (**b**) dorsal seta; (**c**) ventral setae; (**d**) anterior end of sexual worm

and each bundle contain short hair and one bifid crotch with 1.5 times longer upper tooth. (Timm 2009).

Hair chaetae are absent in genera *Amphichaeta, Paranais, Homochaeta, Uncinais, and Ophidonais.* Only one straight, stick-shape chaeta is found in the dorsal bundle of *Ophidonais serpentina*, with rounded or bluntly bifid tip, sometimes lacking in many segments (Plate 34.12). Naididae, without hair chaetae in the dorsal bundles, have dorsal bundles similar in forms to ventral one. All chaetae are similar, bifid and fine with equal teeth in *Amphichaeta spp.* In *paranais frici,* chaetae with upper tooth are at least twice as long as lower, while in other *Paranais* species, the upper tooth is not more than twice as long as lower. Only the hair chaetae of VI in *Slavina appendiculata*, with 800 μm long, is usually considerably longer than those of subsequent segments (279–450 μm). The long setae act as float apparatus and prevent sinking.

Species of this subfamily are characterized by the presence of needle chaetae, always accompanied by hair chaetae. Needle chaetae may be straight, simple pointed

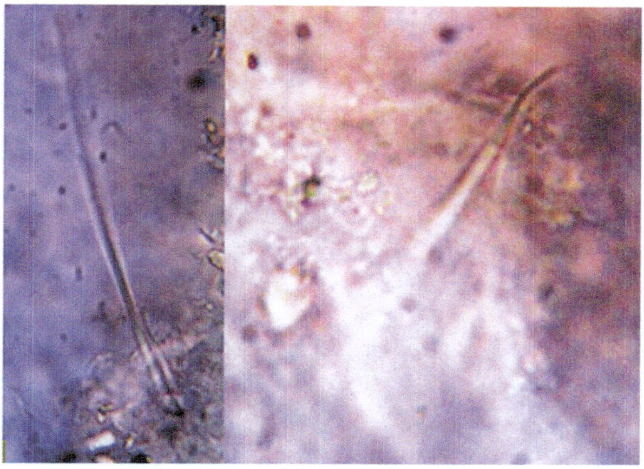

Plate 34.13 Needle seta of *Slavina appendiculata*

as in *Stylaria lucastris.* In *Slavina appendiculata*, the needle chaetae are with tip forming a small knob (Plate 34.13). Bifid needle chaetae found in *Dero(Aulophorus) furcatus and Dero spp.* have shorter upper tooth in D. *(A.) furcatus* and slightly longer upper tooth in *D. dorsalis,* while the upper tooth of needle chaetae in *D. (Dero) digitata* is distinctly longer and straighter,with small intermediate teeth visible under strong magnification. Simple pointed and bifid needle chaetae of different shapes are found within *Nais spp.* Needle chaetae of *N. barbata* resemble in its shape the hair chaetae, while in *N. pseudobtusa*, the needle chaeta bears distinct nodulus and it is obtuse in *N. simplex.* Needle chaetae of *N. pardalis* have short but distinct teeth and long parallel teeth needle chaetae in *N. elinguis* (Plate 34.14a). Very fine bifid needle chaetae are present in *N. cristinae* and *N. varaibilis,* while the teeth in *N. communis* are distinct and diverging (Plate 34.15). In *Allonais inaequalis*, the needle setae are pictinate with upper tooth shorter than the lower one (Plate 34.14b).

Ventral chaetae in Naidinae begin in II. The absolute and relative size of setae is the one of particular importance in the taxonomy of this family. The regional dissimilarities may be exist in size of chaetae. Most species of *Nais* and many of *Dero* whose anterior ventral chaetae, which were formed in a budding zone, are generally longer and straighter than those in the posterior segments (Brinkhurst 1971), which is quite obvious in *N. communis*. The ventral bifid chaetae are thinner in relation to their length than those of other families and in many cases may be more strongly sigmoid or have more localized and sharper curves. Ventral chaetae of the same shape with slightly longer upper teeth are observed in *Specaria* and *Allonais*.

In genus *Chaetogaster,* single pair of ventral bundle is found in II, with next bundle appearing only in VI (plate 34.11). Bundles of II always contain larger and thicker chaetae than the rest, e.g., in *C. diaphanous*, chaetae of II with long 200–350 μm and 4.5 μm thick, 2–20 per bundle; from VI on, 2–10 per bundle, 100–218 μm long (Timm 2009). Ventral chaetae always bifid, with slightly longer upper tooth in *C diaphanus, C. cristallinus, C langi,* and *C. diastrophus*, while in

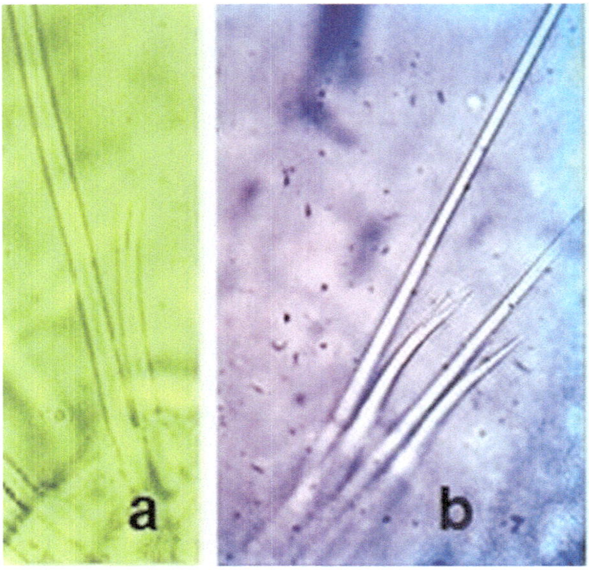

Plate 34.14 Needle setae. (**a**) *Nais elinguis; Allonais inaequalis.* (**b**)

C. limnaei, the ventral chaetae with upper tooth are equal to or slightly shorter than the lower one.

Ventral chaetae also exhibit regional dissimilarities in shape and number. The foremost ventral bundles in *Dero dorsalis* are 3–7 per bundle with teeth of equal thickness, but upper one being 1.5-2x longer, further rearward ventral chaetae becoming gradually shorter, length of their teeth equal but lower tooth growing thicker. In *D. digitata,* ventral chaetae 3–7, with 1.5–2 times longer upper tooth, from VI on and ventral chaetae by 2–5, shorter and more curved, with upper tooth only slightly longer than lower one. Ventral chaetae in II-V in *D. obtuse,* 2–6, slender, with upper tooth twice longer, from VI on, ventral chaetae much shorter, thicker and curved, their teeth at first of equal length, but upper tooth shorter rearward The upper tooth of ventral chaetae of *Stylaria lucastris* is much longer and strongly curved. Chaetae with slightly longer and thinner upper tooth are found in ventral bundles of *Slavina appendiculata.* In *Nais spp.*, single giant chaeta per bundle is found in VI of *N. paradalis* with 80–93 μm long and 4.2 μm thick with upper tooth 2–3 times as long as lower (plate 34.16d). In II-V, ventral chaetae 2–5 are with upper tooth 1.5–2 times as long as lower and from VI on, usually ordinary ventral chaetae 2–5 per bundle, thicker, and more curved, with teeth equally long or upper tooth slightly longer and thinner.

In *N. pseudobtusa,* the chaetae from VI on become slightly thicker. Ventral chaetae of *N. simplex are* with equal thick teeth, but upper one 1.5–2 times as long as lower, and from VI on, ventral chaetae become thicker and more curved than anterior one, with teeth equally long or the upper teeth slightly longer, but always 2–3 times thinner.

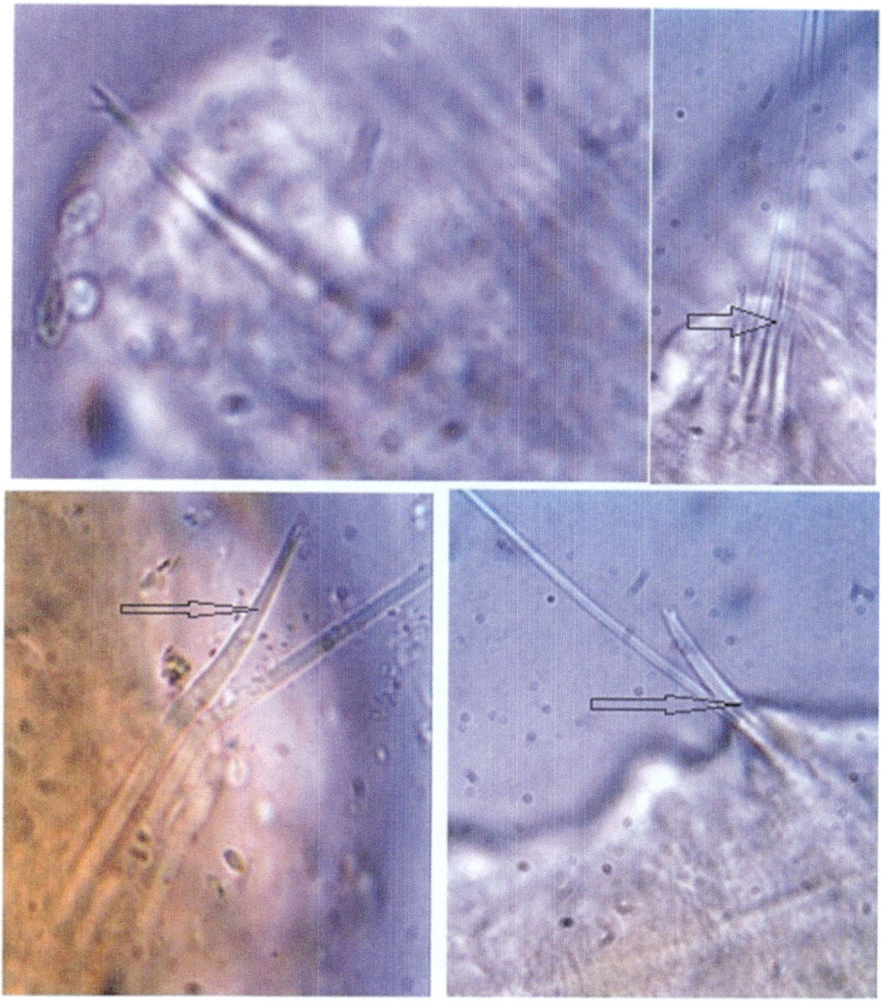

Plate 34.15 Needle setae of *Nais spp.*

All ventral chaetae in *N. elinguis* consist of 2–5 similar chaetae with upper tooth twice as long and strongly curved in its distal end, with lower tooth forming a right angle with the longitudinal axis of chaeta (plate 34.16b).

Beginning from VI, ventral chaetae shorter and thicker than II-V, but with about equally long teeth in *N. christinae; N. varaibilis;* and *N. communis.* In II-V, the teeth are equally thin, but upper tooth is 1.5–2 times as long as lower in *N. christinae,* longer and thinner in *N. varaibilis,* and *N. communis.* Plates (34.17 & 34.18) show different types of ventral setae of Naidid worms.

Penial chaetae are occasionally replaced by spermathecal chaetae; their number ranges from 2–5 per bundle. Penial chaetae may be blunt or with vestigial upper

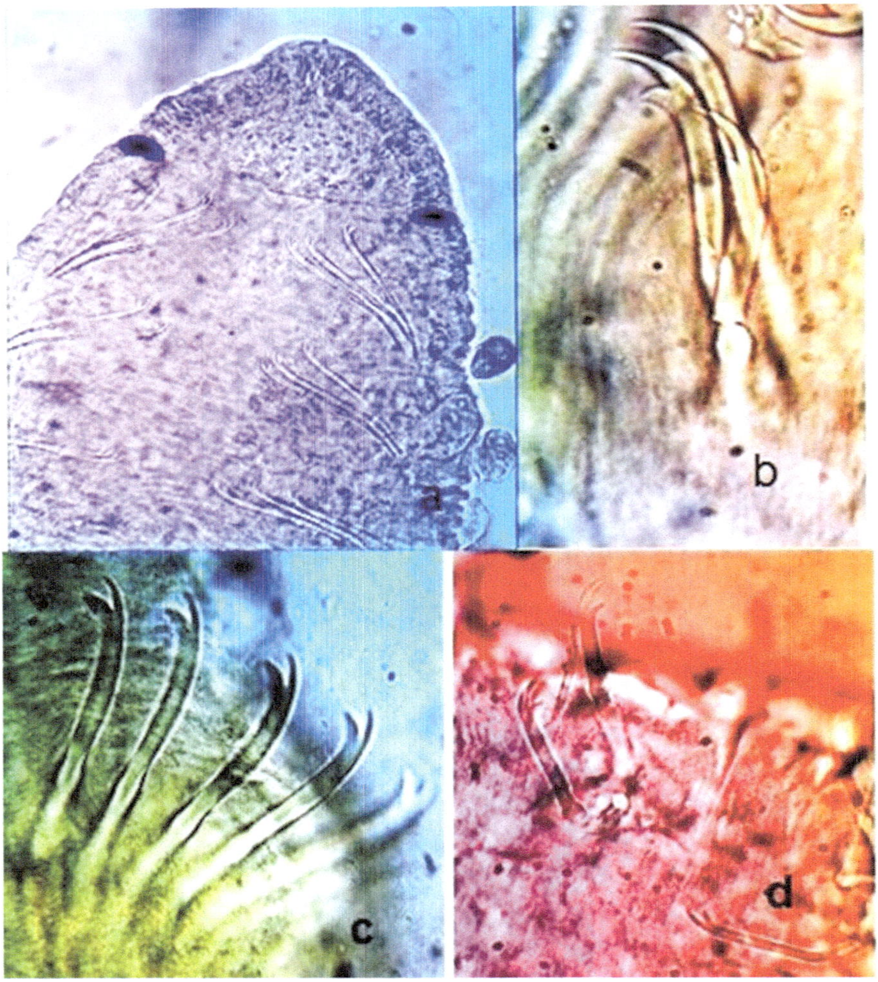

Plate 34.16 : ventral setae. (**a**) Anterior ventral bundles of *Nais.;* (**b**) ventral setae of *N. elinguis;* (**c**) ventral setae of *N. Variabilis;* (**d**) giant ventral setae of *N. paradalis*

tooth as in *C. diphanus,* while it resembles the ordinary ventral chaetae. In *Homochaeta setosa & p. frici*, penial chaetae of *O. serpentine,* with broadened distal end or forming hook (Plate 34.19), aimple hooked and stout penial chaetae are present in *Paranias litoralis*. Two simple penial chaetae per bundle are present with simple hook in *Stylaria lacustris* (Plate 34.19c), while in *Slavina appendiculata*, they are 3 with sharply curved hook. Most *Nias* sp. have 2–3 penial setae (Plate 34.19a); *N. elinguis* have 4–5 per bundle with simple distal hook, and 1–2 per bundle in *N. alpine,* with obtusely simple pointed.

Mature individuals of *Pristina longiseta* have thick spermathecal setae in VI, simple and curved, with prolonged teeth. *P. aequiseta* and *P. amphibiotica* also have

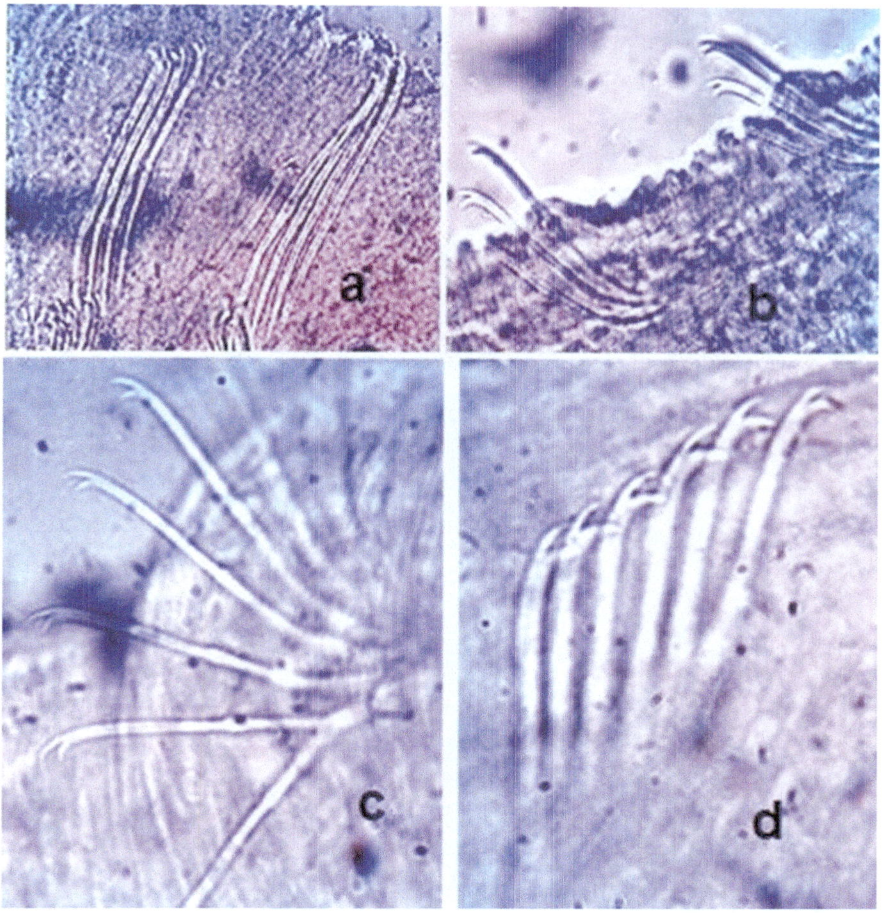

Plate 34.17 Ventral setae of Naidinae genera. (**a**) *Ophidonais*; (**b**) *Dero*; (**c**) *Pristina*; (**d**) *Allonais*

a single modified penial setae in VIII or IX, curved with two distal prongs. No penial setae are found in *Dero (Dero) spp.* and *Dero (Aulophorus) furcatus.*

34.7.2 Subfamily Tubificinae

Species of this subfamily are characterized by

1. Prostomium without proboscis; no eye.
2. Dorsal setae from II; hair setae may be present or absent; bifid, pectinate, palmate,

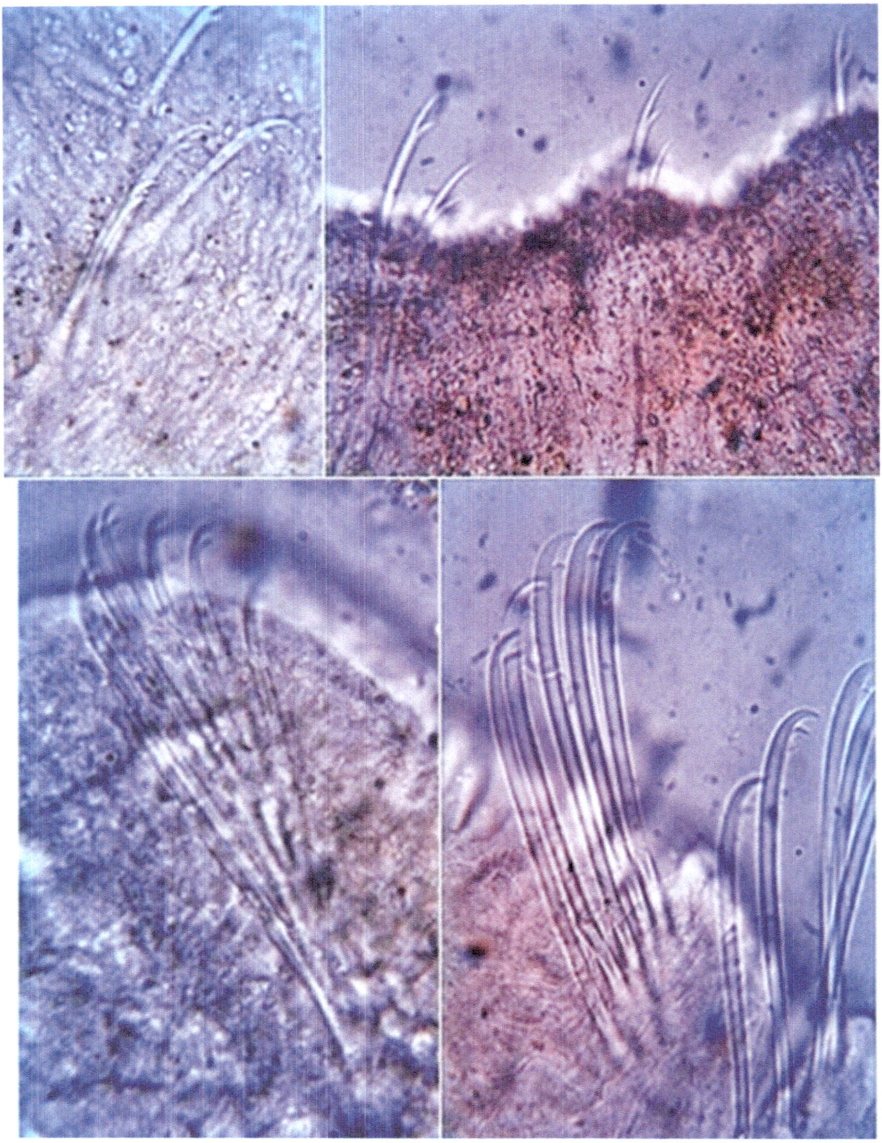

Plate 34.18 Ventral setae of Naidinae genera. (**a**) *Paranais*; (**b**) *Stephensoniana*; (**c**) *Chaetogaster;* (**d**) *Stylaria.* Genital chaetae in Niadidae are usually represented only by hooked

or even simple pointed mostly found in dorsal bundle.

3. Ventral setae indefinite number per bundle beginning in II, bifid or seldom simple pointed.
4. Spermathecal or penial setae or both modified or absent in mature specimen.
5. Clitella in a few segments in the region of gonads.

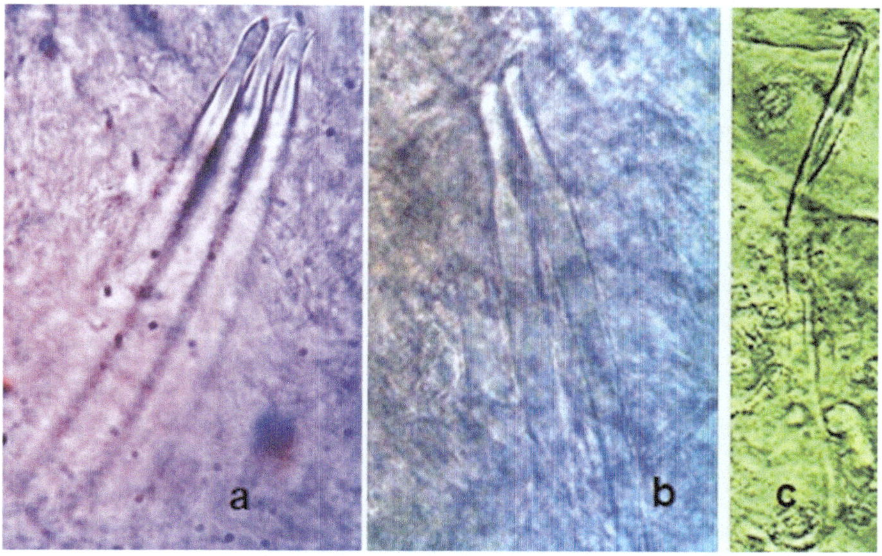

Plate 34.19 Naidinae genital setae. (**a**) Penial setae of *Nais*; (**b**) penial setae of *Ophodonais;* (**c**) spermathecal setae of *Stylaria*

6. Testes and ovaries,one pair each in X and XI; Spermathecae pair in X or single or absent; Spermatophores or free sperm mass in spermathecae; Male funnel in testis segment; Atrium and male pores in succeeding, together with female funnel; Female pores in furrow behind segment bearing male pores; sperm sac and ovisac usually present, unpaired.
7. Asexual reproduction by fragmentation.

Tubificid worms or sometimes called sludge also can be recognized by their chaetae. They are mainly bifid, more often pectinate or palmate. Simple pointed occurs less often (*Embolocephalus vulatinus*). Typical pectinate and palmate chaetae are characteristic of only subfamily Tubificinae, where they are found only in the dorsal. The number of chaetae per bundle ranged from 2–10, sometimes more. Maximum number occurs in the anterior segments. Genital setae are present, either spermathecal or penial or both (Plate 34.20).

Dorsal setae always begin from II. Hair setae may be present or absent, and their presence is useful in identifying species, but not genus. Their number generally ranges from 1–5 per bundle. Hair setae may be serrate as in *Psammoryctides barbatus*, or stiff simply pilose in *P. albicola*. In *T. tubifex,* the hair setae are fine, serrate or nonserrate (Plate 34.21).

Hair setae also differ in size, and they may be short and broad as in *Spirosperma ferox*. Short bayonet-shaped in *Aulodrilus spp.* (Plate 34.22a), straight as in *P. bavaricus* (Plate 34.22c), or may be bent with twisted end as in *Monopylephorus irrotatus* (plate 34.22b).

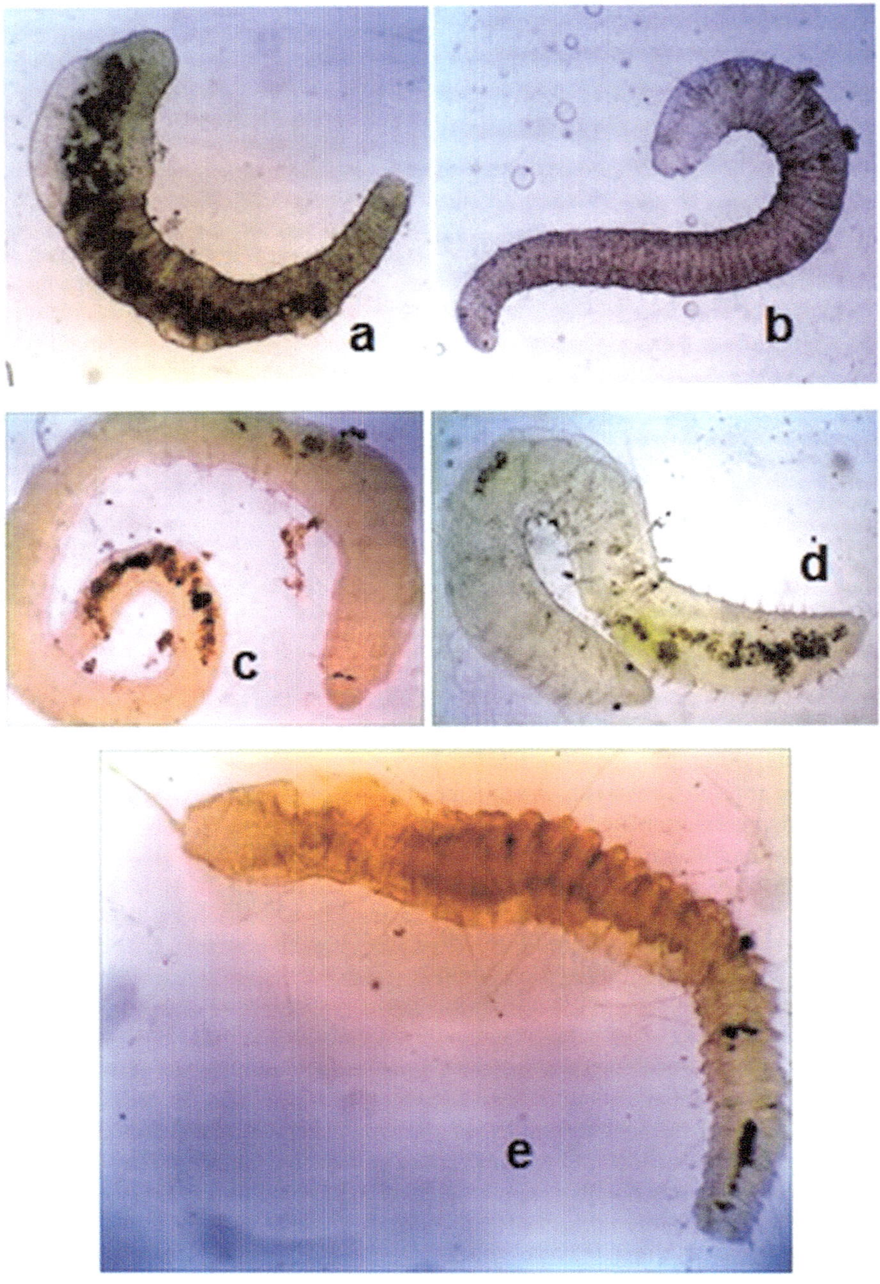

Plate 34.20 Naidid worms. (**a**) *Paranais sp.*; (**b**) *Stephensoniana trivandrana;* (**c** & **d**) *Nais spp.;* (**e**) *Stylaria lacustris*

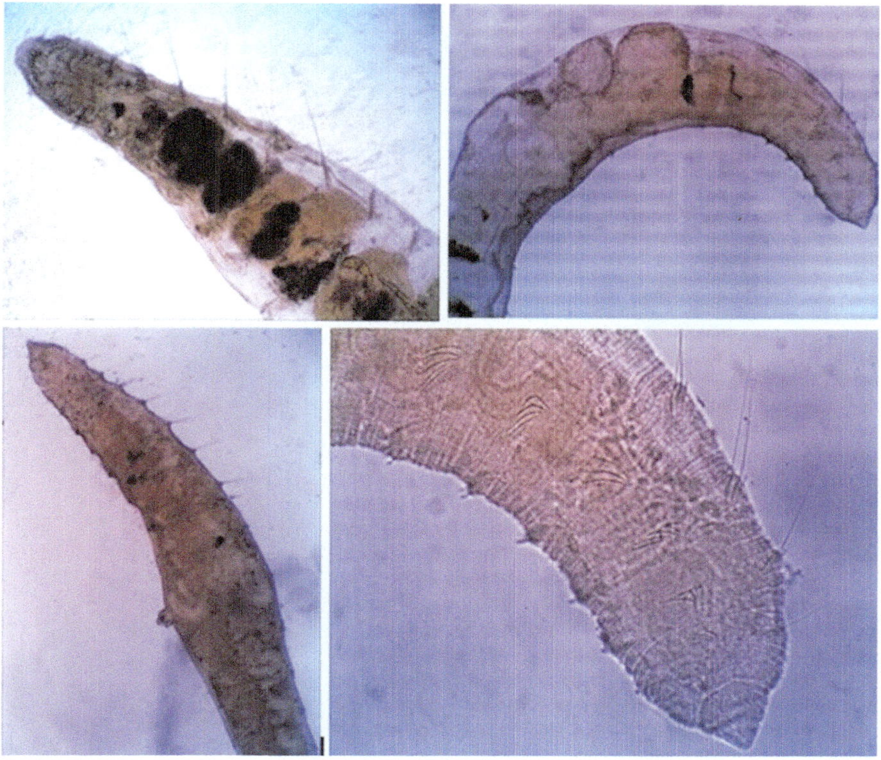

Plate 34.21 Tubifid worms with hair setae

Bifid crotchet with reduced upper tooth is found in dorsal and ventral bundles of *Branchiura sowerbyi,* and shorter and thinner upper tooth is found in the dorsal crotchet of *Aulodrilus pigueti* and *A. pluriseta.* In the case of *A. pluriseta*, distal part of crotchet, immediately under,often slightly dilated, intermediate spines can be visible in dorsal setae, arranged in transversal row (Plate 34.22), while in *A. pigueti*, beginning from VI-XI, dorsal crotchets are replaced by oar-shaped pectinate setae, with their tip rounded or sometimes slightly bifid (plate 34.4d).

Typical pectinate and palmate setae are characterized only in family Tubificicae (Plate 34.3). They are found only in dorsal bundles. They differ in shapes, numbers, and size between different species. Pectinate seta may be with more or less with equal size lateral teeth forming U-shaped tip with well-defined intermediate spines, as *T. nerthus.* In *Rhyacodrilus coccineus,* the pectinate setae are straight proximal with nodulus, teeth equal in length and divergent, with a series of fine spines.

Short intermediate spines are found in the pectinate setae of *Potamothrix spp,* with slightly longer and thinner upper tooth in *P. hammonunsis* and equal teeth in *P. bavaricus. T. tubifex* has pectinate setae with equal main teeth and with shorter but distinct intermediate spines (plate 34.23).

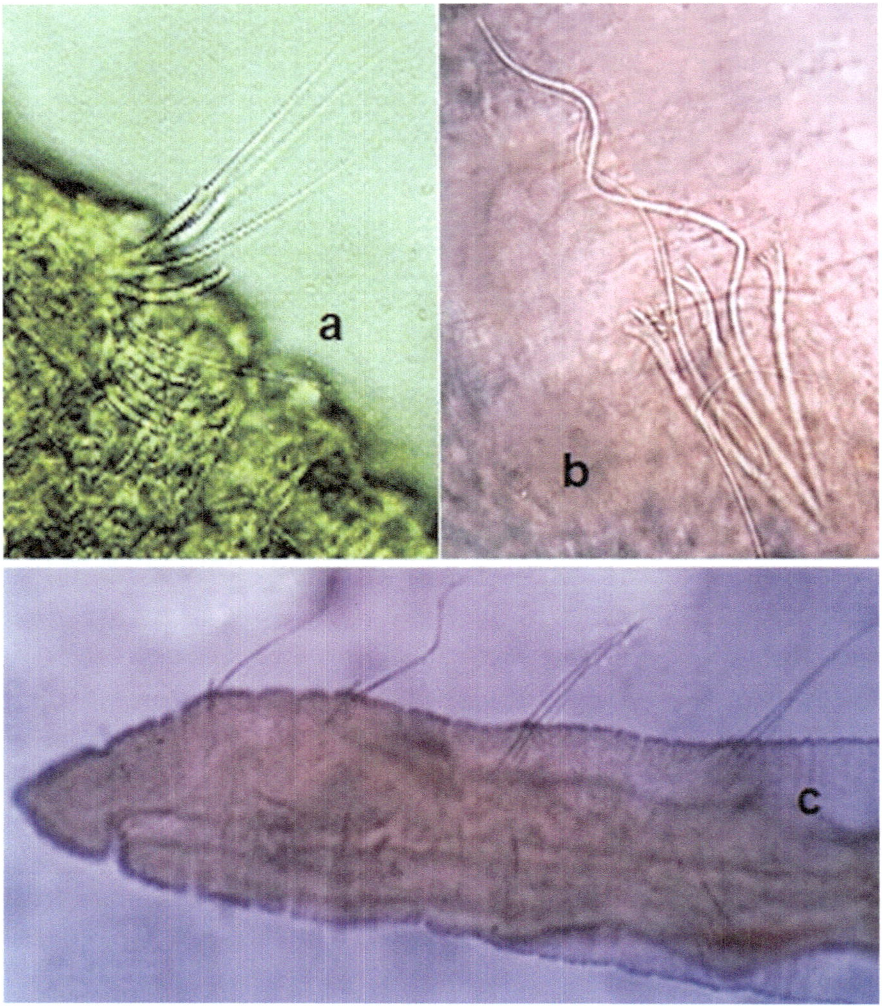

Plate 34.22 hair setae. (**a**) *Aulodrilus pluriseta;* (**b**) *Monopylephorus irrotatus;* (**c**) *Pomatothrix spp.*

In *Psammoryctides barbatus*, anterior dorsal bundles containing 7–8, branched spade-shaped palmate setae, with numerous intermediate denticles, as long as lateral teeth, both fuzing to form common broad, longitudinally folded blade (Plate 34.3g) are present.

Ventral bundles are much uniform in Tubificinae, and all are bifid with different ratios of upper and lower teeth length and thickness There is a number range of 2–20 per bundle. The number and the shape of setae also differ in different regions of the same specimen (Plate 34.24).

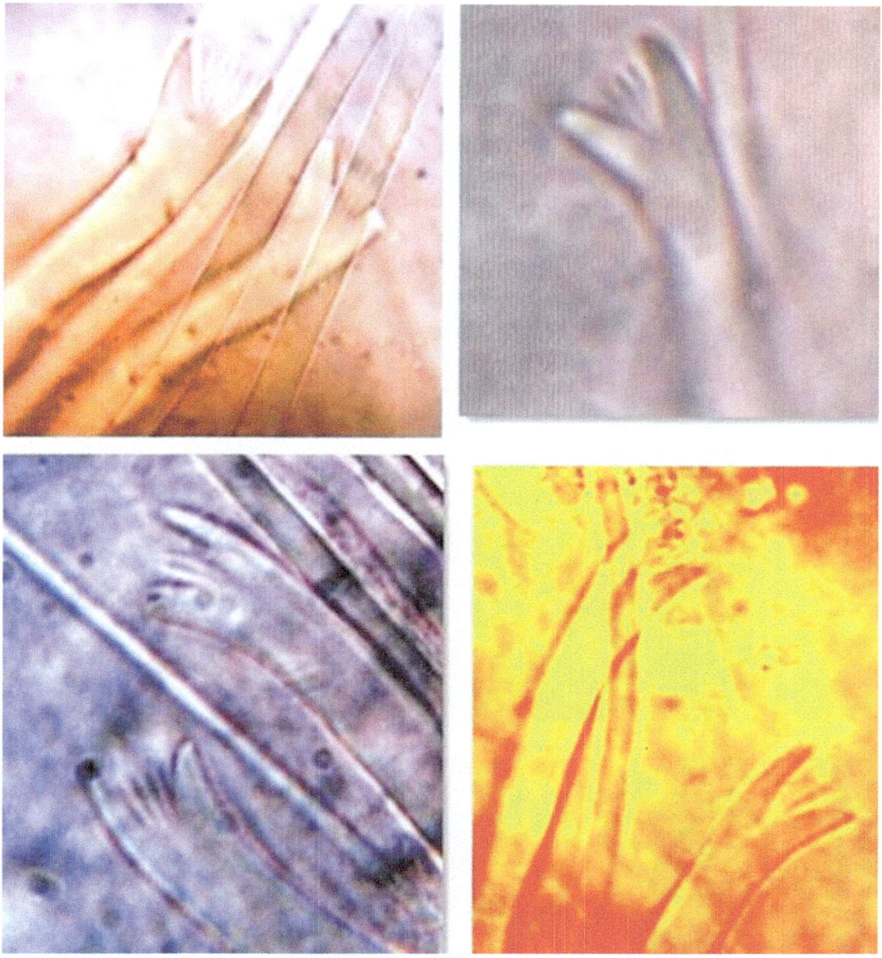

Plate 34.23 pectinate setae

In *Branchiura sowerbyi* (subfamily Branchiurinae), the ventral setae are about 10 in number or more, with reduced upper tooth. Simple pointed setae may be found in *Embolocephalus velutinus*, 1–2 per bundle beside the finely bifid crotchet, which is more curved posteriorly. In *A. pluriseta* and *P. pigueti,* ventral bundles contain up to 10 setae with thinner and shorter upper tooth. The distal part of crotchet in *P. pluriseta* often slightly dilated immediately under teeth. *H. speciosus* has ventral setae with upper tooth in preclitellar segments, twice longer than lower one, but with equal teeth in postclitellar segments. Further, rearward teeth are equally long or the upper one shorter. The same phenomenon is found in *P. barbatus* where the anterior ventral setae 3–5 per bundle anteriorly, with upper tooth longer and thinner than the lower one, only 2–3 posteriorly, with upper tooth shorter and thinner than the broad lower one.

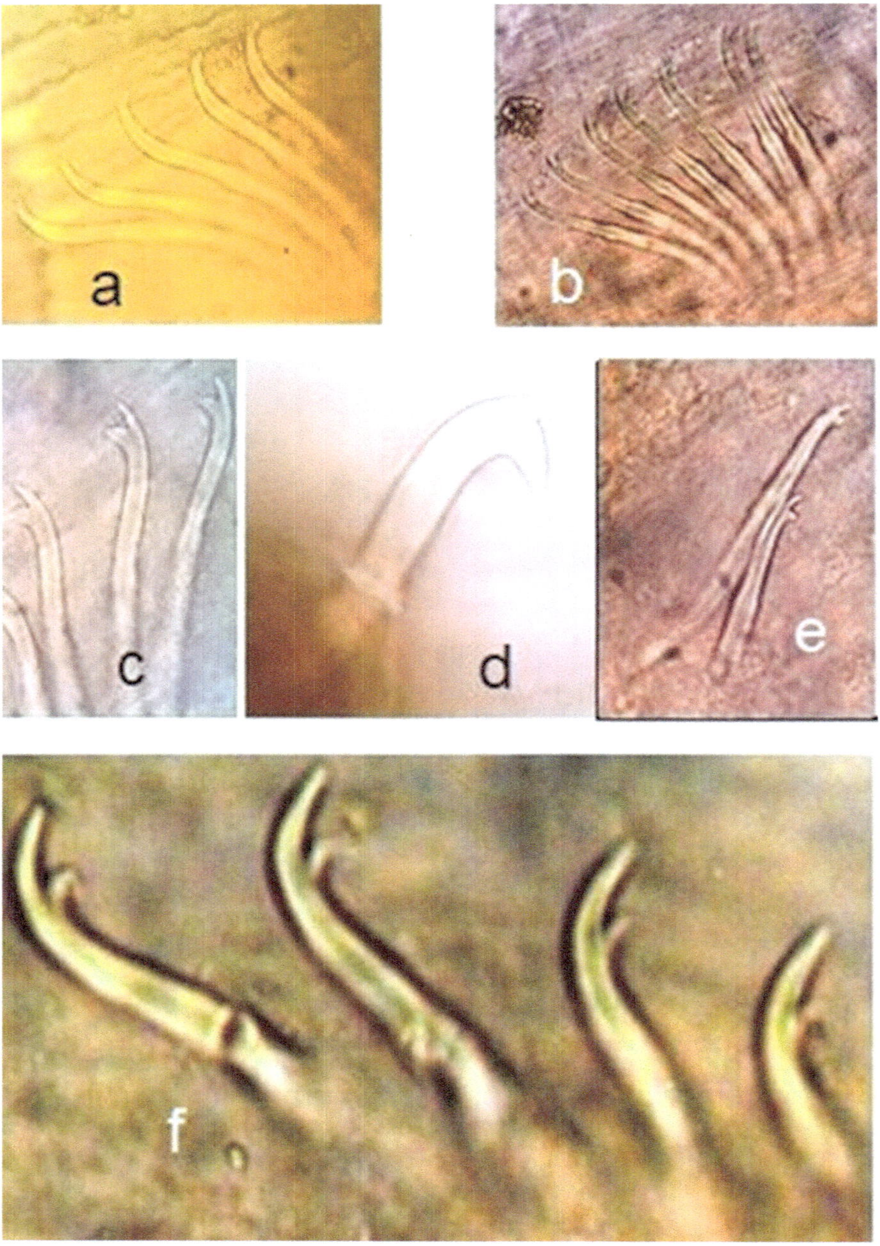

Plate 34.24 Types of ventral setae. (**a**) *Branchiura sowerbyi;* (**b**) *Limnodrilus hoffmeisteri;*(**c** & **d**) anterior & posterior ventral setae of *Psammorhyctides sp.;* (**e** & **f**) posterior *& anterior ventral seta of L. udekemianus*

In all species of *Potamothrix*, ventral bundles contain bifid setae, with slightly longer and thinner upper teeth. *M. irroratus* has 2–7 setae in ventral bundles, less in posterior region, with considerably long and thinner upper tooth. Anterior ventral setae of *R. coccineus,* 4–8 per bundle, with teeth of nearly equal length, become shorter posteriorly. The anterior ventral crotchet of *T. tubifex* has slightly longer upper tooth, while posterior crotchets have short and thinner upper tooth, with lower one being stout and curved.

In the case of Tubificid worms without hair setae, the dorsal and the ventral are quite similar in shape, but less number toward the posterior. The region dissimilarity is noticed in *Limnodrilus udekemianus* (plate 34.24e&f), where the anterior bundles have 3–8 crotchet, with upper tooth thicker and at least twice as long as lower, bent almost at an right angle, posterior setae with equal teeth. All seatae, (dorsal and ventral) are about equal length in *P. moldaviensis*; *L.profundicola; L. hoffmeisteri; L. claparedianus; and L. semiannual* (Plate 34.24b).

Genital setae in Tubificidae are represented either by spermathecal setae in segment IX or X or penial setae in segment XI. Single modified spermathecal setae in X are thin in *Psammoryctides spp.* Mostly broad, with broad or bent distal part, is present in *Potamothrix spp.* (Plate 34.5). It is spearhead-shaped in *P. bavaricus* and gutter-shaped in *P. hammoniensis*, while it is long in *P. heuscheri*, with distal portion straight and narrow, furrowed, with parallel edges and sharp hooked tip, and their proximal end is also hooked ().

Penial setae can occur in VII, 1–2 per bundle ~75 μm long, curved, with longitudinal furrow and sharp tip, in *A. pigueti*, while in *A. pluriseta,* no modified penial setae are present, but male pore is also present in VII.

R. coccineus has penial setae 2–5 per bundle, with blunt bent tip or with reduced upper tooth, often hardly protruding over body surface from aperture shared with male pore ().

Modified penial setae are, often hidden in body wall, near male pore in XI, usually straighter than ordinary setae, often with reduced teeth or with obtuse tip, if several per bundle, then their tips are located close to each other with proximal ends spread fan-like. They may be 1–2 in *T. nerthus* or more than two in *R. coccineus*.

No genital setae found in some Tubificid worms such as *B. sowerbyi*, which have unpaired male pore in XI, *T. tubifex,* have no any ventral setae in XI near the male pores.

References

Al-Abbad MY (2010) New records of *Pristina proboscidea* and *P. aquiseta* (Oligochaeta: Naididae) from Iraq. Marsh Bulletin 5(2):132–142

Al-Abbad MYM (2012) New records of twelve species of Oligochaeta (Naididae and Aeolosomatidae) from southern Iraqi marshes. Jordan J Bio Sci 5(2):105–111

Al-Abbad MYM, Al-Mayah SH (2010) New records of two species of Oligochaetes (Naididae): *Pristina longiseta* and *P. macrochaeta* from Iraq, with notes on their morphology and reproduction. Mesopot J Mar Sci 25(2):57–66
Ali LA (2007a) A study of macroinvertebrates community in the middle sector of Greater Zab River/Iraq. Ph.D. thesis, College of Science for Women, Baghdad Univ., 46 pp.
Ali LA (2007b) First record of Fadejewobdella quinqueannulata (Hirudinea: Erpobdellidae) in greater Zab River-Iraq. Zanco the Scie. J of the Univ of Salahaddin 2(22):47–50
Ali LA, Jaweir HJ (2013) First record of *Dina lineata* (Hirudinea: Erpobdellidae) in greater Zab river Kurdistan region-Iraq. J Univ Zakho (JUOZ) 1A(1):207–209
AL-Lami AA, Jaweir HJ, Nashaat MR (1997) Benthic invertebrates community of river Euphrates up stream and Dawn stream sectors of AL-Qadisia dam, Iraq. Regulated river. Res Manag 14 (4):383–390
Al-Raie HH (2014) Association between Snails and Oligochaees in aquatic habitat in the middle part of Iraq, MSc. Thesis, College of Science for Women, University of Baghdad
Alves R d G, Gorni GR (2007) Naididae species (Oligochaeta) associated aquatic macrophytes in two reservoirs (São Paulo, Brazil). Acta Limnol Bras 19(4):407–413
Arslan N, Ilhan S (2010) Distribution and abundance of Oligochaeta (Annelida) species and environmental variables of Porsuk stream (Sakarya River, Turkey). Review of Hydrobiology 3(1):51–63
Arslan N, Margan D (2018) Check list of freshwater Oligochaeta(Turkey), and oligochaeta fauna of lake Çildir (Adrahan-Kars, Turkey) with emphasis on the differentsamples of *Tubifex* sp.14th. International Symposium on Aquatic Oligochaeta. Abstracts
Arslan N, Sahin Y (2003a) Two new records of *Aulodrilus* Bretscher, 1899(Oligochaeta, Tubificidae) for the Turkish Fauna. Turk J Zool 27:276–280
Arslan N, Sahin Y (2003b) Nine new Naididae(Oligochaeta) species for Sakarya river. Turkey Turk J Zool 27:27–38
Arslan N, Sahin Y (2006) Apreliminarey Study on the Identification of the Littoral Oligochaete. Annelida, Dipter and Chironomidae Fauna Lake Kovada, a National Park in Turkey. Truk J Zool 30:67–72
Bashê SKR (2008) The parasitic fauna of spiny eel *Mastacembelus mastacembelus* (Banks and Solanser, 1794) from Greater Zab River-Kuristan region- Iraq. M. Sc. Thesis, Sci. Edu. Coll., Univ. Salahaddin: 62 pp
Beylich A (2005) A method for preservation and mounting of Enchytraeidae (Oligochaeta: annelid) proceeding of the Estonian Academy of Sciences, biology. Ecology 54(4):352–355
Brinkhurst RO (1971) A guide for the identification of British aquatic oligochaeta. Freshwater Biological Association. Sci.publ. No. 22
Brinkhurst RO, Jamieson BG (1971) The aquatic oligochaeta of the world. Edinburgh, Oliver & Boyd. XI, 860 pp
Chekanovskyana OV (1963) Aquatic oligochaeta of USSR. Key to the fauna of the USSR, Translated to English by Kohli, I. (1981), US. Department of interior and the national science foundation, Washington, D.C. Kothokar, V.S. (ed.) Oxonian Press, Pot. Ldt. Faridabad, India. xxii 513 pp
Corbi JJ, Trivinho-Strixino S, Alves RG (2005) Records of oligochaetes in freshwater sponges, on bryozoarians and on colonial hydrozoans from Brazil. Braz J Biol 65(1):187–188
Envall I, Källersjö M, Ersěus C (2008) Molecular evidence for non-monophyletic status of Naididae (Annelida, Clitellata, Tubificidae). Molecular Phytogenetics and Evolution 40:570–584
Erséus C, Gustarsson L (2002) A proposal to regard the former family Naididae as a subfamily within Tubificidae (Annelida, Clitellata). Hydrobiologia 485:253–256
Erséus C, Gustavsson L, Brinkhurst RO (2005) Case 3305. Tubificidae Vejdovský, 1876 (Annelida, Clitellata): proposed precedence over Naididae Ehrenberg, 1828. Bulletin of Zoological Nomenclature 62:226–231
Glasby CJ, Timm T (2008) Global diversity of polychaetes (Polychaeta; Annelida) in freshwater. Hydrobiologia 595:107–115

Gorni GR, Alves R d G (2006) Naididae (Annelida, Oligochaeta) associated with *Pomacea bridgesii* (reeve) (Gastropoda, Ampullaridae). Revista Brasileira de Zoologia 23(4):1059–1061
Gorni GR, Alves Rd G (2007) Naididae (Annelida, Oligochaeta) associated with briophytes in Brotas, state of Sao Paula, Brazil. Revista Brasileira de Zoologia 24(2):518–519
Gorni GR, Alves Rd G (2008) Naididae species (Annelida: Oligochaeta) associated with the sponge *Metania spinata* (Carter, 1881) (Porifera: Metaniidae) from a southeastern Brazilian reservoir. Acta Limnol. Brassica 20(3):261–263
Herzog PH (1969) Untersuch-ungen über die parasiten der sü Bwasserfische des Irak. Arch Fisch 20 (2/3):132–147
Jaweir HJ (1987) Namalycastis indica (Annelida: Polychaeta) a new species from Iraq. J Bio Sci Res (JBSR) 18(1):229–230
Jaweir HJ (2011) A new records of three tubificid species (Annelida: Oligochaeta) from Al-Hawieza marsh. Iraq Mesopot J Mar Sci 26(2):114–121
Jaweir HJ (2014) Checklist of aquatic oligochaete species in Tigris-Euphrates rivers basin. J Baghdad for Sci 11(3):1397–1404
Jaweir HJ, Albayati SS (2016) A study of soil invertebrates community in a date- palm plantation in Baghdad, Iraq. J Baghdad for Sci 13(1):1–6
Jaweir HJ, Al-Janabi EOS (2012) Biodiversity and abundance of aquatic oligochaetes –Family Naididae- in the middle sector of Euphrates river, at Al-Mussayab City/Iraq
Jaweir HJ, Al-Sarai MH (2016) The aquatic annelid community in Lake Al-Delmage/Iraq. Biologia 71(1):1–6
Jaweir HJ, Alwan AM (2013) Sludge woems species (Oligochaeta: Naididae: Tubificinae) from different aquatic habitat in Baghdad/Iraq. J Baghdad for Sci 10(2):269–281
Jaweir HJ, Rhadi MM (2012) Naididae (Clitellata: Oligochaeta and Aeolosomatidae(Polychaeta: Aphanoneura) species associated with aquatic plants in Tigris River/Baghdad/Iraq). J Baghdad for Sci 10(1):116–125
Jaweir HJ, Sebtie HA, Al-Mukhtar EA (2012) Aquatic Oligochaetes of Iraq's southern marshes. J Baghdad for Sci 9(3):472–480
Kalifa KA (1985) Leeches on freshwater farmed fishes in Iraq. J Wild Dis 21(3):312–313
Kassim TI, Jaweir HJ, AL-Maliky SMSK, Nashaat MR (1997) Benthic fauna in Tigris river. Iraq J Coll Educ For women, Univ Baghdad 8(2):167–170
Martin P, Martinez-Ansemil E, Pinder A, Timm T, Wetzel MJ (2008) Global diversity of oligochaetous clitellates (Oligochaeta: Clitellata). In: Freshwater animal diversity assessment. Development in hydrobiology, vol 198, pp 117–127
Mhaisen FT, Al-Khateeb GH, Balasem AN, Mutar AJ (1997) On a collection of some fish parasites from Euphrates River, Anbar province. Iraq Babylon Univ J, Pure Appl Sci 2(3):267–272
Nashaat MR (2010) Impact of Al-Durah power plant effluents on physical, chemical and invertebrates biodiversity in Tigris River, southern Baghdad, Ph.D. thesis, College of Science, Univ. of Baghdad
Nazarhaghighia F, Timm T, Nadoushanc MR, Shabanipourd N, Mohammad Reza F, Moradia MA (2014) Oligochaetes (Annelida, Clitellata) in the Anzali international wetland, North-Western Iran. Estonian J Ecol 63(3):130–144
Nielson CO, Christensen B (1959) The Enchytraeidae, Critical revision and taxonomy of European species. (Studies on Enchytraeidae VII). Natura Jutlandica, 8–9: 1–160. (Naturhistorisk Museum) Aarhus, Danemark
Parker SP (1982) Synopsis and classification of living organism, vol 2. Macraw-Hill Company, New York
Rahemo ZIF (1989) *Cystibranchus mastacembeli* (Annelida: Hirudinea) from the Iraqi freshwater spinyback, *Mastacembelus simach* (Wabaun, 1792). Riv Parasitol 6(50):121–126
Rodriguez P, Reynoldson TB (2011) The pollution biology of aquatic Oligochaetes. Springer, Dordrocht Heidelberg, p 265
Sabtie HA (2009) An ecological study of the benthic macroinvertebrates community in the Southern Marshes of Iraq. Ph.D.thesis, Colloge of Science for Women, Baghdad Univ., 167 pp

Tas M, Elipek BC, Kirgiz T, Arslan N, Yildiz S (2004) The aquatic and smi-aquatic oligochaeta funa of Turkish Thrace. J Fish Sci. com

Timm T (1981) On the origin and evolution of aquatic oligochaeta. Esti NSV Teuduste Akademia Toimetised Biologia 30:174–181

Timm T (2009) A guide to the freshwater Oligochaeta and Polychaeta of northern and Central Europe. Lauterborina 66:1–235

Yildiz S, Ahiska S (2010) *Nais stolci* Hrabe, 1981. A new Oligochaeta (Annelida: Clitellata: Naididae) species for Turkey. Turk J Zool 34:547–549

Yildiz S, Balik S (2005) The Oligochaeta (Annelida) fauna of the inland waters in the Lake District. E.U. J Fish Aquat Sci 22(1):165–172

Yildiz S, Rusen M, Balik S (2007) The Oligochaeta (Annelida) Fauna of Yuvarlak stream (Koycegiz–Turkey). Turkish J Fish Sc 7:01–06

Chapter 35
A Critical Checklist of the Inland Fishes Native to the Euphrates and Tigris Drainages

Jörg Freyhof, Cüneyt Kaya, and Atheer Ali

The fishes of the Euphrates and Tigris rivers are reviewed and 244 species from 59 families are recognized. From these, 130 are coastal species, many just entering the lower Shatt Al-Arab during salt water intrusion. Additional 116 species are anadromous (*Tenualosa ilisha*) or restricted to freshwaters including one species (*Aphaniops stoliczkanus*) found in freshwater, brackish water, and marine habitats. Three families (Cyprinidae: 32 species, Leuciscidae: 21 species, Nemacheilidae: 39 species) dominate the freshwater fish fauna. The largest genus is *Oxynoemacheilus* with 26 species in the area. Endemic to the Euphrates and Tigris are 90 species (81%) of freshwater species. Most non-endemic freshwater species occur also in the Iranian rivers of the Gulf, while six species are shared with the Mediterranean and one each with the Black Sea and the Caspian Sea basins.

35.1 Introduction

Fishes are the largest group of aquatic vertebrates and they are traditionally valued as an indicator for water quality and a source of protein for human consumption. Fishing remains the largest extractive use of wildlife in the world and provides a

J. Freyhof (✉)
Museum für Naturkunde, Leibniz Institute for Evolution and Biodiversity Science, Berlin, Germany
e-mail: joerg.freyhof@mfn.berlin

C. Kaya
Faculty of Fisheries and Aquatic Sciences, Recep Tayyip Erdogan University, Rize, Turkey

A. Ali
Department of Fisheries and Marine Resources, College of Agriculture, University of Basrah, Basrah, Iraq

L. A. Jawad (ed.), *Tigris and Euphrates Rivers: Their Environment from Headwaters to Mouth*, Aquatic Ecology Series 11,
https://doi.org/10.1007/978-3-030-57570-0_35

lasting vestige of utilizing the resources of a global commons making fishing a serious threat (Youn et al. 2014), especially for large growing fishes as these are often the victim of uncontrolled overfishing (Carrizo et al. 2017).

Fishing is just one driver of biodiversity loss and biodiversity continues to erode globally including in the Euphrates and Tigris drainages. Indeed, biodiversity loss is particularly severe in freshwater ecosystems. Globally, we have lost 69–75% of inland natural wetlands during the twentieth century (Davidson 2014); rates of freshwater species population declines have reached 81% for the period between 1970 and 2012, representing a rate of decline twice that recorded for either marine or terrestrial ecosystems (WWF 2016); and, approximately one in three of the 28,000 species dependent upon freshwater habitats so far assessed for the IUCN Red List of Threatened Species™ is threatened with extinction (IUCN 2017). This general pattern is also true for the Euphrates and Tigris drainages, where overfishing for large species is evident and several fish species are threatened with extinction (Freyhof et al. 2014a).

At the same time, freshwater biodiversity, including the diversity of fishes is still incompletely known and the conservation status has not yet been assessed for all species and has only been assessed more than once for very few species, leaving us poorly informed to measure biodiversity loss and poorly able to detect critical sites or needed action for biodiversity conservation.

Despite that the freshwater fish fauna of the Euphrates and Tigris drainages has been studied for many decades, it remains very incompletely known and new species are recognized regularly from the area. Indeed, there is much progress in understanding the diversity of freshwater fishes in the Euphrates and Tigris drainages, especially in the last 10 years. This is due to the renewal of interest in biodiversity exploration mostly triggered by scientists in the countries of the Euphrates and Tigris drainages and by "new" methodologies and concepts available, especially in the field of molecular taxonomy.

The freshwater fishes of the Euphrates and Tigris drainages have been reviewed comprehensively by Banister (1980) who discusses earlier checklists recognizing 61 species, Coad (1991) who recognized 47 native species, excluding euryhyaline coastal species and Coad (1996), who recognized 52 species of primary freshwater fishes of the orders Cypriniformes, Siluriformes, and Perciformes. From today's viewpoint, it is interesting to see, how much our knowledge has improved in the 38 years since the review by Banister (1980), who's species list cannot at all be compared with the species list we have today, as it is full of small problems resolved by later authors (see also Krupp and Schneider 2008:49). Definitively, we have learned much about the fishes of the Euphrates and Tigris and there is still space to learn more.

Since breakthrough studies by Coad (1991, 1996), authors took a national approach and did compile comprehensive lists for the national parts of the drainages as Coad (2010) for Iraq, Geldiay and Balık (2007) for Turkey, Kaya et al. (2016) for the Turkish part of the Tigris, Krupp and Schneider (2008) for the Syrian Habur River and Esmaeili et al. (2010), Jouladeh-Roudbar et al. (2015) and Esmaeili et al. (2017) for Iran. None of these more recent authors had a dedicated look at the complete watershed of the Euphrates and Tigris and compiled a species list for these rivers. Also, the distribution data presented by these national authors needs to be compiled one by one from their national list.

Here we present a new and up-to-date (March 2021) checklist of the freshwater fishes of the Euphrates and Tigris drainages. This list does not only include all resident freshwater species native to the catchment, but also a large and comprehensive list of all coastal and estuarine species found in the lower parts of the drainage in the last 28 years. This new checklist allows drawing conclusions on endemism and biogeography of the Mesopotamian catchments. It also can function as a backbone for conservation and sustainable use of the highly diverse and interesting fauna of the area.

35.2 Coverage

All species of fishes recorded from the Euphrates and Tigris catchments are included. This includes also the species found in the Qweik River, which was a western tributary of the Euphrates, as well as the basin of Lake Van in eastern Anatolia, which had been connected to the Euphrates and Tigris in the recent geological past. Species of the Karun River in Iran are included but those from the Zoreh River in Iran are excluded as the Zoreh has an own estuary to the Persian Gulf (Arabian Gulf) just east to the Shatt Al-Arab. The Zoreh, as several other rivers in Iran had been tributaries to the Shatt Al-Arab during the last glaciation, but their fauna is ignored here. We follow Tan and Armbruster (2018) in treating the cyprinid subfamilies Danioninae and Leuciscinae as families in their own right, Danionidae and Leuciscidae. The class, order, or any of the intermediate classification levels are not listed and the checklist is restricted to the family, genera, and species level. Synonyms of species are provided only if not provided by the Catalog of Fishes, which should be consulted for actual names of potential synonyms of species not found in this checklist but found in faunal accounts of former publications. Be aware, that some species might not be found in this checklist as they are now restricted to populations outside of the Euphrates and Tigris drainages. Examples are *Capoeta barroisi* and *Luciobarbus pectoralis*, both Cyprinidae, two valid species, both often listed for the fauna of the Euphrates and Tigris. But both species are restricted today to the Mediterranean basin. Furthermore, all knowledge found in FishBase or the Catalog of Fishes is not repeated here, but just regionally relevant information is given on the distribution of species as well as on taxonomic notes and few remarks.

35.3 Check List

The fishes of the Euphrates and Tigris rivers are reviewed and 244 species from 59 families are recognized.

Family: Carcharhinidae

Carcharhinus leucas **(Müller & Henle, 1839)**

Distribution in the area. Not endemic. Enters the Shatt Al-Arab migrating into the Karun and Khowr-e Bahmanshir in Iran. In Iraq, migrated regularly upriver to Baghdad. Today only up to Basrah due to dams.

Rhizoprionodon acutus **(Rüppell, 1837)**

Distribution in the area. Not endemic. Enters rarely the lower of Shatt Al-Arab.

Family: Sphyrnidae

Sphyrna mokarran **(Rüppell, 1837)**

Distribution in the area. Not endemic. Enters rarely the lower Shatt Al-Arab.

Family: Hemiscylliidae

Chiloscyllium arabicum **Goubanov, 1980**

Distribution in the area. Not endemic. Recorded from the Shatt Al-Arab.

Family: Rhinobatidae

Glaucostegus granulatus **(Cuvier, 1829)**

Distribution in the area. Not endemic. Recorded from the lower Shatt Al-Arab and the Khor Al-Zubair.

Family: Dasyatidae

Himantura uarnak **(Gmelin, 1789)**

Distribution in the area. Not endemic. Recorded rarely from the lower Shatt Al-Arab.

Pastinachus sephen **(Fabricius, 1775)**

Distribution in the area. Not endemic. Recorded rarely from the lower Shatt Al-Arab.

Family: Clupeidae

Anodontostoma chacunda **(Hamilton, 1822)**

Distribution in the area. Not endemic. Recorded from the Shatt al-Arab.

Nematalosa nasus **(Bloch, 1795)**

Distribution in the area. Not endemic. Recorded from the Shatt Al-Arab and the marshes. Commonly found on the markets in Basrah.

Nematalosa persara **Nelson & McCarthy, 1995**

Distribution in the area. Not endemic. Recorded from the Shatt Al-Arab.

Sardinella longiceps **Valenciennes, 1847**

Distribution in the area. Not endemic. Recorded from the Shatt Al-Arab.

Tenualosa ilisha **(Hamilton, 1822)**

Distribution in the area. Not endemic. Only anadromous species in the area. Enters the Shatt Al-Arab to spawn upriver in the Tigris and Euphrates. In the Tigris, it migrated as far north as Qal'at Salih and into the East Al-Hammar marshes, as well as in the Euphrates as far north as Yaou (Coad 2010). Also known from the Karun drainage.

Dussumieiriidae

Dussumieria acuta **Valenciennes, 1847**

Distribution in the area. Not endemic. Recorded from the Shatt Al-Arab.

Family: Engraulidae

Thryssa dussumieri **(Valenciennes, 1848)**

Distribution in the area. Not endemic. Recorded from the Shatt Al-Arab.

Thryssa hamiltonii **Gray, 1835**

Distribution in the area. Not endemic. Recorded from the lower Euphrates and Tigris, the Shatt Al-Arab and Al-Hammar marsh.

Thryssa setirostris **(Broussonet, 1782)**

Distribution in the area. Not endemic. Recorded from the Shatt Al-Arab and the marsh area.

Thryssa vitrirostris **(Gilchrist & Thompson, 1908)**

Distribution in the area. Not endemic. Recorded from the Shatt Al-Arab.

Thryssa whiteheadi **Wongratana 1983**

Distribution in the area. Not endemic. Recorded from the lower Euphrates, the Shatt Al-Arab, and Al-Hammar marsh.

Family: Chirocentridae

Chirocentrus dorab **(Fabricius, 1775)**

Distribution in the area. Not endemic. Recorded from the Shatt Al-Arab.

Chirocentrus nudus **Swainson, 1839**

Distribution in the area. Not endemic. Recorded from the Shatt Al-Arab.

Family: Pristigasteridae

Ilisha compressa **Randall, 1994**

Distribution in the area. Not endemic. Recorded from the Shatt Al-Arab and the Al-Hammar marsh.

Ilisha melastoma **(Bloch & Schneider, 1801)**

Distribution in the area. Not endemic. Recorded from the Shatt Al-Arab.

Ilisha sirishai **Seshagiri Rao, 1975**

Distribution in the area. Not endemic. Recorded from the Shatt Al-Arab.

Family: Chanidae

Chanos chanos **(Fabricius, 1775)**

Distribution in the area. Not endemic. Recorded from the Shatt Al-Arab.

Family: Cyprinidae

Arabibarbus grypus **(Heckel, 1843)**

Distribution in the area. Not endemic. Widespread in Euphrates and Tigris drainages from southern Anatolia south to Shatt Al-Arab. Also in the Qweik drainage, from where it is now extirpated.

Remarks. This is a major commercial species and seems to be overfished in large if not all parts of its range. However, data on population trends are very rare and based on fisheries statistics.

Barbus karunensis **Khaefi, Esmaeili, Geiger & Eagderi, 2017**

Distribution in the area. Endemic. Upper Karun drainage.

Taxonomic notes. Based on its description by Khaefi et al. (2017), *B. karunensis* cannot be distinguished from *B. cyri* from the Caspian Sea and Lake Urmia basins and *B. lacerta* from the Lake Van, Euphrates, and Tigris by morphological characters. It can be identified by molecular characters only.

Barbus lacerta **Heckel, 1843**

Distribution in the area. Endemic. Very widespread and usually abundant in headwaters of the Lake Van basin, the Euphrates, and Tigris except of the Karun. Extirpated from the Qweik drainage.

Taxonomic notes. *Barbus kosswigi* described from the upper Greater Zab drainage in Turkey and *B. ercisianus* from the Lake Van basin, both in Eastern Anatolia, are synonyms. Khaefi et al. (2017) reviewed the barbels of the *B. lacerta* species group and recognized four species (*B. lacerta*, *B. cyri*, *B. miliaris*, *B. karunensis*) two of them occur in the area discussed here. *Barbus lacerta* cannot be distinguished from *B. cyri* from the Caspian Sea basin and *B. karunensis* from the Karun drainage by morphological characters. All species of the *B. lacerta* group are well distinguished by molecular characters and are accepted as valid species.

***Caecocypris basimi* Banister & Bunni, 1980**

Distribution in the area. Endemic. Groundwater layer near Haditha in the Iraqi Euphrates drainage.

Taxonomic notes. While no fresh material of *Caecocypris* is available to allow a sound study on its phylogenetic position, it is superficially similar to *Mesopotamichthys* by lacking barbels and having very large scales. If rediscovered, it might turn out to be a species of *Mesopotamichthys*.

Remarks. This species seems not to have been found since the collection of the original materials and might be extinct.

***Capoeta coadi* Alwan, Zareian & Esmaeili, 2016**

Distribution in the area. Endemic. Widespread in the Karun drainage.

***Capoeta damascina* (Valenciennes, 1842)**

Distribution in the area. Not endemic. Very widespread in the Euphrates and Tigris drainage including the Qweik. Absent from Iranian rivers Karun and Karkheh.

Taxonomic notes. *Capoeta kosswigi* and *Capoeta angorae* are synonyms of this species. The relationship of *C. damascina* and *C. umbla* is still unresolved. While recent studies accept both as valid species, they seem to occur in sympatry in the Tigris drainage. Both species show neglectable differences in molecular characters (Alwan 2010; Alwan et al. 2016). Alwan (2010) distinguished both species are by scale counts but Schöter et al. (2009) pointed on a bimodal distribution of lateral line scale numbers in some Anatolian *C. damascina* populations. Kaya (2019) compared *C. damascina* and *C. umbla* populations in Turkey morphologically, however, he did not find any significant differences. We do not exclude, that just one species with a bimodal distribution of scale counts is involved. For the time being, both species are accepted as valid.

***Capoeta macrolepis* (Heckel 1847)**

Distribution in the area. Not endemic. Widespread in the Tigris drainage in Iran. Not yet recorded from Iraq or Turkey.

Taxonomic notes. This species was previously known as *C. aculeata*, which has been restricted to the populations of the Lake Namak and Kavir basins in Iran (Zareian et al. 2018).

***Capoeta pyragyi* Jouladeh-Roudbar, Eagderi, Murillo-Ramos, Ghanavi & Doadrio, 2017**

Distribution in the area. Not Endemic. Tireh and Sezar Rivers, which are headwaters in the Iranian Tigris drainage. Also in Zayandeh drainage. *Capoeta birunii* is a synonym.

Capoeta shajariani **Jouladeh-Roudbar, Eagderi, Murillo-Ramos, Ghanavi & Doadrio, 2017**

Distribution in the area. Endemic to the Gamasiyab, a headwater river in the Iranian Tigris drainage.

Capoeta trutta **(Heckel, 1843)**

Distribution in the area. Endemic. Very widespread in the Euphrates, Tigris, and Qweik drainages.

Taxonomic notes. Fishes with a dorsal spine shorter than the head length are usually identified as *C. barroisi*. This species is endemic to the Orontes River and has not been confirmed from the Euphrates or Tigris drainages so far. *Capoeta trutta* individuals with relatively short last unbranched dorsal-fin ray occur with low frequency in many populations of the species, which normally has a very long last unbranched dorsal-fin ray.

Capoeta umbla **(Heckel, 1843)**

Distribution in the area. Endemic. Very widespread in the Tigris drainage but seems to be absent from the Iranian Karun and Karkheh.

Taxonomic notes. The relationship of *C. umbla* and *C. damascina* is still unresolved. See taxonomic note for *C. damascina*.

Carasobarbus kosswigi **(Ladiges, 1960)**

Distribution in the area. Endemic. Known from few places only: The Botan and Batman Rivers and the Tigris below the confluence with Batman River in Turkey, the Little Zab River in Iraq and the Karkheh and Karun Rivers. In the Euphrates drainage reported from Euphrates at Ḥadithah, Iraq and from the Nahr al Khabur, a tributary of the Euphrates in Syria and Turkey.

Carasobarbus luteus **(Heckel, 1843)**

Distribution in the area. Not endemic. Very common and widespread in the Euphrates and Tigris as well as in the Qweik and marshes.

Carasobarbus sublimus **(Coad & Najafpour, 1997)**

Distribution in the area. Not endemic. Known from the A'la River in wider Karun drainage and from Shatt Al-Arab and Euphrates rivers in Iraq.

Cyprinion kais **Heckel, 1843**

Distribution in the area. Endemic. Widespread in the Qweik, Euphrates and Tigris drainages from southern Anatolia to the Shatt Al-Arab including the Karun River. Records from the Zohreh need confirmation.

Cyprinion macrostomus **Heckel, 1843**

Distribution in the area. Not endemic. Widespread in the Qweik, Euphrates and Tigris drainages from southern Anatolia to the Shatt Al-Arab in Iraq including the Karun.

Taxonomic notes. This species occurs often syntopic with *C. kais*. Mitochondrial data suggest that *C. kais* is fully introgressed by *C. macrostomus*, having lost its own mitochondrial bodies. But position of fins, shape of mouth, patterns of nuptial tubercles as well as color patterns are so different that there is no doubt that both are independent species.

Remark. This species is frequently reported to occur also in the Orontes River. We are not aware of a confirmed record of *Cyprinion* from the Orontes and treat it as endemic to the Gulf basin.

Garra amirhosseini **Esmaeili, Sayyadzadeh, Coad & Eagderi, 2016**

Distribution in the area. Endemic to the Iranian hot spring Sartang-e-Bijar at Mehran.

Remark. This species is very closely related to *G. elegans* and *G. mondica*. While is it very well distinguished from *G. elegans* by morphological characters, it might be a synonym of *G. mondica*.

Garra elegans **(Günther, 1868)**

Distribution in the area. Seems to be endemic to Iraq, where it has been found in the Little Zab River drainage and in the middle Tigris (Freyhof 2016). Might be more widespread in lowland habitats.

Garra gymnothorax **Berg, 1949**

Distribution in the area. Endemic. Karun and Karkheh drainages.

Taxonomic notes. *Garra gymnothorax* might be a real cryptic species. It was described based on populations with a naked breast from the Karun in Iran (Berg 1949). However, most *Garra* from the Karun have a scaled breast and a naked breast is also found in some populations of *G. rufa*. In *G. gymnothorax* as well as in many populations of *G. rufa*, the scales on the breast are often embedded in the skin. In contrast to the diagnosis given by Esmaeili et al. (2016), the breast squamation is variable in *G. gymnothorax* and *G. rufa* and no morphological character has been described to distinguish both species. They are well distinguished by nuclear (Rhodopsin; Behrens-Chapuis et al. 2015) as well as mitochondrial (COI; Hamidan et al. 2014) characters. *Garra gymnothorax* seems to be most closely related to subterranean *G. typhlops* and *G. lorestanensis*, which occur within its range (Hashemzadeh Segherloo et al. 2016).

Garra lorestanensis **Mousavi-Sabet & Eagderi, 2016**

Distribution in the area. Endemic. Only known from a natural well at Kaaje-Ru, in Baq-e-Loveh Oasis, in the Dez system of the Karun drainage.

Garra meymehensis **Zamani-Faradonbe, Keivany, Dorafshan & Zhang,** 2021

Distribution in the area. Endemic. Only known from Meymeh River in Iranian Tigris drainage.

Garra rufa **(Heckel, 1843)**

Distribution in the area. Not endemic. Very widespread in the Euphrates and Tigris drainages from cold headwater in Turkey to brackish Shatt Al-Arab. Extirpated from the Qweik. Absent from Lake Van drainage and Karun and Karkheh drainages.

Garra variabilis **(Heckel, 1843)**

Distribution in the area. Not endemic. Qweik and upper Euphrates and Tigris in Turkey, Syria, and Iraq. Absent from Lake Van drainage. While the species is continuously listed in the fauna of Iran, we are not aware of a record from that area.

Garra tashanensis **Mousavi-Sabet, Vatandoust, Fatemi & Eagderi, 2016**

Distribution in the area. Endemic. Only known from the Tashan Cave at Sarjusher in the Iranian Tigris drainage.

Garra tiam **Zamani-Faradonbe, Keivany, Dorafshan & Zhang,** 2021

Distribution in the area. Endemic. Only known from Abshur River in the Karun drainage.

Garra typhlops **(Bruun & Kaiser, 1944)**

Distribution in the area. Endemic. Only known from a natural well at Kaaje-Ru, in Baq-e-Loveh Oasis, in the Dez system of the Karun drainage.

Garra widdowsoni **(Trewavas, 1955)**

Distribution in the area. Endemic. Groundwater layer near Haditha in the Iraqi Euphrates drainage.

Luciobarbus barbulus **(Heckel, 1847)**

Distribution in the area. Not endemic. Widespread in the Euphrates and Tigris drainages. It was also known from the Qweik River, where it is now extirpated.

Taxonomic notes. This species is often identified as *L. pectoralis* which is endemic to the Mediterranean Sea basin, where it is found in the Orontes, Ceyhan, Seyhan, and Göksu Rivers. There are no confirmed records of *L. pectoralis* from the Euphrates and Tigris drainages. *Luciobarbus barbulus* is identified as *L. mystaceus* in large parts of its range. *Luciobarbus mystaceus* has been described by Pallas (1814) from the Kura River in Georgia and this species is a synonym of *L. capito*. Authors such as Karaman (1971) and Almaça (1983, 1991) refer the species description of *B. mystaceus* to Heckel (1843), who only identified some barbels from Mesopotamia as *B. mystaceus*, but did not describe the species. *Barbus rajanorum* described from "Aleppo" is a hybrid of *L. barbulus* and *Capoeta damascina*.

Luciobarbus esocinus **Heckel, 1843**

Distribution in the area. Endemic. Widespread in the Euphrates and Tigris drainages. It was also known from the Qweik River, where it is now extirpated.

Remarks. *Luciobarbus esocinus* is one of the largest Cyprinids in the world and a small-scale angling tourism has developed around this species in the last years, especially in Turkey and Iran. It is also a major commercial species and seems to be overfished in large if not all parts of its range. However, data on population trends are very rare and based on fisheries statistics.

Luciobarbus subquincunciatus **(Günther, 1868)**

Distribution in the area. Endemic. Euphrates and Tigris drainages including the Karun from southern Anatolia to the Shatt Al-Arab and the Al-Hammar marshes.

Remarks. This species is very rare all over its range and might be at the border of extinction. However, it seems to have been always rare in Turkey and Iran and mostly occurred in Iraq and maybe Syria. Its ecology and the reasons for its decline are unknown as it has never been studied in detail. No captive breeding seems to be in place and no conservation action is taken actually.

Luciobarbus xanthopterus **Heckel, 1843**

Distribution in the area. Endemic. Widespread in the Euphrates and Tigris drainages. It was also known from the Qweik River, where it is now extirpated.

Taxonomic notes. This species often occurs syntopic with *L. esocinus*. Mitochondrial data suggest that *L. xanthopterus* is partly introgressed by *L. esocinus* and many individuals do no longer have the specific mitochondrial bodies, but those of

L. esocinus. There is no doubt that both are independent species. *Luciobarbus kersin* described from Syria and *L. shejch* described from Mossul might to be synonyms of *L. xanthopterus*.

Remarks. This is a major commercial species and seems to be overfished in large if not all parts of its range. However, data on population trends are very rare and based on fisheries statistics.

***Mesopotamichthys sharpeyi* (Günther, 1874)**

Distribution in the area. Not endemic. The southern part of Euphrates and Tigris drainages in Syria, Iraq, and Iran. In Iran in Hawe Al Azim and Shadegan marshes and lower parts of rivers Zohreh, Karkheh and Karun. In Syria in Asad reservoir. In Iraq widespread in the Shatt Al-Arab and the lower part of the Euphrates, Tigris, Diyala including the Al-Hammar marshes and Huwazah marsh, Lakes Saniyah, Habbaniyah, Tharthar and Razzazah, and reservoirs Himreeen and Al Qadisiyah.

Remarks. This is a major commercial species and seems to be overfished in large if not all parts of its range. However, data on population trends are very rare and based on fisheries statistics. Its decline seems to correlate with the destruction of the marshes and wetlands in the lower catchment area as well as the invasion of the alien herbivorous *Coptodon zillii* (Cichlidae) and *Carasssius auratus* and may be other alien species.

Family: Danionidae

***Barilius mesopotamicus* Berg, 1932**

Distribution in the area. Not endemic. Euphrates and Tigris drainages in southern Turkey, Syria, Iraq, and Iran.

Family: Leuciscidae

***Acanthobrama marmid* Heckel, 1843**

Distribution in the area. Endemic. Widely distributed all over the Euphrates and Tigris drainages from northern Anatolia south to the Shatt Al-Arab. Also in the Queik.

***Alburnoides diclensis* Turan, Bektaş, Kaya & Bayçelebi, 2016**

Distribution in the area. Endemic. Restricted to headwater streams in the Great Zab drainage in Turkey and Iraq.

***Alburnoides emineae* Turan, Kaya, Ekmekçi & Doğan, 2014**

Distribution in the area. Endemic. Turkey and most likely adjacent Syria: Beyazsu stream north of the Turkish city of Nusaybin where the species described (Turan et al. 2014). Beyazsu is a headwater stream of the Jagh Jagh River, which flows to the Khabur.

***Alburnoides idignensis* Bogutskaya & Coad, 2009**

Distribution in the area. Endemic. Bid Sorkh River between Sahneh and Kandgavar in the Gav Masiab drainage, a headwater of the Karkheh in the Zagros Mountains.

***Alburnoides nicolausi* Bogutskaya & Coad, 2009**

Distribution in the area. Endemic. Simareh drainage in Lorestan, a tributary of the Karkheh in the Zagros Mountains.

Taxonomic notes. Molecular data characters do not distinguish *A. nicolausi* and *A. idignensis*, which both occur in the Karkheh drainage. For the time being, both species are accepted as valid but it is recommended to restudy the morphological differences between both species described by Bogutskaya and Coad (2009).

Alburnoides velioglui **Turan, Kaya, Ekmekçi & Doğan, 2014**

Distribution in the area. Endemic. Widespread in middle and lower Euphrates drainage in Turkey and also found in Sirvan drainage in Iraq. Might occur in the Iranian part of the Sirvan also.

Alburnus caeruleus **Heckel, 1843**

Distribution in the area. Endemic. Widespread in the Euphrates and Tigris drainages as well as in the Qweik, from southern Anatolia south to Iraq.

Alburnus doriae **De Filippi, 1865**

Distribution in the area. Not endemic. Locally known from the Karun drainage.

Taxonomic notes. This species was previously known as *Petroleuciscus esfahani* and *Alburnus amirkabiri* (Mohammadian-Kalat et al. 2017).

Alburnus heckeli **Battalgil, 1943**

Distribution in the area. Endemic to Lake Hazer in the uppermost Tigris drainage.

Taxonomic notes. *Alburnus heckeli* is very closely related to *A. sellal* but has recently been confirmed to represent a valid species (Freyhof et al. 2018).

Alburnus kurui **(Bogutskaya, 1995)**

Distribution in the area. Endemic. Locally known only from the Yüksekova wetland in Turkey.

Taxonomic notes. Previously known as a species in *Petroleuciscus*. Very closely related to *A. timarensis* from which it is well distinguished by morphological characters (Freyhof et al. 2018).

Alburnus sellal **Heckel, 1843**

Distribution in the area. Not endemic. Very widespread all over the Euphrates and Tigris drainages including the Qweik River.

Taxonomic notes. This species is widely known as *A. mossulensis*, which is now treated as a synonym of *A. sellal* (Mohammadian-Kalat et al. 2017). *Alburnus zagrosensis* from Iran and *A. selcuklui* from Turkey are synonyms (Eagderi et al. 2019, Freyhof et al. 2018).

Alburnus tarichi **(Güldenstädt, 1814)**

Distribution in the area. Endemic to Lake Van drainage in eastern Anatolia.

Taxonomic notes. Closely related to *A. sellal* but has recently been confirmed to represent a valid species (Freyhof et al. 2018).

Alburnus timarensis **Kuru, 1980**

Distribution in the area. Endemic to Lake Van drainage in eastern Anatolia.

Taxonomic notes. *Alburnus timarensis* is very closely related to *A. sellal* but has recently been confirmed to represent a valid species (Elp et al. 2013).

Chondrostoma esmaeilii **Eagderi, Jouladeh-Roudbar, Birecikligil, Çiçek & Coad, 2017**

Distribution in the area. Endemic. Known only from the Sarab-e Ravansar stream in Iran.

Taxonomic notes. Based on its description (Eagderi et al. 2017), *C. esmaeilii* is very different from all other *Chondrostoma* species. However, the authors did not consider that this species might be the hybrid between *Chondrostoma regium* and *Squalius* sp. Hybrids between *Chondrostoma* and *Squalius* are well known from other areas in the Palearctic and are superficially very similar to *C. esmaeilii*. Future research might review the case and test, if these are first-generation hybrids or even a species of hybrid origin. For the time being, *C. esmaeilii* is accepted as a valid species.

***Chondrostoma regium* (Heckel, 1843)**

Distribution in the area. Not endemic. Very widespread all over the Euphrates, Tigris and also in the Qweik and rarely in Shatt Al-Arab.

***Leuciscus vorax* (Heckel, 1843)**

Distribution in the area. Endemic. Euphrates and Tigris drainages from Anatolia south to the Shatt Al-Arab.

Notes. This species is frequently reported to occur also in the Orontes River. We are not aware of a confirmed record of *L. vorax* from the Orontes and treat it as endemic to the Euphrates and Tigris.

***Pseudophoxinus firati* Bogutskaya, Küçük & Atalay, 2006**

Distribution in the area. Not endemic. The species was very locally known from the upper Euphrates drainage. However, the distribution range of the species in the Euphrates has been recently expanded by Saç et al. (2019). Accordingly, the species inhabits from stream Süt Kaynağı, one of the easternmost tributary of upper Murat River (Bitlis); streams Balıklıtohma and Tohma, western Euphrates River tributaries (Sivas). They recorded it also in Ceyhan and Seyhan rivers.

***Pseudophoxinus zeregi* (Heckel, 1843)**

Distribution in the area. Not endemic. Found in the Qweik drainage.

***Squalius berak* Heckel, 1843**

Distribution in the area. Endemic. Very widespread all over the Euphrates, Tigris, and also in the Qweik.

Taxonomic notes. *Squalius berak* hybridizes with *S. lepidus* (Khaefi et al. 2016) and some *Squalius* populations might represent hybrids in most cases, all individuals can be identified without problems.

***Squalius lepidus* Heckel, 1843**

Distribution in the area. Endemic. Euphrates and Tigris drainages from Anatolia south to Shatt Al-Arab where it now absents.

***Squalius semae* Turan, Kottelat & Bayçelebi, 2017**

Distribution in the area. Endemic to uppermost Euphrates drainage in Turkey.

Taxonomic notes. *Squalius semae* seems to be very closely related if not conspecific to unidentified *Squalius* species from the Black Sea basin and more research is needed to clarify the distribution of this species.

Family: Nemacheilidae

***Eidinemacheilus proudlovei* Freyhof, Abdullah, Ararat, Ibrahim & Geiger, 2016**

Distribution in the area. Endemic. Only known from an aquifer close to the village Kanishok in the Tabeen drainage. The Tabeen flows to the Lesser Zab River, a major tributary of the Iraqi Tigris.

***Eidinemacheilus smithi* (Greenwood, 1976)**

Distribution in the area. Endemic. Only known from a natural well at Kaaje-Ru, in Baq-e-Loveh Oasis, in the Dez system of the Iranian Karun drainage.

***Oxynoemacheilus araxensis* (Bănărescu & Nalbant, 1978)**

Distribution in the area. Endemic to the upper Euphrates drainage in Turkey.

Note. Despite its name, this species has never been found in the Arax/Aras drainage, which belongs to the Caspian Sea basin.

***Oxynoemacheilus argyrogramma* (Heckel, 1847)**

Distribution in the area. Endemic. Qweik drainage in Turkey and Syria, as well as in the south-western Euphrates drainage in Turkey and most likely also in the adjacent Euphrates in Syria.

***Oxynoemacheilus arsaniasus* Freyhof, Kaya, Turan & Geiger, 2019**

Distribution in the area. Endemic. The species was described from two localities in the upper Murat River drainages (Freyhof et al. 2019).

***Oxynoemacheilus bergianus* (Derjavin, 1934)**

Distribution in the area. Not endemic. Locally found in the upper Tigris (Kaya et al. 2016). Widespread in the upper Euphrates drainage in Turkey and potentially also in Syria.

Taxonomic notes. *Oxynoemacheilus erdali* was treated as a junior synonym of *O. bergianus* (Freyhof et al. 2019).

***Oxynoemacheilus chomanicus* Kamangar, Prokofiev, Ghaderi, & Nalbant, 2014**

Distribution in the area. Endemic. Seems to be endemic to headwaters of the Choman River, a tributary of the Lesser Zab in Iran and Iraq.

***Oxynoemacheilus ercisianus* (Erk'akan & Kuru, 1986)**

Distribution in the area. Endemic. Northern drainages of Lake Van. It has been recently recorded in microbialites in the lake (Akkuş et al. 2021).

***Oxynoemacheilus euphraticus* (Bănărescu & Nalbant, 1964)**

Distribution in the area. Endemic. Widespread in headwaters of the Euphrates and Tigris in Turkey, Iraq, and Iran.

Taxonomic note. In the upper Tigris drainage, this species seems to have hybridized with *O. kurdistanicus*. Mitochondrial DNA of both species has been found in superficially very similar fish. More research is needed to resolve this case and the distribution of both species in the Tigris drainage. *Oxynoemacheilus freyhofi* from the Karun drainage is a synonym.

***Oxynoemacheilus frenatus* (Heckel, 1843)**

Distribution in the area. Endemic. Described from Mosul but recently only found in the headwaters of the upper Tigris as well as in its tributaries down to the Batman River. Not recorded from rivers downstream of the Batman (Kaya et al. 2016; Freyhof et al. 2017a). Sayyadzadeh & Esmaeili (2020) identified a population of loaches from the Iranian Lesser Zab drainage as *O. frenatus*. Without new materials from Mosul the identity of *O. frenatus* cannot be clarified.

***Oxynoemacheilus gyndes* Freyhof & Abdullah, 2017**

Distribution in the area. Endemic. Only known from one headwater stream in the Iraqi Sirvan drainage.

Oxynoemacheilus hanae **Freyhof & Abdullah, 2017**

Distribution in the area. Endemic. Only known from one headwater stream in the Iraqi Sirvan drainage.

Oxynoemacheilus hazarensis **Freyhof & Özuluğ, 2017**

Distribution in the area. Endemic. Lake Hazar and its tributaries in Turkey.

Oxynoemacheilus karunensis **Freyhof, 2016**

Distribution in the area. Endemic. Karun drainage.

Oxynoemacheilus kaynaki **Erk'akan, Özeren & Nalbant, 2008**

Distribution in the area. Endemic. Euphrates drainage in Turkey.

Oxynoemacheilus kentritensis **Freyhof, Kaya & Turan, 2017**

Distribution in the area. Endemic. Botan, Hezil, and Nerdus Rivers, which are small rivers entering the Tigris in the border area of Turkey, Iraq, and Syria.

Oxynoemacheilus kiabii **Golzarianpour, Abdoli & Freyhof, 2011**

Distribution in the area. Endemic. Karkheh drainage.

Oxynoemacheilus kurdistanicus **Kamangar, Prokofiev, Ghaderi & Nalbant, 2014**

Distribution in the area. Endemic. Widespread in the upper Tigris drainage from headwaters in Turkey down to the Sirvan drainage.

Oxynoemacheilus longipinnis **(Coad & Nalbant, 2005)**

Distribution in the area. Endemic. Meymeh, Doveyrich, and Changooleh drainages in Iran, flowing to the Al-Sannaf marshes near Amarah in Iraq.

Taxonomic notes. Very closely related to *O. bergianus* and might be a synonym.

Oxynoemacheilus marunensis **Sayyadzadeh & Esmaeili,** 2020

Distribution in the area. Endemic. Marun drainage in Iran.

Oxynoemacheilus muefiti **Freyhof, Kaya, Turan & Geiger, 2019**

Distribution in the area. Endemic. The species was described in the upper Murat River drainage as well as in the Eğri, a tributary to Atatürk reservoir (Freyhof et al. 2019).

Oxynoemacheilus paucilepis **(Erk'Akan, Nalbant & Özeren, 2007)**

Distribution in the area. Endemic. Only known from the stream Mancılık, a headwater of the upper Euphrates in Turkey.

Oxynoemacheilus parvinae **Sayyadzadeh, Eagderi & Esmaeili, 2016**

Distribution in the area. Endemic. Only known from the Leileh River, a tributary of the Sirvan in Iran.

Taxonomic notes. Based on its description, this species is indistinguishable by external characters from *O. bergianus*, which is also found in the Sirvan drainage in Iran. Actually, both species can just be distinguished by minor mitochondrial molecular characters and *O. parvinae*, as well as *O. longipinnis*, might be synonyms of *O. bergianus*.

Oxynoemacheilus samanticus **(Bănărescu & Nalbant, 1978)**

Distribution in the area. Not endemic. Found in the stream Mancılık, a headwater stream of the Euphrates in Turkey.

Remark. The Mancılık is close to the watershed with the Kızılırmak River (Black Sea basin), where this species is widely distributed.

Oxynoemacheilus tigris **(Heckel, 1843)**

Distribution in the area. Endemic to the Qweik River and the Merziman, a tributary to the Euphrates.

Oxynoemacheilus zagrosensis **Kamangar, Prokofiev, Ghaderi & Nalbant, 2014**

Distribution in the area. Endemic to headwaters of the Choman River, a tributary of the Lesser Zab in Iran. The Choman flows to Iraq but despite intensive research, this species has not yet been found in Iraq.

Oxynoemacheilus zarzianus **Freyhof & Geiger, 2017**

Distribution in the area. Endemic to the Lesser Zab drainage in Iraq and potentially also in Iran.

Paracobitis basharensis **Freyhof, Esmaeili, Sayyadzadeh & Geiger, 2014**

Distribution in the area. Endemic. Bashar River in the upper Karun drainage.

Paracobitis molavii **Freyhof, Esmaeili, Sayyadzadeh & Geiger, 2014**

Distribution in the area. Endemic. Sirvan and Lesser Zab rivers in Iran and Iraq.

Paracobitis salihae **Kaya, Turan, Kalaycı, Bayçelebi & Freyhof, 2020**

Distribution in the area. Endemic. Distribution of this recently described species is poorly known and restricted in Göksu River (Euphrates) near Gölbaşı Adıyaman (Kaya et al. 2020).

Paracobitis zabgawraensis **Freyhof, Esmaeili, Sayyadzadeh & Geiger, 2014**

Distribution in the area. Endemic. Upper Yanarsu and Botan drainage in Turkey and Greater Zab drainage in Turkey and Iraq (Freyhof et al. 2014b; Kaya et al. 2016).

Paraschistura chrysicristinae **(Nalbant, 1998)**

Distribution in the area. Endemic. Found at two places in the upper Tigris drainage in Turkey.

Note. Despite several attempts to find this species, it has not been found again in the last decades and might be extinct.

Paraschistura ilamensis **Vatandoust & Eagderi, 2015**

Distribution in the area. Endemic. Karun and Karkheh drainages.

Sasanidus kermanshahensis **(Bănărescu & Nalbant, 1966)**

Distribution in the area. Endemic. Headwaters of the Karkheh and Karun Rivers.

Turcinoemacheilus hafezi **Golzarianpour, Abdoli, Patimar & Freyhof, 2013**

Distribution in the area. Endemic. Headwaters of the Karkheh and Karun Rivers.

Turcinoemacheilus kosswigi **Bănărescu & Nalbant, 1964**

Distribution in the area. Endemic. Widespread in headwaters of the Tigris, Greater and Lesser Zab, and Sirvan Rivers.

Turcinoemacheilus minimus **Esmaeili, Sayyadzadeh, Özuluğ, Geiger & Freyhof, 2014**

Distribution in the area. Endemic. Headwaters of the Euphrates in Turkey.

Turcinoemacheilus saadii **Esmaeili, Sayyadzadeh, Özuluğ, Geiger & Freyhof, 2014**

Distribution in the area. Endemic. Bashar River in the Karun drainage.

Family: Cobitidae

Cobitis avicennae **Mousavi-Sabet, Vatandoust, Esmaeili, Geiger & Freyhof, 2015**

Distribution in the area. Endemic. Tributaries of the Karkheh and Karun. A record from the Little Zab River in Iraqi Kurdistan might also belong to this species.

Cobitis elazigensis **Coad & Sarieyyüpoglu, 1988**

Distribution in the area. Endemic. Upper Euphrates in Turkey south to Atatürk reservoir. Expected to be also found in the Syrian Euphrates.

Cobitis kellei **Erk'akan, Atalay-Ekmekçi & Nalbant, 1998**

Distribution in the area. Endemic. Only known from one site in the upper Tigris in Turkey.

Note. Despite several attempts to find this species, it has not been found again in the last decades and might be extinct.

Family: Ariidae

Netuma bilineata **(Valenciennes, 1840)**

Distribution in the area. Not endemic. Recorded from the Shatt Al-Arab and the marsh area.

Netuma thalassina **(Rüppell, 1837)**

Distribution in the area. Not endemic. Recorded from the Shatt Al-Arab and the marsh area.

Plicofollis layardi **(Günther, 1866)**

Distribution in the area. Not endemic. Recorded from the Shatt Al-Arab.

Family: Plotosidae

Plotosus lineatus **(Thunberg, 1787)**

Distribution in the area. Not endemic. Recorded from the Shatt Al-Arab and the marsh area.

Family: Bagridae

Mystus pelusius **(Solander, 1794)**

Distribution in the area. Not endemic. Widespread in the Euphrates and Tigris drainages from southern Turkey to the Shatt Al-Arab. Also in the Qweik.

Family: Sisoridae

The diversity of *Glyptothorax* species in the Euphrates and Tigris drainages seems to have been much underestimated and the genus is awaiting a sound review.

Glyptothorax armeniacus **(Berg, 1918)**

Distribution in the area. Endemic. Until now only been recorded from the Euphrates drainage in Turkey.

Glyptothorax cous **(Linnaeus, 1766)**

Distribution in the area. Endemic. Widespread in the Euphrates and Tigris including the Karun and Qweik (extirpated).

Notes. *Glyptothorax cous* was described from the Qweik in Syria,

Glyptothorax kurdistanicus **(Berg, 1931)**

Distribution in the area. Endemic. Known from the upper Tigris and Lesser Zab drainage.

Glyptothorax daemon Freyhof, Kaya, Abdullah. & Geiger, 2021

Distribution in the area. Endemic. Upper Tigris and Great Zab drainages.

Glyptothorax steindachneri **(Pietschmann, 1913)**

Distribution in the area. Endemic. Widespread in Euphrates and Tigris.

***Glyptothorax silviae* Coad, 1981**

Distribution in the area. Not endemic. Karun and Jarrahi Rivers.

Family: Siluridae

***Silurus triostegus* Heckel, 1843**

Distribution in the area. Endemic. Widespread in the Euphrates and Tigris drainages.

Family: Synodontidae

***Saurida tumbil* (Bloch, 1795)**

Distribution in the area. Not endemic. Recorded from the Shatt Al-Arab.

***Saurida macrolepis* Tanaka, 1917**

Distribution in the area. Not endemic. Recorded from the Shatt Al-Arab.

Taxonomic notes: Reported as *S. undosquamis*, Inoue & Nakabo (2006) found that most specimens so far identified as *S. undosquamis* and collected from the Arabian Gulf are identical with *S. macrolepis*.

Family: Batrachoididae

***Colletteichthys dussumieri* (Valenciennes, 1837)**

Distribution in the area. Not endemic. Recorded from the lower Shatt Al-Arab.

Family: Salmonidae

***Salmo euphrataeus* Turan, Kottelat & Engin, 2014**

Distribution in the area. Endemic to the stream Karasu, which forms part of the headwaters of the northern Euphrates.

***Salmo fahrettini* Turan, Kalayci, Bektaş, Kaya & Bayçelebi 2020**

Distribution in the area. Endemic. The species has recently been described from streams Tekke and Ömertepesuyu, upper Euphrates drainages around Erzurum Province (Turan et al. 2020).

***Salmo munzuricus* Turan, Kottelat & Kaya, 2017**

Distribution in the area. Endemic to the stream Munzur in the upper Euphrates in Turkey. The stream is located in the Munzur Valley National Park, which is known for its rich biodiversity and high endemism (Turan et al. 2017).

***Salmo okumusi* Turan, Kottelat & Engin 2014**

Distribution in the area. Endemic to tributaries of the Tohma and Sürgü streams in the upper Euphrates drainage.

***Salmo tigridis* Turan, Kottelat & Bektaş, 2011**

Distribution in the area. Endemic to the streams Çatak and Müküs in upper the Tigris drainage.

Family: Aphaniidae

Eurasian killifish had been placed in the American family Cyprinodontidae for long but are recognized as an on family: Aphaniidae now (see Freyhof et al. 2017b for discussion). Freyhof and Yoğurtçuoğlu (2020) proposed separation of Aphaniidae into eight monophyletic genera: *Anatolichthys*, *Aphaniops*, *Aphanius*, *Apricaphanius*, *Kosswigichthys*, *Paraphanius*, and *Tellia*, in addition to the new genera *Esmaeilius*. According to their new proposal, in Tigris–Euphrates basin, there are four genera (*Aphaniops, Esmaeilus, Kosswigichthys* and *Paraphanius*) which were listed below.

***Aphaniops stoliczkanus* (Day, 1872)**

Distribution in the area. Not endemic. Lower Shatt Al-Arab and springs around it, southern marsh area, the lower part of Euphrates and Tigris in Iraq and Syria, and lower Karun in Iran.

Taxonomic notes. In the area, this species was previously identified as *A. dispar* (Freyhof et al. 2017c).

***Esmaeilius sophiae* (Heckel, 1847)**

Distribution in the area. Not endemic. Lower Shatt Al-Arab in Iraq and lower Karkheh River.

***Esmaeilius vladykovi* (Coad, 1988)**

Distribution in the area. Endemic. Wetlands in the upper Karun drainage around Boldaji.

***Kosswigichthys asquamatus* Sözer 1942**

Distribution in the area. Endemic to Lake Hazer in Turkey.

***Paraphanius mento* (Heckel, 1843)**

Distribution in the area. Not endemic. Middle to lower Tigris and Shatt Al-Arab.

Taxonomic notes. This species has been described from Mosul, but has not been found again in that area, which has not been studied for fish in the last decades. Populations from the marshes in Iraq and Iran are expected to be conspecific as well as these in the upper Habur River, a tributary to the Euphrates in Syria (Krupp and Schneider 2008).

Family: Mastaembelidae

***Mastacembelus mastacembelus* (Banks & Solander 1794)**

Distribution in the area. Not endemic. Very widespread in the Qweik, Euphrates, and Tigris from Turkey south to Iran and Shatt Al-Arab in Iraq.

Family: Belonidae

***Ablennes hians* (Valenciennes, 1846)**

Distribution in the area. Not endemic. Recorded from the lower Shatt Al-Arab.

***Strongylura leiura* (Bleeker, 1850)**

Distribution in the area. Not endemic. Recorded from the Shatt Al-Arab.

***Strongylura strongylura* (van Hasselt, 1823)**

Distribution in the area. Not endemic. Recorded from the lower Tigris, the Shatt Al-Arab, and the marsh area.

***Tylosurus crocodilus* (Péron & Lesueur, 1821)**

Distribution in the area. Not endemic. Recorded from the lower Shatt Al-Arab.

Family: Hemiramphidae

***Hemiramphus marginatus* (Forsskål, 1775)**

Distribution in the area. Not endemic. Recorded from the lower Shatt Al-Arab.

***Hyporhamphus limbatus* (Valenciennes, 1847)**

Distribution in the area. Not endemic. Recorded from the lower part of the Euphrates, the Al-Hammar marsh, and the Shatt Al-Arab.

***Hyporhamphus unicuspis* Collete & Parin, 1978**

Distribution in the area. Not endemic. Recorded from the lower Shatt Al-Arab.

***Rhynchorhamphus georgii* (Valenciennes, 1847)**

Distribution in the area. Not endemic. Recorded from the Shatt Al-Arab and the Al-Hammar marsh.

Family: Syngnathidae

***Hippocampus kuda* Bleeker, 1852**

Distribution in the area. Not endemic. Recorded from the lower Shatt Al-Arab.

Family: Synanceiidae

***Choridactylus multibarbus* Richardson, 1848**

Distribution in the area. Not endemic. Recorded from the lower Shatt Al-Arab.

***Minous monodactylus* (Bloch & Schneider, 1801)**

Distribution in the area. Not endemic. Recorded from the lower Shatt Al-Arab.

***Pseudosynanceia melanostigma* Day, 1875**

Distribution in the area. Not endemic. Recorded from the lower Shatt Al-Arab.

Family: Platycephalidae

***Grammoplites suppositus* (Troschel, 1840)**

Distribution in the area. Not endemic. Recorded from the lower Shatt Al-Arab.

***Platycephalus indicus* (Linnaeus, 1758)**

Distribution in the area. Not endemic. Recorded from the lower Shatt Al-Arab.

Family: Serranidae

***Epinephelus areolatus* (Forsskål, 1775)**

Distribution in the area. Not endemic. Recorded from the lower Shatt Al-Arab.

***Epinephelus coioides* (Hamilton, 1822)**

Distribution in the area. Not endemic. Recorded from the lower Shatt Al-Arab.

Family: Terapontidae

***Helotes sexlineatus* (Quoy & Gaimard, 1825)**

Distribution in the area. Not endemic. Recorded from the lower Shatt Al-Arab.

***Pelates quadrilineatus* (Bloch, 1790)**

Distribution in the area. Not endemic. Recorded from the lower Shatt Al-Arab.

***Terapon puta* Cuvier, 1829**

Distribution in the area. Not endemic. Recorded from the lower Shatt Al-Arab.

***Terapon theraps* Cuvier, 1829**

Distribution in the area. Not endemic. Recorded from the lower Shatt Al-Arab.

Family: Apogonidae

***Apogonichthyoides taeniatus* (Cuvier, 1828)**

Distribution in the area. Not endemic. Recorded from the lower Shatt Al-Arab.

Family: Sillaginidae

***Sillago arabica* McKay & McCarthy, 1989**

Distribution in the area. Not endemic. Recorded from the Shatt Al-Arab and the East Al-Hammar marsh.

***Sillago attenuata* McKay, 1985**

Distribution in the area. Not endemic. Recorded from the Shatt Al-Arab.

***Sillago sihama* (Fabricius, 1775)**

Distribution in the area. Not endemic. Recorded from the Shatt Al-Arab and the East Al-Hammar marsh.

Family: Rachycentridae

***Rachycentron canadum* (Linnaeus, 1766)**

Distribution in the area. Not endemic. Recorded from the lower Shatt Al-Arab.

Family: Carangidae

Alepes djedaba **(Fabricius, 1775)**

Distribution in the area. Not endemic. Recorded from the Shatt Al-Arab.

Alepes melanoptera **(Swainson, 1839)**

Distribution in the area. Not endemic. Recorded from the lower Shatt Al-Arab.

Alepes vari **(Cuvier, 1833)**

Distribution in the area. Not endemic. Recorded from the lower Shatt Al-Arab.

Atropus atropos **(Bloch & Schneider, 1801)**

Distribution in the area. Not endemic. Recorded from the lower Shatt Al-Arab.

Carangoides chrysophrys **(Cuvier, 1833)**

Distribution in the area. Not endemic. Recorded from the lower Shatt Al-Arab.

Carangoides malabaricus **(Bloch & Schneider, 1801)**

Distribution in the area. Not endemic. Recorded from the Shatt Al-Arab.

Parastromateus niger **(Bloch, 1795)**

Distribution in the area. Not endemic. Recorded from the lower Shatt Al-Arab.

Scomberoides commersonnianus **Lacepède, 1801**

Distribution in the area. Not endemic. Recorded from the Shatt Al-Arab.

Selar crumenophthalmus **(Bloch, 1793)**

Distribution in the area. Not endemic. Recorded from the lower Shatt Al-Arab.

Selaroides leptolepis **(Cuvier, 1833)**

Distribution in the area. Not endemic. Recorded from the lower Shatt Al-Arab.

Trachinotus mookalee **Cuvier, 1832**

Distribution in the area. Not endemic. Recorded from the Shatt Al-Arab.

Uraspis helvola **(Forster, 1801)**

Distribution in the area. Not endemic. Recorded from the Shatt Al-Arab.

Family: Leiognathidae

Photopectoralis bindus **(Valenciennes, 1835)**

Distribution in the area. Not endemic. Recorded from the Shatt Al-Arab.

Family: Gerreidae

Gerres limbatus **Cuvier, 1830**

Distribution in the area. Not endemic. Recorded from the Shatt Al-Arab.

Gerres longirostris **(Lacepède, 1801)**

Distribution in the area. Not endemic. Recorded from the lower Shatt Al-Arab.

Gerres macracanthus **Bleeker, 1854**

Distribution in the area. Not endemic. Recorded from the lower Shatt Al-Arab.

Gerres oyena **(Fabricius, 1775)**

Distribution in the area. Not endemic. Recorded from the lower Shatt Al-Arab.

Family: Haemulidae

Diagramma pictum **(Thunberg, 1792)**

Distribution in the area. Not endemic. Recorded from the lower Shatt Al-Arab.

Pomadasys argenteus **(Forsskål, 1775)**

Distribution in the area. Not endemic. Recorded from the lower Shatt Al-Arab.

Pomadasys kaakan **(Cuvier, 1830)**

Distribution in the area. Not endemic. Recorded from the lower Shatt Al-Arab.

Pomadasys stridens **(Forsskål, 1775)**

Distribution in the area. Not endemic. Recorded from the lower Shatt Al-Arab.

Family: Sparidae

Acanthopagrus arabicus **Iwatsuki, 2013**

Distribution in the area. Not endemic. Recorded from the lower parts of Euphrates and Shatt Al-Arab.

Taxonomic note: Yellow fin seabream populations in the lower Tigris and the marshes most probably belong to this species or to *A. sheim*.

Acanthopagrus berda **(Fabricius, 1775)**

Distribution in the area. Not endemic. Recorded from the Shatt Al-Arab and the East Al-Hammar marsh.

Acanthopagrus sheim **Iwatsuki, 2013**

Distribution in the area. Not endemic. Recorded from the Shatt Al-Arab.

Argyrops spinifer **(Forsskål, 1775)**

Distribution in the area. Not endemic. Recorded from the lower Shatt Al-Arab.

Crenidens crenidens **(Forsskål, 1775)**

Distribution in the area. Not endemic. Recorded from the lower Shatt Al-Arab.

Sparidentex hasta **(Valenciennes, 1830)**

Distribution in the area. Not endemic. Recorded from the Shatt Al-Arab and the East Al-Hammar marsh.

Family: Nemipteridae

Nemipterus bipunctatus **(Valencinnes, 1830)**

Distribution in the area. Not endemic. Recorded from the lower Shatt Al-Arab.

Nemipterus peronii **(Valenciennes, 1830)**

Distribution in the area. Not endemic. Recorded from the lower Shatt Al-Arab.

Scolopsis taeniata **(Cuvier, 1830)**

Distribution in the area. Not endemic. Recorded from the lower Shatt Al-Arab.

Family: Sciaenidae

Johnius belangerii **(Cuvier, 1830)**

Distribution in the area. Not endemic. Recorded from the Shatt Al-Arab and the East Al-Hammar marsh.

Johnius borneensis **(Bleeker, 1851)**

Distribution in the area. Not endemic. Recorded from the lower Shatt Al-Arab.

Johnius dussumieri **(Cuvier, 1830)**

Distribution in the area. Not endemic. Recorded from the Shatt Al-Arab and the East Al-Hammar marsh.

Johnius **sp.**

Distribution in the area. Not endemic. Recorded from the lower Shatt Al-Arab.

Otolithes ruber **(Bloch & Schneider, 1801)**

Distribution in the area. Not endemic. Recorded from the Shatt Al-Arab.

Pennahia aneus **(Bloch, 1793)**

Distribution in the area. Not endemic. Recorded from the Shatt Al-Arab.

Protonibea diacanthus **(Lacepède, 1802)**

Distribution in the area. Not endemic. Recorded from the lower Shatt Al-Arab.

Family: Polynemidae

***Eleutheronema tetradactylum* (Shaw, 1804)**

Distribution in the area. Not endemic. Recorded from the lower Shatt Al-Arab and the East Al-Hammar marsh.

***Polydactylus sextarius* (Bloch & Schneider, 1801)**

Distribution in the area. Not endemic. Recorded from the lower Shatt Al-Arab.

Family: Mullidae

***Upeneus doriae* (Günther, 1869)**

Distribution in the area. Not endemic. Recorded from the lower Shatt Al-Arab.

***Upeneus sundaicus* (Bleeker, 1855)**

Distribution in the area. Not endemic. Recorded from the lower Shatt Al-Arab.

***Upeneus tragula* Richardson, 1846**

Distribution in the area. Not endemic. Recorded from the lower Shatt Al-Arab.

Family: Gobiidae

***Acentrogobius dayi* Koumans, 1941**

Distribution in the area. Not endemic. Recorded from Lake Sawa and Southern marsh.

***Bathygobius fuscus* (Rüppell, 1830)**

Distribution in the area. Not endemic. Recorded from the lower part of Euphrates, Shatt Al-Arab, and the East Al-Hammar marsh.

Family: Oxudercidae

***Boleophthalmus dussumieri* Valenciennes, 1837**

Distribution in the area. Not endemic. Recorded from the Shatt Al-Arab and the East Al-Hammar marsh.

***Periophthalmus waltoni* Koumans, 1941**

Distribution in the area. Not endemic. Recorded from the Shatt Al-Arab.

***Pseudapocryptes elongatus* (Cuvier, 1816)**

Distribution in the area. Not endemic. Recorded from the lower Shatt Al-Arab.

***Scartelaos tenuis* (Day, 1876)**

Distribution in the area. Not endemic. Recorded from the lower Shatt Al-Arab.

***Trypauchen vagina* (Bloch & Schneider, 1801)**

Distribution in the area. Not endemic. Recorded from the lower Shatt Al-Arab.

Family: Ephippidae

***Ephippus orbis* (Bloch, 1787)**

Distribution in the area. Not endemic. Recorded from the lower Shatt Al-Arab.

Family: Scatophagidae

***Scatophagus argus* (Linnaeus, 1766)**

Distribution in the area. Not endemic. Recorded from the lower Shatt Al-Arab and the East Al-Hammar marsh.

Family: Siganidae

***Siganus canaliculatus* (Park, 1797)**

Distribution in the area. Not endemic. Recorded from the lower Shatt Al-Arab

Family: Sphyraenidae

***Sphyraena obtusata* Cuvier, 1829**

Distribution in the area. Not endemic. Recorded from the lower Shatt Al-Arab

Family: Trichiuridae

***Eupleurogrammus glossodon* (Bleeker, 1860)**

Distribution in the area. Not endemic. Recorded from the lower Shatt Al-Arab.

***Eupleurogrammus muticus* (Gray, 1831)**

Distribution in the area. Not endemic. Recorded from the lower Shatt Al-Arab.

***Trichiurus lepturus* Linnaeus, 1758**

Distribution in the area. Not endemic. Recorded from the lower Shatt Al-Arab.

Family: Scomberidae

***Rastrelliger kanagurta* (Cuvier, 1816)**

Distribution in the area. Not endemic. Recorded from the lower Shatt Al-Arab.

***Scomberomorus commerson* (Lacepède, 1800)**

Distribution in the area. Not endemic. Recorded from the lower Shatt Al-Arab.

***Scomberomorus guttatus* (Bloch & Schneider, 1801)**

Distribution in the area. Not endemic. Recorded from the lower Shatt Al-Arab.

Family: Stromateidae

***Pampus argenteus* (Euphrasen, 1788)**

Not endemic. Recorded from the lower Shatt Al-Arab and the East Al-Hammar marsh.

Family: Mugilidae

All except one species of mullets in the Euphrates and Tigris are euryhyaline, widespread species that enter the Shatt Al-Arab from the sea. Their taxonomy is not yet fully resolved. We follow Durand et al. (2012) in the generic assignment of mullets.

***Mugil cephalus* Linnaeus, 1758**

Distribution in the area. Not endemic. Lower Shatt Al-Arab and the marsh areas, where it is very rare.

***Osteomugil speigleri* (Bleeker, 1858)**

Distribution in the area. Not endemic. Recorded from the lower Shatt Al-Arab.

***Planiliza abu* (Heckel, 1843)**

Distribution in the area. Widespread in the Euphrates and Tigris drainages from Turkey downstream to Shatt Al-Arab, also in marshes and reservoirs.

***Planiliza klunzingeri* (Day, 1888)**

Distribution in the area. Not endemic. Lower Euphrates, Shatt Al-Arab, and the East Al-Hammar marsh.

***Planiliza subviridis* (Valenciennes, 1836)**

Distribution in the area. Not endemic. Lower Shatt Al-Arab and the marsh areas.

Family: Paralichthidae

***Pseudorhombus arsius* (Hamilton, 1822)**

Distribution in the area. Not endemic. Recorded from the lower Shatt Al-Arab.

Family: Soleidae

***Brachirus orientalis* (Bloch & Schneider, 1801)**

Distribution in the area. Not endemic. Recorded from the Shatt Al-Arab and the from East Al-Hammar marsh.

***Solea elongata* Day, 1877**

Distribution in the area. Not endemic. Recorded from the lower Shatt Al-Arab.

Zebrias synapturoides **(Jenkins, 1910)**
Distribution in the area. Not endemic. Recorded from the lower Shatt Al-Arab.
Family: Cynoglossidae
Cynoglossus arel **(Bloch & Schneider, 1801)**
Distribution in the area. Not endemic. Recorded from the lower Shatt Al-Arab.
Cynoglossus bilineatus **(Lacepède, 1802)**
Distribution in the area. Not endemic. Recorded from the lower Shatt Al-Arab.
Family: Triacanthidae
Pseudotriacanthus strigilifer **(Cantor, 1849)**
Distribution in the area. Not endemic. Recorded from the lower Shatt Al-Arab.
Triacanthus biaculeatus **(Bloch, 1786)**
Distribution in the area. Not endemic. Recorded from the lower Shatt Al-Arab.
Family: Tetradontidae
Lagocephalus guentheri **Miranda Ribeiro, 1915**
Distribution in the area. Not endemic. Recorded from the Shatt Al-Arab.

35.4 Discussion

35.4.1 Marine and Euryhyaline Fishes in the Shatt Al-Arab

More than half of the fish species found in the inland waters of the Euphrates and Tigris drainages are a marine or euryhyaline. These species mostly are restricted to the lower section of the river, the Shatt Al-Arab, where 128 marine and brackish water species (Table 35.1) are recorded plus one fully euryhyaline species (*Aphaniops stoliczkanus*: found in freshwater, brackish water, and marine habitats) and one anadromous species (*Tenualosa ilisha*). These make the Shatt Al-Arab the most diverse and species-rich part of the drainage. Ali et al. (2018) performed a literature review on the occurrence of the marine and euryhyaline fishes of Iraq. They confirmed the occurrence of 193 species found in the Shatt Al-Arab between 1874 and 2018. In the last 25 years, a total of 129 euryhyaline and marine fish species were found here. Many of these are real marine species, which were found just in the lower Shatt Al-Arab during saltwater intrusion (Mohamed and Abood 2017; Mohamed et al. 2017; Ali et al. 2018). Due to the decreasing amount of freshwater flowing into the Shatt Al-Arab, the salinity increases (about 30 PSU in mid-Shatt Al-Arab in October 2018) and marine and euryhyaline species are and will be found more and more upriver in the future. The marine and euryhyaline species in Table 35.1 are included in the list of fishes of the Euphrates and Tigris, but excluded from all the calculations below. *Tenualosa ilisha* and *Aphaniops stoliczkanus* are kept in the list of freshwater fishes, as they occur in freshwaters during their life cycle or have resident freshwater populations.

Freshwater Species Altogether 116 species are anadromous (*Tenualosa ilisha*) or restricted to freshwaters (112 species) or found in freshwater, brackish water, and

Table 35.1 Euryhaline and marine fish species found in the Shatt Al-Arab

Ablennes hians	*Eupleurogrammus glossodon*	*Pampus argenteus*	*Scomberoides commersonnianus*
Acanthopagrus arabicus	*Eupleurogrammus muticus*	*Parastromateus niger*	*Scomberomorus commerson*
Acanthopagrus berda	*Gerres limbatus*	*Pastinachus sephen*	*Scomberomorus guttatus*
Acanthopagrus sheim	*Gerres longirostris*	*Pelates quadrilineatus*	*Selar crumenophthalmus*
Acentrogobius dayi	*Gerres macracanthus*	*Pennahia anea*	*Selaroides leptolepis*
Alepes djedaba	*Gerres oyena*	*Periophthalmus waltoni*	*Siganus canaliculatus*
Alepes melanoptera	*Glaucostegus granulatus*	*Photopectoralis bindus*	*Sillago arabica*
Alepes vari	*Grammoplites suppositus*		*Sillago attenuata*
Anodontostoma chacunda	*Helotes sexlineatus*	*Planiliza klunzingeri*	*Sillago sihama*
Apogonichthyoides taeniatus	*Hemiramphus marginatus*	*Planiliza subviridis*	*Solea elongata*
Argyrops spinifer	*Himantura uarnak*	*Platycephalus indicus*	*Sparidentex hasta*
Atropus atropos	*Hippocampus kuda*	*Plicofollis layardi*	*Sphyraena obtusata*
Bathygobius fuscus	*Hyporhamphus limbatus*	*Plotosus lineatus*	*Sphyrna mokarran*
Boleophthalmus dussumieri	*Hyporhamphus unicuspis*	*Polydactylus sextarius*	*Strongylura leiura*
Brachirus orientalis	*Ilisha compressa*	*Pomadasys argenteus*	*Strongylura strongylura*
Carangoides chrysophrys	*Ilisha melastoma*	*Pomadasys kaakan*	*Terapon puta*
Carangoides malabaricus	*Ilisha sirishai*	*Pomadasys stridens*	*Eutherapon theraps*
Chanos chanos	*Johnius belangerii*	*Protonibea diacanthus*	*Thryssa dussumieri*
Chiloscyllium arabicum	*Johnius borneensis*	*Pseudapocryptes dentatus*	*Thryssa hamiltonii*
Chirocentrus dorab	*Johnius dussumieri*	*Pseudorhombus arsius*	*Thryssa setirostris*
Chirocentrus nudus	*Johnius* sp.	*Pseudosynanceia melanostigma*	*Thryssa vetrirostris*
Choridactylus multibarbus	*Lagocephalus guentheri*	*Pseudotriacanthus strigilifer*	*Thryssa whiteheadi*
Colletteichthys dussumieri	*Minous monodactylus*	*Rachycentron canadum*	*Trachinotus mookalee*
Crenidens crenidens	*Mugil cephalus*	*Rastrelliger kanagurta*	*Triacanthus biaculeatus*
Cynoglossus arel	*Nematalosa nasus*	*Rhizoprionodon acutus*	*Trichiurus lepturus*

(continued)

Table 35.1 (continued)

Cynoglossus bilineatus	*Nematalosa persara*	*Rhynchorhamphus georgii*	*Trypauchen vagina*
Diagramma picta	*Nemipterus bipunctatus*	*Sardinella longiceps*	*Tylosurus crocodilus*
Dussumieria acuta	*Nemipterus peronii*	*Saurida macrolepis*	*Upeneus sundaicus*
Eleutheronema tetradactylum	*Netuma bilineata*	*Saurida tumbil*	*Upeneus tragula*
Ephippus orbis	*Netuma thalassina*	*Scartelaos tenuis*	*Uraspis helvola*
Epinephelus areolatus	*Osteomugil speigleri*	*Scatophagus argus*	*Zebrias synapturoides*
Epinephelus coioides	*Otolithes ruber*	*Scolopsis taeniata*	

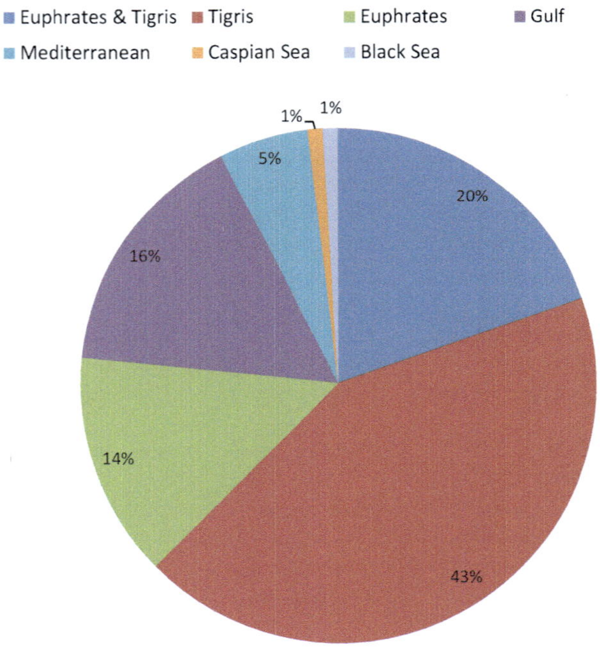

Fig. 35.1 Distribution of the freshwater fish fauna of the Euphrates and Tigris drainages ($n = 107$) in species found in both, the Euphrates and Tigris, endemic species to the Euphrates and Tigris drainages and species shared with the Gulf, Mediterranean, Black, and Caspian Sea basins

marine habitats (*Aphaniops stoliczkanus*). This is about 2.4 times the number of species recorded by Coad (1991) who recognized 47 native species from the area, excluding coastal species and Coad (1996), who recognized 52 species of primary freshwater fishes of the orders Cypriniformes, Siluriformes, and Perciformes. Here, 103 species in these three orders are recognized.

Many New Species Described The increase in numbers of freshwater fish species compared with studies in the twentieth century is mostly due to the description of new species. In total, 49 species have been described after the year 2000 or are

Fig. 35.2 Representative species of freshwater fishes found in the Euphrates and Tigris drainages. Left column from the top: *Barilius mesopotamicus*, *Alburnus caeruleus*, *Pseudophoxinus firati*, *Alburnus sellal*, *Acanthobrama marmid*, and *Chondrostoma regium*. Middle column from the top: *Capoeta trutta*, *Carasobarbus kosswigi*, *Cyprinion kais*, *Garra variabilis*, and *Luciobarbus barbulus*. Right column from the top: *Oxynoemacheilus kurdistanicus*, *Oxynoemacheilus euphraticus*, *Paracobitis zabgawraensis*, *Glyptothorax* sp., *Mastacembelus mastacembelus*, *Planiliza abu*, and *Paraphanius mento*

recognized since as valid from the study area. However, the number of fish species will be reviewed in the coming years, as several genera need revisions and additional new species might be discovered and described. On the other hand, some of the species still recognized as valid here, are likely to represent synonyms of others and such cases need some attention and efforts in the future.

In the twenty-first century, molecular characters have received much attention in the exploration of biodiversity including freshwater fishes. The application of molecular methods has driven the recognition and description of new species but as all sole methodological approaches, there are a number of pit-falls. The literature is full of cases, where molecular trees and morphological species do not fit together (see, e.g., Astrin et al. 2012 and citations herein). Molecular characters, or more correctly molecular distances, must be handled with great care in taxonomy, as there are no agreed molecular distances to recognize the so-called "species level." That means, that "little" or "large" molecular distances are not defined and there is no "species level distance." Naturally, the same is true for morphological characters, and both sets of characters need a certain experience in handling. While there is a trend to reject morphologically "well" distinguished species in the case of "little" molecular distances, the term cryptic species is misused to name molecular lineages as own species without appropriate morphological comparative studies.

Great care has to be taken and we should allow certain molecular differences between populations of geographically widely distributed species occurring in obviously isolated populations. While this is not the place to discuss this topic in depth, we ask not to describe each population making a slightly different clade in a phylogenetic tree as a different, cryptic, species, if no diagnostic morphological characters can be provided.

A good example from the Gulf basin are *Esmaeilius arakensis*, *E. mesopotamicus*, *E. pluristriatus*, and *E. kavirensis*, which are all treated as conspecific with *E. sophiae* due to the absence of morphological characters and the very small molecular differences between these species (Freyhof and Yoğurtçuoğlu 2020). While there are many other, well recognized, fish species having low molecular distances, these are clearly distinguished by morphological characters. This is not the case in these *Esmaeilius*. Naturally, evolution and speciation is usually a process of gradual divergence and there is no discrete point at which populations "mutate into" species. This is true for all character sets, from which morphological and molecular characters are most commonly studied. The grey-zone between populations and species is a challenge in several cases including the *Esmaeilius* discussed here.

Such cases are within the grey-zone between species and populations and need wise treatments. Recognizing, usually young species which are morphologically "well" distinguished but show "little" molecular differences should not open the door to describe just all populations with little molecular differences, also those lacking morphological differences, as cryptic species. But naturally, such grey-zones also open the door for everlasting discussions and disagreements.

Very High Ration of Endemic Species Three families (Cyprinidae: 32 species, Leuciscidae: 21 species, Nemacheilidae: 38 species) dominate the fauna. The largest genus in the area is *Oxynoemacheilus* with 25 species. Looking at the 116 species of freshwater fishes found in the Euphrates and Tigris, it is interesting to understand their distribution patterns. Endemic to the Euphrates and Tigris are 92 species. From these, 19 species are exclusively found in the Euphrates drainage (Table 35.2). Two of them are subterranean species and 14 are strict inhabitants of headwater streams. Three additional species (*Alburnus tarichi*, *A. timarensis*, *Oxynoemacheilus ercisianus*) are endemic to the Lake Van basin, which had faunal input from the Euphrates and the Tigris. Kaya (2020) recently emphasized that one of the headwaters of the Tigris (stream Upper) sourced just 1 km south of the stream Sapur (Lake Van drainage), and there is just 3 m elevation separating both streams. This may be one of the contact zones of Lake Van and Tigris–Euphrates drainage.

In contrast, 51 fish species are endemic to the Tigris drainage including the Karun and Karkheh Rivers in Iran (Table 35.3). Five of these species are subterranean species, three are endemic to Lake Hazar in Turkey, and 41 are strict inhabitants of headwater streams. The number of endemic species and also the total number of freshwater species in the Tigris drainage (87 species) is much higher than in the Euphrates (54 species). This is mostly due to the higher number of headwater stream specialists in the Tigris. It might be speculated, that this difference is just rooted in

Table 35.2 Fish species endemic to the Euphrates and their macrohabitats

Species	Subterranean	Headwaters
Caecocypris basimi		
Garra widdowsoni		
Alburnoides emineae		
Squalius semae		
Oxynoemacheilus araxensis		
Oxynoemacheilus argyrogramma		
Oxynoemacheilus arsaniasus		
Oxynoemacheilus kaynaki		
Oxynoemacheilus muefiti		
Oxynoemacheilus paucilepis		
Oxynoemacheilus tigris		
Paracobitis salihae		
Turcinoemacheilus minimus		
Cobitis elazigensis		
Salmo euphrataeus		
Salmo fahrettini		
Salmo munzuricus		
Salmo okumusi		
Glyptothorax armeniacus		

the much larger area of mountains in the catchment of the Tigris compared to the catchment of the Euphrates.

The most species-rich family of freshwater fishes in the area are Nemacheilid loaches. From the 38 species of nemacheilid loaches native to the Euphrates and Tigris, 36 are endemic and only two of these (*O. bergianus*, *O. euphraticus*) are found in both, the Euphrates and the Tigris drainages, all others are restricted to the Euphrates or to the Tigris (and one species also found in the Black Sea basin). These endemic species are usually restricted to one lake basin or few tributaries of the Euphrates or the Tigris.

This is pointing to very high rate of species that are endemic at small parts of their potential habitat within the drainage system of the Euphrates and Tigris. Theoretically, fish could move from all places of the Euphrates and Tigris drainages to all other places. But indeed, many species do not do this. There might be many environmental filters inhibiting the long-distance dispersal, especially of loaches but also of other fish species and these filters seem to be active since millions of years, as demonstrated by the often deep phylogenetic distances between different loach species within the Tigris. It remains unstudied, why most loaches are locally endemic while others were able to disperse widely thought almost the entire drainage system.

Table 35.3 Fish species endemic to the Tigris and their macrohabitats

Species	Subterranean	Headwaters	Lake Hazer
Barbus karunensis			
Capoeta coadi			
Capoeta pyragyi			
Capoeta shajariani			
Capoeta umbla			
Garra amirhosseini			
Garra elegans			
Garra gymnothorax			
Garra lorestanensis			
Garra tashanensis			
Garra typhlops			
Glyptothorax kurdistanicus			
Glyptothorax steindachneri			
Alburnoides diclensis			
Alburnoides idignensis			
Alburnoides nicolausi			
Alburnus heckeli			
Alburnus kurui			
Chondrostoma esmaeilii			
Eidinemacheilus proudlovei			
Eidinemacheilus smithi			
Oxynoemacheilus chomanicus			
Oxynoemacheilus frenatus			
Oxynoemacheilus gyndes			
Oxynoemacheilus hanae			
Oxynoemacheilus hazarensis			
Oxynoemacheilus kentritensis			
Oxynoemacheilus kiabii			
Oxynoemacheilus karunensis			
Oxynoemacheilus kurdistanicus			
Oxynoemacheilus longipinnis			
Oxynoemacheilus parvinae			
Oxynoemacheilus zagrosensis			
Oxynoemacheilus zarzianus			
Paracobitis basharensis			
Paracobitis molavii			
Paracobitis zabgawraensis			
Paraschistura chrysicristinae			
Paraschistura ilamensis			
Sasanidus kermanshahensis			
Turcinoemacheilus hafezi			
Turcinoemacheilus kosswigi			
Turcinoemacheilus saadii			
Cobitis avicennae			
Cobitis kellei			
Salmo tigridis			
Kosswigichthys asquamatus			
Esmaeilius vladykovi			

Coad (1996) already pointed out, there is no endemism at family level in the catchment and he recognized only endemic genera of subterranean fish (*Caecocypris*, *Iranocypris*, and *Typhlogarra*) from the Euphrates and Tigris. Indeed, there is no river drainage in the Western Palearctic in which an endemic fish family is known and the lack of a family level endemic in the Euphrates and Tigris was and is not expected. There are endemic genera in the area, but these are not those proposed by Coad (1996). *Caecocypris* has never been found again and it might be related to *Mesopotamichthys*. Until this can be resolved, it is still treated as an endemic genus. *Iranocypris* and *Typhlogarra* have been placed in *Garra* by Hamidan et al. (2014). Besides *Caecocypris*, *Sasanidus* and *Eidinemacheilus* are recognized as endemic genera from the Euphrates and Tigris, and *Mesopotamichthys* is almost endemic, being just also found in the lower Zohreh River adjacent to the Shatt Al-Arab.

Country Scale Diversity The high part of fish diversity in the Euphrates and Tigris contributed by mountain fishes is also reflected by the country scale diversity of fishes in Turkey (69 species), Syria (39 species), Iraq (52 species), and Iran (68 species). Here the two countries Iran and Turkey having mostly mountainous rivers in the Euphrates and Tigris catchments show the highest numbers while Syria, with mostly lowland habitats, has the lowest number of species (Table 35.4). But the freshwater fishes of Syria are not well studied and there might be some surprises waiting.

Biogeographic Connections The biogeographic connections of the Euphrates and Tigris drainages had been discussed by Coad (1996) and these connections are meanwhile pretty clear. From the 25 species that are not coastal and not endemic, 17 occur also in the rivers of the Iranian Gulf basin as the Zohreh, Helleh, Mond, and others including the endorheic basins of the Kor and Zayandeh rivers. Only six species are also found in the Mediterranean basin and one species occurs also in the Black Sea basin and one in the Caspian Sea basin. Coad (1996) considered already most (19) species to be shared between the Euphrates and Tigris with other rivers in the Gulf basin of Iran. But he still considered more species of the Euphrates and Tigris to be shared with the Mediterranean (16 species), as well as with the Black and Caspian Sea basins (10 species) but these strong biogeographic connections can no longer be confirmed.

Threats Within such as large area as the catchment of the Euphrates and Tigris there are many different threats. These range from pollution to dams, water abstraction to overfishing and alien species invasion. Pollution is a serious challenge, especially close to cities and towns and wastewater treatment is virtually unknown in the area. Regionally and locally, pollution is massive and "dead" rivers and streams can be found all over the area. But still, there are many clean streams and rivers not or very little impacted by pollution.

Overfishing is a threat especially in the reservoirs as well as in the larger rivers in the Euphrates and Tigris. Several large growing cyprinid species are thought-for commercial species and there is a lack of regulations to protect and manage inland fisheries allow overfishing to be a threat in many areas (Freyhof et al. 2014a). The

Table 35.4 Freshwater fish species in the Turkish, Syrian, Iraqi, and Iranian parts of the Euphrates and Tigris

Species	Turkey	Syria	Iraq	Iran
Tenualosa ilisha				
Arabibarbus grypus				
Barbus karunensis				
Barbus lacerta				
Caecocypris basimi				
Capoeta coadi				
Capoeta damascina				
Capoeta macrolepis				
Capoeta pyragyi				
Capoeta shajariani				
Capoeta trutta				
Capoeta umbla				
Carasobarbus kosswigi				
Carasobarbus luteus				
Carasobarbus sublimus				
Cyprinion kais				
Cyprinion macrostomum				
Garra amirhosseini				
Garra elegans				
Garra gymnothorax				
Garra lorestanensis				
Garra rufa				
Garra variabilis				
Garra tashanensis				
Garra typhlops				
Garra widdowsoni				
Luciobarbus barbulus				
Luciobarbus esocinus				
Luciobarbus subquincunciatus				
Luciobarbus xanthopterus				
Mesopotamichthys sharpeyi				
Barilius mesopotamicus				
Acanthobrama marmid				
Alburnoides diclensis				
Alburnoides emineae				
Alburnoides idignensis				
Alburnoides nicolausi				
Alburnoides velioglui				
Alburnus caeruleus				
Alburnus doriae				
Alburnus heckeli				

(continued)

Table 35.4 (continued)

Alburnus kurui				
Alburnus sellal				
Alburnus tarichi				
Alburnus timarensis				
Chondrostoma esmaeilii				
Chondrostoma regium				
Leuciscus vorax				
Pseudophoxinus firati				
Pseudophoxinus zeregi				
Squalius berak				
Squalius lepidus				
Squalius semae				
Eidinemacheilus proudlovei				
Eidinemacheilus smithi				
Oxynoemacheilus araxensis				
Oxynoemacheilus argyrogramma				
Oxynoemacheilus arsaniasus				
Oxynoemacheilus bergianus				
Oxynoemacheilus chomanicus				
Oxynoemacheilus ercisianus				
Oxynoemacheilus euphraticus				
Oxynoemacheilus frenatus				
Oxynoemacheilus gyndes				
Oxynoemacheilus hanae				
Oxynoemacheilus hazarensis				
Oxynoemacheilus karunensis				
Oxynoemacheilus kaynaki				
Oxynoemacheilus kentritensis				
Oxynoemacheilus kiabii				
Oxynoemacheilus kurdistanicus				
Oxynoemacheilus longipinnis				
Oxynoemacheilus muefiti				
Oxynoemacheilus paucilepis				
Oxynoemacheilus parvinae				
Oxynoemacheilus samanticus				
Oxynoemacheilus tigris				
Oxynoemacheilus zagrosensis				
Oxynoemacheilus zarzianus				
Paracobitis basharensis				
Paracobitis molavii				
Paracobitis salihae				
Paracobitis zabgawraensis				
Paraschistura chrysicristinae				

(continued)

Table 35.4 (continued)

Paraschistura ilamensis				
Sasanidus kermanshahensis				
Turcinoemacheilus hafezi				
Turcinoemacheilus kosswigi				
Turcinoemacheilus minimus				
Turcinoemacheilus saadii				
Cobitis avicennae				
Cobitis elazigensis				
Cobitis kellei				
Mystus pelusius				
Salmo euphrataeus				
Salmo fahrettini				
Salmo munzuricus				
Salmo okumusi				
Salmo tigridis				
Silurus triostegus				
Glyptothorax armeniacus				
Glyptothorax cous				
Glyptothorax kurdistanicus				
Glyptothorax silviae				
Glyptothorax steindachneri				
Aphanius asquamatus				
Paraphanius mento				
Esmaeilius sophiae				
Aphaniops stoliczkanus				
Esmaeilius vladykovi				
Planiliza abu				
Mastacembelus mastacembelus				

absence of government control on fishing activity, e.g., no limited laws on the net mesh size for fishing gears and no prevention legislation during the reproduction period of fishes is a common situation, especially in Iraq.

The construction of dams virtually everywhere in the Euphrates and Tigris drainages has massively alternated the aquatic habitats in the area. This study is not the place to discuss and review the effects of dams and reservoirs in detail. Therefore, just a very short introduction is given. Construction of dams directly affects riverine ecosystems. Dams are mostly perceived as a barrier to the migration of fish, but dam also completely change the river and streams into artificial reservoir ecosystem, usually with strong water level fluctuations and steep shores. Some fish species native to the Tigris and Euphrates, benefit from the construction of reservoirs (Table 35.5) while others are not able to arrange with the new environmental

Table 35.5 Species benefiting from reservoirs

Acanthobrama marmid
Alburnus caeruleus
Alburnus sellal
Arabibarbus grypus
Capoeta damascina
Capoeta umbla
Carasobarbus luteus
Chondrostoma regium
Cyprinion macrostomum
Garra rufa
Garra variabilis
Luciobarbus esocinus
Luciobarbus xanthopterus
Mastacembelus mastacembelus
Mystus pelusius
Planiliza abu
Silurus triostegus
Squalius berak
Squalius lepidus
Squalius semae

conditions (Kuru 1986; Ünlü et al. 1997; Ünlü 2006). Interestingly, while so many dams and reservoirs have been built in the catchment of the Euphrates and Tigris, the effects on fish and other aquatic biodiversity have very rarely been studied. Naturally, for long-distance migrating species as *Tenualosa ilisha* and *Carcharhinus leucas*, coming from the sea and migrating upriver, dams terminate their migration. Anadromous species as *T. ilisha* are strongly impacted, as they can no longer reach their spawning grounds and therefore cannot reproduce anymore. Dams at the lower Euphrates and Tigris have such impacts and limit the spawning migration of *T. ilisha*. However, while the species has strongly declined in the Euphrates and Tigris, a small population is still spawning below the dams.

Riverine fishes which reproduce obligatory in flowing water, usually on clean gravel substrate, might be also limited in their migration patterns by dams. This phenome has not been studied in the Euphrates and Tigris drainages. Some of the gravel spawners migrate to the rivers and streams which flow into the reservoirs and spawn here. In these species, as *Arabibarbus grypus*, *Squalius lepidus*, *Luciobarbus esocinus* and *L. xanthopterus*, it is highly relevant, that reservoirs have headwater rivers in good conditions to allow spawning and natural habitat choice of the juvenile fish. If this is the case, these species might be able to build large populations in reservoirs. Other species are more sensitive and need flowing waters, gravel substrate, and shallow river habitats with high oxygen concentration all year round to complete their life cycle (e.g. *A. grypus* the most dominant occurrence species in Hemreen dam Lake on Tigris river (Lazem and Attee 2016). Ünlü (2006) asserted that lotic species (*Barilus mesopotamicus, Luciobarbus subquincuncinatus, Carasobarbus kosswigi, Arabibarbus grypus, Garra rufa, Cyprinion macrostomum, Mystus pelusius, Glyptothorax cous*, and *Salmo tigridis*) suffer because reservoirs cannot supply their high oxygen requirements and clean shallow gravel beds as habitats. In addition, he emphasized that lentic species such as *Garra variabilis, Silurus triostegus, Mastacembelus mastacembelus*, and *Leuciscus vorax* are strongly impacted by the destruction of the aquatic vegetation used as a spawning substratum by these species. But these lists are far from being complete and we expect, especially many of the nemacheilid, salmonid, and sisorid species as well as the *Alburnoides* species to be very sensitive toward the impacts of dams.

It should also not be forgotten, that dams strongly alternate the sediment flows and the water balance in rivers and reservoirs produce methane and evaporate huge amounts of water, which would otherwise be available for human consumption and irrigation and reservoirs are inevitably stocked by alien fish species, which might have disastrous effects on native species. Two alien species *Carassius auratus* and *Cyprinus carpio* consist the second (20.07%) and third (11.39%) most abundance species in Himreen dam lake (Lazem and Attee 2016).

Strongly increased water needs for human consumption in the area has not only triggered the construction of many reservoirs which evaporate huge amounts of water. Water taken from the rivers, streams, and reservoirs are just no longer available for biodiversity. This leads to a decreasing annual river discharge, which is a considerable reason for the degradation in the diversity and abundance of marine fishes in the Shatt Al-Arab (Mohamed and Abood 2017). More and more the Shatt Al-Arab is an arm of the Gulf inhabited by marine fishes and no longer a river anymore and the border of marine waters is moving more and more inland, interrupted by short periods of high discharges, when freshwater reach the sea again. Indeed, floods have been absent in the Shatt Al-Arab in the last 30 years.

References

Ali AH, Adday TK, Khamees NR (2018) Catalogue of marine fishes of Iraq. Biol Appl Environ Res 2(2):298–368

Almaça C (1983) Remarks on some Heckel's species of *Barbus* from western Asia. Arquivos do Museu Bocage B 2(12):95–102

Almaça C (1991) Evolutionary, biogeographical, and taxonomical remarks on Mesopotamian species of *Barbus s.s.* Arquivos do Museu Bocage, Nova Série 2(4):63–78

Alwan N (2010) Systematics, taxonomy, phylogeny and zoogeography of the *Capoeta damascina* species complex (Pisces: Teleostei: Cyprinidae) inferred from comparative morphology and molecular markers. PhD thesis, Johann Wolfgang Goethe University, Frankfurt, Germany

Alwan N, Esmaeili HR, Krupp F (2016) Molecular phylogeny and zoogeography of the *Capoeta damascina* species complex (Pisces: Teleostei: Cyprinidae). PLoS One 11(6):1–25. https://doi.org/10.1371/journal.pone.0156434

Akkuş M, Sarı M, Ekmekçi FG, Yoğurtçuoğlu B (2021) The discovery of a microbialite-associated freshwater fish in the world's largest saline soda lake, Lake Van (Turkey). Zoosyst Evol 97(1): 181–189. https://doi.org/10.3897/zse.97.62120

Astrin JJ, Stüben PE, Misof B, Wägele JW, Gimnich F, Raupach MJ, Ahrens D (2012) Exploring diversity in cryptorhynchine weevils (Coleoptera) using distance-, character- and tree-based species delineation. Mol Phylogenet Evol 63:1–14. https://doi.org/10.1016/j.ympev.2011.11.018

Banister KE (1980) The fishes of the Tigris and Euphrates rivers. In: Rozoska J (ed) Euphrates and Tigris, Mesopotamian ecology and Destiny. Monographiae Biologicae, vol 38. Springer, Dordrecht, pp 95–108

Behrens-Chapuis S, Herder F, Esmaeili HR, Freyhof J, Hamidan NA, Özuluğ M, Šanda R, Geiger MF (2015) Adding nuclear rhodopsin data where mitochondrial COI indicates discrepancies – can this marker help to explain conflicts in cyprinids? DNA Barcodes 3(1):187–199

Berg LS (1949) Freshwater fishes of the U.S.S.R. and adjacent countries, vol I-II-III. Academy of Sciences of the U.S.S.R. Zoological Institute, Moscow. 341 pp.

Bogutskaya NG, Coad BW (2009) A review of vertebral and fin-ray counts in the genus *Alburnoides* (Teleostei: Cyprinidae) with a description of six new species. Zoosystematica Rossica 18(1):126–173

Carrizo SF, Jahnig SC, Bremerich V, Freyhof J, Harrison I, He F, Langhans SD, Tockner K, Zarfl C, Darwall W (2017) Freshwater megafauna: flagships for freshwater biodiversity under threat. Bioscience 67:919–927

Coad BW (1991) Fishes of the Tigris-Euphrates basin: a critical checklist. Syllogeus 68:1–49

Coad BW (1996) Zoogeography of the fishes of the Tigris-Euphrates Basin. Zool Middle East 13 (1):51–70. https://doi.org/10.1080/09397140.1996.10637706

Coad BW (2009) *Alburnus zagrosensis* n. sp., a new species of fish from the Zagros mountains of Iran (Actinopterygii: Cyprinidae). Zool Middle East 48:63–70

Coad BW (2010) Freshwater fishes of Iraq. Pensoft Publishers, Sofia-Moscow. 274+16pls

Davidson NC (2014) How much wetland has the world lost? Long-term and recent trends in global wetland area. Mar Freshw Res 65:934–941

Durand JD, Chen WJ, Shen KN, Fu C, Borsa P (2012) Genus-level taxonomic changes implied by the mitochondrial phylogeny of grey mullets (Teleostei: Mugilidae). C R Biol 10:687–697

Eagderi S, Jouladeh-Roudbar A, Birecikligil SS, Çiçek E, Coad BW (2017) *Chondrostoma esmaeilii*, a new cyprinid species from the Tigris basin in Iran (Teleostei: Cyprinidae). Vertebr Zool 67(2):125–132

Eagderi S, Nikmehr N, Freyhof J (2019) *Alburnus zagrosensis*, a junior synonym of *Alburnus sellal* (Teleostei: Leuciscidae). Zootaxa 4652(2):367–374. https://doi.org/10.11646/zootaxa.4652.2.9

Elp M, Özuluğ M, Şen F, Freyhof J (2013) Validation of *Alburnus timarensis* from the Lake Van basin, eastern Anatolia (Teleostei: Cyprinidae). Zool Middle East 59:235–244. https://doi.org/10.1080/09397140.2013.841430

Esmaeili HR, Coad BW, Gholamifard A, Nazari N, Teimory A (2010) Annotated checklist of the freshwater fishes of Iran. Zoosystematica Rossica 19:361–386

Esmaeili HR, Sayyadzadeh G, Coad BW, Eagderi S (2016) Review of the genus *Garra* Hamilton, 1822 in Iran with description of a new species: a morpho-molecular approach (Teleostei: Cyprinidae). Iran J Ichthyol 3(2):82–121

Esmaeili HR, Mehraban H, Abbasi K, Keivany Y, Coad BW (2017) Review and updated checklist of freshwater fishes of Iran: taxonomy, distribution and conservation status. Iran J Ichthyol 4 (Suppl.1):1–114

Freyhof J (2016) Redescription of *Garra elegans* (Günther, 1868), a poorly known species from the Tigris River drainage (Teleostei: Cyprinidae). Zootaxa 4173(5):496–500

Freyhof J, Yoğurtçuoğlu B (2020) A proposal for a new generic structure of the killifish family Aphaniidae, with the description of Aphaniops teimorii (Teleostei: Cyprinodontiformes). Zootaxa 4810(3):421–451

Freyhof J, Ekmekçi FG, Ali A, Khamees NR, Özuluğ M, Hamidan N, Küçük F, Smith KG (2014a) Chapter 3: freshwater fishes. In: Smith KG, Barrios V, Darwall WRT, Numa C (eds) The status and distribution of freshwater biodiversity in the eastern Mediterranean. IUCN, Cambridge, pp 19–42. 2014a;xiv+132pp

Freyhof J, Esmaeili HR, Sayyadzadeh G, Geiger M (2014b) Review of the crested loaches of the genus *Paracobitis* from Iran and Iraq with the description of four new species (Teleostei: Nemacheilidae). Ichthyol Explor Fres 25(1):11–38

Freyhof J, Sayyadzadeh G, Esmaeili HR, Geiger M (2015) Review of the genus *Paraschistura* from Iran with description of six new species (Teleostei: Nemacheilidae). Ichthyol Explor Fres 26 (2):1–48

Freyhof J, Kaya C, Turan D (2017a) *Oxynoemacheilus kentritensis*, a new species from the upper Tigris drainage in Turkey with remarks on *O. frenatus* (Teleostei: Nemacheilidae). Zootaxa 4258(6):551–560. https://doi.org/10.11646/zootaxa.4258.6.4

Freyhof J, Özuluğ M, Saç G (2017b) Neotype designation of *Aphanius iconii*, first reviser action to stabilise the usage of *A. fontinalis* and *A. meridionalis* and comments on the family group names of fishes placed in Cyprinodontidae (Teleostei: Cyprinodontiformes). Zootaxa 4294 (5):573–585. https://doi.org/10.11646/zootaxa.4294.5.6

Freyhof J, Weissenbacher A, Geiger MF (2017c) *Aphanius kruppi*, a new killifish from Oman with comments on the *A. dispar* species group (Cyprinodontiformes: Aphaniidae). Zootaxa 4338 (3):557–573. https://doi.org/10.11646/zootaxa.4338.3.10

Freyhof J, Kaya C, Bayçelebi E, Geiger M, Turan D (2018) Generic assignment of *Leuciscus kurui* Bogutskaya from the upper Tigris drainage, and a replacement name for *Alburnus kurui* Mangit & Yerli (Teleostei: Leuciscidae). Zootaxa 4410(1):113–135. https://doi.org/10.11646/zootaxa.4410.1.6

Freyhof J, Kaya C, Turan D, Geiger M (2019) Review of the *Oxynoemacheilus tigris* group with the description of two new species from the Euphrates drainage (Teleostei: Nemacheilidae). Zootaxa 4612(1):29–57. https://doi.org/10.11646/zootaxa.4612.1.2

Freyhof J, Kaya C, Abdullah YS, Geiger MF (2021) The *Glyptothorax* catfishes of the Euphrates and Tigris with the description of a new species (Teleostei: Sisoridae). Zootaxa 4969 (3):453–491. https://doi.org/10.11646/zootaxa.4969.3.2

Geiger MF, Herder F, Monaghan MT, Almada V, Barbieri R, Bariche M, Berrebi P, Bohlen J, Casal-Lopez M, Delmastro GB, Denys GPJ, Dettai A, Doadrio I, Kalogianni E, Kärst H, Kottelat M, Kovačić M, Laporte M, Lorenzoni M, Marčić Z, Özuluğ M, Perdices A, Perea S, Persat H, Porcelotti S, Puzzi C, Robal J, Šanda R, Schneider M, Šlechtová V, Stoumboudi M, Walter S, Freyhof J (2014) Spatial heterogeneity in the Mediterranean biodiversity hotspot affects barcoding accuracy of its freshwater fishes. Mol Ecol Resour 14:1210–1221. https://doi.org/10.1111/1755-0998.12257

Geldiay R, Balık S (2007) Türkiye Tatlısu Balıkları [Freshwater Fishes in Turkey]. In: Ege Üniversitesi Su Ürünleri Fakültesi Yayınları, vol 46, 5th edn. Ege Üniversitesi Basımevi, İzmir. 644 pp.

Hamidan NA, Geiger MF, Freyhof J (2014) *Garra jordanica*, a new species from the Dead Sea basin with remarks on the relationship of *G. ghorensis, G. tibanica* and *G. rufa* (Teleostei: Cyprinidae). Ichthyol Explor Fres 25(3):223–236

Hashemzadeh Segherloo I, Abdoli A, Eagderi E, Esmaeili HR, Sayyadzadeh G, Bernatchez L, Hallerman E, Geiger MF, Özulug M, Freyhof J (2016) Dressing down: multiple reduction of the mental disc in the labeonine genus *Garra* (Teleostei: Cyprinidae) in the Middle East. Hydrobiologia 785(1):47–59. https://doi.org/10.1007/s10750-016-2902-8

Heckel JJ (1843) Ichthyologie. In: Russegger J (ed) Reisen in Europa, Asien und Afrika, mit besonderer Rücksicht auf die naturwissenschaftlichen Verhältnisse der betreffenden Länder, unternommen in den Jahren 1835 bis 1841 von Joseph Russegger. Schweitzerbart'sche Verlagsbuchhandlung, Stuttgart

Inoue T, Nakabo T (2006) The Saurida undosquamis group (Aulopiformes: Synodontidae), with description of a new species from southern Japan. Ichthyol Res 53:379–397. http://dx.doi.org/10.1007/s10228-006-0358-y

International Union for Conservation of Nature (IUCN) (2017). https://www.iucnredlist.org/. 28 Mar 2018

Jouladeh-Roudbar A, Vatandoust S, Eagderi S, Jafari-Kenari S, Mousavi-Sabet H (2015) Freshwater fishes of Iran; an updated checklist. AACL Bioflux 8(6):855–909

Karaman MS (1971) Süßwasserfische der Türkei. 8. Teil. Revision der Barben Europas, Vorderasiens und Nordafrikas. Mitteilungen aus dem hamburgischen Zoologischen Museum und Institut 67:175–254

Kaya C (2019) Taxonomic revision of the species belong to genus *Capoeta* distributed in Turkey. PhD Thesis. Recep Tayyip Erdogan University, Institute of Science and Technology, Rize, Turkey, 126 pp.

Kaya C (2020) The first record and origin of *Salmo trutta* populations established in the upper Tigris River and Lake Van Basin (Teleostei: Salmonidae). J Anatol Environ Animal Sci 5(3):366–372. https://doi.org/10.35229/jaes.777575

Kaya C, Turan D, Ünlü E (2016) The latest status and distribution of fishes in upper Tigris River and two new records for Turkish freshwaters. Turk J Fish Aquat Sc 16(3):545–562. https://doi.org/10.4194/1303-2712-v16_3_07

Kaya C, Turan D, Kalaycı G, Bayçelebi E, Freyhof J (2020) The westernmost known population of *Paracobitis* (Teleostei, Nemacheilidae), with the description of a new species from the Euphrates River in southern Anatolia. Zootaxa 4:525–534. https://doi.org/10.11646/zootaxa.4838.4.6

Khaefi R, Esmaeili HR, Sayyadzadeh G, Geiger MF, Freyhof J (2016) *Squalius namak*, a new chub from Lake Namak basin in Iran (Teleostei: Cyprinidae). Zootaxa 4169(1):145–159

Khaefi R, Esmaeili HR, Geiger MF, Eagderi S (2017) Taxonomic review of the cryptic *Barbus lacerta* species group with description of a new species (Teleostei: Cyprinidae). FishTaxa 2 (2):90–115

Krupp F, Schneider W (2008) Die Fischfauna des Nahr al-Habur, Nordost Syrien. In: Kühne H (ed) Umwelt und Subsistenz der assyrischen Stadt Dur-Katlimmu am unteren Habur, Berichte der Ausgrabung von Tall Seh Hamad / Dur-Katlimmu. (BATSH), Band 8. Harrassowitz, Wiesbaden, pp 41–51

Kuru M (1986) Dicle ve Fırat nehirleri üzerinde kurulacak barajlarla soyu tehlikeye sokulacak balık türleri. 3–5 Eylül 1986, İzmir, VIII. Ulusal Biyoloji Kongresi, 589–598 [in Turkish]

Lazem LF, Attee RS (2016) Structure of fish assemblage in relation to some ecological factors in Himreen Dam Lake, Iraq. Basrah J Agric Sci 29(1):7–16

Mohamed ARM, Abood AN (2017) Compositional change in fish assemblage structure in the Shatt Al-Arab. Iraq Asian J Appl Sci 5(5):944–958

Mohamed ARM, Younis KH, Hameed EK (2017) Status of fish assemblage structure in the Garmat Ali river. Iraq IOSR J Agric Vet Sci 10(2):17–22

Mohammadian-Kalat T, Esmaeili HR, Aliabadian M, Freyhof J (2017) Re-description of *Alburnus doriae*, with comments on the taxonomic status of *A. amirkabiri*, *A. mossulensis*, *A. sellal* and *Petroleuciscus esfahani* (Teleostei: Cyprinidae). Zootaxa 4323(4):487–502

Pallas PS (1814) Zoographia Rosso-Asiatica, sistens omnium animalium in extenso Imperio Rossico et adjacentibus maribus observatorum recensionem, domicilia, mores et descriptiones, anatomen atque icones plurimorum. Tomus III; pp [1], 1–428, I-CXXV[=1–125]. Petropoli

Saç G, Özuluğ M, Elp M, Gaffaroğlu M, Ünal S, Karasu Ayata M, Kaya C, Freyhof J (2019) New records of *Pseudophoxinus firati* from Turkey (Teleostei: Leuciscidae). J Appl Ichthyol 35 (3):769–774. https://doi.org/10.1111/jai.13869

Sayyadzadeh G, Esmaeili HR (2020) *Oxynoemacheilus marunensis*, a new loach species from the Persian Gulf basin with remarks on *O. frenatus* (Teleostei: Nemacheilidae). Zootaxa 4885 (2):189–206

Schöter C, Özuluğ M, Freyhof J (2009) *Capoeta caelestis*, a new species from Goksu River, Turkey (Teleostei: Cyprinidae). Ichthyol Explor Fres 20:229–236

Tan M, Armbruster JW (2018) Phylogenetic classification of extant genera of fishes of the order Cypriniformes (Teleostei: Ostariophysi). Zootaxa 4476(1):6–39. https://doi.org/10.11646/zootaxa.4476.1.4

Turan D, Kaya C, Ekmekci FG, Bayçelebi E (2014) Three new species of Alburnoides (Teleostei: Cyprinidae) from Euphrates River, eastern Anatolia, Turkey. Zootaxa 3754:101–116

Turan D, Kottelat M, Kaya C (2017) *Salmo munzuricus*, a new species of trout from the Euphrates River drainage, Turkey (Teleostei: Salmonidae). Ichthyol Explor Fres 28(1):55–63

Turan D, Kalaycı G, Bektaş Y, Kaya C, Bayçelebi E (2020) A new species of trout from the northern drainages of Euphrates River, Turkey (Salmoniformes: Salmonidae). J Fish Biol 96 (6):1454–1462. https://doi.org/10.1111/jfb.14321

Ünlü E (2006) Tigris River Ichthyological studies in Turkey. A review with regard to the Ilisu Hydroelectric Project. Ilisu Consortium, Turkey; 34pp

Ünlü E, Özbay C, Kılıç A, Coşkun Y, Sesen R (1997) Gap'in Faunaya Etkileri. Turkiye Cevre Vakfi Yayini 125:79–102

Vatandoust S, Eagderi S (2015) *Paraschistura ilamensis*, a new species of loach from the Tigris River drainage (Teleostei: Nemacheiliidae). Int J Aquat Biol 3(3):177–182

World Wide Fund for Nature (WWF) (2016) Living planet report. Risk and resilience in a new era. WWW International, Gland, p 2016

Youn SJ, Taylor WW, Lynch AJ, Cowx IG, Douglas Beard T, Bartley D, Wu F (2014) Inland capture fishery contributions to global food security and threats to their future. Glob Food Sec 3:142–148. https://doi.org/10.1016/j.gfs.2014.09.005

Zamani-Faradonbe M, Keivany Y, Dorafshan S, Zhang E (2021) Two new species of *Garra* (Teleostei: Cyprinidae) from western Iran. Ichthyol Explor Freshw 30(1):249–270. https://doi.org/10.23788/IEF-1137

Zareian H, Esmaeili HR, Gholamhosseini A, Japoshvili B, Özulug M, Mayden RL (2018) Diversity, mitochondrial phylogeny, and ichthyogeography of the *Capoeta capoeta* complex (Teleostei: Cyprinidae). Hydrobiologia 806(1):363–409. https://doi.org/10.1007/s10750-017-3375-0

Chapter 36
Fish Fauna of Shatt al-Arab River, Basrah, Iraq: A More than Quarter a Century of Changes

Laith A. Jawad, Abbas J. Al-Faisal, and Mustafa S. Faddagh Ziyadi

Abstract Shatt Al-Arab River is originated from the convergence of Tigris and Euphrates Rivers, and it is the main surface water source in the region and serves around three million people, the majority living in Basra city. The river is generally used for human consumption, agricultural, trade and industrial activities, transportation, electric power plants and recreation. The decreases in freshwater inflows into the estuary have allowed the salt water to intrude about 80 km upstream the river mouth. A decrease in river discharge into an estuary could increase the tidal range and the wave celerity, and consequent increase in salinity levels.

Across the River of Shatt al-Arab, fish faunas have been highly altered over the past 30 years. Many changes have been noted and summarized in various studies that address the fauna of this river, which generally discuss the status of the fish fauna at the time they have been done. Much of the information is based on general observations and often on data that are not quantitative or comparative in nature. However, long-term studies that are comparative and well quantified will provide a greater knowledge of large-scale faunal changes over long time periods.

This study is based on published records. Although some of the records are unsupported by vouchered specimens, they were able to extend the time period to 31 years. For the fish fauna of Shatt al-Arab River, there is no comprehensive surveys have been published. The aim of this chapter is to determine whether there have been any changes in the species richness (the number of species) of the fish fauna of Shatt al-Arab River 1985–2017. Individual species and overall pattern changes in the fauna also are discussed.

The analysis of the fish fauna of Shatt al-Arab River for the last 3 decades showed the following changes: (1) The number of the native freshwater fish species has been decreased with the time; (2) the number of the introduced species has been increased;

L. A. Jawad (✉)
School of Environmental and Animal Sciences, Unitec Institute of Technology, Auckland, New Zealand

A. J. Al-Faisal · M. S. Faddagh Ziyadi
Marine Science Centre, University of Basrah, Basrah, Iraq

L. A. Jawad (ed.), *Tigris and Euphrates Rivers: Their Environment from Headwaters to Mouth*, Aquatic Ecology Series 11,
https://doi.org/10.1007/978-3-030-57570-0_36

and (3) the presence of the marine species in the area north of Abu al-Khaseeb City locality was noticed in the last few years. There are three possible causes that might behind such changes, these are (1) the decreasing amount of freshwater in Shatt al-Arab River assess the salt front to extend further north; (2) the increase in the number of the introduced freshwater species play a vital role in reducing the number of the native freshwater fish species; (3) changes in the physicochemical factors of the water of Shatt al-Arab River over the 3 decades were intense and can bring changes in the fish species composition of the river in favour of those species that can stand such sever changes.

36.1 Introduction

Among the rich vertebrates groups are Fishes that have been exploited by human through thousands of years of his history. Since the time of Sumerians and Babylonians, they provide food and employment through commercial and traditional fisheries as well as recreation and enjoyment in sport fisheries and as ornamental species in aquaria and ponds (Saggs 1962; Postgate 1994). They are also used extensively in science (Froese and Torres 1999). Regardless of the economic value of these activities, freshwater fishes are probably the most endangered of all aquatic vertebrates. The threat is likely to worsen, as demand for food and conflicts over the use of freshwater fishes endure to increase (Nyman 1991; Maitland 1994).

Although many species are modified to handle chronic, expectable shifts in environmental conditions, the speediness and erratic nature of human impacts on community composition and ecosystem health have created a legacy of disturbance during the human history (Parmesan 2006).

Enumerating observable changes within ecosystems credited to adjustments in resource management is important for assessing the success of conservation initiatives, and a vital step is measuring changes in community structure. Species diversity and functional failure are often signalled as important metrics in identifying the capacity of ecosystems to account for ecological change (Peterson et al. 1999). As such, conservation administration often sponsors actions that lead to the care and increase in species diversity (reviewed by Myers et al. 2000).

The credit can be given to the Sumerian and the Babylonian in studying the fish fauna of Iraq (Saggs 1962). Ancient Mesopotamians succeeded in identifying and naming several freshwater and marine species, which were recorded on clay tablets (Landsberger 1962). However, no further records are available until the nineteenth century, when Heckel (1843) described 17 freshwater fish species from the Tigris River at Mosul City, northern Iraq. The work of Heckel (1843) indicates the actual start of taxonomic work on the fish fauna of Iraq. Previous to that date, the works of Hasselquit (1762–1752) on different parts of the Middle East were considered the early works on fish taxonomy of this part of the world. The authors of these works did not collect the specimens from Iraqi waters in spite of the fact that the species they described were actually present in Iraq later on.

Shatt Al-Arab River which originates from the convergence of Tigris and Euphrates Rivers is the main surface water source in the region and serves around three million people, the majority living in Basra city. The river is widely used for human consumption, agricultural, trade and industrial activities, transportation, electric power plants and recreation. The main agricultural lands extend along the river banks with large date palm plantations. The Shatt Al-Arab River suffered from massive regression in water quality related to the decline in rates of discharge from the Tigris and the Euphrates Rivers (Al-Mahmood et al. 2015) as a result of several hydrological projects constructed in the riparian countries (Partow 2001), and the diversion of the Karun River into Iranian terrene (Hameed and Aljorany 2011). The average rate of discharge in the upstream of the Shatt-al-Arab River was declined from 207 m3/s during 1977–1978 to 60m^3/s during 2014 (Alaidani 2014). The decreases in freshwater inflows into the estuary have allowed the salt water to intrude about 80 km upstream the river mouth (Abdullah et al. 2016). A decrease in river discharge into an estuary could increase the tidal range and the wave celerity and consequent increase in salinity levels (Cai et al. 2011). The modification of water discharge in the Shatt al-Arab River and the saltwater intrusion further upstream have been discussed by several authors (Al-Tawash et al. 2013; Hameed et al. 2013; Yaseen et al. 2017). Several studies have been supportive of the decline of the Shatt al-Arab water quality under these conditions (Al-Tawash et al. 2013; Moyel and Hussain 2015; Yaseen 2016).

Across the River of Shatt al-Arab, fish faunas have been highly altered over the past 30 years. Many changes have been noted and summarized in various studies that address the fauna of this river (Al-Hassan and Hussain 1985; Al-Hassan and Soud 1989; Mohamed et al. 2001, 2008, 2010, 2012, 2013 Lazem 2009; Younis et al. 2010; AL-Shamary 2016; Yaseen, A. T. 2016; Amjed et al. 2016; Mohamed and Abood 2017a, b, c; Mohamed et al. 2017a; b; Nasir and Khalid 2017; Yaseen et al. 2017). Such texts generally discuss the status of the fish fauna at the time they have been done. Much of the information is based on general observations and often on data that are not quantitative or comparative in nature. However, long-term studies that are comparative and well quantified will provide a greater knowledge of large-scale faunal changes over long time periods.

This study relied on published records. Although some of the records are unsupported by vouchered specimens, they were able to extend the time period to 31 years. For the fish fauna of Shatt al-Arab River, there is no comprehensive surveys have been published. The goal of this study is to determine whether there have been any changes in the species richness (the number of species) of the fish fauna of Shatt al-Arab River 1985–2017. Individual species and overall pattern changes in the fauna also are discussed.

36.2 Fish Species Nomenclature Changes

As with the other freshwater fish species in the Middle East and during the last 30 years, the nomenclature of the fish species has faced several periods of changes. In such modifications, species were shifted from genus to another, families were erected, while others were became invalid. Since the members of the family Cyprinidae represent the majority of fish species living in the inland waters of Iraq, then any changes in the nomenclature of its species are important to mention here to remind the readers about the new names of different taxa and their taxonomic status.

Recently, several papers have been published revising the phylogenetic relationship of the family Cyprinidae (Chen and Mayden 2009; Yang et al. 2015; Stout et al. 2016). In these works, several subfamilies taxa were elevated to the family rank such as Danioninae, Sudadanioniae, Xenocypridiniae, Acheilognathinae, Gobioninae, Tanichthyiniae and Leucisciniae. Also, the two cyprinid species, *Ctenopharyngodon idella* and *Hypophthalmichthys molitrix* were removed from Cyprinidae and located in the new upgraded family Xenocyprididae. From the family Cyprinidae, all species of the genus *Alburnus* were removed and located in the newly erected family Leuciscidae. The other important nomenclatural event that happened within the family Cyprinidae was the revision of the genus *Carasobarbus*, which used to contain elements from different groups and the present time includes *Carasobarbus apoensis*, *C. canis*, *C. chantrei*, *C. exulatus*, *C. fritschii*, *C. harterti*, *C. kosswigi*, *C. luteus* and *C. sublimus* (Borkenhagen and Krupp 2013). Still within the family Cyprinidae, the other scientific names change was the erecting of a new genus *Arabibarbus* and inclusion of the cyprinid fish species *Barbus grypus* in that genus so it became *A. grypus* (Borkenhagen 2014, 2017). Berrebi and Tsigenopoulos (2003) moved the large-sized barbus species such as *Barbus xanthopterus* and *B. esocinus* and others to the genus *Luciobarbus*.

For species other than members of the family Cyprinidae, nomenclatural changes have been done within the family Mugilidae. Durand et al. (2012) have performed several movements of species between genera among these the species *Liza abu* has been moved to *Planiliza*. This mugilid species is very well represented in Shatt al-Arab River. On the other hand, Dunz and Schliewen (2013) have moved *Tilapia zilli* to the genus *Coptodon* and Freyhof et al. (2017) have removed members of the genus *Aphanius* from Cyprinodontidae to a new family Aphaniidae.

36.3 Introductions and Range Extensions

Several marine species were shown to have their distributional range extended to freshwater and vice versa, as seen in a number of publications that appeared in the 1980s (Al-Hassan and Naama 1986; Al-Hassan and Al-Badri 1986; Al-Hassan and Muhsin 1986). In addition to their presence in the Shatt al-Arab River, south of

Mesopotamia, some marine species were recorded further north in both the Euphrates and Tigris rivers. Conversely, some freshwater species were recorded in the marine waters of Iraq. In the same period, Al-Hassan and Al-Badri (1986) and Al-Hassan and Miller (1987) recorded an invasive species in Iraq's marine waters that may have been introduced via ballast water.

During the 1990s, Mohammed et al. (1993) recorded the presence of freshwater species in the northwest extension of the Arabian Gulf, while Al-Hassan (1994) reported on the presence of the aquaculture species *Hypophthalmichthys molitrix* in the Shatt al-Arab River. The former record presents the euryhaline characteristic of the cyprinid species *Carasobarbus luteus* that enables it to be present in a marine environment, while the latter was the first alarming incidence of this species in the wild; this species is now well established in Iraqi freshwater systems and has started to compete with the native cyprinid species in the area (Hussein et al. 2000).

The recent work of Coad and Hussain (2007) on the record of the exotic cyprinid species *Hemiculter leucisculus* from the marshes near the Iraq-Iran border is considered significant, as it highlights the invasion of this competitive species. The area where this species was collected is controversial and cannot be considered Iraqi waters, due to several claims by Iran regarding the sovereignty over the marshes.

Scientific records of introduced species (such as some cyprinid species) did not make their appearance until Al-Nasiri and Hoda (1976) mentioned them in their work. This year could mark the start of invasion of the freshwater system of Iraq by species such as the common carp, *Cyprinus carpio*, and the silver carp, *Hypophthalmichthys molitrix*. Comparing lists from the 1960s and 1970s with those of the 1980s, it is clear that the number of *Barbus* species has dropped dramatically in the Shatt al-Arab River and the marsh areas, as they were replaced by introduced species such as *Cyprinus carpio*, *Carassius auratus*, *Ctenopharyngodon idella* and *Hypophthalmichthys molitrix* (Khalaf 1961; Mahdi 1962; Mahdi and George 1969; Al-Nasiri and Hoda 1975, 1976; Al-Hassan et al. 1989). Such major replacements in the species composition are mainly due to significant changes in the environment (Jawad 2003, 2006).

Kennedy (1937) recorded the aquarium fish *Girardinus fosciloides* (= *Poecilia latipinna*) from the Tigris River at Baghdad. He misspelled *fosciloides* as *poeciloides* and *Girardinus* (= *Limia poeciloides*) is a synonym of *Poecilia latipinna*. This species is of South American origin and definitely an escapee from aquarium culture. Recently, Coad (2010) reported the presence of this species in the Shatt al-Arab River and attributed its presence to escape from commercial aquarium fish activity in southern Iraq. Further specimens of this species were collected by the present author from Qarmat Ali (north of Basrah), the Shatt al-Arab River (at Basrah) and from different branches of the Shatt al-Arab River.

The only exotic fish introduction via aquarium trade reported from the marshes of central Iraq is *Pangasianodon hypophthalmus* (Sauvage, 1878) (see Khamees et al. 2013), *Mollienesia latipinna* (Lesueur, 1821) (see Al-Faisal et al. 2014). The same species were also mentioned by Al-Lamy et al. (2012) and Mohamed et al. (2008). Recently, Mutlak et al. (2017) recorded the presence for the first time a single specimen of alligator gar, *Atractosteus spatula*.

36.4 Decadal Analysis of the Fish Fauna of Shatt al-Arab River

In analysing all the literature that dealt with the fish fauna of the River Shatt al-Arab during the last 30 years or so, this chapter will deal also with fish fauna in the estuary of Shatt al-Arab River to show the composition of the marine and freshwater fish species in this estuary.

In Shatt al-Arab River, it meant here in this chapter the waterway stretch starting from the confluent of Euphrates and Tigris Rivers at Qurna north of Basrah City and until its estuary at the City of Fao, south of Basrah City. This stretch has been divided into three regions (Fig. 36.1), upper reaches, which include the confluent point of the Euphrates and Tigris and the formation of the river at the city of Qurna, the Qarma locality represents the confluent of the two arms of Shatt al-Arab River, one emerging from the greater marsh area and the other descending from the north

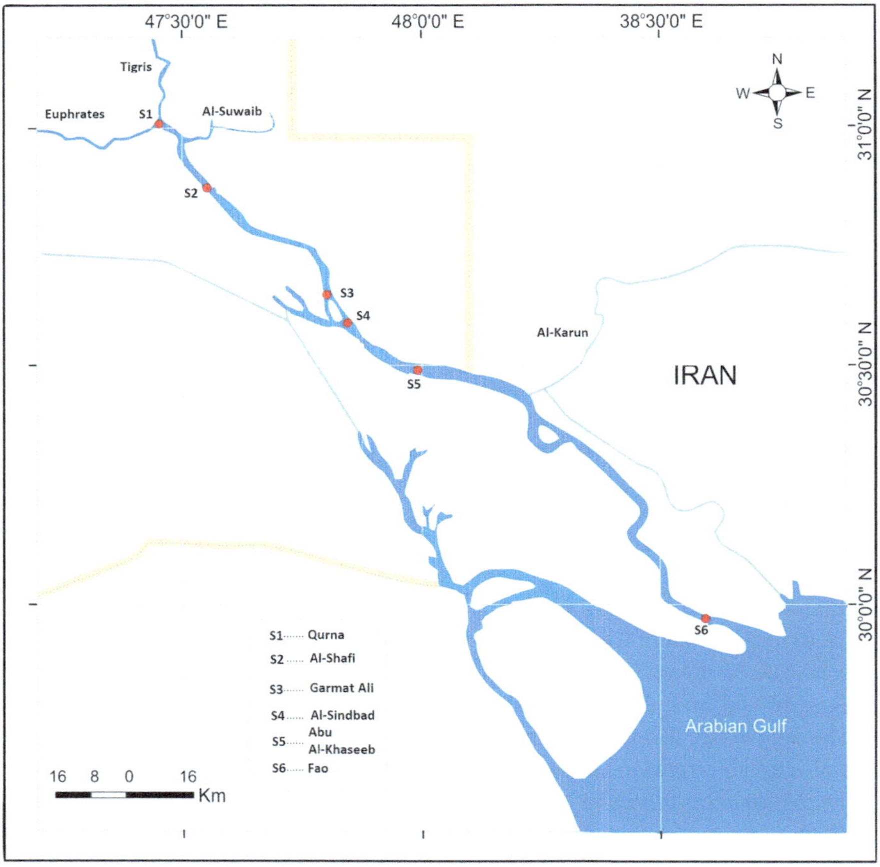

Fig. 36.1 Map showing the waterway stretch of Shatt al-Arab River, Basrah, Iraq

from Qurna locality and, finally, al-Shafi locality, which represents a water body north of Basrah City. The middle reach of the river is the waterway stretching from Qarma locality north to Abu al-Khaseeb City south, and lastly, the lower reaches of the river are Abu al-Khaseeb City locality and the estuary of the river at Fao City.

During the last 3 decades and so, several works have been published on fish assemblages in Shatt al-Arab River with its both components, the freshwater and marine species. In these studies, the total number of freshwater fish species that have dealt with ranges 10–27 belonging to 8–23 genera and contained in 6–11 families. On the other hand, the total number of marine fish species reported was 7–111 contained in 6–92 genera and belong to 5–57 families.

The chronological records of the total number of the freshwater species have shown a trend of decrease in the upper reaches of the river, while it has been slightly increased in the middle reach of the river. On the other hand, the number was slightly lower in Abu al-Khaseeb City locality than the middle reach. The freshwater species were completely disappeared in the estuary of Shatt al-Arab River. The total number of the marine species has been dramatically increased in the 3 regions of the river, and such an increase is clear in the middle and the lower reaches of the river (Table 36.1).

In the upper reaches of the river, it is clear that the number of the marine species has increased since 2008 than in 2017. At present, several species of the family Engraulidae, Mugilidae, Sillaginidae and Gobiidae were recorded to be present in this region. On the other hand, some native species were used to be present in this region during 2008–2010 such as *Arabibarbus grypus*, *Luciobarbus xanthopterus* and *Mesopotamichthys sharpeyi* were disappeared. The introduced species *C. carassius* made their appearance and according to the present analysis in 2016, while individuals of *Cyprinus carpio* where present in the area since 2008.

In the middle reaches of the river, the number of the cartilaginous fish species has cropped up from one species in 1989 to 3 in 2017. This is also true for the number of marine species such as members of the families Clupeidae, Dussumieriidae and Engraulidae and others. It is important here to note that the number of the introduced species has increased towards the present time from being low in the late 1980s.

For the lower reaches of Shatt al-Arab River, there were only 2 cartilaginous fish species reported from Abu al-Khaseeb locality (Yaseen et al. 2017). In the same period, Mohamed and Abood (2017a, b, c) reported 5 species from the middle region of the river, which is located further north of Abu al-Khaseeb City locality. The possible explanation for these discrepancies may be due to the lack of ichthyological survey. On the other hand, the estuary of Shatt al-Arab has shown the presence of 16 cartilaginous fish species. For the freshwater fish species, Abu al-Khaseeb City locality has shown the increase in the number of this group (Mohamed et al. 2013; Al-Shamary 2016; Mohamed and Abood 2017a, b, c).

It is important to note here that the results of the studies used in the decadal analysis of the fish fauna of Shatt al-Arab River are affected by the misidentification factor that could mislead the readers for the presence or absence of a fish species. In addition, the nonexistence of voucher specimens made the process of species verification a difficult task. For example, Mohamed and Abood (2017a, b, c) have

Table 36.1 Chronological distribution of the fish fauna of Shatt al-Arab River, Basrah, Iraq

	Upper reaches					**Middle reaches**							**Lower reaches**	
Fish Family/species	1	2	3	4	5	**6**	**7**	**8**	**9**	**10**	**11**	**12**	**13**	**14**
Hemiscylliidae														
Chiloscyllium arabicum												+		
Chiloscyllium griseum														+
Carcharhinidae														
Carcharhinus dussumieri														+
Carcharhinus falciformis														+
Carcharhinus limbatus										+				+
Carcharhinus sorrah														+
Carcharhinus leucas							+			+			+	+
Rhizoprionodon acutus												+		
Sphyrnidae														
Sphyrna mokarran													+	+
Torpedinidae														
Torpedo panthera														+
Rhinidae														
Rhynchobatus djiddensis														+
Glaucostegidae														
Glaucostegus granulatus												+		+
Dasyatidae														
Brevitrygon imbricata														+
Himantura uarnak													+	+
Maculabatis gerrardi														+
Pastinachus sephen													+	+
Gymnuridae														
Gymnura poecilura														+
Myliobatidae														
Aetomylaeus nichofii														+
Muraenesocidae														
Muraenesox cinereus														
Clupeidae														
Amblygaster sirm									+					+
Anodontostoma chacunda									+	+				+

(continued)

Table 36.1 (continued)

Fish Family/species	Upper reaches					**Middle reaches**							**Lower reaches**	
	1	2	3	4	5	**6**	**7**	**8**	**9**	**10**	**11**	**12**	**13**	**14**
Hilsa kelee														+
Nematalosa nasus		+			+	+	+			+	+	+	+	+
Nematalosa persara												+		
Sardinella albella										+			+	+
Sardinella fimbriata														+
Sardinella longiceps													+	+
Tenualosa ilisha	+	+	+	+	+		+		+	+	+	+	+	+
Dussumieriidae														
Dussumieria acuta												+		
Engraulidae														
Chondrostoma regium														
Thryssa dussumieri					+							+	+	+
Thryssa hamiltoni	+	+		+			+		+	+	+			+
Thryssa malabarica				+		+								
Thryssa mystax				+									+	+
Thryssa vitrirostris					+					+		+		
Thryssa whiteheadi					+				+	+	+	+	+	
Chirocentridae														
Chirocentrus dorab													+	+
Chirocentrus nudus						+	+					+	+	
Pristigasteridae														
Ilisha compressa									+	+	+	+	+	
Ilisha filigera						+								
Ilisha megaloptera						+	+		+					+
Ilisha melastoma												+		+
Chanidae														
Chanos chanos													+	
Cyprinidae														
Arabibarbus grypus	+	+					+	+	+	+		+		
Carasobarbus luteus	+	+		+	+		+	+	+	+	+	+	+	
Carasobarbus sublimus												+		
Carassius auratus			+	+	+			+		+		+	+	
Carassius carassius	+	+												
Carassius gibelio											+			
Cyprinion kais										+	+	+		
Cyprinus carpio	+	+		+	+		+	+	+	+	+	+	+	
Garra rufa	+						+						+	
Garra variabilis												+		
Luciobarbus capito							+							

(continued)

Table 36.1 (continued)

	Upper reaches					**Middle reaches**							**Lower reaches**	
Fish Family/species	1	2	3	4	5	**6**	**7**	**8**	**9**	**10**	**11**	**12**	**13**	**14**
Luciobarbus kersin	+									+				
Luciobarbus subquincunciatus							+							
Luciobarbus xanthopterus	+	+		+			+	+		+	+	+	+	
Mesopotamichthys sharpeyi	+	+					+	+		+	+	+	+	
Leuciscidae														
Acanthobrama marmid	+		+	+	+		+		+	+	+	+	+	
Alburnus caeruleus										+				
Alburnus mossulensis	+		+		+				+	+	+	+	+	
Alburnus sellal							+							
Chondrostoma regium												+		
Leuciscus vorax	+	+		+	+		+	+	+	+	+	+	+	
Xenocyprididae														
Ctenopharyngodon idella		+						+	+	+	+	+	+	
Hemiculter leucisculus				+	+				+	+	+	+	+	
Hypophthalmichthys molitrix								+		+	+	+	+	
Hypophthalmichthys nobilis												+		
Bagridae														
Mystus pelusius	+	+		+	+		+		+	+		+	+	
Siluridae														
Parasilurus triostegus	+	+		+	+		+	+		+	+	+	+	
Heteropneustidae														
Heteropneustes fossilis	+	+		+			+			+		+		
Ariidae														
Netuma bilineata												+		+
Netuma thalassina												+	+	
Plotosidae														
Plotosus lineatus												+	+	
Synodontidae														
Saurida tumbil												+		+
Saurida undosquamis														+

(continued)

Table 36.1 (continued)

	Upper reaches					**Middle reaches**							**Lower reaches**	
Fish Family/species	1	2	3	4	5	**6**	**7**	**8**	**9**	**10**	**11**	**12**	**13**	**14**
Batrachoididae														
Allenbatrachus grunniens														+
Colletteichthys dussumieri												+		
Poeciliidae														
Gambusia holbrooki	+	+		+	+		+			+	+	+	+	
Poecilia sphenops	+		+	+										
Poecilia latipinna					+				+	+	+	+	+	
Aphaniidae														
Aphanius dispar	+			+	+		+			+	+	+	+	
Aphanius mento				+			+			+	+			
Aphanius sophlae							+							
Belonidae														
Strongylura leiura												+		
Strongylura strongylura		+					+			+		+		+
Tylosurus crocodilus												+	+	
Hemiramphidae														
Hemiramphus brasiliensis														+
Hyporhamphus limbatus					+							+		
Hyporhamphus unicuspis													+	
Rhynchorhamphus georgii											+			
Syngnathidae														
Hippocampus kuda														+
Mastacembelidae														
Mastacembelus mastacembelus	+	+		+			+		+	+		+	+	
Scorpaenidae														
Pterois miles													+	
Pterois volitans														+
Synanceiidae														
Choridactylus multibarbus													+	
Minous monodactylus														+
Pseudosynanceia melanostigma												+	+	+

(continued)

Table 36.1 (continued)

	Upper reaches					**Middle reaches**							**Lower reaches**	
Fish Family/species	1	2	3	4	5	**6**	**7**	**8**	**9**	**10**	**11**	**12**	**13**	**14**
Platycephalidae														
Grammoplites suppositus												+		
Platycephalus indicus									+			+	+	+
Terapontidae														
Helotes sexlineatus														+
Terapon puta												+		
Terapon theraps												+		+
Epinephelidae														
Epinephelus areolatus													+	+
Epinephelus coioides												+		
Epinephelus tauvina														+
Apogonidae														
Apogonichthyoides taeniatus												+		
Ostorhinchus aureus					+					+				+
Sillaginidae														
Sillago arabica					+							+		
Sillago attenuata					+							+		
Sillago sihama	+				+	+	+		+	+	+	+	+	+
Rachycentridae														
Rachycentron canadum												+		+
Echeneidae														
Echeneis naucrates														+
Carangidae														
Alepes djedaba						+	+			+			+	+
Alepes kleinii												+		+
Alepes melanoptera												+		
Alepes vari										+		+		
Atropus atropos														+
Atule mate														+
Carangoides chrysophrys												+		+
Carangoides malabaricus										+		+	+	
Parastromateus niger												+		+
Scomberoides commersonnianus									+	+		+	+	+

(continued)

Table 36.1 (continued)

	Upper reaches					**Middle reaches**							**Lower reaches**	
Fish Family/species	1	2	3	4	5	**6**	**7**	**8**	**9**	**10**	**11**	**12**	**13**	**14**
Selar crumenophthalmus												+		
Selaroides leptolepis														
Trachinotus blochii														+
Trachinotus mookalee										+				
Trachurus trachurus														+
Leiognathidae														
Equulites lineolatus														+
Photopectoralis bindus					+					+	+	+	+	+
Menidae														
Mene maculata														+
Lutjanidae														
Lutjanus russellii												+		
Gerreidae														
Gerres limbatus										+		+		
Gerres macracanthus												+		
Gerres oyena									+					
Haemulidae														
Diagramma picta												+		
Plectorhinchus schotaf														+
Pomadasys argenteus												+		+
Pomadasys kaakan												+		
Pomadasys stridens												+		
Sparidae														
Acanthopagrus arabicus					+						+	+		
Acanthopagrus berda						+	+			+				+
Acanthopagrus bifasciatus														+
Acanthopagrus latus	+	+	+				+	+	+	+			+	+
Argyrops spinifer												+		+
Crenidens crenidens												+		+
Diplodus sargus	+			+								+		+
Rhabdosargus sarba														+
Sparidentex hasta						+	+		+	+	+	+	+	+
Lethrinidae														
Lethrinus nebulosus														+

(continued)

Table 36.1 (continued)

	Upper reaches					**Middle reaches**							**Lower reaches**	
Fish Family/species	1	2	3	4	5	**6**	**7**	**8**	**9**	**10**	**11**	**12**	**13**	**14**
Nemipteridae														
Nemipterus japonicus														+
Nemipterus bipunctatus												+		
Scolopsis frenata														+
Scolopsis taeniata												+		
Sciaenidae														
Johnius amblycephalus												+		
Johnius belangerii									+			+	+	+
Johnius borneensis									+					
Johnius carutta						+								
Johnius dussumieri										+				
Johnius sina							+		+				+	+
Otolithes ruber									+			+	+	+
Pennahia anea		+										+		
Protonibea diacanthus												+		+
Polynemidae														
Eleutheronema tetradactylum						+	+		+	+		+	+	+
Polydactylus sextarius												+		+
Mullidae														
Upeneus doriae										+		+		
Upeneus sulphureus														+
Upeneus sundaicus												+		
Upeneus tragula													+	+
Pomacanthidae														
Pomacanthus imperator														+
Cepolidae														
Acanthocepola abbreviata												+		
Mugilidae														
Moolgarda speigleri												+		
Planiliza abu	+	+	+	+	+		+	+	+	+	+	+	+	
Planiliza carinata	+				+							+		+
Planiliza macrolepis						+	+							+
Planiliza subviridis	+	+	+	+	+		+	+	+	+	+	+	+	+
Planiliza klunzingeri				+	+				+	+	+	+	+	

(continued)

Table 36.1 (continued)

	Upper reaches					**Middle reaches**							**Lower reaches**	
Fish Family/species	1	2	3	4	5	**6**	**7**	**8**	**9**	**10**	**11**	**12**	**13**	**14**
Cichlidae														
Coptodon zilli			+		+			+	+	+	+	+	+	
Oreochromis aureus											+	+	+	
Oreochromis niloticus					+						+	+	+	
Scaridae														
Scarus ghobban														+
Gobiidae														
Bathygobius fuscus	+			+	+				+	+	+	+	+	+
Boleophthalmus dussumieri					+				+	+		+	+	
Periophthalmus barbarus							+							
Periophthalmus waltoni	+								+		+	+		+
Pseudapocryptes elongatus														+
Scartelaos tenuis														+
Trypauchen vagina													+	+
Ephippidae														
Ephippus orbis														+
Scatophagidae														
Scatophagus argus					+	+	+			+			+	+
Siganidae														
Siganus canaliculatus												+	+	+
Sphyraenidae														
Sphyraena pinguis														+
Sphyraena obtusata												+	+	
Trichiuridae														
Eupleurogrammus glossodon													+	
Eupleurogrammus muticus													+	
Trichiurus lepturus													+	+
Scombridae														
Auxis thazard														+
Rastrelliger kanagurta												+		
Scomberomorus commerson												+		+
Scomberomorus guttatus												+		

(continued)

Table 36.1 (continued)

Fish Family/species	Upper reaches					**Middle reaches**							**Lower reaches**	
	1	2	3	4	5	**6**	**7**	**8**	**9**	**10**	**11**	**12**	**13**	**14**
Stromateidae														
Pampus argenteus												+		+
Psettodidae														
Psettodes erumei														+
Paralichthyidae														
Pseudorhombus arsius														+
Bothidae														
Bothus pantherinus														+
Soleidae														
Brachirus orientalis					+	+	+		+	+		+	+	+
Pegusa bleekeri														+
Solea elongata												+		+
Solea stanalandi													+	
Zebrias synapturoides														+
Cynoglossidae														
Cynoglossus macrolepidotus						+								
Cynoglossus arel						+	+		+		+	+	+	+
Triacanthidae														
Triacanthus biaculeatus												+		+
Balistidae														
Abalistes stellaris														+
Tetraodontidae														
Lagocephalus guentheri										+				
Lagocephalus lunaris														+
Total number of freshwater species	27	14	10	15	14	0	19	13	18	24	19	26	21	0
Total number of marine species	10	8	7	9	20	16	21	4	25	34	16	85	51	111
Total number of freshwater fish genera	16	14	8	14	13	0	15	14	17	20	17	23	20	0
Total number of marine fish genera	8	8	6	6	16	14	19	3	20	29	12	70	44	92
Total number of freshwater fish families	6	7	7	9	8	0	10	7	11	10	8	10	9	0

(continued)

Table 36.1 (continued)

	Upper reaches					**Middle reaches**							**Lower reaches**	
Fish Family/species	1	2	3	4	5	**6**	**7**	**8**	**9**	**10**	**11**	**12**	**13**	**14**
Total number of marine fish families	7	6	5	5	11	14	16	3	14	19	10	43	32	57

Upper reaches = Qurna, Qarma, Shafi localities; Middle reaches = the waterway stretch of the river from Al-Sindbad Island north to Abu Al-Khaseeb south; Lower reaches = Abu al-Khaseeb area and the estuary f Shatt al-Arab River. 1, Younis et al. 2010; 2, Mohamed et al. 2008; 3, Resen et al. 2016; 4, Mohamed et al. 2010; 5, Mohamed et al. 2017a, b, ; 6, Al-Hassan and Hussain 1985; 7, Al-Hassan and Soud 1989; 8, Nasir and Khalid 2017; 9, Mohamed et al. 2012; 10, Mohamed et al. 2013; 11, Al-Shamary 2016; 12, Mohamed and Abood 2017a, b, c; 13, Yaseen et al. 2017; 14, Mohamed et al. 2001

reported the presence of 85 marine fish species belonging to 70 genera and contained in 32 families. Such large number of marine species in the estuary of Shatt al-Arab River is very high and could be attributed to misidentification of the specimens. Therefore, my suggestions here for those who interested to accomplish any study which deals with the fish assemblages, they need to do the followings in order for the readers after that to verify the identity of the species: (1) keep a voucher specimen/s of each species in an accessible collection and (2) provide a colour image for each species for the same purpose.

From the analysis of the fish fauna of Shatt al-Arab River for the last 3 decades, it is possible to give the following generalizations: (1) The number of the native freshwater fish species has been decreased with the time; (2) the number of the introduced species has been increased; and (3) the presence of the marine species in the area north of Abu al-Khaseeb City locality was noticed in the last few years.

There are two possible causes that might behind such changes, these are (1) the decreasing amount of freshwater in Shatt al-Arab River assesses the salt front to extend further north allowing more marine fish species to move in the river and the same time creates unfavourable conditions for the native freshwater fish species and pushes further north; (2) the increase in the number of the introduced freshwater species plays a vital role in reducing the number of the native freshwater fish species through competition on food, feeding grounds and changing the habitats in different ways and rendering them unsuitable for the life of the eggs and young of the native species; (3) changes in the physicochemical factors of the water of Shatt al-Arab River over the 3 decades (Moyel and Hussain 2015; Abdullah et al. 2015, 2017) were dramatically and can bring changes in the fish species composition of the river in favour of those species that can stand such sever changes.

References

Abdullah A, Karim U, Masih I, Popescu I, van der Zaag P (2016) Anthropogenic and tidal influences on salinity levels and variability of the Shatt al-Arab River, Basra, Iraq. Int J River Basin Manag 14(3):357–366

Abdullah AD, Masih I, van der Zaag P, Karim UF, Popescu I, Al Suhail Q (2015) Shatt al Arab River system under escalating pressure: a preliminary exploration of the issues and options for mitigation. Int J River Basin Manag 13(2):215–227

Abdullah AD, Popescu I, Dastgheib A, van der Zaag P, Masih I, Karim UF (2017) Analysis of possible actions to manage the longitudinal changes of water salinity in a tidal river. Water Resour Manag 31(7):2157–2171

Alaidani MAT (2014) The change in the geographic and agricultural properties impacts in the province of Basra, MSc. Thesis, Basrah University. College of Education Sciences

Al-Faisal AJM, Mutlak FM, Abdullah SA (2014) Exotic freshwater fishes in the southern Iraq. Marsh Bulletin 9(1):65–78

Al-Hassan LAJ (1994) The silver carp, *Hypophthalmichthys molitrix* (Cyprinidae) in the Shatt al Arab River, Iraq. Cybium 18(2):204

Al-Hassan LAJ, Al-Badri MEH (1986) First record of some fishes from Khor al-Zubair, Khor Abdullah and Shatt al-Arab, Basrah, Iraq. Cybium 10(3):295–297

Al-Hassan LAJ, Hussain NA (1985) Hydrological parameters influencing the penetration of Arabian gulf fishes into the Shatt Al Arab River, Iraq. Cybium 9(1):7–16

Al-Hassan LAJ, Miller PJ (1987) *Rhinogobius brunneus* (Gobiidae) in the Arabian gulf. Japanese J Ichthyology 33(4):405–408

Al-Hassan LAJ, Muhsin KA (1986) The presence of *Barbus luteus* and *Heteropneustes fossilis* in Khor al-Zubair, north-west of the Arabian gulf. Zool Middle East 1:116–118

Al-Hassan LAJ, Naama A (1986) New record for some Arab gulf fishes in the freshwater systems of Iraq. Bulletin of Basrah Natural History Museum 6:35–44

Al-Hassan LA, Soud KD (1989) A preliminary, annotated check-list of the fishes of. Pol Arch Hydrobiol 36(2):283–288

Al-Hassan LAJ, Hussain NA, Soud KD (1989) A preliminary annotated check-list of the fishes of Shatt AlArab River River, Basrah, Iraq. Pol Arch Hydrobiol J 36:283–288

Al-Lamy JH, Resen AK, Taher MM, Muhammed SJ (2012) Specialization and diet overlapped between two alien species (Carassius auratus and Tilapia zillii) and native species (Barbus luteus) in east Hammar marsh, southern Iraq. J King Abdulaziz Univ Marine Sci 23(2):25–37

AL-Mahmood H, Hassan W, AL-Hello A, Hammood O, Muhson N (2015) Impact of low discharge and drought on the water quality of the Shatt al-Arab and Shatt al-Basra Rivers (south of Iraq). J Int Academic Res Multidisciplinary 3(1):285–296

Al-Nasiri SK, Hoda SMS (1975) Survey of fish fauna of Shatt al-Arab (from Abu al-Khasib to Karmat Ali). Bulletin of the Basrah Natural History Museum 2:36–46

Al-Nasiri SK, Hoda SMS (1976) A guide to the freshwater fishes of Iraq. Basrah Natural History Museum Publication No. 1: 124pp

AL-Shamary AC (2016) Effect of alien species on assemblage fishes in Shatt Al-Arab River and east Al-Hammar marsh. Mesopotamia Environ J Special Issue A:18–32

Al-Tawash B, Al-Lafta SH, Merkel B (2013) Preliminary assessment of the Shatt al Arab riverine environment, Basra government, southern Iraq. J Nat Sci Res 3(13):120–136

Amjed KR, Jabbar KA, Roaa HA (2016) Some environmental factors and their impact on the fish community composition in the Shatt al-Arab area Shafi, Basrah. *Magazine of Al-Kufa University for Biology*, (العدد الخاص بالمؤتمر العلمي الدولي الثاني لعلوم كلية/الحياة التربية جامعة/للبنات /2016الكوفة), pp.66–75

Berrebi P, Tsigenopoulos CS (2003) Phylogenetic organization of the genus Barbus sensu stricto: a review based on data obtained using molecular markers. In: The freshwater fishes of Europe, vol 5, pp 11–22

Borkenhagen K (2014) A new genus and species of cyprinid fish (Actinopterygii, Cyprinidae) from the Arabian peninsula, and its phylogenetic and zoogeographic affinities. Environ Biol Fish 97 (10):1179–1195

Borkenhagen K (2017) Molecular phylogeny of the tribe Torini Karaman, 1971 (Actinopterygii: Cypriniformes) from the Middle East and North Africa. Zootaxa 4236(2):zootaxa–zoo4236

Borkenhagen K, Krupp F (2013) Taxonomic revision of the genus Carasobarbus Karaman, 1971 (Actinopterygii, Cyprinidae). ZooKeys 339:1

Cai W-J, Hu X, Huang W-J, Murrell MC, Lehrter JC, Lohrenz SE, Chou W-C, Zhai W, Hollibaugh JT, Wang Y, Zhao P, Guo X, Gundersen K, Dai M, Gong G-C (2011) Acidification of subsurface coastal waters enhanced by eutrophication. Nat Geosci 4:766–770

Chen W-J, Mayden RL (2009) Molecular systematics of the Cyprinoidea (Teleostei: Cypriniformes), the world's largest clade of freshwater fishes: further evidence from six nuclear genes. Mol Phylogenet Evol 52:544–549

Coad BW (2010) Freshwater fishes of Iraq. Pensoft series Faunistica no. 93, 274 pp. Pensoft Publishers, Moscow

Coad BW, Hussain NA (2007) First record of the exotic species *Hemiculter leucisculus* (Actinopterygii: Cyprinidae) in Iraq. Zoology in the Middle East 40:107–109

Dunz AR, Schliewen UK (2013) Molecular phylogeny and revised classification of the haplotilapiine cichlid fishes formerly referred to as "tilapia". Mol Phylogenet Evol 68(1):64–80

Durand JD, Chen WJ, Shen KN, Fu C, Borsa P (2012) Genus-level taxonomic changes implied by the mitochondrial phylogeny of grey mullets (Teleostei: Mugilidae). C R Biol 335 (10–11):687–697

Freyhof J, Özuluğ MÜFİT, Saç GÜLŞAH (2017) Neotype designation of Aphanius iconii, first reviser action to stabilise the usage of a. fontinalis and A. meridionalis and comments on the family group names of fishes placed in Cyprinodontidae (Teleostei: Cyprinodontiformes). Zootaxa 4294(5):573–585

Froese R, Torres A, (1999) Fishes under threat: an analysis of the fishes in the 1996 IUCN Red List. In *ICLARM Conference Proceedings (Philippines)*. ICLARM/FAO

Hameed HA, Ali MH, Aljorany YS, Hassan WF, Al-Hello AAZN, (2013) Assessing changes in seawater intrusion and water quality of the Shatt Al-Arab River, Iraq. In *Annales de Limnologie-International Journal of Limnology* (Vol. 49, No. 3, pp. 199–206). EDP Sciences

Hameed AH, Aljorany YS (2011) Investigation on nutrient behavior along Shatt Al-Arab River, Basrah, Iraq. J Appl Sci Res 7:1340–1345

Hasselquist F (1762) Reise nach Palästina in dem Jahren 1749 bis 1762. Herausgegeben von C. Linnaeus. Aus dem Schwedischen [übersetzt von T. H. Gadebusch]. Rostok. Reise nach Palästina in dem Jahren 1749 bis 1762.: i-xviii + 1-606

Heckel JJ (1843) Abbildungen und Beschreibungen der Fische Syriens nebst einer neuen Classification und Characteristik sämmtlicher Gattungen der Cyprinen. Stuttgart, 109 pp.

Hussein SA, Al-Daham NK, Al-Kanaani SM (2000) Dietary overlap between the common carp *Cyprinus carpio* L. and three native cyprinids in Al-Hammar Lake. Southern Iraq J Basrah Res 25:73–83

Jawad LA (2003) Impact of environmental changes on the freshwater fish fauna of Iraq. Int J Environ Stud 60(6):581–593

Jawad LA (2006) Fishing gear and methods of the lower Mesopotamian plain with reference to fisheries management. Marina Mesopotamica 1(1):1–39

Kennedy WP (1937) Some additions to the fauna of Iraq. J Bombay Nat Hist Soc 39:745–749

Khalaf KT (1961) The marine and freshwater fishes of Iraq. Al-Rabitta Press, Baghdad, p 164

Khamees NR, Ali AH, Abed JM, Adday TK (2013) First record of striped catfish *Pangasianodon hypophthalmus* (Sauvage, 1878) (Pisces: Pangasiidae) from inland waters of Iraq. Basrah J Agricultural Sci 26(1):178–183

Landsberger B (1962) The fauna of ancient Mesopotamia. MSL VIII/2, second part. Pontificium Institutum Biblicum, Roma 204, Piazza Pilotta 35. Publié avec le concourse financier de

l'UNESCO et sous less auspices du Conseil International de la Philosophie et des Sciences Humaines, 180pp
Lazem LF (2009) *Ecological evaluation of the Shatt Al-Arab River and fish assemblage by applying geographical information system (GIS)* (Doctoral dissertation, College of Agriculture, University of Basrah). 131pp
Mahdi N (1962) Fishes of Iraq. Ministry of Education, Baghdad, p 82
Mahdi N, George PV (1969) A systematic list of the vertebrates of Iraq. Iraq Natural History Museum Publication, Baghdad 26: 1–104. Maitland, 1994;
Mohamed ARM, Abood AN (2017a) Compositional change in fish assemblage structure in the Shatt Al-Arab River, Iraq. Asian J Appl Sci *(ISSN: 2321–0893)*, *5*(05)
Mohamed ARM, Abood AN (2017b) Dispersal of the exotic fish in the Shatt Al-Arab River, Iraq. J Agriculture Veterinary Sci 10:50–57
Mohamed ARM, Abood AN (2017c) Ecological health assessment of the Shatt Al-Arab River, Iraq. J Agriculture Veterinary Sci 10:1–8
Mohamed ARM, Al-Noor SS, Faris RA (2008) The status of artisanal fisheries in the lower reaches of Mesopotamian rivers, North Basrah, Iraq. Proc 5th Int Con Biol Sci(Zool) 5:126–132
Mohamed ARM, Hussain NA, Ali TS (2001) Estuarine components of the ichthyofauna of the Arabian gulf. Marina Mesopotamica 16(2):209–224
Mohamed ARM, Hussein SA, Lazem LF (2010) Ecological traits on fish assemblage in the Garma River using Canoca program. Basrah J Sci 28(2):92–106
Mohamed ARM, Hussein SA, Lazim LF (2013) Spatiotemporal variability of fish assemblage in the Shatt Al-Arab River, Iraq. Basrah J Agricultural Sci 26(1):43–59
Mohamed ARM, Lazem LF, Sadek A, Hussein SA (2017b) Assessment of water quality and fish species dominance in the Shatt Al-Arab River by GIS technique. J Sci Engg Res 4(8):213–220
Mohamed ARM, Resen AK, Taher MM (2012) Longitudinal patterns of fish community structure in the Shatt Al-Arab River, Iraq. Basrah J Sci 30.(2B Environmental Conference):65–86
Mohamed ARM, Younis KH, Hameed EK (2017a) Status of fish assemblage structure in the Garmat Ali River, Iraq. Global J Biol Agriculture Health Sci 10(2):17–22
Mohammed ARM, Al-Hassan LAJ, Ali TS (1993) The presence of a cyprinid fish, *Barbus luteus* in marine waters of Iraq. Arquivos do Museu Bocage 2(25):415–416
Moyel MS, Hussain NA (2015) Water quality assessment of the Shatt al-Arab river, southern Iraq. J Coastal Life Med 3(6):459–465
Mutlak F, Jawad L, Al-Faisal A (2017) Atractosteus spatula (Actinopterygii: Lepisosteiformes: Lepisosteidae): a deliberate aquarium trade introduction incidence in the Shatt Al-Arab River, Basrah, Iraq. Acta Ichthyol Piscat 47(2):205
Myers N, Mittermeier RA, Mittermeier CG, da Fonseca GAB, Kent J (2000) Biodiversity hotspots for conservation priorities. Nature 403:853–858
Nasir NA, Khalid SA (2017) Fluctuations in the freshwater fish catch of the Basrah province, Iraq during the period from 2005 to 2016. Mesopotamia Environ J 3:15–24
Nyman L (1991) Conservation of freshwater fish: protection of biodiversity and genetic variability in aquatic ecosystems. Fish Devel Series, Gothenburg 50:1–38
Parmesan C (2006) Ecological and evolutionary responses to recent climate change. Annual Review of Ecology and Evolution 37:637–669
Partow H (2001) *The Mesopotamian marshlands: demise of an ecosystem*
Postgate JN (1994) *Text and figure in ancient Mesopotamia: match and mismatch*. Cambridge University Press
Peterson AT, Soberón J, Sánchez-Cordero V (1999) Conservatism of ecological niches in evolutionary time. Science 285(5431):1265–1267
Saggs HWF (1962) The greatness that was Babylon: a sketch of the ancient civilization of the Tigris-Euphrates valley. Hawthorn Books
Stout CC, Tan M, Lemmon AR, Lemmon EM, Armbruster JW (2016) Resolving Cypriniformes relationships using an anchored enrichment approach. BMC Evol Biol 16(1):244

Yang L, Sado T, Hirt MV, Pasco-Viel E, Arunachalam M, Li J, Wang X, Freyhof J, Saitoh K, Simons AM, Miya M (2015) Phylogeny and polyploidy: resolving the classification of cyprinine fishes (Teleostei: Cypriniformes). Mol Phylogenet Evol 85:97–116

Yaseen AT (2016) Effect of some environmental factors on the nature of fish assemblage in stream and estuary of Shatt Al-Arab. A Thesis submitted to the council of the College of Agriculture, The University of Tikrit in partial fulfilment of the requirements for the degree of Master in Agricultural Sciences (Animal Production). 132pp

Yaseen AT, Wahab NK, Abdul Kareem Taher Yesser AKT (2017) Environmental indicators and fish species structure of the Shatt Al-Arab River, Abu Al-Khaseeb, Basra- Iraq. Journal of College of Science, University of Tikrit 17:137–163

Younis KH, Hussain NA, Mohamed ARM (2010) Ecological assessment of fish assemblage in Shatt al-Arab River, Karmat Ali, Basrah using integrated biological index (IBI). Journal of University of Karbala, special issue on the 1st Agriculture Conference of College of Agriculture, 2010

Chapter 37
The Common Carp, *Cyprinus carpio*: Effect on the Environment and the Indigenous Fish Species in Iraq

Laith A. Jawad, Mustafa S. Faddagh Ziyadi, and Abbas J. Al-Faisal

Abstract The common carp (*Cyprinus carpio*) is an extremely invasive species and an ecological destructive. It has been frequently shown to upturn nutrient concentrations and phytoplankton biomass while destroying submerged macrophytes. The effects of the common carp on both the environment and the biota of the freshwater system have discussed briefly in the following pages. The review of the status of C. carpio in the inland Iraqi waters is given in this chapter with emphasis on the population size, reproduction and growth. The results of studies performed on this species about these aspects have shown that the populations of the common carp have flourished in the marsh areas better than the lakes and the main rivers. Such differences in the increase in the population of the common carp could be due to the differences in the environmental and the biological factors between regions.

37.1 Introduction

Due to its impact on the aquatic environment, the common carp (Cyprinus carpio) is considered one of the world's damaging invasive species, which is widely distributed around the world (Florian et al. 2016). This cyprinid species has tough direct consequences on the communities of aquatic invertebrates, fish and waterbirds through predation or competition (Weber and Brown 2009). Additionally, this species of carp is considered ecosystem engineers causing bioturbation and high turbidity through their feeding behaviour (Bajer et al. 2009; Kloskowski 2011). With resulting high turbidity, the cover of submerged vegetation will be reduced and a shift from a clear to turbid state in shallow lakes, which results in a reduction in biodiversity (Bajer et al. 2009). To recover any freshwater body inhabited by this

L. A. Jawad (✉)
School of Environmental and Animal Sciences, Unitec Institute of Technology, Auckland, New Zealand

M. S. Faddagh Ziyadi · A. J. Al-Faisal
Marine Science Centre, University of Basrah, Basrah, Iraq

L. A. Jawad (ed.), *Tigris and Euphrates Rivers: Their Environment from Headwaters to Mouth*, Aquatic Ecology Series 11,
https://doi.org/10.1007/978-3-030-57570-0_37

species needs removal of this invasive species, which result in a recovery of macrophytes together with a drop in turbidity, nutrient concentrations and phytoplankton biomass (Bajer and Sorensen 2015).

The common carp is omnivorous, feeding largely on macrophytes and invertebrates, and up to 25% of the biomass ingested consists of zooplankton (Meijer et al. 1990a, b; Khan 2003; Britton et al. 2007). Therefore, the effect on zooplankton community is eminent by several ways such as: by direct predation (Miller and Crowl 2006), by consuming macroinvertebrates that themselves are zooplankton predators (Khan 2003), through loss of macrophytes that provide shelter, and by increasing phytoplankton biomass and promoting cyanobacterial blooms (Parkos et al. 2003). In addition, the bottom feeding habit of the common carp can cause resuspension of sediment particles, which in turn interfere with the filtering apparatus of cladocerans (Kirk and Gilbert 1990), and bioturbation may also affect the inactive stages in sediments, which has negative effects on emergence patterns (Angeler et al. 2002).

In the warm areas such as the Middle East and the Mediterranean region, the impacts of carp appear to be tougher than in temperate latitudes because relatively high temperatures all year round increase the levels of fish activity. In spite of such information, evidence on how the common carp can affect the zooplankton is not available. Several studies have shown that common carp has a significant effect on zooplankton in part of the semi-arid and temperate lakes (Angeler et al. 2002; Williams and Moss 2003), but a need for integrated studies to be performed on the whole ecosystem scale.

The aim of this chapter is to investigate the changes that may happen in the zooplankton communities due to the presence of the common carp in the freshwater body system. Also, the effects of this species on the diversity and species richness, and how these changes were related to macrophyte cover, phytoplankton abundance. At the end of the chapter, a recommendation about how to deal with such a problem in the Iraqi freshwater system are given in order to contend such a problem.

37.2 The Effects of the Common Carp on the Environmental Factors in the Aquatic Ecosystem

The process of bringing nutrients from the bottom of the aquatic ecosystem to the top is known as 'Bottom-up processes' (McQueen et al. 1986). Fishes in general play a vital role in this process through multiple pathways (Vanni 1996). For the common carp, it links benthic and pelagic food webs through benthic foraging activities and subsequent excretion (Lamarra 1975). It has been found that common carp can increase water column nutrients (Weber and Brown 2009) and may also increase water column phosphorus, nitrogen and ammonia as a result of benthic foraging activities (Parkos et al. 2003), excretion (Lamarra 1975; Qin and Threlkeld 1990) or destruction and subsequent decomposition of aquatic macrophytes (Carpenter and Lodge 1986). Common carp may also interact in collaboration with external nutrient loading to further improve the eutrophication process (Drenner et al. 1998). With the increase in number of individuals of the common carp biomass, this may lead to a proportionately greater effect in water column nutrients and facilitate eutrophication processes (Chumchal et al. 2005).

The bottom feeder fish species may increase water column nutrients and phytoplankton through their excretion (Chumchal and Drenner 2004) and nutrient flux (Carpenter et al. 1992). In the low-productivity freshwater ecosystem, nutrient mobilization may be beneficial as this process will increase the primary productivity. Common carp are among the factors that cause increasing eutrophication of shallow lakes (Chumchal et al. 2005). At least 50% of the phosphorus excreted by common carp may be readily available for phytoplankton production (Lamarra 1975), indicating a potential direct pathway. Food and feeding habits of the common carp can vary due to several causes, and such variations will have a direct effect on the water column nutrients, but the small common carp (<100 cm) that consume primarily zooplankton instead of benthic invertebrates may not have an effect on nutrients (Meijer et al. 1990a, b). On the other hand, the activity of the common carp in mobilizing the nutrients depends on the type of the substrate. The hard substrate can hamper the benthic foraging behaviours of the common carp (Roberts et al. 1995), and this is also true for the sediments composed of large, dense particles (i.e. sand, gravel, cobble), which are not resuspended by common carp foraging behaviours. Because common carp do not increase water column nutrients in systems with hard substrates, bioturbation, not excretion, is likely the major driver behind elevated water column nutrients.

The turbidity in small or shallow ecosystem can be due to the abundance of bottom feeder fish species. (Barton et al. 2000). It has been known that increased turbidity has numerous harmful effects on aquatic ecosystems, including reduced light penetration, primary productivity, aquatic macrophyte growth (Sidorkewicj et al. 1996), zooplankton filtering efficiency (Kirk 1991) and foraging abilities of visual planktivorous (Miner and Stein 1996) and piscivorous fishes (De Robertis et al. 2003). The common carp is one of the main factors in increasing turbidity in the aquatic system and causing resuspended solids (Lammens 1991) and increase turbidity (Chumchal et al. 2005) by search for food and expel non-food items. It has been noted that the large individuals of the common carp have the ability to penetrate up to 12 cm into the substrate while searching for food (Panek 1987). In doing so, such individuals can disturb more than the sediment surface layer and resuspending fine particles.

The effect of mobilization of nutrients in the ecosystem by the common carp can influence the production of the phytoplankton. With the increase in turbidity due to disturbances of the substrate, a shift in the phytoplankton species composition happened. A change from one dominated by chlorophyta (green algae), an important energy source for higher trophic levels, to noxious and often toxic cyanobacteria (blue-green algae) (Williams and Moss 2003) that more commonly compete with macrophytes for nutrients and light can occur.

37.3 The Effects of the Common Carp on the Invertebrate Communities in the Aquatic Ecosystem

The planktivorous fishes have a vital role in structuring the aquatic ecosystems from top to bottom of the freshwater ecosystem. Such ability will be performed by reducing invertebrate communities through predation, which in turn releases algal communities from predatory control by zooplankton (Carpenter et al. 1985). Common carp activity is linked with increases and decreases in zooplankton diversity, size structure and abundance.

Reductions (Lougheed et al. 1998) and increases in zooplankton biomass (Parkos et al. 2003) are associated with the growth of common carp populations. Moreover, size structure of zooplankton communities generally changes from large-bodied to small-bodied taxa in the presence of common carp.

With the variation in the growth of common carp population, variable effects of common carp on zooplankton communities may occur. Indirectly, adult common carp populations may show an effect on zooplankton abundance and size structure through nutrient mobilization or a reduction in zooplankton grazing capabilities through increased turbidity (Parkos et al. 2003). There are two possibilities for increasing the zooplankton biomass. It is either due to the increase phytoplankton biomass (i.e. prey resources) (Khan et al. 2003) or by a reduction in the abundance of benthic invertebrates (Miller and Crowl 2006) that may prey on zooplankton (Johnson and Crowly 1980).

Adults common carp can cause an increase in inorganic turbidity, and they can inhibit ingestion of phytoplankton by large-bodied zooplankton (Kirk 1991), thereby reducing the abundance of these large-bodied grazers. With such changes in the zooplankton size structure from large- to small-bodied species, an effect could happen on phytoplankton density (McQueen et al. 1986). In the balanced freshwater ecosystem, large-bodied zooplankton can regulate phytoplankton biomass (Meijer et al. 1990a, b), whereas small-bodied zooplankton taxa are ineffective grazers. This shift in zooplankton size structure is one of the factors that cause a switch in shallow lakes from a clear to turbid water state (Scheffer et al. 1993). In a direct way and in the early stages of the life of common carp, zooplankton can receive a significant effect as the young of this species of carp ($<$100 mm) feed on zooplankton (Britton et al. 2007).

The bottom feeding behaviour of the common carp, which include foraging on detritus, annelids, chironomids, amphipods and odonates around 100 mm TL (Britton et al. 2007) may reduce benthic invertebrate abundance, diversity, evenness and richness (Stewart and Downing 2008). Change in the population structure of the benthic invertebrate may happen towards large-bodied species due to selective predation by common carp on small-bodied prey (Covich and Knezevic 1978). Annelids, chironomids and odonates face large reductions even at low common carp densities (Zambrano et al. 1999). With the effect on the macrophytes and the periphytic community and their reduction, common carp may affect indirectly the abundance of Decapoda (Hinojosa-Garro and Zambrano 2004). In regard to the native fish species, common carp may be more efficient at consuming benthic invertebrates (Parkos et al. 2003) and therefore enter in completion with those species. At a certain period of time, the water of a freshwater ecosystem will become turbid. This is because the benthic invertebrate populations decline due to the predation of the common carp predation and in turn, common carp increase foraging time and expel additional sediment into the water column (Werner and Anholt 1993) causing the turbid water state.

Aquatic macrophytes are important to the clear water-stable state due to their capability to stabilize sediment, compete with phytoplankton for light and nutrients, provide invertebrates a food resource and refugia, and habitat for fishes (Scheffer et al. 1993; Hanson and Butler 1994). With the increase in turbidity, aquatic macrophytes generally disappear (Lougheed et al. 1998) as a result of increasing wind-driven sediment resuspension (Scheffer 1998). The increase in the biomass of common carp will lead to the reduction of aquatic macrophyte diversity and abundance (Matsuzaki et al. 2009). The common carp also destroy aquatic macrophytes on hard substrates through collisions and increased algal growth that competes with aquatic macrophytes for light and nutrients (Miller and Provenza 2007). On contrary and on soft substrates, common carp directly dislodge aquatic macrophyte roots from the sediment (Parkos et al. 2003) and increase sediment turbidity, nutrient release and algal biomass, attenuating light needed for aquatic macrophyte growth (Skubinna et al. 1995; Sidorkewicj et al. 1999).

In the shallow water system where light penetrates down to the substrate, macrophytes can survive in the presence of the common carp (Lougheed et al. 1998). Furthermore, aquatic macrophyte taxa vary in their susceptibility to common carp uneasiness (Evelsizer and Turner 2006). Aspects such as root system, soil type preference, shade tolerance, tissue strength, and timing of seed dispersal may have a possible influence on macrophyte vulnerability. Physical damage to macrophytes by common carp can be seen in the uprooted of a weak aquatic macrophytes from loose and fine substrate (Parkos et al. 2003), and damaging the soft tissues of aquatic macrophytes in collisions (Zambrano and Hinojosa 1999). The physical damaging has more effects on submerging species than emergent species because they generally have weaker root systems and are more influenced by turbidity (Roberts et al. 1995). The aquatic macrophytes with deeper or stronger root systems can also be eliminated and removed due to an increased turbidity from foraging-induced sediment resuspension.

37.4 The Effects of the Common Carp on the Native Fish Species in the Aquatic Ecosystem

There are several ways that the common carp can affect the life of the native fish species inhabiting the same environment. With the highly eutrophic ecosystems and experience population increases with increasing lake productivity, for example, common carp are more adapted than the native fish and the populations of the later may decline under similar conditions (Egertson and Downing 2004). Common carp may damage spawning habitat or upset spawning behaviour (Panek 1987) and have been shown to eat eggs of some fishes (Miller and Beckman 1996). Fish growth rates are sometimes decreased in the presence of common carp. Examples on such results have seen in the case of several fish species (Wolfe et al. 2006; Egertson and Downing 2004). Fish species that can acclimatize to conditions linked with the presence of common carp may not be affected, whereas those with specific niches for biotic and abiotic conditions may experience reductions in growth or survival, particularly during susceptible early life stages.

37.5 The Status of the Common Carp in the Freshwater System in Iraq

The common carp, Cyprinus carpio, was introduced into the Iraqi inland waters during the period 1960–1972 (Jawad 2003, 2006, 2013). When the number of this species has increased dramatically, it started to be a real danger on the native fish species. By the mid-1980s, the common carp has established very well in the freshwater systems of Iraq and succeeded in eliminating several native fish species such as the Mesopotamichthyes sharpeyi, *Luciobarbus xanthopterus* and *Arabibarbus grypus* (Jawad 2003, 2006, 2013). Several biological features that the common carp has and numerous other environmental factors have assist in the successful of the common carp in the Iraqi freshwater system.

In spite of such a great attainment of this species in Iraqi waters, the studies that dealt with its impact on both the biota and the environment of the freshwater system were very scarce (Jawad 2003). In addition, the studies that are related to the biology of this species, which include detecting the growth and distribution of common carp, were sporadic and did not follow any time interval or any geographical coverage. The studies were mainly performed in the middle and southern of Iraq and have covered certain topics that are related to the stock assessment of the common carp such as reproduction, food, feeding habits and food interaction with the native species, age and growth and population studies. On the other hand, the literature archive of the research on common carp is full of studies that deal mainly on the effects of different pollutants on the common carp, parasitology and basic biology, which contribute less to the issue of controlling the common carp in Iraq.

To know the status of the common carp in the inland waters of Iraq, a short review to the available literature should be given. As mentioned above, the studies published on the common carp that one can come up with an idea about the population status in Iraq are focused on specific subjects, which the following review will deal with it in this section.

37.5.1 Common Carp Composition and Distribution in the Inland Waters of Iraq

In any of the population studies that have been performed on the Iraqi inland waters, *C. carpio* was represented. The available literature for this short review covers the period 2003–2006 and, geographically, covers the north, middle and south of Iraq. There was only one study from North of Iraq and that related to the Dukan Dam Lake, while there are few for the middle and southern reaches of Euphrates and Tigris Rivers. As to the lake, there were five studies that concerned with population study of the Habbaniyah Lake, al-Tharthar Lake (Shakir et al. 2015), Hamrin Dam Lake (Lazem and Attee 2016) and Al-Rathwania Lake (Al-Rudaini et al. 1998; Al-Rudaini 2010). Since the data on the distribution and composition of the common carp in the fish community in the freshwater of Iraq derive from different freshwater ecosystems, i.e. lake, river and marsh, the following review will be constructed on each ecosystem separately.

37.5.2 Lakes

The lakes that studies have been performed in are located in the north of Iraq (Dukan dam Lake on one of the tributaries of Tigris River), Middle part of Iraq and on the Euphrates River, Habbaniyah Lake and Tharthar Lake), on the Tigris River, Al-Rathwania Lake.

37.5.2.1 Al-Rathwania Lake

The earliest study of Al-Rudaini et al. (1998) has shown that the total catch of fish samples was 3884 individuals with a total weight of 1179.79 Kg for a period of 10 months. Out of this number, individuals of *C. carpio* collected were 105 (2.7%) and with total weight of 6.27 Kg (3.5%) of the total catch. During the period of 2000–2001, Al-Rudaini (2010) has shown that the number of individuals that collected from the same lake within the same period of time (10 months) has increased nearly 17 times. The total number of the individuals of *C. carpio* collected was 1758 (36.43%) with total weight of 140.68 Kg (11.42%) out of the total number of fish individuals of 4826 and total weight of 1230.74 Kg. On the other hand, the total number of individuals of the family Cyprinidae other than common carp was 3796 (78.65%) with total weight 1180.58 (95.92%). The individuals of *C. carpio* represent 46.31% with total weight of 11.9% of the same data obtained for the members of the family Cyprinidae.

Habbaniyah Lake

Al-Rudaini et al. (1999) have shown that the number of common carp individuals collected for the period of 12 months in 1997 is 436 (9.46%) with a total weight of 175.24 Kg (43.26%) of the total catch of 4609 and total weight of 410.73 Kg. During this study, the catch of *C. carpio* and *Carassius auratus* was the highest for the catch being made in the period of the study. Also, *C. carpio* was present in all months of the study.

Epler et al. (2001) have collected 134 individuals of common carp (124 and 10 individuals by gill net and trawling respectively) (0.9%) of the total catch. It is clear that there is a significant reduction in the population of the common carp in comparison with the results of Al-Rudaini et al. (1999). Such great decrease in the population within 2–3 years could be explained on the basis that Epler et al. (2001) study was only for 6 months in comparison with that of Al-Rudaini et al. (1999), which last for 12 months.

Comparing the number of individuals and the biomass in Al-Habbaniyah Lake with those in Al-Rathwania Lake for the 1999 period of time, it appears that population of the common carp in the former lake has flourished better than in the latter lake.

Tharthar Lake

In 2001, Epler et al. (2001) have performed an ichthyological investigation for 6 months, but no specimens of common carp were collected. Shakir et al. (2015) have accomplished a study on the ichthyofauna of Al-Tharthar Lake and collected 116 individuals of *C. carpio* (23.6%) with total weight of 80.21 Kg (39.38%) of the total catch. Among the 15 fish species collected in their study, Shakir et al. (2015) have shown that the common carp was present in the catch of every month for the 12-month period of study. Also, the number of individuals of this species was higher than of the 15 species collected. This result is also true for the total weight obtained for the common carp.

Razzazah Lake

Except for the work of Epler et al. (2001), no population study has been performed on the fish fauna of Razzazah Lake. In this study, a total of 9828 fish individuals of 16 species were collected during the period of 6 months. The total number of individuals of common carp was 77 (0.8%) of the total catch.

Hamrin Dam Lake

The only ichthyological survey performed in Hamrin Dam Lake is that of Lazem and Attee (2016). In this study, 12-month collection was accomplished in which the relative abundance of the common carp range 9.54% in summer to 16.67% in winter of the period of the study. With the 19 species studied, C. carpio was in the third place abundance wise after the *Arabibarus grypus* and the goldfish *Carassius auratus*.

Dukan Dam Lake

So far, the only report on the population study of the common carp from north of Iraq is Sediq and Abbas (2013). This study was performed in the period 2007–2008 for 12 months. There were 27 fish species encountered in this study, with total individuals collected of 3006 and total weight of 678.5 Kg. The common carp was not well represented in this lake, where only 237 individuals (7.9%) collected with biomass of 104.7 Kg (15.4%) of the total catch.

37.5.2.2 Euphrates River

The available data on the populations of *C. carpio* of the Euphrates Rivers are only those from the middle and southern reaches of the river. In order to show the fluctuations in the number of individuals and the biomass of the common carp collected from the Euphrates River, the following information is based on the region of the river.

Middle Region

The data on the number of individuals and the biomass of the common carp were obtained from 4 localities in the middle reaches of the Euphrates River, and these are Al-Mussiab area (AI-Rudainy et al. 2006), Al-Hindia area (Al-Amari et al., personal communication; Al-Salman et al. 2014) and Hilla City locality (Al-Amari et al., personal communication).

For the period of 11 years, there was a study increase in the number of individuals of the populations of the common carp collected from the 3 localities at the middle reaches of the Euphrates River. The result of AI-Rudainy et al. (2006) in Al-Mussiab area has shown that the presence of C. carpio formed 11.3% of the total number of individuals collected with total weight of 421.74 (27.2%). During the period of 2010 and 2014, Al-Amari et al., personal communication and Al-Salman et al. (2014) have shown an increase in the number of the individuals collected from Al-Hindia area compared with the result of AI-Rudainy et al. (2006). The catch of individuals from 13.4% and 16% by Al-Amari et al., personal communication and Salman et al. (2014) respectively. The percentage has decreased dramatically to 0.77% at Hilla area in 2010 (Al-Amari et al., personal communication). Such significant drop can be explained on the basis of the short time of survey and probably to the differences in the environmental and the biological factors in Al-Hilla area.

Southern Region

The number of studies performed on the southern reaches of the Euphrates River is very scarce. Hussein et al. (2015) reported very low presence of the common carp with 0.15% of the total catch. On the other hand, Mohamed et al. (2017) showed the relative abundance of *C. carpio* is 1.68% of the total catch of the river for 12 months of study from Al-Diwania River. Mohamed et al. (2017) in their study on the lower reaches of Euphrates River at Qarmat Ali area near Basrah City reported a very low relative abundance of the common carp (0.04).

37.5.2.3 Tigris River

Mohamed et al. (2006) studied the population of the common carp in the Tigris River. They investigated the fish assemblage at the lower reaches of the Tigris River at City of Qurna. They have shown that the relative abundance of the common carp is 2.36% of the total catch of the river for 12 months of study. With one study available, it is not possible to deduce the status of the population of *C. carpio* in the Tigris River.

At the middle region of the Tigris River, Al-Shawi and Wahab (2007) have investigated the fish population in Tuz Jae a tributary of Al-Audhaim River and both are tributary of the Tigris River. They collected 211 individuals of common carp (10.18%), with 60.120 KG total weight (16.32%) of the total catch. In the same region, Wahab and Hassan (2011) have studied this species showed that the individuals of *C. carpio* 65 (1.55%) and 38.8 Kg (7.29%) of the total catch.

37.5.2.4 Shatt al-Arab River

With the start of the flourishing distribution of the common carp in the inland waters of Iraq in the late 1990s, the number of individuals of this species was low as 87 (0.6%) (Hussain et al. 1997) and 26 (0.31%). After nearly two decades, the relative abundance of C. carpio in the Shatt al-Arab River has increased to reach 3.0 and 3.1 reported by Mohamed et al. (2012) and Mohamed et al. (2013), respectively.

37.5.2.5 The Marshes

The studies that took into consideration the populations of C. carpio in the southern marshes of Iraq are scarce. Hussain et al. (2009) reported a relative abundance of 2.9 of this species from Al-Hammar marsh, northwest of Basrah City, while Al-Shamary et al. (2011) have collected 162 individuals for 12 months of study in the same marsh with 3.17% of the total catch. Further to the east at Al-Huwaiza Marsh, Mohamed et al. (2008) gave very low value for the relative abundance of the common carp of 0.72.

37.5.2.6 Shatt al-Basrah Canal

Shatt al-Basrah Canal was dredged in early 1970, but start operating in 1983, to connect the lower reaches of Euphrates River near the Al-Hammar Marsh with Khor al-Zubair, North West of the Arabian Gulf. This canal is under a tidal effect with saltwater intruded during the tide (Al-Daham and Yousif 1990).

An ichthyological survey has been performed in the Shatt al-Basrah Canal, and specimens of C. carpio were collected by Al-Daham and Yousif (1990). The total number of individuals obtained was 26 (0.31%), with total weight of 7539.7 Kg (3.33%) of the total catch. The specimens of C. carpio were only found in the collecting station close to the Euphrates River. No individuals were seen in the middle and southern collecting stations where salinity is high.

37.5.3 Reproduction

So far, there are only three studies on the reproduction of the common carp, two of these were performed in the Middle region of Iraq, while the third study was accomplished at Al-Hammar Marsh, south of Iraq. Hamady (2009) determined the sex ratio of *C. carpio* from AL-Gharraf Canal at Wasit province, Iraq, as 1:0.09 in favourite of males. Male and female fish mature at 213 and 215 mm total length, respectively. Hamady (2010) has calculated the relative fecundity as 177 egg/g.

From the southern part of Iraq and at Al-Hammar marsh, Abdul Retha et al. (2009) have calculated the GIS of the common carp as 0.96–2.20.

37.5.4 Condition Factor

As with the studies on other biological parameters, the condition factor of the common carp has not been investigated very well in Iraq. The only available studies are those for fish samples collected from the Tigris River tributaries in the north and middle of Iraq. At Darbandikhan Dam Lake, the value of condition factor ‘K’ was reported by Rasheed (2012) to be 0.71, while the value has increased dramatically and reached 1.46 in 2012 (Wahab and al-Ani 2012) and range 1.120–1.880 in the eastern Drainage in Balad City (Wahab and Al-Ani 2013).

37.5.5 General Conclusion

In general, the population of the common carp has increased significantly since this species has been introduced to Iraq in the period 1960–1972. Based on the above short review, it is clear that the lakes in Iraq have smaller populations of C. *carpio* than the main rivers and marshes. Similarly, The Euphrates River inhabited by populations of this species smaller than those of the Tigris and Shatt al-Arab Rivers. The population of the common carp seems to be flourished in the marsh areas better than any inland water bodies in Iraq. Having saying this, it is important to mention that such a conclusion is based on what available of data on the populations of the common carp in Iraq which are very scarce and not suitable for any management program to be performed. Because the number of individuals given in the population studies is far less than that given in the reproduction or growth studied, where relatively higher number of individuals were collected from the same localities and at the same period of time.

Reproductive speaking, the common carp has shown better adaptation to the Iraqi new environment and showed good fecundity value. As to the condition factor, those studies on the populations from the Tigris River have shown that individuals of this species have a good well-being. More researches are needed in order to put future plans for the management of controlling the common carp in Iraq.

It is not possible to combine the results of population studies of the common carp performed in the 6 Iraqi lakes due to the differences in their topography, geographical locations, climate, substrate, environmental and biological factors. But on the other hand, it is possible to highlight some points that indicate clearly the increase in the abundance of the common carp in the 6 lakes mentioned above. Future studies on the population of the common carp in all the lakes mentioned above are needed to follow up the increase in the abundance of C. carpio in relation to the native species.

In spite of the experiment of Saleh and Abdul Karim (2013) on the hybridization of the common carp with Mesopotamichthys sharpeyi a native fish species, the future results of such hybridization in the nature will have more ecological impacts than the presence of the alien species *C. carpio* itself.

Conservation and biodiversity can be in great threat in case of the interbreeding that could happen between related species. Endemic species can interbreed with a more common, introduced species occurs frequently (Rhymer and Simberloff 1996). Such dangerous hybridization, if I may call it, cross-species can lead to the development of invasive lineages that are effectively new species (Nolte et al. 2005) and can even replace parental forms (Ellstrand and Schierenbeck 2000). The weak hybridization may produce hybrid generation with low fitness, but can still introduce individual advantageous alleles into populations of non-endemic species and facilitate them becoming invasive in their new environment (Hänfling 2007).

Several factors may affect the inbreeding with the common carp, among these are (1) the potential for hybridization can be difficult to predict; (2) levels of cross-species hybridization can vary greatly between different locations and over time (Kijewska et al. 2009); (3) hybridization can be strongly biased, where the bulk of F1 hybrids happen through the males of one species mating with the females of another (Roberts et al. 2009), or can be bidirectional (Yaakub et al. 2006); (4) hybridization can also be lacking between ecologically similar sympatric sister species (Anderson et al. 2009); (5) the rate and direction of cross-species hybridization can be intensely affected by the spawning times of the species involved (Roberts et al. 2009).

37.6 Recommendation for Common Carp Management

Weber and Brown (2009) have suggested a set of counteractive steps to follow in order to control the presence of the common carp in the freshwater ecosystem. These measures can be summarized as follows:

1. Among the ways to manage the freshwater ecosystem and control the presence of common carp is by removal of individuals of this fish (Schrage and Downing 2004). It is one remedial recovery policy essential to decline internal nutrient cycling, increase water transparency, re-establish aquatic macrophytes and return the ecosystem to the clear water state. Nevertheless, elimination of common carp is problematic, costly and time overwhelming (Schrage and Downing 2004). It has been known that more than 70% of common carp biomass may have to be removed to achieve enhancements in biotic and abiotic variables (Schrage and Downing 2004).

   Besides the plan of removing common carp from an ecosystem, individual common carp size and ecosystem size, substrate and trophic state influence common carp effects on ecosystems should be considered too (Chumchal and Drenner 2004).
2. Moderately regulating plan against the presence of the common carp populations by freshwater body morphometry as the population of the common carp declines with increasing water depth (Egertson and Downing 2004). Consequently, the water body should be selected and decisions must be made to determine where restoration efforts are best directed (Weber and Brown 2009).
3. For example, low abundance of common carp populations may be easy to further reduce, but such efforts might have little ecological effect on aquatic ecosystems.
4. With water bodies that have hard substrate, the effects of the common carp can be reduced by decreasing the sedimentation, nutrient loading and lake eutrophication. For example, common carp have had little effect on suspended solids or

turbidity in lakes with sand, gravel or cobble substrate, even at high densities (Sidorkewicj et al. 1999).

5. Several control methods have been endeavoured to manage common carp populations, including chemical and physical elimination, demolition of common carp spawning habitat, water level manipulation, fish barriers and predator introduction (Parkos et al. 2006).
6. It has been known that individuals of common carp assemble in large quantities deep area of the freshwater body during cold seasons (Penne and Pierce 2008). Therefore, aiming such aggregations may make removal efforts more successful.
7. In spite of the chemical removal of common carp is an accepted removal technique (Bonneau 1999), non-target species are also susceptible (Marking 1992). Chemical spot treatments in areas of common carp aggregations may be an option to minimize undesirable effects on the surrounding lake community and increase removal success on large water bodies (Bonneau 1999).
8. The effectiveness of the physical removal techniques is coupled with habitat types. Such techniques include seining, electrofishing, gill netting and a variety of trap types, and (Pinto et al. 2005).
9. Poison lures have also been recommended to target common carp populations and diminish effects on other species (Rach et al. 1994), but field experiments have been unsuccessful (Bonneau 1999). Even when removal efforts are successful, only short-term benefits may accrue because common carp recolonize quickly (Barton et al. 2000).
10. Precautionary measures (i.e. electrical barriers, fish weirs, traps, etc.) established prior to removal efforts are suggested to lessen the probability of common carp recolonization (Lougheed and Chow-Fraser 2001).
11. Common carp population dynamics may be used in the management plans. It is possible to control the fast-growing, short-lived populations by reducing recruitment (Brown and Walker 2004). The technology that can deal with the sex ratio and biased males over females for multiple generations will ultimately causing population collapse (Grewe 1996) and could also be used to control short-lived populations (Brown and Walker 2004). On the other hand, long-lived, slow-growing populations could be controlled in removing adult population (Brown and Walker 2004) in addition to consideration recruitment.
12. The usual habit of the common carp spawn over aquatic macrophytes in waters generally less than 0.5 m (Edwards and Twomey 1982) can be used in the control efforts. Drawdowns during spawning may reduce spawning habitat and recruitment (Shields 1958). Such a plan has been successfully implemented in Australia (Koehn et al. 2000). Barriers stopping adult common carp contact with the spawning areas may also decrease recruitment (Stuart et al. 2001). Furthermore, physical agitation of water in shallow spawning bays may damage eggs. Nevertheless, consequences of all these techniques on native fishes should be well thought-out when formulating a set of management strategies.

Koehn et al. (2000) have given a strategic management plan, which have been summarized in this chapter and given below. Koehn et al. (2000) have identified the four components of the strategic approach to carp management as defining the problem, developing a management plan, implementing the plan and monitoring and evaluating results. As an editor of this book, I preferred to keep the original recommendations without changes except in certain parts where they are not related to the subject of this chapter. And here I quote,

1. To define the problem. This means determining the nature and scope of the management concern (for example, loss of water quality or aquatic vegetation). A number of factors in addition to the presence of carp can contribute to observed problems, including human interference, habitat type, local influences on water flow and quality and other fish species present. The problems caused by carp may affect many natural resources. Therefore the management of carp as a pest species is a natural resource management issue which extends well beyond the realm of traditional fisheries management.
2. To develop a management plan. This must include clear objectives set in terms of the economic and/or conservation outcomes being sought. These management objectives should include interim and long-term goals. Developing the plan will involve an assessment of the most appropriate control technique(s) and strategy and setting the priorities for management. Best results in pest management are often achieved with a combination of techniques rather than relying on a single technique. Options for carp management include: prevention of further spread, local, small-scale poisoning or removal, exclusion, habitat rehabilitation to enhance native fish species, commercial or recreational removal, and wide-scale control options which may include new technologies. In developing a management plan, one or more of these options need to be selected that will best meet the management objectives. Measurable performance indicators then need to be defined which can be used to measure progress against the management objectives.
3. Implementation, is dependent on an integrated approach for success.

   Although much of the responsibility will rest with a range of government agencies, cooperation and ownership must also be undertaken with other stakeholders and community groups. Ownership of carp management must ultimately reside with many agencies and groups, not just those with fisheries interests.
4. Monitoring and evaluation. This should occur at different levels throughout implementation and on completion of actions. The efficiency of the operation needs to be monitored to ensure that the management plan is executed in the most cost effective manner. Monitoring will help identify inefficiencies so the management strategy can be continually refined. In addition, the effectiveness of the program in achieving the objectives needs to be monitored so that either the program objectives or the management strategy can be modified if necessary, in the light of further knowledge and experience. This may mean modifying the objectives, if they are unrealistic, or adding new objectives. Effectiveness is determined by evaluating achievements and outcomes against the performance

indicators included in the management plan. Different techniques may have different success rates under different circumstances. Economic frameworks are needed to assist in the assessment of the relative cost and value of alternative strategies. Such frameworks require: the definition of the economic problem, data on the relative costs of different carp control strategies, and an understanding of why the actions of individual management agencies may not lead to optimal levels of carp control and how management may be improved.

References

Abdul Retha S, Al-Khafaji BY, Maktoof AA (2009) Seasonal variations in the ovaries of the common carp Cyprinus carpio collected from Al-Hammar marsh, south of Iraq. Thigar J Sci 1:3–13

Al-Rudainy AJ, Mohamed ARM, Abbas LM (2006) Ecology and biodiversity of fish community in Euphrates River at Al-Mussaib Power Station, middle of Iraq. Environmental Conference and Exhibition, Kuwait

Al-Daham NK, Yousif AY (1990) Composition, seasonality and abundance of fishes in the Shatt Al-Basrah Canal, an estuary in southern Iraq. Estuar Coast Shelf Sci 31(4):411–421

Al-Rudaini AJ (2010) Ecology and nature of fish structure in Al-Rathwania Lake, west of Baghdad. Iraqi J Agric 15:1–11

Al-Rudaini AJ, Faisal S, Abu-Al-Hani A, Hassan A, Ahmad S (1998) Some biological aspects of fishes collected from the northern part of Saddam River. Dirasat, Jordan 25:553–563

Al-Rudaini AJ, Ruhaige AM, Qatee AJ, Hussain TS (1999) Some biological aspects of fishes collected from Al-Habbaniya Lake. Iraqi J Agric 4:159–167

Al-Shamary ACH, Younis KH, Al-Zawar GK (2011) Study of fish assemblage in south east Al-Hammar marsh north of Basrah City–Iraq. Basrah J Agric Sci 24:111–125

Al-Shawi SAS, Wahab NK (2007) Structure of fish assemblages for Tuz Jai River basin. Iraqi J Agric, Special issue 12:105–122

Anderson JD, Mcdonald DL, Sutton GR, Karel WJ (2009) Evolutionary associations between sand seatrout (*Cynoscion arenarius*) and silver seatrout (*C. nothus*) inferred from morphological characters, mitochondrial DNA, and microsatellite markers. Fish Bull 107:13–23

Angeler DG, Alvarez-Cobelas M, Sanchez-Carrillo S, Rodrigo MA (2002) Assessment of exotic fish impacts on water quality and zooplankton in a degraded semi-arid floodplain wetland. Aquat Sci 64:76–86

Bajer PG, Sorensen PW (2015) Effects of common carp on phosphorus concentrations, water clarity, and vegetation density: a whole system experiment in a thermally stratified lake. Hydrobiologia 746:303–311

BAJER PG, SULLIVAN G, SORENSEN PW (2009) Effects of rapidly increasing population of common carp on vegetative cover and waterfowl in a recently restored Midwestern shallow lake. Hydrobiologia 632:235–245

Barton DR, Kelton N, Eedy RI (2000) The effects of carp (*Cyprinus carpio* L.) on sediment export from a small urban impoundment. J Aquat Ecosyst Stress Recov 8:155–159

Bonneau JL (1999) Ecology of a fish biomanipulation in a Great Plains reservoir. University of Idaho, Moscow

Britton JR, Boar RR, Grey J, Foster J, Lugonzo J, Harper DM (2007) From introduction to fishery dominance: the initial impacts of the invasive carp *Cyprinus carpio* in Lake Naivasha, Kenya, 1999 to 2006. J Fish Biol 71:239–257

Brown P, Walker TI (2004) CARPSIM: stochastic simulation modeling of wild carp (*Cyprinus carpio* L.) population dynamics, with applications to pest control. Ecol Model 176:83–97

Carpenter SR, Cottingham KL, Schindler DE (1992) Biotic feedbacks in lake phosphorus cycles. Trends Ecol Evol 7:332–336
Carpenter SR, Kitchell JF, Hodgson JR (1985) Cascading trophic interactions and lake productivity. Bioscience 35:634–639
Carpenter SR, Lodge DM (1986) Effects of submersed macrophytes on ecosystem processes. Aquat Bot 26:341–370
Chumchal MM, Drenner RW (2004) Interrelationships between phosphorus loading and carp in the regulation of phytoplankton biomass. Arch Hydrobiol 161:147–158
Chumchal MM, Nowlin WH, Drenner RW (2005) Biomass-dependent effects of carp on water quality in shallow ponds. Hydrobiologia 545:271–277
Covich AP, Knezevic B (1978) Size-selective predation by fish on thin-shelled gastropods (*Lymnaea*): the significance of floating vegetation (*Trapa*) as a physical refuge. Proc Int Assoc Theor Appl Limnol 20:2172–2177
De Robertis A, Ryer CH, Veloza A, Brodeur RD (2003) Differential effects of turbidity on prey consumption of piscivorous and planktivorous fish. Can J Fish Aquat Sci 60:1517–1526
Drenner RW, Gallo KL, Baca RM, Smith JD (1998) Synergistic effects of nutrient loading and omnivorous fish on phytoplankton biomass. Can J Fish Aquat Sci 55:2087–2096
Edwards EA, Twomey KA (1982) Habitat suitability index models: Common carp. U.S. Dept. Int. Fish Wildl. Servo FWS/OBS-82/10.12, 27 pp.
Egertson CJ, Downing JA (2004) Relationship of fish catch and composition to water quality in a suite of agriculturally eutrophic lakes. Can J Fish Aquat Sci 61:1784–1796
Ellstrand CN, Schierenbeck KA (2000) Hybridization as a stimulus for the evolution of invasiveness in plants? PNAS 97:7043–7050
Epler P, Bartel R, Szczerbowski J, Szypula J (2001) The ichthyofauna of lakes Habbaniya, Tharthar, and Razzazah. Archives of Polish Fisheries 9:171–184
Evelsizer VD, Turner AM (2006) Species-specific responses of aquatic macrophytes to fish exclusion in a prairie marsh: a manipulative experiment. Wetlands 26:430–437
Florian N, Lopez-Luque R, Ospina-Alvarez N, Hufnagel L, Green AJ (2016) Influence of a carp invasion on the zooplankton community in Laguna Medina, a Mediterranean shallow lake. Limnetica 35(2):397–412
Grewe P (1996) Review and evaluation of the potential of molecular approaches for the environmentally benign management of the common carp (Cyprinus carpio) in Australian waters. Centre for Research on Introduced Marine Pests, CSIRO Division of Fisheries, Technical Report Number 10 (1996)
Hamady AA (2009) Reproductive biology of common carp (Cyprinus carpio L.) from AL-Gharraf Canal, at Wasit Province–Iraq. I. sex-ratio and maturity. Ibn AL-Haitham J Pure Appl Sci 22:12–18
Hamady AA (2010) Reproductive biology of common carp (Cyprinus carpio L.) from AL-Gharraf Canal at Wasit province–Iraq II. Fecundity of females. Ibn AL-Haitham J Pure Appl Sci 23 (1):105–116
Hänfling B (2007) Understanding the establishment success of non-indigenous fishes: lessons from population genetics. J Fish Biol 71(Suppl. D):115–135
Hanson MA, Butler MG (1994) Responses of plankton, turbidity, and macrophytes to biomanipulation in a shallow prairie lake. Can J Fish Aquat Sci 51:1180–1188
Hinojosa-Garro D, Zambrano L (2004) Interactions of common carp (*Cyprinus carpio*) with benthic crayfish decapods in shallow ponds. Hydrobiologia 515:115–122
Hussain NA, Mohamed ARM, Al-Noo SS, Mutlak FM, Abed IM, Coad BW (2009) Structure and ecological indices of the fish assemblages in the recently restored Al-Hammar marsh, southern Iraq. BIORISK-Biodiversity and Ecosystem Risk Assessment (3):173–186
Hussain NA, Younis KH, Yousif UH (1997) The composition of small fish assemblages in the river Shatt Al-Arab near Basrah [Iraq]. Acta Hydrobiologica 39(1–2)
Hussein SA, Abdulla AM, Abdullah SA (2015) Ecology and fish structure in the southern sector of the Euphrates River, Iraq. Basrah J Agric Sci 28:82–94

Jawad L (2003) Impact of environmental change on the freshwater fish fauna of Iraq. Int J Environ Stud 60:581–593

Jawad LA (2006) Fishing gear and methods of the lower Mesopotamian plain with reference to fishing management. Marina Mesopotamica 1(1):1–37

Jawad L (2013) Threatened freshwater fishes of Iraq, with remarks on their conservation status. Water Res Manag 3(2):27–36

Johnson DM, Crowly PH (1980) Odonate 'hide and seek': habitat-specific rules? In: Kerfoot W (ed) Evolution and ecology of zooplankton communities. University Press of New England, Hanover, NH

Khan TA (2003) Dietary studies on exotic carp (*Cyprinus carpio* L.) from two lakes of western Victoria, Australia. Aquat Sci 65(3):272–286

Khan TA, Wilson ME, Khan MI (2003) Evidence for invasive carp mediated trophic cascade in shallow lakes of western Victoria, Australia. Hydrobiologia 506:465–472

Kijewska A, Burzynski A, Wenne R (2009) Molecular identification of European flounder (*Platichthys flesus*) and its hybrids with European plaice (*Pleuronectes platessa*). ICES J Mar Sci 66:902–906

Kirk KL (1991) Inorganic particles alter competition in grazing plankton: the role of selective feeding. Ecology 72:915–923

Kirk KL, Gilbert JJ (1990) Suspended clay and the population dynamics of planktonic rotifers and cladocerans. Ecology 71:1741–1755

Kloskowski J (2011) Impact of common carp *Cyprinus carpio* on aquatic communities: direct trophic effects versus habitat deterioration. Fundamental and Applied Limnology/Archiv für Hydrobiologie 178:245–255

Koehn J, Brumley A, Gehrke P (2000) Managing the impacts of carp. Bureau of Rural Sciences, Department of Agriculture. Fisheries and Forestry, Canberra, Australia

Lamarra VAJ (1975) Digestive activities of carp as a major contributor to the nutrient loading of lakes. Proc Int Assoc Theor Appl Limnol 19:2461–2468

Lammens EHRR (1991) Diets and feeding behaviour. In: Winfield IJ, Nelson JS (eds) *Cyprinid fishes: Systematics, biology, and exploitation*, pp. 127–155. Chapman and Hall, London

Lazem LF, Attee RS (2016) Structure of fish assemblage in relation to some ecological factors in Himreen dam Lake, Iraq. Basrah J Agric Sci 29(1):7–16

Lougheed VL, Chow-Fraser P (2001) Spatial variability in the response of lower trophic levels after carp exclusion from a freshwater march. J Aquat Ecosyst Stress Recov 9:21–34

Lougheed VL, Crosbie B, Chow-Fraser P (1998) Predictions on the effect of common carp (*Cyprinus carpio*) exclusion on water quality, zooplankton, and submergent macrophytes in a great lakes wetland. Can J Fish Aquat Sci 55:1189–1197

Marking LL (1992) Evaluation of toxicants for the control of carp and other nuisance fishes. Fisheries 17:6–13

Matsuzaki SS, Usio N, Takamura N, Washitani I (2009) Contrasting impacts of invasive engineers on freshwater ecosystems: an experiment and meta-analysis. Oecologia 158:673–686

McQueen DJ, Post JR, Mills EL (1986) Trophic relationships in freshwater pelagic ecosystems. Can J Fish Aquat Sci 43:1571–1581

Meijer M-L, De Hann MW, Breukelaar AW, Buiteveld H (1990a) Is reduction of the benthivorous fish an important cause of high transparency following biomanipulation in shallow lakes? Hydrobiologia 200(201):303–315

Meijer M-L, Lammens EHRR, Raat AJP, Grimm MP, Hosper SH (1990b) Impact of cyprinids on zooplankton and algae in ten drainable ponds. Hydrobiologia 191:275–284

Miller AI, Beckman LG (1996) First record of predation on white sturgeon eggs by sympatric fishes. Trans Am Fish Soc 125:338–340

Miller SA, Crowl TA (2006) Effects of carp (*Cyprinus carpio*) on macrophytes and invertebrate communities in a shallow lake. Freshw Biol 51:85–94

Miller SA, Provenza FD (2007) Mechanisms of resistance of freshwater macrophytes to herbivory by invasive juvenile carp. Freshw Biol 52:39–49

Miner JG, Stein RA (1996) Detection of predators and habitat choice by small bluegills, effects of turbidity and alternative prey. Trans Am Fish Soc 125:97–103
Mohamed ARM, Al-Noor SS, Faris RAK (2006) Analytical study of the fisheries in the lower reaches of the Tigris River, north Qurna, Basrah, Iraq. Iraqi J Aquaculture (1):57–73
Mohamed ARM, Hussain NA, Al-Noor SS, Mutlak FM, Al-Sudani IM, Mojer AM, Toman AJ, Abdad MA (2008) Fish assemblage of restored Al-Hawizeh marsh, southern Iraq. Ecohydrology & Hydrobiology 8(2–4):375–384
Mohamed ARM, Hussein SA, Lazem LF (2013) Spatiotemporal variability of fish assemblage in the Shatt Al-Arab River, Iraq. J Coastal Life Med 3(1):27–34
Mohamed ARM, Kadim H, Younis KH, Entisar K, Hameed EK (2017) Status of fish assemblage structure in the Garmat Ali River, Iraq. IOSR J Agric Veterinary Sci 10:17–22
Mohamed ARM, Resen AK, Taher MM (2012) Longitudinal patterns of fish community structure in the Shatt Al-Arab River, Iraq. Basrah J Sci 30(2):65–86
Nolte AW, Freyhof J, Stemshorn KC, Tautz D (2005) An invasive lineage of sculpins, Cottus sp (Pisces, Teleostei) in the Rhine with new habitat adaptations has originated from hybridization between old phylogeographic groups. Proc Roy Soc Lond Ser B 272:2379–2387
Panek FM (1987) Biology and ecology of carp. In: Cooper EL (ed) *Carp in North America*, pp. 1–15. American Fisheries Society, Bethesda, MD
Parkos JJ, Santucci VJ, Wahl DW (2003) Effects of adult carp (*Cyprinus carpio*) on multiple trophic levels in shallow mesocosms. Can J Fish Aquat Sci 60:182–192
Parkos JJ, Santucci VJ, Wahl DW (2006) Effectiveness of a plastic mesh substrate cover for reducing the effects of carp on aquatic ecosystems. N. Am J Fish Manage 26:861–866
Penne CR, Pierce CL (2008) Seasonal distribution, aggregation, anal habitat selection of common carp in clear Lake, Iowa. Trans Am Fish Soc 137:1050–1062
Pinto L, Chandrasena N, Pera J, Hawkins P, Eccles D, Sim R (2005) Managing invasive carp (*Cyprinus carpio* L.) for habitat enhancement at botany wetlands, Australia. Aquat Conserv: Mar Freshw Ecosyst 15:447–462
Qin J, Threlkeld ST (1990) Experimental comparison of the effects of benthivorous fish and planktivorous fish on plankton community structure. Proc Int Assoc Theor Appl Limnol 119:121–141
Rach JJ, Luoma JA, Marking LL (1994) Development of an antimycin-impregnated bait for controlling common carp. N Am J Fish Manage 14:442–446
Rasheed RO (2012) Length-weight relationships of (9) fish species from Derbendikhan reservoir–Kurdistan region, Iraq. Journal of Babylon University: Pure Appl Sci 5:1434–1440
Rhymer JM, Simberloff D (1996) Extinction by hybridization and introgression. Annu Rev Ecol Syst 27:83–109
Roberts J, Chick A, Oswald L, Thompson P (1995) Effect of carp, *Cyprinus carpio*, an exotic benthivorous fish, on aquatic plants and water quality in experimental ponds. Mar Freshw Res 46:1171–1180
Roberts DG, Gray CA, West RJ, Ayre DJ (2009) Evolutionary impacts of hybridization and interspecific gene flow on an obligately estuarine fish. J Evol Biol 22:27–35
Saleh KI, Abdul Karim MF (2013) Evaluation of inter generic hybridization between fish common carp *Cyprinus carpio* and fish Bunni *Barbus sharpeyi*. Al-Furat J Agric Sci 5:363–370
Salman JM, Salah MM, Dhurgham A, Al-Sultany DA (2014) Spatial variation of composition and distribution of fish in two sites on Euphrates River, middle of Iraq. Magazin of Al-Kufa University for Biology 6:1–9
Scheffer M (1998) Ecology of shallow lakes. Chapman and Hall, London
Scheffer M, Hosper SH, Meijer M-L, Moss B, Jeppesen E (1993) Alternative equilibria in shallow lakes. Trends Ecol Evol 8:275–279
Schrage LJ, Downing JA (2004) Pathways of increased water clarity after fish removal from Ventura marsh; a shallow, eutrophic wetland. Hydrobiologia 511:215–231
Sediq SO, Abbas LM (2013) Fish community structural in Dukan dam Lake, northern Iraq. Iraqi Veterinary J 37:6–12

Shakir HF, Nehad K, Wahab NK (2015) Structure of fish community for south east Al-Tharthar Lake in Salah Alddin Province, Iraq. Tikrit J Agric Sci 25:111–124

Shields JT (1958) Experimental control of carp reproduction through water drawdowns in Fort Randall reservoir, South Dakota. Trans Am Fish Soc 87:23–32

Sidorkewicj NS, Lopez Cazorla AC, Fernandez OA, Mockel GC, Burgos MA (1999) Effects of *Cyprinus carpio* on *Potamogeton pectinatus* in experimental culture: the incidence of the periphyton. Hydrobiologia 415:13–19

Sidorkewicj NS, Lopez-Carzola AC, Fernandez OA (1996) The interaction between *Cyprinus carpio* L. and *Potamogeton pectinatus* L. under aquarium conditions. Hydrobiologia 340:271–275

Skubinna JP, Coon TG, Batterson TR (1995) Increased abundance and depth of submersed macrophytes in response to decreased turbidity in Saginaw Bay, Lake Huron. J Great Lakes Res 21:476–488

Stewart TW, Downing JA (2008) Macroinvertebrate communities and environmental conditions in recently constructed wetlands. Wetlands 28:141–150

Stuart I, Jones M, Koehn J (2001) Targeting spawning habitats to control carp populations. In: Proceedings of the 12th Australian vertebrate Pest conference. Melbourne Department of Natural Resources and Environment, Victoria, Australia

Vanni MJ (1996) Nutrient transport and recycling by consumers in lake food webs: implications for algal communities. In: Polis GA, Winemiller KO (eds) Food webs: integration of patterns and dynamics. Chapman and Hall, New York, pp 81–95

Wahab NK, Al-Ani SM h (2012) Some biological aspects for number of Tigris River fishes, Iraq. Mesopotamia J Agric 40:11–22

Wahab NK, Al-Ani SM h (2013) Some biological aspects of some fishes from the eastern drainage, Balad, Iraq. Diayal J Agric Sci 5:1–15

Wahab NK, Hassan SM (2011) Structure of fish assemblage and food interrelationship in eastern drain (Shtteata) Balad, Tigris River. 5th Scientific Conference of College of agriculture, Tikrit University, 26–27 April, 2011

Weber MJ, Brown ML (2009) Effects of common carp on aquatic ecosystems 80 years after "carp as a dominant": ecological insights for fisheries management. Rev Fish Sci 17(4):524–537

Werner EE, Anholt BR (1993) Ecological consequences of the trade off between growth and mortality rates mediated by foraging activity. Am Nat 142:242–272

Williams AE, Moss B (2003) Effects of different fish species and biomass on plankton interactions in a shallow lake. Hydrobiologia 491:331–346

Wolfe MD, Santucci VJ Jr, Einfalt LM, Wahl DH (2006) Effects of common carp on reproduction, growth, and survival of largemouth bass and bluegills. Trans Am Fish Soc 138:975–983

Yaakub SM, Bellwood DR, Van Herwerden L, Walsh FA (2006) Hybridization in coral reef fishes: introgression and bi-directional gene exchange in Thalassoma (family Labridae). Mol Phylogenet Evol 40:84–100

Zambrano L, Hinojosa D (1999) Direct and indirect effects of carp (*Cyprinus carpio*) on macrophytes and benthic communities in experimental shallow ponds in Central Mexico. Hydrobiologia 408(409):131–138

Zambrano L, Perrow MR, Macias-Garcia C, Aguirre-Hidalgo V (1999) Impact of introduced carp (*Cyprinus carpio*) in subtropical shallow pond in Central Mexico. J Aquatic Ecosyst Stress Recov 6:281–288

Chapter 38
The Good and the Bad in Releasing the Grass Carp *Ctenopharyngodon idella* in the Freshwater System: Recommendations for the Policymakers in Iraq

Laith A. Jawad and Falah Mutlak

Abstract Plants dwelling in water habitats are useful and an essential part of the freshwater system. In addition, a few variety of plants are the main food supply for creatures living in the aquatic areas. Plants are competent to alleviate sediments, recover water clearness and enhance variety in the shallow regions of the freshwater body. Besides, dense plants can cause irrigation problems. The goal of the study at hand chapter is to give a short, but concentrate review on the good and bad points of releasing and introducing grass carp into the freshwater system to control aquatic vegetation. Also, this chapter highlights the status of the grass carp in the Iraqi freshwater system and at the end, a recommendations were given to be taken by policymakers in countries like Iraq to save their freshwater system.

38.1 Introduction

Grass carp (*Ctenopharyngodon idella*) is adaptive to large river systems of eastern Asia. The Amur River on the Russian and southward to the Chinese border is its natural distribution (Pípalova 2006; Bozkurt et al. 2017).

There are contrasting reports about the grass carp, some in a favourite of releasing this species in the freshwater system to control growth of macrophytes (Leslie Jr. et al. 1987; Wells et al. 2003), others contemplate that the grass carp has the impending capacity to produce abundant ecological damage to the freshwater

L. A. Jawad (✉)
School of Environmental and Animal Sciences, Unitec Institute of Technology, Auckland, New Zealand

F. Mutlak
Marine Science Centre, University of Basrah, Basrah, Iraq

L. A. Jawad (ed.), *Tigris and Euphrates Rivers: Their Environment from Headwaters to Mouth*, Aquatic Ecology Series 11,
https://doi.org/10.1007/978-3-030-57570-0_38

systems (Leslie Jr. et al. 1987; Wells et al. 2003; Pípalova 2006; Cudmore and Mandrak 2004; Bozkurt et al. 2017).

Through the literature, this chapter will give a short review about the evaluation of the grass crap in monitoring the aquatic plants in the freshwater systems and itemises the effects and harm that this species can bring to the freshwater habitats (Leslie Jr. et al. 1987; Wells et al. 2003; Pípalova 2006; Cudmore and Mandrak 2004; Bozkurt et al. 2017).

The information given in this chapter is arranged in five sections. In the first section, general information about the grass is given, in the second section, possibility of using grass carp in controlling aquatic vegetation, in the third section, the influences of grass carp on the habitats and biota of the freshwater systems, in the fourth section, the status of grass carp in Iraq is reviewed from the point of view of its biology and establishment in the Iraqi inland waters and finally, in the fifth section, recommendations are given to guide policymakers in countries like Iraq in planning the control of his species (Wells et al. 2003; Cudmore and Mandrak 2004).

38.2 General Information about the Grass Carp

38.2.1 Identification and Short Description

Since it has been described by Valenciennes in 1844 as *Leuciscus idella*, *Ctenopharyngodon idella* is used to be included within the family Cyprinidae. Recently, the phylogenetics of the family Cyprinidae has been revised and some of the subfamilies have been upgraded to the family level. Now, *C. idella* belongs to the new family Xenocyprididae (Pípalova 2006; Chen and Mayden 2009; Stout et al. 2016; Bozkurt et al. 2017).

The grass carp (Fig. 38.1) is large size species and the genus *Ctenopharyngodon* contains only one species (Page and Burr 1991; Chilton III and Muoneke 1992; Billard 1997). This species has the following set of characters: head broad with no scales; mouth located below short snout provided with thin lips and absence of

Fig. 38.1 *Ctenopharyngodon idella*, collected from Shatt al-Arab River, Basrah, Iraq. Courtesy of Mustafa al-Mukhtar, Marine Science Centre, University of Basrah, Iraq

barbels. Body is slim and reasonably compacted with a curved belly and marginally curved lateral line, reaching the middle of the depth of the tail (Skelton 1993; Opuszynski and Shireman 1995; Pípalova 2006; Stout et al. 2016). No spines are present in fins. Body is with dusky grey on the upper part and brighter on the lower, with shade of shiny golden colour (Page and Burr 1991; Billard 1997). This species reaches maximum length of 1500 mm total length (Billard 1997) and reaches a maximum weight of 45 Kg (Skelton 1993; Opuszynski and Shireman 1995; Pípalova 2006; Stout et al. 2016).

38.2.2 Distribution

Grass carp inhabits sub-tropical to temperate regions and natural to large rivers and lakes in eastern Asia. There is a broad range of climatic conditions within the native range of the grass carp and it has been extensively introduced (mainly for macrophyte control) to several countries around the world (Pípalova 2006; Bozkurt et al. 2017).

38.2.3 Biology of the Grass Carp

38.2.3.1 Age and Growth

The life duration of the grass carp in its native country is 11 years, while scales of this species showed various individuals in their 15 years of age (Shireman and Smith 1983). Several factors that affect the growth in this species are feeding, temperature and oxygen (Chilton III and Muoneke 1992; Pípalova 2006; Bozkurt et al. 2017).

38.2.4 Tolerances

Grass carp is of those freshwater species that can tolerate variation in the environmental factors such as water temperature (Federenko and Fraser 1978). On the other hand, factors like variation in dissolved oxygen could initiate stress and Young individuals could face variation in oxygen level in the range of 0.41 to 28 mg/l (Meyer et al. 1975; Shireman and Smith 1983; Pípalova 2006; Bozkurt et al. 2017). Individuals of grass carp could endure salinities of 11 to 12 parts per thousand (ppt) and up to 19 ppt for brief periods (Meyer et al. 1975).

38.2.4.1 Reproductive Biology

Age of maturity depends on the water temperature and obtainable high value diet (Stanely et al. 1978). Developed grass carp needs about 1500 to 2000 degree days within a year for gonadal progress and maturation (Shireman and Smith 1983, Beck 1996; Pípalova 2006). When mature, the individuals of this species reach total length of nearly 50-86 cm, with males reaching maturity before females (Bozkurt et al. 2017).

With the mature individuals, it is possible to separate males and females as the formers have tubercles on the dorsal and medial surfaces of the pectoral fins (Shireman and Smith 1983). On the other hand, deciduous scales are shown in females (Beck 1996; Bozkurt et al. 2017).

Water temperature factor is considered the onset factor in persuading spawning and differs with areas. This species is characterised in several spawning seasons (Shireman and Smith 1983; Beck 1996; Pípalova 2006).

Grass carp prefers to put their eggs in the main branches of rivers during high water (Shireman and Smith 1983; Beck 1996; Pípalova 2006; Bozkurt et al. 2017). Spring to summer is the period chosen by this species for spawning (Shireman and Smith 1983). Individuals usually chose their habitats for laying eggs in turbid, turbulent water at the convergence of rivers or below dams (Federenko and Fraser 1978; Stanely et al. 1978; Bozkurt et al. 2017). In spawning locality, females are frequently more than males by about two to one. For fertilisation of the eggs, each female is regularly trailed by two or more males.

The number of eggs is related to the length, weight and age and ranges from 0.001 to two million eggs, but commonly with mean of 0.5 million for a 5 kg brood stock (Shireman and Smith 1983, Chilton III and Muoneke 1992; Bozkurt et al. 2017). The eggs are 2.0-2.5 mm in diameter when out, but rapidly swell to a diameter of 5-6 mm with absorption of water (from Chilton III and Muoneke 1992). They are semi-buoyant and non-adhesive, requiring water with high oxygen level to assist in their suspension in the water column before hatching (Stanely et al. 1978, Chilton III and Muoneke 1992; Bozkurt et al. 2017).

38.2.4.2 Food and Feeding Habit

There are several factors that may affect the feeding ecology of this species such as age, size, temperature, availability of specific food, depth and density (in pond cultures) (Opuszynski and Shireman 1995). Rigorous feeding occurs precisely when water temperature is at least 20 °C (NatureServe 2003; Pípalova 2006; Bozkurt et al. 2017). Opuszynski and Shireman (1995) have suggested that young individuals immediately after hatching feed on rotifers and protozoans and then with their development, change to larger cladocerans. At the age of 6 months they start to feed on large objects such as *Daphnia* and insect larvae. Later, they feed on plant items more than animal including filamentous algae and macrophytes and they divert to feed on a limited macrophyte when they reach 1 to 1.5 months after hatching (Pípalova 2006; Bozkurt et al. 2017).

Adult grass carps are discriminating in their selection of particular plant species favouring underwater plants with soft leaves (Bailey and Boyd 1971; Pine and Anderson 1991; Bozkurt et al. 2017) and eating the most favoured species first until they become uncommon (Bailey and Boyd 1971; Pípalova 2006; Bozkurt et al. 2017). Less favourite plant species like filamentous algae and firmer-leaved macrophytes are eaten when they are the only species available (Pine and Anderson 1991; Opuszynski and Shireman 1995). Adult individuals of this species can switch to other food items once their favourite food objects become rare (Bailey and Boyd 1971; NatureServe 2003; Pípalova 2006; Bozkurt et al. 2017). The adults of this species also appear to have in their stomachs parts of tree leaves and twigs from banks (Bailey and Boyd 1971; Pine and Anderson 1991; Bozkurt et al. 2017). Such feeding behaviour shows that they inclined to feed on non-aquatic plant objects in the case of lack of submerged plants.

38.3 The Possibility of Using Grass Carp in Controlling Aquatic Vegetation

Plants that inhabit water are important and play a significant role in the freshwater habitats. They contribute noteworthily in energy storing. Furthermore, they act as protection and reproduction sanctuaries for many creatures, and their submerged fragments permit the growth of periphyton groups (Pitelli 1998; Pípalova 2006; Bozkurt et al. 2017).

On the other hand, and in addition to their high benefit, they could simply overgrow and convert to a source of aggravation to the landlord. They easily can destruct the fishing prospective of the water body. The accumulation of a dead plants in any water body could decrease the level of oxygen, which in turn can bring damage to other creatures living in the same environment (Pitelli 1998; Pípalova 2006; Bozkurt et al. 2017).

Monitoring and eradicating aquatic plants from any freshwater system are often unclear and unsatisfying jobs. The choice of a vegetation management plan depends on local settings of the water body (Pitelli 1998). Therefore, there are three approaches that can be followed to get rid of the over-grown macrophytes; these are: mechanical, chemical and biological control. The mechanical control, which involves physical removal of the vegetation and is often more difficult in water than on land, the chemical vegetation control is often unsuccessful, and retreatment may be needed. Biological control has many advantages over the other vegetation control means. For instance, it takes much less human work effort than most of the mechanical control means and does not require using expensive and hazardous aquatic herbicides. In addition, using fish species provides longer-term control than other control mechanisms due to fishes that usually have a lifespan of several years (Pitelli 1998; Zweerde 1990; Pípalova 2006; Bozkurt et al. 2017).

Among fish species that are used to manage the overgrowth of aquatic plants are some species of tilapia (*Tilapia* spp.), silver carp (*Hypophthalmichthys molitrix*) and the grass carp (*Ctenopharyngodon idella*). Among those species, grass carp is the only one to eat large amount of plants (Cross 1969; Zweerde 1990). It has been shown that in some cases, adult individuals of this species can eat more than its own weight of plant material on a day-to-day basis (Cross 1969; Zweerde 1990; Pípalova 2006; Bozkurt et al. 2017).

In order to use grass carp as a method to eradicate vegetation from a freshwater body system, Bozkurt et al. (2017) have suggested the following points that need to be considered in such a process; they wrote the following:

38.4 Stocking of Grass Carp for Controlling Aquatic Vegetation

To manage the overgrowth of the aquatic plants, it is significant to know beforehand the amount of plant groups, plant types, water body sizes and the size of fishes stocked. Then, the number of individuals of grass carp can be made available for this process (Blackwell and Murphy 1996; Pípalova 2006).

Different methods have been used to determine the suitable number of grass carp to stock. The most precise method is to determine the weight of aquatic vegetation in the freshwater body, knowing the ingesting rates of the fish (Blackwell and Murphy 1996; Pípalova 2006).

Despite the studies' investigation on diverse stocking proportions, there is no recommendation that is satisfactory to all circumstances for grass carp. Each aquatic system is dissimilar due to its particular features of fertility, water clarity, shallow water and chemical makeup (Blackwell and Murphy 1996). So, each of these means disturbs the number of grass carp necessary to attain the plant level to the anticipated management level. Supplying rates with individuals of grass carp could show variation ranging between one fish and as many as 20 per acre, liable on the quantity and types of aquatic plants (Adamek et al. 2003; Pípalova 2006).

Supplying rates with individuals of grass carp need to be increased as temperature decreases as grass carp plant ingests and growth reduces. Supplying bulks need to be grounded on the standing crop (biomass) of aquatic plants. This is assessed by multiplying plant distribution by average plant density; consequently, the higher the vegetation biomass, the higher the necessary stocking rate (Blackwell and Murphy 1996; Pípalova 2006).

It has to be noted before the stocking of grass carp started that overstocking will lead to an entire elimination of all aquatic plants, while understocking process will result in a discriminating decrease of aquatic plants (Blackwell and Murphy 1996; Adamek et al. 2003; Pípalova 2006).

It is important to know that grass carp age and size are also significant owing to the conceivable predation on them, which can distinctly reduce their early stocking density. Grass carp should be larger than 30 cm when supplied; elsewhere, they are very susceptible to predators (Adamek et al. 2003).

38.5 Changes in Aquatic Plant Pattern and Plankton Composition

Grass carp can endlessly impact preferred aquatic plant species. Such influences have been noticed for 15–20 years at higher stocking rates. It is presumed that removal of aquatic vegetation favoured by the grass carp leads to decrease of the variety of aquatic macrophyte groups (Richard et al. 1985; Catarino et al. 1997; Bozkurt et al. 2017).

Supplying of grass carp individual density and organised plant area disturbs the expansion of phytoplankton assembly in the freshwater body of any sort. In the case of slow handling of plants by grass carp, the subsidiary effects of grass carp stocking on phytoplankton are trivial (Catarino et al. 1997; Pípalova 2006; Bozkurt et al. 2017).

Primary production of the water body relies on two factors, light and nutrient obtainability. These two issues upset unbalanced equilibrium between macrophytes and phytoplankton. Therefore, the speed and degree of macrophyte elimination by the grass carp disturb the phytoplankton production (Richard et al. 1985; Catarino et al. 1997).

Zooplankton consumption is essential for juvenile and adult grass carp; nonetheless, the injected quantities are minor in case the stocking density is not extremely high (Terrell J W and Terrell 1975). Changes in zooplankton wealth and community structure were owing to an upsurge in phytoplankton and changes in planktivore predation on zooplankton by fish after macrophyte elimination (Richard et al. 1985; Pípalova 2006; Bozkurt et al. 2017).

38.6 Changes in Water Quality and Benthos

The influences of grass carp on plants and water quality are highly flexible and frequently doubtful owing to the lack of appropriate management locations. The amount and rate of plant elimination by the grass carp are imperative. Variations in water quality as a result of plant exclusion by the grass carp frequently happen in minor, non-flowing water bodies and least found when only a small quantity of plants is detached from large, comparatively deep, flowing reservoirs. So that, reductions could be witnessed in oxygen concentration of water subsequent to grass carp supplying, depending on the vanishing of macrophytes (Opuszynski

1972; Pípalova 2006; Bozkurt et al. 2017). Primary producers such as phytoplankton and aquatic macrophytes not only provide oxygen, but also use CO2 during photosynthesis, which outcomes in an upsurge in water pH. Fluctuations in oxygen concentrations subsequent to grass carp supplying were certainly linked with the variations in pH (Opuszynski 1972; Gasaway 1979; Leslie et al. 1983; Pípalova 2006; Bozkurt et al. 2017).

Greater supplying quantities of grass carp or their lengthier effect can upsurge concentrations of nutrients in the water, but this growth chiefly relies on the waterbody features (Gasaway 1979; Opuszynski 1972; Pípalova 2006; Bozkurt et al. 2017). These variations outcome from sediment resuspension throughout feeding and faecal matter deposition by carp as well as failure of systems accountable for upkeep of the vegetated state owing to elimination of macrophytes. Alterations in benthos agreed closely with variations in aquatic plants, which alleviate sediments and deliver extra substrate in the shape of root masses and rotten material. Zoobenthos also reacted to deviations in water quality subsequent to the elimination of aquatic macrophytes (Gasaway 1979).

38.7 The Effects of Grass Carp on the Environment and Biota of the Freshwater Systems

Nevertheless, the collective usage of grass carp (*Ctenopharyngodon idella*) as a biocontrol mediator affects aquatic macrophytes and there has been inevitability about the ecological danger it poses to the ecosystems in some parts of the world (Wittmann et al. 2014; Pípalova 2006; Bozkurt et al. 2017).

Wittmann et al. (2014) re-assessed the ecological threat of grass carp by means of efficient data and prediction tackles. They utilised first questionnaire, where they intend to comprehend whether the fishery specialists in the area where they are performing their assessment discriminate that grass carp generates ecological damage to the area. Second, they compute conceivable ecological danger by means of new investigation tools (eDNA).

The followings are the chief impacts of grass carp on the habitats of the freshwater system and its biota:

38.7.1 Effects on the Sediment Chemistry

In the water system, where grass carp individuals were stocked, the food items from macrophytes were stuck into the sediment (precipitated either by or with organic acids) and consequently were not obtainable to phytoplankton (Terrell 1975; Hestand and Carter 1978; Pípalova 2006; Bozkurt et al. 2017).

38.7.2 Effects on Phytoplankton

Phytoplankton production can be impacted by both the supplying density of grass carp and area of weeds managed. This influence will be insignificant if weed manipulation by grass carp is slow and some aquatic macrophytes are left in a water body. At a low supplying density, divergences in the concentration of chlorophyll-a in the water were trivial (Holdren and Porter 1986; Pípalova 2006; Bozkurt et al. 2017). Though, a high supplying density of grass carp, which can remove all aquatic macrophytes and the nutrients' unconfined yields, augmented phytoplankton. The result of such covering outcome of higher phytoplankton biomass and outstanding aquatic macrophytes are frequently further inhibited. Consequently, wind act can also upsurge water turbidity due to sediment transfer, particularly in the shallow water bodies (Pípalova 2006).

Removal of macrophytes augmented blue-green algae richness in the phytoplankton groups almost 9 times (Holdren and Porter 1986). However, the blue-green algae ruled only during the peak phytoplankton season. Holdren and Porter (1986) stated changes in chief taxa and relative abundances of green and blue-green algae and diatoms, with a general swing to smaller species happening after grass carp supplying.

38.7.3 Effects on Zooplankton

Ingesting of zooplankton is fundamental for both juvenile and adult grass carp, but the quantities eaten are irrelevant if the supplying density is not very high (Terrell and Terrell 1975; Zhang and Chang 1994; Pípalova 2006; Bozkurt et al. 2017). Main influences on zooplankton incline to be ancillary. In lakes supplied with herbivorous fish, the growth of zooplankton and zoobenthos is enhanced through eating macrophytes by the fish and then augmented rates of nutrient remineralization. The entire result can also be inveterate in an upsurge of fish production (Zhang and Chang 1994; Pípalova 2006). Variations in zooplankton wealth and community structure were owing to an upsurge in phytoplankton and changes in planktivore predation on zooplankton by fish after macrophyte exclusion (Richard et al. 1985).

38.7.4 Effects on Zoobenthos

The process of feeding on zoobenthos by grass crap can affect the biomass of these organisms significantly (Terrell and Terrell 1975; Schramm Jr. and Jirka 1989; Bozkurt et al. 2017). Variations in benthos are related mainly to fluctuations in aquatic plants (van der Zweerde 1982; Pípalova 2006), which ease sediments and offer extra substrate in the shape of root masses and decaying material (Schramm

Jr. and Jirka 1989). Zoobenthos also reacted to variations in water quality subsequent exclusion of aquatic macrophytes (Gasaway 1979; Schramm Jr. and Jirka 1989; Pípalova 2006; Bozkurt et al. 2017). It appears clear that the grass carp does not disturb zoobenthos straight. Ancillary variations at the primary level, i.e. phytoplankton, are still not properly enumerated and therefore, variations at the secondary level, i.e. zoobenthos, are not simple to demonstrate. The same looks true for fluctuations in zooplankton.

38.7.5 Effects on Fish Communities

The presence of grass carp for vegetation control decreases the spawning locations for phytophilous fish or lodgings for predatory fish and their prey. It can also incidentally influence the life of some other fish that is dependent on phytophilous animals (Bettoli et al. 1993; Pípalova 2006; Bozkurt et al. 2017).

Food fight between grass carp and other fish can occur in natural water bodies when aquatic macrophytes are detached. In pond polyculture, grass carp prefers commercial food to aquatic plants, which causes competition and decreased growth with common carp (*Cyprinus carpio* L.) (Krupauer 1968). Certain fish species augmented growth, production and survival in the presence of grass carp due to the augmented food supply because of the increased planktonivorous fish (Bettoli et al. 1993; Pípalova 2006; Bozkurt et al. 2017) or because of the fish feeding on faecal pellets (Krupauer 1968; Takamura et al. 1993; Bozkurt et al. 2017). Takamura et al. (1993) proposed that planktonivorous fish did not exploit the faeces of grass carp, but use the involved nutritive microorganisms, which are too small to be eaten directly.

Macrophytes assist as a spawning ground (especially for phytophilous fish species), food (phytophilous animals attached to them) and housing for fish. The number of mainly phytophilous fish species decreased when plants were totally eradicated or their biomass reduced significantly (Krupauer 1968; Takamura et al. 1993; Pípalova 2006; Bozkurt et al. 2017).

38.7.6 Effects on Amphibian and Water Bird Communities

Use of grass carp to control nuisance aquatic plants may reduce environment superiority for waterfowl mainly for the food materials of the grass carp and some species of water birds overlay (Benedict and Hepp 2000; Pípalova 2006).

38.7.7 Effects on Aquatic Macrophyte Community

Plant species succession usually happens when an aquatic plant is eaten from a certain place. The newly comers of plant species must have features to face the grazing impact and its extent.

It is presumed that the consequences of removal of aquatic plant species favoured by the grass carp or in contrary dispersion of disturbing alien species will be decrease of the variety of aquatic macrophyte groups (Catarino et al. 1997; Pípalova 2006).

Variety does indicate not only the number of species but also their relative abundance (biomass). Grass carp favoured the filamentous alga (*Cladophora*) to other macrophyte species and thus, it prohibited the filamentous algae from having massive growth in any aquatic habitat. The influence of the grass carp on variety in aquatic habitats can be influenced by the density and size of fish supplied, presence/absence of favoured and non-favoured species, length of fish effect and definitely on grass carp survival/movement from the system.

38.7.8 Effects on Eutrophication and Water Quality

About one half of the food materials in consumed by grass carp are utilised and digested, the remaining portion go out through the intestine as either partially assimilated or incompletely split material (Stanley 1974; Pípalova 2006).

Hydrochemistry is powerfully influenced by hydrology and morphology of water bodies and weather conditions and therefore, showing statistically important variations in water chemistry instigated by other factors (e.g. grass carp) is difficult. Water quality alteration as a consequence of plant elimination by the grass carp is most influenced in small, non-flowing water bodies and least affected when only small proportion of plants is detached from large, relatively deep, flowing reservoirs (Stanley 1974). The effects of grass carp on plants and water quality are highly adjustable and frequently indecisive due to the absence of proper management sites. The amount and rate of plant exclusion by the grass carp are critical. Higher supplying masses of grass carp or its longer influence can upsurge concentrations of nutrients (especially nitrogen and phosphorus) in the water, but these surges are mostly reliant on the water body features (Stanley 1974; Pípalova 2006).

38.8 Position of the Grass Carp in the Iraqi Inland Waters

In Iraq, the biological status of the grass is unknown, but what is known is that this species has invaded all the freshwater system in Iraq and replaces native fish species in their niches. All the research studies performed so far concentrate mainly on the influences of pollutants on the grass carp, some physiological aspects and breeding experiments. No studies have taken into consideration the impact of grass carp on the habitat or on the biota. Such unawareness will lead definitely to destroying of the freshwater environment including the biota living in.

38.8.1 Recommendations for Management of Stocking and Release of Grass Carp

The following general recommendations are set and so the countries like Iraq can adopt to solve the problem of releasing and culturing grass crap in the inland waters. Such practices are going on in Iraq without any inspection and without any licencing. Policymakers in Iraq in regard to the freshwater fish economy should take a genuine step in the track of supporting the declining freshwater environment in Iraq and solve the problems that the invasive species such the different species carp has created.

1. It is necessary to continue evaluating and update ecological hazard to develop the welfares of introduced species while dropping down unsolicited presence and impacts.
2. It may be recommended that careful evaluation is made to assess the effective uses of the species as a biological weed controller in any freshwater system against the probable impact to the lives of native freshwater species.
3. Investigations of the relationship between the addition of large amounts of undigested faeces and eutrophication should also be made.
4. Studies involving competition of the young grass carp with the young of native non-game species should be performed as well as studies on the direct effects of the adults on non-game species.
5. Barriers need to be located crossways along the emergency spillways (i.e. canals, culverts, etc.) before grass carp can be supplied into any freshwater habitat. Branches and canals of the water body need to be stocked with grass carp need to be supplied with barriers that contain net to stop grass carp from jumping as this species is very good at jumping.
6. Initial valuation of the influence needs to go outside the foundation for introducing exotic species to deliberate influence on aquatic ecology, generally, the consequence on food fishes, on waterfowl and on aquatic plants, the catchability and edibility of the species and the implication to public health. For this purpose, it may be more instructive to study a species that has already been introduced and established in its non-native environment. The species should be well known biologically in its native land.
7. The subject management of the grass carp should be publicised in appropriate newspapers, newsletters, magazines and journals and expert advice should be welcomed. No importation of a species for release purpose is so urgent that its biological implications should not be severely reviewed by a broad panel of experts from representative government and public agencies.

8. The research report should be publicised and an evaluation should be made by a panel of representatives from all involved country agencies together with scientists recommended by a national society or governmental universities. Because animals do not follow political borders, it would seem that regional agencies should be involved at the start.

References

Adamek Z, Kortan D, Lepic P, Andreji J (2003) Impacts of otter (*Lutra lutra* L.) predation on fishponds: a study of fish remains at ponds in the Czech Republic. Aquac Int 11:389–396

Bailey WM, Boyd RL (1971) Some observations on the white amur in Arkansas. Mimeograph. 10 pp. (as cited in: Shireman, J.V. and C.R. Smith. 1983. Synopsis of biological data on the grass carp, *Ctenopharyngodon idella* (Cuvier and Valenciennes, 1844). Food and Aquaculture Organization of the United Nations, Rome)

Beck RH (1996) Risk assessment for the introduction of grass carp (*Ctenopharyngodon idella*). Report to the Canadian Non-Native Species Risk Analysis Committee, Alberta, Canada. 94 p

Benedict RJ, Hepp GR (2000) Wintering waterbird use of two aquatic plant habitats in southern reservoir. J Wildl Manage 64(1):269–278

Bettoli PW, Maceina MJ, Nobel RL, Betsill RK (1993) Response of a reservoir fish community to aquatic vegetation removal. N Am J Fish Manage 13:110–124

Billard R (1997) The freshwater fish of the rivers of France. Identification, inventory and distribution of 83 species. Lausanne, Delachaux & Niestle, 192p

Blackwell BG, Murphy BR (1996) Low-density triploid grass carp stockings for submersed vegetation control in small impoundments. J Freshw Ecol 11:475–484

Bozkurt Y, Yavas İ, Gül A, Balcı BA, Çetin NC (2017) Importance of Grass Carp (Ctenopharyngodon idella) for Controlling of Aquatic Vegetation. In *Grasses-Benefits, Diversities and Functional Roles*. Amjad Almusaed A. and Sammera Mohamed Salih Al-Samaraee S. M. S. (eds.), InTech Publisher, 29-39p

Catarino LF, Ferreira MT, Moreira IS (1997) Preferences of grass carp for macrophytes in Iberian drainage channels. J Aquat Plant Manag 36:79–83

Chen WJ, Mayden RL (2009) Molecular systematics of the Cyprinoidea (Teleostei: Cypriniformes), the world's largest clade of freshwater fishes: further evidence from six nuclear genes. Mol Phylogenet Evol 52(2):544–549

Chilton III EW, Muoneke MI (1992) Biology and management of grass carp (*Ctenopharyngodon idella*, Cyprinidae) for v

Cross DG (1969) Aquatic weed control using grass carp. J Fish Biol 1:27–30

Cudmore BMNE, Mandrak NE (2004) Biological synopsis of grass carp (*Ctenopharyngodon idella*). Can Manuscr Rep Fish Aquat Sci 2705(7):44

Federenko AY, Fraser FJ (1978) Review of grass carp biology. Interagency Committee on Transplants and Introductions of Fish and Aquatic Invertebrates in British Columbia. British Columbia, Department of Fisheries and Environment, Fisheries and Marine Service, Technical Report No. 786. 15 p

Gasaway RD (1979) Benthic macroinvertebrate response to grass carp introduction in three Florida lakes. Proceedings of the Annual Conference on Southeastern Association of Fish and Wildlife Agencies 33:549–562

Hestand RS, Carter CC (1978) Comparative effects of grass carp and selected herbicides on macrophyte and phytoplankton communities. Aquat Plant Manage 16:43–50

Holdren GC, Porter SD (1986) The effects of grass carp on water quality in McNeely Lake. 6. Annual International Symp. Lake and Reservoir Management: Influences of Nonpoint source pollutants and acid precipitation, Portland 5-8 Nov. 1986:20

Krupauer V (1968) Schopnost konzumu rostlin u tˇrílety'ch a ˇctyˇrlety'ch bíly'ch Amuru° [The capacity for plant consumption of three- and four-year old grass carp]. Ž ivoč Výr 13:467–474. (in Czech with English summary
Leslie AJ, Nall LE, Van Dyke JM (1983) Effects of vegetation control by grass carp on selected water-quality variables in 4 Florida lakes. Trans Am Fish Soc 112:777–787
Leslie AJ Jr, Van Dyke JM, Hestand RS III, Thompson BZ (1987) Management of aquatic plants in multi-use lakes with grass carp (*Ctenopharyngodon idella*). Lake and Reservoir Management 3 (1):266–276
Meyer FP, Dupree HK, Muncy RJ, Pardue GB, Stanley JG (1975) Report on the status of the grass carp in North America. 10 p. (As cited in: Federenko, A.Y. and F.J. Fraser. 1978. Review of grass carp biology. Interagency Committee on Transplants and Introductions of Fish and Aquatic Invertebrates in British Columbia. British Columbia, Department of Fisheries and Environment, Fisheries and Marine Service, Technical Report No. 786)
NatureServe (2003) NatureServe Explorer: An online encyclopedia of life. Version 1.8. NatureServe, Arlington, Virginia. Available: http://www.natureserve.org/explorer. Accessed: March 3, 2004
Opuszynski K (1972) Use of phytophagous fish to control aquatic plants. Aquaculture 1:61–74
Opuszynski K, Shireman JV (1995) Herbivorous fishes: culture and use for weed management. In cooperation with James E. Weaver, Director of the United States Fish and Wildlife Service's National Fisheries Research Center. CRC Press, Boca Raton, Florida. Opuszynski *et al.* 1985)
Page LM, Burr BM (1991) A field guide to freshwater fishes. Houghton Mifflin Company, Boston
Pípalova I (2006) A review of grass carp use for aquatic weed control and its impact on water bodies. J Aquat Plant Manag 44(1):1–12
Pine RT, Anderson LWJ (1991) Effect of triploid grass carp on submersed aquatic plants in northern California ponds. Calif Fish Game 77:27–35
Pitelli RA (1998) Macrofitas aquaticas do Brasil, na condiçao de problema. In: Workshop Controle De Plantas Aquaticas. Resumos Brasília, Ibama, p 19
Richard DI, Small JW, Osborne JA (1985) Response of zooplankton to the reduction and elimination of submerged vegetation by grass carp and herbicide in 4 Florida lakes. Hydrobiologia 123:97–108
Schramm HL Jr, Jirka KJ (1989) Effects of aquatic macrophytes on benthic macroinvertebrates in two Florida Lakes. J Freshw Ecol 5:1–12
Shireman JV, Smith CR (1983) Synopsis of biological data on the grass carp, *Ctenopharyngodon idella* (Cuvier and Valenciennes, 1844). Food and Aquaculture Organization Synopsis. 135: 86pp
Skelton PH (1993) A complete guide to the freshwater fishes of southern Africa. Southern Book Publishers. 388 p
Stanley JG (1974) Energy balance of white Amur fed egeria. Hyacinth Control 12:62–66
Stanely JG, Miley WW, Sutton DL (1978) Reproductive requirements and likelihood for naturalization of escaped grass carp in the United States. Trans Am Fish Soc 107:119–128
Stout CC, Tan M, Lemmon AR, Lemmon EM, Armbruster JW (2016) Resolving Cypriniformes relationships using an anchored enrichment approach. BMC Evol Biol 16(1):244
Takamura N, Li JL, Yang HQ, Zhu XB, Miura T (1993) A novelapproach to evaluate feeding by mixed cyprinid species in a Chinese integrated fish culture pond using measurements of chlorophyll derivates and photosynthesis in gut contents. Can J Fish Aquat Sci 50:946–952
Terrell TT (1975) The impact of macrophyte control by white Amur (*Ctenopharyngodon idella*). Verh Internat Verein Limnol 19:2510–2514
Terrell JW, Terrell TT (1975) Macrophyte control and food habits of the grass carp in Georgia ponds. Verhandlungen des Internationalen Verein Limnologie 19:2515–2520
Wells RDS, Bannon HJ, Hicks BJ (2003) Control of macrophytes by grass carp (*Ctenopharyngodon idella*) in a Waikato drain, New Zealand. N Z J Mar Freshw Res 37 (1):85–93

Wittmann ME, Jerde CL, Howeth JG, Maher SP, Deines AM, Jenkins JA, Whitledge GW, Burbank SR, Chadderton WL, Mahon AR, Tyson JT (2014) Grass carp in the Great Lakes region: establishment potential, expert perceptions, and re-evaluation of experimental evidence of ecological impact. Can J Fish Aquat Sci 71(7):992–999

Zhang H, Chang WYB (1994) Management of inland fisheries in shallow eutrophic, mesotrophic and oligotrophic lakes in China. In: Dudgeon D, Lam PKS, editors. Inland Waters of Tropical Asia and Australia: Conservation and Management. Stutgart, Vol. 24. pp. 225-229

van der Zweerde W (1982) The influence of grass carp (*Ctenopharyngodon idella* Val.) on macroinvertebrates in some experimental ditches in the Netherlands. In: Proc. 2nd Int. Symp. Herbivorous fish, pp 158–164

Zweerde VD (1990) Biological control of aquatic weeds by means of phytophagous fish. In: Pieterse AH, Murphy KJ (eds) Aquatic weeds: the ecology and Management of Nuisance Aquatic Vegetation. Oxford University Press, Oxford, pp 201–221

Chapter 39
The Avifauna of Tigris and Euphrates River Basin

Omar F. Al-Sheikhly

Abstract The Avifauna of the Tigris and Euphrates River basin is unique but far to be fully explored. The geographical position at the southernmost coroner of the Western Palearctic realm and within migration routes of many bird species from cold atmospheres in Siberian platform and Europe toward warm climate in Arabia and Africa granted Iraq a remarkable diversity on its avifauna. The earlier contributions to Iraq's ornithology were dated to the early 1920s; however, several additional records and taxonomic revisions were made afterward. In Iraq, a total of 375 bird species were recorded; however, with additional new records, the total number of birds in Iraq has now exceeded 400 species including 246 landbirds, 128 waterbirds, and 25 seabirds. Iraq has 17 bird species with global conservation concern (three Critically Endangered; five Endangered; nine Vulnerable). Moreover, a total of 19 bird species were evaluated as Near Threatened. The Mesopotamian marshes (Ahwar) which occupies the Lower Tigris and Euphrates River basin has been recognized the as one the most important bird areas and endemic bird areas in the Middle East. Among its unique avifauna, several bird species are endemic or known to be restricted-range species to Iraq including Little Grebe *Tachybaptus ruficollis iraquensis*, Basra reed-warbler *Acrocephalus griseldis*, Iraq Babbler *Argya altirostris*, and Mesopotamian crow *Corvus cornix capellanus*. Moreover, isolated breeding populations of Afrotropical Goliath heron *Ardea goliath*, African Darter *Anhinga rufa chantrei*, and Sacred ibis *Threskiornis aethiopicus* are also present. Besides habitat destruction and fragmentation, species persecution through illegal hunting and trapping were highlighted as major threat impacts which were contributed to a significant decline in many bird species populations. However, the current status of the avifauna of northern mountains (Kurdistan), central steppes and aridlands, southern Mesopotamian wetlands, and marine coasts of the Arabian Gulf is comprehensively discussed.

O. F. Al-Sheikhly (✉)
College of Science, University of Baghdad, Baghdad, Iraq

L. A. Jawad (ed.), *Tigris and Euphrates Rivers: Their Environment from Headwaters to Mouth*, Aquatic Ecology Series 11,
https://doi.org/10.1007/978-3-030-57570-0_39

39.1 Introduction

Iraq has remarkable geographical position at the southeastern coroner of the Western Palearctic realm, which includes Europe, Asia (except the southern part), and North Africa. It is situated within routs of many migrant bird species from cold atmospheres in Siberian platform and Europe toward warm climate in Arabia and Africa. Iraq has also remarkable number of resident and breeding birds where they are participating as an important group of its biodiversity. There were four major geographical zones that have been described in Iraq (UNEP 2003). The northeastern highlands occupy approximately 20% of Iraq's territory. This zone extends in the area north of a line between Mosul and Kirkuk (Kurkuk) toward the mountain range reaching up to 3600 m in altitude near Iraq's northern and northeastern borders with Iran and Turkey. The upland region, which occupies approximately 10% of Iraq's territory, extends between the northern highlands and the desert plateau, between the Tigris and Euphrates rivers. The desert and semi-desert plateau, which consists of a broad stony plain and steppes with scattered stretches of sand, is occupying approximately 40% of Iraq's territory and extends to the west and southwest of the Euphrates River. A geological system of seasonal watercourses and valleys (wadis) flows from Iraq's western borders toward the Euphrates River basin. The Alluvial plain occupies approximately 30% of Iraq's territory, which is formed by combined deltas of the Tigris and Euphrates River basins. This zone extends from northern Baghdad toward the head of the Arabian Gulf near Al-Fao (Faw) in the south, covering the Mesopotamian marshes, one of the extensive important wetlands in the region. The wide range of habitats and landscapes gave Iraq marked diversity of its avifauna.

According to Birdlife International (2018), 375 bird species were recorded in Iraq; however, the total number of birds in Iraq has exceeded 400 species including 246 landbirds, 128 waterbirds, and 25 seabirds. Several bird migration routes cross Iraq, including the African-Eurasian Flyways; total of 312 migratory birds were recorded. The large open lakes, waterways, seasonal ponds, and marshlands within Tigris and Euphrates basin contribute as significant stop over and staging sites for migratory birds. Iraq has 17 bird species listed as globally threatened species. Three species were listed as Critically Endangered (Northern bald ibis *Geronticus eremite*; Sociable lapwing *Vanellus gregarious*; and Slender-billed curlew *Numenius tenuirostris*), five Endangered species (White-headed duck *Oxyura leucocephala*; Egyptian vulture *Neophron percnopterus*; Steppe eagle *Aquila nipalensis*; Saker falcon *Falco cherrug*; and Basra reed-warbler *Acrocephalus griseldis*), and nine Vulnerable species (Red-breasted Goose *Branta ruficollis*; Lesser white-fronted goose *Anser erythropus*; Marbled teal *Marmaronetta angustirostris*; Common pochard *Aythya ferina*; European turtle dove *Streptopelia turtur*; Great Bustard *Otis tarda*; Asian houbara *Chlamydotis macqueenii*; Greater spotted eagle *Clanga clanga*; and Eastern imperial eagle *Aquila heliaca*). Moreover, 19 bird species were evaluated as Near Threatened within the avifauna of Iraq (Ferruginous duck *Aythya nyroc*; Little bustard *Tetrax tetrax*; Dalmatian pelican *Pelecanus crispus*; Eurasian oystercatcher *Haematopus ostralegus*; Northern lapwing *Vanellus vanellus*; Eurasian curlew *Numenius arquata;* Bar-tailed godwit *Limosa lapponica*; Black-tailed Godwit *Limosa limosa*; Curlew sandpiper *Calidris ferruginea*; Great Snipe *Gallinago media*; Black-winged pratincole *Glareola nordmanni*; Armenian gull *Larus armenicus*; Bearded vulture *Gypaetus barbatus*; Cinereous vulture *Aegypius monachus*; Pallid harrier *Circus macrourus*; Red-footed falcon *Falco vespertinus*; Redwing *Turdus iliacus*; Meadow pipit *Anthus pratensis*; and Cinereous bunting *Emberiza cineracea*) (Birdlife International 2018).

39.2 The Geographical Landscape of Iraq

Iraq is a part of the Palearctic realm, the largest of the eight terrestrial biogeographic ecozones that have been classified for the earth. The World Wide Fund for Nature (WWF 2018) assigned five terrestrial biomes and nine ecoregions in Iraq, where five (Zagros Mountain Forests-steppe; Middle East Steppe; Mesopotamian Shrub Desert; Tigris-Euphrates alluvial salt marsh; and Arabian Desert and East Sahero-Arabian Xeric Shrublands) account 96% of the total area of Iraq are majorly found. Moreover, three freshwater ecoregions, the Arabian Interior (440), Lower and Upper Tigris, and Euphrates River basins (441,442), and one Arabian Gulf (90) marine ecoregion were also identified in Iraq (Fig. 39.1). However, the Birdlife International ornithological assessment for bird of Iraq especially those listed under "*biome-restricted bird species*" has followed a slightly different list of biomes. The Birdlife International list of biomes includes Mediterranean (ME12, does not constitute a major part in Iraq), Eurasian High-Montane (ME05), Irano-Turanian (ME06), Eurasian Steppe and Desert (ME04), and Sahara-Sindian Biome (ME13). According to the variety of geographical habitats, Iraq is biologically diverse, especially in its avifauna (Nature Iraq 2017).

39.3 Important Bird Areas in Iraq

There were previous continuous efforts to prioritize a list of principal wetlands of ornithological importance in Iraq. A preliminary list of 27 wetlands of special importance for wildfowl (ducks, geese, swans, and coots) was compiled by Savage (1968). A revised version of Savage's list was provided by Georg and Savage (1970). Information on Iraqi wetlands up to 1979 was reviewed by Carp (1980) and produced a list of 19 wetlands of international importance on the basis of the Ramsar criteria in Iraq. This list was progressively revised by Scott and Carp (1982), who provided a list of 32 wetlands known or thought to have been of some importance for waterfowl in Iraq. Scott (1993) included a total of 33 Iraqi wetlands in his provisional list of wetlands of international importance in the Middle East. Total of 42 Important Bird Areas (IBAs) covering a total area of 3,382,410 ha were

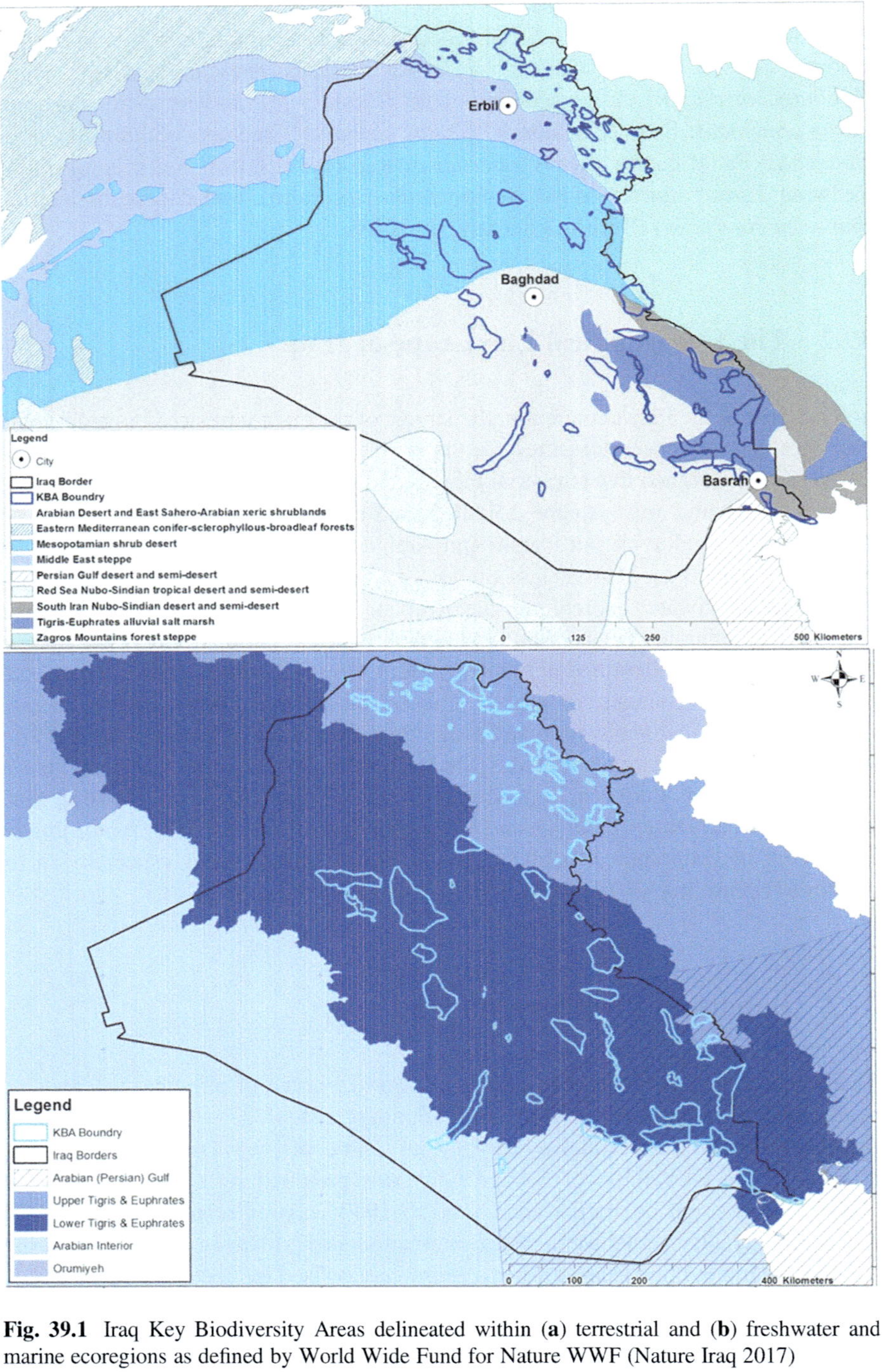

Fig. 39.1 Iraq Key Biodiversity Areas delineated within (**a**) terrestrial and (**b**) freshwater and marine ecoregions as defined by World Wide Fund for Nature WWF (Nature Iraq 2017)

listed in Iraq (Evans 1994). Moreover, The Mesopotamian Marshes was listed as Iraq's only Endemic Bird Area (EBA) (Birdlife International 2018). Resent inventory assessment of Key Biodiversity Areas (KBAs) in Iraq based on Vulnerability and Irreplaceability criteria has designated total of 67 (82%) of the total number of KBA sites covering (*c.* 27,660 km^2) in Iraq. Total of 34 sites were located in northern Iraq (Kurdistan region), 10 sites in central, and 13 sites in the southern Iraq. Total of 19 of these sites situated within the zone of the Mesopotamian marshlands' Endemic Bird Area that encompasses much of the Lower Tigris and Euphrates basin. The ornithological IBAs' criteria of species of global conservation concern (A1), restricted-range species (A2), Biom-restricted species (A3), congregations of waterbirds (A4i), congregations of seabirds/landbirds (A4ii), more than 20,000 waterbirds (A4iii), and Bottleneck sites (A4iv) were evaluated in the assessment. The Key Biodiversity Area assessment has reported total of 12 bird species of global conservation concern (IBA A1), and nine of them are found regularly. A total of 55 Sites (82%; 67 IBAs) are crucial because they support significant numbers of one or more than bird species of global conservation concern. These species are present at IBAs in Iraq in numbers reaching or exceeding the qualifying thresholds; yet, they depend on Iraqi IBAs to a greater or lesser extent for their survival (Nature Iraq 2017).

39.4 Birds of the Upper Tigris and Euphrates Basin

The Zagros Mountain Forest-steppe terrestrial ecoregion extends northwest to southwest and covering the geographical crescent of Iraq-Iran borders along with Zagros mountain chain and falls within the Kurdistan region of northern Iraq. The Birdlife International Irano-Turanian, Mediterranean, with higher elevation parts of Eurasian High-Montane biomes are covering this region. Moreover, the Upper Tigris and Euphrates River basin freshwater ecoregion is covering the total area of northern Iraq. The globally threatened and biome-restricted bird species criteria were mainly used to identify 34 KBAs distributed in northern Iraq (Nature Iraq 2017).

The surrounding mountains of the Upper Tigris and Euphrates basin are providing significant breeding and foraging habitats for many raptor species. One of the flagship bird species in northern Iraq is the endangered Egyptian vulture, a breeding summer visitor to the mountains of northern and western Iraq, and also a winter visitor and passage migrant (Fig. 39.2). Marked population of the Egyptian vulture has been recorded from 27 northern KBAs with highest count of 15 breeding pairs in Mosul Lake in 2009–2010. A large flock consists of 31 vultures, which was observed near Sulaymaniyah city on April 2010. In 2011, a flock consists of 23 adult Egyptian vultures that were observed near Pera Magroon Mountain and other four adult vultures were observed accompanying Eurasian griffon vulture *Gyps fulvus* in QaraDag Mountain where they probably breed. The open steppes and arid plains of the Upper Tigris and Euphrates basin are attracting the wintering and passage migrant large eagles such as the Vulnerable Eastern imperial and the endangered Steppe eagles. Few observations of Steppe eagle were made from the

Fig. 39.2 Egyptian vulture *Neophron percnopterus*, one of the endangered species in Iraq, Barzan-northern Iraq. Courtesy Omar F. Al-Sheikhly

mountains of northern Iraq; a juvenile Steppe Eagle was reported from Chami Rezan in Sulaymaniyah on April 2010. The endangered Saker falcon is another wintering and passage migrant bird in Iraq. Its population is decreasing around the region due to illegal trapping for falconry, impact of agrochemicals, and habitat degradation. It was recorded once in northern Iraq where wintering juvenile *F. c. cherrug* was observed at Pera Magroon Mountain in January 2012 (Al-Sheikhly 2012).

The large freshwater lakes and reservoirs mainly on Tigris River watercourse such as Dukan, Darbandikhan, and Mosul lakes were designated as Important Bird Areas in northern Iraq. Large congregations of wintering waterbirds such as 552 (3.7% of the regional/biogeographical population) of the Greater white-fronted goose *Anser albifrons*, 630 Lesser white-fronted goose, 3200 (3.2%) Great cormorant *Phalacrocorax carbo*, 1200 (1.5%) Common shelduck, 1500 (2%) Armenian gull, and two Red-breasted geese were recorded in Dukan lake. Moreover, 2400 and 3200 breeding pairs of Slender-billed gull *Chroicocephalus genei* were reported from Dukan and Darbandikhan, respectively. In Mosul Lake, 9000 (18%) of the wintering Ruddy shelduck *Tadorna ferruginea*, 1200 (1.2%) breeding summer visitor Collared pratincole *Glareola pratincola*, and Lesser White-fronted goose were recorded.

Several Biome-Restricted bird species were reported from the three biomes covering the area of northern Iraq, the Mediterranean, Eurasian High-Montane,

Fig. 39.3 **Kurdish wheatear** *Oenanthe xanthoprymna* is restricted to Irano-Turanian biome in northern Iraq (Kurdistan), PeraMagroon Mountain-northern Iraq. Courtesy Omar F. Al-Sheikhly

and Irano-Turanian biomes. Despite the fact that the Mediterranean biome does not constitute a major part in Iraq, five bird species were recorded. The northern 34 KBA sites are known or thought to hold a significant component of the group of bird species whose breeding distributions are largely or wholly confined to these biomes. Five Mediterranean-Restricted, six Eurasian High-Montane-Restricted, and 11 of (total 12) Irano-Turanian-Restricted bird species were reported. The Masked shrike *Lanius nubicus*, Somber tit *Poecile lugubris*, Western rock nuthatch *Sitta neumayer*, Eastern black-eared wheatear *Oenanthe melanoleuca*, and Black-headed bunting *Emberiza melanocephala* are restricted to the Mediterranean biome. The Caspian snowcock *Tetraogallus caspius*, Alpine chough *Pyrrhocorax graculus*, White-winged snowfinch *Montifringilla nivalis*, Radde's accentor *Prunella ocularis*, Water pipit *Anthus spinoletta*, and Red-fronted serin *Serinus pusillus* are restricted to the Eurasian High-Montane biome. The See-see partridge *Ammoperdix* griseogularis, Plain leaf warbler *Phylloscopus neglectus*, Upcher's warbler *Hippolais languida*, Ménétries's warbler *Sylvia mystacea*, Eastern rock nuthatch *Sitta tephronota*, White-throated robin *Irania gutturalis*, Kurdish wheatear *Oenanthe xanthoprymna* (Fig. 39.3), Finsch's wheatear *O. finschii*, Hume's wheatear *O. albonigra* (yet not recorded in northern Iraq), Gray-necked bunting *Emberiza buchanani*, Eastern cinereous bunting *E. cineracea semenowi* (Fig. 39.4), and Pale rockfinch *Carpospiza brachydactyla* are restricted to the Irano-Turanian biome in northern Iraq (Nature Iraq 2017).

Fig. 39.4 Eastern cinereous bunting ***Emberiza*** *cineracea* ***semenowi*** is restricted to Irano-Turanian biome in northern Iraq (Kurdistan) PeraMagroon Mountain-northern Iraq. Courtesy Omar F. Al-Sheikhly

39.5 The Avifauna of Central Iraq

The geographical area of central Iraq especially the desert and semidesert habitats was highlighted as sites with ornithological importance by Al-Dabbagh (1998). The key sites for breeding and migratory birds of prey in central Iraq were discussed by Al-Sheikhly (2012). The terrestrial ecoregions of Middle Eastern Steppes, Mesopotamian Shrub Desert, Arabian Desert of East Sahero-Arabian Xeric scrublands, and large portion of the Lower Tigris and Euphrates River basin freshwater ecoregion are largely dominating the geographical landscape of central Iraq. The Lower Tigris and Euphrates River basin in central Iraq is harboring several bird species of conservation concern. In addition, areas distributed within the geographical landscape of central Iraq seem to provide suitable stop-over and/or congregating sites for migratory birds evident by several important ornithological counts. A total of 2000 Greater white-fronted geese in Haweija Marshes to the southeast of the Little Zab River were reported. Records obtained from Tharthar Lake on the plains between the Tigris and Euphrates Rivers basin include about 4000 waterfowl including about 900 ducks of seven species, over 2000 Eurasian coot *Fulica atra*, and 23 Common crane *Grus grus* in December 1972 (Koning and Dijksen 1973). Large numbers of waterfowl were obtained from Tharthar Lake in 1988 and 1989, including up to 10,000 Mallard *Anas platyrhynchos*, 10,000 Common pochard *Aythya ferina*, and 10,000 Eurasian coot. Over 20,000 wintering waterfowl including up to 200 Gadwall *A. strepera*, 5000 Mallard, 1300 Northern pintail *A. acuta*, 300 Northern shoveler *A. clypeata*, 10,000 Common pochard, 250 Tufted duck *A. fuligula*, and 15,000 Eurasian coot were recorded in Samarra Lake on the Tigris River. Baquba Wetlands in the Tigris

River valley were formally considered an important wintering area for migratory waterfowl notably dabbling ducks. Congregating flocks of 24,000 ducks of 10 species and over 39,000 waterfowl were recorded in January 1967 and January 1968, respectively. High counts included 570 White stork *Ciconia ciconia*, 255 Glossy ibis *Plegadis falcinellus*, 5000 Mallard, 14,000 Eurasian teal *A*. crecca, 6000 Gadwall, 5000 Eurasian wigeon *A. penelope*, 5000 Northern pintail, 8650 Northern shoveler, 1500 Common pochard, and 11,000 Eurasian coot. About 180 Marbled teal were found wintering in Baquba Wetlands in January 1968, one of the few areas in Iraq where this threatened species has been found in winter. The globally threatened Greater spotted and Steppe eagles were among the wintering raptors in Baquba Wetlands. However, Baquba Wetlands had lost most of their importance for waterfowl by 1972; only 4850 waterfowl of 18 species were recorded in December 1972.

Baghdad and Attariya plains on the plains to the east of the Tigris River were the only sites in central Iraq that have been covered by the wetland survey in 1979. Some of the large congregation flocks of 412 Ruddy shelduck, 6000 Eurasian teal, 2100 Eurasian coot, and 2500 Black-tailed godwit *Limosa limosa* were recorded. Lake Al Habbaniyah and Ramadi Marshes on the south bank of the Euphrates River were formerly an extremely important staging and wintering area for migratory waterfowl. Large numbers of 200 Eurasian Spoonbill *Platalea leucorodia* have occurred in winter. Moreover, 35 Black-necked grebe *Podiceps nigricollis*, 15 Greater flamingo *Phoenicopterus roseus*, and 130 ducks of different species were recorded. Another extremely important staging and wintering area for a wide variety of waterfowl, notably grebes, cormorants, pelicans, flamingos, ducks, coots, and gulls in central Iraq is Lake Razazah (Bahr Al Milh) on the plains to the west of the Euphrates River. Over 100,000 waterfowl were recorded during very incomplete censuses in January 1975 and January 1979. Some of the high counts included up to 600 Great crested grebe, 1100 Black-necked grebe, 3000 Great cormorant, 600 White pelican *Pelecanus onocrotalus*, three Dalmatian pelican *P. crispus*, 200 Eurasian spoonbill, 3500 Greater flamingo, 730 Common shelduck, 1500 Eurasian wigeon, 3000 Gadwell, 7500 Eurasian teal, 3000 Mallard, 2000 Northern pintail, 5400 Northern pintail, 2000 Common pochard, 1600 Tufted duck, 1000 Smew *Mergellus albellus*, 102,500 Eurasian coot, five Common crane, three Mute swan *Cygnus olor*, 300 Pied avocet *Recurvirostra avosetta*, 100 White-tailed lapwing *Vanellus leucurus*, and 320 Dunlin *Calidris alpina*. The Vulnerable Eastern imperial eagle was recorded in January 1979. The lake is also of considerable importance for breeding waterfowl. In the late 1970s, about 50 pairs of marbled teal were breeding around Lake Razazah and at least another 20 pairs at Lake Usathe (Marchant and Macnab 1963; Georg and Vielliard 1970; Georg and Savage 1970; Koning and Dijksen 1973; Atkinson-Willes 1976; Al-Dabbagh 1998; Ctyroky 1987; Scott and Carp 1982; Scott 1993).

Total of 10 KBA sites were prioritized in central Iraq and consist mainly of large water bodies, lakes, and river watercourse and their tributaries. Five freshwater lakes and dams (Qadissiya, Tharthar, Himreen, Habbaniyah lakes, and Lake Razazah) were recognized as key important sites for migratory and breeding bird populations. Total of 37 individuals of the Vulnerable Lesser white-fronted goose were reported in Himreen Lake in 2011. More than 28,000 congregatory waterbirds were reported from Lake Razazah in 2008–2010. Counts of 100 breeding pairs of Marbled teal with 4350 wintering/on passage in the lake were estimated. Moreover, 300 breeding pairs (2800 wintering/passage) of Red-crested pochard *Netta rufina*, not confirmed 500 breeding pairs (3500) Greater flamingo, 500 pairs (420) White-tailed Lapwing, 400 pairs (1800) Kentish plover *Charadrius alexandrinus*, 1800 pairs (4500) Slender-billed gull, and 1500 wintering/passage Whiskered tern *Chlidonias hybrida* were reported.

The Tigris and Euphrates River surrounding areas such as river tributaries, irrigated fields, cultivated areas, and seasonal wetlands in Mahzam and Al-Alam, Haditha and Khan Al-Baghdadi, and Samara wetlands were highlighted for their ornithological importance. Furthermore, the highlands surrounding the Lower Tigris and Euphrates River basin such as Jabal (Jebel) Makhool and Himreen Foothills have been reported to host many bird species especially birds of prey. The broad range of habitats of central Iraq seems to be important for migratory raptors. The raptor migration in eastern Iraq starts in mid-September when scattered raptors observed soaring with few resident breeding species start to move southward to Arabia and Africa. The Egyptian vulture was recorded from Jabal Makhool and Jabal Himreen on the western banks of Tigris River in central Iraq where it possibly breed in 2011. However, no confirmed record has been made. The Steppe eagle was frequent to the arid plains on the eastern edge of Tharthar Lake in central Iraq. The Greater spotted and Eastern imperial eagles were reported from Jabal Himreen and Tharthar Lake in central Iraq. The Mongolian spp. of Saker falcon *F. c. milvipes* have been reported from the semidesert areas on the eastern bank of Tigris River in 2010, and total 10 falcons of the spp. *F. c. cherrug* were reported from many localities in central Iraq. The Pallid harrier has been observed during its passage migration in central Iraq, and Red-footed falcon was recorded for the first time at one locality in central Iraq (Al-Sheikhly 2012, 2012a; Nature Iraq 2017).

The Critically Endangered Sociable lapwing is known from several localities in Iraq, it was formerly believed to be a common passage migrant and winter visitor in Iraq, but now regarded as a rare passage migrant, some may winter, formerly very common in Iraq (Allouse 1961; Salim et al. 2012). The open plains, steeps, and cultivated fields found along the Lower Tigris and Euphrates basin in central Iraq are designated as key habitats for this species. Large flocks of this species were observed on the cultivated plain on the Tikrit uplands near Tigris River, Fallujah, Aquar Quf (Aqar Quf), Sheikh Said, Hib Hib, Baghdad in central Iraq, with one specimen from Nineveh (Mosul) in northern Iraq. In addition, the species was reported from the semidesert and open steppes in northwestern and western Iraq. Recently, the Sociable Lapwing global population is notably decreasing due to illegal persecution and habitat destruction, which became rare elsewhere in the Middle East. Satellite tracked birds were reported from Haditha wetlands and Baghdadi, and eastern open plains of Qadissiya Lake along the western bank of Euphrates River in central Iraq were reported during 2010–2012.

The Vulnerable Asian (MacQueen's) houbara was previously bred in desert and semidesert areas near Al-Tharthar Lake in central Iraq in the 1980s. It has been recorded at Himreen foothills in central Iraq; however, no recent breeding records have been made. It is likely that isolated scattered population of Asian houbara may breed in suitable habitats in northwestern and northeastern arid plains of Lower Tigris and Euphrates River basin after rainy seasons.

The banks of open freshwater lakes and river islets found along Tigris and Euphrates watercourses are hosting annually large flocks of wintering and migrating waterbirds, especially those belonging to the large White-headed gull complex. These freshwater habitats seem to provide suitable foraging and staging sites for these gulls in central Iraq. Large flocks of 560 Common Black-headed gull *Larus ridibundus* and 70 Caspian gull *L. cachinnans* from Al-Tharthar Lake, 1000 Common Black-headed gull from Baghdad and Attariya plains, 50 Slender-billed gull from Lake Al Habbaniyah, and 200 Black-headed gull, 200 Slender-billed gull, and 80 Caspian gull from Lake Razazah (Bahr Al Milh) were recorded. Moreover, freshwater lakes in central Iraq are attracting migrant gulls from Siberian hemisphere toward Arabia and Africa, which is evident by a recent record of Heuglin's (Siberian) gull *Larus* (*fuscus*) *heuglini*, a new gull species for Iraq reported from Lake Razazah. Recent observation of 156 wintering Armenian gull and 58 juveniles Caspian gulls from Tigris River near Baghdad was made in 2017 (Ticehurst et al. 1921, 1926; Moore and Boswell 1956-1957; Johnson 1958; Allouse 1961; Marchant and Macnab 1963; Scott 1993; Al-Dabbagh 1998; Al-Sheikhly and Al-Barazangy 2015; Nature Iraq 2017).

39.6 Mesopotamian Marshes, an Important Site for Breeding and Migrant Birds

The Mesopotamian marshes (Ahwar) have been recognized as one of the most important bird areas and endemic bird areas in the Middle East. It is hoisting larger numbers of migratory birds from the northern hemisphere toward Africa especially wader (Evans 1994). The Lower Tigris and Euphrates River basin freshwater ecoregion, which covers 227,497 km^2 (67% in Iraq), is dominating the geographical landscape of Mesopotamian marshes. The Mesopotamian marshes are situated within the geographical range of 32.44°29.55' N and 48.30°45.25′ and consist of three main marshlands: (I) Central Marshes; (II) Al-Hammar Marshes (divided in to Eastern and Western parts); (III) Al-Hawizeh Marsh. The area size of the marshes of southern Iraq is varying based on the water level inputs and annual and seasonal rain ratios. The area size estimations were varied from 8000 km^2-20,000 km^2 (Al-Mansori 2008; Hussain 2014). Due its biological and ornithological significance, these main marshlands were delineated and nominated as Ramsar sites on 2007 and 2014 and as UNESCO World Heritage sites on 17th of July 2016.

In the 1990s, the Mesopotamian marshes faced major hydrological draining operations by previous Iraqi political regime, which led to a dramatic decrease in both native and migratory bird species populations. Vast areas of reed beds, which were considered suitable breeding habitats for indigenous avifauna, were drained, and large numbers of migratory birds were enforced to change their migration routes into more prominent wetlands in the region. Prior to the marshland drainage, most of the historical records concerning the birds of the marshes of southern Iraq were restricted on few field observations related to the numbers of migratory waterfowl (Hussain 2014). After 2003, many bird species were returned to the reinundated southern Iraqi marshes along with water flow inputs but with varying levels. Most bird species were reoccupying the reinundated marshes but in small densities compared to the period before the marshland drainage during 2004–2005 (ARDI 2006; IMRP 2006).

In comparison with other biota of the marshes of southern Iraq, birds have attracted attention for many researchers and there are several publications and field notes are available to date. According to Allouse (1960, 1961, 1962), total of 132 bird species are recorded in the geographical range of the Iraqi marshes. Georg (1967) and Kinady (1976) added two species of Bewick's swan *Cygnus columbianus* and the Cotton pygmy goose or Cotton teal *Nettapus coromandelianus* to Iraq's avifauna from the southern marshes. Georg and Vielliard (1970) added further field observations about the winter migration of water birds in central and southern Iraq. Georg and Savage (1970) evaluated the status of migratory waterfowl especially ducks and waterfowl in the Iraqi marshes. During mid-winter surveys in Mesopotamian wetlands in 1979, over 475,000 waterfowl of 81 species were observed. These included almost 4000 pelicans, over 3000 Greater flamingos, 2400 geese, 140,000 ducks, 230,000 coots, 20,000 waders, and 15,000 gulls, and terns were recorded. The surveys were able to cover only a tiny fraction of the total area of wetlands in Iraq, which suggested that vast concentrations of birds were completely overlooked. The total estimated numbers of migratory waterfowl in the overall Iraqi marshes in winter were up to five million individuals (Scott and Carp 1982). Further records to the avifauna of Iraqi marshes were added later to reach more than 134 bird species (Scott 1995). However, after 2003, extensive field surveys concerning the avifauna of southern Iraqi marshes were conducted by Canada-Iraq Marshland Initiative (CIMI) in cooperation with the Iraqi Ministry of Health and Environment (IMoHEn) and other related Iraqi universities and environmental institutes. Salim et al. (2012) published a taxonomic checklist of birds recorded in the Iraqi marshes from 2005–2008. The checklist listed 159 species in which eight globally threatened species were included. After 2003, large numbers of migratory bird species were noticeably returned to the marshes. Total of (57; 40; 29) bird species were counted during 2004–2005 in Hawizeh, Western Al-Hammar, and Eastern Al-Hammar marshes, respectively. These counts were increased (62; 53; 53) during 2005–2006 (IMRP 2006). The highest numbers of waterfowl and other aquatic bird species were recorded in Hawizeh Marsh in comparison with Western and Eastern Al-Hammar marshes. The Pygmy cormorant *Microcarbo pygmaeus* was the resident breeding species in Al-Hawizeh Marsh, while the Little egret *Egretta garzetta* along with gulls and terns were the dominant species in Western and Eastern Al-Hammar marshes (Abed 2007). Total of 57 bird species were recorded in Al-Saffia sanctuary in the southern part of Hawizeh Marsh by Al-Robaee and Habeeb (2011). The highest counts were made in winter, and the total estimated migratory birds were more than 22,000 individuals. Both the Pygmy cormorant and Eurasian coot were the common bird species in Al-Hawizeh. However, the recorded bird species in Al-Safiyyah Reserve were listed as 23 resident species and 32 winter/summer visitor species, and only two species were identified as passage migrants Al-Robaee and Habeeb (2011). The analysis of data obtained from the bird surveys in the Iraqi marshes indicated that wading birds (waders) represent 53.2% from the bird counts followed by Anseriformes (16.8%) and Ciconiiformes (14.3%). The Black-winged stilt *Himantopus himantopus* was the dominant species among the bird communities with 8.6% relative abundance followed by Little tern *Sternula albifrons* with 5.8% relative abundance (ARDI 2006; Hussain 2014). Abed (2008) conducted comprehensive field surveys concerning the waterfowl of Western Al-Hammar Marsh during two succession winters. He recorded total of 14 species and showed that Eurasian wigeon *Mareca penelope* was the highest relative abundance followed by the Gadwall *Mareca strepera*. Al-Robaee (2006a, b) compared the feeding and breeding status of aquatic birds of the marshes before and after the marshland drainage in 2003, which was noticeably varying between the two eras. Moreover, the effect of the environmental factors such as temperature and water level had showed a significant impact on the water bird occurrence, abundance, and dispersal in the reinundated marshes (Abed 2007; Al-Robaee and Habeeb 2011). In addition, recent comprehensive study discussed the bird communities correlated with expected influence of climate change in the Central Marshes, which was conducted by Fazaa et al. (2017).

The Mesopotamian wetlands are situated within routes of migratory raptors and provide extremely important wintering habitat. During the mid-winter surveys in Mesopotamian wetlands in 1979, total of 15 species have been recorded including 305 Black kite *Milvus migrans*, 273 Marsh harrier *Circus aeruginosus*, one Hen harrier *C. cyaneus*,13 Pallid harrier *C. macrourus*, one Montagu's harrier *C. pygargus*, three Eurasian sparrowhawk *Accipiter nisus*, 25 Long-legged Buzzard *Buteo rufinus*, 24 Greater spotted eagle, two Steppe eagle,16 Eastern imperial eagle, two Booted eagle *Hieraaetus pennatus*, 12 Osprey *Pandion haliaetus*, 37 Common kestrel *Falco tinnunculus*, and four Merlin *F. columbarius* and Peregrine falcon *F. peregrines* (Scott and Carp 1982). The Pallas's fish-eagle *Haliaeetus leucoryphus* was formerly a scarce, rare, and uncommon winter visitor to the wetlands of Mesopotamia, but not recorded since 1944 (Moore and Boswell 1956-1957; Salim et al. 2012). The status of the White-tailed eagle *Haliaeetus albicilla* is uncertain. It bred on the cliffs by the Tigris River near Samarra in central Iraq in the 1960s, but none could be found in the late 1980s, despite extensive searches (Scott 1993). Salim et al. (2012) regarded it as former winter visitor in small numbers, not recorded in Iraq since the 1940s. However, it was one of the regular wintering raptors to Mesopotamian wetlands of southern Iraq in 1970s. Five individuals were recorded in Haur Al Shuwaija, near Qurna, and along the Shatt Al-Arab in midwinter surveys in 1972. More recently, White-tailed eagle has been recently recorded from the Central Marshes in southern Iraq (Fazaa et al. 2017). The Eastern imperial, Steppe, and Greater spotted eagles are a fairly widespread winter visitor to the Mesopotamian plains. The Eastern imperial eagle seemed more frequent to the Mesopotamian wetlands, and it is likely that the total wintering population of this species in Mesopotamia at that time exceeded 100 individuals (Scott and Carp 1982; Scott 1993). Ring recoveries of the migrant Steep eagle and Eastern imperial eagle were reported from the ecological zone of Iraqi marshes by Al-Sheikhly et al. (2017). Those recent records highlight the importance of the Iraqi marshes as wintering and/or staging site crucial for migratory raptors.

More recently, a checklist of the diurnal birds of prey was compiled as a first attempt to review all records of raptors obtained from the geographical zone of the Mesopotamian marshes. In addition, a total of 922.7–7689.5 individuals the Marsh harrier were estimated to wintering in the Central Marshes of southern Iraq (Al-Sheikhly and Al-Azawi 2019a, b).

Among the distinctive avifauna of the southern Iraq is the Critically Endangered Slender-billed curlew *Numenius tenuirostris*, which was historically recorded. Ticehurst et al. (1921) mentioned that many individuals with a specimen were recorded in a temporary winter a shallow lake in the bare desert ten miles north of Amara. Later on, a small flock (six birds) on the southern shore of Al-Hammar Marsh was found on 27 January 1979 followed by a record of a single bird found in the Haur Al-Hammar marshes near Nasiriya in the autumn of 1979. Moreover, few birds were reported from Haur Al Shuwaija near Kut and thought to be of this species. There was a major investigation on the status of this species around the region during the past few years, and there were no recent records of this species in Iraq yet; it is believed that it is possibly extinct (Allouse 1961; Scott and Carp 1982; Gretton 1991; Scott 1993; Porter and Aspinall 2010).

The Vulnerable Marbled teal is widely breeding in Iraq and thought to be entirely a breeding summer visitor to Mesopotamian wetlands (Ticehurst et al. 1921; Moore and Boswell 1956-1957; Georg and Savage 1970). It was uncertain that the bulk of the Iraqi population undertakes short migration to winter in Iran where more than 20,000 were observed in winter in south west Iran only a few kilometers from the Iraqi border in the early 1970s. However, it is regarded as local breeding resident in wetlands in central Iraq, more widespread in southern marshes where wintering populations are probably largest in the world (Salim et al. 2012). The total breeding population in Iraq is thought to be at least 4000–6000 pairs, which represents some 40–60% of the world population (Scott and Carp 1982). Recent breeding estimates of 700 pairs were reported from Dalmaj and West Al-Hammar marshes in Southern Iraq. Moreover, counts of 12,000 Marbled teal were reported wintering/on passage in Central Marshes in 2005–2010 (Nature Iraq 2017).

Two passerine species are distinctively related to the wetlands of Lower Tigris and Euphrates basin, the Basra reed-warbler, and Iraq Babbler *Argya altirostris*. The Basra reed-warbler is a restricted-range, biome-restricted, and common breeding summer visitor bird species in the extensive reed beds of central and southern marshlands of Iraq and at one site in western Iraq (Salim et al. 2012). It is along distant migrant to the subtropical or tropical zones of east Africa (Fig. 39.5). It is listed as Endangered due to rapid decrease of its breeding habitats, water scarcity and management, and climate change (Birdlife International 2018). There was a prolonged debate on the breeding ecology of this species in southern Iraqi marshes

Fig. 39.5 Basra reed warbler *Acrocephalus griseldis* on nest in the Central Marshes, restricted-range, biome-restricted, and common breeding summer visitor confined to the Mesopotamian marshes, Central Marshes-southern Iraq. Courtesy Omar F. Al-Sheikhly

made by Al-Sheikhly et al. (2013, 2015) and Porter et al. (2015). The Iraq babbler is another restricted-range and biome-restricted species that inhibits the dense reed beds of the Mesopotamian marshes and found in rural habitats along rivers and irrigation canals (Stattersfield et al. 1998) (Fig. 39.6). The Mesopotamian marshes of Tigris and Euphrates basin in southern Iraq are providing conclusive breeding habitats for both species populations. This fact is evident with an estimation of 2000 breeding pairs of Basra reed-warbler, which was reported form the Central Marshes, and 1000–2000 pairs of Iraq babbler were estimated in the Central and West Al-Hammar marshes, respectively (Nature Iraq 2017).

Another endangered bird species found in the Mesopotamian marshes is the White-headed duck, which appears to be only a very scarce winter visitor. One was shot near Kut in 1920, and one was seen near the west end of Haur Al Hammar in December 1972 (Anstey 1989). In 2005, count of 38 birds was recorded in Al-Hawizeh Marsh and four individuals were reported from Dalmaj in 2011. A single adult bird was recently observed in Shatt Al-Arab in 2017 (Nature Iraq 2017; Ahmad Majed, personal communication 2018).

The populations of two species of waterfowl confined to the Lower Mesopotamian wetlands have been described as distinct subspecies: the African darter *Anhinga rufa chantrei* and Little grebe *Tachybaptus ruficollis iraquensis*. The African darter is confined to the dense reed beds marshes of the Lower Mesopotamia mainly in eastern Hawizeh Marsh. It was formerly bred at Amik Golu (Lake Antioch) in Turkey, but extinct there following drainage in the 1950s. It was common and resident in the huge marshes round Qurna marshes where large numbers were nesting in 1921, commonly found in the Central Marshes and around

Fig. 39.6 Iraq babbler *Argya altirostris* is restricted-range and biome-restricted species that inhibits the dense reed beds of the Mesopotamian marshes, East Al-Hammar Marsh-southern Iraq. Courtesy Omar F. Al-Sheikhly

Hawizeh Marsh. Large breeding colonies were reported near Qurna in 1973, but none was reported in any of the four International Wetlands Research Bureau IWRB waterfowl surveys between 1968 and 1979, and there do not appear any records in Iraq since the early 1980s. The endemic subspecies of Little grebe is confined to southeastern Iraq and southwestern Iran, which is common and widespread breeding in dens reed beds, open freshwater lakes, and even on small temporary pools beside main highways of Mesopotamian wetlands (Fig. 39.7).

Two isolated breeding populations of Afrotropical Goliath heron *Ardea goliath* and Sacred ibis *Threskiornis aethiopicus* are present in the Lower Mesopotamian marshes of southern Iraq. Goliath heron was common breeding resident in the extensive reed beds between Basra and Qurna, from the marshes near Amara and Kut, around the Euphrates Barrage, Central Marshes, Hawizeh Marsh, and Haur Al Shuwaija. The last record was made at Haur Az Zikri in the Central Marshes in 1980. The isolated Mesopotamian population of Sacred ibis was reported from Fao in winter, Amara to Fao, and a breeding colony found in the marshes near Qurna in 1921. It was common in the Central and Hawizeh marshes in spring 1956. The species seems to have become rare in the late 1960s. However, flocks of 36 and four birds were observed in January 1979, at Haur Al Rayan and Qalat Salih, respectively. The recent field surveys in 2005–2010 indicated that Hawizeh marshes are the only wetland in Iraq, which are significant for Goliath heron, Sacred ibis, and African darter breeding populations. However, migrant individuals of African darter were reported from Tigris River in Jadriyah area in Baghdad in 2012–2013 and one adult Goliath heron was shot in Al-Tena area in West Al-Hammar Marsh in 2016. The race of Hooded crow also known as Mesopotamian crow *Corvus cornix capellanus* is an endemic resident to the date palm trees, cultivated areas, and marshlands of the Lower Tigris and Euphrates river basin (Fig. 39.8). It is widely

Fig. 39.7 Little grebe ***Tachybaptus ruficollis iraquensis*** is an endemic resident of the marshlands of Lower Tigris and Euphrates rivers basin, Central Marshes-southern Iraq. Courtesy Omar F. Al-Sheikhly

Fig. 39.8 Mesopotamian crow *Corvus cornix capellanus* is an endemic resident of the marshlands of Lower Tigris and Euphrates rivers basin, Hawizeh Marsh-southern Iraq. Courtesy Omar F. Al-Sheikhly

distributed in central and southern Iraq where it recorded breeding (Cumming 1918; Cheesman 1922; Ticehurst et al. 1921–1922, 1926; Maxwell 1957; Thesiger 1964; Moore and Boswell 1956-1957; Allouse 1961; Scott and Carp 1982; Scott 1993, Nature Iraq 2017).

Several bird species restricted to Sahara-Sindian Desert biome were recorded breeding in southern Iraqi marshes, namely, the Central and Al-Hammar marshes. Assemblage of the breeding bird communities in West Al-Hammar marsh include 1000 pairs of White-tailed lapwing, 40 Spotted sandgrouse *Pterocles senegallus*, 20 Pallid scops owl *Otus brucei*, 300 Egyptian nightjar *Caprimulgus aegyptius*, 500 Gray hypocolius *Hypocolius ampelinus*, 200 White-eared bulbul, and 1300 Dead sea sparrow *Passer moabiticus*. In addition, the marshes of the Lower Tigris and Euphrates basin provide remarkable habitat for 1% or more of biogeographical population of congregatory waterbirds, seabirds, or territorial species. Breeding populations of waterbirds such as 700 Marbled teal, 1500 Kentish plover, 3600 Slender-billed gull, and 1800 Whiskered tern have been observed in West Al-Hammar Marsh during 2005–2010 (Nature Iraq 2017). In 2017, summer surveys were conducted in the three Mesopotamian marshes and significant ornithological observations were made. Breeding colonies consist mainly of late breeding Ardeidae species such as Purple heron *Ardea purpurea*, Black-crowned night-heron *Nycticorax nycticorax*, Little egret *Egretta garzetta*, and Squacco heron *Ardeola ralloides* along with Pygmy cormorant, which were located in the West Al-Hammar marsh. Many hole nests of Pied kingfisher *Ceryle rudis*, a common breeding resident to the Iraqi wetlands, were located in the muddy banks and dicks of Central and West Al-Hammar marsh (Fig. 39.9). In addition, several natively breeding juveniles were observed combining the postbreeding molting adults such as Common moorhen *Gallinula chloropus,* Eurasian coot, Black-winged stilt, Spur-winged lapwing *Vanellus spinosus*, Red-wattled lapwing *Vanellus indicus*, and White-tailed lapwing (Fig. 39.10). More than 173 individuals of the summer breeding Egyptian nightjar were recorded in transect of 5 km on the northern edge of West Al-Hammar marsh (IMoHEn 2017).

Fig. 39.9 Pied kingfisher *Ceryle rudis*, a breeding resident to the Iraqi wetlands. Central Marshes-southern Iraq. Courtesy Omar F. Al-Sheikhly

Fig. 39.10 White-tailed lapwing *Vanellus leucurus*, a breeding resident to the Lower Tigris and Euphrates basin. Central Marshes-southern Iraq. Courtesy Omar F. Al-Sheikhly

39.7 Birds of the Coastal Habitats at the Head of the Arabian Gulf

Al-Fao (Faw) Peninsula is a significant complex of inland and coastal landscapes situated in Basra (Basrah) province in southeastern Iraq at the head of the Arabian Gulf. It comprises a triangle geographical range extending from Sihan to the northeast, Rass Al-Besha, and Al-Fao tongue to the southeast and Umm Qasr to the southwest. Al-Fao Peninsula represents Iraq's southern marine outlet toward the Arabian Gulf. The eastern part of Al-Fao Peninsula is bordered by Shatt Al-Arab River to the south of Tigris and Euphrates confluence at Qurna, which represents the geopolitical borders with Islamic Republic of Iran. There are several marshes that lie along the *c.* 165 km of Shatt Al-Arab waterway toward Rass Al-Besha and Arabian Gulf. Khawr Abd allah represents the south and western parts of Al-Fao Peninsula. There are numerous areas of swampy grasslands (*c.*900 km^2) and intertidal mudflats (*c.* 360 km^2) extending from the region of Rass Al-Besha and Al-Fao tongue at the mouth of the Shatt Al-Arab west along the northern shore of Khawr Abd allah for at least 50 km along the northern edge of Bubiyan Island in Kuwait. The mudflats and swampy flats are backed by a belt of date palms and silt desert with large tidal (>3 m) amplitude (Evans 1994). Al-Fao Peninsula was originally highlighted as a significant area for migratory and breeding birds by Cumming (1918). Shatt Al-Arab Marshes and Khawr Abd allah were listed as wetlands of international importance by Carp (1980) and have been identified as Important Bird Areas (IBA040 and IBA042, respectively) by BirdLife International. In January 1968, 1975, and 1979, a brief and incomplete mid-winter waterfowl survey was undertaken by the International

Waterfowl and Wetlands Research Bureau in collaboration with the Museum of Natural History in Basra. The survey recorded large numbers of migratory birds in Shatt Al-Arab Marshes and Khawr Abd allah of Al-Fao Peninsula (Evans 1994). The marshes of Shatt Al-Arab are important for wintering geese and dabbling ducks (Savage 1968). The muddy shores of Shatt Al-Arab are known to be important feeding area for passage and wintering shorebirds. However, some of the large flocks (>100 individuals) of waterfowl and migratory raptors such as 105 White stork *Ciconia ciconia*; 305 Black kite 100 Common redshank *Tringa totanus*; 400 Little stint *Calidris minuta*; 500 Dunlin *Calidris alpina*; 4000 Common black-headed gull; and 130 Caspian gull *Larus cachinnans* were recorded (Carp 1975a, b; Georg and Vielliard 1968, 1970; Carp and Scott 1979; Scott and Carp 1982; Scott and Evans 1993). The mudflats of Al-Fao Peninsula are believed to be a major staging and wintering area for migratory waterfowl and were listed as such by Summers et al. (1987), but very little information is available. A single visit to tidal mudflats at the head of the Arabian Gulf near Fao was conducted by Georg and Vielliard (1968). Georg and Vielliard (1970) mentioned that the extensive areas of intertidal mudflats around Al-Fao or in the Khor (Haur) Zubair are probably the most important wintering areas for waders in Iraq, and total of 2015 waders were counted. Some of the large migratory flocks (>100 individuals) of waterfowl and shorebirds are 200 White stork*;* 600 Eurasian wigeon; 1200 Mallard; 200 Northern pintail; 230 Kentish plover; 510 Eurasian curlew *Numenius arquata;* 110 Bar-tailed godwit *Limosa lapponica*; 100 Terek sandpiper *Tringa cinerea*; and 1000 Dunlin (Georg and Savage 1970; Georg and Vielliard 1968, 1970; Scott and Carp 1982; Scott 1993). However, the area seems to be more important for migratory gulls and terns. Large flocks of 1100 Slender-billed gull; 700 Caspian gull; and 200 Gull-billed tern *Gelochelidon nilotica* were recorded (Scott and Carp 1982; Scott 1993; Evans 1994). More recently, two Large White-headed gull species, the Steppe gull *Larus (cachinnans) barabensis* and Heuglin's (Siberian) gull, were newly added to the avifauna of Iraq from Al-Fao peninsula by Al-Sheikhly and Al-Barazangy (2015). These two taxa are probably passing and foraging over the entire geographical zone of the southern Iraqi wetlands during their winter migration; however, their status in Iraqi is still uncertain. These observations highlighted the importance of Al-Fao Peninsula for migratory/wintering large white-headed gulls in Iraq and the Middle East (Al-Sheikhly and Al-Barazangy 2015).

The importance of the Al-Fao Peninsula for migratory birds as staging or congregating site is not fully known; but it is thought to be. The variety of inland and coastal areas of Al-Fao-Peninsula seems to be important feeding habitats for wintering and migratory waterfowl and shorebirds. Scott and Carp (1982) indicated that the midwinter surveys were able to visit only a tiny fraction of the total area Iraqi wetlands, while many of the areas that were not visited might hold relatively few birds. It is equally likely that vast concentrations of birds were completely overlooked. Moreover, the area has faced habitat destruction, industrial pollution, and extensive disturbance during the Gulf War I and II. Very few recent mid-winter surveys in Al-Fao Peninsula were conducted; yet, further surveys especially in autumn and spring are required. Surveys may reveal additional important counts of migratory bird populations, which will support further conservation strategies in Iraq and the Middle East.

39.8 Major Threats on the Avifauna of Iraq

The avifauna of Iraq is facing a notable decease in their populations due to broad scale of threat impacts. An assessment of total of 11 threat categories defined by the International Union for Conservation of Nature and Natural Resources (IUCN) on the key biodiversity areas in Iraq was conducted in 2010. Each threat type was evaluated according on its timing, scope, and severity in order to develop an integrated threat status score, an approach adapted from Birdlife International's report on monitoring Important Bird Areas. Not all threats are equally affecting the monitored sites; however, the over-exploitation of species such as hunting and trapping had a greater impact score on biodiversity and avifauna of many sites. In recent study on migratory large eagles in central southern Iraq, threat impacts like species direct persecution (illegal trade, trapping, and shooting) and possible electrocution and pesticide poisoning were identified as threats adversely impacted on migratory eagles in Iraq (Nature Iraq 2017; Al-Sheikhly et al. 2017).

References

Abed JM (2007) Status of water birds in restored southern Iraqi marshes. Marsh Bulletin 2(1):64–79
Abed JM (2008) Restoration rate of waterfowl population in the restored marshes of southern Ira. Marsh bulletin 3(1):67–80
Al-Dabbagh KY (1998) The birds of semi-desert areas of Central Iraq. Sandgrouse 20:135–141
Allouse B (1960, 1961, 1962) Birds of Iraq I III Vols. [Arabic]. Al-Rabita Press. Baghdad
Allouse B (1961) Birds of Iraq. Al-Rabita Press, Baghdad. [Arabic]
Al-Mansori FY (2008) The future expectations for the marshlands restoration. Ph.D. thesis. Basra University. 174pp
Al-Robaee KH (2006a) The breeding of waterbirds in the marshlands of Mesopotamia. Marsh Bulletin 1(1):40–46
Al-Robaee KH (2006b) The effect of the environmental desiccating on water birds of southern Iraqi marshes report. 9 p
Al-Robaee KH, Habeeb MQ (2011) The negative effect of drought on the composition of waterfowls community in Al-Saffia sanctuary. Marsh Bulletin 6(2):98–111
Al-Sheikhly OF (2012) Some Ecological Observations on Lesser Kestrel *Falco naumanni* in Northern and Northern West of Iraq. M.Sc. Thesis. University of Baghdad, Iraq
Al-Sheikhly OF (2012a) Report on the first record of red –footed falcon *Falco vespertinus* in Iraq. Falco 39:10–11
Al-Sheikhly OF, Al-Azawi AJ (2019a) The Diurnal birds of Prey (Raptors) in the Mesopotamian Marshes of Southern Iraq with notes on their conservation status. Bull Iraqi Nat Hist Mus 15(4):381–402
Al-Sheikhly OF, Al-Azawi AJ (2019b) Migration Pattern and wintering population of the Eurasian marsh harrier Circus aeruginosus in the Central Marshes, a wetland of international importance in southern Iraq. Raptor J 13:127–137

Al-Sheikhly OF, Al-Barazangi AN, Mukhtar KH, Fazaa N, Abdulzahra HK, Abou Turab MK, Al-Azawi AJ (2017) Ring recoveries from steppe eagles and eastern Imperial eagles from the Russian and Kazakhstan breeding populations and a review of major threats to eagles in Iraq. Raptors Conservation (35):51–61

Al-Sheikhly OF, Al-Barazangy AN (2015) Occurrence of Heuglin's Gull *Larus heuglini* and steppe Gull *Larus (cachinnans) barabensis* in Iraq. Mar Ornithol 43:245–248

Al-Sheikhly OF, Nader I, Barbanera F (2013) Breeding ecology of the Basra reed warbler, *Acrocephalus griseldis*, in Iraq (Aves: Passeriformes: Acrocephalidae). Zoology in the Middle East 59(2):107–117

Al-Sheikhly OF, Nader I, Barbanera F (2015) A response to the comment of Porter et al. (2014) on "breeding ecology of the Basra reed warbler *Acrocephalus griseldis* (Hartlaub, 1891) in Iraq (Aves: Passeriformes: Acrocephalidae)". Zoology in the Middle East 61(2):193–199

Anstey S (1989) The status and conservation of the white-headed duck, *Oxyura leucocephala*. IWRB spec. Pub!. No. 10. IWRB, Slimbridge, UK

ARDI (Agriculture Reconstruction and Development Programme of Iraq) (2006) Monitor Marsh Ecosystem recovery, Final Report. USAID/Development Alternative, Final report 177pp

Atkinson-Willes GL (1976) The numerical distribution of ducks, swans and coots as a guide in assessing the importance of wetlands in midwinter. In: Smart M (ed) Proc.int. conference on conservation of wetlands and waterfowl, Heiligenhafen, Federal Republic of Germany, December 1974. U.K, IWRB, Slimbridge, pp 199–254

Birdlife International (2018) Data zone: Iraq. Retrieved from: http://datazone.birdlife.org/country/iraq

Carp E (1975a) Report on a mission to Iraq: January/February 1975. Unpublished report in IWRB files. 13 pp.

Carp E (1975b) Waterfowl counts in Iraq. IWRB Bulletin 39(40):51–55

Carp E (1980) A Directory of Western Palearctic Wetlands. UNEP, Nairobi, Kenya, and IUCN, Gland, Switzerland. (Iraq: pp. 187–191)

Carp E, Scott DA (1979) The Wetlands and Waterfowl of Iraq. Report on the Joint Expedition of the International Waterfowl Research Bureau and the University of Basrah, Iraq: 10th January to 3rd February 1979. Unpublished report in IWRB files. 34 pp.

Cheesman RE (1922) Zoological investigations in the Persian Gulf and Iraq. J Bombay Nat Hist Soc 28:1108–1110

Ctyroky P (1987) Ornithological observations in Iraq. Beitr. Vogelkd. 33 (3/4): 141204

Cumming WD (1918) Natural history notes from Fao. J Bombay Nat list Soc 26:292–294

Evans MI (1994) Important bird areas in the Middle East, Bird life conservation series no. 2. Bird Life International, Cambridge, U.K., p 410

Fazaa NA, Dunn JC, Whittingham MJ (2017) Distributions and community composition of birds in Iraq's central marsh. Int J Biodiversity 2017:1–28

Georg P (1967) Cygnus bewickii yarrell Bewick's sawn; an addition to the avifauna of Iraq. Iraq Nat Hist Bull 3(2):11–13

Georg PV, Savage CDW (1970) Status of the Main wildfowl resorts in Iraq. In: Isakov YA (ed) Proc. Int. regional meeting on conservation of wildfowl resources, Leningrad 1968: 328–332. Moscow

Georg PV, Vielliard J (1968) Waterfowl counts in Iraq: 21 December 1967–18 January 1968. Unpublished notes in IWRB files. 7 pp.

Georg PV, Vielliard J (1970) Midwinter observations on birds of central and South Iraq. Bull Iraq Nat Hist Mus IV(4):61–85

Gretton A (1991) The Ecology and Conservation of the Slender-billed Curlew (*Numenius tenuirostris*). ICBP Monograph No. 6. ICBP, Cambridge, U.K.

Hussain NA (2014) Biotopes of the Iraqi marshes, 1st edn. Dhifaf publishing house, Basra, Iraq. 432pp

IMRP (Iraq Marshlands restoration Programme) (2006) Monitor marsh ecosystem recovery. Final report. USAID/development Alternative Inc. 528pp

Iraqi Ministry of Health and Environment (IMoHEn) (2017) Biodiversity Survey Report in the Marshes of Southern Iraq. Baghdad, Iraq pp. 61

Johnson LR (1958) Field notes on some of the birds of Iraq. Iraq Nat Hist Mus 16:1–32

Kinady PVG (1976) The Indian pygmy goose *Nettapus coromandelianus*, in Basrah, Iraq. Basrah Nat Hist Mus 3:107–109

Koning FJ, Dijksen LJ (1973) IWRB Mission to Iraq and Syria, December *1972*. IWRB Bulletin 35:57–62

Marchant S, Macnab JW (1963) Iraq bird notes - 1962. Bull Iraq Nat Hist Inst 2(3):1–48

Maxwell G (1957) A reed shaken by the wind; a journey through the unexplored marshlands of Iraq. Penguin, Harmondsworth, p 223

Moore HJ, Boswell C (1956-1957) Field observations on the birds of Iraq. Parts I and II1956, Part III 1957. Iraq Nat. Hist Mus. Pub!. Nos 9, 10 and 12

Nature Iraq (2017) Key Biodiversity Areas of Iraq. Sulaimaniyah, Iraq. 297 pp.

Porter RF, Aspinall S (2010) Birds of the Middle East. Helm Field Guides, UK. 348 pp.

Porter RF, Batten L, Burton J, Collinson JM, Cowan P, Kennerley P, Kirwan GM, Newell RG, Pearson D, Riddington R, Salim M, Sheldon R, Scott D, Woodcock M (2015) Towards a better understanding of Basra reed warbler *Acrocephalus griseldis* (Aves: Passeriformes: Acrocephalidae) ecology? A comment on Al-Sheikhly et al. (2013). Zoology in the Middle East 61(2):190–192. https://doi.org/10.1080/09397140.2015.1023424

Salim MA, Al-Sheikhly OF, Majeed KA, Porter RF (2012) Annotated checklist of the birds of Iraq. Sandgrouse 34(1):3–44. Sandgrouse 4: 6-76

Savage CDW (1968) The wildfowl and wetland situation in Iraq. In: Elliott HFJ (ed) Proc. technical meeting on wetland conservation, Ankara-Bursa Istanbul, October 1967. IUCN publications new series, vol 12, pp 128–133

Scott DA (1993) Wetlands of West Asia–A regional overview. *In:* Moser, M. and van Vessem, J. (eds), Wetland and Waterfowl Conservation in South and West Asia: 9–22. Proc. International Symposium, Karachi, Pakistan 14–20 December 1991. IWRB Special Publication No. 25. AWB Publication No. 85. IWRB, Slimbridge, U.K. and AWB, Kuala Lumpur, Malaysia

Scott DA (1995) A directory of wetland in the Middle East. IWRP, Slimbridge, UK, pp 223–301

Scott DA, Carp E (1982) A midwinter survey of wetlands in Mesopotamia, Iraq: 1979. Sandgrouse 4:60–76

Scott DA, Evans MI (1993) Wildlife of the Mesopotamian marshlands. Report prepared for Wetlands Ecosystem Research Group, University of Exeter, U.K. 146 pp.

Stattersfield AJ, Crosby MJ, Long AJ, Wege DC (1998) Endemic bird areas of the world: priorities for bird conservation. Bird Life International, Cambridge, U.K

Summers RW, Underhill LG, Pearson DJ, Scott DA (1987) Wader migration systems in southern and eastern Africa and western Asia. In: Davidson, N.C. and Pienkowski, M.W. (eds), The Conservation of International Flyway Populations of Waders: 15–34. Wader Study Group Bull. 49, Supplement. IWRB Special Publication No. 7. IWRB, Slimbridge, U.K.

Thesiger W (1964) The Marsh Arabs. Penguin, Harmondsworth. 233 pp.

Ticehurst CB, Buxton PA, Cheesman RE (1921–1922) The birds of Mesopotamia. J. Bombay Nat. Hist. Soc. 28: 210–250, 381–427, 650–674, 937956

Ticehurst CB, Cox PZ, Cheesman RE (1926) Additional notes on the avifauna of Mesopotamia. J Bombay Nat Hist Soc 31:91–119

United Nations Environmental Programme (UNEP) (2003) Desk study on the environment in Iraq. UNEP, Switzerland

World Wide Fund for Nature (WWF) (2018) Ecoregions. Retrieved from: https://www.worldwildlife.org/biome-categories/terrestrial-ecoregions on 4 June 2018

Chapter 40
The Feasible Approaches to Assist Migratory Birds Visiting the Southern Reaches of Mesopotamia

Laith A. Jawad

Abstract Certain species of birds migrate for several aspects of their biological need. With such migration, birds cross vast numbers of distances in order to locate places with food and other places for reproduction and looking after their young.

The conservation of the migratory birds is important to implement by all the countries that such bird passing through their airspace like Iraq. There are several approaches that can be followed in order to conserve the migratory bird, but the present chapter is too small to accommodate such a comprehensive review.

In the present chapter, information has been given about the importance of the geographical location of Iraq and why Iraq should care about the migratory bird that passing through its airspace. Also, two suggestions were given to assess the migratory birds while they are staying the wintering place.

40.1 Introduction

Birds have been under the attention of human as early as the second millennium B.C. when the ancient Mesopotamians have normally used birds as a means to tell the future. Predictions could be made in two different ways: either observing their flights and behaviours, or their physical peculiarities prior and after the sacrifice (McEwan 1980).

Some species of birds have what is known as a natural process or avian migration. In this process, birds fly over vast areas to find the suitable habitats containing food and appropriate for reproduction and looking after their young.

The issue of bird migration is very spacious and types of migration that birds usually follow are several. The mainstream of birds flies from areas at the northern hemisphere in the summer to southern wintering areas. However, some birds breed in southern regions of Africa and move to northern wintering grounds, or

L. A. Jawad (✉)
School of Environmental and Animal Sciences, Unitec Institute of Technology, Auckland, New Zealand

L. A. Jawad (ed.), *Tigris and Euphrates Rivers: Their Environment from Headwaters to Mouth*, Aquatic Ecology Series 11,
https://doi.org/10.1007/978-3-030-57570-0_40

horizontally, to relish the slighter coastal climates in winter. Further birds fly in terms of altitude, travelling higher up a mountain in summer, and be located in on lowlands throughout the winter months.

Migratory birds infrequently fly to their target non-stop but intrude their journey often to repose and nourish, or to avoid unfavourable weather. Precisely how migrating birds discover their flyways is not entirely implicit. Contemporary studies show that they position along the Earth's magnetic field through distinct light receiver situated in their eyes.

The present chapter looks at the possibility of introducing some ideas that assist the migratory birds while they are visiting the Iraqi lands. This group of birds have received no attention at all in Iraq other than recording their species, but putting conservation measures to protect them have not been implemented yet. Although the suggestions given in this chapter are few, it will be considered a good start for further comprehensive conservation plan to protect migratory birds in Iraq.

40.2 Protection of the Migratory Birds

Migration is a dangerous journey and contains variety of intimidations. Only a trivial number of birds are really endangered by natural measures. Anthropogenic events are the cause for most hazards migratory birds are facing this threat, but the impact varies and depends on the habits of people habits in the different countries and on their activities.

The main threats that face migratory birds are damage of homes owing to contamination or misuse instigated by violation for establishment, agriculture, foraging. Those issues urge migratory birds to discover appropriate areas for reproduction and spending the cold weather in worm places. Contrarily, the damage of any of these places utilized by the birds throughout their yearly sequence might have a noteworthy influence on the birds' likelihoods of existence. Likewise, high-voltage power lines and wind turbines have an intense influence on birds, which are in danger of being exterminated by electrocution or collision. Stealing continues broadly performed in regions where people are highly reliant on biodiversity for their livings.

Flying for long period to cover large areas contains the overpass many international boundaries and arriving not the same political regions with their own habitual politics, regulation, and preservation actions. Without global collaboration, all trials in use to attack the pressures to migratory birds in one country might be ineffective if for instance unmaintainable issue is recognized in another country.

Nonetheless, the further we acquire about birds over scientific studies, the more we are surprised by their unbelievable attainments and their bizarre abilities. Each bird species has its own way, with its own leaving and coming times, its own means to travel (making many stops or flying as far and fast as possible with almost no stops), and its own conducts (flying unaccompanied or with a flock).

Migration is not an easy job, and it is an actual test. Birds fly mostly to pursuit for nourishment. Nonetheless, how do they identify it is time to go? For certain, it is not the deficiency of food that drives them to put away, since they depart long before the food comes to vanish. Literally, it looks that it is the length of the daytime and the temperature that impact the birds' hormones. That is in what way, in springtime, the birds "feel" it is time to depart. They then get ready for the migration: they provide "fuel" in order to have sufficient energy over the entire journey. Actually, throughout two weeks, they are going to eat a lot more than normal and, thanks to the hormone variations, their body is going to stock these assets very fast, acquiring mostly fat.

Many dangers threaten the birds throughout the voyage: storms, predators, vanishing of their usual stop locations where they are used to discover respite and nutrition, etc. The question of how do migratory birds find their way still challenges numerous researchers. It has been exhibited that birds use numerous direction "tools." They can use the sun, for example, which denotes that they always "know" what time it is, in order to know the correct direction on the base of the sun's location. They are also aware of the ultraviolet rays which enter the clouds but are unseen by human beings. Even the night-time birds use the location of the sun at sunset to know their location.

Night-time birds likewise use the stars. This has been shown by letting birds fly in a planetarium and altering the stars' location. Additional means is the earth's magnetic field (earth's north and south magnetic poles). A few birds, like pigeons, have a small zone in their brain made of magnetite (magnetic mineral), just like a small compass. Nonetheless, other investigators deliberate it is somewhat in their eyes that some birds have a system which specifies them where the magnetic north is. In addition, birds also utilize their information of the landscape as they trail rivers, valleys, or roads, or locate themselves with specific mountain tips.

The birds doing the migration trip on their own distinguish their pathway by "instinct." Others, flying in groups, have to acquire the path with their parents during the first flight. That is the situation of geese, cranes, and swans.

40.3 Why Iraq should Care About Migratory Birds Assess them While They are on Transient Visa in its Land

The area that migratory birds visits in Iraq during their stay is a natural freshwater area or wetland rich in vegetation that typically covered the bulk of the water bodies. The habitats are rich in plants such as the common reed (*Phragmites communis*), reed mace (*Typha angustata*) that form excellent nesting sites for migratory birds.

The southern inland water bodies of Iraq form a key area to receive bird species from cold regions to spend their winter in a worm place. These bird species come from various parts of the world. Two-thirds of West Asia's wild bird spent their winter, assessed at some million, and are thought to exist in the marshes of Al Hammar and Al Hawizeh. Insufficient ornithological investigations have specified

134 bird species in important numbers from the area, of which at least 11 are worldwide endangered. Additionally, the marshes have been specified as one of the eleven freshwater marsh areas in the world with Endemic Bird Area status (BirdLife International). They upkeep nearly the whole universal population of two species, the Basrah Reed Warbler and Iraq Babbler, in addition to utmost of the world population of Grey Hypocolius (Maltby 1994).

40.3.1 Recent Iraq Initiatives Toward Bird's Conservation

In the last decades, Iraqi initiatives to conserve the biodiversity have revealed the creation of Key Biodiversity Areas (KBA) (Salim et al. (2009). This project has been guided by Nature Iraq (NI) since 2004. An outline of the KBA Project is accessible in Rubec and Bachmann (2009). In the process of performing the project, availability of historical data about the biodiversity in Iraq has assisted in this aspect considerably. Such historical data helped in assessment with present observations, enabling the studies of habitat variations and evaluating the ecological standing of each location. Widespread waterbird data made over the last 40 years through Europe and Asia—notably by Wetlands International and BirdLife International—are enabling such regional comparisons. Therefore, data on birds were considered the major gauge of the health of Iraq's biological means, mainly as they are also a significant constituent of global struggles to preserve and guarantee wise use of wetlands and other biologically imperative ecosystems. Exhaustive field comments and meetings since 2004 with locals have permitted development of a good sign of the biological health of Iraq's key locations for biodiversity. The KBA investigation places are frequently present within very widespread wetlands (as described in Rubec and Bachmann 2009). Studies on the field investigations for the southern and northern KBA projects have been organized by Nature Iraq (Abdulhasan and Salim 2008; Ararat et al. 2008).

40.4 Recent Bird's Observation in Iraq

Salim et al. (2009) gave an important observations on birds in the northern and southern parts of Iraq made during the 2005 to 2008 surveys.

The list of birds noted during the summer and winter surveys from 2005 to 2008 has shown that 159 species, of which 53 were breeding with an additional 10 perhaps or maybe breeding. A total of 44 species are deliberated to be resident. Additionally, 110 species were perceived as winter visitors from their European and Asian breeding areas. As would be likely, the bulk of species saw in the southern marshes are wetland dependant—107 in total. Of great importance is the fact that eight species reported in the southern surveys are "globally threatened (GT)" and a further 26 species are judged to be of "conservation concern (CC)." Specific care to these two groups of birds was paid during the surveys. Of the 34 species of preservation apprehension recorded during the 2005 to 2008 KBA surveys, eight are globally endangered. They are either resident breeders, winter visitors, or passageway migrants.

40.5 Possible Approaches to Assist Migratory Birds in the Southern Part of Iraq

With the important geographical location of Iraq and with the large number of migratory birds that visiting this land, it is important for the policymakers in Iraq to take action toward the conservation of these birds that passing through airspace. There are several ideas that can be implemented to enhance and protect the visiting migratory birds while they are in Iraq, but the scope of this chapter is not large enough to accommodate explaining all these ideas. Therefore, selected approaches will be described below and for those who are interested in the application of more ideas need to consult references mentioned in this chapter in the future.

40.5.1 Artificial Nests

The use of the imitation nests is to guarantee reproductive achievement of birds. Among the benefits of these nests are that the investigator can regulate the number and distribution of nests in the experiment. These types of nests are modest to place and easier than locating natural nests. Contrarily, the rates of disorder on artificial nests are alike to rates of that on actual nests (Yahner and DeLong 1992).

There are several important points in which the artificial nests differ from natural nests. These differences may affect predation proportions. For example, absence of a hatching adult allied with artificial nests may disturb a predator's skill to find nests. If adults either disguise the natural nest or are competent to energetically protect the nest from predators, rates of predation may be greater on artificial nests. Additional vital dissimilarity between artificial and natural nests is that the former comprises eggs but not nestlings. If predators are fascinated by the pleading calls of nestlings (Haskelld 1994), appraised rates of predation will be low for artificial nests. On contrary, a few predators might be capable to eat eggs but not nestlings. Therefore, predation proportions would be greater for artificial nests than for natural nests. Artificial and natural nests also may be unlike in the size and colour of the eggs and the odour related with the nest; all issues have been shown to impact rates of nest predation.

40.5.1.1 Highlights of the Results of Using Artificial Nests

Wilson et al. (1998) have concluded that rate of predation data obtained from the artificial nests is too close to those collected from the natural nests. On the other hand, the rates are higher for artificial nests than the natural, but these results agree with those of the other investigators (Reitsma et al. 1990). Wilson et al. (1998) have also recommended that higher proportions of nest predation linked with artificial nests might have happened since some predators were competent to consume eggs but not nestlings. Therefore, artificial nests may have been more susceptible as they only held eggs.

Rates of nest predation may be affected by the difference in odour between artificial and natural nests (Wilson et al. 1998). Mammals are customary predators on artificial nests and can signal in on human odours to find nests (Whelan et al. 1994). On the other hand, Roper (1992) proposed higher rates of predation on natural nests than on artificial nests and assign this result to the size of the eggs used in the artificial nest.

As a recommondation, it seems feasible to adopt the suggestions of Wilson et al. (1998), (1) artificial nests must be used mainly for trial studies to test specific theories prior to finding lively nests or in combination with investigations on actual nests to deliver an self-governing basis of data (Ammon and Stacey 1997); (2) data from artificial nests would not be included to measure real proportions of predation or to propose how reproductive achievement differs with landscape and habitat topographies in areas where cowbird parasitism has a large effect on reproductive accomplishment; and (3) in the case of using artificial nests, investigators must endeavour to decrease variances between artificial and natural nests and must increase the number of self-governing places involved in the study at any time possible.

40.5.2 Light to Assess Nocturnally Migrating Birds

Artificial light introduced as a result of human activities has damaging impact on varied animal taxa (Davies et al. 2014). On this aspect, lights straddling over several buildings appeal to night-time flying birds, ensuing in deadly crashes; these accidents subsidize to hundreds of millions of birds passing away yearly in the United States (Horváth et al. 2009; Loss et al. 2014). Moreover, high, set alight constructions confuse birds (Longcore and Rich 2004), which can disturb them to use extra energy throughout flying.

Migratory birds relay in part on astronomic signals for location (Wiltschko and Wiltschko 1996), and birds may become confused when man-made light modifies the observed horizon (Herbert 1970). Diverse sorts of lights may curtail the influence of man-made lighting (Evans et al. 2007; Poot et al. 2008; Doppler et al. 2015), but such bird-friendly lighting is not common.

Most lights produced from human events are at ground level, and a few studies occur on the impact of ground-level lights on migratory birds (Evans et al. 2007). Ground-level human-generated lights affect other features of the performance of birds, such as the timing of nest start, the timing of the dawn chorus, and the occurrence of extra-pair copulations (Kempenaers et al. 2010). Whether prevalent ground-level illumination affects migratory performance of birds has gotten little consideration; nonetheless, utmost migratory birds fly over uncountable man-made lights throughout their journey.

In assessing the impacts of lights produced by human on nocturnal birds, we can assess the influence of these lights on migratory performance (Hüppop and Hilgerloh 2012). There are numerous technical approaches that can be used in following migratory birds and investigate their reactions to man-made disruption. Radar expertise enables capacities of the size, speed, and positioning of migratory bird herds (Gagnon et al. 2010) but cannot determine individual birds or the species contents of migratory groups (Balcomb 1977). Bird banding delivers the capability to study individual birds but does not sample migrants during lively migration. Audio observing of the dialogues made by migratory birds is an auspicious method since it does not hurt from either of these restrictions (Farnsworth 2005). Numerous night-time migrating birds make night flight noises, which are short, high-frequency songs that be unlike in audio construction across species or groups of species (Lanzone et al. 2009).

40.5.2.1 Highlights of the Results on Assessing Nocturnal Migratory Birds

Nocturnal migrants birds frequently fly jointly in herds (Larkin and Szafoni 2008). Inside these groups, night flight songs might permit birds to uphold interaction with other individuals, or they might help in location by keeping herd unity and flight course (Hamilton 1962). Assumed that human-made light has been revealed to confuse night-time migrants (Herbert 1970), the detected upsurge in noises could disclose the birds' requisite for additional alignment signs when flying over well-illuminated locations. The confusion might instigate birds to lesser their altitude, conveying more birds within the range of our recorders, or it could make them to continue in the well-lit recording regions for lengthier periods, leading to an augmented rate of recognition.

Man-made illumination might lead migratory birds to fly at uncommon altitudes, to trail indirect migration tracks, to circle above well-illuminated locations, or to call at higher rates. These lights seemed to cause birds to migrate extravagantly, growing the active strains or time necessities for migration, which consecutively may cut the likelihood of individual birds enduring migration or impact the body state of individuals reaching at the wintering or breeding grounds. These impacts might have an undesirable influence on migratory birds, emphasizing the significance of the investigations of the concerns of man-made modification of the natural habitat.

References

Abdulhasan NA, Salim MA (2008) Key biodiversity survey of southern Iraq. A. Bachmann, J. Davies and L.R. Malone (Eds) Field Report. Sulaimani, Kurdistan, Iraq: Nature Iraq

Ammon EM, Stacey PB (1997) Avian nest success in relation to past grazing regimes in a montane riparian system. Condor 99:7–13. ANGELSTAM, P. 1986. Predation on ground

Ararat K, Abid IM, Rahman SA (2008) Key biodiversity survey of kurdistan, Northern Iraq. A Bachmann, J Davies and LR Malone (eds.). Field Report. Sulaimani, Kurdistan, Iraq: Nature Iraq

Balcomb R (1977) The grouping of nocturnal passerine migrants. Auk 94:479–488

Davies TW, Duffy JP, Bennie J, Gaston KJ (2014) The nature, extent, and ecological implications of marine light pollution. Front Ecol Environ 12:347–355

Doppler M, Blackwell B, DeVault T, Fernández-Juricic E (2015) Cowbird responses to aircraft with lights turned to their eyes: Implications for bird-aircraft collisions. Condor 117:165–177

Evans WR, Akashi Y, Altman NS, Manville AM II (2007) Response of night-migrating songbirds in cloud to colored flashing light. North American Birds 60:476–488

Farnsworth A (2005) Flight calls and their value for future ornithological studies and conservation research. Auk 12:733–746

Gagnon F, elisle MB, Ibarzabal J, Vaillancourt P, Savard JPL (2010) A comparison between nocturnal aural counts of passerines and radar reflectivity from a Canadian weather surveillance radar. Auk 127:119–128

Hamilton WJ (1962) Evidence concerning the function of nocturnal call notes of migratory birds. Condor 64:390–401

Haskelld D (1994) Experimental evidence that nestling begging behaviour incurs a cost due to nest predation. Proc R Sot Lond B 257:161–164

Herbert AD (1970) Spatial disorientation in birds. Wilson Bulletin 82:400–419

Horváth G, Kriska G, Malik P, Robertson B (2009) Polarized light pollution: A new kind of ecological photopollution. Front Ecol Environ 7:317–325

Hüppop O, Hilgerloh G (2012) Flight call rates of migrating thrushes: Effects of wind conditions, humidity and time of day at an illuminated offshore platform. J Avian Biol 43:85–90

Kempenaers B, Borgström P, Löes P, Schlicht E, Valcu M (2010) Artificial night lighting affects dawn song, extra-pair siring success, and lay date in songbirds. Curr Biol 20:1735–1739

Lanzone M, Deleon E, Grove L, Farnsworth A (2009) Revealing undocumented or poorly known flight calls of warblers (Parulidae) using a novel method of recording birds in captivity. Auk 126:511–519

Larkin RP, Szafoni RE (2008) Evidence for widely dispersed birds migrating together at night. Integr Comp Biol 48:40–49

Longcore T, Rich C (2004) Ecological light pollution. Front Ecol Environ 2:191–198

Loss S, Will T, Loss S, Marra P (2014) Bird-building collisions in the United States: Estimates of annual mortality and species vulnerability. Condor 116:8–23

Maltby E (1994) *An Environmental and Ecological Study of the Marshlands of Mesopotamia.* Draft Consultative Bulletin. Wetland Ecosystems Research Group, University of Exeter. Published by The AMAR Appeal Trust, London

McEwan GJP (1980) A Seleucid Augural Request. Zeitschrift für Assyriologie 70:59–69

Poot H, Ens B, de Vries H, Donners M, Wernand M, Marquenie J (2008) Green light for nocturnally migrating birds. Ecol Soc 13(2)

Reitsma LR, Holmes RT, Sherry TW (1990) Effects of removal of red squirrels, Tamiasciurus hudsonicus, and eastern chipmunks, Tamias striatus, on nest predation in a northern hardwood forest: an artificial nest experiment. Oikos 57:375–380

ROPER JJ (1992) Nest predation experiments with quail eggs: too much to swallow? Oikos 65:528–530

Rubec CDA, Bachmann A (2009) The key biodiversity areas program in Iraq: objectives and scope 2004–2008. Project Report. Sulaimani, Kurdistan, Iraq: Nature Iraq

Salim M, Porter R, Rubec C (2009) A summary of birds recorded in the marshes of southern Iraq, 2005–2008. In: Krupp F, Musselman LJ, Kotb MMA, Weidig I (Eds) Environment, Biodiversity and Conservation in the Middle East. Proceedings of the First Middle Eastern Biodiversity Congress, Aqaba, Jordan, 20–23 October 2008. BioRisk 3: 205–219

Wilson GR, Brittingham MC, Goodrich LJ (1998) How well do artificial nests estimate success of real nests? Condor:357–364

Wiltschko W, Wiltschko R (1996) Magnetic orientation in birds. J Exp Biol 199:29–38

Whelan CJ, Dilger ML, Robson D, Hallyn N, Dilger S (1994) Effects of olfactory cues on artificial-nest experiments. Auk 111(4):945–952

Yahner RH, DELONG CL (1992) Avian predation and parasitism on artificial nests and eggs in two fragmented landscapes. Wilson Bull 104:162–168

Chapter 41
A Proposal for Establishing Bird Observatory Centre in the South of Iraq

Laith A. Jawad

Abstract The geographical location of Iraq makes it a country that regularly visited by migratory birds from north in addition to the wealth of bird species present in its north, middle and south regions. For this important location, bird observatory bound to be built in order to integrate with observatories found in the countries of the region. In the present chapter, a brief protocol for monitoring of landbirds is given. The main sort of the data that the observatory needs to have in its library was discussed and itemized. Also, the general rules and policies of the observatory were set in order to be considered by policymakers in Iraq at the time of establishing the bird observatory.

41.1 Introduction

In order to proceed with the current issue of conservation of the wildlife and in particular the bird, Iraq should establish bird observatory centre in its southern part, where large number of migratory birds visiting this area for wintering and breeding. DeSante (2005) has suggested few points that should be taken into consideration if monitoring avian productivity and survivorship are needed to be implemented on any avian fauna. They include management of the data obtained, establishment of centres for better use of the data, standardization of the information, create a central data depository location, provide peer-reviewed procedure after each investigation and provide special budget to cover all the expenses of the project.

In the present chapter, a short, but focused account will be given about the idea of establishing bird observatory in south of Iraq. To know more about why Iraq should have such centre or station, the readers need to consult a section about this subject given in one of the chapters of this book on the migratory birds and written by the same author.

L. A. Jawad (✉)
School of Environmental and Animal Sciences, Unitec Institute of Technology, Auckland, New Zealand

L. A. Jawad (ed.), *Tigris and Euphrates Rivers: Their Environment from Headwaters to Mouth*, Aquatic Ecology Series 11,
https://doi.org/10.1007/978-3-030-57570-0_41

The general rules and policies of the Long Point Bird Observatory Migration Monitoring Protocol, Canada (2005), were adopted in this chapter with modifications so they will match the nature and environment in Iraq.

41.2 How Avian Data that Will Be Available at the Observatory Centre Be Useful?

The next are some of key type of the data that the observatory needs to include in its library accessible by its users.

1. Estimated Totals (ETs), once calculated and amended for several aspects, are used to give an index of species richness. The annual variations in these values will give an impression about which species are growing or declining.
2. The age and sex data gotten throughout banding procedure can be joint with ETs to show dates of migration for each age and sex group within a species.
3. The data will be obtained are also used for verifying migration timing, routes and longevity.
4. There are many matters to be recognized about migratory aspect that needs to be noticed and recorded.
5. Birds have high metabolic proportions, and it is essential to eat regularly to uphold health. Weight is a good indicator of a bird's health status and of the quantity of fat it has stored for a migratory flight.

41.3 General Rules and Policies of the Observatory

The following are the most significant guidelines, principles and procedures should be enforced at the observatory. They are the same as those used by the Long Point Bird Observatory Migration Monitoring Protocol with some amendments so it will suit the status in Iraq. They are about migration observing, day-to-day upkeep, boats, vehicles, radios, safety and general behaviour.

41.3.1 Migration Monitoring Programme

1. Think about the birds' health and well-being over everything else.
2. Do not put an ID ring on bird if the species is in uncertainty.
3. Do not predict at a bird's age or sex.
4. At no time, leave nets or traps under observation for more than 1/2 hour.
5. Do not leave nets unfolded and/or untied at the end of the day.
6. Do not set more nets than you can carefully look after.

7. In case you are tired, remove birds from nets and release them without rings.
8. Birds that are struggling need to be removed first.
9. Always report on cases of injured or dead birds.
10. Do not put ID ring on injured or not well birds.
11. Do not put any bird under strain when you dealing with them through your work.
12. Dogs and bicycles are not allowed in the area where nets are set.
13. Do not cut net to free birds unless there is no option.
14. Nets and traps should be in a place away from people not belong to the working team.
15. Public are not allowed to free birds from the nets at any circumstances.
16. Complete the duty of identification of birds at the end of the working day.
17. The person responsible for putting ID bands should be present at each observatory station.
18. Maintain/repair traps and nets on a regular basis, before they become big problems.

41.3.2 Migration Monitoring Field Operations

41.3.2.1 Overview of the Migration Monitoring Protocol

Every day assessed totals (ETs) are significant for migration information. These are the finest assessments of the numbers of each species present each day and are grounded on three constituents of the migration observing programme: a daily survey, daily ID band putting and other remarks.

It is very cooperative if each station's staff can create a schedule for how the day starts, is showed and finishes. Once the nets are fixed, at least one person should be free to run the survey while the others commence in on the ID rings. It is imperative that these events are corresponding, though the person responsible is eventually accountable for founding the daily schedule.

41.3.2.2 Daily Banding Routine

Preparing ID ring and attaching them to the bird are a job that needs to be done daily. It must begin in the early hours of the day if possible 1/2 hr. before sunrise. Certain species vanish after the first hour or so of light, though others may not show till later. There is frequently a midmorning haste of migrants, so do not give up and close; try to put a good energy in for 6 hours.

Make sure to open as many traps/nets as you can securely manage. This can be influenced by the number of birds in the area, the number and capability of staffs, and weather circumstances. The responsible person will decide on the number of nets that need to be open or closed daily.

41.3.2.3 Bird Catching Effort

Since of day-to-day restrictions passed by weather, bird volume can never function a totally consistent bird gathering. The next offers direction as to what the best/goal effort should be on a daily base for each station.

1. The overall goal of the daily work is catch and mark as much as possible of bird species safely.
2. The condition of the working day controls the netting work.
3. Make sure that the area of the nets and traps should remain the same around the year.
4. Make sure that that each net or trap has a number that the observer can identify it with.
5. After a long day with handling large number of birds, you need to close your net for the next working day.

41.3.2.4 Daily Census

A survey must be completed each day the station is operated, including the rainy days. Maps of the nests should be prepared for the purpose of the census. The survey must begin within 2 hours of sunrise, though weather circumstances might push it to be behind until afterwards in the day. The survey would take about 1 hr. and must be completed as a unit.

The survey every time takes precedence. Furthermore, an Estimated Total cannot be built without a survey. The goal is to grasp a sample of the birds present in the distinct station areas by counting all birds recognized by sight and sound along the survey route. Try not to count the same birds twice and do not guess numbers of birds that were not truly perceived or heard.

Somewhat, one staff on the survey should be skilful at bird ID. If you cannot classify a species, do not predict. Though, attempt to classify it as thoroughly as possible. A survey of the species you do distinguish is better than no survey at all, but write a memo in the comment section of the log sheet if you feel you lost a large proportion of the birds because of special conditions.

Reference

DeSante DF (2005) Suggestions for Planning a Migration-Monitoring Network Based on the Experience of Establishing and Operating the MAPS Program. In: Ralph, C. John; Rich, Terrell D., editors 2005. Bird Conservation Implementation and Integration in the Americas: Proceedings of the Third International Partners in Flight Conference. 2002 March 20–24; Asilomar, California, Volume 2 Gen. Tech. Rep. PSW-GTR-191. Albany, CA: US Dept. of Agriculture, Forest Service, Pacific Southwest Research Station: p. 705–711, vol. 191

Chapter 42
The Potential Role of Waterbirds as a Vector in Dispersing Invertebrates and Plants in the South of Iraq

Laith A. Jawad

Abstract A short review of the ability of the waterfowls to disperse plants and invertebrates is given in the present chapter. For a full dispersal of the aquatic organisms, few steps need to be occurred. In the literature, these steps were given in details and the present chapter has adopted those steps with slight variation so they will match the aquatic environment in Iraq. After that, the effects of the plants and invertebrate dispersal by waterbirds are given and need to be consulted before any future plans for studying the ability of waterfowls to disperse plants and invertebrate. At the end of the chapter, several recommendations were stated to be considered by policymakers in Iraq for the future waterbirds management.

42.1 Introduction

The spatio-temporal patterns of animal disperser movement represent an important component of passive dispersal and are, therefore, a key determinant of seed shadows (the spatial distribution of dispersed seeds and other propagules (Russo et al. 2006; Reynolds 2016)).

In aquatic ecosystems, waterbirds have a significant possibility to disperse aquatic organisms since they are plentiful, widely dispersed through the world's wetlands, and extremely moveable at local, regional, and continental scales. The worth of aquatic species in their diet makes waterbirds a likely route for dispersion by both internal and external transport (Green et al. 2002; Reynolds 2016).

Waterbird-mediated dispersion is an ecological procedure of great preservation cost, chiefly in light of the dilapidation and devastation of aquatic systems taking place worldwide (Ramsar Convention Bureau 1996). The biotic linking among aquatic systems made by waterbirds gives to saving the relationship of wetland

L. A. Jawad (✉)
School of Environmental and Animal Sciences, Unitec Institute of Technology, Auckland, New Zealand

L. A. Jawad (ed.), *Tigris and Euphrates Rivers: Their Environment from Headwaters to Mouth*, Aquatic Ecology Series 11,
https://doi.org/10.1007/978-3-030-57570-0_42

ecosystems, accelerating their restoration after instabilities and rushing the inhabiting of new settlers (Amezaga et al. 2002).

Iraq and especially the southern part, is rich in aquatic avifauna due to the abundance of rivers and marshes. In addition, the geographical location of Iraq makes it the target of the migratory birds from the countries in the north visiting its territories for breeding and overwintering. With such busy avifauna traffic both international and domestic and through the movement of birds, the possibility for plants and invertebrates to disperse from water body to another locally or an invasive species might find new suitable habitats to live in (Reynolds 2016).

The aim of the present chapter is to give a short review about the role of aquatic birds in dispersing plants and invertebrates, and the precautions need to be taken in order to control such an ecological process.

42.2 Steps in Dispersal of Plants and Invertebrates by Waterbirds

Figuerola and Green (2002) in their study on the waterbird as vector for dispersal of plants and invertebrates have underlined some processes intricate and assessing the likely frequency of such dispersal in the field. Figuerola and Green (2002) these processes identified as steps and these are:

1. Internal transport step 1: possible consumption of propagules by waterbirds.
2. Internal transport step 2: possible survive of propagules by digestion.
3. Internal transport step 3: possible staying maximum time for propagules in a waterbird over a long distance.

42.3 External Transport: Can Propagules Adhere to Waterbirds?

Propagules might be dispersed externally by waterbirds trough their hold to the feathers, bills, or legs of waterbirds, but such a transport is very rare (Swanson 1984; Reynolds 2016).

In freshwater habitats, disclosure to dryness through dispersal is probably limiting the ability of several aquatic creatures to transport by grip, but it is not clear how desiccation can affect the viability of the propagules. Various seeds and inactive eggs are probable to be highly resilient to dryness (Bilton et al. 2001; Reynolds 2016). Little evidence is present on the ability of the smaller seeds and those with hook-like configurations are more possible to be fixed to waterbird feathers.

42.4 The Long-Distance Dispersal of Invertebrates

The long-distance dispersal can be defined by several ways. Green and Figuerola (2005) defined this term as, "the invertebrates over dry land dispersion amongst swamps divided by at least 10 km and not associated hydrologically." Comparable type of dispersion cannot be achieved by flows, fish, or other routes functioning within the water and is uncertain by other animal courses. However, small seeds can undergo long-distance spreading using wind or rain. Adjoining flight speeds (Welham 1994) and behaviour (Driscoll and Ueta 2002; Reynolds 2016) of waterbirds with data on retaining times of invertebrate eggs (Charalambidou and Santamaría 2002), it is conceivable to accomplish that maximum dispersion spaces of seeds through endozoochory could exceed 1000 km. Contrarily, how long invertebrates endure fixed to bird will continue is open to guess.

42.5 Waterbirds Versus the Invasive Species

Recently, the world's ecosystems have experienced the most devastating and irreversible impacts through the biological invasions (Strayer 2010; Reynolds 2016). With such mass and global effects, a chain of researches on this issue have been initiated; nevertheless, very little consideration has been given to the clear role of dispersion in species intrusiveness and administration (Westcott and Fletcher 2011). Management non-native species needs an insight of spatial designs of dispersion and the devices that generate those (Levin et al. 2003). Freshwater ecosystems have had the largest impacts of the biological invasion (Green et al. 2005) in addition to its vital role to whole biota (Strayer and Dudgeon 2010).

In freshwater ecosystems, the furthermost significant procedures backing the dispersal of non-native species are those that found across water bodies. However, natural passive dispersion courses across a range of spatial scales also add to the spread of non-native invasive species between water bodies. These comprise activities of prey species between areas by predators (Higgins et al. 2003) and the further usual mechanism of spreading of aquatic organisms by waterbirds (Figuerola and Green 2002; Reynolds 2016).

42.5.1 Methods of Disperse Alien Species by Waterbirds

So the non-native species find its settlement in the new habitat, there should be an effective dispersal mean (Schupp et al. 2010; Reynolds 2016). For dispersal to be fully performed, there are three-stage transfer of creatures and/or their offspring's need to be taken (Bowler and Benton 2005), these are: (1) emigration; (2) transfer and carrying; and (3) settlement.

42.5.1.1 Emigration

This phase includes the passage of foreign species out of their area. Inactively scattering organisms can be taken up by a path and continue in a combination through leaving the location. Feeding investigations can be employed to control the distinctiveness of the strange offspring maybe spread by waterbirds. The number of seeds found in the oesophagus and gizzard is a noteworthy indicator of the existence of whole seeds in the lower gut (Brochet et al. 2010; Reynolds 2016).

42.5.1.2 Movement and Transport

The practicality of the non-native aquatic species can be determined by the transfer distance of the avian path and seed retaining time. Consequently, a rough measure of the maximum distance to which a seed can be disseminated is the maximum time that a seed rests in the gut (or on the feathers) multiplied by the speed at which the bird is able to fly (Reynolds et al. 2015; Reynolds 2016). This information proposes that non-native organisms can continue in the gut of waterbirds and be transported over long distances.

42.5.1.3 Immigration

Immigration is the important phase in the incursion course as it indicates the final dispersion problem before the attack to take place (Kolar and Lodge 2001; Reynolds 2016). The study of biological features that support the existence is serious for approving the dispersal abilities of potentially alien aquatic organisms. To have an actual spreading, alien species must come into appropriate environments where they may become recognized.

Numerous aquatic invertebrates have resilient offspring such as ephippia, cysts, gemmules, or statoblasts (Panov et al. 2004; Reynolds 2016) that not only permit them to live dryness, but these same alterations also help in the spreading via zoochory.

42.6 The Negative Issues in the Dispersal of Plants and Invertebrates by Waterbirds

There are certain negative issues in the dispersal of the plants and invertebrates by waterbirds. These issues incorporate the human health, aquatic environment, and the terrestrial environment. It is important to know these issues before planning any future studies on the subject of dispersal of plant and invertebrates by waterbirds so to include the solutions as far as possible for these issues (Reynolds 2016). With the dispersal of plants and invertebrates by waterbirds, the following possibilities are highly likely to occur:

1. Moving parasites that can infect human and other animals through their flying over a large area.
2. Moving microbes from a contaminated water body to another.
3. Moving zoonotic pathogens for new areas.
4. Dispersing aquatic plant seeds among water bodies.
5. Dispersing unwanted terrestrial plant seeds among fields.
6. Introducing genetically modified plant seeds to non-genetic plant farms.
7. Introducing invasive species with their problems to water bodies.

42.7 Future Studies and Recommendations

The following are future studies suggested by Green and Figuerola (2005) and adopted in here with slight changes to suit the aquatic life in Iraq, but as an author of this chapter, I can see they can be considered as a recommendations to be considered by countries with rich waterbirds fauna such Iraq.

1. The idea of those invertebrate species that are able to fix themselves to birds or to endure going through their gut should be more broadly distributed with less genetic differentiation needs to be tested.
2. Pay special consideration to those species of waterbirds that get sick by many cestodes or acanthocephalans that have aquatic intermediate hosts (typically crustaceans, insects, or molluscs).
3. The necessity for more measurable investigations of long dispersion distance of invertebrates by birds, containing better approximations of dispersal seeds.
4. Comparative experimental examinations of living and retaining times should be united with field investigations of seeds and offspring transport rates and bird movements and of genetic structure in the same organisms.

References

Amezaga JM, Santamaría L, Green AJ (2002) Biotic wetland connectivity—supporting a new approach for wetland management policy. Acta Oecol 23:213–222

Bilton DT, Freeland JR, Okamura B (2001) Dispersal in freshwater invertebrates. Annual Revue of Ecology and Systematics 32:159–181

Bowler DE, Benton TG (2005) Causes and consequences of animal dispersal strategies: relating individual behaviour to spatial dynamics. Biol Rev 80:205–225

Brochet AL, Guillemain M, Fritz H, Gauthier-Clerc M, Green AJ (2010) Plant dispersal by teal (*Anas crecca*) in the Camargue: duck guts are more important than their feet. Freshw Biol 55 (6):1262–1273

Charalambidou I, Santamaría L (2002) Waterbirds as endozoochorous dispersers of aquatic organisms: a review of experimental evidence. Acta Oecol 23:165–176

Driscoll PV, Ueta M (2002) The migration route and behaviour of eastern curlews *Numenius madagascariensis*. Ibis 144:E119–E130
Figuerola J, Green AJ (2002) Dispersal of aquatic organisms by waterbirds: a review of past research and priorities for future studies. Freshw Biol 47(3):483–494
Green AJ, Figuerola J (2005) Recent advances in the study of long-distance dispersal of aquatic invertebrates via birds. Divers Distrib 11(2):149–156
Green AJ, Figuerola J, Sánchez MI (2002) Implications of waterbird ecology for the dispersal of aquatic organisms. Acta Oecol 23:177–189
Green AJ, Sánchez MI, Amat F, Figuerola J, Hontoria F, Ruiz O, Hortas F (2005) Dispersal of invasive and native brine shrimps *Artemia* (Anostraca) via waterbirds. Limnol Oceanogr 50:737–742
Higgins SI et al (2003) Are long-distance dispersal events in plants usually caused by nonstandard means of dispersal? Ecology 84:1945–1956
Kolar CS, Lodge DM (2001) Progress in invasion biology: predicting invaders. Trends in Ecology and Evolution 16:199–204
Levin SA, Muller-Landau HC, Nathan R, Chave J (2003) The ecology and evolution of seed dispersal: a theoretical perspective. Annu Rev Ecol Evol Syst 34:575–604
Panov VE, Piotr I, Riccardi N (2004) Role of diapause in dispersal and invasion success by aquatic invertebrates. J Limnol 63:56–69
Ramsar Convention Bureau (1996) Wetlands and biological diversity: cooperation between the convention of wetlands of international importance especially as waterfowl habitat (Ramsar, Iran, 1971) and the convention on biological diversity. Document UNEP/CBD/COP/3/Inf. 21
Reynolds C (2016) The role of waterbirds in the dispersal of aquatic organisms in southern Africa. Doctoral dissertation, University of Cape Town
Reynolds C, Miranda NA, Cumming GS (2015) The role of waterbirds in the dispersal of aquatic alien and invasive species. Divers Distrib 21(7):744–754
Russo SE, Portnoy S, Augspurger CK (2006) Incorporating animal behavior into seed dispersal models: implications for seed shadows. Ecology 87(12):3160–3174
Schupp EW, Jordano P, Gómez JM (2010) Seed dispersal effectiveness revisited: a conceptual review. New Phytol 188:333–353
Strayer DL (2010) Alien species in fresh waters: ecological effects, interactions with stressors, and prospect for the future. Freshw Biol 55:152–174
Strayer DL, Dudgeon D (2010) Freshwater biodiversity conservation: recent progress and future challenges. J N Am Benthol Soc 29:344–358
Swanson GA (1984) Dissemination of amphipods by waterfowl. J Wildl Manag 48:988–991
Welham CVJ (1994) Flight speeds of migrating birds: a test of maximum range speed predictions from the aerodynamic equations. Behav Ecol 5:1–8
Westcott DA, Fletcher CS (2011) Biological invasions and the study of vertebrate dispersal of plants: opportunities and integration. Acta Oecol 37:650–656

Chapter 43
The Wild Mammals of the Tigris and Euphrates Rivers Basin

Omar F. Al-Sheikhly

Abstract The wild mammal fauna of Tigris and Euphrates River basin is enigmatic and little known. The geographical landscape of Tigris and Euphrates River basin awards Iraq wide range of habitats which significantly influenced its wild mammal biodiversity. The earlier contributions to the wild mammal fauna knowledge of Iraq were made in the early 1950s–1960s; however, several taxonomic revisions were made afterward. The main comprehensive research concerning wild mammals of Iraq was made recently in 2015. In Iraq, there is total of 93 wild mammal species including 3 Erinaceomorphs, 3 Soricomorphs, 20 Bats, 23 Carnivores, 8 Artiodactyls, 2 Lagomorphs, 28 Rodents, and 6 Cetaceans. Their taxonomic revisions, geographical distribution, and conservation status are discussed. Among Iraq's endemic, indigenous, and globally threatened flagship mammal species, the Iraq's Smooth-coated Otter *Lutrogale perspicillata maxwellii* and Long-tailed Nesokia *Nesokia bunnii* which are worth mentioned. Their current geographical distribution and conservation status are under discussion. Several threats impacts were contributed to a significant decline in wild mammal population in Iraq, illegal hunting and trapping were highlighted as major threats. Regardless, the actual distribution and conservation status of many wild mammal species in Iraq required further research and investigation.

43.1 Introduction

The geographical landscape of Iraq is comprised between latitudes 29° - 38° N and longitudes 39° - 49° E (a small area lies west of 39°) and spans over 437,072 km^2 (168,754 sq. mi). It borders with Turkey to the north, Syria to the north-west, Iran to the east, Jordan to the west, Saudi Arabia to the south and south-west, and Kuwait to the south which marked by a narrow ribbon of coastline stretching for 58 km at the head of the Arabian Gulf. The landscape includes high mountains in the north

O. F. Al-Sheikhly (✉)
College of Science, University of Baghdad, Baghdad, Iraq

L. A. Jawad (ed.), *Tigris and Euphrates Rivers: Their Environment from Headwaters to Mouth*, Aquatic Ecology Series 11,
https://doi.org/10.1007/978-3-030-57570-0_43

(Kurdistan), desert, arid lands, and sandy steppes in the western and south-western plateau (Al Badiyah), and the Mesopotamian marshlands in the southern alluvial plain. This wide range of habitats awards Iraq with a noticeable diversity of wild mammals (Al-Sheikhly et al. 2015b; Nature Iraq 2017). Little is known about the wild mammals of Iraq. Most of the field observations and specimens collection available from Iraq were due to Robert T. Hatt and David L. Harrison. In 1959, Robert T. Hatt published "The Mammals of Iraq," where he included data collected on wild and domestic animals during surveys carried out between October 1952 and March 1953 (Hatt 1959). Harrison reviewed Hatt's records and made further field observations especially by surveying the surroundings of western and central Iraq. His observations in Iraq were published in three volumes entitled "The Mammals of Arabia" (1964; 1968; 1972). Further updates and additional observations were included in the second edition of his book that was published in 1991. In his books, Harrison exhaustively addressed the entire information available for the mammalian fauna of Iraq during the period 1953–1991 (Harrison and Bates 1991). A checklist of vertebrate fauna of Iraq was prepared by Nuri Mahdi and P.V. George in 1969. Mahdi and George's checklist included systematic classification of mammals, reptiles, and amphibian taxa reported from Iraq and made according to the systematic arrangement of Ellerman and Morrison-Scott (1951). However, Mahdi & George's checklist failed to provide detailed information on all the listed species (e.g., distribution range and/or collecting localities and dates) (Mahdi and George 1969; Al-Sheikhly et al. 2015b). Presently, the main recent comprehensive research concerning wild mammals and discussing its systematics and geographical distribution throughout Iraq was made by Al-Sheikhly et al. (2015b).

43.2 Wild Mammals of Iraq, Taxonomic Revisions, Distribution, and Conservation Status

In Iraq, there are a total of 93 wild mammal species recorded in Iraq and belong to eight orders, 28 families, and 65 genera, including three Erinaceomorphs, three Soricomorphs, 20 Bats, 23 Carnivores, eight Artiodactyls, two Lagomorphs, 28 Rodents, and six Cetaceans (Al-Sheikhly et al. 2015c). Since Hatt (1959) and Harrison and Bates (1991), several taxonomic revisions on the mammals' fauna of Iraq were made; however, the current taxonomic nomenclature of each taxon exists in Iraq was comprehensively discussed in Al-Sheikhly et al. (2015b).

Order: Erinaceomorpha (one family; three genera; three species).

Family: Erinaceidae.

According to Al-Sheikhly et al. (2015b), three Erinaceomorpha species are present in Iraq, the Southern White-breasted Hedgehog *Erinaceus concolor* which is restricted to the mountains and foothills of upper Tigris and Euphrates rivers basin. The Long-eared Hedgehog *Hemiechinus auritus* and Desert Hedgehog *Paraechinus aethiopicus* are dowelling in the alluvial plain, grasslands, steppes, and desert of central and southern Iraq. However, the historical records of the European Hedgehog *Erinaceus europaeus* from Iraq are referred now to the Southern White-breasted Hedgehog.

Order: Soricomorpha (one family; two genera; three species)

Family: Soricidae

In Iraq, total of three Soricomorpha species are recorded (Al-Sheikhly et al. 2015b). The Asian House Shrew *Suncus murinus* and Etruscan Shrew *Suncus etruscus* are found in urban, cultivated fields, and on the edge of the marshlands in southern Iraq, while the Lesser White-toothed Shrew *Crocidura suaveolens* is confined to the mountain slopes, grasslands, and elevated grounds of northern Iraq. Nader (1969) examined the Iranian specimens from subspecies Bicolored White-toothed Shrew *Crocidura leucodon persica* which were reported by Hatt (1959) and reassigned them to *C. russula*. However, these specimens were provisionally allocated to the Lesser White-toothed Shrew by Hutterer and Harrison (1988).

Order: Chiroptera (seven families; 12 genera; 20 species)

Family: Rhinopomatidae; Emballonuridae; Rhinolophidae; Hipposideridae; Molossidae; Vespertilionidae; Miniopteridae

According to Al-Sheikhly et al. (2015b), a total of 20 Chiropteran (Bats) species were recorded in Iraq. Among the Chiropteran fauna of Iraq, the Mehely's Horseshoe Bat *Rhinolophus mehelyi*, Mediterranean Horseshoe Bat *Rhinolophus euryale*, and Pallid Long-fingered Bat *Miniopterus pallidus* were listed as Vulnerable and Near Threatened respectfully by the International Union for Conservation of Nature (IUCN). Mehely's Horseshoe Bat was recorded from caves of arid areas along the banks of Euphrates River in central and western Iraq. Niazi (1976) reported the occurrence of the Mediterranean Horseshoe Bat in two localities; however, the species was proved to be in fact *R. mehelyi* (Harrison and Bates 1991; Al-Sheikhly et al. 2015b). Mehely's Horseshoe Bat was recorded from *c.* 43 km to the south-west of Ash-Shinaffya and from *c.*130 km west of Ramadi; specimens collected in Al-Alfasia Cave, *c.* 8 km to the south of Haditha (Harrison and Bates 1991). Specimens were collected from El Hashfe and Hadhita areas of Iraq (Matskási 1980) and from Bjil village, Akre District, and Mawat Mountain in northern Iraq (Al-Sheikhly et al. 2015b). Mediterranean Horseshoe Bat is reported from a rocky cave in Bjil village, Akre District, in northern Iraq (Al-Sheikhly et al. 2015b). The current status of the Pallid Long-fingered Bat is uncertain; it is confined to the mountain caves in extreme northern Iraq. It was recorded from one locality in Ser Amadia which is the only known record (Harrison 1964; Harrison and Bates 1991). However, Scott (1995) reported this species from the edge of the Mesopotamian marshes in southern Iraq. Kinnear (1916) reported the presence of "*Rhinopoma arabicum*" (= *R. arabium*) in Iraq, which formerly regarded as subspecies of Lesser Mouse-tailed Bat (*R. hardwickii*) (Ellerman and Morrison-Scott 1951; Simmons 2005). Now, the species *arabicum* is referred to the Levantine and Arabian populations of *R. cystops* (Benda et al. 2009). One of the distinct bat taxa reported from Iraq is the Haditha Mouse-tailed Bat *Rhinopoma hadithaensis* which its occurrence and status require further investigation. It was described as a new bat

Fig. 43.1 Iraq's Smooth-coated Otter *Lutrogale perspicillata maxwellii*, endemic to the Iraqi marshes, Hawizeh Marsh-southern Iraq. Photo Courtesy Omar F. Al-Sheikhly

species from western Iraq by Khajuria (1988); however, this taxon is now considered as junior synonym of Greater Mouse-tailed Bat *R. microphyllum* (Kock et al. 2001).

Order: Carnivora (six families; 17 genera; 23 species).

Family: Canidae; Ursidae; Mustelidae; Herpestidae; Hyaenidae; Felidae

The variations of habitats provided by Iraq' geographical landscape have significantly contributed to the richness of its mammals' diversity especially carnivores. Total of 23 carnivores species were reported from Iraq (Al-Sheikhly et al. 2015b). Among those, four species: Marbled Polecat *Vormela peregusn,* Smooth-coated Otter *Lutrogale perspicillata,* Persian Leopard *Panthera pardus saxicolor,* and Asiatic Cheetah *Acinonyx jubatus venaticus* which are listed as Vulnerable by the IUCN occur in Iraq. Marbled Polecat, the Syrian subspecies *V. p. syriaca* is confined to the high grounds of the upper Tigris and Euphrates rivers basin of northern Iraq. It was recorded from Mousel (Harrison and Bates 1991; Al-Sheikhly et al. 2015b). However, no recent record of this species has been made in recent years.

The Smooth-coated Otter (Iraq's Smooth-coated Otter; Maxwell's Otter *L. p. maxwelli*) is an endemic flagship taxon to Iraq (Fig. 43.1). In 1980s and 1990s, this taxon was guessed to be extinct due to habitat loss and fragmentation as well as to over-hunting, but it has been rediscovered very recently (Omer et al. 2012; Al-Sheikhly and Nader 2013; Al-Sheikhly et al. 2017). It is confined to the dense reed beds of the southern Mesopotamian marshes. It was recorded from Abusakhair and Azair (Hayman 1956; Maxwell 1957, 1960); Azair by (Hatt 1959); Fao-Rass Al-Beisha, Abu Al-Khasib and Umm Al Rassas, Abu Ajaj-Hor Al-Hammar, Umm Al- Na'aj lake-Hawizeh (Al-Sheikhly and Nader 2013; Al-Sheikhly et al. 2014; Al-Sheikhly et al. 2015a, c; Barbanera et al. 2016); and from TaqTaq in northern Iraq (Omer et al. 2012).

Another flagship carnivore species for Iraq is the Persian Leopard, a rare, restricted to the forested mountains of extreme northern Iraq along the Iraqi-Iranian borders. It was recorded from near Rawa (Musil 1927); Aqra and

Rawanduz (Pocock 1930); Jabal Bradost (Felid 1951, 1955); Harir Dagh (Hatt 1959); Bamur (Harrison 1968b); Murdka of Darbandikhan, Mandili, QaraDag, and Suran (Al-Sheikhly 2012b; Raza 2013; Al-Sheikhly et al. 2015b). Since Hatt's (1959), there was no recent report of the Iranian or Asiatic Cheetah in Iraq. Asiatic Cheetah roamed the southern and western deserts of Iraq in 1920s. It was reported from near Busaiya west of Basra in southern desert of Iraq; K3 station and H1and H2 pumping stations in Syrian Desert in western Iraq (Corkill 1929); near Iraq-Saudi Arabia-Jordan borders (Hatt 1959; Harrison and Bates 1991). However, no recent sighting of the Asiatic Cheetah; hence, it is considered as virtually extinct in Iraq (Al-Sheikhly et al. 2015b).

The European Otter *Lutra lutra,* Striped Hyena *Hyaena hyaena,* and Sand Cat *Felis margarita* were listed as Near Threatened mammals' species occur in Iraq. The European Otter is confined to the Lakes, streams, tributaries, and marshes of Tigris and Euphrates River Basin. It was recorded from Amara (Cheesman 1920); Qalat Salih (Sanborn 1940); Baghdad, and Hindiya (Hatt 1959); Abusakhair (South of Hor Al-Hammar Marsh) (Harrison and Bates 1991); Central marshes, vicinity of Al-Maimona, Musharah River, Tarmiya, Al-Alam, Samarra Lake, Himreen, Khan Al-Baghdadi, Haditha, Derbendikhan, Dukan, Little Zab, Barzan, and TaqTaq (Al-Sheikhly and Nader 2013); Mashab in eastern Hammar Marsh by Abass (2013); Mashab River in eastern Hammar Marsh and Salal River; and Taq Taq and Mousl Lake in northern Iraq (Al-Sheikhly et al. 2014).

The Syrian subspecies of Striped Hyena *H. h. syriaca* is a common nomadic in the lowlands of central and southern Iraq, uncommon in the northern mountains. It was recorded from Ur, Basra, and Falluja (Cheesman 1920); between Amara and Sheikh Saad (Capper 1921); Falluja, Nahrwan, and Ad Dawr (Pitman 1922); Tyb River, Jabal Hamrin, and Ur (Pocock 1941); Bradost Mountain (Felid 1955); near Haditha (Lawrence 1956); near Samarra (Harrison 1959); Jezira, *c.* 40 km west of Mousel (Mousl) (Bodenheimer 1959); Mansuriya al Shatt (Hatt 1959); Erbil Liwa, Ctesiphon, Chamchamal, Baghdad, Baquba, Habbaniya, Ramadi, and Sulaymaniyah (Harrison 1968b); near Hammar Marsh (Haba 2009); Hammar Marsh (Abass 2013); Khan Al-Baghdadi; Al-Ga'ara, Rutba (Al-Sheikhly et al. 2015b).

The Arabian Sand Cat *F. m. harrisoni* is confined to the desert and arid steppes of south-western Iraq (Fig. 43.2). It was recently recorded from three specimens obtained from west of Al-Najaf city and Al-Jufaira oasis in the desert of Al-Najaf in south-western Iraq (Mohammad et al. 2013). Additional records in western and south-western deserts of Iraq with evidence of persistence occurrence and breeding were provided by Al-Sheikhly and Haba (2017).

Moreover, the wooded mountains, arid steeps, grasslands, palm tree orchids, and deserts distributed along Tigris and Euphrates River basin are hosting several carnivore species. Packs of the Iranian (*Canis lupus pallipes*) Arabian (*C. l. arabs*) wolves are frequently reported from the wooded mountains and arid steppes of northern and eastern Iraq, respectfully. According to Lawrence (1956), a zone of hybridization between the two subspecies is known in the western part of the country (Harrison and Bates 1991). Very recently, the Arabian Wolf was recorded in the

Fig. 43.2 The IUCN Near Threatened Sand Cat *Felis margarita harrisoni* is confined to the desert and arid steppes of south-western Iraq (In captivity). Photo Courtesy Omar F. Al-Sheikhly

extreme western Iraq by Al-Sheikhly (2012b). Al-Sheikhly (2012) recorded the Syrian Jackal (*Canis aureus syriacus*) for the first time in western Iraq. Similarly, the Indian Grey Mongoose (*Herpestes edwardsi*) was recently observed for the first time in northern Iraq. This record represents a large extension of the species range in northern Arabia (Al-Sheikhly and Mallon 2013).

The Red Fox (*Vulpes vulpes*) observed in the deserts of Iraq and assigned by Hatt (1959) to the Indian subspecies (*V. v. pusillus*) is now listed as Arabian Fox (*V.v. arabica*) according to Harrison and Bates (1991). The presence of Rüppell's Sand Fox (*Vulpes rueppellii*) was investigated in Al-Sheikhly et al. (2015b). However, Rüppell's Sand Fox was reported once at Khan Al-Baghdadi in western Iraq (Al-Sheikhly et al. 2015b).

Mahdi and George (1969) reported the likely occurrence of the Fennec Fox (*Vulpes zerda*) in Iraq. Al-Robaae (1982) collected an individual near Jabal Sanam (*c.* 30 km south-west of Basra, extreme southern Iraq). The same author also reported that an individual was captured alive near Jabal Hafit (United Arab Emirates) and kept at the Al-Ain Zoological Garden. Later on, Gasperetti et al. (1985) proved that the Jabal Hafit fox was in fact *V. rueppellii*. Therefore, the occurrence of Fennec Fox in Iraq is doubtful and required further investigation.

Order: Artiodactyla (three families; seven genera; eight species)

Family: Bovidae; Cervidae; Suidae

According to Al-Sheikhly et al. (2015b), eight Artiodactyls were reported from Iraq, in which five species were evaluated as globally threatened by IUCN. The Mesopotamian Fallow Deer *Dama dama mesopotamica* is listed as Endangered, Wild Goat *Capra aegagrus,* Mouflon *Ovis orientalis,* Goitered Gazelle *Gazella*

Fig. 43.3 Family of Wild Goat *Capra aegagrus* confined to the Zagros Mountain range of north-eastern Iraq. Photo Courtesy Omar F. Al-Sheikhly

subgutturosa marica, and Arabian Oryx *Oryx leucoryx* which listed as Near Threatened by the IUCN are occur in Iraq.

The Mesopotamian Fallow Deer is a rare species confined to the wooded slopes of Zagros Mountain in extreme northern Iraq, where small herds may still survive. It was recorded from Zakho (Ellerman and Morrison-Scott 1951); Bradost Mountain (Felid 1955); and between Maidan and Halabja (Harrison 1968a; Harrison and Bates 1991).

The Wild (Bezoar) Goat is confined to the Zagros Mountain range of north-eastern Iraq: the largest population is in the area of Barzan area in Erbil, northern Iraq (Fig. 43.3). It was recorded from Bradost, Barzan near Aqra, and near Zawitta (Layard 1852); Baradost and Sarsank (Felid 1951); Safin Dagh, Zagarta, and Jabal Sinjar (Hatt 1959); Chamchamal Valley (Hoogstraal and Kaiser 1958); near Shaqlawa, Safin Dagh, Harir Dagh, Erbil Liwa, Nahiya, near Diyana, Rawandoz, and near Darbandikhan (Harrison 1968a; Harrison and Bates1991); Barazan, Qara Dag Mountain; Peramagroon Mountain (Al-Sheikhly 2012b); Qara Dag Mountain and Peramagroon Mountain (Raza 2013). Specimens were collected in Berat Dag, Mousl Liwa, Jabal Zagarta, Sugrimah Dagh, and Kurkuk Liwa (Sanborn 1940); a remarkable range extension was made in Al-Qosh in north-western Mousel (Al-Barazengy et al. 2015). The Mouflon *O. o. gmelini* is uncommon, restricted to the highlands and foothills of eastern and north-eastern Iraq. It was recorded from Zagros Mountain and mountains of Derbendikhan (Harrison 1968a; Nader et al. 1971; Harrison and Bates 1991); Jabal Himreen near Mandli in Diyala (Al-Sheikhly 2012b); Kanispika, Baranand Dagh (Hatt 1959).

Two subspecies of Goitered Gazelle *Gazella subgutturosa* are found in Iraq, the Arabian Sand Gazelle *G. s. marica* which confined to the desert and arid steppes in

Fig. 43.4 Male and female Arabian Sand Gazelle *Gazella subgutturosa marica* confined to the desert and arid steppes in western and southern Iraq. Photo Courtesy Omar F. Al-Sheikhly

western and southern Iraq (Fig. 43.4) and Persian Gazelle *G. s. subgutturosa* which inhabits steppes of northern and eastern Iraq (Al-Sheikhly et al. 2015b). The species is widespread throughout Iraq, but endures in scattered small populations. It was recorded from Euphrates valley (Lydekker and Blaine 1914); Samarra, Basra, Nasiriyah, and Amara, where it was reported as *G. s. marica* (Cheesman 1920; Mallon and Kingswood 2001); Sheikh Saad and to the north-east of Baghdad (Dollman and Burlace 1935); Erbil and Kurkuk (Hatt 1959); Chamchamal, Basra, Kurkuk, Baghdad, and Hinaidia (Hoogstraal and Kaiser 1958; Harrison 1968a); Faidhat Al-Massad, Faidhat Al-Dhaba'a, Al-Dheabeat, Al- Jazera Area near Haditha, Rutba, Al Qa-eem, Jabal Makhool, Al-Sherqat, Baiji, Jabal Sinjar, Rabe'ea.; Makhmoor, Jabal Himreen, Kafri, Kalar, Jalwla, Mandli, Badra, Khanaqeen, and Ser Qella at Garmyan area (Al-Sheikhly 2012a, c); small herds were reported from Garmiyan area (near Kalar), assigned to *G. s. subgutturosa* (Al-Sheikhly et al. 2015b). Details on Gazelles complex and their distribution in Iraq were reported by Al-Sheikhly (2012c). However, further investigation is required to update the distribution of both the Arabian Sand Gazelle and Persian Gazelle in Iraq.

One of the remarkable Artiodactyls species that occurred once in the deserts and aridlands of southern and western Iraq is the Arabian Oryx. This species has a traditional value among Arabic tribes of the Arabian Peninsula especially those in Iraq and the Arabian Gulf states. Several rehabilitation programmes were preformed to re-introduce this species into many places where it was historically abundant, especially in Saudi Arabia, United Arab Emirates, and Jordan (Al-Sheikhy *in litt.* 2019). In Iraq, it is probably extinct; the last individual was shot in Maqil in southern

Iraq in 1914. However, specimens are known from Iraq yet it is unclear whether they have occurred on the eastern side of the Euphrates River (Harrison 1968a, b; Al-Sheikhly 2012c).

Along with the Arabian Oryx, the occurrence of the Syrian Wild Ass "Onager" (*Equus hemionus hemippus*, which belong to Order: Perissodactyla, see Extinct Species section in Al-Sheikhly et al. 2015b), in the western Iraqi plateau is still uncertain. It was abundant in the plains of the upper Tigris and Euphrates River plains in north-western Iraq, but since the middle of the XIXth century, its number has dramatically decreased due to over-hunting. The last herd was reported near Jabal Sinjar in 1927 (Raswan 1935; Hatt 1959).

Order: Lagomorpha (one family; one genus; two species)

Family: Leporidae

With reference to the status of Hare in Iraq, two species occur in Iraq, the Arabian Cape Hare *Lepus capensis arabicus* and European Hare *Lepus* (*Eulagos*) *europaeus connori*. Hatt (1959) regarded *connori* as subspecies of the European Hare (*Lepus europaeus*). However, Harrison and Bates (1991) considered the European and Cape Hare as genetically related taxa and regarded *connori* as subspecies of the Cape Hare (*Lepus capensis*). Sanborn (1940) tentatively identified hares collected in Baghdad and Camp Rashid as European Hare (*L. europaeus connori*), which is now regarded as synonym of *L. c. connori*. Hubbard (1955) reported the occurrence of *Lepus babylonicus* in Baghdad and Hilla, a taxon now considered as synonym of *L. capensis*. The Syrian subspecies of the Cape Hare (*L. c. syriacus*) was recently recorded from *c.*17 km to the north-west of Rutba near the Iraqi-Syrian borders (Al-Sheikhly et al. 2015b).

Order: Rodentia (seven families; 18 genera; 28 species)

Family: Sciuridae; Hystricidae; Dipodidae; Gliridae; Spalacidae; Muridae; Cricetidae

Regarding the order Rodentia of Iraq, a total of 28 rodent species were reported from Iraq (Al-Sheikhly et al. 2015b). One of the distinct and endemic rodent taxa in Iraq is the Long-tailed Nesokia *Nesokia bunnii*. It was originally described under the genus *Erythronesokia* by Khajuria (1980, 1981); later on, it was proved to be a distinctive species of *Nesokia* (Al-Robaae and Felten 1990). This species can be distinguished from the similar Short-tailed Nesokia *Nesokia indica* by its large size, cranial features, long tail, redness of the dorsal pelage, and incipient development of the lower incisor root. Harrison and Bates (1991) carefully examined Khajuria's record and concluded that it could be a large specimen of Short-tailed Bandicoot Rat, with an abnormal long tail possibly reflecting its ecological adaptation to the aquatic habitat. Recent investigation allocated Long-tailed Nesokia into a new neotype was made by Kryštufek et al. (2017). However, the current status of this species is still unknown. It was recorded only in the marshes at the confluence of the Tigris and Euphrates Rivers in south-eastern Iraq (Khajuria 1980, 1981; Al-Robaae and Felten 1990), the species recorded in the northern part of Al-Hawizeh Marsh straddling the Iraq-Iran south-eastern borders (Haba 2009; Kryštufek et al. 2017, 2020).

The Euphrates Jerboa *Allactaga* (*Paralactaga*) *euphratic,* which is listed as Near Threatened by the IUCN, also occurred in Iraq. It has a relatively local distribution in central and southern Iraq. It was recorded along the Euphrates River bank, aridlands with scarce vegetation, alluvial plain, and on the edge of the Mesopotamian marshes (Al-Sheikhly et al. 2015b).

One of the restricted-range rodent species in Iraq is Mesopotamian Gerbil *Gerbillus* (*Hendecapleura*) *mesopotamiae*. It is widespread in arid terrains of central and southern Iraq. It was recorded from surroundings of Amiriya (Harrison 1956); Lake Habbaniya, Baghdad, K-3, Basra; Haur Al Hasa, Al-Jadriyah, Ain Tina, 4 km S Falluja, 10 km to the west of Baghdad, Rashidiya, 4 km south of Abu Ghraib, and Um al Khanazeer Island (Hatt 1959); 5 km to the south-east of Najaf (Nasher 1970); Ramadi Liwa (Nadachowski et al. 1990; Harrison and Bates 1991). However, there were no further records of this rodent species in recent years which may attribute by lack of intensive field surveys targeting the species of this order.

Order: Cetacea (two families; five genera; six species)

Family: Delphinidae; Balaenopteridae

In Iraq, six Cetacean species were recorded, from which the Endangered Blue Whale *Balaenoptera musculus* and Indian Ocean Humpback Dolphin *Sousa plumbea*, the Vulnerable Indo-Pacific Finless Porpoise *Neophocaena phocaenoides*, and the Near Threatened False Killer Whale *Pseudorca crassidens* are found. Most marine mammals' records in Iraq were provided by Al-Robaae (1969, 1970, 1971a, b, 1974, and 1975). However, further investigation is needed to reliably describe the status of this order in Iraq. The occurrence of the Blue Whale is confined to the marine coast of extreme southern Iraq; however, its status is uncertain. It was mentioned by Mahdi and George (1969) without providing further details on its occurrence in Iraq. Al-Robaae (1971a) reports one specimen from Kuwait at the territory of the Arabian Gulf. The Indian Ocean Humpback Dolphin (Regarded as Indo-Pacific Humpbacked Dolphin *Sousa chinensis* in Al-Sheikhly et al. 2015b) and the Indo-Pacific Finless Porpoise (regarded as Finless Porpoise by Al-Sheikhly et al. 2015b) are confined to the marine coasts of southern Iraq. Indo-Pacific Humpbacked Dolphin was recorded from Khor Al Zubeir (Al-Robaae 1970). The Indo-Pacific Finless Porpoise was recorded from the Iraqi territorial water *c.* 37 km to the south of Fao, and another specimen was reported from Fao in extreme southern Iraq (Al-Robaae 1975). The occurrence of the False Killer Whale in Iraq is still uncertain; Al-Robaae (1971a) reported one specimen from offshores of Kuwait where it is abundant in the Arabian Gulf. However, its occurrence in Iraq needs further observations.

43.3 The Wild Mammals of the Lower Tigris and Euphrates Basin (Mesopotamian Marshes)

There were very few records concerning wild mammals of the Mesopotamian marshes in southern Iraq, especially those made during and right after the drainage and re-inundation of the southern marshes in1990–2003. In 1990s, through intensive hydrological projects conducted by previous Iraqi political regime, the marshes were drained and dramatically decreased in their area size. The ecological destruction of the marshes has negatively impacted on the occurrence and distribution of many indigenous and endemic wild mammal species, especially insectivorous, carnivorous, and rodents. The habitats destruction had led to a significant decrease in the area vegetation and prey abundance which sustain the survival of many native mammal species. The loss of foraging, hiding, and nesting and breeding habitats had influenced the zoo-geographical distribution in the marshes where many species were on the brink of extinction (Al-Sheikhly and Haba 2014).

The Mesopotamian marshes are hosting many indigenous and endemic mammal species. However, the status of the wild mammal fauna of southern Iraqi marshes was restricted into few historical field observations and records which have been collected prior the marshes drainage in 1990s. Drower (1949) mentioned that Wild Boars *Sus scrofa* were abundant and heavily roaming the dense Reed *Phragmites australis* beds in Hawizeh marshes. Thesiger (1954) mentioned that he killed 488 wild boars in a two-year period in the area around Al-Hammar Marsh (Fig. 43.5). Scott and Evans (1993) indicated that the drainage of the Mesopotamian marshes of southern Iraq has led to a noticeable decrease in many species population of wild mammals. The habitat loss, fragmentation, and destruction have led to a rapid decline in the populations of the endemic Iraq's Smooth-coated Otter and Long-tailed Nesokia, and possibly 50% of the Mesopotamian Gerbil global population was decreased. Among the large mammals that once abundant in and around the marshes are Golden Jackal *Canis aureus*, Arabian Red Fox *Vulpes vulpes arabica*, and Small Asian Mongoose *Herpestes javanicus pallipes* (Fig. 43.6). Moreover, Honey Badger *Mellivora capensis*, Striped Hyena *Hyaena hyaena*, Wild Cats *Felis silvestris*, Jungle Cat *Felis chaus*, and small packs of Grey Wolf *Canis lupus* were reported from the marshes and still thriving on the terrestrial habitats but in small numbers (Al-Sheikhly *in litt.*).

Fig. 43.5 Adult Wild Boar *Sus scrofa* grazing on the edge of eastern Al-Hammar Marsh. Photo Courtesy Omar F. Al-Sheikhly

Fig. 43.6 Adult Small Asian Mongoose *Herpestes javanicus pallipes* in Hawizeh Marsh. Photo Courtesy Omar F. Al-Sheikhly

The Iraq's Smooth-coated Otter was described as a distinct taxon endemic to the Iraqi marshes (*L. p. maxwelli*) on the basis of a skin from a dead individual and a six-week-old male otter collected by G. Young Maxwell from a tumulus island village called Daub near Al-Azair in Hawizeh Marsh. The newly discovered otter was referred to as Smooth-coated Otter because of both its fur and tail, which were darker and flatter than in the European Otter (Hayman 1956; Young 1977; Al-Sheikhly and Nader 2013; Al-Sheikhly et al. 2014). In 1990s, the endemic-to-Iraq Smooth-coated Otter became very rare due to the ecological destruction the Iraqi marshes. After 2003, the persistence occurrence of Iraq's Smooth-coated Otter in the Mesopotamian marshes was enigmatic; very few records restricted to Hawizeh Marsh were made in 2015 (Al-Sheikhly et al. 2015b). However, the first photographic evidence for this rare taxon in Iraq was made from the northern part of Hawizeh Marsh in 2017. Since 2003, its population faced a dramatic decrease in size due to hunting, trapping, and habitat destruction (Al-Sheikhly et al. 2017). Most recently, ca. 930 individuals of Iraq's Smooth-coated Otter in Hawizeh Marsh was estimated by Al-Sheikhly et al. (2020a).

Another endemic and rare rodent species in the Iraqi marshes is the Long-tailed Nesokia. It was known from five specimens collected on the edge of the Mesopotamian marshes in southern Iraq.

Khajuria (1981) had two specimens (adult male and immature female) from Basra Province, southern Iraq. Further, three specimens (skins and skulls) of Long-tailed Nesokia were collected Bani Mansor, 25 km west of Qurna in Basra Province, southern Iraq by Al-Robaae and Felten (1990). The discovery of Long-tailed Nesokia perhaps was the last discovery of a large and morphologically unique rodent in the Palaearctic region (Khajuria 1981). The animal was originally described as a distinct genus *Erythronesokia khajuria*, but was relegated in the early 1990s to the genus *Nesokia* where it is still placed (Musser and Carleton 2005). In 2017, the morphological uniqueness of Long-tailed Nesokia was documented. A comparison of vouchers specimens available for this species with extensive samples of species from the genera *Nesokia* and *Bandicota* showed that Long-tailed Nesokia has a highly distinct external morphology, which implies a specific ecological niche (Kryštufek et al. 2017). The Euphrates Jerboa and Mesopotamian Gerbil are also among the distractive near endemic rodents to the Mesopotamian marshes. Both species are confined to the alluvial plain and aridlands with scarce vegetation of zoo-geographical zone of southern Iraq and south-western Iran. One of the rare bat species in Iraq is the Pallid Long-fingered Bat which it has very few known records from Iraq (Harrison and Bates 1991). Scott (1995) reported this species from the ruins of kish on the southern edge of the Mesopotamian marshes. This record seems to be overlooked by Al-Sheikhly et al. (2015b); however, no further records for this species were made so far.

After 2003, the study of wild mammals in the Mesopotamian marshes did not take further attention; however, few recent surveys targeting wild mammals were conducted. Haba (2009) indicated the persistence occurrence of Golden Jackal, Wild Boars, Hares, and many rodent species especially the Long-tailed Nesokia in 30 sites distributed on the geographical zone of the Mesopotamian marshes. Moreover, Abass (2013) surveyed the wild mammals' population in eastern Al-Hammar Marsh and documented the occurrence of many carnivore species especially the Jungle Cat and Eurasian Otter. Regardless, the efforts mentioned above are still insufficient to express the actual distribution and conservation status of mammal fauna of the Mesopotamian marshes; therefore, further investigation is required.

43.4 Threats on Wild Mammals of Iraq

Al-Sheikhly et al. (2015b) mentioned that hunting and trapping represent the main threats to the wild mammals of Iraq. Several species of wild mammals are frequently trapped and/or hunted for recreational or traditional illegal practices. Moreover, the hunting and trapping were significantly contributed to reduce many wild mammal populations in Iraq (Al-Sheikhly 2012b). Some mammal species are persecuted by local communities especially in central and western Iraq to be consumed as food items such as Indian Crested Porcupine *Hystrix* (*Hystrix*) *indica*, Hares, and Jerboas. On the contrary, such practices are forbidden in southern Iraq. Gazelle, Wild Goat, and Mouflon are heavily hunted to be consumed as food or trapped to be raised as domesticated pets throughout Iraq. Wild boars are targeted whenever and whatever possible by local farmers for purpose of eradication, as they were considered as a pest. Moreover, foxes, jackals, wolves, striped hyenas, honey badgers, jungle cats, and leopards are persecuted also according to traditional practices and/or myths (Hatt 1959; Harrison 1972). In Iraq as elsewhere in the Arabian Peninsula, wolves have traditional and moral values. The Bedouins are known to chase wolves in order to collect their organs, which are used as symbols of glory and braveness. Bedouins usually trap wolf cubs in order to raise them as domesticated pets (Al-Sheikhly et al. 2020b). Moreover, human-wildlife conflict exists in Iraq. Local people are being constantly engaged to eliminate wolves, leopards, hyenas, and badgers whenever possible. Wolves and leopards are considered dangerous to humans and destructive to cattle herds. A fishermen-wild mammal conflict is also reported from Iraq. The European Otter and the Smooth-coated Otter are widely hunted for their fur and youngsters throughout Iraq. Such a practice led the populations of both species to the edge of the extinction.

Unfortunately, illegal trapping and trafficking of autochthonous mammalian species are widely performed in Iraq. The weakness of the wildlife hunting legislation allows unauthorized local hunters to trap many wild animals (Al-Sheikhly 2012c). It is clear that the conflict between man and wild species is leading to a rapid and irreversible decline of many wild mammal populations in Iraq. Several threat impacts were noticeably contributed to such decline such as increasing hunting pressure, illegal poaching and trade, habitat loss and fragmentation, and lack of scientific knowledge and awareness about the wild fauna among local communities. Despite the huge efforts proposed by the Iraqi authorities in order to protect the wild mammalian fauna, a further enforcement of the present hunting legislation is essential, especially when Iraq became a signatory country to the Convention on International Trade in Endangered Species of Wild Fauna and Flora (CITES) (Al-Sheikhly et al. 2015b).

Acknowledgements I would like to thank Dr. Mukhtar K. Haba, Dr. Nadheer A. Fazaa (College of Science for Women, University of Baghdad-Iraq) for their kind support during the previous and forthcoming wildlife and ecological field surveys in Iraq. I am grateful to the Iraqi Green Climate Organization (IGCO) for supporting the field surveys in Iraq.

References

Abass AF (2013) The relative abundance and biological indicators of mammals' community in east Hammar. M. Sc. Thesis, University of Basra, Iraq

Al-Barazengy AN, Salman AO, Omar BA (2015) New sighting of the wild goat *Capra Aegagrus* (Erxleben, 1777) in Nineveh Province, northern Iraq. Bull Iraq Nat Hist Mus 13(4):21–28

Al-Robaa K (1969) Bryde's whale *Balaenoptera edeni* on the coast of Iraq, new record for Arab gulf. Zeitschrift für Saugtiekunde 34:120–125

Al-Robaae K (1970) First record of the speckled dolphin *Sotalia lentiginosa* of the Arabian gulf. Saugtierkundliche Mitteilungen 18:227–228

Al-Robaae K (1971a) False killer whale *Pseudorca crassidens*, a new record for the Arab gulf. Bulletin of the Iraq Nat Hist Museum 5:31–34

Al-Robaae K (1971b) Notes on a blue whale (Balaenoptera musculus) skeleton in Natural History Museum, Kuwait. Bulletin of the Iraq Nat Hist Museum 5:43–44

Al-Robaae K (1974) *Tursiops aduncus*, bottlenose dolphin: a new record for the Arab gulf, with notes on Cetacea of the region. Bulletin of the Basrah Nat Hist Museum 1:7–16

Al-Robaae K (1975) *Neophocaena phocaenoides*, Asiatic black finless porpoise: a new record for the Arabian gulf. Bulletin of the Basrah Nat Hist Museum 2:47–49

Al-Robaae K (1982) Northeast extension of the geographic distribution of the fennec fox, *Fennecus zerda* Zimmermann in the Arabia Peninsula. Bull Basrah Nat Hist Mus 5:61–64
Al-Robaae K, Felten H (1990) Was ist *Erythronesokia* Khajuria, 1981 (Mammalia: Rodentia: Muridae)? Zeitschrift für Säugetierkunde 55:253–259
Al-Sheikhly OF (2012a) Some ecological observations on the lesser kestrel *Falco naumanni* in north and northern west of Iraq. (M.Sc Thesis submitted to the College of Science-University
Al-Sheikhly OF (2012b) The hunting of the endangered mammals in Iraq. Wildlife of the Middle East 6(2&3):10
Al-Sheikhly OF (2012c) Notes on antelopes in Iraq. Gunsletter 30(2):11–12
Al-Sheikhly OF, Haba MK (2014) The field guide to the wild mammals of Iraq. College of Science for women- University of Baghdad, Al-Frahidi press- Iraq (in Arabic)
Al-Sheikhly OF, Haba MK (2017) Additional records of the Arabian sand cat *Felis margarita harrisoni* in Iraq. Bonn Zoological Bulletin 66(2):135–137
Al-Sheikhly OF, Haba M k, Barbanera F (2014) Otter hunting and trapping, a traditional practice of marsh Arabs in Iraq. IUCN Otter Specialist Group Bulletin 31(2):80–88
Al-Sheikhly OF, Haba MK, Görföl T, Csorba G (2015a) First evidences of the occurrence of two bat species for Iraq. Mammalia. DOI: https://doi.org/10.1515/mammalia-2014-0098
Al-Sheikhly OF, Haba MK, Barbanera F, Csorba G, Harrison DL (2015b) Checklist of the mammals of Iraq (Chordata: Mammalia). Bonn Zoological Bulletin 64(1):33–58
Al-Sheikhly OF, Haba MK, Barbanera F (2015c) Recent sighting of smooth-coated otter *Lutrogale perspicillata maxwelli* in Hawizeh marsh (southern Iraq). IUCN Otter Specialist Group Bulletin 32(1):30–32
Al-Sheikhly OF, Haba MK, Faaza NA, Barbanera F, Yoxon G, McLennan B, Zook DB, Al-Kanani KA, Abdul Gafor SA (2017) First photographic evidence of smooth-coated otter (*Lutrogale perspicillata maxwelli*) and Eurasian otter (Lutra lutra) in Iraq since 1950s. Otter (3):15–20
Al-Sheikhly OF, Haba MK, Mousavi SB, Yoxon GM (2020a) A preliminary population estimate of the Vulnerable smooth-coated otter *Lutrogale perspicillata maxwelli* (Hayman 1956) in the Hawizeh Marsh in Southeastern Iraq with confirmed occurrence in the Hoor Ol-Azim Wetland in Southwestern Iran. OTTER J Int Otter Surviv Fund 6:78–93
Al-Sheikhly OF, Haba MK, Al-Obeidi LA, Abdulzahra HK (2020b) The grey wolf *Canis lupus* in Iraq: notes on its current distribution, major threats and conservation status. Canid Biol Conserv 22(1):1–7
Al-Sheikhly OF, Mallon D (2013) The small Asian mongoose *Herpestes javanicus* and the Indian Grey mongoose *Herpestes edwardsii* in Iraq (Mammalia: Carnivora: Herpestidae). Zoology in the Middle East 59(2):173–175
Al-Sheikhly OF, Nader IA (2013) The status of Iraq smooth-coated otter *Lutrogale perspicillata maxwelli* Hayman 1956 and Eurasian otter *Lutra lutra* Linnaeus 1758 in Iraq. IUCN Otter Specialists Group Bulletin 30(1):18–30
Barbanera F, Moretti B, Guerrini M, Al-Sheikhly OF, Forcina G (2016) Investigation of ancient DNA to enhance natural history museum collections: misidentification of smooth-coated otter (*Lutrogale perspicillata*) specimens across multiple museums. Belg J Zool 146(2):101–112
Benda P, Reiter A, Al-Jumaily M, Nasher AK, Hulva P (2009) A new species of mouse-tailed bat (Chiroptera: Rhinopomatidae) from Yemen. J National Museum (Prague), Nat Hist Series 177 (6):53–68
Bodenheimer FS (1959) A biologist in Israel. Biological Studies Publisher, Jerusalem, p 492
Capper S (1921) Hyaenas in Mesopotamia. J Bombay Nat Hist Soc 27:937
Cheesman RE (1920) Report on the mammals of Mesopotamia collected by members of the Mesopotamian expeditionary force, 1915-1919. J Bombay Nat Hist Soc 27:323–346
Corkill NL (1929) On the occurrences of the cheetah (*Acinonyx jubatus*) in Iraq. The J Bombay Nat Hist Soc 33:700–702
Dollman G, Burlace JB (1935) Rowland War's Records of big Game. African & Asiatic sections, 10th edn. Rowland Ward Publication, London
Drower ES (1949) Arabs of the Hor al Hawiza. Chap. 5, in: Henry Field 1949, op. *cit*

Ellerman JR, Morrison-Scott TCS (1951) Checklist of Palaearctic and Indian mammals; 1758 to 1946. The British Museum of natural History, London
Felid H (1951) Caves and Rockshelters in southwestern Asia. Bulletin of the National Speleological Society 13:14–18
Felid H (1955) Animals of Jabal Baradost, Iraq. American Documentation Institute 4427: 60–62
Gasperetti J, Harrison DL, Büttiker W (1985) The carnivores of Arabia. Fauna Saudi Arab 7:397–346
Haba MK (2009) Mesopotamian Marshland Mammals. Marsh Bulletin 4(2): 179–189
Harrison DL (1956) Gerbils from Iraq, with description of a new gerbil. J Mammal 37:417–422
Harrison DL (1959) Footsteps in the Sand. Benn. London, Publication 254 pp.
Harrison DL (1964) The mammals of Arabia. Insectivora, Chiroptera, primates, vol 1. Ernest Benn Ltd, pp 1–192
Harrison DL (1968a) Mammals of Arabia, vol 2. Carnivora. Artiodactyla. Hyracoidea. Ernest Benn, Ltd, London, pp 195–381
Harrison DL (1968b) Mammals of Arabia, vol 2. Carnivora. Artiodactyla. Hyracoidea. Ernest Benn, Ltd, London, pp 195–381
Harrison DL (1972) The mammals of Arabia. Volume 3: Lagomorpha, Rodentia. Ernest Benn, Ltd. London, pp 385–670
Harrison DL, Bates PJJ (1991)The mammals of Arabia. Harrison Zoological Museum, Sevenoaks, Kent, U.K.
Hatt RT (1959) The mammals of Iraq. Miscellaneous Publications Museum of Zoology University of Michigan 106:1–113
Hayman RW (1956) A new race of the Indian smooth-coated otter from Iraq. Annals and Magazine of Nat Hist 12 9(106):710–712
Hoogstraal HA, Kaiser MN (1958) The Tickes (Ixidoidae) of Iraq: keys, hosts and distribution. J Iraqi Med 6(2–3):7
Hubbard CA (1955) Some ticks from Iraq. Entomol News 66:189–119
Hutterer R, Harrison DL (1988) A new look at the shrews (Soricidae) of Arabia. Bonner Zoologische Beiträge 39(1):59–72
Khajuria H (1980) A new bandicoot rat, *Erythronesokia bunnii* gen. Et sp. nov. (Rodentia: Muridae) from Iraq. Bull Iraq Nat Hist Res Centre 7(4):157164. Mus. Publ. *26: 1-104*
Khajuria H (1981) A new bandicoot rat, *Erythronesokia bunnii* new genus, new species (Rodentia: Muridae), from Iraq. Bulletin of the Nat Hist Res Centre 7:157–164
Khajuria H (1988) A new species of rat-tailed bats (Chiroptera: Rhinopomatidae) from Iraq. Rec Zool Surv India 85(3):391–402
Kinnear NB (1916) Notes on the animals of Mesopotamia. Journal of the Bombay Natural History Society Special Publication 32 pp.
Kock D, Al-Jumaily M, Nasher AK (2001) On the genus *Rhinopoma* E. Geoffroy 1818, and a record of *Rh. muscatellum* Thomas 1903 from Yemen (Mammalia, chiroptera, rhinopomatidae). Senckenberg Biol 81:285–287
Kryštufek B, Al-Sheikhly OF, Hutterer R (2017) A re-description of the long-tailed Nesokia, *Nesokia bunnii*, and designation of a neotype (Rodentia: Muridae). Zootaxa (2):167–187
Kryštufek B, Al-Sheikhly OF, Lazaro J, Haba MK, Hutterer R, Mousavi SB, Ivajnšič D (2020) A forgotten rodent from the Garden of Eden: what really happened to the long-tailed nesokia rat in the Mesopotamian marshes? Mammalia (published online ahead of print 2020), 000010151520200092. https://doi.org/10.1515/mammalia-2020-0092
Lawrence B (1956) Cave Fauna. Appendix E, pp. 80–81, in: Henry Field, 1956 *b: op. cit*
Layard AH (1852) Popular account of discoveries at Nineveh. Harpers, New York, p 360
Lydekker R, Blaine G (1914) Catalogue of the ungulate mammals in the British museum (N. H.). The British Museum, London. Vol. I1 (1914; with Gilbert Blaine), pp. xvi + 295
Mahdi N, George PV (1969) Systematic list of Iraqi vertebrates– mammals. Bulletin of the Iraq Nat Hist Museum 26:34–63

Mallon DP, Kingswood SC (2001) Antelopes. Part 4: North Africa, the Middle East, and Asia. IUCN, Gland, Switzerland, Global Survey and Regional Action Plans
Matskási I (1980) Trematodes of bats in Iraq. Parasitologica Hungarica 13:7–12
Maxwell G (1957) A reed shaken by the wind; a journey through the unexplored marshlands of Iraq. Penguin, Harmondsworth, p 223
Maxwell G (1960) Ring of bright water. Longman's Green and Co. Publication, London
Mohammad MK, Saman RL, Hayder MA (2013) First record of the sand cat, *Felis margarita* Loche, 1858 (Mammalia: Carnivora, Felidae), from Iraq. Zoology in the Middle East 59 (4):358–359
Musil A (1927) The middle Euphrates. American Geographic Society publication, New York, p 426
Musser GG, Carleton MD (2005) Superfamily Muroidae. In: Wilson DE, Reeder DAM (eds) Mammal species of the world. A taxonomic and geographic reference, vol 2, 3rd edn. John Hopkins University Press, Baltimore, pp 894–1531
Nadachowski A, Śmiełowski J, Rzebik-Kowalska B, Daoud A (1990) Mammals from the Near East in Polish collections. Acta Zool Cracov 33(6):91–120
Nader IA (1969) Records of the shrew, *Crocidura russula*, from Iraq. J Mammal 50:614–615
Nader CF, Lay DM, Hassinger JD (1971) Cytogenetic analyses of wild sheep populations in northern Iraq. Cytogenetics 10:137–152
Nasher AKA (1970) Natural history study of the Harrison's gerbil, *Gerbillus dasyurus mesopotamiae* Harrison. (unpublished thesis). Nat Hist Zool Ser 24:49–92
Nature Iraq (2017) Key biodiversity areas of Iraq. Sulaymaniyah, Iraq. 297 pp.
Niazi AD (1976) On the Mediterranean horseshow bat from Iraq. Bulletin of the Iraq Nat Hist Museum 7:167–176
Omer SA, Wronski T, Al Wash A, Elamin MH, Mohammed OB, Lerp H (2012) Evidence for persistence and a major range extension of the smooth-coated otter, *Lutrogale perspicillata maxwelli* (Mustelidae, Carnivora) in Iraq. Folia Zool 61:172–176
Pitman CRS (1922) Notes on Mesopotamian mammals. J Bombay Nat Hist Soc 28:474–480. (Reprinted in: Survey of the Fauna of Iraq Journal of the Bombay Natural History Society 317–23 pp)
Pocock RI (1930) The Panthers and Ounces of Asia. Ibid 64–82:307–360
Pocock RI (1941) Fauna of British India. Vols. I and 11. Mammalia. London: Taylor and Francis. pp. xxxiv + 462. Pp. xii+ 503
Raswan CR (1935) Black tents of Arabia; my life among the Bedouins. Little, Brown and Co., Boston, pp xiii–159
Raza HA (2013) On conserving the wild goat *Capra aegagrus* in Peramagroon and Qara Dagh Mountains, Iraq conservation leadership program. Wildlife of the Middle East 6(4):5
Sanborn CC (1940) Mammals from Iraq. In: Henry field anthropology of Iraq. The Field Museum of Natural History 30:156–162
Scott DA (1995) A directory of Wetland in the Middle East. IWRP, Slimbridge, pp 223–301
Scott DA, Evans MI (1993) Wildlife of the Mesopotamian marshlands. In: Report prepared for wetlands ecosystem research group. University of Exeter, U.K., p 146
Simmons NB (2005) Order Chiroptera. In: Wilson DE, Reeder DM (eds) Mammal 184 species of the world: a taxonomic and geographic reference, 3rd edn. The Johns Hopkins University 185 Press, Baltimore, pp 312–529
Thesiger W (1954) The Marshmen of southern Iraq. J Geogr Syst 120:272–281
Young G (1977) Return to the Marshes. Futura Publications, Great Britain, 186 pp

Chapter 44
The Amphibians and Reptiles of Euphrates and Tigris Basin

Barbod Safaei-Mahroo and Hanyeh Ghaffari

Abstract This chapter focused on Amphibians and Reptiles of Euphrates and Tigris River Basin that is confined to Iraq, Iran, Syria, and Turkey as a main part of Mesopotamia. The Mesopotamian terrestrial biome has an area of approximately 700,000 km^2, divided into ten main ecoregions. Herpetofauna of the basin includes 10 species of amphibians in 8 genera and 4 families in two orders of Caudata and Anura, furthermore 134 species of reptiles in 64 genera and 21 families, 4 turtle species, 71 species of lizards, 57 species of snakes and 2 species of amphisbaenians. Predominant vegetation along the Euphrates and Tigris River and their tributaries provide food, refuge, and breeding ground for many semiaquatic and aquatic species. The main threats to the survival of amphibians and reptiles in this region are human overexploitation, dam construction, water pollution, illegal fishing methods, ongoing habitat loss and fragmentation, wars, and political conflicts.

44.1 Introduction

Historical records show that humans and animals, in this case, amphibians and reptiles, have been living together since antiquity. Archeological investigations have demonstrated these creatures played a significant role as cultural symbols in ancient times (Berthon et al. 2016). This has been investigated in various regions in the Mesopotamia in terms of drawings and carvings on the ceramic pots. Fig. 44.1a displays a tadpole was found in the Jarmoo Plain in the Kurdistan region of Iraq and

B. Safaei-Mahroo
Pars Herpetologists Institute, Tehran, Iran

H. Ghaffari (✉)
Department of Environmental Sciences, Faculty of Natural Resources, University of Kurdistan, Sanandaj, Iran

L. A. Jawad (ed.), *Tigris and Euphrates Rivers: Their Environment from Headwaters to Mouth*, Aquatic Ecology Series 11,
https://doi.org/10.1007/978-3-030-57570-0_44

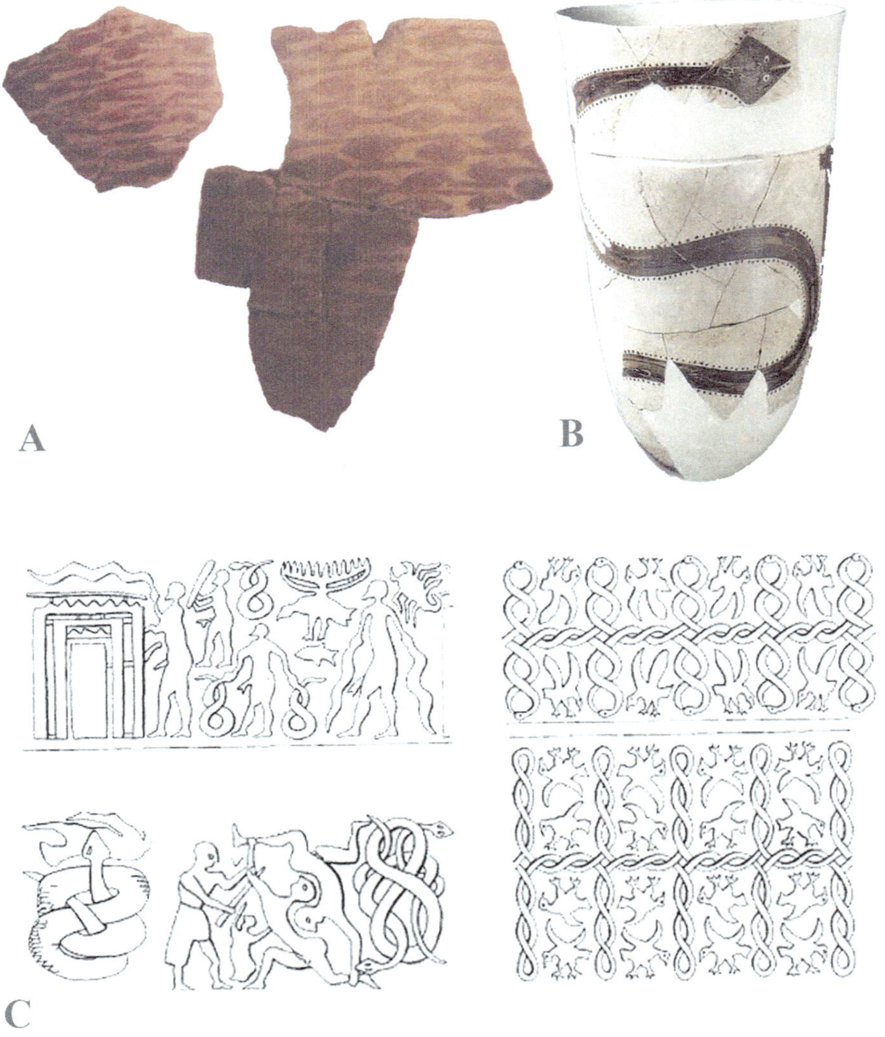

Fig. 44.1 (**a**) Ancient pottery pieces from Mesopotamia showing toadpoles; (**b**) Snake vase from Susa (Harper et al. 1993: 35); (**c**) The snake vase and headless intertwined snakes designs: Middle Susiana (Delougaz and Kantor 1996a, b)

Tamarkan east of Baghdad (Braidwood and Howe 1960; Adams Mc 1983: 215–217; Oates 1968: 4) and other artifacts on snakes (individual or intertwined) (Fig. 44.1b, c) as well as turtles were discovered east of Mesopotamia in Choghamish of Khuzestan Province of Iran (Fig. 44.2) dating back to Middle ages of Susiana (Delougaz and Kantor 1996a, b), which reflects a background of at least 7000 years between humans and reptiles.

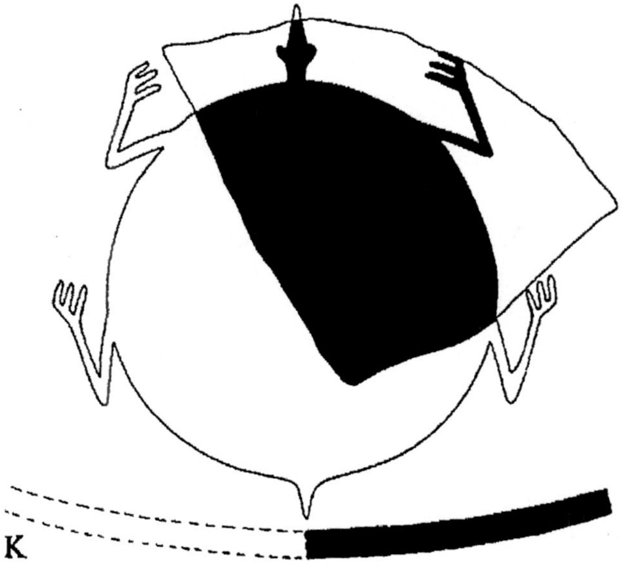

Fig. 44.2 The three-clawed turtle on pottery from Choghamish, Late Middle Susiana (Delougaz and Kantor 1996a, b)

In Middle Susiana 3 (5700–4700 BC), snakes were favored designs compared to earlier ages (Kantor 1976). Furthermore, ceramic pots designed with three-clawed turtles and elongated snout, round and flat shell, have been found, which are very similar to the Euphrates softshell turtles. This species of turtle still lives along the Mesopotamian rivers, including Tigris, Euphrates, and their tributaries in Iran, Iraq, Syria, and Turkey (Fig. 44.2).

44.2 A Short History of Herpetology and Herpetological Publications in Mesopotamia

In terms of new reptiles description on the border of Iraq and Iran from 2015 to 2019, Böhme et al. (2015); Torki (2017); Safaei-Mahroo et al. (2017); Nazarov et al. (2018); and Ablonski et al. (2019) have shown that Mesopotamia is a part of the world that is still herpetologically unexplored.

The first naturalist who visited Mesopotamia was Guillaume-Antoine Olivier, who made extensive expeditions to Persia, Syria, and parts of Turkey from 1792 to 1799. Blanford (1874) described two new lizards from Mesopotamia and Southern Persia. Werner contributed significantly to the herpetology of Mesopotamia; in contributions to the knowledge of the fauna of Syria and Persia paper published in 1929, he described reptiles from Damascus to Baghdad and Basra, which Alfond

Gabriel collected and gave him. Werner (1895) described *Phylodactylus elisae* and *Hemidactylus bornmulleri* from Iraq that were collected by Josef Bornmuller, who was a botanist.

Boulenger (1920a, b) published lists of snakes and lizards collected by members of the Mesopotamian Expeditionary Force from 1915 to 1919 and sent to the Museum of Bombay Natural History Society. Procter (1921) recorded the remaining list of reptiles from Mesopotamia.

Schmidt (1939) was a reptiles and amphibians curator in the Field Museum of Natural History; he reported herpetological specimens deposited in the Field Museum, collected by Henry Field from Iraq during the Marshall Field North Arabian Expedition in 1928 and other collectors from various surveys and investigations. A list of amphibians and reptiles collected by the 1950 and 1955 expeditions from Iran, Iraq, and Syria was published by Henry Field (Anderson 1999).

The specimens of the Austrian Expedition were deposited in the Vienna Natural History Museum were reported by Wettstein (1951), including reptiles and amphibians from Iran and Iraq (Anderson 1999). Bodenheimer and Theodor provided herpetological collections from Baghdad and Addaye; the account of these collections was published by Haas (1952). Hass and Werner (1969) published a report on a collection of reptiles collected by Henry Field from Syria, Iraq, Iran, Saudi Arabia, and Pakistan. The book *Reptiles of Iraq with Some Notes on the Amphibians* was published by Khalaf in 1959, and Nader and Jawdat (1976) published a taxonomic study of the geckos of the Iraq book. By far, the most comprehensive literature on the herpetofauna of the Middle East is Handbook to Middle East Amphibians and Reptiles (Leviton et al. 1992).

44.3 Mesopotamian Biogeography Relative to Reptiles and Amphibians Distribution

In this chapter, we focus on Euphrates and Tigris River Basin that are confined to Iraq, Iran, Syria, and Turkey as a main part of Mesopotamia shown in Fig. 44.3; the boundaries are based on the Euphrates and Tigris River Basin (UN-ESCWA and BGR 2013). The area has a surface of 700,000 km^2 and consists of a diverse range of habitats from complex mountain chains in the north and east to flat, low-lying plains in the center and south.

Mesopotamia is located in Palearctic Zoogeographic Region, the study area located in the western Palearctic (Sindaco et al. 2013). Olson et al. (2001) subdivides the terrestrial world into distinct ecoregions, the main Mesopotamian terrestrial ecoregions including: the Tigris-Euphrates alluvial salt marsh (24,571 km^2, 3.5%), the Arabian Desert, and East Sahero-Arabian xeric shrublands (162,487 km^2, 23.1%), the Eastern Anatolian deciduous forests (67,958 km^2 9.7%), the Eastern Anatolian montane steppe (41,224 km^2 5.9%), the Eastern Mediterranean conifer-sclerophyllous-broadleaf forests (56,334 km^2 8%), the Mesopotamian shrub desert

Fig. 44.3 Ecoregions of Euphrates and Tigris River Basin (based on UN-ESCWA and BGR 2013 and Olson et al. 2001)

(159,554 km^2 22.7%), the Middle East steppe (947,71 km^2 13.5%), the Northern Anatolian conifer and deciduous forests (5131 km^2, 0.7%), the Southern Iran Nubo-Sindian desert and semi-desert (8779 km^2, 1.3%), the Zagros Mountains forest steppe (81,047 km^2, 11.5%).

These species are widely distributed throughout the region: *Bufotes sitibundus, Hyla savignyi, Pelophylax bedriagae, Trapelus ruderatus, Uromastyx aegyptia, Bunopus tuberculatus, Cyrtopodion scabrum, Hemidactylus persicus, Mediodactylus heterocercus, Asaccus elisae, Ophisops elegans, Acanthodactylus boskianus, Heremites septemtaeniatus, Varanus griseus, Dolichophis jugularis, Platyceps chesneii, P. najadum, Malpolon insignitus, Natrix tessellata, Eryx jaculus, Myriopholis macrorhyncha, Macrovipera lebetinus*.

In the northern part of the Euphrates and Tigris River Basin, Common reptiles are found: *Lacerta media, Eumeces schneideri, Dolichophis schmidti, Eirenis (Pediophis) eiselti, E. (P.) punctatolineatus, Hemorrhois nummifer, H. ravergieri, Xerotyphlops vermicularis, Natrix natrix, Testudo graeca*. Species confined to the southern of the Euphrates and Tigris River Basin include *Hemidactylus flaviviridis, Mesalina bernoullii, Eirenis (Pediophis) coronella, Spalerosophis diadema cliffordii, Walterinnesia aegyptia morgani*.

There are similarities between the upper and lower boundaries of the Euphrates and Tigris River Basin in this section and those described by Smith et al. 2014. We corroborate the idea of Smith et al. (2014), who suggested that there are endemic species, which are dependent on freshwater habitats and springs, especially Caudata such as *Salamandra infraimmaculata semenovi, Neurergus crocatus, N. derjugini* and *N. strauchii*, in the upper part of the basin that are not found in the lower part.

The factor of altitude plays a crucial role in the dispersal of amphibians and reptiles of Mesopotamia. The mountains of eastern and northeastern Turkey are the origins of the Tigris and Euphrates Rivers, which prolong from the highland to the low-lying regions of the southern Mesopotamian regions, and the elevation relatively and moderately decreases from the north to the south. This characteristic in controlling the amount of relative humidity, has been the cause of a regulated modification of vegetation throughout Mesopotamia such that the mountainous and forest areas in the north and east have been directly under the impact of Mediterranean rainfalls; and consequently, amphibian and reptile species, which are more dependent on the mountainous and forestry regions, namely, *Salamandra infraimmaculata semenovi, Neurergus crocatus, N. derjugini, N. strauchii, Hemidactylus kurdicus* and *Timon kurdistanica*, have been inhabited here.

The warm and dry climate and little precipitation of lowland and flat areas of the central and southern parts except for the rivers and marshlands shoreline are mainly caused by the climate of northern Arabia. These areas are suitable habitats for reptiles inhabit the north of the Arabian Desert to North Africa like *Acanthodactylus, Stenodactylus, Trigonodactylus, Cerastes, Echis, Platyceps*, and *Rhagerhis* genera.

Taurus-Zagros Mountain range is located between the mountainous area and the Mesopotamian lowlands extending from north to eastern parts of Mesopotamia (across Iran, northern Iraq, and Turkey). This ecotone between forests and grasslands plays an important role in herpetofaunal diversity and speciation. This humid temperate zone has enriched biodiversity of 14 endemic amphibians and reptiles that are as follows: *Neurergus crocatus, N. derjugini, Salamandra infraimmaculata semenovi, Hemidactylus kurdicus, Mediodactylus ilamensis, Microgecko helenae fasciatus, Eremias montana, Timon kurdistanica, Asaccus andersoni, A. kurdistanensis, A. saffinae, Varanus nesterovi, Lytorhynchus levitoni, Pseudocerastes urarachnoide*.

Biogeographical factor is the major driver of herpetofauna patterns in the Euphrates and Tigris River Basin. With 11 ecoregions, each has its own endemic species. Endemic species of amphibians and reptiles of Mesopotamian range in each biome and ecoregion are listed in Table 44.1.

The Zagros Mountain range in northern Iraq that extends to western and southern Iran creates a barrier and a corridor for amphibians and reptiles in Mesopotamia (Anderson 1999).

Tigris and Euphrates Rivers and their tributaries are other major elements of Mesopotamian species diversity. This freshwater ecosystem is titled as factors limiting the dispersion of terrestrials and for riverine reptiles, such as the pond turtle, watersnake, and likewise, for Anura, such as the tree frog and marsh frog, are like the main corridor for the transition and distribution of species.

Table 44.1 Herpetofauna of Mesopotamia in deferent ecoregions

	Biome	*Ecoregion*	Common Species	Endemic species to the Mesopotamia
Terrestrial Biomes	**Desert and Xeric Shrublands**			6 Endemic species
		Arabian Desert and East Sahero-Arabian xeric shraublands	*Uromastyx aegyptia, Hemidactylus flaviviridis, H. persicus, H. robustus, Stenodactylus affinis, S. slevini, , Platyceps rogersi, Cerastes gasperettii, Echis carinatus*	***Diplometopon zarudnyi, Acanthodactylus grandis***
		Mesopotamian shrub desert	*Uromastyx aegyptia, Trapelus pallidus, Stenodactylus grandiceps, Acanthodactylus scutellatus, A. tristrami, Lytorhynchus diadema, L. kennedyi, Platyceps rhodorachis, Telescopus nigriceps*	***Acanthodactylus nilsoni,***
		South Iran Nubo-Sindian desert and semi-desert	*Saara loricate, Stenodactylus affinis, Rhagerhis moilensis, Echis carinatus*	***Phrynocephalus ahvazicus, Xerotyphlops wilsoni, Trigonodactylus persicus***
	Flooded grasslands and savannas			1 Endemic species
		Tigris-Euphrates alluvial salt marsh	*Mauremys capsica, Natrix tessellata*	***Rafetus euphraticus***
	Mediterranean Forests, Woodlands, and Scrub			6 Endemic species
		Eastern Anatolian deciduous forests	*Salamandra infraimmaculata, Stellagama stellio, Elaphe dione, Zamenis hohenackeri, Macrovipera lebetinus*	***Neurergus strauchii, Elaphe urartica, Eirenis (Pseudocyclophis) thospitis***
		Eastern Mediterranean conifer-sclerophyllous-broadleaf forests	*Blanus strauchi, Stellagama stellio, Macrovipera lebetinus*	***Acanthodactylus harranensis, Apathya cappadocica schmidtlerorum, Asaccus barani***
	Temperate broadleaf and mixed forest			14 Endemic species
		Zagros Mountains forest steppe	*Hyla savignyi, Pseudopus apodus, Eublepharis angramainyu, Laudakia nupta, Hemidactylus persicus, Mediodactylus kotschyi, M. heterocercus, M. heteropholis, Apathya cappadocica, Asaccus granularis, Eirenis (Pediophis) collaris, E. (P.) nigrofasciatus, Elaphe sauromates, Hierophis andreanus, Platyceps rhodorachis, Spalerosophis microlepis, Telescopus nigriceps, T. tessellatus, Psammophis schokari, Macrovipera lebetinus, Montivipera raddei*	***Calliopersa luristanica, Neurergus crocatus, N. derjugini, Salamandra infraimmaculata semenovi, Hemidactylus kurdicus, Mediodactylus ilamensis, Microgecko helenae fasciatus, Eremias montana, Timon kurdistanica, Asaccus andersoni, A. kurdistanensis, A. saffinae, Varanus nesterovi, Lytorhynchus levitoni, Pseudocerastes urarachnoides***
	Temperate Grasslands, Savannas, and Shrublands			1 Endemic species
		Eastern Anatolian montane steppe	*Salamandra infraimmaculata, Neurergus strauchii, Paralaudakia caucasia, Eremias suphani, Zamenis hohenackeri, Z. longissimus, Montivipera raddei*	***Darevskia sapphirina***
		Middle East steppe	*Mediodactylus heterocercus*	
Aquatic	**Freshwater**			4 Endemic species
		Lakes and Rivers	*Hyla savignyi, Pelophylax bedriagae, Mauremys capsica, Natrix tessellata*	***Neurergus crocatus, N. derjugini, N. strauchii, Rafetus euphraticus***
Other		**Residential areas**	*Bufotes sitibundus, Cyrtopodion scabrum, Hemidactylus flaviviridis, H. persicus*	

44.4 Amphibians and Reptiles of Euphrates and Tigris River Basin

The Euphrates and the Tigris River Basin is located in a climate transition zone between the desert and humid continental climates; furthermore, high habitat diversity, including mountainous regions, flat and dry plains, marshland, and riverine habitats, has caused herpetofaunal species richness and diversity in this region. The reptiles of Euphrates and Tigris River Basin are 134 species in 64 genera and 21 families, 4 turtle species, 71 species of lizards, 57 species of snakes, and 2 species of amphisbaenians. The *Acanthodactylus* genera with 10 species and relative to the Lacertidae family and the *Asaccus* genus with 7 species from the Phyllodactylidae family are the most diverse groups. Among the snakes' family, Colubridae with 11 species, is the most diverse group. There are also 10 species of amphibians in 8 genera, and 4 families in two orders of Caudata and Anura have been reported in this area. By taking into consideration that half of this basin is dry, there are 3 species and one sub-species of the endemic Salamander, which is extremely significant (Table 44.2). The *Neurergus* genus with 3 species and 3 sub-species from the Salamandridae family, all are endemic to the mountainous area of this basin.

44.5 Checklist of Amphibians in Mesopotamia

Class: Amphibia Gray, 1825
Order: Anura Fischer von Waldheim, 1813
Family: Bufonidae Gray, 1825
Genus: ***Bufotes*** Rafinesque, 1815
Bufotes sitibundus (Pallas, 1771)
Genus: ***Calliopersa*** Safaei-Mahroo & Ghaffari, 2020
Calliopersa luristanica (Schmidt, 1952)
Family: Hylidae Rafinesque, 1815
Subfamily: Hylinae Rafinesque, 1815
Genus: ***Hyla*** Laurenti, 1768
Hyla savignyi Audouin, 1827
Family: Ranidae Rafinesque, 1814
Genus: ***Pelophylax*** Fitzinger, 1843
Pelophylax bedriagae (Camerano, 1882)
Genus: ***Rana*** Linnaeus, 1758
Rana macrocnemis Boulenger, 1885
Order: Caudata Scopoli, 1777
Family: Salamandridae Goldfuss, 1820
Subfamily: Pleurodelinae Tschudi, 1838
Genus: ***Neurergus*** Cope, 1862
Neurergus crocatus Cope, 1862

Table 44.2 Amphibians and reptiles of Euphrates and Tigris River

Class	Order	Suborder	Family	Subfamily: Genus	Total Number
Amphibia					**2 orders, 4 families, 8 genera, 10 species**
	Anura				**3 families, 5 genera, 5 species**
			Bufonidae	*Bufotes, Calliopersa*	2 genus, 2 species
			Hylidae	*Hyla*	1 genus, 1 species
			Ranidae	*Pelophylax, Rana*	2 genera, 2 species
	Caudata				**1 family, 3 genera, 5 species**
			Salamandridae	Pleurodelinae: *Neurergus, Ommatotriton* Salamandrinae: *Salamandra*	3 genera, 5 species
Reptilia					**2 orders, 21 families, 64 genera, 134 species**
	Squmata				**3 suborders, 17 families, 60 genera, 130 species**
		Amphisbaenia			**1 family, 2 genera, 2 species**
			Trogonophiidae	*Blanus, Diplometopon*	2 genera, 2 species
		Sauria			**8 families, 32 genera, 71 species**
			Agamidae	Agaminae: *Laudakia, Paralaudakia, Phrynocephalus, Stellagama, Trapelus* Uromastycinae: *Saara, Uromastyx*	7 genera, 9 species
			Anguidae	Anguinae: *Pseudopus*	1 genus, 1 species
			Eublepharidae	*Eublepharis*	1 genus, 1 species
			Gekkonidae	*Bunopus, Cyrtopodion, Hemidactylus, Mediodactylus, Microgecko, Stenodactylus, Trigonodactylus*	7 genera, 17 species
			Lacertidae	*Acanthodactylus, Apathya, Darevskia, Eremias, Lacerta, Mesalina, Ophisops, Parvilacerta, Timon*	9 genera, 23 species
			Phyllodactylidae	*Asaccus, Ptyodactylus*	2 genera, 9 species
			Scincidae	Eugongylinae: *Ablepharus* Mabuyinae: *Heremites* Scincinae: *Chalcides, Eumeces, Scincus*	5 genera, 9 species
			Varanidae	*Varanus*	1 genus, 2 species
		Serpentes			**8 families, 25 genera, 57 species**
			Colubridae	Colubrinae: *Dolichophis, Eirenis, Elaphe, Hemorrhois, Hierophis, Lytorhynchus, Platyceps, Rhynchocalamus, Spalerosophis, Telescopus, Zamenis* Natricinae: *Natrix*	12 genera, 39 species
			Elapidae	Elapinae: *Walterinnesia*	1 genus, 1 species
			Erycidae	*Eryx*	1 genus, 3 species
			Lamprophiidae	Lamprophiinae: *Malpolon, Rhagerhis*	2 genera, 2 species
			Leptotyphlopidae	Leptotyphlopinae: *Myriopholis*	1 genus, 1 species
			Psammophiidae	*Psammophis*	1 genus, 1 species
			Typhlopidae	*Indotyphlops, Xerotyphlops*	2 genera, 3 species
			Viperidae	Viperinae: *Cerastes, Echis, Macrovipera, Montivipera, Pseudocerastes*	5 genera, 7 species
	Testudines				**4 families, 4 genera, 4 species**
		Cryptodira			**4 families, 4 genera, 4 species**
			Emydidae	Emydinae: *Emys*	1 genus, 1 species
			Geoemydidae	Geoemydinae: *Mauremys*	1 genus, 1 species
			Testudinidae	*Testudo*	1 genus, 1 species
			Trionychidae	Trionychinae: *Rafetus*	1 genus, 1 species

Neurergus derjugini (Nesterov, 1916)
Neurergus strauchii (Steindachner, 1887)
N. s. barani Öz, 1994
N. s. munzurensis Olgun, Avci, Bozkurt, Üzüm, Olgun & Ilgaz, 2016
N. s. strauchii (Steindachner, 1887)
Genus: ***Ommatotriton*** Gray, 1850
Ommatotriton vittatus (Gray, 1835)
Subfamily: Salamandrinae Goldfuss, 1820
Genus: ***Salamandra*** Garsault, 1764
Salamandra infraimmaculata (Martens, 1885)
S. i. semenovi Nesterov, 1916

44.6 Checklist of Reptilians in Mesopotamia

Class: Reptilia Laurenti, 1768
Order: Testudines Batsch, 1788
Suborder: Cryptodira Cope, 1868
Family: Emydidae Gray, 1825
Subfamily: Emydinae Rafinesque, 1815
Genus: ***Emys*** Duméril, 1806
Emys orbicularis (Linnaeus, 1758)
E. o. colchica Fritz, 1994
Family: Geoemydidae Theobald, 1868
Subfamily: Geoemydinae Theobald, 1868
Genus: ***Mauremys*** Gray, 1869
Mauremys caspica (Gmelin, 1774)
Family: Testudinidae Batsch, 1788
Genus: ***Testudo*** Linnaeus, 1758
Testudo graeca Linnaeus, 1758
T. g. boxtoni Boulenger, 1921
T. g. armeniaca Chkhikvadze & Bakradze, 1991
T. g. zarudnyi Nikolsky, 1896
Family: Trionychidae Fitzinger, 1826
Subfamily: Trionychinae Gray, 1825
Genus: ***Rafetus*** Gray, 1864
Rafetus euphraticus (Daudin, 1802)
Order: Squmata Oppel, 1811
Suborder: Amphisbaenia Gray, 1844
Family: Trogonophidae Gray, 1865
Genus: ***Blanus*** Wagler, 1830
Blanus strauchi (Bedriaga, 1884)
B. s. bedriagae Boulenger, 1884
Genus: ***Diplometopon*** Nikolsky, 1907

Diplometopon zarudnyi Nikolsky, 1907
Suborder: Sauria McCarthney, 1822
Family: Agamidae Spix, 1825
Subfamily: Agaminae Spix, 1825
Genus: ***Laudakia*** Gray, 1845
Laudakia nupta (De Filippi, 1843)
L. n. nupta (De Filippi, 1843)
Genus: ***Paralaudakia*** Baig, Wagner, Ananjeva & Böhme, 2012
Paralaudakia caucasia (Eichwald, 1831)
Genus: ***Phrynocephalus*** Kaup, 1825
Phrynocephalus ahvazicus Melnikov, Melnikova, Nazarov, Rajabizadeh, Al-Johany, Amr & Ananjeva, 2014
Genus: ***Stellagama*** Baig, Wagner, Ananjeva & Böhme, 2012
Stellagama stellio (Linnaeus, 1758)
Genus: ***Trapelus*** Cuvier, 1816
Trapelus agilis (Olivier, 1804)
T. a. khuzistanensis Rastegar-Pouyani, 1999
Trapelus mutabilis (Merrem, 1820)
T. m. pallidus (Reuss, 1834)
Trapelus ruderatus (Olivier, 1804)
T. r. ruderatus (Olivier, 1804)
Subfamily: Uromastycinae Theobold, 1868
Genus: ***Saara*** Gray, 1845
Saara loricata (Blanford, 1874)
Genus: ***Uromastyx*** Merrem, 1820
Uromastyx aegyptia (Forskål, 1775)
U. a. microlepis Blanford 1874
Family: Anguidae Gray, 1825
Subfamily: Anguinae Gray, 1825
Genus: ***Pseudopus*** Merrem, 1820
Pseudopus apodus (Pallas, 1775)
Family: Eublepharidae Boulenger, 1883
Genus: ***Eublepharis*** Gray, 1827
Eublepharis angramainyu Anderson & Leviton, 1966
Family: Gekkonidae Gray, 1825
Genus: ***Bunopus*** Blanford, 1874
Bunopus tuberculatus Blanford, 1874
Genus: ***Cyrtopodion*** Fitzinger, 1843
Cyrtopodion scabrum (Heyden, 1827)
Genus: ***Hemidactylus*** Oken, 1817
Hemidactylus flaviviridis Rüppell, 1835
Hemidactylus kurdicus Safaei-Mahroo, Ghaffari, Ghafoor & Amini, 2017
Hemidactylus persicus Anderson, 1872
Hemidactylus robustus Heyden, 1827
Hemidactylus turcicus (Linnaeus, 1758)

Genus: ***Mediodactylus*** Szczerbak & Golubev, 1977
Mediodactylus heterocercus (Blanford, 1874)
M. h. heterocercus (Blanford, 1874)
M. h. mardinensis (Mertens, 1924)
Mediodactylus heteropholis (Minton, Anderson & Anderson, 1970)
Mediodactylus kotschyi (Steindachner, 1870)
M. k. syriacus (Stepánek, 1937)
Mediodactylus ilamensis (Fathinia, Karamiani, Darvishnia, Heidari & Rastegar-Pouyani, 2011)
Genus: ***Microgecko*** Nikolsky, 1907
Microgecko helenae Nikolsky, 1907
M. h. helenae Nikolsky, 1907
M. h. fasciatus (Schmidtler & Schmidtler, 1972)
Genus: ***Stenodactylus*** Fitzinger, 1826
Stenodactylus affinis (Murray, 1884)
Stenodactylus doriae (Blanford, 1874)
Stenodactylus grandiceps Haas, 1952
Stenodactylus slevini Haas, 1957
Genus: ***Trigonodactylus*** Hass, 1957
Trigonodactylus persicus Nazarov, Melnikov, Radjabizadeh & Poyarkov, 2018
Family: Lacertidae Bonaparte, 1831
Genus: ***Acanthodactylus*** Fitzinger, 1834
Acanthodactylus boskianus (Daudin, 1802)
A. b. euphraticus Boulenger, 1919
Acanthodactylus grandis Boulenger, 1919
Acanthodactylus harranensis Baran, Kumlutas, Lanza, Sindaco, Avci & Crucitti, 2005
Acanthodactylus nilsoni Rastegar-Pouyani, 1998
Acanthodactylus opheodurus Arnold, 1980
Acanthodactylus orientalis Angel, 1936
Acanthodactylus robustus Werner, 1929
Acanthodactylus schmidti Haas, 1957
Acanthodactylus scutellatus (Audouin, 1827)
Acanthodactylus tristrami (Günther, 1864)
Genus: ***Apathya*** Mehely, 1907
Apathya cappadocica (Werner, 1902)
A. c. cappadocica (Werner, 1902)
A. c. muhtari (Eiselt, 1979)
A. c. schmidtlerorum (Eiselt, 1979)
A. c. urmiana (Lantz & Suchow, 1934)
A. c. wolteri (Bird, 1936)
Genus: ***Darevskia*** Arribas, 1997
Darevskia raddei (Boettger, 1892)
D. r. vanensis (Eiselt, Schmidtler & Darewsky, 1993)
Darevskia sapphirina (Schmidtler, Eiselt & Darevsky, 1994)

Darevskia valentini (Boettger, 1892)
Genus: ***Eremias*** Fitzinger, 1834
Eremias montana Rastegar-Pouyani & Rastegar-Pouyani, 2001
Eremias suphani Başoğlu & Hellmich, 1968
Genus: ***Lacerta*** Linnaeus, 1758
Lacerta media Lantz & Cyrén, 1920
L. m. media Lantz & Cyrén, 1920
Genus: ***Mesalina*** Gray, 1838
Mesalina bernoullii (Schenkel, 1901)
Mesalina guttulata (Lichtenstein, 1823)
Mesalina microlepis (Angel, 1936)
Genus: ***Ophisops*** Ménétries, 1832
Ophisops elegans Ménétries, 1832
O. e. blanfordi Schmidt, 1939
O. e. ehrenbergii (Wiegmann, 1835)
O. e. elegans Menetries, 1832
O. e. centralanatoliae Bodenheimer, 1944
Genus: ***Parvilacerta*** Arnold, Arribas & Carranza, 2007
Parvilacerta parva (Boulenger, 1887)
Genus: ***Timon*** Tschudi, 1836
Timon kurdistanica Suchow, 1936
Family: Phyllodactylidae Gamble, Bauer, Greenbaum & Jackman, 2008
Genus: ***Asaccus*** Dixon & Anderson, 1973
Asaccus andersoni Torki, Fathinia, Rostami, Gharzi & Nazari-Serenjeh, 2011
Asaccus barani Torki, Ahmadzadeh, Ilgaz, Avci & Kumlutas, 2011
Asaccus elisae (Werner, 1895)
Asaccus granularis Torki, 2010
Asaccus griseonotus Dixon & Anderson, 1973
Asaccus kurdistanensis Rastegar-Pouyani, Nilson & Faizi, 2006
Asaccus saffinae Afrasiab & Mohamad, 2009
Genus: ***Ptyodactylus*** Goldfuss, 1820
Ptyodactylus guttatus Heyden, 1827
Ptyodactylus puiseuxi Boutan, 1893
Family: Scincidae Oppel, 1811
Subfamily: Eugongylinae Welch, 1982
Genus: ***Ablepharus*** Fitzinger, 1823
Ablepharus chernovi Darevsky, 1953
Ablepharus kitaibelii (Bibron & Bory 1833)
Ablepharus pannonicus Fitzinger, 1823
Subfamily: Mabuyinae Mittleman, 1952
Genus: ***Heremites*** Karin, Metallinou, Weinell, Jackman, & Bauer, 2016
Heremites auratus (Linnaeus, 1758)
H. a. transcaucasica (Chernov, 1926)
Heremites septemtaeniatus (Reuss, 1834)
Heremites vittatus (Olivier, 1804)

Subfamily: Scincinae Gray, 1825
Genus: ***Chalcides*** Laurenti, 1768
Chalcides ocellatus (Forskål, 1775)
C. o. ocellatus (Forskål, 1775)
Genus: ***Eumeces*** Wiegmann, 1834
Eumeces schneideri (Daudin, 1802)
E. s. pavimentatus (Geoffroy, 1827)
E. s. princeps (Eichwald, 1839)
Genus: ***Scincus*** Laurenti, 1768
Scincus scincus (Linnaeus, 1758)
S. s. conirostris Blanford, 1881
Family: Varanidae Gray, 1827
Genus: ***Varanus*** Merrem, 1820
Varanus griseus (Daudin, 1803)
V. g. griseus (Daudin, 1803)
Varanus nesterovi Böhme, Ehrlich, Milto, Orlov & Scholz, 2015
Suborder: Serpentes Linnaeus, 1758
Family: Colubridae Oppel, 1811
Subfamily: Colubrinae Oppel, 1811
Genus: ***Dolichophis*** Gistel, 1868
Dolichophis caspius (Gmelin, 1789)
D. c. caspius (Gmelin, 1789)
Dolichophis jugularis (Linnaeus, 1758)
D. j. jugularis (Linnaeus, 1758)
Dolichophis schmidti (Nikolsky, 1909)
Genus: ***Eirenis*** Jan, 1863
Subgenus: ***Eirenis*** Jan, 1863
Eirenis (Eirenis) modestus (Martin, 1838)
E. (*E.*) *m. modestus* (Martin, 1838)
Subgenus: ***Pediophis*** Fitzinger, 1843
Eirenis (***Pediophis***) ***collaris*** (Ménétries, 1832)
E. (*P.*) *c. collaris* (Ménétries, 1832)
Eirenis (***Pediophis***) ***coronella*** (Schlegel, 1837)
E. (*P.*) *c. coronella* (Ménétries, 1832)
Eirenis (***Pediophis***) ***coronlloides*** (Jan, 1862)
Eirenis (***Pediophis***) ***decemlineatus*** (Duméril, Bibron & Duméril, 1854)
Eirenis (***Pediophis***) ***eiselti*** Schmidtler & Schmidtler, 1978
Eirenis (***Pediophis***) ***hakkariensis*** Schmidtler & Eiselt, 1991
Eirenis (***Pediophis***) ***punctatolineatus*** (Boettger, 1892)
E. (*P.*) *p. condoni* (Boulenger, 1920)
E. (*P.*) *p. punctatolineatus* (Boettger, 1892)
Subgenus: ***Pseudocyclophis*** Boettger, 1888
Eirenis (***Pseudocyclophis***) ***nigrofasciatus*** (Nikolsky, 1907)
Eirenis (***Pseudocyclophis***) ***occidentalis*** Rajabizadeh, Nagy, Adriaens, Avci, Masroor, Schmidtler, Nazarov, Esmaeili & Christiaens, 2015

Eirenis* (*Pseudocyclophis*) *persicus (Anderson, 1872)
Eirenis* (*Pseudocyclophis*) *thospitis Schmidtler & Lanza, 1990
Eirenis* (*Pseudocyclophis*) *walteri Boettger, 1888
Genus: ***Elaphe*** Fitzinger, 1833
Elaphe dione (Pallas, 1773)
Elaphe urartica Jablonski, Kukushkin, Avcı, Bunyatova, Ilgaz, Tuniyev & Jandzik, 2019
Genus: ***Hemorrhois*** Boie, 1826
Hemorrhois *nummifer* (Reuss, 1834)
Hemorrhois *ravergieri* (Ménétries, 1832)
Genus: ***Hierophis*** Fitzinger, 1843
Hierophis andreanus (Werner, 1917)
Genus: ***Lytorhynchus*** Peters, 1862
Lytorhynchus diadema (Dumeril, Bibron & Dumeril, 1854)
L. d. gaddi (Nikolsky, 1907)
Lytorhynchus kennedyi Schmidt, 1939
Lytorhynchus levitoni Torki, 2017
Lytorhynchus maynardi Alcock & Finn, 1897
Lytorhynchus ridgewayi Boulenger, 1887
Genus: ***Platyceps*** Blyth, 1826
Platyceps chesneii (Martin, 1838)
Platyceps collaris (Müller, 1878)
Platyceps najadum (Eichwald, 1831)
P. n. najadum (Eichwald, 1831)
P. n. schmidtleri (Schätti & Mccarthy, 2001)
Platyceps rhodorachis (Jan, 1865)
P. r. ladacensis (Anderson, 1871)
P. r. rhodorachis (Jan, 1865)
Platyceps rogersi (Anderson, 1893)
Genus: ***Rhynchocalamus*** Günther, 1864
Rhynchocalamus satunini (Nikolsky, 1899)
Genus: ***Spalerosophis*** Jan, 1843
Spalerosophis diadema (Schlegel, 1837)
S. d. cliffordii (Schlegel, 1837)
Spalerosophis microlepis Jan, 1865
Genus: ***Telescopus*** Wagler, 1830
Telescopus fallax Fleischmann, 1831
T. f. iberus Eichwald, 1831
Telescopus nigriceps (Ahl, 1924)
Telescopus tessellatus (Wall, 1908)
T. t. martini (Schmidt, 1939)
T. t. tessellatus (Wall, 1908)
Genus: ***Zamenis*** Wagler, 1830
Zamenis hohenackeri (Strauch, 1873)
Z. h. tauricus (Werner, 1898)

Zamenis longissimus (Laurenti, 1768)
Subfamily: Natricinae Bonaparte, 1838
Genus: ***Natrix*** Laurenti, 1768
Natrix natrix (Linnaeus, 1758)
Natrix tessellata (Laurenti, 1768)
Family: Elapidae Boie, 1827
Subfamily: Elapinae Boie, 1827
Genus: ***Walterinnesia*** Lataste, 1887
Walterinnesia aegyptia morgani (Mocquard, 1905)
Family: Erycidae Bonaparte, 1831
Genus: ***Eryx*** Daudin, 1803
Eryx elegans (Gray, 1849)
Eryx jaculus (Linnaeus, 1758)
E. j. jaculus (Linnaeus, 1758)
E. j. turcicus (Olivier, 1801)
E. j. familiaris Eichwald, 1831
Eryx jayakari Boulenger, 1888
Family: Lamprophiidae Fitzinger, 1843
Subfamily: Lamprophiinae Fitzinger, 1843
Genus: ***Malpolon*** Fitzinger, 1826
Malpolon insignitus (Geoffroy de St-Hilaire, 1827)
M. i. fuscus (Fleischmann, 1831)
Genus: ***Rhagerhis*** Peters, 1862
Rhagerhis moilensis (Reuss, 1834)
Family: Leptotyphlopidae Stejneger, 1892
Subfamily: Leptotyphlopinae Stejneger, 1892
Genus: ***Myriopholis*** Hedges, Adalsteinsson & Branch, 2009
Myriopholis macrorhyncha (Jan, 1860)
Family: Psammophiidae Bonaparte, 1845
Genus: ***Psammophis*** Fitzinger, 1826
Psammophis schokari (Forskål, 1775)
Family: Typhlopidae Merrem, 1820
Genus: ***Indotyphlops*** Hedges, Marion, Lipp, Marin &Viadal, 2014
Indotyphlops braminus (Daudin, 1803)
Genus: ***Xerotyphlops*** Hedges, Marion, Lipp, Marin &Viadal, 2014
Xerotyphlops vermicularis (Merrem, 1820)
Xerotyphlops wilsoni (Wall, 1908)
Family: Viperidae Laurenti, 1768
Subfamily: Viperinae Oppel, 1811
Genus: ***Cerastes*** Laurenti, 1768
Cerastes gasperettii Leviton & Anderson, 1967
C. g. gasperettii Leviton & Anderson, 1967
Genus: ***Echis*** Merrem, 1820
Echis carinatus (Schneider, 1801)
E. c. sochureki Stemmler, 1969

Genus: ***Macrovipera*** Reuss, 1927
Macrovipera lebetinus (Linnaeus, 1758)
M. l. obtusa (Dwigubsky, 1832)
Genus: ***Montivipera*** Nilson, Tuniyev, Andrén, Orlov, Joger & Herrmann, 1999
Montivipera raddei (Boettger, 1890)
M. r. raddei (Boettger, 1890)
M. r. kurdestanica (Nilson & Andrén, 1986)
Genus: ***Pseudocerastes*** Boulenger, 1896
Pseudocerastes fieldi Schmidt, 1930
Pseudocerastes persicus (Dumeril, Bibron & Dumeril, 1854)
Pseudocerastes urarachnoides Bostanchi, Anderson, Kami & Papenfuss, 2006

44.7 Freshwater Amphibians and Reptiles in the Euphrates and Tigris River Drainages

There are four species of freshwater reptiles in the Euphrates and Tigris River Basin including *Rafetus euphraticus*, *Mauremys capsica*, *Emys orbicularis* and *Natrix tessellate*. Furthermore, ten freshwater amphibians inhabit rivers, lakes, marshlands, and springs of the region including *Bufotes sitibundus*, *Calliopersa luristanica*, *Hyla savignyi*, *Pelophylax bedriagae*, *Rana macrocnemis*, *Neurergus crocatus*, *N. derjugini*, *N. strauchii*, *Ommatotriton vittatus*, and *Salamandra infraimmaculata*, 30% species of which are endemic to the Euphrates and Tigris River Basin.

Lorestan Melodious Toad
Scientific name: *Calliopersa luristanica*
Subspecies: -
Etymology: *luristanicus*: Persian, Lorestan Province of Iran (Fig. 44.4)
Conservation status: IUCN Red List: Least Concern

Fig. 44.4 Lorestan Melodious Toad, *Calliopersa luristanica* (Schmidt 1939) Bina & Bijar Protected Area, Ilam Province, Iran. Photo by Barbod Safaei-Mahroo

Fig. 44.5 Thirsty Toad, *Bufotes sitibundus* (Pallas, 1771) (up female and down male): Qara-Dagh Mountain, SE Sulaymaniah. Photo by Barbod Safaei-Mahroo

Morphology:

It is a small-sized toad (SVL: 7-10 cm). Parotoid glands are almost quadrangular, short, as long as eye diameter, and extending to the beginning of supra-scapula. The dorsal tubercles are mostly same size and homogeneous. Toes of hind-limbs are weakly webbed. The dorsal color is light brown or reddish brown, with reddish to orange warts and scattered pale green spots, the ventral part without blotches or spots.

Distribution in Mesopotamia:

It has been reported from the western slopes of the Zagros region in southeastern Mesopotamia.

Habitat and Ecology:

This species is found in the seasonal rivers with a sandy bed, within hills and mountainous regions. It lays eggs in narrow strings.

Thirsty Toad

Scientific name: *Bufotes sitibundus*

Subspecies: -

Etymology: *sitibundus*: Latin, thirsty (Fig. 44.5)

Conservation status: IUCN Red List: Data Deficient

Morphology:

It is a medium-sized toad (SVL: 7-10 cm). Parotoid glands are medium, elongated, bean-shaped, pass the supra-scapula, and approximately two times larger than eye diameter size. The dorsal skin is covered with heterogeneous tubercles. Toes of hind-limbs are one-third webbed. The dorsal color is cream to light brown with green or olive patches, ventral part is covered with scattered green blotches. Females are larger than males.

Distribution in Mesopotamia:

Throughout Mesopotamia, apart from the desert regions.

Habitat and Ecology:

Fig. 44.6 Savigny's Tree Frog, *Hyla savignyi* Audouin, 1827 Kuna-Masi, Northern Sulaymaniah. Photo by Barbod Safaei-Mahroo

This species mostly inhabits in the vicinity of farms and residential regions, and found in marshlands, ponds, and small streams and can be seen numerous individuals on the roads surface at rainy nights. It lays eggs in wide strings.

Savigny's Tree Frog

Scientific name: *Hyla savignyi*

Subspecies: -

Etymology: *savignyi*: in honor of Marie Jules César Savigny (Fig. 44.6)

Conservation status: IUCN Red List: Least Concern

Morphology:

It is a small-sized tree frog, disc-like digital pads, dorsal surface bright green with a dark brown stripe at the sides with a lemon-colored line from nostril to groin laterally.

Some individuals of this species have gray spots on the dorsum. In the cold season, the dorsal part of the body changes color, ranging from cream to light brown. Males have a vocal sac in the front of the throat.

Distribution in Mesopotamia:

Known from the Euphrates and Tigris Rivers. It has a wide range from the northwest to the southeast and in parts of the central areas of this region.

Habitat and Ecology:

It inhabits marshlands and river shorelines with dense vegetated particularly *Carex, Phragmites,* and *Typha.* It depends intensely on humidity and the canopy cover.

Bedriaga's Marsh Frog

Scientific name: *Pelophylax bedriagae*

Subspecies: -

Etymology: *bedriagae*: in honor of Dr. Jacques Vladimir von Bedriaga (Fig. 44.7)

Conservation status: IUCN Red List: Least Concern

Morphology:

Fig. 44.7 Bedriaga's Frog, *Pelophylax bedriagae* (Camerano, 1882) Ranya, Northern Dukan Lake dam. Photo by Barbod Safaei-Mahroo

Large species with smooth or slightly rough dorsal skin, in a variety of colors from cream to green with irregular brown blotches and some of the individuals have a long green vertebral line. Males have two vocal sacs on either side of their head.

Distribution in Mesopotamia:

Throughout northern and eastern parts of the basin, also the Euphrates and Tigris Rivers.

Habitat and Ecology:

This highly aquatic species mostly inhabits rivers, marshlands, lakes, and small streams. Its breeding season extends from the end of winter to mid-spring. In the mountainous regions, the frogs' calling can be heard at night, in the spring and autumn.

Long-Legged Wood Frog

Scientific name: *Rana macrocnemis*

Subspecies: -

Etymology: *macrocnemis*: Greek, *makrós* = long + Greek, *knḗmē* = Shin or leg (Fig. 44.8)

Conservation status: IUCN Red List: Least Concern

Morphology:

Small species, dorsum is covered with round and small tubercles. Its body is cream or brown color with dark brown blotches. Dark brown triangular temporal marks present. Without external vocal sac opening.

Distribution in Mesopotamia:

Northern Mesopotamia, particularly, in the mountainous regions of the Zagros and Anatolia.

Habitat and Ecology:

Mostly in small lakes or small mountainous springs and slow-flowing streams. Laying eggs in clusters.

Azerbaijan Mountain Newt

Scientific name: *Neurergus crocatus*

Subspecies: -

Fig. 44.8 Long-Legged Wood Frog, *Rana macrocnemis* Boulenger, 1885 Northwest Oshnaviyeh, West Azerbaijan Province, Iran. Photo by Barbod Safaei-Mahroo

Fig. 44.9 Azerbaijan Mountain Newt, *Neurergus crocatus* Cope, 1862 Northwest Oshnaviyeh, West Azerbaijan Province, Iran. Photo by Barbod Safaei-Mahroo

Conservation status: IUCN Red List: Vulnerable

Etymology: *crocatus*: Latin, yellow or saffron-colored (Fig. 44.9)

Morphology:

The species has an elongated and narrow body, tail compressed at the side. The dorsum color ranges from dark brown to black and is covered with yellow spots. The ventral is orange or red.

Distribution in Mesopotamia:

Endemic to the northeast of the Tigris Basin, this species has been reported from the Zagros Mountains only.

Habitat and Ecology:

It prefers the streams and small springs in the mountainous regions with altitude of 1750–2000 meters. Laying eggs and mates in the spring season.

Derjugin's Mountain Newt

Scientific name: *Neurergus derjugini*

Fig. 44.10 Derjugin's Mountain Newt, *Neurergus derjugini* (Nesterov, 1916) Garmab, Western Baneh, Kurdistan Province, Iran. Photo by Barbod Safaei-Mahroo

Subspecies: *N. d. derjugini, N. d. microspilotus*

Conservation status: IUCN Red List: Critically Endangered

Etymology: *derjugini*: in honor of Dr. Konstantin Michailovich Derjugin (Fig. 44.10)

Morphology:

Elongated and narrow body with a compressed tail. Dorsum color ranges from dark brown to black and is covered with yellow spots in a regular manner. Ventral color ranges from orange to red.

Distribution in Mesopotamia:

Endemic to the east of the Tigris Basin and within the limits of the Zagros Forests mountains.

Habitat and Ecology:

Neurergus derjugini inhabits slow-flowing water in springs and small streams in oak forests.

Normally, it mates at the beginning of spring; the female attaches eggs to the stems or roots of aquatic vegetation.

Strauch's Mountain Newt

Scientific name: *Neurergus strauchii*

Subspecies: *N. s. barani*, *N. s. munzurensis*, *N. s. strauchii*

Conservation status: IUCN Red List: Vulnerable

Etymology: *strauchii*: in honor of Dr. Alexander Strauch

Morphology:

Body elongated and narrow, tail compressed. Dorsal surface dark brown or black, with yellowish heterogeneous and irregular small or large dots. In the subspecies, the size of the spots varies, such that in the *N. s. munzurensis* has the smallest size of spots and *N.s. barani* has the largest to be seen.

Distribution in Mesopotamia:

Endemic to the north of Mesopotamia. Its three subspecies have isolated habitat patches that inhabit the northern regions of the Tigris and Euphrates Rivers in the eastern Anatolia deciduous forests and eastern Anatolian montane steppe ecoegions.

N. s. strauchii reported from the eastern part of the Euphrates River also from south to west of the Lake Van.

N. s. barani found in the western part of the Euphrates River.

N. s. munzurensis known from the northern part of the Euphrates River (Olgun et al. 2016).

Habitat and Ecology:

It occurs in the springs and small brooks and prefers the stony shoreline of the rivers.

Banded Newt

Scientific name: *Ommatotriton vittatus*

Subspecies: -

Conservation status: IUCN Red List: Least Concern

Etymology: *vittatus*; Greek, vitta = strip or band.

Morphology:

The dorsal surface milky with small dark gray or dark olive irregular spots, laterally highly compressed tail. The spots on the side of the body are in the form of two parallel lines, with a white-colored center, which runs from the forelimb to the end of the tail. In the breeding season, the males have deeply serrated mid-dorsal and caudal crest.

Distribution in Mesopotamia:

Reported from the northwestern Euphrates River Basin and its easternmost distribution is in Zagros mountains on the border of Turkey and Iraq (Riemsdijk et al. 2018).

Habitat and Ecology:

Ommatotriton vittatus is a highly aquatic species and found in ponds, lakes, ditches, and slow-flowing stream pools within forests (Olgun et al. 2009).

Semenov's Salamander

Scientific name: *Salamandra infraimmaculata*

Subspecies: *S. i. semenovi*

Conservation status: IUCN Red List: Near Threatened

Etymology: *infraimmaculata*; Latin, infra = below, Latin, im = not, Latin, maculate = spotted (Fig. 44.11)

Morphology:

Body robust, parotoid glands distinct and large extending from the posterior edge of the eyes to neck; the dorsum color is dark brown and/or gray, with irregular yellow spots and in adults mostly change into highly irregular. The ventral is brownish-grayish color along with small yellow spots.

Distribution in Mesopotamia:

Salamandra infraimmaculata semenovi subspecies is endemic to the eastern and northern parts of Mesopotamia in the western edge of Zagros Mountains Forest.

Habitat and Ecology:

This viviparous Salamander prefers mountain meadows and brooks; adults can be found under the stones in the vicinity of water.

Euphrates Softshell Turtle

Scientific name: *Rafetus euphraticus*

Fig. 44.11 Semenov's Salamander, *Salamandra infraimmaculata semenovi* Nesterov, 1916 Naw village, Oraman (Hawrāmān) Mountain. Photo by Barbod Safaei-Mahroo

Fig. 44.12 Euphrates Softshell Turtle, *Rafetus euphraticus* (Daudin, 1802) Rofayeh, Southern Bostan, Iran. Photo by Barbod Safaei-Mahroo

Subspecies: -

Conservation status: IUCN Red List: Endangered, CITES: Appendix II, as *Rafetus* spp.

Etymology: *euphraticus*; Greek: Euphrátēs = Euphrates River (Fig. 44.12)

Morphology:

The carapace is flattened, smooth and without tubercles. The carapace color is olive-green with scattered dark blotches and irregular creamy spots. The plastron color is yellowish or light pink with very weak callosities on the hyo- and hypoplastral (Taskavak et al. 2016; Turtle Taxonomy Working Group, 2017).

Distribution:

The distribution of *Rafetus euphraticus* is restricted to the Euphrates and Tigris River Basin and extends from southeastern Turkey to the southwestern Iran, throughout the Euphrates and Tigris Rivers and their tributaries, marshlands, and ponds in Syria, Turkey, Iraq, and Iran (Taskavak et al. 2016).

Habitat and Ecology:

Euphrates softshell turtle is a medium-sized, exclusively riverine and cryptic trionychid turtle that inhabits various freshwater habitats. What we know about *R. euphraticus* home rang size is based on Ghaffari et al. (2014) study that indicated that the mean linear home range size is 2.54 km, and the mean 95% kernel density estimator is 21.75 ha, with a core area of 5.74 ha and prefers vegetated shorelines over open deep water. Ghaffari et al. (2015) feeding ecology study in Iran indicated that the turtle diet predominantly included invertebrates mainly crabs and plants, but also scavenges (Taskavak and Atatür 1998). The reproductive and nesting behavior of *R. euphraticus* is still poorly understood (Biricik and Turğa 2011; Ghaffari et al. 2013; Taskavak et al. 2016). The nesting season in Iran reported from late April to early June and mating observed in March; emerging hatchlings from nest observed in early July (Ghaffari et al. 2013; Taskavak et al. 2016). Published studies (Taskavak and Atatür 1998; Baran and Atatur 1998; Biricik and Turğa 2011; Ghaffari et al. 2013; Taskavak et al. 2016) indicated that the average clutch sizes are 30–40 eggs.

Threats to Survival:

The major threats to *Rafetus euphraticus* are anthropogenic fragmentation, alteration, and destruction of suitable habitat throughout its range; furthermore, drought especially the Mesopotamian Marshes in southeastern Iraq and southwestern Iran, dam construction, unsustainable fishing methods especially electro-fishing, and the use of poisons and explosive materials and pollution are the main concerns for *R. euphraticus* survival (Taskavak et al. 2016).

Besides, in recent years it has been reported that the Chinese soft shell *Pelodiscus sinensis* is imported into Iran illegally; this species has the potential to become a competitor for *R. euphraticus* (Ghaffari et al. 2014).

Caspian Pond Turtle

Scientific name: *Mauremys caspica*

Subspecies: -

Conservation status: -

Etymology: *caspica*; Latin: Caspium = Caspian Sea (Fig. 44.13)

Morphology:

The color of carapace ranges from olive to light brown; the head is olive and has yellow stripes on each side of the head. The plastron is dark brown or black.

Distribution in Mesopotamia:

This species occurs widely across the Middle East, ranges from Central Anatolia east, and extends southeastward Syria and the Caucasus Region to Iraq and Iran. In Mesopotamia, it occurs along Tigris and Euphrates Rivers (Vamberger et al. 2012).

Habitat and Ecology:

Mauremys caspica is a medium-sized freshwater turtle that is found in ponds, creeks, streams, drainage ditches, canals, and marshlands.

European Pond Turtle

Scientific name: *Emys orbicularis*

Subspecies: *E. o. colchica*

Conservation status: IUCN Red List: Near Threatened

Etymology: *orbicularis*; Latin: orbiculāris = circular

Fig. 44.13 Caspian Pond Turtle, *Mauremys caspica* (Gmelin, 1774) Karkheh river. Photo by Barbod Safaei-Mahroo

Morphology:

The carapace color is light or dark olive and occasionally black with yellow radiating lines. The plastron color is yellow or light orange. Head and limb are dark green to black with scattered yellow spots.

Distribution in Mesopotamia:

Although *Emys orbicularis* has a wide distribution from northwestern Africa through Europe to western Asia (Bayrakci and Dinçer 2014); in Mesopotamia this species were reported only from Diyarbakir in southeastern Turkey (Fritz et al. 1998).

Habitat and Ecology:

Emys orbicularis is a semi-aquatic turtle that has variable habitats; it prefers freshwater or slightly brackish water. This turtle inhabits ponds, drainage ditches, canals, creeks, streams, and rivers.

Tessellated Watersnake / Dise Snake

Scientific name: *Natrix tessellata*

Subspecies: -

Conservation status: IUCN Red List: Near Threatened

Etymology: *tessellata*; Greek, tessera = tessellated (Fig. 44.14)

Morphology:

Its dorsal surface brownish to gray color, with dark and irregular blotches. Its ventral surface has a pattern similar to that of a mosaic design, of dark and light-colored squares. In adults this mosaic design can also be observed in an orange color.

Distribution in Mesopotamia:

Occurs extensively throughout the Euphrates and Tigris River Basin, including the entire rivers of this basin.

Habitat and Ecology:

Inhabits in the grasslands, along the rivers and marshlands. Diet includes amphibians, and especially frogs and fishes.

It utilizes two defense mechanisms: one is to emit a strong-smelling fluid, and the other is thanatosis (playing dead).

Fig. 44.14 Dice Snake, ***Natrix tessellata*** (Laurenti, 1768) Hur Al-Azim Wetland. Photo by Barbod Safaei-Mahroo

44.8 Threats to Freshwater Herpetofauna in the Euphrates and Tigris River Basin

The drought of aquatic habitats, especially marshlands and the loss of marginal vegetation; particularly, in the Hoor-Al-Azim marshland in southeastern Mesopotamia, are the main threats to the survival of highly aquatic amphibians and reptiles throughout the Euphrates and Tigris River Basin. Predominant vegetation along the Euphrates and Tigris River and tributaries provide food, refuge, and breeding ground for many semiaquatic and aquatic species. An intense modification in the hydrological regime and the drying up of rivers due to the construction of dams throughout the basin are the main concerns for amphibians and reptiles, which are dependent on freshwater habitat. Furthermore, water pollution, illegal fishing methods, especially electrofishing, and ongoing habitat loss, and fragmentation are accounted as other threats for the survival of amphibians and reptiles in this region. The occurrence of wars and political conflicts in this part of the world has asserted a profound and negative impact on the habitats and wildlife, especially the amphibians and reptiles of Mesopotamia. Other serious threats are habitat alteration, overexploitation of mountainous terrain, and the anthropogenic conveyance of water from springs, which results in the absence of access for species to the most vital source, leading to the decline of the salamanders' population in Mesopotamia. In addition to the abovementioned, the intentional killing in particular of the snakes, Euphrates soft-shell Turtle (*Rafetus euphraticus*), and the Azerbaijan Mountain Newt (*Neurergus crocatus*) by the human is another factor of threat for these creatures in this region. In summer 2018, a wildfire burned the Iraqi part of the Hawr-al-Azim wetlands. An important concern is species that cannot escape from fire easily and move to another suitable habitat may be at risk of local extinction.

44.9 Conclusion

In the Euphrates and Tigris Basins, the Colubridae family has thirty-seven species as the largest reptile family, followed by Lacertidae and Gekkonidae, with twenty-three and seventeen species, respectively. The largest family of amphibians in this region is the Salamandridae, with five species. This basin has 10 species of amphibians and 134 species of reptiles, of which 25 species are endemic to the basin. Of the eleven ecoregions of the Tigris and Euphrates Basin, the Zagros Mountains forest steppe, which accounts for 11.5% of the area, has the most endemic species with 14 endemic amphibians and reptiles. The Zagros Mountains forest steppe ecoregion plays a key role in the speciation of amphibians and reptiles of the region. The speciation level in the Zagros Mountains forest steppe has been high; it has the highest regional endemism level in the Euphrates and Tigris Basins with over half of the endemic amphibians and reptiles of this basin.

Due to the strategic situation of the Euphrates and Tigris Basin and wars, including political conflicts, limited information is available on the wildlife, in particular the amphibians and the reptiles of this region. Therefore, there is no information in access in relevance to the population of amphibians and reptiles. Though, undoubtedly, the threat of dam constructions, habitat degradation and destruction, and increased pollution throughout the catchment area are serious threats to highly aquatic amphibians and reptiles, these threats are more austere in the southern parts, especially in the Mesopotamian Marshlands, extensive parts of which have drained and or else are confronted with salt accumulation.

References

Ablonski D, Kukushkin OV, Avcı A, Bunyatova S, Kumlutaş Y, Ilgaz Ç, Polyakova E, Shiryaev K, Tuniyev B, Jandzik D (2019) The biogeography of *Elaphe sauromates* (Pallas, 1814), with a description of a new rat snake species. Peer J 7:e6944. https://doi.org/10.7717/peerj.6944

Adams Mc R (1983) The Jarmo stone and pottery vessels industries. In: Braiwood LS, Braiwood RJ, Howe B, Reed CA, Watson PJ (eds) Prehistoric archaeology along the Zagros flanks, vol 105. The Oriental institute of the University of Chicago, pp 209–232

Anderson SC (1999) The Lizards of Iran. Society for the Study of Amphibians and Reptiles, Ithaca, New York, USA, 442 p

Angel F (1936) Reptiles Et Batraciens Syrie Et De Mésopotamie Récoltés Par M. P. Pallary. Bulletin De L'institute D'Egypte. 18:107–116

Baran I, Atatur MK (1998) Turkish herpetofauna. ministry of environment, Ankara, Turkey, 214 p

Bayrakci Y, Dinçer A. (2014) Dynamics of a Central Anatolian population of *Emys orbicularis* (Linnaeus, 1758) Herpetozoa 27(1/2):29–37

Berthon R, Erdal YS, Mashkour M, Kozbe G (2016) Buried with turtles: the symbolic role of the Euphrates soft-shelled turtle (*Rafetus euphraticus*) in Mesopotamia. Antiquity 90(349):111–125

Biricik M, Turğa S (2011) Description of an Euphrates softshell turtle (*Rafetus euphraticus*) nest from the Tigris River (SE Turkey). Salamandra 47:99–102

Blanford WT (1874) Descriptions of two Uromasticine lizards from Mesopotamia and southern Persia. Proc Zool Soc London 1874:656–661

Böhme W, Ehrlich K, Milto K, Orlov N, Scholz S (2015) A new species of desert monitor lizard (Varanidae: Varanus: Psammosaurus) from the Western Zagros region (Iraq, Iran). Russ J Herpetol 22(1):41–52

Boulenger GA (1920a) A list of snakes from Mesopotamia, collected by members of the Mesopotamian force 1915-1919. J Bombay Nat Hist Soc 27:347–350

Boulenger GA (1920b) A list of lizards from Mesopotamia, collected by members of the Mesopotamian force 1915-1919. J Bombay Nat Hist Soc 27:351–358

Braidwood JR, Howe B (1960) Prehistoric Investigation in Iraq Kurdistan, Oriental Institute of the University of Chicago, Studies in Ancient Oriental Civilization, No. 31. Chicago University of Chicago Press. 184 pp., + 29 pls

Delougaz P, Kantor HJ (1996a) "Chogha Mish" Volume I, The first five seasons of excavatins 1961–1971, Part 1: Text. The Oriental institute of the University of Chicago. 101:508 pp.

Delougaz P, Kantor HJ (1996b) "Chogha Mish" Volume I, The first five seasons of excavatins 1961–1971, Part 2: Plates. The Oriental institute of the University of Chicago. 101:283 pp.

Fritz U, Baran I, Budak A, Amthauer E (1998) Some notes on the morphology of *Emys orbicularis* in Anatolia, especially on *E. o. luteofusca* and *E. o. colchica*, with the description of a new subspecies from southeastern Turkey. Mertensiella 10:103–121

Ghaffari H, Ihlow F, Plummer MV, Karami M, Khorasani N, Safaei-Mahroo B, Rödder D (2014) Home range and habitat selection of the endangered Euphrates softshell turtle *Rafetus euphraticus* in a fragmented habitat in southwestern Iran. Chelonian Conservation and Biology 13(2):202–215

Ghaffari H, Plummer MV, Karami M, Safaei-Mahroo B, Ahmadzadeh F, Rödder D (2013) Notes on a nest and emergence of hatchlings of the Euphrates softshell turtle (*Rafetus euphraticus*) at the Dez River, Iran. Chelonian Conservation and Biology 12:319–323

Ghaffari H, Safaei-Mahroo B, Plummer MV, Rostami A, Paknejad M (2015) Preliminary data on feeding ecology of the Euphrates Softshell Turtle (*Rafetus euphraticus*) in Iran. The 11th Quasi-annual terrapin, tortoise and freshwater turtle meeting, 35th Annual Symposium on sea turtle biology and conservation.

Haas G (1952) Two collections of reptiles from Iraq, with descriptions of two new forms. Copeia (1):20–22

Hass G, Werner Y (1969) Lizards and snakes from southwestern Asia, collected by Henry field. Bulletin of the Museum of Comparative Zoology 138(6):327–406

Kantor HJ (1976) The prehistoric cultures of chogha mish and boneh fazili. In: Kiani MY (ed) The memorial volume of the sixth international congress of Iranian art and archaeology, Oxford, September 11-16th, 1972, pp 177–193

Khalaf N (1959) Reptiles of Iraq with some notes on the amphibians. Ar-Rabitta Press, Baghdad, p 96

Leviton AE, Anderson SC, Adler K, Minton SA (1992) Handbook to Middle East amphibians and reptiles (Contributions to Herpetology). Society for the Study of Amphibians and Reptiles, St. Louis. USA, 252 p

Mertens R (1924) Ein neuer Gecko aus Mesopotamien. Senckenbergiana 6(1–2):84

Nader IA, Jawdat SZ (1976) Taxonomic study of the geckos of Iraq. Bulletin of the Biological Research Center Baghdad 5:1–41

Nazarov RA, Melnikov DA, Radjabizadeh M, Poyarkov NA (2018) A new species of short-fingered geckos Stenodactylus (Squamata, Geckonidae) from South Iran with taxonomic notes on the validity of the genus Trigonodactylus Hass, 1957. Zootaxa 4457(1):93–113

Nesterov PV 1917 (1916) Tri novykh khvostatykh amfibii iz Kurdistana [Three new forms of amphibians from Kurdistan]. Annuaire du Musée Zoologique de l'Académie des Sciences de Petrograd 21:1–30+6 figs.+3 pls. (In Russian)

Oates J (1968) Prehistoric investigations near Mandali, Iraq. Iraq 30(1):1–20

Olgun K, Arntzen JW, Papenfuss T, Degani G, Ugurtas I, Disi A, Sparreboom M, Anderson S, Sadek R, Hraoui-Bloquet S, Gasith A, Elron E, Gafny S, Werner Y, Avci A, Üzüm N (2009)

Ommatotriton vittatus. The IUCN Red List of Threatened Species. e.T59480A11930635. https://doi.org/10.2305/IUCN.UK.2009.RLTS.T59480A11930635.en. Downloaded on 31 July
Olgun K, Avci A, Bozkurt E, Üzüm N, Olgun H, Ilgaz Ç (2016) A new subspecies of Anatolia newt, *Neurergus strauchii* (Steindachner, 1887) (Urodela: Salamandridae), from Tunceli, eastern Turkey. Russ J Herpetol 23(4):271–277
Olson DM, Dinerstein E, Wikramanayake ED, Burgess ND, Powell GVN, Underwood EC, D'Amico JA, Itoua I, Strand HE, Morrison JC, Loucks CJ, Allnutt TF, Ricketts TH, Kura Y, Lamoreux JF, Wettengel WW, Hedao P, Kassem KR (2001) Terrestrial ecoregions of the world: a new map of life on earth. Bioscience 51(11):933–938
Procter JB (1921) Further lizards and snakes from Persia and Mesopotamia. Bombay Nat Hist Soc 28:251–253
Riemsdijk IV, Nieuwenhuize LV, Martinez-Solano I, Arntzen W, Wielstra B (2018) Molecular data reveal the hybrid nature of an introduced population of banded newts (Ommatotriton) in Spain. Conserv Genet 19:249–254
Safaei-Mahroo B, Ghaffari H (2020) The complete guide to amphibians of Iran, biology, ecology and conservation. Kurdistan University Press, Sanandaj, Iran, 332 p
Safaei-Mahroo B, Ghaffari H, Ghafoor A, Amini S (2017) A new species of *Hemidactylus* (Squamata: Gekkota: Gekkonidae) from Qara Dagh Mountains, Kurdistan region, with a key to the genus in Iraq. Zootaxa 436(3):377–392
Schmidt KP (1939) Reptiles and amphibians from southwestern Asia. Field Museum of natural history. Zoological series 24:49–91
Sindaco R, Venchi A, Grieco C (2013) The reptiles of the Western Palearctic, volume 2: annotated checklist and distributional atlas of the snakes of Europe, North Africa, Middle East and Central Asia, with an update to volume 1. Societas Herpetologica Italica, Latina, Italy. 543 pp
Smith KG, Barrios V, Darwall WRT, Numa C (2014). The status and distribution of freshwater biodiversity in the Eastern Mediterranean. Cambridge, UK, Malaga, Spain and Gland, Switzerland: IUCN. xiv+132pp
Taskavak E, Atatür MK (1998) Distribution and habitats of the Euphrates softshell turtle, *Rafetus euphraticus*, in southeastern Anatolia, with observations on biology and factors endangering its survival. Chelonian Conservation and Biology 3(1):20–30
Taskavak E, Atatur MK, Ghaffari H, Meylan PA. Rafetus euphraticus (Daudin 1801) – Euphrates Softshell turtle. In: A.G.J. Rhodin, J.B. Iverson, P.P. van Dijk, R.A. Saumure, K.A. Buhlmann, P.C.H. Pritchard and R.A. Mittermeier editors. Conservation biology of freshwater turtles and tortoises: a compilation project of the IUCN/SSC tortoise and freshwater turtle specialist group. Chelonian Research Monographs 2016;5(9): 098.1–09811
Torki F (2017) Description of a new species of *Lytorhynchus* (Squamata: Colubridae) from Iran. Zoology in the Middle East 63(2):109–116
Turtle Taxonomy Working Group [Rhodin, AGJ, Iverson JB, Bour R, Fritz U, Georges A, Shaffer HB, van Dijk PP. Turtles of the world: annotated checklist and atlas of taxonomy, synonymy, distribution, and conservation status (8th Ed.). In: Rhodin AGJ, Iverson JB, van Dijk PP, Saumure RA, Buhlmann KA, Pritchard PCH, Mittermeier RA, editors. Conservation biology of freshwater turtles and tortoises: a compilation project of the IUCN/SSC tortoise and freshwater turtle specialist group. Chelonian Research Monographs 2017;7:1–292. doi: https://doi.org/10.3854/crm.7.checklist.atlas.v8.2017
UN-ESCWA & BGR (2013) (United Nations Economic and Social Commission for Western Asia; Bundesanstalt für Geowissenschaften und Rohstoffe). Inventory of Shared Water Resources in Western Asia Beirut. 606 pp.
Vamberger M, Stuckas H, Ayaz D, Graciá E, Aloufi AA, Els J, Mazanaeva LF, Kami HG, Fritz U (2012) Conservation genetics and phylogeography of the poorly known middle eastern terrapin *Mauremys caspica* (Testudines: Geoemydidae). Organisms Diversity & Evolution 13(1):77–85. https://doi.org/10.1007/s13127-012-0102-6

Werner F (1895) Ueber eine Sammlung von Reptilien aus Persien, Mesopotamien und Arabien. Verhandlungen der Kaiserlich-Königlichen Zoologisch-Botanischen Gesellschaft in Wien, 45:13–22

Werner F (1929) Beiträge zur Kenntnis der Fauna von Syrien und Persien. Zool Anz 81 (7/10):238–245

Wettstein OV (1951) Ergebnisse der Österreichischen Iran-Expedition 1949/50 Amphibien und Reptilien. Sitzungsberichte der Österreichische Akademie der Wissenschaften, Wien, Mathematisch naturwissenschaftliche Klasse, 1(5):427–448

Chapter 45
A Preliminary Pictorial Guide to the Herpetofauna of Tigris and Euphrates River Basin

Omar F. Al-Sheikhly

Abstract An extensive review on the taxonomic status, occurrence, geographical distribution, and the IUCN conservation status of 122 species/taxa (10 amphibians and 112 reptiles) belonging to 71 genera (8 genera of amphibians and 63 of reptiles) and to 25 families (5 families of amphibians and 20 families of reptiles) was made. As a part of Iraq's unique biota, the knowledge of the herpetofauna of Iraq is poorly discovered and requires extensive field surveys, further research, and explorations in order to create full frame knowledge of its status and conservation.

45.1 Introduction

As a part of Iraq's unique biota, the knowledge of the herpetofauna of Iraq is poorly discovered and requires extensive field surveys, further research, and explorations. There were contentious attempts from few Iraqi researchers to study the herpetofauna of Iraq; their published notes and observations were significantly contributed to develop the current knowledge on the amphibians and reptiles of Iraq. Some basic specimens collecting have been carried out and checklists of species are available (e.g., Schmidt 1939; Allouse 1955; Khalaf 1959). The Iraqi Biological Research Centre and Museum (nowadays, Iraqi Natural History Research Center and Museum INHM) collected and studied reptiles, and published a monograph on the Gekoniidae of Iraq (Nader and Jawdat 1976); however, the status of many reptiles and amphibian species in Iraq is still poorly known. The recent up-to-date checklist of reptiles and amphibians in Iraq which listed total of 115 species (105 species of reptiles and 10 species of amphibians) belonging to 25 families (20 families of reptiles and five families of amphibians), along with their conservation status in Iraq, was made by Al-Barazengy et al. (2015). This list cited the previous herpetofauna literature in Iraq and neighboring ecoregions such as

O. F. Al-Sheikhly (✉)
College of Science, University of Baghdad, Baghdad, Iraq

L. A. Jawad (ed.), *Tigris and Euphrates Rivers: Their Environment from Headwaters to Mouth*, Aquatic Ecology Series 11,
https://doi.org/10.1007/978-3-030-57570-0_45

(Garstecki and Amr 2011; Arnold 1986) and herpetofauna published records by the Iraqi Natural History Research Center and Museum.

The herpetofauan of Tigris and Euphrates River basin is unique and diverse. Scott (1995) reported many reptile species from Tharthar Lake in central Iraq including Steppe agama *Agama agilis* (*Trapelus agilis agilis*) and Oliver's Agama *A. ruderata* (T. ruderatus), Bosk's Fringe-fingered Lizard *Acanthodactylus boskianus,* Mesopotamian Fringe-toed Lizard *A. grandis*, and several snakes of the genus *Coluber*. The Desert Monitor *Varanus griseus* was present in fairly good numbers in the semi-desert to the east of the lake. The Field's Horned Viper *Pseudocerastes fieldii* has been collected on the shore of the lake. Scott (1995) also reported common reptiles include Steppe Agama, Bosk's Fringe-fingered Lizard, and Mesopotamian Fringe-toed lizard, Desert Monitor, and several snakes of the genus *Coluber* in the surrounding desert of Shari Lake in central Iraq. Scott (1995) also mentioned that were a shortage of information on the amphibians and reptiles of the Mesopotamian marshes. Maxwell (1957) mentioned that were an extreme abundance of frogs in the marshes, and concluded that there were several species. Mahdi and George (1969) listed Toad *Bufo viridis,* a Tree Frog *Hyla arborea*, and two frogs *Rana ridibunda* and *R. esculenta* for Iraq. The common reptiles in the marshes include the Caspian Terrapin *Clemmys caspia,* Euphrates Softshell Turtle *Rafetus* (*Trionyx*) *euphraticus,* geckos of the genus *Hemidaclylus*, two species of skinks *Mabuya aurata* and *M. vittata,* and a variety of snakes including the Spotted Sand Boa

Eryx jaculus, Tessellated Water Snake *Natrix tessellata and* Gray's Desert Racer *Coluber.*

ventromaculatus. The Desert Monitor was formerly common in desert areas adjacent to the marshes, but seemed to be rare at that time. Moreover, Scott indicated that the Spiny-tailed lizard *Uromastix* sp. was to be abundant around the shores of Razazah Lake in central Iraq.

The Tigris and Euphrates River basin is hosting several taxa of globally threatened reptiles and amphibian species. Of course, amphibians are remarkably reflects Iraq's climate and geography. Two amphibian species, the Kurdistan Newt *Neurergus microspilotus* and Lake Urmia Newt *Neurergus crocatus*, were listed as Critically Endangered and Vulnerable by the IUCN respectfully. Both mountain-dwelling newts are allopatric and have restricted patchy distribution in streams and waterfalls of Zagros Mountain in Iraq and Iran (Najafimajd and Kaya 2010). From 2007 to 2012, Kurdistan newt was discovered at seven new localities, while Lake Urmia Newt was located in four new distributed in Zagros Mountain in northern Iraq. The new locations provide a major extension of both species geographical range. However, severe drought of recent years as well as man-made habitat destruction and pollution has been marked as main threat impacts on both species population in northern Iraq (Al-Sheikhly et al. 2013).

A total of nine globally threatened reptile species (seven turtles; one lizard; one viper) was recorded in Iraq. The Critically Endangered Hawksbill Turtle *Eretmochelys imbricata bissa*; the Endangered Green Turtle *Chelonia mydas* and Euphrates Softshell Turtle; the Vulnerable Loggerhead Sea Turtle *Caretta caretta,* Leatherback Sea Turtle *Dermochelys coriacea schlegelii,* Olive Ridley *Lepidochelys olivacea*, Spur-thighed tortoise *Testudo graeca ibera*, and Egyptian Spiny-tailed Lizard Uromastyx aegyptia, and the Near Threatened Caucasus Viper Montivipera (Vipera) raddei kurdistanica were reported from Iraq (Al-Barazengy et al. 2015).

There is scarce information on the occurrence, status, and distribution range of marine turtles in Iraq. In recent years, no dedicated field surveys covering the coastal marine habitats of southern Iraq were conducted. Therefore, up-to-date information on the Iraqi marine herpetofauna is crucially required.

Regarding the Lizard fauna of Iraq, several taxonomic revisions for certain taxa were made. Niazi (1979) considered *Diplometopon shueaibi* as synonym of Zarudnyi's worm lizard *Diplometopon zarudnyi* (Leviton et al. 1992). It was recorded from Rumaila desert by Afrasiab and Ali (1989a). Rudayni et al. (2017) discoursed the variation within the Saudi Arabian Trogonophidae and did not refer it to *D. shueaibi*; however, Afrasiab et al. (2018) considered *D. shueaibi* as a valid name for central Arabia and west Karbala of Iraq. The taxon *Trapelus agnetae* was treated as a synonym of Pale Agama *T. pallidus* until (Disi et al. 1999; Disi and Amr 1998) who assigned it to the subspecies *T. p. haasi;* thus, *T. p. haasi* should be considered a junior synonym of *T. p. agnetae*. Subsequent molecular research found *T. p. pallidus* is synonymous with *T. mutabilis*, while *T. p. agnetae* is related to separate clad (Wagner et al. 2011). However, those authors assigned *T. agnetae* as a full species (Al-Barazengy et al. 2015). Anderson (1999) and Disi (2002) assigned Blandford's Lizard *Trapelus blandfordi* within *T. persicus*, which is now a junior synonym of Oliver's Agama *T. ruderatus* (Rastegar-Pouyani 2000). Moreover, the subspecies *T. ruderatus baluchianus* (Olivier 1804) is now assigned into *Trapelus megalonyx* (Rastegar-Pouyani 2000; Al-Barazengy et al. 2015).

In 2010–2011, during intensive ecological survey in mountains, foothills, desert, semi-deserts, and steppes in north and northwestern Iraq, important records and field observations of amphibians and reptiles were made. These observations and records will significantly contribute to increase the knowledge on the herpetofauna of Iraq (Al-Sheikhly et al. 2013). The Green Toad *Pseudepidalea* (*Bufo*) *viridis*, Long-legged Wood Frog *Rana camerani-macrocnemis,* and Marsh Frog *Rana ridibunda* were found at the muddy banks of Tigris River at Jabal Makhool (Makhool foothills) in northern Iraq. Jabal Makhool seems to be one of the main important sites for reptiles, especially the Lacertidae, where many species have been recorded. The Yellow-headed Rock Agama *Laudakia nupta,* Egyptian Spiny-tailed Lizard, Yellow-belly Gecko *Hemidactylus flaviviridis*, Persian Leaf-toed Gecko *Hemidaciylus persicus,* Rough-tailed Gecko *Cyrtopodion scabrum*, Bosk's Fringe-fingered Lizard *Acanthodactylus boskianus*, Arnold's Fringe-fingered Lizard *Acanthodactylus opheodurus*, Snake-eyed Lizard *Ophisops elegans*, Levant Skink *Trachylepis aurata aurata*, and Desert Monitor were observed. The Endangered Euphrates Soft shell Turtle and Caspian Turtle *Mauremys caspica caspica* were recorded as well. The Dark-headed Dwarf Racer *Pseudocyclophis* (*Eirenis*) *persicus*, Large Whip Snake Dolichophis (Coluber) jugularis Hardwicke's Rat Snake *Coluber ventromaculatus,* Collared Dwarf Snake *Eirenis collaris,* Diademed Sand Snake *Lytorhynchus diadema,* Persian Horned Viper *Pseudocerastes persicus fieldi*, Desert Cobra *Walterinnesia aegyptia*, Dice Water Snake *Natrix tessellate* and Spotted Sand Boa were recorded. In Jabal Himreen (Himreen foothills) in eastern

Iraq, along with western borders of Iran, many reptile species have been observed. The Yellow-headed Rock Agama, Desert agama *Trapelus mutabilis*, Mesopotamian Mastigure *Uromastyx* (*Saara*) *loricatus*, Leaf-toed Gecko *Asaccus elisae*, Yellow-belly Gecko, Persian Leaf-toed Gecko, Rough-tailed Gecko, Snake-eyed Lizard, Levant Skink, the Large Whip Snake, Diademed Sand Snake, Schokari Sand Racer *Psammophis schokari*, Hardwicke's Rat Snake, Persian Horned viper, Black Desert Cobra, Dice Water Snake and Spotted Sand Boa were recorded.

In northern Iraq, Zagros Mountain has a unique diversity of herpetofauna. In Pera Magroon Mountain, Green Toad, Common Tree Frog *Hyla savignyi,* and Long-legged Wood Frog were observed in many freshwater streams and marshy ponds distributed in the mountain. Caspian Pond Turtle, Spur-thighed Tortoise *Testudo graeca,* Yellow-headed Rock Agama, Rough-tail Rock Agama *Laudakia stellio, Trapelus lessonae,* Turkish Gecko *Hemidaciylus turcicus,* Anatolian lizard *Lacerta cappadocica,* Snake-eyed Lizard, Legless lizard *Ophisaurus apodus,* Levant Skink, Large Whip Snake, Collared Dwarf Snake, Persian Horned Viper, Levantine viper *Macrovipera lebetina obtuse,* Montpellier Snake *Malpolon monspessulanus*, and Water Dice Snake were recorded. In Qara Dag Mountain, the Common Tree Frog, Green Toad, Long-legged Wood Frog, Spur-thighed Tortoise, Caspian Pond Turtle, Yellow-headed Rock Agama, Rough-tail Rock Agama, Turkish Gecko, Anatolian Lizard, Snake-eyed lizard, Golden Grass Mabuya *Trachylepis aurata septemtaeniata,* Large Whip Snake, Collared Dwarf Snake, Black Desert Cobra, Persian Horned Viper, Levantine Viper, Montpellier Snake, and Water Dice Snake were recorded.

The survey efforts of Al-Sheikhly et al. (2013) were extended to cover the western bank of Euphrates River in the western plateau of Iraq, where several reptiles and amphibians were recorded. In Faidhatt Al-Massad, to the south of Rutba in extreme western plateau of Iraq (Anbar Province), many terrestrial desert and aridlands reptilian species such as Desert Agama *Trapelus mutabilis,* Brilliant Ground Agama *Trapelus agilis,* Egyptian Spiny-tailed Lizard, Bosk's Fringe-fingered Lizard, Nidua Fringe-fingered Lizard *Acanthodactylus scutellatus,* Giant Fringe-fingered Lizard *Acanthodactylus grandis,* Desert Monitor, Spotted Sand Boa, Black Desert, and Arabian Horned Viper *Cerastes gasperettii* were recorded. In Haditha and Tharthar Lake in western Iraq, Euphrates Softshell Turtle, Caspian Turtle, Bosk's Fringe-fingered lizard, Egyptian Spiny-tailed Lizard, Yellow-belly Gecko, Persian Leaf-toed Gecko, Rough-tailed Gecko, Leaf-toed Gecko, Snake-eyed Lizard, Levant Skink, Desert Monitor, Desert Cobra, Sand Boa, and Dice Water Snake were recorded (Al-Sheikhly et al. 2013).

Two endemic Gekkonidae species are found in Iraq, confined to the mountains of northern Iraqi Kurdistan. The Safin's Mountain Leaf-toed Gecko *Asaccus saffinae*, a cave dowelling species, was discovered in Saffine Mountain in Erbil province in northern Iraq (Afrasiab and Sarbaz 2009). More recently, another gecko species, the Qara Dagh Mountain Gecko (Kurdish Gecko) *Hemidactylus kurdicus*, was discovered in the Mountain of Qara Dagh in Sulymaniyah province in northeastern Iraq (Safaei-Mahroo et al. 2015). More recently, new species of gecko, *Microgecko helenae fasciatus* was recorded near to the Darbandikhan Lake in northeastern Iraq nine globally threatened by Jablonski et al. (2020). However, many areas have never been explored, especially those extending within the foothills of Zagros Mountain geographical range of northeastern and eastern Iraq. Thus, further investigation to the Gekkonidae fauna of Iraq seems crucial to reveal more new discoveries.

Recently, one species of lizards and four species of snakes were recorded for the first time in Iraq (Al-Barazengy et al. 2015). The occurrence of Striated Lizard *Lacerta strigata* in Iraq was based on three specimens collected by Niazi (1976) from Gelki Islam near the Khapoor River and the road to Kanimase, Zakhow district, western Kurdistan. In fact that Niazi's records were restricted to the described areas so that this species may have been introduced from the mountains of eastern turkey by the floods of the Khapoor River (Afrasiab et al. 2013). The Rat Snake *Zamenis hohenackeri* and the Braid Snake *Platyceps ladacensis* were recorded fo the first time from Hawraman Mountain in northeastern Iraq by Afrasiab and Sarbaz (2011). The Zabra Snake *Spalerosophis microlepis* was recorded for the first time in Iraq in Serine Mountain northern of Erbil province by (Afrasiab and Mohamad 2014). The Grass Snake *Natrix natrix persa* was recorded for the first time in Iraq from near Dialah Bridge, 10 km east of Baghdad by (Afrasiab and Mohamad 2011; Afrasiab et al. 2012). In 2016, a new species of whip snakes from the genus *Dolechophis* was described by Afrasiab et al. (2016) from upper Mesopotamia and was named *Dolichophis mesopotamicus* sp. nov. However, based on large morphological variations in color patterns in the mentioned genus and the necessity of genetic confirmation for this record, it was not included in the current list. More recently, the elusive Iranian Spider-tailed Horned Viper *Pseudocerastes urarachnoides* endemic to western Iran was recently discovered in several localities in \the foothills of Zagros Mountains in eastern Iraq.

The occurrence of the Black-tailed Toed Agama *Phrynocephalus maculates longicaudatus* has been confirmed in the south-west of Iraq in Muthanna province near of AL-Khuder city and in semi-desert on the side of Sawa lake by Al-Barazengy (2014).

The Desert Hoodless Cobra *Walterinnesia morgani* (Mocquard 1905) was previously assigned to *W. aegyptia* (Corkill 1932; Khalaf 1959), but later Nilson and Rastegar-Pouyani (2005) decided that the Iraqi and the eastern population belong to *W. morgani*. A juvenile specimen was collected from Taq Taq south eastern Erbil in northern Iraq. Afrasiab et al. (2018) indicated that most probably all of the northern population belongs to *W. aegyptia*, while the south and southwest population belong to *W. morgani*.

In 2013–2014, an investigation on the occurrence of geckos in five Iraqi provinces (Babil, DhiQar, AL-Muthana and Basra) situated in central and southern Iraq was conducted by Mohammed et al. (2015). The survey efforts covered different habitats and landscapes where a total of 111 adults and sub-adult specimens was collected and identified. The collected specimens represented six species belonging to three genera which were one Turkish gecko, two Persian Gecko, 15 Yellow-bellied House Gecko, 53 Rough-tailed Gecko, 16 Iranian Short-fingered Gecko *Stenodactylus affinis*, and 24 Middle Eastern Short-fingered Gecko *Stenodactylus doriae*, the common gecko species recorded in central and southern Iraq.

One of the range-restricted reptiles' species confined to the lakes, rivers, streams, and marshes of Tigris and Euphrates River basin is the Euphrates Softshell Turtle. It occurs across Iraq, Syria, Turkey, and Iran (Ghaffari et al. 2008), although Iraq is thought to contain the largest number of suitable sites for the species (Fazaa et al. 2015). Stadtlander (1992) reported that in 1989, a total of 26 Euphrates Softshell turtles was recorded in the Tigris River in Turkey and a total of at least 55 individuals was recorded in Euphrates River between Al-Fallujah and Hawr Al-Hammar in Iraq. In 2013–2014, a recent survey was conducted in Central Marshes in southern Iraq by Fazaa et al. (2015) and suggested that a maximum population size of Euphrates Softshell Turtle in the Central Marshes is 212–283 individuals/141,615 ha. However, this number may be underestimated. Moreover, Fazaa et al. (2015) suggested that the starting of breeding season for Iraqi Euphrates softshell turtle population is two months earlier than Iranian and Turkish populations.

In concern for the Hydrophiidae sea snakes, Afrasiab et al. (2018) listed *Enhydrina schistosa* (Daudin 1803), *Hydrophis gracilis* (Shaw 1802), *H. spiralis* (Shaw 1802), *H. cyanocinctus* (Daudin 1803), *H. ornatus* (Gray 1842), *H. lapemoides* (Gray 1849), *Pelamis platurus* (Linnaeus 1766), and *Lapemis curtus* (Shaw 1802) from Al-Basrah seashore without providing further details on the voucher specimens cited, collecting dates, and localities. The occurrence of those species required further investigation; therefore, we only referred to this important remark in this section and did not list those species in the current list.

In addition, the reptilian species, the Euphrates Softshell Turtle, Spur-thighed Tortoise and the amphibian species Kurdistan and Lake Urmia newts were used as non-avian criteria to assess, prioritize, and select Key Biodiversity Areas (KBAs) in Iraq. In the KBA site assessment, the Euphrates Softshell Turtle met vulnerability and irreparability KBA criteria while spur-thighed tortoise met vulnerability KBA criteria. The Euphrates Softshell Turtle was found throughout Iraq in wetlands and rivers of Tigris and Euphrates river basin (Central Marshes, Hammar Marsh, Hawizeh Marsh, Dukan and Derbendikhan lakes, etc.). The spur-thighed tortoise was found in the wooded mountains of northern Iraq such as Qara Dagh and Peramagroon mountains. The Kurdistan and Lake Urmia newts both have met the Vulnerability KBA criteria. Kurdistan Newt was recorded at Ahmad Awa, Hawraman area, and Mawat Area in northern Iraq, while Lake Urmia Newt was recorded in Gara and Garagu mountains, Halgured Mountain, and Sakran Mountain (Nature Iraq 2017). Despite what is mentioned above, the herpetofauna of Iraq is unique and diverse, yet, it requires further attention and intensive investigation in order to create a full frame knowledge of its status and conservation.

45.2 Annotated Checklist of Amphibians and Reptiles of Tigris and Euphrates River Basin

The current checklist is initially based on Al-Barazengy et al. (2015) list of amphibians and reptiles of Iraq. An extensive review on the taxonomic status, occurrence, geographical distribution, and the IUCN conservation status of 122 species/taxa (10 amphibians and 112 reptiles) belonging to 71 genera (eight genera of amphibians and 63 of reptiles) and to 25 families (five families of amphibians and 20 families of reptiles) was made. Moreover, several taxonomic updates were made in Al-Barazangy's et al. list. In Amphibia section, the Southern Banded Newt *Ommatotriton* (*Triturus*) *vittatus*, Iranian Earless Toad *Pseudepidalea* (*Bufo*) *surda*, Green Toad *Pseudepidalea* (*Bufo*) *viridis*, and Eurasian Marsh Frog *Pelophylax (Rana) ridibundus ridibundus* were updated amphibian genera.

In Reptilia section, the updated genera of the reptiles were starred agama *Stellagama* (*Laudakia*) *stellio*, Iraqi Keel-scaled Gecko *Mediodactylus* (*Carinatogecko*) *heteropholis*, Blandford's Rough-scaled Gecko *Mediodactylus* (*Cyrtopodion*) *heterocercum,* Kotschy's Gecko *Mediodactylus* (*Cyrtopodion*) *kotschyi syriacus*, Levant Skink *Trachylepis (Mabuya) aurata aurata,* Levant Skink *Trachylepis (Mabuya) septemtaeniata,* Bridled Mabuya *Trachylepis (Mabuya) vittata*, Iraqi Mastigure *Saara* (*Uromastix*) *loricata,* Long-nosed Worm Snake *Myriopholis (Leptotyphlops) macrorhynchus,* Large Whip Snake *Dolichophis* (*Coluber*) *jugularis,* Asian Racer *Hemorrhois* (*Coluber*) *nummifer,* Spotted Wipe Snake *Hemorrhois* (*Coluber*) *ravergieri,* Dahl's Whip Snake *Platyceps* (*Coluber*) *najadum dahlia*, Common Cliff Racer *Platyceps* (*Coluber*) *rhodorhachis,* Rogers' pacer *Platyceps* (*Coluber*) *rogersi,* Gray's Desert Racer *Platyceps* (*Coluber*) *ventromaculatus,* Hooded Malpolon *Rhagerhis* (*Malpolon*) *moilensis,* and Armenian Viper *Montivipera* (*Vipera*) *raddei kurdistanica.* The subspecies *Pseudocerastes persicus fieldi* and *P. p. persicus* were separated into full species Field's Horned Viper and Persian Horned Viper respectfully. Recently the species of the Spiny tailed Lizard [Egyptian spiny-tailed lizard *Uromastyx egyptia* and *Saara* (*Uromastix*) *loricata*] were separated and placed to be under an isolated family Uromastycidae after they were placed under the family Agamidae (Rastegar-Pouyani et al. 2008; Gholamifard et al. 2012).

More recently, a new species of the Desert Monitor Lizard, the Nesterov's Desert Monitor *Varanus nesterovi* sp. nov. was described for the foothills of Zagros Mountains in northeastern Iraq.

In current annotated checklist, races/taxa that taxonomically belong to same species such as Anatolian Rock Lizard *Apathya cappadocica muhtari* and *A. c. urmiana*, Common Sandfish *Scincus scincus conirostris* and *S. s. meccensis*, Javelin Sand Boa *Eryx jaculus familiaris* and *E. j. jaculus*, and Common Leaf-nosed Snake *Lytorhynchus diadema gaddi* and *L. d. kennedyi* were considered as races of one independent species. The European Class Lizard *Pseudopus apodus* was recorded from several localities in northern Iraq, but it somehow was overlooked in Al-Barazengy et al. (2015), and it was added to the current checklist. The scientific nomenclature and taxonomic review of the listed species were based on the International Union for Conservation of Nature (IUCN n.d.), Reptile database (www.reptile-database.org), and on the *Amphibian Species of the World: An Online Reference* (Frost 2019).

A. AMPHIBIA: 10 Species/Taxa (8 Genera; 5 Families)

	Common name	Scientific name	Distribution in Iraq	IUCN Status
A. AMPHIBIA: 10 species/taxa (8 Genera; 5 Families)				
Family: Salamandridae				
1	Mountain Newt, Azerbaijan Newt, Lake Urmia Newt	*Neurergus crocatus* (Cope, 1862)	Mountain streams, waterfalls, and creeks of northern Iraq	VU (Fig. 45.1)
2	Kurdistan Newt	*Neurergus microspilotus* (Nesterov, 1916)	Mountain streams, waterfalls, and creeks of north and northeastern Iraq	CR (Fig. 45.2)
3	Common Fire Salamander, Fire Salamander	*Salamandra salamandra semenovi* (Nesterov, 1916)	Mountain streams, waterfalls, and creeks of northern Iraq. However, the current global distribution map showed that this species is restricted to central and western Europe	LC
4	Southern Banded Newt, Banded Newt, Striped Eft	*Ommatotriton (Triturus) vittatus* (Gray, 1835)	Mountain streams, waterfalls, and creeks of northern Iraq	LC
Family: Bufonidae				
5	Iranian Earless Toad, Iranian Toad, Pakistan Toad	*Pseudepidalea (Bufo) surdus* (Boulenger, 1891)	River banks, seasonal pools, foothills of eastern Iraq	LC
6	Green Toad	*Pseudepidalea (Bufo) viridis* (Laurenti, 1768); Synonym: *Bufotes virids*	Distribution range in uncertain, possibly in wetlands of northern Iraq	DD (Figs. 45.3 and 45.4)
Family: Hylidae				
7	Middle East Tree Frog	*Hyla savignyi* (Audouin, 1829)	River banks, marshlands, and vegetated ponds in northern and central Iraq. Recorded from Hawizeh marsh in southern Iraq.	LC (Fig. 45.5)
Family: Pelobatidae				
8	Eastern Spadefoot, Syrian Spadefoot	*Pelobates syriacus* (Boettger, 1889)	Current status is uncertain, possibly in wetlands of northern and northwestern Iraq	LC
Family: Ranidae				
9	Caucasus frog, Iranian Long-legged Wood frog	*Rana macrocnemis;* Synonym *: Rana camerani* (Boulenger, 1886)	River banks and tributaries, marshy ponds, mountain creeks, seasonal pools throughout Iraq.	LC
10	Eurasian Marsh Frog	*Pelophylax (Rana) ridibundus ridibundus* (Pallas, 1771)	River banks and tributaries, marshy ponds, mountain creeks, seasonal pools throughout Iraq.	LC (Fig. 45.6)

(continued)

Fig. 45.1 The Vulnerable Lake Urmia Newt *Neurergus crocatus* in freshwater streams of Zagros Mountain in northern Iraq. Photo Courtesy Omar F. Al-Sheikhly

Fig. 45.2 The Critically Endangered Kurdistan Newt *Neurergus microspilotus* in freshwater streams of Zagros Mountain in northern Iraq. Photo Courtesy Omar F. Al-Sheikhly

Fig. 45.3 Adult Green Toad *Pseudepidalea* (*Bufo*) *viridis* in Tigris River bank in Salahadin Province in Northern Iraq. Photo Courtesy Omar F. Al-Sheikhly

Fig. 45.4 Adult Green Toad *Pseudepidalea* (*Bufo*) *viridis* in Tigris River in Al-Alam Area-Salahadin Province in Northern Iraq. Photo Courtesy Omar F. Al-Sheikhly

Fig. 45.5 Middle East Tree Frog *Hyla savignyi* in Forests of Qara Daqh Mountain in Northern Iraq. Photo Courtesy Omar F. Al-Sheikhly

Fig. 45.6 Eurasian Marsh Frog *Pelophylax (Rana) ridibundus ridibundus* in a mountain stream of PeraMagroon Mountain in Northern Iraq. Photo Courtesy Omar F. Al-Sheikhly

Common name		Scientific name	Distribution in Iraq	IUCN Status
B. REPTILIA: 112 species/taxa (63 Genera; 20 Families)				
Family: Cheloniidae (Sea Turtles)				
1	Loggerhead Sea Turtle	*Caretta caretta* (Linnaeus, 1758)	Marine shores of Al-Fao peninsula at the head of the Arabian Gulf	VU
2	Green Turtle	*Chelonia mydas* (Linnaeus, 1758)	Marine shores of Al-Fao peninsula at the head of the Arabian Gulf	EN
3	Hawksbill Turtle sub-species	*Eretmochelys imbricata bissa* (Rüppell, 1835)	Marine shores of Al-Fao peninsula at the head of the Arabian Gulf	CR
4	Olive Ridley	*Lepidochelys olivacea* (Eschscholtz, 1829)	Marine shores of Al-Fao peninsula at the head of the Arabian Gulf	VU
Family: Dermochelyidaev (Sea Turtles)				
5	Leather back Sea Turtle sub-species	*Dermochelys coriacea schlegelii* (Garman, 1884)	Marine shores of Al-Fao peninsula at the head of the Arabian Gulf	VU
Family: Emydidae (Freshwater Turtles)				
6	Caspian Pond Turtle, Strip-necked Terrapin	*Mauremys caspica caspica* (Gmelin, 1774)	Rivers , tributaries, ponds, marshlands, and lakes throughout Iraq	NE (Fig. 45.7)
7	Caspian Pond Turtle, Strip-necked Terrapin	*Mauremys caspica siebenrocki* (Wischuf and Fritz, 1997)	Rivers , tributaries, ponds, marshlands, and lakes throughout Iraq	NE
Family: Testudinidae (Tortoises)				
8	Spur-thighed Tortoise, Common Tortoise	*Testudo graeca ibera* Pallas, 1814	Forested mountains in northern Iraq	VU (Fig. 45.8)
Family: Trionychidae (Softshell Turtles)				
9	Euphrates Softshell Turtle	*Rafetus euphraticus* (Daudin, 1801)	Rivers, tributaries, ponds, marshlands, and lakes throughout Iraq	EN (Fig. 45.9)
Family: Amphisbaenidae (Worm Lizards)				
10	Anatolian Worm Lizard, Turkish Worm Lizard	*Blanus strauchi aporus* (Werner1898)	Mountains, foothills, and steppes of northern Iraq. Current distribution range in Iraq is unknown.	LC
11	Zarudnyi's Worm Lizard	*Diplometopon zarudnyi* (Nikolsky, 1907)	Steeps, grasslands, and aridlands with scarce vegetation in southern Iraq	LC

(continued)

	Common name	Scientific name	Distribution in Iraq	IUCN Status
Family: Agamidae (Agama Lizards)				
12	Large scaled-rock Agama	*Laudakia nupta nupta* (De Filippi, 1843)	Mountains with rocky slopes and foothills of northern and eastern Iraq	NE (Fig. 45.10)
13	Starred Agama, Roughtail Rock Agama	*Stellagama (Laudakia) stellio* (Linnaeus, 1758)	Rocky mountains, woods and forests, foothills of northern and northwestern Iraq	LC (Fig. 45.11)
14	Black tailed Toed Agama, Spotted Toad-headed Agama	*Phrynocephalus maculatus longicaudatus* (Haas, 1957)	Steeps, aridlands, semi deserts, and wadies of western and southwestern Iraq	NE
15	Brilliant Agama, Steppe Agama	*Trapelus agilis agilis* (Olivier, 1807)	Steeps, aridlands, semi deserts, and wadies of northern, western and southeastern Iraq	NE (Fig. 45.12)
16	Pale Agama	*Trapelus pallidus haasi* (Y. werner, 1971)	Steeps, aridlands, semi deserts, and wadies of northern, western and southern Iraq	NE (Figs. 45.13 and 45.14)
17	Persian Agama (Field's Agama)	*Trapelus persicus fieldi* (Haas and Werner, 1969)	Steeps, aridlands, semi deserts, and wadies of northern, western and southern Iraq	LC
18	Horny-scaled Agama, Oliver's Agama, Anderson's Agama, Baluch Ground Agama, Persia Agama	*Trapelus ruderatus* (Olivier, 1804)	Steeps, aridlands, semi deserts, and wadies of western and southern Iraq	LC
Family: Gekkonidae (Geckos)				
19	Werner's leaf-toed Gecko	*Asaccus elisae* (Werner, 1895)	Rocky mountains, orchids, cultivated grounds, scrublands, shrublands, ruins and aridlands of northern and southeastern Iraq	LC (Fig. 45.15)
20	Grey spotted leaf-toed gecko, Gray-marked Gecko	*Asaccus griseonotus* (Dixon and Anderson, 1973)	Mountain foothills, orchids, cultivated grounds, scrublands, shrublands, ruins and aridlands of northeastern and southeastern Iraq	LC
21	Safin's Mountain Leaf-toed Gecko	*Asaccus saffinae* (Afrasiab and Mohamed, 2009)	Endemic. Current distribution range is unknown. Restricted to Safin Mountain in Northern Iraq	NE

(continued)

Common name		Scientific name	Distribution in Iraq	IUCN Status
22	Baluch Rock Gecko, Arabian Desert Gecko, Southern Tuberculated Gecko	*Bunopus tuberculatus* (Blanford, 1874)	Mountain foothills, orchids, cultivated grounds, scrublands, shrublands, aridlands, and ruins of central and southeastern Iraq	LC
23	Keeled Rock Gecko, Rough-tailed Gecko, Rough thin-toed Gecko, Rough Bent-toed Gecko	*Cyrtopodion scabrum* (Heyden, 1827)	Wide distribution range; human settlements, cultivated areas, ruins, and in any suitable habitats throughout Iraq	LC (Fig. 45.16)
24	Iraqui eyelid gecko	*Eublepharis angramainyu* (Anderson and Leviton, 1966)	Restricted range to the mountain steppes and foothills of northern Iraq. New localities were recently discovered in eastern Iraq	DD (Fig. 45.17)
25	Yellow-bellied House Gecko, Northern House Gecko	*Hemidactylus flaviviridis* (Rüppell, 1835)	Human settlements, cultivated areas, and ruins in central and southern Iraq	NE (Fig. 45.18)
26	Persian Gecko, Persia Leaf-toed Gecko	*Hemidactylus persicus* (Anderson, 1872)	Human settlements, cultivated areas, cliffs, and ruins in central and southern Iraq, probably in northern Iraq	NE (Fig. 45.19)
27	Mediterranean Gecko, Turkish Gecko, Turkish Warty Gecko	*Hemidactylus turcicus* (Linnaeus, 1758)	The current distribution status is unknown. Rocky mountains, cliffs, foothills, human settlements, and steeps in northern and northwestern Iraq.	LC
28	QaraDagh Mountain Gecko (Kurdish Gecko)	*Hemidactylus kurdicus* (Safaei-Mahroo, Ghaffari, Ghafoor and Amini 2017)	Endemic. Current distribution range is unknown. Restricted to Qara Daqh Mountain in northeastern Iraq	NE
29	Iraqui keel-scaled gecko, Iraqui Gecko	*Mediodactylus (Carinatogecko) heteropholis* (Minton, Anderson and Anderson, 1970)	Status uncertain. Probably restricted to the wooded mountain in northern Iraq	DD
30	Blandford's Rough scaled Gecko, Asia minor thin-toed Gecko	*Mediodactylus (Cyrtopodion) heterocercum mardinensis* (Mertens, 1924)	Status uncertain. Probably restricted to the wooded mountain in northern Iraq	NE

(continued)

Common name		Scientific name	Distribution in Iraq	IUCN Status
31	Kotschy's Gecko	*Mediodactylus (Cyrtopodion) kotschyi syriacus* (Stepanek, 1937)	The current distribution status is unknown. Probably confined to the steppes, rocky areas, and shrublands in northeastern Iraq	LC
32	Common fan-footed Gecko, Yellow Fan-fingered Gecko, Hasselquist's Fan-footed Gecko	*Ptyodactylus hasselquistii* (Donndorff, 1798)	The current distribution status is unknown. Confined to the steppes, rocky areas, and shrublands in northern and eastern Iraq	NE
33	Levante Fan-Fingered Gecko	*Ptyodactylus puiseuxi* (Boutan, 1893)	The current distribution status is unknown. Rocky areas, aridlands, and shrublands in northwestern Iraq	NE
34	Murray's Comb-fingered Gecko, Iranian Short-fingered Gecko	*Stenodactylus affinis* (Murray, 1884)	Marshlands, aridlands and semi desert areas in southeastern Iraq	LC
35	Doriae'scomb-fingered Gecko, Middle Eastern Short-fingered Gecko, Sand Gecko	*Stenodactylus doriae* (Blanford, 1874)	Aridlands and deserts and semi desert areas in central and southern Iraq	LC
36	Stout Gecko, Jordan Short-fingered Gecko	*Stenodactylus grandiceps* (Haas, 1957)	Aridlands and shrunslands, in north-western and western Iraq	LC
37	Slevin's Short-fingered Gecko, Slevin's Sand Gecko	*Stenodactylus slevini* (Haas, 1957)	Aridlands and deserts in western and southern Iraq	LC
38	Dwarf Gecko	*Microgecko helenae fasciatus* (Schmidtler and Schmidtler 1972)	Rocky mountains, woods and forests, foothills of northeastern Iraq	DD
Family: Lacertidae (Fringe-toed Lizards and Lizards)				
39	Bosc's Fringe-toed Lizard, Bosk's Fringe-fingered Lizard	*Acanthodactylus boskianus* (Daudin, 1802)	Aridlands, deserts and semi desert, and wadies in western and southern Iraq	NE (Fig. 45.20)
40	Mesopotamian Fringe-toed lizard, Giant Fringe-toed lizard	*Acanthodactylus grandis* (Boulenger, 1909)	Aridlands, deserts and semi desert, and wadies in western and southern Iraq	LC (Fig. 45.21)
41	Arnold's fringe-fingered lizard, Snake-tailed Fringe-toed Lizard	*Acanthodactylus opheodurus* (Arnold, 1980)	Deserts and semi desert, shrunblands, and on terrestrial edge of the marshlands in southern Iraq	LC (Fig. 45.22)

(continued)

Common name		Scientific name	Distribution in Iraq	IUCN Status
42	Syrian Fringe-fingered Lizard	*Acanthodactylus orientalis* (Angel, 1936)	Grasslands, aridlands, deserts and semi deserts, fields with scarce vegetation, and wadies in central and western Iraq	LC
43	Robust Fringe-fingered Lizard	*Acanthodactylus robustus* (Werner, 1929)	Shrublands, aridlands, desert and semi deserts in western Iraq	LC
44	Schmidt's Fringe-toed lizard	*Acanthodactylus schmidti* (Haas, 1957)	Aridlands, desert, and semi deserts in southern Iraq	LC
45	Nidua Fringe-fingered lizard, Hardy's Fringe-fingered Lizard	*Acanthodactylus scutellatus hardyi* (Haas, 1957)	Aridlands, desert, and semi deserts in western and southern Iraq	NE (Fig. 45.23)
46	Anatolian Rock Lizard	*Apathya cappadocica muhtari* (Eiselt 1979)	Wooded mountains, rocky slopes, forests, and cliffs in northern Iraq	LC (Fig. 45.24)
47	Anatolian Rock Lizard (Lake Uremia Rock Lizard)	*Apathya cappadocica urmiana* (Lantz and Suchow 1934)	Status uncertain. Wooded mountains, rocky slopes, forests, and cliffs in northern Iraq	LC
48	Aralo-Caspian Racerunner	*Eremias persica* (Blanford, 1875)	Status uncertain. wooded mountains and foothills of northern and northeastern Iraq	NE
49	Three-line Lizard, Giant Green lizard	*Lacerta media media* (Lantz and Cyrén, 1920)	Wooded mountains, rocky slopes, and cliffs in northern Iraq; probably occurs in foothills of eastern Iraq	LC
50	Caspian Green Lizard, Striated Lizard	*Lacerta strigata* (Eichwald, 1831)	Occurrence is uncertain; three specimens were collected from northern Iraq which probably introduced by floods. Probably inhabit wooded mountains, rocky slopes, forests, and cliffs in northern Iraq	LC
51	Blandford's short-nosed Desert Lizard	*Mesalina brevirostris* (Blanford, 1874)	Current distribution is unknown. Steppes, grasslands, aridlands, desert and semi desert in southern Iraq. Probably occurs in western Iraq	LC (Fig. 45.25)

(continued)

Common name		Scientific name	Distribution in Iraq	IUCN Status
52	Small-spotted Lizard	*Mesalina guttulata* (Lichtenstein, 1823)	Status uncertain, Aridlands, desert, and semi deserts in western and southern Iraq	NE
53	Olivier's Sand Lizard	*Mesalina olivieri* (Audouin, 1829)	Deserts and semi deserts in southern Iraq	NE
54	Snake-eyed Lizard	*Ophisops elegans* (Ménétriés, 1832)	Wooded mountains, rocky slopes, forests, and cliffs in northern Iraq; probably occurs in eastern Iraq	NE (Fig. 45.26)
55	Zagrosian Lizard, Siirt Lizard	*Timon princeps kurdistanicus* (Suchow, 1936)	Wooded mountains, rocky slopes, and cliffs in northern and north-eastern Iraq	LC (Fig. 45.27)
Family: Anguidae (Legless Lizard Lizards)				
56	European Class Lizard, European Legless Lizard	*Pseudopus apodus* (*Pallas, 1775*)	Wooded mountains, forests, and rocky cliffs in northern Iraq	NE (Fig. 45.28)
Family: Scincidae (Skinks)				
57	Juniper Skink, European Snake-eyed Skink	*Ablepharus kitaibelii kitaibelii* (Bibron and Bory, 1833)	Status uncertain; required further assessment. Probably in wooded mountains northern Iraq	LC
58	Asian Snake-eyed Skink	*Ablepharus pannonicus* (Fitzinger, 1824)	Steppes, aridlands, and foothills in central, eastern, and southern Iraq. see (Afrasiab and Ali, 1989a)	NE
59	Ocellated skink	*Chalcides ocellatus* (Forsskål, 1775)	Steppes, aridlands, shrublands, and foothills in western, eastern, and southern Iraq	NE
60	Schneider's Skink	*Eumeces schneideri princeps* (Eichwld, 1839)	Current distribution is unknown. Steppes, aridlands, shrublands, and foothills in western, eastern, and southern Iraq	NE
61	Common Sandfish	*Scincus scincus conirostris* (Blanford, 1881)	Desert and semi desert of southern and south-western Iraq	NE
62	Common Sandfish	*Scincus scincus meccensis* (Wiegmann, 1837)	Desert and semi desert of southern and south-western Iraq	NE
63	Levant Skink	*Trachylepis (Mabuya) aurata aurata* (Linnaeus, 1758)	Wooded mountains, orchids, cultivated fields, forests, and foothills, and human settlements in northern, central, and southern Iraq	NE

(continued)

Common name		Scientific name	Distribution in Iraq	IUCN Status
64	Levant Skink	*Trachylepis (Mabuya) septemtaeniata* (Reuss, 1834)	Mountains, foothills, forests, cliffs, shrublands in northern, central, and southern Iraq; probably occurs in eastern Iraq	NE (Fig. 45.29)
65	Bridled Mabuya, Bridled Skink	*Trachylepis (Mabuya) vittata* (Olivier, 1804); *Heremites vittatus*	Wooded mountains, rocky slopes, forests, and cliffs in northern, central, and southern Iraq; probably occurs in eastern Iraq	LC
Family: Uromastycidae (Spiny-tailed Lizards)				
66	Egyptian Spiny-tailed Lizard	*Uromastyx aegyptia* (Forskål, 1775)	Desert and semi deserts, aridlands, grasslands, foothills, and marshland edges in central and southern Iraq	VU (Fig. 45.30)
67	Mesopotamian Spiny-tailed Lizard, Small Spiny tailed Lizard, Iraqi Mastigure, Iraqi Spiny-tailed Lizard	*Saara (Uromastix) loricata* (Blanford, 1874)	Steppes, foothills, shrublands, and aridlands of central and southeastern Iraq	LC (Figs. 45.31 and 45.32)
Family: Varanida (Monitors)				
68	Desert Monitor	*Varanus griseus griseus* (Daudin, 1803)	Desert and semi deserts, aridlands, grasslands, foothills, and marshland edges in northern, western, and southeastern Iraq	NE
69	Nesterov's Desert Monitor	*Varanus nesterovi*	Wooded mountains, rocky slopes and cliffs in northern Iraq; probably occurs in eastern Iraq	NE
Family: Leptotyphlopidae (Worm Snakes)				
70	Long-nosed Worm Snake, Hooked Thread Snake	*Myriopholis (Leptotyphlops) macrorhynchus* (Jan, 1860)	Shrublands, aridlands, steppes, foothills of central and southern Iraq	LC
Family: Typhlopidae (Blind Snakes)				
71	Brahminy Blind Snake, Flowerpot Snake	*Ramphotyphlops braminus* (Daudin, 1803)	Aridlands, grasslands, and cultivated fields of central and southern Iraq. see (Afrasiab and Ali 1996)	NE
72	Blind Snake Species	*Typhlops vermicularis* (Merrem, 1820)	Aridlands, grasslands, and cultivated fields of central and southern Iraq	NE

(continued)

Common name		Scientific name	Distribution in Iraq	IUCN Status
Family: Boidae (Boas)				
73	Javelin Sand Boa	*Eryx jaculus familiaris* (Eichwald, 1831)	Desert and semi deserts, shrublands, and aridlands of central and southern Iraq	NE
74	Javelin Sand Boa	*Eryx jaculus jaculus* (Linnaeus, 1758)	Desert and semi deserts, shrublands, and aridlands of central and southern Iraq	NE
75	Saudi Arabian Sand Boa, Arabian Sand Boa	*Eryx jayakari* Boulenger, 1888	Desert and semi deserts, shrublands, and aridlands of southern Iraq	LC
Family: Colubridae (Snakes)				
76	Large Whip Snake	*Dolichophis (Coluber) jugularis* (Linnaeus, 1758)	Rocky foothills, Steppes, orchids, aridlands, shrublands, and deserts of northern, central, and southern Iraq. *Dolichophis* population of southern Iraq requires more collection and more taxonomic study	LC
77	Collared Dwarf Snake	*Eirenis collaris* (Ménétriés, 1832)	Mountain foothills of northeastern Iraq	LC
78	Crowned Dwarf Snake	*Eirenis coronella coronella* (Schlegel, 1837)	Desert and semi deserts, shrublands, and aridlands of western and southern Iraq	LC
79	Sinai Dwarf Snake	*Eirenis coronelloides* (Sivan and Werner, 2003)	Mountain foothills, shrubslands, and steppes of northern and central Iraq	LC
80	Narrow-striped Dwarf Snake	*Eirenis decemlineata* (Duméril, Bibron and Duméril, 1854)	Steppes, grasslands, shrublands of northern and northwestern Iraq	LC
81	Striped Dwarf Snake	*Eirenis lineomaculata* (Schmidt, 1939)	Status uncertain. Steppes, grasslands, shrublands of northern and northwestern Iraq	LC
82	Dark-headed Dwarf Racer	*Eirenis persicus* (Anderson, 1872)	Rocky foothills, steppes, shrublands in central and eastern Iraq	NE
83	Asian Racer, Coin-Marked Snake	*Hemorrhois (Coluber) nummifer* (Reuss, 1834)	Mountains, rocky cliffs, forests, steppes, and grasslands of northern Iraq	NE
84	Spotted Wipe Snake	*Hemorrhois (Coluber) ravergieri* (Ménétriés, 1832)	Mountains, rocky cliffs, forests, steppes, and grasslands of northern and northwestern Iraq	NE

(continued)

Common name		Scientific name	Distribution in Iraq	IUCN Status
85	Common Leaf-nosed Snake	*Lytorhynchus diadema gaddi* (Nikolsky, 1907)	Desert and semi deserts, shrublands, foothills, orchids, cultivated fields, marshlands edges, and and aridlands of central and southern Iraq. *Spalerosophis diadema cliffordii* was reported from southern Iraq by (Afrasiab et al. 2018)	LC
86	Common Leaf-nosed Snake	*Lytorhynchus diadema kennedyi* (K. Schmidt, 1939)	Desert and semi deserts, shrublands, foothills, orchids, cultivated fields, marshlands edges, and and aridlands of western and southern Iraq. See (Afrasiab and Ali,1989b)	LC
87	Montpellier Snake	*Malpolon monspessulanus insignitus* (Geoffroy St-Hilaire, 1809)	Mountains, rocky hills, steppes, grasslands, orchids, cultivated fields, forests, and aridlands in northern, central, and southern Iraq	LC (Fig. 45.33)
88	Tessellated water snake,	*Natrix tessellata tessellata* (Laurenti, 1768)	Wetlands throughout Iraq	LC
89	Grass Snake	*Natrix natrix persa* (Pallas 1814)	Wetlands in central and eastern Iraq; probably has wider distribution range	LC
90	Braid Snake, Jan's Cliff Racer	*Platyceps ladacensis* (Perry, 2012)	Recorded from Hawraman Mountain in northeastern Iraq	NE
91	Dahl's Whip Snake	*Platyceps (Coluber) najadum dahlii* (Fitzinger 1826)	Mountain foothills, shrubslands, and steppes of northern, central, and southeastern Iraq	LC
92	Common Cliff Racer, Wadi Racer, Desert Racer	*Platyceps (Coluber) rhodorhachis* (Jan, 1865)	Deserts and semi deserts, shrublands, rocky areas, steeps, and grasslands of western and southern Iraq	NE
93	Rogers' Racer	*Platyceps (Coluber) rogersi* (Anderson, 1893)	Shrublands, rocky areas, steeps, and grasslands of western and southern Iraq	LC

(continued)

Common name		Scientific name	Distribution in Iraq	IUCN Status
94	Gray's Desert Racer, Hardwicke's Rat Snake, Glossy-bellied Racer	*Platyceps (Coluber) ventromaculatus* (Gray, 1834)	Shrublands, rocky areas, steeps, desert and semi-deserts, and grasslands of central and southern Iraq	NE
95	Hooded Malpolon, Moila Snake, False Cobra	*Rhagerhis (Malpolon) moilensis* (Reuss, 1834)	Mountains, rocky hills, steppes, grasslands, orchids, cultivated fields, forests, and aridlands in northern, central, and southern Iraq	NE
96	Schokari Sand racer, Afro-Asian Sand Snake, Forskal Sand Racer	*Psammophis schokari* (Forsskål, 1775)	Desert and semi deserts, steppes, aridlands, shrublands, cultivated fields, and marshland edges in central and southern Iraq	NE
97	Black-Headed Rhynchocalamus	*Rhynchocalamus melanocephalus satuni* (Nikolsky, 1899)	Mountains, rocky slopes, cliffs, grasslands, and steppes in northern Iraq	LC
98	Diadem snake	*Spalerosophis diadema cliffordii* (Schlegel, 1837)	Grasslands, desert and semi-deserts, shrublands, cultivated fields, human settlements, rural gardens, and marshland edges in central and southern Iraq.	NE
99	Zebra Snake	*Spalerosophis microlepis* (Jan, 1865)	Mountain, foothills, shrublands, and aridlands of eastern Iraq	LC
100	Black Headed Snake	*Telescopus nigriceps* (Ahl, 1924)	Steppes, rocky wadies, and aridlands of western Iraq	LC
101	Soosan Tiger Snake	*Telescopus tessellatus martini* (Schmidt, 1939)	Mountains, rocky slopes, cliffs, grasslands, foothills, and steppes in northern, eastern, and southern Iraq	LC
102	Caucasian Rat Snake, Transcaucasian Rat Snake	*Zamenis hohenackeri* (Strauch, 1873)	Mountains, rocky slopes, and cliffs in northeastern Iraq	LC
Family: Elapidae (Desert Cobras)				
103	Desert Cobra, Desert Black Snake	*Walterinnesia aegyptia* (Lataste, 1887)	Current distribution is unknown, desert and semi deserts, shrublands, grasslands in northern and western Iraq	LC

(continued)

Common name		Scientific name	Distribution in Iraq	IUCN Status
Family: Viperidae (Vipers)				
104	Arabian Horned Viper	*Cerastes gasperettii* (Leviton and Anderson, 1967)	Desret and semi desert, shrublands, aridlands of western and southern Iraq	LC (Fig. 45.34)
105	Saw-scaled Viper	*Echis carinatus sochureki* (Stemmler, 1969)	Desert and semi deserts, aridland, shrublands, scrublands, and on the edges f the marshlands in central and southern Iraq	NE
106	Armenian Viper, Caucasus Viper	*Montivipera (Vipera) raddei kurdistanica* (Nilson and Andren, 1986)	Wooded mountains and rocky cliffs in northern Iraq	NT
107	Blunt-nosed Viper	*Macrovipera lebetina obtuse* (Dwigubsky, 1832)	Rocky foothills, Steppes, orchids, aridlands, shrublands, and deserts of northern and central Iraq	LC
108	Field's Horned Viper	*Pseudocerastes fieldi* (Schmidt, 1930)	Rocky foothills, aridlands, shrublands, and deserts of northwestern and western Iraq	LC
109	Persian Horned Viper	*Pseudocerastes persicus* (Duméril, Bibron and Duméril, 1854)	Wooded moutons, rocky cliffs, and forests of northern Iraq; possibly occurs in the foothills of eastern Iraq	LC
110	Iranian Spider-tailed Horned Viper	*Pseudocerastes urarachnoides* (Bostanchi, S. Ander son, Kami and Papenfuss, 2006)	Rocky foothills of eastern Iraq	DD (Fig. 45.35)
Family: Hydropfiidae (Sea Snakes)				
111	Hook-nosed sea snake, Beaked Sea Snake	*Enhydrina schistosa* (Daudin, 1803)	Marine waters of Al-Fao peninsula at the head of the Arabian Gulf	LC
112	Slender sea snake, Graceful Small Headed Sea snake	*Hydrophis gracilis* (Shaw, 1802)	Marine waters of Al-Fao peninsula at the head of the Arabian Gulf	LC

CR critically endangered, *EN* endangered, *VU* vulnerable, *NT* near threatened, *LC* least concern, *DD* data deficient, *NE* not evaluated

Fig. 45.7 Adult Caspian Pond Turtle *Mauremys caspica caspica* in Tigris River tributary in PeraMagroon Mountain in Northern Iraq. Photo Courtesy Omar F. Al-Sheikhly

Fig. 45.8 The Vulnerable Spur-thighed Tortoise *Testudo graeca ibera* in wooded slpe of Qara Dagh Mountain in Northern Iraq. Photo Courtesy Omar F. Al-Sheikhly

Fig. 45.9 Adult female Euphrates Softshell Turtle *Rafetus* (*Trionyx*) *euphraticus* laying eggs on the banks of Central Marshes in Southern Iraq. Photo Courtesy Omar F. Al-Sheikhly

Fig. 45.10 Adult Large Scaled-rock Agama *Laudakia nupta nupta* in Jabal Himreen foothills in Northeastern Iraq. Photo Courtesy Omar F. Al-Sheikhly

Fig. 45.11 Adult Starred Agama *Stellagama (Laudakia) stellio* in a rocky cliff of PearMagroon Mountain in Northern Iraq. Photo Courtesy Omar F. Al-Sheikhly

Fig. 45.12 Adult Steppe Agama *Trapelus agilis agilis* in the desert of ThiQar Province in Southwestern Iraq. Photo Courtesy Omar F. Al-Sheikhly

Fig. 45.13 Pale Agama *Trapelus pallidus* in rocky desert on Al-Anbar Province in Western Iraq. Photo Courtesy Omar F. Al-Sheikhly

Fig. 45.14 Adult Pale Agama *Trapelus pallidus* in Jabel Himreen foothills in Northeastern Iraq. Photo Courtesy Omar F. Al-Sheikhly

Fig. 45.15 Adult Werner's Leaf-toed Gecko *Asaccus elisae* in Himreen Foothills in Northeastern Iraq. Photo Courtesy Omar F. Al-Sheikhly

Fig. 45.16 Adult gravid female Rough-tailed Gecko *Cyrtopodion scabrum* in human settlements of Al-Alam Area in Salahadin Province in Northern Iraq. Photo Courtesy Omar F. Al-Sheikhly

Fig. 45.17 Adult male Iraqi Eyelid Gecko *Eublepharis angramainyu* in the grasslands of Hirmeen Foothills in Diyala Province in Eastern Iraq. Photo Courtesy Omar F. Al-Sheikhly

Fig. 45.18 Adult male Yellow-bellied House Gecko *Hemidactylus flaviviridis* in a human settlement in Baghdad Province in Central Iraq. Photo Courtesy Omar F. Al-Sheikhly

Fig. 45.19 Adult Persian Gecko *Hemidactylus persicus* in a human settlement in Baghdad Province in Central Iraq. Photo Courtesy Omar F. Al-Sheikhly

Fig. 45.20 Adult Bosc's Fringe-toed Lizard *Acanthodactylus boskianus* in shrublands of Jabel Makhool foothill in Salahadin Province in northern Iraq. Photo Courtesy Omar F. Al-Sheikhly

Fig. 45.21 Adult Mesopotamian Fringe-toed Lizard *Acanthodactylus grandis* in the desert of Faidhatt Al Massad in Anbar Province in Western desert of Iraq. Photo Courtesy Omar F. Al-Sheikhly

Fig. 45.22 Adult Arnold's Fringe- fingered Lizard *Acanthodactylus opheodurus* in the aridlands of Wadi Al-Rodha in Western Desert of Iraq. Photo Courtesy Omar F. Al-Sheikhly

Fig. 45.23 Adult Nidua Fringe-fingered Lizard *Acanthodactylus scutellatus* in semi-desert areas of Rutba in Western Desert of Iraq. Photo Courtesy Omar F. Al-Sheikhly

Fig. 45.24 Adult Anatolian Rock Lizard *Apathya cappadocica* in rocky cliffs of Tabeen Village in Pera Magroon Mountain in Northern Iraq. Photo Courtesy Omar F. Al-Sheikhly

Fig. 45.25 Adult Small-spotted Mesalina guttulata in the desert of Rutba near Faidhatt Al Massad in Anbar Province in Western Iraq. Photo Courtesy Omar F. Al-Sheikhly

Fig. 45.26 Adult male Snake-eyed Lizard *Ophisops elegans* in Rocky cliffs of Peramagroon Mountain in Northern Iraq. Photo Courtesy Omar F. Al-Sheikhly

Fig. 45.27 Zagrosian Lizard *Timon princeps kurdistanicus* in the wooded slopes of Zagros Mountain in Benjween Area in Northeastern Iraq. Photo Courtesy Omar F. Al-Sheikhly

Fig. 45.28 Adult European Class Lizard *Pseudopus apodus* in wooded rocky slopes of Tabeen Village in PeraMagroon Mountain in Northern Iraq. Photo Courtesy Omar F. Al-Sheikhly

Fig. 45.29 Adult Levant Skink *Trachylepis (Mabuya) septemtaeniata* in scarce vegetated steppe near Kahn Baghdadi of Haditha in Anbar Province in Western Desert of Iraq. Photo Courtesy Omar F. Al-Sheikhly

Fig. 45.30 Adult male Vulnerable Egyptian Spiny-tailed Lizard *Uromastyx aegyptius* in the western desert of Anbar Province in Western Iraq. Photo Courtesy Omar F. Al-Sheikhly

Fig. 45.31 Adult breeding male Mesopotamian Mastigure *Uromastyx (Saara) loricatus* in Himreen Foothills in Northeastern Iraq. Photo Courtesy Omar F. Al-Sheikhly

Fig. 45.32 The Mesopotamian Mastigure *Uromastyx (Saara) loricatus* in Al-Sidir Area in Salahadin Province in Northern Iraq. Photo Courtesy Omar F. Al-Sheikhly

Fig. 45.33 Adult Montpellier Snake *Malpolon monspessulanus* in grasslands of Qara Dagh Mountain in Northern Iraq. Photo Courtesy Omar F. Al-Sheikhly

Fig. 45.34 Hornless morph of the Arabian Horned Viper Cerastes gasperettii, a desert dowelling species recorded from the western and southwestern deserts of Iraq. The Iranian Spider-tailed Viper *Pseudocerastes urarachnoides* in Zurbatia Foothills in eastern Iraq. Photo Courtesy Omar F. Al-Sheikhly

Fig. 45.35 The Iranian Spider-tailed Viper *Pseudocerastes urarachnoides* in Zurbatia Foothills in eastern Iraq. Photo Courtesy Omar F. Al-Sheikhly

Acknowledgments I'm are grateful to Ali N. Al-Barazengy for his further comments and personal notes on amphibians and reptiles of Iraq and his permission to allow us to access his preliminary dataset of herpetofauna of Iraq. I would like to thank Mahdi Laith Al-Haidari for his personal observations of some reptile species in Iraq.

References

Afrasiab SR, AL-Ganabi MI, AL-Fartosi K (2012) Snake species new or rare to the herpetofauna of Iraq. Herpetozoa 24(3/4):179–181

Afrasiab SR, Ali HA (1989a) Report on a collection of reptiles from Rumaila desert, south of Iraq. Bulletin of the Iraq Natural History Museum 8(2):65–73

Afrasiab SR, Ali HA (1989b) A new record of the snake *Lytorhynchus kennedyi* Schmidt (Reptilia, Colubridae) from Iraq. Bulletin of the Iraq Natural History Museum 8(2):157–159

Afrasiab SR, Ali HA (1996) Notes on Scolecophidians (blind snakes) Reptilia-Serpentes, of Iraq. Bulletin of the Iraq Natural History Museum 8(4):31–39

Afrasiab SR, Mohamad SI (2009) A study on cave-dwelling geckos in Iraq, with the description of a new species from Saffine mountain. Zoology in the Middle East 47(1):49–56

Afrasiab SR, Mohamad SI (2011) First record of the rat snake, *Zamenis hohenackeri* (Strauch, 1873), from North-Eastern Iraq with notes on other colubrid snakes. Zoology in the Middle East 54(1):19–22

Afrasiab SR, Mohamad SI (2014) New records of snakes from Iraq (Reptilia: Colubridae). Zoology in the Middle East 60(1):92–94

Afrasiab SR, Mohammad MK, Husein AM (2016) Description of a new species of the genus *Dolichophis* Gitstel from the upper Mesopotamian plain-Iraq (Reptilia-Serpentes-Colubridae). J Biodivers Environ Sci 8(4):15–19

Afrasiab SR, Sarbaz IM (2009) A study on cave-dwelling geckos in Iraq, with the description of a new species from *Saffine mountain*. Zoology in the Middle East 47:49–56

Afrasiab SR, Sarbaz IM (2011) Report First record of the Rat Snake, *Zamenis hohenackeri* (Strauch, 1873), from north eastern Iraq with notes on other colubrid snakes, Zoology in the Middle East. 54: 19–22

Afrasiab SR, Sarbaz IM, Raza HH (2013) A review of the Lacertini of Iraq in Iraqi collections. Herpetozoa 25(3/4):93–100

Afrasiab SR, Al-Moussawi AA, Hadi HD (2018) Annotated checklist of reptilian fauna of Basrah, South of Iraq. Bull Iraq Nat Hist Mus 15(1):77–92

Al-Barazengy AN (2014) First observations on *Phrynocephalus maculatus longicaudatus* Haas, 1957 (Squamata: Sauria: Agamidae) in Iraq. 13(2):119–125

Al-Barazengy AN, Salman AO, Abdul hameed FT (2015) Updated list of amphibians and reptiles in Iraq 2014. Bulletin of Iraq Natural History Muisium 13(4):29–40

Allouse BE (1955) A bibliography on the vertebrate Fauna of Iraq and Neighbouring countries. III reptiles and amphibians. Iraq Nat Hist Mus Pubi 6:1–23

Al-Sheikhly OF, Nader IA, Rastegar-Pouyani N, Browne RK (2013) New localities of the Kurdistan newt *Neurergus microspilotus* and Lake uremia newt *Neurergus crocatus* (Caudata: Salamandridae) in Iraq. Amphibian and Reptile Conservation 6(4):42–49 (e68)

Anderson SC (1999) The lizards of Iran, Society for the Study of Amphibians and Reptiles, Oxford, Ohio. 442 pp.

Arnold EN (1986) A key and annotated check list to the lizards and amphibians of Arabia. Fauna Saudi Arabia 8:378–384

Corkill NL (1932) Snakes and snake bite in Iraq. Bailliere, Tindall and Cox, London, Ix+ 51pp

Disi AM (2002) Jordan Country Study on Biological Diversity: Herpetofauna of Jordan. – Withtechnical support from the United Nations Environment Programme (NNEP) and funding from the Global Environment Facility (GEF). Project No. GF/6105-92-65, GF/6105, 92–02 (2991)

Disi AM, Amr ZS (1998) Distribution and ecology of lizards in Jordan (Reptilia: Sauria). p. 43-66. In: Fritz U, Obst FJ, Andreas B (eds) Beiträge zu einer Herpetofauna Arabica. –Faunistische Abhandlungen, Staatliches Museum für Tierkunde Dresden, Supplement, Contributions to a "Herpetologia Arabica", Vol. 21, Suppl, p 6

Disi AM, Modry D, Bunian F, AL-Oran RM, Amr ZS (1999) Amphibians and reptiles of the Badia region of Jordan, Herpetozoa, Wien; 12 (3–4): 135–146

Fazaa NA, Dunn JC, Whittingham MJ (2015) Status of Euphrates softshell turtle *Rafetus euphraticus in the Iraqi Central Marshes*. International Conference on Latest Trends in Food, Biological & Ecological Sciences 44–49
Frost DR (2019) Amphibian specie of the world: an online reference. Version 6.0. Electronic American Museum of Natural History, New York, USA
Garstecki T, Amr Z (2011) Biodiversity and Ecosystem Management in the Iraqi Marshlands-Screening Study on Potential World Heritage Nomination, Amman, Jordan: IUCN. 978–2–8317-1353-3
Ghaffari H, Taskavak H, Karami M (2008) Conservation status of the Euphrates Softshell turtle, *Rafetus euphraticus*, in Iran. Chelonian Conservation and Biology 7(2):223–229
Gholamifard A, Rastegar-Pouyani N, Esmaeili HR (2012) Annotated checklist of reptiles of Fars Province. southern Iran, Iranian journal of Animal biosystematics (IJAB) 2:155–167
IUCN (n.d.) Red list of Iraq: Retrieved from https://www.iucnredlist.org/search/list on Nov. 2018
Jablonski D., Benovics M., Vorel J., Mohammed S., I., Afrasiab S., R.2020. The first record of the genus Microgecko Nikolskii, 1907 forIraq. Herpetozoa 33:149–155
Khalaf KT (1959) Reptiles of Iraq with some notes on the amphibians, Baghdad
Leviton AE, Anderson SC, Adler K, Minton SA (1992) Hand book to the Middle East amphibians and reptiles, Ithaca, NewYork, SSAR, 252 p
Mahdi N, George PV (1969) Systematic list of Iraqi vertebrates– mammals. Bulletin of the Iraq Natural History Museum 26:34–63
Mohammed RG, Rastegar-Pouyani N, Karmiani R, Rhadi FA (2015) Systematics and distribution of geckos in Iraq, systematics and distribution of the Gekkonidae in some provinces of central and southern Iraq. Indian Journal of Natural Science, 7338-7345pp
Nader IA Jawdat SZ (1976) Taxonomic study of the geckos of Iraq (Reptilia: Gekkonidae). Bulletin of Biological Research Center, Baghdad, (5), 1–41
Najafimajd E, Kaya U (2010) A newly found locality of the critically endangered yellow spotted newt *Neurergus microspilotus* (Nesterov, 1916), nourishes hope for its conservation (Salamandridae: Caudata). Zoology in the Middle East 51:51–56
Nature Iraq (2017) Key Biodiversity Areas of Iraq. Sulaimaniyah, Iraq. 297 pp.
Niazi AD (1976) On the Mediterranean horseshow bat from Iraq. Bulletin of the Iraq Natural History Museum 7:167–176
Niazi AD (1979) The Arabian worm lizard: its ecology, morphology with description of a new species. Iraq Journal of Biological Sciences 7:13–29
Nilson G, Rastegar-Pouyani N (2005) *Walterinnesia aegyptia* Lataste, 1887 (Ophidia: Elapidae) and the status of *Naja morgani* Mocquard 1905. Russ J Herpetol 14(1):7–14
Rastegar-Pouyani N (2000) Taxonomic status of *Trapelus ruderatus* (Olivier) and *T. persicus* (Blanford), and validity of *T. lessonae* (De Filippi). Amphibia-Reptilia 21(1):91–102
Rastegar-Pouyani N, Kami HG, Rajabzadeh M, Shafiei S, Anderson SC (2008) Annotated checklist of amphibians and reptiles of Iran. Iranian journal of Animal biosystematics (IJAB) 1:43–66
Rudayni HA, Al-Sadoon MK, Paray B (2017) Morphological characteristics of worm lizard, *Diplometopon zarudnyi* (Squamata: Trogonophidae) in the central rejoin of Saudi Arabia. Saudi Journal of Biological Sciences 24(4):966–971
Safaei-Mahroo B., Ghaffari H., Fahimi H., Broomand S.,Yazdanian M.,NajafiMajd E. Hosseinian, Yousefkhani S. S., Rezazadeh E. H.,Osseinzadeh M. S., Nasrabadi R., Rajabizadeh M., Mashayekhi M.,Moteshareі A., Naderi A., and Kazemi S. M. 2015. The Herpetofauna ofIran: Checklist of taxonomy, distribution and conservationstatus. Asian Herpetological Research 6, 257–290
Schmidt KP (1939) Reptiles and amphibians from southwestern Asia. Chicago field Mus, vol 27, pp 323–346
Scott DA (1995) A directory of wetland in the Middle East. IWRP, Slimbridge, UK, pp 223–301
Stadtlander T (1992) Recent observations of the Euphrates soft-shelled turtle, *Rafetus euphraticus*, in Mesopotamia. Zoology in the Middle East 7(1):55–58
Wagner P, Melville J, Wilms TM, Schmitz A (2011) Opening a box of cryptic taxa – the first review of the north African desert lizards in the *Trapelus mutabilis* Merrem, 1820 complex (Squamata: Agamidae) with descriptions of new taxa. Zool J Linnean Soc 163:884–912

Chapter 46
Policy Guidance for Sustainable Aquaculture in the Inland Waters of Iraq

Laith A. Jawad and Falah Mutlak

Abstract Aquaculture is the fastest rising animal-producing area in the world and is probable to do an imperative part in global food source. The notion of sustainable aquaculture is increasingly known to comprise both the seasonal and geographical ranges of environmental, economic, and social limitations. Sustainable aquaculture growth must be gradual in a manner that is environmentally sustainable and that shelters the features of the environment for other users, while it is equally significant for people to guard the best of the environment for aquaculture. This chapter offers a brief review of the policy and strategies of the increasing sustainable aquaculture so as to be adopted by aquaculture businesses in Iraq, they build an environment-friendly aquaculture commercial. To conclude the chapter, a key suggestions are given, containing what should be the next applied steps.

46.1 Introduction

Aquaculture will be expected to have this important role due to the exploitation of the fisheries to their sustainable limit and beyond. On the negative side, aquaculture has become one of the main factors that can have an influence on the habitat and natural incomes that urge worries of both environmental activists and scientists (Naylor et al. 2000). Boyd (2003) has recognised the following utmost severe impacts, which can be seen through the devastation of swamps, and other subtle aquatic environment by aquaculture plans; changing of agricultural land to ponds; water pollution ensuing from pond wastes; extreme use of ground water and other freshwater provisions for filling ponds, and this concern has appealed to the greatest official attention in most nations (Boyd and Tucker 2000).

L. A. Jawad (✉)
School of Environmental and Animal Sciences, Unitec Institute of Technology, Auckland, New Zealand

F. Mutlak
Marine Science Centre, University of Basrah, Basrah, Iraq

L. A. Jawad (ed.), *Tigris and Euphrates Rivers: Their Environment from Headwaters to Mouth*, Aquatic Ecology Series 11,
https://doi.org/10.1007/978-3-030-57570-0_46

In different parts of the world, the initiation of aquaculture into areas conventionally used mainly for commercial fisheries and a variety of recreational events have occasionally concurred with eager user group struggle. To overwhelm this difference a prearranged, stable and comprehensive community method to rural economic and social development is necessary.

At the present time, there are many aquaculture projects all of which are managed by the private sectors. Such enterprises, when they were created, have no scientific requirements and rules of establishing the proper aquaculture business. These aquaculture corporations did not even follow any guidelines or policies in regard to the environment or the biodiversity. Therefore, giving the correct guidelines and policies for establishing a suitable aquaculture enterprises are essential for the future of the aquaculture industry in Iraq. Based on the status of aquaculture in Iraq, the scope of this chapter includes a short review of how to develop a sustainable aquaculture in Iraq to aid those who are interested in developing such businesses.

46.2 Sustainable Aquaculture

Sustainability signifies to the ability of a people, ecosystem, or any such on-going system to bear working into the infinite future deprived of being pushed into incapacity through exhaustion or overcapacity of key resources on which that system depends. In general, the notion of sustainable development is reasonable and imperative, but translating it into precise standards or measures is hard, frequently particular, and tainted.

46.3 Benefits of Aquaculture

Frankic and Hershner (2003) have stated the following welfares of aquaculture:

1. Upsurge household food source and advance nutrition.
2. Upsurge household income through variation of revenue and food supplies.
3. Reinforce marginal financial status by growing employment and decreasing foodcosts.
4. Enhance water supply and nutrient administration at household or communitystages.
5. Preserve aquatic variety through re-supplying, and improving of endangered species.
6. Decrease impact on fishery assets if done sustainably.
7. Adjusting/augmenting habitats.
8. Inspires research and technology growth.
9. Upsurge education and environmental consciousness.

46.4 Risks of Aquaculture

The following risks that can be resulted from aquaculture have been adopted from Frankic and Hershner (2003) with slight variation to match the aquaculture status in Iraq.

1. Sediment hypoxia/anoxia consequential from organic enhancement.
2. Carbon/nutrient upgrading of the water column and benthos.
3. Decreased rates of dissolved oxygen in a water column.
4. Chemical, pharmaceutical, and toxicant contributions to sediments and water column.
5. Concerns of 'redistributions', containing bioinvasions, pathogens, and disease.
6. Variations in trophic ('food web') relations and productivity.
7. Fluctuations in variety of plants and animals.

46.5 Codes of Conduct

A large number of governmental and non-governmental agencies have put forward guidelines for aquaculture. These strategies are set of regulatory codes comprising of broad statements about how administration and other working events should be engaged. Utmost aquaculture programs refer to those regulations concerning fisheries given by the Food and Agriculture Organization (FAO) of the United Nations (FAO 1997). The majority codes do not have any legal effect, and acceptance usually is intentional. Literally, rules may be recognised in situations where either governmental strategies do not occur or are not compulsory such the case of Iraq. The objective of following rules frequently is to show environmental care for drives of improving the image of an industry. This is surely one of the purposes in aquaculture.

46.6 Best Management Practices

A best management practices (BMP) are revealed to be the top obtainable and practical resources of avoiding a particular environmental influence while still allowing production to be done in an economically active manner. A scheme of several BMPs usually must be mounted to stop water pollution and achieve resource management aims. There has been wide application of BMPs in old-style agriculture to avert soil erosion and ensuing turbidity and sedimentation in streams, and other water bodies.

Amongst the significant BMPs those prerequisites to be stated here are: Erosion of pond watersheds, embankments, bottoms, and discharge canals may be a vital source of suspended soil particles in seepages. These particles can increase turbidity and cause sedimentation in receiptof waters.

The use of BMPs as a foundation for official governmental rules of aquaculture run-offs in developing countries seems more possible than use of effluent standards and permits. Though it would need considerable work to inflict a regulatory system based on BMPs, the volume of assets, manpower, and expertise would be much less than for administration of effluent permits with water quality criteria. It should be conceivable by an annual review by a competent professional to determine if BMPs have been applied on a farm. Certainly, specific guidelines requisite be documented about the actions to be taken in case of non-compliance, just as is necessary in enforcement of effluent permits (Frankic and Hershner (2003).

46.7 Recommendations

Grounded on current national and international determinations to endorse sustainable aquaculture, here is a list of main rules and strategies that should endure to be used, developed, and applied (Frankic and Hershner (2003)):

1. Creating combined coastal and rural/community administration plans (IUCN 2000).
2. Locating appropriateness standards and indicators for aquaculture.
3. Placing observing and assessment tools.
4. Creating quality standards (certifications) for environmentally friendly practice, dispensation and sale in the aquaculture industry.
5. Growing finfish aquaculture industry.
6. Developing united farming systems—integrated aquaculture–agriculture (FAO 2000).
7. Increasing widespread polyculture performs (Naylor et al. 2000).
8. Environmental observing reviews should use the best obtainable methods and technologies for the environmental monitoring of influences and modelling of carrying capacity at farm sites (Frankic 1998).

References

Boyd CE (2003) Guidelines for aquaculture effluent management at the farm-level. Aquaculture 226(1–4):101–112

Boyd CE, Tucker CS (2000) Rule-making for aquaculture effluents in the US. Global Aquaculture Advocate 3(6):81–82

Food and Agriculture Organization of the United Nations (FAO) (1997) Aquaculture development. FAO technical guidelines for responsible fisheries, vol 5. Food and Agriculture Organization of the United Nations, Rome. 40 pp

Food and Agriculture Organization of the United Nations (FAO) (2000) The state of world fisheries and aquaculture 2000. http://www.fao.org/DOCREP/003/X8002E/X8002E00.htm

Frankic A (1998) A framework for planning sustainable development in coastal regions: an island pilot project in Croatia. Dissertation. Virginia institute of Marine Science, College of William & Mary. http://www.vims.edu/library/Frankic/Frankic.pdf

Frankic A, Hershner C (2003) Sustainable aquaculture: developing the promise of aquaculture. Aquac Int 11(6):517–530

IUCN (2000) Coastal and Marine Biodiversity: policy recommendations. Fifth Meeting of the Subsidiary Body on Scientific, Technical and Technological Advice. Montreal, Canada, 31 January–4 February

Naylor RL, Goldburg RJ, Primavera JH (2000) Effect of aquaculture on world fish supplies. Nature 405(6790):1017–1024

Chapter 47
How Possible to Use the Desert Area in Iraq for Aquaculture Industry: Basic Facts and Recommondations

Laith A. Jawad and Saad Mohammed Saleh Abdulsamad

Abstract Iraq is located in an arid or semi-arid. Deserts expand over large part of Iraq's total area, utmost of which are approximately unoccupied. Within the expansion of the Iraqi deserts, groundwater is located in large quantity, each with their sole physical-chemical characters. Numerous of these water supplies may deliver provision for aquaculture purposes. Iraq has not started the practice of the desert aquaculture and in my opinion. It should enter this market so as to create an income besides the oil revenue and to provide protein to the increasing population in Iraq and the neighbouring countries. Unlike other countries, Iraq has no problem of suitable water sources supplying large volumes for the desert aquaculture. In addition to the finfish and crustaceans, Iraq can rear other aquatic organisms such as microalgae. Also, the aquaria food, Artemia and the ornamental fish species are another options for rearing in the desert area.

The present chapter contains a concise overview to desert and arid land aquaculture where the main growing restraints and chances are featured. The narrative of the profits and drawbacks of desert aquaculture along with the choice of appropriate farming species, offers extra data on the disputes confronted in the development of this aquaculture amenities. A number of recommondations have given at the end of this study, which comprise policies by those countries concerned in backup the creation of a desert aquaculture events.

L. A. Jawad (✉)
School of Environmental and Animal Sciences, Unitec Institute of Technology, Auckland, New Zealand

S. M. S. Abdulsamad
Department of Biology, College of Education for Pure Sciences, University of Basrah, Basrah, Iraq

L. A. Jawad (ed.), *Tigris and Euphrates Rivers: Their Environment from Headwaters to Mouth*, Aquatic Ecology Series 11,
https://doi.org/10.1007/978-3-030-57570-0_47

47.1 Introduction

In several aspects, desert areas are considered lands with usage and a source of different sorts of problems such as dust storms. In the last decades, the desert areas were tend to be so valuable and useful for human being. The aquaculture industry has turned these spaces to highly commercial valued areas by using them as desert aquaculture plants. The aquaculture in desert and arid lands has been mounting gradually in the previous years owing to the current technologies and other energy resources that have permitted water in these hard places to be developed more efficiently and more competently, using it for both crop irrigation and production of fish (Crespi and Lovatelli 2011).

With the desert and semi-desert areas, the conventional agriculture and livestock breeding are nearly impossible. Therefore, the availability of protein to human living in these areas is a major problem (FAO Fisheries Department 2011).

During the mid-60s of the last century, the idea of desert fish farming was emerged and tested experimentally, showing the possibility of using salt and brackish waters in the process of aquaculture (Fishelson and Loya 1969). The environmental conditions that are available in the desert areas such as the high mineral substance of these waters, along with high suitable temperatures and solar radiation, in fact backing high primary productivity forming a proper and appropriate food-base for the fish (Pruginin et al. 1988).

The availability of sources of water or transferring water to the sites of aquaculture in the desert areas is the major factor that impacted project of desert aquaculture in several arid and semi-arid countries. With Iraq, these two problems seem nearly solved with little of efforts. The water bodies in Iraq are located not far from any desert area. Therefore, it is possible to transfer water for aquaculture sites inside the desert by digging small branches or by pipes from the nearest main river. If this option is not possible, then transferring water to the aquaculture sites by tankers is possible as the loading points will not be far away from the desert aquaculture sites and the cost of transferring water is high, so the investors will not have problems about this issue.

There are no desert aquaculture plants in Iraq neither the government nor the private sector have thought about this possibility. In the present chapter, the basic aspects of the deserts aquaculture were given based on the experience gained by countries who have invested in this type of aquaculture. These basic information will be a guidelines for both the government and the private sector in Iraq to start such industry in Iraq following the recommondations given at the end of the chapter.

47.2 Appropriate Species for Desert Aquaculture

In the present time and with the advanced technology, large kind of organisms can be cultured in the desert areas providing the availability of the proper technology (Crespi and Lovatelli 2011). Before the step of selection of the organisms starts, there are certain measures need to be considered, these are: the species must be mainly liable to hyper-saline waters and large temperature fluctuations. It is also must be moderately fast growing species to confront water unfavourable conditions that are usually present in such habitats. The choice of the species is also influenced by other matters such as the obtainability of farm gain, market value and volume, local feeding and favourites and feeding behaviours. Presently, the most appropriate fish species for water-confined aquaculture systems contain the tilapias (*Oreochromis* spp.), barramundi or the Asian seabass (*Latescalcarifer*), carps and mullets (*Mugil cephalus* and *Liza ramada*) and numerous catfish's species (*Clarias gariepinus* and *Bagrus* spp.).

Among the best fish species to rear is the tilapias. This fish can be reared in both the mono and the polyculture system and even with other congruent species such as carp and mullet (Crespi and Lovatelli 2011). As to the shrimp, it seems that the Indian white prawn (*Penaeus indicus*) symbolises an explicit case of marine aquaculture production at large profitable level such as in the countries like Saudi Arabia. This shrimp has the ability to tolerate wide range of salinity levels. As to the European fish species used in aquaculture, the European seabass *Dicentrarchus labrax* and the gilthead seabream *Sparus auratus* seem to have a success in certain Middle East countries such as Egypt, where good results have been achieved. In the aquaculture of these marine species, an advanced technology and skills are required to achieve good results (Crespi and Lovatelli 2011).

In addition to finfish, there are several other aquatic organisms that some countries have given a try and go ahead in their aquaculture. Among these are the microalgae and the filamentous *Spirulina*, which are presently being used in aquaculture in numerous states.

With the difficulty in the choice of the species for rearing in a harsh conditions such as those of the arid areas, alien species are often introduced without undertaking adequate risk analysis valuations to evade possible adverse influences on local species and the environment (FAO 2007).

47.3 Benefits and the Drawbacks of Desert Aquaculture

There is a difficulty in drawing a line between the benefits and the drawbacks of aquaculture in the desert areas. In the areas where water supply is available, the aquaculture practices are uncomplicated and this is to a certain degree similar to the desert areas when groundwater is involved in this practices. The overall benefit of desert aquaculture projects is the final price that the reared fish can fetch in the local or the international markets. This final price is the net profit of the project after deducing all the costs that used in transporting materials and equipments prior to the rearing stages and transporting fish at the end of the rearing practices.

Crespi and Lovatelli (2011) have listed some of advantages that are obtainable from the aquaculture in the desert areas.

1. Large storage areas of fresh or brackish waters are usually found in desert and arid territories frequently only partly used for agriculture.
2. High quality water—reduced or low introduction risk of viral diseases and low hazard of pollution owing to lack or restricted industrial events.
3. Production all year round and the likelihood of growing highly priced off-season fish, vegetables and fruits.
4. Growing the competence of water use for the production of high value food products (e.g. fish, vegetables).
5. Plentiful low-cost land.
6. Incorporation of aquaculture with agriculture.
7. Likelihood of culturing aquatic organisms without jeopardising ecological systems or environmental balance.

47.3.1 Strategies Supporting the Desert Aquaculture

Crespi and Lovatelli (2011) have suggested a number of strategies that can support the desert aquaculture and should be considered in any such projects are plan to be done in the arid lands countries:

- Advertising of aquaculture farming schemes.
- Integration of aquaculture events with other present production schemes.
- Inventory and chemical analysis of available surface and subsurface water resources to facilitate selection of suitable farm sites and species to be cultured.
- Running of motivations for the creation, upgrading and modernisation of national feed treating plants.
- Advertising of national plans for the using of renewable energy supplies in isolated locations not aided by the national electricity services.
- Creation of national agendas for minimum data set collection to observe the standing position and tendency of this aquaculture business.

47.4 Types of Farming Systems

There are several types of rearing systems were tested and used by culturists all over the world. The followings are those reviewed by Partridge et al. (2008) and Allan et al. (2009):

Pond-Based Systems
According to Allan et al. (2001), this pond culture might be the utmost economically possible production scheme for inland aquaculture. Ponds are deliberated to be the lowest capital asset system with, frequently, the lowest upkeep expenses. Though, apart from biomass restriction, the possible areas for pond culture are situated in inland areas, where temperatures are high, such as ponds, differ meaningfully around the year (Partridge et al. 2008; Allan et al. 2009; Hutchinson and Flowers 2008).

Enclosed Tanks
This type of tank is designed for the culture *Artemia*. The tanks are totally surrounded, sideways from the manhole at the top. The system comprises an individual filtration system to each tank and an exclusive aeration method to upkeep high-oxygenated water at high salinities (Kolkovski 2010).

47.5 Recommondations

Crespi and Lovatelli (2011) have listed a basic requirements that need to be considered in planning for a desert aquaculture. For the benefit of the government and the private sector in Iraq, I here quote these basic requirements with slight variation to meet the requirements in Iraq and the other similar countries:

1. Education, training and communication. This can be attained over:
    - Improve education institutes with precise aquaculture programmes in their curricula.
    - Support revision courses and mentorship options as means of keeping an unceasing "knowledge" stream of desert and arid lands aquaculture education among investors.
    - Distribute the "how to" information for opening desert and arid lands aquaculture to possible owners as a reassurance of a supportable exercise linked with a positive revenue.
2. Investigation and growth: The necessity for new and/or modified techniques competent to endure the aquaculture circumstances in desert and arid lands is significant.
3. Clear policy and regulations: Desert and arid lands aquaculture is a comparatively new industry and, in numerous circumstances, present rules and guidelines allocating with arid land and obtainable water incomes use for aquaculture dedications are not yet developed or do not occur. Most frequently, certificate for desert and arid lands aquaculture actions are spread among several governmental offices. This situation may significantly restrain and/or even stop the development of this sector.
4. Environmental sustainability: Desert and arid lands aquaculture must be established by accepting strategies that guarantee environmental continual,

particularly through the use of environmentally sound techniques and suitable water administration.

5. Socio-economic aspects: The objective of growing desert and arid lands aquaculture is to protect food supply and protein sources, provide jobs, make business chances and recover incomes in desert and arid lands locations through the development of aquaculture events.
6. Market development and trade: Marketable aquaculture in desert and arid lands in general must be market-oriented for its continuity and well connected with the tendencies and dynamics of world seafood markets.

References

Allan GL, Banens B, Fielder S (2001) Developing commercial inland saline aquaculture in Australia: part 2. Resource inventory and assessment. NSW fisheries final report series 31:116

Allan GL, Fielder DS, Fitzsimmons KM, Applebaum SL, Raizada S, Burnell G, Allan G (2009) Inland saline aquaculture. In: Burnell G, Allan GL (eds) New technologies in aquaculture: improving production efficiency, quality and environmental management. Woodhead Publishing limited, pp 1118–1147

Crespi V, Lovatelli A (2011) Global desert aquaculture at a glance. *Aquaculture in desert and arid lands*, FAO Technical Workshop. 6–9 July 2010, Hermosillo, Mexico. FAO Fisheries and Aquaculture Proceedings No. 20. Rome, FAO. 2011a. 202 pp.

FAO (2007) Aquaculture development. 2. Health management for responsible movement of live aquatic animals. FAO Technical Guidelines for Responsible Fisheries. No. 5, Suppl. 2. Rome. 31 pp. (also available www. fao.org/docrep/010/a1108e/a1108e00.htm)

FAO Fisheries Department (2011) World aquaculture 2010. FAO Fisheries and Aquaculture Technical Paper. No. 500/1. Rome. 105 pp.

Fishelson L, Loya Y (1969) Experiments on rearing tilapia-hybrids in brackish water pond near the Dead Sea. Limnology 17:602–610

Hutchinson WG, Flowers TJ (2008) Research to foster investor attraction and establishment of commercial aquaculture parks aligned to major saline groundwater interception schemes in South Australia: SARDI Aquatic Sciences Report

Kolkovski S (2010) An overview on desert aquaculture in Australia. Paper presented at the aquaculture in desert and arid lands: development constraints and opportunities. FAO Technical Workshop

Partridge GI, Lymbery AJ, George RJ (2008) Finfish mariculture in inland Australia: a review of potential water sources, species, and production systems. J World Aquacult Soc 39:291–310

Pruginin Y, Fishelson L, Koren A (1988) Intensive tilapia farming in brackishwater from an Israeli desert aquifer. pp. 75–81. *In* R.S.V. Pullin, T. Bhukaswan, K. Tonguthai& J.L. Maclean (eds). The Second International Symposium on Tilapia in Aquaculture. ICLARM Conference Proceedings 15, 623 pp. Department of Fisheries, Bangkok, Thailand, and International Centre for Living Aquatic Resources Management, Manila, Philippines

Chapter 48
Aquaculture Industry in Iraq: Past, Present, and Future Perspectives

Nader A. Salman and Khalil I. Saleh

Abstract Recent review of the status of aquaculture in Iraq is presented in terms of previous activities, present practices, and future potential. The chapter described water resources that can be used for aquaculture in Iraq, regarding their availability, technical management, and search for new resources. The cultivated fish species were of special concern. Three types of carp are traditionally raised in Iraq along with limited trials to raise some local cyprinid species. More recently, efforts are made to culture commercial marine species and shrimp due to elevated salinity levels in Southern part of the country. Detailed description of the three main culture systems (Earthen ponds, Floating cages, and Recirculation aquaculture systems) is followed. Production scale and economic feasibility of the culture systems were reviewed in light of some recent research publications. It seems that the floating cages are the most beneficial and profitable projects at present. Two main aquaculture activities, mainly feeding and breeding were discussed under separate titles focusing on the obstacles and development trials to optimize both practices.

Research concerning the use of new feed stuffs which enhance growth and feeding conversion efficiencies are discussed. Trials to improve induced spawning and production of fish seeds are continuing in Iraq by aquaculture researchers, especially young postgraduate students. New hormonal treatments and many management techniques are suggested to maintain supply of fingerlings of cultivable species. Future plans to enhance aquaculture industry and to solve practical difficulties are also discussed in this chapter.

N. A. Salman (✉)
Al-Manara College for Medical Sciences, Maysan, Iraq
e-mail: naderabed@uomanara.edu.iq

K. I. Saleh
Babil Centre for Fisheries Consultancy and Development, Babil, Iraq

L. A. Jawad (ed.), *Tigris and Euphrates Rivers: Their Environment from Headwaters to Mouth*, Aquatic Ecology Series 11,
https://doi.org/10.1007/978-3-030-57570-0_48

48.1 Introduction

Despite richness of Mesopotamian environment which can support fisheries resources, fish production in Iraq has suffered from a progressive decline over the successive years. The inland fish production declined from 30,000 ton in 1984–1990 to 25,500 ton during 1994–1995, and then to 13,200 ton during 1995–2000 (Hussein 2011; AOAD 2013). AOAD (2016) has estimated fish production of Iraqi inland and marine fisheries to be 34,690 metric ton in 2015 with a potential of 40,000 metric ton when managed properly. Such production can contribute to only 15% of the Iraqi people demand which was estimated to be 300,000 metric ton yearly (Hussein 2011). Decline in natural fisheries production can be attributed to many factors such as shortage of water, deterioration of water quality, unsustainable fisheries management, and destruction of nursery grounds, in addition to the use of illegal fishing methods such as electric fishing and use of pesticides.

Per capita consumption has reduced proportionally from about 4 kg/year in 1984 to 1.8–2.0 kg/year during the nineties. It reached nearly 1 kg/year in 1997 (Hussein 1983; Balasim 1999). This rate is comparatively low when compared with international rate of 20 kg/year or even with the fish consumption rate in other Arab countries which averaged 11 kg/year in 2015 (AOAD 2017). This trend has recovered recently during the years (2011, 2012, and 2013) to reach 3–4 kg/year because of extensive fish import policy. Import of fish products increased progressively from 14,780 to 44,830, and then to 100,070 metric ton during the above-mentioned years, respectively (AOAD 2014).

Fish production from aquaculture, however, has improved to reach 34,780 metric ton in 2015 (AOAD 2016) compared with an average low production rate of 2000 metric ton during 1980–2000 (FAO 2009). Aquaculture activities seemed promising and could contribute more to satisfy the market need for fish and to fill the gap of fish production from the natural inland and marine fisheries. All requirements of successful aquaculture activities in Iraq are available. These include land, water, labor, and scientific experts.

The aim of the present chapter is to discuss the history, present status, and future perspectives of aquaculture industry in Iraq. It deals also with the main elements which govern the aquaculture industry requirements and tools of its advancement in the country.

48.2 Water Resources for Aquaculture

Water resources used for fish farming are mainly coming from rivers, lakes, marshes in addition to rain, underground, and marine waters (Fig. 48.1). If we look at the Iraqi map, we can find a wide variation of water resources that can be used to raise fish. Along the country, there exist the two great rivers Tigris and Euphrates and their tributaries, many natural lakes such as Habaniyah and Tharthar, a lot of dam

Fig. 48.1 Water resources that can be used for aquaculture. (**a**) Rivers (Tigris & Euphrates at Basra) (**b**) Lakes (Tharthar) (**c**) Drainage canals (Mid-Iraq)

reservoirs like Dukan, Darbandikhan, Qadisiyah, Mosul, and Audhaim, beside the vast wetland area of the central and southern marshes which occupy 700–800 thousand hectare (Delphy 2008). All these resources are widely used at present in fish farming activities, in addition to the limited use of underground waters. Marine and coastal resources, however, have been partially utilized for mariculture purposes, namely raising Zubaidy (*Pampus argenteus*) in late eighties of the past century and limited shrimp culture projects in the present time (Salman et al. 1991a, b).

These water bodies extending from Northern to Southern Iraq provide wide range of various aquatic ecosystems regarding diversity in water quality and environmental diversity. Cold water temperature, for example, at the mountain area of Northern Iraq near Iran and Turkey provides suitable conditions for raising cold-water species such as salmon and trout (Abdulrahman et al. 2017), while warm water temperature at the southern part is useful for raising warm water species. The same can be said regarding water salinity, which varied from 49.2 mg/l in the upper reaches of Tigris to 312 mg/l where joining Euphrates in the south. As for Euphrates, salinity changes from 109.5 mg/l at the upper reaches in Qaim to 778.9 mg/l at Basra (Central Statistical System CSS 2014). Values recorded by Marsh Research Center MRC (2012) for Euphrates salinity at Dhi Qar Province are higher reaching 2550–5292 mg/l. Salinity changes in Basra, especially in Shatt Al-Arab river, are widely fluctuated during various seasons depending on the water discharge from the Tigris, Euphrates and Karun rivers to the estuary region, in addition to the effect of the Gulf tidal current. Accordingly, fish culture projects need to select the suitable species that can tolerate water salinity at specific areas. Stenohaline freshwater species are raised in northern and middle areas of Iraq while euryhaline and marine species are suitable for the southern areas.

Quantity of water is another point to discuss in terms of rational water availability for fish culture projects. As mentioned above, the amount of freshwater reaching Iraq varied annually and seasonally. The Ministries of Agriculture and Water Resources have put restrictions on permissions for aquaculture projects since 1991 because of the limited and shortage of freshwater resources. They believe that aquaculture activities consume plenty of water, which is necessary for other agricultural, industrial, and humanitarian uses.

Results of a recent study concerning the consumption of water by various activities on Euphrates, Hargreaves (2012) found that aquaculture practices consume as little as 1.4% of Euphrates water (taking into consideration the evaporation and seepage effects). The financial revenue of these projects (Iraqi Dinar ID/cum) exceeds the revenue of rice production by 6–10 turns, 7–8 turns for corn, and 11–12 for wheat revenue. So, it seems that aquaculture projects are more profitable than crops production with limited water consumption compared with other agricultural products.

Adding to all that, another negligible source of water which is the drainage water of agriculture drainage channels which extends for 85,623 km (CSS 2009). The main channels are divided into three parts. The first extends for 130 km in length and 52 m in width from northern Baghdad to Al-Damlag marsh. Water of this part is nearly fresh, with salinity degree of 3 ppt in the main stream and 4–5.4 ppt at tributaries. There have been many fish ponds and hatcheries established on the banks of this part. The second part which extends from Damlag marsh to southern Dhi Qar city at a length of 187 km and 84 m in width, the salinity of this part is higher than the first part, it reaches 3787–5772 mg/l. The third part extends for 174 km with 110 m width from southern Dhi Qar to Shatt Al-Basra drainage canal (38 km) which is connected with Khor Al-Zubair canal, and then to the sea (MRC 2012).

The last source of water is the underground water which is estimated to be around 2 billion cum (Al-Rubaiey 2002). The use of underground waters is limited and only small scale projects of fish raising have been established in northern and mid part of Iraq. Salinity of these waters is low in the north but increased southward.

48.3 Cultivated Species

In the 1950s, carp species were firstly introduced for scientific research purposes. In Iraq, the main aim was to acclimatize this species in the Iraqi inland waters and to establish whether they would be suitable for rearing in the Iraqi environment without interference and without a negative impact on endemic species. However, at that time, this experience was not channeled into commercial activities. Later, significant attention was given to the aquaculture sector, initially with the establishment of hatcheries and the construction of fish farms (FAO 2011).

Freshwater fish production in Iraq consists of pond culture of the common carp, as well as the grass carp and the silver carp. There have been no initiatives to provide opportunities for the development of native fish production due to a limited supply of good quality fish seed, a lack of scientific knowledge and because native species are economically worthless to be produced or cultured in ponds. Such species require 4–5 years to reach marketable size (FAO 2011).

Fish cultivation practices in Iraq concentrate on fish species belonged to family Cyprinidae such as common carp (*Cyprinus carpio*), grass carp

(*Cetenophoryngodon idella*), and silver carp (*Hypophthalmichthys molitrix*) which comprise the main cultivated species in Iraq. Common carp was introduced to Iraq in 1955–1956 for experimental purposes. In 1965, after the establishment of Iraqi Fisheries Directory, the fish cultivation practices started to expand (FAO 1989). According to Al-Hamed (1960), the introduction of common carp started in 1955 by importing carp fry from Indonesia (Runten Carp) and Holland (Japanese Multi-color Carp). They were cultivated in earthen ponds at Zafarania Fish Farm near Baghdad. Upon successful natural and artificial breeding, carp fry were released into natural waters of lakes and marshes such as Hammar, Syneya, Habbania, and Tharthar for restocking purposes and to recruit the exits fish population in these water bodies. Common carp is considered now as the main economic fish species in the inland Iraqi waters and also as the main cultivated species. It is characterized by its fast growth, wide feeding spectrum, toleration to unfavorable environmental conditions, and acceptance by consumers. These characters are considered ideal for any cultivated species.

Grass carp was followed as a cultivable species and was introduced in 1968 from Japan (Shireman and Smith 1983; Al-Shekh et al. 1991). The first record of Silver carp in the Iraqi waters was in 1982 from a private fish farm and then in 1985, few number of brooders were imported for the Central Governmental Hatchery (Al-Wihda) which was established in 1984 (Saleh and Salman 1990). At the same time, brooders of Bighead carp (*Aristichthys nobilis*) were introduced at Alexandria Hatchery in 1986, but unfortunately, artificial spawning was not successful and the species vanished or mixed with the silver carp because of similarity.

Trials were practiced to adapt some local cyprinid species such as Bunni (*Barbussharpeyi*) and Gattan (*Barbus xanthopterus*) to be a cultivated species, but are not successful due to slow growth and high-production costs) Draft 1983; Al-Nasih 1992; Salman et al. 1997; Al-Rudainy et al. 1997). Quite recently, artificial spawning of Bunni was tried in the Marine Science Center for the purpose of restocking the marsh waters with Bunni fingerlings. Trials were successful and nearly 250,000 fingerlings were raised and released in the marshes by the year 2005 (Al-Noor et al. 2012).

As for the marine and euryhaline species, as mentioned above, very few attempts have been tried including raising of Zubaidy in tidal ponds at Khor Al-Zubair Lagoon (Salman et al. 1991a, b). Difficulties of adapting Zubaidy on low salinity levels in the ponds beside the unavailability of the suitable diet (which is mainly a kind of jelly fish) were the main obstacles (Salman 1993a). Shrimp culture has recently being adopted by private sector relying on successful raising of certain species of shrimp in Iran, especially the white-legged shrimp (*Litopenaeus vannamei*).

Due to the salination problem of Shatt Al-Arab and the marshes of Southern Iraq, people are trying these days to adapt certain marine and euryhaline species such as Shanak (*Acanthopagrus latus*) or Biah (*Liza klunzinger*), Sheam (*Acanthopagrus berda*), and Subaity seabream (*Sparidentex hasta*) to be the first marine species for ponds in the area as a replacement for the freshwater species which cannot tolerate the elevation in water salinity. Similar trials were adopted in a salt lake called

Razzaza (near Karbala Province), where fingerlings of Biah and Shanak were released during 1992–1993 (Mohamed et al. 2000, 2001). In the northern part of Iraq, few fish concrete longitudinal ponds have been constructed to suite rainbow trout imported from Iran (Abdulrahman et al. 2017) making use of the cold water of the area.

48.4 Aquaculture Systems and Production

In Iraq, three main aquaculture systems are practiced (Fig. 48.2):

1. Earthen pond culture.
2. Floating cages in rivers and lakes.
3. Recirculation Aquaculture Systems (RAS) or Closed systems.

Table 48.1 shows the registered and actually operating aquaculture projects belonging to the above systems along with their capacities in various Iraqi areas as recorded by the Department of Fisheries, Ministry of Agriculture in December, 2016. Beyond the table, there are a considerable unregistered acting farms established by private sector without governmental permission which contribute to Iraqi aquaculture production.

In 2007, the total production deriving from freshwater and marine aquaculture was estimated to be approximately 16,000 tons. The aquaculture sector is owned by the public and private sector. These two sectors are widely distributed in the middle and southern parts of Iraq. The annual production from aquaculture in Iraq according to FAO statistics for the years 1970–2010 is shown in Fig. 48.3 (FAO 2011). The highest production was recorded in 2010 reaching 20,000 tons. There was a progressive increase from the year 2005 till present. The report of the Arab Organization for Agricultural Development (AOAD 1996) was more optimistic as it divided the

Earthen ponds

Floating cages

Fig. 48.2 Two of the culture systems used in Iraq. (**a**) Earthen ponds (**b**) Floating cages

Table 48.1 Aquaculture projects in Iraq as recorded in 2016 by the Fisheries Department, Ministry of Agriculture, Iraq

Closed systems				Floating cages			Earthen ponds			
Under Const.	Active No.	Total Volume (m^3)	No. of Projects	Active No.	Area (Donum)	No. of Projects	Active No.	Area (Donum)	No. of Projects	Province
—	—	—	—	8	2584	8	92	2301	238	Baghdad Rusafa
2	1	470	1	47	17,532	64	95	2884	242	Baghdad Karkh
—	1	1370	3	73	40,601	102	69	5411	111	Babil
21	—	—	—	14	24,976	59	101	5480	143	Wasit
2	—	21,880	21	11	38,503	215	14	722	76	Misan
—	—	1269	2	11	6685	25	5	32	13	Dewania
1	—	228	—	2	832	7	5	637	22	Muthana
1	—	525	1	7	4926	12	3	119	6	Karbla
—	—	—	1	—	—	—	9	500	31	Kirkuk
—	—	—	—	28	10,054	68	18	255	36	Dhiqar
—	—	—	—	84	21,374	361	4	1005	44	Basra
—	—	—	—	1	1910	11	5	201	10	Najaf
					612	4				Salahldin
—	—	—	—	—	4791	16	—	—	—	Anbar
—	—	—	—	—	15	2	—	—	—	Ninawa
—	1	990	1	1	2178	5	—	371	39	Diyala
27	3	26,792	30	287	177,573	959	420	19,919	1011	Total

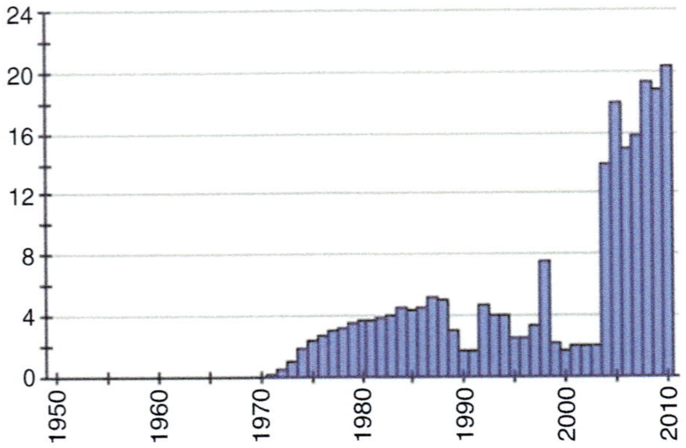

Fig. 48.3 Aquaculture production (thousand tons) in Iraq. Source: Fishery Statistics, Aquaculture production, FAO (2011)

Table 48.2 The estimated annual production of fish farms in Iraq from 2012–2015

Year	Annual production (Ton)
2012	24,070
2013	27,080
2014	24,600
2015	34,780

area into total 33,047 donum (8262 ha) and the aquatic area as 22,049 donum (5512 ha), and the production to be 26,000 ton per year. Table 48.2 shows the estimated annual production of Iraqi fish farms as recorded by AOAD (2016). The Iraqi Society of Fish Culturists ISFC (2018) estimated the production of Iraqi aquaculture projects in 2017 to reach 100,000–150,000 ton due to the establishment of thousands of illegal earthen ponds and fish cages fed by drainage water.

48.4.1 Earthen Pond System

The aquaculture industry started in Iraq as early as 1956 by a scientist called Dr. M. Al-Hamed who introduced carp, tilapia, and trout to Iraq as a cultivated species at Al-Zafaranya fish farm which belonged to the Ministry of Agriculture. After that few governmental farms were established in middle (Latifia) near Baghdad and northern part (Al-Riyadh) near Kirkuk during the seventies of the past century.

Aquaculture activities in earthen ponds continued to spread all over Iraq on small scale basis. It is not until 1982 when private sector started to establish large scale productive fish farms both as individuals or companies. Such development was initiated in the middle part of Iraq and then spread to the southern part. The earthen semi-intensive pond system was the most common system which was followed in all these farms to cultivate the three types of carp until 2010. Number and size of those aquaculture projects varied according to the cited reference and the estimated years. Most sources estimated pond farms in Iraq from 1991 till present by 1600–2700 with

Table 48.3 Annual production of pond farms from 1980–2010 in Iraq

Year	Annual Production (Ton)
1980	4000
1990	2000
2000	2000
2005	17,000
2010	21,000

a total area of 7500 hectare. The governmental record of 1609 farms occupying 31,000 Donum (1 Donum = 2500 m^2), only 886 are active (State Company for Animal Production Services, 2003). Al-Salem (2013), on the other hand, estimated the pond farms by 2712 with only 1509 active farms. The annual production (in Tons) of those farms for the period between 1980 and 2010 is shown in Table 48.3.

According to an old report of FAO (2001), the total area of fish ponds in Iraq was 7500 ha with annual production of 7500 ton. Such production rate (250 kg/donum) is very low and has underestimated Iraqi production from fish ponds which was estimated to be between 400–500 kg/donum or 1600–2000 kg/ha.

48.4.1.1 Pond Specifications

The design of the traditional earthen pond in Iraq includes digging a pond of 0.5 to 200 donum area, supplied with water either by gravity or by pumps. The traditional monoculture of common carp has been replaced by the polyculture technique of three types of carp. Both mono and polyculture practices of carp are tried at present. The polyculture technique includes the release of: 750–1000 common carp fingerlings, 150–200 silver carp fingerlings, and 100–150 grass carp fingerlings at the beginning of the culture season in March and continue cultivation till November or December. Organic fertilizers (cow manure) are usually used during the preparatory period by spreading on the pond bottom inlines at the amount of 400–500 kg/donum. Poultry manure is also used periodically to enrich the natural diet in the pond. Stocking starts in March by releasing a number of fingerlings of 1–2 g in weight and harvest at the end of the rearing season in January or February of 1 kg fish for marketing. Harvest in winter season may face the increased demand for fish due to shortage of fishing harvest from inland fisheries. Cultured fish can then be sold for good prices.

Supplementary feeding is a usual practice in earthen ponds. Fish is offered a 15–18% protein ratio as pellets, consist of barley, wheat, rice bran, corn, and poultry by-products. Some ponds feed only soaked barley grains. Protein ratios of 11–16% are satisfactory for producing 650–700 g fish, to produce fish of higher weights man need to raise the protein ratio gradually to create a balance between protein coming from natural and artificial diets and to fulfill the nutritional requirements of the fish (Al-Aqidy 2010).

48.4.1.2 Multi-Size Polyculture Technique

This technique has been used in the early Seventies by Chinese scientists in multi-grade pond culture system aiming at continuous harvest of fish during the rearing season (WPRIAP 1975). They intend to make use of differences in growth performance of various size-grades of carp to harvest those attained marketable table-size leaving the younger to grow in less crowded environment and bigger chance of feeding opportunities. Similar experimental trials have been practiced in the Fish Research Center (Zafarania, Baghdad) during the Nineties. Several multi-grade earthen ponds were used to grow fish at certain size in a high stocking density. Serial periodical transfer of fish of various sizes was practiced to control optimum stocking density aiming at producing table-size fish at the final stage (Al-Rudainy et al. 1999). Result showed the superiority of multi-size culture upon mono-size culture system in terms of growth performance, feed conversion efficiency, and production. Continuous harvest raised fish production rate to nearly double that of fish produced by the traditional mono-size culture.

Stocking practices have been changed since the year 2000 as farmer started to raise fish at bigger sizes called palm size (Kaffiat 200–250 g) at the aim of harvesting 2 kg fish for the market. They used such practices all over the year to have two harvest seasons. The first catch 300–500 g fish and the second for harvesting fish of more than 1 kg in weight. This technique leads to what is called multi-sized polyculture method (Al-Rudainy et al. 1999), which aimed at using all chains of natural food in the pond. In this technique, common carp is stocked at 100–200 g while silver and grass carp at bigger size of 400–500 g together with small fish of 5 g weight. Results of experimental rearing of polyculture multi-size trial are shown in Table 48.4 (Sabri 2006).

In multi-size culture system, common carp fish with more than 100 g in weight might create a problem of producing newly hatched larva via natural spawning during the spawning season within the rearing period. Such larva would develop into huge number of fingerling (ten of thousands) of nearly 20 g in weight, creating a case of overcrowding in the pond which exceeds the rational stocking density. It will affect the growth of the target fish. To solve such problem, fish culturists prefer to do intermediate fishing of these advanced fingerlings and use them for culture in other ponds or sell them to other fish culturists. This of course may lead to deterioration of the preferable genetic characteristics if repeated so many times.

Table 48.4 Growth of multisize earthen silver and grass carp in pond

Fish species	Initial Wt. (g)	Survival %	Final Wt. (g)	Daily Wt. Incr. (g)
Grass carp	5	61.7	504	2.21
Grass carp	450	84.1	2069	7.75
Silver carp	5	64.4	467	2.39
Silver carp	450	84.8	1623	5.65

48.4.1.3 Economic Feasibility

According to Saleh and Guber (2014), the most costly components of fish culture in Iraq is the artificial feed which reaches 33.76% of the total costs, followed by the cost of fingerling supply (30.84%). Permanent costs comprise 14.45%. Other parameters varied according to the total area of the aquaculture projects. Table 48.5 showed the estimation of these economic factors as recorded for an earthen ponds farm in Babil Province.

48.4.2 Closed Systems

Recirculation Aquaculture Systems (RAS) are recently introduced to Iraq after the crises of freshwater shortage to ensure efficient use of water resources. In 2017, a total of 30 RAS projects have been approved by the Ministry of Agriculture with a total water volume of 26,793 m^3. The projects rely on highly sophisticated instruments and need a continuous power supply which is not available all the time under present Iraqi conditions. Economic feasibility studies for those projects have exhibited negative points. Most of them are not profitable and either were closed or modified to operate as open systems. One of the reasons for failure was bad design which is not suitable to the reared species. Failure to remove waste products or inefficiency of the biological and mechanical filtration processes. It seems that the RAS systems are not suitable for a fish like common carp due to their feeding habits and huge amount of feces produced by the coiling intestine of the fish. RAS systems, however, are successful in Kurdistan (Northern Iraq) where salmonid species are cultivated. Some depend on recirculating well waters for this purpose (Salman and Rasheed 2017).

Table 48.5 Economic factors (million ID) for an earthen ponds farm in Babil Province

Parameters Area (donum)				
	>40	21–40	11–20	< 10
Permanent cost	42,3	–	5,5	3.7
Exchangeable costs	150,1	–	32,7	22.1
Total costs	192,4	–	38,2	25.9
T. Revenue	305,9	–	50,5	31,7
Financial profit	113,5	–	12,3	5,8
Profit of 1 ID	1,59	–	1,31	1,22
Total added value	155,8	–	17,8	9,5

1 USD = 1200 ID

48.4.3 *Floating Cages*

Iraqi water resources can provide the basic requirements for successful fish culture in cages. Environmental conditions may maintain sustainable growth rate for the whole year. This has encouraged the investors and fish culturists to adopt such technique and make use of its advantages. Fish culture in cages has started in Habbaniya Lake early in the eighties, but was not successful for many reasons (Al-Daham 1990). Common carp, Bunni and Gattan fingerlings were cultivated in Razzaza and Tharthar Lakes in cages to monitor their growth rate before releasing into the lakes (Dratf 1982). In 1986, Saleh and Suliman (1988) tried cage culture for common carp in the drainage canals. Salman et al. (1997) used cage culture for common carp in the Northern part of the main drainage canal. The numbers of cage projects in Iraq are shown in Table 48.6 as recorded by the Ministry of Agriculture. Al-Salem (2013) recorded higher number of cages in Babil Province approaching 84 projects with 844 cages. These projects have been established without the permission of the Ministry. The same author summarized the production and economic analysis for those projects as seen in Tables 48.7 and 48.8).

The floating cages culture system is considered now as the main culture system, because the environmental conditions are not suitable for the other two systems. Cage culture industry relies on raw materials that are locally available. They include, frames, bridges, passage ways, nets, floating materials, feeders, and wave breakers. Most of the cage units are locally manufactured from steel or wooden frame in square or rectangular shapes with the dimensions of: $3 \times 3 \times 2$, $2 \times 3 \times 4$, $2 \times 4 \times 4$, $2 \times 3 \times 6$, and $2 \times 4 \times 6$ m. Some circular and octahedral units are imported from Turkey and used in Kurdistan and Mosul. Two kinds of nets are usually used, the external sets for protection and are made of steel BRC covered with plastic. The internal net which accommodates fish is made of nylon of 1 inch diameter. It should be flexible to allow cleaning, sampling, and harvesting. Floating

Table 48.6 Number of cage culture projects in Iraqi Provinces with their total volume

Province	T. Volume m^3	Cages No.	Permissions
Basra	6452	293	130
Misan	16,890	625	24
Wasit	10,992	338	20
Babil	11,040	338	11
Dhiqar	1118	45	10
Qadisya	5750	195	8
Anbar	2900	105	5
Najaf	1028	21	2
Muthana	512	8	2
Salahuldin	522	6	2
Karbala	600	25	1
Mosul	12	1	1
Total	57,816	2000	216

Source: Iraqi Ministry of Agriculture; Department of Fisheries

Table 48.7 Production analysis of cage culture project

Parameters	Average	Range
Initial wt. (g)	124.3	250-50
Rearing period (days)	154	290-90
Production (kg)/cage	1701.78	4500-450
Production kg/m^3	51.47	97.59-12.07
Food conversion ratio	2.68	6.11-1.5
Relative growth (%)	788.31	1775-250
Specific growth (%g/day)	0.655	1,22-0.25
Daily increment (g/day)	6.6	11-2
% mortality	7.67	25-0.5

Al-Salem (2013)

Table 48.8 Economic analysis of the revenue (Million ID) for floating cages of various sizes

Groups Parameters	>200 m^3	120–200 m^3	<120 m^3
T. Capital	43.939	19.62	33.711
T. Revenue	74.128	24.76	49.937
Financial flow	30.189	5.14	16.227
ID profit	1.687	1.26	1.48
Investment profit	0.687	0.26	0.48

Al-Salem (2013)

materials are also used in the form of barrels or polystyrene. Feeding is performed by feeding tables, mechanical feeders or demand feeders. Protection from severe water current is important and is done by several types of wave breakers.

Researches on cage culture continued to optimize parameters such as stocking density, feeding, and stocking sizes (Salman 2013; Al-Shemmary and Saleh 2014; Al-Shemmary 2015; Abbas et al. 2016). Stocking density of fish in the cages was fixed at 60–70 fish per m^3 or 1500–2000 fish per cage of 1.5 × 4 × 4 m (Al-Janabi 2014).

48.4.4 *Fish Farming in the Marshes*

Fish farming represents an additional activity which may increase fish production in the area along with fishing wild fish (Al-Mukhtar et al. 2005). Selected suitable farming practices might be suggested to help the people of the marshes economically and fighting unemployment. Most of these practices are easy to construct besides being very profitable. Aquaculture systems suitable for extensive and semi-intensive fish farming in the marsh were previously suggested by Salman (1993, 2011). They include:

1. Ponds with transect dikes which can maintain adequate water supply for ponds

with larger depth and improve water depth in shallow area, so that it can be used as fish ponds.
2. Ponds with peripheral dikes and ditches are suitable for extensive (large area & low cost) fish culture, by increasing water depth than the surrounding for growing field crops during dry season into an appropriate production cycle between fish and crops water body. Ponds surrounded by peripheral dikes may also be used.
3. Fish cum rice culture style of South East Asia, Production conditions are not encouraging to start such trials but it can be used when the rice cultivation in the marsh returns back as it did in the past.
4. Fish cages and net enclosures which serve as intensive or semi-intensive rearing systems which can be practiced in the marshy area with low cost as a family owned farm. Locally available fishing nets and reeds can be used to build such low cost enclosures which serve both nursing and growing purposes (Salman 2013).

48.5 Fish Feeding and Feed Plants

One of the basic requirements of fish farming industry is the availability of sufficient supplementary feedings to increase fish production. In Iraq, there are more than 160 general feed plants producing 3–25 ton/h and about 17 fish feed plants (10–25 ton/h), three of them produce floating pellets. Most of the feed ingredients are produced locally. They include, corn, barley, wheat, rice bran, broad bean, poultry by-products. Soybean cake and vitamin-mineral premix are also important ingredients that are imported by the government. Research to develop new unconventional feed stuff continued, but little are actually applied (Salman 1993b; Saleh et al. 1995; Al-Shama'a et al. 1999a, b, 1999; Al-Asha'ab 2002; Saleh 2000, 2001).

After 2003, after the expansion of foreign trade in the country, new ingredients became available such as protein concentrates (provimicarp 5‹hammor), vitamin-mineral premix, amino acids additives, antitoxins, enzymes, pro and prebiotics from foreign sources. Due to taxes, transportation costs and merchants speculation, prices of these materials increase by 125–150 $/ton (Inma 2009). For cage culture, floating pellets are imported from Iran and Saudi Arabia, which cost more (750–1400 $/ton) compared with local pellets (500–600 $/ton). Floating pellets, however, enhance better growth reaching 2100 g compared with the local submerged pellets which produce only 1500 g fish after rearing 200 g fish for 125 day (Al-Shemmary and Saleh 2014).

Feed additives proved to be beneficial when added to the basic ingredients, as seen from the study of Al-Jubory and Saleh (2017) who added locally prepared pro and prebiotic along with an enzymatic preparation to the basic ration. Results showed improvement of survival, growth, and feed conversion efficiency as seen in Table 48.9.

Table 48.9 Effects of feed additives on survival, growth and conversion of common carp (*cyprinus carpio* L)

Parameters	Ration with no additives	Ration with additives
Mortality	7.9	4.2
Daily gain wt. (g.)	2.38	4.57
F.C.R.	4.4	2.85
Sp. growth (%g./day)	1.28	1.43

48.6 Hatcheries and Fish Seeds Production

The first attempt for artificial spawning of carp was by using the Kakaban hatching tools at Zafarania Farm in Baghdad in 1956. Then induced spawning using pituitary gland extract was tried in 1976 (Rahim 1999). The large-scale artificial spawning was conducted in Al-Wihda Central Hatchery at Suwera in 1982, Babil Fish Farm (500 ha) and Alexandria Private Farm (500 ha). Soon after, a lot of private and governmental hatcheries were established for the seed production of carp (common, silver, and grass). Some efforts have been done for culture of Bunnei in Iraq (Ali et al. 1985; Ali 1986; Farga and Chabbaq 1988; Al-Nasih 1992; Al-Noor et al. 2012). Number of fish hatcheries has increased rapidly from 25 in 2003 to 50 in 2010 (Hussein 2011). The Arab Organization of Agricultural Development (AOAD 2016) estimated the production to increase from 12.8, 23.5, and 21.6 million fingerlings in 2011, 2012, and 2013, respectively to 55.35 million in 2014. The present number of hatcheries for 2017 is 101 hatchery distributed along Iraqi provinces mainly in Babil (61) and Baghdad (14) (Iraqi Soc. Fish Cult. 2018).

48.6.1 *Hatchery Specifications*

According to Al-Himeary (2011) and Saleh (2015), the total water area of the hatcheries ranged between 23–200 Donum. They consist of external ponds for brooder, buildings for internal tanks, and hatching units. Each consists of 20 jars (7 L) regularly arranged on steel stand. The hatchery usually contains 1–8 units (20–160 hatching jars), in addition to 16–48 incubators (200 L) for larval rearing. The 102 hatcheries in Iraq contain 3780 hatching jars producing around 200 million fingerlings of the three types of carp.

48.6.2 *The Brooder Stock*

The stock of brooders used for induced spawning are Hungarian brooders originally introduced in 1982 which kept its original genetic characteristics for years by internal breeding in the same projects (Naif 2005; Khuleaf 2006; Al-Hilli 2009). It is vital to conserve pure stock of brooders with preferable characters. Failing to do that by some farm owners has resulted in obvious deterioration of spawner's quality. Unfavorable characters such as poor growth, body elongation, reduced disease resistance, reduced feed conversion efficiency, and uncontrolled breeding have separated among certain unprotected farms. As a trial to improve the stock, carp fingerlings of 50 g each have been introduced to Al-Foratand Alexandria Fish Culture Companies by INMA Organization (USAID). They were raised in pond to reach 2 kg in weight and then distributed as brooders to other farms and hatcheries. They had crossbreeding with the old Hungarian brooders to produce new generation with high-quality characters (Hargeaves et al. 2012).

The hybridization between the two strains produces two generations of better quality than the original strains (AL-Jubory and Saleh 2012). Brooders stack is raised in small ponds (0.5–2 ha) with 1.5–2.5 m depth, contributes to only 6–8% of the total hatchery area. They were stocked at the rate of 131–339 fish/donum at biomass of 530–1625 kg/donum. Such stocking rate is high and it is beyond the need of the hatchery for spawning. Brooders need better stocking and feeding rhythm along with suitable environmental conditions. The favorable weight of common carp brooders in Iraqi hatcheries ranges from 3–5 kg for female, 4.5–6.0 kg for grass carp, and 5–7 kg for silver carp. Males are usually smaller weighing 3.5–4.5 kg for common carp, 4.5–6.0 kg for grass carp, and 4.5–6.5 kg for silver carp (Naef, 2005; Al-Himeary 2011).

Brood stock needs special care before and after breeding season regarding feeding and incubation. Feeding at the rate of 0.37–0.6 kg/fish has resulted in poor egg production (13% of body weight) instead of 20–30%. Before spawning, mixed sex and mixed species brood stock are usually kept in the same pond. Silver and grass carp wouldnot breed upon mixing males and females. For common carp, there is a risk of prebreeding in those ponds upon elevated water temperature (>18 °C). This temperature can be achieved during 15–20 March in southern part of Iraq. It could be later than that (April) in the colder area of Kurdistan in Northern Iraq.

On February, they were sexed and transferred to shallow ponds of 1–1.25 m depth for maturation. Care should be taken to prevent spawning in these tanks upon thermal rise. For that, it is better to stock common carp females first in well-circulated deep ponds (3 m depth) and then to maturation pond on batches system. Batches are transferred to the hatchery as soon as possible when fish maturity reaches its optimum level (Saleh and Sulaiman 1992).

48.6.3 Induced Spawning Procedure

Most Iraqi hatcheries followed the technique of Woynarovich and Horvath (1980) of using the pituitary gland P.G extract for induced spawning (Fig. 48.4).The brooders were injected either with the P.G extract or Ovaprime hormonal preparation. The Ovaprime was used, with a dosage of 0.5 ml/kg for female and 0.25 ml/kg for male

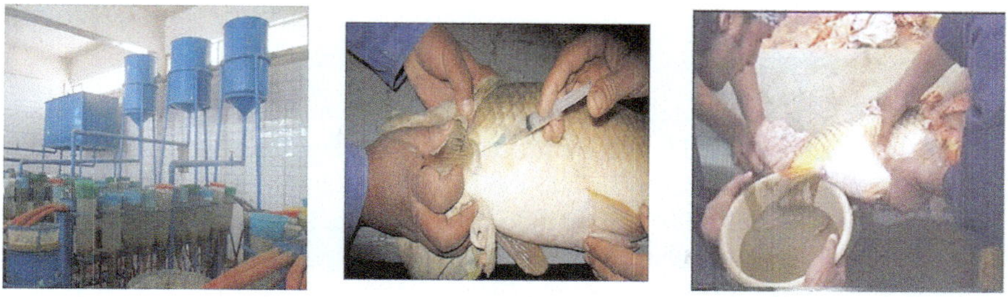

Fig. 48.4 Steps of induced spawning in one of the carp hatcheries in Iraq

in one injection. It has been found that eggs can be obtained after 12–14 hrs of injection, and the smaller size brooder showed more positive response for hormonal administration. Injecting Ovaprime hormones caused 5.05% mortality among fish with more than 1 kg weight. It appeared that the usage of Ovaprime is more easier than PG extract, because it decreased the handling process of the brooders, as well as the time required for each spawning batch. P.G extract, on the other hand, was injected in two dosages. The first (preparatory dosage) was 0.3–0.5 mg/kg for females only. After 12 hr., the second injection (final dosage) was given to common carp female 3–3.5 mg/kg and half dose to male (for Grass & Silver carp, the quantity increases to 5–6 mg/kg for female and half dose to male), depending on water temperature. The behavior of the injected brooders was monitored continuously. After 12–13 hrs from the second injection, the stripping process was conducted to get the sexual products.

It was found that, the most suitable period for getting sex products extended from the first week of March till the last week of May. Dry fertilization method is more suitable, which mean the mixing of the sex products without water. Mixing was done by aid of feather for few seconds, and then the fertilization solution added, just covering the eggs. The fertilization solution had the same composition of the washing solution. The fertilized ova were washed with washing solution (160 NaCl +120 Urea +40 litter water from brooders tanks) for 10 minute with this solution followed by washing with water for 10 minute. This sequence was repeated three times. Then the fertilized ova were washed with tannin solution (10 gm/10 litter water) for 30 seconds followed by washing with water for 30 seconds. The fertilized ova were incubated in Zougar Jars (10 L) with an incubation density of 50–100 g for each jar. Water flow rate was 600 ml/min, just enough for ova circulation. The incubation period extended from 72–96 hr. at water temperature of 22–24°c. Dissolved oxygen concentration in the hatching jars approaching saturation at 10–12 ppm. For hatching, eggs of common carp need thermal aggregation level of 60–72 °C, while silver and grass carp need 24–36 hour for hatching under 22–24 °C. During recent years, successful efforts have been conducted to replace the imported PG extract by other artificial pharmacology materials (Shaker 2006; Khalaf 2009; Al-Hilli 2013; Asal 2015; Al-Tae 2017).

48.6.4 Larviculture

Hatched larva was transported to incubation tanks (0.75 m^3) till complete absorption of yolk sac which extended for 24–36 hr. At postlarval stage, they were transferred to indoor rearing tanks and fed with live food (rotifers nauplii). The total number of larva produced was 600,000 with 40–60% survival rate. For fry production, indoor open water system has been constructed for rearing post larvae which contains 20 tanks (260–300 l. each) rearing density was 100 larvae/l. Live food was used for feeding the post larvae, which includes rotifers (***Brachionus sp.***) and Artemia (***Artemia salina***) nauplii. In the first week, the larval feeding was restricted to rotifers only. From the second week and on Artemia nauplii were added to the feeding regime, with small amount of artificial supplementary food during the 4th week supplied twice daily. The amount depends on number of post larvae in the tanks. In some hatcheries, they used only the liquidized egg yolk to feed carp larva during internal incubation for the first 6–10 days. They were also fed soybean cake and wheat flour as powder soaked with water as a liquid extract distributed in the pond near the banks every 2–3 hours. Survival rates in those ponds are low ranging between 10–20% due to bad preparation of nursery ponds and to high stocking density (Al-Haider 2008).

According to Saleh and Al-Gezali (2014), larva can be reared in external prefertilized earthen ponds (0.5–2 donum) with stocking density of 0.8–1.0 million larva/donum after the absorbance of their air sacs (after 3–4 days in internal incubation). Ponds should be filled with water 5–7 days before introducing larva. They must be treated with 1–2 ppm of Malathion insecticide to eliminate fish enemies (Saleh and Salman 1990) and fertilized with poultry manure at the rate of 250 kg/donum (Al-Amin 2001). After 10–14 days, the seeds are transported to another well-prepared external ponds at bigger size with lower stocking density, to ensure higher survival rate (more than 50%) (Al-Gezali and Saleh 2013).

48.7 Fish Marketing

In Iraq, there are central markets for fish sale in each province (Baghdad 10, Babil 6 and Kurdistan 6), they are locally called “Alawy” (Fig. 48.5). Fish harvested from fish farms is sold in these markets via wholesale system. Prices depend on the arrival time, sizes, culture system, places beside the basic offer, and demand system. The wholesale markets, originally owned by government, are now rented to private sector that controls management and force sale tax of 3–5% of the income. The markets may also be used for retailed sale in some places. Retailed prices are higher than the wholesale prices by 15–25%. Some batches of harvested fish are sold live, as preferred by special consumers with 25% rise in prices than dead fish. Prices of fish fluctuated seasonally from 10,000–12,000 ID/kg to 5000–6000 ID/kg, and to 3500 ID/kg in January 2018. Fluctuations are due to harvest seasons and to fish

Fig. 48.5 Two of the marketing sites for retail and wholesale of cultured live fish

imported from neighboring countries such as I.R. Iran which are sold in competitive prices and affect the income of fish producers in Iraq.

48.7.1 Economical Aspects

Despite being operated by private sector, aquaculture industry in Iraq needs to be planned, controlled, and developed by the governmental authorities, as seen by other countries (Unong 1998; Modandi 2001). This may enhance the production of fish from aquaculture activities by offering loans for initiating aquaculture projects like floating fish cages which are promising and also through monitoring the progress of such projects. Some successful government intervention is seen in Oman and Saudi Arabia (Towers 2014).

48.8 Future Steps

Aquaculture activities in Iraq faced many problems:

1. Local people are not well aware that aquaculture could be a good activity to improve their socioeconomic status.
2. In fish hatcheries, the production is very low due to low survival rate of larva.
3. The majority of feed stuffs used are imported with high prices which increase feed and production costs.
4. The water quality is changing and becoming more saline in the Southern part, which has negative impact on the growth rate of the cultured fresh water species.

To improve aquaculture, especially in the southern part of Iraq, all these parameters should be considered to plan an effective management.

Solutions can be summarized as follows:

1. Making use of the naturally occurred water impoundments in the southern marshes by construction of peripheral and transaction dikes to create extensive fish ponds. Rearing fish in fish rice fields can also be practiced in certain areas in the marshlands.
2. Making use of the lakes and water reservoirs near the dams of Mosul, Tharthar, Audhaim, Dukan and Darbandikhan by releasing fingerlings or increasing fish cages projects.
3. Encouraging the construction of fish cages and other culture systems in the main drainage canals in Iraq in which salinity levels are suitable for certain fish species such as carp and tilapia.
4. Fry and fingerlings supply should be maintained through establishing new local hatcheries and improve hatchery production techniques to increase fingerlings yields of all carp species (common, grass, and silver carp) in addition to some native species such as Bunni (B. sharpey) and Gattan (*Barbus xanthopterus*).
5. The introduction of new fish species that are more tolerable to the present water quality such as euryhaline species Shanak (*Acanthopagrus latus*), Biah (*Liza klunzinger*), Sheam (*Acanthopagrus berda*), and Subaityseabream (*Sparidentex hasta*) or some commercial shrimp species such as the white-legged shrimp (*Litopenaeus vannamei*).

There are big chances of success especially with the availability of sufficient amount of very productive water resources and cheap work power, but research for optimizing stocking and harvesting is essential.

References

Abbas LM, Wahab NK, Salman NA, Abu-Elheni KJ (2016) Effect of stocking density and partitioning of rearing period on growth, feed utilization and production of common carp *Cyprinus carpio* raised in floating cages. Am J Exp Agr 11(4):1–7

Abdulrahman NM, Ahmed VM, Ameen HJ, Salih HA, Latif BM, Ali SS, Ahmed AA (2017) Comparative study of on fish health indices of rainbow trout (Oncorhynchus mykiss) reared in two sites in Sulaimani governorate, north of Iraq. Biol Appl Env Res 1(2):228–236

Al-Amin MA (2001) Use of different technique for production of seeds of grass carp *Ctenopharyngodon idella* val. and silver carp *Hypophthalmichthys molitrix* (val.). PhD Thesis, Univ. of Basrah, Coll. of Agriculture. 96p

Al-Aqidy HG (2010) The effect of gradual replacement of a high protein aliment substitutes another low protein aliment on carp production. MSc. Thesis, Al-Musaib Technical College, Foundation of Technical Education.145 p

Al-Asha'ab MS (2002) The possibility of using corn germ meal as a protein source in common carp *Cyprinus carpio* L. diets. Msc. Thesis Al-Anbar University. 86 p

Al-Daham NK (1990) Fish culture. College of Agriculture, University of Basrah. Dar Alhikmapubl. 481 p

Al-Gezali AR, Saleh KI (2013) Using a mixture of natural food for feeding common carp larvae to obtain a highest survival rate during outside incubation stage. Sci JDhiQar Univ 7:14–29

Al-Haider SM (2008) Studying the biotic and abiotic factors affecting in survival rate of larvae and fingerlings rearing ponds at al-Furat company–Babylon. MSc. Thesis, Al-Musayab Technical College, Foundation of Technical Education. 126 p
Al-Hamed MI (1960) Carp culture in Iraq. Iraqi J Agr Res 1(2):14–23
Al-Hilli AM (2009) Appointing of genetic and phenotypic variation for some traits of common carp *(Cyprinus carpio).*Thesis of high diploma degree. Al-Musayab Technical College, Foundation of Technical Education. 75 p
Al-Hilli AM (2013) Use of Domperidon, Clomifene, Dexamethazone and Human Chorionic Gonadotropin for hormonal induction in Common Carp and Grass Carp. MSc. Thesis, Al-Musayab Technical College, Foundation of Technical Education. 98 p
Al-Himeary KIM (2011) Technical and economical evaluation of fish hatcheries of Babylon Province. MSc. Thesis, Al-Musayab Tech Coll, Found Tech Edu. 148p
Ali MA (1986) Study of propagation of the local fishes in controlled condition. Report submitted to Dept of Fish Studies & Research, Baghdad
Ali MA, Abdul Hussein JS, Dulfi MA, Adul AS, Faried LH (1985) Propagation of the commercial fish of Iraq. Report submitted to Dept. of Fish Studies & Research. Baghdad
Al-Janabi MF (2014) Effect of different densities and aliment protein perventage on growth of common carp *Cyprinus carpio* L. cultured in floating cages, Babilon. MSc. Thesis, Al-Musayab Tech Coll, Found Tech Edu. 122 p
AL-Jubory MO, Saleh KI (2012) Evaluation of some productive and reproductive crossbreeding of two different lines of common carp (*Cyprinus carpio L.*). J Al-Qadisiya Vet Sci 11(3):24–34
Al-Jubory BS, Saleh KI (2017) Effect of enzymatic-bacterial mixture, yeast and Thepax on the productive performance of common carp (*Cyprinus carpio L.*) reared in floating cages. J Babylon Univ Pure & Appl Sci 25(4):16–28
Al-Mukhtar MA, Salman NA, Husunei KH, Al-Janabee AH (2005) Fish farming in Basrah governorate. Iraqi J Aquaculture 2(2):155–164
Al-Nasih MH (1992) Preliminary observations related to the culture of *Barbussharpeyi.* J Aqua Trop 7:69–78
Al-Noor SS, Al-Nukhtar MA, Salman NA, Ali MH (2012) Improvement of induced spawning of Bunnei (Barbussharpeyi, Gunther, 1874) in Iraqi hatcheries. Proceeding of First International Larviculture Conference in Iran, Karaj
Al-Rubaiey H (2002) The underground water of Arab countries. Daar Al-Hikma Pub, Damascus, Syria
Al-Rudainy AJ, Salman NA, Al-Mahdawi GJ, Ruhaig AM, Fadil AA, Farhan RK, Abbas IK (1997) Intensive culture of Gattan (*Barbus xanthopterus*) in earthern ponds. Basrah J Agr Sci 10 (1):25–31
Al-Rudainy AJ, Salman NA, Ruhaig AM, Abul-Hani AJ, Mosa KM (1999) Effect of multi-size culture on growth, food conversion and production of common carp. The Scientific Journal of Iraqi Atomic Energy Commission 4
Al-Salem AF (2013) Technical & economical evaluation of fish cage culture projects in Babylon province MSc. Thesis, Al-Musyab Technical College, Foundation of Technical Education. 181 p
Al-Shama'a AA, Al-Asha'b MH, Saleh KI, Salman AH, Al-Fathly K, Rasheed L, Yousif OS (1999) Use different levels of sunflower meal as a replacement of barley and maize in the feeding of common carp. Al-Techany J Agr Res 56:91–98
Al-Shama'a AA, Salman AH, Saleh KI, Al-Asha'b MH, Adnan MM, Yousif OS (1999a) Use of brewery byproducts in the feeding of common carp. Al-Techany J Agr Res 53:113–121
Al-Shama'a AA, Salman AH, Saleh KI, Al-Asha'b MH, Ahmed AS, Chiad AS (1999b) Use of tomato processing wastes (without seeds) as a replacement of maize in feeding of common carp. Al-Techany J Agr Res 60:200–210
Al-Sheikh MA, Yamur QM, Alloos SB (1991) Fish culture and production. Part II. Al-Hikma Press, Baghdad

Al-Shemmary SM (2015) Effect of initial weight and ration quality on the productive performance of common carp *Cyprinu scarpio* L. reared in floating cages. MSc. Thesis, Al-Musayab Tech Coll, Found Tech Edu. 181p

Al-Shemmary SM, Saleh KI (2014) Technical and economical comparison between sinking and floating aliments used in fish cages projects. Euphrates J Agr Sci 6(4):13–24

Al-Tae BS (2017) Using follicular fluid and some synthetic hormonal materials to induce artificial spawning in the females carp fish. MSc. Thesis, Al-Musayab Technical College, Foundation of Technical Education, 70 p

AOAD (1996) Actual state of aquaculture in Arab countries and its development. Journal of Fish Resources 3:15–17

AOAD (2013) Arab fishery statistics yearbook -Vol 6. Sudan, Khartom

AOAD (2014) Arab fishery statistics yearbook -Vol 9. Sudan, Khartom

AOAD (2016) Arab agricultural statistics yearbook-Vol 36. Sudan, Khartom

AOAD (2017) Arab agricultural statistics yearbook -Vol 37. Sudan, Khartom

Asal ZF (2015) Using of alternatives of pituitary gland hormones in common carp *Cyprinus caprio*& grass carp *Ctenophayngodon idella* artificial reproduction. 72 p

Balasim AN (1999) Fisheries resources of Iraq and role in food security. J Agr Dev 4:43–49

Central Statistical System (CSS) (2009) Agricultural Atlas–the water–water resources report for the year 2009. Agricultural statistic office. Ministry of Planning, Iraq

Central Statistical System (CSS) (2014) Iraqi ecological statistic report. Ministry of Planning, Iraq

Delphy M (2008) Fish production in Iraq. Al-Sabah Newspaper, No. 1410, p. 9

Draft (1982) Detailed report on the development of fisheries in Thartar, Habbaniya and Razzazah lakes. Part. XC: Cage fattening of carp fry lake Razzazah. Inland Fisheries Institute, Poland

Draft (1983) Detailed report of developing fisheries in Tharthar, Habbaniyah and Razzazah. Part X: Production of Bunni and Gattan stocking materials in Tharthar Lake and fattening of carp in Lake Habbaniyah. Inland Fisheries Institute, Poland

FAO (1989) Aquaculture development and coordination programmer (ADCP). Aquaculture Minutes.NO. 4; 1989

FAO (2001) World fish consumption and trade. The consumption in Iraq, FAO, Fishery Information Data and Statistics Unit, Rome. 130p

FAO (2009) The state of world fisheries and aquaculture. FAO Fisheries Department Report, Rome. 176p

FAO (2011) The state of world fisheries and Aquaculture. FAO Fisheries and Aquaculture. Department, Rome

Farga A, Chabbaq FM (1988) Induced spawning technique of three Iraqi fishes Bunnei (*Barbussharpeyi*), Shbout (*Barbus grypus*) and Gattan (*Barbus xanthopterus*). Al-Wihda Fish Hatchery, Departmental Report

Hargeaves J. Stone D, Saleh K (2012) Rebuilding the cultured fish value chain in Iraq. AQUA, global aquaculture,securing and future, Prague, Sep. 1-5

Hargreaves J (2012) Water resources use for aquaculture in Iraq. USAID Inma Agribusiness Program. 20 p

Hussein BJ (1983) Fisheries resources program and role in Arab food security projects. J Fisheries Resources 6-7:14–15

Hussein BJ (2011) Present status and potential of fisheries resources in Iraq. FAO-Iraq Office, Amman

Inma (2009) Competitive evaluation of animal feed plants in Iraq. Agribusiness program, USAID, USA.7 p

Iraqi Society of Fish Culturists (ISFC) (2018) Estimation of aquaculture production from illegal farms and cages. Internal Report

Khalaf KA (2009) Evaluation of three types of hormones in egg production of grass carp. *Ctenopharyngodon idella* Val. 1844. PhD Thesis, Basra Univ. 260p

Khuleaf AH (2006) Management of fish brood stock and the effect of inbreeding. Thesis of high diploma degree. Al-Musayab Technical College, Foundation of Technical Education, 58 p

Marsh Research Center (MRC) (2012) Study of main drainage water. A project submitted by MRC, Dhi Qar University for Marsh Rehabilitation Center

Modandi S (2001) Assessing the competitiveness of India cotton production, a policy analyses matrix approach. The journal of cotton science, India

Mohamed AM, Al-Habeeb FK, Hussein NA, Salman NA, Mutlk FM (2000) Acclimation of small Mugilidae fish from the coasts of South Iraq. Marina Mesopotamica. 16(1):239–256

Mohamed AM, Al-Habeeb FK, Yesser AT, Salman NA, Saleh JH (2001) Culturing of small Mugilidae fish in the Razzaza Lake. Iraq Marina Mesopotamica 16(1):331–346

Naif TS (2005) Some reproduction productive characteristics of fish brood stock flocks in some hatcheries of Babylon Province. MSc. Thesis, Al-Musayab Technical College, Foundation of Technical Education. 98 p

Rahim MF (1999) Fish culture. Zayan Press, Erbil, FAO-Iraq, 47p

Sabri MS (2006) Effect of weight plantation and nutrition in the polyculture on carp fish growth in productive culture ponds. PhD. Thesis, Univ. of Baghdad, Coll. of Agriculture. 129p

Saleh KI (2000) Effect of heat treatment on using of *Sespainasp* seed in common carp (*Cyprinus carpi*) feeding. Seventh Scientific Conference of the Technical Education, Baghdad, Iraq

Saleh KI (2001) Using of starch production by product and dextrin in common carp (*Cyprinus carpi*) feeding. Al-Techeny J Tech Res 71:78–87

Saleh KI (2015) Fish artificial reproduction and hatcheries management. Pub Alfurat al-Awsat Univ Found of Tech Edu:228 p

Saleh KI, Al-Gezali AR (2014) Application of an intensive production program for common carp *Cyprinus carpio* fingerlings in fish hatcheries–larvae production stage. Sci J Karbala Univ 12 (1):75–84

Saleh KI, Guber MA (2014) A view on the nature of the fish culture in Iraq (applied study). Al-Kut J Econ Admin Sci UnivWasit 13:96–116

Saleh KI, Salman MN (1990) Using of Actilic and Malathion with manuring to control the natural foods quality and reducing the mortality of carp larvae. Marine Mesopotamica 5(3):133–141

Saleh KI, Sulaiman MN (1992) A simplified method to prolong common carp *Cyprinus carpio* reproduction season in the climate of middle of Iraq. The third Scientific Conference of Technical Education. Baghdad, Iraq. p 19-28

Saleh KI and Suliman NE (1988) Exploitation of Agriculture drainage channel in fish carp cage culture, to obtain better rearing density. The First Scientific Conference for Technical Education. Baghdad, Iraq. pp. 267–286

Saleh KI, Suliman NO, Al-Azawi RK, Abdelniby K (1995) Effect of using protein concentrate and single cell protein *Saccharomyces carlsbergensis* in common carp (*Cyprinus carpio*) feeding. Al-Techeny J Tech Res. 26:151–160

Salman NA (1993) Study of certain aspects of construction and use of coastal tidal ponds. Proc. scientific symposium on fish farming development. Arab Union of Fish Producers & Arab Union of Sci. Res. Centers:30–42

Salman NA (1993a) Study of certain aspects of construction and use of coastal tidal ponds. Proc. scientific symposium on fish farming development. Arab Union of Fish Producers & Arab Union of Sci Res Centers:30–42

Salman NA (1993b) Practical alterations to the supplementary feeding of common carp in fish ponds. Proc. scientific symposium on fish farming development. Arab Union of Fish Producers & Arab Union of Sci Res Centers:106–119

Salman NA (2011) Suggested methods for enhancement of aquaculture in the southern marshes of Iraq. Iraqi J Aquaculture 8(2):157–177

Salman NA (2013) The use of locally made fish rearing cages in the Iraqi marshes: a case study. Basrah J Agric Sci 26(Special Issue I):198–211

Salman NA, Al-Mahdawi GJ, Al-Azawi AH, Rassoki RH, Abbas LM, Al-Rudainy AJ, Al-Timimi MT, Faisal SM (1997) Acclimation and culture of common carp (*Cyprinus carpio* L.) and Gattan (*Barbus xanthopterus*) in drainage water using floating cages. Marina Mesopotamica 12 (1):169–188

Salman NA, Kittan SA, Habib FK (1991a) Transport of live silver pomfret (*Pampus argenteus*) from fishing area to the tidal ponds of Khor Al-Zubair. IPA J. Agr. Res. 1(1):79–93

Salman NA, Kittan SA, Habib FK (1991b) The use of floating cages and net enclosures for raising silver pomfret (*Pampus argenteus*). IPA J Agr Res 1(2):175–186

Salman NA, Rasheed RM (2017) Economic feasibility of establishing recirculation aquaculture system (RAS) in Kurdistan-Iraq. Project Report:11p

Shaker EF (2006) A comparison of HCG hormone utilization with carp PG for the spawning of silver carp *Hypophthalmichthys molitrix* (val.). Thesis of high diploma degree, Al-Musayab Technical College, Foundation of Technical Education. 79 p

Shireman JV, Smith C (1983) Synopsis of biological data on the grass carp *Ctenopharyngodon idella* Cuvier & Valeneeinnes 1844. Fish Synopsis 135:86p

Towers L (2014) Fisheries and aquaculture markets in Middle East. FAO Report, The Fish Site

Unong IR (1998) Profitability of the broiler industry in Task Malaya. MSC Thesis, Jakarta Univ. Indonesia

Woynarovich E, Horvath L (1980) The artificial propagation of warm water fin fishes a manual for extension. FAO Fisheries Tech. Pap. No 201, Roma. 183 p

WPRIAP (1975) Wangtung Provincial Research Institute of Aquatic products Lecture notes of training course in freshwater fish culture. part. 3: Adult fish culture in pond, Kwangchaw, China. 35pp

Chapter 49
Towards an Improved Tilapia Farming in Iraq: Recommendations for Future Application

Laith A. Jawad and Saad Mohammed Saleh Abdulsamad

Abstract Tilapia is the familiar name used to call three genera of fish in the family Cichlidae: *Oreochromis*, *Sarotherodon*, and *Tilapia*. These species are natural to Africa and the Middle East and became the second most usual farm bred food fish in the world due to numerous positive features and have been cultured in large number of countries. Nonetheless, it has been faced some difficulties in tilapia culture yet in their tropical natural habitats.

Tilapias transported to Iraq in the late 1990s, when the Iraqi government at that time gave permission to introduce individuals for aquaculture purposes and not for stocking in the natural freshwater. Later, tilapia found its way to the rivers and lakes of Iraq and established a sustainable population.

In addition to the success of tilapia culture expansion as a consequence of technical advances linked with the increase of culture activities, several problems have arisen, among these: the development of new strains and hybrids, monosex male culture, formulated diets, a variety of semi-intensive and intensive culture systems (e.g., ponds, cages, tanks, and raceways), and the utilization of greenhouses, geothermal, or industrial waste heat, and advanced water treatment methods.

The present chapter reviews briefly the up-to-date and standard procedures and guidelines of aquaculture of tilapia that are followed globally. These procedures and guidelines together with the specific recommendations need to be followed by both the governmental and the private sectors aquaculturists in Iraq in order to raise the yield of the cultured tilapia in the future.

L. A. Jawad (✉)
School of Environmental and Animal Sciences, Unitec Institute of Technology, Auckland, New Zealand

S. M. S. Abdulsamad
College of Education, University of Basrah, Basrah, Iraq

L. A. Jawad (ed.), *Tigris and Euphrates Rivers: Their Environment from Headwaters to Mouth*, Aquatic Ecology Series 11,
https://doi.org/10.1007/978-3-030-57570-0_49

49.1 Introduction

Among the native fish species of Africa and the Middle East is tilapia. It has been approved to be one of the utmost fecund and worldwide merchandized food fish globally. The history of the farming of tilapia probably goes back to more than 4000 years when the Pharaohs of Egypt have started this industry. In the modern history, the cultivation of tilapias has spread from Africa, where it has been first recorded scientifically oriented culture of tilapia in Kenya in 1924 and rapidly extent through Africa.

The tilapia farming business is deliberated the second most significant farmed fish globally next to carps, but is defined as the best vital aquaculture species of the twenty-first century (Shelton 2002). The fish is being introduced in around 85 countries around the globe (FAO 2002) and about 98% of tilapia produced in these countries is grown outside their original environment (Shelton 2002).

The species of the three genera that belong to the family Cichlidae, *Oreochromis*, *Sarotherodon*, and *Tilapia*, are known in common as “tilapia”. It has been found that the members of the genus Oreochromis are readily hybridized in captivity and can produce several strains.

In the developing countries, there are other than 20 tilapia species that have been cultured in chiefly growing countries, which have shortage of animal protein (Guerrero 2002). The two species (*Oreochromis niloticus* and *O. mossambicus*) and their hybrid are considered the utmost prevalent among tilapias. *Oreochromis niloticus* is the important cultured species and the foundation for the noteworthy increase in global tilapia aquaculture production.

In the mid of 1990s, a permission was given for national Iraqis to import individuals of tilapia species from the neighboring countries such as Jordan. The aim was solely to use these individuals in the closed and secured aquaculture system, so there will be no escape to the natural water systems. The aquaculture business that runs by the private sector was not perfect in their control over the escape of the individuals tilapia to the near rivers. Therefore, in a short period of time, tilapia species has appeared for the first time in the natural freshwater system of Iraq. The first two species of tilapia, *Tilapia* (=*Coptodon*) *zillii* and *Oreochromis aureus* were recorded from the southern part of the general estuary in Basrah Province (Mutlak and Al-Faisal 2009). Later on, *Tilapia* (=*Coptodon*) *zillii* was recorded in Hor Al-Dlameg (AL-Zaidy 2013) and *O. niloticus* in the Shatt Al-Arab south of Iraq (Al-Faisal and Mutlak 2015).

In Iraq, the aquaculture practices and programs that used in the Aquaculture of the common carp are used for the tilapia species. In such practices, several aspects of the normal procedure of the fish aquaculture were not followed. Among these are the control over the environmental factors, rearing the larvae in the hatcheries and collecting the individuals for getting semen and eggs from the field. Therefore, in the present chapter, an attempt is made to gather and review the basic and the regulations that usually followed by tilapia aquacultures in the world in order to be followed by tilapia aquaculturists in Iraq. Such information will certainly enhance and increase the yield tilapia farmed. At the end of the chapter, recommendations are given to assist in the application of the correct and the up-to-date procedure of tilapia aquaculture.

49.2 Synopsis of the Aquaculture Activities in South of Iraq

With the abundant of freshwater quantities in south of Iraq and especially in the vicinity of Basrah Province makes it a good opportunistic future business to develop aquaculture and aquaponics industries to make use of the neglected water bodies and turning them into fish and prawn farms (Salman 2011). Simple aquaculture methods such as fish cages and net enclosures are low cost and would be operative in southern Iraq, mainly for species tolerant of low water quality, such as carp or tilapia (Salman 2011). Such projects can be successful in south of Iraq with more community knowledge and support, as well as gradually improving wetland water quality (Salman 2011).

The fisheries in Iraq, both freshwater and marine, have shown a fluctuation in the fish production. FAO (2004) has estimated the total fish production in Iraq for the years 2001 and 2003 as 22,800 and 32,100 tons live weight, respectively. For the marine fisheries, the fish production in this industry has declined by 12,000–13,000 tons annually from the 1980s through the early 2000s due to the Gulf war and the political instability. Marine fishing now depends completely on the artisanal industry, where the primary species are shad (*Tenualosa ilisha*), pomfret (*Pampus argenteus*), and different species of mullet (family: Mugillidae) (FAO 2004). The fish species that form the main catch is the non-native carp species and to less extents, the different species of the indigenous species of the genus *Barbus* (FAO 2004).

Although there is an increase in the aquaculture production of the fish in the inland waters of Iraq to 16,000 tons in 2007 (Dyck 2017), the procedures and controls are not up to the international standard. As the carp culture is quite easy task and giving less problems through its growth, the majority of the aquaculturists in Iraq have chosen to go with carp farming. The common carp, Cyprinus carpio was introduced into Iraq in the 1950s, as an experiment to study whether carp could be raised in Iraqi waters without negative impact on endemic species (Dyck 2017). Soon after that, the different carp species have become adapted to the Iraqi freshwater and they are now common in the region. Salman (2011) has suggested that the native species Bunni (*Mesopotamichthys sharpeyi)* and Gattan *(Leuciobarbus xanthopterus)* are also possible candidates for aquaculture.

49.2.1 Some Factors Shown to Hinder the Aquaculture Industry in Iraq

In addition to the nonmodern aquaculture practices followed in Iraq, there are several other factors that can hinder the production of fish in this industry. Among these are:

49.2.1.1 Environmental Factors

The main environmental matters that influence the success of aquaculture in southern Iraq are high salinity and poor water quality, high water temperature, and large seasonal variation. Productivity will additionally be determined by stockade level—which is influenced by the availability of fingerlings and food quantity and quality (Dyck 2017).

The studies on the status of water in the fish farm in Iraq in general are few to nil. On the other hand, studies on the water quality of the rivers and marshes are available (Richardson and Hussain 2006; Ewaid and Al-Hamzawi 2017).

49.2.1.2 Availability of Fish Fingerlings and Fish Food

For the availability of fish fingerlings, carp fingerlings are being produced in bigger numbers and are more available to fish farmers, but fish food remains constricted. The raw material for the fish food used for aquaculture in Iraq is an expensive imported. Therefore, it is feasible and commercial to build a fishmeal pellet production in Basrah by growing marsh reeds (mostly *Phragmites*) to be turned into fish food. Local production would reduce costs, boost the economic impact of fish farming, and motivate marsh reed cultivation.

49.2.2 Fish Farms in Iraq

In 1980s, there were 2000 fish farms in Iraq. In Basrah Province, there were 15 fish farm registered in the period 1982 and 2003, with 6.6% produced fish (Salman 2011). Across all of Iraq in 2003, there were 1893 licensed fish farms covering an area of 7500 hectares. The average farm size is 4 hectares; ten farms are larger than 100 hectares; and Babel fish farm is the largest at 500 hectares (Kitto and Tabish 2004). Since 1985, expansion in the aquaculture industry was evident, but, simply reflooding farmland areas and adding fish are hardly successful, because it does not account for the environmental variables.

Among the successful fish farm in Iraq is Alsabah farm (Ali 2013). This farm is 95 hectares in total and 50 of those hectares are used for fish production. The farm contains six ponds that are 1.5 meters deep, 50 meters wide, and 400 meters long, and a feeding canal that is 1.5 meters deep, 10 meters wide, and 800 meters long. The ponds are dried every one to two years and treated with manure and quick lime. Alsabah uses water from the Shatt Al-Arab, so fish are directly affected by the poor water quality and high salinity of the river (Ali 2013). Each of the six ponds began with 10,000, 50–250 gram fingerlings of common carp, 1200 silver carp, and 400 grass carp. A seventh pond is kept for broodstock. Additional fingerlings are placed in feeding and draining canals. The ponds are stocked three times per year (Ali 2013). Fish are fed through floating silos that are stocked with pellets twice daily. Fish food is processed on-site, but the raw material is imported. Alsabah is interested in moving away from soya and meat-protein-based food to algae-based foods (Ali 2013), which would be more sustainable and made of raw materials that could be locally produced.

The other successful, but to less extent of Al-Sabah fish farm is the University of Basrah fish farm. It is investigating at the moment constructing a wastewater treatment system that could support fish farming. In this system, reed marshes would filter effluent from the University. In 2016, Jassim Al-Maliky, then a PhD student at University of Basrah, successfully created test bed marshes that cleaned urban wastewater (Al-Maliky 2016; Pournelle et al. 2017).

The plan of the University of Basrah is to build a water treatment plant and fish farm. This plant can be explained as the University wastewater would flow through a series of treatment ponds and then into an aquaponics system that would grow reeds and fish (mostly carp). The herbivorous fish could then be harvested and used to produce fishmeal for carnivorous fish farms, providing a domestic option for fish food production, rather than importing from abroad, and supporting the larger aquaculture industry. Fish food for the herbivorous fish in the University of Basrah fish farm could come from a local aquatic biomass pellet mill, which would further keep production and support the local economy.

49.3 Tilapia as a Global Important Food Fish

There are several reasons that make tilapia fish as an important food fish globally. Among these, this chapter will enumerate the main and most effective reasons adopted from Fitzsimmons et al. (2011) with slight variation to fit the aquaculture status in Iraq. These are:

49.3.1 Genetics

One of the vital reasons for tilapia''s continued increase of production in future years is grounded on the genetic variety obtainable from which to construct. The farmed tilapias are derivative from many species in the genus *Oreochromis*. Literally, numerous species are simply hybridized and yield large numbers of productive young has enable fish breeders to cross some species and cultivate strains that comprise several characters from each of the parent species. This added support to the argument that the tilapia have been carefully bred and domesticated to an even greater degree than the eatable carps.

49.3.2 Nutrition

The tilapia is omnivores—herbivores, a criterion that endures to make tilapia general with the green movement. They feed chiefly on a very low trophic level. Naturally, the tilapias feed upon algae, fresh and decaying plant substances, and periphyton. In tamed locations, the numerous tilapias remain on the made diet that contains grains and agricultural by-products that aid to keep tilapia diets below the average for generally other farmed fishes. While many of the carps have similar feeding and nutritional patterns, tilapia on the other hand, are smaller and have smaller teeth and mouths, they incline to be even more able at rasping off the finest biofilms and periphyton assemblies.

49.3.3 BioFlocs

The capability of tilapia to flourish in biofloc systems is an additional benefit that tilapia have over many of the other common aquaculture species. Avnimelech and Kochba (2009) describes how tilapia are exclusively modified to flourish under biofloc circumstances that would impact the majority of other fish. This moderately low cost system for making healthy fish and dropping prepared feed charges could be an additional advantage that should keep tilapia prices inexpensive with other wild and farmed species.

49.3.4 Agricultural Plant Wastes

Tilapia have established to be one of the most important fishes used in additional component studies. The most usual goal is to substitute fish meal and fish oils. Whereas these incline to be very slight constituents in tilapia diets, the farmers and researcher still need to more reduce fish products in the diet and use locally obtainable elements (Zerai et al. 2008).

49.3.5 *Production Systems and Locations*

49.3.5.1 Diversity of Production Modes

Tilapia are elite in the selection systems used to rear them in confinement. Commercial procedures contain: ponds, cages, raceways, tanks, net pens, lake ranching, seawater, brackish water, freshwater, aquaponics, plastic drums, and computer-controlled exhaustive recirculation systems. This diversity of production surpasses that of any other farmed fish. Tilapia's treatment with recirculating systems has allowed their production in urban areas, high latitude locations, and even on the international space station (Fitzsimmons 2005).

49.3.5.2 Geographic Distribution

The tilapia production has been introduced to over 100 nations according to the FAO reports (Fitzsimmons et al. 2011). This huge yield and attention in the fish enormously surpasses any other farmed fish. The consumer appeal is likewise prevalent. There are not any accounts of cultural or religious restrictions on eating tilapia.

49.3.5.3 Low Cost Production Costs

Tilapia with their vegetarian grounded feeding habit and capability to collect substantial nutrition from grazing on algae and biofilms, have some of the lowermost feed costs of any farmed fishes. With the high densities achieved on many farms, the infrastructure costs are consequently extent across a larger volume of fish. Lastly, hatchery technique is comparatively modest, letting for fewer hatchery workers.

49.3.6 *Polyculture*

An additional area in which tilapia yield is rapidly growing is polyculture. In several instances, this is for the better market price that tilapia infrequently gets and in others they acquire the different habitats that the tilapia live in related to the carps. Mingling of tilapia and shrimp has been found to be valuable for shrimp health and for financial yield (Yuan et al. 2010). Through many shrimp farming businesses, tilapia are progressively being grown in cages inside shrimp ponds, or are cultivated in supply channels or head ponds. The rising interest in combined multitrophic aquaculture systems for tropical yield is certain to further support to general tilapia yield as most systems consider tilapia to be a main component to the systems.

49.4 General Problems and Solutions in Tilapia Culture: Problems and Solutions

In addition to their positive criteria as an excellent aquaculture fish species, tilapia can show some restrictions toward farming process such as the case in the subtropical region, where it cannot tolerate cold water. Physiological changes occur to tilapia when temperature drops less than 14 °C that could result in the death of the fish.

The other trouble is that tilapias attain sexual maturation in early age and short total length, and then reproduce with the intervals of 4–6 or 8 weeks as reliant on species and habitat circumstances, therefore number of fish start to upsurge, which disturb the care and feeding settings in the ponds. That is not expected in the culture since disarranging the culture scheme instigates high feed conversion rate, low growth rate of the fish, and gathering of immature tilapias having low economic status (Guerrero 2002; Chervinski 1982).

49.5 Extensive Culture

49.5.1 Pond Culture

One of the earliest aquaculture facility to cultivate tilapia was a simple pond in the earth with no inputs and depends on the native plankton and debris present in water and soil (Ramnarine 2000). In most growing countries, large ponds have a low social, cultural, and economic category and restricted admittance to technology, markets, and credit (Alceste-Oliviero 2000). The impact of such ponds is hard to attain, widespread tilapia culture has helped the rural poor to enhance their daily nutrition and increase their lifestyle in some areas of the Americas.

49.5.2 Semi-Intensive Culture (Ponds)

With the market loads increase, growing of industry and enduring technology development, old style widespread culture systems are being substituted by semi-intensive and exhaustive yield schemes.

Semi-intensive pond culture of tilapia is characteristically shared with agricultural or animal husbandry events, because pond fertilization with organic fertilizers can kindle natural pond efficiency in addition to being directly ingested by the tilapia. In this technique of culturing, nutrient contributions fluctuated from inorganic phosphorus to numerous stages of organic fertilizer, mixtures of organic and inorganic fertilizer, and blends of fertilizers and feeds.

49.5.3 *Intensive Culture*

49.5.3.1 Cages

Cultivation of tilapia in cages is mounting in some countries of the Americas. This technique is useful for growers who use public or common waters, containing reservoirs, lakes, bays, irrigation systems, or village ponds. Cages diverge widely in structure, from simple bamboo attainments to complex steel and plastic designs. Capital asset is low related with ponds, and by concentrating fish the farmer has improved control over feeding and yield. Some disadvantages of cage culture contain stealing risk, lack of ability to evade deprived water quality settings, and reliance on nutritionally complete feeds.

49.5.3.2 Integrated Tilapia and Shrimp Production

Consecutive polyculture includes cultivating shrimp and tilapia in detached ponds. In the case of the conditioned farms, water from tilapia ponds is directed to shrimp ponds, while in others the reverse case is true. Tilapia-shrimp polyculture may bring yields in several ways. The behavior of Tilapia in ingesting dead or diseased shrimps can be a disadvantage for the polyculture techniques as the disease will be transferred to the fish from the shrimps. Water from a tilapia culture pond, which tends to contain mainly of Gram-positive bacteria, might restrain the spread of *Vibrio* and other bacterial pathogens in shrimp ponds, which are Gram-negative. Tilapia unsettle bottom sediments when looking for food and in constructing nests, which may improve oxidation of the substrate and interrupt life cycles of shrimp pathogens or release nutrients that could upsurge algal growth (Fitzsimmons 2000).

49.6 Development of Technologies and Tilapia Farming

In tilapia culture, there are particular restrictions regarding the hybridization, masculinization of the whole tilapia groups through hormonal sex reversal. The method, which includes adding of steroids in feeds for a short period during the fry stage, showed to be simply performed, moderately steady in producing nearly all male populations and could be frequent in numerous country situations by farmers. The use of this method though has not been settled upon in some countries owing to environmental and social limitations; for example, the metabolism and the influences on the habitat of the dilapidation yields of synthetic androgen are not yet fully understood in fish (Baroiller 1996). The recently established method for gaining monosex population is by producing "supermales" through genetic management.

49.7 Recommendations

The followings are some main recommendations to enhance the yield of the farmed tilapia. Such recommendations need to be taken on board by both the government and the private sectors who own the tilapia farms in Iraq in order to improve their products.

1. Investigate into the type of plankton/algae that yield chemicals leading to the off-flavors in tilapia.
2. Cautiously management of store feed.
3. Further studies on foods for breeding or nursing stocks are required.
4. The usage of methyltestosterone hormone for sex reversal should be supervised and controlled.
5. Values for accountable culture of tilapia established by the World Wildlife Fund or Best Aquaculture Practices program could be followed.

References

Alceste-Oliviero C (2000) Regional review of aquaculture status and development trends in the Latin America and Caribbean. Book of Synopses. NACA and FAO International Conference on Aquaculture in the Third Millennium, 20–25 February 2000, Bangkok, Thailand

Al-Faisal AJ, Mutlak FM (2015) First record of the Nile tilapia *Oreochromis niloticus* (Linnaeus, 1758), from the Shatt Al-Arab River. Southern Iraq International Journal of Marine Science 5 (38):1–3

Ali S (2013) Alsabah Fish Farm in Basrah, Iraq

Al-Maliky J (2016) Constructed Wetland Systems Station for Treating Wastewater. University of Basrah

Al-Zaidy KJ (2013) First record of *Tilapia zillii* Gewais, 1848 in Al-Delmj marsh west Al-Diwania city middle of Iraq. Diyala Agriculture Science Journal 5:9–16

Avnimelech Y, Kochba M (2009) Evaluation of nitrogen uptake and excretion by tilapia in bio floc tanks, using 15N tracing. Aquaculture 287(1-2):163–168

Baroiller JF (1996) Significant proportions of unexpected males in progenies from single pair mattings with sibling sex reversed males of *Oreochromis niloticus*, p. 229–237. In R. S. V. Pullin, J. Lazard, M. Legendre, J.B. Kothias and D. Pauly (Eds.). The Third International Symposium on Tilapia in Aquaculture. ICLARM Conf. Proc. 41, 575 p. Chervinski 1982)

Chervinski J (1982) Environmental physiology of tilapias. In: Pullin RVS, Lowe-McConnell RH (eds) The biology and culture of tilapias, ICLARM conference proceedings, vol 7. ICLARM, Manila, Philippines, pp 119–128

Dyck M (2017) Potential for aquaculture integration with wastewater treatment wetlands in southern Iraq (2017). *Senior Theses*. 151. https: //scholarcommons. sc.edu/senior theses/151

Ewaid SH, and Al-Hamzawi SA (2017) Water quality index for Al-Gharraf River, southern Iraq. *Egyptian Journal of Aquatic Research*, April

FAO (2002) Fishery statistics. Aquaculture production 90(2)

FAO (2004) Fishery country profile–The republic of Iraq. Food and Agriculture Organization of the United Nations

Fitzsimmons K (2000) Future trends of Tilapia aquaculture in the Americas. In: Costa-Pierce BA, Rakocy JE (eds) Tilapia aquaculture in the Americas, vol 2. The World Aquaculture Society, Baton Rouge, Louisiana, USA, pp 252–264

Fitzsimmons K (2005) Tilapia culture. pp. 563590. In: Kelly A.M. and Silverstein, J. eds. Aquaculture in the twenty-first Century. American Fisheries Society, Symposium 46, Bethesda, Maryland

Fitzsimmons K, Martinez-Garcia R, Gonzalez-Alanis P (2011) Why tilapia is becoming the most important food fish on the planet. *Better science, better fish, better life*, p. 8. Proceedings of the ninth International Symposium on Tilapia in Aquaculture, Shanghai Ocean University, Shanghai, China 22–24 April 2011. Liu Liping and Kevin Fitzsimmons, eds. Published by the Aqua Fish Collaborative Research Support Program

Guerrero RD (2002) Tilapia farming in the Asia-Pacific region. In: Guerrero RDIII, Guerrero del-Castillo MR (eds) Tilapia farming in the twenty-first century. Proceedings of the international forum on Tilapia farming in the twenty-first century (Tilapia forum 2002). Laguna, Philippines, pp 42–49

Kitto MR, Tabish (2004) "Aquaculture and Food Security in Iraq." Food and Agriculture Organization of the United Nations

Mutlak FM, Al-Faisal AJ (2009) A new record of two exotic cichlids fish *Oreochromis aureus* (Steindacher, 1864) and *Tilapia zilli* (Gervais, 1848) from south of the main outfall drain in Basrah city. Mesopotamian Journal of Marine Science 24(2):160–170. (In Arabic)

Pournelle JR, Fawzi NA-M, Ebrahim S, Al-Mukhtar R, Al-Maliky J, Dyck M (2017) Capacity-building in faculty instructional strategies in water resources and environmental engineering. In: Iraq university linkages program small Grants 2015. University of South Carolina-University of, Basrah

Ramnarine IW (2000) Tilapia culture in the eastern Caribbean, Guyana, and Suriname. In: Costa-Pierce BA, Rakocy JE (eds) Tilapia aquaculture in the Americas, vol 2. The World Aquaculture Society, Baton Rouge, Louisiana, pp 245–251

Richardson C, Hussain N (2006) Restoring the garden of Eden: an ecological assessment of the marshes of Iraq. American Institute of Biological Sciences 56(6):477–489

Salman N (2011) Suggested methods for enhancement of aquaculture in the southern marshes of Iraq. Iraqi Journal of Aquaculture 8(2):157–177

Shelton WL (2002) Tilapia culture in the twenty-first century. In: Guerrero RD III, Guerrero-del Castillo MR (eds) Proceedings of the international forum on Tilapia farming in the twenty-first century (Tilapia forum 2002), 184p. Philippine Fisheries Association Inc. Los, Banos, Laguna, Philippines, pp 1–20

Yuan D, Yi Y, Yakupitiyage A, Fitzimmons K, Diana J (2010) Effects of addition of red tilapia (Oreochromis spp.) at different densities and sizes on production, water quality and nutrient recovery of intensive culture of white shrimp (*Litopenaeus vannamei*) in cement tanks. Aquaculture 298:226–238

Zerai DB, Fitzsimmons KM, Collier RJ, Duff GC (2008) Evaluation of brewers waste as partial replacement of fish meal protein in Nile Tilapia (Oreochromis niloticus) diets. J World Aquacult Soc 39(4):556–564

Chapter 50
Information and Recommendations of Aquaculture Stress and its Source in Hatcheries: The Iraqi Aquaculture Industry Investors

Laith A. Jawad

Abstract In both aquaculture and hatcheries, stress is considered among the major factors that cause a serious decline in the aquaculture yield. With the stress, fish can be disposed to conceded growth and health and ultimately promote disease. Studies have shown that the larval stages may be more sensitive to stress as compared with other stages. Although with the increasing interest in fish culture, the study of stress in the hatchery is an important field for future. Proper management practices if followed in the hatchery will reduce the stress and will ultimately lead to the better survival of fish.

The present chapter gives a short review of the stress in fish hatcheries, its types and causes and at the end of the chapter recommendations about how to manage the stress in this important industry. This chapter is written so the managers of the fish hatcheries whether they are from the government or the private sector need to become aware of the issue of the fish stress in the hatcheries and follow the guidelines that teach them how to start and run a healthy and sustainable hatcheries in Iraq.

50.1 Introduction

As in other vertebrates, fish response to stresses through their body physiology by increasing in the level of stress hormones and consequent changes that help maintain the animals normal or homeostatic state (Iwama et al. 2004; Barton 2002). Among the responses that fish can have are increases in plasma cortisol, catecholamines, and glucose levels, increases in branchial blood flow and increases in muscular activity (Barton 2002). Even worse, pathological conditions can be happened leading to

L. A. Jawad (✉)
School of Environmental and Animal Sciences, Unitec Institute of Technology, Auckland, New Zealand

L. A. Jawad (ed.), *Tigris and Euphrates Rivers: Their Environment from Headwaters to Mouth*, Aquatic Ecology Series 11,
https://doi.org/10.1007/978-3-030-57570-0_50

death (Iwama et al. 2004). These physiological stress responses in fish can be triggered by stressors such as environmental and genetic development Barton (2002).

In the aquaculture industry, hatcheries are considered the important part as they are intended and operated for cost-effective rearing of larvae. The growth and sustainability of aquaculture relies on offering suitable conditions for the development of larvae. In hatcheries, the larval fish are reared in conditions which are at significant abnormality from the natural environment. However, the environmental conditions of the culture system must fulfill the physiological and behavioral needs of the fish larvae in order to achieve high survival and growth. Affording the larvae with the environment that best suits its growth and development without compromising the economics of the production system is a big task which the industry is facing all over the world.

Stress is a vital factor, which should be considered in the fruitful rearing of larvae in hatcheries. Stress is outlined as the generic response of the organism to any demand made upon it (Selye 1973). Although, the initial phase of stress, called as eustress, is useful to the organism, distress happens when certain factors cause physiological changes in an organism that ultimately compromise the organism's integrity (Selye 1984). Most of the research on stress is mainly focused on distress phase (Martinez-Porchas et al. 2009). The main aim in the hatchery is to produce a reliable and sufficient good quality of fish seed but certain stressors affect the whole process of production resulting in high mortality rates, and economic loss in aquaculture sector. The intensive culture of larvae exposes them to considerable stressors of chemical, physical, and biological nature and in addition, some of these stressors can occur in an intermittent manner or as pulses, which may further amplify their negative effects. Therefore, the recognition of these stress sources, and their mitigation are critical in the successful operation of hatchery.

In Iraq, there are several hatcheries that belong to the government and the private sector. In all the hatcheries, there are guidelines that follow to study the stress in the fish. Therefore, the yield of these hatcheries is lower than the expected. In the present chapter, the different types of factors that might cause stress in the hatcheries are reviewed in order to draw the attention and assist the managements of these hatcheries for this important issue and raise the yield as much as possible.

50.2 Causes of Stress in Hatchery

Investigators on the issue of the stressors in fish have identified the most common stressors that face fishes in hatchery, they include poor water quality (Tomasso et al. 1981), xenobiotic toxic molecules (Schlenk et al. 1999; Griffin et al. 1999), and handling (Davis et al. 1993). These stressors can be of a short-term acute nature, like handling or may be long-term chronic type, like an extended period of poor water quality. Since the transport, feeding, and treatment are routine activities of a hatchery operation, the larvae are more exposed to both chronic and acute stressors. A similar danger for poor water quality also occurs in the hatcheries because of the high stocking densities, which are common in intensive culture (Kenneth 2006). The following are the main causes of stress in fishes adopted from Rehman et al. (2017).

50.2.1 Water Temperature

It has been much documented that water temperature controls major physiological processes of fish larvae and any slight variation in this factor will have an overwhelming effect on their physiological mechanisms (Blaxter 1991). For example, temperature can disturb the time of hatching, development of larvae, cellular function, muscle ontogeny, and development and ontogeny of internal organs (Herzig and Winkler 1986). In addition to the changes in the external morphology, meristic characters (Torres-Núñez et al. 2014), appearance of skeletal deformities, sex determination, delayed feeding (Dou et al. 2005), changes in the rates of enzymatic reactions, swimming performance, growth, overall survival, and lifespan of fish are affected by temperature (Meeuwig et al. 2013).

In all the hatcheries in Iraq, the temperature of water is not controlled, it follows the great variation in air temperature inside the hatchery during the seasons of the year. The variation in air temperature in Iraq during the year is high. Temperatures drop to less than 10 °C or less during cold months (November–March) and raised to more 50 °C during hot months (May–August).

50.2.2 pH of Water

The pH is an essential variable in water quality as many reactions are pH dependent and normal waters comprise both acids and bases. The biological processes in water tend to increase either acidity or basicity, and the interactions among these contrasting acidic and basic substances determine pH. Carbon dioxide concentration, which is acidic, plays an important role in regulating pH. The alkalinity of water is mainly a consequence of bicarbonate and carbonate ions derived from the reaction of carbon dioxide in water tends to buffer water against extreme pH change (Boyd et al. 2011). Fish larvae are highly sensitive to pH changes than juveniles and adults and at this stage important effects are most likely to be detected. This can partly be understood due to insufficient acid-base regulation prior to the formation of gills (Falk-Petersen 2015). Studies have shown many impacts of predicted CO_2 concentrations on larval fish on sensory capabilities like olfaction (Munday et al. 2009), behavior (Dixson et al. 2012), otoliths (Checkley et al. 2009; Bignami et al. 2013) development, tissue and organ structure (Frommel et al. 2014), survival of eggs specifically the hatching success (Chambers et al. 2013), and survival of very early larval stages (Bromhead et al. 2015).

Especially in the freshwater hatcheries in Iraq, the pH factor is ignored and water never being checked for it. The general idea that all the managers of the hatcheries in Iraq think that freshwater aquaculture does not need a check for the pH value.

50.2.3 Photoperiod

Light plays a significant role as a biological regulator controlling feeding, digestion, and reproduction in fish. Light influences the basic signaling systems by affecting the light sensitive brain tissues including the retina of the eye containing serotonin, an important neurotransmitter. Light comprises different components such as quantity (light intensity), quality (spectral composition), light distribution (point source or evenly), and cycling (photoperiod and season). The outdoor light conditions are quite defined depending on latitude, season, depth, and algal blooms. However, hatchery conditions often provide very different conditions from natural systems like light intensity is lower, spectral composition is different with, e.g., lack of UV radiation, and 24-hour light is commonly used. This may remove any diurnal signals from the fish larvae, with unknown impacts on larval development. Adverse light conditions which do not enable good feeding behavior encourage stress conditions since most marine fish larvae are visual feeders (Tamazouzt et al. 2000).

The control of the intensity of light in all the hatcheries in Iraq is lacking. The lighting in the hatcheries is usually bright and light usually left on all day and night. Beside, all the windows are open and most of the tanks of the hatchery are bathed in the sun light. Therefore, the aquatic organisms have a mixture of natural and artificial light all day and artificial light during the night. Such inconsistency in the light intensity will definitely has direct effects on larvae and juveniles.

50.2.4 Mechanical Stress

Besides the physicochemical aspects of water, several routine practices in the hatchery like overcrowding, handling, and transport can cause stress in fish (Wedemeyer 1996). High stocking density produces a wide variety of effects on fish such as alterations in behavior (Alanärä and Brännäs 1996), poor feed utilization (Jorgensen et al. 1993), behavioral and hormonal changes (Leatherland and Cho 1985) resulting in poor growth (Björnsson 1994), and survival (Sodeberg and Meade 1987). High stocking density also triggers immunosuppression (Tort et al. 1996) and encourages mobilization of energy sources (Vijayan and Leatherland 1988; Vijayan et al. 1990). Capture and handling provoke rapid, marked elevations of plasma cortisol, and glucose level in the fish.

Transportation initiates a physiological stress response by accumulation of carbon dioxide and ammonia in fish hauling tanks, increasing concentrations of circulating cortisol level and reducing survival rate (Schreck et al. 1989).

Behavioral interactions occur between fish in a hatchery may lead to agonistic encounters, interspecific competition, social hierarchies, and territoriality dominance are a significant source of social stress in fish indicated by manifold changes in physiological and neuroendocrine processes (Summers 2002; Blanchard et al. 2001). Fishes show aggressive behaviors and predation causing unnecessary stress and physical injury. Fin nipping, scale loss from ramming, reduced growth, pathological changes in gastrointestinal tissue, and increased susceptibility to diseases can occur in defeated individuals (Earley et al. 2004).

50.3 Consequences of Stress

There are several consequences of stress in fishes. These arise as the stress level has reached a point that can affect other aspects of the fish physiology and anatomy. The scope of this chapter is too small to accommodate a comprehensive list and description of the consequences of stress on fishes. The followings are some of the important of these outcomes suggested by Rehman et al. (2017).

50.3.1 Stress and Immune System

The specific and nonspecific immune system in fish can be conceded in fish due to stress factors present in the hatchery leading to creation of infection. Different kind of stressors induces a variety of immune changes and the outcome of a stress response will depend on the strength of the stressor and its magnitude (Wendelaar Bonga 1997). Majority of stressors, however, produce deleterious effects. Stressful hatchery conditions like crowding (Vazzana et al. 2002), variable water temperature (Varsamos et al. 2006), environmental contaminants, oxidative stress (Franco et al. 2009), and deficient diet, which cause enduring stress normally show suppressive effect.

Most of these effects are reduced immunocompetence, increased level of cortisol, and decreased level of lysozyme.

50.3.2 Stress and Diseases

Microorganisms are usually harmless in the hatcheries, but with stress, they can be infective and can cause infections and epizootic disease (plumb 1994). Fish pathogens are generally classified as obligate or facultative pathogens and the stress-mediated fish diseases are mostly caused by facultative bacterial pathogens (plumb 1994), external fungi, protozoan parasites (Alvarez-Pellitero and Sitja-Bobadilla 1993). Some of the common bacterial diseases occurring in hatcheries can be stress-mediated such as vibiosis (*vibrio* species) and motile *Aeromonas* septicaemia (*Aeromonas hydrophila*), enteric red mouth (*Yersinia ruckeri*), enteric septicemia (Wedemeyer and Goodyear 1984). Furunculosis (*A. salmonicida*) is often correlated with stressful water quality conditions and fish culture procedures (Nomura et al. 1992). Bacterial gill disease (*Flavobacterium branchiophilum*) occurs in hatchery mostly with management errors such as overfeeding, overcrowding, inadequate water flow rates, low DO levels, increased unionized ammonia, and accumulation of suspended particulate matter that causes gill irritation. Stress and cortisol treatment have been shown to increase susceptibility to parasitic infections (Saeij et al. 2003). A number of noninfectious diseases such as skeletal abnormalities, swim bladder stress syndrome, and coagulated yolk disease occur due to improper management of rearing conditions (Soares et al. 1994).

50.3.3 Stress and Reproductive Inhibition

The fish individuals that usually caught in the field and brought to the hatchery as a brooders became stress due to the process of capture or confinement of wild brooders and handling for the collection of gametes (Melotti et al. 1992) influences reproduction by affecting circulatory levels or changing seasonal pattern of reproductive hormones. Such a stress will inhibit their sexual maturation for a certain time (latency period) (Pickering et al. 1987). Effect of stress is mainly due to the increase in plasma cortisol level and catecholamines. Increased plasma cortisol has shown suppressive effects on plasma sex hormones like testosterone, oestradiol and gonadotropin, pituitary level of gonadotropin, circulating vitellogenin, and ovary weight in vitro (Carragher and Sumpter 1990). Repeated acute stress prior to spawning results in delayed ovulation, reduced egg size in females, and lowered sperm counts in males (Pankhurst and Carragher 1992). Progeny from the stressed fish is also known to have a lower survival rate as compared to unstressed fish (Short et al. 1995).

The practice of catching the brooder individuals for the hatchery is not much different when catching fish for commercial use. In this way, the fish individuals will not be treated carefully before they reach the hatchery. They may face a period out of water suffering from lower oxygen level in their blood, kept in small amount of water and in plastic bags, and bad handling that might injure the fish, which in turn creating stress.

50.3.4 Stress and Reduced Feeding Behavior and Growth

Bad handling of the brooders in the process of collecting them from the field is shown to lead to loss of feeding behavior for various period of time depending on the severity of stress and physiological state of fish. Stress disrupts feeding behavior like appetite, visual and chemosensory ability, and restricted area searching, responding to and capturing prey, and handling and ingestion of prey (Beitinger 1990). Low pH of water has shown to depress attraction to food scent (Jones et al. 1987). Natural mortality in fish occurs mostly during larval stages, due to predation and starvation (Hunter 1981). Starvation in larvae can also trigger behavioral abnormalities such as depression and termination of positive phototaxis, abnormalities in gas bladder development leading to lack of buoyancy control and early death (Fletcher 1997).

50.3.5 Stress and Behavioral Response

In fishes, swimming performance, thermoregulation, orientation, avoidance, chemoreception, feeding, predator evasion, and learning can be changed due to stress. These changes may be adaptive and therefore, increase the probability of survival (Olla et al. 1998) and are also sensitive indicator of physiological and biochemical changes occurring in response to stress (Marcucella and Abramson 1978). Pathogenic microorganisms and parasitic infestation can cause behavioral changes in fish, which may include increased activity, flashing at the surface, and ultimately decreased activity and lethargy. Leaping and rolling at the surface can occur as a result of parasitic infestations (Furevik et al. 1993).

50.4 Signs of Stress in Fish

In the process of controlling the stress, the fish is checked for their health, performance, and welfare to investigate the influence by different stressors like handling, rearing, and transport (Portz et al. 2006), as well as anesthesia (Trushenski et al. 2012) affecting captive broodstock health, survival of larvae, and production efficiency. The main aim of assessing is to reduce it and maximize growth and survival. There are three levels of stress, depending on the response to the stress and they know by the name of the responses causing as primary, secondary, and tertiary responses. Physiological indicators of stress include all the responses between the cellular and molecular level and the whole-animal level.

50.4.1 Cellular and Molecular Indicators

Oxidative stress is enumerated by multiple markers through a variety of colorimetric assays measuring ROS, antioxidant levels; or measuring damage to biomolecules (Lesser 2006) in plasma, serum, urine, tissue homogenates, or cell cultures (Valavandis et al. 2006). Heat shock proteins (HSP) sensitive to a range of stressors like rapid temperature and salinity challenges, and handling can be a reliable indicator of stress. In fish, a number of genes like cytochrome c. transcription factor JUNB, NUPR1 have been investigated as potential biomarkers for various stressors (Jeffries et al. 2012). Changes in the gene expression related to inflammation, protein degradation, immune response onset of recovery poststress related to gluconeogenesis, glycogenolysis, and energy metabolism in the liver have been observed (Wiseman et al. 2007).

50.4.2 Primary and Secondary Physiological Indicators

Primary and secondary stress indicators are very useful in assessing responses related to specific aquaculture or handling practices, or acute disturbances in the field. Cortisol and catecholamines are commonly measured as stress indicator (Barton 2002). Cortisol responds more slowly than catecholamines to specific stressors, it can be quantified in laboratory or field settings although the later provides very accurate information about the response to acute stressors (Romero and Reed 2005). Poststress glucocorticoid levels can provide information about specific stimuli (e.g., capture and handling stress, different holding conditions, acute exposures). Secondary stress indicators commonly measured include glucose elevation (Barton 2002); lactate elevation, osmolality or specific ions, leukocytes (Davis et al. 2008).

50.4.3 Whole-Organism Indicators

A number of aspects of fish performance such as changes in growth, disease resistance, metabolism, cardiac activity, swimming performance, behavior, fitness, and even survival are measured in whole-organism (or tertiary) responses to stressors (Sadoul and Vijayan 2016).

50.5 Recommendations

The following recommendations are set to assist hatcheries mangers in both the government and private sector to follow in order to improve the yield in their hatcheries.

1. Proper capturing and handling of brooders.

Handling and grading are highly stressful and more complex because of the aquatic medium. Handling needs a strange element into the water to catch the fish and if not done correctly, excessive stress can risk fish, disrupts the protective mucous coating and fish scales, and increases susceptibility to parasitic or pathogenic infection. To evade upsetting the mucous coat of fish and losses of scales, use of wet hands or soft cotton gloves while handling fish is always recommended. Stress should always be diminished while capturing and handling brooders and this can be partly attained by choosing appropriate gear during the process. The suitable makeup of netting varies with species, and subjects on presence or absence of scales and scale type. Knotted mesh can dislodge fish scales resulting in parasitic and pathogenic invasion. The seine net that is used in ponds and tanks can give good results but should be handled very carefully, while harvesting only a small number of fish per haul. To avoid, the use of knotless stretched mesh nylon nets, as well. Healthy adult fish can also be captured by hook and line, but care should be taken to reject all fish with major injuries in their mouth, gut or, even worse, gills, as well as those which have lost too many scales when hauled on board (Tucker 1985). Temperature affects the stress response to handling and capturing, as water holds less oxygen at higher temperatures (Wheaton 1977). Most of these activities should be performed during the cooler portions of the day or initiated at night so as to finish in the early morning. Temperature-related stress can occur any time when fish are handled, but fish are most susceptible during high temperature resulting in mortality in both the short- and long-term (Piper et al. 1982).

2. Proper transportation.

Several studies have focused on overcoming the challenges of harvesting and transporting fish due to wide interspecific variations in physiological responses to these stressors (Barton 2002). Level of plasma cortisol in different species can vary by as much as two orders of magnitude due to identical stressors. Some fishes can be transported easily other than maintaining proper water quality and temperature, however, other species require specific additional precautionary steps so as to prevent ionic and osmotic imbalances resulting in mortality (Carmichael and Tomasso 1988).

Repeated handling and transport should be avoided and there should be a recovery period. Physiological responses of fish after acute netting, handling and transport can recover after a period of 6 h to1 day, however, it may also vary from 10 days to 2 weeks if they are persisting, but not lethal. Transporting a fish is a multiple-phase operation and should be designed in such a way to minimize stress (Piper et al. 1982). The temporary holding and transport containers should be selected carefully to reduce additional and potentially fatal stress to recently caught fish. They should be heat insulated and filled with water from the same location where the fish have been caught. The containers should be round in shape or square with rounded corners so as to avoid skin abrasions and mechanical shocks and large enough to allow the fish, a fair degree of movement. A correct choice of fish loading density (fish weight per unit volume of water) is of dominant importance in minimizing the stress. The fish density should be inversely proportional to transport time and water temperature. Fish should not be fed for about 24 h prior to harvest and transport to avoid feces and fouling of the transport water. In the container, oxygen saturation should be around 100% and the animals should be brought to the hatchery as soon as possible: speed and care are always recommended. A variety of water additives (anesthetics, hypnotic drugs, mineral salts, and polymers), fasting, hypothermia, and reduced light intensity have been developed to reduce the physiological effects of transportation on fish health (Wedemeyer 1997).

3. Maintaining water quality requirements.

To attain best growth and survival of newly hatched larvae, it is important to have good water quality in the hatchery. The quality of water should be more or less stable without any variation. The hatchery should have both heating and cooling plant to control the temperature fluctuations because natural water may have a suboptimal temperature for growth and survival for the different life stages. Larvae are delicate to high gas saturation in the water and as a result can become diseased quickly and thus aerators should be used properly to avoid such problems in hatchery. Gas bubble disease can occur if the nitrogen gas concentration in the water is above 101–102% (Noga 2010). CO_2-supersaturation is not usual because, its solubility is higher than that of oxygen and nitrogen but the tolerance of aquatic organisms for CO_2 in the water is, however, limited (Good et al. 2010). The ph of water also plays a major role in the hatchery and thus ph should be measured regularly and adjustment should be done to avoid great variations.

4. Control of microorganism.

In the hatchery, diseases can be managed by the disinfection of the inlet water. The most familiar system is ultraviolet (UV), ozone. UV lamp can be used to send out UV radiation having a sufficient dosage to inactivate microorganism. Ozone gas (O_3), a very strong oxidizing agent produced by ozone generator can also be used in the hatchery for disinfection of water. Ozone can also reduce water turbidity, watercolor, organic carbon, metal ions, and algae (Masters et al. 2008).

5. Stabilization of water.

Disinfection of water decreases great quantity of microorganisms but still the water will have dead, inactivated bacteria, and decomposed organic matter which can be a decent food source for the remaining bacteria. Microbial stabilization or maturation is a method where the disinfected water is allowed to pass through a filter where biological organisms growing naturally can stabilize the water quality by inhibiting growth of bacteria. There should also be a retention time for general stabilization of water quality before it is passed into hatchery (Brunvold 2010).

References

Alanärä A, Brännäs E (1996) Dominance-feeding behaviour in Arctic char and rainbow trout: the effect of stocking density. J Fish Biol 48:242–254

Alvarez-Pellitero P, Sitja-Bobadilla A (1993) *Ceratomyxa* spp. (Protozoa: Myxosporea) infections in wild and cultured sea bass, *Dicentrarchus labrax*, from the Spanish Mediterranean area. J Fish Biol 42:889–901

Bignami S, Enochs IC, Manzello DP, Sponaugle S, Cowen R (2013) K. Ocean acidification alters the otoliths of a pantropical fish species with implications for sensory function. Proc Natl Acad Sci U S A 110(18):7366–7370

Barton BA (2002) Stress in fishes: a diversity of responses with particular reference to changes in circulating corticosteroids. Integr Comp Biol 42:517–525

Beitinger TL (1990) Behavioral Reactions for the Assessment of stress in Fishes. J Great Lakes Res Internal Assoc, Great Lakes Res 16(4):495–528

Björnsson B (1994) Effects of stocking density on growth rate of halibut (*Hippoglossus hippoglossus* L.) reared in large circular tanks for three years. Aquaculture 123:259–270

Blaxter JHS (1991) The effect of temperature on larval fishes. Netherlands J Zoology 42 (2):336–357

Blanchard R, McKittrick C, Blanchard D (2001) Animal models of social stress: effects on behavior and brain neurochemical systems. Physiol Behav 73:261–271

Boyd CE, Tucker CS, Viriyatum R (2011) Interpretation of pH, acidity and alkalinity in aquaculture and fisheries. N Am J Aquac 73:403–408

Bromhead D, Scholey V, Nicol S, Margulies D, Wexler J, Stein M et al (2015) The potential impact of ocean acidification upon eggs and larvae of yellow fin tuna (*Thunnus albacares*). Deep Sea Res Part II Mar 113:268–279

Brunvold I (2010) Microbial ecology in the early life stages of intensively reared marine species, PhD Thesis, University of Bergen

Carmichael GJ, Tomasso JR (1988) Survey of fish transportation equipment and techniques. Progr Fish-Cult 50:155–159

Carragher JF, Sumpter JP (1990) The effect of cortisol on the secretion of sex steroids from cultured ovarian follicles of rainbow *Gen*. Comp Endocrinol 77:403–407

Chambers RC, Candelmo AC, Habeck EA, Poach ME, Wieczorek D, Cooper KR et al (2013) Ocean acidification effects in the early life stages of summer flounder. Paralichthys dentatus Biogeosciences Discuss 10:13897–13929

Checkley DM, Dickson AG, Takahashi M, Radich JA, Eisenkolb N, Asch R (2009) Elevated CO enhances otolith growth in young fish. Science 324:1683

Davis AK, Maney DL, Maerz JC (2008) The use of leukocyte profiles to measure stress in vertebrates: a review for ecologists. Funct Ecol 22:760–772

Davis KB, Newsom J, Simco BA (1993) Physiological stress in channel catfish harvested by lift net, vacuum pump, or turbine pump. J Appl Aquac 3:297–309

Dixson DL, Pratchett MS, Munday PL (2012) Reef fishes innately distinguish predators based on olfactory cues associated with recent prey items rather than individual species. Anim Behav 84 (1):45–51

Dou SZ, Masuda R, Tanaka M, Tsukamoto K (2005) Effects of temperature and delayed initial feeding on the survival and growth of Japanese flounder larvae. J Fish Biol 66:362–377

Earley RL, Blumer LS, Grober MS (2004) The gall of subordination: changes in gall bladder function associated with social stress. Proc R Soc Lond B Biol Sci 271:7–13

Falk-Petersen IB (2015) Comparative organ differentiation during early life stages of marine fish. Fish Shellfish Immun 19:397–412

Fletcher TC (1997) Dietary effects on stress and health. In: Iwama GK, Pickering AD, Sumpter JP, Schreck CB (eds) Fish Stress and Health in Aquaculture. Cambridge Univ. Press, Cambridge, UK, pp 223–246

Franco R, Sanchez-Olea R, Reyes-Reyes EM, Panayiotidis MI (2009) Environmental toxicity, oxidative stress and apoptosis: menage a trois. Mutat Res 674(1–2):3–22

Frommel AY, Maneja R, Lowe D, Pascoe CK, Geffen AJ, Folkvord A et al (2014) Organ damage in Atlantic herring larvae as a result of ocean acidification. Ecol Appl 24(5):1131–1143

Furevik DM, Bjordal A, Huse I, Fernoe A (1993) Surface activity of Atlantic salmon (*Salmo salar* L.) in net pens. Aquaculture 110:119–128

Good C, Davidson J et al (2010) The effects of carbon dioxide on performance and histopathology of rainbow trout *Oncorhynchus mykiss* in water recirculation aquaculture systems. Aquac Eng 42(2):51–56

Griffin RJ, Cocker DR III, Flagan RC, Seinfeld JH (1999) Organic aerosol formation from oxidation of biogenic hydrocarbons. J Geophys Res 104

Herzig A, Winkler H (1986) The influence of temperature on the embryonic development of three cyprinid fishes, *Abramis brama*, *Chalcalburnus chalcoidesmento* and *Vimba vimba*. J Fish Biol 28(2):171–181

Hunter JR (1981) Feeding ecology and Predation of marine fish larvae. In: Lasker R (ed) Marine fish Larvae: Morphology, Ecology, and Relation to Fisheries. Washington Sea Grant Program, University of Washington, Seattle, pp 33–77. Distributed by University of Washington Press

Iwama GK, Afonso LO, Todgham A, Ackerman P, Nakano K (2004) Are hsps suitable for indicating stressed states in fish? J Exp Biol 207(1):15–19

Jeffries KM, Hinch SG, Sierocinski T, Clark TD, Eliason EJ, Donaldson MR et al (2012) Consequences of high temperatures and premature mortality on the transcriptome and blood physiology of wild adult sockeye salmon (Oncorhynchus nerka). Ecol Evol 2:1747–1764

Jones KA, Brown SB, Hara TJ (1987) Behavioral and biochemical studies of onset and recovery from acid stress in Arctic char, *Salvelinus alpines*. Can J Fish Aquat Sci 44:373–381

Jorgensen EH, Christiansen JS, Jobling M (1993) Effects of stocking density on food intake, growth performance and oxygen consumption in Arctic char (*Salvelinus alpinus*). Aquaculture 110:191–204

Kenneth BD (2006) Management of Physiological Stress in Finfish Aquaculture North American. J Aquac 68:116–121

Leatherland JF, Cho CY (1985) Effect of rearing density on thyroid and inter-renal gland activity and plasma and hepatic metabolite levels in rainbow trout, *Salmo gairdneri* Richardson. *J Fish Biol* 27:583–592

Lesser MP (2006) Oxidative Stress in Marine Environments. Biochemistry and Physiological Ecology 68:253–278

Marcucella H, Abramson CI (1978) Behavioral toxicology and teleost fish. In: Mostofsky DI (ed) Behavior of fish and Other Aquatic animals. Academic press, New York, pp 33–37

Martinez-Porchas M, Martinez-Cordova LR, Ramos-Enriquez R (2009) Cortisol and Glucose: Reliable indicators of fish stress? Pan-American. J Aquat Sci 4(2):158–178

Masters AL, Vinci BJ et al (2008) Performance characterization of influent and effluent treatment systems: A case study at Craig Brook National Fish Hatchery. Aquac Eng 38(1):66–76

Meeuwig I, Bayer J, Seelye J (2013) Effects of temperature on survival and development of early life stage Pacific and Western brook Lampreys. T Am Fish Soc 134:119–127

Melotti P, Roncarati A, Garella E, Carnevali O, Mosconi G, Polzonetti-Magni A (1992) Effect of handling and capture stress on plasma and androgen levels in brown trout, *Salmo trutta morpho fario*. J Appl Ichthyol 8:234–239

Munday PL, Dixson DL, Donelson JM, Jones GP, Pratchett MS, Devitsina GV (2009) Ocean acidification impairs olfactory discrimination and homing ability of a marine fish. Proc Natl Acad Sci U S A 106(6):1848–1852

Noga EJ (2010) Fish Disease: Diagnosis and Treatment. Wiley, Ames, IA

Nomura T, Yoshimizu M, Kimura T (1992) An epidemiological study of furunculosis in salmon propagation. In: Kimura T (ed) Proceedings of the Oji International Symposium of Salmonid Disease. Hokkaido University Press, Sapporo, pp 187–193. Olla. 1996

Olla BL, Davis MW, Ryer CH (1998) Understanding how the hatchery environment represses or promotes the development of behavioral survival skills. Bull Mar Sci 62(2):531–550

Pankhurst NW, Carragher JF (1992) Oocyte maturation and changes in plasma steroid levels in snapper *Pagrus* (= *Chrysophrys*) *auratus* (Sparidae) following treatment with human chorionic gonadotropin. Aquaculture 101:337–347

Pickering AD, Pottinger TG, Carragher J, Sumpter JP (1987) The effects of acute and chronic stress on the levels of reproductive hormones in the plasma of mature brown trout, Salmo trutta L. Gen Comp Endocrinol 68:249–259

Piper, R.G., McElwain, I.B., Orme, L.E., McCraren, J.P., Fowler, L.G., Leonard, J.R. (1982). Fish Hatchery Management. American Fisheries

Plumb RA (1994) Variations of the least horizontal stress magnitude in sedimentary rocks: 1st North American Rock Mechanics Symposium. North American Rock Mechanics Association Society, Bethesda, MD, USA, p 517

Portz DE, Woodley CM, Cech JJ Jr (2006) Stress-associated impacts of short-term holding on fishes. Rev Fish Biol Fish 16:125–170

Rehman S, Gora AH, Ahmad I, Rasool SI (2017) Stress in Aquaculture Hatcheries: Source, Impact and Mitigation. Int J Current Microbiol Appl Sci 6(10):3030–3045

Romero LM, Reed JM (2005) Collecting baseline corticosterone samples in the field: is under 3 min good enough? Comp Biochem Physiol A 140:73–79

Sadoul B, Vijayan MM (2016) Stress and Growth. In: Schreck CB, Tort L, Farrell AP, Brauner CJ (eds) Fish Physiology-Biology of Stress in Fish, vol 35. Academic Press, San Diego, CA

Saeij JP, Vries BJ, Wiegertjes GF (2003) The immune response of carp to *Trypanoplasma borreli*: kinetics of immune gene expression and polyclonal lymphocyte activation. Dev Comp Immunol 27:859–874

Schlenk D, Davis KB, Griffin BR (1999) Relationship between expression of hepatic metallothionein and sublethal stress in channel catfish following acute exposure to copper sulphate. Aquaculture 177:367–379

Schreck CB, Solazzi MF, Nickelson TE, Johnson ST (1989) Transportation stress affects performance of Coho salmon, *Oncorhynchus kisutch*. Aquaculture 82:15–20

Selye H (1973) The evolution of the stress concept. Am Sci 61:629–699

Selye H (1984) The stress of life. McGraw Hill Publ, 515 pp

Short CE, Crim LW, Morgan MJ (1995) The effects of stress on spawning performance and larval development in Atlantic cod, *Gadus morhua*. In: Goetz FW, Thomas P (eds) *Reproductive Physiology of fish* 1995. University of Texas, Austin

Soares F, Davis MT, Pousae-Ferreira P (1994) Development of the swim bladder of cultured *Sparus aurata* L.: a histological study. Aquaculture Fish. Manage 25:849–854

Sodeberg RW, Meade JW (1987) Effects of rearing density on growth, survival, and fin condition of Atlantic salmon. Progressive Fish-Cult 49:280–283

Summers CH (2002) Social interaction over time, implications for stress responsiveness. Integr Comp Biol 42:591–599

Tamazouzt L, Béatrice C, Pascal F (2000) Tank wall colour and light level affect growth and survival of Eurasian perch larvae (Perca fluviatilis L.). Aquaculture 182:85–90

Tomasso JR, Davis KB, Simco BA (1981) Plasma corticosteroid dynamics in channel catfish (*Ictalurus punctatus*) exposed to ammonia and nitrite. Can J Fish Aquat Sci 38:1106–1112

Torres-Núñez E, Cal R, Rotllant J (2014) Phenotypic plasticity during early ontogeny in cultured turbot (*Scophthalmus maximus*): changes in dorsal and anal fin ray counts by water temperature. J Appl Ichthyol 30:762–766

Tort L, Sunyer JO, Gómez E, Molinero A (1996) Crowding stress induced changes in serum haemolytic and agglutinating activity in the gilthead sea bream *Sparus aurata*. *Vet Immunol Immunopat*h 51: 179–188

Trushenski JT, Bowker JD, Gause BR, Mulligan BL (2012) Chemical and electrical approaches to sedation of hybrid striped bass: induction, recovery, and physiological responses to sedation. Trans Am Fish Soc 141:455–467

Tucker CS (ed) (1985) Channel Catfish Culture. Elsevier, Amsterdam, p 657

Valavandis A, Vlahogianni T, Dassenakis M, Scoullos M (2006) Molecular biomarkers of oxidative stress in aquatic organisms in relation to toxic environmental pollutants. Ecotoxicol Environ Saf 64:178–189

Varsamos S, Flik G, Pepin JF, Bonga SE, Breuil G (2006) Husbandry stress during early life stages affects the stress response and health status of juvenile sea bass Dicentrarchus labrax. Fish Shellfish Immunol 20(1):83–96

Vazzana M, Cammarata M, Cooper EL, Parrinello N (2002) Confinement stress in sea bass (Dicentrarchus labrax) depresses peritoneal leukocyte cytotoxicity. Aquaculture 210:231–243

Vijayan MM, Leatherland JF (1988) Effect of stocking density on the growth and stress response in brook charr, *Salvelinus fontinalis*. Aquaculture 75:159–170

Vijayan MM, Ballantyne JS, Leatherland JF (1990) High stocking density alters the energy metabolism of brook charr, *Salvelinus fontinalis*. Aquaculture 88:371–381

Wedemeyer GA (1996) Physiology of Fish in Intensive Culture Systems. Chapman & Hall, ITP, New York, p 232

Wedemeyer GA, Goodyear CP (1984) Diseases caused by environmental stressors. In. Inne, 0. (ed.) Diseases of mane animals, Vol. IV, Part 1, Pisces. Biologische Anstalt Helgoland, Hamburg, p. 424–434

Wedemeyer GA (1997) Effects of rearing conditions on the health and physiological quality of fish in intensive culture. In: Wama GK, Pickering AD, Sumpter JP, Schreck CB (eds) Fish Stress and Health in Aquaculture. Cambridge University Press, Cambridge, pp 35–71

Wendelaar Bonga SE (1997) The stress response in fish. Physiol Rev 77:591–625

Wheaton FW (1977) Aquaculture Engineering. Wiley, New York, p. 708. Willmer, P., Stone, G., Johnston, I., 2000: Temperature and its effects. In: Willmer P., Stone G., Johnston I. Environmental Physiology of Animals. Blackwell Science Ltd., USA, *pp.* 192–244

Wiseman S, Osachoff H, Bassett E, Malhotra J, Bruno J, Van Aggelen G, et al. (2007) Gene expression pattern in the liver during recovery from an acute stressor in rainbow trout. *Comp Biochem Physiol* D 2, 234–244

Chapter 51
Exotic and Invasive Freshwater Fishes in the Tigris-Euphrates River System

Hamid Reza Esmaeili

Abstract In this paper, 30 exotic fish species belonging to 12 orders and 14 families (Acipenseridae, Atherinidae, Centrarchidae, Cichlidae, Cyprinidae, Esocidae, Gobiidae, Heteropneustidae, Lepisosteidae, Pangasiidae, Percidae, Poecilidae, Salmonidae, Serrasalmidae) are listed from the Tigris-Euphrates River System (TERS) of which four species have not been collected during the last few years. Three species have been recorded in 2019 and their status should be monitored. Some of these fish species (e.g., *Captodon zillii*, *Carassius auratus*, *C. gibelio, Cyprinus carpio*, *Gambusia holbrooki*, *Oreochromis aureus*, *O. niloticus*, *Pesudorasbora parva,* and *Poecilia latipinna*) have established breeding populations, become invasive elements. Some other fishes are frequently stocked by the Department of Fisheries (such as *Hypophthalmichthys molitrix*, *H. nobilis, Ctenopharyngodon idella*) and few other fishes are infrequently recorded from the natural environments/habitats. The main reasons for these fish introductions are activities like aquaculture, recreational or sport fishing, malaria control, ornamental purposes, scientific researches, demonstration in national fairs, and accidental introduction resulted from the activities of aquarium industries. *Pesudorasbora parva* was introduced by accident together with the exotic carps and currently is broadly distributed in inland waters. The deliberate release of exotic species is a complex topic that requires considering several different variables. Suitable management approaches to reduce the impacts of biological invasion, include comparative investigations on the biology and life history of both the introduced fishes, and closely related endemic elements with the similar ecological requirements. Tracking of the invasive fishes, study of their ecological interactions with native fishes, and finally public awareness programs are suggested.

Keywords Exotic · Threats · Diversity · Management · Middle East · Persian Gulf

---

H. R. Esmaeili (✉)
Ichthyology Research Laboratory, Zoology Section, Department of Biology, College of Sciences, Shiraz University, Shiraz, Iran
e-mail: hresmaeili@shirazu.ac.ir

L. A. Jawad (ed.), *Tigris and Euphrates Rivers: Their Environment from Headwaters to Mouth*, Aquatic Ecology Series 11,
https://doi.org/10.1007/978-3-030-57570-0_51

51.1 Introduction

An invasive fish is alien or non-native species (Table 51.1) that has been introduced into areas beyond its natural distributional range and established self-reliant populations. The invasive fish is spreading outside its primary site of introduction, causing damage on the environment structure, economy of the involved countries, and human well-being (Kolar and Lodge 2001; Lymbery et al. 2014; Esmaeili et al. 2014). Biological fish invasion is now a major worldwide environmental issue (Gozlan et al. 2010a, b). Human population growth, increase in transportation, and globalization in economics have assisted introduction rate of exotic organisms all over the map (Vitousek et al. 1997; Sakai et al. 2001; Lymbery et al. 2014). Fish invasive species is an environmental problem in all ecosystems of the world (marine, brackish, and freshwater).

Welcomme (1981, 1988) has reviewed the movement of fishes between countries and basins. Exotic fishes have been primarily introduced into the different tributaries of the Tigris-Euphrates River System (TERS) through deliberately or unintentionally anthropogenic activities. The invasion in the region was started with intentional introductions of *Gambusia* (Poeciliidae) to control malaria and *Cyprinus çarpio* (Cyprinidae) to increase fish production in the 1920s and late 1930s, respectively (Coad 1996a, b, Coad 2017; Esmaeili et al. 2007, 2010a, b). Similar to trends observed in other basins and river systems, the number of introduced and invasive

Table 51.1 Terminology for alien species (after Lymbery et al. (2014) and Gozlan et al. (2010a, b))

Term	Definition
Native species	A species occurring within the range it occupies (or could occupy) naturally, independent of human activity
Alien (exotic, nonindigenous) species	A species that has been transported by human activity into an area outside its natural range
Introduced species	Alien species that has been transported by humans into an area outside its natural range, but has not yet established self-sustaining populations in the wild
Established (naturalized) species	Alien species that has been introduced and established self-sustaining populations in the wild
Feral species	Alien species that has been kept in captivity or domestication after introduction, but has escaped or been released to establish self-sustaining populations in the wild
Invasive species	Alien species that has been introduced, become established and is expanding its range, usually with deleterious consequences for native species
Introduction	The deliberate or accidental release into the wild of a nonnative species
Translocation	The human-assisted movement of species within a specific region (i.e., movement of fish within rivers, lakes, ponds, etc.)
Dispersal	The natural dissemination (i.e., nonhuman assisted) of a species from its point of introduction

fishes in TERS have significantly increased over the last few decades (Coad 1996a, b; Esmaeili et al. 2007, 2010a, b, 2011a, b, 2013, 2017a, b; Khaefi et al. 2014; Kaya et al. 2016; Fatemi et al. 2012; Grave and Ghane 2006; Ivanov et al. 2000; Sadeghi et al. 2013).

The present paper provides a list of exotic and invasive fishes in the Tigris-Euphrates River System. In addition, we discuss their possible environmental impacts, and also effective management plans to conserve the endemic and native ichthyofauna of this vast basin.

51.2 The Tigris-Euphrates Basin

The Tigris-Euphrates basin is one the main drainage of Southwest Asia. It is located between the major basins of the Nile River to the west, the Indus River to the east, and the drainages of the Black and Caspian seas to the north (Coad 1991, 1996b) (Fig. 51.1), hence its ichthyodiversity is interesting in terms of its native, endemic, and exotic fishes. It can be considered as a part of an Irano-Anatolian hotspot of freshwater fish diversity, holding a unique fish fauna containing distinct European, Asian, and African elements and also receiving elements from Nearctic and Neotropical realms (Fig. 51.1).

The Tigris and the Euphrates rivers originating from the Highlands of Armenian in the eastern Turkey, flow through Syria and Iraq and join to form the Arvand River (Shatt-al-Arab) in border of Iran and Iraq. Arvand River enters the head of the Persian Gulf (Fig. 51.1).

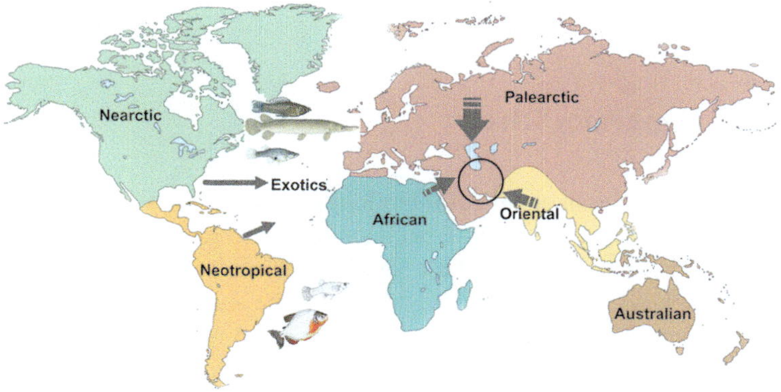

Fig. 51.1 Introduction of exotic fish species from different zoogeographic regions of the world into Iran

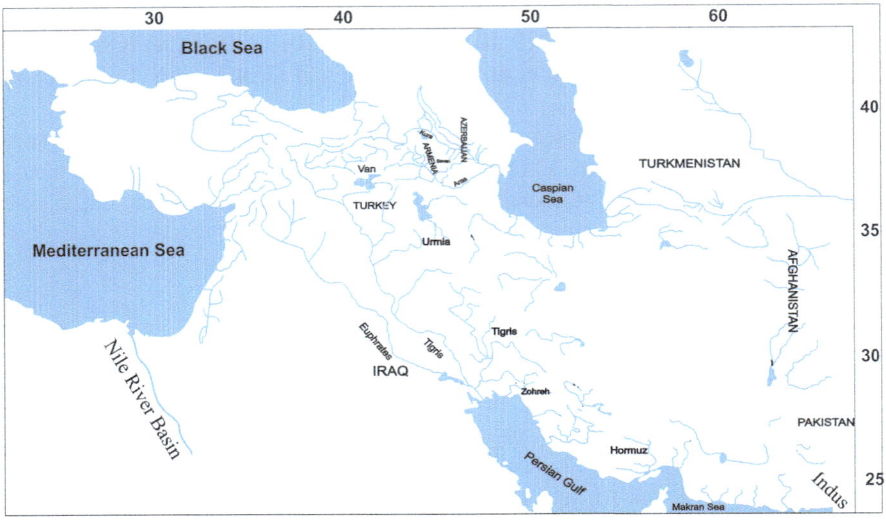

Fig. 51.2 Schematic representation of the Tigris-Euphrates River System in Southwest Asia

51.3 The Tigris River

The Tigris River (Farsi: Dejleh, Tigreh) originates from the Lake Hazar, which is located south of Elazig in Turkey between the Murat River and the Euphrates. It runs southeast, in the border of Syria with Turkey before entering Iraq almost parallel to the Euphrates River (Coad 1991) (Fig. 51.2). Since the Tigris River receives a number of tributaries from the Zagros Mountains, especially the Great Zab River, the Little Zab River, and the Diyala River, it is a swifter and larger river than the Euphrates. The Tigris River which is over 1900 km long (see Coad 1991, 1996a, b), joins the Euphrates River and forms the Arvand River (Shatt-al-Arab) which enters the Persian Gulf.

51.4 The Euphrates River

The Euphrates River (with about 2700 km long) is known as Forat in Iran, Firat in Turkey, and Furat in Iraq and Syria. Two main head streams of the Euphrates River (the Euphrates proper or Karasu and the Murat River) are in Turkey (Fig. 51.2). The sources of these head streams are located between the endorheic Lake Van and the Black Sea. About 22 km north of Erzurum, the Karasu rises and it flows approximately 460 km in a southwesterly direction to join the Murat River. Approximately 65 km southwest of Mount Ararat, the Murat River emerges and it flows nearly 610 km west in a southwesterly direction to join the Karasu. Before connection of the Euphrates and Tigris Rivers in northwest of Basrah in Iraq, the Euphrates runs within 160 km of the Mediterranean Sea in south of Anatolia. In the mountains of Turkey, there are several small tributaries to the Euphrates basin but it receives only two important tributaries, namely the Balikh and the Khabur Rivers in Syria.

51.5 Materials and Methods

This paper is mainly based on field work data since 2004, and also based on Saadati (1977), Armantrout (1990), Holčík and Razavi (1991), Coad and Abdoli (1993), Coad (1991, 1995, 1996a, b, 2017), Al-Sa'adi (2007), Esmaeili et al. (2007, 2010a, b, 2013, 2014, 2017a, b), Mohamed et al. (2008), Al-Sa'adi et al. (2012), Hashemi and Ansary (2012), Khamees et al. (2013), Al-Faisal and Mutlak (2014), Khaefi et al. (2014), Kaya et al. (2016), Mutlak et al. (2017), and Zarei and Rajabi-Maham (2017). Terminology for alien species follows Lymbery et al. (2014) and Gozlan et al. (2010a, b) (Table 51.1). Taxonomic status follows Nelson et al. (2016) and Fricke et al. (2020) (Fig. 51.2).

51.6 Results

Based on the obtained results, exotic fishes reported from the Tigris-Euphrates basin comprise about 30 species in 12 orders, 14 families, and 27 genera (Table 51.2). Cyprinids (the family Cyprinidae) with 11 species are ranked the first, and are followed by Cichlidae and Salmonidae with three species each, Poeciliidae and Acipenseridae with two species each, and nine families each with only one exotic species.

The most widespread exotic fish with invasive behavior is the Eastern Mosquito fish, *Gambusia holbrooki* while some recently introduced fishes, e.g., *Atractosteus spatula* and *Piaractus brachypomus* show restricted distributions. Some of the reported exotic species (e.g., *Cyprinus carpio, Carassius auratus, C. gibelio, Gambusia holbrooki, Oreochromis aureus, O. niloticus, Captodon zillii, Pesudorasbora parva,* and *Poecilia latipinna*) have established sustainable breeding and invader populations. However, several species are frequently stocked by the Fisheries Departments (e.g., the Iranian Fisheries Department) including *Hypophthalmichthys molitrix, H. nobilis, Ctenopharyngodon idella* and a few other fishes are infrequently recorded from natural habitats. Four species (*Esox lucius, Micropterus salmoides, Salvelinus fontinalis,* and *Sander lucioperca*) have not been collected during the last few years.

Order Lepisosteiformes
Family Lepisosteidae
***Atractosteus spatula* (Lacepède, 1803)**
Alligator Gar
(Fig. 51.3)

Table 51.2 Exotic fish species in the Tigris-Euphrates River System

Order	Family	Species
Lepisosteiformes	Lepisosteidae	*Atractosteus spatula* (Lacepède, 1803)
Cypriniformes	Cyprinidae	*Alburnus hohenackeri* Kessler, 1877
		Carassius auratus Linnaeus, 1758
		Carassius gibelio (Bloch, 1782)
		Ctenopharyngodon idella (Valenciennes, 1844)
		Cyprinus carpio Linnaeus, 1758
		Hemiculter leucisculus (Basilewsky, 1855)
		Hypophthalmichthys molitrix (Valenciennes, 1844)
		Hypophthalmichthys nobilis (Richardson, 1845)
		Pseudorasbora parva (Temminck & Schlegel, 1846)
		Rhodeus amarus (Bloch, 1782)
Characiformes	Serrasalmidae	*Piaractus brachypomus* (Cuvier, 1818)
Siluriformes	Heteropneustidae	*Heteropneustes fossilis* (Bloch, 1794)
	Pangasiidae	*Pangasianodon hypophthalmus* (Sauvage, 1878)
Salmoniformes	Salmonidae	*Oncorhynchus mykiss* (Walbaum, 1792)
		Salmo trutta Linnaeus, 1758
		Salvelinus fontinalis (Mitchill, 1814)
Esociformes	Esocidae	*Esox lucius* Linnaeus, 1758
Gobiiformes	Gobiidae	*Rhinogobius lindbergi* Berg, 1933
Cichliformes	Cichlidae	*Oreochromis aureus* (Steindachner, 1864)
		Oreochromis niloticus (Linnaeus, 1758)
		Coptodon zillii (Gervais, 1848)
Atheriniformes	Atherinidae	*Atherina boyeri* Risso, 1810
Cyprinodontiformes	Poeciliidae	*Gambusia holbrooki* Girard, 1859 *Poecilia latipinna* (Lesueur, 1821)
Perciformes	Centrarchidae	*Micropterus salmoides* (Lacepède, 1802)
	Percidae	*Sander lucioperca* (Linnaeus, 1758)

Fig. 51.3 *Atractosteus spatula*, collected from Lake Zarivar, Tigris River drainage, Iran (Chya Green Association, Marivan).

Large size and broad, short snout with two rows of sharp teeth in their mouths are distinguished characters of this species. It is dorsally dark olive-green, and ventrally fading to yellowish white. Juveniles have a mid-dorsal light stripe which is bordered by thin dark lines covering the tip of snout to the dorsal fin. They also have a dark lateral band spreads along the sides with unequal borders. Often, there are oval-shaped dark spots on the unpaired fins of *A. spatula*. Adult fishes lack spots on their body.

Fig. 51.4 *Alburnus hohenackeri*, an exotic fish in the Tigris-Euphrates River System

Introduction of *Atractosteus spatula* to the Tigris-Euphrates basin dates back to 2015 when a single unpreserved specimen (about 65–70 cm) (Fig. 51.3) was collected from the Lake Zarivar/Wetland (Marivan; 35°32′53.32"N, 46°7′19.08″E), a Tigris River tributary of Iran (see Esmaeili et al. 2017b). The identification was made based on digital photographs and videos. Being sold in pet shops, these fishes are kept in aquaria worldwide including Iran. It is also reported that aquarists frequently release unwanted large fishes. Successful reproduction of alligator gar in Zarivar is low which is based on low number of collected specimens. At the same time, the possibility of gradual release of this exotic fish cannot be excluded. As more specimens of alligator gar have been observed by local fishermen in the recent years, monitoring of this wetland is highly recommended (see Esmaeili et al. 2017b).

There is another report of presence of this fish in the Tigris-Euphrates River System. Mutlak et al. (2017) reported a single specimen of alligator gar (TL 90 cm) from southern reaches of the Shatt al-Arab/Arvand River in Iraq (30°30′04.56″N; 47°51′57.18″E).

Although, *Atractosteus spatula* is native to North America, however, it has been reported from Turkmenistan (Salnikov 2010), India, Hong Kong, and Singapore. Due to capability of the alligator gar in creating a sustainable population, it might act as a successful invasive species and becomes a harmful fish. Alligator gar feeds mainly on fishes and it is a successful predator in its native environment. The eggs of *A. spatula* are poisonous, but their consumption by human is unlikely (Boschung and Mayden 2004; Goddard 2009). Additionally, this fish could damage fishing nets due to the shape of its head and the sharp teeth.

Order Cypriniformes
Family Cyprinidae
Alburnus hohenackeri **Kessler, 1877**
North Caucasian Bleak
(Fig. 51.4)

The key distinction characters to distinguish members of the genus *Alburnus* from its closely related genus, *Alburnoides* are the total gill rakers count of 16–29 (usually 20 or more) in this genus vs. 5–12 (usually 7–10) in the genus *Alburnoides*, and longer gill rakers in *Alburnus* vs. shorter in *Alburnoides* (Coad 2017). The other characters are: 7–9 (usually 8) dorsal fin branched rays, 2–4 dorsal fin unbranched rays, 10–21 anal fin branched rays and 3–4 anal fin unbranched rays, 11–16 pectoral fin branched rays, and 6–9 pelvic fin branched rays, lateral line scale count 36–55,

Fig. 51.5 *Carassius auratus*, an exotic fish in the Tigris-Euphrates River System

gill rakers 15–29, elongate reaching the third, or rarely second, below when appressed, vertebrae 36–46 and pharyngeal teeth usually 2,5–5,2 (Coad 2017).

Alburnus hohenackeri has been reported from the Choghakhor Wetland 31°5′ 24.0″ N, 50°56′ 08.0″ E, Alt. 2280 m (Zareian et al. 2013), Zarivar Lake (Tigris River drainage), and Ab-e Sirvan in the upper Diyala River all in the Tigris River drainage (Zareian et al. 2013). It seems that they have established breeding populations.

***Carassius auratus* (Linnaeus, 1758)**

Goldfish

(Fig. 51.5)

Absence of barbels and presence of spines in both anal and dorsal fins are the specific characteristics of gold fish. It is distinguished from other species of *Carassius* in the region by specific coloration of golden-brown or bronze, strongly serrated last simple anal and dorsal rays, usually five branched anal rays, 38–47 total gill rakers, and 26–31 total lateral line scales. Other morphological characters are: 3–4 unbranched and 12–20 branched rays in dorsal fin, anal fin with 2–4, commonly 3, unbranched and 5–6, commonly 5, branched rays, pectoral fin branched rays 11–18, and pelvic fin branched rays 6–9, commonly 8, lateral line scales 21–36, gill rakers counts 35–54, total vertebrae 25–34, and pharyngeal teeth formula 4–4 (Coad 2017).

It is common presumably throughout the Tigris-Euphrates drainage. It is also reported for farm ponds in Iran and Iraq (see Ali 1985; Coad 2017).

Introductions of some populations are definitely the result of discarded aquarium fish.

***Carassius gibelio* (Bloch, 1782)**

Prussian Carp

(Fig. 51.6)

It is distinguished from other species of *Carassius* by specific coloration of silvery-brown, strongly serrated last simple anal and dorsal rays, five branched anal fin rays, 37–52 gill rakers, and 29–33 total lateral line scales (Kottelat and Freyhof 2007).

Fig. 51.6 *Carassius gibelio,* an exotic fish in the Tigris-Euphrates River System

Fig. 51.7 *Ctenopharyngodon idella,* an exotic fish in the Tigris-Euphrates River System

It has been reported from different river drainages of Iran, Iraq, and Turkey (Aydın et al. 2011; Jawad et al. 2012; Esmaeili et al. 2017a). Kaya et al. (2016) reported *C. gibelio* from Batman Stream, Ambar Stream, and Bismil, Salat Stream, all located in the upper Tigris River, Turkey.

Ctenopharyngodon idella (Valenciennes, 1844)

Grass Carp

(Fig. 51.7)

Position of the eyes located low on the side of the head, position of anal fin being far back on the body close to the caudal fin, and presence of large, parallel grooves on the surface of pharyngeal teeth are the main morphological characteristics of this exotic species (Coad 2017).

Other morphological characters include 6–8, commonly 7, branched and three unbranched dorsal fin rays, 7–9, commonly 8, branched and three unbranched anal fin rays, 13–20 pectoral fin branched rays, 7–8 pelvic fin branched rays, lateral line scales 34–47, gill rakers number 15–18, and vertebrae 40–47 (Coad 2017).

The grass carp, has been introduced to reservoirs throughout Iran beginning in the 1950s (Armantrout 1980). This Chinese carp was introduced to Khuzestan Province, Iran in the 1970s to control vegetation in irrigation ditches and to Iraqi fish farms (Anonymous 1970; Saadati 1977; Shireman and Smith 1983; Krasznai 1987; Petr 1987; White 1988). It has also been reported from the Lake Zarivar, Kurdistan (Esmaeili et al. 2011a).

Fig. 51.8 *Cyprinus carpio*, an exotic fish in the Tigris-Euphrates River System

It has been introduced to Iraqi waters, the first record being for 1968 from Japan for use in pond culture. As well as being present in fish ponds throughout Iraq, this species is found in the southern marshes, the Arvand River (the Shatt al Arab River and its tributaries), the Tigris and Diyala rivers, and in Lake Habbaniyah (Iraq).

There is no record of this fish from the upper parts of the Tigris River drainages in Turkey (Kaya et al. 2016).

Cyprinus carpio Linnaeus, 1758

Common Carp

(Fig. 51.8)

This introduced Chinese carp can be easily distinguished from the other species of the Tigris River drainage by having the long dorsal fin, presence of spine in both dorsal and anal fins, and presence of two pairs of barbels. The other main morphological characteristics are: 2–5 unbranched and 14–23, commonly 18–20, branched dorsal fin rays, anal fin with 2–4, commonly 3, unbranched and 3–7, commonly 5, branched rays, pectoral fin branched rays 13–19, pelvic fin branched rays 5–9, commonly 8, lateral line scales 26–41, mostly 36–39, gill rakers 17–29, vertebrae 32–39, and pharyngeal teeth usually 1,1,3-3,1,1 with some variations.

Cyprinus carpio is widely used in aquaculture to enhance fish production and hence, it is transplanted to almost all drainages in the Tigris-Euphrates (Coad 2017). It has been recorded from the Dez and Karun River drainages of Iran. Pond culture of carp was started in Iraq after its introduction in 1955 (Ahmed and Taher 1988). This species is now widespread in Iraq found in all the main rivers and marshes (Al-Faisal and Mutlak 2014). It has been reported from Al-Hawizeh marsh, Iraq (Mohamed et al. 2008) and seems that it has established breeding populations in this wetland.

This species is one of the first transplanted fish species of Turkey. Its introduction started in the1960s leading to a highly productive fishery in this country (Innal and Erk'akan 2006). *Cyprinus carpio* has been reported from the Devegecidi Dam Lake (Kelle 1978), Batman Stream, and Diyarbakır, Yenişehir, Tigris (Kaya et al. 2016).

***Hemiculter leucisculus* (Basilewsky, 1855)**

Sawbelly (Sharpbelly)

(Fig. 51.9)

Fig. 51.9 *Hemiculter leucisculus,* an exotic fish in the Tigris-Euphrates River System

Fig. 51.10 *Hypophthalmichthys molitrix,* an exotic fish in the Tigris-Euphrates River System

Presence of a sharp and scaleless keel, extending from the vent to the throat suddenly downward lateral line at the pectoral fin, and presence of a sharp spine with a flexible tip as a modified unbranched last dorsal fin ray, are the main key characteristics of sawbelly. The other morphological features are: dorsal fin with 2–3 unbranched and 6–8, commonly 7, branched rays, anal fin with three unbranched and 10–18, commonly13–14, branched rays (with some variation in different populations), lateral line scales 43–55, total gill rakers 17–29, and total vertebrae 42.

The sawbelly has been recorded from the Zarivar Lake, Kurdistan (Esmaeili et al. 2011a), Maroon River, Khuzestan Province (Zareian et al. 2015), and from the Hawizeh Marshes in, southern Iraq (Coad and Hussain 2007; Al-Faisal 2008; Al-Faisal and Mutlak 2014). Apparently, the sawbelly has been released accidentally with farmed exotic carps (Coad and Hussain 2007; Zareian et al. 2015). There is no record of this fish from the upper Tigris River of Turkey (Kaya et al. 2016).

It seems that *Hemiculter leucisculus* has established self-reliant populations in different water bodies of the Mesopotamia area due to rapid growth rate, high fecundity, and habitat preference (see Rosenthal in: Welcomme 1988; Serov et al. 2006; Coad and Hussain 2007; Thinh et al. 2012).

***Hypophthalmichthys molitrix* (Valenciennes, 1844)**

Silver Carp

(Fig. 51.10)

Presence of an abdominal compressed keel running from the breast to the vent, low position of the eyes, and minute scales are the main key morphological features of *H. molitrix*. *Hypophthalmichthys molitrix*is distinguished from its congeneric species, *H. nobilis*, by shorter pectoral fins vs. long pectoral fins extending behind the pelvic fin origin in *H. nobilis*, a longer keel running from the breast to the vent vs. a shorter keel (pelvic fins to anus) in *H. nobilis*, and gill raker structure being continuous spongy mass vs. free with no spongy root mass in *H. nobilis*.

Fig. 51.11 *Hypophthalmichthys nobilis,* an exotic fish in the Tigris-Euphrates River System.

The other morphological features are: short dorsal fin with 2–3 unbranched and 6–7 branched rays, anal fin with 2–3, commonly 3, unbranched and 11–15 branched rays, pectoral and pelvic fin branched rays 17 and 7, respectively, lateral line scales 85–125, gill rakers more than 650 making a spongy mass, and total vertebrae 36–40.

This species has been introduced to reservoirs and marshes throughout Iran (Esmaeili et al. 2014; Coad 2017), being cultivated in fish farms in Iraq, Syria, and Iran (Nellen and Ruckes 1975; Ali et al. 1989). It has been introduced to Iraqi waters, the first record being for 1968 from Japan to be used in pond culture. It is known from the Arvand River (Shatt al Arab River) drainage, the Tigris River, and Lake Habbaniyah. There is no record of this fish from the upper Tigris River of Turkey (Kaya et al. 2016).

Hypophthalmichthys nobilis **(Richardson, 1845)**
Bighead Carp
(Fig. 51.11)

Hypophthalmichthys nobilis is distinguished from its congeneric species, *H. molitrix*, by long pectoral fin extending behind the pelvic-fin origin vs. shorter pectoral fin in *H. molitrix*, a short-scaled keel running from pelvic to anal fins vs. long-naked keel extending from breast to vent in *H. molitrix*, gill raker structure being 240–300 long, free with no spongy root mass vs. continuous spongy mass in *H. molitrix*, darker overall coloration vs. lighter in *H. molitrix*, presence of dark and large irregular blotches on the flank vs. absent in *H. molitrix* (Kottelat and Freyhof 2007).

The other morphological features are: short dorsal fin with three unbranched and seven unbranched rays, anal fin with three unbranched and 11–14 branched rays, pectoral and pelvic fin branched rays 16–19 and 7–9, respectively, lateral line scales 92–115, total vertebrae number 36–41, pharyngeal teeth 4–4, gill rakers 240–300.

This fish is native to China, however, it has been introduced to numerous countries in different water bodies outside its native distribution range including the Tigris-Euphrates drainage. It has now almost global distribution. *Hypophthalmichthys nobilis* was reported from fish farms and reservoirs in Khuzestan, Iran (Krasznai 1987; Petr 1987). It has been reported from the Khersan, Karun, Jarrahi, Dez, Kashkan, and Karkheh Rivers in the Tigris River drainage. This species was one of the candidates to be introduced to Syria (Nellen and Ruckes 1975). It was introduced in Iraq to enhance fish production in fish farms. There is no record of this fish from the upper Tigris River of Turkey (Kaya et al. 2016). Bighead

Fig. 51.12 *Pseudorasbora parva,* an exotic fish in the Tigris-Euphrates River System

carp is being used to control plankton bloom in farms. Due to its global distribution, it can effectively transmit parasites and diseases.

***Labeo rohita* (Hamilton, 1822)**
Rohu

It is distinguished by large eyes, not visible from underside of head, small and inferior mouth, thick and fringed lips with distinct inner lobes, presence of a pair of small maxillary barbels, dorsal fin inserted midway between snout tip and base of caudal fin having three unbranched and 12–14 branched rays, anal-fin origin behind distal base of dorsal fin with 2–3 unbranched and five branched rays, pectoral fin shorter than head with one unbranched and 16–18 branched rays, pelvic-fin origin behind dorsal-fin origin with one unbranched and eight branched rays, caudal fin deeply forked, scales cycloid, and lateral line with 40–44 scales.

Occurrence of *L. rohita* in the Karun River drainage of Iran was first recorded and documented by Eagderi et al. (2019). This species probably escaped from rearing farms around the Karun River due to careless management (see Eagderi et al. 2019).

***Pseudorasbora parva* (Temminck & Schlegel, 1846)**
Stone Moroko, Topmouth Gudgeon
(Fig. 51.12)

It is distinguished by having unique structure of mouth, being very small, located entirely before the nostril level. The other morphological features are: dorsal fin with 2–3 unbranched, commonly 3, and 7–8, commonly 7, branched rays, anal fin with 2–3 unbranched, commonly 3, and 5–7 branched rays, commonly 6, pectoral fin branched rays 11–14, pelvic fin branched rays 6–8, usually 7, lateral line scales 29–40, (incomplete in some fishes), gill rakers 6–16, usually 9–13, total vertebrae usually 31–38, pharyngeal teeth formula usually 5–5, rarely 6–5, (see Naseka 1996; Coad 2017).

Introduction of topmouth gudgeon to Iran is incidental, it is now widespread distribution in different basins including Tigris River drainage. *Pseudorasbora parva* has been recorded from several water bodies in the Tigris-Euphrates basin including springs near Kermanshah, the Lake Zarivar in Kurdistan, springs near Kermanshah and in the Yavari, Qarasu, and Niloufar springs (Kermanshah) (Eagderi and Nasri 2012). In Turkey, it was first reported in early 1980s from the Meriç River

Fig. 51.13 *Rhodeus amarus*, an exotic fish in the Tigris-Euphrates River System

and it has expanded its distribution range throughout this country accidentally, mainly through the anthropogenic activities in aquaculture (Özcan and Tarkan 2019). As this invasive species has been considered as a strong competitor (Csorbai et al. 2014) and effective healthy carrier of the rosette agent *Sphaerothecum destruens* (Gozlan et al. 2005), monitoring of its populations is highly recommended. Based on Gozlan et al. (2010a, b), high phenotypic plasticity of the topmouth gudgeon in growth rate, early maturation age, fecundity, paternal care, and the capability to overcome new pathogens, has made it to be a strong invasive fish.

***Rhodeus amarus* (Bloch, 1782)**
European Bitterling
(Fig. 51.13)

It is distinguished by having unique reproductive structure, ovipositor in females, presence of the specific flank stripe, and few number of pored scales (Coad 2017). The other morphological characteristics are: dorsal fin with 2–4, commonly 3, unbranched and 7–11, commonly 9, branched rays, anal fin with 2–4, commonly 3, unbranched and 6–12, commonly 9, branched rays, pectoral fin branched rays 10–13, and pelvic fin branched rays 4–8, usually 6–7, pored lateral line scales 0–11, usually 4–6, along the flank 28–45, usually about 30–32 in some variation (see Abdurakhmanov 1962; Holčík and Jedlička 1994; Pipoyan 1996; Holčík in Bănărescu 1999), gill rakers 9–16, usually 10–14, vertebrae 33–39, and pharyngeal teeth usually 5–5 (Coad 2017).

The bitterling (*Rhodeus amarus*) was reported for the first time from the Iranian part of Tigris-Euphrates drainage (Eagderi and Nasri 2012). Previously, this species was reported from the Caspian Sea and Urmia Lake basins (Esmaeili et al. 2011b; Eagderi and Nasri 2012). Seven individuals of this fish were collected from a tributary of the upper Tigris-Euphrates drainage of Iran, Ghare-sou River (Kermanshah province, Iran) (see Eagderi and Nasri 2012). More specimens were collected from Razavar River, (Kermanshah province, Iran) on 18.8.2012 showing establishment of this fish (HRE).

Fig. 51.14 *Piaractus brachypomus,* an exotic fish in the Tigris-Euphrates River System

Remarks: This species has now been considered as Rhodeus caspius Esmaeili, Sayyadzadeh, Japoshvili, Eagderi, Abbasi & Mousavi-Sabet 2020.

Order Characiformes

Family Serrasalmidae

***Piaractus brachypomus* (Cuvier, 1818)**

Red-bellied Pacu

(Fig. 51.14)

Red-bellied pacu is distinguished from other fishes of the Tigris-Euphrates by the following characters: body deep and laterally compressed, adipose fin small, body silvery belly, chin and pectoral fins red. Other characters are: dorsal fin with 15–18 rays, pectoral fin with 16–19 rays, anal fin with 24–28 rays, and pelvic fin rays 8. The first few rays of the dorsal and anal fins are longer than the remaining rays. The maximum weight and length are 25 kg and 88 cm, respectively (see Innes 1966).

The presence of two individuals of this exotic fish was reported from the Zarivar Lake/Wetland (the Tigris River drainage of Iran) by Zarei and Rajabi-Maham (2017). As, the accidental releases by aquarium hobbyists and aquarium trade have already been reported for the release of *P. brachypomus* into the natural waters of Europe and India, it might be the reason of its introduction to the Zarivar Lake. There is no record of this species from the upper Tigris River of Turkey (Kaya et al. 2016).

Red-bellied pacu, is a tropical species native to the Amazon and Orinoco river basins in South America. It is highly valued as an aquaculture species, and is considered to be one of the most significant and prized species in the aquarium trade.

Predation, competition, behavioral interference, and disease transmission have been considered as possible impacts of introduced red-bellied pacu on indigenous fish species (Leunda 2010).

Siluriformes

Heteropneustidae

***Heteropneustes fossilis* (Bloch, 1794)**

Stinging Catfish

(Fig. 51.15)

*Heteropneustes fossilis*is distinguished from other fishes of the Tigris-Euphrates by the following characters: body elongate, head small and depressed, mouth small and terminal, eyes small, four pairs of well-developed barbels, dorsal fin short, and

Fig. 51.15 *Heteropneustes fossilis*, an exotic fish in the Tigris-Euphrates River System

Fig. 51.16 *Pangasianodon hypophthalmus*, an exotic fish in the Tigris-Euphrates River System (Khamees et al. 2013)

spineless with 6–7 rays, inserted almost above pelvic-fin origin, anal fin long with 60–70 rays, pectoral fin with a strong spine and seven rays, pelvic fin with one spine and five rays, no adipose fin.

The stinging catfish was introduced to the Tigris River drainage in southern Iraq in 1960 (Khalaf 1961; Zakaria 1964; Al-Hassan and Muhsin 1986) and was found the same year in Iran (Coad 2014). It is now widely distributed in all the rivers and marshes of southern Mesopotamia.

In Syria, two specimens of *H. fossilis* were captured from Lake Assad in June 2007 (see Ali et al. 2016).

It is now known from the Turkish part of the Tigris River near Diyarbakır city (Ünlü et al. 2011). It has been reported from Al-Hawizeh marsh, Iraq (Mohamed et al. 2008) which seems that it has established breeding populations in this wetland.

Heteropneustes fossilis, first recorded in Turkish headwaters of Fırat River (Euphrates) in 2011 (Ünlü et al. 2011) putatively from introductions in neighboring countries (Tarkan et al. 2015).

Kaya et al. (2016) reported this catfish from the Tigris River at Diyarbakır and the Göksu Stream at Diyarbakır, Çınar, Tigris River, Turkey.

This species was introduced to Iraq to control the freshwater snail *Bulinus truncatus*. This snail acts as a vector for the human parasite (a trematode known as *Schistosoma*) causing schistosomiasis (Al-Faisal and Mutlak 2014). It was ineffective in this regard (Jawad 2003) and eradicated *Barbus* (= *Arabibarbus*) *grypus* eggs along the Tigris River (see Coad 2014).

Pangasiidae

***Pangasianodon hypophthalmus* (Sauvage, 1878)**

Striped Catfish

(Fig. 51.16)

Pangasiids are large catfishes of the order Siluriformes with distinct morphological features: body laterally compressed, presence of one pair of maxillary and one pair of mandibular barbels, relatively long anal fin with 26–46 rays, short dorsal fin with two spines, a small adipose fin, pelvic fin with 7–8, usually 6, rays, 7–11

Fig. 51.17 *Oncorhynchus mykiss*, an exotic fish in the Tigris-Euphrates River System

branchiostegial rays, 8–9 principal caudal fin rays, and 39–52 vertebrae (see Gustiano and Pouyaud 2008; Khamees et al. 2013).

The striped catfish, *Pangasianodon hypophthalmus* can be distinguished from other catfishes of the Tigris River drainage by following characters: fins dark grey or black fins, dorsal-fin origin anterior to vertical of pelvic-fin origin with 6 branched rays, presence of a dark black stripe along lateral line and a long dark black stripe below lateral line in young individuals, body uniformly gray in large adult specimens, anal fin with dark stripe on its middle part, each caudal lobe with dark stripe (Rainboth 1996; Kottelat 2001).

The striped catfish was recorded for the first time from the Ibn Najim marsh in the middle of Iraq in 2009 and from Shatt Al-Basrah canal, south of Iraq in 2011 by Khamees et al. (2013). Single specimen (photo) of pangasiid caught from Ibn Najim marsh was preliminary identified as *Pangasius* by Brian W. Coad (see Khamees et al. 2013).

Salmoniformes

Family Salmonidae

***Oncorhynchus mykiss* (Walbaum, 1792)**

Rainbow Trout

(Fig. 51.17)

Rainbow trout is distinguished by having 115–130 lateral line scales in midlateral row, 6–9, commonly 8, branched anal fin ray, 11–17 pectoral fin rays, 9–10 pelvic fin rays, 16–17 gill rakers, thick pink to red stripe from head region to caudal base, and males without humps in breeding season.

Breeding populations of *O. mykiss* were reported by Saadati (1977) in the Dez River and the Ab-e Bazuft, Iranian tributaries of the Tigris-Euphrates drainage. There is also record of this salmonid fish from the Karun River drainage in Iran, which is another Tigris-Euphrates tributary (Coad 2017).

This species was probably first introduced in Turkey in 1969 (see Uysal and Alpbaz 2002). Kaya et al. (2016) reported it from the Batman Stream; Diyarbakır, Silvan; Çatak Stream, Van, Çatak and Şetek Stream; Bitlis; upper Tigris River in Turkey.

Competition for food and spawning grounds, piscivorous behavior, and transmission of parasites and diseases (e.g., enteric redmouth disease) can strongly threat native fish fauna (see Timur and Timur 1991; Coad 2017).

***Salmo trutta* (Linnaeus, 1758)**

Brown Trout

Brown trout is distinguished by having emerginated to truncate caudal fin in small and large individuals respectively, presence of 13–16 scales between lateral line and end of adipose fin, 110–36 lateral line scales, 14–16 gill rakers, maxilla reaching or behind posterior margin of eye, 30–60 pyloric caeca, and males with a hooked lower jaw and almost golden body color in breeding season (Kottelat and Freyhof 2007; Coad 2017).

According to Coad (1996a, b), the European brown trout was transplanted the Tigris River drainage. Moreover, five native species, *Salmo euphrataeus, S. fahrettini, S. munzuricus, S. okumusi,* and *S. tigridis* have been recorded or described from the Euphrates and Tigris River drainages (Kaya et al. 2016; Turan et al. 2017, 2020).

Specimens of brown trout were cultivated in the Lake Gahar (the upper Dez River, Tigris River drainage). Viable populations of this fish existed in both the upper and lower lake in the 1970s (Coad 2017).

Possible hybridization and thus gene pollution, introduction of diseases and parasites, and competitor for food and habitat are the main threats.

***Salvelinus fontinalis* (Mitchill, 1814)**

Brook Trout

The brook trout, a North American salmonid species, is characterized by the following characters: body with light-colored spots, anal, pelvic, and pectoral fins with a white edge followed by contrasting black, dorsal fin with 9–14 principal rays, anal fin with 8–13 principal rays, pectoral with 10–15 rays, pelvic fin with 7–10 rays, lateral line scale counts 109–132, gill rakers 13–22, pyloric caeca 20–55, and caudal fin truncate (Coad 2017).

The pelvic axillary scale.

According to Andersskog (1970), the brook trout were cultivated in fish farms in Iran to increase fish production and were accidentally introduced into the Tigris-Euphrates drainage but it needs confirmation (Coad 1996a, b). Competition for food and space, and introduction of parasites and diseases might be the main threats of this piscivorous fish. There is no record of this fish from the upper Tigris River drainage in Turkey (Kaya et al. 2016).

Esociformes

Esocidae

***Esox lucius* (Linnaeus, 1758)**

Northern Pike

The pike is distinguished by the following morphological characters: snout long, broad and flat, large mouth, dorsal and anal fins far back (near the caudal fin), dorsal-fin origin slightly in front of anal-fin origin with 17–25 (total), anal-fin origin behind dorsal-fin origin with 10–22 rays, pectoral rays 11–17, pelvic fin almost in the middle of distance between pectoral and anal fins with 7–13 rays, lateral line with 105–148 scales, and vertebrae 56–65 (Kottelat and Freyhof 2007; Coad 2016a).

Esox lucius stocks were transplanted to the Tigris-Euphrates drainage in Iran from the Caspian Sea basin (Anonymous 1970) and to the Lake Assad (the Syrian Euphrates) in 1977 (Coad 2016a). It has been reported from the Marivan Lake in

Fig. 51.18 *Rhinogobius lindbergi,* an exotic fish in the Tigris-Euphrates River System

Fig. 51.19 *Coptodon zillii,* an exotic fish in the Tigris-Euphrates River System

Kurdistan, Iran (Coad 2016a). There is no record of this fish from the upper Tigris River of Turkey (Kaya et al. 2016).

Gobiiformes

Gobiidae

***Rhinogobius lindbergi* (Berg, 1933)**

Amur Goby

(Fig. 51.18)

Rhinogobius lindbergi is distinguished by a combination of the following characters: elongated body (compressed posteriorly), head depressed with long snout and terminal mouth, absence of predorsal scales, two pores in preopercular canal, absence of a black spot at the upper end of pectoral fin base, presence of a distinct spot on first dorsal fin of male, 19–20 pectoral fin rays, 9–11 transverse scale series, and 27–28 vertebrae (see Sakai et al. 2000; Suzuki et al. 2015; Sadeghi et al. 2019).

The Amur goby is native to the Amur River drainage of Russia and China but it has been introduced to the Lake Zarivar (Tigris River drainage), Kurdistan Province, Iran as an exotic species (Esmaeili et al. 2018; Sadeghi et al. 2019). There is no record of this fish from the upper Tigris River in Turkey (Kaya et al. 2016).

Cichliformes

Family Cichlidae

***Coptodon zillii* (Gervais, 1848)**

Redbelly Tilapia

(Fig. 51.19)

Fig. 51.20 *Oreochromis aureus*, an exotic fish in the Tigris-Euphrates River System (Valikhani et al. 2016)

The Redbelly tilapia is distinguished from other exotic cichlids of the Tigris-Euphrates by having three anal spines, outer teeth on jaws bicuspid, and 8–11 rakers on the lower part of gill arch. Other morphological characters include: dorsal fin with 15–16 spines and 10–14 soft rays, anal fin with three spines and 8–10 soft rays, and presence of 6–7 dark vertical bars cross two horizontal stripes.

The redbelly tilapia has been established in the Syrian Euphrates, and has also been recorded at Al Musayyib (Babil Province) on the Euphrates river and at the main outfall drain in Basrah city (Arvand River), Iraq (Al-Sa'adi 2007; Mutlak and Al-Faisal 2009; Al-Sa'adi et al. 2012). It seems that Egypt was the main source of tilapia introduction to Iraq (JOB 1967).

It was first collected from two sites in the Tigris River drainage of Iran (Shadegan wetland, and a tributary of the Bahman Sheer River, both in Khuzestan Province) in October 2012 by Khaefi et al. (2014). It was also recorded by Roozbhfar et al. (2014) from the Dez River (Dezful city, Khuzestan Province, Iran). There is no record of this species from the upper Tigris River of Turkey (Kaya et al. 2016).

Impacts of *C. zillii* on native species have been well documented. Competition for food and spawning areas with native fishes, effects on native aquatic vegetation, and the organisms that rely on them (Hogg 1976a, b; Molnar et al. 2008) are the main threats.

Oreochromis aureus **(Steindacher, 1864)**
Blue Tilapia
(Fig. 51.20)

The blue tilapia is distinguished from other cichlids of the Tigris-Euphrates River System by having outer teeth on jaws bicuspid, caudal fin without regular bars, breeding males grayish or whitish, mature males without enlarged jaws, lower jaw 29–37% head length, and 18–26 gill rakers (Kottelat and Freyhof 2007). Other characters are: dorsal and anal fins with 15 and three spines, respectively, caudal fin with a wide pink to bright red color distal margin, breeding males show intense bright metallic blue color on their head.

The blue tilapia was first recorded from the Khabour River (a tributary of Euphrates) in Syria (Coad 1996a, b). In Iran, it was captured for the first time from the Arvand and Karun River drainages in Khuzestan province by Valikhani et al. (2016). There is no record of this species from the upper Tigris River of Turkey (Kaya et al. 2016).

Fig. 51.21 *Oreochromis niloticus*, an exotic fish in the Tigris-Euphrates River System (Al-Faisal and Mutlak 2014)

Tilapias have high tolerance for various environmental conditions and can quickly reproduce in new habitats. This characteristic may contribute to their successful reproduction and establishment in the inland waters of Iran in the near future.

***Oreochromis niloticus* (Linnaeus, 1758)**

Nile Tilapia

(Fig. 51.21)

The Nile tilapia, *O. niloticus* is distinguished from other cichlid species of the Tigris-Euphrates System by having outer teeth on jaws bicuspid and regular gray or black bars extending across whole caudal fin. Other characters are: a compressed and deep body, truncate caudal fin with numerous dark vertical stripes, 28–31 gill rakers on the first gill arch, interrupted lateral line with 21–23 and 14–18 scales in the upper and the lower part, respectively, dorsal fin base long with 17–18 spines and 12–13 soft rays, anal fin with three spines and nine rays, and pectoral, dorsal, and caudal fins red.

This fish is endemic to Africa, however it has been released in many parts of the world for aquaculture purposes (Vreven et al. 1998). Its first introduction to a fish pond on the Iraqi Tigris River, occurred in 1969, but the fishes could not tolerate winter condition (Herzog 1969). Lately, Al-Faisal and Mutlak (2014) captured the Nile tilapia from the Arvand River (Shatt Al-Arab River), Iraq and might become established in Iran through the connected water bodies. There is no record of this fish from the upper Tigris River of Turkey (Kaya et al. 2016).

It has been demonstrated that Nile tilapia can change ecosystem function and impact the functional processes of fish populations and communities (Shuai et al. 2019).

Atheriniformes

Atherinidae

***Atherina boyeri* Risso, 1810**

Fig. 51.22 *Atherina boyeri,* an exotic fish in the Tigris-Euphrates River System

Big-scale Sand Smelt

(Fig. 51.22)

Presence of two dorsal fins, position of pectoral fin being located high on the flank, and remote position of vent from anal fin are specific characteristics of this species. It is distinguished from other species of Atherinidae by total scale counts (39–49 in midlateral series), 23–31 gill rakers, and 13–15 anal rays (Kottelat and Freyhof 2007; Coad 2017). The body is compressed with a rounded belly, mouth large and terminal, scales cycloid extending onto the head, short pectoral fins.

The big-scale sand smelt, has a wide native distribution range in the Mediterranean, Black and Caspian sea basins (Kottelat and Freyhof 2007). In Turkey, this species is found in several water bodies (Balık 1979; Altun 1991, 1999; Küçük et al. 2009; Tarkan et al. 2015; Gençoğlu and Ekmekci 2016; Ünlü et al. 2011).

It has been reported from the Lake Devegecidi Dam of Turkey in 2016, which is the first record of this invasive species in the Tigris River drainage (Ünlü et al. 2011). This report might be a sign of further accidental range expansion to the other water bodies in the Tigris-Euphrates (Ünlü et al. 2011).

Order Cyprinodontiformes

Family Poeciliidae

***Gambusia holbrooki* Girard, 1859**

Eastern Mosquitofish

(Fig. 51.23)

It is characterized by having 5–9, commonly 7, dorsal fin rays, presence of gonopodium in males, anal fin with 7–11, usually 10, rays, pectoral fin with 11–14 rays, lateral scale rows number 26–33, and vertebrae 28–34 (Coad 2017).

It has been reported from various waters in the Tigris-Euphrates drainage. Based on historical records, the Iranian introductions occurred in 1922–1930 in order to control malaria (Coad 2017). It has now been established in almost all the Iranian basins/drainages.

In Iraq, it is found in the southern marshes, the Shatt al Arab River and its tributaries, the Tigris and Diyala Rivers, Al-Khawsar River passing through Mosul, and the Lake Razzazah. It is probably more widely distributed in suitable habitats but not always recorded.

Fig. 51.23 Female (up) and male (down) *Gambusia holbrooki,* an exotic fish in the Tigris-Euphrates River System

Apparently, the oldest recorded introduction of *Gambusia holbrooki* in Turkey, was the 1930s using it to control mosquitoes (Geldiay and Balık 1988), but it has been spread over the whole country (Tarkan et al. 2015).

It has been reported from ponds in the Dicle University Campus; Diyarbakır Ponds in the Dicle University Campus; Diyarbakır, Turkey (Kaya et al. 2016).

Possible or demonstrated ecological effects of introduced *G. holbrooki* on native fauna (fishes, amphibians, invertebrates) include predation on eggs and larvae, competition for space and trophic resources, behavioral interference, and transmission of disease (Reynolds 2009; Leunda 2010). Negative impacts are commonly assigned to high fish densities, predatory behaviors, and genetic variability of this small poeciliid fish.

Poecilia latipinna **(Lesueur, 1821)**

Sailfin Molly

(Fig. 51.24)

Male with a single, spineless, large sail-like dorsal fin, anal fin with a gonopodium, and a squared off or rounded caudal fin.

The first evidence of the sailfin molly, *P. latipinna* introduction in Iran comes from the study carried out by Khalaji et al. (2016) who reported it from Zayandehrud River Basin. Later study in 2017, revealed establishment of this fish in the same region (Esmaeili et al. 2017b).

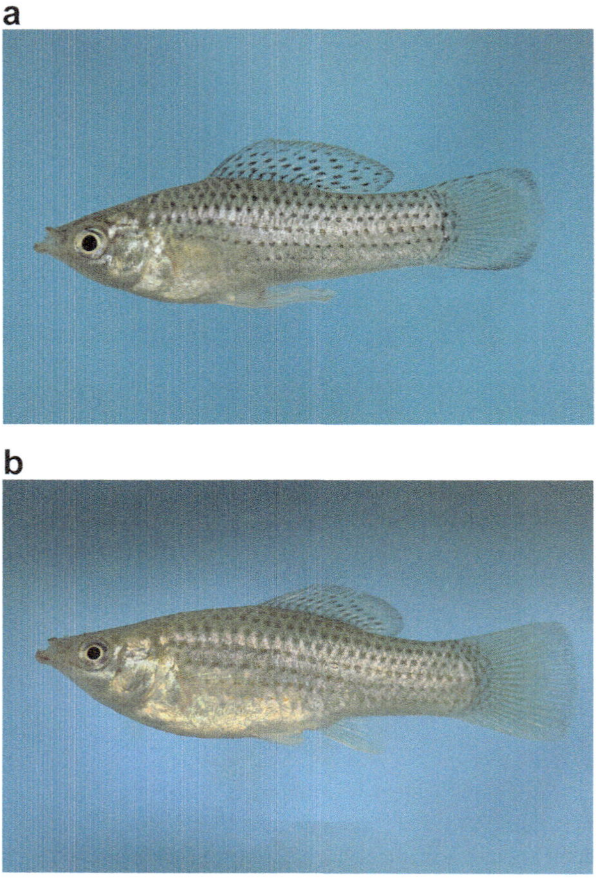

Fig. 51.24 Male (a) and female (b) *Poecilia latipinna,* an exotic fish in the Tigris-Euphrates River System

Presence of *P. latipinna* in the Tigris River tributary in Iran was first reported by Esmaeili et al. (2017b) who further demonstrated evidences for its establishment. There is no record of this species from Iraq and also from the upper Tigris River of Turkey (Kaya et al. 2016).

Native distribution range of *P. latipinna* is southeastern U.S.A. and south to Mexico. However, its high tolerance to broad environmental condition and physiological stress (Nunez et al. 2015), short generation time, and popularity as an aquarium species, have promoted its extensive introduction (Koutsikos et al. 2018) allowing the species to make established breeding populations in many countries. Due to its adverse ecological effects (Smith 1997), especially parasite transmission (e.g., ectoparasites *Ichthyophthirius multifiliis*, *Ichthyobodo* sp., and *Lernaea* sp.) (Khalaji et al. 2016; Esmaeili et al. 2017b), monitoring of this species is highly recommended.

Order Perciformes

Family Centrarchidae

***Micropterus salmoides* (Lacepède, 1802)**

Largemouth Bass

The North American species, largemouth bass, was introduced to the Tigris-Euphrates drainage of Iran (Saadati 1977; Coad 1996a, b). However, there is no further record of this perch in the area including the upper Tigris River of Turkey (Kaya et al. 2016) and Iraq (Al-Faisal and Mutlak 2014).

Family Percidae

***Sander lucioperca* (Linnaeus, 1758)**

Pike-perch

This species is separated from its closely related species, *S. marinum* by dorsal fins being very close together, anal fin spines are not closely joined to soft rays (Coad 2016b).

Other morphological characters are: first dorsal fin spines 11–17, second dorsal spines 1–4, and soft rays 16–27, commonly 19–24, anal fin spines 1–3, soft rays 9–14, pectoral and pelvic fins with 11–18 and five branched rays, respectively, gill rakers number 10–17, lateral line scales 75–150 (scales above and below lateral line 12–17 and 16–24, respectively), and vertebrae usually 45–48 (Coad 2016b).

The pike-perch occurs naturally in the Caspian Sea basin but has been transplanted to the Tigris-Euphrates river drainage (Anonymous 1970). There is no recent record of *S. lucioperca* in the Tigris River tributaries of Turkey (Innal and Erk'akan 2006; Kaya et al. 2016), Iran, and Iraq (Al-Faisal and Mutlak 2014).

Remarks: The first records of the Siberian sturgeon, *Acipenser baerii* (Brandt, 1869), and the beluga, *Huso huso* (Linnaeus, 1758), (Acipenseriformes, Acipenseridae) from the Tigris–Euphrates drainage were documented by Mousavi-Sabet et al. (2019). Status of these sturgeon fishes should be monitored.

51.7 Discussion

Comparable to trends seen in other parts of the world, the number of nonindigenous fish species in the Tigris-Euphrates River system has increased over the last couple of decades, attaining to 30 species. Activities like aquaculture, ornamental purposes, recreational or sport fishing, malaria control, scientific research, and accident release are the significant and major reasons for these introductions (Coad and Abdoli 1993; Esmaeili et al. 2010a, b). Some of these fishes (e.g., *Captodon zillii, Carassius auratus, C. gibelio, Cyprinus carpio, Gambusia holbrooki, Oreochromis aureus, O. niloticus, Poecilia latipinna,* and *Pesudorasbora parva*) have been established in the Tigris-Euphrates River System, become invasion species. Behavior of fish, competition for limited available resources, proper water temperature, water flow regime, suitability of spawning ground, number of release events, species intrinsic parameters (such as food and feeding strategy, generation time, growth rate, body size, genetic variability, and lifespan), and abundant food supply have been proposed to be the key factors promoting establishment of introduced fish species.

The processes involved in species invasions are almost similar in every bioinvasion (Fig. 51.25). An alien species must first overcome geographical barriers,

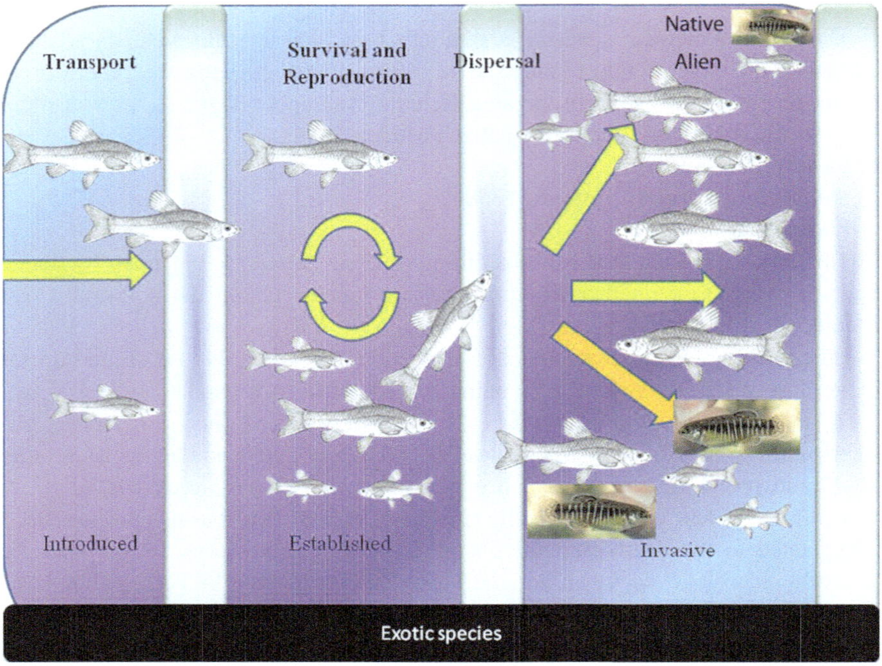

Fig. 51.25 Schematic diagram of processes involved in species invasions (Modified from Lymbery et al. 2014).

survive, reproduce, and then expand its distribution range (disperse) (Blackburn et al. 2011; Lymbery et al. 2014). Due to the nature of the cichlids, *O. aureus*, *O. niloticus,* and *C. zillii*, these species also have surmounted this common framework and act as invasive species in the introduced water body.

Whether through by chance introductions or deliberate releases, numerous species achieve distributions outside their natural distributional ranges and form various novel interactions with indigenous species (Paterson et al. 2013). Introductions always carry risks for the native biota (Gozlan and Newton 2009), resulting in possible detrimental interactions with native species or even on ecosystem function and structure (see Gozlan et al. 2010a, b). The ecological outcomes of freshwater fish invasions have been reviewed and discussed by Cucherousset and Olden (2011) which are given in Fig. 51.26. The effects of invaders appear at different levels including genetic, individual, population, community, and ecosystem.

Our present knowledge about invasion process and invaders is based on information on the introduced populations in other water bodies of the world (see Wilcove et al. 1998; Levine 2000; McNeely 2001; Canonico et al. 2005; Gozlan et al. 2010a, b; Cucherousset and Olden 2011; Blackburn et al. 2011; Lymbery et al. 2014; Koutsikos et al. 2018; Milardi et al. 2019). The effects can be summarized as:

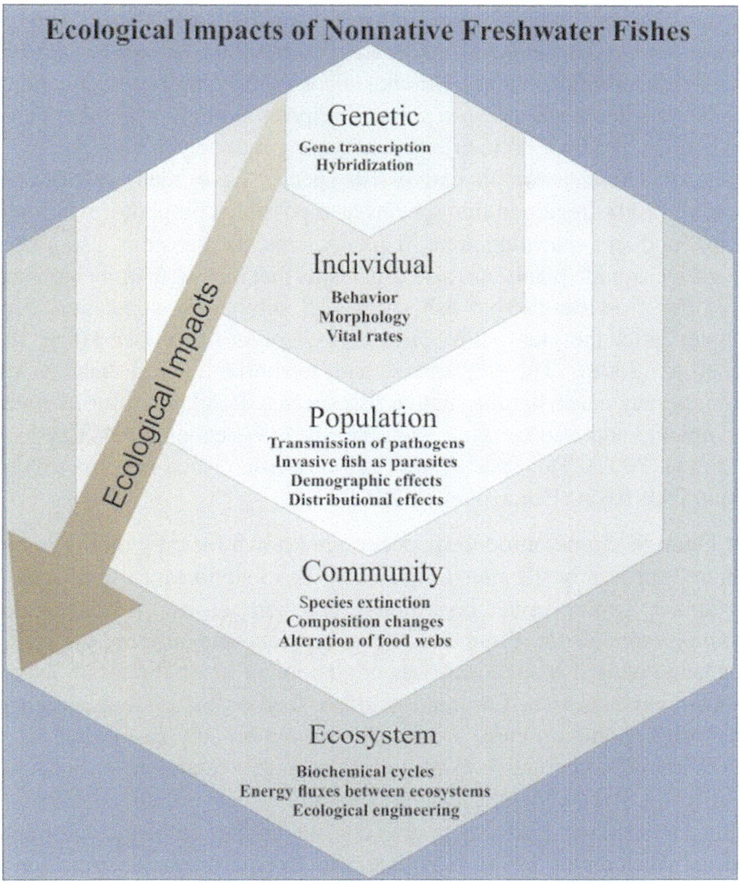

Fig. 51.26 Schematic diagram of ecological impacts of nonnative freshwater fishes (Modified from Cucherousset and Olden 2011).

Competing for food and habitat utilized by native fish species and wildlife, directly preying on native fish species, hybridizing with indigenous fish species leading to genetic introgression and the loss of genetically pure stocks, changing selection pressures transmission of parasites and diseases that were not historically present in those habitats, altering habitat which includes degradation of water quality, modifying energy and nutrient flow and the food web within the aquatic ecosystem, and extirpating endemic and native faunal and floral elements.

Competition Mosquito fish (*Gambusia holbrooki*) has been called "the fish destroyer" by Myers (1965) due to its invasion behavior. They prey on the eggs and larvae of others fishes and also amphibians and invertebrates. They are aggressive, even attacking fish larger than themselves, thus discouraging feeding and reproduction, and compete directly for food with various cyprinid species (see

Armantrout 1990). As in the Tigris-Euphrates, the native and endemic *Aphanius* and *Aphaniops* species (Esmaeili et al. 2020) are present beside *Gambusia,* anywhere the latter has been introduced, hence *Gambusia* competes with these aphaniid species. Another exotic fish, *Pesudorasbora parva*, competes strongly with other fish species for food and is a predator on their eggs and young individuals.

Impacts of *Tilapia zillii* on native fish species have been well documented. Molnar et al. (2008) regarded this species as a possible competitor with native fish species for food and spawning areas. This voracious herbivore is also a significant threat to native aquatic plants and also organisms that rely on them being considered as one of the most destructive fish species to submerged vegetation, especially macrophytes, after the grass carp, *Ctenopharyngodon idella* (see Hogg 1976a, b; Molnar et al. 2008). The aggressive and territorial cichlid fish, *Amatitlania nigrofasciata* may cause harm to native fish species. It is known for its potential to colonize quickly and also carrying alien parasites (Wisenden 1994; Bassleer 1997; Martinez et al. 2002). This species has been now established in some tributaries of the Persian Gulf basin (Esmaeili et al. 2013).

Habitat Change Some introduced species are known for their ability in changing fish habitat. Apparently, *Gambusia holbrooki* have significant impacts on aquatic habitats under experimental conditions (Hurlbert et al. 1972). Grass carp, *Ctenopharyngodon idella* eliminates the vegetation, and reduces the refuge area for larvae and young fish, and decreases production of invertebrates on which many indigenous organisms feed. Carp and goldfish, feed on the bottom and change the water turbidity. Common carp is one of the most broadly distributed fish in the world. It is known to be native to Eastern Europe and central Asia, but it has been successfully translocated/introduced into natural waters of all continents except Antarctica (Welcomme 1988). It is also considered as one of the most cultivated carp species (Welcomme 1988). High tolerance to environmental stress and omnivorous feeding habits make this carp as an attractive subject for commercial aquaculture, hence, these abilities increase its invasive potentials. Non-native carps have been implicated in the degradation of aquatic ecosystems and are considered to belong to the most ecologically harmful freshwater invasive species (Kloskowski 2011). After reaching a critical density in eutrophic shallow lakes and ponds, carp may mediate a transition from clear water, macrophyte-dominated conditions to a turbid water, phytoplankton-dominated condition (Kloskowski 2011). The carp's bottom feeding habit, suctioning food organisms along with the sediment, and ejecting the inappropriate matters (Kloskowski 2011), may cause the uprooting or ingestion of submerged aquatic plants, increase in nutrient and turbidity levels, and also the physical resuspension of settled algal cells from the sediment (Roozen et al. 2007). Furthermore, carp may excrete large amounts of nutrients to the water column, thus accelerating algal production (Kloskowski 2011). The carp is able to achieve large body size, and its effect on aquatic communities seems to rely mainly on traits associated with individual size (Kloskowski 2011). It has been proposed that carp predation and associated impacts may be to a great degree responsible for loss of animal diversity in invaded communities, as they may act earlier than, or independent of, the ecosystem shift to a turbid condition (Kloskowski 2011). Apparently, common carp changes bottom-up processes by altering nutrient and turbidity concentrations, abundance of phytoplankton, and also biodiversity through benthic foraging. However, zooplankton and benthic invertebrates can be affected by top-down processes via predation and reduced foraging efficiency. Common carp also degrades aquatic plants and thus increases the turbidity of stagnant water body (Weber and Brown 2009).

Genetic Changes Numerous endemic and native species live in many water bodies in various environments carrying their own unique genes (Coad 1995). Movement of these species and introduced exotic fishes between the water bodies may well eliminate a diversity yet unrevealed or make gene pollution via hybridization. Some populations of the bighead carp (*Hypophthalmichthys nobilis*) are hybrids with silver carp (*Hypophthalmichthys molitrix*) and hybridization occurs between some populations of *Cyprinus carpio* and *Carassius auratus* in the Caspian Sea basin (Coad 2017), that reveals potential hybridization of closely and distantly related species found in sympatricity in the Tigris-Euphrates. Hybrids of *Aphanius mesopotamicus* and *A. stoliczkanus* (now *Aphaniops stoliczkanus*, Esmaeili et al. 2020) in the Tigris-Euphrates have been documented by Masoudi et al. (2016), showing possibility of hybridization in case of accidental or intentional fish movements between water bodies.

Introduction of Parasites Acting as coinvader, introduced fish species play a key role in the spread of disease or parasites to other species. As numerous parasites have been sampled from invasive and native fish species (e.g., Sayyadzadeh et al. 2016; Coad 2017), it seems that the broad distributional range of invasive fish species may have a notable role in the spread of these parasitic species. A particular example is the carp species with wide distribution range. *Lernaea* spp. (commonly *L. cyprinacea*) has been broadly distributed all over the map, likely through the translocation of cyprinid hosts such as goldfish, *C. auratus* and European carp, *C. carpio* (see Sayyadzadeh et al. 2016). The *Lernaea* parasite was first reported in *Gambusia holbrooki* in 1981 and in common carp and Chinese carps in 1981 and 1982 (see Sayyadzadeh et al. 2016) from north of Iran and now, it has been broadly distributed probably through the translocation of exotic cyprinid hosts (Sayyadzadeh et al. 2016) infesting many native freshwater fish species. Both species of *C. auratus* and *C. carpio* which are supposed to be involved in translocation of *Lernaea* parasite are present in the area under study as exotic cyprinid fishes, and hence, they might have a significant role in the translocation of *Lernaea* during the invasion process illustrated in Fig. 51.27. According to Lymbery et al. (2014), the alien host species carrying an alien parasite element must overcome four major barriers: (i) introduction (transport in alien host and acting as cointroduced species), (ii) establishment (survival and reproduction in alien host, acting as cointroduced species), (iii) spreading with its original host (dispersal), and iv) switching to a native host species to become a coinvader (Fig. 51.27). The same process is suggested for the alien and invader host species of the Tigris-Euphrates River System.

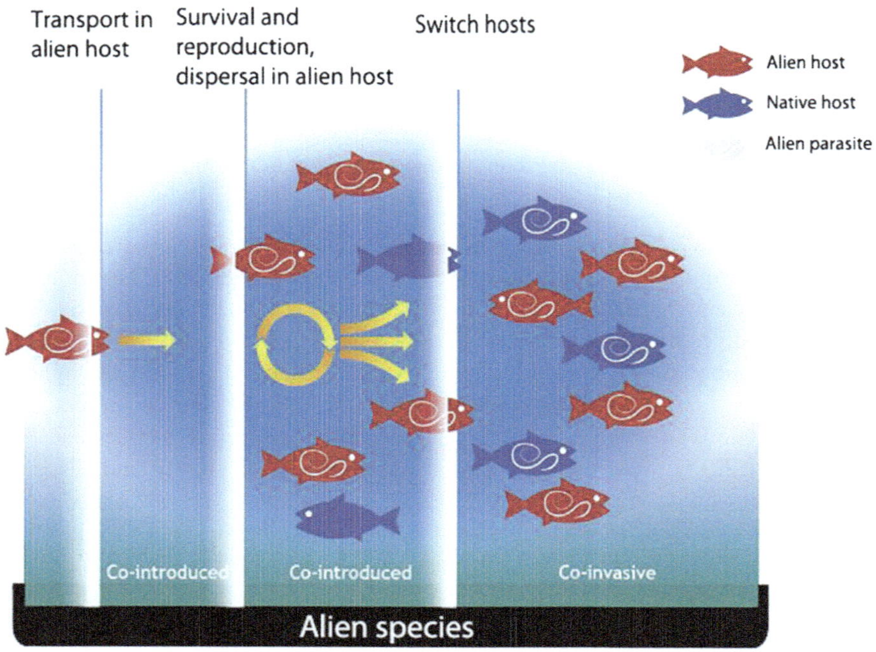

Fig. 51.27 Schematic diagram of processes involved in species invasions and coinvasions (parasitic aliens) (Modified from Lymbery et al. 2014). The alien host species (in red) contains an alien parasite species. The alien parasite goes through the processes of introduction, establishment and spread with its original host and then switches to a native host species (in blue) to become a co-invader

Ornamental Fishes We recorded the occurrences of some aquarium fish species such as *Atractosteus spatula*, *Carassius auratus*, *Pangasianodon hypophthalmus*, *Piaractus brachypomus*, *Poecilia latipinna* in the Tigris–Euphrates River System and *Poecilia reticula, Xiphophorus hellerii,* and *Amatitlania nigrofasciata* from the neighboring basins. Such unintentional releases are unsafe for fish biodiversity. Careful measures and actions are required to observe, monitor, and control the aquarium fish trade to reduce chance of accidental release of ornamental exotic fish species into natural water bodies.

The Convention on International Trade in Endangered Species (CITES) cannot completely protect aquatic habitats from invasions because it deals only with the trade listed as endangered and threatened species. However, under Convention on Biological Diversity (CBD), it is the responsibility of each nation and the state to prevent spread of any invasive alien species which is harmful to the fish biodiversity and aquatic ecosystem. The aquarium trade industry is spreading fast, therefore, it is essential to have specific education programs for involved persons handling ornamental fishes. In addition, presenting specific certificate to show that aquatic ornamental cultivators and large-scale sellers, sterilize their outflows and take necessary steps to prevent the accidental release of species is essential.

Management The deliberate introduction of exotic species is a remarkably complex issue that needs considering a broad number of factors. Understanding the risks that alien species, pose, requires (i) comparative studies on different ecological and life history traits of invasive and their closely related native and endemic species with close ecological requirements, (ii) accurate identification of invasive and native species, (iii) regular checking of fishes for diseases and parasites, (iv), obtaining a more balanced perspective on the ecological impacts of fish invasions across various levels of biological organization, (v) improving our knowledge on how environmental change (e.g., habitat degradation, land-use change, and climate warming) interacts with invasive species, (vi) elucidation of the adaptive and evolutionary responses of native species to the presence of invasive species and vice versa (see Cucherousset and Olden 2011), and (vii) public awareness. All of these actions could be effective management strategies to minimize the bio-invasion impacts. Most important consideration should be given to those alien species that have been established in the nature and on preventing new introductions.

Finally, as pointed out by Gozlan et al. (2010a, b) maps of conservation micro hot spots should be designed based on a combination of predictors of future introductions (e.g., human population density and aquaculture activity) and risk level of incurring losses to local and global biodiversity (e.g., number of critically endangered species). This would accelerate prioritization of the areas at national and global levels where risks of future introductions should be minimized. This includes those where endemic species are at risk and the intensity of introductions of non-native species has been limited.

Acknowledgments We would like to thank H.R. Mehraban, M. Masoudi, G. Sayyadzadeh, R. Khaefi, F. Zarei, R. Sadeghi, H. Darvishnia, H. Zareian, R. Zamanian Nejad, and M. Razbanian for helping in fish collection. This study was financially supported by Shiraz University.

References

Abdurakhmanov YA (1962) Ryby Presnykh vod Azerbaidzhana [Freshwater Fishes of Azerbaidzhan]. Akademii Nauk Azerbaidzhanskoi SSR, Institut Zoologii, Baku, 407 pp

Ahmed AH, Taher MM (1988) Observations on the extensive and intensive cultured carp, *Cyprinus carpio* L. in Basrah, Iraq. Basrah J Agric Sci 1:35–47

Al-Faisal AJ (2008) Taxonomic study for *Alburnus mossulensis*, *Acanthobrama marmid* and *Hemiculter leucisculus* in southern Iraqi marshes. M.Sc. Thesis, University of Basrah. 115 pp (In Arabic)

Al-Faisal AJ, Mutlak FM (2014) First record of the Nile tilapia *Oreochromis niloticus* (Linnaeus, 1758), from the Shatt Al-Arab River, South Iraq. Mesopot J Mar Sci 29(1):45–50

Al-Hassan LAJ, Muhsin KA (1986) The presence of *Carasobarbus luteus* (Heckel, 1843) and *Heteropneustes fossilis* (Bloch, 1797) in Khor Al-Zubair, north west of the Arabian Gulf, Basrah Iraq. Zool Middle East 1:116–118

Ali AM (1985) Morphometric characteristics of the khimri, *Barbus luteus* (Cyprinidae), from Tartar Reservoir and the Tigris River (Iraq). J Ichthyol 25(5):168–171

Ali AS, Jawad LA, Saad AA (2016) Confirmation of the presence of the Indian stinging catfish, *Heteropneustes fossilis* (Bloch, 1794) (Heteropneustidae) in Syrian inland waters. J Appl Ichthyol 32:117–119

Ali NM, Mhaisen FT, Abul-Eis ES, Kadim LS (1989) Parasites of the silver carp *Hypophthalmichthys molitrix* from Babylon fish farm, Hilla, Iraq. Iraq Riv Idrobiol 28(1–2):151–154

Altun Ö (1991) The morphology of sand smelt (*Atherina boyeri* Risso 1810) in Küçük Çekmece Lake. Turk Zool Derg 15:64–75

Altun Ö (1999) Morphological variations observed on the sand smelt (*Atherina boyeri* Risso, 1810) populations. Turk Zool Derg 23:911–918

Al-Sa'adi BA (2007) The parasitic fauna of fishes of Euphrates river: Applied study in Al-Musaib city. M. Tech. Thesis, Al-Musaib Technical College - Foundation Technical Education

Al-Sa'adi BA, Mhaisen FT, Al-Rubae AL (2012) The first parasitological report on the redbelly Tilapia *Tilapia zillii* (Gervais, 1848) in Iraq. In: Proceedings of the Scientific Symposium of Natural History Research Center and Museum, University of Baghdad, Baghdad

Andersskog B (1970) Report to the government of Iran on fisheries activities in Iran with particular reference to the northern fisheries company. Food and Agriculture Organization, Rome, United Nations Development Programme, Technical Assistance, UNDP (TA) 2878: v + 18 pp

Anonymous (1970) Chinese carps in Iran. Food and Agriculture Organization, Rome, Aquaculture Bulletin 3(1):15

Armantrout NB (1980) The freshwater fishes of Iran. Ph.D. Thesis, Oregon State University, Corvallis, OR

Armantrout NB (1990) Fish distribution as influenced by aquatic habitat alteration and species introductions. In: Daniel JC, Serrao JS (eds) Conservation in developing countries: problems and prospects. Proceedings of the Centenary Seminar of the Bombay Natural History Society, Bombay, Delhi, Calcutta, Madras, pp 199–206

Aydın H, Gaygusuz Ö, Tarkan AS, Top N, Emiroğlu Ö, Gürsoy Gaygusuz Ç (2011) Invasion of freshwater bodies in the Marmara region (northwestern Turkey) by nonnative gibel carp, *Carassius gibelio* (Bloch, 1782). Turk J Zool 35(6):829–836. https://doi.org/10.3906/zoo-1007-31

Bassleer G (1997) Color guide of tropical fish diseases: on freshwater fish. Bassleer Biofish, Westmeerbeek, Belgium

Balık S (1979) Taxonomical and ecological investigations upon freshwater fishes of western Anatolia. Fen Fak İlmi Rap Ser 236:61

Bănărescu PM (1999) The freshwater fishes of Europe. Cyprinidae 2. Part I: Rhodeus to Capoeta, vol 5. AULA-Verlag, Wiebelsheim, p 427

Blackburn TM, Pyšek P, Bacher S, Carlton JT, Duncan RP, Jarošík V, Wilson JRU, Richardson DM (2011) A proposed unified framework for biological invasions. Trends Ecol Evol 26:333–339

Boschung HT, Mayden RL (2004) Fishes of Alabama. Smithsonian Institution Press, Washington, DC

Canonico GC, Arthington G, Mccrary GK, Thieme ML (2005) The effects of introduced tilapias on native biodiversity. Aquat Conserv Mar Freshwater Ecosyst 15:463–483

Coad BW (1991) Fishes of the Tigris-Euphrates basin: a critical checklist. Syllogeus 68:21–49

Coad BW (1995) Freshwater fishes of Iran. Acta Scientiarum Naturalium Academiae Scientiarum Bohemicae Brno 29(1):1–64

Coad BW (1996a) Exotic and transplanted fishes in Southwest Asia. Publicaciones Especiales Instituto Español de Oceanografía 21:81–106

Coad BW (1996b) Exotic fish species in the Tigris-Euphrates basin. Zool Middle East 13:71–83

Coad BW (2014) Review of the freshwater catfishes of Iran (Order Siluriformes). Iran J Ichthyol 1 (4):218–257

Coad BW (2016a) Review on the pikes of Iran (Family Esocidae). Iran J Ichthyol 3(3):161–180

Coad BW (2016b) Review of the perches of Iran (Family Percidae). Int J Aquat Biol 4(3):143–170

Coad BW (2017) Freshwater fishes of Iran. www.briancoad.com

Coad BW, Abdoli A (1993) Exotic fish species in the fresh waters of Iran. Zool Middle East 9:65–80

Coad BW, Hussain NA (2007) First record of the exotic species *Hemiculter leucisculus* (Actinopterygii: Cyprinidae) in Iraq. Zool Middle East 40:107–109

Csorbai B, Pereszlényi A, Kovács R, Urbányi B, Horváth L (2014) The habitat use and selectivity by topmouth gudgeon (*Pseudorasbora parva*). Acta Zool Acad Sci Hung 60(4):389–400

Cucherousset J, Olden JD (2011) Ecological impacts of non-native freshwater fishes. Fisheries 36 (5):215–230

Eagderi S, Nasri M (2012) A first record of the Bittrling (*sic*) *Rhodeus amarus* (Bloch, 1782) (Cypriniformes, Cyprinidae) in the Iranian part of Tigris-Euphrates Basin. Int Res J Appl Basic Sci 3(3):639–641

Eagderi S, Mouludi-Saleh A, Nazlabadi SA (2019) First record of the rohu, *Labeo rohita* (Hamilton, 1822) (Cyprinidae) from Karun River, Tigris River drainage, Iran. Fish Taxa 4(1):18–20

Esmaeili HR, Teimori A, Gholamhosseini G (2007) Freshwater Ichthyodiversity and its conservation in Iran. Xii European Congress on Ichthyology. Cavata (Dubrovnik), Croatia, pp 200–201

Esmaeili HR, Coad BW, Gholamifard A, Nazari N, Teimory A (2010a) Annotated checklist of the freshwater fishes of Iran. Zoosyst Rossica 19:361–386

Esmaeili HR, Gholamifard A, Teimori A, Bagh-Bani S, Coad BW (2010b) *Xiphophorus hellerii* (Heckel, 1848) (Cyprinodontiformes, Poeciliidae), a newly introduced fish recorded from natural freshwaters of Iran. J Appl Ichthyol 26:937–938

Esmaeili HR, Gholamifard A, Freyhof J (2011a) Ichthyofauna of Zarivar Lake (Tigris River basin, Iran) with the first records of *Hemiculter leucisculus* and *Alburnus hohenackeri* in the Tigris basin. Electron J Ichthyol 7(1):1–6

Esmaeili HR, Nazari N, Gholamifard A, Gholamhosseini G, Teimori A, Coad BW (2011b) Range extension and translocation for *Rhodeus amarus* (Bloch, 1782) (Actinopterygii: Cyprinidae) in northwest Iran. Turk J Zool 35(6):883–886

Esmaeili HR, Gholamifard A, Sayyadzadeh G, Parsi B, Mirghiyasi S, Ghasemian S (2013) New record of the convict cichlid, *Amatitlania nigrofasciata* (Günther, 1867), from the Middle East (Actinopterygii: Cichlidae), aqua. Int J Ichthyol 19:225–229

Esmaeili HR, Teimori A, Owfi F, Abbasi K, Coad BW (2014) Alien and invasive freshwater fish species in Iran: diversity, environmental impacts and management. Iran J Ichthyol 1(2):62–72

Esmaeili HR, Mehraban H, Abbasi K, Keivany Y, Coad B (2017a) Review and updated checklist of freshwater fishes of Iran: taxonomy, distribution and conservation status. Iran J Ichthyol 4 (Suppl. 1):1–114

Esmaeili HR, Masoudi M, Amini Chermahini M, Esmaeili AH, Zarei F, Ebrahimi M (2017b) Invasion of the neotropical and nearctic fishes to Iran. Fish Taxa 2(3):126–133

Esmaeili HR, Sayyadzadeh G, Eagderi S, Abbasi K (2018) Checklist of freshwater fishes of Iran. FishTaxa 3(3):1–95

Esmaeili HR, Teimori A, Zarei F, Sayyadzadeh G (2020) DNA barcoding and species delimitation of the Old World tooth-carps, family Aphaniidae Hoedeman, 1949 (Teleostei: Cyprinodontiformes). PLoS One 15(4):e0231717. https://doi.org/10.1371/journal.pone.0231717

Fatemi SMR, Nabavi SMB, Vosoghi G, Fallahi M, Mohammadi M (2012) The relation between environmental parameters of Hormuzgan coastline in Persian Gulf and occurrence of the first harmful algal bloom of *Cochlodinium polykrikoides* (Gymnodiniaceae). Iran J Fish Sci 11 (3):475–489

Fricke R, Eschmeyer WN, Van der Laan R (eds) (2020) Eschmeyer's catalog of fishes: genera, species, references. http://researcharchive.calacademy.org/research/ichthyology/catalog/fishcatmain.asp. Accessed 18 Apr 2020

Geldiay R, Balık S (1988) Türkiye Tatlısu Balıkları. Ege Universitesi. Fen Fak. Kitaplar Serisi No. 97, İzmir

Gençoğlu L, Ekmekci FG (2016) Growth and reproduction of a marine fish (*Atherina boyeri*, Risso 1810) in a freshwater ecosystem. Turk J Zool 40:534–542
Goddard N (2009) Alligator gar. Ichthyology at the Florida Museum of Natural History. https://www.floridamuseum.ufl.edu/fish/discover/species-profiles/atractosteus-spatula
Gozlan RE, Newton AC (2009) Biological invasions: benefits versus risks. Science 324:1015–1016
Gozlan RE, St-Hilaire S, Feist SW, Martin P, Kent ML (2005) Disease threat to European fish. Nature 435(7045):1046–1046
Gozlan RE, Britton JR, Cowx I, Copp GH (2010a) Current knowledge on non-native freshwater fish introductions. J Fish Biol 76:751–786
Gozlan RE, Andreou D, Asaeda T, Beyer K, Bouhadad R, Burnard D, Caiola N, Cakic P, Djikanovic V, Esmaeili HR, Falka I, Golicher D, Harka A, Jeney G, Kováč V, Musil J, Povz M, Nocita A, Virbickas T, Wolter C, Tarkan AS, Tricarico E, Trichkova T, Verreycken H, Witkowski A, Zhang C, Zweimueller I, Britton JR (2010b) Pan-continental invasion of *Pseudorasbora parva*: towards a better understand fish invasions. Fish Fish 11:315–340
Grave SD, Ghane A (2006) The establishment of the Oriental River Prawn, *Macrobrachium nipponense* (de Haan, 1849) in Anzali Lagoon, Iran. Aquat Invasions 1(4):204–208
Gustiano R, Pouyaud L (2008) Systematic revision of the genera of Pangasiidae (Silurifomes, Ostariophysi). Indonesian Aquacult J 3(1):13–22
Hashemi SAR, Ansary H (2012) Biomass and production of fish species in the Shadegan wetland, Iran. Res J Recent Sci 1:66–68
Herzog P (1969) Untersuchungen Ilber die Parasiten der SIlBwasserfische des Irak. Archiv fllr Fischereiwissenschaft 20:132–147
Hogg RG (1976a) Established exotic cichlid fishes in Dade County, Florida. Fla Sci 39:97–103
Hogg RG (1976b) Established exotic cichlid fishes in Dade County, Florida. Florida Scientist 39 (2):97–103
Holčík J, Jedlička L (1994) Geographical variation of some taxonomically important characters in fishes: the case of the bitterling *Rhodeus sericeus*. Environ Biol Fish 41(1–4):147–170
Holčík J, Razavi BA (1991) New species of freshwater fish from the Iranian coast of the Caspian Sea. Report prepared for the Anzali Lagoon productivity and fish stocks investigation project. Food and Agriculture Organization, Rome, FI/UNDP/IRA/88/001 Field Document 1. En, 14 pp
Hurlbert S, Zeidler J, Fairbanks D (1972) Ecosystem alteration by mosquito fish (*Gambusia affinis*) predation. Science 175:639–641
Innal D, Erk'akan F (2006) Effects of exotic and translocated fish species in inland waters of Turkey. Rev Fish Biol Fish 16:39–50. https://doi.org/10.1007/s11160-006-9005-y
Innes W (1966) Exotic Aquarium Fishes. T.F.H. Publications, Inc., Neptune, NJ
Ivanov VP, Kamakin AM, Ushivtzev VB, Shiganova T, Zhukova O, Aladin N, Wilson SI, Harbison GR, Dumont HJ (2000) Invasion of the Caspian Sea by the comb jellyfish *Mnemiopsis leidyi* (Ctenophora). Biol Invasions 2:255–258
Jawad LA (2003) Impact of environmental change on the freshwater fish fauna of Iraq. Int J Environ Stud 60(6):581–593
Jawad LA, Al-Faisal A, Al-Mukhtar M (2012) Presence of the non-indigenous fish species, gibel carp, *Carassius gibelio* (Family: Cyprinidae) in Basrah Province freshwater systems, Iraq. Water Res Manag 2(4):41–44
Job TJ (1967) Status of fish culture in the Near East region. In: TVR P (ed) Proceedings of the FAO World Symposium on Warm-Water Pond Fish Culture. Fisheries Report. Food and Agriculture Organization, Rome, pp 54–69
Kaya C, Turan D, Ünlü E (2016) The latest status and distribution of fishes in upper Tigris River and two new records for Turkish freshwaters. Turk J Fish Aquat Sci 16(3):545–562
Kelle A (1978) Taxonomical and ecological investigations on fishes living in Tigris River and its tributaries (pp. 109). PhD Thesis. Dicle University
Khaefi R, Esmaeili HR, Zareian H, Babaei S (2014) The first record of the redbelly tilapia, *Tilapia zillii* Gervais, 1848 in freshwaters of Iran. Turk J Zool 38:96–98

Khalaf KT (1961) The marine and freshwater fishes of Iraq. Al-Rabitta Press, Baghdad, 164 pp

Khalaji M, Sarkhos J, Amini SH, Siyami M, Zangene M, Asadolahi S (2016) The relation between size and parasite load in the Molly fish (*Poecilia latipinna*) of Jarghoyeh qanat, Isfahan Iran. Iran Sci Fish J 25(3):251–256

Khamees NR, Ali AH, Abed JM, Adday TK (2013) First record of striped catfish *Pangasianodon hypophthalmus* (Sauvage, 1878) (Pisces: Pangasiidae) from inland waters of Iraq. Basrah J Agric Sci 26(Special issue 1):178–183

Kloskowski J (2011) Impact of common carp *Cyprinus carpio* on aquatic communities: direct trophic effects versus habitat deterioration. Fundam Appl Limnol 178(3):245–255

Kolar CS, Lodge DM (2001) Progress in invasion biology: predicting invaders. Trends Ecol Evol 16:199–204

Kottelat M (2001) Fishes of Laos. WHT Publications, Colombo, Sri Lanka, p 198

Kottelat M, Freyhof J (2007) Handbook of European Freshwater Fishes. Kottelat, Cornol, Switzerland and Freyhof, Berlin, Germany

Koutsikos N, Vardakas L, Kalogianni E, Economou AN (2018) Global distribution and climatic match of a highly traded ornamental freshwater fish, the sailfin molly *Poecilia latipinna* (Lesueur, 1821). Knowl Manag Aquat Ecosyst 419:23

Krasznai ZL (1987) Mission to Iran to assess the needs in cold water fish culture 14 August - 2 September 1987. A report prepared for the project Training Course in Cold water Fisheries. Food and Agriculture Organization, Rome, TCP/IRA/6755(T), Field Document 1, iii + 23 pp

Küçük F, Sari HM, Demi̇ r, O., & Gülle, I. (2009) Review of the ichthyofaunal changes in Lake Eğirdir between 1915 and 2007. Turk J Zool 33:277–286

Leunda PM (2010) Impacts of non-native fishes on Iberian freshwater ichthyofauna: current knowledge and gaps. Aquat Invasions 5(3):239–262

Levine JM (2000) Species diversity and biological invasions: relating local process to community pattern. Science 288:852–854

Lymbery AJ, Morine M, Kanani HJ, Beatty SJ, Morgan DJ (2014) Co-invaders: the effects of alien parasites on native hosts. Int J Parasitol Parasites Wildlife. https://doi.org/10.1016/j.ijppaw.2014.04.002

Martinez VMV, Macedo MLA, Scholz T, Solis DG, Franco EFM (2002) Atlas de los helmintos parásitos de cíclidos de México. Instituto Politéc-nico Nacional, Dirección de Publicaciones Tresguerras, Mexico

Masoudi M, Esmaeili HR, Teimori A, Gholami Z, Gholamhosseini A, Sayyadzadeh G, Reichenbacher B (2016) Sympatry and possible hybridization among species of the killifish genus *Aphanius* (Nardo, 1827) (Teleostei: Cyprinodontidae) in Southwestern Iran. Limnologica 59:10–20

McNeely JA (2001) Global strategy on invasive alien species. IUCN, Gland, Switzerland

Milardi M, Gavioli A, Soininen J, Castaldelli G (2019) Exotic species invasions undermine regional functional diversity of freshwater fish. Sci Rep 9(1):1–10

Mohamed A-RM, Hussain NA, Al-Noor SS, Mutlak FM, Sudani IM, Mojer AM, Toman AJ, Abdad MA (2008) Fish assemblage of restored Al-Hawizeh marsh, southern Iraq. Ecohydrol Hydrobiol 8(2–4):375–384. https://doi.org/10.2478/v10104-009-0029-5

Molnar JL, Gamboa RL, Revenga C, Spalding MD (2008) Assessing the global threat of invasive species to marine biodiversity. Front Ecol Environ 6:485–492

Mousavi-Sabet H, Salehi M, Sarpanah A, Pirali Kheirabadi E (2019) First records of *Acipenser baerii* and *Huso huso* (Actinopterygii: Acipenseriformes: Acipenseridae) from the Tigris–Euphrates basin, Iran. Acta Ichthyol Piscat 49(3):265–267

Mutlak FM, Al-Faisal AJ (2009) A new record of two exotic cichlids fish *Oreochromis aureus* (Steindacher, 1864) and *Tilapia zilli* (Gervais, 1848) from south of the main outfall drain in Basrah city. Mesopot J Mar Sci 24(2):160–170. (In Arabic)

Mutlak F, Jawad L, Al-Faisal A (2017) *Atractosteus spatula* (Actinopterygii: Lepisosteiformes: Lepisosteidae): A deliberate aquarium trade introduction incidence in the Shatt al-Arab River, Basrah, Iraq. Acta Ichthyol Piscat 47(2):205–207

Myers GS (1965) *Gambusia*, the fish destroyer. Trop Fish Hobbyist 13(5):31–32
Naseka AM (1996) Comparative study on the vertebral column in the Gobioninae (Cyprinidae, Pisces) with special reference to its systematics. Publicaciones Especiales Instituto Espanol de Oceanografia 21:149–167
Nunez JCB, Seale TP, Fraser MA, Burton TL, Fortson TN, Hoover D, Travis J, Oleksiak MF, Crawford DL (2015) Population genomics of the euryhaline teleost *Poecilia latipinna*. PLoS One 10(9):e0137077
Nellen W, Ruckes E (1975) Die Nutzbarmachung des Assadsees, Arabische Republik Syrien, filr die Fischerei und M6glichkeiten einer UnterstUtzung durch die Bundesrepublik Deutschland. - Studie ftlr die Deutsche Gesellschaft ftlr Technische Zusammenarbeit, Projekt-Nr.: 62.2195.6 (75.2007.5), Kiel & Berlin, 124 pp
Nelson JS, Grande TC, Wilson MV (2016) Fishes of the world. Wiley, Hoboken, NJ
Özcan G, Tarkan AS (2019) Distribution revised-fifteen years of changes in the invasion of a freshwater fish, *Pseudorasbora parva* (Temminck and Schlegel, 1846) in Turkey. Transylv Rev Syst Ecol Res 21(2):69–80
Paterson RA, Lal A, Dale M, Townsend CR, Poulin R, Tompkins DM (2013) Relative competence of native and exotic fish hosts for two generalist native trematodes. Int J Parasitol Parasites Wildl 2:136–143
Petr T (1987) Observations on prospects for further inland fisheries development in Iran. A Mission report for the project Training in Coldwater Fish Culture. Technical Cooperative Programme, FAO, FI:TCP/IRA/6675, Field Document 2:vi + 71 pp
Pipoyan SK (1996) Bitterling *Rodeus (sic) sericeus amarus*, a new species in the fauna of Armenia. J Ichthyol 36(8):676–678
Rainboth WJ (1996) Fishes of the Cambodian Mekong. FAO species identification field guide for fishery purposes. FAO, Rome, 265 p
Reynolds SJ (2009) Impact of the introduced poeciliid *Gambusia holbrooki* on amphibians in southwestern Australia. Copeia 2009(2):296–302
Roozbhfar R, Dehestani-Esfandabadi M, Roozbehfar S (2014) First record of the redbelly tilapia, *Tilapia zillii* Gervais, 1848 in Iran. J Appl Ichthyol 30:1–2
Roozen FC, Lürling M, Vlek H, Van Der Pouw Kraan EA, Ibelings BW, Scheffer M (2007) Resuspension of algal cells by benthivorous fish boosts phytoplankton biomass and alters community structure in shallow lakes Freshw Biol 52(6):977–987
Saadati MAG (1977) Taxonomy and distribution of the freshwater fishes of Iran. M.Sc. Thesis, Colorado State University, Fort Collins
Sadeghi R, Zarkami R, Sabetraftar K, Damme PV (2013) A review of some ecological factors affecting the growth of *Azolla* spp. Caspian J Environ Sci 11:65–76
Sadeghi R, Esmaeili HR, Zarei F, Esmaeili A, Abbasi K (2019) The taxonomic status of an introduced freshwater goby of the genus *Rhinogobius* to Iran (Teleostei: Gobiidae). Zool Middle East 65(1):51–58
Salnikov V (2010) First finding of gar *Atractosteus* sp. (Actinopterygii, Lepisosteiformes, Lepisosteidae) in the Caspian Sea near the coast of Turkmenistan. Russ J Biol Invasions 1 (1):17–20
Sakai H, Ikoma K, Frolov SV, Yamazaki Y, Takahashi H, Ida H (2000) Morphological features of a Russian freshwater goby, *Rhinogobius lindbergi* (Pisces: Gobiidae), and its genetic relationships to Japanese species. Biogeogr Int J Biogeogr Phylogeny Taxonomy Ecol Biodivers Evol Conserv Biol 2:51–61
Sakai AK, Allendorf FW, Holt JS, Lodge DM, Molofsky J, With KA, Baughman S, Cabin RJ, Cohen JE, Ellstrand NC, McCauley DE, O'Neil P, Parker IM, Thompson JN, Weller SG (2001) The population biology of invasive species. Annu Rev Ecol Syst 32:305–332
Sayyadzadeh G, Esmaeili HR, Ghasemian S, Mirghiyasi S, Parsi B, Zamanpoore M, Akhlaghi M (2016) Co-invasion of anchor worms *Lernaea cyprinacea* (Copepoda: Lernaeidae) in some freshwater fishes of the Kor River Basin, Southwest of Iran with some remarks on the ecological aspects of lernaeosis in the country. Iran J Fish Sci 15(1):369–389

Serov DV, Nezdoliy VK, Pavlov DS (2006) The Freshwater Fishes of Central Vietnam. KMK Scientific Press, Moscow, Nha Trang
Shireman JV, Smith CR (1983) Synopsis of biological data on the grass carp, *Ctenopharyngodon idella* (Cuvier and Valenciennes, 1844). Food and Agriculture Organization Fisheries Synopsis, 135: iv+86 pp, Rome
Shuai F, Li X, Liu Q, Zhu S, Wu Z, Zhang Y (2019) Nile tilapia (*Oreochromis niloticus*) invasions disrupt the functional patterns of fish community in a large subtropical river in China. Fish Manag Ecol 26(6):578–589
Smith RJF (1997) Avoiding and deterring predators. Behavioural ecology of teleost fishes 1997:163–190
Suzuki T, Shibukawa K, Senou H, Chen I-S (2015) Redescription of *Rhinogobius similis* (Gill, 1859) (Gobiidae: Gobionellinae), the type species of the genus *Rhinogobius* (Gill, 1859), with designation of the neotype. Ichthyol Res 63:227–238
Tarkan AS, Marr SM, Ekmekçi FG (2015) Non-native and translocated freshwater fish species in Turkey. FiSHMED Fishes Mediterranean Environ 003:1–28
Thinh DV, Van NS, Huckstorf V, Nguyen THT (2012) *Hemiculter leucisculus*. The IUCN red list of threatened species. Version 2014.3. www.iucnredlist.org. Accessed 20 December 2014
Timur G, Timur M (1991) An outbreak of enteric redmouth disease in farmed rainbow trout *(0. mykiss)* in Turkey. Bull Eur Assoc Fish Pathol 11(5):182–183
Turan D, Kottelat M, Kaya C (2017) *Salmo munzuricus*, a new species of trout from the Euphrates River drainage, Turkey (Teleostei: Salmonidae). Ichthyol Explor Freshw 28(1):55–63
Turan D, Kalayci G, Bektaş Y, Kaya C, Baycelebi E (2020) A new species of trout from the northern drainages of Euphrates River, Turkey (Salmoniformes: Salmonidae). J Fish Biol. https://doi.org/10.1111/jfb.14321
Ünlü E, Çiçek T, Değer D, Coad BW (2011) Range extension of the exotic Indian stinging catfish, *Heteropneustes fossilis* (Bloch, 1794) (Heteropneustidae) into the Turkish part of the Tigris River watershed. J Appl Ichthyol 27:141–143
Uysal I, Alpbaz A (2002) Comparison of the growth performance and mortality in Abant Trout, *Salmo trutta abanticus* (Tortonese, 1954) and Rainbow Trout, *Oncorhynchus mykiss* (Walbaum, 1792) under farming conditions. Turk J Zool 26:399–403
Valikhani H, Abdoli A, Kiabi BH, Nejat F (2016) First record and distribution of the blue tilapia, *Oreochromis aureus* (Steindachner, 1864) (Perciformes: Cichlidae) in inland waters of Iran. Iran J Ichthyol 3(1):19–24
Vitousek PM, Mooney HA, Lubchenko J, Melillo JM (1997) Human domination of Earth's ecosystems. Science 277:494–499
Vreven EJ, Adepo-Gourene B, Agnèse JF, Teugels GG (1998) Morphometric and allozyme variation in natural populations and cultured strains of the Nile tilapia *Oreochromis niloticus* (Teleostei, Cichlidae). Belg J Zool 128:23–34
Weber MJ, Brown ML (2009) Effects of common carp on aquatic ecosystems 80 years after "carp as a dominant": ecological insights for fisheries management. Rev Fish Sci 17(4):524–537. https://doi.org/10.1080/10641260903189243
Welcomme RL (1981) Register of International Transfers of Inland Fish Species. Food and Agriculture Organization, Rome, Fisheries Technical Paper 213
Welcomme RL (1988) International introductions of inland aquatic species. Food and Agriculture Organization, Rome, Fisheries Technical Paper 294
White PG (1988) A regional survey of the aquaculture sector in eleven Middle East countries (including Bahrain, Iraq, Islamic Republic of Iran, Jordan, Kuwait, Oman, Qatar, Saudi Arabia, United Arab Emirates, Yemen Arab Republic and the People's Democratic Republic of Yemen). Food and Agriculture Organization, Rome, Aquaculture Development and Coordination Programme, ADCP/REP/88/30:v +50 pp
Wilcove DS, Rothstein D, Dubow J, Phillips A, Losos E (1998) Quantifying threats to imperiled species in the United States. Bioscience 48:607–615

Wisenden BD (1994) Factors affecting mate desertion by males in free-ranging convict cichlids (*Cichlasoma nigrofasciatum*). Behav Ecol 5(4):439–447

Zakaria H (1964) A catfish (*Heteropneustes fossilis*) of medical importance invades Iraqi waters. J Faculty Med. Baghdad 6(2):48–55

Zarei F, Rajabi-Maham H (2017) First record of exotic *Piaractus brachypomus* (Cuvier, 1818) (Characiformes: Serrasalmidae) from the Zarivar Lake, Western Iran. J Appl Ichthyol. https://doi.org/10.1111/jai.13383

Zareian H, Esmaeili HR, Gholamhosseini A, Sayyadzadeh G (2013) New records and geographical distribution of *Alburnus hohenackeri* (Kessler, 1870) (Teleostei: Cyprinidae) in Iran. Check List 9:829–831

Zareian H, Esmaeili HR, Zamanian Nejad R, Vatandoust S (2015) *Hemiculter leucisculus* (Basilewsky, 1855) and *Alburnus caeruleus* (Heckel, 1843): new data on their distributions in Iran. CJES 13:11–20

GPSR Compliance

The European Union's (EU) General Product Safety Regulation (GPSR) is a set of rules that requires consumer products to be safe and our obligations to ensure this.

If you have any concerns about our products, you can contact us on ProductSafety@springernature.com

In case Publisher is established outside the EU, the EU authorized representative is:

Springer Nature Customer Service Center GmbH
Europaplatz 3
69115 Heidelberg, Germany

Batch number: 10762843

Printed by Printforce, the Netherlands

Aquatic Ecology Series

Volume 11

Aquatic ecology is an extraordinarily broad and interesting field. It investigates the interplay between aquatic organisms and their physical, chemical, and biological environment. Aquatic ecology encompasses all freshwater and marine ecosystems, including streams, rivers, lakes, wetlands, coastal environments, and the vast expanses of the open ocean. Aquatic ecology studies a wide diversity of different organisms, ranging from tiny bacteria to large whales, facing a myriad of different processes such as biogeochemical cycles, genetic differentiation, and climate change. Fundamental research in aquatic ecology adds new discoveries almost every day. Applied research makes major contributions to biotechnology, fisheries, water management, nature conservation, and environmental policy. Reassessments and syntheses in aquatic ecology are stimulating to the discipline as a whole, as well as enormously useful to students and researchers in ecological sciences.

A series of succinct monographs and specialized evaluations in aquatic ecology has been developed. Subjects covered are topical (e.g., coastal hydrodynamics, microbial loops, alpine lakes) rather than broad and superficial.

The treatments must be comprehensive and state-of-the-art, whether the topic is at the biochemical, mathematical, population, community, or ecosystem level. The objectives are to advance the topics by the development of arguments, with documented support, that generate new insights, concepts, theories to stimulate thought, ideas, directions, controversies. The books are intended for mature as well as emerging scientists to stimulate intellectual leadership in the topics treated.

If you have questions regarding our book series or would like to submit a proposal to Aquatic Ecology, please contact the Publisher Éva Lőrinczi (eva.loerinczi@springer.com).

More information about this series at http://www.springer.com/series/5637

Laith A. Jawad
Editor

Tigris and Euphrates Rivers: Their Environment from Headwaters to Mouth

Volume 2

Springer

Editor
Laith A. Jawad
School of Environmental and Animal Sciences
Unitec Institute of Technology
Auckland, New Zealand

Aquatic Ecology Series
ISBN 978-3-030-57569-4   ISBN 978-3-030-57570-0 (eBook)
https://doi.org/10.1007/978-3-030-57570-0

This Springer imprint is published by the registered company Springer Nature Switzerland AG.
The registered company address is: Gewerbestrasse 11, 6330 Cham, Switzerland

"Once Again to Zainab, my lovely wife"
To my daughters and granddaughters.....
You made me stronger, better and more
fulfilled than I could have ever imagined....
I love you all to the moon and back

Preface

It was a great book edited by Róska (1980), *Euphrates and Tigris, Mesopotamian Ecology and Destiny*. For a period of nearly 40 years, this book was ample in regard to the information that was incorporated into it as there was no flood of reports about the drainages of the twin rivers, Euphrates and Tigris, like what we have now.

Since I saw this book in the early 1980s, an idea came to me to produce a comprehensive book dealing with the drainages of the rivers of Euphrates and Tigris and the landscape they enclosed. This idea remained in my mind in spite of the two Gulf wars and the political turmoil that the region has confronted during the last 30 years. I carried with me through the countries that I settled in temporarily and permanently for the last 26 years the idea of editing a comprehensive publication about Euphrates–Tigris rivers until the opportunity came, nearly 2 and half years ago, when I sent the proposal of the book to Springer, and it has been accepted.

In planning the contents of the book, I took into consideration that of the book of Róska (1980) as a base and started to add and fill in the gaps those chapters that are missing and that contain information about subjects that have been forgotten in spite of being important to the region. I tried my best to make this book containing as much as possible the important aspects of life, the environment and the conservation of the Euphrates–Tigris rivers and their landscapes.

There are several reasons for choosing to edit a book about the Euphrates–Tigris rivers and the landscapes they encompassed. Among these are the resources of this region represented in water, oil, gas, minerals and agriculture. In addition, the historical aspects of the region and the social life of the people living in this area and the strategic geographic location, which is in the middle of the old world and accessible to all continents, are all factors that encourage me to edit such a book and make it available to the readers all over the world.

The fauna and the flora of the Mesopotamia are distinct and peculiar in their distribution and species content. Moreover, the invasion of the different species belonging to different plant and animal groups to this region made the study of these groups an important issue for most scientists.

The other important aspect that this book is dealing with is the issues of the environment and the conservation measures that are taking place at the present time and those that need to be planned in the future. The Mesopotamia is a land that faced over the long years several changes caused naturally and due to the activities of humans in this region for thousands of years. Conflicts of different types between different populations and ethnicities were evident in the historical parts of the region and are still in practice in the present time in some parts of Mesopotamia. Such conflicts and natural disasters have changed and still in the process of changing the resources of the area, which have a negative effect on the people living in the region.

The different chapters of this book were written by scholars who are experts in their field. Thus, they represent the best of their scientific work to educate readers about the status of the environment of the Euphrates–Tigris rivers and the inhabitants of their landscapes and show the important aspects of this region to the Middle East in particular and the world in general. It is important for the people living in Mesopotamia at least to be aware of their land and its resources and how to conserve such wealth as their ancestors in Sumer, Akkad and Assyria have done thousand years ago.

I would like to express my thanks to all the contributors of this book who agreed to share the pieces of their scientific work as chapters are written about different aspects of Mesopotamia in order make it available to the readers all over the world. Also, my sincere thanks should go to Springer that agreed to publish this book and make my dream a reality.

Auckland, New Zealand Laith A. Jawad

Contributors

Abed, Jasim	Iraq
Adriansen, Hanne Kirstine	Denmark
Ahmed, Sumaya M.	Iraq
Akıncı, Zeynep Sıla	Catalonia
Al-Edany, Taha Y.	Iraq
Al-Imarah, Faris J. M.	Iraq
Al-Janabi, Muhammad	Iraq
Al-Saad, Hamid T.	Iraq
Al-Sheikhly, Omar F.	Iraq
Al-Sulami, Amin	Iraq
Al-Taee, Asaad	Iraq
Ali, Atheer	Iraq
Atasağun, Sibel	Turkey
Audai M. Qasim	Iraq
Ayboğa, Ercan	Germany
Aydin, Gazel Burcu	Turkey
Bilgen, Arda	USA
Çamur-Elipek, Belgin	Turkey
Casellas, Antònia	Spain
Çetin, A. Kadri	Turkey
Dawood, Furat	Australia
Ergönül, M. Borga	Turkey
Esmaeili, Hamid Reza	Iran
Freyhof, Jörg	Germany
Genc, Ercument	Turkey
Ghaffari, Hanyeh	Iran
Glöer, Peter	Germany
Halihan, Todd	USA
Hassan, Fikrat M.	Iraq

Hayati, Dariush	Iran
Jajaei, Efat Salmani	Iran
Jawad, Laith A.	New Zealand
Jaweir, Haifa J.	Iraq
Jongerden, Joost	The Netherlands
Kaleli, M. Aydın	Turkey
Kalender, Leyla	Turkey
Kambouzia, Jafar	Iran
Kaya, Cüneyt	Turkey
Kaya, Dogukan	Turkey
Khalfallah, Myriam	Canada
Khayyun Amtair Rahi	Iraq
Koyuncu, Cafer Erkin	Turkey
Malekzadeh-Viayeh, Reza	Iran
Masseti, Marco	Italy
Maulood, Bahram K.	Iraq
Mehrabian, Ahmad Reza	Iran
Mohammed, Ruqayah	Iraq
Mostafavi, Hossein	Iran
Mousavi-Sabet, Hamed	Iran
Muhsin, Tawfik M.	Iraq
Nakayama, Mikiyasu	Japan
Öktener, Ahmet	Turkey
Pauly, Daniel	Canada
Pešić, Vladimir	Montenegro
Pourang, Nima	Iran
Safaei-Mahroo, Barbod	Iran
Sakamoto, Maiko	Japan
Saleh, Khalil I.	Iraq
Salewicz, Kazimierz A.	Austria
Salman, Nadir A.	Iraq
Scholz, Miklas	UK
Scurlock, JoAnn	USA
Secer, Faik Sertel	Turkey
Şen, Bülent	Turkey
Shafizade-Moghadam, Hossein	Iran
Sharif, Adel	UK
Solak, Cüneyt N.	Turkey
Sulaiman, Sadeq Oleiwi	Iraq
Tanaka, Yukio	USA
Taş-Divrik, Menekşe	Turkey
Teimori, Azad	Iran
Valinassab, Tooraj	Iran
Valizadeh, Naser	Iran

Waryani, Baradi	Pakistan
Wojtal, Agata	Poland
Yakhchali, Mohammad	Iran
Yaseen, Zaher Mundher	Vietnam
Yildirim, Pınar	Turkey
Yilmaz, Elif	Turkey
Yilmaz, Harun	Turkey
Zamanpoore, Mehrdad	Iran

Acknowledgements

First, the editor would like to thank each one of the authors, who contributed their time and expertise to this book. Without their support, this book would not have become a reality.

Second, the editor wishes to thank the staff of Springer DE in general about the efforts they put in publishing this book and in particular Éva Loerinczi for her distinguished service, valuable advice and suggestions that she delivered throughout the period of checking the text of this book.

Laith A. Jawad

Contents for Volume 2

Part IV Water Resources

52 The Ecology and Modelling of the Freshwater Ecosystems in Iran 1143
Hossein Mostafavi, Ahmad Reza Mehrabian, Azad Teimori, Hossein Shafizade-Moghadam, and Jafar Kambouzia

53 Enhancing Rural Women's Participation in Fisheries in Iraq 1201
Laith A. Jawad

54 The Impact of Destructive Fishing Gear on the Fish Biodiversity in the Inland Waters of Iraq 1207
Laith A. Jawad

55 Evaluating Variations in Fisheries by Means of Fishers' Information: Suggested Methodology to Improve Small-Scale Fisheries in Rivers in Iraq 1225
Laith A. Jawad

56 A Possibility to Apply a Traditional Fisheries Enforcement Programme in the Inland Waters of Iraq 1231
Laith A. Jawad

57 Market-Resource Relations and Fish Seller Livelihood as Seen in Inland Waters of Iraq 1237
Laith A. Jawad and Audai M. Qasim

58 The Potential Impact of Deformities in Fishes upon Aquatic Production: Case of Iraq 1253
Laith A. Jawad

59 The Impact of some Social Taboos on Fisheries in Iraq 1263
Laith A. Jawad

60 Inland Water Fishes and Fisheries in Iran 1277
Tooraj Valinassab and Hamid Reza Esmaeili

61 The Possibility of Introducing an Inland Fisheries Education in Iraq 1291
Laith A. Jawad

62 The Marine and Diadromous Fisheries of Iraq 1301
Myriam Khalfallah and Daniel Pauly

Part V Stress of the Environment of the Two Rivers

63 Fish Deformities in the Freshwater Fishes of Iraq: A Short Review and a Study Case on the Indian Catfish *Heteropneustes fossilis* 1311
Laith A. Jawad and Jasim Abed

64 The Phenomenon of Fluctuating Asymmetry: As Fish Welfare Indicator Represented by Case Study from the Freshwater Fishes of Iraq 1321
Laith A. Jawad and Jasim Abed

65 The Studies on Sediments Pollution by Different Types of Metals in Turkey 1339
Leyla Kalender

66 Macroplastic and Microplastic in the Freshwater Environment of Southern Iraq: Evidences Obtained from Freshwater Fish Species 1353
Laith A. Jawad

67 Heavy Metals in Freshwater Invertebrates of Iran: A Review on the Bioaccumulation and Effects 1375
Nima Pourang

68 Fish Parasites of Tigris and Euphrates River Systems 1389
Cafer Erkin Koyuncu, Harun Yilmaz, Faik Sertel Secer, Dogukan Kaya, and Ercument Genc

69 The Parasites of Fishes of the Euphrates and Tigris Rivers: Iraq and Turkey 1419
Ahmet Öktener

70 Ornamental Fishes: A Looming Danger for Inland Fish Diversity of Iraq 1445
Laith A. Jawad, Audai Qasim, and Muhammad Al-Janabi

Part VI The Health of the Human Community Inhabiting Freshwater Zones

71 Aquatic Snails as a Vector of Diseases to the Human in Iran 1459
Mohammad Yakhchali

72 Death by Drowning in Rivers in Iraq 1479
Laith A. Jawad

73 The Dangerous Catfish Species in the Freshwater System of Iraq: First Time Reports on Cases of Envenomation 1487
Laith A. Jawad

74 Fish Species of the Order Cypriniformes as a Source of Ichthyootoxin and Ichthyogallotoxin in Iraq: Cases Reports 1493
Laith A. Jawad

75 First Reports on Cases of Hallucinatory Fish Poisoning (Ichthyoallyeinotoxism) and Scombrotoxic Fish Poisoning in Iraq 1499
Laith A. Jawad

Part VII Conservation

76 Freshwater Management and Conservation in Iran: Past, Present, and Future 1507
Dariush Hayati and Naser Valizadeh

77 The Need of Biodiversity Conservation Strategies in Iraq: The State of Protected Areas 1535
Laith A. Jawad

78 Benthic Macroinvertebrates of the Tigris and Euphrates Rivers in Turkey 1539
Belgin Çamur-Elipek, Bülent Şen, Gazel Burcu Aydın, Menekşe Taş-Divrik, and Pınar Yıldırım

79 Freshwater Ecosystem Conservation in Iraq: Recommendations for Management 1561
Laith A. Jawad

80 Utilising Phenotypic Difference to Regulate Protection Value: A Scheme for Application of a Novel Approach to the Inland Water of Iraq 1565
Laith A. Jawad

81 Hatchery-Reared Fish Stocks Released into the Wild: A Conservation Problem as Seen in a Case Study from Iraq 1571
Laith A. Jawad

Part VIII Social Perspectives

82 Is the Glass Half Empty or Half Full? An Appraisal of the Four Decades of Turkey's Southeastern Anatolia Project (GAP) 1581
Arda Bilgen, Zeynep Sıla Akıncı, Antònia Casellas, and Joost Jongerden

Part IX Food Security

83 How Possible to Build Rice–Fish Farming in Iraq in order to Support Food Security Plan: Positive and the Negative Aspects 1601
Laith A. Jawad

84 The Importance of Non-commercial and Small-Sized Fish Species: A Proposal for an Additional Revenue to Iraq 1611
Laith A. Jawad

85 Sociocultural Aspects Influence Food Consumption Habits in Iraq: Management of Food Security 1625
Laith A. Jawad

86 A Preliminary Investigation of Determinants of Food Security in Rural Areas of Basrah Province, Iraq 1635
Laith A. Jawad

Contents for Volume 1

1 **Introduction** 1
Laith A. Jawad

Part I Historical Perspectives

2 **Vertebrates of Upper Mesopotamia: Present Evidence and Archaeological Data** 13
Marco Masseti

3 **Fishing Gears and Methods: A Comparison of Ancient Mesopotamia and Other Ancient Worlds** 73
Laith A. Jawad

4 **Ichthyological Characteristics Available in the Fish Images Existed in the Art of the Ancient Mesopotamia** 85
Laith A. Jawad

5 **The Effectiveness of Ancient Mesopotamian Medical Practices: The Example of *šūšu*-Licorice** 127
JoAnn Scurlock

Part II The Abiotic Aspects of the Tigris–Euphrates River System

6 **Management of Water Resources Using Storage Reservoirs** 171
Kazimierz A. Salewicz, Maiko Sakamoto, and Mikiyasu Nakayama

7 **Estimation of Irrigated Agricultural Area and Water Consumption in Iraq** 205
Yukio Tanaka

8 **The Nature of Tigris–Euphrates Rivers Flow: Current Status and Future Prospective** 229
Zaher Mundher Yaseen, Sadeq Oleiwi Sulaiman, and Adel Sharif

9 Streamflow Alteration Impacts with Particular Reference to the Lower Zab River, Tributary of the Tigris River 243
Ruqayah Mohammed and Miklas Scholz

10 Ecohydrology in Iraq: Challenges and its Future Pathways 275
Laith A. Jawad

11 Oil Pollution in the Shatt Al-Arab River and its Estuary 1980–2018 285
Faris J. M. Al-Imarah and Hamid T. Al-Saad

12 Pesticides in the Waters, Sediments, and Biota of the Shatt Al-Arab River for the Period 1980–2017 299
Hamid T. Al-Saad and Faris J. M. Al-Imarah

13 Surface Water Salinity of the Euphrates, Tigris, and Shatt al-Arab Rivers 309
Khayyun Amtair Rahi and Todd Halihan

14 Renewable Energy for Water–Energy Nexus in Euphrates and Tigris River Basin: A Literature Review 337
Furat Dawood

15 Impacts of Dams on Aquatic Biodiversity, Fisheries, Fishes and Their Environment: Problems that Could Be Present in Iraq with Recommendations 369
Laith A. Jawad

16 Water, Politics and Dams in the Mesopotamia Basin of the Northern Middle East: How Turkey Instrumentalises the South-Eastern Anatolia Project for Political, Military and Strategic Interests 383
Joost Jongerden, Zeynep Sıla Akıncı, and Ercan Ayboğa

17 Dams and their Impacts on Fishes in Iran 401
Hamed Mousavi-Sabet

Part III The Biotic Aspects of the Tigris–Euphrates River System

18 Preliminary Review of the Aquatic Biodiversity in Al-Kahla River, Missan Province, Iraq 423
Laith A. Jawad

19 Freshwater Fish Biodiversity in Iraq: Importance, Threats, Status, and Conservation Challenges 479
Laith A. Jawad

20 Effect of Climate Changes on the Freshwater Biodiversity in the Mesopotamian Plain: Recommendations for Avoidance and Plans for the Future 499
Laith A. Jawad and Baradi Waryani

21 River Corridors as a Refuge for Freshwater Biodiversity: Basic Information and Recommendations to the Policymakers for Possible Implications in Iraq 515
Laith A. Jawad and Baradi Waryani

22 Biodiversity of Fungi in Aquatic Environment of Iraq 525
Tawfik M. Muhsin

23 Potability of Drinking Water in Basra-Iraq 541
Amin Al-Sulami and Asaad Al-Taee

24 Algal Studies in Iraqi Inland Waters. A Review 553
Bahram K. Maulood and Fikrat M. Hassan

25 The Distribution of Epilitic Diatoms in the Turkish Part of the Tigris-Euphrates River Basin 571
Cüneyt N. Solak, Agata Wojtal, Elif Yılmaz, M. Borga Ergönül, and A. Kadri Çetin

26 Plant Biodiversity in Shatt Al-Arab Estuary and Ecological Variations 621
Taha Y. Al-Edany

27 Medicinal Plants of Shatt al-Arab River and Adjacent Area 643
Taha Y. Al-Edany

28 The Role of Plants as a Canopy in the Inland Waters: Basic Information for Application in Iraq 663
Laith A. Jawad and Audai M. Qasim

29 The Zooplankton Fauna of the Turkish Part of the Euphrates-Tigris River Basin 679
Mehmet Borga Ergönül and Sibel Atasağun

30 Kinds and Distribution of Icthyoplankton in Shatt Al-Arab River 693
Sumaya M. Ahmed

31 Rotifer Diversity in Iranian Waters: A Review 705
Reza Malekzadeh-Viayeh

32 Biodiversity of the Freshwater Amphipods in Iran 743
Mehrdad Zamanpoore

33 The Freshwater Molluscs of the Mesopotamian Plain 763
Vladimir Pešić and Peter Glöer

34 Freshwater Annelida of Iraq 779
Haifa J. Jaweir

35 A Critical Checklist of the Inland Fishes Native to the Euphrates and Tigris Drainages 815
Jörg Freyhof, Cüneyt Kaya, and Atheer Ali

36 Fish Fauna of Shatt al-Arab River, Basrah, Iraq: A More than Quarter a Century of Changes 855
Laith A. Jawad, Abbas J. Al-Faisal, and Mustafa S. Faddagh Ziyadi

37 The Common Carp, *Cyprinus carpio*: Effect on the Environment and the Indigenous Fish Species in Iraq 877
Laith A. Jawad, Mustafa S. Faddagh Ziyadi, and Abbas J. Al-Faisal

38 The Good and the Bad in Releasing the Grass Carp *Ctenopharyngodon idella* in the Freshwater System: Recommendations for the Policymakers in Iraq 897
Laith A. Jawad and Falah Mutlak

39 The Avifauna of Tigris and Euphrates River Basin 913
Omar F. Al-Sheikhly

40 The Feasible Approaches to Assist Migratory Birds Visiting the Southern Reaches of Mesopotamia 937
Laith A. Jawad

41 A Proposal for Establishing Bird Observatory Centre in the South of Iraq 947
Laith A. Jawad

42 The Potential Role of Waterbirds as a Vector in Dispersing Invertebrates and Plants in the South of Iraq 951
Laith A. Jawad

43 The Wild Mammals of the Tigris and Euphrates Rivers Basin . . . 957
Omar F. Al-Sheikhly

44 The Amphibians and Reptiles of Euphrates and Tigris Basin 975
Barbod Safaei-Mahroo and Hanyeh Ghaffari

45 A Preliminary Pictorial Guide to the Herpetofauna of Tigris and Euphrates River Basin 1007
Omar F. Al-Sheikhly

46 Policy Guidance for Sustainable Aquaculture in the Inland Waters of Iraq 1041
Laith A. Jawad and Falah Mutlak

47 How Possible to Use the Desert Area in Iraq for Aquaculture Industry: Basic Facts and Recommondations 1047
Laith A. Jawad and Saad Mohammed Saleh Abdulsamad

48 Aquaculture Industry in Iraq: Past, Present, and Future Perspectives 1053
Nader A. Salman and Khalil I. Saleh

49 Towards an Improved Tilapia Farming in Iraq: Recommendations for Future Application 1077
Laith A. Jawad and Saad Mohammed Saleh Abdulsamad

50 Information and Recommendations of Aquaculture Stress and its Source in Hatcheries: The Iraqi Aquaculture Industry Investors 1089
Laith A. Jawad

51 Exotic and Invasive Freshwater Fishes in the Tigris-Euphrates River System 1103
Hamid Reza Esmaeili

About the Editor

Laith A. Jawad obtained a degree (MSc) in fish taxonomy from the Zoology Department, University of Bristol, UK, in 1980. He continued as fish taxonomist at Basrah University, Iraq, where he worked for more than 20 years before he immigrated to New Zealand in 1997. During this time, he started the biochemical taxonomy of fishes of Iraq and published over 420 scientific papers and book reviews in leading scientific journals. He is the author and co-author of several textbooks in biology published in Arabic. Recently, he contributed five chapters to a book about coastal fishes, *Coastal Fishes: Habitat, Behavior and Conservation*, published by Nova Publishers, Canada, and authored a book about *Dangerous Fishes of the Eastern and Southern Arabian Peninsula* published by Springer in 2017. He served as fish biodiversity expert and consultant at the Ministry of Agriculture and Fisheries in Oman for the period 2008–2012 during which he co-authored two papers describing a new fish species from the Omani waters and reported over 80 fish species as a new record to the Omani waters. He authored a guide to the fishes of the southern coasts of Oman published by the Ministry of Agriculture and Fisheries in Oman in 2018. He also published over 90 papers on fish fauna of Oman, Iraq, Kuwait and Saudi Arabia. In 2013, he broadened his scientific contact and started to collaborate with over 60 scientists from more than 50 countries around the world in researches dealing with different aspects of fish taxonomy and ichthyology.

Part IV
Water Resources

Chapter 52
The Ecology and Modelling of the Freshwater Ecosystems in Iran

Hossein Mostafavi, Ahmad Reza Mehrabian, Azad Teimori, Hossein Shafizade-Moghadam, and Jafar Kambouzia

52.1 Introduction

Iran has an area of 1,623,779 km^2 and, in terms of size, is believed to be the second and fourteenth country in South-west Asia and in the world, respectively (Mostafavi et al. 2015; Coad 2018). The largest proportion of Iran is located in an average altitude of about 1000 m. The main highlands consist of four distinct mountainous regions: Alborz in the north; Kopet-Dagh and north Khorasan ranges in the north-

H. Mostafavi (✉)
Department of Biodiversity and Ecosystem Management, Environmental Sciences Research Institute, Shahid Beheshti University, Tehran, Iran
e-mail: hmostafaviw@gmail.com

A. R. Mehrabian
Department of Plant Sciences and Biotechnology, Faculty of Life Sciences and Biotechnology, Shahid Beheshti University, Tehran, Iran
e-mail: A_Mehrabian@sbu.ac.ir

A. Teimori
Department of Biology, Faculty of Sciences, Shahid Bahonar University of Kerman, Kerman, Iran
e-mail: a.teimori@uk.ac.ir

H. Shafizade-Moghadam
Department of Water Resources Engineering and Management, Tarbiat Modares University, Tehran, Iran
e-mail: h_shafizadeh@modares.ac.ir

J. Kambouzia
Department of Agroecology, Environmental Sciences Research Institute, Shahid Beheshti University, Tehran, Iran
e-mail: j_kambouzia@sbu.ac.ir

L. A. Jawad (ed.), *Tigris and Euphrates Rivers: Their Environment from Headwaters to Mouth*, Aquatic Ecology Series 11,
https://doi.org/10.1007/978-3-030-57570-0_52

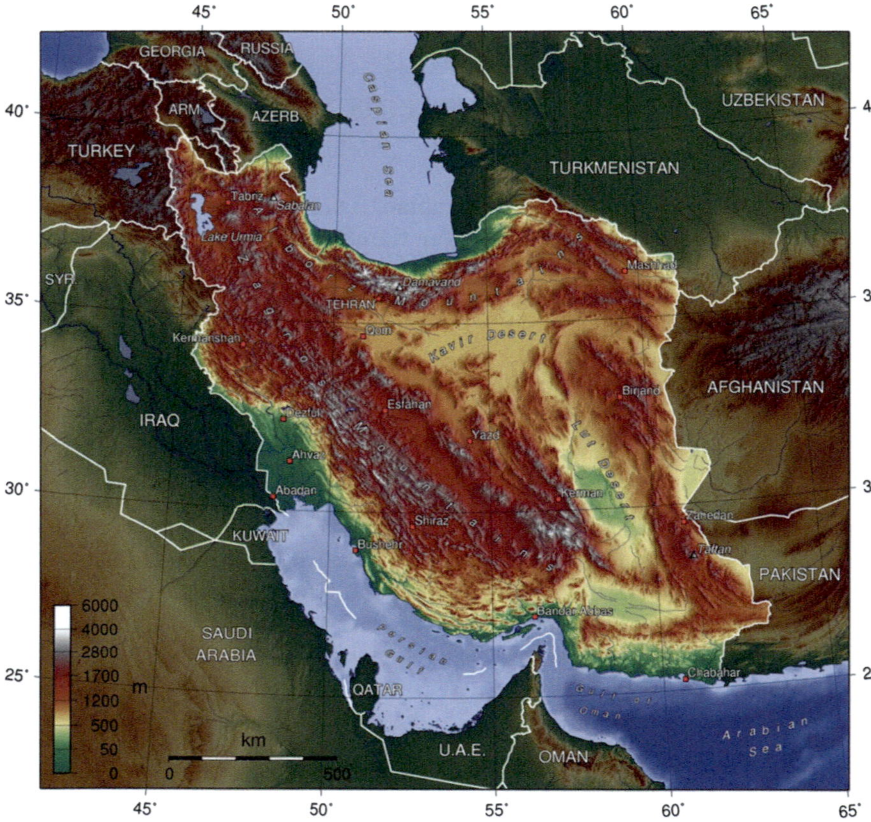

Fig. 52.1 Topographic map of Iran shows various physiographic regions (source: www.worldofmaps.net)

east; Zagros in the west and south-west; and Jebal Barez and Baluchestan mountains in the central to south-east (Fig. 52.1) (Zehzad et al. 2002; Rahimi 2011).

Seven desert plains and depressions, which totally change the appearance of the landscape, are Dasht-e Kavir in central Iran, Dasht-e-Lut desert, Sistan, and Jazmurian depressions in the south-east, Khuzestan plain in the south-west, Moghan steppe in the north-west, and the Turkmen Sahra steppe in the north-east. The country encompasses six major basins; nonetheless, different authors have divided the central and southern basins into 21. Iran's landscape is also formed by over 20 larger lakes, out of which the saline Urmia Lake in the north-west is the largest one, covering about 4868 km^2 area (Zehzad et al. 2002; Majnonian et al. 2005; Coad 2018) (Fig. 52.2).

From a biogeographic point of view, Iran is situated in the mid-latitude belt of arid and semi-arid regions of the Earth, with such areas covering more than 60% of the country. In this region, precipitations vary with regard to time, space, amount, and duration, and water is the most critical restricting element for biological and agricultural activities (Modarres and de Paulo Rodrigues da Silva 2007). The slopes located in the northern part of the Alborz Mountains and the Caspian lowland have

Fig. 52.2 Photographs show different landscapes of Iran

an annual rainfall of 800–2000 mm constituting the country's most humid regions. The Dasht-e-Kavir and Dasht-e-Lut deserts are the driest parts of the country experiencing an annual rainfall of less than 150 mm. The highlands typically receive between 250 and 800 mm. The hydrological cycle and environmental processes are altered by seasonal changes in rainfall pattern as well as the vegetation and the entire ecosystem (Zehzad et al. 2002; Majnonian et al. 2005; Akhani et al. 2010; Coad 2018).

52.2 Biodiversity of Plants and Animals

Plants' diversity and richness in Iran are highly affected by environmental characteristics. For example, there are deciduous forests in the northern flanks of Alborz, while mangroves forests cover the coastal areas in the Oman Sea and the Persian

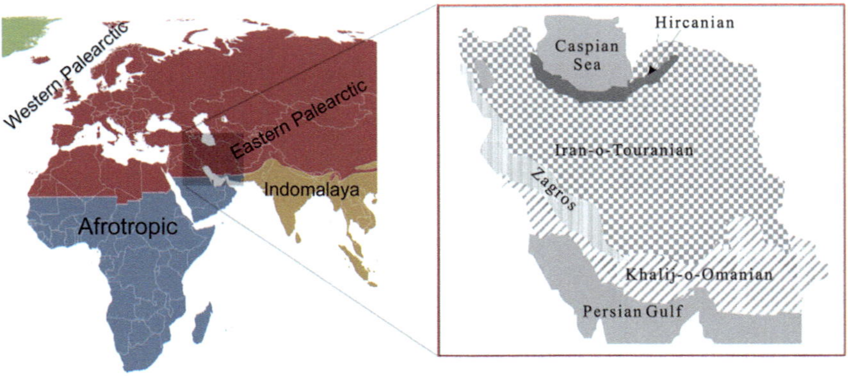

Fig. 52.3 Distribution of four ecological zones in Iran (Heshmati 2012)

Gulf (Darehshouri and Kasraian 1988). One of the most significant features of the habitats in Iran is that they can support a wide range of flowering plant species (belonging to 167 families and 1200 genera), of which almost 1700 are endemic (Majnonian et al. 2005). The Iranian vascular plant flora is composed of around 10,000 species, with approximately 20% of them being endemic (Zehzad et al. 2002).

Considering environmental factors, the country was divided into four ecological zones with specific plant richness. These zones ranging from the lowest area to the highest area, respectively, include Hyrcanian, Khalij-o-Omani, Zagross, and Iran-o-Touranian zones (Heshmati 2012).

With respect to its biodiversity, northern Iran is rich in accommodating 8000 plant species that represent many diverse life forms (Herb, Grass, Shrub and tree). The smallest degree of plant diversity is found in the southern area of the country (Khalij-o-Omanian zone), which is a flat region (Heshmati 2012) (Fig. 52.3).

Also, an extensive spectrum of animal species is living in Iran, a phenomenon that is mainly attributed to high habitat diversity of the country (Figs. 52.2, 52.3, 52.4). Accordingly, the vertebrate fauna of the Iranian ecosystems contains about 1912 species including 197 mammals, 535 bird, 227 reptiles, 21 amphibians, more than 222 freshwater fish, and 710 marine fish (Farashi and Shariati 2017). However, more species are expected to be discovered from this country.

Irano-Anatolian region has been recognised as one of the known 34 hot spots in the world (http://www.thecropsite.com/focus/5m/52/biodiversity-protecting-the-natural-world). In spite of its high diversity, many plant and animal species are currently under serious threats. Therefore, it is in high priority for the conservation management programmes (Fig. 52.5).

Fig. 52.4 Biodiversity of Iran at a glance

52.3 Freshwaters of Iran

As previously mentioned, both mountainous areas and vast deserts can be found in Iran. Furthermore, land-locked inland areas contain numerous tiny springs and streams, which have no recent or present association with other water bodies. Iran has been categorised into 19 major basins (Fig. 52.6) based on fish distributions, hydrography of the different water bodies, and the areas that are important for the understanding of zoogeography (Coad 2018). Following this classification, Iran is land to two major basin types, i.e. exorheic basins in which the rivers and lakes end up in the sea and endorheic, where rivers flow to an internal basin (e.g. a lake), or are discharged into the desert; hence, this second class of basins is not connected to the sea. The exorheic basins are mainly located in the southern parts of the Caspian Sea in the north of Iran, and all the southern, south-western, and south-eastern regions of the country. The endorheic basins (15 in numbers) are mainly confined to the inland regions and cover about 78.1% of the country's area. Two main mountains ranges (i.e. the Alborz and Zagros mountain chains) are the major water sources and guarantee a constant and appreciable flow for major rivers in the north and south/ south-west parts of the country (Coad 2018).

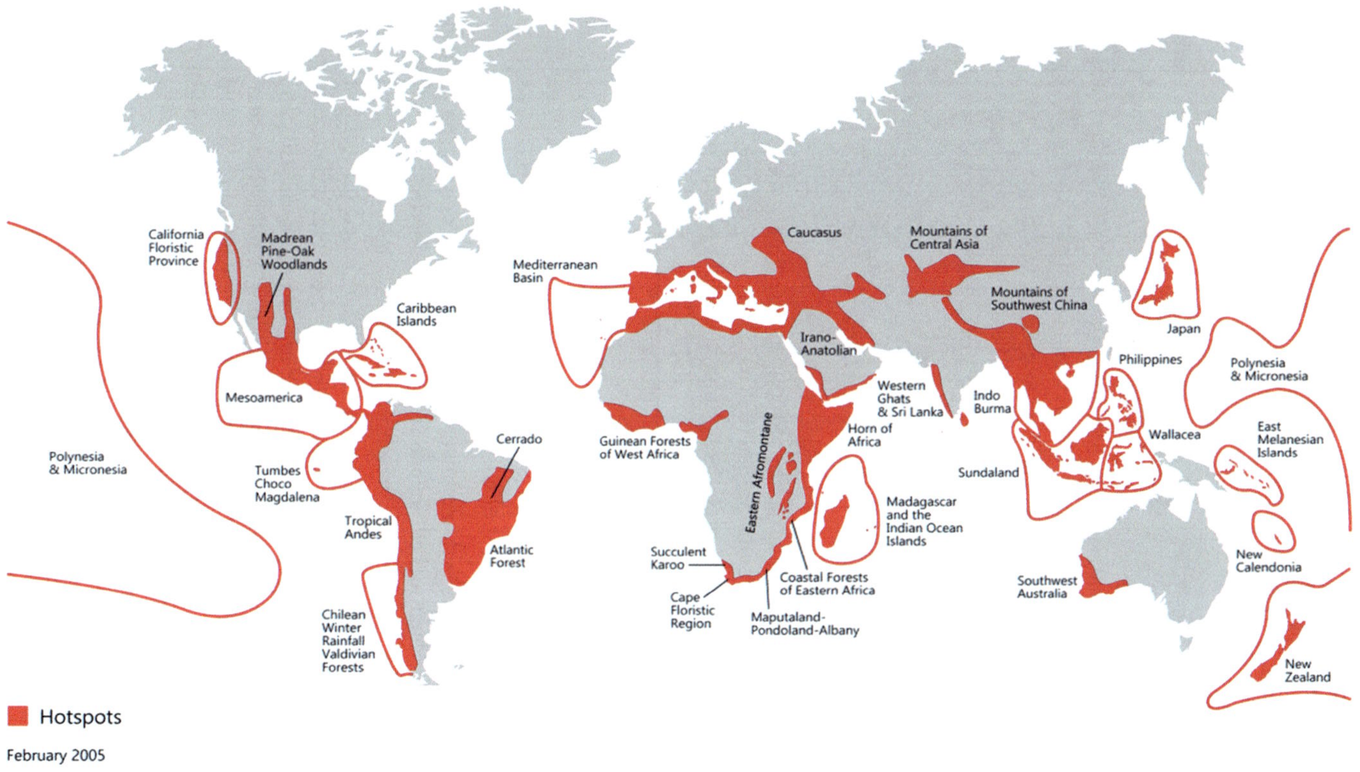

Fig. 52.5 Hot spots of the world showing the Irano-Anatolian hot spot (Source: http://www.thecropsite.com/focus/5m/52/biodiversity-protecting-the-natural-world)

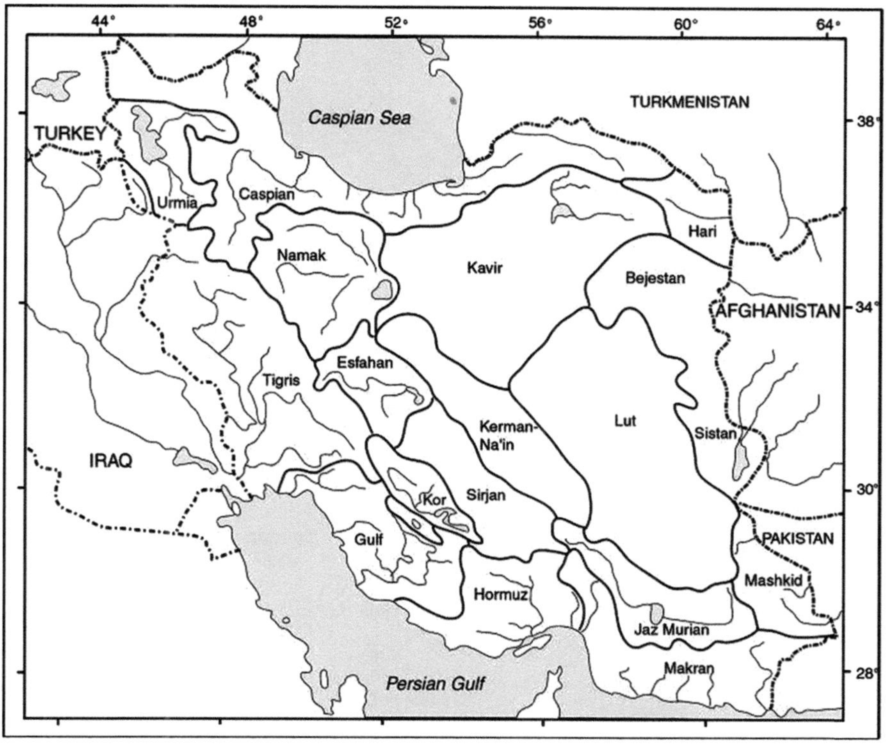

Fig. 52.6 Nineteen basins of Iran (www.briancoad.com)

52.4 Usual Habitats

As mentioned above, adequate amounts of snow and rainfall in the Zagros and Alborz mountain ranges are the primary sources of water for the rivers which causes a continual and noticeable flow. The outstanding characteristic of most rivers in Iran is that shallow and clear water runs through them. However, there are several rivers with considerable amounts of water, which are home to a variety of fish species. Figure 52.7 illustrates Iran's pivotal rivers drainages. Most large lakes are salty (save for the ones in Sistan, the Southern Caspian Sea basin, some lakes in the Persian Gulf basin, and the plains of Khuzestan, all of which are freshwater lakes) (Coad 2018; Teimori et al. 2016).

52.4.1 Freshwater Spring-Stream Systems

The spring-stream systems in Iran typically originate from mountains, underground waters, or Qantas systems and are usually identified by the rather small or medium

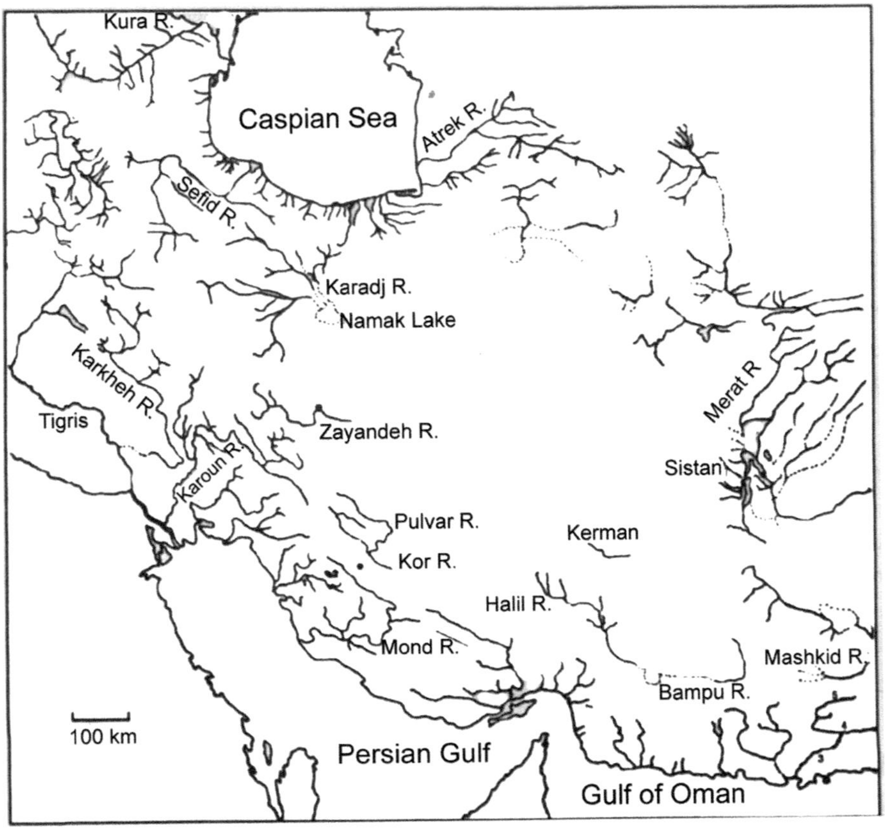

Fig. 52.7 Main river drainages of Iran (Coad 1976)

sizes (one or few metres wide and few centimetres/or metres deep). The diversity of the Iranian inland fish types, especially in arid and semi-arid regions, could be particularly attributed to these spring-stream systems (Teimori et al. 2016; Coad 2018).

52.5 Fish Diversity of Iran

From a biogeographic perspective, the Iranian plateau is positioned in Palearctic region neighbouring the Oriental and African zones (Coad and Vilenkin 2004). Nevertheless, based on its ichthyofauna formation, the Iranian plateau borders the Eastern Mediterranean (Western-Palearctic), the Southern Asian (Indo-Oriental), and the Ethiopian regions (Coad 1998; Nalbant and Bianco 1998). Consequently, it represents diverse kinds of freshwater fish species, i.e. over 222 species (Abdoli 2000; Coad and Vilenkin 2004; Teimori 2006; Esmaeili et al. 2007, 2010, 2013,

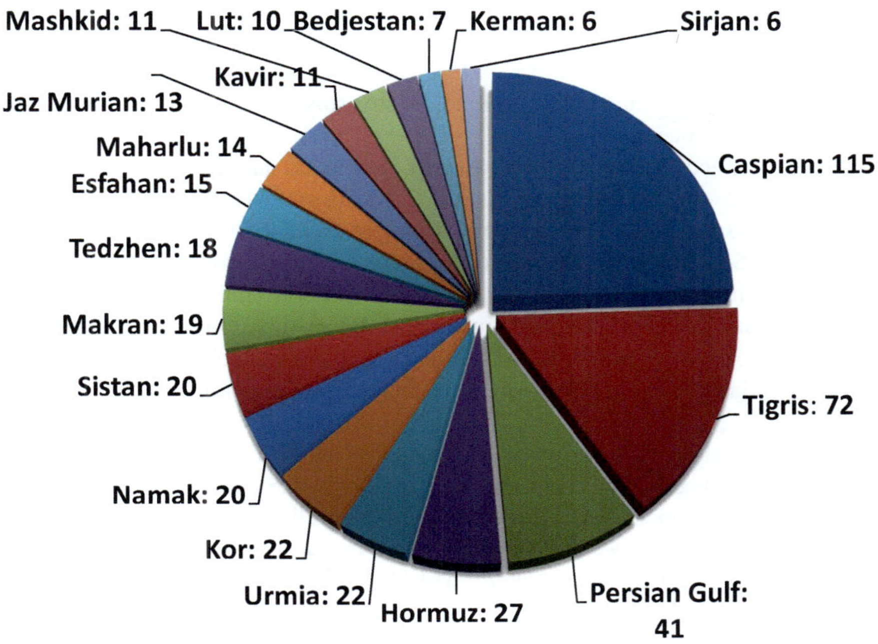

Fig. 52.8 Fish diversity of Iran at different basins (Teimori et al. 2016)

2015; Kamangar et al. 2014; Teimori et al. 2011a, b, 2012a, b, 2014, 2016; Abdoli 2017). As a result, the Iranian plateau may be regarded as the original source of many native species (Esmaeili et al. 2007, 2010; Hrbek and Meyer 2003; Teimori et al. 2012b). Figure 52.8 demonstrates that, in Iran, the highest fish diversity rates are listed for the Caspian Sea basin with 112 species, Tigris basin with 72 and the Persian Gulf with 41 species in that order.

52.6 Environmental Concerns of Iranian Freshwater

Freshwater ecosystems are vital for human beings. They offer a wide variety of services such as provisioning, regulatory, cultural, and supporting facilities. The provision of such services largely depends on the existence of freshwater ecosystems. Therefore, it is of prime importance to preserve, increase, and recover all surface waters to achieve a good ecological status, an issue that is demanded by the EU-Water Framework Directive (Hering et al. 2010).

As a result of development in societies, rivers are manipulated to meet humans' needs. In Iran, for example, dams are constructed and straightened on rivers, exotic species having rivalry with native biota are introduced, and water is extracted to irrigate crops and diverted to establish urban, recreational, and agricultural areas.

The rise of human requirements has jeopardised freshwater ecosystem health and ecological integrity in many parts in Iran and around the world (Schinegger et al. 2012, 2013; Mostafavi et al. 2014, 2015). Therefore, human activities make freshwater ecosystems encounter various kinds of stressors which endanger biodiversity and ecosystem process.

Freshwater ecosystems in Iran have been profoundly and negatively affected by man's activities (Coad 1980; Esmaeili et al. 2007; Mostafavi et al. 2015), which lead to undue pressures on the biota and biological processes (Schmutz et al. 2000; Pont et al. 2005; Schmutz et al. 2007). According to previous literature (Esmaeili et al. 2010; Abdoli 2017) and following the IUCN criteria (International Union for Conservation of Nature (http://www.iucnredlist.org/.)), there exist seven endangered (EN) and five vulnerable fish species in Iran. Yet, it is very likely that, due to the use of inappropriate data collection procedures, many fishes may not have been counted in this list.

In Iran, ecological assessment is missing given that the aquatic ecosystem monitoring has been mainly accomplished in light of the existence of the physical and chemical parameters. On the contrary, almost all freshwater ecosystems are influenced by several types of human pressures such as land-use change, connectivity, hydro-morphology, water quality, and biology (Mostafavi et al. 2004, 2014, 2015). It is critical to identify proper procedures to analyse all types of human pressures and model them for the suitable assessment towards restoration and integrated river basin management. Hence, this chapter primarily concentrates on identifying different human pressure types in Iran (here, it is shown that water quality is not the only pressure threatening the freshwater ecosystems). Second, it aims to carry out an ecological assessment by modelling and analysing human pressures at different scales (here, how the human pressures are quantified at different spatial scales and fish indicators are used for the ecological assessment by modelling). The third part of the chapter revolves around modelling the potential spatial distribution of brown trout (*Salmo trutta*) at Iranian scale and the effects of climate change on its distribution (Mostafavi et al. 2014, 2015).

52.6.1 Human Pressures

In this section, all categories of human pressure types influencing freshwater ecosystems (particularly rivers) in Iran are explicated.

52.6.1.1 Land-Use Pressures

Land-use change is a substantial driver on limiting of ecosystem services. Rivers including their catchments (also known as rivers capes) are considered as an ecosystem profoundly affected by human activities in the landscape (Allan 2004; Trautwein et al. 2012). In fact, land use along stream margins and throughout catchments can measurably impact instream physical, chemical, and biological habitats (Trautwein et al. 2012).

Deforestation

Deforestation has an intense effect on the ecology and hydrology of rivers, occasionally over great spaces (Allan and Flecker 1993). Some of the outstanding influences entail wide variation in temperature and discharge rate, significant rise in nutrient loss from terrestrial sources, considerable reduction in the contribution of leaf litter to energy flow, and alteration in the time and quantity of run-off altered. Due to increased erosion, deforestation can also significantly increase the loading of fine sediments to rivers (Harrison et al. 1995). Additionally, deforestation drives climate change.

No accurate data are available about forest areas in Iran; however, on the basis of some estimations, different national and international organisations argue that Iran is facing a considerable loss in its forest areas. According to a report, which was published by Iran's natural resources engineering office in 1964, 18-m hectares of Iran is covered by natural forests. 3.4-m hectares of the entire forests are situated in Northern slopes of Alborz mountains, alongside the Caspian Sea. But, the current report reveals a decline of 14.2 million hectares in overall forest areas and a decrease of 1.8 m hectares in the three Northern provinces. Cutting and extracting trees from protected and unprotected areas, agricultural activities, civil projects, excessive grazing, and massive fires are believed to be the major culprits for deforestation (Akhani et al. 2010; Majidi et al. 2011; http://www.tehrantimes.com/news/422244/The-significant-deforestation-trend-in-Iran-during-last-7-decades). Figure 52.9 clearly displays land-use change and deforestation in the Caspian Sea basin during past decades.

De-vegetation has severely affected the hydrological regime with consequent destruction of fish habitat in Iran. For a long time, it has reduced rainfall, resulting in stream flow and desert encroachment. The outcome has been continuous torrential flow and flash floods. Moreover, a shift happens in the flow of lower reaches of rivers, which are not even situated in de-vegetated areas, to seasonal round instead of year round (Coad 1980). Many species carry out spawning in the spring when there is a high unexpected overflow. In this time of the year, the eggs are also washed away and covered with silt, or in some cases, they are desiccated due to high water levels (Coad 1980).

Overgrazing

For a long time, overgrazing by cattle has been a big challenge in Iran. Coad (1980) indicated that so many sheep, goats, and cows have influenced the country's vegetation (Fig. 52.10). Moreover, their waste is usually released into the water, hence lowering its quality (see Fig. 52.10).

Fig. 52.9 Land use change in the north of Iran during the last decades (map from Google Earth)

Fig. 52.10 Overgrazing as well as releasing of waste into the water

Fig. 52.11 Agricultural land use

Agricultural Land Use (Fig. 52.11)

Using the land for agricultural purposes will add pollutants to the water which has a negative impact on riparian and river habitat and leads to a shift in stream flows and river damages.

The growing use of agricultural land use will increase the amount of deposits, nutrients, and pesticides (Osborne 1988). The stream habitat in highly agricultural areas is not in good condition, and it is not of high quality (Wang et al. 1997; Mostafavi et al. 2014). Water is extracted from rivers to such a large extent that the rivers become devoid of water for at least some months of the year, or there is a significant decline in flows in comparison with their original capacity (Mostafavi et al. 2015). Due to this significant reduction, current waters sometimes turn to stagnant water.

The disturbance in spawning and early life stages of stream fishes are caused due to some alternations in flow system (Harvey 1987; Schlosser 1985). In addition, fish assemblage generally changes as rheophilic fishes are replaced by stagnant fishes (Melcher et al. 2007; Mostafavi et al. 2015). Although agricultural land use induces a rise in annual and storm flows, it reduces the base flow by decreasing infiltration and more transportation of water (Poff et al. 1997). As a consequence of this reduction, the area of shallow water with structure deficiency which gets warm easily increases (Richards et al. 1996). The more agriculture or other human activities spread to stream margin and eradicate natural riparian forest, the warmer will be the streams during summer and the fewer energy input will be gained, and it leads to a rise in the main production. This may lead to bank stability decrease. It should, however, be noted that the stability of banks increases by planting deep-rooting grasses (Lyons et al. 2000), and the amount of large wood in the stream dwindles significantly (Johnson et al. 1997).

Fig. 52.12 Urban land use

Urban Land Use (Fig. 52.12)

The principal changes are caused by manipulations of the urban land area including the rise in size and variety of toxic waste in drainage, more irregular hydrology due to increased impenetrable surface area, increase of water temperatures originating from the lack of riparian vegetation, and decline in channel structure because of sediments, destabilisation of bank, channelisation, and restrained connections between the river and its banks (Paul and Meyer 2001; Allan 2004).

52.6.1.2 Hydrological Pressures

Hydrological pressures typically take place in the areas with low precipitation and high population mass and also in the areas with intensive agricultural or industrial activities. Dams and artificial channels/pumps are used to abstract water for the purpose of agricultural irrigation (Figs. 52.13, 52.14, 52.15 and 52.16). Water courses and wetland regions in Iran have dwindled due to overexploitation of water. As a matter of fact, manipulations of the normal flow may increase threat to the ecological integrity of such ecosystems (Poff et al. 1997; McKay and King 2006; Mostafavi et al. 2015). Overuse of water resources reduces the availability of water for human uses downstream and undermines the capacity of streams in supporting native biota. There is a coincidence between the rise in water extraction and naturally low flows in drier months of summer (NRE 2002). Consequently, total water capacity will decrease while the frequency and persistence of ecological flow will increase and summer flow patterns will change (McKay and King 2006).

In some months of the year, some rivers and streams face either lack of running water or a small portion of their real capacity. Indeed, our observations indicated that even the pace of the flow in some areas were zero during excessive water abstraction (Figs. 52.13, 52.14, 52.15 and 52.16). It was also discovered that the water quality of rivers as well as habitat structure was affected by water abstraction (Figs. 52.13, 52.14, 52.15 and 52.16) (Mostafavi et al. 2019).

Fig. 52.13 Water abstraction and its effects in different areas in Iran

Flow dynamic alterations may raise sedimentation rate which have influence on species sensitive to sediment loadings, like fish, as well as spawning success. Colonisation of the invasive plant species usually leads to the accumulation of river sediments such as gravel bars and temporary islands (Figs. 52.13, 52.14, 52.15 and 52.16).

52.6.1.3 Connectivity Pressures

Despite the efficient functions of dams, such as managing flows, generating electricity, and supply water for different human uses, they constitute one of the major causes of running water ecosystems alternation (Wang et al. 2011). A huge number of dams (around 647) have been constructed in Iran over past decades (Mostafavi et al. 2019).

a

b

Fig. 52.14 Water abstraction and its effects in different areas in Iran

Fig. 52.15 Water abstraction and dis-connectivity of rivers

The continuity of longitudinal rivers is interrupted by dams and, in particular, by ground sills and weirs (there are numerous weirs and ground sills in most basins), which hinder fish migration (Fig. 52.17). These barriers have blocked long-distance migrations to up-streams in Iran (e.g. Sturgeon fishes) despite the fact that fish species have been reported in many up-streams in the past (e.g. Holčík and Oláh

Fig. 52.16 Water abstraction and its effects on the other biota related to the freshwater ecosystems (Photo from Rafaela Schinegger lecture at BOKU University (Austria, Vienna))

Fig. 52.17 Connectivity pressures influencing fish migration

1992; Mostafavi 2007; Abdoli and Naderi 2009; Mostafavi et al. 2019; Coad 2018). Nearly none of the dams in Iran have fish-ways or fish ladders.

Moreover, dams change lotic river sections to lentic ones and thus have a profound effect on the ecosystems. Indeed, the fish assemblage is altered under this condition (Mostafavi et al. 2015, 2019). Dams further change sediment and nutrients proportion to downstream, alter fluvial thermal regimes, and disconnect

Fig. 52.18 Water quality pressures on rivers

river segments (Wang et al. 2011). Due to these mentioned reasons, the majority of downstream natural habitats, which are widespread in Iran, have been changed and also fragmented (Fig. 52.18).

52.6.1.4 Water Quality Pressures

In recent decades, water quality in Iran has been mostly affected by untreated sewage of urban and agriculture as well as industrial sectors (Fig. 52.19) (Mostafavi et al. 2015). The aquatic ecosystem pollution especially running waters pose the most obvious threats to the ecological health of ecosystem and biota survival. Nevertheless, Mostafavi et al. (2015) have demonstrated that species living in rivers are under the influence of multiple pressures not by chemical pollution only.

Fig. 52.19 Fish mortality due to water quality pressures (eutrophication) on rivers

In addition to habitat deterioration, chemical pollutants are severe threat for species of highly limited ranges. Although all rises are attributed to growing land-use alternation or human progress, pinpointing the culprit factors accurately is dubious, as there may exist various primary causes. The pollution could alter organisms' behavioural decision in natural populations. Making a wrong decision impacts species' conservation at least in two ways. The first one is known as the "ecological traps" or "attractive sink" (Battin 2004), and the second one could be interpreted as making wrong decision by individuals, who are consequently unable to manage the quality of their habitats. On the other hand, water contamination favours wrong decisions. More precisely, because animals (especially freshwater organisms like fish) cannot identify cues about the existence of pollutions, their breeding is reduced and, therefore, mortality goes up.

In Iran, a chain of channels is used for transferring water throughout the city or towns (this system is named "Jube" in Iran) which can lead to contamination. This system functions as source of irrigation for roadside trees; however, it has a negative effect on nearby rivers or streams by releasing the pollutants to them (Coad 1980; Mostafavi et al. 2014, 2015). As this effluent contains nutrient elements, the richness and abundance of sensitive or intolerant species may be reduced or even disappear entirely in the ecosystem (Fig. 52.20). On the contrary, a number of tolerant species like invasive may become abundant (Mostafavi et al. 2014 and 2015). Such a

Fig. 52.20 Eutrophication due to an increase of nutrients

situation leads to Eutrophication (Fig. 52.21), which refers to a considerable growth in algae (microscopic organisms similar to plants) as a consequence of higher accessibility to one or more growth elements essential for photosynthesis, such as sunlight, carbon dioxide, and nutrients (nitrogen and phosphorus).

The uncontrolled growth of algae results in the formation of a large biomass, which certainly will be deteriorated. In deep water, a great mass of organic substance compiles, as a result of culminating the algae's life cycle. A large amount oxygen is consumed by microorganisms to destroy all the dead algae. An anoxic (oxygen-free) environment is consequently established on the lower part of the lake, paving the way for the development of organisms which can survive even without oxygen (anaerobic), a phenomenon that is blamed for the degradation of the biomass. Due to lack of oxygen, the microorganisms disintegrate the organic substance which leads to releasing toxic compounds such as ammonia and hydrogen sulphide (H2S). The death of animal and plant species is another outcome of lack of oxygen as a result of a decrease in biodiversity. The above-mentioned events occur as a result of higher algae degradation compared with oxygen reconstruction, which is already low in summer (Smith et al. 1996).

52.6.1.5 Morphological Pressures

Channelisation causes stream habitat loss and degradation (Lau et al. 2006; Cowell and Stoudt 2002). Based on our observations, the reasons for channelisation in Iran are in general connected with agriculture activities, constructions within the rivers for establishment of bridges, urban development, flood protection, and river bank erosion control (Mostafavi et al. 2015, 2019) (Fig. 52.22).

Channelisation actually refers to the river restriction to the chief channel and its detachment from nearby riparian zone. Habitat heterogeneity that is required for a variety of organisms is declined due to this action (Kennedy and Turner. 2011) (Figs. 52.23 and 52.24). When habitats degrade, resistant species also raise and altered communities become highly vulnerable to other threats, such as population fragmentation and invasion species. Moreover, one of the prevalent issues is that

Fig. 52.21 River channelisation

almost all road bridges constructed on the rivers often have a weir underneath them that hinder the migration of fish and specially during the reproduction seasons; therefore, migrating fish become delay and leads to degeneration of eggs, so that a fish pass could reduce this occurrence markedly (Coad 1980; Paukert et al. 2011; Mostafavi et al. 2015, 2019).

Moreover, other following services which are generally supported by a natural channel can be lost by channelisation according to some studies (e.g. Kingsford 2000; Poff and Hart 2002; Tockner and Stanford 2002; Padmalal et al. 2008) are as follows: control of soils wet and aquifers by flood waters naturally, regulating of water temperature, retention of flow, sediment transport and nutrient, ecological health of streams.

Mining and extraction of sand and gravel are other major reasons for morphological pressure type. This activity has been considerably distributed to the most

Fig. 52.22 Gravel mining and sand extraction

Fig. 52.22 (continued)

parts of Iran and similar to dams are dramatically increasing along the river network. In addition to this, in some basins, they may reach to more or less 20 points along the only one river, which intensify the impacts (Mostafavi et al. 2004). These activities actually result in serious changing the stream's features as well as water quality (Mostafavi et al. 2015) (Fig. 52.22).

52.6.1.6 Biological Pressures

Considering our recent studies, and previous investigations, overexploitation and abnormal fishing methods such as cast net, electricity, toxics can threaten ecosystems seriously (Teimori 2006; Mostafavi et al. 2019) (Fig. 52.25). Moreover, using of unusual fishing methods such as toxins not only destroy fish species but also affect all ecosystem components (Gholami et al. 2009).

Alien organisms which are known as invasive species spread to parts other than their natural ranges and establish self-sustaining populations. They also cross their limits, with detrimental effects on the environment, economy, and human health (Kolar and Lodge 2001; Lymbery et al. 2014). Parallel with trends detected in other

Fig. 52.23 Habitat heterogeneity and fish life cycle at different stages with preferred habitats (Photo from Rafaela Schinegger lecture at BOKU University (Austria, Vienna))

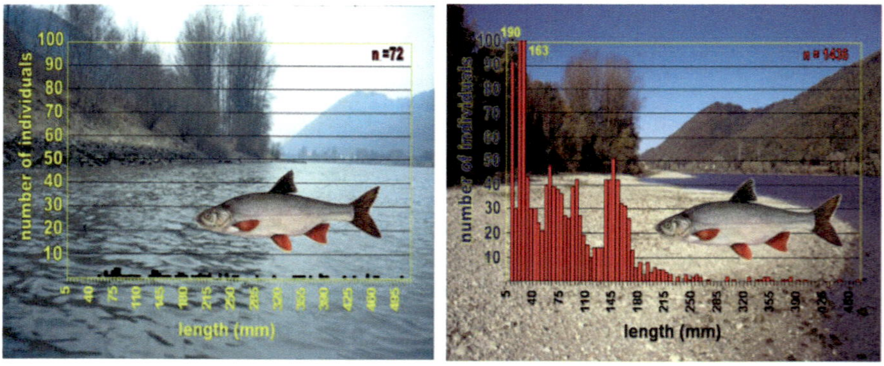

Fig. 52.24 The effect of channelisation on the abundance of fish species (Photo from Stefan Schmutz lecture at BOKU University (Austria, Vienna))

countries, the number of aquatic invasive species in Iran has considerably increased over recent decades. These include harmful algal blooms by *Cochlodinium polykrikoides* (Gymnodiniaceae) in the Persian Gulf, southern Iran (Fatemi et al. 2012), *Mnemiopsis leidyi* blooms in the Caspian Sea (Ivanov et al. 2000), *Azolla filiculoides* blooms in Anzali wetland, and invasion of the Oriental River Prawn, *Macrobrachium nipponense* in Anzali wetland (Grave and Ghane 2006). Fishes are one of the aquatic groups which have been widely introduced and translocated. The movement of fishes between countries was reviewed by Welcomme (1988).

Fig. 52.25 Over fishing

Non-native species were initially introduced into ecosystems in Iran through human-related activities, either deliberately or unintentionally. In this regard, around 30 fish species were recorded from different Iranian freshwater ecosystems ever since (Esmaeili et al. 2011; Khaefi et al. 2014; Esmaeili et al. 2015). They have multiple synergistic impacts particularly on the freshwater ecosystem through various ways. The species diversity and faunistic structure within water bodies can be changed.

52.6.1.7 Global Climate Change

Climate change is one of multiple interacting stresses on ecosystems. Nearly all regions of the world will encounter the effects of climate change on aquatic ecosystems (Intergovernmental Panel on Climate Change (IPCC)). The negative effects of global warming on freshwaters are critical because many species' ability for distribution will be limited by environment alteration (Woodward et al. 2010). In addition, temperature and accessibility of water are affected by climate change while diverse anthropogenic stressors are already affecting majority of the aquatic ecosystems. The effects of climate change on biodiversity of freshwater ecosystems have already been observed and demonstrated in the past. Also and according to predictive climate models and bioclimatic modelling, it is expected to be increased in the future (e.g. Thuiller 2003, 2004; Araújo and New 2007; Abbaspour et al. 2009; Mostafavi et al. 2017;). Usually, species can respond to climate change in several ways, i.e. move to track climatic conditions, stay in place, and adapt to the new climate, or they become extinct (Berteaux et al. 2004; Lovejoy and Hannah 2006). However, it is clear that pattern of their response mostly depends on the regions, where they are distributed. Some particular parts are more dangerous for the biota due to their available temperature regimes and the insufficient escape paths to more suitable habitats. Iran is a country with large climatic variability. Over 82% of Iran's area is situated in arid and semi-arid world regions. The mean value for annual

Fig. 52.26 some exotic fish species from the Iranian drainage basins

Fig. 52.27 Aquaculture along the rivers and introducing exotic species to the ecosystem

rainfall in this country is about 250 mm, which is less than 1/3 of the world's average rainfall, which is 860 mm. Furthermore, with respect to time and location, this sparse precipitation is not favourable (Abbaspour et al. 2009; Coad 2018) (Fig. 52.26 and 52.27).

Severe alteration in temperature which sometimes varies from −20 °C to +50 °C and drastic drought are other important elements and features of Iran's climate. In the few past years, the country has experienced pervasive losses as a result of desiccation (Fig. 52.28). It is estimated that if CO_2 amount is doubled by the year 2100, there will be a rise up to 1.5 °C–4.5 °C in Iran's temperature (Amiri and Eslamian 2010).

Alteration in precipitation and temperature patterns, water resources, agriculture and food production, sea level rise and coastal zone, forestry, drought intensity, and frequency and human health are considered as direct unfavourable effects of climate change. However, the developed countries' responses measures are counted as

Fig. 52.28 Drought and climate change effects in some areas of Iran

indirect disadvantageous impact (Abbaspour et al. 2009; Amiri and Eslamian 2010; Mostafavi et al. 2017).

Similar to the other regions of the world, different human pressure types in Iran have influenced freshwater ecosystems. Figure 52.29 simply shows the DPSIR framework for water management, advocated by the European Environment Agency to assess water problems.

52.7 Case Study 1

Ecological assessment of Salmonid and Cyprinid running waters of Caspian Sea basin by quantifying of human pressures and modelling techniques.

52.7.1 Study Area and Site Selection

The area under study was situated in two ecoregions (Kura-South Caspian Drainages and Caspian Highlands) of southern Caspian Sea basin, north of Iran (Fig. 52.30, case study 1). Salmonid and cyprinid streams where were, respectively, dominant by brown trout (*Salmo trutta*) fish and cyprinid species have been selected for this study (see further information in Mostafavi et al. 2015, 2019). In total, we selected 130 medium-sized streams (up to 20 m) involving 44 reference and 44 impacted sites on salmonid zone (SZ) as well as 50 reference and 52 impacted sites on cyprinid zone (CZ). All reference and impacted sites were selected according to Table 52.1, case study 1.

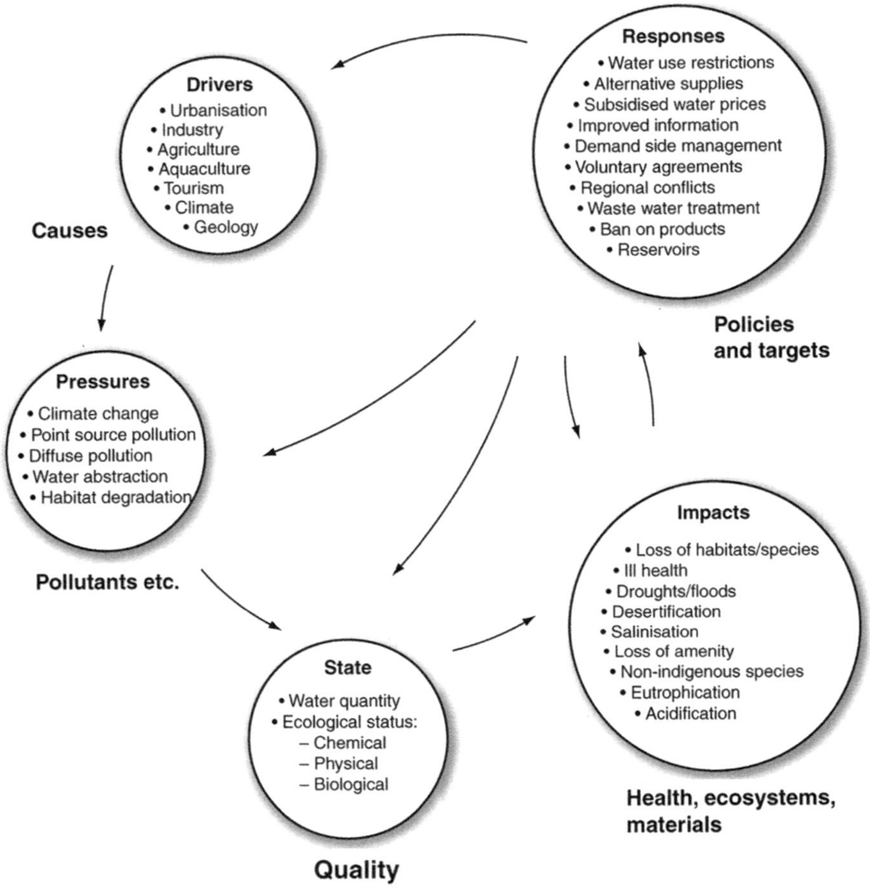

Fig. 52.29 The generic DPSIR framework for water, promoted by the European Environment Agency as an analytical framework to assess water issues. The framework allows a comprehensive assessment of issues through an examination of relevant Driving forces and Pressures on the environment, the consequent State and its Impacts, and the resulting Response, and of the linkages between each element in the framework. Modified from Kristensen (2004)

52.7.2 Fish Sampling As Well as Environmental and Human Pressures Data Measurements

Fish sampling was undertaken in autumns 2011 (for salmonid zone) and 2012 (for cyprinid zone) due to low flow conditions according to CEN (2003). The length of sampling size was defined 10–20 times of stream width based on Langdon (2001) to cover all habitat types (Figs. 52.31, 52.32 (just to show other fish survey and sampling), 4, case study 1). Fishes were caught by electrofishing equipment (see more in EFI+ Consortium 2009; Mostafavi et al. 2015, 2019) (Fig. 52.31, case study 1). Then, the stunned fish were first recognised according to some key identification

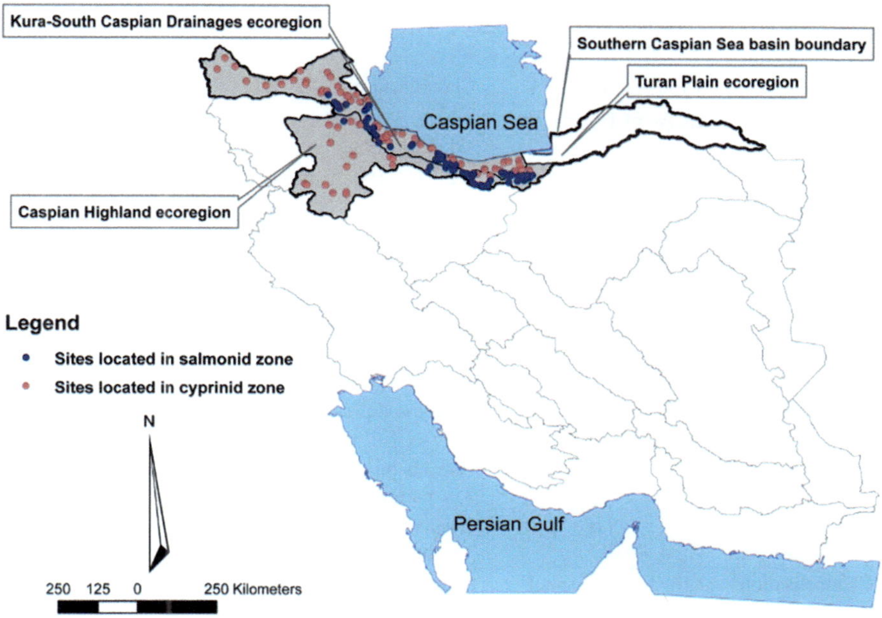

Fig. 52.30 Map of fish sampling sites in salmonid and cyprinid zones in the studied area of the Southern Caspian Sea basin of Iran

books and articles (e.g. Abdoli 2000; Abdoli and Naderi 2009; Esmaeili et al. 2010) and finally were released back into the water (Fig. 52.32).

We subsequently measured twelve environmental parameters at each sampling site including elevation, average bankfull width, average wetted width, flow velocity, discharge, water temperature, dissolved oxygen (DO), pH, conductivity (EC), turbidity, NO^{3-}, NO^{2-}, and PO_4^{3-}. Elevation was measured by Garmin GPS map 62 s. A digital water velocity meter (Global Water Flow Probe, FP111) was utilised to gauge flow velocity (Mostafavi et al. 2015, 2019). Furthermore, we measured water temperature, pH, and EC by multi-parameter water analyser portable (HANNA HI 9828); DO by Oxygen meter portable (HACH HQ30D); turbidity by Turbiditimeter portable (HACH 2100Qis); NO^{3-}, NO^{2-}, and PO_4^{3-} by Multi-parameter analyser portable (HACH DR/890) (Mostafavi et al. 2015, 2019) (see Table 52.2, case study 1). Then, for every single sampling site, diverse anthropogenic pressures were gathered (Table 52.1, case study 1) based on Degerman et al. (2007); EFI+ consortium (2009); and Schinegger et al. (2012). Actually, the data set involved 30 pressure variables relevant to seven pressure types: (1) land use, (2) connectivity, (3) hydrology, (4) morphology, (5) water quality, (6) biological pressures, and (7) other pressures (Table 52.2, case study 1). Human pressures were measured at four spatial levels: drainage, primary catchment, segment, and site (see detailed information in Mostafavi et al. 2015, 2019). Afterwards, all pressure variables were considered in five classes, i.e. (1) high, (2) good, (3) moderate, (4) poor, and (5) bad status. Then, in order to identify redundant variables, a Spearman rank

Table 52.1 Human pressure classification into seven human pressure types

Human pressure variable	Type	Code	Classification
Agriculture	LUP	LU_agri_sit	Range: 50 m from stream; 1 = none, 3 = along one side, 5 = along both sides
Urbanisation	LUP	LU_urb_sit	Range: 100 m from stream; 1 = <5%, 3 = ≥5% & < 10%, 5 = ≥10%
[a]Agriculture	LUP	LU_agri_pc	Extent and pressure of agriculture and silviculture; 1 = <10%, 3 = ≥10% & < 40%, 5 = ≥40%
[a]Urbanisation	LUP	LU_urb_pc	Extent and pressure of urban areas; 1 = <1%, 3 = ≥1% & < 15%, 5 = ≥15%
[a]Agriculture	LUP	LU_agri_dr	Extent and pressure of agriculture and silviculture; 1 = <10%, 3 = ≥10% & < 40%, 5 = ≥40%
[a]Urbanisation	LUP	LU_urb_dr	Extent and pressure of urban areas; 1 = <1%, 3 = ≥1% & < 15%, 5 = ≥15%
Migration barrier upstream	CP	C_B_s_up	Barriers on the segment level upstream; 1 = no, 3 = partial, 3 = yes
Migration barrier downstream	CP	C_B_s_do	Barriers on the segment level downstream; 1 = no, 4 = partial, 4 = yes
Channelisation	MP	M_channel	Alteration of natural morphological channel plan form; 1 = no, 3 = intermediate, 5 = straightened
Channelisation	MP	M_crosssec	Alteration of cross-section; 1 = no, 3 = intermediate, 5 = technical cross-section /U-profile
Channelisation	MP	M_instrhab	Alteration of instream habitat condition; 1 = no, 3 = intermediate, 5 = high
[a]Channelisation	MP	M_embankm	Artificial embankment; 1 = no (natural status), 2 = slight (local presence of artificial material for embankment), 3 = intermediate (continuous embankment but permeable), 5 = high (continuous, no permeability)
Channelisation	MP	M_ripveg	Alteration of riparian vegetation close to shoreline; 1 = no, 2 = slight, 3 = intermediate, 5 = high (no vegetation)
Flood protection	MP	M_floodpr	Presence of dykes for flood protection; 1 = no, 3 = yes
[a]Flood protection	MP	M_remfloodpl	If the river has a former floodplainProportion of connected floodplain still remaining. Floodplain = area connected during the flood; 1 = >50%, 2 = 10–50%, 3 = <10%, 5 = some water bodies remaining or no
Sedimentation	MP	M_sediment	Input of fine sediment (mainly mineral input; bank erosion, erosion from agricultural land); 1 = no, 3 = yes
[a]Flow velocity increase	HP	H_veloincr	Pressure on flow conditions (mean velocity) Due to channelisation, flood protection, etc.; 1 = no, 3 = yes
Impoundment	HP	H_imp	Natural flow velocity reduction on site because of impoundment; 1 = no (no impoundment), 3 = intermediate, 5 = strong

(continued)

Table 52.1 (continued)

Human pressure variable	Type	Code	Classification
Hydropeaking	HP	H_hydrop	Site affected by hydropeaking; 1 = no (no hydropeaking), 3 = partial, 3 = yes
Water abstraction	HP	H_waterabstr	Site affected by water flow alteration/minimum flow; 1 = no (no water Abstraction), 3 = intermediate (less than half of the mean annual flow), 5 = strong (more than half of mean annual flow)
[a]Reservoir flushing	HP	H_reflush	Fish fauna affected by flushing of reservoir upstream of site; 1 = no, 3 = yes
Temperature pressure	HP	H_tempimp	Water temperature pressure; 1 = no, 3 = yes
Eutrophication	WQP	W_eutroph	Artificial eutrophication; 1 = no, 3 = low, 4 = intermediate (occurrence of green algae), 5 = extreme (oxygen depletion)
Acidification	WQP	W_aci	Acidification; 1 = no, 3 = yes
Organic siltation	WQP	W_osilt	Siltation; 1 = no, 3 = yes
[a]Organic pollution	WQP	W_opoll	Is organic pollution observed; 1 = no, 3 = intermediate, 5 = strong
Toxicity	WQP	W_toxic	Toxic priority substances (organic and nutrient appearance); 1 = no or very minor, 3 = weak (important risk, link to particular substance), 5 = high concentration (a clearly known input)
Pressure of exploitation	BP	B_explo	Fishing, at site affecting fauna, information based on Local fishermen; 1 = no, 3 = intermediate, 5 = strong
Introduction of fish	BP	B_intro	New fish species to river basin; 1 = no introduction, 2 = introduction, but no reproduction and low density, 3 = not reproduction, high density, 4 = reproducing, low density, 5 = reproducing, high density
Other pressures	OP	O_imp	e.g. explosion of oil pipe 1 = no, 3 = weak, 5 = strong (expert judgment)

LUP Land-Use Pressure, *CP* Connectivity Pressure, *MP* Morphological Pressure, *HP* Hydrological Pressure, *WQP* Water Quality Pressure, *BP* Biological Pressure, *OP* Other Pressures
[a]Excluded variables after correlation test

correlation test was applied (EFI+ consortium 2009). Finally, variables with high co-linearity ($\rho > |0.70|$) were excluded (Schinegger et al. 2012; Mostafavi et al. 2015, 2019). Human pressure photographs have been already shown in previous section; therefore, they have not brought here again.

Fig. 52.31 Fish sampling sites in salmonid and cyprinid zones in the studied area of the Southern Caspian Sea basin of Iran

52.7.3 Regional Pressure Index (RPI) Calculation

In this part, first, M_morph_instr was figured for each salmonid and cyprinid zone according to Schinegger et al. (2012):

$$M\_\text{morph\_instr} = \frac{M\_\text{channel} + M\_\text{crosssec} + M\_\text{instrhab}}{3} \quad (52.1)$$

Second, for each dominant pressure groups in each salmonid and cyprinid zone, an index was computed (Schinegger et al. 2012). These indices (i.e. land-use pressures (LUP), connectivity pressures (CP), morphological pressures (MP), hydrological pressures (HP), water quality pressures (WQP), biological pressures (BP), and other pressures (OP)) were computed by averaging the single pressure parameter

Fig. 52.32 Other methods of fish survey and sampling by authors in Iran

values of classes 3, 4, and 5 for avoiding values <3 recompensing for values ≥3 (Schinegger et al. 2012; Mostafavi et al. 2015, 2019).

$$\mathrm{LUP} = \frac{\mathrm{LU\_agri\_sit} + \mathrm{LU\_urb\_sit}}{2} \tag{52.2}$$

$$\mathrm{CP} = \frac{C\_B\_s\_up + C\_B\_s\_do}{2} \tag{52.3}$$

$$\mathrm{MP} = \frac{M\_\mathrm{Morph\_instr} + M\_\mathrm{ripveg} + M\_\mathrm{floodpr} + M\_\mathrm{sediment}}{4} \tag{52.4}$$

$$\mathrm{HP} = \frac{H\_\mathrm{imp} + H\_\mathrm{hydrop} + H\_\mathrm{waterabstr} + H\_\mathrm{tempimp}}{4} \tag{52.5}$$

$$\mathrm{WQP} = \frac{W\_\mathrm{eutroph} + W\_\mathrm{aci} + W\_\mathrm{osilt} + W\_\mathrm{toxic}}{4} \tag{52.6}$$

Table 52.2 Environmental characteristics measured at the sampling sites

	Total sites of salmonid zone					Total sites of cyprinid zone				
Parameters	N	Mean	S.D.	Min.	Max.	N	Mean	S. D.	Min.	Max.
Elevation (m)	88	1694	495	118	2638	102	745	552	18	1675
Average bankfull width (m)	88	7.1	4.1	2.1	22.2	102	15.1	11.6	4.2	52.0
Average wetted width (m)	88	4.7	2.0	1.0	9.7	102	7.4	4.6	1.0	20.0
Average flow velocity (m/s)	88	0.7	0.2	0.1	1.2	102	0.7	0.4	0.0	1.8
Water temperature (°C)	88	6.1	2.5	0.0	14.0	102	17.0	2.0	13.0	24.0
DO (mg/l)	88	9.7	2.2	4.1	12.5	102	8.0	1.0	3.0	14.0
pH	88	8.0	1.2	5.5	8.8	102	8.0	0.0	7.0	9.0
EC (μSiemens/cm)	88	551	395	122	1149	102	537	357	151	1780
Turbidity (NTU)	88	26	41	2	163	102	19	192	2	1185
NO^{3-} (mg/l)	88	6.224	3.126	0.007	17.333	102	1.44	1.30	0.01	9.60
NO^{2-} (mg/l)	88	0.027	0.022	0.002	0.094	102	0.06	0.03	0.00	0.17
PO_4^{3-}(mg/l)	88	1.292	0.244	0.001	3.227	102	1.68	1.10	0.00	5.70

DO dissolved oxygen, *N* number of sites, *S.D.* standard deviation, *max.* Maximum and *min.* Minimum

$$BP = \frac{B\_explo + B\_intro}{2} \tag{52.7}$$

$$OP = O - Pr \tag{52.8}$$

The number of pressure types (LUP, CP, MP, HP, WQP, BP, and OP) affected was considered as "affected group." This value ranged from one to four for salmonid zone and one to five for cyprinid zone. Finally, for calculation of regional pressure index (RPI) for each site in salmonid and cyprinid zone, the following formula was used (Schinegger et al. 2012; Mostafavi et al. 2015, 2019):

$$RPI = \frac{LUP + CP + MP + HP + WQP + BP + OP}{7} \times \text{affected group} \tag{52.9}$$

The RPI contained 0 to 25 which was finally rescaled into five classes as follows:

class one (1)—values between 3 and 5 (single pressure affected by one group);
class two (2)—values between 6 and 8 (double pressures affected by two groups);
class three (3)—values between 9 and 11 (triple pressures affected by three groups);
class four (4)—values higher than 11 (multiple pressures affected by four or five groups).

Overall, among all impacted sites, seven pressure types with 30 variables for salmonid zone (SZ) and six pressure types with 29 for cyprinid zone (CZ) were recognised which are influencing the streams of these areas. Actually, "other pressure" type (OP) which was related to the oil pipe explosion was only observed in SZ. After excluding redundant variables, 21 pressures for SZ and 20 for CZ were remained (see Table 52.1, case study 1). The most common human pressure among all human pressure types was land use (LUP) in both zones, taking place at 37.50% sites in SZ followed by water quality (32.29%) and morphological pressures (30.21%) respectively and 44.79% in CZ followed by hydrological and morphological pressures (36.46 and 32.29% correspondingly) (Fig. 52.33, case study 1). Overall, the most frequent pressure in both zones was land use (82.29%) followed by morphological and water quality pressures, respectively (62.50 and 61.46%) (Fig. 52.34, case study 1). Most sites in SZ were affected by double and triple pressures (14.58% each one), while in CZ, most sites were affected by multiple pressures (27.08%). However, in both zones least sites were only under the impact of a single pressure (5.21% for SZ and 2.08% for CZ) (Fig. 52.34). Overall, most sites in both zones were affected by multiple pressures and least by single (38.54 and 7.29%, respectively) (Fig. 52.34, case study 1). Moreover, the regional pressure index (RPI) of each sampling sites represented in Fig. 52.35, case study 1.

52.7.4 Fish Metrics Description

Five types of metrics (i.e. biodiversity, reproduction, habitat, water quality sensitivity, and trophic level) were assigned for each species (EFI+ Consortium 2009). These attributes were inferred from the available studies and fish experts in Iran (see Mostafavi et al. 2015, 2019) (Table 52.3, case study 1). Most important ecological characters of fish communities are actually reflected by these selected metrics as stated in Noble et al. (2007) and Schinegger et al. (2013) and Mostafavi et al. (2015).

Overall, 23 taxa out of six families (including 19 native and four alien) were found at both zones of which in the reference sites of SZ, brown trout (*S. trutta*) happened as sole species; however, at the reference sites of CZ, 19 taxa were observed and cyprinid species were dominant (Table 52.3, case study 1). In the impacted sites of SZ, some other species were also observed and called non-type specific species for this zone. In the impacted sites of CZ, not only 19 taxa of reference sites could be found but also three additional alien species were observed (Table 52.3, case study 1).

52.7.5 Determination and Assessment of Predictor Variables

A restricted number of environmental predictors (i.e. drainage size, annual mean air temperature (Tmean), January mean air temperature (Tmin), July mean air temperature (Tmax), the thermal range between January and July (Trange), annual mean precipitation, and average slope) for predicting of reference metric values at the

Fig. 52.33 Different meso- and micro-habitats (a: meso, b: micro)

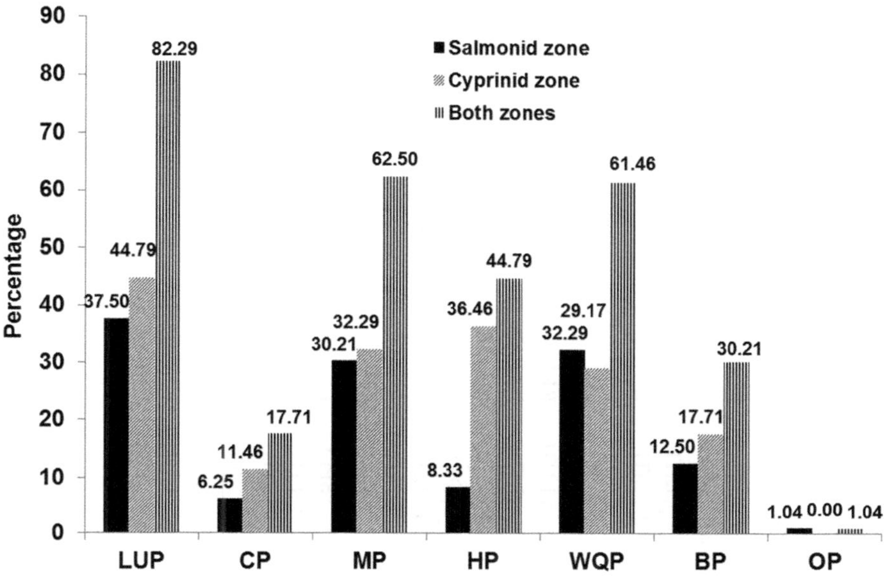

Fig. 52.34 Percentage of sites affected by no/slight pressure, land-use pressures (LUP), connectivity pressures (CP), morphological pressures (MP), hydrological pressures (HP), water quality pressures (WQP), biological pressures (BP), and other pressures (OP) at salmonid and cyprinid zones

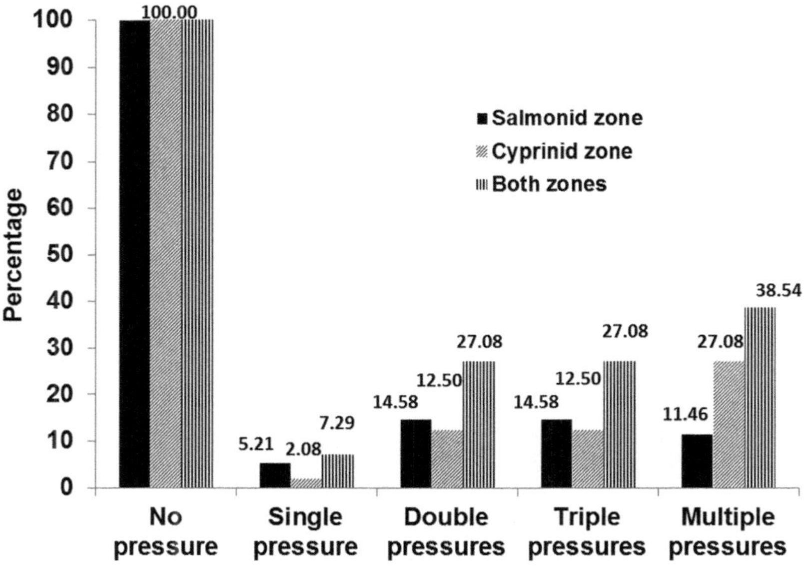

Fig. 52.35 Number of sites affected with no, single, double, triple, and multiple pressures in salmonid and cyprinid zones

Table 52.3 List of 23 fish species sampled in salmonid and cyprinid streams

Species	Family	WQgen	WQO2	HTOL	Hab	Atroph	Repro	HabSp	Native/Alien	Reference	Impact
Acanthalburnus microlepis	Cyprinidae	IM	O2IM	HIM	EURY	INSV	LITH	EUPAR	Na	CZ	CZ
Alburnus chalcoides	Cyprinidae	TOL	O2IM	HINTOL	EURY	OMNI	LITH	LIPAR	Na	CZ	CZ
Alburnoides eichwaldii	Cyprinidae	INTOL	O2INTOL	HINTOL	RH	INSV	LITH	RHPAR	Na	CZ	SZ/CZ
Alburnus filippii	Cyprinidae	INTOL	O2INTOL	HINTOL	RH	INSV	LITH	EUPAR	Na	CZ	CZ
Alburnus hohenackeri	Cyprinidae	TOL	O2IM	HTOL	EURY	PLAN	PHLI	EUPAR	Na	CZ	CZ
Barbus lacerta	Cyprinidae	INTOL	O2INTOL	HINTOL	RH	INSV	LITH	RHPAR	Na	CZ	SZ/CZ
Capoeta capoeta	Cyprinidae	IM	O2IM	HIM	RH	HERB	LITH	RHPAR	Na	CZ	SZ/CZ
Carassius carassius	Cyprinidae	TOL	O2TOL	HTOL	LIMNO	OMNI	PHYT	LIPAR	Na	–	CZ
Hemiculter leucisculus	Cyprinidae	TOL	O2TOL	HTOL	EURY	OMNI	PELA	EUPAR	Al	–	CZ
Luciobarbus capito	Cyprinidae	INTOL	O2INTOL	HINTOL	RH	INSV	LITH	RHPAR	Na	CZ	CZ
Luciobarbus mursa	Cyprinidae	INTOL	O2INTOL	HINTOL	RH	INSV	LITH	RHPAR	Na	CZ	CZ
Pseudorasbora parva	Cyprinidae	TOL	O2TOL	HTOL	EURY	OMNI	PHLI	EUPAR	Al	–	CZ
Rhodeus amarus	Cyprinidae	INTOL	O2IM	HINTOL	LIMNO	OMNI	OSTRA	LIPAR	Na	CZ	CZ
Squalius cephalus	Cyprinidae	TOL	O2IM	HTOL	RH	OMNI	LITH	RHPAR	Na	CZ	SZ/CZ
Cobitis sp.	Cobitidae	IM	O2IM	HIM	RH	INSV	PHYT	EUPAR	Na	CZ	CZ
Sabanejewia aurata	Cobitidae	IM	O2IM	HIM	RH	INSV	PHYT	EUPAR	Na	CZ	CZ
Neogobius pallasi	Gobiidae	TOL	O2IM	HTOL	EURY	INSV	SPEL	EUPAR	Na	CZ	SZ/CZ
Neogobius melanostomus	Gobiidae	TOL	O2IM	HTOL	EURY	INSV	LITH	EUPAR	Na	CZ	CZ
Paracobitis malapterura	Nemacheilidae	INTOL	O2IM	HIM	RH	INSV	LITH	EUPAR	Na	CZ	SZ/CZ
Oxynoemacheilus sp.	Nemacheilidae	IM	O2IM	HIM	RH	INSV	LITH	EUPAR	Na	CZ	CZ
Gambusia holbrooki	Poeciliidae	TOL	O2TOL	HTOL	LIMNO	INSV	VIVI	LIPAR	Al	–	CZ
Salmo trutta	Salmonidae	INTOL	O2INTOL	HINTOL	RH	INSV	LITH	RHPAR	Na	SZ/CZ	SZ
Oncorhynchus mykiss	Salmonidae	IM	O2INTOL	HIM	RH	INSV	LITH	RHPAR	Al	–	CZ

Codes for assignments are as follows: *WQgen* water quality tolerance general, *IM* intermediate, *TOL* tolerant, *INTOL* intolerant, *WQO2* water quality tolerance O2, *O2IM* intermediate, *O2INTOL* intolerant, *O2tolerant* tolerant, *HTOL* Habitat degradation tolerance, *HIM* intermediate, *HINTOL* intolerant, *HTOL* tolerant, *HAB* Habitat, *EURY* eurytopic, *RH* rheophilic, *LIMNO* limnophilic, *Atroph* Adult trophic guild, *INSV* insectivorous, *OMNI* omnivorous, *PLAN* planktivorous, *HERB* herbivorous, *Mig* Migration guild, *RESID* resident, *POTAD* potamodrom, *Repro* Reproductive guild, *LITH* lithophilic, *PHLI* phyto-lithophilic, *PHYT* phytophilic, *PELA* pelagophilic, *OSTRA* ostracophilic, *VIVI* viviparous, *SPEL* speleophilic, *HabSp* Habitat spawning preferences, *EUPAR* euryoparous, *LIPAR* limnoporous, *RHPAR* rheoparous, *Na* native, *Al* alien, *SZ* salmonid zone, *CZ* cyprinid zone reach and regional scale were selected (Oberdorff et al. 2001, 2002; Pont et al. 2009; Filipe et al. 2013). The climatic variables were extracted from the WorldClim website (https://www.worldclim.org/, provided by Hijmans et al. 2005, 2007) and Slope as well as drainage size from CCM2 (Catchment Characterisation and Modelling database provided by Vogt et al. 2003, 2007; de Jager and Vogt 2010). All mentioned variables were tested for co-linearity by Spearman rank correlation (ρ). Redundant variable was excluded if two variables were highly correlated ($\rho > |0.70|$) (Mostafavi et al. 2015, 2019).

52.7.6 Fish Metrics Prediction

A Generalised Linear Model (GLM) for the modelling was applied (see Mostafavi et al. 2015, 2019). A stepwise procedure according to Akaike's information criterion (AIC; Logez and Pont 2011) was used to choose predictor variables. The model performance was evaluated according to R square, leverage values, and standardised residuals. The normality of residuals (using Q-Q plot and histogram), the heteroscedasticity of residuals (graph of standardised residuals versus standardised expected values), the influence of leverage values (graph of residual values versus leverage values), and the relationship between observed and expected values (a linear relation of the form $y = x$ was expected) were visually checked (see Mostafavi et al. 2015, 2019). Furthermore, this procedure was confirmed by internal validation based on bootstrapping (Efron and Tibshirani 1993). After mentioned procedures, some metrics were selected for further steps.

After fitting of models, residuals were computed according to following equation (Pont et al. 2009):

$$R_i = \log(O_i + 1) - \log(E_i + 1) \tag{52.10}$$

where,

R_i is the residual; O_i is the observed; and E_i is the expected value.

Then, by standardising the residuals of the model, the score of each metric (M_i) was acquired according to following way (Mostafavi et al. 2015, 2019):

$$M_i = \frac{(R_i - M)}{S_q} \tag{52.11}$$

where

- R_i: Is the residual value (difference between observed and expected metric) from sites i to n.
- M: Is the median value of the residuals; i to n.
- S_q: Is the standard deviation of the residuals in the whole undisturbed dataset.

Considering the fact that standardised residuals vary from $-\infty$ to $+\infty$, two transformations were adopted in order to ensure that all metrics fluctuate within a finite interval from 0 to 1.

This maximum (Max) and this minimum (Min) were two substitutions for all values over a maximum (percentile 95) and below a minimum (percentile 5). Afterwards, the transformation below was applied to all metric scores (Mostafavi et al. 2015):

$$\text{Rescaled } M_i = \frac{(M_i - \text{Min})}{\text{Max} - \text{Min.}} \tag{52.12}$$

Finally, the redundant metrics were excluded by using Spearman rank correlation test ($\rho > |0.80|$).

52.7.7 Testing Sensitivity of Fish Metrics to Human Pressures

The sensitivity of metrics against human pressures between impacted and reference sites was examined by Mann-Whitney U-test as well as regression test (i.e. testing metrics against regional pressure index (RPI)). As can be seen in Table 52.4, case study 1, some metrics, i.e. Nsp-all, Nsp-native, and Atroph-HERB, revealed conflicting reaction in two zones ($p < 0.0001$, R square > 60), some similar, i.e., Wqgen-INTOL, WQO2-O2INTOL, HTOL-HINTOL, Hab-RH, Hab-EURY, HabSp-RHPAR, HabSp-EUPAR, Repro-LITH, Atroph-INSV ($p < 0.0001$, R square > 60), and some had unclear reaction ($P > 0.0001$ and R square < 10) but observed only in CZ, i.e. Wqgen-IMTOL, WQO2-O2IMTOL, HTOL-HIMTOL, Repro-PHLI, Repro-PHYT, Repro-PELA, Repro-OSTRA, Repro-SPEL, Atroph-PLAN. Finally, metrics with high co-linearity were excluded so that more variants would remain.

52.7.8 Index Calculation, Scoring, and Validation

An arithmetic mean of the standardised and transformed metric scores was used to develop multi-metric fish index. Then, this index was classified into five categories according to Mostafavi et al. (2015, 2019). Figure 52.36, case study 1 shows the general process of modelling.

Table 52.4 Name and definition of the candidate metrics as well as their reaction to the human pressures

Trait	Definition	Type	Reaction to pressure in SZ	Reaction to pressure in CZ
Nsp-all	Total number of fish species, including native and alien species.	Biodiv	Incr	Decr
Nsp-native	Number of native species.	Biodiv	Incr	Decr
Nsp-alien	Number of alien species.	Biodiv	Incr	Incr
Wqgen-INTOL Wqgen-IMTOL Wqgen-TOL	In general, intolerant, intolerant/tolerant, and tolerant to usual (national) water quality parameters, respectively.	Wq	Decr – –	Decr Unclear Incr
WQO2- O2INTOL WQO2-O2IMTOL WQO2- O2TOL	Intolerant, intolerant/tolerant, and tolerant to low oxygen concentration (O_2), respectively, more than 6 mg/l in water.	Wq	Decr – –	Decr Unclear Incr
HTOL-HINTOL HTOL-HIMTOL HTOL-HTOL	Intolerant, intolerant/tolerant, and tolerant to habitat degradation, respectively.	Hab	Decr – –	Decr Unclear Incr
Hab-RH	Degree of rheophily (habitat). Fish prefer to live in a habitat with high flow conditions and clear water.	Hab	Decr	Decr
Hab-EURY	Degree of rheophily (habitat). Fish that exhibit a wide tolerance of flow conditions, although generally not considered to be rheophilic.	Hab	Incr	Incr
Hab-LIMNO	Degree of rheophily (habitat). Fish prefer to live, feed, and reproduce in a habitat with slow flowing to stagnant conditions.	Hab	–	Incr
HabSp-RHPAR	Preference to spawn in running waters.	Hab	Decr	Decr
HabSp-EUPAR	No clear spawning habitat preferences.	Hab	Incr	Incr
HabSp-LIPAR	Preference to spawn in stagnant waters.	Hab	–	Decr
Repro-LITH	Fish spawn exclusively on gravel, rocks, or pebbles.	Repro	Decr	Decr
Repro-PHLI	Fish deposit eggs in clear water habitats on submerged plants or on other submerged items such as logs, gravel, and rocks. Larvae are photophobic.	Repro	–	Unclear
Repro-PHYT	Fish deposit eggs in clear water habitats on submerged plants.	Repro	–	Unclear
Repro-PELA	Fish spawn into the pelagic zone.	Repro	–	Unclear
Repro-OSTRA	Spawning in shells of bivalve molluscs.	Repro	–	Unclear
Repro-SPEL	Fish spawn in interstitial spaces, crevices, or caves.	Repro	Incr	Unclear
Repro-VIVI	Live bearers.	Repro		Incr
Atroph-OMNI	Adult consists of more than 25% plant material and more than 25% animal material. Generalists.	Troph	–	Incr

(continued)

Table 52.4 (continued)

Trait	Definition	Type	Reaction to pressure in SZ	Reaction to pressure in CZ
Atroph-INSV	Insectivorous species.	Troph	Decr	Decr
Atroph-PLAN	Plantivorous species.	Troph	–	Unclear
Atroph-HERB	Herbivores species.	Troph	Incr	Decr

Biodiv biodiversity, *hab* habitat, *repro* reproduction, *troph* trophic level, *wq* water quality, *decr* metric decreases with increasing human pressure, *incr* metric increases with increasing human pressure, *SZ* salmonid zone, *CZ* cyprinid zone

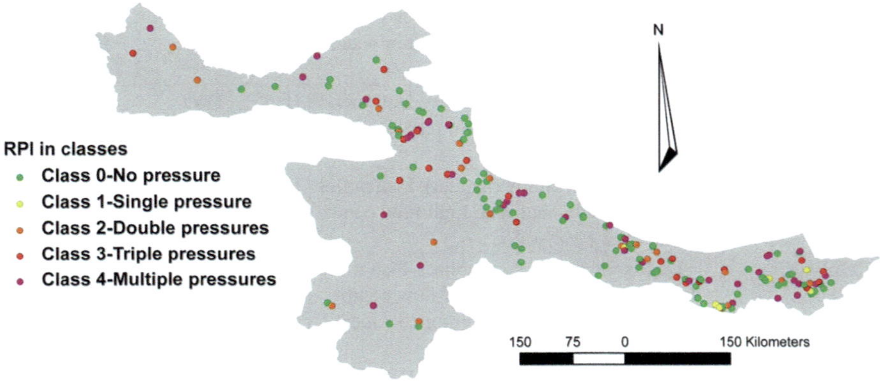

Fig. 52.36 Distribution of the regional pressure index (RPI) for sampling sites

52.7.9 Results for Salmonid Zone

Finally, only two metrics (i.e. density of brown trout and biomass of sub-adult brown trout) remained for the fish index development. This multi-metric fish index was classified into five classes (Fig. 52.37, case study 1). Sites in moderate, poor, or bad status need an urgent action for restoration.

52.7.10 Results for Cyprinid Zone

Finally, seven fish metrics were identified as main metrics for development of multi-metric fish index (MMICS). These metrics were included number of native species, density of intolerant species to oxygen depletion, biomass of intolerant species to water quality degradation, biomass of intolerant species to habitat degradation, density of rheophilic species, biomass of lithophilic species, and percentage biomass of insectivorous species. Further, this index was classified into five categories

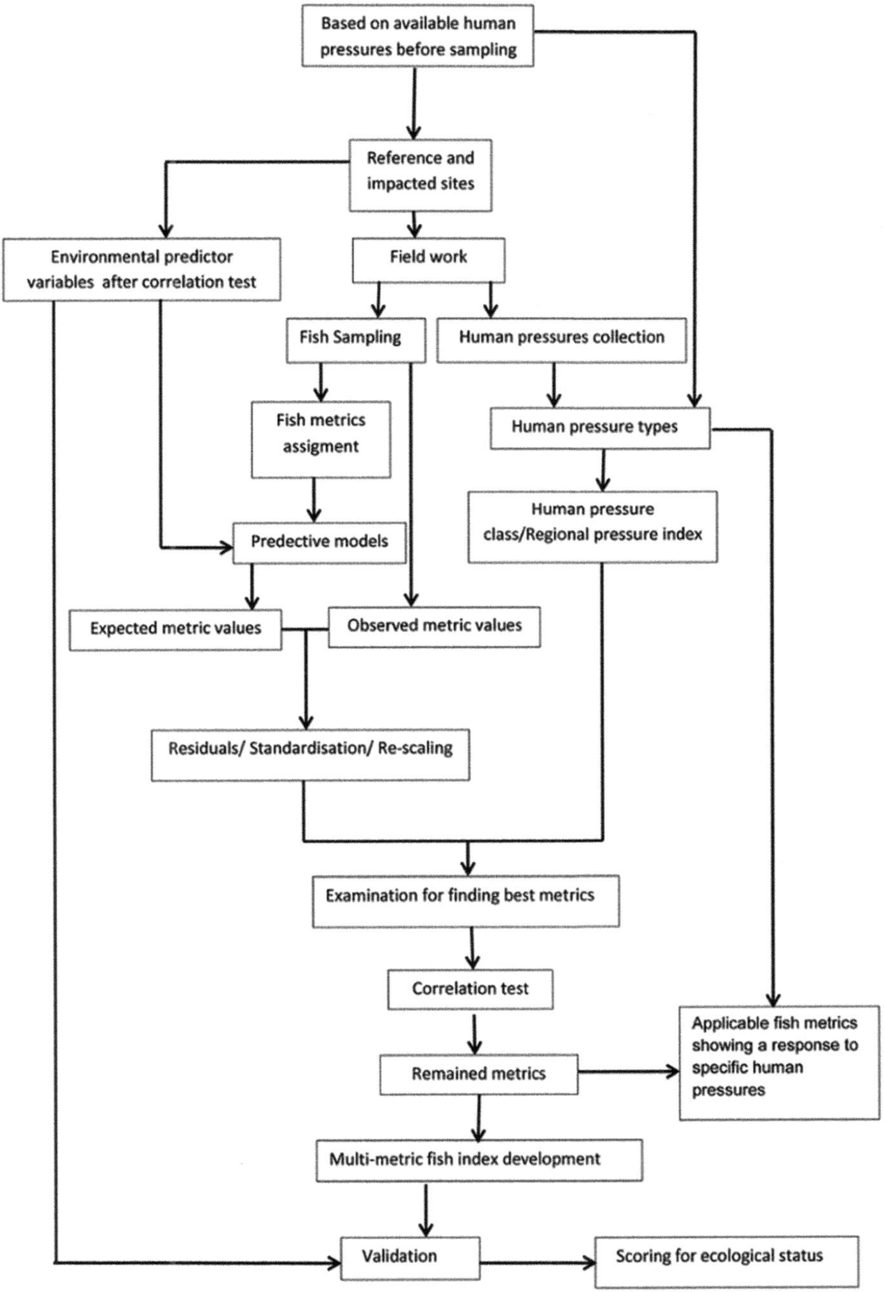

Fig. 52.37 Flow chart describing the procedure of multi-metric fish index development for ecological assessment

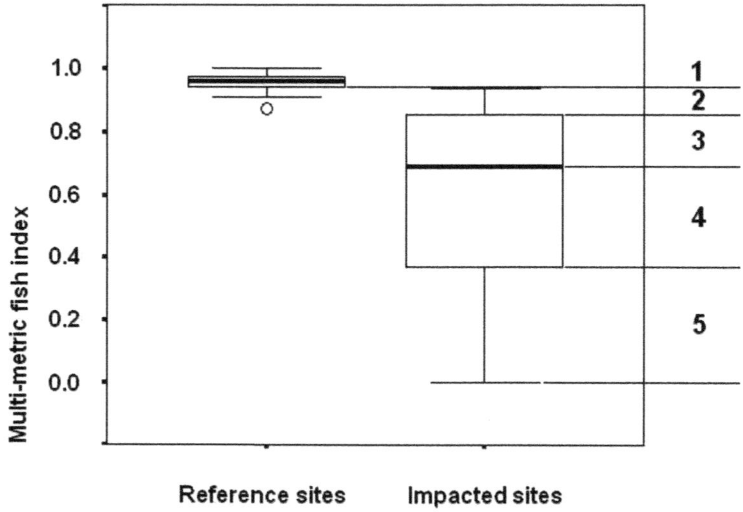

Fig. 52.38 The quartiles of both reference and impacted sites by classification into five categories in the salmonid zone

(Fig. 52.38, case study 1). Sites in moderate, poor, or bad status need an urgent action for restoration.

52.7.11 Conclusion

According to the general concept of WFD (Water Framework Directive) (see Fig. 52.39, case study 1), it is possible to act for the restoration of sites which categorised in the classes 3, 4, and 5 to achieve good ecological status. Therefore, by quantifying of all human pressure types and modelling approach, we are able to restore and manage freshwater ecosystems ecologically. Generally, the advance in such an index provides an opportunity to make an improvement in national bio-monitoring projects in Iran and elsewhere. For uncertainty, it is necessary to be indicated that our fish indices must be used for wadeable streams in the salmonid and cyprinid zone with characters which were indicated above in Table 52.2, case study 1. These fish indices are solely appropriate for data gathered in autumn. Moreover, due to the crucial impact of sampling efficiency and sampling effort on the fish index scores, sampling method is of prime importance (e.g. Simon and Sanders 1999; Mostafavi et al. 2015). There is currently a paucity of standardised fish sampling method in Iran, and thus, European standards, as in CEN (2003) are preferable.

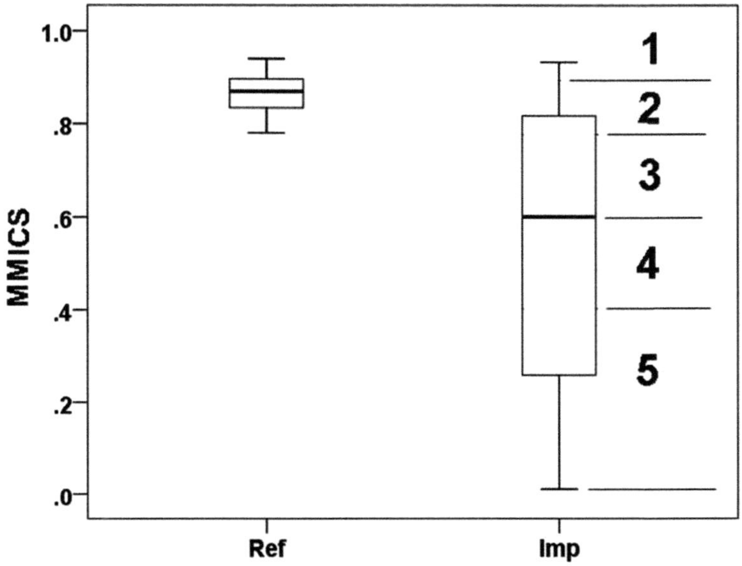

Fig. 52.39 The quartiles of both reference and impacted sites by classification into five categories in the cyprinid zone

52.8 Case Study 2

Modelling of the potential spatial distribution of brown trout (*Samo trutta*) at Iranian scale and the effects of climate change on its distribution (Mostafavi et al. 2014).

Brown trout (*Samo trutta*) is an indicator species because it is sensitive to many human pressures (e.g. water quality, habitat alteration) (Abdoli 2000; Coad 2018; Mostafavi et al. 2014, 2019). It is usually found in headwaters in which high oxygen saturation, steep slope, fast flow, suitable temperature regimes, and adequate food are dominant (Mostafavi et al. 2014, 2019). Over the past decades, human impacts declined its population in Iran in some areas (Coad 2018). Our main goal in this study was to predict potential spatial distribution of this species at the scale of Iran with predicting the climate change effects as well by species distribution model (SDM) tool. According to available literature, this species was recorded in the Urmia Lake, Caspian Sea, and Namak basins (Abdoli 2000; Abdoli and Naderi 2009; Coad 2018; Mostafavi et al. 2014).

Species distribution modelling has been recognised as a critical issue in ecology, biogeography, and conservation biology in order to conserve and manage ecosystems (Guisan and Thuiller 2005). In this regard, many studies have been carried out such as habitat shift impacts assessment (Logez and Pont 2011), evaluating of habitat suitability for species re-introductions (Lek et al. 1996), forecasting the probability of invasive species distribution (Poulos et al. 2012), investigating environmental variables impacts on fish species distribution (Pont et al. 2006), and predicting areas of persistence for endangered species (Dauwalter and Rahel 2008).

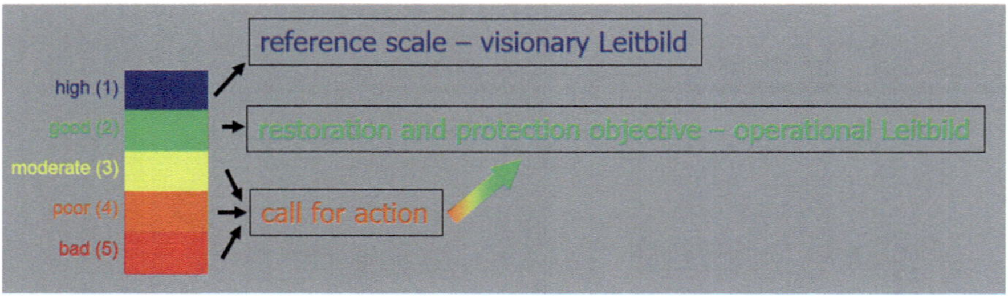

Fig. 52.40 General concept of WFD (Water Framework Directive) for the restoration and management of freshwater ecosystems (Photo from Stefan Schmutz lecture at BOKU University (Austria, Vienna))

52.8.1 Study Area and Fish Data

The whole of country (Iran) has been selected for this study (Fig. 52.40, case study 2). Real data (i.e. recorded data) regarding brown trout spatial distribution were collected from different periods (i.e. prior samplings, museums, and historical information (e.g. Saadati 1977; Abdoli 2000, personal database). After data quality check by authors, 1090 sites (presence of brown trout was only recorded in 63 sites) were prepared for further analysis (see Mostafavi et al. 2014) (Fig. 52.41, case study 2).

52.8.2 Environmental Predictor Variables

Environmental predictor variables such as elevation (ELE), stream slope (SLO), wetted width (W_WID), bankfull width (B_WID), maximum air temperature (Max_TEM), minimum air temperature (Min_TEM), mean air temperature (A_TEM), the range of air temperature (R_TEM), and annual precipitation (PRE) were defined for modelling according to some studies (see Pont et al. 2005; Logez and Pont 2011; Mostafavi et al. 2014). Some variables such as ELE, W_WID and B_WID were extracted on Google Earth because catchment layer similar to CCM2 is not available for the whole of Iran. Climatic variables (i.e. Max_TEM, Min_TEM, A_TEM and PRE) with a circular buffer (5 km) around each sampling site were extracted in WorldClim website (https://www.worldclim.org/) provided by Hijmans et al. (2005) and Hijmans et al. (2007). Afterwards, redundant variables with high correlation ($\rho > |0.75|$) were excluded by Spearman rank correlation test (Mostafavi et al. 2014). Finally, B_WID, W_WID, SLO, ELE, A_TEM, R_TEM, and PRE were remained as independent environmental predictor variables for the modelling.

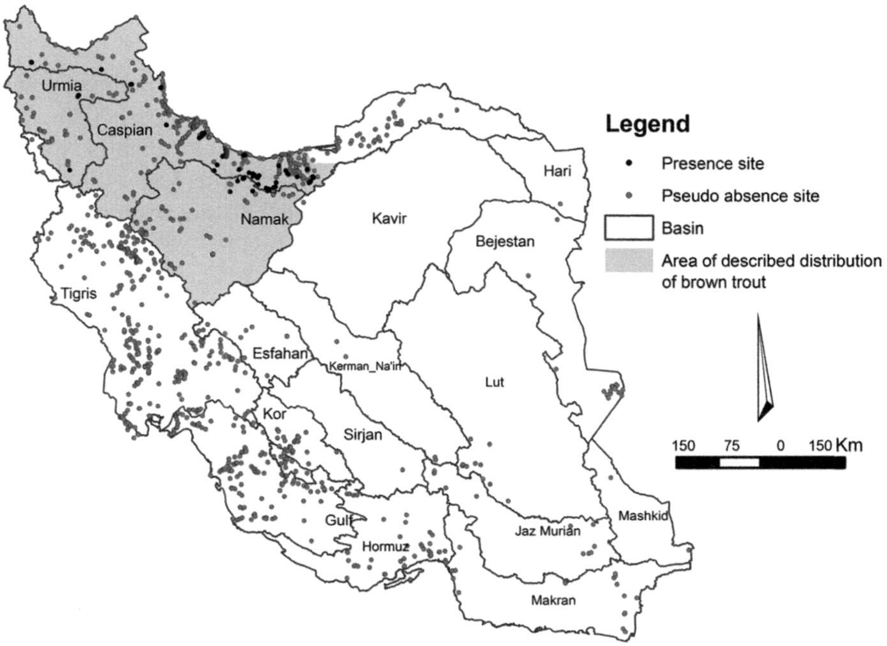

Fig. 52.41 Distribution of the studied sites with occurrence data used in the modelling of brown trout in different freshwater basins plus the distribution of brown trout in Iran as described in the literature

52.8.3 Modelling by SDM Tool

BIOMOD2 (BIOdiversity MODelling) package within R software environment was applied for modelling by SDM (Thuiller et al. 2009). We preferred using "pseudo-absence"-approach since our database involved heterogeneous data (Thuiller et al. 2009; Barbet-Massin et al. 2012; Mostafavi et al. 2014). First, our dataset was randomly divided into two parts (i.e. calibration (80% of the data) and validation (20% of the data)) via a cross-validation procedure and 10 repetition runs (Thuiller et al. 2009; Mostafavi et al. 2014). The ratio of presence data to pseudo-absence data was 0.33 in the model calibration.

Five models (i.e. GLM (Generalised linear model), GAM (Generalised Additive Model), CTA (Classification Tree Analysis), GBM (Generalised Boosting Models), and RF (Random Forests)) were applied for modelling and in order to reduce the uncertainty, all mentioned models were converged by an ensemble approach finally (according to Araújo and New 2007; Mostafavi et al. 2014).

Different criteria were applied for the assessment of modelling performance and accuracy which are as follows: TSS (True Skill Statistic) according to Lobo et al. (2008) classification, sensitivity ("true positives") and specificity ("true negatives") according to Thuiller et al. (2010); Barbet-Massin et al. (2012) as well as using an independent data (15 absences and 15 presences) which were gathered in the filed by sampling.

Hereafter, for prediction of climate change effects at different optimistic and pessimistic climatic scenarios (RCP2.6 and RCP8.5 for 2050 and 2080), we implemented the fitted model to appropriate climate and habitat variables, so that generate climate suitability maps for species in each model. Afterwards, all models were converged by an ensemble approach. Finally, spatial distribution maps of the species at different climatic scenarios were produced.

52.8.4 Results of Modelling

All mentioned criteria for modelling assessment were "excellent" (i.e. $\geq$0.8 for TSS and $>$ 89% for sensitivity and specificity) (Table 52.5, case study 2). GLM showed lowest performance compared to other applied techniques (i.e. TSS =0.81) while the performance of RF was highest.

Models mostly predicted brown trout in the Caspian Sea, Urmia Lake, and Namak Lake basins which are similar to the described distribution areas according to previous recorded information (Fig. 52.42, case study 2). Nevertheless, models predicted some potential sites out of the known distribution areas. These sites were predicted in the Tigris basin as well as in the eastern region of the Caspian Sea basin (Fig. 52.42, case study 2). Based on the information of some local fisheries experts, it is assumed that most likely this species living in the eastern region of the Caspian Sea basin although evidence is missing to date. Regarding prediction of brown trout in Tigris basin, it seems to be rational as this species was recorded in the upstream of Tigris in Turkey (Turan et al. 2011). Furthermore, from a biogeographical viewpoint, the Tigris basin was the migration path of brown trout to the Namak Lake basin in palaeo-historic times before the mountains between the basins lifted up (Boulenger 1896; Mostafavi et al. 2014). As a matter of fact, fish sampling for these regions are not adequate particularly in upstream areas. This is a crucial issue since new species are discovered nearly every year in remote and mountainous areas of Iran (e.g. Abdoli 2017; Teimori et al. 2012a; Mostafavi et al. 2014). In addition, we cannot ignore the effects of human activities on fish species distribution and occurrence in Iran.

Table 52.5 Prediction accuracy measured using sensitivity, specificity, and TSS

Model	Sensitivity (%)	Specificity (%)	TSS
CTA	95.1	91.3	0.86
GAM	97.5	85.3	0.83
GBM	95.9	88.6	0.85
GLM	97.9	82.4	0.80
RF	99.0	98.9	0.98
Average	**97.1**	**89.3**	**0.86**

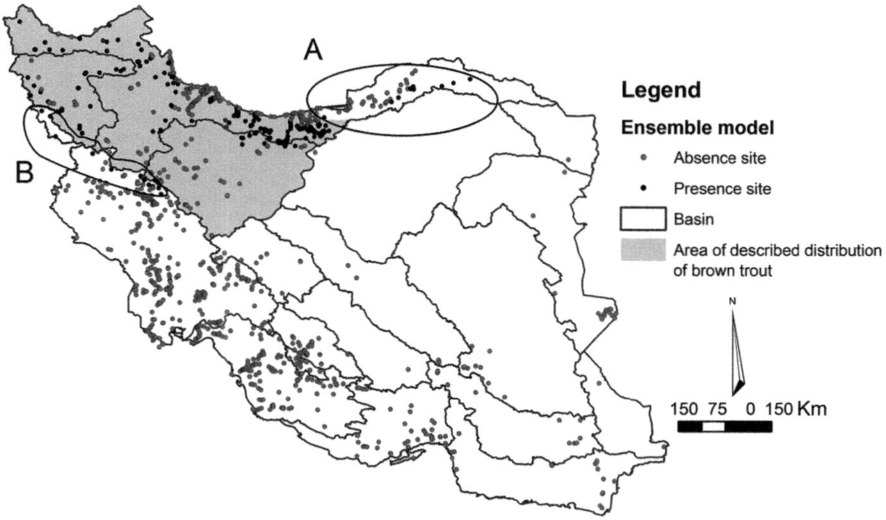

Fig. 52.42 Predicted distribution of brown trout according to the ensemble model: (**a**) predicted sites outside of the described distribution area in the eastern part of Caspian basin and (**b**) predicted sites outside of the described distribution area in Tigris basin

Table 52.6 Relative importance (in percentage) of environmental variables for each extent and all models

Model type	B_WID	W_WID	SLO	ELE	A_TEM	R_TEM	PRE
Ensemble	1.9	0.3	27.4	0.6	44.6	19.7	5.6
CTA	5.9	0.0	35.4	0.0	23.0	22.3	13.5
GAM	0.0	0.0	12.8	0.0	53.0	23.2	11.0
GBM	0.4	0.4	40.4	0.4	45.8	12.0	0.5
GLM	0.0	0.0	2.3	0.0	67.6	30.1	0.0
RF	3.2	1.1	45.9	2.4	33.4	10.8	3.2

B_WID bankfull width, *W_WID* wetted width, *SLO* stream slope, *A_TEM* mean air temperature, *PRE* annual precipitation, *ELE* elevation, and *R_TEM* range of air temperature

The environmental predictor variables, i.e. SLO, A_TEM, and R_TEM, showed the most importance values (>18%), but B_WID, W_WID, ELE, and PRE displayed the lowest values (<6%) (Table 52.6, case study 2), variable significance for all models is presented as well (Table 52.6, case study 2).

Finally, as can be seen in Table 52.7, case study 2 and Fig. 52.43, case study 2, climate change may significantly influence on the potential habitats of brown trout therefore decision-makers should plan for its conservation in early future.

Table 52.7 Significant reduction of brown trout from its potential habitats because of climate change effects at different climatic scenarios

RCP 8.5		RCP 2.6		Scenario
2080	2050	2080	2050	Year
1481	1481	1481	1481	Total sites
1318	1318	1318	1318	Unsuitable
0	0	0	0	No change
163	163	163	163	Reduction
0	0	0	0	Expansion
100.00	100.00	100.00	100.00	Reduction percentage
0.00	0.00	0.00	0.00	Expansion percentage
−100.00	−100.00	−100.00	−100.00	Species range changes

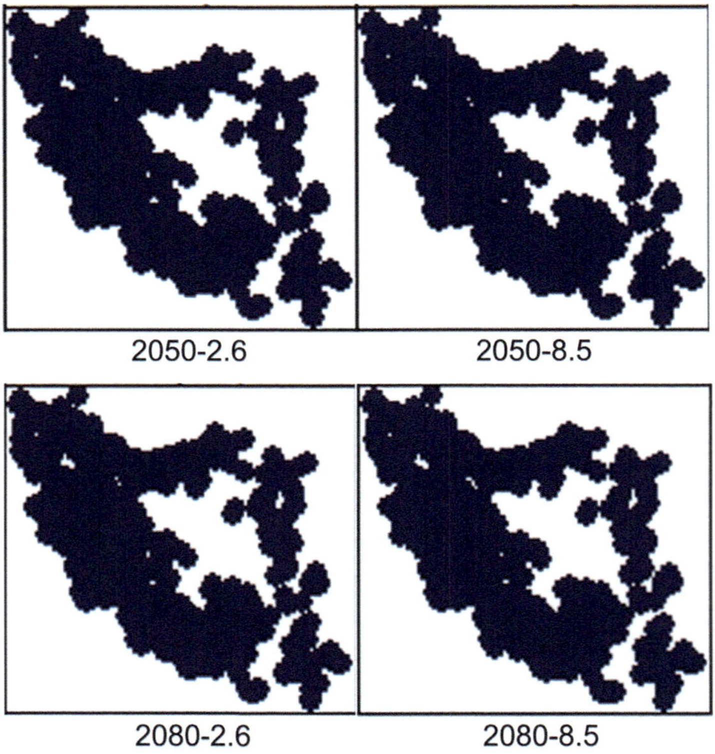

Fig. 52.43 Prediction of climate change effects on the spatial distribution at different climatic scenarios

52.9 General Conclusion

To conclude, some fundamental steps are introduced in the current chapter to develop a monitoring system for Iran representing rivers' ecological status. In fact, as an essential action for a sustainable use of aquatic resources, river managers can develop and utilise this monitoring system which is based on the modelling approach to apply more useful measures so that they can preserve or re-achieve a good ecological status in the future. Despite the existence of a large amount of water in Iran, many parts are to a large extent dry. Additionally, based on the results of this chapter, various human pressures with negative impact on running waters were identified.

Multiple pressures are more pervasive than single pressure occurrence. As a consequence, using water wisely is vital for Iran's development. Accordingly, there is a call for such modelling tools in order to stabilise the diverse needs in different riverine systems. In the current condition, there exist some constraints in regard with hydrology, morphology, water quality, connectivity, etc., which can be identified by developed fish indices. Some questions that may be responded in future studies include the following: how much water has to be kept in the river, what has to be the water quality, what kind of framework has to be preserved, etc. Future researchers can further improve the method developed in this study since many other questions can be answered with such indices. Besides, SDM tool enables us to anticipate the effects of some pressure like climate change which provide very useful information for the conservation of our biodiversity.

Acknowledgments Our special thanks are given to Prof. Dr. Stefan Schmutz and Dipl.-Ing. Dr. Rafaela Schinegger at BOKU University for their support.

References

Abbaspour KC, Faramarzi M, Ghasemi SS, Yang H (2009) Assessing the impact of climate change on water resources in Iran. Water Resour Res 45:1–16

Abdoli A (2000) The inland water fishes of Iran. Iranian Museum of Nature and Wildlife, Tehran

Abdoli A (2017) The field guide of inland water fishes of Iran. Iran-shenasi, Tehran

Abdoli A, Naderi M (2009) Biodiversity of fishes of the Southern Basin of the Caspian Sea. Abzian Scientific Publications, Tehran

Akhani H, Djamali M, Ghorbanalizadeh A, Ramezani E (2010) Plant biodiversity of Hyrcanian relict forests, N Iran: an overview of the flora, vegetation, palaeoecology and conservation. Pak J Bot 42:231–258

Allan JD (2004) Landscapes and riverscapes: The influence of land use on stream ecosystems. Annual Rev Ecol Evol Systematics 35(1):257–284

Allan JD, Flecker AS (1993) Biodiversity conservation in running waters. Bioscience 43(1):32–43

Amiri M, Eslamian S (2010) Investigation of climate change in Iran. J Environ Sci Technol 3:208–216

Araújo MB, New M (2007) Ensemble forecasting of species distributions. Trends Ecol Evol 22:42–47

Barbet-Massin M, Jiguet F, Albert CH, Thuiller W (2012) Selecting pseudo-absences for species distribution models: how, where and how many? Methods Ecol Evol 3:327–338
Battin J (2004) When good animals love bad habitats: ecological traps and the conservation of animal populations. Conserv Biol 18:1482–1491
Berteaux D, Réale D, McAdam AG, Boutin S (2004) Keeping pace with fast climate change: can arctic life count on evolution? Integr Comp Biol 44:140–151
Boulenger G (1896) XXII.—on freshwater fishes from Smyrna. J Nat Hist 18:153–154
CEN (2003) Water quality–sampling of fish with electricity. European Committee for Standardization
Coad BW (1976) Freshwater fishes of Iran: a check-list and bibliography. Syllogeus, 57 p
Coad BW (1980) Environmental change and its impact on the freshwater fishes of Iran. Biol Conserv 19:51–80
Coad BW (1998) Systematic biodiversity in the freshwater fishes of Iran. Italian J Zool 65:101–108
Coad BW (2018) Freshwater fishes of Iran. http://www.briancoad.com
Coad BW, Vilenkin BY (2004) Co-occurrence and zoogeography of the freshwater fishes of Iran. Zool Middle East 31:53–62
Cowell CM, Stoudt RT (2002) Dam-induced modifications to upper Allegheny River stream flow patterns and their biodiversity implications 1. JAWRA J Am Water Res Assoc 38:187–196
Darehshouri BF, Kasraian N (1988) Nature of Iran. Rowzaneh Kar Publication, Tehran
Dauwalter DC, Rahel FJ (2008) Distribution modelling to guide stream fish conservation: an example using the mountain sucker in the Black Hills National Forest, USA. Aquat Conserv Mar Freshwat Ecosyst 18:1263–1276
De Jager A, Vogt J (2010) Development and demonstration of a structured hydrological feature coding system for Europe. Hydrol Sci J 55:661–675
Degerman E, Beier U, Breine J, Melcher A, Quataert P, Rogers C et al (2007) Classification and assessment of degradation in European running waters. Fish Manag Ecol 14:417–426
EFI+Consortium (2009) Manual for the application of the new European Fish Index EFI+. A fish-based method to assess the ecological status of European running waters in support of the Water Framework Directive
Efron B, Tibshirani R (1993) An introduction to the bootstrap chapman and hall. Google Scholar, New York
Esmaeili HR, Teimori A, Khosravi AR (2007) A note on the biodiversity of Ghadamgah spring–stream system in Fars province, Southwest Iran. Iranian J Animal Biosystematics 3:15–23
Esmaeili H, Coad B, Gholamifard A, Nazari N, Teimory A (2010) Annotated checklist of the freshwater fishes of Iran. Zoosystematica Rossica 19:361–386
Esmaeili HR, Nazari N, Gholamifard A, Gholamhosseini G, Teimory A, Coad BW (2011) Range extension and translocation for *Rhodeus amarus* (Bloch, 1782) (Actinopterygii: Cyprinidae) in northwestern Iran. Turkish J Zool 35:882–885
Esmaeili HR, Gholamifard A, Sayyadzadeh G, Parsi B, Mirghiyasi S, Ghasemian S (2013) New record of the convict cichlid, *Amatitlania nigrofasciata* (Günther, 1867), from the Middle East (Actinopterygii: Cichlidae). Aqua, Int J Ichthyol 19:225–229
Esmaeili HR, Brian WC, Mehraban HR, Masoudi M, Khaefi R, Abbasi K et al (2015) An updated checklist of fishes of the Caspian Sea basin of Iran with a note on their zoogeography. Iranian J Ichthyol 1:152–184
Farashi A, Shariati M (2017) Biodiversity hotspots and conservation gaps in Iran. J Nat Conserv 39:37–57
Fatemi SMR, Nabavi SMB, Vosoghi G, Fallahi M, Mohammadi M (2012) The relation between environmental parameters of Hormuzgan coastline in Persian gulf and occurrence of the first harmful algal bloom of *Cochlodinium polykrikoides* (Gymnodiniaceae). Iran J Fish Sci 11 (3):475–489
Filipe AF, Markovic D, Pletterbauer F, Tisseuil C, De Wever A, Schmutz S et al (2013) Forecasting fish distribution along stream networks: brown trout (*Salmo trutta*) in Europe. Divers Distrib 19:1059–1071

Gholami Z, Akhlaghi M, Esmaeili H (2009) Infection of the Isfahan tooth-carp, Aphanius isfahanensis (Actinopterygii: Cyprinodontidae) with *Diplostomum spathaceum* (Trematoda: Diplostomidae): a case report. Int J Vet Res 3:129–131

Grave SD, Ghane A (2006) The establishment of the oriental river prawn, *Macrobrachium nipponense* (de Haan, 1849) in Anzali lagoon. Iran Aquatic Invasions 1(4):204–208

Guisan A, Thuiller W (2005) Predicting species distribution: offering more than simple habitat models. Ecol Lett 8:993–1009

Harrison AF, Howard PJA, Howard DM, Howard DC, Hornung M (1995) Carbon storage in forest soils. Forestry 68:335–348

Harvey BC (1987) Susceptibility of young-of-the-year fishes to downstream displacement by flooding. Trans Am Fish Soc 116:851–855

Hering D, Borja A, Carstensen J, Carvalho L, Elliott M, Feld CK et al (2010) The European water framework directive at the age of 10: a critical review of the achievements with recommendations for the future. Sci Total Environ 408:4007–4019

Heshmati G (2012) Vegetation characteristics of four ecological zones of Iran. Int J Plant Production 1:215–224

Hijmans RJ, Cameron SE, Parra JL, Jones PG, Jarvis A (2005) Very high resolution interpolated climate surfaces for global land areas. Int J Climatol 25:1965–1978

Hijmans R, Cameron S, Parra J (2007) WorldClim Version 1.4. Museum of Vertebrate Zoology, University of California 2007: http://www.worldclim.org

Holčík J, Oláh J (1992) Fish, fisheries and water quality in Anzali Lagoon and its watershed. Report prepared for the project – Anzali Lagoon productivity and fish stock investigations. Food and Agriculture Organization, Rome, FI:UNDP/IRA/88/001

Hrbek T, Meyer A (2003) Closing of the Tethys Sea and the phylogeny of Eurasian killifishes (Cyprinodontiformes: Cyprinodontidae). J Evol Biol 16:17–36

Ivanov VP, Kamakin AM, Ushivtzev VB, Shiganova T, Zhukova O, Aladin N, Wilson SI, Harbison GR, Dumont HJ (2000) Invasion of the Caspian Sea by the comb jellyfish *Mnemiopsis leidyi* (Ctenophora). Biol Invasions 2:255–258

Johnson L, Richards C, Host G, Arthur J (1997) Landscape influences on water chemistry in Midwestern stream ecosystems. Freshw Biol 37:193–208

Kamangar BB, Prokofiev AM, Ghaderi E, Nalbant TT (2014) Stone loaches of Choman River system, Kurdistan, Iran (Teleostei: Cypriniformes: Nemacheilidae). Zootaxa 3755:33–61

Kennedy TL, Turner TF (2011) River channelization reduces nutrient flow and macroinvertebrate diversity at the aquatic terrestrial transition zone. Ecosphere 2(3):art35

Khaefi R, Esmaeili HR, Zareian H, Babaei S (2014) The first record of the redbelly tilapia, *Tilapia zillii* (Gervais, 1848), in freshwaters of Iran. Turkish J Zool 38:96–98

Kingsford RT (2000) Ecological impacts of dams, water diversions and river management on floodplain wetlands in Australia. Austral Ecol 25:109–127

Kolar CS, Lodge DM (2001) Progress in invasion biology: predicting invaders. Trends Ecol Evol 16:199–204

Kristensen P (2004) The DPSIR framework. In: Proceedings of the Workshop on a Comprehensive/ Detailed Assessment of the Vulnerability of Water Resources to Environmental Change in Africa Using River Basin Approach, UNEP Headquarters, Nairobi, Kenya, 27–29 September 2004

Langdon RW (2001) A preliminary index of biological integrity for fish assemblages of small Coldwater streams in Vermont. Northeast Nat 8:219–232

Lau JK, Lauer TE, Weinman ML (2006) Impacts of channelization on stream habitats and associated fish assemblages in east Central Indiana. Am Midl Nat 156:319–330

Lek S, Belaud A, Baran P, Dimopoulos I, Delacoste M (1996) Role of some environmental variables in trout abundance models using neural networks. Aquat Living Resour 9:23–29

Lobo J, Jiménez-valverde A, Real R (2008) AUC: erratum: predicting species distribution: offering more than simple habitat models. Glob Ecol Biogeogr 17:145–151

Logez M, Pont D (2011) Development of metrics based on fish body size and species traits to assess European Coldwater streams. Ecol Indic 11:1204–1215
Lovejoy TE, Hannah LJ (2006) Climate change and biodiversity. Yale University Press, New Haven, CT
Lymbery AJ, Morine M, Kanani HG, Beatty SJ, Morgan DL (2014) Co-invaders: the effects of alien parasites on native hosts. Int J Parasitol: Parasites and Wildlife 3:171–177
Lyons J, Thimble SW, Paine LK (2000) Grass versus trees: managing riparian areas to benefit streams of Central North America 1. JAWRA J Am Water Res Assoc 36:919–930
Majidi E, Sobhani K, Rezaei MR (2011) Deforestation of forests in Iran. J Appl Environ Biol Sci 8:184–186
Majnonian H, Kiabi BH, Danesh M (2005) Reading in Zoogeography of Iran (part1). Sazman hefazat mohit zist Iran (Department of environment of Iran). Dayereh Sabz, Tehran
McKay SF, King AJ (2006) Potential ecological effects of water extraction in small, unregulated streams. River Res Appl 22(9):1023–1037
Melcher A, Schmutz S, Haidvogl G, Moder K (2007) Spatially based methods to assess theecological status of European fish assemblage types. Fish Manag Ecol 14:453–463
Modarres R, de Paulo Rodrigues da Silva V (2007) Rainfall trends in arid and semi-arid regions of Iran. J Arid Environ 70(2):344–355
Mostafavi H (2007) Fish biodiversity in Talar River, Mazandaran province. Majale olom mohity (Iranian J Environ Stud) 32:127–135
Mostafavi H, Babaii S, Mahmoudi H (2004) Assembling of preliminary report on natural potentialities in Mazandaran Province at its present condition. Environmental Sciences Research Institute, Shahid Beheshti University, Tehran
Mostafavi H, Pletterbauer F, Coad BW, Mahini AS, Schinegger R, Unfer G et al (2014) Predicting presence and absence of trout (*Salmo trutta*) in Iran. Limnologica-Ecol Manage Inland Waters 46:1–8
Mostafavi H, Schinegger R, Melcher A, Moder K, Mielach C, Schmutz S (2015) A new fish-based multi-metric assessment index for cyprinid streams in the Iranian Caspian Sea Basin. Limnologica-Ecol Manage Inland Waters 51:37–52
Mostafavi H, Valavi R, Rashidian M, Makki T (2017) Prediction climate change effects on the Iranian sturgeon fishes (Acipenseridae) distribution under different climatic scenarios. In: The fifth Iranian conference of ichthyology. Islamic Azad University of Babol, Iran
Mostafavi H, Teimori A, Schinegger R, Schmutz S (2019) A new fish based multi-metric assessment index for cold-water streams of the southern Caspian Sea Basin in Iran. Environ Biol Fish 102:645–662
Nalbant TT, Bianco PG (1998) The loaches of Iran and adjacent regions with description of six new species (Cobitoidea). Italian J Zool 65:109–123
Noble RAA, Cowx IG, Starkie A (2007) Development of fish-based methods for the assessment of ecological status in English and Welsh rivers. Fish Manage Ecol 14:495–508
NRE (2002) Sustainable Diversions Limit Project, Recommendations for sustainable diversions limits over winter fill periods in unregulated Victorian Catchments. A report prepared by Sinclair Knight Merz and the Cooperative Research Centre for Freshwater Ecology
Oberdorff T, Pont D, Hugueny B, Chessel D (2001) A probabilistic model characterizing fish assemblages of French rivers: a framework for environmental assessment. Freshw Biol 46:399–415
Oberdorff T, Pont D, Hugueny B, Porcher JP (2002) Development and validation of a fish-based index for the assessment of 'river health' in France. Freshw Biol 47:1720–1734
Osborne LL (1988) Empirical relationship between land use/cover and stream water quality in an agricultural watershed. J Environ Manag 26:9–27
Padmalal D, Maya K, Sreebha S, Sreeja R (2008) Environmental effects of river sand mining: a case from the river catchments of Vembanad lake, southwest coast of India. Environ Geol 54:879–889

Paukert CP, Pitts KL, Whittier JB, Olden JD (2011) Development and assessment of a landscape-scale ecological threat index for the lower Colorado River basin. Ecol Indic 11:304–310
Paul MJ, Meyer JL (2001) Streams in the urban landscape. Annu Rev Ecol Syst 32:333–365
Poff NL, Hart DD (2002) How dams vary and why it matters for the emerging science of dam removal: an ecological classification of dams is needed to characterize how the tremendous variation in the size, operational mode, age, and number of dams in a river basin influences the potential for restoring regulated rivers via dam removal. AIBS Bull 52:659–668
Poff NL, Allan JD, Bain MB, Karr JR, Prestegaard KL, Richter BD et al (1997) The natural flow regime. Bioscience 47:769–784
Pont D, Hugueny B, Oberdorff T (2005) Modelling habitat requirement of European fishes: do species have similar responses to local and regional environmental constraints? Can J Fish Aquat Sci 62:163–173
Pont D, Hugueny B, Beier U, Goffaux D, Melcher A, Noble R et al (2006) Assessing river biotic condition at a continental scale: a European approach using functional metrics and fish assemblages. J Appl Ecol 43:70–80
Pont D, Hughes RM, Whittier TR, Schmutz S (2009) A predictive index of biotic integrity model for aquatic-vertebrate assemblages of western US streams. Trans Am Fish Soc 138:292–305
Poulos HM, Chernoff B, Fuller PL, Butman D (2012) Ensemble forecasting of potential habitat for three invasive fishes. Aquat Invasions 7:59–72
Rahimi M (2011) Analyzing the temporal and spatial variation of fog days in Iran. Pure Appl Geophys 169(5–6):1165–1172
Richards C, Johnson LB, Host GE (1996) Landscape-scale influences on stream habitats and biota. Can J Fish Aquat Sci 53:295–311
Saadati MAG (1977) Taxonomy and distribution of the freshwater fishes of Iran: Colorado State University, Fort Collins
Schinegger R, Trautwein C, Melcher A, Schmutz S (2012) Multiple human pressures and their spatial patterns in European running waters. Water Environ J 26:261–273
Schinegger R, Trautwein C, Schmutz S (2013) Pressure-specific and multiple pressure response of fish assemblages in European running waters. Limnologica-Ecol Manage Inland Waters 43:348–361
Schlosser IJ (1985) Flow regime, juvenile abundance, and the assemblage structure of stream fishes. Ecology 66:1484–1490
Schmutz S, Kaufmann M, Vogel B, Jungwirth M, Muhar S (2000) A multi-level concept for fish-based, river-type-specific assessment of ecological integrity. In: Assessing the ecological integrity of running waters. Springer, New York, pp 279–289
Schmutz S, Cowx I, Haidvogl G, Pont D (2007) Fish-based methods for assessing European running waters: a synthesis. Fish Manag Ecol 14:369–380
Simon TP, Sanders RE (1999) Applying an index of biotic integrity based on great river fish communities: considerations in sampling and interpretation. In: Simon TP (ed) Assessment approaches for estimating biological integrity using fish assemblages. CRC Press, Boca Raton, pp 475–505
Smith BM, Desvigne LD, Slade JB, Dooley JW, Warren DC (1996) Transcutaneous oxygen measurements predict healing of leg wounds with hyperbaric therapy. Wound Repair Regen 4:224–229
Teimori A (2006) Preliminary study of freshwater fish diversity in Fars province. MSc, Shiraz University, Shiraz, Iran
Teimori A, Esmaeili H, Gholamhosseini A (2011a) The ichthyofauna of Kor and Helleh River basins in southwest of Iran with reference to taxonomic and zoogeographic features of native fishes. Iran J Anim Biosystematics 6:1–8
Teimori A, Esmaeili HR, Reichenbacher B (2011b) Aphanius farsicus, a replacement name for *A. persicus* (Jenkins, 1910) (Teleostei, Cyprinodontidae). Zootaxa 3096:53–58

Teimori A, Esmaeili HR, Gholami Z, Zarei N, Reichenbacher B (2012a) Aphanius arakensis, a new species of tooth-carp (Actinopterygii, Cyprinodontidae) from the endorheic Namak Lake basin in Iran. ZooKeys 215:55–76
Teimori A, Schulz-Mirbach T, Esmaeili HR, Reichenbacher B (2012b) Geographical differentiation of Aphanius dispar (Teleostei: Cyprinodontidae) from southern Iran. J Zool Syst Evol Res 50:289–304
Teimori A, Esmaeili HR, Erpenbeck D, Reichenbacher B (2014) A new and unique species of the genus Aphanius Nardo, 1827 (Teleostei: Cyprinodontidae) from southern Iran: a case of regressive evolution. Zoologischer Anzeiger-A J Comparative Zool 253:327–337
Teimori A, Mostafavi H, Esmaeili H (2016) An update note on diversity and conservation of the endemicfishes in Iranian inland waters. Turk J Zool 40:87–102
Thuiller W (2003) BIOMOD–optimizing predictions of species distributions and projecting potential future shifts under global change. Glob Chang Biol 9:1353–1362
Thuiller W (2004) Patterns and uncertainties of species' range shifts under climate change. Glob Chang Biol 10:2020–2027
Thuiller W, Lafourcade B, Engler R, Araújo MB (2009) BIOMOD–a platform for ensemble forecasting of species distributions. Ecography 32:369–373
Thuiller W, Lafourcade B, Araujo M. ModOperating manual for BIOMOD Thuiller W, Lafourcade B (2010) BIOMOD: species/climate modelling functions. R package version 2009a:1.-5
Tockner K, Stanford JA (2002) Riverine flood plains: present state and future trends. Environ Conserv 29:308–330
Trautwein C, Schinegger R, Schmutz S (2012) Cumulative effects of land use on fish metrics in different types of running waters in Austria. Aquat Sci 74:329–341
Turan D, Kottelat M, Bektas Y (2011) Salmo tigridis, a new species of trout from the Tigris River, Turkey(Teleostei: Salmonidae). Zootaxa 2993:23–33
Vogt J, Colombo R, Paracchini ML, de Jager A, Soille P (2003) CMM river and catchment database. EUR 20756 EN, EC-JRC, Inspra
Vogt J, Soille P, de Jaeger A, Rimaviciute E, Mehl W, Foisneau S, et al. (2007) A Pan-European river and catchment database. EC-JRC (Report EUR 22920 EN) p 120, Luxembourg
Wang L, Lyons J, Kanehl P, Gatti R (1997) Influences of watershed land use on habitat quality and biotic integrity in Wisconsin streams. Fisheries 22:6–12
Wang L, Infante D, Lyons J, Stewart J, Cooper A (2011) Effects of dams in river networks on fish assemblages in non-impoundment sections of rivers in Michigan and Wisconsin. USA River Res Appl 27(4):473–487
Welcomme RL (1988) International introductions of inland aquatic species. Food and Agriculture Organization, Rome, Fisheries Technical Paper 294
Woodward G, Perkins DM, Brown LE (2010) Climate change and freshwater ecosystems: impacts across multiple levels of organization. Philos Trans R Soc of Lond B: Biol Sci 365:2093–2106
Zehzad B, Kiabi BH, Madjnoonian H (2002) The natural areas and landscape of Iran: an overview. Zool Middle East 26:7–10

Chapter 53
Enhancing Rural Women's Participation in Fisheries in Iraq

Laith A. Jawad

Abstract Men are believed to have a control in the fishery area, and women are not deliberated to have a notable role. Nevertheless, numerous women are tangled up in the fishery business. Several cases from around the world showed that women are involved straight in a fishery by involving in fishing one or the other exclusively or by associated with their husbands, and by gathering seaweed. They could also get intricate in a fishery implicitly by involving in food and feeding events.

In Iraq, the part of women in fisheries has been discounted, and to some extent, it has been not accounted for at all. The present chapter deals with the idea of establishing a project that can recover the incomes of women and to improve the aids of women in the fisheries sector. This project should be established by the government of Iraq and probably in collaboration of the Food and Agriculture Organization (FAO) to train women in order to grow their skills, add denomination to the additional yield and generate jobs and upsurge profits. Such project should start to operate in the rural areas, where water bodies are accessible such as the southern part of Iraq and the coastal area at Fao City, south of Iraq. At the end of the chapter, several recommendations were given on what is desired to tolerate and recover these women-centred events.

53.1 Introduction

It is known for everyone that the term 'fisherman' means fishing is achieved by men, but closer look into the fisheries business around the world, it is clear that specific fishing events are more normally undertaken by men, others are controlled by women. Women are intricate in the collecting, handling and sale, as well as money facets of fisheries, yet (Weeratunge et al. 2010). Men usually considered as earners (i.e. hunters, fishers), while women remain at home and support the family.

L. A. Jawad (✉)
School of Environmental and Animal Sciences, Unitec Institute of Technology, Auckland, New Zealand

L. A. Jawad (ed.), *Tigris and Euphrates Rivers: Their Environment from Headwaters to Mouth*, Aquatic Ecology Series 11,
https://doi.org/10.1007/978-3-030-57570-0_53

In the rural areas of Iraq, women form a big part of the agricultural workers and are concerned in the arrangement, handling and marketing of agricultural supplies. Women comprise a significant monetary source, yet their role in this area is low. Commonly, present policies for refining the status of women incline to highlight their parts as wives, mothers, caregivers and homemakers. Gathering sign displays that women also have wider parts in development, applying and assessing as Iraqi women increasingly take part in the labour force, containing in the fisheries and in group administrations.

Despite of their participation, women and young girls in the rural areas of Iraq still often have very low incomes or no income at all, including in the fisheries sector. While men are chiefly intricate in fishing activities, women are involved in pre- and post-fishing actions, but sometimes, women do catch fish. As with the other women in the fishers' communities, they are also engaged in repairing nets and looking after fishing gear, and other doings (FAO 2005; Raquiza 2005). Seeing the elementary job and tasks of women in fishing populations, distinct attempts should be made to deliver for the requirements of women workers and their families.

Fish is a very subtle product and people conducting it require awareness and talent to extend its morals to the next generations in order to enhance value to capture higher prices in markets. Therefore, the government in Iraq needs to mandate to uphold the rights and privileges of fisher folk with special attention to women for priority development assistance. Government should introduce programmes to assist in this aspect and should provide equipment and materials to cover training on the activities of the fisheries sector (Gaerlan et al. 2012).

The present chapter proposes an idea of establishing a training centre to enhance the parts of women and their aids through this government involvement to the growth of the fisheries business.

53.2 How to Achieve the Goal

The project intended to advance talents of women in the country agricultural areas communities in contact with fisheries practices. The training includes fish handling indispensable for enhancing value to fish yields that upsurge their likelihoods of receiving a reasonable income. Precisely, the idea of the project required to advance women's talents in handling fish. The long-term aim was to produce earnings and jobs, improve women's abilities, free enterprise and business insight. The project needs to be located in the southern Iraq, where are many villages that located along the side of the different sizes of water bodies.

The project should start with small number of participants and once it proves success, then it can be extended to involve larger number and can be applied in several localities. The scheme should adopt and support relations among the different connotations. Over the exercise courses and meetings where the members and officers were present, individual members and the organizations swapped ideas and actions and more progressive relations became styles for the others.

53.3 Women and Technology

The use of labour-saving techniques looks to be gender linked and has been obliging to women in some circumstances but not suitable in others.

With the development of fishery production, men have taken control of the mechanized operations and left the non-mechanized and more labour-intensive jobs, such as feeding, and net repairing, to women. The usage of machines has unrestricted women from many uninteresting tasks, and it has also replaced a noteworthy fund of revenue for many women who be determined by these professions to yield extra revenue for their families. In numerous country areas, though, such modern techniques are remain not obtainable. In such areas, families do not have clean water and rely on women and young girls to fetch water from the nearby spring for daily usage.

A vital issue remains to be solved in the issue of educating women to perform better in the fisheries sector. Women are usually with less educational status in the fisheries sector than men due to the absence of teaching plans in technological enhancements. However, the number of women looking for higher schooling in fisheries is on the growth in several countries including Iraq, but those women are from the urban part of the country. With any training programme that aim to upgrade the talent pf women in the fisheries sector, women should have knowledge in the following subjects: fish technology and nutrition, biology and management. With the problem of age, old women are obviously not suitable for such a programme. Therefore, special intention should be paid for young girls to enrol for the subjects mentioned through the training programme.

53.4 Inferences for Women

Suwanrangsi (1998) has recommended some implications for women. Due the importance of these inferences, I thought it is worth give it below with slight variation to fit the status of women in Iraq:

1. Education is vital and proved significantly family health and nourishment (Telefood 1998).
2. Fewer work hours: using new techniques might decrease women's work hours by generating their work a little simpler and less time-consuming.
3. Creation of an extra income: Development in technology will offer women with selections for producing revenue.
4. Substitute of the female labour force: In some circumstances, technology could unenviably disturb women by removing their occupations.

53.5 Areas of Women's Participation

There are several areas that women can be trained in so she will assist in the fisheries sector. Some of these aspects are related to marine fisheries, and others are general and can be applied for any fisheries industries.

In Iraq, the marine fisheries industry is very limited in its usage of the natural marine resources in spite of the availability of several marine organisms that have been marked as not commercial for the local market such different species of shells, shellfish, crustaceans other than shrimps and prawns, octopus, sea cucumber and squids. If training courses are given to women in processing these marine products for export, then this project will form an additional revenue to Iraq and at the same time involve in such business large number of women and young girls who they are usually either work for free or sit at home do nothing. To bring such a product to life, a complete proposal needs to be put in hand so it will contains several parts such as the building of the aquatic product plants fully equipped with modern machinery, vessels and boats for catching aquatic organisms that have no commercial value in the local markets in Iraq, training centre that give short courses in handling aquatic products and a marketing department.

Among the areas that women can be trained for such a project are the followings, which include both freshwater and marine resources. These types of training issues were suggested by Srinath (1987):

1. Prawn shedding: removing skins of prawn is a fundamental element of getting foreign currency to the country. This is a periodic action concurring with prawn fishery season. In this job, there are two kinds of events are usually in progress. This job is referable for women. It does not need any special talent; however, some difference is seen in the percentage of meat retrieval from person to person.
2. Net making: Net manufacture is the customary talent in Iraq, and women can learn this job simply. Net making is a relaxation time action.
3. Collection shells: This job can be performed in southern of Iraq at the city of Fao, where people can go at the low tide to collect shell and bring them back to the factory.
4. Processing: Handling is the chief fisheries action can be directed by women. Main treating takes place in the form of gutting, scaling, cleaning of fish and shellfish for family feeding, and sale.
5. Marketing: In Iraq and in the rural areas, there is no fish market. Instead, the women sell their yields to the main road users by sitting on the main road leading to their villages. Or taking out to nearest large centre by boats or by cars early in the morning.

53.6 Women in Fisheries in Asia as an Example

Women in Asia share with men much of the agricultural and fisheries works (Sharma 2004). In doing so, women go through different types jobs in fisheries sector that can be taken as a role model to follow in other countries that would like to upgrade the status of their women in the fisheries sector. Sharma (2004) has suggested types of jobs that related to fisheries sector and that women can do, and these are as follows:

1. Involving in pre-harvest work such as net-making.
2. Engaging in fish processing and marketing activities.
3. Participating in gleaning and collection activities in inshore areas and intertidal zones.
4. Women as workers in processing plants.

53.7 Recommendations

So to recover the place of women in Iraq, it is essential that the following difficulties be adopted:

1. Combat Illiteracy among women especially in the rural areas.
2. Involve women in all programmes set by the government.
3. Improve and enhance the conservative nature of women in the country side of Iraq and teach them to how to become vital components of the society.
4. Women in the rural areas and due to their illiteracy, they fall in poverty, which can be cause of several social problems in addition to the health ones.

References

FAO (Food and Agriculture Organization) (2005) Fact Sheet Philippines - Women in agriculture, environment and rural production. Accessed at: http://www.fao.org/docrep/008/ae946e/ae946e01.htm#TopOfPage

Gaerlan RSP, Lachica RB, Dumol MB (2012) Enhancing rural Women's participation in fisheries: post-harvest livelihoods, Ilocos region, Philippines. Gender in aquaculture and fisheries: moving the agenda forward. Asian Fish Sci Spec Issue 25S:199–205

Raquiza MVR (2005) A condensed report on Beijing +10: celebrating gains, facing new challenges: a report of Philippine NGOs. United Nations Population Fund, Manila, Philippines, 12 p. Accessed at: http://www.socialwatch.org/sites/default/files/pdf/en/beijing2005_phi.pdf

Sharma C (2004) The impact of fisheries development and globalization processes on women of fishing communities in the Asian region. Asia-Pacific Research Network J 8:1–12. SPC Women Fish Inf Bull 14:27–29

Srinath K (1987) Role of women in small scale fisheries-A case study. Mar Fish Inf Serv Tech Ext Ser 72:12–17

Suwanrangsi S (1998, November) Technological changes and their implications for women in fisheries. In: 2001. International Symposium on women in Asian Fisheries: Fifth Asian Fisheries Forum (Vol. 13).
Telefood (1998) Women Feed the World. Telefood of FAO of the United Nations
Weeratunge N, Snyder KA, Choo PS (2010) Gleaner, fisher, trader, and processor: underestimated gendered employment in fisheries and aquaculture. Fish Fish 11:405–420

Chapter 54
The Impact of Destructive Fishing Gear on the Fish Biodiversity in the Inland Waters of Iraq

Laith A. Jawad

Abstract Human behaviour is the chief component of all major hazards to biodiversity. In order to move from authenticating damages and recognizing reasons for failure to undertaking the underlying drivers and applying answers, we need to know that preservation is not only about animals and plants but similarly concerning people and their performance.

The objective of this chapter is to assess recognized features shaping fishers' conduct heading to unmaintainable supply use, by using the sample of damaging fishing gear such as electro-, poison and blast fishing in Iraq. Factors influencing the fishers to use a destructive fishing gear were discussed in addition to their acceptance and economic aptitude to modify less damaging gears. Therefore, the significance of attending all of these issues, toning to the diverse backgrounds, somewhat than concentrating on fast-moving routine measures, is highlighted to upsurge probabilities of administration achievement. Additional favourable strategies would be supply necessities to more sustainable fishing means, well-managed gear interchange plans, as well as consciousness rising on the benefits of more supportable fishing methods, their normal and cultural values, information on the actual income they make, as well as education and an interchange of traditional information on how to use them.

54.1 Introduction

Small-scale fisheries are frequently not well-achieved (Berkes et al. 2001) and are under substantial stress on both national and international levels (Kittinger et al. 2013). This can have particularly overwhelming concerns for family finance and food safety in areas where the dependence on fisheries capitals is high, poverty widespread and living substitutes inadequate (Béné et al. 2007). To meet the short-term food reassurance deliveries of inland fisheries societies with fishing means

---

L. A. Jawad (✉)
School of Environmental and Animal Sciences, Unitec Institute of Technology, Auckland, New Zealand

L. A. Jawad (ed.), *Tigris and Euphrates Rivers: Their Environment from Headwaters to Mouth*, Aquatic Ecology Series 11,
https://doi.org/10.1007/978-3-030-57570-0_54

which permit a responsible and sustainable use of freshwater incomes, bearings a great contest.

One stern problem in small-scale fisheries is the use of damaging fishing methods like electro, poison and blast fishing. Despite the availability of comprehensive guidelines and total ban of these negative fishing means, they are usually used in developing countries, which can cause environment dilapidation as well as overexploitation and removes of juvenile fish that deteriorate assets.

In circumstances where damaging methods are prohibited but the level of compliance is low, efforts to reinforce jurisdictions, law implementation and penalties are often seen as quick solutions. But in a situation of hardship as well as political and institutional shortcomings, the emphasis cannot exclusively lounge on rules and regulations (Jentoft 2004) as this would likely to have poor long-term achievement and could instead have devastating social costs for supply users. A much more suitable answer to the problem of non-maintainable and presently miss-managed fisheries is to create better institutional events that in turn lead to a more tolerable use of assets (De la Torre-Castro and Lindström 2010). To be able to achieve such behavioural modification, it is vital to comprehend the social, economic and institutional mechanics changing the current behaviour of non-sustainable resource removal like fishing with damaging gears, as well as to discover means to a behavioural alteration.

The present chapter gives a short review of a selective destructive and most effective fishing methods that have been used by the fishers in the inland fisher's communities in Iraq. The chapter is also focus on the effects of these methods on the aquatic biodiversity and the inland fisheries in Iraq and finally on the health of both those who are using them and eating the fish caught by these methods.

54.2 Poverty and the Use of Destructive Fishing Gear

The administration events such as gear limitations are obligatory outside of endangered areas to help withstand aquatic habitats (McClanahan et al. 2008).

Particular fishing methods have a higher propensity to harm the ecosystem, capture a high ratio of juvenile fish (Mangi et al. 2007) and aim for species that have no profitable values, but imperative for the ecosystem.

Efficiently lessening damaging fishing methods used will be determined by at least in part on a well appreciation of these fishers and the socio-economic motives behind their performs. Several investigations have recommended that poverty together with declining catch linked with environmental dilapidation might generate circumstances that push fishers to use damaging fishing methods (Pauly 1990; Tobey and Torell 2006). The notion of poverty is with several trajectories and can include facets of income or expenses, access to infrastructure, education, the variety of living sets and social capital (Narayan 1997).

54.3 Review of the Destructive Fishing Methods

There are several destructive fishing methods that are used by fishers around the world both in the marine and in the freshwater environments. From the perspectives of the consumers, the fishers, the biodiversity and the inland fisheries, the followings are the most effective methods that can bring a drastic change to both the environment and the socio-economic status of the fishers' communities.

54.3.1 Electrofishing

Electrofishing is a method of fishing where electric fields are used in water to capture fish. This fishing gear has been an appreciated sampling method for over half a century, but it contains several difficulties and poorly assumed mix of physics, physiology and behaviour. The electric field in the water needs to be adequately strong at suitable distances from the electrodes to incite the expected reactions by aimed fish. Numerous issues can disturb the performance of this method such as the size, shape and nature of the electric field, the delivery and design of electrical strength which is controlled by the peak electrical potential, kind of current and waveform produced between and around the electrodes; the location, size and shape of those electrodes; the conductivity of the water and hurdling and enclosed media; and the size and extents of the water body (Snyder 2003a).

The overall reaction of the fish to the electric field contains twitches, taxis, narcosis and tetany, are fundamental features of the same stages of epilepsy (automatism, petit mal and grand mal). Such influences are comparable to those saw in human and other animals subjected to electroconvulsive therapy (Sharber and Black, 1999).

54.3.1.1 Harmful Effects of Electrofishing on Fishes

Possible destructive impacts of electrofishing on individual fish contain cardiac or respiratory failure, injury, stress and fatigue. Death can be instant or late.

Small fish whose normal behavioural retorts are slowed or repressed might be more vulnerable to predation. Fish that endure notwithstanding electrofishing damage or other opposing influences may grieve short-term, long-term or lifetime deformities that disturb their biological life in general. Such stayers may unfavourably impacted fish groups structure, population size, quality of the fishery supply and management rules.

It is conceivable to track down the harmful impacts of electrofishing to one of two causes-excessive contacts to high-intensity parts of electric fields causing in tetany or features of electric fields that result in rapid and powerful, but not continued, contractions of the body musculature, occasionally referred to as myoclonic jerks or seizures. Damages due to such seizures are usually categorized as spinal injuries but may contain injury to tissues or organs not linked with the vertebral column (or notochord in cartilaginous fishes).

Types of Injuries Caused by Electric Shock

Injuries Additional to the Spinal-Associated Injuries

The most dangerous examples of this type of injuries are electrocution when fish are adequately exposed to very high voltage grades. Unlike humans and other mammals where fibrillation of the heart and death by cardiac seizure are familiar results of contact with strong electric currents, electrofishing deaths are normally rare.

Snyder (2003b) has discussed several cases of the non-spinal-related injuries and here a summary of this description is given in the following points:

1. Northrop (1967) suggested that "temporary" cardiac seizure might happen in electrically narcotized (perhaps tetanized) fish, whereas Kolz and Reynolds (1990) stated that such incidence is infrequently issued in fish death.
2. In tests by Schreck et al. (1976), improvement of normal heart activity needed much more time. Their fish also showed uneven cardiac activity directly after being shocked (probably tetanized).
3. The visceral organs of fish may also be upset by electric fields. Shparkovsky and Vataev (1985) stimulated the lateral areas of hindbrain and midbrain, and peristalsis of the stomach, and gut was subdued.
4. Bleeding from the gills was possibly first stated as an electrofishing injury by Hauck (1949) in his description of injuries to rainbow trout.
5. Walker et al. (1994) stated that haemorrhages in both paired and median fins of juvenile northern pike exposed to electric current is occurred.
6. Respiratory malfunction is perhaps the eventual reason for death in most electrically shocked fish.
7. Stress and fatigue due to electric shock are physiological reactions that upset physicochemical balance and other metabolic activities in fishes.
8. Schreck et al. (1976) reported speedy upsurges in blood concentrations of plasma corticoid, lactate or lactic acid, and thrombocytes in young rainbow trout.
9. Electrofishing also affects subsequent fish behaviour.

Spinal Related Injuries

Snyder (2003b) has given a detailed description of electrofishing injuries that originally described by Hauck (1949). It is of interest to let the readers of this chapter know about this description. Fish showing incomplete paralysis swam in an arc around the electrode (oscillotaxis), while those showing total paralysis (probably tetany, including cessation of respiratory movements) would float briefly on their sides then sink slowly to the bottom. Hauck (1949) designated the damages in captured fish as follows: "A number of fish haemorrhaged from the gills or vent, or both. Others displayed widened and haemorrhaged blood vessels in the skin near the vent. Several were observed with the intestine bulging from the vent. Physical contact with the electrode caused the appearance of dark vertical bars on that area of the fish which touched the electrode". Hauck (1949) noted that "Paralysis of swimming muscles continued in some fish for several days. This damage, or partial loss, of movement would designate a damage to the nervous system.

54.3.1.2 Effects of Electrofishing on Invertebrate Fauna

Electrofishing can affect the invertebrate fauna in the freshwater system where the electrofishing is used. Such an effect can be seen on different levels of the invertebrate's community. Woolmer et al. (2011) have studied different invertebrate communities and showed short-term effects of the electric field such as stupefaction and disorientation, which suggest an effect of the electrical field on the nervous system of these animals. These effects were observed to be temporary and short-term and can take more than 5 min to resume normal behaviour. Some species took a maximum of 10 min before a return to normal behaviour, i.e. extension of foot and reburial, took place.

It may be relevant to focus that there was no physical damage to the epifauna meeting the gear, but on the other hand, a risk factor of increasing predation on these animals is increased due to the electric shock (Woolmer et al. 2011). The long-term effects of the electrofishing gear on epifaunal community can be seen in changes in the relative abundance of component species.

Shentyakova et al. (1970) studied the effects of electric shock on benthic insects and concluded that there are direct hazardous, but instead, it modifies their normal diel drift patterns. Many specialists consider drift is of the major importance for regulating population densities and reducing competition of stream invertebrates (Walton et al. 1977; Corkum 1978). The dislocation rates of species during electrofishing are directly related to the shape and size of the insect as the shape of some species prevents large displacement, while others cannot help drifting.

The long-term reductions in benthic populations will be similar among most species except that populations of large-bodied species and those with high growth rates (Mesick 1979) will be reduced to relatively lower levels. Reductions in populations will persist only as long as electrofishing is so frequent that rates of insect displacement are greater than rates of insect recolonization; endangered invertebrate species, especially those of low productivity, could be eliminated from heavily shocked areas.

54.3.1.3 Use of Electrofishing in Iraq

Recently and in the last two decades, the use of electrofishing in the inland waters of Iraq has gone off the leash especially with the absence of any governmental enforcements and policing. Salim et al. (2009) have reported the wide range of use of the electrofishing as a fishing gear in all the inland water bodies in Iraq. In some cases, the use of this gear was heavy and prolonged for long time. In addition, fishers using electrofishing in Iraq are using AC current instead of DC current, which in turn has more destructive effects on the biodiversity, environment and the fishers themselves.

The level of unawareness and absence of knowledge of using the electrofishing gear have added on the top of the problems that the electrofishing gear might cause. The Iraqi fishers are aiming in using electric shock gear is to collect as much as possible of large-sized fish. They do not care about the small-sized fish or the fish species that not have local commercial values even they were large sized. In addition, they have no knowledge at all about aquatic animals other than fishes that might get trapped in the electric field of their gear and damaged. On the top of all these unawareness, they are ignorant of the health and safety issues of using such deadly fishing gear.

Experimentally, the effect of the electric shock on fishes was studied on the physiology of some commercial fish species (Al-Dubaikel et al. 1999) and also on the chemical composition of some freshwater fish species in southern Iraq (Al-Mhnawi 2006). Al-Mukhtar et al. (2006) have described several cases of vertebral anomalies and haemorrhage in several places of the fish body of several fish species collected from the marshes in the vicinity of Basrah, Iraq. This study represents the only study on electrofishing in Iraq and showed the drastic effects of this gear on the fish body.

Reports were collected about the bycatch resulted from using electrofishing in the inland waters of Iraq. This bycatch contains small-sized commercial fish individuals, large non-commercial fish individuals and large number of invertebrates of different species. Although the number of the fish or invertebrates that have died and wasted is not large, but it represents the general attitude that followed by the fishers using the electrofishing gear in the inland waters of Iraq.

The effect of the misuse of electrofishing on the fishers in Iraq cannot be overlooked. The fishers are using AC current which is dangerous to human. In several reports, incidents of death have happened due to mistake in using the electrofishing gear. Nearly all the electrofishing gear use in Iraq are home-made, and there are several malfunctions due to poor-quality manufacturing. In addition, the age of the fishers can also be a problem. A large number of young fishers' even boys of under 10 years of age use the home-made electrofishing without adult supervision and end up having death incidents. Cases of death resulted from fight over the fishing areas are also reported. In such fight, electrofishing gears are used as weapon.

54.3.1.4 Recommendations

Snyder (2003b) has suggested several recommendations for users of electrofishing.

These guidelines are given below with the amendment to fit the status of users in Iraq.

1. Used inaccurately or incorrectly, any collection gear and method can be injurious to fish, other aquatic organisms or their habitat.
2. Notwithstanding of target species, if electrofishing is the least harmful of applied and actual methods for gaining needed data or specimens, it must always be steered in such a way and with such currents as to minimize potential for strain and damage as much as possible while upholding sufficient efficiency.
3. Use the minimum harmful current obtainable for actual capture of target fish.
4. Where practical, use DC.
5. Spherical, circular or dropper array anodes are normally suggested somewhat than cables (especially single or paired, small-diameter cables).
6. Adjust the electrofishing gear in such a way as to net and remove fish from the electric field as soon as possible.
7. Select and position anodes such that fish are brought as near to the surface and as close to the netters as possible before narcosis.
8. Position netters, and lighting at night, such that fish are more simply perceived and seized.
9. Evade electrofishing when and where waters are rough, too fast for effective netting or disproportionately turbid.
10. Improve fish handling and holding amenities for fast recovery and least possible stress.
11. Evade overcrowding captured specimens.
12. Handle fish as gently and as little as possible.
13. Team leaders should be appropriately trained and certified in electrofishing theory and practice. An appropriate course and certification program should be introduced in Iraq.

54.3.2 Poison Fishing

Poison used in fishing can be obtained either from pure chemicals such the different compounds of the cyanide or the several types of the insecticides that usually are available at the disposal of fishermen or natural poison attained from certain plants that known for their poisonous effects on the aquatic fauna.

In the followings, a short review of both types of poisons will be given together with their effects on the biodiversity, the fish collected and the human health.

54.3.2.1 Poison Fishing with Chemicals

Poison Fishing with Cyanide

Probably, the use of the different compounds of cyanide is the famous poison used for fishing around the world especially in the coral reef areas, where the ornamental fish trade is flourishing. The date of 1962 marks the start of using cyanide in fishing operations (Robinson 1985). This method is simple as the fishermen first squash sodium cyanide (NaCN) pellets into squirt bottles filled with water. Then, they dive down to coral reef areas and squirt the concentrated cyanide solution into the cracks where fish conceal. The cyanide shocks the fish briefly, making them simpler to catch (Rubec 1988). Live fish are transported to the ship and put into fresh seawater. This is what happened at the coral reef areas, but it is easier and more effective in the freshwater bodies, where the area is confined and with less motion. Cyanide fishing is deliberated a quick method to shock and collect fish in all environment.

Even though the harmful influences of cyanide are well recognized, its use continues, although unlawfully, in many countries (Barber and Pratt 1997). Cyanide is very destructive since it can kill both target fish and non-target organisms (Johannes and Riepen 1995).

It appears that the poisoning method of fishing frequently applies to get big fishes which will raise higher price. The fishes caught by this way will be sold in the close areas, where consumers buy them and not alert because it is very difficult to differentiate the variations between the fish caught by usual and poisoned caught.

Rao and Girijavallabhan (1984) have suggested that the pathogenic effects found in the poisoning incidence resulted from eating fish cached by cyanide poison fishing are similar to the suicide cases using potassium cyanide. The cyanide fishing technique will generate health difficulties even to those fishermen in the long run by their constant dealing with these chemicals. They are facing to the danger of getting into their system tiny amounts of cyanide which may abolish their health. As such, this type of fishing should be totally forbidden directly by an appropriate regulation by the authorities apprehensive.

Effects of Insecticides on the Aquatic Life

Probably DDT, which is discovered in 1939, is the first well-known insecticide that has been used in wide range. This compound has noted to have drastic effects on the aquatic fauna through the irradiation of the disease vectors such as insect larvae and molluscs. Later, several other organic insecticides were developed which include chlordane and others with fatal effect on the aquatic fauna including fishes.

From their introduction, the use of insecticides has increased tremendously since the time when they were effectively employed in policies to increase crop productivity (Wijngaarden et al. 2005).

The previously published works review of freshwater (model) ecosystem findings with neurotoxic insecticides were achieved to evaluate ecological verge levels, to match these levels with the first level approach within European Union (EU) administration measures and to appraise the ecological effects of surpassing these thresholds (Wijngaarden et al. 2005). The most subtle taxa were species of crustaceans, insects and fish. Conferring to the toxic units, start values were matching to mixtures with a similar mode of act (Leeuwangh 1994).

The best subtle endpoints for direct influences of the insecticides investigated were structural ecosystem features and usually related population densities of the aquatic organisms such as crustaceans and insects. These straight influences can normally be well forecast on the basis of laboratory examinations with similar species (Schroer et al. 2004).

54.3.2.2 Insecticides Fishing in Iraq

Due to the availability and the variety of insecticides, these chemicals became the target of fishers in Iraq for use as a fishing gear. This method of catching fishes does not involve equipment and less efforts in obtaining fishes, but on the other hand, it encounters unawareness about the destructive effects of catching fishes by this method on the environment, biodiversity and the human health.

Chlordane is the well-known insecticide that usually used by fishers in the inland waters of Iraq. This chemical is normally sold for those people building houses so they can use it in the foundations of the house to prevent termites and other harmful insects from entering the house. The chlordane is usually dissolved in water from the river and sprinkled in small and closed areas to accomplish high fish catch. It has been known to the public that the cached fishes with red eyes, due to internal haemorrhage in the eye, mean these were cached using chlordane. Even the eyes were not red, there is a strong chemical test in the muscle that cannot be mistaken. As an author of this chapter, I have passed through the experience of eating BBQ fish collected from the marsh areas near Basrah City, Iraq. The body of the fishes looks normal and healthy. Once the fish has been put on BBQ, a distinguished chemical smell can be traced in the muscles with unbearable chemical test. I stop eating the fish, but other members of the family have continue their lunch, but later on, they experienced stomach pain and strong diarrhoea has started. Later they were taken to the hospital for stomach flush to remove the effect of the insecticides. Similar incidents of chlordane poisoning due to eating fish cached with this chemical have been reported from the rural areas in the southern marshes of Iraq, where the use of chlordane fishing is widely used. In some of these incidences, the patients have died due to the consumption of large portion of fish muscles contaminated with chlordane.

The effects of insecticides on the aquatic biodiversity are evident as you visit the inner villages of the rural areas of the southern marshes and you can see dead unwanted small fishes and other aquatic organism floating on the surface of the water. In several occasions, the author has collected and examined these fishes and organisms and found that they suffer from a severe haemorrhage in different part of their body. In addition, a strong chlordane smell ascended from their bodies.

The insecticides as a fishing gear is not a selective method and can kill all the aquatic life within certain range without discrimination. Early life stages of different aquatic animals and microorganisms are the main distressed groups. In the rural areas, the use of insecticides and other destructive fishing gear is a daily business that its continuation will definitely lead to drastic changes in the environment due to the death and disappearance of certain ecological animal groups.

54.3.2.3 Plant Poison Fishing

Using plant poisons as fishing gear has been a part of human civilization and still in use in many parts of the world. The natives usually discover certain plant species and experiment its extract as poison to kill fish and then cultivate it in their village as a source for their fishing activities. The good side of this poison has no side effects after eating the fish caught by this method (Neuwinger 2004). This type of fishing has been forbidden worldwide as it destructs the aquatic environment, but it is still in use in the rural and the remote areas in Africa, East Asia and the Middle East. In order to get an efficient catch result, this method of fishing is usually used in the dry season.

Neuwinger (2004) described the basic method for the preparation of the poison, which is simple as follows: it is a simple method and includes scraping the plant into small fragments or pounded and thrown into the water. The poison dissolves quickly. Others use a concentrated plant suspension, or the plant material is soaked overnight. If latex (milk-sap) is used (e.g. of *Euphorbia* species), it is first obtained into small vessels which are then released into small pools. Others cut open the succulent *Euphorbia* branches, weight them with stones and let them down into the water.

The general action of the poison is to paralyse the fish first and follow by the death.

The time required to distress the fish varies significantly conferring to the species of plant and fish and the kind and concentration of poison. Liable to numerous issues, it may take 10 min to several hours for the fish to be distressed and rise, still breathing but seemingly paralysed, to the surface, where it can then readily be taken by hand. For best results, one must look for slow-flowing waters or rivers with a low water level. The method of poison fishing in blocked water pools is mainly common. The plant amount used to be influenced by the size of the pool or dammed section of a stream; it is always a large quantity (Neuwinger (2004)).

The effect on the person eating the fish caught by plant poison depends on the species of plant used. Neuwinger (2004) found that fish poisoned with the highly effective roots of the *Gnidia kraussiana* (Family: Thymelaeaceae) cause diarrhoea if eaten.

Plant Species Used in Poison Fishing

There is a large number of plant species that can be used in poison fishing. There is a countless diversity of plants used in poison fishing. Neuwinger (2004) has numerated these species as follows: the quantified 325 fishing poisons belong to 71 plant families with 183 genera. Five families give with 10 and more genera, 12 families with 3–9 genera, 13 families with 2 genera and 41 families each with 1 genus. Caesalpiniaceae, Mimosaceae and Papilionaceae are considered the most affected plants in poison fishing. The members of the genus *Tephrosia*, Family: Papilionaceae is one of the utmost interesting genera among these plants, not only on account of the large number of species of a toxic nature which it comprises, but also on the issue of the extensive range of the genus through both hemispheres. Exceptional also is the great quantity of Euphorbiaceae that are fishing poisons of high strength, especially *Euphorbia tirucalli*, which is an exceptional fishing poison—easy to obtain and highly effective.

The Structure and the Mechanism of the Plant Poison Used for Fishing Purposes

In the present section, the account is given by Neuwinger (2004) on the structure and the mechanism of the plant poison is adopted and presented with slight variation so to put in a generic form. The poisonous part of the plants that are used for fishing purposes having several chemical structures, e.g. triterpene saponins, diterpene esters, rotenoids, sesquiterpenes, lignans, proanthocyanidin polymeres, polyacetylenic compounds.

The triterpene saponins, diterpene esters and rotenoids are the utmost significant and most communal components in ichthyotoxic plants. Saponins found broadly in plants and are the best-known toxic ingredients in fishing poisons. Esters of alcohols based on the tigliane, daphnane and ingenane group of diterpenes present mainly in Euphorbiaceae and Thymelaeaceae. Rotenoids present chiefly in Papilionaceae. Cyanogenic glycosides are particularly located in Passifloraceae, for example *Adenia* species (Nishizawa et al. 1983).

Most poisons affect the respiratory organs of the fish, creating a stupefying and paralysing effect, later death. After a short time, the fish starts to rise to the surface of water breathless for oxygen. The fishermen gather the shocked fish rapidly as they became to the surface. There are some influences of plant poison on human. Nwude (1982) reported an anaesthetic influence on limbs and roughness of skin of people who wade into streams to collect fish poisoned with *Tephrosia vogelii*; its toxic agents are rotenoids. Li et al. (2003) have recommended the paralysis of these respiratory chains of the cell consequences of asphyxia of the tissue and paralysis of the organs. Some current indication displayed that rotenone could encourage cell death in a diversity of cells, and the mechanism is still indefinable. Additionally, Gaudin and Vacherat (1938) suggested that twenty grams of the leaves of *Tephrosia vogelii* paralysed the fish within 1 h and killed them in 3 h later.

Plant Poison Fishing in Iraq

In Iraq, the use of poison obtained from certain plants is a practice that has been performed since ancient time. Locally, the plant poison used in fishing is known as "Zahar= Iraqi dialect for poison" and it is usually mixed with flower and water in a form of dough and it is widely used in the southern marshes of Iraq.

There are several plant species that have been recognized having potential poison that can be used to catch fishes. The following list of plant species was found to grow in different parts of Africa (Neuwinger 2004) and they also appear to be grown in different regions in Iraq too: *Jacaranda mimosifolia* (Family: Bignoniaceae), *Carica papaya* (family: Caricaceae), *Erigeron eanandensis* (Family: Compositae), *Erigeron bonariensis* (Family: Cucurbitaceae), *Luffa acutangula*, *Momordica charantia* (Family: Euphorbiaceae), *Jatropha curcas* (Family: Gramineae), *Dactyloctenium aegyptium* (Family: Malvaceae), *Hibiscus cannabinus* (Family: Sapindaceae), *Dodonaea viscosa* (Family: Tiliaceae) and *Grewia subinaequalis* (Family: Malvaceae).

In the southern part of Iraq, natives used the plant *Anagallis arvensis* (Fig. 54.1) to extract the poison for fishing. This plant is usually grown in the vicinity Basrah Province, south of Iraq (Taha Al-Edani, personal communication).

Fig. 54.1 *Anagallis arvensis*. (**a**) Scarlet pimpernel (curtsey of Joaquim Alves Gaspar, Lisboa, Portugal); (**b**) Azure-blue *Anagallis arvensis*, a blue form (Curtsey of Zachi Evenor, Israel; (**c**) *Anagallis arvensis* f. *azurea* together with a colour variant closer to f. *carnea* (Curtsey of Wikimedia); (**d**) *A. arvensis* f. *azurea*. The glandular hairs on the petal margins (Curtsey of Wikimedia)

The plant *Anagallis arvensis* is also known by its several common names as scarlet pimpernel, blue-scarlet pimpernel (Korea National Arboretum 2015), red pimpernel, red chickweed, poor man's barometer, poor man's weather-glass, shepherd's weather glass or shepherd's clock (Fig. 54.1) (Fact Sheet 2018).

In numerous cultures and countries, the plant material has been spread outwardly to slow-healing ulcers and wounds. It also has been used as a therapy for pruritus, rheumatism, haemorrhoids, rabies, leprosy and snake-bite. *Anagallis* has been used in cure of non-specified types of phthisis, and of kidney-related conditions such as dropsy and chronic nephritis. *Anagallis arvensis* is an insecticidal, or at least is revolting to some insects, possibly by benefit of its strong vital oil which has a distinctive smell. An absorbing experimental quantities of the liquid in humans produced 24 h of powerful nausea, headache and bodily pain. Some people also showed dermatitis from interaction with the leaf (Watt and Breyer-Brandwijk 1962).

The additional plant that frequently utilized to excerpt poison from for fishing dedications in parts of Iraq particularly the northern region is the common mullein, also known as woolly Mullein (*Verbascum thapsus* L., Scrophulariaceae) (Fig. 54.2). It is a biennial plant, low growing rosette of bluish grey-green colour in the first year, while mature second-year flowering plants grow to 600–1200 mm in height, containing the visible flowering stalk. It is natural to Europe and Asia (Semenza et al. 1978).

Through the human life, the common mullein has been used as a medicinal herb. The leaves and flowers are noted to have linctus and demulcent properties which are used to cure respiratory difficulties such as bronchitis, dry coughs, whooping cough, tuberculosis, asthma and hoarseness (Tyler 1994). The mucilaginous constituents are mainly accountable for the calming actions on mucous membranes, and saponins may be liable for the linctus actions of mullein (Tyler 1993).

Experimentally, Husein and Rasheed (2011) have shown that the plant *Verbascum thapsus* can cause histopathological effects on the freshwater fish species *Cyprinus carpio*. A clear destruction was clear in the tissue of the brain. A congestion of blood vessels, infiltration of glial cells. They showed also effect of this plant on the heart tissue, where congestion of blood vessel and haemorrhage were present. The liver tissues are as well have shown to be affected by this plant, where swelling of hepatocyte oedema, inflammatory infiltration and haemorrhage were observed.

Shoot part of this plant is used for fishing since of their saponin substances which are toxic to fish (Baytop 1999). Haematological features of fishes are very subtle to strain like oil crude, toxicant plant. The reduction in the values of these features could be accredited to haemolysis resulting in the mechanism for diluting the concentration of the pollutant in the circulatory system (Smith et al. 1979). *Erythropaenia* may also be responsible for by swelling of the erythrocytes (Annune and Ahuma 1998), damages to haematopoietic tissues in the kidneys and accumulation of cells at the gills, thus initiating a reduction in the number of circulating cells of strained fish (Singh and Singh 1982).

In the use of the poison of both scarlet pimpernel and mullein and any other pants for fishing, the aim of the fishers is to collect as much as possible of fish and preferably large fish individuals. The conservation of the biodiversity and the

Fig. 54.2 *Verbascum thapsus*. (**a**) habitat (Curtesy of Forest Starr & Kim Starr, Hawaii, USA); (**b**) A close-up of the flowers (Curtesy of Forest Starr & Kim Starr, Hawaii, USA); (**c**) *Verbascum thapsus* grows best where there is little competition (Curtesy of Forest Starr & Kim Starr, Hawaii, USA); (**d**) The fruit of great mullein contains large numbers of minute seed (Curtsy of Krzysztof Ziarnek, botanist from Poland)

environment awareness are very low in the fishermen who are using the poison fishing. Particularly, in the southern Iraq, the usage of poison plant fishing is a daily practice. This means that the level of toxicant substances from the used plant will be increased to high level enough to kill or to erode vast number of non-targeted aquatic animal species that found in the fishing area. The concerns are mainly on the early life stages of the aquatic animals that if they were affected will have serious negative results on the stock of fish or any other commercial aquatic animals that the natives are using in their daily life as a food.

54.3.2.4 Blast or Bomb Fishing

No matter the disruption is severe or long-lasting, it has important inferences for the troubled ecosystem's time frame for recovery, with lower probabilities for recovery after chronic, long-term disturbances (Connell et al. 1997).

The Blast fishing is a man-made disturbance that actually changes any aquatic environment that is used in. The explosion of domestic bombs not only kills fish but destruct the nearby environment such as the coral reefs (Alcala and Gomez 1987). The extensive use and shocking influences of blast fishing have been well recognized (McManus et al.1997), and some researchers deliberate damaging fishing events to be the largest prompt threat to coral reef ecosystems in a few countries (Fox and Erdmann 2000).

Explosives are obtained from quarries and companies engaged in mining, demolition and road construction, but are also progressively made from ammonium fertilizers and diesel. More classy bombs, supposedly obtained from the army, are also used.

Dynamite fishing has other influences. There are results showing that people have died or lost limbs as a result of blasting events. A zero tolerance for blasting needs to be created at both society and official level so that it converts an intolerable action.

Blast or Bomb Fishing in Iraq

Fishing using bombs of all sorts is the usual method of catching fish in Iraq and accounts designate that it has been in use here since 2003. Notwithstanding being made unlawful by the Iraqi Government years ago, it is presently deliberated to be more extensively used now than at any other point in history, to the degree that on some parts of Iraq such as the southern part is becoming "fishing as usual". The history of usage of dynamite in fishing in Iraq was verified by various elders in the rural villages of the south of Iraq and in the remote parts of the southern marshes of Iraq. According to those elderlies, fishermen from marsh areas were the first to use dynamite fishing in their waters using home-made dynamites. Later, the practice spreads further to the other marshes and regions in Iraq. The practice was encouraged by the collaboration of individuals from the nearby cities who came with the idea of using the dynamite for collecting fish.

Explosive gears no longer depend on the classical sticks of dynamite; home-made bombs are easily made using plastic bottles, granular fertilizer, petrol, detonator caps and small quantities of explosive gel. Hand grenades and army explosive have found their way to the fishers in Iraq mainly after the political unrest period in the last two decades. Fishers are unaware of the detrimental effects of blast fishing on the environment. Local fishers use dynamite due to the pressures of poverty and the quick returns afforded by the use of explosives—they realize that they are often operating as pawns in a larger game.

There has always been frequent and widespread fisher migration within the southern freshwater system. As boat engines have become more accessible, people travel further and can reach the market more quickly.

In Iraq, there is a need for projects that can halt blast fishing. Such projects were implemented in some developing countries. In those countries, there have been numerous advantages to control the use of dynamite in their aquatic habitats, government and donor subsidized such schemes. Some of these plans confirmed some achievement and for a few years. Another income-producing events and delivery of boats, revolving funds, etc. have not shown to be operative. With such a large and complicated project, funds from other countries are required in order to accomplish the prerequisite steps that should go ahead of the implementation of the practical steps of the project. These preliminary steps include exercise lectures, participating video and film shows and technical help, the project needs to mobilize several fishers' communities to start the projects' idea. In order to perform the application of the project smoothly, an establishment of a local umbrella organization should be achieved. The involvement of the government is important in such a project to provide police force or may be the army to control the application of the ban of the blast fishing and capture those who were taking part in the event. This may look contentious but it might effectively decrease blast fishing to the minimum or to zero level.

Recommendations

1. Outlining the supply chain of bomb-making materials.
2. Consciousness educating and politicization of decision-makers at the national level.
3. Trial of key instigators in the dynamite fishing networks.

Moreover, longer term but similarly vital resolutions can be useful:

1. Evaluation of legislation to remove uncertainties and upsurge fines.
2. Official, implementation and inspection officer training.
3. Upgrade of good practice at the village level.

References

Alcala AC, Gomez ED (1987) Dynamiting coral reefs for fish: a resource-destructive fishing method. In: Salvat B (ed.), Human impacts on coral reefs: facts and recommendations, pp 52–60

Al-Dubaikel AY, Ahmed SM, Jasim AA (1999) The physiological influences of electric current on the ionic balance of common carp (*Cyprinus carpio*) and mullet (Liza abu). Marina Mesopotamica 14:339–349

Al-Mhnawi BHH (2006) Impact of electrical current (AC) on chemical composition of four fish species and required lethal voltages. J Basrah Res Sci 2(32):38–41

Al-Mukhtar MA, Al-Noor SS, Fadag MS, Ali RA, Faris RA (2006) The effect of electrofishing on some fishes in Basrah marshes, Iraq. Marina Mesopotamica 21:95–111

Annune PA, Ahuma FTA (1998) Changes in mudfish exposed concentrations of copper and lead. Acta drochem Hydrobiol 1(3):33–36

Barber CV, Pratt VR (1997) Sullied seas: strategies for combating cyanide fishing in Southeast Asia and beyond. World Resources Institute and International Marinelife Alliance-Philippines, Washington DC

Baytop T (1999) Therapy with medicinal plants in Turkey (past and present). (in Turkish), 2nd edn. Nobel Tıp Kitabevleri, İstanbul, pp 334–335

Béné C, Macfadyen G, Allison EH (2007). Increasing the contribution of small-scale fisheries to poverty alleviation and food security (no. 481). Food & Agriculture org.

Berkes F, Mahon R, McConney P, Pollnac R, Pomeroy R (2001) Managing small-scale fisheries-alternative directions and methods. International Development Research Centre, Ottawa, Canadá, p 350

Connell JH, Hughes TP, Wallace CC (1997) A 30-year study of coral abundance, recruitment, and disturbance at several scales in space and time. Ecol Monogr 67:461–488

Corkum LD (1978) The influence of density and behavioural type on the active entry of two mayfly species (Ephemeroptera) into the water column. Can J Zool 56(5):1201–1206

De la Torre-Castro M, Lindström L (2010) Fishing institutions: addressing regulative, normative and cultural-cognitive elements to enhance fisheries management. Mar Policy 34(1):77–84

Fact Sheet (2018) *Anagallis arvensis*. http://www.iewf.org/weedid/Anagallis_arvensis.htm. Retrieved 29 June 2018

Fox HE, Erdmann MV (2000) Fish yields from blast fishing in Indonesia. Coral Reefs 19(2):114–114

Gaudin O, Vacherat R (1938) Recherches de la rotenone et du pouvoir ichthyotoxique chez quelques plantes du Soudan franc¸ais. Bull Soc Pharmacol 40:385–394

Hauck FR (1949) Some harmful effects of the electric shocker on large rainbow trout. Trans Am Fish Soc 77:61–64

Husein RH, Rasheed RO (2011) Ecophysiological effects of Ichthyotoxic (Verbascum sp.) on some local fishes in Sulaimani City-Iraq. J Univ Babylon 18(1–2):418–426

Jentoft S (2004) Institutions in fisheries: what they are, what they do, and how they change. Mar Policy 28(2):137–149

Johannes RE, Riepen M (1995) Environmental economic and social implications of the live reef fish trade in Asia and the Western Pacific. The Nature Conservancy, Jakarta, p 81

Kittinger JN, Finkbeiner EM, Ban NC, Broad K, Carr MH, Cinner JE, Gelcich S, Cornwell ML, Koehn JZ, Basurto X, Fujita R (2013) Emerging frontiers in social-ecological systems research for sustainability of small-scale fisheries. Curr Opin Environ Sustain 5(3–4):352–357

Kolz AL, Reynolds JB (1990) Principles and techniques of electrofishing [outline manual for course]. Office of Technical Fisheries Training, U.S. Fish and Wildlife Service fisheries academy—Leetown, Kearneysville, West Virginia

Korea National Arboretum (2015) English names for Korean native plants. Korea National Arboretum, Pocheon, 778pp

Leeuwangh P (1994) Comparison of chlorpyrifos fate and effects in outdoor aquatic micro- and mesocosms of various scale and construction. In: Hill IR, Heimbach F, Leeuwangh P, Matthiessen P (eds) Freshwater field tests for Hazard assessment of chemicals. Lewis Publishers, Boca Raton, FL, pp 217–248

Li N, Ragheb K, Lawler G, Sturgis J, Rajwa B, Melendez JA, Robinson JP (2003) Mitochondrial complex I inhibitor rotenone induces apoptosis through enhancing mitochondrial reactive oxygen species production. J Biol Chem 278:8516–8525

Mangi SC, Roberts CM, Rodwell LD (2007) Financial comparisons of fishing gear used in Kenya's coral reef lagoons. Ambio 36:671–676

McClanahan TR, Cinner J, Maina J, Graham NAJ, Daw TM, Stead SM, Wamukota A, Brown K, Ateweberhan M, Venus V, Polunin NVC (2008) Conservation action in a changing climate. Conserv Lett 1:53–59

McManus JW, Reyes RB Jr, Nanola CL Jr (1997) Effects of some destructive fishing methods on coral cover and potential rates of recovery. Environ Manag 21:69–78

Mesick CF (1979) The effects of electricity on some aquatic invertebrates. M.Sc. Thesis, the University of Arizona, p 50

Narayan D (1997) Voices of the Poor: Poverty and Social Capital in Tanzania. Environmentally and socially sustainable development studies and monograph series no 20. World Bank, Washington, DC

Neuwinger HD (2004) Plants used for poison fishing in tropical Africa. Toxicon 44(4):417–430

Nishizawa M, Inoue A, Sastrapradja S, Hayashi Y (1983) (þ)-8-Hyroxycalmenene: a fish-poison principle of Dysoxylum acutangulum and D. alliaceum. Phytochemistry 22:2083–2085
Northrop RB (1967) Electrofishing. IEEE Trans Biomed Eng 3:191–200
Nwude N (1982) Plants poisonous to man in Nigeria. Vet Hum Toxicol 24:101–106
Pauly D (1990) On Malthusian overfishing. Naga ICLARM Q 13:3–4
Rao DS, Girijavallabhan KG (1984) An undesirable method of fishing using poisonous and toxic chemicals. Mar Fish Inf Serv Tech Ext Ser 60:13–14
Robinson S (1985) Collecting tropical marines—an interview with Eral Kennedy: the founding father of the Philippine fish trade speaks out, part 2. Freshw Mar Aquar 8(3):27–28. 30–31
Rubec PJ (1988) The need for conservation and management of Philippine coral reefs. Environ Biol Fish 23(1/2):141–154
Salim MA, Abd IM, Abdulhassan NA, Minjal MS (2009) Key biodiversity survey of southern Iraq. Nature
Schreck CB, Whaley RA, Bass ML, Maughan OE, Solazzi M (1976) Physiological responses of rainbow trout *(Salmo gairdneri)* to electroshock. J Fish Res Board Can 33:76–84
Schroer AFW, Belgers D, Brock TCM, Matser A, Maund SJ, Van den Brink PJ (2004) Comparison of laboratory single species and field population-level effects of the pyrethroid insecticide lambda-cyhalothrin on freshwater invertebrates. Arch Environ Contaminat Toxicol 46:324–335
Semenza RJ, Young JA, Evans RA (1978) Influence of light and temperature on the germination and seedbed ecology of common mullein (*Verbascum thapsus*). Weed Sci 26(6):577–581
Sharber NG, Black JS (1999) Epilepsy as a unifying principle in electrofishing theory: a proposal. Trans Am Fish Soc 128:666–671
Shentyakova LF, Shentyakov VA, Stepanov V'S, Smirnova NF, Kuz'mina VV, Baronkin OF, Bayeva GD, Zaytsev KN (1970) The effect of an alternating electric current on fishes and aquatic invertebrates. J Ichthy 10(3):363–372
Shparkovsky IA, Vataev SI (1985) The effect of electrical stimulation of different brain structures on the motor fiinction of the cod digestive tract [Russian, English summary]. Fiziologicheskii Zhumal (Kiev) 71(10):1265–1270
Singh A, Singh B (1982) Effects of copper and zinc sulphate on the blood parameters of *Mystus vittatus*. Matsya 8:1–6
Smith GL, Hattingh J, Burger AP (1979) Haematological assessment of anaesthesia, MS 222 in natural and neutral forming three freshwater fish species: interspecific differences. J Fish Bio 15:633–643
Snyder DE (2003a) Invited overview: conclusions from a review of electrofishing and its harmful effects on fish. Rev Fish Biol Fish 13:445–453
Snyder DE (2003b) Electrofishing and its harmful effects on fish, information and technology report USGS/BRD/ITR-2003-0002: U.S. government printing office, Denver, CO, 149 p
Tobey J, Torell E (2006) Coastal poverty and MPA management in mainland Tanzania and Zanzibar. Ocean Coast Manag 49:834–854
Tyler VE (1993) The honest herbal. Pharmaceutical Products Press, New York, pp 219–220
Tyler VE (1994) Herbs of choice: the therapeutic use of phytomedicinals. Pharmaceutical Products Press, New York, p 92
Walker MK, Yanke EA, Gingerich WH (1994) Use of electronarcosis to immobilize juvenile and adult northern pike. Prog Fish Cult 56:237–243
Walton OE Jr, Reice SR, Andrews RW (1977) The effects of density, sediment particle size and velocity on drift of *Acroneuria abnormis* (Plecoptera). Oikos 28:291–298
Watt, JM; Breyer-Brandwijk, MG (1962) The Medicinal and Poisonous Plants of Southern and Eastern Africa 2nd ed Pub. E & S Livingstone
Wijngaarden RPV, Brock TC, Van Den Brink PJ (2005) Threshold levels for effects of insecticides in freshwater ecosystems: a review. Ecotoxicology 14(3):355
Woolmer A, Maxwell E, Lart W (2011) SIPF C0083-effects of electrofishing for Ensis spp. on benthic macrofauna, epifauna and fish species. Seafish Report SR652

Chapter 55
Evaluating Variations in Fisheries by Means of Fishers' Information: Suggested Methodology to Improve Small-Scale Fisheries in Rivers in Iraq

Laith A. Jawad

Abstract To apply a quantitative fishery study, a broad range and amount of data need to be prepared and available from a wide spectrum of fields such as biological and socioeconomic. Nowadays, there are several attempts to develop collecting such data among these is the fishers' interview method. The information available with the fishers can be used to supplement other data-recording systems or used independently to document the changes that have occurred in the resource base over a lifetime of fishing. The results can be utilised to guide the assessment and management of assets to conserve habitats and incomes.

Fishers' interview appeared also important in the incorporation of scientific and indigenous information, particularly the first-hand skills and insights of means users such as fishers. The present chapter aims to present a cheap, but efficient methodology to assist the fisheries management in the lower reaches of Mesopotamia. This methodology is based on fishers' interview. The chapter also sheds light on how profitable riverine fishermen operating in the southern Iraq recall great climatic measures (flood and drought) and how they confront these effects on fish obtainability.

55.1 Introduction

The contact and link between ecological and social subsystems are a significant mediator of the consequence of environmental alteration on natural properties. In fisheries scenery, the fishermen ecological information indicates a substantial facts base and is frequently utilised to evaluate fisheries source levels (Berkes et al. 2000), plan fisheries maintenance (Shackeroff and Campbell 2007; Murray et al. 2011) and

L. A. Jawad (✉)
School of Environmental and Animal Sciences, Unitec Institute of Technology, Auckland, New Zealand

L. A. Jawad (ed.), *Tigris and Euphrates Rivers: Their Environment from Headwaters to Mouth*, Aquatic Ecology Series 11,
https://doi.org/10.1007/978-3-030-57570-0_55

to backing elementary research on indigenous fish group (Santos and Nóbrega Alves 2016). Such data are increasingly being inserted to the inquiries of climate difference through ethno-climatological methods (Peterson and Broad 2009), which can be involved to control the conceivable worries of climate change on burdened societies across different seasonal and geographical levels. Data from such modules can be utilised for assistance administration (Mackinson and Nøttestad 1998) and to control upcoming research (Pinto et al. 2013). Additionally, the combination of scientific and traditional information enables and backings social collaboration during the growth and use of supply management processes (Wilson et al. 2006).

Artisanal fishing communities in the lower reaches of Mesopotamia have historically been exposed to the influences of climatic irregularities, such as draught, flood, and dust storms, and they have established policies to deal with exciting occurrences. Though, new challenges in the outline of present climate variation may interrupt their economy and food safe keeping as they have done with fishermen societies in other parts of the world (Guerreiro et al. 2016). Climate alteration may, therefore, rush fishermen requirement on having well-known data of climate and its influence to foster active decision-making. In Iraq, investigations on ethnoichthyology *do* not exist and attention should be given to communities operating in both coastal and inland fishing areas.

Fisheries have a chief influence on the aquatic environment (Worm et al. 2013), and an appropriate valuation of the supply and handlers will have a critical responsibility in operative conservation. Socioeconomic factors can influence the limited fisheries, which are related for more than 95% of the world's fishermen (Pauly 2006; Andrew et al. 2007). Such type of fisheries does not receive the adequate attention and care. Therefore, researches should be performed on enhancing the small-scale fisheries and bring it to the courtesy of the fisheries managers seems to be commanding (Berkes et al. 2001).

The present chapter aims to build up a methodology for using the knowledge available with the local fishers and that have been followed in different countries and presented as suggestion for the fisheries management in Iraq to follow in order to get use of this buried information in enhancing the inland fisheries in rivers in Iraq.

55.2 The General Plan of the Methodology

Interview of fishers system was used a basic tool in developing the methodology. This methodology is similar to that suggested by Tesfamichael et al. (2014) and adopted in this study with slight changes to fit the inland fisheries in Iraq.

There are two main issues need to base on in generating the enhancing methodology; these are (1) perform a series of an interviews with the fishers to compute the variation in catch rate at diverse times and by means of the top catch they remembered having made; and (2) enumerate deviations in the normal catch rates of fishers between the time they taking place fishing and the time of the interviews. To perform such analysis, a calculation to estimate the rate of alteration in catch rate to inspect the major changes in the fisheries.

The main concern through the interviews is the methodical collection of information that includes data obtained through meetings and figure out the information in such a way that the statistics can help. In addition, some information should be included about the fisheries biology, sociology, and pragmatic instances of the use of meetings in fishery investigation information collection.

55.3 How Useful the Fishers Knowledge about the Small-Scale Fisheries

Usually, the *data about the small-scale fisheries are not available* when scientists need them, but this does not mean they do not exist (Johannes et al. 2000). Particularly, information can be attained from the remembrance of source workers, which accomplished more care in fisheries research. However, how these data are best used argued (Haggan et al. 2007).

Fishers' knowledge about the small-scale fisheries has been reviewed by Soto (2006) and Hind (2012), and they decided on their usefulness or their discard inside mainstream fisheries research and management. There is wide range of variation in the opinions on the practicality of fishermen information. It differs among "useless" and does not increase to fisheries science and administration (Walley 2002); others deliberate it can be valuable but hard to counterpart with fisheries science (Jentoft et al. 1998) or that it can assist fisheries science (Neis et al. 1999).

Among the information that can be acquired by the interview-based methods is the state of ecosystems in the past (Bunce et al. 2008; Lozano-Montes et al. 2008). Besides, interview-based approaches have been utilised for present analyses, e.g. initial valuation of the ecological and socioeconomic uphold of fisheries (Teh et al. 2005), or to get data on the bycatch that is absent in landing recording systems (Moore et al. 2010).

55.4 Characteristics of the Methodology

1. The methods of meeting fishers depend on the knowledgeable ability of candidates and have been used for gathering information over wide parts and periods of time at comparatively low cost (Moore et al. 2010).
2. There are investigations that have utilised meeting with fishers simultaneously with other means to assess fisheries. Some of the studies, when examining the same fisheries, discovered similar styles conclusions (Lozano-Montes et al. 2008).
3. Provoking sole memories needs work. Through experimental meeting with fishermen, fishers were requested a direct question: "What is your best catch ever?" Nearly always, their answer was that "the catch differs as the water gives"

or similar words. Later, the identical question was knitted into a story: "When you go for fishing you do not always catch the same quantity, when you are fortunate you catch a lot, and other days you may come back empty and lose money. But if you look back, there must be one day where you caught a lot of fish and came back happy." The last approach almost always instigated an alteration in the meeting expression.
4. The assessment collected over meeting with fishermen must be verified alongside autonomously measured features, which can be utilised as a recommendation point to interpret the interview data to an absolute scale.

55.5 Example on the Expected Results of Fishers' Interview

In order to show the readers the results of the interview of fishers, it is helpful to draw an example showing these results so the idea behind suggesting the methodology of fishers' interview will be clear. Tesfamichael et al. (2014) have performed similar study in fishers' communities in 3 countries on the coast of the Red Sea. Adopting these results here in this chapter with some changes to accommodate the status and the culture of the fishers in southern Iraq has been preferred.

In this example, a total of 423 different fishers were used with age fluctuating from 12 to 83 years and with fishing knowledge of 1–65 years. In choosing such a wide range of fishers' age, a good representation of all age groups will be included. The fishers of less than 30 years were better embodied in the example than the other age sets. The analysis of the best catch per unit effort (CPUE) fishers remembered was calculated out by fishing method type as this diagnose the fisheries specifically.

Tesfamichael et al. (2014) concluded that systematic tactic during the meeting with the fishermen is important. They discovered was that asking for an special knowledge, e.g. the best catch they ever made, and comparing diverse skills, e.g. the typical catch at altered times, permitted fishermen to answer to the questions more easily than by posing more general or unclear questions, e.g., how much the catch has altered. These results were confirmed by others (e.g. Sáenz-Arroyo et al. 2005; Daw 2008). In a conclusion, they found that the information obtained through the fishers' interview can be valuable to accompaniment data breaks for resource valuation; else, it can be utilised easily for a rapid, low-cost valuation of resources without past information.

55.6 Use of Fishers' Knowledge in Predicting Climate Changes

Climate is not a constant weather phenomenon and its altering events can affect people and their resources. Acclimatising to such influences will necessitate the incorporation of scientific and local (folk) information, particularly the first-hand skills and insights of supply workers such as fishermen. With the knowledge available at the disposal of the fishers, it is possible through their interview to evaluate how commercial riverine fishermen recall dangerous climatic occasions upsetting the local aquatic habitats (drought and flood); to recognise how fishermen reaction to the results of severe actions on fish obtainability; to conclude the bases of information about dangerous occasions utilised by fishermen, seeing their education and skill; and to evaluate fishermen data of intense occasion years (Guerreiro et al. 2016).

The cultural events and landmarks have similar effects in recognising past extreme occasions. Fishers in particular had thorough insight about drought impacts on fisheries incomes. In countries in South America, investigation has only been performed on fishermen remarks of bird's biology (Andrade et al. 2016), the communication among birds and fishing (Suazo et al. 2013), and territorial use rights (Gelcich et al. 2016). Similar studies need to be accomplished in the southern Iraq in order to attain the knowledge of the fishers for better understanding of the fisheries management.

Indication from other parts of the world such as the Amazon River establishes that fishermen with different stages of information use diverse data sources. However, fishermen skill, the number of fishing grounds, and educational stage did not interrupt their memory. These results are backing those with the historical dynamics of environmental, social, and economic circumstances in Central Amazon (Guerreiro et al. 2016). This is also true for the fishers in the southern Iraq as they live in the area for whole their life and have not involved in jobs other than fishing.

References

Andrade LP, Silva-Andrade HML, Lyra-Neves RM, Albuquerque UP, Telino- Júnior WR (2016) Do artisanal fishers perceive declining migratory shorebird populations? J Ethnobiol Ethnomed. 12:16

Andrew NL, Béné C, Hall SJ, Allison EH, Heck S, Ratner BD (2007) Diagnosis and management of small-scale fisheries in developing countries. Fish Fish 8:227–240

Berkes F, Colding J, Folke C (2000) Rediscovery of traditional ecological knowledge as adaptive management. Ecol Appl. 10:1251–1262

Berkes F, Mahon R, McConney P, Pollnac R, Pomeroy R (2001) Managing smallscale fisheries: alternative directions and methods. International Development Research Centre, Ottawa, Ontario, Canada

Bunce M, Rodwell LD, Gibb R, Mee L (2008) Shifting baselines in fishers' perceptions of island reef fishery degradation. Ocean Coastal Manage 51:285–302

Daw TM (2008) How fishers count: engaging with fishers' knowledge in fisheries science and management. Dissertation, Newcastle University, Newcastle upon Tyne, UK

Gelcich S, Cinner J, Donlan C, Tapia-Lewin S, Godoy N, Castilla J (2016) Fishers' perceptions on the Chilean coastal TURF system after two decades: problems, benefits, and emerging needs. Bull Mar Sci 92:000–000

Guerreiro AIC, Ladle RJ, da Silva Batista V (2016) Riverine fishers' knowledge of extreme climatic events in the Brazilian Amazonia. J Ethnobiol Ethnomed 12(1):50

Haggan N, Neis B, Baird I (eds) (2007) Fishers' knowledge in fisheries science and management. UNESCO, Paris
Hind EJ (2012) Last of the hunters or the next scientists? Arguments for and against the inclusion of fishers and their knowledge in mainstream fisheries management. Dissertation, National University of Ireland, Galway, Ireland
Jentoft S, McCay BJ, Wilson DC (1998) Social theory and fisheries co-management. Marine Policy 22:423–436
Johannes RE, Freeman MMR, Hamilton RJ (2000) Ignore fishers' knowledge and miss the boat. Fish Fish 1:257–271
Lozano-Montes HM, Pitcher TJ, Haggan N (2008) Shifting environmental and cognitive baselines in the upper Gulf of California. Front Ecol Environ 6:75–80
Mackinson S, Nøttestad L (1998) Combining local and scientific knowledge. Rev Fish Biol Fish 8:481–490
Moore JE, Cox TM, Lewison RL, Read AJ, Bjorkland R, McDonald SL, Crowder LB, Aruna E, Ayissi I, Espeut P, Joynson-Hicks C, Pilcher N, Poonian CNS, Solarin B, Kiszka J (2010) An interview-based approach to assess marine mammal and sea turtle captures in artisanal fisheries. Biological Conserv 143:795–805
Murray C, Wieckowski K, Hurlburt D, Soto C, Johnnie K (2011) Incorporation of traditional and local ecological knowledge and values in fisheries management. ESSA Technologies Ltd, Canada
Neis B, Schneider DC, Felt L, Haedrich RL, Fischer J, Hutchings JA (1999) Fisheries assessment: what can be learned from interviewing resource users? Can J Fish Aquat Sci 56:1949–1963
Pauly D (2006) Major trends in small-scale marine fisheries, with emphasis on developing countries, and some implications for the social sciences. Mar Stud 4:7–22
Peterson N, Broad K (2009) Climate and weather discourse in anthropology: from determinism to uncertain futures. In: Crate S, Nutall M (eds) Anthropology and climate change: from encounters to actions. Left Coast Press, Walnut Creek, CA, pp 70–86
Pinto M, Mourão J, Alves R (2013) Ethnotaxonomical considerations and usage of ichthyofauna in a fishing community in Ceará State, Northeast Brazil. J Ethnobiol Ethnomed. 9:17
Sáenz-Arroyo A, Roberts CM, Torre J, Cariño-Olvera M, Enríquez-Andrade RR (2005) Rapidly shifting environmental baselines among fishers of the Gulf of California. Proc Royal Soc B: Biol Sci 272:1957–1962
Santos CAB, Nóbrega Alves RR (2016) Ethnoichthyology of the indigenous Truká people, Northeast Brazil. J Ethnobiol Ethnomed 12:1
Shackeroff JM, Campbell LM (2007) Traditional Ecological Knowledge in Conservation Research: Problems and Prospects for their Constructive Engagement. Conserv Soc. 5:343–360
Soto CG (2006) Socio-cultural barriers to applying fishers' knowledge in fisheries management: an evaluation of literature cases. Dissertation, Simon Fraser University, Burnaby, British Columbia, Canada
Suazo CG, Schlatter RP, Arriagada AM, Cabezas LA, Ojeda J (2013) Fishermen's perceptions of interactions between seabirds and artisanal fisheries in the Chonos archipelago, Chilean Patagonia. Oryx 47:184–189
Teh L, Cabanban AS, Sumaila UR (2005) The reef fisheries of Pulau Banggi, Sabah: a preliminary profile and assessment of ecological and socio-economic sustainability. Fish Res 76:359–367
Tesfamichael D, Pitcher TJ, Pauly D (2014) Assessing changes in fisheries using fishers' knowledge to generate long time series of catch rates: a case study from the Red Sea. Ecol Soc 19(1)
Walley CJ (2002) They scorn us because we are uneducated': knowledge and power in a Tanzanian marine park. Ethnography 3:265–298
Wilson DC, Raakjær J, Degnbol P (2006) Local ecological knowledge and practical fisheries management in the tropics: a policy brief. Mar Policy 30:794–801
Worm B, Davis B, Kettemer L, Ward-Paige CA, Chapman D, Heithaus MR, Kessel ST, Gruber SH (2013) Global catches, exploitation rates, and rebuilding options for sharks. Mar Policy 40:194–204

Chapter 56
A Possibility to Apply a Traditional Fisheries Enforcement Programme in the Inland Waters of Iraq

Laith A. Jawad

Abstract The present chapter presents a methodology for implementation of fisheries enforcement in rivers and the coastal waters of Iraq. The principles for assessing prosecution usefulness were consequent from the traditional economic opinion of controlling enforcement that focuses hindrance through legal penalties. The use of this procedure in Iraq might certainly be reliable to a degree as it could stand on the observed opinions of all local implementation officials, and more remarkably, in utmost cases, the apparent opinions were found to be similar to fishermen. The administration consultant can move onward in this knowledge-building study of depicting prosecution certainty over the repetition of this study which may also help guarantee rationality of the matters given in this chapter and therefore make balanced deliberate conclusions on fisheries law execution.

56.1 Introduction

Literatures on citizens' law-abidingness have frequently stressed the preventive effects of formal law implementation (Meier and Johnson 1977; Tyler 1990). This view can be tracked to the legacy of Thomas Hobbes (1984) (in Norrie 1984), who regards law as a symbol of the shared benefits of citizens who are unable of making social order through moral community. Hobbes' state ratifies external rules upon individuals and guarantees their compliance "by the terror of some punishment, greater than the benefit they expect by the breach of their Covenant. . . ." The punishment points out the divergent and reproachable nature of the crime and thus approves the ethical unity that has been attacked. Punishment's main purpose is thus to emblematically restore the moral order of society.

In the literature on acquiescence in fisheries, Hønneland (1998) has claimed that a conversational enforcement style has proven efficient to encourage fishermen to

L. A. Jawad (✉)
School of Environmental and Animal Sciences, Unitec Institute of Technology, Auckland, New Zealand

L. A. Jawad (ed.), *Tigris and Euphrates Rivers: Their Environment from Headwaters to Mouth*, Aquatic Ecology Series 11,
https://doi.org/10.1007/978-3-030-57570-0_56

fulfil with measures meant to protect fish stocks, and Randall (2004) has approved the view in a set of endorsements for fisheries law enforcement. These viewpoints characterize a premeditated approach to compliance because they emphasize the importance of agreement between citizens and the state reached through communication.

The problem of not to go along with the guidelines ascends since the launch of instructions and directions in the fishery does not routinely denote that the agents' motivations to disrupt regulations are removed, although the fishermen's inducements certainly can be changed. Law implementation and obedience have been investigated by researchers demonstrating a wide range of issues, e.g., economists, sociologists, criminologists, and psychologists, and these diverse areas do not essentially approve with each other in every regard.

Sighted the significant part of fisheries law implementation and with specific mention to the traditional fisheries subdivision, this chapter aims to give a short review about the fisheries law enforcement and the criteria used to evaluate them in order to apply such enforcement principles in both the inland and coastal fisheries in Iraq.

Comparable to other countries rich in rivers, Iraq is opposing both working and strategy contests with respect to non-obedience in its fisheries area. Iraq has about 34 km^2 of surface water accessible per year (FAO 2003) and short coastal on the Arabian Gulf with many landing places located inside an unapproachable areas in the southern Iraq and the marsh areas. This working constraint together with restricted financial and well-trained human means it is puzzling to inspect fishing actions in a normal operation. Additionally, from a tactical view, Iraq is one of the most significant fishing nations in the Middle East and could be a net exporter of fisheries foodstuffs in the Arabian Gulf region, the Arab and other parts of world. This shows a future sector's dependence on the export market as well as its possible in foreign exchange incomes.

56.2 Historical Background

Fisheries enforcement is not a new issue in human life. It has been in use since the early civilizations that appeared thousands of years ago in the land between the two rivers Euphrates and Tigris and that known as Mesopotamia.

Fish have been the main food component in Mesopotamia over the last 5000 years separately from milk and grain. Fish were thought as one of the goods of early human development in the Euphrates and Tigris Rivers with their tributaries and coastal waters of the Arabian Gulf (Sahrhage 1999). Many kinds of fish are specified in managerial records from the third millennium B. C. up to the period of the first dynasty in Babylon. A Sumerian text from about 2000 B. C. entitles the behaviours and presence of many species of fish in some detail (Saggs 1962). The peoples of Babylonia were no less involved in the watching of animals, and fish were as exactly and noticeably portrayed on the bas-relief as were present mammals (Jawad 2006).

Ancient society of Mesopotamia had competent ways of catering fish for food that comprised eating it fresh, salted, smoked, and dried in the sun (Maspero 1896). Fish was not just deliberated a food for the deprived people, but was also involved in the food list for the palace of kings of Mesopotamia (Postgate 1994). Additionally, it was intricate among the food that was presented to the gods in temples (Saggs 1987). Fishermen were fundamentally found in large numbers among the temple staffs, being divided into "freshwater fishers," "sea fishers," and "fishers in salt waters." The last group signifies fishermen working in the tidal lagoons of the delta of the Tigris and Euphrates Rivers. Bestowing to the cipher of Hammurabi, fishermen were given some rights similar to those given to certain classes of society like ministers and traders (Postgate1994). Fishing methods such as Lower Mesopotamian fishing gear and methods in regard to fishing administration spearing, harpooning, angling, and the use of baskets and numerous other sorts of nets were by this time well established during the time of ancient Mesopotamia (Sahrhage 1999).

With such an activity, the code of Hammurabi has included specific laws to regulate fishing in the Euphrates and Tigris Rivers and control fishers in their daily work so they will not damage the environment and to not over fish. These laws seem to be important to protect the fish stocks as the ancient Mesopotamians have involved in trade and exporting commodities, where fish in its different conditions are included in the list of goods that usually exported from southern Mesopotamia and in particular from city of Lagash (Crawford 1973).

While fishing unquestionably started as an individual, periodic action, it advanced in some areas into a talented one that was, in some circumstances, highly prepared. This was chiefly true in Mesopotamia, where we have detailed records of fisheries run by government and temple foundations (Potts 2012). These data emerge as early as the Uruk III period at Uruk where the archaic occupations list comprises two sign groupings that may assign a foreman of fishermen and a fisheries manager (Englund 1991). In the Early Dynastic period, the temple of the goddess Bau at Girsu (modern Telloh) in the territory of Lagash (Bauer 1998) ruled over a number of fisheries managers who, in turn, controlled units of both freshwater and saltwater fishermen, the former restricted to work on the canals and lakes and the latter, a numerically larger group, subdivided into brackish water and coastal fishermen (Bauer 1998). It seems that the temple has a set of rules in providing type and number of fish to the temple, which represents a sort of fisheries enforcement (Englund 1991).

Environment and animals conservation including fish have been mentioned in the wholly books of Judaism, Christianity, and Islam. In Judaism, the basis for humanity's duty to defend nature may be found in the biblical expression, "For the earth is Mine" (Lev. 25:23). The Earth is not subject to man's absolute ownership, but is rather given to us "to use and protect" (Gen. 2:15).

There can be no uncertainty to any rational or solicitous person that the "dominion" stated in the book of Judaism in the phrase, "and have dominion over the fish of the sea, and over the fowl of the air, and over every living thing that creeps upon the earth," is not the dominion of a dictator who deals severely with his people and servants in order to attain his own personal supplies and needs (Finlay and Palmer 2003).

In Christianity, all of establishment is the loving action of God, who not only wanted the creation but also endures to care for all facets of life (Finlay and Palmer 2003). It has been recognized by the World Council of Churches in 1988 that *the effort to have "mastery" over creation has ensued in the ridiculous mistreatment of natural resources, the unfriendliness of the land from people and the obliteration of indigenous cultures.... Creation came into being by the will and love of God, and as such it owns an inner unity and goodness. Though human eyes may not always distinguish it and accord with which creation is gifted. And when our human eyes are opened and our tongues released, we too learn to admiration and contribute in the life, love, power and freedom that is God's ongoing gift and grace.*

In Islam, the care and conservation of the nature and the creatures were more obvious and came in details. In the chapter on Islam and the conservation, Finlay and Palmer (2003) narrated, "There are a number of Qur'anic values that have a clear linking with preservation. They were given in clear terms that Allah, the One True God, is the Universal God and the Creator of the Universe and, indeed, the Owner of the Universe. To Him belong all the animate and inanimate objects, all of whom should or do submit themselves to Him. Allah, in His Wisdom, appointed us, the creatures that He has conferred with the faculty of reason and with free will, to be His vice regents on earth."

It was given clearly in Quran the enforcements and punishments of people that ill responsible toward the environment and the creature. Quran asked Muslims to not harm or abuse animal or overconsume them while they are on the pilgrimage, where thousands of people attending this wholly events. If each person hunts one bird or any animal for food, then the stock of prey will be affected and drop down (Bagader et al. 1994; Brammer et al. 2007).

56.3 The Traditional Fisheries Sector in Iraq

When concepts of natural incomes come to mind in the Arab world, most people deliberate of oil, gas and phosphates. In an area where most climates are arid, it would be hard to envisage aquaculture even occurs in the region, let unaccompanied has expansion Potential (US-Arab Tradeline 2003). In Mesopotamia, one of the supplies of early human civilization, fish, crustacean, molluscs, and turtles in the Euphrates and Tigris Rivers with their tributaries and the coastal waters of the Arabian Gulf was a chief food varieties already 5000 years ago (Sahrhage 1999).

Iraq's inland fishery is based on the Tigris-Euphrates riverine system, its lakes, and it has an important part in the country's wealth. There is an opportunity to improve these assets through administration, supplying, and development of extensive culture applies. Fisheries in numerous small water bodies and reservoirs can be improved through providing and management systems that take consider particular configurations of the individual fishery. The inland fisheries are grounded mainly on carps *Cyprinus* spp., whereas the most significant Iraqi native fishes are barbs belonging to the family Cyprinidae.

The old-style fisheries sector in Iraq is mostly contained of small-scale fishermen who are intricate in fishing actions with a low level of money inputs and technique. In spite of being considered the lowest revenue to the country, this sector provides direct employment to large number of workers mainly young people. Although the fishing operations in the inland waters of Iraq have been ruled by law documented in the Iraqi constitution, controlling this sector is left open without any guardian. There are no licensing system, no fishing season limitation, and no fishing grounds specificity. With such situation, enforcement of the fishing laws is eminent and must so the fisheries resources of Iraq can be protected.

56.4 Suggestions for the Iraqi Government to Introduce Fisheries Lawmaking and Implementation Actions

The Ministry of Agriculture or the Ministry of environment in Iraq can control fishing linked actions through a set of rules postulated in the specific law on fisheries in the Iraqi constitution. A certain department belonging to the responsible ministry about the fisheries enforcements should follow up the implementation of the law of enforcement. The means of fisheries implementation should include air, sea, and land-based observation at landing locations and of fisheries corporations and handling plants. The air and land investigations should be led by the Air and Police forces, while the sea scrutiny is directed by the Iraqi Navy and the Coast Guards.

Categories of fisheries abuses conferring to their possible impact on the status of fisheries assets can be stated by the ministry of Agriculture if they are not described in the fisheries law given in the Iraqi constitution. The high jeopardy abuses could be mostly related with high-value fisheries, while the medium hazard desecrations contain the usage of unregistered boat, illegal and delimited fishing gear, and illegal service of foreign labour. Actions such as fishing without authorization, not carrying lawful license, fishing in constrained zones can be considered as low risk class.

56.5 Key Points of the Enforcement Programme

There are some main features such as type of fishermen intricate, period of participation, the degree of fishers' data, and expert skill; age group is significant for scheduling forthcoming growth in the fisheries area. Additionally, permanent fishermen low levels of educational achievement may restrict their work-related flexibility and upsurge fishermen reliance on fisheries for their living. With such situation, sensible fishermen and the consultant should have optimistic approaches and are predictable to be practical to upkeep and long-term endurance of fisheries incomes.

New techniques such as the handheld Personal Digital Assistant or palmtop computers can be utilized in delivering infringement bills and speed up the procedure of sending violation reports with indication and action. It will aid upholding reliability in the cataloguing of breaches and creating an electronic database for forthcoming appraisal and scheduling.

References

Bagader A, El-Sabbagh A, Al-Gayand M, Samarrai M, Llewellyn O (1994) 'Environmental protection in Islam.' IUCN environmental policy and law paper no. 20. http://cmsdata.iucn.org/downloads/eplp_020reven.pdf
Bauer J (1998) Die vorsargonische Abschnitt der mesopotamischen Geschichte. In: Attinger P, Wäfler M (eds) Mesopotamien: Späturuk-Zeit und Frühdynastische Zeit. Universitätsverlag, Freiburg
Brammer S, Williams G, Zinkin J (2007) Religion and attitudes to corporate social responsibility in a large cross-country sample. J Bus Ethics 71:229–243
Crawford HEW (1973) Mesopotamia's invisible exports in the third millennium BC. World Archaeol 5(2):232–241
Englund RK (1991) Hard work-where will it get you? Labor management in Ur III Mesopotamia. Journal of Near Eastern Studies 50(4):255–280
FAO (Food and Agriculture Organization of the United Nations) (2003) Review of world water resources by country, Water reports number 23. Available at: ftp://ftp.fao.org/agl/aglw/docs/wr23e.pdf
Finlay V, Palmer M (2003) Faith in conservation: new approaches to religions and the environment. The International Bank for Reconstruction and Development, The World Bank, Washington, DC, p 166
Hønneland G (1998) Enforcement and legitimacy in the Barents Sea fisheries. Northern Waters: Management Issues and Practice. Blackwell Science, London
Jawad LA (2006) Fishing gear and methods of the lower Mesopotamian plain with reference to fishing management. Marina Mesopotamica 1(1):1–37
Maspero G (1896) The dawn of civilization. Society for Promoting Christian Knowledge, London, p 800
Meier RF, Johnson WT (1977) Deterrence as social control: the legal and extralegal production of conformity. Am Sociol Rev 42:292–304
Norrie A (1984) Thomas Hobbes and the philosophy of punishment. Law Philos 3(2):299–320
Postgate JN (1994) Early Mesopotamia. Routledge, London, p 567
Potts DT (2012) Fish and fishing. In: Potts DT (ed) A companion to the archaeology of the ancient Near East. John Wiley & Sons, Chichester, pp 220–235
Randall JK (2004) Improving compliance in US federal fisheries: an enforcement agency perspective. Ocean Dev Int Law 35(4):287–317
Saggs HWF (1962) Babylon. Sidgwick and Jackson, London, p 562
Saggs HWF (1987) Everyday life in Babylonia and Assyria. Dorset Press, New York, p 207
Sahrhage D (1999) Fishing and fish culture in ancient Mesopotamia. Peter Lang, Frankfurt-am-Main, p 341
Tyler TR (1990) Why people obey the law. Yale University Press, New Haven
US-Arab Tradeline (2003) Arab Aquaculture investment deepens- Scott Bortot. 1st May 2003

Chapter 57
Market-Resource Relations and Fish Seller Livelihood as Seen in Inland Waters of Iraq

Laith A. Jawad and Audai M. Qasim

Abstract This chapter discusses the small-scale fish selling in Iraq in regard of two viewpoints: (1) as a linking between natural reserve use and eating, (2) and as a living in itself. Utmost fishmongers come from fishing societies, but their supply is frequently a collection of buys from other fishermen. There is minor sign of proper preparations between fishermen and vendors, however most adapt to the highly variable natural and social surroundings of the region. This chapter is considered a pilot and preliminary study to encourage stakeholders and government to pursue an advanced and comprehensive studies on the small-scale fisheries in both riverine and coastal areas of Iraq.

57.1 Introduction

The importance of small-scale fisheries within the fisheries management and development is progress and gaining wide recognition (Allison 2001). Limited fisheries are often considered as "the profession of last resort" and fishermen societies as "the poorest of the poor" (Christy 1986).

Investigates of limited number of fishmongers and the place of selling their commodities have typically put enquiries such as how this action determines contacts between supply use and depletion (Mendelson et al. 2003) and how limited occupation can be expected as a upkeep itself (e.g., Brockington 2001). The link between utilization and eaten, numerous authors have suggested that the advertisement of an open-access supply, yet at a restricted level, can maybe lead to unsustainable degrees of handling (Wilkie and Carpenter 1999).

L. A. Jawad (✉)
School of Environmental and Animal Sciences, Unitec Institute of Technology, Auckland, New Zealand

A. M. Qasim
Department of marine vertebrates, Marine Science Centre, University of Basrah, Basrah, Iraq

L. A. Jawad (ed.), *Tigris and Euphrates Rivers: Their Environment from Headwaters to Mouth*, Aquatic Ecology Series 11,
https://doi.org/10.1007/978-3-030-57570-0_57

The value of the catches in the artisanal fisheries are both eating and product values, which causing a difference between upkeep and marketable actions problematic (Allison and Ellis 2001; Berkes et al. 2001).

Through city and country side of the southern Iraq, a large quantity of the interchange in locally manufactured food takes place through small-scale marketing. Such activities are not uncommon and similar to what happened in other parts of the world (Fafchamps 1997). The present chapter gives a short review of limited selling and themes of resource consumption and livelihoods. It also focuses on the fish market and the product sequences and the demographic and living features of the salespersons in southern of Iraq. The chapter ends with a conclusion about the effect of small-scale vending on the fisheries management policies.

57.2 Small-Scale Retailing and Artisanal Fishing

The relationship between market supply and livings implication also has an effect on small-scale selling connected with non-industrial fisheries. Andersson and Ngazy (1998), in their investigation of the fishermen of this type of fisheries at coastal Tanzania, recognized that those living in communities nearby to tourist zones augmented the amount of time consumed fishing in order to supply the commercial claim for fresh seafood. This case is similar to that in lower reaches of Mesopotamia, where fishers' villages near the main market and near main road spent more time in fishing to provide commodity to main markets and selling fish in stalls on the side of the main highways. On the other hand, those villages located far away from the main cities or the main road spent lesser time in fishing, so the catch will cover their personal need and the small market of the village they live in. The situation is reversed at the time of fishing ban period, where fishers' villages near the main cities usually under the surveillance of the officers of fisheries to locate any infringements, the far away villages spent more time than usual in collecting fish for people visiting them by boats to purchase their catch with slightly higher prices.

The relationship between limited fish selling implication as a living, Medard et al. (2001) displayed that the mongers in Lake Victoria are reigned by single, divorced and widowed women, a demographic subsection often linked with deficiency in income. In Uganda, the situation is different, where fish trading is considered significant part of the life among women (Dolan 2002). In the Indian port of Goa, Rubinoff (1999) shows that fish trading is associated with lower castes. In southern Iraq, fish selling in villages is a mixed profession between men and women. On the sides of the main highway between large cities, women are usually sitting with their baskets full of fish offered for sale (Fig. 57.1). They usually brought to the site by men, who will collect them at the end of the day. Fishes in Iraq can be sold in a proper fish market or any other market, i.e. spices market or cloth market (Fig. 57.2).

Nevertheless, bigger holds due to new techniques indicate that a dealer can be the centre of fish trading and hereafter increase status if she can profit from this new supply of fish. The introduction of new technology as electrofishing in the rivers in

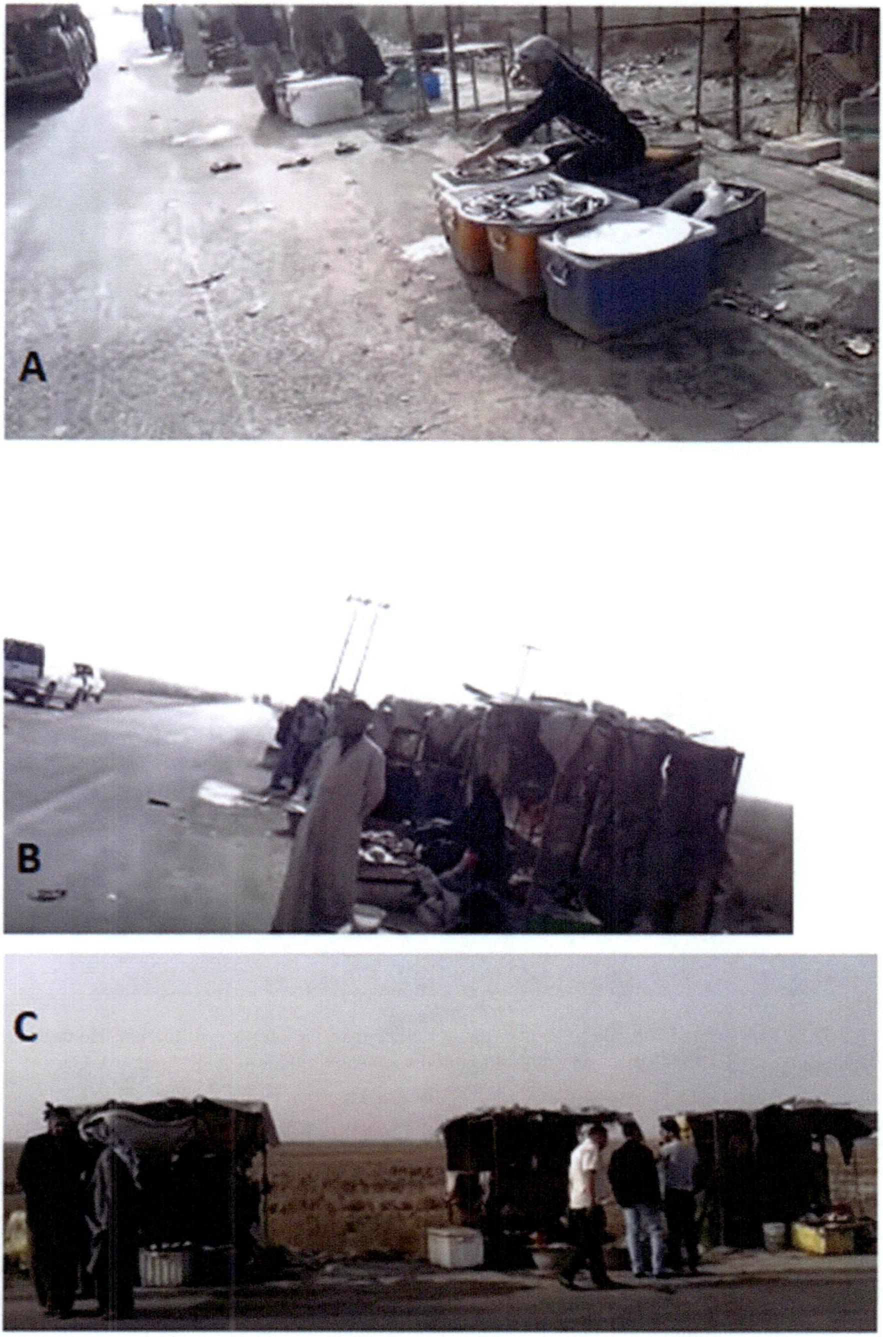

Fig. 57.1 Fish offered for sale on the side of the highway between big cities. (**a**) fisher preparing his catch for display; (**b**) customer seems to have an idea about the offered fish and women vendors shown in the background; (**c**) temporary stalls and customers discussing price of offered fish (Courtesy of Laith Jawad)

Fig. 57.2 Fish offered for sale in stall in market other than the proper fish market. Here, in the background shop selling spices and other food items are shown (Courtesy of Audai M. Qasim)

Iraq has changed the demography of the fishers. Therefore, those fishers that use this technology usually come to the market with large number of fish in comparison with those using different types of nets. It is important to note here that another group of fishers that are using illegal fishing gear such blast fishing or poison are also bring their catch to the market as small-scale vendors.

As in other parts of the world, the common of the investigations inspect limited fish selling in marine or lake habitats. For instance, less consideration has been directed to limited fish selling in sub-Saharan Africa's floodplains (FAO 1996; Geheb and Sarch 2002). In Iraq, although fisheries studies are not many in comparison with other countries, they are centred on the coastal fishing activities. Though, significant biological, social and economic features discriminate the riverine fisheries from those in marine or lake regions. Such dissimilarities disquiet the efficiency and variability of the fishery, as well as the fishers that exploit it.

57.3 Commodity Chains

The differences between the riverine and coastal fisheries may encompass after catch activities such as offering the catch for sale. In case of marine fisheries, landing areas for catches are well distinct and often stable, with fishers and sellers focused in the same place. Thus, the level of opposition is possible to be great. Otherwise, fishers and salespersons may choose to arrive into engagements in order to limit jeopardy and protected credit and supply, respectively (Walker 2001). In contrast, landing sites of the riverine catch are both extensive and possible to change position over time, occasionally at a comparatively fast rate, conferring to the best site of catch and presence of buyers. Therefore, with only small numbers of fishermen utilizing each landing area, it may be hard for fishmongers to know where, when and if fishermen will land their catch. Generally, the degree to which the intricate actions seen in the riverine fisheries in Iraq is largely indefinite for instance of commercial level manipulation and distribution in this area.

57.4 Types of Fish Markets in Iraq

Southern Iraq probably differs from the middle and north of Iraq in offering fish for sale. This difference could be attributed to two main factors, the location of the market, (i.e. in a main city, village or in a remote area of marsh region) and type of market (i.e. stalls in open area, a series of shops, wood carriages filled with ice and fish on top).

(a) Fish markets at main cities (Fig. 57.3)

Fish in this type of market are usually offered for sale in small shops lined in two rows opposite to each other and the whole area is covered. Outside the main market area, some fishers are sitting selling their catch fish with prices less than that offered in the shops. Fishers deliver their catch early morning hours for auction to take place (Fig. 57.4).

In Baghdad in particular and in the different markets such as the gold market, vegetable and meat markets or even on the sides of the main streets, wooden carriages are usually seen filled with ice and large fish lie on the top (Fig. 57.5). While there are other markets located in the outskirt of the big cities, such markets are restricted in the range and volume of commodity sold, as well as the number of sellers. Fish markets in Iraq highlight the role seller has to meet the living and feeding objectives of all social sectors of people. Though, customers who visit the

Fig. 57.3 Old Basrah City fish market. (**a**) customers viewing the offered fish for sale; (**b**) fish market at the mid-day when number of customers drop down; (**c**) fish market at busy time (Courtesy of Laith Jawad and Audai M. Qasim)

Fig. 57.4 Fishers deliver their catch early morning hours for auction to take place in Basrah City (Courtesy of Laith Jawad)

fish markets of the main cities do not seem to be restricted to a definite revenue bracket or social group. While the common of the customers are "average" (i.e. with limited disposable income), comparatively well-paid civil servants and consultants

Fig. 57.5 Fish for sale in wooden carriage in market in Baghdad City. (**a**) wooden carriage specialized in large fish individuals; (**b**) vendors offering fish for sale in wooden carriage in one of non-fish market place in Baghdad City (Courtesy of Omar Al-Sheikhly)

often visit the market, mainly to buy fish. Folks journeying to other cities by road or by plane will often take a cooler full of large fresh fish for friends and family.

In Basrah City specifically, the main fish markets offer for sale both marine and freshwater fish species (Fig. 57.6). Also, imported frozen fish from the neighbouring countries such as Iran and the Gulf States are also present among the fish commodity offered for sale.

(b) Fish market at the outskirt of main cities

This type of fish market is usually located at the outskirt of the main cities and at the same time considered a landing sites for fishers from villages that their inhabitants practise fishing for their daily life (Fig. 57.7).These type of markets are small and fish offered for sale on plastic sheets cover small areas of the ground. They are usually located near the river, where the fishers are able to deliver their catch by their traditional boats. In these markets, men and women are sharing the trade of selling fish (Fig. 57.8). No frozen fish are usually sold in this type of market due to the unavailability of ice. On the other hand, fresh fish is usually obtained from these markets especially in the early hours of the morning, where fishers bring their catch. Better prices can get if wholesale is aimed by the customers. Just after the mid-day, these markets are usually finished selling all the fish offered for sale and the vendors go home preparing for another day. These markets are open 7 days a week and prices of fish usually more during holidays.

(c) Fish markets or gathering at rural areas

It is not possible to call the area where the fish offered sales in the rural villages as fish market, because they have no infrastructure of the market (Fig. 57.9). It is possible to call them a gathering area for selling fish. Here, women acting as vendors are more than men. No tables or stalls are used to put the fish on. Instead, fish are

Fig. 57.6 Freshwater and marine fish species offered for sale at Basrah City fish Market. (**a**) Bunni fish, *Mesopotamichthys sharpeyi*; (**b**) Mixture of crabs and marine fish species; C, Mixture of freshwater and marine fish species offered for sale (Courtesy of Audai M. Qasim)

usually put in a large open metal containers covered with water. Number of vendors are very small, also the variety of fish is very limited depends on the local fishing area nearby. In these fish gathering, fish are usually sold per individual and no weight is used due to the unavailability of weighing scales.

(d) Fish offered for sale in fishing boat

In southern Iraq and at the marsh area, fish is usually bought directly from the traditional fishing boats. The trade is based on selling the fish indivually or by sacs. The total price will be negotiated between the fishers and the customers, who are approaching the fishers' boat in the middle of the marsh. Those trading fishers, if I may call them, are not abundant and rarely seen in the area.

(e) Fish market and landings in coastal areas of Iraq

In the coastal area of Iraq and precisely at the City of Fao, which is located at the estuary of Shatt al-Arab River, fishers usually bring their catch early morning hours (Fig. 57.10) for auction. Here, in this market, marine fish species are usually dominating the catch (Fig. 57.11). The auction usually lasts for 1–2 h after which, the vendors start collecting their commodity and directed to the retail fish market in the centre of Fao City. Fishing processes take place in different types of boats including fast boats and Dhows (Fig. 57.12). In these boats, several fishing gears are used such as half-spherical wire baskets (Fig. 57.13).

Fig. 57.7 Fish market at the outskirt of main cities. (**a**) Fish landing and market at Abu Al-Khaseeb City, south of Basrah City; (**b**) Fish market at the outskirt of north Basrah City (Courtesy of Laith Jawad and Audai M. Qasim)

Fig. 57.8 Women vendor offering fish for sale in the open air. Fishes are placed in a metal tries. In the background, a fan operated by car battery is installed to aerate the stall area (Courtesy of Laith Jawad)

Fig. 57.9 Women vendor offering fish for sale in a rural area (Courtesy of Laith Jawad)

Fig. 57.10 Fisher's activities after catch. (**a**) Fishers preparing catch to be taken for the auction; (**b**) In the auction ground, fishers separating catch according to size and species (Courtesy of Laith Jawad and Audai M. Qasim)

Fig. 57.11 Marine fish species offered for sale in the coastal city of Fao, south of Basrah City. (**a**) Mixture of a marine species offered for sale in metal try; (**b**) vendor showing yellowbar angelfish, *Pomacanthus maculosus* (Courtesy of Audai M. Qasim)

57.5 Important Questions Need to be Asked in Relation to the Market and the Riverine Livelihoods in Southern Iraq

Due to the high stages of potential human and natural inconsistency present in riverine fisheries communities and livelihoods, the following three questions concerning limited fish selling in these environments come important to have their answers. These questions were adopted with slight changes from Abbott et al. (2007).

1. *What is the product series connecting fish to market in the riverine fisheries community?*

   Usually, there are arrangements between fishers and salespersons for the buildup, selection, buying, planning and transference of catch for marketing. Likewise there are actions within the market, such as sharing of stalls, intra-seller communication and extra-regional distribution of catch.
2. *Is market source mainly influenced directly or indirectly by periodic issues, such as the natural difference in fish richness?*

   There a numbers of salespersons occupying stalls or places to sell fish at the market for a long time. Also it is interesting to know the perceptions of the seller about the best and worst periods for selling fish, as well as the foundations behind these insights.
3. *How do limited fish sellers reply to high-deceptive grade of doubt in fish stock?*

   The riverine fisheries naturally vary in the volume and composition of catches, it is expected that small-scale fish sellers would have to be very flexible.

Fig. 57.12 Different types of fishing boats at the harbour of the coastal city of Fao, south of Basrah City. (**a**) Fast boats; (**b**) Dhows (Courtesy of Audai M. Qasim)

Fig. 57.13 Hemispherical wire basket used for catching fish in the coastal area (Courtesy of Audai M. Qasim)

Therefore, it should be considered where salespersons normally got their fish (i.e. from family, non-related fishermen or middle-man) and the occurrence and ranking of diverse limits to selling, such as moving, receiving enough fish, getting the right size of fish or obtaining the right kind of fish.

57.6 Livelihoods Perspectives and Small-Scale Fisheries Management and Development

Allison and Ellis (2001) emphasized some issues for additional conversation regarding the livings viewpoint relative to limited fisheries administration and growth. At this point, these issues have been accepted with slight variations to fit the riverine fisheries status in Iraq:

1. Income diversification is a characteristic of numerous fishing societies.
2. Growth in country side areas where fishing is significant might not be top-aided by interference to upsurge fishing revenues, but rather to backing matching domestic actions.
3. Geographical movement is essential to endure catches on movable or changeable fish supply.
4. The payment economy is significant in country side locations, and whether or not such payment is participated in fishing, can act to control capitalization in fisheries.

References

Abbott JG, Campbell LM, Hay CJ, Næsje TF, Purvis J (2007) Market-resource links and fish vendor livelihoods in the upper Zambezi river floodplains. Hum Ecol 35(5):559–574
Allison EH (2001) Big laws, small catches: global ocean governance and the fisheries crisis. J Int Dev 13:933–950
Allison EH, Ellis F (2001) The livelihoods approach and management of small-scale fisheries. Mar Policy 25(5):377–388
Andersson J, Ngazy Z (1998) Coastal communities production choices, risk diversification, and subsistence behavior: responses in periods of transition. Ambio 27(8):686–693
Berkes F, Mahnon R, McConney P, Pollnac R, Pomeroy R (2001) Managing small-scale fisheries: alternative directions and methods. International Development Research Centre, Ottawa
Brockington D (2001) Women's income and the livelihood strategies of dispossessed pastoralists near the Mkomazi game reserve. Tanzania Human Ecol 29(3):307–338
Christy FT Jr (1986) Special characteristics and problems of small-scale fisheries management in developing countries. In: Miles E, Pealy R, Stokes R (eds) Natural resource economics and policy applications. University of Washington Press, Seattle, pp 118–151
Dolan C (2002) Gender and diverse livelihoods in Uganda. LADDER working paper. No. 10. Overseas Development Group, University of East Anglia, Norwich, p 30
Fafchamps M (1997) Introduction: Markets in sub-Saharan Africa. World Dev 25(5):733–734
FAO (1996) Fisheries and aquaculture in sub-Saharan Africa: situation and outlook in 1996. Food and Agricultural Organization, Rome
Geheb K, Sarch M (2002) Africa's inland fisheries: the management challenge. Fountain, Kampala
Medard M, Sobo F, Ngatunga N, Chirwa S (2001) Women and gender participation in the fisheries sector in Lake Victoria. Paper presented to the global symposium on women in fisheries, Kaohsiung, Taiwan
Mendelson S, Cowlishaw G, Rowcliffe J (2003) Anatomy of a Bushmeat commodity chain in Takoradi, Ghana. J Peasant Stud 31(1):73–100
Rubinoff J (1999) Fishing for status: impact of development on Goa's fisherwomen. Women's Stud Int Forum 22(6):631–644

Walker BLE (2001) Sisterhood and seine nets: endangering development and conservation in Ghana's marine fishery. Prof Geogr 53(2):160–177
Wilkie D, Carpenter J (1999) Bushmeat hunting in the Congo Basin: an assessment of impacts and options for mitigation. Biodivers Conserv 8(7):927–955

Chapter 58
The Potential Impact of Deformities in Fishes upon Aquatic Production: Case of Iraq

Laith A. Jawad

Abstract Fish can be the receivers of abundant injuries that are possibly lethal to aquacultural production performance and welfare. A short review was given in the present chapter of the publications on fish deformities in freshwater and marine species of Iraq. This field of study is new and only started 36 years ago. A selected common fish anomalies that could happen in fish of Iraq were also discussed in this chapter such as deformities in the mouth, eye and fins as these structures are the most eminent to receive injuries and deformities in fish. The chapter then accomplishes with recommendations and suggestions for aquaculture and aquatic managements in Iraq to follow in order to enhance the aquatic production in Iraq.

58.1 Introduction

Abnormalities can be present in both wild (Slooff 1982) and farmed fish (Matsuoka 2003; Noble et al. 2012) and can occur at any time from the larval (López-Albors et al. 1995) to the adult phase (Korsøen et al. 2009) of the life of the animal. In the present study, the expression deformity is expressed as 'an attained or congenital alteration of an organ or part' (Collins English Dictionary 1998). Such body aberrations may cause performing damage, which can be threateneing to fish health (Huntingford et al. 2006) and production functioning (Noble et al. 2012).

From production viewpoint, abnormalities can cause a reduction in development (Miyashita et al. 2000), diminished feeding (Dyková et al. 1998), lessened feeding capability (Kurokawa et al. 2008), augmented vulnerability to infection (Turnbull et al. 1996) and enhanced degrees of death (Cobcroft and Battaglene 2009; Noble et al. 2012). Abnormalities can also lessen the market price of both farmed and collected fish from the wild (Michie 2001; Jawad and Ibrahim 2018).

L. A. Jawad (✉)
School of Environmental and Animal Sciences, Unitec Institute of Technology, Auckland, New Zealand

L. A. Jawad (ed.), *Tigris and Euphrates Rivers: Their Environment from Headwaters to Mouth*, Aquatic Ecology Series 11,
https://doi.org/10.1007/978-3-030-57570-0_58

The present chapter provides a review of the cases of abnormalities that reported in both freshwater and marine fish species collected from Iraq. In addition, a review to other types of anomalies have not reported to present in Iraqi fish species, but they include a large number of exterior characters of the fish that can be injured during the life of the fish or during the aquaculture activities. Most of these deformities are easy to locate and identify by eye or simple veterinary measures. At the end of the chapter, a conclusion is given to summarize a possible collaboration between hazard reasons and anomaly/damage type, before recognizing possible links in both working farming and fisheries performs and extenuation policies.

58.2 Review of the Fish Deformities so Far Reported from Freshwater and Marine Fish Species of Iraq

The subject of the study of fish abnormalities is new to Iraq. Publications started to appear in this field 36 years ago when Al-Hassan and Basrah (1982) has published his first report on some cases of vertebral deformities observed in marine and freshwater fish species collected from United Arab Emirates and Iraq, respectively (Table 58.1). Since then, only few papers have been published on anomalies in fishes. These publications described mainly several cases of vertebral deformities that include vertebral coalescence and deformation (Al-Hassan 1983; Al Hassan and Naama 1986, 1988; Jawad et al. 2015, 2018c), lordosis (Jawad et al. 2014) and other minor vertebral anomalies (Al-Hassan 1985, 1987). The case of pugheadness is described from only one marine species, *Johnius aneus* (Family: Sciaenidae). This specimen was collected from Khor al-Zubair area, north-west of the Arabian Gulf, Iraq. Whether the fish species is freshwater or marine, they are all commercially important species and obtained from the commercial catch in the inland waters of Iraq and its marine areas.

To analyse the published works on the fish anomalies in the freshwater and marine species, the following dialogue will be centred on the severity of the case of the anomaly and on the fish body part that has been affected.

From the published works on the fish anomalies in Iraq, it seems that *Mesopotamichthys sharpeyi* showed the larger incidences of abnormalities followed by *Luciobarbus xanthopterus* and *Carasobarbus luteus*. This does not mean that those species are more vulnerable than the other species, but the result is based on the availability of published reports. Within the marine species, the following three species were only reported for abnormalities, *Johnius aneus*, *Tenualosa ilisha* and *Platax teira* (Al-Hassan and Náama 1988; Jawad and Bannai 2014; Jawad et al. 2014).

As to the severity of the anomaly incidence, all the vertebral deformities that have been reported in the fish species of Iraq have been classified as severe cases due to the level of damage that has produced as result of either coalescence of the vertebrae, the deformation of the vertebral centra and unusual curve of the vertebral column to

Table 58.1 Chronological arrangement for the publication of fish anomalies from both freshwater and marine habitats of Iraq

Fish species		Area of deformity on vertebral column	Type of deformity	Level of deformity	References
Mesopotamichthyes sharpeyi	Freshwater	Caudal vertebrae	Coalescence of vertebrae, deformed centra	Minor and severe	Al-Hassan and Basrah (1982)
Carasobarbus luteus		Thorasic and caudal vertebrae	Deformed centra	Severe	
Luciobarbus xanthopterus		Caudal vertebrae		Minor and severe	
Aphanius dispar					
Mesopotamichthyes sharpeyi	Freshwater	Thorasic and caudal vertebrae	Coalescence of vertebrae, deformed centra	Minor and severe	Al-Hassan (1983)
Carasobarbus luteus		Caudal vertebrae			
Gambusia affinis	Freshwater	Caudal vertebrae	Coalescence of vertebrae	Minor and severe	Al-Hassan (1985)
Mesopotamichthyes sharpeyi	Freshwater	Caudal vertebrae	Coalescence of vertebrae, deformed centra	Severe	Al Hassan and Naama (1986)
Arabibarbus grypus					
Planiliza abu					
Mesopotamichthyes sharpeyi	Freshwater	Caudal vertebrae	Coalescence of vertebrae, deformed centra	Severe	Al-Hassan (1987)
Luciobarbus xanthopterus					
Johnius aneus	Marine	Head	Pugheadness	Severe	Al Hassan and Naama (1988)
Platax teira	Marine	Caudal vertebrae	Hyperostosis	Severe	Jawad and Bannai (2014)
Carasobarbus luteus	Freshwater	Caudal vertebrae	Lordosis	Severe	Jawad et al. (2014)
Tenualosa ilisha	Marine				
Luciobarbus xanthopterus	Freshwater	Caudal vertebrae	Kyphosis and coalescence	Severe	Jawad et al. (2015)
Mastacembelus mastacembelus	Freshwater	Thorasic and caudal vertebrae	Coalescence of vertebrae, compact centra	Severe	Jawad et al. (2018c)

give lordosis incident. The case of pugheadness is also considered severe as it affects the anterior part of the head that includes mouth and bones of the head. Such deformity in the bones of the anterior part of the head could bring damage to the brain and other sense centres located in the anterior part of the skull. Pugheadness and hyperostosis are cases that only reported in marine species and no such reports were obtained for freshwater species. As to the region of deformity on the vertebral column of the fish, it appears that the caudal region of the vertebral column is the most affected area. This is not unusual as such cases were reported for other fish species around the world (Jawad et al. 2018a; Jawad and Akyol 2018).

58.3 Selected Major Incidences of Injuries and Deformities in Fishes in General

(a) Mouth injuries and deformities

Mouth aberration and deformities happen in both wild and aquaculture fish populations (Slooff 1982), but are more frequently seen in cultured fish (Sadler et al. 2001; Noble et al. 2012). Their presence has been recognized through the different stages of life of the fish starting from yolk sac larvae to adults (Miyashita et al. 2000; Noble et al. 2012).

There are several types of deformities that can affect the mouth of the fish, these comprise the following: (1) pug-headedness (Matsuoka 2003; Jawad and Hosie 2007; Jawad et al. 2018b), (2) lower jaw abnormality syndrome (Sadler et al. 2001; Noble et al. 2012), (3) double mouth (Swan 1968), (4) cross bite (Barahona-Fernandes 1982) and (5) gaping jaws syndrome (Pittman et al. 1990; Noble et al. 2012). Mouth can have somewhat fewer anomalies than those stated above and they can be: elongation or extension of either the upper or lower jaw (Matsuoka 2003; Noble et al. 2012), contracting of the entire snout (Yamauchi et al. 2006), the bending down or oblique of the lower jaw (Okamura et al. 2007; Noble et al. 2012), total nonexistence of an upper jaw (Sumagaysay et al. 1999), the union of jaws in an open mouthed position (Roberts et al. 2001), twisting or winding of the upper and lower jaws (Fraser and de Nys 2005) and the occurrence of additional bones or deficiency of bones (Martinez et al. 2007). In extreme cases, mouth anomalies present in link with abnormalities to the operculum and spine of fish (Sadler et al. 2001; Noble et al. 2012).

There are two ways in which anomalies and injuries of the mouth can affect the production and welfare of the fish. As to the production, decreased market values owing to the unacceptable look of fish with mouth deformities showed a main problem for industry (Sadler et al. 2001; Noble et al. 2012). Assumed that mouth anomalies frequently hinder a fish's ability to take up food (Branson and Turnbull 2008; Noble et al. 2012), development rates can also be unfavourably upset and endurance rates diminished (Cobcroft and Battaglene 2009). From a practical standpoint, fish with mouth deformity and anomaly face two main issues which may result from the inability to correctly open and close their mouths.

The use of inappropriate raising temperatures throughout the early development of fish is also a well-known hazard cause in the growth of mouth deformities. Throughout egg development and larval growth, many fish are mainly susceptible to inappropriate culturing temperatures. Temperatures that are too high, too low or that vary too much can possibly upsurge the prevalence of jaw anomaly in these early stages (Okamura et al. 2007; Noble et al. 2012). Moreover, unsuitable salinities (Okamoto et al. 2009) and light intensities can also strongly upsurge the rate of occurrence of mouth abnormalities in these early life stages.

In aquaculture, it is possible to change conditions of the rearing fishes so to reduce the possibility of mouth injuries and deformities and enhance the fish welfare as a result. These measures include confirming the diets of fish comprise suitable levels of phosphorus, niacin, HUFA, vitamins A, C, D, K (Martinez et al. 2007; Noble et al. 2012). Raising temperatures should be attuned to escape the ranges where each species has proven susceptibility to augmented mouth deformities (Lein et al. 1997). Fish should also be grown in water free from pollutants such as dioxins and heavy metals (Yamauchi et al. 2006), and where severe contact to these impurities happens, rearing constructions should be quickly flooded with contaminant free water (Noble et al. 2012).

(b) Eye injuries and deformities

The causative agent of the usual eye deformity case of 'pop eye' or exophthalmia can be due to many reasons (Bouck 1980; Noble et al. 2012), but occasionally, it can be instigated by gas bubble disease. Additional anomalies and damages can be a consequence of eye nipping. This case is an unusual problem that occurs when rays of light enter a tank and reflect off the eyes of fish. The reflected flash hits the eye directly and resulted in one-sided periocular dermatitis, ulceration of the cornea and even loss of the eye.

Additional common eye deformity is cataracts, which are opaqueness or clouding of the eye lens. Numerous investigators suggested the causes of fish cataracts containing nutritional deficiencies, toxic agents, parasites, contact to ultraviolet light, hereditary factors, disparity in water temperature and rapid growth (reviewed by Björnsson 2004).

Mechanical causes for eye injuries and deformity are available through the practices of husbandry and aquaculture management. Therefore, great care should be taken in the design of the infrastructure of aquaculture plant. For example, fish pumps, nets, sorting and grading gadget come in a range of materials and patterns. Therefore, there has been much saleable concern in recognizing the approaches that decrease the likelihoods of impairment when moving fish (Noble et al. 2012).

(c) Fins injuries and deformities

Fin injury has been reported in wild fish, particularly in ruined habitats (Latremouille 2003), but is more common under aquaculture circumstances (Bosakowski and Wagner 1994). The main groups of fin harm can be divided into three classes: (1) splitting, (2) erosion and (3) thickening (MacLean et al. 2000;

Noble et al. 2012). An extra type of fin damage can also be deformed fins (Turnbull et al. 1996; Noble et al. 2012).

Fin damage may have a harmful consequence upon two common production routine features, growth and survival and may also interrupt product excellence. It may also affect fish health due to being a direct damage to living tissue (Ellis et al. 2008; Noble et al. 2012) and may upsurge susceptibility to infection and possibly lessens swimming capability. As with other morphological anomalies, fin damage reduces the market price of the deformed fish from both table fish processors and live fish buyers for re-stocking drives (Hoyle et al. 2007; Noble et al. 2012).

Fin injury could decrease fin job and have a damaging consequence upon a fish's capacity to regulate its body or reduce its swimming ability (Barthel et al. 2003; Noble et al. 2012) during the functions of swimming or feeding.

58.4 Conclusions and Suggestions

The cases and incidences of fish deformities that appeared in the fisheries catch are happened in the wild and there is no control to reduce them in the fish environment. On the other hand, the injuries and deformities happened during the aquaculture processes can be reduced and control to certain extent for the sake of the production and welfare of the fish.

Noble et al. (2012) have suggested a number of recommendations regarding the reduction of cases of injuries and deformities in aquaculture procedures that appeared to be important to adopt in this chapter with slight changes to fit the application in Iraq. These are as follows:

1. To accept a mitigation approach, recognizing husbandry performs and working policies for enhancing the health of farmed fish.
2. Aquaculturists need to adopt risk factors to lessen the incidence and harshness of the damages and abnormalities.
3. Controlling food class and feed administration performs can also diminish the incidence of wounds.
4. Justification policies can be equally severe and long lasting. For instance, egg development temperature is a hazard issue for pug-headedness, moreover to gill anomalies and spinal wounds (Ornsrud et al. 2004; Noble et al. 2012), and this can be treated by having information of the optimal egg incubating temperatures for a specified species.

References

Al Hassan LAJ, Naama AK (1986) Some vertebral abnormalities in fishes collected from the Euphrates in Iraq. Indian J Fish 33(2):237–241

Al Hassan LAJ, Naama AK (1988) A case of pugheadedness in the croaker *Johnius aneus* from Khor Al-Zubair northwest of the Arab gulf, Basrah, Iraq. Indian J Fish 35(1):68–69

Al-Hassan LAJ (1983) Vertebral anomalies in Mesopotamichthys sharpeyi and Carasobarbus luteus (Pisces: Cyprinidae) from Basrah, Iraq. Cybium 7:7–10

Al-Hassan LAJ (1985) Vertebral anomalies in *Gambusia affinis* (Teleostei: Poeciliidae) collected from Basrah and Baghdad, Iraq. Pak J Zool 17:169–171

Al-Hassan LAJ (1987) Vertebral anomalies in fishes collected from Iraq and Arabistan, Iran. Pak J Zool 19(1):95–97

Al-Hassan LAJ, Basrah E (1982) Vertebral abnormalities in fishes from Iraq and UAE, Arabian Gulf Iraqi. J Mar Sci 1:13–23

Barahona-Fernandes MH (1982) Body deformation in hatchery reared European sea bass *Dicentrarchus labrax* (L.)— types, prevalence and effect on fish survival. J Fish Biol 21:239–249

Barthel BL, Cooke SJ, Suski CD, Philipp DP (2003) Effects of landing net mesh type on injury and mortality in a freshwater recreational fishery. Fish Res 63:275–282

Björnsson B (2004) Can UV-treated seawater cause cataract in juvenile cod (*Gadus morhua* L.)? Aquaculture 240:187–199

Bosakowski T, Wagner EJ (1994) Assessment of fin erosion by comparison of relative fin length in hatchery and wild trout in Utah. Can J Fish Aquat Sci 51:636–641

Bouck GR (1980) Etiology of gas bubble disease. Trans Am Fish Soc 109:703–707

Branson EJ, Turnbull T (2008) Welfare and deformities in fish. In: Branson EJ (ed) Fish welfare. Blackwell, Oxford, pp 202–216

Cobcroft JM, Battaglene SC (2009) Jaw malformation in striped trumpeter *Latris lineata* larvae linked to walling behaviour and tank colour. Aquaculture 289:274–282

Collins English Dictionary (1998) Millennium edition. HarperCollins Publishers, Glasgow

Dyková I, Figueras A, Novoa B, Casal JF (1998) Paramoeba sp., an agent of amoebic gill disease of turbot *Scophthalmus maximus*. Dis Aquat Org 33:137–141

Ellis T, Oidtmann B, St-Hilaire S, Turnbull J, North B, Mac-Intyre C, Nikolaidis J, Hoyle I, Kestin S, Knowles T (2008) Fin erosion in farmed fish. In: Branson E (ed) Fish welfare. Blackwell, Oxford, pp 121–149

Fraser MR, de Nys R (2005) The morphology and occurrence of jaw and operculum deformities in cultured barramundi (*Lates calcarifer*) larvae. Aquaculture 250:496–503

Hoyle I, Oidtmann B, Ellis T, Turnbull J, North B, Nikolaidis J, Knowles TG (2007) A validated macroscopic key to assess fin damage in farmed rainbow trout (*Oncorhynchus mykiss*). Aquaculture 270:142–148

Huntingford FA, Adams C, Braithwaite VA, Kadri S, Pottinger TG, Sandoe P, Turnbull JF (2006) Current issues in fish welfare. J Fish Biol 68:332–372

Jawad LA, Akyol O (2018) Vertebral anomalies in *Mullus barbatus* (Actinopterygiidae: Osteichthyes: Mullidae), collected from Izmir Bay, North-Eastern Aegean Sea, Turkey. Int J Mar Sci 8(7):59–65

Jawad L, Bannai M (2014) Characterization of hyperostosis in *Platax teira* (Forsskål, 1775) collected from marine water of Iraq, North west Arabian gulf. Sky J Agric Res 3(6):109–111

Jawad LA, Hosie A (2007) On the record of pug-headedness in snapper, *Pagrus auratus* (Forster, 1801) (Perciformes, Sparidae) from New Zealand. Acta Adriat 48(2):205–210

Jawad LA, Ibrahim M (2018) Saddleback deformities in fish species collected from the Arabian Gulf Coast of Jubail City, Saudi Arabia. J Ichthyol 58(3):401–409

Jawad LA, Al-Faisal AJ, Al-Mutlak FM (2014) Incidence of Lordosis in the cyprinid fish, Carasobarbus luteus and the Shad, *Tenualosa ilisha* collected from Barash waters, Iraq. Int J Mar Sci 4:1–4

Jawad LA, Al-Faisal AJ, Al-Mukhtar M (2015) A case of vertebral coalescence in Luciobarbus Xanthopterus (Heckel, 1843) (Pisces: Cyprinidae) obtained from the lower reaches of Mesopotamia. Travaux du Muséum National d'Histoire Naturelle "Grigore Antipa" 57(2):127–132

Jawad LA, Akyol O, Aydin İ (2018a) Severe case of Lordosis-kyphosis-Ankylosis in *Mullus barbatus* Linnaeus, 1758 (Teleostei: Mullidae) collected from the northern Aegean Sea, Turkey. Int J Mar Sci 8(12):101–105

Jawad LA, Aydin I, Akyol O (2018b) The first record of pughead deformity in the European hake, *Merluccius merluccius* (Linnaeus, 1758) collected from the northern Aegean Sea, Turkey. Cah Biol Mar 59:137–141

Jawad L, Mutlak F, Abbas AF (2018c) On the record of vertebral deformities in Mastacembelus mastacembelus collected from the lower reaches of Euphrates River, Iraq. Boletim do Instituto de Pesca, 42(1):216–220

Kurokawa T, Okamoto T, Gen K, Uji S, Murashita K, Unuma T, Nomura K, Matsubara H, Kim SK, Ohta H, Tanaka H (2008) Influence of water temperature on morphological deformities in cultured larvae of Japanese eel *Anguilla japonica* at completion of yolk resorption. J World Aquacult Soc 39:726–735

Korsøen Ø, Dempster T, Fjelldal, PG, Folkedal O, Kristiansen, T, Oppedal F (2009) Longterm submergence of Atlantic salmon (Salmo salar L.) during winter affects behaviour, growth and condition. Aquaculture 296, 373–381

Latremouille DN (2003) Fin erosion in aquaculture and natural environments. Res Fish Sci 11:315–335

Lein I, Holmefjord I, Rye M (1997) Effects of temperature on yolk sac larvae of Atlantic halibut (*Hippoglossus hippoglossus* L.). Aquaculture 157:123–135

López-Albors O, Gil F, Ramirez-Zarzosa G, Latorre R, Garcia-Alcazar A, Abellan E, Blanco A, Vazquez JM, Moreno F (1995) Early muscle injuries in a standard reared stock of sea bass *Dicentrarchus labrax* (L.). Aquaculture 138:69–76

MacLean A, Metcalfe NB, Mitchell DF (2000) Alternative competitive strategies in juvenile Atlantic salmon (*Salmo salar*): evidence from fin damage. Aquaculture 184:291–302

Martínez GM, Baron MP, Bolker JA (2007) Skeletal and pigmentation defects following retinoic acid exposure in larval summer flounder *Paralichthys dentatus*. J World Aquacult Soc 38:353–366

Matsuoka M (2003) Comparison of meristic variations and bone abnormalities between wild and laboratory-reared red sea bream. Jpn Agric Res Q 37:21–30

Michie I (2001) Causes of downgrading in the salmon farming industry. In: Kestin SC, Warriss PD (eds) Farmed fish quality. Fishing News Books, London, pp 129–136

Miyashita S, Sawada Y, Hattori N, Nakatsukasa H, Okada T, Murata O, Kumai H (2000) Mortality of northern bluefin tuna *Thunnus thynnus* due to trauma caused by collision during grow out culture. J World Aquacult Soc 31:632–639

Noble C, Jones HAC, Damsgård B, Flood MJ, Midling KØ, Roque A, Sæther BS, Cottee SY (2012) Injuries and deformities in fish: their potential impacts upon aquacultural production and welfare. Fish Physiol Biochem 38(1):61–83

Okamoto T, Kurokawa T, Gen K, Murashita K, Nomura K, Kim SK, Matsubara H, Ohta H, Tanaka H (2009) Influence of salinity on morphological deformities in cultured larvae of Japanese eel *Anguilla japonica* at completion of yolk resorption. Aquaculture 293:113–118

Okamura A, Yamada Y, Horie N, Utoh T, Mikawa N, Tanaka S, Tsukamoto K (2007) Effects of water temperature on early development of Japanese eel *Anguilla japonica*. Fish Sci 73:1241–1248

Ornsrud R, Gil L, Waagsbø R (2004) Teratogenicity of elevated egg incubation temperature and egg vitamin a status in Atlantis salmon, *Salmo salar* L. J Fish Dis 27:213–223

Pittman K, Berg Ø, Opstad I, Skiftesvik AB, Skjolddal L, Strand H (1990) Development of eggs and yolk-sac larvae of halibut (*Hippoglossus hippoglossus* L.). J Appl Ichthyol 6:142–160

Roberts RJ, Hardy RW, Sugiura SH (2001) Screamer disease in Atlantic salmon *Salmo salar* L. in Chile. J Fish Dis 24:543–549

Sadler J, Pankhurst PM, King HR (2001) High prevalence of skeletal deformity and reduced gill surface area in triploid Atlantic salmon (*Salmo salar* L.). Aquaculture 198:369–386

Slooff W (1982) Skeletal anomalies in fish from polluted surface waters. Aquat Toxicol 2:157–173

Sumagaysay NS, Hilomen-Garcia GV, Garcia LMB (1999) Growth and production of deformed and non-deformed hatchery–bred milkfish (*Chanos chanos*) in brackishwater ponds. Isr J Aquac Bamidgeh 51:106–113

Swan MA (1968) Double mouth deformity in a trout (*Salmo trutta*) and its cause. J Zool 156:449–455

Turnbull JF, Richards RH, Robertson DA (1996) Gross, histological and scanning electron microscopic appearance of dorsal fin rot in farmed Atlantic salmon, *Salmo salar* L., parr. J Fish Dis 19:415–427

Yamauchi M, Kim EY, Iwata H, Shima Y, Tanabe S (2006) Toxic effects of 2378—tetrachlorodibenzo-p-dioxin (TCDD) in developing red seabream (*Pagrus major*) embryo: an association of morphological deformities with AHR1 AHR2 and CYP1A expressions. Aquat Toxicol 80:166–179

Chapter 59
The Impact of some Social Taboos on Fisheries in Iraq

Laith A. Jawad

Abstract Food taboos are recognized from almost all human cultures. Most religions proclaim definite food items fit and others unfit for human consumption. Nutritional instructions and rules may direct specific stages of the human life cycle and may be related with special actions such as pregnancy, childbirth, lactation, and—in traditional societies—preparation for the hunt, battle, wedding, funeral, etc. On a comparative basis, many food taboos seem to make no sense at all, as to what may be acknowledged unhealthy by one group may be perfectly suitable to another. On the other hand, food taboos have a long history and one should expect a sound justification for the presence of certain nutritive customs in a given culture. Food taboos, whether scientifically correct or not, are frequently intended to guard the human individual and the observation, for example, that certain allergies and depression are associated with each other could have led to declaring food items taboo that was identified as causal agents for the allergies.

The present chapter revives interest in fish-related taboos that were practised by Iraqi people and attempts a functionalist's explanation. However, to illustrate some of the complexity of possible reasons for food taboo, two examples have been selected, namely traditional or social and religious-based taboos. These taboos are mainly practised by the majority Shia Muslim sector in Iraq. The paper showed how the taboos can affect the socio-economic status of the fisher's communities. At the end of the chapter, an assessment to the economic status of the Iraqi government is given and suggestions to use the available aquatic resources in both the freshwater and marine environments in Iraq instead of continuing loss of millions of dollars through the usage of large number of fish, shellfish, and other aquatic organisms for food or for export to increase national revenue.

L. A. Jawad (✉)
School of Environmental and Animal Sciences, Unitec Institute of Technology, Auckland, New Zealand

L. A. Jawad (ed.), *Tigris and Euphrates Rivers: Their Environment from Headwaters to Mouth*, Aquatic Ecology Series 11,
https://doi.org/10.1007/978-3-030-57570-0_59

59.1 Introduction

Taboos signify unwritten social guidelines that control human performance. Such restrictions not only may govern human social life, but also may disturb, and sometimes even directly manage, many elements of the local natural environment (Colding and Folke 1997). Whatsoever the reason for such limitations, taboos may, at least locally, play a main role for the conservation of natural resources, species, and ecosystems (Gadgil et al. 1993). There are criticizers who view the rehearsal of taboos as irrational and a hindrance towards development (Edgerton 1992), who dismiss any ecological reasons behind them (Rea 1981), or who claim that the taboos may not be observed to by some groups and, therefore, may be of no value in nature conservation (Alvard 1994).

Taboos associated with the natural environment originally may not have been envisioned for nature conservation. For example, species may be excluded because of their behavioural arrays and morphological characteristics (Zann 1983), or in the belief that they are toxic (Begossi and de Souza Braga 1992). Human insight of, and relation to, species may govern which ones to avoid. Species may be avoided simply because they occur in creation myths, because they represent religious representations, or because they are pet animals. For example, several species associated with different gods and goddesses in the Hindu scripture are sacred among Hindus.

In many circumstances, accidental nature conservation may be the result of such an escaping, which may be highly adaptive in ecological terms. Good examples are taboos associated with particular habitat patches, known as *sacred groves*. Sacred groves are smaller or larger ecosystems, set aside for religious purposes (Gadgil and Vartak 1974). These habitat patches, once widespread throughout India, Africa, and Europe (Frazer 1922), may be the only primary forests remaining locally (Wilson 1993).

Taboos may appear as mechanisms for the protection of species and habitats in modern society, but they have other social rules and consents, rooted in the traditional belief systems. Ecologically oriented anthropologists have revealed the complex ecological adaptations behind taboos (Balée 1985). For example, Harris (1971) suggested that the taboo on the Indian cattle is ecologically adaptive, in that it increases rather than decreases the capacity of the present Indian system of food production to support human life, through the production of milk, dung, and bullocks.

Taboos have also been a social device in the administration of natural assets, as recognized in several studies. Among traditional societies in Oceania, it was customary to impose taboos on the use of subsistence crops to prevent their being harvested at inappropriate times (Child and Child 1993). Taboos were also imposed on certain marine locations in order to avoid overexploitation of aquatic resources (Chapman 1985). These taboos were not always permanent in time and space, but could be removed when food resources were plentiful. Such taboos that directly manage nature are reportedly found among traditional groups from various parts of the world (Sankhala 1993). They may have been the outcome of a trial-and-error process of resource management strategy resembling the contemporary practice of adaptive management (Walters 1986). In the same means, taboos may be engaged as a social appliance for the implementation of ecologically adaptive behaviour, even though different cultural settings are devoted to them.

Definite food sources may be prohibited to women during childbirth, children under a particular age, or parents of newborns. Thus, stresses on a target resource may be greatly reduced. There is perhaps no single theory explaining why people employ specific food taboos. Conceivable ecological associations behind specific food taboos are often neglected by anthropologists, but should not be ruled out.

In this chapter, a review of the taboos that currently and since ancient time running throw the human society mainly in south of Iraq is given. These taboos are related to certain fisheries and precisely to certain freshwater and marine species. As far as the author is concerned, no previous study has provided such a review and draws such a relationship previously.

59.2 Species-Specific Taboos

General food taboos, on the other hand, comprise a total prohibition of species as food at all times. Here, we call such taboos *specific-species taboos*, stressing that species may be avoided not only as foods, but also for a variety of other plausible reasons. Such taboos totally protect animals and plants, both in time and in space, by eliminating their killing and harmful use by all members of a human community. At least in theory, this total prohibition may be of direct value in protecting threatened, and perhaps ecologically critical, species. Taboos may be an important factor for some vertebrates not yet extinct. In fact, no one knows how many of the world's species owe their survival to human groups that have avoided using them.

Although several specific-species taboos may be a consequence of the symbolical, mythological, or religious qualities that certain groups assign to animals (Child and Child 1993), conservation causes behind such taboos should not be ruled out. For example, Reichel-Dolmatoff (1976), studying the Tukanos of the Colombian north-western Amazon, states that the shaman in this group may interfere directly with hunting, fishing, gathering, and most other harvesting activities, and may personally control the number of animals to be killed. He might even completely prohibit the killing of certain animals in a restricted area whenever he thinks that a certain species is too low in abundance. Thus, the shaman becomes a powerful agent in the control and management of resources. Among other ethnic groups, the medicine man, the elders, or other prominent figures may hold the same position with a similar responsibility (Ntiamoa-Baidu 1991; Wilson 1993). This situation is also present in the tribes in south of Iraq, where the head of tribe or the head religious man declares the prohibition of hunting certain bird species or catching certain fish species.

The main point at this stage is that species conservation may exist in traditional societies as a means of human survival even they do not mean to employ. Even though such practices may have been diluted or even lost in many traditional societies (Alvard 1994) or might not be expressed in conservation terms, species protection by taboos seems to have existed over millennia (Gadgil et al. 1993).

Some specific-species taboos may have been established in the earlier in order to endure natural resources when local human groups confronted resource shortage and disasters of different kinds and extents (Gadgil and Berkes 1991). Thus, taboos may have evolved to increase the buffering capacity, or flexibility, of local ecosystems (Holling et al. 1997). Why a specific-species taboo continues to exist in a culture is difficult to determine. It may be that the species is still low in numbers or that the species is recognized, through traditional ecological knowledge, as keystone in the ecosystem. There may be institutional and cultural inertia to changing conditions.

An alternative hypothesis is that species are protected through specific-species taboos because they play a role in religious and cultural belief systems (Ingold 1994). In reality, it may be almost impossible to separate the belief system from performs and traditional ecological knowledge systems. Knowledge, practices, and beliefs tend to intermix among traditional peoples (Folke et al. 1997). No single theory can explain specific-species taboos; species may be avoided for many reasons, depending on regional conditions and the world view of different peoples.

59.3 The Historical and Religious Aspects of Fish Taboos

Through the history of human civilization, it seems that Sumerians the first world civilization that settled in the south of Tigris and Euphrates rivers around 5000 or 4000 B.C. have used fish as a symbol of Sun God. In the ancient Mesopotamian culture, people thought that the sun downed in the water in the sunset and rose in the next day again. Consequently, they associate the pictures of sun and fish together. In Sumerian culture, fish is the symbol of Ea. In Sami, language (Akkadian and Babylonian culture) Ea is the God of water and in Sumerian language Enki is the other name for Ea. It was the symbol of underwater and the death world, the third great God in the council of Mesopotamia, the God of wisdom and witchcraft, illustrated, and carved in some amulets (Cooper 2007).

Ancient Mesopotamian with the first civilization on earth distinguished the kind and measure of necessary fishes like fresh, fried, or dried fishes accurately. They commonly offered abundant spices of fishes wholesale and rare ones were ordered in retail. It means people were aware of fish diversity, but probably in a different way that we understand it at the present time. Sometimes, Assyrian Kings offered 10,000 species of fishes in their feasts. They used cheaper kind of fishes in these occasions, but Sumerian kings preferred plant-eating crabs that usually lived at the bottom of pits and muddy pools (Sahrhage and Lundbeck 1992).

The fish in the ancient Mesopotamian culture and as it has been shown by literature that fish was considered among the respected food and there are no taboos about eating them (Liverani 1995). On contrary, there were rituals about how to offer the fish to gods in temples (Hooke 1961). No taboos against eating any kind of fish were known in the ancient history of Mesopotamia.

With the descended of the three heavenly religions, Judaism, Christianity, and Islam, and with the subsectioning of each religion to several groups that have slight changed their faith and such changes have affected their fish consumption habits.

Unlike other heavenly religions, Christianity has never to this day established any generally or even commonly practised food taboos (Barclay 2010). Other two major world religions (Judaism and Islam) have more or less clearly clear rules in relation to diet—for instance, refraining from certain kinds of meat, from meat of animals slaughtered in the wrong way, or even from all meats. In the case of Christianity, the absence of food taboos is all the more remarkable given that its roots lie in Judaism, whose very precise food rules are preserved in divine laws still retained within the Christian Scriptures (Barclay 2010). This fact has been confirmed by Islam, where Allah (God) said that Jesus son of Merry came to free human being from the food restrictions that Jews have created. The Jewish food rules, and in particular self-restraint from pork, were very well known in ancient times, a frequent subject of question, delight or disdain by non-Jews, who wondered what on earth pigs were for if they were not meant to be eaten. This pork taboo, and the Jewish anxiety about contamination by Gentile "idolatry," combined to make it very difficult for Jews to share meals with non-Jews: they could be hosts, perhaps, but given the requirement of exchange in giving and sharing hospitality, it was difficult to make profound friendships across ethnic boundaries by meal fellowship, and the resulting autonomy caused Jews to be regarded at times as unfriendly and anti-social (Barclay 2010).

There are certain rules in Islam in relation to kind of food that human being should eat and offer policies and recommendations as responses to the concurrent health and social problems (Kamarulzaman and Saifuddeen 2010). Muslims contemplate the divine rules in every aspect of life. For the followers of Islam, there is a complete code of nutritional rules in the Holy Quran (Khattak et al. 2011). The recommendations on nutrition offered by Islam not only lead to physical health, but guarantee the mental health. One of the most important determinants of health is following the health teachings of Islam on eating and drinking (Avari et al. 2007). Moreover, a lot of consideration has been paid to food so that the Quran says: "Then let man look to his food" (Marwat et al. 2009). The "all-provider" term as one of the attributes of Allah, Allah's swearing on foods (swearing on figs and olives), and requests of manna and foods by Isa (Jesus) and Musa (Moses) each confirm the importance of food and nutrition (Akhondi 2008). There are many verses in the Quran about foods. In the 118 verses, Allah has spoken to us about foods and eating. This indicates the importance of nutrition from the perspective of Islam since everything we eat or drink influences not only the body but also the soul (Gorbani 2011). In this holy book, an emphasis has been placed on some nutrients, and their benefits are occasionally described (Marwat et al. 2009).

59.4 Fish Taboos in Iraq

In Iraq, the Shia sector of Muslims follows certain taboos that related to fish. These taboos have developed through the time and became as a rules and principles for Shia Muslim sector in Iraq and in the whole world. Abandon some fish species or aquatic organisms are affecting the socio-economic of the fishers in the first place and the country accordingly. In addition, the results of the fish taboos, amount of protein becomes available to the individuals in the fishers' communities, will be decreased and this will affect the well-being of the fishers populations.

The types of taboos that the people of Iraq have against eating or handling certain fishes covers both freshwater and marine species. They also cover fish consumption habits and routine. It is possible to list those taboos as follows:

Taboo no. 1: *"Eat fish every Wednesday of the week."* This taboo has a religious base. It originates from the fish eating routine that cousin of Prophet Mohammed "Ali" used to have. He used to eat fish every Wednesday of the week. In addition to Wednesday, Friday is another day for eating fish.

Taboo no. 2: *"Do not eat fish at night."* This taboo has a cultural origin and it has been developed through the practice. People believe that fish meal is heavy on the stomach and it needs sometime to be digested and since there is not much enough time between the last meal of the day and the time to go to bed, therefore, pain resulted from indigestion will occur. To cure this incident, people usually eat dates and drink black tea after eating fish even during the day as both dates and tea will help digest fish in the stomach.

Taboo no. 3: *"Do not drink yogurt after eating fish and mainly if you are eating carp fish 'Shaboot' (Arabibarbus grypus)."* This taboo has a cultural base and became to practise after people experienced some health problem represented in having absence of colour from their skin. Also, relate this case if the person eating fish and drink yoghurt and became angry, the skin health incidence will display. As the relation of this taboo to the carp fish "Shaboot," it may relate to a story that usually told by people. Shaboot fish is one of the cyprinid fish species that has been known for its great taste, and it has been on high demand by all sectors of people living in the Euphrates-Tigris Rivers basin. Zivotofsky and Amar (2006) wrote on this fish species from the Jewish religion point of views and here I quote, "The *Babylonian Talmud is a 2711 page encyclopaedic work of Jewish law and lore that was redacted in Persia approximately 1500 years ago and has served as the focus of Jewish religious study and the foundation of Jewish jurisprudence ever since. Among the intricate legal debates and anecdotes, the Talmud student spy's myriad glimpses of the daily life of the Jews living in Persia during the period of the Sassanian Empire. Because the Jews in Persia lived clustered around the Tigris and Euphrates Rivers, it is reasonable to assume that fish were an important part of their diet. Probably the most frequently mentioned fish Talmudic literature is the shibuta (Shabout), a species that appears in a variety of contexts and is described as a popular and tasty fish."* They also narrated," *...The Talmud records that the shibuta, a permitted food, tastes like prohibited* *pork (Chullin 109b), that it was commonly prepared for the festive Sabbath meal (Shabbat 119a), that salted shibuta was recommended as a treatment for 'yarkuna' (perhaps jaundice) (Shabbat 110b), and that eating shibuta in the spring can lead to specific skin ailment (Pesachim 112b).* Zivotofsky and Amar (2006) suggested and here I quote," *Iraqi Jews in the early medieval period (*e.g. *Rav Hai Gaon, 969–1038 C.E) referred to it by its Arabic name, shabout. . .An important clue to the identity of the shibuta is the statement of the Jerusalem Talmud, redacted in Palestine around 1600 years ago, (Ta'anit ch. 4) that the species was not available in the Land of Israel,* i.e. *Roman-occupied Palestine. Thus, the shibuta is a species that existed in Persia, but not along the eastern shore of the Mediterranean".* Zivotofsky and Amar (2006) suggested that the identity of Shabout fish for all Iraqi is very simple and no two person can miss the identity of this fish. In this aspect, Zivotofsky and Amar (2006) suggested and her I quote," *In the context of discussing medieval Babylonian cuisine,* Nasrallah (2003) *states: "Fish from the river Tigris was highly valued by the medieval Baghdadis ... Top quality fishes were shabout (carp), ... Of the commercially important fishes that are still swimming in the two rivers the medieval favourites still hold their ground, especially the shabout (carp).*"

With this religious importance of this fish to Jewish people, the story of relating eating this fish and rink yoghurt can cause skin disease might be created by Jewish people to keep away others from buying this fish.

Taboo no. 4: *"Do not eat fish without scales."* This taboo is very well known and very practised among Shia sector of Islam. It has mixed cultural and religious bases. The religious base derives from the following story that Shia Muslim sector usually narrated. The story says that 1 day cousin of the Prophet Mohamed was washing his hands, face, and feet prior to prey. Once he start taking the water from the river and start washing, a catfish came and disturb the bottom of the river and render it muddy not suitable for washing. Ali condemned this fish and since then the followers of Ali (Shia sector of Islam) do not eat catfish species. Individuals of catfish species usually have no scales on their bodies, and in the addition to the above-mentioned story of condemnation by Ali cousin of the prophet Mohamed, any scaleless fish is forbidden to be eaten. Here, we have cultural base, where in Shia believe the fish without scales is easily to get injured and consequently get infected rendering the fish not suitable as a food.

Taboo no. 5: *"Do not eat fish that can eat human such as sharks and relatives."* This taboo has a religious base and could be related to a verse in Quran that prohibited eating dead animals and drinking blood. The verse says, "*Say: I find not in that which is revealed to me aught forbidden for an eater to eat thereof, except that it be what dies of itself, or blood poured forth, or flesh of swine- for that surely is unclean- or what is a transgression, other than (the name of) Allah having been invoked on it, But whoever is driven to necessity, not desiring nor exceeding the limit, then surely thy Lord is Forgiving, Merciful.*" The taboo is also may related to the say of the Prophet Mohamed that he suggested that all the beast of the forest with canine and birds of prey with claws that can prey on human their meat is forbidden for Muslims.

In the present time and with the import of frozen fillets of fishes from around the world and due to its low prices, many Muslims contradict their belief and eat imported frozen fillet of unknown species.

There are other types of taboos that usually practised by people in Iraq and not in other Arab or Muslims countries. These taboos have cultural bases rather than religious base grounds. People in Iraq do not eat any crustaceans other than shrimps, molluscs, shellfishes, and any other aquatic animals present in the vicinity of the fresh and coastal waters. Moreover, if fishers collected any of these animals, they throw them back to water (in case of sea fishing) and leave it on the river bank (in case of freshwater catch).

59.5 Aquatic Organism Included in the Food Taboos Practised in Iraq

There are several animal groups that are classified as prohibited from being used as food by Iraqi people for either religious or cultural reasons or both. Species of these groups will be listed and discussed briefly in this section.

59.5.1 Species of Scaleless Fish

59.5.1.1 Catfish Species

The catfishes (Order Siluriformes) encompass 35 families and over 2867 species live worldwide in freshwaters although two families are primarily marine. These fishes originated in the Late Cretaceous. The greatest diversity is found in South America (Nelson 2006). Iraq has four families with four genera and five species, one of which is an exotic. Their biodiversity in relation to other Iraqi fishes is given in Coad (2014).

Members of the catfish groups range in size from under 10 cm to over 3 m, one of the largest species, *Silurus glanis*, being found in Iran. Among their characteristic features are naked body and lacking scales. There are 1–4 pairs of barbels around the mouth (one nasal, one maxillary, and two on the chin), eyes are usually small since the barbels are used to find food, the premaxillae bones of the upper jaw usually have teeth, the vomer bone in the roof of the mouth is toothed as are the pterygoid palatine bones, the preopercle and interopercle bones of the gill cover are small, an adipose fin is present, principal caudal fin rays are 18 or fewer (usually 17), serrate spines frequently present at the front of the dorsal and pectoral fins which can be locked erect, there are no pelvic fin spines, and some have an air-breathing apparatus (e.g. Heteropneustidae in Iraq).

In many parts of the world, catfishes are important food, but in the countries with Shia Muslims majority, catfishes do not have an economic value due to the food taboos against them (see above section). The pectoral fin spines can be venomous and can cause death in humans. Individuals of this group feed on a wide range of invertebrates and fishes. They prefer open water area for spawning, build nests to protect the young, or brood eggs in the mouth.

59.5.1.2 Species Account of the Catfishes

(a) **Family Sisoridae**

The sisorid are catfishes that have the ability to attach to items in the environment with aid of suckers. This group of catfish comprises 18 genera with about 196 species (Eschmeyer and Fong 2011). Five nominal species are reported from the Tigris-Euphrates basin in Southwest Asia but the diversity is very limited, compared to India for example (Coad 2014). This family contains one genus *Glyptothorax* including two species *G. kurdistanicus* and *G. steindachneri.*

(b) **Family Siluridae**

This family comprises 12 genera and about 99 species (Nelson 2006; Eschmeyer and Fong 2011) with one reported from Iraq so far. These family members are distinguished in having large, scaleless and elongate body, a moderately compressed head, a non-protractile mouth, depressible teeth on the jaws. The only species of this family in Iraq is Silurus triostegus, but in neighbouring Iran, there is another species of the genus *Silurus*, *S. glanis*.

(c) **Heteropneustidae**

Individuals of this family are known also as stinging or airsac catfishes. They belong to a genus with four species found naturally from Pakistan through India to Thailand (Eschmeyer and Fong 2011). In Iraq, this family contains only one species, *H. fossilis*, which is introduced to Iraq to control malaria vector (Jawad 2003). They are diagnosed in having an elongated and compressed body with a flattened head, presence of 4 pairs of barbels present (nasal, maxillary and 2 mandibular), presence of an air sacs, very long anal fin confluent with the caudal or separated from it by a notch, pectoral fin with a strong and venomous spine. These fishes can live in standing water by breathing air. They are dangerous to man since the pectoral spine harbours strong venom (Jawad 2015). Stinging catfishes nonetheless are an important food in their native range except in the Shia Muslims majority areas such as Iraq and Iran.

59.5.2 *Species of eel Group*

59.5.2.1 Freshwater eel Species

The only family found in freshwater system of Iraq that belongs to the Order Synbranchiformes is Mastacembelidae (Nelson 2006). This family comprises 84 species (Eschmeyer and Fong 2015), among which Iraq has only one species.

(a) **Family Mastacembelidae**

Mastacembelids, the spiny eel, or spinyback family is mainly freshwater riverine family with an Old World distribution throughout tropical Africa and eastwards to Korea and Malaysia in Asia although the majority of species occur in Africa. Members of this family reach a maximum length which is less than 1.0 m. The family is diagnosed by a very elongate, compressed or subcylindrical body, minute cycloid scales or body naked, a non-protractile mouth, gill opening a slit, an elongate snout with a sensitive tip flanked by tubular anterior nostrils; the posterior nostrils are hence far to the rear, long dorsal, and anal soft fins (Coad 2015). *Mastacembelus mastacembelus* is the only species of this family living in Iraq.

59.5.2.2 Marine eel Species

There are several species of eel living in the marine waters of Iraq, but still there are several species to be reported or described from this part of the world due to the lack of an ichthyological exploration (Jawad et al. 2014). The list of the species is the common species found in the Iraqi marine waters:

(a) **Family: Congridae**
    Conger cinereus
(b) **Family: Muraenesocidae**
    Muraenesox cinereus

59.5.3 *Cartilaginous Fishes*

In the marine waters of Iraq, there is certain number of sharks and rays species as water depth and other water physical characters do not allow for the presence of a wide variety of cartilaginous fish species. In general, The Arabian Gulf area is poor in regard to the species composition of sharks and their allies (Almojil et al. 2015).

Sharks and rays are among the sea commodity that has a high market demand all over the world except in the countries where Shia Muslims are majority such as Iraq and Iran. Such global demand has led to some species to be listed in the International Union for Conservation of Nature (IUCN) Red List of Threatened Species, and many regional fisheries management organizations consider the implementation of shark conservation a priority (Jawad 2013).

59.5.3.1 Sharks and Rays Species Content of Iraq

The shark and rays species content found in the Iraqi marine waters is not very well known due to the fish taboos that restrict their catch and reaching samples to the local fish markets as the main source of obtaining samples in the absence of major ichthyological surveys in Iraq. The followings are the most common species of sharks reported from the marine waters of Iraq.

(a) **Family: Hemiscylliidae**
  Chiloscyllium arabicum
(b) **Family: Stegostomatidae**
  Stegostoma fasciatum
(c) **Family: Odontaspididae**
  Carcharias taurus
(d) **Family: Triakidae**
  Mustelus mosis
(e) **Carcharhinidae**
  Carcharhinus amboinensis.
  Carcharhinus dussumieri.
  Carcharhinus leucas.
  Carcharhinus limbatus.
  Carcharhinus macloti.
  Carcharhinus sorrah.
  Rhizoprionodon acutus.
  Rhizoprionodon oligolinx
(f) **Family: Sphyrnidae**
  Sphyrna lewini.
  Sphyrna mokarran
(g) **Family Pristidae**
  Pristis zijsron
(h) **Family: Rhinidae**
  Rhina ancylostoma
(i) **Family: Rhynchobatidae**
  Rhynchobatus djiddensis
(j) **Family: Torpedinidae**
  Torpedo sinuspersici
(k) **Dasyatidae**
  Himantura imbricate
(l) **Family: Gymnuridae.**
  Gymnura poecilura

(m) **Family: Myliobatidae**
Aetobatus flagellum.
Aetobatus ocellatus.
Aetomylaeus milvus.
Aetomylaeus nichofii
(n) **Family: Rhinopteridae**
Rhinoptera jayakari.

59.6 The Effects of Fish Taboos on the Socio-Economic of the Fishers and the Iraq

From the above-mentioned list of freshwater and marine organisms that are affected by the fish taboos in Iraq, it is clear that the socio-economic status of the fishers' communities are facing dilemma. For example, in the lower reaches of Mesopotamia more than half of the daily catch of the fishers is composed of catfish species and freshwater eels. This part of the catch will throw back to the river or leave it on the river's bank for birds to eat. Such a decline in the catch will have a direct impact on the income of the fishers and consequently on their social life. As to the government, this loss in the catch represents a national harm to the revenue of the country. Those aquatic resources both freshwater and marine that have no local market can be a worthy export commodity that can have a good revenue. In addition to the undesirable fish species locally, there is large number of crustaceans and shellfish species that are considered food delicacy in other countries. The government should commence aid investors to utilize the aquatic resources of Iraq and build factories for exporting aquatic resources, i.e. fish, shellfish, sharks, mollusc, seaweeds in different forms and categories, i.e. frozen, canned, smoked. The value of such tremendous aquatic resources could rise to millions of dollars and can be a major revenue to the country.

References

Akhondi M (2008) An overview of Islamic perspective about food and nutrition. Hason 15:175

Almojil DK, Moore AB, White WT (2015) Sharks & Rays of the Arabian/Persian gulf. MGB (INT) Limited, London, p 178

Alvard MS (1994) Conservation by native peoples – prey choice in a depleted habitat. Hum Nat An Interdisciplinary Biosocial Perspective 5:127–154

Avari M, Azizi M, Ghanee Ezabadi B, Rigazadeh M (2007) Knowledge attitude and practice of people aged 25 and older in Yazd regarding foods mentioned in Quran and hadiths. Res Q J Public Health School Yazd Univ Med Sci 6(1):26–35

Balée W (1985) Ka'apor ritual hunting. Hum Ecol 13(4):485–510

Barclay JM (2010) Food, Christian identity and global warming: a Pauline call for a Christian food taboo. Expository Times 121(12):585–593

Begossi A, de Souza Braga FM (1992) Food taboos and folk medicine among fishermen from the Tocantins River (Brazil). Amazoniana XII:101–118
Chapman M (1985) Environmental influences on the development of traditional conservation in the South Pacific region. Environ Conserv 12:217–230
Child AB, Child IL (1993) Religion and magic in the life of traditional peoples. Prentice-Hall, Englewood Cliffs, NJ
Coad BW (2014) Review of the freshwater catfishes of Iran (order Siluriformes). Iran J Ichthyol 1 (4):218–257
Coad BW (2015) Native fish biodiversity in Afghanistan. Iran J Ichthyol, 2(4):227–234
Colding J, Folke C (1997) The relations among threatened species, their protection, and taboos. Conserv Ecol 1:1–6
Cooper JC (2007) Illustrated dictionary of traditional symbols. (Karbassian M Trans.) Tehran: Nashre-Nou Publication
Edgerton RB (1992) Sick societies: challenging the myth of primitive harmony. The Free Press, New York
Eschmeyer WN, Fong JD (2011) Pisces. In: Zhang, Z.-Q. (ed.). Animal biodiversity: an outline of higher level classification and survey of taxonomic richness. Zootaxa 3148:26–38
Eschmeyer WN, Fong JD (2015) Catalog of fishes: genera, species, references. California Academy of Sciences, San Francisco
Folke C, Berkes F, Colding J (1997) Ecological practices and social mechanisms for building resilience and sustainability. In: Berkes F, Folke C (eds) Linking social and ecological systems: management practices and social mechanisms for building resilience. Cambridge University Press, Cambridge, p 30
Frazer JG (1922) *The golden bough.* From the 1992 Swedish reprint "Den gyllene grenen: studier i magi och religion". In: Bokförlaget Natur och Kultur. Sweden, Stockholm
Gadgil M, Berkes F (1991) Traditional resource management systems. Res Manage Optimization 8:127–141
Gadgil M, Vartak VD (1974) The sacred groves of Western Ghats in India. Econ Bot 30:152–160
Gadgil M, Berkes F, Folke C (1993) Indigenous knowledge for biodiversity conservation. Ambio 22:151–156
Gorbani Z (2011) Fruit Paradise in Quran. Electronic J Borhan 1:16–24
Harris M (1971) Culture, man, and nature. An introduction to general anthropology. Thomas Y. Crowell, New York
Holling CS, Berkes F, Folke C (1997) Science, sustainability, and resource management. In: Berkes F, Folke C (eds) Linking social and ecological systems: management practices and social mechanisms for building resilience. Cambridge University Press, Cambridge, p 30
Hooke SH (1961) Fish symbolism. Folklore 72(3):535–538
Ingold T (1994) Food taboos and prohibitions. Companion encyclopedia of anthropology. Humanity, culture, and social life. Routledge, London, pp 250–256
Jawad L (2003) Impact of environmental change on the freshwater fish fauna of Iraq. Int J Environ Stud 60(6):581–593
Jawad LA (2013) Sharks of the world: a fully illustrated guide-edited by DA Ebert, S. Fowler & L. Compagno. J Fish Biol 83(5):1487–1488
Jawad LA (2015) The Indian catfish, *Heteropneustes fossilis* (Bloch, 1794) envenomation-a report of four cases from Basrah and Maysan, south of Iraq. TOFIQ J Med Sci 2(2):52–61
Jawad LA, Al-Mukkhtar M, Faddagh MS (2014) Confirmation of the presence of *Heniochus acuminatus* (Linnaeus, 1758)(Chaetodontidae) and *Pomacanthus maculosus* (Forsskål, 1775) (Pomacanthidae) in Iraqi marine waters, Arabian gulf. Arxius de Miscel· lània Zoològica 12:124–129
Kamarulzaman A, Saifuddeen SM (2010) Islam and harm reduction. Int J Drug Policy 21 (2):115–118
Khattak JZK, Mir A, Anwar Z, Wahedi HM, Abbas G, Khattak HZK et al (2011) Concept of halal food and biotechnology. Adv J Food Sci Technol 3(5)

Liverani M (1995) The deeds of ancient Mesopotamian kings. Civilizations Anc Near East 4:2352–2366
Marwat SK, Khan MA, Khan MA, Ahmad M, Zafar M, REHMAN F et al (2009) Fruit plant species mentioned in the holy Qura'n and Ahadith and their Ethnomedicinal importance. Am Eurasian J Agric Environ Sci 5(2):284–295
Nasrallah N (2003) Delights from the garden of Eden: a cookbook and a history of the Iraqi cuisine. Author House Publishing, Bloomington
Nelson JS (2006) Fishes of the World, 4th edn. John Wiley & Sons, New York, p xix+601
Ntiamoa-Baidu Y (1991) Conservation of coastal lagoons in Ghana: the traditional approach. Landsc Urban Plan 20:41–46
Rea AM (1981) Resource utilization and food taboos of Sonoran desert peoples. J Ethnobiol, 1 (1):69–83
Reichel-Dolmatoff G (1976) Cosmology as ecological analysis: a view from the rainforest. Man 11:307–318
Sahrhage D, Lundbeck J (1992) A history of fishing. Springer-Verlag, Berlin, pp 1–348
Sankhala K (1993) Prospering from the desert. In: Kemf E (ed) Indigenous peoples and protected areas. Earthscan, London, pp 18–23
Walters CJ (1986) Adaptive management of renewable resources. McGraw-Hill, New York
Wilson A (1993) Sacred forests and the elders. In: Kemf E (ed) Indigenous peoples and protected areas. Earthscan, London, UK, pp 244–248
Zann LP (1983) Traditional management and conservation of fisheries in Kiribati and Tuvalu atolls. In: K. Ruddle and R.E. Johannes, editors. Contending with global change: Study Number 2. Traditional marine resource management in the Pacific basin: an anthology. UNESCO/ROSTSEA, pp 79–101
Zivotofsky AZ, Amar Z (2006) Identifying the ancient shibuta fish. Environ Biol Fish 75 (3):361–363

Chapter 60
Inland Water Fishes and Fisheries in Iran

Tooraj Valinassab and Hamid Reza Esmaeili

60.1 Inland Fishing and Fisheries in Iran

Fisheries in Iran are classified as three geographical groups of:

1. Southern part of Iran fisheries in the Persian Gulf and Oman Sea with focus on fishing of different marine fishes. More than 900 species are inhabited in these two ecosystems (Valinassab 2013) of which about 70 species are considered as commercial catch such as tunas, mackerels, seabreams, snappers, croakers, groupers, silver pomfret, carangids, sardines, anchovies, and so on with a total catch of 6,00,802 tons (IFO Programming and Budget Office 2017).
2. Northern part of Iran fisheries in the Caspian Sea as a brackish water ecosystem (salinity ~12 ppt) inhabits 114 fish species (Valinassab 2013) of which only a few species are considered as commercial ones with emphasize on sturgeons (endangered species and new rule as protected species), bony fishes (kutum, mullets, carps, . . .), and kilka fishes (clupeids) with an annual total catch of 33,396 tons (IFO Programming and Budget Office 2017).
3. Inland water fisheries from rivers, lakes, and dam reservoirs as freshwater ecosystems inhabit about 300 fish species (Table 60.3). The main parts of freshwater fishes catch as commercial ones are: different species of barbs, trout, salmon, and carps with an annual total catch of 61,000 tons (IFO Programming and Budget Office 2017). The main fishing methods are gillnet, using electricity, hook, Shill as a traditional method, beach seine, and cast net.

T. Valinassab (✉)
Iranian Fisheries Science Research Institute, Agricultural Research, Education and Extension Organization, Tehran, Iran

H. R. Esmaeili
Ichthyology Research Laboratory, Zoology Section, Department of Biology, College of Sciences, Shiraz University, Shiraz, Iran

L. A. Jawad (ed.), *Tigris and Euphrates Rivers: Their Environment from Headwaters to Mouth*, Aquatic Ecology Series 11,
https://doi.org/10.1007/978-3-030-57570-0_60

Table 60.1 Fisheries and Aquaculture statistics in Inland waters of Iran

Year	2012	2013	2014	2015	2016
Catch (tons)	42,363	45,551	51,666	55,430	61,391
Cold-water fishes culture (tons)	131,000	143,917	126,000	140,244	163,325
Warm-water fishes culture (tons)	155,021	168,447	170,991	185,135	203,243
Ornamental fishes (N*1000)	147,854	186,301	203,670	213,901	232,416
Annual per capita fish consumption	8.1	8.5	9.2	10.0	10.6

Table 60.2 A glance on different facilities for fisheries and Aquaculture in inland waters of Iran

Year	2012	2013	2014	2015	2016	2017
Number of farms for warm-water fishes	14,295	14,615	16,254	17,193	18,382	1931
Number of farms for cold-water fishes	1907	1923	1595	1776	1995	1894
Number of fishermen and Aquaculturist in inland waters	60,761	65,173	68,287	71,976	82,428	87,502

The fishing and aquaculture trend during last 5 years from 2012 to 2016 have been tabulated in Table 60.1. Also, some added informations about number of fishermen and aquaculturist in inland waters and number of farms for both cold-water and warm-water fishes are presented (Table 60.2).

A total of 3300 rivers distributed in Iran and the main of them are: Karoon, Sefid-rud, Arvand, Tajan, Gorgan-rud, Aras, Bahu kalat, Sarbaz, Mond, Dalaki, Gamasiab, Bashar, Sardab-rud, Bazoft, Zayandeh-rud, and Jajrud.

- Cold-water fishes are mostly trout.
- Warm-water fishes are mostly carps.

Different researchers have studied and reported the Inland water fishes of Iran (Armantrout 1980; Abbasi et al. 1998; Coad, 1998, 1999; Abdoli 2000; Abell et al. 2008; Valinassab 2013; Ramin et al. 2018; Berg 1949; Abdoli 2016; Naderi Jolodar & Abdoli 2010), but the last and the best one was done by Esmaeili (2017) that he did a new review and presented an updated checklist of freshwater fishes of Iran with focus on taxonomy, distribution, and conservation status based on literature, and taxon occurrence data obtained from natural history and new fish collections.

Esmaeili (2017) reports that Iran is a mountainous country and much of it is desert and there are thousands of small springs and streams with no present or recent connection to other water bodies. For a long time, the Iranian drainage basins have been divided into 19 major basins based on field work, maps, fish distributions, history of research, works on hydrography, and areas deemed important for an understanding of zoogeography (see Esmaeili et al. 2010, Esmaeili et al. 2014a, b; Coad 2017) (Fig. 60.1). However, they divide the Persian Gulf basin into four main sub-basins including the Tigris, Zohreh, Persis (previously known as

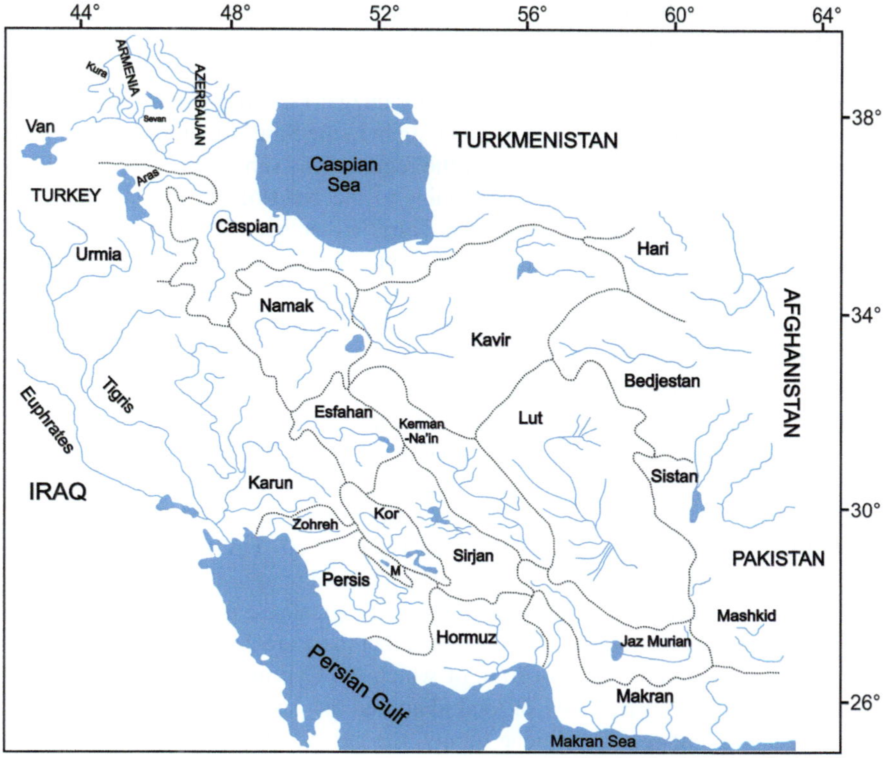

Fig. 60.1 Map of Iran showing different basins, (Derived from Esmaeili 2017)

Gulf basin), and Hormuz (Hormuzgan) basins which all drain to the Persian Gulf (Fig. 60.1).

There are two main types of basin, exorheic where the rivers and lakes drain to the sea and endorheic, where rivers drain to an internal basin such as a lake, or are lost in the desert, and have no connection with the sea. The exorheic basins all fringe the southern part of Iran. The bulk of the basins, in number (15) and area (about 78.1% of Iran), are endorheic. These plateau basins lie at an average altitude of 800 m, alternating with mountains ridges at an average of 2000 m. The salt lakes and flats of these basins are fed primarily by groundwater rather than rain and water is lost by evaporation.

According to Esmaeili (2017) report, the *endorheic basins* are: Bejestan, Caspian Sea, Dasht-e Kavir, Dasht-e Lut, Esfahan, Hamun-e Mashkid, Hamun-e Jaz Murian, Kor River, Lake Maharlu, Lake Urmia, Namak Lake, Sirjan, Sistan, Hari (Tedzhen) River, Kerman-Na′in, and *exorheic basins,* in which are divided into two large basins, (a) the Makran and (b) the Persian Gulf.

The Makran basin is the coastal region of southeastern Iran between the Strait of Hormuz and the Pakistan border that drains to the Oman Sea.

The Persian Gulf basin is subdivided to four sub-basins of: *(1) Hormuz (Hormuzgan) sub-basin*: The Hormuz (or Hormuzgan) sub-basin comprises a number of intermittent streams and rivers which drain to the Straits of Hormuz. *(2) Persis sub-basin:* Persis is an ancient name for the region. This sub-basin comprises rivers which drain the southern Zagros Mountains to the head of the Persian Gulf, but which are neither now tributaries of the Tigris River nor are the salt streams of Hormuz. *(3) Tigris sub-basin:* The Tigris-Euphrates basin is the largest and most important river system between the Nile and the Indus. *(4) Zohreh sub-basin:* At its northern edge, the Zohreh River flows across the Khuzestan plains and is close to Tigris River tributaries. Shesh Peer, Fahlian, and Kheir Abad are the main tributaries of the Zohreh sub-basin.

The checklist of each basin is also provided in recent years. Esmaeili et al. (2014b) reported a total of 119 species belonging to 63 genera, 18 families, 16 orders, and two classes from southern Caspian Sea basin (SCSB), including 19 exotic fish species in seven families. The number of species was higher than two published checklists in 1988 (74 species and 42 genera) and in 2010 (116 confirmed species belonging to 61 genera and 18 families). The ichthyofauna of the southern Caspian Sea has been of great importance for modern biological research.

On the other hand, Abdoli (2000) reported that there are 23 families and 139 species of fishes in inland waters of Iran, of which the Cyprinidae family as the most abundant family comprises 58% of total identified fish species till 20 years ago. On the other hand, amongst Cyprinidae family, the genus *Capoeta* with 20 species, is the most abundant and widespread taxon inhabiting in nearly all water basins. Two families of Namacheilidae and Cyprinidae are found in all water basins of inland waters of Iran. From point of distribution pattern, the highest fish species diversity is found in the Caspian basin in north of Iran and Tigris Euphrates basin in the west and southwest of Iran; meanwhile the lowest fish species diversity is observed in other basins far from two above mentioned ones (Table 60.3).

Also, 27 exotic species have been introduced to Iran's inland waters and also some native species have been transplanted from one basin to other basins. The exotic and transplanted species have been tabulated in Table 60.4.

Table 60.3 The checklist of Inland water fishes of Iran

No.	Family	Species
1	Petromyzontidae	*Caspiomyzon wagneri*
2	Carcharhinidae	*Carcharhinus leucas*
3	Acipenseridae	*Acipenser gueldenstaedtii*
		Acipenser nudiventris
		Acipenser persicus
		Acipenser ruthenus
		Acipenser stellatus
		Huso huso
4	Lepisosteidae	*Atractosteus spatula*
5	Anguillidae	*Anguilla anguilla*
6	Clupeidae	*Alosa braschnikowi*
		Alosa caspia
		Alosa curensis
		Alosa kessleri
		Alosa saposchnikowii
		Alosa sphaerocephala
		Alosa volgensis
		Clupeonella caspia
		Clupeonella engrauliformis
		Clupeonella grimmi
		Tenualosa ilisha
7	Chanidae	*Chanos chanos*
8	Acheilognathidae	*Rhodeus caspius*
9	Cyprinidae	*Arabibarbus grypus*
		Bangana dero
		Barbus cyri
		Barbus karunensis
		Barbus lacerta
		Barbus miliaris
		Barbus urmianus
		Capoeta aculeate
		Capoeta alborzensis
		Capoeta anamisensis
		Capoeta birunii
		Capoeta buhsei
		Capoeta capoeta
		Capoeta coadi
		Capoeta ferdowsii
		Capoeta fusca
		Capoeta gracilis
		Capoeta heratensis
		Capoeta macrolepis
		Capoeta mandica

(continued)

Table 60.3 (continued)

No.	Family	Species
		Capoeta pyragyi
		Capoeta razii
		Capoeta saadii
		Capoeta sevangi
		Capoeta shajariani
		Capoeta trutta
		Capoeta umbla
		Carasobarbus kosswigi
		Carasobarbus luteus
		Carasobarbus sublimus
		Carassius auratus
		Carassius gibelio
		Carassius langsdorfii
		Cyprinion kais
		Cyprinion macrostomum
		Cyprinion milesi
		Cyprinion tenuiradius
		Cyprinion watsoni
		Cyprinus carpio
		Garra amirhosseini
		Garra gymnthorax
		Garra lorestanensis
		Garra mondica
		Garra nudirentris
		Garra persica
		Garra roseae
		Garra rossica
		Garra rufa
		Garra tashanensis
		Garra typhlops
		Labeo rohita
		Luciobarbus barbulus
		Luciobarbus brachycephalus
		Luciobarbus capito
		Luciobarbus caspius
		Luciobarbus esocinus
		Luciobarbus kersin
		Luciobarbus mursa
		Luciobarbus subquincunciatus

(continued)

Table 60.3 (continued)

No.	Family	Species
		Luciobarbus xanthopterus
		Mesopotamichthys sharpeyi
		Schizocyprisalti dorsalis
		Schizopygopsis stolickai
		Schizothorax intermedius
		Schizothorax pelzami
		Schizothorax zarudnyi
		Tariqilabeo adiscus
		Tariqilabeo diplochilus
10	Danionidae	*Barilius mesopotamicus*
		Cabdio morar
11	Gobionidae	*Gobio nigrescens*
		Pseudorasbora parva
		Romanogobio macropterus
		Romanogobio persus
12	Leuciscidae	*Abramis brama*
		Acanthobrama marmid
		Acanthobrama microlepis
		Acanthobrama persidis
		Acanthobrama urmianus
		Alburnoides coadi
		Alburnoides damghani
		Alburnoides eichwaldii
		Alburnoides holciki
		Alburnoides idignensis
		Alburnoides namaki
		Alburnoides nicolausi
		Alburnoides parhami
		Alburnoides petrubanarescui
		Alburnoides qanati
		Alburnoides samiii
		Alburnoides tabarestanensis
		Alburnus atropatenae
		Alburnus caeruleus
		Alburnus chalcoides
		Alburnus doriae
		Alburnus filippii
		Alburnus hohenackeri
		Alburnus sellal
		Alburnus taeniatus
		Ballerus sapa

(continued)

Table 60.3 (continued)

No.	Family	Species
		Blicca bjoerkna
		Chondrostoma cyri
		Chondrostoma esmaeilii
		Chondrostoma orientale
		Chondrostoma regium
		Leucaspius delineates
		Leuciscus aspius
		Leuciscus latus
		Leuciscus vorax
		Pelecus cultratus
		Pimephales promelas
		Petroleuciscus ulanus
		Rutilus kutum
		Rutilus lacustris
		Scardinius erythrophthalmus
		Squalius break
		Squalius lepidus
		Squalius namak
		Squalius turcicus
		Vimba persa
13	Tincidae	*Tinca tinca*
14	Xenocypridae	*Ctenopharyngodon idella*
		Hemiculter leucisculus
		Hypophthalmichthys molitrix
		Hypophthalmichthys nobilis
		Mylopharyngodon piceus
15	Cobitidae	*Cobitis avicennae*
		Cobitis faridpaki
		Cobitis linea
		Cobitis saniae
		Sabanejewia aurata
		Sabanejewia caspia
16	Nemacheilidae	*Eidinemacheilus smithi*
		Oxynoemacheilus bergianus
		Oxynoemacheilus brandtii
		Oxynoemacheilus chomanicus
		Oxynoemacheilus elsae
		Oxynoemacheilus euphraticus
		Oxynoemacheilus frenatus
		Oxynoemacheilus karunensis
		Oxynoemacheilus kiabii
		Oxynoemacheilus kurdistanicus

(continued)

Table 60.3 (continued)

No.	Family	Species
		Oxynoemacheilus longipinnis
		Oxynoemacheilus marunensis
		Oxynoemacheilus parvinae
		Oxynoemacheilus persa
		Oxynoemacheilus tongiorgii
		Oxynoemacheilus zagrosensis
		Paracobitis abrishamchiani
		Paracobitis atrakensis
		Paracobitis basharensis
		Paracobitis hircanica
		Paracobitis longicauda
		Paracobitis malapterura
		Paracobitis molavii
		Paracobitis persa
		Paracobitis rhadinaea
		Paraschistura abdolii
		Paraschistura alta
		Paraschistura aredvii
		Paraschistura bampurensis
		Paraschistura cristata
		Paraschistura delvarii
		Paraschistura hormuzensis
		Paraschistura ilamensis
		Paraschistura kermanensis
		Paraschistura kessleri
		Paraschistura makranensis
		Paraschistura naumanni
		Paraschistura nielseni
		Paraschistura susiani
		Paraschistura turcmenica
		Paraschistura turcomana
		Sasanidus kermanshahensis
		Turcinoemacheilus bahaii
		Turcinoemacheilus hafezi
		Turcinoemacheilus kosswigi
		Turcinoemacheilus saadii
17	Serrasalmidae	*Piaractus brachypomus*
18	Bagridae	*Mystus pelusius*
19	Siluridae	*Silurus glanis*
		Silurus triostegus
20	Sisoridae	*Glyptothorax kurdistanicus*
		Glyptothorax silviae

(continued)

Table 60.3 (continued)

No.	Family	Species
21	Salmonidae	*Coregonus lavaretus*
		Oncorhynchus keta
		Oncorhynchus mykiss
		Salmo caspius
		Salmo trutta
		Salvelinus fontinalis
		Stenodus leucichthys
22	Esocidae	*Esox lucius*
23	Gadidae	*Lota lota*
24	Gobiidae	*Anatirostrum profundorum*
		Babka gymnotrachelus
		Babka macrophthalma
		Benthophiloides brauneri
		Benthophiloides turcomanus
		Benthophilus abdurahmanovi
		Benthophilus baeri
		Benthophilus casachicus
		Benthophilus ctenolepidus
		Benthophilus granulosus
		Benthophilus grimmi
		Benthophilus kessleri
		Benthophilus leobergius
		Benthophilus leoptocephalus
		Benthophilus leptorhynchus
		Benthophilus macrocephalus
		Benthophilus mahmudbejovi
		Benthophilus pinchuki
		Benthophilus ragimovi
		Benthophilus spinosus
		Benthophilus svetovidovi
		Boleophthalmus dussumieri
		Glossogobius giuris
		Hyrcanogobius bergi
		Knipowitschia caucasica
		Knipowitschia iljini
		Knipowitschia longecaudata
		Mesogobius nigronotarus
		Mesogobius nonultimus
		Neogobius caspius
		Neogobius melanostomus
		Neogobius pallasi
		Periophthalmus waltoni

(continued)

Table 60.3 (continued)

No.	Family	Species
		Ponticola bathybius
		Ponticola cyrius
		Ponticola goebelii
		Ponticola gorlap
		Ponticola iranicus
		Ponticola patimari
		Ponticola syrman
		Proterorhinus nasalis
		Rhinogobius lindbergi
		Scartelaos tenuis
25	Mugilidae	*Chelon aurata*
		Chelon saliens
		Ellochelon vaigiensis
		Mugil cephalus
		Planiliza abu
		Planiliza subviridis
26	Cichlidae	*Amatitlania nigrofasciata*
		Coptodon zillii
		Iranocichla hormuzensis
		Iranocichla persa
		Oreochromis aureus
		Oreochromis niloticus
27	Atherinidae	*Atherina caspia*
28	Aphaniidae	*Aphaniops furcatus*
		Aphaniops ginaonis
		Aphaniops hormuzensis
		Aphaniops stoliczkanus
		Aphanius arakensis
		Aphanius darabensis
		Aphanius farsicus
		Aphanius isfahanensis
		Aphanius kavirensis
		Aphanius mesopotamicus
		Aphanius pluristriatus
		Aphanius shirini
		Aphanius sophiae
		Aphanius vladykovi
		Paraphanius mento
29	Poeciliidae	*Gambusia holbrooki*
		Poecilia latipinna
		Poecilia reticulata
		Xiphophorus hellerii

(continued)

Table 60.3 (continued)

No.	Family	Species
30	Mastacembelidae	*Mastacembelus mastacembelus*
31	Channidae	*Channa gachua*
32	Syngnathidae	*Syngnathus caspius*
33	Percidae	*Perca fluviatilis*
		Sander lucioperca
		Sander marinus
34	Gasterosteidae	*Gasterosteus aculeatus*
		Pungitius platygaster
35	Sparidae	*Acanthopagrus arabicus*

Derived from (Esmaeili 2017; Abdoli 2000; Abbasi et al. 1998; Ramin et al. 2018)

Table 60.4 Exotic or transplanted fish species in the inland waters of Iran

	Family	Species	Exotic (E) or Transplanted (T)	Result
1	Anguillidae	*Anguiella anguiella*	E	–
2	Cyprinidae	*Alburnus charusini*	T	+
3		*Carassius auratus*	E	+
4		*Carassius carassius*	E	–
5		*Ctenopharyngodon idella*	E	–
6		*Cyprinus carpio*	T	+
7		*Hemiculter leuciosculus*	E	+
8		*Hypophthalmichthys molitrix*	E	–
9		*Hypophthalmichthys nobilis*	E	–
10		*Pimephales promelas*	E	–
11		*Pseudorasbora parva*	E	+
12	Heteropneustidae	*Heteropneustes fossilis*	E	+
13	Esocidae	*Esox lucius*	T	+
14	Salmonidae	*Coregonus lavaretus*	E	+
15		*Oncorhynchus keta*	E	–
16		*Oncorhynchus mykiss*	E	+
17		*Salmo trutta*	T	+
18		*Salvelinus fontinalis*	E	–
19	Mugilidae	*Chelon aurata*	E	+
20		*Chelon saliens*	E	+
21		*Mugil cephalus*	E	–
22	Poecillidae	*Gambusia holbrooki*	E	+
23	Gasterosteidae	*Gasterosteus aculeatus*	E	+
24	Centrarchidae	*Lepomis macrochirus*	E	–
25		*Micropterus salmoides*	E	–
26	Persidae	*Stizostedion lucioperca*	T	–
27	Pleuronectidae	*Platchthys flesus*	E	–

Derived from (Coad and Abdoli 1993; Abdoli 2000)

References

Abbasi K, Valipour AR, Talebi-Haghighi D, Sarpanah AN, Nezami S (1998) Atlas of fishes of Iran. Inland water of Guilan province. Guilan Fisheries Research Center, Bandar Anzali, Iran, p 126. In Persian

Abdoli A (2000) Inland water fishes of Iran. Naghsh Mana Press, Tehran, p 377. In Persian

Abdoli A (2016) The field guide of the inland water fishes of Iran. Iranshenasi Publisher, Tehran, p 272. In Persian

Abell R, Thieme ML, Revenga C, Bryer M, Kottelat M, Bogutskaya N, Coad BW, Mandrak N, Contreras Balderas S, Bussing W, Stiassny MLJ, Skelton P, Allen GR, Unmack P, Naseka A, Ng R, Sindorf N, Robertson J, Armijo E, Higgins JV, Heibel TJ, Wikramanayake E, Olson D, López HL, Reis RE, Lundberg JG, Sabaj Pérez MH, Petry P (2008) Freshwater ecoregions of the world: a new map of biogeographic units for freshwater biodiversity conservation. Bioscience 58(5):403–413

Armantrout NB (1980) The freshwater fishes of Iran. Ph.D. Thesis, Oregon State University, Corvallis, Oregon, USA

Berg LS (1949) Freshwater fishes of Iran and adjacent countries. Trudy Zoologicheckogo Instituta Akademii Neuk SSSR 8(In Russian):783–858

Coad BW (1998) Systematic biodiversity in the freshwater fishes of Iran. Ital J Zool 65 (Supplement):101–108

Coad BW (1999) Freshwater fishes. In: Yarshater E (ed) EncyclopædiaIranica. Bibliotheca Persica Press, New York, pp 655–669

Coad BW (2017) Freshwater Fishes of Iran. www.briancoad.com

Coad BW, Abdoli A (1993) Exotic fish species in the freshwaters of Iran. Zool Middle East 9:65–80

Esmaeili H (2017) Review and updated checklist of freshwater fishes of Iran: taxonomy, distribution and conservation status. Iran J Ichthyol 4:1–114. https://doi.org/10.7508/iji.2017

Esmaeili HR, Coad BW, Gholamifard A, Nazari N, Teimory A (2010) Annotated checklist of the freshwater fishes of Iran. Zoosystematica Rossica 19:361–386

Esmaeili HR, Teimory A, Owfi F, Abbasi K, Coad BW (2014a) Alien and invasive freshwater fish species in Iran: diversity, environmental impacts and management. Iran J Ichthyol 1(2):62–72

Esmaeili HR, Coad BW, Mehraban HR, Masoudi M, Khaefi R, Abbasi K, Mostavavi H, Vatandoust S (2014b) An updated checklist of fishes of the Caspian Sea basin of Iran with a note on their zoogeography. Iran J Ichthyol 1(3):152–184

Iran Fisheries Organization (IFO) Programming and Budget Office (2017) Annual catch statistics of Iran. Iran Fisheries Organization Press, Tehran, p 64

Naderi Jolodar M, Abdoli A (2010) The fish species atlas of Caspian Sea Basin (Iranian waters). Iranian Fisheries Research Organization Press, Tehran, p 89

Ramin M, Doostadar M, Valinassab T (2018) The diversity of inland water fishes of Iran. The final report of research project, Iranian Fisheries Science Research Institute, p 58

Valinassab T (2013) List of fishes of the Persian Gulf, Oman Sea and Caspian Sea. Mooj-e-Sabz Press, Tehran, p 273

Chapter 61
The Possibility of Introducing an Inland Fisheries Education in Iraq

Laith A. Jawad

Abstract Maintainable operation of the fishery resources along with manpower is one of the most important features of fisheries science. While an enormous manpower is available in the fisheries sector, gaps in knowledge also widespread in various subsections of fishery-related occupations. In this situation, this chapter attempts to give suggestions to initiate various levels of fisheries education in Iraq. It has been observed that clear demarcation of the level of fisheries education such as undergraduate, postgraduate and basic level exists in Iraq.

Although a low level of student enrolment, university education of subjects related to fisheries and aquaculture are under a satisfactory level. Expansion of fisheries education for basic levels such as education of fisherwomen need to be improved. Scientific distribution of knowledge by published works such as journals should further be expanded, and indexing of those journals in reputable and reliable databases is needed. Several constraints might stand against the progress of fisheries education in Iraq such as inadequate funding, low level of student enrolment, quality of the students and less cooperation with industries were also identified. Therefore, finding possible solutions to these issues is essential to ensure the quality of the education in Iraq. A set of recommendations is given at the end of this chapter to assist policymakers to start such education in Iraq.

61.1 Introduction

Throughout 2011–2050, the world population is likely to increase from 7.0 to 9.6 billion (Population Reference Bureau 2011). Further than 95% of this increase will happen in developing countries, and in sub-Saharan Africa, the population is expected to increase by 134%. To meet increased request, overall world food production will need to increase by 70%, and in the developing countries, production

L. A. Jawad (✉)
School of Environmental and Animal Sciences, Unitec Institute of Technology, Auckland, New Zealand

L. A. Jawad (ed.), *Tigris and Euphrates Rivers: Their Environment from Headwaters to Mouth*, Aquatic Ecology Series 11,
https://doi.org/10.1007/978-3-030-57570-0_61

will need to almost double (Bruinsma 2011). Fisheries and aquaculture are vital to food security and deficiency alleviation. Millions of the world's poor depend on fish as an essential source of protein and micronutrients (FAO 2010). Likewise, millions of people depend on catching, processing and trading fish to provide their main source of income, so it is serious to withstand the productivity of the fish stocks upon which they depend and to participate in developing sustainable aquaculture.

Higher education is a significant aspect of the development process. It builds the human capital needed to compete in a global economy, advances and maintains the institutions that generate new knowledge and technology, and those that conduct business, and generates knowledge that has economic influence (Bloom et al. 2006; McMahon 2009). But higher-education institutions in developing countries now face growing external tasks, including globalization, increasing numbers of students, and the information and communication technology (ICT) revolution. At the same time, their ability to regulate to these challenges is restricted by internal limitations, such as the questionable quality and importance of the education they provide, problems related to recruiting, developing and retaining staff, and their limited external linkages with the private sector (Saint 2009).

Numerous fisheries specialists and others concerned in fish, fishing and fisheries management issues believe youth education as a vital step in developing an informed, responsible community (Kelso and Murphy 1988). This awareness has resulted in backing for fisheries education, which in turn has led to the expansion of youth education materials and programmes that can be used in school (formal education) and out-of or away-from-school (nonformal education) settings. However, these fisheries education materials have not been inventoried and reviewed to determine whether existing instructional resources meet fisheries education needs. Consequently, it has been challenging for most fisheries professionals and educators to identify appropriate materials and for curriculum developers to decide what content new materials should include.

Among the issues in the future plan to establish the fisheries education are (1) develop objectives and an outline for fisheries education, (2) recognize fisheries education materials, (3) evaluate the content covered by these materials based on the framework, (4) summarize results in a user-friendly guide to fisheries education resources and (5) develop recommendations for youth fisheries education materials.

Being a nation in the Middle East, where huge resources of freshwater comparing to its total land area, the importance of conservation of the natural resources arises as a key issue and needs to be developed through time and through the generations. All the evidences for the presence of water resources in Iraq and especially in the southern part of this country reveal that there is a huge potential in developing fisheries sector in terms livelihood development, earning significant foreign exchange and food security. Fish is the most important source of animal protein for people of Iraq particularly for those living in the middle and south of Iraq (Morgan 2006).

Human resource constituent in the fisheries sector of Iraq indicates that the number of active fishermen involved in the fishery industry is gradually increasing despite the flourishing job possibilities in other sectors (Kitto and Tabish 2004).

While the quantity of manpower is increasing, the quality of well-trained and knowledgeable fishermen in the fishery sector is gradually decreasing. De Santisteban (2005) showed that the average education level in the rural areas, which include fishers operating in Iraq over 95% have attended only secondary schools. This phenomenon acts as one of the major impediments for developing fisheries sector in Iraq. However, the situation is more or less the same in other South Asian countries like Bangladesh where low level of education of fishermen is well-documented (e.g. Islam et al. 2013; Galib et al. 2016).

As to the aquaculture sector, although some forms of novel approaches are still underway, inland capture-based culture fishery is well practised. Apart from labour, educated manpower to successfully handle those two sectors is essential. Lack of knowledge is a major impairment to the biosecurity measures aquaculture as in other countries such as Sri Lanka (Munasinghe et al. 2010). Furthermore, human ability development in fish post-harvest industry is very significant for additional development of fisheries sector in Iraq. In Iraqi context, it was noted that decline of fish products happens before final processing due to gaps in the fish distribution channels, rough handling procedures and poor storage conditions, a case noted in other countries (NSC 1980). Such problems originate mostly due to insufficient knowledge of proper handling of raw and processed fish (NSC 1980).

These issues have emphasized the significance of consistent and quality fisheries education for supportable utilization of the fisheries resources. In return, viable utilization of fisheries resources will guarantee food security and important economic benefits. In that context, this paper attempts to discuss the possibility of introducing fisheries education in Iraq with possible improvements of institutional set-up and future outlook for the major constraints.

61.2 Summary of the History of the Higher Education in Iraq

Harb (2008) gave a brief history of the higher education in general in Iraq. In this chapter, a summary of that history will be given for its relevance to the subject of the chapter.

1. Iraqis have conventionally placed a high value on education. The country has a long intellectual history dating back to the ancient Mesopotamian civilizations, and it appreciated a period of remarkable activities during the early Arab-Islamic empires of the ninth century. Iraq's modern higher education sector dates back a century, when a college of law was established in Baghdad in 1908.
2. The higher education sector thrived in the 1960s and early 1970s. Advances were made in both the arts and the sciences. The Iraqi Academy of Sciences became a link for research in language, history and literature. The University of Baghdad's medicine and science faculties attracted students from throughout the Arab world. Fuelled by oil wealth, the universities' research helped support an aggressive

import-substitution drive that produced household items, construction materials and agricultural machinery.

3. The political unbalanced destabilized the vitality of Iraqi higher education. As the Baathists gradually came to dominate public life after 1968—especially after Saddam Hussein became president in 1979—the higher education sector became a venue for political correctness, cronyism, corruption and manipulation of resources to advance the regime's ideology and policies.
4. The sector became under the control of the Ministry of Higher Education and Scientific Research (MHESR), in the process losing any impression of academic independence. The higher education sector quickly found itself obliged to dedicate its research and talents to the political agenda of the Baath party.
5. One of the most destructive consequences of Baathist control was the brain drain inspired by the persecution, imprisonment and assassination of faculty members and students. Thousands of university professors left to work in neighbouring countries or in the West. Those who remained in Iraq had to contend with a lack of new research materials and a lack of contacts with the outside world. This situation became especially acute after the imposition of UN sanctions following the first Gulf War in 1991.
6. Overseas travel was soon deprived of the most academics except those with close ties to the regime. Meanwhile, retention and advancement of faculty members became more politicized. Those who were close to the regime found lectureships quickly and advanced rapidly. Some universities also received preferential treatment. Most, however, had to cope with poorly maintained facilities and buildings, old and dilapidated equipment, a lack of technology, outdated journals and books and low faculty salaries. Corruption and isolation affected morale, further damaging performance at all levels of education.

61.3 Fisheries Education in Iraq

61.3.1 Undergraduate Education

In Iraq, there are 35 universities run by the Ministry of Higher Education and Scientific Research. All these universities grant BSc, MSc and PhD in biological sciences and different aspects of environmental science. Out of this number, there are only 6 universities that have specialized centres in the environment and fisheries sciences. University of Basrah located at City of Basrah in south of Iraq recognized among the other universities in Iraq in granting undergraduate degrees in aquatic science in both marine and freshwater sectors. At the College of Science, University of Basrah, Department of Biology grant BSc degree in biology with special degree in marine science), College of Agriculture awards BSc in Fisheries and Marine Resources, and recently, College of Marine Sciences has been established to grant BSc in different branches of marine aspects.

61.3.2 Postgraduate Education

As with the undergraduate degree, postgraduate qualifications are granted by all universities in Iraq. The higher degrees include MSc and PhD in all branches of knowledge including aquatic and environmental sciences. The students who register for higher degrees in aquatic, fisheries and environmental sciences usually receive their supervision from either academics at the appropriate departments or those serve in research centres such as the Marine Science Centre at University of Basrah and the Marsh Research Centre at Thiqar University.

Duration of these courses is normally 2 years for MSc degree consisting of two academic semesters which cover both theoretical and practical components in addition to one full year devoted to research and writing a thesis. For PhD degree, the duration lasts for 3 years, where the student spends first year in having theoretical and practical courses and an additional 2 year confined for research and writing a thesis. The outcomes of the master and doctoral courses geared towards the common approach. The primary objective of these programmes is to develop the aquaculture and fisheries sectors of Iraq by producing experienced, talented scientists and managerial personnel. In addition, conservation of fishery resources in inland and marine waters is also highlighted.

61.4 Dissemination of the Knowledge

61.4.1 Scientific Journals

Every college of the 35 official universities in Iraq publishes at least 2 journals deal with discipline of these colleges. Some of the colleges publish more than 2 journals that are regular in having at least 4 issues per year. In total and according to the records of the ministry of Higher Education and Research, there are 261 scientific journals published in Iraq by the universities run by the government. Aquatic, fisheries and environment subjects are usually appeared in all the journals of the colleges of sciences and education of all Iraqi universities. Specialized scientific journals such as those published by Marine Science Centre, Mesopotamian Journal of Marine Science and the Iraqi Journal of Aquaculture publish specialized article in the fields of aquatic sciences and aquaculture. In addition, University of Basrah publishes special scientific journal, Marsh Bulletin that is dedicated to the different aspects of life in the southern marshes of Iraq. On the other hand, The Natural History Research Centre, Baghdad University, publishes his bulletin under the same name to include studies on different aspects of natural history in Iraq. All the above-mentioned journals receive and publish articles in various fields of fisheries and aquaculture including marine biology, post-harvest technology, aquaculture, aquatic plants and ornamental fish from all parts of the world. In addition to these two journals, a number of other journals also available in the field of agriculture and zoology. Most of these journals are adequately indexed in reputable databases (e.g. Web of Science, Scopus).

61.4.2 *Other Publications by Ministry of Higher Education and Scientific Research (MHESR) in Iraq*

Apart from journal, the MHESR encourages the academics in the different colleges and in the different discipline of knowledge to publish specialized books for both under and postgraduate studies and for researches. Most of the books are in Arabic Language in an aim to enrich the Arabic science library in general. Broad subjects including brackish and marine water fisheries, ornamental fish breeding and culture of various aquatic organism are focused in these publications, and these are available at a reasonable price.

61.5 Fisheries Education Needs

It is always known that better-qualified employees the better education is important in the graduate education programme and while more education is important, better education is even more important (Hester 1979). The present fisheries and aquatic sciences education in Iraq is falling short of the following need as the other institutions in other parts of the world have gone through. Hester (1979) recognized the following short of need for the fisheries education in USA, which looks comparable to that in Iraq.

1. If there are still universities that endure to consider of the BSc degree as adequate for making professional fisheries biologists, they are out of step with the need. This method produces two problems. First, the student is incompetently organized for the complex problems, both biological and sociological, that he must face. Second, it demands requiring too many fishery and related courses at the undergraduate level, at the expense of a broader education. Even if the student later goes to graduate school, his undergraduate education has been too dedicated to scientific and technical courses. The Master's degree is the minimum education necessary for most fisheries positions, and the content of this education is critically important.
2. Both students and faculty appear enthusiastic for students to take every fishery and related biological science course that the university can suggest. While on superficial examination this may appear needed, it must be weighed against what must be given up in the sense of courses that were 'traded-off'. Students, however, on several occasions had shown about their inabilities to deal effectively with the public. Even after graduation, a biologist has a great chance to learn biological information through books, meetings, workshops, as well as

other forms of on-the-job training, and he or she is keen and hungry for this knowledge. But a biologist who has never been exposed to public administration, environmental law, public speaking, human psychology, legislative process, planning, economics, technical writing, policy formulation or political science is unprepared in these areas and is unlikely to practice any of these after graduation.

3. Students seem too often to be open to a philosophy that says 'manage the resource, but don't let the public have anything to do with the decisions and above all, don't let politicians get involved'. We need the students to understand and hopefully appreciate that the public has a right to be involved in the decision-making process. And, interestingly, the strongest and most successful programmes are usually those that are understood and consequently supported by the public.
4. Universities seem to do an outstanding job at providing education and training for doing individual research, sometimes on one aspect of the biology of one species. Within a short period of time after graduation, most of the students will have managerial, not individual research, positions. Faculty members may do their best at training students to become faculty members, whereas in actuality, the greatest demand is not to provide new faculty members, but rather to provide fisheries biologists to do researches on fisheries in the field.
5. Universities are too oriented to production of degrees, with too little importance on other types of education, including continuing education. Regrettably, the little bit that is done inclines to be basically a repeat of the formal courses and thus is more of the same. For example, short courses on what every fisheries biologist should know about the ecological effects of environmental contaminants on fish populations? Or how most effectively to obtain public input into natural resource decisions?
6. Major fisheries-related degree programmes in Iraq are currently being carried out mainly by the government universities. Enrolment in higher education is highly competitive in Iraq, which results in high dropout during the selection process. Despite this, very few students are being enrolled in these courses. One of the reasons for low enrolment could be the people's attitude towards the fisheries education. Fishing is considered as a job of poor people and young people seeking jobs that have high ranks in the society such as doctors, engineers and lawyers. In general, the majority of the people also believe that fisheries-based jobs are not as profitable as other occupations. Therefore, building the trust in fisheries occupations among people and provide proper professional respect for fisheries occupations are commanding.
7. Iraqi education system primarily depends on the government funding. However, more than 80% of the annual allocation for the universities is spent for the recurrent expenditure and remuneration. Also, Gross domestic expenditure on the research is particularly low when compared to other countries (Herath and Radampola 2017). Lack of funding removes basic infrastructure facilities needed for the continuation of the quality education. In most cases, additional funding to universities/ institutions may be obtained by projects that are in cooperation with

NGOs/other collaborative partners (Harb 2008). Since Iraq is rich in water resources with many exploitable fishery capitals, it would be wise to increase the fund allocation to fisheries education for maximizing the future benefits.

61.6 Recommendations to Set and Improve Fisheries Education in Iraq

Hester (1979) recognized the following basic recommendations to enhance the fisheries education and they can be applied in the case of Iraq.

1. Consider the faculty members themselves. Their own education, experience and attitude are of paramount importance. In addition to some obvious aspects about selection, there are sometimes opportunities for additional training and experience. A regular programme needs to be set for the training of Academics at the universities and research centres for a year or two.
2. Provide B.S. degrees in Biological Science, not in Fisheries or emphasizing fisheries at the undergraduate level include an adequate exposure to the political sciences, economics. But the specialized courses in Fisheries at the graduate level.
3. Get students involved with summer jobs with governmental agencies or industry. Let them find out early what these agencies actually do.
4. Open up curriculum requirements to allow and even more important-to require courses in sanitary engineering, water resource planning and similar courses, especially at the graduate level.
5. Invite seminar speakers to communicate with students on the methods by which policy is established, the importance of communicating with the public, legislation and legislative authorities and what they mean, as well as a host of other such subjects.
6. Help students to understand equally the aquatic environment of the fish and the political and sociological environment of the biologist use elective courses wisely.
7. Provide meaningful continuing education that will be helpful to those who are already professional fisheries biologists. Ask the biologists and administrators what their needs are and then work towards meeting those needs.

Olmsted (1979) in his study on the fisheries education questioned whether the responsibility for reform in this sector lies in industry or in the educational system? He suggested it needs to coordinate by both sectors if better-trained individuals are be insured. He proposed the followings: (1) that industry can do much to improve the situation. Since the demands of industry are so diverse, universities cannot be expected to tailor fisheries people to individual positions; (2) industry must be willing to provide on-the-job training and allow sufficient time for the biologist to grow into his position; (3) industry must identify the weaknesses of new graduates and clearly state its needs to universities; (4) industry should become actively involved with the educational process, perhaps in a seminar atmosphere where representatives can clearly present a picture of the requirements and qualifications for a job in industry; (5) summer intern jobs or co-op programmes for students, industry can allow students to get first-hand experience in industry; (6) industry must expect fisheries scientists to perform their work in a scientific and professional manner.

Olmsted (1979) recommended that much of the responsibility for accommodating changes in the job market lies with the educational system. Initial responsibility lies in honesty to the potential fisheries student. He gave an example on how his suggestion can be met by arranging an introductory seminar ought to be offered to familiarize the student with types of employment, job opportunities, expected salaries, etc. A thorough comparison of government, academic and industry employment would greatly benefit the student in educational and career decisions. Employment in industry must be presented as a viable alternative to traditional employment in academia or government.

References

Bloom D, Canning D, Chan K (2006) Higher education and economic development in Africa. World Bank, Washington, DC

Bruinsma J (2011) The resources outlook: by how much do land, water and crop yields need to increase by 2050? In: Conforti P (ed) Looking ahead in world food and agriculture: perspectives to 2050. Rome, Food and Agriculture Organization, pp 233–278

De Santisteban AV (2005) Sanctions, war, occupation and the de-development of education in Iraq. Int Rev Educ 51(1):59–71

FAO (2010) The state of world fisheries and aquaculture 2010. Food and Agriculture Organization, Fisheries and Aquaculture Department, Rome

Galib SM, Hoque MN, Akter S, Chaki N, Mohsin ABM (2016) Livelihood, climate change and fisheries: a case study of three fishing communities of northwestern Bangladesh. Int Res J Soc Sci 5(8):18–25

Harb I (2008) Higher education and the future of Iraq. United States Institute of Peace (USIP), Washington, DC

Herath TN, Radampola K (2017) Fisheries education in Sri Lanka: current status, constraints and future outlook. J Fish 5(3):535–540

Hester FE (1979) Fisheries education from the federal perspective. Fisheries 4(2):22–24

Islam MR, Hoque MN, Galib SM, Rahman MA (2013) Livelihood of the fishermen in Monirampur Upazila of Jessore district, Bangladesh. J Fish 1(1):37–41

Kelso WE, Murphy BR (1988) University fisheries programs in the United States: structure and needs. Fisheries 13(2):9–15

Kitto MR, Tabish M (2004) Aquaculture and food security in Iraq. Aquaculture Asia 9:31–31

McMahon WW (2009) Higher learning, greater good: the private and social benefits of higher education. John Hopkins Press, Baltimore

Morgan G (2006) Country review: Iraq. Review of the state of world marine capture fisheries management: Indian Ocean, 488:247

Munasinghe M, Stephen C, Abeynayake P, Abeygunawardena I (2010) Shrimp farming practices in the Puttalam district of Sri Lanka: implications for disease control, industry sustainability, and rural development. Vet Med Int 2010:679130. https://doi.org/10.4061/2010/679130

NSC (1980) Post-harvest food losses in Sri Lanka. National Science Council. http://thakshana.nsf.ac.lk. Accessed 10 Oct 2018

Olmsted LL (1979) Fisheries education from the private sector perspective. Fisheries 4(2):26–28

Population Reference Bureau (2011) 2011 World population data sheet. Population Reference Bureau, Washington, DC

Saint W (2009) The worldwide context for higher education: demands, constraints and trends. Discussion Paper, Higher Education for Development (HED). Available at http://www.glp.net/c/document_library/get_file?folderId=219997&name=DLFE-7 684.pdf Accessed 21 May 2018

Chapter 62
The Marine and Diadromous Fisheries of Iraq

Myriam Khalfallah and Daniel Pauly

Iraq has very rich freshwater resources, among the most abundant in the Middle East, due to the Tigris and Euphrates Rivers, the Basrah River and the Mesopotamian marshes, which serve as nursery grounds for a number of diadromous fish. Jointly, these rivers and the marshlands provide crucial nutrients to the fisheries of the northern Gulf, particularly through the Shatt al-Arab River (Jawad 2006), which is formed by the confluence of the Euphrates and Tigris river about 200 km upstream (Abdullah et al. 2015). However, Iraq's maritime Exclusive Economic Zone (EEZ) is tiny (Fig. 62.1) and so strongly influenced by the Shatt-al-Arab's outflow that it can be considered an extension of its estuary. Thus, one could argue that Iraq's marine waters are an extension of its river system, and hence, their inclusion in a book devoted to Iraq's rivers.

Given the smallness of its EEZ, Iraq's estuarine and marine fisheries (henceforth 'marine fisheries') generate catches that are 3–4 times smaller than its freshwater fisheries. However, while the fish fauna of Iraq has been studied by a number of authors (Table 62.1), there are currently no management plans in place for any of Iraq's fisheries. Detailed catch statistics have not been properly collected since the early 1990s with missing statistics for some years, and few stock assessments have been performed. Thus, the numbers presented here will remain tentative until this situation is addressed. The marine fisheries are predominantly artisanal in nature, with gillnetting for the diadromous hilsa shad (*Tenualosa ilisha*), grey mullet (*Liza* spp.) and pomfret (*Pampus* spp.) being dominant fishing activities. Fish supply is relatively low throughout Iraq's coastal region and does not meet local demand (Jawad 2006). Since 1950, the first year considered here, and especially during the twenty-first century, large areas of the Mesopotamian marshes were drained, at

M. Khalfallah (✉) · D. Pauly
Institute for the Ocean and Fisheries, the University of British Columbia, Vancouver, BC, Canada
e-mail: m.khalfallah@oceans.ubc.ca

L. A. Jawad (ed.), *Tigris and Euphrates Rivers: Their Environment from Headwaters to Mouth*, Aquatic Ecology Series 11,
https://doi.org/10.1007/978-3-030-57570-0_62

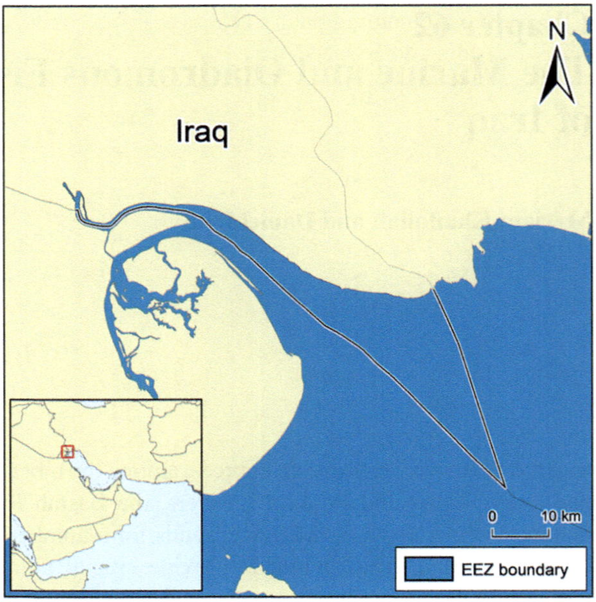

Fig. 62.1 Map of the Exclusive Economic Zone of Iraq, comprising 540 km^2 of shallow water heavily influenced by the Shatt-al-Arab outflow

Table 62.1 Notes on the ichthyofaunal diversity of Iraq

Taxa	Remarks	Source
Marine and freshwater fishes	A thorough quantitative account of the ichthyology of Iraq, from Heckel (1843) to Sultan and Hassan (2009)	Jawad (2012)
Marine and freshwater fishes	First monograph on the fishes of Iraq	Khalaf (1962)
Marine and freshwater fishes	Fishes of the Tigris-Euphrates Basin, consisting of at least 58 freshwater and 53 marine species	Coad (1991)
Freshwater fishes	65[a] species recorded, based on literature published until early 2018.	FishBase (February 2018; www.fishbase.org)
Marine and brackish water fishes	269[a] species recorded, based on the literature published until early 2018.	FishBase (February 2018; www.fishbase.org)

[a]The numbers of fish Iraqi fish species in FishBase add up to only 331, as some species occur in marine, brackish- and freshwater

different times and for different reasons (Jawad 2003; Al-Yamani et al. 2007). Although the initial draining of the central marshes was intended for land reclamation for agricultural purposes, it later became a political attempt to force the 'Marsh Arabs' (i.e. the Ma'dan people) out of the area through water diversions. The marshes, which have been reduced in extent by over 90%, have long been considered as a refuge for people persecuted by the government at that time. Not surprisingly, millions of fish and waterfowl died as the waters receded (North 1994).

In addition, damming naturally flowing rivers reduced freshwater discharge into the sea, leading to reduced nutrient input to coastal waters, which consequently diminished plankton productivity, and in turn, fish landings (Al-Yamani et al. 2007). The damming also increased the salinity of the northwestern Gulf, raising concerns about jellyfish outbreaks and changes in plankton (and hence fish) community density and distribution (Al-Yamani et al. 2007).

Major wars have greatly shaped the country's marine fisheries. The Iran-Iraq War, which lasted from 1980 to 1988, led to decreased marine fisheries catch, though the lack of detailed records precludes detailed inferences (Ali et al. 2000). In addition, the first Gulf War in 1991 led to sanctions by the United Nations which implied that areas in the Northern Persian Gulf traditionally exploited by Iraqi fishers could no longer be accessed. This, combined with other factors (general insecurity and perhaps oil pollution; Linden et al. 2004), meant that Iraq's catches dropped to near zero in 1991 and 1992.

A previously completed historical catch reconstruction of the Iraqi marine fisheries for 1950–2015 completed by Khalfallah (2020) is summarized here. This summary illustrates the need to establish management infrastructure for fisheries monitoring and regulation enforcement, especially in light of the many marine and estuarine fish populations that are shared with other Persian Gulf countries.

62.1 The Large-Scale Marine Fisheries of Iraq

Information on the marine Iraqi large-scale fishery is sparse; however, an account on the introduction of the trawl fishery in what is now the Iraqi EEZ is available. As a national initiative to develop the fishing industry, authorities introduced an experimental and exploratory trawler '*Zubeidi*' in what is now the Iraqi EEZ. This initiative, however, did not meet expectations leading to a cessation of the trawling experiments in 1952 (Amodeo 1956). In his 'Report to the Government of Iraq' on Iraqi fisheries in the 1960s, Andersskog (1966) did not mention any trawlers or trawl nets. During the 1970s, small Iraqi trawlers began to be deployed (Ali et al. 2000; Mohamed et al. 2002b; Jawad 2012). This trawl fishery was government-owned and was privatized only in the early 1990s (Ali et al. 1998; Mohamed et al. 2002b). National laws and regulations of the marine fishing industry were improved in 1976 (Al-Assadi and Al-Matouri 2014). In the late 1980s, Iraq initiated a war against Iran, and marine fisheries ceased operation for almost 8 years (Ali et al. 2000; Mohamed et al. 2002b; Mohamed and Qasim 2014). Following the first Gulf War in the early 1990s, the United Nations imposed sanctions on Iraq which lasted until early 2000s. This, however, played a role in the increase of the marine fishing activities (Mohamed 1993; Mohamed and Qasim 2014). Fishing, however, stopped again for almost a year in 2003 due to the US-led coalition invasion of Iraq that same year (Nasir and Khalid 2013). The period following the US forces withdrawal from the Iraqi territories in 2011 was marked by violence and instability.

The annual catches by Iraqi trawlers were estimated by Khalfallah (2020) using a variety of sources and estimates of the CPUE or catch/effort of different classes of trawlers, whose numbers were provided by Mohamed et al. (2002a), Mohamed et al. (2002b) and Ali et al. (1998). According to Khalfallah et al. (2019), reported large-scale catches represented 6% of total reconstructed catches, while the unreported catches of this sector represented 35%. Industrial marine fisheries catches started with an average of 700 t·year^{-1} in the 1970s and increased to around 8000 t·year^{-1} between 2010 and 2015. Catches have, however, dropped to merely 100 t·year^{-1} during the 1980s then increased to almost 9900 t·year^{-1} and 12,600 t·year^{-1} during the 1990s and 2000s, respectively. Hilsa shads and greenback mullet (*Planiliza subviridis*) made up most of the catch by the industrial Iraqi marine fishery over the past 65 years (1950–2015). Catches of hilsa shads have dropped drastically in the past decade, while industrial landings of the low-value Klunzinger's ponyfish (*Equulites klunzingeri*) increased tremendously since the early 2000s (20% of total industrial catch in 2015).

62.2 The Small-Scale Marine Fisheries of Iraq

During the 1950s, the artisanal Iraqi fishery consisted mainly of few dozen boats, deploying various traps (Amodeo 1956). To improve fish yield, the FAO introduced Iraqi fishers to the use of hooks and lines, and floating lines (Amodeo 1956). Since then, the small-scale fishery has not developed much. It still consists mainly of wooden launches and small boats deploying traps, gillnets, and hooks and lines (Mohamed et al. 2002a; Mohamed et al. 2002b; Al-Assadi and Al-Matouri 2014). Al-Abdulrazzak and Pauly (2014) did not detect any weirs ('*hadra'*) from Google Earth maps of the Iraqi coast, although such traps were detected in neighbouring Kuwait and Iran.

Khalfallah (2020) estimated marine landings to be about 41% and 6% for reported and unreported artisanal commercial fisheries, respectively. Reported and unreported commercial small-scale landings increased from about 1000 t·year^{-1} in the 1950s to around 7700 t·year^{-1} in the 1970s. Following a drastic drop in the 1980s, i.e. 1900 t·year^{-1}, artisanal landings increased to just over 7000 t·year^{-1} in the 1990s and 8800 t·year^{-1} in the 2000s. Between 2010 and 2015, artisanal landings were about 5700 t·year^{-1}. Hilsa shads were the most caught by the artisanal fishery but their landing had decreased dramatically by 2015.

As it appears unlikely that any marine recreational fishing occurs in Iraq (Morgan 2006), the only other subsector of the small-scale fishery is the subsistence fishery. Subsistence fishing is the activity performed by people who gather fish and invertebrates for their personal and family consumption. Khalfallah (2020) estimated subsistence catches to be 3% of total reconstructed catches, increasing from an average of 200 t·year^{-1} in the 1950s and 1960s to almost 500 t·year^{-1} between 2000 and 2015. In the 1980s, subsistence catches decreased considerably to 65 t·year^{-1} due to the Iran-Iraq war.

62.3 Discarded Catch

There is almost no information on discards in the Iraqi marine waters despite some attempts to estimate the Iraqi marine fisheries bycatch (Mohamed 1993; Mohamed et al. 2002a). In many studies, 'bycatch' is frequently (and erroneously) used as a synonym for discards. At other times, 'bycatch' includes both discards and what is referred to as 'non-commercial species'. This category refers to the catch destined for fishmeal production. In the study completed by Khalfallah (2020), the term bycatch was not considered as a fishery component. Instead, landed by-catch was considered commercial if sold and subsistence if used for personal consumption. Only rejected by-catch was classified in the 'discards' category.

Discards by the artisanal fishery were usually small, primarily due to the fact that for religious reasons most of the population around the Iraqi coast does not consume fishes without visible scales, e.g. sea catfish of the Ariidae family (Ye et al. 2000; Jawad 2006). According to Khalfallah (2020), total discards increased from around 65 t·year^{-1} in the 1950s and 1960s to almost 400 t·year^{-1} in the 1970s. This is likely a direct consequence of the introduction of the trawl fishery in the Iraqi fishery. Discards were estimated at around 8500 t·year^{-1} between 2010 and 2015.

62.4 Discussion

This chapter summarizes the results of the historical catch reconstruction completed by Khalfallah (2020) for Iraq for 1950–2015. Despite being based on far more sources than the preliminary study of Al-Abdulrazzak and Pauly (2013), the resulting catch time series presented here as Fig. 62.2 is still tentative, as nothing can fully replace a non-existent catch monitoring programme. This also applies to the catch composition data in Fig. 62.3.

However, clear messages appear from these two figures. One, very obvious, is that wars are generally good for fish (as they were in the North Sea during two World Wars; see Beverton and Holt 1957), because fishing cannot proceed, and fishing kills fish.

The other clear message is that the anadromous hilsa, which made up most of the marine catch of Iraq (Al-Dubakel 2011) and is also important in neighbouring Iran (Hashemi et al. 2010), has now severely declined, in both relative (i.e. as component of total catch) and absolute term. This is probably due to the water quality in the Shatt-al-Arab and to its low quantity. For information on this, see Jawad (2003); Salman et al. (2009); Al-Said et al. (2017); Ben-Hasan et al. (2018).

Needless to say, both of these items require attention and could be remedied by good applied science and resource management.

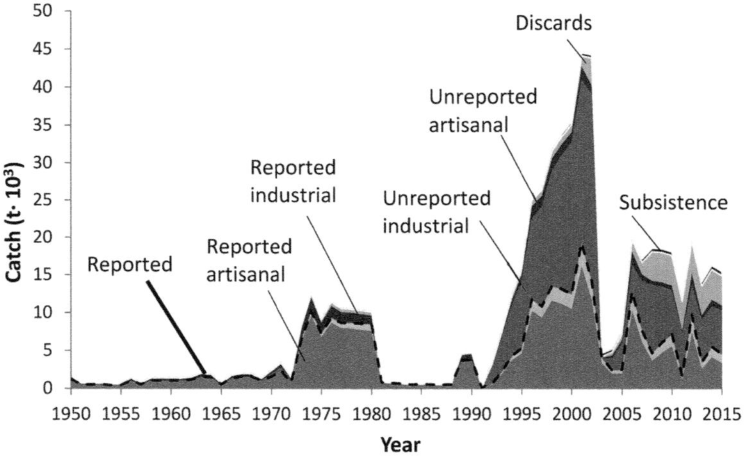

Fig. 62.2 Time series of the marine and diadromous catch of Iraq, 1950 to 2015, distinguishing industrial and artisanal landings, along with (mainly industrial) discards and subsistence catch. The black dotted line refers to the catch level reported by Iraq to the FAO (Khalfallah 2020)

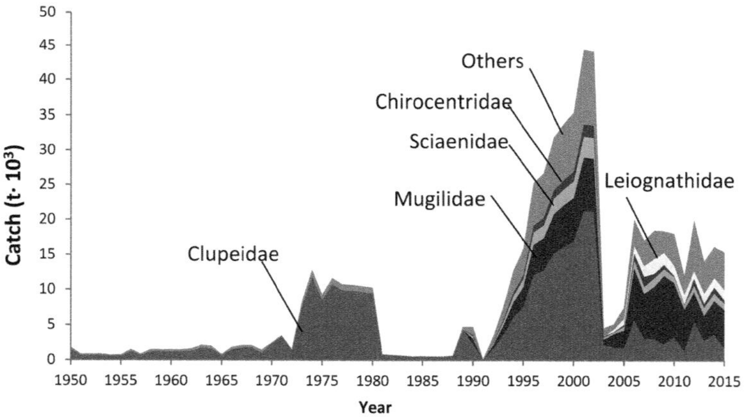

Fig. 62.3 Time series of the marine and diadromous catch of Iraq, 1950–2015, by taxa. Note the diminishing importance of Clupeidae mainly comprising hilsa (*Tenualosa ilisha*) both in absolute and relative terms (see www.seaaroundus.org for catch by taxon and by fishery sector) (Khalfallah et al. 2019)

Acknowledgements We acknowledge the support of the *Sea Around Us*, a research initiative at The University of British Columbia which receives funding from a number of philanthropic foundations, notably the Oak and Marisla Foundations. We also thank Laith Jawad (Auckland, New Zealand) and Abdul Ben Hasan (IOF, UBC) for useful discussions and references. Also, Abdul-Razak M. Mohamed and Nadir A. Salman (University of Basrah) provided us with several useful articles and reports.

References

Abdullah AD, Masih I, Pvd Z, Karim UFA, Popescu I, Suhai QA (2015) Shatt al Arab system under escalating pressure: a preliminary exploration of the issues and options for mitigation. Int J River Basin Manage 13(2):215–222

Al-Abdulrazzak D, Pauly D (2013) Reconstructing Iraq's fisheries: 1950–2010. In: Al-Abdulrazzak D, Pauly D (eds) From dhows to trawlers: a recent history of fisheries in the Gulf countries, 1950 to 2010, fisheries Centre research reports 21(2). Fisheries Centre, University of British Columbia, Vancouver, pp 17–22

Al-Abdulrazzak D, Pauly D (2014) Managing fisheries from space: Google earth improves estimates of distant fish catches. ICES J Mar Sci 71(3):450–454

Al-Assadi SA, Al-Matouri SC (2014) Evaluation of marine fishing in Iraq. J Geograph Res of Iraqi Fish 20:261–276

Al-Dubakel AY (2011) Commercial fishing and Marketing of Hilsa Shad *Tenualosa ilisha* (Hamilton-Buchanon, 1822) in Basrah -southern Iraq. Emirate J Food Agric 23(2):178–186

Ali TS, Mohamed A-RM, Hussain NA (1998) The status of the Iraqi marine fisheries during 1990–1994. Marina Mesopotamica 33(1):129–147

Ali TS, Mohamed A-RM, Hussain NA (2000) Catch per unit of effort of Iraqi marine fisheries northwest Arabian gulf. Marina Mesopotamica 15(2):285–301

Al-Said T, Al-Ghunaim A, Rao DVS, Al-Yamani F, Al-Rifaie K, Al-Baz A (2017) Salinity-driven decadal changes in phytoplankton community in the NW Arabian gulf of Kuwait. Environ Monit Assess 189(268):17

Al-Yamani FY, Bishop JM, Al-Rifaie K, Ismail W (2007) The effects of the river diversion, Mesopotamian marsh drainage and restoration, and river damming on the marine environment of the northwestern Arabian gulf. Aquat Ecosyst Health Manage 10(3):277–289

Amodeo G (1956) Iraq – the possibilities of developing the sea fishing industry – Report to the government. 438, FAO, p 30

Andersskog B (1966) Report to the government of Iraq on a preliminary fishery survey. FAO/UN, p 8

Ben-Hasan A, Walters C, Christensen V, Al-Husaini M, Al-Foudari H (2018) Is reduced freshwater flow in Tigris-Euphrates rivers driving fish recruitment changes in the Northwestern Arabian Gulf? Marine Pollution Bulletin 129(1):1–7

Beverton R, Holt S (1957) On the dynamics of exploited fish populations. Great Britain Ministry of Agriculture. Fisheries and Food, London, p 538

Coad BW (1991) Fishes of the Tigris-Euphrates basin: a critical list. Sillogeus 68(1):1–49

Hashemi S, Mohammadi G, Eskandary G (2010) Population dynamics and stock assessment of Hilsa shad (*Tenualosa ilisha* Hamilton-Buchanan) in coastal waters of Iran (northwest of Persian gulf). Aust J Basic Appl Sci 4(12):5780–5786

Heckel JJ (1843) Abbildungen und Beschreibungen der Fische Syriens nebst einer neuen Classification und Characteristik sämmtlicher Gattungen der Cyprinen. Stuttgart, p 109

Jawad LA (2003) Impact of environmental change on the freshwater fish fauna of Iraq. Int J Environ Stud 60(6):581–593

Jawad LA (2006) Fishing gear and methods of the lower Mesopotamian plain with reference to fishing management. Marina Mesopotamica 1(1):1–37

Jawad LA (2012) History of the study of the fish Fauna of Iraq. Water Res Manage 2(3):11–20

Khalaf KT (1962) The marine and freshwater fishes of Iraq. Ar-Rabitta Press, Baghdad

Khalfallah M (2020) Data-poor fisheries: case studies from the Southern Mediterranean and the Arabian Peninsula, Ph.D. thesis, the University of British Columbia, Vancouver, Canada, p 349. https://doi.org/10.14288/1.0389951

Linden O, Jernelov A, Egerup J (2004) The environmental impacts of the Gulf War 1991. Interim Reports, International Institute for Applied Systems Analysis (IIASA), pp xi+82

Mohamed ARM (1993) Seasonal fluctuations in the fish catches of the northwestern Arabian gulf , Iraq. Marina Mesopotamica 8(1):63–78

Mohamed A-RM, Qasim AMH (2014) Trends in the artisanal fishery in Iraqi marine waters, Arabian gulf (1965–2011). Asian J Appl Sci 2(2):209–217
Mohamed A-RM, Ali TS, Hussain NA (2002a) The assessment of fisheries in the Iraqi marine waters during 1995–1999. Iraqi J Agric Sci 7(1):127–136
Mohamed ARM, Ali TS, Hussain NA (2002b) The physical oceanography and fisheries of the Iraqi marine waters, Northwest Arabian Gulf. Regional Seminar on Utilization of Marine Resources, Karachi, Pakistan, pp 47–56
Morgan G (2006) Country review: Iraq. In: DeYoung C (ed.) review of the state of world marine capture fisheries management: Indian Ocean. FAO fisheries technical paper. No. 488. Rome, pp 247–253
Nasir NA-N, Khalid SA-R (2013) A statistic survey of marine and freshwater fish catch in Basrah, Iraq 1990–2011. Arab Gulf J Sci Res 31(1):1–9
North A (1994) Saddam drains life from Arab marshes: scientists fear Iraq's historic wetlands face destruction in 10 to 20 years, says Andrew North. The Independent, edition of May 17, 1994, London
Sultan FAM, Ahmed SM (2009) Effect of salinity acclimation on growth in *Acanthopagrus latus* (Houttyn, 1782) juveniles. The 6th. Scientific conference of fisheries resources, 3–4 March 2009, Basrah, Iraq (Abstract)
Salman JM, Salah MM, Hassan FM (2009) Used of some bioindicators to study the heavy metals pollution in Euphrates River, Iraq. The 6th. Scientific conference of fisheries resources, 3–4 March 2009, Basrah, Iraq (Abstract)
Ye Y, Alsaffar AH, Mohammed HMA (2000) Bycatch and discards of the Kuwait shrimp fishery. Fish Res 45(1):9–19

Part V
Stress of the Environment of the Two Rivers

Chapter 63
Fish Deformities in the Freshwater Fishes of Iraq: A Short Review and a Study Case on the Indian Catfish *Heteropneustes fossilis*

Laith A. Jawad and Jasim Abed

Abstract Situations fish meeting during embryogenesis and early life history can cause lasting influences not only on morphology, but also on development rate, life history and behavioural appears.

As a case study, an incidence of absence of pelvic fins in the Indian catfish *H. fossilis* collected from Al-Hammar Marsh, south of Iraq was described. The specimen is a mature male of 148 mm in total length. Comparison of the abnormal specimen with a normal *H. fossilis*, shows considerable changes in the posterior part of the axial skeleton,the course of the intestine and especially the extent and shape of the body cavity. Comments are also made to account for such an abnormality. At the end of the chapter, a future studies are suggested.

63.1 Introduction

Morphological characters appear as a result of a complex relations between genes and environment (Gilbert and Epel 2008), and development influences are produced by circumstances that organisms undergo throughout embryogenesis (Martin et al. 2013). The early the environmental chaos, the sturdier its long-lasting impacts can be (Lindström 1999). Developmental impacts are frequently overlooked when studying life-history difference, and morphological changes in features, such as growth rate, age at maturity, can simply be misunderstood for genetic diversity. Consequently, likely influences of early damage should be thought about when studying ecological difference along environmental grades or between conflicting environments (Jonsson and Jonsson 2014).

L. A. Jawad (✉)
School of Environmental and Animal Sciences, Unitec Institute of Technology, Auckland, New Zealand

J. Abed
Department of Fisheries and Marine Resources, College of Agriculture, University of Basrah, Basrah, Iraq

L. A. Jawad (ed.), *Tigris and Euphrates Rivers: Their Environment from Headwaters to Mouth*, Aquatic Ecology Series 11,
https://doi.org/10.1007/978-3-030-57570-0_63

The present chapter sheds light on the freshwater fish abnormalities studies of Iraq that published during the last 40 years. This short review will give an idea about what has been made and what needs to be done in the field of fish abnormalities study in Iraq in order to have an idea about the health of the environment and plan for the future environment. At the end of the chapter, a study case of anomaly in the Indian catfish was given as an example of the several cases of abnormalities that may occur in the freshwater system of Iraq.

63.2 Fish Abnormality Studies in Iraq.

In Iraq, studies reporting on cases of morphological anomalies in the freshwater fish species have not started until 1982, when Al-Hassan (1982) published the first scientific account reporting on some abnormal vertebral cases obtained in some freshwater fish species collected from Basrah Province waters. During the last 40 years, nine scientific reports were published on Iraqi freshwater fish species. These reports described different cases of abnormalities in both the external morphology of the fish and its skeletal parts. Due to the lack of expertise in Iraq, the publication on the cases of anomalies in fishes has delayed until the last two decades of the twentieth century and due to the political unsettle in this country, there was a big gap of no publication on records for the period 1987–2014.

Table 63.1 shows the types of abnormalities observed in some freshwater fish species collected from the southern parts of Iraq. It is clear that the vertebral anomalies were the dominant type of deformities in the cases observed. Within the vertebral abnormalities, coalescence of vertebrae was the common aberration that affecting the vertebral column of the fish species, where there were six species reported this type of vertebral deformity. Within the minor anomalies, the bifurcation of the neural and the haemal spines came was second common anomaly. *Mesopotamichthyes sharpeyi* was the most species with large number of cases of abnormalities. Except for the loss of vertebrae, fusion of neural spines, lordosis and eye abnormalities, this species has shared all other types of abnormalities so far observed and described. *Luciobarbus xanthopterus* was second in having abnormality varieties. This species shared *M. sharpeyi* in both coalescence of vertebrae and bifurcation of neural and haemal spines. The remaining species were observed with less variety than *M. sharpeyi* and *L. xanthopterus*. The studies showed that the type of the abnormality was not confined to a certain locality. This means that the causes of these deformities were present in all the localities studied from Baghdad at the Middle of Iraq and to Basrah Province in the south.

Table 63.1 Types of fish abnormalities observed in freshwater fish species collected from Iraq

Type of abnormality	Species	Part of the fish body	Reference	Locality
Vertebral anomalies				
Coalescence of vertebrae				
	Mesopotamichthys sharpeyi	Caudal vertebrae	Al-Hassan (1982, 1983)	Shatt al-Arab River, Basrah
	M. sharpeyi	Caudal vertebrae	Al-Hassan and Na'ma (1986)	Euphrates River, Nasiria
	M. sharpeyi	Caudal vertebrae	Al-Hassan and Na'ma (1986)	Shatt al-Arab River, Basrah
	M. sharpeyi	Caudal vertebrae	Al-Hassan (1987)	Shatt al-Arab River, Basrah
	Luciobarbus xanthopterus	Thoracic vertebrae	Al-Hassan (1982)	Shatt al-Arab River, Basrah
	L. xanthopterus	Caudal vertebrae	Al-Hassan (1987)	Shatt al-Arab River, Basrah
	L. xanthopterus	Caudal and thoracic vertebrae	Jawad et al. (2015)	Al-Huwaiza marsh
	Carasobarbus luteus	=	Al-Hassan (1982, 1983)	Shatt al-Arab River, Basrah
	Aphanius dispar	Caudal vertebrae	=	Shatt al-Arab River, Basrah
	Gambusia affinis	Caudal vertebrae	Al-Hassan (1985)	Tigris River, Baghdad
	Planiliza abu	Caudal vertebrae	Al-Hassan (1985)	Shatt al-Arab River, Basrah
Compact vertebrae				
	Mesopotamichthys sharpeyi	Thoracic vertebrae	Al-Hassan (1982, 1983)	Shatt al-Arab River, Basrah
	Mastacembelus mastacembelus	Caudal and thoracic vertebrae	Jawad et al. (2016)	Euphrates River, Nasiria
Bifurcation of neural and haemal spines				
	M. sharpeyi	Caudal vertebrae	=	Shatt al-Arab River, Basrah
	C. luteus	=	=	Shatt al-Arab River, Basrah
	L. xanthopterus	=	=	Shatt al-Arab River, Basrah
	L. xanthopterus	Caudal vertebrae	Al-Hassan (1987)	Shatt al-Arab River, Basrah
Undulation of haemal spines				
	M. sharpeyi	Caudal vertebrae	=	Shatt al-Arab River, Basrah
Abnormal calcification				
	M. sharpeyi	Caudal vertebrae	=	Shatt al-Arab River, Basrah

(continued)

Table 63.1 (continued)

Type of abnormality	Species	Part of the fish body	Reference	Locality
Loss of neural spines				
	M. sharpeyi	Thoracic vertebrae	Al-Hassan (1983)	Shatt al-Arab River, Basrah
Loss of vertebrae	*C. luteus*	Thoracic vertebrae	=	Shatt al-Arab River, Basrah
Fusion of neural spines	*G. affinis*	Caudal vertebrae	=	Tigris River, Baghdad
Lordosis	*C. luteus*	Caudal vertebrae	Jawad et al. (2014)	Al-Hammar marsh
Eye abnormalities	*Pampus argenteus*	Eye	Jawad et al. (2018)	Shatt al-Arab River, Basrah

63.3 Effect of Heavy Metals on Fish

Among the common pollutants that stand behind the fish deformities are heavy metals. Since the responsible for all the anomalies so far being described from the inland fish species of Iraq are the heavy metals, it is appropriate to devote a section of this chapter to talk briefly on the role of the heavy metals in causing deformities in fishes in general. Such an information will be a short guide to those who plan for a healthy environment in Iraq in the future.

Heavy metals were behind many fish abnormalities in natural groups as well as in laboratory-cultivated specimens (Cheng et al. 2000).The occurrence of contaminants and mainly heavy metals in the aquatic habitats of fish evidences of harsh hostile impacts on the organisms and has been a matter of fear for many years (Muramoto 1981). In the 1970s, numerous efforts were made to recognise the impacts of heavy metals on organisms such as the influence of cadmium on vertebral deformities of *Cyprinus carpio* (Muramoto 1981) and *Pimephales promelas* (Pickering and Gast 1972) or the effect of zinc on the vertebral column of *Phoxinus phoxinus* (Bengtsson 1974). Generally speaking, metals can be deliberated as biologically important and supplementary.The lack of an essential metal can therefore lead to contrasting health consequences, while its high intensity can also result in negative effects which are similar to or destructive than those initiated by non-essential metals (Kennedy 2011). The most usually present heavy metals in fish organisms are cadmium, lead, mercury, zinc, copper, nickel, cobalt, molybdenum, chromium and tin. Among them, the most frequently investigated, relative to fish abnormalities, include cadmium, copper, lead, zinc, mercury and chromium. Heavy metals accumulate in the tissues of aquatic animals and convert to toxic when levels reach particular toxicity verges, values which differ meaningfully between metals, metal species, taxonomic species and organism life stages. Fish consumption of metals is mainly through the gills and the digestive track, and to a smallerextent, via the skin (Kennedy 2011). Also, fish anomalies have uncontrollable results on fish groups since they distract their survival, growth rate, health and shape (Boglione et al. 2013). Some of the most usual anomalies can be found in the vertebral column (Koumoundouros et al. 2002; Sfakianakis et al. 2006), the swim bladder (Chatain 1994), the cephalic region (Georgakopoulou et al. 2007), the fins (Favaloro and Mazzola 2003) and the lateral line (Sfakianakis et al. 2013). The highest continuing of the mare, the ones in the vertebral column and mainly lordosis (V-shaped dorsal–ventral curvature), kyphosis (Λ-shaped dorsal–ventral curvature) and scoliosis (lateral curvature).

Heavy metals may destructively distract many metabolic events in developing fish, ensuing in growing interruption, morphological and functional deformities, or death of the most subtle individuals. Furthermore, heavy metals trigger energy-expending detoxification procedures; therefore, in fish that are heavily affected by heavy metals have less energy to be utilised for development. The majority of investigations on heavy metals in relation to the growth of fish on growing fish showed great cases of death, late hatching, deformation in the external features changed and skeletal deformities (Jezierska et al. 2009).

In numerous instances, fish anomalies are simple to discriminate, even macroscopically, and therefore present a great benefit over other approaches particularly for scientists working on the field away from the laboratory equipment. Toxicants are often joining the food chain through water, micro-organisms, plants, fish, and then arrive into human bodies by drinking water and fishery products. Several current investigations show the build-up of many metals on animals' livers and kidneys particularly in closed seas such as the Mediterranean and address the need for a careful observations (Storelli et al. 2005).

63.4 Loss of Pelvic Fins in the Indian Catfish, *Heteropneustes fossilis* Collected from Al-Hammar Marsh, Basrah, Iraq–a Study Case

Absence of pelvic fins in teleost fish has been previously identified and accredited to inherited or post-natal anomalies, in addition to chemical pollution (Slooff 1982). Deformities in growth or the whole non-appearance of this pair of fins promote queries on additional advanced useful variations in the shape of the fish as a consequence of habitat modification. Pelvic fins are normally deliberated as the part responsible for turning the fish body, whereas their hydrodynamic role has got minor care corresponds to the pectoral and median fins (Lauder and Drucker 2004).

One specimen of the Indian catfish *Heteropneustes fossilis* (total length = 148 mm, standard length = 132 mm, weight = 156.3 g) lacking the pelvic fins was found in the shallow area of Al-Hammar marsh east of Basrah City, Iraq on November 2018 (Fig. 63.1a, b). In addition, one normal specimen (Fig. 63.1c) with total length = 149 mm, standard length = 135 mm, weight = 160 g was obtained from the same locality for comparison. Eutrophication and town, industrialised and agricultural contamination are deliberated to be the chief human impacts affecting the marsh areas in Iraq (Yaseen et al. 2016; Mahdi and Fawzi 2014).

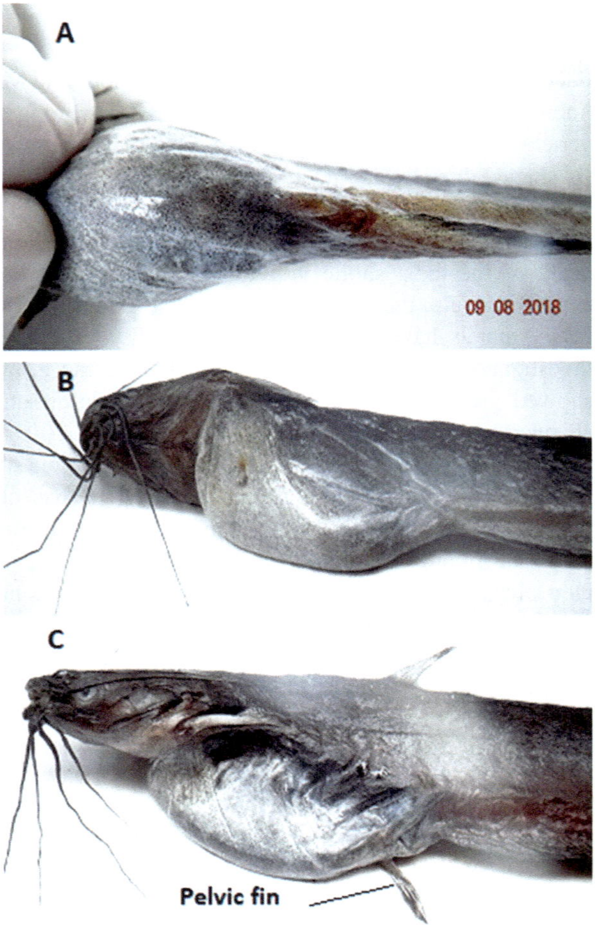

Fig. 63.1 The Indian catfish *Heteropneustes fossilis*. (**a**), Abnormal specimen, 148 mm TL showing absence of pelvic fins (a close up view); (**b**) general view; (**c**) normal specimen, 135 mm TL

So to ratify the lack of the entire pelvic assembly, the abnormal specimen captured in the marsh area was dissected and related with a normal specimen of the same standard length (Fig. 63.1c).The studied materials were kept in 70% ethanol and deposited at the fish collection of the Fisheries and Marine Resources Department, College of Science, University of Basra, Basrah, Iraq.

Exterior inspection of the area where the origins of the pelvic fins must be occur (underneath the pectoral fins) has shown no mark or excised and was normal looking, demonstrating that the specimens never established the exact structures. This was further inveterate by dissection of the specimen. Though, the pterygiophores of the dorsal and anal fins, the vertebrae and the air bladder in the examined specimen seemed normally grown.

Since only one deformed specimen was captured and no genetic study was conceivable, it might be expected that the perceived anomaly may have been a consequence of a genetic mutation, or could be accredited to the dilapidation of the water class of the marsh area.

63.5 Future Research

Before we proceed to put a plan to improve the environment where the fishes are living in, it is important to ask the following questions of what grade parental and early habitat impacts disturb upcoming acts of fishes and what the genetic, epigenetic, neural and hormonal contrivances are? For instances, environmental influences throughout development might have influence on the adult morphology that is hard to understand when testing acts onthe organism in the field**.** Deprived of this data, climate modification influences could be incorrect. Since there is a time gap between the reason and consequence, influences of such early impressions are mainly exposed through long-lasting experimental studies.The relations between early impacts, behaviour, growth and life-history features are not very well understood, and new data about these events will be valued for accepting the population and species influences in relation to natural and human-made habitat modifications and climate modification.

Early habitat influences on later functioning in fishes are as yet a poorly studied field. New investigations must be intended at establishing the contrivances of this plasticity, and how the norms of reaction are and copy changes in capability of the fishes to react to habitat modifications? By growing the appreciation of relations between habitats and genotypes, the aptitude to imagine population reactions to early influences on later morphologic plasticity will be established.

References

Al-Hassan LAJ (1982) Vertebral abnormalities in fishes from Iraq and UAE. Arabian Gulf Iraqi J Mar Sci 1:13–23

Al-Hassan LAJ (1983) Vertebral abnormalities in Mesopotamichthys sharpeyi and Carasobarbus luteus (Pisces: Cyprinidae) from Basrah, Iraq. Cybium 7(1):7–10

Al-Hassan LAJ (1985) Vertebral anomalies in *Gambusia affinis* (Bairdand Girard) (Teleostei: Poeciliidae) collected from Basrah and Baghdad. Iraq Pakistan J Zool 17(2):169–171

Al-Hassan LAJ (1987) Vertebral anomalies in fishes collected from Iraq and Arabistan. Iran Pakistan J Zool 19(1):93–95

Al-Hassan LAJ, Na'ama AK (1986) More vertebral anomalies in fishes from Euphrates River. Iraq Indian J Fisheries 33(2):237–241

Bengtsson BE (1974) Vertebral damage to minnows *Phoxinus phoxinus* exposed to zinc. Oikos 25:134–139

Boglione C, Gisbert E, Gavaia P, Witten E, Moren P, Fontagné M, Koumoundouros GS (2013) Skeletal anomalies in reared European fish larvae and juveniles.Part2: main typologies, occurrences and causative factors. Rev Aquac 5:S121–S167

Chatain B (1994) Abnormal swim bladder development and lordosis in seabass (*Dicentrarchus labrax*) and seabream (*Sparus auratus*). Aquaculture 119:371–379

Cheng SH, Wai AWK, So CH, Wu RSS (2000) Cellular and molecular basis of cadmium-induced deformities in zebrafish embryos. Environ Toxicol Chem 19:3024–3031

Favaloro E, Mazzola A (2003) Meristic variation and skeletal anomalies of wild and reared sharp snout seabream juveniles (*Diplodus puntazzo*, Cetti 1777) off coastal Sicily, Mediterranean Sea. Aquac Res 34:575–579

Georgakopoulou E, Angelopoulou A, Kaspiris P, Divanach P, Koumoundouros G (2007) Temperature effects on cranial deformities in European seabass, *Dicentrarchus labrax* (L.). J Appl Ichthyol 23:99–103

Gilbert SF, Epel D (2008) Ecological developmental biology. Sinauer Associates, Inc., Sunderland, MA

Jawad L, Al-Faisal AJ, Al-Mukhtar M (2015) A case of vertebral coalescence in Luciobarbus xanthopterus (Heckel, 1843) (Pisces: Cyprinidae) obtained from the lower reaches of Mesopotamia. Travaux du Muséum national d'Histoire Naturelle «Grigore Antipa» 57:127–132

Jawad LA, Al-Faisal AJ, Al-Mutlak FM (2014) Incidence of lordosis in the cyprinid fish, Carasobarbus luteus and the Shad, *Tenualosa ilisha* collected from Basrah waters, Iraq. International Journal of Marine science 4(45):1–5

Jawad LA, Mutlak F, Al-Faisal A (2016) On the record of vertebral deformities in *Mastacembelus mastacembelus* collected from the lower reaches of Euphrates River, Iraq. Bol. Inst. Pesca, São Paulo 42(1):216–220

Jawad LA, Jayabalan N, Jawad HL (2018) Unilateral microphthalmia and slow growth rate in a natural population of the marine teleost *Pampus argenteus* (Actinopterygii: Perciformes: Stromateidae). Bull Fish Biol 17(1/2):39–44

Jezierska B, Sarnowski P, Witeska M, Ługowska K (2009) Disturbances of early development of fish caused by heavy metals (a review). Electron J Ichthyol 2:76–96

Jonsson B, Jonsson N (2014) Early environment influences later performance in fishes. J Fish Biol 85(2):151–188

Kennedy CJ (2011) The toxicology of metals in fishes, in Encyclopaedia of fish physiology: from genome to environment. In: Farrell AP (ed) , 3rd edn. Academic Press, San Diego, CA, pp 2061–2068

Koumoundouros G, Maingot E, Divanach P, Kentouri M (2002) Kyphosis in reared seabass (*Dicentrarchus labrax* L.): ontogeny and effects on mortality. Aquaculture 209:49–58

Lauder GV, Drucker EG (2004) Morphology and experimental hydrodynamics of fish fin control surfaces. IEEE J Ocean Eng 29(3):556–571

Lindström J (1999) Early development and fitness in birds and mammals. Trends Ecol Evol 14:343–348

Mahdi BA, Fawzi NAM (2014) Iraq's inland water quality and their impact on the North-Western Arabian gulf. Marsh Bulletin 9(1):1–22

Martin TE, Ton R, Niklison A (2013) Intrinsic vs. extrinsic influences on life history expression: metabolism and parentally induced temperature influences on embryo development rate. Ecol Lett 16:738–745

Muramoto S (1981) Vertebral column damage and decrease of calcium concentration in fish exposed experimentally to cadmium. Environ Poll Ser A: Ecol Biol 24:125–133

Pickering QH, Gast MH (1972) Acute and chronic toxicity of cadmium to the fathead minnow (*Pimephales promelas*). J Fish Res Board Can 29:1099–1106

Sfakianakis DG, Georgakopoulou E, Papadakis IE, Divanach P, Kentouri A, Koumoundouros G (2006) Environmental determinants of haemal lordosis in European seabass, *Dicentrarchus labrax* (Linnaeus, 1758). Aquaculture 254:54–64

Sfakianakis DG, Katharios P, Tsirigotakis N, Doxa CK, Kentouri M (2013) Lateral line deformities in wild and farmed seabass (*Dicentrarchus labrax*, L.) and seabream (*Sparus aurata*, L.). J Appl Ichthyol. 29:1015–1021

Slooff W (1982) Skeletal anomalies in fish from polluted surface waters. Aquat Toxicol 2 (3):157–173

Storelli MM, Giacominelli-Stuffler R, Storelli A, Marcotrigiano GO (2005) Accumulation of mercury, cadmium, lead and arsenic in swordfish and bluefin tuna from the Mediterranean Sea: a comparative study. Mar Poll Bull 50:1004–1007

Yaseen BR, Al Asaady KA, Kazem AA, Chaichan MT (2016) Environmental impacts of salt tide in Shatt al-Arab-Basra/Iraq. IOSR J Environ Sci Toxicol Food Technol 10(1–2):35–43

Chapter 64
The Phenomenon of Fluctuating Asymmetry: As Fish Welfare Indicator Represented by Case Study from the Freshwater Fishes of Iraq

Laith A. Jawad and Jasim Abed

Abstract It has been suggested that fluctuating asymmetry shows an animal's capability to manage with the amount of sufferings throughout its developing period and, consequently, is likely health indicator. Fluctuating asymmetry is a satisfactory gauge of animal fitness, notwithstanding a great number of open queries, e.g., linking to the development of FA or its delicateness to numerous impacts. The substantial impending of fluctuating asymmetry as a health marker renders it valuable to follow more influentially endorsement studies as well as applied studies. These investigations should pay specific consideration to an appropriate procedural method.

In the case study, an investigation of the asymmetry analysis was carried out for eight bilateral characters of the Indian catfish, *Heteropneustes fossilis* collected from 8 localities along the Tigris, Euphrates, and Shatt al-Arab Rivers, Iraq. The lowest bilateral asymmetry value was observed in fishes collected from the Tigris River at Mosul City, while the higher value was obtained from fishes collected from Tigris River at Baghdad City. Fishes obtained from the Tigris River are shown to have higher asymmetry value than those collected from Euphrates River for all the characters studied. The grade of asymmetry rate for the features of fishes obtained from the Shatt al-Arab River was inside the range of those fishes attained from Tigris and Euphrates Rivers. Bilateral asymmetry augmented with the fish size at all regions for the features, preorbital length, upper and lower jaws length, and pectoral fin length. The conceivable reasons for bilateral asymmetry in this species are deliberated relative to diverse contaminants.

L. A. Jawad (✉)
School of Environmental and Animal Sciences, Unitec Institute of Technology, Auckland, New Zealand

J. Abed
Department of Fisheries and Marine Resources, College of Agriculture, University of Basrah, Basrah, Iraq

L. A. Jawad (ed.), *Tigris and Euphrates Rivers: Their Environment from Headwaters to Mouth*, Aquatic Ecology Series 11,
https://doi.org/10.1007/978-3-030-57570-0_64

64.1 Introduction

Among the external features of fishes that its variation has been related to the changes of the habitats is fluctuating asymmetry [the haphazard deviancy from perfect bilateral symmetry (Van Valen 1962)]. The homeostatic regulator of the external trait growth, or developmental constancy, may be upset when naturally occurring or man-made impacts are accomplished throughout the growth period. Consequently, growth does not pursue its pre-programmed course, and deformed morphological feature, containing bilateral asymmetry, may be the consequence.

Studies of fluctuating asymmetry have come to be a widespread exercise for quickly and rationally testing before death impact, and the consequences of such investigations are frequently concluded to group health (Graham et al. 1993; Lens et al. 2002). However, in the wider scientific society, settlement on fluctuating asymmetry expediency is absent may be due to fluctuating asymmetry's link to impact is fragile or owing to the connection converts to be mixed up during frequent phases of the analytical procedure (Lens et al. 2002; Graham et al. 1993).

Apparently, it is appropriate to enquire if fluctuating asymmetry investigators have observed this risk avoidance. A number of studies scrutinize the connection between FA and habitat impacts crosswise the broad phylogenetic scales (Lens et al. 2002; Leung et al. 2003; Graham et al. 2010), but none exactly mention this aspect in fishes. Fishes are an exceptional group for such an appraisal. As outside habitat circumstances considerably affecting their growth, fish are frequently used as examination species for inspection the health of the aquatic habitats (Graham et al. 1993; Ranaldi and Gagnon 2010) as they are closely related with their niches (Helfman et al. 1997; Lens et al. 2002; Leung et al. 2003).

The phenomena of fluctuating asymmetry investigation in fish are analytically studied with an aim to conclude overall subjects, recognize faults, and reprise some basic thoughts for the study of this phenomenon (Leung et al. 2003).

The aim of the present chapter to provide a short review of the research of fluctuating asymmetry on fishes, give suggestions to recover the class and make-up of field and laboratory tests, thus permitting for general and strong deductions about the reasons and results of FA. At the end of the chapter, a case study on fluctuating asymmetry in the Indian catfish *Heteropneustes fossilis* collected from several localities in the north, middle, and south of Iraq was given as an example on the health of the environment in Iraq and urge the policy makers in this country to go ahead and plan for a better environment for the people of Iraq.

64.2 Fish Fluctuating Asymmetry Studies

The investigations on the fluctuating asymmetry in fishes evidently showed that this occurrence initiated by external impacts does occur in one or more morphological features or as a complex of characters. However, attempts to attain indications from these studies are gained by the diversity of biological or environmental settings in which fluctuating asymmetry was assessed, as well as by the diversity of species used and characters measured (Leung et al. 2003; Allenbach 2011).

Fluctuating asymmetry is perhaps most often examined in relation to the man-made impacts. In fish, aggravations vary from general (physical habitat deviations, urban effect, undefined, or mixed stressors) to exact (heavy metals, temperature, radiation, acidification, and pesticides) (Allenbach 2011).

Several investigations have considered the use of fluctuating asymmetry in fish, but significant question leftovers: what do those rates and parameters really mean? One assertion is that fluctuating asymmetry can be a marker of health, that stable during the growth of the fish have higher metabolic effectiveness and therefore have more energy for growth, reproduction, and endurance (Allenbach 2011).

Several fish features were involved in the process of studying fluctuating asymmetry valuations, and such traits are frequently divided into two groups: meristic (counted) and morphometric (discrete). These characters vary in their level of fluctuating asymmetry in any fish species. Therefore, it is important to pay much care in choosing the features for the fluctuating asymmetry (Allenbach 2011).

Concerning influence nature, the hazards evaluated in such investigations and it is possible to assemble them into three general classes: physical (e.g., temperature, water staining), biological (e.g., infection, rearing density), or chemical (e.g., acidification, pesticides) (Allenbach 2011).

64.3 Fluctuating Asymmetry Studies in Iraq

The studies on the fluctuating asymmetry in fishes of Iraq, both freshwater and marine, are scarce. The first study published on fish asymmetry was in 1994 by Al-Hassan and Hassan on the freshwater catfish, *Mystus pelusius* and later in 2004, Jawad reported on the fluctuating asymmetry in the freshwater mullet *Planiliza abu* from Shatt al-Arab River, Basrah, Iraq. In those papers, the authors have used both meristic and morphometric characters to show the level of asymmetry in these characters as an indication of the health of the environment. Therefore, more research are needed to be done in this field on a large number of fish species inhabiting different localities in order to put baseline studies for further environmental monitoring.

64.4 Asymmetry Measurement

There are specific limitations about placing the fish part that needs to be measured (Palmer 1994; Helm and Albrecht 2000). This is particularly correct if structures with stretchy body features are measured. Once starting to take measurements, the resolution of measurements must be considered as it must counterpart the predictable average variances between sides. The repetitions of the dimensions need to be taken without referring to those already done. Consequently, duplicate trials should happen during not at the same settings (Palmer 1994; Helm and Albrecht 2000).

The conceivable sum of features in each organism is extremely high, but there are definite limitations that changing some attributes more valuable than others. These parameters need to have sights that are simple to identify with any mistake. They should be well reachable, if for example a caliper is used, and allow duplicable placing (Palmer 1994; Helm and Albrecht 2000).

64.5 Bilateral Asymmetry in the Indian Catfish Collected from Several Localities in Iraq: A Case Study

64.5.1 Overview

The Indian stinging catfish, *Heteropneustes fossilis* is belonging to the family Heteropneustidae contained in the genus *Heteropneustes* namely *H. fossilis* Bloch (1794).

It distributed over the south and Southeast Asian countries including Bangladesh, India, Laos, Myanmar, Nepal, Pakistan, Sri Lanka, and Thailand (Talwar and Jhingran 1991). It has been brought into Iraq in the late 1950s and documented its presence in the freshwaters of southern Iraq in 1960 by Khalaf (1961) and from the marine waters of Khor Al-Zubair, northwest Arabian Gulf (Al-Hassan and Muhsin 1986). The objective of getting into Iraq this Indian stinging catfish is to combat the snail *Bulinus truncatus*, which is considered the causative agent for disease of schistosomiasis (Jawad 2003).

Whereas it has not been taken as food species in Iraq and the Middle East, *H. fossilis* is commercially and aquaculturally is a significant fish in several Asian countries (Hossain et al. 2015). It is extensively caught for its activating meat class that contains taste, nutritional, and medicinal values (Jha and Rayamajhi 2010).

This chapter portrait an incidence of fluctuating asymmetry examined in a familiar fish species in the Tigris, Euphrates, and Shatt al-Arab Rivers to disclose the range of bilateral asymmetry in eight morphological characters and to then use those traits to show the uppermost levels of bilateral asymmetry in additional progressive investigations. No preceding study on bilateral asymmetry has been done on *H. fossilis*. Hence, this chapter is considered the first information on the bilateral asymmetry in one of the freshwater fishes of Iraq grounded on a wide range of geographical samples selection.

64.5.2 *Methodology*

64.5.2.1 Study Area

In Iraq, Tigris and Euphrates are the two main rivers. The former is 1850 Km long, rising from the Taurus Mountains in eastern Turkey. The river runs over the Turkish territories before entering the Iraqi territories from the north. In the Middle of Iraq and near Mysan Province, Majar al-Kabir branch splits from this river to feed the central marsh area (Isaev and Mikhailova 2009). The Euphrates River is the sister river to Tigris River that runs to its right. It is the most historically important and longest river in western Asia. It is 450-Km long running from its origin in south-eastern Turkey and enter Syrian territories before running the Iraqi planes (IM 2006). Shatt al-Arab River established by the meeting of the Tigris and Euphrates Rivers at al-Qurnah, Basrah Governorate, south of Iraq, with 200-Km length (Country data.com 2015).

The eight sampling sites that are situated on the three great rivers of Mesopotamia, Tigris, Euphrates, and Shatt al-Arab, were nominated since they are profoundly inhabited and are impacted by huge human-made waste discarding. The geographic location and description are given in Table 64.1 and Fig. 64.1.

64.5.2.2 Chemical Analyses of Organochlorine Pesticides and Heavy Metals in Surficial Sediments

Since *H. fossilis* feeds mainly on detritus in addition to fish larvae and eggs, the level of concentration of organochlorine pesticides and heavy metals were targeted in the sediments rather than the water. The method of chemical analyses was followed that given by Jawad and Ibrahim (2018).

Table 64.1 Coordinates and nature of bottom of the eight localities on the Tigris and Euphrates Rivers, Iraq

Localities	Coordinates	Description of the area
Mosul	36.34°N 43.13°E	Mainly silt and rocky bottom with plenty of aquatic plants
Baghdad	33°20′N 44°26′E	Mainly silt and peoples with plenty of aquatic plants
Al-Amarah	1°54′N 47°2′E	Mainly silt rich with large aquatic plants
Ar Ramadi	3°25′11″N 43°18′45″E	Mainly rocky with fine silt and fewer aquatic plants
As Samawah	31°19′N 45°17′E	Mainly silt with small rocks and plenty of aquatic plants
An Nasiria	31°03′N 46°16′E	Mainly silt and plenty of aquatic plants
Al-Hammar Marsh	30°46′0″N 47°3′0″E	Mainly silty and rich with large aquatic plants and algae
Basrah	30°30′N 47°49′E	Mainly silty and rich with large aquatic plants and algae

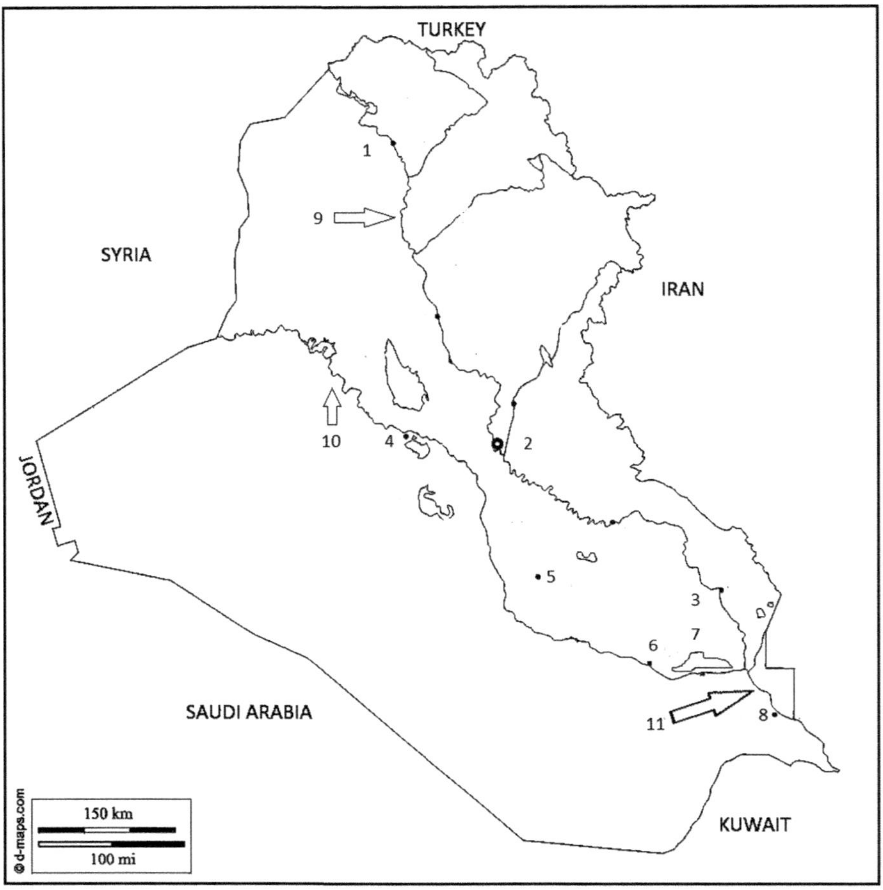

Fig. 64.1 Map showing sampling localities through Iraq. (1) Tigris River at Mosul City; (2) Tigris River at Baghdad City; (3) Tigris River at Al-Amarah City; (4) Euphrates River at As Samawah City; (5) Euphrates River at An Nasiria City; (6) Al-Hammar Marsh area; (8) Shatt al-Arab River at Basrah City; (9) Tigris River; (10) Euphrates River; and (11) Shatt al-Arab River

64.5.3 Data analysis

64.5.3.1 Probability of Toxicity

The procedures of Long et al. (1995) for probability of toxicity were pursued. The recommendations contain two parametric concentrations for each of the chemicals given in these plans. The lower level (effects range low, or ERL) indicates the concentration below which opposing biological impacts are infrequently witnessed and the influences range median (ERM) level differentiates concentrations above which contrasting biological concerns are foreseeable to happen frequently. Any value falls between ERL and ERM parameters specify a midway, frequently uneven biological reaction. Individual chemicals were evaluated for conceivable opposing biological impacts by associating concentrations at each site to the respective ERL and ERM values. Sediments were categorized into four risk classes, low, moderate, high, and severe after the system of Long et al. (1995).

64.5.3.2 Fluctuating Asymmetry

Fluctuating asymmetry levels and size errors are in many instances small and usually dispersed around a mean of zero. Morphological traits might be disturbed by a number of factors, such as circumstances of measuring and operator's knowledge (e.g., Lajus et al. 2003). Consequently, haphazardly nominated samples were measured irrespective of the order of sample location to avoid the result of these causes on outcomes of sample assessments.

To calculate the value of the fluctuating asymmetry, the square coefficient of the following equation was used conferring to Valentine et al. (1973) as:

$$\mathrm{CV}_a^2 = (S_{l-r}X100/X_{l+r})$$

where S_{l-r} is the standard deviation of the signed difference, X_{l+r} is the mean of the character, which is calculated by adding the absolute scores for both sides and dividing by the sample size. For each locality, sampled individuals were categorized into classes based on their total length. Specimens from the eight localities were divided into three size groups to calculate the fluctuating asymmetry value for each size class.

64.5.3.3 Heavy Metals Data Analysis

Statistical analyses were achieved using SAS software. Analysis of variance (ANOVA) was done for each of the six metals distinctly to examine the alterations amongst the four sites. Significant differences confirmed by ANOVA *F*-test were additionally tested using Tukey's honest significant difference (HSD) multiple comparison method (alpha = 0.05). As a measure of link between asymmetry level and the concentration of both heavy metals and pesticides, the Pearson correlation coefficient was computed using the data from the eight sites for each of the 12 pollutants (6 heavy metals and 6 pesticides). The normality and equal variance molds of ANOVA were tested using residual plots.

64.5.4 Fish Samples

The eight sampling sites that are situated on the three great rivers of Mesopotamia, Tigris, Euphrates, and Shatt al-Arab, were chosen since they are profoundly inhabited and are impacted by enormous man-made discarded removal. The geographic location and description are given in Table 64.1 and Fig. 64.1.

Fish samples were attained from fishers working in the sampling sites. Gillnets (200 m × 1.30 m, 25, 40, and 50 mm mesh) and cast nets (6 diameter, 20 mm mesh) were utilized by fishers to catch the fish. Fish samples were originated from Tigris River locations in August 2004 and from Euphrates and Shatt al-Arab Rivers locations in July 2005. The depth at all sampling sites ranged from 0.5 to 2.4 m. The total number of specimens of the Indian catfish, *H. fossilis* studied was 1500 ranging in total length between 200 and 320 mm.

The eight metric traits taken for bilateral asymmetry analysis (Fig. 64.2) were earlier used in fish studies (Jawad et al. 2010; Hechter et al. 2000; Lucentini et al. 1998). These are as follows: (1) preorbital length (ProL), measured from the tip of the mouth to the anterior edge of the orbit; (2) postorbital length (PosL), measured from the posterior edge of eye to the edge of operculum; (3) eye diameter (ED), measured from the anterior edge of the orbit to the posterior edge the orbit; (4) head length (HL), measured from the anterior tip of the mouth to the edge of the operculum; (5) upper jaw length (UJL), measured from the anterior symphysis of the left and right upper jaws to the posterior bony edge of the upper jaw; (6) lower jaw length (LJL), measured from the anterior symphysis of the left and right parts of the lower jaw to the posterior bony edge of the lower jaw; (7) pectoral fin length (PFL), measured from the origin of the fin to the tip of the longest ray; and (8) pelvic fin length (PVFL), measured from the origin of the fin to the tip of the longest ray (Fig. 64.1). Traits were measured to the nearest 0.1 cm using a digital calliper.

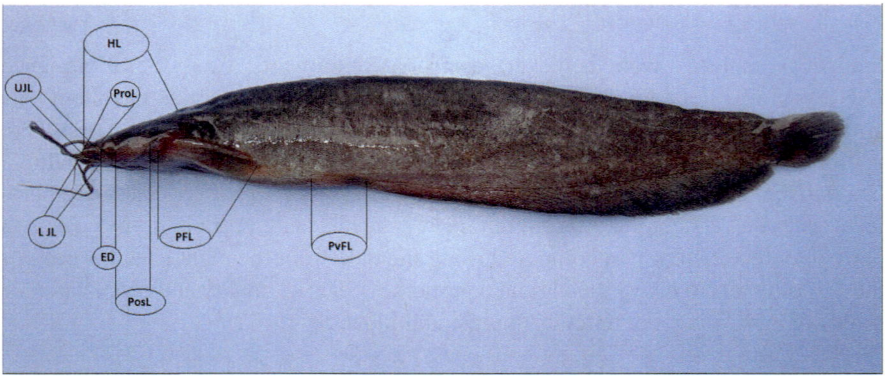

Fig. 64.2 Bilateral body characters of *Heteropneustes fossilis*. *ED* eye diameter, *HL* Head length, *LJL* Lower jaw length, *POSL* postorbital length, Preorbital length, *PVFL* pelvic fin length, *UJL* upper jaw length

64.6 Results and Discussion

The results of the bilateral asymmetry calculations of the eight morphological traits of *H. fossilis* are shown in Table 64.2. Excluding for Mosul locality, the asymmetry levels of these features were greater in localities from the Tigris River than in Euphrates River, while those for the Shatt al-Arab River were within the range of the levels attained for Tigris and Euphrates Rivers. The eight morphological traits exhibited greater values in fishes collected from the Baghdad area and the lowest values in fishes obtained from Mosul city (Table 64.2).

Analysis of variance showed that the values of asymmetry for the eight morphological traits varied between the groups of the *H. fossilis* studied along the Tigris, Euphrates, and Shatt al-Arab Rivers ($P < 0.001$).

All eight morphometric characters exhibit highest and lowest percentage of bilateral asymmetry Baghdad (90–96%) and Mosul (32–46%) localities, respectively.

Individuals of *H. fossilis* from the eight localities were assembled into length classes (Table 64.2). A growing trend in asymmetry levels was observed for preorbital length, postorbital length, upper jaw length, lower jaw length, and pectoral fin length in the eight populations of *H. fossilis*.

High bilateral asymmetry values for morphometric characters have formerly been reported in numerous freshwater and marine fish species (Al-Hassan et al. 1990; Al-Hassan and Hassan 1994; Jawad et al. 2012a, b; Jawad 2013; Mabrouk et al. 2014). The eight morphometric characters studied showed variable values of bilateral asymmetry range from 24.56 in preorbital length of fishes collected from Mosul locality to 120.3 in pelvic fin and upper jaw length of Baghdad locality. This concords in results of bilateral asymmetry could specify the susceptibility of these traits to instant modifications in the habitat. Hence, they could be used as an effective biomarker of stress in the environment. Although the value of bilateral asymmetry of some morphometric characters is not high enough in fishes from certain localities (e.g., Mosul), they considered high in comparison with the value obtained for other morphometric characters of the same area. Alternatively, the lower bilateral asymmetry values, which proposes that this trait may be less subtle to habitat impact issues containing pollution. The lower bilateral asymmetry levels attained for morphometric traits in fishes from Mosul could be elucidated on the ground that the growth period of these traits may not concur with the occurrence of opposing habitat measures (Jawad 2003).

The low bilateral asymmetry level at Mosul locality can be considered as normal since biological systems cannot have a faultless bilateral symmetry even in an unspoiled habitat. Minor inconsistencies throughout growth can diverge from normal growing procedures (Palmer and Strobeck 1992). Such indiscretions may be owing to the quality and quantity of food, thrilling temperatures, parasites, disease, and behavioral impact forced by connections with members of the same species (Markow 1995).

Table 64.2 Squared coefficient asymmetry (CV^2a) values and character means (X_{r+1}) of *Hetropneustes fossilis* collected from different localities at the Tigris and Euphrates Rivers, Iraq. Length group no.1 = 80–100 mm; length group no.2 = 101–150 mm; length group no.3= 151–180 mm

		Characters																							
		Preorbital length			Postorbital length			Eye diameter			Head length			Upper jaw length			Lower jaw length			Pectoral fin length			Pelvic fin length		
	Locality	1	2	3	1	2	3	1	2	3	1	2	3	1	2	3	1	2	3	1	2	3	1	2	3
CV^2a	Mosul	21.1	26.3	29.4	30.5	32.5	36.5	44.2	46.2	44.9	45.4	43.3	48.6	33.3	34.5	39.4	24.1	25.4	29.8	51.9	52.3	54.6	43.2	42.3	43.2
	Baghdad	75.2	88.4	94.3	92.5	98.4	98.4	99.8	87.9	99.9	111.2	99.9	99.3	91.4	96.3	121.0	92.8	92.7	97.2	99.5	99.8	114.2	120.0	99.4	98.5
	Al-Amarah	74.2	75.8	79.4	87.1	83.6	88.6	95.4	92.8	98.1	97.4	96.6	98.6	92.1	95.3	98.4	75.1	74.5	78.4	93.4	96.4	99.2	95.4	97.5	93.6
	Ar Ramadi	46.7	48.8	56.1	65.2	66.4	63.3	74.2	73.4	78.8	76.4	79.9	80.6	83.1	85.3	88.3	73.3	75.3	78.1	81.3	85.8	88.3	89.7	89.9	84.3
	As Samawah	70.5	70.4	77.5	86.8	82.5	88.2	94.4	92.4	99.4	96.1	96.4	96.4	90.8	91.3	94.8	83.1	86.3	89.0	86.5	88.4	91.4	86.0	86.4	85.4
	An Nasiria	60.9	63.5	68.5	73.2	70.1	76.4	86.3	82.7	89.0	86.4	84.3	87.7	92.1	95.3	99.3	80.1	83.4	87.2	83.4	85.4	90.0	89.3	90.8	90.4
	Al-Hammar Marsh	62.4	63.5	69.6	87.3	83.2	87.8	98.3	94.2	98.7	95.3	92.6	94.7	90.6	90.4	92.3	72.4	73.9	78.1	74.4	75.3	78.4	89.4	87.6	89.3
	Basrah	93.5	97.4	97.6	99.3	96.3	99.7	96.6	94.8	99.4	99.9	99.3	99.0	95.2	98.4	99.5	91.4	94.2	95.2	93.4	98.4	99.4	100	99.3	99.5
N	Mosul	60	80	50	60	80	50	60	80	50	60	80	50	60	90	40	60	80	50	60	80	50	60	80	40
	Baghdad	50	80	50	50	80	50	50	80	50	50	80	50	50	80	50	50	80	50	50	80	50	50	90	50
	Al-Amarah	0	60	80	0	60	80	0	60	80	0	60	80	0	60	80	0	60	80	0	60	80	0	60	80
	Ar Ramadi	60	80	45	60	80	45	60	80	45	60	80	45	60	80	45	60	80	45	60	80	45	70	80	45
	As Samawah	65	60	16	65	60	16	65	60	16	65	60	16	65	60	16	65	60	16	65	60	16	65	60	15
	An Nasiria	70	80	37	70	80	37	70	80	37	70	80	37	70	80	37	70	80	37	70	80	37	70	80	37
	Al-Hammar Marsh	90	80	40	90	80	40	90	80	40	90	80	40	90	80	40	90	80	40	90	80	40	90	80	40
	Basrah	80	75	43	80	75	43	80	75	43	80	75	43	80	75	43	80	75	43	80	75	43	80	75	43
Character mean (X_{r+1})	Mosul	0.8	0.8	0.9	5.2	5.3	5.4	1.5	1.6	1.7	9.4	9.5	9.6	2.4	2.5	2.6	3.1	3.2	3.4	4.8	4.9	5.0	2.8	2.9	3.9
	Baghdad	0.9	1.0	1.0	5.0	5.1	5.1	0.3	0.4	0.4	9.6	9.7	9.8	2.4	2.5	2.6	2.8	2.9	3.0	5.0	5.1	5.2	0.2	0.3	0.4
	Al-Amarah	1.1	1.2	1.3	5.5	5.6	5.7	0.5	0.5	0.6	9.0	9.5	10	2.4	2.5	2.6	3.2	3.4	3.5	5.0	5.1	5.2	0.2	0.3	0.4
	Ar Ramadi	0.9	1.0	1.0	5.4	5.5	5.6	0.4	0.5	0.5	9.6	9.7	9.8	2.4	2.5	2.6	3.2	3.3	3.4	3.3	3.4	3.5	0.3	0.4	0.5
	As Samawah	1.3	1.4	1.6	5.5	5.5	5.6	0.2	0.3	0.3	7.4	7.5	7.6	1.3	1.4	1.5	2.0	2.1	2.1	4.4	4.5	4.6	0.3	0.4	0.5
	An Nasiria	0.9	1.0	1.0	5.6	5.7	5.7	0.4	0.5	0.5	10.3	9.9	10.4	2.4	2.5	2.6	3.3	3.4	3.5	3.0	3.1	3.2	0.3	0.4	0.5
	Al-Hammar Marsh	0.9	0.9	0.9	5.3	5.4	5.5	0.3	0.4	0.4	9.1	9.2	9.3	2.4	2.5	2.6	3.3	3.4	3.5	4.8	4.9	5.0	0.2	0.3	0.4
	Basrah	0.9	1.0	1.0	5.2	5.3	5.3	0.5	0.6	0.6	9.1	9.2	9.3	2.4	2.5	2.6	3.3	3.4	3.5	4.8	4.9	5.0	0.2	0.3	0.4

Percentage of individuals with asymmetry	Mosul	30	35	50	35	25	45	34	43	52	34	45	48	23	35	36	10	16	19	30	34	40	32	33	32
	Baghdad	53	67	89	76	85	86	87	90	92	89	90	94	87	88	86	80	83	87	87	86	80	98	99	97
	Al-Amarah	52	63	89	78	86	90	86	88	90	86	78	77	79	85	89	54	53	60	87	78	79	85	83	87
	Ar Ramadi	23	35	78	53	65	67	67	75	74	64	63	54	65	67	78	65	67	69	32	37	40	76	74	79
	As Samawah	45	55	86	78	80	87	64	53	78	78	80	83	67	69	79	67	76	87	78	88	89	76	78	79
	An Nasiria	43	62	77	65	63	70	75	76	63	65	67	76	87	88	78	76	77	79	78	75	79	87	86	89
	Al-Hammar Marsh	54	44	78	76	77	80	87	80	79	87	79	90	67	78	80	65	66	70	65	66	69	62	66	69
	Basrah	87	89	96	89	87	85	87	88	90	88	78	98	85	86	89	87	85	88	87	88	94	98	99	99

The results also showed that bilateral asymmetry levels for the eight morphological traits examined varied significantly between the eight groups of *H. fossilis* ($P < 0.001$). There is a noteworthy level of pollution by diverse kinds of contaminants such organic pollutants and heavy metals in the areas where fishes were obtained at all localities from north to the south of Iraq (Kassim et al. 1997; Al-Lami and Al-Jaberi 2002; Al-Juboury 2009; Al-Noor et al. 2013; Al-Obaidy et al. 2014). It is possible that pollution might be the cause of the high bilateral asymmetry levels in these areas. Certainly, many investigations have revealed that pollution was blamed for instances related to high bilateral asymmetry values (e.g., Franco et al. 2002; Mabrouk et al. 2014). Generally speaking, the toxicity of some trace metals and other chemicals is revealed to upsurge with the rise in both temperature and salinity (Sogorb et al. 1988; Wright 1995; Rainbow 1997; Kwok and Leung 2005; Franco et al. 2002). Water temperature at Mosul locality is usually lower (11.9–20 °C) (Al-Sanjari and Al-Tamimi 2009) than the remaining localities at the middle, west, and south of Iraq (15 °C to 40 °C)(Al-Noor et al. 2013; Hassan et al. 2014; Abbas et al. 2015). Therefore, the role of low temperature at Mosul locality in increasing toxicity is clear with low bilateral asymmetry value at this locality. On the other hand, and due to the presence of certain level of salinity in the lower reaches of Tigris and Euphrates Rivers and in Shatt al-Arab River (0.5–1.0%) (Kadhim 2014; Hassan et al. 2014), the salinity as a factor enhancing the toxicity to pollutants can be considered in these localities and the high values of bilateral asymmetry might be correlated to this factor, but it is not possible to take into consideration the salinity factor to increase toxicity in the middle and west region's localities due to low water salinities (Jehad 1984; Al-Tamimi and Al-Gafily 2009).

Growing disarrays bring about by chemical and organic pollution can cause severe morphological deformities (Bengtsson et al. 1988; Sogorb et al. 1988; Rainbow 1997; Franco et al. 2002; Kwok and Leung 2005). In some localities studied, skeletal abnormalities were reported for several fish species, which were connected to heavy metal and organic pollution (Al-Hassan 1983, 1985; Al-Hassan and Naama 1986; Jawad et al. 2014).

As revealed in preceding investigations (Al-Mamry et al. 2011; Jawad et al. 2012a, b; Mabrouk et al. 2014), ANOVA test analysis in the present study showed that large size specimens of *H. fossilis* from all eight localities had greater bilateral asymmetry levels than smaller the young specimens for preorbital length, postorbital length, upper jaw length, lower jaw length, and pectoral fin length ($P < 0.001$). It was obvious that these levels of the five morphological traits augmented with fish size in this species (Table 64.2). This tendency is perhaps the consequence of partial growth; trait means are always lowest in smaller size classes (Valentine et al. 1973). Comparable outcomes were reached by Valentine et al. (1973) in selected fish species collected from California, USA. They recommended two conceivable theories that may lead to such a tendency; developmental variations linked to an upsurge in bilateral asymmetry with size (age), and likely historical procedures which results in a secular rise in bilateral asymmetry. Otherwise, Thiam (2004) proposed that a growing tendency in bilateral asymmetry levels with fish size might be owing to the fact that large size individuals had longer periods of contact to opposing habitat circumstances and consequently lost their constancy in such niches.

Sediment pollution in the Baghdad area was notably higher than in the other locations studied (Table 64.3). Average MERMQ levels in waters in the vicinity of Baghdad City exceeded those in the vicinity of Mosul City by nearly 2.5 folds and those in the vicinity of Ramadi City by 2.7 folds in heavy metals. For organic pesticides, average MERMQ levels in waters in the vicinity of Baghdad City are higher than those of Mosul and Ramadi City by 5.1 folds. Except for fishes from the vicinity of Mosul City, the cities located on the Tigris River, i.e., Baghdad and Mysan, the average MERMQ for heavy metals and organic pesticides are higher than those of fishes from cities located on Euphrates and Shatt al-Arab Rivers (Table 64.3).

The use of MERMQ methods displays that heavy metals have a low hazard of toxicity in locations situated in Mosul and Ramadi Cities, whereas the other locations were shown a modest danger (Table 64.3). Hazard of toxicity linked with organic pollution was significantly higher and reaching a severe level in Baghdad and Mysan sites, while it was low in the remaining sites.

The overall ANOVA exhibited that there were significant differences in sediment concentrations among sites for all 12 pollutants. Tukey's HSD (alpha = 0.05) indicated that sites on the Tigris River were significantly different from those on the Euphrates River for all pollutants studied. Correlation analyses showed a positive correlation between concentrations of all metals and pesticides and level of asymmetry (r-values >0.7).

An administration strategy is immediately necessary in order to reinstate a healthy habitat in Tigris, Euphrates, and Shatt al-Arab Rivers. Numerous aquatic plant and animal species within these water bodies have already been harmfully impacted (Kassim et al. 1997; Salman 2011).

64.7 Conclusions and Recommendations

Numerous environmental biologists are worried about whether morphological asymmetry is a marker of niche destruction giving a developmental impact. A number of studies are available checking this association in fish, and close by investigation of these studies have shown not only the causative agents may causing a significant fluctuating asymmetry result but likewise ways to improve study design and analysis. Allenbach (2011) has recommended the subsequent trials:

1. Character selection—Detailed care must be specified for picking traits in fluctuating asymmetry studies.
2. Measurement mistake—Since fluctuating asymmetry itself is such a small part of within-individual feature dissimilarity, extreme caution needs be given to keep measurement mistakes at a lower level.

Table 64.3 Concentration (μg/ml) (upper line), quotient (lower line), and mean effects range median quotients (MERMQs) of six heavy metals and six organochlorine pesticides in sample sites in the Euphrates, Tigris, and Shatt al-Arab Rivers, Iraq

Locality		Heavy metals								Organochlorine pesticides								Overall toxicity
		Cd	Cr	Cu	Pb	Ni	Zn	MERMQ	Risk of toxicity	Lindane	DDE	DDD	DDT	ttChlordane	Dieldrin	MERMQ	Risk of toxicity	
Mosul	Mean	10.3	8	16	28	4	82	0.264	Low	1.9	2.0	3.0	2.6	1.1	2.0	1.132	Low	0.102
	Quotient	0.02	0.03	0.08	0.14	0.02	0.20			0.07	0.07	0.15	0.07	0.210	0.28			
Baghdad	Mean	0.87	164	99.6	198	20.8	627	0.6282	Moderate	76.5	77.2	57.9	22	135.7	46.0	5.72	Severe	3.35
	Quotient	0.08	0.43	0.37	0.92	0.41	1.53			2.61	2.66	2.99	0.52	21.9	5.83			
Al-Amarah	Mean	0.742	160	97	190	19.5	617	0.6191	Moderate	70.1	72.2	565	20	129	45.7	5.61	Severe	3.27
	Quotient	0.096	0.45	0.34	0.86	0.37	1.38			2.47	2.48	2.85	0.47	20.4	5.33			
Ar Ramadi	Mean	8	8	17	25	4	81	0.230	Low	1.8	1.9	2.8	2.4	1.1	1.8	1.122	Low	0.100
	Quotient	0.300	0.02	0.07	0.13	0.01	0.21			0.07	0.07	0.15	0.07	0.21	0.27			
As Samawah	Mean	0.652	143	87.5	181	17.8	605	0.571	Moderate	67.0	65.1	556	19	123.5	34.8	3.62	Low	0.980
	Quotient	0.081	0.43	0.35	0.89	0.40	1.45			2.50	2.51	2.88	0.48	20.82	5.42			
An Nasiria	Mean	0.647	141	86.2	187	17	604	0.550	Moderate	58.1	58.2	558	19.9	120.2	34.6	4.51	Low	0.9600
	Quotient	0.077	0.42	0.33	0.88	0.38	1.43			2.37	2.38	2.79	0.47	20.63	5.32			
Al-Hammar Marsh	Mean	0.638	140	84.1	176	17.6	608	0.531	Moderate	57.1	58.0	546	19.2	119.9	40.0	4.41	Low	0.7900
	Quotient	0.068	0.42	0.31	0.88	0.39	1.40			2.36	2.37	2.69	0.45	2.42	5.21			
Basrah	Mean	0.696	153	88.5	187	19.5	616	0.527	Moderate	66.5	66.8	567	20.9	124.5	35.7	4.67	Low	0.2100
	Quotient	0.088	0.44	0.37	0.91	0.40	1.52			2.61	2.65	2.97	0.52	21.8	5.81			

3. Laboratory experiments—Additional laboratory trials are wanted not only to confirm field observations but also to better elucidate precisely, which impactor may be producing fluctuating asymmetry since natural environments frequently involved in a mixture of man-made impacts.
4. Repetition of examinations—If a particular study is only done once, its result could have been got purely by chance.

References

Abbas MF, Salman SD, Al-Mayahy SH (2015) Diversity and seasonal changes of zooplankton communities in the Shatt Al-Arab River, Basrah, Iraq, with a special reference to Cladocera. Intern J Mar Sci 5(24):1–14

Al-Hassan LAJ (1983) Vertebral abnormalities in *Mesopotamichthys sharpeyi* and Carasobarbus luteus (Pisces: Cyprinidae) from Basrah, Iraq. Cybium 7:7–10

Al-Hassan LAJ (1985) Vertebral anomalies in *Gambusia affinis* (Bairdand Girard) (Teleostei: Poeciliidae) collected from Basrah and Baghdad. Iraq Pakistan J Zool 17:169–171

Al-Hassan LAJ, Hassan SS (1994) Asymmetry study in *Mystus pelusius* collected from Shatt al-Arab River, Basrah. Iraq Pakistan J Zool 26(3):276–278

Al-Hassan LAJ, Muhsin KA (1986) The presence of Carasobarbus luteus (Heckel, 1843) and *Heteropneustes fossilis* (Bloch, 1797) in Khor Al-Zubair, north west of the Arabian gulf, Basrah, Iraq. Zool Middle East 1:116–118

Al-Hassan LAJ, Naama AK (1986) More vertebral anomalies in fishes from Euphrates River. Iraq Indian J Fish 33:237–241

Al-Hassan LAJ, Al-Doubaikel AY, Wahab NK, Al-Daham NK (1990) Asymmetry analysis in the catfish, *Heteropneustes fossilis* collected from Shatt al-Arab River, Basrah. Iraq Rev Idrobiol 29:775–780

Al-Juboury AI (2009) Natural pollution by some heavy metals in the Tigris River, northern Iraq. Intern J Environ Res 3:189–198

Al-Lami AA, Al-Jaberi HH (2002) Heavy metals in water, suspended particles and sediment of the upper-mid region of Tigris River, Iraq. In: Proceedings of international symposium on environmental pollution control and waste management, 7–10 January 2002, Tunis (EPCOWM'2002), pp 97–102

Allenbach DM (2011) Fluctuating asymmetry and exogenous stress in fishes: a review. Rev Fish Biol Fish 21(3):355–376

Al-Mamry J, Jawad LA, Al-Bimani S, Al-Ghafari F, Al-Mamry D, Al-Marzouqi M (2011) Asymmetry in some morphological characters of *Leiognathus equulus* (Forsskål) (Leiognathidae) collected from the sea of Oman. Arch Polish Fish 19:51–55

Al-Noor TA, Arslan LK, Abd Ali LJ (2013) Studying the effects of industrial wastes on Tigris water in Al- Great City-Baghdad-Iraq. Advan Phys Theories Appl 25:48–57

Al-Obaidy AMJ, Al-Janabi ZZ, Al-Kubaisi AR (2014) Distribution of some heavy metals in Tigris River middle of Iraq. Baghdad J Sci 11:983–990

Al-Sanjari MN, Al-Tamimi MA (2009) Interpretation of water quality parameters for Tigris River within Mosul City by using principal components analysis. Tikrit J Pure Sci 14:1–6

Al-Tamimi AAM, Al-Gafily AAG (2009) The effect of Ramadi sewage treatment plant on the phytoplankton and some physico-chemical characters in Euphrates River. Iraq Baghdad J Sci 6:673–682

Bengtsson ÅB, Bengtsson E, Lithner G (1988) Vertebral defects in four horn sculpin, *Myoxocephalus quadricornis* L., exposed to heavy metal pollution in the Gulf of Bothnia. J. Fish Biol 33:517–529

Country data.com (2015). Iraq – Major Geographical Features. country-data.com. Retrieved 12 Dec 2015

Franco A, Malavasi S, Pranovi F, Nasci C, Torricelli P (2002) Ethoxyresorufin O-deethylase (EROD) activity and fluctuating asymmetry (FA) in *Zosterisessor ophiocephalus* (Teleostei, Gobiidae) as indicators of environmental stress in the Venice lagoon. J Aquat Ecosyst Stress Recover 9(4):239–247

Graham JH, Emlen JM, Freeman DC (1993) Antisymmetry, directional asymmetry, and dynamic morphogenesis. Genetica 89:121–137

Graham JH, Raz S, Hel-Or H, Nevo E (2010) Fluctuating asymmetry: methods, theory, and applications. Symmetry 2(2):466–540

Hassan FM, Salman JM, Abdulameer SH (2014) Seasonal variation of environmental properties and phytoplankton Community in Al- Hussainya River, Holly Karbala. Iraq Mesopotamia Environ J 1:56–82

Hechter RP, Moodie PF, Moodie GEE (2000) Pectoral fin asymmetry, dimorphism and fecundity in brook stickleback, *Culaea inionstans*. Behavior 137:999–1009

Helfman G, Collette B, Facey D (1997) The diversity of fishes. Blackwell, Malden

Helm B, Albrecht H (2000) Human handedness causes directional asymmetry in avian wing length measurements. Anim Behav 60:899–902

Hossain Y, Islam R, Ahmed ZF, Rahman M, Hossen A, Naser SM, Rasel RI (2015) Threatened fishes of the world: *Heteropneustes fossilis* (Bloch, 1794) (Siluriformes: Heteropneustidae). Croatian J Fish 73:77–79

IM (Iraqi Ministries of Environment, Water Resources and Municipalities and Public Works) (2006) Overview of present conditions and current use of the water in the marshlands area/ book 1: water resources. New Eden master plan for integrated water resources management in the Marshlands areas, New Eden Group (No doi is available)

Isaev VA, Mikhailova MV (2009) The hydrology, evolution, and hydrological regime of the mouth area of the Shatt al-Arab River. Water Res 36:380–395

Jawad LA (2003) Asymmetry in some morphological characters of four sparid fishes from Benghazi, Libya. Oceanol Hydrobiol Stud 32:83–88

Jawad LA (2013) On the asymmetry of some morphological characters of the silver-cheeked toadfish *Lagocephalus sceleratus* (Gmelin, 1789) collected from the sea of Oman. Water Res Manag 3:25–30

Jawad LA, Ibrahim M (2018) Environmental oil pollution: a possible cause for the incidence of ankylosis, kyphosis, lordosis and scoliosis in five fish species collected from the vicinity of Jubail City, Saudi Arabia, Arabian gulf. Int J Environ Stud 75(3):425–442

Jawad LA, Al-Mamry JM, Al-Kharusi AA, Al-Habsi S, H. (2010) Asymmetry in some morphological characters of the carangid species, *Decapterus russelli* collected from Lema coastal area, north of the sea of Oman. Oceanol Hydrol Stud 39:55–62

Jawad L, Al-Mamry J, Al-Bimani S, Al-Ghafri F, Al-Marzouqi M (2012a) Asymmetry of some morphological characters of *Upeneus doriae* (Osteichthyes: Mullidae) collected from the sea of Oman. Thalassia Salentina 34:3–10

Jawad LA, Al-Mamry JM, Al-Shuaily S (2012b) Bilateral asymmetry in some morphological characters of *Parapercis alboguttata* (Günther, 1872) (family: Piguipedidae) collected from the Arabian Sea coasts of Oman. Rom J Biol Zool 57:51–62

Jawad LA, Al-Faisal AJ, Al-Mutlak FM (2014) Incidence of lordosis in the cyprinid fish, Carasobarbus luteus and the Shad, *Tenualosa ilisha* collected from Basrah waters, Iraq. Intern J Mar Sci 4:1–5

Jehad AK (1984) Effect of Tharthar Canal salty water on the quality of Euphrates water. M.Sc. Thesis University of Technology, Baghdad, Iraq

Jha BR, Rayamajhi A (2010) *Heteropneustes fossilis*. The IUCN red list of threatened species. Version 2014.2. www.iucnredlist.org. Accessed 10 Dec 2015

Kadhim NF (2014) Monthly variations of Physico-chemical characteristics and phytoplankton species diversity as index of water quality in Euphrates River in Al-Hindiya barrage and Kifil City region of Iraq. J Biol Agri Health 4:105–119

Kassim TI, Al-Saadi HA, Al-Lami AA, Al-Jaberi HH (1997) Heavy metals in water, suspended particles, sediment and aquatic plants of the upper region of Euphrates River. Iraq J Environ Sci Health A32(9&10):2497–2506

Khalaf KT (1961) The marine and freshwater fishes of Iraq. Ar-Rabitta Press, Baghdad, p 164

Kwok KWH, Leung KMY (2005) Toxicity of antifouling biocides to the intertidal harpacticoid copepod *Tigriopus japonicus* (Crustacea, Copepoda): effects of temperature and salinity. Mar Poll Bull 51:830–837

Lajus D, Knust R, Brix O (2003) Fluctuating asymmetry and other parameters of morphological variation of eelpout *Zoarces viviparus* (Zoarcidae, Teleostei) from different parts of its distributional range. Sarsia: North Atlantic Marine Sci 88(4):247–260

Lens L, Van Dongen S, Kark S, Matthysen E (2002) Fluctuating asymmetry as an indicator of fitness: can we bridge the gap between studies? Biol Rev 77(1):27–38

Leung B, Knopper L, Mineau P (2003) A critical assessment of the utility of fluctuating asymmetry as a biomarker of anthropogenic stress. In: Polak M (ed) Developmental instability: causes and consequences. Oxford University Press, Oxford, pp 415–426

Long ER, Macdonald DD, Smith SL, Calder FD (1995) Incidence of adverse biological effects within ranges of chemical concentrations in marine and estuarine sediments. Environ Manag 19 (1):81–97

Lucentini L, Caroti A, Ezra R, Giovinazzo F, Lorenzoni M, Mearelli M (1998) Fluctuating asymmetry in perch, *Perca fluviatilis* (Percidae) from three lakes of the region Umbria (Italy) as a tool to demonstrate the impact of manmade lakes on developmental stability. Ital J Zool 65 (Supplement):445–447

Mabrouk L, Guarred T, Hamza A, Messaoudi I, Hellal AN (2014) Fluctuating asymmetry in grass goby, *Zosterisessor ophiocephalus* Pallas, 1811 inhabiting polluted and unpolluted area in Tunisia. Mar Poll Bull 85:248–251

Markow TA (1995) Evolutionary ecology and developmental instability. Annu Rev Entomol 40 (1):105–120

Palmer AR (1994) Fluctuating asymmetry analyses: a primer. In: Markow TA (ed) Developmental instability: its origins and evolutionary implications. Kluwer Academic Publishers, Dordrecht, Netherlands, pp 335–364

Palmer AR, Strobeck C (1992) Fluctuating asymmetry as a measure of developmental stability: implications of non-normal distribution and power of stability tests. Acta Zool Finnica 191:57–72

Rainbow PS (1997) Trace metal accumulation in marine invertebrates: marine biology or marine chemistry? J Mar Biol Ass UK 77:195–210

Ranaldi MM, Gagnon MM (2010) Trace metal incorporation in otoliths of pink snapper (*Pagrus auratus*) as an environmental monitor. Comp Biochem Physiol C 152(3):248–255

Salman NA (2011) Assessment of environmental toxicity in Iraqi Southern marshes using fish as bioindicators. Ekologija 57(1)

Sogorb A, Andreu-Moliner ES, Almar MM, del Rame J, Múñez A (1988) Temperature-toxicity relationship of fluvalinate (synthetic pyrethroid) on Procanbarus clarkii (Girard) under laboratory conditions. Bull Environ Cont Toxicol 40:13–17

Talwar PK, Jhingran AG (1991) Inland fishes of India and adjacent countries, vol 2. Oxford & IBH Publishing, New Delhi, Bombay, Calcutta, p 1158

Thiam N (2004) Ecomorphologie de *Trisopterus luscus* (Linnaeus, 1758) tacaud, adaptation a la température at l'asymmétric fluctuante. Ph.D. Thesis, Faculté des Sciences, Université de Vigo, p 190

Valentine DW, Soule ME, Samollow P (1973) Asymmetry in fishes: a possible statistical indicator of environmental stress. Fish Bull 71:357–370

Van Valen L (1962) A study of fluctuating asymmetry. Evolution 16:125–142

Wright DA (1995) Trace metal and major ion interactions in aquatic organisms. Mar Poll Bull 31:8–18

Chapter 65
The Studies on Sediments Pollution by Different Types of Metals in Turkey

Leyla Kalender

The Euphrates and Tigris Rivers are located at the Mesopotamian area, and the most important rivers in the southwestern Asian district. The Euphrates ends up mixing with the Persian Gulf. Both of the rivers contributed greatly to the use of water resources. They also played a major role in the development of ancient Mesopotamian culture.

65.1 The Previous Studies on Sediments Pollution of Euphrates–Tigris Rivers, Turkey with Metals

65.1.1 The Euphrates River Sediments

The Euphrates River, which is commonly used name Firat River in the region, is 2800-km long. It covers an area of 580,000 square kilometers. Therefore, it is the larger river in the southwestern Asian (Frenken 2008). Euphrates River was formed by mixing waters of the Karasu and Murat Rivers along the east Anatolian fault zone in the eastern Anatolian region in Turkey. This region is located 10 kilometers north of the Palu township, and upstream of the Keban Dam. There are many dams built on the Euphrates River until now. These are respectively the Keban Dam downstream of the Karasu/Murat Rivers, the Karakaya Dam, the huge Atatürk, the Birecik, and Karkamiş Dams. There are few scientific studies on the Euphrates River. Keban township is located at the Euphrates River side. Keban township has been an important mining area since 2000 B.C. Archaeological studies have been conducted

L. Kalender (✉)
Firat University, Department of Geological Engineering, Elazığ, Turkey
e-mail: leylakalender@firat.edu.tr

L. A. Jawad (ed.), *Tigris and Euphrates Rivers: Their Environment from Headwaters to Mouth*, Aquatic Ecology Series 11,
https://doi.org/10.1007/978-3-030-57570-0_65

on the history of ancient mining. These studies have revealed that ancient miners mined the Keban area for Au, Cu, Pb, Ag, Zn, Mn, and Fe as a polymetallic mine deposit. During the twentieth century, mining activities were limited for Fe, Pb, Zn, Ag, and F. However, mining ceased in 1988 due to economically exhausted, and the site has been completely abandoned. Nevertheless, the studies of the wastes in the east and west Euphrates slag sites shown that metal contents are still high (average of 2.16 ppm Au, 66.41 ppm Ag, 2.38% Pb, and 0.52% Zn; Kalender and Hanelçi 2001). Kalender and Bölücek (2004), Bölücek (2007), and Kalender (2012) studied the geochemical properties of the stream sediments on the Euphrates River and its west tributaries (Karamagara Stream) upside of the Keban Dam Lake. They found an increase in heavy metal concentrations due to previous Keban Pb and Ag mining. The same researchers focused on the environmental geochemical studies on the Euphrates. They stated that the high metal concentrations of As, Mo, Se, Cu, Zn, and Pb in the Karamağara Stream were dispersed both hydromorphically and mechanically. The mechanical distribution has been found to be close to wastes because of ancient mining. In the same study, metal contents of the algae and moss samples have been investigated from the Euphrates River bed.

Calculated total BAF (bioaccumulation factor) values for *algae species are much less than the moss samples.* The spatial distributions of metals in the Euphrates sediment, algae, and moss suggest that the causes of the contamination are due to abandoned Keban Pb–Zn and Ag deposits and its wastes. In subtropical climatic conditions, natural weathering was effective on metal enrichment. The geochemical results indicate that algae, moss, and river bed sediments have positive effects on ecosystem in terms of metal pollution within river water. Kalender and Çiçek Uçar (2013) have explained the source and effect of contamination of the metals according to obtained the trace element analysis results in Gelidere sediment. The calculated enrichment factor values of the heavy REE (rare earth element) are more than those of light REE in the Geli stream sediments, which is a tributary of the Euphrates River. The same study pointed out that the median values of As, Cu, Pb, Zn, Ag, Ni, Co, Mn, Cd, Sb, Cr, Ti, and Al exceeded the concentrations in normal/background granodioritic host rocks. Normalization was done according to Fe for calculated EF (enrichment factor) values. The highest EF (enrichment factor) values were estimated rather than background values for Sb, As, and Cr, which values showed a significant increase in concentration. Sediment quality guideline (SQG) values known as ecotoxicological indices are used to determine pollution levels of the concentration of metals in the sediments. Estimated SQG values show heavy polluted (37.95 ppm and 85.12 ppm for As and Cu, and 64.50 ppm moderately polluted for Cr). In the Geli stream sediments, As, Ba, Cr, Cu, Pb, Ni, and Zn are higher than acceptable background values. In the studied sediments, considering the calculated threshold effect levels [TEL (threshold effect level), LEL (lowest effect level), MET (minimal effect threshold), and TEC (threshold effect concentrations), and PEC (probable effect concentrations)], for all metals and semimetals show that the most important anomaly values are for As, Cu, and Ba. The anomaly map drawn on the drainage system considering the calculated threshold values indicated that the possible source of As, Cu, and Ba pollution is located southeast of the Keban polymetallic mining area in which closes to waste storage area. The distribution of metals and semimetals varies depending on many factors. The statistically calculated factor analysis (using correlation coefficients between metals and semimetals) indicated that source of metals and semimetals are derived from both natural local lithologic units, natural geogenic mineralizations, and anthropogenic pollutants, which are wastes and ancient mining activities. In order to determine the source rock features of the Euphrates sediments, the first isotogenic isotope research was studied by Kalender and Aytimur (2016). Nd and Sr isotope compositions were selected for this study due to interpreting source of both basic and acidic rocks. They also used REE contents to make comments of the radiogenic isotope compositions in the sediments. Samples for REE and $^{87}Sr/^{86}Sr$ and $^{143}Nd/^{144}Nd$ isotopic analysis were selected along the Euphrates River bank sediments in the north-west of the Keban Dam, and between the Keban Dam and the Karakaya Dam lakes. Distribution of the radiogenic isotopic ratios shows that $^{87}Sr/^{86}Sr$ and $^{143}Nd/^{144}Nd$ isotopic compositions in the Euphrates River bank sediments are 0.7053, 0.7048, and 0.7057 and 0.512654, 0.512836, and 0.512775, respectively. The study shows that the weathering of Elazığ Magmatites and shallow marine sediments (Keban Metamorphic rocks) has been effective on the chemical composition of the bank sediments of Euphrates. The positive εNd (0) values (0.35, 3.9, and 2.7) are higher downstream in the studied sediments due to weathering of the mafic volcanic rocks. The chondrite, NAS (North American Shale), and UCC (Upper Continental Crust) normalized patterns show that the REE (Rare Earth Element) compositions of the Euphrates River sediments are higher than chondrite composition but close to NAS and UCC due to origin of host rocks. Hereby, it can be said that the host rocks are contaminated by the upper continental crust. The same study explained that the tectonic zone and the weathered granodioritic rocks of the Elazig Magmatites have been effective on the Euphrates River water compositions, too. Irmak et al. (2017) studied chemical composition of the Euphrates River bank sediments, which are located between Euphrates River starting point and the Keban Dam Lake. The study showed that the Euphrates River sediments are named sub-litharenite and feldspar litharenite. In terms of the major oxide contents in the Holocene Euphrates River sediments, the concentrations of Fe_2O_3, MgO, CaO, and TiO_2 are higher than UCC and SiO_2, Al_2O_3, Na_2O, and P_2O_5 concentrations are depleted. Firstly, the paper shows the Pb isotope composition ratios of the Euphrates River sediment samples that the natural weathering of the local lithologic unites is to be more effective on the river sediment chemistry than anthropogenic effects. The river sediments are mostly between natural lead values which come from the mafic volcano-sedimentary rocks due to $^{206}Pb/^{207}Pb$ (1.205–1.22) ratio. The lithological units which contribute to the chemical composition of the Euphrates River sediments by natural weathering have characterization of Plio-Quaternary Palu Formation, Late Miocene Çaybağı Formation, and Late Cretaceous Elazığ Magmatic rocks. The study has been carried out to investigate the usability of sediments as aggregates. Physico-mechanical experiments show that the sediments of the Fırat River have 2.6 gr/cm^3 specific weight and sediments optimum bitumen is 4.6% and their mixture type stability value is 2300 kg. The physical–mechanical features of the sediments at the western East

Anatolian Fault Zone show that clay size fractions are higher than the updown river sediment samples.

65.1.2 The Tigris River Sediments

Tigris River is popularly known as the Dicle River. The Tigris (Dicle) River originates in the morphological structure of the Eastern Anatolia of Turkey and flows southeast into Iraq after briefly forming the extreme eastern portion of the border between Syria and Turkey. The river is approximately 1900 kilometers long traversing a stretch of 523 kilometers through Turkey (Akbulut et al. 2009). Diyarbakir is the largest city on the river. Maden copper deposit is located in the region, and wastes of the plant are being discharged in the tributaries of the Tigris River. Maden copper deposit is an important Cyprus-type ore deposit of Turkey. Mining activity had been operated from 4000 B.C. up to the recent (Seeliger et al. 1985). Mining activities have continued since Rome, Seljuk, and Ottoman Empires (1860–1915) and the Republic of Turkey. Many geological, biogeochemical anomalies for Zn and Fe along the Maden Stream, analytical methods and metal dispersion patterns studies have been conducted around the Maden copper deposit area (Erdoğan 1977; Özkaya 1978; Özdemir and Sağıroğlu 1998, 2000; Kırat et al. 2008). Some researchers have started investigating the Tigris River sediments to investigate heavy metal pollution from the copper mine at the upper Tigris River (Ünlü and Gümgüm 1993; Gümgüm et al. 1994; Karadede Akın and Ünlü 2007; Varol 2011). The concentration of some metals, such as Cd, Co, Cu, Fe, Mn, Ni, Pb, and Zn, are determined in Tigris River water, sediment, muscle, liver, and gill of fish by Karadede Akın and Ünlü (2007). The average Cd, Cu, Mn, Ni, Zn, and Fe values are found to be high in different seasons (spring and summer). The results showed that Cu levels have gradually decreased during recent years due to dilution. The average content of Cd, Cu, Mn, Ni, Zn, and Fe show seasonal variation (spring and summer). Because the metals can precipitate in the sediments close to the source. They cannot be carried too far along the flowing water. The time interval during the migration increases the dilution. Al-Juboury (2009) indicates a contamination level of heavy metals such as Cr, Cu, Ni, Pb, and Zn in the Miocene and Quaternary sediment samples from the Tigris River and certain of its tributaries of northern Iraq because of natural weathering of the main sources. The researcher indicates that mineralogical features of the Tigris River sediments form more than one-third of the heavy fraction, and most of the opaque minerals are well-developed crystals as they may be originated from the relatively nearby solid rocks in the highlands of southern Turkey. Varol (2011) used different pollution indices to assess the degree of metal contamination in sediments of the Tigris River. According to the sediment quality standards, it is stated that contents of Cr, Cu, Ni, and Pb can cause harmful effect for organisms living on the sediment. In addition to As, Cd, Co, Cr, Cu, Mn, Ni, and Zn might be derived from the anthropogenic sources. Particularly metals should be enriched in sediments by leaching with metallic discharges from the copper mine waste. Varol and Şen (2012) studied the concentrations and distributions of total nitrogen, total phosphorus, and As, Cd, Co, Cr, Fe, Mn, Ni, Pb, and Zn in both surface water and sediments of the upper Tigris River. The researchers were determined that Total nitrogen (TN) and total phosphorus (TP) and metal concentrations in sediment samples at the downstream of Maden Copper Plant are higher than upstream. However, the study indicates a significant decrease in the concentrations of heavy metals in sediment at the downstream of the Dicle Dam Lake.

65.1.3 Investigation on the Other Surface Waters and Sediments in Turkey

Gemici and Oyman (2003) investigated the impacts on stream sediment and waters of the ancient Kalecik Hg mine. The researchers point out high Hg values in studied water samples. The measured Ni values are higher than the maximum allowable level (20 mg/l) for drinking water in the abandoned Kalecik Hg mining area in the western Turkey. The mine waters show a notable concentration of SO_4 according to the drinking water standards. The same study shows that the studied stream sediment samples have severe positive anomaly values for Hg As, Sb, Ni, and Cr. The calculated pollution inside values vary between 69 and 82 for stream sediment samples. These values show that pollution level is quite high. Işık et al. (2008) studied that the effects of anthropogenic activities on the Lower Sakarya River basin by data from 1965, 2003, and 2006 at the different sample sets along with the river profile. The researchers briefly stated that human activities or anthropogenic effects have changed the hydrogeochemical and morphological characteristics of the Upper and Middle Sakarya Rivers during the past 50 years. Öztürk et al. (2009) studied seasonal concentration changes of certain heavy metals (Cd, Cr, Cu, Fe, Ni, and Pb) in sediment, waters, and some tissues of *Cyprinus carpio* from Avsar Dam Lake in western Turkey. The researchers thought that sediment and water interactions were effective in the remobilization of the metals and semimetals such as pesticides and heavy metals. The contaminants in the aquatic system cause pollution. The study indicated that the contents of heavy metal in the sediments are Fe maximally accumulated and followed by the amount of abundance Ni, Cu, Cr, Pb, and Cd. Manav et al. (2016) studied the pollution level ^{210}Po and ^{210}Pb natural radionuclides concentrations and certain heavy metals (Cr, Ni, Pb, Cd, Mn, Fe, and Zn) in the fish, water, and sediment (surface and core) samples in the Lake Bafa. The paper indicates that the sediment mass accumulation rate was found to be 3.27 $g.m^{-2}.day^{-1}$ (0.119 $g.cm^{-2}.y^{-1}$) from a core sample. The highest ^{210}Pb concentrations were observed in the winter season because of tropical climate and the main source of ^{210}Pb is from atmospheric deposition through precipitation. The Lake Bafa is the delta progradation of the Büyük Menderes River, so it could be considered a sedimentation environment (Manav et al. 2016).

The determined Cr, Cu, Pb, and Zn pollution levels show the effect of sediment metal content on bioleaching. (Güven and Akıncı, 2012). They indicated that the metal pollution in the studied sediments after bioleaching using by Pollution Index (PI) are provided low PI values as 0.88 and 0.5. As it is known, if the pollution index value is >1, it poses a risk in terms of environmental pollution.

65.2 Material and Methods Used in the Pollution of Sediments of Euphrates–Tigris Rivers, with Metals in Turkey

Until now, studies have been carried out to investigate the effect of sediment and water chemistry on aquaculture. However, in recent years, detailed studies on the environmental effects of sediment chemistry of the major rivers in Turkey have been developed using different pollution indices. Below, only the used analytical methods for determining the metal contents of the Euphrates and Tigris Rivers sediments around of mine deposit areas in Turkey will be explained. Firstly, an orientation for the Euphrates River sediments survey has been conducted in the western Elazığ area by Bölücek and Kalender (2009). The survey includes a geochemical orientation study used to compare the analytical methods of the river sediments on the semiarid climate zone. Thus, optimize grain size selection and analytical procedures for delineating mineralization and differentiating bedrock types have been determined (Bölücek and Kalender 2009). Thirteen stream sediment samples were sieved into −180 μm (80 mesh) and −75 μm (200 mesh) size fractions. The analyzed were both digested using aqua regia, and weak cyanide leach. All samples were analyzed for trace elements by ICP–MS after both digestions. Higher concentrations of nearly all trace elements are typically found in the finest grain size fraction (silt and clay, <75 μm), which usually represents less than 3 wt% of the stream sediment samples. Geochemical anomalies were determined to extend several hundreds of meters downstream from the known Pb, Zn, and Cu-bearing veins, which could be delineated by element dispersion patterns obtained by both extractions. Previously unknown Au anomalies were detected in the southwestern part of the study area, which was shown only by the aqua-regia extractable Au anomalies. The study shows that weak cyanide leaching may offer an acceptable alternative to aqua regia analysis for Cu, Zn, Mo, Co, Ag, and Hg in the geochemical orientation prospecting studies except for Au. Gemici and Oyman (2003) collected from a sediment sample in the abandoned Hg mining area. They selected analytical method from Herr and Gray (1995), stream sediment was sieved to −10 mesh (2 mm) in the field. In the laboratory, the sediment samples were air dried and sieved to −80 mesh (0.18 mm). 2–2 HCl–HNO_3–H_2O 95 °C for 1 h and diluted to 20 ml. Stream sediment samples were analyzed for Hg, As, Sb, Fe, Cu, Pb, Zn, Co, Ni, Cr, and Cd using by ICP–MS. Karadede Akın and Ünlü (2007) chose a different method that the dried sediments were passed through a − 60 mesh stainless screen to remove larger particles, and 0.5 g sediment was weighed into a 100-ml beaker. The sediment samples were prepared by the procedure described earlier by using concentrated HNO_3/ HCI (1:3 v/v). The beaker was cooled, and 10 ml HNO_3 was added. The solution was transferred into a 50-ml standard flask and then diluted with triple distilled water. Varol (2011), Varol et al. (2012) used similar to analytical methods for determining the concentration of heavy metal of the Tigris River sediment samples. The papers were showed that 0.25 g and - 500 μm sieved sediment samples were digested in Teflon vessels with 12 ml HNO_3 (65%): HCl (37%)(3:1) mixture in a microwave oven for heavy metal content determinations. After microwave digestion, the sample solutions were filtered, adjusted to a suitable volume with double deionized water. The sediment extracts were analyzed for Co, Cr, Cu, Fe, Mn, Ni, and Zn by a flame atomic absorption spectrometry (FAAS) equipped with deuterium background correction (Bağ et al. 2000). As, Cd, and Pb in extracts were measured by using a graphite furnace atomic absorption spectrometry (GFAAS) with Zeeman background correction (Bağ et al. 2000). The sediment samples were prepared using routine analytical techniques around of Avşar Dam Lake such as powdered after air-dried, and passed through 160-μm sieve (Öztürk et al. 2009). The sediment samples were digested with 10 mL of HNO_3/HCl (1:3 v/v) in a microwave digestion method for Cd, Cr, Cu, Fe, Ni, and Pb. When all these experimental studies are evaluated carefully; The best analysis method for the determination of metal contents of stream sediment samples is to be dissolved with aqua regia. In addition, grain size of sediment is an important factor and also microwave method can be tried.

Quality Control Data (QCD) The accuracy and precision of analytical data is the most important criterion for the healthy interpretation of the data. The obtained analytical data quality is very important in the evaluation of the results and also need to guarantee through the implementation of laboratory quality assurance and quality control methods, including the use of standard operating procedures, calibration with standards, analysis of reagent blanks, recovery of spiked samples and analysis of replicates. The accuracy and precision of the analytical procedures are tested by recovery measurements on spiked sediment samples. The precision of the analytical procedures, expressed as the relative standard deviation (RSD), ranged within approximately 5%. Parametric statistical evaluations (sample number > 30) have a positive effect on the precision of data distribution. Reanalysis of one sample in every 10 samples will eliminate the analytical error.

The obtained geochemical data from the studied river sediments are interpreted using both geostatistical methods (factor analysis, correlation coefficient, anomaly contrast and threshold values, etc.) and different background values. The data from the Tigris River and Euphrates River sediments have been interpreted mostly the average shale, average continental crustal values, and sometimes the data from local uncontaminated lithological units as reference baselines (Kalender and Çiçek Uçar 2013). The selected background values were used to calculate different indices (CF, PLI, Igeo, EF). The indices were used to evaluate the degree of metal contamination in sediments of both the Tigris River and Euphrates River (Varol 2011; Kalender and Çiçek Uçar 2013).

Contaminant Factor The Contaminant factor (CF) is obtained by dividing the concentration of each metal in the studied sediment samples by the background value (chemical concentrations of the uncontaminated lithological units outcropping in the region).

$CF = C_{heavy\ metal} = C_{background}$ CF values were interpreted as suggested by Hakanson (1980), where: $CF < 1$ indicates low contamination; $1 < CF < 3$ is moderate contamination; $3 < CF < 6$ is considerable contamination; and $CF > 6$ is very high contamination.

Pollution Load Index Pollution load index (PLI) was determined as the *n*th root of the product of the *n* CF:

$$PLI = (CF1 \times CF2 \times CF3 \times \ldots.. \times CFn)^{1/n}.$$

This empirical index is used for evaluating the level of heavy metal pollution. When $PLI > 1$, it means that pollution exists; otherwise if it is <1, there is no metal pollution (Tomlinson et al. 1980).

Geo-Accumulation Index The geoaccumulation index (Igeo) is defined by the following equation:

$$\text{Igeo} = \text{Log}_2(Cn)/1.5(Bn).$$

where *Cn* is the concentration of metals examined in sediment samples, and *Bn* is the geochemical background concentration of the metal (*n*). The mentioned background concentration may be local lithogeochemical rock compositions. Factor 1.5 is the back ground matrix correction factor due to lithospheric effects. The geoaccumulation index consists of seven classes (Muller 1969, 1981). Class 0 (practically unpolluted): Igeo ≤ 0; Class 1 (unpolluted tomoderately polluted): $0 <$ Igeo <1; Class 2 (moderately polluted): $1 <$ Igeo <2; Class 3 (moderately to heavily polluted): $2 <$ Igeo <3; Class 4 (heavily polluted): $3 <$ Igeo <4; Class 5 (heavily to extremely polluted): $4 <$ Igeo <5; and Class 6 (extremely polluted): >5 Igeo (Buhiyan et al. 2010).

Enrichment Factor Enrichment factor (EF) is used in determining the degree assessment of heavy metal pollutants accumulation in the river sediments (Sakan et al. 2009). The EF is computed using the relationship below:

$$EF = (Cx/C_{\text{ref sediment}})/\left(Cx/C_{\text{refbackground}}\right).$$

where *Cx* is the concentration of element *x*, and Cref is the concentration (Fe, Ti, etc.) of the reference element in sediment(s) and Cref background is the concentration of the Earth's crust or local lithogeochemical Fe/Ti content. Normalization to Fe/Ti is an important factor for estimating EF values.

The used methods and sediment pollution assessment indices showed that the Tigris and Euphrates rivers; Arsenic, Cd, Co, Cr, Cu, Mn, Ni, and Zn are derived from the anthropogenic sources. Anthropogenic pollution is more important than natural weathering due to particularly metallic discharges to the Tigris River of the copper mine plant. High As, Ba, Cr, Cu, Pb, Ni, and Zn concentrations in the Geli Stream sediments, which are tributaries of the Euphrates River are derived from abandoned Keban Pb, Zn, and Ag deposit and its wastes. The sample location sites are important in these studies. If the sample sites close to abandoned mine deposits, the heavy metal contents reach to the highest values. However, the highest heavy metal contents decrease along the downstream due to dilution. Climate change is another important factor to consider in metal concentrations. The preparation of geochemical anomaly maps is important during the prospecting process. Geochemical anomaly maps are the easiest way to reach the area where the indicator element is enriched. Descriptive statistical values of geochemical data are used to make geochemical anomaly maps. For an easier understanding of the geochemical anomaly map drawing, an example is described below.

Kalender and Çiçek Uçar (2013) described potential metal pollution on the Euphrates River sediments. The determination of the mineralization source was calculated to the threshold values according to the equation below (Fig. 65.1). Logarithmic values and median can be used when the difference between the maximum and the minimum values is large.

$$\text{Threshold values} = \mu(\text{arithmetic mean}) \pm 2\sigma(\text{standard deviation})$$

The possible mineralization locations for As, Cd, Cu, Pb, Sb, Zn, and Mn are drawn on the drainage network anomaly map according to the calculated threshold values. The method is an effective method to determine the metal anomaly field in mine exploration stages (Fig. 65.2).

A number of researchers have studied Nd–Sr isotopic and trace element geochemistry of river sediments and soils as tracers of clastic sources. The geochemical characterizations and Sr–Nd isotopic fingerprinting of sediments in any fluvial system can be done using radiogenic isotopic compositions (Burke et al. 1982; Martin and McCulloch 1999; Gingele and De Deckker 2005; Revel-Rolland et al. 2005). Isotopic data are an important factor to determine the provenance of sediments. A common approach to use Pb isotopes for source identification is to use cross-plots of the isotope ratios, e.g., $^{206}Pb/^{207}Pb$ versus $^{208}Pb/^{207}Pb$, or $^{206}Pb/^{207}Pb$ versus $^{208}Pb/^{206}Pb$ (Reimann et al. 2012). Natural Pb comprises four stable isotopes (natural abundance in brackets): ^{204}Pb (1.4%), ^{206}Pb (24.1%), ^{207}Pb (22.1%), and ^{208}Pb (52.4%) in varying proportions, uniquely defined by the three ratios $^{206}Pb/^{207}Pb$, $^{208}Pb/^{206}Pb$, and $^{206}Pb/^{204}Pb$ (Reimann et al. 2012).

Kalender and Aytimur (2016) determined the Sr and Nd isotope compositions of the Euphrates River sediment samples for the first time. An 80-mg aliquot was taken for analysis of Sr and Nd isotope ratios. The samples were dissolved in beakers in a

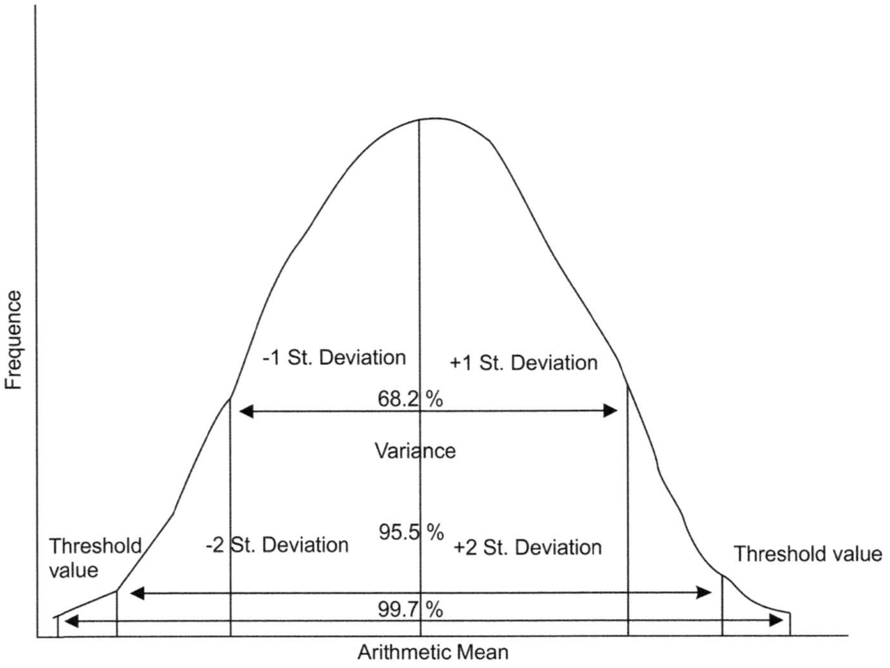

Fig. 65.1 The distribution lies within two standard deviations of the mean

4-mL 52% HF at 160 °C on the hotplate along 4 days. The samples were dried on the hotplate using 2.5 N HCl and 2 mL bis (ethylhexyl) phosphate using Bio-Rad AG50 W-X8, 100–200 mesh, and chemical separation of Sr ionic chromatographic columns was prepared. After the chemical separation of Sr, REE fractionation was collected using 6NHCl. Sr isotopes were measured using a single Taactivator with Re filament and 0.005 N H_3PO_4. $^{87}Sr/^{86}Sr$ ratios were corrected for mass fractionation by normalizing to $^{86}Sr/^{88}Sr = 0.1194$, and strontium standard (NBS 987) was measured more than 2 times. The chemical separation of Nd from REE was made in a Teflon column using 0.22 N HCl and 2 mL bis (ethylhexyl) phosphate. $^{143}Nd/^{144}Nd$ data were normalized by $^{146}Nd/^{144}Nd = 0.7219$, and the neodymium standard (0.511848 ± 5) was measured more than two times. Irmak et al. (2017) made a lead isotopic analysis of the Euphrates River sediments for the first time. 130 mg and 80 mesh sediment samples were dissolved in 4 ml HF, 1 ml HNO_3, 15 M HNO_3 along 7 days at 140 °C. The lead isotope analysis were made using by ICP–MS (Inductively Coupled Plasma-Mass Spectrometer), and NIST (National Institute of Standards and Technology) 981-1Y, 983-1Y, and STD DS 10.

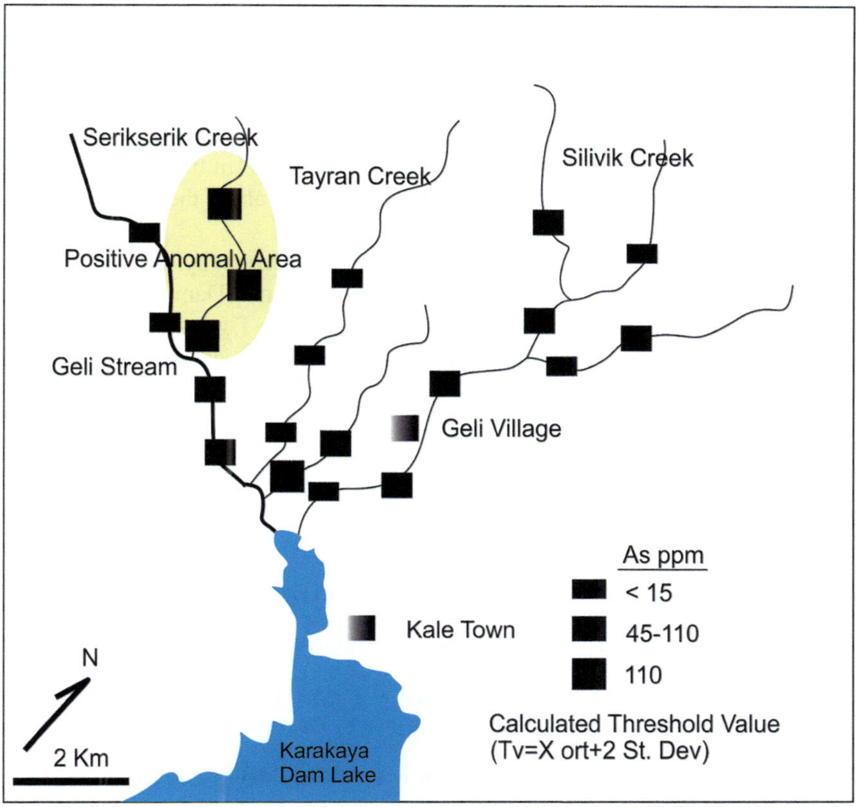

Fig. 65.2 Drawn anomaly map for As according to calculated threshold values from Kalender and Çiçek Uçar (2013). Vertical lines show anomaly fields

65.3 Review of the Euphrates–Tigris River Sediment's Pollution by Metals in Turkey

As can be seen from the above information, the metal contents of the Euphrates and Tigris River sediments show anomalous values at the important ancient mine deposit locations. One of the ancient mine deposits is the Keban polymetallic Au, Cu, Pb, Ag, Zn, Mn, and Fe deposit. The other is the Maden copper deposit. The Keban polymetalic mine deposit has been abandoned since 1988. In the Cu deposit around the Tigris River, abandoned mining wastes have been operated since 2007. The operation of ancient copper mining wastes is important in terms of reducing environmental pollution. Because the weathering of the wastes by surface water and groundwater cause to dissolve and transport of the metals both hydromorphically and mechanically. The previous studies have been carried on the metal pollution since 2004 shows that As, Cd, Cu, Pb, Sb, Zn, and Mn values are higher in the

Euphrates River sediments close to abandoned Keban polymetallic mine deposit area. During the last 25 years of studies on pollution in the Tigris River sediments show that generally contents of Cd, Co, Cu, Fe, Mn, Ni, Pb, and Zn are higher at the close to the ancient mining area. However, the sediment content of both streams suggests that as the rivers move along the southward direction of flow, the metal content of the river waters decreases as they move away from the ore deposits. The reason for this is that freshwaters from different sources along the current direction mix into the waters of both rivers.

The fine size fraction sediments in the dam reservoirs by the Keban Dam, Karakaya Dam, Atatürk Dam, Birecik Dam, and Karkamış Dam on the Euphrates River, Kralkızı Dam, Batman Dam, and Dicle Dam on the Tigris River, which cause to absorb onto clay minerals of the metals and do not allow metal transport along the river's current direction because they have high ion exchange capacity.

References

Akbulut EN, Bayari S, Akbulut A, Şahin Y (2009) Rivers of Turkey. In: Tockner K, Robinson CT, Uehlinger U (eds) Rivers of Europa. Academic Press, New York, pp 643–672

Al-Juboury AI (2009) Natural pollution by some heavy metals in the Tigris River northern Iraq. Int J Environ Res 3(2):189–198

Bağ H, Türker AR, Lale M (2000) Determination of Cu, Zn, Fe, Ni and Cd by flame atomic absorption spectrophotometry after preconcentration by Escherichia coli immobilized on sepiolite. Europe PMC 51(5):1035–1043

Bölücek C (2007) Environmental contamination in the Keban Mining District eastern Turkey. AJSE 32(1a):3–18

Bölücek C, Kalender L (2009) A drainage sediment geochemical orientation study Keban, Elazığ, Eastern Turkey. Arab J Sci Eng 34(1A):91–102

Buhiyan MAH, Parvez L, Islam MA, Dampare SB, Suzuki S (2010) Heavy metal pollution of coal mine effected agricultural soils in the northern part of Bangladesh. J Hazard Mater 173:384–392

Burke WH, Denison RE, Hetherington EA, Koepnick RB, Nelson HF, Otto JB (1982) Variation of seawater $^{87}Sr/^{86}Sr$ throughout Phanerozoic time. Geol 10(10):516–519

Erdoğan B. (1977) Geology geochemistry and genesis of the sulphide deposits of the ergani-maden region SE Turkey. Ph.D. Thesis the University of New Burnswick, p. 289 (unpublished)

Frenken K (2008) Irrigation in the Middle East region in figures. Water reports 34, FAO, Rome, Italy. AQUASTAT Survey, p. 402

Gemici Ü, Oyman T (2003) 2003. The influence of the abandoned Kalecik Hg mine on water and stream sediments (Karaburun Izmir Turkey). Sci Total Environ 312:155–166

Gingele FX, De Deckker P (2005) Late quaternary fluctuations of palaeoproductivity in the Murray canyons area south Australian continental margin. Palaeogeogr Palaeoclimat Palaeoecol 220:361–373

Gümgüm B, Ünlü E, Tez Z, Gülsün Z (1994) Heavy metal pollution in water sediment and fish from the Tigris River in Turkey. Chemosphere 29:111–116

Güven D, Akıncı G (2012) Assessment of metal pollution in contaminated sediment after bioleaching with different solid contents. GU J Sci 25(2):363–370

Hakanson L (1980) Ecological risk index for aquatic pollution control a sediment logical approach. Water Res 14:975–1001

Herr C, Gray NF (1995) Environmental impact of acid mine drainageon the Avoca River: metal fluxes in water and sediment. Part II. Metal contamination of riverine sediments. Technical Report 14 Water Technology Research Trinity College University of Dublin, Dublin, p. 36
Irmak A, Kalender L, Yılmaz M, Yamaç ÖE (2017) Geochemical features and study of the industrial raw materials usability of the Euphrates River Bank sediments between Palu and Keban dam Lake. Geol Bull Turkey 60:283–311
Işık S, Dogan E, Kalin L, Sasal L, Agiralioglu N (2008) 2008. Effects of anthropogenic activities on the lower Sakarya River. Catena 75:172–181
Kalender L (2012) Environmental impact and drainage geochemistry of the abandoned Keban ag Pb Zn deposit. In: Panagiotaras D (ed) Working Maden cu deposit and alpine type Cr deposit in the eastern Anatolia Turkey. Geochemistry. In Tech, Croatia, pp 345–370. www.intechopen.com
Kalender L, Aytimur G (2016) REE geochemistry of Euphrates River sediments. J Chem 2016:1–13. Article-id: 1012021
Kalender L, Bölücek C (2004, March) Major and trace element contamination of groundwaters stream sediments and plants of the abandoned mines in Keban district (Elazığ) of eastern Anatolia Turkey. In: Proceedings of the 57th geological congress of Turkey. MTA Culture Sites, Ankara, Turkey, p 187
Kalender L, Çiçek Uçar S (2013) Assessment of metal contamination in sediments in the tributaries of the Euphrates River using pollution indices and the determination of the pollution source Turkey. J Geochem Explor 134:73–84
Kalender L, Hanelçi Ş (2001) Mineralogical and geochemical features of au, ag, Pb and Zn mineralization in Keban (Elazığ) wastes. Geol Bull Turkey 44:91–104
Karadede Akın H, Ünlü E (2007) Heavy metal concentrations in water sediment fish and some benthic organisms from Tigris River Turkey. Environ Monit Assess 131:323–337
Kırat G, Bölücek C, Kalender L (2008) Distribution of cu Pb Zn as cd and Fe in stream sediments around of Maden copper deposit. Geosound/Yerbilimleri 53:203–217
Manav R, Ugur Görgün A, Filizok I (2016) Radionuclides (^{210}Po and ^{210}Pb) and some heavy metals in fish and sediments in Lake Bafa Turkey and the contribution of ^{210}Po to the radiation dose. Int J Environ Res Public Health 13:11–13
Martin CE, McCulloch MT (1999) Nd-Sr isotopic and trace element geochemistry of river sediments and soils in a fertilized catchment New South Wales. Aust Geochim Cosmochim Act 63(2):287–305
Muller G (1969) Index of geoaccumulation in sediments of the Rhine River. Geo J 2:108–118
Muller G (1981) Die Schwermetallbelstung der sedimente des Neckarsund seiner Nebenflusse: eine Bestandsaufnahme. Chem Zeit 105:157–164
Özdemir Z, Sağıroğlu A (1998) The study of biogeochemical anomalies for Fe along the Maden River (Maden-Elazig). Geol Bull Turkey 41(1):49–54
Özdemir Z, Sağıroğlu A (2000) The study of biogeochemical anomalies for Zn along the Maden River (Maden-Elazig). MU Series Essay 4:93–100
Özkaya İ (1978) Stratigraphy of Ergani-Maden vicinity. Geol Bull Turkey 21:129–139
Öztürk M, Özözen Ö, Minareci O, Minareci E (2009) Determınatıon of heavy metals in fish water and sedıments of Avsar dam Lake in Turkey. J Environ Health Sci Eng 6(2):73–80
Reimann C, Flem B, Fabian K, Birke M, Ladenberger A, Négrel P, Demetriades A, Hoogewerff J (2012) GEMAS project team. Lead and lead isotopes in agricultural soils of Europe – the continental perspective. Appl Geochem 27:532–542
Revel-Rolland M, Arnaud F, Chapron E Sr (2005) Nd isotopes as tracers of clastic sources in Lake Le Bourget sediment (NW Alps France) during the little ice age: palaeohydrology implications. Chem Geol 224(4):183–200
Sakan SM, Djordjevic DS, Manojlovic DD, Polic PS (2009) Assessment of heavy metal pollutants accumulation in the Tisza river sediments. J Environ Manag 90:3382–3390
Seeliger TC, Pernicka E, Wagner GA, Begemann F, Schmitt-Strecker S, Eibner C, Oztunali O, Baranyi I (1985) Archaometallurgische untersuchungen in nord- und ostanatolien. Jahrbuch des Romisch-Germanisches Zentralmuseums Mainz 32:597–659

Tomlinson DC, Wilson JG, Harris CR, Jeffery DW (1980) Problems in the assessment of heavy metals levels in estuaries and the formation of a pollution index. Helgoländer Meeresun 33:566–575
Ünlü E, Gümgüm B (1993) Concentrations of copper and zinc in fish and sediments from the Tigris River in Turkey. Chemospher 26(11):2055–2061
Varol M (2011) Assessment of heavy metal contamination in sediments of the Tigris River (Turkey) using pollution indices and multivariate statistical techniques. J Hazard Mater 195:355–364
Varol M, Şen B (2012) Assessment of nutrient and heavy metal contamination in surface water and sediments of the upper Tigris River, Turkey. Catena 92:1–10
Varol M, Bekleyen A, Gökot B, Şen B (2012) Spatial and temporal variations in surface water quality of the dam reservoirs in the Tigris River Basin, Turkey. Catena 92:11–21

Chapter 66
Macroplastic and Microplastic in the Freshwater Environment of Southern Iraq: Evidences Obtained from Freshwater Fish Species

Laith A. Jawad

Abstract The objective of this chapter was to report on the presence of plastics ingested by fish species collected from the freshwater system in southern Iraq. The identification of ingested fragments of plastics might help as an opening stage in evaluating freshwater biota contact to the plastic ingredients and eventually deciding the possible influences of such materials on fish. It is approved that in additional taxonomic groups such as birds, sea turtles, and cetaceans, swallowed plastics may cause numerous difficulties containing inner wounds, ulcers, and tumors or obstructions of the digestive tract. The latter yields an untrue feeling of fullness, which decreases eating action and may cause several digesting problems. The evidence on ingested macro-, meso-, and microplastic fragments by a selected freshwater fish species presented in the present study represents an early phase to examine pollutant transmission crossways the food web, containing humans in Iraq. Fish species studied showed variation in the amount and sort of plastic materials in their stomachs. Such variation could be related to the feeding habit that the fish has.

66.1 Introduction

It has been recognized that plastic is manufactured from a wide variety of artificial or semisynthetic organic materials that are lenient and can be converted into solid substances of diverse shapes. They are frequently artificial, most usually resulting from petrochemicals and numerous are partly natural (Verma et al. 2016).

The increase in production and consumption of plastic materials results in a constant plastic waste increase. Plastic materials are predominantly not biodegradable and having a low density makes them unfit for disposal in the rubbish collecting

L. A. Jawad (✉)
School of Environmental and Animal Sciences, Unitec Institute of Technology, Auckland, New Zealand

L. A. Jawad (ed.), *Tigris and Euphrates Rivers: Their Environment from Headwaters to Mouth*, Aquatic Ecology Series 11,
https://doi.org/10.1007/978-3-030-57570-0_66

area (Aguado et al. 2007). Deprived of suitable waste administration, augmented plastic left-over, will add to the backlog of plastic surplus now in presence.

Increase in temperature and habitat circumstances could interrupt the deprivation of plastic into secondary microplastics or the release of chemicals enclosed or changed on plastic waste. Secondary microplastics are those shaped from breakdown of larger plastic constituents (Arthur et al. 2009).

The volume of data on the buildup and impacts of plastics in freshwater and terrestrial systems is much fewer than in marine systems (Wagner et al. 2014). In oceans, the small size and low density of microplastics add to their prevalent transport across large distances chiefly by ocean currents (Cole et al. 2011). Till lately the spreading of microplastics in freshwater systems as in marine systems was unidentified. Even large plastic items (e.g., fragments >5 mm, line, films, and polystyrene) have only lately been noted in lakes, rivers (e.g., Moore et al. 2011), and estuaries. The microplastics perceived in these investigations are of diverse origins containing primary and secondary origins and are of diverse configurations.

Fishes are susceptible to consumption of the plastic constituents and their size allows ingesting macro- or microplastics (Arthur et al. 2009) obtainable in the habitat. Absorption can happen straight, through resemblances with the typical prey of these fish (Hoss and Settle 1990), or incidentally, through trophic transfer, when the prey has earlier swallowed plastic remains (Eriksson and Burton 2003).

This chapter sheds light on the concealed fact that the most commercial freshwater fish species in the southern Mesopotamian plane were contaminated with plastic materials (macro- and microplastics). Through the food investigations of several commercial freshwater fish species from the lower reaches of Mesopotamia, it became obvious that the stomach contents of 10 of these species were contained plastic fragments of different sizes. Description of the plastic materials was given together with a future policy to protect the environment from this human-made monster, the plastic.

66.2 Features of the Microplastics

Wu et al. (2018) have demonstrated the characteristics of the macroplastic are very well known as it represents a large piece of plastic (>5 mm). Therefore, the present section will deal with features of the microplastic particles as they are widely distributed and commonly found in stomach of diverse animal groups. The small size of the microplastic fragments makes them not noticeable to the observer.

66.2.1 Shape

According to their shapes, microplastics are characteristically considered as follows: sheet, film, line/fiber, fragment, pellet/granule, and foam. Nevertheless, there is no set technique, and different preparations could be used by diverse researchers. This morphological knowledge from the microplastic examples can be used to specify their conceivable origins (Boerger et al. 2010; Collignon et al. 2014; Eerkes-Medrano et al. 2015).

66.2.2 Particle Size

Size is additional issue frequently dignify microplastics, but no combined criteria are currently available. Different size groups were stated by different researchers, which make it hard to relate the data from diverse works. Small microplastics can be studied for sediment and biota samples as density separation united with filtration are used. While the inspection of microplastics <0.05 mm will get progressively tough, unconventional tools such as Raman microscopy, micro-Fourier transform infrared spectroscopy (μ-FTIR), or scanning electron microscope (SEM) with energy dispersive spectroscopy (EDS) should be used. Normally, microplastic richness upsurges with declining size (Hoss and Settle 1990; Kukulka et al. 2012; Verma et al. 2016). The tendency of microplastic size distribution can be linked to the origin of microplastics and could also reveal the level of enduring. A higher degree of change might result in a higher richness of smaller segments. Biofouling and hydrodynamic circumstances were also thought to disturb the size distribution of microplastics (Phillips and Bonner 2015; Romeo et al. 2016; Wu et al. 2018).

66.2.3 Color

Some investigations have designated the colors of the microplastics. Microplastics can receive their colors from their parent plastic produces, but their colors can change due to enduring. Preceding studies conclude that predators may rather swallow microplastics with colors like their prey. It may be nosy to discover additional how color disturbs the habitat destiny and ecological impacts of microplastics. For instance, colorants can frequently affect the final thermal and UV steadiness of a plastic material (Boerger et al. 2010; Collignon et al. 2014; Eerkes-Medrano et al. 2015).

66.2.4 Surface Texture

The first characteristic that microplastic acquire in association is the enduring process, and this process will affect the surface of the microplastics. Limited surface textures on microplastics can be used to designate the courses of mechanical and oxidative changes. Surface textures are frequently inspected using SEM.

Topographies such as grooves, fractures, and mechanical pits are thought of as consequence from mechanical enduring, while flakes, granules, and solution pits are deliberated as oxidative changes characters. The surface oxidation of plastics can be established using FTIR as shown by the exterior of peaks for carbonyl groups (Hoss and Settle 1990; Kukulka et al. 2012; Verma et al. 2016).

66.3 Freshwater Fish Species as an Evidence for Plastic Contamination in Iraq

The information given in this chapter represents the first evidence on the contamination of the freshwater system in Iraq with plastic materials. Such contamination might happened a long time ago with no report to disclose it. Due to the small size and coloration, the plastic debris can be easily missed during the examination of the feeding studies of the aquatic animals.

In this chapter, 10 commercial freshwater fish species were selected to undergo stomach examination for food in general and for detecting the presence of plastic materials. They were collected from different sites along Shatt al-Arab River, Basrah, Iraq, and some branches of this river (Figs. 66.1 and 66.2).

Before discussing the presence of the polluted plastic debris, a short note on the feeding biology of those species will be given. The 10 freshwater fish species fall into 6 families, these are: Clupeidae, *Tenualosa ilisha*; Cyprinidae, *Carasobarbus luteus*, *Carassius auratus*, *Cyprinus carpio*; Leuciscidae, *Alburnus mossulensis*, *Leuciscus vorax*; Bagridae, *Mystus pelusius*; Siluridae, *Parasilurus triostegus*; Heteropneustidae, *Heteropneustes fossilis*; Mugillidae, *Planiliza abu*.

Family Clupeidae
Tenualosa ilisha

Hilsa Shad (*Tenualosa ilisha*) (Fig. 66.3a) lives coastal shelf, estuaries, and freshwater rivers in Indonesia, Sumatra, Myanmar, Bangladesh, India, Pakistan, Kuwait, Iraq, and Iran. It is an anadromous species, recounted to ascend rivers as far as 1200 km inland for breeding (Pillay and Rosa 1963) and after spawning in freshwater goes back to marine habitats. It is the most important fish species from the socioeconomic point of view along its geographical range. The overall diet of hilsa larvae consisted mainly of adult and larvae copepod (55.7%), cladocera (27.3%), and organic detritus (17.0%). According to Castello plot (Costello 1990), the larvae of *T. ilisha* were a generalist feeder.

Family: Cyprinidae
Carasobarbus luteus

The cyprinid fish *Carasobarbus luteus* (Heckel 1843) (Fig. 66.4a), which is called himri in Iraq, widely distributed in the rivers of Tigris and Euphrates and adjacent drainage basins (Coad 2010). This fish is considered one of the most

Fig. 66.1 Shatt al-Arab River, Basrah, Iraq. (**a**) photo taken at the southern part of the river; (**b**) photo was taken at the north part of the river (Photo courtesy of Laith Jawad)

important species for artisanal fisheries and is consumed domestically as fresh fish. Mohamed et al. (2008) mentioned that the *C. luteus* represented 8.6% of the total fish landing in the artisanal fishery of Swab River which is part of Al-Huwazah marsh, Iraq during 2005. This species is an omnivorous and diet is primarily composed of algae (24.3%), aquatic insects (23.0%), macrophytes (21.6%), detritus (10.9%), diatoms (9.9%), and snails (8.0%) (Mohamed and Abood 2008).

Carassius auratus

The silver crucian carp *Carassius auratus* (Linnaeus, 1758) (Fig. 66.5a), a native cyprinid species to Eastern Asia. It is a model of a nonindigenous fish species that

Fig. 66.2 One of the main branches of Shatt al-Arab River that pass through the City of Basrah. (**a**) southbound of the branch; (**b**) northbound of the branch

has positively sustained populations through Europe, North and South America, New Zealand, and Australia (Lorenzoni et al. 2010). Al-Nasiri and Shamsul Hoda (1976) listed the freshwater fish species of Iraq and referred to the presence of *C. auratus*, this species is now well established in the Shatt Al-Arab River and dominated the fish assemblage in the river, constituting 20.8, 23.7, and 13.24% of the total catch during 2010, 2012, and 2016, respectively (Mohamed et al. 2012, 2015; Mohamed and Abood 2017).

Carassius auratus is an omnivorous fish species feeds on aquatic insects (28.9%), followed by macrophytes (26.2%), algae (12.8%), detritus (12.4%), zooplankton (6.0%), diatoms (5.8%), and snails (5.5%). The diet of this species was more close to that of *C. luteus* (S% = 0.75) than to *C. carpio* (S% = 0.50). It means that these fish species are in competition for food items in the Shatt Al-Arab River (Mohamed and Abood 2008).

Cyprinus carpio

The common carp (*Cyprinus carpio* Linnaeus, 1758) (Fig. 66.6a) is native to temperate portions of Europe and Asia. It has been broadly introduced to other parts of the world; North America, southern Africa, New Zealand, Australia, and Asia (Kottelat and Freyhof 2007). Carp is a nonnative species in Iraq and was first brought in 1955 from Indonesia and Netherland to be cultivated in fish ponds in Baghdad, later in 1960 this species was introduced into Tharthar, Habbaniya, and Hammar lakes (Al-Hamed 1966). Since then, this species became one of the most widely distributed fish species in various water bodies of Iraq (Coad 2010). Al-Hassan et al.

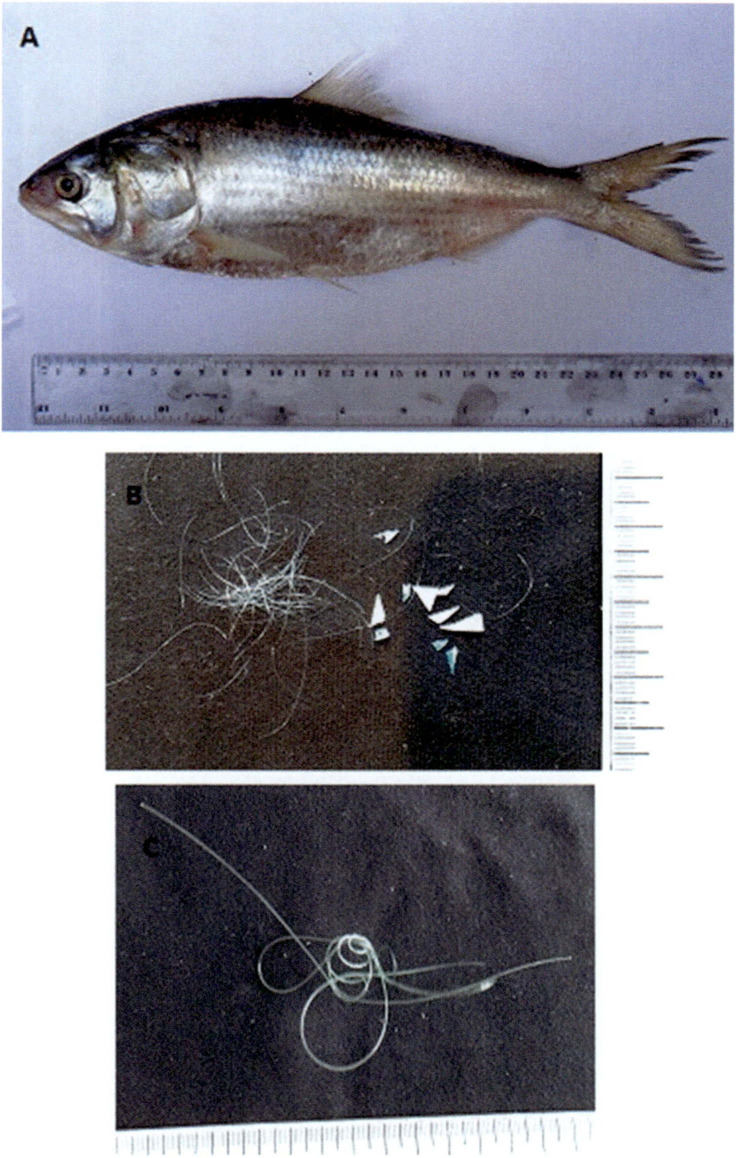

Fig. 66.3 *Tenualosa ilisha* (Family: Clupeidae). (**a**) general body aspect; (**b**) samples of meso- and microplastic materials found inside the stomach; (**c**) sample of macroplastic fragment obtained from the stomach of the fish

(1989) reported the presence of *C. carpio* in the Shatt Al-Arab River. The diet of this species is contained of aquatic insects (37.8%), followed by macrophytes (19.0%), snails (17.2%), detritus (9.8%), and fish (7.1%).

Fig. 66.4 *Carasobarbus luteus* (Family: Cyprinidae). (**a**) general body aspect; (**b**) samples of meso- and microplastic materials found inside the stomach; (**c**) sample of macroplastic fragment obtained from the stomach of the fish

Family: Leuciscidae
Alburnus mossulensis

The Mosul bleak (Fig. 66.7a) is a cyprinid fish widely distributed in the Middle East. (Coad, 2017; Esmaeili et al. 2017), however, the biology of this species is not well documented in these areas, partly because of its slow growth rate and small size (Keivany et al. 2017). Nevertheless, it is used as an edible fish at a local scale. The

Fig. 66.5 *Carassius auratus* (Family: Cyprinidae). (**a**) general body aspect; (**b**) samples of meso- and microplastic materials found inside the stomach; (**c**) sample of macroplastic fragment obtained from the stomach of the fish

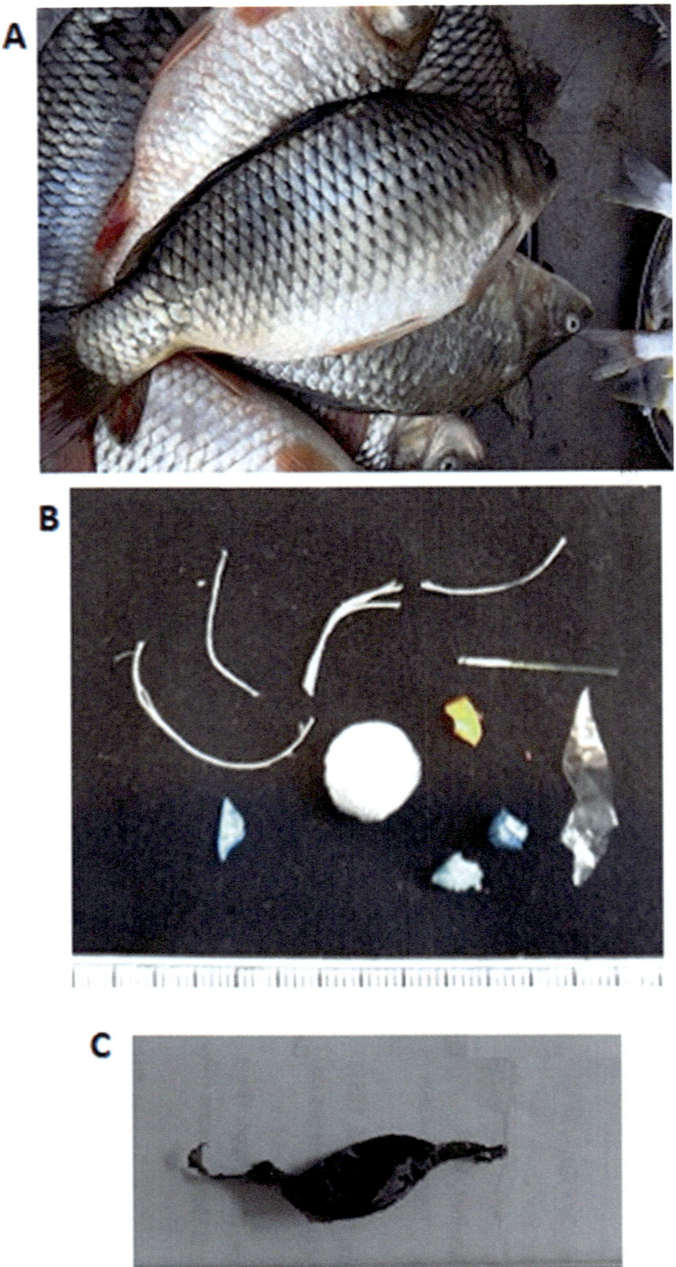

Fig. 66.6 *Cyprinus carpio* (Family: Cyprinidae). (**a**) general body aspect; (**b**) samples of meso- and microplastic materials found inside the stomach; (**c**) sample of macroplastic fragment obtained from the stomach of the fish

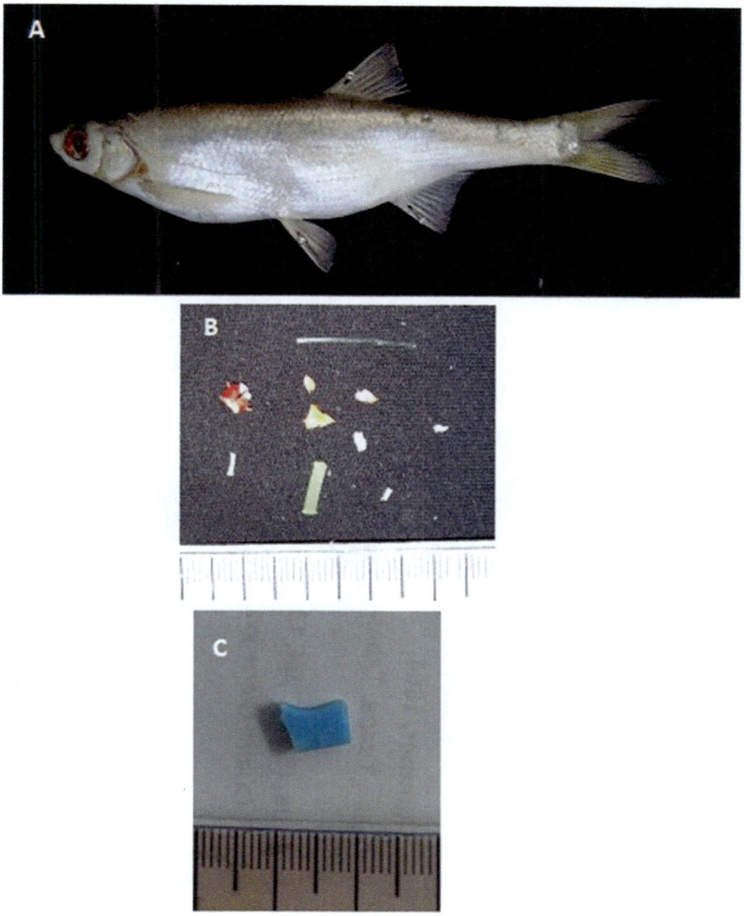

Fig. 66.7 *Alburnus mossulensis* (Family: Leuciscidae). (**a**) general body aspect; (**b**) samples of meso- and microplastic materials found inside the stomach; (**c**) sample of macroplastic fragment obtained from the stomach of the fish

feeding activity of the species ranged from 60.5% in March to 87.7% in October and the feeding intensity varied from 6.08 point/fish in December to 8.7 point/fish in July (Mohamed et al. 2016). This species is omnivorous, feeds mainly on insects (41.2%), algae (28.74%), diatoms (15.94%), aquatic plants (12.36%), fishes (3.74%), and snails (0.3%) (Mohamed et al. 2016).

Leuciscus vorax

The freshwater fish *L. vorax* Heckel 1843 (Fig. 66.8a), is a cyprinid species inhabits the Euphrates and Tigris Rivers in Turkey, Syria, and Iraq (Bogutskaya 1997). The biological features of this species have been examined in some studies at diverse regions in Iraq (Epler et al. 2001; Szczerbowski et al. 2001 and Szypula et al.

Fig. 66.8 *Leuciscus vorax* (Family: Leuciscidae). (**a**) general body aspect; (**b**) samples of meso- and microplastic materials found inside the stomach; (**c**) sample of macroplastic fragment obtained from the stomach of the fish

2001). It is a carnivorous species, with fish representing 81.67% of its food items (Mohamed et al. 2016).

Family: Bagridae
Mystus pelusius

In Iraq, this species (Fig. 66.9a) is known for its common name as Abu-zummair or abu zumir. It can reach the age of 3 years based on the eye lens diameter and vertebrae (Al-Hassan et al. 1991).

Among the food items that Roberts (1994) has found are eggs in the branchial chamber, while stomach seemingly identical with those from the ovary. Other stomach objects were fish fin fragments and cyprinid fish scales. Aquatic insects, crustaceans, debris, and plant remains are also found in stomach subjects of fish inspected by me and Al-Rudainy (2008) also mentions fish. Al-Shamma'a (2005)

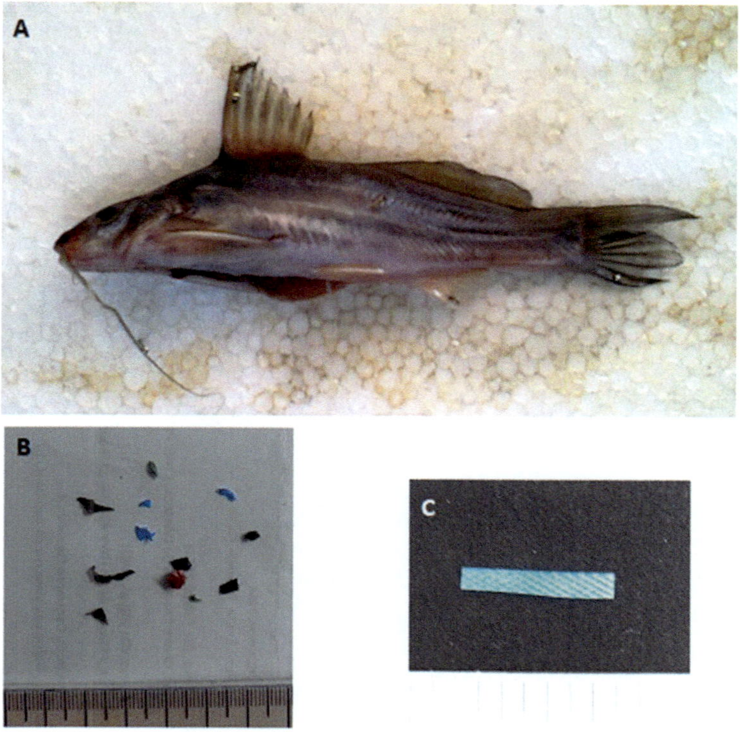

Fig. 66.9 *Mystus pelusius* (Family: Bagridae). (**a**) general body aspect; (**b**) samples of meso- and microplastic materials found inside the stomach; (**c**) sample of macroplastic fragment obtained from the stomach of the fish

reported the presence of shrimp and insects to form 47% by volume of the diet of this fish at Al-Fuhoud, Hawr al Hammar, Iraq.

Family: Siluridae
Parasilurus triostegus

The common name of this species is *Silurus triostegus* (Fig. 66.10a) distributed in the Tigris–Euphrates drainages (Esmaeili et al. 2010). Van den Eelaart (1954) records this species from open and lakes rich in vegetation and marshes and rivers in Iraq.

The chief food article of this species is the mugillid species *Planiliza abu* (=*Chelon abu*) and *Acanthobrama marmid* (Al-Shamma'a and Jasim 1993; Ünlü and Bozkurt 1996; Dawood 1997). Aquatic insects are also consumed. The food in Hawr al Hammar, Iraq was mainly fish (*Planiliza abu*, *Aphanius* spp., *Leuciscus vorax*, and *Thryssa* spp.).

Family: Heteropneustidae
Heteropneustes fossilis

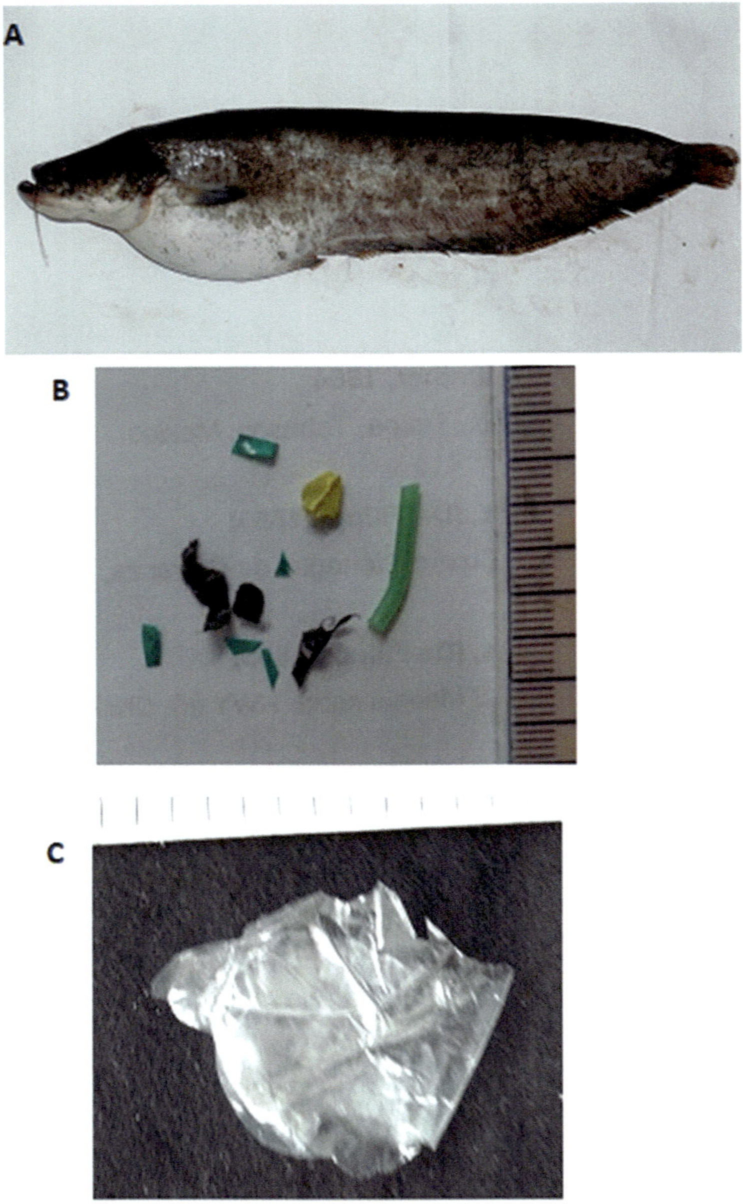

Fig. 66.10 *Parasilurus triostegus* (Family: Siluridae). (**a**) general body aspect; (**b**) samples of meso- and microplastic materials found inside the stomach; (**c**) sample of macroplastic fragment obtained from the stomach of the fish

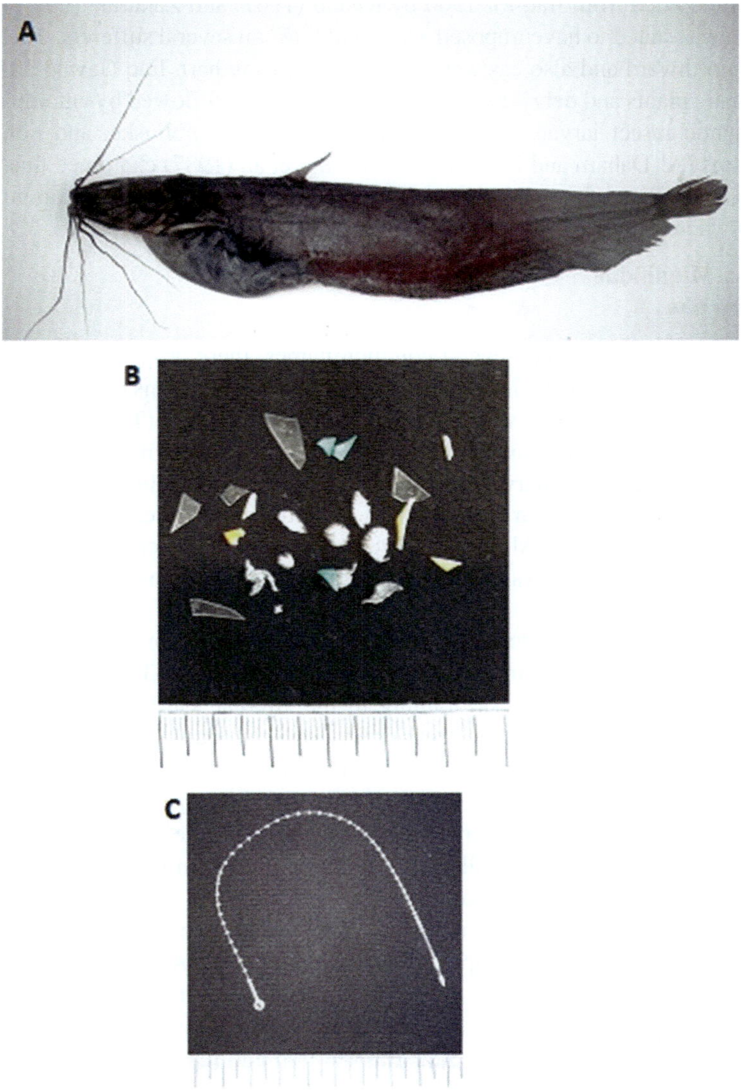

Fig. 66.11 *Heteropneustes fossilis* (Family: Heteropneustidae). (**a**) general body aspect; (**b**) samples of meso- and microplastic materials found inside the stomach; (**c**) sample of macroplastic fragment obtained from the stomach of the fish

The common name of this species (Fig. 66.11a) in Iraq are abu-al-hukum, abu al-hakim, jamhoori, samak al-za'em (named after a former president who had this species introduced for mosquito control). *Silurus fossilis* was originally described from Tranquebar, Tamil Nadu, India.

First recorded from Iraq for 1960 by Khalaf (1961) and Zakaria (1964) when an odd fish was stated to have imposed a "painful bite" on several sufferers. The species spread northward and also eastward into Iran from southern Iraq (Jawad 2015).

Aquatic plants and debris are the majority of its food, followed by entomostracans and aquatic insect larvae. Also present are fish parts, mollusks, and nonaquatic organisms (Al-Daham and Bhatti (1977). Khalaf et al. (1987) examined this species in the Diyala River, Iraq and observed that young fish consuming chironomids and worms whereas larger ones ate fish.

Family: Mugillidae
Planiliza abu

In Iraq, this species has several common names, these are zuri, khishni, hishni, and abu-khraiza (Coad 2017) (Fig. 66.12a). Diverse populations occur in the Tigris, Euphrates, and Orontes rivers as shown by morphology (Turan et al. 2004). Khayyami et al. (2014) examined the morphometry of fish from the Karun River (fresh), Arvand River estuary (brackish), and Mousa Creek (saline) and found a high degree of overlap between the first two localities while these two were highly different from the third. Al-Hassan (1984) displayed variances in the number of vertebrae and dorsal fin rays between fish from Basrah, Iraq, and the Karkheh River in Iran.

Planiliza abu is distributed in the Tigris–Euphrates and Orontes (= Asi) (Özdilek 2003) river drainage, and in Pakistan. In Iran, found in the Tigris River basin in rivers and marshes and creeks draining to the Arabian Gulf such as the Helleh, Mond, and Zohreh in the Gulf or Persis drainage (Najafpour 1997; Jawad et al. 2010; Hashemi et al. 2011; Maghtouie et al. 2011; Pazira et al. 2016; Keivany and Zamani-Faradonbe 2017).

Food of this species comprises phytoplankton as the most important food followed by organic detritus. Silt is a significant food item of the gut in all fishes studied. Eighteen species of Bacillariophyceae, 6 Chlorophyceae, and 4 Cyanophyceae species were reported from gut analyses (Coad 2017). Abdoli (2000) recorded Daphnia, Hemiptera, Navicula, Nitzschia, Amphora, and Cymbella species for Iranian specimens. Al-Nasiri et al. (1977) investigated the feeding ecology of this species in the Shatt al Arab, Iraq. Ahmad and Hussain (1982) inspected the food of young fish near Basrah, Iraq and reported on the presence of organic debris to be a vital food followed by phytoplankton.

66.4 The Evidence of the Presence Macro- and Microplastic in the Stomach of Iraqi Fish Species

The macro-, meso-, and microplastics materials obtained from the stomachs of selected fish species living in the vicinity of Shatt al-Arab River are shown in Figs. 66.3, 66.4, 66.5, 66.6, 66.7, 66.8, 66.9, 66.10, 66.11, 66.12b, c. The

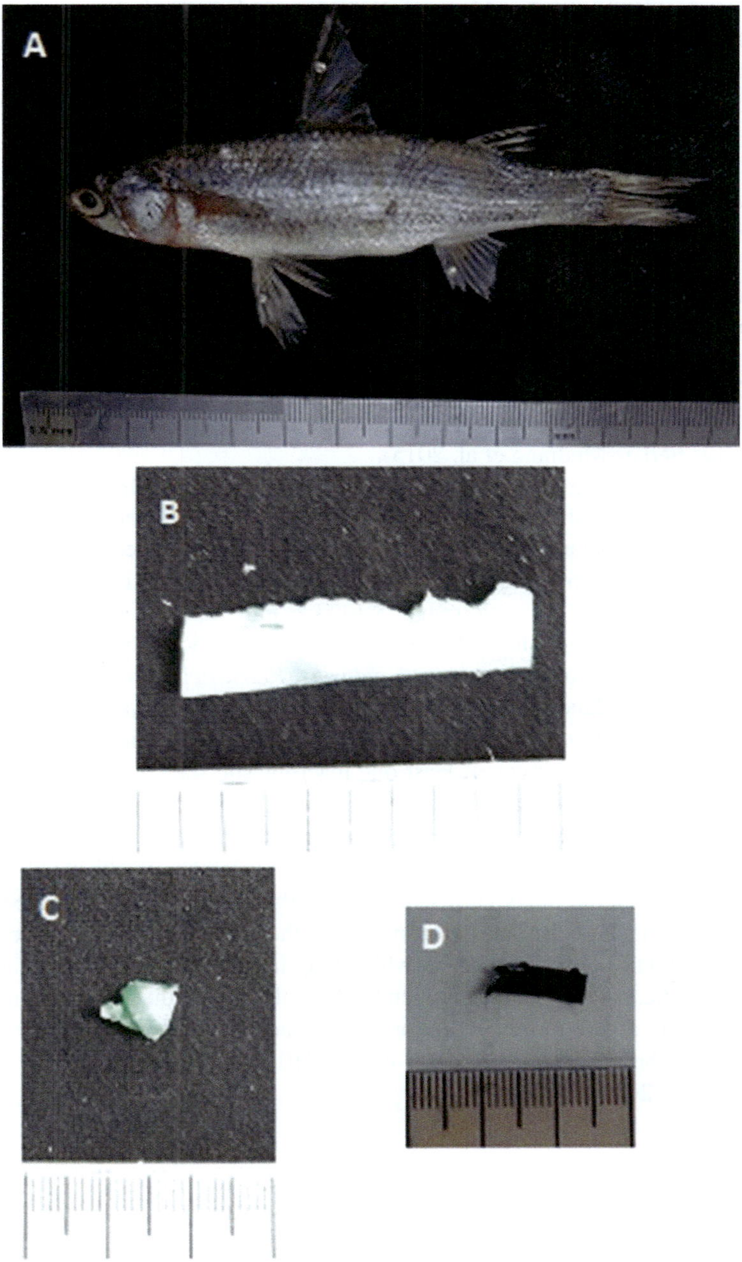

Fig. 66.12 *Planiliza abu* (Family: Mugillidae). (**a**) general body aspect; (**b**) samples of meso- and microplastic materials found inside the stomach; (**c**) sample of macroplastic fragment obtained from the stomach of the fish

examination, identification, and proof of macro- and microplastics were attained under a Stereomicroscope, and images of plastic items would be taken with a digital camera at different magnifications. In the procedure of evaluating the plastic fragments, visual cataloging was used first (Hidalgo-Ruz and Thiel 2013; Hoss and Settle 1990; Kukulka et al. 2012; Verma et al. 2016).). Then plastics were categorized conferring to their physical features into fibers (elongated), fragments (small angular pieces), pellets (spherical, ovoid), sheets (irregular flat, flexible), and films (thin, soft, and transparent) (Li et al. 2016). The longest or widest sizes of each fragment was measured to the nearest millimeters (Phillips and Bonner 2015). When maximum piece size was smaller than 5 mm, the plastic sample was deliberated as microplastic; particles with 5.1 mm–24.8 mm were taken as mesoplastic and macroplastic fragments having >25 mm in length (Collignon et al. 2014; Romeo et al. 2016). It is important to take special care to avert sample pollution during dissection, removal, sorting, and visual recognition (Boerger et al. 2010; Collignon et al. 2014; Eerkes-Medrano et al. 2015).

The evidence of the presence of plastics in the freshwater of Iraq is based on the examination of 10 fish species that are considered commercially important. All species examined have plastic fragments of one sort or another. The size of the plastic pieces range between macroplastics of size larger than 25 mm length, mesoplastic fragments of size range 5.1–24.8 mm, and microplastics of size <5 mm. The fish species showed variation in the abundance of plastic materials in their stomach. The 3 catfish species, *M. pelusius*, *P. triostegus*, and *H. fossilis* have shown greater abundance of plastic materials in their stomachs, while the stomach of the mugilid species *P. abu* was with the least abundance of plastic materials. All ten species examined have shown at least one macroplastic fragment in their stomach and all 5 types of plastic materials, i.e., fiber (elongated), fragments (small angular piece), pellets (spherical, ovoid), sheets (angular flat), and films (thin, soft, transparent) were observed to be found in the stomach of at least one species. Species of catfish feed on benthic preys gulp sediment jointly with the prey, after identifying them with their barbels, and ejecting then the sediment through their gills (Labropoulou and Eleftheriou 1997), which may upsurge the hazard of by mistake ingesting plastic. As has been recommended for pelagic fish, the active consumption of plastic fibers owing to misperception with some of their prey, like polychaetes, must also not be abandoned (Boerger et al. 2010; Collignon et al. 2014; Eerkes-Medrano et al. 2015).

66.5 The Future Next Steps and Opportunities

Eerkes-Medrano et al. (2015) recommended the following to be done: (1) advance best practice for observing microplastics in freshwater systems; (2) enumerate all features causing occurrence, richness, and distribution of microplastics in the habitat; (3) comprehend the dilapidation conduct containing particle lifetimes and eventual destiny in freshwater; (4) evaluate the ability of rivers to be a supply of microplastic to the oceans; (5) evaluate and comprehend microplastic relations with biota; (6) measure microplastic effects on ecosystem functions; and (7) assess the costs of microplastic for humans.

References

Abdoli A (2000) The inland water fishes of Iran. Iranian Museum of Nature & Wildlife, Tehran, p 378. In Farsi

Aguado J, Serrano DP, San Miguel G (2007) European trends in the feedstock recycling of plastic wastes. Global NEST J 9(1):12–19

Ahmad TA, Hussain NA (1982) Observations on the food of young *Liza abu* (Heckel) from Salihyia River (Basrah, Iraq). Iraqi J Mar Sci Basrah 1(1):79–88

Al-Daham NK, Bhatti MN (1977) Salinity tolerance of *Gambusia affinis* (Baird & Girard) and *Heteropneustes fossilis* (Bloch). J Fish Biol 11:309–313

AL-Hamed MI (1966) Carp culture in the Republic of Iraq. In: Pillay, T.V.R. (Ed.). Proceedings of the FAO world symposium on warm- water pond fish culture. FAO Fish. Rep. No. 44(2), Rome, Italy, p 174

Al-Hassan LAJ (1984) Meristic comparison of *Liza abu* from Basrah, Iraq and Karkhah River, Arabistan, Iran. Cybium 8(3):107–108

Al-Hassan LAJ, Hussain NA, Soud KD (1989) A preliminary annotated checklist of the fishes of Shatt Al-Arab River, Basrah, Iraq. Polskie Archiwum Hydrobiologii 36:283–288

Al-Hassan LAJ, Al-Daham NK, Hassan SS (1991) Eye lens as an age indicator in *Mystus pelusius* (Bagridae). Cybium 15(2):171–172

Al-Nasiri SK Shamsul Hoda SM (1976) A guide to the freshwater fishes of Iraq. Bull Basrah Nat Hist Mus Publ, No.1: 126 p

Al-Nasiri SK, Sarker AL, Hoda SMS (1977) Feeding ecology of a mugilid fish, Liza abu (Heckel) in Basrah, Iraq. Bull Basrah Nat Hist Museum 4:27–40

Al-Rudainy AJ (2008) Atlas of Iraqi fresh water fishes. Ministry of the Environment, Baghdad, p 107. In English and Arabic

Al-Shamma'a AA (2005) The diet composition of three catfishes from Al-Hammar Marsh, Al-Fuhoud, Iraq. First Scientific Conference on the Rehabilitation of the Southern Iraq Marshes, 11–12 April 2005, Marine Science Centre, University of Basrah (abstract)

Al-Shamma'a AA, Jasim ZM (1993) The natural food of *Liza abu* during the flood in Al-Hammar marsh, South Iraq. Zool Middle East 9:59–64

Arthur C, Baker J, Bamford H (2009) Proceedings of the international research workshop on the occurrence, effects and fate of microplastic marine debris, in NOAA technical, Memorandum

Boerger CM, Lattin GL, Moore SL, Moore CJ (2010) Plastic ingestion by planktivorous fishes in the North Pacific central gyre. Mar Pollut Bull 60:2275–2278

Bogutskaya NG (1997) Contribution to the knowledge of leuciscine fishes of Asia minor. Part 2. An annotated check-list of leuciscine fishes (Leuciscinae, Cyprinidae) of Turkey with descriptions of a new species and two new subspecies. Mitt Hamb Zool Mus Inst 94:161–186

Coad BW (2010) Freshwater fishes of Iraq. Pensoft Publishers, Moscow

Coad BW (2017) Review of the freshwater mullets of Iran (family Mugilidae). Iranian J Ichthyol 4 (2):75–130

Cole M, Lindeque P, Halsband C, Galloway TS (2011) Microplastics as contaminants in the marine environment: a review. Mar Pollut Bull 62(12):2588–2597

Collignon A, Hecq JH, Galgani F, Collard F, Goffart A (2014) Annual variation in neustonic micro- and meso-plastic particles and zooplankton in the bay of Calvi (Mediterranean–Corsica). Mar Pollut Bull 79(1–2):293–298

Costello MJ (1990) Predator feeding strategy and prey importance: new graphical analysis. J Fish Biol 36:261–263
Dawood AH (1997) Food habit of *Silurus triostegus* Heckel, 1843 (Pisces, Siluriformes) from Al-Hammar marsh, southern Iraq. Marina Mesopotamica 12(1):75–85
Eerkes-Medrano D, Thompson RC, Aldridge DC (2015) Microplastics in freshwater systems: a review of the emerging threats, identification of knowledge gaps and prioritisation of research needs. Water Res 75:63–82
Epler P, Bartel R, Szczerbowski JA, Syzpuła J (2001) The ichthyofauna of lakes Habbaniya, Tharthar and Razzazah. Archiwum Rybactwa Polskiego 9(supplement 1):171–184
Eriksson C, Burton H (2003) Origins and biological accumulation of small plastic particles in fur seals from Macquarie Island. Ambio 32:380–384
Esmaeili HR, Coad BW, Gholamifard A, Nazari N, Teimori A (2010) Annotated checklist of the freshwater fishes of Iran. Zoosystematica Rossica 19:361–386
Esmaeili HR, Mehraban H, Abbasi K, Keivany Y, Coad BW (2017) Review and updated checklist of freshwater fishes of Iran: taxonomy, distribution and conservation status. Iranian J Ichthyol 4:1–114
Hashemi S, Eskandary G, Ansary H, Yooneszadeh M (2011) Stock assessment and production of fish species in the Shadegan wetland, Iran. World J Fish Marine Sci 3(6):502–508
Heckel JJ (1843) Ichthyologie. In: Russegger, J. Reisen in Europa, Asien und Afrika, mit besonderer Rücksicht auf die naturwissenschaftlichen Verhältnisse der betreffenden Länder, unternommen in den Jahren 1835 bis 1841 von Joseph Russegger. Schweitzerbart'sche Verlagsbuchhandlung, Stuttgart 1(2): 991–1099, Taf. II-XIII
Hidalgo-Ruz V, Thiel M (2013) Distribution and abundance of small plastic debris on beaches in the SE Pacific (Chile): a study supported by a citizen science project. Mar Environ Res 87:12–18
Hoss DE, Settle LR (1990) Ingestion of plastics by teleost fishes. In: Proceedings of the Second International Conference on Marine Debris. NOAA Technical Memorandum. NOAA-TM-NMFS-SWFSC-154. Miami, FL, pp 693–709
Li WC, Tse HF, Fok L (2016) Plastic waste in the marine environment: a review of sources, occurrence and effects. Sci Total Environ 566:333–349
Jawad LA (2015) The Indian catfish, *Heteropneustes fossilis* (Bloch, 1794) envenomation-a report of four cases from basrah and maysan, south of Iraq. TOFIQ J Med Sci 2(2):52–61
Jawad LA, Sadighzadeh Z, Valinassab T (2010) Malformation of the caudal fin in the freshwater mullet, *Liza abu* (Actinopterygii: Mugilidae) collected from Karkhe River Iran. An Biol 32:11–14
Keivany Y, Zamani-Faradonbe M (2017) Length-weight and length-length relationships for seven fish species from the Zohreh River, Iran. J Appl Ichthyol 33:625–627
Keivany Y, Ghorbani M, Paykan-Heyrati F (2017) Reproductive biology of Mossul bleak (*Alburnus mossulensis*) in Bibi-Sayyedan River of Tigris basin in Iran. Caspian J Environ Sci 15(2):135–145
Khalaf KT (1961) The marine and freshwater fishes of Iraq. Ar-Rabitta Press, Baghdad, p 164
Khalaf AN, Allouse SB, Al-Yamour KY, Al-Jafary AR, Sadek SE (1987) Food and feeding of the catfish *Heteropneustes fossilis* (Bloch) from a polluted river. J Environ Sci Health A22 (5):397–410
Khayyami H, Movahedinia A, Zolgharnein H, Salamat N (2014) Morphological variability of *Liza abu* (Heckel, 1843) from northwestern part of Persian gulf. World J Fish Marine Sciences 6 (5):386–394
Kottelat M, Freyhof J (2007) Handbook of European freshwater fishes. Publications Kottelat, Cornol and Freyhof, Berlin, p 646
Kukulka T, Proskurowski G, Morйt-Ferguson S, Meyer DW, Law KL (2012) The effect of wind mixing on the vertical distribution of buoyant plastic debris. Geophys Res Lett 39:L07601
Labropoulou M, Eleftheriou A (1997) The foraging ecology of two pairs of congeneric demersal fish species: importance of morphological characteristics in prey selection. J Fish Biol 50:324–340

Lorenzoni M, Ghetti L, Pedicillo G, Carosi A (2010) Analysis of the biological features of the goldfish *Carassius auratus auratus* in Lake Trasimeno (Umbria, Italy) with a view to drawing up plans for population control. Folia Zool 59(2):142–156

Maghtouie AH, Neissi G, Nasseri S, Gholaminezhad E, Shalamzari N, Nikpour Y (2011) Determination of mercury in mullet fish (*Liza abu*) from Arvand River, Iran. World journal of fish and marine. Sciences 3(6):514–517

Mohamed ARM, Abood AN (2008) Diet and trophic status of three cyprinids fish in the Shatt Al-Arab River, Iraq. J Agric Vet Sci 11:49–57

Mohamed ARM, Abood AN (2017) Compositional change in fish assemblage structure in the Shatt Al-Arab River, Iraq. Asian J Appl Sci 5(5):944–958

Mohamed ARM, Al-Noor SS, Faris RAK (2008) The status of artisanal fisheries in the lower reaches of Mesopotamian rivers, North Basrah, Iraq. Proc 5th Int Con Biol Sci (Zool) 5:126–132

Mohamed MAR, Ahmed SM, Al-Okailee MTK (2012) Variations in occurrence, abundance and diet of Hilsa, *Tenualosa ilisha* larvae in the north of Shatt Al-Arab River, Basrah, Iraq. Basrah J Agri Sci 25:40–52

Mohamed ARM, Hussein SA, Lazem LF (2015) Spatiotemporal variability of fish assemblage in the Shatt Al-Arab River, Iraq. J Coastal Life Med 3(1):27–34

Mohamed ARM, Aufy LA, Jasim BM (2016) Food habit of Mosul bleak, *Alburnus mossulensis* (Heckel, 1843) in the southern reaches of Euphrates River, Iraq. Basrah J Agric Sci 29:587–593

Moore CJ, Lattin GL, Zellers AF (2011) Quantity and type of plastic debris flowing from two urban rivers to coastal waters and beaches of Southern California. J Integr Coast Zone Manag 11 (1):65–73

Najafpour N (1997) Identification of some freshwater fishes of Khuzestan Province. Fisheries Research Centre of Khuzestan Province, Iranian Fisheries Research and Training Organisation, Tehran, p 96. (In Farsi)

Özdilek ŞY (2003) Occurrence of the abu mullet, *Liza abu* (Heckel, 1843) (Pisces, Mugilidae), in the Orontes River. Zool Middle East 30:111–113

Pazira A, Abdoli A, Ghanbari F, Moghdani S (2016) Comparison of fish species diversity in Dalaki and Helleh Rivers of the Persis basin in Bushehr Province. Iran J Ichthyol 3(3):222–228

Phillips MB, Bonner TH (2015) Occurrence and amount of microplastic ingested by fishes in watersheds of the Gulf of Mexico. Mar Pollut Bull 100:264e269

Pillay TVR, Rosa H Jr (1963) Synopsis on biological dataon Hilsa, *Hilsa ilisha* (Hamilton) 1882. FAO Fish Biol Synop 25:1–6

Roberts TR (1994) Systematic revision of Asian bagrid catfishes of the genus Mystus sensu stricto, with a new species from Thailand and Cambodia. Ichthyol Explor Fresh 5(3):241–256

Romeo T, Peda C, Fossi MC, Andaloro F, Battaglia P (2016) First record of plastic debris in the stomach of Mediterranean lantern fishes. Acta Adriat 57(1):115–123

Szczerbowski JA, Bartel R, Epler P (2001) Fishing gear selectivity, fish survival rates and resources in lakes Tharthar, Habbaniya and Razzazah. Arch Polish Fish 9:225–233

Szypula J, Epler P, Bartel R, Szczerbowski JA (2001) Age and growth of fish in lakes Tharthar, Razzazah and Habbaniya. Arch Polish Fish 9:185–197

Turan C, Ergüden D, Turan F, Gürlek M (2004) Genetic and morphologic structure of *Liza abu* (Heckel, 1843) populations from the rivers Orontes, Euphrates and Tigris. Turkish J Vet Ani Sci 28:729–734

Ünlü E, Bozkurt R (1996) Notes on the catfish, *Silurus triostegus* (Siluridae) from the Euphrates River in Turkey. Cybium 20(3):315–317

van den Eelaart A (1954) Report to the government of Iraq on the development of inland fisheries. Food and agriculture organization, Rome, report of the expanded technical assistance. Program 270:42

Verma R, Vinoda KS, Papireddy M, Gowda ANS (2016) Toxic pollutants from plastic waste-a review. Procedia Environ Sci 35:701–708

Wagner M, Scherer C, Alvarez-Muñoz D, Brennholt N, Bourrain X, Buchinger S, Fries E, Grosbois C, Klasmeier J, Marti T, Rodriguez-Mozaz S (2014) Microplastics in freshwater ecosystems: what we know and what we need to know. Environ Sci Eur 26(1):1–9

Wu C, Zhang K, Xiong X (2018) Microplastic pollution in inland waters focusing on Asia. In: *Freshwater microplastics*. Springer, Cham, pp 85–99

Zakaria H (1964) A catfish (*Heteropneustes fossilis*) of medical importance invades Iraqi waters. J Faculty Med Baghdad 6(2):48–55

Chapter 67
Heavy Metals in Freshwater Invertebrates of Iran: A Review on the Bioaccumulation and Effects

Nima Pourang

Abstract Inputs of heavy metals into the aquatic environment have increased their levels to large extents within past few decades, which are severely affecting and endangering the ecological equilibrium of the relevant ecosystems. Some of the aquatic organisms, including freshwater invertebrates, are widely recognized for their capacity to accumulate heavy metals in their tissues, and hence can be used for biomonitoring studies. Considering the importance of freshwater ecosystems of Iran from different points of view, and at the same time lack of any published review on heavy metals accumulation in the freshwater invertebrates, this chapter aims to present a brief overview of the issue, based on available literature.

67.1 Introduction

Heavy metals are released from both natural sources and human activity. The impact of the metals on the environment is an increasing problem worldwide (Walker et al. 2006). Some of these metals, such as Cd and Pb, are toxic to living organisms even at quite low concentrations, whereas others, such as Zn and Cu, are biologically essential and natural constituents of aquatic ecosystems, and generally only become toxic at very high concentrations (Bahnasawy et al. 2011). Heavy metals have also been shown to bioaccumulate in a range of organisms, and can be toxic to a number of aquatic organisms, including invertebrates (Jezierska et al. 2009; Bonanno and Giudice 2010). The extent of bioaccumulation of metals is dependent on the total amount, the bioavailability of each metal in the environmental medium and the route of uptake, storage, and excretion mechanisms (Chiarelli and Roccheri 2014). Some of the freshwater organisms are widely recognized for their capacity to accumulate a variety of environmental pollutants, including metals, in their tissues (Cairns and

N. Pourang (✉)
Iranian Fisheries Science Research Institute (IFSRI), Agricultural Research, Education and Extension Organization (AREEO), Tehran, Iran
e-mail: n.pourang@areeo.ac.ir

L. A. Jawad (ed.), *Tigris and Euphrates Rivers: Their Environment from Headwaters to Mouth*, Aquatic Ecology Series 11,
https://doi.org/10.1007/978-3-030-57570-0_67

Pratt 1995), among them, some of the freshwater invertebrates (e.g., bivalve mollusks) have an ability to accumulate heavy metals to various orders of magnitude with respect to the levels found in their environment. These organisms accumulate most of the contaminants at much higher levels than those found in the water column and they are representative of the pollution of an area (Pourang et al. 2010).

There are no comprehensive studies published on the identification of freshwater invertebrates of Iran, but based on the results of several related studies (Zamanpoore et al. 2011; Glöer and Pešić 2012; Keikhosravi 2013; Rhmanipanah et al. 2015; Sharifinia 2015; Bojková et al. 2018), more than 145 species belonging to 4 phyla have been identified in the freshwater ecosystems of the country.

With regard to the great importance of freshwater ecosystems from different points of view, especially in developing and semiarid countries such as Iran (Aazami et al. 2015), on the one hand, and the potential of using these invertebrates as heavy metal bioindicators (Li et al. 2010) on the other hand, this chapter provides a brief overview of the results of the relevant researches in Iran.

67.2 Brief Review of the Relevant Studies

Considering the identified taxa of freshwater invertebrates of the country, only relatively limited studies concerning heavy metals (accumulation or effect) were performed on the taxa among them, which are briefly reviewed below. As shown in Table 67.1, most of the studies have been conducted in the northern part of the country.

The first published research on the effects of heavy metals on freshwater invertebrates of Iran was conducted by Nehring (1976). He evaluated the toxicity of heavy metals (copper, lead, zinc, and silver) in two species of aquatic insects, a mayfly (Ephemeroptera): *Ephemerella grandis*, and a stonefly (Plecoptera): *Pteronarcys californica*, collected from the Chalus River (northern part of Iran), as well as their potential use as biological monitors of heavy metal pollution. The insects tested were found to be more tolerant of the heavy metals than the most fish. The results indicated that the aquatic insects may serve as effective biological monitors of heavy metal pollution where fish-kills are involved. They accumulate metals in relative proportion to the metal concentration in the water, and they concentrate the metal by some predictable, reproducible factor.

Nehring et al. (1979) evaluated the accuracy of the concentration factor method in determining the amount of lead pollution in the Chalus River. Initial laboratory experiments indicated that the mechanism of lead uptake and loss varies with the type of exposure, making field determination of concentration factors essential. Subsequently, the field concentration factors were determined for six groups of aquatic insects. These concentration factors were used to determine an estimated lead level in the river during subsequent sampling periods. The estimated lead levels rarely differed by more than 0.1 mg/L. The range of concentration factors was between 2488 and 8076.

Table 67.1 Studies conducted on heavy metals effects on (and accumulation by) freshwater invertebrates from different areas of Iran

Taxon	Metal	Geographical area	Type of studies		References
			F	E	
Ephemerella grandis (Ephemeroptera), *Pteronarcys californica* (Plecoptera)	Pb, Zn, cu, and ag	Chalus River and labora-tory study-N		√	Nehring (1976)
Ephemeroptera (Heptageniidae), Diptera (Tipulidae: Craneflies), TrichopteraNehring et al. (1979)(*Macronemum*), Odonata: (Anisoptera: Drag-onflies), Plecoptera (*Paragnetina*, *Nemoura*), (a single genus or species of each)	Pb	Chalus River and labora-tory study-N	Nehring et al. (1979)	√	Nehring et al. (1979)
Culiseta subochrea and *Aedes caspius* (Diptera: Culicidae)	As, cd, co, Cr, cu, Fe, hg, Mn, Pb, and Zn	Shadegan international Wetland-SW	√		Nasirian et al. (2014)
Astacus leptodactylus	Cu, Pb, cd, Cr, Zn, Ba, and Fe	Abbasa River-N	√		Hosseini et al. (2004)
A. leptodactylus	Zn, cu, Cr, and Pb	Tehran market	√		Ahmadi Kordestani et al. (2015)
A. leptodactylus	Fe, cu, Zn, Mn, Pb, Ni, co, and Cr	Aras dam-NW			Naghshbandi et al. (2007)
Chironomidae and *Gammarus pulex*	Cu and Zn	Namrood River-N	√		Javanshir (2013)
Chironomidae and *G. pulex*	Cd and Pb	Namrood River-N	√		Rezaei et al. (2011)
Mytilaster lineatus and *Cor-bicula fiuminalis, Tubifex tubifex,* Chironomidae	Mn, cu, Zn, and Pb	Anzali wet-land-N	√		Pourang (1996)
C. fluminalis	Cd, Zn, cu, and Pb	Zayandeh-Roud River-N	√		Movahedi et al. (2016)
Anodonta cygnea	Cd, cu, and Pb	Anzali wet-land-N	√		Pourang et al. (2010)
A. cygnea	Zn	Tajan River estuary and laboratory study-N		√	Moëzzi et al. (2013a)
A. cygnea	Zn and Cr	Tajan River estuary and laboratory study-N		√	Moëzzi et al. (2013b)

(continued)

Table 67.1 (continued)

Taxon	Metal	Geographical area	Type of studies F	E	References
Dreissena polymorpha	Pb	Anzali wetland and laboratory study-N		√	Rahnama et al. (2011)
M. lineatus	Cd and Pb	Sisangan, Mazandaran-N		√	Salahinejad et al. (2017)
Physa sp.	Cd	Sarab Chehr stream and laboratory study-W		√	Sharifi et al. (2007)

Types of studies: Field study (F); Experimental study (E). *N* North, *W* West, *SW* Southwest, *NW* Northwest

The accumulation of heavy metals in crayfish, *Astacus leptodactylus* has been investigated in three separate studies, which were conducted by Hosseini et al. (2004), Naghshbandi et al. (2007), and Ahmadi Kordestani et al. (2015).

Hosseini et al. (2004) studied heavy metals (Ba, Ca, Cr, Cu, Fe, Pb, and Zn) concentrations in the edible tissues of *A. leptodactylus,* water and sediments from Abbasa River (flowing into the Caspian Sea). The following metal accumulation pattern was observed in the crayfish: Zn > Ba > Cu > Fe > Cr > Ca > Pb, which was similar to the order metals concentration in the sediments. Mean contents of heavy metals in the muscle of *A. leptodactylus* were 45.73, 2.44, 3.27, 184.20, 145.12, and 9.66 for Cu, Pb, Cd, Cr, Zn, Ba, and Fe, respectively (μg/g dry weight). There were significant correlations between heavy metals contents in water, sediments, and crayfish. Ahmadi Kordestani et al. (2015) measured concentrations of Pb, Cr, Cu, and Zn in muscle tissues of *A. leptodactylus*, which were randomly purchased from the Tehran market. No correlation was found between weight and length with the metal concentrations. There was also no correlation between sex and concentration of toxic elements. However, sex was found to have a significant correlation with the concentration of essential elements. The pattern of accumulation of the metals was Zn > Cu > Cr ≥ Pb. Comparison of the metal concentrations with standard values indicated that Cu and Zn concentrations were lower than concentrations allowed by WHO (World Health Organization) and ANHMRC (Australian National Health and Medical Research Council), while Pb concentration was higher than the allowable level. Naghshbandi et al. (2007) determined concentration of Fe, Cu, Zn, Mn, Pb, Ni, Co, and Cr in the tissues (exoskeleton, gills, hepatopancreas, and muscle) of *A. leptodactylus* from the Aras Dam reservoir. The highest concentration of Cu and Fe was found in the gills and the highest level of Zn was found in hepatopancreas. The highest Mn concentration was found in exoskeleton. No significant differences were found between the male and the female in the level of heavy metals in their tissues, except for the concentration of Fe in hepatopancreas.

Three relevant studies (Pourang 1996; Pourang et al. 2010; Rahnama et al. 2011) have been performed on invertebrates from Anzali Wetland, one of the most important water bodies in northern Iran, connected to the Caspian Sea. In one of the researches, Pourang (1996) measured heavy metals (lead, copper, zinc, and manganese) in chironomid larvae, tubificid worms, and two species of bivalve molluscs (*Mytilaster lineatus* and *Corbicula fiuminalis*) from the Wetland (Pourang 1996). Significant differences were found in bioaccumulation of Pb and Zn in three size categories of chironomid larvae. Pb was higher in smaller *M. lineatus*, while the reverse was observed for Cu. Only Pb and Zn body burdens in chironomid larvae were significantly correlated with levels in sediments. An inverse relationship occurs between size and metal levels in the larvae, probably due to reduction of body surface ratio in larger individuals. The mean concentration of Pb was higher in smaller molluscs while a reverse case was detected for Cu. The relatively higher level of Cu in the larger specimens may reflect their pumping rate, because the pumping rate of some filter-feeder species (e.g., *Mytilus edulis*) is positively correlated with shell length. The highest concentrations of Zn and Pb were measured in *M. lineatus* and the order of these metals for other investigated benthos were: tubificid worms > chironomid larvae > *C. fluminalis*. Mn and Cu concentrations showed a different pattern in decreasing order, as follows: tubificid worms > chironomid larvae > *M. lineatus* > *C. fluminalis*. With regard to above-mentioned results, it can be concluded that in the case of Mn and Cu, detritivorous benthos showed higher metal concentrations than filter feeders whereas no definite trend was found for Zn and Pb. Pourang et al. (2010) determined concentrations of Cd, Cu, and Pb in the soft tissues (foot and gills) of swan mussel *Anodonta cygnea* from two sampling sites in Anzali Wetland. The metal contents in the mussel species from the studied region were comparable to other world areas. Clear and significant differences could be detected between the tissues for the accumulation of Cd and Pb. Age-related correlations were found in the case of Cu accumulation in the foot and Cd levels in gills. There was no weight-dependent trend for the accumulation of the three elements. The pattern of metal occurrence in gills and foot exhibited the following order, respectively: Pb, Cu > Cd and Cu > Pb, Cd. Rahnama et al. (2011) investigated the effect of Pb on condition indices of zebra mussel (*Dreissena polymorpha*) from Anzali Wetland. No significant differences were observed in condition indices (TCI and SCI) among the mussels in control and Pb treatments. Mussels' condition indices decreased during the experimental period, that it caused by function of time, not by metal accumulation. Besides the metal pollution, other environmental factors, such as food availability, could affect the mussel's condition. The results suggest that the accumulation of Pb in the soft tissue of mussels was affected by exposure time and Pb concentration in the ambient water.

The results of several other related researches, conducted in different areas of the country, are presented below.

Histopathological effects of Zn and Cr and their incidence time on *A. cygnea*, collected from the Tajan River estuary, were studied by Moëzzi et al. (2013a, b). The accumulated level of Zn was significantly higher than Cr and the cumulative capacity of both metals in gill was higher than mantle. Mantle and gill histopathology showed remarkable changes in mussels exposed to metals, but no difference was observed between changes caused by each metal. Moreover, the results indicated that heavy metal accumulation potential of gill is higher than mantle. The results showed that sensitivity of digestive gland is lesser than mantle and foot in exposure to Zn. The histopathological alterations in the organs of swan mussel can be considered as reliable biomarkers in biomonitoring of heavy metal pollution in aquatic ecosystems. Javanshir (2013) measured Cu and Zn concentration in benthos (Chironomidae and *Gammarus pulex*) and sediment from Namrood River located in Firoozkooh (Tehran Province). The Namrood River is situated by the main road being by pollutants from tourism and recreational centers, gas stations, sewage of villages, agricultural wastewater, and fish culture effluent. The water was extremely contaminated in some parts and possibly contains heavy metals. The comparison of the results concerning Cu and Zn concentrations in up and down stream suggests that the amount of Cu in sediments was higher than that of benthos. In the case of Cu concentration in the body, gammarids showed higher concentrations than chironomids. The researcher explained that chironomids are filter feeders and feed on plankton and fine particulate organic materials. The low concentrations of Cu in chironomid bodies may be due to low quantities of this metal in their food. The concentration of Zn in chironomids was higher than gammarids. It is possible that natural sources of zinc were transferred from upstream to downstream. As mentioned before, chironomids are filter feeders and thus may assimilate more Zn content than gammarids (shredders of dead leaves). Copper and zinc concentrations ranged 0.170–0.966 and 0.187–3.846 ppm, respectively. Sediments of the upstream station had the highest copper concentration among the samples in both cold and warm seasons of the year. Rezaei et al. (2011) also selected two sampling sites in upstream and downstream of Namrood River. Cadmium and lead concentrations were measured in Chironomidae, *G. pulex*, and sediment samples. The results showed that the concentration of lead in the samples varied between 0.010 and 0.2033 ppm, while the range of lead levels was between 0.11 and 2.16 ppm.

It was concluded that the selected invertebrates can be used for biomonitoring of the metals in the river.

The accumulation of heavy metals (Cd, Zn, Cu, and Pb) in the soft tissues and shells of *C. fluminalis* from in Zayandeh-Roud River (Isfahan Province) and its potential use for biomonitoring was investigated by Movahedi et al. (2016). They concluded that concentrations of Cd, Pb, and Zn in the shells indicate their concentrations in the river sediments.

Salahinejad et al. (2017) studied the filtration rate of *M. lineatus* exposed to Cd and Pb and its ability to remove these metals from water. The experiments were carried out in a specially designed mesocosm system. The results showed a significant difference in the filtration rate of *M. lineatus* exposed Cd and Pb. *M. lineatus* revealed a higher filtration rate exposed to Pb concentrations in comparison to that of Cd.

The quantity of some heavy metals (As, Cd, Co, Cr, Cu, Fe, Hg, Mn, Pb, and Zn) of mosquito larvae (*Culiseta subochrea* and *Aedes caspius*) in Shadegan International Wetland (Khuzestan Province, southwest of Iran) was evaluated by Nasirian et al. (2014). The results showed that the sediments and *C. subochrea* larvae are polluted with all the metals investigated except As and Hg. The heavy metals bioaccumulated in the *C. subochrea* larvae range from 31.78 μg/g dry weight at the lowest level for Cr to 3822.7 at the highest level for Cd. The researchers concluded that *C. subochrea* will likely be used as a bioindicator to heavy metal pollution in aquatic ecosystems such as wetlands.

Sharifi et al. (2007) determined the concentration of cadmium in the freshwater snail *Physa* sp., collected from the Sarab Chehr Stream (Kermanshah Province) under seminatural conditions provided by artificial streams. The mean Cd level was more than three orders of magnitude greater than the nominal concentrations in the artificial streams. The average rate of cadmium uptake was 0.14 mg g^{-1} per day. Both uptake and depuration were influenced by biological active processes. No mortality or other overt signs of toxic effects were noticed during this study.

Tables 67.2, 67.3, 67.4, 67.5, 67.6, 67.7 and 67.8 summarize the results of researches concerning the concentrations of the heavy metals in different taxa of freshwater invertebrates of Iran. With regard to the tables the following points can be drawn:

(a) The minimum and maximum of lead concentrations were observed in *Aedes caspius* and *Ephemerella grandis*, respectively (0.56 and 104,700 ppm dry weight). The oldest and the most recent related researches were conducted by Nehring (1976) and Ahmadi Kordestani et al. (2015), respectively (Table 67.2).
(b) The range of copper concentrations was between 0.62 (for *Aedes caspius*) and 9125 (for *Ephemerella grandis*) ppm dry weight. The first relevant study was also conducted by Nehring (1976) and the most recent one by Ahmadi Kordestani et al. (2015) (Table 67.3).
(c) The minimum and maximum of zinc concentrations can be seen in *Ephemerella grandis* and *Aedes caspius*, respectively (0.895 and 2361 ppm dry weight). The oldest and the most recent related studies were carried out by Nehring (1976) and Ahmadi Kordestani et al. (2015), respectively (Table 67.4).
(d) The minimum and maximum of cadmium levels can be observed in Chironomidae and *Culiseta subochrea*, respectively (0.0204 and 114.7 ppm dry weight). To our knowledge, the first relevant research was conducted by Hosseini et al. (2004) and the most recent one by Movahedi et al. (2016), respectively (Table 67.5).
(e) The range of iron concentrations was between 9.66 (for *Astacus leptodactylus*) and 881.8 (for *Aedes caspius*) ppm dry weight. The oldest and the most recent related studies were carried out by Hosseini et al. (2004) and Nasirian et al. (2014), respectively (Table 67.6).

Table 67.2 Mean lead levels in different taxa of freshwater invertebrates of Iran. In cases where two or more tissues/organs were analyzed, their means are presented

Taxon	Concentration (ppm dry weight)	Year of sample collection	References
Ephemerella grandis	104700[a]	1976	Nehring (1976)
Pteronarcys californica	8172[a]	1976	Nehring (1976)
Tipulidae (one species)	4037.8	1976–1977	Nehring et al. (1979)
Perlidae (Paragnetina)	669.75	1976–1977	Nehring et al. (1979)
Heptageniidae (one species)	2835.4	1976–1977	Nehring et al. (1979)
Nemouridae (Nemoura)	1013.2	1976–1977	Nehring et al. (1979)
Hydropsychidae (Macronemum)	1787.4	1976–1977	Nehring et al. (1979)
Anisoptera (one species)	980.38	1976–1977	Nehring et al. (1979)
Anisoptera (one species)	802.07	1976–1977	Nehring et al. (1979)
Mytilaster lineatus	24.3	1993–1994	Pourang (1996)
Corbicula fiuminalis	5.8	1993–1994	Pourang (1996)
Tubifex tubifex,	19.2	1993–1994	Pourang (1996)
Chironomidae	13.9	1993–1994	Pourang (1996)
Astacus leptodactylus	2.44	2003	Hosseini et al. (2004)
Anodonta cygnea	0.144	2006	Pourang et al. (2010)
Astacus leptodactylus	79.71	2006	Naghshbandi et al. (2007)
Chironomidae	0.3775	2010	Rezaei et al. (2011)
Gammarus pulex	0.7100	2010	Rezaei et al. (2011)
Culiseta subochrea	803.5	2011	Nasirian et al. (2014)
Aedes caspius	0.56	2011	Nasirian et al. (2014)
Corbicula fiuminalis	7.78	2014	Movahedi et al. (2016)
Astacus leptodactylus	10.18	2015	Ahmadi Kordestani et al. (2015)

[a]Maximum of the average accumulation levels after exposure to different levels of the metal

(f) The range of manganese levels was between 0.26 (for *Aedes caspius*) and 31931.6 (for *Culiseta subochrea*) ppm dry weight. The oldest and the most recent related studies were carried out by Pourang (1996) and Nasirian et al. (2014), respectively (Table 67.7).

(g) The minimum and maximum of chromium concentrations can be found in *Astacus leptodactylus* and *Culiseta subochrea*, respectively (3.40 and 1578 ppm dry weight). The first relevant study was conducted by Hosseini et al. (2004) and the most recent one by Ahmadi Kordestani et al. (2015), respectively (Table 67.8).

Table 67.3 Mean copper levels in different taxa of freshwater invertebrates of Iran. In cases where two or more tissues/organs were analyzed, their means are presented

Taxon	Concentration (ppm dry weight)	Year of sample collection	References
Ephemerella grandis	9125[a]	1976	Nehring (1976)
Pteronarcys californica	2540[a]	1976	Nehring (1976)
Mytilaster lineatus	35.6	1993–1994	Pourang (1996)
Corbicula fiuminalis	24.8	1993–1994	Pourang (1996)
Tubifex tubifex,	77.4	1993–1994	Pourang (1996)
Chironomidae	49.9	1993–1994	Pourang (1996)
Astacus leptodactylus	45.73	2003	Hosseini et al. (2004)
Astacus leptodactylus	666.80	2006	Naghshbandi et al. (2007)
Anodonta cygnea	0.209	2006	Pourang et al. (2010)
Culiseta subochrea	7223.04	2011	Nasirian et al. (2014)
Aedes caspius	0.62	2011	Nasirian et al. (2014)
Chironomidae	0.216	2012	Javanshir (2013)
Gammarus pulex	0.480	2012	Javanshir (2013)
Corbicula fiuminalis	3.43	2014	Movahedi et al. (2016)
Astacus leptodactylus	58.11	2015	Ahmadi Kordestani et al. (2015)

[a]Maximum of the average accumulation levels after exposure to different levels of the metal

67.3 Conclusions

In general, with regard to the identified taxa of freshwater invertebrates of the country (more than 145 species), only relatively limited studies concerning their potential for biomonitoring of heavy metals have been carried out. According to Table 67.1, most of the studies have been conducted in the northern part of the country. On the other hand, almost half of the studies (7 of 15) are related to the laboratory experiments (Table 67.1). Based on the results presented in Tables 67.2, 67.3, 67.4, 67.5, 67.6, 67.7 and 67.8, it can be concluded that the widest and narrowest range of variations can be observed in the cases of Pb and Cd, respectively.

Considering the presence of numerous species with heavy metals bioaccumulation potential in the list of identified freshwater invertebrates of the country, further relevant studies need to be undertaken.

Table 67.4 Mean zinc levels in different taxa of freshwater invertebrates of Iran. In cases where two or more tissues/organs were analyzed, their means are presented

Taxon	Concentration (ppm dry weight)	Year of sample collection	References
Ephemerella grandis	2361[a]	1976	Nehring (1976)
Pteronarcys californica	561.2[a]	1976	Nehring (1976)
Mytilaster lineatus	262.9	1993–1994	Pourang (1996)
Corbicula fiuminalis	53.1	1993–1994	Pourang (1996)
Tubifex tubifex,	154.3	1993–1994	Pourang (1996)
Chironomidae	79.5	1993–1994	Pourang (1996)
Astacus leptodactylus	184.20	2003	Hosseini et al. (2004)
Aedes caspius	112.2	2011	Nasirian et al. (2014)
Chironomidae	2.112	2012	Javanshir (2013)
Gammarus pulex	0.895	2012	Javanshir (2013)
Corbicula fiuminalis	18.39	2014	Movahedi et al. (2016)
Astacus leptodactylus	136.89	2015	Ahmadi Kordestani et al. (2015)

[a]Maximum of the average accumulation levels after exposure to different levels of the metal

Table 67.5 Mean cadmium levels in different taxa of freshwater invertebrates of Iran. In cases where two or more tissues/organs were analyzed, their means are presented

Taxon	Concentration (ppm dry weight)	Year of sample collection	References
Astacus leptodactylus	3.27	2003	Hosseini et al. (2004)
Astacus leptodactylus	5.85	2006	Naghshbandi et al. (2007)
Anodonta cygnea	0.064	2006	Pourang et al. (2010)
Chironomidae	0.0204	2010	Rezaei et al. (2011)
Gammarus pulex	0.0337	2010	Rezaei et al. (2011)
Culiseta subochrea	114.7	2011	Nasirian et al. (2014)
Aedes caspius	0.03	2011	Nasirian et al. (2014)
Corbicula fiuminalis	1.35	2014	Movahedi et al. (2016)

Table 67.6 Mean iron levels in different taxa of freshwater invertebrates of Iran. In cases where two or more tissues/organs were analyzed, their means are presented

Taxon	Concentration (ppm dry weight)	Year of sample collection	References
Astacus leptodactylus	9.66	2003	Hosseini et al. (2004)
Astacus leptodactylus	485.00	2006	Naghshbandi et al. (2007)
Culiseta subochrea	464,808	2011	Nasirian et al. (2014)
Aedes caspius	881.8	2011	Nasirian et al. (2014)

Table 67.7 Mean manganese levels in different taxa of freshwater invertebrates of Iran. In cases where two or more tissues/organs were analyzed, their means are presented

Taxon	Concentration (ppm dry weight)	Year of sample collection	References
Mytilaster lineatus	6.2	1993–1994	Pourang (1996)
Corbicula fiuminalis	4.3	1993–1994	Pourang (1996)
Tubifex tubifex,	8.4	1993–1994	Pourang (1996)
Chironomidae	3.3	1993–1994	Pourang (1996)
Astacus leptodactylus	159.60	2006	Naghshbandi et al. (2007)
Culiseta subochrea	31931.6	2011	Nasirian et al. (2014)
Aedes caspius	0.26	2011	Nasirian et al. (2014)

Table 67.8 Mean chromium levels in different taxa of freshwater invertebrates of Iran. In cases where two or more tissues/organs were analyzed, their means are presented

Taxon	Concentration (ppm dry weight)	Year of sample collection	References
Astacus leptodactylus	3.40	2003	Hosseini et al. (2004)
Culiseta subochrea	1578	2011	Nasirian et al. (2014)
Aedes caspius	107.6	2011	Nasirian et al. (2014)
Astacus leptodactylus	16.28	2015	Ahmadi Kordestani et al. (2015)

References

Aazami J, Esmaili-Sari A, Abdoli A, Sohrabi H, Van den Brink PJ (2015) Monitoring and assessment of water health quality in the Tajan River, Iran using physicochemical, fish and macroinvertebrates indices. J Environ Health Sci Eng 13(29):1–12

Ahmadi Kordestani Z, Hamidian AH, Hosseini SV, Ashrafi S (2015) Determination of heavy metals concentration in muscle tissue of crayfish (*Astacus leptodactylus*). J Nat Environ (Iranian Journal of Natural Resources) 68(3):345–351. (In Persian)

Bahnasawy M, Khidr A, Dheina N (2011) 2011. Assessment of heavy metal concentrations in water, plankton, and fish of Lake Manzala, Egypt. Turkish J Zool 35(2):271–280

Bojková J, Sroka P, Soldán T, Namin JI, Staniczek AH, Polášek M, Hrivniak Ľ, Abdoli A, Godunko RJ (2018) Initial commented checklist of Iranian mayflies, with new area records and description of *Procloeon caspicum* sp. n. (Insecta, Ephemeroptera, Baetidae). Zookeys 10 (749):87–123

Bonanno G, Giudice RL (2010) Heavy metal bioaccumulation by the organs of *Phragmites australis* (common reed) and their potential use as contamination indicators. Ecol Indic 10:639–645

Cairns J Jr, Pratt JR (1995) The relationship between ecosystem health and the delivery of ecosystem services. In: Rapport DJ, Gaudet CL, Calow P (eds) Evaluating and monitoring the health of large-scale ecosystems. Springer-Verlag, Berlin, pp 63–76

Chiarelli R, Roccheri MC (2014) Marine invertebrates as bioindicators of heavy metal pollution. Open J Metal 4:93–106

Glöer P, Pešić V (2012) The freshwater snails (Gastropoda) of Iran, with descriptions of two new genera and eight new species. ZooKeys 219:11–61

Hosseini SV, Amininasab SM, Tahergorabi R, Sari AE, Bor S (2004) Determination of heavy metals content in water, sediments and muscle of crayfish, *Astacus leptodactylus* in Southern Coasts of Caspian Sea (Abbasa River of Nour city). Proceedings of the Fourth International Iran & Russia Conference, pp 1463–1467

Javanshir A (2013) Effects of season and altitude on copper and zinc concentrations in benthos (Chironomids and Gammarids) and sediment. Int J Aquatic Biol 1(1):33–35

Jezierska B, Ługowska K, Witeska M (2009) The effects of heavy metals on embryonic development of fish (a review). Fish Physiol Biochem 35:625–640

Keikhosravi A (2013) Taxonomic and phylogenetic relationships of freshwater crabs of the genus Potamon (Crustacea: Brachyura: Potamidae) from Iran. PhD Thesis, University of Regensburg

Li L, Zheng B, Liu L (2010) Biomonitoring and bioindicators used for river ecosystems: definitions, approaches and trends. Procedia Environ Sci 2:1510–1524

Moëzzi F, Javanshir A, Eagderi S, Poorbagher H (2013a) Histopathological effects of zinc (Zn) on mantle, digestive gland and foot in freshwater mussel, *Anodonta cygnea* (Linea, 1876). Int J Aquatic Biol 1(2):61–67

Moëzzi F, Javanshir A, Eagderi S, Poorbagher H (2013b) Tissue specific accumulation and histopathological alterations of zinc and chromium and their effects on clearance rate in swan mussel, *Anodonta cygnea*. Am Eurasian J Agric Environ Sci 13(6):740–746

Movahedi H, Fattollahi M, Pir Ali E (2016) Zamani Ahmad Mahmoudi R. the bivalve *Corbicula fluminalis* (Müller, 1774): an indicator for heavy metals accumulation in the habitat of Zayandeh-Roud River. J Aquatic Ecol 6(3):33–44. (In Persian)

Naghshbandi N, Zare S, Heidari R, Razzaghzadeh S (2007) Concentration of heavy metals in different tissues of *Astacus leptodactylus* from Aras dam of Iran. Pak J Biol Sci 10 (21):3956–3959

Nasirian H, Vazirianzadeh B, Taghi Sadeghi SM, Nazmara S (2014) *Culiseta subochrea* as a bioindicator of metal contamination in Shadegan international wetland, Iran (Diptera: Culicidae). J Insect Sci 14:258. https://doi.org/10.1093/jisesa/ieu120.

Nehring RB (1976) Aquatic insects as biological monitors of heavy metal pollution. Bull Environ Contam Toxicol 15:147–154

Nehring RB, Nisson R, Minasian G (1979) Reliability of aquatic insects versus water samples as measures of aquatic lead pollution. Bull Environ Contam Toxicol 22:103–108

Pourang N (1996) Heavy metal concentrations in surficial sediments and benthic macroinvertebrates from Anzali wetland, Iran. Hydrobiologia 331:53–61

Pourang N, Richardson CA, Mortazavi MS (2010) Heavy metal concentrations in the soft tissues of swan mussel (*Anodonta cygnea*) and surficial sediments from Anzali wetland, Iran. Environ Monit Assess 163:195–213

Rahnama R, Javanshir A, Mashinchian A (2011) The effect of lead bioaccumulation on condition indices of zebra mussel (*Dreissena polymorpha*) from Anzali wetland-Caspian Sea. Turk J Fish Aquat Sci 11:561–568

Rezaei M, Kamali A, Shapoori M (2011) Investigation of heavy metals (cadmium, lead) in *Chironomidae* and *Gammarus pulex* Namrood River – Tehran Province. J Chem Health Risks 1(2):39–44

Rhmanipanah S, Belgheiszadeh H, Darabi Darestani K, Jalali TH (2015) Freshwater leeches (Clitellata: Hirudinida) of South-Western Iran: small area - high diversity. Ecol Montenegrina 4:29–32

Salahinejad A, Javanshir Khoei A, Poorbagher H, Eagderi S, Attaran A (2017) Study of cd and Pb absorption and filtration rates in *Mytilaster lineatus* of the Caspian Sea. J Fish (Iranian Journal of Natural Resources) 70(3):284–294. (In Persian)

Sharifi M, Esmaeili Nezhad M, Zereshki S (2007) Different patterns of uptake and depuration of cadmium by periphyton community and a grazer species (*Physa* sp.): a mesocosm evaluation. Freshwater. Forum 27:15–26

Sharifinia M (2015) Macroinvertebrates of the Iranian running waters: a review. Acta Limnol Bras 27(4):356–369

Walker CH, Hopkin SP, Silby RM, Peakall DB (2006) Principles of ecotoxicology, 3rd edn. CRC Press, Boca Raton

Zamanpoore M, Grabowski M, Poeckl M, Schiemer F (2011) Taxonomic review of freshwater *Gammarus* (Crustacea: Amphipoda) from Iran. Zootaxa 3140(1):14

Chapter 68
Fish Parasites of Tigris and Euphrates River Systems

Cafer Erkin Koyuncu, Harun Yilmaz, Faik Sertel Secer, Dogukan Kaya, and Ercument Genc

The Tigris and Euphrates are the eastern enrollees of the two old rivers that characterize Mesopotamia. The River Tigris flows to the south, and from the mountains of southeastern Turkey through Iraq. The River Euphrates flows through Syria–Iraq lines (basin), which is born in Turkey. Both rivers join in the Shatt al-Arab point, which flow itself into the Persian Gulf.

The Tigris River is 1900 km long, rising in the Taurus Mountains of Elazig, Turkey. The Euphrates River is rising to the surface (from the northwest part of Lake Van) and then it is running to the confluence area of the Karasu River and the Murat River. It is approximately 2900 km long and also known that the longest river of western Asia. Both rivers have numerous branches, and the entire river system drains mostly in a mountainous area.

This river system is one of the great hot points for the species diversification between the west and the east part of the Anatolia (Fig. 68.1).

The word of parasite, originating from two different Greek words; para = alongside and sitos = food. The parasite is a kind of symbiotic organism. Parasitic mode of life depends on host availability. In this context, the parasite is described as

C. E. Koyuncu (✉)
Department of Aquaculture (Programme of Fish Disease), Faculty of Fisheries, Mersin University, Mersin, Turkey
e-mail: ekoyuncu@mersin.edu.tr

H. Yilmaz · F. S. Secer · E. Genc
Department of Fisheries and Aquaculture Engineering, Faculty of Agriculture, Ankara University, Ankara, Turkey

D. Kaya
Department of Fisheries and Aquaculture Engineering, Faculty of Agriculture, Agricultural Applications and Research Centre, Tokat Gaziosmanpasa University, Taslıciftlik Yerleskesi, Tokat, Turkey
e-mail: dogukan.kaya@gop.edu.tr

L. A. Jawad (ed.), *Tigris and Euphrates Rivers: Their Environment from Headwaters to Mouth*, Aquatic Ecology Series 11,
https://doi.org/10.1007/978-3-030-57570-0_68

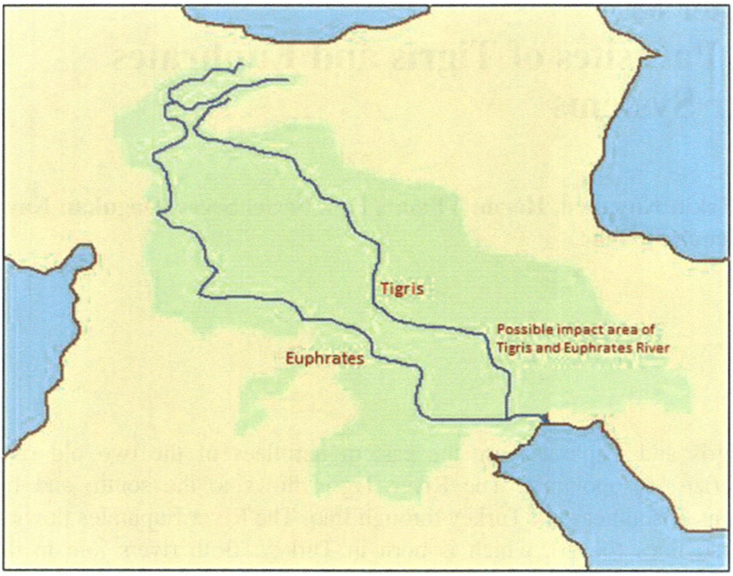

Fig. 68.1 Map of the combined Tigris–Euphrates Rivers and their drainage basin (Anonymous 2018)

temporally or compensatory lives with another organism (live in the internal or external part of the host) and sucks or assimilates by deriving nutrients. Parasites living on the outer surface, such as skin or gill of the host are called "ecto-parasites," and those living in the internal cavities, such as tissues of the host are called "endo-parasites." Parasites are developing and reproducing themselves mostly by harming the organism they hosted in.

The parasites are not only dependent on the host species but also dependent on the water quality, pH, and temperature of the environment, which makes their ecological networks both fragile and complex (Anderson 2003).

68.1 The Impact of Parasites on the Host

The host is in a mutual and close relationship with the parasite, which is on or in the host, and both the parasite and the host influence each other. Parasite–host interaction/relationship is particular and different forms. On the other hand, the host shows a reaction to these effects, and in some cases this reaction has a serious effect on the parasite.

The severity of parasites pathological effects of the host depends on many factors; the type, the amount, the morphological and physiological characteristics, the virulence of parasite, the specific sensitivity of the host which exposed to the

Table 68.1 Basic taxonomic groups of fish parasites (Arda et al. 2017)

Protozoa	Flagellata Sporozoa Ciliata
Helminthes	Monogenean Cestoda Trematoda Nematoda Acanthocephala Hirudinea (Leech)
Parasitic crustaceans	

infection, which organ/s are infected and the severity of reactions occurring in the host organism against parasites are at the head of these factors (Table 68.1).

It is possible to summarize the effects of parasites on host in five fields, these are:

(a) Peeler or exploiter effects: Parasite takes such as vitamins, minerals and salts, and deprives the host of these nutrients.
(b) Mechanical and functional effects: Such as wounds, ruptures and the so on disorders, made by parasite, in muscles and other internal organs are mechanical. These disorders are completely local lesions due to the parasite and the condition of the host. Functional effects are also caused when parasites block vital activities by plugging or piercing the stomach, intestines, and other similar organs.
(c) Toxic effects: Some parasites have toxic secretions which affects the resistance of the host and becoming susceptible to various diseases.
(d) The effects of appetite of the host: The anorexia in host, caused by parasites, brings the absence of normal nutrition and the deterioration of metabolic events.
(e) Irritative and inflammatory effects (parasites block respiration by locating in the gill plates of fish): The tissue proliferation occurs in the region where the parasite is located/attached. The gill lamellas and filaments cling to each other (lamellar fusion) and prevent breathing. In this case, the oxygen, cannot pass through the capillary veins in the filaments and the host suffers from respiratory difficulties which will result in mortality by anorexia. This is one of the most significant effects of certain parasite on host.

68.2 Recorded Parasites from Tigris and Euphrates River Systems

Fish parasitic groups recorded from Tigris and Euphrates Rivers (between 1977 and 2018) are listed and summarized in Table 68.2.

Masoumian et al. (1994) collected cyprinid species; *Arabibarbus grypus*, *Mesopotamichthys sharpeyi*, and *Carasobarbus luteus*, monthly from seven different stations of River Karun in Khuzestan Province situated in Southwest Iran. *Myxobolus* (Myxosporea, Myxozoa) species (*M. karuni* sp. n. and *M. persicus*

Table 68.2 Group of parasites and host's infection sites in Tigris and Euphrates

Group	Parasite	Host	Site of infection[a]	References
Protozoan	*Apiosoma amoeba*	*Poecilia latipinna*	N	Kadhim (2009)
	Apiosoma sp.	*Mystus pelusius*	N	Adday (2001)
		Silurus triostegus (Syn. *Parasilurus triostegus*)		
	Balantidium sp.	*Cyprinus carpio*	N	Al-Daraji et al. (1999)
		Planiliza abu (Syn. *Liza abu*)		
		Planiliza carinata (Syn. *Liza carinata*)		
	Chilodonella hexasticha	*Silurus triostegus* (*Syn. Parasilurus triostegus*)	N	Abbas (2007)
	Entamoeba sp. (Amoebozoa)	*Cyprinus carpio*	N	Al-Daraji et al. (1999)
	Ichthyobodo necator (Syn. *Costia necatrix*)	*Heteropneustes fossilis*	N	Bhatti (1979)
	Ichthyophthirius multifiliis	*Acanthobrama marmid*	G	Abdullah (2005)
		Carasobarbus luteus (Syn. *Barbus luteus)*		Al-Saadi et al. (2010)
		Mesopotamichthys sharpeyi (Syn.*Barbus sharpeyi*)		Al-Awadi et al. (2010)
		Carasobarbus luteus (Syn. *Barbus luteus)*	N	Al-Daraji (1986)
				Khamees (1983)
				Mhaisen et al. (1986)
				Mhaisen (1986)
		Carassius auratus	N	Al-Janae'e (2010)
		Chondostoma regium	G	Abdullah (2005)
		Ctenopharyngodon idella	N	Jassim (2007)
		Cyprinus carpio	G	Abdullah (2005)
				Ahmed and Ali (2013)

				Al-Daraji et al. (2002)
				Al-Janae'e (2010)
				Eassa et al. (2014)
				Jassim (2007)
		Gambusia holbrooki (misidentified as *Gambusia affinis*)	N	Kadhim (2009)
		Garra rufa	N	Mhaisen (1986)
		Heteropneustes fossilis	G	Al-Awadi et al. (2010)
		Hypophthalmichthys molitrix		Abdullah (2005)
		Leuciscus vorax (Syn. *Aspius vorax)*	N	Al-Daraji and Al-Salim (1990)
				Al-Daraji (1986)
				Mhaisen (1986)
		Planiliza abu		Al-Saadi et al. (2010)
		Mystus pelusius	G	Al-Awadi et al. (2010)
		Planiliza abu (Syn. *Liza abu*)	N	Al-Dosary (1999)
				Jori (1998)
		Poecilia latipinna	N	Kadhim (2009)
		Silurus triostegus (Syn. Parasilurus triostegus)	N	Abbas (2007)
			N	Jori (2006)
	Ichthyophthirius sp. (Ciliophora)	*Mystus pelusius*	N	Al-Dosary (1999)
		Planiliza abu (Syn. *Liza abu*)	N	Sharma (1977)
	Opalina ranarum (Ciliophora)	*Cyprinus carpio*	N	Al-Daraji et al. (1999)
		Planiliza abu (Syn. *Liza abu*)		
	Tetrahymena corlissi (Ciliophora)	*Cyprinus carpio*	N	Al-Ali et al. (2013)
			N	Al-Daraji et al. (2006)

(continued)

Table 68.2 (continued)

Group	Parasite	Host	Site of infection[a]	References
		Gambusia holbrooki (misidentified as *Gambusia affinis*)	N	Al-Ali et al. (2013)
		Mesopotamichthys sharpeyi (Syn. *Barbus sharpeyi*)	N	Al-Daraji et al. (2006)
	Trichodina domerguei (Ciliophora)	*Acanthobrama marmid*	S, G	Al-Janae'e (2010)
				Abdullah (2005)
		Acanthopagrus arabicus (Syn. *Acanthopagrus latus*)	S, G	Al-Janae'e (2010)
		Alburnus mossulensis		
		Arabibarbus grypus (Syn. *Barbus grypus*)		
		Carasobarbus luteus (Syn. *Barbus luteus*)	S, G	Al-Saadi et al. (2010)
		Mesopotamichthys sharpeyi (Syn. *Barbus sharpeyi*)	S, G	Al-Awadi et al. (2010)
		Carassius auratus	S, G	Al-Janae'e (2010)
		Ctenopharyngodon idella	S, G	Jassim (2007)
		Cyprinus carpio	S, G	Al-Saadi et al. (2010)
		Heteropneustes fossilis	S, G	Al-Awadi et al. (2010)
				Al-Salim and Mohamad (1995)
				Mohamad (1989)
		Hypophthalmichthys molitrix	S, G	Jassim (2007)
		Planiliza abu(Syn. *Liza abu*)	S, G	Al-Saadi et al. (2010)
			S, G	Al-Awadi et al. (2010)
		Luciobarbus xanthopterus (Syn. *Barbus xanthopterus*)	S, G	Al-Janae'e (2010)

		Mastacembelus mastacembelus	S, G	Adday (2001)
		Mystus pelusius	S, G	Al-Janae'e (2010)
		Planiliza abu (Syn. *Liza abu*)	S, G	Al-Janae'e (2010)
				Jori (1998)
		Planiliza subviridis (Syn. *Liza dussumieri and Liza Subviridis*)	S, G	Al-Janae'e (2010)
				Al-Salim (1992a)
		Poecilia latipinna	S, G	Al-Janae'e (2010)
		Silurus triostegus (Syn. *Parasilurus triostegus*)	S, G	Adday (2001)
				Al-Janae'e (2010)
				Jori (2006)
		Sparidentex hasta	S, G	Al-Janae'e (2010)
		Carasobarbus luteus (Syn. *Barbus luteus*)		
		Cyprinus carpio	S, G	Al-Janae'e (2010)
				Eassa et al. (2014)
				Jassim (2007)
				Mhaisen (1986)
		Gambusia holbrooki (misidentified as *Gambusia affinis*)	S, G	Kadhim (2009)
	Trichodina nigra (Ciliophora)	*Silurus triostegus* (Syn. *Parasilurus triostegus*)	S, G	Jori (2006)
	Trichodina prowazeki (Ciliophora)			
	Trichodina reticulata (Ciliophora)			
	Trichodina sp. (Ciliophora)	*Aphanius dispar*	S, G	Kadhim (2009)
		Cyprinus carpio	S, G	Ahmed and Ali (2013)
				Al-Dosary (1999)
		Planiliza abu (Syn. *Liza abu*)	S, G	Al-Dosary (1999)
		Poecilia latipinna	S, G	Kadhim (2009)

(continued)

Table 68.2 (continued)

Group	Parasite	Host	Site of infection[a]	References
		Silurus triostegus (Syn. *Parasilurus triostegus*)	S, G	Al-Dosary (1999)
				Jori (2006)
	Trichodina cottidarum (Ciliophora)	*Arabibarbus grypus* (Syn. *Barbus grypus*)	N	Al-Nasiri and Mhaisen (2009a)
		Carasobarbus luteus (Syn. *Barbus luteus*)		
		Carasobarbus xanthopterus		
		Cyprinion macrostomum		
		Cyprinus carpio		
		Planiliza abu (Syn. *Liza abu*)		
	Trichodina elegini (Ciliophora)	*Arabibarbus grypus* (Syn. *Barbus grypus*)		
		Carasobarbus luteus (Syn. *Barbus luteus*)		
		Luciobarbus xanthopterus (Syn. *Barbus xanthopterus*)		
		Cyprinion macrostomum		
		Cyprinus carpio		
		Planiliza abu (Syn. *Liza abu*)		
	Trichodina murmanica (Ciliophora)	*Arabibarbus grypus* (Syn. *Barbus grypus*)		
		Carasobarbus luteus(Syn. *Barbus luteus*)		
		Luciobarbus xanthopterus (Syn. *Barbus xanthopterus*)		
		Cyprinion macrostomum		
		Cyprinus carpio		
		Planiliza abu(Syn. *Liza abu*)		
	Trypanosoma arabica (Flagellata, Kinetoplastida)	*Silurus triostegus* (Syn. *Parasilurus triostegus*)	N	Al-Salim and Al-Daraji (1990)
				Jori (2006)

	Trypanosoma basrensis (Flagellata, Kinetoplastida)	*Leuciscus vorax* (Syn. *Aspius vorax*)	N	Al-Salim and Al-Daraji (1992)
	Trypanosoma carasobarbi (Flagellata, Kinetoplastida)	*Carasobarbus luteus* (Syn. *Barbus luteus*)	N	Al-Daraji and Al-Salim (1990)
				Al-Daraji (1986)
				Al-Salim (1985)
			N	Mhaisen et al. (1986)
	Trypanosoma mystuii (Flagellata, Kinetoplastida)	*Mystus pelusius*	N	Al-Dosary (1999)
	Trypanosoma sp. (Flagellata, Kinetoplastida)	*Planiliza abu* (Syn. *Liza abu*)	N	Al-Dosary (1999)
		Silurus triostegus (Syn. *Parasilurus triostegus*)	N	Al-Dosary (1999)
			N	Jori (2006)
	Unidentified protozoan cyst	*Coptodon zillii* (Syn. *Tilapia zillii*)	N	Al-Maliki et al. (2015)
	Haemogregarina majeedi (Sporozoa)	*Mesopotamichthys sharpeyi* (Syn. *Barbus sharpeyi*)	N	Al-Salim (1989a)
			N	Al-Salim (1993)
	Haemogregarina cyprinid (Sporozoa)	*Cyprinus carpio*	N	Al-Salim (1992b)
	Haemogregarina meridianus (Sporozoa)	*Planiliza abu* (Syn. *Liza abu*)	N	Al-Salim (1989b)
	Haemogregarina sp. (Sporozoa)	*Planiliza subviridis* (Syn. *Liza dussumieri and Liza Subviridis*)	N	Al-Salim (1992a)
	Henneguya tachysuri (Sporozoa, Myxozoa)	*Silurus triostegus* (Syn. *Parasilurus triostegus*)	N	Abbas (2007)
			N	Jori (2006)
	Myxobolus cyprinicola (Sporozoa, Myxozoa)	*Silurus triostegus* (Syn. *Parasilurus triostegus*)	N	Abbas (2007)
	Myxobolus dispar (Sporozoa, Myxozoa)	*Leuciscus vorax* (Syn. *Aspius vorax*)	K, Sp	Al-Jawda et al. (2000)
		Luciobarbus barbulus(Syn. *Barbus barbulus*)	G	
		Mesopotamichthys sharpeyi (Syn. *Barbus sharpeyi*)	G, K, Sp	

(continued)

Table 68.2 (continued)

Group	Parasite	Host	Site of infection[a]	References
		Chondostoma regium	L	
	Myxobolus diversus (Sporozoa, Myxozoa)	*Planiliza subviridis* (Syn. *Liza dussumieri and Liza Subviridis*)	N	Al-Salim (1992a)
	Myxobolus dogieli (Sporozoa, Myxozoa)	*Arabibarbus grypus* (Syn. *Barbus grypus*)	G, K, Sp	Al-Jawda et al. (2000)
	Myxobolus intrachondrealis (Sporozoa, Myxozoa)	*Cyprinus carpio*	N	Bannai et al. (2005)
	Myxobolus mesopotamiae (Sporozoa, Myxozoa)	*Silurus triostegus* (Syn. *Parasilurus triostegus*)	N	Jori (2006)
	Myxobolus oviformis (Sporozoa, Myxozoa)	*Leuciscus vorax* (Syn. *Aspius vorax*)	G, H, K, Sp	Al-Jawda et al. (2000)
		Luciobarbus barbulus(Syn. *Barbus barbulus*)	G, Gb, H, K, Sp	
		Capoeta damascina	G, K, S	Al-Jawda et al. (2000)
		Arabibarbus grypus (Syn. *Barbus grypus*)	Ab, G, H, K, L, Sp	
		Carasobarbus luteus (Syn. *Barbus luteus*)	G, K, L, S, Sp	
		Mesopotamichthys sharpeyi (Syn.Barbus sharpeyi)	G, K, L, Sp	
		Carassius carassius	N	Abdul-Rahman (1999)
		Chondostoma regium	Sp	Al-Jawda et al. (2000)
		Cyprinus carpio	N	Abdul-Rahman (1999)
		Leuciscus lepidus	G, K, Sp	Al-Jawda et al. (2000)
		Leuciscus vorax (Syn. *Aspius vorax*)	N	Abdul-Rahman (1999)
		Planiliza abu (Syn. *Liza abu*)	H, K, L, Sp	Al-Jawda et al. (2000)
		Mesopotamichthys sharpeyi (Syn. *Barbus sharpeyi*)	N	Al-Salim (1989a)

		Planiliza abu (Syn. *Liza abu*)	N	Abdul-Rahman (1999)
		Planiliza subviridis (Syn. *Liza dussumieri and Liza Subviridis*)	N	
		Varicorhinus trutta	H, K, SP	Al-Jawda et al. (2000)
		Carasobarbus luteus (Syn. *Barbus luteus*)	N	Abdul-Rahman (1999)
	Myxobolus pfeifferi (Sporozoa, Myxozoa)	*Alburnus sellal* (Syn. *Chalcalburnus sellal*)	N	
		Leuciscus vorax (Syn. *Aspius vorax*)	H, K, L, Sp	Al-Jawda et al. (2000)
		Luciobarbus barbulus(Syn. *Barbus barbulus*)	G, K	
		Arabibarbus grypus (Syn. *Barbus grypus*)	G, K, L, S, Sp	
		Capoeta damascina	S	
		Carasobarbus luteus(Syn. *Barbus luteus*)	S, G	Al-Awadi et al. (2010)
			Gb, H, K, Sp	Al-Jawda et al. (2000)
		Mesopotamichthys sharpeyi (Syn.Barbus sharpeyi)	Gb, K, L	Al-Jawda et al. (2000)
		Carassius carassius	N	Abdul-Rahman (1999)
		Ctenopharyngodon idella		
		Cyprinus carpio	N	
			N	Eassa et al. (2014)
		Heteropneustes fossilis	N	Abdul-Rahman (1999)
		Squalius lepidus	H, Sp	Al-Jawda et al. (2000)
		Leuciscus vorax (Syn. *Aspius vorax*)	N	Abdul-Rahman (1999)
			N	Al-Daraji and Al-Salim (1990)
			N	Al-Daraji (1986)
		Planiliza abu (Syn. *Liza abu*)	S, G	Al-Awadi et al. (2010)
		Mastacembelus mastacembelus	N	Abdul-Rahman (1999)
		Planiliza abu (Syn. *Liza abu*)	N	Abdul-Rahman (1999)

(continued)

Table 68.2 (continued)

Group	Parasite	Host	Site of infection[a]	References
				Al-Daraji and Al-Salim (1990)
				Al-Daraji (1986)
				Al-Janae'e (2010)
				Al-Niaeem (2002)
				Al-Niaeem (2006a)
				Al-Niaeem (2006b)
				Al-Salim (1989b)
				Jori (1998)
				Khamees (1983)
				Mhaisen et al. (1986)
		Planiliza subviridis (Syn. *Liza dussumieri and Liza Subviridis*)	N	Abdul-Rahman (1999)
				Al-Salim (1992a)
		Silurus triostegus (Syn. *Parasilurus triostegus*)	N	Abbas (2007)
				Abdul-Rahman (1999)
				Al-Daraji and Al-Salim (1990)
				Al-Daraji (1986)
				Jori (2006)
		Acanthopagrus arabicus (Syn. *Acanthopagrus latus*)	N	Al-Janae'e (2010)
		Carasobarbus luteus (Syn. *Barbus luteus*)	N	Abdul-Rahman (1999)
				Al-Daraji and Al-Salim (1990)
				Al-Daraji (1986)

				Al-Salim (1986)
				Khamees (1983)
				Mhaisen et al. (1986)
		Mesopotamichthys sharpeyi (Syn. *Barbus sharpeyi*)	N	Abdul-Rahman (1999)
				Al-Daraji and Al-Salim (1990)
				Al-Daraji (1986)
		Periophthalmus waltoni	N	Mhaisen and Al-Maliki (1996)
	Myxobolus punctatus (Sporozoa, Myxozoa)	*Cyprinus carpio*	N	Al-Daraji et al. (1999)
	Myxobolus karuni	*Carasobarbus luteus* (Syn. *Barbus luteus*)	G	Masoumian et al. (1994)
		Arabibarbus grypus (Syn. *Barbus grypus*)		
		Mesopotamichthys sharpeyi (Syn.Barbus sharpeyi)		
	Myxobolus persicus	*Carasobarbus luteus* (Syn. *Barbus luteus*)	G	
		Arabibarbus grypus (Syn. *Barbus grypus*)		
		Mesopotamichthys sharpeyi (Syn.Barbus sharpeyi)		
	Myxobolus mülleri (Sporozoa, Myxozoa)	*Cyprinus carpio*	N	Al-Nasiri and Mhaisen (2009a)
		Arabibarbus grypus (Syn. *Barbus grypus*)		
		Carasobarbus luteus (Syn. *Barbus luteus*)		
		Luciobarbus xanthopterus (Syn. *Barbus xanthopterus*)		
		Cyprinion macrostomum		
		Cyprinus carpio		
		Planiliza abu (Syn. *Liza abu*)		

(continued)

Table 68.2 (continued)

Group	Parasite	Host	Site of infection[a]	References
	Myxobolus sp. (Sporozoa, Myxozoa)	*Planiliza abu* (Syn. *Liza abu*)	N	Al-Dosary (1999)
		Silurus triostegus (Syn. *Parasilurus triostegus*)	N	Jori (2006)
	Thelohanellus sp. (Sporozoa, Myxozoa)	*Cyprinus carpio*	N	Bannai et al. (2005)
	Dermocystidium percae	*Arabibarbus grypus* (Syn. *Barbus grypus*)	N	Al-Nasiri and Mhaisen (2009a)
		Carasobarbus luteus(Syn. *Barbus luteus*)		
		Luciobarbus xanthopterus (Syn. *Barbus xanthopterus*)		
		Cyprinion macrostomum		
		Cyprinus carpio		
		Planiliza abu (Syn. *Liza abu*)		
Monogenean	*Microcotyle donavini*	*Arabibarbus grypus* (Syn. *Barbus grypus*)	N	Al-Nasiri and Mhaisen (2009a)
		Carasobarbus luteus(Syn. *Barbus luteus*)		
		Luciobarbus xanthopterus		
		Cyprinion macrostomum		
		Cyprinus carpio		
		Planiliza abu (Syn. *Liza abu*)	N	Al-Saadi et al. (2010)
	Dactylogyrus barbuli	*Luciobarbus barbulus*(Syn. *Barbus barbulus*)	G	Abdullah (2005)
	Dactylogyrus cornu	*Capoeta damascina*	G	Al-Jawda et al. (2000)
		Carasobarbus luteus(Syn. *Barbus luteus*)	G	Al-Awadi et al. (2010)
		Planiliza abu (Syn. *Liza abu*)	G	
	Dactylogyrus extensus	*Arabibarbus grypus* (Syn. *Barbus grypus*)	N	Al-Saadi et al. (2010)
		Luciobarbus xanthopterus (Syn. *Barbus xanthopterus*)		

		Cyprinus carpio	G	Abdullah (2005)
	Dactylogyrus hypophthalmichthys	*Hypophthalmichthys molitrix*	G	
	Dactylogyrus pavlovski	*Arabibarbus grypus* (Syn. *Barbus grypus)*	G	
	Dactylogyrus vastator	*Luciobarbus barbulus*(Syn. *Barbus barbulus*)	G	Al-Jawda et al. (2000)
		Luciobarbus xanthopterus (Syn. *Barbus xanthopterus*)		
		Cyprinus carpio		Abdullah (2005)
		Arabibarbus grypus (Syn. *Barbus grypus)*		Al-Nasiri and Mhaisen (2009a)
		Carasobarbus luteus(Syn. *Barbus luteus*)		
		Luciobarbus xanthopterus (Syn. *Barbus xanthopterus*)		
		Cyprinion macrostomum		
		Cyprinus carpio		
		Planiliza abu (Syn. *Liza abu*)		
	Dactylogyrus vistulae	*Squalius lepidus*	G	Abdullah (2005)
		Alburnus mossulensis		Tunç and Koyun (2018)
	Dactylogyrus holciki	*Alburnus mossulensis*	G	Tunç and Koyun (2018)
	Dactylogyrus lenkorani			
	Dactylogyrus carassobarbi	*Carasobarbus luteus*(Syn. *Barbus luteus*)	N	Al-Saadi et al. (2010)
	Dactylogyrus rohdeianus	*Carasobarbus luteus*(Syn. *Barbus luteus*)	G	Al-Saadi et al. (2009)
		Mesopotamichthys sharpeyi (Syn.Barbus sharpeyi)		Al-Saadi et al. (2010)
	Dactylogyrus skrjabini	*Arabibarbus grypus* (Syn. *Barbus grypus)*	N	Al-Saadi et al. (2010)
	Dactylogyrus tuba	*Mesopotamichthys sharpeyi* (*Syn.Barbus sharpeyi*)	N	Al-Saadi et al. (2010)
	Dactylogyrus varicorhini	*Luciobarbus xanthopterus* (Syn. *Barbus xanthopterus)*	N	Al-Saadi et al. (2010)
	Dactylogyrus alatus	*Alburnus mossulensis*	G	Tunç and Koyun (2018)

(continued)

Table 68.2 (continued)

Group	Parasite	Host	Site of infection[a]	References
	Diplozoon kasimi	*Cyprinion macrostomum*	N	Rahemo (1980)
		Carasobarbus luteus (Syn. *Barbus luteus)*	N	Al-Saadi et al. (2010)
	Diplozoon paradoxum		G	Al-Saadi et al. (2009)
			N	Al-Saadi (2007)
			N	Al-Saadi et al. (2010)
	Dogieius planus		N	
			G	Abdullah (2005)
	Diplozoon barbi	*Carasobarbus luteus* (Syn. *Barbus luteus)*	N	Al-Saadi et al. (2010)
		Cyprinus carpio		
	Eudiplozoon nipponicum	*Cyprinus carpio*	N	Al-Nasiri (2003)
		Mesopotamichthys sharpeyi (Syn.Barbus sharpeyi)	N	Al-Saadi et al. (2010)
	Gyrodactylus sp.	*Alburnus mossulensis*	N	Abdul-Ameer and Atwan (2017)
	Gyrodactylus elegans	*Capoeta damascina*	G	Al-Jawda et al. (2000)
		Carasobarbus luteus (Syn. *Barbus luteus)*	S	Al-Awadi et al. (2010)
		Cyprinus carpio	G	Abdullah (2005)
		Heteropneustes fossilis	S	Al-Awadi et al. (2010)
		Planiliza abu (Syn. *Liza abu*)	S	
	Gyrodactylus paralatus	*Cyprinus carpio*	S	Abdullah (2005)
		Hypophthalmichthys molitrix	S	
	Paradiplozoon amurensis	*Cyprinion macrostomum*	N	Al-Nasiri (2010)
	Paradiplozoon barbi	*Chondrostuma nasus*	N	Rasheed (1989)
		Chondostoma regium	N	
		Cyprinus carpio	N	

	Paradiplozoon bliccae	*Cyprinion macrostomum*	N	Al-Nasiri (2009)
	Paradiplozoon cyprini	*Arabibarbus grypus* (Syn. *Barbus grypus*)	N	Al-Nasiri and Mhaisen (2009a, b)
		Arabibarbus grypus (Syn. *Barbus grypus*)		Al-Nasiri and Mhaisen (2009a)
		Carasobarbus luteus(Syn. *Barbus luteus*)		
		Luciobarbus xanthopterus (Syn. *Barbus xanthopterus*)		
		Cyprinion macrostomum		
		Cyprinus carpio		
		Planiliza abu (Syn. *Liza abu*)		
	Paradiplozoon ergensi	*Leuciscus vorax* (Syn. *Aspius vorax*)	N	Al-Jubori and Al-Nasiri (2014)
	Paradiplozoon homoion	*Luciobarbus xanthopterus* (Syn. *Barbus xanthopterus*)	N	Al-Saadi et al. (2010)
				Al-Saadi (2007)
			G	Al-Saadi et al. (2009)
		Alburnus mossulensis	G	Tunç and Koyun (2018)
	Paradiplozoon megan	*Leuciscus vorax* (Syn. *Aspius vorax*)	N	Al-Saadi et al. (2010)
		Luciobarbus xanthopterus (Syn. *Barbus xanthopterus*)	G	Al-Saadi et al. (2009)
			N	Tunç and Koyun (2018)
	Paradiplozoon pavlovskii	*Carasobarbus luteus* (Syn. *Barbus luteus*)	N	Al-Saadi et al. (2010)
		Luciobarbus xanthopterus (Syn. *Barbus xanthopterus*)		
		Leuciscus vorax (Syn. *Aspius vorax*)	N	Khamees (1983)
	Paradiplozoon rutili	*Leuciscus vorax* (Syn. *Aspius vorax*)	N	Al-Jubori and Al-Nasiri (2014)
		Cyprinion macrostomum	N	Al-Jubori and Al-Nasiri (2014)

(continued)

Table 68.2 (continued)

Group	Parasite	Host	Site of infection[a]	References
	Paradiplozoon vojteki	*Luciobarbus xanthopterus* (Syn. *Barbus xanthopterus*)	G	Al-Saadi et al. (2009)
			N	Al-Saadi et al. (2010)
			N	Al-Saadi (2007)
	Paradiplozoon magnum	*Capoeta trutta*	N	Al-Nasiri (2017)
		Hemigrammocapoeta elegans		
Trematoda	*Ascocotyle coleostoma* (Digenea)	*Alburnus capito*	S	Al-Jawda et al. (2000)
		Leuciscus vorax (Syn. *Aspius vorax*)	G, S	
		Capoeta damascina	G, S	
		Carasobarbus luteus (Syn. *Barbus luteus*)	S	
		Mesopotamichthys sharpeyi (Syn.*Barbus sharpeyi*)	S	
		Chondostoma regium	G, S	
		Heteropneustes fossilis	S	Al-Awadi et al. (2010)
		Squalius lepidus	S	Al-Jawda et al. (2000)
		Planiliza abu (Syn. *Liza abu*)	G, S	
	Clinostomum complanatum (Digenea)	*Aphanius dispar*	S	Al-Awadi et al. (2010)
		Carasobarbus luteus (Syn. *Barbus luteus*)	S	
		Gambusia affinis	S	
		Heteropneustes fossilis	S	
		Planiliza abu(Syn. *Liza abu*)	S	
	Clinostomum complanatum (metacercariae)	*Carasobarbus luteus* (Syn. *Barbus luteus*)	S	Al-Saadi et al. (2010)
		Planiliza abu (Syn. *Liza abu*)		
	Diplostomum spathaceum	*Alburnus mossulensis*	N	Tunç and Koyun (2018)
	Diplostomum sp. (Digenea)	*Capoeta damascina*	E	Al-Jawda et al. (2000)
		Carasobarbus luteus (Syn. *Barbus luteus*)	E	

		Chondostoma regium	E	
	Pseudochetosoma salmonicola (Digenea)	*Leuciscus vorax* (Syn. *Aspius vorax*)	Gb	
		Luciobarbus barbulus (Syn. *Barbus barbulus*)	Gb	
		Arabibarbus grypus (Syn. *Barbus grypus*)	Gb	
	Allocreadium isoporum (Digenea)	*Oxynoemacheilus tigris*	N	Koyun et al. (2016)
Nematoda	*Contracaecum* sp.	*Acanthobrama marmid*	I	Abdullah (2005)
		Aphanius dispar	K, L, M	Al-Awadi et al. (2010)
		Leuciscus vorax (Syn. *Aspius vorax*)	L	Al-Jawda et al. (2000)
		Luciobarbus barbulus (Syn. *Barbus barbulus*)	L	
		Luciobarbus esocinus (*Syn.Barbus esocinus*)	I	Abdullah (2005)
		Arabibarbus grypus (Syn. *Barbus grypus*)	K, L, M	Al-Awadi et al. (2010)
		Chondostoma regium	I	Abdullah (2005)
		Gambusia affinis	BC	Al-Awadi et al. (2010)
		Heteropneustes fossilis	I	Abdullah (2005)
		Leuciscus cephalus	I	Abdullah (2005)
		Planiliza abu (Syn. *Liza abu*)	I, K, L, M, Sp	Al-Awadi et al. (2010)
		Mastacembelus mastacembelus	I	Abdullah (2005)
	Spyroxis sp.	*Capoeta damascina*	H	Al-Jawda et al. (2000)
Cestoda	*Bothriocephalus acheilognathi*	*Cyprinus carpio*	I	Abdullah (2005)
	Khawia sinensis	*Arabibarbus grypus* (Syn. *Barbus grypus*)		
	Paracaryophyllaeus gotoi	*Carasobarbus luteus* (Syn. *Barbus luteus*)		
	Caryophyllaeus laticeps	*Alburnus mossulensis*	I	Tunç and Koyun (2018)
	Ligula intestinalis			
	Proteocephalus osculatus	*Leuciscus vorax* (Syn. *Aspius vorax*)		Al-Jawda et al. (2000)
Acanthocephala	*Neoechinorhynchus iraqensis*	*Barbus xanthopterus*	I	Al-Awadi et al. (2010)
		Planiliza abu(Syn. *Liza abu*)	I	Al-Jawda et al. (2000)

(continued)

Table 68.2 (continued)

Group	Parasite	Host	Site of infection[a]	References
	Nyctotheroides cordiformis (Syn. *Nyctotherus cordiformis*)	*Cyprinus carpio*	N	Al-Daraji et al. (1999)
		Planiliza abu (Syn. *Liza abu*)		
		Planiliza carinata (Syn. *Liza carinata*)		
	Paulisentus fractus	*Luciobarbus barbulus*(Syn. *Barbus barbulus*)	I	Al-Jawda et al. (2000)
Hirudinea	*Piscicola geometra*	*Arabibarbus grypus* (Syn. *Barbus grypus*)	S	Al-Jawda et al. (2000)
		Capoeta capoeta		Koyun et al. (2015)
		Alburnus mossulensis		
		Cyprinion macrostomum		
		Alburnus mossulensis		Tunç and Koyun (2018)
Crustacea	*Argulus foliaceus* (Maxillopoda, Branchiura)	*Cyprinus carpio*	S	Al-Jawda et al. (2000)
	Dermoergasilus varicoleus	*Carasobarbus luteus* (Syn. *Barbus luteus*)	N	Al-Saadi et al. (2010)
		Planiliza abu (Syn. *Liza abu*)		
	Ergasilus barbi	*Leuciscus vorax* (Syn. *Aspius vorax*)		
		Carasobarbus luteus (Syn. *Barbus luteus*)		
		Cyprinus carpio		
		Planiliza abu (Syn. *Liza abu*)		
	Ergasilus mosulensis	*Leuciscus vorax* (Syn. *Aspius vorax*)		
		Carasobarbus luteus (Syn. *Barbus luteus*)		
		Planiliza abu (Syn. *Liza abu*)		
	Ergasilus peregrinus	*Leuciscus vorax* (Syn. *Aspius vorax*)		
		Carasobarbus luteus (Syn. *Barbus luteus*)		
		Mesopotamichthys sharpeyi (Syn.Barbus sharpeyi)		
	Ergasilus rostralis	*Leuciscus vorax* (Syn. *Aspius vorax*)		
		Arabibarbus grypus (Syn. *Barbus grypus*)		

		Carasobarbus luteus (Syn. *Barbus luteus*)		
		Mesopotamichthys sharpeyi (Syn.Barbus sharpeyi)		
	Ergasilus sieboldi (Maxillopoda, Cyclopoida)	*Leuciscus vorax* (Syn. *Aspius vorax*)	N	Al-Saadi et al. (2010)
		Arabibarbus grypus (Syn. *Barbus grypus*)		
		Mesopotamichthys sharpeyi (Syn.Barbus sharpeyi)		
		Cyprinus carpio		
		Planiliza abu(Syn. *Liza abu*)		
		Carasobarbus luteus(Syn. *Barbus luteus*)	G	Al-Jawda et al. (2000)
	Lamproglena pulchella (Maxillopoda, Cyclopoida)	*Carasobarbus luteus* (Syn. *Barbus luteus)*		Abdullah (2005)
		Leuciscus vorax (Syn. *Aspius vorax*)	N	Al-Saadi et al. (2010)
	Lernaea cyprinacea (Maxillopoda, Cyclopoida)	*Capoeta damascina*	S	Al-Jawda et al. (2000)
		Cyprinus carpio	G, S, F	Abdullah (2005)
	Pseudolamproglena annulata (Maxillopoda, Cyclopoida)	*Carasobarbus luteus* (Syn. *Barbus luteus*)	G	Al-Jawda et al. (2000)
	Tracheliastes domergueri (Maxillopoda, Siphonostomatoida)	*Carasobarbus luteus* (Syn. *Barbus luteus*)	G, S	Abdullah (2005)
		Chondostoma regium	G, S	
		Hypophthalmichthys molitrix	G, S	
		Squalius cephalus	G, S	
		Squalius spurius	G, S	
	Tracheliastes polycolpus (Maxillopoda, Siphonostomatoida)	*Capoeta damascina*	F, S	Al-Jawda et al. (2000)
		Luciobarbus kersin (Syn.*Barbus kersin*)	F	Abdullah (2005)
	Copepoda: *Ergasilus* sp.	*Alburnus mossulensis*	N	Tunç and Koyun (2018)

[a]Site of infection: Ab, air bladder; BC, body cavity; B, blood; E, eye; F, fin; G, gills; Gb, gall bladder; H, hearth; I, intestine; K, kidney; L, liver; M, mesenteries; N, not specified from the author; S, skin; Sp, spleen

sp. n.) were identified from the gills of host fish of *Arabibarbus grypus*. A total of 75% of *Arabibarbus grypus*, 16% of *Mesopotamichthys sharpeyi* and 14% of *Carassobarbus luteus* were infected by *Myxobolus* spp. cysts on the gills.

Al-Jawda et al. (2000) collected 12 cyprinids and 1 mugilid fish species from the Tigris River at Salah Al-Deen, Iraq. The authors identified a total of 21 parasite species, including 4 sporozoans (*Myxobolus dispar*, *M. dogieli*, *M. oviformis*, *M. pfeifferi*), 3 monogenic trematodes (*Dactylogyrus cornu*, *D. vastator*, *Gyrodactylus elegans*), 3 trematodas (*Ascocotyle coleostoma*, *Diplostomum* sp., *Pseudochetosoma salmonicola*), 1 cestoda (*Proteocephalus osculatus*), 2 nematodas (*Contracaecum* sp., *Spiroxys* sp.), 2 acanthocephalans (*Neoechinorhynchus iraqensis*, *Paulisentus fractus*), 1 leech (*Piscicola geometra*), and 5 crustaceans (*Argulus foliaceus*, *Ergasilus sieboldi*, *Lernaea cyprinacea Pseudolamproglena annulata*, *Tracheliastes polycolpus*).

Abdullah (2005) collected 313 freshwater fish specimens, belonging to 17 species, from Darbandikhan lake, Sulaimaniya province, Iraq. Fish were surveyed for ecto- and endo-parasites. His study revealed the presence of two species of protozoa (*Trichodina domerguei*, *Ichthyophthirius multifiliis*), nine monogenetic trematodas (*Dactylogyrus barbuli*, *D. extensus*, *D. hypophthalmichthys*, *D. pavlovskyi*, *D. vastator*, *D. vistulae*, *Gyrodactylus elegans*, *G. paralatus*, *Dogielius planus*), three adult cestoda (*Bothriocephalus acheilognathi*, *Paracaryophyllaeus gotoi*, *Khawia sinensis*), one nematoda larvae (*Contracaecum* spp.), and four crustaceans (*Lamproglena pulchella*, *Pseudolamproglena annulata*, *Tracheliastes polycolpus*, *Lernaea cyprinacea*) in different sites of the host bodies.

Al-Saadi et al. (2009) investigated 2615 fish collected from Al-Husainia creek, Karbala province, Iraq. Fish were infected with 33 species of ecto- and endo-parasites. Among the ecto-parasites monogenetic trematodas, *Dactylogyrus rohdeianus*, *Diplozoon paradoxum*, *Paradiplozoon homoion*, *Paradiplozoon megan*, and *Paradiplozoon vojteki* were recorded for the first time in Iraq. *Dactylogyrus rohdeianus* identified from gill filaments of *Carassobarbus luteus* and *Mesopotamichthys sharpeyi* with a prevalence of 0.3 and 0.4%, respectively. *Diplozoon paradoxum* was recorded from the gill tissues of *Carassobarbus luteus* with a prevalence of 0.3%. *Paradiplozoon homoion* was recorded from gills of *Barbus xanthopterus* with a prevalence of 0.6%. *Paradiplozoon megan* was recorded on gills of both *Aspius vorax* and *Luciobarbus xanthopterus* with a prevalence of 0.5% and 0.3%, respectively. *Paradiplozoon vojteki* was recorded in the present study on gills of *Barbus xanthopterus* with a prevalence of 0.6%.

Al-Awadi et al. (2010) inspected 6992 fish for external and internal parasites, belonging to 11 species, in Bahr Al-Najaf depression, mid Iraq. Fish were infected with three protozoans (*Ichthyophthirius multifiliis, Myxobolus pfeifferi*, and *Trichodina domerguei*), two monogeneans (*Dactylogyrus cornu* and *Gyrodactylus elegans*), two digenetic trematodas (*Ascocotyle coleostoma* and *Clinostomum complanatum*), one nematoda (*Contracaecum* sp.), and one acanthocephalan (*Neoechinorhynchus iraqensis*).

Al-Saadi et al. (2010) investigated a total of 2615 fish specimens belonging to seven fish species (*Aspius vorax*, *Arabibarbus grypus*, *Mesopotamichthys sharpeyi*,

Carasobarbus luteus, *Luciobarbus xanthopterus*, *Cyprinus carpio*, and *Planiliza abu*) of two families (Cyprinidae and Mugilidae) were collected from Al-Husainia creek, Karbala province, Iraq. Fish were infected with 26 ecto-parasitic species, which included two ciliated protozoans (*Ichthyophthirius multifiliis* and *Trichodina domerguei*), sixteen monogeneans (*Dactylogyrus carassobarbi*, *D. extensus*, *D. rohdeianus*, *D. skrjabini*, *D. tuba*, *D. varicorhini*, *Dogielius planus*, *Diplozoon barbi*, *D. kasimii*, *D. paradoxum*, *Eudiplozoon nipponicum*, *Paradiplozoon homoion*, *P. megan*, *P. pavlovskii*, *P. vojteki*, and *Microcotyle donavini*), one digenetic trematode metacercariae (*Clinostomum complanatum*), and seven crustaceans (*Dermoergasilus varicoleus*, *Ergasilus barbi*, *E. mosulensis*, *E. peregrinus*, *E. rostralis*, *E. sieboldi*, and *Lamproglena pulchella*).

Mansoor and Al-Shaikh (2010) collected 255 cyprinid fish *Cyprinus carpio* (Linnaeus, 1758) from fish markets east of Baghdad city, Iraq. Fish were examined with a microscope and found that the fish were infected with five species of ciliated protozoans. The percentage of infestation were; 1.6% *Chilodonella cyprini*, 3.5% *Ichthyophthirius multifiliis*, 21.6% *Trichodina domerguei*, 0.4% *Apiosoma piscicola*, and 1.2% *Epistylis solidus*.

Al-Sa'adi et al. (2013) studied a total of 471 fish specimens belonging to 24 species from the Euphrates River at Al-Musaib city, Iraq. Among these fishes, 65 monogenean parasites were recorded from gills. Authors categorized infected fish species and the percentage of incidence for monogenean. *Dactylogyrus dogieli* were reported from *Carassobarbus luteus* (5.2%), *Mesopotamichthys sharpeyi* (50%), *Alburnus sellal* (4.3%), *Cyprinion kais* (1.7%), and *Ctenopharyngodon idella* (50%). And the other monogenean species of *Octomacrum europaeum* were collected from *Cyprinion kais* (6.7%), *Cyprinion macrostomum* (1.7%) and *Garra rufa* (9.1%), respectively.

Al-Jubori and Al-Nasiri (2014) collected a total of 92 specimens of cyprinid fishes (15 *Aspius vorax*, 77 *Cyprinion macrostomum*) from the Tigris River, Tikrit city, Salah Al-Deen province, Iraq. They declared five specimens of *Paradiplozoon ergensi*, one specimen of *Paradiplozoon rutile*, and two specimens of *Paradiplozoon rutili* as monogenean parasites for sampled host fishes.

Koyun et al. (2015) identified the parasitic infections of *Piscicola geometra* (Hirudinea), on three different cyprinid species (*Capoeta capoeta*, *Alburnus mossulensis*, and *Cyprinion macrostomum*) in two reservoirs (May–November 2013 Dumlu creek and September 2014–April 2015 Goynuk stream) of the Euphrates–Tigris basin, Turkey. They recorded fourteen hirudin specimens from 7 of the 94 examined fishes.

Koyun et al. (2016) determined the seasonal prevalence, and intensity of *Allocreadium isoporum* (Digenea: Allocreadiidae), it was founded as an endoparasite in Tigris loach *Oxynoemacheilus tigris* (Cypriniformes, Nemacheilidae). They declared from December 2012 to November 2013, 136 fish samples (209 parasites) obtained from Murat River (Euphrates–Tigris basin), Bingol, Turkey. Additionally, the seasonal parasite prevalence, number of collected parasites and the mean intensity results were recorded by the authors (prevalence: Winter: 35.3%, Spring: 50%, Summer: 69.4%, Autumn: 52.9%, number of parasites: Winter: 21, Spring:

59, Summer: 84, Autumn: 45, and intensity: Winter: 1.8, Spring: 3.7, Summer: 3.4, Autumn: 2.5).

Al-Nasiri and Mhaisen (2009a) sampled 109 fish specimens (*Arabibarbus grypus*, *Carasobarbus luteus*, *Luciobarbus xanthopterus*, *Cyprinion macrostomum*, *Cyprinus carpio*, and *Planiliza abu*) from the Tigris River passing through Salah Al-Deen province, Iraq during September 2008–January 2009 period. They identified five different protozoans including in 2 sporozoans: *Myxobolus mülleri*, *Dermocystidium percae* and 3 ciliophorans: *Trichodina cottidarum*, *T. elegini*, and *T. murmanica* and 4 monogenean trematodes which were *Dactylogyrus rhodeianus*, *D. vastator*, *Paradiplozoon cyprini*, and *Microcotyle donavini* from the host fishes.

Al-Nasiri (2017) collected 87 cyprinid specimens (June 2013–April 2014) (13 *Capoeta trutta*, 41 *Carasobarbus luteus*, 17 *Cyprinion macrostomum*, 7 *Cyprinus carpio*, and 9 *Hemigrammocapoeta elegans*) from the Tigris River passing through Tikrit City, Salah Al-Deen Province, Iraq. The author reported one diplozoid parasite *Paradiplozoon magnum* (Monogenean trematoda) from the gills of two infected fish species/hosts. The prevalence and mean intensity values of *Paradiplozoon magnum* for *Capoeta trutta* 15.38%, 1.5 and for *Hemigrammocapoeta elegans* 44.44%, 2.25, respectively.

Tunç and Koyun (2018) studied on the frequency of metazoan parasitic infections of Mosul bleak (*Alburnus mossulensis*) in Murat River (January 2015 to March 2016) in Eastern Anatolia (Euphrates–Tigris basin), Turkey. They isolated 972 metazoan parasites from 182 fish samples. At the end of the investigation, authors reported parasites and the prevalence as Monogenea: *Dactylogyrus holciki* (12.1%), *Dactylogyrus lenkorani* (4.9%), *Dactylogyrus alatus* (6.6%), *Dactylogyrus vistulae* (1.1%), *Gyrodactylus* sp. (3.8%), and *Paradiplozoon homoion* (19.8%); Digenea: *Diplostomum spathaceum* (33%), Cestoda: *Caryophyllaeus laticeps* (12.1%), *Ligula intestinalis* (0.5%); Nematoda: *Rhabdochona denudata* (4.9%); Hirudinea: *Piscicola geometra* (1.6%); and Copepoda: *Ergasilus* sp. (34.6%), respectively.

68.3 Major Groups of Parasites Detected from Tigris and Euphrates River Systems

In this review, the literature on fish parasites in the Tigris and Euphrates River systems has been evaluated (published between 1997 and 2018). In the literature, it has been found that the two rivers are generally assessed together. Furthermore, it has been understood that the work areas in the present literature are generally referred to as "Tigris-Euphrates basin." In this context, data recorded for "Tigris-Euphrates basin" in the literature are 45 protozoans, 35 monogeneans, 7 digenean, 2 nematoda, 6 cestoda, 3 achantocephalan, 1 hirudin, and 13 crustacean parasites. It was also determined that these parasites were reported as 42.86% protozoan, 33.33% monogenean, 6.67% digenean, 1.90% nematoda, 5.71% cestoda, 2.86%

achantocephalan, 0.95% hirudin, and 12.38% crustacean. It is understood that the protozoan group is a common agent group for fish (as a causative parasitic infection agent) from the presented references.

In the southern part of the region (especially Iraq and Iran) is made numerous scientific studies, whereas in the northern part (Turkey) shows that a limited number of studies reached. In this context, it is considered necessary to conduct more extensive, quantitative, and qualitative (based on species-specific/related gene region—DNA based/barcoding—molecularly confirmed identifications) research from the northern parts of the Tigris–Euphrates basin (Turkey) in the future.

68.4 Conclusion

Kotob et al. (2017) claimed that parasites usually exist in a unique equilibrium with their hosts. Environmental damages may trigger outbreaks of parasitic disease. Parasites can cause physical and physiological pathologies such as proliferation and fusion in lamellar tissue of gills, cell proliferation, immune-depression, loss of weight, loss of appetite, and also behavioral and reproductive changes/anomalies (Buchmann and Lindenstrøm 2002; Knudsen et al. 2008; Al-Jahdali and Hassanine 2010; Iwanowicz 2011).

This chapter summarized with a list of fish parasites from Tigris and Euphrates River systems and consists of table group by group, parasite species were listed with their hosts, and the original reference. In the wild populations; isolation, identification, and quantifying of the fish parasite is not easy. In the river systems of Tigris and Euphrates, more than 114 parasite species were reported in approximately 52 host fish.

In conclusion, it is understood that protozoan parasites constitute a significant part of the listed parasites obtained from these rivers system. As it is known, the parasites will be the agents or the vectors of other diseases. Some parasites that presented in this list has known as zoonotic characteristics (Nematoda: *Contracoceum* sp. and *Spyroxis* sp.). For this reason, as a precaution, public awareness should be created in terms of environment, animal, and human health subject. As far as there were no notified cases of human infections by fish nematodes in this region.

Parasites and parasitic infections of fish in river systems are important for the identification and understanding of the pathways of host–parasite relationships. These assessments are important in terms of future ecological studies and protection measures to be applied for the sustainability of species.

References

Abbas AAK (2007) Histopathological studies of some parasites of the Asian catfish, *Silurus triostegus* (Heckel, 1843) and potassium permanganate on the black molly, *Poecilia sphenops* (Valenciennes, 1846). Ph.D. thesis, College of Education, University of Basrah, 144pp

Abdul-Ameer KN, Atwan FK (2017) First record of four species of the genus Gyrodactylus Nordmann, 1832 (Monogenea: Gyrodactylidae) from some Iraqi freshwater fishes. In: Journal of Kerbala University. Third scientific conference of the Faculty of Veterinary Medicine, pp 289–297

Abdullah SM (2005) Parasitic fauna of some freshwater fishes from Darbandikhan Lake, north of Iraq. Skin 18(6.8):9–1

Abdul-Rahman NM (1999) Parasites infection in fish from Garmat Ali River and its relation with food items. M. Sc. Thesis, College of Agriculture, University of Basrah, 103pp

Adday TK (2001) Biology of the crustacean *Ergasilus ogawai* Kabata, 1992 which parasitized on some Iraqi fishes. M.Sc. thesis, College of Agriculture, University of Basrah, 117pp

Ahmed SM, Ali AH (2013) Serum proteins and leucocytes differential count in common carp (*Cyprinus carpio* L.) infested with ectoparasites. Mesopotamian J Mar Sci 28(2):151–162

Al-Ali MFM, Atte RS, Banay MA (2013) Activity of the pesticide sumcidin on the copepods *Lernaea cyprinacea* and *Tetrahymena corlissi* infecting of *Cyprinus carpio* and *Gambusia affinis* fishes. Diyala Agric Sci J 5(1):24–31

Al-Awadi HM, Mhaisen FT, Al-Joborae FF (2010) Parasitic fauna of fishes in Bahr Al-Najaf depression, mid Iraq. Bull Iraq Natl Hist Mus 11:1–9

Al-Daraji SAM (1986) Survey of parasites from five species of fishes found in Al-Hammar Marsh. M.Sc. thesis, College of Agriculture, University of Basrah, 130pp

Al-Daraji SAM, Al-Salim NK (1990) Parasitic fauna of five species of fishes from Al-Hammar Marsh, Iraq. I: Protozoa and Monogenea. Marina Mesopotamica 5(2):275–282

Al-Daraji SAM, Salim YAK, Jori MM, Nasir AM (1999) Endoparasites of some cultured fishes from Basrah. Basrah J Sci 17(1):81–86

Al-Daraji SAM, Jori MM, Salim YAK (2002) Chemotherapeutic control of some ectoparasitic diseases infesting some Iraqi fishes. Basrah J Sci 20(1):53–62

Al-Daraji SAM, Bannai MAA, Jarallah HM, Bannai JAA (2006) The effect of aqueous extract of fruit peel of *Punica granatum* on a newly recorded ciliate parasite *Tetrahymena corlissi* (Hoffman and Glenn, 1978). Mesopotamian J Mar Sci 21(2):155–161

Al-Dosary SH (1999) A study on some protozoan parasites of six species of freshwater fishes of Qarmat Ali River, Basrah. M.Sc. thesis, Univ Basrah, 61pp

Al-Jahdali MO, Hassanine RMES (2010) Ovarian abnormality in a pathological case caused by *Myxidium* sp. (Myxozoa, Myxosporea) in onespot snapper fish *Lutjanus monostigma* (Teleostei, Lutjanidae) from the Red Sea. Acta Parasitol 55:1–7

Al-Janae'e AMS (2010) Parasites of some Iraqi fishes in two localities varied in their trophic levels in inland water of Basrah. M.Sc. thesis, College of Agriculture, University of Basrah, 228pp

Al-Jawda JM, Balasem AN, Mhaisen FT, Al-Khateeb GH (2000) Parasitic fauna of fishes from Tigris river at Salah Al-Deen province, Iraq. Iraqi J Biol Sci 19:16–24

Al-Jubori MI, Al-Nasiri FS (2014) First record of two *Paradiplozoon* (Monogenea) from cyprinid fishes in Iraq. Jordan J Agric Sci 10:673–680

Al-Maliki GM, Al-Khafaji KK, Al-Shemary AJ (2015) Incidence of parasites in *Tilapia zillii* from Tigris at north of Qurna with some environmental parameters of the river. J Basrah Res Sci 41(2) A:86–92

Al-Nasiri FS (2003) First occurrence of the monogenetic trematoda *Diplozoon nipponicum* Goto, 1891 in Iraq from common carp *Cyprinus carpio* (Pisces). Iraqi J Agric (Spec Issue) 8(6):95–99

Al-Nasiri FS (2009) Diplozoid species (Monogenea) parasitizing gills of some cyprinid fishes from *Tigris river* passing through Tikreet city, Salah Al-Deen province. Iraqi J Agric 14(5):182–186

Al-Nasiri FS (2010) First record of *Paradiplozoon amurensis* (Monogenea: Diplozoidae) in Iraq from gills of the cyprinid fish *Cyprinion macrostomum*. Parassitologia 52:439–440

Al-Nasiri FS (2017) *Paradiplozoon magnum* (Monogenea: Diplozoidae) from some species of cyprinid fish in Tigris River, Iraq. J Tikrit Univ Agric Sci Proc Sixth Sci Conf Agric Sci 17:129–133

Al-Nasiri FS, Mhaisen FT (2009a) Parasites of fishes collected from Tigris River, Salah Al-Deen Province, Iraq. Ibn AL-Haitham J Pure Appl Sci 22(2)

Al-Nasiri FS, Mhaisen FT (2009b) First record of Paradiplozoon cyprini Khotenovsky, 1982 (Monogenea: Diplozoidae) in Iraq, from gills of the cyprinid fish Barbus grypus. J Tikreet Univ Agric Sci 9(1):535–540

Al-Niaeem KSK (2002) Biochemical changes in the ovary of *Liza abu* due to infection with *Myxobolus pfeifferi* (Cnidosporidia: Myxosporida) from Basrah-Iraq. 1- Cholesterol and gonado somatic index. Basrah J Vet Res 2(1):10–16

Al-Niaeem KSK (2006a) Biochemical changes in the ovary of *Liza abu* due to infection with *Myxobolus pfeifferi* (Cnidosporidia: Myxosporida) from Basrah, Iraq. 2- Moisture, total protein, fat and glycogen. Basrah J Vet Res 5(2):1–5

Al-Niaeem KSK (2006b) Infection distribution of fish parasites in Basrah Province and pathological effects of *Saprolegnia* sp. and its susceptibility to some plant extracts. Ph.D. thesis, College of Agriculture, University of Basrah, 172pp

Al-Saadi AAJJ (2007) Ecology and taxonomy of parasites of some fishes and biology of *Liza abu* from Al-Husainia creek in Karbala province, Iraq. Ph.D. thesis, Coll. Educ. (Ibn Al-Haitham), Univ. Baghdad, 155pp

Al-Saadi AA, Mhaisen FT, Hasan HR (2009) Description of five monogenetic trematodes for the first time from fishes of Iraq. Iraqi J Agric (Spec Issue) 14(1):187–193

Al-Saadi AAJJ, Mhaisen FT, Hasan HR (2010) Endoparasites of seven fish species from Al-Husainia creek, Karbala province, mid Iraq. Basrah J Agric Sci 23:139–146

Al-Sa'adi BA, Mhaisen FT, Al-Rubaie ARL (2013) The first record of two monogeneans: *Dactylogyrus dogieli* Gussev, 1953 and *Octomacrum europaeum* Roman & Bykhovskii, 1956 from fishes of Iraq. Basrah J Agric Sci 26(1):1–7

Al-Salim NK (1985) *Trypanosoma carasobarbi* sp. n. from a freshwater fish, *Carasobarbus luteus* (Heckel, 1843) (family Cyprinidae) from Shatt Al-Arab River, Basrah, Iraq. J Biol Sci Res 16 (2):205–215

Al-Salim NK (1986) *Myxobolus pfeifferi* Thélohan, 1894 a new record from *Carasobarbus luteus* (Heckel, 1843) (family Cyprinidae) from Shatt Al-Arab River, Basrah, Iraq. Dirasat 13 (7):163–166

Al-Salim NK (1989a) Pathology of the freshwater fish *Barbus sharpeyi* infested by four diseases including *Haemogregarina majeedi* n. sp. In: Second scientific conference, fisheries and marine resources, Basrah, 16–17 May 1989

Al-Salim NK (1989b) The occurrence of *Myxobolus pfeifferi* Thelohan, and Haemogregarina meridianus n. sp. in freshwater fish *Liza abu* Heckel (family Mugilidae) from Shatt Al-Arab River, Basrah, Iraq. Bollettino del Museo Regionale di Scienze Naturali, Torino 7(2):287–295

Al-Salim NK (1992a) First record of four blood diseases in Liza subviridis. In: Third scientific conference, fisheries and marine resources, Basrah, 14–15 Apr 1992

Al-Salim NK (1992b) First record and pathology of *Cyprinus carpio* L. infected by *Haemogregarina cyprini* in Iraq. Marina Mesopotamica 7(2):195–201

Al-Salim NK (1993) Haemogregarina majeedi n. sp. from the freshwater fish *Barbus sharpeyi* Günther (Cyprinidae). Zool Middle East 9:97–101

Al-Salim NK, Al-Daraji SAM (1990) *Trypanosoma arabica* sp. n. from the freshwater fish *Silurus triostegus* Heckel in Iraq. Acta Ichthyologica et Piscatoria 20(2):85–89

Al-Salim NK, Al-Daraji SAM (1992) *Trypanosoma basrensis* n. sp. from the freshwater fish, *Aspius vorax* Heckel, 1843 in Iraq. Basrah J Agric Sci 5(2):151–155

Al-Salim NK, Mohamad ET (1995) Studies on some parasites of the stinging catfish (*Heteropneustis fossilis*) from Al-Hammar Marsh- Basrah. A) Infection of *Heteropneustes fossilis* by four parasites and their effects on the fish infected. Marina Mesopotamica 10 (1):155–170

Anderson DP (2003) Text book of fish immunology. Narendra, New Delhi
Anonymous (2018) Tigris. https://en.wikipedia.org/wiki/Tigris
Arda M, Seçer S, Sarıeyyüpoğlu M (2017) Fish disease, 3rd edn. Medisan, Ankara
Bannai MAA, Al-Daraji SAM, Jarallah HM, Hussian NA (2005) Three new parasites recorded from the common carp, *Cyprinus carpio* L., 1758, captured from southern part of Al-Hammar marshes. In: The first scientific conference to rehabilitation of marshes of South Iraq, Basrah, 11–12 Apr 2005
Bhatti MN (1979) A note on the occurrence of costiasis disease in the stinging catfish, *Heteropneustes fossilis* (Bloch) from Basrah waters. Arab Gulf 11(1):216
Buchmann K, Lindenstrøm T (2002) Interactions between monogenean parasites and their fish hosts. Int J Parasitol 32:309–319
Eassa AM, Al-Jenaei AM, Abdul-Nabi ZA, Abood MA, Kzaal RS, Aliwy YJ (2014) Comparative ecological study of pathogens structure between wild and cultured common carp *Cyprinus carpio* L. in Basrah. Marsh Bull 9(2):107–123
Iwanowicz DD (2011) Overview on the effects of parasites on fish health. In: Proceedings of the third bilateral conference between Russia and the United States, 12–20 July 2009, held in Shepherdstown, West Virginia, pp 176–184
Jassim AAR (2007) Parasites of cultured fishes for three localities in Basrah, Iraq. M. Sc. thesis, College of Agriculture, University of Basrah, 95pp
Jori MM (1998) Study of the parasites of two mugilid fish species and the effect of some on the blood parameters. M.Sc. thesis, College of Agriculture, University of Basrah, 136pp
Jori MM (2006) Parasitic study on the Asian catfish *Silurus triostegus* (Heckel, 1843) from Al-Hammar marshes, Basrah, Iraq. Ph.D. thesis, College of Agriculture, University of Basrah, 192 pp
Kadhim AH (2009) Disease agents of four fish species in Basrah Province. M.Sc. Thesis, College of Agriculture, University of Basrah, 102pp
Khamees NR (1983) A study of the parasites of *Carasobarbus luteus* (Heckel), Liza abu (Heckel) and *Aspius vorax* Heckel from Mehaijeran Canal, south of Basrah. M. Sc. thesis, College of Agriculture, University of Basrah, 148pp
Knudsen R, Amundsen PA, Jobling M, Klemetsen A (2008) Differences in pyloric caeca morphology between Arctic charr *Salvelinus alpinus* ecotypes: adaptation to trophic specialization or parasite induced phenotypic modifications. J Fish Biol 73:275–287
Kotob MH, Menanteau-Ledouble S, Kumar G, Abdelzaher M, El-Matbouli M (2017) The impact of co-infections on fish: a review. Vet Res 47(1):98
Koyun M, Tepe Y, Mart A (2015) First record of *Piscicola geometra* (Annelida, Hirudinea) on some species of Cyprinidae from Euphrates-Tigris Basin in Turkey. J Fish Aquat Sci 10 (6):575–580
Koyun M, Ulupınar M, Mart A, Tepe Y (2016) Seasonal prevalence of *Allocreadium isoporum* (Loos, 1894) (Digenea: Allecreadiidae) in *Oxynoemacheilus tigris* (Osteichthyes: Balitoridae) (Steindachner, 1897) from Murat River, Eastern Anatolia, Turkey. Biharean Biol 10(1):1–3
Mansoor NT, Al-Shaikh SMJ (2010) Protozoans infection of *Cyprinus carpio* L. from Bab Al-Muatham fish markets, Baghdad City. Iraqi J Vet Sci 34(1):158–164
Masoumian M, Baska F, Molnar K (1994) Description of *Myxobolus karuni* sp. n. and *Myxobolus persicus* sp. n. (Myxosporea, Myxozoa) from *Barbus grypus* of the River Karun, Iran. Parasitologia Hungarica 27:21–26
Mhaisen FT (1986) Records of some fish parasites from Shatt Al-Arab River and the north west of the Arab Gulf. Bull Basrah Nat Hist Mus 6(1):111–124
Mhaisen FT, Al-Maliki NS (1996) Parasites, diseases and food of the dark-blotched mudskipper *Periophthalmus waltoni* (Perciformes: Gobiidae) in the Khor Al-Zubair Estuary (Iraq). Zool Middle East 13:85–87
Mhaisen FT, Al-Salim NK, Khamees NR (1986) The parasitic fauna of two cyprinids and a mugilid fish from Mehaijeran Creek, Basrah. J Biol Sci Res 17(3):63–73

Mohamad ET (1989) Study on parasites of the stinging catfish *Heteropneustes fossilis* (Bloch, 1797) from Al-Hammar Marsh, Basrah. M. Sc. thesis, College of Agriculture, University of Basrah, 101pp

Rahemo ZIF (1980) Diplozoon kasimii new species from a freshwater teleost fish, *Cyprinion macrostomum* Heckel. Bull Biol Res Cent Baghdad 12(1):109–114

Rasheed ARAM (1989) First record of *Diplozoon barbi* Reichenbach-Klinke, 1951 from some freshwater fishes from Tigris river, Baghdad, Iraq. Zanco 2(3):5–15

Sharma KP (1977) Occurrence of *Ichthyophthirius* in khishni, *Liza abu* (Heckel) fingerlings of Shatt Al-Arab. Arab Gulf 7:35–36

Tunç A, Koyun M (2018) Seasonal Infection of Metazoan Parasites on Mosul Bleak (Alburnus mossulensis) Inhabiting Murat River and Its Tributaries in Eastern Anatolia, Turkey. Turk J Agric Nat Sci 5(2):153–162

Chapter 69
The Parasites of Fishes of the Euphrates and Tigris Rivers: Iraq and Turkey

Ahmet Öktener

Abstract The preliminary study aims to present and compare the parasites of the freshwater fish from the Euphrates and Tigris Rivers of Iraq and Turkey. This study is also the first to provide a collective look at the parasites of freshwater fish from both these rivers. The comparison is mainly based on the checklists studied from the published papers on the fish parasites of these rivers. The parasite diversity is examined according to Phyla Platyhelminthes (Monogenea, Digenea, Cestoda), Acanthocephala, Nematoda, Annelida, Arthropoda, Cnidaria, and Ciliophora. The number of parasites reported in the freshwater fish from the Tigris River (241 species) is higher than the Euphrates River (170 species). In addition, the parasite diversity of the Euphrates River Part in Iraq (129 species) is higher than the Euphrates River Part in Turkey (62 species). The fish parasite diversity from the Tigris River of both countries was not compared due to insufficient information. It is recommended to do parasitological study as detailed in both rivers in future, especially the Tigris River.

69.1 Introduction

The preservation of biodiversity is vital for continuation of humanity. For this, the diversity of flora and fauna of the environment must be determined. The Euphrates and Tigris River Systems is one of the important wetlands in the world. It is also a rich biodiversity. The environmental factors including domestic, industrial, and agricultural pollution, the recharge of river flow, the accumulation of salt, the establishment dam, the primitive irrigation activities are affecting the Euphrates and Tigris Rivers. The biodiversity of these rivers is threatened by these environmental problems. The diversity of fish parasite is one of the important elements of biodiversity. The prominence of fish parasites in the biological and environmental studies is discussed below. Until today, the parasite diversity of the fishes living in these rivers has not been revealed.

A. Öktener (✉)
Department of Fisheries, Sheep Research Institute, Bandirma, Balikesir, Turkey

L. A. Jawad (ed.), *Tigris and Euphrates Rivers: Their Environment from Headwaters to Mouth*, Aquatic Ecology Series 11,
https://doi.org/10.1007/978-3-030-57570-0_69

69.2 The Importance of Fish Parasites

Fish parasites can be found in different sites of the host fish body. The gill filaments and the digestive tract are the sites where the parasites are most abundant in fish. They are also found in other organs such as the body surface, fins, oral cavity, eye liquor, blood, gonads, and liver.

Fish parasites have indirect and direct harmful effects on the host fish. There are several examples of direct effects. Ectoparasites are localized to the epidermis, mouth, eyes, nostrils, and gill filaments of the fish and they can cause extreme flaking, pigmentation changing in the body, excessive mucus accumulation in the gill tissues and filaments, adherence, swelling, and mechanical lesions, anemia, and exophthalmos.

Fish louse as known, *Argulus* is one of the dangerous parasites of fish. It often causes serious mortalities of fish. The feeding sites of Argulids become hemorrhagic and ulcerated, also cause to access to secondary infections by other parasites, bacteria, fungi, and viruses (Hoffman 1977). Sahoo et al. (2013) emphasized that comprehensive report on economic loss in carp culture ponds due to an argulosis.

Vinobaba (2010) examined the effects of parasitic copepod, *Ergasilus* on several hosts in Batticaloa Lagoon. He found that the atrophy on the secondary lamellae of the gill filaments, the hyperplasia and mucous cell proliferation of the gill epithelium, the high numbers of eosinophilic granular cells and rodlet cells in the gills of the infected fish.

Endoparasites lead to the formation of hemorrhagic lesions in the internal organs, to postnatal malnutrition of the gastrointestinal tract, especially to the intestinal tract, to secrete toxic substance to the tissue of host organisms as direct effects. All these negative effects cause the decrease of growth rate and condition of fish, the deterioration in functions of digestion, circulation, respiration, reproductive system, and body resistance. Arme and Wynne Owen (1968) searched the infection of *Ligula intestinalis* in the fish of British freshwater. They determined that the reduction of liver weight, the count of erythrocytes, and hemoglobin in roach infected by Ligula. Barson and Marshall (2003) found that the atrophy in the gonads of all fish infected by *Ligula intestinalis*.

Indirectly, parasites can reduce the commercial and nutritional value of fish with other factors such as the color changes in the outer appearance, damaged areas, and deep wounds on the body of the fish. In addition, the serious economic losses can be inevitable as they can cause mass mortality.

Fish parasites have also great ecological, medical, and aquaculture importance. The relationship between parasites and fish is balanced in natural ecosystems generally. When the ecological balance deteriorates (for reasons such as domestic or industrial pollution), the effects and damage of the parasites on the fish are increasing. Parasites can be more severe damage in fish farms. Shinn et al. (2015) indicated that the major impact of fish parasites on aquaculture and their significant effects. They examined the potential economic damages of parasite-originated disease in aquaculture based on 498 specific events.

There are a lot of scientific papers concerning the use of parasites as a bioindicator in monitoring aquatic pollution, environmental changes, and interrelations between fish parasites and pollution in the world (MacKenzie et al. 1995; Overstreet 1997; MacKenzie 1999; Sures 2004).

Fish parasites have also been used as an ecological indicator or tag, determining biodiversity, comparing, and obtaining identification of fish stocks (MacKenzie 1983; Pascual and Hochberg 1996; MacKenzie and Abaunza 1998; Yamaguchi et al. 2003).

Fish parasites have medical importance. They can cause serious clinical symptoms. In humans, fish parasites cause zoonosis by consuming fish and aquatic invertebrates raw or undercooked. Anisakiasis is a disease that humans are infected by larval worms of Anisakidae in squids and marine fish. Acute gastric infections cause gastric pain, nausea, and vomiting 4–6 h after ingesting raw infected seafood. During the chronic phase, vague epigastric pain, nausea, and vomiting may last from several weeks to 2 years (Ko 2006).

69.3 The Euphrates and Tigris Rivers

The Euphrates (Fırat in Turkish; Nahr Al-Furāt in Arabic) originated from Karasu and Murat Rivers in Turkey. It runs approximately 3000 km through Turkey, Syria, and Iraq. The Euphrates river is divided between Turkey (1230 km), the Syrian Arab Republic (710 km), and Iraq (1060 km). The Tigris River (Tigra in Persian; Dicle in Turkish) rises from the Hazar Lake at an altitude of 1150 m near Elazig province in eastern Anatolia, Turkey. It runs approximately 1850 km; 400 km with Turkey; 1418 km with Iraq; 32 km on the border between Turkey and Syria (Fig. 69.1). These two big rivers constitute the Tigris–Euphrates River System. The Euphrates and Tigris Rivers combine at Al-Qurnah and form the Shatt al-Arab and discharge into the Persian Gulf (FAO 2016) (Fig 69.1).

The region between the Tigris–Euphrates River System is called as the Mesopotamia, historically. This region is also known as the region in which many ancient civilizations are established such as Sumerian, Assyrian, Akkadian, and Babylonian civilizations. The Region has fertile soil for agriculture. The Tigris–Euphrates River System forms important and fertile areas for agriculture in the Mesopotamia due to its continuous flow. These rivers supply water to irrigate for agriculture. Even, passenger and freight transports were made in this river system via large and small boats many years ago (Lloyd et al. 2018).

Fig. 69.1 The Euphrates and Tigris Rivers (https://en.wikipedia.org)

69.4 Freshwater Fish Fauna of Euphrates and Tigris Rivers of Iraq and Turkey

The presence of 74 species of fish in Turkey and 53 species of fish in Iraq have been reported from the Euphrates and Tigris Rivers according to the available resources can be accessed. Several taxonomical studies have been used for the current situation of freshwater fish of these two rivers for Turkey (Kuru 1980; Ünlü 1991, 1999, 2012; Ünlü et al. 1994, 1997, 2000, 2017; Erk'akan et al. 1998, 2007, 2008; Tsigenopoulos et al. 2003; Bogutskaya et al. 2006; Fricke et al. 2007; Turan et al. 2009, 2011, 2013, 2016, 2017; Kara et al. 2011, 2016; Liao et al. 2011; Çiçek et al. 2015; Freyhof et al. 2017; Freyhof and Özuluğ 2017; Küçük et al. 2017; Pers. commun with Prof. Ünlü) and for Iraq (Coad 2010; Jawad 2012) (Table 69.1).

Table 69.1 Freshwater fish species living in Euphrates and Tigris Rivers of Iraq and Turkey[a]

Fish family	Fish species	
	Turkey	Iraq
Cyprinidae	*Acanthobrama marmid, Alburnoides velioglui, Alburnoides emineae, Alburnoides recepi, Alburnoides diclensis, Alburnus mossulensis, Alburnus caeruleus, Alburnus heckeli, Alburnus selcuklui, Alburnus sellal, Arabibarbus grypus, Barbus lacerta, Barilius mesopotamicus, Capoeta barroisi, Capoeta trutta, Capoeta umbla, Carasobarbus luteus, Carasobarbus kosswigi, Chondrostoma regium, Cyprinion kais, Cyprinion macrostomum, Cyprinus carpio, Garra rufa, Garra variabilis, Leuciscus vorax, Luciobarbus barbulus, Luciobarbus esocinus, Luciobarbus mystaceus, Luciobarbus xanthopterus, Luciobarbus kosswigi, Luciobarbus subquincunciatus, Luciobarbus kersin, Pseudophoxinus firati, Petreleuciscus kurui, Squalius berak, Squalius lepidus, Squalius semae, Squalius seyhanensis, Squalius kottelati, Ctenopharyngodon idella*	*Acanthobrama marmid, Alburnoides bipunctatus, Alburnus caeruleus, Alburnus mossulensis, Arabibarbus grypus, Barbus lacerta, Barilius mesopotamicus, Caecocypris basimi, Capoeta aculeata, Capoeta barroisi, Capoeta damascina, Capoeta trutta, Carasobarbus kosswigi, Carasobarbus luteus, Carassius auratus, Chondrostoma regium, Ctenopharyngodon idella, Cyprinion kais, Cyprinion macrostomum, Cyprinus carpio, Garra rufa, Garra variabilis, Garra elegans, Hemiculter leucisculus, Hypophthalmichthys molitrix, H. nobilis, Leuciscus vorax, Luciobarbus barbulus, Luciobarbus esocinus, Luciobarbus kersin, Luciobarbus pectoralis, Luciobarbus subquincunciatus, Luciobarbus xanthopterus, Mesopotamichthys sharpeyi, Squalius cephalus, Squalius lepidus, Typhlogarra widdowsoni*
Nemacheilidae	*Oxynoemacheilus euphraticus, Oxynoemacheilus frenatus, Oxynoemacheilus kaynaki, Oxynoemacheilus erdali, Oxynoemacheilus kurdistanicus, Oxynoemacheilus bergianus, Oxynoemacheilus chomanicus, Oxynoemacheilus kentritensis, Oxynoemacheilus hazarensis, Oxynoemacheilus araxensis, Oxynoemacheilus argyrogramma, Oxynoemacheilus tigris, Oxynoemacheilus panthera, Paracobitis zabgawraensis, Paracobitis malapterura, Schistura chrysicristinae, Turcinemacheilus kosswigi*	*Oxynoemacheilus frenatus, Oxynoemacheilus argyrogramma, Paracobitis malapterura*
Cobitidae	*Cobitis elazigensis, Cobitis kellei*	*Cobitis taenia*
Sisoridae	*Glyptothorax armeniacus, Glyptothorax kurdistanicus, Glyptothorax cous, Glyptothorax steindachneri*	*Glyptothorax kurdistanicus, Glyptothorax steindachneri*

(continued)

Table 69.1 (continued)

Fish family	Fish species	
	Turkey	Iraq
Heteropneustidae	*Heteropneustes fossilis*	*Heteropneustes fossilis*
Mastacembelidae	*Mastacembelus mastacembelus*	*Mastacembelus mastacembelus*
Bagridae	*Mystus pelusius*	*Mystus pelusius*
Salmonidae	*Oncorhynchus mykiss, Salmo okumusi, Salmo euphrataeus, Salmo tigridis*	
Mugilidae	*Planiliza abu*	*Planiliza abu*
Siluridae	*Silurus triostegus*	*Silurus triostegus*
Cyprinodontidae		*Aphanius dispar, Aphanius mento, Aphanius* sp.
Ariidae	*Arius cous*	
Poeciliidae		*Gambusia holbrooki, Poecilia latipinna*
Atherinidae	*Atherina boyeri*	

[a]Freshwater fish species living in Euphrates and Tigris Rivers of Iraq and Turkey are listed together

69.5 The Parasite Diversity of the Euphrates and Tigris Rivers of Iraq and Turkey

Fish parasite checklists studies are important taxonomic documents obtaining the fish–parasite relationships, host selectivity, and geographic distribution of fish parasites. They may contribute as baseline data in the disciplines of parasitology, zoology, medicine, environmental science in terms of determining biological diversity, treatment and control of parasites, identifying the parasite, determining host selectivity, understanding the geographic distribution of fish parasites, comparing the fish parasite fauna of local, regional, and worldwide.

It is known that valid names and synonyms of the parasite, host species may be changed. Parasite findings may be published in different or same dates and regions by different researchers. Therefore, checklists are also useful in minimizing synonyms, doubtful, error reports, and notifications of both the parasite–host. Although checklists are important, they need to revise and update and it must be delivered to many more readers. In this sense, checklists may contain the little restrictive and outdated information. Hence, these constitute disadvantages of checklists (Alaş and Öktener 2015).

The checklists studies on parasite diversity of fish in freshwater and marine water habitats have been prepared in the world (Hewitt and Hine 1972; Kennedy 1974; Holland and Kennedy 1997; Arthur and Lumanlan-Mayo 1997; Moravec 2001; Arthur and Ahmed 2002; Afonso-Dias and MacKenzie 2004; MacKenzie et al. 2004; Salgado-Maldonado 2006; Arthur and Te 2006; Salgado-Maldonado et al. 2006; Kirjušina and Vismanis 2007; Luque and Tavares 2007; Moles 2007; Pazooki and Masoumian 2012).

Aquatic environments include same or different ecological characteristics and flora, fauna. Therefore, the information about parasite and host obtained from different or same regions, countries cannot be compared together at the same time. In this process, checklists are important in comparing and obtaining all information about parasite and hosts among the different countries at a glance.

A significant number of studies (articles, unpublished MSc and PhD thesis, conference, symposium oral abstracts) concerning parasites of freshwater and marine fish have been published in Turkey and Iraq. Several checklists concerning fish parasites in both countries have also been done according to these studies. They have been prepared for Iraq by Mhaisen et al. (1991, 2011, 2013a, b, c, 2016, 2017a, b), Mhaisen (1993a, 1993b, 1995, 2002), Mhaisen and Khamees (1995), Mhaisen and Al-Nasiri (2012), Mhaisen and Abdul-Ameer (2013, 2014), Mhaisen and Abdullah (2016, 2017), Mhaisen and Al-Rubaie (2016), and for Turkey by Öktener (2003, 2005, 2014a, b, 2015), Öktener et al. (2004, 2009, 2010), Arslan and Öktener (2012), Alaş et al. (2015), Alaş and Öktener (2015), Vlizzi et al. (2015), İnnal et al. (2016).

The checklist of the parasites of fishes from the Euphrates and Tigris Rivers of Iraq and Turkey were prepared according to the above-indicated checklists.

The main literature were based on Mhaisen et al. (2017a) for the Euphrates River; Mhaisen and Al-Nasiri (2012), Mhaisen and Abdullah (2017) for Tigris River of Iraq; Öktener (2003, 2014b), Öktener et al. (2010), Öktener (2014b), Alaş et al. (2015), Alaş and Öktener (2015) for Turkey. The information obtained in these literature were controlled with an online database (WoRMS Editorial Board 2018; Global Cestode Database, 2018). In addition, the valid names and synonymies of the freshwater mentioned in these literature were checked with Froese and Pauly (2018).

The parasite checklists are arranged by phylum. Each parasite species is shown in separate rows and columns in the table according to the river (Tigris and Euphrates) and country (Iraq and Turkey) in which they are reported. Against each parasite are listed the country and river of that parasite in tables. The purpose of these tables are to facilitate the comparison of parasite diversity from the concerned country and river. These checklists are present in ten groups such as Cestoda, Monogenea, Digenea, Acanthocephala, Nematoda, Annelida, Cnidaria, Ciliophora, Mollusca, and other groups (Table 69.6, 69.7, 69.8, 69.9, 69.10, 69.11, 69.12, 69.13, 69.14, 69.15, and 69.16). This review includes only parasite species lists reported from the Euphrates and Tigris Rivers of both two countries. In this study, only a list of parasites reported from fish of the Euphrates and Tigris Rivers of Iraq and Turkey was given.

It is known that the Euphrates and Tigris Rivers System runs from Turkey to Iraq, Syria. Although there are several checklists to be published about the fish parasite diversity from these rivers of Iraq and Turkey, a checklist including a comparative analysis of these papers of both countries together has not been carried out until today. If the parasite checklist or the comparative analysis about the parasites reported from the fishes from the Euphrates and Tigris Rivers are arranged, it allows scientists to compare easily the parasite diversity according to countries and rivers at a glance and to aid in creating hypotheses about ecological and zoogeographical topics.

Table 69.2 Parasite diversity reported in the fish from the Euphrates and Tigris Rivers of Iraq and Turkey

	Euphrates River	Tigris River
Parasite phyla/class	Number of species	Number of species
Monogenea	52	99
Digenea	12	16
Cestoda	15	23
Acanthocephala	11	7
Nematoda	11	22
Parasitic annelida	5	4
Parasitic arthropods	13	11
Ciliophora	30	26
Cnidaria	21	33
Total	170	241

Although I think that all literature on the parasites reported from freshwater fishes from the Euphrates and Tigris Rivers have been presented here, it is certain that some literature will be overlooked. I have been informed by Prof. Mhaisen continuing to prepare several checklists and the existence of 1000 articles approximately about this subject in Iraq (personal communication). However, it should be accepted that this review represents a preliminary study or a first step or a sight of a picture of the parasite diversity reported from freshwater fishes from the Euphrates and Tigris Rivers in general terms. In addition, this preliminary checklist will be useful in revealing the differences and similarities of the parasite diversity reported from freshwater fishes of the Euphrates and Tigris Rivers of both the countries.

The number of parasite species belonging to each river of both Iraq and Turkey is given together. The parasite diversity of fishes from two rivers mainly belong to the Phyla Platyhelminthes (Monogenea, Digenea, Cestoda), Acanthocephala, Nematoda, Annelida, Arthropoda, Cnidaria, and Ciliophora. Compared the parasite diversity of the fishes from two rivers, it can be said that the parasite diversity of the Tigris River is higher than the Euphrates River, except Acanthocephala, Nematoda, and Ciliophora phyla (Tables 69.2 and 69.3).

Compared the parasite diversity of the fishes from only the Euphrates River in both Iraq and Turkey, it can be said that the parasite diversity of the Euphrates River Part in Iraq is higher than the Euphrates River Part in Turkey except for Acanthocephala and Annelida (Tables 69.4 and 69.5).

The number of parasite species reported in the freshwater fish from the Euphrates River of Turkey is only about one-third of the total parasite species recorded from Iraq. This big difference in the number of parasites between the two countries may be attributed to several reasons as follows.

Polluted waters may impair the immunological structure of fish that this status causes parasites to be effective and dominant in fish. The fish parasite diversity may be affected by the density or size of the host population. The presence of the populations of a fish species with the populations of other species together and the relationships between them may affect the fish parasite biodiversity. The feeding habits and food preference of fish species are very different. Fish can be omnivorous,

Table 69.3 Parasite diversity reported in the fish from the Euphrates and Tigris Rivers of Iraq and Turkey

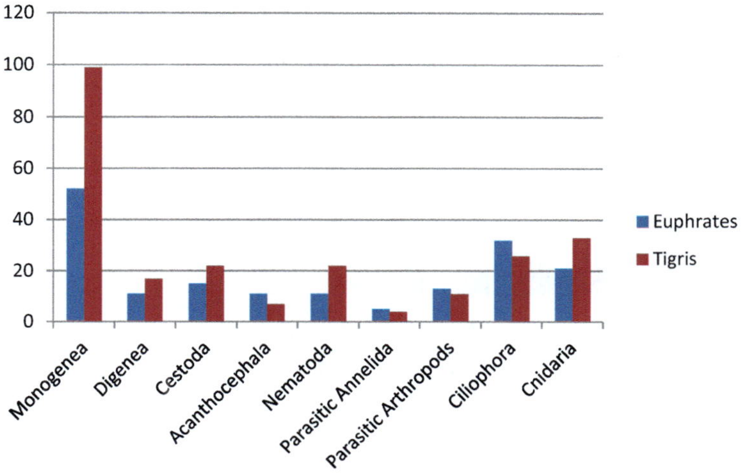

Table 69.4 Parasite diversity reported in the fish from only the Euphrates River of Iraq and Turkey

	Turkey	Iraq
Parasite phyla/class	Number of species	Number of species
Monogenea	25	31
Digenea	2	11
Cestoda	9	10
Acanthocephala	7	6
Nematoda	5	8
Parasitic annelida	3	3
Parasitic arthropods	7	11
Ciliophora	3	29
Cnidaria	1	20
Total	62	129

herbivorous, carnivorous according to their feeding patterns. It is known that the species belonging to digenea, cestoda, Acanthocephala use an invertebrate such as a mollusc or an arthropod for development in their life cycle. Thus, the lack of an intermediate hosts in the habitat may limit the parasites to colonize in fish. The spawning and feeding migratory abilities (from the river to the lake or from the lake to tributaries), the predator–prey interrelationships, the presence of invasive species, the large or small size of fish, the school-solitary life of fish, the morphological characters of the fish may be attributed to increasing or decreasing of fish parasite diversity. Euphrates and Tigris Rivers display very different limnological features to mouth from headwaters. These ecological properties are also effective in the parasite diversity.

Table 69.5 Parasite diversity reported in the fish from only the Euphrates River of Iraq and Turkey

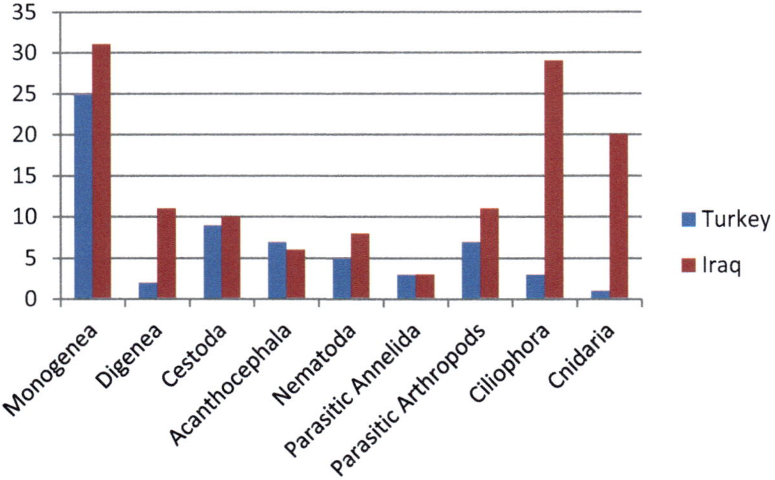

One or more of these reasons above indicated may cause the geographical isolation or the colonization, results also contribute to fish parasite diversity significantly.

In addition to, compared the number of Turkish and Iraqish parasitologists' studies, it may be said that the fish parasitologists in Iraq have carried out many more studies of parasites of freshwater fish of the Euphrates and Tigris Rivers, especially Tigris River.

There is a significant difference in species diversity between the monogenea fauna in the freshwater fish from the Euphrates and Tigris Rivers System in both Iraq and Turkey (Table 69.6). Comparing the diversity of monogenea in the freshwater fish from the Euphrates and Tigris Rivers System of Iraq has the richest community with 108 species than Turkey with 25 species. Thirteen monogenea species listed freshwater fish from the Euphrates and Tigris Rivers System were common in Iraq and Turkey, e.g., *Dactylogyrus affinis*, *Dactylogyrus alatus*, *Dactylogyrus elegantis*, *Dactylogyrus extensus*, *Dactylogyrus lenkorani*, *Dactylogyrus minutus*, *Dactylogyrus pulcher*, *Dactylogyrus rectotrabus*, *Dactylogyrus sphyrna*, *Dactylogyrus vistulae*, *Mastacembelocleidus heteranchorus*, *Dogielius mokhayeri*, and *Paradiplozoon barbi*. Four monogenea species (*Dactylogyrus affinis*, *Dactylogyrus extensus*, *Dactylogyrus minutus*, and *Mastacembelocleidus heteranchorus*) are also common in the freshwater fish from the Euphrates River of Turkey and Iraq. The monogenea fauna of the freshwater fish from the Tigris River of both countries could not be compared as there are no parasites in Turkey. Twenty one (*Dactylogyrus achmerowi*, *Dactylogyrus affinis*, *Dactylogyrus anchoratus*, *Dactylogyrus carassobarbi*, *Dactylogyrus cornu*,

Table 69.6 Monogenea reported from fishes in the Euphrates and Tigris Rivers of Iraq and Turkey

Rivers	Turkey	Iraq
Euphrates	*Dactylogyrus affinis*, *Dactylogyrus alatus*, *Dactylogyrus ancylostylus*, *Dactylogyrus asper*, *Dactylogyrus auriculatus*, *Dactylogyrus elegantis*, *Dactylogyrus extensus*, *Dactylogyrus malleus*, *Dactylogyrus minutus*, *Dactylogyrus prostae*, *Dactylogyrus pulcher*, *Dactylogyrus rectotrabus*, *Dactylogyrus sphyrna*, *Dactylogyrus vistulae*, *Dogielius forceps*, *Mastacembelocleidus heteranchorus*, *Diclybothrium hamulatum*, *Paradiplozoon barbi*, *Paradiplozoon megan* (Öktener 2003, 2014b), *Dactylogyrus goktschaicus*, *Dactylogyrus lenkorani*, *Gyrodactylus* sp, *Dogielius mokhayeri* (Koyun et al. 2015), *Paradiplozoon bingolensis* (Civáňová et al. 2013), *Solostamenides paucitesticulatus* (Kritsky and Öktener 2015)	*Dactylogyrus achmerowi*, *Dactylogyrus affinis*, *Dactylogyrus anchoratus*, *Dactylogyrus carassobarbi*, *Dactylogyrus cornu*, *Dactylogyrus extensus*, *Dactylogyrus jamansajensis*, *Dactylogyrus lamellatus*, *Dactylogyrus minutus*, *Dactylogyrus varicorhini*, *Dactylogyrus vastator*, *Discocotyle sagittata*, *Discocotyle* sp, *Gyrodactylus baicalensis*, *Gyrodactylus barbi*, *Gyrodactylus elegans*, *Gyrodactylus kherulensis*, *Gyrodactylus markevitschi*, *Gyrodactylus tincae*, *Gyrodactylus umbrae*, *Gyrodactylus varicorhini*, *Paradiplozoon ergensi*, *Paradiplozoon kasimii*, *Paradiplozoon pavlovskii*, *Paradiplozoon rutili*, *Paradiplozoon skrjabini*, *Microcotyle donavini* (Mhaisen et al. 2017a), *Thaparocleidus gomitus*, *Thaparocleidus vistulensis*, *Ligophorus vanbenedenii*, *Mastacembelocleidus heteranchorus* (Mhaisen et al. 2015a)
Tigris		*Thaparocleidus vistulensis*, *Dactylogyrus cornu*, *Dactylogyrus kulwieci*, *Dactylogyrus pulcher*, *Dactylogyrus rhodeianus*, *Dactylogyrus varicorhini*, *Dactylogyrus vastator*, *Gyrodactylus elegans*, *Gyrodactylus markewitschi*, *Paradiplozoon kasimii*, *Diplozoon paradoxum*, *Paradiplozoon amurensis*, *Paradiplozoon barbi*, *Paradiplozoon bliccae*, *Paradiplozoon cyprini*, *Paradiplozoon pavlovskii*, *Microcotyle donavini* (Mhaisen and Al-Nasiri 2012); *Dactylogyrus achmerowi*, *Dactylogyrus acinacus*, *Dactylogyrus affinis*, *Dactylogyrus alatus*, *Dactylogyrus anchoratus*, *Dactylogyrus arcuatus*, *Dactylogyrus barbioides*, *Dactylogyrus barbuli*, *Dactylogyrus baueri*, *Dactylogyrus carassobarbi*, *Dactylogyrus carpathicus*, *Dactylogyrus charbinensis*, *Dactylogyrus cyprinioni*, *Dactylogyrus deziensioides*, *Dactylogyrus deziensis*, *Dactylogyrus dulkeiti*, *Dactylogyrus dyki*,

(continued)

Table 69.6 (continued)

Rivers	Turkey	Iraq
		Dactylogyrus elegantis, *Dactylogyrus extensus*, *Dactylogyrus fallax*, *Dactylogyrus formosus*, *Dactylogyrus hypophthalmichthys*, *Dactylogyrus inexpectatus*, *Dactylogyrus inutilis*, *Dactylogyrus kersini*, *Dactylogyrus lenkorani*, *Dactylogyrus macracanthus*, *Dactylogyrus macrostomi*, *Dactylogyrus mascomai*, *Dactylogyrus microcirrus*, *Dactylogyrus minutus*, *Dactylogyrus molnari*, *Dactylogyrus orbus*, *Dactylogyrus pavlovskyi*, *Dactylogyrus persis*, *Dactylogyrus polylepidis*, *Dactylogyrus rectotrabus*, *Dactylogyrus sahuensis*, *Dactylogyrus skrjabinensis*, *Dactylogyrus skrjabini*, *Dactylogyrus suchengtaii*, *Dactylogyrus vistulae*, *Dactylogyrus* sp, *Dogielius mokhayeri*, *Dogielius molnari*, *Dogielius persicus*, *Dogielius planus*, *Mastacembelocleidus heteranchorus*, *Gyrodactylus baicalensis*, *Gyrodactylus barbi*, *Gyrodactylus cyprini*, *Gyrodactylus gobioninum*, *Gyrodactylus gussevi*, *Gyrodactylus katharineri*, *Gyrodactylus kherulensis*, *Gyrodactylus longoacuminatus*, *Gyrodactylus macracanthus*, *Gyrodactylus medius*, *Gyrodactylus molnari*, *Gyrodactylus shulmani*, *Gyrodactylus sprostonae*, *Gyrodactylus vicinus*, *Diplozoon* sp, *Paradiplozoon bingolensis*, *Paradiplozoon cyprini*, *Paradiplozoon homoion*, *Paradiplozoon leucisci*, *Paradiplozoon tadjikistanicum*, *Paradiplozoon vojteki*, *Mazocraes alosae* (Mhaisen and Abdullah 2017); *Daclytogyrus phoxini*, *Dactylogyrus sphyrna* (Balasem et al. 2009); *Ligophorus vanbenedenii, Dactylogyrus jamansajensis, Thaparocleidus gomitus* (Adday et al. 1999); *Gyrodactylus branchicus*, *Gyrodactylus kearni*, *Gyrodactylus rarus*, *Gyrodactylus seravshani* (Rasheed and Al-Saadi 2018); *Gyrodactylus cotti*, *Gyrodactylus mikailovi*, *Gyrodactylus prostae* (Al-Saadi and Rasheed 2017)

Table 69.7 Digenea reported from fishes in the Euphrates and Tigris Rivers of Iraq and Turkey

Rivers	Turkey	Iraq
Euphrates	*Diplostomum* sp, *Allocreadium isoporum* (Öktener 2003, 2014b)	*Aspidogaster limacoides*, *Diplostomum spathaceum*, *Diplostomum* sp, *Ascocotyle coleostoma*, *Asymphylodora tincae*, *Asymphylotrema macracetabulum* (Mhaisen et al. 2017a); *Pseudozoogonoides subaequiporus*, *Azygia lucii*, *Asymphylodora demeli*, *Asymphylodora markewitschi*, *Orientocreadium pseudobagri* (Mhaisen et al. 2015b)
Tigris		*Ascocotyle coleostoma*, *Distomum globiporum*, *Sphaerostoma bramae*, *Pseudochetosoma salmonicola*, *Aspidogaster limacoides*, *Sanguinicola* sp, *Diplostomum* sp (Mhaisen and Al-Nasiri 2012), *Azygia robusta*, *Clinostomum complanatum*, *Diplostomum flexicaudum*, *Diplostomum spathaceum*, *Diplostomum* sp, *Paracoenogonimus ovatus*, *Allocreadium transversale*, *Megamonostomella rashediansis*, *Asymphylotrema macracetabulum*, *Orientocreadium siluri* (Mhaisen and Abdullah 2017)

Dactylogyrus extensus, *Dactylogyrus jamansajensis*, *Dactylogyrus minutus*, *Dactylogyrus varicorhini*, *Dactylogyrus vastator*, *Gyrodactylus baicalensis*, *Gyrodactylus barbi*, *Gyrodactylus elegans*, *Gyrodactylus kherulensis*, *Gyrodactylus markevitschi*, *Paradiplozoon kasimii*, *Paradiplozoon pavlovskii*, *Microcotyle donavini*, *Thaparocleidus gomitus*, *Thaparocleidus vistulensis*, *Mastacembelocleidus heteranchorus*) are common from between the Euphrates River and the Tigris River of Iraq.

Comparing the diversity of digenean in the freshwater fish from the Euphrates and Tigris Rivers System, Iraq has the richest community with 22 species than Turkey with 2 species (Table 69.7). Only one digenea species listed freshwater fish from the Euphrates and Tigris Rivers System was common in Iraq and Turkey, e.g., *Diplostomum* sp. The same parasite is also common in the freshwater fish from the Euphrates River of Turkey and Iraq. The digenea fauna of the freshwater fish from the Tigris River of both countries could not be compared as there are no parasites in Turkey. Five species (*Aspidogaster limacoides*, *Diplostomum spathaceum*, *Diplostomum* sp., *Ascocotyle coleostoma*, and *Asymphylotrema macracetabulum*) are common between the Euphrates River and the Tigris River of Iraq.

Comparing the diversity of cestoda in the freshwater fish from the Euphrates and Tigris Rivers System, Iraq has the richest community with 28 species than Turkey with 9 species (Table 69.8). Seven cestoda species listed from freshwater fish of the Euphrates and Tigris Rivers System were common in Iraq and Turkey, e.g., *Ligula intestinalis*, *Khawia sinensis*, *Khawia armeniaca*, *Schyzocotyle acheilognathi*,

Table 69.8 Cestoda reported in the fishes from the Euphrates and Tigris Rivers of Iraq and Turkey

Rivers	Turkey	Iraq
Euphrates	*Ligula intestinalis*, *Khawia armeniaca*, *Schyzocotyle acheilognathi*, *Caryophylleous laticeps*, *Senga mastacembeli*, *Postgangesia inarmata*, *Caryophyllaeus auriculatus*, *Diphyllobothrium* sp (Öktener 2003, 2014b), *Khawia sinensis* (Barata 2012)	*Caryophyllaeus laticeps*, *Schyzocotyle acheilognathi*, *Ligula intestinalis*, *Schistocephalus solidus*, *Valipora campylancristrota*, *Valipora mutabilis*, *Glanitaenia osculata* (Mhaisen et al. 2017a), *Senga ophiocephalina*, *Senga magnum*, *Caryophyllaeus auriculatus* (Mhaisen et al. 2015c)
Tigris	*Ligula intestinalis* (Öktener 2003, 2014b)	*Schyzocotyle acheilognathi*, *Caryophyllaeus* sp, *Khawia armeniaca*, *Khawia rossittensis*, *Caryophyllaeus auriculatus*, *Ligula intestinalis*, *Postgangesia hemispherous*, *Glanitaenia osculata*, *Proteocephalus torulosus* (Mhaisen and Al-Nasiri 2012), *Senga* sp, *Tetracampos ciliotheca*, *Caryophyllaeides fennica*, *Caryophyllaeus fimbriceps*, *Paracaryophyllaeus gotoi*, *Caryophyllaeus laticeps*, *Monobothrium wageneri*, *Caryophyllid* sp, *Khawia sinensis*, *Dibothriocephalus latus*, *Neogryporhynchus cheilancristrotus*, *Postgangesia inarmata*, *Proteocephalus longicollis*, *Proteocephalus* sp (Mhaisen and Abdullah 2017)

Caryophylleous laticeps, *Postgangesia inarmata*, *Caryophyllaeus auriculatus*. Four species (*Ligula intestinalis*, *Schyzocotyle acheilognathi*, *Caryophylleous laticeps*, *Caryophyllaeus auriculatus*) are also common in the freshwater fish from the Euphrates River of Turkey and Iraq. The cestoda fauna of the freshwater fish from the Tigris River of both countries could not be compared as there are no parasites in Turkey. Five species (*Caryophyllaeus laticeps*, *Schyzocotyle acheilognathi*, *Ligula intestinalis*, *Glanitaenia osculata*, and *Caryophyllaeus auriculatus*) are common between the Euphrates and the Tigris Rivers of Iraq.

The number of Acanthocephala species reported in the freshwater fish from the Euphrates and Tigris Rivers System of Iraq and Turkey are close to each other, Turkey with seven species and Iraq with eight species (Table 69.9). But the species diversity is different. Only three acanthocephalans species listed in the freshwater fish from the Euphrates and Tigris Rivers System were common in Iraq and Turkey, e.g., *Neoechinorhynchus rutili*, *Neoechinorhynchus zabensis*, and *Pomphorhynchus spindletrancatus.* Two species (*Neoechinorhynchus rutili*, *Neoechinorhynchus zabensis*) are also common in the freshwater fish from the Euphrates River of Turkey and Iraq. The Acanthocephala of the freshwater fish from the Tigris River of both countries could not be compared as there are no parasites in Turkey. Five species

Table 69.9 Acanthocephala reported in the fishes from the Euphrates and Tigris Rivers of Iraq and Turkey

Rivers	Turkey	Iraq
Euphrates	*Neoechinorhynchus rutili*, *Neoechinorhynchus zabensis*, *Pseudoechinorhynchus clavula*, *Neoechinorhynchus* sp, *Pomphorhynchus* sp (Öktener 2003, 2014b); *Pomphorhynchus spindletrancatus* (Heckmann et al. 2010); *Echinorhynchus baeri* (Amin et al. 2016)	*Neoechinorhynchus chilkaensis*, *Neoechinorhynchus cristatus*, *Neoechinorhynchus iraqensis*, *Neoechinorhynchus rutili*, *Neoechinorhynchus zabensis* (Mhaisen et al. 2017a), *Paulisentis fractus* (Mhaisen et al. 2015d)
Tigris		*Neoechinorhynchus cristatus*, *Neoechinorhynchus iraqensis*, *Neoechinorhynchus rutili*, *Neoechinorhynchus zabensis*, *Paulisentis fractus* (Mhaisen and Al-Nasiri 2012), *Pomphorhynchus laevis*, *Pomphorhynchus spindletruncatus* (Mhaisen and Abdullah 2017)

(*Neoechinorhynchus cristatus*, *Neoechinorhynchus iraqensis*, *Neoechinorhynchus rutili*, *Neoechinorhynchus zabensis*, and *Paulisentis fractus*) are common between the Euphrates and the Tigris Rivers of Iraq.

Comparing the diversity of parasitic nematods in the freshwater fish from the Euphrates and Tigris Rivers System of Iraq has the richest community with 26 species than Turkey with 5 species (Table 69.10). Only two nematod species listed in freshwater fish from the Euphrates and Tigris Rivers System were common in Iraq and Turkey, e.g., *Rhabdochona denudata*, *Contracaecum* sp. Two same species are also common in the freshwater fish from the Euphrates River of Turkey and Iraq. The parasitic nematod fauna of the freshwater fish from the Tigris River of both countries could not be compared as there are no parasites in Turkey. Four species (*Contracaecum* sp., *Cucullanus cyprini*, *Spiroxys* sp., and *Rhabdochona denudata*) are common between the Euphrates and the Tigris Rivers of Iraq.

Comparing the diversity of parasitic annelids on the freshwater fish from the Euphrates and Tigris Rivers System, Iraq has the richest community with five species than Turkey with three species (Table 69.11). Only one leech species listed from the freshwater fish of the Euphrates and Tigris Rivers System was common in Iraq and Turkey, e.g., *Piscicola geometra*. This leech species is also common in the freshwater fish from the Euphrates River of Turkey and Iraq. The leech fauna of the freshwater fish from the Tigris River of both countries could not be compared as there are no parasites in Turkey. *Piscicola geometra* is also common in the freshwater fish between the Euphrates River and the Tigris River of Iraq.

Comparing the diversity of parasitic arthropod listed in the freshwater fish from the Euphrates and Tigris Rivers System, Iraq has the richest with 14 species than

Table 69.10 Parasitic nematodes reported in the fishes from the Euphrates and Tigris Rivers of Iraq and Turkey

Rivers	Turkey	Iraq
Euphrates	*Rhabdochona denudata*, *Molnaria intestinalis*, *Contracaecum* sp, *Philometra ovata*, *Philometra rischta* (Öktener 2003, 2014b)	*Dioctophyma* sp, *Contracaecum* sp, *Cucullanus cyprini*, *Camallanus lacustris*, *Echinocephalus uncinatus*, *Spinitectus armatus*, *Spiroxys* sp, *Rhabdochona denudata* (Mhaisen et al. 2017a)
Tigris		*Capillaria* sp, *Contracaecum* sp, *Porrocaecum* sp, *Cucullanus cyprini*, *Cucullanus pseudeutropi*, *Procamallanus viviparus*, *Spiroxys* sp, *Rhabdochona grandipapillata*, *Rhabdochona hellichi*, *Rhabdochona denudata* (Mhaisen and Al-Nasiri 2012), *Anisakis* sp, *Cucullanus* sp, *Procamallanus siluri*, *Philometra* sp, *Rhabdochona chodukini*, *Rhabdochona kurdistanensis*, *Rhabdochona* sp, *Rhabdochona gnedini*, *Rhabdochona similis*, *Rhabdochona tigridis*, *Agamospirura* sp (Mhaisen and Abdullah 2017); *Rhabdochona* (*Rhabdochona*) *longispicula* (Bilal et al. 2017)

Table 69.11 Parasitic Annelids reported on the fishes from the Euphrates and Tigris Rivers of Iraq and Turkey

Rivers	Turkey	Iraq
Euphrates	*Piscicola geometra*, *Illinobdella patzcuarensis*, *Actinopdella* sp (Öktener 2003, 2014b)	*Acipenserobdella volgensis*, *Piscicola geometra*, *Piscicola* spp (Mhaisen et al. 2017a)
Tigris		*Hemiclepsis marginata*, *Piscicola geometra* (Mhaisen and Al-Nasiri 2012); *Cystobranchus mammillatus*, *Piscicola* sp (Mhaisen and Abdullah 2017)

Turkey with 7 species (Table 69.12). Six parasitic arthropod species on the freshwater fish from the Euphrates and Tigris Rivers System were reported in Iraq and Turkey, e.g., *Argulus foliaceus*, *Ergasilus mosulensis*, *Ergasilus sieboldi*, *Lernaea cyprinacea*, *Lamproglena pulchella*, *Tracheliastes polycolpus*. Five parasitic arthropod species (*Argulus foliaceus*, *Ergasilus mosulensis*, *Ergasilus sieboldi*, *Lernaea cyprinacea*, and *Lamproglena pulchella*) are common in the freshwater fish from the Euphrates River of Turkey and Iraq. The parasitic arthropod fauna of the freshwater fish from the Tigris River of both countries could not be compared as there are no

Table 69.12 Parasitic Arthropods reported in the fishes from the Euphrates and Tigris Rivers of Iraq and Turkey

Rivers	Turkey	Iraq
Euphrates	*Argulus foliaceus* (Öktener 2003; Öktener et al. 2010), *Ergasilus mosulensis*, *Ergasilus sieboldi*, *Ergasilus briani*, *Lernaea cyprinacea*, *Lamproglena pulchella*, *Tracheliastes polycolpus* (Öktener 2003; Alaş et al. 2015)	*Argulus foliaceus*, *Dermoergasilus varicoleus*, *Ergasilus barbi*, *Ergasilus mosulensis*, *Ergasilus peregrinus*, *Ergasilus rostralis*, *Ergasilus sieboldi*, *Lamproglena pulchella*, *Lernaea cyprinacea*, *Pseudolamproglena annulata*, *Pseudolamproglena boxshalli* (Mhaisen et al. 2017a)
Tigris		*Argulus foliaceus*, *Ergasilus mosulensis*, *Ergasilus peregrinus*, *Ergasilus sieboldi*, *Lamproglena pulchella*, *Lernaea cyprinacea*, *Pseudolamproglena annulata*, *Tracheliastes polycolpus* (Mhaisen and Al-Nasiri 2012), *Ergasilus barbi*, *Ergasilus* sp, *Arrenurus* sp (Mhaisen and Abdullah 2017)

parasites in Turkey. Why is the diversity of parasitic arthropod listed in the freshwater fish from the Euphrates and Tigris Rivers System of Iraq so rich? The differences in the parasite diversity in fish could be attributed to biotic and abiotic factors. The water qualities such as temperature, pH, salinity, or pollution, the climate and the existence of migratory fish are sufficient explanations for the rich of parasite biodiversity listed in Iraq. For example, migratory fish such as *Liza abu* may have transferred parasites such as *Ergasilus peregrinus* and *Dermoergasilus varicoleus* from the Shattul-Arab to inner parts of Iraq. More stable water parameters, polluted environments, or other factors in Iraq may be prepared as appropriate for the formation of colony for parasites. It is obvious that Iraqi scientists make more publications, when the current state of other parasite phylas (Ciliophorian, Cnidarian, Myzozoa, Euglenozoa, Choanozoa, Microsporidia, Mollusca) reported in the fishes from the Euphrates and Tigris Rivers of Iraq and Turkey is analyzed (Tables 69.13, 69.14, 69.15, and 69.16).

*"Finally, I want to tell a story about the fish parasites of the Euphrates River. Ergasilus mosulensis was described in Iraq waters by Rahemo (*1982*). I learned the presence of this parasite in the freshwater fish (Planiza abu) of Iraq, while I read Mhaisen (*1993b*)'s article. I decided to go to Atatürk Dam Lake for a parasitological survey. I also founded the samples of Ergasilus mosulensis and the deformed samples belonging to Planiza abu (Öktener et al.* 2007*; Jawad and Öktener* 2007*). We also discovered a monogenea, Solostamenides paucitesticulatus (Kritsky and Öktener* 2015*) (Fig. 69.2)." In this respect, the checklists represent a guide for zoologists, biologists, and parasitologists.* I think that the detailed parasitological studies must be carried out on the parasites of freshwater fish from these two rivers

Table 69.13 Ciliophorian parasites in the fishes from the Euphrates and Tigris Rivers of Iraq and Turkey

Rivers	Turkey	Iraq
Euphrates	*Trichodina nigra*, *Trichodina* sp, *Chilodonella cyprini* (Alaş and Öktener 2015)	*Balantidium ctenopharyngodoni*, *Balantidium polyvacuolum*, *Chilodonella cyprini*, *Chilodonella hexasticha*, *Ichthyophthirius multifiliis*, *Apiosoma amoeba*, *Apiosoma campanulata*, *Apiosoma cylindriformis*, *Apiosoma miniciliatum*, *Apiosoma minuta*, *Apiosoma piscicola*, *Apiosoma poteriformis*, *Epistylis kronwerci*, *Epistylis solida*, *Ambiphrya ameiuri*, *Scyphidia globularis*, *Vorticella* sp, *Trichodina cottidarum*, *Trichodina domerguei*, *Trichodina elegini*, *Trichodina gracilis*, *Trichodina lepsii*, *Trichodina murmanica*, *Trichodina nigra*, *Trichodina reticulata*, *Trichodina schulmani*, *Trichodina spatulata*, *Trichodina sphaeroides*, *Tetrahymena pyriformis* (Mhaisen et al. 2017a)
Tigris		*Trichodina cottidarum*, *Trichodina domerguei*, *Trichodina elegeni*, *Trichodina murmaniaca*, *Trichodina* sp, *Ichthyophthirius multifiliis* (Mhaisen and Al-Nasiri 2012); *Balantidium polyvacuolum*, *Chilodonella cyprini*, *Apiosoma amoebae*, *Apiosoma robusta*, *Apiosoma* sp, *Riboscyphidia arctica*, *Trichodina acuta*, *Trichodina anguilli*, *Trichodina erbilensis*, *Trichodina heterodentata*, *Trichodina kurdistani*, *Trichodina mutabilis*, *Trichodina nobilis*, *Trichodina pediculus*, *Trichodina ranae*, *Trichodina reticulata*, *Trichodina* sp, *Tetrahymena pyriformis*, *Tetrahymena* sp (Mhaisen and Abdullah 2017); *Trichodina urinaria* (Abdul-Ameer and Atwan 2018)

Rivers having mysterious future. As long as these two important rivers of the Mesopotamia Region, which unites the Turkish, Iraqi, and Syrian peoples, exist, the Euphrates and Tigris Rivers will continue to bring fertility, peace, brotherhood, and happiness for the peoples of the region. Therefore, we must protect these natural wonder rivers, which have been home to various civilizations for centuries and have significant biodiversity, for the continuation of humanity.

Table 69.14 Cnidarian parasites in fishes from the Euphrates and Tigris Rivers of Iraq and Turkey

Rivers	Turkey	Iraq
Euphrates	*Myxobolus cyprinicola* (Alaş and Öktener 2015)	*Myxidium pfeifferi*, *Myxidium rhodei*, *Myxobolus acutus*, *Myxobolus alienus*, *Myxobolus bliccae*, *Myxobolus bramae*, *Myxobolus dispar*, *Myxobolus dogieli*, *Myxobolus ellipsoides*, *Myxobolus follius*, *Myxobolus macrocapsularis*, *Myxobolus musculi*, *Myxobolus nemacheili*, *Myxobolus orientalis*, *Myxobolus oviformis*, *Myxobolus pseudodispar*, *Myxobolus schulmani*, *Myxobolus spatulatus*, *Myxobolus sphaericus*, *Thelohanellus catlae* (Mhaisen et al. 2017a)
Tigris		*Myxobolus chondrostomi*, *Myxobolus cyprinicola*, *Myxobolus dispar*, *Myxobolus dogieli*, *Myxobolus ellipsoides*, *Myxobolus karelicus*, *Myxobolus koi*, *Myxobolus macrocapsularis*, *Myxobolus mülleri*, *Myxobolus nemachili*, *Myxobolus orientalis*, *Myxobolus oviformis*, *Myxobolus parvus*, *Myxobolus pfeifferi*, *Myxobolus pseudodispar*, *Myxobolus sandrae*, *Myxobolus schulmani*, *Myxobolus sphaericus*, *Thelohanellus catlae*, *Myxosoma acuta* (Mhaisen and Al-Nasiri 2012); *Myxobolus amurensis*, *Myxobolus bulbocordis*, *Myxobolus iranicus*, *Myxobolus karuni*, *Myxobolus mesopotamiae*, *Myxobolus molnari*, *Myxobolus persicus*, *Myxobolus poljanski*, *Myxobolus rotundus*, *Myxobolus shadgani*, *Myxobolus sharpeyi*, *Myxobolus* sp (Mhaisen and Abdullah 2017); *Myxidium rhodei* (Adday et al. 1999)

Table 69.15 Parasites belonging to different Phyla from the fishes of the Euphrates and Tigris Rivers of Iraq and Turkey

Rivers	Turkey	Iraq
Euphrates	**Myzozoa** *Eimeria truttae*, *Oodinium pillularis* (Alaş and Öktener 2015)	**Euglenozoa:** *Trypanosoma carasobarbi*, *Trypanosoma mystuii*, *Trypanosoma* sp., *Cryptobia* sp. (Mhaisen et al. 2017a)
Tigris		**Euglenozoa:** *Trypanosoma* sp (Mhaisen and Abdullah 2017) **Choanozoa:** *Dermocystidium percae* (Mhaisen and Al-Nasiri 2012) **Microsporidia:** *Glugea anomala* (Mhaisen and Al-Nasiri 2012), *Pleistophora longifilis* (Mhaisen and Abdullah 2017)

Table 69.16 Glochidium (Mollusca) parasites reported from fishes in the Euphrates and Tigris Rivers according to Iraq and Turkey

Rivers	Turkey	Iraq
Euphrates		*Unio pictorum* (Mhaisen et al. 2017a)
Tigris		*Unio pictorum* (Mhaisen and Abdullah 2017)

Fig. 69.2 (**a**) Atatürk Dam Lake, (**b**) the Euphrates River, (**c**, **d**) the concrete boat showing increase and decrease in the amount of water in the Atatürk Dam Lake, (**e**) deformed mullet samples, and (**f**) scanning electron image of *Ergasilus mosulensis* from Turkish samples

Acknowledgments I would like to thank Laith A. Jawad for offering me chapter authorship in this book; Erhan Ünlü for providing me literature about fishes of Euphrates and Tigris Rivers of Turkey; to Furhan Mhaisen for inform me literature about fish parasites of Iraq.

References

Abdul-Ameer KN, Atwan FK (2018) First Record of *Trichodina urinaria* Dogiel, 1940 (Ciliophora: Trichodinidae) in Iraq from Gills of Mugilid fish *Planiliza abu*. Biol Appl Environ Res 2:44–48

Adday TK, Balasem AN, Mhaisen FT, Al-Khateeb GH (1999) A second survey of fish parasites from Tigris river at AL-Zaafaraniya, South of Baghdad. Ibn Al-Haitham J Pure App Sci 12:22–31

Afonso-Dias I, MacKenzie K (2004) A checklist of the protozoan and metazoan parasites reported from the anglerfishes *Lophius piscatorius* L. and *L. budegassa* Spinola. Bull Eur Ass Fish Pathol 24:246–252

Alaş A, Öktener A (2015) Different parasitic phyla of fish from Turkey excluding Helminths and Crustacea. J Zool Stud 2:24–41

Alaş A, Öktener A, Türker ÇD (2015) Review of parasitic copepods recorded in fish from Turkey. Transylv Rev Syst Ecol Res 17:39–62

Al-Saadi AAJJ, Rasheed R (2017) The first record of three Monogenean species of gyrodactylus in Iraq from gills of some fishes in Tigris river, Al-Shawwaka Region, Baghdad Province. Ibn Al-Haitham J Pure Appl Sci 30:306–314

Amin OM, Heckmann RA, Evans RP, Tepe Y (2016) A description of *Echinorhynchus baeri* Kostylew, 1928 (Acanthocephala: Echinorhynchidae) from *Salmo trutta* in Turkey, with notes on synonymy, geographical origins, geological history, molecular profile, and X-ray microanalysis. Parasite 23:1–14

Arme C, Wynne Owen R (1968) Occurrence and Pathology of *Ligula intestinalis* Infections in British Fishes. J Parasitol 54:272–280

Arslan N, Öktener A (2012) A general review of parasitic Annelida (Hirudinea) recorded from different habitats and hosts in Turkey. T J Zool 36:141–145

Arthur JR, Ahmed ATA (2002) Checklist of the parasites of fishes of Bangladesh. FAO Fisher. Tech. Pap. No 369/1. FAO, Rome, 77p

Arthur JR, Lumanlan-Mayo S (1997) Checklist of the parasites of fishes of the Philippines. FAO Fisher. Tech. Pap. No 369, FAO, Rome. 102p.

Arthur JR, Te BQ (2006) Checklist of the parasites of fishes of Vietnam. FAO Fish. Tech. Pap. No. 369/2. FAO, Rome, 133p

Balasem AN, Mhaisen FT, Asmar KR, Al-Jawda JM, Adday TK (2009) Record of two species of the monogenetic trematodes fenus Dactylogyrus for the first timein Iraq on Gills of the cyprinid fish *Alburnus caeruleus*. Bull Iraq Nat Hist Mus 10:11–16

Barata S (2012) Study of endohelminthes of some fishes caught in Komurhan region of Karakaya Dam Lake. Fırat University, Science Institution, Master Thesis, 73p

Barson M, Marshall BE (2003) The occurrence of the tapeworm, *Ligula intestinalis* (L.), in *Barbus paludinosus* from a small dam in Zimbabwe. Afr J Aquat Sci 28:1–4

Bilal SJ, Rahman SK, Abdullah YS, Abdullah SMA (2017) First record of *Rhabdochona* (*Rhabdochona*) *longispicula* (Nematoda) in *Glyptothorax kurdistanicus* from Greater Zab River, Kurdistan Region/ Iraq. ZANCO J Pure Appl Sci 29(1):48–53

Bogutskaya NG, Küçük F, Atalay MA (2006) A description of three new species of the genus *Pseudophoxinus* from Turkey (Teleostei: Cyprinidae: Leuciscinae). Zoosyst Ross 15:335–341

Çiçek E, Birecikligil SS, Fricke R (2015) Freshwater fishes of Turkey: a revised and updated annotated checklist. Biharean Biol 9:141–157

Civáňová K, Koyun M, Koubková B (2013) The molecular and morphometrical description of a new diplozoid species from the gills of the *Garra rufa* (Heckel, 1843) (Cyprinidae) from Turkey-including a commentary on taxonomic division of Diplozoidae. Parasitol Res 112:3053–3062

Coad BW (2010) Freshwater fishes of Iraq. Pensoft, Sofia, Bulgaria. 275p

Erk'akan F, Atalayay A, Ekmekçi FG, Nalbant TT (1998) Four new species and one new subspecies of the genus *Cobitis* (Pisces: Ostariophysi: Cobitidae) from Turkey. Turk J Zool 22:9–16

Erk'akan F, Nalbant TT, Özeren SC (2007) Seven new species of Barbatula, three new species of Schistura and a new species of Seminemacheilus (Ostariophysi: Balitoridae: Nemacheilinae) from Turkey. J Fish Int 2:69–85

Erk'akan F, Özeren SC, Nalbant TT (2008) Two new species of stone loaches from Turkey (Teleostei: Nemacheilidae). J Fish Int 3:115–119

FAO (2016) AQUASTAT website. Food and Agriculture Organization of the United Nations (FAO). http://www.fao.org/nr/water/aquastat/basins/euphrates-tigris/index.stm. Accessed 03/2018

Freyhof J, Özuluğ M (2017) *Oxynoemacheilus hazarensis*, a new species from Lake Hazar in Turkey, with remarks on *O. euphraticus* (Teleostei: Nemacheilidae). Zootaxa 4247:378–390

Freyhof J, Kaya C, Turan D (2017) *Oxynoemacheilus kentritensis*, a new species from the upper Tigris drainage in Turkey with remarks on *O. frenatus* (Teleostei: Nemacheilidae). Zootaxa 4258:551–560

Fricke R, Bilecenoğlu M, Sarı HM (2007) Annotated checklist of fish and lamprey species (Gnathostomata and Petromyzontomorphi) of Turkey, including a Red List of threatened and declining species. Stutt Beitr Naturkd 706:1–169

Froese R, Pauly D (eds) (2018) FishBase. World Wide Web electronic publication. www.fishbase.org. Accessed 02/2018

Global Cestode Database (2018) A survey of the tapeworms (Cestoda: Platyhelminthes) from vertebrate bowels of the earth. http://tapewormdb.uconn.edu. Accessed 3/2018

Heckmann RA, Oguz MC, Amin OM, Dusen S, Tepe Y, Aslan B (2010) Host and geographical distribution of *Pomphorhynchus spindletrancatus* (Acanthocephala: Pomphorhynchidae) in Turkey, with enhanced description from new fish and amphibian hosts using SEM, and histopathological notes. Sci Parasitol 11:129–139

Hewitt GC, Hine PM (1972) Checklist of parasites of New Zealand fishes and of their hosts. N Z J Mar Fresh Res 6:69–114

Hoffman GL (1977) Argulus, a branchiuran parasite of freshwater fishes. US Fish and Wildlife Publications. Paper 137, 9p

Holland CV, Kennedy CR (1997) A checklist of parasitic helminth and crustacean species recorded in freshwater fish from Ireland. Biol Environ Proc R Ir Acad 97B:225–243

İnnal D, Aydoğdu A, Güçlü SS (2016) Invasion of *Schyzocotyle acheilognathi* (Yamaguti, 1934) (Cestoda: Bothriocephalidea) in Turkey. Acta Biol Turc 29:20–25

Jawad LA (2012) History of the study of the fish fauna of Iraq. Water Res Manag 2:11–20

Jawad LA, Öktener A (2007) Incidence of lordosis in the freshwater mullet, *Liza abu* (Heckel, 1843) collected from Atatürk Dam Lake, Turkey. Ann Biol 29:105–113

Kara C, Alp A, Gürlek ME (2011) Morphological variations of the trouts (*Salmo trutta* and *Salmo platycephalus*) in the rivers of Ceyhan, Seyhan and Euphrates, Turkey. TrJFAS 11:77–85

Kaya C, Turan D, Ünlü E (2016) The latest status and distribution of fishes in upper Tigris River and two new records for Turkish freshwaters. TrJFAS 16:545–562

Kennedy CR (1974) A checklist of British and Irish freshwater fish parasites with notes on their distribution. J Fish Biol 6:613–644

Kirjušina M, Vismanis K (2007) Checklist of the parasites of fishes of Latvia. FAO Fish. Tech. Pap. No. 369/3. FAO, Rome, 106p

Ko RC (2006) Fish-borne parasitic zoonoses. In: Woo PTK (ed) Fish diseases and disorders, Vol. 1: Protozoan and metazoan infections. CAB, Singapore, pp 592–628

Koyun M, Ulupınar M, Gül A (2015) Seasonal Distribution of Metazoan Parasites on Kura Barbell (*Barbus lacerta*) in Eastern Anatolia, Turkey. Pak J Zool 47:1253–1261
Kritsky DC, Öktener A (2015) *Solostamenides paucitesticulatus* n.sp. (Monogenoidea: Mazocraeidea: Microcotylidae) from the freshwater mullet *Liza abu* (Heckel) (Mugiliformes: Mugilidae) from Atatürk Reservoir on the Euphrates River in southern Turkey. Syst Parasitol 91:139–144
Küçük F, Turan D, Güçlü SS, Mutlu AG, Çiftci Y (2017) Two new species of Chondrostoma Agassiz, 1832 (Teleostei: Cyprinidae) from the Ceyhan, Seyhan and Göksu Rivers in the East Mediterranean Region of Turkey. TrJFAS 17:795–803
Kuru M (1980) Key to inland water fishes of Turkey. Hacettepe Bull Nat Sci Eng 9:103–133
Liao TY, Ünlü E, Kullander SO (2011) Western boundary of the subfamily Danioninae in Asia (Teleostei, Cyprinidae): derived from the systematic position of *Barilius mesopotamicus* based on molecular and morphological data. Zootaxa 2880:31–40
Lloyd SHF, Owen L, Gibson M, Lotha G, Cunningham JM, Pletcher K, Tikkanen A (2018) Tigris-Euphrates river system. Encyclopædia Britannica. https://www.britannica.com/place/Tigris-Euphrates-river-system. 3/2018
Luque JL, Tavares LR (2007) Checklist of Copepoda associated with fishes from Brazil. Zootaxa 1579:1–39
MacKenzie K (1983) Parasites as biological tags in fish population studies. Adv Appl Biol 7:251–331
MacKenzie K (1999) Parasites as pollution indicators in marine ecosystems: a proposed early warning system. Mar Pollut Bull 38:955–959
MacKenzie K, Abaunza P (1998) Parasites as biological tags for stock discrimination of marine fish: a guide to procedures and methods. Fish Res 38:45–56
MacKenzie K, Williams HH, Williams B, McVicar AH, Siddall R (1995) Parasites as indicators of water quality and the potential use of helminth transmission in marine pollution studies. Adv Parasit 35:85–144
MacKenzie K, Campbell N, Mattiucci S, Ramos P, Pereira A, Abaunza P (2004) A checklist of the protozoan and metazoan parasites reported from the Atlantic horse mackerel, *Trachurus trachurus* (L.). Bull Eur Ass Fish Pathol 24:180–184
Mhaisen FT (1993a) Check-lists of parasites and disease agents of Barbus spp (Teleostei: Cyprinidae) of Iraq. Marina Mesopot 8:236–253
Mhaisen FT (1993b) Fish Parasites of Neinava Province: a review and check-lists. Iraqi J Vet Med (1st Sci Conf Coll Vet Med Baghdad) 17:110–125
Mhaisen FT (1995) Parasites of fishes of Basrah Province Marshy Area, Iraq. J Environ Sci Health A30:41–49
Mhaisen FT (2002) Literature review and check lists of acanthocephalans of fishes of Iraq. Al-Mustansiriya J Sci 13:13–25
Mhaisen FT, Abdul-Ameer KN (2013) Checklists of *Gyrodactylus* species (Monogenea) from fishes of Iraq. Basrah J Agric Sci 26:8–25
Mhaisen FT, Abdul-Ameer KN (2014) Checklists of diplozoid species (Monogenea) from fishes of Iraq. Bull Iraq Nat Hist Mus 13:95–111
Mhaisen FT, Abdullah SMA (2016) Checklists of parasites of farm fishes of Kurdistan Region, Iraq. Iraqi J Agric Res 21:204–216
Mhaisen FT, Abdullah SMA (2017) Parasites of fishes of Kurdistan Region, Iraq. Checklists Biol Appl Environ Res 1:131–218
Mhaisen FT, Al-Nasiri FS (2012) Literature review on the parasites of fishes of Salah Al-Deen Province, Iraq. Tikrit Univ J Agric Sci 12:209–219
Mhaisen FT, Al-Rubaie ARL (2016) Checklists of fish parasites of Al-Najaf Al-Ashraf Province, Iraq. Al-Kufa Univ J Biol (Spec Second Int Sci Conf Life Sci Fac Educ Women Univ Kufa) 1:86–95
Mhaisen FT, Khamees NR (1995) Check-list of the parasites and disease agents of the Siluriformes of Iraq. Zool Mid East 11:113–120

Mhaisen FT, Khamees NR, Al-Daraji SAM (1991) Parasites and disease agents of carps in Iraq: a check-list. Basrah J Agric Sci 4:133–139
Mhaisen FT, Al-Niaeem KS, Al-Zubaidy AB (2011) Literature review on fish parasites of Al-Furat fish farm, Babylon province, Iraq. Iraqi J Aquacult 9:95–122
Mhaisen FT, Ali AH, Khamees NR (2013a) Checklists of Monogeneans of freshwater and marine fishes of Basrah Province, Iraq. Basrah J Agric Sci 26:26–49
Mhaisen FT, Khamees NR, Ali AH (2013b) Checklists of trematodes of freshwater and marine fishes of Basrah Province, Iraq. Basrah J Agric Sci 26:50–77
Mhaisen FT, Khamees NR, Ali AH (2013c) Checklists of cestodes of freshwater and marine fishes of Basrah Province, Iraq. Basrah J Agric Sci 26:78–98
Mhaisen FT, Al-Rubae AL, Al-Sa'adi BA (2015a) Monogenean parasites of fishes from the Euphrates River at Al-Musaib City, Mid Iraq. Am J Biol Life Sci 3:50–57
Mhaisen FT, Al-Rubae AL, Al-Sa'adi BA (2015b) Trematodes of fishes from the Euphrates River at Al-Musaib City, Mid Iraq. Am J Biol Life Sci 3:91–95
Mhaisen FT, Al-Rubae AL, Al-Sa'adi BA (2015c) Cestodes of fishes from the Euphrates River at Al-Musaib City, Mid Iraq. Am J Biol Life Sci 3:96–101
Mhaisen FT, Al-Rubae AL, Al-Sa'adi BA (2015d) Nematodes and acanthocephalans of fishes from the Euphrates River at Al-Musaib City, Mid Iraq. Am J Biol Life Sci 3:108–115
Mhaisen FT, Ali AH, Khamees NR (2016) Checklists of protozoans and myxozoans of freshwater and marine fishes of Basrah Province, Iraq. Mesopot J Mar Sci 31:29–52
Mhaisen FT, Al-Jawda J, Asmar KR, Ali MH (2017a) Checklists of fish parasites of Al-Anbar Province, Iraq. Biol Appl Environ Res 1:17–56
Mhaisen FT, Ali AH, Khamees NR (2017b) Checklists of fish parasites of Basrah Marshlands, Iraq. Biol Appl Environ Res 1:237–277
Moles A (2007) Parasites of the fishes of Alaska and surrounding waters. Alaska Fish Res Bull 12:197–226
Moravec F (2001) Checklist of the Metazoan parasites of fishes of the Czech Republic and the Slovak Republic. Academia, Praha. 168 p
Öktener A (2003) A checklist of Metazoon parasites recorded in freshwater fish from Turkey. Zootaxa 394:1–28
Öktener A (2005) A checklist of parasitic helminths reported from sixty-five species of marine fish from Turkey including two new records of monogeneans. Zootaxa 1063:33–52
Öktener A (2014a) Diversity of parasitic helminths of marine and freshwater fish from Turkey. J Zool Stud 1:12–15
Öktener A (2014b) Revision of parasitic helminths reported in freshwater fish from Turkey with new records. Transylv Rev Syst Ecol Res 16.1:1–56, 27
Öktener A (2015) An updated checklist of parasitic helminths of marine fish from Turkey. Transylv Rev Syst Ecol Res 16:55–96
Öktener A, Yalçın M, Koçyiğit E (2004) Protozoon parasites recorded in fish from Turkey. Anadolu Univ Fen Bil Der 5:297–307
Öktener A, Trilles JP, Leonardos I (2007) Five ectoparasites from Turkish fishes. Acta Parasitol Turc 31:154–157
Öktener A, Koç HT, Erdoğan Z (2009) Fish/parasites interrelationships in Turkey: a review. Ittiopatologia 6:121–139
Öktener A, Alaş A, Solak K (2010) Findings of fish lice, *Argulus foliaceus* (Crustacea, Branchiura) in Turkey. E J Ichthyol. 1:9–14
Overstreet RM (1997) Parasitological data as monitors of environmental health. Parassitologia 39:169–175
Pascual S, Hochberg FG (1996) Marine parasites as biological tags of cephalopod hosts. Parasitol Today 12:324–327
Pazooki J, Masoumian M (2012) Synopsis of the parasites in Iranian freshwater fishes. Iran J Fish Sci 11:570–589

Rahemo ZIF (1982) Two new species of Ergasilus (Copepoda: Cyclopoida) from the gills of two Iraqi freshwater fishes. Bull Basrah Nat Hist Mus 5:39–59

Rasheed R, Al-Saadi AAJJ (2018) The first record of four species of Gyrodactylus (Monogenea) from gills of four freshwater fish species in Iraq. Biol Appl Environ Res 2:1–11

Sahoo PK, Mohanty J, Garnayak SK, Kar B, Prasanth H, Jena JK (2013) Estimation of loss due to argulosis in carp culture ponds in India. Indian J Fish 60:99–102

Salgado-Maldonado G (2006) Checklist of helminth parasites of freshwater fishes from Mexico. Zootaxa 1324:1–357

Salgado-Maldonado G, Cohen SC, Kohn A (2006) Checklist of Monogenea parasites of freshwater and marine fishes, amphibians and reptiles from Mexico, Central America and Caribbean. Zootaxa 1289:1–114

Shinn AP, Pratoomyot J, Bron JE, Paladini G, Brooker EE, Brooker AJ (2015) Economic costs of protistan and metazoan parasites to global mariculture. Parasitology 42:196–270

Sures B (2004) Environmental parasitology: relevancy of parasites in monitoring environmental pollution. Trends Parasitol 20:170–177

Tsigenopoulos C, Durand JD, Ünlü E, Berrebi P (2003) Rapid radiation of the Mediterranean *Luciobarbus* species (Cyprinidae) after the Messinian salinity crisis of the Mediterranean Sea, inferred from mitochondrial phylogenetic analysis. Biol J Linnean Soc 80:207–222

Turan D, Yilmaz BT, Kaya C (2009) *Squalius kottelati*, a new cyprinid species (Teleostei: Cyprinidae) from Orontes River, Turkey. Zootaxa 2270:53–62

Turan D, Kottelat M, Bektaş Y (2011) *Salmo tigridis*, a new species of trout from the Tigris River, Turkey (Teleostei: Salmonidae). Zootaxa 2993:23–33

Turan D, Kaya C, Ekmekçi FG, Doğan E (2013) Three new species of Alburnoides (Teleostei: Cyprinidae) from the Euphrates River, Eastern Anatolia, Turkey. Zootaxa 3754:101–116

Turan D, Bektaş Y, Kaya C, Baycelebi E (2016) *Alburnoides diclensis* (Actinopterygii: Cyprinidae), a new species of cyprinid fish from the upper Tigris River, Turkey. Zootaxa 4067:79–87

Turan D, Kottelat M, Bayçelebi E (2017) *Squalius semae*, a new species of chub from the Euphrates River, Eastern Anatolia (Teleostei: Cyprinidae). Zool Middle East 63:33–42

Ünlü E (1991) Investigation on the biological characteristics of *Capoeta trutta* (Heckel, 1843) living in Tigris River. Turk J Zool 15:12–38

Ünlü E (1999) *Cyprinion macrostomus* Heckel, 1843 ve *Cyprinion kais* Heckel, 1843 (Cyprinidae) türlerinin taksonomisi ve dağılışı üzerine bir araştırma, X. Ulusal Su Ürünleri Sempozyumu:688–697

Ünlü E (2012) Fisheries and aquaculture in the Euphrates and Tigris Basin in Turkey. FAO. 2014. Report of the expert meeting on the review of fisheries and aquaculture activities in the Tigris Euphrates Basin, Erbil, Iraq, 11–12 Nov 2012. FAO Fisheries and Aquaculture Report No. 1079. Rome. 125 p

Ünlü E, Balcı K, Akbayın H (1994) Some biological characteristics of the *Acanthobrama marmid* Heckel,1843 in the Tigris River (Turkey). Tr J Zool 18:131–139

Ünlü E, Özbay C, Kılıç A, Coşkun Y, Şeşen R (1997) Gap'ın Faunaya Etkileri. Türkiye Çevre Vakfı Yayını 125:79–102

Ünlü E, Balcı K, Meriç N (2000) Aspects of biology of *Liza abu* (Mugilidae) in the Tigris River (Turkey). Cybium 24:27–43

Ünlü E, Gaygusuz Ö, Çiçek T, Bilici S, Coad BW (2017) New record and range extension of the big-scale sand smelt *Atherina boyeri* Risso, 1810 (Atherinidae) in the Devegecidi Dam Lake, Tigris River basin, Turkey. J Appl Ichthyol 33:63–68

Vinobaba P (2010) Histopathological changes induced by ergasilid copepod infections on the gills of food fish from Batticaloa lagoon, Sri Lanka. Sri Lanka J Aquat Sci 12:77–87

Vlizzi L, Tarkan AS, Ekmekçi FG (2015) Parasites of the common carp *Cyprinus carpio* L., 1758 (Teleostei: Cyprinidae) from water bodies of Turkey: updated checklist and review for the 1964–2014 period. Turk J Zool 39:545–554

WoRMS Editorial Board (2018) World register of marine species. http://www.marinespecies.org. Accessed 03/2018

Yamaguchi A, Yokoyama H, Ogawa K, Taniuchi T (2003) Use of parasites as biological tags for separating stocks of the starspotted dogfish *Mustelus manazo* in Japan and Taiwan. Fish Sci 69:337–342

Chapter 70
Ornamental Fishes: A Looming Danger for Inland Fish Diversity of Iraq

Laith A. Jawad, Audai Qasim, and Muhammad Al-Janabi

Abstract The aquarium trade has been renowned as a significant passageway for the introduction of offensive species everywhere in the world. Ornamental fishes are nice-looking colourful fishes of nonviolent nature, which are retained as pets in aquarium for recreation purposes. The process of keeping these fishes is the second most popular hobby next to photography. And the ornamental fish industry is one of the most booming ones among the World.

Invasions of an exotic species in the freshwater habitats can be ecologically destructing and economically expensive. Identifying 'hot-spots' of the exotic species and their origin of introduction is important to decide the utmost of the usefulness of invasion isolation plans.

Management efforts for the eradication of the invasive fish species must emphasis on (a) enhancing cataloguing and recognition of fish species in pet shops, (b) teaching buyers of pet fish and store staffs about the hazards caused by pet discharge, and (c) offering improved choices for accountable discarding of surplus fish.

In conclusion, the aquarium fish sector in Iraq is practically uncontrolled given the presence of threatened species, species potentially harmful to humans and species capable of establishing non-indigenous populations, if released into the wild. Therefore, a set of recommendations is given at the end of this chapter for the policyholders in Iraq to adopt and curb the invasiveness of the ornamental and fish food species in the freshwater system of this country.

L. A. Jawad (✉)
School of Environmental and Animal Sciences, Unitec Institute of Technology, Auckland, New Zealand

A. Qasim
Department of Marine Vertebrates, Marine Science Centre, University of Basrah, Basrah, Iraq

M. Al-Janabi
Iraq Natural History Research Centre and Museum, University of Baghdad, Baghdad, Iraq

L. A. Jawad (ed.), *Tigris and Euphrates Rivers: Their Environment from Headwaters to Mouth*, Aquatic Ecology Series 11,
https://doi.org/10.1007/978-3-030-57570-0_70

70.1 Introduction

Non-native fishes that found external to their natural home zones and have dispersion capability. These fish species are frequently introduced to any area either deliberately or else for the devotions of aquaculture, aquarium business, healing value, research, and biological regulator (Singh and Lakra 2011).

Normally, invasion attainment might be motivated by the features of the trespasser, such as inherent developmental rates, food-related grade, and reproductive life history (Fausch et al. 2001). However, investigational and comparative indications show that all being identical, the likelihood of achievement increases with growing impact of the stages of life history of the organism (Kolar and Lodge 2001).

This chapter a short review is given about the potential influences of the exotic fish species in general and the ornamental species in particular in regard to both welfares and hazards. This data will convey a scientific foundation for assessments of the introduction of a certain exotic species into Iraq, limiting the extent of possibly intrusive species and constructing controlling devices to deal with safe aquaculture events. The chapter is also dealt with the issues of conservation and continuous efforts to improve any assessments and interventions.

70.2 Invasive Fish Species of Iraq

Over the last few decades, the process of introduction of an alien fish species has been increased and it continues to be increased (Coad and Hussain 2007; Mutlak and Al-Faisal 2009; Coad 2010; Al-Sa'adi et al. 2012; Jawad et al. 2012; Al-Faisal et al. 2014; Khamees et al. 2013, Mutlak et al. 2017), with the appearance of species that have never thought to be accommodated in the Aquatic environment of Iraq such as freshwater species of neotropical origin (Jawad and Qasim 2019).

In this section, a short review of the invasive fish species has been recorded is given. Those species can be divided into groups depending on their nature and use for the society.

70.2.1 Food Fish Species

70.2.1.1 Family Cyprinidae

Carasobarbus sublimus (Coad and Najafpour, 1997)

This species was initially designated from Iran by Coad and Najafpour (1997). The distribution of *C. sublimus* is confined to the A'la River in Khuzestan Province in Iran (Coad and Najafpour 1997). Recently, Mohamed et al. (2017) reported on specimens of cyprinid fish have been obtained from the higher reaches of the Shatt al-Arab River, and later appeared to be *C. sublimus*. This is the first record of this species from the freshwater system of Iraq.

Gibel Carp, *Carassius gibelio* (Bloch 1782)

The natural distributional range of the gibel carp, *C. gibelio* is from northern Europe to Asia (Tarkan et al. 2012). Jawad et al. (2012) recorded *C. gibelio* from Basrah City, Iraq, differentiating it from *C. auratus* and *C. carassius*. This species found in the Basrah fish market when an upsurge in the Tigris River release, which decreases the salinity of the Shatt al-Arab River level enabling for *C. gibelio* to survive (Coad 1996). The carp farming events in Iran contain many species of carp that are well sustained and fruitful (Salehi 2007; Yousefian, 2011). Such actions could be the reason for the carp species entering freshwater systems in Iran (Nash 1997), which may also be the supply for the adjoining freshwater systems of Iraq. *Carassius gibelio* is now ascertained in Iraq, subsequently being misrecognized with the other two *Carassius* species brought into Iraq: goldfish, *C. auratus* and crucian carp, *C. carassius,* but their influences on the local species have not been known and established yet.

Common Carp, *Cyprinus carpio*

Cyprinus carpio was brought into Iraqi waters over the period 1960–1972. By the mid-1980s the common carp had reached full setting up in the habitats of Iraq and accomplished in eliminating several fish species that had previously controlled the area.

Hemiculter leucisculus (Basilewsky, 1855)

The natural distribution of this fish is from Maritime Russia south through China to Korea and Vietnam. It is brought into Iran, perhaps by coincidence together with commercial consignments of Chinese major carps from central Asia in the former U.S.S.R. and/or Romania in 1967 (Al-Faisal et al. 2014). It was reported from the Hawizah marsh Coad and Hussain (2007) that recorded this species for the first time from the Hawizah marsh and latter from southern marshes of Iraq by Al-Faisal (2008). This species can be taken as a food fish, but it expresses high feeding competition with the native species in Iraq (Coad 2010).

70.2.1.2 Family Cichlidae

Coptodon zillii (Gervais, 1848)

In Iraq, the Redbelly tilapia is recorded from Al Musayyib on the Euphrates River in Iraq (Al-Sa'adi et al. 2012), and recorded at the chief outfall drain in Basrah city (Mutlak and Al-Faisal 2009). It is also reported from the Euphrates River in Syria. The populations of this species inhabiting Shatt al-Arab River have shown segregation so there are two groups, one positioned at the north and middle parts of the river, while the second group is located in the south (Jawad et al. 2018).

Oreochromis aureus (Steindacher, 1864)

This species has been recorded from the main outfall drain in Basrah city (Mutlak and Al-Faisal 2009). It is also recorded from the Euphrates River and other rivers in Syria (Coad 1996). The morphometric and meristic characters of the populations of *O. aureus* inhabiting Shatt al-Arab River showed significant variation indicating the separation of the stock of this species into two groups, the northern group located at the middle and north part of the river and a second group situated at the southern part of River Shatt al-Arab (Jawad et al. 2018).

The Nile Tilapia, *Oreochromis niloticus* (Linnaeus, 1758)

The Nile tilapia is native to Africa, but it has been brought into several countries for aquaculture business (Vreven et al. 1998). Conferring to Herzog (1969), the Nile tilapia was brought into fish ponds on the Tigris River near Baghdad but did not endure during winter (Coad 1996). Another report of this species from southern part of Iraq, Basrah was by Al-Faisal and Mutlak (2015).

70.2.1.3 Family: Xenocyrididae

Ctenopharyngodon idella (Valenciennes in Cuvier and Valenciennes, 1844)

Grass carp was introduced to the freshwater system of Iraq in 1968 from Japan for use in pond culture (Coad 2010). It has a wide distribution range within Iraq. In addition, this species is found in the ponds and marshes of neighbouring Iran (Shireman and Smith 1983). In Iraq, this species is favoured due to its large size and can grow faster, which could be preferred by farmers. In Iran, grass carp used to control weeds in the ponds and marshes as they can get rid of huge amounts of vegetation or destroy them in a short time. In doing so, human diseases that such as schistosomiasis can be kerb down since the snail carried their early stages of these worms live on water plantation (Coad 2010).

Silver Carp, *Hypophthalmichthys molitrix*

This species was brought into the Middle East area such as the marshes and reservoirs in Iran. The record of the silver carp in Iraq goes back to 1992, when Al-Hassan (1994) recorded this species for the first time from the lower part of the Shatt al-Arab River near Abu al-Khaseeb (Al-Hassan 1994). This record was a result of an escape from aquaculture facilitates in the Middle part of Iraq. An additional option for its record is that it has come from the Iranian freshwater system via the Karoon River. This species was reported as being brought into Iran in 1974 and formed a sustained population there (Kamaly 1991; Coad 1992).

Hypophthalmichthys nobilis (Richardson, 1844)

This species is recorded from fish farms and basins in Khuzestan, Iran, it has been also brought into Syrian reservoirs (Coad 1996). It is used in fish farms in Iraq. It is a significant profitable species for farm ponds, feeding on phytoplankton when adult. The record of this species in Iraq could go back to 2005, when a foreign soldier caught one specimen from one of Saddam Palaces in Baghdad. Its effect on the aquatic ecosystem can be seen through the specialized feed of this species on plankton and remove them drastically from the environment. In addition, individuals of this species can turn the water of the habitats living in turbid and not suitable for living of eggs and larvae of other native fish species of Iraq.

70.2.2 Medical Purposes Fish Species

70.2.2.1 Family: Poeciliidae

Mosquito Fish, *Gambusia holbrooki* Girard, 1859

Individuals of this species have been disseminated dynamically through Iraq through a programme to combat malarial disease. This species is native to North and Central America. No specific date is known for the bringing of *Gambusia* to Iraq has been reported (Jawad 2003). *Gambusia* must not be introduced to newly created basins because it contests for food with various cyprinids of commercial status.

The Indian Catfish, *Heteropneustes fossilis* (Bloch, 1794)

The precise date for bringing in this species to Iraq is not known, but it exhibited in large numbers in the inland waters towards the end of the 1950s. This species was introduced intentionally to get rid of the gastropod mollusc, *Bulinus truncatus*, which plays as an intermediate host for the Bilharzia parasite (*Schistosoma haematobium*). Unluckily, this aim was incorrect and as a substitute, this fish displayed an ability to feed frequently on the aquatic plants and organic detritus in addition to crustaceans, aquatic insects and mollusks (Menon 1965). The feeding habit is alike to that of the cyprinid *M. sharpeyi*, but the struggle between the two species is less than that between *M. sharpeyi* and the common carp, probably since of the great variance in their ecological niches. This species is well ascertained in the freshwater system in Iraq and it would be very hard at this stage to get rid of it. Coad (2010) reported that R. Beckman has reported this species from localities at the Euphrates River in Syria and Ali et al. (2016) have recorded this species from the Syrian Inland waters recently.

70.2.3 Ornamental Fish Species

70.2.3.1 Family: Poeciliidae

Sailfin Mollie, *Poecilia latipinna* (Lesueur, 1821)

The distribution of the sailfin molly is confined to the coastal waters of the Gulf of Mexico and the Atlantic Ocean, from southeast North Carolina to the Yucatan. No particular date has been allocated for the first introduction to Iraq, but Coad (2010) thinks and based on personal communication that the date 2006 marks the first introduction of this species into Iraq. Individuals of this species have been reported from shallow marsh areas, and large groups inhabit areas were water flow has been changed. The presence of the sailfin mollie in the freshwater scheme of Iraq is through the deliberate introduction of the ornamental fish trade.

70.2.3.2 Family: Pangassidae

The Striped Catfish, *Pangasianodon hypophthalmus* (Sauvage, 1878)

The original locality of this species is the Mekong River and Maeklong basins. Individuals of the striped catfish usually inhabit large rivers (Rainboth 1996). They are omnivorous (Ukkatawewat 2005), feeding on fish and crustaceans as well as on vegetable debris (Rainboth 1996). The formal date of introduction of this species is 2013, when Khamees et al. (2013) have recorded this species from the Shatt

Al-Basrah canal south of Iraq and concluded that their record of the striped catfish is due to aquarium trade introduction.

70.2.3.3 Family: Lepisosteidae

The Alligator Gar, *Atractosteus spatula* (Lacepède, 1803)

The native locality of this species is North America as it is distributed from the Mississippi River basin from southwestern Ohio and southern Illinois in the USA south to the Gulf of Mexico; Gulf of Mexico Coastal Plain from Enconfina River in Florida, USA to Veracruz, Mexico (Etnier and Starnes 1993). This species feeds mostly on fishes and it is an effective predator in its natural habitat. The eggs of alligator gar are toxic, but their consumption by people is unlikely (Goddard 2009). Furthermore, this fish could damage fishing nets owing to the form of its head and the piercing teeth.

Mutlak et al. (2017) recorded on the capture of a single specimen of alligator gar, from the inland waters of Iraq. The specimen was obtained from the lower reaches of the Shatt al-Arab River, Iraq. This report symbols the first sighting a member of the family Lepisosteidae in the freshwater system of Iraq. Aquarium profession path is the as an alleged agent for the occurrence of this species in the Iraqi freshwater area.

70.2.3.4 Family: Serrasalmidae

The Speckled Piranha, *Serrasalmus maculatus* Kner, 1858

The distribution of this species is confined to the Amazon and Paraguay-Paraná River basins in South America (Jégu 2003). Jawad and Qasim (2019) reported on the presence of this species in the open waters of the Tigris River near the capital city of Baghdad in Iraq. This appearance suggests an introduction of a potentially dangerous species into the inland waters of Iraq. Furthermore, it looks to be well sustained in its new habitat, so that it can be considered an invasive species. Moreover, a new maximum size was also measured with the specimens.

The only Neotropical fish species that has been recorded from the natural waters in the Middle East is the red-bellied pacu, *Piaractus brachypomus*. One specimen of this species has been collected from the freshwater system in Iran (Esmaeili et al. 2017). No other records of piranha have been reported from neighbouring areas. Therefore, the possibility of the speckled piranha to invade the inland waters of Iraq in a natural way is nil.

The speckled piranha may be injurious in several ways to both the humans and the freshwater environment of Iraq, particularly because it is now established in this new habitat.

70.3 Negative Side of the Aquarium Trade

70.3.1 The Disease Risks

Matched to the land animals greater numbers of individual aquatic animals of a greater number of species are transacted. These animals will form a source of diseases if each species and each country of origin is hypothetically linked with disease causative agent abundance. Linked possibilities contain the formation of species which may develop pests that contend with the exotic species. On the other hand, the launch of aquarium plants and the setting up of invertebrates, some of which could play as a middle hosts for pathogens of further animals. Also, the introduction of farmed ornamental fish does not fundamentally diminish the hazard of disease attack.

70.3.2 Ecological Impact

Often, non-native species inclusion to an area is related to other reasons of habitat alteration (McDowall 2006) and administration measures (Lewin et al. 2006). It appears generally that the modest action of bringing an exotic species into an area is a source of ecological influence (Simberloff 2006).

There are many significant aspects frequently deliberated in relation with the non-native fish bring to any area like predation, niche dilapidation, struggle for foods, hybridization, and disease spread. Conferring to a current study to the European Commission in 2008, and support the work of Gozlan (2009), negotiation of these matters in departure is insufficient to define the ecological impact of bringing into an area a fish species. As a substitute, there is a responsibility for the above significant aspects to be linked with a calculable and vital miscarriage of biological or genetic diversity threatening long-term reliability of local species. Consequently, in this instance fish species introduction can be deliberated destructive.

A difference in the ecosystem, and mostly macrophytes or phytoplankton community or biomass, could usually give an outcome such as a widespread alteration of ecosystem utility, with long-term implications for several species and the general integrity of the habitat. This is well shown by the worldwide introduction of *C. carpio* (Pinto et al. 2005), a species that alters aquatic ecosystems via its feeding habitat, causing in displaced plants and the re-suspension of sediments.

70.3.3 Economic Impacts

Economic drive, chiefly globalization via line of work, is the essential human reason for non-native organism forwarding's (Pimentel et al. 2005). The overall aspect is that the non-native species has an unwanted economic impact. It is often demanded that banning, regulation and abolition of non-native species must be a worldwide action that all the countries should join in (Perrings et al. 2002). Consequently, a necessity for best strategies concerning non-native species is desirable that lessen the likelihood of their entrance, bearing in mind the costs of avoidance, and policies that decreasing the harm they therefore perform, in addition to the costs of control or abolition.

70.4 Recommendations

The followings are recommendations selected from research from the literature and summarized and altered to suit the status of aquarium trade in Iraq. These recommendations are set for policy makers to adopt and put in action in order to kerb the process of invasion of freshwater fish species through the aquarium trade.

1. Obtainability of scientific values for taking hazard of bringing ornamental fish species.
2. Treatment is the procedure by which the non-native species is completely eradicated from the receiver habitat.
3. Distinguishing resolution for the problem of intrusiveness and justification responses (as summarized in Britton et al. 2008).
4. The physical removal of the non-native species from infested waters by means of procedures such as seining, electrofishing and gillnetting decreases their richness in the site and permits all non-target species to be reverted (Britton et al. 2009).
5. Producing regulations to forbid the discharge of surplus ornamental pet fish into natural waters.
6. Announcing a plan whereby undesirable pets could be reverted to stores or local zoos and make it possible with operative policy.
7. Public awareness plans for aquarists and sellers are essential and perhaps have the success (Strecker et al. 2011).
8. Hazard valuations should be performed to recognize species that indicate the most hazards of conquering natural waters and creating damaging effects.

References

Al-Faisal A (2008) Taxonomic study for Alburnus mossulensis, Acanthobrama marmid and Hemiculter leucisculus in southern Iraqi marshes. Doctoral dissertation, M. Sc. thesis (in Arabic). College of Agriculture, University of Basra

Al-Faisal AJM, Mutlak FM (2015) First record of the Nile tilapia *Oreochromis niloticus* (Linnaeus, 1758), from the Shatt Al-Arab River, Southern Iraq. Int J Mar Sci 5(38):1–3

Al-Faisal AJM, Mutlak FM, Abdullah SA (2014) Exotic freshwater fishes in the Southern Iraq. Marsh Bull 9(1):65–78

Al-Hassan LAJ (1994) The silver carp, *Hypophthalmichthys molitrix* (Cyprinidae) in Shatt al-Arab River. Cybium 18(2):204

Ali AS, Jawad LA, Saad AA (2016) Confirmation of the presence of the Indian stinging catfish, Heteropneustes fossilis (Bloch, 1794) (Heteropneustidae) in Syrian inland waters. J Appl Ichthyol 32(1):117–119

Al-Sa'adi BA, Mhaisen FT, Al-Rubae AL (2012) The first parasitological report on the redbelly Tilapia *Tilapia zillii*(Gervais, 1848) in Iraq. In: Proceedings of the scientific symposium of Natural History Research Center and Museum, Univ. Baghdad, Baghdad, 20 June 2012, pp 1–6

Britton JR, Brazier M, Davies GD, Chare SI (2008) Case studies on eradicating the Asiatic cyprinid Pseudorasbora parva from fishing lakes in England to prevent their riverine dispersal. Aquat Conserv Mar Freshw Ecosyst 18(6):867–876

Britton JR, Davies GD, Brazier M (2009) Eradication of the invasive *Pseudorasbora parva* results in increased growth and production of native fishes. Ecol Freshw Fish 18:8–14

Coad BW (1992) Freshwater fishes from Iran. A checklist and Bibliography, Ichthyology section, Canadian museum of Nature

Coad BW (1996) Exotic fish species in the Tigris-Euphrates basin. Zool Middle East 13:71–83

Coad BW (2010) Freshwater fishes of Iraq. Pensoft

Coad BW, Hussain NA (2007) First record of the exotic species Hemiculter leucisculus (Actinopterygii: Cyprinidae) in Iraq. Zool Middle East 40(1):107–109

Coad BW, Najafpour N (1997) Barbus sublimus, a new species of cyprinid fish from Khuzestan Province, Iran. Ichthyol Explor Freshw 7:273–278

Esmaeili HR, Masoudi M, Chermahini MA, Esmaeili AH, Zarei F, Ebrahimi M (2017) Invasion of the Neotropical and Nearctic fishes to Iran. FishTaxa 2(3):126–133

Etnier DA, Starnes WC (1993) The fishes of Tennessee. University of Tennessee Press, Knoxville, TN

Fausch KD, Taniguchi Y, Nakano S, Grossman GD, Townsend CR (2001) Flood disturbance regimes influence rainbow trout invasion success among five holarctic regions. Ecol Appl 11 (5):1438–1455

Goddard N (2009) Alligator gar. Ichthyology at the Florida Museum of Natural History. https://www.floridamuseum.ufl.edu/fish/discover/species-profiles/atractosteus-spatula

Gozlan RE (2009) Response by R Gozlan biodiversity crisis and the introduction of non-native fish: solutions, not scapegoats. Fish Fish 10(1):109–110

Herzog P (1969) Untersuchungen über die Parasiten der Süßwasserfische des Irak. Archiv für Fischereiwissenschaft 20:132–147

Jawad L (2003) Impact of environmental change on the freshwater fish fauna of Iraq. Int J Environ Stud 60(6):581–593

Jawad LA, Qasim AM (2019) Neotropical piranha, *Serrasalmus maculatus* (Actinopterygii: Characiformes: Serrasalmidae), in the Tigris River, Baghdad, Iraq—a case of deliberate introduction by the aquarium trade. Acta Ichthyol Piscat 49(2):199–201

Jawad LA, Al-Faisal A, Al-Mukhtar M (2012) Presence of the non-indigenous fish species, Gibel Carp (Carassius gibelio)(Family: Cyprinidae) in Basrah Province Freshwater Systems, Iraq. Water Res Manag 2(4):41–44

Jawad LA, Habbeb FS, Al-Mukhtar MA (2018) Morphometric and meristic characters of two cichlids, *Coptodon zillii* and *Oreochromis aureus* collected from Shatt al-Arab River, Basrah, Iraq. Int J Mar Sci 8(2):12–24

Jégu M (2003) Serrasalminae. In: Reis RE, Kullander SO, Ferraris CJ (eds) Check list of freshwater fishes of South and Central America. Edipucrs, Porto Alegre, pp 182–196

Kamaly A (1991) Effect of cold weather on the growth of mixed carp species in ponds. In: Seventh international ichthyol. congress, Den Haag, 26—30 August. Abstract

Khamees NR, Ali AH, Abed JM, Adday TK (2013) First record of striped catfish Pangasianodon hypophthalmus (Sauvage, 1878)(Pisces: Pangasiidae) from inland waters of Iraq. Basrah J Agric Sci 1:184–197

Kolar CS, Lodge DM (2001) Progress in invasion biology: predicting invaders. Trends Ecol Evol 16(4):199–204

Lewin I, Smoliński A (2006) Rare and vulnerable species in the mollusc communities in the mining subsidence reservoirs of an industrial area (The Katowicka Upland, Upper Silesia, Southern Poland). Limnol Ecol Manag Inland Waters 36(3):181–191

McDowall RM (2006) Crying wolf, crying foul, or crying shame: alien salmonids and a biodiversity crisis in the southern cool-temperate galaxioid fishes? Rev Fish Biol Fish 16(3-4):233–422

Menon AG (1965) Gambusia and mosquito control. J Bonibay Nat Hist Soc 62:567–568

Mohamed ARM, Abood AN, Jawad LA (2017) Presence of Carasobarbus sublimus (Coad & Najafpour, 1997) in the upper reaches of Shatt al-Arab River, Basrah, Iraq. Zool Ecol 27 (1):30–34

Mutlak FM, Al-Faisal AJ (2009) A new record of two exotic cichlids fish *Oreochromis aureus* (Steindacher, 1864) and *Tilapia zilli* (Gervais, 1848) from south of the main outfall drain in Basrah city. Mesop J Mar Sci 24(2):160–170. (In Arabic)

Mutlak F, Jawad L, Al-Faisal A (2017) Atractosteus spatula (Actinopterygii: Lepisosteiformes: Lepisosteidae): a deliberate aquarium trade introduction incidence in the Shatt Al-Arab River, Basrah, Iraq. Acta Ichthyologica et Piscatoria 47(2):205

Nash CE (1997) Iran develops farm skills to meet fishing needs. Fish Farm Int 24:26–28

Perrings C, Williamson M, Barbier EB, Delfino D, Dalmazzone S, Shogren J, Simmons P, Watkinson A (2002) Biological invasion risks and the public good: an economic perspective. Conserv Ecol 6(1)

Pimentel D, Zuniga R, Morrison D (2005) Update on the environmental and economic costs associated with alien-invasive species in the United States. Ecol Econ 52:273–288

Pinto L, Chandrasena N, Pera J, Hawkins P, Eccles D, Sim R (2005) Managing invasive carp (Cyprinus carpio L.) for habitat enhancement at Botany Wetlands, Australia. Aquat Conserv Mar Freshw Ecosyst 15(5):447–462

Rainboth WJ (1996) Fishes of the Cambodian Mekong. FAO species identification field guide for fishery purposes. FAO, Rome. 265 p

Salehi H (2007) An analysis of inputs cost for carp farming sector in 2001 in Iran. Pak J Biol Sci 10 (21):3808–3814

Shireman JV, Smith CR (1983) Synopsis of biological data on the grass carp, Ctenopharyngodon idella (Cuvier and Valenciennes, 1844) (No. 135). Food & Agriculture Organization

Simberloff D (2006) Invasional meltdown 6 years later: important phenomenon, unfortunate metaphor, or both? Ecol Lett 9(8):912–919

Singh AK, Lakra WS (2011) Risk and benefit assessment of alien fish species of the aquaculture and aquarium trade into India. Rev Aquac 3(1):3–18

Strecker AL, Campbell PM, Olden JD (2011) The aquarium trade as an invasion pathway in the Pacific Northwest. Fisheries 36(2):74–85

Tarkan AS, Copp GH, Top N, Özdemir N, Önsoy B, Bilge G, Filiz H, Yapici S, Ekmekci FG, Kirankaya SG, Emiroğlu Ö, Gaygusuz Ö, Oymak A, Özcan GÖ (2012) Are introduced gibel carp Carassius gibelio in Turkey more invasive in artificial than in natural waters? Fish Manag Ecol 19:178–187

Ukkatawewat S (2005) The taxonomic characters and biology of some important freshwater fishes in Thailand. Manuscript. National Inland Fisheries Institute, Department of Fisheries, Ministry of Agriculture, Bangkok, Thailand, 55 p

Vreven EJ, Adépo-Gourène B, Agnèse JF, Teugels GG (1998) Morphometric and allozyme variation in natural populations and cultured strains of the Nile tilapia Oreochromis niloticus (Teleostei, Cichlidae). Genetics and Aquaculture in Africa. ORSTOM, France, pp 175–182

Yousefian M (2011) Genetic variations of common carp (Cyprinus carpio L.) in south-eastern part of Caspian Sea using five microsatellite loci. World J Zool 6:56–60

Part VI
The Health of the Human Community Inhabiting Freshwater Zones

Chapter 71
Aquatic Snails as a Vector of Diseases to the Human in Iran

Mohammad Yakhchali

71.1 Introduction

Iran is the second largest country in the Middle East with an area of about 1,650,000 km^2 divided among 30 Provinces. It is bounded to the north by the Caspian Sea, to the east by Afghanistan and Pakistan, to the south by the Persian Gulf and Sea of Oman, and to the west by Iraq and Turkey. The center and east of the country are largely barren undulating deserts, while there are highlands of Zagros Mountains in the west, along the Turkish and Iraqi borders. In the north, the Alborz mountain ranges rise steeply from a fertile belt around the Caspian Sea (Rokni 2008) (Fig. 71.1).

Human intestinal parasites are one of the important health problems in almost all communities, especially in tropical and subtropical areas. The pattern and frequency of intestinal parasitic infections vary over the time, due to changes in population's behavior and lifestyle (Rokni 2008). Diversity and prevalence of each parasitic infection are influenced by such factors as social, cultural, and economic attributes of the societies, environmental conditions, and life cycle of the parasites (Mehran et al. 2017). Factors that contribute to a significant decrease in intensity and prevalence of the intestinal helminths in a given area, include human activities, such as improved toileting behavior of the children, mass awareness of disease transmission, improvement of individual hygiene, and environmental health conditions, such as improvement in sanitary conditions of the environment, access to quality drinking water sources, construction of toilets in rural areas, and sanitary disposal of human excreta. Most of these contributing factors in prevalence of

M. Yakhchali (✉)
Department of Pathobiology, Parasitology Division, Faculty of Veterinary Medicine, Urmia University, Urmia, West Azerbaijan Province, Iran
e-mail: m.yakhchali@urmia.ac.ir

L. A. Jawad (ed.), *Tigris and Euphrates Rivers: Their Environment from Headwaters to Mouth*, Aquatic Ecology Series 11,
https://doi.org/10.1007/978-3-030-57570-0_71

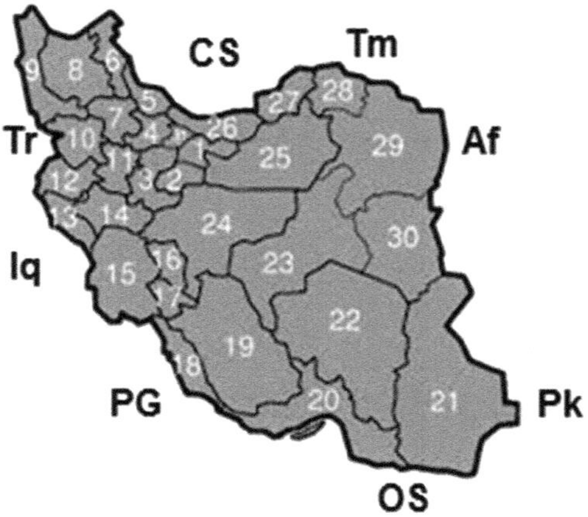

Fig. 71.1 Map of geographical location of the Provinces and neighboring countries of Iran. (1) Tehran, (2) Markazi, (3) Qume, (4) Alborz, (5) Gilan, (6) Ardabil, (7) Zanjan, (8) East Azerbaijan, (9) West Azerbaijan, (10) Kurdistan, (11) Hamadan, (12) Kermanshah, (13) Ilam, (14) Lorestan, (15) Khuzestan, (16) Chaharmahal-va-Bakhtiari, (17) Kohgyluyeh-va-Boyerahmad, (18) Bushehr, (19) Fars, (20) Hormozgan, (21) Sistan-va-Baluchestan, (22) Kerman, (23) Yazd, (24) Isfahan, (25) Semnan, (26) Mazandaran, (27) Golestan, (28) Khorasan Shomali, (29) Khorasan Razavi, (30) Khorasan Junubi. CS, Caspian Sea; PG, Persian Golf; OS, Oman Sea; Tr, Turkey; Iq, Iraq; Tm, Turkmenistan; Af, Afghanistan; Pk, Pakistan

helminthic infection have drastically been improved in Iran in recent years, not only in urban but also in rural sites. Support for this view comes from the results of recent studies that have been conducted in different parts of the country (Akhlaghi et al. 2009; Sharif et al. 2010b; Daryani et al. 2012; Hosseini et al. 2015; Zebardast et al. 2015). Ashtiani et al. (2011) reported that the prevalence of parasitic infections in the children referred to Tehran medical centers has dropped from 8 to 1% during the years 1991–2008. The prevalence of intestinal parasitic infection in the students of South Khorasan Province, eastern Iran, was 0.55% (Taheri et al. 2011). However, the infection rates displayed considerable differences in varying locations. High rates of the infection have been documented in other Iranian Provinces, such as 4.7% in Alborzthe prevalence was 47.2% in Kerman Province (Nasser and Jafar 1997), 22.4% in Chaharmahal-va-Bakhtiari Province (Koroosh 1997), 21.9% and 65.5% in Mazandaran Province (Azam 1995), 32.2% in Khorasan Junubi Province (Saied 1999), 27.3% in Tehran Province (Akhlaghi et al. 2009), 21.2% in Sistan-va-Balouchestan Province (Haghighi et al. 2009), 11.9% in Lorestan Province (Kheirandish et al. 2011), 4.7% in Alborz Province (Nasiri et al. 2009), and 13.35% in Khuzestan Province (Khoshnood et al. 2015). In a recent study by Mehran et al. (2017), it was found than 31.2% of the people in Hormozgan Province, southern Iran, were infected by one or more intestinal parasites. With regard to Veterinary Medicine, many domesticated herbivorous animals have been suffering from metabolic and liver disorders due to the infection with the parasites, losing their economic efficiency (Ghobadi and Yakhchali 2003; Yakhchali and Ghobadi 2005).

Trematodes are a diverse group of endoparasites requiring molluscan and vertebrate animals as intermediate and definitive hosts in their life cycle (Sharif et al. 2010a). As an old age-group being found among the early fossils, mollusks have great diversity in size, distribution, habitat, and utility throughout the world (Yakhchali et al. 2013). Freshwater snails have been important for human livelihood across the history as a food source, tools, and even pets (Sharma et al. 2013). They also play a key role, as intermediate hosts, for part of the developmental stages of the digenean trematodes. In this regard, snails of the order Basomatophorahave substantial contribution to the growth and transmission of parasitic flukes (Imani-Baran et al. 2013). They are intermediate hosts of the foodborne fluke infecting the liver, lungs, urogenital system, and the intestines of humans or animals. The snails engage in the life cycle of at least 71 trematode species belonging to 13 different families (Imani-Baran et al. 2011a). Snail-mediated diseases are among the major categories of helminthic diseases caused by trematode parasites. However, the main snail species involving in the transmission of flukes vary in different geographic regions of Iran (Farahnak and Essalat 2003; Karimi et al. 2004; Imani-Baran et al. 2011b, 2013). Several malacological studies on the freshwater snail fauna of Iran have been performed (Issel 1866; Forcart 1935; Starmühlner and Edlauer 1957; Starmühlener 1965; Eliazian et al. 1979; Mansoorian 1994, 2000; Moghaddam et al. 2004b; Imani-Baran et al. 2011a). The most comprehensive study on the Iranian snails of medical importance was conducted by Gloer and Pesicwho reported 73 snail species from different regions of the country (Bargues et al. 2001). Mansoorian (2000) identified 13 species of mollusks, including 5 operculated shell-bearing and 8 pulmonate snails, in the lotic and lentic freshwater bodies of the northern Provinces of Iran, i.e., Gilan, Golestan and Mazandaran, in the summer of 1992. The radular teeth morphology and morphometry have been found to be the character sets useful for classification of lymnaeid snails (Yakhchali and Jamshidi-Deilamy 2012).

The distribution of freshwater mollusks varies with physical, chemical, and biological attributes of the environments, and their susceptibility to trematodes is highly specific (Sharif et al. 2010a). In Iran, lymnaeid snails play a crucial role in public health and are mostly reported from the lentic waters or slow-moving streams with muddy beds (Imani-Baran et al. 2015). For instance, *Lymnaea* (*Radix*) *gedrosiana* (Annandale and Prashad 1919) has been reported to be a preferred intermediate host for *Fasciolagigantica* (Cobbold 1855) (Mansoorian 2000; Ashrafi et al. 2004; Imani-Baran et al. 2012) and *Trichobilharzia* species (Farahnak and Essalat 2003). It also has a notable participation in the transmission of zoonotic diseases, such as cercarial dermatitis (with infection rates of 0.05% in North and 1.1% in South-West of Iran), fascioliasis (0.35%) (Farahnak and Essalat 2003), Plagiorchids infection (0.1%), *Ornitobilharziaturkestanicum* (Motamedi et al. 2008; Yakhchali et al. 2013), and *Clinostomum* infection (0.2%) (Farahnak and Essalat 2003; Ashrafi et al. 2004; Ghobadi and Farahnak 2004; Athari et al. 2006; Sharif et al. 2010b) in Iran. Furthermore, cercariae infections, i.e., by the trematode genera *Heterophys*, *Echinostoma*, *Cyathocotyle*, and *Phylophtalmus*, in *Melanoides tuberculata* were reported from Khuzestan Province, southwestern Iran (Farahnak et al. 2005a). The heterophyid trematodes were also found to be responsible for human infection in a swampy area located in center of the latter Province (Massoud et al. 1981; Farahnak et al. 2005a). Their cercariae together with those of echinostomatids were detected in *Melanopsis* species in the same Province (Farahnak et al. 2006; Karamian et al. 2011). The contaminated raw vegetables with the eggs of *Heterophyes heterophyes* was also reported in Zahedan, Sistan-va-BaluchestanProvince, Southeast Iran (Ebrahimzadeh et al. 2013). Sharif et al. (2010a, b) also noted the lymnaeid snail's infection with the cercariae of Plagiorchidae, Diplostomatidae, Clinostomatidae, and Echinostomatidae families in Mazandaran.

The infections by parasitic helminths are an important issue throughout the world, mainly in developing countries. They are severe problems for human and veterinary medicine, while also affecting economy, agriculture, and wildlife management (Yakhchali et al. 2014). However, during the last decades, a significant decrease has been noticed in the prevalence of intestinal helminthiasis in Iran (Rokni 2008). In accordance with this, there appears to have also been a marked decline in prevalence and absolute numbers of intestinal helminths in several other Asian countries. This is mostly because of nationwide controlling measures, together with social and economic development (De Silva et al. 2003). In this chapter, common trematode infections, mediated by aquatic snails, in Iranian individuals, including their geographic distribution, epidemiology, and clinical aspects of disease will be discussed.

71.2 Human Fascioliasis

Fascioliasis is recognized by World Health Organization (WHO) as one of the "neglected tropical diseases" with an estimated 2.4–17 million people infected and 180 million at-risk of its infection across the globe (WHO 2010). The public health concern on human fascioliasis increased following the record of 2594 infected individuals in 42 countries of all continents during 1970–1990 (Chen and Mott 1990). Several areas have been described as endemic to human fascioliasis with low to very high intensities and prevalence rates, switching state of the disease from an almost exclusive veterinary problem to of both veterinary and medical importance (Mas-Coma et al. 2005). In Asia, human fascioliasis occurs mainly in Iran and at a lower level in Vietnam (Ashrafi et al. 2015). In Iran, fascioliasis first came to the attention some 50 years ago; while afterward it has impacted several Iranian Provinces (Yadegari et al. 1999; Ashrafi et al. 2004; Moghaddam et al. 2004a; Salahimoghaddam 2009; Imani-Baran et al. 2012; Sarkari et al. 2012; Ashrafi et al. 2015; Hosseini et al. 2015).

Fascioliasis is a cosmopolitan issue being traditionally considered as an important human disease caused by the taxonomically valid digenean trematodes of the two liver fluke species of *Fasciola hepatica* (Linnaeus 1758) and *F. gigantica* (Cobbold 1856) (Trematoda: Fasciolidae) (Mas-Coma and Bargues 1997). *Fasciola hepatica* has succeeded in expanding from the European original geographical areas to form actual colonies in 51 countries of the five continents (Mas-Coma et al. 2005).

Characterization of the definite cause of fascioliasis in each geographic region has been a major global concern from public health points of view. Earlier studies suggested that *F. gigantica* may be the predominant fasciolid species in Gilan Province and some other Provinces, a fact supported by morphology of the liver flukes found in livers of the slaughtered livestock and the widespread distribution of *L.* (*Radix*) *gedrosiana* throughout the endemic areas (Ashrafi et al. 2004). In a recent study by Yakhchali et al. (2015b), the identities of two morphologically described fasciolid species were verified with no host-specific restriction pattern for both the species.

Fascioliasis is estimated to be responsible for the annual loss of thousands of dollars in the Iranian agriculture industries (Imani-Baran et al. 2012). Human cases have also been reported throughout the country over a long time period (Sahba et al. 1972). In Iran, human fascioliasis has been sporadic and several cases of the infection have been registered from different parts of the country until 1989, when an epidemics of human fascioliasis occurred in Gilan Province (Massoud 1989; Assmar et al. 1991; Rokni 2008). Thousands of individuals with clinical signs, manifested mainly by epigastric and right upper-quadrant pains, fever, chill, sweating, weight loss, urticarial, and chest signs accompanied with high eosinophilia, attended in medical centers of the Province during a short period (Ashrafi 2015). The second epidemics of human fascioliasis that happened in 1997 left some 5000 infected people in Gilan Province. Hundreds of the infections have also been recorded following the two outbreaks and thereafter in local health centers. Two northern cities of Rasht and Bandar-Anzali appeared to be the most important endemic zones hosting most of the human cases during the epidemics and inter-epidemics periods (Yadegari et al. 1999; Ashrafi et al. 2004; Moghaddam et al. 2004a; Ashrafi 2015). The prevalence of human fascioliasis in western Mazandaran was higher than that of its eastern parts and 107 human cases were reported from western Mazandaran during 1997–2002 (Moghaddam et al. 2004a; Ashrafi 2015).

Mollusks of the *Galba/Fossaria* (*Lymnaea* Lamarck, 1799, *Galba* Schrank, 1803) and *Radix* (Montfort 1810) groups act as intermediate hosts of numerous trematode parasites, many of which receive concern for their interference with human health. Experimental infections, field observations, and molecular analyses have demonstrated the competence of Iranian lymnaeids nails in transmission of human and animal fascioliasis (Yakhchali et al. 2014, 2015a, b; Ashrafi et al. 2015). To date, seven species of the snails, i.e., *Lymnaea* (*Radix*) *auricularia* (Linnaeus, 1785), *L.* (*Stagnicola*) *palustris*, *L. schiraziensis*, *L. pregra*, *L. stagnalis*, *L.* (*Radix*) *gedrosiana* and *L.* (*Galba*) *truncatula* are reported from different parts of Iran, most of which were found to be the intermediate hosts for *F. hepatica* or *F. gigantica* (Mansoorian 1986, 1994, 2000; Ashrafi et al. 2004, 2007; Hosseini et al. 2004; Karimi et al. 2004; Moghaddam et al. 2004a; Salahimoghaddam 2009; Imani-Baran et al. 2011a,b; Yakhchali et al. 2014, 2015a, b). The susceptibility of different Iranian *Lymnaea* species to *F. hepatica* and *F. gigantica* has been investigated by many researchers (Arfaa et al. 1969; Massoud and Sadjadi 1980; Rohani and Massoud 1990; Shahlapour et al. 1994). In endemic zone of Gilan, *L.* (*Radix*) *gedrosiana*, *L. palustris* and *L. schiraziensis* were prevalent, while *G. truncatula* was rarely observed. Close to the mountainous areas, at foothills, *G. truncatula* appears in superficial clean and cold waters of the small ponds originating from the adjacent mountains. It seems to be the only prevalent intermediate host snail of the trematodes in mountainous regions (Ashrafi 2005). Water pH has also been found to be a crucial factor affecting the life cycle of *Fasciola* in Iran (Yakhchali and Bahramnejad 2016).

In Iran, the role of *L.* (*Galba*) *truncatula* and *L.* (*Radix*) *gedrosiana* in transmission of *F. hepatica* has experimentally been explored (Arfaa et al. 1969). Massoud and Sadjadi (1980) confirmed the susceptibility of *L.* (*Radix*) *gedrosiana* and *L. pregra* to the infection with *F. gigantica* (Arfaa et al. 1969; Massoud and Sadjadi 1980). Rohani and Massoud (1990) also noticed the vulnerability of *L.* (*Radix*) *gedrosiana* to be infected by various numbers of *F. gigantica* miracidia. Shahlapour (1994) indicated that young *L. palustris* snails were susceptible to *F. gigantica*, while its adults were prone to the infection with both *F. hepatica* and *F. gigantica* miracidia. Ashrafi and Mas-Coma (2014) following comprehensive experimental studies, verified *L.* (*Radix*) *gedrosiana* as a potent intermediate host for the transmission of *F. gigantica* in the endemic zone of Gilan Province. *Lymnaeastagnalis*, which has been shown to play a role in the transmission of both the fasciolid species, was found mainly in Khuzestan, Lorestan, Charmahal-va-Bakhtiari, and west Azerbaijan Provinces (Yakhchali et al. 2015a, b; Ashrafi et al. 2015).

In recent years, molecular tools have been employed to clarify the role of lymnaeid snails in life cycle of liver flukes. Molecular studies suggested that *F. hepatica* and *L.* (*Galba*) *truncatula* as well as *F. gigantica* and *L.* (*Radix*) *gedrosiana* are the main synergistic fasciolids and lymnaeids involving in fascioliasis in Iran (Massoud and Sadjadi 1980; Ashrafi et al. 2004, 2007; Yakhchali et al. 2014, 2015a, b). In accordance with this finding, Yakhchali et al. (2014) has molecularly confirmed the presence of the developmental stages of *F. gigantica* in *L. (Radix) auricularia*. Salahimoghaddam et al. (2004) proved the presence of *L. (Stagnicola) palustris* (a secondary intermediate host of *F. hepatica*), *G. truncatula* (the main intermediate host of *F. hepatica*), and *L.* (*Radix*) *gedrosiana* (the main intermediate host of *F. gigantica*) in northern Province of Mazandaran by genetic analyses. *Lymnaea(Radix) auricularia* is, reportedly, the second most common lymnaeid snail, after *L. gedrosiana*, in northwestern Iran (Mansoorian 1986; Imani-Baran et al. 2011b). Ashrafi et al. (Ashrafi et al. 2007, 2015) disclosed the in situ infections of *G. truncatula* and *L.* (*Radix*) *gedrosiana* with the larval stages of *F. hepatica* and *F. gigantica* in endemic zone of Bandar-Anzali, Gilan Province.

Humans are only the accidental host for *F. hepatica* through the ingestion of uncooked and unwashed vegetables. *Fasciola* infection in human may also happen following the ingestion of encysted infective metacercariae attached to the aquatic plants. Several species of wild aquatic or semiaquatic plants are associated with human fascioliasis in Iran (Emami Al-Agha and Athari 1995; Hatami et al. 2000; Ashrafi et al. 2004; Moghaddam et al. 2004a; Sarkari et al. 2012). Many species of freshwater plants were responsible for human infection in different geographical zones, some of which were not necessarily parts of the usual human diet

(Moghaddam et al. 2004a). Ashrafi et al. (2014) highlighted various sources of the infection, exposing travelers and migrants at risk of human fascioliasis in different parts of the world, including Iran. In endemic regions of Iran, such as Gilan Province, *Mentha pulegium*, *Mentha piperita,* and *Eryngium caucasicum* were the main species implicated in the transmission of human fascioliasis. The people in Gilanuse fresh or ground local veggies, olives mixed with walnuts, garlic, and various spices for the preparation of an appetizer called "Zeitoon-e-Parvarde." They also use the veggies to make a herbal paste called "Dalar" to which an excessive amount of salt is added, giving it the name "Green Salt" (Ashrafi et al. 2015). These traditional side dishes could transmit the disease if prepared with fresh aquatic plants containing the stuck metacercariae (Ashrafi et al. 2006). In other parts of the country, *Nasturtium* spp. and *Falcaria vulgaris* (locally named "Paghaze") in Kermanshah Province (Emami Al-Agha and Athari 1995), western Iran, *Eryngium* and *Mentha* species in Mazandaran Province, on the southern coasts of the Caspian Sea (Moghaddam et al. 2004a), *Nasturtium microphyllum* (locally named "Boolaghotti") and *Mentha Longifolia* in Yasuj district, Southwest Iran (Sarkari et al. 2012), and *Nasturtium officinalis* (locally named "Balmak") in Lorestan Province (Kheirandish et al. 2016) were noticed as the main carrier of the infection to the people.

Due to the potential for reoccurrence of large-scale human fascioliasis outbreaks, WHO considers Iran among the six countries, which were known to having a serious problem with the disease(WHO 2007). In spite of the low disease prevalence, a specific pattern of transmission has been proposed for the Caspian Sea areas, namely "Caspian Pattern," defined as a hypoendemic area with large-scale epidemics sometimes affecting more than 10,000 people (Marcos et al. 2006; Mas-Coma 2007). Although the world's largest ever outbreaks of human fascioliasis have occurred in Iran (Massoud 1989; Assmar et al. 1991), recent surveys reveal the disease prevalence rates of 0.4 and 1.2% using coprological and serological methods, respectively (Ashrafi et al. 2012). In the northern Province of Gilan, the reports of local health clinics documented 68–223 cases per year. The first observed case of human fascioliasis in Kermanshah Province belonged to a female patient from Sahneh district (Emami Al-Agha and Athari 1995). There also was a minor emergence of fascioliasis Province with 17 nonfatal cases from different localities of the latter Province, especially in its capital, Kermanshah City, in a 10-year follow up (1998–2008) (Hatami et al. 2012). Recently, a new focus on human fascioliasis in Boyer-Ahmad Township and Yasuj district, Kohkyluyeh-va-Boyerahmad Province, showed anti-*Fasciola* antibodies in the sera of 0.2% and 1.8% of the residents, respectively, with the history of consuming wild freshwater plants (Sarkari et al. 2012; Hosseini et al. 2015). In 2010 and 2011, two cases of human fascioliasis were reported from Ardabil Province, northwestern Iran, following the isolation of *Fasciola* eggs from the liver sections and peritoneal mass (Mohammadi-Ghalehbin et al. 2012; Mowlavi et al. 2015). Later, a supplementary seroprevalence study of human fascioliasis also showed that 1.96% of the examined individuals in the Province were seropositive Province (Asadian et al. 2013). In Lorestan Province,

1.03% of the people were seropositive during the years 2015–2016 (Heydarian et al. 2017).

Clinical signs of *Fasciola* infection in humans begin 1–3 months after the ingestion of metacercariae in two clinical phases; acute and chronic. In 1989, thousands of individuals in Gilan had classical symptoms of *F. hepatica* infection, urticaria, chest pain, and hypereosinophilia (Ashrafi et al. 2015). Fascioliasis caused by *F. hepatica* was also reported in an Iranian man with colicky abdominal pain and bile duct obstruction and dilatation (Bafandeh et al. 2003). Although the number of reported cases infected with *F. hepatica* has increased significantly in recent decades, the removal of living *F. hepatica* from biliary tract was rare (Aminian et al. 2012). However, Niknam et al. (2015) reported the acute cholangitis with three living *F. hepatica* in the biliary tract of a woman with a basket via endoscopic retrograde cholangiopancreatography (ERCP).

Although rare, a few ectopic cases of Iranian human fascioliasis with flukes in non-hepatic sites, such as the thyroid, eye, hepatobiliary, hepatic abscess, ascites, cutaneous painful mass, peritoneal or spleen, pancreas, and kidney involvement, have been reported (Mohsenin and Ebrahimi 1969; Farid-Moayer 1971; Dowlati et al. 1987; Asefzadeh 1998; Riazi 2001; Mansouri and Jamshidi-Fard 2002; Esmaeili et al. 2002; Zali et al. 2004; Dalimi and Jabarvand 2005; Sanei-Taheri et al. 2007; Moghadami and Mardani 2008; Aminian et al. 2012; Mohammadi-Ghalehbin et al. 2012; Behzad et al. 2014). In most of the cases, eosinophilia higher than 20% was also detectable (Ashrafi et al. 2015). Emami Al-Agha and Athari (1995) described a case of ectopic fascioliasis due to cutaneous migration of an immature *F. hepatica* fluke in a woman from western Iran. An allergic fascioliasis due to *F. hepatica* infection, with the main clinical signs of urticaria and asthmatic attacks, was reported from Tehran City, capital Iran (Hanjani et al. 1971).

Various approaches have been used for the diagnosis of different stages of fascioliasis in Iran. The parasitological methods, such as Kato-Katz and Cup-sedimentation, have the highest specificity, but acute and obstructive infections reduce the sensitivity of these methods. Serological tests such as ELISAs (e.g., Fas2-ELISA, CL1-ELISA) have routinely been employed for the detection of anti-*Fasciola* antibodies in the sera in acute phases of the ectopic fascioliasis (Rokni et al. 2002; Rahimi et al. 2011; Ashrafi et al. 2015). Immunological assays were also found suitable for diagnosis of chronic fascioliasis by detecting specific antigens in stool samples and using antibodies against *Fasciola* Excretory–Secretory antigens (Fas2 and CL1) secreted by juvenile and adult *Fasciola* in the sera (Ashrafi 2015). In Iran, during the first outbreak of fascioliasis in Gilan, serological diagnosis was carried out using ELISA and Counter-Current Immune Electrophoresis (CCIE). Rahimi et al. (2011) established a Fast-ELISA method and evaluated its efficacy versus Standard-ELISA for the diagnosis of human fascioliasis in Iran. Mohammad-Alizadeh et al. (2011) suggested Endoscopic Ultrasonography (EUS) as an excellent noninvasive tool for the detection of biliary fascioliasis and showed that it was as accurate as magnetic resonance cholangiopancreatography (MRCP) and endoscopic retrograde cholangiopancreatography (ERCP).

Due to ineffectiveness of praziquantel (PZQ), in Iran, the drug of choice to treat human cases of fascioliasis was triclabendazole (TCBZ, a single dose of 10 mg/kg for 1–3 days). Yadegari et al. (1991) administered PZQ (70 mg/kg) for the treatment of one hundred infected cases. Bithionol (40 mg/kg for 15 days) was also used during the first outbreak of human fascioliasis in the endemic regions of the disease in Iran (Sarshad et al. 1990). Metronidazole (1.5 g/day for 3 weeks) has been regarded as another alternative for the treatment of the disease in the country (Nik-Akhtar and Tabibi 1977; Mansour-Ghanaei et al. 2003).

For proper implementation of control strategies, understanding where the at-risk populations live is essential. Due to the important role of pond snails in life cycle of *Fasciola* species, estimation of the geographic distribution and abundance of the snails and the rate of their infection with different trematode cercariae is pivotal for establishment of the control programs in each region (Imani-Baran et al. 2013). Results of a community-based epidemiological study in the endemic region of Bandar-Anzali in Gilan Province, where the largest ever outbreaks of human fascioliasis have occurred, revealed that human fascioliasis was hypoendemic in the region (Ashrafi et al. 2015). The measures performed for controlling human fascioliasis in Gilan Province included a passive case-finding approach, establishment of effective veterinary and public health standards, awareness of people living in the endemic areas about the danger of consuming raw and uncooked aquatic and semiaquatic plants, and encouraging them to culture the required culinary herbs in safe places (home gardens) (Ashrafi 2015).

71.3 Human Schistosomiasis

Schistosomiasis, also called bilharziasis, has affected millions of people in 76 countries and is the second most common parasitic cause of human death after malaria (Farahnak et al. 2005b). The Global Burden of Disease study in 2013 estimated that almost 300 million people suffer from varying forms of schistosomiasis, with approximately 90% of the cases found in Africa with up to 200,000 deaths each year (Rollinson et al. 2013).

Urinary, or vesical, schistosomiasis reported from Africa, Asia, and the Americas is caused by *Schistosoma haematobium* which is found to also be the causative agent of urinary schistosomiasis in Iran (Sahba and Malek 1979; Massoud et al. 1982; Mansoorian 1994; Farahnak and Essalat 2003). The endemic area for the disease is Khuzestan Province with a wide range of geographic areas and mostly a tropical climate. The Province has been spotted as a suitable locale for propagation of the intermediate host snails, in particular, *L.*(*Radix*) *gedrosiana* (Chu et al. 1968; Arfaa et al. 1976; Massoud et al. 1982; Noorpisheh and Farahnak 2011). In addition, seasonal variation in the population density of *Bulinus truncatus* was also found to affect the transmission of the urinary schistosomiasis in the region (Arfaa et al. 1967; Chu et al. 1968). Urinary schistosomiasis is also prevalent in southeastern Iraq, a neighboring territory of Khuzestan Province. In Khuzestan Province, Arfaa et al.

(1967, 1970) found that 10–11.3% of the patients were infected with this trematode. Since 1976, however, the prevalence of human infection has decreased to 0.7% (Massoud et al. 1982), and by 1995, it was only 0.02% (Kejbafzadeh et al. 1995). This downward trend in the prevalence rate was also observed during a 10-year (1980–1989) field survey in Khuzestan Province by Kejbafzadeh et al. (1995). So that, the prevalence decreased from 0.65% in 1980 to much lower values in 1988 (0.021%) and 1989 (0.042%). Most of the cases were diagnosed by urine examination, but a few were detected by bladder biopsy. Over this period, strict control measures, including mass chemotherapy and mollusciciding, were employed to help decrease the prevalence of urinary schistosomiasis. One age-group of patients aged 20–29 years was found to have the greatest risk of infection (Kejbafzadeh et al. 1995). Similarly, Mombeini and Kheradmand (2005) accounted 245 cases of infection with the disease in 1981 but only 16 in 1993 in Khuzestan Province. Although a total of 1158 cases were recorded throughout the Province between 1981 and 1990, number of the impacted individuals dropped to only 98 during 1991–2000. While no cases were encountered in 2000 and 2001, there was an uncertainty on elimination of the disease and eradication of *S. haematobium* (Mombeini and Kheradmand 2005). Despite the fact that thousands of the evacuees entering the country from the surrounding regions have probably added to the reservoir of infection and complicated the attempts for effective control, no Iranian cases of human schistosomiasis have been reported since 2001 (Gooya 2004; Mombeni and Kheradmand 2005). Farahnak et al. (2008) because of the lack of *S. haematobium* furcocercariae in the examined *Bulinustruncatus*snails concluded that there were no cases of urinary schistosomiasis in Khuzestan Province. Nevertheless, the construction of new water routes, as potential snail habitats, and the proximity of the disease-endemic regions (e.g., Iraq) both pose threats to the effective control of schistosomiasis, and the Iranian Ministry of Health is therefore maintaining its monitoring of the disease. Efficient control has been achieved by molluscstep, the training of healthcare workers, screening for the infection, and the prompt treatment of the confirmed cases. In this regard, Farahnak et al. (2005b) suggested that analyzing the interaction of *Biomphalaria* snail hemocytes with miracidia of *Schistosoma* could be useful for evaluation of cellular immunity of this snail against the infection, and ultimately, for the schistosomiasis control programs in Iran.

71.4 Human Cercarial Dermatitis

Human Cercarial Dermatitis, or swimmer's itch, is a human disorder caused by the schistosome parasites of the genus *Trichobilharzia* living initially in animals, mainly birds, with a worldwide distribution if they enter into a nonspecific host for completion of their life cycle. It is also a potential economic hazard to persons who have invested in fisheries and tourism industries. Cercarial dermatitis should be considered a potential risk whenever warm-blooded and molluskan hosts share a water resource that is also used by humans. More emphasis has been given to the temporal occurrence of this disease in humans (Jauhari and Nongthomban 2014). *Trichobilharzia* is the only genus of the family Schistosomatidae whose furcocercariae have morphologically been described from Iran (Athari et al. 2006). Adult helminths of the genera *Ornithobilharzia* and *Trichobilharzia* have been isolated from birds and other animals in south, southwest, northwest, and north of Iran (Maleki et al. 1994; Athari et al. 2006; Yakhchali et al. 2013, 2016). Some of the geographic regions in Iran, such as north and northwest of the country, possess the climates and habitats well-suited for the completion of the life cycle of the trematodes such as *Trichobilharzia* species (Yakhchali et al. 2015a, b). Sahba and Malek (1979) documented several cases of cercarial dermatitis in the Iranian coastal areas of the Caspian Sea. The major snail families which are known as the intermediate hosts for *Trichobilharzia* species are Lymnaeidae (Rafinesque 1815), Planorbidae (Rafinesque 1815), Pleuroceridae (Fischer 1885), and Physidae (Fitzinger 1833; Brant and Loker 2009). In Iran, snails of the genera *Lymnaea* and *Planorbis* have been designated as the main intermediate hosts of the causative agents of cercarial dermatitis. Various cercariae groups including xiphidiocercariae, strigeidcercariae, echinostome cercariae, and animal schistosome cercariae were isolated from *L. (Radix) gedrosiana* (Farahnak and Essalat 2003). Yakhchali et al. (2016), using molecular analysis, discovered infection of the natural populations of the snail *L. (Radix) auricularia* with two trichobilharzian species of *T. szidati* and *T. franki* in northwest Iran. These snails, chiefly *L. (Radix) gedrosiana*, were abundant in the water canals and observed mainly in the oxygenated marginal waters. They were also living in different ecological habitats, such as streams, swamps, ponds, springs (standing waters), ditches of agriculture canals and drains (flowing waters). Many water canals and ponds while are suitable for the living snails, have been used for swimming, drinking, and washing by the local people. Upon the presence of the infected *Lymnaea* species, water resources could be contaminated by the emerging furcocercariae, including bird schistosomes, which could consequently attack the swimmer's skin in summer (Farahnak and Essalat 2003). There were a few reports of cercarial dermatitis caused by *Trichobilharzia* in Iran, typically in the areas where people swam with uncovered bodies (Farahnak and Essalat 2003). The furcocercariae of animal schistosomes generating swimmer's itch was also found in the farmers working in the rice fields in northern Iran (Sahba and Malek 1979). In Khuzestan Province, 1.1% of the examined people had the clinical signs of cercarial dermatitis, and the furcocercariae of *Trichobilharzia* species were detected in 2.4% of the inspected *L.(Radix) gedrosiana* snails. Majority of the cercarial dermatitis cases were observed as maculopapular rashes in the children. According to their claims, a burning or itchy sensation appeared after water drops had dried on their skin. Symptoms started with erythema (reddening of the skin) and itching, and several hours later, maculopapular rashes have emerged (Farahnak and Essalat 2003).

Diagnosis of cercarial dermatitis is based on a history of exposure to the water that may be contaminated with furcocercariae. Infected people use saline water for the treatment of itching and dermatitis. While the itching is reduced by saline water, the patients are usually referred to physicians for lessening the local swelling and itching by prescribed topical cortisone (steroid) cream (Farahnak and Essalat 2003). Control of the disease in natural waters is, however, difficult, requiring strict maintenance of bodies of water and if necessary, the use of molluscicides such as niclosamide (Byluscide) for ponds, springs, canals, and small swamps (Farahnak and Essalat 2003).

71.5 Human Dicrocoeliasis

Human dicrocoeliasis is a zoonotic liver fluke infection caused by the trematode helminth *Dicrocoelium dendriticum* (Dujardin 1845), also called "lancet liver fluke" (Bizhani et al. 2017). The parasite has been reported from most parts of Europe and Asia, while also has foci in North America and Australia. Paleo-parasitological evidence reveals the antiquity of *Dicrocoelium* species since 550,000 years BP in Europe (Jouy-Avantin et al. 1999). Review of the literature showed the records of these species in several archeological sites of different regions in the ancient world (Le Bailly and Bouchet 2010). Study of the parasites of ancient times is a new line of research in Iran initiated by the finding of several helminths, including *D. dendriticum* eggs in a cemetery of the Bronze Age in southwestern Iran (Mowlavi et al. 2015). Bizhani et al. (2017) announced the isolation of *D. dendriticum* eggs in Kiasar archeological site in the Caspian Sea shore in northern Iran, dating back to 250 BC.

The lancet fluke, *D. dendriticum*, is principally in the bile ducts of different herbivorous animals, i.e., cattle, sheep, goats, and pigs, and is a less common liver parasite in human. Ingestion of the infected ants harboring metacercarial stages of the fluke is fundamental in transmission of the infection to humans (Manga-González et al. 2001). Humans may, infrequently, acquire the infection through the ingestion of raw vegetables containing infective metacercariae anchored in ants (Cengiz et al. 2010). True cases of human infection with *D. dendriticum* arise when people accidentally or deliberately eat ants. The infection with *D. dendriticum* and co-infection with *F. hepatica* were previously reported from Gonbad-e-KavoosCity in northeast and Isfahan Province in central Iran (Farid-Moayer 1971; Arfaa et al. 1977). Infection with *D. dendriticum* has also been reported in three residents of the southern Caspian Sea coast in 1982 (Sohrabi 1982). Mahmoodi et al. (2010) disclosed the infection of a male individual with *D. dendriticum* upon a trip to southern regions of Iran, with severe diarrhea after controlling other symptoms and signs. A gluten-free diet could depress the symptoms and signs in 2 months, while diarrhea began again. The patient denied consumption of liver or liver products within the next several weeks. This eliminated the chance of spurious infection through the infected liver. The stool examination (three times) revealed *Dicrocoelium* ova and infection was proved. Treatment was restarted with Bithionol (PZQ was not accessible) thereafter further stool examinations for parasites were negative, and diarrhea was disappeared.

References

Akhlaghi L, Shamseddin J, Meamar AR, Razmjou E, Oormazdi H (2009) Frequency of intestinal parasites in Tehran. Iran J Parasitol 4(2):44–47

Aminian K, Rezayat KA, Shafaghi A, Tanhaeevash R (2012) Living *Fasciola hepatica* in biliary tree: a case report. Ann Hepatol 11:395–398

Arfaa F, Bijan H, Farahmandian I (1967) Present status of urinary bilharziasis in Iran. Trans R Soc Trop Med Hyg 61:358–367

Arfaa F, Movafagh K, Mahdavi M (1969) *Lymnaea gedrosiana*, an intermediate host of *Fasciola hepatica* in Iran. J Parasitol 55(1):134–135

Arfaa F, Farahmandian I, Sahba GH, Bijan H (1970) Progress towards the control of bilharzizsis in Iran. Trans R Soc Trop Med Hyg 64:912–917

Arfaa F, Sahba GH, Massoud J (1976) Studies on the biology of *Bulinus truncatus* under laboratory conditions. Iran J Publ Health 5(1):1–4

Arfaa F, Keivan E, Ghadirian E, Missaghian GH (1977) Report of one human *Dicrocoelium* from Gonbad region, Iran. Iran J Publ Health 6(1):3–4

Asadian S, Mohebali M, Mahmoudi M, Kia EB, Heidari Z, Asgari M, Aryaeipour M, Moradi S, Rokni MB (2013) Seroprevalence of human fascioliasis in Meshkin-Shahr district, Ardabil Province, northwestern Iran in 2012. Iran J Parasitol 8(4):516–521

Asefzadeh M (1998) A case of fascioliasis in Bu-Ali-Sina hospital in Qazvin city, Iran. J Qazvin Univ Med Sci 2(1):71–74

Ashrafi K (2005) Study of human and animal fascioliasis and determination of genotype and phenotype of fasciolids and their intermediate hosts in Gilan Province. PhD. thesis, University of Tehran Medical Sciences

Ashrafi K (2015) The status of human and animal fascioliasis in Iran: a narrative review article. Iran J Parasitol 10(3):306–328

Ashrafi K, Mas-Coma S (2014) *Fasciola gigantica* transmission in the zoonotic fascioliasis endemic lowlands of Gilan, Iran: experimental assessment. Vet Parasitol 205:96–106

Ashrafi K, Massoud J, Holakouie Naieni K, Mahmoodi M, Jo-Afshani MA, Valero MA, Fuentes MV, Khoubbane M, Artigas P, Bargues MD, Mas-Coma S (2004) Evidence suggesting that *Fasciola gigantica* may be the most prevalent causal agent of fascioliasis in the endemic province of Gilan, northern Iran. Iran J Pub Health 33(4):31–37

Ashrafi K, Valero MA, Panova M, Periago MV, Massoud J, Mas-Coma S (2006) Phenotypic analysis of adults of *Fasciola hepatica*, *F. gigantica* and intermediate forms from the endemic region of Gilan, Iran. Parasitol Int 55(4):249–260

Ashrafi K, Massoud J, Holakouie Naieni K, Jo-Afshani MA, Mahmoodi M, Ebadati N, Rezvani SM, Artigas P, Bargues MD, Mas-Coma S (2007) Nuclear Ribosomal DNA ITS-2 sequence characterization of *Fasciola hepatica* and *Galba truncatula*. Iran J Publ Health 36(4):42–49

Ashrafi K, Saadat F, Rahmati B (2012) Evaluating the level of endemicity of human fascioliasis in rural communities of Bandar-e-anzali district; an important step for implementation of disease control programs. In: First international and 8thnational congress of parasitology and parasitic diseases in Iran, Tehran, p 121

Ashrafi K, Bargues MD, O'Neill S, Mas-Coma S (2014) Fascioliasis: a worldwide parasitic disease of importance in travel medicine. Travel Med Infect Dis 12:636–649

Ashrafi K, Saadat F, O'Neill S, Rahmati B, Tahmasebi HA, Dalton JP, Nadim A, Asadinezhad M, Rezvani SM (2015) The endemicity of human fascioliasis in Gilan province, northern Iran: the baseline for implementation of control strategies. Iran J Public Health 44(4):501–511

Ashtiani MT, Monajemzadeh M, Saghi B, Shams S, Mortazavi SH, Khaki S, Mohseni N, Kashi L, Nikmanesh B (2011) Prevalence of intestinal parasites among children referred to Children's Medical Center during 18 years (1991-2008), Tehran, Iran. Ann Trop Med Parasitol 105 (7):507–512

Assmar M, Milaninia A, Amir-Khani A, Yadegari D, Forghan-Parast K, Nahravanian H, Piazak N, Esmayli A, Hovanesian A, Valadkhani Z (1991) Seroepidemiological investigation of fascioliasis in northern Iran. Med J Islamic Republic Iran 5(1-2):23–27
Athari A, Gohar-Dehi SH, Rostami M, Jalilian MD (2006) Determination of definitive and intermediate hosts of cercarial dermatitis-producing agents in Northern Iran. Arch Iran Med 9:11–15
Azam S (1995) Frequency distribution of intestinal worms in Behshar and Nour cities. School of Health Sciences and Research Institute, Tehran University
Bafandeh Y, Daghestani D, Rad S (2003) Biliary Tract obstruction due to *Fasciola hepatica* managed by ERCP. Iran J Med Sci 28(1):43–45
Bargues MD, Vigo M, Horak P, Dvorak J, Patzner RA, Pointier JP, Jackiewicz M, Meier-Brook C, Mas-Coma S (2001) European Lymnaeidae (Mollusca: Gastropoda), intermediate hosts of trematodiases, based on nuclear ribosomal DNA ITS-2 sequences. Inf Gen Evol 1(2):85–107
Behzad C, Lahmi F, Iranshahi M, Mohammad-Alizadeh AH (2014) Finding of biliary fascioliasis by endoscopic ultrasonography in a patient with eosinophilic liver abscess. Case Rep Gastroenterol 8(2):310–318
Bizhani N, Sharifi AM, Rokni MB, Dupouy Camet J, Rezaeian M, Fallah Kiapi M, Paknezhd N, Najafi F, Mowlavi G (2017) *Dicrocoelium* egg identified in an ancient cemetery in Kiasar Archeological Site, Northern Iran, Dated Back 247 BC–224 AD. Iran J Public Health 46 (6):792–795
Brant SV, Loker ES (2009) Molecular systematic of the avian schistosome genus *Trichobilharzia* (Trematoda: Schistosomatidae) in North America. J Parasitol 95(4):941–963
Cengiz ZT, Yilmaz H, Dulger AC, Cicek M (2010) Human infection with *Dicrocoelium dendriticum* in Turkey. Ann Saudi Med 30:159–161
Chen MG, Mott KE (1990) Progress in assessment of morbidity due to *Fasciola hepatica* infection: a review of recent literature. Trop Dis Bull 87(4):1–38
Chu KY, Massoud J, Arfaa F (1968) Distribution and ecology of *Bulinus truncatus* in Khuzestan. Bull World Health Organ 39:607–637
Dalimi A, Jabarvand M (2005) *Fasciolahepatica* in the human eye. Trans R Soc Trop Med Hyg 99 (10):798–800
Daryani A, Sharif M, Nasrolahei M, Khalilian A, Mohammadi A, Barzegar G (2012) Epidemiological survey of the prevalence of intestinal parasites among schoolchildren in Sari, northern Iran. Trans R Soc Trop Med Hyg 106(8):455–459
De Silva NR, Brooker S, Hotez PJ, Montresor A, Engels D, Savioli L (2003) Soil-transmitted helminths infections: updating the global picture. Trends Parasitol 19(12):547–551
Dowlati Y, Dowlati A, Seyyedi BV (1987) Cutaneous fascioliasis. Med J Islamic Republic Iran 1:62–65
Ebrahimzadeh A, Jamshidi A, Mohammadi S (2013) The parasitic contamination of raw vegetables consumed in Zahedan, Iran. Health Scope 1(4):205–209
Eliazian M, Tamiji Y, Akbar-Zadeh M, Hagh-Nazari J (1979) Snails from the northern parts of Iran (Caspian Sea). Arch Inst Razi 31:29–36
Emami Al-Agha M, Athari A (1995) Subcutaneous fascioliasis: a rare complication. Med J Islamic Republic Iran 8(4):269–270
Esmaeili M, Khatami G, Najafi M, Khoda-dad A (2002) Fascioliasis in a child with ascites treated with triclabendazole. J Babol Univ Med Sci 4(2):44–47
Farahnak A, Essalat M (2003) A study on cercarial dermatitis in Khuzestan province, south western Iran. BMC Public Health 3:1–4
Farahnak A, Setodeh S, Moebedi I (2005a) A faunistic survey on the cercariae (Larva;Trematode; Parasites) of fresh water snails:I. *Melanoidestuberculata* (Thiaridae; Snails)and their role on transmission diseasesin Khuzestan Province of South WesternIran. Arch Razi Inst 59:113–119
Farahnak A, Ghobadi H, Bagheri F (2005b) Morphology of *Biomphalaria glabrata* hemocytes and their interaction with miracidium of *Schistosoma mansoni*. Acta Medica Iranica 43(3):209–2011

Farahnak A, Vafaie-Darian R, Mobedi I (2006) A faunistic survey of cercariae from fresh water snails: *Melanopsis* spp. and their role in transmission diseases. Iran J Public Health 35(4):70–74

Farahnak A, Mobedi I, Eshraghian MR (2008) The use of cercariae infection of the *Bulinus truncatus* snail for evaluation of schistosomiasis control in Iran. Ann Saudi Med 28(1):58

Farid-Moayer H (1971) Human infection with *Fasciola hepatica* and *Dicrocoelium dendriticum* in Isfahan area, central Iran. J Parasitol 2:160

Forcart L (1935) Die Mollusken der nordpersischen Provinz Masenderan und ihre tiergeographische Bedeutung. Archive für Naturgeschichte Leipzig NF 4(3):404–447

Ghobadi H, Farahnak A (2004) A faunistic survey on the cercariae of *Bellamya* (*Viviparus*) *bengalensis* snails and their zoonotic importance. Iran J Publ Health 33(2):38–42

Ghobadi K, Yakhchali M (2003) Survey of helminthes infection rate of liver and economical losses in Urmia slaughterhouse, Iran. J Fac Vet Med Tehran Univ 58(1):19–20

Gooya M (2004) Risks of travel to Iraq. Mon J Minist Health Med Educ Iran 8:74–75

Haghighi A, Khorashad AS, Nazemalhosseini Mojarad E, Kazemi B, Rostami Nejad M, Rasti S (2009) Frequency of enteric protozoan parasites among patients with gastrointestinal complaints in medical centers of Zahedan, Iran. Trans R Soc Trop Med Hyg 103(5):452–454

Hanjani AA, Nikakhatar F, Arfaa F, Khakpour M, Rashed MA (1971) A case of infection with *Fasciola hepatica* with allergic manifestations. Acta Medica Iranica 14:149–151

Hatami H, Asmar M, Massoud J, Aryanifar S, Mansori F, Fatemi S, Shahrezaei A, Rezaei R, Namdaritabar H (2000) Report of the first outbreak of human fasciolosis in Kermanshah Province. Moddares J 3:79–87

Hatami H, Asmar M, Massoud J, Mansouri F, Namdaritabar H, Ramazankhani A (2012) The first epidemic and new emerging human fascioliasis in Kermanshah, western Iran, and a ten-year follow up, 1998-2008. Int J Prev Med 3(4):266–272

Heydarian P, Ashrafi K, Mohebali M, Kia EB, Aryaeipour M, Sharafi AC, Mokayeri H, Bozorgomid A, Rokni MB (2017) Seroprevalence of human fasciolosis in Lorestan Province, western Iran, in 2015-2016. Iran J Parasitol 12(3):389–397

Hosseini SH, Vaezi V, Jafari G (2004) Epidemiology of fasciolosis in Kermanshah Province. J Vet Fac Tehran Univ 59(3):202–206

Hosseini G, Sarkari B, Moshfe A, Motazedian MH, Khabisi S (2015) Epidemiology of human fascioliasis and intestinal helminths in rural areas of Boyer-Ahmad township, southwest Iran; a population based study. Iran J Public Health 44(11):1520–1525

Imani-Baran A, Yakhchali M, Malekzadeh-Viayeh R (2011a) A study on the geographical distribution and diversity of Lymnaeidae snails in West Azerbaijan Province, Iran. Pajouhesh Sazandegi 82(4):53–63

Imani-Baran A, Yakhchali M, Malekzadeh-Viayeh R, Farhangpajuh F (2011b) Prevalence of cercariae infection in *Lymnaea auricularia* (Linnaeus, 1758) in North West of Iran. Vet Res Forum 2(2):121–127

Imani-Baran A, Yakhchali M, Malekzadeh-Viayeh R, Paktarmani R (2012) Molecular study for detecting the prevalence of *Fasciola gigantica* in field-collected snails of *Radix gedrosiana* (Pulmonata: Lymnaeidae) in northwestern Iran. Vet Parasitol 189:374–377

Imani-Baran A, Yakhchali M, Malekzadeh-Viayeh R, Farahnak A (2013) Seasonal and geographic distribution of cercarial infection in *Lymnaeagedrosiana* (Pulmunata: Lymnaeidae) in North West Iran. Iran J Parasitol 8(3):423–429

Imani-Baran A, Yakhchali M, Malekzadeh-Viayeh R, Sehhatnia B, Darvishzadeh R (2015) Ecology of snail family Lymnaeidae and effects of certain chemical components on their distribution in aquatic habitats of West Azerbaijan, Iran. J Vet Res 70(4):433–440

Issel A (1866) Catalogo dei molluschi raccoltidalla Missione Italiana in Persia aggiuntavi la descrizione delle specie nuove o poco note. Memorie della RealeAccademia delle Scienze di Torino, Serie Seconda 23:387–442

Jauhari RK, Nongthomban PD (2014) Occurrence of a snail borne disease, cercarial dermatitis (Swimmer Itch) in Doon Valley (Uttarakhand), India. Iranian J Publ Health 43(2):162–167

Jouy-Avantin F, Combes C, Lumley H, Miskovsky JC, Mone H (1999) Helminth eggs in animal coprolites from a Middle Pleistocene site in Europe. J Parasitol 85:376–379
Karamian M, Aldhoun JA, Maraghi S, Hatam G, Farhangmehr B, Sadjadi SM (2011) Parasitological and molecular study of the furcocercariae from *Melanoides tuberculata* as a probable agent of cercarial dermatitis. Parasitol Res 108(4):955–962
Karimi GR, Derakhshanfar M, Paykari H (2004) Population density, trematode infection and ecology of *Lymnaea* snails in Shadegan, Iran. Arch Razi Ins 58:125–129
Kejbafzadeh AM, Hoghooghi-Rad N, Shenyari I, Nemat R (1995) Progress in urinary schistosomiasis control measures in Iran. J Trop Med Hyg 98:131–135
Kheirandish F, Tarahi M, Haghighi A, Nazemalhosseini ME, Kheirandish M (2011) Prevalence of intestinal parasites in bakery workers in khorramabad, lorestan Iran. Iran J Parasitol 6(4):76–83
Kheirandish F, Kayed MH, Ezatpour B, Anbari K, Karimi-Rouzbahani HR, Chegenisharafi A, Zendehdel A, Bizhani N, Rokni MB (2016) Seroprevalence of human fascioliasis in Pirabad, Lorestan Province, Western Iran. Iran J Parasitol 11(1):24–29
Khoshnood S, Saki J, Rafiei A, Alizadeh K (2015) Prevalence of intestinal parasitic infections among people in Baghmalek during 2013-2014. Jentashapir J Health Res 6(2):e24792
Koroosh MN (1997) Prevalence of intestinal parasitic infestations in patients attending the parasitology laboratory in Shahrekord. Iran J Parasitol 131a:11
Le Bailly M, Bouchet F (2010) Ancient dicrocoeliosis: occurrence, distribution and migration. Acta Trop 115:175–180
Mahmoodi M, Zahraei Ramazani AR, Izadi S, Najafian J (2010) Dicrocoeliiasis with signs of chronic diarrhea. Acta Medica Iranica 48(3):198–199
Maleki M, Khodokaram A, Oryan A, Aslani M, Housseinzadeh SA, Sadjjadi SM (1994) Pathological findings in ornithobilharziasis in the herds of sheep and goat of Assyrians in Fars Province, Iran. Res Reconstruct 24:143
Manga-González MY, Gonzalez-Lanza C, Cabanas E, Campo R (2001) Contributions to and review of dicrocoeliosis, with special reference to the intermediate hosts of *Dicrocoelium dendriticum*. Parasitology 123:91–114
Mansoorian AB (1986) A practical guideline for identification of Iranian freshwater snails. Iran J Public Health 15(1-2):41–54
Mansoorian A (1994) Final report entitled: freshwater snails of Iran. Technical Series No. 2145/ 1374,Tehran:School of Public Health & Institute of Public Health Research (Publ.)
Mansoorian A (2000) Some freshwater snails from Northern Iran. Iranian J Pub Health 29 (1-4):77–82
Mansour-Ghanaei F, Shafaghi A, Fallah M (2003) The effect of metronidazole in treating human fascioliasis. Med Sci Monit 9(10):127–130
Mansouri D, Jamshidi-Fard A (2002) A case re-port of fascioliasis with rare manifestations of recurrent fever and ascites. J Babol Univ Med Sci 4(4):50–54
Marcos LA, Maco L, Salmavides F, Tereshima A, Espinoza JR, Gotuzzo E (2006) Risk factors for *Fasciola hepatica* infection in children: a case-control study. Trans R Soc Trop Med Hyg 100 (2):158–166
Mas-Coma S (2007) Parasitic diseases, global change and the developing world: the example of emerging fascioliasis. Scientia Parasitologica 1:10–21
Mas-Coma S, Bargues MD (1997) Human liver flukes: a review. Res Rev Parasitol 57 (3-4):145–218
Mas-Coma S, Bargues MD, Valero MA (2005) Fascioliasis and other plant-borne trematode zoonoses. Int J Parasitol 35:1255–1278
Massoud J (1989) Fasciolosis outbreak in man and drug test (Triclabandazole) in Caspian littoral, northern part of Iran. Bull Soc Fr Parasitol 8(Suppl 1):438
Massoud J, Sadjadi S (1980) Susceptibility of different species of *Lymnaea* snails to miracidia of *Fasciola gigantica* and *F. hepatica* in Iran. J Helminthol 54(3):201–202

Massoud J, Jalali H, Reza M (1981) Studies on trematodes of the family Heterophyidae (Odhner, 1914) in Iran: 1. Preliminary epidemiological surveys in man and carnivores in Khuzestan. J Helminthol 55(4):255–260
Massoud J, Arfaa F, Farahmandian I, Ardalan A, Mansoorian A (1982) Progress in national schistosomiasis control program in Iran. Bull World Health Organ 60:577–582
Mehran S, Haghighi A, Khazan H, Azargashb E, Ghasemian-Moghadam HR (2017) Epidemiological study of intestinal parasites in referred individuals to the medical centers' laboratories of Haji-Abad City, Hormozgan Province, Iran, 2015. Novelty Biomed 3:119–126
Moghadami M, Mardani M (2008) *Fasciola hepatica*: a cause of obstructive jaundice in an elderly man from Iran. Saudi J Gastroenterol 14(4):208–210
Moghaddam AS, Massoud J, Mahmoodi M, Mahvi AH, Periago MV, Artigas P, Fuentes MV, Bargues MD, Mas-Coma S (2004a) Human and animal fascioliasis in Mazandaran province, northern Iran. Parasitol Res 94(1):61–69
Moghaddam AS, Massoud J, Mahmoodi M, Khoubbane M, Artigas P, Periago MV, Fuentes MV, Bargues MD, Mas-Coma S (2004b) Distributional outline of lymnaeid snails (Gastropoda) in the fascioliasis endemic area of Mazandaran, Iran. Acta Parasitol 49(2):145–152
Mohammad-Alizadeh AH, Roshani M, Lahmi F, Davoodi NA, Rostami Nejad M, Seyyedmajidi MR, Zali MR (2011) Cholangiocarcinoma in magnetic resonance cholangiopancreatography and fascioliasis in endoscopic Ultrasonography. Case Rep Gastroenterol 5(3):569–577
Mohammadi-Ghalehbin B, Chinifroush-Asl MM, Ramzi F (2012) Extra-hepatic fascioliasis with peritoneal malignancy tumor feature. J Parasit Dis 36(1):78–80
Mohsenin H, Ebrahimi MA (1969) Human fascioliasis in Iran, report of a case with *Fasciola hepatica* in biliary ducts. Bulletin de la Société de Pathologie Exotique et de ses Filiales 62:871–874
Mombeni H, Kheradmand A (2005) *Schistosoma haematobium* control in Khuzestan province in Iran: prosperities and failures. Med J Islamic Republic Iran 19:19–22
Motamedi GR, Ghorashi SA, Paykari H, Dalimi AH, Salehi Tabar R, Motamedi N, Karimi GR (2008) Detection of *Ornithobilharzia turkestanikum* cercariae (Trematoda) by nested-PCR in intermediate host snail, *Lymnaea gedrosiana*. Arch Razi Inst 63(2):35–40
Mowlavi G, Mokhtarian K, Makki MS, Mobedi I, Masoumian M, Naseri R, Hoseini G, Nekouei P, Mas-Coma S (2015) *Dicrocoelium dendriticum* found in a Bronze Age cemetery in western Iran in the pre-Persepolis period: the oldest Asian palaeofinding in the present human infection hottest spot region. Parasitol Int 64:251–255
Nasiri V, Esmailnia K, Karim G, Nasir M, Akhavan O (2009) Intestinal parasitic infections among inhabitants of Karaj City, Tehran province, Iran in 2006-2008. Korean J Parasitol 47 (3):265–268
Nasser ZA, Jafar M (1997) Prevalence of intestinal parasites in the city of Kerman. Iran J Parasitol 129a:11
Nik-Akhtar B, Tabibi V (1977) Metronidazole in fascioliasis: report of four cases. J Trop Med and Hyg 80(8):179–180
Niknam R, Kazemi MH, Mahmoudi L (2015) Three living *Fasciola hepatica* in the biliary tract of a woman. Iran J Med Sci 40(5):465–468
Noorpisheh S, Farahnak A (2011) Determination of the infection of *Lcmnaea gedrosiana* to trematodes larval in the water region of Khuzcstan. J Fasa Univ Med Sci 1(l):9–12
Rahimi MT, Ashrafi K, Koosha S, Abdi J, Rokni MB (2011) Evaluation of fast-ELISA versus standard-ELISA to diagnose human fasciolosis. Arch Iran Med 14(1):18–21
Riazi H (2001) Large liver abscess: unusual presentation of fascioliasis: case report and literature review. Saudi J Gastroenterol 7(3):113–115
Rohani S, Massoud J (1990) The effect of variation in miracidial exposure dose on experimental infection of *Fasciola gigantica* in *Lymnaea gedrosiana*. Iran J Public Health 19:1–4
Rokni MB (2008) The present status of human helminthic diseases in Iran. Ann Trop Med Parasitol 102(4):283–295

Rokni MB, Massoud J, O'Neill SM, Parkin-Son M, Dalton JP (2002) Diagnosis of human fasciolosis in the Gilan province of northern Iran: application of cathepsin l-ELISA. Diag Microbiol Infect Dis 44:175–179

Rollinson D, Knopp S, Levitz S, Stothard JR, Tchuem Tchuente LA, Garba A, Mohammed KA, Schur N, Person B, Colley DG, Utzinger J (2013) Time to set the agenda for schistosomiasis elimination. Acta Trop 128(2):423–440

Sahba GH, Malek EA (1979) Dermatitis caused by cercariae of *Orientobilharzia turkestanicum* in the Caspian Sea area of Iran. Am J Trop Med Hyg 28:912–913

Sahba GH, Arfaa F, Farahmandian I, Jalali H (1972) Animal fascioliasis in Khuzestan, southwestern Iran. J Parasitol 58:712–716

Saied HM (1999) Prevalence of intestinal parasites in Tabas city. Tehran University: School of Health Sciences and Research Institute (Publ.)

Salahimoghaddam A (2009) Epidemiology of human fasciolosis in Iran. J Kerman Univ Med Sci 16 (4):385–398

Salahimoghaddam A, Massoud J, Mahmoodi M, Khoubbane M, Artigas P, Periago MV, Fuentes MV, Bargues MD, Mas-Coma S (2004) Distributional outline of lymnaeid snails (Gastropoda) in the fascioliasis endemic area of Mazandaran, Iran. Acta Parasitol 49(2):134–141

Sanei-Taheri M, Aminzade Z, Shokohi S, Aghazade K (2007) Hepatobiliary fascioliasis: clinical and radiological features. Iran J Parasitol 2(4):48–55

Sarkari B, Ghobakhloo N, Moshfea A, Eilami O (2012) Seroprevalence of human fasciolosis in a new-emerging focus of fasciolosis in Yasuj district, southwest of Iran. Iran J Parasitol 7 (2):15–20

Sarshad A, Yadegari D, Forghan-Parast K, Haji-Abdolbaghi M (1990) A report on treatment of 31 cases of human fascioliasis with bithionol. In: 1st National Iranian congress of parasitology and parasitic, rasht: diseases. Guilan University of Medical Sciences

Shahlapour AA, Massoud J, Nazary JH, Rahnou MN (1994) Further observations on the susceptibility of different species of *Lymnaea* snails of Iran to miracidia of *Fasciola hepatica* and *Fasciola gigantica*. Arch Razi Inst 44:11–17

Sharif M, Daryani A, Karimi A (2010a) A faunistic survey of cercariae isolated from lymnaeid snails in central areas of Mazandaran, Iran. Pak J Biol Sci 13(4):158–163

Sharif M, Daryani A, Asgarian F, Nasrolahei M (2010b) Intestinal parasitic infections among intellectual disability children in rehabilitation centers of northern Iran. Res Dev Disabil 31 (4):924–892

Sharma KK, Bangotra K, Saini M (2013) Diversity and distribution of Mollusca in relation to the physico-chemical profile of Gho-Manhasan stream, Jammu (J & K). Int J Biodivers Conserv 5(4):240–249

Sohrabi A (1982) Human dicrocoeliasis: 2 case reports from Iran. Iran J Public Health 11(1):55–56

Starmühlener F (1965) Ein weiterer Beitrag zur Wassermolluskenfauna des Iran. Sitzberichte der Oesterreichischen Akademie derWissenschaften,Wien, Mathem.-Naturwiss. Klasse, Abteilung I 174(5–6):171–184

Starmühlner F, Edlauer Ä (1957) Ergebnisse des österreichischen Iran-Expedition 1949/50. Beiträge zur Kenntnis der Molluskenfauna des Iran. Knochyliologische Bestimmungen und Beschreibungen. Sitzberichte der Oesterreichischen Akademie der Wissenschaften, Wien, Mathem. Naturwiss. Klasse, Abteilung I 166(9–10):435–494

Taheri F, Namakin K, Zarban A, Sharifzadeh G (2011) Intestinal parasitic infection among school children in South Khorasan Province, Iran. J Res Health Sci 11(1):45–50

World Health Organization (2007) The neglected worms: action against worms 10:1–8

World Health Organization (2010) Working to overcome the global impact of neglected tropical diseases. First WHO report on neglected tropical diseases, p 172

Yadegari D, Forghan-Parast K, Assmar M (1991) Survey of praziquantel effect on fascioliasis. Med J Islamic Repub Iran 5(1-2):43–44

Yadegari D, Talaie H, Massoud J (1999) Clinical trial of triclabendazole on human fascioliasis: long term follow up. Med J Islamic Repub Iran 13:89–91

Yakhchali M, Bahramnejad K (2016) Inhibition effect of pH on the hatchability of *Fasciola* miracidia under laboratory conditions. Iran J Parasitol 11(1):30–34

Yakhchali M, Ghobadi K (2005) A survey on helminths infection of liver and economic loss in slaughtered sheep of Urmia slaughterhouse, Iran. Iran Vet J 11:60–65

Yakhchali M, Jamshidi-Deilamy L (2012) Radular variations in freshwater snails of the family Lymnaeidae (Mollusca: Gastropoda: Basommatophora) from northwestern Iran. Am Malacol Bull 30(2):323–326

Yakhchali M, Mirrajaei SY, Malekzadeh-Viayeh R (2013) Detection of infection with larval stages of *Ornithobilharzia turkestanicum* using PCR in field-collected snails of *Lymnaea gedrosiana* from northwestern Iran. Iran J Parasitol 8(4):627–633

Yakhchali M, Malekzadeh-Viayeh R, Imani BA (2014) PCR-RFLP analysis of 28SrDNA for specification of *Fasciola gigantica* (Cobbold, 1855) in the infected *Lymnaea auricularia* (Linnaeus, 1785) snails from northwestern Iran. Iran J Parasitol 9(3):358–364

Yakhchali M, Imani Baran A, Malekzadeh-Viayeh R (2015a) Molecular detection of the infection with *Fasciola hepatica* in field-collected snails of *Galba truncatula* and *Lymnaea stagnalis* from West Azerbaijan, Iran. Arch Razi Inst 70(3):195–202

Yakhchali M, Malekzadeh-Viayeh R, Imani Baran A, Mardani K (2015b) Morphological and molecular discrimination of *Fasciola* species isolated from domestic ruminants of Urmia city, Iran. Iran J Parasitol 10(1):46–55

Yakhchali M, Hosseinpanahi A, Malekzadeh-Viayeh R (2016) Molecular evidence of *Trichobilharzia* species (Digenea: Schistosomatidae) in the snails of *Lymnaea auricularia* from Urmia suburb, North West Iran. Iran J Parasitol 11(3):296–302

Zali MR, Ghaziani T, Shahraz S, Hek-matdoost A, Radmehr A (2004) Liver, spleen, pancreas and kidney involvement by human fascioliasis: imaging findings. BMC Gastroenterol 4:15

Zebardast N, Gharavi MJ, Abadi AR, Seyyed-Tabaei SJ, Yeganeh F, Khazan H, Fallahi S, Kohansal K, Salehi N, Farideh F, Haghighi A (2015) Frequency of intestinal parasites in patients with gastrointestinal disorders, in different parts of Iran during 2012-2013. Int J Enteric Pathog 3(1):1–5

Chapter 72
Death by Drowning in Rivers in Iraq

Laith A. Jawad

Abstract Drowning incidents may happen when people have different activities in or around water areas, with water depth allows for human body to drown. The drowning events happened when people do not follow the particular instructions about the safety in the aquatic habitats.

For Iraq, there is a need to raise public awareness about the issue of drowning. Therefore, the present chapter is designed to review briefly the concept of drowning, estimation of the cost of drowning and put possible recommendations to prevent drowning. Such information is important to country like Iraq, where there are no clear legislation about health and safety for people to follow when they become in contact with aquatic environment.

72.1 Introduction

There are several explanations for death by drowning in the literature, among these the old-style categories delivered by Roll: "death by drowning is the consequence of a obstructing of the respiration by blockading the mouth and nose by a fluid medium (usually water)" (Roll 1918). Drowning is a major cause of morbidity and mortality worldwide, predominately affecting low- and middle-income countries (LMICs). Conferring to the World Health Organization (WHO), drowning caused about 372,000 deaths in 2012, with 91% of these deaths happening in LMICs (WHO 2014). Children (1–18 years of age) (UNICEF 2018) are particularly vulnerable with over 450 children drowning each day worldwide and thousands experiencing devastating injuries, containing brain injury, as a result of drowning consequences (WHO 2008).

A question should be answered here, why including a chapter about drowning in this book? There are several reasons that impel me as an author of this chapter to raise public awareness about an important issue that is drowning. These reasons are:

L. A. Jawad (✉)
School of Environmental and Animal Sciences, Unitec Institute of Technology, Auckland, New Zealand

L. A. Jawad (ed.), *Tigris and Euphrates Rivers: Their Environment from Headwaters to Mouth*, Aquatic Ecology Series 11,
https://doi.org/10.1007/978-3-030-57570-0_72

(1) lack of public awareness about the risk in dealing with aquatic environment, i.e., boating, fishing, and swimming; (2) Iraq is one the countries that has a variety of aquatic environments that people can get drown easily in; (3) Lack of a clear legislations regulating the use of the aquatic environments. Therefore, the aim of this chapter is to review the state of death by drowning in the rivers, to give a model for estimating the cost of drowning. Such information will assist in promoting better drowning prevention policies. At the end of the chapter, recommendations are given so the policy makers in Iraq can adopt and decrease the number of death by drowning.

72.2 The Risk of the Rivers

The aquatic environment in which drowning can take place includes numerous aquatic locations. More precisely, research has shown that people drown in open water (Lifesaving Society 2000); in inland water (Bierens et al. 1996); in public or home swimming pools (Mackie 1999; in water parks (Raging Waters 2001); in waterholes, dams, and garden ponds (Pearn et al. 1976); and others (Petridou and Klimentopoulou 2006).

River currents can be fairly complex (Hey and Thorne 1975), worsened by flooding. Strong currents produce important swirls that can take swimmers under rapidly. Once a swimmer begins to sink, the action of going down can occur very swiftly (Hunsacker and Davison 2013; Dilen 1984). This explains stories of speciously strong swimmers unexpectedly vanishing underneath the surface. Swimmers in freshwater rivers also are lack of buoyancy that they may be used as in the case of swimming in the sea.

Flooding and/or strong currents happening and upheld by heavy rains may also upsurge underwater rubbish such as logs or trees, which create difficulties that can capture or wound swimmers as they pass fast downstream, or tangle bodies and so stop flotation to the surface. With current, it is difficult to look for bodies as the perceptibility is very low due to high level of turbidity in the water. These factors together with the possibility of attack by fish and other organisms might produce problems at the time of autopsy (Byard et al. 2002) in the valuation of injuries or conceivable fundamental organic circumstances that may have contributed to.

72.3 Factors Affecting and Assisting in Drowning

72.3.1 Geography and Shape of Aquatic Environments

The geographical features of a country might be a second cause of drowning. For example, countries such as New Zealand and Greece are surrounded by and contain a wide variety of aquatic environments that provide outstanding opportunities for aquatic recreational and sporting activities (Avramidis 1998; Water Safety New Zealand 2000). Similarly, in Iraq, the three large rivers, Euphrates, Tigris, and Shatt al-Arab and their tributaries and the great marsh areas were considered one of the great freshwater resources in the Middle East. Living anywhere in Iraq, in the main cities such as Baghdad, Mosul, or Basrah, people are vulnerable for drowning in the river if they come in contact with. In north of Iraq, where there are waterfalls, the risk of drowning in these waterfalls is very high. The problem gets enlarged in the wetlands at the marsh area, where people living on huts made of reeds floating on water of depth, not more than 2–3 m. The more water in an environment the greater the likelihood of increased number of drowning incidents. No research has yet related the number of drowning incidents to the number of the kilometers of a country's coastline, the total diameter or area of inland lakes, or the total length of all rivers (Avramidis et al. 2009a, b).

The size and shape of the aquatic environment might be related to the incidence of drowning. People supposedly *can drown in as little as 2 cm of water depth* (Ridder et al. 2002). Many drownings occur in water only just deeper than the person's height (Stallman 2008). In terms of the water density, people can drown in freshwater (Orlowski 1987) and even in the extremely buoyant water of the Dead Sea that has high density due to the salt content (Yagil et al. 1983).

72.3.2 Additional Environmental Factors

Lighting, heating, glare, and reflection are indoor causes also likely to lead to drowning. Lighting can influence the safety of bathers. There must be adequate visibility for bathers and lifeguards to move around and supervise the aquatic activities (Schwartz 1998). Heating *plays a major role* in relation to the levels of attention of lifeguards and other supervisors, as it can influence the level of watchfulness. Anecdotal information discloses that the lifeguards are less attentive when the environment is very warm. Glare and reflection can affect the view of lifeguards, and therefore it should be minimized so a lifeguard can see a submerged swimmer on the pool bottom (Schwartz 1998). The same applies to the sea or inland bodies of water when the *sun shines from the south in the afternoon and creates a glare that shines in the eyes of the lifeguard who is looking in that direction* (Brons 2006). These additional factors seem to affect the lifeguards more than the swimmers themselves. Talking about the lifeguards, there is no such job designation in Iraq whether in the inland water bodies (rivers, lakes, dams, and marshes) or in the coastal areas at south of Iraq.

72.3.3 Open-Air Aquatic Environmental Factors

A number of outdoor reasons are also expected to lead to drowning. First, inland flooding is the number one weather-related killer worldwide (Ray 2006), but it has been underestimated because drownings resultant from floods and natural tragedies frequently are not regarded as drownings but as natural catastrophes (Smith 2006) and in 1998 alone, nearly 100,000 died in floods worldwide (Greenhalgh 1997). The second source, extreme and heavy rain, relates to floods and therefore can kill many people at the same time. A large wave (e.g., tsunami) can be the cause of drowning deaths on a very large scale. The earthquake of 2004 that generated the great Indian Ocean tsunami led to more than 150,000 deaths (National Geographic 2005). In tidal waters at midtide, the water level will be rising at its fastest and could rise 1.5 m in 30 min, which would be critical for those unable to move away, such as unattended children on a beach (Whatling 1994). Unsuspecting poor or weak swimmers can suddenly find themselves in difficulty in the same spot they safely played in earlier (Lifesaving Society 1999). The third cause, heat, has been found to negatively influence both bathers and lifeguards, increasing the chances of drowning for different reasons. A high number of drowning incidents have been reported during warm months (Tapadinhas et al. 2002) where thousands of people go swimming.

The fourth cause, many mysterious drownings may be caused by carbon monoxide emissions from boat engines and generators (United Press International 2003). Coldwater can cause immediate death due to the cold shock response (Golden and Tipton 2002) and speed the process of drowning because it reduces the period during which a person is aware (Groneng 2006). Finally, offshore winds can push inflatables and their passengers away from beaches (Avramidis 1998), and wind speed and direction are therefore critical to safety (World Health Organization 2003).

72.4 Future Method to Investigate Drowning-Related Deaths in Iraq

72.4.1 The Protocol

A general view on the drownings is significant to planning and employing suitable national directed interferences that greatest use the restricted means obtainable for drowning stoppage. This section offers the approaches that used in defining the epidemiology of drowning-related events in Iraq.

There are several steps that need to be taken in order to set up an epidemiological study of drowning-related incidents in Iraq. First, a database should be present to provide the information about the incidences of drowning for a number of years. In Iraq, there is no specialized health information center that can provide previous data on people drown for different reasons. Instead, governmental hospitals represent the only source of information in this respect.

The fundamental reasons of death prerequisite to be defined conferring to the International Classification of Diseases Ninth Revision (ICD-9) (ICD 1975). In instances in which multiple reasons back the death, the ICD-9 guidelines show which reason is nominated as the causative agent of death in order to include it cortically in the statistical database. Deaths should be incorporated in the database if the death is recorded in Iraq using an Iraqi death certificate and contains deaths to overseas visitors.

Classification of a drowning conferring to single action is difficult. For instance: a duck shooter drowns crossing a river. This case could be coded as "duck shooting" or "river crossing." With this incidence, the ranking should be the generic action the victim was engaged in (e.g., duck shooting) rather than the specific action or activity directly before death. For sites of drowning a special code should be used.

72.4.2 The Expected Results of the Survey

If the above-mentioned protocol is used, then the following type of results can be obtained:

1. The total number of incidences of drowning related.
2. Causes of drowning such as road traffic crashes, external causes, and natural reasons such as epilepsy or ischemic heart disease. The drowning can be separated into two main groups according to the psychological state of the victim. These are:

   (a) Intentional drownings include mainly the cases of suicides, which have a rate increased with age. With the available data set, it is possible to group ages according to the number of suicide incidences. Also to separate the cases of drowning into homicide or non-homicide.
   (b) Unintentional drownings, which contain drowning while boating. This type of drowning is considered causing an unintentional drowning pursued by swimming and other water activity, motor vehicle drownings, and falls or slips. Drownings happened while involved in swimming and other water events such as scuba and snorkeling. In some instances, it was obvious that the casualty was walking or playing near the edge of the water and in others the precise action was indistinct but they were positioned near the body of water. Fishing-linked drownings should be contained persons aged 15 years or older, but in Iraq, young people go fishing alone without adult supervision.

72.4.3 The Outcome of the Survey Analysis

This plan of study of epidemiologic study of drownings if it will apply to data of Iraq it will connect both important statistical data and data from hospitals and other institutions. While water contact might be high in Iraq, the latest publication of an in-depth review of drownings in other countries such as Australia, using analogous means to recognize all drownings, offers a significant evaluation for defining where the drowning rates of Iraq are extreme and therefore propose where inhibition attempts should be directed (Mackie 1999). In Iraq, a high percentage of unintentional drownings could be due to fishing from a boat or sitting on the river bank and swimming. The latter cause might be higher than fishing as there is a habit for young people to swim in the rivers, streams, and creeks with adult supervision.

For Iraq to begin making considerable improvement in showing its general drowning proportion, it is essential to mention drowning threats in a number of areas. It is a significant requisite to endure and reinforce the influences in precedence areas, precisely those due to motor vehicle crashes, boating, and among pre-schoolers, adolescents and young adults. General concern with various phases of water safety should be established in Iraq to draw attention of the public for the dangerous of aquatic habitats and to show how to become safe using these environments.

72.5 The Impact of Lack of Legislation Governing Water Safety and Drowning Prevention

In any country like Iraq, it is needed to know exactly what is the next step to control and manage the drowning in freshwater bodies. Also, to legislate specific guidelines and rules for this purpose.

In this section, the status of absence of clear legislations regarding drowning will be discussed and solutions are put forward. In this respect, Sewduth (2006) has proposed the case of South Africa to cover this state.

Sewduth (2006) promoted the parts and duties (terms of references) for the Water Safety Councils will be envisioned as including the following:

1. Put forward regulations that govern the using water areas and showing the safety measures to prevent drowning.
2. Arrange for a national plan to include neighboring countries in the safety of swimming and stopping drowning.
3. Create a public awareness program at schools to make assurance the application of the guidelines.
4. Assist to improve and uphold class levels on lifeguard facility and first aid training.
5. Control the function of all organizations accountable for guaranteeing lifesaving amenities.

72.6 Solicitation to Prevention

Risk vision can be distinct as a set of psychosocial issues that regulate whether specific circumstances or behavior are regarded as hazardous or risk-free (Glik et al. 1991). Efforts to adjust risk awareness, then, need to include both the individual perception eccentricities and the socioeconomic settings of the populations aimed. Interaction and administration of hazard may be better organized as a procedure that contains assigning risk valuation data, emphasizing the significance of the hazard deliberated to the society afraid, adopting self-responsibility, admitting individual worries and personal principles, and motivated for broad thoughtful through consciousness and understanding.

References

Avramidis S (1998) The manual of the Greek lifeguard. Anavasi, Athens

Avramidis S, Butterly R, Llewellyn D (2009a) Where do people drown? Encoding the third component of the 4W model. Int J Aquat Res Educ 3(3):4

Avramidis S, Butterly R, Llewellyn D (2009b) Who rescues? Encoding the first component of the 4W model. Int J Aquat Res Educ 3(1):66–82

Bierens J, van Stiphout WAHJ, van Ingen SMME, Knape JTA (1996) 2944 submersion victims: An analysis of external causes, concomitant risk factors, complications and prognosis. In: Bierens J (ed) Drowning in the Netherlands, pathophysiology, epidemiology and clinical studies. Author, Utrecht, pp 81–99

Brons R (2006) Quality assessment and risk monitoring of lifesaving. In: Bierens J (ed) The handbook of drowning. Springer, Heidelberg, pp 148–151

Byard RW, James RA, Gilbert JD (2002) Diagnostic problems associated with cadaveric trauma from animal activity. Am J Forensic Med Pathol 23:238–244

Dilen DR (1984) The motion of floating and submerged objects in the Chattahoochee River, Atlanta. GA. J Forensic Sci. 29:1027–1037

Glik D, Kronenfeld J, Jackson K (1991) Predictors of risk perceptions of childhood injury among parents of preschoolers. Health Educ Q 18:285–301

Golden FSC, Tipton MG (2002) Essentials of sea survival. Human Kinetics, Champaign, IL

Greenhalgh JR (1997) Special rescue teams flood/river rescue. Available at: http://www.usla.org/PublicInfo/library/Flood_Rescue.pdf. Retrieved 28 Oct 2018

Groneng I (2006) Personal lifesaving appliances other than lifejackets. In: Bierens J (ed) The handbook of drowning. Springer, Heidelberg, pp 229–232

Hey RD, Thorne CR (1975) Secondary flows in river channels. Roy Geo Soc 7:191–195

Hunsacker JL, Davison SJ (2013) Time required for a drowning victim to reach bottom. J Search Rescue 1:19–28

International Classification of Diseases (ICD) (1975) 9th revision. World Health Organization, Geneva

Lifesaving Society (1999) Alert, lifeguarding in action. Lifesaving Society, The Lifeguarding Experts, Regina, Saskatchewan

Lifesaving Society (2000) The national drowning report. Lifesaving Society The Lifeguarding Experts, Regina, Saskatchewan

Mackie IJ (1999) Patterns of drowning in Australia, 1992–1997. Med J Aust 171(11-12):587–590

National Geographic (2005) The deadliest tsunami in history? Available at: http://news.nationalgeographic.com/news/2004/12/1227_041226_tsunami_2.html. Retrieved 28 Oct 2018

Orlowski JP (1987) Drowning, near-drowning, and ice-water submersions. Pediatr Clin North Am 34(10):75–92
Pearn J, Nixon J, Wilkey I (1976) Freshwater drowning and near–drowning accidents involving children: a five–year total population study. Med J Aust 2(25-26):942–946
Petridou E, Klimentopoulou A (2006) Risk factors for drowning. In: Bierens J (ed) The handbook of drowning. Springer, Heidelberg, pp 63–69
Raging Waters (2001) 2001 rescue statistics. Raging Waters 2(1):1–2
Ray S (2006) Recommendation 2-training personnel for flood rescue. In: Bierens J (ed) The handbook of drowning. Springer, Heidelberg, p 140
Ridder M, Hoofwijk M, Dijk B, Hemmes B (2002) Drowning risk for children at home. In: Bierens J, Knape H (eds) World drowning congress. Stichting Foundation Drownin, Amsterdam, p 35
Roll HF (1918) Leerboek der Gerechtelijke Geneeskunde voor de scholen tot opleiding van Ind. artsen, 's-Gravenhage, Martinus Nijhoff
Schwartz D (1998) Managing risk in the physical environment. Parks Recreat 33(2):68–73
Sewduth D (2006) South Africa: The impact of lack of legislation governing water safety and drowning prevention. Lifesaving South Africa, Durban, p 11
Smith G (2006) The global burden of drowning. In: Bierens J (ed) The handbook of drowning. Springer, Heidelberg, pp 56–61
Stallman KR (2008) Reflections on philosophical issues on lifesaving. In: Avramidis S (ed) Handbook on safety and lifesaving. European Lifeguard Academy, Piraeus, Greece, pp 444–450
Tapadinhas F, Anselmo M, Rocha E, Barros MF, Maio J (2002) Near-drowning and drowning in children—analysis of 90 cases (Portugal-Faro 1991-2001). In: Bierens J, Knape H (eds) World drowning congress. Stichting Foundation Drowning, Amsterdam, p 116
UNICEF (2018) Convention on the rights of the child. 1989. http://www.unicef-irc.org/portfolios/crc.html. Accessed 24 Oct 2018
United Press International (2003) Pollution in water leads to drowning. United Press International, 4, 10.
Water Safety New Zealand (2000) Water safety across the curriculum 2000. Water Safety New Zealand, Wellington, New Zealand
Whatling S (1994) Beach lifeguarding. Royal Life Saving Society UK, Warwickshire
World Health Organization (WHO) (2003) The injury fact book: A graphical overview of the global burden of injuries. World Health Organization, Geneva
World Health Organization (WHO) (2008) World report on child injury prevention. World Health Organization, Geneva
World Health Organization (WHO) (2014) Global report on drowning: preventing a leading killer. World Health Organization, Geneva
Yagil R, Etzion Z, Oren A (1983) The physiology of drowning. Comp Biochem Physiol 74 (2):189–193

Chapter 73
The Dangerous Catfish Species in the Freshwater System of Iraq: First Time Reports on Cases of Envenomation

Laith A. Jawad

Abstract Among the dangerous fish are the catfish species, those of which are poisonous are a customary habitat threat worldwide. Though such stings are frequently harmless, important injuries may be the outcome of these incidences, containing harsh pain, engaged foreign bodies, infection, respiratory constrained, arterial hypotension, and cardiac dysrhythmias. Cure included hot water immersion, analgesia, injury examination, and prophylactic antibiotics. In the present chapter, attack on human by three common catfish species that inhabit the freshwater system of southern Iraq were reported in this chapter. The incidences of both *Mystus pelusius* and Parasilurus triostegus were considered the first-ever fish attack reports on humans.

73.1 Introduction

The freshwater fish groups of Iraq comprise an assorted set of families and species. These create a vital element of the aquatic ecosystem and a number of species are of commercial or other significance. The Catfishes (Order Siluriformes) contain 35 families and well over 2867 species found worldwide in freshwaters though two families are primarily marine. These fishes were evolved sometimes in the Late Cretaceous. Iraq has three families with three genera and three species, one of which is an exotic. These fishes range in size from under 10 cm to over 3 m, one of the largest species, *Parasilurus triostegus*, being found in Iraq and the neighboring countries. Catfish species of Iraq distinguished by a naked body, lacking scales. There are 1–4 pairs of barbels around the mouth (one nasal, one maxillary, and two on the chin), eyes are usually small since the barbels are used to find food, a Weberian apparatus is present (fused and modified 5 anterior vertebrae for transmission of sound from the gas bladder, used as a sounding board to the inner ear), an adipose fin is usually present, serrate spines frequently found anterior to the dorsal and pectoral which can be

L. A. Jawad (✉)
School of Environmental and Animal Sciences, Unitec Institute of Technology, Auckland, New Zealand

L. A. Jawad (ed.), *Tigris and Euphrates Rivers: Their Environment from Headwaters to Mouth*, Aquatic Ecology Series 11,
https://doi.org/10.1007/978-3-030-57570-0_73

locked erect, there are no pelvic fin spines, and some have an air-breathing apparatus (e.g., Heteropneustidae in Iraq).

Catfishes are chief food and sport fishes in many countries of the world and smaller species are popular in the aquarium trade. The larger species are significant predators on commercially important fishes. The pectoral fin spines can carry venomous tissue and can cause death in humans. Food comprises a wide range of invertebrates and fishes. Catfishes spawn in open water, build nests to protect the young, or brood eggs in the mouth.

The venom glands of catfishes are linked with their pectoral fins and dorsal fins (Satora et al. 2008). When flattened, the series of sharp curving spicules in the fins break in the casualty's skin and, successively, the toxin found in the sac connected to the spine ruptured and toxin envenomed in the body of the causality (Huang et al. 2013). This event causes inflammation in the area of attack, which exposes to bacteria and a secondary infection resulted (Baack et al. 1991). People get injured by the spine of this fish as the person walked upon (Huang et al. 2013).

The machineries of getting toxin inside the body of the victim can fall into two categories: a sting penetration and rupture of the venom glandular tissue surrounding the sting, and crinotoxicity, which associated with making of poison in the entire fish skin (Cameron and Endean 1973; Jawad 2015, 2018). It has been known that the poison of the catfish has function composed of hemolytic, dermonecrotic, edema-producing, and vasospastic actions (Blomkalns and Otten 1999; Burnett et al. 1985).

This chapter reports on cases of envenomation occurred to people while they got in contact with the fishes. Such cases were reported previously from Iraq for the Indian catfish, *Heteropneustes fossilis* (Jawad 2015), but not for the other two catfish species, *Mystus pelusius* and *Parasilurus triostegus*. Therefore, the instances of attack by *M. pelusius* and *P. triostegus* catfish species reported in this chapter are first to be publicized. The aim of this chapter is to draw the attention of the policy makers in Iraq to take action and run an awareness campaign to educate people especially those living in south of Iraq, where the three species of catfishes are found about the danger of these fishes.

73.2 Case Reports

In this section, reports on cases of envenomation incidents occurred to people due to contact with different species of catfishes. These cases were arranged according to the species of the catfish.

73.2.1 *Family Bagridae*

73.2.1.1 *Mystus pelusius*

This case about a young man of 21-years old. This man has received a severe wound resulting from a catfish *Mystus pelusius* while he was fishing in the Tigris River, Mysan, South of Iraq. The wound occurred when the spine of the fish pierced through the rubber shoes the man used to wear. A friend accompanying the man in the fishing trip took him to the nearest clinic, where first aids were applied. The man started to show some symptoms such as severe pain in the foot after 4 days of the incident.

When the man examined, the wound was closed and there is no indication of infection at the wound location, but intense erythema was prominent on the dorsal side of the foot. A superficial search was made to the wound for the presence of a catfish barb, but nothing was found, but when the lesion opened for drainage, a small part of the catfish spine was located. Several clinical complications were revealed in the tissues of the backside of the foot. Presence of bacteria causing such clinical problems was shown in the wound after doing laboratory investigations. This is also associated to a type of fungus that rendered the case of the wound very bad.

The wound remains active and giving problems due to the noncompliance of the patient, but after a few weeks, the wound cured and the man went back to his normal performances.

73.2.1.2 Conclusion

As far as the author is known, there are no previous reports on record about attack cases by the catfish *M. pelusius* on humans. Therefore, this case report is the first of its kind on this species of catfish from Iraq. Incidences of catfish attacks are not unusual and they have been reported from different parts of the world (Roth and Geller 2010).

73.2.2 *Family Siluridae*

73.2.2.1 *Parasilurus triostegus*

A 50-year-old man was injured while clearing catfish specimens, *P. triostegus* from his fishing net. He was operating in the northern part of Shatt al-Arab River, Basrah, Iraq. He had grabbed a live catfish with his bare right hand, after which he was throbbed in the right nail groove of his thumb by the spine of the fish. The patient felt an instant and harsh pain at the location of the wound. Later, he established pain, erythema, and growth throughout his right thumb. In the next few days, the patient showed a ruthless pain in the abdomen uprising to his right forearm, and advanced erythema and swelling which protracted proximally up his right arm. Then, medical care was given to him and found the patient has infection and prescribed amoxicillin-clavulanate to treat cellulitis and ibuprofen as needed for pain control. The victim saw his doctor 3 days later and at this time an abscess has developed in the area. A small operation suggested by his physician to open the wound and clean out the abscess. Upon admission, the patient reported a stable case, with clean wound.

73.2.2.2 Conclusion

Almost all catfish have the ability to inflict extremely painful wounds with their pectoral and dorsal spines. The case reported in this chapter is considered the first-ever record of a catfish attack on human by *P. triostegus*. This species has never reported to attack human and it has been handled by fishermen regularly. On the other hand and with the presence of a strong pectoral spine, *P. triostegus* looks able of instigating major wound with its stings.

Envenomations usually happen when the catfish are being touched. They respond to being grabbed by slamming from side to side and locking their dorsal and pectoral spines, which are surrounded in an integumentary sheath comprising poison glands, into a rigid and protracted position. These razor-sharp spines may pierce skin during which the toxin gets into the body of the victim after the tissue of the poison gland gets broken.

Generally, the poison of the catfish containing two sources of venoms, components that act as hemolytic, dermonecrotic, edema-producing, and vasospastic agents (Venkaiah and Lakshmipathi 2000). The other type of venom has crinotoxins entering the body of the victim at the time of penetration by the spine.

While an uncommon incidence, the most crucial long-term obstacles of catfish stinging include infections. The body of the catfish is usually covered with debris of different sorts as this species inhabits bottom of the freshwater body. Therefore, their spines are also covered with different kinds of bacteria that can find their way inside the body of the victim, once the spine penetrates into and causing a secondary infection in addition to the problem that the venom of the fish is causing (Pacy 1966). Among the pathogenic microorganisms that may found on the spines of the catfish and could enter the body of the victim through the wound are: *Klebsiella*, *Erysipelothrix*, *Nocardia*, *Chromobacterium*, *Sporothrix*, *Actinomyces*, *Pseudomonas*, *Staphylococcus*, *Morganella*, *Edwardsiella* (Baack et al. 1991), *Mycobacterium*, Aeromonas, and Vibrio species (Pacy 1966).

Live catfish should be touched sensibly with gloves to evade unintentional run into with spines. The best way to grip a live catfish out of water is to hold it behind the pectoral fins, maintaining the dorsal spine pressed down with the palm of the hand (Pacy 1966).

73.2.3 *Family Heteropneustidae*

73.2.3.1 *Heteropneustes fossilis*

An incidence involves a 33-year-old woman introduced to the Clinic with a tender wound instigated by a Stinging Catfish (*Heteropneustes fossilis*). The woman was washing dishes in the river and sitting on a stone with her feet immersed in the water. This is a normal habit for people living in villages around Iraq, especially in the south. The woman felt severe pain at her foot and she was able to recognize the stinging catfish was the causative agent. The victim was rushed to the hospital, where the wound was cleaned with excessive clean water to remove the effect of the venom. Later, the patient was given tetanus anatoxin and antibiotic course. Later and in 2 weeks' time, the wound was healed and the victim returned her daily functions.

The other case of envenomation of *H. fossilis* to man is the incident of 35-year-old fisherman operate at the northern part of Basrah City. The man was introduced to the hospital after 2 h of the accident. The patient got involved in this incidence as he tried to remove the stinging catfish from the net, when the fish stung him in his hand.

Among the medical features that wound shown are persistent cutaneous edema, erythema, intense burning, or throbbing pain that appeared at the wound site. Associated with these events, the victim showed paresthesias, weakness, localized sweating, and muscular fibrillation accompanied by cyanosis and inflammation around the injury location. The patient advised that ten fishesinvolved in the attack on him. The injuries of the hand were complicated as the spines were broken inside the wound. Other medical attributes were experienced by the patient like numbness to the elbow, and slight dizziness.

The evaluation of the attack of the Indian catfish, *H. fossilis*, and the structure of the poison gland and the pectoral fin spine of this species were given by Jawad (2015). The mechanism of the inflection of spine in the body of a victim is generally the same in all freshwater catfish species. It comes as defense response acquired by the fish.

All the incidences of attack by *H. fossilis* on a man described in this section are common in one issue in that the violence of the fish that happened as a result of the fish has been distressed or bothered. The fishes were alarmed by the action of the fisherman to catch them and remove them from the fishing net.

73.2.3.2 Conclusions

There is a crucial necessity for guidelines regarding venomous and poisonous animals which will be given the opportunity for local to get eduacated about the dangerous of venomous and poisonous animals in the Regional Poison Control Centres should be obligatory.

Dangerous organisms including fishes are living in the freshwater system of Iraq in addition to several other species that enter the country through the ornamental fish trade will develop a dilemma for the locals that they need guidance about the environment they living in.

References

Baack BR, Kucan JO, Zook EG, Russell RC (1991) Hand infections secondary to catfish spines: case reports and literature review. J Trauma 31(10):1432–1436
Blomkalns AL, Otten EJ (1999) Catfish spine envenomation: a case report and literature review. Wilderness Environ Med 10:242–246
Burnett JW, Calton GJ, Morgan R (1985) Catfi sh poisoning. Cutis 35:208
Cameron AM, Endean R (1973) Epidermal secretions and the evolution of venom glands in fishes. Toxicon 11:401–406
Huang G, Goldstein R, Mildvan D (2013) Catfish spine envenomation and bacterial abscess with Proteus and Morganella: a case report. J Med Case Rep 7:122
Jawad LA (2015) The Indian catfish, Heteropneustes fossilis (Bloch, 1794) envenomation—a report of four cases from Basrah and Maysan, South of Iraq. TOFIQ J Med Sci 2(2):52–61
Jawad LA (2018) Dangerous fishes of the Eastern and Southern Arabian Peninsula. Springer, Cham
Pacy H (1966) Australian catfish injuries with report of a typical case. Med J Aust 2(2):63–65
Roth BJ, Geller SM (2010) Deep soft-tissue necrosis of the foot and ankle caused by catfish envenomation: a case report. J Am Podiatr Med Assoc 100(6):493–496
Satora L, Kuciel M, Gawlikowski T (2008) Catfish stings and the venom apparatus of the African catfish Clarias gariepinus (Burchell, 1822), and stinging catfish *Heteropneustes fossilis* (Bloch, 1794). Ann Agric Environ Med 15:163–166
Venkaiah Y, Lakshmipathi V (2000) Biochemical composition of epidermal secretions and poisonous spine of two freshwater catfishes. Asian Fish Sci 13:183–189

Chapter 74
Fish Species of the Order Cypriniformes as a Source of Ichthyootoxin and Ichthyogallotoxin in Iraq: Cases Reports

Laith A. Jawad

Abstract This chapter reports for the first time incidence of ichthyootoxication and ichthyogallotoxication in Iraq. The former case was resulted from ingestion of roe of the cyprinid fish *Carasobarbus luteus*, while the former caused by eating gallbladder of two carp species, Cyprinus carpio, and *Ctenopharyngodon idella*. These cases have happened in the southern part of Iraq at Basrah and Mysan Provinces. In incidences of both types of intoxications, signs are shown later on after ingestion. Patient experienced abdominal pain, vomiting for 6 h, diarrhea, intense thirst, cyanosis, and faintness, among milder symptoms.

74.1 Introduction

Ichthyootoxication and ichthyogallotoxication are two types of intoxications that humans can get when ingesting fish or parts of their body. The former is related to ingestion of the gonads of the fish, where there is a specific link between gonad function and toxin production. The latter is linked to drinking the bile of the gallbladder of the fish, where a certain poison is confined. People are getting to know the ichthyogallotoxin over and done with their traditional medicine practices, where they consider that the bile of these fish species can remedy many illnesses.

The ichthyootoxism is one from the past known intoxications. Since 1491 it has been known in Europe that the eating of the roe of the cyprinid fish *Barbus barbus* will lead to an incidence of poisoning (Halstead 1967). Toxic materials were removed from the milt of some fish species such as salmon and called "salmin," from Acipenser sturio, "sturin," and from *Cyprinus carpio*, "cyprinin" (Knox 1888). A trial on the effect of pike and barbel roe was done on dogs and rabbits by Kossel

L. A. Jawad (✉)
School of Environmental and Animal Sciences, Unitec Institute of Technology, Auckland, New Zealand

L. A. Jawad (ed.), *Tigris and Euphrates Rivers: Their Environment from Headwaters to Mouth*, Aquatic Ecology Series 11,
https://doi.org/10.1007/978-3-030-57570-0_74

(1896) and revealed sensory turbulences, respiratory and muscular paralysis, and finally death. Ichthyootoxin was known to contain a lipoprotein called "dinogrunellin" as a result of studies done by Asano and Itoh (1962).

On the other hand, ichthyogallotoxin originated through consuming fish gallbladder, which is regularly done in India and other East Asian countries including China (Pandey et al. 2014). People of the Assam region, India usually eat the roe and gallbladder of the cyprinid fish Rohu (*Labeo rohita*) thinking that it is benefit for health, recovers vision, and remedies rheumatism (Bhattacharyya et al. 2009; Pandey et al. 2014) and some young people, particularly in Barak valley, eat the gallbladder for fun and for contest (Das et al. 2015). In the northeastern part of India, it is a normal habit that people take fish gallbladder as therapies for fever and as a food addition to recover strength (Jamil et al. 2013).

The present chapter sheds light on the results of some feeding and traditional medicine adopted by some locals in the remote villages in southern Iraq. Reports on cases of intoxication by eating the roe and gallbladder of the fish are documented. The causative agents of such intoxications were given besides the treatment.

74.2 Cases Reports

74.2.1 Ichthyootoxication

Six members of a family including parents and seven children were introduced to the emergency department of a hospital in Basrah, Iraq showing symptoms of food poisoning. They admitted that they ate grilled roe of the cyprinid fish *Carasobarbus luteus* known by its local common name as "Himri." All the family members were shown the following symptoms: faintness, abdominal discomfort, nausea, diarrhea, bitter taste, dryness of mouth, and weakness. All members of this family were hospitalized for 2 days and their stomach was pumped. An hour after ingestion, the patients suffered cold sweats, chills, dizziness, headache, cyanosis, and vomiting for 6 h but recovered after 1 day's rest. No other accounts of poisoning by this species have been noted in Khuzestan.

The habit of eating grilled roe of fish is common in Iraq in general and in its southern part in particular. Therefore, people need to differentiate and educated about roe of what species they need to eat and which they need to avoid.

The fish species accountable for the ichthyootoxism are mostly freshwater. Since these species contain high food usefulness, frozen or even fresh specimens are frequently found in the different shops in the market. *Carasobarbus luteus* (Heckel 1843), common name "Himri" is the suspected fish species that causing ichthyootoxication. It is distributed in the rivers and lakes in Asia. In Iraq, it is found in all rivers and other water bodies (Froese and Pauly 2016).

The polysaturated fatty acids present in aquatic organisms come mostly from the food chain that is related to the plankton-type organisms (Mancini et al. 2011). Such polyunsaturated fatty acids are utilized by aquatic organisms to acclimatize themselves in cold habitats by upsurging membrane fluidity and stopping any tendency to crystallize (Jütter 2001). In species of the genera *Barbus* and *Carasobarbus* (Family: Cyprinidae), there are high dilutions of polyunsaturated fatty acids (u3 and u6) coming from the food chain (Aras et al. 2009). Irrespective of their beneficial input to human health, high additives of free unsaturated fatty acids are known to provoke toxic influences in higher invertebrates and humans as some of them instigate hemolysis, which could be responsible for icthyotoxicity by storing in fish (Fu et al. 2004). It is known that the poisonousness of unsaturated fatty acids was noted to reach to eight times more than that of saturated ones, the greatest poisonous being linolenic acid (C18:3 u3), AA (20:4 u6), and EPA (C20:5. u3). Arachidonic acid, which is available in large quantities in extricate of roe of the barbel has hemolytic and cytotoxic actions (Mancini et al. 2011).

74.2.2 Ichthyogallotoxication

Documentation of using natural and traditional remedies has been in practice since 8000 BC (Dugo et al. 2010). The approach of traditional medicine to diseases is through ceremonial performs intended to reinstate social, mystical, and psychological accord (Ross 2008; Cocks and Moller 2002). Currently, substitute medicines can frequently be gained easily from pharmacies, but in several areas, medications are still controlled by traditional physicians or family members and maybe "home made" conferring to local ethnicities (Hegde et al. 2007).

The kidney is highly vulnerable to poisonous actions, high blood flow, high metabolic motion, and probable energetic reabsorption of toxins (Colson De Broe 2005). As such, varied kinds of renal damage, both severe and long-lasting diseases, have been designated with the use of traditional medications from around the world. Moreover, sick people with or at danger of renal malfunction may have augmented vulnerability to opposing impacts of such therapies.

In the countryside of Asia, people ingest gallbladders from numerous fish species for the cure of different illnesses, containing arthritis and decreased visual acuity (Chen et al. 1976). The usual fish species from which gallbladders are eaten is the different species of carp, ostariophysan families Cyprinidae and Xenocyprididae. Once linked with toxic impacts, it signifies a form of ichthyotoxism resultant in severe renal failure. The toxin is thought to be cyprinol sulfate or cyprinol, a C27 bile acid (Chan et al. 1985). Gallbladders of particular freshwater fish are progressively known as a significant source of ichthyosarcotoxism in Asian societies.

The custom of ingesting raw gallbladders of carp, *Cyprinus carpio*, and other cyprinid fish, such as grass carp, *Ctenopharyngodon idella* is well known in Asia. In Iraq, and in the rural villages, the gallbladder of the fish is used as a traditional medication with a customary conviction for bettering rheumatism, arthritis, hypotensive, improvement of visual insight, cough and shortness of breath, detoxification, and for upholding body equilibrium.

In the following three cases of fish poisoning due to eating gallbladder of two carp species *C. carpio* and *C. idella* were reported. These cases reflect the decline level of education of the locals in these rural villages located in Basrah and Mysan Provinces, south of Iraq.

Case No. 1
As a practice in the *alternative* medicine, a 57-year-old woman living in a remote village at the edge of the southern marshes *consumed the grass carp gallbladders stewed with vinegar as part of a meal to improve general sickness. She developed nausea and epigastric pain two hours after ingestion and arrived at an emergency department eight hours after ingestion. Her liver tests indicated severe parenchymal injury. Her alanine aminotransferase (ALT) (formerly glutamic pyruvic transaminase (GPT)) level was 5227 IU/L at 8 h, and peaked at 7340 IU/L thirteen hours after ingestion. Despite adequate hydration, she developed oliguria (2 mL/h) on day three. The diagnosis was carp C. idella gallbladder poisoning. Hemodialysis was performed for persistent fluid retention. On day 26, her renal function showed continued improvement. She was stable and followed up with renal function monitoring.*

Case No. 2
A 47-year-old housewife living in Mysan Province, south Iraq. She consumed the raw gallbladder of a grass carp and six hours later had central colicky abdominal pain, vomiting, and watery diarrhea for several hours. She was admitted to the clinic 3 days later due to oliguria. Physical inspection exhibited only jaundice. There was no dehydration. Results of the medical examination were not normal. The serum creatinine dilution summited on the 10th day and then improved. Renal tissue sample at 4 weeks revealed indication of new interstitial edema with interstitial fibrosis and acute tubular necrosis with regeneration. Both renal and liver functions go back to normal on discharge.

Case No. 3
A 50-year-old male living in a remote village north of Basrah was admitted in the emergency department. Patient was diagnosed with severe kidney wound and changed sensorium with history of consumption of large amounts of gallbladder of carp (*Cyprinus carpio*) fish with bile, in dinner a few days ago. The patient started to experience vomiting and abdomen discomfort. He became unconscious later in the day. Patient was found to be nervous and twitchy suffering from pain in the abdomen. Medical investigations showed reasonable dehydration, icterus was present, but no pallor and edema. The case of the patient was steady except for tenderness over the upper abdomen.

Further medical examinations revealed an augmented echo texture of both kidneys with normal collecting system. Later the patient was managed conservatively for all clinical problems and was cleared after 8 days of hospital.

74.3 Conclusion

The freshwater carp fish species is considered among the daily food in Asia. Eating the gallbladder raw of this species is thought by majority to help in advancing health, particularly for rheumatism. Incidences of poisonousness due to eating raw gallbladder of particularly common and grass carps have been stated only infrequently. The victims of such intoxication revealed gastrointestinal distressed containing abdominal pain, nausea, vomiting, and watery diarrhea several hours after consumption. All of the victims have reported having hepatotoxicity and nephrotoxicity. The medical examination of the liver of these victims' revealed hepatotoxins alternatively to an infective factor in the raw bile.

More severe is the nephrotoxicity, which ends in either the oliguric or the non-oliguric form of severe renal failure, usually within 48–72 h after ingestion while liver malfunction is solving. The tissue sample examination of acute tubular necrosis revealed similar to that shown by other common nephrotoxins.

Toxin in fish gallbladder is believed to damage lysosomes, meanwhile inhibiting cytochrome oxidase and blocking cellular energy metabolism, so as to cause necrosis of the proximal tubular epithelial cells. This case of liver malfunction is not usually happened in every gallbladder consumption, but it occurs when large amount of the later has been ingested.

For cure, patients usually take tablets of sodium bicarbonate three times a day to remove fish bile toxins from the urethra and oral lactulose is given thrice a day to remove fish bile toxins from the intestinal tract. Customary Chinese drug (Niaoduqing) may be given to guard renal function. Hemodialysis may be made several times and dextrose and sodium chloride injections could be given for adequate rehydration.

To make a strategy to stop intoxication with gallbladder toxin, public awareness is important to educate people, especially those living in rural areas about the intoxication action hidden in the habit of eating fish gallbladder. Pharmacokinetics and potential interactions of alternative medicines with many conventional "western" medications are being increasingly recognized, although limited knowledge of the composition of many remedies makes interactions difficult to predict (De Smet 2002).

References

Aras NM, Hasiloglu MA, Haliloglu HI, Bayir A, Sirkecioglu AN (2009) Comparison of fatty acid composition of some tissues and conversion ratios of stomach containing fatty acids to tissue fatty acids in Barbus capito capito Güldenstaed, 1773. Asian J Chem 21:6969–6974

Asano M, Itoh M (1962) Toxicity of a lipoprotein and lipids from the roe of a blenny, *Dinogunellus grigorjewi herzenstein*. Tohoku J Agric Res 13(2):151–167

Bhattacharyya PC, Nayak M, Barkataky A (2009) Acute renal failure following consumption of fish gall bladder. Indian J Nephrol 19(4):161–162

Chan DWS, Yeung CK, Chan MK (1985) Acute renal failure after eating raw fish gall bladder. Br Med J 290:PMC1417670

Chen WY, Yen TS, Cheng JT, Hsieh BS, Hsu HC (1976) Acute renal failure due to ingestion of raw bile of grass carp (*Ctenopharyngodon idellus*). J Formos Med Assoc 75:149–157

Cocks M, Moller V (2002) Use of indigenous and indigenised medicines to enhance personal well-being: a South African case study. Soc Sci Med 54:387–397

Colson CR, De Broe ME (2005) Kidney injury from alternative medicines. Adv Chronic Kidney Dis 12:261–275

Das D, Bhattacharjee K, Kalwar AK, Debnath B (2015) A case series on fish bile toxicity. J Evid Based Med Healthc 2(33):5073–5076

De Smet PAGM (2002) Herbal Remedies. N Engl J Med 347:2046–2056

Dugo M, Gatto R, Zagatti R, Gatti P, Cascone C (2010) Erbe: nefrotossicita e interazioni farmacologiche. G Ital Neprhol 27(suppl 52):S5–S9

Froese R, Pauly D (eds) (2016) FishBase. World Wide Web electronic publication. www.fishbase.org, version (01/2016)

Fu M, Koulman A, van Rijssel M, Lützen A, de Boer MK, Tyl MR, Liebezeit G (2004) Chemical characterization of three hemolytic compounds from the microalgal species Fibrocapsa japonica (Raphidophyceae). Toxicon 43:355–363

Halstead BW (1967) Poisonous and venomous marine animals of the world, Vertebrates, vol 2. U.S. Government Printing Office, Washington, DC

Hegde HV, Hegde GR, Kholkute SD (2007) Herbal care for reproductive health: ethno medicobotany from Uttara Kannada district in Karnataka, India. Complement Ther Clin Pract 13:38–45

Jamil M, Lyngdoh M, Lynrah KG (2013) Multiorgan dysfunction syndrome following consumption of fish gall bladder: a case report. J Med Health Sci NEIGRIHMS 5(1):43–44

Jütter F (2001) Liberation of 5,8,11,14,17-eicosapentaenoic acid and other polyunsaturated fatty acids from lipids as a grazer defense reaction in epilithic diatom biofilms. J Phycol 37:744–755

Knox G (1888) Poisonous fishes and methods of preventing cases of poisoning by them. Voen Med Zh 161(399):464

Kossel A (1896) Yber die basischen Stoffe des Zellkerns. S-B Kgl preuss Akad Wiss 18:403

Mancini I, Defant A, Mesarič T, Potočnik F, Batista U, Guella G, Turk T, Sepčić K (2011) Fatty acid composition of common barbel (Barbus barbus) roe and evaluation of its haemolytic and cytotoxic activities. Toxicon 57(7):1017–1022

Pandey NR, Yu Yao B, Khakurel S (2014) Acute renal failure after consumption of fish gall bladder. Case Rep Emerg Med. https://doi.org/10.1155/2014/194129

Ross E (2008) Traditional healing in South Africa: ethical implications for social work. Soc Work Health Care 46:15–33

Chapter 75
First Reports on Cases of Hallucinatory Fish Poisoning (Ichthyoallyeinotoxism) and Scombrotoxic Fish Poisoning in Iraq

Laith A. Jawad

Abstract This chapter reports for the first time on incidences of Ichthyoallyeinotoxism scombrotoxic fish poisoning in Iraq. All cases were described and causes of such intoxications were given and discussed. It is vital for the policy makers Iraqi to introduce plans to educate people about the dangerous fish species inhabit their environment especially those living in the rural areas.

75.1 Introduction

Numerous psychotropic elements, extensively dispersed in plants and animals, were revealed by early hunter–gatherers previous to the Neolithic agricultural uprising (Guerra-Doce 2015; Nesher et al. 2013). Furthermore, some psychoactive constituents, naturally arising in "fauna," utilized for ceremonial and religious drives over thousands of years, could be considered by humans as the benefits in taking them would surpass the disadvantages, from an evolutionary viewpoint. Consequently, the amount of these psychoactive constituents, naturally present in "fauna," may be permitted by some "cultural" procedures.

The "Hallucinogenic" fish are particular species of fish have been proved to comprise hallucinogenic constituents which may give a "fishing trip" like that made by lysergic acid diethylamide (LSD) intake (Halstead 2000; Samorini 2002).

Numerous hallucinatory and onyroid knowledges, also called "Ichthyoallyeinotoxic or hallucinatory mullet poisoning" or "ichthyoallyeinotoxism," have been stated after consumption of the above-mentioned fish as raw (Helfrich and Banner 1960; De Haro and Pommier 2006). The effects of eating Ichthyoallyeinotoxic raw fish might include vivid/frightening auditory and visual hallucinations, dizziness, loss of equilibrium, lack of motor coordination and mental depression, terror and nightmares, itching, burning of the

L. A. Jawad (✉)
School of Environmental and Animal Sciences, Unitec Institute of Technology, Auckland, New Zealand

L. A. Jawad (ed.), *Tigris and Euphrates Rivers: Their Environment from Headwaters to Mouth*, Aquatic Ecology Series 11,
https://doi.org/10.1007/978-3-030-57570-0_75

throat, muscular weakness, infrequently abdominal distress (Halstead 1988; De Haro and Pommier 2006). The indications might happen within a few minutes to 2 h after eating, and may last for up to 24 h (De Haro and Pommier 2006). The first signs frequently include loss of balance, decreasing of direction and general illness, followed by fever, visual and/or auditory hallucinations, despair and terrible dreams (Froese 2011).

Scombroid or histamine intoxication is a kind of taking food polluted with toxin coupled with signs and treatment similar to those related with seafood allergies. Scombroid intoxication is a consequence of consumption of mismanaged fish (Hungerford 2010). Histamine and other putrefaction yields are produced in time-temperature affecting raw fish by bacterial, enzymatic change of free histidine (Rawles et al. 1996; Hungerford 2010). The term "scombroid" derives from the type of fish (i.e., Scombridae) first associated, such as tuna and mackerel. These fish species have in common is upraised grade of histidine in their muscle tissues (Ruiz-Capillas and Moral 2004). Scombroid intoxication is obviously related with elevated histamine levels in the outbreak-associated specimens (Taylor 1986). Though there is not a clear cut dose-response association between oral administration of histamine, and histamine levels consumed in the disintegrated fish, with scombrotoxic fish displaying higher toxicity than an equal oral dose of pure histamine (Taylor et al. 1989; Lehane and Olley 2000; Hungerford 2010). Therefore scombroid intoxication is simple histamine poisoning (Lehane and Olley 2000).

This chapter reports on cases of hallucinatory fish Poisoning (Ichthyoallyeinotoxism) and scombrotoxic fish poisoning for the first time in Iraq after victims eating mullet fish species *Mugil cephalus* and scombrid fish species *Scomberomorus commerson*, respectively. These cases were observed in Basrah Province, south of Iraq (Hungerford 2010; Taylor et al. 1989).

75.2 Hallucinatory Fish Poisoning (Ichthyoallyeinotoxism)

75.2.1 Case Report No. 1

A healthy 35-year-old man living in a village south of Basrah City, Iraq. He felt feeble and exhausted within 2 h after eating freshly cached mullet (*Mugil cephalus*). In addition to the flesh, he ate the flesh in the head including the brain of the fish. Nausea and vomiting were established fast during the night. Medical indications continued and were complemented by noticeable muscle faintness. The patient began to have distorting of dreams and hallucinations involving violent and screaming animals and seeing giant arthropods around him. Nervousness and awkwardness made him to see a physician. Physical investigation showed: no fever, no indication of focalization or sensory motor deficit excluding sinusal tachycardia linked directly to the mental disorders. Throughout his stay at the clinic, the man improved quickly with whole resolve of signs within 36 h post consumption.

75.2.2 Case Report No. 2

A 65-year-old household woman started having auditory hallucinations 2 h after eating a whole fish specimen of mullet (*Mugil cephalus*) that she had bought from a fisherman in a village north of Basrah, Iraq. She regularly bought and consumed this species of fish. Hallucinations were of a mainly frightening nature (human screams and bird squealing), and she had many bad dreams for the next two nights. Be afraid of that these signs could indicate the start of a chief mental disease. The appearances decreased 3 days after she had eaten the fish. Later, and after her health deteriorated, her husband took her to the hospital for care.

75.2.3 Remarks

Medical signs of ichthyoallyeinotoxism occur within minutes to 2 h after consumption of poisonous fish. The first indications look like inebriety with imbalance and direction and general sickness (Halstead 1988; De Haro et al. 1998). Additional signs such as pain in the throat and heartburn were also reported. After some hours, indications of the toxin start to appear such as delirium, visual, and/or auditory hallucinations (often involving animals), depression. Patients classically report terrifying nightmares if they manage to sleep (Halstead 1988). Other clinical signs like nausea, abdominal pain, and diarrhea have been noted in patients experiencing this incidence (Helfrich and Banner 1960). Suitable supervision is significant at this stage to stop self-hurting or other damage. Medical indications usually subside within 24–36 h, but faintness may continue for several days (Halstead 1988). Ichthyoallyeinotoxism is prevalent in tropical and temperate areas of the Indian and Pacific Oceans as well as in the Mediterranean Sea. The majority of the species of the toxic fish are herbivores or scavengers living in freshwater water or coastal areas. It should be suggested that toxicity in all species differs conferring to fishing location, season, and way of cooking the fish (Helfrich and Banner 1960; Davis and Weil 1992). A well-known stated intensifying agent is eating fish prepared without removing the head, and/or not immediately removing the viscera after being cached (Helfrich 1963). Though the toxic factors are still unidentified, some investigators have concerned about a toxic macroalgae (Caulerpaceae family) that are consumed and polluted the flesh of fish (Chevaldonne 1990). These poisons are perhaps heat resistant since a case stated described after eating of fried, boiled, steamed, or raw fish (Davis and Weil 1992; Raikhlin-Eisenkraft and Bentur 2002).

75.3 Scombrotoxic Fish Poisoning

75.3.1 Case Report No. 1

A fit and well 30-year-old male living in Basrah City introduced to the hospital suffering from prevalent erythema, dizziness, profuse sweating, and chest tension two hours after eating cooked scombrid fish species *Scomberomorus commerson*. The prepared fish was cached at Fao City, South of Iraq. The fish had not been chilled and was uncovered under the sun for a certain period of time. The victim did not show facial, lip, or tongue swelling.

The preliminary clinical signs include tachycardia and substantial hypotension, which then enhanced with intravenous fluid rehydration. The patient was given antihistamines medication. Specifically, the chest tension and global ST-segment depression were quickly fixed with glycerol trinitrate. Treatment for the heart was done after getting clinical data from the patient. Later, the victim showed fatigue after given an exercise for 14 min, where myocardial ischemia was not shown.

75.3.2 Remarks

With the clinical investigations taken from the patient, the incidence was reported under scombroid fish intoxication. This kind of toxication is owing to consuming of inappropriately chilled fish of the family Scombridae (mackerel, bluefin and yellowfin tuna, bonito, skipjack), as in the designated incidence (McLauchlin et al. 2006). These fish comprise histidine within muscle tissue. Bacteria within the fish include an enzyme histidine decarboxylase, which can change histidine to histamine at temperatures 20–30 °C (Codori and Marinopoulos 2010; Bjeldanes et al. 1978). These bacteria belong to the following species: *Pseudomonas*, *Klebsiella*, *Enterobacter*, *Escherichia,* or *Clostridium* and can be originated through handling the fish specimens. Since the fish was not chilled and left for sometimes in the sun, such factors enable the histamine to be formed in the tissue of the fish.

In addition, another agent could also contribute to the intoxication incident mentioned above. This factor is bioamines that are also considered as toxins and can be enhanced due to the formation of histamine (Hungerford 2010; Bjeldanes et al. 1978).

Histamine, a potential causative factor of vasoconstriction, may deposit coronary vasospasm over direct stimulation of vascular smooth muscle cells in the setting of endothelial malfunction. In the present incidence, the histamine resulting from histidine within the fish is a likely hastening factor for coronary vasospasm. This type of intoxication is also linked with self-limiting signs of skin flushing, pruritus, throbbing headache, dizziness, nausea, vomiting, abdominal cramps, and diarrhea (Codori and Marinopoulos 2010).

Scombroid toxication sickness can stand for few hours and the treatment, in this case, is supportive, comprising the use of antihistamines, antipyretics, and intravenous fluid rehydration (Morrow et al. 1991; Kusama et al. 2011). The described incidence is significant as it shows a serious cardiac display of scombroid fish toxic sickness. It is also imperative for the Iraqi fishing industry to encourage precise fish management in their procedure of protection fishing industry guidelines and thus growing consciousness of the industry rules for best usage.

References

Bjeldanes LF, Schutz DE, Morris MM (1978) On the aetiology of scombroid poisoning: cadaverine potentiation of histamine toxicity in the guinea-pig. Food Cosmet Toxicol 16(2):157–159

Chevaldonne P (1990) Ciguatera and the saupe, *Sarpa salpa*, in the Mediterranean: a possible misinterpretation. J Fish Biol 37:503–504

Codori N, Marinopoulos S (2010) Scombroid fish poisoning after eating seared tuna. South Med J. 103:382–384

Davis W, Weil AT (1992) Identity of a new world psychoactive toad. Ancient Mesoamerica 3 (1):51–59

De Haro L, Pommier P (2006) Hallucinatory fish poisoning (Ichthyoallyeinotoxism): two case reports from the Western. Clin Toxicol 44:185–188

De Haro L, Prost N, Arditti J, David JM, Jouglard J (1998) Ichthyoallyeinotoxism: a rare pathology. Marseille Poison Centre Experience and literature. Toxicon 12(36):1738–1739

Froese R (2011) The science in Fishbase. In: Christensen V, Maclean J (eds) Ecosystem approaches to fisheries: a global perspective. Cambridge University Press, Cambridge, pp 47–54

Guerra-Doce E (2015) Psychoactive substances in prehistoric times: examining the archaeological evidence. J Archeol Conscious Cult 8:91–112

Halstead BW (1988) Poisonous and venomous marine animals. Darwin, Princeton, NJ, pp 683–686

Halstead BW (2000) Fish toxins. In: Foodborne disease handbook. CRC, pp 43–70

Helfrich P (1963) Fish poisoning in Hawaii. Hawaii Med J 22(5):361–372

Helfrich P, Banner A (1960) Hallucinatory mullet poisoning. J Trop Med Hyg 1:86–89

Hungerford JM (2010) Scombroid poisoning: a review. Toxicon 56:231–243

Kusama Y, Kodani E, Nakagomi A, Otsuka T, Atarashi H, Kishida H, Mizuno K (2011) Variant angina and coronary artery spasm: 10. The clinical spectrum, pathophysiology, and management. J Nippon Med Sch 78:4–12

Lehane L, Olley J (2000) Histamine fish poisoning revisited. Int J Food Microbiol 58:1–37

McLauchlin J, Little CL, Grant KA, Mithani V (2006) Scombrotoxic fish poisoning. J Public Health 28(1):61–62

Morrow JD, Margolies GR, Rowland J, Roberts LJ (1991) Evidence that histamine is the causative toxin of scombroid-fish poisoning. N Engl J Med 324(11):716–720

Nesher E, Gross M, Lisson S, Tikhonov T, Yadid G, Pinhasov A (2013) Differential responses to distinct psychotropic agents of selectively bred dominant and submissive animals. Behav Brain Res 236:225–235

Raikhlin-Eisenkraft B, Bentur Y (2002) Rabbitfish ("Aras"). An unusual source of ciguatera poisoning. Isr Med Assoc J 4:28–30

Rawles DD, Flick GJ, Martin RE (1996) Biogenic amines in fish and shellfish. Adv Food Nutr Res 39:329–364

Ruiz-Capillas C, Moral A (2004) Free amino acids and biogenic amines in red and white muscle of tuna stored in controlled atmospheres. Amino Acids 26:125–132

Samorini G (2002) Animals and psychedelics. The natural world and the instinct to alter consciousness. Park Street Press, Rochester, VT

Taylor SL (1986) Histamine food poisoning: toxicology and clinical aspects. Crit Rev Toxicol 17:91–128

Taylor SL, Stratton JE, Nordlee JA (1989) Histamine poisoning (scombroid fish poisoning): an allergy-like intoxication. J Toxicol Clin Toxicol *27*(4–5):225–240

Part VII
Conservation

Chapter 76
Freshwater Management and Conservation in Iran: Past, Present, and Future

Dariush Hayati and Naser Valizadeh

Abstract This chapter provides an overview of the status of Iran's freshwater management and conservation in three periods of time (past, present, and future). In order to figure out the past situation of freshwater management and conservation in Iran, first the characteristics of Qanats and their social structure (Buneh) as the backbones of traditional water management system of the country before the Land Reforms Law were discussed. Emergence of modernity and land reforms destroyed Qanat-based water management system and thus new modern water management system prevailed in the country which has been followed by the Western world's water management systems and has also been dominant up to now. The process of water management modernization has been explained in continuation of traditional water management system. In the second section, the current situation of water management has been explained and some of the most important water-related problems of the country have been discussed. The last part of the chapter which was entitled "future of the water management in Iran" presents some future-oriented and exit strategies for water management of Iran in order to solve its water problems.

Keywords Iran · Freshwater management · Water crisis · Agriculture · Modernity

76.1 Introduction

Iran has an area of 1,648,195 km^2, with a population of about 80 million based on the latest census in 2016, about 59 million of that are living in the urban areas and 21 million are living in the rural areas (Statistical Centre of Iran 2018). Geographically, Iran is located between 44° 05′ and 63° 18′ east longitude and 25° 03′ and 39° 47′ north latitude and is bordered to the northwest by Armenia and Azerbaijan; to the

D. Hayati (✉) · N. Valizadeh
Department of Agricultural Extension and Education, School of Agriculture, Shiraz University, Shiraz, Iran
e-mail: hayati@shirazu.ac.ir; n.valizadeh@shirazu.ac.ir

L. A. Jawad (ed.), *Tigris and Euphrates Rivers: Their Environment from Headwaters to Mouth*, Aquatic Ecology Series 11,
https://doi.org/10.1007/978-3-030-57570-0_76

north by the Caspian Sea; to the northeast by Turkmenistan; to the east by Afghanistan and Pakistan; to the south by the Persian Gulf and the Oman Sea; and to the west by Turkey and Iraq. The country's central location in Eurasia and Western Asia, and its proximity to the Strait of Hormuz give it geostrategic importance. The elevation ranges from below sea level to more than 5000 m above sea level. The temperature is variable between –30 and 50 °C.

Approximately 90% of land area in Iran falls within the Iranian Plateau. Deserts cover one-fourth of the country and only one-fourth of it is related to arable lands. The rest of the country's land area is covered with mountains (Madani 2014). The annual precipitation in Iran is 250 mm/year (Hayati et al. 2010; Hayati and Karami 2005; Madani 2014). An important thing in this context is that the distribution of precipitation is uneven around the country (Spatial distribution). In a way that it is variable between 50 mm in the central parts of the country to 1000 mm in the northern parts (Caspian Sea coast). The average annual rainfalls is greatly less than the average of global annual rainfalls (one-third of average at global level). Regarding spatial distribution of rainfalls, it should be mentioned that most of the land area in Iran receives less than 100 mm of annual rainfalls. However, 25% of the country receives 75% of total precipitation. Temporal distribution of annual rainfalls has also indicated that most of precipitation (75%) falls when agricultural sector does not need it. It is even possible that these unexpected and offseason rainfalls damage the agricultural sector and livelihood of farmers and/or villagers (Madani 2014). It should be noted that about 88.9% of freshwater resources are used in agriculture sector, about 8.3% in municipal (sanitation), and about 2.8% in industry sector in Iran (Mazaheri and Abdolmanafi 2018).

Climate has varied over the centuries in Iran and most of the country's surface area has always been lain within what is now termed arid region (65%). That is while 20% of the country's surface area is semi-arid and 15% has humid or semi-humid climate (Lehane 2014; Madani 2014). Streams are seasonal, causing flooding during spring and drying during summer, leading to significant variability in freshwater access for those reliant on surface water resources. The overall renewable freshwater is about 130 bcm annually and the overall returned water from consumption is about 29 bcm annually. According to the United Nation Development Program, the level of Iran's per capita freshwater resources are anticipated to fall to as little as 816 m^3 in 2025, down from 2025 m^3 in 1990 (currently Iran's freshwater resources per capita is less than 1700 m^3). That is while the average per capita freshwater at global level is about 7000 m^3 (Lehane 2014).

According to Lehane (2014), groundwater levels have dropped two meters in recent years across 70 plains, affecting as much as 100 million hectares. Iran is fragmented into six catchment areas. The main catchment areas include the Central Plateau (Markazi) in the middle of the country (831,000 km^2), the Lake Urmia basin in the north-west (53,000 km^2), the Persian Gulf and the Gulf of Oman basin in the west and south (430,000 km^2), the Lake Hamun basin in the east (Mashkil Hirmand, 106,000 km^2), the Kara-Kum basin (Sarakhs) in the north-east (44,000 km^2), and the Caspian Sea basin in the north (Khazar, 170,000 km^2). Except the Persian Gulf and Gulf of Oman Basins, all of Iran's basins are located where renewable freshwater is limited. Approximately half of overall renewable water of the country is located in the Persian Gulf and Gulf of Oman Basins, representing one-fourth of landmass in Iran. The important thing is that the Markazi Basin covers more than half of the country's land area, but holds less than one-third of the available freshwater.

Ninety-nine percent of the population in urban areas and 75% of rural population have access to clean tap water. The average water consumption in a day is estimated at 250 l/person which is twice the global average and can reach up to 400 l in some cities like Tehran (Madani et al. 2016). There are several large rivers throughout the country. Only one river is navigable, and the others are too steep and irregular. Streams are seasonal and variable. They normally flood in spring (with the ability to create some damage), but have little or no water in summer (Madani 2014).

This chapter provides an overview of the status of Iran's freshwater management and conservation in the three periods of time (past, present, and future) that seems to be salient in the history of water management in this country. Each period of time has its own distinct characteristics which were discussed in different sections. In the beginning of the first section, the characteristics of Qanats and their social structure (Buneh) as the traditional water management system of Iran were discussed. But regarding that emergence of modernity and especially White Revolution policies destroyed this water management system, the modern water management system of Iran has been explained in the second part of this section. In the second section which is related to the present situation of water management in Iran and even some scholars call it the crisis period, the authors tried to figure out the current trends in water management system of the country. The last part of the chapter which was entitled "future of the water management in Iran" presents some future-oriented and exit strategies for water management of Iran in order to solve its water problems.

76.2 Freshwater Management and Conservation in the Past: Traditional Water Management System

76.2.1 Before Modernization

As it was previously mentioned, most of the land area in Iran is arid and semi-arid. But, Iran has an international reputation for its significant achievements in managing freshwater resources overs thousands of years (Madani 2014). In the past years, Iranian people managed their water resources in a way that many scholars still call it the most sustainable water management system for arid regions like Iran (Yazdanpanah et al. 2013a). More than 3000 years ago, Iranians created an amazing and interesting system to conduct snowmelt through underground channels. This system is called Qanat that starts from the mountains (which has a lot of snow and ice masses) and carries water through the channels to the downstream areas using gravity (Foltz 2002; Balali 2009). In other words, Qanat is a pro-environmental and sustainable method of groundwater use and extraction. In the plateau regions of

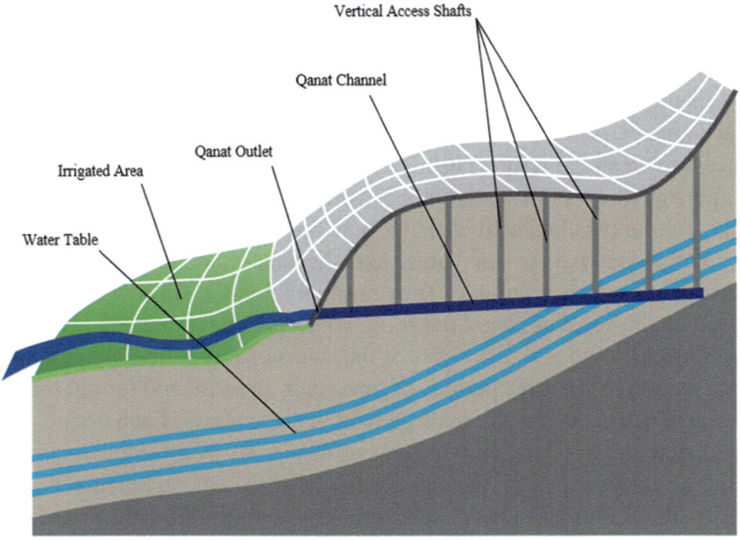

Fig. 76.1 The Qanat system: traditional water management system of Iran (Emadodin et al. 2012)

the country, water extracted from the underground in this way is used for domestic uses and agricultural activities. As a man-made tunnel, the Qanat has created an opportunity for inhabitants to find new pastures and cultivable lands in the deserts and remote areas without sufficient water sources. There is a strong belief that the earlier inhabitants might have incrementally developed the idea of a Qanat to convert the arid but fertile lands in deserts and remote regions under cultivation and make them livable and/or habitable (Ahmadi et al. 2010; Papoli Yazdi 2000).

Iranians had and/or have plausible reasons for using Qanat irrigation system. The first reason is that water can be transferred to remote regions through these subterranean conduits with minimal evaporation losses and little possibility of contamination. The second reason is that unlike other old irrigation methods, such as the counterpoised sweep, Qanats do not need a power source other than gravity to maintain the flow of water. In other words, it is a technique for tapping water without the use of lifting devices. By digging a line of wells and then connecting them with a gently sloping conduit (Fig. 76.1), the groundwater is brought from the higher ground until (after sometimes more than tens of kilometers) it reaches the surface (Yazdanpanah et al. 2013a; Lambton 1953; Ahmadi et al. 2010). Finally, the flow of water in Qanats is proportionate to the current supply in the aquifer and, if correctly protected, these irrigation canals can provide a reliable supply of freshwater for many years (Balali 2009; Haeri 2003).

One of the most important features of traditional freshwater management in Iran is that we can separate the technological systems of this management system from human activities and social entities. In other words, technological systems have an interwoven relationship with the social, ethical, and religious and governance systems (Davarpanah 2005; Kalantari 2011; Balali 2009; Lambton 1953). Ancient

Iranians had two fundamental tools in their water management approach which included valuable accumulated technical knowledge and a social–ethical–cultural system. These factors highlighted the ecological realities of the Iran's desert climate and the social imperative of conserving and distributing water in a way that ensured its availability to all individuals (Yazdanpanah et al. 2013a). From this perspective, Qanat seems to be a socio-technical system (Balali 2009). Similar to all traditional resource management systems, Qanats also has a wide sociocultural dimensions. In fact, Qanat is a substantial social phenomenon and should not be viewed only as an engineering wonder (Habashiani 2011; Balali 2009; Yazdanpanah et al. 2013a). Qanat system has to be closely linked to the local communities' capacity to plan and manage their own water resources, especially for agriculture. The social institution on which Qanat depends to operate properly is called Buneh. It is worth mentioning that Buneh is known by different names in different areas. For example, Sahra and Harasseh are other names that refer to Buneh system (Habashiani 2011).

Buneh is defined as a social system in Qanat regions (Kalantari 2011) and has been the product of adaptive reaction of human to water shortages in which individuals who are subject of the water scarcity share the managerial and ownership activities of water resources using a complex sharing ethics among local stakeholders (Habashiani 2011; Balali 2009). There are different definitions for this Qanat-based cooperative Buneh system which each of them reflects slightly different functions and roles for Buneh. In Table 76.1 some of most important definitions and functions of Buneh system were presented.

The important thing about Buneh is that before radical changes in Iran's land and water structures or modernization of agriculture and water management (Land reform and White Revolution took place in 1962), it has been as a collective or cooperative production unit that has a crucial role in determining the rural social structure. Qanat and Buneh as technical and social phenomena, along with relevant value systems, were the main characteristics of the land and water management in pre-modern Iranian rural society (Balali 2009). In Iran's traditional water management system, at the state level, the ancient water authority had an obligation to adjudicate on water ownership, record shares of water, and calculate how much tax individuals (farmers) had to pay for their shares. Water management duties were not limited to these activities and included a broader range of actions (Yazdanpanah et al. 2013a). For example, the allocation and division of the waters according to land ownership, irrigation rights, timeshares, and so on had to be scheduled based on the possible variations in the volume of water throughout the year (Emadodin et al. 2012; Lahsaeizadeh 2002). In addition, precise and legitimate adjudications on that sort of matching could only be obtained at the local level using farmers, landlords (and/or their representatives/agents), Qantas diggers, blacksmiths, carpenters, and so on—whose various contributions were crucial to the sustainable functioning of this complex socio-technical system. In fact, Buneh was a Key to this grassroots element within the system of water governance.

As it formerly mentioned, Buneh can be articulated as a social structure that had been created to cooperate in construction and maintaining Qanat and also using Qanat's water for irrigating agricultural farms (Habashiani 2011; Jomehpour 2009).

Table 76.1 definitions of Buneh system

Definition	References	Functions
The Buneh was a multifamily collective: a farming cooperative, whose main function was to reconcile the efficient exploitation of productive land with the careful use of the available water (that availability itself being, in large measure, a function of the socio-technical skills inherent in the Buneh)	Yazdanpanah et al. (2013a)	To reconcile the efficient exploitation of productive land with the careful use of the available water (that availability itself being, in large measure, a function of the sociotechnical skills inherent in the Buneh)
A traditional cooperative organization of sharecropping, in which people cultivated the land cooperatively	Rezaei-Moghaddam et al. (2005) Lahsaeizadeh (2002)	Organizing cooperative sharecropping
A communal system under which the arable lands of a village was organized into units farmed cooperatively by teams of sharecroppers	Hooglund (1982)	Organizing and dividing shared lands between villagers and farmers
Collective organization of production management, common in the central and eastern provinces, in the Qanat regions	Jomehpour (2009)	Organizing production management in a collective method
An agricultural unit on which some farmers have the right to work cooperatively	Balali (2009)	Division of labor between farmers
A cooperative lifestyle more than just a management system	Farshad and Zinck (1997)	As a lifestyle for villagers which not only is a resource management system, but also influences all aspects of their life
A social hierarchy that defines roles and responsibilities	Jomehpour (2009)	Division of labor/responsibilities among individuals
As a system of collective ownership and management of Qanat water along with some participatory practices	Habashiani (2011)	Collective management of Gant water and agricultural activities

All the Buneh's members usually being peasants that were from the same social status, but there was an exact division of labor among them. A Buneh, normally, included six peasant members: the Buneh head (or irrigator), two assistants, and three sharecroppers (Yazdanpanah et al. 2013a; Rezaei-Moghaddam et al. 2005). It should be mentioned that Buneh was not independent and each Buneh was tied into a broader network that included the landlord (or his representative/agent), other Buneh within the village and many of vital specialists: the Qantas diggers (muqani), blacksmiths, carpenters and village-level service providers such as barbers and bath keepers. This sort of network, with all its diversity and mutuality, constituted the production system (Yazdanpanah et al. 2013a). Salmanzadeh and Jones (1981), call this an "agrarian structure," shaped over many centuries by a complex set of interrelated physical and cultural factors and replicated again and again across the Iranian landscape. All this came to an abrupt end with the 1962 Land Reform Act, an intervention that was derived directly from a modern theoretical framework.

76.2.2 After Modernization

The beginning of modernization in Iran is related to around the middle of nineteenth century. Nevertheless, the trend and process of penetrating modernity in Iran progressed very slowly until ascension of Reza Shah Pahlavi to the throne. He created comprehensive economic, cultural, educational, industrial, and public facility programs based on the modernization approach of European countries. But modernization programs of Reza Shah coincided with the outbreak of World War II in which his government collapsed and his son Mohammad-Reza Shah Pahlavi ascended to the throne. After the World War II, Iran's newly arrived government attempted to follow the modernization process more quickly than the past. This government was a secular and dependent on world capitalism (but it was peripheral for the capitalism world). These fundamental changes in political structure of the country made a basis for the White Revolution and Land Reforms in 1962. This revolution was inspired by the United States (one of the most important allies of Iran government before the Islamic Revolution in 1979) in order to transition to the sort of free world stance (Yazdanpanah et al. 2013a; Asaesh 1994).

Basically, White Revolution and Land Reforms were provided as a social background for the modernization of water management system (Habashiani 2011). Because, these policies facilitated destruction of feudalism and then replacement of capitalism in agricultural sector of Iran's economy. Progress (using industrialization, urbanization, and Westernization) was the primary aim of these radical changes in Iran, but the agriculture and water sectors were also dramatically changed (Yazdanpanah et al. 2013a). Industrial growth was the main tool to achieve progress. So, the agricultural and/or rural areas were considered as the main source of labor for the industrial sectors in urban areas (Habashiani 2011). In other words, rural and agricultural sectors were as a backward part of the economy, which had no considerable connection with the industrialized and urbanized state that was about to be put in place (Yazdanpanah et al. 2013a; Hayati et al. 2006; Asaesh 1994). In order to achieve these goals, a range of social and economic reforms were conducted in the form of White Revolution (Lahsaeizadeh 2002).

Along with modernization that was one of the most important triggers of White Revolution, there were other reasons for conducting White Revolution policies. For example, landlords were significantly strong in terms of economic and political influence in pre-modernization feudalism system. The government tried to decrease its influence and develop new peasant-based support among deprived classes in rural and agricultural areas (Azkia 2012). Land reform law was one of the focuses of White Revolution and was considered as a prerequisite for modernization in rural and agricultural areas. Furthermore, it is worth mentioning that this law was the most brilliant part of White Revolution which had interwoven relationships with water management in Iran. There was a strong belief that land tenure system and land ownership in Iran were considered as the biggest constrains of modernization in feudalism system of Iran (Habashiani 2011; Lahsaeizadeh 2002). To surmount these basic obstacles, approving and implementation of Land reform law were the first options (Habashiani 2011). Through the land reforms, the symptoms of capitalism emerged in rural and agricultural sectors of Iran. Independent peasantry agricultural system and capitalist agricultural units were the most impressive results of capitalization and modernization of agriculture in Iran (Lahsaeizadeh 2002).

Independent peasants who could own the high-quality lands, had strong control over rural life, while they were supporting the government's goals in rural areas. Capitalist agricultural units were another result of penetrating capitalism in rural and agricultural areas, which exerted their authority just using economic tools; despite landlords in feudalism system who sometimes used violent coercion over farmers and rural people. These capitalist units had so many features of modernized units in Western countries. For example, they benefited from the public and private facilities, advanced agricultural machineries, modern irrigation systems, wage-earner workers, and so on (Ehsani 2006; Habashiani 2011; Safinejad 1980). Land reforms granted water and land as the two important production factors (sources) to the farmers who did not already have them. Peasants allowed to have a dominant role in the production of agricultural products and power distributed between Buneh members and they were entitled to equal rights. Buneh members previously only had the workforce factor of production, but land reforms gave them the ownership of all production factors (Safinejad 1989). In other words, land reforms took the lands from landlords and gave to the farmers and individuals who had the right of cultivation to specific parts of land and water (Habashiani 2011). These actions under the land reform law eliminated the feudal class (landlords) form the agricultural and rural areas of Iran and helped to the emergence of small-scale family farms, independent peasants, capitalistic production units, large-scale capitalist farmers, and also industrial projects in agricultural sectors (Habashiani 2011; Ehsani 2006).

These transformations in structure of agricultural sector affected the social, physical, and technical systems of agriculture and production. Consequently, water management system changed. The changing of water management included transformation of Buneh-pre-modern social institution of water management to new cooperatives. The technological change in this process was replacement of Qanat with deep and semi-deep wells (Habashiani 2011). On the other hand, it is worth mentioning that from the late nineteenth century to 1970s approximately all of the water management policies and practices around the world influenced by "hydraulic mission" paradigm (Molle et al. 2008; Molle 2006; Balali 2009). Inspired by this paradigm, Iranian water management policy-makers and water authorities thought that new and advanced hydrological technologies, borrowed from the West, can satisfy the country's increasing water needs (Yazdanpanah et al. 2013a, b; Madani 2014). It was supposed that arid and semi-arid areas could be industrialized by making the required water resources accessible by constructing gigantic dams, extracting and exploiting groundwater and building canals to bring water from remote sources to make the deserts bloom (Yazdanpanah et al. 2013a; Allan 2005). Hydraulic mission paradigm used the nature as a tool for the scientific, technological, and industrial goals like construction of new and large-scale water systems comprising reservoirs, canals, gigantic irrigation systems, and hydroelectric stations. The focus of this mission was on technical solutions to narrowly defined environmental challenges, with regulatory authorities implementing those solutions on the basis of specialists' advice (Yazdanpanah et al. 2013a; Abdullaev et al. 2009). Very evident signs of hydraulic mission in Middle East and North Africa countries and other developing countries emerged from the second half of the twentieth century (Habashiani 2011; Balali 2009; Yazdanpanah et al. 2013a). The introduction of this mission in developing countries coincided with the White Revolution. Regarding that the goals hydraulic mission were in line with the goals of the White Revolution, the policy-makers and decision-makers used the hydraulic mission as a tool for modernizing water management system in Iran.

Through this new water management system many of values and practices changed (from traditional to mechanistic values and worldviews). To facilitate the process of industrial modernity, the Iranian officials and authorities tried to eliminate the traditional irrigation and agricultural production systems from technical and social aspects (Habashiani 2011; Balali 2009). This contextualization for eliminating previous social and technical water management systems (Buneh and Qanat) was in the midst of the White Revolution and Land Reforms (Kardovani 2007). Most Iranian experts and politicians started to exaggeration of the deficiencies and shortcomings of the Qanat to justify their own programs and to convince farmers to use pump extraction instead of Qanats (Yazdanpanah et al. 2013a). One of the consequences of these efforts was destruction of Qanat and the social ties of it (Habashiani 2011). For example, as it was previously mentioned, Buneh was a social body of Qanat water management system which was replaced with governmental farm corporations and cooperatives (Habashiani 2011; Safinejad 1989).

During the land reforms, land and water were distributed among farmers. Sharecroppers in Buneh became owner of water and land. In other words, the functions of Buneh and its division of labor destroyed (Habashiani 2011; Balali 2009). But it soon became clear that the productivity of the fragmented lands and small-scale farmers is lower than the Buneh system. In this regard, land reforms' policy-makers offered the governmental farm cooperatives as the suitable option for improvement in productivity of distinct production lands (Hooglund 1982). The main roles of the cooperatives included adoption of communal cropping patterns, joint cultivation, marketing, production of inputs for agriculture, granting credit, training, and education of farmers. In addition, at the beginning of land reforms, these modernized cooperatives were independent, but they incrementally became dependent on the national government. Many experts believe that creating these cooperatives as an alternative for the Buneh was not successful (Habashiani 2011; Ajami 1993). The land reform had thrown the farmers into the unsuitable situation of not knowing where he stands with the new group. With the old landlord, they always knew when he could push and how he could get what he needed (Balali 2009; Bill 1970).

In a nutshell, land reforms in Iran had some negative impacts on water management. Because it did not consider the interwoven and complex associations among water management, environment, and agricultural systems. Most of these impacts originated from destruction of Buneh and Qanat systems (Farshad and Zinck 1997). Buneh was well adapted to the optimal use of the Qant system. However, it seems that the traditional sense water resources management for the benefit of the community was replaced with an individualistic perspective or every man for himself mentality (Balali 2009; Habashiani 2011).

Generally, the main agricultural policy during and after modernization was the green revolution with its focus on high payoff input pattern (Forouzani and Karami 2010; Kalantari 2011). Consequently, so many land areas were brought into cultivation (Rezvani 2005), the use of fertilizer dramatically increased (Deihimfard et al. 2007) along with diversification of crops and particularly cash cropping (Emadodin et al. 2012; Rezaei-Moghaddam et al. 2005; Faramarzi et al. 2009). Based on Habashiani (2011), declining the irrigation efficiency in agricultural and rural areas is one of the most important negative impacts of modernizing water management system, which this diminution originated from some specific causes:

1. Dependency on governmental services: Before modernization policies, the utilization of freshwater had been governed by traditional and local customs. After elimination of Buneh system and replacement of landlords with governmental agencies, small-scale farmers, and villagers were no longer able to collectively maintain the rural infrastructures. Therefore, they became dependent on the government for the necessary services. It was while that the national government was not able to supply all of their requirements effectively and new water management alienated people from the source of supply.
2. Dramatic increase in the number of deep wells: Maintaining Qanats in pre-modern water management system was the responsibility of landlords. But after land reforms, the farmers could not meet the costs of maintaining the Qanats. As a result, they preferred to use pumped wells. The depletion of groundwater by deep wells meant that many of Qanats fell into decay. On the other hand, some of the landlords were forced to substitute the Qanat with pump extraction to save their lands (under the land reform laws, landlords could save land from being transferred if they equipped their agricultural land with modern irrigation systems). Thus, the number of deep wells remarkably increased. But, owning private deep wells was one of the most important constrains of collective actions to better and efficient water management.
3. Migration of Qanat practitioners: Implementation of land reforms did not have benefit for the Qanat practitioners and forced them to leave the rural and agricultural areas. Qanat practitioners included a professional community. But, based on the traditional class structure, they did not have the right to work on the lord's lands as some permanent farmers had. During land reforms, the lands were distributed among people who worked on the lord's fields. In this regard, they were encouraged to migrate to the cities. In absence of the Qanat practitioners, unskilled individuals in irrigation activities wasted a large proportion of the water

and thus water use dramatically increased and its demand quickly surpassed the supply of Qanats.

It is obvious that Iran experienced a double setback, with an industrialized style of agriculture and modern agriculture's package of high-yield varieties, fertilizers, pesticides, and heavy irrigation being lumped on top of the White Revolution's land reforms. Inevitably, water demand increased way beyond what the Qantas system could provide (Karimi 2009). As a result, many deep, and semi-deep wells were sunk (about 500,000 deep and shallow wells and the same number of manually excavated open wells), and water was pumped with scant concern about the environmental impacts of what is now called competitive deepening (Beck et al. 2011). The green revolution, centralized, expert approach, in line with the then-current goal of modernization, opened the way to an industrial style of agriculture. Based on Nelson et al. (2008); Brunner and Steelman (2005), this approach was both positivist and reductionist in the sense that it provided the instruments that enabled the conquest of nature for the improvement of human welfare and, in the process, separated society from nature. Within that approach, water management was seen as primarily an engineering problem, that is, one in which the efficient solution lay in technological fixes. The discourse of this new Modernization continued after the Islamic Revolution, but in a different direction.

White Revolution and especially land reform law caused to incremental marginalization of agricultural sector (Lahsaeizadeh 2002; Halliday 1981; Hooglund 1982). On the other hand, petrolization of the country's economy following the rise in oil prices in global markets, increased the blindness of government to the agricultural and rural areas (Ehsani 2006). This carelessness toward agricultural and rural sectors and also unsatisfactory results of modern agricultural development programs led to criticism and then 1979 Islamic Revolution (Hooglund 1973; Habashiani 2011). The revolution has so many impacts. Increasing solidarity with other Islamic and revolutionary nations (Tajik and Darvishi 2004), together with a marked antipathy toward capitalist West, led to more rapid population growth (Fozi 2004), to a brake on the previous regime's factory building and also to a questioning of the regime's efforts to industrialize the agricultural sector (Foltz 2002). As a result, revitalization of rural and agricultural sectors became one of the most important economic goals of new Islamic government (Wosooghi 2007; Ghadiri Masoum and Najafi kani 2003; Yazdanpanah et al. 2013a) and there was also a marked expansion of mono-cropping and incentive-based grain production through interventions, such as guaranteed purchase, subsidies, crop insurance, and so on (Deihimfard et al. 2007).

After the revolution 1979, traditional agricultural and irrigation methods received renewed support along with reconstruction and surviving of some the Qanats (Habashiani 2011). Some institutional changes also took place after the revolution that has significant impacts on water management and agricultural sector. For example, establishment of Jihad-e-Sazandehgi (a revolutionary organization for agricultural and rural development) was one of the most significant actions after the revolution (Ghaffari 2006; Habashiani 2011). The purposes of Jihad-e-

Sazandehgi included moving toward self-sufficiency and independence; creating a basis for the growth of dignity of rural people; and encouraging the participation of local people in the construction of villages (Balali 2009). The Jihad-e-Sazandehgi performed a range of activities for improvement of socioeconomic conditions in the rural and agricultural areas. These practices included water supply, building of rural baths, schools, health centers, and many other activities. It can be concluded that, after the Islamic Revolution, the process of modernization in the rural areas was continued with an emphasis on physical development (Yazdanpanah et al. 2013a).

Due to the post-revolutionary socio-political situations, some of the previous capitalist enterprises were dissolved and some new management organizations were established (Ghaffari 2006). One of these organizations was Mosha which has emerged as a result of the revolutionary measures (construction of Moshas was carried out by Jihad-e-Sazandehgi) and was one of the most important agricultural organizations in the early years of the revolution. Mosha was a new collective ownership of agricultural land and production system. These new production systems were societies of farmers (5–15 farmers) who jointly own a part of the land and work in partnership. Cultivation plans for them were based on the priorities of the government. Moshas had three potentialities: (1) land that was a sum of all the members' shares; (2) water resources, all the costs of which should be paid by members; and (3) agricultural machinery which all members can equally use (Habashiani 2011). Establishing Mosha was a response to overcome the objections of fragmentation of the lands through land reforms in 1960. Land reforms separated the integrated lands of the landlords into some small parts and also distributed them between farmers. But this action led to problems such as reduction of efficiency and increasing water consumption. In order to avoid such issues, construction Mosha was proposed. However, after a while the farmers and peasants rejected these conditions in practice. As a result, lands were more fragmented. Meanwhile, the new government was also vigorously pursuing the policies of land allocation for farmers, without thinking about increasing water consumption in agricultural sector. The result was a new, and distinctly Iranian, modern agriculture, with more and more land (even land with marginal agricultural potential) being brought into cultivation. Between 1973 and 1998, approximately 483,000 ha went under the plough, and many deep boreholes were sunk to provide those hectares with pumped-water irrigation (Yazdanpanah et al. 2013a).

At present, there are more than 350,000 deep and semi-deep water wells throughout the country, which are being exploited without permit. The excessive exploitation of water through these wells has resulted in a negative water balance in most areas, and has accelerated the trend toward desertification. Moreover, the existence of high chain of mountains, sediment plains, and so on has led people to use both surface and underground water resources. Also, the ease of digging in the sediment plains of the country (due to particular conditions) has resulted in major damages to these areas. Likewise, control of the land and water resources has been transferred from religious endowments to government bodies. These changes have been brought about as a result of the adoption of development models imported from the West (Foltz 2002). Furthermore, it is worth mentioning that overexploitation of underground water has some negative impacts on surface water resources. In other words, past and present intense modernization activities in Iran have led to side effects like reduction and salinity of surface water resources, eradication of rivers all over the country, and destruction of wetlands and their valuable ecosystems. In this regard, the next part of this chapter (Sect. 76.3) figures out the present conditions of water management in Iran which some scholars describe it as the crisis period.

76.3 Present Situation of Iran Water Management System: Crisis Period

Currently, Iran is facing with different water-related issues like drying lakes and rivers, declining groundwater resources, land degradation, water pollution, water supply rationing and disruptions, forced migration, agricultural losses, salt and sand storms, and ecosystem damages, which most of them are the results of modernization policies in previous periods (Madani 2014). In this section, in order to figure out the situation of water resource management, some of the most important representations of present water crisis are offered. Lake Urmia with approximately 5000–6000 km^2 is the largest lake in Iran and is located in northwest of the country. In recent decades, the lake's water level has been decreased by up to 6 m (Hassanzadeh et al. 2012). This lake has a crucial socioeconomic and ecological role in the Northwestern part of the country. However, this lake's level has been decreased and most of the lake has changed to saline area in the last 16 years (Water Research Institute 2015). There are different ideas about drying this lake. But it is obvious that the roles of frequent droughts (Fathian et al. 2015; Zarghami 2011) water use (Madani 2014; Valizadeh et al. 2018), mismanagement (Hassanzadeh et al. 2012; Fathian et al. 2015), and storage are not deniable in exaggerating the situation (Norouzi et al. 2013; Sima and Tajrishy 2013). Decreasing the lake's water have too many negative impacts on livelihood of the inhabitants and exceptional ecosystem of the lake (Hassanzadeh et al. 2012; Madani 2014; Valizadeh et al. 2016). Furthermore, salt storm is expected to destroy the agriculture and to have catastrophic health impacts as well (Madani 2014). In the next step, these impacts can lead to other crises such as migration and slum spreading around cities.

Lake Hamun which is located in east of Iran is destroying as a result of the everlasting trans-boundary contradictions with the Afghanistan government over the Hirmand River as well as water mismanagement in the Iranian portion of the watershed (Madani and Hipel 2011; Najafi and Vatanfada 2011; Madani 2014). The Hirmand River is a trans-boundary river, which flows from Afghanistan to Iran and is important for agriculture in both countries and the survival of Hamun Lake in Iran's Sistan-va-Balouchestan Province (Madani and Hipel 2011). Changing the process of the birds' migrations, destroying habitats of immigrated animals and fishes, destroying unique reeds and land covers of Hamun Lake and threatened animals of region such as Sistan Cow, changing quality of wetlands' freshwater and salination of water, developing sand storm and 120 days winds of Sistan, changing people living ways through increasing fields for fuel and goods smuggled from Iran to Afghanistan and opium smuggled from Afghanistan to Iran, migration of the people are only some objective impacts of dying Hamun Lake and water crisis in this area of Iran (Najafi and Vatanfada 2011).

The Zayandeh-Rud River is the highest-volume river in semi-arid central Iran and forms one of the most strategic and important river basins of Iran with large agricultural, industrial, and domestic water uses (Madani and Mariño 2009). But, this river dries in the hot–dry seasons due to disturbing pressure of agricultural and industrial activities and also high population growth (Madani 2014) and thus water management in the river basin of this river is currently challenged with providing sufficient water supply for various user groups in the face of increasing demands. The Gav-Khuni wetland (which is located at the end of the river) receives something less than half of the amount of water which is necessary to sustain the wetland habitat for fauna and flora (Salemi et al. 2000) and some of scholars (Gohari et al. 2014; Madani 2014) believe that there is almost no hope in the recovery of the Gav-Khuni wetland and its valuable ecosystem. Water consumption by different stakeholders has exceeded available surface supply and resulted in decreasing groundwater tables. If overusing the groundwater is continued, available groundwater resources will most likely be exhausted. Of course, it should be mentioned that the arid and semi-arid climatic conditions of the region which Zayandeh-Rud River located and also frequent droughts make the situation harder and more complex for water authorities to supply the region with sustainable water resources.

The Lake Parishan is in Kazerun, Fars Province and is the largest freshwater lake in the country. It receives only a very small amount of water from feeder rivers and the whole lake is a protected area, as it is considered a globally significant wetland ecosystem. The lake has been the subject of intense pressure during the last years and is currently in a state of ecological crisis, with major negative effects on biodiversity and socioeconomic conditions (Djamali et al. 2016). The present state of water inflows and outflows into and from the lake is unbalanced, and inflows are much less than annual evaporation. The main cause of this problem consists of reduced precipitation (because of long-term climate change), increased water abstraction from groundwater resources, and trends of expanding cash crop cultivation which consume more groundwater (local farmers increasingly tend to expand early season cash crop production under plastic cover, which generates higher income). Contamination of the water is another important threat to the lake. Sewage outflows from villages around the lake enter the lake directly or through seepage. Also, residual fertilizers and pesticides dissolved in irrigation return flow eventually discharge into the lake and increase its eutrophic level (Sharifi Moghaddam 2013; Djamali et al. 2016).

Iran's water management tragedy is no limited to these cases. Drying and vanishing other lakes (Bakhtegan Lake, Shadegan Lake, etc.) and wetlands (Anzali Wetland, Amir-Kalayeh Wetland, etc.) should be added to this list. Madani (2014) stated that not only the end-closed water bodies have been the victims of Iranians' thirsty for modernization, but also rivers have been the victims of offensive human manipulations. As one of the most important results of the Iranians' hydraulic mission, too many dams are constructed to capture water in order to support agricultural practices, increase power generation, and secure urban water supplies. Iran has the third rank in the world with respect to the number of dams it has under construction.

The groundwater resources have also the same situation and it is believed that the country has already used most of its groundwater resources (Karami and Hayati 2005). The government has not much control over groundwater abstraction. Energy and water are highly subsidized by the Ministry of Energy (The country's highest regulatory authority in the field of water and energy management), leaving no motivators for farmers and other users to raise their water consumption efficiency (Foltz 2002; Madani 2014). In the previous section, we introduced Qanat as a sustainable and pro-environmental system of groundwater resources management. But it should be kept in mind that this system is approximately useless in present situation (at least in most areas). Because, most of them have dried up also the social, economic, and structural changes following the land reforms and Islamic Revolution have dramatically changed the situation of water management (Balali 2009; Habashiani 2011; Madani 2014; Madani 2008). Increasing the discrepancy of current groundwater resources' level from the standard level has also made the situation more dramatic to reuse of Qanat systems.

The quality of water in Iran is another concern, which attracted an increasing attention in recent water research literature (Esmaeili et al. 2014; Farhadinejad et al. 2014; Razmkhah et al. 2010). Generally, freshwater quality in Iran is better than its neighboring countries and most of the people have access to improved drinking water (Madani 2014). However, in some parts of the country (especially in mega-cities and regions with intensive agricultural and industrial activities) contamination of freshwater is becoming a serious challenge for water management. The most important sources of water pollution include agricultural activities (i.e., fertilizer and manure application, leguminous crops, irrigation with groundwater containing nitrogenous compounds), decaying vegetation, atmospheric depositions, septic tank, sewage systems, industrial wastewaters, municipal waste, and dairy and poultry farming which facilitate the penetration of the contaminants into the fresh-waters (Masetti et al. 2008; Anayah and Almasri 2009).

In the recent past, Iran experienced a water crisis so severe that the government made a paradigm shift from its supply side-driven water management approach to a more demand-driven one (Valizadeh et al. 2018; Salehi et al. 2018). Supply side-driven strategy is unsustainable in the long run and water conservation moves to the forefront for tackling the problem of water shortages. However, to be successful in the macro (country) level, water conservation has to be implemented on the micro (household) level. Accordingly, from a demand side-driven perspective, the question for the policy-makers was how to encourage farmers' behavior to apply water conservation strategies, which is vital for the success of any overall water management plan (Malekian et al. 2017; Yazdanpanah et al. 2014a; Bijani et al. 2017; Shiri et al. 2011; Abbasian et al. 2017). The important point in this context is that although the water policy-makers shifted their perspective from a supply side-driven approach toward the demand-driven approach, but it seems that they do not really believe in behavioral approaches in practice. A review of their water plans is a clear crystal proof of this claim. Looking for new water resources, building new dams, transferring water from one location to another, desalination of seas' water, extracting deepwater resources, etc. are some of the strategies that Iranians are following now. These plans are not in line with the paradigm shift which we explained it before. In other words, the government does not sufficiently invest in improving the knowledge, belief, and behavior of freshwater beneficiaries.

Population growth, depletion of surface and groundwater resources, frequent droughts and so many other reasons in the country have created a competitive situation for water consumption among different stakeholders (Bijani and Hayati 2015; Bijani and Hayati 2011). As a result, water conflict is raising among them. Bijani and Hayati (2015) stated that water conflict in Iran occurs mostly between the government (as owner and manager of water) and farmers (as one of the users). On the other hand, the environment has been damaged by these conflicts. Generally in Iran, there are three turning points in trend of water conflict (Table 76.2). Water conflicts in Iran are taking place in different levels. Conflicts between farmers are becoming a common phenomenon in different regions of the country. However, conflicts over water are not limited to conflicts between farmers. There are so many conflicts between villages, regions, and even provinces over trans-boundary rivers and water systems (Madani 2014; Najafi et al. 2013). But this is not the end of the story and at the international level, conflicts continue over sharing the surface water resources like the traditional conflict over the Hirmand River with the Afghanistan Government (Najafi and Vatanfada 2011; Madani 2014) and over the ownership of the Caspian Sea in the north. According to Madani (2014) failure in reaching an agreement over sharing the Caspian Sea with Azerbaijan, Kazakhstan, Russia, and Turkmenistan since the collapse of the Soviet Union has resulted in an international tragedy of the commons, associated with high level of water pollution and ecosystem damages. The situation is less tragic in other trans-boundary surface water systems in which Iran has better access to upstream water sources such as the Tigris–Euphrates system, shared with Turkey, Iraq, and Syria; and Aras system, shared with Armenia, Azerbaijan, and Turkey. However, Iran's attempts to increase its use of outgoing surface flows can increase the international tensions with the neighbors in the future.

The current water crisis of Iran, exacerbated by severe droughts which affected approximately all sectors in the country and caused huge costs both in human and in socioeconomic terms (Hayati et al. 2010; Foltz 2002; Yazdanpanah et al. 2013b; Maleksaeidi et al. 2016). To give some indications of the severity some examples are presented. Hayati (2009) estimated the total costs due to droughts between 1998 and 2001 to be more than US$4.2 billion. In 2001, more than 100,000 farmworkers in the Esfahan Province lost their jobs because of a severe drought event. In 2003, more than 100 villages in southeast Iran were evacuated due to lack of water, and more than one million head of livestock perished throughout the country (Beik Mohammadi et al. 2006). Although, Iran has always had cycles of droughts in its history there are strong indications that Iran now faces not only a periodic dry spell, but a severe water crisis (Balali 2009), which is further exacerbated by recent high

Table 76.2 Turning points in trend of water conflict in Iran (Bijani and Hayati 2015)

Stages	Characteristics of each stage
Before land reform (Latent water conflict: Before 1963)	Before land reform in Iran, landlords were known as owners of water resources. They were managers of water and controlled consumption and distribution of water. Consumption and distribution of water were controlled by them under a specific discipline. Therefore, there was no particular water conflict in that period
After land reform up to Islamic Revolution of Iran (The start of water conflict:1963–1979)	After land reform, landlords became weak and the system of feudal and peasant was overthrown. In the continuation of land reform, water was declared as national resource. The government was recognized as responsible for control and managing of water in all sectors. The government was assigned to administrate distribution of water among stakeholders. In that way, some rules were registered for better control of water distribution. Forcefully, water conflicts started from this point, because the government had not needed control on water resources and water beneficiaries, especially in regard to landlords. For example, unauthorized revenue from water resources increased after land reform. Also, digging deep wells developed. We can say in this period not only the government had insufficient power to control water conflict, but that was a factor causing conflict between itself and stakeholders
After the Islamic Revolution up to now (Culmination of water conflict: After 1979)	After political revolution in 1979, the government decreased its control over water resources. In that condition, there was no anticipation and legal mechanism for controlling water conflicts. Furthermore, in the last decade, climate change, especially drought added to this trend and increased water conflicts. With the condition of drought and water scarcity, managing water conflict is more complex. The main part of conflict is between the government and stakeholders, especially in districts that are confronting with drought. In the meantime, urban and industrial water consumption has increased in recent years and government has allocated most of the water savings to these sectors. It is one of the reason for conflict between the government and stakeholders. It should be noted that increasing demands, climate change, agricultural development, and so on have caused some international level water conflicts between Iran and neighboring countries

population growth rates (FAO 2006). Furthermore, climate change is expected to raise the pressure on Iran's water resources by making the region hotter and drier, having major implications for agricultural production, hydroelectricity production, and reliability of water supply and reservoir operations (Madani et al. 2016; Madani 2014; Yazdanpanah et al. 2014a).

Decreasing water resources and increasing tensions over water between different stakeholders like farmers, villages, cities, provinces, economic sectors (industry, agriculture, etc.) have forced these stakeholders to increase their share from water resources (Bijani and Hayati 2015; Valizadeh et al. 2016). Trying to increase the share of water has developed an egoistic-individualistic perspective among the stakeholders (Bijani and Hayati 2013; Valizadeh et al. 2016) and this situation has intensified in recent years due to economic insecurity and high inflation rates following international sanctions. High inflation rates and economic insecurity develops the short-term benefit attitude and egoistic-individualistic values of water management. Such an attitude among the decision makers and policy makers of water sector have encouraged a reactive and symptom-based water management (which usually follows short term benefits) rather than a preventive water management system (which follows the long-term goals) that tries to find and improve the roots and main causes of the water problem (Madani and Dinar 2012; Madani 2014). Studies show that Iran is currently facing severe water and environmental crises and if suitable practical and preventive strategies are not taken to resolve the problem, it should prepare itself to deal with larger and more tragic water problems which will take place in the near future. In this regard, in the next section some managerial strategies were offered.

76.4 Future of the Water Management in Iran

Current water crisis of Iran has been formed over several decades and it is not expected to be solved immediately. But, it should be kept in mind that Iran's water sectors need urgent exit strategies in order to solve and mitigate the negative impacts of the problem. Iran's water management system should remarkably change its intervention paradigms if it wants to be able to mitigate similar problems and their negative consequences in the future.

There is a necessary need for some revisions in water management sector of Iran in terms of institutional structure. Agriculture lands and/or farmers are the main users of freshwater resources and agricultural sector of the country is significantly inefficient in terms of water consumption. As a result, the Ministry of Jihad-Agriculture has always been the subject of criticisms. But in fact, the Ministry of Jihad-Agriculture is not the "owner" of water resources and it is only one of several "users" of water. As it was previously mentioned, the Ministry of Energy has the ownership of water resources in Iran and is responsible for management of water resources. Therefore, many authorities of the Ministry of Jihad-Agriculture claim that water management is not related to them and this ministry cannot engage in water management effectively. In other words, policies, strategies, objectives, and programs of water management are out of the control of the Ministry of Jihad-Agriculture. On the other hand, the Ministry of Energy which is directly responsible for the water management claims that inefficiency of water consumption in agricultural sector is not the responsibility of this ministry. Such institutional arguments are not limited to the Ministry of Jihad-Agriculture and the Ministry of Energy. Department of Environment is another water-related institution that is politically weaker than the other government sectors such as the Ministry of Energy and the Ministry of Jihad-Agriculture (Madani 2014). This arguments and conflict of interests are originated from "institutional perturbation," "lack of coordination system," and "lack of common vision" among water-related organizations (Bijani and Hayati 2015). One of the first steps, in order to fill this gap, is a comprehensive revision in the structure of these governmental bodies and establishment of a Monitoring Board that can be able to exert a direct influence on all water-related organizations. This Board can facilitate the process of coordination and common vision building among water-related institutions. On the other hand, Madani (2014) states that removing the unnecessary hierarchies and minimize the number of stakeholders can be another option in revising the structure of water management system.

As well as revisiting the institutional structure of water management and facilitating the interconnections among water-related bodies, involving communities (especially the farmers) in adopting new strategies and technologies can increase acceptance of new water management systems. One of the important issues for Iran is the change in cropping systems. Paillard et al. (2011) stated that the Middle East countries will not be able to produce as much food as they will presumably consume under standard demographic assumptions and it may be better to rely (again) on cropping systems that were compatible with their climate in combination with other productive activities. For example, these countries could focus on ecological modernization with an emphasis on multifunctional agriculture rather than productivism. This could include the development of alternative sources for income (such as tourism or the development of small-scale industries) in arid regions where groundwater-based irrigation is not sustainable in the long run anymore; for other parts, incentives to change cultivation patterns or to produce more tolerant crops could be initiated, more specifically cultivation of rice, one of the most water consumers of all crops could be limited, and raising barley instead which would make more ecological sense (Yazdanpanah et al. 2013a). Other options include the support in rainwater harvesting via the provision of capital and knowledge, introducing indirect subsidies that are promoting less water-intensive cropping patterns compared to the intensive ones.

Although environmental awareness in Iran has increased in recent years, but it is generally low (Bijani et al. 2017; Keshavarz and Karami 2016; Valizadeh et al. 2016; Madani 2014; Madani et al. 2016). In this regard, Institutions and organizations in charge of water resources management (i.e., The Ministry of Energy in Iran) must prepare an extensive planning in order to raise awareness about the negative impacts of losing water resources. Because, many Iranians (especially farmers) do not have a high education level and this in many cases can hinder their access to different sources of information on the current status of water resources and their relevant problems. Of course, it should be kept in mind that the educational level of farmers is not the only criterion to create positive changes in their awareness, and water authority has also a task to do here. Water resources monitoring system can play an important role in creating awareness and enlightening farmers. Monitoring surface and groundwater resources not only can provide required information for understanding current status of water resources, but also can construct a strong foundation for conducting preventive measures and farmers' access to reliable sources of information. In fact, accountability of governmental organizations will reinforce farmers' personal obligation and accountability. Thus, water authority and other governmental organizations in charge of water resources management should try to set some of their water-related boundaries with respect to farmer communities' boundaries. It means that farmers are one of the undeniable stakeholders in water sector and their needs and problems must be reflected in public sector policies and practices. An appropriate water accounting system can be very important in this process, as well. Because, reporting and assuring information about water are tasks that are done by water accounting. This system provides a basis for processing and interpreting different data that have resulted from organizing hydrological, environmental, economic, and social contexts.

On the other hand, many farmers are not concerned and committed to participating in water conservation activities because of low cost of agricultural water use in Iran as well as due to the responsibility of the government to meet their needs. Regarding that agricultural water has low price in Iran (it has a low price compared to other agriculture inputs and is often not considered as a cost by farmers) and its use is subsidized, with the result that many farmers have had no qualms about taking full advantage of what they see to be a plentiful supply (Yazdanpanah et al. 2014a, b; Esmaeili and Vazirzadeh 2009). In doing this, they become both the victims and the cause of the ever-worsening shortfall in that supply. As a result, it seems that water pricing by the government can be applied as an instrument for awareness and obligation among farmers (Madani 2014). Water prices should be raised significantly to be reflective of the true cost of water in each region across the country. This, of course, can have undeniable negative effects on the socioeconomic conditions of farmers in the short-run and is tied with a high political cost for the government. In order to prevent such impacts, Madani (2014) stated that the government can finance the modernization of agricultural practices that help farmers cut water usage effectively. Although, this strategy requires large initial investments, in the long term, it is expected to cost less than the current government policy, which heavily subsidizes the increasing water and energy use in the agricultural sector.

The mechanisms to monitor and control freshwater use should also be revised. Water management in Iran is centralized. Significant participation of local stakeholders would shift the decision-making power from central agencies to local governments and/or local civil society groups. However, one would need to go beyond simply broadening democratic participation to a new process, which includes continuous open dialogues and engagement on various levels (Yazdanpanah et al. 2013a). These longer-term strategies include implementing legal, institutional reforms, water rights, land rights, social and civic institutions, and legal regimes. Additionally, it is argued (Timmerman 2013) to strengthen the management, coordination, transparency, accountability, and overall technological know-how of the water governance authorities. We argue that this can only be achieved via the stronger involvement of different stakeholders and the empowerment of water users.

As it was previously mentioned, shortage and degradation of water resources in Iran have forced the water authorities to find solutions for increasing rate of water demand. Iran's government usually opted two alternative strategies for responding to water demand. First alternative is economic and technological approach to find the new water resources and increasing the water supply. But the second alternative that has recently approached is a behavioral approach with focusing on the users of fresh water to decrease the water demand (Yazdanpanah et al. 2014b; Bijani et al. 2017; Shiri et al. 2011; Abbasian et al. 2017). This paradigm change has been much more intense, especially in the domain of agriculture because the agricultural sector is the largest consumer of fresh water (Forouzani and Karami 2010; Yazdanpanah et al. 2014a; Bijani and Hayati 2015; Valizadeh et al. 2016). But in fact, these paradigm shift has not been a real change. Because, although decision makers claim that their policy is creating fundamental changes in the behavior of Iranian society members, but in practice their activities (building dams, using deep waters, Water diversion, etc.) are opposite of this policy and thus they following policies that have immediate benefits rather than long term benefits. Some scholars (Yazdanpanah et al. 2013a, 2014a; Madani 2014; Balali 2009) claim that this orientation appears that Iran's water management style is a reactive management. If Iran's water management wants to be sustainable, it should shift its management style from crisis/re-active management type to a pro-active/preventive management. Preventive approach of water management considers the dynamics of different water-related sectors and its goals are identifying and eliminating the roots of water problems, shifting from water governmentality to water governance, and using soft solutions like regulations, institutions, taxation, and so on alongside the hard solutions (Madani 2014; Yazdanpanah et al. 2013c).

It seems that lack of attention to the population issues among decision makers and even being interested in its growth will soon dramatically raise Iran's water management problems (see Balali 2009; Madani and Hipel 2011; Madani et al. 2016; Hashemi 2015; Madani 2014). Given the support that is being made by the policy-makers, Iran's population is expected to continue to grow in next years. But it should be noted that this unwise population growth can encounter the country with too many challenges. One of the first impacts of rapid population growth is significant increase in water demand and a decreasing per capita water availability. That is while Iran's per capita water availability is lower than the global average. On the other hand, inadequate water distribution infrastructures will make the situation more problematic if the population growth continues. In this regard, Iran needs urgent revision of population growth policies if it wants to apply a preventive water resource management style to solve current water problems.

Though the countless ecological and water-related crises in Iran (tragedies of Lake Urmia, Lake Parishan, Bakhtegan Lake, and so many other cases) are very unfortunate, these tragedies can be considered a milestone point for creating positive changes. Because, they have made the Iranians more responsive to the environmental issues and it seems that the Iranian community members are now feeling more sense of responsibility about water and environmental problems than the past (Madani 2014). But, the important point in this regard is the way of orienting this environmentalist trend. Iran's media can play a key role in this regard. Because, the media can facilitate the long-term and positive outcomes of this environmentalist movement. The media can enlighten the people that their pro-environmental behaviors (for example, water-saving behavior) have a very positive impact on reducing current environmental crises. Media can also focus on the fact that water and environmental crises are not just about now and in order to prevent a similar crises in the future, all members of the community must be responsive. This process can gradually lead to the institutionalization of a culture of water conservation in the society.

Another issue that needs to be addressed is having an integrated perspective in water resource management planning. Managing complex water resources systems using unidirectional, mechanistic models may be doomed to provide unrealistic, or at least, questionable results (Hjorth and Bagheri 2006). Many of Iran's water problems are the products of unidirectional focus on one sector without paying attention to the dynamic relationships of the water with the other sectors (economy, agriculture, and infrastructure) (Mirchi et al. 2010) and also are the product of the absence of an integrated view of the complex human-natural system of systems (Hjorth and Madani 2014). But it should be noted that the principles of systems thinking are critical to solving problems in water resources systems, which inevitably consist of interrelated subsystems.

76.5 Conclusion

Reviewing the water management literature of Iran reveals that it is facing with a serious water crisis. Decreasing groundwater levels, drying rivers, drying lakes, freshwater pollution, soil erosion, and desertification are some of water-related problems in Iran. However, Iran was one of the pioneers of sustainable water resources management in the not too distant past. Water crisis management in Iran raised questions which are really significant for both practitioners and scholars dealing with water crisis and water management around the world. Specifically, in Iran, there is a growing consensus among key water decision makers and researchers that current water management methods and practices need to make the transition to more sustainable ones involving an iterative process of continuous improvement on all scales. Thus, in future, water management system of Iran should include characteristics like participatory management, collaborative action, decentralization, attention to both hard and soft systems, networking, integrated perspective, information sharing and learning, and having a common vision about water problems among water-related bodies. But it should be noted that Iran is a vast country and it has specific and different demographic, climatic, agricultural, topographic, cultural, and ethnic characteristics; so there is no one-size-fits-all water management solution for all regions and water problems of Iran. In other words, the water problems of each region should be addressed by examining the strengths, weaknesses, opportunities, threats, as well as the appropriateness of solutions to specific regional conditions.

References

Abbasian AR, Chizari M, Bijani M (2017) Farmers' views on the factors inhibiting the implementation of soil conservation practices (the case of Koohdasht township, Iran). J Agric Sci Technol (JAST) 19(4):797–807

Abdullaev I, Kazbekov J, Manthritilake H, Jumaboev K (2009) Participatory water management at the main canal: a case from south Ferghana canal in Uzbekistan. Agric Water Manag 96 (2):317–329

Ahmadi H, Samani AN, Malekian A (2010) The qanat: a living history in Iran. In: Schneier-Madanee G, Courel MF (eds) Water and sustainability in arid regions. Springer, Dordrecht, pp 125–138

Ajami A (1993) Cooperatives. Encyclopaedia Iranica. http://www.iranicaonline.org/articles/cooperatives-serkat-e-taawoni-economic-organizations-owned-jointly-by-and-operated-for-the-benefit-of-groups-of-individu

Allan JA (2005) Water in the environment/socio-economic development discourse: sustainability, changing management paradigms and policy responses in a global system. Govern Opposition 40(2):181–199

Anayah FM, Almasri MN (2009) Trends and occurrences of nitrate in the groundwater of the West Bank, Palestine. Appl Geogr 29(4):588–601

Asaesh H (1994) Rural economy. Payam Noor University Press, Tehran

Azkia M (2012) Development sociology and rural underdevelopment in Iran. Ettelaat, Tehran

Balali MR (2009) Towards reflexive land and water management in Iran: linking technology, governance and culture. Doctoral thesis, Wageningen University, Wageningen, The Netherlands

Beck MB, Thompson M, Ney S, Gyawali D, Jeffrey P (2011) On enabling and disabling governance for re-engineering city infrastructure. Eng Sustain 164(ES2):129–142

Beik Mohammadi H, Noori H, Bazrafshan J (2006) The effects of drought of 1999-2005 on Sistan rural economy and coping strategies. J Geogr Dev 3:53–72

Bijani M, Hayati D (2011) Water conflict in agricultural system in Iran: a human ecological analysis. J Ecol Environ Sci 2(2):27–40

Bijani M, Hayati D (2013) Application of environmental attitudes toward analyzing water conflict: the case of Doroodzan dam irrigation network. Iran Agric Ext Educ J 9(1):83–102

Bijani M, Hayati D (2015) Farmers' perceptions toward agricultural water conflict: the case of Doroodzan dam irrigation network, Iran. J Agric Sci Technol 17(3):561–575

Bijani M, Ghazani E, Valizadeh N, Fallah Haghighi N (2017) Pro-environmental analysis of farmers' concerns and behaviors towards soil conservation in central district of Sari County, Iran. Int J Soil Water Conserv Res 5(1):43–49

Bill JA (1970) Modernization and reform from above: the case of Iran. J Polit 32(1):19–40

Brunner R, Steelman T (2005) Beyond scientific management. In: Brunner R, Steelman T, Coe-Juell L, Cromley C, Edwards C, Tucker D (eds) Adaptive governance: integrating science, policy and decision making. Columbia University Press, New York, pp 1–32

Davarpanah G (2005) Comparable survey of benefits and disadvantages from ground water to the method of wells and qanats. In: Proceedings of the 2nd international conference on Qanat, Kerman, Iran
Deihimfard R, Zand E, Damghani AM, Soufizadeh S (2007) Herbicide risk assessment during the wheat self-sufficiency project in Iran. Pest Manag Sci 63(10):1036–1045
Djamali M, Jones MD, Migliore J, Balatti S, Fader M, Contreras D, Condect S, Hosseini Z, Lahijani H, Naderi A, Shumilovshikh LS, Tengberg M, Weeks L (2016) Olive cultivation in the heart of the Persian Achaemenid Empire: new insights into agricultural practices and environmental changes reflected in a late Holocene pollen record from Lake Parishan, SW Iran. Veg Hist Archaeobot 25(3):255–269
Ehsani K (2006) Rural society and agricultural development in post-revolution Iran: the first two decades. Middle East Critique 15(1):79–96
Emadodin I, Narita D, Bork HR (2012) Soil degradation and agricultural sustainability: an overview from Iran. Environ Dev Sustain 14(5):611–625
Esmaeili A, Vazirzadeh S (2009) Water pricing for agricultural production in the South of Iran. Water Resour Manag 23(5):957–964
Esmaeili A, Moore F, Keshavarzi B (2014) Nitrate contamination in irrigation groundwater, Isfahan, Iran. Environ Earth Sci 72(7):2511–2522
Faramarzi M, Abbaspour KC, Schulin R, Yang H (2009) Modelling blue and green water resources availability in Iran. Hydrol Process 23:486–501
Farhadinejad T, Khakzad A, Jafari M, Shoaee Z, Khosrotehrani K, Nobari R, Shahrokhi V (2014) The study of environmental effects of chemical fertilizers and domestic sewage on water quality of Taft region, Central Iran. Arabian J Geosci 7(1):221–229
Farshad A, Zinck JA (1997) Indigenous knowledge and agricultural sustainability: a case study in semi-arid regions of Iran. In: Beek KJ, de Bie CA, Driessen P (eds) CD-ROM proceedings of the international conference on geo-information for sustainable land management: Enschede, ITC, 17–21 Aug 1997. ITC, Enschede. 10 p
Fathian F, Morid S, Kahya E (2015) Identification of trends in hydrological and climatic variables in Urmia Lake basin, Iran. Theor Appl Climatol 119(3–4):443–464
Foltz RC (2002) Iran's water crisis: cultural, political and ethical dimension. J Agric Environ Ethics 15(4):357–380
Food and Agricultural Organization of the United Nations (FAO) (2006) National strategy and action plan on drought preparedness, management and mitigation in the agricultural sector: Iran. Terminal statement prepared for the government of the Islamic Republic of Iran by the Food and Agricultural Organization of the United Nations. FAO, Cairo
Forouzani M, Karami E (2010) Agricultural water poverty index and sustainability. J Agron Sustain Dev 31(2):415–431
Fozi Y (2004) Social challenge of Islamic republic of Iran. Sociol J Iran 5(3):172–190
Ghadiri Masoum M, Najafi kani AA (2003) Development program after the Islamic revolution and its impact on rural areas. Geogr Res 44:11–121
Ghaffari G (2006) The state and rural development. Crit Middle East Stud 15(3):283–293
Gohari A, Madani K, Mirchi A, Bavani AM (2014) System-dynamics approach to evaluate climate change adaptation strategies for Iran's Zayandeh-Rud water system. In: Proceedings of the world environmental and water resources congress 2014. ASCE, Portland, OR, pp 1598–1607
Habashiani R (2011) Qanat: a sustainable groundwater supply system. Master's thesis, School of Arts and Social Science, James Cook University, Queensland, Australia
Haeri MR (2003) Kariz (Qanat): an eternal friendly system for harvesting groundwater. In: Adaptation workshop, Cenesta, New Delhi
Halliday F (1981) Response to Ghorayshi on capitalism in rural Iran. MERIP reports, no. 98, Iran two years after. pp 28–30
Hashemi H (2015) Climate change and the future of water management in Iran. Middle East Critique 24(3):307–323

Hassanzadeh E, Zarghami M, Hassanzadeh Y (2012) Determining the main factors in declining the Urmia Lake level by using system dynamics modeling. Water Resour Manag 26(1):129–145
Hayati D (2009) Responsibility of extension institution in planning, implementing, monitoring and evaluating of a drought action oriented system. In: Paper presented at national drought seminar issues and mitigation, 13–14 May, Shiraz, Iran
Hayati D, Karami E (2005) Typology of causes of poverty: the perception of Iranian farmers. J Econ Psychol 26(6):884–901
Hayati D, Karami E, Slee B (2006) Combining qualitative and quantitative methods in the measurement of rural poverty: the case of Iran. Soc Indicat Res 75(3):361–394
Hayati D, Yazdanpanah M, Karbalaee F (2010) Coping with drought: the case of poor farmers of south Iran. Psychol Dev Soc 22(2):361–383
Hjorth P, Bagheri A (2006) Navigating towards sustainable development: a system dynamics approach. Futures 38(1):74–92
Hjorth P, Madani K (2014) Sustainability monitoring and assessment: new challenges require new thinking. J Water Resour Plann Manag 140:133–135
Hooglund E (1973) The khwushnishin population of Iran. Iran Stud 6(4):229–245
Hooglund E (1982) Land and revolution in Iran. University of Texas Press, Austin, TX
Jomehpour M (2009) Qanat irrigation systems as important and ingenious agricultural heritage: case study of the qanats of Kashan, Iran. Int J Environ Stud 66(3):297–315
Kalantari K (2011) Rural sociology. Payame Noor University Press, Tehran
Karami E, Hayati D (2005) Rural poverty and sustainability: the case of groundwater depletion in Iran. Asian J Water Environ Pollut 2(2):51–61
Kardovani P (2007) Water resources and issues in Iran. Agah Press, Tehran
Karimi AA (2009) Water situation in Iran: challenges and achievements. In: European water and sanitation services vs. sustainable development, potable water services in Morocco-China-Austria-Iran, Proc. November, 2009, Athens
Keshavarz M, Karami E (2016) Farmers' pro-environmental behavior under drought: application of protection motivation theory. J Arid Environ 127:128–136
Lahsaeizadeh A (2002) Social change in rural Iran. Navid Press, Shiraz
Lambton AKS (1953) Landlord and peasant in Persia. Oxford University Press, London, p 459
Lehane S (2014) The Iranian water crisis. Strategic Analysis Paper, Future Directions International International Pty Ltd.: Perth, Australia, 11.
Madani K (2008) Reasons behind failure of Qanats in the 20th century. In: Paper presented at the world environmental and water resources congress 2008, Honolulu, HI, ASCE
Madani K (2014) Water management in Iran: what is causing the looming crisis? J Environ Stud Sci 4(4):315–328
Madani K, Dinar A (2012) Non-cooperative institutions for sustainable common pool resource management: application to groundwater. Ecol Econ 74:34–45
Madani K, Hipel K (2011) Non-cooperative stability definitions for strategic analysis of generic water resources conflicts. Water Resour Manag 25:1949–1977
Madani K, Mariño MA (2009) System dynamics analysis for managing Iran's Zayandeh-Rud river basin. Water Resour Manag 23(11):2163–2187
Madani K, AghaKouchak A, Mirchi A (2016) Iran's socio-economic drought: challenges of a water-bankrupt nation. Iran Stud 49(6):997–1016
Malekian A, Hayati D, Aarts N (2017) Conceptualizations of water security in the agricultural sector: perceptions, practices, and paradigms. J Hydrol 544:224–232
Maleksaeidi H, Karami E, Zamani GH, Rezaei-Moghaddam K, Hayati D, Masoudi M (2016) Discovering and characterizing farm households' resilience under water scarcity. Environ Dev Sustain 18(2):499–525
Masetti M, Poli S, Sterlacchini S, Beretta GP, Facchi A (2008) Spatial and statistical assessment of factors influencing nitrate contamination in groundwater. J Environ Manag 86(1):272–281
Mazaheri M, Abdolmanafi N (2018) Water crisis and its consequences in Iran. Iran parliament research center report no. 15608, Tehran, Iran

Mirchi A, Watkins D Jr, Madani K (2010) Modeling for watershed planning, management, and decision making. Chapter 6. In: Vaughn JC (ed) Watersheds: management, restoration and environmental impact. Nova, Hauppauge, NY, pp 221–244
Molle F (2006) Planning and managing water resources at the river basin level: emergence and evolution of a concept. Comprehensive assessment research report 16. International Water Management Institute, Colombo, Sri Lanka. www.Iwmi.cgiar.org/assessment/publications/researchreports.htm
Molle F, Mollinga PP, Meinzen-Dick R (2008) Water, politics and development: introduction water alternatives. Water Altern 1(1):1–16
Najafi A, Vatanfada J (2011) Environmental challenges in trans-boundary waters, case study: Hamoon Hirmand Wetland (Iran and Afghanistan). Int J Water Resour Arid Environ 1(1):16–24
Najafi H, Bagheri A, Madani K (2013) The topology of generic shared water resources games: insights for the Lake Urmia Disaster. In: 6th international conference on water resources and environment research, Koblenz, Germany
Nelson R, Howden M, Smith MS (2008) Using adaptive governance to rethink the way science supports Australian drought policy. Environ Sci Policy 11(7):588–601
Norouzi H, AghaKouchak A, Madani K, Mirchi A, Farahmand A, Conway C (2013) Monitoring changes in water resources systems using high resolution satellite observations: application to Lake Urmia. American geophysical union fall meeting 2013, AGU fall meeting, San Francisco, CA, Abstract # H43O-04
Paillard S, Treyer S, Dorin B (2011) Agrimonde: scenarios and challenges for feeding the World in 2050 Editions Quae
Papoli Yazdi MH (2000) Qanats and its economic value. In: Proceedings of the first international conference on Qanat, Yazd, Iran, 9–22 (in Persian)
Razmkhah H, Abrishamchi A, Torkian A (2010) Evaluation of spatial and temporal variation in water quality by pattern recognition techniques: a case study on Jajrood River (Tehran, Iran). J Environ Manag 91(4):852–860
Rezaei-Moghaddam K, Karami E, Gibson J (2005) Conceptualizing sustainable agriculture Iran as an illustrative case. J Sustain Agric 27(3):25–56
Rezvani MR (2005) An introduction to rural development planning at Iran. Ghoomes, Tehran
Safinejad J (1980) Traditional irrigation system in Iran. University of Tehran Press, Tehran
Safinejad J (1989) Buneh: traditional agricultural system in Iran. Amirkabir, Tehran
Salehi S, Chizari M, Sadighi H, Bijani M (2018) Assessment of agricultural groundwater users in Iran: a cultural environmental bias. Hydrgeol J 26(1):285–295
Salemi HR, Mamanpoush A, Miranzadeh M, Akbari M, Torabi M, Toomanian N, Murray-Rust H, Droogers P, Sally H, Gieske A (2000) Water management for sustainable irrigated agriculture in the Zayandeh Rud Basin, Esfahan Province, Iran. IAERI-IWMI research reports 1
Salmanzadeh S, Jones GE (1981) Transformations in the agrarian structure in Southwestern Iran. J Dev Areas 15:199–214
Sharifi Moghaddam M (2013) Development of sustainable agricultural for applying ecosystem approach in wetlands management, case studies on Lake Parishan, Conservation of Iranian wetlands project. (in Farsi)
Shiri S, Bijani M, Chaharsoughi Amin H, Noori H, Soleymanifard A (2011) Effectiveness evaluation of the axial plan of wheat from expert supervisors' view in Ilam Province. World Appl Sci J 14(11):1724–1729
Sima S, Tajrishy M (2013) Using satellite data to extract volume–area–elevation relationships for Urmia Lake, Iran. J Great Lakes Res 39:90–99
Statistical Centre of Iran (2018) Iran population information based on 2016 census. Retrieved from: https://www.amar.org.ir/
Tajik M, Darvishi F (2004) Aspirations of the Islamic revolution in the age of globalization: Challenges and reactions. Sociol J Iran 5(3):140–171
Timmerman M (2013) Water consumption in the Middle East and North Africa (Mena) region: a virtual water approach

Valizadeh N, Bijani M, Abbasi E (2016) Pro-environmental analysis of farmers' participatory behavior toward conservation of surface water resources in southern sector of Urmia Lake's catchment area. Iran Agric Ext Educ J 11(2):183–201. (In Persian)

Valizadeh N, Bijani M, Abbasi E (2018) Farmers' active participation in water conservation: insights from a survey among farmers in southern regions of west Azerbaijan Province, Iran. J Agric Sci Technol (JAST) 20(5):895–910

Water Research Institute (2015) Data report. Integrated water resources management of Urmia Lake, Tehran

Wosooghi M (2007) Rural sociology. Keyhan, Tehran

Yazdanpanah M, Hayati D, Zamani GH, Karbalaee F, Hochrainer-Stigler S (2013a) Water management from tradition to second modernity: an analysis of the water crisis in Iran. Environ Dev Sustain 15(6):1605–1621

Yazdanpanah M, Monfared N, Hochrainer-Stigler S (2013b) Inter-related effects due to droughts for rural populations: a qualitative field study for farmers in Iran. Int J Mass Emerg Disast (IJMED) 31(2):106–129

Yazdanpanah M, Thompson M, Hayati D, Zamani GH (2013c) A new enemy at the gate: tackling Iran's water super-crisis by way of a transition from government to governance. Prog Dev Stud 13(3):177–194

Yazdanpanah M, Hayati D, Hochrainer-Stigler S, Zamani GH (2014a) Understanding farmers' intention and behavior regarding water conservation in the Middle-East and North Africa: a case study in Iran. J Environ Manag 135:63–72

Yazdanpanah M, Hayati D, Thompson M, Zamani GH, Monfared N (2014b) Policy and plural responsiveness: taking constructive account of the ways in which Iranian farmers think about and behave in relation to water. J Hydrol 514:347–357

Zarghami M (2011) Effective watershed management; case study of Urmia Lake, Iran. Lake Reserv Manag 27(1):87–94

Chapter 77
The Need of Biodiversity Conservation Strategies in Iraq: The State of Protected Areas

Laith A. Jawad

Abstract The significance of developing instructions is studied for Bangladesh in order to assist in deciding future co-administration urgencies for the protected areas in Iraq. Co-organization tactic is chiefly set up appropriate for many countries that are closely spread with local societies, who exercise agriculture and rely on forests for their income. The guard of protected areas and the included biota in contrast to biotic impact cannot be efficiently tested deprived of creating profitable companies with local investors. Additional natural forest zones should be put under protected areas where co-administration performs for included biota upkeep should be applied.

Protected areas are a new subject to Iraq and the policy makers in Iraq. In order to learn a lesson from other countries, Iraq needs to take several steps to implement what other countries have done in this field.

77.1 Introduction

It has been thought that during this century restoration to the environment will be developed further (Sunding 2011). Though, natural, self-supporting, and well-protected rivers and swamps almost vanished from the map of inner-city areas world producing the regeneration of the strong habitats a puzzling objective due to the absence of known circumstances.

It is well known that it is the humblest individuals in the societies of most developing countries, who rely most on native habitats for their livings, that will be most impacted by the impacts of this biota loss (CBD 2006). Therefore, essential to conserve this threatened biodiversity for the well-being of these people.

Iraq, one of the countries of the Middle East that is rich in water resources, both freshwater and marine. A special chapter in this book has been allocated for water

L. A. Jawad (✉)
School of Environmental and Animal Sciences, Unitec Institute of Technology, Auckland, New Zealand

L. A. Jawad (ed.), *Tigris and Euphrates Rivers: Their Environment from Headwaters to Mouth*, Aquatic Ecology Series 11,
https://doi.org/10.1007/978-3-030-57570-0_77

resources in Iraq. Therefore, identification of water resources in Iraq will not be given in this chapter.

As in other areas around the world, the biota of Iraq is likewise ingoing within a serious era. The number of species both on land in water that has become extinct or near extinction is unknown due to the lack of studies and research in this field.

As mentioned above, the biodiversity research and studies are scarce in Iraq. Probably, freshwater fish groups have gained slight of these studies in the last few decades.

The Main Biodiversity Areas plan, based on a quick valuation method by Nature Iraq from 2005 to 2007, has recognized important data on the position and dispersal of ecologically and economically serious fish species for Iraqis. In 3 years of studies, a range of 41–52 species (including freshwater and marine entrant species) were reported in swamps in southern Iraq (Abed et al. 2009). Numerous approvals are given for additional scientific and administration investigations. Sixteen fish species are projected as possible "Species of Special Concern" in deliberation of the plan of a fisheries administration policy for Iraq grounded on ecological and economic issues.

Iraq needs to practice both in situ and ex situ preservation approaches to uphold the fragments of the country's biological biota. The announcement of endangered areas, ecologically serious areas (ECAs), World Heritage Sites, and Ramsar sites are some samples of in situ conservation; otherwise, ex situ conservation events contains botanical gardens, protection plots, gene banks, arboretums, etc. The country wants to support the international agreements and resolutions associated to protection of biota in order to develop numerous judicial strategies and developments to preserve its outstanding biota. Such activities were achieved by other countries such as Bangladesh (Ali and Ahmed 2001; Kothari et al. 2000).

In the present chapter, a short review of the major legislative policies and approaches that Iraq needs to take to provide necessities for biodiversity conservation in the country.

77.2 Creating Protected Areas in Iraq

Announcing sheltered zones has long been the maximum actual and prevalent issue for saving nature and natural assets throughout the world; though, the statement of such areas for biota protection is somewhat a new idea in Iraq. Before this step, Iraq needs to create Wildlife Preservation Act to define the three types of the protected areas to be included within IUCN sheltered area administration classes in the country, namely:

1. *Natural Reserve:* A zone preserved as an undisturbed upbringing ground for wild fauna and where the habitat is protected for the continued health of the living or migratory fauna.
2. *Nationwide Estate:* A moderately large area of natural beauty to which the members of the public have admission for leisure, tutoring and investigation, and in which the wildlife is secure.

3. *Disposed Zone:* Normally comprises a relatively isolated area meant for protection of wildlife in general and to increase the group of definite species. Presently, there are no such areas in Iraq.

77.3 Threats to Protected Areas of Iraq

In case sheltered zones have been built in Iraq, they will be a part of the Reserved Forests. The following are some major challenges that might face the future protected areas in Iraq, thus a managing plans need to be put forward at the same time of the plans for building protected areas are put forward:

1. Countryside scarcity and the high joblessness proportion in the nearby sheltered areas.
2. Struggles between the Forest Department and local forest-dependent inhabitants owing to overlooking local people's usual supply use performs.
3. Wood as fuel, prohibited stealing and over misuse of other non-timber forest assets.
4. Incompetent administration services owing to a lack of qualified staffs, up-to-date facilities, and financial plan shortages.
5. The lack of an appropriate observing organization and inappropriate application of laws.
6. Absence of people's consciousness regarding the significance of biota.

77.4 Conclusions and Recommendations

In Iraq, throughout the last few periods and it continues now a noteworthy area of the country agricultural lands and forests have been converted to other land use practices (e.g., agricultural crop fields) owing to country's huge population impact. It is nowadays essential to conserve the country's remaining natural forest patches by bringing them under a well-defined protected area system, guaranteeing a reasonable representation of all vegetation varieties. However, in Iraq, simply declaring a protected area under the provision of the law, but excluding the requirements of the people at the countrysides, cannot halt the rapid loss of biodiversity or secure the future of protected areas, meanwhile the local residents will consider such efforts to be ignoring their old-style privileges to the forest and lands, which they have liked for age band. The Government has to accomplish protected areas through an acceptable cooperative administration scheme and should directly deliberate the following in order to attain a sustainable and active protected area administrative scheme:

1. Applying deficiency lessening plans about sheltered zones by creating substitute revenue events.
2. Generate chances to grow the tourist business grounded on sheltered zones.

3. Restore and manage fitter zones as an alternative resource exploitation area, also fixing a permissible supply misuse limit from the protected areas.
4. Frame a distinct official body for the administration and watching of protected areas.
5. Encourage bulk construction of protected area administrative personnel by scheming particular studies in the maintenance and administration of protected areas.
6. Create internal revenue supply (e.g., selling entry tickets, permissions for photographs, souvenir, etc.) for the maintainable funding of sheltered zones.

References

Abed I, Rubec C, Coad B (2009) Key biodiversity areas: Rapid assessment of fish fauna in southern Iraq. BioRisk 3:161–171

Ali MO, Ahmed M (2001) Bio-diversity conservation: vision for Bangladesh. In: Chowdhury QI (ed) Bangladesh: State of biodiversity. Forum of Environmental Journalists of Bangladesh (FEJB), Dhaka, pp 33–49

CBD (2006) Global biodiversity outlook 2. Secretariat of the Convention on Biological Diversity (SCBD), Montreal, 81 pp

Kothari A, Pathak N, Vania F (eds) (2000) Where communities care: community based wildlife and ecosystem management in South Asia. Kalpavriksh, Pune and International Institute of Environment and Development (IIED), London, pp 17–23

Sunding KN (2011) Toward an era of restoration in ecology: successes, failures, and opportunities ahead. Annu Rev Ecol Evol Syst 42:465–487

Chapter 78
Benthic Macroinvertebrates of the Tigris and Euphrates Rivers in Turkey

Belgin Çamur-Elipek, Bülent Şen, Gazel Burcu Aydın, Menekşe Taş-Divrik, and Pınar Yıldırım

78.1 Introduction

Tigris–Euphrates River Basin is one of the great river systems of south-western Asia. Tigris (Dicle) River and Euphrates (Fırat) River are born in Turkey and they follow roughly parallel courses from their sources to the upper part of the Persian Gulf. The Lake Hazar located in the southeast of Elazığ Province is the main source of the Tigris River while the Murat and Karasu streams originated in northeastern Turkey are major tributaries of the Euphrates. A total of ten dam lakes or reservoirs are located on Tigris and Euphrates Rivers in Turkey. A lot of them are used to provide hydroelectric power besides recreational and fisheries purposes. In this chapter, benthic macroinvertebrate communities in this area reported at previous studies were submitted.

In the present chapter, the findings/data obtained through previous studies carried out on benthic macroinvertebrate communities in the rivers and dam lakes present in the Tigris–Euphrates River Basin in Turkey are reviewed and evaluated. In addition,

B. Çamur-Elipek · G. B. Aydın
Department of Biology, Faculty of Science, Trakya University, Edirne, Turkey
e-mail: belginelipekcamur@trakya.edu.tr; gburcuaydin@trakya.edu.tr

B. Şen (✉)
Faculty of Fisheries, Fırat University, Elazığ, Turkey
e-mail: bulentsen23@gmail.com

M. Taş-Divrik
Şarkışla Aşık Veysel Vocational School, Sivas Cumhuriyet University, Sivas, Turkey
e-mail: menekse.tas@cumhuriyet.edu.tr

P. Yıldırım
Faculty of Çanakkale Applied Sciences, Department of Fisheries Technology, Çanakkale Onsekiz Mart University, Çanakkale, Turkey
e-mail: pinaryildirim@comu.edu.tr

L. A. Jawad (ed.), *Tigris and Euphrates Rivers: Their Environment from Headwaters to Mouth*, Aquatic Ecology Series 11,
https://doi.org/10.1007/978-3-030-57570-0_78

environmental effects on benthic macroinvertebrates and their role in the aquatic habitats of the two rivers are discussed in relation to their occurrence.

78.2 Streams and Dam Lakes of the Two Rivers

There are four dam lakes on the Tigris (Dicle) River namely Kralkızı, Batman, Ilısu, and Cizre. The Euphrates (Fırat) River has two main streams: Murat and Karasu, and five dam lakes: Keban, Karakaya, Atatürk, Birecik, and Karkamış.

78.2.1 Tigris (Dicle) River

The Tigris River originates from a tectonic lake called Lake Hazar located south of Elazığ Province in Eastern Anatolia. The lake fed by Behremaz, Kürk, and Zıkkım streams is the rising source of the Tigris River. Numerous tributaries that rise in the Zagros Mountains in Iran, Iraq, and Turkey are known to feed the Tigris River. Thus, the rivers' basin is expanded in four countries: Iran, Iraq, Syria, and Turkey. Tigris River is known to be the second largest river in Western Asia and carries a higher volume of water than the Euphrates River (UN-ESCWA and BGR 2013). The most important tributaries of the river are Batman Stream, Botan Stream, Habur Brook, Big Zap Brook, and Little Zap Brook (Zeybek et al. 2016). The Tigris has been an important river throughout history as it was one of the main water sources for the ancient Mesopotamian civilizations (Varol et al. 2010).

78.2.2 Euphrates (Fırat) River

The Euphrates River has two main tributaries, the Murat and Karasu rivers, which drain from the high plateau to the north-west of Lake Van (Kibaroğlu and Scheumann 2011). Three largest dams of Turkey named Keban (the 1970s), Karakaya and Atatürk (1980s) were built on the Euphrates River (Brismar 2002; Kibaroğlu and Scheumann 2011). They are the largest dam lakes and also the biggest electric generation plants in Turkey (Gökçe and Özhan 2011).

78.3 Benthic Macroinvertebrate Communities in the Two Rivers of Turkey

Macroinvertebrates are known useful biological monitors because as they are found in all aquatic environments. In addition, they are less mobile than many other groups of aquatic organisms, and are of a size, which makes them easily collectible (Gökçe and Özhan 2011). Therefore, this group has a very important value for researchers, both taxonomically and ecologically. Although the studies on benthic macroinvertebrates have increased in Turkey, there are still a few studies performed on the benthos of Tigris–Euphrates River Basin. It is worthwhile to mention that the information on macroinvertebrate communities of the main stretch of the Dicle River is still incomplete (Çetinkaya and Bekleyen 2017) although it is one of the most important transboundary rivers in Turkey.

In this chapter, the benthic macroinvertebrate populations determined at some previous studies performed in the basin were examined. The list of the taxa recorded at the previous studies in the area were presented at Table 78.1. The present chapter also aimed to emphasize the ecological roles of benthic macroinvertebrates in the aquatic environments.

78.3.1 Group Mollusca

Lymnaeidae family (Gastropoda) was reported from Hazar lake by Şahin and Baysal (1972). The species belonging to ***Radix* sp.**, ***Menetus* sp.**, ***Promenetus* sp.**, and ***Probythinella* sp.** from Gastropoda, ***Pisidium* sp.** from Bivalvia were also reported from Hazar Lake by Türkgülü (2010).

Dreissena polymorpha (Pallas, 1771) also known as Zebra Mussels is reported from the Euphrates River and all the reservoirs (Keban, Karakaya, Atatürk, Birecik, and Karkamış) in the Euphrates Basin (Bobat et al. 2004; Aksu et al. 2017). This species was also reported in Atatürk Dam Lake since 1997 and their accumulation problems have also been known in Birecik Dam since 2000 (Bobat et al. 2004; Aksu et al. 2017). The occurrence of ***Unio mancus*** (Lamarck, 1819) in Atatürk Dam Lake and the subspecies ***Unio elongatus eucirrs*** (Bourguignat, 1860) in Karakaya Dam Lake are known (Alkan-Uçkun et al. 2017; Küçükyılmaz and Şahin 2017). The species ***D. polymorpha*** was reported as the most abundant taxa in Karakaya Dam Lake (Gökçe and Özhan 2011). ***Dreissena siouffi*** Locard, 1893 was also reported from Euphrates River (Ekin et al. 2008). However, the species ***Physa fontinalis*** (Linnaeus, 1758), ***Radix ovata*** (Drapparnaud, 1805), and ***Valvata pulchella*** Studer, 1820 were reported from Karakaya Dam Lake (Gökçe and Özhan 2011). Some specimens belonging to group Mollusca are shown in Figs. 78.1 and 78.2.

78.3.2 Group Oligochaeta

Oligochaeta fauna of the Tigris River was studied by Zeybek et al. (2016) who reported a total of 25 taxa (13 from Naidinae, 6 taxa from Tubificinae, and 6 taxa from Enchytraeidae): ***Ophidonais serpentina*** Müller, 1774; ***Stylaria lacustris*** Linnaeus, 1767; ***Nais behningi*** Michaelsen, 1923; ***Nais barbata*** Müller, 1773; ***Nais elinguis*** Müller, 1774; ***Nais bretscheri*** Michaelsen, 1899; ***Nais pardalis*** Piguet, 1906; ***Nais variabilis*** Piguet, 1906; ***Nais communis*** Piguet, 1906; ***Nais simplex*** Piguet, 1906; ***Nais pseudobtusa*** Piguet, 1906; ***Pristina aequiseta*** Bourne, 1891; ***Pristina osborni*** Walton, 1906; ***Tubifex tubifex*** Müller, 1774; ***Limnodrilus***

Table 78.1 Benthic macroinvertebrate taxa recorded at previous studies in the Tigris–Euphrates Basin of Turkey

Group	Taxa	Tigris basin	Euphrates basin
Mollusca			
Gastropoda			
	Menetus sp.	√	
	Physa fontinalis		√
	Probythinella sp.	√	
	Promenetus sp.	√	
	Radix ovata		√
	Radix sp.	√	
	Valvata pulchella		√
Bivalvia			
	Dreissena polymorpha		√
	Dreissena siouffi		√
	Pisidium sp.	√	
	Unio elongatus eucirrs		√
	Unio mancus		√
Annelida			
Oligochaeta			
	Aulodrilus pluriseta	√	
	Cognettia glandulosa	√	
	Cognettia sphagnetorum	√	
	Dugesia sp.	√	
	Haplotaxis sp.	√	
	Henlea nasuta	√	
	Henlea perpusilla	√	
	Henlea ventriculosa	√	
	Limnodrilus claparedeianus	√	
	Limnodrilus hoffmeisteri	√	
	Limnodrilus profundicola	√	
	Mesenchytraeus sp.	√	
	Nais barbata	√	
	Nais behningi	√	
	Nais bretscheri	√	
	Nais communis	√	
	Nais elinguis	√	
	Nais pardalis	√	
	Nais pseudobtusa	√	
	Nais simplex	√	
	Nais variabilis	√	
	Ophidonais serpentina	√	
	Potamothrix alatus hazaricus	√	

(continued)

Table 78.1 (continued)

Group	Taxa	Tigris basin	Euphrates basin
	Potamothrix hammoniensis		√
	Pristina aequiseta	√	
	Pristina osborni	√	
	Psammoryctides deserticola	√	
	Rhynchelmis sp.	√	
	Stylaria lacustris	√	
	Tubifex sp.		√
	Tubifex tubifex	√	
Ostracoda			
	Candona angulata		√
	Candona neglecta		√
	Cryptocandona sp.		√
	Cypris sp.	√	
	Darwinula stevensoni		√
	Fabaeformiscandona sp.		√
	Heterocypris salina		√
	Ilyocypris sp.		√
	Kovalevskiella sp.		√
	Limnocythere inopinata		√
	Paralimnocythere relicta	√	
	Potamocypris sp.		√
	Pseudocandona cf albicans		√
Hydrachnidia			
	Arrenurus (s.str.) abbreviator	√	
	A (s.str.) cuspidifer	√	
	A. (s.str.) albator	√	
	A. (s.str.) ayyildizi	√	
	A. (s.str.) bicuspidator	√	
	A. (s.str.) bruzelii	√	
	A. (s.str.) cuspidator	√	
	A. (s.str.) demirsoyi	√	
	A. (s.str.) distans	√	
	A. (s.str.) radiatus	√	
	A. (s.str.) robustus	√	
	A. (s.str.) rodrigensis	√	
	A. (s.str.) tricuspidator	√	
	A. (Megalura.) globator	√	
	A. (Micrura.) novus	√	
	A. (Micrura.) octagonus	√	
	A. (Micrura.) sinuator	√	
	A.(Rhino.) hazarensis	√	
	Atractides nodipalpis	√	

(continued)

Table 78.1 (continued)

Group	Taxa	Tigris basin	Euphrates basin
	Eylais extendens	√	
	E. megalostoma	√	
	E. rahmi	√	
	E. setosa	√	
	Forelia variegator	√	
	Georgella helvetica	√	
	Hydrachna (s.str.) skorikowi	√	
	Hydrodroma despiciens	√	
	Hydryphantes dispar	√	
	Hygrobates (s.str.) fluviatilis	√	
	Lebertia (Pilo.) leioderma	√	
	L.(Pseudolebertia) tuberosa	√	
	Limnesia (s.str.) undulata	√	
	L. (s.str.) walteri	√	
	Mideopsis orbicularis	√	
	Neumania deltoides	√	
	Piona carnea	√	
	Pionopsis lutescens	√	
	Sperchon setiger	√	
	Tiphys (s.str.) ornatus	√	
Malacostraca			
Isopoda			
	Asellidae	√	
Amphipoda			
	Gammarus spp.	√	
Insecta			
Ephemeroptera			
	Baetis alpinus	√	√
	B. bisri	√	
	B. buceratus		√
	B. lutheri		√
	B. lutheri georgiensis		√
	B. macrospinosus	√	
	B. pseudogemellus	√	
	B. tricolor		√
	Caenis macrura	√	
	Centroptilum sp.	√	
	Cinygma caucasica	√	√
	Cloeon sp.	√	
	Ecdyonurus dispar	√	
	Ecdyonurus venosus		√

(continued)

Table 78.1 (continued)

Group	Taxa	Tigris basin	Euphrates basin
	Electrogena ressli		√
	Epeiron amseli	√	√
	Epeorus zaitzevi	√	√
	Ephemera vulgata	√	
	Heptagenia coerulans	√	
	H. perflava	√	
	Iron nigripilosus	√	√
	Oligoneurella tskhomelidzei	√	
	Potamanthus luteus		√
	Rhithrogena sp.	√	
	R. caucasica		√
	R. iridina	√	√
	R. iranica	√	
	Siphlonurus sp.	√	
	Torleya major		√
Trichoptera			
	Nyctiophylax sp.	√	
	Psychomyia sp.	√	
Diptera			
Culicidae			
	Culicoides sp.	√	
	Ochlerotatus(Rustic.)rusticus	√	
Ephydridae			
	Scatella sp.	√	
Ceratopogonidae			
	Sphaeromias sp.	√	
Chironomidae			
	Cardiocladius capucinus	√	
	Chironomus anthracinus		√
	Chironomus holsatus	√	
	Chironomus plumosus	√	√
	Chironomus thummi	√	
	Cladotanytarsus mancus	√	
	Cladotanytarsus sp.	√	
	Cricotopus bicinctus	√	
	Cricotopus sp.	√	
	Cricotopus triannulatus	√	
	Cricotopus trifascia	√	
	Cricotopus vierriensis	√	
	Cryptochironomus defectus	√	
	Dicrotendipes sp.	√	

(continued)

Table 78.1 (continued)

Group	Taxa	Tigris basin	Euphrates basin
	Eukiefferiella coerulescens	√	
	Eukiefferiella fuldensis	√	
	Eukiefferiella sp	√	
	Euryhapsis sp.	√	
	Larsia curticalcar	√	
	Nanocladius spiniplenus	√	
	Orthocladius (O.) sp.	√	
	Orthocladius (E.) thienemanni	√	
	Orthocladius (E.) rivicola	√	
	Orthocladius (O.) oblidens	√	
	Orthocladius (S.) holsatus	√	
	Parakiefferiella sp.	√	
	Paratanytarsus lauterborni	√	
	Paratanytarsus sp.	√	
	Polypedilum sp.	√	
	Procladius sp.	√	
	Psectrocladius sordidellus	√	
	Psectrocladius sp.	√	
	Rheotanytarsus sp.	√	
	Stictochironomus histrio	√	
	Tanypus punctipennis	√	
	Tanypus sp.	√	
	Tanytarsus sp.	√	
	Thienemanniella sp.	√	
	Thienemannimyia sp.	√	
	Tvetenia calvescens	√	
Simuliidae			
	Metacnephia subalpina		√
	Prosimulium rachiliense		√
	Simulium (Mont.)alizadei		√
	S.(Nevermannia) costatum		√
	S.(N.) vernum		√
	S.(Simulium) bezzii		√
	S. (S.) ornatum		√
	S. (S.) alajense		√
	S. (S.) kerisorum		√
	S. (S.) kiritshenkoi		√
	S. (S.) morsitans		√
	S. (S.) niha		√
	S. (S.) variegatum	√	
	S. (Trciho.) margaritae		√
	S.(Wilhel.) pseudequinum		√
	Sulcicnephia znoikoi		√

(continued)

Table 78.1 (continued)

Group	Taxa	Tigris basin	Euphrates basin
Coleoptera			
	Donacia sp.	√	
Hemiptera			
	Micronecta sp.	√	
Odonata			
	Enallagma sp.	√	
	Libellula sp.	√	
	Ophiogomphus sp.	√	

Fig. 78.1 *Physa* sp. (Gastropoda: Mollusca)

hoffmeisteri Claparede, 1862; ***Limnodrilus profundicola*** Verrill, 1871; ***Limnodrilus claparedeianus*** Ratzel, 1868; ***Aulodrilus pluriseta*** Piguet, 1906; ***Psammoryctides deserticola*** Grimm, 1877; ***Mesenchytraeus* sp.**; ***Henlea nasuta*** Eisen, 1878; ***Henlea perpusilla*** Friend, 1911; ***Henlea ventriculosa*** Udekem, 1854; ***Cognettia glandulosa*** Michaelsen, 1889; and ***Cognettia sphagnetorum*** Vejdovsky, 1878. Although Tubificidae and Planariidae families were reported from Hazar Lake by Şahin and Baysal (1972), new endemic Oligochaeta species (***Potamothrix alatus hazaricus* n. ssp**.) in Hazar Lake which is the main source of Tigris River was also determined by Timm et al. (2013). The species belonging to genera ***Rhynchelmis, Haplotaxis, Dugesia,*** and Tubificidae were also reported from Hazar Lake by Türkgülü (2010).

The species of ***Tubifex*** was reported as the most abundant taxa in Karakaya Dam Lake in the Euphrates Basin (Gökçe and Özhan 2011). More recently Aras and Fındık (2016) investigated the Oligochaeta fauna in the eight dam lakes at the

Fig. 78.2 *Dreissena* sp. (Bivalvia: Mollusca)

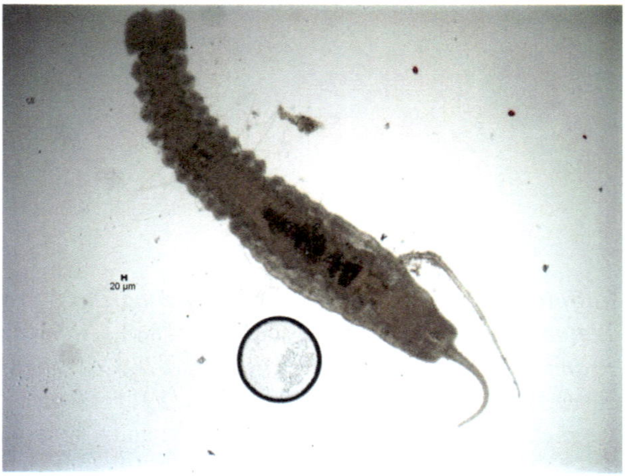

Fig. 78.3 *Stylaria lacustris*

Euphrates Basin: Hacıhıdır lake, Atatürk lake, Uçöz lake, Dumluca lake, Seve lake, Çat lake, Karakaya lake, and Birecik lake. They found a total of ten species of two subfamilies within the family Naididae and reported ***Limnodrilus hoffmeisteri***, ***Potamothrix hammoniensis*** (Michaelsen, 1901), ***Tubifex tubifex***, ***Nais simplex***, and ***Nais communis*** as the most abundant taxa in those lakes. Some species belonging to group Oligochaeta are shown in Figs. 78.3 and 78.4.

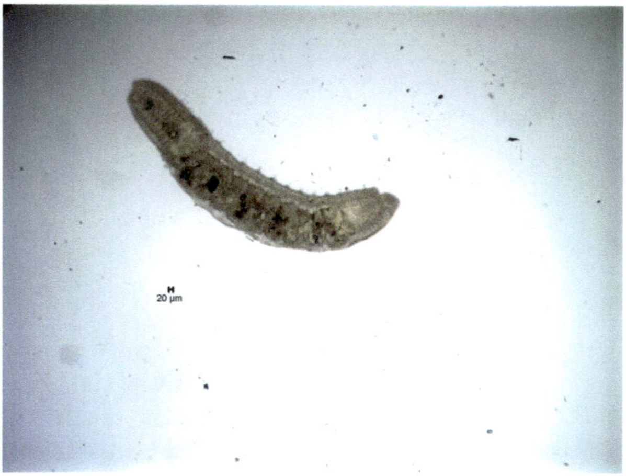

Fig. 78.4 *Nais elinguis*

78.3.3 Group Ostracoda

Members of Ostracoda were reported by Şahin and Baysal (1972) and the species belonging to genus ***Cypris*** were determined by Türkgülü (2010) in Hazar Lake. Akdemir and Külköylüoğlu (2011) reported presence of ***Paralimnocythere relicta*** from Kral Kızı Dam Lake on the Tigris River.

Özuluğ and Dökümcü (2014) reported a total of 11 species from Birecik Reservoir in Euphrates Basin: ***Darwinula stevensoni*** (Brady and Robertson, 1870); ***Candona neglecta*** Sars, 1887; ***Candona angulata*** Müller, 1900; ***Fabaeformiscandona* sp**.; ***Pseudocandona cf albicans***; ***Cryptocandona* sp**.; ***Ilyocypris* sp.; *Heterocypris salina*** (Brady, 1868); ***Potamocypris* sp**.; ***Limnocythere inopinata*** (Baird,1843); and ***Kovalevskiella* sp**. An Ostracoda sample is shown in Fig. 78.5.

78.3.4 Group Hydrachnidia

Erman et al. (2006) reported a total of 26 species belonging the water mites from Hazar Lake: ***Hydrachna (s.str.) skorikowi*** Zachvatkin 1941; ***Eylais extendens*** (O.F. Müller, 1776); ***Hydryphantes dispar*** (Schaub, 1888); ***Georgella helvetica*** (Haller, 1882); ***Hydrodroma despiciens*** (O.F. Müller, 1776); ***Limnesia (s.str.) undulata*** (O.F. Müller, 1776); ***Atractides nodipalpis*** Thor, 1899; ***Neumania deltoides*** (Piersig, 1894); ***Piona carnea*** (Koch, 1836); ***Pionopsis lutescens*** (Herrmann, 1804); ***Tiphys (s.str.) ornatus*** (Koch, 1836); ***Arrenurus (s.str.) abbreviator*** Berlese, 1888; ***A. (s.str.) albator*** (Müller, 1776); ***A. (s.str.) bicuspidator*** Berlese, 1885; , ***A. (s.str.) bruzelii*** Koenike, 1885; ***A. (s.str.) cuspidator*** (Müller, 1776); ***A (s. str.) cuspidifer*** Piersig, 1896; ***A. (s.str.) distans*** Walter, 1927; ***A. (s.str.) radiatus*** Piersig, 1894; ***A. (s.str.) robustus*** Koenike, 1894; ***A. (s.str.) rodrigensis*** Lundblad,.

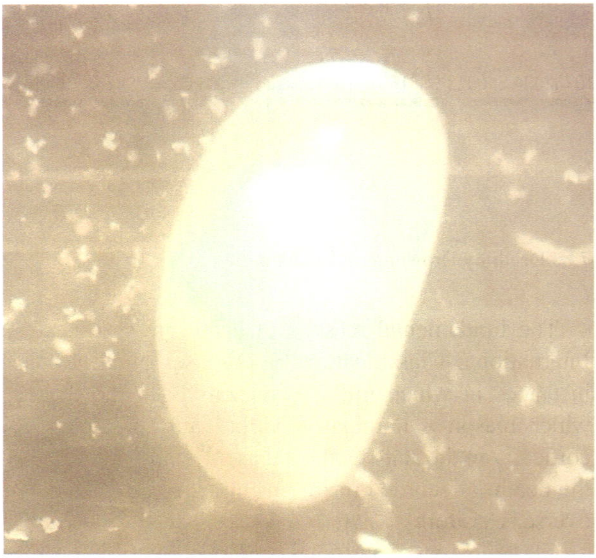

Fig. 78.5 An Ostracoda sample

1954.; ***A. (Megaluracarus) globator*** Müller, 1776; ***A. (Micruracarus) novus*** (George, 1884); ***A. (Micruracarus) octagonus*** Halbert, 1996; ***A. (Micruracarus) sinuator*** Müller, 1776; ***A. (Rhinophoracarus) hazarensis*** Ozkan&Erman, 1990. Numerous taxa were also reported in Behremaz stream that flows into the Hazar Lake: *Eylais extendens*, ***E. megalostoma*** Koenike, 1897; ***E. setosa*** Koenike, 1897; ***E. rahmi*** Ozkan, 1982; *Hydrodroma despiciens*; ***Sperchon setiger*** Thor, 1898; ***Lebertia (Pilolebertia) leioderma*** Vietz, 1940; ***L. (Pseudolebertia) tuberosa*** Vietz, 1926; ***Limnesia (s.str.) undulata*** (Müller, 1776); ***L. (s.str.) walteri*** Migot, 1926; ***Hygrobates (s.str.) fluviatilis*** (Ström, 1768); ***Neumania deltoides*** (Piersig, 1894); ***Pionopsis lutescens*** (Herman, 1804); ***Forelia variegator*** (Koch, 1837); ***Mideopsis orbicularis*** (O.F. Müller, 1776); *Arrenurus (s.str.) albator*; ***A. (s.str.) ayyildizi*** Erman, 1993; *A. (s.str.) bruzelii*; *A. (s.str.) cuspidator*, *A. (s.str.) cuspidifer*, ***A. (s. str.) demirsoyi*** Erman, 1993; ***A. (s.str.) tricuspidator*** (Müller, 1776); ***A. (Micruracarus) octagonus*** Halbert, 1906; ***A. (Micruracarus) sinuator*** Müller, 1776 (Erman et al. 2006). A Hydrachnidia sample is shown in Fig. 78.6.

78.3.5 Group Malacostraca

Asellidae family (Order: Isopoda) was reported from Hazar lake by Şahin and Baysal (1972). The species belonging to the genus ***Gammarus*** (Order: Amphipoda) were reported in Hazar Lake by Türkgülü (2010). Some specimens belonging to group Malacostraca are shown in Figs. 78.7 and 78.8.

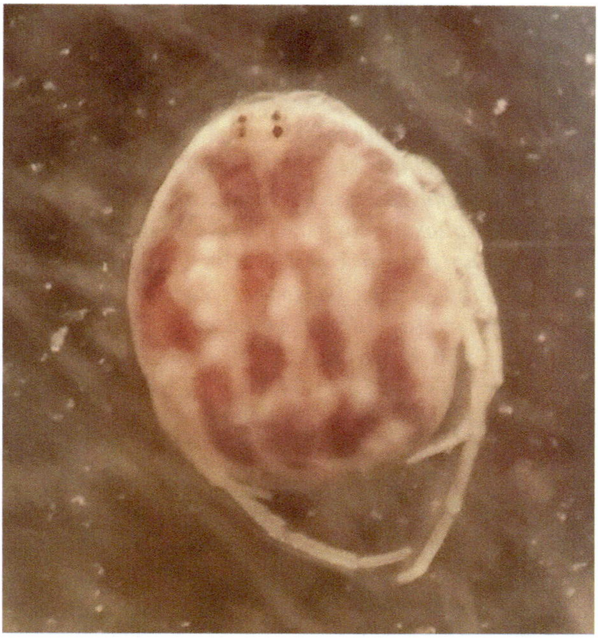

Fig. 78.6 A water mite sample (Hydrachnidia)

Fig. 78.7 Asellidae (Isopoda: Malacostraca)

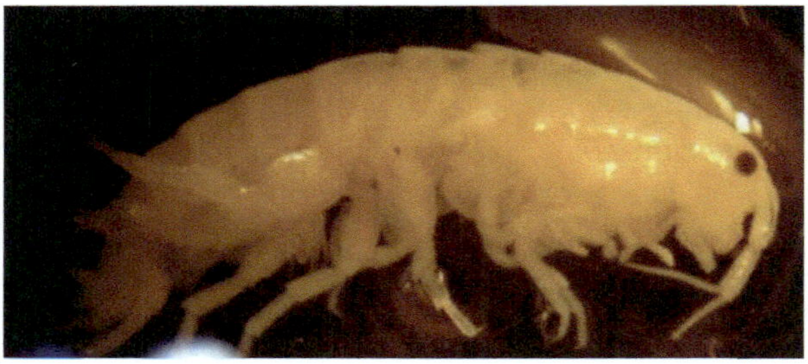

Fig. 78.8 Amphipoda (Malacostraca)

78.3.6 Group Aquatic Insecta

Kazancı (2009) reported a total of 14 Ephemeropteran species in the Tigris River Basin: ***Baetis alpinus*** (Pictet, 1843); ***B. bisri*** Thomas & Dia, 1983; ***B. pseudogemellus*** Soldan, 1977; ***Oligoneurella tskhomelidzei*** Sowa & Zosidze, 1973; ***Cinygma caucasica*** Thsernova, 1938; ***Epeorus zaitzevi*** Thsernova, 1981; ***Iron nigripilosus*** Sinitshenkova, 1976; ***Epeiron amseli*** Demovlin, 1964; ***Rhithrogena iranica*** Braasch, 1983; ***R. iridina*** Kownackii, 1979; ***Ecdyonurus dispar*** (Curtis, 1834); ***Heptagenia coerulans*** Rostoc, 1877; ***H. perflava*** Brodsky, 1930; ***Ephemera vulgata*** Linnaeus, 1758. In the recent study performed by Çetinkaya and Bekleyen (2017), a total of 35 taxa from the orders Trichoptera (***Psychomyia* sp.**), Ephemeroptera (***Baetis macrospinosus*** Koch, 1905; ***Caenis* macrura** Stephens, 1835; ***Rhithrogena* sp.**), and Diptera (***Culicoides* sp.**, ***Chironomus plumosus*** (Linnaeus, 1758), ***Cladotanytarsus* sp.*;* *Dicrotendipes* sp.*;* *Paratanytarsus* sp.*;* *Polypedilum* sp.*;* *Rheotanytarsus* sp.*;* *Tanytarsus* sp.*;* *Cardiocladius capucinus*** (Zetterstedt, 1850), ***Cricotopus* sp.*; Cricotopus bicinctus*** (Meigen, 1818), ***Cricotopus triannulatus*** (Macquart, 1826); ***Cricotopus trifascia*** Edwards, 1829; ***Cricotopus vierriensis*** Goetghebuer, 1935; ***Eukiefferiella coerulescens*** (Kieffer, 1926), ***Eukiefferiella fuldensis*** Lehmann, 1972; ***Euryhapsis* sp.*; Nanocladius spiniplenus*** Saether, 1977; ***Orthocladius (E.) rivicola*** Langton, 1984; ***O. (E.) thienemanni*** Kieffer, 1906; ***Orthocladius (O.)* sp.*; Orthocladius (O.) oblidens*** (Walker, 1856); ***Orthocladius (S.) holsatus*** Goetghebuer, 1937 ***;*** ***Parakiefferiella* sp.*;* *Psectrocladius sordidellus*** (Zetterstedt, 1838); ***Thienemanniella* sp.*; Tvetenia calvescens*** (edwards, 1929)***; Thienemannimyia* sp.*; Larsia curticalcar*** (Kieffer, 1918)***; Ochlerotatus (Rusticoidus) cf. rusticus***; and ***Scatella* sp.** were identified in the Tigris River. The species *Cricotopus bicinctus* and *Orthocladius (S.) holsatus* were determined as the most common taxa (Çetinkaya and Bekleyen 2017). Şahin (1984) reported 55 taxa of larval chironomids (Diptera) from the Tigris River: ***Larsia curticalcar*** Kieffer, 1906; ***Thienemannimyia* sp.**, *Cardiocladius capucinus, Cricotopus bicinctus, C. triannulatus, C. vierriensis,* ***Eukiefferiella* sp.**, *Orthocladius (E.) thienemanni,* ***Psectrocladius* sp.**, *Thienemanniella* sp., *Chironomus gr. plumosus, Cladotanytarsus* sp., *Paratanytarsus* sp., *Polypedilum* sp., *Rheotanytarsus* sp., and *Tanytarsus* sp. The species ***Simulium (Simulium) variegatum*** Meigen, 1818 was reported from Tigris River by Başören and Kazancı (2016) while members of subfamilies Tanypodinae and Chironominae were reported from Hazar lake (Şahin and Baysal 1972). Tellioğlu et al. (2008) reported the larval Chironomidae fauna from Hazar Lake. Tellioğlu et al. (2008) also reported the occurrence of ***Tanypus* sp.**; ***Tanypus punctipennis*** Meigen, 1818; ***Procladius* sp.**; *Chironomus plumosus*; ***Cryptochironomus defectus*** (Kieffer, 1913); ***Stictochironomus histrio*** (Fabr., 1794); ***Paratanytarsus lauterborni*** (Kieffer, 1909); ***Cladotanytarsus mancus*** (Walker, 1856); ***Chironomus thummi*** (Kieffer, 1911); ***Chironomus holsatus*** (Lenz, 1959) in Lake Hazar. The species belonging to Chironomidae, Ceratopogonidae (***Sphaeromias* sp.**), Coleoptera (***Donacia* sp.**), Hemiptera

Fig. 78.9 *Chironomus plumosus* (Chironomidae: Diptera) (larvae)

(***Micronecta* sp.**), Trichoptera (***Nyctiophylax* sp.**), Odonata (***Libellula* sp.**, ***Ophiogomphus* sp.**, and ***Enallagma* sp.**), Ephemeroptera (***Centroptilum* sp.**, ***Cloeon* sp.**, and ***Siphlonurus* sp.**) were reported from the same lake by Türkgülü (2010).

Kazancı (2009) reported a total of 15 Ephemeroptera species in Euphrates River Basin: *Baetis alpinus* (Pictet, 1843); ***B. buceratus*** Eaton, 1870; ***B. lutheri*** species group; ***B. lutheri georgiensis*** Zimmermann, 1981; ***B. tricolor*** Thsernova, 1928; *Cinygma caucasica* Thsernova, 1938; *Epeorus zaitzevi* Thsernova, 1981; *Iron nigripilosus* Sinitshenkova, 1976; *Epeiron amseli* Demovlin, 1964; ***Rhithrogena caucasica*** Braasch, 1979; *R. iridina* Kownackii, 1979; ***Ecdyonurus venosus*** (Fabricius, 1775); ***Electrogena ressli*** (Braasch, 1981); ***Torleya major*** (Klapalek, 1905); ***Potamanthus luteus*** Linnaeus, 1767. Also the chironomid species ***Chironomus anthracinus*** Zetterstedt, 1860; *C. plumosus* were found in Karakaya Dam Lake (Euphrates Basin) by Gökçe and Özhan (2011). ***Simulium (S.) bezzii*** (Corti, 1914) and ***Simulium (S.) ornatum*** Meigen, 1818; ***Prosimulium rachiliense*** Djafarov, 1954 (*P. pronevitschae* Rubtsov, 1955); ***Metacnephia subalpina*** Rubtsov, 1956; ***Simulium (Montisimulium) alizadei*** Djafarov, 1954; ***Simulium (Nevermannia) costatum*** Friederichs, 1920; ***Simulium (Nevermannia) vernum*** Macquart, 1826 (*S. aestivale* (Rubtsov, 1962)); ***Simulium (Simulium) alajense*** Rubtsov, 1956; ***Simulium (Simulium) kerisorum*** Rubtsov, 1956; ***Simulium (Simulium) kiritshenkoi*** Rubtsov, 1940 (*S. caucasicum* Rubtsov, 1940); ***Simulium (Simulium) niha*** Giudicelli and Dia, 1986; *Simulium (Simulium) variegatum* Meigen, 1818; ***Simulium (Simulium) morsitans*** Edwards, 1915; ***Simulium (Trcihodagmia) margaritae*** (Rubtsov, 1956); ***Simulium (Wilhelmia) pseudequinum*** Seguy, 1921 (*S. mediterraneum* Puri, 1925); ***Sulcicnephia znoikoi*** Rubtsov, 1940 were reported to occur in lakes and rivers present in Euphrates River Basin (Başören and Kazancı 2016). Some specimens belonging to group Aquatic Insecta are shown in Figs. 78.9, 78.10, 78.11, and 78.12.

Fig. 78.10 *Libellula* sp. (Odonata) (nymph)

Fig. 78.11 *Cloeon* sp. (Ephemeroptera) (nymph)

78.4 The Roles of Benthic Macroinvertebrates and Environmental Effects on Their Distribution in the Basin

Benthic macroinvertebrates are one of the most important group in aquatic environments. They are functional groups, i.e., grazers, shredders, gatherers, filterers, and predators, in stream ecosystem processes (Wallace and Webster 1996). Zoobenthos in an aquatic ecosystem has very critical roles such as energy translocation, nutrient flow, and detrital decomposition (Hellmann et al. 2015). Thus, they can have an

Fig. 78.12 Coleoptera (adult)

important influence on nutrient cycles, primary productivity, decomposition, and translocation of materials (Wallace and Webster 1996). In addition, the considerable number of species belonging to these groups are used as bioindicators of pollution and water quality in environmental studies (Kazancı and Dügel 2000; Türkmen and Kazancı 2008; Kalyoncu and Gülboy 2009; Girgin et al. 2010). They can answer the ecological changes in many ways as the species richness of freshwater benthic invertebrates can decrease, increase, or exist from the aquatic ecosystems due to changes in their own environment. Therefore, information related to the species composition and their growth dynamics in the ecosystem can be expected to provide more important knowledge than physicochemical analyses.

Benthic macroinvertebrates serve detrital composition in freshwater food webs. Especially crustaceans, influence both energy flow through freshwater food webs and nutrient cycling (Covich et al. 1999). It is reported that the presence or absence of a single species can dramatically alter ecological processes such as rates of grazing and decomposition (Covich et al. 1999). Dissolved nutrients are transformed from organic detritus by benthic invertebrates and they can be mixed into overlying waters to use by rooted plants (macrophytes) and algae (phytoplankton). This increased growth of algae and rooted macrophytes is in turn consumed by herbivorous and omnivorous benthic invertebrates. Thus, they can also control the primer productivity and biomass in an aquatic environment. While benthic species feed on macrophytes, algae, and zooplankton, they can also serve as food for fishes or other animals. All the above explanations clearly show that benthic invertebrates are significant organisms for aquatic life.

The structure of macroinvertebrate communities in an aquatic ecosystem depends on abiotic and biotic factors that vary across spatial scales from regional to habitat specific (Kenney et al. 2009). Not only the natural features of a stream but also the geographic and anthropogenic characters of terrestrial habitats surrounding the stream can affect macroinvertebrate assemblage structure.

To understand the effects of benthic invertebrates in an aquatic ecosystem, firstly we have to know the the ecological characteristics of the system. As seen in the chapter, the investigations on species composition of benthic macroinvertebrates and their distribution in the Tigris–Euphrates Basin in Turkey has not yet completed. While some group of benthic macroinvertebrates (e.g., Chironomidae and Oligochaeta) were studied in detail in some parts of the Basin (only in some dam or streams), however, the rest of the basin still needs to be studied.

The Tigris and Euphrates Rivers serve for many vital purposes such as irrigation, fishing, recreation, and drinking water source for this huge area (Varol et al. 2012). It is reported that the basin has been particularly affected by anthropogenic influences, from damming, impoundments, and other hydromorphological alterations (Varol et al. 2012). In addition, Çetinkaya and Bekleyen (2017) reported that the rivers have been directly receiving the partially treated domestic wastewater from many urban settlements (Diyarbakır, Bismil, Hasankeyf, and Cizre) and effluents from several industries along their course. The Tigris River has the highest water temperature among all the Eastern Anatolian rivers (Akbulut et al. 2009). Çetinkaya and Bekleyen (2017) reported that when the construction of the Ilısu Dam is completed, Tigris will lose a large part of its basin and as a result of this many organisms will be endangered.

Numerous springs carrying different types and amount of discharges join the Euphrates River that carries all these pollutants into the Keban Dam Lake. Similarly Tohma River, Kahta River, and Nizip tributary are known to carry pollution to Karakaya, Birecik, and Atatürk dam lakes (Kibaroğlu and Scheumann 2011). In fact, the direct discharge of sewage and industrial wastes as well as pollutants coming from agricultural areas have been constituting great threats for habitat degradations and biodiversity loss in these reservoirs.

Oligochaets and chironomids are among the important components of the aquatic communities in terms of taxa richness and their role in the food web (Çamur-Elipek et al. 2010). It is reported that *N. variabilis, N. communis, N. elinguis, Tubifex tubifex, Limnodrilus hoffmeisteri, L. profundicola*, and *Mesenchytraeus* spp. belonging to group Oligochaeta are common in areas polluted with organic wastes (Timm 1999; Timm and Veldhuijzen van Zanten 2002; Zeybek et al. 2016). These species are usually considered the signs of the eutrophic conditions in aquatic environments (Sperber 1948; Timm 1970; Learner et al. 1978). These species are reported from the Tigris River by Zeybek et al. (2016). However, *N. pardalis* and *Stylaria lacustris* found in the Tigris River are known for their low tolerance to organically polluted waters (Timm 1970).

Certain benthic invertebrate species (e.g., *Tubifex tubifex*) serve as parasite transmitting vectors; if these species increase in abundance in the aquatic environments, they may spread a lethal disease to trout (Brinkhurst 1997). The species *Tubifex* sp. was reported the most abundant benthic macroinvertebrate species with invasive mollusc *Dreissena polymorpha* in Karakaya Dam Lake (Oğuzkurt and Özhan 2008). Although Zebra mussels that are also present in the Tigris–Euphrates

Basin (originally lived in the lakes of Southeast Russia) are able to disperse rapidly reaching high densities and shortly after colonizing a new habitat (Aksu et al. 2017). For this reason, once zebra mussels enter an ecosystem, it is really difficult to control them.

Wallace and Webster (1996) informed that aquatic insects, as well as other components of the macroinvertebrates, have been used to evaluate the degree of anthropogenic disturbance to both lotic and lentic ecosystems. Some species like *Simulium (S.) bezzii* and *Simulium (S.) ornatum* belonging Diptera can also live in eutrophic conditions and they can survive in physically damaged environments (Kazancı 2006; Kazancı and Ertunç 2010; Başören et al. 2013). Especially it was reported that the substrate structure, water temperature, and water discharge are not important for the distribution of *S. ornatum* (Bernotiene 2006). These species were recorded from the Euphrates River (Başören and Kazancı 2016).

The endemic species from the Tigris–Euphrates Basins are the oligochaet species *Potamothrix alatus hazaricus* Timm et al. 2013 and the mite species *Arrenurus (Rhinophoracarus) hazarensis* Ozkan and Erman, 1990 from Hazar lake. In addition, the water mite species *A. (s.str.) ayyildizi* Erman, 1993 and *A. (s.str.) demirsoyi* Erman, 1993 are also the endemic species reported from Behremaz Stream. However, environmental changes affect the distribution of the endemic species in the basin. The endemic species are also important as food organisms for endemic fish species *Aphanius asquamatus, Kosswigichthys asquamatus*, and *Orthrias angorae eregliensis*.

Covich et al. (1999) informed that native species are generally well adapted to local conditions, movements of additional species into aquatic assemblages can sometimes alter energy flow and change nutrient cycling. The arrival of an additional species is often associated with the loss of one or more resident species and it is also important to ask about the effects on ecosystem processes (Lodge et al. 1998; Covich et al. 1999). In addition, loss of some species will change the ecosystem process and populations of some remaining species would increase or decrease (Covich et al. 1999). This also holds true for the Tigris and Euphrates Rivers.

In conclusion, the many roles performed by macroinvertebrates in aquatic ecosystems are related to giving importance to their conservation. They have also served as valuable indicators of degradation of aquatic environments. Thus macroinvertebrates and other organisms should be protected to prevent biodiversity loss in the two largest river systems of Turkey.

Tigris–Euphrates River Basin is a great river system of south-western Asia and they follow roughly parallel courses that occur in Mesopotamia from their sources to the head of the Persian Gulf. Their ecological roles are very important for the area. To sustainable management of the Basin, we have to protect their ecological balance.

References

Akbulut N, Bayarı S, Akbulut A, Şahin Y (2009) Rivers of Turkey. In: Tockner K, Robinson CT, Uehlinger U (eds) Rivers of Europa. Elsevier, Academic, New York, pp 643–772

Akdemir D, Külköylüoğlu O (2011) Freshwater Ostracoda (Crustacea) of Diyarbakır Province, including a new report for Turkey. Turk J Zool 35(5):671–675

Aksu S, Yildiz D, Güngör AP (2017) How zebra mussels threaten to water supply security and effects of preventive measures in Turkey. World Sci News 64:99–126

Alkan-Uçkun A, Yoloğlu E, Uçkun M (2017) Seasonal monitoring metals in water, sediment and mussel (*unio mancus*) from Atatürk Dam Lake (Euphrates River). Van Vet J 28(2):75–83

Aras S, Fındık Ö (2016) The aquatic oligochaetes (Annelida: Clitellata) of eight lakes in the Aşği Firat River Basin (Lower Euphrates, Turkey). Biologia 71(1):38–43

Başören Ö, Kazancı N (2016) A checklist of Simuliidae (Insecta, Diptera) species of Turkey. Rev Hydrobiol 9(2):153–164

Başören Ö, Kazancı N, Girgin S (2013) The distribution of Simuliidae (Insecta, Diptera) Species İn Ankara Stream (Turkey). Rev Hydrobiol 6(2):157–167

Bernotiene R (2006) Bloodsucking dipterous insects (Ceratopogonidae, Culicidae) on the Curonian spit. Žmogaus ir gamtos sauga. In: Proceedings of the XIIth international scientific—practical conference "human and nature safety", Akademija, 1–2 June, 2006, pp 123–126

Bobat A, Hengirmen MO, Zapletal W (2004) Zebra mussel and fouling problems in the euphrates basin. Turk J Zool 28:161–177

Brinkhurst RO (1997) On the role of tubificid oligochaetes in relation to fish disease with special reference to the Myxozoa. Annu Rev Fish Dis 6:29–40

Brismar A (2002) The Atatürk Dam project in south-east Turkey: changes in objectives and planning over time. Nat Resour Forum 26:101–112

Çamur-Elipek B, Arslan N, Kırgız T, Öterler B, Güher H, Özkan N (2010) Analysis of benthic macroinvertebrates in relation to environmental variables of Lake Gala, a National Park of Turkey. Turk J Fish Aquat Sci 10:235–243

Çetinkaya F, Bekleyen A (2017) Spatial and temporal distribution of aquatic insects in the Dicle (Tigris) River Basin, Turkey, with new records. Turk J Zool 41:102–112

Covich AP, Palmer MA, Crowl TA (1999) The role of benthic invertebrate species in freshwater ecosystems: zoobenthic species influence energy flows and nutrient cycling. BioScience 49 (2):119–127

Ekin İ, Başhan M, ve Şeşen R (2008) Fırat Nehrinden Toplanan *Dreissena siouffi* (Locard 1893) (Bivalvia: Dreissenidae)' nin Yağ Asiti İçeriği. Fırat Üniv Fen ve Müh Bil Dergisi 20 (2):243–250

Erman O, Tellioğlu A, Orhan O, Çitil C, ve Özkan M (2006) Water mites (Hydrachnidia: Acari) Fauna of Hazar Lake and Behremaz stream (Elazıg) and their seasonal distribution. Fırat Üniv Fen ve Müh Bil Der 18(1):1–10

Girgin S, Kazanci N, Dügel M (2010) Relatinship between aquatic insects and heavy metals in an urban stream using multivariate Techniques. Int J Environ Sci Technol 7:653–664

Gökçe D, Özhan D (2011) Limno-ecological properties of deep reservoir, Karakaya HEPP. Turkey GU J Sci 24(4):663–669

Hellmann JK, Erikson JS, Queenborough SA (2015) Evaluating macroinvertebrate community shifts in the confluence of freestone and limnostone streams. J Limnol 74:64–74

Kalyoncu H, Gülboy H (2009) Benthic macroinvertebrates from Darıören and Isparta streams (Isparta/Turkey)-Biotic indices and multivariate analysis. J Appl Biol Sci 3:85–92

Kazancı N (2006) Ordination of Simuliidae and climate change impact. Acta Entomologica Serbica suppl, 69–76

Kazancı N (2009) Türkiye Ephemeroptera (Insecta) Faunası: Doğu Anadolu (Türkiye)'dan kayıtlar. Rev Hydrobiol 2:187–195

Kazancı N, Dügel M (2000) An evaluation of the water quality of Yuvarlakçay Stream, in the Köyceğiz-Dalyan protected area, SW Turkey. Turk J Zool 24:69–80

Kazancı N, Ertunç Ö (2010) Use of Simuliidae (Insecta: Diptera) species as indicators of aquatic habitat quality of Yeşilırmak River Basin (Turkey). Rev Hydrobiol 3(1):27–36

Kenney MA, Sutton-Grier AE, Smith RF, Gresens SE (2009) Benthic macroinvertebrates as indicators of water quality: The intersection of science and policy. Terr Arthropod Rev 2:99–128

Kibaroğlu A, Scheumann W (2011) Euphrates-Tigris rivers system: political rapprochement and transboundary water cooperation. In: Kibaroglu A et al (eds) Turkey's water policy. Springer, Heidelberg, pp 277–299

Küçükyılmaz M, Şahin AG (2017) Determination of the conchological features of Unio elongatulus eucirrs (Bourguignat, 1860) (Mollusca: Bivalvia) living in Karakay Dam Lake. Int J Pure Appl Sci 3(1):1–7

Learner MA, Lochhead G, Hughes BD (1978) A review of the biology of British Naididae (Oligochaeta) with emphasis on the lotic environment. Freshw Biol 8(4):357–375

Lodge DM, Stein RA, Brown KM, Covich AP, Bronmark C, Garvey JE, Klosiewski SP (1998) Predicting impact of freshwater ex-otic species on native biodiversity: Chal-lenges in spatial and temporal scaling. Aust J Ecol 23:53–67

Oğuzkurt D, Özhan D (2008) Bioindicator benthic macroinvertebrate for assesing water quality: a case study on Karakaya Dam Lake. In: Proceedings of Taal 2007: the 12th world lake conference, pp 2148–2153

Özuluğ O, Dökümcü N (2014) Birecik Baraj Gölü Havzası (Şanlıurfa) Ve Hazar Gölü(Elazığ) Ostracodları (Crustacea) Hakkında Ön Çalışma. İstanbul Üniversitesi Su Ürünleri Dergisi 29-2:1–8

Şahin Y (1984) Doğu ve Güneydoğu Anadolu bölgeleri akarsu ve göllerindeki chironomidae (Diptera) larvalarının teşhisi ve dağılışları. Anadolu University Press, Eskişehir, Turkey (in Turkish)

Şahin Y & Baysal A (1972) Hazar Gölü Dip Faunası ve Yayılışları, İstanbul Üniversitesi Fen Fakültesi Hidrobiyoloji Araştırma Enstitüsü Yayınları, 9, İstanbul

Sperber C (1948) A taxonomical study of the Naididae. Zool Bidrag Uppsala 28:1–296

Tellioğlu A, Çitil C, Şahin İ (2008) Distribution of Chironomidae (Diptera) larvae in Hazar lake, Turkey. J Appl Biol Sci 2(1):77–80

Timm T (1970) On the fauna of Estonian Oligochaeta. Pedobiologia 10:52–78

Timm T (1999) A guide to the Estonian Annelida. In: Naturnalist's handbooks 1, Tart-Tallin, 208 pp

Timm T, Veldhuijzen van Zanten HH (2002) Freshwater Oligochaeta of North-West Europe. World Biodiversity Database, CD-ROM Series. University of Amsterdam: Expert Center for Taxonomic Identification

Timm T, Arslan N, Rüzgar M, Martinsson S, Erséus C (2013) Oligochaeta (Annelida) of the profundal of Lake Hazar (Turkey), with description of *Potamothrix alatus hazaricus* n. ssp. Zootaxa 3716(2):144–156

Türkgülü İ (2010) Hazar Gölü litoral bölge bentik makroomurgasızları ve dağılımları, Fırat Üniversitesi Fen Bilimleri Enstitüsü Doktora Tezi, 245 s

Türkmen G, Kazancı N (2008) Water quality evaluation of reference sites by using saprobic index in some running waters in the province of Bolu. Rev Hydrobiol 2:93–118

UN-ESCWA and BGR (United Nations Economic and Social Commission for Western Asia; Bundesanstalt für Geowissenschaften und Rohstoffe) (2013) Inventory of shared water resources in Western Asia. Beirut. Part I Chapter 3: Tigris River Basin, pp 99–125

Varol M, Gökot M, Bekleyen A (2010) Assesment of water pollution in the Tigris river in Diyarbakır. Turk Water Pract Technol 5(1):1–13

Varol M, Gökot B, Bekleyen A, Şen B (2012) Water quality assessment and apportionment of pollution sources of Tigris River (Turkey) using multivariate statistical techniques—a case study. River Res Applic 28:1428–1438

Wallace JB, Webster JR (1996) The role of macroinvertebrates in stream ecosystem function. Annu Rev Entomol 41:115–139

Zeybek M, Ahıska S, Yıldız S (2016) A preliminary taxonomical investigation on the Oligochaeta (Annelida) fauna of Tigris River (Turkey). Ege J Fish Aquat Sci 33(1):47–53

Chapter 79
Freshwater Ecosystem Conservation in Iraq: Recommendations for Management

Laith A. Jawad

Abstract The freshwater system in Iraq contains several troubles for years. Such problems were originated from accumulated local competitions and struggles between societies used to inhabit this region. As a result of such challenging have influenced the nation of Iraq, their structure and the people themselves. The important question that needs to be answered is what we determine to do about the freshwater system in Iraq in order to improve these waterways and keep it for the next generations.

As with the other freshwater systems in the other parts of the world, the social, economic and environmental outcomes in the freshwater water system valley in Iraq could all be improved if certain amount of money is allocated to such a project. Also, if Iraq adopts one of the mandates that put forward by several international agencies related to the biodiversity conservation such as Ramsar and implement exactly such a mandate, the results certainly will be positive in favourite of conserving the aquatic areas of Iraq.

79.1 Introduction

There are several policies that have been put forward by international agencies such as Ramsar. The Agreement on Wetlands and the Resolution on Biological Diversity of Ramsar offer much of the legitimate order for Iraq to adopt. Such mandate was adopted by country such as Australia in 2011 (Pittock and Finlayson 2011). In the plan of this mandate, an effect will be given to these global covenants by decreasing extreme water alterations to habitat maintainable states, using environmental movements to preserve key ecological assets such as wetlands, accomplish developing hazards to water deliveries and then optimize socioeconomic benefits. Other required progress contain a hard valuation of ecosystem amenities in the main ecological benefits, preserving the remaining unregulated water flows, refunctioning

L. A. Jawad (✉)
School of Environmental and Animal Sciences, Unitec Institute of Technology, Auckland, New Zealand

L. A. Jawad (ed.), *Tigris and Euphrates Rivers: Their Environment from Headwaters to Mouth*, Aquatic Ecology Series 11,
https://doi.org/10.1007/978-3-030-57570-0_79

water substructure and recognizing urgencies for research to progress the basin of any freshwater body.

In this chapter, a lesson by Australian freshwater management for the upkeep of a growing number of river basins was suggested for the policy makers in Iraq to adopt. Within this lesson, a number of recommendations were suggested and given in this chapter with slight changes to suit the environment in Iraq.

79.2 General Overview of the Conservation Plan

Chief results are present that related to the conservation project to accomplish the freshwater environments in Iraq with a need for reallocation of an amount of water for the habitat, these are:

- Vast regions of the freshwater systems in Iraq cannot be sustained extensive and maintenance of diverse swamps would be essential.
- Progressively, preservation of the main zones would depend on continuous interference within environmental—water demand—administration actions when government experts in Iraq adopt freshwater-conservation measures by suspending any previous legislations regarding the freshwater system.
- If Iraq, adopts the mandate of Ramsar, then it would be compulsory to continue to maintain the ecological features of all of the chosen Ramsar sites. Linked with the present, terrible state, there is no reluctance that the freshwater habitats will be in improved state with a rearrangement of water to the environment and with new guidelines for sharing future water losses between the heavy users and the environment.

79.3 Recommendations

1. Recognizing and assessment of the full range of ecosystem amenities for recovering improvement.
2. Appraisal of freshwater areas alongside the variety of lands established by the Ramsar Agreement, to guarantee that a satisfactory range of marsh areas biota will be continued present in the region.
3. Rearrangement of adequate water quantities to uphold the ecological feature of all localities of the 16 chosen Ramsar wetlands.
4. Documentation and securing the continuing free-flowing tributary rivers and of river reaches that obtain net inflows from groundwater ('gaining reaches') as added methods of saving freshwater biota.
5. More healthy administration of climate change-induced risks to water obtainability, containing larger reorganization of water to the environment.

6. Practical administration of current water substructure to guarantee that water quality and habitat performance are enhanced.
7. Recognizing an extra method for studies such as maintenance of groundwater-dependent habitats and climate change alteration choices, to guarantee that the following restatement of the upkeep plan is further upgraded.

Reference

Pittock J, Finlayson CM (2011) Freshwater ecosystem conservation: principles versus policy. In: Connell D, Grafton RQ (eds) Basin futures: water reform in the Murray-Darling basin. ANU E Press, pp 39–57

Chapter 80
Utilising Phenotypic Difference to Regulate Protection Value: A Scheme for Application of a Novel Approach to the Inland Water of Iraq

Laith A. Jawad

Abstract In general, fish populations showed high grade of distinct disparity in phenotype. This takes the form of variation not only in morphometric and meristic features conventionally in practice to distinguish species, but also specific of life history, behaviour, coloration and ecology. This inconsistency has a number of concerns, one of these is that there is a robust case for the maintenance of the groups with risky morphological characters. Though, if disparity is distinct between groups but unceasing across many other groups, this postures problems in splitting those groups of high protection value from those of lower protection rate. In this chapter, a description of a statistical method was given, which allows scientists in Iraq to identify freshwater fish groups on the outermost limits of the range of phenotypic difference and apply this to the morphometric characters of other freshwater fish species in Iraq. The method permits the recognition of any amount of the most variable morphological characters. It also can possibly be used on any species and on any set of features as an impartial mean of protection unit of a group in a discrete phenotypic range.

80.1 Introduction

Differences in the morphological characters across the geographic range of a species are not unusual criterion. Where the habitats engaged by a species are spatially fragmented or where movement of the species is restricted, then limitations on free gene flow and/or local environmental variation working upon morphological flexibility can give important disruptions in the phenotype through the series of a species (Jenkins 1997). In fishes, it is supposed that microevolutionary routes functioning

L. A. Jawad (✉)
School of Environmental and Animal Sciences, Unitec Institute of Technology, Auckland, New Zealand

L. A. Jawad (ed.), *Tigris and Euphrates Rivers: Their Environment from Headwaters to Mouth*, Aquatic Ecology Series 11,
https://doi.org/10.1007/978-3-030-57570-0_80

over small scales joint with limited gene flow between groups reproducing in diverse freshwater systems have steered to the variety of form in several species.

The component of the problem of trying to poise distribution of assets besides the need for administration is that some groups might be considerably more different within the species than others and therefore may be deliberated as of possible larger protection mean. Consequently, some methods of enumerating the grades of deviation of groups from each other; a degree of their peculiarities, could be used to highlight protection activities.

Freshwater fish species of Iraq showed a distribution from north to the south, but it is unknown for most of these species whether such distribution is uniformed and the populations of each species are morphologically and genetically unique. The recent studies on morphometrics and meristic characters of the Asian catfish, *Silurus triostegus* collected from several localities along its national distribution within Iraq have shown that these populations are morphologically distinct (Jawad and Al-Janabi 2016; Jawad et al. 2017). Therefore, genetic studies need to be allocated to reveal whether those populations are unique and do not overlap. With the results of such studies, it only possible to set a plan for management of the fisheries of this species.

The aim of this chapter to make available for fisheries biology scientists in Iraq a reliable and cheap method leads to a measure of population 'uniqueness' based on variability in phenotypic characteristics. Such method is already in use in the population studies for other fish species around the world (Bush and Adams 2007).

80.2 Methodology

To accomplish the analysis, fish samples of not less than 100 should be collected from as much as possible of population of any fish species that needs to be studied across its national geographical in Iraq. Similar standard survey gill nets should be used in all localities. After catch, fish should be taken from the nets, kept on ice and photographed in a lateral view for morphometric analysis within 24 h. The morphological features that going to be chosen for this study chosen should be linked to the shape of the head and body as the fish species selected might have a practical part in feeding such a case is found in Arctic charr (Adams and Huntingford 2002) and has been revealed to be linked to ecological changes between groups (Skúlason and Smith 1995).

More than eleven morphological features linked to the body and head shape together with the fin lengths, were measured straight from photographs.

Since these linear features of the fish shape differ with fish size, to develop a measure of morphology that was free of the effect of the size of individual fish, each of morphologic measure was regressed upon fish fork length and regression residuals used as a measure of that variable, free from the effect of body size (Reist 1986; Adams et al. 2003). Principal Component scores were resulting for all fish from all populations from the first and second principal components (PC1 and PC2, respectively) (for a more complete explanation of the technique see Adams et al. 2003). To calculate the degree of phenotypic deviancy between groups, the population mean principal component scores for PC1 and for PC2 were computed for each population distinctly. A mean of population was used as the morphologic centroid position of the complete data set and Pythagoras' theorem was used to compute the two-dimensional vector distance of each individual population from the phenotype centroid for all groups.

80.3 Expected Results

If the individuals of fish examined were different and vary in their morphological traits, then there will be a substantial difference in stated head and body morphology between groups from through the studied range of the fish species. Such differences can be shown when two dimensions principle component analysis is plotted. The size amended morphometric measures may display a broad dispersed outline of distribution across both dimensions. The dispersal of the scatter of data should not vary meaningfully from that of the PC1 dimension nor in the PC2 dimension. In both cases, there should be a significant difference in body and head morphology between populations for both PC1 and for PC2 ($P < 0.001$).

The z distribution offers a device to derive thresholds above which one would envisage a specific percentile of the population would lie within a normal distribution of a measure of known mean and standard deviation. Thus *Y*, the variable threshold is given by:

$$\mathrm{Y} = (\mathrm{z\,x\,SD}) + \mathrm{x}$$

where *z* is the standard normal SD, the variable standard deviation and x the variable mean (Moore and McCabe 2002).

80.4 Concluding Remarks

The technique that usually used is based on the hypothesis that the grade to which groups vary in their features can be calculated numerically and that this alteration reveals the peculiarities of the group's characters. This presentation undertakes that the level of difference between groups itself diverges and that this disparity is normally distributed; as it is in this case study (however, this approach could be adjusted if an alternative distribution were more appropriate).

It is possible to apply a simple statistical method to calculate the level of peculiarities of each group and to isolate populations with the most extreme levels of 'uniqueness'. This technique can be employed to any numerical morphologic feature and also to genetic data. One postulation of this method is that all the features that will be encompassed in the study should be of an equal value in evaluating individuality. One benefit of this method is that it permits a threshold for individuality to be set at any level.

The usefulness of stock separation methods should be revealed on a case-by-case basis relying likewise on the level of purpose requisite. Tagging and parasitic information usually offer broad-scale stock separation information, but maybe insufficient for defining more complex multi-stock assemblies, except better importance is add on gaining more detailed recapture information than is typical (MacKenzie 1983; Begg et al. 1997). Morphometrics, meristics and life history features have been used effectively for stock separation at a range of different levels (Casselman et al. 1981; Elliott et al. 1995; Cadrin and Friedland 1999), but are frequently restricted by their likely change by environmental disparity (Lindsey 1964; Todd et al. 1981). Chemical methods enable elemental 'fingerprints' to be renowned for multiple stock complexes (Campana et al. 1995; Begg et al. 1998; Thresher 1999), though habitat difference may happen within the dispersal of a single genetic stock.

The procedure of separating fish stocks is indispensable for active fisheries administration, and will endure to go through alteration as new gears and techniques are established (Kutkuhn 1981). The dilemma of separating fish groups will stay forming a contest until a better method to recognise such groups will be in place. Group separation must thus be acknowledged as an ongoing procedure, developing as administration prerequisites for stock valuation alteration, but continuously seen alongside the background of a serious inspection of all available information and new investigations as read out by altering reserve state and experimental techniques (Brown et al. 1987).

The above-mentioned discussion about fish stock and their discrimination is only a short review for the future studies need to be accomplished in the freshwater system of Iraq in order to solve the problem of fish population identification and then introducing a plan for managing the fish stock and conserve them from depletion as the deterioration of the aquatic habitats was already started in the Iraqi inland waters.

References

Adams CE, Fraser D, McCarthy I, Sheilds S, Waldron S, Alexander G (2003) Stable isotopes analysis reveals ecological segregation in a bimodal size polymorphism in Arctic charr from loch Tay, Scotland. J Fish Biol 78:43–49

Adams CE, Huntingford FA (2002) The functional significance of inherited differences in feeding morphology in a sympatric polymorphic population of Arctic charr. Evol Ecol 16:15–25

Begg GA, Cameron DS, Sawynok W (1997) Movements and stock structure of school mackerel (*Scomberomorus queenslandicus*) and spotted mackerel (*S. munroi*) in Australian east coast waters. Mar Freshw Res 48:295–301

Begg GA, Cappo M, Cameron DS, Boyle S, Sellin MJ (1998) Stock discrimination of school mackerel (Scomberomorus queenslandicus) and spotted mackerel (S. munroi) in coastal waters of eastern Australia using analysis of minor and trace elements in whole otoliths. Fish Bull 96:653–666

Brown BE, Darcy GH, Overholtz W (1987) Stock assessment/stock identification: an interactive process. In: Kumpf, H.E., Vaught, R. N., Grimes, C. B., Johnson, A.G., Nakamura, E. L. (Eds.), Proceedings of the Stock Identification Workshop, 5±7 November 1985, Panama City Beach, FL. NOAA Technical Memorandum NMFS±SEFC-199, US Government Printing Office, pp. 1±24

Bush V, Adams CE (2007) Using phenotypic variation to determine conservation value: application of a novel approach to Arctic charr. Ecol Freshw Fish 16(1):29–33

Cadrin SX, Friedland KD (1999) The utility of image processing techniques for morphometric analysis and stock identification. Fish Res 43(1-3):129–139

Campana SE, Gagné JA, McLaren JW (1995) Elemental fingerprinting of fish otoliths using ID-ICPMS. Mar Ecol Prog Ser 122:115–120

Casselman JM, Collins JJ, Grossman EJ, Ihssen PE, Spangler GR (1981) Lake whitefish (*Coregonus clupeaformis*) stocks of the Ontario waters of Lake Huron. Can J Fish Aquat Sci 38(12):1772–1789

Elliott NG, Haskard K, Koslow JA (1995) Morphometric analysis of orange roughy (Hoplostethus atlanticus) off the continental slope of southern Australia. J Fish Biol 46:202–220

Jawad LA, Al-Janabi MI (2016) Morphometric characteristics of catfish *Silurus triostegus* Heckel, 1843 from the tigris and Shatt al-Arab rivers, Iraq. Croatian Journal of Fisheries: Ribarstvo 74 (4):179–185

Jawad LA, Ligas A, Al-Janabi MI (2017) Meristic character variability among populations of *Silurus triostegus* Heckel, 1843 from the Euphrates, Tigris, and Shatt al-Arab Rivers, Iraq. Archives of Polish Fisheries 25(1):21–31

Jenkins SH (1997) Perspectives on individual variation in mammals. J Mammal 78:271–273

Kutkuhn JH (1981) Stock definition as a necessary basis for cooperative management of Great Lakes fish resources. Can J Fish Aquat Sci 38:1476–1478

Lindsey CC (1964) Temperature-controlled meristic variation in the paradise fish, Macropodus opercularis (L.). Can J Zool 32:87–98

MacKenzie K (1983) Parasites as biological tags in fish population studies. Adv Appl Biol 7:251–331

Moore DS, McCabe GP (2002) Introduction to the practice of statistics, 4th edn. Freeman & Co., New York

Reist JD (1986) An empirical evaluation of the coefficients used in residual and allometric adjustment of size covariation. Can J Zool 64:1363–1368

Skúlason S, Smith TB (1995) Resource polymorphism in vertebrates. Trends in Ecology and Evolution 10:366–370

Thresher RE (1999) Elemental composition of otoliths as a stock delineator in fishes. Fish Res 43:165–204

Todd TN, Smith GR, Cable LE (1981) Environmental and genetic contributions to morphological differentiation in ciscoes (Coregoninae) of the Great Lakes. Can J Fish Aquat Sci 38:59–67

Chapter 81
Hatchery-Reared Fish Stocks Released into the Wild: A Conservation Problem as Seen in a Case Study from Iraq

Laith A. Jawad

Abstract The corporate administration procedure of presenting aquaculture produced fish into wild groups has elevated worries among fishery scientists. Such anxieties ascend from the comment that hatchery-produced fish commonly change from wild fish in ways that may impact ecological connections between them. The present chapter a short revise of the subject of the relationship between reared fish stock and wild population was given. The variances between hatchery-reared and wild fish might have adverse inferences for the success of supplying plans. A number of reports stating group reactions to supplying support this, signifying that the enactment of hatchery fish and their exchanges with wild fish is of such a feature that many of the present supplying performs might be harmful to the receiver population.

At the end of the chapter, a case study was presented to determine meristic traits of two populations (wild and cultured) of *Cyprinus carpio* and to launch whether groups could be distinguished grounded on meristic inconsistency. The five meristic characters chosen for this case study have shown to be significantly different between the wild population and the reared stock of the common carp studied from a locality at the Euphrates River, Iraq.

81.1 Introduction

Accidental or intentional discharges of aquaculture created fish into wild populations have initiated worry among fishery scientists (Ryman et al. 1995; Youngson and Verspoor 1998). While such discharges (deliberate) are frequently carried out to balance the decrease production caused by human encouraged habitat dilapidation, a range of possible ecological harms might be linked with this exercise. First, supplying large numbers of fish into a restricted area will certainly disturb population mass, at least initially. Therefore, any density-dependent features of the habitat or of the

L. A. Jawad (✉)
School of Environmental and Animal Sciences, Unitec Institute of Technology, Auckland, New Zealand

L. A. Jawad (ed.), *Tigris and Euphrates Rivers: Their Environment from Headwaters to Mouth*, Aquatic Ecology Series 11,
https://doi.org/10.1007/978-3-030-57570-0_81

fish itself are possibly disturbed (Elliott 1990). This mathematical impact of supplying might, for instance, contain variations in the occurrence of competitive relations, degrees of food obtainability, or a practical reaction of predators, and therefore affect development and survival of the wild fish. Hypothetical thoughts propose that this may lead to hatchery discharges to upsurge temporal alterations of group strength (Fagen and Smoker 1989). Second, aquaculture fish might vary from the genetic point of view or morphologically from wild fish. Such variances might disturb how supplied and wild fish cooperate, and therefore lead to an impact of supplying outside those due to pure density dependence.

Morphometric and meristic morphological traits are used broadly to distinguish fish stocks (Turan et al. 2004), and they continue the easiest, major direct approaches of species recognition. From preceding studies (Creech 1992; Mamuris et al. 1998; Bronte et al. 1999; Hockaday et al. 2000), it is assumed that the analysis of the morphological variation in the discrete or meristic traits is the most frequently used technique to define stocks of fish. It has regularly been used in separating and classification investigations by statistical methods (Agnew 1988; Avsar 1994). Notwithstanding the usage of methods which straight inspect biochemical or molecular genetic difference, these conformist approaches endure to have a significant part in population discrimination till now (Swain and Foote 1999). Changes in the morphometric and meristic traits of a species from different localities can result alterations in genotypes, habitat factors functioning on one genotype, or both of these acting together (Parish and Sharman 1958). While both morphometric and meristic traits react to variations in environmental causes, their reactions are diverse in some circumstances and can fluctuate from species to species.

The examination of dissimilarities and inconsistency in morphometric and meristic features of fish populations is significant in phylogenetics and in delivering information for following studies on the genetic development of groups. Conferring to Turan et al. (2006), decades of introduction and domestication of a fish species (especially those from the wild) clues to high acclimatization to a wide range of geographical areas, which cause phenotypic differences with regard to the pure population (strains) of the brood stock. This is perhaps because of the impacts of the habitat or hybrids evolving through widespread inbreeding (El-Serafy et al. 2007).

In the present chapter, a short review of the issue of the reared-wild fish populations was given. In addition, a study case showing the effect of the reared stockfish on the wild population of the common carpe *Cyprinus carpio* collected from Euphrates River, Babil Province, Iraq was provided. Such an example will be an evidence on the effect of reared-fish stocks in the freshwater system of Iraq on the native freshwater fish species. Recently the activities of aquaculture in cages built in rivers have been increased dramatically in Iraq (Salman 2011). Therefore, the aim of this chapter is to draw the attention of the policy makers in Iraq to the problems of accidental and intentional release of reared-fish individuals in the wild on the wild fish populations. Alteration among hatchery and wild populations grounded on morphological traits have not yet been investigated, and to the information of the author, this is the first such study attentive on inspecting the degree of morphological differences of the common carp, *Cyprinus carpio* in cultured and wild environments.

81.2 The Reasons why Do Hatchery and Wild Fish Differ

Fish grew in aquaculture conveniences may vary from their wild individuals of similar species for three reasons. First, fish are highly morphologic plasticity and hence their shape may be followed significantly by the culturing system (Pakkasmaa 2000). Second, why fish from aquaculture might diverge from wild one is that the strength and course of selection vary between the two habitats. Third, why reared fish might vary from wild one is the use of non-native fish for supplying. Luckily, the possible significance of local adjustments is being progressively recognized (reviewed by Taylor 1991), and the process of freeing non-native fish has consequently reduced in occurrence. Deliberate unnatural selection may also produce genetic modification in reared fish groups, as has happened with commercially reared fish (Einum and Fleming 1997; Fleming and Einum 1997).

81.3 Study Case from the Freshwaters System of Iraq

The plasticity of the fish body shape is very great, with larger variances in morphological characters both within and between groups than any other vertebrates. The reason for such differences can be partially accredited to intraspecific inconsistency, which is under the effect of habitat factors (Wimberger 1992). Fish are very subtle to habitat fluctuations and rapidly acclimate by altering essential morphometric traits (Hossain et al. 2010). Shape disparity between groups may be appropriate for investigating short-term habitat prompted disparity (Pinheiro et al. 2005). Moreover, whereas both morphometric and meristic traits react to variations in habitat features, their reactions are diverse in some cases and can vary between species. Lastly, is significant to aquaculturists to be acquainted with the alterations between reared and wild fish of diverse species; this might cause them to comprehend the chemical, physical, nutritional, and sensorial outlines of the wild animal and try to reproduce these in their cultured yields (Orban et al. 2003).

Difference among cultured and wild *Cyprinus carpio* populations built on morphological features have not yet been investigated in Iraq as with the populations of other commercial fish species living freshwater system. Consequently, this is the first such examination that has dedicated on studying the level of their morphological differences in reared and natural habitats. Meanwhile, this data is imperative for the appropriate administration of the fisheries and for best use of the assets, the objective of the present case study was to evaluate the morphological and meristic features of *C. carpio* collected in different habitats (cultured and wild). This will aid in development of additional breeding and protection policies for this fish and enhancing yield.

81.3.1 Data, Sampling, and Statistical Analysis

The study comprised two localities of the Euphrates River and a fish rearing facility farm in Babil Province, Iraq. One hundred matured fish samples of *C. carpio*, encompassing 50 individuals from natural habitat (wild population) and 50 from a reared environment (private fish farms, cultured stock), were obtained with the help of typical fishing gears such as cast and hand nets. As males and females cannot be differentiated from their outside features, separating sex was not performed. Specimen collection was achieved weekly by purchasing samples of the two selected populations from local fishermen (wild fish) or a fish farm (cultured fish). After approval of their death, the fish were identified and weighed, and then body measurements and meristic counts were done.

In all, 100 carp (*C. carpio*) were examined for the five meristic characters as given by Hubbs and Lagler (1964). Meristic counts were of: number of lateral line scales, dorsal fin rays, caudal fin rays, anal fin rays, and number of vertebrae. As meristic traits were not linked to fish size and did not vary during growth (Murta 2000), the raw meristic data were not transformed. Canonical discriminant analysis (CDA) was used to evaluate the efficacy in the meristic characters in classification by location. Cross validated discriminant analysis was used to assess the extent to which meristic characters allow identification by locations. Percentages of correct classifications were recorded. For these analyses, the only characters that showed no overlapping values between the two localities were used. The analyses were performed using SPSS ver 13.

81.4 Results

The calculation for each of the meristic and morphometric traits is given in Table 81.1. Relations between meristic features utilized in the study and sex were not significant ($P < 0.05$), representing a trivial consequence of sex on detected differences. Data for both the sexes were consequently united for all the following analyses. From Table 81.1, it is obvious that the five meristic characters examined were variable. The level of variation in all five meristic characters was the same, but

Table 81.1 Definition of meristic counts, range mean ± standard deviation (in parentheses), and F values (derived from analysis of variance) examined in wild and farm conditions of *Cyprinus carpio*

Meristic characters	Euphrates River	Aquaculture farm	*F* value
Number of lateral line scales	34–37 (35.4 ± 1.65)	32–33 (32.4 ± 1.54)	10.478**
Dorsal fin rays	17–23 (19 ± 1.84)	24–25 (24.3 ± 1.75)	23.502**
Caudal fin rays	17–19 (17.8 ± 0.711)	15–16 (15.4 ± 1.83)	10.004**
Anal fin rays	5–6 (5.7 ± 0.020)	7–8 (7.4 ± 2.34)	16.891**
Total number of vertebrae	36–37 (36.4 ± 0.123)	38–39 (38.3 ± 2.01)	37.481**

** Significant at 1 % level

all the five characters have shown a significant differences between wild and reared populations of *C. carpio* ($P < 0{\cdot}05$).

81.5 Conclusions

In both wild and reared individuals, the meristic features used were of main significance in differentiating the two populations as revealed by the statistical analysis. The farmed specimens were characterized in having less number of: scales on their lateral line, number of caudal fin rays, and number of vertebrae. On the other hand, they showed to have more number of dorsal and anal fin rays. These dissimilarities might also be accredited to numerous agents linked to the culturing settings such as stock density, violence, strain, and quality of food (Favaloro and Mazzola 2003), as well as altered swimming ability (Basaran et al. 2007) and fish flexibility (Hanson et al. 2007). Phenotypic variations in meristic traits might not only be genetic but may be habitat tempted (Swain and Foote 1999). In some investigations, habitat setting, mainly temperature, which win out during some subtle growing phases, have been revealed to have the highest impact on meristic features (Quilang et al. 2007). Bhatt et al. (1998) observed similar types of differences in both morphometric as well as meristic traits of *T. putitora* between the populations of the Ganga and Gobindsagar reservoir group. Alternatively, morphologic changes are not essentially symptomatic of genetic diversity between groups (Ihssen et al. 1981; Allendorf 1988), and therefore the finding of morphological alterations among groups cannot frequently considered as a sign of genetic variation (Turan 1999). Morphologic flexibility of fish permits them to react flexibly to habitat variations by change in their physiology and behavior, which cause modifications in their morphology, reproduction, or endurance that alleviate the impacts of habitat variations (Mostafa et al. 2010).

In assumption, the discussed case offers indication that body meristic features of *C. carpio* are diverse conferring to their source at diverse levels, which might be accredited to both culturing setting and genetic diversity. Further, the usage of meristic trait looks to have widespread applications in the recognition of wild and reared fish. These features contain no pronounced scientific knowledge and they have been revealed to be a valued means for defining variations in shape attributes. The usage of the joint method, such as morphometry, genetic, and other biological signs (e.g., growth pattern of scales and otoliths, fatty acids, and trace elements), must be deliberated for the more exact outcomes. This will not only underwrite significantly to biological and ecological explanation of the species but will also aid in the growth of a policy for natural populations for protection and refining the aquaculture endurance.

References

Agnew DJ (1988) Evidence for the existence of two populations of Irish Sea cod (*Gadus morhua* L.) from consideration of growth rates. ICES Committee Meeting Papers 1988/G 65:20
Allendorf FW (1988) Conservation biology of fishes. Conserv Biol 2:145–148
Avsar D (1994) Stock differentiation study of the sprat off the southern coast of the Black Sea. Fish Res 19:363–378
Basaran F, Ozbilgin H, Ozbilgin YD (2007) Comparison of the swimming performance of farmed and wild gilthead sea bream, *Sparus aurata*. Aquac Res 38:452–456
Bhatt JP, Nautiyal P, Singh HR (1998) Comparative study of morphometric characters of Himalayan mahseer Tor putitora (ham.) between ganga and Gobindsagar reservoir stocks. Indian J Fish 45(1):85–87
Bronte CR, Fleischer GW, Maistrenko SG, Pronin NM (1999) Stock structure of Lake Baikal omul as determined by whole body morphology. J Fish Biol 54:787–798
Creech S (1992) A multivariate morphometric investigation of *Atherina boyeri* Risso, 1810 and *A. presbyter* Cuvier, 1829 (Teleostei: Atherinidae): morphometric evidence in support of the two species. J Fish Biol 41:341–353
Einum S, Fleming IA (1997) Genetic divergence and interactions in the wild among native, farmed and hybrid Atlantic salmon. J Fish Biol 50:634–651
Elliott JM (1990) Mechanisms responsible for population regulation in young migratory trout, *Salmo trutta*. II. Fish growth and size variation. J Anim Ecol 59:171–185
El-Serafy SS, Abdel-Hameid N-AH, Awwad MH, Azab MS (2007) DNA riboprinting analysis of *Tilapia* species and their hybrids using restriction fragment length polymorphisms of the small subunit ribosomal DNA. Aquac Res 38:295–303
Fagen R, Smoker WW (1989) How large-capacity hatcheries can alter interannual variability of salmon production. Fish Res 8:1–11
Favaloro E, Mazzola A (2003) Shape change during the growth of sharpsnout seabream reared under different conditions in a fish farm of the southern Tyrrhenian Sea. Aquac Eng 29:57–63
Fleming IA, Einum S (1997) Experimental test of genetic divergence of farmed from wild Atlantic salmon due to domestication. ICES J Mar Sci 54:1051–1063
Hanson KC, Hasler CT, Suski CD, Cooke SJ (2007) Morphological correlates of swimming (*Micropterus salmoides*) in their natural environment. Comp Biochem Physiol Part A 148:913–920
Hockaday S, Beddow TA, Stone M, Hancock P, Ross LG (2000) Using truss networks to estimate the biomass of *Oreochromis niloticus* and to investigate shape characters. J Fish Biol 57:981–1000
Hossain MAR, Nahiduzzaman M, Saha D, Habiba Khanam MU, Alam MS (2010) Landmark-based morphometric and meristic variations of the endangered carp, Kalibaus *Labeo calbasu*, from stocks of two isolated Rivers, the Jamuna and Halda, and a hatchery. Zool Stud 49:556–563
Hubbs CL, Lagler KF (1964) Fishes of the Great Lakes region. Univ Mich Press *Ann Arbor* 213
Ihssen PE, Booke HE, McGlade JM, Payne NR, Utter FM (1981) Stock identification: materials and methods. Can J Fish Aquat Sci 38:1838–1855
Mamuris Z, Apostolidis AP, Panagiotaki P, Theodorou AJ, Triantaphllidis C (1998) Morphological variation between red mullet populations in Greece. J Fish Biol 52:107–117
Mostafa HAR, Mohammad N, Saha D, Habiba Mst KU, Md. Alam S (2010) Landmark-based morphometric and meristic variations of the endangered carp, Kalibasu, *Labeo calbasu*, from stocks of two isolated rivers, the Jamuna and Halda, and a hatchery. Zoolog Stud 49(4):556–563
Murta AG (2000) Morphological variation of horse macherel (*Trachurus trachurus*) in the Iberian and north African Atlantic: implication for stock identification. ICES J Mar Sci 57:1240–1248
Orban, E., Nevigato, T., Di Lena, G., Casini, I., & Marzetti, A. (2003). Differentiation in the lipid quality of wild and farmed sea bass (Dicentrarchus labrax) and Gilthead Sea bream (Sparus aurata). J Food Sci, 68, 128–132

Pakkasmaa S (2000) Morphological and early life history variation in salmonid fishes. Ph.D. Dissertation, Department of Ecology and Systematics, University of Helsinki, Finland
Parish BB, Sharman DP (1958) Some remark on method used in herring 'racial' investigations, with special reference to otolith studies. Rapp P-V Reun Cons Int Explor Mer 143:66–80
Pinheiro A, Teixeira CM, Rego AL, Marques JF, Cabral HN (2005) Genetic and morphological variation of Solea lascaris (Risso, 1810) along the Portuguese coast. Fish Res 73(1-2):67–78
Quilang JP, Basiao ZU, Pagulayan RC, Roderos RR, Barrios EB (2007) Meristic and morphometric variation in the silver perch, *Leiopotherapon plumbeus* (Kner, 1864), from three lakes in the Philippines. J Appl Ichthyol 23:561–567
Ryman N, Utter F, Hindar K (1995) Introgression, supportive breeding, and genetic conservation. In: Ballou JD, Gilpin M, Foose TJ (eds) Population management for survival and recovery. Analytical methods and strategies in small population conservation. Columbia University Press, New York, pp 341–365
Salman NA (2011) Suggested methods for enhancement of aquaculture in the southern marshes of Iraq. Iraqi Journal of Aquaculture 8(2):157–177
Swain DP, Foote CJ (1999) Stocks and chameleons: the use of phenotypic variation in stock identification. Fish Res 43:1123–1128
Turan C (1999) A note on the examination of morphometric differentiation among fish populations: the truss system. Turk J Zool 23:259–263
Turan C, Erguden D, Turan F, Gurlek M (2004) Genetic and morphologic structure of *Liza abu* (Heckel, 1843) populations from the Rivers Orontes, Euphrates and Tigris. Turk J Vet Anim Sci 28:729–734
Turan C, Oral M, Ozturk B, Duzgunes E (2006) Morphometric and meristic variation between stocks of bluefish (*Pomatomus saltatrix*) in the black, Marmara, Aegean and northeastern Mediterranean seas. Fish Res 79:139–147
Wimberger PH (1992) Plasticity of fish body shape, the effects of diet, development, family and age in two species of *Geophagus* (Pisces: Cichlidae). Biol J Linn Soc 45:197–218
Youngson AF, Verspoor E (1998) Interactions between wild and introduced Atlantic salmon (*Salmo salar*). Can J Fish Aquat Sci 55(Suppl. 1):153–160

Part VIII
Social Perspectives

Chapter 82
Is the Glass Half Empty or Half Full? An Appraisal of the Four Decades of Turkey's Southeastern Anatolia Project (GAP)

Arda Bilgen, Zeynep Sıla Akıncı, Antònia Casellas, and Joost Jongerden

Abstract The Southeastern Anatolia Project (*Güneydoğu Anadolu Projesi*, GAP) was launched as a massive package of water and land resources development project in 1977. Over time, the project was redefined as a multisectoral, integrated, and sustainable regional and human development project. This chapter seeks to make a critical appraisal of GAP on the basis of its formal objectives. To this end, the chapter explores GAP-induced changes and challenges specifically in the (i) infrastructural and economic and (ii) sociocultural dimensions of the project. Our main conclusion is that the state's claim about the project's success is not supported by the state's own data.

Keywords Appraisal · Development · Modernization · The Southeastern Anatolia Project · GAP · Turkey

> "[T]he will to improve [...] is both benevolent and stubborn. [...] For vast numbers of people, it falls short of the promise to make the world better than it is. Yet, [...] it deeply shapes the conditions of their lives."—Tania Murray Li.[1]

---

[1]Li 2007, p. 283.

A. Bilgen (✉)
Department of International Development, Community, and Environment, Clark University, Worcester, MA, USA
e-mail: abilgen@clarku.edu

Z. S. Akıncı · A. Casellas
Department of Geography, Universitat Autònoma de Barcelona (UAB), Barcelona, Spain
e-mail: zeynepsila.akinci@uab.cat; antonia.casellas@uab.cat

J. Jongerden
Rural Sociology, Wageningen University, Wageningen, The Netherlands
e-mail: joost.jongerden@wur.nl

L. A. Jawad (ed.), *Tigris and Euphrates Rivers: Their Environment from Headwaters to Mouth*, Aquatic Ecology Series 11,
https://doi.org/10.1007/978-3-030-57570-0_82

82.1 Introduction

The year 2017 marked the 40th anniversary of the official initiation of the Southeastern Anatolia Project (*Güneydoğu Anadolu Projesi* in Turkish, GAP in its acronym) in southeastern Turkey. Launched as a massive water and land resources development project, GAP included seven groups of energy production and irrigation projects on the Euphrates and six on the Tigris. As one of the largest regional development projects in the world, the construction of 22 dams, 19 hydroelectric power plants (HPPs), and extensive irrigation networks was expected to utilize the available water resources of the Southeastern Anatolia Region (henceforth, the GAP region),[2] generate 27 billion kWh of hydroelectric energy each year, and water 1.8 million hectares (ha) of land (Kibaroğlu 2006, p. 178). In the 1980s, GAP was expanded to include additional sectors such as agriculture, transportation, telecommunications, healthcare, education, and infrastructure and, thus, transformed to a multisectoral and integrated regional development project (Devlet Planlama Teşkilatı [DPT] 1989a, pp. 1.1–1.5). Following the "sustainability" and "human development" turn in development research and policy in the 1990s, GAP was once again redefined, this time as a "sustainable human development project" (Kibaroğlu 2006, p. 178). From the 2000s onward, in parallel with the increasing neoliberal character of the Turkish political economy, the project has made a market- and private sector-friendly turn (Bilgen 2018c, p. 151). In other words, the scope, mechanisms, and even definitions of GAP have undergone significant changes over the decades. The overall objectives of the project, however, have remained as to reduce inter- and intra-regional disparities and to modernize and transform the socioeconomic, sociopolitical, and sociocultural structure of the GAP region.

The state has framed GAP as a panacea for the GAP region's political, economic, social, and ecological issues such as water scarcity, poverty, insurgency, or poor infrastructure. However, even official figures fail to present a convincing argument regarding the ultimate success of the project. Even though GAP has been the subject of critical studies and appraisals (Bilgen 2018b; Glass 2017; Harris 2008; Jongerden 2010; Konak 2013; Warner 2012) in the past, only a few have made an appraisal of the project on basis of its formal objectives (see Bilgen 2018d for a comprehensive review of the literature on GAP). This chapter, therefore, seeks to fill this gap and present a critical and most up-to-date appraisal of GAP within the contours of its own claims.

[2]This value-laden label varies according to the positionality of a person or a political entity. Alternative labels include, but not limited to, the East, the Southeast, the region, or a part of Northern Kurdistan (*Bakur*).

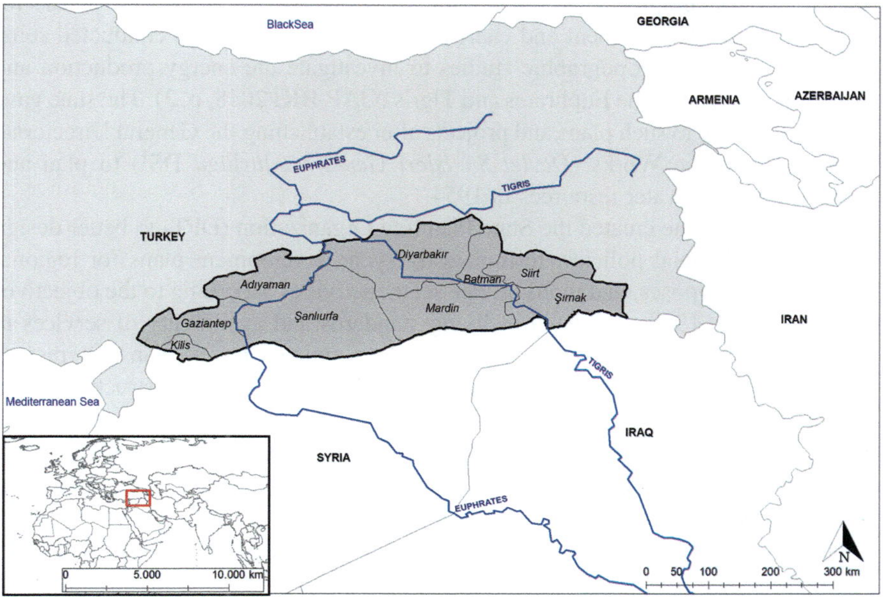

Fig. 82.1 The GAP region. Source: Akıncı et al. 2020, p. 182

82.2 GAP

The GAP region (Fig. 82.1) lies on the northern basin of the Euphrates and Tigris and covers around 75,000 km^2 surface area and 8.7 million people, corresponding to around 10% of the total surface and population in Turkey (GAP Bölge Kalkınma İdaresi [GAP-BKİ] 2018, p. 8). It includes the provinces of Adıyaman, Batman, Diyarbakır, Gaziantep, Kilis, Mardin, Siirt, Şanlıurfa, and Şırnak. Around 20% of the total irrigable land and 30% of the water potential of Turkey are located in the region (p. 23). The region is mostly populated by the Kurds and Kurdish is the dominant language, especially in rural settings (Gökçe et al. 2010, pp. 37–38).

After the establishment of modern Turkey in 1923, the overarching goals of the state were to build a (modern) Turkish nation in Anatolia (Bozdoğan and Kasaba 1997). To this end, the state, inter alia, extended the railroad system to connect different regions and to have greater control over society, and created new industries in small and remote cities in central Anatolia (Tekeli 2008, pp. 53–54). In the 1930s, Mustafa Kemal Atatürk introduced the idea of constructing a dam and an HPP on the Upper Euphrates after his fascination with the Soviet Union's plans for the Dnieper (Turgut 2000, p. 47).[3] In 1935, the state established the Electrical Power Resources

[3]Indeed, there were earlier attempts to develop the water and land resources of the region. For instance, in the 1910s, Sir William Willcocks (1910) was appointed by the Ottoman Empire as an advisor to, among others, "survey and level the rivers and canals of the Tigris–Euphrates delta and devise projects for the rehabilitation of the country" (p. 1).

Survey and Development Administration (*Elektrik İşleri Etüt İdaresi*, EİEİ) for water resources development and energy planning purposes. EİEİ conducted some early geological and topographic studies to investigate the energy production and irrigation potential of the Euphrates and Tigris (GAP-BKİ 2018, p. 2). The state gave more importance to such plans and projects after establishing the General Directorate of State Hydraulic Works (*Devlet Su İşleri Genel Müdürlüğü* DSİ) to plan and manage Turkey's water resources in 1954.

In 1960, the state created the State Planning Organization (DPT) to better design economic and social policies, formulate five-year development plans for regional development purposes, and mold private sector activities according to the objectives of plans. In the 1960s, the level of living standards and availability of services in eastern and southeastern Turkey continued to remain lower than the rest of the country (Jafar 1976, p. 124). In this context, DSİ began the construction of the Keban Dam in 1966. Similarly, DPT began to work on a plan that focused on the Keban region, considered as the basis of GAP. Throughout the 1960s, DSİ continued to conduct feasibility studies and formulate various reconnaissance reports to construct more dams and HPPs on the Euphrates and Tigris. Because the gap between western and eastern Turkey further widened in the 1970s, the state designated provinces in southeastern and eastern Turkey as 'Priority Localities in Development' in 1971. The Keban Dam was completed in 1974. Finally, 13 projects for the Tigris and Euphrates were combined and renamed GAP in 1977.

In 1986, DPT replaced DSİ as the new coordinator of GAP until the establishment of the GAP Regional Development Administration (GAP-BKİ) in 1989. GAP-BKİ was tasked to engage in planning, implementing, monitoring, and evaluating activities for regional development purposes as well as to ensure coordination among state institutions involved in the project. In the same year, *the GAP Master Plan* was formulated, aiming at increasing the gross regional product (GRP) by 445% and per capita income by 209% while creating 3.8 million new jobs by 2005 (GAP-BKİ 2016, p. 4). A turning point for GAP took place in the mid-1990s when the concepts of sustainability, participation, social development, and human development became the fundamental principles of GAP. Also, after becoming a European Union (EU) candidate in 1999, Turkey became obliged to change its conception of regional development and formulate a national policy aimed at reducing regional disparities. To this goal, in 2002, the government formulated *the GAP Regional Development Plan* to "increase income and welfare through protecting and enhancing environment and resources based on the principles of equity and fairness; to consider and integrate disadvantaged groups into development; and to ensure sustainability and private sector and public participation at all stages" (GAP-BKİ 2002a, p. 13). In 2008, to accelerate the project's schedule, the government formulated *the GAP Action Plan of 2008–2012,* which included four major development axes: economic development, social development, infrastructure building, and institutional capacity building (GAP-BKİ 2008, p. 9). Finally, in 2014, the government formulated *the GAP Action Plan of 2014–2018* and added "increasing the livability of cities" as the fifth development axis (GAP-BKİ 2014, p. 15). To date, past governments allocated an average of 8.5% of public funds to GAP from 1989 to 2018 (GAP-BKİ 2018, p. 19).

82.3 Is the Glass Half Empty or Half Full?

To advance in the analysis and to provide evidences that would allow to establish the level of success of GAP, the sections below explore GAP-induced changes and challenges organized under two dimensions: i) infrastructural and economic and ii) sociocultural.

82.3.1 Infrastructural and Economic Dimension

GAP has accorded great importance to the production of hydro-energy, generation of income by hydro-power, and electrification of urban and rural areas. There has been a significant increase in each field over the decades. By 1987, only three dams—Karakaya, Tahtaköprü, and Hancağız—were completed (DPT 1989a, p. 5.1). In 1992, 40% of energy projects were in operation, 31% were under construction, and the remaining 29% were under planning stage (GAP-BKİ 1993, p. 74). In 2001, 70% of these projects were in operation, 4% were under construction, 19% were under bidding stage, and the remaining 7% were under planning stage (GAP-BKİ 2001, p. 11). By 2018, 74% of the energy projects under GAP have been completed (GAP-BKİ 2018, p. 35).[4]

The electricity access has notably increased too. In 1980, only 16.3% of villages located in the region had access to electricity, while 50.7% of villages located in all of Turkey were electrified (DPT 1989a, p. 5.8). The access rates increased to 84% in the region and to 94% in Turkey in 1986. Despite this expansion, however, per capita electricity consumption in the region (350 kWh) continued to be lower than the national average (650 kWh) in this period (p. 5.8). By 2018, all villages in the region had electricity access.[5] The region's electricity consumption has gradually increased as well. In 1990, 2.2 billion kWh of electricity was consumed in the region (GAP-BKİ 1993, p. 13). In 2006, this amount was quadrupled to 8.8 billion kWh (GAP-BKİ 2007, p. 46). By 2016, the Dicle electricity distribution area covering all provinces in the region except Adıyaman, Gaziantep, and Kilis consumed 21.2 billion kWh of electricity alone (Türkiye Elektrik İletim A.Ş. Genel Müdürlüğü 2017, pp. 24–25).

The second objective to highlight was the construction of irrigation canals and networks. The completion rate and pace of irrigation projects have been lower than

[4]The dams and HPPs in operation generated 9.6 billion kWh of energy in 1993 (GAP-BKİ 1994, p. 11). The amount generated from the beginning of the project until 2001 was as high as 184 billion kWh of energy, delivering $11 billion in financial value to the state (GAP-BKİ 2001, p. 12). By 2018, the amount of energy produced increased to 443.8 billion kWh, delivering $26.63 billion in financial value to the state (GAP-BKİ 2018, p. 35).

[5]This information was obtained on January 29, 2019, through a personal email correspondence between one of the authors and a corporate relations manager at Enerjisa, a company engaged in the production, distribution, and sale of electricity in Turkey.

those of energy projects. Only 4% of irrigation projects were in operation in 1992, irrigating 70,049 ha then. In this period, 14% of irrigation projects covering 225,306 ha were under construction and 81% of irrigation projects covering 1.3 million ha were under planning stage (GAP-BKİ 1993, p. 74). The portion of completed irrigation projects increased to 12% in the early 2000s; they provided irrigation to 215,080 ha. The remaining 8% were under construction, 25% were under bidding stage, and 55% were under planning stage (GAP-BKİ 2001, p. 14). Today, 30.4% of irrigation projects have been completed and 546,345 ha has been opened to irrigation. Still, new projects to water 197,969 ha are under construction and the remaining projects to further water 1.05 million ha are under planning stage (GAP-BKİ 2018, p. 29).

Many improvements took place in urban and rural infrastructure. Of the 145 municipalities in the region in 1990, around 80% lacked a sewage system and drinking water, only 9% had a partially functioning sewage system and 24% had a drinking water network (GAP-BKİ 1993, p. 11). In 2002, the number of municipalities in the region increased to 196. Then, 60–65% of these municipalities had a functioning sewage system and access to drinking water, while around 20% had no or partially operating sewage systems and drinking water infrastructure (GAP-BKİ 2002b, p. 50). By the 2010s, the number of municipalities decreased to 122. In this period, 118 out of 122 municipalities—96.7% of all—had a functioning sewage system (GAP-BKİ 2016, p. 47) and all municipalities had functioning drinking water networks (GAP-BKİ 2018, p. 57).

As for the economy, two of the most articulated goals of the project were to increase GRP and per capita income in the region. While both figures have improved, the improvement has failed to elevate the region to the level of other regions in Turkey. The region's GRP was 3,365,559 billion TL in 1985. This figure formed only 4% of Turkey's gross domestic product then. In this period, per capita GRP in the region was 47% of the national average (DPT 1990, p. 1). In 2000, while the share of the region's economy in the national economy increased to 5%, per capita GRP in the region decreased to 46% of the national average (GAP-BKİ 2002b, p. 14). By 2016, the regional average increased to 51.3% of the national average (GAP-BKİ 2018, p. 16). According to household income quintiles, however, the first 20% in the region (the lowest income group) earned 3.486 TL, whereas the last 20% (the highest income group) earned 22,001 TL. In this period, the first 20% in Turkey earned 5880 TL and the last 20% earned 45,173 TL (p. 16).

Unemployment has been a serious problem in the region. In the early 1990s, the region's unemployment rate was 22%, whereas Turkey's unemployment rate was 11% (GAP-BKİ 1996, p. 26). In this period, the labor participation rate in the region ranged between 35–40%, while nationwide labor participation ranged between 55–60% (DPT 1989c, p. G50). Toward the late 2000s, the region's unemployment declined to 15.8%. It was, however, still higher than the national average, which remained unchanged at 11% (GAP-BKİ 2018, pp. 44–45). Labor participation in the region continued to be lower than labor participation in all of Turkey too. In 2008, participation rates for the region and country were 35.7% and 46.9%, respectively (pp. 44–45). In 2016, the region's unemployment increased to 18.7%, while

Turkey's unemployment declined to 10.9%. Labor participation in the region continued to be lower than the national average—45.7% as opposed to 52%—too (pp. 44–45).

The distribution of working population has also undergone changes over time. In the 1980s, around 70% of economically active population in the region was employed in the agriculture sector (GAP-BKİ 1993, p. 22), around 5% was employed in the industry sector (p. 49), and the rest was employed in the service sector. In 2006, however, 22.1% of the labor force worked in agriculture, 24.7% worked in industry, and 53.3% worked in service sectors (GAP-BKİ 2018, p. 46). This distribution remained largely intact more recently in 2016: 22.7%, 24.5%, and 52.8% of the labor force worked in the agriculture, industry, and service sectors, respectively (pp. 46–47).

Agriculture has been a key pillar of GAP and the region's economy. In the early stages of the project, 36% of the cultivated land was used as dry farming land, 1.7% as irrigated farming land, and the rest for horticulture and other activities (DPT 1989a, p. 4.5). Between 1984 and 2004 alone, the cultivation of some water-intensive crops such as maize and sunflower increased by around 1500% and 7500%, respectively (Benek 2009, p. 63). By 2018, around 60% of the cultivated land was used as dry farming land, while around 15% was used as irrigated farming land (Dikme 2017). The cultivated land occupied 43.6%, pastures occupied 29.4%, and forests and bushes occupied 19.2% of the total land area in this period (GAP-BKİ 2018, p. 23).

Traditionally, dry farming included the production of grains, red lentil, chickpea, sesame, and tobacco as well as horticultural crops like pistachio and grape (GAP-BKİ 1993, p. 22). Crops such as wheat, cotton, rice, sorghum, sunflower, melon, watermelon, tomato, eggplant, pepper, and cucumber were also produced through irrigated farming techniques (p. 22). The shift from dry to irrigated farming has had an impact on both the variety of crops and their portions in nationwide agricultural production. The increase in the production of grains, specifically wheat, was particularly noteworthy. In 1989, the region produced 98% of lentils, 90.8% of pistachios, 44% of sesame, 30% of tobacco, 20.3% grapes, 15.5% of chickpeas, 14% of garlic, 13% of onions, 14% of barley, and 7% of wheat produced in Turkey (GAP-BKİ 1993, p. 24). In 2016, the region produced 96.94% of lentils, 92.2% of pistachios, 4.1% of sesame, 12.6% of tobacco, 25.2% of table grapes, 28.2% of wine grapes, 10% of chickpeas, 57.7% of garlic, 12.5% of onions, 21.7% of barley, and 35.7% of wheat produced nationwide (GAP-BKİ 2018, pp. 25–27).

The increase in cotton production in the region has been remarkably high. In 1985, the region produced 13.2% of cotton produced in Turkey. In 1994, this portion increased to 22.8% (GAP-BKİ 1996, p. 7). After the Harran Plain was opened to irrigation in 1995, around 200,000 out of 300,000 ha of irrigated land was allocated to cotton production. Thus, more than 25% of national cotton production originated from the region (p. 4). The rise was huge between 1995 and 2015: the region produced 55% of cotton produced in Turkey in 2016, making the region the cotton production hub of the country (GAP-BKİ 2018, p. 23).

The region witnessed a significant surge in the number of tractors and agricultural equipment since the 1980s. The region had 31,475 tractors in 1989. This figure constituted 4.7% of all tractors in Turkey (GAP-BKİ 2007, p. 24). In 2017, the region had 65,321 tractors, making up 5.3% of all tractors nationwide (GAP-BKİ 2018, p. 32). The number of agricultural equipment increased in a similar manner. In 1988, the region had 213,592 equipment that included combine harvesters, seed drill-planters, cultivators, tractor hoes, tractor plows, fertilizer distributors, movers, threshers, sprayers, motor pumps, trailers, wooden plows, and livestock plows (DPT 1989b, p. A33). By 2017, the region had 483,493 equipment (GAP-BKİ 2018, pp. 32–33). Fertilizer use has also increased. The region used 162,422 tons of fertilizer in 1990 (GAP-BKİ 1993, p. 34) and 230,717 tons of fertilizer in 2003 (Çolakoğlu et al. 2005, p. 17). By 2013, fertilizer use in some provinces exceeded even the world average: the amount used in Mardin, Şanlıurfa, and Diyarbakır was 192.7 kg/ha, 148 kg/ha, and 145.5 kg/ha, respectively (Şahin 2016, p. 24).

Landownership has long been skewed in the region. In 1985, for instance, 38.4% of rural families in the region possessed no land or animals, while 8% of farmers possessed 51% of agricultural lands (DPT 1989a, p. 6.10). In the 2000s, the skewed ownership pattern and the number of landless farmers remained almost the same: 61.4% of agricultural enterprises continued to cultivate only 10.5% of the land, while 6.2% of enterprises continued to cultivate almost 50% of the land (Gülçubuk, 2005, as cited in Görgü 2006). By 2010, 70.1% of 1227 families studied in the region were landless (Gökçe et al. 2010, p. 96), suggesting that GAP had little or no impact in terms of improving the conditions of landless people.

Animal husbandry sector has also undergone significant changes. Over the decades, the number of livestock population has fluctuated and lands for grazing have shrunk due especially to the submerging of lands and security measures imposed under the State of Emergency Law between the 1980s and 2000s. Between 1980 and 1998, the number of sheep, goats, and cattle decreased by 36%, 89%, and 44%, respectively (GAP-BKİ 2002b, p. 26). By 2018, however, the number of sheep, goats, and cattle increased by 25.4%, 52.9%, and 126%, while the number of mohair goats decreased by 72.3% (GAP-BKİ 2018, p. 34). Unlike husbandry activities, fishery activities have followed a steady upward trend: the total production of freshwater fish increased from 805 tons (DPT 1989b, p. A19) to 6600 tons (GAP-BKİ 2018, p. 34) between 1980 and 2016.

The industry sector has also been a crucial pillar of GAP. A wide array of actions has been undertaken to increase industrial production and efficiency, especially in Gaziantep. As a result, the region has become more industrialized and relied more on a market-based form of economy. It provides evidence of this shift that the number of "incentive certificates" given to investors so that they can promote industrialization, create jobs, and boost further investments have steadily increased since the 1980s. In 1980, only 14 certificates were given and, eventually, 3501 jobs were created and 1.8 billion TL of investment was encouraged (DPT 1989b, p. B11). In 2000, 209 certificates helped the creation of 9396 jobs and the attraction of 302 billion TL of investment (GAP-BKİ 2001, p. 23). By 2017, the number of certificates given reached 970, creating 46,571 jobs, and attracting 8.3 billion (Y)TL of investment (GAP-BKİ 2018, p. 36).

Another indicator of industrialization is the number and capacity of organized industrial zones (*Organize Sanayi Bölgeleri*, OSB) and small industrial zones (*Küçük Sanayi Siteleri*, KSS) located in the region. OSB "are designed to allow companies to operate within an investor-friendly environment with ready-to-use infrastructure and social facilities" and enable investors to benefit from investment and employment incentives, research and development support, and tax and real estate duty exemptions (Presidency of the Republic of Turkey Investment Office 2019). In the 1980s, only Gaziantep had an organized industrial zone. Between 2000 and 2018, the number of OSB went up from 8 to 24, increasing the number of jobs created by OSB from 43,500 (GAP-BKİ 2001, p. 19) to 174,170 (GAP-BKİ 2018, p. 41). Also between 2000 and 2018, the number of KSS increased from 18 to 38 and the number of businesses in these zones went up from 5514 to 9643 (GAP-BKİ 2001, p. 19, 2018, p. 43).

There has been an increase in the value of exports from the region as well as the region's share in Turkey's total exports too. In 1996, the region exported $405 million worth of goods and services, constituting 1.7% of total exports of the country (GAP-BKİ 2002b, p. 31). By 2017, 3759 companies exported $8.6 billion worth of goods and services which formed 5.5% of Turkey's total exports (GAP-BKİ 2018, p. 38). The pattern of exports has changed too: 60% of exported goods consisted of textile goods and raw material, 15% of grains and legumes, 8% of dried fruit, 3% of livestock and fishery products, and 14% of other products in 2001 (GAP-BKİ 2002b, p. 31). Yet, in 2017, 94% of exported goods consisted of manufactured goods, 5% of agricultural and forestry goods, and 1% of mining products (GAP-BKİ 2018, p. 38). Before 2000, 45% of exports went to Germany, the United States, and the United Kingdom. After 2000, however, 45% of exports went to Iraq alone. In 2017, Iraq continued to be the top export destination for receiving 36% of exports (p. 40).

The increase in the number of domestic and foreign tourists visiting the region and/or staying overnight is considered another indicator of economic and industrial development. In 1986, around 4.5 million tourists visited Turkey and around 10.8 million stayed overnight in Turkey. Yet, only 2.8% of those visiting (130,758 people) and only 1.8% of those staying (192,776 people) traveled to the region then (DPT 1989b, p. B16). By 2016, 453,139 tourists lodged at municipally approved facilities, while 733,998 tourists lodged at a facility holding a tourism operation license. They formed 2% and 1.8% of all tourists lodging in such facilities in Turkey in that year (GAP-BKİ 2018, p. 49). Concurrently, around one million stayed overnight at municipal-related facilities, while more than 1.5 million stayed at a facility with a tourism license, constituting 2.7% and 1.4% of all tourists staying in these facilities in Turkey overall (p. 49). In other words, while the number of tourists visiting the region has increased, the region has not been able to capture a higher percentage of total tourism in Turkey over time.

82.3.2 Sociocultural Dimension

Turning our attention to the social dimension of GAP, it is relevant to stress that a wide array of sociological studies was conducted by, or through the support of, GAP-BKİ to better understand the sociocultural characteristics of the region's population and obtain control over their progress, development, and movements in the 1990s. These include, but not limited to, *the Management, Operation, and Maintenance Project Socioeconomic Studies*; *the Survey on the Trends of Social Change in the GAP Region*; *Population Movements in the GAP Region*; *the Survey on the Problems of Employment and Resettlement in Areas Affected by Dam Lakes in the GAP Region*; and *Women's Status in the GAP Region and Their Integration into the Process of Development*. This took place against the background of the rise of the Kurdistan Workers' Party (*Partiya Karkerên Kurdistan*, PKK) and the wide support it garnered from the local population. From the state's perspective, political problems were understood as deriving from economics in a very broad sense: because the region was relatively poor and dissatisfaction and disaffection among the local population were pervasive, the PKK was able to channel these sentiments to its own ends. Therefore, the logic went, developing the Kurdish region economically would lead to the automatic solution of the Kurdish issue. GAP-employed social scientists argued that the introduction of modern, irrigation-based agriculture, and market integration would bring a new lifestyle to the region. The importance of tribal relations and extended families would diminish, while dependency on state institutions such as the Ministry of Agriculture would increase. In the course of this process, tribal Kurds would turn into modern Turks (Jongerden 2010).

To begin with the demographics, the region's population has increased steadily but moderately since the 1980s. In 1985, the region had 4.3 million people who formed 8.5% of 50.6 million people in Turkey then (DPT 1989a, p. 2.3). In 2000, the region's population increased to 6.6 million people, constituting 8.8% of 67.8 million people living in Turkey (GAP-BKİ 2002b, p. 12). In 2017, with 8.7 million inhabitants, the region constituted 10.7% of 80.8 million inhabitants of Turkey (GAP-BKİ 2018, p. 8). The region's annual population growth rate has long been higher than Turkey's average. In 1990, the region's population growth was around 3.5% (GAP-BKİ 1993, p. 5), while Turkey's population grew by 2.3% (Kasarcı 1996, p. 248). Population growth in the region and in all of Turkey was respectively 2.47% and 1.82% between 1990 and 2000 (GAP-BKİ 2018, pp. 10–11). In the same order, these figures declined to 1.38% and 0.8% between 2000 and 2010. Between 2010 and 2017, however, the region's average rate increased to 1.88%, while Turkey's average rate increased to 1.3% (pp. 10–11).

The region's population density was lower than the density in all Turkey in the early phases of GAP. In 1990, the number of people per km^2 in the region was 68, while it was 72 in all of Turkey (GAP-BKİ 1993, p. 5). This changed in the project's later stages. In 2000, the region's population density was slightly higher than the national average, but grew to 115 people per km^2 as opposed to 105 in Turkey by 2018 (GAP-BKİ 2018, p. 10). Just as population density, the average number of household members in the region has been higher than the average in all Turkey. In 1990, the region's average was 6.8, while the national average was 5.0 (p. 11). As of 2018, the average number of households was 5.05 in the region and 3.4 in all of Turkey. Six out of ten cities with the highest number of household members remain located in the region (p. 11).

As for population movements, emigration has outpaced immigration for decades. Even before GAP, the region's net migration rate was −3.23% between 1965 and 1970, −3.15% between 1970 and 1975, and −2.90% between 1975 and 1980 (DPT 1989c, p. G3). The same trend remained in the 1990s, when a military depopulation campaign roamed through the countryside, displacing 2–4 million people (Jongerden 2007). In 2010, 151,585 people immigrated to the region, while 183,752 people emigrated from the region. The net migration rate was −0.41% then (GAP-BKİ 2012). More recently in 2016, 149,022 people immigrated to the region. However, the emigration of 222,997 people from the region kept the net migration rate negative, at −0.86% (GAP-BKİ 2018, p. 13).

The destinations of emigrants have not changed much over time. From 1975 to 1980, more than half emigrated to metropolitan cities such as İstanbul, Ankara, İzmir, or richer regions such as Çukurova (DPT 1989c, p. G5). Between 1995 and 2000, the Mediterranean, Aegean, and Marmara regions continued to be the top emigrant destinations (GAP-BKİ 2007, p. 17). As of 2018, too, metropolitan cities in these regions such as İstanbul, Ankara, İzmir, Kocaeli, Bursa, and Antalya continued to receive the most emigrants from southeastern Turkey (GAP-BKİ 2018, p. 14). Also, the region's population has become more urbanized over time. In 1990, 56% of people in the region and 59% of people in all Turkey lived in urban settings (GAP-BKİ 1993, p. 5). In 2000, urbanization increased to 62.7% in the region and to 64.9% in Turkey. By 2017, 92.3% of people in the region and 92.51% of people in Turkey were urbanized (GAP-BKİ 2018, p. 10).

Education is considered a crucial element of social and human development. GAP has emphasized the importance of improving Turkish literacy and schooling among children. In this context, in parallel to the increase nationwide, the region's literacy rate has increased since the implementation of the project. In 1980, 55% of adults in the region and 67% of adults in all Turkey were literate (DPT 1989c, p. G11). In 1990, the region's literacy increased to 60.4% (75.5% in men, 44.8% in women). In this period, the literacy rate in Turkey was 78.4% (89.8% in men and 67.4% in women). In the ensuing years, the region's literacy remained lower than the national average. By 2017, for instance, the region's literacy was 78% (82% in men, 74% in women), whereas the country's literacy was more than 95% (Türkiye İstatistik Kurumu 2017).

The schooling rates in preschool, primary school, middle school, and high school education in the region have also increased over the decades. To illustrate, the schooling rate among preschool children was 2.36% between 1990 and 1991 (GAP-BKİ 1996, p. 36), 3% between 2001 and 2002 (GAP-BKİ 2002b, p. 44), and 52.08% between 2016 and 2017 (GAP-BKİ 2018, p. 51). In 2017, all provinces in the region except Kilis ranked below 52.08%, yet their schooling rates were closer to the national average compared to the past (p. 51). Similarly, schooling among high school children was 19.12% between 1990 and 1991 (GAP-BKİ 1996, p. 36) and 27.3% between 2001 and 2002 (GAP-BKİ 2002b, p. 44). Between 2016 and 2017, the schooling in GAP provinces ranged from 79.68% to 99.53%. However, except Adıyaman (107.14%), they failed to reach the national average, which was 106.94% then (GAP-BKİ 2018, p. 51). In addition, the number of universities in the region increased from three (GAP-BKİ 1993, p. 57) to twelve (Yükseköğretim Kurulu 2018) between 1992 and 2018, constituting 5.8% of 206 universities in Turkey.

Health, too, is seen as a vital element of social and human development. Therefore, great importance has been given to improve mortality rate, fertility rate, and the number of hospitals and health personnel in the region. In 2000, infant mortality in the region and all of Turkey was 42‰ and 43‰, respectively (GAP-BKİ 2018, p. 12). In 2010, the region's average declined to 15.8‰, while Turkey's average declined to 12‰. Recently in 2016, infant mortality was 14.5‰ in the region and 10‰ in all of Turkey (p. 12). Fertility rate has similarly decreased. In the region, the number of children per woman was 7.0 in 1980 (DPT 1989c, p. G1). In 2000, fertility in the region and in all of Turkey declined to 4.6 and 2.5 children per woman, respectively (GAP-BKİ 2018, p. 12). By 2016, they further declined to 3.37 for the region and 2.1 for Turkey (p. 12). The importance given to teach family planning methods to the local population played a significant role in this decline.

As for hospitals, 53 hospitals out of 844 hospitals in Turkey, or 6.2% of all hospitals in the country, were located in the region in 1990 (GAP-BKİ 1993, pp. 60–63). By 2015, the number of hospitals in the region increased to 128, constituting 8.3% of 1533 hospitals located in all of Turkey then (GAP-BKİ 2018, p. 53). The number of beds per 10,000 people has increased too. While the number was 11.4 regionwide and 21.4 nationwide in 1990, it increased to 20.9 in the region and 26.6 in the country in 2015 (GAP-BKİ 1993, pp. 60–63; 2018, p. 53). The number of specialist doctors and general practitioners in the region has also increased, yet failed to draw near the national average. In 1990, the region had 1.7 specialists and 2.6 practitioners per 10,000 people as opposed to 4.4 specialists and 4.2 practitioners in all Turkey (GAP-BKİ 1993, pp. 63–64). In 2015, the region had 5241 specialists and 4832 practitioners. In other words, 6.7% of specialists and 11.6% of practitioners in Turkey were located in the region (GAP-BKİ 2018, p. 54).

Last but not least, the number of Multi-Purpose Community Centers (*Çok Amaçlı Toplum Merkezleri,* ÇATOM) has increased since their establishment in 1995. ÇATOM are designed as community-based centers where local, mostly Kurdish women aged between 14–50 years can be trained in the fields of Turkish literacy, computer literacy, birth control, maternal-child health, hygiene, nutrition, home economics, and income-generating activities. Officially, the objectives of ÇATOM are raising women's awareness about their own problems and their capability to address them, increasing their participation in the public sphere, enhancing their employment and entrepreneurship, and empowering them toward a gender-balanced development (GAP-BKİ 2017, p. 56). First ÇATOM were opened in Şanlıurfa thanks to the cooperation and coordination among GAP-BKİ, United Nations Children's Emergency Fund, and involved governmental organizations. Even though initially women were the primary target group of ÇATOM, children, adult males, or other household members were also included into programs later on. In 1996, the region had seven ÇATOM in Şanlıurfa and Mardin, providing service to 1430 participants (GAP-BKİ 1997, p. 14). In 2007, the number of ÇATOM increased to 29 and the number of ÇATOM participants increased to 125,000 (GAP-BKİ 2007, p. 79). In 2017, there were 44 ÇATOM in the region. These centers provided services to 1,115,985 people since their establishment (GAP-BKİ 2017, p. 56).

82.4 Discussion

As a project of modernization, GAP was erected upon a strong belief in the possibility of linear and limitless progress with an elite-driven and top-down scheme (see Akıncı et al. 2020). In this process, the state has instrumentalized the project to reorder the political, economic, social, cultural, and ecological spheres in the region and to increase its own political reach, control, and legitimacy. GAP policies and practices can thus be located in the broader context of the state's efforts to make the GAP region's nature and society more legible, homogenized, and standardized (see Scott 1998). One can argue that GAP has provided more opportunities to the state than it has provided to the people of the GAP region and that what is referred to by the state as the development of the region has been ephemeral, fragile, and unsustainable rather than long-term, inclusive, and sustainable.

GAP has been highly "selective" and "exclusive" due to failing to provide benefits to all communities and equal benefits to different community groups. The distribution of income and resources is still highly uneven: only a limited number of groups—those already possessing large amounts of land, capital, resources, and power—in certain locations have reaped the benefits of improvements. In that sense, the project has widened the existing inter- and intra-regional disparities and created further inequalities and exclusions among local communities in certain contexts.

Similarly, the shift from dry to irrigated farming has increased agro-industrial production and efficiency, yet the commodification of the land has decreased the quality of the very same products. Over-irrigation has caused salinization and waterlogging problems that significantly decrease soil quality. Increased agricultural mechanization as well as diminished animal husbandry activities cost many small landowners as well as landless, unqualified, and seasonal workers their jobs and other sources of income. For example, only 20% of the landless could find jobs more easily and 26% had higher incomes after the project (Kudat and Bayram 2000, p. 247).

From a cultural and political reading, massive dams and HPPs are often presented as the symbols of the nation's techno-economic power. However, they have also caused the erasure of historical and cultural heritages in the region. For instance, Nevali Çori in Şanlıurfa was inundated by the Atatürk Dam in the early 1990s. Zeugma, a once Hellenistic and Roman city with several well-preserved mosaics and inscriptions located in today's Gaziantep, was also inundated by the Birecik Dam in the 2000s. Recently, Hasankeyf in Batman, an ancient settlement dating back to more than 10,000 years ago, was flooded by the Ilısu Dam. The unique heritages of the town such as the historical commercial center, multi-layered cave dwellings, Roman ruins, Medieval monuments, and several Islamic and Christian holy sites were destroyed.

Regarding population shifts, GAP has caused the displacement of hundreds of thousands of people. Around 3000 settlements were already destroyed and 400,000 to 1.5 million people were already forced to migrate between 1990 and 2000, mostly when the State of Emergency Law was in effect (Jongerden 2007). In addition, estimates show that 181,000 (GAP-BKİ 1994) to 400,000 people were displaced due to all dams built within GAP (see Jongerden et al. 2021 in this volume). The Ilısu Dam is estimated to displace as high as 78,000 people, excluding up to 30,000 nomads living in the close vicinity of the dam (Eberlein and Walicki 2017, pp. 2–5). Displaced persons face many challenges. For example, 72% of the people who were displaced due to the Atatürk Dam described their economic conditions as worse compared to before, while only 20.5% said they were better off (Güler and Çırakoğlu 2014, p. 12). It is widely argued that especially the Kurds with no choice but to migrate to western cities face more challenges because they also face discrimination and often lack choices except working in construction, tourism, and textile sectors (Koç 2015). To put it differently, "the promise that modernization would provide a pathway from country to city, and from farm to factory, has proven to be a mirage" for thousands of people because their old set of relations that enabled them to live and work were gone and the new ones were far from providing a viable livelihood (Li 2014, p. 3).

From a gender perspective, despite the efforts to include women in the development process, women continue to be socio-spatially excluded. Their empowerment remain limited; they largely continue to be under the "decision-making authority of the husbands, brothers, or fathers" (Harris 2006, p. 21). They are still expected to carry out unpaid domestic tasks such as preparing food, making handicraft, and looking after children as well as to work in informal and labor-intensive jobs such as weeding and picking crops in the fields. Furthermore, irrigated farming has brought more work burden for women, especially in the fields. Modernization paradigm admits that some groups lose when agriculture intensifies and becomes competitive, yet also claims that former farmers will become workers who sell their labor (Li 2014, p. 2). In this case, some landless women believe that they are "left more backward" after the shift to irrigated farming (Harris 2006, p. 7). Even though some women, especially seasonal workers travel less to other cities to find temporary jobs during harvest seasons, some others like those in Adıyaman continue to travel for jobs, often without their consent (p. 8). In addition, the already low number of women holding land titles remain steady. The number of women who take on roles within the Water Users Associations, established to engage farmers into the water management process in a democratic and participatory manner in the 1990s, also remain very limited. In other words, the participation of women in the local decision-making mechanisms remain low and local political, economic, and social institutions continue to be dominated by powerful [male] elites (Harris 2006). Also, child marriages remain widespread in the region; the region's average was still higher than the country's average in the 2010s (T.C. Aile ve Sosyal Politikalar Bakanlığı 2015, p. 36). ÇATOM activities become perceived as the state's assimilation and containment efforts through birth control programs or literacy courses offered only in Turkish language (Özok-Gündoğan 2005). Rather than GAP-BKİ or ÇATOM arguably the Kurdish women's movement has empowered women in the region (Çağlayan 2012).

Since the early stages of GAP, many high ranking military officials as well as top-level politicians and bureaucrats have been vocal about their expectation of GAP to bring order to the region and eventually end the Kurdish conflict (Bilgen 2018a). However, apparently the policy of "doing politics via development" has failed to provide a quick fix to the conflict, and even paved the way for the (re-)politicization of conflictual issues and processes inherent in the insurgency. The "Turkishness" and so-called homogenized characteristics of the region have been emphasized throughout the project. Anatolia had become the imagined geography of Turkishness after 1923, yet its peoples unaware of its consequences. In 1933, the architect Abdullah Ziya had remarked that "there are brothers who have forgotten their language and talk another language. There are brothers who consider it an insult if you called them Turk. It is our responsibility to construct their villages and to make our brothers talk, dress, and live like us" (Ziya 1933). Designing for the nation, through village plans in the 1930s and 1940s, or the construction of dams from the 1970s onward, did not yield the results attributed to it, and the social projects did not manage to turn Kurds into Turks. As a social engineering project, GAP has simply failed.

82.5 Conclusion

For the past 40 years, GAP has functioned as an essential instrument for the Turkish state to utilize the water and land resources of the GAP region toward achieving the desired level of national development. Following the expansion of scope and coverage of GAP in the ensuing decades, the project has been given a wider set of goals. The project administration became responsible for an erasure of the "problematic" aspects of the region and been assigned a grandioso role to "change the ill-fate of the region."

Even after 40 years, the question remains: is GAP a success or failure story? Depending on where one sits, one can conclude that the glass is half full. In the state's discourse, the project is represented as a success story in terms of radically reorganizing the political, social, economic, and cultural landscape of the region and redefining the characteristics of the ideal citizen model in the region. From another perspective, one can also conclude that the glass is half empty. It might look like not only the impact of the project on the development of the region and betterment of its local population has been quite limited, but also the project has rather been a disappointment for millions of people because the delivery of its promises was based on a negation of their (Kurdish) identity.

Beyond these opposing views, though, a third way to assess the past, present, and future implications of GAP would be to question whether the glass has actually been *malformed* from the beginning and whether such comprehensive development packages emblematize a colonial project of resource extraction and assimilation in the guise of a positive, benign, and transformative power (see Akıncı et al. 2020). Thus, unpacking GAP and similar development projects and understanding what kind of—overlooked—impacts they have on the voiceless and disempowered non-state actors would not only be possible, but also open up new theoretical, practical, and methodological avenues that would enable a more inclusive, empowering, and emancipatory development process for all in the future.

References

Akıncı ZS, Bilgen A, Casellas A, Jongerden J (2020) Development through design: knowledge, power, and absences in the making of southeastern Turkey. Geoforum 114:181–188

Benek S (2009) Ortaya çıkışı, gelişme seyri ve bölgeye etkileri bakımından Güneydoğu Anadolu Projesi (GAP). Ankara Üniversitesi SBF Dergisi 64(3):45–71

Bilgen A (2018a) A project of destruction, peace, or techno-science? Untangling the relationship between the southeastern Anatolia project (GAP) and the Kurdish question in Turkey. Middle East Stud 54(1):94–113

Bilgen A (2018b) The southeastern Anatolia project (GAP) in Turkey: an alternative perspective on the major rationales of GAP. J Balkan Near East Stud. https://doi.org/10.1080/19448953.2018.1506287

Bilgen A (2018c) The southeastern Anatolia project (GAP) revisited: the evolution of GAP over forty years. New Perspect Turk 58:125–154

Bilgen A (2018d) Turkey's southeastern Anatolia project (GAP): a qualitative review of the literature. British Journal of Middle Eastern Studies. https://doi.org/10.1080/13530194.2018.1549978

Bozdoğan S, Kasaba R (1997) Rethinking modernity and national identity in Turkey. University of Washington Press, Seattle & London, pp 15–37

Çağlayan H (2012) From Kawa the blacksmith to Ishtar the goddess: gender constructions in ideological-political discourses of the Kurdish movement in post-1980 Turkey. Eur J Turk Stud 14

Çolakoğlu H, Çokuysal B, Çakıcı H (2005) Türkiye'de gübre üretimi ve tüketimi. Tarımda Potasyumun Yeri ve Önemi Çalıştayı, Eskişehir, Türkiye, pp 13–19

Dikme Y (2017, December 27) GAP çiftçisinin kuraklık endişesi. *Milliyet*. http://www.milliyet.com.tr/gap-ciftcisinin-kuraklik-endisesi%2D%2Dyerelhaber-2492964/

DPT (1989a) Güneydoğu Anadolu Projesi Master Plan Çalışması: Master Plan Nihanı Raporu (Cilt 2)

DPT (1989b) Güneydoğu Anadolu Projesi Master Plan Çalışması: Master Plan Nihanı Raporu (Cilt 3)

DPT (1989c) Güneydoğu Anadolu Projesi Master Plan Çalışması: Master Plan Nihanı Raporu (Cilt 4)

DPT (1990) Güneydoğu Anadolu Projesi Master Plan Çalışması: Master Plan Nihanı Raporu (Yöneticiler İçin Özet)

Eberlein C, Walicki N (2017) Lessons not learned: Turkey's Ilısu Dam. International Displacement Monitoring Centre (IDMC)
GAP-BKİ (1993) GAP: 1992 Durum Raporu. GAP-BKİ, Ankara
GAP-BKİ (1994) GAP: 1993 Durum Raporu. GAP-BKİ, Ankara
GAP-BKİ (1996) GAP: 1996 Yarı Yıl Durum Raporu. GAP-BKİ, Ankara
GAP-BKİ (1997) Güneydoğu Anadolu Projesi'nde Son Durum (Nisan 1997). GAP-BKİ, Ankara
GAP-BKİ (2001) GAP: Son Durum 2001. GAP-BKİ, Ankara
GAP-BKİ (2002a) *GAP Bölge Kalkınma Planı: Ana Rapor* (Cilt 2). GAP-BKİ, Ankara
GAP-BKİ (2002b) GAP'ta Son Durum. GAP-BKİ, Ankara
GAP-BKİ (2007) GAP: Son Durum. GAP-BKİ, Ankara
GAP-BKİ (2008) GAP Eylem Planı (2008–2012). GAP-BKİ, Ankara
GAP-BKİ (2012) GAP'ta Son Durum. GAP-BKİ, Ankara
GAP-BKİ (2014) GAP Eylem Planı (2014–2018). GAP-BKİ, Ankara
GAP-BKİ (2016) GAP Son Durum. GAP-BKİ, Ankara
GAP-BKİ (2017) Performans Programı:2017
GAP-BKİ (2018) GAP Son Durum. GAP-BKİ, Ankara
Glass S (2017) Twisting the tap: water scarcity and conflict in the Euphrates-Tigris River basin. Independent Study Project (ISP) Collection 2594
Gökçe B, Kasapoğlu A, Kaya NC, Güler Z (2010) Bölgesel kalkınmanın can suyu GAP: Karşılaştırmalı sosyal ve ekonomik yapı araştırması. GAP-BKİ & Sosyoloji Derneği, Ankara
Görgü E (2006, January 10). 30 yılda GAP gerçeği. *Evrensel*. https://www.evrensel.net/haber/168524/30-yilda-gap-gercegi
Güler Z, Çırakoğlu D (2014) Barajlar ve yeniden iskanın tarıma etkileri: Samsat Örneği. In: XI. Ulusal Tarım Ekonomisi Kongresi Doğal Kaynaklar. Enerji Ekonomisi ve Politikaları, Çevre, pp 10–15
Harris LM (2006) Irrigation, gender, and social geographies of the changing waterscapes of southeastern Anatolia. Environment and Planning D: Society and Space 24(2):1–35
Harris LM (2008) Modernizing the nation: Postcolonialism, postdevelopmentalism, and ambivalent spaces of difference in southeastern Turkey. Geoforum 39:1698–1708
Jafar MR (1976) Under-underdevelopment: a regional case study of the Kurdish area in Turkey. Helsinki, Painoprint Oy
Jongerden J (2007) The settlement issue in Turkey and the Kurds. Brill, Leiden & Boston
Jongerden J (2010) Dams and politics in Turkey: utilizing water, developing conflict. Middle East Policy 17(1):137–143
Jongerden J, Akıncı Z, Ayboğa E (2021) Water, politics and dams in the Mesopotamia Basin of the northern Middle East: how Turkey instrumentalises the southeastern Anatolia project for political, military and strategic interests. In: Jawad LA (ed) Tigris and Euphrates Rivers: their environment from headwaters to mouth. Springer, New York
Kasarcı R (1996) Türkiye'de nüfus gelişimi. Ankara Üniversitesi Türkiye Coğrafyası Araştırma ve Uygulama Merkezi Dergisi 5:247–266
Kibaroğlu A (2006) GAP: a grand design for sustainable development. Economy and Territory 1:178–180
Koç F (2015) 1985–1995 Kürt göçü ve Türkiye'nin büyük proleterleştirme süreci. DİSKAR 4:74–85
Konak N (2013) Development and the southeastern Anatolia project (GAP): an ecofeminist inquiry. In: Dönmez RÖ, Özmen FA (eds) Gendered identities: criticizing patriarchy in Turkey. Lexington Books, Plymouth, pp 51–66
Kudat A, Bayram M (2000) Şanlıurfa-Harran Plains on-farm and village development project. In: Kudat A, Peabody S, Keyder Ç (eds) Social assessment and agricultural reform in Central Asia and Turkey. The World Bank, Washington, DC, pp 255–302
Li TM (2007) The will to improve: Governmentality, development, and the practice of politics. Duke University Press, Durham

Li TM (2014) Land's end: capitalist relations on an indigenous frontier. Duke University Press, Durham

Özok-Gündoğan N (2005) "Social development" as a governmental strategy in the southeastern Anatolia project. New Perspect Turk 32(32):93–111

Presidency of the Republic of Turkey Investment Office (2019) *Investment Zones*. http://www.invest.gov.tr/en-US/investmentguide/investorsguide/Pages/SpecialInvestmentZones.aspx

Scott J (1998) Seeing like a state: how certain schemes to improve the human condition have failed. Yale University Press, New Haven & London

Şahin G (2016) Türkiye'de gübre kullanım durumu ve gübreleme konusunda yaşanan problemler. Tarım Ekonomisi Dergisi 22(1):19–32

T. C. Aile ve Sosyal Politikalar Bakanlığı (2015) *Türkiye'de evlilik tercihleri*

Tekeli İ (2008) Türkiye'de bölgesel eşitsizlik ve bölge planlama yazıları. Tarih Vakfı Yurt Yayınları, İstanbul

Turgut H (2000) GAP ve Demirel. ABC, İstanbul

Türkiye Elektrik İletim A.Ş. Genel Müdürlüğü (2017) 10 yıllık talep tahminleri raporu (2018–2027)

Türkiye İstatistik Kurumu (2017) *Eğitim istatistikleri*. www.tuik.gov.tr

Warner J (2012) The struggle over Turkey's Ilısu dam: domestic and international security linkages. International Environmental Agreements: Politics, Law and Economics 12(3):231–250

Willcocks SW (1910) Mesopotamia: past, present, and future. Geogr J 35(1):1–15

Yükseköğretim Kurulu (2018) Üniversitelerimiz. *Yükseköğrenim Kurulu*. http://www.yok.gov.tr/web/guest/universitelerimiz

Ziya A (1933) Köy mimarisi. Ülke 9:37–41

Part IX
Food Security

Chapter 83
How Possible to Build Rice–Fish Farming in Iraq in order to Support Food Security Plan: Positive and the Negative Aspects

Laith A. Jawad

Abstract Growing rice as only product in the field cannot offer a maintainable food source devoid of a cost to long-term habitat durability. Alternatively, combined rice–fish farming has shown to play a vital part in growing food yield as the joined farming system is better than culturing rice alone in terms of resource utilization, variety, yield, and both the quality and quantity of the food produced. Integrated rice–fish rearing too provides various socioeconomic and habitat benefits. In Iraq, unified rice–fish farming has not been introduced and government should develop such program to learn lesson from other countries who tried it before such as Bangladesh. Many studies have shown that combined rice–fish culturing can aid any country keeping speed with the current request for food through rice and fish yield but entails larger reassurance if it is to understand its full prospective.

As with any other project, there are a number of factors that influencing the adoption of the project. In case of the integrated rice–fish farming, the socioeconomic issues such as family size, number of associates with an extension agent, participation in extension-education activities, membership in social institutes, and the existence of farmworkers were the most imperative agents for the acceptance of rice–fish farming system.

Iraq has the potentials to develop the united rice–fish farming correctly if all the mistakes that other countries fall in are avoided. There are plenty of water resources in Iraq, fields of rice farming, and high possibility of obtaining fingerlings. In addition, Iraq has over 20 universities that graduate hundreds of scholars in different aspects of environmental science and fisheries who can educate the farmers about the integrated rice–fish farming. With such aim, Iraq easily can secure food for future generations.

L. A. Jawad (✉)
School of Environmental and Animal Sciences, Unitec Institute of Technology, Auckland, New Zealand

L. A. Jawad (ed.), *Tigris and Euphrates Rivers: Their Environment from Headwaters to Mouth*, Aquatic Ecology Series 11,
https://doi.org/10.1007/978-3-030-57570-0_83

83.1 Introduction

Culturing rice in combination with rearing fish is an ancient exercise in Asia (Rabbani et al. 2004; Tasnoova et al. 2008). The mixing of fish into rice farming offers vital protein, particularly for existence farmers who achieve rain-fed agricultural systems. The proposal of rice–fish culturing taking place to being prevalent in the 1980s (Nabi 2008). The new technique was assumed to have probable for manifold environmental profits in Asia. Several studies recommended that united rice—fish culturing is ecologically inclusive because fish recover soil fertility by producing nitrogen and phosphorus (Dugan et al. 2006). Looking for food by fish in rice fields leads to aerate the water. Fish also involve in regulating pests by eating aquatic weeds and algae that carry diseases, act as hosts for pests, and compete with rice for nutrients. Furthermore, fish consume flies, snails, insects, and can help control malaria mosquitoes and waterborne diseases (Matteson 2000). Alternatively, rice fields deliver fish with planktonic, periphytic, and benthic food (Mustow 2002). Covering by rice plants also upholds the water temperature suitable to fish during the summer (Kunda et al. 2008). Several fish species favor the rice fields for their reproduction (Halwart 1998). The natural combination of culturing rice and fish has encouraged rice–fish farming to increase yield (Gurung and Wagle 2005).

At the farm level, rice–fish integration reduces the use of fertilizers, pesticides, and herbicides in fields. This decrease in charges drops the farmer's budget load and upsurges their extra revenue from the sale of fish (Ahmed and Garnett 2011). Likewise, united rice–fish culturing offers greater rice product and raises a higher gross margin than a sole rice culturing system (Das 2002).

No attention has been given to the ecological and economical features of the rice–fish scheme in Iraq. Even though the possibility for rice–fish culturing in Iraq is available, it has not been assumed by farmers. Noorhosseini-Niyaki and Allahyari (2012) suggested that owing to socioeconomic, habitat, technological, and institutional restraints, culturing only rice persists the primary culturing system in countries such as Iran in spite of the united rice–fish culturing signifies a more well-organized use of resources. Assumed this, it is serious that attempts should be made to upsurge the consciousness among the farmers regarding the role of fish in biocontrol, the upkeep of soil health, and nutrient recovering (Saikia and Das 2008).

The chapter in hand discusses the applicability of developing rice–fish culturing scheme in Iraq and what chances and restraints to upsurge food source for the country. As with other countries that adopted this technology, it should be hypothesized that rice–fish culture, in specific united, rice–fish farming can make improved use of obtainable contributions and that it can offer socioeconomic and nutritional welfares to the households of poor farmers and, more generally, food security in Iraq. This can be tested by evaluating yield productivity of culturing rice alone and rice–fish culturing households through a two-stage tactic.

83.2 Rice Agriculture in Iraq

The rice (*Oryza sativa* L.) is one of the highly significant cereal crops and helps as the primary foundation of important food for more than half of the global population (Emani et al. 2008; Jiang et al. 2013).

Rice as the main food crop in Iraq comes after wheat in agricultural area and yield. The area for planting rice in 2006 was 125,641 ha, giving yield of 363,338 tons of paddy with an average yield 2892 kg ha^{-1} (MOP 2007). Iraqi rice farmers grew rice (conventional method) according to the cultural performs passed through generations. This method is the dry method with large areas, using large amounts of seed (about 160 kg ha^{-1}) and growing rice in interchange with wheat, which has caused a reduction in soil fertility (El-Hakim et al. 2017).

The conservative technique of rice irrigation in Iraq is nonstop submergence throughout the rice cycle, which needs large amounts of water (Saleh et al. 1999) assessed at 70,000 m^{-3} ha^{-1} (Ito 1965). This also has an undesirable impact on the habitat since the discarding of large quantities of drainage water that comprise residual agrochemicals and herbicides, which lead to pollution of water and soil (Willingham et al. 2008).

The conservative approval for rice irrigation in Iraq is to keep a 10-cm layer of water on the field throughout the crop cycle. This needs large amounts of water, besides having an adverse impact on the habitat due to the large quantities of water are removed from natural ecosystems, and large quantities of drainage water are thrown of carrying agrochemicals that cause environmental pollution (Saleh et al. 1999).

Water for the purpose of agricultural use is a restrictive agent for rice culturing and growth in Iraq. Increased water is obligatory in the summer season, and there is a growing water shortage in Iraq, as most of the country is frequently situated in semi-arid zone (Hameed et al. 2011). The agricultural area in Iraq lies sideways and between the Tigris and Euphrates Rivers ranging from the country's central region south-eastward to the marshlands of the Tigris–Euphrates Delta. Small amount of rain falls in the center-south zone of Iraq and agriculture in this region is dependent on irrigation.

83.3 Aquaculture in Iraq

In this section, only a short review of the aquaculture industry in Iraq will be given because there is a separate chapter devoted to this subject included in this book.

The total area allocated for aquaculture production in Iraq is estimated to be 7500 ha. The mean annual yield for 1986–1997 was 4000 t. In 1998, production is reported to have been increased to about 7500 t. A total of 1893 fish farms are approved for aquaculture, all operated by the private sector. The only system of fish culture is in earth ponds. Aquaculture in Iraq rests on freshwater supply, with no marine aquaculture performed (Kitto and Tabish 2004).

Adequate hatcheries are obtainable, although most yield is of common carp. Culture cages extended in the early 1980s Habania Lake, but was ultimately left for commercial production, restraining its use for research (El Gamal 2001). The latest information available indicates that the total area belong to fish culturing is estimated at 7500 ha consisting of about 1900 fields. These are earthen fishponds without appropriate insert or insulation (Kitto and Tabish 2004). Only the Babel fish farm, a government-owned venture, is an integrated field that is fully insulated and well equipped, built on a 500-ha area. All other fields are smaller, operated by private companies and individuals.

Water resources used for fish farming are mainly originated from rivers, lakes, marshes in addition to rain, underground and marine waters (Delphy 2008). All these resources are widely used at present in fish farming activities, in addition to the limited use of underground waters. The mountain areas of Northern Iraq at the border with Iran and Turkey and due to the low temperature provide suitable conditions for raising cold-water species such as salmon and trout (Abdulrahman et al. 2017).

Recent study showed that the consumption of water by various activities on Euphrates, Hargreaves (2012) as little as 1.4% of Euphrates water (taking into consideration the evaporation and seepage effects). The financial revenue of these projects (Iraqi Dinar ID/cum) exceeds the revenue of rice production by 6–10 turns, 7–8 turns for corn, and 11–12 for wheat revenue. So, it seems that aquaculture projects are more profitable than crop production with limited water consumption compared with other agricultural products.

Fish cultivation activities in Iraq focus on fish species, common carp (*Cyprinus carpio*), grass carp (*Cetenopharyngodon idella*), and silver carp (*Hypophthalmichthys molitrix*) which comprise the main cultivated species in Iraq. Nowadays, common carp is considered among the chief economic fish species in the inland Iraqi waters and also as the main cultivated species. On the other hand, grass carp was introduced in 1968 from Japan (Shireman and Smith 1983; Al-Sheikh et al. 1990). Experiments to adapt some local cyprinid species such as Bunni (*Mesopotamichthys sharpeyi*) and Gattan (*Luciobarbus xanthopterus*) to be a cultivated species were made, but without success due to slow growth and high production costs) Draft 1983; Al-Nasih 1992; Salman et al. 1997; Al-Rudainy et al. 1997).

83.4 Socioeconomic Benefits of the Rice–Fish Farming within the Food Security Process

Ahmed and Garnett (2011) in their research discussed the socioeconomic impacts of the rice–fish combined culture. Here, I thought it is a good idea to summarize what they suggested as what they have said will encourage farmers to change to rice–fish culture. Fish culturing in rice fields has also been related to considerable social welfares, liable on yield systems. Affirmative influences seem to be maximum among farmers intricate in combined rice–fish culturing who incline to eat more food, have higher standards of living and better buying ability than other farmers. Due to the mixing, the culture of fish has augmented rice production. Such an upsurge has also augmented obtainability of paddy straw, using for construction houses, as cooking fuel, and as food for cattle. As a result of silage obtainability, milk and cow dung are accessible and inexpensive in the study area. Cow dung is utilized as a cooking fuel and as fertilizer, while milk is viewed as appropriate food for children. Farmers have also enhanced their housing situations as revenue from rice–fish culturing has been utilized to substitute bamboo roofs and walls with tin sheets and wood.

83.5 The Negative Side of the Rice–Fish Farming within the Food Security Process

Each project has its positive and negative sides and the rice–fish farming is not different. In the section above, the positive side which is represented by the socio-economic factors was discussed briefly and examples from the trial of Bangladesh were given. In this section, the negative side of the project will be discussed and what issues and cares need to be taken will be briefly discussed and pointed out.

83.5.1 Economic Problems in Rice–Fish Farming

83.5.1.1 Initial Financial Problems

As in any other new projects, some financial problems will be formed. In order to overcome these problems, assistance and help from other agencies for farmers that have used rice–fish culture should be available and fishery organizations should support producers.

83.5.1.2 Diet Difficulties in Combined Rice and Fish Culturing Fields

Among the problems that might arise with the rice–fish culture is providing diet for the fish. The new technique and the facility it needs to play an important role for the farmers to adapt with. These farmers need to know exactly how to raise fish and what needed to be done quickly and at low cost. Among these problems is delivering nourishment for the fish in the farm. Experts from the fisheries organizations should show the farmers how to take care of the fish in the rice farms.

83.5.1.3 Loss of Fish in Combined Rice and Fish Culturing Fields

To the majority of farmers, the problem of losing fish is trivial. Losing fish through transporting fingerlings or through diseases are the main two ways the fish can be lost with by the farmers due to lack of knowledge in handling fishes.

83.5.1.4 Difficulties of Fish Seed in Rice–Fish Culturing

The supply of fingerlings to farmers in good and healthy status are among the factors that affect the production of fish in the farm. The supply of the fingerlings should be continuous and regular in the time they ordered by the farmers. Such regularity will ensure that the farm is continuously having fish to grow. Kapanda et al. (2005) observed that the farmers in the Mchinji rural area of Malawi did not have admittance to the accessible fish seeds. This demonstrates to be costly in consideration of transport and death is great owing to the strain that the Youngs are exposed to. This situation is not maintainable. The fishery department need to teach some farmers in the area to yield Youngs of the fish, who in turn sell to other farmers.

83.5.1.5 Aquatic Difficulties in Rice–Fish Culturing

Matters with water supply and quality may face farmers of rice–fish as sometimes shortage of water, unsuitable water temperatures, and pollution of water from upstream farms by pesticides were giving difficulties to their farms.

83.5.1.6 Biota in Rice–Fish Culturing

Problems generated from the wildlife are not unusual for the farmers of rice–fish. The occurrence of snakes, turtles, and birds eating fish in their paddy fields, which generate glitches. To avoid such difficulties, some farmers may surround the ponds with plastic bags, others sheltered the top of the pond with a plastic grid or used a scarecrow. Others used an electric fence to stop wild animal's incoming to the pond. Generally, the enclosure of ponds with a fence is a suitable solution.

83.5.1.7 Deficiency of Information and Skill in Rice–Fish Culturing

The lack of expert staffs that give directions to the farmers will generate problems. Most farmers may not have the experience to deal with the technique of culturing rice and fish at the same time and therefore they fall in mistakes that could be large enough to cost a large amount of money. It has been recommended that some educational films and courses should be delivered.

83.5.1.8 Deficiency of Oxygen in Combined Rice and Fish Culturing Scheme

One of the major problems that face the farmers in the field is the lack of oxygen or hypoxia. One of the solutions the farmers have used is use ducks to recover oxygen content. Islam et al. (2004) in Bangladesh stated that diverse constituents of rice, duck, and fish in a polyculture scheme were profited each other. Such a structure is linked with the oxidation and reduction. Consequently, the duck dropping was found very active for better growth and development of fishes and rice.

83.6 Conclusions

In some parts of Iraq, from the Southwestern Bedouins to the Northeastern Kurds, old-style self-adequate food schemes have been sustained. However, in furthest of the country, food security has desirable importations both carbohydrates and proteins. Vitamins and micronutrients have all also frequently been in short source.

Located in what is traditionally known as the Fertile Crescent, Iraq is provided with plentiful amounts of water from the Tigris and Euphrates. Adding to this, there are water assets in the form of lakes and ponds, particularly in the northern part of the country and the Shatt Al-Arab delta area in the south. Proof to this is the quantity of growth that took place in Iraqi fish farming before 1990. Nowadays, more than ever, as the country looks for ways to feed itself, aquaculture may be an area of solemn development.

In order to meet the mounting demand for food, there is a need for the start and improve rice and fish production in Iraq as in other countries such as Bangladesh. Several studies have concluded that rice–fish mixing might be a feasible choice for variation. These field variations will improve food safekeeping.

The estimates for combined rice–fish farm growth in Iraq are substantial but such effects need to have full support from different parties. A variety of public and private sector reserves and enterprises are desirable to comprehend the possibility for growth of this mixing. These associations have already underwritten to food security in many growing countries (Spielman and von Grebmer 2004). Moreover, additional studies would be compulsory on social, economic, habitat, and living matters for the implementation of rice–fish culturing in rural Iraq.

References

Abdulrahman NM, Ahmed VM, Ameen HJ, Salih HA, Latif BM, Ali SS, Ahmed AA (2017) Comparative study of on fish health indices of rainbow trout (Oncorhynchus mykiss) reared in two sites in Sulaimani governorate, north of Iraq. Biol Appl Env Res 1(2):228–236

Ahmed N, Garnett ST (2011) Integrated rice-fish farming in Bangladesh: meeting the challenges of food security. Food Security 3(1):81–92
Al-Nasih MH (1992) Preliminary observations related to the culture of *Barbus sharpeyi*. J Aqua Trop 7:69–78
Al-Rudainy AJ, Salman NA, Al-Mahdawi GJ, Ruhaig AM, Fadil AA, Farhan RK, Abbas IK (1997) Intensive culture of Gattan (*Barbus xanthopterus*) in earthen ponds. Basrah J Agr Sci 10 (1):25–31
Al-Sheikh MA, Yamur QM, Alloos SB (1990) Fish culture and production. Part II. Al-Hikma Press, Baghdad
Das DN (2002) Fish farming in rice environments of north eastern India. Aquaculture Asia 7 (2):43–47
Delphy M (2008) Fish production in Iraq. Al-Sabah Newspaper 1410:9
Draft (1983) Detailed report of developing fisheries in, Tharthar, Habbaniyah and Razzazah. Part X: Production of Bunni and Gattan stocking materials in Tharthar Lake and fattening of carp in Lake Habbaniyah. Inland Fisheries Institute, Poland
Dugan P, Dey MM, Sugunan VV (2006) Fisheries and water productivity in tropical river basins: enhancing food security and livelihoods by managing water for fish. Agric Water Manag 80 (1–3):262–275
El Gamal AR (2001) Status and development trends of aquaculture in the near east. In: Subhasinghe RP, Bueno P, Phillips MJ, Hough C, McGladdery SE, Arthur JR (eds) Aquaculture in the third millennium. Technical proceedings of the conference on aquaculture in the third millennium, Bangkok, Thailand, 20–25 Feb 2000. NACA, Bangkok and FAO, Rome, pp 357–376
El-Hakim A, Ahmed M, El-Sherbiny AE, Dahdouh SM, Galal YGM (2017) Role of bioeffectors and soil ameliorates on cowpea yield grown on saline soil with aid of 15N isotope dilution technique. Zagazig J Agric Res 44(6):2105–2115
Emani C, Jiang, Miro B, Hall TC, Kohli A (2008) Transgenic cereals and forage grasses. In: Kole C, Hall TC (eds) Compendium of transgenic crop plants, pp 1–234
Gurung TB, Wagle SK (2005) Revisiting underlying ecological principles of rice-fish integrated farming for environmental, economic and social benefits. Our Nature 3:1–12
Halwart M (1998) Trends in rice-fish farming. FAO Aquacult Newsl 18(3):C11
Hameed KA, Mosa AKJ, Jaber FA (2011) Irrigation water reduction using system of rice intensification compared with conventional cultivation methods in Iraq. Paddy Water Environ 9 (1):121–127
Hargreaves J (2012) Water resources use for aquaculture in Iraq. USAID Inma Agribusiness Program:20
Islam SS, Azam MG, Adhikary SK, Wickramarachchi KS (2004) Efficiency of integrated rice, fish and duck polyculture as compared to rice and fish culture in a selective area of Khulna District, Bangladesh. Pak J Biol Sci 7(4):468–471
Jiang SH, Zhou H, Lin DZ, Dong YJ, Ye SH, Zhang XM (2013) Identification and gene mapping of a thermo-sensitive leaf-color mutant at seedling stage in Rice. Chin J Rice Sci 27:359–364
Kapanda K, Matiya G, N'Gong'Ola DH, Jamu D, Kaunda EK (2005) A logit analysis of factors affecting adoption of fish farming in Malawi: a case study of Mchinji rural development program. J Appl Sci 5:1514–1517
Kitto MR, Tabish M (2004) Aquaculture and food security in Iraq. Aquaculture Asia 9:31–31
Kunda M, Azim ME, Wahab MA, Dewan S, Roos N, Thilsted SH (2008) Potential of mixed culture of freshwater prawn (Macrobrachium rosenbergii) and self-recruiting small species mola (Amblypharyngodon mola) in rotational rice–fish/prawn culture systems in Bangladesh. Aquac Res 39(5):506–517
Matteson PC (2000) Insect pest management in tropical Asian irrigated rice. Annu Rev Entomol 45 (1):549–574
MOP (2007) Annual statistical abstract 2007. Central Bureau of Statistics and Information Technology. Directorate of Agricultural Statistics, Ministry of Planning and Development Cooperation, Baghdad

Mustow SE (2002) The effects of shading on phytoplankton photosynthesis in rice-fish fields in Bangladesh. Agric Ecosyst Environ 90:89–96

Nabi R (2008) Constraints to the adoption of rice-fish farming by smallholders in Bangladesh: a farming systems analysis. Aquaculture Econ Manag 12:145–153

Noorhosseini-Niyaki SA, Allahyari MS (2012) Logistic regression analysis on factors affecting adoption of rice-fish farming in North Iran. Rice Sci 19(2):153–160

Rabbani MG, Islam MS, Hossain MI, Hossain TMB, Begum MEA (2004) An economic study on alternate rice-fish culture in selected areas of Mymensingh District in Bangladesh. Pak J Biol Sci 7(5):685–688

Saikia SK, Das DN (2008) Rice-fish culture and its potential in rural development: a lesson from Apatani farmers, Arunachal Pradesh, India. J Agriculture Rural Develop 6(1):125–131

Saleh RO, Jadooa KA, Hameed KA (1999) Water use efficiency under intermittent irrigation for rice (Oryza sativa L.) in Iraq. In: *Paper for international conference on water resources management, use and policy in dry areas* (pp. 1–3). Amman, Jordan, pp 1–3

Salman NA, Al-Mahdawi GJ, Al-Azawi AH, Rassoki RH, Abbas LM, Al-Rudainy AJ, Al-Timimi MT, Faisal SM (1997) Acclimation and culture of common carp (*Cyprinus carpio* L.) and Gattan (*Barbus xanthopterus*) in drainage water using floating cages. Marina Mesopotamica 12 (1):169–188

Shireman JV, Smith C (1983) Synopsis of biological data on the grass carp *Ctenopharyngodon idella* Cuvier & Valenciennes 1844. Fish Synopsis 135:86

Spielman DJ, von Grebmer K (2004) Public-private partnerships in agricultural research: an analysis of challenges facing industry and the consultative group on international agricultural research. Int Food Policy Res Inst 113

Tasnoova S, Iqbal KM, Iwamoto I, Haque MM (2008) Economic performance of fish based farming systems in Bangladesh. J Fish Aquat Sci 3(3):206–212

Willingham SD, Falkenberg NR, McCaulem GN, Chandler M (2008) Early post-emergence clomazone tank mixes on coarse-textured soil. Weed Technol 22(4):565–570

Chapter 84
The Importance of Non-commercial and Small-Sized Fish Species: A Proposal for an Additional Revenue to Iraq

Laith A. Jawad

Abstract Iraq is one of the fortunate countries in the Middle East that its territories encompass large quantities of freshwater represented by the three rivers, Euphrates, Tigris and Shatt al-Arab and their tributaries. In addition, Iraq has a limited coastal area that can supply good marine resources.

With such aquatic wealth, a number of non-commercial species of fish and other aquatic organisms are high and goes to waste sector during the fisheries management. This amount of fish and other aquatic organism's waste can be turned into valuable products and can be considered an additional income to Iraq besides oil and vegetation.

This chapter has shown the usefulness of the non-commercial species and gave examples of the approaches that can be reached in order to make use of the wealth of this type of aquatic resource and change them to profits that will enhance the economy of the country.

84.1 Introduction

Iraq is one of oil-producing and exporting countries. The oil revenue is considered the main income for this country. Besides the oil, Iraq is also rich in its water resources, cultivation and aquatic life. Until recently, Iraq was one of the countries that export different agricultural products to the countries of the Middle East region. This commercial activity has declined due to the decrease in amount of water in the rivers, change in climate and factor of desertification.

For a certain degree, the fisheries' wealth, both freshwater and marine, remained to be one of the revenue to Iraq at the local market, where a large number of freshwater and marine fish and other aquatic organisms are available in the Iraqi waters.

L. A. Jawad (✉)
School of Environmental and Animal Sciences, Unitec Institute of Technology, Auckland, New Zealand

L. A. Jawad (ed.), *Tigris and Euphrates Rivers: Their Environment from Headwaters to Mouth*, Aquatic Ecology Series 11,
https://doi.org/10.1007/978-3-030-57570-0_84

The Iraqi people are considered among the people that have a selective test towards the freshwater and marine products. For example, all shellfishes have no local market in addition to several other fish species that are considered delicacy and can be considered a national revenue if good management for this sector is performed.

This chapter overview fish resources that are not subject to trade and thus are not accounted for in economic statistics and make use of them in Iraq. The finfish and shellfish resources of both marine and freshwater will be dealt with to show their importance as an effective or potential food supply. Also, the importance of non-commercial fish for human consumption is pointed out. In developing countries, population expansion combined with poor economic conditions and limited employment opportunities lead to the development of low-cost subsistence fisheries (Blaber 1997).

This chapter also highlighted some industrial ideas that can be taken by private enterprises or governmental agencies for the use of non-commercial species. If establishing the factories and other required structures, the export of the non-commercial species to those countries that can make use of them is always an alternative solution. In both ways, an additional revenue will be added to Iraq in addition to those already present and at the same time, no waste of the national aquatic resources was implemented.

84.2 The General Benefit of the Non-commercial Fish

Larval and juvenile fish usually have no commercial, but they are important as they provide the recruits for future harvests. Though several species of commercial interest spawn and grow at sea, the bulk of fish species collected on continental shelves have larval and juvenile phases in estuarine and littoral areas.

From an ecological perspective, the importance of certain non-commercial species in the functioning of coastal and reef systems should be stressed. Thus, they should play a consistent role in the diet of several commercial species. The contribution of the non-commercial taxa to trophic webs leading to harvestable fish resources has probably been notably underestimated. The early life history of many non-commercial species are considered an essential diet for important commercial species (Baran 2002).

The content of the most by-catch contains mostly of immature individuals of commercially valuable species, and of non-commercial species that are a food source for targeted ones. These juvenile fish could contribute to replenish stocks if allowed to mature; however, a detailed impact assessment of by-catch requires more information on natural mortality, growth rates and proportions of juveniles discarded relative to adult stocks (Baran 2002).

84.3 The Benefit of Non-commercial Fish for Humans

Fish is composed of the main commercial issue of humans since the early civilizations of Sumer, Babylon, and Assyria in the Mesopotamia (Saggs 1962). This situation continues to be the same at the present time. For several countries with high population and with limited financial resources and employment opportunities, fish gathering can be expected to be an important activity at the family level in these countries. Most of the studies have concentrated on the harvest of the commercial species by human, but studies of fish gathers of non-commercial fisheries are rare, and most of them deal with the Pacific or the Indian Ocean reef systems.

The picture of human dependence of fisheries can vary between locations around the world. Baran (2002) has reviewed this subject and came up with figures on the human dependence of sea products in some countries around the world. A summary of his review is given below: the American Samoa islands, gleaning accounted for 32% of annual fishing effort and provided 9.7% of annual finfish catches (and 78% of invertebrate's catches); this activity dominantly involved women and children. In Viti Levu Island (Fiji) in 1996, the annual catch of subsistence fisheries was amounted to 3500 tons, when the artisanal catch reached 6200 tons per year. In the Philippines, on Pamilacan Island.

In south-western Madagascar, subsistence fisheries provide 3 to 4 kg of fish per day per family, and total gathering was assessed to reach 15 to 16 tons per year.

For Freshwater non-commercial fish resources, Baran (2002) has reported that the trends are similar: scarcity of quantitative data in general and surprising amounts in the few available figures, particularly in Asia. In Cambodia, subsistence fisheries production amounts to 115–140,000 tons a year. In developed countries, studies on subsistence fisheries are not abundant and very few of them have quantified fish captures.

The need for protein input is vital in several developing countries. In Vanuatu in 1983–1984, "small-scale unstructured village fishing" provided an average of 3750 tons of fish, when artisanal fishery provided only 90 tons. This subsistence fishery provided between 16 and 18% of the population's yearly protein requirement. In the Mekong river basin, proteins of aquatic origin make up 50–75% of the diet of rural populations, and the consumption of aquatic animals (mostly fish) ranges between 20 and 64 kg per person per year. This protein input is also significant along African coasts and rivers.

There is another benefit of the non-commercial fisheries to human. This is related to the social aspects of the population. In the Philippines, the global fishing and gathering activity related to the Bolinao reef flat provided about 230 jobs per km^2. Such activity is also important to women and children. In Oceania, the contribution of women and children to total fish catches ranges from 16 to 50%. In Africa, several studies demonstrated the importance of invertebrates gathering, but apparently no study enumerated family finfishing, although mosquito net fishing and harvests were often supposed to be far from insignificant.

84.4 The Negligence of the Non-commercial Fish by Human

Scientists usually concentrate on doing researches on commercial fish species and ignoring the non-commercial ones. There were several reasons for this neglect:

1. The non-commercial species are in most time out of the scene of the research due to their small number or size. They only attract attention when an important species is present.
2. They show much variation in their geographical variation. Therefore, one species might important to another area, but not important to another.
3. The interference among fisheries was less intense.
4. The ecosystem concerns were not a management priority. The emphasis on commercial species management models and schemes did not leave much room for consideration of non-commercial species.

84.5 Non-commercial Species Utilization

Among the issue of the management of the non-commercial species is to encourage their utmost utilization. This is, of course, a sensitive and difficult issue given the history of over-exploitation of fisheries and the very real danger that adding new groups of species to the list of acceptable targets will lead to yet further depletion and other ecological consequences. This issue is further complicated by the fact that, as species become serially depleted, what was once a non-commercial soon becomes a targeted resource.

Conversely, however, the failure to make use of fish that are already killed is considered in many cultures as highly unwanted, particularly in developing countries where the supply of adequate protein to the public is a trial. In such countries, the concept of non-commercial species is often rather weak, particularly in most of the poorer fishing communities in developing countries where these species may provide an important food subsidy to the poor.

In order to reduce waste, a legislation might be introduced binding the trawlers operating in both marine and freshwater systems to land 1 tonne of non-commercial species per trip. Such legislation was applied by some countries on the by-catch (Gordon 1981 cited in Clucas 1997).

84.5.1 The Effects of the Social System on the Utilization of Non-commercial Species

In order to conserve the aquatic resources, an alteration of the feeding attitudes of the people so they can try to eat different species than they usually do. It is important to keep in mind that economic incentives are aligned with those for conserving aquatic ecosystems and communities. It is certainly true that without such an alignment and shift in values to drive changes in feeding behaviour, the effectiveness of the technical and legislative systems will be diminished.

The following examples on cases of food preference conservation are currently in practice by people in Iraq. Until recently, people living in the capital Baghdad City and further north of Iraq do not eat at all kinds of crustaceans or shellfish including shrimps and prawn. On the contrary, people who belong to "Shiat" sect of Islam do not eat any species of catfish or any species of fish without scales including sharks and rays. In general, all Iraqis do not eat octopus, squids, sea turtles, dolphins, and whales.

Perhaps the first step towards achieving correction in the taste of the people towards food is to let the public aware about the cost of catching and discarding the non-commercial species and prepare for a campaign to show the people how tasty, for example to eat species that they have never try before and even teach them through the different channels of the media about to prepare tasty meals of these species.

The waste in the economics resulted from discarding the non-commercial catch can be classified into three categories: (i) foregone income associated with discarding juvenile and adult target species; (ii) interfishery costs associated with discarding juvenile by-catch species; (iii) costs associated with measuring/estimating the levels of discards. It is also important to keep in mind that costs can also be suffered when markets are closed because of non-commercial species considerations. While economic losses might be expected to be fundamental promoters for changing the feeding behaviour of people, this does not appear to always be the case.

84.5.2 Processing the Non-commercial Species for the Local and Export Purposes

Processing the non-commercial fishes is one technique of their utilization in the proper way that will have a positive impact on the economy of the country that uses it. The processing of the non-commercial fishes can follow the need of the country that going to be exported to. For fish, South East Asia is the best market for processed food. Therefore, some suggestions of non-commercial fish species processing will be given below in order to give an idea for small enterprises to be open in Iraq to run such business.

Among the South East Asian countries that form a promised market for the processed non-commercial fish is Malaysia. In this country, traditional methods of preserving non-commercial fish include drying, smoking and fermentation. In Iraq, the possibility to obtain non-commercial fish species of both marine and freshwater origins is high and they can get for a low price and good condition (Yeoh and Merican 1978). For most dishes of the South East Asians, prawn and shrimps are the main ingredients. For this type of food, the small-sized shrimp that lives in the freshwaters of Iraq is suitable. This species of small prawn is abundant in large quantities and easy to collect.

84.5.3 Production of Fish Protein Concentrates

Fish is a source of extremely nutritive protein consumed by people around the globe. Virtually all fish species have the same nutritional quality in terms of proteins; still, only a few of them are frequently consumed, due to cultural factors, habits and other aspects such as poor appearance, unusual size and unattractive taste. Many species that are caught have no commercial value due to these deliberations, but are equally as nutritive as the desired ones. To increase commercialization they are frequently used for the making of fish meal, fish protein concentrate or fish protein hydrolyzates.

Additional supply of fish proteins is found in the waste created during the processing of commercial fish species to obtain fillets or during canning operations. In order to assist the poor people to have enough protein supplement, a great attention in the use of fish proteins from non-commercial species and/or processing wastes arose during the 1950s.

It has been known that protein sources are not consumed for their nutritive value but rather by their organoleptic properties and, unfortunately, in reduced amounts.

As has been mentioned above, the social rituals of food habits around the world are the main obstructions to the introduction of a new food as part of the everyday diet; therefore, the possibility of introducing a new source of protein, outside the economic determinants, be influenced by on its incorporation in pre-existing foods, making every effort to make its addition non-evident, or as an additional of some products with sufficient appeal. Nonetheless, fish proteins are very well employed for animal feed which lastly adds to the human food supply, either supporting the growth of animal species for food or releasing foodstuffs for human consumption.

The use of proteolytic enzymes to yield fish protein hydrolyzates as a process of utilizing non-commercial species or wastes from the fish industry released a new and striking possibility to the sectors allocating with fish protein concentrates. Presently, there is an increasing request for protein sources with satisfactory functional properties as food additives.

84.5.3.1 Fishmeal and Fish Protein Concentrate

Fishmeal is the crude flour obtained after milling and drying fish flesh (Barzana and Garía-Garibay 1994). Due to its reduced sensory quality and its absence of functional properties, its use is nearly limited to animal feed. The first technical progress for fishmeal production were attained at the end of the last century (Mackie 1983). The protein concentration is increased by removal of water and, in certain cases, of oil, bones and other materials (Finch 1977). Accordingly, the simple removal of water from fish flesh produces a product that can be considered a fish protein concentrate.

84.5.3.2 Utilization of Fish Products

Fishmeal is broadly used as animal feed and, due to its lack of functional properties and to its flavour, is scarcely used as human food. Concerning fishmeal protein concentrate is basically designed for human consumption, there is now a slight interest in its production on an industrial scale. The main problem with this product is the economic factor as the process of producing it very costly (Mackie 1983). Therefore, the development and use of this kind of product by governmental agencies or international organizations require funding and political will, which are not always available. However, intensive research has been done to include fishmeal protein concentrate in human food, in addition to the utilization of fishmeal in animal feed. The use of this product in the human diet has been mainly as a cereal supplement in bakery commodities, breakfast cereals, and macaroni, in vegetable dishes or soups, and as a meat extender. Among the suggested products that contain fishmeal protein concentrate include high-protein beverages, soups, snacks, desserts, sauces, meat extenders, and dairy. The product can also be in the form of a powder and use dry spray (Anonymous 1987).

The other product that contains fishmeal protein concentrate is succinylated fishmeal protein concentrate, which had an interesting whipping property, this form has been used successfully in dessert toppings, a souffle and a whipped gelatin dessert (Spinelli et al. 1975).

Phosphate–protein complex isolates from fish protein have been tested as egg albumin substitutes in angel cake formulations, as meat extenders in frankfurters and for the preparations of milk-like beverages (Spinelli et al. 1975).

84.5.4 Food Hydrocolloids as Additives

Polysaccharides and proteins are food hydrocolloids with a vital role in the structure, stability and functional properties of several processed foods. Protein–carbohydrate interactions regulate the functional properties in foods where proteins are the major ingredients, such as meat and fish processed products. Different hydrocolloids have been recommended to improve the mechanical and functional characteristics of surimi and restructured fish gels (Park 2000; Ramírez et al. 2002).

Fish proteins solubilized by salt and water form a continuous matrix. Some additives can be entrapped within this matrix, filling the gel and exerting their functional effects in the restructured products by a) influencing the formation of the continuous surimi gel matrix during thermal-inducing gelation; b) modifying the viscosity, mobility and other properties of the liquid phase; c) influencing texture and appearance of the gel, i.e. particle size, distribution, rheological (textural) properties, and relative volume fraction of the gel (Lee et al. 1992).

With the increasing demand for fish and fish products by consumers and the drop down drastically the fishery resources worldwide, the need for the non-commercial fish and other marine organisms has aroused. These underutilized organisms were neglected due to their size, flavour, odour, colour or texture and proved to be transformed into high-value products by surimi technology and restructuring technology. These technologies are used to obtain new products using a collection of additives to improve the mechanical and functional properties (Ramírez et al. 2011).

The restructuring process permits the gaining of products with high commercial value. This process infers the milling of fish muscle, solubilization of fish proteins with salt, formatting of fish paste and induction of the gelling phenomenon, usually by heat. Numerous restructured fish products have been established, including the following: vacuum-tumbling processing of trimmed salmonid fish with posterior canning and retorting; and tumbling of channel catfish using egg white as binder (Yetim and Ockerman 1995; Zimmerman et al. 1998). The restructuring process allows for the commercialization of some low-value fish species with higher profits: non-commercial fish species, smaller fishes than commercial size (such as caught as shrimp by-catch), and trimmings from filleting of commercial fish species (Pacheco-Aguilar et al. 2010).

The mechanical and functional properties of restructured products depend on the biochemical and physicochemical properties of muscle proteins, mainly myosin and actomyosin. In this regard, the biochemistry of fish muscle is different from that of mammals and birds. Therefore, fish products must be processed in a different way from red meat or poultry. The main fish products are surimi and restructured products. Fish products can be improved or modified by using hydrocolloids (carbohydrates and protein) as additives.

84.5.4.1 Surimi

Surimi technology suggests a great chance to transform different fish species into high commercial value products. Therefore, the need for surimi and kamaboko is increasing worldwide. Nevertheless, surimi technology has several negative environmental effects, which are necessary to reduce. One such impact is the overexploitation of white fish stocks, which has compromised the supply of these species. Policies for using non-commercial species and fisheries by-catch, together with the maximum utilization of fish, are being proposed (Martin-Sanchez et al. 2009).

Surimi is not a finishing product; it is a wet concentrate of high-quality myofibrillar proteins, obtained by washing the minced fish meat with cold water several times. The washing operation removes the water-soluble sarcoplasmic proteins responsible for organoleptic properties and concentrates the water-insoluble myofibrillar proteins, such as myosin, actin, and the actomyosin complex (Ramírez et al. 2000). Although it is technically feasible to obtain surimi from any fish species, the functional properties are highly dependent on fish species. Surimi is also considered an ingredient in the establishment of many products. Recently, the use of surimi as a protein-based carrier in developing high omega-3 fatty acid-containing seafood products has been proposed because it allows for uniform dispersion and oxidative stability of omega-3 fatty acid oil in the highly cohesive gel system without the use of antioxidants (Tolasa et al. 2010).

84.5.5 Low-Salt Restructured Products

Among the other methods to obtain fish paste is to use restructuring products.

Restructuring yields fish products with high commercial value from different sources: non-commercial fish species, fish with size smaller than commercial. All these techniques use salt to solubilize and extract myofibrillar proteins which form a sticky exudate, responsible for the binding in these products (Zimmerman et al. 1996). It was demonstrated, in other fish species, that it is not feasible to obtain restructured products in the absence of salt (Uresti et al. 2004). On the other hand, consumers are demanding healthier foods and there is a great interest in lightly salted products to prevent and control adverse blood pressure levels in humans. Among the products that have been replaced salt in the restructuring process is microbial transglutaminase (MTGase) or whey protein concentrate (WPC) as binding agents to prepare low-salted restructured fish products from non-commercial fish species (Uresti et al. 2004).

84.5.6 Non-commercial Small Fish as a Source of Micronutrients

Non-commercial small fish species issue has attracted little attention within the general subject of the role of fish as a source of micronutrients (Kawarazuka and Béné 2011). Recent studies have shown that small fish species that are consumed whole with bones, heads and viscera play a vital role in micronutrient intakes, as these parts are where most micronutrients are concentrated. Small fish also offer other nutritional advantages: they can be processed and stored for a long period; they are more affordable for the poor as they can be purchased in small quantities; and they can also be more evenly divided among household members (Thilsted et al. 1997). The contribution of fish to micronutrient intakes is therefore determined not only by the nutrient content of the species but also by the local processing methods and eating patterns. As a consequence, several studies have indicated the actual nutrient content of the edible part by reflecting the local methods used to clean and prepare the fish for the meal (e.g. leaving or cutting off the head, removing a part of the viscera).

Although small fish are rich in Ca and some forms of marine fish are rich in iodine. In addition vitamin A, Fe and Zn can also be obtained from the small fish species. Recently, several methods were introduced to extract the compounds mentioned above and for the readers who are interested to read more about this subject, they should consult (Kawarazuka and Béné 2011).

84.6 What Fish Species in Iraq Is a Small-Sized Fish Species and Non-commercial?

The Tigris–Euphrates drainage is regarded as one of the most separated foremost freshwater areas in the world. Though, endemism is only at the species level and diversity is low with only about 50 primary division species in 27 genera and 7 families. Out of this number, there are about 34 species are Cyprinidae. States for freshwater extinctions are due to climate change and water withdrawal that show the combined effect of these two influences in leading to the loss of 30% of the Tigris River fish species and 54% of the Euphrates River species by the year 2070 (Coad 2010).

Here, it is important to distinguish between what is meant by small-sized fish species that are naturally small in size and not more than few centimetres in total length such as members of the Aphaniidae, Poeciliidae and Cyrinodontidae and small size individuals of any of the large-sized fish species.

In Iraq, the number of the main small-sized fish species is 25 belonging to 8 families, with the family Nemacheilidae having the largest number of species (10 species). Next are the families Cyprinidae and Leuciscidae, with 4 species each. Among this number of species, there are 2 introduced species, *Heteropneustes fossilis* and *Gambusia holbrooki* and one species of marine origin *Planiliza abu*. The list of the small-sized species is:

Family: Nemacheilidae.
Oxynoemacheilus chomanicus.
Oxynoemacheilus euphraticus.
Oxynoemacheilus frenatus.
Oxynoemacheilus gyndes.
Oxynoemacheilus hanae.
Oxynoemacheilus kurdistanicus.
Oxynoemacheilus zarzianus.
Paracobitis molavii.
Paracobitis zabgawraensis.
Turcinoemacheilus kosswigi.
Family: Cyprinidae.
Garra elegans.
Garra rufa.
Garra variabilis.

Garra widdowsoni.
Family: Leuciscidae.
Alburnoides diclensis.
Alburnoides velioglui.
Alburnus caeruleus.
Alburnus sellal.
Family: Bagridae.
Mystus pelusius.
Family: Heteropneustidae.
Heteropneustes fossilis.
Family: Aphaniidae.
Aphanius mento.
Aphanius sophiae.
Aphanius stoliczkanus.
Family Poeciliidae.
Gambusia holbrooki Girard, 1859.
Family: Mugilidae.
Planiliza abu.

84.7 Conclusion

In the above sections of this chapter, a short review of the importance of the small-sized fish species and non-commercials in the plan of food security in any country. These fish species are usually neglected in most countries and they are included in the by-catch that return to the sea or to the river. In this chapter, it has been made clear the nutritional value of the small-sized fish species in addition to their availability in distribution and in abundance. The chapter is also listed the small-sized fish species that inhabit the freshwater system of Iraq. It seems that Iraq has an excellent potential to invest in the industry small-sized fish species in the form of manufacturing them for sale in the local and international markets as whole specimens. It is also possible for Iraq to build factories for extracting several pharmaceuticals, making protein concentrate for chicken and aquaculture.

On account of its high nutrient content, fish could be used as a key component in strategies aimed at reducing micronutrient deficiencies in Iraq. A few studies have been published in relation to this issue; however, evidence is still sparse and fragmented and the overall potential contribution of fish to nutritional security is yet to be fully understood.

The present study showed that the non-commercial fish species can be caught and/or traded locally, which in turn represent a major source of micronutrients in the everyday diet of the poor in Iraq and other developing countries due to the presence of high nutrient content and bioavailability in their bodies. Existing data suggest that adding fish to plant-based diets boosts protein absorption from food staples. Small fish, frequently consumed by the poor, are therefore nutritionally important, even in small quantities. Currently, in most developing countries the production of small fish species is highly dependent on the activities of small-scale fisheries and to a lesser extent on aquaculture.

A sustainable supply of these species should be prioritized and the production and consumption of nutrient-dense fish should be promoted in order to make full use of the capacity of these species in reducing micronutrient deficiencies. For this, conservation of wild stocks and distribution of aquaculture techniques producing these small-sized species are needed. Further research is required in areas such as nutrient composition, cleaning/cooking methods, impact of declining fish catches on the fish consumption patterns of the poor and seasonal availability in order to derive appropriate strategies and actual food-based methods. More technically, research protocols need to be developed to better quantify the outcomes of increased nutritional intake.

References

Anonymous (1987) Enzyme process creates creamy fish protein. Food Eng 59(9):44–45

Baran E (2002) The importance of non-commercial fish. Fisheries and aquaculture: towards sustainable aquatic living resources management. UNESCO Encyclopaedia of Life Support Systems. EOLSS Publishers, Oxford, pp. 2–11

Barzana E, Garía-Garibay M (1994) Production of fish protein concentrates. In: Fisheries processing. Springer, Boston, MA, pp 206–222

Blaber SJM (1997) Fish and fisheries of tropical estuaries. In: Fish and fisheries series. Chapman and Hall, London

Clucas I (1997) A study of the options for utilization of bycatch and discards from marine capture fisheries. FAO Fisheries Circular No 928:59

Coad BW (2010) Freshwater fishes of Iraq. Pensoft, Sofia, Bulgaria, p 294

Finch R (1977) Whatever happened to fish protein concentrate? Food Technol 31(5):44–53

Kawarazuka N, Béné C (2011) The potential role of small fish species in improving micronutrient deficiencies in developing countries: building evidence. Public Health Nutr 14(11):1927–1938

Lee CM, Wu MM, Okada M (1992) Ingredient and formulation technology for surimi-based products. In: Lanier TC, Lee CM (eds) Surimi technology. Marcel Dekker, New York, p 273e302

Mackie LM (1983) New approaches in the use of fish proteins. In: Hudson BJF (ed) *Developments in Food Proteins* - 2. Appl Sci, London, pp 215–262

Martin-Sanchez AM, Navarro C, Perez-Alvarez JA, Kuri V (2009) Alternatives for efficient and sustainable production of surimi: a review. Compr Rev Food Sci Food Saf 8:359e374

Pacheco-Aguilar R, Ocano-Higuera VM, Ezquerra-Brauer JM, Castillo-Yaez FJ, García-Sánchez G, Marquez-Rios E (2010) Partial characterization of 50- nucleotidase from giant squid (Dosidicus gigas) mantle. CyTA e J Food 7:173e179

Park JW (2000) Ingredient technology and formulation development. In: Park JW (ed) Surimi and surimi seafood. Marcel Dekker, New York, NY, p 343e391

Ramírez JA, Rodríguez-Sosa R, Morales OG, Vázquez M (2000) Surimi gels from striped mullet (Mugil cephalus) employing microbial transglutaminase. Food Chem 70:443–449

Ramírez JA, Uresti RM, Téllez S, Vázquez M (2002) Using salt and microbial transglutaminase as binding agents in restructured fish products resembling hams. J Food Sci 67(5):1778–1784

Ramírez JA, Uresti RM, Velazquez G, Vázquez M (2011) Food hydrocolloids as additives to improve the mechanical and functional properties of fish products: a review. Food Hydrocoll 25 (8):1842–1852

Saggs HWF (1962) The greatness that was Babylon. Sidgwick and Jackson, London, p 562

Spinelli J, Groninger H, Koury B, Miller R (1975) Functional protein isolates and derivatives from fish muscle. Process Biochem 10(10):31–36. 42

Thilsted SH, Roos N, Hassan N (1997) The role of small indigenous fish species in food and nutrition security in Bangladesh. World Fish Centre Q 22:82–84

Tolasa S, Lee CM, Cakli S (2010) Physical and oxidative stabilization of omega-3 fatty acids in surimi gels. J Food Sci 75:C305eC310

Uresti RM, Téllez S, Ramírez JA, Vázquez M (2004) Use of dairy proteins and microbial transglutaminase to obtain lowsalt fish products from filleting waste from silver carp (Hypophthalmichthys molitrix). Food Chem 86:257–262

Yeoh QL, Merican Z (1978) Processing of non-commercial and low-cost fish in Malaysia. Proc IPFC 18(3):572–580

Yetim H, Ockerman W (1995) The effects of egg white, tumbling and storage time on proximate composition and protein fractions of restructured fish product. J Aquatic Food Prod Tech 4:65–77

Zimmerman PA., Bissel HM, McIntosh GS (1996) Method of processing salmonoid fish. Arctic Alaska Seafoods Inc., US patent 5,846,594

Zimmerman PA., Bissel HM, McIntosh GS (1998) Arctic Alaska Seafoods Inc., issued on December 8, 1998. Method of processing salmonoid fish. U.S. patent 5,846,594

Chapter 85
Sociocultural Aspects Influence Food Consumption Habits in Iraq: Management of Food Security

Laith A. Jawad

Abstract Several factors have been found to determine the food habits and the consequent food security of the people in Iraq. Such factors are similar to those present in other Arab countries in the Middle East. Food consumption pattern has radically changed in Iraq as a result of a sudden increase in income from oil revenue. It is believed that food subsidy policy has adversely affected the food habits and consequently food security in Iraq by encouraging the intake of fat, sugar, rice, wheat flour and meat in large quantities and create a large amount of food waste. The factors that are affecting the food consumption in Iraq are leading in one way or another to create food waste that the government pays a large amount of money to make available for the people of Iraq. The following three main habits are related to food that the Iraqis have used in their daily life. These habits are present in both the urban and the rural societies in Iraq: (1) the habit of obtaining food items in large quantities more than what the family needs. Such large quantity of food items will end up in the trash before it has been used due to their date of expiry; (2) the habit of cooking more than a variety of food and with a large amount. This habit can affect the income of the family and at the same time produce more food waste and consequently more assets and (3) the habit of cooking huge amount of food during the special days and ceremonies, which can be a burden the budget of the family and increase the food waste that leads to influence food security plans of the country. Broad studies on social, cultural and economic factors associated with food consumption patterns in Iraq are highly recommended and needed.

85.1 Introduction

Contemporary Iraq overlays almost the same area as ancient Mesopotamia, which centred on the land between the Tigris and the Euphrates Rivers. Mesopotamia, also referred to as the Fertile Crescent, was a significant centre of early civilization and

L. A. Jawad (✉)
School of Environmental and Animal Sciences, Unitec Institute of Technology, Auckland, New Zealand

L. A. Jawad (ed.), *Tigris and Euphrates Rivers: Their Environment from Headwaters to Mouth*, Aquatic Ecology Series 11,
https://doi.org/10.1007/978-3-030-57570-0_85

saw the rise and fall of many cultures and settlements. In the medieval era, Iraq was the name of an Arab province that made up the southern half of the modern-day country. In today's Republic of Iraq, where Islam is the state religion and claims the beliefs of 95 percent of the population, the majority of Iraqis identify with Arab culture. The second-largest cultural group is the Kurds, who are in the highlands and mountain valleys of the north in a politically autonomous settlement. The Kurds occupy the provinces of As Sulaymaniyah, Dahuk and Irbil, the area of which is commonly referred to as Kurdistan (Every Culture 2019).

Iraq is located in the Middle East, with an area of 168,754 square miles (437,073 square kilometres), which is equal to twice the size of Idaho. Baghdad was the name of a village that the Arabs chose to develop as their capital and is in the central plains. The northern border areas near Iran and Turkey are mountainous and experience cold, harsh winters, while the west is mostly desert. The differences in climate have influenced the economies of the various areas and ethnic groups, especially since a large part of the economy used to be agriculturally based.

Among the population of Iraq, Arabs comprise about three-fourths and Kurds compose about one-fifth. The remaining people are divided into several ethnic groups, including Assyrian, Turkoman, Chaldean, Armenian, Yazidi, and Jewish. Almost all Iraqis speak and understand their official language, Arabic. Kurdish is the official language in Kurdistan, and serves to distinguish Kurds from other Iraqis. It is not of Semitic origin nor an Arab or Persian dialect, but a distinct language from the Indo-European family. Other minority languages include Aramaic, Turkic, Armenian, and Persian (Every culture 2019).

As with the people of other Arab countries, Iraqis face 2 kinds of nutritional problems; those associated with underdevelopment such as iron deficiency anaemia and undernutrition, and those associated with affluence such as obesity, diabetes, hypertension and heart diseases. Current studies have shown that sociocultural and economic environments play an important role in the prevalence of malnutrition in developing countries (Bhuiya et al. 1986; Hassan and Ahmad 1986). A fundamental transformation in food consumption patterns in the Arab countries in general and in Iraq, in particular, has occurred and affected the health and nutritional status of the people.

Prior to the United Nations economic sanctions, the traditional diet of the Iraqis included rice with soup or sauce, accompanied by lamb and vegetables. During the sanction food was tightly rationed, most people eat rice or another grain sometimes with sauce. Both vegetables and meat are hard to come by. In rural areas, it is customary for families to eat together out of a common bowl, while in urban areas individuals eat with plates and utensils. It is traditional to sacrifice a lamb or a goat to celebrate holidays. The common dish between the different ethnicities in Iraq is the rice with sauce of a sort of vegetables, but for each ethnicity and religion there are certain kinds of dishes and sweets that they are recognized with.

The present chapter highlights the influence of some of the social, cultural and economic issues on food consumption patterns in Iraq. The chapter discusses the habits of consumption of the Iraqis and it relates to the food security of their country.

85.2 Factors Affecting Food Consumption

85.2.1 Income

Obtaining and consumption of food in Iraq and as in other parts of the world depend mainly on the income of the individual. During the last few decades, the trend of food consumption was changed due to the increase in the income of the Iraqi individual. This coincides with the increase in the income per capita for individuals in the other Arab countries during the period 1973–1980 (Khaldi 1984; Musaiger 1993).

As with the other Arab high-income countries, Iraqis showed a marked rise in the consumption of meat, milk, eggs and cereals during the last ten decades. The change in meat consumption exceeded 200% during this period. Milk showed a similar trend, but the percentage change was lower than that of meat. This is also true for egg consumption, Iraq has shown growth rates similar to other high-income countries not due to the high oil revenue, but also for investments in the poultry industry.

The consumption of meat, milk, eggs and cereals varies between the social levels in Iraq. These four food items may not available for the poor people every day, but probably once a week. The availability of these items depends on the income of the family. On contrary to this theory, we can find the marsh Arab people, who are very poor, but they can eat two types of meat daily (fish and aquatic birds), drink fresh milk obtained from water buffalo that lives with them, and eat eggs from chickens they raise in their small living area. On the other hand, in the large cities of Iraq, cheap foods such as potatoes, bread, sugar and rice are the main source of energy for the low-income people.

Whether the income of the family high or low and whatever kind of food they consume, it is clear that the amount of food thrown in the trash is high at the different levels of the Iraqi society. For example, if the income of any family increased, the members of which will switch to eat more variety of food and obtain them in large quantity a case that results in large amount of food waste. The case is the same even with the low-income families that depend on their food consumption on a limited variety of food items. These families, for example buy a large number of bread more than what they need aiming to keep everybody in the family full, but they do not estimate the actual number of bread that needed for their families, which resulted in a waste of bread that goes to the trash.

85.2.2 Food Price

Prices of food items and the subsidies to consumers by the government have long been a part of social and economic policy in Arab countries including Iraq. Such subsidies help the Arab individuals to consume more food or in other words obtain food even they do not need it (Scobie 1983). In such a way, food items will be accumulated in the family stores and end up in the trash or as a food for the livestock (Alderman and Von Braun 1986). In the Arabian Gulf countries, for example the consumption of meat, rice, wheat flour, sugar, fat and oil has increased dramatically during the past 15 years, mainly due to lower subsidized pricing of these foods and increased income (Musaiger 1993). With such an increase in consumption the nutritional status of the population, as the increased intake of the energy-rich foods such as rice, fat and sugar have also increased and at the same increased the risk of obesity, heart diseases and dental caries (Musaiger 1991).

85.2.3 Local Differences

In the rural areas of Iraq and as in other Arab countries, the variety of food is not wide and the rural people depend largely on food locally produced. The difference between the dietary habits of rural and urban populations can be seen in many parts of the region. In Iraq, for example wheat bread is consumed mostly in urban areas, while bread made from a mixture of wheat and any other cereals is consumed in rural areas. This is the case in Egypt for example (Ramadan 1986). In the rural areas of Iraq, no bread can be purchased from the market. All bread consumed is baked by the household. Such a case is similar to what happened in Lebanon, where 98% of the rural households made their own bread compared to only 10% of the urban ones. In all Arab countries including Iraq, rural families ate meals together and all meals started earlier in the day as compared to those of urban dwellers (Al-Isi et al. 1975). In the urban areas of Iraq, nearly every member of the family eats his meal separately from the rest of the family. In this way, extra food will be purchased and consumed while the weekly or the monthly ration remain as it is in store and get expired after a short period of time. Fresh vegetables, fruits and milk are less taken by urban people in Iraq due to the long working hours. In the rural areas, there is a kind of food management scheme. This system is created because of the short-time storage of some food items such as milk and meat. Not all the rural areas have electricity and refrigerators to keep food in for a longer period. Farmers sell milk while it is still fresh or process it domestically for conversion into products before the milk spoils. In most rural areas, milk is utilized for making butter and soft cheese. The skimmed milk remaining after removal of butterfat is used for cheese preparation. In this way, milk has been transformed from one state to another and new food items will be created that might fulfill the need of the members of the families of the rural areas.

85.2.4 Geographical Differences

The geography of Iraq is variable and contains different habitats such as mountains, deserts, flood plains and coastal areas. Such habitats have a great influence on the dietary patterns of the Iraqi people. The desert area is inhabited by nomadic animal breeding tribes, who live mainly on cereal, dates, milk and milk products. The settled nomads consume cereal, pulses, some vegetables and relatively less milk and fewer milk products. Meat is rarely consumed in the desert, being reserved particularly for occasions when guests have been invited or for feasts (Moore 1970). However, the food consumption patterns may differ slightly from area to area based on the available food resources. The nomads in Iraq consume wheat bread, dates, ghee, milk and yoghurt. Rice is sometimes substituted for bread, and the only vegetable consumed is the onion (Al-Ani 1985). In the mountainous area of Iraq and due to the cold weather, people eat less rice and replace it with barley as the later give more heat calories and keep them warm. Also, they eat less fish, but more meat. Milk products such as cheese and yoghurt are made in large quantities. In the north of Iraq, people are famous in drying fruit of different kinds. In the big cities, the food of the people is similar to that of any other city in the world, where fast food is dominating the daily life of the Iraqi individuals. Meat is usually consumed more by people living in the cities by eating it as grilled or in the vegetable sauce. In the southern Iraq and due to the availability of fishes, people eat fish at least once a week. This increased probably everyday in the marsh areas, where fish and aquatic birds are abundant. In the south of Iraq, high percentage of households bake their bread traditionally. This is also true for the marsh Arab. In coastal areas and south of Basrah City, marine fish and crustaceans such as shrimps play an important role in the diet. The consumption of fish in these areas is higher than in other parts of Iraq and nearly similar to what is present in the neighbouring Arabian Gulf countries (Feidi 1986).

It is clear from the wide variation in the diet of the people of Iraq, that different food item are consumed and in large quantities reaching to the level of more than the need of one family. Iraqis are known to obtain food items in large quantities whenever it becomes available even the nomads in the desert they following suit. For example, when the nomads reaching an Oasis or a small village at the edge of the desert, they shop from the local market for what they need of food items. They usually take more than what they need to cover their daily food and the food that they offer in the meals during the ceremonies.

85.2.5 Influence of Religious

Iraq is a country with people of mixed religions, Muslims, Christians, Jewish, Yazidis and Mandeism. For people of those religions certain food items to eat and food habits. For example, pork and ham are not eaten by Muslims, but they have eaten by Christians. Within Muslims, there are some minor restrictions about some food items such as fish. The Shiite sect of Muslim do not eat any species of catfish or shark, while people of the Sunni sect do eat catfish and shark fish species. Food beliefs are rooted in folk beliefs, religion and sociocultural traditions (Leininger 1970). Despite the rapid changes in food habits which have taken place, food beliefs are enduring (Todhunter 1973). The taboos about the food in different sectors of Muslims and other religions have been discussed in another chapter of this book.

Fig. 85.1 (**a**, **b**), People throwing food waste in the trash after a celebration; (**c**, **d**), people looking for food in the trash. Both cases were taken from Middle east countries

There are special days and ceremonies that members of each religion perform during their own calendar. In spite of the differences in the food items each group of religion used in their special days, the common feature remains the same between the members of all religions and that is the amount of food prepared and the amount of food that remain after the celebrations and waiting to throw in the trash.

During the Ramadan month, Muslims stop eating and drinking from the sunrise to sunset. Once they started to eat, they do not stop eating until the sunrise. In the month of Ramadan, Muslims usually put on extra weight due to the habit of eating food with high level of fat and desserts in large quantities. The Christians during Christmas days and the New Year eat large amount of food that high in fat, consume large amount of sweet, and drink a large amount of alcohol.

The food and feeding habits of the different sectors of Iraq lead in one way or another to health problems. For example, Christians eat pork and ham, which are characterized by high levels of fat that cause high blood pressure and heart diseases. Muslims on the other hand, do not eat pork and ham, but eat meat with level of fat in it, which in turn causes health problems (Fig. 85.1).

85.2.6 Food Distribution within the Family

In some parts of Iraq, especially in the rural areas, there are some discrimination in offering meals to the different members of each family. In general, on any occasion and celebration and if food is served, men usually eat first and then women and children. Also, in some areas the food is served to the fathers and then wife and children will eat latter (Abdulla 1979). Such habits are no longer followed in the urban and even in the rural areas that are located at the edge of the cities. The most important issue in distributing food among the members of the family that each member will have enough food to eat. Gender discrimination in the rural areas during food serving is considered the cause of the lower health and nutritional status of girls in these areas (UNICEF 1985).

85.2.7 Education

Illiteracy and unawareness of sound eating habits greatly influence food consumption. The rapid industrialization of some Arab countries has led to a sharp increase in income, but not in the educational level (Musaiger et al. 1986). In all the countries in the region, illiteracy is relatively high, especially among the women, who are mainly responsible for food preparation and infant feeding. Studies in other countries showed that the quality and quantity of food improved as education of the mothers improved (Delfina et al. 1984; Chaudhury 1984).

The illiteracy can lead to health problems by choosing the wrong food items and the method of cooking them. Al-Shawi (1985) reported that the involvement of the housewife in food preparation decreased as education increased. This leads to employing a cook to do the job. Such cook will prepare the food in the way she/he knows previously, where the level of fat, sugar and other ingredients might be high or low enough to cause health problems such as heart diseases and blood pressure. In addition, the cook will not be careful about the number of food items she/he is using. In many cases, the cook who employed to serve a family shoed to waste a large amount of food through cooking a large amount of rice in one day and in the next day she/he cook the same amount of rice and throw the quantity of rice that she/he prepared in the day before in the trash. This an example of wasting food and should be taken into consideration in food security measures in Iraq. However, education is not always connected with good food habits. For example, educated people are aware that obesity ill adversely associated with health, but this awareness does not necessarily mean that they change their eating habits in order to lose weight or that they choose low-calorie foods. The reason: for the lack of success of education possibly lies behind a lack of awareness of the severity of health problems and a suitable solution for overcoming these problems (McKenzie 1964).

85.3 Conclusion

Factors determining food consumption patterns in Iraq are also considered limiting factors for the food security of this country. In the above discussion, several issues have been elucidated that might cause health problems and at the same time are regarded as concerns that affecting the future food security plans in Iraq.

From the discussion of the factors that are affecting the food consumption in Iraq, it is clear that these factors lead in one way or another to create food waste that the government pay a large amount of money to make available for the people of Iraq. It is possible to distinguish three main habits related to food that the Iraqis have used in their daily life. These habits are present in both the urban and the rural societies in Iraq:

1. The habit of obtaining food items in large quantities more than what the family needs. Such a large quantity of food items will end up in the trash before it has been used due to their date of expiry.
2. The habit of cooking more than variety of food and with a large amount. This can affect the income of the family and at the same time produce more food waste and consequently more assets.
3. The habit of cooking a high quantity of food during the special days and ceremonies, which can be a burden to the budget of the family and increase the food waste that leads to influence food security plans of the country.

However, the shortage of literature dealing with this topic makes an in-depth discussion of the impact of such and other factors difficult. There are many other factors that are not mentioned and have some influence on dietary habits and ultimately on food security in Iraq. Population growth, for example has affected not only food demand in the region, but also food exports. The political situation, the Gulf Crisis and the decline in the price of oil are all important factors which have recently been varying food consumption patterns and food security plans. In-depth studies on the role of the sociocultural and economic environments in the food habits and security of Iraq are highly recommended.

References

Abdulla SM (1979) Food habits in Iraq. Master thesis, high Institute of Public Health, University of Alexandria, Egypt AL-ANI M (1980). Diet and dietary habits of nomads in Iraq. Ecol Food Nutr 9:55–58

Alderman H, Von Braun J (1986) Egypt's food subsidy policy, lessons and options. Food Policy 11:223–236

Al-Ani A, Bruzau F, Raymond P, Saint-Ges V, Leblanc JM, Pradet A (1985) Germination, respiration, and adenylate energy charge of seeds at various oxygen partial pressures. Plant Physiol 79(3):885–890

Al-Isi IJ, Kanawati AA, McLaren DS (1975) Formal education of mothers and their nutritional behaviour. J Nutr Ed 7:22–24

Al-Shawi A (1985) The dietary patterns and food habits of Kuwaiti housewives of 3 educational levels. Educ J Kuwait 2:9–19

Bhuiya A, Mojtyniak B, D'Souza S, Zimicki S (1986) Socio-economic determinants of child nutritional status. Food Nutr Bull 8:3–7

Chaudhury RH (1984) Determinants of dietary intake and dietary adequacy for preschool children in Bangladesh. Food Nutr Bull 6:24–33

Delfina BA, Caedo MM, Arnold JC, Engel RW (1984) The relationship of family characteristics to the nutritional status of preschool children in Bangladesh. Food Nutr Bull 6:24–33

Every culture (2019). https://www.everyculture.com/Ge-It/Iraq.html (Retrieved 24 February 2019)

Feidi I (1986) Consumer preferences of fishery products in the Arab countries. Infofish Marketing Digest 2:10–12

Hassan N, Ahmad K (1986) Household distribution of energy intake and its relationship to socio-economic and anthropometric variables. Food Nutr Bull 8:3–6

Khaldi N (1984) Evolving food gaps in the Middle East/North Africa: prospects and policy implications. International Food Policy Research Institute, Washington DC

Leininger M (1970) Some cross-cultural universal and non-universal functions, beliefs and practices of food. In: Dupont J (ed) Dimension of nutrition. Colorado Associated University Press, Colorado, pp 153–172

McKenzie JC (1964) Food trends: the dynamics of accomplished change. In: Yudkin J, McKenzie J (eds) Changing food habits. Macgibbon and Kee, London, pp 25–45

Moore FW (1970) Food habits in non-industrial societies. In: Dupont J (ed) Dimension of nutrition. Colorado Associated University Press, Colorado, pp 181–222

Musaiger AO (1991) Towards food and nutrition policy in Bahrain. Ministry of Health, Bahrain

Musaiger AO (1993) Socio-cultural and economic factors affecting food consumption patterns in the Arab countries. J R Soc Health 113(2):68–74

Musaiger AO, Al-Sherbini AF, Elsayed NA, Amine EK, Darwish OA (1986) Children's response to television food advertisements in Bahrain. Int J Hlth Educ (Hygie) V:30–35

Ramadan AAS (1986) Some characteristics of Egyptian baladly bread as affected by partial substitution of maize flour in relation to economic state. Egypt J Food Sci 14:237–240

Scobie GM (1983) Food subsidies in Egypt: their impact on foreign exchange and trade. International Food Policy Research Institute, Washington DC

Todhunter EN (1973) Food habits, food faddism and nutrition. Wrld Rev Nutr Diet 16:286–317

UNICEF (1985) Girl's adolescence, UNICEF/MENA, Amman UNICEF (1991). The girl child

Chapter 86
A Preliminary Investigation of Determinants of Food Security in Rural Areas of Basrah Province, Iraq

Laith A. Jawad

Abstract Food security occurs when all individuals at all the occasions have bodily and economic admittance to adequate, safe, and healthful food to encounter their nutritional requirements and diet favorite for an energetic and well life. Diet security might be examined four elements at diverse theoretical stages: areas, states, family, and individual. Iraq in general and Basrah Province people have not sufficient admittance to food for their appropriate diet. The present investigation was taken to explore the causes of diet security and key matters upsetting diet security in rural areas of Basrah Province, Iraq. A sample of 200 family participants was designated methodically from countryside of Basrah Province through a multistage sampling technique. Studies have shown that 63.3% of questionaries stated that they had confronted the difficulties concerning the obtainability of all kinds of diet and 50.3% of contributors were content about the quality of food. Nearly two-third of the questionaries, i.e., 63.3% suggested that they had no admittance to all types of food goods which they need to acquire. The results of Chi-Square and Gamma showed an important and strong meaning between age, schooling, and revenue of the questionaries and their consciousness regarding diet security.

86.1 Introduction

Chakiso and Emana (2012) suggested that there are two types of food insecurity: long-lasting and passing. The former type is a provisional weakening in a family's admittance to enough nutrition, while the latter is a constantly insufficient intake instigated by the incapability to obtain diet. In Iraq and in the southern province of Basrah, food uncertainty continues an unsatisfied vision for presently around more than one million people have not admittance to sufficient diet.

L. A. Jawad (✉)
School of Environmental and Animal Sciences, Unitec Institute of Technology, Auckland, New Zealand

L. A. Jawad (ed.), *Tigris and Euphrates Rivers: Their Environment from Headwaters to Mouth*, Aquatic Ecology Series 11,
https://doi.org/10.1007/978-3-030-57570-0_86

The wealth that Basrah delivers to Iraq in general, the standard of living in Basrah is still under the international standard that other similar cities in the world living in. The Inter-Agency Information and Analysis Unit in its report retrieved on February 2019 showed that: in the education section, the percentage of prevalence of illiteracy (aged 10+ years) is 14.6% compared to the total percentage of Iraq of 17.5%; in the health section, the prevalence of chronic disease showed 8.5% vs 9.3 for whole Iraq, and diarrhea (aged 0–5 years) of 4.2% vs 3.4% of whole Iraq; in poverty section, the families in the bottom per capita spending part was 9.8% vs 20% for the whole Iraq; in the water and sanitation, the proportion of households with a toilet is 99.6% vs 91.3% for the whole Iraq; in the protection section, the percentage of post-2006 IDPs is 2.5% vs 100% for whole Iraq. In the outskirt of City of Basrah, the values of the previously mentioned components vary significantly and some areas, the illiteracy and the health sectors dropped well below the whole Iraq standard.

Diet uncertainty and therefore food deficiency has been on upsurge during time in Basrah Province. As to the status of scarcity in Basrah Province is the same as that suggested by the United Nations, which stated that the occurrence of diet scarcity is greater in the countryside regions than in city zone (Tawfeek and Salom 2001).

Food security in Basrah Province is not different from that of the whole of Iraq. In general, the food security in Iraq has been intensely impacted by Iraq's oil-based budget, over the last 30 years of fight and its policies that have been changed by controlling investors.

This chapter signifies out that the number of hungry, and underfed individuals are living in the rural areas of the province of Basrah, Iraq. This study further reported on the percentages of households in the area of investigation that suffered food insecurity.

86.2 The Method of Obtaining Data

The investigation at hand was intended to evaluate the factors of food safety in countryside regions of Basrah Province, Iraq. The data for the analysis was based on 250 random respondents living in rural areas in the north, south, west, and east of Basrah Province. Simple random and proportional techniques were used in dealing with the questioners.

86.3 Analysis of the Data Collected

The social and economic life and inhabitant's analytical feature of the participants in the survey were analyzed and the results are shown in Table 86.1. The information showed that 24.0% of the participants fall in the age category 18 to 25 years, specified that the good contribution of adolescence in this survey. Mainstream of

Table 86.1 Socio economic and demographic characteristic of the respondents

Age of the respondents (year)	Frequency	Percentage
18–25	48	24.0
26–40	50	25.0
41 and above	27	13.5
Sex of the respondents		
Male	21	10.5
Female	134	67.0
Educational level		
Illiterate	76	38.0
1–5 group	32	16.0
6–10 group	41	20.5
11 and above group	56	28.0
Household monthly income (Iraqi Dinar)		
Up to 600,000	56	28.0
600,001–2,500,000	76	38.0
>2,500,000	21	10.5

the participants 25.0% in the age range 26 to 40 years, whereas the rest of 13.5% participants related to 42 years and above. The analysis showed that mainstream of the participants 67.0% were female. Information obtained also verified that bulk of the participants 38.0% were uneducated, whereas the 16.0% of participants had 1–5 grade schooling. 20.5% and 28.0% of participants had 6 to 10 level and 11 and above level schooling correspondingly. As to the income section, 28.0% participant's revenue was up to 600,000 Iraqi Dinars, whereas bulk 38.0% of the participant's wages were falling in the range of 600,000 to 2,500,000 Iraqi Dinars and extra than one–third, i.e., 10.5% had above 2,500,000 Iraqi Dinars household monthly income.

One-third of the participants, i.e., 31.3% confirmed this accounts that the "The entire sort of food goods obtainable in their region," whereas 43.2% of the participants stated that there were not admission on the appropriate obtainability of food items in their regions. Around 10.0% and 15.5% of the participants expressed that food items were accessible frequently and sometimes correspondingly. Information collected indicates that bulk of the participants 73.3% testified that they have not admission on all sort of food stuffs which we need to acquire in our regions. Less than one-third of the participants, i.e., 26.6% were in the view that, they had admittance on the accessibility of all sorts of food items in their regions. Statistics display that 43.3% of participants were content with the condition of food which they purchased from their nearby areas, while 56.7% of participants were not content about the condition of food. Similar results were obtained by Arene and Anyaeji (2011) in their study on Nsukka metropolis of Enugu State, Nigeria. They establish that about 60 percent of the families were food uncertain. Likewise, Mahmood et al. (2014) in their study on the rural areas of Faisalabad City, Pakistan.

Regarding the eating habits of the participants, information show 14.3% were eating two times in a day, whereas, a large bulk 82.7% of participants ate meal

Table 86.2 Distribution of the respondents with regard to availability, accessibility and satisfaction with the quality of food products

	Response categories	Frequency	Percentage
1	All types of food products available in your area		
i	Yes	62	31.3
ii	No	87	43.2
iii	Often	20	10.0
iv	Sometime	31	15.5
2	Accessibility of food products which you want to get		
i	Yes	147	73.3
ii	No	53	26.7
3	Satisfaction with the quality of food		
i	Yes	86	43.3
ii	No	114	56.7
4	How many time eat meal in a day		
i	Two	28	14.3
ii	Three	166	82.7
iii	More than three	6	3.0
5	Family members take milk regularly		
i	Yes	65	32.8
ii	No	135	67.2
6	Income spent on buying their food		
i	Up to 150,000	43	21.5
ii	150,000–200,000	32	16.0
iii	>200,000	12	6.0
7	Knowledge about balance diet		
i	Yes	46	23.0
ii	No	154	77.0

3 times a day (Table 86.2). A small part of the participants 3.0% ate meal more than three times in a day. Mahmood et al. (2014) found that in city zones in Pakistan, about 48% were food protected whereas 52% were food unprotected, compared with rural areas where 39.4% of households were food secure and 60.6% were food insecure. It is important to note here that the present information is based on the rural areas of Basrah province where the natural foods, i.e., milk, butter, and eggs are effortlessly obtainable but only 32.8 participants were in the view that family associates took milk frequently and bulk 67.2% of the participant's family fellow had not admittance to milk frequently. From the questionaries given to the households, it seems that 50% of the women and children were found to be underfed. The data displays that the maternal and child undernourishment is the main question for Basrah people and it should be rectified.

Information obtained from the participants displayed that 43% of participants used their salary on food up to 150,000 Iraqi Dinars per month, whereas the 32.0% of participants spent 150,000–200,000 Iraqi Dinars per month on buying food. Such results indicate that a large portion of monthly family revenue was assigned for acquiring the foodstuffs in the countryside regions. The not unexpected results that in the countryside zones 77.0% of participants had no/any idea regarding the stability of diet, while 23.0% of participants had knowledge about it (Table 86.2).

There is an exceedingly important connotation between age of the participants and their acquaintance about food safety established (Table 86.3). Gamma analysis result displays a robust adverse link among data. Statistics discloses that young adults had more familiarity and consciousness about food safekeeping as related to older variables. Arene and Anyaeji (2011) they indicated out that working rank and older family heads incline to be food more food protected. Table 86.5 displays a greatly important link between the revenue of the variables and their information about food safekeeping.

Table 86.4 indicates that Chi-square result (78.04) extremely noteworthy reflecting link between schooling of the variables and their information about food safety. It signifies uneducated individuals had less information as related to taught variables. Sultana and Kiani (2011) also designated that the improved enlightening level of family's head had also a vital and helpful influence on food safety. Place of living (Urban) had a vital and undesirable influence on family's food safety. In Table 86.5 gamma result (0.47) indicates another robust positive link between reliant on and non-reliant on data. The result of Bashir et al. (2012) intensely agrees with the current results as they suggested that monthly revenue, cattle, and better learning status definitely have impact on rural family food safekeeping. Alternatively, larger family heads' age and family size had negative influences on family food safekeeping.

Table 86.3 Association between age of the respondents and their knowledge about food security. Chi-square = 13.71; degree of freedom = 4; *P*-value = .008; Gamma value = −.394

	Knowledge about food security			
Age of the respondents (years)	Low	Medium	High	Total
Up to 25	17 (21.8%)	18 (23.1%)	43 (55.1%)	78
26–40	25 (35.2%)	18 (25.4%)	28 (39.4%)	71
40 and above	20 (39.2%)	19 (37.3%)	12 (23.6%)	51
Total	62	55	83	200

Table 86.4 Association between education of the respondents and their knowledge about food security. Chi-square = 78.04; degree of freedom = 6; *P*-value = .000**; Gamma value = 0.781

	Knowledge about food security			
Education of the respondents	Low	Medium	High	Total
Illiterate	12 (13.9%)	27 (31.4%)	47 (54.7%)	86
Up to 8 grade	12 (26.1%)	16 (34.8%)	18 (39.1%)	46
12 grade	9 (26.5%)	18 (52.9%)	7 (20.6%)	34
14 grade and above	1 (2.9%)	3 (8.8%)	30 (88.2%)	34
Total	34	64	102	200

Table 86.5 Association between income of the respondents and their knowledge about food security. Chi-square = 21.08; degree of freedom = 4; *P*-value = .000**; Gamma value = 0.475

	Knowledge about food security			
Income of the respondents in Iraqi Dinars	Low	Medium	High	Total
Up to 150,000	16 (30.8%)	16 (30.8%)	20 (38.5%)	52
150,000–200,000	10 (16.7%)	10 (16.7%)	40 (66.7%)	60
>200,000	2 (2.3%)	20 (22.7%)	66 (75.0%)	88
Total	28	46	126	200

86.4 Conclusion

The majority of family food uncertainty in this investigation was great and perturbing. In countryside regions, where this survey was performed, the bulk of the inhabitants had not appropriate admission to food and frequent participants stated that the unavailability of diet items in their region. It was set up that mainstream of the inhabitants had low purchasing ability. Owing to this agent, inhabitants particularly children were grieving from undernourishment and additional communicable diseases. It is recommended that administration should pay an appropriate care for physical and economic admittance to adequate food for satisfactory nutritional requirements for each resident.

References

Arene CJ, Anyaeji RC (2011) Determinants of food security among households in Nsukka Metropolis of Enugu state, Nigeria. Pak J Soc Sci 30(1):9–16

Bashir MK, Schilizzi S, Pandit R (2012) The determinants of rural household food security in the Punjab, Pakistan: an econometric analysis, working paper 1203, School of Agricultural and Resource Economics. University of Western Australia, Crawley, Australia

Chakiso KT, Emana B (2012) Household food insecurity, coping strategies and policy options: food insecurity causes and coping strategies among farm households in Dodota-Sire District, Arsi zone. Lambert Academic Publishing, Germany

Mahmood S, Tabassum A, Khan SA, Jabeen N (2014) An investigation of determinants of food security in rural areas of Faisalabad, Pakistan. Mediterr J Soc Sci 5(4):676

Sultana A, Kiani A (2011) Determinants of food security at household level in Pakistan. Afr J Bus Manag 5(34):12972–12979

Tawfeek HI, Salom AH (2001) Iraqi national survey data on malnutrition and breastfeeding practices among children under five years of age. Food Nutr Bull 22(1):19–22

GPSR Compliance

The European Union's (EU) General Product Safety Regulation (GPSR) is a set of rules that requires consumer products to be safe and our obligations to ensure this.

If you have any concerns about our products, you can contact us on ProductSafety@springernature.com

In case Publisher is established outside the EU, the EU authorized representative is:

Springer Nature Customer Service Center GmbH
Europaplatz 3
69115 Heidelberg, Germany

Batch number: 10732594

Printed by Printforce, the Netherlands

Aquatic Ecology Series

Volume 11

Aquatic ecology is an extraordinarily broad and interesting field. It investigates the interplay between aquatic organisms and their physical, chemical, and biological environment. Aquatic ecology encompasses all freshwater and marine ecosystems, including streams, rivers, lakes, wetlands, coastal environments, and the vast expanses of the open ocean. Aquatic ecology studies a wide diversity of different organisms, ranging from tiny bacteria to large whales, facing a myriad of different processes such as biogeochemical cycles, genetic differentiation, and climate change. Fundamental research in aquatic ecology adds new discoveries almost every day. Applied research makes major contributions to biotechnology, fisheries, water management, nature conservation, and environmental policy. Reassessments and syntheses in aquatic ecology are stimulating to the discipline as a whole, as well as enormously useful to students and researchers in ecological sciences.

A series of succinct monographs and specialized evaluations in aquatic ecology has been developed. Subjects covered are topical (e.g., coastal hydrodynamics, microbial loops, alpine lakes) rather than broad and superficial.

The treatments must be comprehensive and state-of-the-art, whether the topic is at the biochemical, mathematical, population, community, or ecosystem level. The objectives are to advance the topics by the development of arguments, with documented support, that generate new insights, concepts, theories to stimulate thought, ideas, directions, controversies. The books are intended for mature as well as emerging scientists to stimulate intellectual leadership in the topics treated.

If you have questions regarding our book series or would like to submit a proposal to Aquatic Ecology, please contact the Publisher Éva Lőrinczi (eva.loerinczi@springer.com).

More information about this series at http://www.springer.com/series/5637